SIXTH EDITION

HEAVY DUTY TRUCK SYSTEMS

SIXTH EDITION

HEAVY DUTY TRUCK SYSTEMS

SEAN BENNETT

CENGAGE Learning

Australia • Brazil • Mexico • Singapore • United Kingdom • United States

CENGAGE Learning

Heavy Duty Truck Systems: Sixth Edition

Senior Manager, Custom Production:
Donna Brown
Linda deStefano
Julie Dierig

Manager, Custom Production:
Terri Daley
Louis Schwartz

Marketing Manager:
Rachael Kloos

Manager, Premedia:
Kim Fry

Manager, Intellectual Property Project Management:
Brian Methe

Manager, Manufacturing & Inventory:
Spring Stephens

Printed in Mexico

Heavy Duty Trucks: Sixth Edition
Sean Bennett

For product information and technology assistance, contact us at
Cengage Learning Customer & Sales Support, 1-800-354-9706
For permission to use material from this text or product,
submit all requests online at **cengage.com/permissions**
Further permissions questions can be emailed to
permissionrequest@cengage.com

This book contains select works from existing Cengage Learning resources and was produced by Cengage Learning Custom Solutions for collegiate use. As such, those adopting and/or contributing to this work are responsible for editorial content accuracy, continuity and completeness.

Compilation © 2015 Cengage Learning

ISBN: 978-1-3057-5569-7

WCN: 01-100-101

Cengage Learning
20 Channel Center Street
Boston, MA 02210
USA

Cengage Learning is a leading provider of customized learning solutions with office locations around the globe, including Singapore, the United Kingdom, Australia, Mexico, Brazil, and Japan. Locate your local office at: **www.international.cengage.com/region.**

Cengage Learning products are represented in Canada by Nelson Education, Ltd.

For your lifelong learning solutions, visit **www.cengage.com/custom.**

Visit our corporate website at **www.cengage.com.**

CONTENTS

Contents for Photo Sequences xi
Preface xv
About the 6th Edition xv
Changes in Technology xv
New to This Edition xv
Feedback from Educators xvi
Features of the Text xvii
Supplements xix
Acknowledgments xxi

Chapter 1 **Introduction to Servicing Heavy-Duty Trucks 1**

1.1 Truck Classifications • **2** 1.2 Heavy-Duty Trucks • **5**
1.3 Career Opportunities for Heavy-Duty Truck Technicians • **11**
1.4 Job Classifications • **14** 1.5 Advancement in the Profession • **19**

Chapter 2 **Shop Safety and Operations 23**

2.1 Personal Protective Equipment • **24** 2.2 Work Area Safety • **28**
2.3 Oxyacetylene Equipment • **33** 2.4 Electric Arc Welding • **35**
2.5 Shop Tool Safety • **35** 2.6 Hazardous Materials and WHMIS • **35**
2.7 Handling and Disposal of Hazardous Waste • **36** 2.8 Shop Records • **40**

Chapter 3 **Tools and Fasteners 45**

3.1 Hand Tools • **46** 3.2 Power Tools • **56** 3.3 Measuring Tools • **62**
3.4 Manufacturers' Service Literature • **70** 3.5 Fasteners • **76**

Chapter 4 **Maintenance Programs 88**

4.1 Setting up a Maintenance Program • **92**
4.2 Out-of-Service or Deadlining a Vehicle • **99** 4.3 PM Scheduling • **100**
4.4 Lubricants • **108** 4.5 Winterizing • **115** 4.6 Performing a Lube Job • **121**
4.7 Fleet Management Software • **123**
4.8 Maintaining Emissions Control Equipment • **124**

Chapter 5 **Fundamentals of Electricity 127**

5.1 Atomic Structure and Electron Movement • **128**
5.2 Conductors and Insulators • **133** 5.3 Current Flow • **134**
5.4 Magnetism • **135** 5.5 Electromagnetism • **136**
5.6 Electrical Current Characteristics • **138** 5.7 Sources of Electricity • **139**
5.8 Electrical Circuits and Ohm's Law • **141**
5.9 Electric Motor and Generator Principle • **148**
5.10 Coils, Transformers, and Solenoids • **151**

Chapter 6 **Fundamentals of Electronics and Computers** 156

6.1 Using Electronic Signals • **157** 6.2 Semiconductors • **159**
6.3 Diodes • **161** 6.4 Transistors • **163**
6.5 Photonic Semiconductors • **166** 6.6 Testing Semiconductors • **167**
6.7 Integrated Circuits (I/Cs) • **169** 6.8 Gates and Truth Tables • **170**
6.9 Binary System Basics • **171** 6.10 Microprocessors • **173**
6.11 Data Retention in a Vehicle ECU • **182**
6.12 Auxiliary Electronic Systems • **183**
6.13 Accident Avoidance Systems • **184**

Chapter 7 **Batteries** 192

7.1 Battery-Operating Principles • **193** 7.2 Battery Ratings • **200**
7.3 Battery Maintenance • **201** 7.4 Battery Testing • **205**
7.5 Charging the Battery • **210** 7.6 Conductance Testing • **216**
7.7 Alternative Battery Technology • **219**
7.8 Ultracapacitors and Supercapacitors • **222**
7.9 Battery Management Systems • **223**

Chapter 8 **Charging Systems** 227

8.1 Alternator Construction • **228** 8.2 Alternator Operation • **231**
8.3 Charging System Failures and Testing • **235**
8.4 Alternator Rebuild • **240**

Chapter 9 **Cranking Systems** 249

9.1 Cranking Circuit Components • **250** 9.2 Starter Motors • **252**
9.3 Cranking Circuit Testing • **260** 9.4 Starter Rebuild • **266**

Chapter 10 **Chassis Electrical Circuits** 270

10.1 Lighting Systems • **271** 10.2 Auxiliary Electrical Equipment • **279**
10.3 Rapid Checking of a Truck Electrical Circuit • **290**

Chapter 11 **Diagnosis and Repair of Electronic Circuits** 297

11.1 Types of EST • **299** 11.2 Using Digital Multimeters • **302**
11.3 Multifunctional Handheld Diagnostic Tools • **313**
11.4 Handheld Diagnostic Instruments • **313**
11.5 PCs and OEM Software • **318** 11.6 Bus Codes and Protocols • **326**
11.7 Troubleshooting CWS and CMS • **328**
11.8 Electrical Wiring, Connector, and Terminal Repair • **329**
11.9 Splicing Guidelines • **334** 11.10 Supplemental Restraint System • **335**

Chapter 12 **Multiplexing** 343

12.1 Multiplexing, Clients, and Servers • **345** 12.2 Multiplexing Basics • **345**
12.3 Multiplex Switching • **353**
12.4 Multiplex Transactions • **354** 12.5 Accessing the Data Bus • **356**

Chapter 13 **Hydraulics** 365

13.1 Fundamentals • **366** 13.2 Hydraulic Components • **371**
13.3 Pumps • **372** 13.4 Actuators • **380** 13.5 Hydraulic Motors • **382**
13.6 Conductors and Connectors • **384** 13.7 Hydraulic Fluids • **391**
13.8 Symbols and Schematics • **392** 13.9 Maintenance • **392**

Chapter 14 **Clutches** 400

14.1 Clutch Function • **401** 14.2 Clutch Components • **401**
14.3 Troubleshooting • **413** 14.4 Maintenance • **415**
14.5 Clutch Servicing • **425**

Chapter 15 **Standard Transmissions** 448

15.1 Main and Auxiliary Gearing • **449** 15.2 Geartrain Configurations • **456**
15.3 Mechanical Shift Mechanisms • **456** 15.4 Shift Housing Assembly • **466**
15.5 Twin-Countershaft Transmissions • **469** 15.6 Auxiliary Gear Sections • **470**
15.7 Transfer Cases • **486**

Chapter 16 **Standard Transmission Servicing** 494

16.1 Lubrication • **494** 16.2 Preventive Maintenance Inspections • **497**
16.3 Troubleshooting • **501** 16.4 Transmission Overhaul • **507**
16.5 General Overhaul Guidelines • **511** 16.6 Step-By-Step Overhaul • **511**
16.7 Troubleshooting Air Shift Systems • **527**

Chapter 17 **Torque Converters** 535

17.1 Torque Converter Construction • **536** 17.2 Principles of Operation • **540**
17.3 Maintenance and Service • **546**

Chapter 18 **Automatic Transmissions** 556

18.1 Planetary Gearset Components • **557** 18.2 Compound Planetary Gearsets • **564**
18.3 Transmission Hydraulic Systems • **573** 18.4 Hydraulic Retarders • **581**

Chapter 19 **Automatic Transmission Maintenance** 584

19.1 Inspection and Maintenance • **585** 19.2 Shift Point Adjustments • **594**
19.3 Power Take-Offs (PTOs) • **600**

Chapter 20 **Automated Manual and Hybrid Transmissions** 602

20.1 Transmission Identification • **604** 20.2 Ultrashift Components • **604**
20.3 Transmission ECU • **613** 20.4 Fault Detection and Data Links • **617**
20.5 AMT clutch actuators • **619** 20.6 Hybrid transmissions • **621**

Chapter 21 **Electronically Controlled Automatic Transmissions (ECATs)** 628

21.1 ECAT Management • **630** 21.2 Allison DOC • **633**
21.3 WT Transmissions • **635** 21.4 Allison TC10 • **665**
21.5 Caterpillar CX Truck Transmissions • **683**
21.6 Voith Diwa Transmission • **692**

Chapter 22 **Driveshaft Assemblies** 696

22.1 Driveshaft Construction • **697** 22.2 Driveshaft Inspection • **704**
22.3 Lubrication • **705** 22.4 U-Joint Replacement • **707**
22.5 Chassis Vibration Diagnosis • **713**

Chapter 23 **Heavy-Duty Truck Axles** 722

23.1 Drive Axles • **724** 23.2 Power Dividers • **734**
23.3 Two-Speed Drive Axles • **740** 23.4 Drive Axle Shaft Configurations • **747**
23.5 Nondriving Axles • **747**

Chapter 24 **Heavy-Duty Truck Axle Service and Repair** 752

24.1 Drive Axle Lubrication • **753** 24.2 Truck Axle Cleaning and Inspection • **756**
24.3 Carrier, Axle, and Gearing Identification • **757** 24.4 Failure Analysis • **757**
24.5 Axle Adjustments and Checks • **774**

Chapter 25 **Steering and Alignment** 785

25.1 Steering System Components • **786** 25.2 Front-End Alignment • **791**
25.3 Electronic Alignment Systems • **802** 25.4 Steering Axle Inspection • **805**
25.5 Manual Steering Gears • **816** 25.6 Power-Assist Steering Systems • **822**
25.7 Troubleshooting Power Steering Systems • **826**
25.8 Rack and Pinion Steering • **829**
25.9 Electronic and Load-Sensing Steering • **831**
25.10 Commercial Vehicle Safety Alliance (CVSA) OOS Steering Criteria • **833**

Chapter 26 **Suspension Systems** 837

26.1 Leaf Spring Suspensions • **838** 26.2 Cushion and Torsion Suspensions • **845**
26.3 Air Spring Suspensions • **847** 26.4 Servicing Mechanical Suspensions • **852**
26.5 Servicing Equalizing Beams • **858** 26.6 Servicing Air Suspensions • **862**
26.7 Suspension Alignment • **867** 26.8 Roll Stability Control • **872**
26.9 Cab Air Suspensions and Driver Seats • **872** 26.10 Ride Analysis • **874**

Chapter 27 **Wheels and Tires** 878

27.1 Wheels and Rims • **879** 27.2 Tires • **884** 27.3 Tire Maintenance • **889**
27.4 Tire, Rim, and Wheel Service • **894** 27.5 Wheel Balancing • **908**
27.6 Chassis Tire Pressure Management • **912**
27.7 Wheel Hubs, Bearings, and Seals • **914** 27.8 Wheel Bearing Adjustment • **921**
27.9 Wheel Seal Service • **925** 27.10 Wheel End Conclusion • **928**

Chapter 28 **Truck Brake Systems** 931

28.1 Air Brakes: a Driver's Point of View • **933** 28.2 Air Brake Subsystems • **935**
28.3 The Air Supply Circuit • **937** 28.4 Primary Circuit • **947**
28.5 Secondary Circuit • **950** 28.6 Trailer Circuit • **954**
28.7 Foundation Brakes • **958** 28.8 Air Brake System Components • **964**
28.9 Brake System Balance • **981** 28.10 Air Start Systems • **985**

Chapter 29 **Hydraulic Brakes and Air-over-Hydraulic Brake Systems** 989

29.1 Hydraulic Brake System Components • **990**
29.2 Air-over-Hydraulic Brake Systems • **1002**
29.3 Hydraulic Brake Service Procedures • **1005**
29.4 Hydraulic ABS • 1015 29.5 CVSA OOS • **1018**

Chapter 30 **ABS and EBS** 1023

30.1 Basics of ABS Operation • **1024** 30.2 ABS Components • **1026**
30.3 ABS Configurations • **1033**
30.4 Automatic Traction Control (ATC) Systems • **1035** 30.5 Trailer ABS • **1035**
30.6 ABS Diagnostics • **1037** 30.7 ABS Service Procedure • **1040**
30.8 ABS Add-Ons • **1044** 30.9 Electronic Braking Systems (EBS) • **1047**

Chapter 31 **Air Brake Servicing** 1052

31.1 Maintenance and Safety • **1053** 31.2 Test Equipment • **1056**
31.3 Assessing Brake System Performance • **1057**

31.4 Supply Circuit Service • 1063 31.5 Service Application Circuit • 1069
31.6 Parking/Emergency Circuit • 1075 31.7 Slack Adjusters • 1080
31.8 Trailer Brake System • 1083 31.9 Foundation Brake Service • 1084
31.10 Brake Adjustment • 1091 31.11 General Brake Valve Troubleshooting • 1097
31.12 Service Procedures on Air Disc Brakes • 1099
31.13 Servicing Wedge Brakes • 1102
31.14 Brake Certification, Inspection, and Testing • 1102
31.15 Commercial Vehicle Stopping Distances • 1104

Chapter 32 **Vehicle Chassis Frame** **1107**

32.1 Basic Frame Terms • 1108 32.2 Frame Theory • 1109
32.3 Frame Construction • 1113 32.4 Frame Damage • 1116
32.5 Frame Alignment • 1118
32.6 Repair of Frame, Side Rails, and Cross-Members • 1121

Chapter 33 **Heavy-Duty Truck Trailers** **1131**

33.1 Tractor/Trailer Combinations • 1132 33.2 Freight Van Trailers • 1136
33.3 Refrigerated and Insulated Vans • 1140
33.4 Specialty Trailer Designs • 1144
33.5 Trailer Undersides • 1146 33.6 Trailer Maintenance • 1153
33.7 Mobile Universal Trailer Tester • 1157

Chapter 34 **Fifth Wheels and Coupling Systems** **1160**

34.1 Types of Fifth Wheels • 1161 34.2 Fifth Wheel Height and Location • 1164
34.3 Principles of Fifth Wheel Operation • 1165
34.4 Fifth Wheel Maintenance • 1171 34.5 Coupling • 1174
34.6 Mounting Fifth Wheels • 1179 34.7 Sliding Fifth Wheels • 1180
34.8 Kingpins and Upper Couplers • 1181
34.9 Pintle Hooks/Ball Hitches • 1184
34.10 Coupling Device CVSA OOS • 1186

Chapter 35 **Heavy-Duty Heating, Ventilation, and Air Conditioning Systems** **1190**

35.1 Principles of Refrigeration • 1191 35.2 Refrigerant • 1193
35.3 The Refrigeration Cycle • 1197
35.4 Air Conditioning System Components • 1199
35.5 Safety Precautions • 1212
35.6 Performance Testing an Air Conditioning System • 1213
35.7 Air Conditioning Service Equipment • 1216
35.8 Air Conditioning Service Procedures • 1220
35.9 Common Air Conditioning Problems • 1227
35.10 Converting an R-12 System to an R-134a System • 1229
35.11 Cab Ventilating and Heating Systems • 1230
35.12 Liquid-Cooled Heating System • 1231
35.13 Electronically Managed Climate Control • 1234
35.14 Bluecool Truck Thermal Storage System • 1239

Appendix A Acronyms **1243**
Appendix B Truck Nominal Class Designations **1248**
Appendix C Brake Shoe Identification Chart **1249**
Appendix D Vibration Diagnosis **1250**
Appendix E Hard Steering Troubleshooting Flow Chart **1252**
Appendix F Dual Circuit Brake System Troubleshooting **1253**

Appendix G Driveline Configurations **1257**
Appendix H Measuring Driveline Angles **1258**
Appendix I Vehicle Configurations **1259**
Appendix J Quick Conversion Formula **1260**
Appendix K J1939 Message Decoder **1261**
Appendix L Tire Code Interpretation and Tread Pattern Designations . . . **1262**

Glossary **1263**
Index **1303**

Photo Sequences

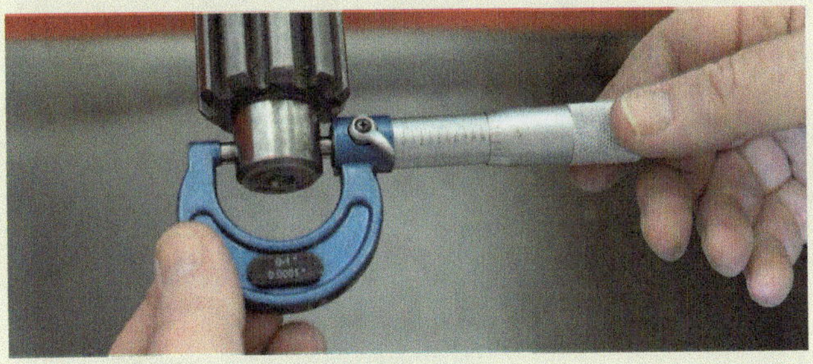

1 Micrometer Familiarization 67

2 Repairing Damaged Threads 83

3 Testing Truck Batteries 220

4 Performing a Rapid Assessment of a Truck Electrical Circuit 291

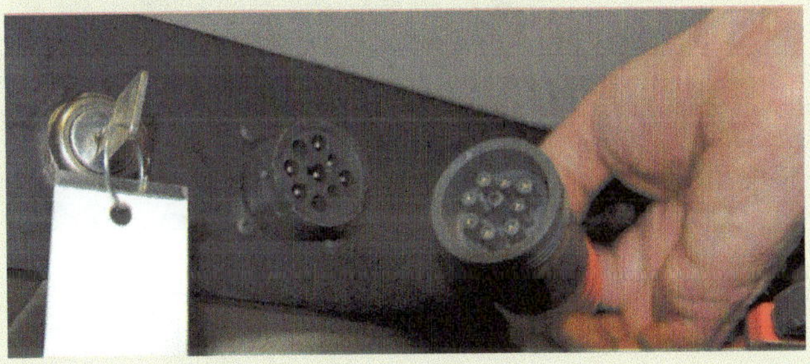

5 Accessing a Truck Data Bus with an EST 360

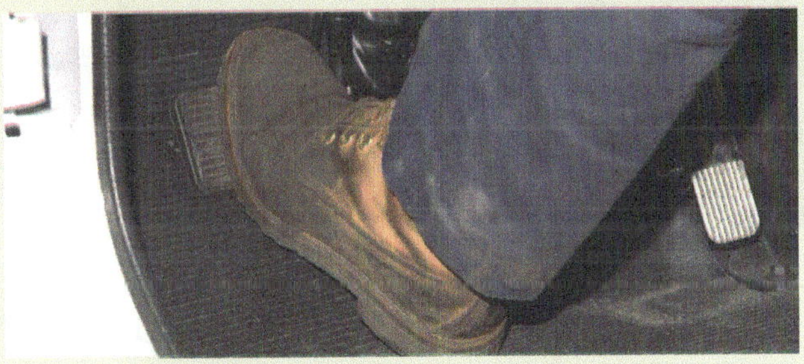

6 Clutch Adjustment 419

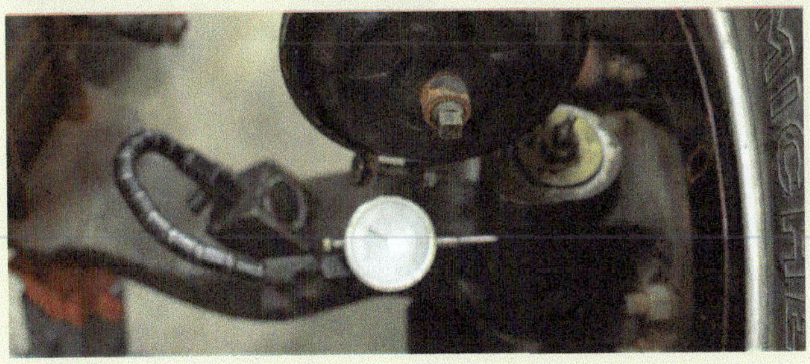

7 Measuring Kingpin Wear and Vertical Play 811

8 Flow and Pressure Test a Hydraulic Assist Steering Gear System 827

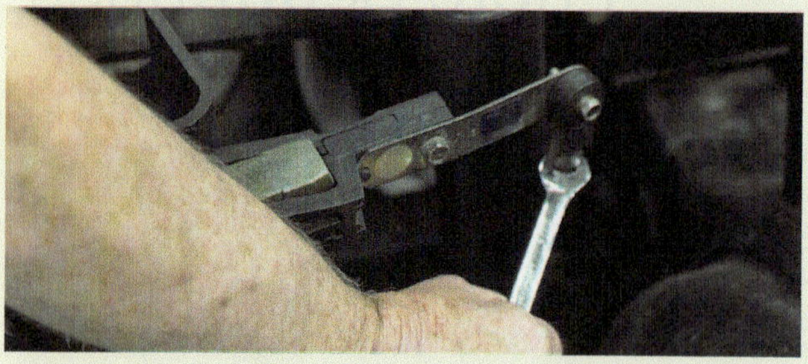

9 Air Suspension Height Control Adjustment....................864

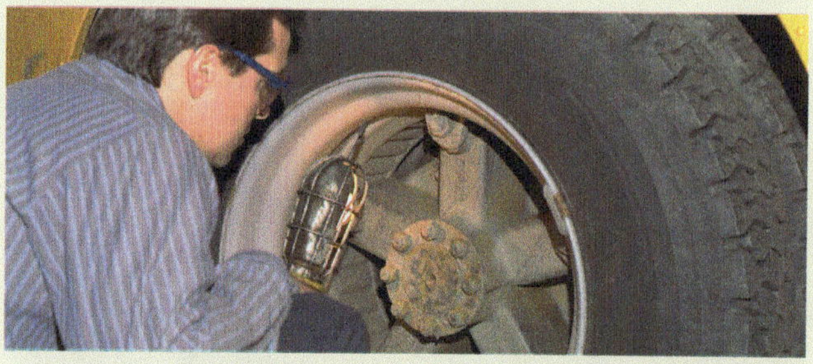

10 Mount a Set of Duals to a Cast Spoke Wheel.................896

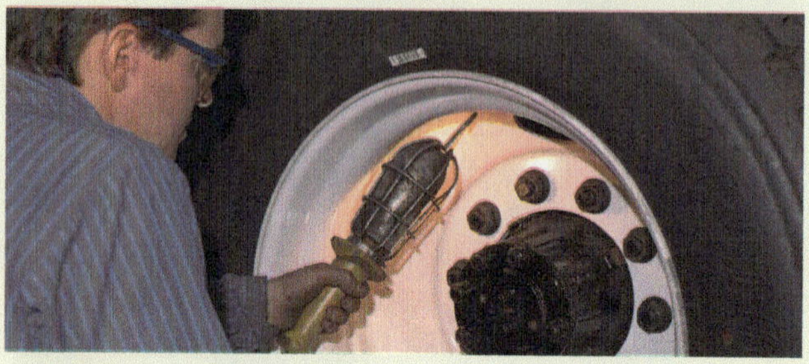

11 Install a Set of Hub-Piloted Duals to a Wheel Assembly..904

12 Wheel End Procedure: TMC Method of Bearing Adjustment 922

13 Wheel Speed Sensor Testing 1043

14 Governor Adjustment 1060

15 Check Freestroke on S-CAM Foundation Brakes........ 1092

16 Wheel-Down Brake Adjustment Procedure on a Tractor/ Trailer.. 1095

17 Wheel-Up Brake Adjustment Procedure on a Tractor/Trailer 1096

18 Measuring a Brake Rotor 1101

19 Tractor to Trailer Coupling Procedure 1176

PREFACE

ABOUT THE SIXTH EDITION

The sixth edition of *Heavy Duty Truck Systems* adheres to the basic structure of the previous two editions, a model that appears to work for most users. However, preserving this structure was not without its challenges and one result is that there are three or four overly lengthy chapters, along with retention of content that might appear to be a generation out of date. In balancing which older technology is retained and which is eliminated in the textbook, I am made constantly aware that there is a difference between the technology operated by our largest fleets and repaired by dealerships, and that which is used in teaching and learning. The reality is that there are a large number of legacy vehicles (more than ten years old) operated daily on our roads: Anyone who doubts this should take a look at the online HD forum of the International Automotive Technicians Network (iATN).

CHANGES IN TECHNOLOGY

As we progress into the electronic age of vehicle technology, multiplexing has advanced beyond networking powertrain components and, in the latest equipment, multiple data bus are often used. Telematics and condition based maintenance (CBM) have now become commonplace and it is only a matter of time before enforcement and safety regulators can wirelessly access vehicle communications systems. Perhaps the biggest question posed to the trucking industry and highway safety regulators today is who has ownership and right-of-access to all that information traded between the modules on vehicle data buses.

Much of the new content in this edition is either directly or indirectly connected with the Administration's drive toward improved fuel economy and reduced CO_2 emission targets scheduled for 2017. Better fuel efficiency and reduced CO_2 dump are closely connected, because they are about reducing the fuel consumed per ton of freight hauled. This has led to the term *freight efficiency*. You can witness the evidence of freight efficiency initiatives on our highways in the form of longer trailers, heavy loads, double-trailer trains, and vehicle aerodynamics.

NEW TO THIS EDITION

Most chapters in the book have been updated, and some have been significantly revised. Some of the new content that has been integrated into this edition includes:

- Full color and updating of the images.
- Fifty-eight short video sequences demonstrating everyday service shop practices are included within the MindTap offering. Look for select videos within the instructor resource material.
- Emerging trends in trucking heading for EPA 2017 freight efficiency and CO_2 reduction requirements.
- Impact of CSA 2010 on truck record keeping and maintenance facilities.
- Aligned with the 2014 NATEF Medium/Heavy Duty Truck Program Accreditation Standards.
- Expanded coverage of hydraulics, fastener identification, and cutting and welding safety.
- Updating of electrical chapters to include Delco Remy 33/34 SI alternator diagnostics and reconditioning, smart ignition switches, internal mag solenoid switches (IMSS), overcrank protection, and highway grid, hotel load stops for trucks and trailers.
- Updating of electronics chapters including J1939 electronics (2014 bus connectors), Wingman radar, collision mitigation systems (CMS), new electronic service tools (ESTs), ultracapacitors, and battery information management systems.

- Updating of content on centrifugal and wet clutches.
- Expansion of content on Automated Manual Transmissions (AMTs).
- Detailed study of Allison's new Class 8 truck TC10 automatic transmission.
- Revision of tire and wheel end chapters including new balancing methods, expanded coverage of air disc brakes and a significantly revised chapter on hydraulic/hydraulic power-boost brakes.
- New generation trailers (freight efficiency), pintle and ball hooks, and an updated HVAC chapter to include hermetic seal and vane compressors.

FEEDBACK FROM EDUCATORS

The number of individuals and companies who help me as collaborators increases with every new edition. These contributors are mostly educators working in truck, heavy equipment and diesel programs in the United States, Canada, Australia, and New Zealand, but I want to acknowledge the role played by students. Students are perhaps more inclined to think a little outside the box defined by curriculum and view technology from the perspective of a learner, so their feedback can be invaluable. Of course, it is communications technology that has made this kind of direct connection to readers possible. Collating and organizing input from a wide range of sources so that it works on the printed page is just as important to an author today, as researching and writing.

While I am indebted to all of my reviewers, some contributors deserve a special mention. In this respect, I would like to single out John Murphy of Centennial College, Bernie Andringa of Skagit Valley College, and James Mack of Berks Career. Special thanks are also due to Steve Belitsos of Vermont Technical College for the use of their facility for video and photography sessions.

Sean Bennett
January 2015
email@seanbennett.org
http://www.seanbennett.org

Features of the Text

Learning how to maintain and repair heavy-duty truck systems can be a daunting endeavor. To guide the readers through this complex material, we have built in a series of features that will ease the teaching and learning processes.

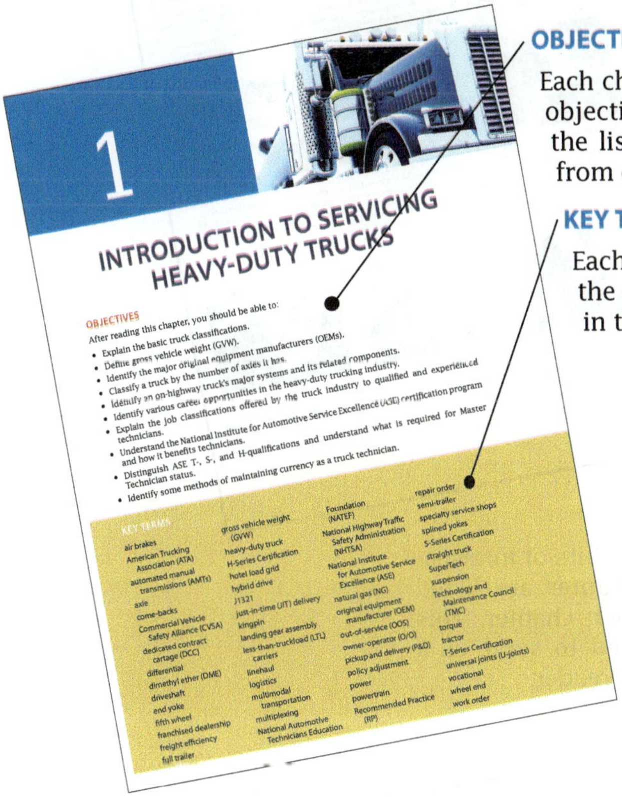

OBJECTIVES

Each chapter begins with the purpose of the chapter, stated in a list of objectives. Both cognitive and performance objectives are included in the lists. The objectives state the expected outcome that will result from completing a thorough study of the contents of the chapters.

KEY TERMS

Each chapter also includes a list of the terms that are introduced in the chapter. These terms are defined in the glossary and highlighted in the text when they are first used.

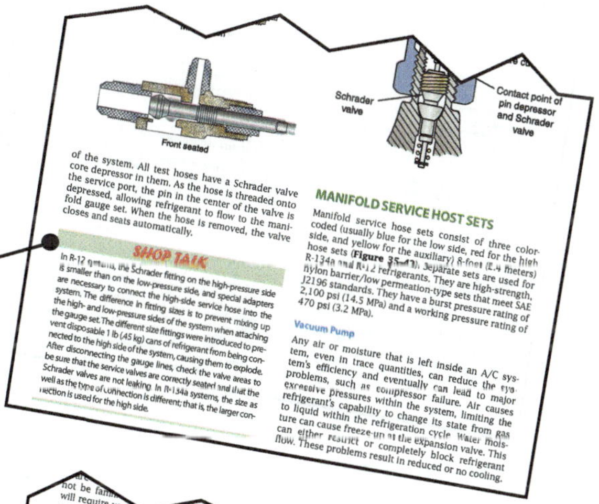

SHOP TALK

These features are sprinkled throughout each chapter to give practical, common-sense advice on service and maintenance procedures.

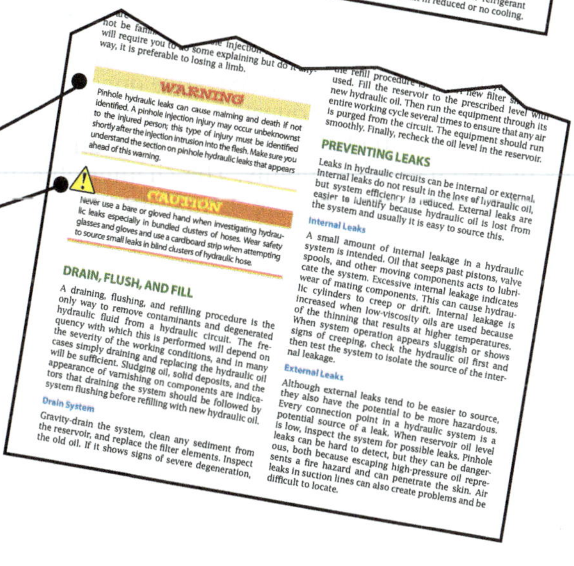

WARNING and CAUTION

Since shop safety is the most important concern among instructors, cautions and warnings appear frequently to alert students of safety concerns.

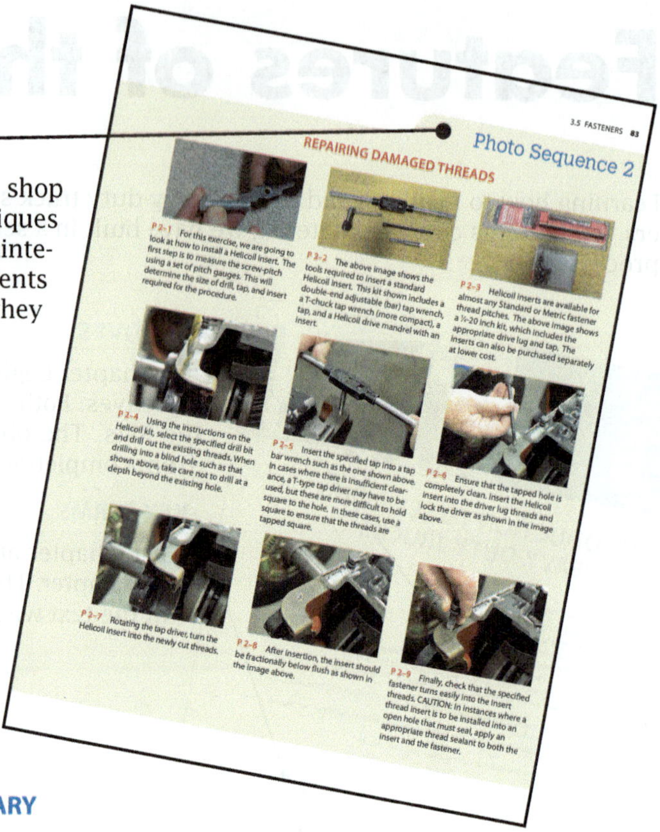

PHOTO SEQUENCE

Step-by-step photo sequences illustrate practical shop techniques. The photo sequences focus on techniques that are common, need-to-know service and maintenance procedures. These photo sequences give students a clean, detailed image of what to look for when they perform these procedures.

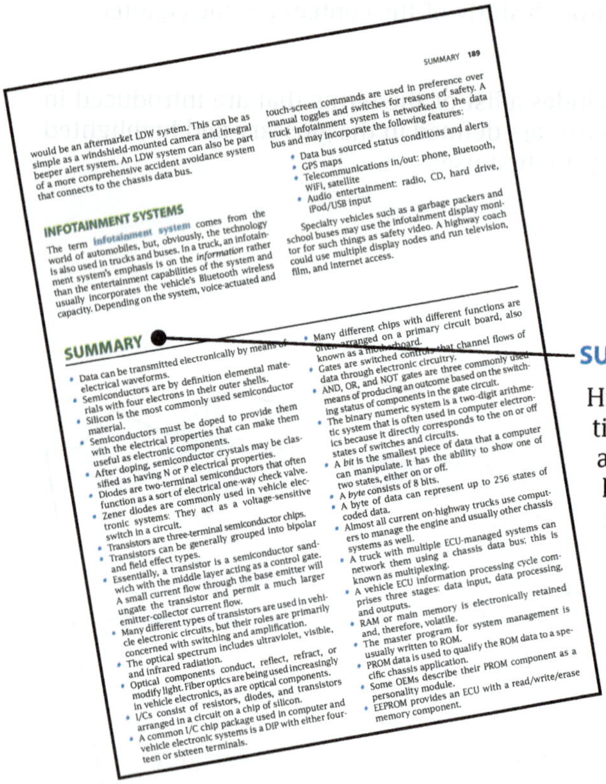

SUMMARY

Highlights and key bits of information from the chapter are listed at the end of each chapter. This listing is designed to serve as a refresher for the reader.

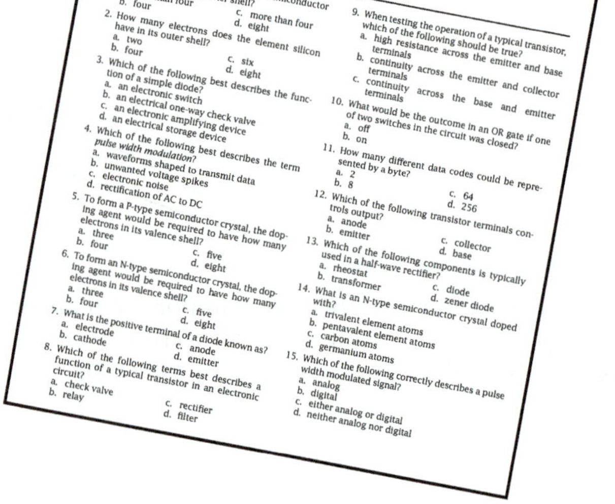

REVIEW QUESTIONS

A combination of short-answer essay, fill-in-the-blank, multiple-choice, and ASE-style questions make up the end-of-chapter questions. Different question types are used to challenge the reader's understanding of the chapter's contents. The chapter objectives are used as the basis for the review questions.

Supplements

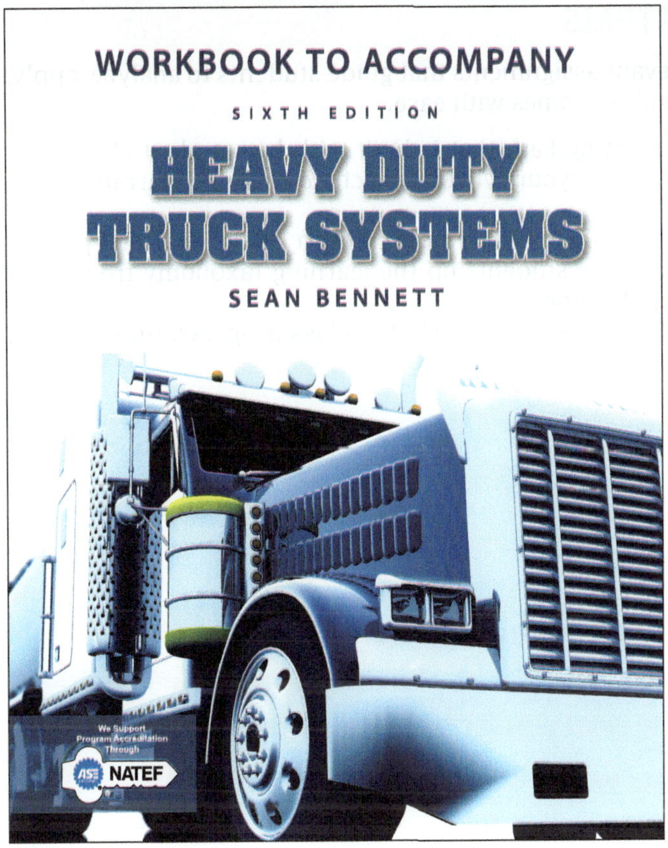

WORKBOOK

The Student Workbook reinforces the foundations provided by the textbook. In addition to special emphasis on the use of precision measuring tools, the Workbook includes study tips, practice questions, and online tasks to address the increasing importance of networking skills to truck technicians. Finally, in keeping with the objective reinforcing performance skills, the job sheets in the Workbook make the connection between the theoretical concepts in the textbook and the hands-on application of that knowledge, bridging theory with practice. Each job sheet has been correlated with all relevant 2014 NATEF tasks.

INSTRUCTOR RESOURCES CD

Carefully prepared, the Instructor Resources CD brings together several time-saving tools that allow for effective, efficient instruction. The Instructor Resources CD contains the following components:

- **PowerPoint®** lecture slides, which present the highlights of each chapter.
- An **Image Gallery**, which offers a database of hundreds of images in the text. These can easily be imported into the PowerPoint® presentations.
- An **Answer Key** file, which provides the answers to all end-of-chapter questions and the Practice Questions in the Student Workbook.
- **NATEF Correlations** in which the current NATEF Medium/Heavy Truck Standards are correlated to the chapter of the core text and all relevant Workbook Job Sheets.
- **End-of-Chapter Review Questions**, which are provided in MS Word format.

INSTRUCTOR COMPANION WEBSITE

The Instructor Companion Website, found on cengagebrain.com, includes the following components to help minimize instructor preparation time and engage students:

- **PowerPoint®** lecture slides, which present the highlights of each chapter.
- An **Image Gallery**, which offers a database of hundreds of images in the text. These can easily be imported into the PowerPoint® presentations.
- An **Answer Key** file, which provides the answers to all end-of-chapter questions and the Practice Questions in the Student Workbook.
- **NATEF Correlations** in which the current NATEF Medium/Heavy Truck Standards are correlated to the chapter of the core text and all relevant Workbook Job Sheets.
- **End-of-Chapter Review Questions**, which are provided in MS Word format.

Cengage Learning Testing Powered by Cognero is a flexible, online system that allows you to:

- Author, edit, and manage test bank content from multiple Cengage Learning solutions.
- Create multiple test versions in an instant.
- Deliver tests from your LMS, your classroom, or wherever you want.

MINDTAP FOR HEAVY DUTY TRUCK SYSTEMS

MindTap is a personalized teaching experience with relevant assignments that guide students to analyze, apply, and improve thinking, allowing you to measure skills and outcomes with ease.

- *Personalized Teaching*: Becomes YOURS with a Learning Path that is built with key student objectives. Control what students see and when they see it—match your syllabus exactly by hiding, rearranging, or adding your own content.
- *Guide Students*: Goes beyond the traditional "lift and shift" model by creating a unique learning path of relevant readings, multimedia and activities that move students up the learning taxonomy from basic knowledge and comprehension to analysis and application.
- *Measure Skills and Outcomes*: Analytics and reports provide a snapshot of class progress, time on task, engagement and completion rates.

Acknowledgments

REVIEWERS

We would like to acknowledge and thank the following educators for their insightful suggestions and comments on way to improve the text for the sixth edition:

David Conant
Lincoln College of Technology
Nashville, TN

Casey Eglinton
Western Technical College
LaCrosse, WI

Kevin Heimbach
Berks Career and Technology Center East Campus
Leesport, PA

David Lund
Memphis, TN

James Mack
Berks Career and Technology Center East Campus
Leesport, PA

Martin Wilmer
Universal Technical Institute
Exton, PA

INDIVIDUALS

Tim Allan, Centennial College
Doug Anderson
Bernie Andringa, Skagit Valley College
Dave Archibald, Navistar International
Kathi Barringer, Allison Transmission
Stephen Belitsos, Vermont Technical College
Joe Bell, John Deere, Iowa
John Bissonnette, Centennial College
Douglas Bradley, Utah Valley State College
Lawrence Brett, Tallman International Trucks
Walter Brueggeman, Tidewater Community College
Darrin Bruneau, Automotive Resources International
Mike Cerato, Centennial College
Dennis Chapin, Roque Community College
Alan B. Clark, Lane Community College
Dave Coffey, Toromont Caterpillar
David Conant, NADC, Lincoln South
Pedro Correiea, Harper DDC-Allison, Toronto
Cameron Cox, SAIT
Jeff Curtis, Bellingham Technical College
John Dixon, Centennial College
Bob Drabecki, Navistar International
Ken DeGrant, Dearborn Group
Dave Drummond, Mack Trucks Inc. (Ret)
Owen Duffy, Centennial College
Dave Embury, Eaton Dana Corporation
Craig Fedder, Navistar International
Lou Gilbert, Allison Transmission
Len Gonzalez, MGM Corporation
Terry Harkness, Toromont Caterpillar
Scott Heard, Fleming College
Jason Hedges, Buckham Transport
Michael Henich, Linn-Benton Community College
Sergio Hernandez, Palomar College
Rick Higinbotham, Ohio Auto-Diesel Technical College
Ted Hrdlicka, Denver Automotive and Diesel College
Kyle Hutchinson, Toromont Caterpillar
Robert Huzij, Cambrian College
Winston Ingraham, University College of Cape Breton
John Kay, Allison Transmission
Kenneth W. Kephart, Central Texas College
Mark Koslan, Texas State Technical College
John Kramer, Centennial College
Bobby Leatherman, NADC, Lincoln South
Roger LeBlanc, Counterbalance Beads
Wayne Lehnert, Universal Technical Institute
Pat Leitner, Freightliner Corporation
Sam Lightowler, Toronto Transit Commission
George Liidemann, Centennial College
Terryl Lindsey, Oklahoma State University, Okmulgee
Rolf Lockwood, *Today's Trucking* Magazine
Alan McClelland,Centennial College
Dale McPherson, East Idaho Technical College
John Montgomery, Volvo-Mack Trucks
Dave Morgan, Centennial College
John Murphy, Freightliner Training
Steve Musser, Wyoming Technical Institute
Gary Nederynan, Orion Bus
John Overing, Michelin Canada
Jim Park, *Today's Trucking*
George Parsons, Sault College
Fred Pedler, Haldex Midland
Roger Perry, Harper, Toronto
Allan Pritchard, Toronto Transit Commission
T. Grant Ralston, Fairview College
Dexter Rammage, Santa Rosa Junior College
Leland Redding, Texas State Technical Institute
Robin Reed, Deutsch Corporation
Robert D. Ressler, Lebannon Co., CTC
Martin Restoule, Algonquin College
Ken Riley, Toromont Caterpillar
Ovi Robotin, Fluke Corporation
Bob Rutherford, Auburn University
Sara Saplin, Wards Forest Media, NY

Ron Scoville, Sargeant Reynolds CC
Wayne Scott, Loblaws Logistics
Martin Sissons, Centennial College
Darren Smith, Centennial College
Angelo Spano, Centennial College
John Stone, Washington State Community College
Brian Strach, Hendrickson Corporation
Dan Sullivan, Sullivan Training Solutions
Gino Tamburro, Centennial College
Robert Taylor, Voith Incorporated
Michael Thomas, Arizona Automotive Institute
Alan Thompson, Centennial College
Michael Tunnel, American Trucking Associations
Pierre Valley, Centennial College
Robert Van Dyke, Denver Automotive and Diesel College
Claude Williams, Toromont Caterpillar
Gus Wright, Centennial College
Gilles Ybarro, Université Laval

ORGANIZATIONS

The Truck Faculty Team, Lincoln College of Technology, Nashville
The Truck Faculty Team University of Northwestern Ohio
The Automotive Faculty Team, WelTech, Wellington, New Zealand
The Heavy Vehicle Instructor Team, Hunter TAFE, Kurri Kurri, NSW, Australia
North American Conference of Automotive Teachers (NACAT)
Technology and Maintenance Council of the ATA
Vermont Technical College
Centennial College, Toronto

CONTRIBUTING COMPANIES

We would like to thank the following companies that provided technical information and art for this edition.

APW Engineered Solutions (Power-Packer)
Accuride Corporation
Aidco International, Inc.
Alcoa Wheel Products
Allison Transmission
American Trailer Industries, Inc.
Arvin Meritor
ASE
Battery Council International
Bee Line Co., Bettendorf, IA
Bendix Commercial Vehicle Systems LLC
Bosch GmbH
Bostrom Seating, Inc.
Bridgestone Americas Tire Operations
Bridgestone / Firestone Inc.
Buckham Transport, Millbrook
The Budd Company
Carrier Refrigeration Operations
Caterpillar Inc.
Central Tools, Inc.
Chalmers Suspensions International Inc.
Chicago Rawhide
Consolidated Metco, Inc.
Counterbalance Beads Corporation
Daimler Trucks North America
Dake
Dalloz Safety
Dana Corporation
Dearborn Group
Deere & Company
Detroit Diesel Corporation
Dorsey Trailers, Inc.
DuPont Automotive Finishes
Eaton Corporation
Espar Heater Systems
Firestone Industrial Products Co.
Fontaine International
Goodson Tools & Supplies
Gray USA Corp.
Haldex Brake Products Corporation
Harper Freightliner and Detroit Diesel
Heavy Duty Trucking
Hendrickson International
Hennessy Industries, Inc.
Hunter Engineering Company, Inc.
Industrial Seats
Innovative Products of America
Intel Corporation
International Business Machines
International Truck and Engine Corporation
J.J. Keller & Associates, Inc.
JOST International
Kenworth Trucks of Toronto
Kim Hotstart Mfg. Co.
Knorr-Bremse Group
Lubriquip, Inc.
Meritor WABCO
Michelin Tires (Le groupe Michelin)
Midtronics, Inc.
MPSI (Micro Processor Systems, Inc.)
National Seating
NEXIQ Technologies
Paccar Trucks
Penske Truck Leasing Corp.

Prestolite Electric Inc.
Protectoseal Company
Remy International, Inc.
Robinair, SPX Corporation
RTI Technologies, Inc.
SAF-HOLLAND, Inc.
SKF USA Inc.
Snap-On Tools Company
Stanley Tools, New Britain, CT
Tallman Navistar Truck Center, Oshawa
Toromont Caterpillar Industries
Trailmobile Trailer LLC
Truck Trailer Manufacturers Association
TRW Automotive
Utah Valley State College
Voith GmbH
Vulcan Materials Company
Wagner Brake, Division of Federal-Mogul Corp.
Wards Forest Media
Webasto Product North America, Inc.
White Industries, Division of K-Whit Tools, Inc.
Winslow Gerolamy International Trucks

1

INTRODUCTION TO SERVICING HEAVY-DUTY TRUCKS

OBJECTIVES

After reading this chapter, you should be able to:

- Explain the basic truck classifications.
- Define gross vehicle weight (GVW).
- Identify the major original equipment manufacturers (OEMs).
- Classify a truck by the number of axles it has.
- Identify an on-highway truck's major systems and its related components.
- Identify various career opportunities in the heavy-duty trucking industry.
- Explain the job classifications offered by the truck industry to qualified and experienced technicians.
- Understand the National Institute for Automotive Service Excellence (ASE) certification program and how it benefits technicians.
- Distinguish ASE T-, S-, and H-qualifications and understand what is required for Master Technician status.
- Identify some methods of maintaining currency as a truck technician.

KEY TERMS

air brakes
American Trucking Association (ATA)
automated manual transmissions (AMTs)
axle
come-backs
Commercial Vehicle Safety Alliance (CVSA)
dedicated contract cartage (DCC)
differential
dimethyl ether (DME)
driveshaft
end yoke
fifth wheel
franchised dealership
freight efficiency
full trailer
gross vehicle weight (GVW)
heavy-duty truck
H-Series Certification
hotel load grid
hybrid drive
J1321
just-in-time (JIT) delivery
kingpin
landing gear assembly
less-than-truckload (LTL) carriers
linehaul
logistics
multimodal transportation
multiplexing
National Automotive Technicians Education Foundation (NATEF)
National Highway Traffic Safety Administration (NHTSA)
National Institute for Automotive Service Excellence (ASE)
natural gas (NG)
original equipment manufacturer (OEM)
out-of-service (OOS)
owner-operator (O/O)
pickup and delivery (P&D)
policy adjustment
power
powertrain
Recommended Practice (RP)
repair order
semi-trailer
specialty service shops
splined yokes
S-Series Certification
straight truck
SuperTech
suspension
Technology and Maintenance Council (TMC)
torque
tractor
T-Series Certification
universal joints (U-joints)
vocational
wheel end
work order

INTRODUCTION

According to the U.S. Department of Transportation (DOT) the total number of registered highway vehicles exceeded 253 million vehicles in 2012. More than 15 percent of this total is made up of trucks and trailers used primarily for commercial purposes. That represents close to 40 million vehicles. In 2014, total registrations for Class 8 trucks alone exceeded 3.66 million units. For most of the past three decades, through periods of boom and recession, a shortage of truck technicians has existed throughout the continent. This shortage of technicians is worsening as the median age of those employed in the trucking industry increases. For a number of years, the trucking industry has retired more personnel than it recruits and in some areas of the country there are significant shortages. Although this may be bad news for freight transportation managers, it is good news for anyone wanting to get into the industry especially for those willing to move to those areas in which there are jobs available. Job opportunities, rates of pay, and potential for advancement in the trucking industry have never been better. Although the modern highway truck requires much less frequent service work to keep it in top mechanical condition than the truck of a generation ago, most of that service work performed by a truck technician today requires a higher level of skills.

Good truck technicians are in high demand. Today a good truck technician is required not just to diagnose and repair trucks but also to be computer literate, to regularly update technical knowledge, and to practice customer service skills. Perhaps more than any other skill set, the skill most required of the modern truck technician is that of being a lifelong learner to keep abreast of the fast-changing technology of this industry. One of the objectives of this chapter is to outline some of the strategies both student and certified technicians can use to maintain technical currency.

1.1 TRUCK CLASSIFICATIONS

For purposes of registering commercial vehicles for highway use, trucks are classified by their **gross vehicle weight (GVW)**. GVW is the maximum allowable weight of the vehicle plus the weight of the load it can safely carry. There are three classes of "light-duty" trucks, three classes of "medium-duty" trucks, and two classes of "heavy-duty" trucks (**Table 1–1**). A **heavy-duty truck** has a GVW of 26,001 pounds (11,794 kg) or more.

ORIGINAL EQUIPMENT MANUFACTURERS

There are four major truck manufacturers of medium- and heavy-duty trucks in North America. We refer to these manufacturers by the term **original equipment manufacturer (OEM)**. In addition, the four major OEMs are being joined by import OEMs such as Toyota/Hino. None of the import OEMs as yet offers the full range of weight class classifications but this is expected to change. **Table 1–2** lists the major truck manufacturers of Class 8 heavy-duty trucks and their respective market share.

AXLE CLASSIFICATIONS

The term **tractor** is used to describe a highway truck that is designed to haul a trailer. The term **straight truck** is often used to describe a highway truck not designed to pull a trailer. Trucks are also classified by the number of axles, the number of wheels, and the number of drive wheels. For example, a tractor with a tandem (close-coupled pairs) rear axle will be either a 6 × 2 or a 6 × 4. The first number refers to the total number of wheels (or sets of wheels in the case of dual wheels). The second number indicates the number of wheels that are driven by the vehicle **powertrain**. The powertrain consists of the engine,

TABLE 1–1 Truck Weight Classifications

General Designation	Weight Category	Standard GVW (pounds)	Metric GVW (kilograms)
Light-duty	Class 1 Class 2 Class 3	up to 6,000 GVW 6,001–10,000 GVW 10,001–14,000 GVW	up to 2,722 GVW 2,722–4,535 GVW 4,535–6,350 GVW
Medium-duty	Class 4 Class 5 Class 6	14,001–16,000 GVW 16,001–19,500 GVW 19,501–26,000 GVW	6,350–7,257 GVW 7,257–8,845 GVW 8,845–11,793 GVW
Heavy-duty	Class 7 Class 8	26,001–33,000 GVW 33,001 GVW and over	11,794–14,970 GVW 14,970 GVW and over

TABLE 1–2 Market Share by OEM: Class 8 Trucks

North American Market Sales Percentage for November 2012 to October 2013 Reported by Transport Topics Magazine.			
OEM Corporation	% Total	Brand	% Total
Daimler AG	35.2%	Freightliner Western Star	32.5% 1.7%
DAF-Paccar	28.6%	Peterbilt Kenworth	11.8% 16.7%
International Trucks	15.5%	Navistar	15.5%
Volvo-Mack	20.6%	Mack Trucks Volvo Trucks	9.5% 11.1%

clutch, transmission, drive shafts, differentials, and drive wheels).

A tractor with tandem rear axles (total of 3 axles) where only one axle is driven by the powertrain would be classified as a 6 × 2; it has six wheels, but only two wheels (one axle) drive the vehicle. This classification method is no different from that used to describe smaller vehicles. Most of us are familiar with the term "4 × 4," which is the common means of describing four-wheel drive in an automobile or pickup truck. On the highways the most common heavy truck axle classification is the 6 × 4 Class 8 truck, as shown in the example in **Figure 1–1**. A 6 × 4 vehicle consists of three axles: a forward located steering axle and a pair of rear drive axles. The truck shown in Figure 1–1 is a highway tractor while **Figure 1–2** shows some types of vocational trucks. A tractor is designed to haul trailers. **Table 1–3** lists the common axle–wheel configurations and their driven wheels and axles.

FIGURE 1–1 A 2013 Class 8 tandem drive tractor

6 × 4 versus 6 × 2 versus 4 × 2

Up to the present day the most common highway tractor is the 6 × 4 configuration, but a shift toward

FIGURE 1–2 Examples of vocational trucks

TABLE 1–3 Truck Classification by Wheel Number

Motor Vehicle	Total Wheels	Driven Wheels	Total Axles	Drive Axles
4 × 2	4	2	2	1
4 × 4	4	4	2	2
6 × 2	6	2	3	1
6 × 4	6	4	3	2
6 × 6	6	6	3	3
8 × 4	8	4	4	2
8 × 6	8	6	4	3

improved fuel economy is beginning to change this. Outside of North America there has been wider acceptance of 4 × 2 tractor units and this is related to lower cargo payload limits and better fuel economy. Comparative fuel economy usually references the SAE **J1321** test, which attempts to ensure that strict rules are used to evaluate fuel consumption tests in commercial trucks. Comparative advantages and disadvantages of each configuration based on J1321 testing and simplified are listed in **Table 1–4**.

Trailers

There are many trailer designs, sizes, and applications. A majority of the trailers we see on our highways are **semi-trailers** such as that shown in **Figure 1–3**. A semi-trailer is one that depends on the tractor to support at least some of its weight. The semi combination shown in Figure 1–3 is the most common truck configuration seen on our roads. In this figure, a dual axle semi-trailer is being hauled by a 6 × 4 tractor in an arrangement we know as an "18-wheeler." The less common **full trailer** fully supports its load; in other words, a full trailer does not rest a portion of its weight on the tractor (or trailer in a train combination) hauling it.

TABLE 1–4 Comparative Advantages of Highway Tractor Configurations

	6 × 4	6 × 2	4 × 4
Fuel efficiency	Lowest	Medium	Best
Traction	Best	Medium	Lowest
Maneuverability	Least	Least	Best
Payload	Highest	Medium	Lowest

FIGURE 1–3 Tractor-trailer combination: a dual axle semi-trailer coupled to a 6 × 4 tractor. This is the most common rig observed on our highways.

Freight Efficiency

The most common type of trailer is the semi van trailer. For some time, 48-foot trailers were dominant on our highways. However, because of the recent acceptance of 53-foot trailers by most jurisdictions, companies have rapidly renewed their trailer fleets to take advantage of the increased cargo volume and potential to reduce fuel consumption per ton hauled. The term **freight efficiency** is being used to justify heavier loads, longer trailers, and multiple trailer combinations. Most of the trucking industry favors replacing the term *fuel economy* by *freight efficiency*, because it is more accurate. Freight efficiency factors the volume and weight of cargo hauled to fuel consumed, rather than the more vague miles-per-gallon equation of vehicle fuel economy. Also due to its potential to reduce costs, multimodal transportation systems are rapidly increasing in popularity.

Multimodal Containers

Multimodal transportation refers to containers that can be moved by truck, railway, ship, and even aircraft while keeping load and unload costs to a minimum. Containers are ideal for stacking on ships and railway flatbeds but a primary design consideration is adaptability to haulage by transport truck. Following the acceptance by most jurisdictions of 53-foot highway trailers, there has been a shift toward 53-foot multimodal containers. There is a wide range of specialty trailers, including dry bulk carriers, tank trailers, and refrigerated trailers (reefers), as shown in **Figure 1–4**. We take a closer look at the many different types of trailers in Chapter 33.

Rig Dimensions

Tractor and semi-trailer dimensions are provided in **Figure 1–5**. An important consideration in determining tractor/semi-trailer dimensions is the distance between the two vehicles when coupled. There must be sufficient clearance between the tractor cab and

FIGURE 1–4 A triaxle reefer semi coupled to a tractor: weight-over-axle limitations by some jurisdictions requires multiple trailer axles.

semi-trailer front (nose) as well as between the rear of the tractor and semi-trailer **landing gear assembly** to allow for sharp turns and the effect of grade changes.

1.2 HEAVY-DUTY TRUCKS

Heavy-duty truck technicians need to understand the systems and components that power or move, slow and stop, control, direct, support, and stabilize a tractor/trailer combination. **Figure 1–6** shows some of the components discussed in this book. The following major systems are found in on-highway trucks. More information on the classification of trailers can be found in Chapter 33.

ENGINES

Current commercial heavy-duty trucks are powered almost exclusively by diesel engines. In fact, most commercial medium- and light-duty trucks also use diesel power today. However, in recent years, the abundance of less costly, domestically sourced **natural gas (NG)** is making some fleets consider NG-fueled engines: the downside is a less developed fueling and repair infrastructure. For this reason, it has been adopted as an ideal fuel for city transit applications, especially on the West Coast. Volvo recently announced that it will make **dimethyl ether (DME)** fueled trucks available in 2015. Like NG, DME is

FIGURE 1–5 Truck/trailer dimensions and terms

Dimensions

A	Distance from centerline of rear axle to centerline of body and/or payload. Centerline of body (as 1/2 body length)
AF	Center of rear axle to end of frame
BA	Bumper to centerline of front axle
BBC	Bumper to back of cab
BL	Body length
CA	Back of cab to centerline of rear axle or tandem suspension
CE	Back of cab to end of frame
CFW	Back of cab to centerpoint of kingpin hole in 5th wheel
CT	Back of cab to front of semi-trailer in straight-ahead relationship
FH	Frame height
FW	Centerline of rear axle or tandem to centerpoint of 5th wheel
KP	Kingpin setting—front of semi-trailer to centerpoint of kingpin on semi-trailer
LGC	Landing gear clearance—center point to nearest interference point
OAL	Overall length
OWB	Overall wheel base
TL	Semi-trailer length
WB	Wheel base—distance between centerline of front and rear axle or tandem suspension

Terms

Chassis: Basic vehicle-cab, frame, and running gear
Body: Container in which the load is carried
Payload: Commodity to be carried
Curb weight: Finished vehicle weight, excluding cargo or occupants
Body weight: Weight of complete body to be installed on chassis
Payload weight: Weight of commodity to be carried
Gross vehicle weight (GVW): Total or curb, body, and payload weight

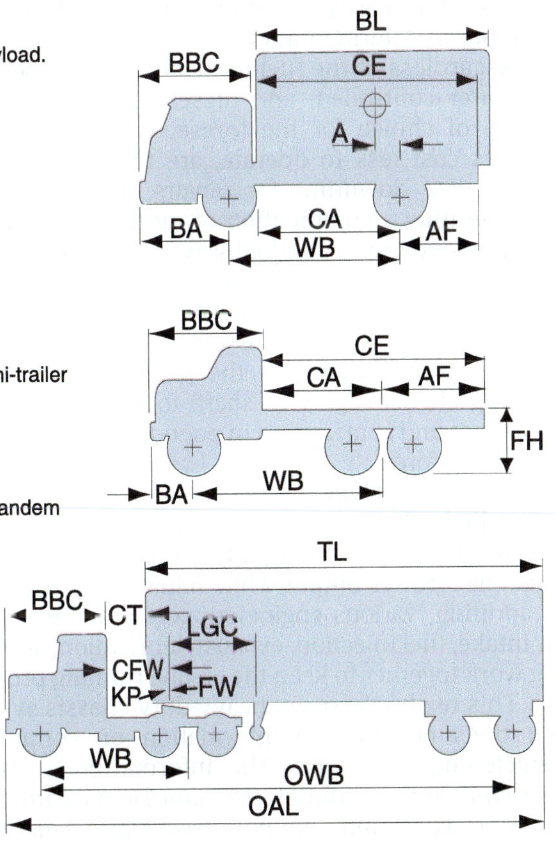

FIGURE 1–6 Some of the components found on a Class 8 heavy-duty truck.

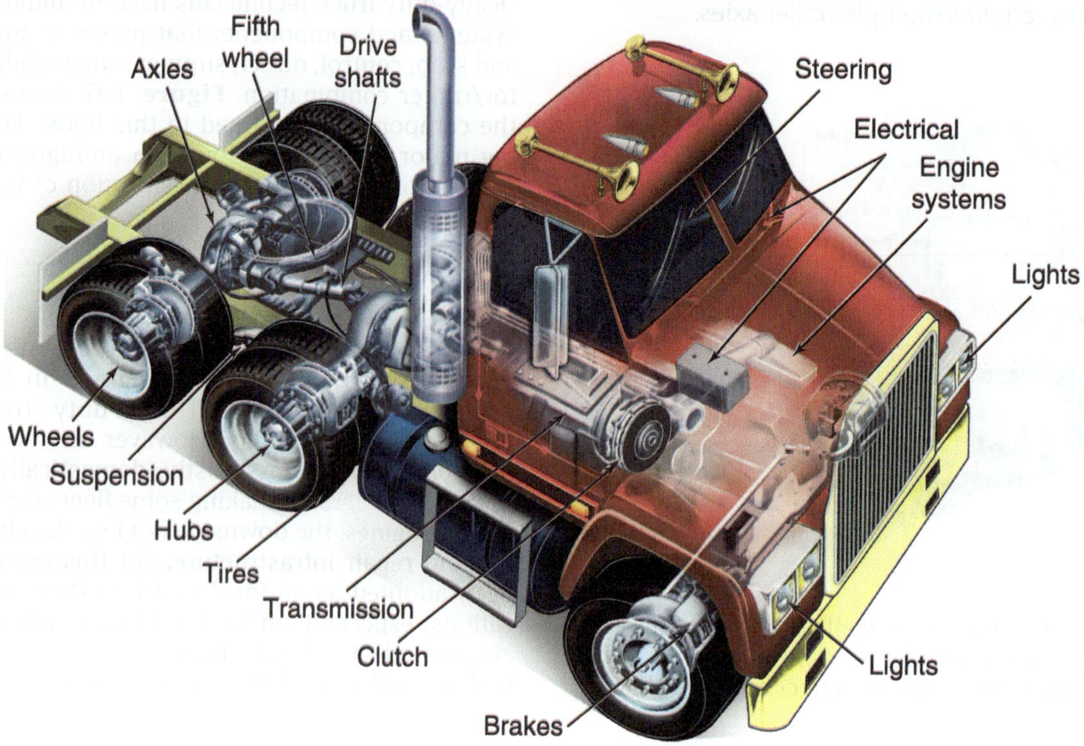

composed primarily of methane, but it can be sourced from biomass fermentation.

Regardless of the fuel used, all engines today are computer controlled, and diesel power will be the engine of choice for the foreseeable future. Diesel engines cost less to operate, are more dependable, require less downtime for repairs, and are capable of generating a combination of more **power** (140 to 600 horsepower for on-highway trucks) and **torque** (180 to 2,000 lb-ft.) than equivalently sized NG or gasoline engines. *Power* specifies the engine's ability to move, or propel, the vehicle and is measured in horsepower. *Torque* is the potential ability of a rotating element (gear or shaft) to overcome turning resistance and is measured in pounds per linear foot. It is twisting force.

Current highway diesel engines are electronically controlled—that is, computer controlled—such as the example shown in **Figure 1–7**. Electronic controls maximize power output while minimizing emissions. In addition, various engine support systems such as air intake, fuel injection, exhaust, lubrication, and cooling work together to keep the engine running properly.

This textbook covers heavy-duty chassis systems but does not include information on commercial vehicle engine systems. This information is covered in a companion textbook (*Medium/Heavy Duty Truck Engines, Fuel, and Computerized Management Systems, Fourth Edition*).

Hybrid Drive

Hybrid drive technologies have been introduced recently. Hybrid drive systems make a lot of sense in specialized vocational trucks used in inner city, stop-start applications. They also have been embraced by urban transit authorities interested in reducing emissions and lowering fuel costs. Hybrid drive systems used today can be divided as follows:

- Diesel electric series drive: consists of a small diesel engine powering a generator/motor and battery bank. Vehicle can only move under electric motive power.
- Diesel electric parallel drive: consists of a small diesel engine powering a generator/motor/battery bank and mechanical drivetrain. Vehicle can option electric or engine drive. It is the most common technology used in hybrid urban transit buses.
- Diesel hydraulic series drive: consists of a small diesel engine powering a hydraulic motor/pump and accumulator. Vehicle can only move under hydraulic motive power.
- Diesel hydraulic parallel drive: consists of a small diesel engine powering a hydraulic motor/pump and mechanical drivetrain. Vehicle can option hydraulic or engine drive. It is the most common technology used in hybrid urban pickup and delivery vehicles.

FIGURE 1–7 Current heavy-duty, Class 8 trucks are powered by computer-controlled diesel engines such as this Detroit Diesel DD15.

ELECTRICAL SYSTEMS

The batteries, alternator, and starter must be sized to match the operating requirements of the engine and chassis electrical systems. Coverage of truck electrical and electronic systems appears in Chapters 5 through 11 in this textbook. Most heavy-duty trucks have between two and four batteries to supply current for the starter motor. Some older trucks have 24-volt starters that require a series-parallel switch in the system. In addition, the electrical systems of heavy-duty trucks provide electricity to operate such safety components as lighting, windshield wiper motors, and gauges plus the operation of the computers, accessories, and tractor cab amenities.

CLUTCHES

Most current heavy-duty trucks use either a push- or pull-type clutch, but more recently, centrifugal and wet clutches have been integrated with some types of transmission. Trucks equipped with high-torque engines and designed to haul heavy payloads use a two-plate clutch. Two-plate clutches double the friction contact area of the clutch compared with a single-plate clutch. The additional friction contact surface area is necessary to transmit high torque to the transmission without slippage. Lighter-duty trucks may use a single-plate clutch assembly. Clutches used with automated manual transmissions are managed electronically but most use the same fundamental principles as mechanical clutches.

TORQUE CONVERTERS

Torque converters are used to transmit engine torque to fully automatic transmissions. Automatic transmissions are not as widely used in commercial trucks as they would be in automotive applications but they are more common in transit bus and vocational truck applications. Torque converters are fluid couplings that use hydrodynamics to multiply the input to output torque ratio.

TRANSMISSIONS

Heavy-duty truck transmissions can be classified as:

- Conventional
- Semiautomated
- Automated
- Fully automatic

Transmission gears, shafts, bearings, forks, and other internal components must perform for thousands of hours, hundreds of thousands of miles, year after year. It is not unusual for a truck transmission to run for a half million or even a million miles with little maintenance other than checking lubrication levels and observing drain intervals.

A majority of heavy-duty trucks are equipped with standard (manual) transmissions (see Chapter 15) but many of these are computer-controlled, automated units known as **automated manual transmissions (AMTs)**. In addition, the market share of electronically controlled fully automatic transmissions continues to increase. Depending on engine output, the intended application of the vehicle, and the terrain over which it must operate, a commercial truck transmission might have from six to twenty forward gears.

Conventional Transmissions

Conventional transmissions have two or three countershafts that transmit engine torque from the input shaft to the output shaft. This divides the torque two or three ways so that there is less stress on individual gears, extending the service life of the transmission. Conventional transmissions are ratio shifted directly by the driver, sometimes assisted by pneumatic (air) pressure. For this reason they are sometimes known as manual transmissions.

Automated Manual Transmissions

Automated manual transmissions (AMTs) adapt a conventional standard transmission platform to electronic controls. The guts of the conventional transmission remain unchanged. However, the unit is adapted so that the responsibility of selecting and shifting ratios is electronically controlled. Automated transmissions are categorized as:

- Semiautomated: also known as three-pedal automated transmissions because the driver is required to use the clutch under some circumstances. These were the first generation of AMTs.
- Automated: commonly known as two-pedal automated transmissions because the clutch is actuated automatically, eliminating the clutch pedal.

Automatic Transmissions

Most automatic transmissions use planetary gearsets to transmit drive torque through different ratios. They are coupled to the engine by a type of fluid coupling known as a torque converter. Fully automatic transmissions are initially high-cost components, but in vocational applications, which require constant shifting, they are often cost effective over time. In 2014 to address the gearing requirements of Class 8 tractors, Allison introduced an innovative power-shift automatic transmission that uses a combination of conventional and planetary gearing.

Computer-controlled, automated, and automatic transmissions are becoming increasingly popular, in part due to a shortage of truck drivers and retaining existing drivers. Electronically managed shifting tends to be popular with new generation, younger drivers. Computer-controlled transmissions require a lower level of driver skills and reduce driver fatigue. In addition, they are not vulnerable to shifting abuse practices that often shorten the life of conventional transmissions. **Figure 1–8** shows an electronically controlled, fully automatic Allison transmission.

FIGURE 1–8 An Allison computer-controlled transmission

DRIVE SHAFTS

A flange or **end yoke** splined to the output shaft of the transmission transfers engine torque to the drive shaft (see Chapter 22). The **driveshaft** is a hollow tube with end yokes welded or splined to each end. **Splined yokes** allow the drive shaft to increase and decrease in length while rotating to accommodate movements of the drive axles. Sections of the drive shaft are connected to each other and to the transmission and differentials with **universal joints** ("U-joints"). The U-joints allow torque to be transmitted to components that are operating on different planes.

AXLES

Axles provide a mounting point for the suspension system components, wheels, and steering components. The drive axles also carry the **differential** gearing and axle shafts (**Figure 1–9**). The differential transfers the motion of the drive shaft, which is turning perpendicular (at a right angle) to the rotation of the axle shaft, into motion that is the same as the direction the vehicle is moving. A differential carrier assembly also provides a gear reduction, increasing the torque delivered to the drive wheels. A differential divides torque between the left and right wheels. Axles and drive axles are covered in Chapters 23 and 24.

FIGURE 1–9 Axle and drive shaft components

FIGURE 1–10 Truck air suspension assembly

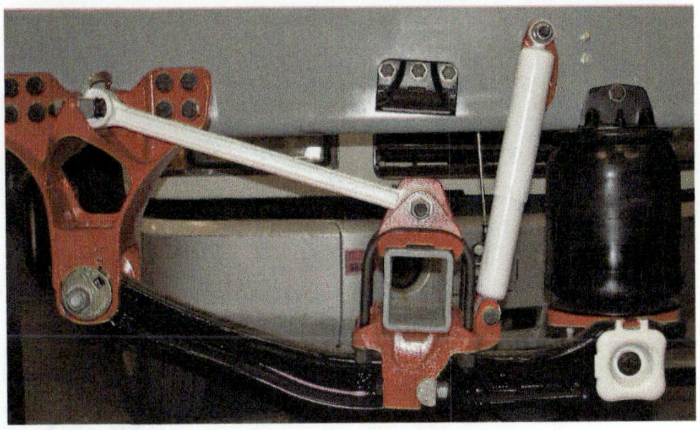

STEERING

Both manual and power steering assist systems are installed on heavy-duty trucks (see Chapter 25). Two types of manual steering gears are used: worm roller and recirculating ball. Power steering systems use a hydraulic pump to provide steering assist. In older trucks, steering assist can also be provided by an air-powered cylinder installed in the steering linkage. More recently, rack and pinion steering has appeared on trucks, providing the driver with improved steering response and better road feel.

SUSPENSION SYSTEMS

Although some trucks continue to use steel and rubber spring suspensions, smooth-riding air **suspensions** are popular for reasons of comfort and cargo protection. But while truckers and the cargo benefit, so does the vehicle itself. Trucks and trailers equipped with air suspensions absorb less road shock than conventional spring suspensions and require less maintenance. Another advantage of an air suspension is that ride and handling are improved when running empty because an air suspension maintains the same ride height regardless of GVW. Cab air suspension systems are becoming standard for similar reasons. By minimizing vibration transmitted to the cab, they reduce repairs to cab hardware and electrical components. They also reduce driver fatigue and can increase vehicle resale value. **Figure 1–10** shows a typical truck air suspension assembly.

WHEELS AND TIRES

There are four basic types of wheel systems:

- Cast spoke
- Steel disc
- Aluminum disc
- Wide-base disc

Cast spoke wheels, used almost universally through the 1970s, are much less popular today than disc wheels (both steel and aluminum). Wide-base disc wheels have been principally used on front axles of some extra-heavy-duty applications such as construction trucks. Their popularity is slowly increasing, though, and they are occasionally observed as super-singles replacing sets of duals. Wide-base-singles can offer slightly improved fuel economy so their use is predicted to increase over the next few years. Truck wheels are mounted to axles by what is known as the **wheel end**. Understanding wheel end procedure is critical for truck technicians because so many vehicle safety issues depend on this. **Figure 1–11** shows a cutaway view of a truck wheel end assembly. Truck and trailer tires are available in a variety of tread patterns to suit different driving conditions. Wheels, tires, and wheel end procedure are studied in Chapter 27.

BRAKES

Along with wheel assemblies, service brakes are one of the most important systems on a vehicle. If they should fail, the consequences can be fatal. Even when brakes perform well, they can be expensive to repair when they wear out prematurely. Brakes are a key preventive maintenance item on trucks. Any truck inspected during routine **Commercial Vehicle Safety Alliance (CVSA)** found to have defective brakes can be issued an **out-of-service (OOS)** infraction, resulting in fines and downtime. Four chapters in this textbook are devoted to truck brake systems (Chapters 28–31), and these are among the most important for rookie technicians.

AIR BRAKES

Highway heavy-duty trucks use **air brakes**, almost exclusively. A complex circuit of pneumatic lines, valves, and cylinders controls the delivery of

FIGURE 1–11 Cutaway, sectional view of wheel end and foundation brake assembly

compressed air to the brakes. Trucks are equipped with drum brakes (see Figure 1–11) and/or disc brakes. Air brakes provide tractors with the versatility to couple and uncouple to air brake–equipped trailers without the intervention of skilled mechanical labor. Some lighter-duty trucks are equipped with hydraulic or air-over-hydraulic brake systems.

Since 1998 **National Highway Traffic Safety Administration (NHTSA)** regulations have required that all current tractors and trailers be equipped with antilock brake systems (ABS). Chapter 30 describes the operating principles of truck ABS. However, technicians should not assume that every rig is ABS equipped. Many highway trailers have a life expectancy that greatly exceeds that of the tractors that haul them. This means that there are still many trailers in service that are not ABS equipped.

Hydraulic Brakes

Hydraulic brakes tend to be primarily used in light- and some medium-duty on-highway trucks. However, they are extensively used in applications such as school buses and off-road equipment, so the chances are that a commercial vehicle technician will work on hydraulic brakes from time to time.

VEHICLE RETARDERS

Vehicle retarders refer to auxiliary brake systems. An auxiliary brake system is an alternative means of slowing (retarding) a moving tractor/trailer without engaging the service brakes.

The most popular vehicle retarder is the internal engine brake; it turns the engine into a power-absorbing compressor to slow the vehicle. Similarly, an exhaust brake creates a restriction in the exhaust system to slow the truck drivetrain. Some vehicles, particularly those equipped with some types of automatic transmission, use a hydraulic retarder to reduce vehicle speed. A fourth type of retarder is the electrical retarder, which uses electromagnetism to resist the rotation of the vehicle powertrain. When used, it is usually located in the driveline behind the transmission. Electric retarders are more likely to be encountered in off-road equipment.

CHASSIS FRAME

The function of a truck is to carry a load, and the strength of its frame plays a large role in defining how heavy a load can be carried. The frame is the backbone of a truck, and like a human backbone it must be capable of flexing to accommodate movement as well as be able to sustain the weight of its cargo.

Most tractor frames are shaped like a ladder. Although a ladder's function is far different, its two main components—rails and steps—can be compared to the truck frame's rails and crossmembers. The cross sections of truck frame rails resemble a "C" or "I," and as they are increased in size, the "duty rating" of the ladder or truck frame is upgraded. Rails and crossmembers also get stronger as the anticipated workload increases. Frames are studied in Chapter 32. Not all truck frames are the same. Differences in frame design show up within one truck manufacturer's model lineup and to a greater extent among the different OEMs.

FIFTH WHEEL

A **fifth wheel** (**Figure 1–12**) is used as a coupling device on a tractor/trailer combination. The tractor-mounted fifth wheel connects to the trailer **kingpin**. A percentage of the weight of a semi-trailer is supported by the fifth wheel plate. The coupling formed between the fifth wheel and kingpin allows the tractor/trailer combination to pivot (articulate) when turning corners. Fifth wheels seldom fail but technicians must understand how they function and the consequences of a failure. A key to understanding fifth wheels is knowing how the various locking mechanisms function. Most current fifth wheels can be slid backward or forward to properly position the weight of the trailer on the tractor (see Chapter 34).

HEATING/AIR CONDITIONING

Most truck and tractor cabs are equipped with heating, ventilation, and air-conditioning systems, known as HVAC. A typical truck has an HVAC or climate control system used to condition the air in the driver cab and the sleeper unit if so equipped. Many current trucks complement climate control systems with auxiliary systems that automatically maintain a comfortable cab and/or engine temperature with engine shutdown. Because many jurisdictions enforce anti-idling legislation, auxiliary climate control systems are becoming more common. Such systems allow a driver to sleep without the need to keep the engine running.

FIGURE 1–12 Typical fifth wheel

ELECTRONIC CONTROLS

Most trucks on the roads today use computers to manage systems such as engines, transmissions, brakes, climate control, suspension, and communications systems. When we say that a system is electronically controlled, it is managed by computer. In addition, most current trucks enable all of the chassis computers to communicate with each other using a data backbone network and a technology known as **multiplexing**. Understanding "smart" or electronic systems is required knowledge for the technician, and this textbook describes the basics of computer-controlled systems.

HOTEL LOAD GRID

The **hotel load grid** has grown exponentially in recent years. It provides a means of supplying trucks and trailers with a mains AC-electrical supply enabling the shutdown of the vehicle (or reefer) engine. Many truck stops and DOT inspection stations along interstates in the United States and Canada are equipped with a hotel load electrical supply.

The hotel load grid can support on-vehicle AC-electrical equipment when parked and plugged in, saving fuel ... and the environment. Truck and trailer hotel loads can include battery chargers, air conditioning, heating, lighting, and entertainment systems. Chapter 10 takes a closer look at this subject.

1.3 CAREER OPPORTUNITIES FOR HEAVY-DUTY TRUCK TECHNICIANS

Qualified heavy-duty service technicians are needed in the different sectors of the trucking industry. Trucking accounts for nearly 80 percent of all domestic freight revenues. Percentage breakdown for other domestic revenues are as follows: less than 10 percent by rail; 5 percent by air; 3 percent by pipeline; and 7 percent by water, according to the U.S. Department of Commerce.

FLEET OPERATIONS

The trucking industry was deregulated in the United States by the Motor Carrier Act of 1980, and this changed the structure of the trucking industry. Truck haulage used to be an industry of small businesses, but this has changed due to intense competition and cost undercutting, which makes it difficult for small operations to compete. The owner-operator today retains an ever-decreasing portion of the freight market. To survive, the owner-operator must usually be affiliated with one of the large operators. As each year goes by, the top 100 carriers increase their percentage of the total haulage market revenues, helped along by rising fuel costs. Large operations can function on much smaller margins while they squeeze out smaller operations.

Fleets operate in different ways; some have specialties of some kind, whereas the largest keep as much of the operation in-house as possible. The following section explains some of the terminology used to describe fleet operations.

- **Logistics**. **Logistics** is the organization of tracking, moving, storing, and delivering cargo.
- **Linehaul**. Fleets manage **linehaul** carriers to operate terminal to terminal. The terminals are usually located in the outer limits of large cities. This enables linehaul trucks to run mostly on highways. A large fleet uses linehaul trucks to run goods terminal to terminal and pickup and delivery (P&D) trucks to link its customers to the terminals.
- **Just-in-time delivery**. Computerized management of trucking activity has enabled **just-in-time (JIT) delivery** operations, usually known simply as JIT. The objective is to time the delivery of a cargo to a major assembly plant so that a load can be unloaded and phased directly into a production line. The result is that the cargo does not have to be stored as inventory by the assembly plant.
- **Less-than-truckload**. **Less-than-truckload (LTL)** carriers specialize in collecting smaller cargoes in terminals, then bulk shipping to another major terminal before redistributing the load for customer delivery. This works a little like a courier service but usually is more economical and less fussy about the type of cargo.
- **Pickup and delivery**. **Pickup and delivery (P&D)** trucks are usually lighter-duty units used to deliver loads to a linehaul or LTL operator terminal—or disperse them for final delivery.
- **Courier fleets**. The largest operator of a truck fleet (measured by revenues and number of power units) in the country is United Parcel Service (UPS). A courier operation consists of both air and ground vehicles that operate between customers and operational hubs. The emphasis is on speed and reliability. UPS operates a combination of light- to heavy-duty vehicles in its ground fleet. In other words, it combines P&D, LTL, and linehaul operations and enhances them with air freight capability and comprehensive logistics.
- **Dedicated contract cartage**. **Dedicated contract cartage (DCC)** is managed to accommodate a specific customer's cargo requirements, and in addition to trucking it often provides warehousing and logistics planning.
- **Vocational**. **Vocational** trucking operations cover a broad range of applications of a specialty nature and include such operations as aggregate haulers, concrete trucks, fire trucks, milk tankers, livestock carriers, crane chassis, and others.
- **Renting and leasing**. Leasing operations continue to be popular with large companies who wish to maintain a hands-off distance between their shipping/trucking requirements and their main business activity. For instance, a food services provider has a primary mandate of getting food from growers and suppliers to its customers, and because of this it has considerable truck haulage requirements. In some cases it makes sense to make a third party responsible for the trucking aspect by leasing both trucks and logistical services.
- **Owner-operator**. The **owner-operator (O/O)** is an American tradition in trucking, and despite diminishing numbers, they still account for a little over 10 percent of heavy-duty trucks on our highways. O/Os may be retainer contracted by a large carrier, operate on specialty runs, or make themselves available on a general for-hire basis.

SHOP TALK

Because it involves long-distance highway driving, linehaul mileage is considered by chassis and engine manufacturers as an indicator of vehicle longevity. The life of a chassis or chassis components is often rated in "linehaul" miles. To convert linehaul miles into engine hours, divide the mileage value by 50.

FLEET SHOPS

A company that owns or operates vehicles has ongoing vehicle service and maintenance requirements. Whereas small fleets often use the services of a dealership or an independent shop to do this work, larger companies usually have their own service and repair facility whose servicing capacity is determined by the age and size of the fleet. A large fleet will often bundle vehicle purchases with OEM technician training packages so that they can perform warranty repairs within their own service facilities.

Employment in a medium-to-large fleet is one way of beginning a career as a heavy-duty truck technician. Most provide training opportunities and are usually run with the latest equipment. Large fleets with 500 or more vehicles probably have repair facilities located in their terminals. A high percentage of the repairs undertaken in a fleet service shop are *running repairs*. Running repairs are executed quickly. Repairing failed lights, adjusting clutches, checking brakes, and adjusting suspension ride height are examples of running repairs. The remainder of the work is usually related to preventive maintenance. Large fleets tend to avoid major repairs such as engine overhauls by negotiating warranties from OEMs that may last for the entire period of the planned ownership of a truck.

DEALERSHIP SHOPS

Heavy-duty truck **franchised dealership** shops are major employers of truck technicians. Dealerships are privately owned businesses. A franchised dealership is one that has a contractual agreement with a particular manufacturer to sell and service a particular line of vehicles. Some dealerships have contracts with more than one manufacturer. A dealership may also handle a line of trailers under an arrangement similar to one with a truck manufacturer. **Figure 1–13** shows a mid-sized Paccar Kenworth dealership service shop specializing in repairs to a single brand: such a facility will ensure that its service personnel are factory trained to handle repairs to the most recent technology.

The sales and service policies of the dealership are usually set by the manufacturer. Service performed while the vehicle is under warranty is usually undertaken by dealerships or authorized service centers because truck manufacturers have been aggressive in securing service business for their dealerships. Extended warranties and service plans are designed to channel repair and maintenance work to the dealership shop(s). Manufacturers provide special diagnostic equipment designed specifically for their vehicles. They stress the superiority of their replacement parts over aftermarket components and promote their service personnel as the most qualified to work on their products. **Figure 1–14** shows an independently owned Freightliner dealership service facility.

FIGURE 1–13 A mid-sized Kenworth truck dealership.

FIGURE 1–14 Independently owned Freightliner dealership service facility

Working for a major OEM dealership has advantages. Technical support, equipment, and the opportunity for ongoing training are usually excellent. When working for an auto dealership, the service technician's scope of service expertise may be limited to one or two particular model lines. This is not true in heavy-duty tractor and trailer dealerships because each truck tends to be custom built for customers to a much greater extent. However, this trend is undergoing a gradual shift due to widespread European influence (Mercedes/Freightliner, DAF/Paccar, Volvo/Mack) on U.S. trucking OEMs. In offshore markets, trucks tend not to be customized during manufacture.

INDEPENDENT TRUCK SERVICE SHOPS

Independent heavy-duty service shops are not associated with any specific manufacturer or trucking fleet, although they may service both segments of the industry. Some shops are authorized under agreement with the manufacturer to make warranty repairs and replacements. Today some small and mid-sized fleets that at one time performed their own servicing now hire independent shops. In most cases, the hourly rates charged to customers are more competitive than OEM dealerships. This type of shop has its limitations and most will not undertake major repair work requiring specialized training and tooling.

SPECIALTY SERVICE SHOPS

Specialty service shops are shops that specialize in areas such as engine rebuilding, transmission/axle overhauling, brake, air-conditioning/heating repairs, and suspension work. Because maintaining expertise in one specific chassis system is easier, specialty service facilities survive by establishing a good reputation that can make them favored over a dealership option. Service technicians employed by such shops have the opportunity to become highly skilled in one specific area of vehicle service and repair.

FIGURE 1–15 Truckstop service facility

OTHER TRUCK SHOPS

Truck leasing/rental companies, construction/mining/refuse haulers, van truckers, buses, agriculture haulers, and private and for-hire carriers usually operate their own service shops. A recent trend led by one major OEM has been to run lube and running-repair facilities at the rest stops on major interstate highways. These shops are usually equipped to handle no more than basic vehicle servicing, brake adjustments, and minor electrical repairs, but it should be noted that this type of work represents a significant segment of the truck service industry. **Figure 1–15** shows a TravelCenters of America (TA) truck stop service facility where basic services and limited Daimler Trucks warranty repairs can be undertaken.

1.4 JOB CLASSIFICATIONS

The trucking industry offers many varied employment opportunities, and it is interesting to note that most technicians who begin a career in a truck service facility will spend their career associated with trucking, though not necessarily as a technician. The certified truck technician can use shop floor experience to progress to a wide range of occupations that include service management, equipment sales, field technical support, customer service, technical teaching, and logistics, to mention a few.

SERVICE TECHNICIANS

There is no better way to begin a career in the trucking industry than to train as a truck technician. In most cases, aspiring truck technicians will be required to undertake a 1- or 2-year college training program before obtaining a position as a trainee or apprentice technician. Many colleges run training programs in conjunction with the major truck manufacturers. The OEM provides equipment and technical support to the college program and may define the exit level skills required for graduation or certification. The objective of most in-school programs is to provide students with a sound set of theoretical skills and some exposure to the hands-on challenges of functioning effectively on the shop floor.

Today's technician needs to have a good mechanical understanding of truck chassis systems combined with some knowledge of machine shop processes, hydraulics, pneumatics, electronics, and computer skills. Although a sound theoretical understanding of the technology of the modern truck is required, the technician's ability to succeed is primarily determined by the ability to perform hands-on work on the shop floor. Because the trucking industry is one of the most highly competitive, most repairs are required to be performed accurately, within a specified time, and without "**come-backs**." Come-backs occur when a "repair" is returned to the service facility, requiring further attention. Accountants call this type of repair a **policy adjustment**. Come-back repairs usually have to be performed without additional charge and are costly to:

- The service facility: The shop bay is tied up and the technician has to be paid.
- The customer: Equipment downtime means that the truck is not earning money and its operator still has to be paid.

Certification

Technicians planning to spend a career in the truck industry should be certified. The **National Institute for Automotive Service Excellence** (now known by the acronym **ASE**) manages certification throughout the United States by holding hard copy and online tests twice yearly in the spring and fall. Passing an ASE test in a given subject certifies the technician for a period of 5 years. To qualify for ASE testing, a technician must be able to prove a minimum of 2 years' experience in working with the subject matter of each test. The test certification areas that connect with the objectives of this textbook are:

- Medium/Heavy Duty Truck Technicians. T-series
- School Bus Technicians. S-series
- Transit Bus Technicians. H-series

In Canada, certification is known as licensing and is mandatory. Licensing usually requires enrollment in an apprenticeship program that includes college training and testing. To work on commercial vehicles in Canada, technicians must either be registered as apprentices or be licensed by the jurisdiction in which they are located.

T-Series Certification

If you are planning to make a career in the trucking industry, it makes sense to aim to achieve Master Medium/Heavy Duty Technician status. This requires passing tests in the following **T-Series Certification** areas: T2, T3, T4, T5, T6, and T7. Refer to **Table 1–5** for a complete list of the Truck ASE tests and task lists.

S- and H-Series Certification

More recently, the ASE has added **S-Series Certification** (school bus) and **H-Series Certification** (transit bus) to its qualification categories. The ASE has designed all three heavy-duty certification fields so that the numeric value of each test corresponds to the same subject matter while the letter indicates the vehicle-type classification. For instance, the number 4 relates

TABLE 1–5 Task Lists for Medium/HD Truck Certification

Gasoline Engines (Test T1)		
Content Area	**Questions in Test**	**Percentage of Test**
A. General Engine Diagnosis	14	28%
B. Cylinder Head and Valve Train Diagnosis and Repair	4	8%
C. Engine Block Diagnosis and Repair	4	8%
D. Lubrication and Cooling Systems Diagnosis and Repair	3	6%
E. Ignition System Diagnosis and Repair	6	12%
F. Fuel, Air Induction, and Exhaust Systems Diagnosis and Repair	6	12%
G. Emissions Control Systems Diagnosis and Repair	5	10%
H. Computerized Engine Controls Diagnosis and Repair	8	16%
Total	**60**	**100%**
Diesel Engines (Test T2)		
Content Area	**Questions in Test**	**Percentage of Test**
A. General Engine Diagnosis	11	20%
B. Cylinder Head and Valve Train Diagnosis and Repair	4	7%
C. Engine Block Diagnosis and Repair	5	9%
D. Lubrication and Cooling Systems Diagnosis and Repair	6	11%
E. Air Induction and Exhaust Systems Diagnosis and Repair	6	11%
F. Fuel System Diagnosis and Repair 1. Mechanical Components (8) 2. Electronic Components (12)	16	29%
G. Starting and Charging System Diagnosis and Repair	4	7%
H. Engine Brakes	3	5%
Total	**55**	**100%**
Drive Train (Test T3)		
Content Area	**Questions in Test**	**Percentage of Test**
A. Clutch Diagnosis and Repair	11	28%
B. Transmission Diagnosis and Repair	13	33%
C. Driveshaft and Universal Joint Diagnosis and Repair	7	18%
D. Drive Axle Diagnosis and Repair	9	23%
Total	**40**	**100%**

(continued)

Brakes (Test T4)		
Content Area	Questions in Test	Percentage of Test
A. Air Brakes Diagnosis and Repair 1. Air Supply and Service Systems (16) 2. Mechanical/Foundation and Wheel Hub (13) 3. Parking Brakes (4)	33	66%
B. Hydraulic Brakes Diagnosis and Repair	12	24%
C. Air and Hydraulic Antilock Brake Systems (ABS), and Automatic Traction Control (ATC), and Electronic Stability Control Systems	5	10%
Total	**50**	**100%**

Suspension and Steering (Test T5)		
Content Area	Questions in Test	Percentage of Test
A. Steering System Diagnosis and Repair 1. Steering Column (3) 2. Steering Units (6) 3. Steering Linkage (3)	12	24%
B. Suspension, Frame and Fifth Wheel Diagnosis and Repair	16	32%
C. Wheel Alignment Diagnosis, Adjustment, and Repair	13	26%
D. Wheels, Tires, and Hub Diagnosis and Repair	11	18%
Total	**50**	**100%**

Electrical/Electronic Systems (Test T6)		
Content Area	Questions in Test	Percentage of Test
A. General Electrical Diagnosis	14	28%
B. Battery Diagnosis and Starting System Diagnosis and Repair	11	22%
C. Charging System Diagnosis and Repair	7	14%
D. Lighting Systems Diagnosis and Repair	6	12%
E. Related Vehicle Systems Diagnosis and Repair	12	24%
Total	**50**	**100%**

Heating, Ventilation, and Air Conditioning (HVAC) Systems (Test T7)		
Content Area	Questions in Test	Percentage of Test
A. HVAC Systems Diagnosis, Service, and Repair	6	15%
B. A/C System and Component Diagnosis, Service, and Repair	20	50%
C. Heating and Engine Cooling Systems Diagnosis, Service, and Repair	6	15%
D. Operating Systems and Related Controls Diagnosis and Repair	8	20%
Total	**40**	**100%**

Preventative Maintenance and Inspection (PMI) (Test T8)		
Content Area	Questions in Test	Percentage of Test
A. Engine Systems	10	20%
B. Cab and Hood	5	10%
C. Electrical/Electronics	10	20%
D. Frame and Chassis	22	44%
E. Road/Operational Test	3	6%
Total	**50**	**100%**

to Brakes systems, so the S-4, H-4, and T-4 test subject matter is brakes, with the letter prefix indicating the type of equipment. Some distinct S- and H-tests exist or are planned. These include:

- H1 certification: compressed natural gas (CNG) fueled bus engines
- H8 certification: preventative maintenance
- S1 certification: school bus body systems

Transit bus technicians are in high demand. Buses move people. This means that passenger and equipment safety are of high importance. **Figure 1–16** shows some examples of urban transit buses. See if you can identify the features of a hybrid drive city bus.

Master Technician Status

Certification as an ASE Master Technician may be attained specifically in one of the T-, H-, or S-series fields by passing certification tests as follows:

- T-Series Master Tech: certification in tests T2 through T7 inclusive
- S-Series Master Tech: certification in tests S1 through S6 inclusive
- H-Series Master Tech: certification in tests H2 through H7 inclusive

It is also possible for technicians to qualify as a Master Medium/Heavy Vehicle technician by holding certifications in tests 2 through 7 regardless of which specific vehicle field each test is in. For instance, a technician certified in T2, S3, S4, H5, H6, and H7 tests would qualify as a Master Medium/Heavy Vehicle Technician.

SPECIALTY TECHNICIAN

The heavy-duty truck specialty technician specializes in servicing and repairing a single vehicle system such as electrical (and/or electronics), engines, brakes, transmission, drivetrain, suspension/steering, trailers, heating/air conditioning, or tire/wheel. These specialties often require advanced and continuous training in that particular field.

There has been a shift toward specialization in recent years due to the increased complexity of modern chassis systems. Specialty technicians will often only maintain certification in their area of specialization. **Figure 1–17** is an image of a Class 3 school bus and **Figure 1–18** shows a Class 6 straight truck used as a utility vehicle.

SELF-EMPLOYMENT

The skills required to manage a business are quite different from those required to repair equipment. It is a fact that talented mechanical brains have

FIGURE 1–16 Urban transit buses

FIGURE 1–17 Class 3 school bus

FIGURE 1–18 Class 6 utility vehicle

often failed miserably as self-employed technicians, whereas others possessing much lower levels of skills have succeeded. Good truck and diesel technicians are much sought after, so surely they should succeed in self-employment. It is, however, essential to remember that the skills required of a good business person and those required of a good technician are often in conflict. Working for yourself rather than others can sound like a great option, but caution and some self-examination are required.

Self-employed technicians who have succeeded have done so because they understood the challenges required of running a business before getting into it. Successful self-employed technicians often owe that success to a partner who looks after the financial aspects of running a business—or to having had the foresight to have taken business administration courses before starting into the venture. Remember this: Understand what you are getting into upfront because it is almost impossible to learn business practices on the fly.

SERVICE WRITER

When a customer enters a large truck service facility, the service writer is usually the first and often the only person he or she speaks to. A good service writer should have a thorough understanding of truck chassis technology and of how the shop facilities are organized. A key attribute is the ability to communicate with people in a friendly manner. In many operations, the service writer will key-in or write the repair order and direct the truck to the appropriate area of the shop for repair. The ability to understand a customer problem and then express it clearly in the shop information systems is critical for a service writer. It is critically important to express anything written on a **repair order** or **work order** in the simplest terms. In the event of a disputed bill, the repair order assumes a legal status and the language used on it will be interpreted by lawyers. A typical work order is shown in Chapter 4.

PARTS MANAGER

The parts manager manages the parts inventory in a service repair facility. The ordering and timely delivery of parts is important for a shop to operate smoothly and on schedule. Delays in obtaining parts or omitting a critical component from the initial parts order can cause frustrating holdups for both the service technicians and customers or fleet operators.

Most fleets and large independent service shops maintain an inventory of commonly used parts such as filters, belts, hoses, and gaskets. Most modern truck shops manage parts inventories by networked computers. These networks can also identify the nearest location of a required component and the means required to get it to the shop. ASE certifies medium/heavy duty parts specialists.

SHOP SUPERVISOR

The shop supervisor or foreperson is directly in charge of the service technicians, including directing, routing, and scheduling service and repair work. The supervisor often helps hire, transfer, promote, and discharge technicians to meet the needs of the service department. The supervisor also instructs and oversees the technicians in their work procedures, inspects completed repairs, and is responsible for satisfactory shop operation.

SERVICE MANAGER

The service manager oversees the service operation of a large dealership, fleet, or independent shop, usually from a business perspective. Customer concerns and complaints are usually handled through the service manager. Good communications skills, business savvy, and a sound technical background are essential job requirements.

In a franchised or company dealership, the service manager ensures that OEM policies concerning warranties, service procedures, and customer relations are adhered to. The service manager coordinates training programs and keeps all shop personnel informed and working together.

1.5 ADVANCEMENT IN THE PROFESSION

The most common sources of training are:

- Vocational/technical schools
- Community colleges
- Fleet training programs
- Manufacturer training programs

Heavy-duty truck service courses are offered at various training levels—secondary, postsecondary, vocational/technical, or community colleges, both private and public. To help schools keep pace with rapidly changing technology and maintain a curriculum that meets the service industry's needs, many truck manufacturers and fleets assist schools by running cooperative programs. A sister organization to the ASE, the **National Automotive Technicians Education Foundation (NATEF** [http://www.natef.org]) certifies secondary and postsecondary heavy-duty truck training programs.

Apprenticeship programs offered by some large dealerships and fleets are another good way to receive training. In such a program, the trainee receives job training under supervision. For information on available apprenticeship programs, contact the U.S. Department of Labor, Bureau of Apprenticeships and Training, Washington, D.C. 20006 (http://www.dol.gov/dol/topic/training).

SELF-EDUCATION

The professional heavy-duty truck technician must constantly learn. Truck manufacturers, aftermarket parts manufacturers, and independent publishers are always producing new training materials to keep technicians informed on how to service the next generation of trucks. In addition, technical clinics are sponsored by truck manufacturers, aftermarket parts manufacturers, and parts dealers. Reading trade magazines and publications is also an excellent way to stay informed and up to date. A competent technician should take advantage of every opportunity to maintain currency with the latest technology. There are many hard copy and online magazines, most of which can be subscribed to at no cost. Examples of domestic and international (Australia/NZ and UK) publications are:

- *American Trucker*
- *Truck and Bus Builder*
- *Fleet Owner*
- *Heavy Duty Trucking*
- *Transport Topics* (requires TMC membership)
- *Commercial Carrier Journal*
- *Diesel Progress*
- *Land Line and Road King Magazine*
- *Today's Trucking*
- *Truckin' Life*
- *Big Rigs Magazine*
- *Commercial Motor Magazine*
- *Trucking Mag*

PROFESSIONAL ASSOCIATIONS

The **Technology and Maintenance Council (TMC)** division of the **American Trucking Association (ATA)** sets standards and practices in the industry. Membership in the TMC is an excellent value and is one of the best means of staying on top of truck technology. There are many membership categories determined by job function and including a student member (currently registered in truck technology college program) and technician member (requires 2 years' documented experience as a technician). The current (2014) costs are:

- Student member: $30 annually
- Technician member: $75 annually

As a TMC technician member, you receive continual technical updates and are eligible to participate in feedback on its development of Recommended Practices (RPs).

Since being formed in 2003, the technician's wing of the TMC has supported technician professional development and hosted technical skills competitions at its annual meetings. A hard copy of the TMC *Recommended Practices (RP) Manual* should be available in every truck shop and electronic versions should be loaded onto at least one general access shop computer. Most of the **Recommended Practices (RPs)** outlined in the TMC manual are consensually agreed to by OEM and member experts. The TMC newsletter *Fleet Advisor* is sent to all TMC members monthly. Check out the TMC at http://tmc.truckline.com. The address is: American Trucking Association, Technology and Maintenance Council, 950 N. Glebe Road, Suite 210, Arlington, VA 22203-4181.

TMC SUPERTECH

A technician rodeo is one in which technicians get to challenge their skills against each other. Some states, fleets, and OEMs have run this type of competition

FIGURE 1–19 TMC SuperTech electrical station

FIGURE 1–20 TMC SuperTech axle-end station

in-house for a number of years, but since 2005, the TMC has been running its **SuperTech** competition on a national scale and it takes place at the TMC fall meetings. The competition is divided into stations, each specializing in a truck chassis subsystem such as brakes, suspensions, transmissions, engines, and so on. Technicians compete in each specialty station and there is a high emphasis on technical literacy and computer diagnostic skills. Winners are recognized for each specialty station and the overall results are computed to award first, second, and third prizes overall. The overall first-place technician is identified as the grand champion of the event. **Figures 1–19** and **1–20** show a couple of stations at a TMC SuperTech competition.

Many of the country's largest fleets have sponsored their technicians in the TMC SuperTech. The contest is open to registered TMC technician members at no charge though it helps to have the backing of your employer to help with travel and accommodation costs. Companies sponsoring technicians in the event have recognized that there is a return beyond the costs of sponsorship, most notably that of promoting technical excellence and profiling in the trucking media.

DEGREE PROGRAMS

Some technicians may wish to consider using their technical certification to return to higher education. Most colleges and universities today are required to practice prior learning assessment recognition (PLAR) that can help shorten the time required to achieve degree status. Assuming you want to remain in the trucking industry, areas that you may want to consider for more advanced studies are business administration, marketing, and engineering.

SUMMARY

- Although the number of trucks and automobiles in America is increasing, the number of technicians available to service and maintain them is decreasing.
- Trucks are classified by their gross vehicle weight (GVW), the weight of the vehicle and maximum load, and by the number of axles they have. Heavy-duty trucks have a GVW of 26,001 pounds (11,794 kg) or more.
- The major systems in on-highway trucks are engines, electrical systems, clutches, transmissions, drive shafts, axles, steering, suspension systems, wheels and tires, brakes, vehicle retarders, chassis frame, fifth wheel, heating and air conditioning, electronic controls, and accessories.
- Heavy-duty truck technicians are employed by fleet operations, fleet shops, dealership shops, independent truck service shops, specialty service shops, and other types of truck shops such as truck leasing and refuse haulers.
- Job classifications in the heavy-duty truck industry include the service technician, who maintains and repairs all systems; the specialty technician, who maintains and repairs a single system; the service writer, who deals directly with drivers and communicates truck problems to the service technician; the parts manager, who maintains the inventory of parts needed for maintenance and repair; the shop supervisor, who is in charge of the service technicians; and the service manager, who oversees the entire service operation of a large dealership, fleet, or independent shop.
- A successful heavy-duty truck technician must be able to maintain good customer relations and

working relations, use effective communications skills, maintain a safe work environment, perform preventive maintenance, use tools and equipment properly, troubleshoot, correct problems by repairing or replacing, and upgrade skills and knowledge continuously.

- Training for heavy-duty truck technicians is offered by vocational/technical schools, community colleges, fleet training programs, and manufacturer training programs.
- Heavy-duty truck technicians with at least 2 years of hands-on experience can obtain certification from the National Institute for Automotive Service Excellence (ASE) by passing written exams.
- Technicians should consider joining a professional association such as the Technology and Maintenance Council (TMC) to maintain technical currency.
- SuperTech is a national truck technician rodeo held at the TMC annual fall meeting during which technicians compete for grand champion status.

REVIEW QUESTIONS

1. Which of the following is a correct definition of gross vehicle weight?
   a. The mean weight of a vehicle.
   b. The weight of a vehicle plus its freight.
   c. The maximum allowable weight of a vehicle plus its cargo.
   d. The minimum allowable weight of a vehicle minus its cargo.
2. How many major OEMs manufacture Class 8 trucks for the North American market?
   a. two
   b. three
   c. four
   d. six
3. Which classification group of highway trucks would be called "heavy duty"?
   a. Class 1 and 2
   b. Class 3 and 4
   c. Class 5 and 6
   d. Class 7 and 8
4. What is the GVW range of a Class 6 truck?
   a. 16,001 to 19,500 pounds (7,257–8,845 kg)
   b. 19,501 to 26,000 pounds (8,846–11,793 kg)
   c. 26,001 to 33,000 pounds (11,794–14,970 kg)
   d. over 33,001 pounds (14,970 kg)
5. What is the GVW range of a Class 8 truck?
   a. 16,001 to 19,500 pounds
   b. 19,501 to 26,000 pounds
   c. 26,001 to 33,000 pounds
   d. over 33,001 pounds
6. What is used to power most heavy-duty trucks on our highways?
   a. propane-fueled engines
   b. gasoline-fueled engines
   c. hydromechanical diesel engines
   d. computer-controlled diesel engines
7. A truck equipped with six wheels, four of which are driven, would be classified as:
   a. 4 × 6
   b. 2 × 8
   c. 10 × 4
   d. 6 × 4
8. What is used to provide steering power assist in most heavy-duty highway truck systems?
   a. compressed air
   b. hydraulic pump
   c. electric pump
   d. drag link
9. What do fully automatic transmissions use to transmit drivetrain torque?
   a. planetary gearsets
   b. computer controls
   c. twin countershafts
   d. triple countershafts
10. Which of the following does most to reduce driver fatigue in highway trucks?
    a. 20-speed transmissions
    b. 6-speed transmissions
    c. computer-controlled automated manual transmissions
    d. hydromechanical power-shift transmissions
11. What type of frame is used on most heavy-duty trucks?
    a. monocoque
    b. unibody
    c. ladder
    d. aluminum
12. What percentage of all transportation revenues is accounted for by the trucking industry in the United States?
    a. 5 percent
    b. 10 percent
    c. 20 percent
    d. 80 percent
13. How many ASE truck technician tests are currently available?
    a. five
    b. six
    c. seven
    d. eight
14. In order to obtain Master Truck Technician certification, how many ASE tests must a technician pass?
    a. five
    b. six
    c. seven
    d. eight
15. What type of truck suspension is most commonly found on highway trucks today?
    a. air spring
    b. solid rubber spring
    c. steel coil spring
    d. steel leaf spring
16. A truck that operates long distance, terminal to terminal, would be classified as:
    a. LTL
    b. P&D
    c. vocational
    d. linehaul

17. Which of the following is an appropriate classification for a concrete truck?
    a. P&D
    b. LTL
    c. JIT
    d. vocational
18. What term is used to describe the organization of tracking, moving, storing, and delivering cargo in the trucking industry?
    a. P&D
    b. logistics
    c. JIT
    d. LTL
19. Convert 500,000 linehaul miles to engine hours.
    a. 5,000
    b. 10,000
    c. 50,000
    d. 100,000
20. Which of the following is the accounting term for a *come-back* repair?
    a. freebie
    b. policy adjustment
    c. OOS repair
    d. catastrophe

2

Prerequisite: Chapter 1

SHOP SAFETY AND OPERATIONS

OBJECTIVES

After reading this chapter, you should be able to:

- Explain the special notations in the text labeled SHOP TALK, CAUTION, and WARNING.
- Identify the basic procedures for lifting and carrying heavy objects and materials.
- Explain how to use personal protective equipment (PPE).
- Identify the UL requirements of shop safety boots and ESR footwear.
- Describe safety warnings as they relate to work area safety.
- Identify the different classifications of fires and the proper procedures for extinguishing each.
- Operate the various types of fire extinguishers based on the type of extinguishing agent each uses.
- Identify the four categories of hazardous waste and their respective hazards to health and the environment.
- Explain laws regulating hazardous materials, including both the "Right-to-Know" and employee/employer obligations.
- Identify which types of records are required by law to be maintained on trucks involved in interstate shipping.
- Outline the precautions required to work on hybrid hydraulic, hybrid electric, and gaseous fueled vehicles.
- Explain what hydraulic pinhole injection is and the action required if you suspect it.
- List the safety requirements required to work around high-voltage electrical equipment.
- Identify the precautions required to work safely with oxyacetylene equipment.
- Identify the precautions required to work safely with electric welding stations.
- Discuss the role of computers in the administration, logistics, and maintenance management of transport truck operations.

KEY TERMS

corrosive
dispatch sheet
electric shock resistant (ESR) footwear
electrocution
Federal Motor Vehicle Safety Standard (FMVSS)
flammable
hazardous materials
material safety data sheet (MSDS)
mechanic's gloves
Occupational Safety and Health Administration (OSHA)
parts requisition
personal protective equipment (PPE)
pinhole injection
reactive
Resource Conservation and Recovery Act (RCRA)
Right-to-Know Law
single-phase main
solvents
spontaneous combustion
static charge
static discharge
toxic
vehicle identification number (VIN)
Workplace Hazardous Materials Information System (WHMIS)

INTRODUCTION

Shop safety is a concern for all technicians, forepersons, managers, and shop and fleet owners. Safety rules and regulations should be adhered to, to prevent injuries in the workplace. Carelessness and a lack of safety awareness result in accidents. Accidents have far-reaching consequences, not only on the victim, but also on the victim's family and society in general. It is an obligation of all shop employees and the employer to develop a safety program to protect the health and welfare of both employees and the general public.

Throughout this book, the text contains special notations labeled **SHOP TALK, CAUTION**, and **WARNING**. Each one has a specific purpose. **SHOP TALK** gives added information that will help the technician to complete a particular procedure or make a task easier. A **CAUTION** message is given to prevent the technician from making an error that could damage the vehicle. A **WARNING** message reminds the technician to be especially careful of those areas where carelessness can cause personal injury. The following text contains some general warnings that should be followed when working in a truck service facility.

2.1 PERSONAL PROTECTIVE EQUIPMENT

Personal protective equipment (PPE) refers to the equipment you wear so that you can safely do your job and protect yourself from injury. It involves wearing protective gear, dressing for safety, and the measures an individual takes to handle tools, hazardous materials, and equipment safely. Perhaps just as important, it requires that individuals develop an awareness of how long-term health can be protected by being aware of potential hazards in the workplace such as air quality, noise levels, or carcinogens.

EYE PROTECTION

Most shops today require persons entering a service facility to wear safety glasses every moment they are on the shop floor. This is a no-brainer. Eyes are sensitive to dust, vapors, metal shavings, and liquids. Grinding and machining generate tiny particles that are thrown off at high speeds. Gases and liquids escaping a broken hose or fuel line fitting can be sprayed great distances under massive force. Dirt and sharp bits of corroded metal can easily fall into your eyes when working under a vehicle.

Whether or not it is mandated, eye protection should be worn whenever you are exposed to a shop floor environment. There are many types of eye protection available; some of which are shown in (**Figure 2–1**). The lenses must be made of safety glass or plastic and offer some sort of side protection.

FIGURE 2–1 The various eye protection available include safety glasses, splash goggles, and face shields.

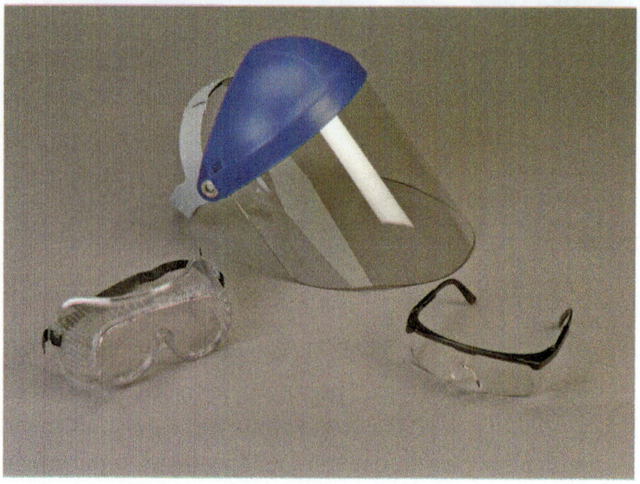

FIGURE 2–2 Eyewash station

Regular prescription glasses will not suffice. Select safety glasses that fit well and feel comfortable. Make a habit of putting on eye protection and leaving it on. A full face shield such as that shown in Figure 2–1 should be used when handling corrosive liquids. If chemicals such as battery acid, fuel, or solvents enter your eyes, flush them continuously with clean water until you can get medical help. **Figure 2–2** shows an eyewash station which has become mandatory in most service facilities.

SHOP TALK

Truck technicians who invest $100 in a single wrench they might use once a year balk at paying a fraction of that for a good quality pair of safety glasses. The freebie safety glasses provided by shops may be fundamentally safe but they are also uncomfortable and usually impair vision. Don't be cheap. Invest in a good quality pair of safety glasses. After 3 days of wearing them continuously, you will forget you are wearing them. It will be one of the best investments you will ever make!

CLOTHING

Clothing should be durable, comfortable, and well fitted. Loose, baggy clothing can get caught on moving parts and machinery. Neck ties should not be worn. Even in shops where it is not mandatory, many service technicians prefer to wear coveralls or shop coats to protect their personal clothing. Cutoffs and short pants are not satisfactory for shop work and appear unprofessional. As a consequence some employers implement strict dress codes for employees.

Ideally, coveralls should be manufactured from 100% cotton for safety and personal comfort. In cases where shops provide artificial fabric, coveralls exercise extreme caution when working near open flame or when welding: even when polyester clothing is treated with fire retardant, prolonged exposure to high temperatures can make them ignitable. When polyester and other petrochemical fibers burn, they melt and fuse to the skin resulting in severe burns.

FOOTWEAR

Even when it is not mandated, technicians should wear properly certified safety footwear in service facility or performing field work. Because safety footwear standards are slightly different in the United States and Canada, they are briefly outlined here. Australian and New Zealand standards conform to the U.S. ASTM standard.

ASTM Class 75

ASTM Class 75–approved footwear sold in the United States must meet an internal clearance height of 0.5 in. (12 mm) when impacted with a force of 75 lb-ft. (102 joules): this is equivalent to having a 40-lb object dropped from a height of 2 feet onto the toe of the boot. The crush standard must be met whether the toe protective cap is of steel or composite construction.

CSA Class 1

CSA Class 1–approved footwear sold in Canada must meet an internal clearance height of 12 mm (0.5 in) when impacted with a force of 74 lb-ft. (100 joules): this is equivalent to having a 20-kilogram weight dropped from a height of 0.5 meter onto the toe of the boot. The crush standard must be met whether the toe protective cap is of steel or composite construction.

AS/NZ

The Australia and New Zealand standard pertaining to safety boots is known as AS/NZ 2210.3 and conforms to ASTM Class 75. This requires that the footwear meet an internal clearance height of 0.5 in. (12 mm) when impacted with a force of 75 lb-ft. (102 joules):

Electric Shock Resistant (ESR) Footwear

Any mechanical technician working around high-voltage potential chassis equipment should wear **electric shock resistant (ESR) footwear**. ESR footwear is designated by an orange omega on a white rectangle in both the United States and Canada. The standards are as follows:

- ASTM ESR footwear: must withstand the application of 14,000 volts at 60 hertz for 1 minute with zero current flow or leakage exceeding 3.0 milliamperes under dry conditions.
- CSA ESR footwear: must withstand the application of 18,000 volts at 60 hertz for 1 minute with zero current flow or leakage exceeding 1.0 milliamperes under dry conditions.

ESR footwear should be worn when working on hybrid AC-DC-AC drivetrains. High-potential electrical discharge from battery or capacitor banks are capable of **electrocution**. To emphasize the definition of *electrocution* it means *death* by electrical energy.

GLOVES

Good hand protection is often overlooked. A scrape, cut, or burn can seriously impair your ability to work for many days. **Mechanic's gloves** are artificial fiber, stretchable gloves that have become popular in recent years. They are thin and fit tight to the hand to provide the technician with some tactile sense. They are suitable for working on light-duty jobs.

A well-fitted pair of heavy work gloves should be worn during operations such as cutting and welding or working on suspensions; uncured leather works best. Do not wear uncured leather gloves when working on refrigeration or AC systems because these can wick liquid refrigerant into the hands. When handling caustic chemicals or high-temperature components, specialty gloves are required.

WARNING

Never wear gloves when operating a stationary grinding wheel. The glove may catch and pull the hand into the wheel.

EAR PROTECTION

Exposure to high noise levels for extended periods damages hearing. Air wrenches, engines run on dynamometers, and vehicles running in enclosed areas can all generate annoying and, with prolonged exposure, harmful levels of noise. Buck-riveting panels on a trailer produce noise levels sufficient to damage hearing in a short period. Simple disposable wax or sponge

FIGURE 2–3 Hearing protection

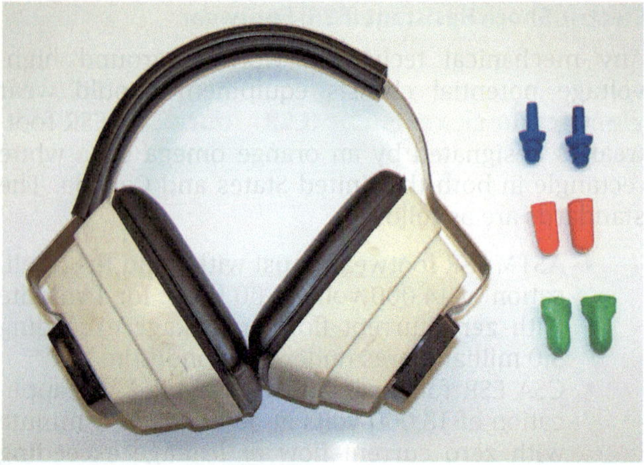

earplugs are good enough for temporary use. Hearing muffs or earphone-type protectors (**Figure 2–3**) can be worn in constantly noisy environments.

WARNING

Electronic noise-cancelling headphones or audio headphones should not be worn on the shop floor. It can be dangerous to wear electronic noise-cancelling devices when doing anything but sitting still. These and audio headphones may isolate the wearers from certain noises that they may want to be aware of such as a verbal warning or running machinery.

HAIR AND JEWELRY

Long hair and hanging jewelry can create the same type of hazard as loose-fitting clothing. They can become caught on moving engine components and machinery.

Tie up long hair securely behind your head or cover it with a cap. Bump caps (similar to construction helmets) are recommended when working in pits or under overhead hoists.

Remove all rings, watches, bracelets, and neck chains. These items can get caught on moving parts, causing serious injury or electrically arc on live circuits.

LIFTING AND CARRYING

Knowing the proper way to lift heavy materials is important. You should always lift and work within your ability and seek help from others when you are not sure you can handle the size or weight of the material or object. Even small, compact auto parts can be surprisingly heavy or unbalanced. Always examine the lifting task before beginning. When lifting any object, follow these steps:

FIGURE 2–4 Use your leg muscles, never your back, when lifting any heavy load.

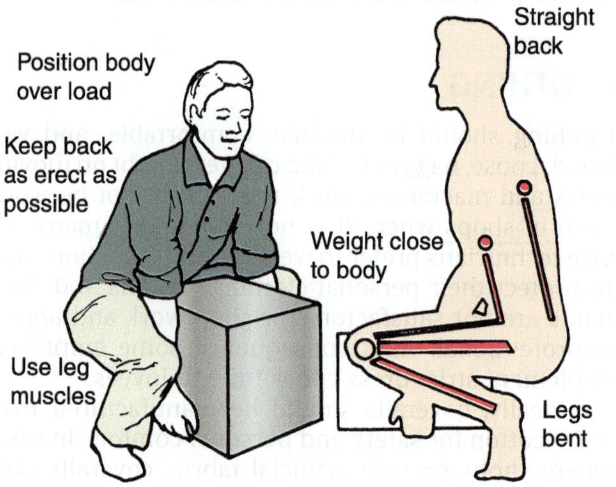

1. Place your feet close to the load and properly positioned for balance.
2. Keep your back and elbows as straight as possible. Bend your knees until your hands reach the best place for getting a strong grip on the load (**Figure 2–4**).
3. If a part or component is stored in a cardboard box, be certain that the box is in good condition. Old, damp, or poorly sealed boxes can tear or otherwise fail. A heavy object could rip through the side or bottom of the container, causing injury or damage.
4. Grasp the object close to your body and lift by straightening your legs. Use your leg muscles, not back muscles.
5. When changing direction of travel, do not twist your body. Turn your whole body, including your feet.
6. When placing an object on a shelf or counter, do not bend forward. Place the edge of the load on the surface and slide it forward. Be careful not to pinch your fingers.
7. When lowering a load, bend your knees and keep your back straight. Do not bend forward—this strains the back muscles.
8. Use wood blocks to protect your fingers when picking up or lowering heavy objects to the floor.

OTHER PERSONAL SAFETY WARNINGS

Never smoke while working on any vehicle or machine in the shop. Tilt cab-over-engine (COE) style cabs and engine compartment hoods with care (**Figure 2–5**). Proper conduct can help prevent accidents. Horseplay is not fun when it sends someone to the hospital. Such things as air nozzle fights, creeper races, or practical jokes have no place in a truck shop.

FIGURE 2–5 The tilting of a hood must be undertaken with care. Make sure there is adequate clearance in front of the vehicle and that the area is free of people and all objects. Do not tilt a cab with the engine running. Tilting the cab could engage the transmission. If the engine is running, the vehicle could move, causing an accident that could result in personal injury or property damage.

Welding and Cutting

A welding helmet or goggles with the proper shade lens must be worn when welding or cutting. Check that the filter lens is suited to the actual work being performed. When available, it makes sense to introduce welding and cutting techniques using a virtual welding station that simulates the experience in the safest possible way, eliminating personal danger and material waste. **Figure 2–6A** shows the helmet used on a virtual welder that can simulate electric welding techniques. Welding filters protect the eyes and face from airborne molten steel and harmful light rays: **Figure 2–6B** shows a typical set of oxyacetylene goggles. Selecting the appropriate filter for the welding procedure you are using is essential. Use optical filters. Reference **Table 2–1** as a guide and use the highest filter rating that allows you to properly see the work. Never use welding equipment unless you have received instruction in how to safely use it. To prevent serious burns, avoid contact with hot metals such as radiators, exhaust manifolds, turbochargers, tailpipes, and mufflers. Welding and cutting should never be performed near flammable substances, and welders should be mindful that the process can produce harmful fumes. Snorkel extractors, fresh air extractors, and fresh air replacement systems are now mandatory in some work environments but they have to be used in order to be effective.

FIGURE 2–6 (A) Virtual welding helmet used to introduce electric welding techniques to students and (B) oxyacetylene goggles.

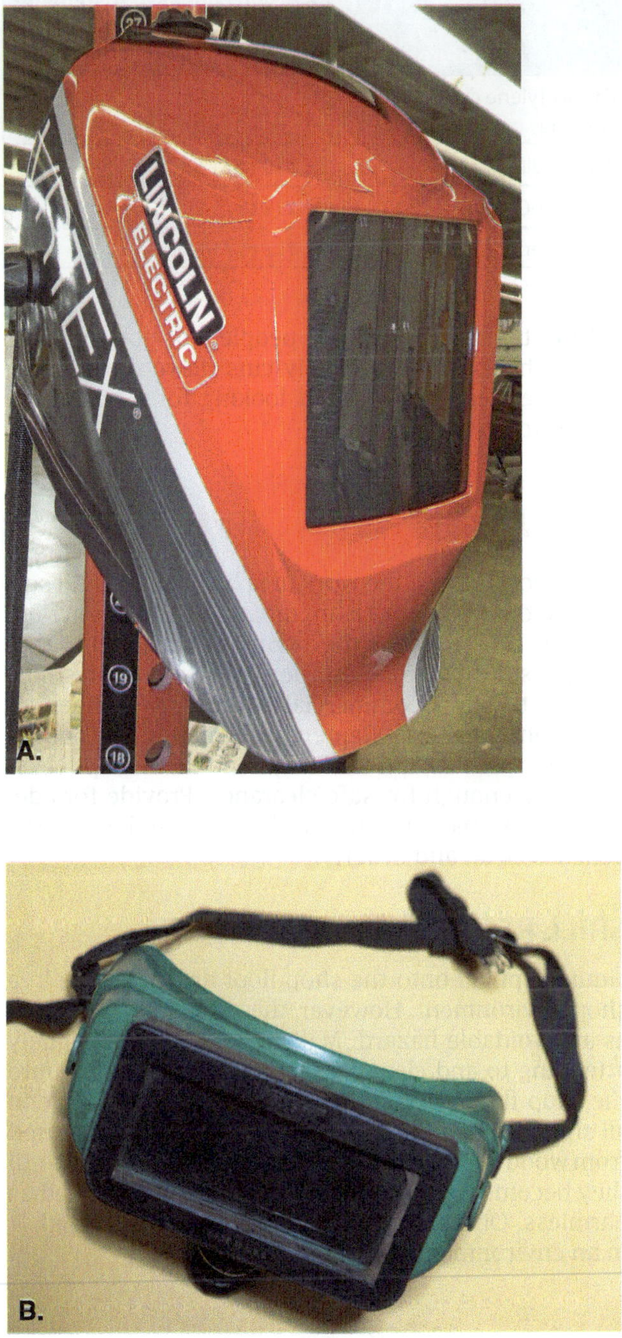

Hydraulic Presses and Housekeeping

When working with a hydraulic press, make sure that hydraulic pressure is applied in a safe manner. It is generally smart to stand to the side when operating a press. Make use of safety cages and always wear safety glasses when using a press. Sloppy housekeeping causes injuries. Store parts and tools properly by

TABLE 2–1 Filter Lens Selection for Welding Procedures

Procedure	Light Filter Lens Required (Use the highest rating that allows you to see the work.)
Oxyacetylene cutting and welding	#4 to #6
Arc welding	#10 to #13
MIG or TIG	#11 to #14
Carbon and plasma arc cutting	#12 to #14

putting them away so that people will not trip over them. This practice not only cuts down on injuries, it also reduces time wasted looking for a misplaced part or tool.

2.2 WORK AREA SAFETY

It is important that the work area be kept safe. All surfaces should be kept clean, dry, and orderly. Any oil, coolant, or grease on the floor can cause slips that could result in serious injuries. To clean up oil spills, be sure to use a commercial oil absorbent. Keep all water off the floor. Remember, water can conduct electricity. Aisles and walkways should be kept clean and wide enough for safe clearance. Provide for adequate work space around machinery. Also keep workbenches clean and orderly.

SPILL RESPONSE

Liquids spilled onto the shop floor are inevitable in a shop environment. However, they should be regarded as an avoidable hazard. Make a point of immediately attending to and cleaning up any liquids spilled onto the shop floor. Ground up absorbents are available in all shops and should be used. They are manufactured from wood bark, clay, and artificial substances and until they become saturated with liquid, can be regarded as harmless. Once used, absorbent must be disposed of in an environmentally responsible manner.

WARNING

Absorbents soaked with liquid fuel, oil, and solvents represent a fire hazard. Take great care when disposing of used absorbent.

VENTILATION

Another important safety requirement for any work area is good ventilation. Although diesel engines produce less carbon monoxide (CO) fumes than gasoline engines, no engine exhaust fumes can be safely inhaled. Truck exhaust pipes should be connected to a shop exhaust extraction system any time an engine is run in the shop. Oil-fueled space heaters used in some shops can also be a source of deadly CO and, therefore, must be periodically inspected to make sure they are adequately vented and do not become blocked. Proper ventilation is also important in areas where flammable chemicals are used.

WARNING

Diesel engine exhaust fumes have been designated by the State of California to cause respiratory problems, cancer, birth defects, and other reproductive harm in humans. Avoid operating diesel engines unless in a well-ventilated area. When starting an engine outside a shop, warm the engine before driving it into the shop to reduce the contaminants emitted directly into the shop while parking the unit.

EMERGENCY TELEPHONE NUMBERS

Keep a list of up-to-date emergency telephone numbers clearly posted next to the telephone. These numbers should include a doctor, a hospital, and the fire and police departments. Also, the work area should have a first-aid kit for treating minor injuries. Facilities for flushing the eyes should also be in or near the shop area.

FLAMMABLE SUBSTANCES

Diesel fuel, gasoline, and solvents are flammable liquids. **Flammable** liquids are easily ignited. For this reason, keep solvents or diesel fuel in an approved safety container and never use it to wash hands or tools. Oily rags should also be stored in an approved metal container. When oily, greasy, or paint-soaked rags are left lying about or are stored improperly, they are possible sources of **spontaneous combustion**—that is, fires that start by themselves.

Check that all drain covers are snugly in place. Open drains can cause toe, ankle, and leg injuries.

Handle all **solvents** (solvents are substances that dissolve other substances) with care to avoid spillage. Keep solvent containers closed, except when pouring. Extra caution should be used when moving flammable materials from bulk storage (**Figure 2–7**). Static electricity can accumulate sufficient charge to create a spark that could cause an explosion. Discard or clean all empty solvent containers. Solvent fumes in empty containers can start a fire or explosion. Do not strike matches or smoke near flammable solvents and chemicals, including battery acids. Solvents and other combustible materials must be stored in approved and designated storage cabinets or rooms (**Figure 2–8**). Storage rooms should have adequate ventilation. Storage of flammable material and combustible liquids should never be near exits or stairways.

FIGURE 2–7 Safe methods of transferring flammable materials from bulk storage.

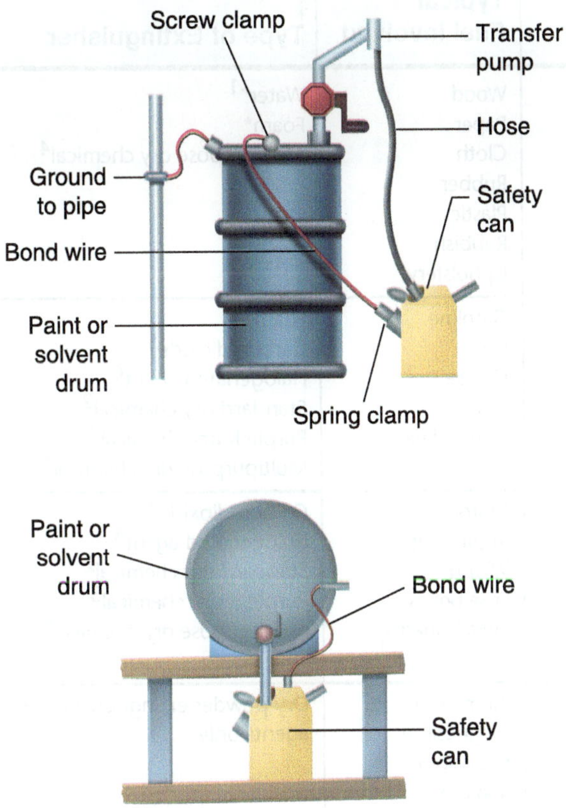

FIGURE 2–8 Store combustible materials in approved safety cabinets.

Where natural gas (NG)-, dimethyl ether (DME)-, and propane-fueled vehicles are serviced, special care must be taken to identify and repair gas leaks. Technicians should take precautions when working in the vicinity of gaseous-fueled vehicles, especially where welding and high-potential electricity is close-by.

FIRE SAFETY

Familiarize yourself with the location and operation of the firefighting equipment in the work area. All fires are classified in one or more of the following categories.

- **Class A**—Fires in which the burning materials are ordinary combustibles, such as paper, wood, cloth, or trash. Putting out this type of fire requires drowning with water or foam solutions containing a high percentage of water, or a multipurpose dry chemical extinguisher.
- **Class B**—Fires in which the burning material is a liquid, such as gasoline, diesel fuel, oil, grease, or solvents. To extinguish this type of fire requires a smothering action from foam, carbon dioxide, or dry chemical-type extinguisher. *Do not use water on this type of fire.* It can cause the fire to spread.
- **Class C**—Fires in which the burning material is "live" electrical equipment such as motors, switches, generators, transformers, or general wiring. To extinguish this type of fire requires a nonconductive smothering action, such as carbon dioxide or dry chemical extinguisher. Do not use water on this type of fire.
- **Class D**—Fires in which the burning materials are combustible metals. Special extinguishing agents are required to put out this type of fire.

The following are some general tips for operating the various types of portable extinguishers based on the type of extinguishing agent they use (**Figure 2–9**):

- **Foam**—Do not spray the jet directly into the burning liquid. Foam works by smothering to allow the foam to fall lightly onto the fire.
- **Carbon dioxide**—Direct discharge as close to the fire as possible, first at the edge of the flames and gradually forward and upward.
- **Soda-acid, gas cartridge**—Direct the stream at the base of the flame.
- **Pump tank**—Place the foot on the footrest and direct the stream at the base of the flame.
- **Dry chemical**—Direct the nozzle at the base of the flames. In the case of Class A fires, follow up by directing the dry chemicals at the remaining material that is burning.

FIGURE 2–9 Guide to fire extinguisher selection

Class of Fire	Description of Fire	Typical Fuel Involved	Type of Extinguisher
Class A Fires	**For Ordinary Combustibles** Put out a Class A fire by lowering its temperature or by coating the burning combustibles.	Wood Paper Cloth Rubber Plastics Rubbish Upholstery	Water*[1] Foam* Multipurpose dry chemical[4]
Class B Fires	**For Flammable Liquids** Put out a Class B fire by smothering it. Use an extinguisher that gives a blanketing, flame-interrupting effect; cover whole flaming liquid surface.	Gasoline Oil Grease Paint Lighter fluid	Foam* Carbon dioxide[5] Halogenated agent[6] Standard dry chemical[2] Purple K dry chemical[3] Multipurpose dry chemical[4]
Class C Fires	**For Electrical Equipment** Put out a Class C fire by shutting off power as quickly as possible and by always using a nonconducting extinguishing agent to prevent electric shock.	Motors Appliances Wiring Fuse boxes Switchboards	Carbon dioxide[5] Halogenated agent[6] Standard dry chemical[2] Purple K dry chemical[3] Multipurpose dry chemical[4]
Class D Fires	**For Combustible Metals** Put out a Class D fire of metal chips, turnings, or shavings by smothering or coating with a specially designed extinguishing agent.	Aluminum Magnesium Potassium Sodium Titanium Zirconium	Dry powder extinguishers and agents only

**Cartridge-operated water, foam, and soda-acid types of extinguishers are no longer manufactured. These extinguishers should be removed from service when they become due for their next hydrostatic pressure test.*

Notes:

(1) Freezes in low temperatures unless treated with antifreeze solution, usually weighs over 20 pounds (9 kg), and is heavier than any other extinguisher mentioned.

(2) Also called ordinary or regular dry chemical (sodium bicarbonate).

(3) Has the greatest initial fire-stopping power of the extinguishers mentioned for Class B fires. Be sure to clean residue immediately after using the extinguisher so sprayed surfaces will not be damaged (potassium bicarbonate).

(4) The only extinguishers that fight A, B, and C classes of fires. However, they should not be used on fires in liquefied fat or oil of appreciable depth. Be sure to clean residue immediately after using the extinguisher so sprayed surfaces will not be damaged (ammonium phosphates).

(5) Use with caution in unventilated, confined spaces.

(6) May cause injury to the operator if the extinguishing agent (a gas) or the gases produced when the agent is applied to a fire is inhaled.

If a fire extinguisher is used, report it to your instructor or service manager so that it can be immediately recharged.

HYBRID DRIVE AND NG PRECAUTIONS

Although the subject matter of this textbook does not deal with alternative fuels and hybrid drive technology, both hybrid drive and gaseous-fueled (NG, propane, and DME) powertrains are becoming increasingly common in commercial heavy vehicle shops. Specialized training is required to work on the powertrains or fuel systems of these vehicles. And if you are asked to do a lube job on one, you should clearly understand the potential dangers. That begins with the ability to define them.

Hybrid Electric

Hybrid-electric drive vehicles have become increasingly common, especially in inner-city transit operations. A hybrid-electric drive vehicle typically uses a diesel- or NG-fueled engine and a generator system to produce high-voltage electricity. Once produced this electricity is stored on chassis in battery and capacitor banks. High-voltage potential points are clearly marked on new vehicles but these cautions are not so easy to see once the vehicle has been in service for a few years. If you have not been trained to work on hybrid-electric vehicles, exercise extreme caution when working on those chassis systems where you would not normally expect high-voltage impulses.

Hybrid Hydraulic

Hybrid-hydraulic drive vehicles are most often seen in light-duty commercial vehicles, especially in inner-city pick-up and delivery operations. A hybrid-hydraulic drive vehicle typically uses a gasoline-, diesel-, or NG-fueled engine and hydraulic pump combined with accumulators. Hydraulic fluid can be stored in accumulators at extremely high pressures so great caution should be exercised when working on these vehicles. Minor, barely visible leaks can result in severe injuries. Never check for hydraulic leaks using your hand because high pressure hydraulic fluid can penetrate the glove. Use cardboard held with a gloved hand.

WARNING

Pinhole hydraulic leaks can cause maiming and death if not identified. A **pinhole injection** injury may occur unbeknownst to the injured person: this type of injury must be identified shortly after the injection intrusion into the flesh. If you suspect a hydraulic pinhole injection injury, get to a hospital immediately.

Gaseous-Fueled Vehicles

These have been a presence on the West Coast of North America for a generation, but the low cost of NG has now made this a fueling option for Class 8 linehaul trucks. In addition, propane continues to be used in some light-duty commercial trucks and vans and Volvo-Mack are in the process of introducing DME-fueled trucks to North America.

DME-fueled vehicles are clearly identified with warnings and cautions around the fuel tanks and fuel system components. NG-fueled engines use the terms liquefied natural gas (LNG) and compressed natural gas (CNG) for identification purposes and technicians should always exercise extreme caution when working on or in the vicinity of these vehicles. Special training is required to work on the engines and fuel systems of propane, DME, CNG, and LNG vehicles.

CHASSIS AND SHOP ELECTRICAL SAFETY

Trucks today can use numerous computers, all networked to a central data backbone using multiplexing technology (Chapter 12). The computers used to control chassis subsystem components function on low-voltage electrical signals and use thousands of solid-state components. While some of these electronic subcircuits are protected against transient voltage spikes, others are not. An unwanted high-voltage spike caused by static discharge or careless placement of electric welding grounds can cause thousands of dollars worth of damage.

Some trucks are equipped with electrical isolation switches. These should be opened any time major service or repair work is performed on a vehicle.

STATIC DISCHARGE

When you walk across a plush carpet, your shoes "steal" electrons from the floor. This charge of electrons accumulates in your body, and when you go to grab a door handle, the excess of electrons discharges itself into the door handle, creating an arc as it does so. This is known as **static discharge**. Accumulation of a **static charge** is influenced by factors such as relative humidity and the type of footwear you are wearing. Getting a little zap from the static charge that can accumulate in the human body is seldom going to produce any adverse effects to human health but it can damage sensitive solid-state circuits.

Picture a fuel tanker transport running down an interstate. In the same way your body steals electrons from a carpet, so does the tanker steal electrons from the atmospheric air it is forcing itself through. However, the charge differential that can be accumulated by the tanker is much greater and can exceed 50,000 volts. This type of charge differential can be highly dangerous and produce a spark that can easily ignite fuel vapors. This potential danger accounts for the legal requirement to ground out a tanker chassis before undertaking any load or unload operation.

Static Discharge and Computers

Static charge accumulation in the human body can easily damage computer circuits. Because some pieces of equipment today have a dozen, sometimes more, computer-controlled circuits, it is important for technicians to understand the effects of static discharge. The reason that static discharge has not caused more problems than it has in the service repair industry is due primarily to:

- Technicians' footwear of choice, usually rubber-soled boots.
- The tendency of shop floors to be concrete-surfaced rather than carpeted.

Neither of the above factors is conducive to static charge accumulation, but technicians should remember that any carpeted flooring is conductive, and due precautions should be taken. That said, it is good practice when troubleshooting requires you to access electronic circuits, to use a ground strap before separating sealed connectors, connecting breakout Tees/boxes, or accessing the data bus beyond just connecting an electronic service tool (EST) to it. Special care should be taken when working with modules that require you physically remove and replace solid-state components such as PROM chips from a motherboard.

Chassis Wiring and Connectors

Every year, millions of dollars of damage to mobile equipment is created by service technicians who ignore original equipment manufacturer (OEM) precautions regarding working with chassis wiring systems. Perhaps the most common abuse is puncturing wiring insulation with test lights and digital multimeter (DMM) leads. When you puncture the insulation on copper wiring, in an instant that wiring becomes exposed to both oxygen (in the air) and moisture (relative humidity). The chemical reaction almost immediately produces copper oxides, which then react with moisture to form corrosive cupric acid. The acid begins to eat away the wiring, first creating high resistance, and ultimately consuming the wire. The effect is accelerated when copper-stranded wiring is used because the surface area over which the corrosion can act is so much greater.

CAUTION

Never puncture the insulation on chassis wiring. Read the section titled "Chassis Wiring and Connectors" if you want to know why!

The sad thing about this type of abuse is that it is so easily avoided. There are so many ways that a repair technician can access wiring circuits using the correct tools; it is just stupidity not to use them. Use breakout Tees, breakout boxes, and test lead spoons.

Breakout Tees. A breakout Tee allows you to access an energized circuit. Most breakout Tees allow you to access a single-wire circuit: to "Tee" into the circuit, you separate it at a connector, then connect the breakout Tee. This allows you to test the circuit status while it is electrically active.

Breakout Box. A breakout box is a troubleshooting device required by some OEMs. A breakout box does the same thing as a breakout Tee except that, when it is connected into the circuit, you can check the status of a large number of circuit wires and/or terminals. A troubleshooting tree in interactive diagnostic software will often direct a technician to install a breakout box into a circuit, usually at a connector block, then perform a sequence of DMM driven tests, the results of which may have to be entered into the diagnostic computer.

Test Lead Spoons. Test lead spoons have the potential to save so much unnecessary expenditure, it is surprising that they are not distributed free of charge to anyone working on mobile equipment. Test lead spoons are designed for insertion into weatherproof connectors and terminal blocks without creating damage to either the terminal or the wiring. Get a set from your tool supplier.

Circuit Test Lights. Test lights have few uses on modern equipment and have the potential to inflict damage on electrical wiring when misused. A circuit test light consists of a sharp spike intended to abuse wiring insulation, a light bulb, a wire, and a ground clamp. If you connect the ground clamp on the tool to a good ground, then any time the spike contacts a voltage differential, the bulb illuminates. Some are specified for working on electronic circuits, and these will illuminate with much smaller voltage potentials. Although this category of circuit test light will not create a current overload in a sensitive circuit, the spike may damage connector blocks and wiring insulation.

CAUTION

Be aware of the damage you can create using a circuit test light. Many can cause current overloads in sensitive electronic circuit, and even those rated for working on electronics can damage insulation, insulated connectors, and connector blocks.

Mains Electrical Equipment

Mains electrical circuits, unlike vehicle electrical circuits, operate at pressures that can be lethal, so you have to be careful when working around any electrical equipment. Electrical pressures may be **single-phase main** operating at pressure values between 110 and 120 volts *or* three-phase operating at pressures between 360 and 600 volts. In some jurisdictions, repairs to mains electrical equipment and circuits are required to be undertaken by certified personnel. If you undertake to repair electrical equipment, make sure you know what you are doing!

Take extra care when using electrically powered equipment when the area you are working in is wet. And remember that a transient spike of AC voltage driven through a chassis data bus can knock out electronic equipment networked to it. Electrical equipment can also be dangerous around vehicles because of its potential to arc and initiate a fire or explosion.

WARNING

Do not undertake to repair mains electrical circuit and equipment problems unless you are qualified to do so.

Some examples of shop equipment using single-phase mains electricity are:

- Electric hand tools
- Portable electric lights
- Computer stations
- Drill presses
- Burnishing and broaching tools

CAUTION

Take care when using trouble lights with incandescent bulbs around volatile liquids and flammable gases: these are capable of creating sufficient heat to ignite flammables. Many jurisdictions have banned the use of this type of trouble light, and they should be never used in garages in which gasoline-, propane-, and natural gas-fueled vehicles are present. Best bet: use fluorescent or LED trouble lights in rubber-insulated housing!

Some examples of shop equipment using high voltage, three-phase electricity are:

- Most welding equipment
- Dynamometers
- Lathes and mills
- Large shop air compressors

2.3 OXYACETYLENE EQUIPMENT

Technicians use oxyacetylene for heating and cutting on a daily basis. This equipment is used less commonly for braising and welding. Some basic instruction in the techniques of oxyacetylene equipment safety and handling is required. The following information should be understood by anyone working in close proximity with oxyacetylene equipment.

ACETYLENE CYLINDERS

Acetylene regulators and hose couplings use a left-hand thread. The regulator gauge working pressure should *never* be set at a value exceeding 15 psi (100 kPa): acetylene becomes extremely unstable at pressures higher than 15 psi (100 kPa). The acetylene cylinder should always be used in the upright position. Using an acetylene cylinder in a horizontal position will result in the acetone draining into the hoses.

It is not possible to determine with any accuracy the quantity of acetylene in a cylinder by observing the pressure gauge because it is in a dissolved condition. The only really accurate way of determining the quantity of gas in the cylinder is to weigh it and subtract this from the weight of the full cylinder, often stamped on the side of the cylinder.

CAUTION

It is a common malpractice to set acetylene pressure at high values. Check a welder's manual for the correct pressure values to set for the equipment and procedure you are using.

CAUTION

Never operate an acetylene cylinder in anything but an upright condition. Using acetylene when the cylinder is horizontal results in acetone exiting with the acetylene can destabilize the remaining contents of the cylinder.

OXYGEN CYLINDERS

It should be noted that oxygen cylinders tend to pose more problems than acetylene when exposed to fire. They should always be stored in the same location in a service shop when not in use (this location should be identified to the Fire Department during an inspection) and not left randomly on the shop floor.

Oxygen regulators and hose fittings use a right-hand thread. The cylinder pressure gauge will accurately indicate the oxygen quantity in the cylinder, meaning that the volume of oxygen in the cylinder is approximately proportional to the pressure.

Oxygen is stored in the cylinders at a pressure of 2,200 psi (15 MPa) and the hand wheel actuated valve "forward-seats" to close the flow from the cylinder and "back-seats" when the cylinder is opened. Therefore, it is important to ensure that the valve is fully opened when in use. The consequence of not fully opening the valve is leakage past the valve threads.

WARNING

Never use oxygen as a substitute for compressed air when cleaning components in a shop environment. Oxygen can combine with solvents, oils, and grease, resulting in an explosion.

Regulators and Gauges

A regulator is a device used to reduce the pressure at which gas is delivered: it sets the working pressure of the oxygen or fuel. Both oxygen and fuel regulators function similarly in that they increase the working pressure when turned clockwise (CW). They close off the pressure when backed out counterclockwise (CCW).

Pressure regulator assemblies are usually equipped with two gauges. The cylinder pressure gauge indicates the actual pressure in the cylinder. The working pressure gauge indicates the working pressure, and this should be trimmed using the regulator valve to the required value while under flow.

Hoses and Fittings

The hoses used with oxyacetylene equipment are usually color-coded. Green is used to identify the oxygen hose, and red identifies the fuel hose. Each hose connects the cylinder regulator assembly with the torch. Hoses may be single or paired (Siamese). Hoses should be routinely inspected and replaced when defective. A leaking hose should never be repaired by being wrapped with tape. In fact, it is generally bad practice to consider repairing welding gas hoses by any method; the safe solution is to replace them when they fail.

Fittings couple the hoses to the regulators and the torch. Each fitting consists of a nut and gland. Oxygen fittings use a right-hand thread, and fuel fittings use a left-hand thread. The fittings are machined out of brass that has a self-lubricating characteristic. Never lubricate the threads on oxyacetylene fittings.

TORCHES AND TIPS

Torches should be ignited by first setting the working pressure under flow for both gases, then opening the fuel valve *only* and igniting the torch using a flint spark lighter. Set the acetylene flame to a clean burn (no soot), then open the oxygen valve to set the appropriate flame. When setting a cutting torch, set the cutting oxygen last. When extinguishing the torch, close the fuel valve first, then the oxygen: lastly, the cylinders should then be shut down and the hoses purged.

Welding, cutting, and heating tips may be used with oxyacetylene equipment. Consult a welder's manual to determine the appropriate working pressures for the tip/process to be used. There is a tendency to set gas working pressure high. Even when using a large heating tip often described as a rosebud, the working pressure of both the acetylene and the oxygen should be set at 7 psi (50 kPa)—and that is probably the highest you will ever have to set the acetylene pressure in normal shop use. Oxygen pressures are set higher when using a cutting torch.

Backfire

Backfire is a condition where the fuel ignites within the nozzle of the torch, producing a popping or squealing noise: it often occurs when the torch nozzle overheats. Extinguish the torch and clean the nozzle with tip cleaners. Torches may be cooled by immersing in water briefly with the oxygen valve open.

Flashback

Flashback is a much more severe condition than backfire: it takes place when the flame travels backward into the torch to the gas-mixing chamber and beyond. Causes of flashback are inappropriate pressure settings (especially low-pressure settings) and leaking hoses/fittings. When a backfire or flashback condition is suspected, close the cylinder valves immediately beginning with the fuel valve. Flashback arresters are usually fitted to the torch and will limit the extent of damage when a flashback occurs.

Eye Protection

Safety requires that a #4- to #6-grade filter be used whenever using an oxyacetylene torch. The flame radiates ultraviolet light that can damage eyesight. UV-rated sunglasses are not sufficient protection.

OXYACETYLENE PRECAUTIONS

Take the following precautions when working with oxyacetylene:

- Store oxygen and acetylene upright in a well-ventilated, fireproof room.
- Protect cylinders from snow, ice, and direct sunshine.
- Remember that oil and grease may ignite spontaneously in the presence of oxygen.
- Never use oxygen in place of compressed air.
- Avoid bumping and dropping cylinders.
- Keep cylinders away from electrical equipment where there is a danger of arcing.
- Never lubricate the regulator, gauge, cylinder, and hose fittings with oil or grease.
- Blow out cylinder fittings before connecting regulators: make sure the gas jet is directed away from equipment and other people.
- Use soapy water to check for leaks: NEVER use a flame to check for leaks.
- Thaw frozen spindle valves with warm water: NEVER use a flame.

ADJUSTMENT OF THE OXYACETYLENE FLAME

To adjust an oxyacetylene flame, the torch acetylene valve is first turned on and the gas ignited. At the point of ignition, the flame will be yellow and produce black smoke. Next, the acetylene pressure should be increased by using the torch fuel valve. This will increase the brightness and reduce the smoking. At the point the smoking disappears, the acetylene working pressure can be assumed to be correct for the nozzle jet size used. Now, the torch oxygen valve is turned on. This will cause the flame to become generally less luminous, and an inner blue luminous cone surrounded by a white-colored plume is formed at the tip of the nozzle. The white-colored plume indicates excess acetylene. As more oxygen is supplied, this plume reduces until there is a clearly defined blue cone with no white plume visible. This indicates the *neutral* flame used for most welding and cutting operations.

Oxidizing Flame

If after setting a neutral flame, the oxygen supply is increased, the blue cone will become smaller and sharper in definition and the outer envelope will become streaky. This is known as an *oxidizing flame,* which indicates excess oxygen. For most welding procedures an oxidizing flame should be avoided, but in some special applications, a slightly oxidizing flame is required.

Carburizing Flame

A *carburizing flame* is indicated by the presence of a white plume surrounding an inner blue cone. A carburizing flame should also be avoided in most welding operations, although a very slight carburizing flame is used in certain special applications.

2.4 ELECTRIC ARC WELDING

Some trucks are equipped with electrical isolation switches: make sure you open this switch before using an electric welder on a vehicle. When any type of electric welding is performed on a chassis, ensure that the ground clamp is placed close to the work. Placing a welding ground clamp on the front bumper when you are welding at the rear of the chassis is not only capable of causing electronic damage but also of taking out bearings and journals in the major powertrain components of the vehicle. Although electricity can be relied on to take the shortest path to complete a circuit, sometimes it will experiment with determining which is the shortest path. Pulsing electricity through crankshaft journals and transmission bearings causes arcing that results in costly damage that will probably result in someone getting fired.

CAUTION

Whenever performing electric arc welding or cutting on a chassis, make sure you place the ground clamp as close to the work area as possible to avoid creating chassis electronic or arcing damage.

Electric arc welding and cutting processes are used extensively in truck and heavy equipment service garages. Arc welding stations either receive or generate high-voltage charges, then transform them to lower voltage, high-current circuits. Before attempting to use any type of arc welding equipment, make sure you receive some basic instruction and training. The following types of welding station are nonspecialized in application and are found in many repair shops:

- Arc welding station: uses a flux-coated, consumable electrode, often known as *stick welding*.
- Metal inert gas (MIG): uses a continuous reel of wire that acts as the electrode, around which inert gas is fed to shield the weld from air and ambient moisture.
- Tungsten inert gas (TIG): uses a nonconsumable tungsten electrode surrounded by inert gas, and filler rods are dipped into the welding puddle created.
- Carbon-arc cutting: arc is ignited using carbon electrodes to melt base metal while a jet of compressed air blows through the puddle to make the cut.
- Plasma-cutting: electric arc is shrouded by an inert gas blown through the arc-created molten puddle turning the material into plasma: neater than carbon arc cutting.

Because less training and faster weld speeds are possible with MIG welding processes, stick arc welding is less commonly used in today's shops. However, when specialty alloy and tempered materials are required to be welded (truck frame rails), arc welding electrodes (sticks) are commonly selected.

2.5 SHOP TOOL SAFETY

Understanding the proper use of non-power-driven hand tools, portable electric power tools, pneumatic power tools, and stationary equipment will help eliminate accidents. Observe the following:

- Select the proper size and type of tool for the job.
- Use tools only for the purpose for which they are designed.
- Keep tools in safe working condition.
- Store tools safely when not in use.
- Report any breakage or malfunctions to your instructor or service manager.
- Make sure that cutting tools are properly sharpened and in good condition.
- Do not use tools with loose or cracked handles.
- Never use tools unless you know how to operate them.

Shop tool safety depends mainly on the person who uses the tool. Knowing what a tool is designed to do and how to use it correctly is the key. Because of the importance of this, Chapter 3 is devoted to safe use of the tools used by the heavy-duty truck service technician.

2.6 HAZARDOUS MATERIALS AND WHMIS

Heavy-duty truck repair work involves use of many materials classified as hazardous by both state and federal governments. These materials include such items as solvent and cleaners, paint and body repair products, adhesives, acids, coolants, and refrigerant products.

Hazardous materials are those that could cause harm to a person's well-being. Hazardous materials can also damage and pollute land, air, or water. There are four types of hazardous waste:

- **Flammable—Flammable** materials easily catch fire or explode.
- **Corrosive—Corrosive** materials are so caustic that they can dissolve metals and burn the skin and eyes.
- **Reactive—Reactive** materials will become unstable (burn, explode, or give off **toxic** vapors) if mixed with air, water, heat, or other materials.

- **Toxic**—Toxic materials can cause illness or death after being inhaled or upon contacting the skin.

LAWS REGULATING HAZARDOUS MATERIALS

The Hazard Communication Regulation—commonly called the **Right-to-Know Law**—was passed by the federal government and is administered by **Occupational Safety and Health Administration (OSHA)**. This law mandates that any company that uses or produces hazardous chemicals or substances must inform its employees, customers, and vendors of any potential hazards that may exist in the workplace as a result of using the products. All employers today are required to provide **Workplace Hazardous Materials Information System (WHMIS)** training to their employees: this training underscores the type of potentially dangerous materials a worker might encounter in a specific workplace. This requires that each workplace provide site-specific training to employees that teaches them how to interpret **material safety data sheet (MSDS)** labels.

Most important is that you keep yourself informed. You are the only person who can keep yourself and those with whom you work protected from the dangers of hazardous materials. These are some of the highlights of the Right-to-Know Law:

- You have a right to know what hazards you may face on the job.
- You have a right to learn about these materials and how to protect yourself from them.
- You cannot be fired or discriminated against for requesting information and training on how to handle hazardous materials.
- You have the right for your doctor to receive the same hazardous material information that you receive.

Employee/Employer Obligations

Familiarize yourself with the material safety data sheet (MSDS) shown in **Figure 2–10**. This MSDS is for the substance we know as caustic soda. It is used as a cleaning agent and is commonly dissolved into water-based solutions in shop hot soak tanks. Knowing something about the MSDS helps prevent injuries because it teaches you to work safely with potentially dangerous substances. An employer or school that uses hazardous materials must:

- Provide a safe workplace.
- Educate employees about the hazardous materials they will encounter while on the job.
- Recognize, understand, and use warning labels and MSDS (Workplace Hazardous Material Information Systems—WHMISs).
- Provide personal protective clothing and equipment and train employees to use them properly.

You, the employee or student, must:

- Read the warning labels on the materials.
- Follow the instructions and warnings on the MSDS or WHMIS.
- Take the time to learn to use protective equipment and clothing.
- Use common sense when working with hazardous materials.
- Ask the service manager if you have any questions about a hazardous material.

Personal Protection

One of the greatest concerns to personal safety is the effect of long-term exposure to hazardous materials, particularly solvents, cleaning agents, and paint products.

The importance of this is easily seen in concerns over asbestos. When first introduced on the market, asbestos was widely used in brake pads, brake shoes, clutches, and other automotive applications. We now know that asbestos fibers pose a health risk and that long-term exposure to small amounts of asbestos can cause lethal health problems. For this reason, asbestos has gradually been phased out of the automotive and truck components market.

To handle hazardous materials properly, you need to:

- Know what the material is.
- Know that the material is dangerous.
- Know the correct safety equipment needed for working with that material.
- Know how to use the safety equipment properly and properly dispose of flammable materials in a fireproof container (**Figure 2–11**).
- Make sure that the safety equipment fits properly and is in working order.

Good personal hygiene practices are important in minimizing exposure to asbestos dust and other hazardous wastes:

- Do not smoke.
- Wash before eating.
- Shower after work.
- Change to work clothes upon arrival at work, and change from work clothes after work. Work clothing should not be taken home.

2.7 HANDLING AND DISPOSAL OF HAZARDOUS WASTE

Specific laws govern the disposal of hazardous wastes. You and your shop must be aware of how these laws affect shop operation. These laws include the **Resource Conservation and Recovery Act (RCRA)**. This law states that after you have used hazardous materials, they must be properly stored

FIGURE 2–10 Typical material safety data sheet

ABC CHEMICALS

24-Hour Emergency Phone (212) 987-XXXX

Division of ABC Materials Company / P. O. Box 12345 • New York, NY

I – IDENTIFICATION

CHEMICAL NAME	CHEMICAL FORMULA	MOLECULAR WEIGHT
Sodium Hydroxide Solution	NaOH	40.00

TRADE NAME
Caustic Soda, 73%, 50% and Weaker Solutions

SYNONYMS	DOT IDENTIFICATION NO.
Liquid Caustic, Lye Solution, Caustic, Lye, Soda Lye	UN 1824

II – PRODUCT AND COMPONENT DATA

COMPONENT(S) CHEMICAL NAME	CAS REGISTRY NO.	% (wt.) Approx.	OSHA PEL
Sodium Hydroxide	1310-73-2	73, 50 and less	2 mg/m^3 Ceiling

Note: This material safety data sheet is also valid for caustic soda solutions weaker than 50%. The boiling point, vapor pressure, and specific gravity will be different from those listed.

* Denotes chemical subject to reporting requirements of Section 313 of Title III of the 1986 Superfund Amendments and Reauthorization Act (SARA) and 40 CFR Part 372

III – PHYSICAL DATA

APPEARANCE AND ODOR	SPECIFIC GRAVITY
Colorless or slightly colored, clear or opaque; odorless	50% Solution: 1.53 @ 60°F/60°F 73% Solution: 1.72 @ 140°F/4°F
BOILING POINT	**VAPOR DENSITY IN AIR (Air = 1)**
50% Solution: 293°F (145°C) 73% Solution: 379°F (192.8°C)	N/A
VAPOR PRESSURE	**% VOLATILE, BY VOLUME**
50% = 6.3 mm Hg @ 104°F 73% = 6.0 mm Hg @ 158°F	0
EVAPORATION RATE	**SOLUBILITY IN WATER**
0	100%

IV – REACTIVITY DATA

STABILITY	CONDITIONS TO AVOID
Stable	Mixture with water, acid or incompatible materials can cause splattering and release of large amounts of heat (Refer to Section VIII). Will react with Some metals forming flammable hydrogen gas.

INCOMPATIBILITY (Materials to avoid)
Chlorinated and fluorinated hydrocarbons (i.e., chloroform, difluoroethane), acetaldehyde, acrolein, aluminum, chlorine trifluoride, hydroquinone, maleic anhydride, phosphorous pentoxide, and tetrahydrofuran.

HAZARDOUS DECOMPOSITION PRODUCTS Will not decompose.

HAZARDOUS POLYMERIZATION Will not occur.

V – FIRE AND EXPLOSION HAZARD DATA

FLASHPOINT (Method used)	FLAMMABLE LIMITS IN AIR
None	None

EXTINGUISHING AGENTS
N/A NFPA Hazard Ratings: Health 3; Flammability 0; Reactivity 1

UNUSUAL FIRE AND EXPLOSION HAZARDS
Firefighters should wear self-contained positive pressure breathing apparatus, and avoid skin contact. Refer to Reactivity Data, Section IV.

EXPOSURE LIMITS (When exposure to this product and other chemicals is concurrent, the exposure limit must be defined in the workplace.)

ACGIH: 2 mg/m^3 Ceiling
OSHA 2 mg/m^3 Ceiling

IDLH: 250 mg/m^3

Effects described in this section are believed not to occur if exposures are maintained at or below appropriate TLVs.
Because of the wide variation in individual susceptibility, these exposure limits may not be applicable to all persons and those with medical conditions listed below.

(continued)

FIGURE 2–10 (continued)

VI – TOXICITY AND FIRST AID

MEDICAL CONDITIONS AGGRAVATED BY EXPOSURE

May aggravate existing skin and/or eye conditions on contact.

ACUTE TOXICITY **Primary route(s) of exposure:** ☒ **Inhalation** ☒ **Skin Absorption** ☐ **Ingestion**

Inhalation: Inhalation of solution mist can cause mild irritation at 2 mg/m^3. More severe burns and tissue damage at the upper respiratory tract, can occur at higher concentrations. Pneumonitis can result from severe exposures.

Skin: Major potential hazard–contact with the skin can cause severe burns with deep ulcerations. Contact with solution or mist can cause multiple burns with temporary loss of hair at burn site. Solutions of 4% may not cause irritation and burning for several hours, while 25% to 50% solutions can cause these effects in less than 3 minutes.

Eyes: Major potential hazard–Liquid in the eye can cause severe destruction and blindness. These effects can occur rapidly effecting all parts of the eye. Mist or dust can cause irritation with high concentrations causing destructive burns.

Ingestion: Ingestion of sodium hydroxide can cause severe burning and pain in lips, mouth, tongue, throat, and stomach. Severe scarring of the throat can occur after swallowing. Death can result from ingestion.

FIRST AID

Inhalation: Move person to fresh air. If breathing stops, administer artificial respiration. Get medical attention immediately.

Skin: Remove contaminated clothing immediately and wash skin thoroughly for a minimum of 15 minutes with large quantities of water (preferably a safety shower). Get medical attention immediately.

Eyes: Wash eyes immediately with large amounts of water (preferably eyewash fountain), lifting the upper and lower eyelids and rotating eyeball. Continue washing for a minimum of 15 minutes. Get medical attention immediately.

Ingestion: If person is conscious, give large quantities of water to dilute caustic. Do not induce vomiting. Get medical attention immediately. Do not give anything by mouth to an unconscious person.

CHRONIC TOXICITY

No known chronic effects

Carcinogenicity: No studies were identified relative to sodium hydroxide and carcinogenicity. Sodium hydroxide is not listed on the IARC, NTP, or OSHA carcinogen lists.

Reproductive Toxicity: No studies were identified relative to sodium hydroxide and reproductive toxicity.

VII – PERSONAL PROTECTION AND CONTROLS

RESPIRATORY PROTECTION

Where concentrations exceed or are likely to exceed 2mg/m^3 use a NIOSH/MSHA approved high-efficiency particulate filter with full facepiece or self-contained breathing apparatus. Follow any applicable respirator use standards and regulations.

VENTILATION

As necessary to maintain concentration in air below 2mg/m^3 at all times.

SKIN PROTECTION

Wear neoprene, PVC, or rubber gloves; PVC rain suit; rubber boots with pant legs over boots.

EYE PROTECTION

Chemical goggles which are splashproof and faceshield.

HYGIENE

Avoid contact with skin and avoid breathing mist. Do not eat, drink, or smoke in work area. Wash hands prior to eating, drinking, or using restroom. Any protective clothing or shoes which become contaminated with caustic should be removed immediately and thoroughly laundered before wearing again.

FIGURE 2–10 (continued)

OTHER CONTROL MEASURES

Safety shower and eyewash station must be located in immediate work area. To determine the exposure level(s), monitoring should be performed regularly.
NOTE: Protective equipment and clothing should be selected, used, and maintained according to applicable standards and regulations. for further information, contact the clothing or equipment manufacturer or the ABC Chemicals Technical Service Department.

VIII – STORAGE AND HANDLING PRECAUTIONS

Follow protective controls set forth in Section VII when handling this product.
Store in closed, properly labeled tanks or containers. Do not remove or deface labels or tags.
When diluting with water, slowly add caustic solution to the water. Heat will be produced during dilution. Full protective clothing, goggles, and faceshield should be worn. Do not add water to caustic because excessive heat formation will cause boiling and spattering.
Contact of caustic soda cleaning solutions with food and beverage products (in enclosed vessels or spaces) can produce lethal concentrations of carbon monoxide gas. Do not enter confined spaces such as tanks or pits without following proper entry procedures as required by 29 CFR 1910.146.

SARA Title III Hazard Categories: Immediate Health.

IX – SPILL, LEAK AND DISPOSAL PRACTICES

STEPS TO BE TAKEN IN CASE MATERIAL IS RELEASED OR SPILLED

Cleanup personnel must wear proper protective equipment (refer to Section VII). Completely contain spilled material with dikes, sandbags, etc., and prevent run-off into ground or surface waters or sewers. Recover as much material as possible into containers for disposal. Remaining material may be diluted with water and neutralized with dilute hydrochloric acid. Neutralization products, both liquid and solid, must be recovered for disposal. Reportable Quantity (RQ) is 1000 lb. Notify National Response Center (800/424-8802) of uncontained releases to the environment in excess of the RQ.

WASTE DISPOSAL METHOD

Recovered solids or liquids may be sent to a licensed reclaimer or disposed of in a permitted waste management facility. Consult federal, state, or local disposal authorities for approved procedures.

X – TRANSPORTATION

DOT HAZARD CLASSIFICATION

Sodium Hydroxide Solution, 8, UN 1824, PG II, RQ

PLACARD REQUIRED

Corrosive, 1824, Class 8

LABEL REQUIRED

Corrosive, Class 8. Label as required by OSHA Hazard Communication Standard, and any applicable state and local regulations.

Medical Emergencies

Call collect 24 hours a day for emergency toxicological information 212 987 YYYY

Other Emergency information

Call 212 987 XXXX (24 hours)

For any other information contact:

ABC Chemicals
Technical Service Department
P.O. Box 12345
New York, NY
8 AM to 5 PM Eastern Time
Monday Through Friday

DATE OF PREPARATION: November 1, 2014

NOTICE: ABC Chemicals believes that the information contained on this material safety data sheet is accurate. The suggested procedures are based on experience as of the date of publication. They are not necessarily all-inclusive nor fully adequate in every circumstance. Also, the suggestions should not be confused with nor followed in violation of applicable laws, regulation, rules, or insurance requirements.

NO WARRANTY IS MADE, EXPRESS OR IMPLIED, OF MERCHANTABILITY, FITNESS FOR A PARTICULAR PURPOSE OR OTHERWISE.

FIGURE 2–11 Wear proper safety equipment when handling waste and dispose of flammable materials in a fireproof container.

DuPont Performance Coatings

until an approved hazardous waste hauler arrives to take them to the disposal site. In addition, your responsibility continues until the materials arrive at an approved disposal site and are processed in accordance with the law.

When dealing with hazardous wastes:

1. Consult the MSDS or WHMIS under the "Waste Disposal Method" category.
2. Check with your instructor or service manager for the exact method for correct storage and disposal.
3. Follow their recommendations exactly.

WARNING

A service facility is ultimately responsible for the safe disposal of hazardous wastes, even after the waste leaves the shop. In the event of an emergency hazardous waste spill, contact the National Response Center (1-800-424-8802) immediately. Failure to do so can result in a fine or a year in jail, or both.

Never, under any circumstance, do any of the following:

- Throw hazardous materials into a dumpster.
- Dump waste anywhere but into a collection site of a licensed facility.
- Pour waste down drains, toilets, sinks, or floor drains.
- Use hazardous waste to kill weeds or to suppress dust in gravel.

2.8 SHOP RECORDS

It is the "law" that certain records must be kept by service facilities when trucks are involved in any kind of cargo haulage in North America. The legal requirement to do this has been recently reinforced by the introduction of CSA, which is discussed in more detail in Chapter 4 of this book. Shop records include the following:

- Identification of each vehicle, including the company unit number, model, serial number, year, and tire size
- A schedule showing the nature and due date of the various inspections and maintenance to be performed
- A record of the nature and date of inspections, maintenance, and repairs made
- Lubrication record

Maintaining shop records is important because:

- They are required by the Department of Transportation (DOT), the Federal Motor Carrier Safety Alliance (FMCSA), and Compliance Safety Accountability (CSA).
- A study of component failures can alert the manager to problems or highlight components that have performed well.
- A service manager cannot develop a good preventive maintenance (PM) program without the use of these records (see Chapter 4).
- In the event of a serious accident, an up-to-date vehicle maintenance file is usually beneficial in the defense of a lawsuit.

SHOP TALK

Federal Motor Vehicle Safety Standard (FMVSS) specifies that all vehicles in the United States be assigned a **vehicle identification number (VIN)**. The VIN is located on the left frame rail over the front axle and on the Vehicle Specification Decal (see the driver's manual for the location of the decal). All heavy-duty trucks are assigned a seventeen-character VIN (**Figure 2–12**). Using a combination of letters and numerals, the VIN codes the vehicle make, series or type, application, chassis, cab, axle configuration, gross vehicle weight rating (GVWR), engine type, model year, manufacturing plant location, and production serial number. A check digit (ninth position) is determined by assignment of weighted values to the other sixteen characters. These weighted values are processed through a series of equations designed to check the validity of the VIN and to detect VIN alteration. The VIN can also be accessed off the chassis data bus in current trucks.

FIGURE 2–12 Typical VIN

Description	Typical Identification Number								
	1FU	P	D	CY	B	2	J	P	345678
Decoding Table Number*	1	2	3	4	5		6	7	
Manufacturer, Make, Type of Vehicle	1FU								
Chassis, Front Axle Position, Brakes		P							
Model Series, Cab			D						
Engine Type				CY					
Gross Vehicle Weight Rating (GVWR)					B				
Check Digit						2			
Vehicle Model Year							J		
Plant of Manufacture								P	
Production Number									345678

WORK OR REPAIR ORDER

A work or repair order initiates the repair process. It is usually a "soft" (electronic) form with set fields requiring a minimum of keystroke data entry. Ideally, a work order results from a collaboration among the service writer, truck driver, and diagnostic technician. It may result from a condition noted in a driver's pre- and post-trip reports. The service writer usually quarterbacks a work order with input from the truck driver and the diagnostic technician. In today's truck shop, a work order is not finalized until a diagnostic scan has been performed by connecting the chassis data bus to an electronic service tool (EST). After the diagnostic routine has been run, the work order is produced and passed on to the repair technician either electronically (soft format) or in hard copy. **Figure 2–13** shows a hard copy example of a typical repair order.

Another form that is used in some shops that concerns the technician is the **parts requisition**. To order new parts, the technician creates a parts list and identifies the vehicle VIN or company identification folder. In many shops, a **dispatch sheet**, or work schedule, keeps track of dates when repair work is completed. Some dispatch sheets follow the job through each step of the repair process. This information is usually tracked by computer and helps the truck dispatcher route units for dispatch.

Most fleets, large or small, maintain a file on every vehicle, whether it be a tractor or a trailer. This file includes all the vehicle maintenance and repair records, schedule, and PM inspection results. Work orders performed on that particular vehicle are generally kept in its history folder. Federal Motor Carrier Safety Regulations require the mandatory records of a vehicle file to be retained where the vehicle is housed or maintained for a period of 1 year. They also require these records to be kept on hand for 6 months after the vehicle leaves the carrier's control. The same regulations require that the driver's vehicle condition report be retained for at least 3 months from the date of the report.

COMPUTERS IN THE SHOP

Almost every aspect of life on the service facility shop floor is computerized to some degree. Computers are used by the technician as tools for a wide range of diagnostic and data tracking routines. Shop computers (**Figure 2–14**) equipped with bar code wands help relieve technicians of the task of writing legible repair orders, assist shop managers with service scheduling, and minimize parts department errors. With the right software, computers can turn the repair/maintenance process into a "paperless" operation. Software packages are available to track every aspect of a vehicle. Increasingly, tracking and diagnostic software is being designed to be driven from compact, mobile phones that are fast becoming shop floor tools. Here are some examples:

- **Vehicle Maintenance Reporting Standards (VMRS)**. Developed by the American Trucking Association, this is a coding system that can be used at an individual part level up to total operating systems. Used within a software package, it can provide information on items such as part inventory, part usage, part wear rates, warranty information, and overall cost control.
- **Parts Inventory Control**. A basic inventory control system should maintain at least the following information on each part: part number, description, current quantity, and minimum (reorder) quantity. The goal is to balance maximum part availability against minimum investment in inventory.
- **Bar Coding**. One of the weaknesses of a computer inventory system is human input error. Bar coding systems reduce the error rate from 1 in 100 to 1 in 100 million. Bar coding used in maintenance shops is similar to that used in grocery stores, except it uses the 3-of-9 code instead of the universal product code (UPC).
- **Replace/Repair Analysis**. The ability to keep track of which parts are being replaced and repaired on fleet vehicles. This ability to analyze component failures can be a big help when spec'ing out future vehicles.
- **Bokodes**. A design to replace bar codes. A Bokode is a 3 mm (0.1 in.) dot that is read optically and can be embedded with much more data.
- **Preventive Maintenance**. A schedule set up in the computer tells you which vehicle needs servicing at what time. Fixed station hubs can interact wirelessly with vehicles in a fleet to ensure that service intervals do not get exceeded.
- **Electronic Data Interchange (EDI)**. A technology that allows computers to network and exchange data about business transactions that formerly had to be transmitted on paper or by telephone.

FIGURE 2–13 A typical work or repair order.

REPAIR ORDER INFORMATION

ABC FREIGHTLINER

REPAIR ORDER NO. ....

WRITTEN BY	DELIVERY DATE ....
SERIAL NO.	DELIVERY MILES ....
CUSTOMER ACCOUNT NO.	YR/MAKE/MODEL ....
NAME	UNIT NO. ....
ADDRESS	UNIT DATA ....
CITY	ENGINE MODEL ....
PHONE	ENGINE SERIAL NO. ....
P.O. NO.	TRANS. MODEL ....
ESTIMATE	TRANS. SERIAL NO. ....
MILEAGE	R. AXLE MODEL ....
MEMO - 1	R. AXLE SERIAL NO. ....
MEMO -2	R. AXLE MODEL ....
	R. AXLE SERIAL NO. ....

1	CONDITION		
	TYPE		
2	CONDITION		
	TYPE		
3	CONDITION		
	TYPE		
4	CONDITION		
	TYPE		
5	CONDITION		
	TYPE		
6	CONDITION		
	TYPE		
7	CONDITION		
	TYPE		
8	CONDITION		
	TYPE		
9	CONDITION		
	TYPE		
10	CONDITION		
	TYPE		

TERMS: STRICTLY CASH, APPROVED ACCOUNT OR CREDIT CARD

I the undersigned authorize you to perform the repairs and furnish the necessary materials. I understand any costs verbally quoted are an estimate only and not binding. Your employees may operate vehicle for inspecting, testing, and delivery at my risk. You will not be responsible for loss or damage to vehicle or articles left in it. I agree to pay reasonable storage on vehicle left more than 48 hours after notification that repairs are completed.

I AGREE THAT YOU HAVE AN EXPRESS LIEN ON THE DESCRIBED VEHICLE FOR THE CHARGES FOR PARTS AND LABOR FURNISHED UNDER THIS REPAIR ORDER INCLUDING THOSE FROM ANY PRIOR REPAIR ORDERS ON THE VEHICLE. IF I FAIL TO PAY SUCH CHARGES, I AGREE THAT THE VEHICLE MAY BE HELD UNTIL CHARGES ARE PAID IN FULL. IN THE EVENT OF LEGAL ACTION TO COLLECT ANY SUMS DUE, I AGREE TO PAY COSTS OF COLLECTION AND FEES INCLUDING REASONABLE ATTORNEY FEES.

If charges are handled on an approved open account, they are due on the 5th of the month following the purchase. A FINANCE CHARGE of 2% per month will be added to all balances 30 days past due.

AUTHORIZE BY: ____________________

FIGURE 2–14 Shop record keeping is made easier by specialty software programs.

- **Vehicle History**. A database that can provide a detailed profile on any vehicle in the fleet.
- **Work Order Generating**. The ability to automatically generate a work order when a vehicle needs servicing.
- **Cost Tracking**. Using databases on parts, vehicle history, PM scheduling, and warranty information to keep track of overall costs.
- **Warranty Information**. A database linked with vehicle history that provides specific information on parts warranty. Helpful in obtaining prompt warranty payments from vendors.
- **Vehicle and Driver Performance Analysis**. Trip data can be wirelessly downloaded during (telematics) or after each trip (microwave transponder).

SUMMARY

- Personal safety on the job may require eye or ear protection, or both, plus protective clothing and shoes. Long hair and loose jewelry are hazards.
- Lifting and carrying heavy materials the correct way will protect against injury.
- Tilt hoods and COE cabs with care.
- Do not smoke or engage in horseplay in the shop.
- Take care when oxyacetylene cutting or electric welding and always use protective eyewear with the correct filter shades. Avoid contact with hot metal components, and practice caution when working around flammable substances and potentially toxic gases.
- Take care when using a hydraulic press. Use protective eyewear. Make sure you are protected by safety cages or barriers.
- The work area should be kept clean, dry, and organized. Flammable liquids and solvents should be handled and stored carefully.
- Emergency telephone numbers and a first-aid kit should be handy.
- Use firefighting equipment appropriately: water or foam on ordinary combustibles; foam, carbon dioxide, or dry chemicals on burning liquids; carbon dioxide or dry chemicals on burning "live" electrical equipment; and special extinguishing agents on burning metals.
- Select, store, use, and maintain shop tools properly.
- Hazardous materials used in heavy-duty truck repair include flammable, corrosive, reactive, and toxic materials. Your employer is obligated to inform you of potential hazards in your workplace, and you have a right to protect yourself from them.
- Specific laws govern the disposal of hazardous wastes, including oil, antifreeze/coolants, refrigerants, batteries, battery acids, acids and solvents used for cleaning, and paint and body repair product wastes. Hazardous wastes may be recycled in the shop or removed by a licensed disposal hauler.
- By law, records must be kept by each shop of the repair and maintenance of trucks involved in interstate shipping.
- Computers have become part of the way of life in truck service shops. They are used for diagnostic routines, maintenance tracking, parts inventory control, work order generating, personnel management, and cost tracking.

REVIEW QUESTIONS

1. In this book, the sections that contain cautions about situations that might result in personal injury are labeled with what special notation?
   a. SHOP TALK
   b. SAFETY RULES
   c. CAUTION
   d. WARNING
2. The best way to prevent eye injury is to:
   a. take care during grinding or other operations that throw off particles.
   b. always wear safety glasses.
   c. always wear a bump cap.
   d. make sure a source of running water is available to flush foreign matter out of the eyes.
3. Which of the following describes a safe lifting and carrying practice?
   a. Twist your body when changing your direction of travel while carrying a heavy object.
   b. Bend forward to place a heavy object on a shelf or counter.

c. Lift by bending and then straightening your legs, rather than by using your back.
d. Position your feet as far as possible from the load when you begin to lift.

4. Oil on the floor of the work area can be cleaned up using a ______.

5. The exhaust pipe of a diesel engine must be connected to the shop exhaust system to protect against:
   a. breathing harmful emissions.
   b. carbon monoxide.
   c. carbon dioxide.
   d. fire.

6. Do NOT attempt to put out a Class B fire using:
   a. foam.
   b. carbon dioxide.
   c. a dry chemical type extinguisher.
   d. water.

7. A Class C fire involves:
   a. ordinary combustibles, such as paper or cloth.
   b. a flammable liquid.
   c. live electrical equipment.
   d. combustible metals.

8. To use carbon dioxide to extinguish a fire, direct the discharge:
   a. at the top of the flames.
   b. first at the edge of the fire, then forward and upward.
   c. at the base of the flames.
   d. several feet over the top of the flames.

9. What are the four types of hazardous wastes?

10. The "Right-to-Know Law" was passed by the government to:
    a. require any company that disposes of hazardous materials to inform their community.
    b. protect employees, customers, or vendors from hazards in the workplace caused by hazardous chemicals.
    c. require industries to compensate employees injured by contact with hazardous materials.
    d. require chemical industries to reveal complete information about the chemicals they produce.

11. Which of the following is covered under the Resource Conservation and Recovery Act?
    a. waste water
    b. waste oil
    c. cleaning solvents
    d. all of the above

12. Which of the following is an approved way of disposing of hazardous wastes?
    a. Washing them down the drain with plenty of water.
    b. Using them as weed killer.
    c. Recycling them by reusing them in the shop.
    d. Placing them in leak-proof containers and disposing of them in an RCRA-approved method.

13. What record must be kept by the shop on trucks involved in interstate shipping?
    a. out-of-service times
    b. names of all drivers
    c. names of all service technicians
    d. nature and date of inspections

14. What information is provided by the first digit of the VIN of a heavy-duty truck?
    a. model year
    b. axle configuration
    c. manufacturer
    d. gross weight rating

15. Which of the following is bar coding more likely to be used for in a truck service facility?
    a. parts inventory control
    b. tracking preventive maintenance
    c. profiling specifications on a vehicle
    d. generating work orders

16. When shutting down an oxyacetylene torch, which valve should be closed off first?
    a. oxygen
    b. acetylene
    c. time it so they are both shut off together
    d. neither: immerse torch head in water

17. When arc welding on a truck, where should the ground clamp be placed on the chassis?
    a. on the engine ECM
    b. on the engine cylinder block
    c. as far from the weld area as possible
    d. as close to the weld area as possible

18. Which of the following oxyacetylene conditions is potentially more dangerous?
    a. oxidizing flame
    b. flashback
    c. carburizing flame
    d. backfire

19. What footwear should be worn when working on high-voltage vehicle systems?
    a. flip flops
    b. UL safety boots
    c. sturdy leather-soled shoes
    d. ESR footwear

20. Technician A says that Right-to-Know legislation requires employers to provide WHMIS training to their employees. Technician B says that WHMIS training teaches employees how to interpret a MSDS. Who is correct?
    a. Technician A only
    b. Technician B only
    c. both A and B
    d. neither A nor B

3

Prerequisite: Chapters 1 and 2

TOOLS AND FASTENERS

OBJECTIVES

After reading this chapter, you should be able to:

- List some of the common hand tools used in heavy-duty truck repair.
- Describe how to use common pneumatic, electrical, and hydraulic power tools used in heavy-duty truck repair.
- Identify the mechanical and electronic measuring tools used in the heavy-duty truck shop.
- Describe the proper procedure for measuring with a micrometer.
- Identify the types of manufacturer service literature used in truck repair facilities and describe the type of information each provides.
- Explain the principles and precautions of working with various heavy-duty truck fasteners.
- Identify SAE Grade 2, Grade 5, and Grade 8 fasteners.
- Outline the process for removing and installing rivets.
- Identify some common rivets.
- Describe the application of sealants and adhesives used in the trucking industry.

KEY TERMS

adjustable pliers
adjustable wrench
air ratchet wrench
Allen wrench
bench grinder
block diagnosis chart
blowgun
box-end wrench
chisel
combination pliers
combination wrench
deburring
diagonal cutting pliers
dial caliper
driver's manual
extractor
feeler gauge
flaring
frame machine
graphical user interface (GUI)
hacksaw
hand tap
hand-threading die
helical
hex
Huck fastener
impact sockets
impact wrench
Industrial Fastener Institute (IFI)
locking pliers
machinist's rule
maintenance logs
maintenance manual
micrometer
needle nose pliers
open-end wrench
Phillips screwdriver
pliers
Posi-Drive™ screwdriver
press
punch
recall bulletin
rivet
screw pitch
screw pitch gauge
service bulletin
service information system (SIS)
service manual
sidecutter
socket wrench
stud remover
swaging
technical service bulletin (TSB)
thickness gauge
time guide
torque wrench
Torx® screwdriver
tree diagnosis chart
Unified National Coarse (UNC)
Unified National Fine (UNF)
Vise Grips™
Wrench

INTRODUCTION

It has been said that the mechanical technician's most important tool is the brain. However, most service repair work requires the use of tools. Tools allow the technician to convert thought into action. Knowing how to undertake a repair is essential, but not even the best technician can effect a repair without having the proper tools at hand.

Many of the tools that the heavy-duty truck technician uses daily are general-purpose hand and power tools. For instance, a complete collection of wrenches is indispensable. Most truck components and shop equipment use common fasteners. Depending on the manufacturer of the vehicle, the fasteners can be standard Society of Automotive Engineers (SAE) or metric size fasteners. A well-equipped technician should own both metric and standard wrenches in a range of sizes and styles. The proper use of the appropriate hand and power tools by the technician is important for performing quality heavy-duty truck service.

3.1 HAND TOOLS

There are many repair tasks that do not require the use of power tools. For those jobs, the heavy-duty truck technician must have suitable hand tools. Most service departments and garages require their technicians to buy their own hand tools. Technicians assemble a complete set of tools over a number of years: No technician is expected to own a complete set of tools on the first day of employment. A typical set of technician's hand tools and tool chest are shown in **Figure 3–1**.

HAMMERS

Hammers are identified by the material and weight of the head. There are two groups of hammer heads: steel and soft face. The heads of steel-face hammers (**Figure 3–2**) are made of high-grade alloy steel. The steel is deep forged and heat treated to a suitable degree of hardness. Soft-face hammers (**Figure 3–3**) have a surface that yields when it strikes an object. Soft-face hammers are preferred when machined surfaces and precision are involved or when marring a finish is undesirable. For example, a brass hammer can be used to drive in gears or shafts. Technicians should also consider purchasing ball peen hammers in which the head and haft (handle) are forged as one piece as shown in **Figure 3–4**; the insulation on the handle of these hammers is better at absorbing shock and they last longer in a truck shop environment.

FIGURE 3–1 A professional set of heavy-duty truck hand tools in a typical tool cabinet.

FIGURE 3–2 Types of steel hammers: (A) brass soft face; (B) ball peen; and (C) cross peen or blacksmith.

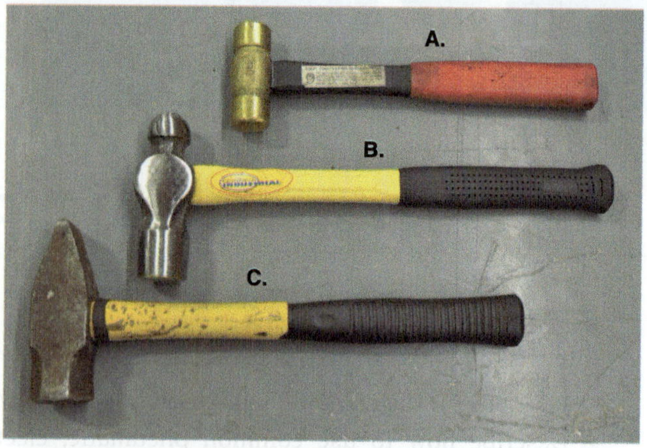

FIGURE 3–3 Soft-face mallets

HAMMER SAFETY

- Wear eye protection. Always wear eye protection when striking tempered tools and hardened metal surfaces. This will protect your

FIGURE 3–4 Heavy-duty ball peen hammers

eyes from flying chips of fractured steel. When possible, use soft-face hammers (plastic, wood, or rawhide) when striking hardened surfaces.

- Never strike one hammer against another. A hammer can chip if struck against another hammer or hardened surface, resulting not only in damage to the hammer but possibly in bodily injury.
- Check the fit and condition of the handle. Keep handles securely wedged in hammer heads to prevent injury to yourself and others nearby.
- Replace cracked or splintered handles and do not use the handle for prying or bumping. Handles are easily damaged and broken this way.
- Select the right size for the job. A light hammer tends to bounce off the work. One that is too heavy is hard to control.
- Grip the handle close to the end. This increases leverage for harder, less tiring blows. It also reduces the possibility of crushing your fingers between the handle and the protruding parts of the work piece if you should miss.
- Prevent injuries to others. Swing in a direction that will not let your hammer strike someone if it slips from your hand. Keep the handle dry and free of grease and oil.
- Keep the hammer face parallel with your work. Force is then distributed over the entire hammer face, reducing the tendency of the edges of the hammer head to chip or slip off the object.

SAWS AND KNIVES

The **hacksaw** is a much-used tool by truck technicians. It is required for cutting bolts, angle iron, tubing, and so on. Hacksaw blades are generally available with 14, 18, 24, or 32 cutting teeth per inch. As shown in **Figure 3–5**, the 14-tooth blade is used for mild steels. Whereas an 18-tooth blade is required to cut harder steels, the 24-tooth blade is usually used on heavy sheet metal, copper, brass, and medium tubing. For thin sheet metal and thinwall tubing, a 32-tooth blade is required. Although the blades may be made of a variety

FIGURE 3–5 The material to be cut determines the best blade tooth-per-inch selection.

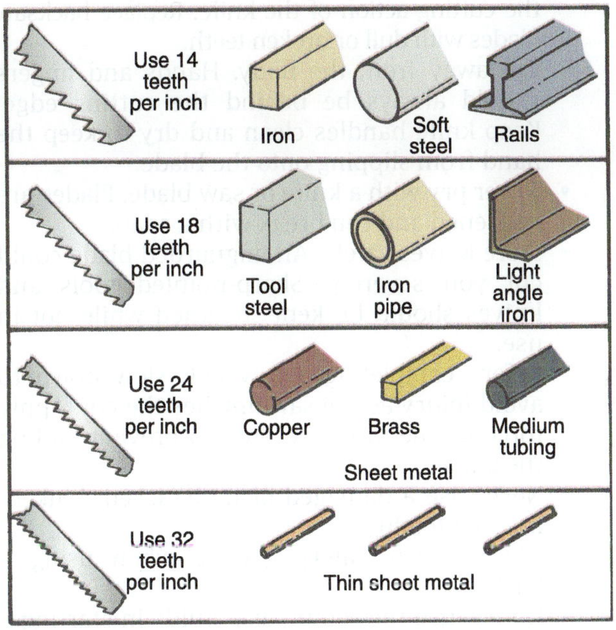

of materials, high-speed and tungsten alloy steels are best for cutting alloy steels.

Hacksaw Technique

When hacksawing, ensure that the blade is held taut in the saw frame with the cutting teeth edges facing away from the handle. Securing the work firmly to prevent slippage, grasp the top of the frame in one hand while holding the handle in the other (**Figure 3–6**). Apply enough pressure on the forward stroke to ensure cutting, and then relieve the blade pressure slightly on the return stroke. Cut at 90 degrees to the work. Keep the blade level and do not rock the saw. Avoid starting the cut on a sharp edge; this may chip the saw teeth. Maintain a smooth cutting action.

FIGURE 3–6 The proper way to hold a hacksaw when cutting. Note that the hacksaw is level.

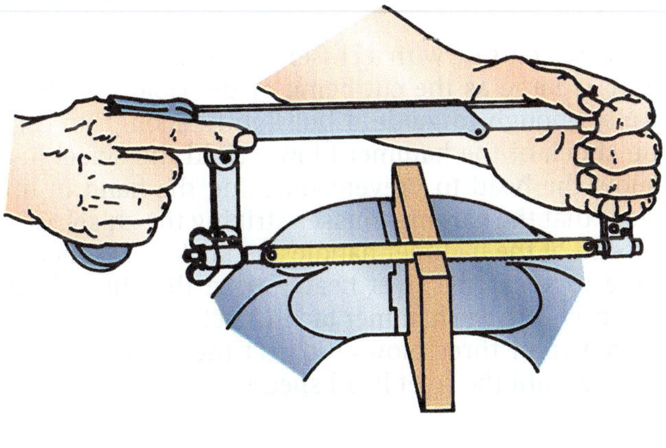

Saw and Knife Safety

- Keep knife blades sharp. The greater the force you have to apply, the less control you have over the cutting action of the knife. Replace hacksaw blades with dull or broken teeth.
- Cut away from the body. Hands and fingers should always be behind the cutting edge. Keep knife handles clean and dry to keep the hand from slipping onto the blade.
- Never pry with a knife or saw blade. Blades are hardened and can break with a snap.
- Store knives safely. An unguarded blade could cut you severely. Sharp-pointed tools and knives should be kept sheathed while not in use.
- Before completing a saw cut, slow down to avoid injury as the saw finishes the cut. Apply force in one direction only to prevent dulling the blade.
- Never use a damaged blade (cracked, kinked, missing teeth).
- Always wear safety glasses when using a hacksaw.
- Do not use the thumb as a guide in starting a hacksaw. If starting is a problem, file a starting notch in the work.
- Use full travel, even strokes to maximize the life of the blade.

Tube Cutting Tools

Tubing made from steel, copper, aluminum, and plastic is used frequently on trucks. During service work, tubing often needs to be repaired or replaced. Tubing tools are made for such tasks as cutting, **deburring** (removing sharp edges from a cut), **flaring** (spreading gradually outward), **swaging** (reducing or tapering), bending, and removing fittings. **Figure 3–7** illustrates some of these tools.

FIGURE 3–7 Common tubing tools: (A) tube or pipe cutter and (B) a flaring tool kit.

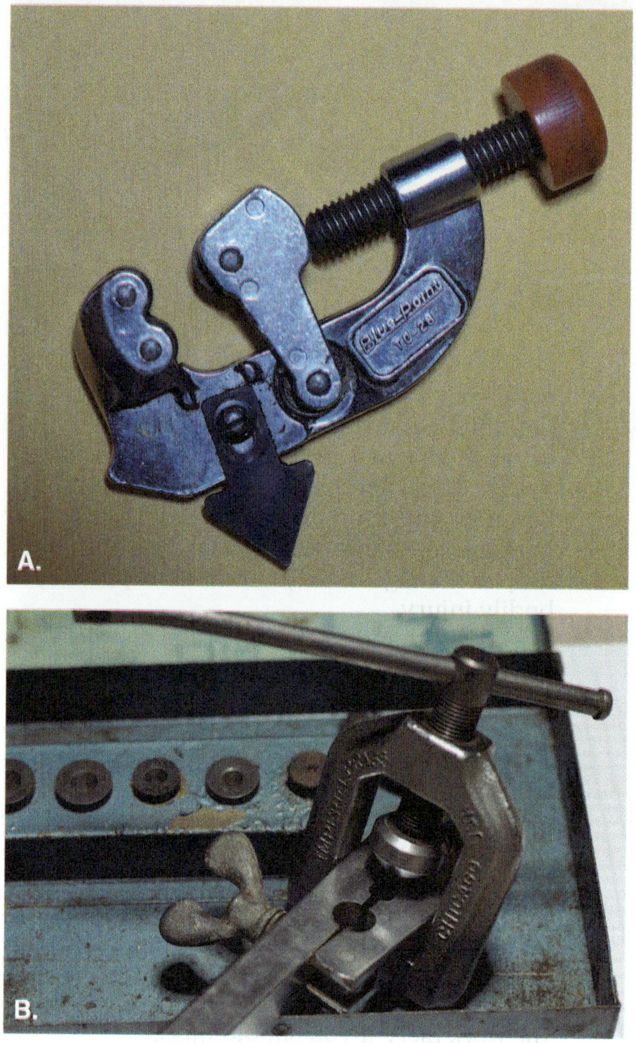

CHISELS AND PUNCHES

Chisels (**Figure 3–8**) are used to cut metal by driving them with a hammer. Technicians use a selection of chisels for cutting sheet metal, shearing off rivet and bolt heads, splitting rusted nuts, and chipping metal.

When cutting with a chisel, the blade should be at least as large as the cut being made. Hold the chisel firmly enough to guide it but lightly enough to ease the shock of the hammer blows. Hold the chisel just below the head to prevent pinching the hand in the event that the hammer misses striking the chisel. Grip the end of the hammer handle and strike with enough force to shear the rivet head. Strike the chisel head squarely with the hammer face. Check your work after every two or three blows. Correct the chisel angle as needed until the rivet head shears.

Punches (**Figure 3–9**) are used for driving out pins, rivets, or shafts; aligning holes in mating components during assembly; and scribing the starting point for drilling a hole. Punches are designated by point diameter and shank shape.

Chisel and Punch Safety

- Wear eye protection when cutting with a chisel or using a punch.
- Avoid using a punch or chisel on hardened metal.
- Grind off mushroom heads (**Figure 3–10**) before using a chisel or punch. This will help prevent cutting your hand and will keep chips from flying into your eyes.
- Do not drive a punch too deep into a bore or it may become wedged due to its taper.

FIGURE 3–8 Chisels classified by their cutting edges.

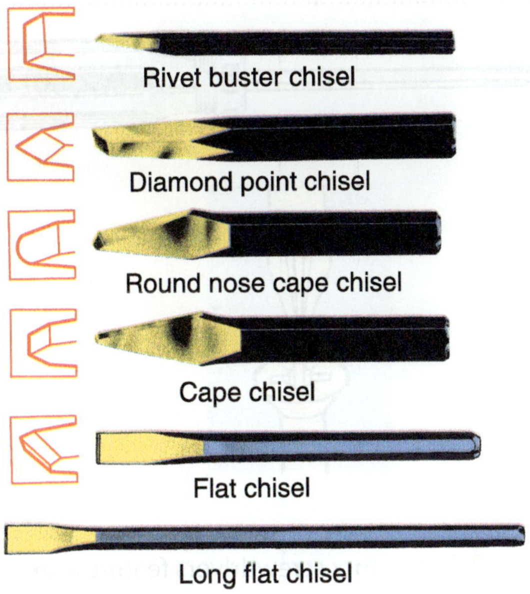

FIGURE 3–9 Punches are designated by point diameter and punch shape.

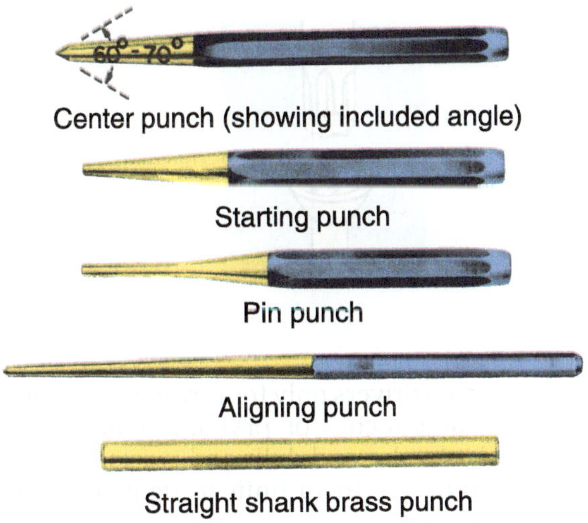

FIGURE 3–10 The proper method for dressing a punch.

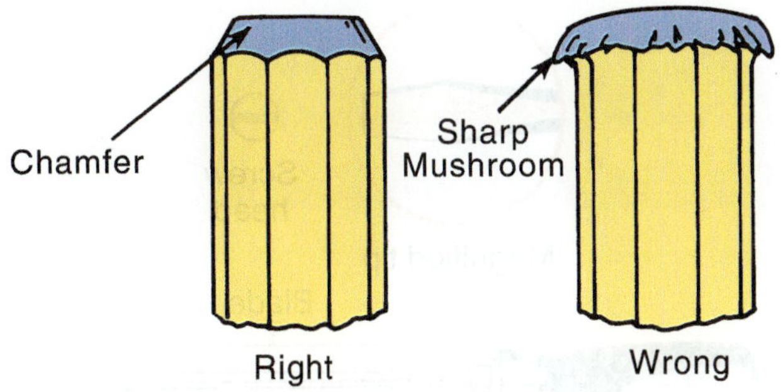

FIGURE 3–11 Selection of common screwdrivers

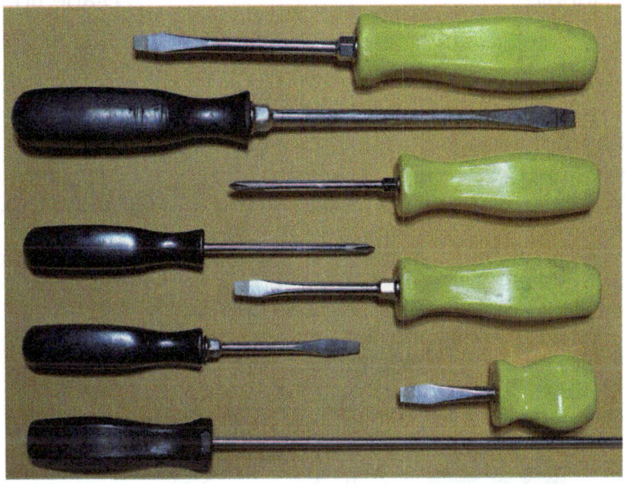

SCREWDRIVERS

A variety of threaded fasteners used in the trucking industry are driven by a screwdriver. Each fastener requires a specific kind of screwdriver, and a well-equipped technician will have several sizes of each (**Figure 3–11**). All screwdrivers, regardless of the type of fastener they were designed for, have several things in common. The size of a screwdriver is determined by the length of the shank and size of the tip. The size of the handle is important, too. The larger the handle diameter, the better its grip and the more torque can be applied to it when turned. Screwdriver handles should also be insulated from the blade and manufactured of a material that does not conduct electricity.

Standard Tip Screwdrivers

A slotted screw accepts a screwdriver with a standard tip. The standard tip screwdriver is probably the most common type (**Figure 3–12**). It is used for turning carriage bolts, machine screws, and sheet metal screws.

Phillips Screwdrivers

The tip of a **Phillips screwdriver** has four prongs that fit to the four slots in a Phillips head screw (**Figure 3–13**). This type of fastener is commonly used in truck and bus chassis. It looks and grips better than a slot head screw and is also easier to install and remove. The four recessed slots engage the screwdriver tip so that there is less likelihood of slippage or damage.

FIGURE 3–12 A standard tip screwdriver fits slotted head screws.

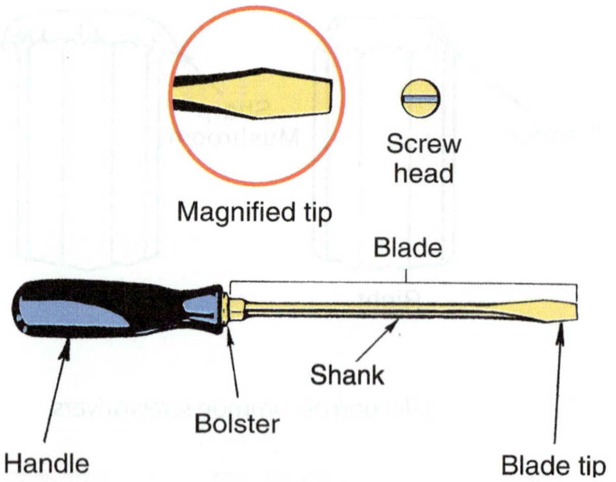

FIGURE 3–13 The tip of a Phillips screwdriver has four prongs that help prevent slippage of the fastener.

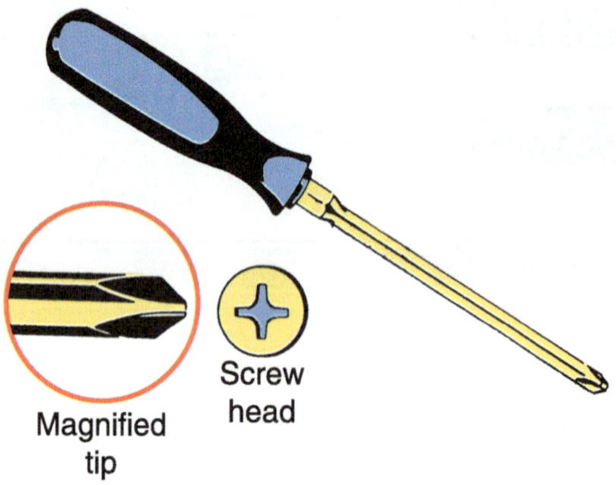

FIGURE 3–14 The Posi-Drive screwdriver is similar in appearance to a Phillips.

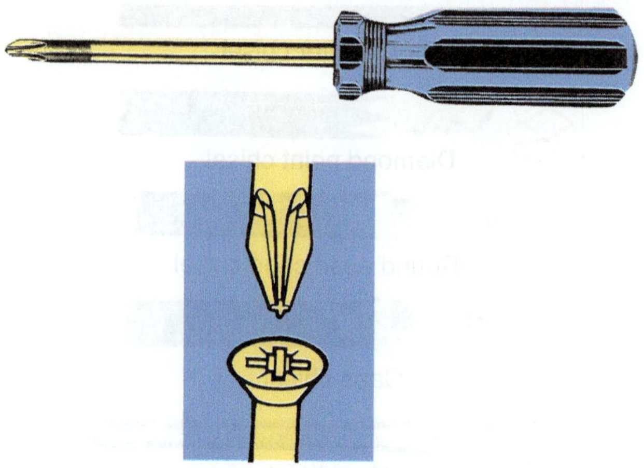

FIGURE 3–15 Torx screwdrivers feature a six-prong tip.

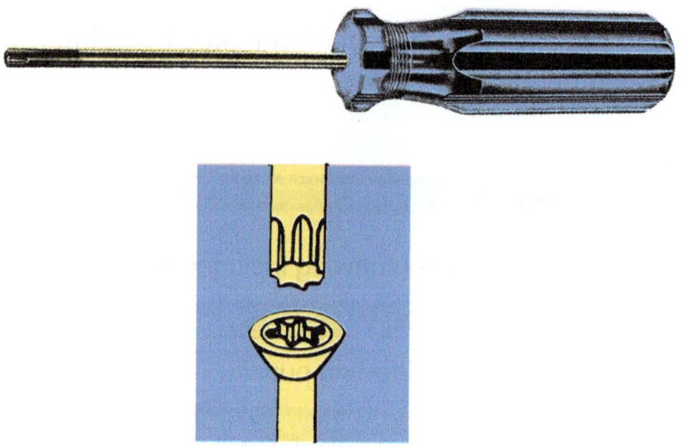

Specialty Screwdrivers

A number of specialty fasteners can be used in place of slot and Phillips head screws. The objective is to improve transfer of torque from the screwdriver to the fastener, minimize slip, and enable installation on robotic assembly lines.

The **Posi-Drive screwdriver** is similar to a Phillips but with a tip that is flatter and blunter (**Figure 3–14**). The squared tip grips the screw head and slips less than an equivalent Phillips screwdriver. A Torx fastener is often used to secure headlight assemblies, mirrors, and brackets. Not only does its six-prong tip sustain greater torque and less slippage (**Figure 3–15**), but it also provides a measure of tamper resistance. It requires a **Torx screwdriver** or bit to remove and install them.

- Use screwdrivers only for turning screws. Using them as punches or prybars breaks handles, bends shanks, and dulls and twists the tips.
- Abuse makes them unfit to tighten or loosen screws safely.
- A slotted screwdriver tip can easily be dressed to its original shape. Never grind the tip, because excess heat will destroy the temper. Always file by hand; you will have more control on the shape of the tip and not remove the temper.
- If the screwdriver blade fits the screw slot properly, you will produce maximum torque with minimum effort. A blade tip that does not fit properly will damage not only the screw slot but perhaps the tip itself.
- It is a good rule to keep your other hand clear when applying force to any type of screwdriver.

- Always have the screwdriver and the screw correctly lined up. You cannot get a good grip in the slot if the tip is held at an angle.
- Screwdrivers designed for use with wrenches (for high torque application) have either a square shank or a hex bolster at the handle to withstand the application of extra force.
- Do not hold components in your hand while turning fasteners with a screwdriver. Put the work on a bench or in a vise to avoid the possibility of piercing your hand with the screwdriver tip.
- When working around anything electrical, use a screwdriver with an insulated handle and shank to avoid shock and short circuits.

WRENCHES

A complete collection of wrenches is indispensable for a truck technician. A well-equipped technician will have both metric and standard wrenches in a variety of sizes and styles (**Figure 3–16**). But remember that metric and standard wrenches are not interchangeable. For example, a 9⁄16-inch wrench is 0.02 inch larger than a 14-millimeter nut. If a 9⁄16-inch wrench is used to turn or hold a 14-millimeter nut, the wrench will probably slip. This can cause rounding of the hex points of the nut and possibly skinned knuckles as well.

The word **wrench** means twist. A wrench is a tool for twisting and/or holding bolt heads or nuts. The width of the jaw opening determines its size. For example, a ½-inch wrench has a jaw opening (from face to face) of ½ inch. The size is actually slightly larger than its nominal size so that the wrench easily fits around a fastener head of equal nominal size.

FIGURE 3–16 Types of wrench: (A) combination wrench; (B) open-end wrench; (C) box-end wrench; and (D) line-wrench, also known as flare-nut or crowsfoot wrench.

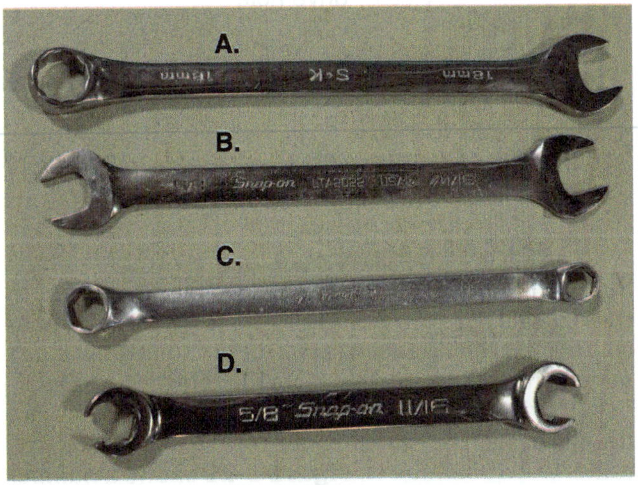

FIGURE 3–17 The open-end wrench grips only two faces of a fastener.

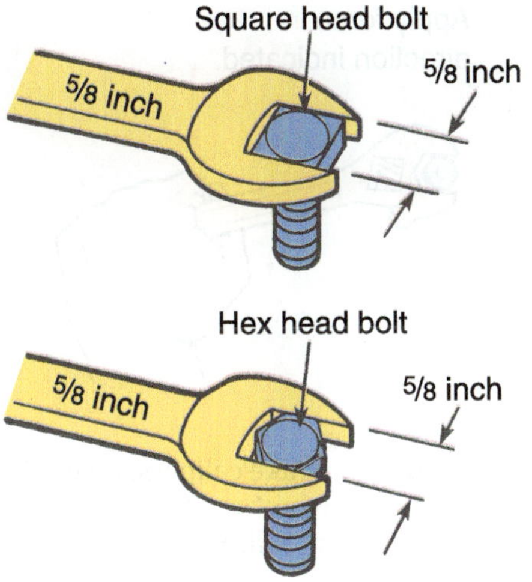

Open-End Wrenches

The jaws of an **open-end wrench** (**Figure 3–17**) allow the wrench to slide around bolts or nuts where there might be insufficient clearance above a fastener to enclose it with a box wrench.

Box-End Wrenches

The end of a **box-end wrench** is boxed or closed rather than open. The jaws of the wrench completely surround a bolt or nut **hex**, gripping on all six points of the fastener. A box-end wrench is not likely to slip off a nut or bolt. It is therefore safer than an open-end wrench.

Combination Wrenches

The **combination wrench** has an open-end jaw on one end and a box-end jaw on the other. Both ends are the same nominal size. Every truck technician should have at least two sets of combination wrenches, one for holding and one for turning. A combination wrench set is an essential toolbox item because it complements open-end, box-end, and socket wrench sets.

Adjustable Wrenches

An **adjustable wrench** has one fixed jaw and one movable jaw. Wrench jaw opening can be adjusted by rotating a **helical** adjusting screw that is mated to teeth in the sliding jaw. Because this type of wrench does not firmly grip the fastener, it is likely to slip. Adjustable wrenches should be used carefully and only when necessary. Be sure to apply all of the turning pressure on the fixed jaw (**Figure 3–18**).

Allen Wrenches

Allen-type set screws and fasteners with a recessed hex head are used to fasten door handles, instrument

FIGURE 3–18 Pull the adjustable wrench so that the force bears against the fixed jaw.

panel knobs, and even brake calipers. A set of hex head wrenches, or **Allen wrenches** (**Figure 3–19**), should be in every technician's toolbox.

Socket Wrenches

In many situations, a socket wrench is safer, faster, and easier to use than an open-end or box-end wrench. Some applications require the use of a socket. A basic **socket wrench** set consists of a ratchet handle and several barrel-shaped sockets. The socket fits over and around a given size bolt or hex head (**Figure 3–20**). Inside, it is machined as either a hex or double hex box-end wrench. Sockets are available with 6, 8, or 12 points. The upper side of a socket has a square hole that engages with a square lug on the socket handle (**Figure 3–21**). One handle fits all the sockets in a set. The size of the lug ($\frac{3}{8}$ inch, $\frac{1}{2}$ inch, and so on) indicates the drive size of the socket wrench. On better quality handles, a spring-loaded ball in the square lug fits into a depression in the socket. This ball holds the socket to the handle drive lug.

Sockets are available in various sizes, lengths, and bore depths. Both standard SAE and metric socket wrench sets are required to work on current trucks. Normally, the larger the socket size, the deeper the well. Deep-well sockets (**Figure 3–22**) are made extra long to fit over bolt ends or studs. They are also useful for reaching nuts or bolts in limited access areas. Deep-well sockets should not be used when a regular size socket will do the job. The longer socket develops more torque twist and tends to slip off the fastener.

Heavier walled sockets made of softer steel are designed for use with an **impact wrench** and are called **impact sockets**. **Figure 3–23** shows a number of socket wrench set accessories. Accessories increase

FIGURE 3–19 Allen key set

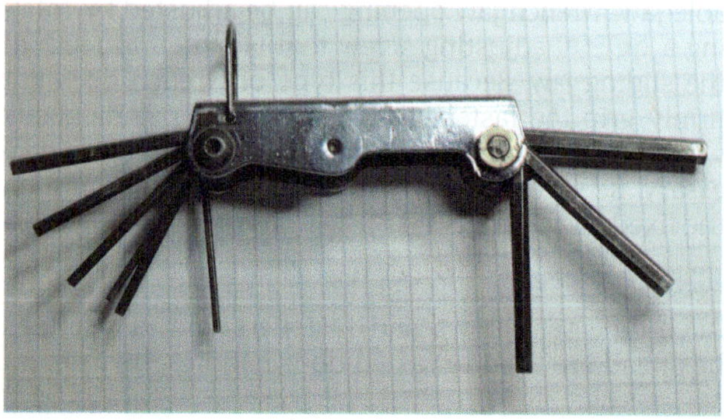

FIGURE 3–20 The specified size of a socket is the same as the bolt head or nut nominal size.

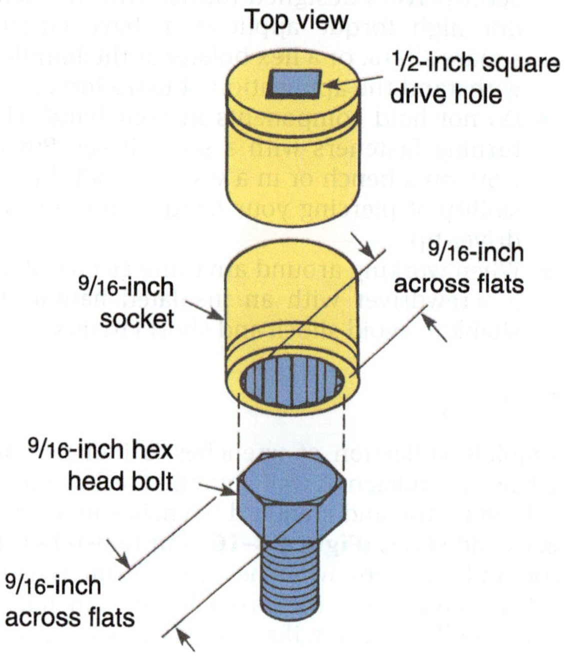

FIGURE 3–21 A socket square drive lug recess is the same size as the handle drive lug.

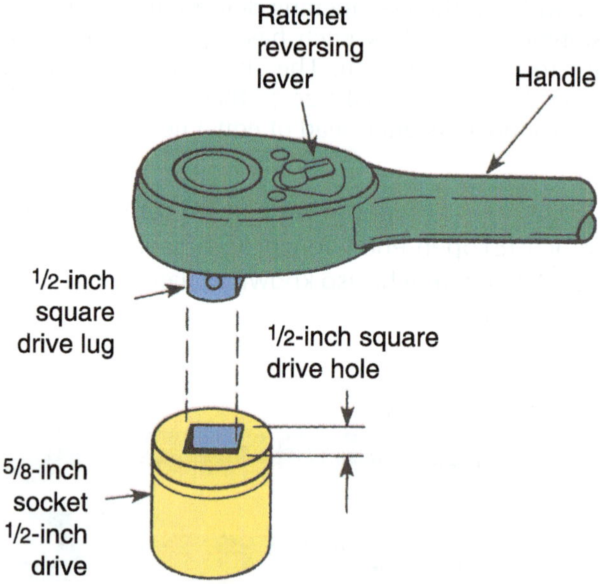

FIGURE 3–22 Deep-well sockets fit over the bolt hex and long studs.

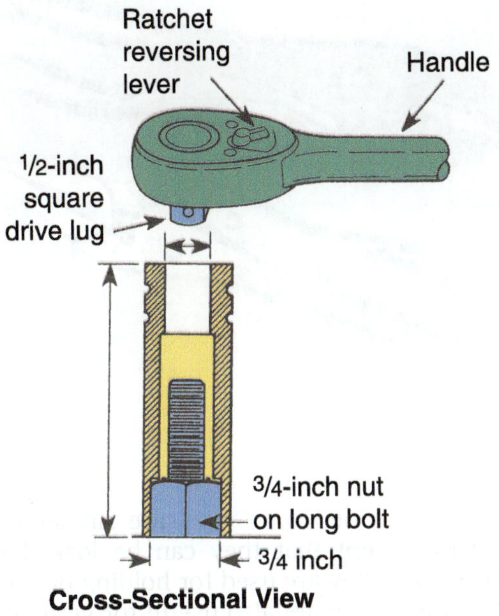

FIGURE 3–23 Typical socket wrench sets

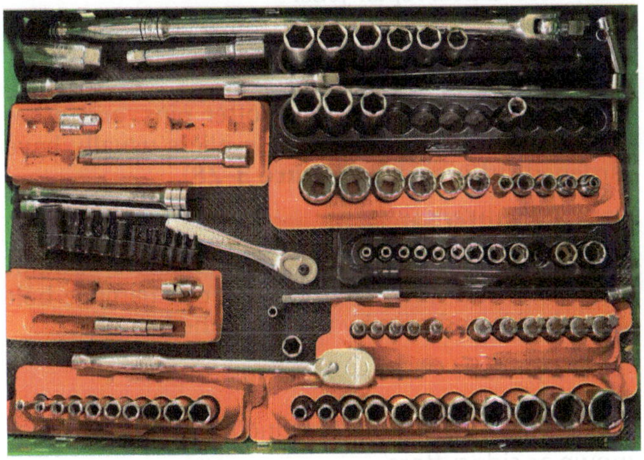

the usefulness of a socket wrench set. A good socket wrench set has a variety of accessories. Screwdriver attachments are also available for use with a socket wrench. These socket wrench attachments are very handy when a fastener cannot be loosened with a regular screwdriver. The extra leverage provided by the ratchet handle is often just what it takes to break loose a stubborn screw.

Wrench Safety

- Use wrenches that fit. Wrenches that slip damage bolt heads and nuts, skin knuckles, and cause the user to fall.
- Use the proper wrench to get the job done—the one that gives you the surest grip and a straight clean pull. Cocking a wrench puts concentrated stress on the points of contact, a frequent cause of tool failure under pressure. (Other types of wrenches, such as the angle head, offset, and socket type, give you the ability to work in difficult-to-access places and get a clean pull.)
- Do not extend the length of a wrench. Do not use a pipe to increase the leverage of the wrench.
- The handle length is made to apply the maximum safe force the socket can sustain. Excessive force may break the wrench or bolt unexpectedly, or the wrench may slip, rounding off hex corners; skinned knuckles, a fall, or a broken wrench may result.
- Do not use a hammer on wrenches unless they are designed for that type of use.
- Pull on the wrench. This is not always possible, but if you push, you take the risk that if the wrench slips or if the nut suddenly breaks loose, you may skin your knuckles or cut yourself on a sharp edge. Use an open palm to push on a wrench when you cannot pull it toward you.
- Replace damaged wrenches. Straightening a bent wrench weakens it. Cracked and worn wrenches are dangerous to use, because they could break or slip at any time.
- The adjustable wrench is a multipurpose tool, but it should never be used if a properly fitting combination wrench is at hand.

PLIERS

Pliers are an all-around gripping tool used for working with wires, clips, and pins. The truck technician must own several types: standard pliers; needle nose pliers for small, difficult-to-access components; and large adjustable pliers for heavy-duty work and pipes.

Combination Pliers

Combination pliers (**Figure 3–24**) are one of the most common types of pliers and are used in many

FIGURE 3–24 Linesman pliers sometimes known as combination pliers.

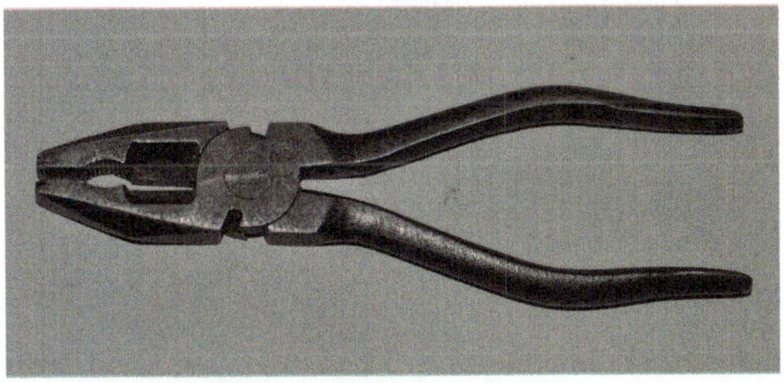

kinds of mechanical repair. The jaws have both straight and curved surfaces for holding flat or round objects. Also called *slip-joint* pliers, combination pliers have multiple jaw-opening sizes. One jaw can be moved up or down on a pivot pin attached to the other jaw to change the size of the opening.

Adjustable Pliers

Adjustable pliers, also known as *pipe pliers* and *channel locks* (**Figure 3–25**), have a multiposition slip joint that allows for many jaw-opening sizes. They may be used to grip pipes, shafts, and filters.

Needle Nose Pliers

Needle nose pliers have long tapered jaws (**Figure 3–26**). They are indispensable for grasping small components or for reaching into tight spots. Most needle nose pliers also have wire cutting edges and a wire stripper. They are indispensable for electrical work.

FIGURE 3–25 Adjustable pliers often known as waterpump pliers.

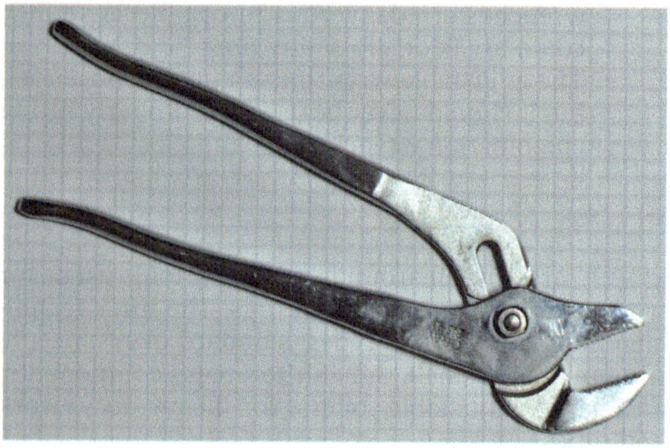

FIGURE 3–26 Needle nose pliers

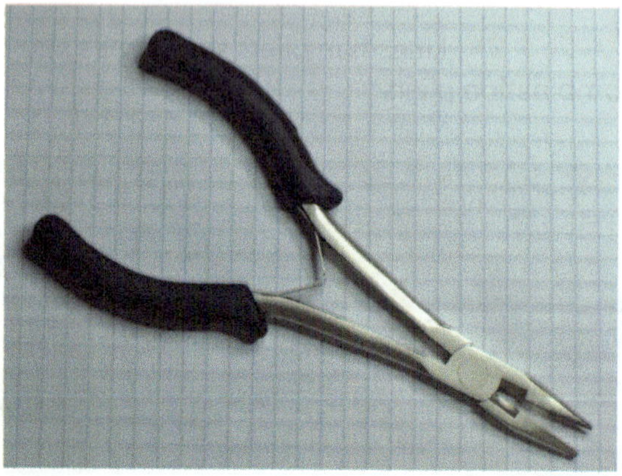

FIGURE 3–27 Two types of Vice Grips.

Locking Pliers

Locking pliers, or **Vise Grips**, are similar to standard pliers, except that they can be locked closed (**Figure 3–27**). They are used for holding or clamping parts together. They are also useful for getting a firm grip on a badly rounded fastener on which wrenches and sockets are no longer effective. Locking pliers come in several sizes and jaw configurations for use in many mechanical repair jobs.

Diagonal Cutting Pliers

Diagonal cutting pliers, or **sidecutters**, are used to cut electrical connections, cotter pins, and other wires on vehicles. Jaws on these pliers usually have hardened cutting edges (**Figure 3–28**).

Pliers and Cutter Safety

- Do not use pliers as a wrench. They cannot hold the work securely and can damage bolt heads and nuts by rounding the hex.

FIGURE 3–28 Diagonal cutting pliers—often known as sidecutters

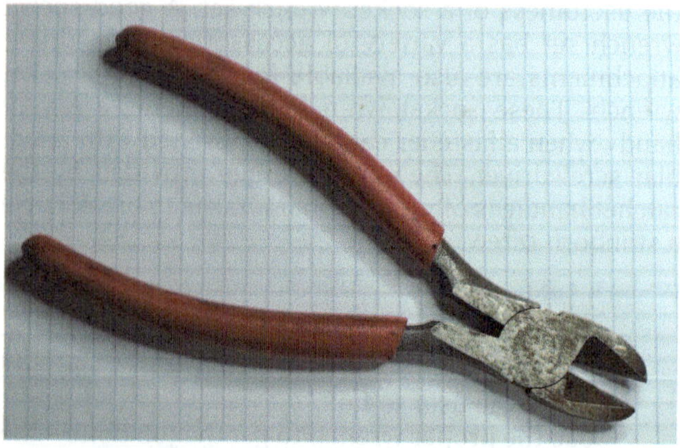

- Guard against eye injuries when cutting with pliers or cutters. Short and long strands of wire often fly through the air when cut. Wear eye protection and cup a hand over the pliers to protect yourself.
- Observe the following sidecutter precautions:
  - Select a cutter big enough for the job.
  - Keep the blades at right angles to the stock.
  - Do not rock the cutter to get a faster cut.
  - Adjust the cutters to maintain a small clearance between the blades. This prevents the hardened blades from striking each other when the handles are closed.
- Pliers are made for holding, pinching, squeezing, and cutting—not usually for turning.

FILES

Files are designed to remove small amounts of metal, for smoothing, or sharpening surfaces. They are classified as single cut, double cut, fine, and coarse. These are easily identified by the file tooth pattern. **Figure 3–29** shows various file cuts and shapes. The abrasive teeth on a file cut in only one direction—on the forward stroke. Choose a file that is appropriate for the job at hand. Use a coarse-cut file on soft material such as aluminum to prevent clogging.

Using Files

Drive the file in a forward direction and apply only a slight drag on the return stroke. When filing soft material, applying a slight drag on the return stroke should prevent clogging of the cutting teeth by filings. When filing, grasp the file handle in one hand, with the thumb on top. Hold the point of the file with the thumb and first two fingers of the other hand (**Figure 3–30**). File only on the push strokes. Do not attempt to cut on the return stroke. Cross the stroke at short intervals.

FIGURE 3–29 Various types of files

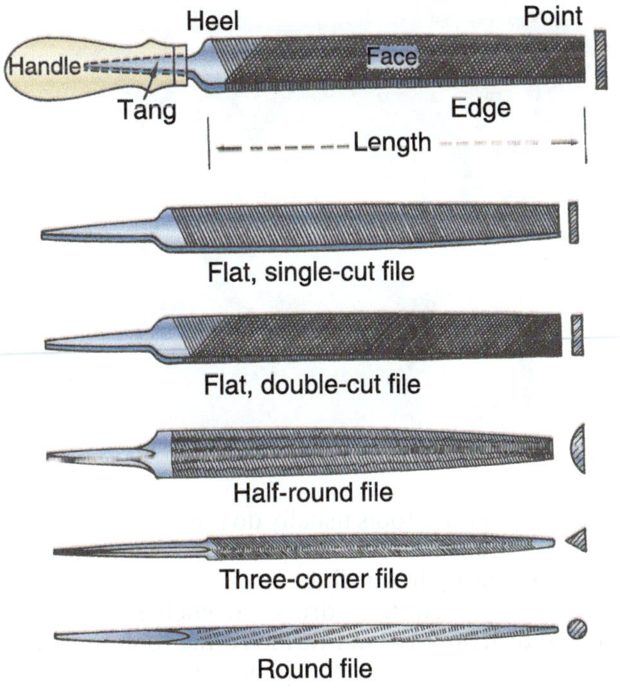

FIGURE 3–30 The correct way to hold a file

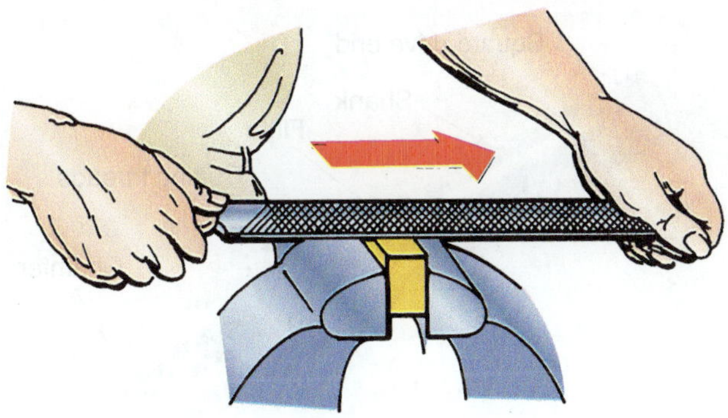

File Safety

- Wear eye protection when filing.
- Never strike a file with a hammer. The file may shatter, resulting in serious injury.
- Always cut away from the body.
- Never use a file without a securely attached handle.
- Do not use worn (dull) files; replace them.

SPECIAL TOOLS

There are many other tools that can be found in a well-equipped truck technician's tool chest. Among them are the following:

Taps and Dies

The **hand tap** is a small tool used for hand cutting internal threads (**Figure 3–31**). An internal thread is cut on the inside of a hole, such as a thread on a nut. This tap is also used for cleaning and restoring previously cut threads known as thread chasing.

Hand-threading dies (**Figure 3–32**) function oppositely to taps because they cut external (outside) threads on bolts, rods, and pipes. Dies are made in various sizes and shapes, depending on the particular work for which they are intended. Dies may be solid (fixed size), split on one side to permit adjustment, or have two halves held together in a collet that provides for individual adjustments. Dies fit into holders called *die stocks.*

Gear and Bearing Pullers

Many precision gears and bearings have a slight interference fit (press- or shrink-fit) when installed on a shaft or housing. An interference fit allows no movement

FIGURE 3–31 Tap terminology

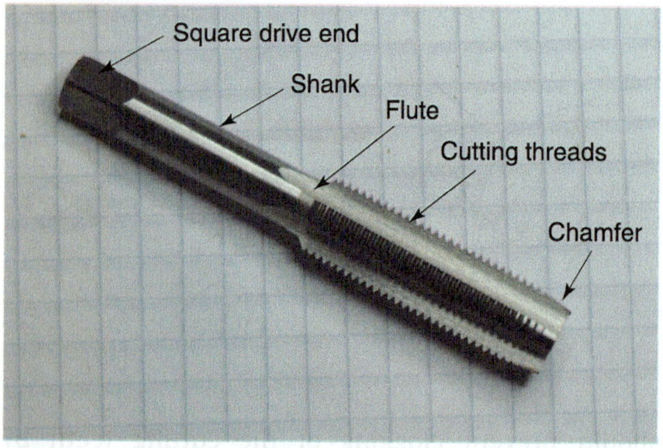

FIGURE 3–32 A standard/metric tap and die set

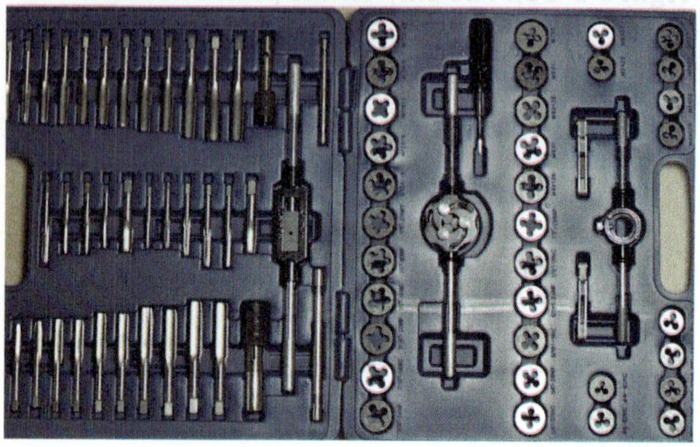

between the mating components and eliminates wear. The removal of gears and bearings must be done carefully. Prying or hammering can fracture or bind the mating components. A puller with the proper jaws and adapters should be used when applying force to remove gears and bearings. Force can be applied with gradually increasing intensity. Various gears and bearing puller styles and sizes are shown in **Figure 3–33**.

Other hand tools that come in handy when servicing trucks include prybars, wire brushes, scrapers, strap wrenches, propane torch, tire gauges, grease guns, tire irons, and trouble lights. Some types of special hand tools used are discussed in the appropriate chapters in this book. There are also manufacturers' special tools (factory shop tools) that are required to repair their equipment. These are usually assigned a part number by the manufacturer (**Figure 3–34**).

3.2 POWER TOOLS

Power tools make a technician's job easier. They operate faster and with more torque than hand tools. However, using power tools requires greater safety awareness. Power tools usually do not stop unless they are turned off. Power is supplied by air (pneumatic), electricity, or hydraulic fluid.

FIGURE 3–33 Various gear, sleeve, and bearing pullers

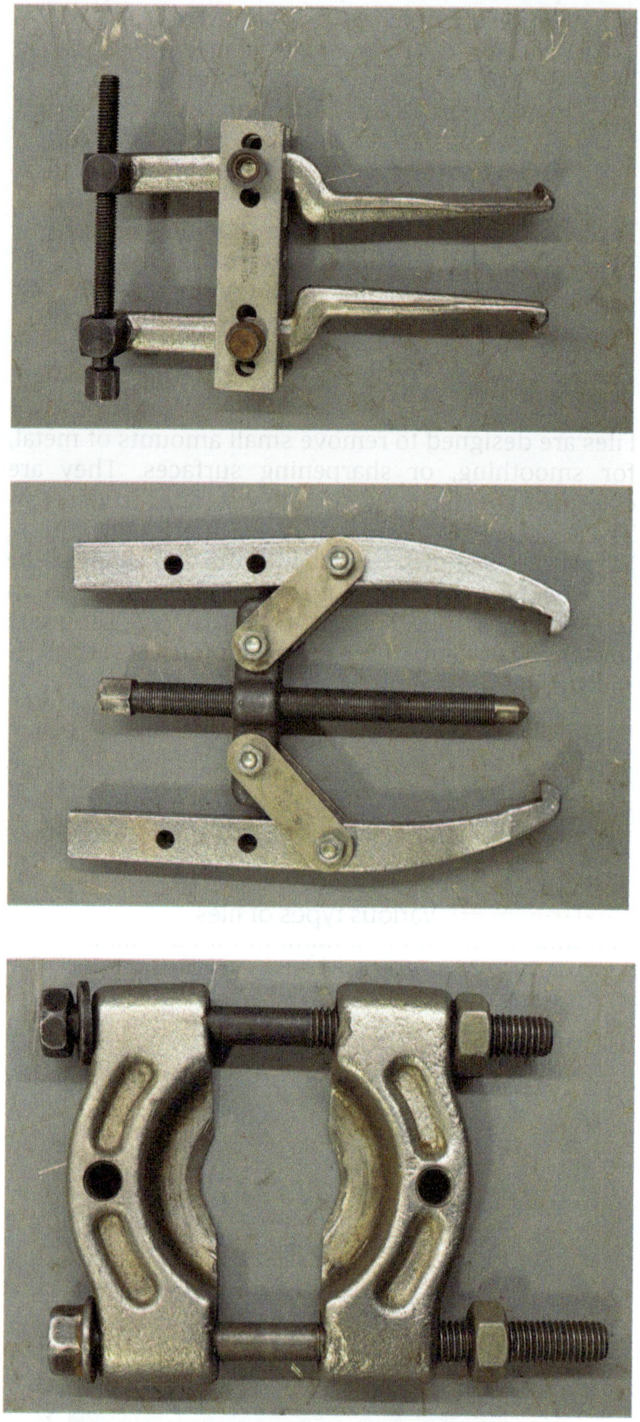

Although electric drills, wrenches, grinders, chisels, drill presses, and various other tools are found in shops, pneumatic (air) tools are used more frequently. Pneumatic tools have four major advantages

FIGURE 3–34 Manufacturers' special tools make jobs easier and are usually assigned a part number that can be identified in service literature.

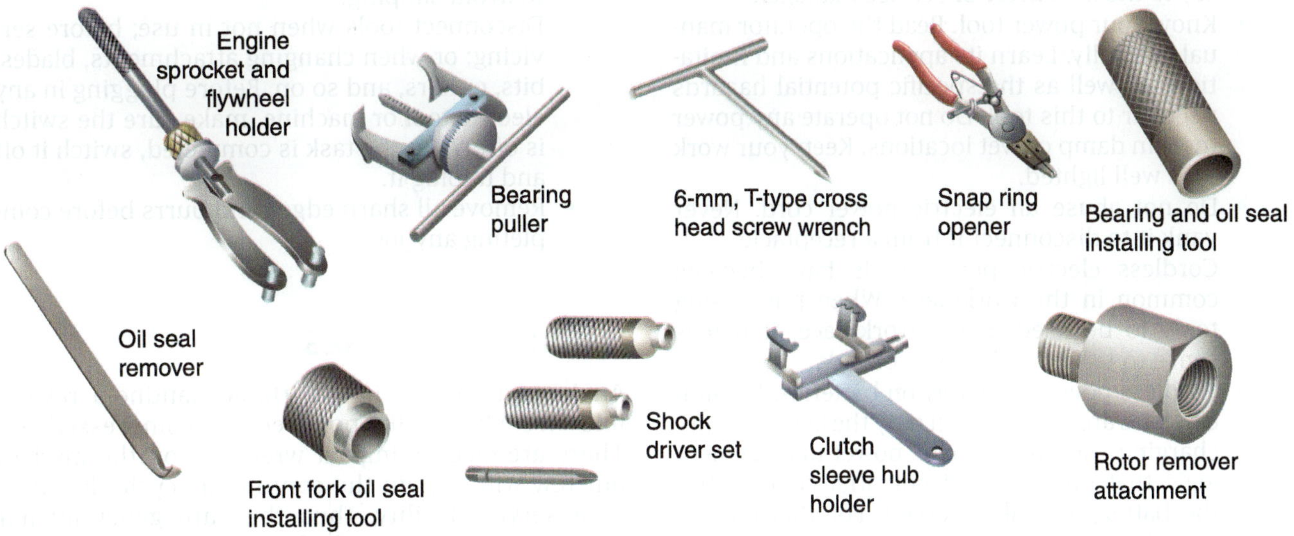

over electrically powered equipment in an engine rebuilding shop:

- **Flexibility**. Air tools run cooler and have the advantage of variable speed and torque; damage from overload or stall is minimized. They can fit in tight spaces.
- **Lightweight**. Air tools are lighter in weight and permit faster work with less fatigue (**Figure 3–35**).

FIGURE 3–35 Comparison of electric and air impact guns: Note that the air gun has a ½-inch and the electric gun a ⅜-inch drive lug.

- **Safety**. Air equipment reduces the danger of fire and shock hazards in some environments where arcing of electric power tools can be a problem.
- **Low-cost operation and maintenance**. Due to fewer parts, air tools require fewer repairs and less preventive maintenance. Also, the purchase cost of air-driven tools is usually less than equivalent electric tools.

The mechanical repair industry was one of the first industries to recognize the advantages of air-powered tools. Today they are essential tools for the professional truck technician. However, a major disadvantage of air tools is excessive noise.

POWER TOOL SAFETY

Safety is critical when using power tools, regardless of how they are powered. Carelessness or mishandling of power tools can cause serious injury. Here are some general power tool safety rules that must always be followed:

- Return all equipment to its proper place when finished.
- Wear eye protection.
- Noise may be a hazard with some portable power tools, especially pneumatic tools. Wear hearing protection whenever noise is excessive.
- Wear gloves when operating air chisels or air hammers.
- All electrical equipment should be grounded, unless it is the double insulated type.
- Never make adjustments, lubricate, or clean a machine while it is running.

- Do not clean yourself or anyone else with compressed air.
- Report any suspect or malfunctioning machinery to the instructor or service manager.
- Know your power tool. Read the operator manual carefully. Learn its applications and limitations as well as the specific potential hazards peculiar to this tool. Do not operate any power tools in damp or wet locations. Keep your work area well lighted.
- Do not abuse an electric power cord. Never yank it to disconnect it from a receptacle.
- Cordless electric power tools have become common in the workplace. When purchasing tools to be used in the workplace, it makes sense to target well-known brands.
- Cordless power tools rely on batteries. Lithium batteries are common but only the better quality charging stations have power-management logic that prevents full potential charge after the battery is fully charged. For this reason, it makes sense not to leave batteries under perpetual charge because it can limit battery life.
- No machine should be started unless guards are in place and in good condition. Defective or missing guards should be reported to the instructor or service manager immediately.
- Check and make all adjustments before applying power.
- Give the machine your undivided attention while you are using it. Do not look away or talk to others.
- Inspect all equipment for safety and for apparent defects before using.
- Whenever safeguards or devices are removed to make repairs or adjustments, equipment should be turned off and the main switch locked and tagged.
- Start and stop your own machine and remain with it until it has come to a complete stop.
- Always allow any machine to reach operating speed before loading it.
- No attempt should be made to retard rotation of the tool or work.
- Do not try to strip broken belts or other debris from a pulley in motion or reach between belts and pulleys.
- Do not use loose rags around operating machinery.
- Use the right tool. Do not use too small a tool or attachment to do the job of a heavy-duty tool.
- Maintain tools with care. Keep tools sharp and clean at all times for best and safest performance. Follow instructions for lubricating and changing accessories.
- Remove adjusting keys and wrenches. Make a habit of checking that keys and adjusting wrenches are removed from chucks and tools before switching them on.
- Do not overreach. Maintain a balanced stance to avoid slipping.
- Disconnect tools when not in use; before servicing; or when changing attachments, blades, bits, cutters, and so on. Before plugging in any electric tool or machine, make sure the switch is off. When the task is completed, switch it off and unplug it.
- Remove all sharp edges and burrs before completing any job.

IMPACT WRENCHES

An impact wrench is a portable handheld reversible wrench usually powered by compressed air. There are electric impact wrenches on the market, but few will sustain the continuous cycle demands of a service facility, thus they are generally not recommended. A ½-inch drive air-driven, medium-duty model such as that shown in **Figure 3–36** can deliver up to 400 lb-ft. of torque using a 100-psi air supply. In fact, the ½-inch drive pneumatic impact gun is the workhorse of the truck shop floor. When triggered, the output shaft, onto which the impact socket is fastened, spins freely at 2,000 to 14,000 revolutions per minute (rpm), depending on the wrench make and model. When the impact wrench meets resistance, a small spring-loaded hammer, located on the drive end of the tool, strikes an anvil attached to the drive shaft onto which the socket is mounted. Each impact moves the socket around a little until torque equilibrium is reached, the fastener breaks, or the trigger is released. **Figure 3–37** shows a technician removing wheel nuts with an impact gun.

FIGURE 3–36 Workhorse of the truck shop floor: the half-inch drive, air impact gun.

FIGURE 3–37 An half-inch impact wrench used to remove wheel nuts.

Using Air Impact Wrenches

When using an impact wrench, keep the following safety rules in mind:

- Sockets designed for hand tool use should not be used on impact wrenches. They can break, causing damage or injury. Use soft steel impact sockets (these are unchromed) only.
- Make sure the socket is securely snapped into the drive lug before using the wrench.
- Hold the wrench firmly with both hands.
- Keep hands clear of moving parts.
- Hearing protection is recommended for use of more than 15 minutes.
- Keep your face away from work when using an impact wrench.
- Wear eye protection.

Air Ratchet Wrenches

An **air ratchet wrench**, like the hand ratchet, has a special ability to work in hard-to-reach places. Its right-angle drive loosens or tightens fasteners that other hand or power wrenches cannot access (**Figure 3–38**). The air wrench looks like an ordinary ratchet but has a broad handgrip that contains an air vane motor and drive mechanism.

Air Drills

Air drills are usually available in ¼-, ⅜-, and ½-inch chuck sizes and operate in much the same manner as an electric drill but have the advantage of being smaller and lighter. This compactness makes them easier to use for drilling operations on truck chassis.

Air Chisels and Hammers

The air chisel or hammer (**Figure 3–39**) is one of the most useful tools in the truck technician's toolbox. Used with the accessories illustrated in **Figure 3–40**, this tool will perform many different operations, including the following.

- **Universal joint and tie-rod tool**. This tool helps to vibrate loose stubborn universal joints and tie-rod ends.
- **Ball joint separator**. The wedge action breaks apart frozen ball joints.
- **Shock absorber chisel**. Designed to crack frozen shock absorber nuts.
- **Exhaust pipe cutter**. A cutter designed to slice through mufflers and exhaust pipes.
- **Tapered punch**. Used for driving frozen bolts, installing pins, and punching or aligning holes.
- **Bushing removal**. Designed to remove all types of bushings. The blunt edge hammers but does not cut.
- **Bushing installer**. Drives all types of bushings to the correct depth. A pilot prevents the tool from sliding.

FIGURE 3–38 A typical ⅜-inch drive, air ratchet wrench.

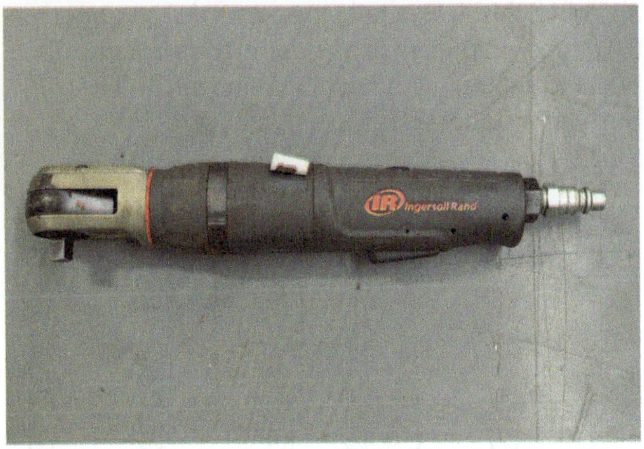

FIGURE 3–39 A typical air hammer or chisel with accessory kit

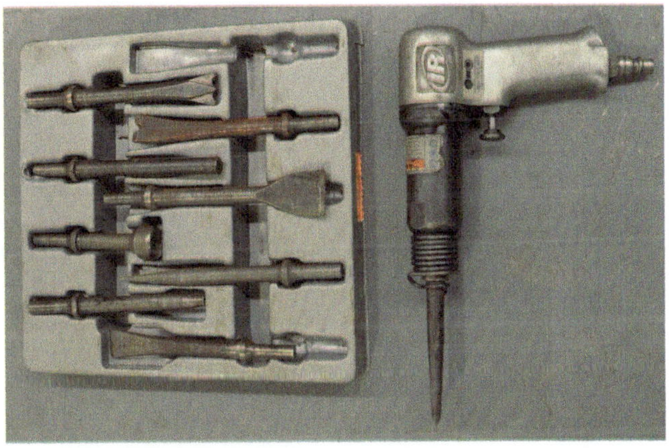

FIGURE 3–40 Air chisel accessories: (A) a universal joint and tie-rod tool; (B) a smoothing hammer; (C) a ball joint separator; (D) a panel crimper; (E) a shock absorber chisel; (F) a tail pipe cutter; (G) a scraper; (H) a tapered punch; (I) an edging tool; (J) a rubber bushing splitter; (K) a bushing remover; and (L) a bushing installer.

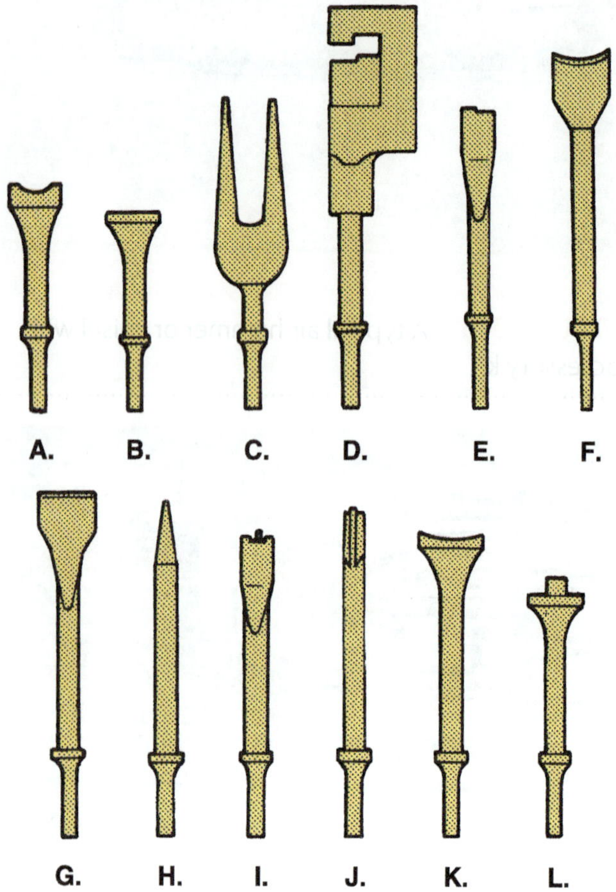

Blowgun

One way to use compressed air from a pneumatic hose is with a **blowgun** (**Figure 3–41**). A blowgun snaps into the air hose coupler and directs airflow when a trigger is depressed. Before using a blowgun, be sure that it has not been modified to eliminate air-bleed holes on the side. Blowguns are used for blowing off parts during cleaning. Never point a blowgun at yourself or someone else.

Torque Measuring Wheel Nut Gun

A welcome recent addition to the tireman's toolbox is the torque measuring wheel nut pneumatic wrench, more commonly known as a wheel nut gun. Because tire technicians are usually remunerated on a piecework basis, they are more likely to take shortcuts than other techs. The result of taking shortcuts is that tire lug nuts are frequently over-torqued to a substantial degree. The torque measuring wheel nut wrench shown in **Figure 3–42** is fast and accurate because it only requires the tech to set the torque value prior to pulling the trigger.

FIGURE 3–41 A typical blowgun

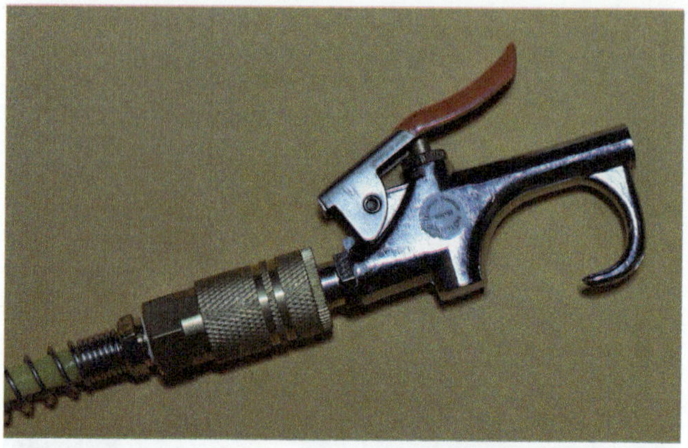

More Air Tool Safety

Other safety rules to remember when using pneumatic powered tools are:

- Air hose and hose couplers used for conducting compressed air to equipment must be used only for the pressure and service application for which they are rated.
- Do not use compressed air for cleaning, unless it is reduced (regulated) to less than 30 psi by a blowgun.
- Ear protection is required when using pneumatic tools for more than 15 minutes, or whenever noise is excessive.

FIGURE 3–42 Torque setting, wheel nut pneumatic wrench: It enables a fast, accurate means of installing wheel lug nuts.

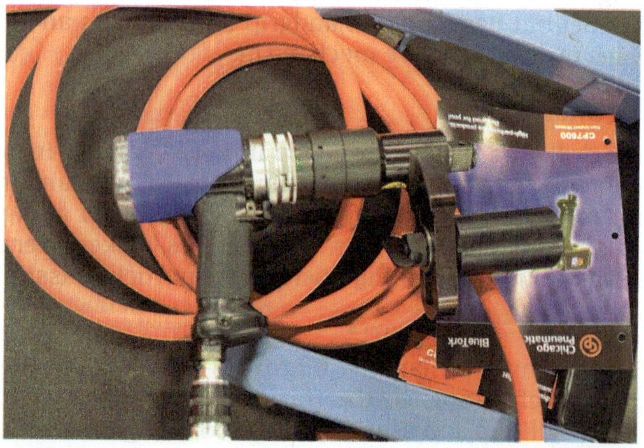

OTHER POWER TOOLS

There are several other power tools that the truck technician may use, including the following:

Bench Grinders

These electric power tools are generally bolted to a workbench. A **bench grinder** is classified by wheel size. The most common in auto repair shops are the 6- to 10-inch wheels. Three types of wheels are available for use with this bench tool.

- **Grinding wheel**. For a wide variety of grinding jobs from sharpening cutting tools to deburring.
- **Wire wheel brush**. Used for general cleaning and coarse buffing, removing rust and scale, removing paint, deburring, and so forth.
- **Buffing wheel**. For light-duty buffing, polishing, light cutting, and finish coloring operations.

WARNING

Never wear gloves when operating a bench grinder.

LIFTS AND HOISTS

Raising a heavy-duty truck trailer on a lift system or **frame machine** (**Figure 3–43**) requires special care. Adapters and hoist plates must be positioned correctly on multiple post- and rail-type lifts to prevent damage to the underbody of the vehicle. There are specific lift points to use where the weight of the vehicle is evenly supported by the adapters or hoist plates. The correct lift points can be found in the vehicle service literature. Before operating any lift or alignment machine, carefully read the manufacturer's literature and understand all the operating and maintenance instructions.

FIGURE 3–43 A typical frame-straightening deck

Courtesy of Bee Line

Independent Post Lift Systems

Independent post lift systems (**Figure 3–44**) have become commonplace in truck shops in recent years. These consist of "portable" hydraulic lift posts that can be positioned under a truck, bus, or trailer chassis and phased to lift the vehicle level. Various systems are available using between two and six posts. Lift and lower of each post is phased using a hardwired or wireless system. Great care should be exercised when using these systems! Make sure that you receive some basic training in the use of post lift devices before using one yourself. An example is the wireless Gray four post hoist shown in Figure 3–44; post lift and lower phasing is mastered by a computerized-control console.

Other Lift Devices

Heavy subcomponents, such as engines and transmissions, should be removed using chain hoists or cranes. To prevent serious injury, chain hoists and cranes must be properly attached to the parts being lifted. Use bolts and lift shackles of sufficient strength rating to attach the hoist to the object being lifted. Secure the lifting chain or cable to the component that is to be lifted.

Hoisting Safety

The following are some general rules for using jacks, lifts, frame machines, and hoists:

- Do not allow anyone to remain in a vehicle when it is being raised.

FIGURE 3–44 A typical truck hoist system

- Make certain that you know how to operate the equipment and know its limitations.
- Never overload a lift, hoist, or jack.
- Chain hoists and cranes must be properly attached to the parts being lifted. Always use bolts and shackles of sufficient strength rating to attach the hoist to the object being lifted.
- Independent post lift systems must be properly set and aligned. Always check the user's literature before using. Check phasing of wired and wireless devices before raising a chassis to a height that could be dangerous.
- Mechanical locks or stands must be engaged after lifting a truck on any kind of hoist.
- Do not use any lift, hoist, or jack that you believe to be defective or not operating properly. Tag it and report it to your instructor or service manager immediately.
- Make sure all persons and obstructions are clear before raising or lowering an engine or vehicle.
- Avoid working, walking, or standing under suspended objects that are not mechanically supported.

Presses

Many truck repair jobs require the use of considerable force to assemble or disassemble components that are press fit or seized. Servicing rear axle bearings, pressing brake drum and rotor studs, transmission assembly, and frame work are just a few examples.

Presses can be hydraulic, electric, air, or hand actuated.

Capacities range up to 150 tons of pressing force, depending on the size and design of the press. Smaller arbor and C-frame presses can be bench or pedestal mounted (**Figure 3–45**), whereas high-capacity units are freestanding or floor mounted (**Figure 3–46**).

FIGURE 3–45 A bench-mounted, manually actuated arbor press

3.3 MEASURING TOOLS

A number of precision-measuring tools are essential to the truck technician. Knowing how to use both mechanical and digital measuring tools is important. The modern truck technician must be capable of working in both standard and metric units of measurement.

MACHINIST'S RULE

The **machinist's rule** looks like an ordinary ruler (**Figure 3–47**) made out of hardened alloy steel. However, unlike a common ruler, it is precisely divided into small (1⁄16 inch/1.6 mm) increments. A typical standard machinist's rule is marked on both sides. One side is marked off at 1⁄16-, 1⁄8-, 1⁄4-, 1⁄2-, and 1-inch intervals. The other side is marked at 1⁄32- and 1⁄64-inch intervals.

Machinist's rules are also available with metric or decimal graduations. Metric rules are usually divided into 0.5 mm and 1 mm increments. Standard decimal rules are typically divided into 1⁄10-, 1⁄50-, and 1⁄100-inch (0.10-, 0.50-, and 0.01-in.) increments. Decimal machinist's rules are convenient when measuring component dimensions that are specified in decimals.

DIAL CALIPERS

The **dial caliper** (**Figure 3–48**) is a versatile measuring instrument. It is capable of taking inside, outside, depth, and step measurements. It can measure these dimensions from 0 to 150 mm in increments of 0.02 mm, or in standard units up to 0.0005 inch. The standard dial caliper features a depth scale, bar scale, dial indicator, inside measurement jaws, and outside measurement jaws. The bar scale is divided into one-tenth (0.10) of an inch graduations. The dial indicator is divided into one-thousandth (0.001) of an inch graduations. Therefore, one revolution of the dial indicator needle

FIGURE 3–46 Typical heavy-duty hydraulic press

FIGURE 3–47 Machinist's rule graduations

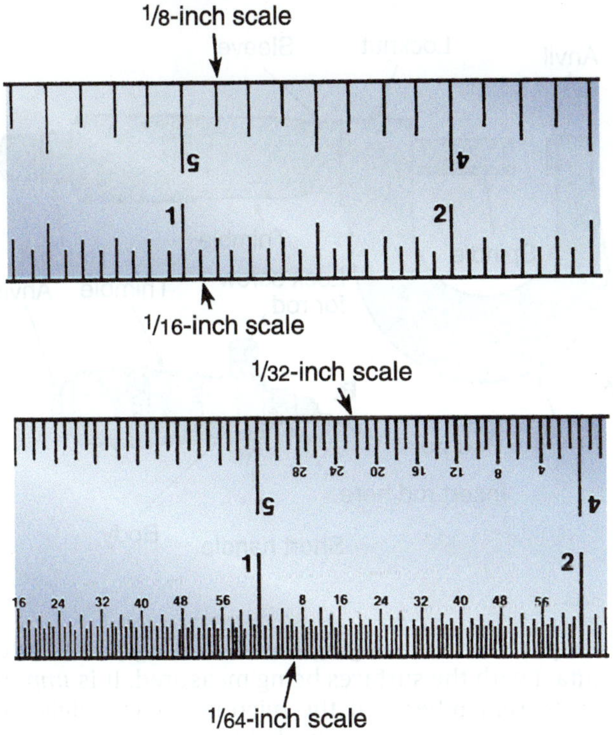

FIGURE 3–48 A typical digital caliper

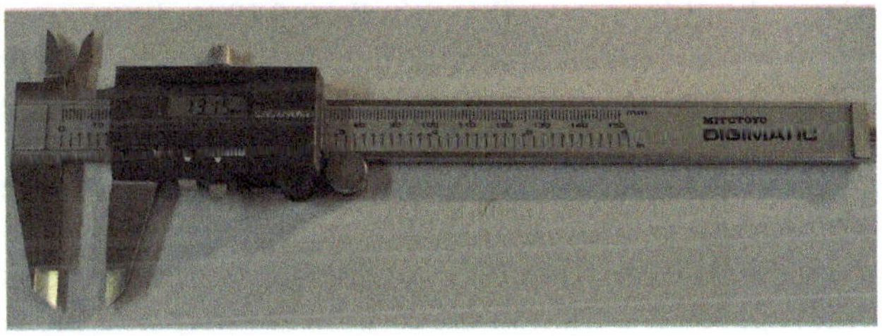

equals one-tenth of an inch on the bar scale (one hundred thousandths of an inch equals one-tenth of an inch or 0.10-inch).

The metric dial caliper is similar in appearance to the standard (inch-reading) model. However, the bar scale is divided into 0.02 mm increments. Additionally, on the metric dial caliper, one revolution of the dial indicator needle equals 2 mm. Both standard and metric dial calipers use a thumb-operated roll knob for fine adjustment. When you use a dial caliper, always move the measuring jaws backward and forward to center the jaws in the work. Make sure that the caliper jaws lay flat on the work. If the jaws or the work are tilted in any way, you will not obtain an accurate measurement. However, although dial calipers are precision-measuring instruments, they truly are only accurate to within ±0.002 inch. The factors that limit dial caliper accuracy include jaw flatness and "feel." Micrometers are better suited to measuring tasks that require high precision.

MICROMETERS

Some truck servicing operations require precise measurements of both outside and inside diameters, or the diameter of a shaft and the bore of a hole. The **micrometer** is the common instrument for taking these measurements. Both outside and inside micrometers are calibrated and read in the same manner and are both operated so that the measuring points exactly contact the surfaces being measured.

The components of a micrometer include the frame, anvil, spindle, locknut, sleeve, sleeve numbers, sleeve long line, thimble marks, thimble, and ratchet (**Figure 3–49**). Micrometers are calibrated in either inch or metric graduations. On both outside and inside micrometers, the thimble is revolved between the thumb and forefinger. Very light pressure

FIGURE 3–49 Nomenclature and components of (A) an outside and (B) an inside micrometer.

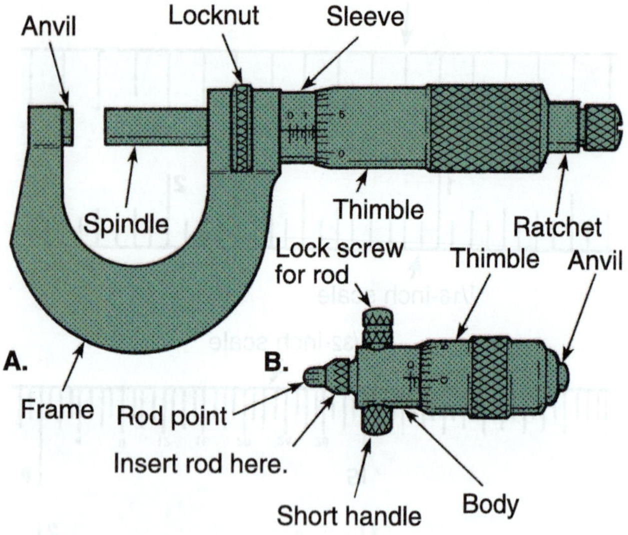

is required when bringing the measuring points into contact with the surfaces being measured. It is *important* to remember that the micrometer is a delicate instrument and that even slight excessive pressure will result in an incorrect reading.

Reading an Inch-Graduated Outside Micrometer

Standard micrometers are made so that each full turn of the thimble moves the spindle 0.025 inch (twenty-five thousandths of an inch). This is accomplished by using 40 threads per inch on the thimble. The sleeve index line is marked with sleeve numbers 1, 2, 3, and so on up to 10. Each sleeve number represents 0.100 inch, 0.200 inch, 0.300 inch, and so on. The sleeve on the micrometer contains sleeve graduations that represent 1 inch in 0.025-inch (twenty-five thousandths of an inch) increments. Each of the thimble graduations represents 0.001 inch (one thousandth of an inch). In one complete turn, the spindle moves through 25 graduations or 0.025 inch (twenty-five thousandths of an inch). Inch-graduated micrometers come in a range of sizes—0 to 1 inch, 1 inch to 2, 2 inches to 3, 3 inches to 4, and so on. The most commonly used micrometers are calibrated in one thousandth of an inch increments.

To read a micrometer, first read the last whole sleeve number visible on the sleeve index line. Next, count the number of full sleeve graduations past the number. Finally, count the number of thimble graduations past the sleeve graduations. Add these together for the measurement. These three readings indicate tenths, hundredths, and thousandths of an inch, respectively. For example, a 2- to 3-inch micrometer that has taken a measurement is described as follows (**Figure 3–50**):

FIGURE 3–50 The three steps in reading a micrometer: (A) measuring tenths of an inch; (B) measuring hundredths of an inch; and (C) measuring thousandths of an inch.

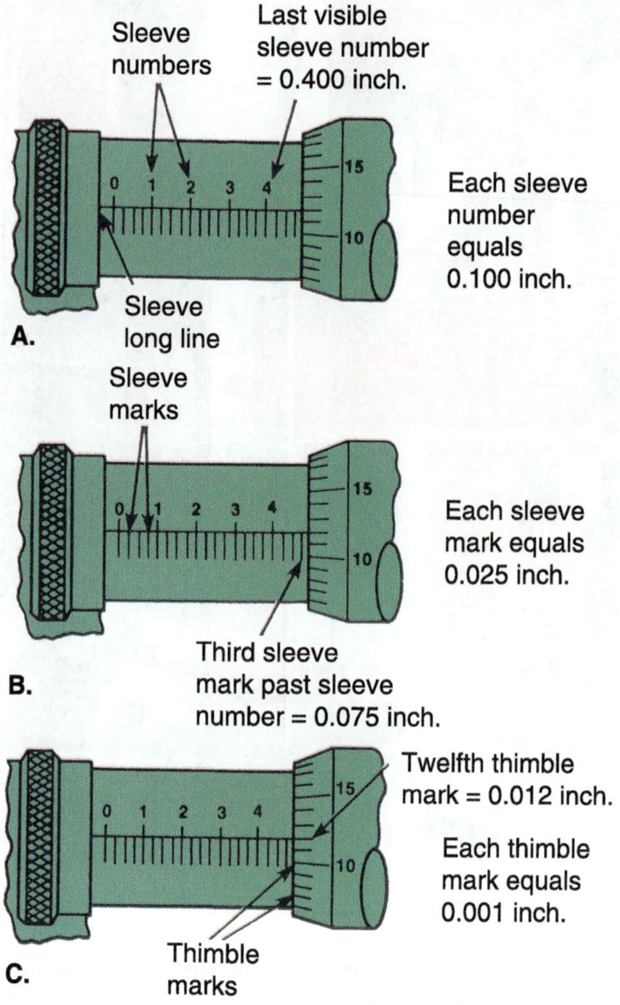

1. The largest sleeve number visible is 4, indicating 0.400 inch (four-tenths of an inch).
2. The thimble is three full sleeve graduations past the sleeve number. Each sleeve graduation indicates 0.025 inch, so this indicates 0.075 inch (seventy-five thousandths of an inch).
3. The number 12 thimble graduation is lined up with the sleeve index line. This indicates 0.012 inch (twelve thousandths of an inch).
4. Add the readings from Steps 1, 2, and 3. The total of the three is the correct reading. In our example:

i.	Sleeve	0.400 inch
ii.	Sleeve graduations	0.075 inch
iii.	Thimble graduations	0.012 inch
iv.	Total =	0.487 inch

5. Now add 2 inches to the measurement, because this is a 2- to 3-inch micrometer. The final reading is 2.487 inches.

Photo Sequence 1 covers some of the basics of using a micrometer. Obtain a micrometer and practice using it.

Reading an Outside Micrometer with a Vernier Scale

In cases in which a measurement must be within 0.0001 inch (one ten-thousandth of an inch), a micrometer with a Vernier scale should be used. This micrometer is read in the same way as a standard micrometer. However, in addition to the three scales found on the typical micrometer, this type has a Vernier scale on the sleeve. When taking measurements with this micrometer (sometimes called a *mike*), read it in the same way as you would a standard mike. Then, locate the thimble graduation that aligns precisely with one of the Vernier scale lines (**Figure 3–51**). Only one of these lines will align exactly. The other lines will be misaligned. The Vernier scale number that aligns with the thimble graduation is the 0.0001-inch (one ten-thousandth of an inch) measurement.

Reading a Metric Outside Micrometer

A metric micrometer is read in the same manner as the inch-graduated micrometer except that the graduations are in the metric system of measurement.

Readings are obtained as follows:

- Each number on the sleeve of the micrometer represents 5 mm or 5/1,000 of a meter (**Figure 3–52A**).
- Each of the ten equal spaces between each number, with index lines alternating above and below the horizontal line, represents 0.5 mm or five-tenths of a millimeter. One revolution of the thimble changes the reading one graduation on the sleeve scale or 0.5 mm (**Figure 3–52B**).

FIGURE 3–51 Measuring ten-thousandths of an inch using a micrometer with Vernier scale

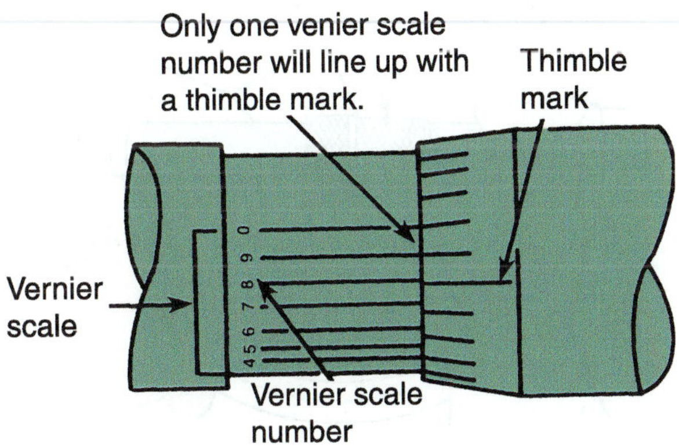

FIGURE 3–52 Reading a metric micrometer: (A) 5 mm; (B) 0.5 mm; and (C) 0.01 mm.

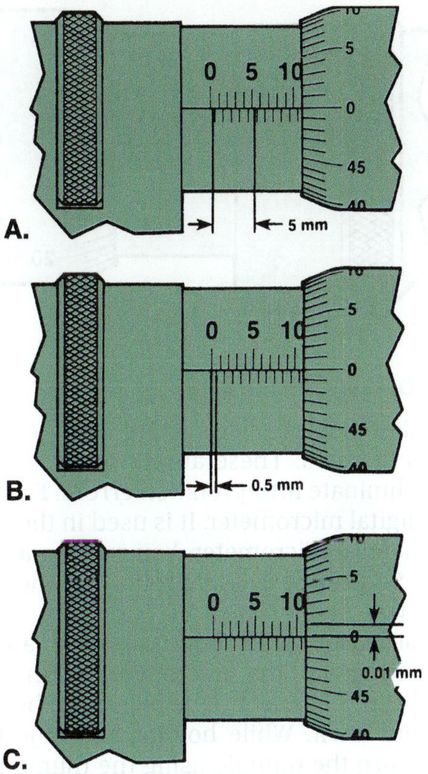

- The beveled edge of the thimble is divided into fifty equal divisions with every fifth line numbered 0, 5, 10...45. Because one complete revolution of the thimble advances the spindle 0.5 mm, each graduation on the thimble advances the spindle 0.5; each graduation on the thimble is equal to 1/50 of 0.5 mm or one hundredth of a millimeter (**Figure 3–52C**).

As with the inch-graduated micrometer, the three separate readings are added together to obtain the total reading (**Figure 3–53**):

1. Read the largest number on the sleeve that has been exposed by the thimble. In the illustration it is 5, which means the first number in the series is 5 mm.
2. Count the number of lines past the number 5 that the thimble has exposed. In the example, this is 4, and because each graduation is equal to 0.5 mm, 4 graduations equal 4 × 0.5 or 2 mm. This, added to the figure obtained in Step 1, gives 7 mm.
3. Read the graduation line on the thimble that coincides with the horizontal line of the sleeve scale and add this to the total obtained in Step 2. In the example, the thimble scale reads 28 or 0.28 mm. This, added to the 7 mm from Step 2, gives a total reading of 7.28 mm.

FIGURE 3–53 The reading shown on this metric micrometer is 7.28 mm.

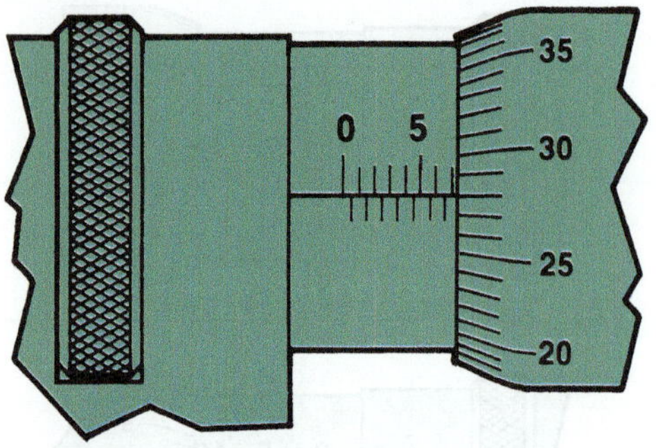

Using an Outside Micrometer

Using digital display micrometers has become popular in recent years. These are no more accurate but virtually eliminate interpretation errors. **Figure 3–54** shows a digital micrometer. It is used in the same way as a mechanical micrometer. You might want to refer to **Photo Sequence 1** while following the sequence that follows:

To measure small objects using an outside micrometer, grasp the micrometer with the right hand and slip the object to be measured between the spindle and anvil. While holding the object against the anvil, turn the thimble using the thumb and forefinger until the spindle contacts the object. Never clamp the micrometer tightly. Use only enough pressure on the thimble to allow the component to just fit between the anvil and spindle. If the micrometer is equipped with a ratchet screw, use it to tighten the micrometer around the object for final adjustment. For a correct measurement, the object must just slip while adjusting the thimble. It is important to slide the mike back and forth over the work until you feel a very light resistance, while at the same time rocking the mike from side to side to ensure that the spindle cannot be closed any further (**Figure 3–55**). These steps should be taken with any precision-measuring device to ensure accurate measurements.

Measurements will be reliable if the mike is calibrated correctly. To calibrate a micrometer, close the mike over a micrometer standard. If the reading

FIGURE 3–54 An outside micrometer with a digital display

Courtesy of Mitutoyo American Corporation

FIGURE 3–55 Slip the micrometer and adjust the thimble until it just begins to grab.

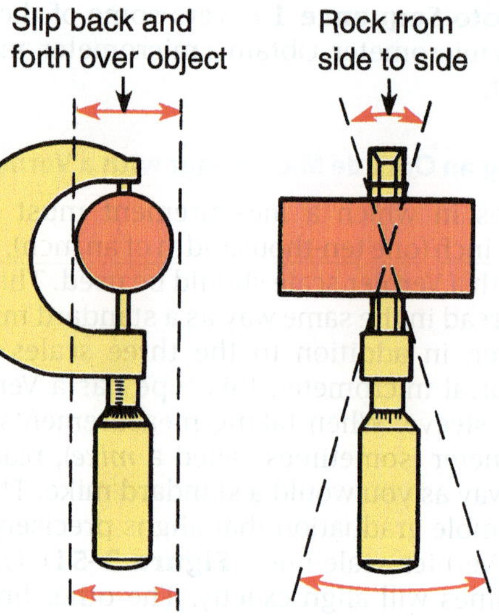

differs from that of the known micrometer standard, then the mike will need adjustment.

Reading an Inside Micrometer

Inside mikes (**Figure 3–56**) are used to measure bore sizes. They are frequently used with outside mikes to reduce the chance of error. To use an inside mike, place it inside the bore or hole and extend the measuring surfaces until each end touches the bore surface. If the bore is large, it might be necessary to use an extension rod to increase the measuring range. Extension rods come in various lengths. The inside micrometer is read in the same manner as an outside micrometer.

FIGURE 3–56 Obtaining a precise measurement with an inside micrometer

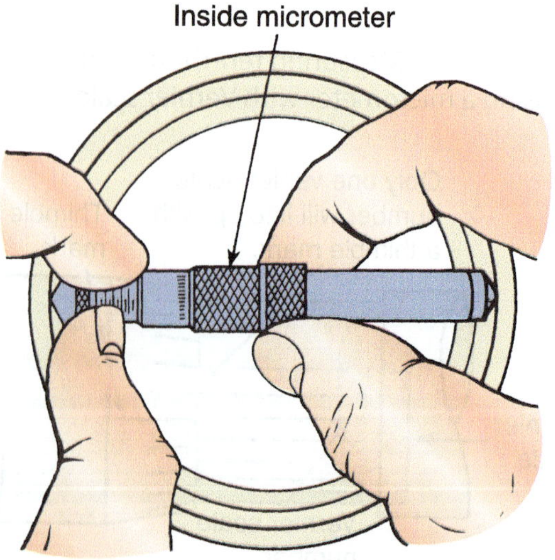

Photo Sequence 1

MICROMETER FAMILIARIZATION

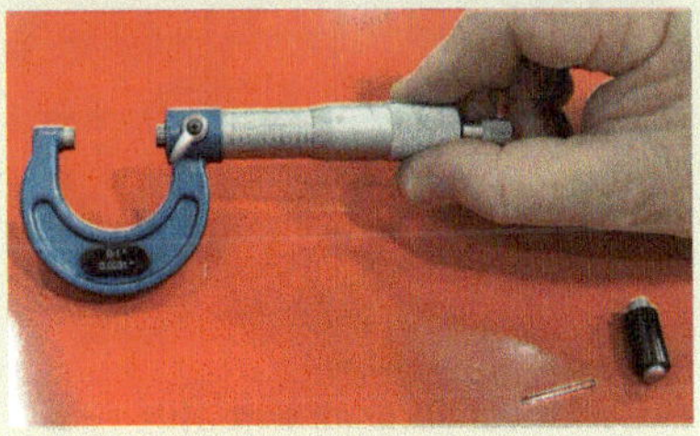

P 1–1 For this exercise, a standard 1-inch micrometer will be used to check out a transmission clutch shaft pilot for wear. Even though most U.S. components are engineered using the metric system, many OEMs continue to convert and publish specifications using English standard values.

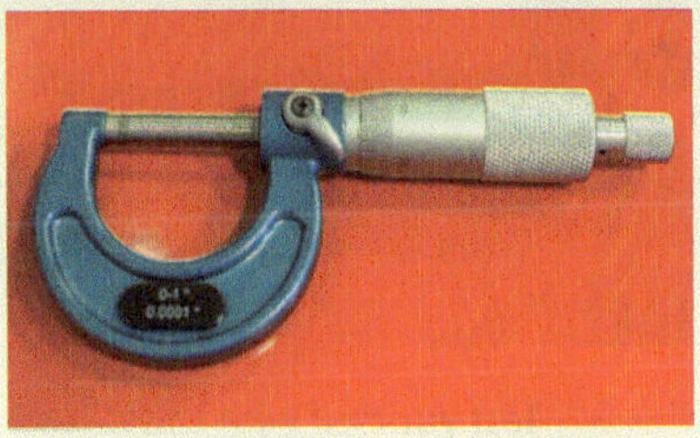

P 1–2 The first step is to check the micrometer calibration. This is important if a shop micrometer is to be used: shop-owned instruments are subject to more abuse than those personally owned. Use the thimble to close the spindle so that it contacts the anvil and check that the reading is zero.

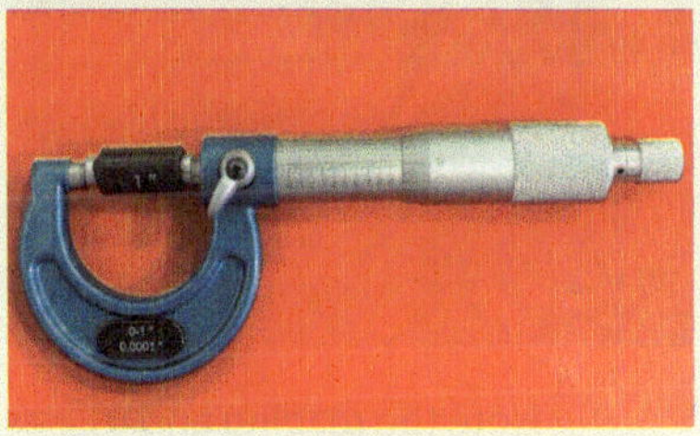

P 1–3 Next, use the thimble to open the spindle past the 1-inch setting. Insert a 1-inch micrometer tram and use the ratchet to turn the spindle into the anvil so that it produces light clearance drag. The reading should be exactly 1.000 inches.

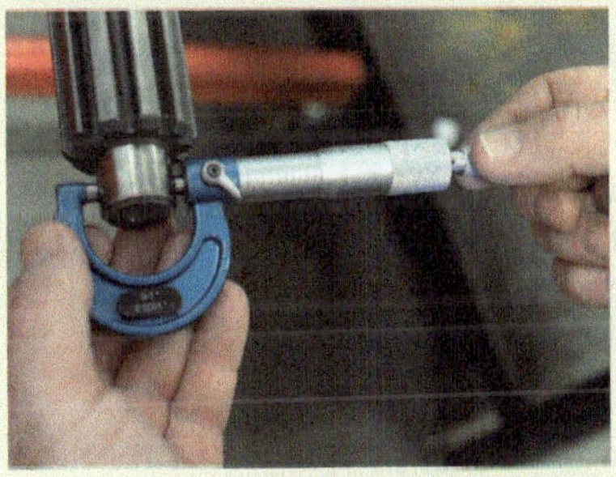

P 1–4 Open the micrometer sufficiently so that it clears the item to be measured. In our example, this is the pilot race of a clutch shaft which measures just under 1-inch. Use the micrometer ratchet to close the spindle to contact the pilot race.

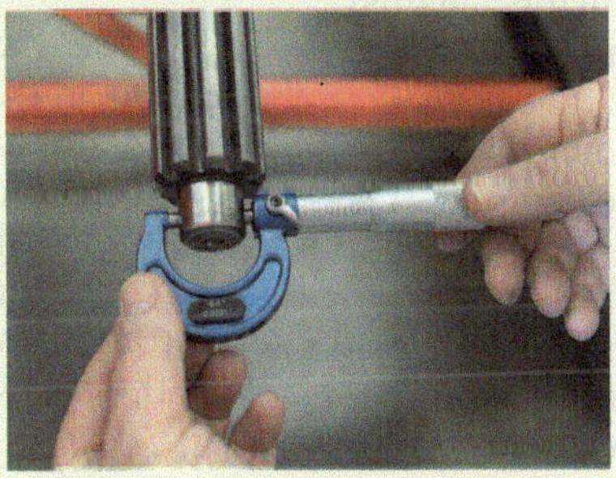

P 1–5 Next use light pressure on the thimble to establish minimal drag clearance between the pilot and the anvil-spindle surfaces of the micrometer. Ensure that the micrometer is supported in a horizontal position so that its weight does not interfere with the reading.

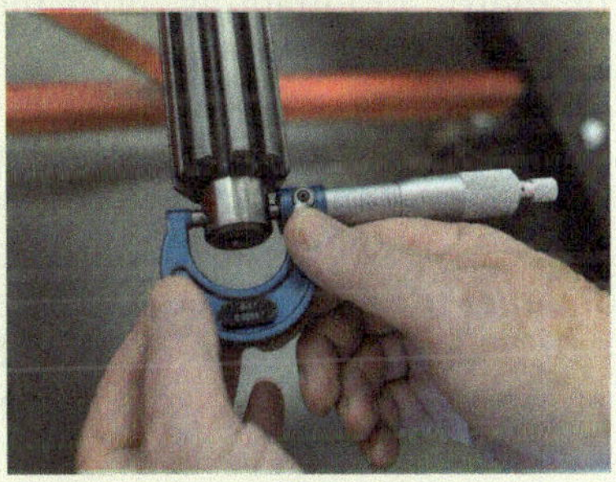

P 1–6 When you are satisfied that you have set the correct dimension, lock the micrometer setting as shown in the above image.

(continued)

Photo Sequence 1 (Continued)

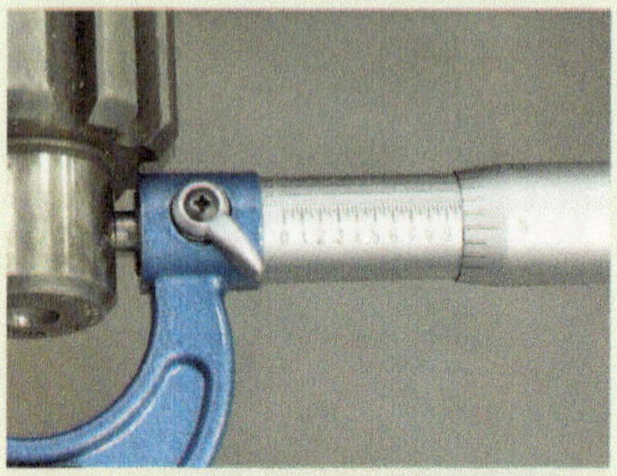

P 1–7 Note the reading and compare to OEM specification. The reading shown here is 0.982-inch. Another way of saying this is 982 thousandths of an inch. The manufacturer specification indicates that this clutch shaft pilot is acceptable for re-use providing it not worn below 0.980-inch. This clutch shaft pilot measurement is therefore within specification.

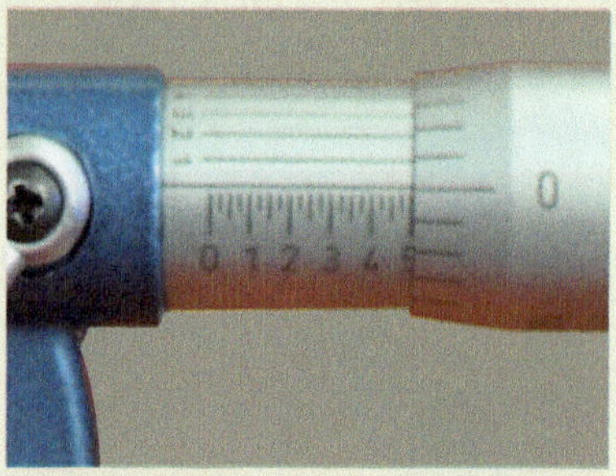

P 1–8 Test yourself with this simple 1-inch micrometer reading: ________ inch.

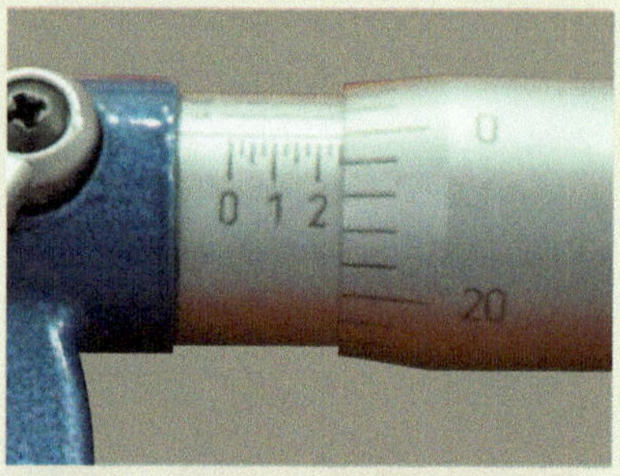

P 1–9 Test yourself with this simple 1-inch micrometer reading: ________ inch.

To obtain a precise measurement in either inch or metric graduations, hold the anvil firmly against one side of the bore and rock the inside mike back and forth and side to side. This ensures that the mike fits in the center of the work with the correct amount of resistance. As with the outside micrometer, this procedure will require a little practice until you get the feel for the correct resistance and fit of the mike. After taking a measurement with the inside mike, use an outside mike to take a comparison measurement. This reduces the chance of errors and helps ensure an accurate measurement.

SHOP TALK

Follow these tips for taking care of a micrometer:

- Always clean a micrometer before using it.
- Do not touch the measuring surfaces.
- Store the micrometer properly. The spindle face should not contact the anvil face, or a change in temperature might spring the micrometer.
- Clean the mike after use. Wipe it clean of any oil, dirt, or dust using a lint-free cloth.
- Do not drop the mike. It is a sensitive instrument and must be handled with care.
- Check the calibration weekly. If it drops at any time, check it immediately.

DIGITAL MEASURING TOOLS

There are many digital precision-measuring tools available in the market today that can make the life of truck technicians a lot easier, especially in the engine rebuilding and machine shops. They can be used to measure anything—from a brake drum inside diameter to a cylinder head valve guide bore—and just about anything in between. Digital measuring tools have the following advantages:

- They are as accurate as or more accurate than their mechanical counterparts.
- They eliminate technician reading/interpretation errors.
- They perform accurate measurements more quickly than their mechanical counterparts.
- They convert standard to metric measurements (and vice versa) at the push of a button.
- They can be zero'd at any location.
- They are more versatile than mechanical measuring tools: One quality digital caliper can replace a full set of 1-inch through 6-inch micrometers.
- They are often lower in cost than their equivalent mechanical counterparts.

Digital measuring instruments perform precise linear and bore measurements to a greater accuracy than their mechanical predecessors. In the same way that the scientific calculator rendered obsolete the slide rule, digital measuring instruments will make the need to interpret the calibrations on a micrometer or Vernier scale a thing of the past. The only puzzle is why so many technicians continue to use mechanical measuring instruments.

CAUTION

Although no one who has worked with machine shop procedures would question the efficacy of digital measuring instruments, remember that at this moment in time the ability to interpret standard, metric, and Vernier micrometer scales is a required learning component in most technician certification programs. Do not throw away your mechanical micrometer yet!

OTHER MEASURING GAUGES

There are several measuring gauges that truck technicians may use. They are:

Thickness Gauges

The **thickness gauge** is a strip of metal of a known and precisely manufactured thickness. It is often referred to as a **feeler gauge**. Several of these metal strips are often stacked into a multiple measuring instrument that pivots in a manner similar to a pocket knife (**Figure 3–57**). The desired thickness gauge leaf can be pivoted away from others for convenient use. A standard steel thickness gauge pack contains leaves of 0.002- to 0.010-inch thickness (in steps of 0.001 in.) and leaves of 0.012- to 0.024-inch thickness (in steps of 0.002 in.). Thickness gauges up to 0.050 inch are available.

Screw Pitch Gauges

Sometimes it is necessary to determine the **screw pitch** (threads per inch) of a fastener. The use of a **screw pitch gauge** (**Figure 3–58**) provides a quick and accurate method of checking the threads per inch. The toothed leaves of this measuring tool are cut at various pitch angles. Just match the teeth of the gauge blade with the threads of a fastener and the correct pitch can be read directly from the leaf.

Screw pitch gauges are available for the various types of fastener threads used in the truck industry: Unified National coarse and fine threads, metric threads, and International Standard threads.

FIGURE 3–57 A typical thickness gauge pack

FIGURE 3–58 A screw pitch gauge pack

Bolt and Nut Dimension Gauges

Bolt and nut dimension gauges are a useful addition to any toolbox. Using one in conjunction with a screw pitch gauge can help identify metric and standard fasteners. A bolt and nut dimension gauge is shown in **Figure 3–59**.

Torque-Indicating Wrenches

A **torque wrench** measures the amount of turning force being applied to a fastener (bolt or nut). Conventional torque wrench scales are usually read in pound-feet (lb-ft.) and metric scales in Newton-meters (N·m). The fact that every truck manufacturer publishes torque recommendations for every fastener proves the importance of using the specified torque when tightening nuts or cap screws.

FIGURE 3–59 A fastener dimension gauge

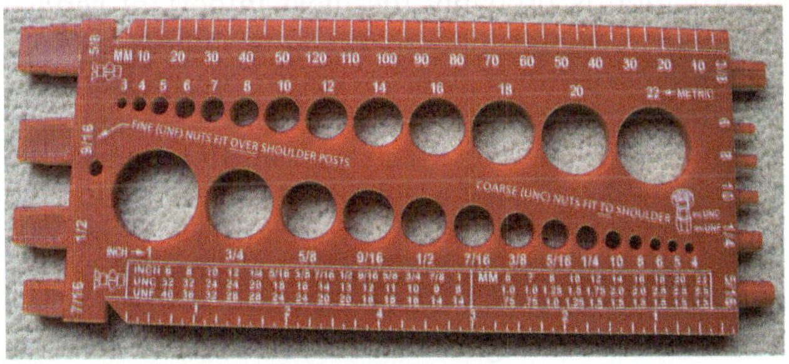

FIGURE 3–60 Three types of torque wrenches: (A) a flex bar; (B) a dial indicator; and (C) a click or sound-indicating wrench.

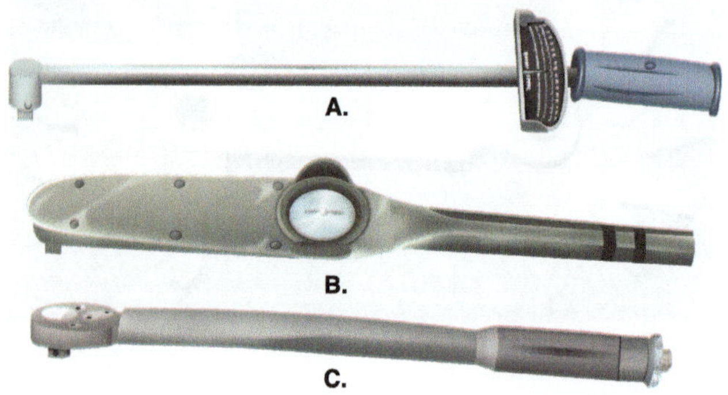

The three general types of manual torque wrenches are the flex bar, dial indicator, and sound-indicating (click) types. These are shown in **Figure 3–60**. The flex bar torque wrench is inexpensive and accurate. The dial indicator torque wrench is very accurate but can be hard to read in tight quarters. The sound-indicating type torque wrench is fast and easy to use. It makes a "pop," "click," or digital "beep" when a preselected torque value is reached. With this type of torque wrench it is not necessary to watch an indicating needle while torquing.

As described later in this chapter, a torque wrench makes it possible to apply the correct clamping force without overstressing either the tool or the fastener. Do not use a torque wrench, however, to break fasteners loose. A torque wrench should be used for tightening purposes only. In addition to manual torque wrenches, pneumatic torque wrenches are used on the shop floor. Because pneumatic torque wrenches have lower accuracy than their manual counterparts, they are best used for fasteners requiring lower clamping precision such as for wheel lug nuts in the example shown earlier in Figure 3–42.

3.4 MANUFACTURERS' SERVICE LITERATURE

Information is probably the most important service instrument in the modern trucking industry. Rapid change makes up-to-date access to the most recent information essential to the service technician. The means we use to access the latest information in this industry has undergone significant change during the past decade. In general, but not always, electronic media has replaced hard copy. Service information in the truck service and repair industry is made available in the following formats.

SERVICE INFORMATION SYSTEMS

Online **service information systems (SIS)** have generally replaced hard copy **service manuals**. For those persons who dislike working from soft copy, the contents of SIS can be printed out. An original equipment manufacturer (OEM) SIS uses a data hub to store service and repair information for all its vehicle systems and components. Any updates are automatically tagged to a downloaded repair procedure. Some examples of OEM SIS are:

- Allison Data Bus Viewer (**DBV**)
- Caterpillar Service Information System or **SIS**
- Detroit Diesel/Mercedes Benz **DDDL** (Detroit Diesel Diagnostic Link)
- Freightliner Daimler **AccessFreightliner**
- Mack Trucks Electrical Information System (**EIS**)
- Navistar International **OnCommand** Service Information System or **ISIS**.
- Paccar **ServiceNet**
- Volvo **IMPACT**

In some cases, OEMs prefer that their chassis electronics and SIS be accessed by dedicated computer. This lowers the likelihood of software file clashes. **Figure 3–61** shows the **graphical user interface (GUI)** of Navistar International's ISIS on the display screen of a dedicated EZ-Tech notebook computer.

SIS Organization

SIS data is organized much like the information in service manuals. It is menu driven and compiled of separate sections dealing with individual systems but makes use of such features as hyperlinks and interactive wiring schematics. This makes navigation

FIGURE 3–61 The GUI of the Navistar service information system

through repair information much easier, making the technician's life easier. In most cases, the SIS works in concert with the OEM diagnostic software so that a typical troubleshooting routine would connect the vehicle electronics with both the OEM SIS and diagnostic software. There is nothing magical about SIS data; it is simply a soft copy service manual information. A downloaded page from Freightliner's ServiceLink SIS is shown in **Figure 3–62**.

Using SIS and Software Diagnostics

The real advantage of online SIS is its ability to interface with the OEM diagnostic software package. Troubleshooting can be interactive or suggestion driven. If the source of a problem is difficult to locate, a troubleshooting chart can be used. The objective of a troubleshooting chart is to guide the technician in identifying the root cause of a specific problem. The best SIS software does this interactively. Whether soft copy or hard copy, there are usually two basic formats of troubleshooting charts:

- A **tree diagnosis chart** (**Figure 3–63**) provides a step-by-step sequence for what should be inspected or tested when trying to solve a repair problem. The routing through to the source of a problem is determined by tests and conditions. Some OEM diagnostic routines can make a tree diagnostic chart interactive. When interactive, progressing from step to step depends on test data entered at each step.
- A **block diagnosis chart** (**Figure 3–64**) lists conditions (problem symptoms), causes (problem sources), and remedies (needed repairs) in columns. This is used for more general types of problem and is based on suggestion. It is useful to technicians when more precise electronic diagnostic routines are unavailable or outside the mapping capability of the software.

Maintenance Logs

Maintenance manuals (hard copy) or **maintenance logs** (soft, electronically retained) contain checklists of routine maintenance procedures and service intervals. Built in to the checklists is a log that is retained in the chassis management electronics and can be downloaded to land-base tracking and analysis software. Maintenance logs are important resources for preventive maintenance (PM) service technicians because they contain information on OEM-required lubrication procedures and charts, service intervals, fluid capacities, specifications, adjustment procedures, checking fasteners, and torque specification charts.

Driver's Manuals

Driver's manuals contain information required by the driver to operate and care for the vehicle and its components. At this moment in time, they are almost always produced in hard copy format. Each driver manual contains a chapter that covers pretrip inspection and daily maintenance of vehicle components. Driver's manuals seldom contain detailed repair or service information.

Parts Literature

Today, parts manuals have been integrated into OEM SIS. This makes sense because it results in service technicians and parts personnel working with information from the same root source. Parts-specific information contains illustrations and part numbers and is designed to aid in the identification of serviceable replacement components for heavy-duty vehicles. They are organized by service group number. Access to the OEM SIS is essential for ordering the correct parts for today's vehicles.

Service Bulletins

Hard copy **service bulletins** have also become a thing of the past. They have been integrated into SIS and although the name is still used, technicians using SIS are guaranteed to have the most recent service information at their disposal. Service bulletins provide the latest service tips, field repairs, product improvements, and related information of benefit to service personnel. **Figure 3–65** is an example of a service bulletin concerning a Freightliner torque arm. Service bulletins are also known as **technical service bulletins (TSBs)**. Service bulletins are usually available only to OEM dealers. When servicing a vehicle system or component, check for valid service bulletins for the latest information on the subject.

Recall Bulletins

Recall bulletins apply to required service work or replacement of parts resulting from documented series of failures. Recall bulletins are often related to vehicle safety. All bulletins are distributed to dealers; customers receive notices that apply to their vehicles.

Field Service Modifications

The Field Service Modification publication is concerned with nonsafety–related service work or replacement of parts. Field service modifications are distributed to OEM dealers; the procedure is usually only accessible by OEM levels of access to SIS and not to customers with SIS access. Customers receive notices that apply to their vehicles.

Time Guides

A **time guide** is used for computing compensation payable by the truck manufacturer for repairs or service work to vehicles under warranty or for other special conditions authorized by the company. A time guide usually covers all vehicle models of the manufacturer. This software contains operation numbers and time allowances for various procedures on

FIGURE 3–62 A downloaded page from Freightliner's ServiceLink SIS

Heater and Air Conditioner Removal and Installation

Removal

1. Review the information under "Safety Precautions, 100."

WARNING: Failure to review the safety precautions, and to be aware of the danger involved when working with refrigerant, could result in serious personal injury.

2. Depending on the type of temperature control, open the water shutoff valve.

   If equipped with a toggle switch control, set the heater water-temperature toggle switch to HOT (**Fig. 1**). When in this position, there is no air pressure to the water shutoff valve.

   If equipped with a cable control, pull the temperature control cable up (**Fig. 2**).

   If equipped with C.T.C., turn the temperature control knob clockwise to HEAT (**Fig. 3**).

3. Disconnect the batteries.

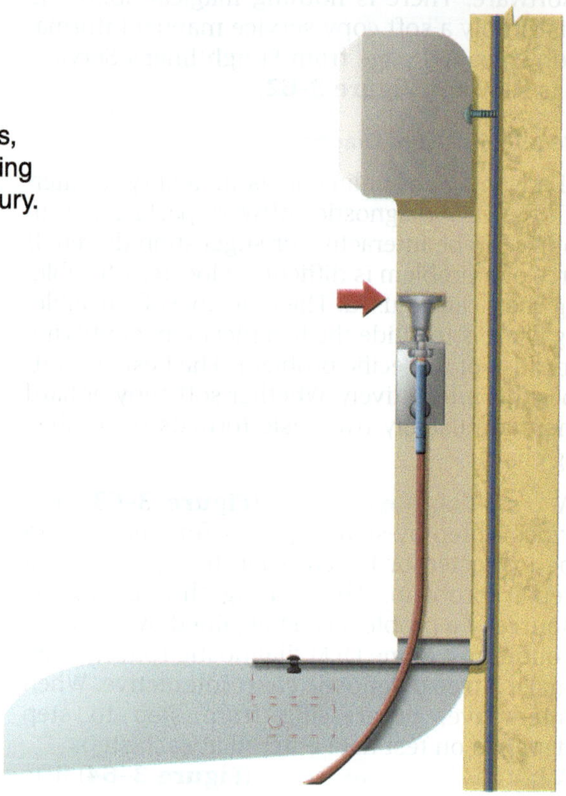

Fig. 2 Manual Temperature Control Cable Handle

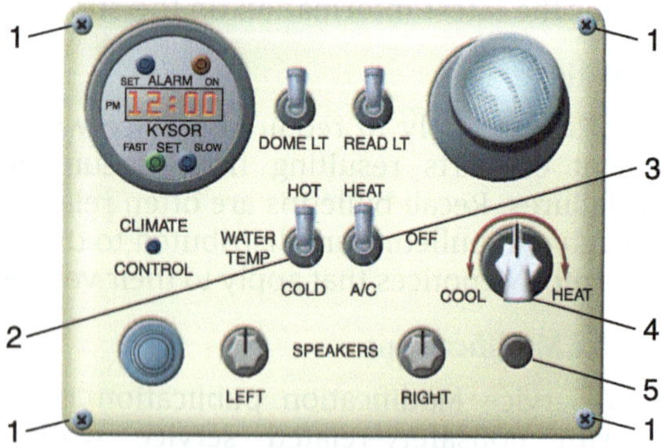

1. Control Panel Screw
2. Water Temperature Switch
3. Mode Switch
4. Temperature Control Knob
5. Temperature Sensor

Fig. 1 Control Panel for Electronic Thermostat Control System

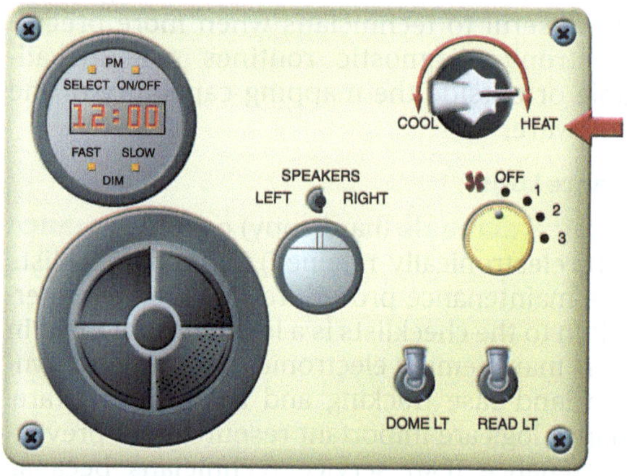

Fig. 3 Temperature Control Knob

Courtesy of Daimler Trucks North America

FIGURE 3–63 A tree diagnosis chart layout

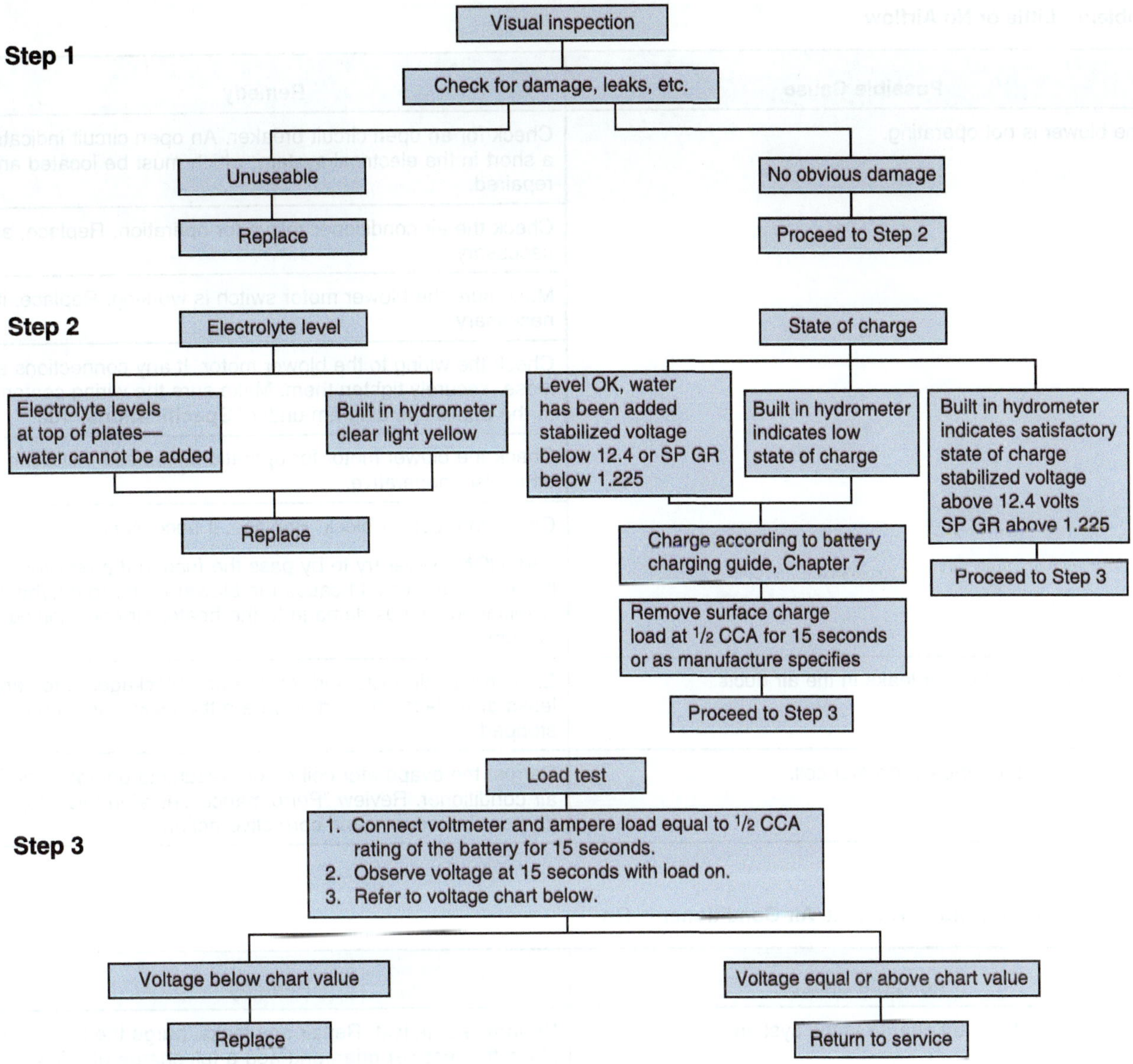

Voltage Chart

Estimated electrolyte temperature		Minimum required voltage is under 15 second load use 1/2 these values for 6 volt batteries
70°F	(21°C) & above	9.6
60°F	(16°C)	9.5
50°F	(10°C)	9.4
40°F	(4°C)	9.3
30°F	(–1°C)	9.1
20°F	(–7°C)	8.9
10°F	(–12°C)	8.7
0°F	(–18°C)	8.5

FIGURE 3–64 A block diagnosis chart layout

Problem—Little or No Airflow

Possible Cause	Remedy
The blower is not operating.	Check for an open circuit breaker. An open circuit indicates a short in the electrical system, which must be located and repaired.
	Check the air conditioner relays for operation. Replace, as necessary.
	Make sure the blower motor switch is working. Replace, if necessary.
	Check the wiring to the blower motor. If any connections are loose, securely tighten them. Make sure the wiring conforms to the applicable diagram under **"Specifications, 400."**
	Check the blower motor for operation. Replace if sticking or otherwise inoperative.
	Check the resistor block. Replace, if necessary. **CAUTION:** Never try to by pass the fuse in the resistor block. To do so could cause the blower motor to overheat, resulting in serious damage to the heater/air-conditioning system.
There are restrictions or leaks in the air ducts.	Examine all air ducts and remove any blockages. Stop any leaks or replace any portion where the leaks cannot be stopped.
Ice has formed on the evaporator coil.	Defrost the evaporator coil before resuming operation of the air conditioner. Review "Performance Tests" in this subject for possible causes and corrective action.

Problem—Warm Airflow When the Air Conditioner Is On

Possible Cause	Remedy
There is no refrigerant charge in the system.	Perform a leak test. Repair any leaks, purge the system, replace the receiver-drier, and add a full charge of refrigerant.
Moisture in the system.	If moisture is in the system, ice crystals may form at the expansion valve, blocking the flow of refrigerant (off and on). Discharge the refrigerant charge, purge the system, replace the receiver-drier, and add a full charge of refrigerant.
The refrigerant compressor is not operating.	If the refrigerant charge is low, charge and leak test the system. Repair any leaks.
	The refrigerant compressor clutch or drive belt needs repair or replacement. For instructions, refer to the refrigerant compressor section elsewhere in this group.

FIGURE 3–65 A typical service bulletin

Tightening the Torque-Arm Clamp Bolts, Reyco® Model 102W Suspensions

32—1

Service Bulletin

> COE
FLA CaE
Medium Trucks

> 120 Conventional
FLO Conventional
FLC 112 Conventional

The Reyco Model 102W suspension now uses an adjustable cast-end torque arm (**Fig. 1**). On the cast-end torque arm, the clamp and torque-arm end are a one-piece casting. The torque for the clamp bolts on the cast-end torque arm is 150 lbf·ft (200 N·m).

On Reyco Model 102W suspensions equipped with the previous style two-piece end assembly (**Fig. 2**), tighten the clamp bolts 80 lbf·ft (108 .N·m).

NOTE: On the two-piece end assembly, the clamp must be positioned correctly to maintain the torque value. Be sure that the small dimple stamped into the clamp is against (not on) the torque-arm end.

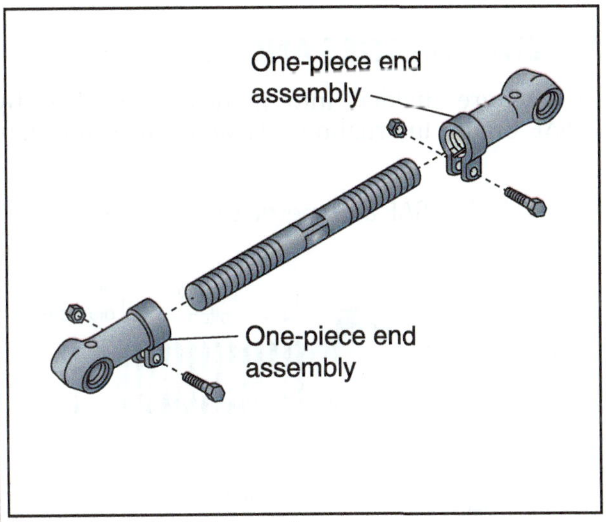

Fig. 1 Cast-End Torque Arm

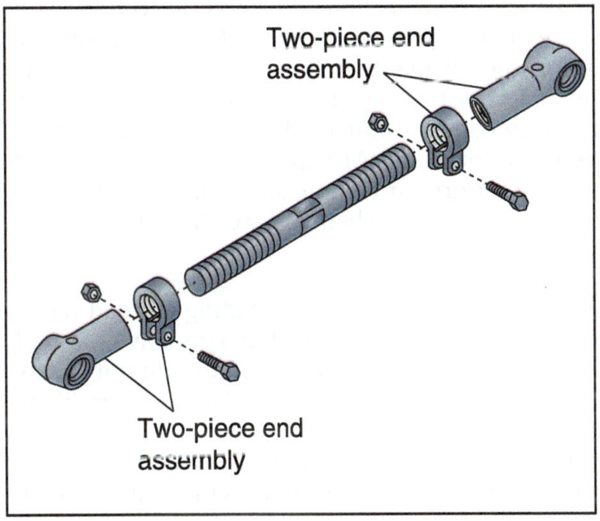

Fig. 2 Two-Piece End Torque Arm

Courtesy of Daimler Trucks North America

vehicle systems and parts. Time guide fields in SIS software are usually available only to dealers.

OTHER SERVICE LITERATURE

There are other sources of servicing information available to truck technicians.

Supplier Manufacturers' Guides and Catalogs

Many of the larger parts manufacturers publish guides on the original equipment and aftermarket parts that they manufacture or supply for truck repairers and builders. Many of these contain detailed installation guides and service bulletins so they can be useful to repair technicians. They can also be a useful resource for truck body builders.

Training Literature

It is in the interest of truck and component manufacturers to make service repair procedures and training programs available to the truck service repair industry. For instance, Eaton manufactures transmissions that can be found in any OEM chassis. Eaton has established its own in-house certification program, which is required of technicians if they wish to perform warranty work on Eaton products. Most OEM training is either online or computer based.

General Service Repair Information

This source of service information is produced by independent publishers rather than the manufacturers. They contain component information, diagnostic steps, repair procedures, and specifications for several brands of trucks in one online accessed data hub. This source of service information is usually in condensed format and is more general in nature. It may also be out of date because truck OEMs often stall the release of recent service data to third party software providers to protect their dealer networks.

General repair information providers sell their services to shops with no dealership affiliation. The provider offers the service by subscription. This enables access to the provider's data hub. Examples are Mitchell's Online or Motor Truck publications. Technicians should be wary about relying exclusively on this type of repair information when working with some of the more complex systems found on truck chassis.

Trade Magazines

Other valuable sources of up-to-date technical information are trade magazines and newsletters. More often than not, these publications are provided free of charge to anyone working in the trucking industry. They can help technicians keep up to date with the pace of change in the industry. Because most subscriptions are available online, they are an easy way for technicians to keep up to date. The Technology and Maintenance Council (TMC) of the American Trucking Association (ATA) produces some excellent support literature for technicians; consider joining as a technician member when you are eligible.

Make Reading a Habit

As we suggested in Chapter 1, reading a broad range of technical literature helps technicians stay on top of technology and this respect, it is important not to rely exclusively on phone downloads. You are more likely to read an article from start to finish if you are doing so on a tablet, computer screen, or in hard copy. **Figure 3–66** shows some examples of technical reading material.

FIGURE 3–66 Reading a broad range of publications is one of the best ways of keeping up with technology.

3.5 FASTENERS

Trucks and trailers on our roads today use both standard and metric threads; metric fasteners are common but not universal even in late model applications (**Figure 3–67**). For a number of years, engine and major chassis components have used metric fasteners almost exclusively (diameter and pitch are measured in millimeters).

Most fasteners used on a truck are hex head fasteners (usually nonflanged); most metric fasteners are also nonflanged. Hardened flat washers are used under the bolt head, between the clamped components and the hex nut, to distribute the load: this prevents localized stress. The washers are cadmium- or zinc-plated and have a hardness rating of 38 to 45 Rockwell "C" hardness (HRC). Some fasteners, often those smaller than ½-inch diameter, have integral flanges that fit against the clamped surfaces. These flanges eliminate the requirement for washers.

FASTENER GRADES AND CLASSES

Fasteners are divided into grades established by the SAE or the International Fastener Institute (IFI).

FIGURE 3–67 SAE and metric thread bolts

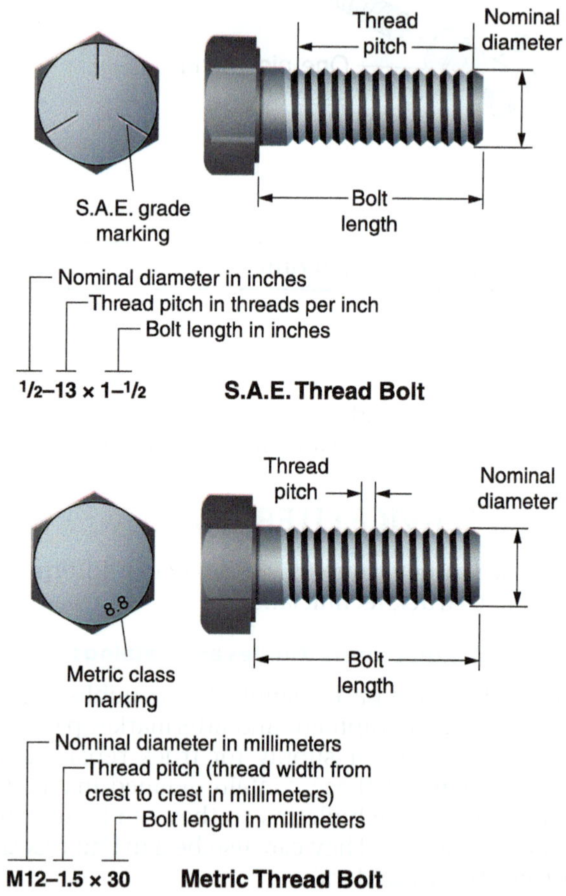

FIGURE 3–68 Bolt (capscrew) identification

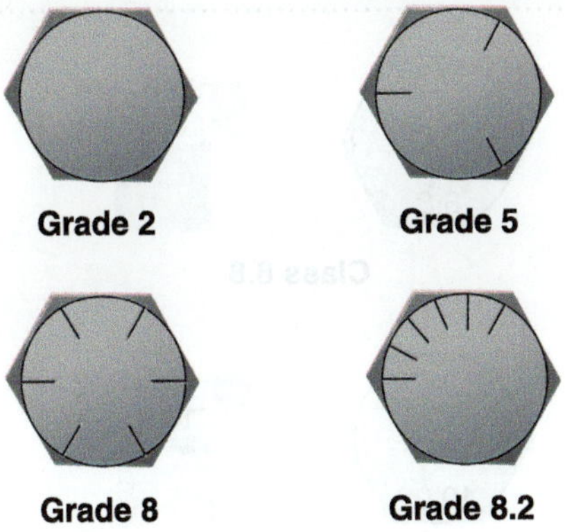

Note: Grade 2 bolts have no grade markings; grade 2 bolts are rarely used on trucks.

These grade markings are used on plain hex-type and flanged bolts (capscrews). In addition to the grade markings, the bolt head must also carry the manufacturer's trademark or identification.

Fastener grades indicate the tensile strength of the fastener; the higher the number (or letter), the stronger the fastener. Bolt (capscrew) grades can be identified by the number and pattern of radial lines/dashes/dots forged on the bolt head (**Figure 3–68**).

Hex nut (and locknut) grades can also be identified by the number and pattern of axial lines and dots on various surfaces of the nut (**Figure 3–69**). Nearly all of the bolts used on the heavy-duty vehicle are grades 5, and 8, but some OEMs use special spec variants of both grades, that are identified using a numeric coding qualifier, such as grade 9. Matching grades of hex nuts are used with grade 5 bolts; grade 8, grade C, or grade G (flanged) hex nuts are used with grade 8 bolts.

Every fastener manufacturer is required by law to register their headmarking logo with the **Industrial Fastener Institute (IFI)** for purposes of identification and maintaining manufacturing standards. All graded fasteners are also required by law to meet certification standards for size, tensile strength, chemical composition, and hardness.

FIGURE 3–69 Hex nut (and lock washer) identification

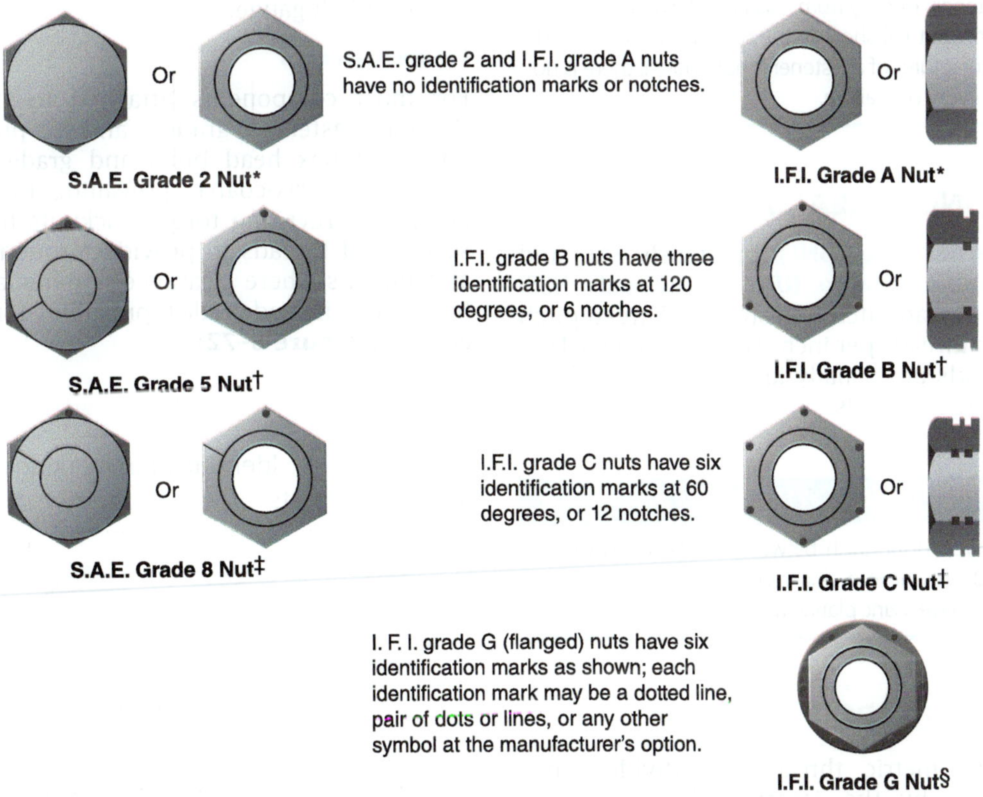

* Strength compatible with grade 2 bolt.
† Strength compatible with grade 5 bolt.
‡ Strength compatible with grade 8 or grade 8.2 bolt.
§ Flanged locknut, strength compatible with grade 8 or grade 8.2 bolt.

Tech Tip

A grade 5 bolt or nut is identified when the radial marks or dots are more than 60 degrees (1 hex flat). A grade 8 bolt or nut is identified when the radial marks or dots are spaced 60 degrees (1 hex flat) or less apart. Study Figure 3-72 and make sure you can interpret fastener gradings at a glance. An absence of radial marks or dots identifies a grade 2 bolt which is generally unsuitable for use in truck applications. In the case of SAE grade 5 or grade 8 fasteners, only two marks or dots may be used depending on the manufacturer.

CAUTION

Although most technicians have no difficulty identifying grade 5 and grade 8 bolts/capscrews, many get confused when attempting to identify their equivalent nuts: remember the 60-degrees or less (grade 8), MORE than 60-degrees rule explained in the preceding tech tip. Be especially careful not to confuse the crimp marks used on locknuts with grade ID marks.

Tech Tip

Avoid using fasteners with no obvious manufacturer's logo or icon. Over the past decade, the market has been flooded with bogus fasteners (mostly from Asia), which sometimes use SAE grade marks but fall short of the specifications required. When the manufacturer of a fastener is not known, there is no recourse in the event of a failure.

UNF AND UNC THREADS

Unified National Fine (UNF) (previously SAE) and **Unified National Coarse (UNC)** (previously USS) designate the thread pitch of fasteners. Thread pitch is classified as threads per inch (TPI) in fasteners. UNF and UNC threads are usually known respectively as "fine" and "coarse" threads.

Tech Tip

Color can help you distinguish between Grade 8 and Grade 5 nuts: Grade 8 nuts have a zinc yellow dichromate plating whereas Grade 5 have a zinc plating finish.

METRIC FASTENERS

Fasteners with metric threads are divided into classes adopted by the American National Standards Institute (ANSI). The higher the class number, the stronger the fastener. Bolt classes can be identified by the numbers forged on the head of the bolt

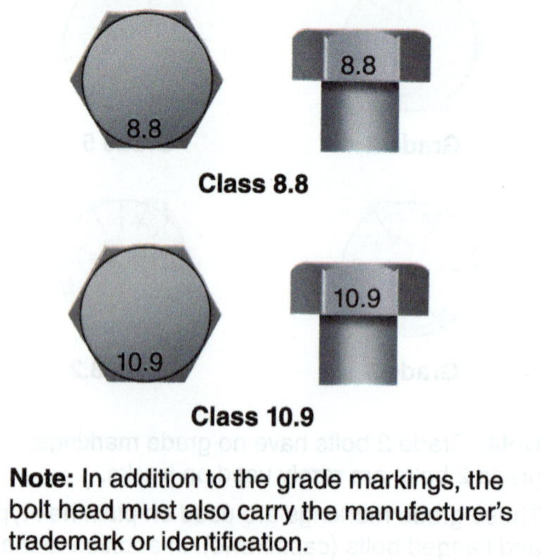

FIGURE 3–70 Bolt classes can be identified by the numbers forged on the head of the bolt.

(**Figure 3–70**). Hex nut (and locknut) classes can be identified by the lines or numbers on various surfaces of the nut (**Figure 3–71**). Class 8 hex nuts are always used with class 8.8 bolts; class 10 hex nuts with class 10.9 bolts. Threads can be measured with a screw pitch gauge.

Frame Fasteners

For most components attached to the frame by threaded fasteners, grade 8 and 8.2 phosphate- and oil-coated hex head bolts and grade C cadmium-plated and wax-coated prevailing torque locknuts are used. Prevailing torque locknuts have distorted sections of threads to provide torque retention. For attachments where clearance is minimal, low-profile hex head bolts and grade C prevailing torque locknuts are used (**Figure 3–72**).

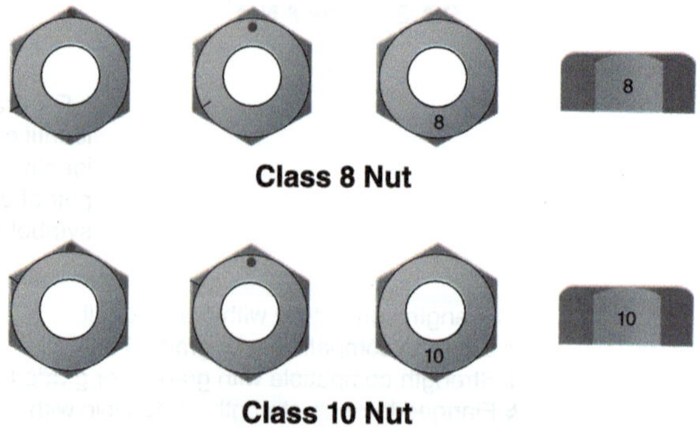

FIGURE 3–71 Identification markings on Class 8 and Class 10 nuts.

FIGURE 3–72 A Grade C fastener assembly with a prevailing torque locknut

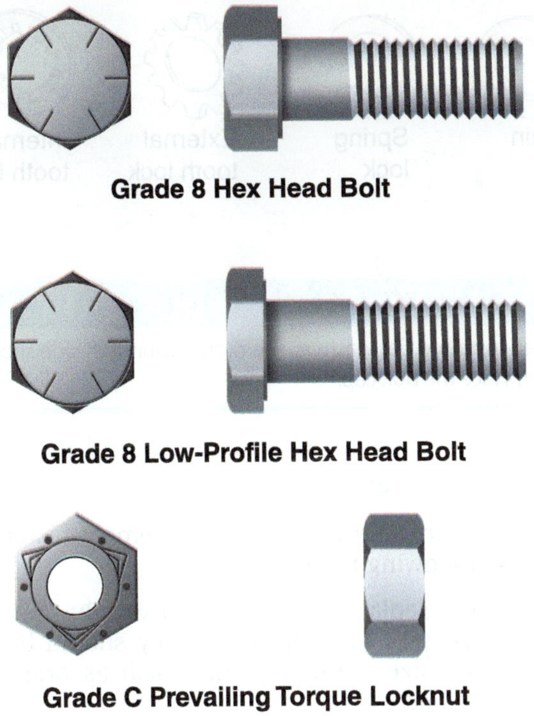

Many trucks today are manufactured with constant clamping force fasteners such as **Huck fasteners**. These fasteners are often installed by robotic assembly processes in new truck assembly but are usually destroyed on removal. Replace with either Huck fasteners (access with a bulky Huck tool can be a problem) or a similar grade of bolt.

TIGHTENING FASTENERS

When a capscrew or bolt is tightened to its specified torque, or a nut is tightened to its torque value on a bolt, the shank of the capscrew or bolt stretches slightly. This stretching (tensioning) results in a preload. OEM-specific torque values are calculated to provide enough clamping force on bolted components and the correct tension on the fastener to maintain the clamping force.

Friction and Clamping Force

Use of a torque wrench to tighten fasteners will help prevent overtensioning. Overtensioning causes permanent stretching of the fasteners, which can result in breakage of components or fasteners. When torquing a fastener, typically 80 to 90 percent of the turning force is used to overcome thread, cap, and nut face friction; only 10 to 20 percent results in capscrew or bolt clamping force. About 40 to 50 percent of the turning force is needed to overcome the friction between the underside of the capscrew head or nut and the washer. Another 30 to 40 percent is needed to overcome the friction between the threads of the capscrew and the threaded hole or the friction between the threads of the nut and bolt.

All metals are marginally elastic, which means they can be stretched and compressed to a certain point. This elastic, spring-like property is what provides the clamping force when a bolt is threaded into a tapped hole or when a nut is tightened. As the bolt is stretched, clamping force is created due to bolt tension. Like a spring, the more a bolt is stretched, the tighter it becomes. However, a bolt can be stretched too far, which will result in shear. At this point, the bolt can no longer safely clamp the load it was designed to support.

Fastener Elasticity

Elasticity means that a bolt can be stretched a certain amount, and each time the stretching load is reduced, the bolt will return to its original, normal size. In other words, it is reusable. However, if a bolt is stretched beyond its yield point it permanently deforms. A bolt will continue to stretch a little more each time it is used, just like a piece of taffy that is stretched until it breaks (**Figure 3–73**).

Importance of Correct Torque Procedure

Proper use of torque avoids exceeding the yield point of a bolt. Torque values are calculated with at least a 25 percent safety factor below yield point. Some fasteners, however, are intentionally torqued just barely into a yield condition, although not quite enough to create the classic coke bottle shape of a necked-out bolt. This type of fastener, which is known as a torque-to-yield (TTY) bolt, uses close to 100 percent of its tensile strength, compared to around 75 percent on a regular fastener when both are torqued to specification. TTY bolts are normally required to be template torqued; this requires using incrementally increasing torque values using a torque wrench, with final torque turning through a specified number of degrees. TTY fasteners, however, should not be reused unless otherwise specified.

FIGURE 3–73 This bolt has been torqued beyond its yield point, resulting in shear.

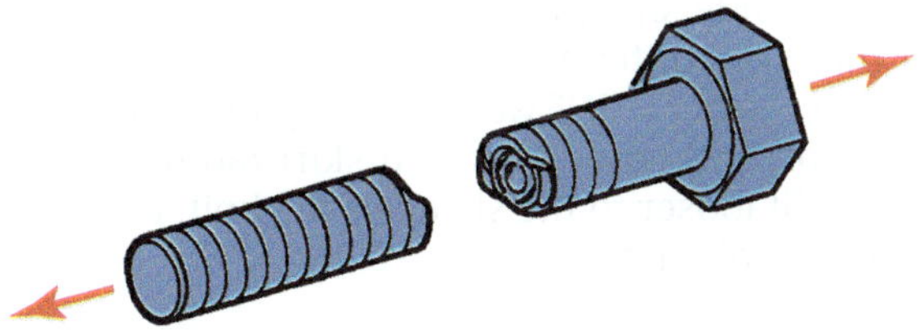

Effect of Lubricants

The torque required to tighten a fastener is reduced when friction is reduced. If a fastener is dry (unlubricated) and plain (unplated), thread friction is high. If a fastener is wax coated or oiled or has a zinc phosphate coating or cadmium plating, friction forces are reduced. Each of these coatings and combinations of coatings has a different effect. Using zinc-plated hardened flat washers under the bolt (capscrew) head and nut reduces friction. Dirt or other foreign material on the threads or clamping surfaces of the fastener or component can increase friction to the point that the torque specification is met before any clamping force is produced.

Even though varying conditions affect the amount of friction, a different torque value cannot be given for each. To ensure that they are always torqued accurately, most OEMs recommend that all fasteners be lubricated with oil (unless specifically instructed to install them dry), and then torqued to the values for lubricated- and plated-thread fasteners. When locking compound or antiseize compound is recommended for a fastener, the compound acts as a lubricant, and oil is not required.

Overtorquing Fasteners

Be careful not to strip bolt threads when using power wrenches. It is easy to turn a bolt beyond its yield point within a split second. Impact wrenches are the worst offenders. Some friction is required to prevent a nut from spinning. When a nut is lubricated, there is insufficient friction to stop the impact wrench from hammering the nut beyond the bolt yield point and/or stripping the threads.

Do not run a nut full speed onto the bolt threads with an impact gun. Instead, run it up slowly until it contacts the work, and then note the socket position and observe how far it turns. Smaller air-powered speed wrenches do not produce the aggressive turning force of impact wrenches and are safer to use. Follow this procedure with a torque-modulated air wrench as well.

Washers and Lock Washers

A rule of thumb on lock washers is that if the fastener assembly did not come with one, do not add one. Lock washers are extremely hard and tend to break under severe pressure. Use locknuts with hard, flat washers. Properly torqued, this type of fastener should never loosen—even when lubricated (**Figure 3–74**). As a general rule, when using flat washers, the radius (rounded) side should face the head of the bolt. Note that there is a difference between UNF (SAE) and UNC (USS) flat washers: UNF washers fit tighter to the shank of the bolt and have a reduced skirt radius. UNC flat washers fit looser to the shank of the bolt, and have a larger skirt diameter.

FIGURE 3–74 Washers are sometimes used to lock fasteners to keep them from coming loose.

> **Tech Tip**
>
> When installing flat washers, the radius (rounded) side should always face the bolt cap.

FASTENER REPLACEMENT

When selecting and installing replacement fasteners, keep the following points in mind:

- When replacing fasteners, use only identical bolts, washers, and nuts; they should be the same size, strength, and finish as originally specified.
- When replacing graded (or metric class) bolts and capscrews, use only fasteners that have the manufacturer's trademark or identification on the bolt head; do not use substandard bolts.
- When using nuts with bolts, use a grade (or class) of nut that matches the bolt.
- When installing nonflanged fasteners, use hardened steel flat washers under the bolt (capscrew) head and under the hex nut or locknut.
- For bolts 4 inches (100 mm) or less in length, make sure that at least 1½ threads and no more than ⅝-inch (16 mm) bolt length extends through the nut after it has been tightened. For bolts longer than 4 inches (100 mm), allow a minimum of 1½ threads and a maximum of ¾-inch (19 mm) bolt length protrusion.
- Never hammer or screw bolts into place. Align the holes of the mating components so that the fastener surfaces are flush with the washers, and the washers are flush with the clamped surfaces.
- When installing fasteners into threaded aluminum or plastic components, start the fasteners by hand to ensure that cross-threading does not damage the threads.
- Do not use lock washers (split or toothed) next to aluminum surfaces.
- When installing studs that do not have an interference fit, install them with thread locking compound.

- When installing components mounted on studs, use free-spinning (nonlocking) nuts and helical spring (split) lock washers or internal-tooth lock washers. Avoid using locknuts because they tend to loosen the studs during removal. Do not use flat washers.
- Do not use lock washers and flat washers in combination (against each other); each defeats the other's purpose.
- Use stainless steel fasteners against chrome plating, unpainted aluminum, or stainless steel.

SHOP TALK

If a torque-to-yield bolt is replaced with a new bolt of identical grade but torqued to a value found in a standard torque chart, the clamping force produced will be at least 25 percent less.

CAUTION

A fastener without strength markings must be assumed to be at the lowest common denominator of ratings (Grade 2) and not suitable for use in vehicle applications. A fastener with no manufacturer's logo is probably from Asia and likely to be of lower quality: there are many of these in circulation so technicians should be very aware!

Fastener Tightening

When tightening fasteners, remember the following procedures:

- Clean all fasteners, threads, and all surfaces before installing them.
- To ensure they are torqued accurately, fasteners should be lubricated with oil (unless specifically instructed to install them dry), and then torqued to the values for lubricated- and plated-thread fasteners. When locking compound or antiseize compound is recommended for a fastener, the compound acts as a lubricant, and oil is not needed.
- Hand turn fasteners so they contact before using a torque wrench to tighten them to their final torque values.
- Tighten the nut, not the bolt head, when possible. This gives a truer torque reading by eliminating bolt body friction.
- Always use a torque wrench to tighten fasteners, and use a slow, smooth, even pull on the wrench. Do not use a short, jerky motion or inaccurate readings can result.
- When reading a bar-type torque wrench, look straight down at the scale. Viewing from an angle can give a false reading.
- Only pull on the handle of the torque wrench.
- Do not allow the beam of the wrench to touch anything.
- Tighten bolts and nuts incrementally. Typically, this should be to one-half specified torque, to three-fourths torque, to full torque, and then to full torque a second time.
- Never overtorque fasteners; overtightening causes permanent stretching of fasteners, which can result in breakage of parts or fasteners.
- If specific torque values are not given for countersunk bolts, use the torque value for the corresponding size and grade of regular bolt.
- Follow the torque sequence when provided to ensure that clamping forces are even and mating parts and fasteners are not distorted.

FASTENER FAILURES

Most fastener failures can be attributed to human error concerning application and/or assembly. The consequence of fastener failure on transportation equipment can be fatal. The following are the most common reasons:

- Overtorquing
- Mismatched graded fasteners
- Re-use of a fastener (especially nuts)
- Use of an excessively long bolt
- Compression of clamped materials
- Improperly installed washers

Re-Use of Fasteners

The Industrial Fastener Institute has conducted detailed studies on the loss of strength in re-used fasteners. A key conclusion is that a re-used nut is more likely to fail. Threads on a nut are manufactured to be slightly softer than those on the bolt so that deform slightly to form into the contour of the bolt threads when torqued. Although the IFI recommends that both the nut and bolt be replaced in any critical application, it emphasizes that nuts are a poor re-use risk. Some facts:

- The first thread of a UNC nut supports 38 percent of the total load on a bolt.
- The second thread supports 25 percent of the total bolt load.
- The third thread supports 18 percent of the total bolt load.
- Therefore over 80 percent of the total bolt load is supported by the first three threads on the nut.

Thread Repair

A common fastening problem is stripped threads. This is usually caused by high torque or by cross-threading.

FIGURE 3–75 Steps in the installation of a helical screw repair coil: (A) Drill the damaged threads using the correct size drill bit. Clean all metal chips out of the hole. (B) Tap new threads in the hole using the specified tap. The thread depth should exceed the length of the bolt. (C) Install the proper size coil insert on the mandrel provided in the installation kit. Bottom it against the tang. (D) Lubricate the insert with oil and thread it into the hole until it is flush with the surface. Use a punch or sidecutter to break off the tang.

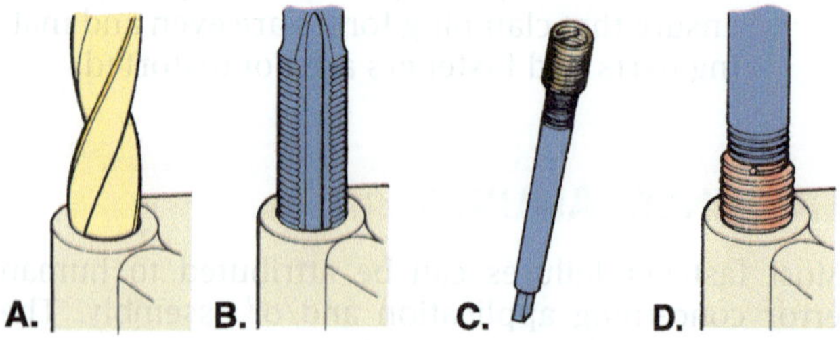

Threads can sometimes be replaced by using threaded inserts. Several types of threaded inserts are available; the helically coiled insert is the most popular (**Figure 3–75**). To install this and similar thread reconditioning inserts, proceed as follows:

1. Establish the size, pitch, and length of the thread required. Refer to the insert manufacturer's instructions for correct size drill for the thread tap to be used for the repair.
2. Drill out the damaged threads with the specified drill. Clean out the drill swath and chips from the hole.
3. Tap new threads in the hole using the specified tap. Lubricate the tap while threading the hole. Back out the tap every quarter turn or so. When the hole is threaded to the required depth, remove the tap and all metal cuttings from the hole.
4. Select the appropriate size insert and screw it onto the special installing mandrel or tool. Make sure that the tool engages with the tang of the insert. Screw the insert in the hole by turning the installing tool clockwise. Lubricate the thread insert with engine oil if it is installed in cast iron (do not lubricate if installing into aluminum). Turn the thread insert into the tapped hole until it is flush with the surface or one turn below. Remove the installer.

Photo Sequence 2 shows a thread repair procedure.

Screw/Stud Removers and Extractors

Two types of **stud removers** are shown in **Figure 3–76**. These tools are also used to install

FIGURE 3–76 A collar-type stud remover

and remove studs. Stud removers have a hardened, grooved eccentric roller or jaws that grip the stud tightly when turned. Stud removers/installers may be turned by a socket wrench drive handle.

Extractors are used on screws and bolts that have sheared below a surface. Twist drills, fluted extractors, and hex nuts are included in a screw extractor set (**Figure 3–77**). This type of extractor lessens the tendency to expand a screw or stud that has been drilled out by providing gripping contact through the full length of the stud.

FIGURE 3–77 A screw extractor set

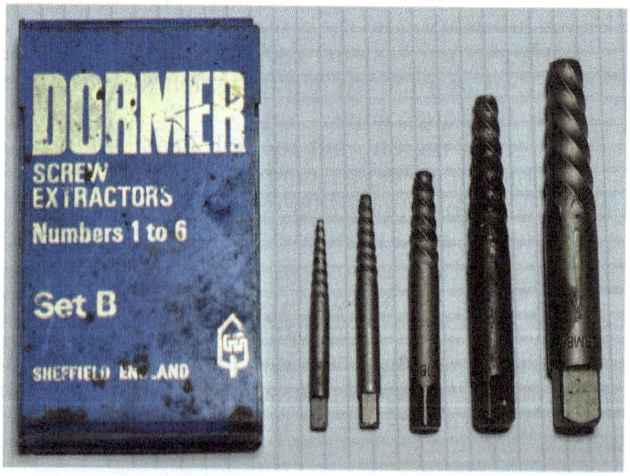

Photo Sequence 2

REPAIRING DAMAGED THREADS

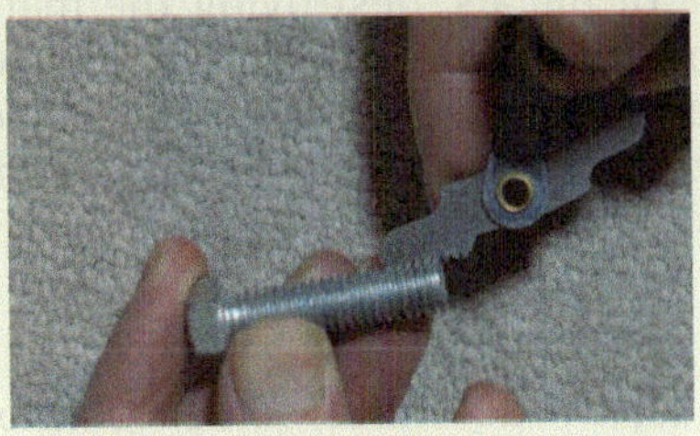

P 2-1 For this exercise, we are going to look at how to install a Helicoil insert. The first step is to measure the screw-pitch using a set of pitch gauges. This will determine the size of drill, tap, and insert required for the procedure.

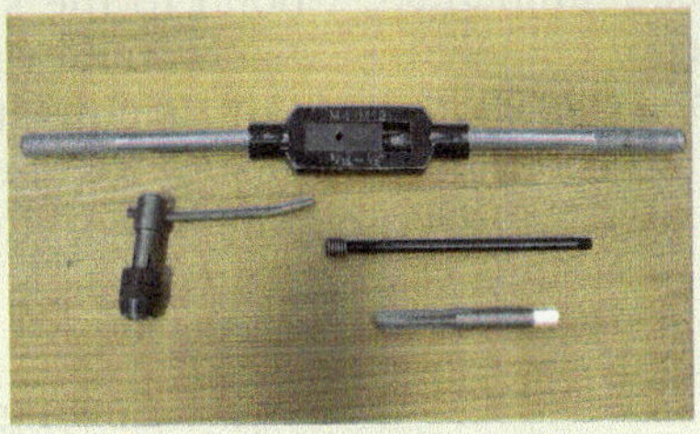

P 2-2 The above image shows the tools required to insert a standard Helicoil insert. This kit shown includes a double-end adjustable (bar) tap wrench, a T-chuck tap wrench (more compact), a tap, and a Helicoil drive mandrel with an insert.

P 2-3 Helicoil inserts are available for almost any Standard or Metric fastener thread pitches. The above image shows a ¼-20 inch kit, which includes the appropriate drive lug and tap. The inserts can also be purchased separately at lower cost.

P 2-4 Using the instructions on the Helicoil kit, select the specified drill bit and drill out the existing threads. When drilling into a blind hole such as that shown above, take care not to drill at a depth beyond the existing hole.

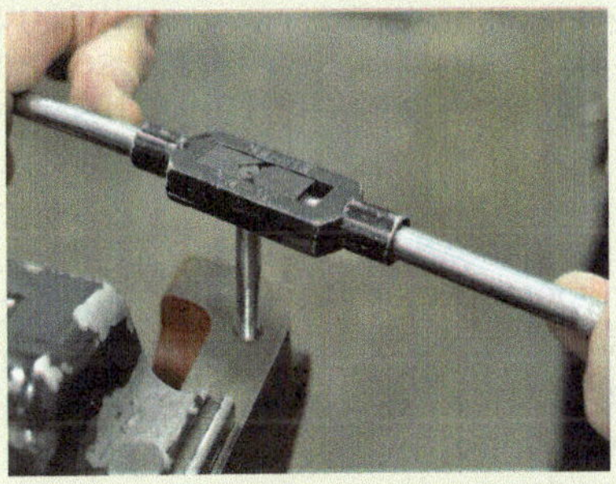

P 2-5 Insert the specified tap into a tap bar wrench such as the one shown above. In cases where there is insufficient clearance, a T-type tap driver may have to be used, but these are more difficult to hold square to the hole. In these cases, use a square to ensure that the threads are tapped square.

P 2-6 Ensure that the tapped hole is completely clean. Insert the Helicoil insert into the driver lug threads and lock the driver as shown in the image above.

P 2-7 Rotating the tap driver, turn the Helicoil insert into the newly cut threads.

P 2-8 After insertion, the insert should be fractionally below flush as shown in the image above.

P 2-9 Finally, check that the specified fastener turns easily into the insert threads. CAUTION: In instances where a thread insert is to be installed into an open hole that must seal, apply an appropriate thread sealant to both the insert and the fastener.

Thread Locking Compound Application

When applying a thread locking compound, follow the safety precautions given on the locking compound container. Then proceed as follows:

1. Clean the male and female threads of the fasteners, removing dirt, oil, and other contaminants. If the area around the fastener is contaminated, clean with solvent, and then allow everything to air dry for 10 minutes. Be sure solvent is completely evaporated before applying thread adhesive.
2. Apply a small amount of locking compound from the container to the circumference of three or four fastener threads.
3. Install and immediately torque the nut. Retorquing the nut is not possible after installation without destroying the adhesive locking bond.

CAUTION

Thread locking compounds are powerful adhesives. They are color coded. Use only the color code recommended by the OEM.

SHOP TALK

To disassemble fasteners that have been held together with a thread locking compound, it may be necessary to heat the bond line to 400°F (205°C) before removing the nut. Every time the fasteners held by locking compounds/adhesives are disassembled, replace them. If mating components are damaged by overheating, replace them.

Riveting

Body skin panels on the tractor cab, buses, and some trailers are fastened using aluminum alloy or mild steel rivets. A **rivet** is a clamping type of fastening device. It must have two heads to clamp the material together. One head is preformed by the manufacturer and is referred to as the manufactured head. The other end of the rivet is formed after the rivet has been driven through the material to be clamped. This head is referred to as the bucked head and is shaped when driven against a bucking bar.

Although truck cabs are usually repaired by a body shop, truck trailer repair technicians are often required to become expert riveters. For this reason rivet removal, replacement, and repair practices are covered in some detail in Chapter 33 of this book.

ADHESIVES AND CHEMICAL SEALING MATERIALS

Chemical adhesives and sealants may provide added holding force and sealing ability when two components are joined. Sealants are applied to threads where fluid contact is frequent. Chemical thread retainers are either aerobic (cures in the presence of air) or anaerobic (cures in the absence of air). When using a chemical adhesive or sealant, follow the manufacturer's instructions. Note that some adhesives molecularly bond to the surface of metals, destroying the material on removal. Some can be harmful on contact with skin. In almost all cases, the material safety data sheet (MSDS) for these products is published on the company's Web site.

Sealants

The following list briefly describes some common shop sealants:

- High tack. Available as a paste or aerosol spray. Rapidly cures and can be used to hold gaskets in place while clamping two components together. Every toolbox should be equipped with some.
- RTV silicone. Room temperature vulcanized silicone. Cheap and outlasts paper gaskets. Uses acid to cure, which is corrosive, and for this reason should not really be used in any chassis applications. Having said this, it is more commonly used than any other shop sealant.
- Nonacidic vehicle silicone. RTV silicone that uses a nonacidic desiccant. Designed to replace gaskets, not to glue them into position. These take longer to cure and may be system specific, usually coded by color—for example, high temperature for exhaust manifolds and turbos, transmissions, wheel ends, etc. Apply a single bead about ¼ inch or less dead center along one side of the mating face. Circle around the bolt holes. Do not apply too much. Mate up wet; do not allow cure time.
- Rubber gasket dressing. This should be applied to reuse rubber gaskets. It tacks them into position while mating up components.
- Rubber gasket maker. Rubber compound that cures slowly but can actually be put into service immediately due to its high integrity. When fully cured it is very difficult to remove and can actually destroy thinner steels. High temperature rated and may be good for problem leaks.
- Weatherstrip adhesive. Designed to adhere weatherstripping. Should not be used for anything else. Can be extremely difficult to remove, especially when applied to aluminum or carbon fibers.

Thread Lock Compounds

Thread lock compounds are usually manufactured in liquid and gel forms. Several manufacturers produce thread sealants. Although there are no rules about this, most manufacturers usually abide by the following color codes:

- Red. The highest strength adhesive and sealant. Usually red locking compound functions effectively at high temperatures. Powerful enough to damage some metals on removal. Use in small quantities.
- Blue. A good multipurpose adhesive and sealant rated at medium strength.
- Green. A medium strength adhesive and sealant with wicking capability; this allows it to run down threads and create a better seal.
- Purple. A low-strength thread adhesive that seals threads effectively.

SUMMARY

- Hand tools are used in many repair tasks. Proper selection and use of hand tools by the technician are important requirements for performing quality heavy-duty truck service.
- Power tools make a technician's job easier. Pneumatic (air) tools have four advantages over electrically powered equipment: flexibility, light weight, safety, and low cost operation and maintenance. The disadvantage is the noise produced. Power tool safety rules must be followed, and that includes wearing hearing protection.
- The power tools used in heavy-duty truck repair include impact wrenches, air ratchet wrenches, air drills, air chisels and hammers, blowguns, bench grinders, grinding wheels, wire wheel brushes, buffing wheels, presses, and lifts and hoists.
- Using jacks, lifts, frame machines, and hoists to raise trucks, trailers, or heavy parts on a truck chassis requires careful adherence to safety rules.
- The machinist's rule looks like an ordinary ruler, but it is precisely divided into small increments, either in metric or decimal graduations.
- Dial or digital calipers are used for taking inside, outside, depth, and step measurements and are calibrated to read in either metric or standard scales.
- A micrometer is used for measuring the inside or outside diameter of a shaft or the bore of a hole, either in metric or standard units.
- Electronic digital measuring instruments are accurate, easy to use, and reduce interpretation errors on the part of the technician.
- Gauges are used to measure thickness and screw pitch. A torque wrench is used to measure the amount of twisting force applied in tightening a fastener.
- The main source of repair and specification information for heavy-duty trucks is the online OEM SIS. Other service literature may be available in service manuals, CDs, or DVDs.
- Troubleshooting routines are provided by linking diagnostic software and online SIS; this provides a way to systematically, and often interactively, track the root causes of problems.
- Fasteners used on heavy-duty trucks come in a range of grades and classes. Proper use of torque wrenches is necessary when installing threaded fasteners. The correct fastener is important. In some cases, threads must be repaired with thread inserts or by thread chasing.

REVIEW QUESTIONS

1. Which of the following is NOT a requirement for using a hammer safely?
   a. Wear eye protection.
   b. Grip the hammer close to the head.
   c. Select the correct size hammer for the job.
   d. Keep the hammer face parallel with the surface being struck.
2. What is the reason for slowing down just before completing a hacksaw cut?
   a. To keep from dulling the blade.
   b. To prevent the blade from breaking.
   c. To prevent the surface being sawed from slipping.
   d. To prevent injury to hands.
3. You should never use a punch or chisel on what material? ___
4. Which of the following is a specialty fastener often used to secure mirrors, headlight assemblies, and often racks?
   a. Torx
   b. Posi-Drive
   c. Phillips
   d. Robertson
5. Which of the following is recommended for the maintenance and care of a screwdriver?
   a. Periodically harden with a torch.
   b. File by hand to dress the tip.
   c. Shape the tip with a grinder.
   d. Reshape the tip with a hammer.

6. Which type of hand wrench is fastest for turning?
   a. combination wrench
   b. box-end wrench
   c. open-end wrench
   d. socket wrench and ratchet
7. Which of the following is NOT good practice when using wrenches?
   a. Use wrenches that fit snugly.
   b. Do not extend the length of a wrench.
   c. Never hammer on wrenches.
   d. Straighten a bent wrench before using.
8. Which type of pliers is best suited for grasping small components?
   a. combination pliers
   b. needle nose pliers
   c. locking pliers
   d. diagonal cutting pliers
9. Using a coarse cut file on soft material will prevent:
   a. dragging
   b. clogging
   c. injury
   d. damage to the file
10. How do hand-threading dies differ from taps?
    a. They remove, rather than cut, threads.
    b. They increase thread pitch.
    c. They cut external threads.
    d. They cut internal threads.
11. Precision bearings and gears should be removed with:
    a. a hammer
    b. a prybar
    c. a puller
    d. gear and bearing wrenches
12. Which of the following is NOT true of pneumatic tools compared to electrically powered equipment?
    a. They run cooler.
    b. They are lighter in weight.
    c. There is less risk of fire and electric shock.
    d. They are not as noisy.
13. Which of the following accessories, used with an air chisel or hammer, can be used to install pins and drive seized bolts?
    a. universal joint and tie-rod tool
    b. ball joint separator
    c. exhaust pipe cutter
    d. tapered punch
14. What must always be used with any type of hydraulic hoist to prevent serious injury? ___
15. Which of the following is NOT a safe practice when using truck lifts and hoists?
    a. Allowing the driver to remain in the vehicle.
    b. Locating the recommended lift points.
    c. Using cables or chains to secure the object being lifted.
    d. Checking that the attachments are secure.
16. Which of the following measuring tools has the most precision?
    a. a metric machinist's rule
    b. a decimal machinist's rule
    c. a thickness gauge
    d. a micrometer
17. What part of the micrometer is revolved between the thumb and forefinger to bring the measuring points into contact with surfaces being measured?
    a. thimble
    b. spindle
    c. sleeve
    d. frame
18. When should the calibration of a micrometer be checked?
19. Which of the following would a screw pitch gauge be used for?
    a. to determine the number of threads per inch
    b. to measure thickness
    c. to measure turning force
    d. to determine clamping force
20. Technician A says that most OEM SIS require an online connection to function at full potential. Technician B says that OEMs have designed their SIS and troubleshooting software to work together. Who is correct?
    a. Technician A only
    b. Technician B only
    c. both A and B
    d. neither A nor B
21. Hardened flat washers are used with:
    a. flanged fasteners
    b. hex-type fasteners
    c. all fasteners with metric threads
    d. rivets
22. Tightening a capscrew or bolt beyond its torque value may result in:
    a. reducing the amount of friction
    b. reducing the elasticity of the metal
    c. permanent stretching or breakage
    d. increasing the elasticity of the metal
23. What is required to cure an anaerobic sealant?
    a. heat
    b. freezing
    c. presence of air
    d. absence of air
24. Threads can be repaired using:
    a. inserts
    b. studs
    c. extractors
    d. red-code thread locking compound
25. Which of the following colors would be used to identify a thread locking compound of the highest available strength?
    a. purple
    b. blue
    c. green
    d. red

26. Technician A says that a bolt with six radial dashes is rated as an SAE grade 8 fastener. Technician B says that a nut with two identification dots 60 degrees apart is a grade 8 nut. Who is correct?
    a. Technician A only
    b. Technician B only
    c. both A and B
    d. neither A nor B

27. Technician A says that a flat washer should be installed with the flat side toward the head of the bolt. Technician B says that the re-use of nuts in a fastener assembly should be avoided Who is correct?
    a. Technician A only
    b. Technician B only
    c. both A and B
    d. neither A nor B

28. Technician A says that UNC fasteners have coarse threads. Technician B says that every domestic fastener manufacturer is required by law to register the headmarking logo of their fasteners. Who is correct?
    a. Technician A only
    b. Technician B only
    c. both A and B
    d. neither A nor B

4

Prerequisite: Chapters 1, 2, and 3

MAINTENANCE PROGRAMS

OBJECTIVES

After reading this chapter, you should be able to:

- Explain the characteristics and benefits of a well-planned maintenance program.
- Define the terms *preventive maintenance* (PM), *condition-based maintenance* (CBM), and *predictive-based maintenance* (PBM).
- Explain how trucking companies are increasingly using prognostics to reduce maintenance and service costs.
- List and describe the steps of the pretrip inspection procedure.
- Identify the different categories of preventive maintenance schedules.
- Describe the criteria for deadlining or out-of-service (OOS) tagging a vehicle.
- Implement a preventive maintenance schedule that conforms to federal inspection regulations.
- Outline inspector qualifications and record-keeping requirements.
- Select the correct lubricants for servicing trucks on preventive maintenance schedules.
- Describe the operation of on-board chassis lube systems.
- Prepare trucks and trailers for cold weather by winterizing.

KEY TERMS

American Trucking Association (ATA)
anticorrosion agents
antirust agents
automated maintenance environment (AME)
Behavioral Analysis Safety Improvement Categories (BASIC)
Commercial Vehicle Safety Alliance (CVSA)
Compliance, Safety, Accountability (CSA)
condition-based maintenance (CBM)
coolant hydrometer
CSA 2010
CVSA L1 inspection
deadline
detergent additives
EP1DM
Federal Motor Carrier Safety Administration (FMCSA)
gladhands
hours of service (HOS)
model year (MY)
National Highway Traffic Safety Administration (NHTSA)
out-of-service (OOS)
oxidation inhibitors
performance-based maintenance (PBM)
predictive-based maintenance (PBM)
preventive maintenance (PM)
prognostics
pull circuit
push circuit
refractometer
repair order
Safety Management System (SMS)
Technology and Maintenance Council (TMC)
thermoplastic hose clamps
total base number (TBN) additives
triage
viscosity
work order
zerk fitting

Note on Maintenance and Out-of-Service Standards: The Commercial Vehicle Safety Alliance (CVSA) publishes commercial truck out-of-service (OOS) standards every year, usually in February. Although both OOS standards and maintenance standards are referenced in this textbook, they should be regarded as no more than references. For up-to-date OOS enforcement standards, the CVSA guidelines must be referenced in their OOS book. It is also necessary to underline that OOS standards are not maintenance standards. CVSA OOS citations issued by the Department of Transportation (DOT) define the point at which a truck becomes dangerous and therefore illegal to operate.

INTRODUCTION

It has been conservatively estimated that in the United States alone, more than $25 billion is spent annually on performing unnecessary preventive maintenance (PM). This is not to say that PM is unnecessary, but it does mean that many trucking companies are looking at maintenance practices in a more scientific way with the objective of cutting costs. When a skilled technician is paid to inspect a component that passes 99 percent of inspections, that represents money lost.

In recent years, terms such as *predictive maintenance* have become commonplace in the trucking industry. Although PM will always be an important feature in the management of any fleet, because of advances in software monitoring and analysis of equipment performance, the trend over the next generation of trucks will be to spend less on PM and invest more on predictive maintenance. The U.S. military uses the term **prognostics** to describe the science of using performance analysis to gauge component life cycles. The idea is to calculate the best moment to intervene within the life cycle with either a repair or replacement before a failure occurs. Predictive maintenance practice should be based on good prognostics. In this chapter, we introduce some of the service, PM, and predictive maintenance practices used by the trucking industry. If you are just entering this industry, you can expect to hear a lot about predictive maintenance practices over the next few years.

UNDERSTANDING MAINTENANCE PRACTICES

Although some of the knowledge required to properly service a truck is not covered until the later chapters in this textbook, apprentice technicians are usually expected to perform routine maintenance almost from their first day on the job. This means that having some idea about maintenance practices is essential from the start. For this reason, this chapter appears toward the beginning of this book rather than at the end. In some cases, the student will need to refer to later chapters in the book during the reading of this one.

DEFINING MAINTENANCE CATEGORIES

The objective of maintenance is to minimize the cost of vehicle ownership and eliminate costly on-the-road breakdowns. An effective protective maintenance program is one that enables the operator to perform the least number of unscheduled repairs while owning a vehicle. Maintenance practices today can be divided as follows:

- Unscheduled repairs. Usually the result of a breakdown; costly.
- **Preventive maintenance (PM)**. Scheduled maintenance routines usually platformed on original equipment manufacturer (OEM) recommendations and adapted to fleet-/operator-specific requirements.
- **Condition-based maintenance (CBM)**. Also known as **predictive-based maintenance (PBM)** and **performance-based maintenance (PBM)** depending on who is using the term. Interventionary maintenance based on software analysis of component and system aging.
- **Automated maintenance environment (AME)**. Maintenance based on CBM profiles applied in such a way that PM can be eliminated.

A good maintenance program should meet the truck OEM requirements and the specific requirements of the fleet that manages the vehicle. Any well-planned maintenance program is designed to offer the following advantages:

- The lowest overall maintenance cost
- Maximum vehicle availability (least downtime)
- Better fuel economy
- Minimum number of road failures
- Minimization of out-of-service citations
- Reduced possibilities of accidents due to defective equipment
- Fewer driver complaints

Many factors are involved in the success of a maintenance program. The program's success depends on careful planning and equipment monitoring. Although this is primarily the responsibility of a maintenance manager, it cannot be accomplished unless every player on a service team is working toward the same goal. Inspections have to be performed conscientiously and on schedule. Accurate records have to be logged and maintained. Increasingly the appropriate maintenance software must be used to profile equipment performance and trends. Finally, when maintenance trends have been profiled, they must be used to adjust a maintenance program for maximum effectiveness.

DRIVER COMMUNICATIONS

The first key to a maintenance program is the driver. The driver is responsible for understanding how equipment performs and that means identifying some of the tattletales of imminent failure. Drivers are required to be relied on to be able to communicate suspect conditions into a report that a technician can understand. The purpose of a driver inspection is to detect at least some failures before they become breakdowns. Drivers have to perform daily pre- and post-trip inspections and they should be taught how to conduct them.

LEGAL REQUIREMENTS

Highway safety is mandated by federal and state legislation. There are a number of bodies responsible for ensuring that highway trucks are maintained in a safe condition, and we address some of them in this chapter. However, it is important to understand that there is a significant difference between a maintenance standard and a maintenance defect so severe that it breaks the law. Maintenance inspections are managed in-house. Their objective is to minimize maintenance costs. Minimum safety standards are managed by the government. Their objective is to protect the public. **Figure 4–1** lists some of the concerns that fleet operators cite when asked to name their top five maintenance problems.

PENCIL INSPECTIONS

A major cause of maintenance program failure is what is known as a "pencil inspection." This happens every time a driver or technician checks off an "OK" without bothering to actually inspect the system or component. This type of inspection wastes time and resources. It is also dishonest and can endanger others. They have a way of coming back to bite you. Remember the cautionary phrase "what goes around, comes around."

FIGURE 4–1 Brakes top the list of maintenance concerns that fleets cite when asked to name the top five trailer maintenance problems.

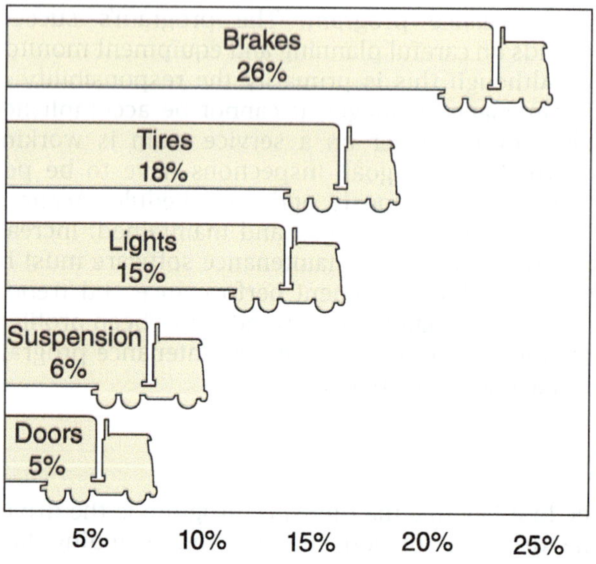

KEEPING RECORDS

Maintaining accurate records is another essential component of a maintenance program. Sometimes it is referred to as a paper trail, though today it is more often than not managed electronically. Records begin with driver inspection reports. From there they move to the technician's ability to manage data in the form of PM forms, repair orders, vehicle files, and chassis component histories. Computer-managed data records make management of these tasks easier for both the driver and technician. Records are important for several reasons:

- In most jurisdictions they are required by the DOT.
- Permanent records are invaluable from a performance standpoint. Analysis of component failures should be used for purposes of predictive maintenance and future equipment spec'ing. On the positive side, it can highlight components that do perform well in an operation.
- A maintenance program cannot be developed without reference to chassis repair history files and comparative analysis of fleet breakdowns.
- In the event of a serious accident vehicle repair files, along with documented evidence of maintenance inspections, may be required by investigators. In addition, they can be of some benefit to defend a lawsuit if they were actually done.

REPAIR ORDERS

Repair orders or **work orders** are basically blank documents that are filled in by service writers or technicians (see Figure 2–13). It is important to take some care about how these are completed because the language used can come back to bite you if there is a dispute. The function of the work order when it is initiated is to communicate to the service technician an outline of the repair required: this is because in most cases a technician does not write the work order. Once in the hands of the technician, the work order becomes the only means of fully describing the exact nature of the repair along with any additional problems or complications. If a customer bill is disputed, then what that technician writes on the repair order (whether in soft or hard copy format) can make the difference between a shop getting paid for the completed work...or writing off some of the labor.

When a bill is under dispute, the work order assumes the status of a legal document emphasizing the importance of accurately describing the repairs completed. All technicians must think carefully about what they write in these documents. When a bill is being contested, it can be weeks or months after the work has been completed and after that period of time, a technician is unlikely to recall the specific job unless there is a detailed written account of why a projected two hour-job took three hours.

The "Ideal" Repair Order

In an ideal world, we would not have to worry about litigation and disputed work orders. So the type of information written onto a work order—and the sequence in which diagnostics take place—would run along these lines:

- Collect information (this is known as **triage** in some dealerships): talk to driver, review previous service history of the vehicle, connect to the chassis data bus and note any fault codes or abnormalities in audit trails. Record the relevant information on the repair order.
- Analyze the information: this is done by the technician on receipt of the repair order. This requires looking at the complaint that initiated the repair order and evaluating other chassis systems that may have contributed to the problem.
- Run diagnostic routines: these may be software-guided depending on the system. Record the information on the repair order because this is the point at which a repair cost estimate is made.
- Repair the problem and its root cause: document this on the repair order making a note of anything noted in the previous step.
- Verify the repair: this requires the system that caused the initial complaint to be tested. May require a chassis dynamometer or road test.

Keep repair order language brief. Use bullet points and avoid the use of slang: this is a business document!

Automated Maintenance Environment

Earlier we mentioned that an estimated $25 billion dollars are spent every year on "unnecessary" PM and that this expenditure has driven a more scientific approach to maintenance practices known as prognostics. Because of the enormous range of advanced equipment operated by the defense department and the expense required to maintain it, the U.S. military has led the initiative to eliminate service procedures that are not cost effective. We use the acronym "CBM" to describe the predictive maintenance program described a little earlier in this chapter. CBM software is capable of analyzing real time equipment profiles and calculating such things as:

- Reliable life cycles
- Predicted wear rates
- Component risk management

FIGURE 4–2 Component life-cycle algorithm and Y factor

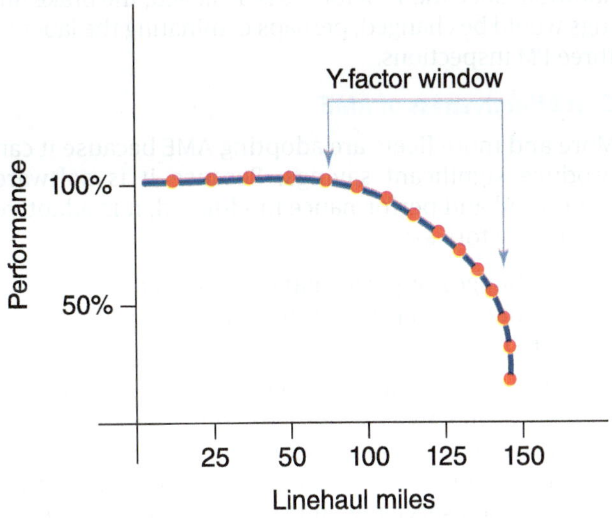

Once these factors have been calculated, they are built into a software map we call an algorithm.

Using the CBM Algorithm

Basically what this algorithm profiles is an expected life cycle of a component or a system such as that shown in **Figure 4–2**. This graph plots the incidence of failure versus the operational hours of service. Once the algorithm has been established, some human decision making is required. We could call the human a maintenance manager. That human then decides at what point on the graph the Y should take place. The Y factor indicates the point at which some maintenance intervention should be done. This could mean servicing, reconditioning, or replacement.

If the graph shown in Figure 4–2 was created to represent the life-cycle algorithm of the friction facings of the brake linings versus the linehaul miles, then you can observe a direct relationship between an increase in linehaul miles and the point at which the friction facings wear to a point of outright failure. If the position of Y on the graph indicates the point at which those linings are replaced, then you can see that the greater the mileage on a vehicle, the closer you get to the moment when outright failure occurs. Determining the Y factor intervention point is a human decision. Risk can be managed to a minimum by moving the Y factor to the left on the graph (lower level of the linehaul miles), or you can live dangerously by moving the Y factor to the right, deeper into the linehaul mile count.

Managing AME

When fleet equipment management is ceded to CBM software the role of the maintenance manager is reduced into determining what level of risk should be undertaken. Using the example of brake linings cited earlier, wheel-off brake inspections would not be undertaken on PM routines. The Y factor on the algorithm would be determined by the CBM software along with a human decision on what level of risk should be undertaken. In addition, once the Y factor life is achieved, the brake linings would be changed, perhaps eliminating the labor of three PM inspections.

Cost Effectiveness of AME

More and more fleets are adopting AME because it can produce significant savings. Because it is software monitored and performance modulated, it is adaptive to such factors as:

- Changes of performance conditions
- Changes in product quality
- Driver abuse

Like any software we use today, AME systems can only be as effective as those persons interacting with the software. Accurate data entry is essential. Get used to it; this is fast becoming the way in which we will approach servicing, at least in the near future.

4.1 SETTING UP A MAINTENANCE PROGRAM

No single maintenance program applies to all operations. There are simply too many variables. Factors such as the age and type of equipment, percentage of units under warranty, and distances traveled make it essential to create fleet-specific programs.

Adhering to the service recommendations outlined in each major OEM's PM checklist is a safe foundation for any program. Another source of help is government guidelines. The DOT for years has influenced the maintenance practices and record-keeping procedures of those involved in interstate trucking. Rules and guidelines are covered in a manual called *Federal Motor Carrier Safety Regulations*. Every truck operation may or may not be governed by these regulations. However, whether or not compliance is mandatory, the guidelines and requirements can be used as a basis for a planned, controlled maintenance program on most types of equipment.

PREVENTIVE MAINTENANCE

A key to any successful PM program is the PM form (**Figure 4–3**). This form provides the technician with a list of items to be checked or inspected and instructions on those items that should be cleaned, lubricated, tightened, adjusted, and replaced. It is the technician's responsibility to know specifically what to do at each step; this is why many of the major fleets run in-house PM training programs. The following is a list of some types of inspections in A, B, C, D order:

- Schedule "A" is a light inspection.
- Schedule "B" is a more detailed check.
- Schedule "C" is a detailed inspection, service, and adjustment.
- Schedule "D" is a comprehensive inspection and adjustment and often includes component rebuilds and replacements.
- Schedule "L" is a chassis lubrication: it is often incorporated into an A inspection.

MAKING SENSE OF PM FORMS

It makes sense either to perform a road test or use a driver's inspection report before performing a scheduled PM. Sometimes the instructions on a PM form are too general. For instance, what should the technician do in response to the instruction "Inspect the cooling system"? When instructions are more specific, the technician is left in no doubt as to what is required. Detailed instructions would be clear cut and laid out in steps such as:

1. Pressure-test the system to its specified maximum pressure (15 psi is typical).
2. Hold under pressure 10 minutes while checking for leaks at the:
   a. Cab heater
   b. Coolant hoses
   c. Water pump
   d. Surge tanks
   e. Radiator core
   f. Head gasket
3. Record pressure drop-off after 10 minutes.

When PM instructions are produced in this kind of detail, there is less room for error. The technician should be persuaded to perform each step conscientiously. The goal is to identify and eliminate potential problems.

DRIVER INSPECTION

It is often the driver who first identifies the need for a repair during a pretrip or a post-trip inspection. Apart from being required by federal legislation, a "Driver Inspection Report" is a key to sound maintenance management. Part of a driver's post-trip responsibilities, as outlined by the DOT's Federal Motor Carrier Safety Regulations and now reinforced by the Compliance, Safety, and Accountability program, is completing a post-trip vehicle inspection report after each day's work and each vehicle driven. The form is almost always printed on the back of the driver hours-of-service log and completed before drivers turn in their daily logs. **Figure 4–4** is a copy of the driver's report form that is issued by the **American Trucking Association (ATA)**. Items the regulations say must be addressed include:

- Service brakes
- Frames
- Parking brakes
- Sliding subframes (trailers)

FIGURE 4–3 An example of a PM form

PM Inspection Linehaul Tractor

Unit # ________ ☐ "A" PM ☐ "B" PM ☐ "C" PM Mechanic ________
Date ________ Check- High Oil/Fuel Checklist ☐ Supervisor ________
Mileage ________

On D/L, Check and Inspect

	Ck	Remarks
Verify PM on History Card		
Start Engine		
Gauges, Warning Lights		
Low Air Buzzer & Light on at 60#		
Air Buildup & Cutoff		
Air Dryer & Cutoff		
Air Loss Test, 3# Per Min.		
Heater/Defroster Operation		
A/C Operation		
Horns, Air & Electric		

Moving to Shop, Check and Inspect

	Ck	Remarks
Park Brake Application		
Clutch Operation, Free Travel		
Transmission Shift, Hi-Lo		
Steering Free Travel, Bind, Pull		
Speedometer Operation		
Any Unusual Noise or Vibration		
Windshield Wiper/Washer Operation		
Foot Brake Application		
Headlights & Driving Lights		

In Shop, Check and Inspect

	Ck	Remarks
Turn Signals, Marker Lights		
Brake Lights-Reflectors		
Dome Light (Rear Light)		
Floor Mats, Boots, Coat Hook, DH Seal		
Sun Visors, Dash Screws		
All Glass and Mirrors		
Door Locks, Regulators		
Safety Equipment		
Seat Belts, Retractors		
(Sleeper Compartment Items)		
Trailer Cord Test		
Air Hoses and Hangers		
Lube: Door Latches, Hinges		
Seat Rails, Pivot Points		
Brake/Clutch Peddle Pivot		
Accelerator Peddle Pivots		
Heater/Defroster Cables		
Clean & Lube Floor Mtd. Foot Valve		

Engine Compartment, Check and Inspect

	Ck	Remarks
Hood Condition, Cracks, Damage		
Hinges, Bug Screen, Brackets, Wiring		
(Cab Over, Ck Jack Lift Cyl and Lines)		
Air Cleaner Ducts, Hoses, Brackets		
Air Restriction, Repl. Element @ + 25 in./H_2O		
All Belts, Tension-Condition		
A/C Compressor, Condenser,		
Receiver-Dryer, Wiring, Pressure Lines		
Pressure Test Cooling System		
Radiator, Leaks, Mounts, & Brackets		
Cooling Hoses, Any Leaks		
Water Pump Leaks		
Antifreeze protection		
DCA Check		
Change Water Filter if Equipped		
Alternator Mounting, Wiring		
Starter Mounting, Wiring		
Spray Protectant on All Wiring Term.		
Fuel System, Leaks, Lines Rubbing		
Air Comp. Lines, Mtgs., Air Gov. Screws		
Lube Hood Linkage		
Drain Fuel Tank Sumps of Water		
Drain Fuel Heater of Water		
Exhaust System, Clamps, Leaks		
Inspect Fan Hub		
Engine and Transmission Mounts		
Cab Mounts		
Drain Air Tanks of Water		

Under Vehicle

	Ck	Remarks
Change Oil, Oil Filters and Fuel Filters		
Check Trans Lube ("C" PM Change)		
Check Diff(s) Lube ("C" PM Change)		
Inspect Axle Breather		
Inspect Pinion Seals, Wheel Seals		
Inspect Dr. Shaft, U-Joints Yokes, Lube		
Inspect Clutch Linkage, Lube as Req.		
Insp. Exh. Pipe, Muffler, Hangers, Leaks		
Insp. Fuel Tanks, Lines, Hangers, Leaks		
Insp. Air Tanks, Hangers, Lines Rubbing		
Record Oil Pressure Hi-Lo		
Check Dipstick Full Mark		
Lube All Required Points		

(continued)

FIGURE 4–3 (continued)

PM Inspection Linehaul Tractor

Unit # ________ Date ________

☐ "A" PM ☐ "B" PM ☐ "C" PM

Check- High Oil/Fuel Checklist ☐

Mechanic ________ Supervisor ________ Mileage ________

Axles and Chassis

	Ck	Remarks
Vehicle Jacked Up		
Insp. Frt. Spgs., Pins. Hanger, U-Bolts		
Ins. Frt. Brake Chamber, Slacks, Lining, Air Lines, Brake Adjustment		
Insp. Frt. End. Tie-Rod Ends, Drag Link		
Pitman Arm, Kingpin, Steering Box, Steering Shaft, Hub Lube Level, Lube All Points.		
Lower Vehicle		
Insp. Rear Spgs., Pins, Hangers, U-Bolts		
Torque Rods, Equalizer Wear		
Insp. Rear Brake Chambers, Slacks, Lining, Air Lines, Brake Adjustment		
Insp. Frame and Cross Members		
Insp. 5th Wheel, Legs, Brackets, Ground		
Strap, Lube Slider Assembly		
Insp. Battery Box Mounts		
Clean Battery and Post		
Check Battery		
Insp. Battery Cables, Spray Prot.		
Insp. All Chassis Wiring, Spray Prot. on All Terminals		
Check Lic., Regis, Permits, State Insp.		
Clean Glass & Cab Interior		

Add for "C" PM

	Ck	Remarks
See Foreman for Compuchek		
Replace Gladhand Rubbers		
Check Air Shield Angle		
Check Toe-in		
Check Tandem Alignment		
Change Trans and Diff Lube		
Change Rear Center Pump Filter/Screen		
Check U-Bolt Torque		
Change Steering Box Oil, As Required (Delvac SHC Synthetic)		
Clean Heater Core & A/C Evap.		
Clean Trans. Air Valve Filter		
Air System Check		
Clean & Check 7 Way Cord		
Use UYK Comp. Reverse Ends		

"A" PM ONLY

on D/L & Moving to Shop, Check & Insp.

	Ck	Remarks
Verify PM on History Card		
Start Eng.: Ck Gauges, Warning Lights		
Low Air Buzzer and Light on @ 60#		
Air Build Up & Cut Off		
Air Dryer Cut Off		
Heater, Defroster and A/C Operation		
Horns: Air & Electric		
Clutch Operation, Free Travel		
Steering Free Travel, Bind, Pull		
Foot & Park Brake Application		

In Shop, Check and Inspection

	Ck	Remarks
All Lights		
Floor Mats, Boots, Coat Hook		
Safety Equipment		
All Glass and Mirrors		
Trailer Cord Test		
Check Engine Oil, Coolant Level		
Change Fuel Filter		
Check Front End, Lube All Points		
Check Trans and Diff Lube Levels		
Lube Driveshaft U-Joints		
Lube 5th Wheel and Sliders		
Drain Air Tanks		
Check Batteries		
Clean All Glass, Inside Cab		

Wheels and Tires, Check and Inspect

Air Pressure	LF	RF	RFI	RFO	RRI	RRO	LRI	LRO	LFI	LFO
Tread Depth										
Loose Lugs										
Cracked Rims										
Valve Caps										

FIGURE 4–4 Driver's report form as issued by the American Trucking Association

Vehicle Inspection Report

Date: ____________

Company: ____________

Terminal: ____________

CHECK ANY DEFECTIVE ITEM WITH AN X AND GIVE DETAILS

Tractor Number: ____________ **Odometer:** ____________

- ☐ Air Compressor
- ☐ Air Lines
- ☐ Battery
- ☐ Body
- ☐ Brake Accessories
- ☐ Brakes, Service
- ☐ Clutch
- ☐ Coupling Lines
- ☐ Defroster/Heater
- ☐ Drive Line
- ☐ Engine
- ☐ Exhaust
- ☐ Fifth Wheel
- ☐ Frame and Assembly
- ☐ Front Axle
- ☐ Fuel Tanks

- ☐ Horn
- ☐ Lights
  - Head - Stop
  - Tail - Dash
  - Turn Indicators
- ☐ Mirrors
- ☐ Muffler
- ☐ Oil Pressure
- ☐ Radiator
- ☐ Rear End
- ☐ Reflectors
- ☐ Safety Equipment
  - Fire Extinguisher
  - Reflective Triangles
  - Flags - Flares - Fusees
  - Spare Bulbs & Fuses
  - Spare Seal Beam

- ☐ Suspension System
- ☐ Starter
- ☐ Steering
- ☐ Tachograph
- ☐ Tires
- ☐ Tire Chains
- ☐ Transmission
- ☐ Wheels and Rims
- ☐ Windows
- ☐ Windshield Wipers
- ☐ Other

Trailer Number: ____________ **Hubdometer:** ____________

- ☐ Brake Connections
- ☐ Brakes
- ☐ Coupling Devices
- ☐ Kingpin
- ☐ Doors

- ☐ Hitch
- ☐ Landing
- ☐ Lights - All
- ☐ Roof
- ☐ Suspension System

- ☐ Tarpaulin
- ☐ Tires
- ☐ Wheels and Rims
- ☐ Other

Remarks: ____________

☐ Condition of the above vehicles is satisfactory

Driver's Signature: ____________

☐ Above defects corrected

☐ Above defects need not be corrected for safe operation of vehicle

Technician's Signature: ____________ Date: ____________

Driver Reviewing Repairs: Signature: ____________ Date: ____________

ORIGINAL

Courtesy of American Trucking Association

- Brake drums
- Tire and wheel clearance
- Tire and wheel condition
- Brake hoses and tubing
- Low air pressure warning
- Wheels and rims
- Tractor protection valve
- Windshield glazing
- Air compressor
- Wipers
- Hydraulic brake systems
- Vacuum brake systems
- Fifth wheels
- Exhaust system
- Fuel system
- Lighting
- Cargo securement
- Steering componentry
- Suspensions

In addition, drivers are required to note any defects or problems that could affect safe operation and then sign the report.

Pretrip Inspection

Pretrip inspection guidelines are less specific. In addition to ensuring that the vehicle is safe, drivers are required to review the previous vehicle inspection report. If any defect(s) noted by the previous driver have not been repaired, both the technician and the new driver must sign off on the work before the vehicle can be dispatched. If the problem noted did not require repair, then that too should be noted. In addition, a copy of the latest inspection report must remain with the vehicle. (Carriers are required to keep the original copy of each report and the certification of repairs for at least 3 months.)

Although the regulations seem to place greater emphasis on the post-trip inspection, most operators agree that the pretrip inspection is more important. Because the driver performing the inspection will soon be operating that particular vehicle, the incentive to ensure that the vehicle is safe is probably greater than if he or she had just returned from a long day on the road. Repairs are cheaper and less time consuming if the driver identifies defects on the pretrip inspection before an on-the-road breakdown or safety citation takes place.

INSPECTION PROCEDURE

The driver (and technician) should perform a pretrip inspection in the same way every time to learn all the steps and to reduce the possibility of forgetting something (**Figure 4–5**). The following procedure is an adaptation of the **Commercial Vehicle Safety Alliance (CVSA)** standard inspection and should be a useful guide. This can be obtained on DVD or online from http://www.cvsa.org.

FIGURE 4–5 Make inspection a habit by following a circular, walk-around sequence (numbers correspond with the text). Most jurisdictions require the use of this, or variation of this, pre-trip circle inspection.

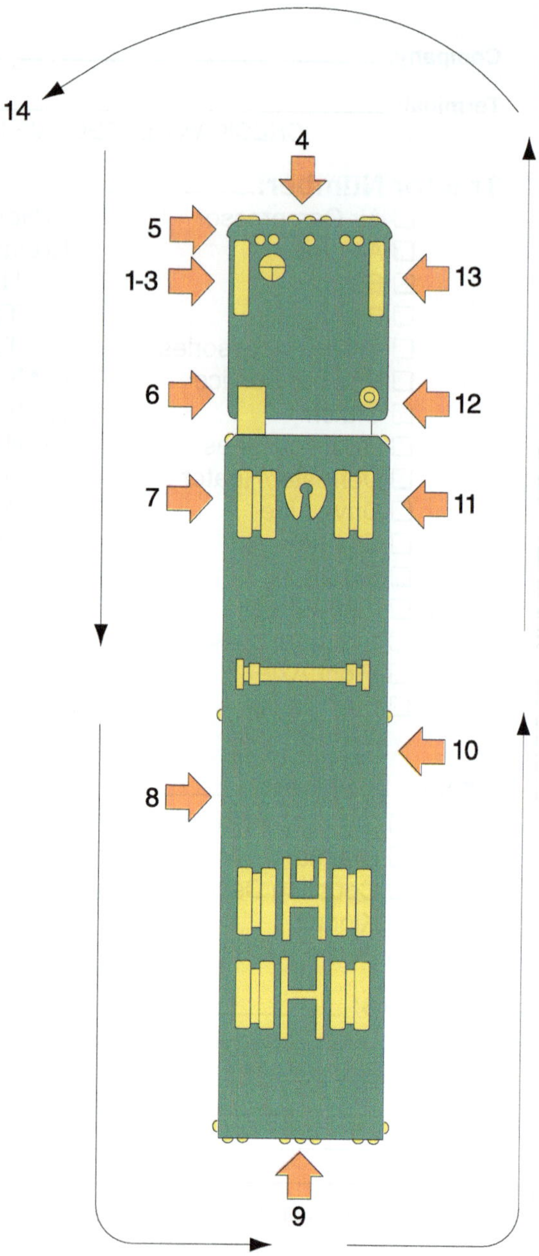

Step 1 (Vehicle Overview). Observe the general condition of the vehicle. If the unit is leaning to one side, this might indicate a broken spring, poor load distribution, or a flat tire. Look under the vehicle for signs of fuel, oil, grease, or coolant leaks. Check the area around the vehicle for hazards to vehicle movement (people, other vehicles, objects, low-hanging wires or tree limbs, and so forth).

Review the previous vehicle inspection report. Note any defects reported by the previous driver.

Inspect the vehicle and confirm that any necessary repairs were made.

Step 2 (Engine Compartment). Check that the parking brakes are applied and/or the wheels are chocked. Raise the hood, tilt the cab (secure loose items so they do not fall and break something), or open the engine compartment door, and check the following:

- Coolant level in the radiator; condition of hoses
- Power steering fluid level; hose condition (if so equipped)
- Windshield washer fluid level
- Battery electrolyte level, connections, and tie-downs (battery may be located elsewhere)
- Automatic transmission fluid level (may require engine to be running)
- Belts for tightness and excessive wear (alternator, water pump, air compressor)
- Leaks in the engine compartment (fuel, coolant, oil, power steering fluid, hydraulic fluid, battery fluid, and air lines)
- Cracked, worn electrical wiring insulation

Lower and secure the hood, cab, or engine compartment door.

Step 3 (Inside the Cab). Make sure the parking brakes are applied. Put the gearshift in neutral (or park if automatic) and then start the engine and listen for unusual noises. Check the dash gauges for the following readings:

- Oil pressure should rise to normal within seconds after the engine is started.
- Ammeter and/or voltmeter should indicate in the normal range(s).
- Coolant temperature should begin a gradual rise to the normal operating range.
- Engine oil temperature should begin a gradual rise to the normal operating range.
- Warning lights and buzzers for the oil, coolant, and charging circuit should go out right away.

Check the seats and the seat belts. Inspect the windshield and mirrors for cracks, dirt, illegal stickers, or other obstructions to vision. Clean and adjust as necessary. Check for excessive play in the steering wheel. Manual steering play should not normally exceed 2 inches, and power steering play should not exceed 2½ inches. Check all of the following for looseness, sticking, damage, or improper setting:

- Steering wheel
- Clutch
- Accelerator pedal
- Brake controls
- Foot brake
- Trailer brake
- Parking brake
- Retarder controls (if vehicle is equipped)
- Transmission controls
- Interaxle differential lock (if vehicle is equipped)
- Horn(s) (air and electric)
- Windshield wiper/washer
- Lights
- Headlights
- Dimmer switch
- Turn signal
- Four-way flashers
- Clearance, identification, marker light switch(s)

SHOP TALK

The **National Highway Traffic Safety Administration (NHTSA)** standardizes the buckle-release mechanism on seat belts that requires them to have either an emergency locking retractor or an automatic locking retractor. The regulation also requires that retractors be attached to the seat so they will move with the air suspension system.

Check that the truck is equipped with all the necessary emergency equipment. There should be spare fuses (for circuits not protected by electronic or mechanical breakers), three red reflective triangles, and a properly charged and rated fire extinguisher within arm's reach of the driver's seat. Flares, lanterns, and flags are optional. Other options include tire chains (where winter conditions require them), tire changing equipment, a list of emergency phone numbers, and an accident reporting kit.

Step 4 (Front of Cab). Check the steering system for defects. Look for missing nuts, bolts, cotter pins, or other parts. Check for bent, loose, or broken parts such as the steering column, steering gearbox, or tie-rods.

Inspect the headlights, turn signals, and emergency flashers for proper color and operation.

Check for the following on both sides of the suspension system:

- Spring hangers that allow movement of the axle from proper position.
- Cracked or broken spring hangers.
- Missing or broken leaves in any leaf spring. If one-fourth or more are missing, it will put the vehicle out of service, but any defect could be dangerous.
- Broken or shifted leaves in a multileaf spring or pack might contact a tire or other moving component.
- Leaking shock absorbers.
- Torque rod or arm, U-bolts, spring hangers, or other axle-positioning components that are cracked, damaged, loose, or missing.
- Air suspension systems that are damaged and/or leaking.
- Any loose, cracked, broken, or missing frame members.

Check the front brakes on both sides to make sure that all components are attached and operational. Check the brake lines for leaks or damage and the chambers (if visible) for cracks or insecure mounting. Check the brake linings. They should be free of large cracks or missing pieces. No grease or oil should be on the linings or drums. Make certain that the pushrod and slack adjuster are mechanically operational. Check for audible air leaks. If possible, ask a helper to apply the brakes, hold them, and then release them when signaled. Check for excessive slack adjuster travel. If visible, check the brake drums for external cracks that open upon application.

Step 5 (Left Side of Cab). Inspect the left front wheel. Check for defective welds, cracks, or breaks (especially between handholds or stud holes); unseated locking rings; broken, missing, or loose lugs, studs, or clamps; and bent or cracked rims. Check for "bleeding" rust stains, defective nuts, or elongated stud holes. Check cast spoke wheels for cracks across the spokes. Inspect for scrubbed or polished areas on either side of the lugs, which indicates a slipped rim. Also check the rims for cracks or bends. The valve stem should be straight and equidistant from the wheel spokes.

Check the left front tire for bulges, leaks, sidewall separation, cuts, exposed fabric, and worn spots. Check for proper inflation. Measure the tread depth.

Check for tire contact with any part of the vehicle.

Inspect the frame for cracked, sagging rails. Check for broken or loose bolts or brackets.

Step 6 (Left Fuel Tank Area). Check the fuel level. Check for unsecured mounting, leaks, or other damage. Check for an unsecured cap or loose connections. Verify that the fuel crossover line is secure.

Check the air and electrical lines between the tractor and the trailer for tangles, crimps, chafing, or dragging. Check the connections. Listen for air leaks.

Check the mounting of the hose couplers (**gladhands**). Look for leaks or other damage.

Inspect the frame for cracked, sagging rails. Check for broken or loose bolts or brackets.

Step 7 (Left Rear Tractor Area). Inspect the wheels, rims, and tires as described in step 5. Examine the inside tire and make sure that both tires are the same height. Check between the tires for debris or contact.

Inspect the suspension and brakes as described in step 4.

Check for cracks along the fifth wheel plate and mounting area. Make sure that the locking jaws are properly engaged. Check for loose or missing nuts and bolts.

For sliding fifth wheels, make sure that the slider is locked.

Check to see that the operating handle is closed and latched. Check the tractor stoplights and turn signals for proper color and operation. Inspect the frame for cracked, sagging rails and check for broken or loose bolts or brackets.

Step 8 (Left Side of Trailer). Inspect the wheels, rims, and tires as described in steps 5 and 7. Inspect the visible suspension and brake components as described in step 4. Check for burned-out or missing marker lights or reflectors.

For flatbeds, check the header board for proper type and mounting. Check the blocking and bracing, chains, straps, and side posts. Check for shifted cargo. Inspect the tarp.

Step 9 (Rear of Trailer). Check for proper operation, color, and cleanliness of the stoplights, taillights, turn signals, emergency flashers, reflectors, and clearance and marker lights.

Check the suspension and brakes as described in step 4. Inspect the wheels, tires, and rims as described in steps 5 and 7. Check the rear bumper for damage or missing pieces. Verify that the doors are locked/latched. For flatbeds, inspect as described in step 8. Check for proper placarding. Make sure that the license plate is visible and that the license plate light is operable. Inspect the frame for cracked, sagging rails and check for broken or loose bolts or brackets.

Step 10 (Right Side of Trailer). Check the same items that were checked on the left side. In addition, check the spare tire for secure mounting and proper inflation. Make sure that the landing gear or dollies are fully raised and that the crank handle is secured. Check for missing, bent, or damaged parts.

Step 11 (Right Rear Tractor Area). Check the same items that were checked on the left side.

Step 12 (Right Fuel Tank Area). Check the same items that were checked on the left side. In addition, inspect the exhaust system. Check it for secure mounting and leaks (under the cab). Make sure that the exhaust system is not contacted by fuel or air lines or electrical wires. Look for carbon deposits around seams and clamps, indicating exhaust leaks.

Step 13 (Right Side of Cab). Check the same items that were checked on the left side.

Step 14 (Cab). Bleed down the air system and check the air pressure gauge. Check the low air pressure warning device by dumping air (pump the foot brake valve). The warning light/buzzer should activate at about 55 psi or above. Check the system cutout pressure.

With the seat belt fastened, release the brakes. As the vehicle begins to move, activate the parking brakes to check their operation.

At about 5 mph, apply the service brakes. Note any unusual pulling, delay, or play in the brake pedal.

If any problem is noted during the pretrip inspection, the technician should record it on the pretrip form. This allows the technician to make a judgment call on whether a questionable item is to be repaired or allowed to be placed back in service. The technician is usually responsible for verifying that the vehicle is safe and for the consequences if it is not.

4.2 OUT-OF-SERVICE OR DEADLINING A VEHICLE

The decision to take a vehicle **out-of-service (OOS)** can be a tough one for technicians because it can place them in conflict with dispatchers and the companies that employ them. If the fault is a CVSA OOS condition, the technician's decision is easy, the vehicle cannot be put into service because under current CSA regulations, the technician, the company, and the driver will be all at fault. If the problem is not of sufficient severity to be classified as a CVSA OOS then the technician must evaluate the risk that a mechanical problem could result in:

- An accident
- A breakdown
- A safety citation

So how does the dispatcher—or anyone else—determine when to **deadline** a vehicle?

DEFINING OOS

Out-of-service (OOS) has a fleet or operator interpretation and a legal one and it is important to understand the difference. Every fleet has its definition of a deadline or OOS standard but this should not be confused with the legal standard. Federal Motor Carrier Safety Regulations state that a mechanical system that can either cause or prevent an accident is classified as a safety item. Potential safety problems should be repaired even when minor in nature. For example, drivers could be injured by sticking trailer van doors or slippery tractor steps. These are concerns that should be addressed. But they are not official OOS standards, though they may be designated as such by fleet operating standards. For our purposes, OOS is defined as follows:

- In-house, fleet OOS and deadline standards
- Federally mandated minimum safe operation standards defined by the CVSA

In-House OOS

It does not make business sense to dispatch a vehicle that is likely to break down. A dead vehicle on a highway is a traffic hazard. Even if the driver makes it to the shoulder, parking it there can be dangerous. Adding up the costs of driver downtime, unhappy freight customers, and a towing and/or road repair bill testifies that dispatching a truck likely to break down is shortsighted. In-house and CVSA OOS standards should never be the same and if they are, the fleet is probably not going to be in business for very long. In other words, in-house OOS standards should set a minimum maintenance standard, not a minimum safety standard. Vehicles with mechanical or safety defects should be tagged immediately (**Figure 4–6**) because the consequences of not doing so can be costly.

FIGURE 4–6 Vehicles with serious defects should be tagged immediately.

CVSA OOS

Every April, the CVSA publishes a new set of OOS criteria to be used by inspectors while undertaking vehicle inspections. In other words, they are the guidelines used by law enforcement as a basis to cite an unsafe vehicle with a fine. A CVSA OOS standard is not a maintenance standard and should never be regarded as such. It is actually the absolute minimum standard at which a vehicle should be operated. An OOS citation is issued when there is an imminent hazard. A federal standard is the minimum maintenance standard—unfortunately, many fleets confuse the difference between the two, meaning that they are maintaining equipment too close to the lowest standards permitted.

CVSA Inspections

The mandate of the CVSA is to establish uniformity for truck inspections across North America. There are seven types of inspections, classified as L1 through L7, but the one that truck technicians should understand is known as a **CVSA L1 inspection**. It consists of thirty-seven steps and can take an inspector between 15 and 45 minutes to perform. Other L inspections involve such things as driver inspections and driver credential inspections and a driver can be cited for an **hours-of-service (HOS)** violations. All CVSA inspections are logged electronically by the inspector who must upload them to the CVSA database within 21 days. An access program known as Query Central enables access to this record-keeping database to maintain transparency.

DOT annual inspections have been mandatory since 1988. "Random" roadside inspections are not necessarily random. Electronic screening is available in some areas, allowing trucks with transponders to bypass inspection stations. In the year 2013 the CVSA published the following data:

- Of vehicles inspected, 21.2 percent are issued with OOS citations.

- Of drivers checked, 4.9 percent were issued HOS citations.
- Of drivers checked, 1.5 percent were issued seatbelt violations.
- The average repair time required to correct an OOS citation is 8.5 hours.
- Of truck crashes, 30 percent are equipment related.

Drivers and technicians should familiarize themselves with the CVSA OOS criteria (**Table 4–1**). It makes sense for service facilities to ensure that a current copy of the CVSA OOS criteria is available as a technician reference.

CAUTION

It is dangerous to regard CVSA OOS specifications as maintenance standards. The OOS limit indicates that a system has become potentially dangerous and merits a citation.

CSA 2010

December 2010 marked the introduction of the **Compliance, Safety, Accountability (CSA)** program and this has become known in the trucking community as **CSA 2010**. CSA is a **Federal Motor Carrier Safety Administration (FMCSA)** initiative designed to provide motor carriers and drivers with accountability for potential safety problems. CSA is a **Safety Management System (SMS)** that tracks carrier performance in seven areas identified as **Behavioral Analysis Safety Improvement Categories (BASIC)**. In a short period of time, these acronyms have become commonplace in today's trucking world. The seven BASIC fields are:

1. Unsafe driving
2. Fatigued driving (HOS)
3. Driver fitness
4. Controlled substance/alcohol infractions
5. Vehicle maintenance (OOS)
6. Cargo handling
7. Crash indicators

The objective of SMS is reduce commercial truck and bus crashes, injuries, and fatalities by making proven shortcomings in the seven BASIC fields transparent to consumers and enforcement agencies.

4.3 PM SCHEDULING

While maintenance strategies are undergoing some changes, an important phase of any maintenance program is *scheduled* periodic PM. Implementing a PM model requires good scheduling to ensure that vehicles are serviced on a set cycle measured in miles, hours, gallons of fuel consumed, or days/months in service (**Figure 4–7**).

SHOP TALK

Off-highway vehicles rarely have odometers. Most use engine or vehicle hours of service to set PM intervals rather than distance traveled. Some operators of vocational vehicles bases PM timetabling on the amount of fuel consumed.

The PM form illustrated in Figure 4–3 shows three categories of PM:

- A inspection
- B inspection
- C inspection

A recent **Technology and Maintenance Council (TMC)** report indicated that:

- A inspections averaged 2.4 hours but ranged from 45 minutes to 5.5 hours.
- B inspections averaged 5 hours but ranged from 2 to 8 hours.
- C inspections averaged about 1 day depending on the equipment.

Some fleets use an even more detailed inspection, called D inspection, usually those running specialized equipment where safety is paramount.

Over the years, linehaul miles between unscheduled repairs have averaged about 20,000 miles (32,000 km). A wide range of different PM forms are included in the Student Guide that accompanies this textbook.

TRAILER PM

Like the tractor, trailer components require PM inspections and servicing at periodic intervals. Maintenance will help ensure maximum service life from the trailer. **Table 4–2** is a typical trailer inspection guide.

FEDERAL INSPECTION REGULATIONS

The NHTSA has set up a periodic minimum inspection standards program under which the following vehicles must be inspected:

- Any vehicle involved in interstate commerce with a gross vehicle weight over 10,000 pounds with or without power, including trucks, buses, tractor/trailers, full trailers, semi-trailers, converter dollies, container chassis, booster axles, and jeep axles
- Any vehicle, regardless of weight, designed to carry more than fifteen passengers, including the driver
- Any vehicle, regardless of weight, carrying hazardous materials in a quantity requiring placarding

Each vehicle must carry proof that an inspection was completed. Proof can be either a copy of the inspection form kept on the vehicle or a decal. If using

TABLE 4–1 CVSA Vehicle Out-of-Service Criteria

Inspected Item	Out-of-Service If:
Brake System Defective Brakes	20 percent or more of the brakes on a vehicle are defective. A brake is defective if:

1. Brakes are out of adjustment. (Measured with engine off and reservoir pressure at 80–90 psi with brakes fully applied.)

CLAMP-TYPE BRAKE CHAMBER			ROTOCHAMBER		
Type	Outside Dia.	Max. Stroke	Type	Outside Dia.	Max. Stroke
6	4 1/2"	1 1/4"	9	4 9/32"	1 1/2"
9	5 1/4"	1 3/8"	12	4 13/16"	1 1/2"
12	5 11/16"	1 3/8"	16	5 13/32"	2"
16	6 3/8"	1 3/4"	20	5 15/16"	2"
20	6 25/32"	1 3/4"	24	6 13/32"	2"
24	7 7/32"	1 3/4" (2" for long stroke)	30	7 1/16"	2 1/4"
30	8 3/32"	2"	36	7 5/8"	2 3/4"
36	9"	2 1/4"	50	8 7/8"	3"

BOLT-TYPE BRAKE CHAMBER			
Type	Effective Area	Outside Dia.	Max. Stroke
A	12 sq. in.	6 15/16"	1 3/8"
B	24 sq. in.	9 3/16"	1 3/4"
C	16 sq. in.	8 1/16"	1 3/4"
D	6 sq. in.	5 1/4"	1 1/4"
E	9 sq. in.	6 3/16"	1 3/8"
F	36 sq. in.	11"	2 1/4"
G	30 sq. in.	9 7/8"	2"

WEDGE BRAKE
Movement of the scribe mark on the lining shall not exceed 1/16".

2. On application of service brakes, there is no braking action (such as brake shoe(s) failing to move on a wedge, S-cam, cam, or disc brake).
3. Mechanical components are missing, broken, or loose (such as shoes, linings, pads, springs, anchor pins, spiders, cam rollers, pushrods, and air chamber mounting bolts).
4. Audible air leak at brake chamber is present.
5. Brake linings or pads are not firmly attached or are saturated with oil, grease, or brake fluid.
6. Linings show excessive wear.
7. Required brake(s) are missing.

Inspected Item	Out-of-Service If:
Steering Axle Brakes	1. On vehicles required to have steering axle brakes, there is no braking action on application (includes the dolly and front axle of a full trailer). 2. Air chamber sizes or slack adjuster lengths are mismatched on tractor steering axles. 3. Brake linings or pads on tractor steering axles are not firmly attached to the shoe; are saturated with oil, grease, or brake fluid; or lining thickness is insufficient.
Parking Brakes	Upon actuation, no brakes are applied, including driveline hand controlled parking brake.*
Brake Drums or Rotors	1. External crack(s) on drums open on brake application. 2. A portion of the drum or rotor is missing or in danger of falling away.
Brake Hose	1. Hose damage extends through the outer reinforcement ply. 2. Hose bulges/swells when air pressure is applied or has audible leak at other than proper connection. 3. Two hoses are improperly joined and can be moved or separated by hand at splice. 4. Hoses are improperly joined but cannot be moved or separated by hand.* 5. Air hose is cracked, broken, or crimped and airflow is restricted.

*Indicates a restricted service waiver may apply.

(*continued*)

TABLE 4–1 (continued)

Inspected Item	Out-of-Service If:
Brake Tubing	Tubing has audible leak at other than proper connection or is cracked, heat-damaged, broken, or crimped.
Low-Pressure Warning Device	Device is missing, inoperative, or does not operate at 55 psi and below, or half the governor cutout pressure, whichever is less.*
Air Loss Rate	If air leak is discovered and the reservoir pressure is not maintained when governor is cut in, pressure is between 80 and 90 psi, engine is at idle, and service brakes are fully applied.
Tractor Protection	Protection valve on tractor is missing or inoperable.
Air Reservoir	Mounting bolts or brackets are broken, missing, or loose.*
Air Compressor	1. Drive belt condition indicates impending or probable failure.* 2. Mounting bolts are loose. Pulley is cracked, broken, or loose. Mounting brackets, braces, and adapters are cracked or broken.
Electric Brakes	1. Absence of braking on 20 percent or more of the braked wheels. 2. Breakaway braking device is missing or inoperable.
Hydraulic Brakes	1. No pedal reserve with engine running, except by pumping the pedal. 2. Master cylinder is less than one-quarter full. 3. Power-assist unit fails. 4. Brake hoses seep or swell under application. 5. Hydraulic fluid is visibly leaking from brake system and master cylinder is less than one-quarter full. 6. Check valve is missing or inoperative. 7. Hydraulic fluid is visibly leaking from brake system. 8. Hydraulic hoses are worn through outer cover-to-fabric layer. 9. Fluid lines or connections are restricted, crimped, cracked, or broken. 10. Brake failure light/low fluid warning light is on and/or inoperative.*
Vacuum System	1. Vacuum reserve is insufficient to permit one full brake application after engine is shut off. 2. Hoses or lines are restricted, excessively worn, crimped, cracked, broken, or collapsed under pressure. 3. Low-vacuum warning light is missing or inoperative.*
Coupling Devices Fifth Wheels	1. Mounting to frame—more than 20 percent of frame mounting fasteners are missing or ineffective. 2. Movement between mounting components is observed. Mounting angle iron is cracked or broken. 3. Mounting plates and pivot brackets, more than 20 percent of fasteners on either side are missing or ineffective. Movement between pivot bracket pin and bracket exceeds 3/8". Pivot bracket pin is missing or not secured. 4. Cracks in any weld(s) or parent metal are observed on mounting plates or pivot brackets.* 5. Sliders—more than 25 percent of latching fasteners, per side, are ineffective. Any fore or aft stop is missing or not secured. Movement between slider bracket and base exceeds 3/8". 6. Cracks are observed in any slider component parent metal or weld.* 7. Lower coupler—horizontal movement between the upper and lower fifth wheel halves exceeds 1/2". Operating handle is not closed or locked in position. Kingpin is not properly engaged. Cracks are observed in fifth wheel plate. Locking mechanism parts are missing, broken, or deformed to the extent the kingpin is not securely held. 8. Space between upper and lower coupler allows light to show through from side to side.*
Pintle Hooks	1. Mounting to frame—fasteners are missing or ineffective. Frame is cracked at mounting bolt holes. Loose mounting. Frame cross member providing pintle hook attachment is cracked. 2. Integrity—cracks are anywhere in the pintle hook assembly. Section reduction is visible when coupled. Latch is insecure. 3. Any welded repairs to the pintle hook are visible.*
Drawbar Eye	1. Mounting—any cracks are visible in attachment welds. Fasteners are missing or ineffective. 2. Integrity—any cracks are visible. Section reduction is visible when coupled. Note: No part of the eye should have any section reduced by more than 20 percent.

*Indicates a restricted service waiver may apply.

TABLE 4–1 (continued)

<table>
<tr><th>Inspected Item</th><th>Out-of-Service If:</th></tr>
<tr><td>Drawbar/Tongue</td><td>1. Slider (power/manual)—latching mechanism is ineffective. Stop is missing or ineffective. Movement of more than $\frac{1}{4}$" between slider and housing. Any leaking air or hydraulic cylinders, hoses, or chambers.
2. Integrity—any cracks. Movement of $\frac{1}{4}$" between subframe and drawbar at point of attachment.</td></tr>
<tr><td>Safety Devices</td><td>1. Missing or unattached, or incapable of secure attachment.
2. Chains and hooks worn to the extent of a measurable reduction in link cross section.*
3. Improper repairs to chains and hooks including welding, wire, small bolts, rope, and tape.*
4. Kinks or broken cable strands. Improper clamps or clamping on cables.*</td></tr>
<tr><td>Saddlemounts (method of attachment)</td><td>1. Any missing or ineffective fasteners, loose mountings, or cracks in a stress or load-bearing member.
2. Horizontal movement between upper and saddlemount halves exceeds $\frac{1}{4}$".</td></tr>
<tr><td>Exhaust System</td><td>1. Any exhaust system leaking in front of or below the driver/sleeper compartment and the floor pan permits entry of exhaust fumes.*
2. Location of exhaust system is likely to result in burning, charring, or damaging wiring, fuel supply, or combustible parts.</td></tr>
<tr><td>Fuel System</td><td>1. System with visible leak at any point.
2. A fuel tank filter cap missing.*
3. Fuel tank not securely attached due to loose, broken, or missing mounting bolts or brackets.*</td></tr>
<tr><td>Lighting Devices
When Lights Are Required</td><td>1. Single vehicle without at least one headlight operative on low beam and without a stoplight on the rearmost vehicle.
2. Vehicle that does not have at least one steady burning red light on the rear of the rearmost vehicle (visible at 500 feet).
3. Projecting loads without at least one operative red or amber light on the rear of loads projecting more than 4 feet beyond vehicle body and visible from 500 feet.</td></tr>
<tr><td>At All Times</td><td>1. No stop light on the rearmost vehicle is operative.
2. Rearmost turn signal(s) do not work.*</td></tr>
<tr><td>Safe Loading</td><td>1. Spare tire or part of the load is in danger of falling onto the roadway.
2. Protection against shifting cargo—any vehicle without front-end or equivalent structure as required.*</td></tr>
<tr><td>Steering Mechanism
Steering Wheel Free Play</td><td>When any of the following values are met or exceeded, the vehicle will be taken out of service.
<table>
<tr><th>Steering Wheel Diameter</th><th>Manual System Movement</th><th>Power System Movement</th></tr>
<tr><td></td><td>30 Degrees or</td><td>45 Degrees or</td></tr>
<tr><td>16"</td><td>$4\frac{1}{2}$" (or more)</td><td>$6\frac{3}{4}$" (or more)</td></tr>
<tr><td>18"</td><td>$4\frac{3}{4}$" (or more)</td><td>$7\frac{1}{8}$" (or more)</td></tr>
<tr><td>20"</td><td>$5\frac{1}{4}$" (or more)</td><td>$7\frac{7}{8}$" (or more)</td></tr>
<tr><td>21"</td><td>$5\frac{1}{2}$" (or more)</td><td>$8\frac{1}{4}$" (or more)</td></tr>
<tr><td>22"</td><td>$5\frac{3}{4}$" (or more)</td><td>$8\frac{5}{8}$" (or more)</td></tr>
</table></td></tr>
<tr><td>Steering Columns</td><td>1. Any absence or looseness of U-bolts or positioning parts.
2. Worn, faulty, or obviously repair-welded universal joints.
3. Steering wheel not properly secured.</td></tr>
<tr><td>Front Axle Beam</td><td>1. Includes all steering components other than steering column, including hub.
2. Any cracks.
3. Any obvious welded repair(s).</td></tr>
<tr><td>Steering Gear Box</td><td>1. Any mounting bolt(s) loose or missing.
2. Any cracks in gear box or mounting brackets.</td></tr>
<tr><td colspan="2">*Indicates a restricted service waiver may apply.</td></tr>
</table>

(continued)

TABLE 4–1 (continued)

Inspected Item	Out-of-Service If:
Pitman Arm	Any looseness of the arm or steering gear output shaft.
Power Steering	Auxiliary power-assist cylinder loose.
Ball and Socket Joints	1. Movement under steering load of a stud nut. 2. Any motion (other than rotational) between linkage member and its attachment point over 1/4".
Tie-Rods and Drag Links	1. Loose clamp(s) or clamp bolt(s) on tie-rods or drag links. 2. Any looseness in any threaded joint.
Nuts	Loose or missing nuts in tie-rods, pitman arm, drag link, steering arm, or tie-rod arm.
Steering System	Any modification or other condition that interferes with free movement of any steering component.
Suspension Axle Parts/Members	Any U-bolt(s), spring hanger(s), or other axle positioning part(s) cracked, broken, loose, or missing resulting in shifting of an axle from its normal position.
Spring Assembly	1. One-fourth of the leaves in any spring assembly broken or missing. 2. Any broken main leaf in a leaf spring. 3. Coil spring broken. 4. Rubber spring missing. 5. Leaf displacement that could result in contact with a tire, rim, brake drum, or frame. 6. Broken torsion bar spring in torsion bar suspension. 7. Deflated air suspension.
Torque, Radius, or Tracking	Any part of a torque, radius, or tracking component assembly that is cracked, loose, Components broken, or missing.
Frame Frame Members	1. Any cracked, loose, or sagging frame member permitting shifting of the body onto moving parts or other condition indicating an imminent collapse of the frame. 2. Any cracked, loose, or broken frame member adversely affecting support of functional components such as steering gear, fifth wheel, engine, transmission, or suspension. 3. Any crack 1 1/2" or longer in the frame web that is directed toward the bottom flange. 4. Any crack extending from the frame web around the radius and into the bottom flange. 5. Any crack 1" or longer in bottom flange.
Tire and Wheel Clearance	Any condition, including loading, causes the body or frame to be in contact with a tire or any part of the wheel assemblies at the time of inspection.
Adjustable Axle (sliding subframe)	1. Adjustable axle assembly with more than one-quarter of the locking pins missing or not engaged. 2. Locking bar not closed or not in the locked position.
Tires Any Tire on Any Steering Axle of a Power Unit	1. With less than 2/32" tread when measured in any two adjacent major tread grooves at any location on the tire. 2. Any part of the breaker strip or casing ply is showing in the tread. 3. The sidewall is cut, worn, or damaged so that ply cord is exposed. 4. Labeled "Not for Highway Use" or carrying markings that would exclude use on steering axle. 5. A tube-type radial tire without the stem markings. These include a red band around the tube stem, the word "radial" embossed in metal stems, or the word "radial" molded in rubber stems.* 6. Mixing bias and radial tires on the same axle.* 7. Tire flap protrudes through valve slot rim and touches stem.* 8. Regrooved tire except on motor vehicles used solely in urban or suburban service.* 9. Visually observable bump, bulge, or knot related to tread or sidewall separation. 10. Boot, blowout patch, or other ply repair.* 11. Weight carried exceeds tire load limit. This includes overloaded tire resulting from low air pressure.* 12. Tire is flat or has a noticeable leak. 13. So mounted or inflated that it comes in contact with any part of the vehicle.

*Indicates a restricted service waiver may apply.

TABLE 4–1 (continued)

Inspected Item	Out-of-Service If:
All Tires Other Than Those Found on the Steering Axle of a Powered Vehicle	1. Weight carried exceeds tire load limit. This includes overloaded tires resulting from low air pressure.* 2. Tire is flat or has noticeable leak (that is, one that can be heard or felt). 3. Bias ply tire—when more than one ply is exposed in the tread area or sidewall, or when the exposed area of the top ply exceeds 2 sq. in. Note: On duals, both tires must meet this condition. 4. Bias ply tire—when more than one ply is exposed in the tread area or sidewall, or when the exposed area of the top ply exceeds 2 sq. in.* 5. Radial ply tire—when two or more plies are exposed in the tread area or damaged cords are evident in the sidewall, or when the exposed area exceeds 2 sq. in., tread or sidewall. Note: On dual wheels, both tires must meet this condition. 6. Radial ply tire—when two or more plies are exposed in the tread area or damaged cords are evident in the sidewall, or when the exposed area exceeds 2 sq. in., tread or sidewall.* 7. Any tire with visually observable bump or knot apparently related to tread or sidewall separation. 8. So mounted or inflated that it comes in contact with any part of the vehicle. (This includes any tire contacting its mate in a dual set.) 9. Is marked "Not for Highway Use" or otherwise marked and having similar meaning.* 10. So worn that less than $\frac{1}{32}$" tread remains when measured in any two adjacent major tread grooves at any location on the tire. Exception: On duals, both tires must be so worn.*
Wheels and Rims	1. Lock or side ring is bent, broken, cracked, improperly sealed, sprung, or mismatched. 2. Rim cracks—any circumferential crack except at valve hole. 3. Disc wheel cracks—extending between any two holes. 4. Stud holes (disc wheels)—50 percent or more elongated stud holes (fasteners tight). 5. Spoke wheel cracks—two or more cracks more than 1" long across a spoke or hub section. Two or more web areas with cracks. 6. Tubeless demountable adapter cracks—cracks at three or more spokes. 7. Fasteners—loose, missing, broken, cracked, or stripped (both spoke and disc wheels) ineffective as follows: For 10 fastener positions: 3 anywhere, 2 adjacent. For 8 fastener positions or less (including spoke wheel and hub bolts): 2 anywhere. 8. Welds—any cracks in welds attaching disc wheel disc to rim. Any crack in welds attaching tubeless demountable rim to adapter. Any welded repair on aluminum wheel(s) on a steering axle. Any welded repair other than disc to rim attachment on steel disc wheel(s) mounted on the steering axle.
Windshield Glazing	Any crack over $\frac{1}{4}$" wide, intersecting cracks, discoloration not applied in manufacture, or other vision-distorting matter in the sweep of the wiper on the driver's side.
Windshield Wipers	Any power unit that has inoperable parts or missing wipers that render the system ineffective on the driver's side.

*Indicates a restricted service waiver may apply.

a decal, a copy of the inspection form must be kept on file and indicate where an inspector can call or write to get a copy of it.

Size and shape of the decal is not specified. It may be purchased from a supplier or even be made by the PM shop. The only requirement is that it remains legible. Each "vehicle" must be inspected separately. This means a tractor/trailer is two vehicles, each requiring a decal; a converter dolly is a separate vehicle. The decal (**Figure 4–8**) must show the following information:

- Date (month and year) the vehicle passed the inspection
- Name and address to contact about the inspection record. The record can be stored anywhere as long as the decal states where to contact the fleet.
- Certification that the vehicle has passed the inspection
- Vehicle identification information (fleet unit number or serial number) sufficient to identify the vehicle

You can check out the NHTSA Web site at http://www.nhtsa.dot.gov.

Record-Keeping Requirements

Many fleets have modified their "B" PM forms to include all the federal safety checks. This way whenever a "B" service is performed, the vehicle also passes the safety inspection, and a new decal can be affixed. This procedure simplifies the task of tracking each vehicle to perform the annual inspection. Most fleets stagger inspections through the year to prevent shop scheduling congestion.

FIGURE 4–7 Typical monthly PM schedule

Vehicle Maintenance Record and Schedule

Truck # ________ **Make** ________ **Model** ________ **Serial #** ________

Month or mileage	Due date	Service due	Mileage serviced	Date serviced	Repairs performed	RPO#
January		(1) O&F				
February		(1)				
March		(1) O&F				
April		(1) WF				
May		(1) O&F				
June		(1)				
July		(1) O&F				
August		(1) WF				
September		(1&2) O&F				
October		(1)				
November		(1) O&F				
December		(1) WF				

O = Oil
F = Oil Filter
WF = Water Filter

Experience has shown that it is a good idea to keep work orders in the vehicle file folder, and current fleet practice indicates that maintaining electronic vehicle records on the chassis data bus is effective. The same regulations require that the driver's vehicle condition report be retained for at least 3 months from the date of the report.

Today, computerized records play an important role in maintenance programs. Increasingly sophisticated maintenance software with automated data entry facilitates tracking and scheduling. Wireless data exchangers at fuel islands and the use of telematics enable the chassis data bus to be continuously read and in conjunction with diagnostic recorders help manage drivers, vehicle maintenance, and trip information. Today maintenance managers are expected to be computer literate, which means they have better knowledge of equipment condition than ever before. That knowledge is important when determining warranty claims, spec'ing decisions, and cost-per-mile reduction.

SHOP TALK

When undertaking PM checks and servicing trucks, it is usually necessary to raise a COE cab. Check the operator manual for the correct procedure, but be sure to follow the safety tips given in Chapter 2. With most hydraulic cab-lift systems, there are two circuits: the **push circuit** that raises the cab from the lowered position to the desired tilt position, and the **pull circuit** that brings the cab from a fully tilted position up and over the center (**Figure 4–9**). Remember that in most systems, whenever raising or lowering the cab, stop working the hydraulic pump once the cab goes over center. The cab falls at a controlled rate and continued pumping could lock up the tilt cylinders.

MAINTENANCE SOFTWARE

Most medium to large fleets manage maintenance using software programs that can be adapted to their specific needs. There are a wide range of programs available; some examples include:

- FastMaint
- FleetMatics
- Manager Plus
- Manger Pro
- EmDecs
- eMaint

Today, maintenance software is integrated into comprehensive fleet management programs. The potential of these programs is addressed in a section at the end of this chapter.

TABLE: 4–2 PM Trailer Inspection Guide

Inspection after the first 500 miles	Conduct pretrip inspection. Check axle alignment. Tighten axle U-bolts. Adjust brakes. Retorque inner and outer cap nuts on disc wheels or wheel nuts if trailer is equipped with spoke wheels.
Monthly inspection	Conduct pretrip inspection. Inspect brakes to ensure that the lining is not worn excessively. When lining is replaced, check drums for excessive wear, heat cracks, and grooving. Check travel of brake chamber pushrods and adjust brakes as necessary. Check and retorque wheel nuts. Inspect tires for uneven wear. Check axle alignment. Check for loose or missing fasteners, and replace or tighten as necessary. Check and repair any cracked welds. Inspect all areas where sealant is used. Reseal as necessary. Check body integrity and subframe finish, particularly in areas above trailer tires. Replace as necessary. Check optional items (including refrigeration and heating equipment) according to manufacturer's inspection guide. Check other items or special type trailers or construction (FRP trailers, dump trailers, etc.). Lubricate trailer according to the service manual.
Semiannual inspection	Complete the 500-mile and monthly inspection checklists. Inspect brake linkage, cams, and shoes. Inspect wheel bearings. Set wheel ends. Refill hubs to proper oil level. Inspect body integrity, doors, ventilators, roof, floor, sides, bulkheads, and scuff plates. Replace or repair unserviceable parts. Inspect suspension components, springs, U-bolts, landing gear, and tire carrier. Repair or replace worn or damaged parts as needed. Inspect kingpin for excessive wear and damage. Inspect upper coupler plate for excessive wear, distortion, and cracks. Inspect interior and exterior finish. Replace as necessary. Lubricate according to service manual recommendations.

Note: If the trailer was delivered directly from the manufacturer, check axle alignment, tighten axle U-bolts, and retorque inner and outer cap nuts or wheel nuts on arrival.

FIGURE 4–8 Make your own inspection sticker, but be sure it meets the federal requirements shown.

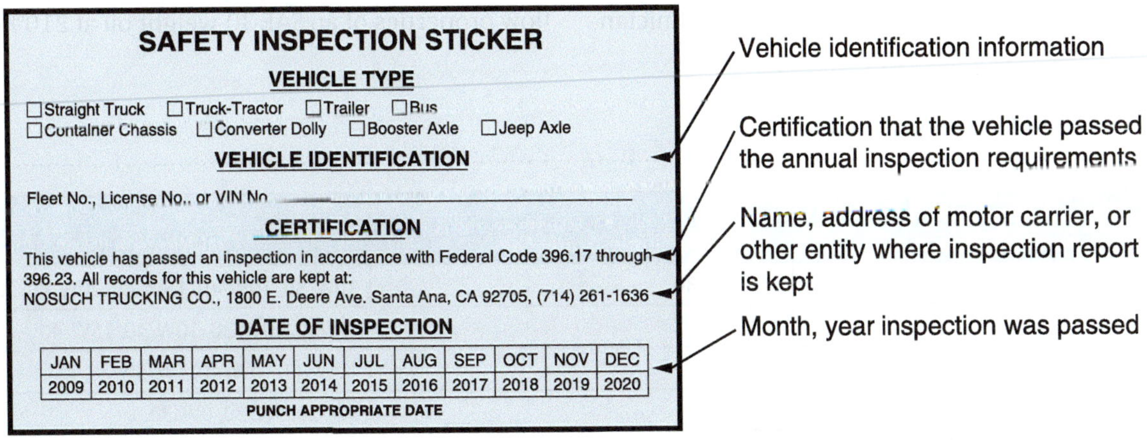

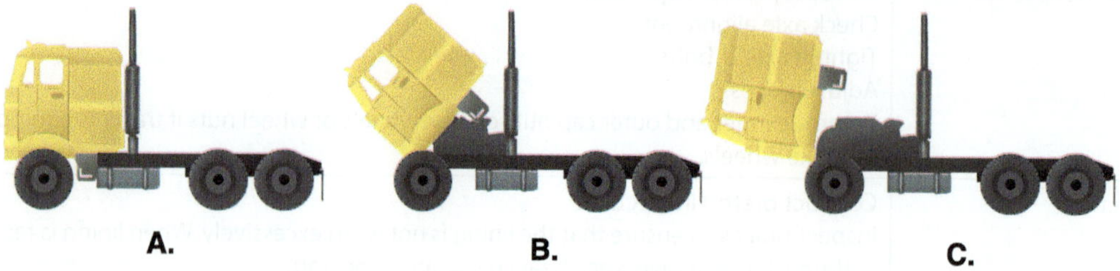

FIGURE 4–9 Cab-lift positions: (A) lowered or in operating position; (B) 45-degree tilt position; and (C) full-tilt position (80 degrees).

4.4 LUBRICANTS

Proper lubrication is an important factor in reducing wear and preventing premature failure of truck components. The oil film provided by lubricants between moving parts reduces friction, prevents excessive heat, and holds dirt particles in suspension. Lubricants also contain additives that inhibit corrosion and aid in reducing component wear. Heat and use alter the chemical properties of lubricants, making regular lubricant change intervals necessary.

In later chapters of this book, lubricants and lubrication are discussed in the context of specific systems. Actually, the proper use of lubricants is the backbone of most good fleet PM programs. Meeting the lubrication requirements of a vehicle is not difficult. It requires adhering to the manufacturer's recommendations provided in the service literature. However, there are some other factors to consider—especially in fleets made up of different OEM chassis and mixed component brands.

In some cases, manufacturers approve several types of lubricants (such as engine oil or gear lubricants in transmissions and final drive carriers).

Synthetic lubricants are tending to replace mineral-based lubes in transmissions, differential carriers, and wheel ends because service intervals can be greatly extended. Also critical is the lube **viscosity**. When should a 90-weight, 140-weight, or multiviscosity lubricant be used? How does gear lube viscosity compare with engine oil viscosity? To perform a proper lubrication job, the service technician should have some understanding of lubricant properties and the lubrication requirements of chassis components.

ENGINE OILS

The PM technician should have a basic understanding of the properties of engine oils. The viscosity of engine oil describes its resistance to fluid shear, but this is best understood as a rating of oil's resistance to flow. Because oil thins as temperature rises and thickens as temperature drops, the viscosity of engine oil should be matched to both ambient temperature and the operating temperatures of the equipment being lubricated.

The Society of Automotive Engineers (SAE) has established an oil viscosity classification system accepted throughout the trucking industry. This system grades heavier weight oils with a higher number or rating. For example, oil with an SAE rating of 50-weight is heavier and flows more slowly than an SAE 10-weight oil. Heavier weight oils tend to be used in warmer climates, whereas the lighter weight oils are used in low-temperature areas.

Multiviscosity oils are formulated to cover a wider range of operational temperatures than the fixed standard weight lubes. Multiviscosity oils are classified by the range of temperature conditions they meet; examples are 10W/30, 15W/40, and so on (**Table 4–3**). A 10W/30 oil, for example, has the same viscosity characteristics as a 10-weight oil at 0°F (−17°C) but has the flow properties of an SAE 30-weight oil at 210°F (99°C).

TABLE 4–3 SAE Grades of Engine Oil

Lowest Atmospheric Temperature Expected	Single-Grade Oils	Multigrade Oils
32°F (0°C)	20, 20W, 30	10W/30, 10W/40, 15W/40, 20W/40, 20W/50
0°F (−18°C)	10W	5W/30, 10W/30, 10W/40, 15W/40
−15°F (−26°C)	10W	10W/30, 10W/40, 5W/30
Below −15°F	5W	5W/20, 5W/30, 0W-30

This oil provides both easy starting in cold weather and good protection at all operating temperatures.

Resistance to oxidation is an important characteristic governing the service life of lubricating oil. Oxidation occurs when the hydrocarbons in the oil chemically combine with oxygen. Although most oils contain antioxidants, they are gradually used up through service. Heat, pressure, and aeration speed up the oxidation process. Other common causes of oxidation include extended oil drain intervals, the wrong oil for the application, excessive combustion gas blowby, and high sulfur fuels. As oil oxidizes, corrosive acids can form and deposits may accumulate on critical engine parts, inhibiting operation and accelerating wear.

SHOP TALK

Both the SAE viscosity grade and American Petroleum Institute (API) classification are clearly displayed on the label of any oil container.

ADDITIVES

To enable a lubricating oil achieve maximum performance and service life, additives are blended into oil by the manufacturers. The following are some of the more common additives used with lubricating oils.

- **Oxidation inhibitors** keep oil from oxidizing even at very high temperatures. Because they prevent the oil from being oxidized they also help limit varnish and sludge formations.
- **Antirust agents** prevent corrosion of metals when the engine is not running. They assist in forming a protective coating that repels water and neutralizes harmful acids.
- **Detergent additives** help keep metal surfaces clean and prevent sludge deposits. These additives suspend carbonized soot and particulates in the oil. These contaminants are removed from the system when the oil is drained so they cannot accumulate as sludge.
- **Anticorrosion agents** protect metal surfaces from corrosion. These agents will work with oxidation inhibitors.
- **Total base number (TBN) additives** are acid molecule neutralizers that are blended into the oil to reduce the harmful effects of acid compounds that are produced as a by-product of the combustion process.

Additives gradually lose their effectiveness. The engine oil should be drained and filters serviced before the additives are totally depleted. Engine oils are blended to a precise chemical formula, and the addition of aftermarket additives can upset the balance of the additives and often adversely affect engine performance. Follow the manufacturer's recommendations when using engine oil additives and generally avoid using them.

SHOP TALK

Using the incorrect grade or type of oil in an engine can cause a variety of problems. For example, the wrong viscosity oil can cause either an oil consumption or low oil pressure problem, depending on the weather and driving conditions. Using an oil with the incorrect service rating can result in inadequate protection for engine bearings and other moving engine parts. Consult the engine manufacturer's recommendations to ensure that the engine is using the correct grade and type of oil.

The frequency of oil changes as well as oil, fuel, and coolant filters depend on the PM schedule. Most PM programs follow the OEM service recommendations but some of the largest fleets use their own.

DIESEL ENGINE OIL CLASSIFICATIONS

Diesel engine oils are classified by the API based on the type of service operation under which an engine is to be operated. There are two basic classifications:

- **C**: appropriate for diesel engine (compression-ignited) applications
- **S or G**: appropriate for gasoline engine (spark-ignited) applications

The C or S/G classification is followed by a second alpha rating that describes the generation of the specification. For instance, it can be assumed that a CF-rated oil supersedes a CC-rated oil. This is important. As engine technology changes so do the lubricants required. For instance, 2004 emissions standards forced most manufacturers to adopt cooled exhaust gas recirculation (C-EGR) systems, resulting in higher soot loads delivered into engine intake. From 2007, diesel particulate filters (DPFs) were added to diesel engines, requiring the use of engine oils with low ash. As each new engine oil classification is introduced, it is designed to be backward compatible. Today most highway diesel engines require the use of one of the following oil classifications:

- **CI-4**—Introduced in September 2002 for use in high-speed, highway diesel engines meeting 2004 exhaust emissions standards, implemented in October 2002 by the Environmental Protection Agency (EPA) agreement with engine OEMs. CI-4 oil is formulated for engines using fuels containing less than 0.05 percent sulfur and maximizes lube oil and engine longevity when exhaust gas recirculation (EGR) devices are used. CI-4 engine oil supersedes CD, CE, CF, CG, and CH category oils. It is not designed for use in post-2007 diesel engines, although a couple of OEMs have authorized it. CI-4 should

be available until around 2010 and will probably be used until then by some operators due to its lower cost than CJ-4.
- **CJ-4** is formulated for use in 2007 diesel engines equipped with cooled-EGR and DPFs but is also backward compatible. The main difference between CJ-4 and CI-4 is a change in the additive package designed to reduce ash generated when the oil is either combusted in the cylinder or through the closed crankcase ventilation system. This ash in the exhaust gas can produce a negative effect on DPFs. This negative effect could be by plugging or poisoning the catalyst.
- **PC-11** has been undergoing testing since 2011 and will probably replace CJ-4 in either 2016 or 2017 likely renamed CK. The current formulation of this oil has better high temperature shear resistance and viscosity properties that lower internal drag of engine components. When it replaces CJ-4 it is expected to be backward compatible.

2017 Engine Oils

Because of the stringent fuel economy improvements scheduled for implementation in 2017, the oil currently known as PC-11 may not be available as 15W-40. Some engine OEMs are stating that their requirement will be for oils specified as 5W-30 or 10W-30.

Engine Oils for Gasoline-Fueled Vehicles

Oils formulated for use in gasoline-fueled engines are categorized in much the same way. The most recent engine oils formulated for service in gasoline engines are as follows:

SH	1993
GF-2	1996
GF-3	2001
GF-4	2004

There still are some engine oils formulated claiming to meet the needs for both spark-ignited (SI) gasoline-fueled and diesel-fueled engines, but this is not so common as it was a decade ago. In most cases, it is a suspect claim.

NG Engine Oils

Commercial engines fueled with natural gas (NG) require an NG-specific lubricant and shorter service intervals. These oils are usually distinguished by an NGEO or GEO suffix. Always consult the OEM service literature before selecting oil for NG engines.

Used Oil

Most states and provinces consider used oil as hazardous waste. For this reason, check environmental agencies for the regulations concerning the disposal of used oils. Federal regulations state that used oil cannot be stored in unlined containers. Tanks and storage containers must be properly maintained and have a "used oil" label. Spills must be cleaned up. If you have used oil recycled, make sure that both the transporter and recycler are in compliance with federal and state regulations. If not drained of free-flowing oil, used oil filters are regulated as used oil. Even after oil is drained from filters, state or local recycling rules may apply. Transporters and recyclers who specialize in used oil often also take filters. Crushing may be required.

BELTS AND TENSIONERS

Belt materials used on truck engines changed during the mid-1990s from neoprene to **EPDM** (ethylene, propylene, diene monomer) construction. EPDM greatly outlasts neoprene and does not crack. This means that when engine drive belts are inspected, the technician is not looking for cracks but checking for wear. Loss of engine accessory drive belt tension is caused by:

- Belt wear (as little as 5% can cause problems)
- Tensioner failure

The function of the tensioner is to:

- Maintain the correct belt tension
- Dampen engine torsionals

Like tires, belts wear and when EPDM belts are used, they usually last about as long as the tensioners so it is good practice to replace both. To source belt noise, spray water onto the belt:

- If noise increases, the cause is tension related
- If noise disappears, the cause is alignment related

COOLANT HOSES

Like drive belts, coolant hoses are a key maintenance inspection item because inability to identify an imminent failure can cripple a vehicle on the road. Coolant hoses should be physically inspected for cracks, collapsing, and rub points, along with being checked for leaks. It is regarded as good practice to eliminate worm screw and spring-tension hose clamps with silicone-specific or thermoplastic clamps. Heat-sensitive, **thermoplastic hose clamps** (Gates PowerGribSB is an example) dynamically adjust to changes in temperature to automatically maintain tension. This is important with today's silicone coolant hoses that are sensitive to excessive hose clamp tension.

GEAR LUBRICANTS

The lubricants and lubrication of transmissions and drive axles are covered in Chapters 15 through 24. Fleet maintenance personnel do not pay enough attention to axle and transmission lubes because, unlike engine crankcase oils, these oils are contained in an isolated system. There is no potential for contamination from antifreeze or fuel leaking into the lube.

Cold temperature operation can cause transmission and axle lubrication problems known as *channeling*. High-viscosity oils used in some transmissions and axles thicken and may not flow at all when the vehicle is first started. Rotating gears can displace lubricant films, leaving voids or channels on gear contact surfaces. With transmission lubes, a 50-weight oil has a channeling temperature of 0°F (−17°C). In fleet operations where trucks are parked in cold temperatures, this could cause transmission gears to lack proper lubrication for the first couple of miles of operation.

Extreme pressure (EP) additives were developed to meet the increased stresses of hypoid and amboid axle gearing. Some earlier EP lubes broke down under high temperatures produced in standard transmissions. For this reason, many transmission manufacturers option a 90-weight straight mineral oil or an SAE 50 engine oil.

Synthetic Gear Lubes

Synthetic lubricants for transmissions and axles introduced in recent years offer extended oil life. Synthetics do not contain the readily oxidized ingredients found in mineral-based lubes, so they have better oxidation stability, can handle higher operating temperatures, and can double or even triple oil change intervals. Channeling temperatures are as low as −65°F (−54°C). Synthetics can be used in both transmissions and axles. Some are compatible with mineral oil-based lubes and with seals but combining synthetic and mineral oils may reduce service life. Certain synthetic oils should never be mixed with mineral oils (especially those formulated for automatic transmissions), so always check with the manufacturer's information. Synthetic gear lubes may help improve fuel economy (1 to 3%) by reducing gear drag in colder temperatures. Most fleets using synthetics in transmissions report easier gear shifting in cold weather, too. Synthetic gear lubes can cost three times as much as equivalent petroleum-based products but tend to be cost effective because of extended service intervals. Some transmission OEM specifically require the use of synthetic gear lubes.

Automatic Transmission Fluids

Automatic transmission fluids (ATF) have various specs that comply with automatic transmission manufacturers' requirements. ATFs are also available as synthetic gear lubricants with longer service intervals and superior cold/hot temperature operations. Some synthetics are functionally compatible with conventional ATFs (see Chapter 19) but when mixed, the shorter projected service life of the mineral base oil should be observed. Some transmission OEMs require the use of synthetic ATF.

TABLE 4–4 Classification of Gear Lubricants

Classification	Description
GL-3	Originally developed for spiral bevel gear lubricant requirements, it is not for use with hypoid gearing.
GL-5	Hypoid gearing, severe service. This is the top gear lube spec today. MIL-L-2105D (a common military spec) is closely related to GL-5.
GL-6	Not much used. Designed for gears with extremely high pinion offset.
In 1988, the SAE identified specific needs for two additional gear lubricant categories as follows:	
PG-1	Lubricant spec for truck manual transmission with high oil temperature (over 300°F). It must also have synchronizer and oil seal compatibility.
PG-2	Heavy-duty gear lubricant with specifications to meet improved gear spalling resistance standards, better oil seal compatibility, and improved thermal stability/ component cleanliness. It must also have improved corrosion protection with copper alloys under both wet and dry conditions and it must have improved gear score protection over MIL-L-2105D spec.

Gear Lube Classifications

True gear lubes designed for truck and heavy equipment gearing applications use the prefix GL. The prefix MT (manual transmission) is used by a couple of lubricant suppliers addressing the light-duty market. A listing of some current GL classifications appears in **Table 4–4**.

CHASSIS LUBRICANTS

Chassis lubricants are usually solidified greases, although several on-board automatic lube (chassis centralized) systems require gear oil or liquefied greases. These systems are covered later in this chapter.

Grease is formulated to lubricate just as oil would but without the cooling and cleansing functions normally associated with oil. Chemists call the oil-holding gel characteristic of grease *soap*, and it may have a calcium, sodium, or lithium base. Each soap provides its particular characteristics to the grease product. Qualities such as appearance, stability, pumpability,

TABLE 4–5 NLGI Grades and Method of Application

NLGI Grade No.	Penetration Worked, 77°F. Typical Methods of Application
000	445–475 Semifluid; used in centralized systems
00	400–430 Semifluid; used in centralized systems
0	355–385 Used in centralized systems
1	310–340 Used in grease guns, centralized systems
2	265–295 Used in grease guns, centralized systems
3	220–250 Used in grease guns
4	175–205 Used in grease guns
5	130–160 Used in grease guns
6	85–115 Block grease, very hard; for open grease cellars

and heat and water resistance depend largely on the type of soap base used in grease.

In addition to soap and liquid lubricant, nearly all greases have additive packages that bring additional properties such as corrosion resistance, adhesiveness, and EP capability. Sulfur, phosphorus, zinc, molybdenum disulfide, and graphite are some common additives. The most frequently measured property of grease is consistency. This can be compared to viscosity in engine oil.

The consistency of grease, a measure of its relative hardness, is commonly expressed in terms of the American Society for Testing and Materials (ASTM) penetration or National Lubricating Grease Institute (NLGI) consistency number (**Table 4–5**).

The penetration is the depth of penetration of a grease sample, in tenths of a millimeter, by a standard test done under stated conditions (ASTM D 217): the greater that penetration, the softer the grease. The corresponding NLGI grades range from a semifluid 000 ("triple ought") to a very hard 6. The latter grade has been described as "almost hard enough to walk on." The properties of grease constituents are outlined in **Table 4–6**.

Dust and dirt do not settle to the bottom of a grease reservoir as they would in oil. If airborne contaminants are allowed to enter the grease, they will remain in suspension, acting like a lapping compound on metal surfaces or possibly blocking a centralized system pump or injector.

Greases are not all the same, so avoid mixing them. Refill a reservoir only with a grease using the same thickener, or soap, such as calcium, lithium, or aluminum. Sodium-soap greases are water soluble and are not generally recommended for heavy-duty truck applications.

ON-BOARD CHASSIS LUBRICATING SYSTEM

One of the greatest benefits to PM chassis lubricating is the on-board, centralized lube system. These systems are manually or automatically operated. Manual systems are used on vehicles such as trailers that have no electrical or air power to operate the lubricant pump used on automatic systems.

Manual Systems

A manual system consists of a manifold, or distribution block, with twelve to twenty-four grease lines connected to critical lube points. To lube the chassis, a grease gun is connected to one central zerk fitting mounted on the block, through which grease is routed to all connected points. These have become less common in recent years but are still available.

Automatic Chassis Lube Systems (ACLS)

A vehicle with ACLS delivers a periodic, uniform shot of grease to critical points as often as every few minutes. That helps offset incomplete or inadequate lubrication during PMs and in some fleets has reportedly helped components such as steering kingpins and other components last through the vehicle's life.

An advantage of the mobile system versus stationary lubing is that the continuous addition of fresh lubricant tends to force out old, contaminated lubricant.

To distribute grease, on-board lube systems use the same basic components:

- A reservoir holds the supply of grease to be distributed (**Figure 4–10**).
- A pump delivers the lubricant through a network of grease lines. Metering valves dispense the grease.
- An automatic timer mechanism tells the pump-and-valve system when to pump grease.
- An electrical motor or other power source (the truck's air system is the most widely used).

Air-Driven Systems. In air-driven systems (**Figure 4–11**), progressive feeders, piston distributors, or metering valves at the end of the dispensing lines are strategically located at the chassis' multiple grease points; major suspension points; shackle pins; brake camshafts; slack adjusters; clutch releases; steering linkages (tie-rod ends, drag links); kingpins, and the fifth wheel (pivot points, fifth wheel plate,

TABLE 4–6 Properties of Greases

Type Grease Base	Usual Appearance	Mechanical Stability	Cold Weather Pumpability	Heat Resistance	High Temp Life	Water Resistance	Compatibility with Other Greases
Calcium (lime soap)	Buttery	Good	Fair	Fair	NA	Excellent	Excellent
Sodium (soda soap)	Fibrous or spongy	Fair	Poor	Good to excellent	Fair	Poor	Good
Barium	Fibrous	Good	Poor	Excellent	Fair	Excellent	Fair
Lithium	Buttery	Excellent	Good to excellent	Good to excellent	Good	Excellent	Excellent
Lithium complex	Buttery	Excellent	Good to excellent	Excellent	Excellent	Excellent	Excellent
Calcium complex	Buttery to grainy	Good	Fair	Good	Fair	Good to excellent	Fair
Aluminum complex	Buttery to grainy	Good to excellent	Good	Excellent	Fair	Excellent	Poor
Nonsoap (Bentone)	Buttery	Good	Excellent	Excellent	Good	Good	Poor
Urea	Buttery	Good	Good	Excellent	Excellent	Excellent	Excellent

slider rails). The pump in air-driven systems is activated by a solenoid valve.

Motor-Driven Systems. Air or electric motor-driven systems use a multioutlet pump and no secondary distribution parts. The grease is metered directly at the pump. A third type of system, using a single outlet, electric-driven gear pump, uses a secondary system of metering valves, piston distributors or progressive feeders, like the truck air system-driven setup. Most U-joints must still be lubed manually. This creates a convenient maintenance interval for the system itself. While technicians are manually lubing the U-joints, they can fill the automatic system reservoir, which should last for 2 to 4 months, and at the same time check system lines to ensure they are in place and functioning properly.

FIGURE 4–10 Key components of a centralized, on-chassis lube delivery system: The reservoirs are usually mounted to the frame, where they may be viewed and refilled at PM intervals.

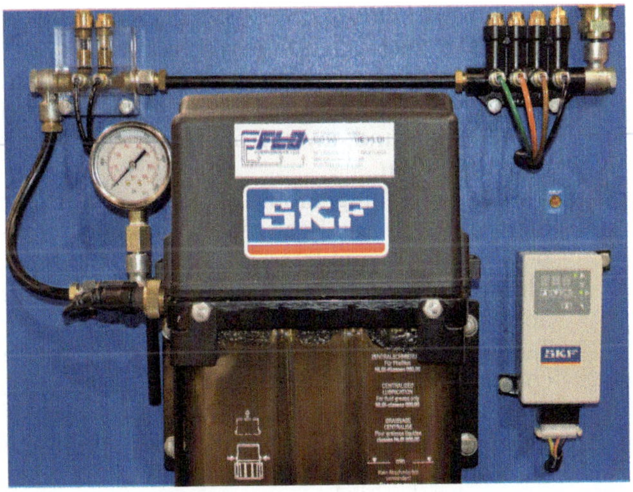

Distribution. One system can be configured to lubricate more than thirty-two points, including steering kingpins, spring pins, brake "S"-cams, brake slack adjusters, the clutch, shift shaft linkage, and fifth wheel plates. The unit's integrated pump, powered by the truck air supply circuit, feeds grease through a main line to modules that serve groups of lube points (left and right front, rear axles, fifth wheel, and suspension). The modules are fitted with different-sized meters that feed varying amounts of grease to individual points through distribution lines. (Nonmodule systems require a separate line for each individual lube point.)

System Cycling. A 12- or 24-volt DC solenoid controller on this particular system, working in conjunction with the pump, initiates lube cycles at regular

FIGURE 4–11 Layout of an automatic on-board chassis lubrication system. Key to number for Class 8 vehicle lube points: 1. Front spring pin; 2. Front spring shackles (upper and lower); 3. Kingpin (upper and lower); 4. Brake camshaft; 5. Brake slack adjuster; 6. Tie rod; 7. Drag link; 8. Fifth wheel pivot; 9. Fifth wheel plate; 10. Clutch crossshaft.

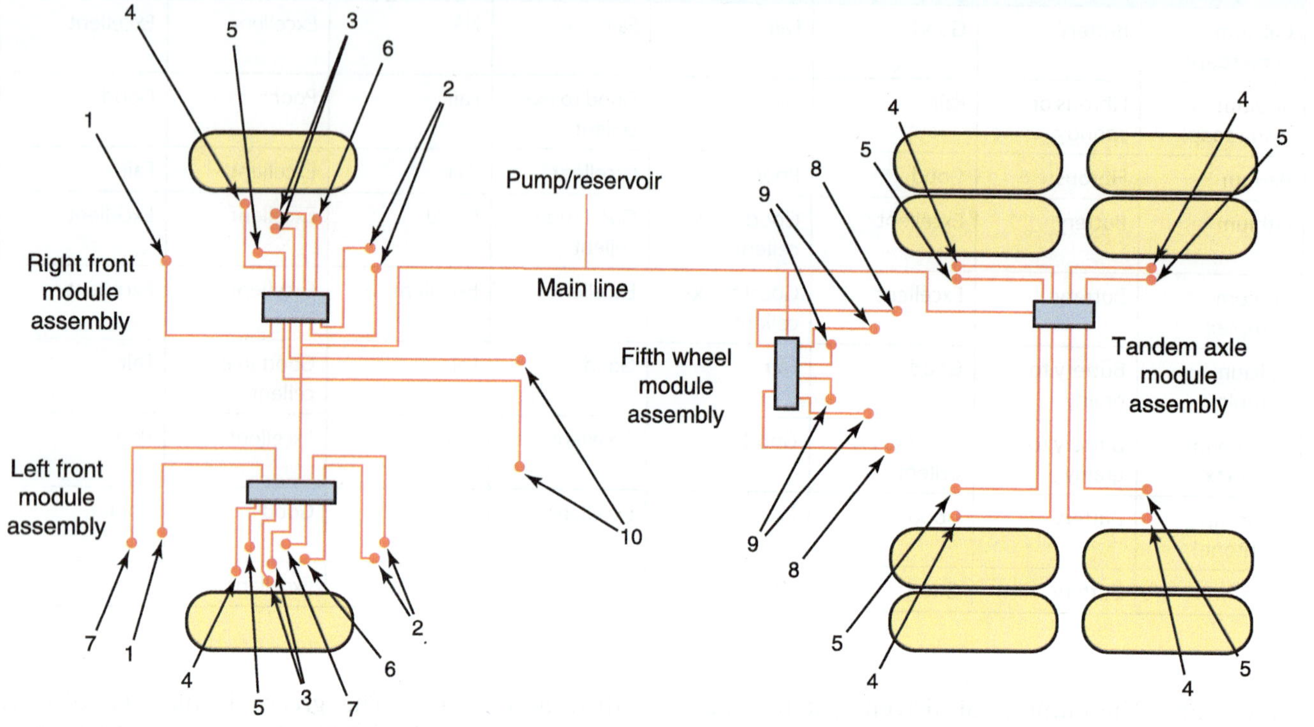

intervals. Adjustment of a simple knob on the face of a small, solid-state timer programs the controller to trigger a lube cycle anywhere from once every 6 hours to as often as every ½ hour. The timer's built-in memory retains elapsed cycle time when the ignition is switched off and resumes when it is switched back on. The grease reservoir typically holds 10 pounds of lubricant, or enough to last a truck in an average line-haul operation from 14 to 16 weeks.

PM Checks. The routing and securing of the grease lines from the pumping system to the individual lube points are critical (**Figure 4–12**). Check that the lines are not loose and flopping around; route them away from areas where the line could be pinched or cut. Also include the system in PM checks. Technicians should cycle the automatic lube system each time a tractor is serviced. The objective is to verify that the solenoids are functioning and to check if the proper amount of grease is being pumped into a component. If not, the controller time cycle can be adjusted. Being able to adjust individual grease amounts is important because some components require more grease than others.

Both manual and automatic chassis lubricating systems are, in simplest terms, hydraulic systems for pumping grease. Some automatic systems use standard shop greases rated #1 or #2 by the NLGI; others require liquid greases with NLGI ratings of 000 and 00. These greases tend to stay fluid over a wide range of operating temperatures. Using these two ratings of grease, one manufacturer reports that its systems are dependable in temperatures ranging from −25°F to 100°F (−32°C to + 38°C).

Automatic chassis lube systems can be added to older trucks that do not have it. When retrofitting a system, there are several installation tips to keep in mind:

- In tilt cab vehicles, do not tilt the cab any further than the truck builder recommends.
- Metering valves, which distribute lubricant, should be placed alongside the vehicle, and if on the ground, away from any dirt or other contamination.
- All lines or tubing should be fixed a sufficient distance from any heat sources (exhaust components, for example).
- Clip/trim all tubing well to avoid chafing and abrasion.
- Take care when drilling to avoid damage to the compressed air or electrical lines.
- It is important to make the system a part of the truck, but separate from the air and brake lines, and chassis wiring.

FIGURE 4–12 Some automatic-lube systems run a single tube to each lubrication point; others bundle tubes together into "groups" of lube points, fanning them one per point.

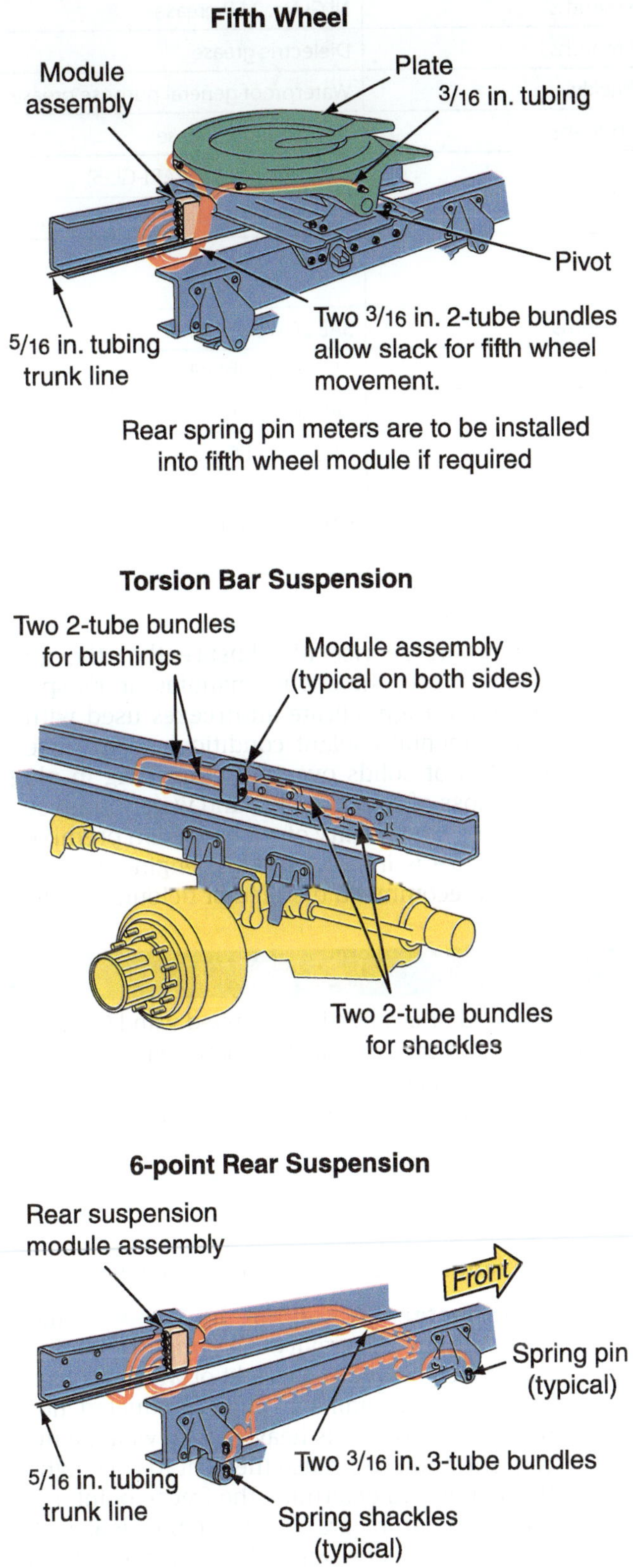

TRAILER LUBRICATION

Table 4–7 is a typical PM trailer lubrication chart recommended by many trailer manufacturers. A trailer PM checklist is also included in the Student Guide that accompanies this book.

4.5 WINTERIZING

Winterizing trucks is an important part of any PM program. A cold weather (+40°F to −40°F; +4°C down to −40°C) operation maintenance checklist should be integrated into the PM schedule.

ENGINE COOLANT

The engine cooling system must be properly winterized for cold weather operation. Key to maintaining and testing antifreeze protection is identifying the coolant. In today's trucks, five types of antifreeze solution are used:

- Ethylene glycol (EG): mixed with water
- Propylene glycol (PG): mixed with water
- Premix extended life coolant (ELC): never mixed with water; used as premix only
- Straight ELC: can be mixed with water
- Waterless engine coolant (WEC): cannot be mixed with anything.

Coolant Color

At this point in time, engine coolants cannot be identified by color. The color added to antifreeze is simply a dye. Although attempts have been made over the years to standardize antifreeze colors by type, so far these attempts have not produced any results. The only really significant coolant color tattletale is the color brown. When you observe a muddy brown color in antifreeze you can assume that one of the following has occurred:

- Two incompatible antifreezes have been mixed.
- Severe corrosion has occurred in the cooling system.

Coolant Maintenance Schedule

The maintenance schedule outlined here is a winterizing routine that should be performed in this sequence.

- **Inspection**. A visual check of the entire cooling system (radiator, water pump, hoses, and so on). Look for leaks or dried inhibitor deposits (gel or crystals). Heater and water hoses should be sound, soft, and pliable. If they are hard, cracked, or brittle, replace them. Check all hose clamps. Pressure-test the cooling system at the specified system pressure after turning on heat control valves. Repair any leaks. Also

TABLE 4–7 Trailer Lubrication Chart

Item	Check or Service	Frequency	Lubricant or Equivalent
Brake camshaft support bearings	Lubricate	3 months	NLGI No. 2 EP grease
Brake slack adjuster	Lubricate	3 months	NLGI No. 2 EP grease
Electrical connections	Add to void areas	3 months	Dielectric grease
Fifth wheel base and ramps	Lubricate	Weekly	Waterproof general purpose grease
Landing gear with grease fittings	Lubricate	6 months	NLGI No. 2 EP grease
Oil-type wheel bearings	Fill to full level	Daily	SAE 90 weight oil (API-GL-5)
Grease-type wheel bearings	Clean and repack	6 months	NLGI No. 2 EP wheel bearing grease
Sliding suspension: linkages of locking mechanism	Lubricate	3 months	SAE 20 weight oil
Door and ventilator hinge pins	Lubricate	3 months	SAE 20 weight oil
Door handle linkage	Lubricate	3 months	SAE 20 weight oil
Roll-up door rollers	Lubricate	3 months	SAE 20 weight oil
Door lock rod retaining points with grease fittings	Lubricate	3 months	NLGI No. 2 EP grease
Miscellaneous moving parts and linkages	Lubricate	3 months or less, check to be sure	SAE 20 weight oil

pressure-test the radiator cap at its rated pressure. Best results from these tests are obtained when the engine is cold.

- **Clean the Cooling System**. If inspection of the coolant reveals rust or cloudy appearance or if deposits have formed on the top header of the radiator, flush the cooling system. Also clean the cooling system if the truck history indicates a history of cooling system complaints. Follow the recommended cleaning procedure to remove rust, corrosion by-products, solder bloom, scale, and other deposits. Cleaning opens coolant passages and removes insulating scale and deposits that can cause overheating and possible engine block damage.
- **Antifreeze**. The fluid used as coolant in heavy-duty vehicles is a mixture of water and usually EG- or PG-based antifreeze coolant, though WECs are rapidly becoming accepted by fleets. The normal mixture used is a 50/50 solution of water and antifreeze/coolant. A 60 percent (EG or straight ELC to water) mixture (never higher) can be used for additional protection. PG can be used in extreme weather conditions. It provides maximum antifreeze protection at 100 percent, but as water is reduced in the solution, so is its ability to transfer heat. In engines using ELC, make sure you know whether premix ELC (should be topped up only with premix) or straight ELC (may be topped up with distilled water) is being used. Premix ELC does not need to be tested but it may require a recharge to extend its service life. Ensure that the antifreeze meets the engine manufacturer's specifications. High silicate antifreezes used with a supplemental coolant conditioner can cause a buildup of solids over time, resulting in plugging, loss of heat transfer, and water pump seal damage. The quality of water used in premixing antifreeze is important. Most engine manufacturers recommend distilled or deionized water.

Tech Tip

Use of distilled or deionized water mixed with EG, PG, or straight ELC can save thousands of dollars in repair costs over the life of a single truck. Tap water destroys engines and engine cooling systems. Better still, use premix-ELC or WEC for even greater overall savings.

Checking Coolant Condition

Engine coolant is checked in the following ways:

- **Hydrometers**. A **coolant hydrometer** measures the specific gravity of a sample by comparing its weight with the weight of water. A typical hydrometer contains a single graduated float. Because antifreeze is heavier than water, a high concentration will cause the float to ride high in the sample. To determine the freeze point, read the temperature graduation at the float water line. Hydrometer readings MUST be temperature corrected to obtain an accurate reading.

- **Refractometers**. **Refractometers** test the refractive index of coolant. They do this by measuring the angle at which light bends (refracts) when passing through a liquid droplet. The refractometer compares a known angle for pure water with the angle produced as light passes through a sample. Freeze point is noted on a scale viewed through an eyepiece. They are recommended because the reading is not temperature dependent.
- **WEC**. WEC does not have to be tested providing it is never mixed with anything but itself. The cooling system fill point should be labeled with a caution advising that it should be topped up only with WEC.
- **Litmus/Chemical Tests**. The basis for litmus tests is the chemical reactions affecting the color of a strip of test paper. Often, the sample must be mixed with plain water and/or a chemical reactant before the test strip is inserted into it. After the paper has been moistened, comparing its color with a chart indicates the acidity and/or additive strength of the solution. Litmus tests measure nitrate concentration and pH. Coolant conditioner protects metals by coating them with a chemical barrier against corrosion and cavitation. Nitrates are a component of that barrier. Measuring nitrate concentration, therefore, gives an indication of the conditioner charge level.
- **Laboratory Sample Analysis**. Laboratory analysis provides the most information about coolant at a moderate cost. For example, coolant may be analyzed for the freeze point, silicate concentration, nitrate level, and reserve alkalinity. The sample also may be tested for traces of metals such as iron, copper, aluminum, and solder (lead and tin). Laboratory testing gives a more thorough picture of the cooling system condition.
- **Supplemental Coolant Additive (SCA) or Diesel Coolant Additive (DCA)**. Use of an SCA or a DCA (term depends on manufacturer) is important to proper engine and cooling system protection. If the system has been cleaned, a precharge of SCA or DCA may have to be added per manufacturer's recommendations and coolant volume. WEC requires no SCA testing or addition during its service life so topping up should only be necessary if there has been system leakage. Perform the necessary adjustments to the coolant if the inhibitor levels are under or over the proper concentrations.

Disposal of Antifreeze

Antifreeze is biodegradable and is currently classified by the EPA as a substance of "unknown" toxicity. The EPA classifies used antifreeze as "toxic waste" due to the amount of heavy metals it can absorb from the engine cooling system. Lead from soldered radiator and bundle cores is the metal that poses the most environmental danger. There is legislation that regulates the disposal of used antifreeze. A number of aftermarket companies use antifreeze recycling equipment that permits used antifreeze to be cleaned and replenished, minimizing the disposal problem. The other alternative is to pay for the disposal of the antifreeze.

FIGURE 4–13 On-board air filter restriction gauge

Air Intake System

Check the air filter restriction gauge (**Figure 4–13**) to ensure it is within specification. Most fleets prefer air filters to be changed on either a calendar time period or when the restriction value is exceeded: remember that a filter is functioning at its peak efficiency just before it fails so do not be over-eager to change out air filters out of schedule.

AIR BRAKE SYSTEM

The air brake system requires special attention during cold weather. Contaminants such as road dirt, salt, oil, and so on should be removed. The air system must be protected against freezing. Drain the air tanks daily and inspect the system to determine if any repairs are necessary. The inspection should include the following:

- Remove and clean the air dryer purge line and air governor signal line.
- Check the vent on the exhaust port of the air governor.
- Check air dryer heater operation; a 4- to 5-amp draw with the element cold is normal.

There are several methods of moisture and freeze-up protection used in an air brake system: alcohol injectors, which release vaporized methyl alcohol into the system; both automatic and manual moisture ejectors mounted at the reservoir; and air dryers mounted between the compressor and air reservoir.

Avoid dumping methyl alcohols into air systems as a matter of routine.

STARTING AIDS

There are different types of starting aids available. Starting aids operation should be checked each fall.

1. Check the condition of the block heater with an ohmmeter. A 10- to 15-ohm element resistance is normal for 1,000- to 1,500-watt heaters. Check the condition of the electrical cord and plug. Check for coolant leakage at the mounting points.
2. Check the fuel heater mounting, condition of hoses, and hose routing. Heaters with manual shutoff valves should be turned on before cold weather to ensure proper operation.
3. Check the water separator mounting, condition of hoses, and hose routing. Drain any collected water.
4. Check the ether starting system by removing the cylinder and activating the valve. Weigh the cylinder for contents. Replace it if it is low.
5. For oil pan heaters and battery warmers, check the element resistance with an ohmmeter. (Oil pan heaters have a normal resistance of about 50 ohms, battery warmers 75 to 125 ohms.) Check the condition of the electrical cord and plug.
6. Fuel-fired heaters and auxiliary power units (APUs) should be tested. Run the unit through at least two thermostatically controlled cycles. Check hoses and wiring for proper routing. Test the ignitor if required. Consult the manufacturer's recommendations for any other tests.

ETHER STARTING SYSTEMS

Diesel fuel will not ignite in an engine until it reaches its ignition temperature—typically about 550°F (288°C). Using ether-based starting fluid aids helps start diesel engines because it ignites at about 350°F (177°C) and pilots the ignition of diesel fuel in the cylinder.

Typically, starting fluid systems vary from the handheld aerosol can (total operator control) to "automatic" and electronic systems that actuate with the engine cranking (no operator control).

Dispensing Ether

A spray aerosol dispenses about 12 cubic centimeters of starting fluid per second. An operator typically delivers at least 2 to 3 seconds of spray or approximately six times the amount of ether required. The resulting explosion in the engine cylinder can damage pistons and rings. A measured shot or metering flow valve limits the amount of ether injected per activation and is recommended by most diesel engine manufacturers. Measured shot valves are made in three or four shot sizes to match engine displacements. This valve can be remotely operated by a switch on the dashboard. A thermal switch can be wired in series with the ether valve to prevent ether from being injected into a hot engine.

In operation, when the valve is energized (pushing the button), ether is injected into the measured shot chamber. De-energizing the valve (releasing the button) dispenses the measured shot of ether into the engine intake through the atomizer. Push-pull cable operating systems that operate in the same manner are available; pulling the cable charges the shot chamber and pushing the cable dispenses the ether to the engine.

Automatic Cold Start

Automatic ether systems are wired into the cranking circuit and dispense starting fluid when the cranking circuit is energized (**Figure 4–14**). Flow of the ether is controlled by a valve orifice or rapid cycling of the valve, resulting in continuous atomizer flow. An engine temperature switch prevents operation when the engine is warm. The atomizer or nozzle directs the flow of starting fluid into the airstream and determines the rate at which it is injected. The recommended location for the atomizer is in the intake air system, upstream of the inlet of the intake manifold. The direction of spray should be toward the incoming

FIGURE 4–14 Typical ether injection kit

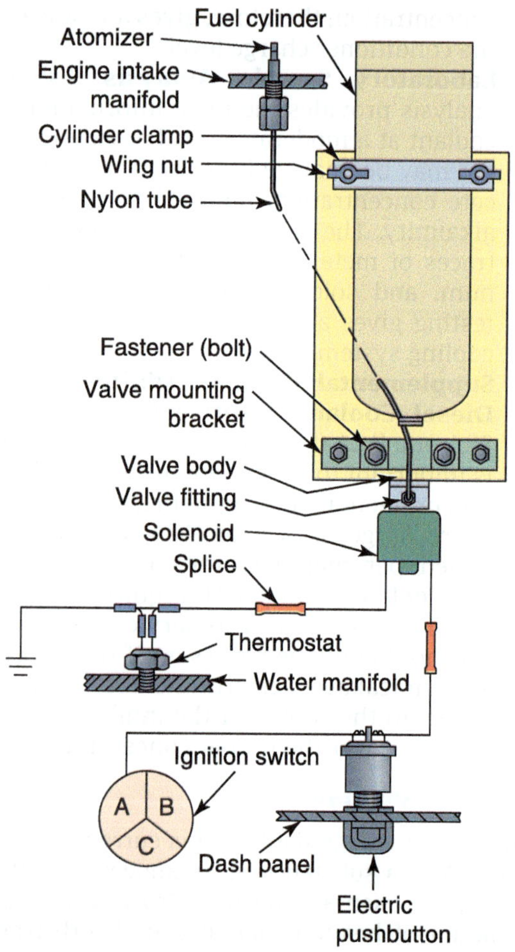

FIGURE 4–15 Electronic ether system control system, networked to chassis data bus

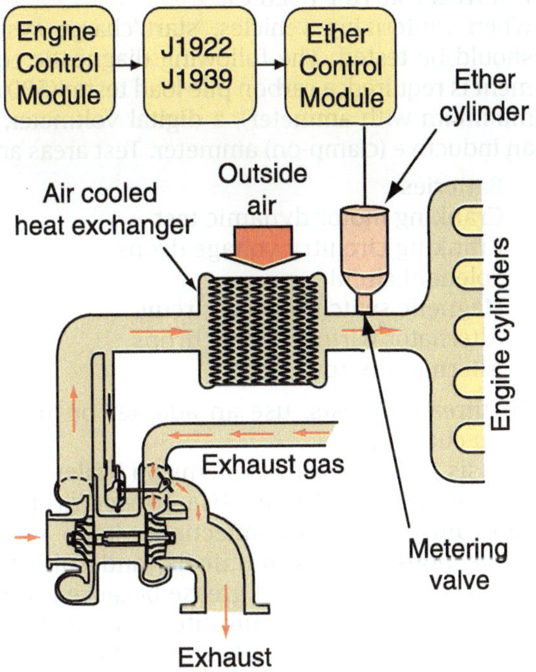

airstream to maximize mixing with the air. The valves and atomizers are connected with nylon or metal tubing of ⅛-inch diameter.

An electronic system operates in the same manner as the automatic, except the ether is controlled, measured, and metered by the electronic control module, which is networked to the chassis data bus (**Figure 4–15**). The module collects engine temperature data and if the engine starter is engaged, it supplies ether as needed. Most alert the operator when the fluid is low.

WARNING

Avoid using ether with glow plug-equipped engines. On engines with an air preheater, consult the engine owner's manual.

With automatic and electronic systems, the atomizer container must be checked by the technician on a regular basis to be sure that there is always ether in it.

CAB COMFORTS

To keep the driver happy, check the following:

1. Clean out the heater screens and check for any heater core obstructions.
2. Check heater fan operation using an ammeter to check for current draw. It should be no more than the rated value (see the service literature).
3. Check bunk heater operation and switch condition.
4. Check heater controls for proper function. Operate in the defrost and floor heat modes. Cab heaters with manual controls should be turned on before cold weather to ensure proper operation.
5. Check the cab floor grommets for holes and for sealing capability (especially around the clutch, throttle, wiring, and hoses).
6. On cab-over-engine (COE) vehicles, make sure that the transmission tower seals with the cab down. Replace if it is not sealing properly against the outside air.
7. Check the weather stripping around the doors and door adjustment.

ELECTRICAL SYSTEM

Night driving during winter months can overload a heavy-duty truck's batteries and electric system (**Table 4–8**). The reserve capacity rating (RCR) is important if long-term parking with electrical loads is required. A battery at 80°F (27°C) has 100 percent of its power. That same battery at 14°F (−10°C) loses 50 percent of its available power. As battery available power is reduced with low temperature,

TABLE 4–8 Tractor/Trailer Power Usage*

Device	Amps	Hours	Amp Hours
Key on	2.0	10.0	20.0
Starter	1,500.0	0.01	15.0
Wipers	7.5	10.0	75.0
Headlights	8.0	10.0	80.0
Panel lights	3.0	10.0	30.0
Taillights	1.5	10.0	15.0
Marker lights	12.0	10.0	120.0
Brake lights	8.0	0.25	2.0
Turn signals	5.0	0.15	0.75
Engine electronics	6.25	10.0	62.5
Heater fan	15.0	10.0	150.0
Heated mirrors	5.0	10.0	150.0
CB radio	1.5	10.0	15.0
AM/FM radio	1.0	10.0	10.0
Radar detector	0.5	10.0	5.0
Subtotal			650.25
+ 25 percent safety factor			162.56
TOTAL			812.81

*Calculated for 10-hour shift on a winter night.

the mechanical effort required to turn the engine over increases. An engine at −20°F (−29°C) is over three times as difficult to crank as an engine at 80°F (27°C).

Battery Heaters

Battery heaters can help solve the problem. The most common battery heaters are the blanket type and the plate type. The plate-type battery heater usually is placed under the battery, but it can also be placed between the batteries. The blanket type of heater comes in a variety of lengths and wattages and incorporates its own insulation. It is best to heat batteries slowly. A well-insulated battery box can help batteries maintain heat.

Electrical System Checklist

To minimize battery and electric system problems, prepare them for winter as follows:

1. Look for the accumulation of dirt or corrosion on top of the battery, corroded terminals and cables, broken or loose terminal posts, and containers or covers that are broken or cracked. All affect the battery's ability to perform. Also check the condition and tension of the alternator belts. Inspect all wiring and cables for frays and breaks in insulation.
2. Avoid damaging a battery when installing or removing it. Truck batteries are heavy and their lightweight plastic casings make them vulnerable to handling abuse.
3. Check the voltage regulator setting at every PM inspection. Overcharging is a leading cause of battery and electrical component failures. For trucks operating linehaul, the voltage setting should be 13.6 to 14.2 volts using flooded cell batteries. For vehicles involved in pickup and delivery, set voltage should be no more than 14.5 volts. An additional benefit besides longer battery life is increased lamp life—high chassis voltage settings damage electrical components.
4. On conventional batteries, maintain the electrolyte level above the plates and below the bottom of the vent well. Never allow the electrolyte level to drop below the top of the plates. Low electrolyte level can cause reduced battery capacity. If water must be added, use distilled water. Overfilling can cause acid spewing and corrosion.
5. Ensure that battery hold-downs are properly attached to keep the batteries from bouncing. Vibration damages batteries. Vibration also breaks alternators and brackets and elongates mounting holes. Alternators should be mounted using grade 5 or grade 8 fasteners.
6. Keep all terminals and connections bright and tight. Use a water-resistant grease or lubricant such as petroleum jelly to seal out moisture on connections that are not totally encapsulated.
7. Keep batteries from freezing by maintaining a full charge. This includes parked vehicles and batteries on the shelf. A completely discharged battery will freeze at 18°F (−28°C).
8. When winterizing vehicles, start/charge systems should be tested. The following diagnostic equipment is required: a carbon pile load tester (500 amp minimum with ammeter), a digital voltmeter, and an inductive (clamp-on) ammeter. Test areas are:
   - Batteries
   - Cranking motor dynamic test
   - Cranking circuitry voltage drops
   - Solenoid circuit
   - Magnetic switch control circuit
   - Alternator wiring voltage drops
   - Alternator output test
9. On threaded posts, use an adapter on the post when charging.
10. Chassis ground wiring systems on trailers should also be bright and tight. Use an electrical grade lubricant on pigtail connections, lamp sockets, connections, battery terminals, and splices. Do not use a sodium-based grease because it causes corrosion. For longer bulb life, clean lenses and housings frequently and keep the chassis voltage regulated below 14 volts.
11. All light bulbs should illuminate brightly. Locating the cause of a dull or failed bulb is usually easier during a PM inspection than outside in the middle of winter.

DIESEL FUEL

In preparing for cold weather operations, fuel plays an important role. There are three areas relating to fuel that should be considered. They are fuel storage, tank configuration, and chemical properties of the fuel. The following factors are critical when dealing with the winterization of fuel and fuel systems.

- **Cloud Point**. The temperature at which a haze of wax crystals is formed. Cloud point is more significant than pour point when determining whether a fuel can be pumped through filters in cold weather. Cloud point has not been shown to be significantly lowered by the additives used in general purpose fuel conditioners: the problem is most noticeable when a summer fuel is run in winter temperature conditions.
- **Pour Point**. This is the lowest temperature at which a diesel fuel will flow. This temperature is lower than the cloud point and is easier to modify or lower using additives than cloud point. Long before the temperature has dropped to the pour point of a fuel, the engine has ceased to run.
- **Fuel Storage**. Storage tanks can form condensation. Water found in tanks must be removed. Water removal filters should be installed on

fuel-dispensing pumps. Microbiological contamination can also be a problem. These living organisms produce slime-like metabolic waste that can plug fuel filters. Fill areas should be raised to eliminate the possibility of water, snow, or debris entering fill and vent ports. Vents should be checked to ensure that only air can enter them.

- **Tank Configuration**. Vehicle fuel tanks should be checked for water contamination, and any water or debris should be removed if found. Crossover lines should be checked for both correct size and low spots that can collect water. Any exposed fuel lines can be subject to freeze-up during severely cold temperatures.
- **Fuel**. Although winter diesel fuel is prepared by refineries on a regional basis, it should be recognized that all fuel is not the same. ASTM specifications for winter fuel are based on a 10 percent mean low temperature for the specific month it is sold. Fuel refiners, therefore, adjust the ignition temperature of fuel (cetane number [CN]) and pour point inhibitors seasonally. The use of aftermarket fuel additives by the user is not generally recommended.

Some operators use fuel heaters. The two most practical heat sources for heating fuel during highway operation are electricity and engine coolant. The problems caused when wax or ice crystals form at the fuel tank upper or lower outlet can be solved by an in-tank heater. The fuel temperature should be regulated below 80°F (27°C) and never exceed 100°F (38°C).

FUEL/WATER SEPARATORS

There are many types and sizes of fuel/water separators available on the market. To select the proper unit, it is important to know the requirements in terms of fuel flow rate. There are three types: fine mesh filters, coalescer screens, and centrifugal fuel/water separators.

When using fuel/water separators during cold weather operation, the following must be considered:

1. Mount the fuel/water separator in a protected area. Some fleets wrap them with insulation for cold weather operation.
2. If a diesel fuel warmer is used in conjunction with a fuel/water separator, mount the separator downstream from the fuel warmer so it receives warm fuel.
3. Do not use 90-degree elbow fittings; these may freeze in cold weather.
4. Drain the fuel/water separator frequently so water does not accumulate in the separator bowl. This water can freeze in operation, especially when exposed to ram air wind chills. Ice crystals can plug a fuel filter in the same way wax crystals can.

GENERAL WINTERIZING TIPS

The following are some other checks while winterizing a vehicle:

1. Check all engine and chassis hoses (fuel, oil, and air) for wear and deterioration. Replace as necessary.
2. Check the operation of the windshield washers and ensure that the wipers remove solution without streaking.
3. Check the windshield washer reservoir for the proper type and amount of solution.
4. Check the exhaust system for leakage, for example, soot around connections.
5. Clean and grease the electrical connections to the headlights, taillights, light cord receptacle, and clearance lights.
6. Check the fan blade and fan belt condition and tension.
7. Check the fan clutch for play with the fan clutch released. Play should not be more than 3/16 inch maximum at the fan blade tip.
8. Lubricate the door latches with a Teflon™ lubricant.
9. Lubricate the door locks with dry graphite or silicone lubricant.
10. Inspect the condition of the tires and wheels. Replace any damaged or worn tires. It may be fleet policy to install lug-type treads on drive wheels to enhance traction on snow and ice. Inflate all tires to proper pressures. Check the wheels and rims for bends and cracks and at the lug nuts for signs of looseness or breakage. Repair any problems.
11. Make sure that the foundation brakes and the brake cams and slack adjusters are properly lubed. Check the brake shoes, linings, and drums for wear, cracks, and breakage. Check the gladhands and grommets. Ensure that the air tanks are securely mounted and that the air compressor and air dryer are performing properly. All necessary repairs should be made before winter sets in.
12. Check that the fire extinguishers are charged, that safety triangles are available, and that tire chains are in good condition.
13. Run the truck outside until it reaches operating temperature. Check the operation of the cooling fan and cab heater operation.
14. Road test the vehicle and check for leaks (coolant, oil, and air).

4.6 PERFORMING A LUBE JOB

A lube job can be known as either an A- or L-type inspection. A lube job is the simplest and most frequently performed service procedure. For that reason, it is a task frequently assigned entry-level technicians.

If an aspiring technician cannot perform a lube job effectively, that person should probably seek an alternative career. It consists of applying grease to every **zerk fitting** (grease nipple) on the chassis, checking fluid levels, checking lights, and checking brake adjustment. The smart technician will be giving the chassis a thorough visual and physical inspection while performing this service and recording anything else that might require repair. All the procedures of this basic inspection are repeated in the higher-level inspections—B, C, and so on.

GREASING THE CHASSIS

Most shops use pneumatic grease guns. These enable a chassis to be lubricated at a faster speed than would be possible with a hand-actuated grease gun. However, you should remember that in the event a zerk does not take grease, a hand-actuated grease gun develops higher pressure than an air grease gun and may open a partially seized nipple. The following are some tips that may help you perform this job to the standard required by most service facilities:

- Pump grease into zerks only until grease begins to exit the seal. Do not make a mess by overgreasing.
- Steering knuckle kingpins: The design of some kingpins requires that you jack the front axle up, relieving the weight from the steering knuckle before attempting to pump grease into the assembly.

Tech Tip

When preparing to perform a lube job, place a few zerk fittings in your pocket before beginning. Zerk (spring-loaded ball) seals may seize and prevent grease from entering; replace a failed zerk fitting with a new one.

- When greasing U-joint assemblies, grease should exit each of the four journal seals. If this does not happen, you may have to loosen the journal cap. This is important on any U-joint not "greased for life." Failure to do this will result in premature failure of the U-joint assembly.
- When greasing driveshaft slip splines, place a finger over the breather hole and stop pumping grease the moment you feel grease attempt to exit.
- You may have to take the weight off the spring pins before they will accept grease. Do this by jacking the frame up evenly. In some cases, heat may have to be applied, but when this is required the pin and/or bushing are probably already damaged.

CAUTION

Some spring hangers are manufactured of cast aluminum alloy. Never apply heat anywhere near cast aluminum suspension components.

- Do not overgrease fifth wheels. If the grease on the fifth wheel plate is dirty or oxidized, scrape it away with a wooden board and apply new grease. If you are servicing a fifth wheel while it is coupled to the trailer, relieve the weight from the pivot bushings before attempting to grease them.
- If the chassis is equipped with an automatic lube system, check that it is actually getting the oil to each component and replenish the reservoir level with the appropriate lube.

CAUTION

Remember that grease is both difficult to remove and may permanently stain surfaces: There is probably nothing that will enrage a driver more quickly than grease on any part of a truck chassis other than where it is supposed to be. Work cleanly! Use floor mats and change coveralls before driving a vehicle after a lube job.

Tech Tip

Remember that a hand-actuated grease gun actually develops higher pressure than an air grease gun and may open a zerk nipple that a pneumatic gun failed to open.

CHECKING FLUIDS

Make sure you check the chassis fluid levels properly. They may be correct, but if you skip checking a level and there is a subsequent lube-related failure, your employer will be paying for the repair, which could result in some inconvenience for you.

- Engine coolant: Check the fluid level, test additive protection, and record antifreeze protection with a refractometer. Before adding coolant, make sure you know what type it is: EG, PG, premixed ELC, and WEC should never be mixed with each other. Premixed solution-type ELC should only be replenished with ELC premix.
- Engine oil: Check the level using the dipstick. If this has to be topped up, make sure you correctly identify the oil in the engine first: this may be recorded electronically in the vehicle maintenance log so check using an EST. Never overfill.

FIGURE 4–16 Correct method of checking lube oil in transmissions and differential carriers with no sight glass or electronic gauge: Oil must be level with bottom of fill hole as shown.

- Transmission oil/fluid: To check the automatic transmission fluid level, check to see whether this should be performed cold or at operating temperature. When checking fluid level in a standard transmission, the oil level should be exactly level with the bottom of the oil fill hole, as shown in **Figure 4–16**. If you have to top up, ensure that you use the correct oil, remembering that standard transmissions may use either petroleum- or synthetic-based gear lubes or engine oils.
- Differential carrier: Check the oil level in exactly the same manner as with the standard transmission as shown in Figure 4–15. If you have to top up, ensure that you use the specified gear oil. Again, check to see if the lubricant is synthetic or petroleum based.
- Wheel ends: The wheel hub oil level should be level with the high level index on the hub inspection sight glass. Check that the specified lubricant is used when topping up.

CAUTION

Be wary of dents in disposable filter canisters. Dents create stress risers in the canister that can easily develop into cracks. Experience has shown that a filter is most likely to be damaged during installation. If this happens, replace the filter. The cost of a filter is small compared to the consequences of an over-the-road failure of a filter.

BRAKE STROKE AND LIGHTING INSPECTION

You will find detailed information elsewhere in this textbook that explains the procedure for checking brake stroke and lighting. It does not take much to identify a failed lamp, but repairing the circuit that supplies it requires a higher level of skills that are covered in the electrical chapters of this book. Similarly, anyone working on trucks should know how to identify an out-of-adjustment brake stroke. Remember that the CVSA overstroke limits shown earlier in this chapter are not the required inspection performance standards. Correcting an overstroke condition is covered in greater detail in dedicated chapters later in this textbook so make sure you reference Chapter 31.

CAUTION

When an automatic slack adjuster overstrokes, the TMC recommends not to manually adjust it. The cause is usually a defect in the adjusting mechanism. The recommendation is that the cause of the overstroke be identified; that often requires the replacement of the slack adjuster.

4.7 FLEET MANAGEMENT SOFTWARE

Today there are dozens of fleet management software packages some of which are identified earlier in this chapter. They are designed for the owner-operator who runs a single truck, for small fleets, and for the largest fleets running our highways. In fact, many of the largest fleets use software packages that have been specifically created for their operation. Fleet management software is often designed to interface with other software packages, especially when specifically designed for a major fleet. This allows vehicle data buses to transfer chassis data to the management software for analysis. Some of the features of a comprehensive fleet management software package are as follows:

- Accident tracking
- Accounting integration
- Audit trail
- Bill of lading generation
- Billing and invoicing
- Costs prediction
- Dispatch
- Driver compliance management
- Driver data analysis
- Driver logs
- Expense tracking
- Freight ratings
- Fuel optimization
- Inventory management
- Parts ordering
- PM scheduling
- Payroll management
- Maintenance costs

- Mileage tracking
- Scheduling
- Secure against hacking
- Tire tracking
- Trip reporting

The best fleet management software packages usually interface with the AME software described earlier in this chapter. The objective is to produce key decisions that are not muddled by conflicting human decisions sourced from multiple departments of an operation.

4.8 MAINTAINING EMISSIONS CONTROL EQUIPMENT

Any maintenance program today must take into account the emissions control devices on the vehicle. For this reason, part of every maintenance procedure involves connecting an EST to the chassis data bus and checking for emissions related diagnostic fault codes (DTCs). This step will become even more important when heavy-duty, on-board diagnostics (HD-OBD) finally become enacted: at the time of writing this is scheduled for **model year (MY)** 2015, though this could be delayed again.

DIESEL PARTICULATE FILTER

The diesel particulate filter (DPF) should be checked for status codes and a stationary regeneration should be performed if required. Almost all highway diesel engines of MY 2007 and later are equipped with DPFs. It is not advisable to perform a DPF regeneration using a shop exhaust extraction system, so move the truck outside and away from any flammable materials. Never perform a DPF regeneration close to a fuel station, near agricultural grain chaff, or any other flammable substances.

SELECTIVE CATALYTIC REDUCTION

Almost all highway diesel engines from MY 2010 and later are equipped with selective catalytic reduction systems (SCR) so any maintenance schedule requires checking for DTCs and the level of diesel exhaust fluid (DEF). Outside of North America, DEF is known as AdBlue: regardless of name, it is aqueous urea composed of 2.5 percent urea and 7.5 percent distilled water. SCR operation is described in more detail in the companion textbook to his one,

DEF can be replenished using 1-gallon (4-liter) jugs available truck service facilities or by using a DEF refill station such as that shown in **Figure 4–17**. Failure to maintain DEF levels can result in crippling a vehicle on the road which can be costly.

FIGURE 4–17 DEF refill station

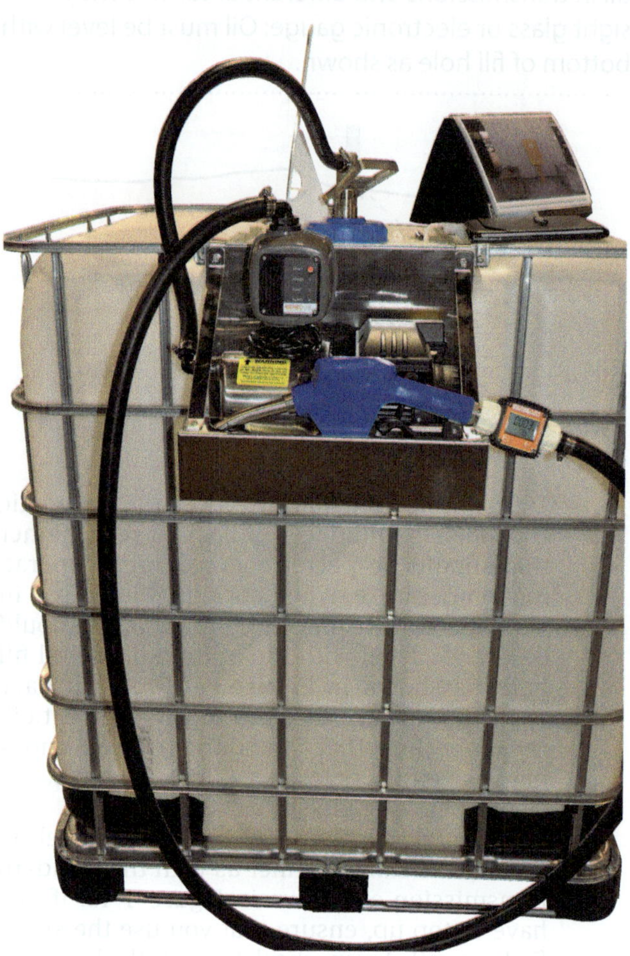

CLEAN IDLE CERTIFICATION

Clean idle certification is mandatory in some jurisdictions: the sticker indicates that idle shutdown mode has been programmed into a protected field. Use an EST to access the chassis data bus and check the customer data programming fields to verify that the idle shutdown programming is enabled. **Figure 4–18** shows an idle shutdown sticker that is displayed on the truck/bus in jurisdictions that require it.

FIGURE 4–18 Certified clean air idle certificate: mandatory in some jurisdictions

SUMMARY

- A proper maintenance program involves the inspection and servicing of the vehicle as a whole. The program's success depends on careful planning, often using predictive maintenance software.
- CBM is reducing the amount of time spent on PM in the modern truck shop. The objective is to achieve an AME in which maintenance technology becomes more of an exact science.
- A good maintenance program should benchmark OEM maintenance standards tailored to the requirements of each individual operation.
- The driver is always a key identifier of maintenance shortfalls when performing pretrip and post-trip inspections.
- The term *out-of-service* (OOS) when used in-house by a fleet defines minimum acceptable maintenance standards.
- The term *out-of-service* (OOS) when used by the CVSA is NOT a maintenance standard. CVSA OOS criteria define legal minimum safety standards. Falling short of these standards can result in a citation.
- A vehicle considered likely to cause an accident or breakdown because of mechanical conditions or improper loading should be written up by the technician or driver as OOS.
- It is the fleet's responsibility to ensure that all personnel performing annual inspections are qualified.
- Maintenance managers and truck technicians are expected to be computer literate because most service tracking is computerized.
- Proper lubrication is important in reducing wear and preventing premature failure of truck components.
- Winterizing a heavy-duty truck is an important part of any PM program.
- Performing a lube job properly is a requirement of an entry-level truck technician from day 1 on the job.

REVIEW QUESTIONS

1. Pretrip and post-trip inspections are a primary responsibility of the:
   a. technician.
   b. maintenance manager.
   c. driver.
   d. vehicle inspector.
2. Which of the following helps reduce the amount of PM without compromising equipment safety?
   a. CBM software
   b. predictive maintenance software
   c. AME
   d. all of the above
3. Which body sets the OOS standards that can legally deadline a vehicle?
   a. ATA
   b. CVSA
   c. TMC
   d. PMI
4. Which type of PM service on a vehicle requires the most hours to perform?
   a. category A
   b. category B
   c. category C
   d. category L
5. When carrying out an inspection of a trailer after the first 500 miles, Technician A checks brake stroke travel. When inspecting trailers, Technician B checks all areas where sealant is used and reseals as necessary. Who is correct?
   a. Technician A only
   b. Technician B only
   c. both A and B
   d. neither A nor B
6. How many separate inspections are required for a tractor-double combination with a converter dolly?
   a. two
   b. three
   c. four
   d. five
7. Which of the following is affected by an oil's viscosity?
   a. ability to flow
   b. lubricity
   c. oxidation stability
   d. resistance to sulfur contamination
8. Which engine oil additive neutralizes harmful acids?
   a. oxidation inhibitors
   b. antirust agents
   c. detergent additives
   d. TBN additives
9. Which of the following grades of grease can be used in a grease gun?
   a. 000
   b. 00
   c. 0
   d. none of the above
10. What damage can be caused by overgreasing?
    a. channeling
    b. blowing out seals
    c. scoring
    d. grit contamination
11. SAE 20-weight oil is used to lubricate what part of a trailer every 3 months?
    a. brake camshaft support bearings
    b. fifth wheel base and ramps
    c. brake slack adjuster
    d. roll up door rollers

12. In winterizing a truck, which of the following should be done first?
    a. Check coolant pH.
    b. Visually inspect the cooling system.
    c. Add antifreeze.
    d. Test the antifreeze protection of the coolant.
13. Which of the following would help reduce power loss in batteries during winter driving?
    a. battery heaters
    b. overcharging
    c. slight overfilling with electrolyte
    d. deep cycling
14. In winterizing a vehicle, Technician A uses a Teflon lubricant on the door latches and a dry graphite lubricant on the door locks. Technician B mounts the fuel/water separator downstream from the fuel heater. Who is correct?
    a. Technician A only
    b. Technician B only
    c. both A and B
    d. neither A nor B
15. Which of the following is good practice when greasing the kingpin steering knuckle on a heavy-duty truck?
    a. Heat with a rosebud.
    b. Jack the front axle off the ground.
    c. Remove the zerk fittings.
    d. Flush all the old grease out of the knuckle.
16. When checking the oil level in a standard transmission with no sight glass, where should the oil level be?
    a. above the fill hole level
    b. exactly even with the fill hole
    c. one finger joint below the fill hole
    d. within reach of a finger inserted into the fill hole
17. When checking the oil level in a differential carrier, where should the oil level be?
    a. Above the fill hole level.
    b. Exactly even with the fill hole.
    c. One finger joint below the fill hole.
    d. Within reach of a finger inserted into the fill hole.
18. Which of the following should be done first when the zerk fitting on a suspension spring pin does not accept grease?
    a. Replace the zerk fitting.
    b. Lift the chassis weight off the pin.
    c. Heat the shackle with a rosebud.
    d. Depress the zerk nipple ball seal with a screwdriver.
19. What is the recommended tool used to check the antifreeze protection of coolant?
    a. litmus paper
    b. a battery hydrometer
    c. spectrographic analysis
    d. a refractometer
20. When greasing a U-joint on a driveshaft, grease only exits from three of the four journal seals on the trunnion. What should you do first?
    a. Leave it; three of four journals are good enough.
    b. Loosen the journal cap and try again.
    c. Replace the U-joint.
    d. Heat the trunnion with a rosebud.

5

Prerequisite: Chapters 1, 2, 3, and 4

FUNDAMENTALS OF ELECTRICITY

OBJECTIVES

After reading this chapter, you should be able to:

- Define the terms *electricity* and *electronics.*
- Describe atomic structure.
- Outline how some of the chemical and electrical properties of atoms are defined by the number of electrons in their outer shells.
- Outline the properties of conductors, insulators, and semiconductors.
- Describe the characteristics of *static electricity.*
- Define what is meant by the *conventional* and *electron theories* of current flow.
- Describe the characteristics of magnetism and the relationship between electricity and magnetism.
- Describe how electromagnetic field strength is measured in common electromagnetic devices.
- Define what is meant by an electrical circuit and the terms *voltage*, *resistance*, and *current flow.*
- Outline the components required to construct a typical electrical circuit.
- Perform electrical circuit calculations using Ohm's Law.
- Identify the characteristics of DC and AC.
- Describe some methods of generating a current flow in an electrical circuit.
- Describe and apply Kirchhoff's First and Second Laws.

KEY TERMS

ampere
anode
armature
capacitor
cathode
circuit breakers
closed circuit
conductor
conventional theory
current flow
dielectric
electricity
electromotive force (EMF)
electronics
electron theory
electron
ground
hotel loads
insulator
ion
Kirchhoff's Laws
magnetomotive force (mmf)
negative temperature coefficient (NTC)
ohm
Ohm's Law
open circuit
parallel circuit
positive temperature coefficient (PTC)
resistance
semiconductors
series circuit
series-parallel circuits
shorepower
short circuit
solenoid
static electricity
step-down transformers
step-up transformer
thermocouple
transformer
valence
voltage
watt

INTRODUCTION

This chapter introduces basic electrical principles. These principles are applied in later chapters that deal with chassis electrical components. Any vehicle technician must have a good understanding of basic electricity, because almost every subcomponent on a modern truck is managed and monitored by electrical and electronic devices.

By definition, electricity is a form of energy that results from charged particles such as **electrons** and protons. These electrically charged particles can be static (at rest), when they would be described as accumulated charge. They may also be dynamic or moving, such as **current flow** in an electrical circuit.

ORIGINS

All matter (anything that has mass and occupies space) has some electrical properties. However, only in the last couple of hundred years have humans been able to make electricity work for them. The first type of electricity to be identified was **static electricity**. Over 2,000 years ago, the Greeks observed that amber rubbed with fur would attract lightweight objects such as feathers. The Greek word for *amber* (a translucent, yellowish resin that comes from fossilized trees) is "electron," from which we get the word **electricity**. Probably the next significant step forward occurred toward the end of the sixteenth century when the English physicist William Gilbert (1544–1603) made the connection between electricity and magnetism. More than 100 years later, Benjamin Franklin (1706–1790) proved the electrical nature of thunderstorms in his famous kite experiment. He coined the terms *positive* and *negative* and established the conventional theory of current flow in a circuit. The **conventional theory** of current flow says that current flows from positive to negative in an electrical circuit. From this point in time, progress accelerated. In 1767, Joseph Priestley established that electrical charges attract each other. He saw that this attraction of opposites decreased as the distance between the forces increased. Around 1800, Alessandro Volta invented the first battery. A battery converts chemical energy into electrical potential. Electrical potential means electrical energy. Volta's battery was the forerunner of the modern vehicle battery.

Michael Faraday (1791–1867) opened the doors of the science we now know as electromagnetism. His law of induction states that a magnetic field induces (produces) an **electromotive force (EMF)**, electrical potential or **voltage** in a moving **conductor**. A conductor is something that conducts or transmits heat or electricity. Vehicle alternators and cranking motors each use electromagnetic induction principles. Thomas Edison (1847–1931) invented the incandescent lamp (the electric light bulb) in 1879, but perhaps even more important he built the first central power station and distribution system in New York City in 1881. New York City's power grid provided a means of introducing electrical power into industry and the home.

ELECTRONICS

Electronics is a branch of electricity. Electronics deals with the flow of electrons through solids, liquids, and gases and across vacuums. An introduction to electronics theory is provided in Chapter 6. Many electrical components in modern vehicles, such as alternators, have electronic subcircuits. Increasingly, chassis systems are managed by electronic components. It should be noted that terms like *electronics* and *electronic engine management* are often used in vehicle technology to describe any system that is managed by a computer.

Many types of vehicle computers exist. Managing a function such as the pulse wiper mode on windshield wipers would be a simple use of a vehicle computer. A more complex application of a vehicle computer is monitoring and controlling the multiple subsystems of engines and transmissions. Because of the ever-increasing use of computers on modern trucks, developing an understanding of electricity and electronics is essential.

This understanding begins with learning about electrical and electronics theory, which form the building blocks of mastering practical skills. The practical skills are required to diagnose and repair electrical and electronic systems and circuits. It is important to understand the basic theory of electricity and electronics before advancing to the chapters that address applications of this theory.

The days when many truck technicians could avoid working on an electrical circuit through an entire career are long past. A good course of electrical study will always begin with the study of atomic structure. Understanding a little about the behavior of atoms and electrons can help students visualize some of the elements of "invisible" electricity.

5.1 ATOMIC STRUCTURE AND ELECTRON MOVEMENT

An atom is the smallest particle of a chemical element that can take part in a chemical reaction. An atom is usually made up of protons, neutrons, and electrons. Protons and neutrons form the center or nucleus of each atom, whereas the electrons orbit the nucleus in a manner similar to that of planets orbiting the sun in our solar system. The simplest atom is hydrogen, which has a single electron orbiting a nucleus consisting of a single proton; it has no neutron. **Figure 5–1** shows how an atom of hydrogen is represented in diagram form.

Electrons orbit the nucleus, the center point of an atom, in a concentric ring known as a *shell*. Figure 5–1 is not unlike a picture of a solar system, with the nucleus representing the sun and the orbiting electron representing a single planet. Electrons have a negative electrical charge. Protons and neutrons are clustered into and form the nuclei of all atoms. The electrons orbit around the nucleus in their shells. Protons have

FIGURE 5–1 Hydrogen atom

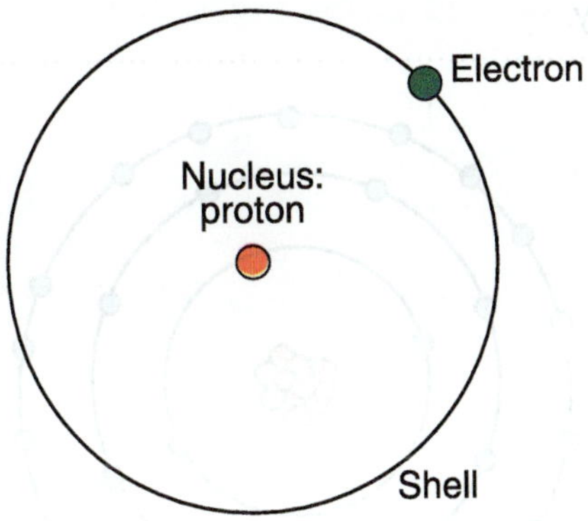

a positive electrical charge, whereas neutrons have no electrical charge. The nucleus of an atom makes up 99.9 percent of its mass. This means that most of the weight of any atom is concentrated in its nucleus. The number of protons in the nucleus is the atomic number of any given element. The sum of the neutrons and protons is the atomic mass number. In a balanced or electrically neutral atom, the nucleus is surrounded by as many electrons as there are protons.

ELECTRONS AND PROTONS

All electrons are alike. All protons are alike. The number of protons associated with the nucleus of an atom identifies it as a specific element. Electrons have a tiny fraction of the mass of a proton. Under normal conditions, electrons are bound—that is, held in their orbital shells—to the positively charged nuclei of atoms by the attraction between opposite electrical charges.

Any atom may possess more or fewer electrons than protons. An atom with an excess or deficit of electrons retains the character of the element. Such an atom is described as *negatively* (an excess of electrons) or *positively* (a net deficit of electrons) charged and is known as an **ion**. For instance, a copper atom with a shortage or deficit of electrons could be called a positive copper ion, meaning that it would be inclined to steal electrons from other substances.

ATOMIC SHELLS

Each shell within the structure of an atom is an orbital path. The concentric shells of an atom proceed outward from the nucleus of an atom. The electrons in the shells closest to the nucleus of an atom are held most tightly. Those in the outermost shell are held more loosely and tend to be more inclined to move. As we have seen, the simplest element, *hydrogen*, has a single shell containing one electron. The most complex atoms may have seven shells. The maximum number of electrons that can occupy shells 1 through 7 are in sequence, progressing away from the nucleus: 2 (closest to the nucleus), 8, 18, 32, 50, 72, 98 (furthest from the nucleus). The heaviest elements in their normal states have only the first four shells fully occupied with electrons. The outer three shells are only partially occupied. The outermost shell in any atom is known as its **valence** shell. The number of electrons in the valence shell will dictate some basic electrical and physical characteristics of an element.

Chemical Properties of Atoms

The chemical properties of all the elements are defined by how their atoms' shells are occupied with electrons. Atomic elements with similar orbital structures will probably have other similarities in the way they behave. For instance, an atom of the element helium with an atomic number of 2 has a full inner shell, which also happens to be its valence or outermost shell. An atom of the element neon with an atomic number of 10 has both a full first and second shell (2 and 8). Its second shell is its valence. **Figure 5–2** shows a neon atom. Both helium and neon are relatively inert elements. "Inert" means that these elements are unlikely to participate in chemical reactions. Similarly, other more complex atoms that have eight electrons in their outermost shell (although this shell might not be full) will resemble neon in terms of their chemical inertness. In other words, we can say that atoms with a full valence shell are likely to have inert characteristics.

The atomic structure of an atom defines the chemical characteristics of an element. For instance, a copper atom has twenty-nine protons and twenty-nine electrons. This means that only one electron orbits the outer or valence shell (**Figure 5–3**). Unlike helium and neon, copper is reactive. It oxidizes easily and is also an excellent conductor of electricity. In fact, copper is often used as material for electrical wiring. Its hold on the electron in its valence shell is a light one; it both readily gives it up and acquires additional electrons.

FIGURE 5–2 Structure of a neon atom: The valence shell is full.

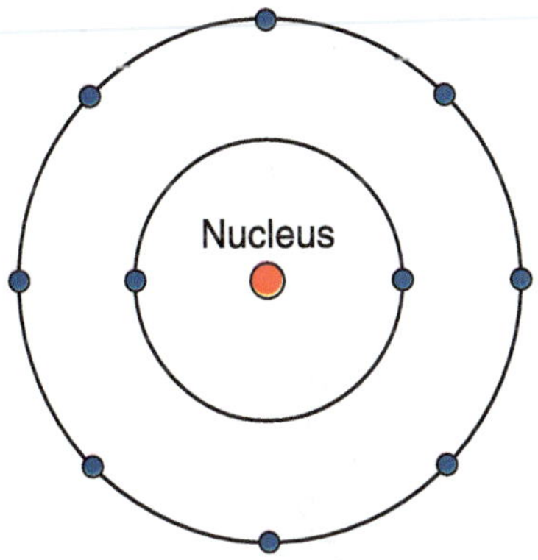

FIGURE 5–3 Atomic structure of a copper atom: The valence shell has one electron, making it a good conductor of electricity.

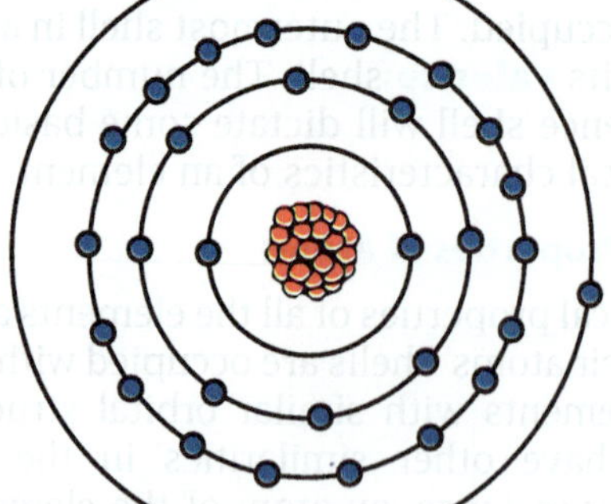

The atomic structure of the element nickel is very similar in appearance to copper when represented in diagram form. It has an atomic structure with twenty-eight protons and twenty-eight electrons (**Figure 5–4**) and is just one proton short of an atom of the element copper. Nickel has a full valence or outer shell. It therefore has characteristics that are quite different from those of copper. Nickel is a poor conductor and does not readily react with oxygen. In industry, nickel is an alloyed ingredient of stainless steel that reduces the reactivity of steel, especially its tendency to rust (oxidize).

FIGURE 5–4 Atomic structure of a nickel atom: The valence shell is full, making it a poor conductor of electricity.

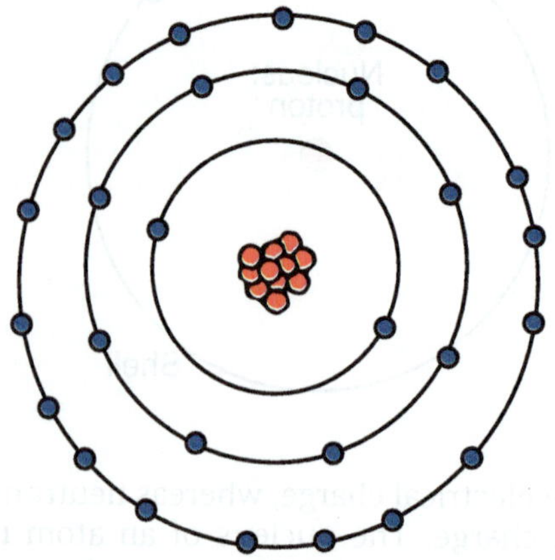

FIGURE 5–5 Balanced atom: The number of protons and electrons is equal.

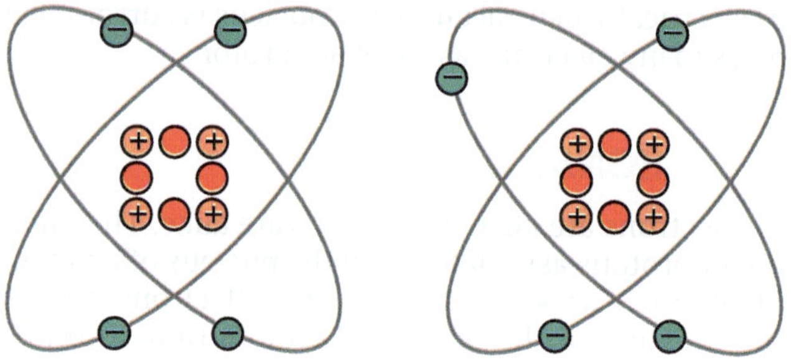

IONS

An electrically balanced atom is one in which the number of electrons equals the number of protons. **Figure 5–5** shows an electrically balanced atom. When atoms of the chemical elements are diagrammed they are almost always shown in a balanced state. Remember that an *ion* is defined as any atom with either a surplus or a deficit of electrons. Free electrons can rest on the surface of a material. Electrons resting on a surface will cause that surface to be negatively charged. Because the electrons are not moving, that surface can be described as having a negative static electrical charge. The extent of the charge is measured in voltage or charge differential.

Electrons are also capable of transferring EMF through matter (or a vacuum) at close to the speed of light. This rate of electron transfer is what is known as an electrical current. For instance, if a group of positive ions passes in close proximity to electrons resting on a surface, they will attract the electrons (negatively charged) by causing them to fill the “holes” left by the missing electrons in the positive ions; the greater the number of electrons subjected to EMF, the greater the current flow. Electron transfer of EMF through a conductor is shown in **Figure 5–6**. The unit of measurement of current flow is the **ampere** (amp).

FIGURE 5–6 Electron flow through a conductor

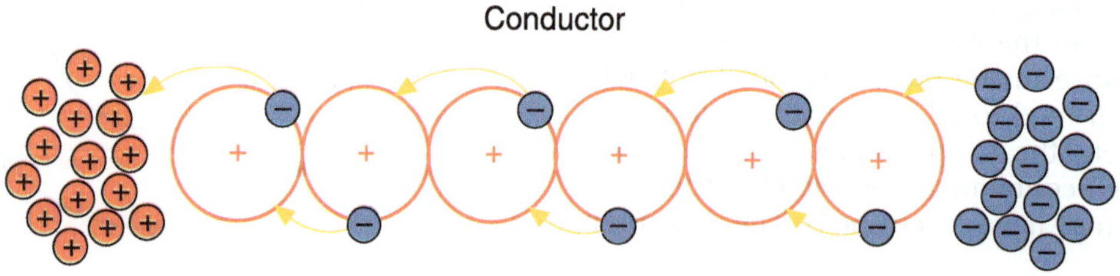

FIGURE 5–7 The rate of electron flow is measured in amperes.

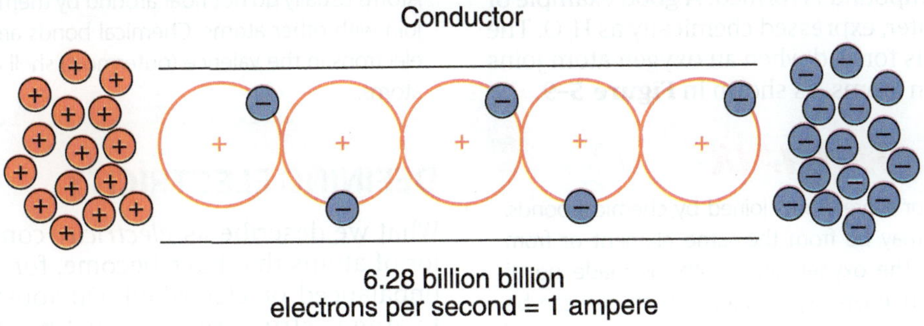

An ampere, therefore, describes the number of electrons passing through a circuit over a specified time period. Specifically, 1 ampere is 6.28 billion billion (6.28×10^{18}) electrons passing a given point per second. This is shown in **Figure 5–7**. The ampere is the quantitative measurement of current. One ampere or 6.28 billion billion electrons passing a given point in a circuit in 1 second can be described as 1 *coulomb*. It is not especially important to remember these values other than to acknowledge that it is a large number.

A number of factors such as friction, heat, light, and chemical reactions can "steal" electrons from the surface of a material. When this happens, the surface becomes positively charged, meaning that it has a deficit of electrons. Atoms with a deficit of electrons are known as positive ions. Providing positive ions remain at rest, the surface will have a positive static electrical charge differential. Let us take a look at how this works in an everyday example. Every time someone walks across a carpet, electrons are "stolen" from the carpet surface. This has an electrifying effect (*electrification*) on both the substance from which electrons are stolen (the carpet) and the moving body (the person) that steals the electrons.

When this moving body has accumulated a sufficient charge differential (measured in voltage), the excess electrons will be discharged through an arc. This arc, seen as a spark, balances the charge. When the charge has been balanced, an electrically neutral state has been reestablished.

ATTRACTION AND REPULSION

Electrification results in both attractive and repulsive forces. In electricity, like charges repel and unlike charges attract. When a plastic comb is run through hair electrons will be stolen by the comb, giving it a negative charge. This means that the comb can now attract small pieces of paper, as shown in **Figure 5–8**. The attraction will continue until the transfer of electrons results in balancing the charge differential. This experiment will always work better on a dry day. Electrons can travel much more easily through humid air, and any accumulated charge will dissipate (be lost) rapidly. Two balloons rubbed on a woolen fiber will both acquire a negative charge. This enables the two balloons to "stick" to a wall, but at the same time they will tend to repel each other. The attraction of unlike charges and repulsion of like charges holds true in both electricity and magnetism.

An atom is held together because of the electrical tendency of unlike charges attracting and like charges repelling each other. Positively charged protons hold the negatively charged electrons in their orbital shells. Also, the electrons do not collide with each other because like electrical charges repel each other. All matter is composed of atoms. Electrical charge is a component of all atoms. When an atom is balanced (the number of protons match the number of electrons; see Figure 5–5), the atom can be described as being in an electrically neutral state. Given that atoms are the building blocks of all matter, it can be said that all matter is electrical in essence.

MOLECULES

Atoms seldom just float around by themselves. More often, they join with other atoms; that is, they bond chemically. Chemical bonding of atoms occurs when electrons in the valence shell of atoms are transferred or shared by a companion atom or atoms. When two

FIGURE 5–8 Unlike electrical charges attract

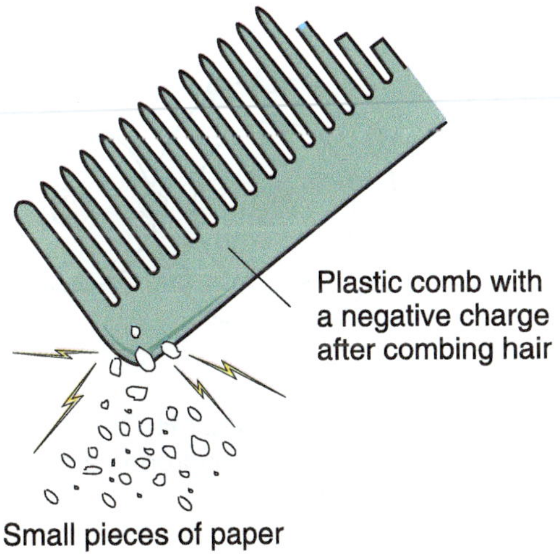

or more atoms are held together by chemical bonding, they form a molecule. When the atoms are from different elements a compound is formed. A good example of a compound is water, expressed chemically as H_2O. The compound water is formed when an oxygen atom joins with two hydrogen atoms, as shown in **Figure 5–9**.

SHOP TALK

A molecule is two or more atoms joined by chemical bonds. The atoms joined may be from the same element or from different elements. The oxygen we breathe is made up of pairs of chemically bonded oxygen atoms. Each pair can be described as a molecule. A single water molecule is made up of two hydrogen atoms chemically bonded to an oxygen atom to form a compound.

A molecule does not necessarily have to be made up of atoms of separate elements. For example, ground-level oxygen is oxygen bonded in pairs known as O_2, whereas oxygen in the upper stratosphere is ozone, chemically bonded in threes and known as O_3. Although just a few more than 100 elements exist, millions of different compounds can be formed from these. Many occur naturally and many are man-made.

FIGURE 5–9 A molecule of the compound water

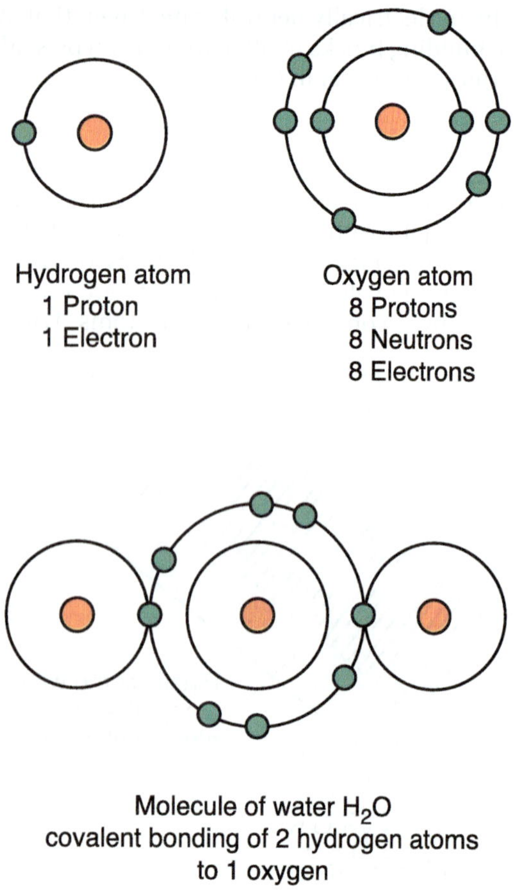

SHOP TALK

Atoms usually do not float around by themselves. They tend to join with other atoms. Chemical bonds are formed when the electrons in the valence (outermost) shell are shared by other atoms.

DEFINING ELECTRICITY

What we describe as *electricity* concerns the behavior of atoms that have become, for whatever reason, unbalanced or electrified. Obviously, there is more to atomic structure than what has been outlined so far. However, understanding a little about atomic structure and behavior can help the technician "see" electricity in a different way when it comes to figuring out failures in electrical circuits. Here is a summary of what we have said about atomic structure so far:

- In the center of every atom is a nucleus.
- The nucleus is made up of positively charged matter called protons.
- The nucleus contains matter with no charge, called neutrons.
- Negatively charged particles, called electrons, are orbiting each atomic nucleus.
- Electrons orbit the nucleus in concentric paths called shells.
- All electrons are alike.
- All protons are alike.
- Every chemical element has a distinct identity and is made up of distinct atoms. That is, each has a different *number* of protons and electrons.
- In an electrically balanced atom, the number of protons equals the number of electrons. This means that the atom is in what is described as a neutral state of electrical charge.
- An atom with either a deficit or an excess of electrons is known as an ion.
- Charge can move from one point to another.
- Like charges repel.
- Unlike charges attract.
- Electrons (negative charge) are held in their orbital shells by the nucleus (positive charge) of an atom.
- Electrons are prevented from colliding with each other because they all have similar negative charges that tend to repel each other.
- A molecule is a chemically bonded union of two or more atoms.
- A compound is a chemically bonded union of atoms of two or more dissimilar elements.

The preceding statements are reinforced as we continue to study electricity and magnetism throughout this chapter. In fact, they are repeated many times, although in slightly different formats.

5.2 CONDUCTORS AND INSULATORS

The ease with which an electron moves from one atom to another determines the conductivity of a material. *Conductance* is the ability of a material to carry electric current. To produce current flow, electrons must move from atom to atom as shown in Figure 5–6. Materials that easily permit this flow of electrons from atom to atom are classified as conductors.

Some examples of good conductors are copper, aluminum, gold, silver, iron, and platinum. Materials that do not easily permit, or perhaps prevent, a flow of electrons are classified as insulators. Some examples of good insulators are glass, mica, rubber, and plastic. Sometimes it can be more difficult to determine. For instance, pure (distilled) water is a good insulator, whereas tap water (because of its mineral load) is a conductor of electricity.

- A *conductor* is generally a metallic element that contains fewer than four electrons in its outer shell or valence. The electrons in atoms of elements with one to three electrons in the valence shell are not so tightly bound by the nucleus.
- An **insulator** is a nonmetallic substance that contains five or more electrons in its outer shell or valence. When an atom has five or more electrons in the valence they are more tightly bound to the nucleus, meaning that they are less likely to be given up.
- **Semiconductors** are a group of materials that cannot be classified either as conductors or insulators. They have exactly four electrons in their outer shell. Silicon is an example of a semiconductor.

Any material that can conduct electricity, even when in an electrically neutral state, contains vast numbers of moving electrons that move from atom to atom at random. When a battery is placed at either end of a conductor such as copper wire and a complete circuit is formed, electrons are pumped from the more negative terminal to the more positive. The battery provides the charge differential or potential difference, and the conductive wire provides a path for the flow of electrons. The transfer of electrons continues until either the charge differential ceases to exist or the circuit path is opened. The number of electrons does not change. In order to have a continuous transfer of charge (that is, have a continuous flow of current) between two points in an electrical circuit, it is necessary to produce a new supply of electrons at a negative point in the circuit as fast as this supply is consumed at a positive point in the circuit. A variety of methods of achieving this exist, and they are covered in a later section in this chapter under the heading "Sources of Electricity" (5–7).

FACTORS AFFECTING RESISTANCE

Copper is a commonly used conductor in vehicle wiring systems. Copper wiring may consist of a single extruded wire coated with plastic insulation or multiple thin strands of braided and bound wire coated with insulation. Braided or stranded wire has greater flexibility and due to something known as Litz effect can conduct slightly more current. Flexibility is a desirable characteristic in vehicle electrical systems so braided wire is used almost exclusively. The **resistance** to current flow in a section of copper wire will depend on its sectional area (wire gauge), temperature, and overall length; the greater the sectional area of copper wire, the less resistance to current flow. Quite simply, thicker electrical wire provides a greater area for flow and is, therefore, less restrictive. This is because more free electrons are available in heavy-duty wire gauge than in a thin gauge. For instance, 8-gauge wire is thicker than 16-gauge wire. So 8-gauge wire is capable of permitting greater current flow than 16-gauge wire. In most vehicle applications, wiring is designed to be as thin as possible while being capable of handling the electrical load it is designed for. Wire gauge is usually **original equipment manufacturer (OEM)** specified to handle the intended electrical load plus a small safety margin.

Temperature and Resistance

The resistance of materials also depends on temperature. In most conductors, the higher the temperature, the greater is the resistance to current flow. When this is the case, the conductor has a **positive temperature coefficient (PTC)**. Copper wire has a PTC electrical characteristic. When a conductor subjected to an increase in temperature has *lower* resistance to current flow it can be said to possess a **negative temperature coefficient (NTC)**. The thermistors used to signal temperature conditions commonly have an NTC electrical characteristic. Thermistors are studied in detail in Chapter 6.

When electrons move through a conductor, they often find that the atoms of the substance limit their motion to some extent. Some electrical energy is lost in overcoming this resistance to motion in the conductor. If the temperature of a PTC conductor rises, the atoms of the conductor vibrate more vigorously and will further inhibit the movement of electrons through the conductor. In other words, the resistance of the conductor increases with temperature rise.

Insulators

Insulators are substances with atomic structures that tightly hold their electrons in the valence shell.

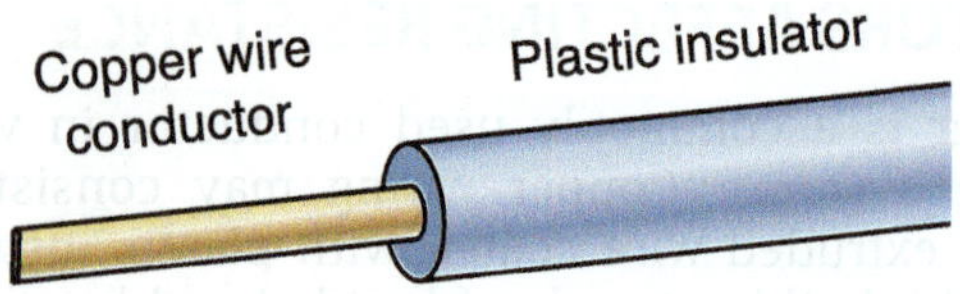

FIGURE 5–10 A copper wire conductor protected by a plastic insulator.

It is possible to flow very little current through such substances, even when they are subjected to massive charge differentials (very high voltages). Rubber, glass, and some plastics have extremely high resistivity. Both are commonly used as insulators. For instance, plastic is used to protect copper wiring in most vehicle electrical systems, as shown in **Figure 5–10**.

Between low- and high-resistance conductors and insulators is a group of materials known as semiconductors. All semiconductors have four electrons in their valence shell. The most commonly used semiconductors are silicon and germanium. These will be studied when electronics theory is introduced in Chapter 6.

ELECTRON MOVEMENT

Electrons do not actually move through an electrical circuit at high velocity, although the force that moves them does. If a row of pool balls is lined up so that each kiss-contacts the other and the row is struck by a rapidly moving ball, the moving ball stops, imparting a force to the stationary balls that is transmitted through them at high speed but does not move any but the final ball in the row. Most of the force of the moving ball is transferred to the final ball in the row, which separates and moves off from the row. Electron movement through a conductor is somewhat similar. Although actual electron movement through a conductor is not rapid, the electromotive force travels through the conductor at nearly the speed of light.

5.3 CURRENT FLOW

Current flow will occur only when there is a path and a difference in electrical potential. This difference is known as *charge differential* or *potential difference* and is measured in *voltage*. Charge differential exists when the electrical source has a deficit of electrons and therefore is positively charged. Because electrons are negatively charged and unlike charges attract, electrons will flow toward a positive source when provided with a path. Charge differential is electrical pressure (**Figure 5–11**), and it is measured in voltage.

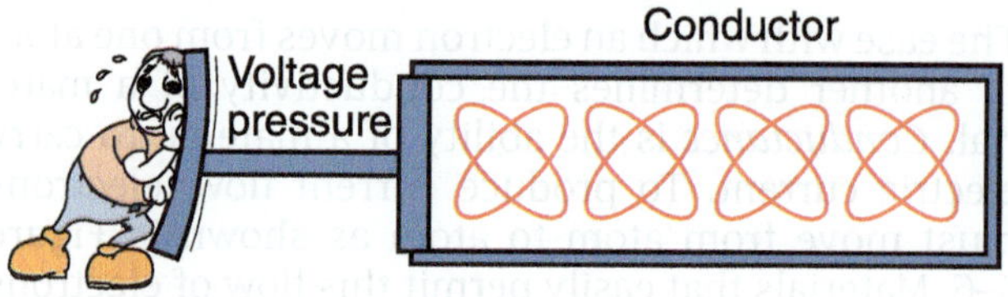

FIGURE 5–11 Voltage or charge differential is the pressure that causes electrons to move.

Initially many scientists thought that current flow in an electrical circuit had one direction of flow—that is, from positive to negative. This is known as the *conventional theory* of current flow. It originated from Ben Franklin's observations and conclusions from his kite in the electrical storm experiment. When the electron was discovered, scientists revised the theory of current flow and called it **electron theory**. In studying electricity, the technician should be acquainted with both the conventional and electron theories of current flow. However, vehicle schematics and voltmeters use the conventional theory almost exclusively. Conventional and electron theories of electrical circuit flow are shown in **Figure 5–12**.

A conductor such as a piece of copper wire contains billions of neutral atoms whose electrons move randomly from atom to atom, vibrating at high frequencies. When an external power source

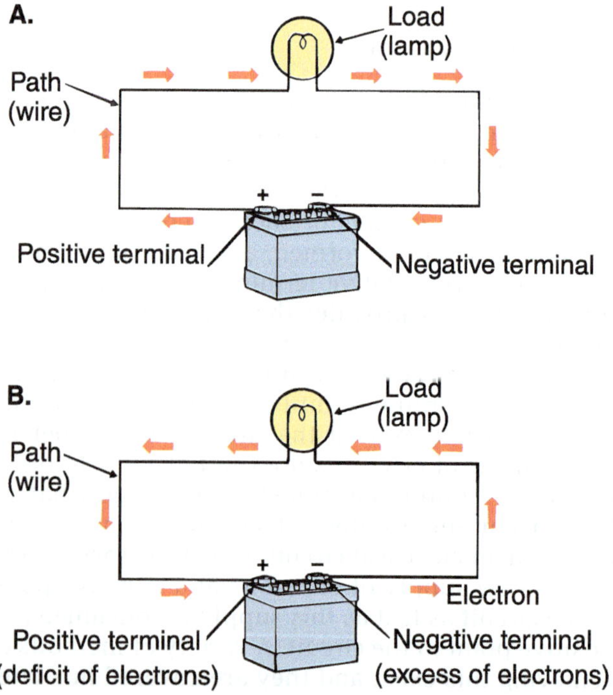

FIGURE 5–12 Electrical circuit flow: (A) conventional theory; (B) electron theory.

such as a battery is connected to the conductor a deficit (shortage) of electrons occurs at one end of the conductor and an excess of electrons occurs at the other end. The negative terminal will have the effect of repelling free electrons from the conductor's atoms, whereas the positive terminal will attract free electrons. This results in a flow of electrons through the conductor from the negative charge point to the positive charge point in a circuit. The rate of flow will depend on the charge differential (or potential difference/voltage). The greater the voltage (pressure) or charge differential, the greater the rate of flow. Voltage is the force that pushes current through a circuit.

The charge differential or voltage is a measure of electrical *pressure*. The role of a battery, for instance, is to act as a sort of electron pump. A 12V battery will pump electrons through a circuit faster than a 6V battery of similar amperage capability. In a closed electrical circuit, electrons move through a conductor, producing a displacement effect close to the speed of light. The higher the voltage, the higher the force potential or pressure available, so the more electrons get pumped through the circuit. Charge differential can also be expressed as *EMF* or *potential difference* (PD). In fact, because all of the following terms can be used to describe electrical pressure, becoming familiar with them is a good idea:

- Charge differential
- Voltage
- EMF
- PD

As mentioned a little earlier, the physical dimensions of a conductor are also a factor. The larger the sectional area (measured by wire gauge size), the more atoms there are over a given sectional area, so there are more free electrons available. Therefore, as wire size increases, so does the ability to flow more electrical current through the wire. Wire size is specified to the amount of current it is expected to carry.

The higher the expected current requirement, the thicker the wire must be; therefore, wire gauge size is matched to the expected current load plus a small safety margin in vehicles. The rate of electron flow is called current and it is measured in amperes. A current flow of 6.28 billion billion electrons per second is equal to 1 ampere, as shown in Figure 5–7.

Terminals

The terminal from which electrons exit in an electrical device connected in a circuit is known as the **anode** or positive terminal. The terminal through which electrons enter an electrical component is known as the **cathode** or negative terminal. Understanding something about the nature of current flow and exactly what is meant by the terms *conventional* and *electron theories* of current flow is crucial because both are used in vehicle technology.

Conventional and Electron Theories

In any electrical circuit, electrons will flow from a more negative point in a circuit toward a more positive point. That is, they will flow from a point in the circuit with an excess of electrons toward a point with a deficit of electrons. So the actual direction of current flow is always from negative to positive. However, many electrical diagrams and schematics use conventional current flow theory. Conventional current flow theory originated from Ben Franklin's observations and mistaken conclusions in his kite experiment. Although the conventional theory of current flow is technically incorrect, it is the basis of all standard vehicle wiring schematics. The conventional theory must be understood so an electrical wiring schematic can be accurately interpreted. The electron theory must be understood so the technician knows electrically what is happening in a circuit for purposes of troubleshooting it.

5.4 MAGNETISM

Magnetism was first observed in lodestone (magnetite) and the way ferrous (iron-based) metals reacted to it. When a bar of lodestone was suspended by string, the same end would always rotate to point toward the earth's North Pole. Even today, we do not fully understand magnetism. By observation, we can certainly say quite a bit about magnetism. The molecular theory of magnetism tends to be the most widely accepted. In most materials, the magnetic poles of the composite molecules are arranged randomly so no evident magnetic force exists. However, in certain metals such as iron, nickel, and cobalt, the molecules can be aligned so their north or N poles all point in one direction and their south or S poles point in the opposite direction. In one known material, lodestone, the molecules align themselves in this manner naturally.

Some materials have better magnetic retention than others. This means that when they are magnetized by having their north and south poles aligned, they are able to hold that molecular alignment longer. Other materials are only capable of maintaining their molecular alignment when positioned within a magnetic field. When the field is removed, the molecules disarrange themselves randomly and the substance's magnetic properties are lost. All magnetism is essentially electromagnetism in that it results from the kinetic energy (energy of movement) of electrons. Whenever an electric current is flowed through a conductor, a magnetic field is created. When a bar-shaped permanent magnet is cut in two, each piece will assume the magnetic properties of the

FIGURE 5–13 Magnetic principles: (A) all magnets have poles; (B) unlike poles attract; and (C) like poles repel.

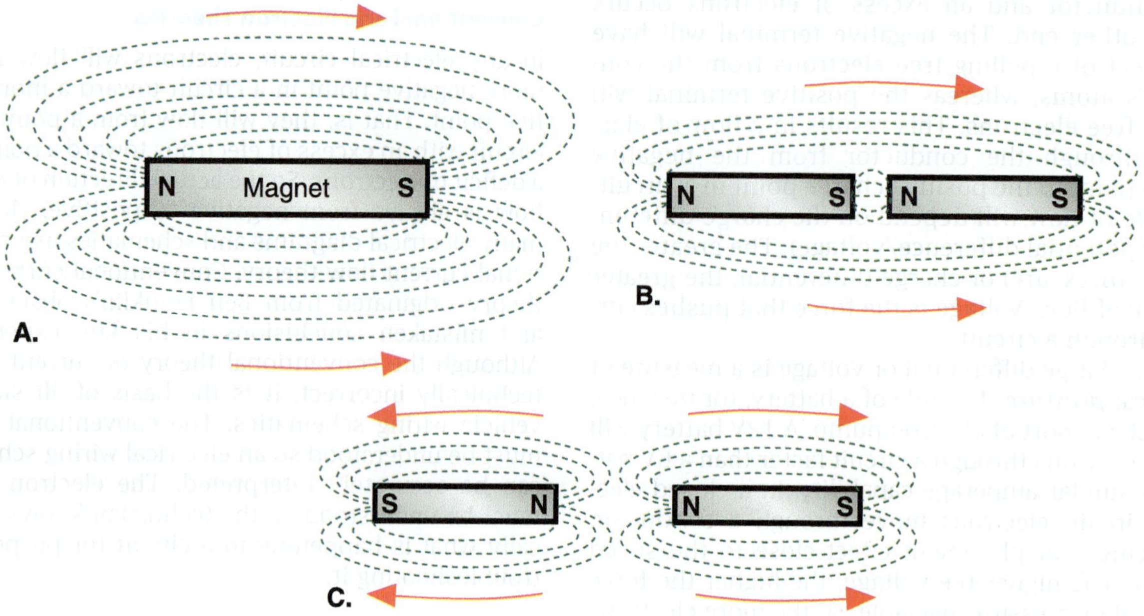

parent magnet, with individual north and south poles. **Figure 5–13** shows some basic magnetic principles.

RELUCTANCE

The term *reluctance* is used to describe resistance to the movement of magnetic lines of force. Reluctance can be reduced by using permeable (susceptible to penetration) materials within magnetic fields. The permeability of matter is rated by giving a rating of 1 for air. Air is generally considered to be a poor conductor of magnetic lines of force. In contrast, iron would be given a permeability factor of 2,000. This means that in comparison with air, iron has 2,000 times better permeability. In other words, air has high reluctance and iron has low reluctance. Certain ferrous (iron-based) alloys may have permeability ratings exceeding 50,000.

The force field existing in the space around a magnet can be demonstrated when a piece of thin cardboard is placed over a magnet and iron filings are sprinkled on top of the cardboard. The pattern produced as the filings arrange themselves on the cardboard is referred to as *flux lines*. By observing the behavior of magnetic lines of flux, we can say the following:

- Flux lines flow in one direction.
- Flux lines exit from the magnet's north pole and enter through the south pole.
- The flux density (concentration) determines the magnetic force. A powerful magnetic field has a dense flux field, whereas a weak magnetic field has a low-density flux field.
- The flux density is always greatest at the poles of a magnet.
- Flux lines do not cross each other in a permanent magnet.
- Flux lines facing the same direction attract.
- Flux lines facing opposite directions repel.

ATOMIC THEORY AND MAGNETISM

In an atom, all of the electrons in their orbital shells also spin on their own axes. This is much the same as the planets in our solar system orbiting the sun. Each rotates axially and each produces magnetic fields. Because of the orbiting electrons' axial rotation, each electron can be regarded as a minute permanent magnet. In most atoms, pairs of electrons spinning in opposite directions will produce magnetic fields that cancel each other out. However, an atom of iron has twenty-six electrons, only twenty-two of which are paired. This means that in the second from the outermost shell, four of the eight electrons are not paired, resulting in these electrons rotating in the same direction and not canceling each other out. This fact accounts for the magnetic character of iron.

A large number of vehicle electrical components use magnetic and electromagnetic principles in some way. Uses for permanent magnets include the AC pulse generators used as shaft speed sensors (rpm sensors would be a good example) and some electric motors. Electromagnetic principles are used extensively in such devices as motors, generators, solenoids, magnetic switches, and coils.

5.5 ELECTROMAGNETISM

Current flow through any conductor creates a magnetic field. Whenever electrical current is flowed through copper wire, a magnetic field is created surrounding

FIGURE 5–14 Electromagnetic field characteristics

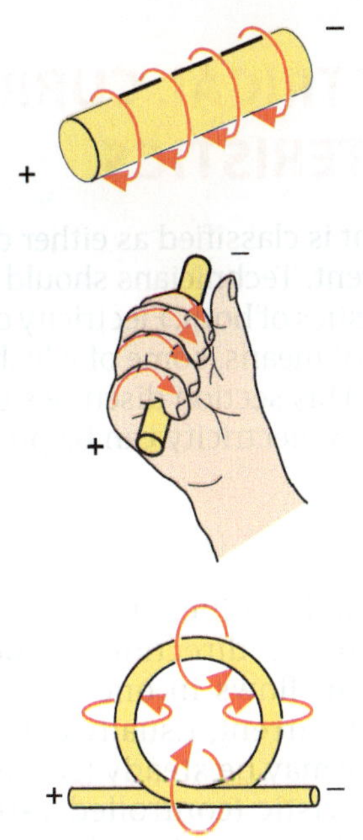

the wire, as shown in **Figure 5–14**. This effect can be observed by passing a copper wire through which current is flowing lengthwise over a compass needle. The compass needle will deflect from its north-south alignment when this occurs. Any magnetic field created by electrical current flow is known as *electromagnetism*. Study of the behavior of electromagnetic fields has proven the following:

- Magnetic lines of force do not move when the current flowing through a conductor remains at a constant. When current flowed through the conductor increases, the magnetic lines of force will extend further away from the conductor.
- The intensity and strength of magnetic lines of force increase proportionally with an increase in current flow through a conductor. Similarly, they diminish proportionally with a decrease in current flow through the conductor.
- A rule called the right-hand rule is used to indicate the direction of the magnetic lines of force: The right hand should enclose the wire with the thumb pointing in the direction of conventional current flow (positive to negative), and the fingertips will then point in the direction of the magnetic lines of force, as shown in **Figure 5–15**.

USING ELECTROMAGNETISM

We have already said that when an electric current is flowed through a straight wire, a magnetic field is produced around the wire. When the wire is coiled and electric current is flowed through it, the magnetic field combines to form a larger and more intense magnetic field. Just as in a straight wire, this is identified by north and south poles. This effect can be further amplified by placing an iron core (high permeability) through the center of the coil (**Figure 5–16**), which reduces the reluctance of the magnetic field.

FIGURE 5–15 (A) Magnetic lines of force join together and attract each other, and (B) the right-hand rule

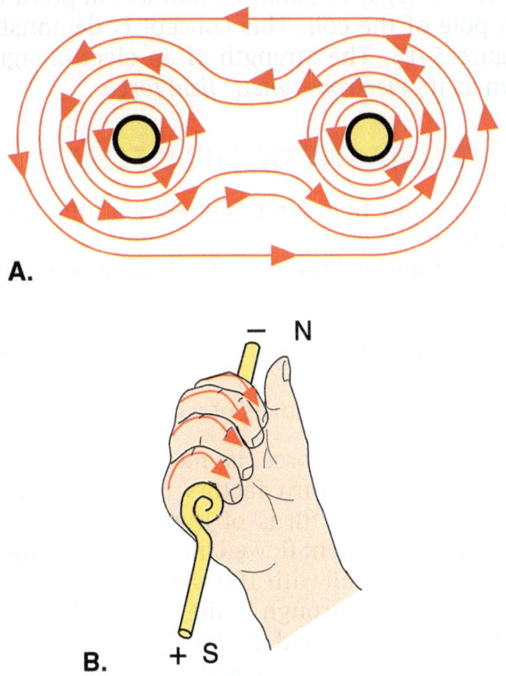

FIGURE 5–16 Magnetic field characteristics of a coiled conductor

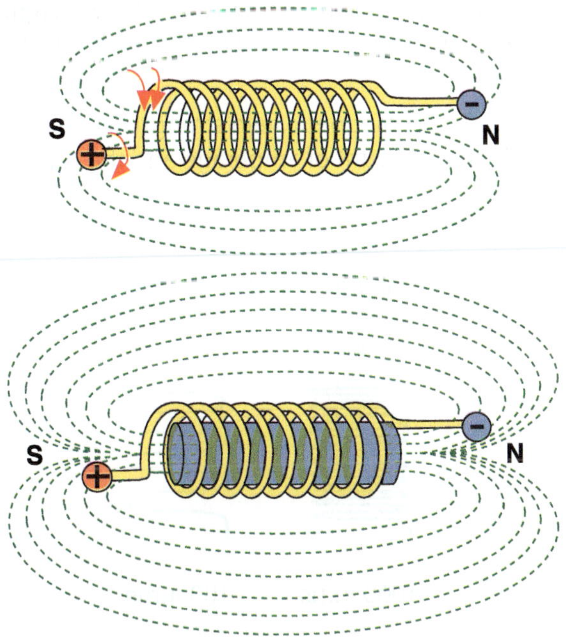

The polarity of the electromagnet created can be determined by the right-hand rule for coils. The coiled wire should be held with the fingers pointed in the direction of conventional current flow (positive to negative), and the thumb will point to the north pole of the coil. This concept is demonstrated in Figure 5–15. The strength of an electromagnet is known as its *electromagnetic field force.*

ELECTROMAGNETIC FIELD FORCE

Electromagnetic field force is often quantified or measured as **magnetomotive force (mmf)**.

Magnetomotive force (mmf) is determined by two factors:

- The amount of current flowed through the conductor
- The number of turns of wire in a coil

Magnetomotive force is measured in ampere-turns (At). Ampere-turn factors are the number of windings (complete turns of a wire conductor) and the quantity of current flowed (measured in amperes). For instance, if a coil with 100 windings has 1 ampere of current flowed through it, the result will be a magnetic field strength rated at 100 At. An identical magnetic field strength rating could be produced by a coil with 10 windings with a current flow by 10 amperes. The actual field strength must factor in reluctance. In other words, the actual field strength of both coils would be increased if the coil windings were to be wrapped around an iron core. **Figure 5–17** shows a pair of coils of equal field strength, but because of its larger number of windings, the one on the left requires much less current flow.

A common use of an electromagnet would be that used in an automobile salvage yard crane. By switching the current to the lift magnet on and off, the operator can lift and release scrap cars. Electromagnetic principles are used extensively in vehicle electrical systems. They are the basis of every solenoid, relay, coil, generator, and electric motor.

FIGURE 5–17 Magnetic field strength is determined by the amount of amperage and the number of coils.

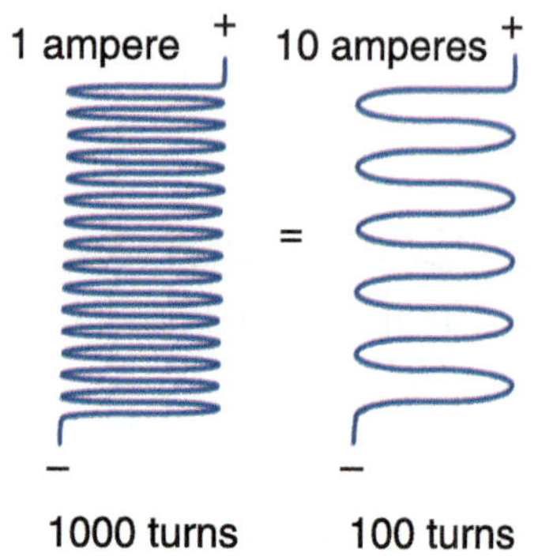

5.6 ELECTRICAL CURRENT CHARACTERISTICS

Electrical current is classified as either direct current or alternating current. Technicians should understand the basic characteristics of both. Electricity can be produced by any number of means, some of which have been discussed already. This section discusses the types of current flow and how electricity can be produced.

DIRECT CURRENT (DC)

Like water through a pipe, electrical current can be made to flow in two directions through a conductor. If the current flows in one direction only, it is known as direct current, usually abbreviated as DC. The current flow may be steady (continuous) or have a pulse characteristic (controlled variable flow). DC can be produced in a number of ways outlined later in this section. DC has many applications and is extensively if not quite exclusively used in highway vehicles throughout most of the chassis electrical circuits.

ALTERNATING CURRENT (AC)

Alternating current, or AC, describes a flow of electrical charge that cyclically reverses at high speed due to a reversal in polarity at the voltage source. AC is often shown in graph form (**Figure 5–18**). It is usually

FIGURE 5–18 Graphed AC voltage signal typical of that produced by an inductive pulse generator sensor

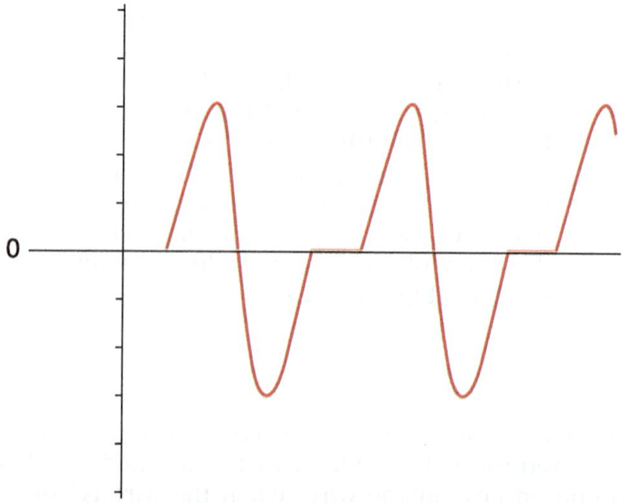

FIGURE 5–19 Principle of an AC inductive pulse generator

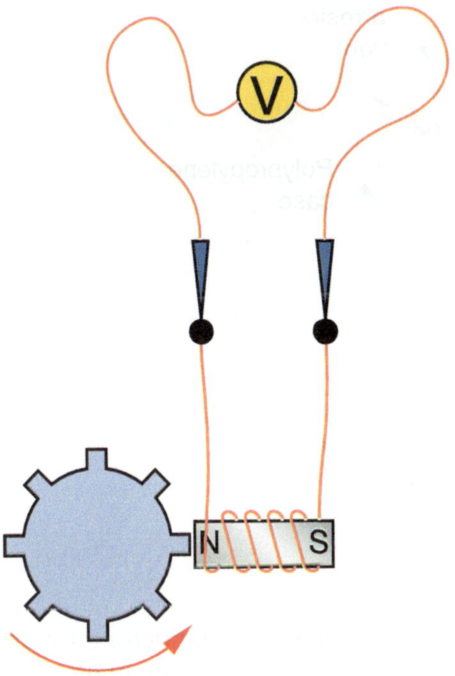

produced by rotating a coil in a magnetic field. AC is used in vehicle AC generators (perhaps more often called *alternators*) and in certain sensors on modern vehicles. **Figure 5–19** shows a typical shaft speed sensor, also called an inductive pulse generator. AC is better suited than DC for transmission through power lines. The frequency at which the current alternates is measured in cycles. A *cycle* is one complete reversal of current from zero through positive to negative and back to the zero point. Frequency is usually measured in *hertz*. One hertz is one complete cycle per second. For instance, the frequency of household electrical power supply in North America is 60 Hz or 60 cycles per second.

Shorepower

Many recent trucks accommodate what is known as **shorepower** or **hotel load** hook-ups. Shorepower is used to refer to a truck designed to be connected to the AC mains grid while parked and stationary. Shorepower can be supplied highway rigs as 110 V-AC, 220 V-AC, or three-phase voltages, typically between 360 and 600 V-AC. Technicians must be mindful that working on high-voltage shorepower systems requires specialized training.

Shorepower can be used on a truck in the form of a battery charger to run driver comfort systems such as microwaves, HVAC, and entertainment systems. High-voltage AC is used to run a trailer reefer units while parked: This is mandatory in many offshore jurisdictions and there is no doubt it will arrive here, probably sooner rather than later. Currently, shorepower supply stations are operated across major truck routes across the United States and Canada and many more are planned.

5.7 SOURCES OF ELECTRICITY

Electricity is an energy form. As stated previously, electrical energy may be produced by a variety of methods. The production of electricity involves changing one energy form, such as mechanical or chemical energy, into electrical energy. Some of these methods are outlined in this section.

CHEMICAL

Batteries are a means of producing DC from a chemical reaction. In the lead-acid battery, a *potential difference* is created by the chemical interaction of lead and lead peroxide submerged in sulfuric acid electrolyte. When a circuit is connected to a charged battery (one in which there is a charge differential), the battery will pump electrons through the circuit from the negative terminal through the circuit load to the positive terminal. This electrochemical process will continue until there is no difference in potential between the two posts of the battery. If a voltmeter were to be placed across the battery terminals, the reading would be zero. When the charge differential ceases to exist, no difference exists between the potential at either terminal and the battery is discharged. The operating principles of lead-acid batteries are explored in much greater depth in Chapter 7. A typical lead-acid battery is shown in **Figure 5–20**.

STATIC ELECTRICITY

The term *static electricity* is somewhat misleading because it implies that it is unmoving. Perhaps it is more accurately expressed as *frictional electricity* because it results from the contact of two surfaces. Chemical bonds are formed when any surfaces contact and the atoms on one surface tend to hold electrons more tightly. The result is the "theft" of electrons. Such contact produces a charge imbalance by pulling electrons from one surface to another. As electrons are pulled away from a surface, the result is an excess of electrons (the result is a negative charge) and a deficit in the other (the result is a positive charge).

The extent of the charge differential is, of course, measured in voltage. Although the surfaces with opposite charges remain separate, the charge differential will exist. When the two polarities of charge are united, the charge imbalance will be canceled. Static electricity is an everyday phenomenon, as described

FIGURE 5–20 A lead-acid battery

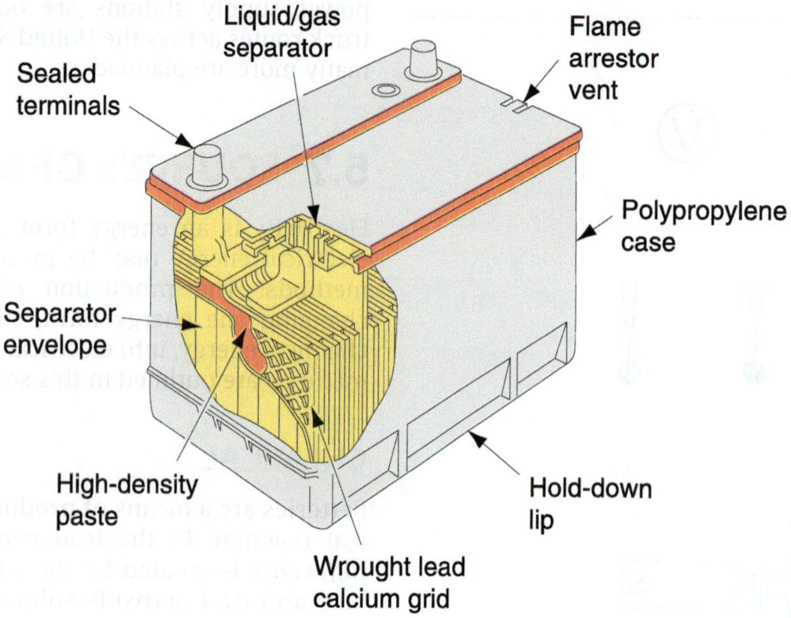

in examples in the opening to this chapter. It usually involves voltages of over 1,000V and perhaps rising to as much as 50,000V. A fuel tanker trailer towed by a highway tractor steals electrons from the air as it is hauled down the highway (as does any moving vehicle) and can accumulate a significant and potentially dangerous charge differential. This charge differential, which can be as high as 40,000V, must be neutralized by grounding before any attempt is made to load or unload fuel. Failure to ground a fuel tanker before loading or unloading could result in an explosion ignited by an electrostatic arc.

ELECTROMAGNETIC INDUCTION

Current flow can be created in any conductor that is moved through a magnetic field or alternatively by a mobile magnetic field and a stationary conductor. The voltage induced increases both with speed of movement and the number of conductors. Densely wound conductors will tend to produce higher voltage values. Generators, alternators, and cranking motors all use the principle of electromagnetic induction to either produce or use electricity. Electromagnetic induction is a means of converting mechanical energy into electrical energy—it is much used in the production of electricity. **Figure 5–21** and **Figure 5–22** demonstrate the principle of electromagnetic induction.

THERMOELECTRIC

Electron flow can be created by applying heat to the connection point of two dissimilar metals.

FIGURE 5–21 Moving a conductor through a magnetic field induces an electric current flow.

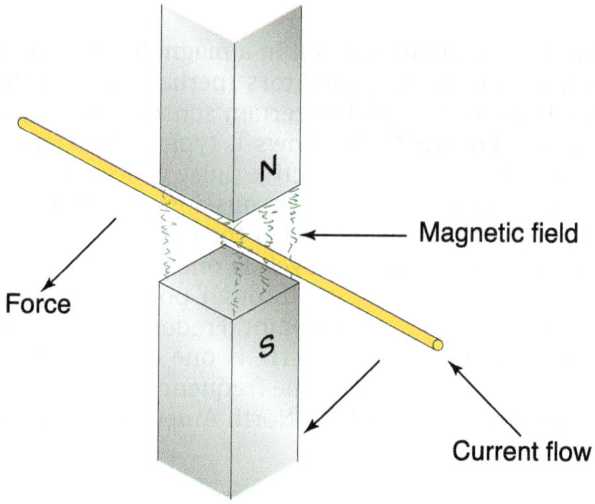

FIGURE 5–22 Rotating a looped conductor in a magnetic field induces current flow.

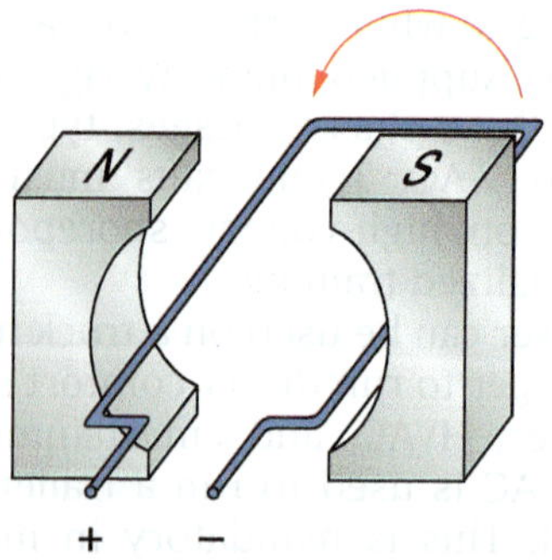

A **thermocouple** used to measure high temperatures consists of two dissimilar metals (typically, iron and constantin, a copper-tin alloy) joined at the "hot" end and connected to a sensitive voltmeter at the gauge end. As temperature increases at the hot end, a potential difference is created at the gauge end. The greater the potential difference, the greater the reading on the display gauge (a millivolt meter). In this way, heat energy is converted into electrical energy. Pyrometers used in truck diesel engine exhaust systems use this thermocouple principle.

PHOTOELECTRIC

When light contacts certain materials such as selenium and cesium, electron flow is stimulated. In this way, light energy or photons are converted into electrical energy. Photoelectric cells are used as sensors that can control headlight beams and automatic daylight/night mirrors.

PIEZOELECTRIC

Some crystals, notably quartz, become electrified when subjected to direct pressure, the potential difference or voltage increasing with pressure increase. *Piezoelectric sensors* are used as detonation (knock) sensors used on spark-ignited (SI) electronically controlled engines. Piezoelectricity works both ways; in other words it is reversible. Mechanical movement results when a quartz crystal is hit with a charge of electrical potential. The reversibility of piezoelectricity is the principle behind the actuators used in bubble jet printers and some types of diesel engine injector actuators.

5.8 ELECTRICAL CIRCUITS AND OHM'S LAW

The German physicist Georg Simon Ohm (1789–1854) proved the relationship between electrical potential (pressure or voltage), electrical current flow (measured in amperes), and the resistance to the current flow (measured in **ohms**, the symbol of which is the Greek letter Ω). Ohm proved the relationship among potential difference, circuit resistance, and current intensity in mathematical terms. To better understand the behavior of electricity in an electrical circuit, you should be familiar with the terminology used to describe its characteristics. When introducing electricity, comparisons are often made between electrical circuits and hydraulic circuits.

These comparisons are used in the following explanations, but it is probably not advisable to think of electricity in terms of hydraulic circuits after having mastered the basics. A simple electric circuit must have three basic elements:

- **Power source**. This provides the charge differential that will drive the flow of electrons through the circuit. A battery or a generator would be typical power source devices.
- **Path**. A feed path for electron flow to the load and a return path from the load. The wires that connect the power source to the load would provide the circuit path.
- **Load**. The load in an electric circuit acts as a current limiting restriction and as a means of converting electrical energy into another energy form. For instance, a light bulb as a load in an electric circuit converts electrical energy into light energy: An electric motor converts electrical energy into kinetic energy (the energy of movement).

Figure 5–23 shows a simple electrical circuit. The power source is a battery and the load is a light bulb. The switch is closed, indicating that there is current flow. Next, we should take a look at some terms we have already defined in the context of an actual electric circuit.

VOLTAGE

We said previously that voltage is a measure of charge differential. Charge differential can also be defined as electrical pressure or potential difference. Using a hydraulic comparison, voltage in an electrical circuit can be compared to fluid pressure. Just as in the hydraulic circuit, voltage may be present in an electrical circuit without any current flow taking place. The higher the voltage, the faster the potential for current flow. Voltage is measured in parallel with a circuit. A voltmeter can be used to check circuit voltage or to check one location in a circuit for higher potential than another.

FIGURE 5–23 Simple series light bulb circuit

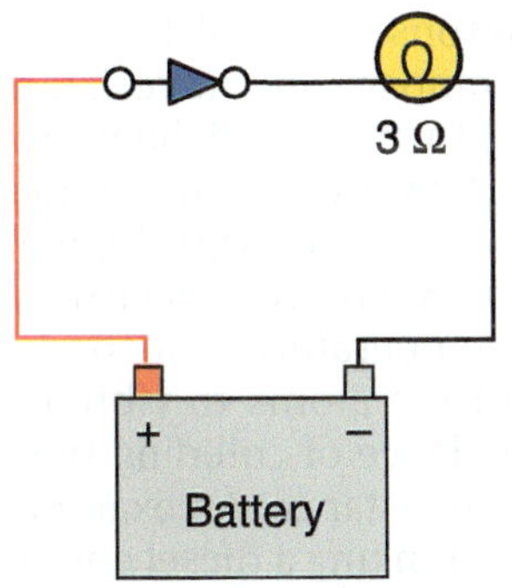

FIGURE 5–24 Connecting meters into circuits

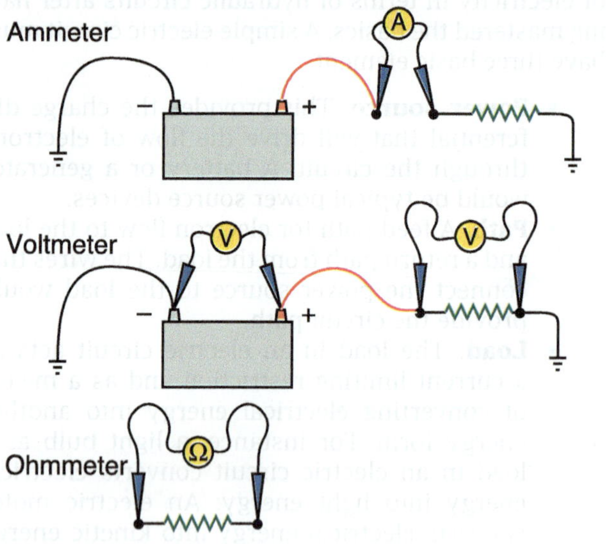

A couple of simple applications of a voltmeter are shown in **Figure 5–24**.

CURRENT

Current intensity or the flow of electrons is measured in amperes. One ampere is equal to 6.28 billion billion electrons passing a given point in an electrical circuit in 1 second—or 1 coulomb. If electrical current is compared to what happens in a hydraulic circuit, amps can be compared to gpm (gallons per minute). If current flow is to be measured in an electric circuit, the circuit must be electrically active or closed—that is, actively flowing current. Current flow is usually measured with an ammeter as shown in Figure 5–24.

RESISTANCE

Resistance to the flow of electrons through a circuit is measured in *ohms.* The resistance to the flow of free electrons through a conductor results from the innumerable collisions that occur. Generally, the greater the sectional area of the conductor (wire gauge size), the less resistance to current flow, simply because more free electrons are available. Once again, using the hydraulic comparison, the resistance to fluid flow through a circuit would be defined by the pipe internal diameter or flow area. In an electrical circuit, resistance generally increases with temperature because of collisions between free electrons and vibrating atoms. As the temperature of a conductor increases, the tendency of the atoms to vibrate also increases, as does the incidence of colliding free electrons. The result is higher resistance. An example of this can be observed when cranking a diesel engine: As the cranking circuit heats up, the less efficient electrical energy is converted into the mechanical energy required to rotate the engine. The resistance in an electrical circuit must be checked with the circuit open—that is, not energized. Resistance checks are often performed on components with an ohmmeter after they have been isolated from their electrical circuit, as shown in Figure 5–24.

CIRCUIT COMPONENTS

We said before that a complete electrical circuit is an arrangement of a power source, path, and load that permits electrical current to flow. Such a circuit can be described as **closed** whenever current is flowing. The same circuit would be described as **open** when no current is flowing. **Figure 5–25** shows a simple series light bulb circuit in both a closed state and an open state.

A switch is used to open and close a circuit in much the same way a valve acts in a hydraulic circuit. Figure 5–23 shows an electrical circuit consisting of a power source, a switch, a load, and wiring to complete the circuit. In most vehicle electrical circuits, a circuit protection device such as a fuse or circuit breaker would also be used. The following are vehicle circuit components:

- **Power source**. The energy source of the circuit and the means used to pump electrons through the circuit when it is closed. A battery or a generator is the usual power source in a vehicle electrical system.

FIGURE 5–25 (A) Closed circuit; and (B) open circuit

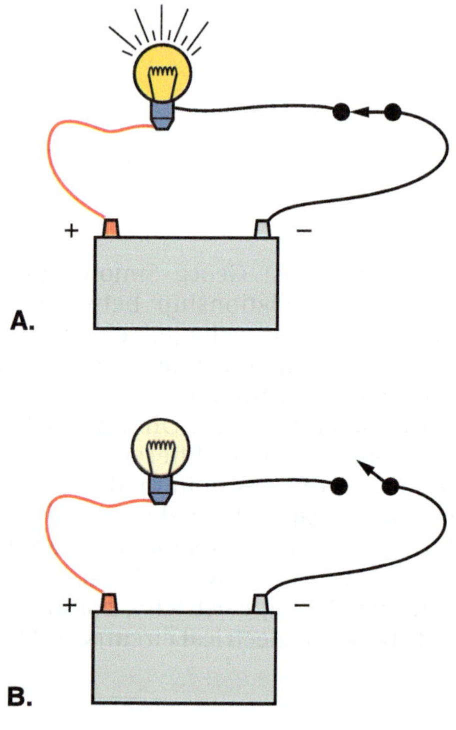

- **Conductors**. The means used to provide a path for current flow in an electrical circuit. This would include electrical wiring and the vehicle chassis in a typical common ground system.
- **Switches**. The means used to open and close an electrical circuit or subcircuits within the main circuit. The ignition switch and dash toggle switches are examples. Note how a switch appears in open and closed states in a circuit diagram by referencing Figure 5–25.
- **Circuit protection devices**. **Circuit breakers** and fuses protect electrical circuits and subcircuits from current overloads by opening the circuit. A fuse permanently fails when overloaded with current and it must be replaced. **Figure 5–26** shows a common blade-type fuse in good and failed states. Circuit breakers are widely used in truck chassis in preference to fuses. When a circuit breaker trips (opens) due to a circuit overload, it may reset automatically (cycling) or the circuit may have to be switched open (noncycling) before current flow can resume.
- **Load**. The objective of the electrical circuit. The function of the load is to convert electrical energy into another energy form, such as light or mechanical energy.

SERIES CIRCUITS

A **series circuit** may have several components such as switches, resistors, and lamps but connected in such a manner that only one path for current flow through the circuit exists, as shown in Figure 5–23. In other words, the loads in the circuit are connected one after another. In a series circuit such as the one in Figure 5–23, the circuit would open if the element in the light bulb were to fail. Because the path for current flow includes the light bulb element, no current would flow. Series circuits have the following characteristics:

- The resistances placed in the circuit are always added. Total resistance in the circuit is always the sum of all resistances in the circuit because each added resistance represents an added hindrance to current flow.
- Current flow is the same at any point in the circuit when it is closed. That means the amperage through each resistor in the circuit will be identical.
- The voltage drop through each resistor in the circuit will mathematically relate to its actual resistance value. This can be calculated using **Ohm's Law**, which is covered a little later in this chapter.
- The sum of the voltage drops through each resistor in the circuit will equal the source voltage.

SHOP TALK

A series circuit is one in which all the components are connected through a single path. For instance, if ten light bulbs were wired in series and the filament in one bulb failed, no current could flow because each load in the circuit represents a portion of the current path.

FIGURE 5–26 Blade-type fuses in good and failed (open) states

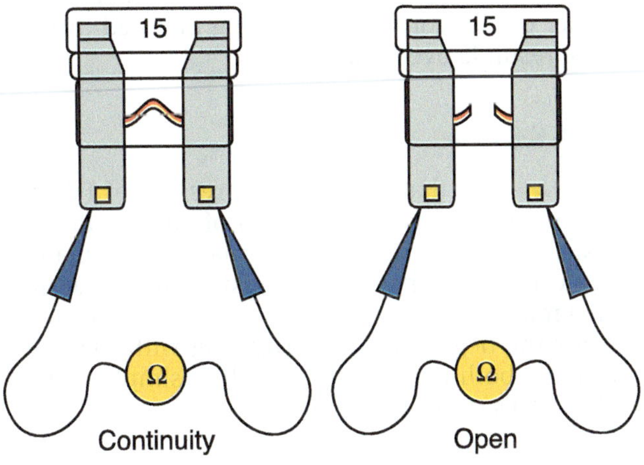

PARALLEL CIRCUITS

A **parallel circuit** is one with multiple paths for current flow, meaning that the components in the circuit are connected so current flow can flow through a component without having first flowed through other components in the circuit (**Figure 5–27**). In other words, the circuit elements are connected side by side. Parallel circuits have the following characteristics:

FIGURE 5–27 A parallel circuit with different resistances in each branch

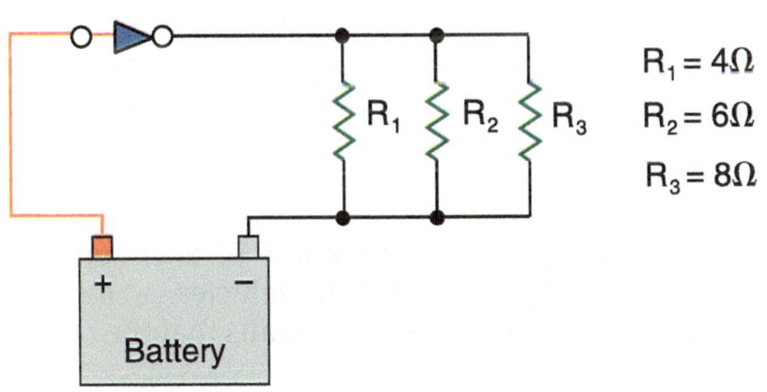

- Total circuit resistance will always be less than the lowest resistance load device in the circuit. As resistances are added in parallel, they provide additional paths for current flow; therefore, total circuit resistance must drop.
- The current flow through each load device will depend on its actual resistance. This may be calculated using Ohm's Law.
- Applied voltage is the same on each branch of a parallel circuit.
- The voltage drop through each path in a parallel circuit is the same and equals the source voltage.
- The sum of the current flows measured in amperage in the separate branches of a parallel circuit must equal the total amperage flowed through the circuit.

SERIES-PARALLEL CIRCUIT

Many circuits are constructed using the principles of both series and parallel circuits. These are known as **series-parallel circuits**. The principles used to calculate circuit values in series and parallel circuits are used to calculate circuit behavior in series-parallel circuits.

ELECTRICAL CIRCUIT TERMINOLOGY

The following terminology is used to describe both normal and abnormal behaviors in an electrical circuit.

Short Circuit

A **short circuit** describes what occurs in an electrical circuit when a conductor is placed across the connections of a component and some or all of the circuit current flow takes a shortcut. Short circuits are generally undesirable and can quickly overheat electrical circuits. Electricity will almost always choose to flow through the shortest possible path—that is, the path of least resistance—in order to complete a circuit. Short circuits result in excessive current flow, which can rapidly overheat wiring harnesses and cause vehicle fires.

Open Circuit

The term *open circuit* describes any electrical circuit through which no current flow occurs. A switch is used in electrical circuits to intentionally open them. An electrical circuit may also be opened unintentionally. This might occur when a fuse fails, a breaker opens, a wire breaks, or connections corrode.

Ground

The ground represents the point of a circuit with the lowest voltage potential. In vehicles, **ground** or *chassis ground* is integral. This means that the chassis forms the electrical path that acts to supply electrons for all components in the circuit. It has an excess of electrons because it is directly connected to the negative terminal of the battery. For instance, a clearance light has two terminals. One terminal is connected to the chassis ground, which has a negative potential (an excess of electrons). The second terminal is connected to a switch. When the switch is open, there is no path for current flow, so none takes place. The moment the switch is closed, a path for electrical current flow is provided. Electrons can now flow from the chassis ground (negative potential) through the light bulb filament, exiting at the positive terminal, passing through the closed switch, and returning to the positive terminal of the power source—that is, the battery.

This light bulb circuit can now be described as *energized*. Technicians used to working on vehicles always use the term *ground* to mean *chassis ground*. The voltage potential at true ground should measure close to zero.

SHOP TALK

In vehicles, one battery terminal is connected to ground, that is, the chassis. In today's vehicles, this is usually the negative terminal. So to complete a circuit a load must be connected by a current path only from the nongrounded terminal. This greatly reduces the amount of wiring required.

Short to Ground

The term *grounded circuit* is sometimes used to describe a short to ground. This means that when a circuit is closed, the electron flow bypasses the intended load by taking a shorter route to the positive point or terminal of the power source. Shorts to ground have almost no resistance. The result is excessive current flow and overheating.

High-Resistance Circuits

A high-resistance circuit is caused by loose or corroded terminals. In a high-resistance circuit, a path for current flow exists, but the path is too small for the number of electrons that are being pumped through it. The result is overheating.

OHM'S LAW

Ohm's Law says that an electrical pressure of 1V (volt) is required to move 1A (ampere) of current through a resistance of 1 Ω. It is a mathematical formula that technicians must know. In fact, just by playing around with a multimeter, you can prove it. This can be more fun for the beginner than struggling with the math, but doing both is probably best. It is simple to work with, so let's take a look at it:

I = "intensity" = current in amperes
E = EMF = pressure in voltage
R = resistance = resistance to current flow in Ω

$$E = I \times R \qquad R = \frac{E}{T} \qquad I = \frac{E}{R}$$

Ohm's Law can also be expressed in the units of measurement used in the following formulae:

$$V = A \times \Omega \qquad \Omega = \frac{V}{A} \qquad A = \frac{V}{\Omega}$$

Or reference **Figure 5–28**, which shows the application of Ohm's Law formulae. Better still, memorize it—it will prove to be invaluable later when we perform electronic troubleshooting.

WORKING WITH OHM'S LAW

We can solve Ohm's Law with the circular diagram in Figure 5–28 by simply covering the value we wish to find and using simple algebra to calculate it. For instance, to calculate the amount of voltage required to push 4A through a resistance of 3 V, we would do the following: The unknown value is voltage represented by the letter *E* in the Ohm's Law formula. Cover the E. The result is that the I and R values are side by side. Construct an equation as follows:

$$E = I \cdot R$$

$$E = 4\,(\text{amps}) \cdot 3\,(\text{ohms}) = 12\,(\text{volts})$$

FIGURE 5–28 Ohm's Law graphic

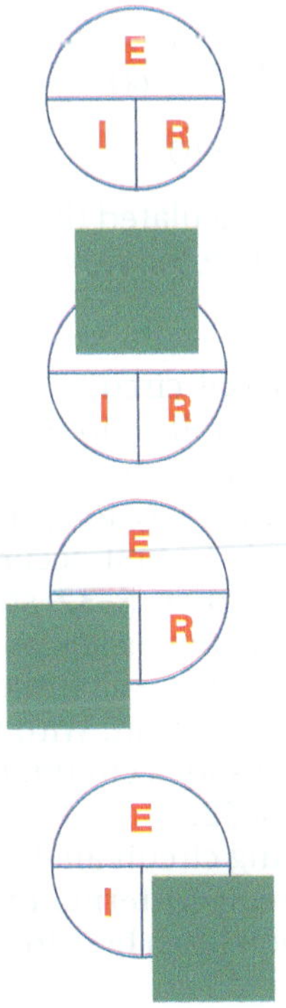

Now we use another set of values and calculate current flow. To determine how much current will flow in a circuit in which there is a charge differential of 12V acting on a resistance of 4 Ω, first we cover the I, which represents current in the formula. Now we see E over R, so an equation can be constructed as follows:

$$I = \frac{E}{R} \qquad I = \frac{12V}{4\Omega} = 3A$$

Ohm's Law Applied to Series Circuits

Now we will use some of this theory in some actual circuits. In a series circuit all of the current flows through all of the resistances in that circuit. The sum of the resistances in the circuit defines the total circuit resistance. If a series circuit were to be constructed with 1Ω and 2Ω resistances, as shown in **Figure 5–29**, total circuit resistance (R_t) would be calculated as follows:

$$R_t = R_1 + R_2$$

$$R_t = 1\Omega + 2\Omega$$

$$R_t = 3\Omega$$

Notice that to construct the formula, each resistance has been identified with a number. Also, the letter *t* is often used to mean *total* when constructing formulae.

If the power source in the series circuit shown is a 12V battery, when the circuit is closed, current flow can be calculated using Ohm's Law as follows:

$$I = \frac{E}{R_t}$$

$$I = \frac{12V}{3\Omega}$$

$$I = 4\,A$$

Ohm's Law Applied to Parallel Circuits

According to Kirchhoff's Law of Current (see the next section), current flowed through a parallel circuit divides into each path in the circuit. When the current flow in each path is added, the total current equals the current flow leaving the power source. When calculating the current flow in parallel circuits, each

FIGURE 5–29 Series-circuit calculation

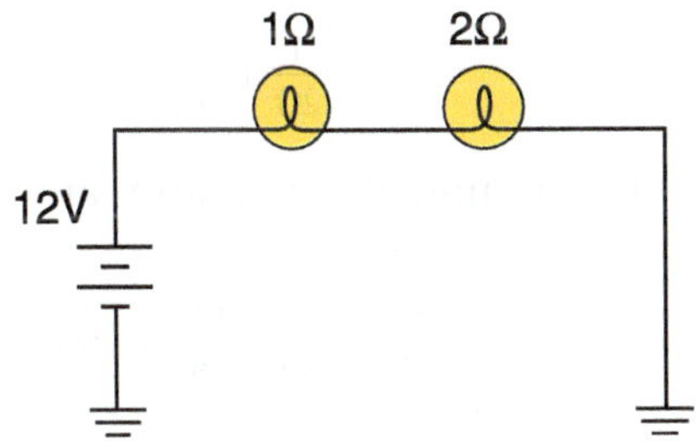

FIGURE 5–30 Parallel-circuit calculation

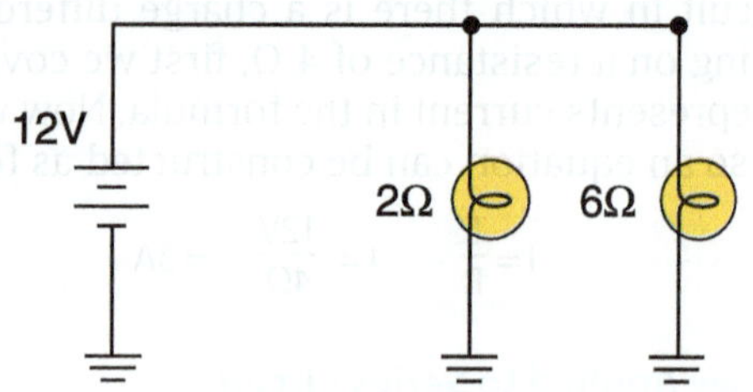

FIGURE 5–31 Multiple branch parallel circuit

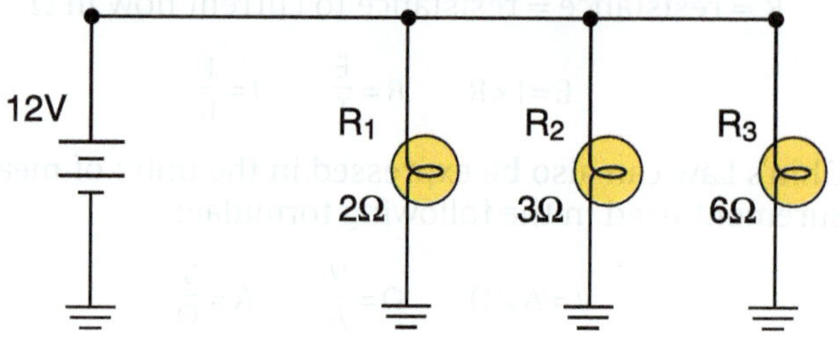

current flow path must be treated as a series circuit, or the total resistance of the circuit must be calculated before calculating total current. When performing calculation on a parallel circuit, remember that more current will always flow through the path with the lowest resistance. This confirms what we said earlier in this chapter about electricity always choosing to flow through the path of least resistance. If a parallel circuit is constructed with 2Ω and 6Ω resistors in separate paths and supplied by a 12V power source as shown in **Figure 5–30**, total current flow can be calculated by treating each current flow path separately, as follows:

$$I_1 = \frac{12V}{2\Omega} = 6\,A$$

$$I_2 = \frac{12V}{6\Omega} = 2\,A$$

$$I_t = 8\,A$$

Note how more current flows through the path with the least resistance. Another means of accomplishing the same result can be used to calculate R_t:

$$R_t = \frac{R_1 \times R_2}{R_1 + R_2}$$

$$R_t = \frac{6\Omega \times 2\Omega}{6\Omega + 2\Omega} = \frac{12V}{8} = 1.5\Omega$$

$$I = \frac{E}{R}$$

$$I = \frac{12V}{1.5\Omega} = 8\,A$$

You may also use the following formula known as the reciprocal method to produce the same result:

$$\frac{1}{R_t} = \frac{1}{R_1} + \frac{1}{R_2}$$

$$\frac{1}{R_t} = \frac{1}{2\Omega} + \frac{1}{6\Omega}$$

Find the common denominator and solve as follows:

$$\frac{1}{R_t} = \frac{3+1}{6\Omega} = \frac{4}{6\Omega} = \frac{6\Omega}{4} = 1.5\Omega$$

Once you know the total resistance value, you use Ohm's Law to calculate current flow:

$$I = \frac{E}{R}$$

$$I = \frac{12V}{1.5\Omega} = 8\,A$$

R_t in Multiple Branch Parallel Circuits

The challenge of applying Ohm's Law to multiple branch parallel circuits is in calculating the total circuit resistance before using Ohm's formula. You should use the reciprocal method we used before. So in a parallel circuit with three branches, the reciprocal formula is as follows:

$$\frac{1}{R_t} = \frac{1}{R_1} + \frac{1}{R_2} + \frac{1}{R_3}$$

Take a look at **Figure 5–31** and apply the data to the foregoing formula as follows:

$$\frac{1}{R_t} = \frac{1}{2\Omega} + \frac{1}{3\Omega} + \frac{1}{6\Omega}$$

Find the common denominator:

$$R_t = \frac{3+2+1}{6\Omega} = \frac{6}{6} = 1\Omega$$

$$R_t = 1\Omega$$

Once you have calculated the R_t value, you can use Ohm's Law to calculate the other circuit conditions.

Series-Parallel Circuit

Some vehicle electrical circuits are of the series-parallel type. A series-parallel circuit combines the characteristics of both the series circuit and the parallel circuit. Once again, the key to applying Ohm's Law to this type of circuit is to first accurately calculate the circuit resistance. **Figure 5–32** shows how a series-parallel circuit calculation is achieved.

The first step is to resolve the resistance of the parallel branch of the circuit. With that done, you can treat the circuit as a series circuit. Follow the steps as outlined in Figure 5–32.

When performing circuit analysis and calculation, visualizing the circuit in terms of paths for current flow helps. When you learn how to visualize electricity

FIGURE 5–32 Series-parallel circuit calculation.

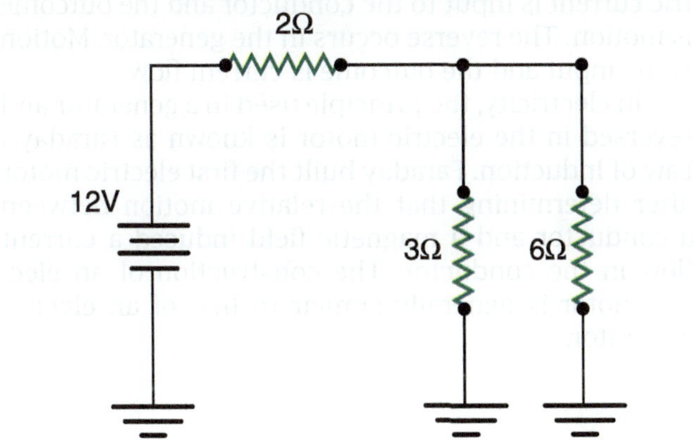

in these terms, it becomes common sense rather than a mystery.

Kirchhoff's Law of Current (Kirchhoff's First Law)

Kirchhoff's Law of Current states that the current flowing into a junction or point in an electrical circuit must equal the current flowing out. If you use a hydraulic comparison and visualize a water wheel being rotated by a flow of water through its paddles, this makes some sense. In a parallel circuit, the current flow through each path of the circuit adds up to the total current flow. Basically, what flows into a circuit must exit it. The calculations that accompany Figure 5–29 and Figure 5–30 prove this.

Kirchhoff's Law of Voltage Drops (Kirchhoff's Second Law)

Kirchhoff's Law of Voltage Drops states that voltage will drop in exact proportion to the resistance, and that the sum of the voltage drops must equal the voltage applied to the circuit. This is a critical law that has an everyday application for technicians working on electrical circuits. Calculation of voltage drop is frequently performed by technicians troubleshooting electrical circuits. Let's take a closer look. If a series circuit is constructed with a 12V power source and 2Ω and 6Ω resistors, as shown in **Figure 5–33**, the voltage drop across each resistor when the circuit is closed and subject to a current flow of 1.5A can be calculated: Voltage drop for R_1:

$$E_1 = I \times R_1$$

$$E_1 = 1.5A \times 2\Omega$$

$$E_1 = 3V$$

Voltage drop for R_2:

$$E_2 = I \times R$$

$$E_2 = 1.5A \times 6\Omega$$

$$E_2 = 9V$$

Total voltage drop through the circuit should equal that of the two preceding calculations.

$$E_1 + E_2 = \text{source voltage}$$

$$3V + 9V = 12V$$

In every case, the sum of the voltage drops in a circuit equals the source voltage. Voltage drop testing is an important diagnostic tool for technicians because it is an easy way to locate unwanted resistances caused by corroded connections, damaged wiring, and failed terminals. It must always be performed on a closed or energized circuit.

POWER

Just as in the internal combustion engine, the unit for measuring power is the **watt**, named after James Watt (1736–1819), usually represented by the letter *P*. In engine technology, power is defined as the *rate* of accomplishing work and, therefore, is always factored by time. This is also true in an electrical circuit. Remember, the definition of an ampere is 6.28 billion billion electrons (1 coulomb) passing a point in a circuit per *second*. The formula for calculating electrical power is:

$$P = I \times E \text{ (spells "pie")}$$

Using the data from the previous formula in which the circuit voltage was 12V and the current flow was 1.5A:

$$P = 1.5 \times 12 = 18W = \text{power consumed}$$

One HP equals 746W, so calculated values can be converted into standard values used to rate some components.

FIGURE 5–33 Voltage drop calculation

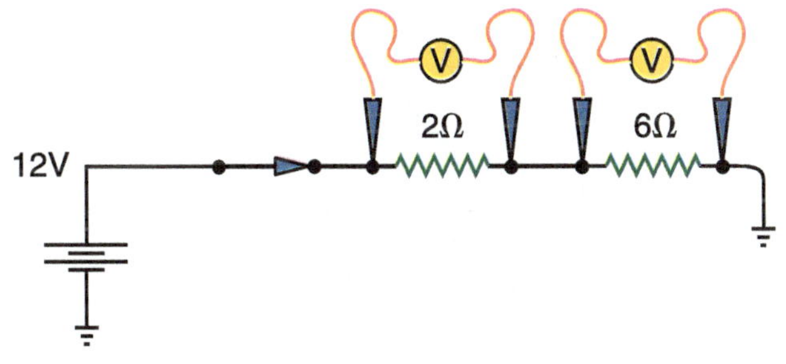

SHOP TALK

How does a light bulb operate? Resistance increases in a conductor with a decrease in cross-sectional area. The light bulb uses a filament of high resistance but conductive alloy that is both long and thin (low cross-sectional area). This filament, when subjected to a current flow, will produce temperatures high enough to radiate photons. Photons are visible light. To ensure that the filament does not rapidly fail by overheating and burning out, it is manufactured from tungsten or alloys capable of sustaining high temperatures and contained in a sealed glass container filled with inert gas under moderate pressure. The inert gas prevents reactions between the filament and the surrounding gas (exposed to oxygen, the filament would rapidly oxidize and burn out) and the pressure inhibits filament evaporation. Incandescent light bulbs are an inefficient method of producing light. This is why light-emitting diodes (LEDs) (explained in Chapter 6) are becoming more common.

5.9 ELECTRIC MOTOR AND GENERATOR PRINCIPLE

Electric motors are used extensively in vehicles. They are required to drive blowers, power windows, and wipers and to crank the engine. After the engine is started, an alternator, which is a type of generator, is required to keep the batteries charged and provide electricity for the entire rig. In this section, we take a look at the operating principles of both the electric motor and the generator. They are looked at together because the process by which an electric motor converts electrical energy into mechanical energy can be reversed. The function of a generator is to convert mechanical energy into electrical energy, and the alternator is the type of generator used in all current vehicles.

Both the electric motor and the generator put the principle of electromagnetic induction (discussed earlier in this chapter) to work, so they both require a conductor and a magnetic field, as shown in **Figure 5–34**.

FIGURE 5–34 DC motor principle. When current is flowed through the wire loop, torque (twisting force) is created.

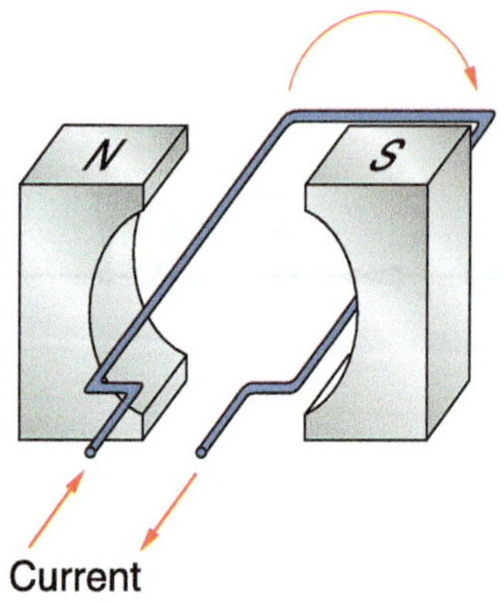

One of these two components must rotate while the other is held stationary. In the case of the motor, electric current is input to the conductor and the outcome is motion. The reverse occurs in the generator. Motion is the input and the outcome is current flow.

In electricity, the principle used in a generator and reversed in the electric motor is known as Faraday's Law of Induction. Faraday built the first electric motor after determining that the relative motion between a conductor and a magnetic field induced a current flow in the conductor. The construction of an electric motor is generally similar to that of an electric generator.

DC MOTORS

The objective of an electric motor is to convert electrical energy into mechanical motion. The current-carrying conductors on a DC motor are loops of wire arranged on an **armature**. The armature is the rotating member of the motor. It is supported on bearings, usually bronze bushings. The loops of wire begin and end at copper terminals called *segments* on the armature shank. These segments on the armature rotate between stationary brushes. The brushes are used to create current flow through the loops of wire. Surrounding the armature are stationary magnets called *fields*. The magnets can be either permanent or electromagnets.

When an electrical current is flowed through the loops of wire on the armature, a force is exerted on the wire. Because the current is flowing in opposite directions on either side of the loop (it passes up on one side, down on the other), this creates torque, causing the armature to rotate. This principle is shown in Figure 5–34 and **Figure 5–35**.

Electric motors can produce high torque at low speeds. In fact, some types of DC motor are designed

FIGURE 5–35 Construction of a simple electric motor

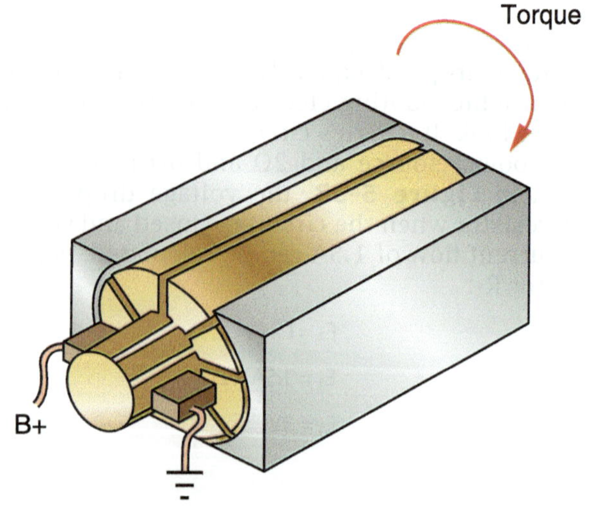

to produce peak torque at 0 rpm. This type of DC motor is used as the cranking motor for internal combustion engines, which are unable to produce working torque at low rpm and need considerable help to get them rotating at a speed at which they can run unassisted.

GENERATORS

Until the 1960s, DC generators were used to meet the electrical needs of vehicles. As we have already discovered, the DC generator functions exactly as a DC motor except in reverse. Mechanical energy, usually from the vehicle engine transmitted by a pulley and drive belts, drives the armature through the stationary fields. This results in current flow in the armature wire loops, which flows through the brushes to the vehicle electrical circuit. Because a generator is simply an electric motor with its function reversed, it can be made to function as a motor simply by connecting it to a power source. This concept is shown in **Figure 5–36**. A disadvantage of the DC generator is its inability to function efficiently at low rpm. This means that under conditions of low rpm operation such as city driving, they are incapable of meeting the electrical requirements of the vehicle.

An alternator is an AC generator. Because an alternator produces AC, this must be rectified to DC so it can be used by the chassis electrical system. In an alternator, the rotor is an electromagnet. A small current is flowed into the rotor field windings by means of brushes and copper slip rings. This produces a magnetic field. As the rotor turns, it induces current flow in stationary conductors called the *stator windings*. An alternator produces AC because the current is induced in opposite directions when opposite magnetic poles are driven through the stator conductor loops. Diodes (electrical one-way valves, described in Chapter 6) are used to change or rectify the AC to DC so it can be used by the vehicle electrical system.

Reluctor-Type Generators

A common type of shaft-speed sensor used in today's vehicles is really a miniature AC generator. It consists of a permanent magnet and a coil of wire mounted to the toothed reluctor. The reluctor is located on the rotating shaft. As the teeth in the reluctor cut through the magnetic field, voltage is induced in the pickup coil. This voltage (and the frequency) increases proportionally with an increase in shaft speed. This type of sensor is commonly used to signal wheel speed, transmission tailshaft speed, engine speed, and so on, to electronic control modules (ECMs).

CAPACITANCE

The term *capacitance* is used to describe the electron storage capability of a commonly used electrical component known as a **capacitor**. *Capacitors*, which are also called condensers, all do the same thing: They store electrons. The simplest type of capacitor consists of two conductors separated by some insulating material called **dielectric**. The conductor plates could be aluminum and the dielectric may be mica (silicate mineral). The greater the dielectric (insulating) properties of the material, the greater the resistance to voltage leakage. When a capacitor is connected to an electrical power source, it becomes capable of storing electrons from that power source. When the capacitor's charge capability is reached, it will cease to accept electrons from the power source. The charge is retained in the capacitor until the plates are connected to a lower voltage electrical circuit. At this point, the stored electrons are discharged from the capacitor into the lower potential (voltage) electrical circuit.

As a capacitor is electrified, for every electron removed from one plate, one is loaded onto the other plate. The total number of electrons in the capacitor is identical when it is in either the electrified or completely neutral states. What changes is the location of the electrons. The electrons in a fully charged capacitor will in time "leak" through the dielectric until both conductor plates have an equal charge. At this point, the capacitor is in a discharged condition. Capacitance, the ability to store electrons, is measured in *farads* (named after Michael Faraday, the discoverer of the principle). One farad is the ability to store 6.28 billion billion electrons at a 1-volt charge differential.

FIGURE 5–36 Electric motors and generators convert one energy form to another.

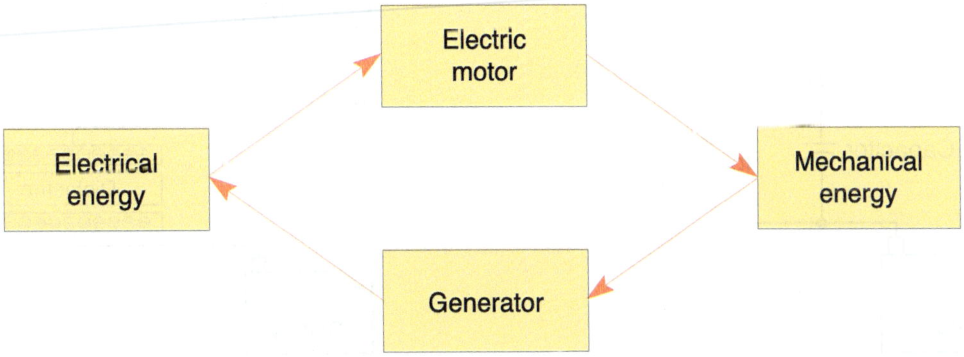

Most capacitors have much less capacitance than this, so they are rated in picofarads (trillionths of a farad) and microfarads (millionths of a farad).

$$1\ \text{farad} = 1\text{F}$$

$$1\ \text{microfarad} = 1\mu\text{F} = 0.000001\text{F}$$

$$1\ \text{picofarad} = 1\text{pF} = 0.000000000001\text{F}$$

Types of Capacitors

Many capacitors used in electric and electronic circuits are fixed-value capacitors. Fixed-value capacitors are coded by capacitance and voltage rating. Some capacitors have a variable capacity. Variable capacitors have a combination of fixed and moving conductor plates and the capacitance is varied by a shaft that rotates the moving plates. In some variable capacitance-type capacitors, the dielectric may be air.

Electrolytic-type capacitors tend to have much higher capacitance ratings than nonelectrolytic types. They are polarized, so they must be connected accordingly in a circuit. The dielectric in this type of capacitor is the oxide formed on an aluminum conductor plate, but the operating principle is identical.

Figure 5–37 and **Figure 5–38** show how a capacitor functions in an electrical circuit. In Figure 5–37, a capacitor is located parallel to the circuit load to smooth voltage changes in the circuit. When the circuit switch is open, no electron flow will occur. When the circuit has been closed by the switch, the battery's positive terminal pulls electrons from the capacitor, leaving its positive plate with a deficit of electrons. This will attract electrons to its negative plate: The dielectric keeps the electrons from being pulled through to the capacitor's positive plate. In Figure 5–38, the capacitor is shown with a stored charge differential that can be unloaded when the circuit switch is opened.

When working on electrical circuits, note that capacitors can retain a charge for a considerable time after the circuit has been opened and current flow has ceased. Accidental discharge can damage circuit components and may represent an electric shock hazard.

Capacitors are used extensively in electrical and electronic circuits performing the following roles:

- **Power supply filter.** Smoothes a pulsating voltage supply into a steady DC voltage form.

FIGURE 5–37 Operating principles of a capacitor

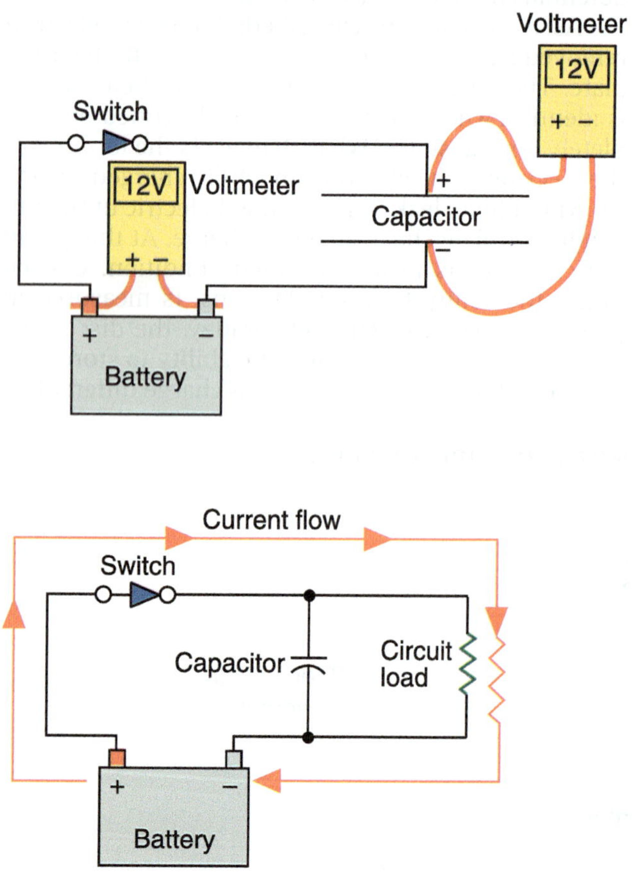

FIGURE 5–38 Current flow in a fully charged capacitor

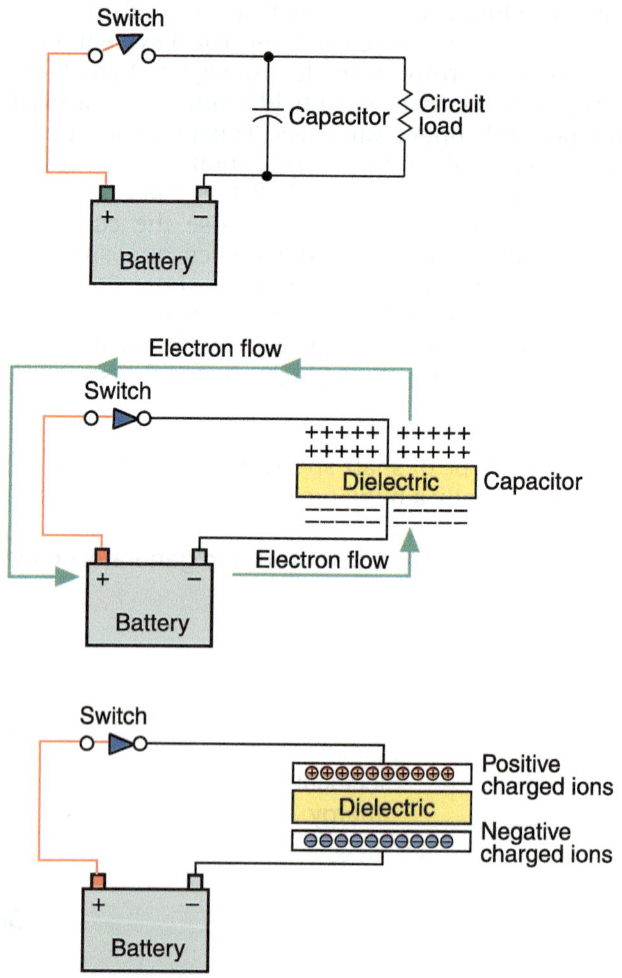

- **Spike suppressant.** When digital circuitry is switched at high speed, transient (very brief) voltage inductions may occur. Capacitors can eliminate these spikes or glitches by compensating for them.
- **Resistor-capacitor circuits (R-C circuits).** R-C circuits incorporate a resistor and a capacitor. These are used to reshape a voltage wave or pulse pattern from square wave to sawtooth shaping or modify a wave to an alternating pattern. These concepts are discussed in more detail in Chapter 6.

5.10 COILS, TRANSFORMERS, AND SOLENOIDS

When electrons are moved through a conductor, an electromagnetic field is created surrounding the conductor. When the conductor is a wire wound into a coil, the electromagnetic field created is stronger. As discussed previously, such coils are the basis of electric **solenoids** and motors. Coils are used in electronic circuits to shape voltage waves because they tend to resist rapid fluctuations in current flow. Like capacitors, coils can be used in electrical circuits to reshape voltage waves. Also, the energy in the electromagnetic field surrounding a coil can be induced in any nearby conductors. If the nearby conductor happens to be a second coil, a current flow can be induced in it. The principle of a transformer is essentially that of flowing current through a primary coil and inducing a current flow in a secondary or output coil. Variations on this principle are coils that are constructed with a movable core that permits their inductance to be varied, thereby altering frequency.

TRANSFORMERS

The basic **transformer** functions by having two coils located beside each other. When an electric current is flowed through either of the coils, the resulting electromagnetic field induces a current flow in the other coil. This principle is shown in **Figure 5–39**.

FIGURE 5–39 Transformer operating principle

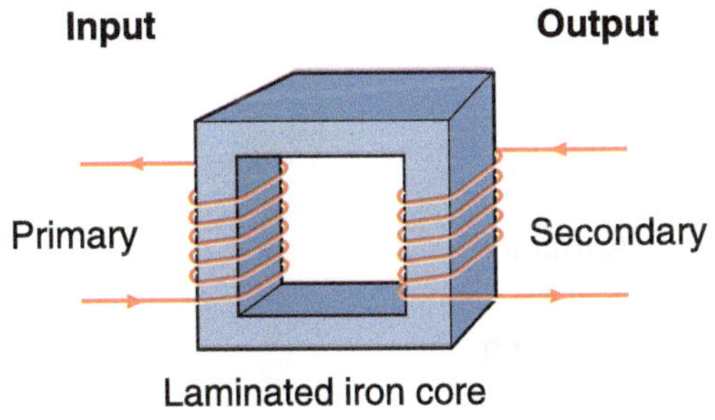

The coil through which current is flowed is known as the *primary coil.* The coil through which current is induced by the electromagnetic effect of the primary coil is known as the *secondary coil.* A transformer requires a changing current characteristic to operate. AC current constantly changes, so when it is flowed through the primary coil, it creates a changing magnetic field required to induce current flow in the secondary coil. The operating principle of a transformer is often described as *mutual inductance.* Transformers of various types are used in vehicle electrical circuits, but they all fall into three categories of function: stepping up voltage, stepping down voltage, and isolating portions of a circuit. Therefore, we can say in a general sense that transformers have three functions in truck electrical circuits: to step up voltage, to step down voltage, and circuit isolation.

Isolation Transformers

In an isolation transformer, the primary and secondary coils have the same number of windings, producing a 1:1 input to output ratio similar to that shown in Figure 5–39. Their objective is to "isolate" one portion of an electrical circuit from another. Secondary voltage and current equal that of the primary coil. Isolation transformers are used in electronic circuits to isolate or protect subcircuits against voltage surges.

Step-Up Transformer

The objective of a **step-up transformer** is to multiply the primary coil voltage by the winding ratio of the primary coil versus that of the secondary coil. For instance, if the primary to secondary coil winding ratio is 1:10, 12V through the primary coil will produce 120V through the secondary coil with a similarly proportional drop in current flow. With electricity, you never get something for nothing. When voltage is stepped up by a transformer, current drops by a corresponding amount. Examples of step-up transformers are automotive ignition coils and some types of injector driver units used on diesel electronic unit injection (EUI) systems.

Step-Down Transformer

Step-down transformers function in an opposite way from step-up transformers. The primary coil winding ratio exceeds that of the secondary coil, resulting in a diminished output voltage and increased output current. This results in reduced voltage and amplified current. A common shop use of a step-down transformer is an AC welding station. Electric arc welding requires high current at a relatively low voltage and the AC welding unit uses transformers to achieve this.

Ignition Coils

SI engines require high voltage, up to 50,000V, to fire spark plugs. The available voltage in a vehicle's electrical system is somewhere between 12 and 14V DC. Direct current will not operate in a transformer

FIGURE 5–40 Operating principle of an ignition coil.

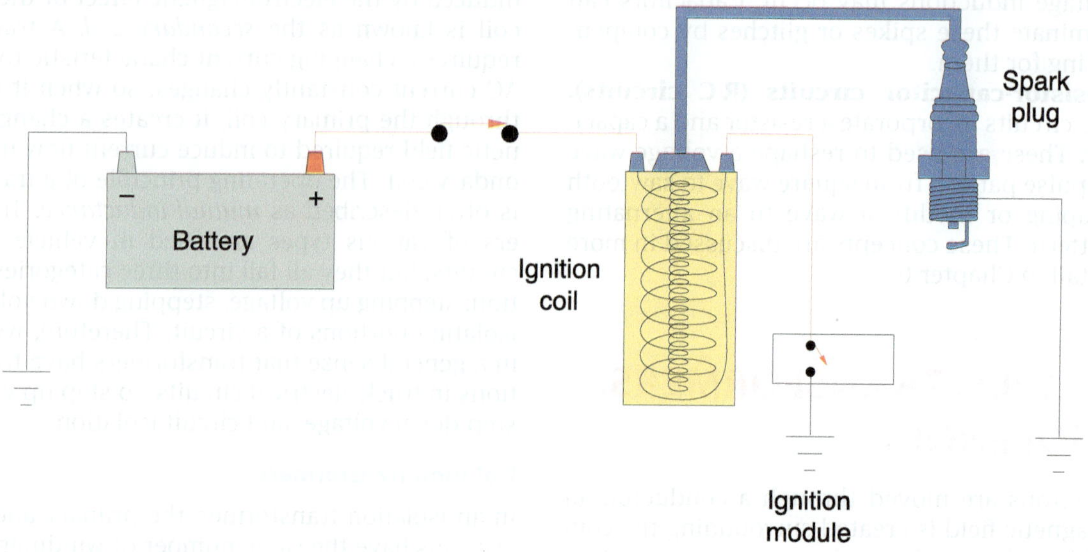

because it will not produce a changing electromagnetic field. To make an ignition coil operate, the low-voltage primary coil must be pulsed, that is, switched on and off at high speed. The ignition coil uses a high-current, low-voltage primary coil with a low number of windings to induce a low-current, high-voltage output in a secondary coil with a large number of windings. Each time the primary current is pulsed, a magnetic field is created and then collapsed, inducing current flow in the secondary coil. The high-speed switching is performed by a transistor located in and controlled by the ignition module. **Figure 5–40** shows the coil's role in firing a spark plug.

Transformer Summary

- Two coils are arranged so one is subject to a magnetic field created in the other.
- The input coil is the primary coil.
- The output coil is the secondary coil.
- Step-up transformers have secondary coils with a greater number of windings.
- Step-down transformers have secondary coils with a lower number of windings.

SOLENOIDS AND MAGNETIC SWITCHES

Electromagnetic switches are used in vehicle electrical circuits so a low-current circuit can control a high-current circuit. A typical example would be in the cranking circuit, as shown in **Figure 5–41**. An electromagnetic switch opens or closes an electrical subcircuit by using an electromagnet to pull on movable contacts when energized.

A solenoid is not much different in its operating principle from a magnetic switch except that it functions specifically to convert electrical energy into mechanical movement. Solenoids are used extensively on vehicles in such applications as starter engage mechanisms, diesel electronic unit injector control, gasoline injector switching, pilot switches, automatic transmission clutch controls, and so on. They may be small or large, but they all operate on the same principle of converting an electrical signal into mechanical movement. A solenoid consists of two components: a coil and an armature. The coil is generally stationary. When the coil is energized, the armature moves either into or out of the coil, depending on polarity. The action of a typical solenoid is shown in **Figure 5–42** and **Figure 5–43**.

FIGURE 5–41 Magnetic switch shown open

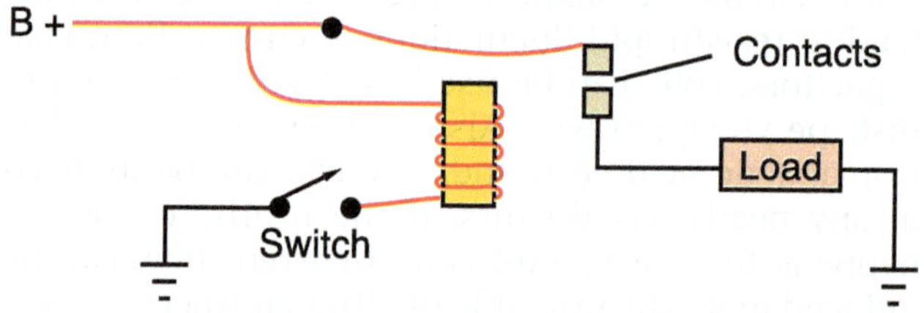

FIGURE 5–42 A solenoid when not energized

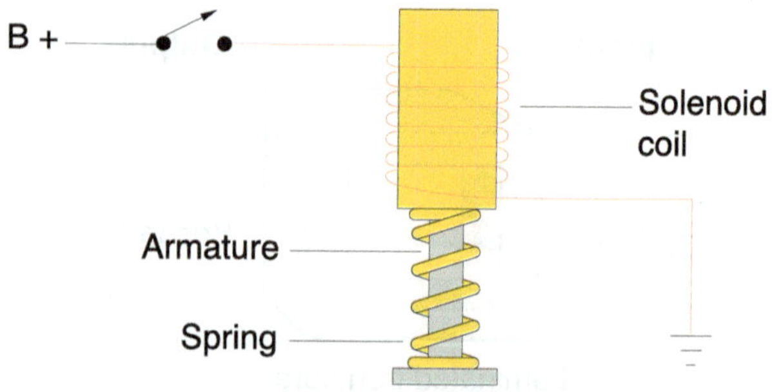

FIGURE 5-43 An energized solenoid

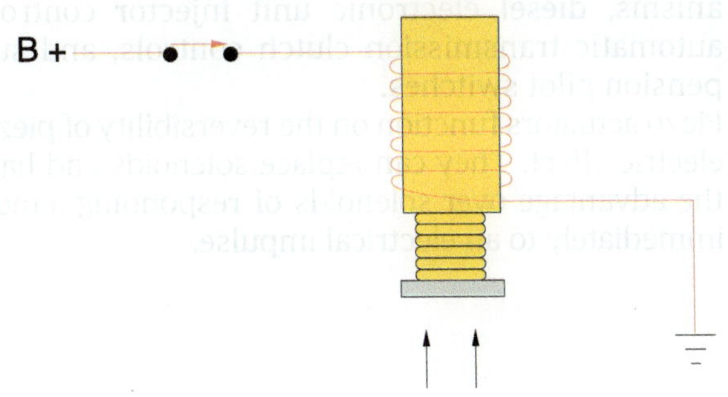

PIEZO ACTUATORS

Earlier in the chapter we described piezoelectricity and said that its effect is reversible. When the opposite faces of a piezo crystal have a voltage applied to them, the crystal lattice (the way the crystals structurally bond to each other) changes almost instantly, resulting in either expansion or contraction. This change of shape produces a powerful force that can be used for triggering mechanical movement in much the same way as a solenoid but at much higher speeds. Solenoids use electromagnetism and respond more slowly due to the time required to build and collapse electromagnetic fields.

Uses of Piezo Actuators

Reverse piezoelectric effect is becoming common in diesel engine actuators, replacing solenoids as fuel injector switches. The reason for adopting piezoelectric actuators is much faster response times along with lower electrical power draw. However, the use of piezo actuators is not limited to the fuel injection equipment. As time makes this type of actuator more compact and less costly, you can expect to see them increasingly used as replacements for solenoids.

SUMMARY

This concludes the introduction to electrical principles. It is not necessary that you understand everything in this chapter before progressing to the next few that deal mostly with the application of these electrical principles, but it will help. Knowing something about electrical theory can greatly increase your ability to visualize electrical circuit behavior and make you more effective at mastering the challenges of electrical troubleshooting. Make sure you reference this chapter again when you have acquired some knowledge of applied electricity—or when an electrical principle is referred to but not explained in later chapters.

- All matter is composed of atoms.
- All atoms have an electrical charge. When an atom is balanced (the number of protons match the number of electrons), the atom can be described as being in an electrically neutral state.
- All matter is electrical in essence. *Electricity* concerns the behavior of atoms that have become, for whatever reason, unbalanced or electrified.
- Electricity may be defined as the movement of free electrons from one atom to another.
- Current flow is measured by the number of free electrons passing a given point in an electrical circuit per second.
- Electrical pressure or charge differential is measured in *volts*, resistance in *ohms*, and current in *amperes*.
- The magnetic properties of some metals such as iron are due to electron motion within the atomic structure.
- A direct relationship exists between electricity and magnetism. Electromagnetic devices are used extensively on vehicles.
- Magnetomotive force (mmf) is a measure of electromagnetic field strength: Its unit is ampere-turns.
- Ohm's Law is used to perform circuit calculations on series, parallel, and series-parallel circuits.
- In a series circuit, there is a single path for current flow and all of the current flows through each resistor in the circuit.
- A parallel circuit has multiple paths for current flow: The resistance in each path determines the current flow through it.
- Kirchhoff's Law of Voltage Drops states that the sum of voltage drops through resistors in a circuit must equal the source voltage.
- When current is flowed through a conductor, a magnetic field is created.
- Reluctance is resistance to the movement of magnetic lines of force. Iron cores have permeability and are used to reduce reluctance in electromagnetic fields.
- Capacitors are used to store electrons. They consist of conductor plates separated by a dielectric.
- Capacitance is measured in farads. Capacitors are rated by voltage and by capacitance.
- When current is flowed through a wire conductor, an electromagnetic field is created. When the wire is wound into a coil, the electromagnetic field strength is intensified.
- The principle of a transformer can be summarized by describing it as flowing current through a primary coil and inducing a current flow in a secondary or output coil.
- Transformers can be grouped into three categories: isolation, step-up, and step-down.

- An electromagnetic switch is used in a truck electrical circuit to enable a low-current circuit to control a high-current circuit.
- A relay is an example of an electromagnetic switch.
- A solenoid uses similar operating principles to an electromagnetic switch except that it converts electromagnetic energy into mechanical movement.
- Solenoids are used extensively in truck electrical circuits for functions such as starter engage mechanisms, diesel electronic unit injector controls, automatic transmission clutch controls, and suspension pilot switches.
- Piezo actuators function on the reversibility of piezoelectric effect. They can replace solenoids and have the advantage over solenoids of responding almost immediately to an electrical impulse.

REVIEW QUESTIONS

1. A material described as an insulator has how many electrons in its outer shell?
   a. less than 4
   b. 4
   c. more than 4
2. Which of the following is a measure of electrical pressure?
   a. amperes
   b. ohms
   c. voltage
   d. watts
3. Which of following units of measurement expresses electron flow in a circuit?
   a. volts
   b. watts
   c. farads
   d. amps
4. How many electrons does the element silicon have in its outer shell?
   a. 2
   b. 4
   c. 6
   d. 8
5. Who originated the branch of electricity generally described as *electromagnetism*?
   a. Franklin
   b. Gilbert
   c. Thomson
   d. Faraday
6. Which of the following elements could be described as being electrically inert?
   a. oxygen
   b. neon
   c. carbon
   d. iron
7. Which of the following is a measure of charge differential?
   a. voltage
   b. wattage
   c. amperage
   d. ohms
8. An element classified as a semiconductor would have how many electrons in its outer shell?
   a. less than 4
   b. 4
   c. more than 4
   d. 8
9. Which of the following describes resistance to movement to magnetic lines of force?
   a. reluctance
   b. inductance
   c. counter electromotive force
   d. capacitance
10. Use Ohm's Law to calculate the current flow in a series circuit with a 12V power source and a total circuit resistance of 6 Ω.
11. Calculate the power consumed in a circuit through which 3A are flowed at a potential difference of 24V.
12. A farad is a measure of:
    a. inductance.
    b. reluctance.
    c. charge differential.
    d. capacitance.
13. Technician A says that magnetism is a source of electrical energy in truck electrical systems. Technician B says that a chemical reaction is a source of electrical energy in truck electrical systems. Who is correct?
    a. Technician A only
    b. Technician B only
    c. both A and B
    d. neither A nor B
14. Which of the following is a measure of electrical circuit current flow?
    a. watts
    b. volts
    c. ohms
    d. amps
15. Resistance to current flow is measured in:
    a. amps
    b. volts
    c. watts
    d. ohms
16. Which of the following is the best insulator of electricity?
    a. copper
    b. aluminum
    c. distilled water
    d. tap water
17. Which type of circuit offers only one path for current flow?
    a. series
    b. parallel
    c. series-parallel
    d. open
18. A break in a circuit that causes a loss of continuity is called a:
    a. short.
    b. ground.
    c. open.
    d. overload.
19. Which of the following would cause high resistance in a closed electrical circuit?
    a. short to ground
    b. corroded terminals
    c. broken wire
    d. heavy gauge electrical wiring

20. If one branch of a parallel circuit has high resistance, how will total circuit current be affected?
    a. Current will increase.
    b. Current will decrease.
    c. Voltage will decrease.
    d. Voltage will increase.
21. When describing the negative post of a battery terminal in an energized electrical circuit, which of the following applies?
    a. It is the source of electrons flowing through the circuit.
    b. It has a deficit of electrons.
    c. It is more positive than the first load in the circuit.
    d. It is electrically neutral.
22. If an electrical connection in a closed electrical circuit heats up, which of the following would be the more likely cause?
    a. a blown fuse
    b. wire gauge thickness too large
    c. corroded connectors
    d. overcharged battery
23. When a dead short occurs, which of the following would be true?
    a. Circuit resistance increases.
    b. Current flow increases.
    c. Voltage increases.
    d. All of the above are true.
24. Which of the following components is used in a truck electrical circuit to convert electromagnetic energy into mechanical movement?
    a. an electromagnetic switch
    b. a relay
    c. a transformer
    d. a solenoid
25. A circuit protection device that opens when overloaded and then automatically resets is known as a:
    a. fuse
    b. ballast resistor
    c. cycling circuit breaker
    d. noncycling circuit breaker
26. According to Ohm's Law, what should happen to current flow if circuit resistance increases and voltage remains the same?
    a. It ceases.
    b. It increases.
    c. It decreases.
27. If you add up all the voltage drops (Kirchhoff's Second Law) in any closed electrical circuit, what should they add up to?
    a. 12 volts
    b. source voltage
    c. zero
    d. voltage at the ground terminal
28. What is the moving electrical component in a solenoid known as?
    a. the armature
    b. the coil
    c. the actuator
    d. the pinion
29. Which of the following devices is used to open and close electrical circuits?
    a. a solenoid
    b. a magnetic switch
    c. a transformer
    d. a capacitor
30. According to Watt's Law, if current flow through an electrical circuit increases, what is likely to happen to the power consumed?
    a. It increases.
    b. It decreases.
    c. It has no effect.

6

Prerequisite: Chapter 5

FUNDAMENTALS OF ELECTRONICS AND COMPUTERS

OBJECTIVES

After reading this chapter, you should be able to:

- Outline some of the developmental history of electronics.
- Describe how an electrical signal can be used to transmit information.
- Define the term *pulse width modulation.*
- Define the principle of operation of N- and P-type semiconductors.
- Outline the operating principles and applications of diodes.
- Describe the construction and operation of a typical transistor.
- Describe what is meant by the *optical spectrum.*
- Identify some commonly used optical components used in electronic circuitry.
- Explain what is meant by an *integrated circuit* and outline its application in on-board vehicle electronics.
- Define the role of gates in electronic circuits.
- Describe the operating modes of some common gates used in electrical circuits, including AND, OR, and NOT gates.
- Interpret a *truth table* that defines the outcomes of gates in an electronic circuit.
- Explain why the binary numeric system is used in computer electronics.
- Define the role of an electronic control module in an electronic management system.
- Outline the distinct stages of a computer processing cycle.
- Describe the data retention media used in vehicle ECUs.
- Demonstrate an understanding of input circuits on a vehicle electronic system.
- Troubleshoot a potentiometer-type TPS.
- Describe the operating principles of some accident avoidance systems.

KEY TERMS

active cruise with braking (ACB)
AND gate
anode
bias
binary system
bipolar
bit
byte
cathode
central processing unit (CPU)
chopper wheel
collision mitigation system (CMS)
collision warning system (CWS)
controller
Darlington pair
data processing
doping
driver display unit (DDU)
driver interface unit (DIU)
duty cycle
electronic control module (ECM)
electronically erasable programmable read-only memory (EEPROM)
electron
fiber optics
gateway module
global positioning satellites (GPS)
Hall effect
infotainment system
input circuit
integrated circuit (I/C)

lane departure warning (LDW) system
light-emitting diode (LED)
memory banks
NOT gate
OR gate
output circuit
photonic semiconductor
potentiometer
programmable logic controller (PLC)
programmable read-only memory (PROM)
protons
pulse wheel
pulse width (PW)
pulse width modulation (PWM)
random access memory (RAM)
ratiometric analog
read-only memory (ROM)
reluctor
semiconductor
solid state
thermistor
thyristor
tone wheel
transistor
triangulation
trilateration
truth table
VORAD
Wingman
zener diode

INTRODUCTION

Basic electrical principles were introduced in Chapter 5. We discovered in that chapter that electricity is the form of energy that results from charged particles such as **electrons** or **protons** being either static (not flowing) or dynamic (flowing) as current flow in a circuit. Electronics is the branch of electricity that addresses the behavior of flows of electrons through solids, liquids, gases, and across vacuums. It is probably essential to have a sound understanding of basic electricity before attempting to understand basic electronics theory.

The science of electronics began with the discovery of the *electron* by J. J. Thomson (1856–1940) in 1897. This discovery quickly resulted in the invention of the diode (1904), the triode (1907), and, perhaps most importantly, the **transistor** in 1947 by three Bell Laboratories scientists: Bardeen, Brattain, and Shockley. A diagram of the world's first transistor is shown in **Figure 6–1**. Diodes and transistors are fundamental components in any electronic circuit. They are defined as solid-state components.

FIGURE 6–1 Cutaway view of the world's first transistor, produced by Bell Laboratories.

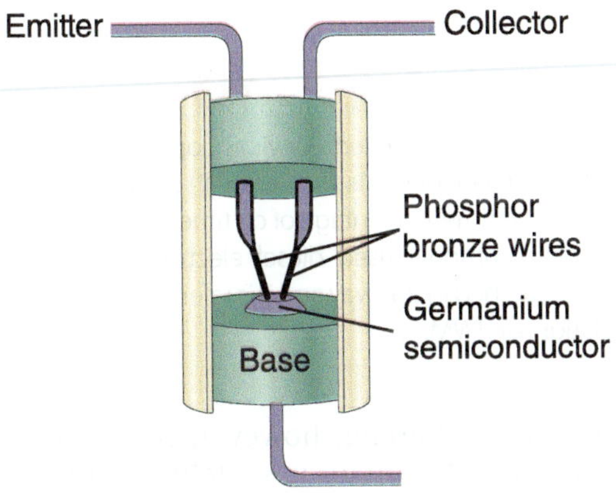

SOLID-STATE DEVICES

Solid state devices are the building blocks of electronic circuits. They are called solid-state devices because they are manufactured from solid materials called **semiconductors**. A number of different types of semiconductor devices exist, but the previously mentioned diode and transistor are probably the most important. Though they are never repaired, the technician will benefit from having a fundamental knowledge of these devices, especially when troubleshooting.

SCOPE

The objective of this chapter is to deliver a foundation-level understanding of electronics. It should be recognized that most of the concepts introduced can be studied in much greater depth, but this would probably exceed what the average truck technician is required to know at this time. The terms *electronics* and *electronic engine management* are generally used in automotive technology to describe any systems managed by a computer. Basic computers are introduced in terms of their operating principles in the final portion of this chapter. The application of electronics theory is explored in Chapters 8, 10, and 11.

6.1 USING ELECTRONIC SIGNALS

Electricity can be used to transmit signals or information as well as current. An electric doorbell is designed to deliver a simple electrical signal. If it is functioning properly, the circuit remains unenergized, or inactive, for most of its service life. A visitor signals his or her presence by mechanically depressing a switch that closes the electrical circuit and produces an audible chime. Electronic signals often operate in a somewhat similar manner using on/off signals at high speeds and of different durations. When an electronic circuit is constructed to manage information, low-current and low-voltage circuits are generally used. Electrical signals can be classified as

FIGURE 6–2 Analog signal

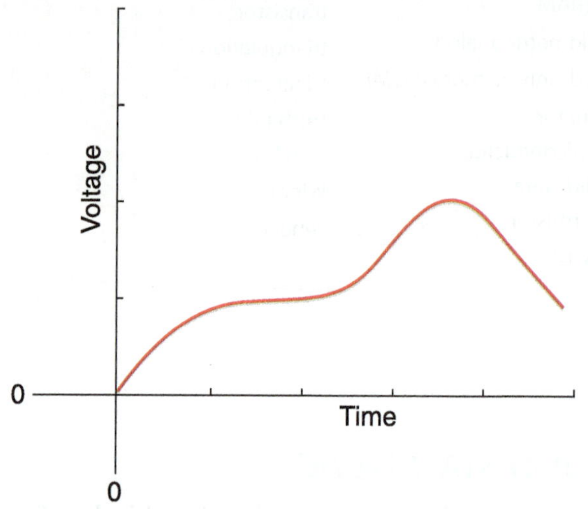

analog or digital. An analog signal operates on variable voltage values. A graphic representation of an analog signal is shown in **Figure 6–2**.

DIGITAL BASICS

Digital signals operate on specific voltage values, usually the presence or lack of voltage. Alternatively, it could be a low (LO) voltage and high (HI) voltage. Simple electronic circuits can be designed to transmit relatively complex data by using digital signals. A digital signal produces a square wave, which is shown in diagram form in **Figure 6–3**.

SHOP TALK

A square wave is a train of high and low/no voltage pulses. Some types of digital sensors use varying square wave frequencies to transmit information.

FIGURE 6–3 Square wave digital signal

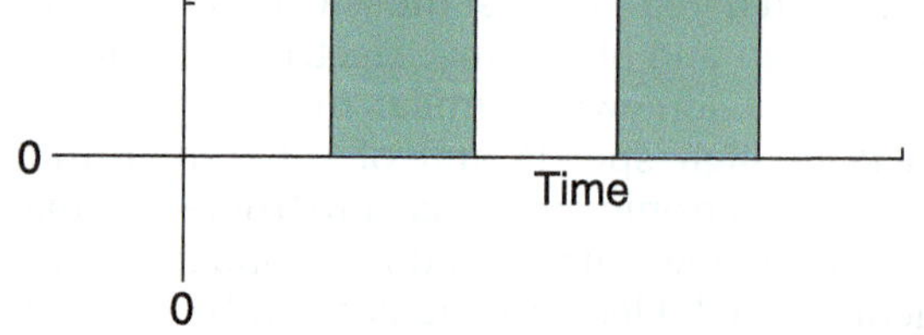

FIGURE 6–4 Varying frequency: The number of pulses per second is measured in hertz

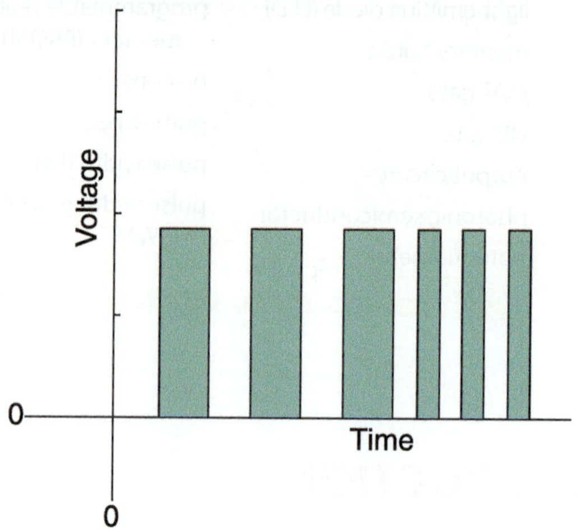

A square wave is a train of voltage pulses with specific high and low values. When the high and low values do not change from the predetermined specific values, the elements that may be changed are *frequency* and **duty cycle**. Frequency is simply the number of pulses per second. It is expressed in *hertz* (from H. R. Hertz, a German physicist). Information can be transmitted by varying the frequency of a signal, as in **Figure 6–4**.

FREQUENCY

Varying duty cycle can also be used to transmit information. A square wave of fixed frequency but variable duty cycle is achieved by changing the percentage of on time. This is known as **pulse width modulation (PWM)**. This term is used extensively in digital electronics. If a circuit consisting of a power source, light bulb, and switch is constructed, the switch can be used to "pulse" the on/off time of the bulb. This pulsing can be coded into many types of data such as alpha or numeric values. Pulses are controlled, immediate variations in current flow in a circuit. In fact, the increase or decrease in current would ideally be instantaneous. If this were so, the pulse could be represented graphically as in **Figure 6–5**.

SHOP TALK

Some computer-controlled devices operate on a fixed frequency but variable **pulse width (PW)** duration—that is, on time versus off time. Percentage of on time is referred to as duty cycle. The commonly used diesel electrohydraulic injectors (EHIs) are switched in this way and it is referred to as pulse width modulation, or PWM.

True pulse shaping, however, results in graduated rise when the circuit is switched to the on state

FIGURE 6–5 Pulse width modulation: Note how the on time of the pulses is varied while the frequency remains the same. On time is referred to as duty cycle.

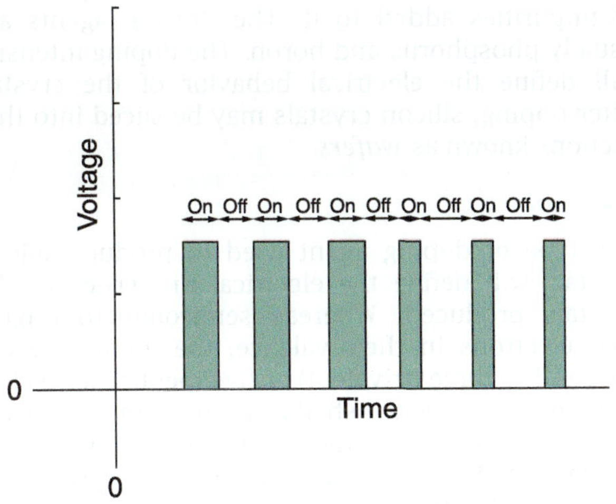

and in graduated fall when the circuit is switched to the off state. This explains why a square wave is often not truly square when displayed on a labscope. We call the deviation from the "square" appearance *ringing*. Most true square waves usually display some ringing when viewed on a scope. **Figure 6–6** compares a theoretical square sine wave with an actual wave.

Waves are rhythmic fluctuations in circuit current or voltage. They are often represented graphically and are described by their graphic shapes. **Figure 6–7** shows some typical waveforms.

SIGNALS

The term *signals* is therefore used to describe electrical pulses and waveforms that are spaced or shaped to transmit data. The processes that are used to shape data signals are known as *modulation*. In vehicle electronics, the term *modulation* is more commonly used in reference to digital signaling. An example would be the signal that a diesel engine **electronic control module (ECM)** uses to control the actuator in an EHI.

FIGURE 6–6 Diagram of theoretical square sine wave compared with actual square sine wave shown with ringing.

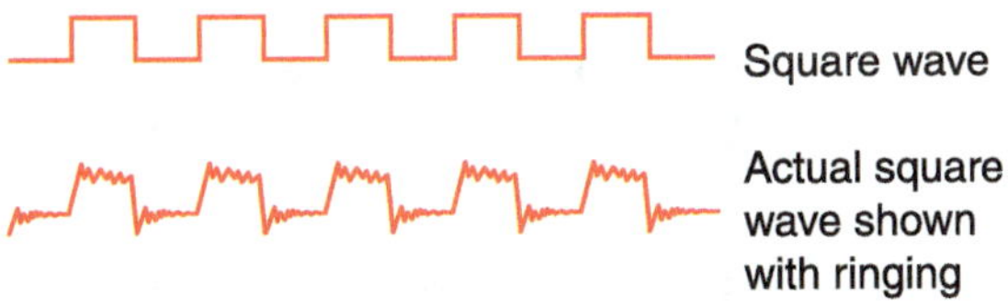

FIGURE 6–7 Waveform characteristics

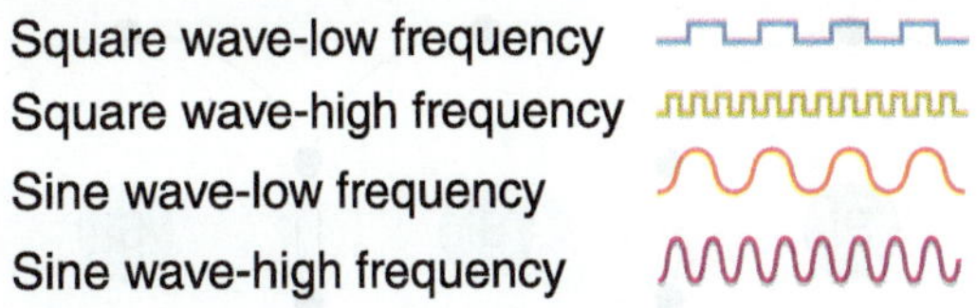

MODULATION

A PWM signal can be divided into primary modulation, which controls the amount of on time, and secondary or submodulation, which controls the current flow. When we study Allison electronic transmissions in Chapter 21 we will take a closer look at primary and submodulation. Frequency is another element of an electronic signal. A signal modulated at a frequency of 50 Hz completes 50 cycles per second. If voltage is the electrical signal that is being modulated at a frequency of 50 Hz, then each second is divided into 50 segments within which the voltage will be in the on state for a fraction of time. The percentage of on time is expressed as duty cycle or PW. A 100 percent duty cycle would indicate the maximum on time signal. A 0 percent duty cycle would indicate no signal or maximum off time.

Noise

Electronic *noise* is usually unwanted pulse or waveform interference that can scramble signals but it should be noted that all electrical and electronic components produce electromagnetic fields that may generate noise. Note also that all electronic circuits are vulnerable to magnetic and electromagnetic fields. The ringing shown in the actual square sine wave in Figure 6–6 is what electronic noise can look like in diagram form. When electronic noise becomes excessive, signals can be corrupted.

A corrupted signal means that the information intended for relay is inaccurate. Most electronic circuits need protection against electronic noise. This means they must be shielded from interference from low-level radiation such as other vehicle electrical systems, high-tension electrical wiring, and radar.

6.2 SEMICONDUCTORS

Semiconductors are a group of materials with exactly four electrons in their outer shell. As such, they cannot be classified as either insulators or conductors. Silicon is the most commonly used semiconductor in the manufacture of electronic components, but other substances, such as germanium, are also used. The first transistor shown in Figure 6–1 used a germanium semiconductor. Silicon and germanium both have four electrons in their valence shells, but they would

FIGURE 6–8 Semiconductor crystal formation

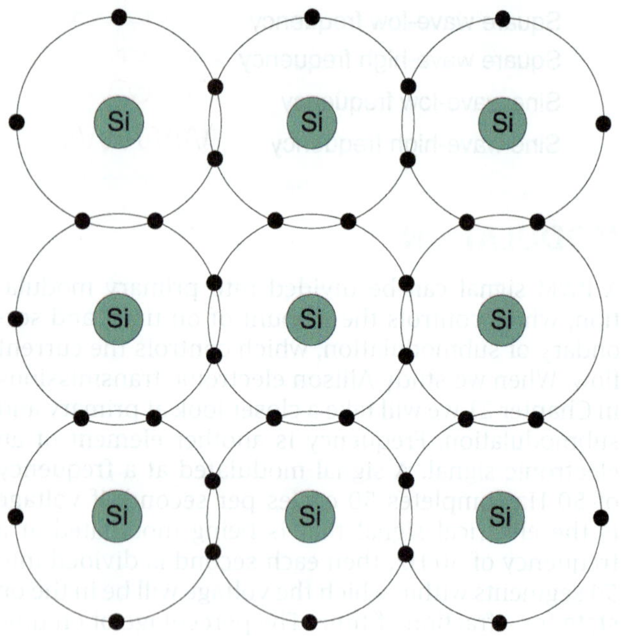

"prefer" to have eight. This means that semiconductor atoms readily unite in clusters called crystals, sharing electrons in their outer shells. Silicon and germanium can be grown into large crystals by heating the elements to their melting temperature, followed by a period of cooling. **Figure 6–8** shows how electrons in the valence shell are shared in a semiconductor crystal, and **Figure 6–9** shows the crystal structure of a crystallized cluster of germanium atoms.

FIGURE 6–9 A crystallized cluster of germanium atoms.

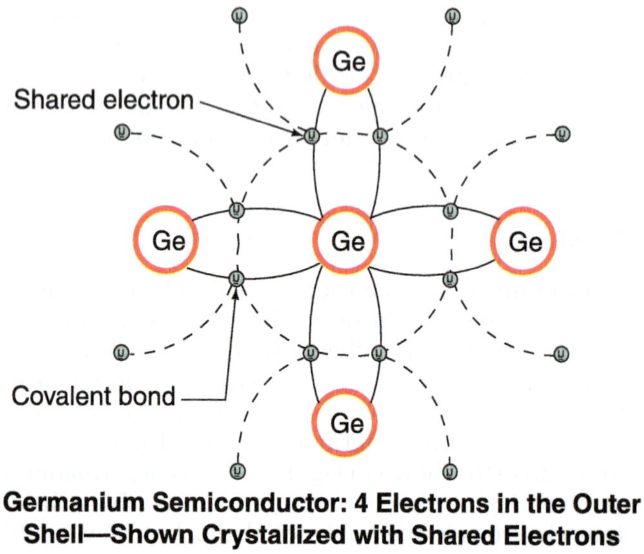

DOPING

Pure silicon and pure germanium are of little use in electronic components. For a semiconductor to be useful it must be doped, that is, have small quantities of impurities added to it. The **doping** agents are usually phosphorus and boron. The doping intensity will define the electrical behavior of the crystal. After doping, silicon crystals may be sliced into thin sections known as *wafers*.

Bias

The type of doping agent used to produce silicon crystals will define the electrical properties of the crystals produced. Whereas semiconductors have four electrons in their valence, the doping agents have either three (trivalent) or five (pentavalent) electrons in their valence shells. When a semiconductor crystal is constructed, the doping agent will produce a "bias" in the electrical character of the semiconductor crystal produced. For instance, a boron atom (common doping element) has three electrons in its outer shell. A boron atom in a crystallized cluster of silicon atoms will produce a valence shell with seven electrons instead of eight. This "vacant" electron opening is known as a *hole*. The hole makes it possible for an electron from a nearby atom to fall into the hole. In other words, the holes can move, permitting a flow of electrons. Silicon crystals doped with boron (or other trivalent elements) form P-type (positively doped) silicon. **Figure 6–10** shows the hole created by doping a semiconductor crystal with a trivalent element such as boron.

A semiconductor crystal can also be produced with an N or negative electrical characteristic. A phosphorus atom has five electrons in its outer shell. It is *pentavalent*. In the bonding between the

FIGURE 6–10 P-type semiconductor crystal

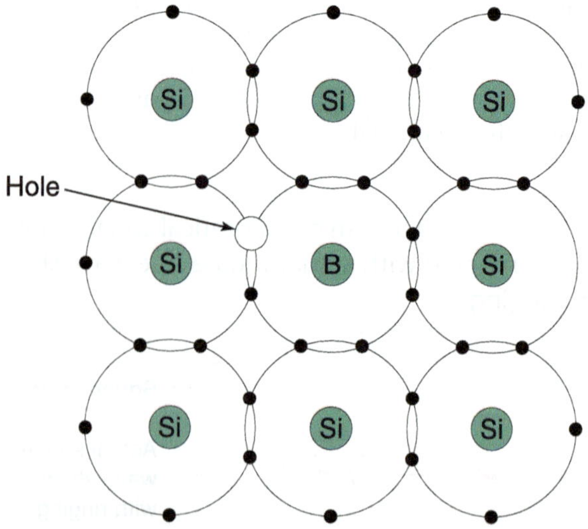

FIGURE 6–11 N-type semiconductor crystal

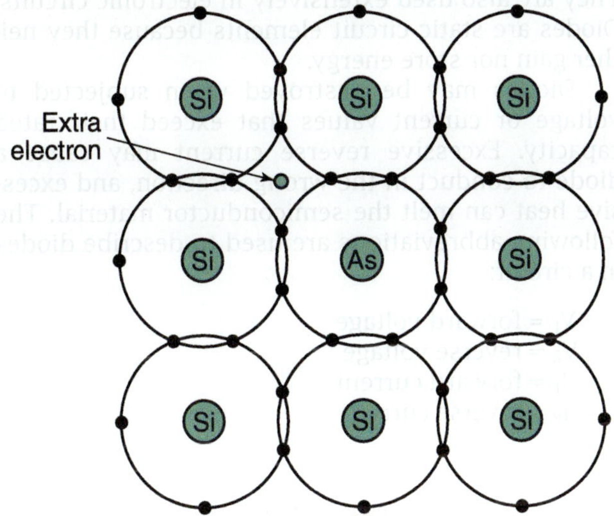

FIGURE 6–12 Doped germanium semiconductor crystals of the P and N types

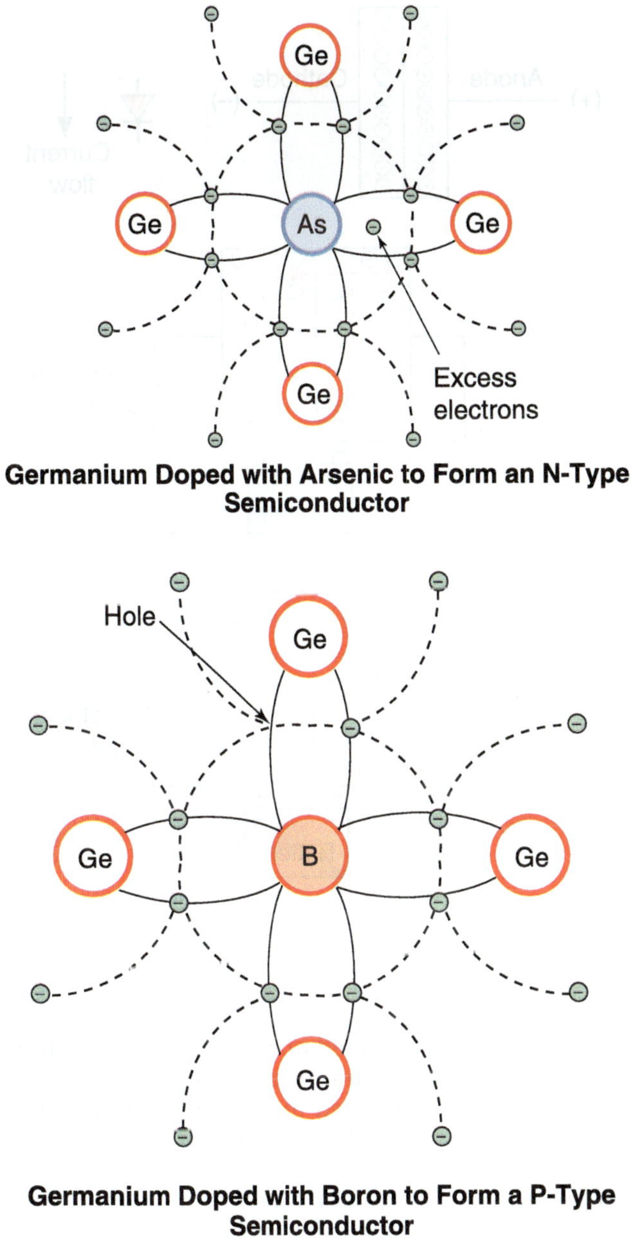

semiconductors and the doping material there is room for only eight electrons in the center shell. Even when the material is in an electrically neutral state, the extra electron can move through the crystal. When a silicon crystal is manufactured using a doping material with five electrons in the valence shell it forms N-type (negatively doped) silicon. The first transistor manufactured used a germanium semiconductor. Because the element germanium is a semiconductor, just like silicon, it also has four electrons in its valence shell. **Figure 6–11** shows a silicon crystallized cluster doped with arsenic to form an N-type semiconductor.

Doping a semiconductor crystal always defines its electrical characteristics. We describe the electrical characteristics of a semiconductor as P type or N type. In **Figure 6–12**, germanium is used once again to show how the element forms a P-type semiconductor when doped with arsenic and an N-type semiconductor when doped with boron. P-type and N-type semiconductor crystals permit an electrical current to flow in different ways. In the P-type semiconductor, current flow is made possible by a deficit of electrons, whereas in the N-type semiconductor, current flow is made possible by an excess of electrons. Whenever a voltage is applied to a semiconductor, electrons will flow toward the positive terminal and the holes will move toward the negative terminal.

6.3 DIODES

The suffix *-ode* literally means "terminal." For instance, it is used as the suffix for the words *cathode* and *anode*. The word *diode* means literally "having two terminals." Diodes are constructed of semiconductor materials. A little earlier in this chapter, we discovered how both P-type and N-type semiconductor crystals can conduct electricity. The actual resistance of each type is determined by either the proportion of holes or surplus of electrons. When a chip is manufactured using both P- and N-type semiconductors, electrons will flow in only one direction. The diode is used in electronic circuitry as a sort of one-way check valve that will conduct electricity in one direction (forward) and block it in the other (reverse). Diodes will not conduct current in forward bias until the turn-on voltage is exceeded. Turn-on voltage for a silicon diode is between 0.6 and 0.7 V-DC.

FIGURE 6–13 Diode operation and diodes in forward and reverse bias

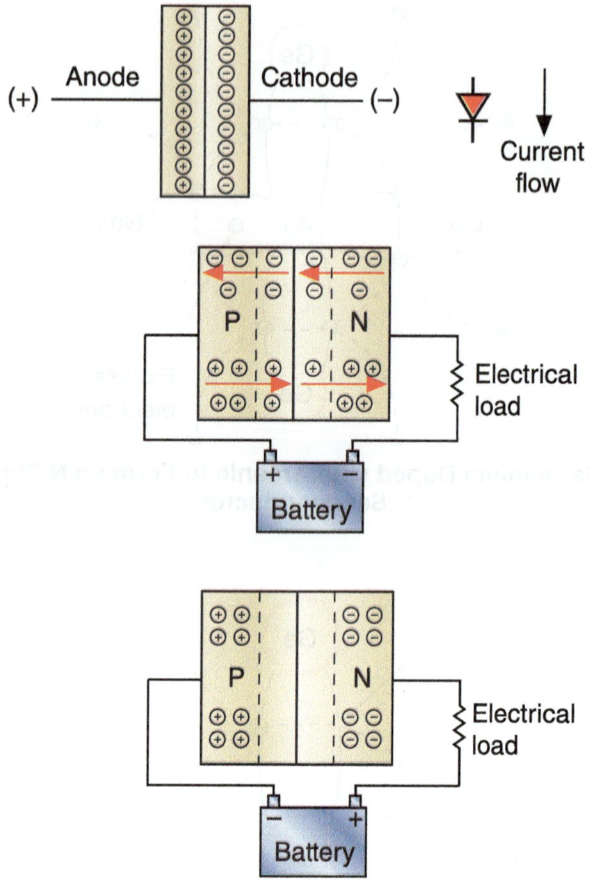

DIODE BIAS

When a diode is connected in its proper polarity, that is, to permit current flow, it is said to be in forward bias. When it is connected in opposite polarity, that is, to block current flow, it is said to be in reverse bias. The word **bias** in vehicle electrical systems simply means voltage potential at input referenced to ground. In electronics, the word *bias* can be also be used to refer to voltage potential at input referenced to another point in a circuit: In other words, the bias can be referenced to values other than zero volts. In a diagram of a diode, the arrow always points in the direction of conventional current flow when forward biased (**Figure 6–13**).

ANODE AND CATHODE

The positive terminal (1) is called the **anode** and the negative terminal (2) the **cathode**. As an electrical one-way check valve, diodes will permit current flow only when correctly polarized. For instance, diodes are used in alternating current (AC) generators (alternators) to produce a direct current (DC) characteristic from AC in a process known as *rectification*. They are also used extensively in electronic circuits. Diodes are static circuit elements because they neither gain nor store energy.

Diodes may be destroyed when subjected to voltage or current values that exceed their rated capacity. Excessive reverse current may cause a diode to conduct in the wrong direction, and excessive heat can melt the semiconductor material. The following abbreviations are used to describe diodes in a circuit:

V_F = forward voltage
V_R = reverse voltage
I_F = forward current
I_R = reverse current

SHOP TALK

A diode is simply a one-way electrical check valve. When it is connected in the correct polarity to allow current flow, it is described as forward biased. When connected in opposite polarity, that is, to block current flow, it is described as reverse biased.

TYPES OF DIODES

Numerous types of diodes exist, and they play a variety of roles in electronic circuits. The following is a sample of some of the more common types.

Small Signal Diodes

Small signal diodes are used to transform low-current AC to DC (using a rectifier), perform logic data flows, and absorb voltage spikes. Small signal diodes are used to rectify the AC current produced by an AC generator's (alternator's) stator to the DC current required by the vehicle electrical system.

Power Rectifier Diodes

These function in the same manner as small signal diodes but are designed to permit much greater current flow. They are used in multiples and often mounted on a heat sink to dissipate excess heat caused by high-current flow.

Zener Diodes

A **zener diode** (**Figure 6–14**) functions as a voltage-sensitive switch. It is named after its inventor, Clarence Zener, in 1934. The zener diode is designed to block reverse bias current but only up to a specific voltage value. When this reverse breakdown voltage is attained, it will conduct the reverse bias current flow without damage to the semiconductor material. Zener diodes are manufactured from heavily doped semiconductor crystals. They are used in electronic

FIGURE 6–14 Zener diode

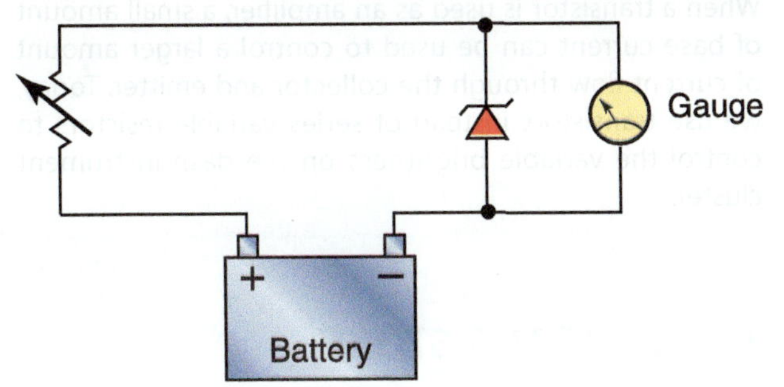

voltage regulators and in other automotive electronic circuitry. Zener diodes are rated by their breakdown or threshold voltage (V_Z), and this can range from 2 to 200V.

Light-Emitting Diodes (LEDs)

All diodes emit electromagnetic radiation when forward biased, but diodes manufactured from some semiconductors (notably gallium arsenide phosphide) emit it at much higher levels. **Light-emitting diodes (LEDs)** may be constructed to produce a variety of colors and are commonly used for digital data displays. For instance, a digital display with seven linear LED bars arranged in a bridged rectangle could display any single numeric digit by energizing or not energizing each of the seven LEDs. LED arrangements constructed to display alpha characters are only slightly more complex.

LEDs convert electrical current directly into light (photons) and therefore are highly efficient because no heat losses occur. LEDs have been used for a number of years as tail and clearance lights in trucks and trailers, and more recently in clusters as headlights. They consume a fraction of the power of conventional bulbs, illuminate at faster speeds, and significantly outlast them. The only disadvantage is that they do not produce heat when operated and are therefore more susceptible to ice and snow buildup, when used as external lighting, during winter operation. **Figure 6–15** shows a diagram of an LED and the LED in schematic form.

FIGURE 6–15 (A) Light-emitting diode (LED) and (B) LED in schematic form

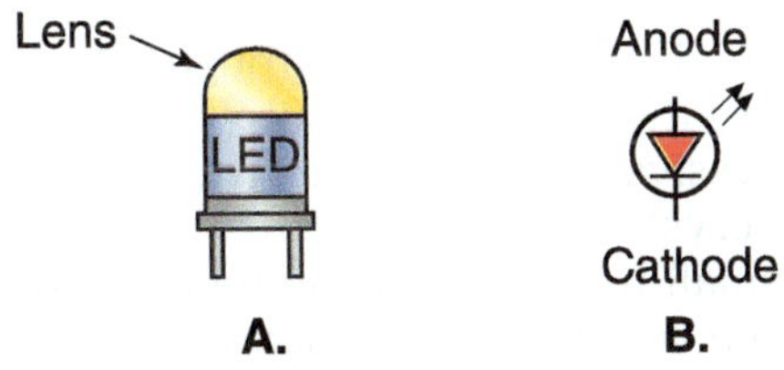

PHOTO DIODES

All diodes produce some electrical response when subjected to light. A photo diode is designed to detect light and therefore has a clear window through which light can enter. Silicon is the usual semiconductor crystal medium used in photo diodes.

Summary of Diodes

The following list summarizes this discussion of diodes:

- Diodes are two-terminal devices.
- A diode always has an anode (positive terminal) and a cathode (negative terminal).
- Diodes act like one-way check valves, allowing current flow in one direction (forward bias) and blocking it in the other (reverse bias).
- Diodes can be regarded as static circuit elements in that they neither gain nor store energy.
- For a diode to conduct in forward bias, its turn-on voltage threshold of 0.6 to 0.7 V-DC must be met.

6.4 TRANSISTORS

Transistors are three-terminal semiconductor chips that are used extensively in electronic circuits. Unlike diodes, which are static elements in electronic circuits, transistors are active circuit elements capable of amplifying or transforming a signal level. A transistor consists of two P-N junctions. As with the diodes, P and N describe the semiconductors that define the transistor's electrical characteristics. That means two basic types of transistor exist, NPN and PNP. A terminal connects each of the semiconductor segments. A transistor's three terminals are known as *base, collector*, and *emitter*. To some extent, the names used to describe the terminals explain their functions. A small base current is used to control a much larger current flow through the collector and emitter. The base can be regarded as a "switch." The collector can be regarded as the input and the emitter as the output of the transistor.

AMPLIFICATION

In electricity, a relay is used to enable a small electrical current to control a much larger electrical current. In electronics, a transistor can be used in much the same way, that is, as a pilot relay, to switch a circuit on or off. However, the transistor is more versatile because it can be used to switch and amplify. Amplifying means that by varying the amount of base current, the collector-emitter current flow through the transistor can be controlled. An everyday use of an amplifying transistor is replacing the variable resistor that controls the brightness of display lighting. Audio amplifiers in stereo systems have used transistors for many years now.

FIGURE 6–16 Transistor operation

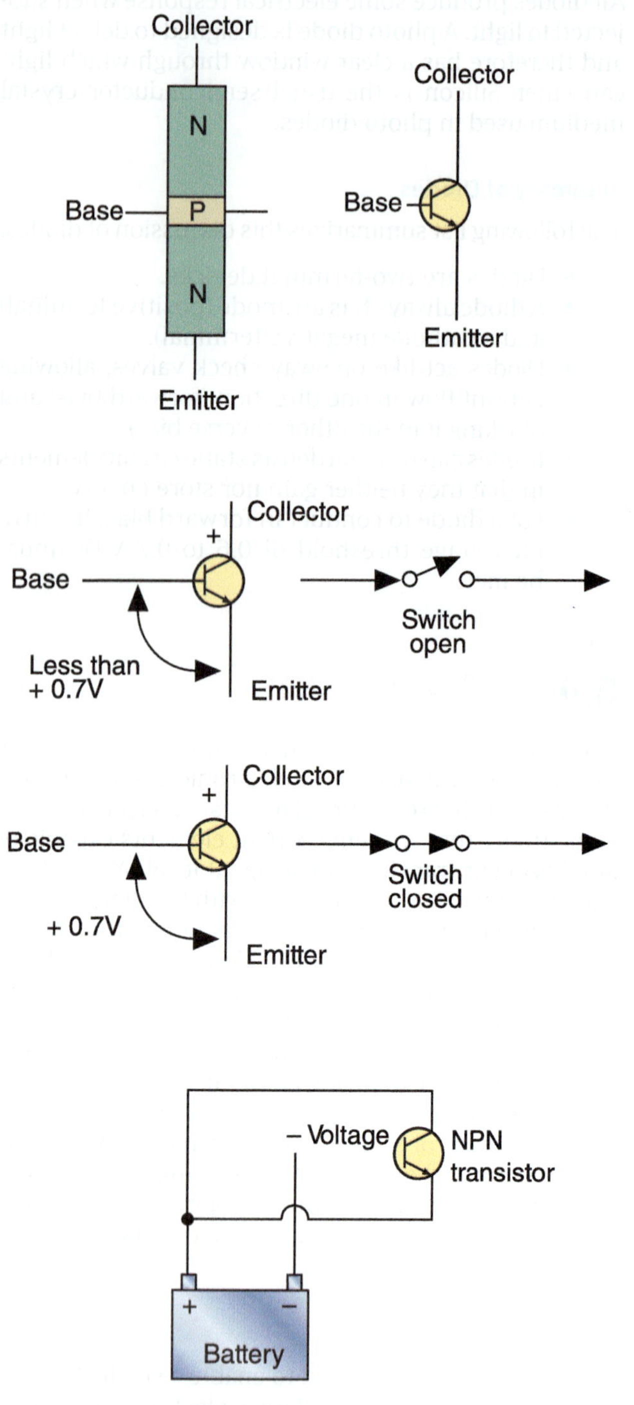

Types of Transistors

Transistors can be generally grouped into **bipolar** and *field effect* categories. (These are explained later in this chapter.) Transistors offer many advantages over the electrical devices they replace, such as the now-obsolete glass tubes. They are faster, much more compact, and have a significantly greater life span. **Figure 6–16** shows how transistors are represented in schematics and how they operate.

SHOP TALK

When a transistor is used as an amplifier, a small amount of base current can be used to control a larger amount of current flow through the collector and emitter. Today, we use transistors instead of series variable resistors to control the variable brightness on the dash instrument cluster.

BIPOLAR TRANSISTORS

A bipolar transistor functions very much like a sort of switched diode. It has three terminals known as collector, emitter, and base. Each terminal is connected to semiconductor media in what is known as a silicon sandwich. The sandwich may either be NPN or PNP.

Bipolar Construction

An NPN sandwich is constructed of a layer of P-type semiconductor material sandwiched between two layers of N-type semiconductor. The middle of the sandwich always acts as a *gate* capable of controlling the current flow through all three layers. The base is fairly thin and has comparatively fewer doping atoms than that of the semiconductor material on either side of it. They are designed so a small emitter-base current will ungate the transistor and permit a larger emitter-collector current flow.

Bipolar Operation

Looking at the operation of a typical NPN transistor, the base-emitter junction acts as a diode. Current flows forward into the base and out of the emitter but cannot flow in the reverse direction. The collector base also acts as a sort of diode, but supply voltage is applied to it in the reverse direction. However, when the transistor is used as a switch in the on state, the collector-base junction may become forward biased. Under normal operation, current flow through the base-emitter terminals can be controlled by the current flowing through the base terminal. In this way, small base currents are used to control large collector currents. In fact, the amplification capability of transistors is typically 100 times (although it may be much less or much more). This allows a base current to control a collector current 100 times greater.

Facts about Bipolar Transistor Operation

- The base-emitter junctions will not conduct until the forward-bias voltage exceeds 0.6V.
- Excessive current flow through a transistor will cause it to overheat or fail.
- Excessive voltage can destroy the semiconductor crystal media.
- A small base current can be used to control a much larger collector current.

Examples of Bipolar Transistors

Many different types of bipolar transistors exist; notable are small signal switching and power transistors. These two types are used extensively in vehicle electronic systems. Small signal transistors are used to amplify signals. Some may be designed to fully gate current flow in their off position and others may both amplify and switch. Power transistors may conduct high-current loads and are often mounted on heat sinks to enable them to dissipate heat. They are sometimes known as *drivers* because they serve as the final or output switch in an electronic circuit used to control a component such as a solenoid, piezo actuator, or pilot switch. For instance, the actuators in injectors and transmission shift controls are switched by power transistors located in injector driver units built into the ECMs.

FIELD EFFECT TRANSISTORS (FETs)

FETs are more commonly used than bipolar transistors largely because they are cheaper to manufacture. They may be divided into *junction*-type and *metal-oxide semiconductors.* Both types are controlled by a very small input (base/gate) voltage.

JUNCTION FETS (JFETs)

There are two types of JFETs: N-channel and P-channel. The channel behaves as a resistor that conducts current from the source side to the drain side. Voltage at the gate will act to increase the channel resistance, thereby reducing drain to source current flow. This allows the FET to be used as either an amplifier or a switch. When gate voltage is high, high-resistance fields are created, narrowing the channel available for conductivity current to the drain. In fact, if the gate voltage is high enough, the fields created in the channel can join and completely block current flow. FETs are used in the smart switch multiplex circuits described in Chapter 12.

Facts about JFETs

You should be aware of a couple of facts regarding JFETs:

- JFET gate-channel resistance is very high, so the device has almost no effect on external components connected to the gate.
- Almost no current flows in the gate circuit because the gate-channel resistance is high. The gate and channel form a "diode," and as long as the input signal "reverse biases" this diode, the gate will show high resistance.

METAL-OXIDE SEMICONDUCTOR FIELD EFFECT TRANSISTORS (MOSFETs)

The MOSFET has become the most important type of transistor in microcomputer applications in which thousands can be photo-infused onto minute silicon wafers. They are easy to manufacture and consume fractional amounts of power. These are also used in truck multiplex circuits described in Chapter 12.

MOSFETs are classified as P type or N type. However, in a MOSFET, there is no direct electrical contact between the gate with the source and the drain. The gate's aluminum contact is separated by a silicon oxide insulator from the remainder of the transistor material.

MOSFET OPERATION

When the gate voltage is positive, electrons are attracted to the region around the insulation in the P-type semiconductor medium. This produces a path between the source and the drain in the N-type semiconductor material, permitting current flow. The gate voltage will define the resistance of the path or channel created through the transistor, permitting them to both switch and act as variable resistors. MOSFETs can be switched at very high speeds and, because the gate-channel resistance is high, almost no current is drawn from external circuits. **Figure 6–17** shows the operating principle of a transistor: Note how the device is switched by a voltage applied to the gate.

FIGURE 6–17 PNP and FET operating principles

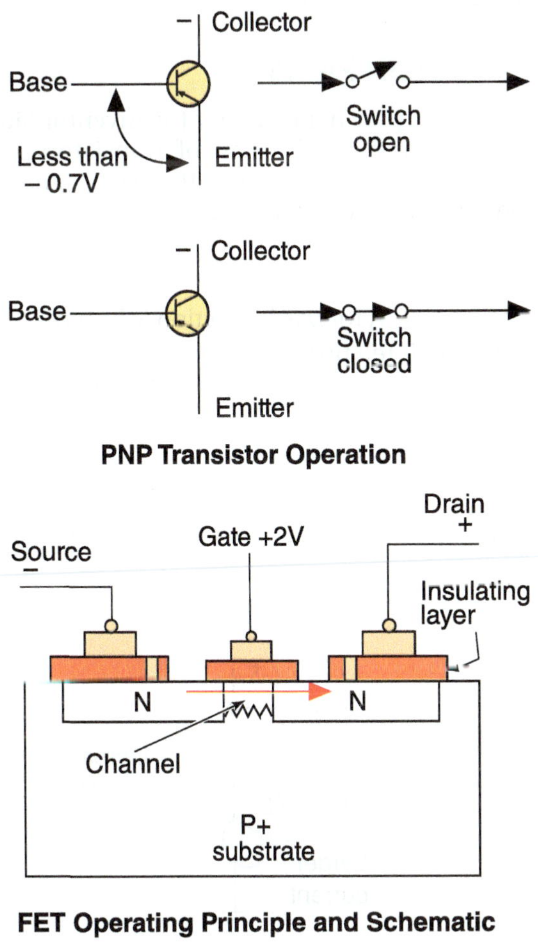

THYRISTOR

Thyristors are three-terminal, solid-state switches. They are different from transistors in that they are only capable of switching. A small current flow through one of the terminals will switch the thyristor on and permit a larger current flow between the other two terminals. Thyristors are switches, so they are either in an on or off condition. They fall into two classifications depending on whether they switch AC or DC current. Some thyristors are designed with only two terminals and they will conduct current when a specific trigger or breakdown voltage is achieved.

Silicon-Controlled Rectifiers (SCRs)

SCRs are similar to a bipolar transistor with a fourth semiconductor layer added. They are used to switch DC. When the anode of an SCR is made more positive than the cathode, the outer two P-N junctions become forward biased: The middle P-N junction is reverse biased and will block current flow. A small gate current, however, will forward bias the middle P-N junction, enabling a large current to flow through the thyristor. SCRs will remain on even when the gate current is removed. The on condition will remain until the anode-cathode circuit is opened or reversed biased. SCRs are used for switching circuits in vehicle electronic and ignition systems. **Figure 6–18** shows a forward-direction SCR.

DARLINGTON PAIRS

A **Darlington pair** (named after the inventor Sidney Darlington) consists of a pair of transistors wired so the emitter of one supplies the base signal to a second, through which a large current flows. The objective once again is to use a very small current to switch a much larger current. This type of application is known as amplification. Darlington pairs are used extensively in vehicle computer control systems and in ignition modules. They are the switching component used in the ECMs in some diesel injector drivers. Figure 6–18 shows a Darlington pair relationship.

FIGURE 6–18 A forward-direction SCR and Darlington pair relationship

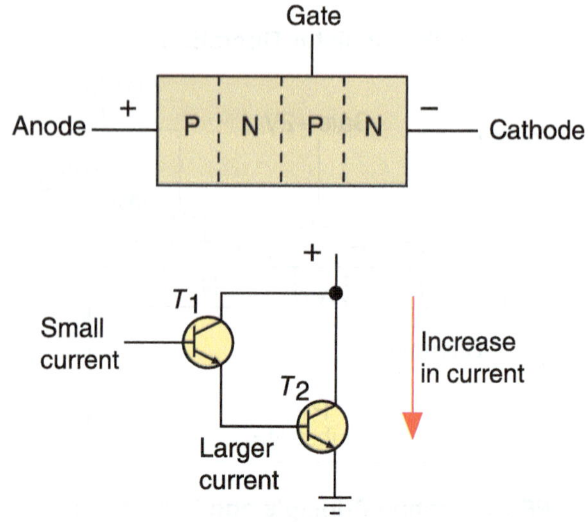

6.5 PHOTONIC SEMICONDUCTORS

Photonic semiconductors emit and detect light or *photons.* A photon is a unit of light energy. Photons are produced electrically when certain electrons excited to a higher-than-normal energy level return to a more normal level. Photons behave like waves. The distance between the wave nodes and antinodes (wave crests and valleys) is known as *wavelength.* Electrons excited to higher energy levels emit photons with shorter wavelengths than electrons excited to lower levels. Photons are not necessarily visible and it is perhaps important to note that they may truly be described as *light* only when they are visible.

All visible light is classified as electromagnetic radiation. The specific wavelength of light rays will define its characteristics. Light wavelengths are specified in nanometers, that is, billionths of a meter.

THE OPTICAL LIGHT SPECTRUM

The optical light spectrum includes ultraviolet, visible, and infrared radiation. **Figure 6–19** shows a graphic representation of the optical light spectrum. Photonic semiconductors either emit or can detect near-infrared radiation frequencies. Near-infrared means that the frequency is slightly greater than the red end of the visible light spectrum (find this in **Figure 6–20**) and is therefore usually referred to as

FIGURE 6–19 The optical spectrum

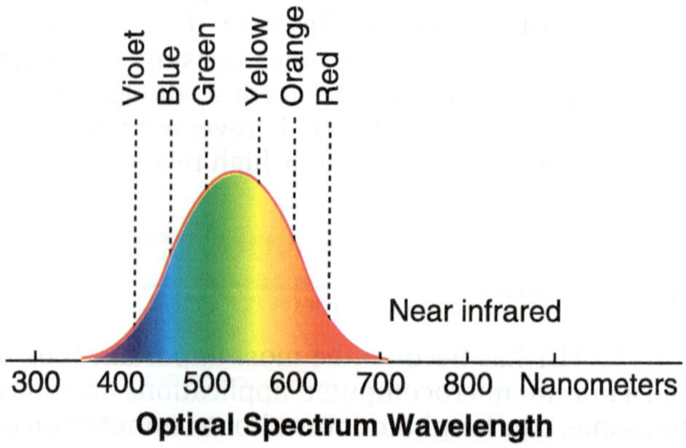

FIGURE 6–20 The electromagnetic spectrum

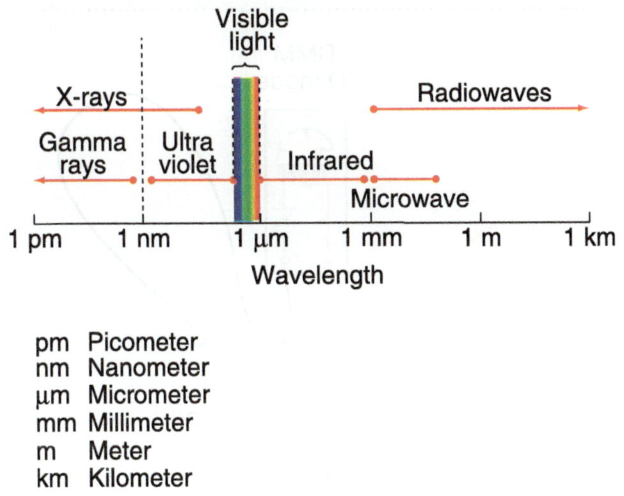

light. Figure 6–20 shows the full optical light, or electromagnetic, spectrum. Note the narrow portion of the spectrum that is classified as visible light.

OPTICAL COMPONENTS

As we progress into the electronic age, the importance of optical components will increase: Most signaling functions will be removed from hard wire buses and performed using fiber optics. The advantage of fiber optics is faster transaction and vastly increased bandwidth or data volume capability. Fiber optic buses are already used extensively in off-highway equipment with complex powertrain management systems such as rock trucks.

Optical components may conduct, refract, or modify light. In the same way electrical signals can relay information by modulating duty cycle and frequency, light waves can be similarly used. The use of optics in vehicle technology is rapidly increasing. Some of the more common optical components are discussed in the following sections.

Filters

Filters transmit only a narrow band of the spectrum and block the remainder. They operate much like the filter lens used in a welding helmet.

Reflectors

Reflectors reflect a light beam, or at least most of it, in much the same way a mirror functions.

Beam Splitters

Beam splitters transmit some of the optical wavelength and reflect back the rest of it.

Lenses

Lenses bend light waves. They are often used in conjunction with semiconductor light sources and detectors. They are often used to collect and focus light onto a detector.

Fiber Optics

Optical fibers are thin, flexible strands of glass or plastic that conduct light. The light travels through a core surrounded by conduit or cladding. Optical fibers are increasingly used to transmit digital data by pulsing light. The use of **fiber optics** is rapidly increasing, and in the coming years some truck chassis multiplexing functions will be driven optically rather than by hard wire. Fiber optics are efficient, especially when large data volumes are required to be transferred with a high degree of accuracy. Because of the standardization of core multiplexing (see J1939 in Chapter 12), when fiber optics are currently used in trucks and transit buses, this is done outside the core bus. In automobiles, fiber optics are also used as auxiliary buses.

Solar Cells

A solar cell consists of a P-N or N-P silicon semiconductor junction built onto contact plates. A single silicon solar cell may generate up to 0.5V in ideal light conditions (bright sunlight), but output values are usually lower. Like battery cells, solar cells are normally arranged in series groups, in which case the output voltage would be the sum of cell voltages. They can also be arranged in parallel, in which case the output current would be the sum of the cell currents. They are sometimes used as battery chargers on vehicles.

6.6 TESTING SEMICONDUCTORS

The technician is seldom required to test an individual diode or transistor in a vehicle electronic circuit, but it is an activity that is often taught in technical training programs because it increases awareness of how electronic systems work and fail. Diodes and transistors are normally tested using an ohmmeter. The semiconductor to be tested should be isolated from the circuit. The digital multimeter (DMM) should be set to the diode test mode.

TESTING DIODES

Diodes should produce a low-resistance reading with the DMM test leads in the forward-bias direction and high resistance when the leads are reversed. Low-resistance readings both ways indicate a shorted diode. High-resistance readings both ways indicate an open diode. **Figure 6–21** and **Figure 6–22** introduce the basic testing of a diode. This is explored in a little more detail in Chapter 11. To perform the tests outlined in Figure 6–21 through Figure 6-24 a digital multimeter (DMM) in resistance mode must

FIGURE 6–21 Testing a diode in forward bias

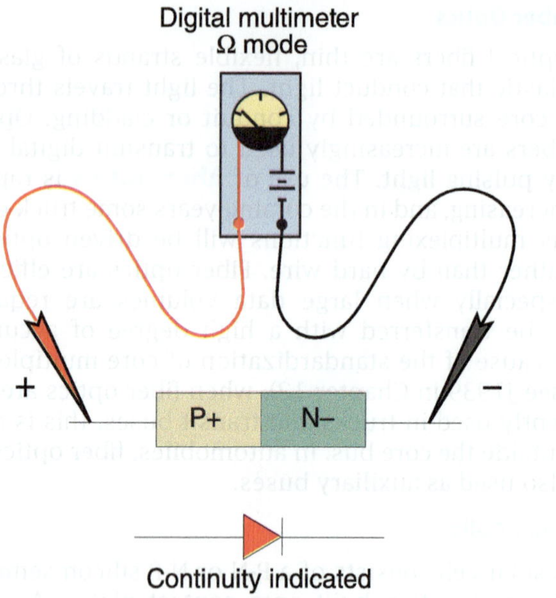

FIGURE 6–22 Testing a diode in reverse bias

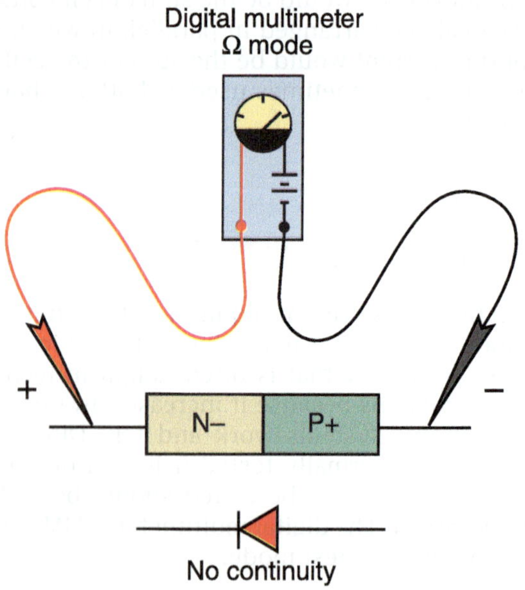

be used; however, we have used an analog display face to indicate the difference between high and low readings.

TESTING TRANSISTORS

Again, the instrument used to bench test a transistor is the DMM in ohmmeter mode. The testing of transistors is covered in a little more detail in Chapter 11, but this should give you some idea of how to go about

FIGURE 6–23 Testing transistor base to collector continuity

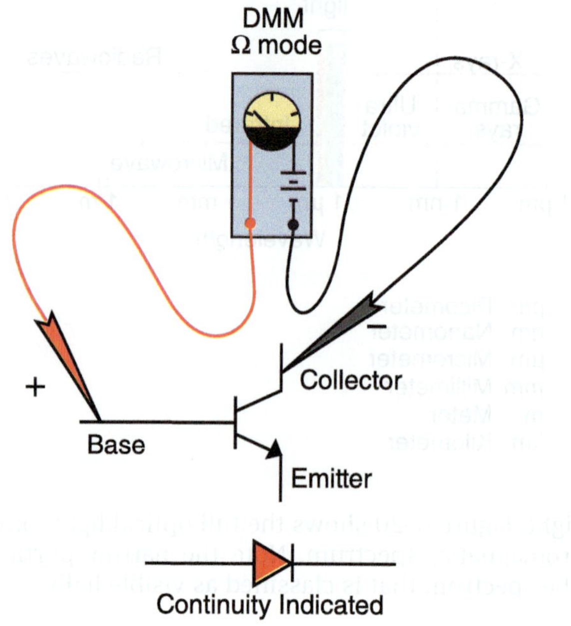

FIGURE 6–24 Testing a transistor for high resistance across the emitter-collector junction

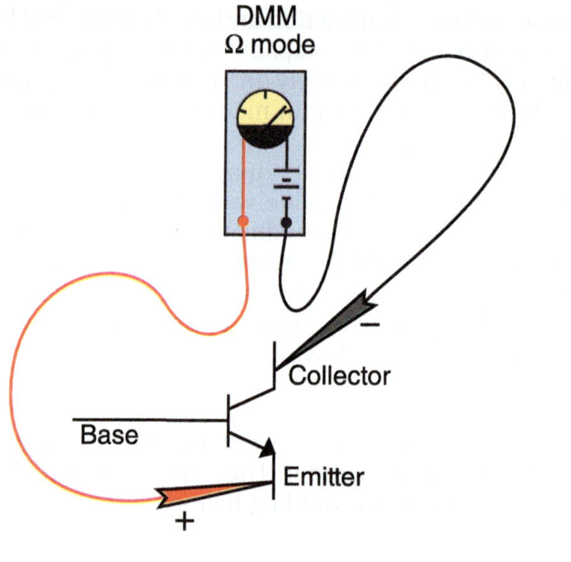

the procedure. Use **Figure 6–23** and **Figure 6–24**, a DMM, and some transistors to practice on. A functional transistor should test as follows:

- Continuity between the emitter and base
- Continuity between the base and the collector when tested one way, and high resistance when DMM test leads are reversed
- High resistance in either direction when tested across the emitter and collector terminals.

FIGURE 6–25 An electronic breadboard that can be helpful in enabling students to understand electronics by building circuits from scratch.

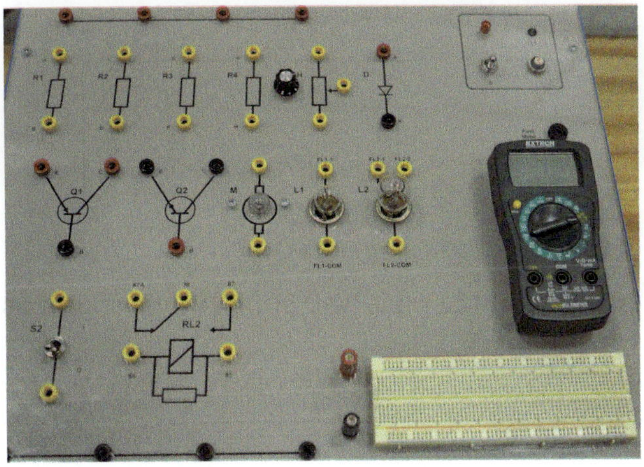

Circuit Board Trainers

An effective way for students to familiarize themselves with the basics of electronics is to work on electronic breadboards and build electronic circuits from scratch. You can find good quality circuit boards in Radio Shack outlets or use one that has been specifically manufactured for the requirements of motive power techs such as the example shown in **Figure 6–25**.

6.7 INTEGRATED CIRCUITS (I/Cs)

Integrated circuits (I/Cs), consist of resistors, diodes, and transistors arranged in a circuit on a *chip* of silicon (**Figure 6–26**). The number of electronic components that comprise the I/C vary from a handful to hundreds of thousands, depending on the function of the chip. I/Cs have innumerable household, industrial, and automotive applications and are the basis of digital watches, electronic pulse wipers, and all computer systems.

SHOP TALK

A chip is an I/C. One chip can be smaller than a fingernail and contain many thousands of resistors, diodes, and transistors.

ANALOG AND DIGITAL CIRCUITS

I/Cs fall into two general categories. Analog I/Cs operate on variable voltage values. Electronic voltage regulators are a good vehicle example of an analog I/C. Digital I/Cs operate on two voltage values only, usually *presence of voltage* and *no voltage*. Digital I/Cs are the basis of most computer hardware, including processing units, main memory, and data retention chips. I/C chips can be fused into a motherboard (main circuit) or socketed. The latter has the advantage of easy removal and replacement.

FIGURE 6–26 Integrated circuit

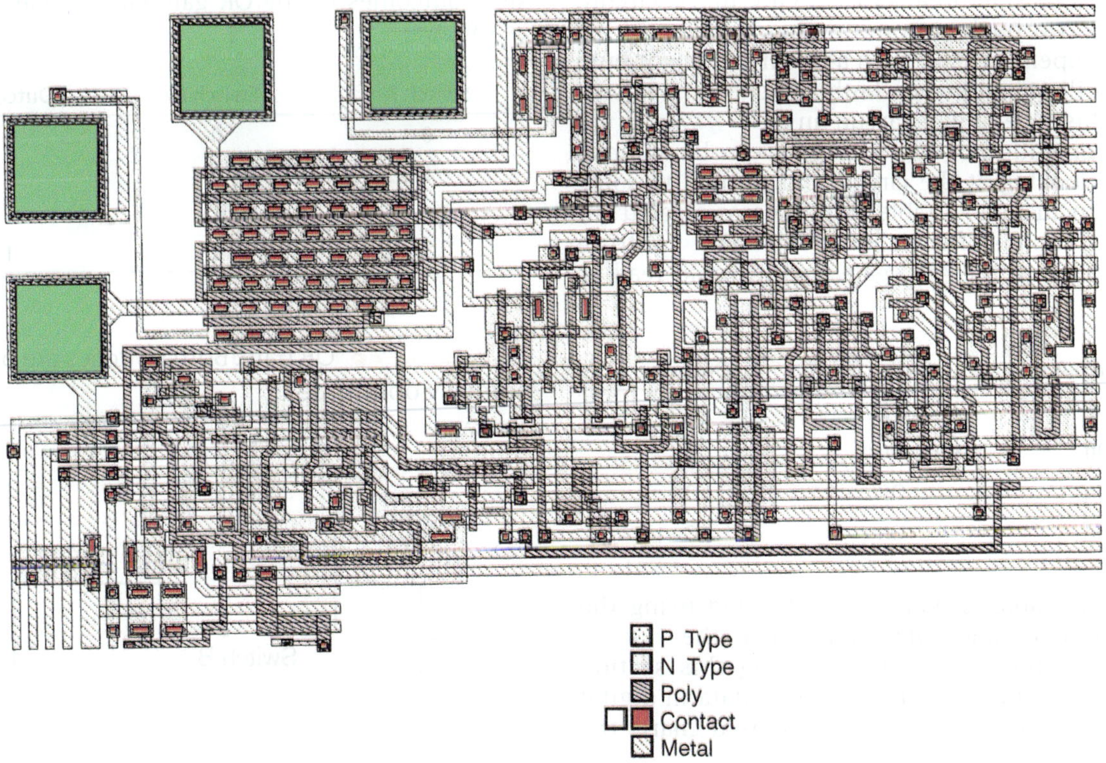

DIPs

A common chip package used in computer and vehicle engine/electronic control modules (ECMs) is the dual in-line package (DIP). This package consists of a rectangular plastic-enclosed I/C with usually fourteen to sixteen pins arranged evenly on either side. DIPs may be fused (not removable) or socketed to the motherboard.

6.8 GATES AND TRUTH TABLES

The importance of gates was emphasized in the process of outlining the operation of transistors in one of the previous sections. These are the electronically controlled switching mechanisms that manage the operating mode of a transistor. Digital I/Cs are constructed by using thousands of gates. In most areas of electronics, gates can be either open or closed; in other words, in-between states do not exist. Just like the toggle switch that controls the lights in a room, the switch is either on or off. The terms used to describe the state of a gate are *on* and *off*. In a circuit, these states are identified as *presence of voltage* or *no voltage* (or high-/low-voltage states).

AND GATES

The best way to learn about the operation of gates in a digital circuit is to observe the operation of some electromechanical switches in some simple electrical circuits. In **Figure 6–27**, a power source is used to supply a light bulb in a series circuit. In the circuit, there are two push-button switches that are in the normally open state. In such a circuit, the light bulb will only illuminate when both switches are closed. We call this type of gate circuit an **AND gate**.

The operation of the AND gate can be summarized by looking at the circuit and coming to some logical conclusions. A table that assesses a gated circuit's operation is often called a **truth table**. A truth table applied to the AND gate circuit would read as follows:

Switch A	Switch B	Outcome
Off	Off	Off
Off	On	Off
On	Off	Off
On	On	On

A truth table is usually constructed using the digits *zero* (0) and *one* (1) because the **binary system** (outlined in detail following this section on gates) is commonly used to code data in digital electronics. Therefore, a truth table that charts the outcomes of the same AND switch would appear as follows:

Switch A	Switch B	Outcome
0	0	0
0	1	0
1	0	0
1	1	1

FIGURE 6–27 AND gate: Both switches are of the normally open type.

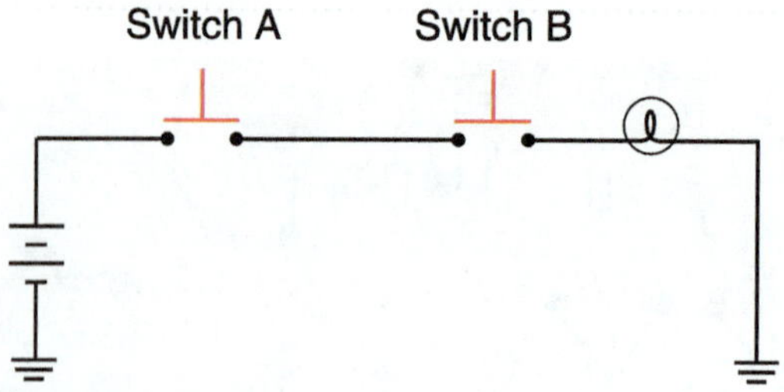

OR GATES

If another circuit is constructed using two normally open electromechanical switches, this time arranged in parallel, the circuit would appear as in **Figure 6–28**. This kind of gate is called an **OR gate**. Both switches are of the normally open type: If either one is closed, the gate outcome is on.

A truth table constructed to represent the possible outcomes of the OR gate circuit would appear as follows:

Switch A	Switch B	Outcome
0	0	0
1	0	1
0	1	1
1	1	1

FIGURE 6–28 OR gate: Both switches are of the normally open type.

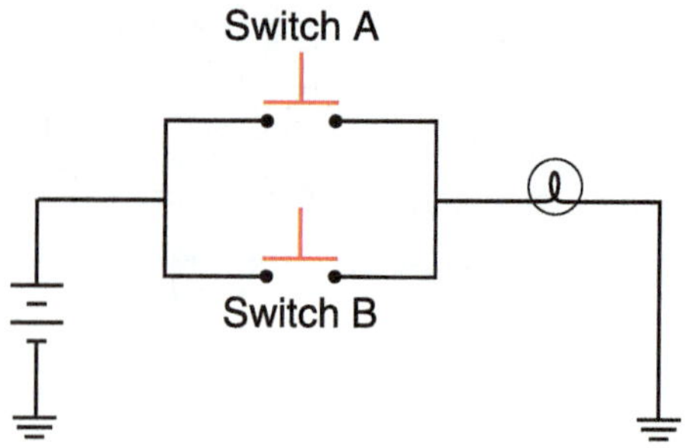

FIGURE 6–29 NOT gate: The single switch is of the normally closed type.

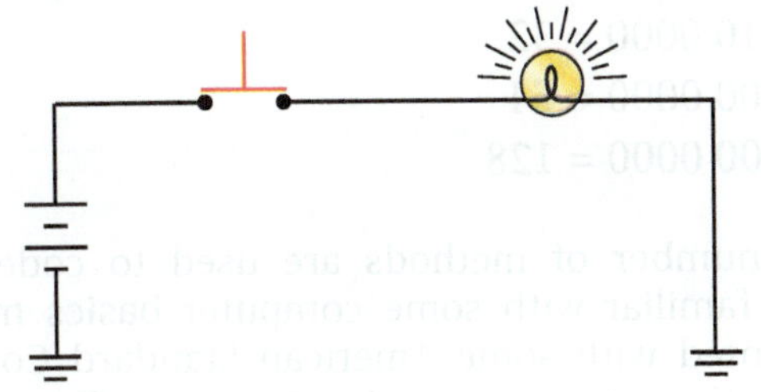

NOT OR INVERTER GATES

A **NOT gate** circuit switch can be constructed by using a push-button switch that is in the normally closed state. In other words, circuit current flow is interrupted when the button is pushed. In the series circuit shown in **Figure 6–29**, current will flow until the switch opens the circuit.

If a truth table is constructed to graphically represent the outcomes of a NOT gated circuit, this is what it would look like:

In	Out
0	1
1	0

Although the examples presented here used electromechanical switches for ease of understanding, in digital electronics, circuit switching is performed electronically using diode and transistor gates. AND, OR, and NOT gates are three commonly used means of producing an outcome based on the switching status of components in the gate circuit.

GATES, TRUTH TABLES, AND BASIC DATA PROCESSING

Figure 6–30 shows some simple input logic gates and the outcomes that can be produced from them. When the number of inputs to a logic gate is increased, its outcome processing becomes more complex. Gates are normally used in complex networks in which they are connected by buses, or connection points, to form a logic circuit. **Bits** of data can be moved through the logic circuit or highway to produce outcomes that evolve from the logical processing of inputs. By massing hundreds of thousands of logic circuits, the processing of information becomes possible.

A computer works simply by processing input data and generating logical outcomes based on the input data. Although it may not be required that a vehicle technician understand the precise operation of a digital computer, these have become so much part of the workplace that it is certainly desirable to have some idea of the basics. Computer operation is explored in a little more detail later in this chapter under the heading Microprocessors.

FIGURE 6–30 Logic gates and truth tables showing the switching outcomes

AND gate

A	B	Out
0	0	0
0	1	0
1	0	0
1	1	1

NAND gate

A	B	Out
0	0	1
0	1	1
1	0	1
1	1	0

OR gate

A	B	Out
0	0	0
0	1	1
1	0	1
1	1	1

NOR gate

A	B	Out
0	0	1
0	1	0
1	0	0
1	1	0

6.9 BINARY SYSTEM BASICS

The binary system is an arithmetic numeric system using two digits and, therefore, has a base number of 2. The base number of a numeric system is the number of digits used in it. The decimal system is, therefore, a 10-base number system. The binary numeric system is often used in computer electronics because it directly corresponds to the *on* or *off* states of switches and circuits. In computer electronics, the binary system is the primary means of coding data using the digits 0 and 1 to represent alpha, numeric, and any other data. Numeric data is normally represented using the decimal system. Again, this is simple coding. The digit 3 is a representation of a quantitative value that most of us have been accustomed to decoding since early childhood.

If decimal system values were coded into binary values, they would look like this:

Decimal digit	Binary digit
0	0
1	1
2	10
3	11
4	100
5	101
6	110
7	111
8	1000
9	1001
10	1010

In digital electronics, a *bit* is the smallest piece of data that a computer can manipulate. It is simply the ability to represent one of two values, either *on* (1) or *off* (0). When groups of bits are arranged in patterns, they are collectively described as follows:

4 bits = nibble
8 bits = byte

Most digital electronic data is referred to quantitatively as **bytes**. Computer systems are capable of processing and retaining vast quantities of data. In most cases, millions (megabytes) and billions (gigabytes) of bytes are described. A byte is 8 bits of data. It has the capability of representing up to 256 data possibilities. For instance, if a byte were to be used to code numeric data, it would appear as follows:

Decimal digit	Binary coded digit
0	0000 0000
1	0000 0001
2	0000 0010
3	0000 0011
4	0000 0100
5	0000 0101
6	0000 0110
7	0000 0111
8	0000 1000

and so on up to 256. Try it. Note that each time a one/on value progresses one column to the left, it doubles the value of the column to its right.

0000 0001 = 1
0000 0010 = 2
0000 0100 = 4
0000 1000 = 8
0001 0000 = 16
0010 0000 = 32
0100 0000 = 64
1000 0000 = 128

A number of methods are used to code data. Those familiar with some computer basics may be acquainted with some American Standard Code for Information Interchange (ASCII) codes. This coding system has its own distinct method of coding values and would not be compatible with other coding systems without some kind of translation. Because on/off states can so easily be used to represent data, most digital computers and communications use this technology. It is also used in optical data processing, data retention, and communications. Digital signals may be transmitted in series 1 bit at a time, which tends to be slower, or in parallel, which is much faster. If the numbers 0 through 3 had to be transmitted through a serial link, they would be signaled sequentially, as shown in **Figure 6–31**. If the same numbers were to be transmitted through a parallel link, they could be outputted simultaneously as shown in **Figure 6–32**, which would be much faster.

However, in critical vehicle communications systems, data is usually streamed through series circuits. Serial communications require every message to be

FIGURE 6–31 Sequential switching of binary coded numbers through a serial link

Decimal number	Binary number
0	0 0 0 0
1	0 0 0 1
2	0 0 1 0
3	0 0 1 1

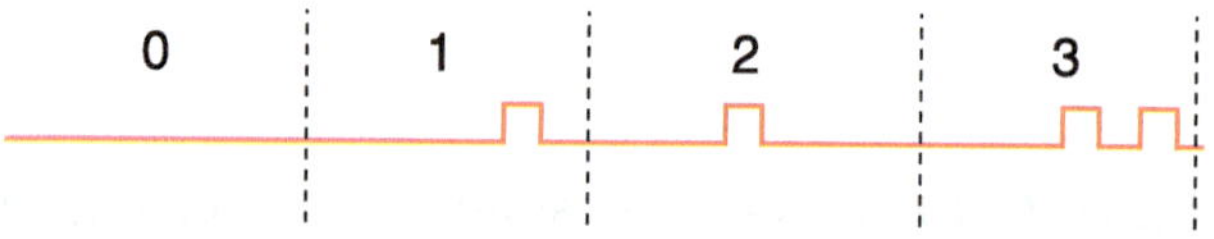

FIGURE 6–32 Parallel link: Data transmission time is greatly reduced

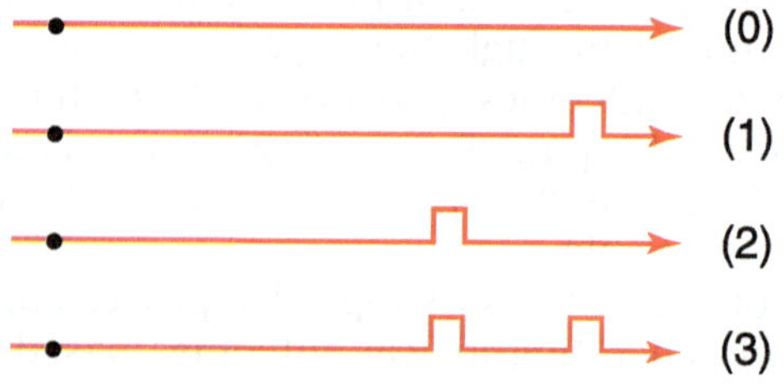

assigned priority status. This helps prevent conflicts. For instance, when a driver effects a panic brake stop you can bet that the braking electronics have very high priority status on the communications network, making other less important messages wait.

6.10 MICROPROCESSORS

A microprocessor is a solid-state chip that contains many hundreds of thousands of gates per square inch. The microcomputer is the operational core of any personal or vehicle computer system. All current highway trucks, buses, and off-highway equipment, use computers to manage powertain and other on-board system functions. All highway trucks manufactured in North America today are "drive by wire" with full authority chassis management. On-board vehicle computers are referred to as engine/electronic control modules (ECMs) or electronic/engine control units (ECUs). ECMs and ECUs are system **controllers**, that is, they are the computers used to manage a system. ECM is the SAE acronym of choice when referring to engine controllers while ECU tends to be used for other system controllers. However, each original equipment manufacturer (OEM) has its own preferences; so in this book, the acronym ECU will be generally used, except when discussing components manufactured by an OEM that prefers a different term.

ECU CONSTRUCTION AND NETWORKING

ECUs normally contain a microprocessor, data retention media, and often the output or switching apparatus. The ECU can be mounted on the component to be managed (a transmission management ECU is often located on the transmission) or, alternatively, inside the vehicle cab. Increasingly as this technology develops, several system ECUs can be networked by connecting them to a data bus. This means that a vehicle equipped with computerized engine management, electronic transmission, ABS/traction control, and any other electronically controlled systems with an address on the communications bus would be capable of exchanging data and command requests to optimize performance. The term *multiplexing* is used to refer to a vehicle management system using multiple, interconnected ECM/ECUs on a data bus network.

CAUTION

Static discharge can destroy sensitive computer equipment. Wear a ground strap or use common sense to avoid damaging computer equipment by careless static discharge.

FIGURE 6–33 Information processing cycle

PROCESSING

A truck technician must have a basic understanding of both vehicle and personal computers to interact effectively with today's technology. This section introduces the essentials of electronic management of vehicle systems. *Computerized system management* is simply summarized as a set of electronically connected components that enable an information processing cycle comprising three distinct stages:

1. Data input
2. Data processing
3. Outputs

The information processing cycle is shown in diagram form in **Figure 6–33**.

DATA INPUT

Data is simply raw information. Most of the data that has to be signaled in a truck ECU comes from monitoring sensors and command sensors. Most data input devices are sensors. An example of a monitoring sensor would be a tailshaft speed sensor that signals road speed information to the various ECUs that need to know the vehicle road speed. An example of a command sensor would be the throttle position sensor, whose signal is used by the engine and transmission electronics. In fact, anything that signals input data to an electronic system can be described as a sensor. This means that sensors may be simple switches that an operator toggles open or closed, devices that ground or modulate a reference voltage (V-Ref), or devices that are powered up either by V-Ref or battery voltage (V-Bat).

V-Ref is voltage conditioned by the system ECM or ECU. It is almost always 5V. V-Ref is the supply voltage for the most common category of sensors known as **ratiometric analog**. Ratiometric analog sensors receive V-Ref and output a proportion of that voltage as a signal in "ratio" with change in status. Some components used in the **input circuit** of a vehicle computer system are discussed in the following sections.

Thermistors

Thermistors precisely measure temperature. There are two types. Resistance through a thermistor may decrease as temperature increases, in which case it is known as a negative temperature coefficient (NTC) thermistor. A thermistor in which the resistance increases as temperature rises is known as a positive temperature coefficient (PTC) thermistor. The ECU receives temperature data from thermistors in the

FIGURE 6–34 An NTC-type thermistor. The chart indicates resistance of an NTC thermistor decreases as temperature increases. Output of thermistor is not linear.

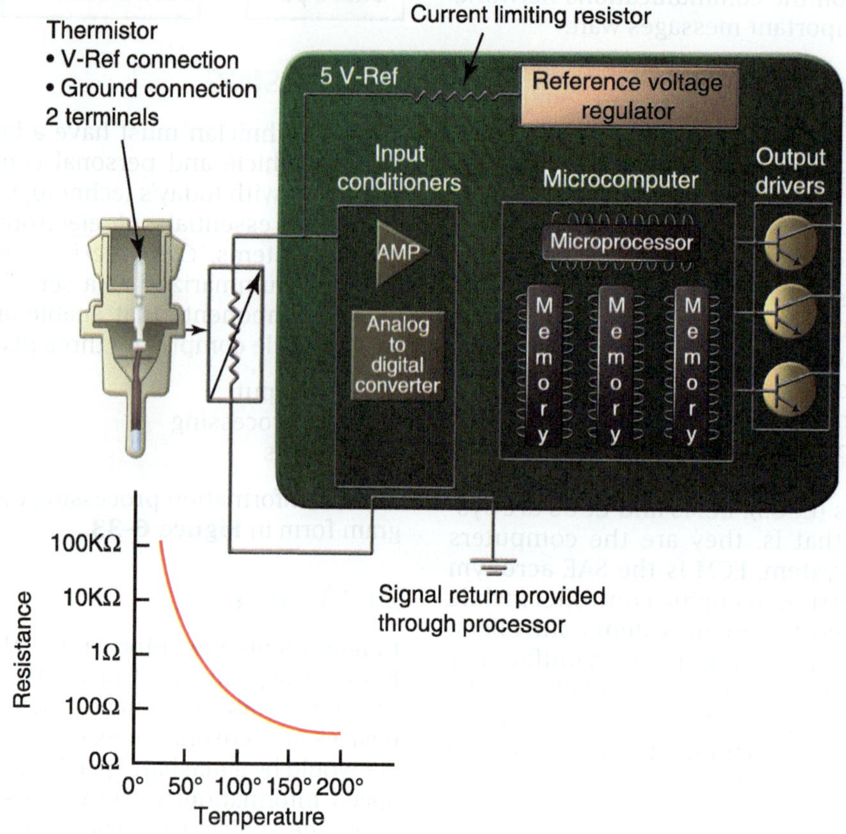

form of analog voltage values. A transmission oil temperature sensor is usually a thermistor. **Figure 6–34** shows a schematic of a thermistor.

Variable Capacitance (Pressure) Sensor

These are supplied with reference voltage and usually designed to measure pressure values. The medium whose pressure is to be measured acts on a ceramic disc connected to a flat wound coil and moves it either closer to or farther away from a steel disc. This varies the capacitance of the device and, therefore, the voltage value returned to the ECU.

Variable capacitance-type sensors are used for oil-pressure sensing, MAP sensing (turbo boost pressure), barometric pressure sensing (BARO), and fuel pressure sensing. **Figure 6–35** shows an engine oil–pressure sensor that uses a variable capacitance electrical operating principle.

Piezo-Resistive Pressure Sensor

This type of sensor is often used to measure manifold pressure and can be known as a Wheatstone bridge sensor. A doped silicon chip is formed in a diaphragm shape so it measures 250 microns around its periphery and only 25 microns at its center; in other words, it is about 10 times thinner in the center. This permits the silicon diaphragm to flex at the center when subjected to pressure. A set of sensing resistors is located around the edge of a vacuum chamber behind the diaphragm, so when pressure causes the diaphragm to deflect, the resistance of the sensing resistors changes in direct proportion to the increase in pressure.

An electrical signal proportional to pressure is obtained by connecting the sensing resistors into a Wheatstone bridge circuit in which a voltage regulator holds a constant DC voltage across the bridge. When there is no pressure acting on the silicon diaphragm, all the sensing resistance will be equal and the bridge will be balanced. When pressure causes the silicon diaphragm to deflect, the resistance across the sensing resistors increases, unbalancing the bridge and creating a net voltage differential that can be relayed to the ECU as a signal.

Potentiometers

The **potentiometer** is a three-wire sensor (the wires used are V-Ref, ground, and return signal). The potentiometer is designed to vary its resistance in proportion to mechanical travel. The potentiometer receives a V-Ref or reference voltage from the system ECM. It outputs a voltage signal exactly proportional to the motion of a mechanical device. The signal returned

FIGURE 6–35 Variable capacitance-type sensor

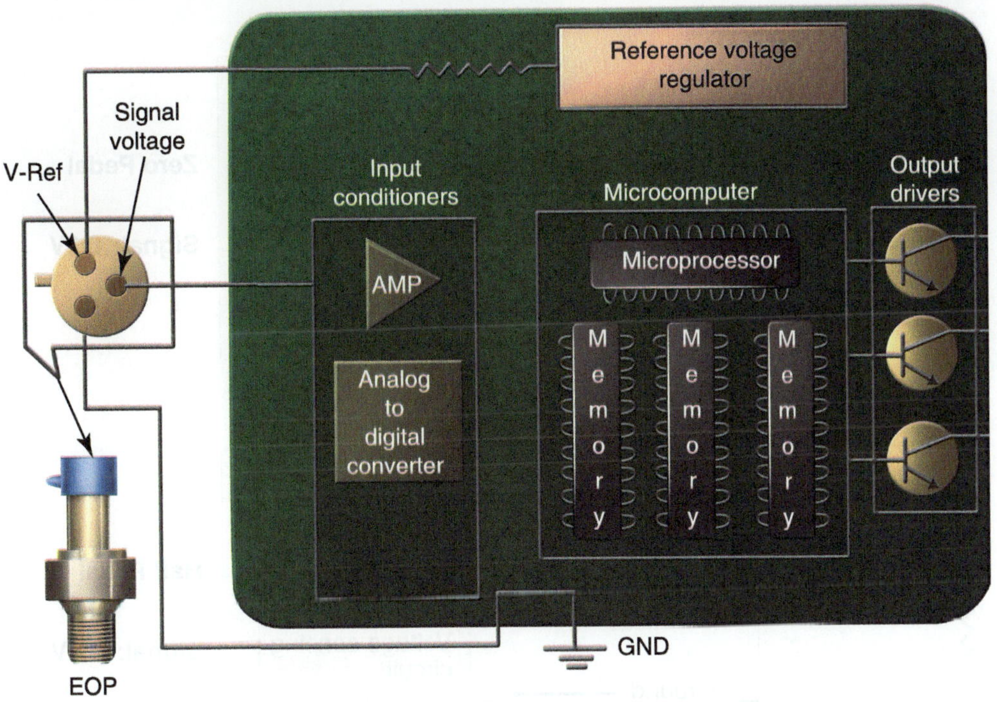

to the system ECU is always less than V-Ref, so sometimes they can be referred to as voltage dividers. As the mechanical device moves, the resistance is altered within the potentiometer. Most throttle position sensors (TPSs) used in trucks before 2007 used a potentiometer operating principle as demonstrated in **Figure 6–36**.

Rheostats and Voltage Dividers

A rheostat is a two-terminal variable resistor incorporating a resistive track and slider, meaning that there is a single path for electrical current flow: The mechanical position of the slider will determine how much current flow is choked down by the resistor. The key

FIGURE 6–36 Potentiometer as used in a throttle position sensor

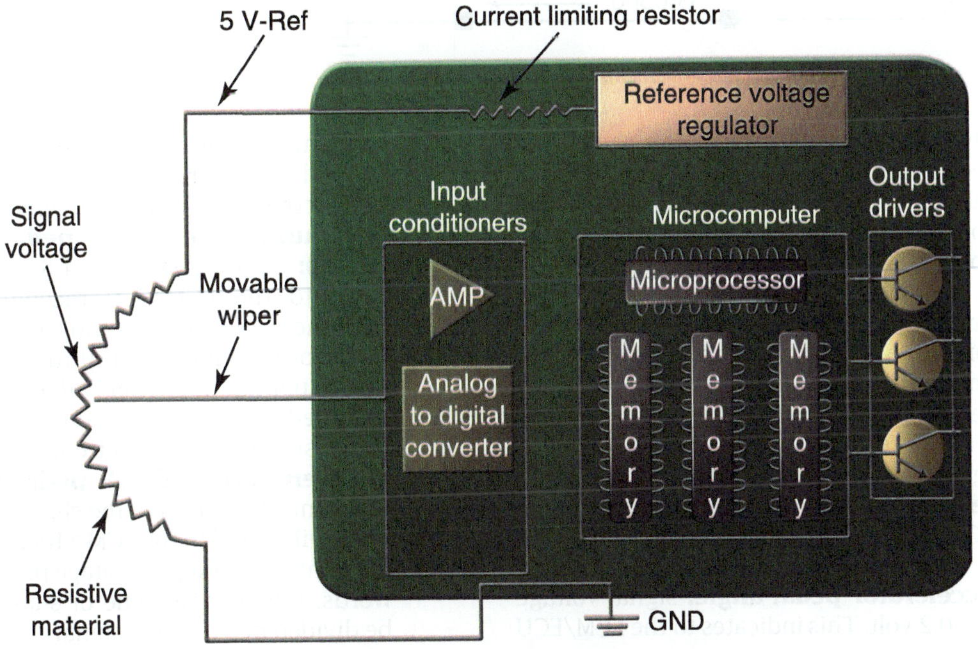

FIGURE 6–37 TPS functioning properly

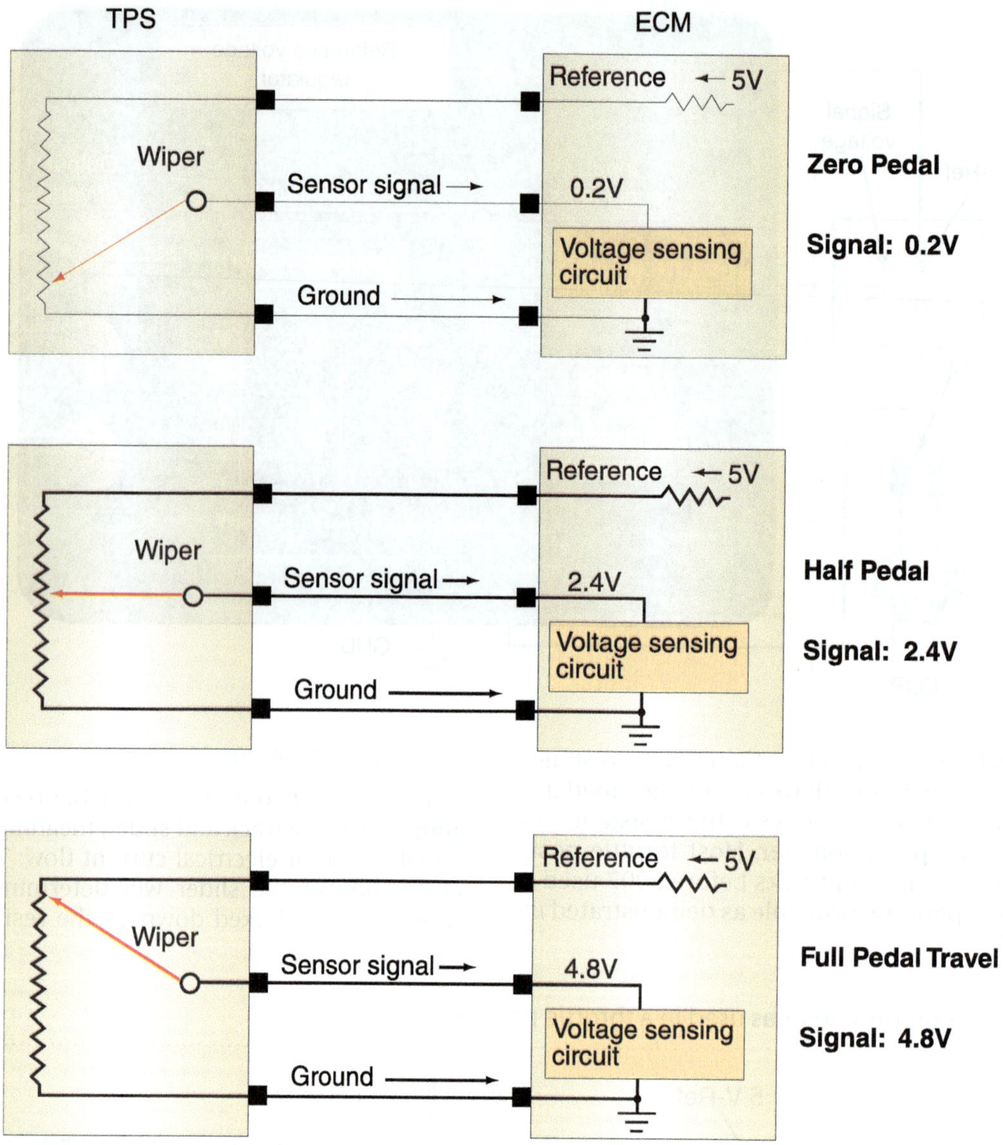

difference between a potentiometer and a rheostat is there are three terminals on a potentiometer: supply voltage, signal voltage, and ground. This means that the position of the slider on the resistive track determines how voltage is divided between the signal and ground paths. It is for this reason that we often call a potentiometer a voltage divider.

It is important to understand the voltage divider concept when troubleshooting a TPS. This requires that you understand the signal voltage produced and how the ECM/ECU interprets those signals when a TPS is functioning normally (**Figure 6–37**) and in failure mode (**Figure 6–38**). Typically a TPS supplied with a V-Ref of 5 VDC performs as follows:

- **Zero accelerator pedal angle:** Signal voltage output is 0.2 volt. This indicates to the ECM/ECU receiving the signal that the TPS is functioning properly. If a zero voltage signal was returned to the controller, it would diagnose an open.
- **Maximum accelerator pedal angle (full travel):** Signal voltage is 4.8 volts. This indicates to the ECM/ECU receiving the signal that the TPS is functioning properly and the operator is requesting full fuel. If a signal voltage equaling the V-Ref value of 5 volts was returned, the ECM/ECU would diagnose a short in the supply to signal circuit.
- **Between zero and full pedal travel:** Actual mechanical position of the slider on the resistive track will signal to the ECM a fueling request proportional to the signal voltage produced. In other words, the voltage value of supply voltage will be divided between the signal and ground paths.

FIGURE 6–38 TPS malfunctions. Pay special attention to the bad ground schematic—this is a common occurrence in a truck that produces an unexpected fault code.

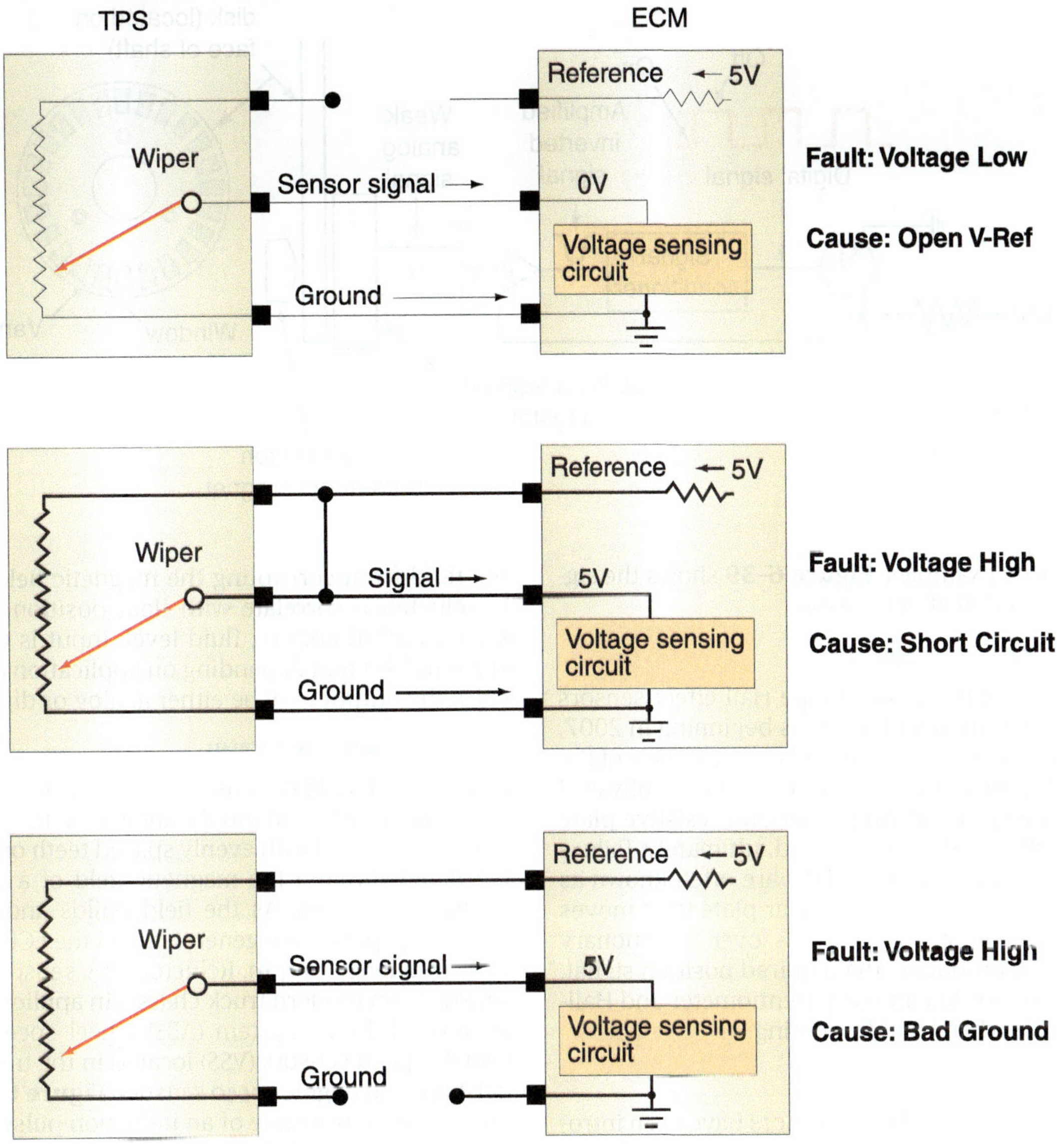

- **Loss of potentiometer ground:** This is a common problem because of the location of the TPS on the floor of a truck. The potentiometer is a voltage divider so in the event that the ground path is lost, no voltage will be dropped on through the ground circuit. This means that signal voltage will equal supply voltage; this is read as a "short" but was actually caused by the bad ground.

Figure 6–37 and Figure 6–38 show some typical TPS normal and failure mode operating conditions.

Hall Effect Sensors

Hall effect sensors generate a digital signal. They may be rotary or linear. Timing windows or vanes on either a rotating disc or linear plate pass through a magnetic field. When a rotating disc is used it is known as a **pulse wheel** or **tone wheel**, terms that are both used to describe the rotating member of the induction pulse generator, so care should be taken to avoid confusion. The frequency and width of the signal provides the ECU with speed and position data. The disc or linear plate incorporates a narrow window or vane for accurately relaying position data.

Rotary-Type Hall Effect Sensors

Rotary-type Hall effect sensors are used to input shaft speed and position data for purposes of event timing computations such as the beginning and duration of the pulse width. The camshaft position sensor (CPS), timing reference sensor (TRS), and engine position

FIGURE 6–39 Rotary Hall effect sensor

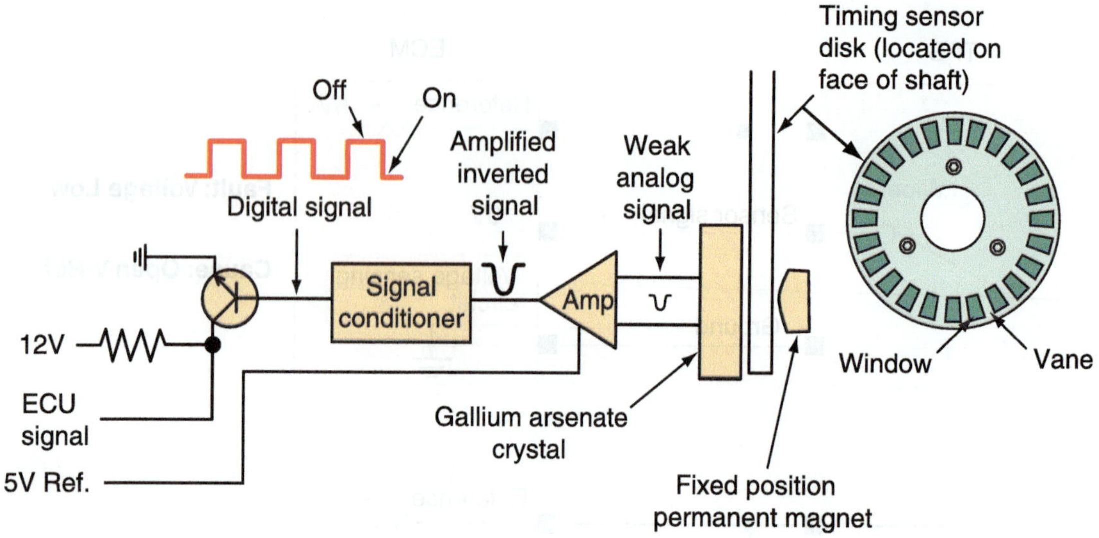

sensor (EPS) are examples. **Figure 6–39** shows the signal output of a Hall effect sensor.

Linear-Type Hall Effect Sensors

Some OEMs introduced linear-type Hall effect sensors to replace potentiometer-type TPSs beginning in 2007, and these have become commonplace today. One of the problems of potentiometer-type TPS was the physical contact between the moving wiper and resistive plate that inevitably resulted in wear and, ultimately, failure of the component. Hall effect TPSs are often known as noncontact TPSs. A windowed linear plate (that moves with the accelerator pedal) moves over a stationary magnetic field, producing the required position signal. It is important not to confuse potentiometer and Hall-effect type TPS when troubleshooting.

Hall Effect Fluid Level Sensors

Recently, Hall effect fluid level sensors have been introduced. When properly calibrated, this category of fluid level sensor can indicate the extent of underfill (or overfill) and display it to the operator on request. Fluid levels can be written to an audit trail, which can help in analyzing warranty culpability. An example of a Hall effect fluid level sensor is found in the oil sump of a Caterpillar CX transmission studied later in Chapter 21.

The sensor uses a float suspended by a pair of coil springs on the top and bottom, as shown in **Figure 6–40**. A coil spring has the characteristic of being very flexible axially but fairly rigid radially. Fluid (for instance, transmission oil) enters the bottom of the sensor assembly by means of inlet ports and as the float moves either up or down, air is allowed in and out through breathing ports at the top of the sensor assembly. The breathing ports at the top of the sensor are located low enough to maintain a constant air bubble around the magnet; this prevents magnetic particles in the fluid from corrupting the magnetic field. Changes in reluctance correlate with float position within the sensor and, therefore, fluid level. Input is usually the system V-Ref and depending on application (sensor circuit), the output may be either analog or digital.

Induction Pulse Generator

A toothed disc known as a **reluctor**, pulse wheel, or tone wheel (also commonly known by the slang term **chopper wheel**) with evenly spaced teeth or serrations is rotated through the magnetic field of a permanent stationary magnet. As the field builds and collapses, AC voltage pulses are generated and the ECU correlates their frequency to rpm. Reluctor-type sensors are used variously on modern truck chassis in applications such as antilock brake system (ABS) wheel speed sensors, vehicle speed sensors (VSS) located in the transmission tailshaft, and engine speed sensors. **Figure 6–41** shows the operating principle of an induction pulse generator while **Figure 6–42** shows what one actually looks like when used as a transmission tailshaft speed sensor.

Switches

Switches can either be electromechanical or "smart." Electromechanical switches can either open or close a circuit by toggling (driver command) or ground-out (such as a radiator coolant level sensor). Smart switches use either ladder logic or an integral processor and are discussed in Chapter 12.

Data Processing

Data processing is the "thinking" function of a computer or microprocessor. This thinking function involves receiving inputs, consulting program instructions and memory, and then generating outputs. The term *data processing* is used to describe both the simple and complex processes that take place within the computer.

FIGURE 6–40 Linear Hall effect fluid level sensor

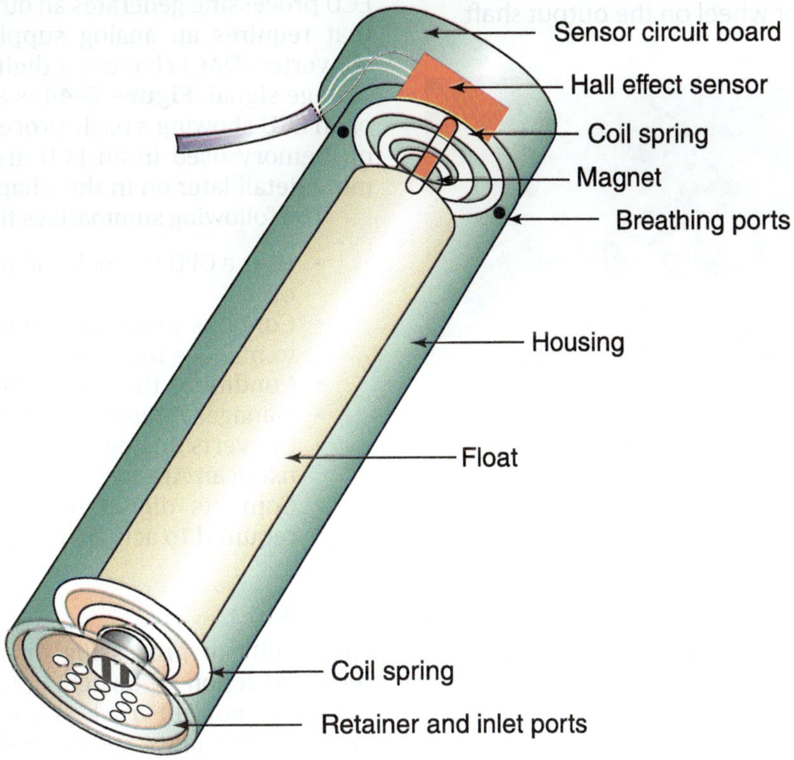

CAUTION

Chassis electrical voltage pressure can destroy vehicle computers and sensors. When testing vehicle computer circuits, exercise caution when connecting break-out boxes and Ts. Never use jumper wires and diagnostic forks across terminals unless the test procedure specifically requires this.

FIGURE 6–41 Induction pulse generator

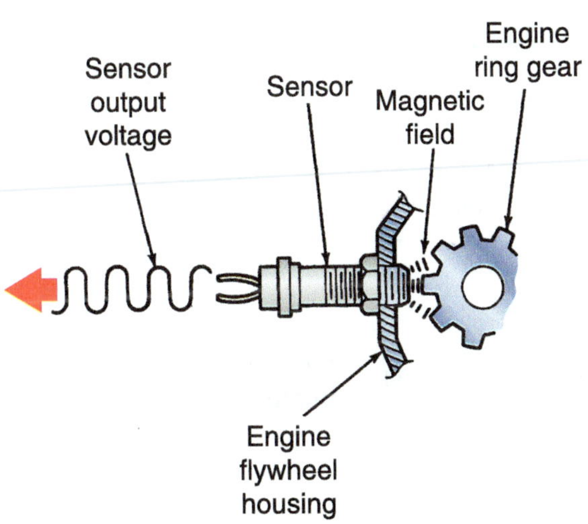

Probably the most important component in the microprocessor is the **central processing unit (CPU)**. The CPU contains a control unit that executes program instructions and an arithmetic logic unit (ALU) to perform numeric calculations and logic processing such as comparing data. It also clocks the processing frequency: the higher the frequency, the faster the processing speed of the computer. Most vehicle microprocessors run on voltage values lower than the 12V chassis voltage in much the same way that a home computer runs on much lower voltage than household supply voltage. Typically, this is 5V or less. It is a function of the ECU to transform the chassis voltage to that required to run the microprocessor circuit. Additionally, a lower voltage value is required as reference voltage for the input circuit. Again, this is commonly 5V.

Memory Management

The role of the CPU is to manage the processing cycle. This requires receiving input data and locating it in the processing cycle. The input conditioning discussed previously are shown in **Figure 6–43**. The objective of Figure 6–43 is to show how typical inputs to a vehicle computer are conditioned prior to processing. The CPU also fetches and carries information from the ECU memory compartments (usually known as **memory banks**) and loads this into the processing cycle.

FIGURE 6–42 Transmission tailshaft sensor that uses an induction pulse generator principle: Note the location of the reluctor wheel on the output shaft.

Random access memory (RAM) is data that is electronically retained in the ECU. Only this data can be manipulated by the CPU. Input data and magnetically retained data in **read-only memory (ROM)**, **programmable read-only memory (PROM)**, and **electronically erasable programmable read-only memory (EEPROM)** are transferred to RAM for processing.

RAM can be called primary storage or main memory. Because RAM data is electronically retained, it is lost when its electrical power circuit is shut off. Because many of the signals into the ECU processing cycle are in analog format, these signals have to be converted to digital signals. The component required to perform this is known as an analog-to-digital converter (ADC). The same is true in reverse when the ECU processing generates an outcome to a component that requires an analog supply. A digital-to-analog converter (DAC) changes a digital signal to an analog voltage signal. **Figure 6–44** is a simplified schematic of an ECU showing a basic processing cycle. The types of memory used in an ECU are explained in a little more detail later on in this chapter.

The following summarizes the functions of an ECU:

- Uses a CPU to clock and manage the processing cycle.
- Contains in memory banks the data required to manage the system.
- Conditions the processor circuit voltage.
- Manages reference voltage.
- Converts analog input data to a digital format using an ADC.
- Converts digital outputs to analog voltages required to actuate electrical components.

Outputs

The results of processing operations must be converted to action by switching units and actuators. In most (but not all) truck system management ECUs, the switching units are integral with the ECU. Using the example of an ECU managing a typical automatic transmission, clutch actuators or solenoid valves are among the primary output devices to be switched. ECU commands are converted to an electrical signal that is used to energize clutch solenoids so the outcomes of the processing cycle are effected by shifting range ratios. The **output circuit** of any computing

FIGURE 6–43 Input conditioning to a typical ECU

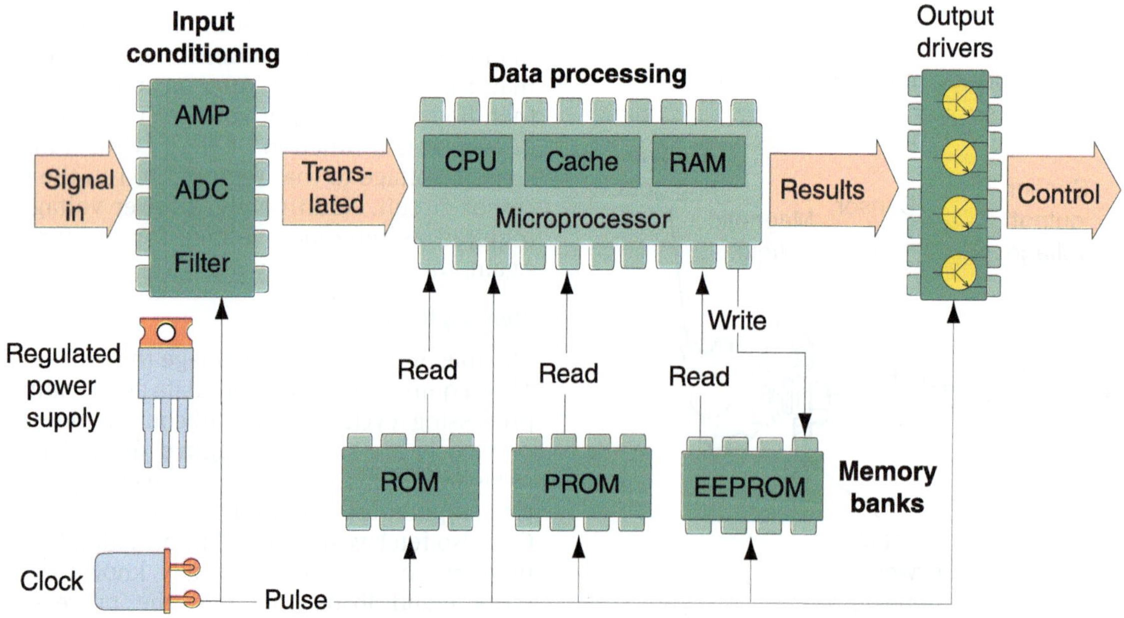

FIGURE 6–44 The ECU processing cycle

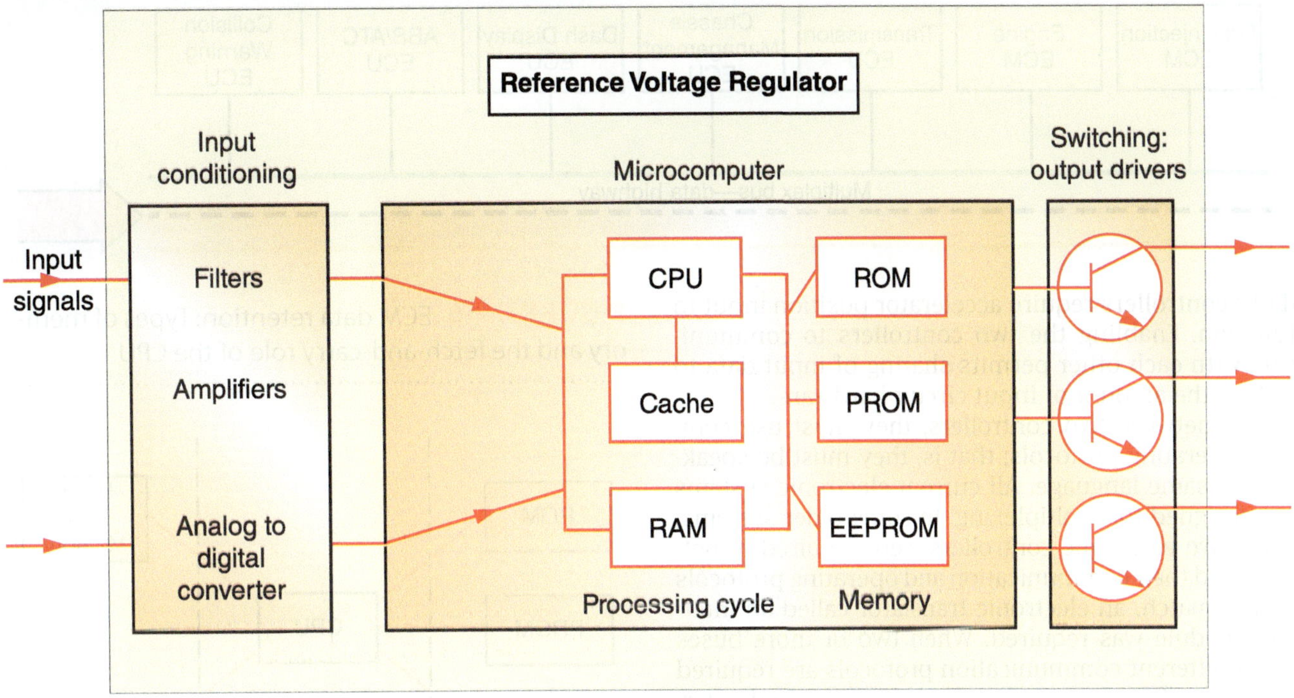

system simply effects the result of the processing cycle. Output devices on a home computing system would be the display monitor and the printer.

SAE Hardware and Software Protocols

Among the OEMs of truck and bus engines in the United States, there has been a generally higher degree of cooperation in establishing shared electronics hardware and software protocols than in the automobile manufacturing segment of the industry. To some extent this cooperation has been orchestrated by the Society of Automotive Engineers (SAE) and American Trucking Association's (ATA) Technology and Maintenance Council (TMC). It has been necessary because most North American trucks are engineered by a truck manufacturer using major component subsystems supplied by an assortment of component OEMs. For instance, a Cummins engine could be specified to any truck chassis and be required to electronically interact with a Fuller UltraShift transmission and Wabco ABS/ATC (antilock brake system/automatic traction control).

In the first generation of truck communication buses, separate SAE J standards (surface vehicle recommended practices) controlled the hardware and software multiplexing protocols that enabled a TPS on a Cummins engine to provide the throttle position data required to manage an Allison WT transmission. The current J1939 standard incorporates the software and hardware standards covered in J1587 and J1708 into a single standard. All current trucks use the J1939 standard, but note that some J1939 hardware currently integrates a J1587/1708 bus so that earlier generation electronics can be accommodated depending on the OEM.

SAE J1587	First-generation data exchange protocols used for data transactions on heavy-duty multiplexed chassis. Uses J1708 hardware.
SAE J1708	Serial multiplexing hardware compatibility protocols used on the J1587 data bus. For example, a data connector enabling communication with the J1587 data bus is a J1708 connector.
SAE J1939	The faster serial communications data bus used on all current multiplexed heavy commercial vehicles. J1939 addresses both software and hardware protocols and compatibilities and functions at much higher speeds than the J1587/1708 bus. J1939 standards are updated by simply adding a suffix. J1939-compatible systems share common hardware and communicate using a common "language."

Multiplexing

When a truck chassis uses multiple electronically managed circuits, multiplexing the system ECUs on a data bus makes sense. Networking the system controllers avoids duplication of hardware and serves to synergize (a synergized system's components work in harmony) the operation of each system. For instance, both the engine management (ECM) and transmission

FIGURE 6–45 Multiplex bus

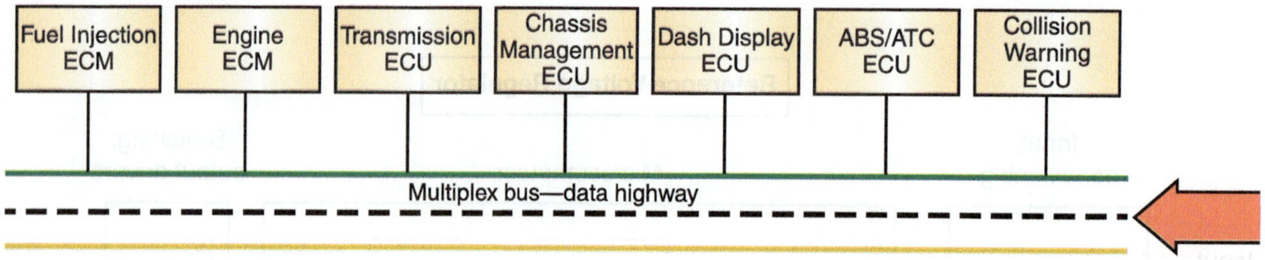

(ECU) controllers require accelerator position input to function. Enabling the two controllers to communicate with each other permits sharing of input data to reduce the amount of input circuit hardware.

To network two controllers, they must use common operating protocols; that is, they must be speaking the same language. All current electronic systems are designed for multiplexing. In some older systems, when two electronic controllers were required to network and their communication and operating protocols do not match, an electronic translator called an interface module was required. When two or more buses using different communication protocols are required to communicate, a **gateway module** is required. Interface modules are often not manufactured by either of the system OEMs. **Figure 6–45** shows how multiplexing is used in a modern truck integrated electronic system, and a more detailed introduction is provided in Chapter 12.

FIGURE 6–46 ECM data retention: Types of memory and the fetch-and-carry role of the CPU

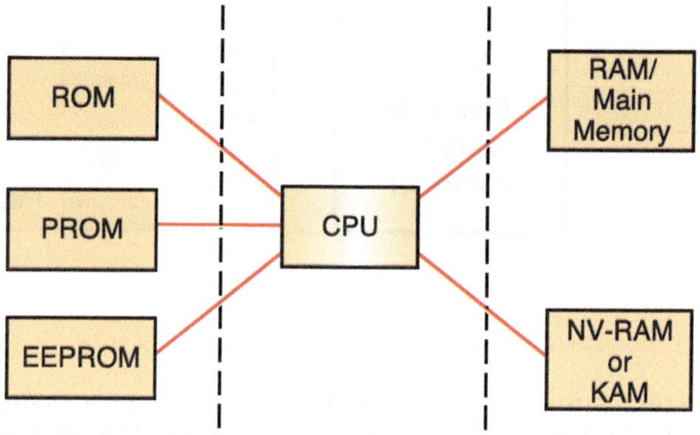

6.11 DATA RETENTION IN A VEHICLE ECU

Data retention means memory storage. Data is retained both electronically and magnetically in current-generation truck and bus ECUs; however, optical data-retention, laser-read systems are beginning to appear and will be used with increasing frequency as costs diminish. The data categories discussed in the following sections are used by contemporary ECUs (**Figure 6–46**).

RANDOM ACCESS MEMORY (RAM)

The amount of data that may be retained in RAM is a primary factor in measuring the computing power of a computer system. It is also known as *main memory* because the CPU can only manipulate data when it is retained electronically. At startup, RAM is electronically loaded with the management operating system instructions and all necessary running data retained in other memory categories (ROM/PROM/EEPROM).

RAM data is electronically retained, which means that it is always to some extent volatile (lost when the circuit is not electrically energized). In other words, data storage could be described as temporary—if the circuit is opened, RAM data is lost. It should be noted that most truck ECUs use only fully volatile RAM; therefore, when the ignition key circuit is turned off, all RAM data is lost.

NV-RAM

However, a second type of RAM is used in some truck and many automobile systems, often those that have no EEPROM capability. This is nonvolatile RAM (NV-RAM) or keep alive memory (KAM) in which data is retained until either the battery is disconnected or the ECU is reset. The ECU is usually reset by depressing a computer reset button that temporarily opens the circuit. Codes and failure strategy (action sequence) would be written to NV-RAM and retained until reset.

READ-ONLY MEMORY (ROM)

ROM data is magnetically or optically retained and is designed not to be overwritten. It is permanent, although it can be corrupted (rare) and, when magnetically retained, it is susceptible to damage when exposed to powerful magnetic fields. Low-level radiation such as that encountered routinely (from police radar and high-tension electrical wiring while driving on a highway) will not affect any current ECUs. A majority of the data retained in the ECU is logged in ROM. The *master program* for the system management is loaded into ROM. Production standardization is permitted by constructing ROM architecture so it contains common necessary data for a number of different systems. For example, identical ROM chips can be manufactured to run a group of different transmissions in a series. However, to actually make a transmission operate in a specific drivetrain application, the ROM data would require further qualification from data loaded into PROM and EEPROM.

PROGRAMMABLE ROM (PROM)

PROM is magnetically retained data. It is usually a chip, a set of chips, or a card socketed into the ECU motherboard that can often be removed and replaced. PROM's function is to qualify ROM to a specific chassis application. In the earliest truck engine management systems, programming options such as idle shutdown time could only be altered by replacing the PROM chip. Today customer programmable options are often written to EEPROM where they can be easily altered without changing any hardware so PROM chips are no longer used for this function. Some OEMs describe the PROM chip as a personality module, an appropriate description of its actual function of trimming or fine-tuning the ROM data to a specific application. Newer personality modules also contain the EEPROM data, explained in the next section.

ELECTRONICALLY ERASABLE PROM (EEPROM)

The EEPROM data category contains customer data programmable options and proprietary data (programming data owned and controlled by the OEM) that can be altered and modified using a variety of tools ranging from a generic electronic service tool (EST) up to a mainframe computer. A EST such as Nexiq iQ programmed with the correct software can be used to rewrite customer options such as tire rolling radius, progressive shifting, road speed limit, and so on. Usually, only the owner password is required to access the ECU and make any required changes to customer data options.

Programming

Proprietary data is more complex both in character and methods of alteration. It contains data such as the fuel map in an engine system. The procedure here normally requires accessing a centrally located mainframe computer, usually at the OEM central location, via a modem. The appropriate files are then downloaded, from the Web, to a personal computer (PC) or EST, and logged in memory. The ECU can then be reprogrammed (rewriting the original data) from the logged memory. Some proprietary ESTs—manufactured by one OEM for use only on its equipment—can act as the interface link between the vehicle ECU and the mainframe, but the procedure is essentially the same.

6.12 AUXILIARY ELECTRONIC SYSTEMS

Vehicle navigation, collision warning, lane departure warning, and infotainment systems have been embraced by the trucking industry. In particular, vehicle navigation systems have become commonplace in today's vehicles. Our objective is to provide a brief introduction to each technology in this section.

VEHICLE NAVIGATION

We commonly use the term *global positioning satellite* (GPS) *systems* to describe vehicle navigation systems. These systems vary in complexity (and cost) but a fully loaded system can consist of:

- Geographic positioning
- Target destination
- Route mapping, including real time updates
- Route guidance, including real time updates
- Map display

The detail that a navigation system offers depends on:

- Stored data capacity
- Telecommunications capability

Details such as real time traffic reports and detour negotiation tend to be expensive due to the cost of maintaining and updating the necessary information. **Global positioning satellites (GPS)** are the key to vehicle navigation technology. In taking a look at GPS communications, the student should remember that this is a one-way communications technology. To achieve two-way communications, GPS must be additionally supported by satellite communications or other wireless technologies such as WiFi or Bluetooth.

Trilateration

To understand how GPS functions, you have to understand a simple geometric principle called **trilateration**. Suppose you are completely lost somewhere in America. You pull into a truck stop to ask someone where you are, and the answer is that he or she is not sure. However, the person does know for sure that the location is exactly 490 miles from Cleveland. At this point you can get a map of the United States, find Cleveland, and draw a circle with a radius on a scale that equals 490 miles. You ask the next trucker, and you get the answer that he does not know exactly, but he has come directly from Houston and has traveled 710 miles. Now you can go back to the map and draw a circle with a radius representing 710 miles around Houston. At this point you still cannot say exactly where you are until you approach a third trucker and ask him the same

FIGURE 6–47 Position location by trilateration

question. He informs you that he is exactly 520 miles out of Jacksonville. Again, you draw a circle with a radius representing 520 miles around Jacksonville. Now if you look at your map (**Figure 6–47**), you will see that the three circles you have drawn all intersect at one point and you can determine your exact position.

GPS Location

If the world were flat, trilateration could be used to determine position. For the principle to work in three-dimensional space you have to use spheres instead of circles, so in order to pinpoint exact position, four GPS satellites are required. If the GPS receiver can only locate three satellites, then its software creates a *virtual* sphere, which is a dummy sphere that allows it to use the coordinates established by the three satellites it has located. This means that your position can be calculated but not your altitude.

Measuring Distance. GPS satellites broadcast radio signals that a GPS receiver can detect. The receiver then calculates how long the signal has taken to travel from the satellite to the receiver. For a GPS system to function, twenty-four satellites are required so that there is a minimum of four on the horizon at any time or location during the day. The signals received tend to be weak, so this technology works better driving through the countryside than it does in the middle of a large city. The term **triangulation** can also be used in place of trilateration.

Mapping. The raw positional data produced by the GPS receiver unit on your car or truck can next be correlated to geographic information retained in the memory of the GPS ECU. The retained data may be in the controller's memory, but the system works best with a telecom link found on some trucks that can provide live data on such things as traffic congestion and road construction.

Navigational Aids to GPS. Depending on the system, the effectiveness of a GPS unit can be improved by inputting data from the chassis electronics (road speed) and dedicated sensors such as a gyrometer to signal change of direction data. A top-of-the-line system can be integrated with the collision warning electronics discussed next.

6.13 ACCIDENT AVOIDANCE SYSTEMS

Accident avoidance systems are all about using electronics to make vehicle operation safer. They can be divided into systems that alert the driver to act to avoid an accident (collision warning), and more comprehensive collision mitigation which makes use of the chassis electronic communication systems to act independently (from the driver) to avoid an accident by doing such things as applying brakes and derating the engine. Truck accident avoidance systems are covered in theory in this chapter and in practice in Chapter 11. The systems currently used can be divided into:

- Normal driving assists
- Conflict alerts
- Accident and damage control response

NORMAL DRIVING

Examples of normal driving assists can be grouped as follows:

- Adaptive (smart) cruise control
- Hill descent control
- Active headlamp controls
- Alcohol impairment lockout

Conflict Alerts and Intervention

Examples of conflict alerts and intervention strategies can be grouped as follows:

- Stability control electronics (SCE)
- Antilock brake system (ABS)
- Automatic traction control (ATC)
- Collision warning systems (CWS)
- Collision mitigation systems (CMS)
- Trailer stability assist
- Antirollover electronics
- Lane departure warning (LDW) systems

Accident and Damage Control Response

Examples of accident and damage control response strategies can be grouped as follows:

- Supplemental restraint systems (SRS)
- Rollover protection (RollTech)
- Accident analysis audit trails

COLLISION DETECTION SYSTEMS

A **collision warning system (CWS)** or **collision mitigation system (CMS)** is designed to alert the driver to imminent danger and, when programmed to do so, take evasive action. CWS and CMS are common options on today's commercial trucks, and they use a combination of radar, microwave, and optical (camera) sensing to function. The first CWS was Eaton **VORAD** system, which in 2009 was purchased by Bendix and renamed **Wingman**. At the same time, Bendix X-vision was phased out. Bendix Wingman is a CMS because it offers **active cruise with braking (ACB)**, meaning that the software can be programmed to actuate engine and service braking when vehicle velocity and closing distance exceed a programmed threshold value. Wingman broadcasts command requests to the chassis data bus, which may result in actuating up to 100 percent engine braking potential and as much as one-third of the service braking potential of the vehicle.

Because VORAD and now Wingman have been used by the trucking industry for a generation, we are going to base our description of CWS/CMS primarily on these two systems. Troubleshooting of both systems will be covered in Chapter 11.

VORAD is a loose acronym for Vehicle On-board RADar. The word *radar* comes from the initial letters of RAdio Detecting And Ranging. Radar is capable of precisely measuring both distance and relative speed. VORAD and Wingman CMS use radar, laser, and microwave technology to sense some specific vehicle danger conditions and alert the driver using sound and warning lights. It is integrated into the cruise control (CC) electronics but has features that can alert the driver whether in CC mode or not. Wingman CMS uses the same detection technology as VORAD but it can broadcast commands into the chassis data bus so that it can intervene or mitigate an imminent collision event by applying the engine and vehicle brakes. Both VORAD and Wingman sense straight-ahead closing velocities and proximity as well as blind side proximity and alert the driver when danger is detected. CWS and CMS have been proven to reduce accident rates and have a special appeal to fleet operations with higher-than-average accident rates because they can reduce insurance payments. Radar systems were first used because they required simple processors. In more sophisticated systems, a radar CWS platform may be complemented by video.

Radar versus Video

Cameras (video) produce much more dense input data so powerful **programmable logic controllers (PLCs)** are required to process the information. The advantage of video/camera over laser or microwave radar is that distinctions can be made as to the type of image profile detected; that is, a motorcycle versus a truck. Video has the advantage of being adapted for a **lane departure warning (LDW) system**. Radar has the advantage of "seeing" in the dark or through fog. So the best systems use a combination of both technologies.

Doppler Effect

Christian Doppler (1803–1853) was an Austrian physicist who theorized that because the pitch of sound is highest at a sound source and gets progressively lower with distance, the same should apply to other waves such as light waves. Change in frequency can be used to determine relative speed. For instance, if the light waves produced by a star were observed to shift toward the red end (lower frequency/longer wavelength) of the spectrum (see Figure 6–20), the star is becoming more distant from the observer; also, the rate of change in the frequency can be used to calculate the velocity of the star. In the case of an observed star accelerating toward an observer on Earth, the reverse would be true: The light wave produced would shift toward the violet (higher frequency/shorter wavelength) end of the spectrum. Again, the velocity of the star can be calculated by determining the rate of shift. Doppler effect is used extensively by astronomers and is the basis of radar systems.

Pulse Radar

Radar operates by transmitting electromagnetic waves toward an object and then monitoring the return wave bounced back by the object. The properties of the returned wave are then analyzed by a signal processor and converted to a form usable by the operator or system. Microwaves and lasers are commonly used.

For instance, an air traffic controller would have a video display that would have a map-like image of the area scanned, indicating the location and velocity of aircraft in the area. The objectives of VORAD are more basic because its return signals are processed as warning signals. The more recent Wingman is capable of mitigating what it detects (using program logic) as imminent danger so in addition to providing driver warning alerts, it can also intervene to apply the engine retarder and service brakes.

Pulse radar requires that a transmitter produce short pulses of radiated energy and that a receiver monitor the return signals as they are bounced off objects. Close objects would produce an *echo* from the pulse sooner than distant objects. Microwaves travel at the speed of light, about 1,000 feet (300 meters) per microsecond, so the delay between the transmission of a pulse and the echo can be used to calculate distance. The Doppler effect, the change in observed frequency produced by motion described in the previous section, is used by the processor to determine speed.

CWS and CMS Detection Electronics

Wingman and VORAD provide object detection on the front and right (blind) side of a truck and may optionally provide left (driver) side object detection. The systems are electronically managed and consist of relatively few components. Both Wingman and

FIGURE 6–48 CWS and CMS components

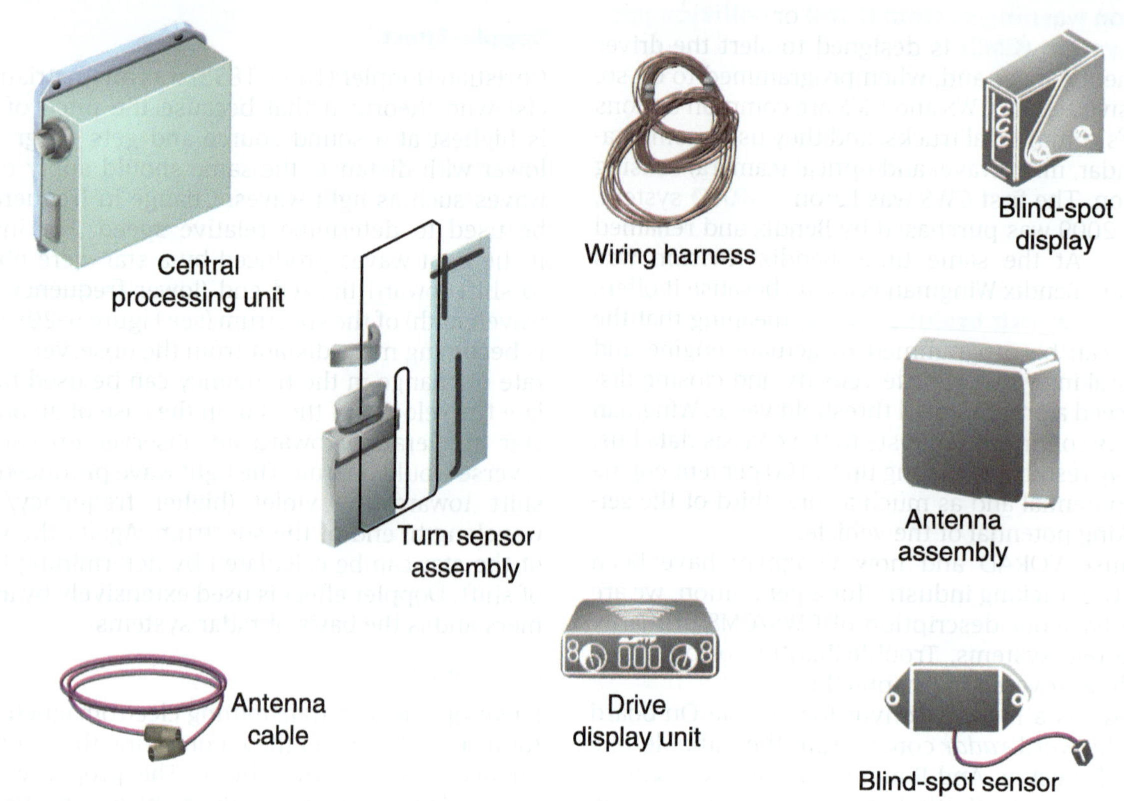

VORAD are connected into the vehicle electrical and electronic system in order to share sensor signals such as the input from the vehicle road speed sensor (tailshaft sensor). But only Wingman broadcasts command data to the bus, with the objective of mitigating a potential collision. The components used by Wingman and VORAD system are shown in **Figure 6–48**.

Antenna Assembly (AA)

The forward antenna is mounted directly to the center of the bumper and is used to transmit and receive microwave pulses. Transmitted radar signals are bounced off objects in a straight line path from the transmitting antenna. Return signals are received by the antenna and the Doppler effect (frequency shift) of the transmitted and received signals is calculated. This information is then converted to a digital format and sent to a CPU that performs additional processing.

The mounting location ensures that the radar beam is aimed directly in front of the truck. The AA has a range of about 350 feet (107 meters) and is capable of monitoring up to 20 vehicles (the 20 closest to the front of the truck) in its scanned range. **Figure 6–49** shows the location of the antenna on a typical tractor application.

Blind-Spot Sensors (BSS)

The blind-spot sensor (BSS) is a simple motion detector that senses objects in the driver's blind spot on the right side of the vehicle. It is mounted not more than 3 feet (0.9 meter) ft. above road level and about 4 feet (1.2 meters). behind the mirror assemblies. BSS information is sent to the ECU, which will output the appropriate driver alerts. Figure 6–49 shows the required location of a BSS.

Turn Sensor Assembly (TSA)

The turn sensor monitors the movement of the steering wheel shaft. Because the antenna outputs its microwave signal in a straight line, during a turn it will sense objects located on the side of the road such as buildings, billboards, stationary vehicles, and so on. The turn sensor unit signals to the ECU that a turn is being made, and the ECU will disable the audible driver warnings for the duration of the turn. The turn sensor consists of a shaft-mounted magnet that rotates past sensors when the shaft rotates more than 2 to 4 degrees. When the steering wheel is turned and the magnet passes across either sensor, the ECU receives a signal indicating that a turn is being made. **Figure 6–50** shows the TSA components.

Central Processing Unit (CPU)

It is the system CPU that receives CWS and CMS input signals from the antenna and the turn sensor and BSS, processes the input data and programmed instructions, and switches the system outputs, meaning the lighting of appropriate indicator lamps and

FIGURE 6–49 Antenna and blind-spot sensor location

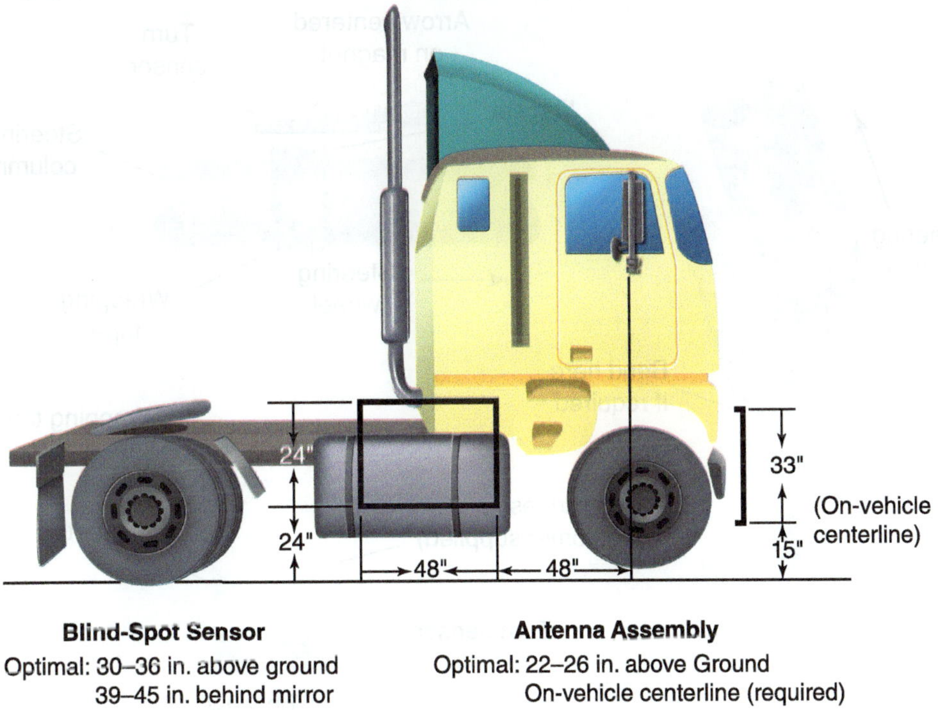

the sounding of audible driver alerts. The CPU can be located in a number of places on the truck, but it is most commonly found inside the firewall (under the dash) or behind the driver's seat. The CPU contains a slot for an optional driver's card. The driver card has two functions:

- It can log up to 10 minutes of system data. This feature is typically used to reconstruct the events leading up to an accident.
- It can function as an ID card. It may be programmed as personal electronic ID, identifying the name and employee number of the driver.

The CPU has write-to-self memory and a clock. This permits Wingman and VORAD to store trip recorder information, crash times, engine idle time, miles traveled, average speed, average following time, and distances. Like other electronic driver monitoring devices, Wingman and VORAD can play a role in driver education. For instance, truck drivers who are inclined to tailgate will be less likely to do so if they know their driving habits are being electronically monitored. Note the location of the Wingman or VORAD ECU in a typical highway tractor.

Driver Interface Unit (DIU)

The Wingman **driver interface unit (DIU)** or VORAD **driver display unit (DDU)** is mounted to the top of the dash in a location that is both easily visible and reachable by the driver. The unit houses the controls and indicators used by the system. The DIU controls system power-up, speaker volume, the range thresholds of the vehicle alerts, and writing of data to the driver card. The indicator lights are used to signal system power, system failure, three stages of distance alerts, and the detection of stationary or slow-moving objects.

A light sensor on the DIU automatically adjusts the intensity of the display lights. Display light intensity is greatest in bright sunlight and is dimmed for night driving. The DIU also contains a small speaker used to warn the driver when second- or third-stage intervals to impact are detected, or if a vehicle is detected in the blind spot when a left turn is indicated. Current versions of Wingman can broadcast to the driver information display to show status conditions as shown in Chapter 11.

Blind-Spot Display (BSD)

The BSD is mounted to the right-side, inside-front cab pillar in a close-to-direct sight line from the driver and the right-side mirror. The unit contains red and yellow lights that are switched by the ECU, indicating whether or not a vehicle is detected in the blind spot. As with the DIU, a light sensor in the BSD unit automatically adjusts the light intensity of the display lights as light conditions vary.

When the blind-spot motion sensor detects a vehicle in the blind spot, the red light is illuminated. When the blind spot is detected to be clear, the yellow lamp is illuminated. When left-side sensors are used on a CMS, the setup is identical to the BSD, except that BSS are used on both sides of the vehicle and display

FIGURE 6–50 CWS and CMS steering sensor and magnet mounted on the steering shaft

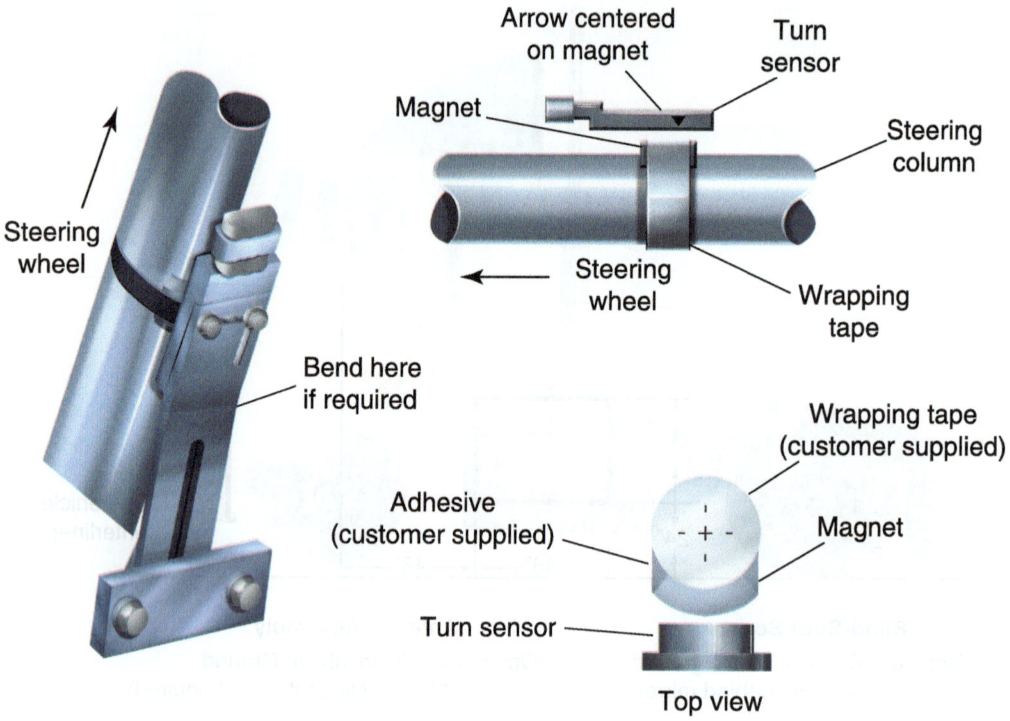

units are located on both sides of the cab as shown in Figure 6–49.

The CWS/CMS Cab

Figure 6–51 shows the location of the Wingman and VORAD components in a typical truck cab. Note that there are alternatives to this arrangement, but the location of the driver alerts shown here is optimum. **Figure 6–52** shows what the Wingman radar antenna/sensor actually looks like on the front bumper of a truck.

VIDEO

Video has been used in specialty truck, school bus, and transit bus applications for a couple of decades. Because the processing of video signals requires more computing power, it has been only recently that its use has become economical and, therefore, more widespread. Compact (and now less costly) PLCs have enabled the increased use of video on commercial vehicles. A simple application of video technology

FIGURE 6–51 Location of CWS and CMS components in a typical truck cab

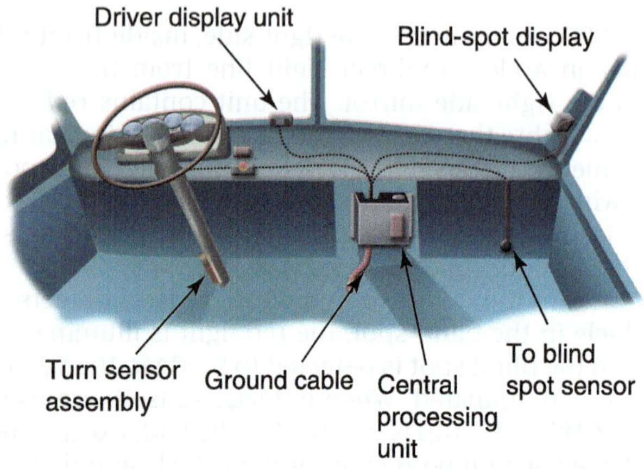

FIGURE 6–52 Wingman radar sensor/antenna on the front bumper of a truck

would be an aftermarket LDW system. This can be as simple as a windshield-mounted camera and integral beeper alert system. An LDW system can also be part of a more comprehensive accident avoidance system that connects to the chassis data bus.

INFOTAINMENT SYSTEMS

The term **infotainment system** comes from the world of automobiles, but, obviously, the technology is also used in trucks and buses. In a truck, an infotainment system's emphasis is on the *information* rather than the entertainment capabilities of the system and usually incorporates the vehicle's Bluetooth wireless capacity. Depending on the system, voice-actuated and touch-screen commands are used in preference over manual toggles and switches for reasons of safety. A truck infotainment system is networked to the data bus and may incorporate the following features:

- Data bus sourced status conditions and alerts
- GPS maps
- Telecommunications in/out: phone, Bluetooth, WiFi, satellite
- Audio entertainment: radio, CD, hard drive, iPod/USB input

Specialty vehicles such as a garbage packers and school buses may use the infotainment display monitor for such things as safety video. A highway coach could use multiple display nodes and run television, film, and Internet access.

SUMMARY

- Data can be transmitted electronically by means of electrical waveforms.
- Semiconductors are by definition elemental materials with four electrons in their outer shells.
- Silicon is the most commonly used semiconductor material.
- Semiconductors must be doped to provide them with the electrical properties that can make them useful as electronic components.
- After doping, semiconductor crystals may be classified as having N or P electrical properties.
- Diodes are two-terminal semiconductors that often function as a sort of electrical one-way check valve.
- Zener diodes are commonly used in vehicle electronic systems: They act as a voltage sensitive switch in a circuit.
- Transistors are three-terminal semiconductor chips.
- Transistors can be generally grouped into bipolar and field effect types.
- Essentially, a transistor is a semiconductor sandwich with the middle layer acting as a control gate. A small current flow through the base emitter will ungate the transistor and permit a much larger emitter-collector current flow.
- Many different types of transistors are used in vehicle electronic circuits, but their roles are primarily concerned with switching and amplification.
- The optical spectrum includes ultraviolet, visible, and infrared radiation.
- Optical components conduct, reflect, refract, or modify light. Fiber optics are being used increasingly in vehicle electronics, as are optical components.
- I/Cs consist of resistors, diodes, and transistors arranged in a circuit on a chip of silicon.
- A common I/C chip package used in computer and vehicle electronic systems is a DIP with either fourteen or sixteen terminals.
- Many different chips with different functions are often arranged on a primary circuit board, also known as a motherboard.
- Gates are switched controls that channel flows of data through electronic circuitry.
- AND, OR, and NOT gates are three commonly used means of producing an outcome based on the switching status of components in the gate circuit.
- The binary numeric system is a two-digit arithmetic system that is often used in computer electronics because it directly corresponds to the on or off states of switches and circuits.
- A *bit* is the smallest piece of data that a computer can manipulate. It has the ability to show one of two states, either on or off.
- A *byte* consists of 8 bits.
- A byte of data can represent up to 256 states of coded data.
- Almost all current on-highway trucks use computers to manage the engine and usually other chassis systems as well.
- A truck with multiple ECU-managed systems can network them using a chassis data bus; this is known as multiplexing.
- A vehicle ECU information processing cycle comprises three stages: data input, data processing, and outputs.
- RAM or main memory is electronically retained and, therefore, volatile.
- The master program for system management is usually written to ROM.
- PROM data is used to qualify the ROM data to a specific chassis application.
- Some OEMs describe their PROM component as a personality module.
- EEPROM provides an ECU with a read/write/erase memory component.

- *Multiplexing* is the term used to describe a system where two or more ECUs are networked on a data bus to reduce input hardware and optimize vehicle operation.
- Input data may be categorized as command data and system monitoring data.
- A potentiometer is a common input component; It is a three-terminal voltage divider.
- GPS technology uses trilateration to geographically position vehicles to help them navigate road routes.
- CMS and CWS use a combination of Doppler radar, camera, and motion sensors to alert drivers to imminent collision hazards.
- CMSs such as Wingman and VORAD process data received from radar and microwave motion/proximity sensors into driver alerts and can launch electronic intervention to avoid accidents.
- When programmed, a CMS can broadcast data commands to the chassis data bus meaning that the engine retarder and vehicle service brakes can be applied when a potential collision is detected.
- Accident avoidance systems may use video (cameras) for such things as object identification and can launch LDW systems.
- Most current trucks are equipped with infotainment systems.

REVIEW QUESTIONS

1. How many electrons does a semiconductor element have in its outer shell?
   a. less than four
   b. four
   c. more than four
   d. eight
2. How many electrons does the element silicon have in its outer shell?
   a. two
   b. four
   c. six
   d. eight
3. Which of the following best describes the function of a simple diode?
   a. an electronic switch
   b. an electrical one-way check valve
   c. an electronic amplifying device
   d. an electrical storage device
4. Which of the following best describes the term *pulse width modulation*?
   a. waveforms shaped to transmit data
   b. unwanted voltage spikes
   c. electronic noise
   d. rectification of AC to DC
5. To form a P-type semiconductor crystal, the doping agent would be required to have how many electrons in its valence shell?
   a. three
   b. four
   c. five
   d. eight
6. To form an N-type semiconductor crystal, the doping agent would be required to have how many electrons in its valence shell?
   a. three
   b. four
   c. five
   d. eight
7. What is the positive terminal of a diode known as?
   a. electrode
   b. cathode
   c. anode
   d. emitter
8. Which of the following terms best describes a function of a typical transistor in an electronic circuit?
   a. check valve
   b. relay
   c. rectifier
   d. filter
9. When testing the operation of a typical transistor, which of the following should be true?
   a. high resistance across the emitter and base terminals
   b. continuity across the emitter and collector terminals
   c. continuity across the base and emitter terminals
10. What would be the outcome in an OR gate if one of two switches in the circuit was closed?
    a. off
    b. on
11. How many different data codes could be represented by a byte?
    a. 2
    b. 8
    c. 64
    d. 256
12. Which of the following transistor terminals controls output?
    a. anode
    b. emitter
    c. collector
    d. base
13. Which of the following components is typically used in a half-wave rectifier?
    a. rheostat
    b. transformer
    c. diode
    d. zener diode
14. What is an N-type semiconductor crystal doped with?
    a. trivalent element atoms
    b. pentavalent element atoms
    c. carbon atoms
    d. germanium atoms
15. Which of the following correctly describes a pulse width modulated signal?
    a. analog
    b. digital
    c. either analog or digital
    d. neither analog nor digital

16. Which of the following is a three-terminal device?
    a. zener diode
    b. transistor
    c. capacitor
    d. diode
17. The acronym that describes the component in a computer that executes program instructions is the:
    a. CRT
    b. RAM
    c. CPU
    d. ROM
18. Which of the following data retention media is volatile?
    a. RAM
    b. ROM
    c. PROM
    d. EEPROM
19. Which of the following memory categories would the master program for system management be written to in a typical truck chassis ECU?
    a. RAM
    b. ROM
    c. PROM
    d. EEPROM
20. Which of the following memory categories would customer data programming be written to from an electronic service tool (EST)?
    a. RAM
    b. ROM
    c. PROM
    d. EEPROM
21. Technician A says that a CWS system is designed to alert the truck driver when a potential collision is detected. Technician B says that the Wingman CMS uses video technology to signal the proximity of straight-ahead objects. Who is correct?
    a. Technician A only
    b. Technician B only
    c. both A and B
    d. neither A nor B
22. What is the key difference between a CWS and a CMS?
    a. The CWS uses radar.
    b. The CMS uses microwave.
    c. The CWS can derate the engine.
    d. The CMS can actuate the service brakes.
23. Which of the following principles is used by GPS to determine positional data?
    a. radar
    b. trilateration
    c. microwaves
    d. geosynchronous satellites
24. Technician A says that trilateration requires three known positional reference points to identify position on a flat surface. Technician B says that because the earth is a sphere, signals from a minimum of four satellites are required by a GPS to reckon position and altitude. Who is correct?
    a. Technician A only
    b. Technician B only
    c. both A and B
    d. neither A nor B

7

Prerequisite: Chapter 5

BATTERIES

OBJECTIVES

After reading this chapter, you should be able to:

- Define the role of a battery in a vehicle electrical system.
- Outline the construction of standard, maintenance-free, gelled electrolyte, and absorbed glass mat (AGM) batteries.
- Describe the chemical action within the battery during the charging and discharging cycles.
- Outline how batteries are arranged in multiple battery banks in truck chassis.
- Verify the performance of a lead-acid battery using a voltmeter, hydrometer, refractometer, and carbon pile tester.
- Analyze maintenance-free battery condition using an integral hydrometer sight glass.
- Describe the procedure required to charge different types of batteries.
- Jump-start vehicles with dead batteries using another vehicle and generator methods.
- Outline how batteries should be safely stored out of chassis.
- Describe the operating principles of ultracapacitors and how they can be used in hybrid electric vehicles.
- Outline the features of typical vehicle battery management information systems.

KEY TERMS

absorbed glass mat (AGM) battery
AC conductance testing
ampere-hour rating
anode
Battery Council International (BCI)
battery management information system (BMIS)
cathode
charging system
cold cranking amperes (CCA)
conductance testing
deep cycling
electrochemical impedance spectroscopy (EIS)
electrolyte
flooded cell battery
galvanic
gel cell
hybrid electric vehicle (HEV)
hydrometer
lead-acid battery
lithium-ion (Li-On)
lithium-ion polymer (Li-Po)
nickel-metal-hydride (NiMH)
NiMH battery
recombinant battery
refractive index
refractometer
reserve capacity rating
specific gravity (SG)
sulfation
supercapacitors
ultracapacitors
wet cell battery

INTRODUCTION

The principles of battery operation are discussed in this chapter, but their application in truck electrical circuits is discussed in Chapters 8, 9, and 10. Almost all of the focus on batteries in this chapter is on flooded cell, lead-acid batteries. However, because of the rapid emergence of hybrid drivetrain technologies in commercial vehicles, it is essential that a little text be devoted to the battery packs in vehicles that use battery-sourced electricity for part-time motive power. Over the next decade, electric hybrid drivetrains have the potential to increase in sales in commercial vehicles that operate within large cities.

7.1 BATTERY-OPERATING PRINCIPLES

A battery is a **galvanic** device—that is, a device that produces a charge differential by chemical action. Any galvanic device requires two electrodes, one positive and one negative, plus some **electrolyte**, to interact between the electrodes. The word *galvanic* is sourced from Luigi Galvani (Italy, 1737–1798) who invented the voltaic pile, an early type of battery. The battery stores electrical energy not as electrons but as chemical energy. In terms of its operation, it is probably best described as an electrical energy storage device. During the charge cycle, electrical energy is converted into chemical energy and during the discharge cycle, chemical energy is converted into electrical energy.

LEAD-ACID BATTERY

The lead-acid battery is the electrical energy storage device used on most vehicles. Its function is to store the electrical energy generated by the alternator. It performs a secondary role as a stabilizer for the voltage in the vehicle electrical system. The battery or batteries on a vehicle must be capable of supplying a high-current load to the vehicle cranking circuit and support the operation of other vehicle electrical system loads, especially when the alternator is operating at lower efficiency—such as when idling. Most truck electrical systems use multiple lead-acid battery units connected in series and series-parallel arrangements that are the basis of a 12-volt electrical system. In some cases, a 12V chassis electrical system may have a 24-volt cranking circuit, in which case multiple batteries are arranged so that they can be switched between the 12V chassis electrical circuit and the 24V cranking circuit. The automotive and trucking industries are in a slow process of introducing 42-volt systems but there are problems with this, which are discussed later in the chapter.

Lead-acid batteries used in trucks today can be grouped into six categories:

- Conventional
- Low maintenance
- Maintenance free
- Sealed maintenance free
- Gel cell
- Absorbed glass mat (AGM)

The fundamental operating principles of all six classes of batteries are identical. The most commonly used battery in truck applications at this moment in time is the flooded cell, Group 31 battery. For this reason, the main focus of this chapter is directed at a Group 31 battery classification.

Isolated Hybrid Electrical Circuits

At the end of this chapter we will take a brief look at isolated hybrid electric circuits and the battery and capacitor banks used to store electrical potential much higher than chassis voltage. High-voltage hybrid electrical circuits are always isolated from the chassis electrical circuit.

BATTERY CONSTRUCTION

Initially, we will describe the most commonly used batteries in truck applications. These lead-acid batteries are known as **flooded cell batteries** or **wet cell batteries**. There are variations of lead-acid batteries in use and these are described later in the chapter. A current 12V lead-acid battery consists of 6 series connected cells arranged in a polypropylene casing as shown in **Figure 7–1**. Each cell has positive and negative plates—or **anode** (positive) and **cathode** (negative) plates. Each plate is constructed of a lead grid. Due to the relative softness of lead, the grids are toughened by adding antimony (a metal alloying element) or calcium. The plates must then be coated or "pasted" with chemically active matter. The positive plates are pasted with lead peroxide (lead dioxide) and the negative plates are pasted with sponge lead, which is a pure, porous form of lead.

Within each cell in the battery housing the plates are arranged so that positive and negative plates are located next to each other and alternate as shown in **Figure 7–2**. They must not contact each other so separators are used to prevent short circuits. The separators (see Figure 7–2) are manufactured of nonconductive materials such as plastics, fiberglass, and resin-coated paper. These separators are porous to permit the chemical interaction of the liquid battery electrolyte with plates. Below the plates in each cell in the battery housing is a series of ribs that help support and separate the plates but also provide a space to store spent sediment from the pasting on the plates. This spent sediment is capable of causing a short circuit between the plates. Some batteries prevent the spent sediment from settling in the bottom of the battery case by enclosing each plate in a microporous plastic envelope as shown in **Figure 7–3**. This method prevents the sediment from getting to the bottom of the battery case.

Each cell in a battery is separated from the other cells by partitions as shown in **Figure 7–4**. The partitions isolate the chemical interactivity of each cell, with the result that they are only connected electrically. The connectors used are lead and are designed to minimize resistance.

FIGURE 7–1 Battery construction and terminology

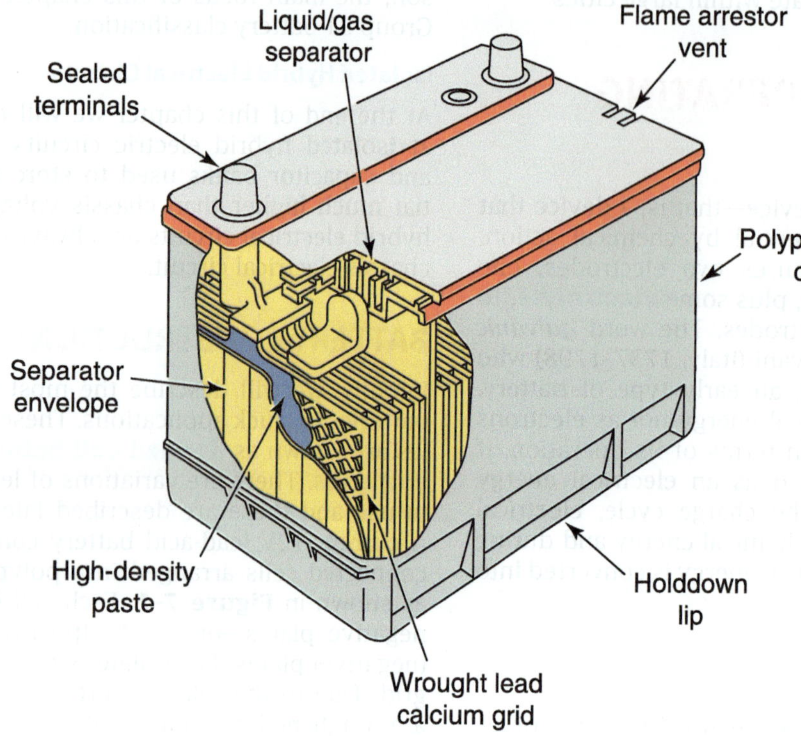

FIGURE 7–2 Positive and negative plates with separator

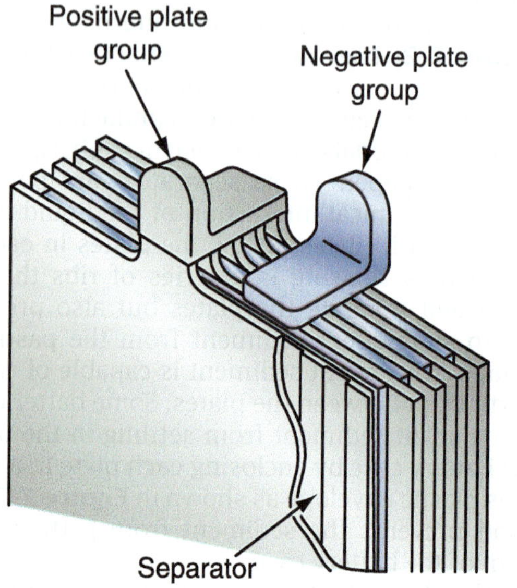

FIGURE 7–3 Envelope separator

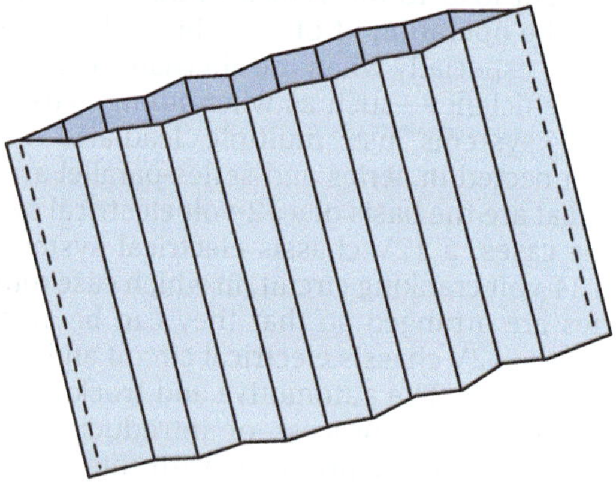

ELECTROLYTE

By definition, electrolyte is any substance that conducts electricity. In the lead-acid battery it is the liquid solution that enables the galvanic or chemical action of the battery. In both standard and maintenance-free batteries, the electrolyte used is a solution (mixture) of sulfuric acid (H_2SO_4) and pure water (H_2O). The solution proportions should be 36 percent sulfuric acid and 64 percent distilled H_2O. This solution will produce a **specific gravity (SG)**—the weight of a liquid or solid versus that of the same volume of H_2O—of 1.265 at 80°F (27°C). The acid-to-water proportions should never be tampered with. When a new

FIGURE 7–4 Maintenance-free battery case showing partitions

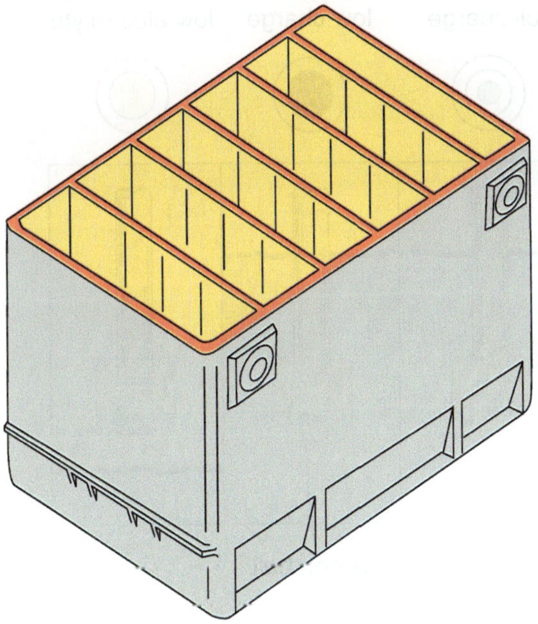

battery is activated, premix electrolyte is introduced to the cells. The electrolyte solution in a lead-acid battery is both conductive (capable of conducting electricity) and reactive (able to participate in chemical reactions).

In batteries that permit addition to the electrolyte when reduced by gassing, only distilled H_2O should be used. It should be noted that pure H_2O or distilled H_2O does not conduct electricity; in fact, it can be used as an insulator. The conductivity of H_2O depends on its content of total dissolved solids (TDS) or minerals. The minerals are conductive. This means that the conductivity of the electrolyte mixture in a **lead-acid battery** is dependent on the sulfuric acid that makes up the solution. Using tap H_2O to replenish battery electrolyte can in time cause battery cells to short out.

Discharge Effect on Electrolyte

The chemical action of a battery during discharge reduces the proportional ratio of sulfuric acid to that of H_2O. This reduces the density (SG) of the solution—something that can be measured using a hydrometer or refractometer. A **hydrometer** is a device that directly measures SG by floating a sealed bulb in the solution. A **refractometer** measures the refractive index of a solution. The **refractive index** is essentially the amount a light ray is slowed when it passes through a solution in comparison with pure H_2O. The higher the density of the solution, the more the ray is slowed.

CAUTION

Great care should be exercised when testing battery electrolyte with a refractometer. The process involves placing a drop of electrolyte on the hydrometer window close to the eye. Battery electrolyte is highly corrosive and is capable of causing permanent eye damage. Always execute the test instructions precisely.

SPECIFIC GRAVITY AND TEMPERATURE

Temperature directly affects the SG of a substance. SG specifications are based on a temperature of 80°F (27°C). When a hydrometer is used to test SG, the reading must always be temperature adjusted. The usual method of adjusting SG to temperature is to add 0.004 for every 10°F (5°C) above 80°F (27°C) and to subtract 0.004 for every 10°F (5°C) below 80°F (27°C). As we have said, the sulfuric acid concentration in the electrolyte may also be tested using a refractometer and these readings do not require temperature correction. However, today a large percentage of batter ies used in truck applications are maintenance free. In a maintenance-free battery, there is no means of measuring the SG of the electrolyte solution other than with a built-in hydrometer. The built-in hydrometer has all of the disadvantages of any other type of hydrometer in terms of temperature accurate readings, plus the fact that it does not produce an actual reading. Instead it produces one of three color readings as shown in **Figure 7–5**. At best, this produces a ballpark estimate of the electrolyte SG. In addition, it is limited to displaying the SG status of just one cell. However, it is the only means of reading electrolyte SG in a maintenance-free battery.

84 Factor

Actual SG of the electrolyte in any cell in a battery relates directly to its voltage. The *84 factor* enables cell voltage to be calculated by adding 0.840 to the SG measured. If the SG in a cell is measured to be 1.260, cell voltage would be calculated as follows:

$$1.260 + 0.840 = 2.100\,V$$

In addition if all 6 cells in the battery were measured with the same SG, then battery voltage could be calculated as follows:

$$2.100 \times 6 = 12.6\,V$$

The 84 factor is a "ballpark" method of relating cell SG with voltage. It does not apply when the battery electrolyte is in a completely discharged state.

BATTERY CHARGE/DISCHARGE CYCLES

Next, we take a look at the chemical interactions that occur within a lead-acid battery during the charge

FIGURE 7–5 Built-in hydrometer operating principle: Note what caused the color in the eye to change.

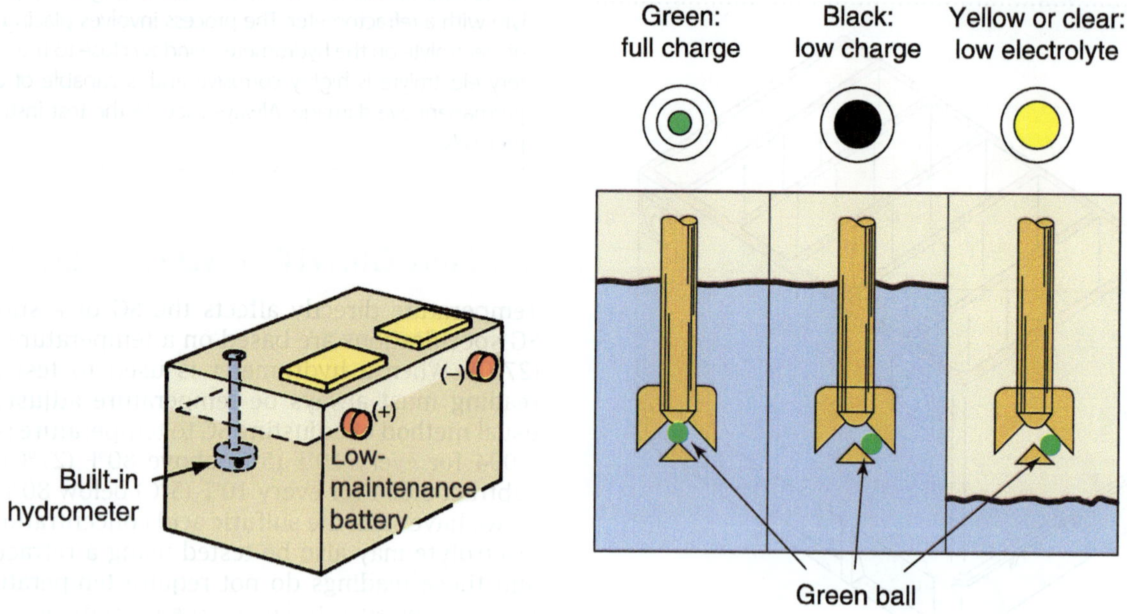

and discharge cycles. A charge cycle occurs when the battery is being regenerated by the alternator. A discharge cycle occurs when battery power is being consumed by the vehicle electrical system such as when cranking the engine.

Discharge Cycle

When a circuit is constructed connecting the positive and negative battery terminals with a load, current will flow through the circuit. The lead peroxide (PbO_2) on the positive plate reacts with the sulfuric acid solution electrolyte (H_2SO_4), with the result that its oxygen molecule (O_2) is released to the electrolyte, forming water (H_2O). In the reaction, the negative plate (Pb) reacts with the electrolyte to form lead sulfate ($PbSO_4$). **Figure 7–6** and **Figure 7–7** show the chemical action that takes place during the discharge cycle.

This chemical action will continue until both the positive and negative plates are coated with lead sulfate ($PbSO_4$) and the electrolyte has been chemically reduced to water (H_2O). It is just about impossible to reduce battery electrolyte to a state of pure H_2O that would produce an SG reading of 1.000. In other words, batteries seldom become discharged to the point where they can be regarded as being electrically neutral. Typically, what is described as a "dead" battery contains electrolyte with an SG in the region of 1.140 to 1.160. It is also important to note that a fully charged battery, one with the correct ratio of sulfuric acid to H_2O, provides a good measure of antifreeze protection. H_2O freezes at 32°F (0°C) and the higher the concentration of sulfuric acid, the lower the freeze temperature. As the electrolyte solution degrades to H_2O during discharge, the electrolyte freeze point rises. Batteries that freeze up in winter conditions are almost always in a discharged state.

Charge Cycle

When a generating device such as an alternator is connected in a circuit between the battery terminals, the battery can be recharged. This essentially reverses the chemical reaction that takes place in the charging cycle. During the charge cycle, the sulfate coatings that have formed on both the positive and negative plates are reacted to return them to the liquid electrolyte. This process corrects the ratio of sulfuric acid (H_2SO_4) and water (H_2O). During the charge process H_2O molecules in the electrolyte reduce to hydrogen and oxygen. The hydrogen combines with the sulfate in the electrolyte to form sulfuric acid while the oxygen is drawn to the positive plate to reconstruct the lead peroxide coating.

Gassing. When a battery is in charge cycle, gassing is caused by electrolysis. *Gassing* is the conversion of the H_2O in the electrolyte to hydrogen and oxygen gas. In older batteries, these gases were vented from the battery housing, forming an explosive combination that, given a small spark, could destroy a battery or battery bank (two or more interconnected batteries) and could be highly dangerous. Current lead-acid batteries have

FIGURE 7–6 Chemical status during battery discharge and charge cycles

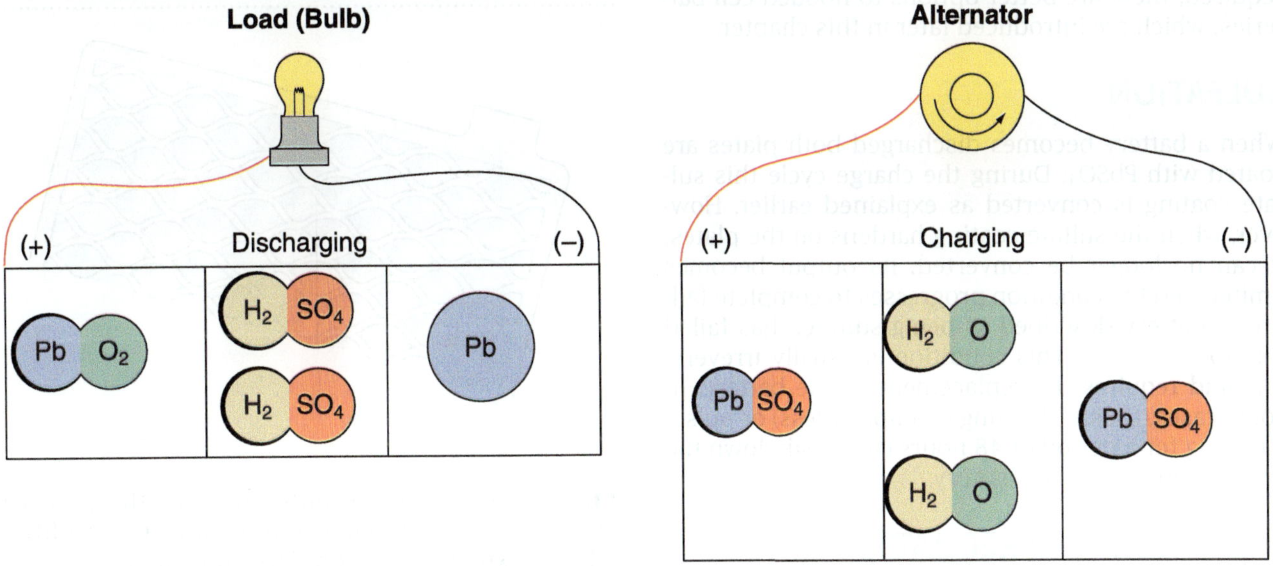

FIGURE 7–7 Changes to the chemical composition during charge and discharge cycles

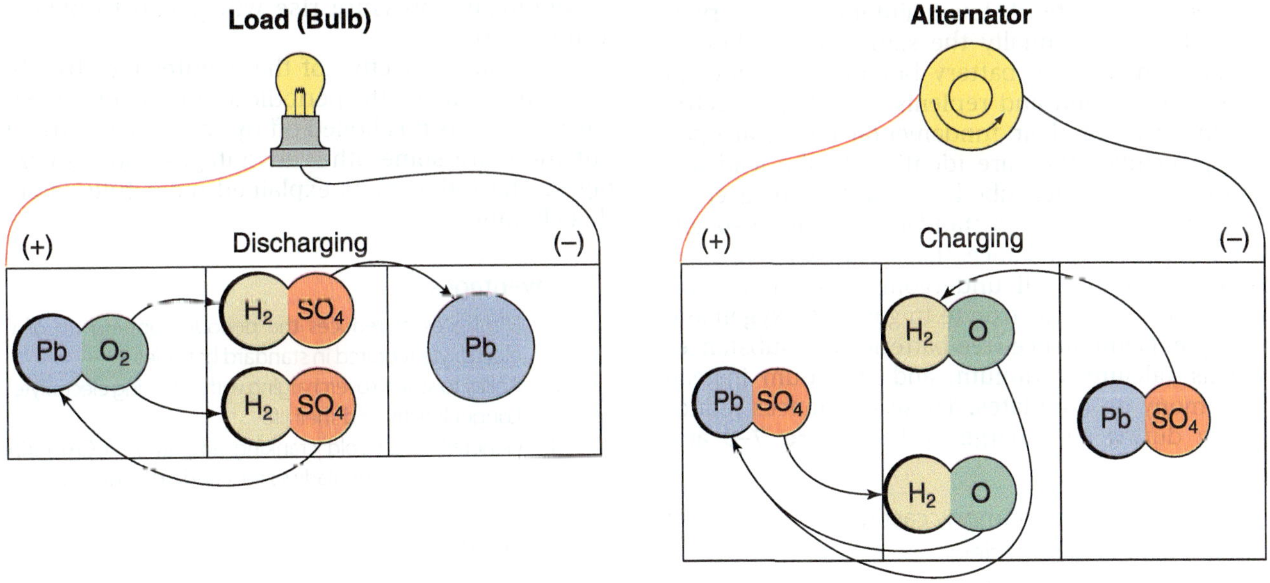

expansion or condensation chambers that prevent the escape of charging gases.

CAUTION

Hydrogen gas may be discharged during charging and care should always be taken when connecting or disconnecting battery terminals so as not to create a spark that could result in a battery explosion. Always disconnect the ground cable first . . . and reconnect it last.

Deep Cycling Precautions

Most flooded cell batteries are not designed for **deep cycling**. Deep cycling occurs when a battery is brought close to a completely discharged state and then recharged. Recharging a fully discharged battery tends to be hard on the battery due to the heat created by internal resistance. This can cause plate buckling. Applications that use deep-cycle (cycling by complete discharge followed by full recharge) batteries, such as electric fork lift trucks using lead-acid batteries, do so with batteries constructed with much thicker plates. Repeated deep cycling will fail most batteries prematurely. In applications in which deep cycling is required, there are better options to flooded cell batteries, which are introduced later in this chapter.

SULFATION

When a battery becomes discharged both plates are coated with $PbSO_4$. During the charge cycle this sulfate coating is converted as explained earlier. However, when the sulfate coating hardens on the plates, it can no longer be converted. Its output becomes limited and the condition progresses to complete failure. A battery described as being sulfated has failed due to **sulfation**. This condition is usually irreversible and requires the replacement of the battery. In some cases, trickle-charging (2 amps or less) or pulse-charging for a period of 48 hours can break down the sulfate and restore the battery.

MAINTENANCE-FREE BATTERIES

Maintenance-free and low-maintenance batteries have become widely used in the trucking industry in recent years. When a battery is described as maintenance free, no provision is made for adding H_2O to depleted electrolyte. A low-maintenance battery is one that uses essentially the same construction as the maintenance-free battery but does provide for periodic inspection and replenishing of the electrolyte. In terms of their fundamental chemical operating principles, they are identical to the lead-acid batteries already described. The antimony used to strengthen lead plates in standard batteries also acts as a catalyst (a substance that enables a chemical reaction without itself undergoing any change) for the electrolysis of electrolyte to gaseous oxygen and hydrogen. Maintenance-free batteries use substances such as calcium, cadmium, and strontium instead of antimony on the plates. These substances reduce gassing during the charge cycle. **Figure 7–8** and **Figure 7–9** show the differences in the plates in standard and maintenance-free batteries. Additionally, an expansion or condensation chamber is used to contain the gassing, condensing the hydrogen and oxygen to water and allowing it to drain back into the electrolyte. A vent provides a measure of safety to the battery housing, but only in the event of significant pressure rise will it trip to bleed to atmosphere.

FIGURE 7–8 Lead antimony cast grid used on most standard lead-acid batteries.

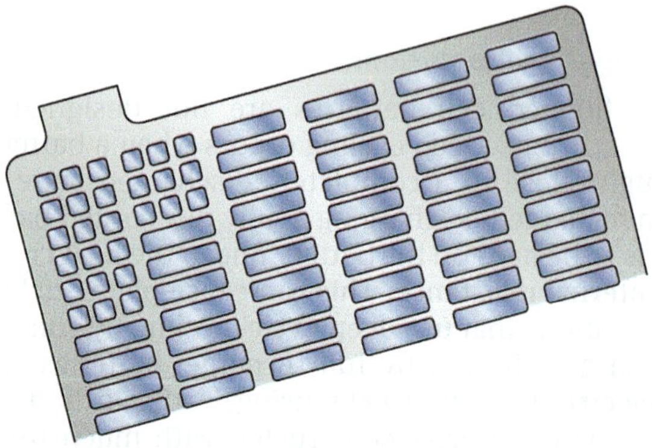

FIGURE 7–9 Lead calcium wrought grid used on many maintenance-free batteries.

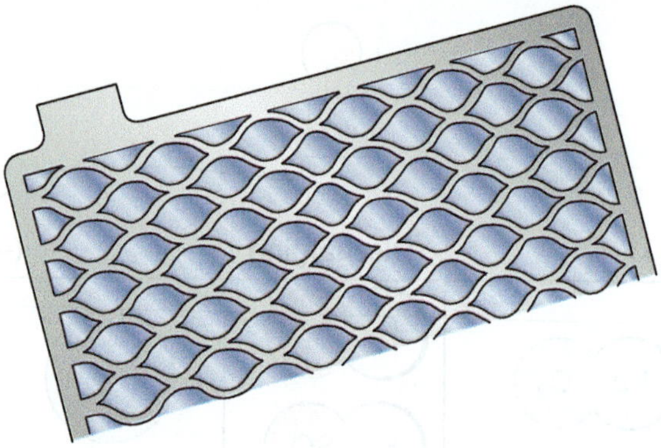

The main objective of the maintenance-free battery is to eliminate the periodic addition of H_2O to the battery cells as it is boiled off by charging electrolyte, but there are some other advantages and disadvantages. The following is explained more fully later in this chapter.

Advantages:

1. Eliminates or reduces the periodic topping up of the electrolyte required in standard batteries
2. Ability to sustain overcharging without losing electrolyte
3. Longer inactivated shelf life
4. Generally high cold cranking amperage (CCA) ratings compared with similarly sized standard batteries

Disadvantages:

1. Less ability to sustain deep cycling
2. Generally lower reserve capacity
3. Shorter total service life expectancy
4. Faster discharge rates

Some maintenance-free batteries have a built-in hydrometer. This is usually called a visual state of charge indicator. This was described a little earlier in this chapter in the section on electrolyte and is shown in Figure 7–5. **Figure 7–10** shows a comparison of the rate of self-discharge of standard lead-acid batteries versus maintenance-free batteries. The electrolyte degrades at a faster rate in standard batteries, especially at higher ambient temperatures.

FIGURE 7–10 Self-discharge comparison graph

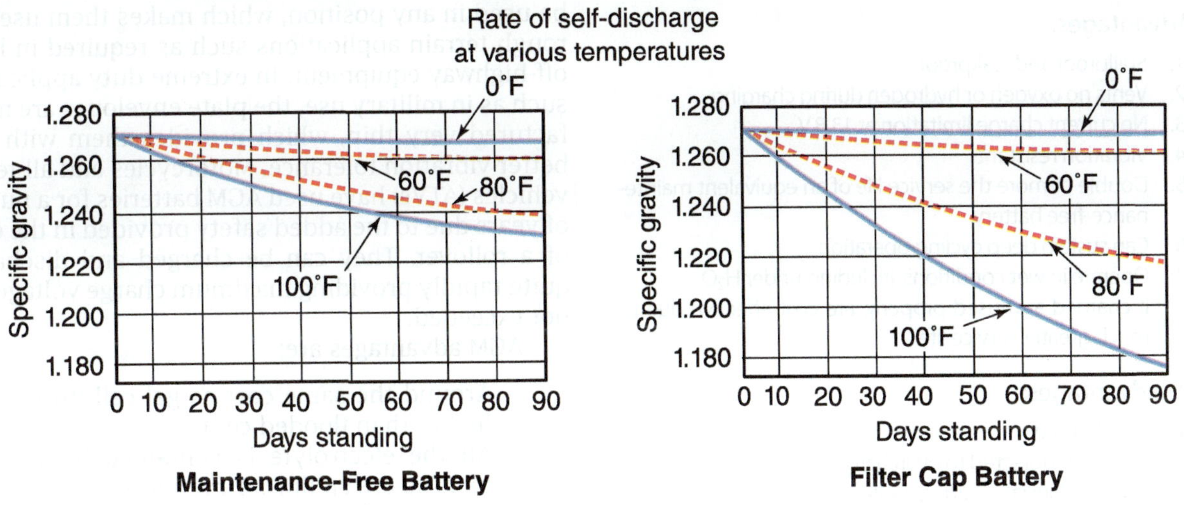

SHOP TALK

Technicians should note that high CCA-rated batteries do not sustain deep-cycling as well as lower rated CCA batteries. Use the OEM-recommended CCA-rated batteries for the application.

RECOMBINANT BATTERIES

The next two types of battery are both classified as **recombinant batteries**. The two types of recombinant batteries used in truck and heavy off-road equipment are:

- Gelled cell
- Absorbed glass mat (AGM)

Gelled cell batteries were introduced to the trucking industry in the early 1990s because they had the advantage of sustaining deep cycling. More recently, AGM batteries have been introduced and these tend to be replacing gel cells. In the charge cycle of a recombinant lead-acid battery, oxygen released at the positive plate recombines within the cell with the hydrogen released at the negative plate. The recombination of hydrogen and oxygen produces H_2O that is then reabsorbed to the electrolyte. Recombinant batteries are entirely maintenance free and do not cause corrosion around the terminals.

GELLED ELECTROLYTE BATTERIES

A gelled electrolyte battery is usually known as a **gel cell** battery. The gel cell is a type of lead-acid battery and uses all the fundamental operating principles of any other type of lead-acid battery. The main difference of the gel cell is the electrolyte. The gel cell battery uses a special thixotropic (a solid or gel that liquefies under certain conditions) electrolyte that liquefies when stirred or shaken but when left at rest returns to the gelled state. The battery is pressurized and sealed using special vents. For this reason, it never requires filling. Gel cells, therefore, have the advantage of being operated in almost any position, although upside-down installation is not recommended. They were commonly used where isolated-battery, deep-cycle duty was called for before the introduction of AGM batteries. In a vehicle with multiple battery banks, an isolated battery is used for the chassis electrical requirements when the engine is not running.

Gel Cell Operating Principle

Gel cells are designed to be acid limited. This means that the electrolyte is diluted to a discharge condition before the plates sulfate. This protects the plates from ultra-deep discharging that can cause plate shedding and accelerated positive grid corrosion that can, ultimately, destroy the battery.

If a gel cell battery is subjected to electrical charging voltages that exceed 14.1V, they will release more hydrogen and oxygen than can be recombined. This excess of gas raises the pressure in the battery housing to the point at which it can trip the pressure-sensitive vents. Once the excess of oxygen and hydrogen has been released, it is lost and cannot be recombined to H_2O. When a gel cell is subjected to overcharging, the electrolyte dries out and the battery fails prematurely. Because gel cell batteries operate under pressure they may appear to bulge slightly even during appropriate charging. However, extreme bulging usually indicates overcharging and/or blocked bleed vents and the battery should be immediately taken out of service. When used as an isolated battery in multibattery truck electrical systems the gel cell must be connected directly to the battery charger.

Advantages:

1. Spillproof and leakproof
2. Vents no oxygen or hydrogen during charging
3. No current charge limitation at 13.8 V
4. Vibration resistant
5. Double or more the service life of an equivalent maintenance-free battery
6. Can sustain deep cycling operation
7. Operate in wet conditions, including under H_2O
8. If charged and used properly, are cost effective due to much greater service life

Disadvantages:

1. Weigh more
2. Will fail if subjected to overcharging
3. Require special chargers (automatic, temperature sensing, voltage regulated) that regulate charge pressure within a narrow range: 13.8 to 14.1V
4. Vulnerable to abuse

ABSORBED GLASS MAT BATTERIES

Absorbed glass mat (AGM) batteries are a type of lead-acid battery in which the electrolyte is absorbed into a very thin boron silicate fiberglass mat. The plates in an AGM battery are usually flat like flooded cell lead-acid batteries, or they may be wound into a spiral. Their construction permits the lead plates to be purer in composition because they do not have to support their weight as they would in a flooded cell battery. The internal resistance of AGM batteries is lower than wet cells. This is accounted for by the close plate proximity of the plates and the fact that they are manufactured from pure lead. This provides AGM batteries with the following advantages:

- Ability to function at higher temperatures
- Ability to function at lower temperatures
- Much slower self-discharge rates
- Lighter in weight than equivalent flooded cell batteries
- Better tolerance of vibration

AGM batteries are not nearly as sensitive to voltage overload during charging as are gel cell batteries. Having said that, generally, charging voltage should not exceed 13.8 volts. In the event of overcharge, AGM batteries are equipped with a pressure relief valve. This pressure relief valve is usually designed to trip at 14 volts, permitting bleed down of oxygen and hydrogen gas. Should bleed down occur, active material (oxygen and hydrogen) is dumped, which results in reduced battery capacity. The battery covers incorporate gas diffusers designed to safely disperse hydrogen and oxygen dumped during overcharge. AGM batteries are maintenance free.

Applications

Unlike flooded cell batteries, AGM batteries can also be used in any position, which makes them useful in rough terrain applications such as required in heavy off-highway equipment. In extreme duty applications such as in military use, the plate envelopes are manufactured very thin, which provides them with even better vibration tolerance. Motorcycles and all-terrain vehicles (ATVs) have used AGM batteries for a number of years due to the added safety provided in the event of a rollover. They can be charged and discharged quite rapidly providing maximum charge voltages are not exceeded.

AGM advantages are:

- Around the same cost as gel cell (three times higher than flooded cell)
- All the electrolyte is contained in the glass mats so are spillproof even if broken
- Nonhazardous, meaning that shipping and storage costs are lower
- Immune from freezing due to no liquid to freeze and expand
- Low internal resistance, so almost no heating of the battery takes place even under heavy charge and discharge currents
- Store well because of low self-discharge typically from 1 to 3 percent per month
- Withstand shock and vibration better than flooded cell batteries
- Produce no fumes or leakage
- Can withstand arctic-type temperatures without damage

7.2 BATTERY RATINGS

All vehicle batteries are rated to standards established by the **Battery Council International (BCI)** in conjunction with the Society of Automotive Engineers (SAE). In terms of design, the current capacity of any battery depends primarily on the size of the plates and the quantity of chemically active material coated to the plates. The larger the surface area of the plate in contact with the electrolyte, the more chemical action can take place to produce current flow. Battery rating classification can be defined as follows:

COLD CRANKING RATING

Cold cranking amperes (CCA) is the primary method of rating a vehicle battery. It specifies the current load a battery is capable of delivering for 30 seconds at a temperature of 0°F (−18°C) without dropping the voltage of an individual cell below 1.2V, or 7.2V for a 6-cell battery across the terminals. Truck batteries may have CCA ratings exceeding 1,000 amperes. The electrical energy required to crank an engine tends to

FIGURE 7–11 Relationship of battery power available and power required

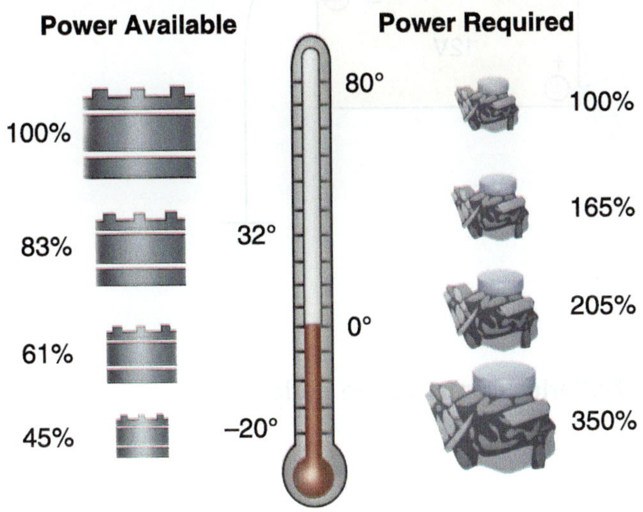

go up as temperature drops while available battery power goes down. **Figure 7–11** shows the relationship of power available and cranking power required. Engines that have to be started in subzero conditions require batteries with high CCA ratings.

AMPERE-HOUR RATING

The **ampere-hour rating** of a battery is the amount of current that a fully charged battery can feed through a circuit before the cell voltage drops to 1.75V. In a typical 6-cell, 12V battery, this would be equal to a battery voltage of 10.5V. Ampere-hour rating is used by some fleets when spec'ing batteries.

RESERVE CAPACITY RATING

The **reserve capacity rating** of a battery system determines the amount of time a vehicle can be driven with its headlights on in the event of a total charging system failure. The current output varies with the type of vehicle electrical system, but the critical low-voltage value is 10.5V.

Table 7–1 shows some common battery size groups with their typical ampere-hour ratings; for truck technicians today, the Group 31 data is most important because these are so commonly used.

BATTERY BANKS

Current trucks tend to use multiple batteries to fulfill the engine cranking and electrical chassis requirements of the vehicle. Most highway trucks use 12V electrical systems supplied by batteries arranged in banks. Although arrangements of both 6V and 12V batteries have been used in the past, today it is most common for arrangements of 12V batteries to be used. A small number of older highway trucks use a 12/24 system, but the introduction of compact, increasingly high torque–output cranking motors has made such systems unnecessary. A 12/24 system is a 12V chassis electrical system with a 24V cranking capability. The switching may be performed electromechanically (rare today) or electronically. **Figure 7–12** shows how 12V batteries can be arranged in series or in parallel in a battery bank.

TABLE 7–1 Ampere-Hour Ratings for Common Battery Classifications

Battery Classification	Ampere-Hour Rating	Voltage
Group L-16	350–415	6 V
Group 4D	180–215	12 V
Group 8D	225–250	12 V
Group 24	70–85	12 V
Group 27	85–105	12 V
Group 31	95–125	12 V

7.3 BATTERY MAINTENANCE

A lead-acid battery has a finite service life and requires maintenance to achieve it. With a reasonable amount of care, the life of a battery can be appreciably extended. Neglect and abuse will invariably cause shorter battery service life. The battery should be inspected at each chassis lubrication (known as an A or L schedule preventive maintenance inspection [PMI]) or during other periodic services. Battery maintenance includes inspecting the battery and its mounting for corrosion, loose mounting hardware, case cracks, and deformation (**Figure 7–13**). Loose holddown straps or covers allow the battery to vibrate or bounce during vehicle operation. This vibration can shake the active materials off the grid plates, considerably shortening battery life. It can also loosen the plate connections to the plate strap, loosen cable connections, and even crack the battery case. In heavy-duty trucks, the physical mounting and location of the batteries is important to ensure a reasonable service life. **Figure 7–14** illustrates some mounting recommendations of the Technology and Maintenance Council (TMC) of the American Trucking Association.

Battery corrosion is caused by spilled electrolyte or electrolyte condensation from gassing. Acidic corrosion attacks and damages battery components such as connectors and terminals. Corroded connections

FIGURE 7–12 Batteries arranged in series and in parallel

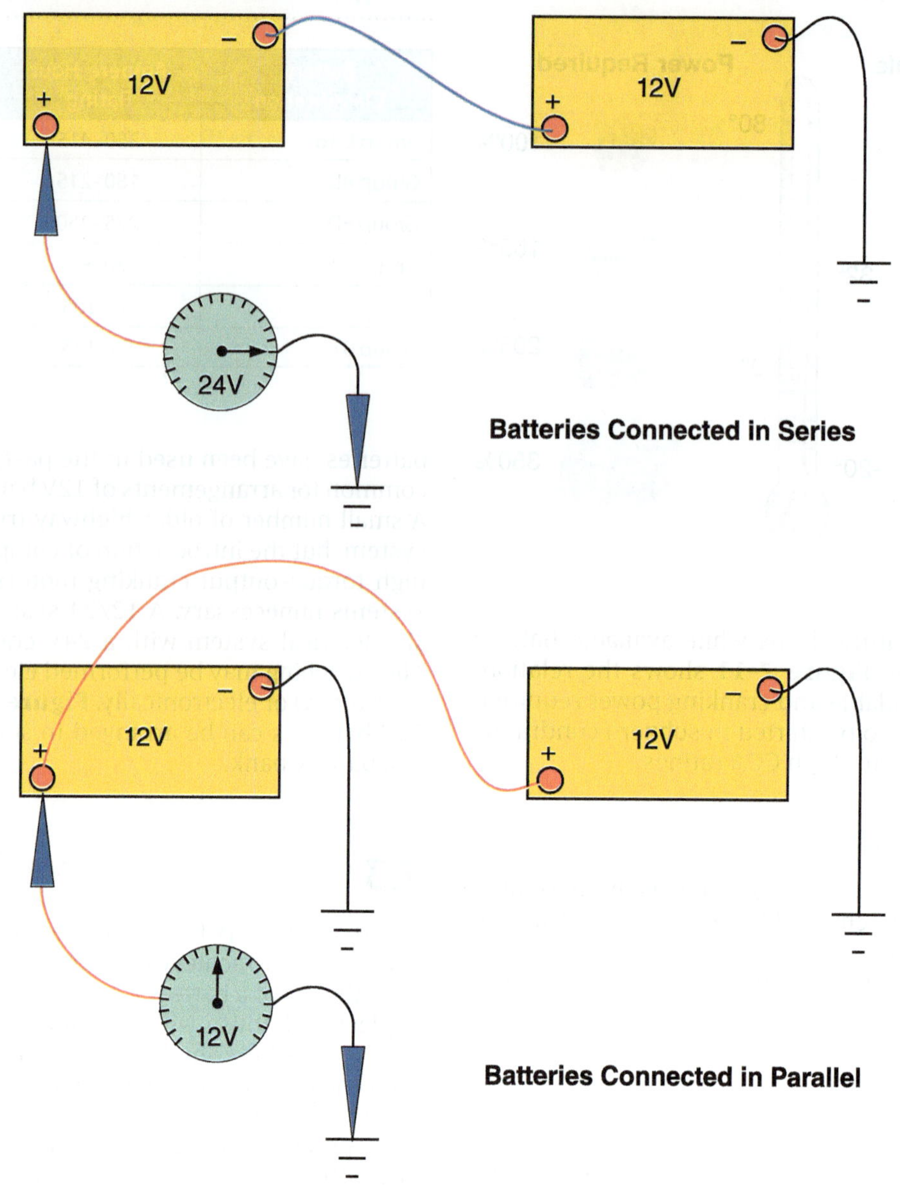

increase resistance at the battery terminals, reducing the voltage available to the chassis electrical system. Built-up corrosion on the battery housing can create current leak paths that can result in slow discharge. Left unchecked, battery corrosion can spread beyond connectors and terminals to the holddown straps and carrier box, resulting in physical damage to the battery.

BATTERY MAINTENANCE PROCEDURE

If corrosion is found on the terminal posts, remove the cable terminals from the battery and clean them. Always remove the ground cable first using the proper-sized wrench and a terminal puller. A terminal cleaning brush can be used to clean tapered posts and the mating surfaces of the cable clamps. Other types of terminals can be cleaned with a wire brush, taking care not to create sparks. The cable terminals can then be cleaned with a solution of baking soda and H_2O to neutralize any remaining battery acid. Do not allow the baking soda and H_2O solution to enter the cells. Clean dirt from the top of the battery first with a cloth moistened with baking soda and H_2O, then wipe it with a cloth moistened with clean H_2O. A wire brush can be used to remove dirt, corrosion, or rust from the battery tray or holddown components. When the rust and dirt have been removed, pressure wash the battery with clean water, allow to air dry, and repaint.

After cleaning, reinstall the battery and holddown components. Coat the battery and cable terminals

FIGURE 7–13 Battery maintenance inspection points

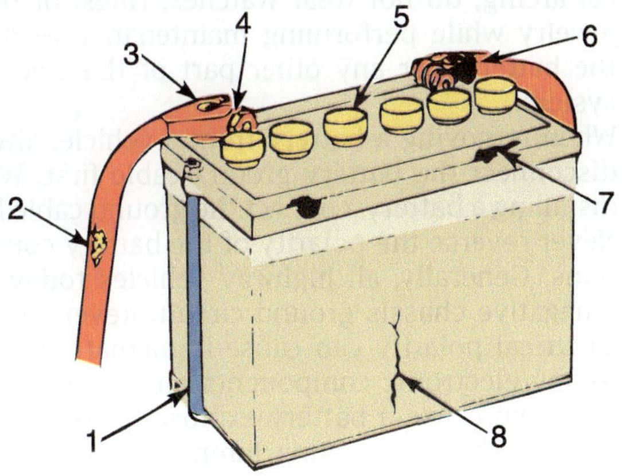

Inspect for:

1. Loose holddowns
2. Defective cables
3. Damaged terminal posts
4. Loose connections
5. Clogged vents
6. Corrosion
7. Dirt or moisture
8. Cracked case

with terminal grease and connect the cables to the battery terminals. It is best to use grease specially formulated for terminal protection. Note that any grease described as being dielectric is nonconductive and should never be used on the contact surfaces of terminals.

Chassis grease is conductive. Connect ground cables last when reconnecting the batteries. Maintaining the correct electrolyte level in nonsealed batteries is essential to keeping them in good operating condition. In any nonsealed battery, electrolyte is boiled off during charging. When the electrolyte level is low, fill each cell with distilled H_2O to bring the liquid level to the full level indicator. If the battery does not have a level indicator, bring the level to ½ inch above the top of the separators. Do not overfill any cell. This weakens the concentration of the acid (reducing the electrolyte SG), which decreases battery efficiency. Also, when a cell is overfilled, the excess liquid electrolyte can be forced from the cell by the gas formed in the battery during charging. This type of leakage corrodes any metals it comes into contact with and reduces both battery performance and life. Low electrolyte levels in batteries results in higher acid concentration and may deteriorate the plate grids at an accelerated rate.

Batteries can be overcharged by either the vehicle **charging system** or by battery chargers. In either case, the result is an accelerated chemical reaction within the battery that causes gassing and a loss of electrolyte in the cells. This can remove the active materials from the plates, permanently reducing the capacity of the battery. Overcharging can also cause excessive heat, which can oxidize the positive plate grid material and even buckle the plates, resulting in a loss of cell capacity and early battery failure.

FIGURE 7–14 Proper battery mounting recommendations

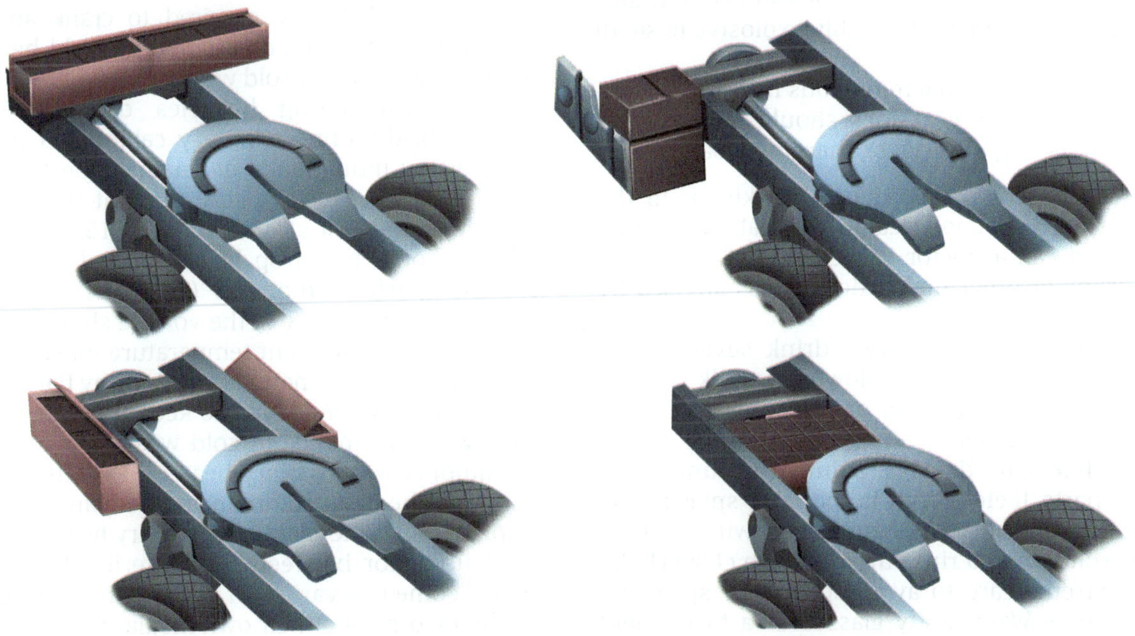

SHOP TALK

Frequent need for refilling battery cells is a typical indication that the batteries are being overcharged. Test the charging system (see Chapter 8) and adjust the voltage regulator as needed.

Undercharged batteries are operating in a partially discharged condition. Undercharging can be caused by excessive system electrical loads, stop-and-go driving, or problems in the charging system. An undercharged battery will eventually become sulfated when the sulfate normally formed on the plates becomes dense, hard, and chemically irreversible. A *sulfation* condition occurs when sulfate remains on the battery plates for a prolonged time. Sulfation causes two problems. First, it lowers the SG levels and increases the danger of freezing at low temperatures. Second, in cold weather, a sulfated battery may not have the reserve power needed to crank the engine. Rapid continuous discharging and recharging of batteries is known as *deep cycling*. Repeated deep cycling can cause the positive plate material to break away from its grids and fall into the sediment chambers at the base of the battery case. Deep cycling can reduce battery capacity and eventually short-circuit the battery plates.

BATTERY SERVICING PRECAUTIONS

The electrolyte in a battery contains sulfuric acid and is very corrosive. If splashed into eyes or onto the skin, it can cause blindness and serious burns. Battery electrolyte is also poisonous as well as corrosive; if swallowed, immediate medical attention is required.

Batteries produce hydrogen and oxygen gases during normal operation. These gases may vent during charge cycles and can be highly explosive in small quantities.

To avoid the potential for serious injury and damage, the following precautions should be observed whenever handling, testing, or charging a battery:

1. If electrolyte is splashed on the skin or in the eyes, rinse with cool, clean H_2O for about 15 minutes and call a doctor immediately. Do not add eyedrops or other medication unless advised by a doctor.
2. If electrolyte is swallowed, drink several large glasses of milk or H_2O followed by milk of magnesia, a beaten raw egg, or vegetable oil. Call a doctor immediately.
3. Electrolyte can damage painted or unpainted metal parts. If electrolyte is spilled or splashed on painted metal surfaces, neutralize it with a baking soda solution and rinse the area with clean H_2O.
4. Use extreme care to avoid spilling or splashing electrolyte. Wear safety glasses or a face shield when working with batteries. Wear rubber gloves and an apron when handling or carrying batteries.
5. To prevent possible skin burns and electrical arcing, do not wear watches, rings, or other jewelry while performing maintenance work on the batteries or any other part of the electrical system.
6. When removing a battery from a vehicle, always disconnect the battery ground cable first. When installing a battery, connect the ground cable last.
7. Never reverse the polarity of the battery connections. Generally, all highway vehicles today use a negative chassis ground circuit. Reversing the electrical polarity can cause immediate damage to any electronic components on the chassis. It may also cause a battery explosion, so be extra careful when connecting jumper cables. Always connect negative to negative and positive to positive.
8. A battery that has lead-calcium cells or one that is operated in hot weather conditions may have a shorted cell. Shorted cells occur when the grid plates expand to the point where they contact, resulting in a short path. Sometimes a shorted cell can be identified by visual inspection: When all cells have normal electrolyte levels but one is almost dry, a short condition is indicated.

WINTERIZING BATTERIES

As mentioned earlier, available battery energy decreases as temperature drops. A fully charged battery at 80°F (27°C) has 100 percent of its potential energy. The same battery at 32°F (0°C) has only 83 percent of that energy available. At 0°F (−18°C), battery potential has dropped to 61 percent, and further decreases to a mere 10 percent at −30°F (−34°C). Also, as temperature drops the mechanical energy required to crank an engine is much greater, so much attention should be paid to the cranking circuit in cold weather.

When winterizing batteries during PMs (see Chapter 4), disconnect battery cables and load test each battery individually at one-half the CCA rating for 15 seconds. For a battery rated at 900 CCA, carbon pile B load test at 450 amps for 15 seconds. The voltage should not drop below 9.6 volts during the test when performed at 70°F (21°C). When the test is performed at 0°F (−18°C) the voltage should not drop below 8.5 volts. Ambient temperature must always be taken into account when performing any battery tests.

Because it is important to keep batteries warm to optimize performance in cold weather, many trucks are equipped with battery heaters. Various styles are available. The most common are the blanket type and the plate type. The plate-type battery heater is placed either under or between the batteries. Blanket-type heaters come in a variety of lengths and wattages and usually incorporate their own insulation.

If you suspect a battery is frozen do not attempt to heat it rapidly. Trickle charge and observe for charge response.

7.4 BATTERY TESTING

Batteries should be tested during PMIs and whenever performance is suspect. Battery testing should indicate:

- If the battery is satisfactory and can remain in service
- If the battery should be recharged before placing it back in service
- If the battery must be replaced

A complete battery test, as outlined here, includes these steps:

1. Visual inspection
2. State of charge test
3. Battery capacity (load) test

CAUTION

Be sure to follow all instructions and safety procedures suggested by the test equipment manufacturer.

VISUAL INSPECTION

Begin by inspecting the outside of the battery for obvious damage such as a cracked or broken case or cover. Check for damage to the battery terminals. If obvious physical damage is found, replace the battery. When possible, determine the cause of damage and correct it. Next, check the electrolyte level in the battery. On low-maintenance batteries, simply unscrew the vent caps or pull off the vent plugs and observe the level of electrolyte in each cell. You can check electrolyte levels on maintenance-free batteries, but be aware that you could be voiding any kind of warranty offered. Use a knife blade to slice through the decal on top of the battery and peel it back to expose the vent caps or vent manifold as shown in **Figure 7–15**. Remove the caps or use a screwdriver to pry up the vent manifold.

If the electrolyte level is below the plates, add distilled H_2O until the electrolyte level just covers the separators. After topping up the electrolyte, charge the battery for at least 15 minutes at around 15 to 25 amps to mix the H_2O with the electrolyte.

STATE OF CHARGE TEST

Battery state of charge can be determined either by measuring SG or by using a stabilized open-circuit voltage test. Checking SG on low-maintenance or

FIGURE 7–15 The electrolyte level can be checked in a maintenance-free battery by removing the decal to expose the vent manifold or caps.

conventional batteries requires the use of a hydrometer or refractometer. Maintenance-free batteries normally have a built-in hydrometer.

Specific Gravity (Hydrometer) Test

Perform the SG (hydrometer) test on all cells (**Figure 7–16**). Measure and record the SG, corrected to 80°F (27°C), of each cell. Determine the battery state of charge. If the SG readings are 1.225 or higher

FIGURE 7–16 Reading a hydrometer

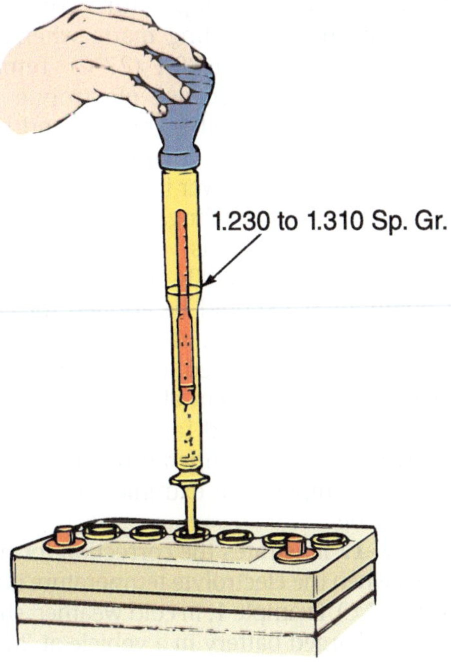

and are within 50 points (0.050 SG) between the highest and lowest cells, proceed to the load test. If the SG readings are low (below 1.225) or vary more than 50 points between highest and lowest cells, recharge the battery as instructed under "Charging the Battery" and be sure to inspect the vehicle electrical system to determine the cause for the low state of charge. If, after charging, the SG readings are within 50 points between highest and lowest cells, proceed to step 3. If readings still vary more than 50 points after charging, replace the battery.

How to Use a Hydrometer

The correct method of reading a hydrometer requires that the barrel be held vertically so the float is not contacting the sides of the barrel. Electrolyte should be drawn in and out of the hydrometer barrel a few times to balance the temperature of the hydrometer float and that of the electrolyte. Draw a full charge of electrolyte into the hydrometer by fully expanding the suction bulb, and ensure that the hydrometer float is lifted free, touching neither side.

When reading the hydrometer, eyeball the float level with the surface of the liquid in the hydrometer barrel. Disregard any surface tension curvature of the liquid in the barrel. Keep the float clean and ensure it is not cracked. Never take a hydrometer reading immediately after H_2O has been added to a cell. Any added distilled H_2O must be mixed with the battery electrolyte, something that is accomplished by a short period of charging. There are different types of battery hydrometers available, so follow the manufacturer's instructions.

Temperature Correction

Hydrometers are calibrated to give a true reading at a fixed temperature. A temperature correction factor applies to any SG reading made when the electrolyte temperature is not 80°F (27°C). Temperature correction is required because electrolyte expands and becomes less dense when heated and contracts and increases in density as it cools. Accordingly, the hydrometer float will sit lower in less dense solutions, producing lower SG readings—with the opposite occurring in colder solutions.

A correction factor of 0.004 SG (sometimes referred to as 4 points of SG) should be applied for each 10°F (5°C) change in temperature. That means that 4 points of SG (0.004) should be added to the indicated reading for each 10°F (5°C) increment above 80°F (27°C) and 4 points should be subtracted for each 10°F (5°C) increment below 80°F (27°C). Temperature correction is critical because at temperature extremes it can become a substantial value.

Figure 7–17 illustrates the correction for hydrometer readings when the electrolyte temperature is above or below 80°F (27°C). Example 1, in cold weather, shows that a partially discharged battery in a vehicle at 20°F (−7°C)

FIGURE 7–17 Temperatures above 80°F (27°C) require subtracting from the recorded specific gravity; temperatures below 80°F (27°C) require adding to the specific gravity reading.

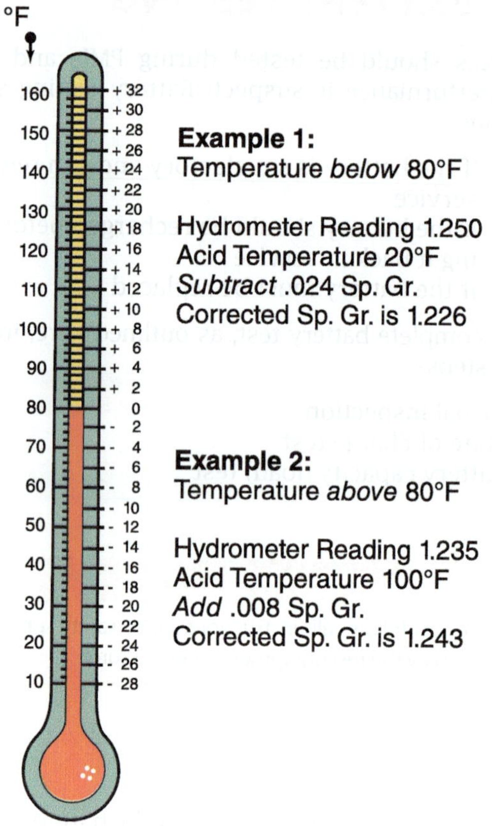

might read 1.250, indicating it is close to being fully charged. However, when the correction factor is applied, the actual SG measures only 1.226.

Example 2 in Figure 7–17 shows what might be encountered in a battery exposed to the sun in hot weather. Note also that the operating temperature of electrolyte is typically 100°F (38°C) during summer. The uncorrected SG reading of 1.235 indicates a low enough state of charge to warrant replacing the battery. However, a corrected reading of 1.246 tells a different story and may not be unreasonably low, depending on the length of storage of the battery or the type of service that it has been experiencing in the vehicle.

Integral Hydrometers

On many sealed maintenance-free batteries a temperature-compensated hydrometer is built into the battery cover. A quick visual check will indicate the battery state of charge. The integral or built-in hydrometer has a colored ball within a cage that is attached to a clear plastic rod. The colored ball will float at a predetermined SG of the electrolyte that represents about a 65 percent state of charge. When the colored ball floats, it rises within the cage and positions itself under the rod. A colored dot then shows in the center of the hydrometer (**Figure 7–18A**).

The built-in hydrometer provides a guide for battery testing and charging. In testing, the colored dot means the battery is charged enough for testing. If the colored dot is not visible and the hydrometer has a dark appearance (**Figure 7–18B**), it means the battery must be charged before the test procedure is performed.

The hydrometer on some batteries may appear clear or light in color (**Figure 7–18C**). This usually means the electrolyte level is below the bottom of the rod and attached cage. This might have been caused by excessive or prolonged charging, a broken case, excessive tipping, or normal battery wearout. Whenever this clear or light color appearance is present while looking straight down on the hydrometer, always tap the hydrometer lightly with a small screwdriver to dislodge any gas bubbles that might be giving a false indication of low electrolyte level. If the clear or light color appearance remains, and if a cranking complaint exists caused by the battery, replace it. It is important when observing the hydrometer that the battery sight glass be clean. A flashlight might be required in poorly lit areas. Always look straight down when viewing the hydrometer sight glass.

FIGURE 7–18 Reading a built-in hydrometer: Although this is the most common color code used with truck batteries, keep in mind that a few battery manufacturers use a slightly different one. Check the top of the battery case to find out the color codes for a given battery.

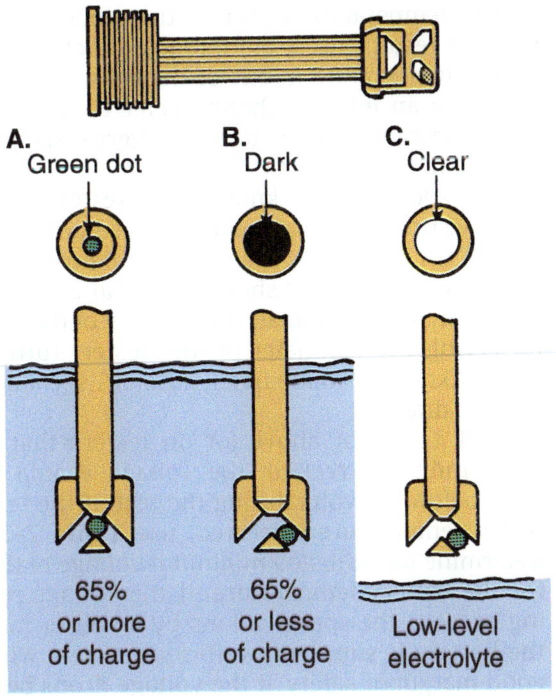

WARNING

Never jump-start or attempt to recharge a fully discharged maintenance-free battery. Jump-starting and charging can create an explosion hazard. These batteries have limited means to vent gas buildup and, therefore, should be replaced if fully discharged.

Reading a Refractometer

This is probably the most reliable means of testing electrolyte, primarily because no temperature compensation is required. Dip the test probe ¼ inch into the electrolyte; surface tension should pull a drop of the liquid onto the probe. Carefully contact the test probe onto the refractive glass of the hydrometer, depositing the drop of electrolyte. Close the refractive window and move the scope to your dominant eye to view the reading. The reading is displayed as dark shadow on the refractive scale. This test is covered in Photo Sequence 3 toward the end of this chapter.

Tech Tip

To obtain an accurate reading on a refractometer it is necessary to view the reading in good light; attempting to read a refractometer in poor light will make it difficult to read the refractive scale.

OPEN-CIRCUIT VOLTAGE TEST

The open-circuit voltage of the battery can also be used to determine its state of charge. Follow these steps to test voltage:

1. If the battery has just been recharged or has recently been in vehicle service, the surface charge must be removed before an accurate voltage measurement can be made. To remove surface charge, crank the engine for 15 seconds. Do not allow the engine to start. To prevent the engine from starting, apply the engine stop control or disconnect the fuel solenoid valve lead wire as required.
2. After cranking the engine, allow the battery to rest for 15 minutes.
3. Connect the voltmeter across the battery terminals (**Figure 7–19**) and observe the reading. Compare the reading obtained with **Table 7–2** to determine the battery state of charge.
4. If the stabilized voltage is below 12.4 volts, the battery should be recharged. Also, inspect the vehicle's electrical system to determine the cause of the low state of charge. After charging the battery, proceed to the load test.

FIGURE 7–19 Open-circuit voltage test with a digital multimeter and minimum load test specs

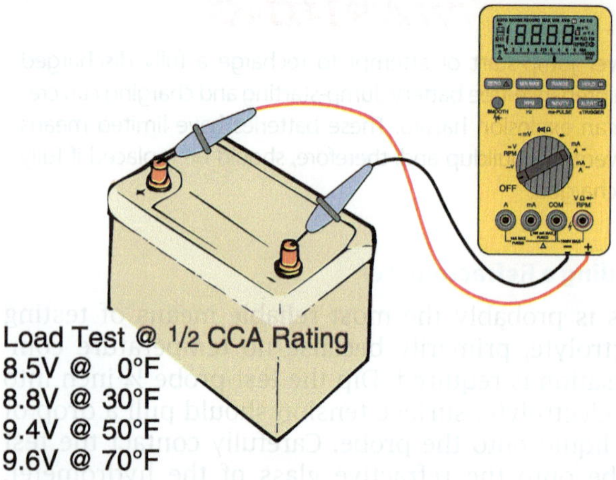

TABLE 7–2 State of Charge Determined by Open-Circuit Voltage Test

Stabilized Open-Circuit Voltage	State of Charge
12.6 volts or more	Fully charged
12.4 volts	75% charged
12.2 volts	50% charged
12.0 volts	25% charged
11.7 volts or less	Discharged

LOAD TEST

The *load*, or *capacity*, test determines how well any type of battery, sealed or unsealed, functions under dynamic load.

A load test requires the use of an adjustable carbon pile. Load testing effectively deep cycles the battery so it should not be performed frequently. Most original equipment manufacturers (OEMs) prefer the capacitance testing outlined later in this chapter. Carbon pile testing can be performed with the batteries either in or out of the vehicle, but each battery should be at or close to a full state of charge for the result to have validity. First use either the SG or open-circuit voltage test to determine state of charge, making sure that you recharge the battery if required. When load testing, the electrolyte should be as close to 80°F (27°C) as possible. Cold batteries will test at lower capacity. Never load test a sealed battery if its temperature is below 60°F (15°C).

After attaching the carbon pile (**Figure 7–20**), use the following steps to load test a battery:

FIGURE 7–20 Performing a capacity test on a battery. Adaptors are attached to the terminals.

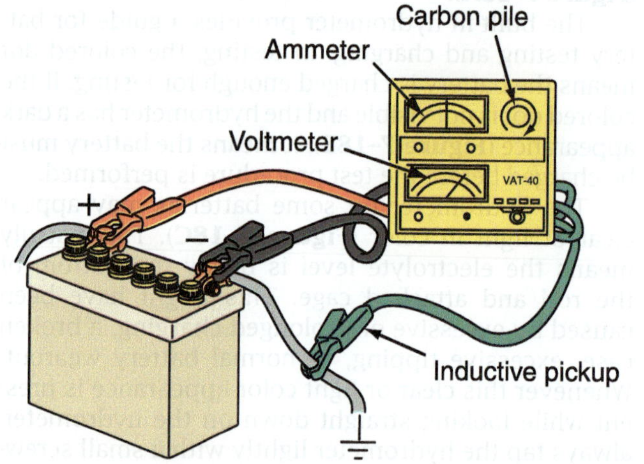

1. First, put on safety glasses. If the battery was recently charged or in service, remove the surface charge of the battery by applying a 300-amp load for 15 seconds or by disabling the engine and cranking it for that length of time.
2. Adjust all mechanical settings on the load tester to zero. Rotate the load control knob fully counterclockwise to the off position.
3. Connect the load tester to the battery. An inductive pickup must surround the wires from the ground terminal if used in the test. Observe correct polarity and be sure the test leads contact the battery posts. Batteries with sealed terminals require adapters to provide a place for attaching the test leads (**Figure 7–21**).
4. If the tester is equipped with an adjustment for battery temperature, turn it to the proper setting. For best results, use a thermometer to check electrolyte temperature in one of the cells. On sealed units use an infrared thermometer to determine the temperature. Refer to the battery's specifications to determine its CCA rating.
5. Turn the load control knob on the tester to draw battery current at a rate equal to one-half the battery's CCA rating. For example, if a battery's CCA rating is 440, the load should be set at 220 amps.
6. Maintain specified load for 15 seconds while observing the voltmeter on the tester. Turn the control knob off immediately after 15 seconds of current draw.
7. At 70°F (21°C) or above (or on testers that are temperature corrected), the voltage should not drop below 9.6 volts during the test. If the tester is not temperature corrected, use **Table 7–3** to determine the adjusted minimum voltage reading for the battery temperature. If the voltage reading exceeds the specifications by a volt or more, the battery is supplying sufficient current with a good margin of safety. If the voltage drops below

FIGURE 7–21 Batteries with threaded terminals require the use of adapters for testing or charging.

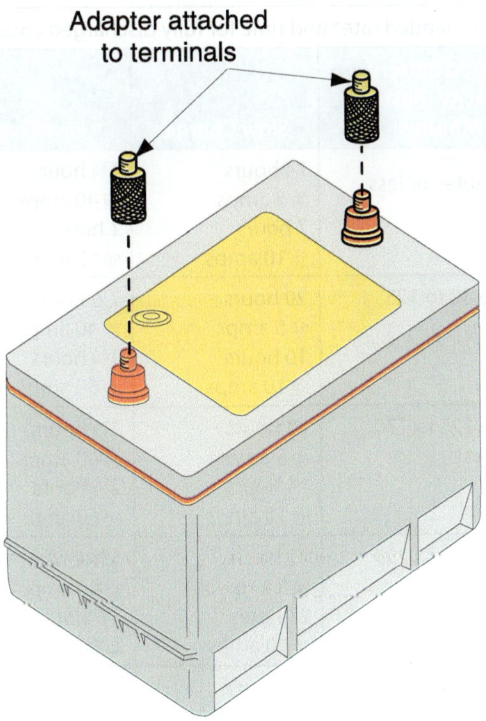

TABLE 7–3 Minimum Load Test Voltages as Affected by Temperature

Battery Minimum Temperature (F)	Test Voltage
70°	9.6 volts
60°	9.5 volts
50°	9.4 volts
40°	9.3 volts
30°	9.1 volts
20°	8.9 volts
10°	8.7 volts
0°	8.5 volts

the minimum specification, the battery should be replaced. If the reading is right on the spec, the battery might not have the reserve necessary to handle cranking under demanding low temperature conditions. Keep in mind that this varies with the state of charge determined before the load test was made. For instance, a battery at 75 percent charge whose voltage dropped to the minimum acceptable load specification is probably in good shape.

ELECTROCHEMICAL IMPEDANCE SPECTROSCOPY (EIS)

Electrochemical impedance spectroscopy (EIS) is widely used in automated battery testers, to verify battery performance and most OEMs use it to evaluate warranty claims. An EIS shop floor battery tester enables technicians to rapidly assess and report battery condition and performance in easy to understand terms such as OK and Not OK.

EIS operates by evaluating the electrochemical characteristics of a battery by applying alternating current (AC) at varying frequencies and measuring the current response of each battery cell. Frequencies range from 100 microhertz (mHz) to 100 kilohertz (kHz). 100 mHz is such a low frequency that it takes more than 2 hours to complete a full cycle. By contrast, applying a 100 kHz frequency pulse completes 100,000 cycles in 1 second.

The idea of applying various frequencies is to test the different layers of the battery and examine its characteristics at all levels. Battery resistance falls into three categories: pure Ohmic resistance, inductance, and capacitance. *Capacitance* is responsible for the capacitor effect, whereas the *inductance* is accountable for so-called magnetic field, or coil effect. While the voltage on a capacitor lags behind the current, the opposite is true on a magnetic coil, where the current lags behind the voltage. When applying a sine wave to a battery, the reactive resistance produces a phase shift between voltage and current. This is the basis used by EIS to evaluate battery performance.

EIS is not new and was used originally to perform in-flight analysis of satellite batteries, providing an estimate of such factors as grid corrosion and H_2O loss. Because this type of testing provides the ability to observe the kinetic reaction of the electrodes, it has gained acceptance as the most reliable means of evaluating performance and expected remaining battery life. Increases in impedance readings can suggest intrusion of corrosion and other deficiencies. Impedance studies using the EIS methods have been used for a number of years on lead-acid batteries and they have generated a good database for reliably determining battery capacity and predicting life span. When a warranty claim for a supposedly defective battery is rejected, EIS testing by the battery manufacturer is usually responsible. This has increased the use of EIS testers by dealerships.

CONDUCTANCE TESTING

Less costly automated battery testers that use a variation of EIS technology known as **conductance testing** or **AC conductance testing** are often used in service facilities. These are not as accurate as EIS testers but they can help reduce the incidence of incorrectly diagnosed battery problems. The advantages of AC conductance testing are that it is noninvasive and quick, and requires no more skill on the part of the technician other than the ability to connect test leads and read a data display screen. AC conductance test instruments are best at identifying batteries with definite deficiencies.

Midtronics Conductance Testing

Many fleets and dealerships are using EIS with the objective of reliably predicting the remaining life span of batteries in their vehicles. Most battery suppliers are adopting EIS to assess warranty claims. Commonly used EIS or conductance testing instruments are manufactured by Midtronics. The first generation of Midtronics electrical test instruments were battery specific, but they have evolved in comprehensive electrical diagnostic tools. A Midtronics battery test procedure is outlined at the end of this chapter, and a Midtronics electronic service tool is introduced in Chapter 11.

7.5 CHARGING THE BATTERY

There are two methods of recharging a battery: the slow-charge method and the fast-charge method. Either method can be used to recharge flooded cell, lead-acid batteries. However, there are some batteries (recombinant types) that should be charged slowly. Both methods are explained next. The charge that a battery receives is equal to the charge rate in amperes multiplied by the time in hours. Accordingly, a 5-ampere rate applied to a battery for 10 hours would translate to a 50 ampere-hour charge to the battery. A 60 ampere-hour charge applied for 1 hour would be a 60-ampere charge.

The battery charging guides given in **Table 7–4** and **Table 7–5** show approximately how much recharge a fully discharged battery requires. For partially discharged batteries, the charging current or charging time should be reduced accordingly. For example, if the battery is 25 percent charged (75 percent discharged), reduce charging current or time by one-fourth. If the battery is 50 percent charged, reduce charging current or time by one-half. If time is available, lowering the charging rates in amperes is recommended. While the battery is being charged, periodically measure the temperature of the electrolyte. If the temperature exceeds 125°F (52°C) or if violent gassing or spewing of electrolyte occurs, the charging rate should be reduced or temporarily halted. This must be done to avoid damage to the battery.

CAUTION

Do not overcharge batteries, particularly maintenance-free type batteries. Overcharging causes excessive boil-off of water from the electrolyte. Overcharging also causes the battery to produce explosive combinations of hydrogen and oxygen.

TABLE 7–4 Battery Charging Guide—6- and 12-Volt Low-Maintenance Batteries

(Recommended rate* and time for fully discharged condition)

Rated Battery Capacity (Reserve Minutes)	Slow Charge	Fast Charge
80 minutes or less	14 hours @ 5 amps 7 hours @ 10 amps	1¾ hours @ 40 amps 1 hour @ 60 amps
Above 80 to 125 minutes	20 hours @ 5 amps 10 hours @ 10 amps	2½ hours @ 40 amps 1¾ hours @ 60 amps
Above 125 to 170 minutes	28 hours @ 5 amps 14 hours @ 10 amps	3½ hours @ 40 amps 2½ hours @ 60 amps
Above 170 to 250 minutes	42 hours @ 5 amps 21 hours @ 10 amps	5 hours @ 40 amps 3½ hours @ 60 amps
Above 250 minutes	33 hours @ 10 amps	8 hours @ 40 amps 5½ hours @ 60 amps

*Initial rate for standard taper charger.

SLOW CHARGING

The slow-charging method uses a low-charging rate for a relatively long period. The recommended rate for slow charging is 1 ampere per positive plate per cell. If the battery has nine plates per cell, normally four of the nine will be positive plates. Therefore, the slow-charge rate would be 4 amperes. Charging periods as long as 24 hours might be needed to bring a battery to full charge.

The best method of making certain that a battery is fully charged, but not overcharged, is to measure the SG of a cell once per hour. The battery is fully charged when no change in SG occurs over a 3-hour period or when charging current stabilizes (constant voltage-type charger). The problem is that this type of labor-intensive procedure does not get much acceptance in the trucking industry.

Batteries with charge indicators are sufficiently charged when the colored dot in the hydrometer is visible. Gently shake or tilt the battery at hourly intervals during charging to mix the electrolyte and to see if the colored dot appears. Do not tilt the battery beyond a 45-degree angle. If the colored dot does not appear after a 75 ampere-hour charge, continue

TABLE 7–5 Battery Charging Guide—12-Volt Maintenance-Free Batteries

(Recommended rate* and time for fully discharged condition)

Rated Battery Capacity (Reserve Minutes)	Slow Charge	Fast Charge
80 minutes or less	10 hours @ 5 amps 5 hours @ 10 amps	2½ hours @ 20 amps 1½ hours @ 30 amps 1 hour @ 45 amps
Above 80 to 125 minutes	15 hours @ 5 amps 7½ hours @ 10 amps	3¾ hours @ 20 amps 2½ hours @ 30 amps 1¾ hours @ 45 amps
Above 125 to 170 minutes	20 hours @ 5 amps 10 hours @ 10 amps	5 hours @ 20 amps 3 hours @ 30 amps 2¼ hours @ 45 amps
Above 170 to 250 minutes	30 hours @ 5 amps 15 hours @ 10 amps	7½ hours @ 20 amps 5 hours @ 30 amps 2½ hours @ 45 amps
Above 250 minutes	20 hours @ 10 amps	10 hours @ 20 amps 6½ hours @ 30 amps 4½ hours @ 45 amps

*Initial rate for standard taper charger.

charging for another 50 to 75 ampere-hours. If the colored dot still does not appear, replace the battery.

SHOP TALK

Batteries with charge indicators cannot be charged if the indicator is clear or light in color; replace these batteries.

If a low-maintenance (conventional) battery is to be charged overnight (10–16 hours), use the specified slow-charge rate in Table 7–4. Maintenance-free batteries must not be charged at rates greater than specified in the maintenance-free, battery charging guide (see Table 7–5). If a maintenance-free battery is to be recharged overnight (16 hours), a timer or voltage-controlled charger is recommended to avoid overcharging. If the charger does not have such controls, a 3-ampere rate should be used for batteries of 80-minute capacity or less and 5 amperes for 80- to 125-minute reserve capacity batteries. Batteries charged for over 125 minutes should be charged at the specified slow-charge rate (see Table 7–5).

Batteries that have remained in a discharged condition for long periods without a recharge have probably become sulfated; sulfur in the electrolyte combines with the lead in the plates, hardening plates and making them inactive. A sulfated battery can sometimes be rescued by recharging at a low rate to avoid overheating and excessive gassing. It can require 2 or 3 days of slow charging to bring a sulfated battery to a fully charged condition. Again, care should be taken not to overcharge maintenance-free-type batteries.

Batteries can become so badly sulfated that they cannot be restored to a normal operating condition, regardless of the rate of charge or the length of time the charge is applied. If a battery cannot be restored to a fully charged condition by slow charging, it should be rejected. Sulfated batteries after charging will typically produce abnormally high voltage readings across the terminals. This voltage drops the moment a load is placed on the battery.

FAST CHARGING

The fast-charge method provides a high-charging rate for a short period. The charging rate should be limited to 60 amperes for 12-volt batteries. The maximum charging rate for 6-volt batteries (above 180 reserve capacity minutes) can be approximately doubled.

Ideally, fast charges should be limited to the charging times shown under "Fast Charge" in the battery charging guides (see Table 7–4 and Table 7–5). The battery generally cannot be fully charged within these time periods, but it will receive sufficient charge (70 to 90%) for practical service. To completely recharge a battery, follow the fast charge with a slow charge until no change in SG occurs over a 3-hour period (or until the colored dot is visible in the built-in hydrometer). When fast charging a battery in the vehicle, always disconnect the battery ground cable before applying the charge.

A battery with an SG of 1.225 or above should never be charged at a high rate. If the charger has not tapered to a low rate, adjust to a slow charge, preferably at a rate of 1 ampere per positive plate per cell.

CHARGING INSTRUCTIONS

Follow these steps to charge a battery:

1. Before placing a battery on charge, clean the battery terminals if necessary.
2. Add distilled H_2O sufficient to cover the plates. Fill to the proper level near the end of charge.

If the battery is extremely cold, allow it to warm before adding distilled H_2O because the level will rise as it warms. In fact, an extremely cold battery will not accept a normal charge until it becomes warm.

3. Connect the charger to the battery, following the instructions of the charger manufacturer (**Figure 7–22**). Connect the positive (+) charger lead to the positive battery terminal and the negative (–) lead to the negative terminal. If the battery is in the vehicle, only connect the negative lead to the chassis or engine block ground in older truck chassis; newer trucks with data bus-driven electronics require that ground clamps be connected to the battery ground posts. Move the charger lead clamps back and forth to ensure a good connection.
4. Switch the charger on and gradually increase the charging rate until the recommended ampere value is reached.

CAUTION

If smoke or dense vapor comes from the battery, shut off the charger and reject the battery. If violent gassing or spewing of electrolyte occurs, reduce or temporarily halt the charging.

5. When the battery is charged, turn the charger off and disconnect it.
6. Install the battery in the test truck. If the engine does not crank satisfactorily when a recharged battery is installed, capacitance test or load test the battery. If the battery passes the load test, the vehicle's fuel, ignition, cranking, and charging systems should be checked to locate and correct the no-start problem. If it does not pass the load test, the battery should be replaced.

CHARGING SAFETY

When charging the batteries, gas forms in each cell and can bleed through the vents. In poorly ventilated areas, explosive gas may remain around the battery for some time after charging. The hydrogen gas vented by batteries is easily ignited by sparks, flame, or heat and can cause an explosion. Follow these precautions when charging the batteries:

1. Make sure that the area is well ventilated.
2. Do not separate energized circuits at the terminals: A spark can result. Use great care when connecting or disconnecting booster leads or cable clamps on chargers. Poor connections are common causes of electrical arcs, which cause explosions.
3. Do not smoke near batteries. Keep the batteries away from open flames or sparks.
4. If a battery is frozen, warm it to room temperature before attempting to charge it. A dead battery can freeze at temperatures near 0°F (−18°C). Never attempt to charge a battery that has visible ice in the cells. Passing current through a frozen battery can cause it to rupture or explode. If ice or slush is visible or the electrolyte level cannot be seen, allow the battery to thaw at room temperature before servicing it. Do not take chances with sealed batteries; if there is any doubt, allow them to warm to room temperature.
5. Do not place or drop tools or metal objects on batteries. This could short the batteries, causing arcing and possible damage.
6. Never lean over a battery during charging, testing, or jump-start operations. When a battery explodes, the case cover and electrolyte tend to blow upward.
7. When handling a plastic-cased battery, pressure placed on the end walls may cause electrolyte to spew through the vents. Use a battery carrier strap to lift batteries. When lifting by hand, place hands at opposite corners.

JUMP-STARTING

The procedure outlined here should be followed when it becomes necessary to boost a set of discharged batteries on a vehicle—a relatively common practice.

Take a good look at the electrical systems and battery configuration on both vehicles before making any connections. Both 6- and 12-volt batteries are used on trucks and they can be arranged in 12V- and 12/24V-systems.

SHOP TALK

Ensure that the vehicle with the dead batteries and the boost vehicle do not directly contact. If the two vehicles are in contact, a ground connection could be established, which could cause sparking when jumper cables are being connected.

1. Set the parking brake on both vehicles. Place both transmissions in neutral or park. Turn off lights, heater, and other electrical loads. The ignition key should be turned off in both vehicles. Put on safety glasses.
2. Connect the clamp of the positive (red) jumper cable to the insulated (positive) terminal of the booster batteries that is closest to the starter. Attach the other end of the positive jumper cable to the insulated terminal of the discharged battery bank, preferably the one that is closest to the starter (**Figure 7–23**).
3. Connect the clamp of the ground jumper cable to the ground terminal of the booster batteries. Clamp the other end of the ground jumper cable to the battery ground post closest to that of the chassis ground (current truck chassis with a data

FIGURE 7–22 Battery charger connection options

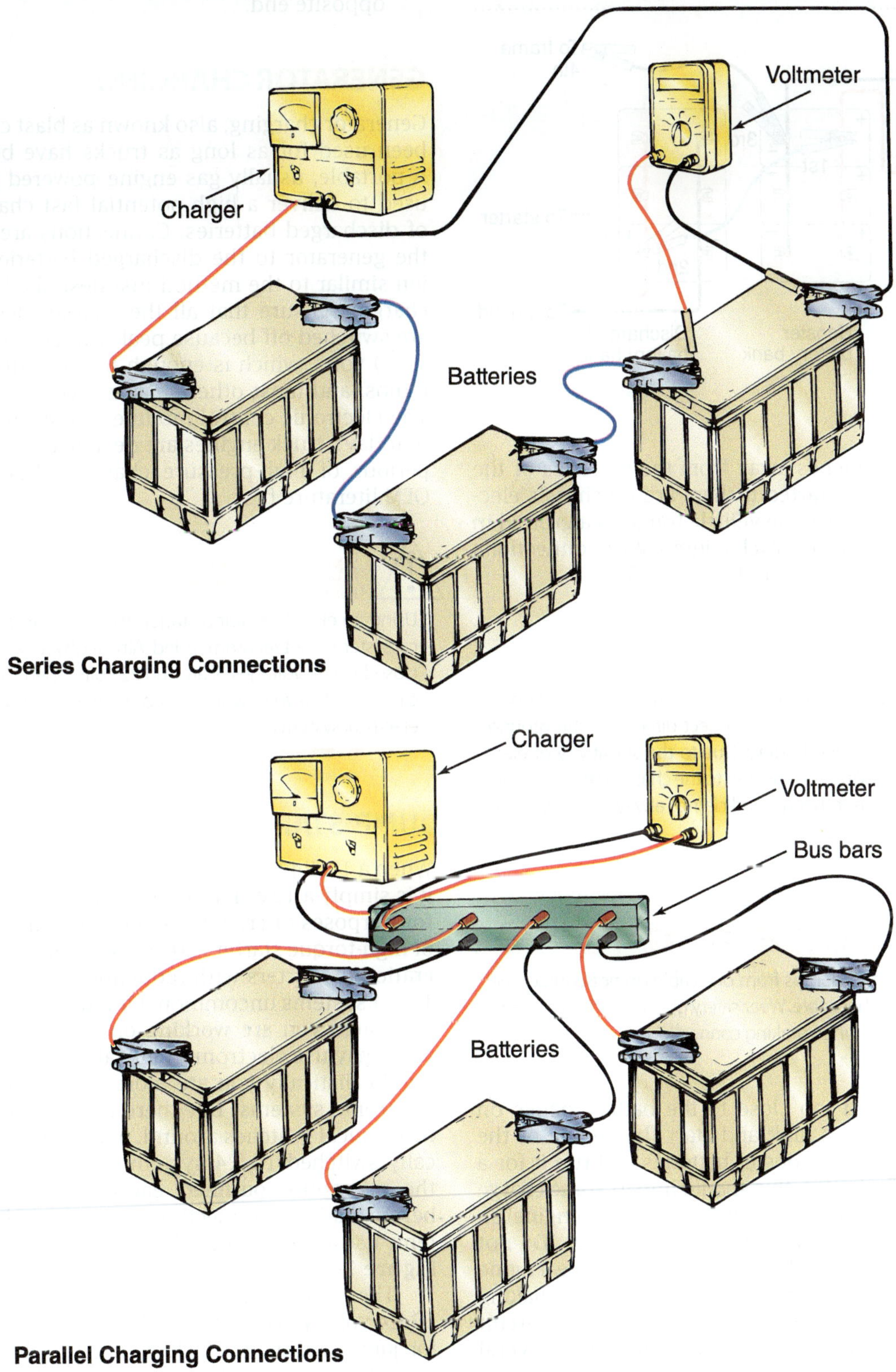

FIGURE 7–23 Correct method of attaching jumper cables to a 12-volt system

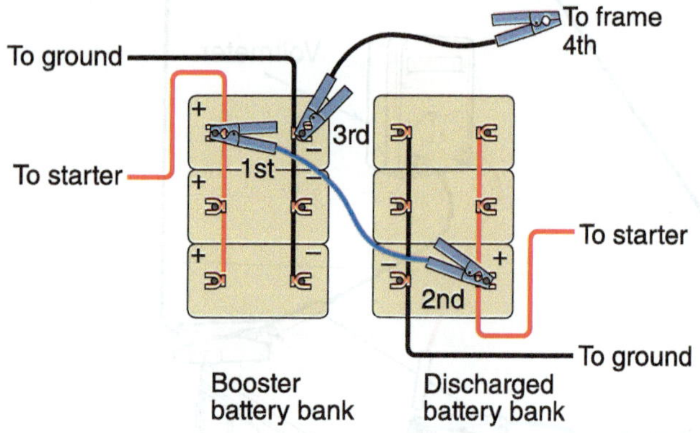

bus) or 12 inches away from the batteries of the vehicle to be started if there are no chassis electronics. Only use the vehicle frame as a ground in pre-electronic era trucks; jump-start connections for these are shown in Figure 7–23.

SHOP TALK

The ground connection must be sound. It used to be recommended that you should not connect directly to the ground post of the discharged battery. Due to the sensitivity of electronic control modules connected to the chassis data bus, most OEMs today suggest that ground clamps always be connected directly to battery ground.

WARNING

Make sure that the clamps from one cable do not contact the clamps on the other cable. Wear safety glasses, and do not lean over the batteries when making connections.

4. Do not stand too close to the battery banks on either vehicle. Crank and start the engine of the vehicle with the booster batteries and run it for a couple of minutes; this should impart some charge to the dead batteries. Next, crank the engine on the vehicle with the discharged batteries. Do not crank for longer than 30 seconds. If the engine fails to start, wait at least 2 minutes before making another attempt: This allows the cranking circuit to cool. If the engine does not start after several attempts, check for the cause.
5. After starting allow the engine to idle. First, disconnect the ground connection from the vehicle with the discharged battery and then disconnect the opposite end of the ground cable.
6. Disconnect the positive clamp from the discharged battery first and, finally, disconnect the opposite end.

GENERATOR CHARGING

Generator charging, also known as blast charging, has been used for as long as trucks have been around. A portable, usually gas engine–powered generator is used to deliver a high-potential fast charge to a set of discharged batteries. Connections are made from the generator to the discharged batteries in a fashion similar to the method just described. When blast charging ensure that all the chassis electrical loads are switched off because peak voltage can be as high as 20 volts, which is enough to take out headlamps, radios, and most other chassis electrical equipment. The electronic control modules (ECMs) used on most of today's truck engines are designed to sustain short periods of high-pressure charging, but check with OEM literature first.

CAUTION

Using an electric welding station to jump-start a dead electrical system is not recommended. Although arc welders produce closed-circuit voltages of around 22 volts, open-circuit voltages can exceed 70 volts, which is enough to severely damage truck electrical systems.

JUMP-STARTING A 12/24-VOLT SYSTEM

When a 12/24-volt system is used on a highway truck it is simply a 12V chassis system that switches to 24V for purposes of cranking the engine only. The advent of high-torque starters 20 years ago and a current generation of starters with reduction gearing have made 12/24 systems uncommon. First determine what type of system you are working on; 12/24 cranking systems may use electromechanical or electronic switching. Realistically you are more likely to come across electronic systems, but there are still some electromechanical switches around. In most cases, electronically switched 12/24 systems can be boosted using the procedure outlined earlier, but always check before making connections. Electromechanical systems use a series-parallel (SP) switch that is shown in **Figure 7–24.**

The procedure for jump-starting a truck with a 12/24-volt system with an SP switch is similar to that for jump-starting a vehicle with a 12-volt system. The main difference is in how you connect the cables to the batteries. You will have to closely observe how the batteries are arranged in the battery bank. The procedure is outlined here and you should reference the illustration in Figure 7–24 for the correct sequence. You will need

FIGURE 7–24 Sequence of attaching jumper cables to an older 12/24-volt system

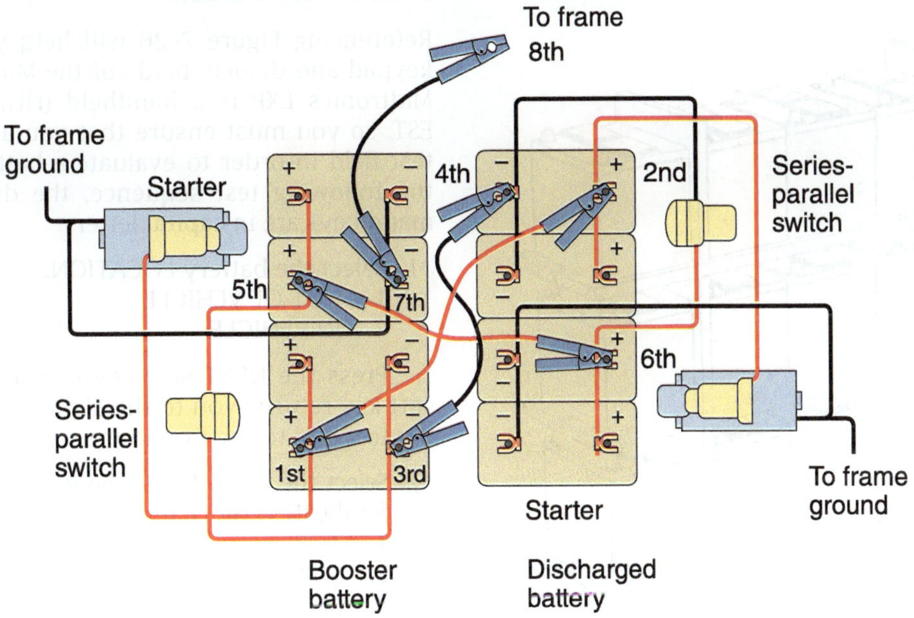

a couple of sets of booster cables to use this method. Wear safety glasses.

1. Connect the clamp of one positive (red) jumper cable to the insulated (positive) terminal of the booster batteries that is closest to the starter. Attach the other end of this positive jumper cable to the insulated terminal of the discharged battery bank, preferably the one that is closest to the starter. This is shown as steps 1 and 2 in Figure 7–24.
2. Clamp one end of a second jumper cable to the ground terminal of the booster battery that connects the SP switch. Attach the other end of this jumper cable to the ground terminal of the discharged battery that connects to the SP switch. This is shown as steps 3 and 4 in Figure 7–24.
3. Using a third jumper cable, clamp one end to the insulated terminal of the booster battery that connects to the SP switch. Attach the other end of this jumper cable to the insulated terminal of the discharged battery that connects to the SP switch. This is shown as steps 5 and 6 in Figure 7–24.
4. With the fourth jumper cable, clamp one end to the ground terminal of the booster battery that connects to the starter. Attach the other end of this cable to the battery ground post (electronic truck chassis) or a good ground at least 12 inches from the battery (vehicles with no chassis electronics) of the vehicle being started. The vehicle frame is usually a good ground. This is shown as steps 7 and 8 in Figure 7–24.

After starting, allow the engine to idle. Disconnect the ground clamp from the vehicle with the discharged batteries. Next, disconnect the opposite end from the booster vehicle chassis. Disconnect the remaining cables from the vehicle with discharged batteries first and, finally, those from the booster vehicle chassis.

BATTERY STORAGE AND RECYCLING

The Environmental Protection Agency (EPA) recommends that lead-acid batteries be stored properly to prevent contamination or injury resulting from spills or leaks. Batteries should be stored indoors because this will reduce the possibility of damage caused by extreme temperatures. Asphalt or concrete floors can be protected by applying acid-resistant coatings; epoxy emulsions are commonly used. Acid-resistant curbing should be constructed around the storage area to contain spills.

Batteries should be stored in a cool, well-ventilated area. They should be placed upright on pallets and stacked on shelves in the manner shown in **Figure 7–25**. Battery casings are not designed to support the weight of another battery, so do not stack batteries directly on top of other batteries. Inspect stored batteries regularly. Because used batteries may leak, acid-resistant, leakproof material should be placed below the storage shelves. Battery acid has to be considered as hazardous waste because it is corrosive and may also contain toxic levels of lead. Spills that cannot be contained should be reported to the Department of Environmental Resources (DER).

FIGURE 7–25 Proper way to store batteries when not in use

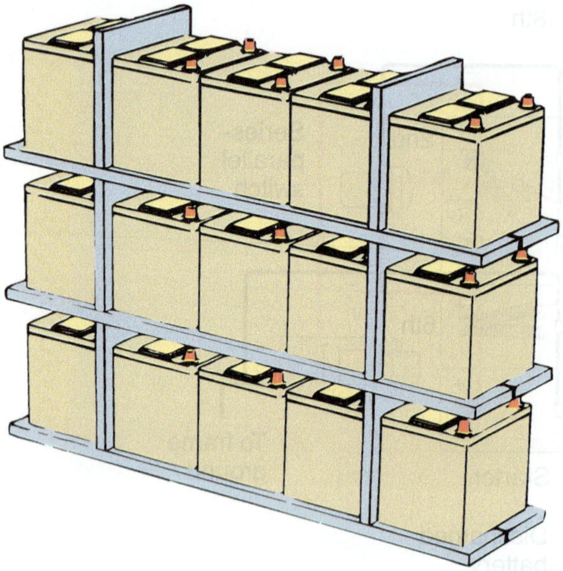

Most jurisdictions have specific laws in effect for safe disposal of lead-acid batteries, so you should be aware of those that apply to your area. The BCI sets these recycling guidelines that have been converted to regulations by many states and provinces:

1. Spent lead-acid batteries must not be disposed of in the garbage. They should be returned to a battery retailer or wholesaler or delivered to an approved recycling center.
2. Battery retailers and wholesalers are required to accept at least as many used batteries as the total number they sell from any outlet.
3. Retailers of batteries should post a 5 × 7 inch (13 × 18 cm) sign that displays the universal recycling logo and provide instruction on proper battery disposal.
4. Warn technicians that violation of approved battery disposal regulations could result in fines and/or imprisonment.

7.6 CONDUCTANCE TESTING

The principles of EIS and AC conductance testing are explained earlier in this chapter. The test outlined here is based on a current-generation Midtronics electronic service tool (EST) switched to battery test mode. There are a couple of different models of this family of very useful, multifunctional test instruments. For the basis of this introduction we reference the Midtronics inTELLECT EXP model shown in **Figure 7–26**.

BATTERY TEST PROCEDURE MIDTRONICS EXP

Referencing Figure 7–26 will help you navigate the keypad and display fields of the Midtronics EXP. The Midtronics EXP is a handheld (HH) multifunctional EST, so you must ensure that you are in the battery test field in order to evaluate a battery. In outlining the following test sequence, the display and command keys are in capital letters.

1. Select the battery LOCATION.
   1. OUT OF VEHICLE
   2. IN VEHICLE

   Press the NEXT soft key to continue. The BACK soft key returns you to the Main Menu at the start of the test and to the previous screen as you progress.
2. Select the POST TYPE. The REMOTE option appears for the IN VEHICLE test.
   1. TOP
   2. SIDE
   3. REMOTE

   Press the NEXT soft key to continue.
3. Select the BATTERY TYPE.
   1. REGULAR/AUTO
   2. AGM
   3. AGM/SPIRAL
   4. GEL

   Press the NEXT soft key to continue.
4. Select the battery capacity rating standard. This information is printed on the battery specification label. If the information cannot be interpreted the battery manufacturer should be contacted.
   1. CCA
   2. CA
   3. MCA
   4. JIS
   5. DIN
   6. SAE
   7. IEC
   8. EN

   Table 7–6 explains some battery acronyms used on the Midtronics instrument, including the foregoing menu options acronyms. Press the NEXT soft key to continue.
5. Press the UP or DOWN arrow keys or use the keypad to select the RATING UNITS or in the case of JIS, the part number. You can increase the scrolling speed by depressing the UP or DOWN arrow keys. The default selection is 500. The entry range is 100 to 3,000 except for DIN and IEC, in which case the range runs between 100 and 1,000. The display values increase or decrease by 5 units at a time. Press the NEXT soft key to start the test.

FIGURE 7–26 Main menu and keypad orientation of the Midtronics inTELLECT EXP

The **Internal Batteries Status Indicator**, which appears in the screen's top left corner, lets you know the status and charge level of the analyzer's 6 1.5 V batteries. The X shown in the figure shows that the EXP is powered by the battery you're testing to conserve the internal batteries.

Press the two **Soft Keys** linked to the bottom of the screen to perform the functions displayed above them. The functions change depending on the menu or test process. So it may be helpful to think of the words appearing above them as part of the keys. Some of the more common soft-key functions are SELECT, BACK, and END.

When you first connect the EXP to a battery it functions as a voltmeter. The voltage reading appears above the left 12.31V soft key until you move to other menus or functions.

In some cases, you can use the **Alphanumeric Keypad** to enter numerical test parameters instead of scrolling to them with the **ARROW** keys.

You'll also use the Alphanumeric Keys to create and edit customer coupons. The keypad includes characters for punctuation. To add a space, press the **RIGHT** and **LEFT ARROW** keys simultaneously.

The **Title Bar** shows you the name of the current menu, test tool, utility, or function.

Press the **POWER** button to turn the EXP on and off. The EXP also turns on automatically when you connect its test leads to a battery.

Whichever way you turn on the EXP, it always highlights the icon and setting you last used for your convenience.

The **Selection Area** below the **Title Bar** contains items you select or into which you enter information. The area also displays instructions and warnings.

The **Directional Arrows** on the display show you which **Arrow Keys** to press to move to other icons or screens. The Up and Down Directional Arrows, for example, let you know to press the **UP** and **DOWN ARROW** keys to display the screens that are above and below the current screen.

The Left and Right Directional Arrows let you know to use the **LEFT** or **RIGHT ARROW** keys to highlight an icon for selection.

Another navigational aid is the **Scroll Bar** along the right side of the screen. The position of its scroll box tells you which menu screen you're viewing.

Scroll Bar

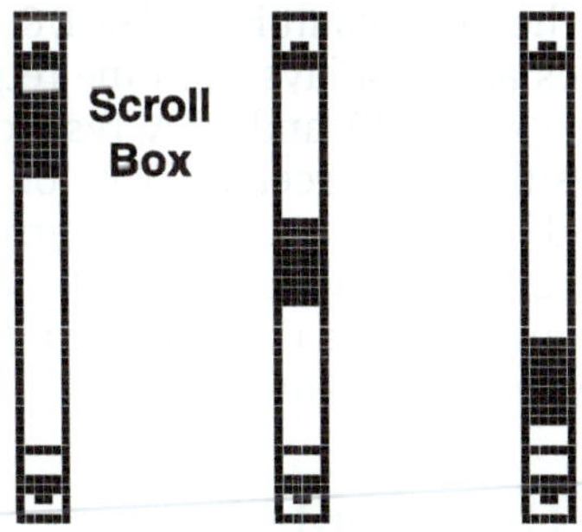

Top or only screen | **Middle screen** | **Last screen**

Courtesy of Midtronics, Inc.

TABLE 7–6 Battery Acronyms

Acronym	Decode
CCA	Cold cranking amps
CA	Cranking amps
MCA	Marine cranking amps
JIS	Japanese Industrial Standard
DIN	Deutsches Institut fur Normung
SAE	Society of Automotive Engineers
IEC	International Electrochemical Commission
EN	Euro-standards (European normung)
AH	Ampere hours
CCA	Cold cranking amperes
RC	Reserve capacity
CCV	Closed-circuit voltage
OCV	Open-circuit voltage
EMF or E	Electromotive force (voltage)
CEMF	Counter-electromotive force

TABLE 7–7 Midtronics Exp Battery Test Results

Decision	Recommended Action
GOOD BATTERY	Return battery to service.
GOOD-RECHARGE	Fully charge the battery and return it to service.
CHARGE & RETEST	Fully charge the battery and retest. *Failure to fully charge the battery before retesting may cause false readings.* If CHARGE & RETEST appears again, the battery should be replaced.
REPLACE BATTERY	Replace the battery and retest. This may be caused by a poor connection between the battery cables and the battery. After disconnecting the battery cables, retest the battery using the out-of-vehicle test before replacing it.
BAD CELL–REPLACE	Replace the battery and retest.

For the next few seconds the EXP will display the word TESTING and a stopwatch while the battery is evaluated. In addition, the analyzer displays the selected rating standard and units.

Interpreting Test Results

After evaluating the battery, the EXP displays one of five possible test results known as DECISIONS. The five possible results from the test procedure are shown in **Table 7–7**.

When the result is CHARGE & RETEST, the EXP calculates and displays the time required to charge the battery at 10, 20, and 40 A, respectively. **Figure 7–27** shows the EXP screen display following the testing of a good battery. The UP/DOWN arrow keys can be used to scroll through each result. The results can also be printed; some OEMs require this when replacing under-warranty batteries.

BATTERY CABLES

Battery cables that have insufficient conductor sectional area will overheat. When replacing battery cables, ensure that the original specification is met or exceeded. SAE battery cables are coded using both standard gauge and metric systems, but in the trucking industry the tendency is to reference battery cables using the standard gauge system. You may find old timers on the shop floor referring to the 0 gauge as "odd." This term is sourced from *ought*, meaning zero. For instance, a standard gauge 4/0 can be referred to as 0000 or 4-odd cable. The total length of cable used in the circuit is also a factor; the longer the cable, the greater the voltage losses that will occur. **Table 7–8** correlates high-current chassis cable standard gauge to the nearest metric equivalent.

PMI BATTERY TESTING

When verifying battery performance during a PMI, most maintenance forms require both a refractomer test (if electrolyte can be accessed) and amp/volt/resistance (AVR) test to be performed at minimum. One of the most popular AVR testers is shown in **Figure 7–28**. The VAT 33 tester is popular for its simplicity and you will find more recent digital units capable of more precise analysis introduced in later chapters of this book.

Photo Sequence 3 illustrates the procedure required to perform a refractometer measurement and connect a digital AVR. The results of a digital AVR test are displayed electronically and can usually be printed for a hard-copy record.

New-Generation Test Equipment

Batteries account for more than their share of truck downtime and at least part of the reason is that technicians often do not take the trouble to understand how they work, and what makes them fail. As a consequence, there has been a surge of more sophisticated diagnostic equipment available in the market

FIGURE 7–27 GOOD BATTERY display after performing a Midtronics EXP battery test

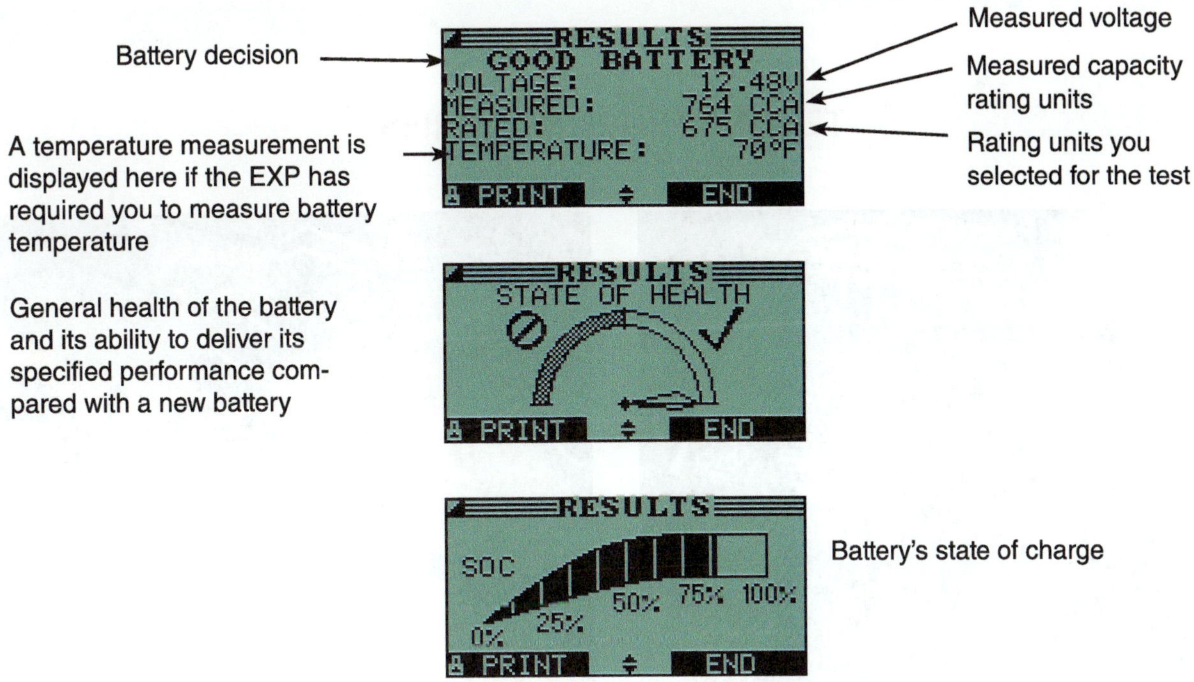

Courtesy of Midtronics, Inc.

TABLE 7–8 Chassis High-Current (Battery) Cables

SAE Standard Gauge	SAE Metric
4 gauge	20 mm
2 gauge	32 mm
1 gauge	40 mm
0 gauge	50 mm
2/0 or 00	60 mm
3/0 or 000	80 mm
4/0 or 0000	100 mm

designed to eliminate some of the guesswork in diagnosing battery-related problems. These automated battery testers use EIS and AC conductance testing described earlier in this chapter.

Figure 7–29 shows an OTC HH diagnostic battery tester while **Figure 7–30** shows a not so portable unit, the latest offering from Midtronics, which does everything their HH EXP unit does, along with being able to charge the batteries and produce status printouts. This newer equipment is designed to dumb-down diagnostics for technicians by producing an "OK" or "Not OK" conclusion to troubleshooting.

FIGURE 7–28 A Sun VAT-33 AVR tester. This unit is known for its simplicity, ease of use, and reliability.

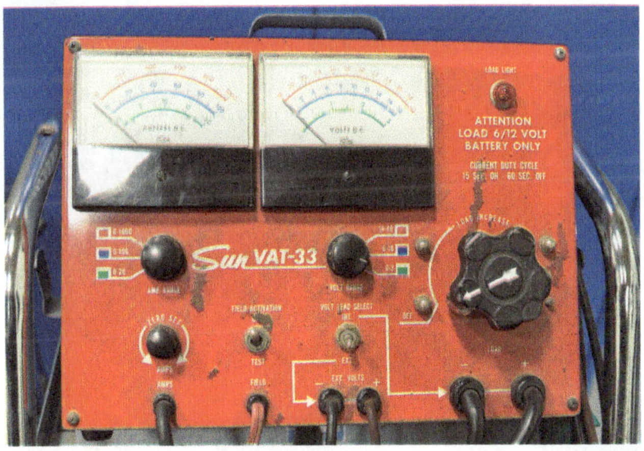

7.7 ALTERNATIVE BATTERY TECHNOLOGY

Up to this point, we have examined the various types of lead-acid batteries used in truck, bus, and heavy equipment applications. Alternative battery technology refers to all categories of batteries that do not

Photo Sequence 3

TESTING TRUCK BATTERIES

P 3–1 Dip the refractometer probe into the batter cell, wetting just the tip of the probe. Throughout this procedure, remember that battery electrolyte is corrosive.

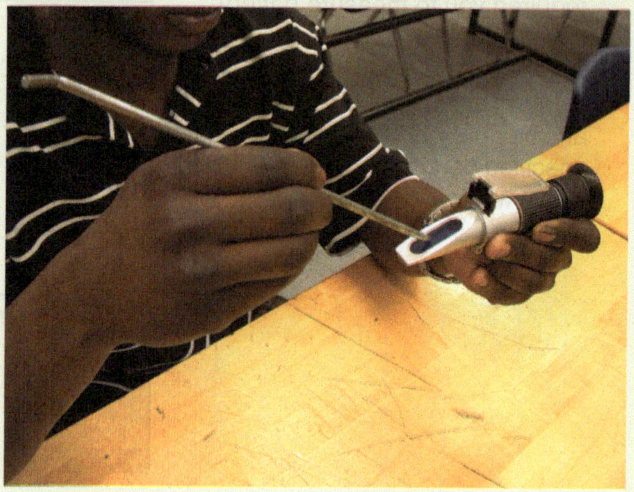

P 3–2 Deposit a drop of electrolyte onto the refractometer read-lens as shown. Close the refractive lid.

P 3–3 Raise the refractometer view scope to your eye and point the refractive window toward a light source, preferably natural. The shaded area in the view-finder correlates to a specific gravity reading.

P 3–4 Connect a digital AVR to test a battery(s) by connecting the polarized clamps as shown. Then connect the amp pickup lead with its arrow pointing in the direction of current flow.

P 3–5 A digital AVR with inductive pickup connected to a battery bank ready for a load test.

P 3–6 Turn the load test knob CW as shown to load the batteries. Typically, you will load to ½ CCA for 15 seconds and observe the voltmeter reading, which should not drop below 9.6 volts.

use a lead-acid principle. These alternative battery technologies have become important because of the increasing popularity of electric hybrid drive systems. Probably the most limiting factor to the growth of **hybrid electric vehicles (HEVs)** is the fact that current battery technology falls short of industry expectations of reliability and longevity. Nevertheless, it is becoming increasingly important for truck technicians working on pickup-and-delivery units and transit bus technicians to know something about the batteries used in electric hybrid drivetrains. Some of the first-generation commercial HEVs used banks of lead-acid batteries but these have proven to be costly. Today HEVs are usually spec'd with banks of one of the following battery categories.

FIGURE 7–29 An OTC HH battery and circuit status diagnostic tool

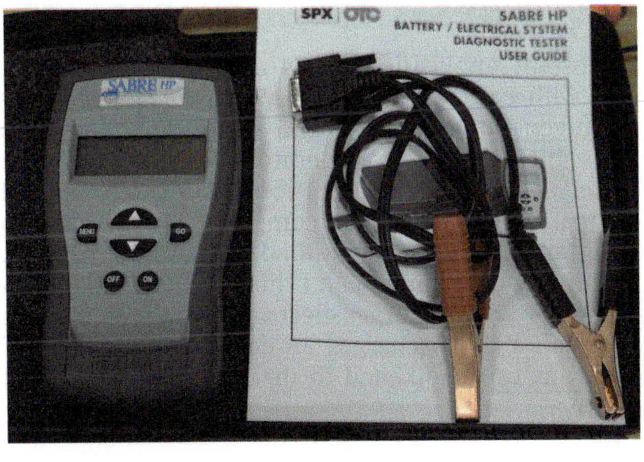

NiMH BATTERIES

At the time of writing, HEVs continue to use **nickel-metal-hydride (NiMH)** batteries, although there has been a recent shift toward lithium batteries. In an NiMH, an aqueous (H_2O base) solution of potassium hydroxide is used as the electrolyte. The electrodes used in these batteries are:

- Positive electrode: nickel oxide
- Negative electrode: a metal hydride

FIGURE 7–30 The latest battery diagnostic station manufactured by Midtronics. This unit does everything the Midtronics HH EXP does, plus charge the batteries and produce status printouts.

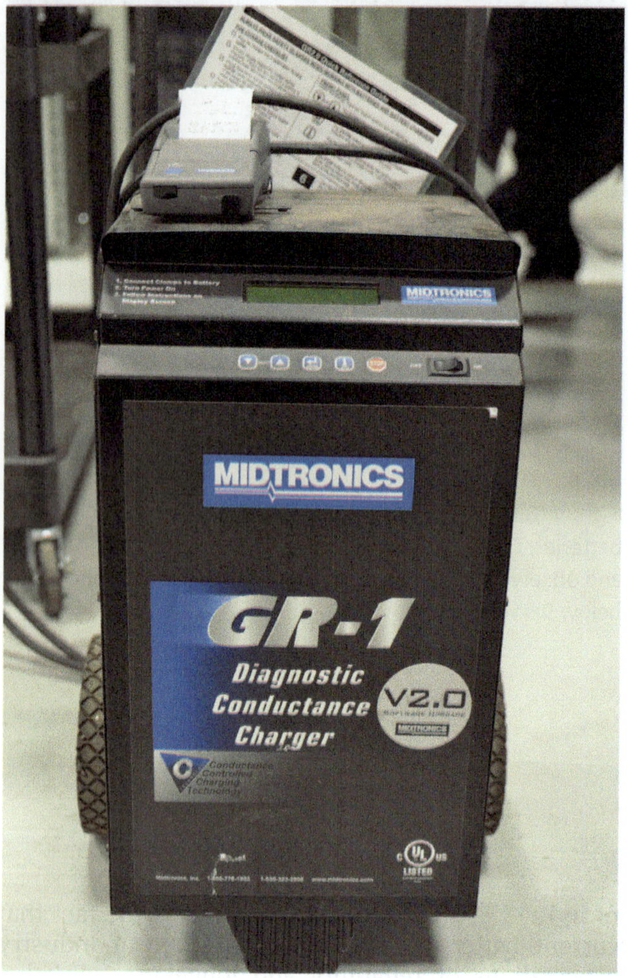

The active element in the NiMH battery is hydrogen. This is absorbed as the battery is discharged and de-absorbed as the battery is charged. NiMH batteries produce a per-cell voltage of 1.2 volts. Banks of these batteries are used in HEV transit buses.

The problem with NiMH batteries is initial high cost and longevity (life cycle) that has fallen way short of expectations. While diesel electric, hybrid drive buses have been embraced by inner city transit operators for their low emissions, quiet operation, and performance, the fact remains that their batteries fail sometimes when only 30 percent of their projected life has been attained. Replacing banks of NiMH batteries is costly.

LITHIUM-BASED BATTERIES

Lithium has the highest electrochemical potential of any current batteries and they are becoming more common in HEVs. Typically they produce a per-cell potential of between 2 and 5 volts. Their storage and power density features make them attractive for HEV applications where they are in trial development. Manufacturers of commercial HEVs are beginning to adopt lithium-based batteries in preference to NiMH. It is possible to retrofit lithium-based batteries into a unit that was originally spec'd for NiMH but this must be undertaken by the OEM. The common categories are cobalt based, **lithium-ion (Li-On)**, and **lithium-ion polymer (Li-Po)**. Lithium-base batteries must be tested in exact accordance with OEM procedure for two reasons:

- They are more volatile than lead-acid batteries and can represent a fire/explosion hazard if not properly handled.
- Most have a processor chip that can be destroyed if an incorrect diagnostic or charging procedure is applied.

7.8 ULTRACAPACITORS AND SUPERCAPACITORS

Ultracapacitors (UCs) and **supercapacitors** (the terms are synonymous) are used to store electrical potential. They are currently used in conjunction with battery storage systems in some electrically powered and hybrid electric vehicles. UCs function by physically separating the positive and negative charges, holding the charge in a manner that can be compared to static electrical buildup. The UC's major advantage is their superfast rate of charge and discharge. Another advantage is that they do not chemically and physically degrade as do batteries. On the down side, they have lower energy density than most batteries.

UC VERSUS BATTERY DISCHARGE RATES

In current applications, UCs are used in conjunction with batteries where they can be charged during regenerative braking, and rapidly discharged as the vehicle needs to accelerate. However, the way a battery can pump out electrons is severely limited compared to a UC. A good comparison is the bathtub drain analogy: if a tub is filled with water, it drains when the plug is pulled and this can be compared to how a battery discharges electrons. However, the rate at which a UC discharges an equivalent quantity of electrons can be likened to the spill rate that would

FIGURE 7–31 Operating principle of an ultracapacitor

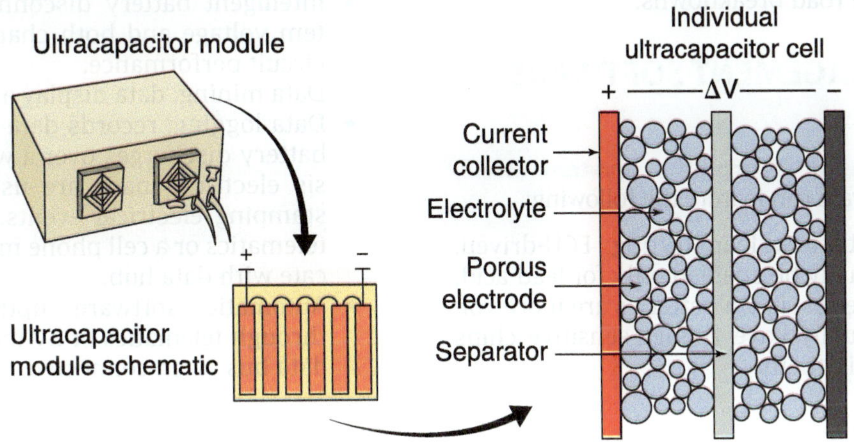

result if a basketball-sized hole is punched into the same tub. UCs can be used in parallel to batteries with the objective of prolonging battery life. This is how they are being used in hybrid electric transit buses and straddle carriers utilized in ports because they work really well in stop-start type applications.

UC Operation

An ultracapacitor functions by polarizing an electrolytic solution to store charge energy electrostatically. Although it is an electrochemical device, no chemical reactions take place. The charge accumulation is highly reversible, and ultracapacitors are said to be capable of up to 1,000,000 charge/discharge cycles providing them with extra long service life.

After a UC is loaded with a charge, a load (such as an electric vehicle's motor) can use this energy. The amount of energy stored is much greater than an equivalent standard capacitor, because the charge storage surface area is much larger. A UC can be charged in as little as 1 second with electrical storage volume that would take an equivalent lithium battery 1 hour, and a lead-acid battery 10 hours. In addition, a lead-acid battery is capable of around 3,000 load cycles versus over 1,000,000 load cycles for the UC. Modern UCs have low leakage rates, typically 1 farad per 24 hours for a 3,000-farad battery.

UC Construction

The basic material of an ultracapacitor (UC) is carbon sourced from coconut shells. The coconut shells are ground into a dust and then the activated carbon is then rolled into shape by kneading. The double-sided, plate-type cassette is the most common. A UC the size of a standard 12-ounce soda could be rated at up to 3,000 farads.

UC Operating Principle

A UC uses a pair of nonreactive porous plates, or collectors, that are suspended within electrolyte. In each UC, the applied voltage on its positive electrode attracts negative ions in the electrolyte, while the potential on the negative electrode attracts the positive ions. A dielectric separator between the two electrodes prevents the charge from moving between the two electrodes.

UCs actually store their electrical charge on an electrode. Devices such as batteries use the electrode to create—by chemical reaction—an electrical charge at the electrodes. This means that a battery's ability to store or create electrical charge is a function of the surface area of the electrode. UCs have vastly greater electrode surface area than battery electrodes to increase both the capacitance or energy storage capacity.

As a storage device, a UC depends on the microscopic charge separation at an electrochemical interface to store electrical energy. Because UC capacitance is proportional to the active electrode area, increasing that surface area increases capacitance. The electrode surface area in UCs uses activated carbon or sintered metal powder coatings. **Figure 7–31** shows the operating principle of an ultracapacitor.

7.9 BATTERY MANAGEMENT SYSTEMS

Battery and vehicle electrical power management systems are being introduced to on-highway trucks. This is regarded as good news given the disproportionate number of electrically related problems on trucks today. There are several systems available but the objectives are all similar—to reduce the incidence of no-start and on the road breakdowns.

BATTERY MANAGEMENT SOFTWARE FEATURES

Battery management information system (BMIS) software and hardware can feature the following:

- Automated battery load testing: ECU-driven. This can usually be enabled only for lead-acid, wet cell batteries. Li-On batteries are more volatile and usually have voltage-sensitive chips located within them.
- Seven-wire trailer cord tester (verifies performance in each circuit).
- Intelligent battery disconnect: monitors system voltage and both charging and cranking circuit performance.
- Data mining: data display and analysis.
- Data logging: records data such as how much battery discharges over a weekend, how chassis electrical loads are used, time and date stamping electrical events. May make use of telematics or a cell phone modem to communicate with data hub.
- Automatic software updates: may occur through telematics or by means of the chassis data bus when connected.

SUMMARY

- A battery converts chemical energy into electrical energy.
- A battery is a galvanic device.
- The storage device for electrical energy on current trucks is the lead-acid battery.
- A lead-acid battery acts as a sort of electron pump in a truck electrical circuit.
- A typical battery contains anode (positive) and cathode (negative) plates arranged in cells that are grouped in series within the battery housing. The electrolyte used in lead-acid batteries is a solution of distilled water and sulfuric acid (H_2SO_4).
- The electrolyte in a lead-acid battery is both conductive and reactive.
- During the discharge cycle of the battery, lead peroxide on the anode (positive plate) combines with electrolyte, releasing its oxygen (O_2) into the electrolyte to form H_2O, or water. Meanwhile, lead on the cathode (negative plate) reacts with electrolyte to form lead sulfate ($PbSO_4$). The result of discharging a battery is $PbSO_4$ formation on both the anode and cathode plates.
- During the charge cycle, the sulfate coating on the anode and cathode plates is drawn off and recombined into the electrolyte. Charging reestablishes the correct proportions of sulfuric acid and water in the electrolyte.
- A fully charged battery has an electrolyte solution that consists of 36 percent sulfuric acid and 64 percent pure water.
- A fully charged battery should produce a specific gravity reading of 1.260 at 80°F (27°C).
- During the charge cycle, both oxygen and hydrogen are released from electrolyte in a process known as gassing. In maintenance-free and low-maintenance batteries, the results of charge gassing are contained in a condensation chamber.
- Gel cell batteries are designed to sustain deep cycling and are used in truck electrical systems requiring an isolated battery to power auxiliary accessories during shutdown.
- Gel cell batteries should be recharged by direct connection to the battery charger.
- Only approved chargers may be used to charge gel cell batteries: These regulate the charging voltage to between 13.8V and 14.1V. Most battery chargers use a charging voltage of about 16V, which can destroy a gel cell battery.
- On trucks equipped with gel cell batteries, the voltage regulator must be set at 14.1V or less. Voltages higher than 14.2V can destroy gel cell batteries.
- Batteries are performance rated by cold cranking amps, reserve capacity, and ampere-hour rating.
- Most truck batteries are specified to a chassis electrical system by their CCA rating.
- Temperature has to be considered when evaluating the output capacity of any battery: Available cranking power significantly drops as temperature falls.
- When removing a battery from a vehicle, always disconnect the ground cable first.
- Battery electrolyte should be tested with either a refractometer or a hydrometer.
- When jump-starting vehicles, be sure to study the battery configuration on both vehicles before attempting to make electrical connections. Also switch off all the chassis electrical loads before connecting jumper cables.
- Generator charging is popular and effective, but ensure that the correct procedure is followed before using this method of fast charging.
- Conductance testing using instruments such as the Midtronics inTELLECT EXP is commonly used to assess battery serviceability.

- Battery performance can be accurately tested using a digital AVR.
- Early commercial HEVs used banks of lead-acid batteries.
- Today, the HEV battery of choice is the NiMH.
- Battery management software and hardware is becoming more common today. BMIS can test batteries, automatically disconnect batteries, test chassis electrical systems, and mine data for analysis.

REVIEW QUESTIONS

1. What is an anode?
   a. a diode
   b. a transistor
   c. a positive terminal
   d. a negative terminal
2. What is a cathode?
   a. a diode
   b. a triode
   c. a positive terminal
   d. a negative terminal
3. If two 12-volt batteries are connected in parallel, the resulting voltage would be:
   a. 6 volts
   b. 12 volts
   c. 18 volts
   d. 24 volts
4. Which of the following would result when two batteries are connected in series?
   a. Voltage doubles.
   b. Current capability doubles.
   c. Voltage and current both double.
   d. Nothing, if both batteries have the same CCA and nominal voltage.
5. Which of the following would be associated with the negative plate in a fully charged battery?
   a. H_2SO_4
   b. $PbSO_4$
   c. PbO_2
   d. Pb
6. Which of the following would be associated with the electrolyte in a lead-acid battery?
   a. H_2SO_4
   b. $PbSO_4$
   c. PbO_2
   d. Pb
7. When the term *gassing* is applied to battery performance, which gas(es) is/are being referred to?
   a. oxygen
   b. hydrogen
   c. both oxygen and hydrogen
   d. neither oxygen nor hydrogen
8. Which of the following terms describes the amount of time that a vehicle can be operated with its headlights on before battery voltage drops to 10.5V?
   a. ampere-hours rating
   b. cold cranking amps
   c. reserve capacity
   d. wattage
9. The term *deep cycling* in the context of a battery means:
   a. complete discharge followed by full recharge
   b. prolonged overcharging
   c. sulfation
   d. recombination cycling
10. Which of the following terms describes a *recombinant* battery?
    a. standard
    b. low maintenance
    c. maintenance free
    d. gel cell
11. Which of the following *must* be done when using a portable generator to blast charge discharged batteries on a vehicle?
    a. Switch on all the chassis electrical system loads.
    b. Switch off all the chassis electrical system loads.
    c. Disconnect the engine ECM.
    d. Remove the discharged batteries before connecting to the generator leads.
12. What is the recommended maximum charging rate for a heavy-duty, 12-volt battery?
    a. 12 amps
    b. 30 amps
    c. 60 amps
    d. 120 amps
13. What is the recommended maximum charging voltage for a 12-volt gel cell battery?
    a. 12 volts
    b. 12.6 volts
    c. 14.1 volts
    d. 14.8 volts
14. When observing the sight glass in the integral hydrometer in a maintenance-free battery, what should be done if it appears clear or light color?
    a. Charge the battery for 3 hours and retest.
    b. Load test the battery.
    c. Replace the hydrometer.
    d. Replace the battery.
15. What is the recommended procedure when cranking a vehicle that is reluctant to start?
    a. Crank for 30 seconds and then wait for 2 minutes before cranking again.
    b. Crank for 2 minutes and then wait for 30 seconds before cranking again.
    c. Crank for 15 seconds and then wait for 30 minutes before cranking again.
    d. Crank for 2 minutes and then wait 20 minutes before cranking again.

16. When jump-starting a truck with electronically controlled chassis components, where should the ground cable clamp be connected on the truck with the dead batteries?
    a. On the frame rail 12. inches from the ground post.
    b. On a battery ground post.
    c. On the engine cylinder block.
    d. On the alternator housing bracket.
17. Why should frequent carbon pile load testing of batteries be avoided?
    a. It deep cycles batteries, reducing service life.
    b. It produces severe gassing, which can cause explosions.
    c. It overloads the alternator.
    d. It heats up the insulated circuit wiring.
18. Technician A says that conductance testing is considered to be an accurate method of assessing battery service life. Technician B says that a disadvantage of conductance testing is that the battery electrolyte covers must be removed. Who is correct?
    a. Technician A only
    b. Technician B only
    c. both A and B
    d. neither A nor B
19. Which of the following data *must* be input for a typical automated battery tester to evaluate battery performance?
    a. electrolyte specific gravity
    b. ambient temperature
    c. battery capacity rating
    d. starter motor type
20. Technician A says that a REPLACE BATTERY prompt on an automated battery tester could be caused by poor battery terminal connections. Technician B says that a BAD CELL–REPLACE prompt always requires battery replacement. Who is correct?
    a. Technician A only
    b. Technician B only
    c. both A and B
    d. neither A nor B

8

Prerequisite: Chapter 7

CHARGING SYSTEMS

OBJECTIVES

After reading this chapter, you should be able to:

- Identify charging circuit components.
- Navigate a charging circuit schematic.
- Voltage drop test charging circuit wiring and components.
- Describe the construction of an alternator.
- Explain full-wave rectification.
- Full-field an alternator.
- Measure AC leakage in the charging circuit.
- Verify the performance of an alternator.
- Use Intelli-check™ to assess charging circuit performance.
- Disassemble and reassemble a Delco Remy 33/40 SI alternator.

KEY TERMS

A-circuit voltage regulator	diode	remote sensing technology (RST)	transformer rectifier (TR) 12/24 alternator
B-circuit voltage regulator	full-wave rectification	rotor	voltage regulator
brushes	half-wave rectification	sensing voltage	windings
delta	Intelli-check	slip rings	wye
	load dump	stator	
	remote sensing		

INTRODUCTION

A truck charging system consists of the batteries, alternator, **voltage regulator**, associated wiring, and the electrical loads of the truck. The purpose of the system is to recharge the batteries whenever necessary and to provide the current required to power the electrical components of the truck. It does this by converting a portion of the mechanical energy of the engine into electricity. A charging system can be compared to a compressed air system on a truck (**Figure 8–1**). The compressor forces air into a reservoir or tank. When a pneumatic component needs air, air is drawn from the reservoir. Eventually, the air pressure in the tank will drop below a certain level and a pressure-sensitive switch will activate the compressor. Air will be pumped into the reservoir again until cutout pressure is reached. Then the switch (governor) will unload the compressor.

POTENTIAL

In a charging system (refer to Figure 8–1 again), the battery acts as a reservoir of electrical energy. Electrical energy is drawn from it to crank the engine and to power electrical systems. The electrical "pressure" in a battery is called voltage. You can also call this "potential." When fully charged, a heavy-duty 12-volt battery has 12.6 volts potential available across its terminals. As the battery is discharged, the voltage level drops, just as the pressure of compressed air in a tank as it is used. When battery potential reaches a preset voltage level, an

FIGURE 8–1 A charging system can be compared to a compressed air system. Just as a compressor keeps air tanks full of compressed air, an alternator keeps a battery "full" of voltage or potential energy.

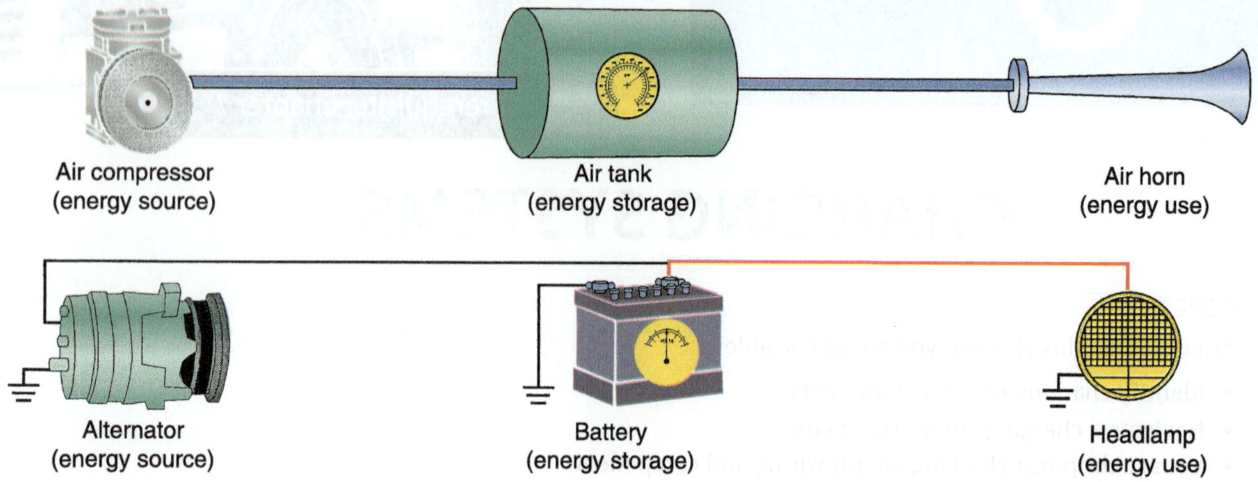

electrical switch, called a *voltage regulator*, activates the alternator.

ELECTRON PUMP

The alternator acts as an electron pump. It sends electrical current into the battery to restore the voltage level. On 12V systems, charging is typically at 14.0 to 14.2V to get the battery up to 12.6V. Just as the air compressor in our example has to push against air already in the tank, the alternator has to push against the voltage in the batteries. When the batteries are discharged, it does not take much voltage to start current flowing into the battery. A regulator might turn the alternator on and off thousands of times per second. When demand is high, the alternator will stay on longer; when low, it will freewheel. This helps fuel economy because a 100-amp alternator can consume 6 to 7 horsepower.

8.1 ALTERNATOR CONSTRUCTION

To generate electricity, the alternator uses this basic law of physics: *When magnetic lines of force move across a conductor* (such as a wire or bundle of wires), *an electrical current is produced in the conductor.* This principle is illustrated in **Figure 8–2**. Electricity can be induced in an electrical circuit either by moving the circuit wiring through the magnetic field of a magnet or by moving the magnetic field past the wiring. In both situations, the direction of electrical current flow is determined by the polarity of the magnetic field. The magnetic forces of the north pole of a magnet will force electrons to flow in one direction and the south pole of a magnet will force electrons to flow in the opposite direction. If the wiring is exposed alternately to both north and south poles, an alternating current will be produced.

Actual current flow induced depends on several factors: the strength of the magnetic field, the speed of the wire passing through the field, and the size and number of wires. In an alternator, the **rotor** provides the magnetic fields necessary to induce a current flow. The rotor spins inside a stationary coil of wires called

FIGURE 8–2 A magnetic field moving past a wire will produce or induce current flow in the wire.

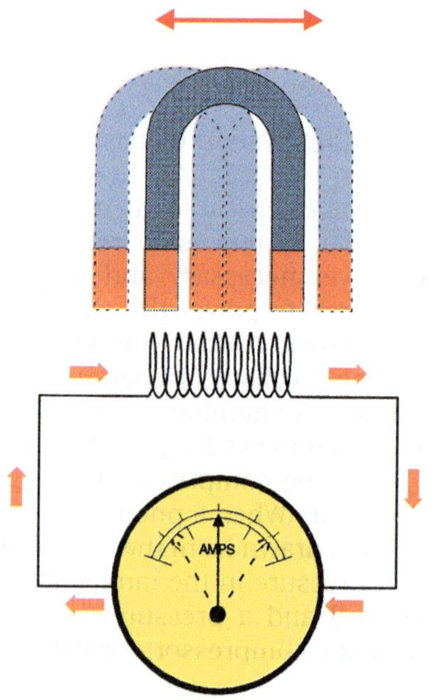

FIGURE 8–3 Components of a heavy-duty truck alternator

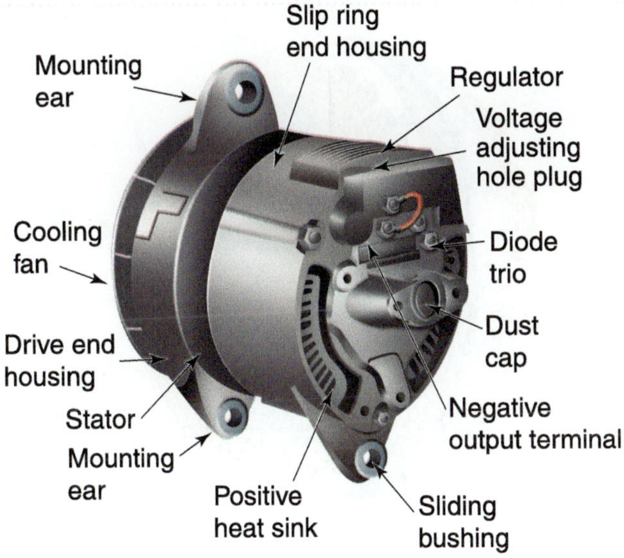

a **stator**. The moving magnetic fields induce a flow of electrons through the stator wiring that is pumped to the batteries when they are in need of recharging. The following paragraphs explain how the rotor and stator and other alternator components work together to provide charging current to the battery. The components of an alternator are shown in **Figure 8–3**.

ROTOR

The rotor is the only moving component within the alternator. It is responsible for producing the rotating magnetic field. The rotor (**Figure 8–4**) consists of a coil, two pole pieces, and a shaft. The magnetic field is produced when current flows through the coil; this coil is simply a series of **windings** wrapped around an iron core. Increasing or decreasing the current flow through the coil varies the strength of the magnetic field, which in turn defines alternator output. The current passing through the coil is called the *field current*. It is usually 3 amperes or less.

The coil is located between the interlocking pole pieces. As current flows through the coil, the core essentially becomes a magnet. The pole pieces assume the magnetic polarity of the end of the core that they touch. Thus, one pole piece has a north polarity and the other has a south polarity. The extensions on the pole pieces, known as the *fingers*, form interlacing magnetic poles. A typical rotor has fourteen poles, seven north and seven south, with the magnetic field between the pole pieces moving from the north poles to the adjacent south poles (**Figure 8–5**). The more poles a rotor has, the higher the alternator output will be.

SLIP RINGS AND BRUSHES

The wiring of the rotor coil is connected to **slip rings**. The slip rings and **brushes** conduct current to the rotor. Most alternators have two slip rings mounted

FIGURE 8–4 Cutaway view of an alternator

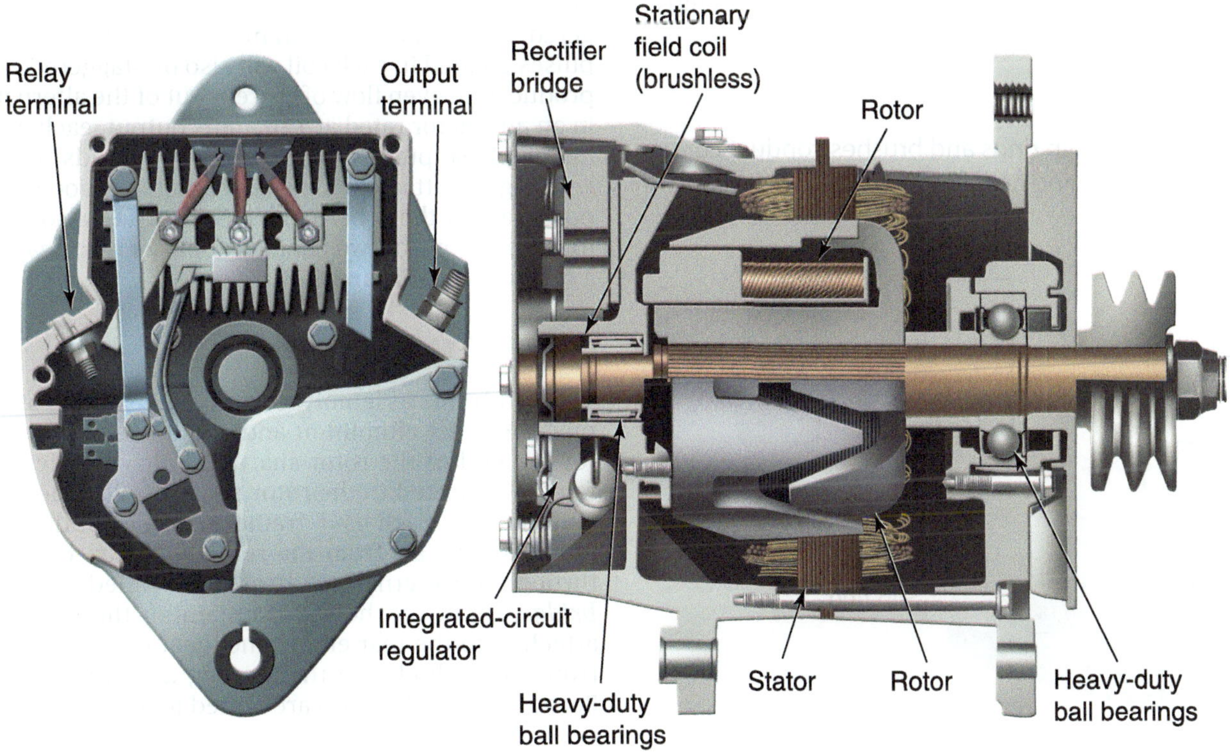

FIGURE 8–5 The interlacing fingers of the pole pieces create alternating north and south poles.

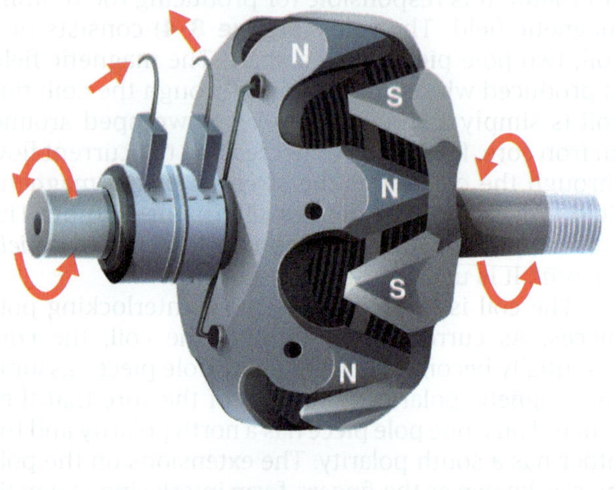

directly on the rotor shaft; they are insulated from the shaft and from each other. A spring-loaded carbon brush is located on each slip ring to carry the current to and from the rotor windings (**Figure 8–6**). Because the brushes carry only 1.5 to 3.0 amperes of current, they do not require frequent maintenance. This is in direct contrast to generator brushes, which conduct all of the generator's output current and consequently wear rapidly. Common brushed alternators are:

- Delco 24 SI with ratings up to 200 amps
- Delco 28 SI with ratings up to 200 amps
- Leece-Neville 4900 series with ratings up to 185 amps
- Bosch 9963 series with ratings up to 200 amps

FIGURE 8–6 Slip rings and brushes conduct current to the spinning field coil.

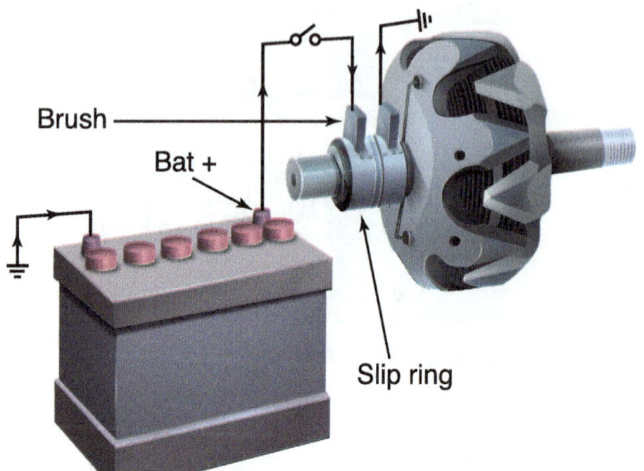

FIGURE 8–7 A stator consists of three windings looped around the inside perimeter of the alternator frame or case.

STATOR

The stator is made up of many conductors, or wires, into which the spinning rotor induces voltage. The wires are wound into slots in the alternator frame, with each wire forming several coils spaced evenly around the frame (**Figure 8–7**). There are as many coils in each wire as there are pairs of north and south rotor poles.

Windings

The wires are grouped into three separate bundles, or windings. The coils of the three windings are staggered in the alternator frame so that the electrical pulses created in each coil will also be staggered. This produces an even flow of current out of the alternator. In an alternator rated at 120 amps output, each winding or phase pumps out 40 amps. The ends of each winding are attached to separate pairs of diodes, one positive and the other negative. They are also wired together in one of two configurations: a **wye** shape or a **delta** shape (**Figure 8–8**).

END FRAME ASSEMBLY

The end frame assembly, or housing, is made of two pieces of cast aluminum and contains the bearings for the end of the rotor shaft. The drive pulley and fan are mounted to the rotor shaft outside the drive end frames. Each end frame also has built-in air ducts, so the air from the rotor shaft fan can pass through the alternator. A heat sink—called a *rectifier bridge*, or **diode** holder—containing the diodes is attached to the rear end frame; heat can pass easily from these diodes to the moving air (**Figure 8–9**). Because the end frames are bolted together and then

FIGURE 8–8 The stator windings are wired together in one of two arrangements: (A) a wye configuration; or (B) a delta configuration.

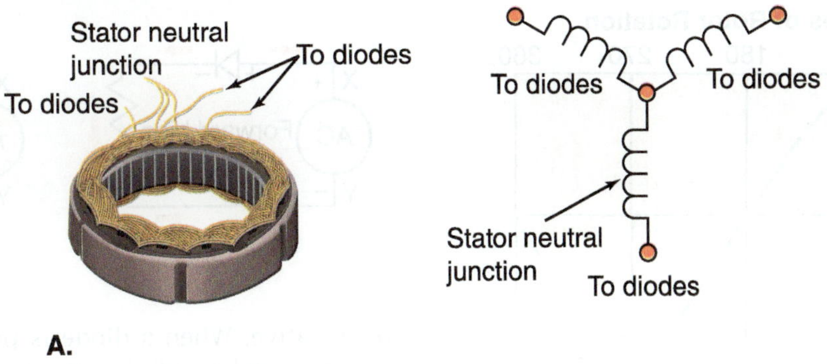

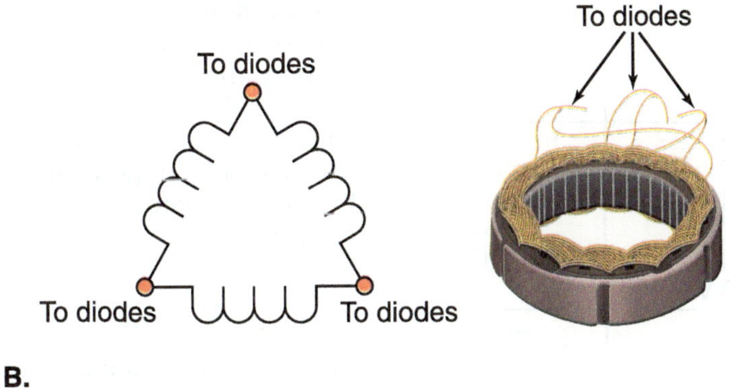

bolted directly to the engine, the end frame assembly provides the electrical ground path for many alternators. This means that anything connected to the housing that is not insulated from the housing is grounded.

FIGURE 8–9 Diodes are mounted in a heat sink to keep them cool.

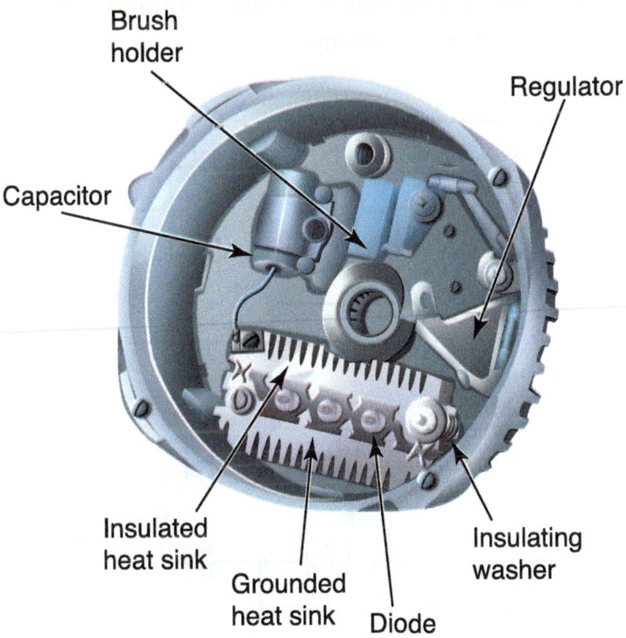

8.2 ALTERNATOR OPERATION

As the rotor, driven by a belt and pulley arrangement, rotates inside the alternator, the north and south poles (fingers) of the rotor alternately pass by the coiled windings in the stator. Only a very small air gap separates the rotor from the stator so that the windings are subjected to a maximum intensity of the magnetic force fields. As each pole alternately passes by the coils, magnetic lines of force cause electrons to flow in the wires. As a north pole passes by a coil in the winding, electrons first flow in one direction. As the next pole, having a south polarity, passes by the coil, the flow of electrons changes direction. In this way, an alternating current (AC) is produced. When viewed on an oscilloscope, the alternating pulses of current are seen as a sine wave (**Figure 8–10**). The positive side of the wave is produced by a north pole and the negative side of the wave is produced by a south pole. Because a truck alternator has three windings staggered around the rotor, a three-phase sine wave is produced.

FIGURE 8–10 Alternating current is seen as a sine wave on an oscilloscope.

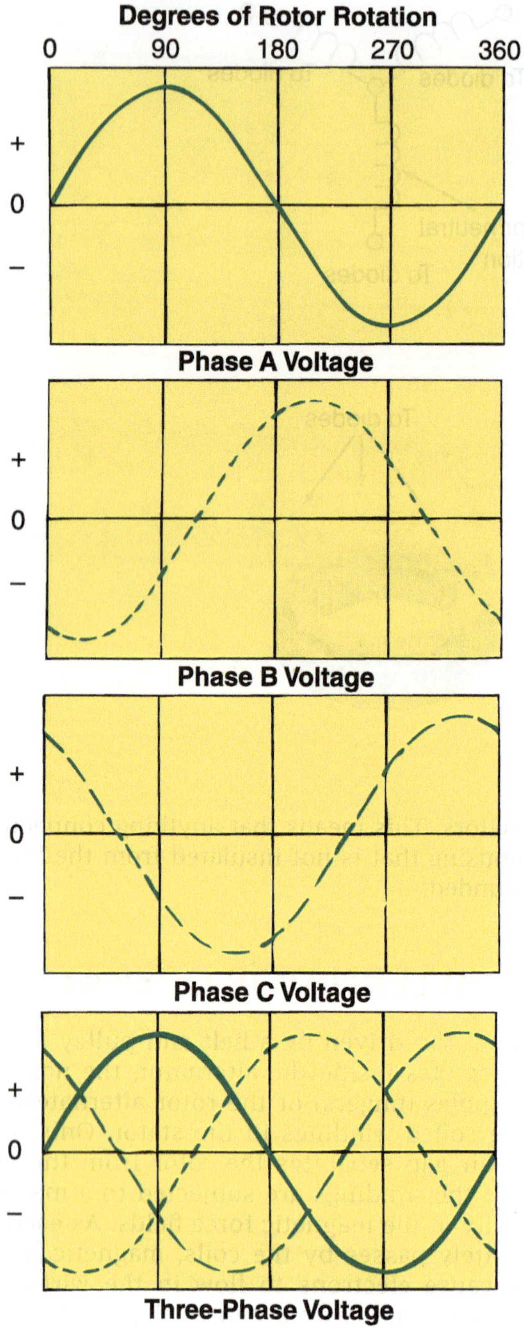

AC TO DC

A battery can only be charged using direct current (DC). Truck accessories also are powered by DC. In order for the alternator to provide the electricity required, the AC it produces must be converted, or rectified, to DC. While generators accomplish this using a mechanical commutator and brushes, alternators achieve this electronically with diodes. A *diode* can be thought of as an electrical check valve. It permits current to flow in only one direction—positive

FIGURE 8–11 One diode in an AC circuit results in half-wave rectification. Only the positive half of the sine wave is allowed to pass.

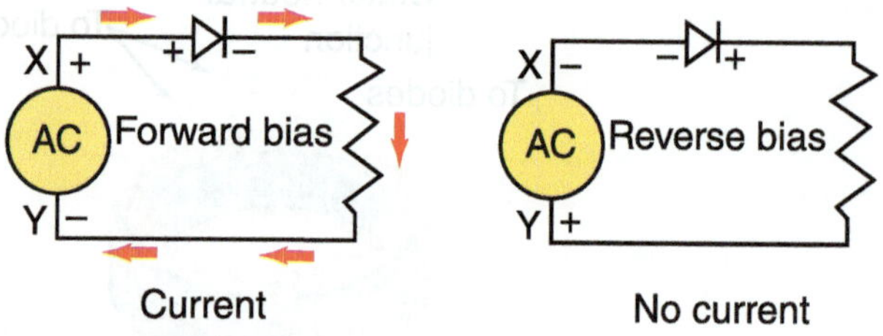

or negative. When a diode is placed in a simple AC circuit, one-half of the current is blocked. In other words, the current can flow from X to Y as shown in **Figure 8–11**, but it cannot flow from Y to X. When the voltage reverses at the start of the next rotor revolution, the current again can pass from X to Y, but not back. This type of output is not efficient because current would be available only half of the time. Because only 50 percent of the AC voltage produced by the alternator is being converted to DC, this is known as **half-wave rectification**.

FULL-WAVE RECTIFICATION

By adding more diodes to the circuit, the full wave can be rectified to DC. In **Figure 8–12A**, current flows from X to Y. It flows from X, through diode 2, through the load, through diode 3, and then to Y. In **Figure 8–12B**, current flows from Y to X. It flows

FIGURE 8–12 By adding more diodes to the circuit, current can now flow (A) from X to Y, and (B) from Y to X, resulting in rectifying both halves of the sine wave to DC current.

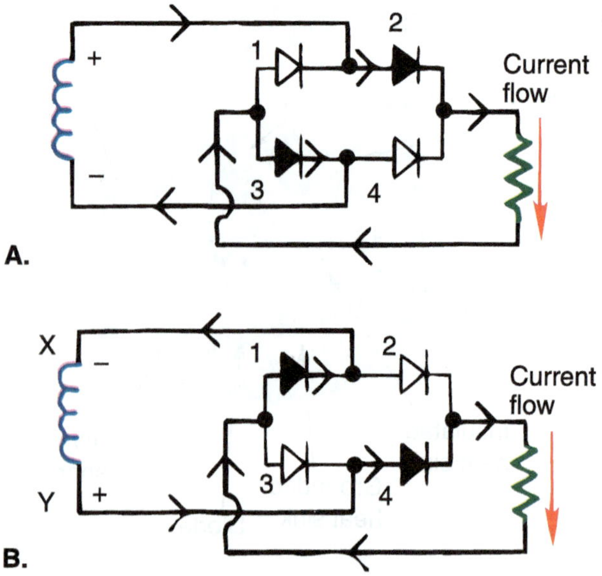

FIGURE 8–13 A three-phrase alternator has six diodes to provide full rectification.

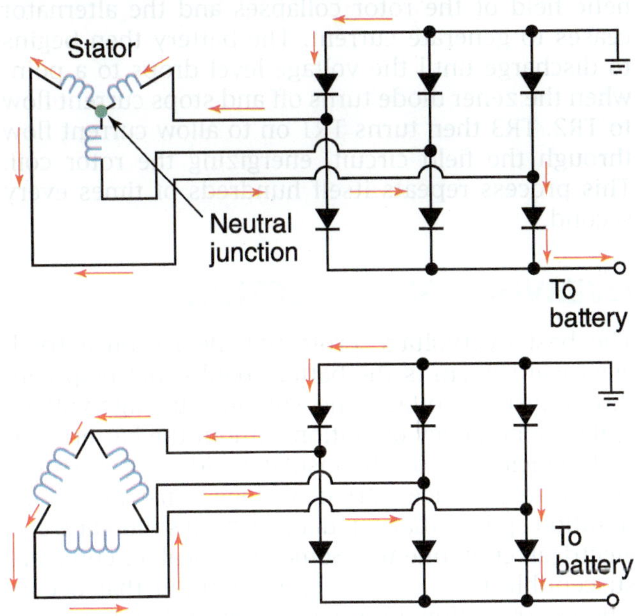

from Y, through diode 4, through the load, through diode 1, and back to X. In both cases, the current flows through the load in the same direction. Because all of the AC is now rectified, this is known as **full-wave rectification**. However, there remain brief moments when the current flow is zero; therefore, most alternators use three windings and six diodes to produce overlapping current pulses, thereby ensuring that the output is never zero.

THREE-PHASE ALTERNATOR

In a heavy-duty three-phase alternator, there are three coils and six diodes (**Figure 8–13**). At any time, two of the windings will be in series and the third winding will be neutral, doing nothing. Depending on the combination of coils and the direction of current flow, one positive diode will rectify current flowing in one direction and a negative diode will rectify the current flowing in the opposite direction. This is true for both wye and delta configurations. Many heavy-duty truck alternators use a delta arrangement. The windings in a delta stator are arranged in parallel rather than in series circuits. The parallel paths permit more current to flow through the diodes, thereby increasing the output of the alternator.

BRUSHLESS ALTERNATORS

Many alternators do not use brushes to deliver voltage to the field windings of the rotor. In this type of alternator, the field windings do not rotate with the rotor. They are held stationary while the rotor turns around them. The rotor itself retains sufficient residual magnetism to energize the stator when the truck engine is first started. Part of the voltage induced in the stator windings is then diverted to the field windings, which energizes the electromagnetic field. As the strength of the stationary field windings increases, the magnetic field of the rotor also increases and stator output reaches its specified potential.

Many current alternators use a brushless design because they produce better longevity. Examples are the Delco 33 SI, 34 SI, 35 SI, 36 SI, and 40 SI alternators and the high-capacity 50DN used in many motor coaches. Examples of competing brushless alternators are the Leece-Neville 320, the Bosch T1 series, and Denso 130A. Truck brushless alternators can be rated up to 300 amperes output in 12V systems. The output ratings in current trucks are necessarily higher because of the more complex electrical and electronic circuits. The heavy-duty, oil-cooled 50DN alternator is designed to supply the electrical charging requirements of motor coaches.

VOLTAGE REGULATORS

For more than 50 years prior to the introduction of solid-state voltage regulators in the early 1970s, mechanical regulators were used to control alternator (or generator) output. Like so many other mechanical components that have been replaced with electronic parts, the mechanical voltage regulator is slow and imprecise when compared to today's transistorized regulators. A wiring schematic of a solid-state voltage regulator is shown in **Figure 8–14**.

A voltage regulator defines the amount of current produced by the alternator and the voltage level in the charging circuit. It does this by turning the field circuit off and on. Without a voltage regulator, the batteries would be overcharged and the voltage level in the electrical system would rise to a point where lights would burn out and fuses and fusible links would blow. Controlling the voltage level is particularly important as electronic components are added to heavy-duty trucks. Microprocessors and electronic sensors and switches are easily damaged by voltage spikes and high-voltage levels.

Sensing Voltage

The voltage regulator receives battery voltage as an input. This is called the **sensing voltage**; it allows the regulator to sense and monitor the battery voltage level. When the battery voltage rises to a specified level (approximately 13.5 volts), the regulator will turn the field current off. When the battery voltage drops below a set level, the regulator will again turn the field circuit back on. This on/off cycling of the field circuit takes place hundreds of times each second.

Remote Sensing

Due to resistance in the battery cables, the voltage output by the alternator drops by the time it arrives at the battery. This voltage drop can be as much as 0.5 volt. When an alternator is equipped with **remote sensing** a second sensing wire reads actual voltage at the battery pack. This second wire signals the alternator to compensate for any voltage drop. The result is that battery charge times are reduced by half. Remy uses the term **remote sensing technology (RMT)** to describe their specific system but almost all alternators use this.

Temperature Factors

The ability of a battery to accept a charge varies with temperature. A cold battery needs a greater charging current to bring it up to capacity. In warm or hot weather, less current is needed. Therefore, the voltage regulator uses a thermistor to vary the voltage level at which the regulator will switch the field circuit on and off.

Types of Field Circuits

The field circuit, which is controlled by the voltage regulator, might be one of two types. If it is an **A-circuit voltage regulator**, it is positioned on the ground side of the rotor. Battery voltage is picked up by the field circuit inside the alternator. The regulator turns the field circuit off and on by controlling a ground. A **B-circuit voltage regulator** is positioned on the feed side to the alternator. Battery voltage is fed through the regulator to the field circuit, which is then grounded in the alternator. The regulator turns the field circuit off and on by controlling the current flow from the battery to the field circuit. Most voltage regulators are mounted either on or in the alternator. Some charging systems might have a regulator mounted separately from the alternator. A typical electronic regulator is an arrangement of transistors, diodes, zener diodes, capacitors, resistors, and thermistors.

Most alternators with internal regulators use current generated by the alternator stator to energize the field circuit. The regulator controls a ground to turn the field circuit off or on. Because the stator generates AC current, a set of three diodes is used to rectify the current to DC. These three diodes are often referred to as a *diode trio.* A separate sensing circuit delivers battery voltage to the regulator to control the on/off cycle of the alternator.

Figure 8–14 demonstrates how current flows through a typical voltage regulator. Battery current is delivered to the regulator through the sensing circuit. While the battery voltage is low, it turns on transistor 3 (TR3), which then turns on transistor 1 (TR1) (see Figure 8–14A). When TR1 is on, current flows from the stator, through the diode trio, through the field coil, through TR1, and to ground.

When battery current builds to approximately 13.5 to 14 volts, the *zener diode* (D1) trips to switch transistor 2 (TR2) on. This diverts current away from TR3, which then turns off. When TR3 turns off, TR1 must turn off also. This blocks the flow of current through the field circuit (see Figure 8–14B). The magnetic field of the rotor collapses and the alternator ceases to generate current. The battery then begins to discharge until the voltage level drops to a point when the zener diode turns off and stops current flow to TR2. TR3 then turns TR1 on to allow current flow through the field circuit, energizing the rotor coil. This process repeats itself hundreds of times every second.

OVERVOLTAGE PROTECTION

The best overvoltage protection device on a truck electrical system is the battery banks that help condition system voltage and smooth out high-voltage spikes. Emergency operation without the batteries or with deficiencies on the insulated side of the battery circuit can produce extreme voltage transients (short-duration spikes) that can damage both solid-state and electrical components. Some electrical circuits are susceptible to **load dump**, a condition that occurs when current flow to a major electrical consumer is opened, creating a voltage spike.

Zener diodes can be installed in the rectifier bridge to limit high-energy voltage spikes. These can be used elsewhere in the electrical system to protect voltage-sensitive components. Other solid-state overvoltage protection devices can be used to protect the alternator ground circuit for voltage spikes. These are designed to short the alternator to ground at the excitation winding.

CAUTION

Truck alternators are not provided with reverse-polarity protection. Reverse-polarity connections such as those caused by jump-starting can destroy alternator diodes and numerous other chassis solid-state devices.

TR ALTERNATORS

Transformer rectifier (TR) 12/24 alternators have been used since the 1980s on trucks requiring the use of 24V cranking systems. They were introduced to replace the electromechanical series-parallel switch used on early 12/24V systems. Transformer rectifier alternators are identified by the suffix TR appearing after the alternator numeric code. A TR alternator has both 12V- and 24V-terminal studs located on the TR module, which is bolted onto the alternator.

In a 12/24 TR system, the batteries are arranged in banks—commonly two for chassis voltage requirements and another pair for cranking

FIGURE 8–14 Schematic of voltage regulator operation

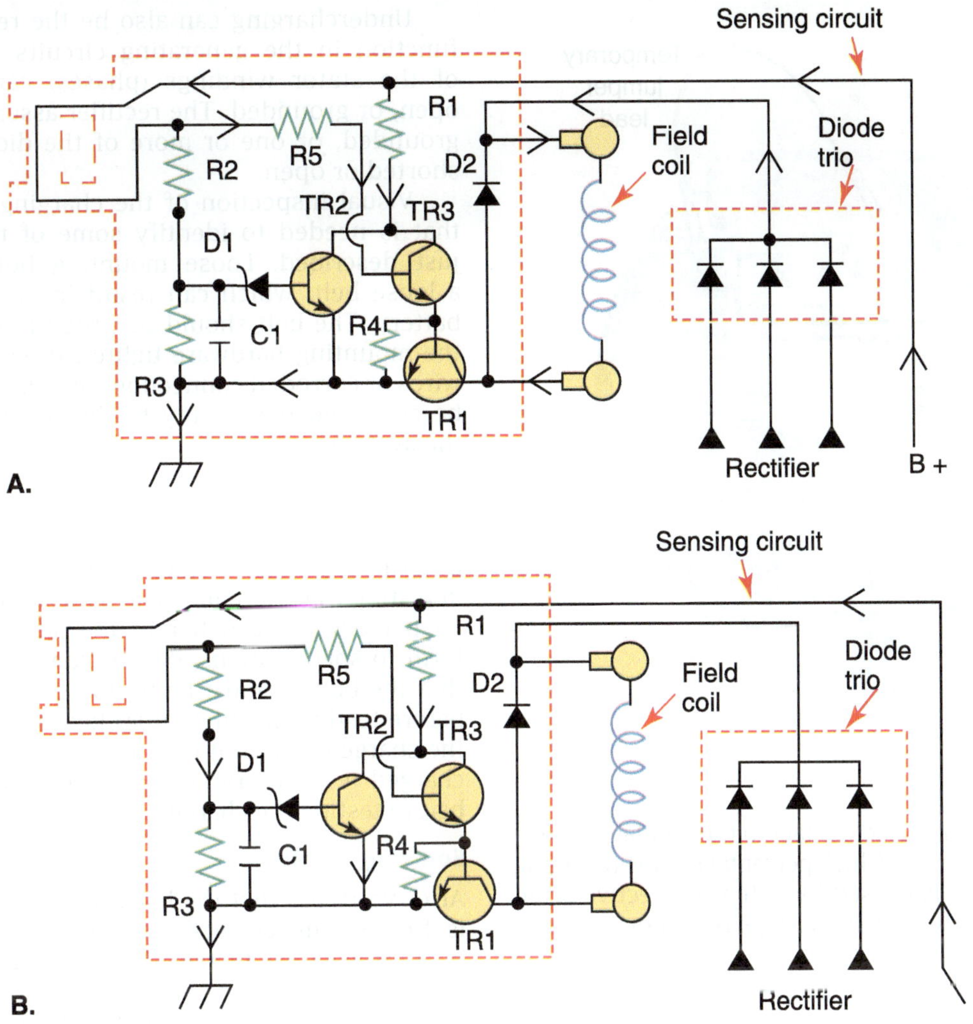

only; the TR unit switches both banks in series for cranking the engine. After the engine is running, the battery bank that handles the chassis voltage requirements is charged by the alternator as it would be with any other truck electrical system. The bank that is used solely for cranking is trickle-charged at low amperage during normal operation. The need for TR alternators has been largely eliminated by the introduction of high-torque, 12V cranking motors.

8.3 CHARGING SYSTEM FAILURES AND TESTING

A malfunction in the charging system results in either an overcharged battery or an undercharged battery.

SHOP TALK

Alternators that energize the field circuit with current produced by the stator windings rely on residual magnetism in the rotor to initially energize the stator when the engine is being started. During handling or repair, this residual magnetism can be lost. It must be restored before testing the system. This is done by connecting a jumper wire between the diode trio terminal and the alternator output terminal as shown in **Figure 8–15**.

OVERCHARGING

An overcharged battery will produce water loss, eventually resulting in hardened plates and the inability to accept a charge. Overcharging can be caused by one or a combination of the following:

- Defective battery
- Defective or improperly adjusted regulator

FIGURE 8–15 Restoring residual magnetism to the rotor

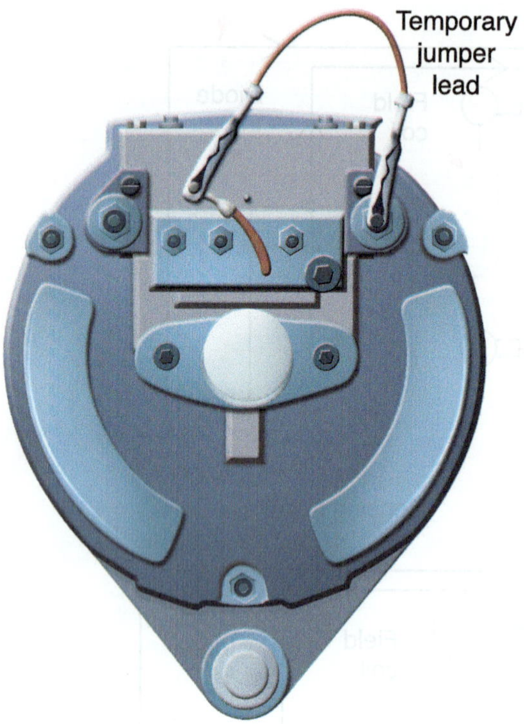

- Poor sensing lead contact to the regulator or rectifier assembly. A high percentage of overcharging problems is caused by either a defective or an improperly adjusted voltage regulator.

UNDERCHARGING

An undercharged battery will result in slow cranking speeds and a low specific gravity of the electrolyte. Undercharging is not always caused by a defect in the alternator. Undercharging has various causes: a loose drive belt; loose, broken, corroded, or dirty terminals on either the battery or alternator; undersize wiring between the alternator and the battery; or a defective battery that will not accept a charge.

Undercharging can also be caused by one or a combination of the following defects in the alternator field circuit:

- Poor contacts between the regulator and carbon brushes
- Defective diode trio
- No residual magnetism in the rotor
- Defective or improperly adjusted regulator
- Damaged or worn brushes
- Damaged or worn slip rings
- Poor connection between the slip ring assembly and the field coil leads
- Shorted, open, or grounded rotor coil
- Open remote sensing wire
- Problems with alternator drive pulleys or belts.

Undercharging can also be the result of a malfunction in the generating circuits. One or more of the stator windings (phases) can be shorted, open, or grounded. The rectifier assembly might be grounded, or one or more of the diodes might be shorted or open.

Visual inspection of the charging system is all that is needed to identify some of the conditions just described. Loose mounting bolts will cause a loose belt, which can result in an undercharged battery. The belt should be properly tensioned and the mounting hardware tightened. Loose or broken wires or corroded connections are visible and should be corrected before any further testing is done to the system.

CHARGING SYSTEM TESTING

Determining battery state of charge is the first step in testing the alternator. The battery must be at least 75 percent charged before the alternator will perform to specifications. (Some manufacturers specify that the battery must be 95 to 100 percent charged before testing the alternator.) Recharge the battery if the charge is low; if it will not accept a charge, a fully charged test battery should be installed in its place before testing the charging system.

Sourcing Problems

After verifying that the battery is fully charged and is functioning correctly, the charging system can be performance tested. There are basically three tests, although some manufacturers might specify more. First, the output of the alternator is tested. If the output is below specifications, the voltage regulator is bypassed and battery current is wired directly to the field circuit of the rotor. This is called *full-fielding* the rotor. If this corrects the problem, the fault is in the regulator; if the output remains low with the regulator bypassed by full-fielding, the alternator might be defective. However, before condemning the alternator or regulator, voltage drop tests should be performed on the system wiring to determine if high resistance could be the cause of the charging problem.

The following procedures are general in nature and will not apply to every alternator. The major differences from model to model are the meter test points and the specifications. Keep in mind that not all vehicle manufacturers require all of these tests to be performed, and others may suggest even more. When undertaking any test, refer to the vehicle manufacturer's specifications; even the most accurate test results are meaningless if they are not correlated to the correct specs. Before testing an alternator, either go online and obtain a printout of the original equipment manufacturer (OEM) test specifications or locate the specifications in the service manual.

ALTERNATOR-OUTPUT TESTING

To test the maximum output of an alternator, follow these steps:

1. Disconnect the ground battery cable from the alternator.
2. Using a digital multimeter (DMM) equipped with an inductive pickup (amp clamp), position the clamp jaws over the alternator battery wire. Ensure that the amp clamp directional arrow conforms with the direction of current flow from the alternator to the battery. Set the DMM to current test mode. Next install a voltmeter and a carbon pile across the terminals of the battery (**Figure 8–16**).
3. Start the engine and run it fast enough to obtain maximum output from the alternator (typically above 1,500 rpm).
4. Turn on all accessories and increase the load on the carbon pile until the voltage in the system drops to 12.7V. This will cause the regulator to send full-fielded voltage to the rotor, maximizing alternator output.

FIGURE 8–16 Test connections for an ammeter, voltmeter, and carbon pile.

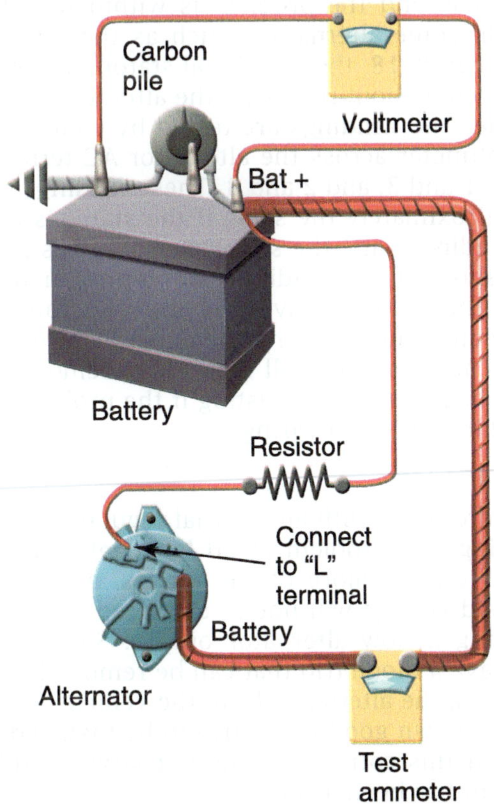

5. Compare the amperage reading with the manufacturer's specifications. If the output is within 10 percent of rated amperage (15 amps for a 34 SI rated at 150 amps), the alternator is good. If the output is more than 10 percent below specs, full-field test the alternator to identify the cause of the problem.

SHOP TALK

For two reasons, an inductive (amp) clamp ammeter should be used instead of a series ammeter. First, alternator current greatly exceeds the DMM maximum amperage measurement. Second, it is clamped over the battery cable instead of connected in series with it, meaning that the cable is not required to be disconnected.

FULL-FIELD TESTING THE ALTERNATOR

By applying full battery voltage directly to the field windings in the rotor, it can be determined whether or not the regulator is the cause of an undercharging condition. There are two variations of this procedure that apply to alternators with external regulators. If the field circuit is grounded through the regulator (an A circuit), the regulator is disconnected from the field terminal on the alternator and a jumper is connected between the terminal and a ground. If the alternator receives battery voltage through the regulator (a B circuit), the regulator is disconnected from the field (F) terminal and a jumper is connected to the terminal and to the insulated battery terminal. In either case, the regulator circuit is bypassed completely and full battery voltage is available to the rotor.

Now repeat the procedure outlined for testing the alternator output. With the load applied, observe whether or not the alternator output rises to rated amperage. If it does, the alternator is functioning correctly and the regulator must be replaced. If it does not, the alternator must be further tested to determine the cause of undercharging.

CAUTION

When testing the output of a full-fielded alternator, carefully observe the rise in system voltage. Because the current output is not regulated, battery voltage can quickly rise to an excessive level that is sufficient to overheat the batteries, causing electrolyte to spew from the vent holes and possibly damage sensitive electronic components. Do not allow system voltage to rise above 15V.

SHOP TALK

Some alternators with remote mounted electronic regulators are connected to the regulators by a wiring harness and a multipin connector. Full-fielding the field circuit is accomplished by removing the connector from the regulator and connecting a jumper wire between two pins (terminals) in the harness connector (consult the OEM service literature to correctly identify the pin assignments). Doing so bypasses the regulator, sending battery current directly to the field circuit.

FULL-FIELDING INTERNALLY REGULATED ALTERNATORS

Full-field testing is not possible on some alternators with internal voltage regulators. To isolate and test the regulator on these types of alternators, the alternator must be removed from the truck and disassembled. Other internally regulated alternators have a hole through which the field circuit can be tested. The test will require the use of a short jumper with insulated clips and a stiff paper clip wire or a $^{1}/_{32}$-inch drill bit (**Figure 8–17**).

1. With the engine off and all electrical accessories turned off, measure the voltage across the battery terminals. Make a note of the reading.
2. Start the engine and run it at the speed necessary to generate full alternator output.
3. Connect a short jumper wire to the alternator negative output terminal and to the shank of the straightened clip wire or drill bit.
4. Insert the wire or bit into the full-field access hole as far as it will go and make a note of the voltage reading. If a fault in the regulator or diode trio is causing the undercharged condition, the alternator output should climb to within 10 percent of its rated output.

FIGURE 8–17 Full-fielding an alternator with an integral voltage regulator

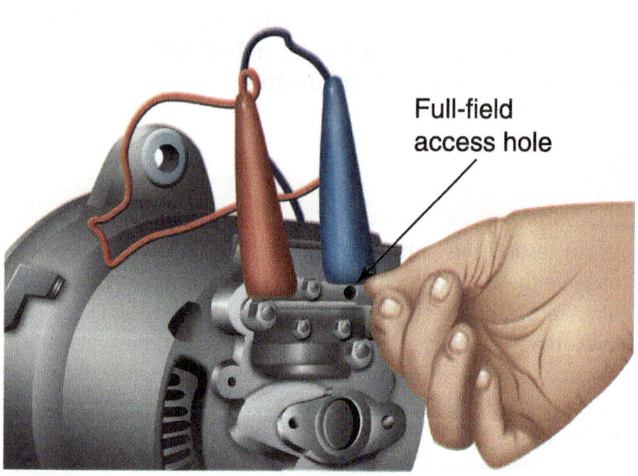

FIGURE 8–18 AC terminals of the stator windings

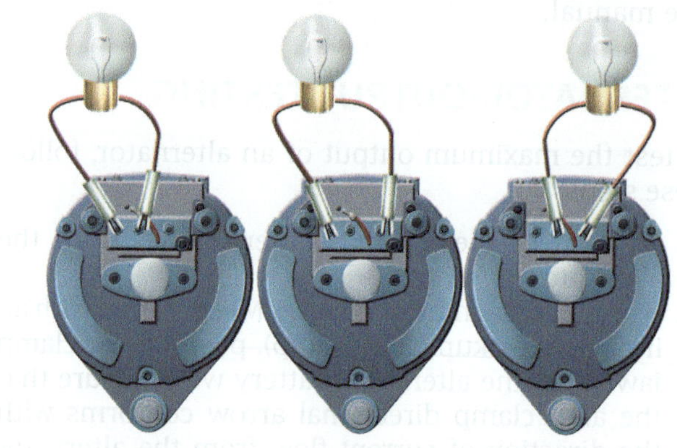

CAUTION

Never full-field an alternator without applying an electrical load to the batteries. Whenever the voltage regulator is bypassed, there is nothing to control peak alternator output; this can cause voltage spikes that can damage both electrical and electronic components.

STATOR WINDING TESTING

If full-fielding the alternator does not solve the undercharging condition, the regulator is probably okay and the problem is within the alternator. On some alternators, such as the one shown in **Figure 8–18**, the individual stator windings can be tested without removing the alternator from the vehicle. The windings are tested by connecting an AC voltmeter across the alternator AC terminals 1 and 2, 1 and 3, and 2 and 3. These readings should be approximately the same if the stator is okay. If the readings vary, the stator probably has an open or a short. If the readings are approximately the same, the stator is okay and some other fault exists in the alternator, causing the undercharging condition. The alternator will have to be removed from the vehicle for further testing if the problem is not found in the circuit wiring.

Diode Trio Test

If an alternator with an internal regulator is undercharging, the problem could be a defective diode trio. If one or more of the diodes are defective, full-field voltage will not be delivered to the rotor, resulting in low alternator output. Some alternators have a diode trio that can be removed without removing the alternator from the vehicle. This permits a known good diode trio to be installed in its place. If this corrects the undercharging problem, the fault has been located.

Worn Brushes

Undercharging might also be caused by worn or corroded brushes or contact pads in the regulator. On some alternators, the regulator can be removed to provide access to the brushes and the contact pads. If the brushes appear burned, cracked, or broken, or if they are worn to a length of $^3/_{16}$-inch or less, they should be replaced. Brushes should also be replaced if the shunt lead inside the brush spring is broken. If the brushes are okay, the contact caps on the brushes and the contact pads on the regulator should be cleaned using 600-grit (or finer) grade sandpaper or crocus cloth.

REGULATOR CIRCUIT TESTING

In addition to providing current to the field windings of the rotor, the voltage regulator must also keep the system voltage within a predetermined range, typically 13.5 to 14.5 on a 12V system. Note that on systems using gel cell batteries, this should never exceed 14.1V. The regulator must reduce the output of the alternator when necessary to keep system voltage below the set level. If it does not, the battery will be overcharged. To test system voltage regulation, follow this procedure:

1. Connect a volt-amp tester to the batteries as described earlier under "Alternator-Output Testing."
2. Start the engine and run it at the speed necessary to achieve full output from the alternator.
3. Observe the voltage readings and current output. As voltage approaches 13.8 to 14.5 volts, alternator output should slowly decrease. When the batteries are fully charged, the output of the alternator should be low. If alternator output remains high after system voltage reaches its specified peak, the voltage in the system will rise above 15 volts. Batteries will overheat, begin to gas, and might even start spewing electrolyte through vent holes. The problem might be a defective regulator, a short in the field circuit, or high resistance in the wiring. For example, unwanted resistance in the sensing wire will cause the voltage regulator to read system voltage at a lower value than it actually is. This would result in the alternator generating current to fully charged batteries.

To test the ground of the regulator, install a voltmeter between the case of the regulator (if externally mounted) and the battery ground. Start the engine and measure the voltage difference between the two points. With the engine running, there should not be any difference between the two ground points. The voltmeter should read zero.

If the ground side tests okay, measure the voltage supplied to the regulator by the sensing circuit. It should be equal to battery voltage. If resistance is high on the ground side or if voltage is low on the sensing (supply) side of the regulator, the problem might be loose or corroded connections or a partial open in the circuit. If resistance and supply voltage meet specifications, the overcharging condition is the fault of either the alternator or the regulator.

CHARGING CIRCUIT TESTING

If there is excessive resistance in the charging circuit, the alternator might not fully charge the battery during peak load periods. The design of the alternator limits current output to a maximum level. High resistance in the circuit will prevent full current from reaching the battery. During peak load demand, the batteries will not be fully charged and, over time, can be damaged.

To test voltage drop in the charging circuit, connect the volt-amp tester as explained earlier and load the battery to full alternator output. Then connect a voltmeter from the output terminal of the alternator and the insulated terminal of the battery. The voltage drop between the two points should be low (0.2 volt on alternators rated at 14V and 0.5 volt on alternators rated at 28V are typical maximums). If the voltage drop is higher than the manufacturer's specifications, there might be loose or corroded connections in the circuit or a fusible link might be degenerating.

The ground side of the circuit should also be tested. Move the voltmeter leads to the alternator casing and the battery ground pole. The voltage drop in this case should be zero. If not, the alternator mounting might be loose or corrosion might be built up between the casing and mounting bracket.

AC Leakage Test

An important test of the diode rectifier bridge is the AC leakage test. Over time, the diode bridge can start to "leak" AC current to ground. Typically AC leakage should not exceed 0.3-volt AC, but check the OEM maximum specification. Use a DMM and switch to AC voltage. Place one test lead on the alternator insulated terminal and connect the other lead to chassis ground. Record the voltage specification produced when the engine is run at above 1,500 rpm. A within-specification-but-high reading can be an indication of an alternator that will fail sooner rather than later.

INTELLI-CHECK

In an effort to both simplify and improve the accuracy of electrical system diagnostics, manufacturers have recently introduced troubleshooting instruments that enable reasonably accurate diagnoses (see the description of battery conductance testing in the Chapter 7). One such instrument is the Delco Remy **Intelli-check**. This handheld diagnostic tool is inexpensive and can be used by technicians short on electrical diagnostic skills because it can only display one of five status conditions. The tool is easy to use and can be used as first-level diagnosis by technicians. Experience has shown, however, that if one of Intelli-check's four failure status conditions is displayed, the system should be checked over by a technician with good electrical diagnostic skills rather than replacing components on the basis of the status lights.

Intelli-check Procedure

Intelli-check should first be plugged into the two-way connector attached to the alternator output and ground terminal. Next, the engine should be started and run with no chassis electrical loads turned on. Rev up the engine once to high idle. Then turn on the lights and the heater blower to high, and once again rev up the engine again to high idle. This completes the test, and the charging circuit condition will be displayed by illuminating one of five light-emitting diodes (LEDs). Status conditions displayed are:

- Good
- Overcharge
- Partial charge
- No charge
- Low battery voltage

Figure 8–19 shows a Delco Remy Intelli-check being connected for a charging system test. Note the status lights. Each is illuminated by LEDs to display charging circuit status.

FIGURE 8–19 Intelli-check connected for a charging circuit test

FIGURE 8–20 Alternator diagnosis: scope graphics identification.

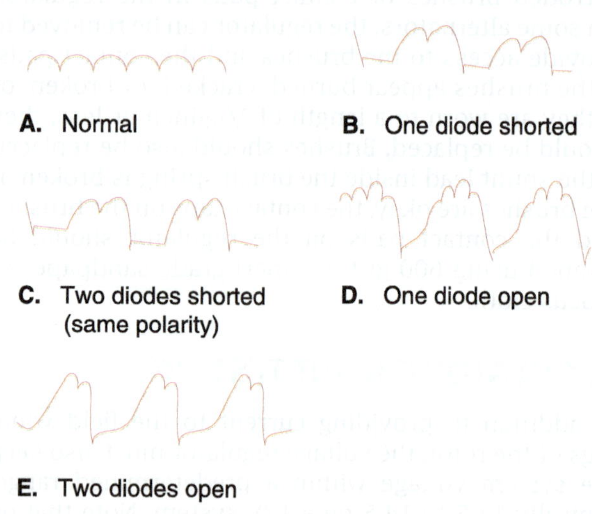

SCOPE TESTING

Oscilloscopes can be used to diagnose alternator conditions. The scope used may either be a full-sized oscilloscope or a good-quality, handheld lab scope. Use the scope manufacturer's instructions to make the connections. **Figure 8–20** shows some typical wave patterns produced.

8.4 ALTERNATOR REBUILD

The reality of today's truck shop is that truck technicians seldom are called upon to rebuild alternators. It is our task to diagnose a failure and then, if necessary, replace a defective alternator with a rebuilt unit. However, it is good learning practice for trainee technicians to understand what is required to overhaul an alternator because it can greatly improve diagnostic skills. As we go through the disassembly procedure of a commonly used alternator, the procedure used to test each subcomponent will be covered in detail, with the objective of improving diagnostic skills.

The alternator we are going to use as our model for overhaul is the brushless heavy-duty Delco 33/34 SI unit. The 33/34 SI is used for electrical systems requiring up to 300-amp alternators but is more commonly rated at half that amperage. **Figure 8–21** is an internal wiring schematic of this alternator that can be used for troubleshooting purposes.

Alternator Disassembly and Testing

The procedure for disassembling a Delco Remy 33/34 SI alternator is outlined in sequential steps here and

FIGURE 8–21 Electronics wiring diagram for a Delco Remy 33/34 SI brushless alternator.

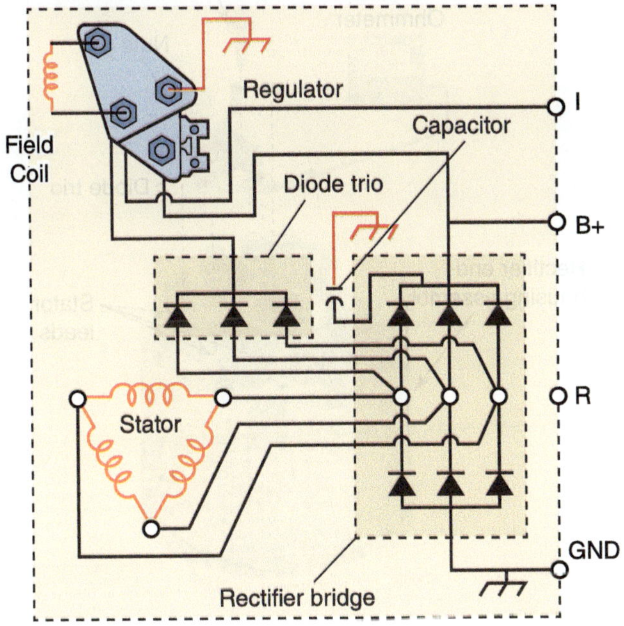

includes instructions on how to test each subcomponent. An exploded view of this alternator is shown in **Figure 8–22**.

1. Remove the endplate screws to expose the electronics compartment (see **Figure 8–23**).
2. Check for evidence of contamination and corrosion. Look for any obviously shorted or grounded wires, or any loose connections. Clean and dry the electronics compartment. When undertaking the tests that follow, remember that the protective coating on the electronic components and connectors is an insulator that must be scraped away to make an electrical contact.
3. Test field coil: Disconnect the field leads from the regulator. Use an ohmmeter to check the field coil resistance to specification as shown in **Figure 8–24**; the usual spec window is 1.4 to 2.1 ohms at 80°F (27°C) and the field coil should be replaced if this specification is not met.
4. Test diode trio: Remove the diode trio by removing the three nuts and the stator. Use DMM in diode test mode placing its negative lead on the regulator bridge strap, and the positive lead on

FIGURE 8–22 Exploded view of a Delco Remy 33/34 SI alternator

1. Housing, rectifier end
2. Bearing, outer race
3. Bearing, inner race
4. Frame, drive end
5. Bearing, ball
6. Regulator
7. Diode trio
8. Rectifier bridge
9. Field coil & support
10. Rotor assembly
11. Stator assembly
12. Capacitor
13. Fan
14. Pulley (not shown)
15. Cover plate
16. Retainer, bearing
17. Bushing, rectifier end
18. Package, output terminal
19. Package, relay terminal
20. Package, indicator light terminal
21. Package, hardware

A. Field coil leads (2)
B. Stator leads (3)
C. Insulator, grommet

18. Output Terminal Package
181. Nut, hex head (2)
182. Washer, flat (2)
183. Insulator
184. Insulator, housing
185. Insulator, connector
186. Strap, output
187. Stud, output
19. Relay Terminal Package
191. Nut, hex head
192. Washer, lock
193. Stud, relay

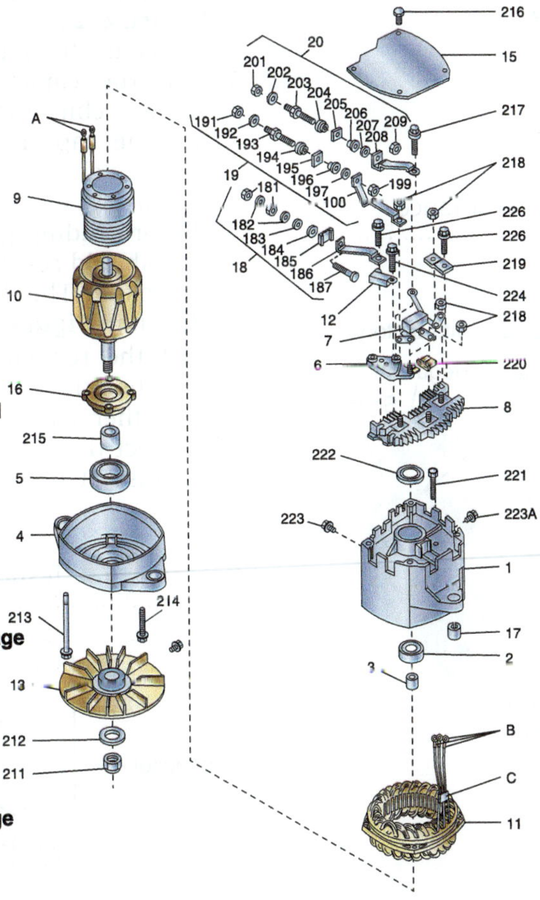

194. Insulator, terminal
195. Washer (square)
196. Insulator, housing
197. Washer, flat
198. Strap, relay
199. Nut, hex flanged
20. Indicator Light Terminal Package
201. Nut, Hex Head
202. Washer, lock
203. Stud, indicator light
204. Insulator, terminal
205. Washer (square)
206. Insulator, housing
207. Washer, flat
208. Strap, indicator
209. Nut, hex flanged
21. Hardware Package
211. Nut, lock
212. Washer, flat
213. Bolt, thru (4)
214. Screw, retainer (4)
215. Collar, shaft
216. Screw, cover plate (4)
217. Screw, insulated hex head (regulator) (2)
218. Nut, hex head (4)
219. Connector, bridge to regulator stud
220. Cover, terminal
221. Screw, hex head, field coil attch (4)
222. Plug, rectifier housing
223. Screw and lockwasher assembly
224. Screw, ground (3)
225. Cap, pin terminal
226. Screw, insulated hex head for rectifier (2).

FIGURE 8–23 Electronics compartment of a 33/34 SI alternator

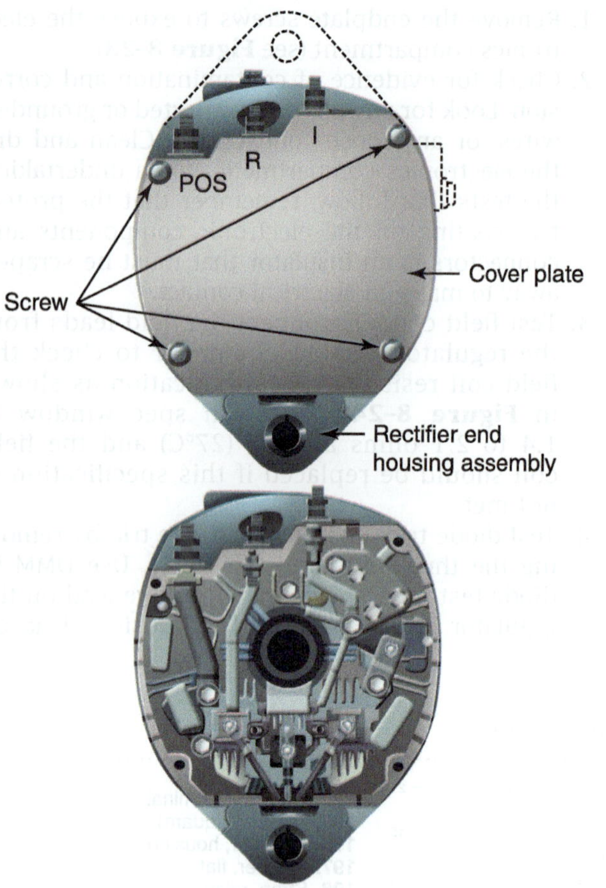

FIGURE 8–24 Checking field coil resistance

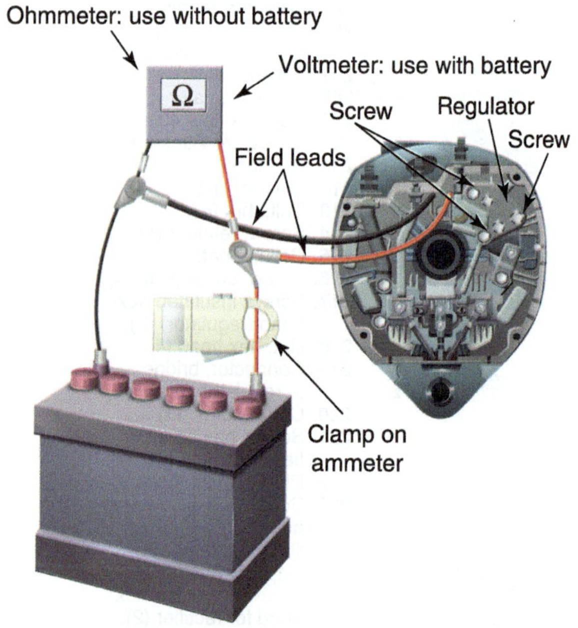

FIGURE 8–25 Diode trio test

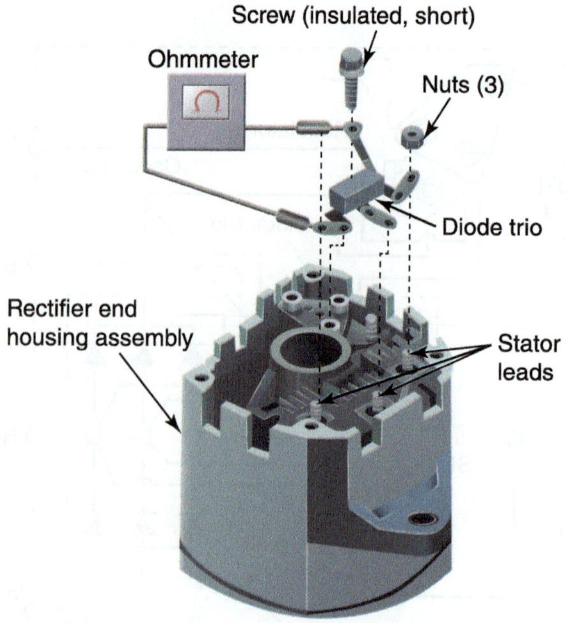

each rectifier bridge strap in turn; all should indicate continuity. Next, reverse the ohmmeter leads and repeat the tests; all should indicate an open condition. The test lead locations are shown in **Figure 8–25**. If the readings are good, the diode trio is functioning; replace if it fails.

5. Test the capacitor: Remove the attachment screws, chip away the varnish seal and slide out the capacitor. Use a DMM in diode test mode to check for a shorted capacitor by touching the DMM leads on to the capacitor lead and the protruding grounding tab; an open condition should result. Continuity indicates a failed capacitor. The DMM test lead locations are shown in **Figure 8–26**.
6. Test the regulator: Remove the sensing lead nut, the bridge to regulator stud connector, and any other screws not previously removed. Disconnect the "I" terminal sensing strap. Discard the insulated screw if the insulation fractures.

FIGURE 8–26 Testing the capacitor

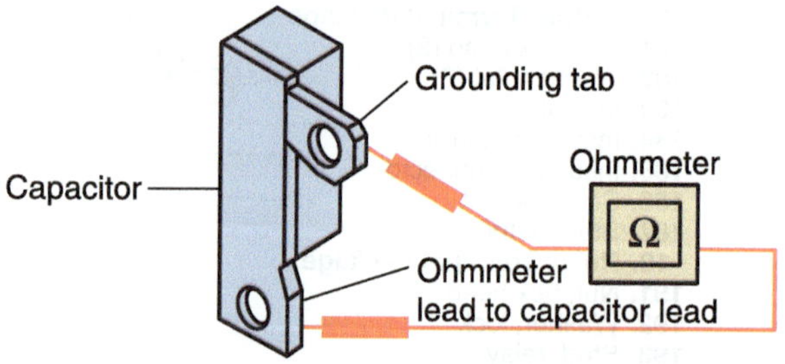

FIGURE 8–27 Removing the voltage regulator

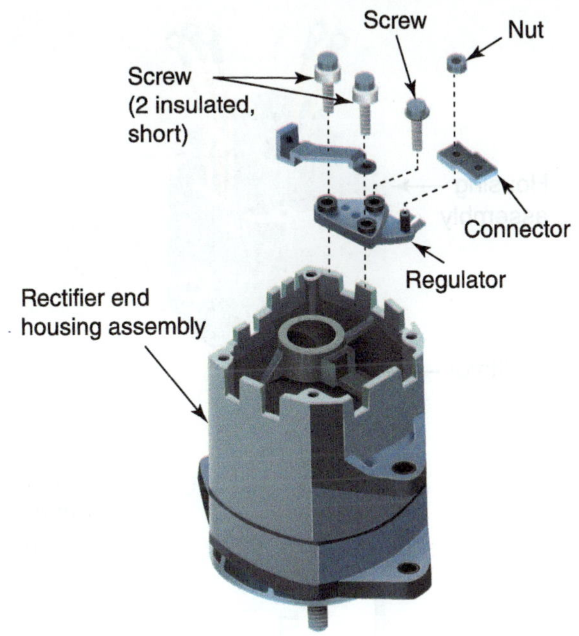

Check the regulator on an approved SI regulator tester; this will indicate one of two conditions, "good" or "bad." Replace the regulator if necessary referencing **Figure 8–27**.

7. Test the rectifier bridge: This can be checked without removing from the housing using a DMM in diode test mode and referencing **Figure 8–28**. Disconnect the terminals from the phase studs; the "R" terminal and output terminal straps can be raised about an inch (2.5 mm) above the rectifier to facilitate removal. Place the DMM negative test lead on the grounded heat sink; next contact the DMM positive test lead to the metal diode clips that surround the three threaded studs. All three readings should indicate continuity. Next, repeat the checks using the insulated (positive) heat sink in place of the grounded heat sink; all three readings should indicate continuity. Finally, switch the DMM test leads and repeat the tests; all three readings should indicate an open condition. If any of the preceding tests fail, replace the rectifier bridge.
8. Test stator continuity: Disconnect the three stator phase leads from the diode bridge studs. Use the DMM in diode test mode to test the stator windings by placing one DMM test lead onto one of the stator phase connections as shown in **Figure 8–29** and check for continuity to the other two stator phase connectors: there should be continuity to both. If there is no continuity, replace the stator.
9. Check for stator grounds: Use the DMM in diode test mode and touch one lead to a stator phase connector and the other to a clean metal ground on the alternator housing; it should indicate an open condition. Repeat the test for each of the stator phase contacts and replace the stator if a phase is grounded.
10. Shorted stator windings: It is not possible to detect shorted stator windings with standard shop equipment. However, if the alternator is being checked for low output and all the preceding checks are as specified, shorted stator windings may be the cause.

FIGURE 8–28 Testing the rectifier bridge

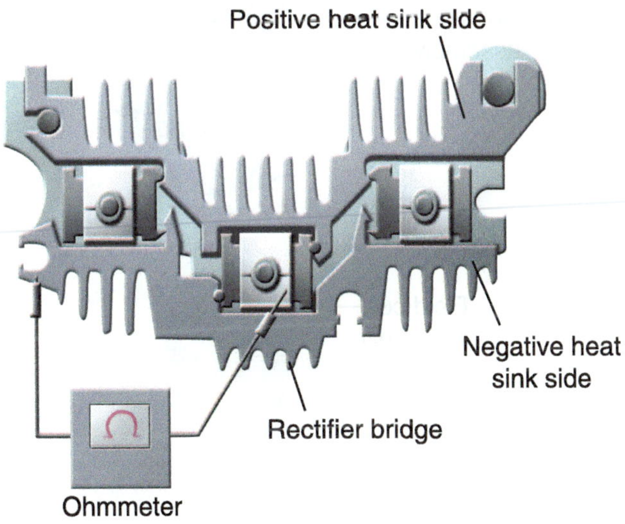

FIGURE 8–29 Checking the stator windings

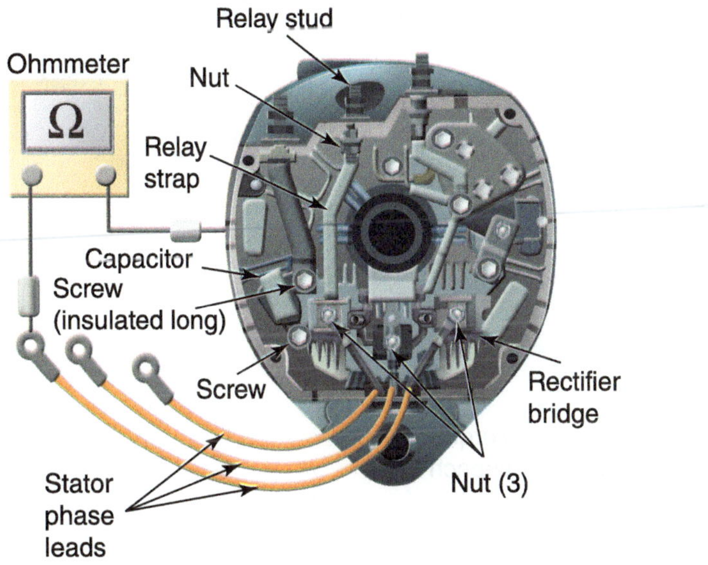

11. During final disassembly, take care not to damage exposed stator and field windings by scratching or scraping the insulation or a short or ground problem could result.
12. Separate the end frame assembly from the rectifier end housing using a 5⁄16-inch socket to hold the end of the shaft while removing the shaft nut. Then separate the washer, pulley, and fan as shown in **Figure 8–30**.
13. Remove the four through bolts. Then carefully separate the drive end housing from the stator and rectifier end housing as shown in **Figure 8–31**. Note that the rectifier bridge must always be removed before attempting to remove the stator. Ensure that the three stator phase leads are disconnected from the diode bridge studs in the electronics compartment. Carefully separate the stator and housing, guiding the stator phase leads and grommet through the housing.
14. Remove the field coil from the rectifier end housing by separating the four attaching screws. Lift the coil and its support from the housing, guiding the field leads through the hole as shown in **Figure 8–32**.
15. Remove the rectifier end bearing using a small screwdriver to pry the plug from the housing. Use an arbor press (do not attempt to hammer it out) to push the bearing from the housing.

FIGURE 8–30 Removing the shaft nut, pulley, and fan

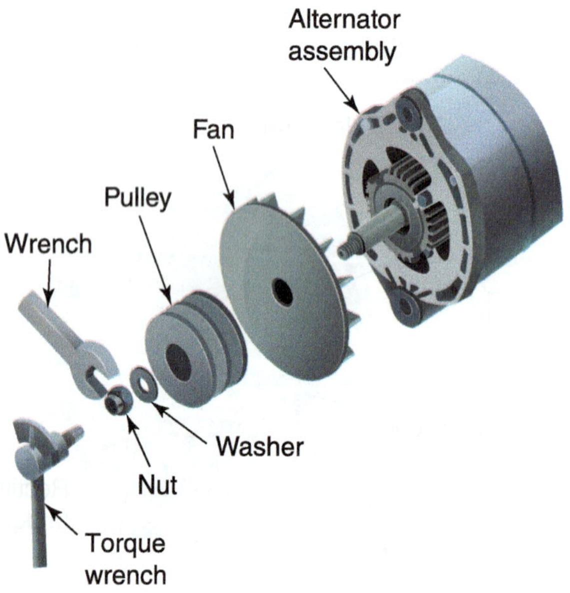

FIGURE 8–31 Separating the housing sections

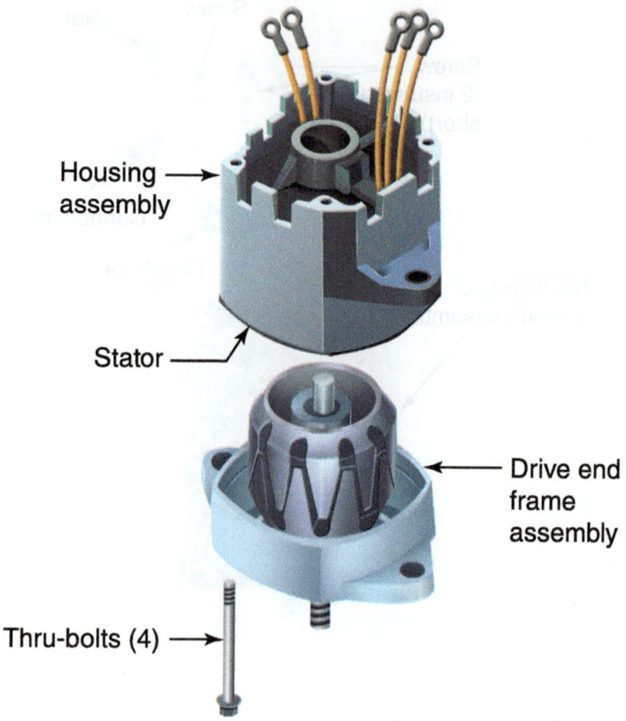

FIGURE 8–32 Removing the field coil

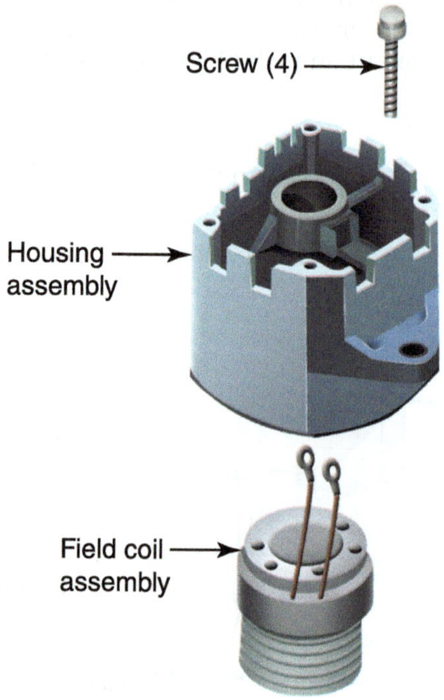

16. If the inner race appears to be worn, use a puller to separate it from the rotor shaft.
17. If the rotor or drive end bearing require replacement, remove the bearing retainer screws from the outside of the housing. Lift the rotor and bearing from the housing, then separate the bearing from the rotor shaft. If the inside collar is worn or damaged, separate this from the rotor shaft.

CAUTION

Avoid using a hammer to separate alternator components because they are easily damaged: use pullers and arbor presses.

FIGURE 8–33 Restoring residual magnetism.

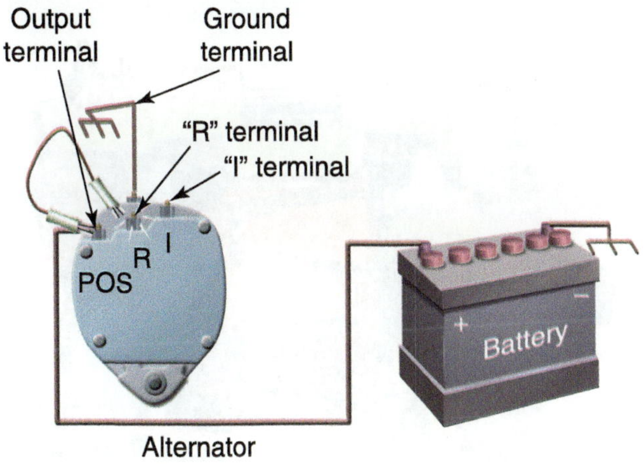

ALTERNATOR REASSEMBLY

Alternator reassembly essentially reverses the disassembly procedure but the OEM service literature should be consulted in order to avoid damaging the components and identifying the torque specifications. In the case of the family of Delco Remy 33/34 SI alternators we used as our example for disassembly, the appropriate service manual is Delco Remy 1G-500.

Alternator Bench Test Procedure

The bench test procedure outlined in this section requires the use of an alternator test stand rated at 5 HP and capable of 5000 rpm. The alternator to be tested is specified for a 12-volt system.

1. Mount the alternator to the test bench and connect it to a fully charged battery.
2. Make the electrical connections as shown in **Figure 8–33**; ensure that the carbon pile is switched off before connecting it to the battery posts. The connections shown in Figure 8–33 assume that the alternator is specified for a negative ground system.
3. Run the alternator rpm up to the specified cold output speed, typically 5,000 rpm, and observe the voltage reading which should rise above battery voltage (V-Bat). If it does not, there is no alternator output; check to see if residual magnetism in the rotor has been lost. If the voltage rises above 15.5 volts, the voltage is uncontrolled and the test should be halted and the voltage regulator checked.
4. If the alternator has passed step #3, continue to run it at 5,000 rpm and turn on the carbon pile load; adjust the load to obtain the highest current output reading on the ammeter. If the ammeter reading is within 15 amps of its specified rating (this is recorded on the alternator ID plate, not in the service literature), it can be regarded as functioning properly. If not, the alternator should be disassembled and the subcomponents retested as outlined in the disassembly section.

Restoring Residual Magnetism

To restore the residual magnetism in an alternator rotor with an "R" or "I" terminal, the alternator ground terminal should be connected to battery ground, either directly or through the test stand wiring circuit, referencing Figure 8–33. If the alternator is on the test bench, ensure the carbon pile leads are disconnected. Connect a jumper lead to battery positive and momentarily make contact (like striking a match) with the alternator "R" or "I" terminal. This should be sufficient to restore the residual magnetism in the rotor. The bench test procedure can then be repeated.

33/34 SI SPECIFICATIONS

When performance testing 33/34 SI alternators, the following data should be close to the specifications:

Drive rotation: can be either CW or CCW
Field current draw at 80°F (27°C):
amps .... 5.5–9.0 amps
volts 12.0
Output check for 135 amp rated version at 80°F (27°C):
amps @ 1,800 rpm .... 52 amps
amps @ 5,000 rpm ...... 135 amps

FIGURE 8–34 Alternator mounting: (A) hinge mount; and (B) pad mount

A.

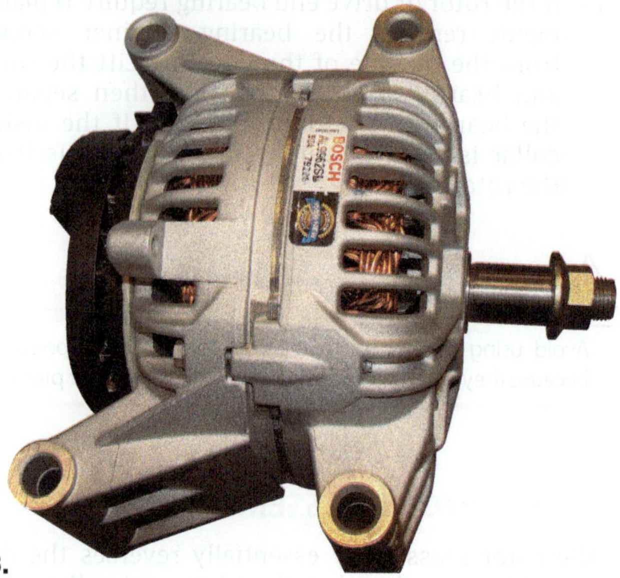

B.

ALTERNATOR MOUNTINGS

There are two common methods of mounting alternators to the engine frame:

- Hinge mount
- Pad mount

Pad mount alternators are relatively new. The pad mount provides more balanced support for the alternator frame as the torque loads change. **Figure 8–34** shows hinge and pad mount versions of a Delco 28 SI alternator.

SUMMARY

- A truck charging system consists of batteries, an alternator, a voltage regulator, associated wiring, and the electrical loads of the chassis.
- The purpose of the charging system is to recharge the batteries whenever necessary and to provide the current required to power the electrical components on the truck chassis.
- A malfunction in the charging system results in either undercharged or overcharged batteries.
- The subcomponents of an alternator consist of a stator, a rotor, slip rings, brushes, and a rectifier.
- A magnetic field is established in the rotor windings, and this is used to induce current flow in a stationary stator.
- Slip rings are used to conduct current to the rotor to establish a magnetic field.
- Because the brushes in slip rings conduct very low current to the rotor windings, they significantly outlast the brushes used in now-obsolete generators.
- The alternator rectifier is located in the end frame assembly.
- Most current truck alternators use a delta configuration of diodes in the rectifier to achieve full-wave rectification.
- Current truck alternators use solid-state, electronic voltage regulators.
- TR-type alternators use electronic switching in 12/24V systems, eliminating the need for electromechanical series-parallel switches.
- When the voltage regulator is shorted out of the circuit, the alternator is full-fielded and will produce maximum output.
- The *sensing voltage* of the charging system should be battery voltage at any given moment of operation.
- AC leakage testing can verify the performance of the diodes in the rectifier and help predict imminent alternator failures.
- Scope testing can graphically identify rectifier solid-state component failures.
- Delco Remy's Intelli-check tool enables technicians to make a rapid test of charging system performance.

- When disassembling any alternator it is good practice to test each subcomponent as it is removed; this will avoid any surprises when reassembling it.
- After reassembling an alternator, it may be necessary to reestablish its residual magnetism before it will function.
- It is good practice to bench test an alternator before returning it to service.

REVIEW QUESTIONS

1. What part of an alternator produces the magnetic field?
   a. stator
   b. rotor
   c. brushes
   d. poles
2. A typical rotor contains:
   a. fourteen north poles and fourteen south poles.
   b. fourteen north poles or fourteen south poles.
   c. seven north poles and seven south poles.
   d. a pair of poles, one north and one south.
3. What is used in an alternator to rectify AC to DC?
   a. a stator
   b. a regulator
   c. diodes
   d. a rotor
4. What instrument should be used to measure alternator current output?
   a. an ammeter
   b. an ohmmeter
   c. a diode tester
   d. a voltmeter
5. How much horsepower is required to drive a 100-amp alternator?
   a. None, the engine is turning anyway.
   b. 4 horsepower.
   c. 6–7 horsepower.
   d. 25–30 horsepower.
6. What is the function of slip rings in an alternator?
   a. to conduct current to the rotor
   b. to conduct current to the stator
   c. to conduct current to the armature
   d. to act as bearings to support the rotor
7. What is the moving component in an alternator?
   a. the stator
   b. slip rings
   c. the rotor
   d. the diode trio
8. Why do the brushes in an alternator last much longer than brushes in a DC generator?
   a. They conduct much less current.
   b. They conduct much more current.
   c. They rotate at slower speeds.
   d. They rotate at faster speeds.
9. How many windings are in a typical alternator stator?
   a. one
   b. two
   c. three
   d. six
10. What is used to control the current output of an alternator?
    a. a voltage regulator
    b. a storage battery
    c. a resistor
    d. a two-pole switch
11. How many diodes are used in a typical alternator rectifier bridge?
    a. one
    b. three
    c. six
    d. twelve
12. When a diode is in reverse bias, what is happening to current flow through it?
    a. Current is blocked.
    b. Current is converted to AC.
    c. Current passes through it.
    d. Only the positive sine wave passes through it.
13. What type of diode bridge arrangement is more common in alternators used in heavy truck applications?
    a. delta
    b. wye
    c. half-wave rectification
    d. zener diodes
14. What component is used to switch field current on and off in an alternator?
    a. a diode bridge
    b. a stator
    c. an armature
    d. a voltage regulator
15. When the term *sensing voltage* is referred to in alternator operation, which of the following best describes it?
    a. 5 volts DC
    b. 5 volts AC
    c. battery voltage
    d. diode leakage voltage
16. What is used to perform switching functions in an electronic voltage regulator?
    a. diodes
    b. transistors
    c. a two-pole, two-throw switch
    d. breaker points
17. When an alternator is full-fielded, which of the following components is bypassed?
    a. the stator
    b. the rotor
    c. the slip rings
    d. the voltage regulator

18. When a carbon pile is used to load the electrical system to test the charging circuit, how much load should be applied?
    a. enough to drop system voltage to 12.7 volts
    b. one-half the CCA rating of the batteries
    c. the rated amperage specification
    d. enough to drop system voltage to 9.6 volts
19. An alternator rated at 130 amps is carbon pile load tested and system voltage drops to 12.6 volts at a current load of 120 amps. Technician A says that the alternator is okay and can be returned to service. Technician B says that alternator output should test within 10 percent of the rated specification. Who is correct?
    a. Technician A only
    b. Technician B only
    c. both A and B
    d. neither A nor B
20. Technician A says that Intelli-check is an industry acceptable means of testing charging system performance. Technician B says that when Intelli-check produces a *good* status rating, the charging system should still be further tested. Who is correct?
    a. Technician A only
    b. Technician B only
    c. both A and B
    d. neither A nor B

9

Prerequisite: Chapter 7

CRANKING SYSTEMS

OBJECTIVES

After reading this chapter, you should be able to:

- Identify the components in a truck cranking circuit.
- Explain the operating principles of magnetic switches, solenoids, and starter motors.
- Identify the control circuit components in a cranking circuit and define the role of the ECM in providing overcrank protection.
- Describe the operating principles of lightweight, planetary gear reduction starter motors.
- Test and troubleshoot a cranking circuit using voltage drop testing.
- Disassemble a heavy-duty truck starter motor.
- Test an armature for shorts using a growler.
- Test an armature for grounds and opens.
- Use a testlight to check out field coils.
- Outline the procedure required to rebuild a Remy 42MT starter motor.

KEY TERMS

armature
brushes
commutator
control circuit
counter-electromotive force (CEMF)
electromechanical ignition switch
external ground
field coils
ground circuit
insulated circuit
internal ground
internal magnetic solenoid switch (IMSS)
neutral safety switch
overcrank protection (OCP)
smart ignition switch
starter circuit
starter motor
starter relay
thermostat
windings

INTRODUCTION

The cranking system in any vehicle is designed to turn the engine over until it can operate under its own power. A cranking system can be divided into two subcircuits known as the **control circuit** and the **starter circuit**. The objective of the cranking system is to energize a **starter motor** using energy from the vehicle batteries. The control circuit is activated either directly from the vehicle ignition key or by a dedicated starter button. The control circuit uses low current to switch the starter circuit. The switch used to bridge the control circuit and the starter circuit is a magnetic switch. The magnetic switch receives a low-current command signal that switches the high-current starter circuit. Once energized, the starter circuit uses full battery power to energize the starter motor. The arrangement of a typical truck electrical system is shown in **Figure 9–1**. Note how the battery connects with the starter motor and identify the cranking system control circuit.

STARTING CIRCUIT COMPONENTS

There are five basic components and two distinct electrical circuits (**Figure 9–2**) in a typical cranking system. The components are:

- Battery
- Key switch (or starter button)

FIGURE 9–1 Typical heavy-duty truck cranking system

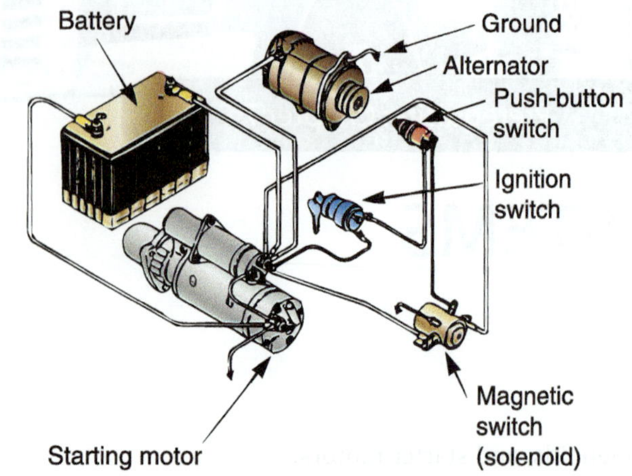

FIGURE 9–2 (A) A basic cranking circuit and (B) a cranking circuit with a thermostat.

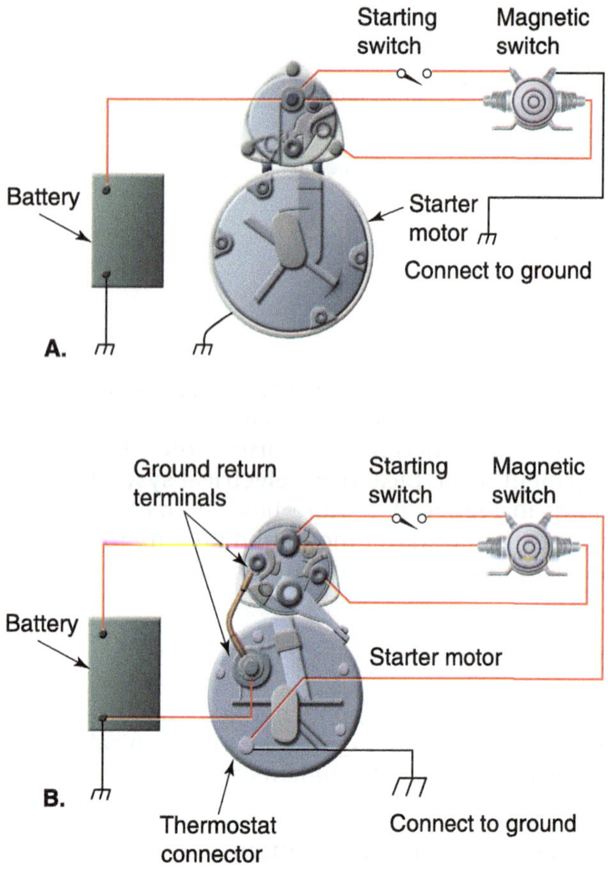

- Battery cables
- Magnetic switch
- Starter motor

A starter motor draws high current from the batteries, typically 300 to 400 amperes. This current flows through the heavy-gauge cables that directly connect the battery to the starter. The driver switches the current flow using the ignition key. However, if the high-current cables were to be routed from the battery to the ignition key and then on to the starter motor, the voltage drop caused by resistance in the cables would be too great.

The starter motor is energized directly by the vehicle batteries. The function of the starter motor is to convert electrical energy into sufficient mechanical torque to turn a high-compression diesel engine. The output torque from the starter motor pinion is imparted to ring gear teeth on the engine flywheel. A cranking circuit must be capable of starting an engine even under the most demanding conditions such as cold weather extremes.

GROUND AND INSULATED CIRCUITS

Almost all current highway vehicles use negative chassis ground electrical circuits. In such a system, the entire chassis is used as the ground or negative path for current flow. This is known as the **ground circuit**. An advantage of chassis negative ground systems is that only the positive side of the current path to components has to be insulated. When the term **insulated circuit** is referred to in a vehicle system, it refers to the electrically positive portion of the circuit.

A majority of highway trucks in North America use the electric cranking circuits described in this chapter. Those that do not use air start systems, which is covered briefly in Chapter 28. Air start systems are limited in cranking duration by the amount of compressed air available, so these have fallen out of favor recently, especially in northern states and Canada.

Dividing the cranking system into a low-current carrying switching (control) circuit and high-current carrying starter circuit minimizes the length of heavy-duty cables that conduct the high-current load directly from the batteries to the starter motor. It also reduces the incidence of voltage drop–related problems in the circuit.

9.1 CRANKING CIRCUIT COMPONENTS

The vehicle batteries are a fundamental component of the cranking circuit. Any study of cranking systems must begin with a thorough understanding of battery operation, performance, and testing. Batteries are covered in detail in Chapter 7 of this book and their operation must be understood before proceeding with this chapter. This section addresses the key cranking system components.

The control circuit incorporates the ignition switch, starter relay (magnetic switch), associated wiring, batteries, and safety starting devices such as the clutch switch and neutral starting switch. The starter circuit consists of the starter relay, starter motor solenoid, batteries, and high-current cables. The control circuit uses low current to energize the high-current starting circuit.

BATTERY TERMINALS

Battery terminals connect the vehicle battery cables to battery posts. They are the source of more cranking circuit malfunctions than all other cranking circuit problems combined, yet experienced technicians somehow find ways of skipping this key step in cranking circuit troubleshooting. Battery terminals can either be lead collets that are bolted to threaded or recessed battery posts, or lead clamps that are bolted to lead battery posts. In most cases, the copper battery cable is soldered into a socket integral with the lead terminal. Terminals that are crimped or clamped to the socket tend to have higher failure rates. Battery terminals are susceptible to oxidation and corrosion, both of which are insulators that can create high resistance. High resistance anywhere in the cranking circuit will result in cranking circuit problems.

Cleaning Terminals

If any evidence of corrosion is observed on visual inspection, a battery terminal should be removed from the post and cleaned. Cleaning a battery post requires that acidic corrosion be neutralized and washed away using a solution of water and baking soda. Next, both the terminal clamp inside bore and the battery post can be prepped using battery wire brushes. The objective of any battery terminal is to maximize the surface contact area between the connections—that is, provide the largest possible path for electron flow. When reinstalling a battery clamp, the bolt should be sufficiently torqued so that the clamp is moderately deformed to achieve this. When installing threaded terminals, the fastener should be torqued to the original equipment manufacturer (OEM) specification.

Testing Terminals

The only way to determine whether a battery terminal is doing its job is to test it by measuring voltage drop. Voltage drop testing is explained many times over in this book because it is a dynamic test. Voltage drop testing is performed with a circuit energized, so when you are testing battery terminals it should be done while attempting to crank the engine. To test the insulated side of the circuit, set an autoranging digital multimeter (DMM) to V-DC and place the positive test lead on the battery positive post and the negative test lead on the cable clamp. When the engine is cranked, the voltage reading should not exceed 0.1 V-DC; if it does, there is high resistance at the connection. To test the ground side of the circuit, set an autoranging DMM to V-DC and place the negative test lead on the battery negative post and the positive test lead on the cable clamp. When the engine is cranked, the voltage reading should not exceed 0.1 V-DC; if it does, there is high resistance at the connection. High-resistance connections cause excessive voltage drops, resulting in reduced circuit performance and heat.

CABLES

The cranking circuit in Figure 9–2 requires two or more heavy-gauge cables. Two of these cables attach directly to the batteries. One cable connects between the battery negative terminal and a good ground or a ground stud on the starter housing. The other cable connects the battery positive terminal with the starter relay. On vehicles where the starter relay does not mount directly on the starter motor, two cables are needed. One runs from the positive battery terminal to the relay and the second from the relay to the starter motor terminal. These cables conduct the heavy current load from the battery to the starter and from the starter back to the battery. All cables must be in good condition. Cables can be corroded by battery acid. Corrosion will cause a voltage drop and decrease circuit amperage, reducing power available to the starter. Contact with the engine and other metal surfaces can fray the cable insulation. Deteriorated insulation can result in a dead short that can damage electrical components. A short to ground is the cause of many dead batteries and can result in fire. Cables must be heavy enough to carry the required current load.

When checking cables and wiring, check any fusible links in the wiring. Some vehicles are equipped with fusible links to protect wiring from overloads. Fusible links are different in construction from a fuse but operate in much the same way. The most common type consists of a wire with a special nonflammable insulation. Wire used to make fusible links is ordinarily two sizes smaller than the wire in the circuit they are designed to protect. When a fusible link is subjected to a current overload, the insulation becomes charred.

The largest fusible link is usually located at the starter solenoid battery terminal. From this terminal, current is distributed to the remainder of the vehicle electrical system. A second fusible link joins this battery terminal to the main body harness and

TABLE 9–1 Maximum Voltage Drop Specifications

Insulated circuit voltage drop (VD) maximums:	
Cranking Circuit Component	**Maximum VD**
Starter cable	0.2 volt
Each cable connection	0.1 volt
Starter solenoid	0.3 volt
Maximum total permissible voltage loss	0.5 volt

protects the vehicle wiring. This link may take several forms—from a wire to a small piece of metal with terminal connections on each end. When a link fails, troubleshoot the system and locate the cause before replacing the link.

Because battery cables conduct the current required to crank the engine, they must be of sufficient size to conduct the current load. In Chapter 7, dealing with batteries, Table 7–5 lists both the standard and metric cable size dimensions used on current trucks. **Table 9–1** shows truck industry standard maximum voltage drop specifications. Note that the maximum total voltage drop should not exceed 0.5 volt.

IGNITION SWITCH

The ignition switch gets its name from the automobile key circuit. The term ignition switch is commonly used to describe the switch that energizes the control circuit in the cranking system. There are two types. The first type is integral with the ignition key: To close the control circuit to initiate cranking, the key is turned. The second type is a push-button and requires that the ignition key circuit be closed before the push-button starter is powered. In both cases, the switch is spring-loaded to the open position. This means that it must be either turned (by key) or pushed (by thumb) against the spring pressure to close the starter control circuit and initiate cranking.

When the ignition switch is closed, the coil in the **starter relay** (magnetic switch) is energized, enabling current flow in the high-current starter circuit. In some newer truck chassis, ignition switch status is broadcast on the chassis data bus. Always observe the OEM service procedures when troubleshooting and replacing ignition switches.

STARTER RELAY

The starter relay is a magnetic switch that enables the control circuit to open and close the high-current cranking circuit. It is either mounted at a remote location from the starter or built into the front end of the starter solenoid assembly. Remote-mounted starter relays are located close to the battery or starter to

FIGURE 9–3 A magnetic switch, or starter relay, mounted on a firewall near a windshield wiper motor.

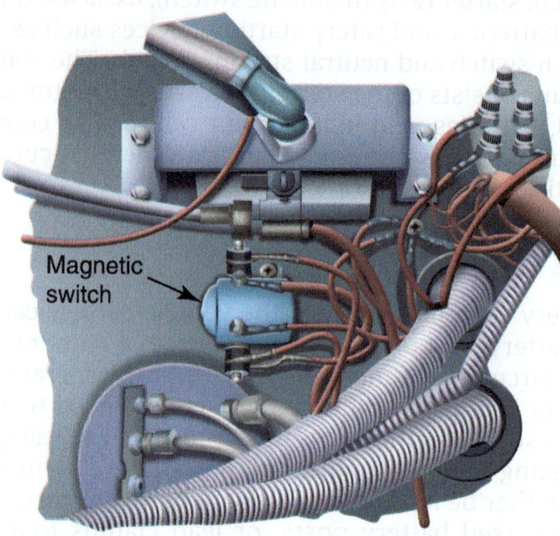

keep the cables as short as possible. **Figure 9–3** shows a starter relay mounted on the firewall of a conventional truck near the windshield wiper motor on the driver's side. On cab-over-engine models, the starter relay is often mounted on the front frame cross member.

The starter relay consists of an electromagnet plunger, contact disc, and two springs (**Figure 9–4**). The electromagnet has a hollow core in which the plunger moves.

Starter Relay Operation

When the operator turns the ignition switch to the crank position, battery current flows through the starting switch to a control circuit terminal on the starter relay. Control circuit current flows through the windings in the electromagnet, creating a magnetic field that pulls the plunger into the hollow core. Spring pressure then forces the contact disc against the starter circuit terminals, closing the circuit. High amperage current flows through contacts to the starter motor.

When the engine starts, the ignition switch is released, and the control circuit is opened. This deactivates the electromagnet in the starter relay. The return spring forces the plunger out of the hollow core, which moves the contact disc away from the starter circuit terminals, interrupting current flow to the starter motor.

9.2 STARTER MOTORS

The starter motor (**Figure 9–5**) converts electrical energy from the battery into mechanical energy for cranking the engine. The starter is an electric motor

FIGURE 9–4 Component parts of a magnetic switch

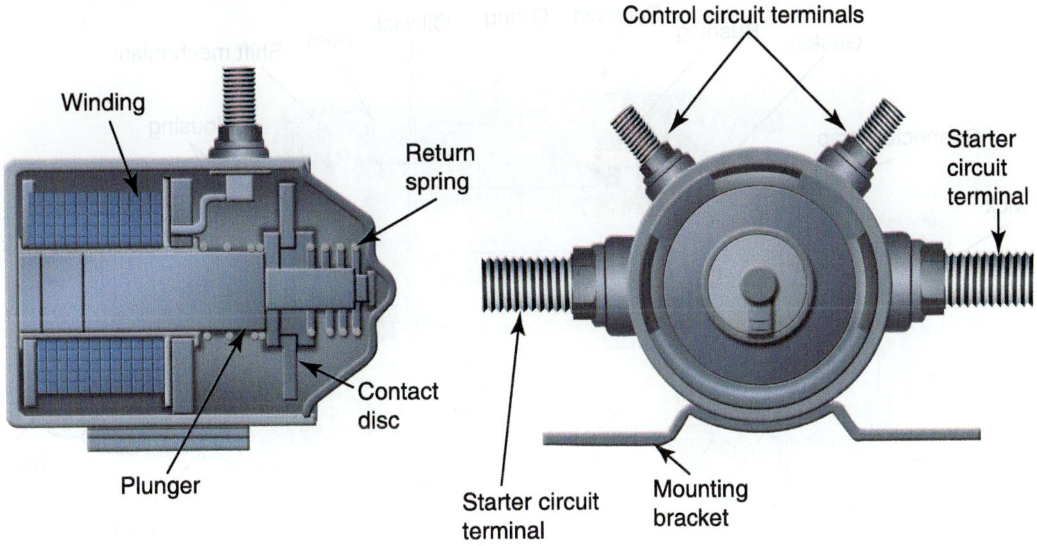

designed to operate under great electrical loads and produce high torque.

SHOP TALK

Starter motors can only operate for short periods without rest. The high current needed to operate the starter creates considerable heat, and continuous operation will cause overheating. A starter motor should never operate for more than 30 seconds at a time and should rest for 2 minutes between cranking cycles. This permits the heat to dissipate without damage to the unit.

STARTER MOTOR CONSTRUCTION

Starter motors do not differ much in design and operation. A starter motor consists of a housing, **field coils**, an **armature**, a **commutator**, **brushes**, end frames, and a solenoid-operated shift mechanism. **Figure 9–6** shows a cutaway of a typical starter used on heavy-duty trucks. The starter housing or frame encloses the internal starter components and protects them from damage, moisture, and foreign materials. The housing supports the field coils (**Figure 9–7**) and forms a conducting path for the magnetism produced by the current passing through the coils. Some starter motors use an integral magnetic switch.

FIGURE 9–5 (A) Front view and (B) rear view of a starter motor

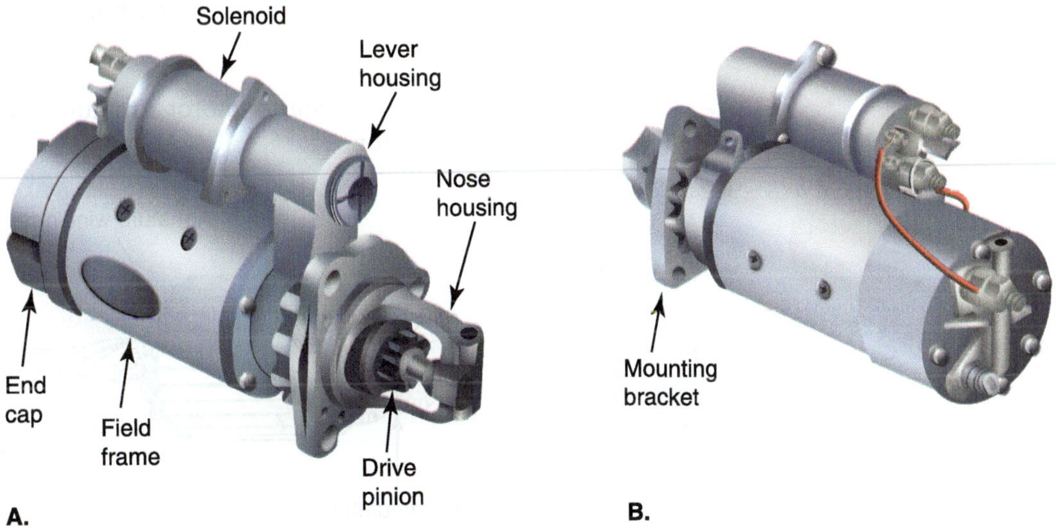

FIGURE 9–6 Cutaway of a cranking motor

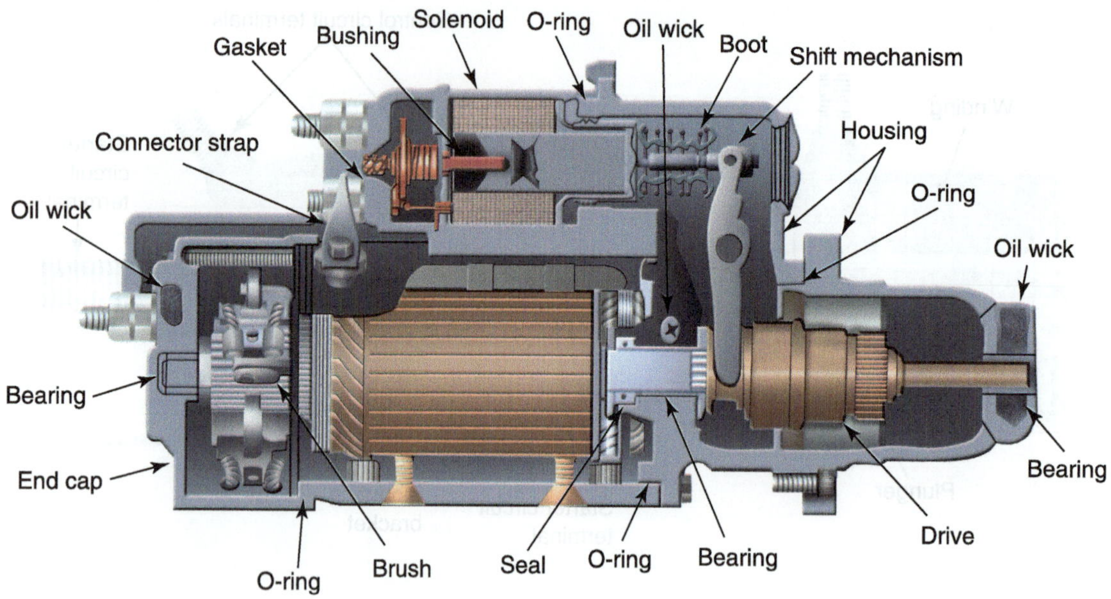

Field Coils

The field coils and their pole shoes are securely attached to the inside of the iron housing. They are insulated from the housing but are connected to a terminal that protrudes through the outer surface of the housing. The field coils and pole shoes are designed to produce strong stationary electromagnetic fields within the starter body as current is passed through the starter. These magnetic fields are concentrated at the pole shoes. Field coils will have an N or S magnetic polarity depending on the direction the current flows. The coils are wound around respective pole shoes in opposite directions to generate opposing magnetic fields. The field coils connect in series with the armature winding through the starter brushes. This design permits all current passing through the field coil circuit to also pass through the armature windings.

FIGURE 9–7 Four field coils used in a starter motor

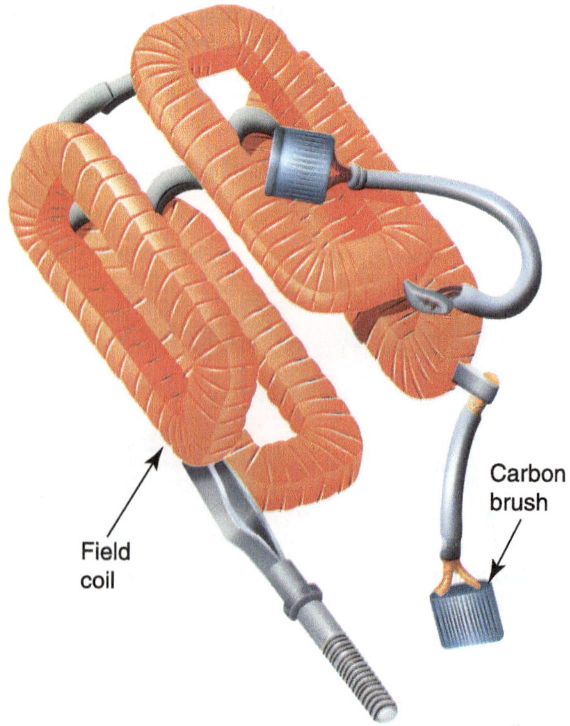

Armature

The armature is the rotating component of the starter. It is located between the drive and commutator end frames and the field windings (**Figure 9–8**). When the starter motor operates, the current passing through

FIGURE 9–8 Starter motor armature and commutator

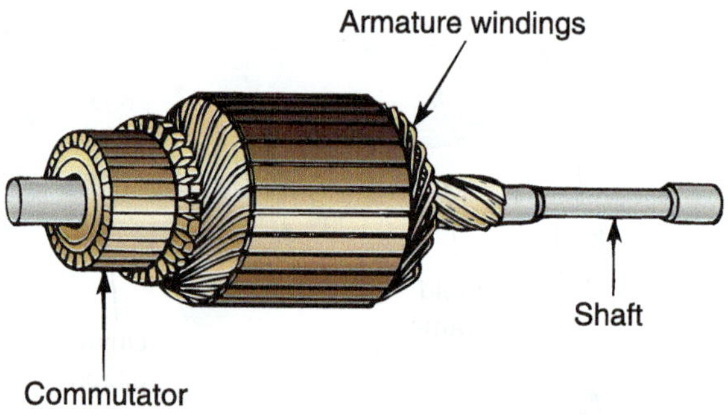

the armature produces a magnetic field in each of its conductors. The reaction between the armature's magnetic field and that of the field coils causes the armature to rotate. This creates the torque required to crank the engine.

Armature Windings

The armature has two main components: the armature windings and the commutator. Both mount to the armature shaft. The armature **windings** are not made of wire. Instead, heavy flat copper strips that can handle heavy current flow are used. The windings are constructed of several coils of a single loop each. The sides of these loops fit into slots in the armature core or shaft but are insulated from it. Each slot contains the side of two of the coils. The coils connect to each other and to the commutator so that current from the field coils flows through all of the armature windings at the same time. This action generates a magnetic field around each armature winding, resulting in a repulsion force all around the conductor. This repulsion force causes the armature to turn.

Commutator

The commutator assembly (**Figure 9–9**) presses onto the armature shaft. It is made up of heavy copper segments separated from each other and the armature shaft by insulation. The commutator segments connect to the ends of the armature windings. Starter motors have four to twelve brushes that ride on the commutator segments and carry the heavy current flow from the stationary field coils to the rotating armature windings via the commutator segments. The brushes are held in position by a brush holder.

FIGURE 9–9 Different starter motor configurations

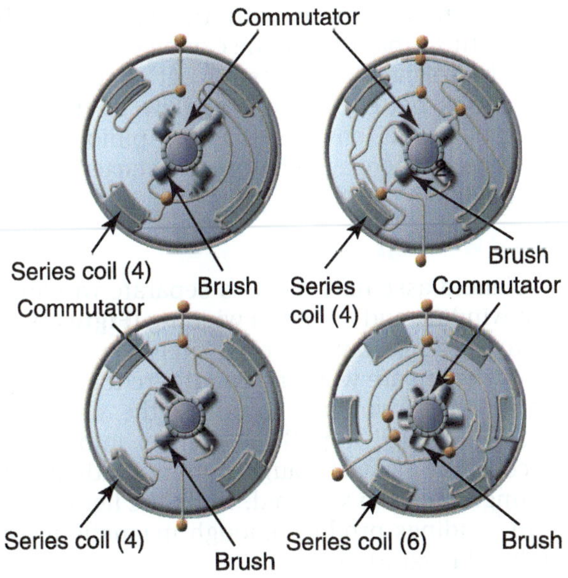

OPERATION

The starter motor converts electric current into torque or twisting force through the interaction of magnetic fields. It has a stationary magnetic field (created by passing current through the field coils) and a current-carrying conductor (the armature windings). When the armature windings are placed in this stationary magnetic field and current is passed through the windings, a second magnetic field is generated with its lines of force wrapping around the wire (**Figure 9–10**). Because the lines of force in the stationary magnetic field flow in one direction across the winding, they combine on one side of the wire, increasing the field strength, but are opposed on the other side, weakening the field strength. This creates an unbalanced magnetic force, pushing the wire in the direction of the weaker field.

Current Flow

Because the armature windings are formed in loops or coils, current flows outward in one direction and returns in the opposite direction. This means

FIGURE 9–10 (A) When a current-carrying conductor passes through a magnetic force field, the field is deflected to one side of the conductor. This creates pressure on one side of the wire and forces the conductor to move away from the strong force field. (B) When a loop of wire is placed between two magnets, the intersection between the force fields causes the loop to rotate around its axis.

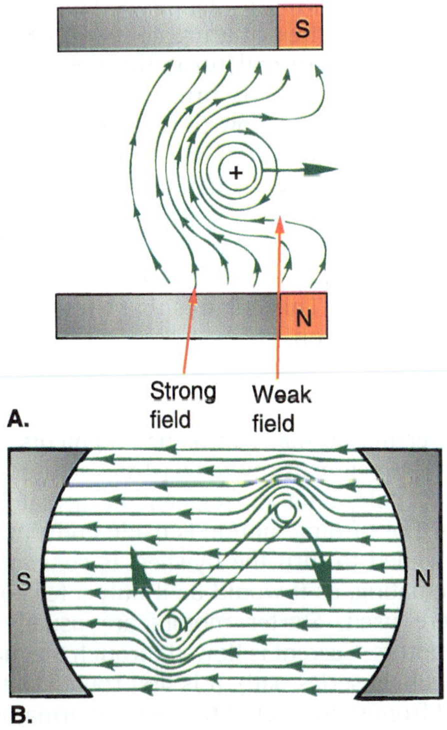

FIGURE 9–11 Current flow in the conductor must be reversed every 180 degrees of rotation so that the armature will continue to rotate in the same direction.

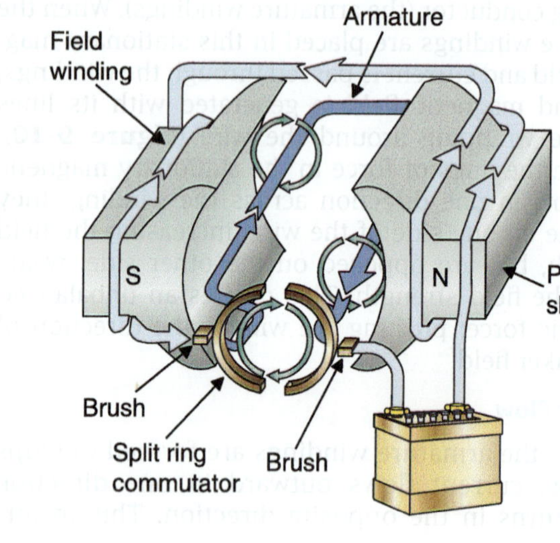

the magnetic lines of force are oriented in opposite directions in each of the two segments of the loop. When placed in the stationary magnetic field of the field coils, one part of the armature coil is pushed in one direction while the other part is pushed in the opposite direction. This causes the coil, and the shaft to which it is mounted, to rotate. Each end of the armature windings is connected to one segment of the commutator. Two carbon brushes are connected to one terminal of the power supply. The brushes contact the commutator segments, conducting current to and from the armature coils (**Figure 9–11**).

As the armature coil turns through a half revolution, the contact of the brushes on the commutator causes the current flow to reverse in the coil. The commutator segment attached to each coil end will have traveled past one brush and is now in contact with the other. In this way, current flow is maintained constantly in one direction while allowing the segment of the rotating armature coils to reverse polarity as they rotate. This ensures that the armature will turn in one direction.

Armature Segments

In a starter motor, many armature segments must be used. As one segment rotates past the stationary magnetic field pole, another segment immediately takes its place. The turning motion is made uniform and the torque needed to turn the flywheel is constant rather than fluctuating, as it would be if only a few armature coils were used. Starter motors are usually series-wound motors. Current is flowed through the motor circuit in a single series path. It is first routed to the field coils and from there routed through the armature. This is one factor that enables cranking motors to produce maximum torque at zero rpm. All series-wound motors produce maximum torque at low rpm close to stall.

Counter-Electromotive Force

In taking a look at the starter motor as it has been described so far, it should be noted that all the conditions for *generating* a voltage have been met. When a conductor cuts through magnetic lines of force, a voltage is induced in the conductor. In the starter motor, this induced voltage acts in opposition to battery current. The induced voltage is known as **counter-electromotive force (CEMF)**. The effect of CEMF reduces the current supplied by the battery.

Because voltage induced by a conductor is proportional to the speed at which it cuts through the magnetic field, CEMF increases proportionally with armature speed. As the armature is wound out to its maximum speed, CEMF gets close to but cannot exceed battery voltage. The ability of an armature to produce torque is determined by the current flowed through it. The absence of CEMF is one factor that determines that peak torque is produced at zero rpm in a starter motor. As the cranking motor rpm rises, CEMF rises proportionally, reducing torque potential. In other words, CEMF acts a bit like a *magnetic brake* to help limit starter motor overspeed.

STARTER SOLENOIDS

The solenoid-operated shift mechanism (**Figure 9–12**) is mounted in a solenoid housing that is sealed to keep out oil and road splash. The case is flange mounted to the starter motor housing. It contains an electromagnet with a hollow core. A plunger is installed in the hollow core, much like the starter relay described earlier. The solenoid performs two functions:

- When energized, plunger movement acts on a shift lever that shifts the starter motor drive pinion into mesh with the engine flywheel teeth so that the engine can be cranked.
- The solenoid also closes a set of contacts that allows battery current to flow to the starter motor; this ensures that the pinion and flywheel are engaged before the starter armature begins to rotate.

Solenoid Windings

The solenoid assembly has two separate windings: a pull-in winding and a hold-in winding (**Figure 9–13**). The two windings have approximately the same number of turns but are wound from different gauge wire. Together these windings produce the electromagnetic force needed to pull the plunger into the solenoid coil. The heavier gauge pull-in windings draw the plunger into the solenoid, while the lighter gauge hold-in windings produce enough magnetic force to hold the plunger in this position.

FIGURE 9–12 The solenoid engages the drive mechanism when the contact disc closes the circuit across the terminals.

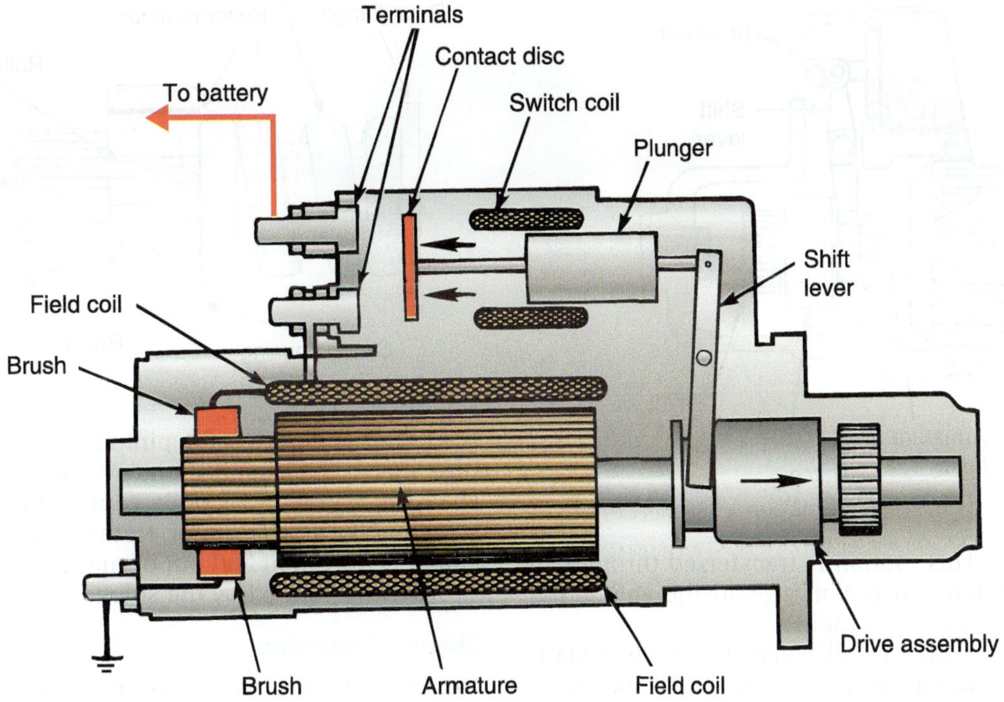

Current Routing

Both windings are energized when the starting switch is turned to the start position. If the system has a remote starter relay, battery current passes through the relay to the "F" terminal on the solenoid. When the plunger contact disc touches the solenoid terminals, the pull-in winding is deactivated. At the same time, the plunger contact disc makes the motor feed connection between the battery and the starting motor, directing full battery current to the field coils and starter motor armature for cranking power.

FIGURE 9–13 The starter solenoid has two windings: a pull-in winding and a hold-in winding.

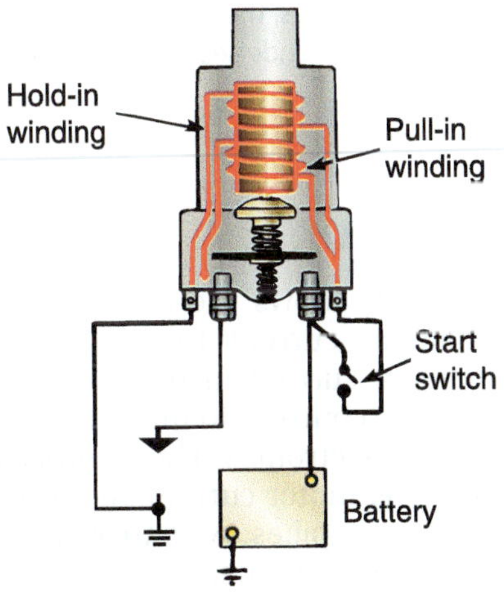

SHOP TALK

In almost all diesel engine starters, the solenoid performs the function of a relay. The control circuit is wired to the windings of the solenoid. Battery current is routed directly to the starter motor through the drive solenoid.

Internal Solenoid

Many current starter motors integrate the solenoid switch into the housing. In most cases a one-wire design is used with the solenoid grounded into the motor housing. Internal switches eliminate the corrosion damage that shortens the life of external solenoids and allow ECM control of cranking. We will take a closer look at internal solenoids later in this chapter.

Initiating Cranking

As this electrical connection is being made at the terminal end of the solenoid, the mechanical motion of the solenoid plunger is being transferred to the drive pinion through the shift lever, bringing the pinion gear into mesh with the flywheel ring gear (**Figure 9–14**). When the starter motor receives current, its armature

FIGURE 9–14 Drive mechanism

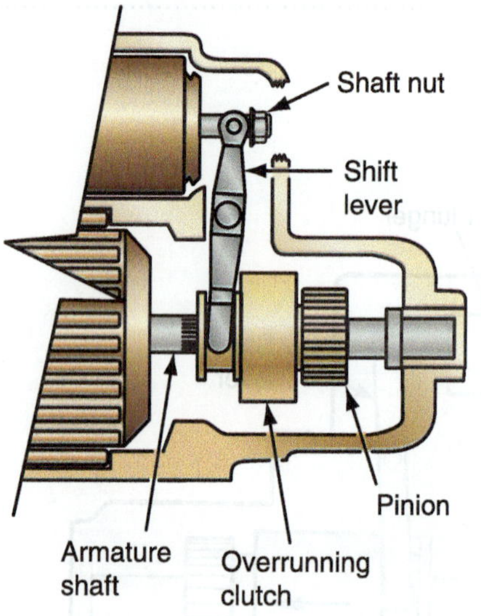

FIGURE 9–15 Override clutch operation

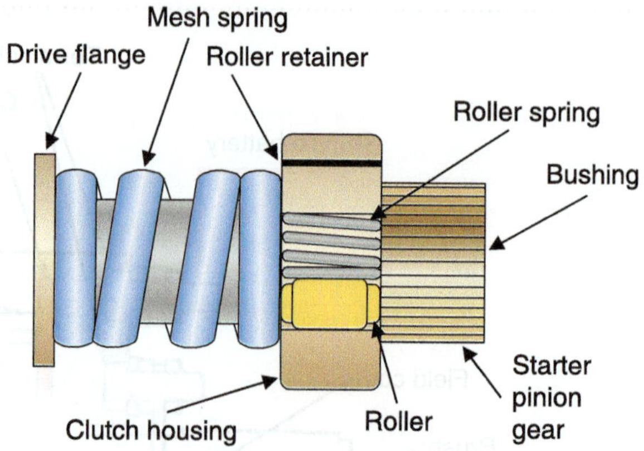

starts to turn. This motion is transferred through an overriding clutch and pinion gear to the engine flywheel and the engine is cranked.

With this type of solenoid-actuated direct-drive starting system, teeth on the pinion gear might not immediately mesh with the flywheel ring gear. If this occurs, a spring located behind the pinion compresses so that the solenoid plunger can complete its stroke. When the starter motor armature begins to turn, the pinion teeth will mesh with the flywheel teeth under spring pressure.

Overrunning Clutches

The overrunning clutch performs an important job in protecting the starter motor. When the engine starts and runs, its speed increases. If the starter motor remained connected to the engine through the flywheel, the starter motor would spin at very high speeds, destroying the armature windings. To prevent this, the starter must be disengaged from the engine as soon as the engine turns more rapidly than the starter. But with a solenoid-actuated drive mechanism, the pinion remains engaged with the flywheel until current stops flowing to the starter. In these cases, an overunning clutch is used to disengage the starter. A typical overriding clutch is shown in **Figure 9–15**.

The clutch housing is internally splined to the starting motor armature shaft. The drive pinion turns freely on the armature shaft within the clutch housing. When the clutch housing is driven by the armature, the spring-loaded rollers are forced into the small ends of their tapered slots and wedged tightly against the pinion barrel. This locks the pinion and clutch housing solidly together, permitting the pinion to turn the flywheel and, thus, crank the engine.

When the engine starts, the flywheel spins the pinion faster than the armature. This action releases the rollers, unlocking the pinion gear from the armature shaft. The pinion then "overruns" the armature shaft freely until being pulled out of the mesh without stressing the starter motor. Note that the overrunning clutch is moved in and out of mesh with the flywheel by linkage operated by the solenoid.

Thermal Protection

Most modern starter motors are equipped with some form of **overcrank protection (OCP)** device. An OCP circuit consists of a thermostat (see Figure 9–2B) and internal circuit breaker. The **thermostat** monitors the temperature of the motor. If prolonged cranking causes the motor temperature to exceed a safe threshold, the thermostat causes a cycling-type circuit breaker to open and the starter current is interrupted. The starter motor will not operate until the motor cools and the thermostat allows the breaker to close.

SHOP TALK

Heavy-duty truck electric starter motors are sometimes used to power hydraulic pumps on applications such as automobile carrier trailers; in these cases, the motor can safely operate for periods longer than 30 seconds because there is almost no load on the motor.

CONTROL CIRCUIT

The control circuit allows the driver to use a small amount of battery current to control the flow of a large amount of current in the starting circuit. In some chassis, the control circuit is managed by the engine controller electronics. This allows engine logic to control cranking motor current and protect the circuit against voltage spikes. A control circuit consists of the following components.

Ignition Switch

The control circuit begins at an ignition switch. The ignition switch can be:

- Electromechanical
- Smart (virtual: signals start request)

An **electromechanical ignition switch** is connected by light gauge wire to the batteries and the starter relay. When the key is turned to the start position, a small current flows through the coil of the starter relay, closing it and switching high-current flow directly to the starter motor. A **smart ignition switch** is designed to signal the engine control module (ECM) with a start engine request. ECM logic then determines whether that is going to happen or not, and assuming that it is, activates the starter control circuit. The ignition switch performs other jobs besides controlling the starting circuit. It normally has at least four separate positions: accessory, off, on (run), and start.

Some trucks have a push-button starter switch. Battery voltage is available to the switch when the ignition switch is in the on position. When the push-button is depressed, current flows through the control circuit to the starter relay coil. Electronic push-button starter switches may require that data bus wake-up be complete and the clutch or brake pedal fully depressed before the control circuit closes to initiate cranking.

Neutral Safety Switch

The **neutral safety switch** prevents vehicles with automated and automatic transmissions from being started in gear. Neutral safety switches are located in either of two places in the control circuit. One position is between the ignition switch and the starter relay. Placing the transmission in park or neutral will close the switch so current can flow to the starter relay. The neutral safety switch can also be connected between the starter relay and ground so that the switch must close before current can flow from the starter relay to ground. The neutral safety switch can also use smart technology in which case, the transmission ECU broadcasts a signal over the data bus network.

Starter Relays

The starter relay is the point in the cranking circuit where the low-current control circuit and high-current starter circuit come together. Low current in the control circuit passes from the ignition switch or ECM to the starting switch and neutral safety switch to energize the starter relay and activate the cranking circuit.

Many current starter motors use an **internal magnetic solenoid switch (IMSS)**, which is ECM-managed. Most IMSS use a single wire design with grounding into the motor housing. Physically locating the IMSS within the starter motor eliminates many of the corrosion problems that failed external solenoids. In addition, the IMSS allows the ECM to monitor and control input current to the starter motor and protect the circuit against voltage spikes.

REDUCTION-GEARING STARTER MOTORS

In recent years, reduction-gearing starters capable of cranking diesel engines up to 16 liters displacement have become common. This has been good news for truck technicians familiar with struggling with the weight of a 42MT starter. Gear-reduction heavy-duty starters include the Remy 29MT, 38MT, and 39MT, but there are competitor models that are similar. The 39MT has the appearance and weight of the starter motor used on a typical automobile but effortlessly cranks a 15-liter Cummins ISX engine (**Figure 9–16**). Gear-reduction, soft-start, positive engagement starter motors with integral magnetic switches now outsell direct-drive starter motors in truck applications. However, direct-drive starter motors have some advantages that we will take a look at later in this chapter.

Planetary Gearset

The key to enabling a lightweight starter motor such as the 39MT to crank large-bore, high-compression engines is a planetary gearset. A full description of the principles of planetary gearsets and input to output ratios appears in Chapter 18. **Figure 9–17** shows a 39MT armature assembly, where the drive pinion on the armature shaft engages with the four planetary gears. This means that the role of the armature pinion is that of the sun gear in the planetary set. In the 39MT planetary set, the sun gear is the input, the planetary carrier is the output, and the ring gear that surrounds the four planetary gears is held stationary. This arrangement provides for a reduction ratio of about 3½:1. This means that for every output rotation of the output carrier, the sun gear has to turn over three

FIGURE 9–16 39MT starter motor on a 15-liter Cummins ISX engine

FIGURE 9–17 Armature assembly from a Remy 39MT: The pinion stub engages with the planetary gears shown in Figure 9–18.

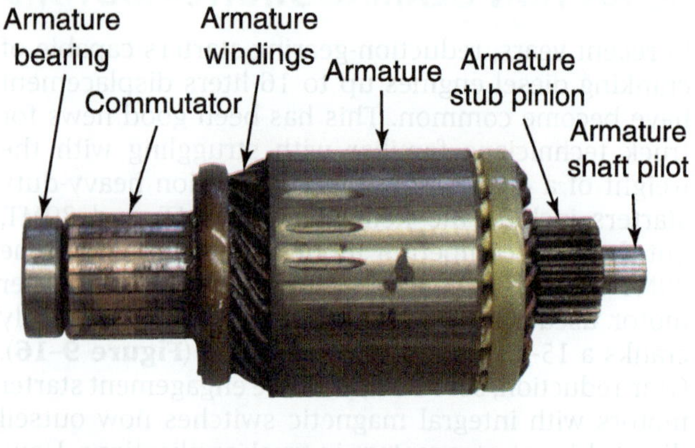

FIGURE 9–18 The 39MT planetary gearset: The armature pinion stub meshes with the planetary gears to provide input torque.

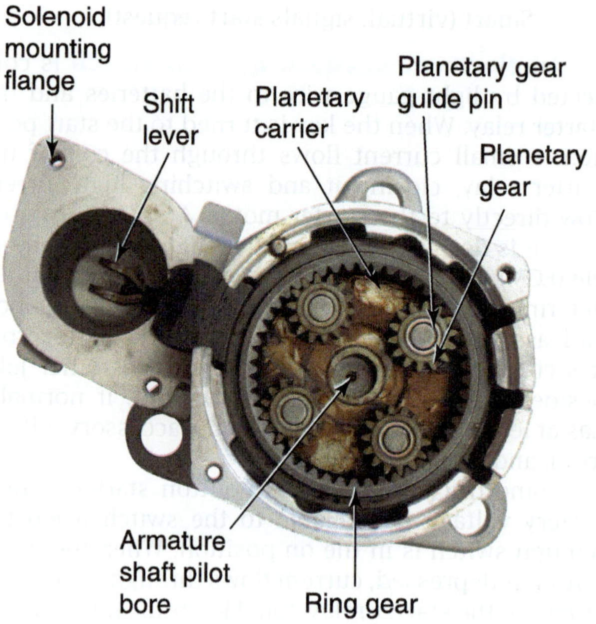

revolutions. A planetary gearset arrangement such as this provides for maximum rpm reduction and maximum output torque.

The output from the 39MT planetary gearset is the starter pinion. The starter pinion is solenoid/shift lever engaged to the ring gear on the flywheel for cranking. The 39MT uses a noseless drive snout and soft start engagement technology. **Figure 9–18** shows a split 39MT starter motor: The armature pinion shown in Figure 9–17 engages to the planetary carrier gears.

9.3 CRANKING CIRCUIT TESTING

The cranking circuit requires testing when the engine will not crank, when the engine cranks slowly, or when the starter motor will not turn.

PRELIMINARY CHECKS

Cranking output obtained from the motor is affected by the condition and charge of the battery, the wiring circuit, and the engine cranking requirements. The battery should be checked and charged as needed before testing. Ensure that the batteries are rated to meet or exceed the vehicle manufacturer's recommendations. The voltage rating of the batteries must also match the voltage rating of the starter motor.

SHOP TALK

The starter should not be operated if the voltage at the battery is below 9.6 volts. Some leasing companies now use a voltage sensing module to prevent starter operation if voltage is below 9.6 volts to prevent drivers from destroying starters.

Check the wiring for clean, tight connections. Loose or dirty connections will cause excessive voltage drop. Clean and tighten all connections as necessary. The cranking system cannot operate properly with excessive resistance in the circuit.

The engine crankcase should be filled with the proper weight oil as recommended by the engine manufacturer. Heavier-than-specified oil combined with low temperatures lowers cranking speed to the point where the engine will not start.

Check the starting switch for loose mounting, damaged wiring, sticking contacts, and loose connections. Check the wiring and mounting of the safety switch, if so equipped, and make certain that the switch is properly adjusted. Check the mounting, wiring, and connections of the starter relay and starter motor.

In starters equipped with a thermostat, you must check the thermostat circuit if the starter does not crank. Check the resistance between the two thermostat terminals on the motor. The ohmmeter should read close to zero. If it does not, the thermostat is open circuited. Do not check the thermostat when the starter motor is hot. The thermostat operates by opening the circuit when a specified temperature is exceeded. This makes cranking impossible until the temperature has dropped below the threshold.

TROUBLESHOOTING

Systematic troubleshooting is essential when servicing the starting system. Half of "defective" starters returned on warranty claims function to specification when tested. This results from poor or incomplete diagnosis of the starting circuit. **Table 9–2** itemizes a systematic approach to starting circuit diagnosis. Testing the starting system can be divided into area tests, which check voltage and current in the entire system, and more detailed pinpoint tests, which target one particular component or segment of the wiring circuit.

Starter Relay Testing

The starter relay bypass test is a simple method of determining if the relay is operational. This test should be performed when the starter motor does not activate when the ignition key is in the start position (or when the starter button is depressed).

Ensure that the transmission is in neutral. Connect a jumper cable around the starter relay as

TABLE 9–2 Troubleshooting a Cranking Circuit

Problem	Possible Cause	Tests and Checks	Remedy
Engine cranks slowly or unevenly	1. Weak battery	1. Perform battery open-circuit voltage and load voltage tests. Perform battery load tests (capacity). Check capacity and voltage ratings against engine requirements.	1. Service, recharge, or replace defective battery.
	2. Undersized or damaged cables	2. Perform visual inspection.	2. Replace as needed.
	3. Poor starter circuit connections	3. Perform visual inspection for corrosion and damage.	3. Clean and tighten. Replace worn parts.
	4. Defective starter motor caused by high internal resistance	4. Perform cranking current test and no-load test.	4. If cranking current is under specs, proceed with no-load bench testing.
	5. Engine oil too heavy for application	5. Check oil grade.	5. Change oil to proper specs.
	6. Seized pistons or bearings	6. Check compression and cranking torque.	6. Repair as needed.
	7. Overheated solenoid or starter motor	7. Check for missing or damaged heat shields.	7. Replace shield. Service as needed.
	8. High resistance in starter circuit	8. Use cranking current test, insulated circuit test, and ground circuit tests to pinpoint area of high resistance.	8. Replace defective components.
	9. Poor starter drive/flywheel engagement	9. Perform visual inspection of drive and flywheel components.	9. Replace damaged components.
	10. Loose starter mounting	10. Perform visual inspection.	10. Tighten as needed.
Engine does not crank	1. Discharged battery	1. As listed above.	1. As listed above.
	2. Poor or broken cable connections	2. As listed above.	2. As listed above.
	3. Seized engine components	3. As listed above.	3. As listed above.
	4. Loose starter mounting	4. As listed above.	4. As listed above.
	5. Open in control circuit	5. Perform control circuit test to locate "open" or high resistance.	5. Repair or replace components as needed.
	6. Defective starter relay	6. Perform starter relay bypass test.	6. Replace starter relay if engine cranks when bypassed.

(continued)

TABLE 9–2 (continued)

Problem	Possible Cause	Tests and Checks	Remedy
	7. Defective starter motor caused by internal motor malfunction	7. Perform starter relay bypass test.	7. Replace starter motor if engine will not crank when starter relay is bypassed.
Starter motor spins but does not crank engine	1. Defective starter drive	1. Perform starter drive test.	1. Replace starter drive.
	2. Worn or damaged pinion gear	2. Perform visual inspection of components.	2. Replace starter drive.
	3. Worn or damaged flywheel gears	3. Perform visual inspection of flywheel.	3. Replace as needed.
Starter does not operate or movable pole shoe starter chatters or disengages before engine has started	1. Battery discharged	1. Perform battery load test.	1. Recharge or replace.
	2. High resistance in starting circuit	2. Perform cranking current test, insulated circuit test, and ground circuit tests to pinpoint area of high resistance.	2. Replace defective components.
	3. Open in solenoid or movable pole shoe hold-in winding		3. Replace solenoid or movable pole shoe starter.
	4. Worn solenoid unable to overcome return spring pressure		4. Replace solenoid. Install lighter return spring.
	5. Defective starter motor.	5. Perform visual inspection.	5. Replace starter motor.
Noisy starter cranking	Loose mounting		Tighten mounts. Correct alignment.

shown in **Figure 9–19**, bypassing the relay. If the engine cranks with the jumper installed, the starter relay or the control circuit may be defective. A starter relay also can be checked by connecting a voltmeter across the winding from the push-button key start connection to one of the mounting bolts that attaches the switch to the bulkhead. Have someone hold the starting switch closed. If the voltmeter reading is 0 volts, check for an open circuit. If the voltmeter reading is less than 11 volts, check for

FIGURE 9–19 Check the starter relay operation by bypassing the switch with a jumper wire.

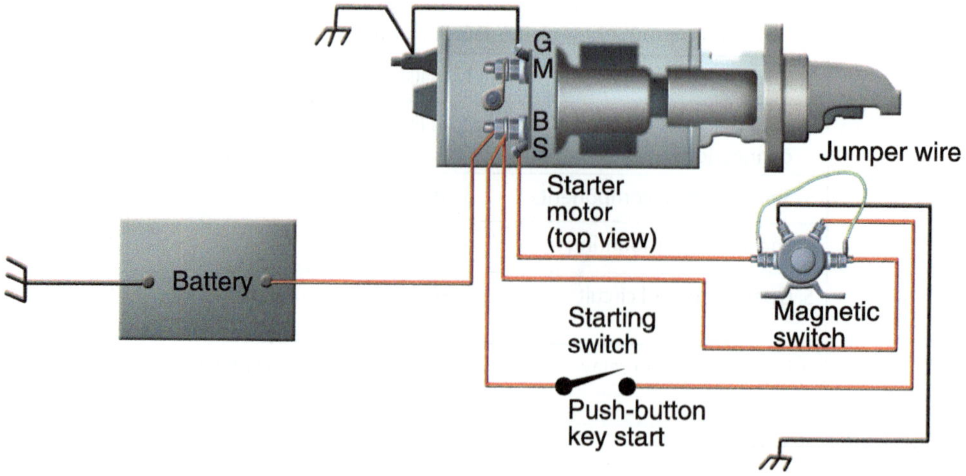

corroded or loose connections (refer to "Control Circuit Testing"). Repair or replace any damaged wires. Then check the voltage to the starter relay again. If the voltmeter reading is now 11 volts or more, the relay and control circuit are okay. If less than 11 volts, perform a voltage drop test across the power terminals.

Cranking Current Testing

This on-engine, cranking current test measures the amperage that the starter circuit draws to crank the engine. This amperage reading can be useful in isolating the source of certain types of starter problems. Before beginning, verify that the batteries are fully charged.

1. Connect the leads of a volt-amp tester (**Figure 9–20**).
2. Set the carbon pile to its maximum resistance (open). Install an inductive pickup clamp if equipped.
3. Crank the engine and observe the voltmeter reading.
4. Stop cranking. Adjust the carbon pile until the voltmeter reading matches the reading taken in step 3.
5. Note the ammeter reading.

SHOP TALK

If the analyzer uses an inductive pickup (amp clamp), ensure that the arrow on the inductive pickup is pointing in the right direction as specified on the ammeter clamp. Then crank the engine for 15 seconds and observe the ammeter reading.

TABLE 9–3 Result of Cranking Current Testing

Problem	Possible Cause
Low-current draw	Undercharged or defective battery
	Excessive resistance in circuit due to faulty components or connections
High-current draw	Short in starter motor
	Mechanical resistance due to binding engine or starter system component failure or misalignment

Compare the reading obtained during testing to the manufacturer's specifications. **Table 9–3** summarizes the most probable causes of high- or low-current draw. If the problem appears to be caused by high resistance in the circuit, test as shown in the next section.

Control Circuit Testing

The control circuit test verifies the performance of the wiring and components used to activate the starter relay. The control circuit can be checked by connecting a voltmeter across the coil terminals of the solenoid or starter relay. Remove the positive battery cable from the solenoid/starter when performing the test. This allows both windings of the solenoid to energize while preventing the cranking motor from turning. If there are other leads connected to the main positive terminal of the solenoid, you will have

FIGURE 9–20 Test connections for a current draw test

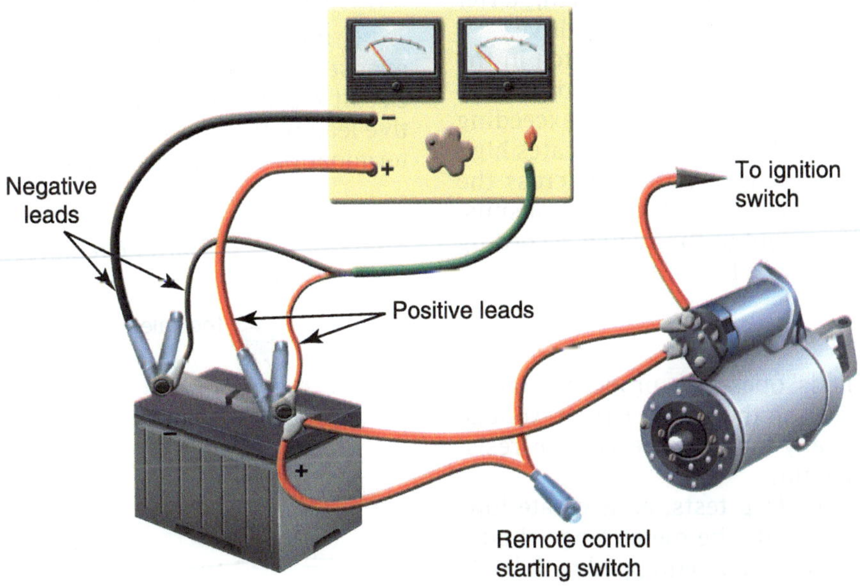

to remove these and temporarily reconnect them to the battery cable. Operate the starting circuit and read the voltage across the solenoid coil terminals. This should be at least 10 volts for a 12V system and 20 volts for a 24V system.

WARNING

When performing this test, do not operate the solenoid for extended periods because severe overheating will result.

ECM-Managed Control Circuit

The cranking control circuit in some current engines is ECM managed. This requires technicians to adhere to the OEM troubleshooting and testing procedure because the chassis electronics must be protected from inductive spikes that could destroy the electronics. When the cranking circuit uses an ECM-managed IMSS, technicians should connect an electronic service tool (EST) as a first step in diagnosing cranking circuit problems. That said, most starter circuit problems in current trucks are usually related to electrical rather than electronic problems.

If the voltage available at the relay terminals is lower than specifications, check the control circuit wiring and components for high resistance.

High resistance in the solenoid switch circuit will reduce current flow through the solenoid windings, which can cause improper functioning of the solenoid. In some cases of high resistance, it might not function at all. Improper functioning of the solenoid switch will generally result in the burning of the solenoid switch contacts, causing high resistance in the starter motor circuit.

Check the vehicle wiring diagram, if possible, to identify all control circuit components. These normally include the starting switch, safety switch, starter drive solenoid winding, or a separate relay drive.

While someone holds the starting switch in the start position, connect the voltmeter leads across each wire or component. A voltage drop exceeding 0.1 volt across any one wire or switch indicates high resistance. If a high reading is obtained across the neutral safety switch used on automatic transmissions, check the adjustment of the switch according to the manufacturer service literature.

Starter Circuit Testing

If the control circuit is operating properly, voltage drop tests can be made on the starter circuit. Voltage drop testing locates any source of excessive resistance in the starter circuit.

To perform voltage drop tests, an accurate low-range voltmeter is required. The meter range should be 2 or 3 volts full scale and be equipped with leads that are long enough to reach the various points being checked. One of the leads should be equipped with a sharp probe so that when battery cables are being checked, it can be jabbed into the battery post. This will allow the drop across the clamp and post to be measured along with the cable drop. When connecting the voltmeter to the switch or motor, connect it to the terminal stud rather than to the terminal so that the drop across the connection will also be measured. Also, connect the positive voltmeter lead to the part of the circuit that is more positive and the negative lead to the more negative point.

The first test to be performed on starters is the ground circuit resistance check. Determine whether the starter uses an **external ground** circuit (grounded through starter frame to engine) or **internal ground** circuit (dedicated negative cable). When working with an internally ground starter motor, one lead of the voltmeter is connected to the ground terminal of the battery and the other lead is connected to the ground post of the starter (**Figure 9–21**). Read the voltmeter while the engine is cranked. If the voltage reading exceeds 0.2 volt, excessive resistance exists in the ground circuit. Further voltage drop checks must be made between the battery ground terminal and the starter ground to pinpoint the source of the unwanted resistance. The same problem can be isolated in externally grounded starters by measuring the voltage drop across the following connections:

- Ground cable connection at the battery
- Ground cable connection at the engine
- Starter bolt connection to the engine
- Contact between the starter main frame and the end frames
- Any connection between the ground terminal and the starter base. Dirt, acid corrosion, loose connections, or any other contaminant can cause excessive resistance.

If the ground circuit resistance check does not expose any problems with resistance, the insulated circuit resistance check can be performed. The positive lead of the voltmeter is connected to the positive terminal of the battery. Then the engine is cranked.

FIGURE 9–21 Starter circuit testing

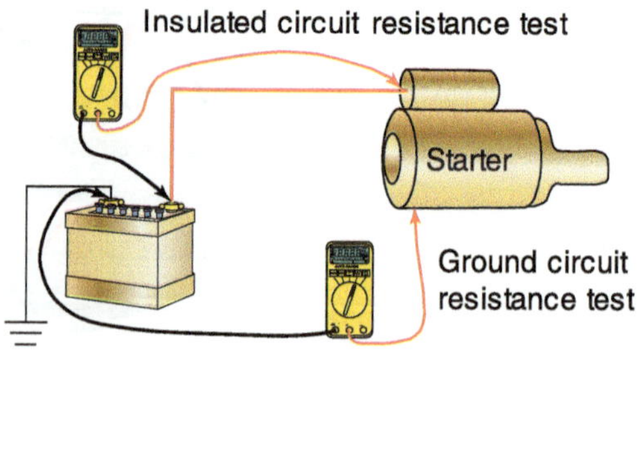

While the engine is being cranked, the other voltmeter lead is brought into contact with the starter input terminal.

Tech Tip

Note that direct-drive and gear reduction starter motors behave the same when subjected to voltage drop testing. With a current draw of up to 500 amps, the maximum permissible voltage drop is 0.5 V-DC.

A voltmeter reading higher than 0.5 volt indicates high resistance in one of the following components or connections of the insulated circuit:

- Cable connection at the battery
- Cable connection at the solenoid
- Cable
- Starter solenoid

Isolate the cause of excessive resistance by performing additional voltage checks across these possible sources. Repair or replace any damaged wiring or faulty connections.

SHOP TALK

When testing starter circuits use the OEM-recommended method of preventing the engine from starting. When performing voltage drop tests, make sure you record the results to two decimal places. Record the results by writing them down so you can easily check them to spec afterward.

NO-LOAD TESTS

When testing indicates that a starter malfunction is the cause of the no-start or hard-start condition, the starter must be removed from the vehicle for additional testing, the first of which should be a no-load test. The no-load test is used to identify specific defects in the starter that can be verified with tests when disassembled. Also, the no-load test can identify open or shorted fields, which are difficult to check when the starter is disassembled. The no-load test also can be used to indicate normal operation of a repaired motor before installation.

To perform a no-load test, first clamp the starter motor in a bench vise. Then, connect the test equipment as shown in **Figure 9–22**. Connect a voltmeter from the motor terminal to the motor frame and use an rpm indicator to measure armature speed. Connect the motor and an ammeter in series with a fully charged battery of the specified voltage and a

FIGURE 9–22 No-load test connections

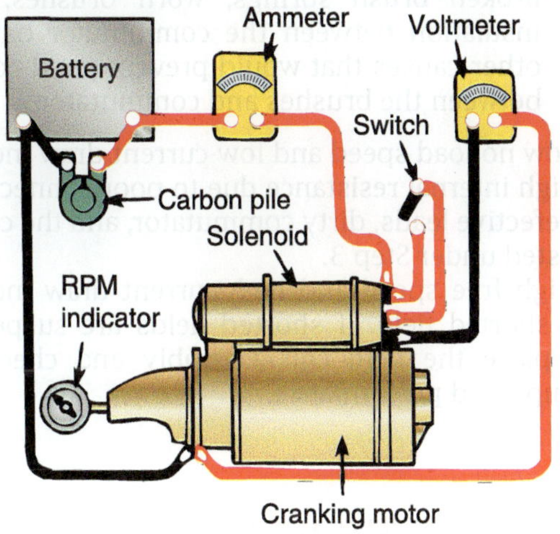

switch in the open position from the solenoid battery terminal to the solenoid switch terminal. Close the switch and compare the rpm, current, and voltage reading with specifications. It is not necessary to obtain the exact voltage specified, because an accurate interpretation can be made by recognizing that if the voltage is slightly higher, the rpm will be proportionately higher, with the current remaining essentially unchanged. However, if the exact voltage is desired, a carbon pile connected across the battery can be used to reduce the voltage to the specified value. If more than one 12V battery is used, connect the carbon pile to only one of the 12V batteries. If the specified current draw does not include the solenoid, deduct from the ammeter reading the specified current draw of the solenoid hold-in winding. Disconnect only with the switch open. Interpret the test results as follows:

1. Rated current draw and no-load speed indicate normal condition of the cranking motor.
2. Low free speed and high-current draw indicate:
   - Too much friction. Tight, dirty, worn bearings, bent armature shaft, or loose pole shoes, allowing the armature to drag.
   - Shorted armature. This can be checked further after disassembly.
   - Grounded armature or fields. Check further after disassembly.
3. Failure to operate with high-current draw indicates:
   - A direct ground in the terminal or fields.
   - Seized bearings. This can be determined by turning the armature by hand.
   - Failure to operate with no-current draw indicates:
   - Open field circuit. This can be checked after disassembly by inspecting terminal connections and tracing the circuit with a test lamp.

- Open armature coils. Inspect the commutator for burned bars after disassembly.
- Broken brush springs, worn brushes, high insulation between the commutator bars, or other causes that would prevent good contact between the brushes and commutator.

4. Low no-load speed and low-current draw indicate high internal resistance due to poor connections, defective leads, dirty commutator, and the causes listed under Step 3.
5. High free speed and high-current draw indicate a shorted field. If shorted fields are suspected, replace the field coil assembly and check for improved performance.

9.4 STARTER REBUILD

The rebuilding of starter motors is a less common practice today than it used to be, but it is still performed. The procedure outlined here references that for a Remy 42MT, which continues to be one of the most commonly used in the truck industry because it can better sustain prolonged cranking than the new-generation gear-reduction units. As usual, you should consult the OEM service literature when reconditioning starter motors in the field.

DISASSEMBLY

1. Scribe the starter housing so that the locations of the solenoid, lever housing, nose plate, and endplate are indicated for reassembly.
2. Disconnect the field coil connector from the solenoid to motor terminal and the solenoid ground lead where applicable.
3. Remove the brush inspection caps and then remove the brush lead caps. This separates the field leads from the brush holders.
4. Remove the solenoid and then remove the through bolts, separating the commutator end cap from the field frame body. Separate the brush plate assembly from the field frame body.
5. Next, separate the nose housing and lever housing. Remove the pinion stop.
6. Remove the clutch assembly retainer and snapring from the armature shaft by driving the collar toward the armature core.
7. Separate the clutch assembly from the armature shaft—and the armature from the lever housing. Press the bearings from both the gear housing and the drive housing using a mandrel.
8. Clean all of the disassembled components with a solvent with no residual oil and a firm bristle brush. Allow the components to air dry. If the commutator is dirty, clean it with #100 or finer sandpaper. If the commutator cannot be cleaned the armature assembly should be replaced.

INSPECTION

1. Check the brushes for wear and replace if necessary: The brush contact face should have maximum contact with the commutator. The brush holders should be checked for binding and the springs visually inspected. If the springs are discolored or distorted, replace them.
2. Inspect the armature assembly. Check for loose commutator connections. These can be resoldered to the riser bars, but it is not recommended. Test the armature for shorts, grounds, and opens.
3. A growler is used for testing shorts. Place the armature on the growler and hold a steel strip on the armature; it will vibrate over a short.
4. Using a test lamp, check for armature grounds; one lead should be placed on the commutator and the other on the armature shaft. If the test lamp illuminates, the armature is grounded and should be replaced.
5. Locate opens by checking for continuity any place that conductors are joined to the commutator. Either resolder connections or replace the armature if there are opens.
6. Test the field coils for grounds by using a 110-volt test lamp and connecting one lead to the field frame and the other to the field connector. If the lamp illuminates, the field coils are grounded.
7. Check the field coils for opens with the testlight: If it does not illuminate, the field coil is open. If you have to replace the field coils, use a pole shoe spreader and pole screw screwdriver.
8. Check out the bearings by rotating and noting any roughness.

REASSEMBLY

1. Install the field coils in the housing. Immerse the bearings and both tangent wicks in engine oil prior to installation.
2. Press the bearings into the housing using a mandrel.
3. Next, install the oil seal in the lever housing followed by the field frame. Coat the spacer washer with the appropriate lubricant and fit it to the armature shaft. Then insert the armature into the field frame, making sure you align the scribe marks made on disassembly, and leave the commutator exposed.
4. Run the through bolts through the assembly and insert the brush inspection caps.

5. Now install the clutch assembly onto the armature shaft and install the solenoid plunger and shift lever mechanism. Ensure that the shift lever assembly operates freely. Place the clutch assembly retainer onto the armature shaft, ensuring that the cupped surface faces the snapring groove. Then slide the snapring into position.
6. Install the solenoid and field coil connection. Install the ground lead securely. Finally, check pinion clearance as outlined in the next section.

ADJUSTING PINION CLEARANCE

1. Disconnect the field coil connector (the M terminal) from the solenoid motor terminal and insulate it. Connect a battery (same voltage as solenoid) from the solenoid S terminal to the solenoid housing.
2. Flash a jumper lead from the solenoid Mtr terminal and the ground return terminal; this should kick the starter shift pinion into cranking position, where it will remain while the battery is connected.
3. Apply a little pressure on the pinion, forcing it back toward the commutator, and measure clearance between the pinion and nose housing as shown in **Figure 9–23**. The specification should be typically around 0.40 inch but check the Remy data.

FIGURE 9–23 Pinion clearance on a 42MT starter motor: Measure with feeler gauges—adjust at shaft nut if required.

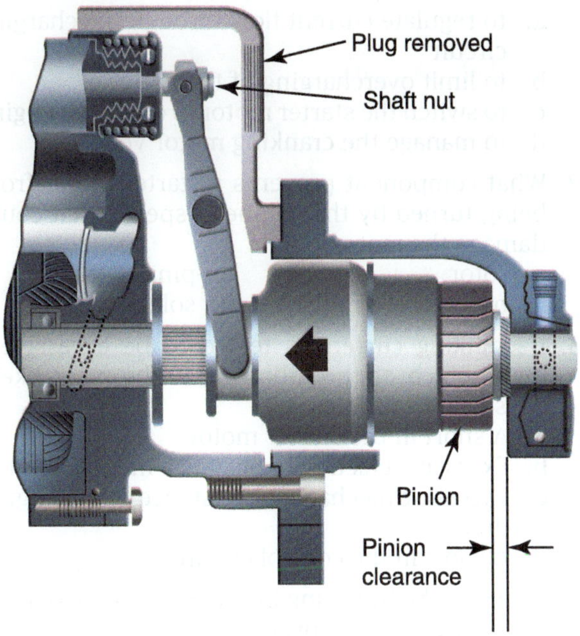

4. To adjust pinion clearance, remove the plug on the rear of the solenoid and turn the shaft nut that connects the shift lever to the solenoid as shown in Figure 9–23.

SUMMARY

- The vehicle cranking circuit functions to crank the engine until it can operate under its own power.
- A cranking circuit is managed by a control circuit that uses low current to switch and energize a high-current starter motor circuit.
- The starter circuit in some trucks is managed electronically by an ECM. This allows the ECM to monitor and control input current to the starter motor and protect the circuit from voltage spikes.
- A starter motor converts the electrical energy of the vehicle batteries into mechanical energy for cranking the engine.
- A powerful electromagnetic field in the starter motor field coils is used to rotate the armature assembly in a typical starter motor.
- Most starter motors are series wound, so there is only one path for current flow through the assembly. This means that all of the current flowed through the field coils also flows through the armature, providing the conditions that along with the absence of CEMF produce peak torque at close to stall speeds.
- CEMF works in opposition to battery current in an armature. It rises proportionally with armature speed. For this reason, CEMF helps limit starter overspeed.
- A new generation of lightweight, high-torque starter motors uses planetary gear reduction to multiply cranking torque and greatly reduce motor weight.
- Voltage drop testing should be used to troubleshoot a cranking circuit malfunction because it does so dynamically.
- A heavy-duty truck starter motor found to be defective should be removed from the engine and either rebuilt or replaced.
- A growler is used to test an armature for shorts: A steel strip such as a hacksaw blade is placed on top of the armature during testing and if it vibrates, a short is indicated.
- A testlight is used to test an armature for grounds and opens.
- Field coils should be checked using a testlight.
- After reassembling a starter motor, the pinion clearance should be checked with thickness gauges and adjusted if necessary.

REVIEW QUESTIONS

1. What is the function of the control circuit in a truck cranking circuit?
   a. to regulate current flow through the charging circuit
   b. to limit overcharging of the batteries
   c. to switch the starter motor to crank the engine
   d. to manage the cranking motor voltage
2. What component prevents a starter motor from being turned by the engine at speeds that could damage the motor?
   a. rotor
   b. overrunning clutch
   c. pinion gear
   d. solenoid gear
3. If cranking current is tested to be excessively high, which of the following is the most likely cause of the problem?
   a. A short in the starter motor.
   b. Excessive electrical resistance in the circuit.
   c. Excessive mechanical resistance due to engine drag.
   d. A short in the control circuit.
4. Which of the following best describes the operating principle of a typical truck starter motor?
   a. AC motor
   b. permanent-magnet motor
   c. series-wound motor
   d. constant-torque motor
5. When will output torque be greatest in a truck starter motor?
   a. when energized at full stall
   b. when energized at between 200 and 250 rpm
   c. when energized at its highest rpm
   d. when open and stationary
6. What is the usual maximum permitted voltage drop through the insulated side of the starter circuit?
   a. 0.05 volt
   b. 0.1 volt
   c. 0.5 volt
   d. 1.0 volt
7. What is the usual maximum permitted voltage drop through the ground side of the starter circuit?
   a. 0.05 volt
   b. 0.1 volt
   c. 0.5 volt
   d. 1.0 volt
8. What device is used on some cranking circuits to prevent starter motor operation if the voltage drops below 9.6 volts?
   a. a voltage regulator
   b. a voltage sensing module
   c. an isolation switch
   d. a neutral safety switch
9. What is the reason for using a planetary gearset in modern starters such as the Remy 39MT?
   a. high torque output
   b. reduced weight
   c. more compact size
   d. all of the above
10. Which cranking system component is common to both the control circuit and the high-current starter circuit?
    a. an ignition switch
    b. a fusible link
    c. a starter relay
    d. a solenoid
11. When checking voltage drop through a negative battery clamp, where should the positive voltmeter lead be placed?
    a. on the negative battery post
    b. on the negative battery clamp
    c. on the chassis ground
    d. on the positive battery post
12. When growler testing an armature, if a steel strip vibrates when placed on the armature, which of the following conditions is indicated?
    a. a short
    b. an open
    c. high resistance
    d. no residual magnetism
13. Technician A says that a voltage drop of 0.4 volt measured over a battery terminal and post indicates a high-resistance connection. Technician B says that ground circuit specifications for voltage drops tend to be higher than on insulated circuits. Who is correct?
    a. Technician A only
    b. Technician B only
    c. both A and B
    d. neither A nor B
14. Technician A says that a starter relay is a magnetic switch. Technician B says that the control circuit in a starter relay conducts the starter motor current load. Who is correct?
    a. Technician A only
    b. Technician B only
    c. both A and B
    d. neither A nor B
15. Which of the following produces the stationary magnetic fields in a typical truck starter motor?
    a. brushes
    b. armature
    c. commutator
    d. field coils and pole shoes
16. Technician A says that most heavy-duty starter motors are series wound. Technician B says that most heavy-duty starters produce peak torque when driven at the highest rpm. Who is correct?
    a. Technician A only
    b. Technician B only
    c. both A and B
    d. neither A nor B
17. What has to be adjusted to change the pinion clearance specification in a 42MT starter motor?
    a. the solenoid pinion shaft and shift lever assembly
    b. the pinion pilot bushing
    c. the armature endplay
    d. the shift lever clevis on the armature shaft

18. Technician A uses a spring tension gauge to set the starter motor pinion clearance specification. Technician B uses thickness gauges to measure pinion clearance. Who is correct?
    a. Technician A only
    b. Technician B only
    c. both A and B
    d. neither A nor B
19. What is the function of a neutral safety switch in the cranking circuit?
    a. to prevent accidental engine startup in neutral gear
    b. to prevent engine startup when an automatic transmission is in gear
    c. to prevent the engine from being over-revved in neutral
    d. to prevent the starter motor from overspeeding after engine start
20. Technician A says that the sun gear pinion is integral with the armature shaft on a Remy 39MT starter motor. Technician B says that the ring gear is the output member of the planetary gearset on a 39MT starter motor. Who is correct?
    a. Technician A only
    b. Technician B only
    c. both A and B
    d. neither A nor B

Prerequisite: Chapters 5, 6, 7, 8, and 9

10 CHASSIS ELECTRICAL CIRCUITS

OBJECTIVES

After reading this chapter, you should be able to:

- Describe how a light bulb functions.
- Explain the operating principles of halogen, LED, and high-intensity discharge (HID) lamps.
- Describe the function of the reflector and lens in a headlamp assembly.
- Aim truck headlights.
- Troubleshoot lighting circuit malfunctions.
- Describe the operation of typical truck auxiliary equipment.
- Explain how a trailer electrical plug and connector are connected.
- Outline the operating principles of truck instrument cluster components.
- Diagnose and repair some typical truck instrument cluster failures.
- Explain the function and operation of warning and shutdown systems.
- Identify the types of circuit protection used in truck electrical systems, including fuses and cycling and noncycling circuit breakers.
- Describe the procedure and material required to solder a pair of copper wires.
- Outline the procedure required to quickly check out a truck electrical system.

KEY TERMS

cycling circuit breaker
data display unit (DDU)
daylight running lights (DRLs)
Federal Motor Vehicle Safety Standard 108 (FMVSS 108)
filament
halogen infrared (HIR)
halogen lamp
heads-up display (HUD)
high-intensity discharge (HID)
hotel load
hotel load grid
instrument cluster unit (ICU)
light-emitting diodes (LEDs)
luminosity
noncycling circuit breakers
parabolic
relay
stalk switch
tungsten filament
virtual breaker
virtual circuit protection
virtual fuse
voltage limiter
xenon headlamp
XVision

INTRODUCTION

As each model year passes, the reliance on chassis electrical system components increases in trucks. The type of electrical circuits used on trucks is also changing and the expanding role of chassis data buses (see Chapter 12) is resulting in the replacement of electromechanical switches by "smart switches." Electromechanical switches function by changing the electrical status in a circuit—that is, by directly energizing or opening circuits to control electrical components. Smart switches simply send low-potential messages to the chassis data bus. These messages can then be interpreted and acted on by one or more of the various modules connected to the data backbone of the chassis.

This chapter examines the chassis electrical equipment that is managed off-network (at this moment in time) from the vehicle multiplex system. We begin with chassis lighting systems and then move on to some of the other auxiliary electrical components.

10.1 LIGHTING SYSTEMS

The lighting systems used on heavy-duty highway trucks can be generally divided into the interior and exterior circuits. Technicians will learn a lot more about the exterior circuits because these are more exposed to road and weather-related hazards and require more frequent repairs. All U.S.-built trucks must comply with lighting standards identified in **Federal Motor Vehicle Safety Standard 108 (FMVSS 108)**. **Figure 10–1** shows the location of some of the exterior lights on a Class 8 highway tractor.

Because of the negative-ground chassis system used on trucks and trailers, the insulated or positive side of the circuit routes the electrical circuit to the load and the circuit is completed by grounding it at, or close to, the load. Although some light circuits are equipped with dedicated two-wire circuits, truck and trailer electrical lighting circuits consist of the insulated portion of the circuit and use local chassis ground to complete the circuit.

Exterior lighting consists of headlamp, marker, clearance, fog lights, **daylight running lights (DRL)**, stoplights, and turnlights. Interior lighting includes dome lights, reading lights, gauge and switch illumination, and warning lights. The means used to provide the light can be generally categorized as incandescent, gaseous discharge, or light-emitting diode (LED).

FIGURE 10–1 Exterior lighting locations on a Class 8 tractor

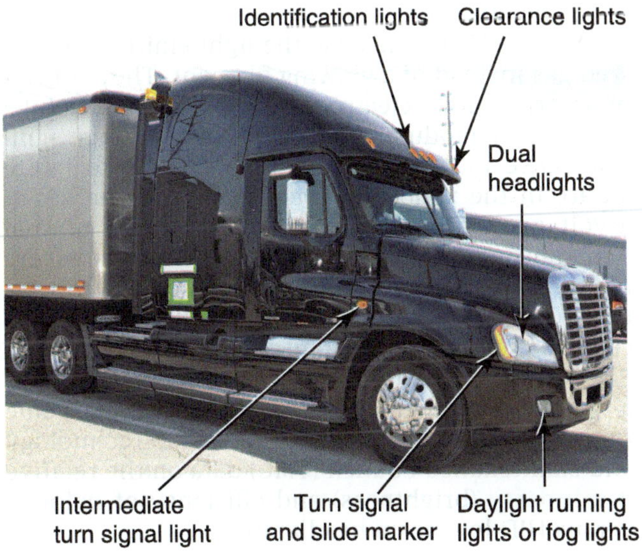

LIGHT BULB OPERATION

Most devices we call light bulbs use electrical energy to excite atoms to produce photons. Photons are light energy. A typical incandescent light bulb is constructed of a bulbous, thin-walled, glass envelope. Sealed inside the glass is an inert gas such as argon or nitrogen. Also inside the glass is an electrical circuit. It consists of a resistor called a **filament**. The filament allows current flow through it by means of a pair of contact terminals on the socket of the bulb. The filament is typically made out of tungsten. Current flow heats the **tungsten filament** up to a white-hot temperature. Like any heated metal, the tungsten becomes white hot, something in the region of 4,500°F (2,500°C). A metal heated to these kinds of temperatures emits both heat and visible light in a reaction known as incandescence. Visible light energy is radiated as photons. **Figure 10–2** shows a typical vehicle incandescent light bulb.

Incandescent light bulbs use a thermal radiation principle. Thermal radiation light bulbs are not especially efficient because much of the electrical energy consumed is radiated as heat as well as photons. A typical light bulb lasts until the tungsten in the filament vaporizes, creating a high-resistance spot, following which the bulb burns out.

FLUORESCENT LIGHTS

Fluorescent lights are used in many truck sleepers. The tubes are filled with mercury vapor that, when used to conduct current, converts the electrical energy into ultraviolet (UV) light. A coating on the interior of the tubing converts the UV light into lower-frequency visible light. Fluorescent lights require a ballast to function. The ballast provides a high-voltage ignition pulse to ignite the arc through the mercury vapor and then maintains the arc with a low-voltage pulse. The advantage of fluorescent lights is that they provide good light while consuming little electrical energy, so they are ideal for illuminating a truck sleeper.

FIGURE 10–2 Incandescent bulb

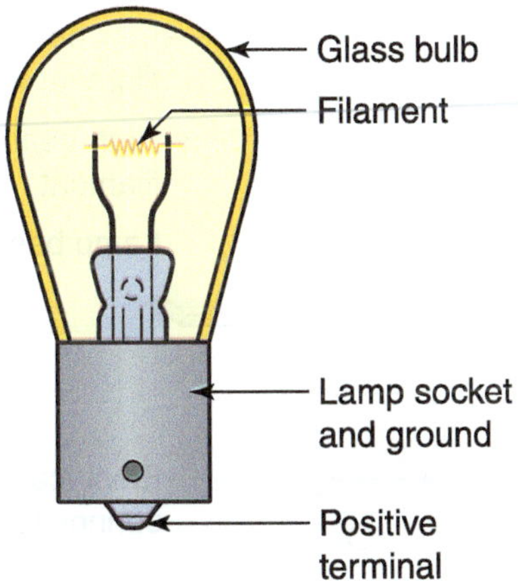

HALOGEN LAMPS

Halogen lamps use tungsten filaments. Halogen lamps use a thermal radiation principle. However, in a halogen light the tungsten filament is encased inside a small quartz envelope. Because of the small size of the quartz envelope it is necessarily located so close to the filament that it would melt if it were made from glass. Inside the quartz envelope are halogen gases that have the characteristic of combining with tungsten vapor. This means that when the bulb is at operating temperature, the halogen gas combines with tungsten atoms as they vaporize and redeposits them on the filament. This recycling of vaporized tungsten allows the filament to outlast typical comparative headlamp bulbs by a lot. Also, in the halogen bulb it becomes possible to run the filament at higher temperatures—meaning you get more light per unit of energy consumed. Halogen bulbs still produce a lot of heat; in fact, they run at much higher temperatures compared to a normal light bulb. **Figure 10–3** shows a typical halogen bulb and **Figure 10–4** shows its operating principle.

HALOGEN INFRARED (HIR) HEADLAMPS

Halogen infrared (HIR) bulbs use a multilayer, thin-film technology that allows light to pass through the bulb glass while reflecting some infrared energy back to a tungsten filament. This redirecting of energy back toward the filament allows bulb manufacturers to:

- Reduce power (wattage)
- Increase light emitted
- Increase headlamp service life

FIGURE 10–3 Halogen bulb

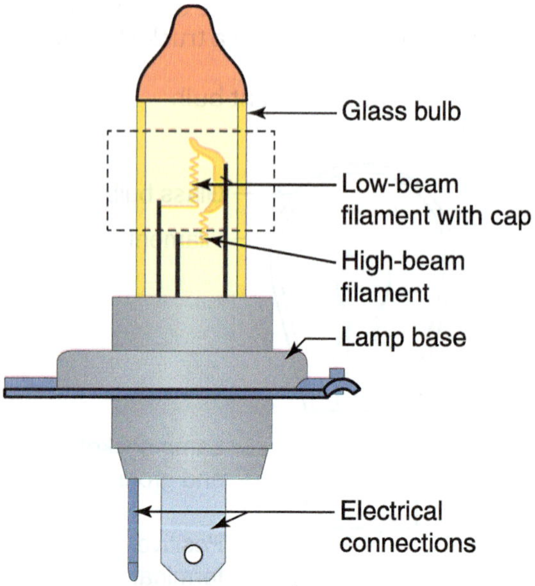

FIGURE 10–4 Operating principle of a halogen bulb: Vaporized tungsten is returned to the tungsten filament.

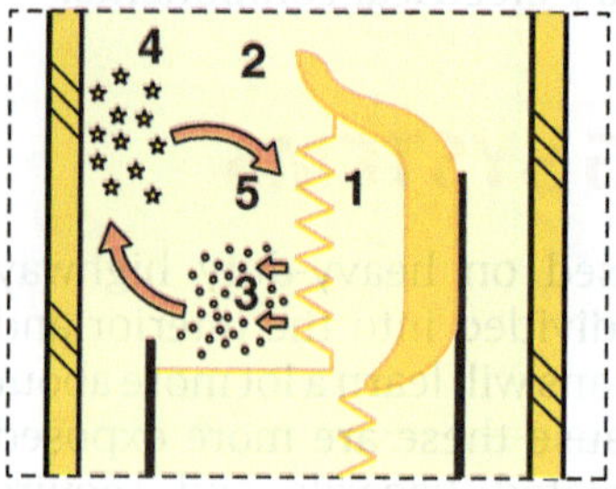

or any combination of the three to allow the units to significantly outperform standard halogen lights. HIR lamps are only available in single-filament design because the filament coil has to be centered in the elliptical filament tube to optimize the capture of reflected infrared light.

XENON HEADLAMPS

Xenon headlamps, sometimes known as **high-intensity discharge (HID)** lamps, have gained in popularity despite their initial high cost because of increased light output and reduced power consumption compared with halogen headlamps. When they are observed on trucks today, they are usually on owner-operator commercial trucks. They use a gaseous-discharge operating principle. A xenon light works a little like a sodium streetlight that produces light using an arc between electrodes through a gas. Sodium streetlights have the disadvantage of taking up to 5 minutes before producing normal light levels—not an acceptable shortcoming in a vehicle headlamp. **Figure 10–5** shows a typical xenon headlamp.

Vehicle HID lamps use the light emitted by ionized gas instead of a glowing filament. They consist of an arc source, electrodes, and a ballast module. The ballast module produces a high-voltage ignition pulse in the region of 20,000 volts to ignite an arc in the xenon gas then maintains the arc at a voltage of around 85 volts. In this way, normal light output is achieved within 4 seconds of arc ignition, which is satisfactory for highway operation. The efficiency with which HID headlamps convert electrical energy into light energy exceeds 80 percent compared to around 20 percent for a halogen headlamp. **Table 10–1** shows some American Trucking Association (ATA) Technology and Maintenance Council (TMC) data on the relative **luminosity** (brightness) and efficiency of halogen versus HID low-beam headlamps.

FIGURE 10–5 Xenon headlamp components

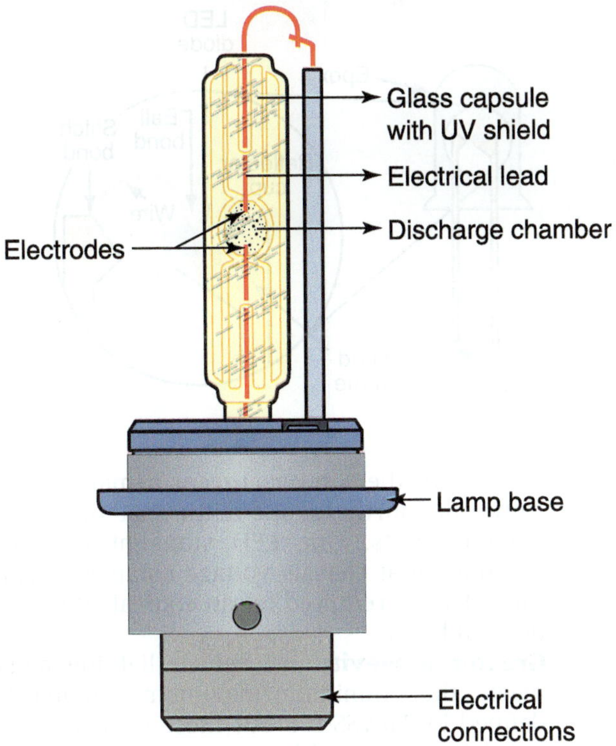

TABLE 10–1 HID and Halogen Luminosity and Efficiency

Factor	Halogen 9006	HID
Wattage (W)	55	35
Lumens (lm)	1,000	3,000
Efficiency (lm/W)	18	86

HID OPTICS

Xenon headlamps require specially designed optics and reflectors to ensure that they meet FMVSS 108 standards in trucks. At this moment in time it is illegal to retrofit HID lamps to existing chassis using halogen lamps. HID lamps must be fitted 33.5 inches (85 cm) or lower from the ground. Despite creating some complaints due to their increased brightness, HID lamps meet all current U.S. and Canadian standards for highway use. Xenon headlamps allow the driver to see the road better, and the "blue" light they reflect is closer to natural sunlight and better stimulates reflective signs and road markers. HID headlights are becoming commonplace but the additional cost has been a limiting factor.

Reflectors. When a beam of light is directed at a mirror, the angle of incidence and reflection should be equal. Vehicle headlight reflectors are designed to be **parabolic** (even conical recess) so that light beams leaving the focal point at the source (bulb unit) are emitted parallel to the reflector axis as shown in **Figure 10–6**. The objective is to concentrate the focus of the emitted light from the headlamp assembly in such a fashion that it will effectively illuminate the road area in front of the driver.

Lens. A light assembly lens is used to disperse the light rays either directly from the bulb or reflected by the parabolic reflector. In other words, the lens refracts light from the reflector to achieve the desired distribution pattern on the road ahead of the vehicle. The lens is designed to act in conjunction with the reflector assembly to properly direct or refract the light emitted from the bulb. **Figure 10–7** shows the optics of a typical lens.

Service Performance

Road shock and vibration, the culprits in defining the service life of halogen bulbs, have no effect on xenon elements because an arc is used in place of a filament. Most HID failures are accounted for by oxidation of the electrodes used to conduct the arc. Oxidation of the

FIGURE 10–6 Reflector operation: Light is reflected in parallel rays.

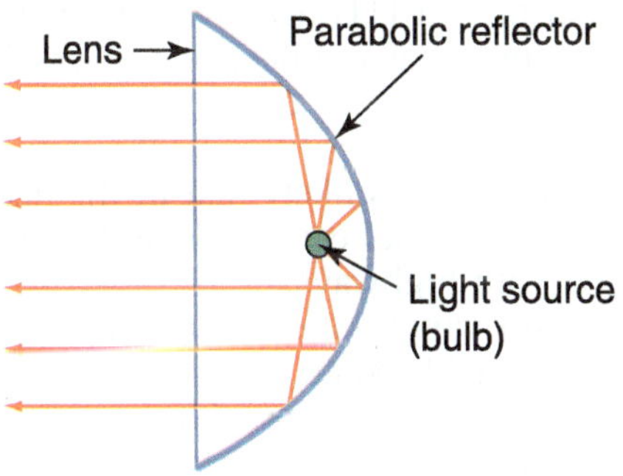

FIGURE 10–7 Lens optics: Light is refracted from the reflector to the desired distribution pattern.

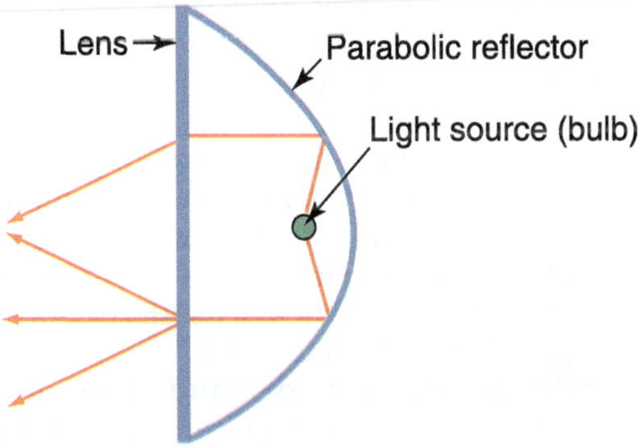

electrodes is accelerated by frequent cycling, which is more of a problem in passenger cars than in heavy truck applications. Nevertheless, drivers should be discouraged from using HID headlamps to flash signals.

MANAGED HEADLAMP VOLTAGE

Voltage spikes and current rushes both shock-load the tungsten filaments used in standard and halogen headlamp bulbs. This type of damage is most common at initial startup when the lamp is cold. Additionally, headlamps are engineered for operation at 12.6 volts, and bulb life is reduced if chassis voltage exceeds this value. One fleet determined that a 1 percent increase in system voltage correlated to a 14 percent reduction in bulb life. Some fleets are using electronic voltage-stabilizing devices to condition chassis system voltage and suppress voltage spikes. At this moment in time, such devices are only available from aftermarket suppliers.

LIGHT-EMITTING DIODES (LEDS)

A **light-emitting diode (LED)** is a solid-state component. The operating principles of LEDs are covered in Chapter 6; there we said that all diodes emit some electromagnetic light when in forward bias. Certain diodes such as those constructed of gallium arsenide phosphide emit high levels of light energy, and because they have greater longevity and consume less electrical energy than any other current light bulbs, they are widely used in commercial truck lighting systems.

The adoption of LED headlights in medium- and heavy-duty trucks has not been as universal as it has in light-duty trucks and automobiles, but in 2013 Penske Truck leasing retrofitted 5,000 units for a trial evaluation period. It is likely that in the near future, LED commercial truck headlights will become a common sight on our highways.

Because no vaporization takes place in an LED, they can have a life span that exceeds that of the vehicle, although some dimming may take place. **Figure 10–8** shows a typical LED. The TMC of the ATA has identified the following advantages of using LEDs on truck lighting systems:

- **Safety.** LEDs illuminate in less than 0.1 second compared with about 0.25 second for incandescent bulbs. This translates to 20 feet of extra stopping distance running at a speed of 60 mph and can reduce the incidence of rear-end collisions. The low power consumption of LEDs benefits antilock brake system (ABS) operation. ABS systems require a minimum voltage of 9.5 volts to meet certification standards, so the reduced electrical requirement for the lighting circuit assists ABS performance required during panic stops.
- **Reduced current demand.** Use of LEDs can alter the wiring requirements for trailers because of their lower current demand, typically one-fourth to one-tenth that of incandescent lights. One LED stoplight requires 0.5 ampere at chassis voltage compared with the 2.0 amps required for an equivalent incandescent lamp.
- **Greater longevity.** All vehicle lighting must meet the minimum/maximum standards defined by FMVSS-108 but these standards do not define expected longevity. Some manufacturers rate LEDs to a 100,000-hour lifetime but it should be noted that LEDs seldom outright fail. However, subjected to prolonged usage they do dim and require replacement for that reason.

FIGURE 10–8 LED construction

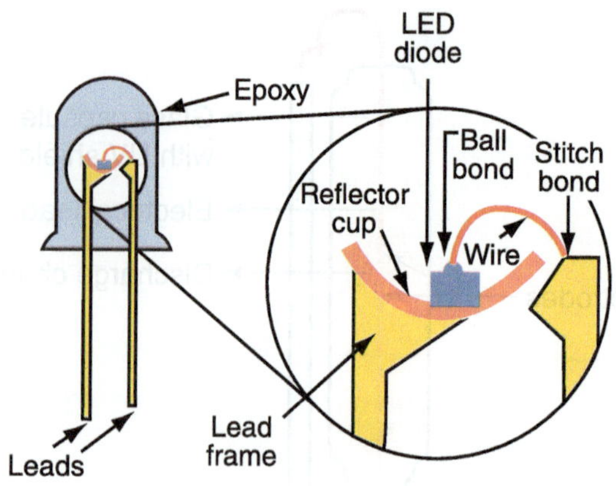

LED Precautions

LEDs operate at cooler temperatures than incandescent bulbs and some operators have experienced problems with snow and ice buildup that would not be a factor with incandescent bulbs. Mounting location of LED assemblies is critical for trailer original equipment manufacturers (OEMs), but the bottom line is that operators are responsible for ensuring that light units are not compromised by snow and ice. Theft of LED light assemblies is a factor in some operations. The TMC recommends the use of theft-resistant brackets and one-way mounting screws to minimize theft.

LEDs Conclusion

The trucking industry has embraced the use of LED lighting systems for the aforementioned reasons. LED lighting systems are commonly specified on new highway transport equipment, and many fleets have retrofit LEDs to older equipment because of their benefits.

NIGHT VISION SYSTEMS

Although a typical truck operates only 30 percent of its operational life at night, more than half of truck accidents occur when it is dark. The cause is reduced visibility. A driver with 20/20 vision sees with no better than 20/50 vision when driving at night with halogen headlamps. This has led to the emergence of infrared thermal imaging systems such as the Bendix **XVision** system.

XVision consists of a front-mounted infrared camera that senses heat from the environment and processes the signal electronically to produce a virtual display as an output on the vehicle dash. Its ability to detect heat is effective to differential values of as little as 0.4°F (0.2°C). This enables it to distinguish between a concrete barrier and painted stripes on the barrier that absorb and retain more heat during daylight. The driver display consists of a black-and-white image. Warmer objects such as people, animals, and headlights appear in shades of white, whereas cooler objects such as guardrails, trees, and abutments appear in shades of gray and black.

MAINTENANCE

In addition to replacing defective lamps and bulbs, periodically check to see that all wiring connections are clean and tight. In addition, lamp units should be tightly mounted, should provide a good ground, and be properly adjusted. Loose or corroded connections can discharge batteries, dim lights, and damage charging circuit components.

Wires and/or harnesses must be replaced if insulation becomes burned, cracked, or deteriorated. Whenever it is necessary to splice a wire or repair one that is broken, use rosin flux solder to bond the splice and shrink tube to seal splices or bare wires. A repair alternative is to use crimped butt splices. This can be done with connectors that come with tubing and sealant built-in. Do not attempt to make a splice by twisting the wires together and taping or by using wire nuts. These methods make high-resistance connections that deteriorate quickly.

SHOP TALK

Lights should be turned off when cranking the engine to avoid damage by transient voltage spikes. As the engine is cranked, 650 to 1,200 amps are drawn through the system. At the precise moment when the starter is disengaged, electricity may surge into any active electrical circuits. This random surge can shorten the life of the lights, causing them to burn out prematurely.

HEADLIGHT FIXTURES

A typical current headlight system consists of either two or four sealed-beam tungsten headlight bulbs. In a two-headlight system, each light has a pair of filaments, one for high beam and the other for beam operation. In a typical four-headlight arrangement, the two outer lights have a double filament, whereas the inner pair of lights has only the high-beam filament. Where the four headlights are vertically arranged, the upper lights have the double filament. Headlamps may be either round or rectangular in shape.

Replacing Headlamp Bulbs

Halogen bulbs may be replaced by unplugging the electrical connector, twisting the retaining ring about one-eighth turn counterclockwise, and removing the bulb from its socket. To install a new bulb, simply reverse this procedure.

Figure 10–9 shows the procedure for changing the element from a typical halogen bulb assembly.

CAUTION

When replacing halogen bulbs, take care not to contact the bulb with fingers because this can cause a rapid failure. When installing a bulb, handle it by the base only.

When replacing HID lights always observe the OEM-recommended procedure. Although HID lamps generally outlast others, failure of the light assembly and exciter unit can be caused by frequent cycling. Advise drivers not to use HID headlamps for signaling. First, inform them that HID units take up to 4 seconds to reach full luminosity so they are not ideal for flashing other drivers; next, point out that they fail by frequent rapid cycling.

Headlight Adjustment

Headlights must be maintained in adjustment to obtain maximum illumination. Out-of-adjustment headlights can temporarily blind oncoming drivers and create hazardous conditions. To adjust headlights, headlight aim must be checked first. Various types of headlight-aiming devices are available commercially. When using headlight-aiming equipment, follow the instructions provided by the equipment manufacturer. Where headlight-aiming equipment is

FIGURE 10–9 Replacing a halogen light element

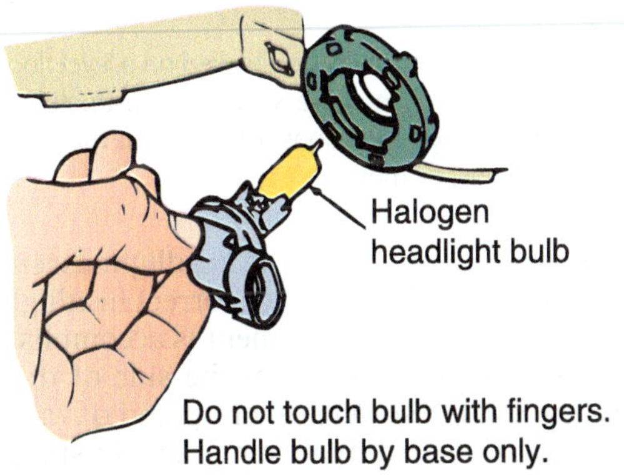

FIGURE 10–10 Headlight aiming pattern

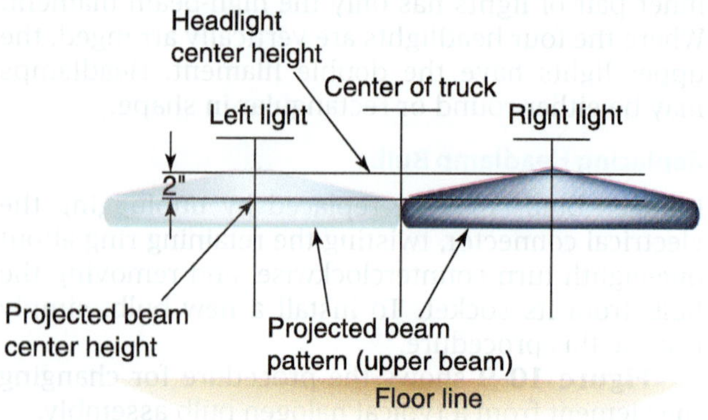

not available, headlight aim can be checked by projecting the high beam of each lamp onto a screen or chart at a distance of 25 feet. The truck should be exactly perpendicular to the chart.

The chart should be marked in the following manner (**Figure 10–10**). First, measure the distance between the centers of the matching headlights. Use this measurement to draw two vertical lines on the screen with each line corresponding to the center of a headlight. Then draw a vertical centerline halfway between the two vertical lines. Next, measure the distance from the floor to the centers of the headlights. Subtract 2 inches (50 mm) from this height and then draw a horizontal line on the screen at this new height. With headlights on high beam, the focal point of each projected beam pattern should be centered on the point of intersection of the vertical and horizontal lines on the chart. If necessary, adjust headlights vertically and/or laterally to obtain proper aim.

Headlight Aim Station

A larger service shop will probably have a headlight aiming station, which simplifies the procedure previously outlined. There are a number of different headlight aim stations available and that shown in **Figure 10–11** is typical; this unit also has a headlight intensity gauge shown in **Figure 10–12**.

SHOP TALK

Headlight aim should usually be checked on a level floor with the vehicle unloaded. In some states, this instruction might conflict with existing regulations. Where this is the case, legal requirements must be met.

Adjusting screws move the headlight assembly in relation to the hood (fender) to correct headlight aim. Lateral or side-to-side adjustment is accomplished by turning the adjusting screw at the side of the headlight (**Figure 10–13**). Vertical or up-and-down adjustment is accomplished by turning the adjusting screw

FIGURE 10–11 Typical headlight aiming station

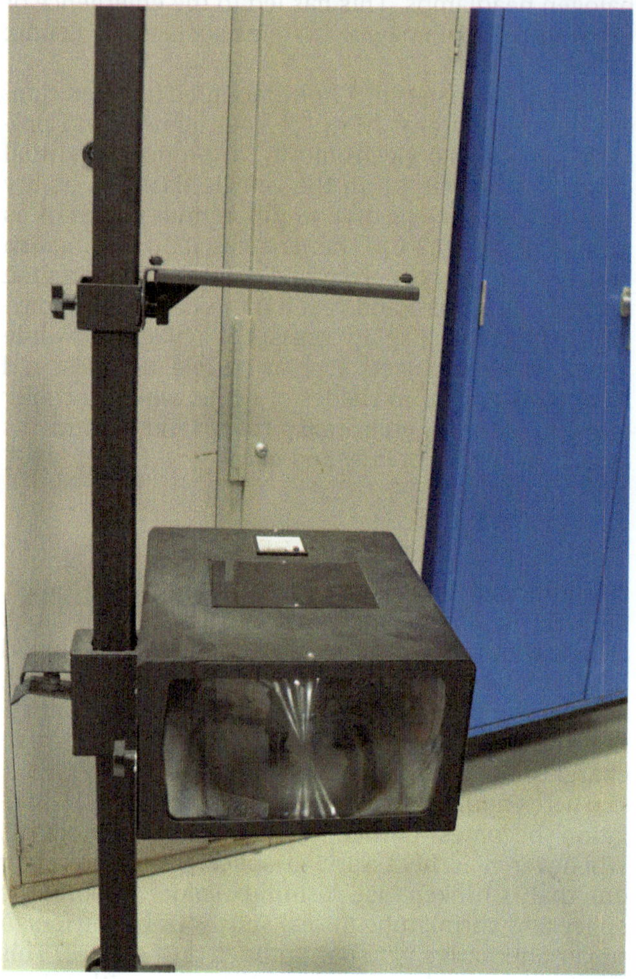

FIGURE 10–12 Headlight intensity gauge used on some headlight aiming stations.

at the top of the headlight. Adjustments can be made without removing headlight bezels.

Headlight Replacement

There are variations in procedures from one model to another when replacing headlights. On some models, the turn signal light assembly must be removed before the headlight can be replaced. Overall, however, the

FIGURE 10–13 Typical location of headlight adjusting screws

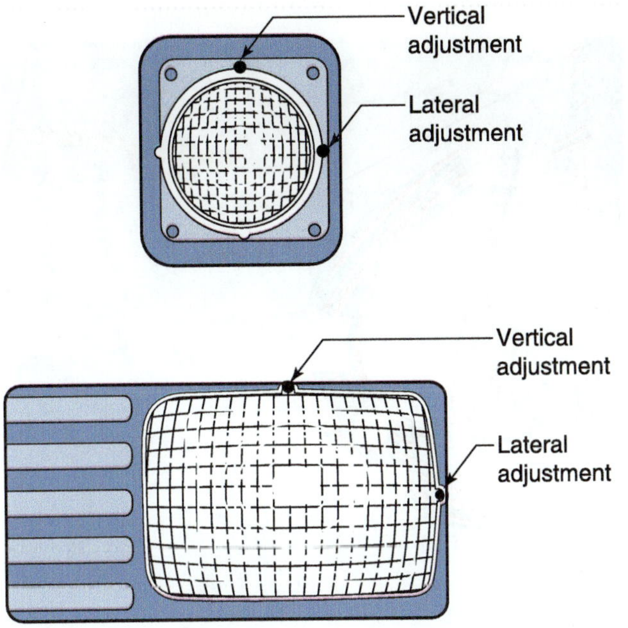

procedure will not differ much from the following typical instructions:

1. Remove the headlight bezel retaining screws. Remove the bezel. If necessary, disconnect the turn signal plug.
2. Remove the retaining ring screws from one or both lamps.
3. Remove the retaining rings.
4. Remove the light from the housing; disconnect the connector behind the light.
5. Insert the connector onto the terminal prongs of the new light.
6. Place the new light in the headlight housing. Position it so that the embossed number in the lamp lens is on top.
7. Place the retaining ring over the lamp and install the retaining ring screws; tighten them securely.
8. Check the aim of the headlight and adjust it in the manner described earlier.
9. Install the headlight bezel; secure it with the retaining screws. Connect the turn signal wiring (if it was disconnected). Do not overtighten bezel retaining screws. Overtightening could cause damage (stripping) of threads in the hood (fender).

SHOP TALK

Some manufacturers recommend coating the prongs and base of the new sealed beam with terminal grease for corrosion protection. Use an electrical terminal protective approved by the manufacturer.

FIGURE 10–14 Side marker/turn signal light

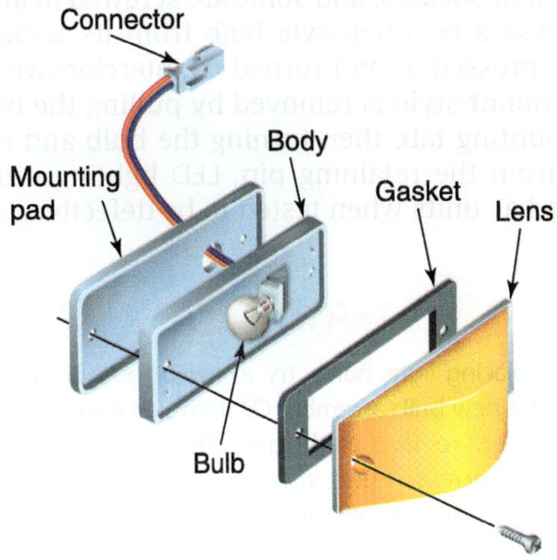

Bulb Replacement

When replacing a bulb, inspect the bulb socket. If the socket is rusty or corroded, the socket or light assembly base should be replaced. Also, inspect the lens and gasket for damage when the lens is removed and replace any damaged part.

There are several types of construction designs for exterior lights: those in which the lens is removed and the bulb is removed from the front (**Figure 10–14**), those in which the light assembly must be removed as a unit (**Figure 10–15**), and LED lighting units. Removing the lens from the assembly of some types could cause damage, and wiping the reflector surface to clean it can reduce brightness. Never attempt to remove the lens from the type in which the socket and bulb are removed from the back of the assembly.

FIGURE 10–15 Front turn signal/marker light assembly located in a hood pod

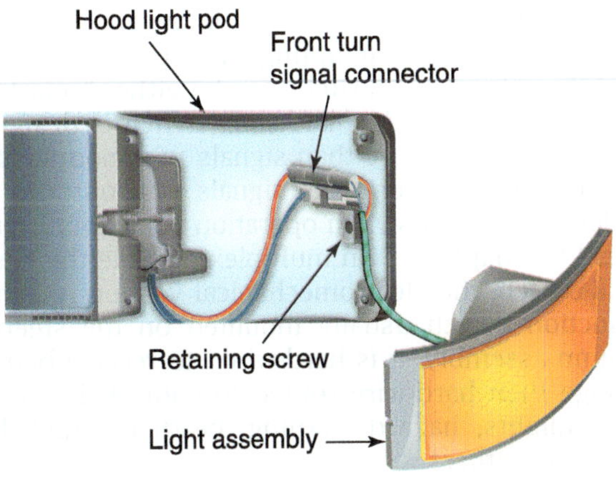

Bulbs are retained in their sockets in a number of ways. Some bulbs are simply pushed into and pulled out of their sockets, and some are screwed in and out. To release a bayonet-style bulb from its socket, the bulb is pressed in and turned counterclockwise. The blade mount style is removed by pulling the bulb off the mounting tab, then turning the bulb and removing it from the retaining pin. LED lighting units are replaced as units when tested to be defective.

SHOP TALK

When replacing light bulbs, try to avoid touching the glass part of the new bulb assembly. Oil from skin even on recently washed hands remains on the glass. This oil inhibits heat from dissipating. Increased heat inside the bulb can cause the filament to burn out prematurely.

Headlight Dimmer Switch

The headlight dimmer switch, or *courtesy switch*, can either be mounted on the floor or it can be a part of the turn signal assembly. The switch may be electromechanical or smart. To replace a floor-mounted electromechanical unit, use the following procedure:

1. Remove the two screws that hold the dimmer switch to the mounting bracket (**Figure 10–16**).
2. Disconnect the dimmer switch wiring connector from the main wiring harness.
3. Remove the dimmer switch and pigtail as an assembly.
4. Position the new switch on the mounting bracket. Install and tighten the screws until snug.
5. Connect the dimmer switch wiring connector to the main wiring harness.
6. Check the dimmer switch operation.

FIGURE 10–16 Typical components of an electromechanical dimmer switch designed to be foot-actuated.

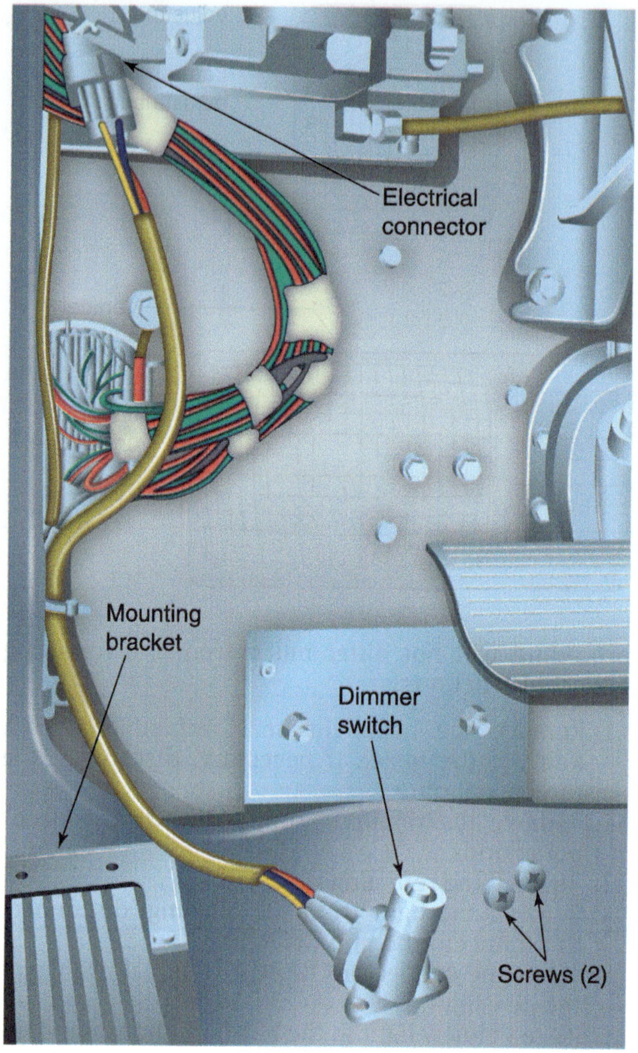

STALK SWITCH

Never forget that trucks today may use either electromechanical switches or smart switches. The procedures you use for diagnosing and repairing switches differ vastly so make sure you know what you are working on before attempting any repairs. The **stalk switch** or turnlight switch may be either "smart" or electromechanical. A smart stalk switch is designed to send nothing more than signals to a module: The module then converts the signals into outputs. To understand smart switch operation you should jump ahead to Chapter 12 on multiplexing. Like the smart stalk switch, the electromechanical version is a multifunction switch usually mounted on the steering column assembly. It is hard-wired to receive battery voltage then hard-wire routed to control the status of turnlights, hazard warning, headlamp high/low beam, and others.

Replacing a Turn Signal Switch

The following procedure describes how to replace an electromechanical turn signal switch. The type described was in common use in truck chassis until the introduction of smart stalk switches. Typical instructions for replacing a switch located in the turn signal switch are as follows:

1. Turn the ignition key switch off and disconnect the battery negative cable.
2. Disconnect the switch wiring connector from the cab wiring harness connector (**Figure 10–17**).
3. Remove the switch mounting screws and the switch assembly from the steering column.
4. Remove the three screws holding the plastic grip together (**Figure 10–18**). Lift off the grip section with the function symbols embossed on it.

FIGURE 10–17 A multifunction turn switch assembly: This unit is electromechanical.

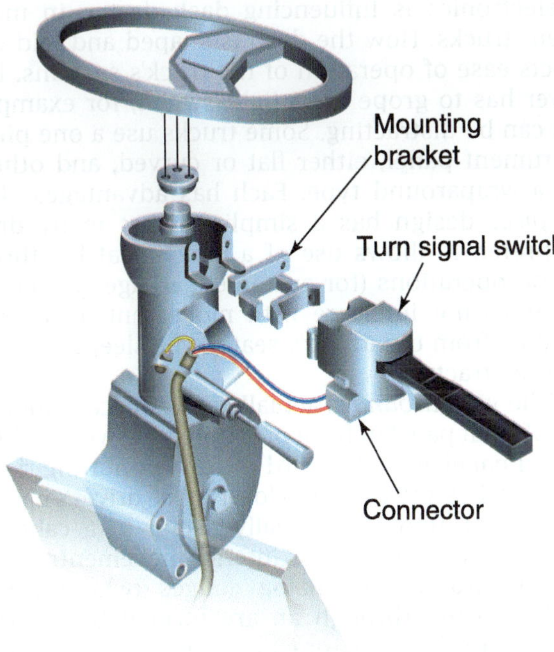

5. Lift out the courtesy switch, disconnecting the edge terminals of the yellow and red wires (**Figure 10–19**).
6. Carefully reassemble the edge terminals to the new switch.
7. Position the new courtesy switch inside the handle.
8. Attach the grip with the embossed symbols to the other grip and fasten it with the three screws. Tighten until snug.
9. Replace the battery negative cable and check the switch operation.

FIGURE 10–18 Three screws holding a turn signal grip together

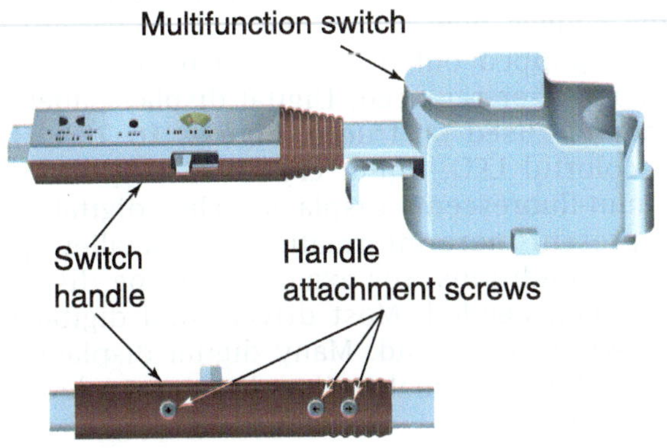

FIGURE 10–19 Courtesy switch (hard wire) detail

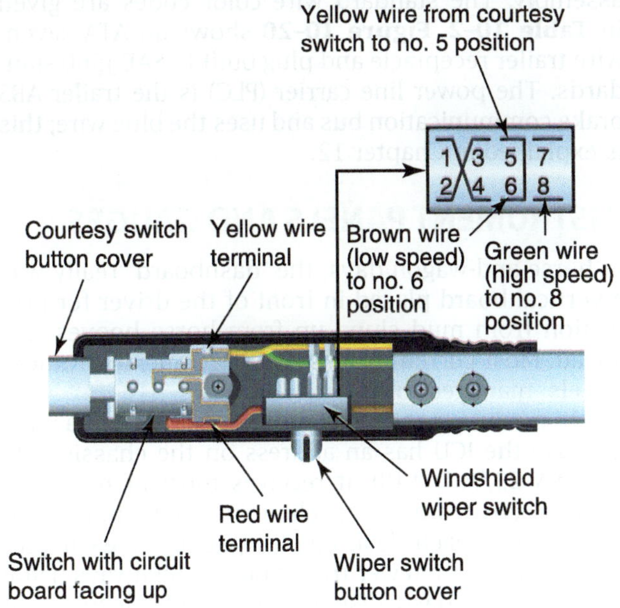

Turn Signal Switch Diagnosis

Before replacing the turn signal switch, ensure that the trouble is in the switch and not elsewhere in the circuit. Check that the circuit breaker and fuse are functional, and inspect the signal light bulbs for broken filaments. Also, check the flasher relay, and replace it if necessary. If the turn signal switch must be replaced, make sure that the key is off and that the battery negative cable is disconnected.

10.2 AUXILIARY ELECTRICAL EQUIPMENT

This section takes a look at some of the auxiliary chassis electrical circuits. Every technician should be familiar with the operating principles and maintenance of circuit breakers, relays, junction blocks, trailer plugs, and the dozens of chassis electrical components that fail and keep truck technicians busy coast to coast. Truck electrical equipment tends to be operated on a run-until-fail basis and is seldom the focus of preventive maintenance programs.

TRAILER CIRCUIT CONNECTOR

The wiring terminal block for the trailer is usually located inside the tractor cab, directly behind the driver's seat. Access is gained by removing the plastic cover held in place by screws. In sleeper models, the connector is in the luggage compartment.

A seven-wire core leads back from the terminal block to the trailer electrical connector cord and plug assembly. The standard wire color codes are given in **Table 10–2**. **Figure 10–20** shows an ATA seven-wire trailer receptacle and plug built to SAE J560 standards. The power line carrier (PLC) is the trailer ABS brake communication bus and uses the blue wire; this is explained in Chapter 12.

INSTRUMENT PANELS AND GAUGES

In horse-and-wagon days, the "dashboard" really was a vertical board placed in front of the driver for protection from mud slung up from horse hooves just ahead. Most current trucks use electronic instrument panels managed by an **instrument cluster unit (ICU)** module networked to the chassis data bus. Because the ICU has an address on the chassis data bus (SA 23/MID 140), it receives most of its status data directly from it. This eliminates the requirement to hard-wire each dash display gauge. As each year passes, there are fewer trucks on our highways using nonelectronic dash displays. The multiplexing that enables ICU operation is explained in Chapter 12.

TABLE 10–2 Trailer Cord Color Codes

Wire Color	Light and Signal Circuits
White	Ground return to towing vehicle
Black	Clearance, side marker, and identification lights
Yellow	Left-hand turn signal and hazard signal
Red	Stoplights and antilock brake circuit
Green	Right-hand turn signal and hazard signal
Brown	Tail and license plate lights
Blue	Auxiliary circuit and PLC

FIGURE 10–20 Tractor-trailer 7-pin light cord plug connector (SAE J560)

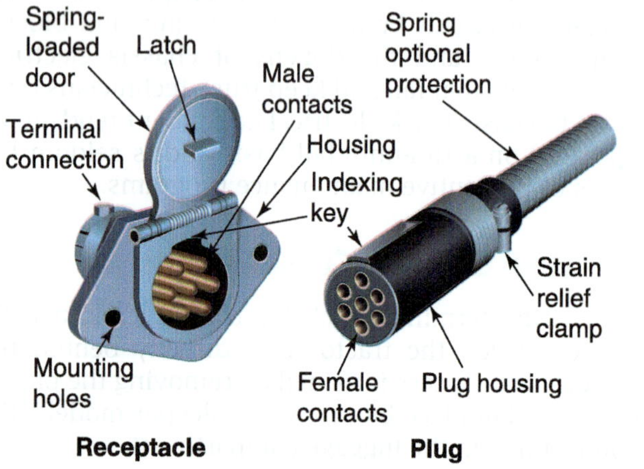

Dash Design

Figure 10–21 shows some examples of instrument panel design and layout, but the increased use of electronics is influencing dash design in more recent trucks. How the dash is shaped and laid out affects ease of operation of the truck's systems. If a driver has to grope for a light switch, for example, this can be distracting. Some trucks use a one-piece instrument panel, either flat or curved, and others use a wraparound type. Each has advantages. The *one-piece* design has a simplicity that many drivers prefer. It allows use of a bench seat for three-person operations (for example, garbage collection) and does not interfere with movement inside the cab (like from the driver's seat to the sleeper area in highway tractors).

The *wraparound* is actually a two-piece combination of a flat panel in front of the steering wheel with a second panel angled toward the driver, putting right-side switches and controls closer to the driver's hands. The wraparound is especially useful in a cabover, where the doghouse limits interior movement.

With traditional analog gauges (where gauge needles swing through an arc formed by a circle of numerals), field practice has indicated that white numerals on black backgrounds are the easiest to read (**Figure 10–22**). Most truck gauge faces are white-on-black, with orange or red highlights. Because red lighting aids the operator's night vision, many OEMs now illuminate their instruments in red or orange rather than white. Most gauges indicate normal conditions when the indicator needles are somewhere between 10 and 2 o'clock, but this can vary.

As indicated earlier, truck manufacturers have mostly converted from mechanically and electrically actuated gauges to electronic instrument displays. These are more reliable and easier to repair. An electronic speedometer can be reprogrammed to compensate for changes in tire and wheel size and not fail due to a broken cable. However, electromechanical gauge devices continue to be used due to the fact that analog readouts are better suited to displaying some types of information.

Easy removal of display gauges from the panel is made possible by electronic instrument clusters. Some gauges now are of the plug-in variety and can be popped out of the panel to be fixed on a workbench or replaced. Digital display panels are commonly used on trucks (**Figure 10–23**). These use colorful LED, liquid crystal display (LCD), or vacuum-fluorescent displays. The digital dash can present more information in a smaller space. Engine conditions, for example, can be displayed only when needed. Most drivers find digital data displays easy to read. Many digital displays now feature both an analog-like climbing scale and a numerical readout.

FIGURE 10–21 Two typical dash layouts

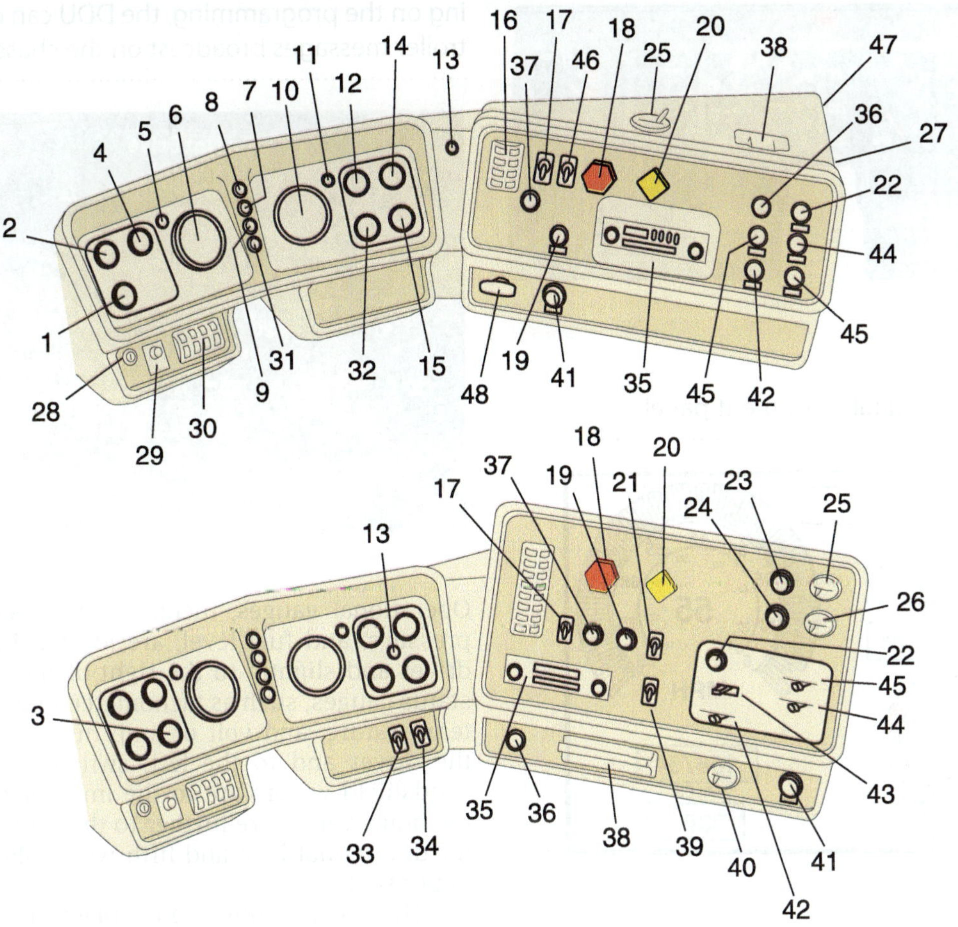

1. Voltmeter gauge
2. Engine oil pressure gauge
3. Optional gauge
4. Engine coolant temperature
5. Left-hand turn signal indicator
6. Engine tachometer
7. Hi temp/low water indicator
8. Hi beam indicator
9. Low air pressure indicator
10. Speedometer
11. Right-hand turn indicator
12. Fuel gauge
13. Preheater indicator
14. Dual air pressure gauge
15. Optional gauge
16. Heater/AC diffuser
17. Headlight switch
18. Trailer air supply valve
19. Wiper/washer control
20. System park brake
21. Spot light switch
22. Heater/AC switch
23. Interdifferential Indicator
24. Interwheel indicator
25. Interdifferential control
26. Interwheel control
27. Heater/AC diffuser
28. Ignition switch
29. Starter button
30. Heater/AC diffuser
31. Low oil pressure
32. Optional gauge
33. Cargo lamp
34. Heated mirrors
35. Radio
36. DC power outlet
37. Panel rheostat
38. Ash tray
39. Driving lights
40. Air slide 5th wheel
41. Hand throttle control
42. Floor/defrost lever
43. AC on/off switch
44. Fresh air lever
45. Heat/cold lever
46. Marker Interrupter
47. Radio power supply
48. Engine stop knob

Dash Components

Most OEMs prefer to make the dash appearance distinctive to their chassis but the ATA has guidelines outlined in RP 401. **Figure 10–24** shows an RP 401 dash face. These are merely guidelines that may or may not be observed by the OEM and **Figure 10–25** shows a 2013 electronic dash interface with a **data display unit (DDU)** just below the center of the driver's field of vision. A decade ago, some OEMs optioned **heads-up display (HUD)** visuals projected onto the driver's side lower windshield; these are seldom used today because they are thought to distract a driver.

FIGURE 10–22 Analog gauge instrument panel

FIGURE 10–23 Digital instrument panel

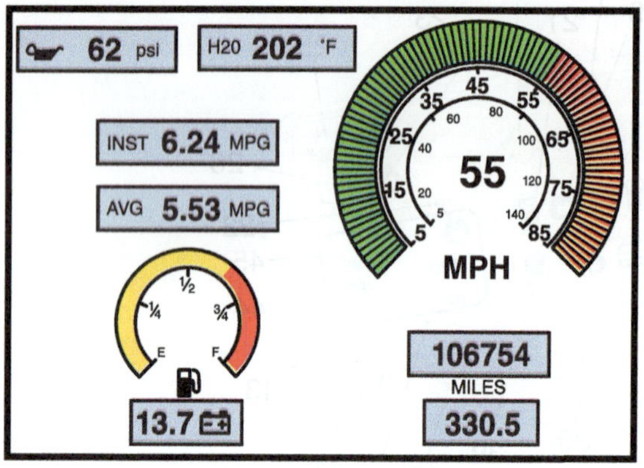

FIGURE 10–24 The Technology and Maintenance Council's RP 401 maps location for engine and safety gauges, plus recommended locations for switches and controls. Until data display units (DDUs) became common on trucks, most OEMs followed this layout.

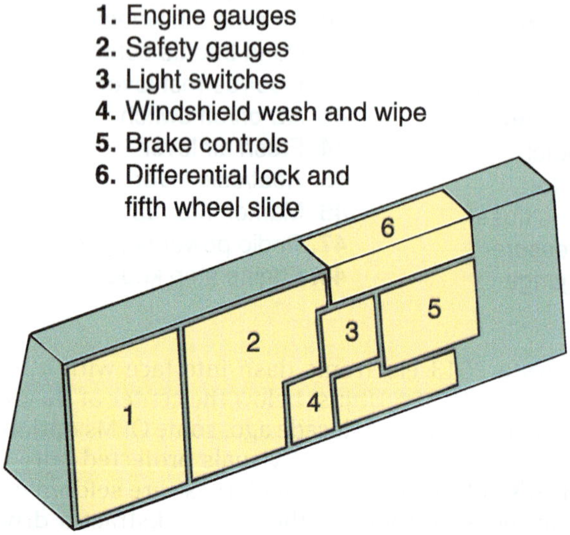

FIGURE 10–25 A typical electronic dash display featuring a data display unit (DDU) window. Depending on the programming, the DDU can display controller messages broadcast on the chassis data bus.

Operational gauges, including the speedometer, air pressure, and fuel level, are located in front of the driver and slightly to the right of the driver; engine status gauges, such as tachometer, oil pressure, water temperature, and volt or amp meter, are in front of the driver and to the left; switches for lights and windshield wipers are to the immediate right; parking brake valves are further to the right; and controls for differential lock and fifth wheel slider are to the upper right.

The dash layout should place the most critical displays and switches (for example, speedometer and light switches) where they can be easily seen and operated by the driver. Using the RP guidelines shown in Figure 10–24 helps truck drivers in fleets that are not assigned to a specific truck. Standard instrumentation in heavy trucks includes the following:

- **Oil Pressure Gauge.** This gauge indicates engine oil pressure. Normal oil pressure values should be between 30 and 90 psi (207 and 620 kPa) when the engine is running at rated engine speed at normal operating temperature. A lower pressure is normal at low idle speed. In most systems, a low oil pressure fault code is not produced until oil psi drops to well under the indicated normal range. The signal originates from a variable capacitance sensor. The dash may also display engine oil temperature.
- **Coolant Temperature Gauge.** This gauge indicates engine coolant temperature. It should normally indicate between 170°F and 195°F (77°C and 91°C). Higher temperatures may occur under certain conditions. Maximum temperatures range between 200°F and 220°F (93°C and 104°C) with the cooling system pressurized. The signal originates from a thermistor.

- **Tachometer.** The tachometer indicates engine rpm (engine speed). The engine can be operated at high idle for short periods without damage but should not be allowed to overspeed. The signal originates from an inductive pulse generator.
- **Fuel Level Gauge.** This gauge indicates the fuel level from multiple tanks or an overall fuel level reading. Usually electrically operated, it registers only when the key is switched on. The signal originates from a float and sending unit assembly.
- **Service Hour Meter.** This gauge indicates the total number of clock hours that the engine has operated. Some dashes have a separate time-of-day clock and an elapsed timer. One engine hour is equivalent to 50 highway miles.
- **Pyrometer.** The pyrometer monitors the exhaust gas temperature. The signal originates from a bimetallic thermocouple.
- **Voltmeter.** This gauge measures the battery voltage using a standard voltmeter.
- **Odometer.** An odometer records the distance traveled, calculated by rotations of an inductive pulse generator on the transmission tailshaft.
- **Speedometer.** This gauge indicates road speed. In addition, some instrument panels are equipped with an audible speed alarm, cruise control set speed, and cruise status. The signal originates from an inductive pulse generator on the transmission.
- **Air Pressure Gauges.** These gauges monitor the air pressure in the primary and secondary air reservoirs. Service brake application pressure is often also displayed.
- **Ammeter.** This gauge indicates the amount of charge or discharge in the battery charging circuit. Normal operation of the indicator should be slightly to the positive side of zero. During normal operation with the engine running, if the indicator is constantly to the negative side of zero or shows excessive charge, the electrical system should be checked for malfunction.
- **Fuel Pressure Gauge.** This gauge indicates fuel pressure on the pressure side of the transfer pump or the fuel rail pressure. Check fuel pressure data with values specified by the engine OEM. The signal originates from a variable capacitance sensor.
- **Diagnostic Data.** Diagnostic information relating to the instrument panel itself or relating to the chassis data bus is often displayed by a DDU on the instrument panel, though some fleets opt not to use a DDU believing that it can distract drivers.

Gauges should be properly functional because they display the status of the truck major components and safety systems during operation. Most ICUs are networked to the chassis data bus and have self-diagnostic capability. Diagnosing ICU malfunctions should be accomplished by connecting to the chassis data bus through the chassis data connector and accessing the ICU source address (SA) or message identifier (MID 140). This process is explained in more detail in Chapter 11 and Chapter 12.

Voltage Limiter

Some instrument panel gauges require protection against voltage fluctuations that could damage the gauges or cause them to give incorrect readings. A **voltage limiter** provides this protection by limiting voltage to the gauges to a preset value.

The voltage limiter can either be electromechanical (functions like a circuit breaker) or solid state. Most current trucks use solid-state voltage limiters. Dash electronic systems commonly function at 5 volts but some use lower potential. Crossing system voltage into a dash-integrated circuit board can destroy the dash, so always observe OEM diagnostic repair instructions. Never be tempted to use some of the older troubleshooting techniques, such as jumping voltage using jumper leads on electronic dashes.

AIR AND OIL PRESSURE GAUGE FAILURE

If pressure gauges do not show a reading, or the reading is suspected of being in error, a cluster gauge reading and master shop gauge reading should be taken by placing a Y or T connector in the circuit. If air pressure is evident at either of the two air lines, check the air filter inside the air pressure gauge to make sure it is not plugged. If the filter is not plugged, and the cluster air gauge pressure reading exceeds a 65 psi (450 kPa) master gauge reading, replace the cluster air gauge.

ELECTRICAL TEMPERATURE AND FUEL DISPLAY FAILURE

If the displays on electrical temperature gauges or the fuel gauge indicate a problem, the vehicle wiring that connects the gauge in the circuit must be checked. If the wiring and connector are okay, negative temperature coefficient (NTC) thermistor-type temperature sensors may be tested using either a resistance or voltage drop test. Some OEMs provide voltage drop specifications for testing NTC thermistor performance with greater accuracy. Check for engine electronic (engine) control module (ECM) fault codes before testing circuit components.

Thermistor Resistance Check

To check thermistor-type temperature sensors, the approximate temperature of the component being checked should be known. To test NTC thermistors at the cluster harness connector, use the ohmmeter on a digital multimeter (DMM) and consult a chart such as the one shown in **Table 10–3**.

TABLE 10–3 NTC Thermistor Resistance Check

Component Temperature (°F)	Thermistor Resistance (Ohms)
32	8,335
50	4,945
68	3,010
86	1,871
104	1,185
140	505
158	341
194	169.8
212	125.6
230	95.2
248	73.5
266	57.5

Thermistors used in truck chassis are usually NTC type; this is fully explained in Chapter 6.

Restricted Needle Movement

If the sensor electrical resistance checks okay, the gauge should be checked for restricted pointer movement. Very gently move the pointer between the minimum and maximum dial positions. If the pointer moves freely, the gauge should be checked for an open or shorted winding.

Open or Shorted Winding Check

Remove the gauge and check for an open or shorted winding with an ohmmeter. All resistance values below 20 ohms should be considered as short. All resistance values above 1,000 ohms should be considered as open. All readings between 20 and 1,000 ohms should be considered within an acceptable range. If the resistance reading is not within the acceptable range, replace the gauge. This test cannot be performed accurately on a voltmeter because there is a semiconductor inside the gauge that will make resistance readings inaccurate.

CAUTION

Static electricity can cause permanent damage to the cluster. Before working on the cluster, be sure to remove all static electricity from your body by touching grounded metal. Do not touch pin connectors during removal and installation of gauges.

Internal Connection Check

If gauge winding resistance is within range, check the internal connections between the cluster's harness pins and the gauge pins for a good connection. Refer to the service literature for the proper gauge pins to use for this test procedure. Measure the resistance on the circuits that connect the cluster harness ground and sensor line to the gauge pin ground and sensor line for the gauge in question. They should read under 1 ohm.

Voltage Check

Check voltage with a voltmeter across the power pin and ground pin in the cluster for the gauge being checked. Accessory voltage is applied and all other gauges should be installed in the cluster. The voltage reading should be just below accessory voltage. If either the voltage measurement or the two electrical resistances are incorrect, replace the cluster with a new cluster without gauges. The cluster voltmeter should have about the same reading as the voltmeter used to make the gauge voltage check. If the cluster voltmeter has a different reading, replace it.

Fuel Gauge

If an electric fuel gauge shows an inaccurate reading, the fuel level sensor must be checked. The fuel level sensor used is a float/rheostat type. **Table 10–4** shows typical resistance values at different fuel levels.

SHOP TALK

Before checking the fuel gauge, be sure that the cab interior is warmed up (during cold weather), and that the vehicle has been sitting still long enough to allow the fuel to settle.

When a sending unit sensor is shorted to ground, the gauge should read empty. Locate the fuel sensor gauge input in the cluster harness. If resistance for the fuel tank level matches that in Table 10–4 or the OEM spec and the gauge reading do not correspond, replace the fuel gauge.

VOLTMETER DISPLAY FAILURE

If either the voltmeter or pyrometer displays are indicating a problem, the connection wiring should be checked for clean, tight, unbroken, and nonshorted connections. If the wiring is okay, measure the appropriate

TABLE 10–4 Sending Unit Resistance

Tank Fuel Levels	Sensor Resistance (Ohms)
Full	88
¾	66
½	44
¼	22
Empty	01

voltage on the vehicle's harness connector that mates to the cluster for the gauge in question. The key switch must be on during this test. The voltmeter reads across accessory voltage and ground. Replace the voltmeter if it does not read close to accessory voltage.

SPEEDOMETER/ODOMETER AND TACHOMETER/HOURMETER DISPLAY FAILURE

If the speedometer, odometer, tachometer, or hourmeter displays are indicating a problem, the wiring must be checked. Also check engine ECM fault codes. Make sure the accessory voltage to the cluster is correct as specified.

If all four meters are not functional and the voltmeter and fuel gauge are, try replacing the cluster with a new gaugeless cluster. Also check for proper dip switch settings inside the cluster. The speedometer and odometer use the road speed sensor; the tachometer and hourmeter use the engine speed sensor. If either pair of meters does not operate, but the other pair does, the sensor that signals the failed pair should be checked. Both the correct signal amplitude and frequency must be as specified at the harness pins of the vehicle connector.

If the appropriate sensor input into the cluster checks okay and both of the paired functions still do not work, the cluster must be replaced with a new cluster without gauges.

Mechanical Engine Coolant Temperature Display Failure

If a mechanical water temperature gauge indicates a reading, but the reading is suspect, a test to check gauge accuracy must be performed. Place the sensing bulb in boiling water and check the gauge reading against a master temperature gauge inserted in the water. If the cluster gauge reading and the master gauge readings differ by more than 5°F (3°C), the cluster gauge must be replaced. No cluster voltage is required for this test.

COIL-TYPE DISPLAY GAUGES

A variety of analog coil-type gauges are used in vehicle dashes to provide a user friendly display that operators tend to find easier to interpret. Coil-type gauges use electromagnetic principles of operation and, therefore, their operating principles are closely related. The operation of a balancing coil display gauge is described here. Most coil-type display gauges use variations of the balancing-type display gauge.

Balancing Coil

A pair of coils is arranged in 90-degree angled vectors off the pivot point of a magnetically responsive display needle. The coil on the left (low) side of the display is angled so that if the display needle were to exactly align with it, this would represent the lowest gauge display reading. Similarly, the coil on the right (high) side of the display is angled so that if the display needle were to precisely align with it, this would represent the highest gauge display reading. When both coils are supplied with the same voltage and fully grounded, the high-side coil is designed to produce the stronger magnetic field, biasing the display needle toward it. Both coils are supplied with a nonvariable voltage value such as V-Bat or V-Ref. The low-side coil is direct grounded. This means that its electromagnetic field intensity is constant. While the high-side coil is provided with the same voltage as the low side, it is grounded by means of a variable resistor.

To explain the operation of a balancing coil gauge, the example we use is that of an NTC thermistor connected to the ground circuit of the high-side coil. When the NTC thermistor is cold, resistance in the high-side coil is high, choking down on current flow. The result is low electromagnetic field intensity in the high-side coil biasing the display needle to the low side of the display gauge. As the NTC thermistor is heated, its internal resistance goes down, allowing more current to flow through the high-side coil. More current flowing through the high-side coil intensifies its magnetic field, pulling the display needle toward it. At the highest expected temperature, the NTC thermistor produces almost no resistance on the ground side of the high-side coil. Because the high-side coil is designed to produce the stronger magnetic field, when the NTC internal resistance is lowest, the display needle is pulled fully over to the high side of the gauge. **Figure 10–26** is an example of the gauge we have just described.

Although we have used the example of an NTC thermistor here, balancing coil gauges can be used for analog display of a range of status signals. The only requirement is that a means of varying the resistance on one of the two coils is provided.

MOTORIZED CAB CIRCUITS

Current trucks use a number of motorized cab electrical circuits. These would include electric window regulator, window washer, and electric wiper circuits.

FIGURE 10–26 Balancing coil operating principle

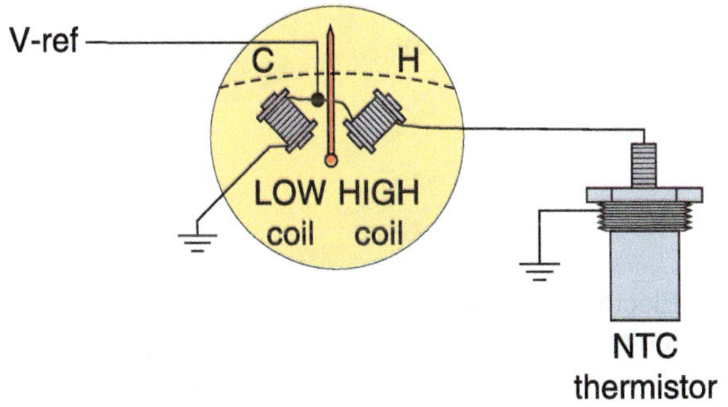

These motorized circuits are switched closed and open by the driver. When a motor circuit is closed, a pinion on the electric motor drives a mechanical assembly that creates the desired movement, such as opening/closing a window or wiping the windshield. Problems in this type of circuit may have either an electrical or mechanical cause and it is important to remember this when troubleshooting.

Windshield Wipers

Windshield wipers may be electric or air operated. One motor may be used to drive both wipers, or in a cab-over-engine (COE) each wiper may be operated separately by its own motor. The components of a typical electric windshield wiper system are shown in **Figure 10-27**. The two-speed electric motor is activated by turning a control knob on the instrument panel to the first detent for low speed and to a second detent for high speed. Electric wiper motors may also be multi- or variable-speed, but in practice they are used less that air-driven wiper motors on commercial vehicles.

Sometimes an intermittent control is available to operate the wipers in the delay mode. When the wiper control knob is switched to low or high speed, the intermittent control ceases to function. When air brake equipment is specified for trucks, air-operated windshield wipers are a preferred alternative to electric wipers.

FIGURE 10-27 Electric windshield wiper components

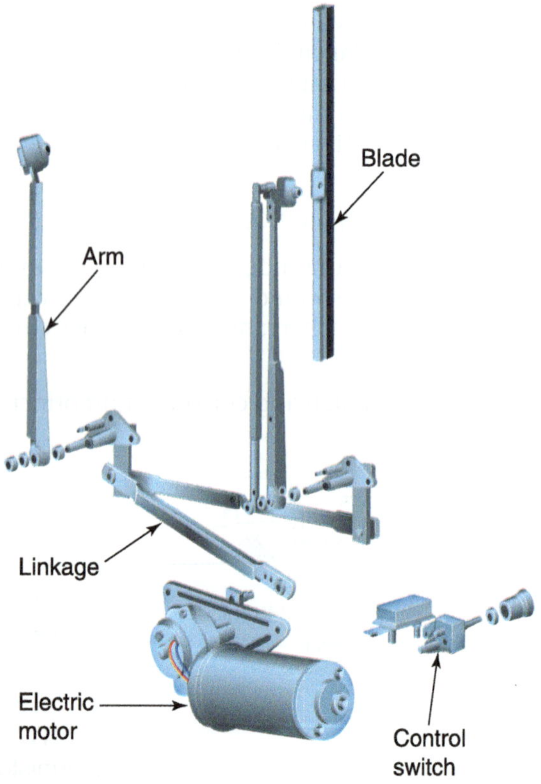

Alarm and Shutdown Systems

Alarm and shutdown systems are integrated into all engine management systems. They monitor parameters such as coolant temperature and engine-mounted control devices and can warn the driver of potentially hazardous conditions such as low oil pressure. The objective is to alert the driver to any condition that could result in engine or drivetrain damage, or become potentially dangerous highway hazard. These systems are usually programmed with three levels of alert and intervention:

1. Level one: display alert only.
2. Level two: display alert and derate engine and shift logic to a limp home mode.
3. Level three: display alert, warn imminent shutdown, shutdown engine.

The actual response action depends on how the system has been programmed. Fleet trucks would normally be programmed for level three intervention to safeguard the equipment. On the other hand, a fire truck may be programmed to display alerts only. Typical conditions monitored include:

- Air pressure in brake systems
- Hydraulic fluid level in brake systems
- rpm
- Lubricant level in the transmission or rear axle
- Transmission or rear axle temperature
- Coolant level in the radiator
- Coolant temperature
- Turbo temperature
- Oil temperature
- Oil level in the engine

Depending on the type and programming of the alarm/shutdown system, the out-of-normal range signal from a sensor can result in either a driver alert or an engine shutdown.

SOLDERING

When electrical wiring in trucks fails, it is sometimes possible to repair sections by soldering. When considering a soldering repair, first determine whether this type of repair is acceptable for the failed circuit. Some electrical circuits should not be repaired by soldering, for instance, the high bus circuits used for data transmission (explained in Chapter 12). When soldering electrical circuits, it is important that a soldering iron of around 100 watts be used: Too much heat, such as that produced by a propane torch, can damage wiring and adjacent circuits. Insufficient wattage can also result in too much heat being produced because it will take much longer to heat the joint wires to solder temperature. The idea of soldering is to heat the surface of the wires to be connected so that the solder molecularly bonds to the surface and creates the conductive path. It is a degree of bonding lesser than fusion that we would refer to as *welding*. Soldering can achieve a good conductive path in an electrical joint without applying excessive heat. Most OEMs publish soldering guides in their service literature.

CAUTION

Soldering unloads a lot of heat into the circuit being repaired. Before beginning, make sure that you are not going to melt other soldered connections in the circuit or scorch wiring conduit.

CIRCUIT PROTECTION DEVICES

Circuit protection components are designed to limit current flow through a circuit. Almost all truck electrical circuits are protected by:

- Fuses
- Circuit breakers (electromechanical)
- Virtual fuses (electronic)
- Virtual breakers (electronic)

Fuses

An electromechanical fuse is a device designed to fail once. However, we can also use **virtual fuses** in truck electrical circuits. Virtual fuses (and breakers) are covered in the section that immediately follows this. After a failure an electromechanical fuse has to be physically replaced. A typical fuse consists of a wire bridge through which circuit current is routed. The bridge is rated for specific maximum amperage. If the specified amperage is exceeded, the bridge overheats and melts. The symbol for a fuse is shown in **Figure 10–28**.

Circuit Breakers

There are two types of circuit breakers:

- Virtual
- Electromechanical

When a circuit draws excessive amperage, the circuit breaker is designed to open the circuit, interrupting current flow. The objective is that the breaker can be reset, either automatically or otherwise, without any physical damage being created in the circuit. Virtual breakers are covered in the section that follows this. A typical electromechanical circuit breaker consists of an armature and contacts. The contacts "break," that is, *open*, when the circuit is overloaded. Two classifications of breakers are used on trucks whether they function electromechanically or virtually. The description that follows describes the operation of the electromechanical versions of breakers.

Cycling Circuit Breakers. **Cycling circuit breakers** are known as Society of Automotive Engineers (SAE) type-1 circuit breakers. The symbol for a type-1 circuit is shown in Figure 10–28. When current flowed through the armature exceeds the specified rating, the armature heats up and bends away from the contact

FIGURE 10–28 Circuit protection devices: The units shown are electromechanical but some OEMs use virtual equivalents.

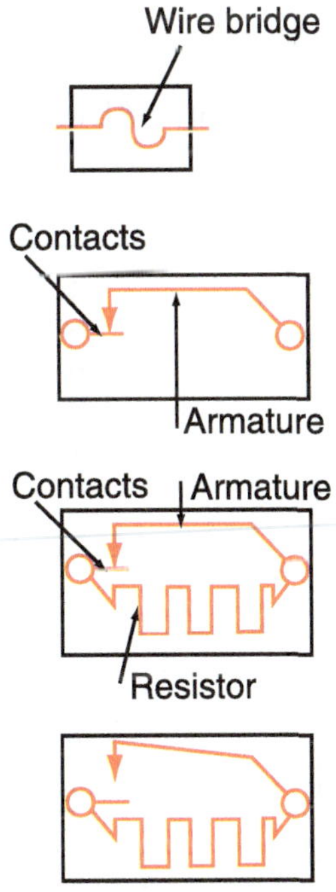

Fuse	The wire bridge overheats and permanently fails when amperage rating is exceeded.
Type 1 circuit breaker closed	Current overload heats the armature, which bends, opening the contacts. The breaker automatically resets when the armature cools and closes the contacts.
Type 2 circuit breaker closed	Current flows through the armature with the contacts closed.
Type 2 circuit breaker open	When current exceeds rating, the armature heats and bends, opening the contacts. Current is now forced to flow through the resistor, the heat from which holds the armature open.

point shown by the arrow; this immediately opens the circuit. With the circuit open no current flows through the armature, so it cools and as it straightens, it once again closes the circuit as the contacts connect. Typical circuits that use SAE type-1 breakers are headlights, fog lights, electric windshield wiper circuit, and so on.

Noncycling Circuit Breakers. **Noncycling circuit breakers** are known as SAE type-2 circuit breakers. The symbol for a type-2 circuit breaker is shown in **Figure 10–28**. In this figure, you can see the breaker in both closed and tripped conditions. A type-2 circuit breaker functions similarly to the type 1 except that it has a resistor below the armature. When the circuit that it is protecting is closed (energized) there are two possible current paths through the breaker, but because electricity will use the path of least resistance most of the current will flow through the armature. If the specified amperage of the circuit breaker is exceeded, the armature heats up and bends away from the contacts, opening the circuit; at this point, circuit current has no choice but to flow through the resistor. The current flowed through the resistor is converted to heat that acts to keep the armature warm enough so that the contacts do not meet. When a noncycling circuit breaker trips the current load can only pass through the resistor, resulting in greatly reduced current and voltage loss to the circuit.

So long as the circuit is energized, the only path for current flow through a noncycling circuit breaker is through the resistor. To "reset" a noncycling circuit breaker, the circuit must be switched open. This allows the circuit breaker resistor and armature to cool and close the breaker contacts. Noncycling circuit breakers are used extensively in truck electrical circuits.

VIRTUAL FUSES AND BREAKERS

The term *virtual* is used when a logic circuit replicates an electromechanical event without any actual physical action or damage taking place. When **virtual circuit protection** devices are tripped, the performance outcome is the same as if a physical fuse or breaker had been overloaded. OEMs use virtual fuses and **virtual breakers** in a wide range of chassis circuits. When a virtual device is tripped, the technician is required to diagnose the condition using OEM diagnostic routines. Virtual circuit protection devices can be programmed to perform exactly as their electromechanical counterparts, so terms such as *cycling* and *noncycling* virtual breakers or fuses are used.

RELAYS

A **relay** is an electromechanical switch. Relays are used extensively on truck electrical circuits and every technician should know exactly how one functions. A relay consists of two electrically isolated circuits. One of the circuits is used to change the switch status of the other. The other is a coil circuit used to electromagnetically switch the circuit. Relays are used so that a low-current circuit can control high-current circuits. Take a look at **Figure 10–29** showing an SAE standard relay when following this explanation.

Coil Circuit

The relay shown in Figure 10–29 is used in all OEM electrical systems, and the terminal numeric codes do not change. The control or coil circuit is indicated by terminals 85 (2) and 86 (1). The polarity usually does not matter, but it might. For instance, Volvo Trucks use a diode on some of its control circuits, so on these terminal 85 (2) must always be chassis ground.

When no current flows through the coil circuit, the status of switching circuit is in NC, or normally closed. In most cases (depending on how the relay is being used) this will mean that the output is open. When current flows through the control coil, an electromagnetic field is created and the movable switch is pulled toward the coil, opening the 30-87a (3–4) circuit and closing the 30-87 (3–5) circuit.

CAUTION

The wiring designations outlined here are the most common. If you take a look at Figure 10–29 you can see that other wiring options are available depending on how the coil circuit is switched. Take nothing for granted and use the OEM wiring schematics.

Switched Circuit

Referencing Figure 10–29 again, the switched circuit of the relay is represented by terminals 30, 87, and

FIGURE 10–29 Standard SAE relay terminal assignments

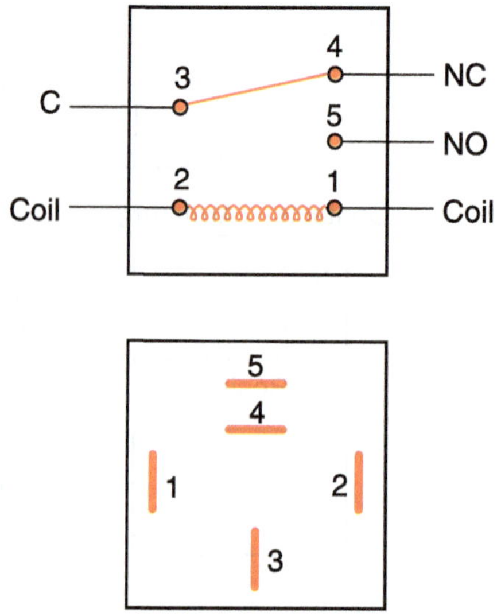

87a. Terminal 30 is common. This means that a voltage reading could normally be measured at terminal 30 regardless of the status of the switch. When the coil circuit of the switch is not energized, 30 would close on the NC terminal 87a. In most cases this would mean that the switching status of the relay is open, so no current flows through the switch circuit. When the control coil is energized, the switch is moved from the 87a NC pole, to the 87 normally open (NO) coil. This action permits current to flow from common terminal 30 to the NO terminal 87.

Terminal Assignments

Standard SAE designations have been recently revised by the SAE with the result that two numbering systems are in use. These methods of numbering relays are standard through the industry, so commit the terminal designations to memory.

Terminal Designation	Former Number	Revised Number
Coil (control current positive)	86	1
Coil (control current ground)	85	2 (direct or switched)
Common (supply current in)	30	3
Normally closed (NC)	87a	4
Normally open (NO)	87	5

Check these terminal assignments by referencing Figure 10–29 again. Relays are simple switches that only cause confusion if you do not know how one operates. Take a little time to memorize the terminal assignments and it will pay off in troubleshooting time saved many times over.

FETS

Field effect transistors (FETs) have replaced relays in many circuits. For this reason it is always necessary to consult OEM circuit schematics and diagnostic routines so that you know exactly what type of circuit you are working with. FETs are covered in Chapter 12 of this textbook.

TEST LEAD SPOONS

Most manufacturers of DMMs today supply test lead spoons. These are designed to access sealed electronic circuits for testing while the circuit is energized without damaging either the wiring or connector seals. Although they do not access all types of circuit connectors (you will have to continue to use breakout tees for these) they can be a useful addition to your test equipment.

CIRCUIT ANALYSIS

As you start to make yourself *think* electrically, you will find that tackling electrical problems becomes a cinch. There are certain rules about circuit behavior that really do not change. **Figure 10–30** captures some of the key rules about how circuit problems can be zoned in and quickly identified. Treat your DMM as a friend, and become very familiar with **Figure 10–30A**. Knowing how a circuit behaves when it is functioning properly is the first step. Next, note how your DMM can immediately identify circuit malfunctions by studying **Figure 10–30B**, **Figure 10–30C**, and **Figure 10–30D**.

FIGURE 10–30 Circuit analysis

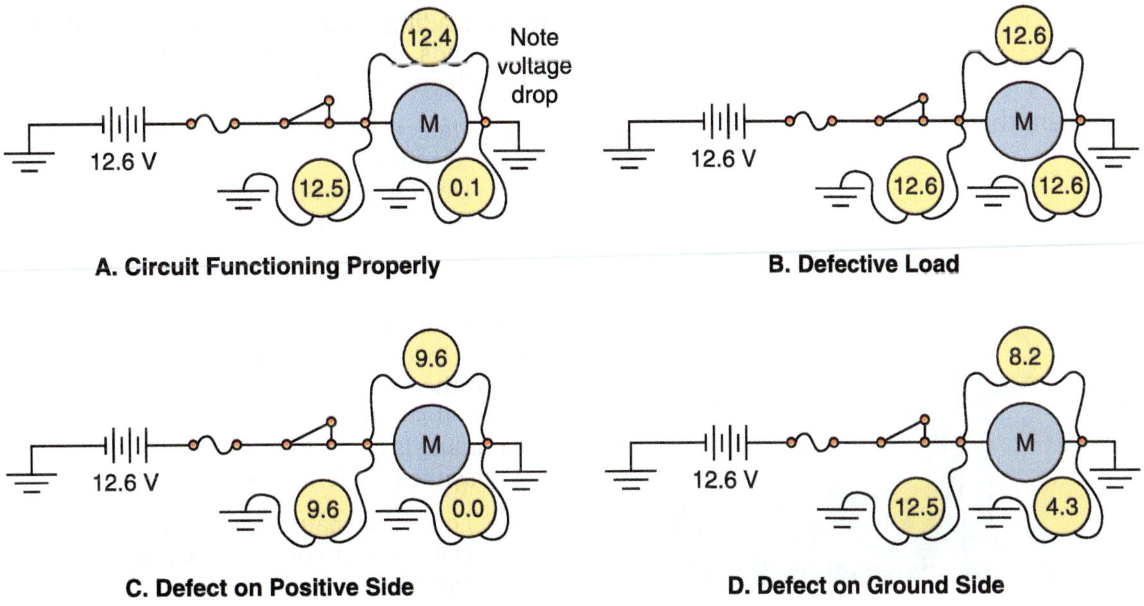

10.3 RAPID CHECKING OF A TRUCK ELECTRICAL CIRCUIT

The following sequence of tests is designed to produce a quick assessment of the truck electrical circuit. This type of sequential check is often performed as part of routine preventive maintenance. The essential diagnostic tool is your DMM but you should also have a portable AVR instrument such as the Sun VAT shown in **Figure 10–31**. In addition, you might make use of a new generation, handheld (HH) circuit diagnostic instrument such as that shown in **Figure 10–32**. **Photo Sequence 4** shows some of the key steps electrical circuit problems.

FIGURE 10–31 A Sun VAT 40 AVR commonly used in truck service facilities.

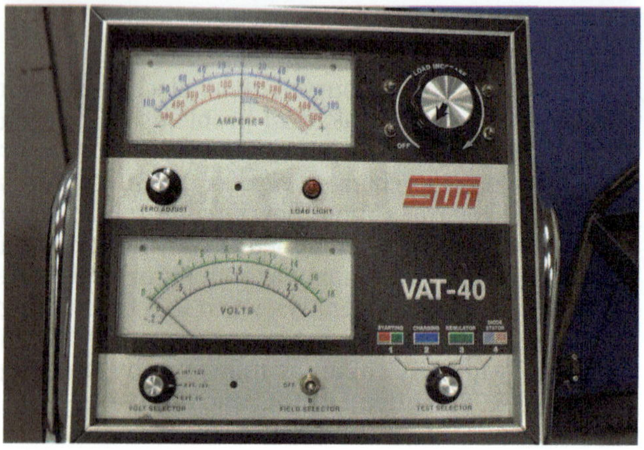

FIGURE 10–32 A handheld electric circuit diagnosis instrument that can assist in finding short and open circuits.

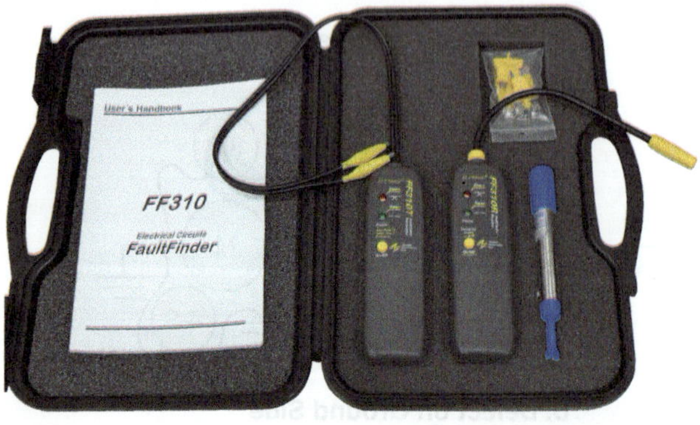

GENERAL VOLTMETER TEST

These tests are used to determine the general condition of a vehicle electrical system. The idea is to produce a report card on the battery, the cranking circuit, and the charging circuit. In fact, the test is so fast and easy to perform that it should become part of routine service procedure.

BATTERY VOLTAGE

Battery voltage can only be accurately measured when there is no surface charge; this may be removed by applying a load on the battery for a 1-minute period. (Turning on the headlamps will usually suffice.) Now electrically isolate each battery by disconnecting the terminals. If the voltmeter is not autoranging, set the scale to read up to 18 volts and connect the positive lead to each positive battery post and the negative lead to each negative battery post. The voltmeter readings produced at 70°F (21°C) may be interpreted as follows:

12.6 volts or higher	100 percent charged
12.4 volts	75 percent charged
12.2 volts	50 percent charged
12.0 volts	25 percent charged
11.9 volts or lower	fully discharged

A fully charged lead-acid battery should produce 2.1 volts per cell: 6 cells in series will produce a reading of 12.6 volts.

CRANKING VOLTAGE

Ensure that the engine is not capable of starting by no-fueling it either electronically or mechanically. Connect the voltmeter leads across the cranking motor terminals. Crank the engine for 15 continuous seconds. The voltmeter reading should read above 9.6 volts for the full 15 seconds. If the reading falls below 9.6 volts, a problem with one of the following is indicated:

- Defective or corroded battery cables or terminals
- Defective or discharged batteries
- Defective cranking motor, solenoid, or relay

Charging Voltage

Start the engine and run it at 75 percent of rated speed with no load. Now turn on all the electrical accessories on the vehicle. Use the voltmeter to test battery voltage. It should read between 13.5 and 14.5 volts. If the reading is lower than 13.5 volts, one of the following possible problems is indicated:

- Loose alternator drive belt(s)
- Defective or corroded electrical connections
- Defective voltage regulator

Photo Sequence 4

PERFORMING A RAPID ASSESSMENT OF A TRUCK ELECTRICAL CIRCUIT

P 4–1 Check the battery voltage. Turn the headlights on for a minute to remove any surface charge. Now isolate each battery by disconnecting it from the chassis electrical circuit (remove the terminals). Use a voltmeter and select V-DC. If it is not auto-ranging, set it in the 18V scale. Connect the positive lead to the positive battery post and the negative lead to the negative battery post.

P 4–2 Record the reading for each battery tested. Make sure that the voltage values meet the specifications after the temperature adjustment. Any battery out of specification requires further testing. Reconnect the batteries.

P 4–3 Check cranking voltage. Ensure that the engine will not start during this test. Set the DMM to V-DC and connect the test leads across the cranking motor terminals, observing the correct polarity. Now crank the engine for 15 continuous seconds. The reading should remain above 9.6V for the test period. The DMM should then read at least 10.5V.

P 4–4 Charging voltage test. Start the engine and run at 75 percent of its rated speed (1,500 rpm if rated speed is 2,000 rpm). Turn on all the electrical accessories and use a DMM on the V-DC scale to check the battery voltage. This should read between 13.5 and 14.5 volts. Note the DMM reading.

P 4–5 If the voltage reading is lower than 14.5V, a problem with the alternator, alternator drive mechanism, voltage regulator, or electrical connections is indicated.

P 4–6 If the voltage is higher than 14.5V, a problem with the voltage regulator or the electrical connections is indicated. Note that the maximum charging pressure in some systems should not exceed 14.1 V.

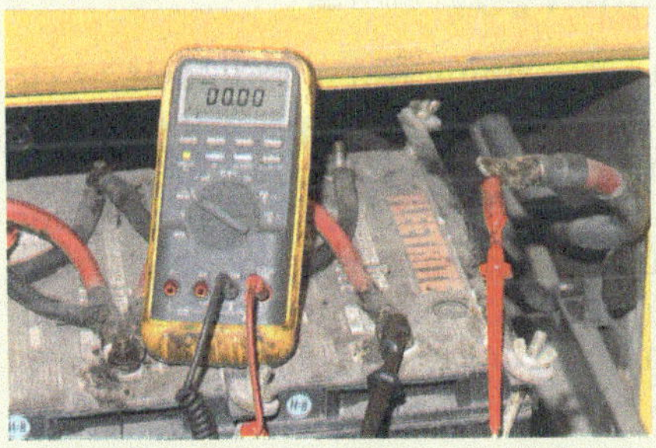

P 4–7 Problems with the wiring or connections can be located by performing a voltage drop test. Set the DMM to V-DC on a scale so that the readings of less than 1 volt can be read. To voltage drop test a starter cable, crank the engine with the positive test lead placed at the end of the cable closest to the positive terminal of the battery and the negative test lead placed at the other end. The DMM should read 0.1 volt or less. Voltage drop testing can also help identify defective terminals. Just remember that the circuit must be energized and the positive test lead placed at the more positive location in the circuit.

- Defective alternator
- If the reading exceeds 14.5 volts, the problem is likely one of the following:
  - Defective voltage regulator
  - Poor electrical connections

Note that maximum charging system voltage should be set at 14.1 volts on some systems; gel cell batteries should never exceed a 14.1-volt charge.

Voltage Drop Test

Voltage drop testing is a means of testing the condition of wires, cables, connections, and loads in the electrical circuit. It is a much better method of testing electrical circuit resistances because the testing is performed in a fully energized electrical circuit as opposed to checks with an ohmmeter, which must be performed on inactive circuits. The ohmmeter uses very low current flow and voltage and, therefore, does not test a circuit under active circuit conditions. In any electrical circuit, the cables should be of a gauge and length that represent minimal circuit resistance.

A voltage drop test is a quick and easy way to determine the resistance of electrical wires and cables. Cables and connections in good condition should test at zero or very low voltage drop values. When the voltage drop value exceeds 0.2 volt across a cable, the cable is indicating high resistance. Electrical connections should not produce voltage drops greater than 0.1 volt and, perhaps most important, total voltage drop of all the cables and terminals in a cranking circuit should not usually exceed 0.5 volt.

Performing a Voltage Drop Test

The best instrument for performing a voltage drop test is a DMM. If the DMM is not autoranging, select a low scale such as 0 to 3 volts for the test. The most common circuit tested for voltage drops is the cranking circuit. For instance, if an engine cranks slowly, two possible causes are the circuit cables and terminals. In fact, many cranking motors and batteries are needlessly changed every day in the trucking industry due to defective cables and terminals. Voltage drop tests are a quick method of verifying the integrity of any connections in the cranking circuit.

1. Set the DMM on the V-DC scale so readings of 1 volt or less can be easily read. Ensure that the DMM test leads and probes are inserted into the correct sockets. If they are color-code polarized, ensure that the red lead is identified as the positive and the black lead is identified as the negative.
2. The positive probe should always be connected to the most positive section of the circuit being tested. This is simply the section closest to the positive post of the battery. The black probe should always be connected to the most negative section of the circuit. This is always that closest to the negative post of the battery (chassis ground).
3. No-fuel the engine and crank for up to 30 continuous seconds. Test each terminal and wire sequentially through the circuit from the positive battery post, through the positive cables and connections, from the cranking motor to chassis ground, and from chassis ground back to the negative post of the battery.
4. While performing the voltage drop tests, use a remote starter switch to make the job easier. Also, crank for 30 seconds maximum, ensuring at least a 2-minute interval between each cranking cycle. Remember, a voltage drop value exceeding 0.2 volt through any individual terminal or cable indicates a possible problem. A voltage drop exceeding 0.5 volt in the cranking circuit is out of specification and requires repair.

Visual Inspection

Never overlook the importance of simply observing the condition of cranking circuit components. When voltage drop tests are performed on a cranking circuit, observe the condition of cables and terminals, taking note of corrosion, insulation damage, and signs of overheating. After the voltage drop test has been performed, touch sections of the cranking circuit to check for heat; high resistances in electrical circuits produce heat, which is easily detected. Any terminal that overheats should either be cleaned or replaced. A cable that overheats should always be replaced.

Testing a Solenoid or Relay

1. Initially set the voltmeter range to a value that exceeds the battery voltage value: 12 volts for most truck electrical systems and 24 volts on 12/24 systems and some buses.
2. Connect the voltmeter probes in sequence through the circuit so that the positive probe is in the most positive location and the negative at the most negative. See Figure 10–30 for the required test sequence.
3. While the engine is being cranked, switch the voltmeter to a low scale in order to read the voltage drop value.
4. A voltage drop that exceeds 0.2 volt through a relay or solenoid indicates that the component is defective and must be replaced.

BATTERY-DRAIN TEST

The objective is to determine if a component or circuit in the vehicle is causing a drain on the battery when the ignition circuit is off. It is good practice to perform this test each time a battery or set of batteries is replaced, but bear in mind that modern trucks have a wide range of on-board modules that will continue to draw some current even when in shutdown or "sleep" mode.

1. Ensure that the ignition circuit is open, all accessories are switched off, and the cab door and courtesy lights are off.
2. Disconnect the main negative terminal from the battery pack. If a significant drain is suspected, you can check this out using an ammeter. Disconnect the negative battery cable from the battery (bank). Set the DMM to ammeter mode and position the test leads appropriately. Next, place one test lead on the negative battery post and the other on the (disconnected) battery cable. This places the DMM in series with the negative cable terminal and the negative battery terminal. Measure the amperage and compare to specification. The test can be performed using a testlight; if the testlight illuminates you can assume that there is a significant battery drain.
3. To identify low level battery drains, an ohmmeter is required. Ensure that the positive battery cable is removed from the battery pack. Place one DMM test probe on the now disconnected positive battery terminal and the other on a good chassis ground location. A resistance of 100 ohms or less will result in draining the battery. To calculate the extent of current drain, use Ohm's Law. Assuming a 12-volt chassis electrical system:

$$I = E/R$$
$$A = 12V \div 100 \text{ ohms}$$
$$A = 0.12 \text{ amperes or } 120 \text{ milliamps}$$

SHOP TALK

When performing a resistance battery-drain test, the lower the resistance reading, the greater the amperage drain on the battery.

Note: When performing battery-drain tests, it should be remembered that a number of electronic components draw some battery power when the ignition circuit is open, including some system ECMs/ECUs, electronically tuned radios, and digital clocks. In addition, diode leakage at the alternator rectifier can also account for battery drain-off.

SOURCING A BATTERY DRAIN

Obtain a vehicle wiring schematic and become familiar with the circuits. Check for the presence of any circuits controlled by mercury switches (these can remain closed) and any electronic components that require a closed circuit to function. To identify the circuit responsible for the current draw, disconnect the circuit breakers (or fuses) from each subcircuit until the ammeter readings fall within specification. If, after performing this test, the subcircuit causing the drain has not been identified, the drain is probably in the alternator or cranking motor circuits.

First, isolate the alternator. Electrically disconnect the alternator and observe the ammeter readings (or whether the testlight illuminates). If the current flow drops off or the testlight extinguishes, the problem is probably in the alternator rectifier. Remove the alternator and recondition. If the current drain remains, reconnect the alternator. Next, remove the positive terminal connection at the starter solenoid. If the current drain ceases, the cranking motor/solenoid is defective and must be reconditioned.

TYPICAL ELECTRICAL SYSTEM VALUES

It can help when you are troubleshooting electrical circuits if you have a rough idea of what normal operating values should be. Here are some approximate examples:

• Battery voltage	12.6 V-DC
• Charging voltage	14.0 V-DC
• Cranking motor current draw	± 400 amps
• Halogen headlight draw per unit	4 amps
• HID headlight current per unit	3 amps draw
• Typical bower motor on high	15 amps
• LED stoplight current draw	300 ma
• LED taillight current draw per unit	40 ma
• V-Ref	5 V-DC
• Alternator output	around 130 amps

Note that the above values are typical. In older trailers using incandescent bulbs, current draw is much greater.

The following lists TMC RP maximum continuous operation amperage ratings for on-highway electrical cable. The cable gauge section sizes use American wire gauge dimensions.

4/0	400 amps
3/0	300 amps
2/0	250 amps
1/0	200 amps
1 AWG	175 amps
2 AWG	100 amps

SHORT CIRCUITS

A vehicle electrical short circuit usually results in tripping the circuit breaker or blowing the fuse, protecting the subcircuit. A cycling circuit breaker will break the circuit at the point current flow through the circuit exceeds the amperage rating. When it cools, it will close the circuit. This on-off sequence will recur until the cause of the problem is located. A noncycling breaker will not automatically reset. The circuit must be opened before the breaker can close the circuit again. A fuse, when subjected to current overload, will blow. When a fuse blows immediately following replacement, the cause is often a short circuit.

The best method of locating a short circuit is to use a DMM in ohmmeter mode. Once again, locate the vehicle wiring schematic and use it as a guide through each subcircuit. Disconnect the positive wire to the breaker power strip. Connect one of the ohmmeter test probes to a good ground and the other to the circuit side of the circuit breaker. If the circuit is shorted to ground, the ohmmeter will read zero or close to zero (indicating low resistance). Next, using the vehicle wiring schematic as a guide, follow through the circuit, disconnecting one component at a time while observing the ohmmeter. If a component is unplugged and the resistance reading jumps to a high or infinity value, it is the likely source of the short.

FEEDBACK PROBLEMS

When current supplied to an electrical component lacks an adequate path to ground, it can seek an alternative route to complete the circuit. The dual filament incandescent bulbs used in older truck stop/turn/taillight circuits are a common cause of feedback. These dual filament bulbs share a common ground in the socket. When all of the current fed to the bright filament (brake/turn light) exceeds the ground's ability to handle it (due to high resistance), some of the current backfeeds through the other filament (taillight). This can energize both the taillight and the clearance light circuit, resulting in all clearance lights flashing when a turn signal is actuated or clearance lights illuminated at braking. When troubleshooting this condition, remember that it is caused by a high-resistance ground connection. Restore the ground integrity of the component and the condition will be repaired.

HIGH-VOLTAGE HYBRID CIRCUITS

High-voltage hybrid electrical vehicle (HEV) circuits are completely isolated from the 12-volt chassis electrical system. For obvious reasons, technicians working on HEVs should at least be aware of the potential danger of high-voltage electrical circuits. Current practice has been to use a fluorescent orange color on connectors and cables to indicate high-voltage potential. In addition, battery banks and high-voltage components are fully sealed and usually inaccessible without special tools. Most high-voltage components on HEVs are designed to have redundancy. This means that when there are problems with the high-voltage circuit components, the system defaults to limp home on mechanical power.

CAUTION

Do not attempt to work on HEV high-voltage circuits without having received the required OEM training. You should also be in possession of the special tools.

HOTEL LOADS

The term **hotel loads** is used to describe trucks and trailers fitted with an alternate mains electrical supply capability when parked and connected to the grid. Hotel load feed is used to run battery chargers, air conditioning, cab heating, block heaters, lighting, and entertainments systems. The **hotel load grid** has emerged and grown very rapidly over the past 5 years; stations are located at truck stops and DOT inspection stations along key interstate routes in the United States and Canada. The stations are equipped with AC electrical supply that allows the truck and/or reefer engine to be shut down while the rig is plugged in and parked.

The advantage of the hotel load grid is considerable. The fuel cost savings (typically about 10% of running a diesel engine), and wear and tear on the engine are significant to the vehicle owner. In addition, elimination of the emissions of running a diesel for hours while a driver eats or sleeps helps the environment. The hotel load grid program is expected to be expanded as the trucking industry moves toward 2017 CO_2 reduction and fuel efficiency compliance. When working on a vehicle connected to the grid, technicians must be mindful that unlike 12V-DC circuits, mains AC voltages can cause electrocution.

WARNING

Mains AC electricity can be fatal. Exercise caution when working on any vehicle connected to a hotel load AC-supply grid.

SUMMARY

- Truck and trailer lighting systems can be broadly divided into interior and exterior.
- Circuit wiring usually proceeds from a power source to a circuit breaker device (often virtual), to a switch, to a junction block, and then finishing at the load. The circuit is completed by grounding it at or close to the load.
- Headlamp assemblies consist of either a pair (a single lamp on either side of the chassis) or double units (two lamps on either side of the chassis).
- Truck chassis lighting systems use incandescent, gaseous-discharge, or LED operating principles.
- Halogen headlamps use a tungsten filament within a gas envelope.
- Xenon or HID headlamps provide longer service life and a brighter, whiter light. They are a more costly option than halogen headlights.
- LED light units are commonly used on truck chassis marker and taillight assemblies because they last longer and illuminate faster. Faster illumination makes them ideal as brake lights. LED headlights are beginning to be adopted in trucks and will likely become common in the near future.
- The wiring output junction block for the trailer is usually located inside the tractor cab, often directly behind the driver seat.
- Most current trucks use computer-managed instrument clusters that are networked with the chassis data bus; for this reason, the OEM service literature should be consulted before attempting troubleshooting.
- Relays are a means of using a low-current control circuit to switch a high-current circuit. Standard SAE relays use standard numeric coding to indicate the terminal assignments. FETs replace electromechanical relays in some more recent truck chassis.
- Fuses and cycling and noncycling circuit breakers are all circuit protection devices. They may be electromechanical or virtual.
- Electromechanical fuses fail and have to be replaced when overloaded. Circuit breakers trip (either virtually or electromechanically) when overloaded, opening the circuit; they reset either automatically, when the circuit is switched open, or using an EST depending on the type of circuit.
- Virtual fuses and virtual breakers do not physically fail. When they trip, the technician must use electronic diagnostic routines to locate the root cause.
- SAE type-1 circuit breakers are cycling and automatically reset. SAE type-2 circuit breakers are noncycling and the circuit must be switched open to allow them to cool after tripping. In some cases, a virtual type-2 circuit breaker must be reset by a technician using an appropriate EST and OEM software.
- Virtual fuses and breakers may be programmed by the OEM as SAE type 1 and type 2.
- Voltage drop testing is a key to diagnosing truck chassis electrical systems. Note the procedures for performing a fast electrical system assessment on a truck in this chapter, and study voltage drop testing with a DMM in the next.
- When connected to a hotel load grid, some vehicle components and systems are powered by AC mains electricity. Technicians must exercise extreme caution when working on a vehicle connected to the grid.

REVIEW QUESTIONS

1. When an incandescent light bulb is energized in an electrical circuit, what is electrical energy converted to?
   a. light only
   b. heat only
   c. heat and light
   d. gas
2. What is the function of a headlamp dimmer switch?
   a. applies resistance to the headlamps to reduce brightness
   b. switches headlamps between separate low- and high-beam circuits
   c. increases voltage for high-beam operation
   d. reduces voltage for low-beam operation
3. What type of headlight fires an arc through gas to produce light energy?
   a. incandescent
   b. halogen
   c. LED
   d. HID
4. Which of the following best explains the construction of a halogen headlamp element?
   a. inert gas in a tube
   b. tungsten filament encased in a quartz envelope
   c. tungsten filament encased in a fluorescent tube
   d. electrodes and arc igniter
5. How much voltage is required to sustain the arc in a typical HID headlamp assembly?
   a. 5 volts
   b. 12 volts
   c. around 85 volts
   d. more than 110 volts
6. Which of the following best defines light energy?
   a. electrons
   b. photons
   c. protons
   d. heat

7. Which of the following would be used to switch turnlights on a truck chassis?
   a. stalk switch
   b. toggle switch
   c. headlamp switch
   d. brake pressure switch
8. How many contacts are in a typical tractor-to-trailer electrical connector?
   a. one
   b. three
   c. five
   d. seven
9. What color code is used for the brake light circuit in a truck trailer electrical connector?
   a. green
   b. red
   c. blue
   d. white
10. When truck and trailer clearance lights and headlamps seem to burn out at a high rate, which of the following would be a required step in troubleshooting?
    a. battery state of charge
    b. battery surface charge
    c. voltage losses in ground circuit
    d. charging system voltage regulator
11. Which of the following governs highway truck exterior lighting standards in North American trucks?
    a. FMVSS 108
    b. FMVSS 121
    c. RP 401
    d. RP 532
12. In a standard SAE relay, which of the following is normally open (NO)?
    a. 3
    b. 5
    c. 4
    d. 7
13. In an older standard SAE relay, which pair of terminals energizes the coil circuit?
    a. 87 and 87a
    b. 30 and 86
    c. 85 and 86
    d. 85 and 87
14. The headlights on a truck with electromechanical controls and circuit breaker protection continuously illuminates for a short period and then cut out. Technician A says that if the circuit breaker is a type 1, replacing it with a type 2 should repair the problem. Technician B says that the problem is likely caused by a malfunction in a type-2 circuit breaker. Who is correct?
    a. Technician A only
    b. Technician B only
    c. both A and B
    d. neither A nor B
15. Technician A says that on a standard SAE relay, terminal 3 is always the common terminal. Technician B says that in a standard SAE relay, the coil circuit must be energized to close the normally open 5 terminal. Who is correct?
    a. Technician A only
    b. Technician B only
    c. both A and B
    d. neither A nor B
16. What is the maximum acceptable total voltage drop across all of the terminals and cables on the insulated side of a cranking circuit?
    a. 12.6 volts
    b. 12.0 volts
    c. 9.6 volts
    d. 0.5 volt
17. Technician A says that a corroded light unit terminal could result in an increased voltage drop across the terminal. Technician B says that a corroded light unit terminal could result in heat at the terminal. Who is correct?
    a. Technician A only
    b. Technician B only
    c. both A and B
    d. neither A nor B
18. Technician A says that a disadvantage of using LEDs as stoplights is that when the circuit is switched on, they illuminate at a slower speed than incandescent bulbs. Technician B says that xenon headlamps are a disadvantage in trucks because they have a shorter service life than halogen bulbs. Who is correct?
    a. Technician A only
    b. Technician B only
    c. both A and B
    d. neither A nor B
19. What type of filament is used in a halogen headlight?
    a. halogen
    b. tungsten
    c. xenon
    d. argon
20. Technician A says that HID headlamps consume more electrical power than equivalent halogen headlamps. Technician B says that HID headlamps are brighter and their blue light is more similar to natural sunlight than that of halogen lamps. Who is correct?
    a. Technician A only
    b. Technician B only
    c. both A and B
    d. neither A nor B

11

Prerequisites: Chapters 5, 6, and 10

DIAGNOSIS AND REPAIR OF ELECTRONIC CIRCUITS

OBJECTIVES

After reading this chapter, you should be able to:

- Explain what is meant by sequential electronic troubleshooting.
- Perform tests on some key electronic components, including diodes and transistors.
- Define the acronym EST.
- Identify some types of EST in current use.
- Identify the levels of access and programming capabilities of each EST.
- Explain why electronic damage may be caused by electrostatic discharge and by using inappropriate circuit analysis tools.
- Describe the type of data that can be accessed by each EST.
- Identify what type of data may be read using the onboard flash codes.
- Perform some basic electrical circuit diagnosis using a DMM.
- Identify the function codes on a typical DMM.
- Test some common input circuit components such as thermistors and potentiometers.
- Test semiconductor components such as diodes and transistors.
- Describe the full range of uses of handheld reader/programmer ESTs.
- Connect an EST to a vehicle data bus via the data connector and scroll through the display windows.
- Define the objectives of snapshot test.
- Outline the procedure required to use a PC and OEM software to read, diagnose, and reprogram vehicle electronic systems.
- Understand the importance of precisely completing each step when performing sequential troubleshooting testing of electronic circuits.
- Identify the SAE J1587/1939 codes for SAs, SPNs, PGNs, MIDs, PIDs, SIDs, and FMIs.
- Repair the sealed electrical connectors used in most electronic wiring harnesses.
- Describe the operating principles and work safely around truck SRS and rollover protection systems.

KEY TERMS

active codes
active cruise with braking (ACB)
air bag
air bag control unit (ACU)
blink codes
Brake-Link
breakout box
breakout T
butt splice
collision mitigation system (CMS)
collision warning system (CWS)
communications adapter (CA)
continuity
controller
crash sensor
data connector
Deutsch connector
diagnostic link connector (DLC)
data link connector (DLC)

- diagnostic trouble code (DTC)
- digital multimeter (DMM)
- driver interface unit (DUI)
- electronic control unit (ECU)
- electronic service tools (ESTs)
- failure mode indicator (FMI)
- flash codes
- ground strap
- handheld (HH)
- handshake
- historic codes
- heavy-duty onboard diagnostics (HD-OBD)
- inactive codes
- inductive amp clamp
- labscope
- locking tang
- message identifier (MID)
- meter resolution
- multiprotocol cartridge (MPC)
- open circuit
- parameter identifiers (PIDs)
- personal computer (PC)
- Pro-Link GRAPHIQ
- Pro-Link iQ
- reader/programmer
- RollTek
- root mean square (rms)
- RP 1210
- SAE J1587
- SAE J1708
- SAE J1939
- safing sensor
- scopemeter
- service information system (SIS)
- snapshot test
- source address (SA)
- subsystem identifiers (SIDs)
- supplemental restraint system (SRS)
- suspect parameter number (SPN)
- tang release tool
- thermistor
- three-way splice
- triage tool
- Wingman

INTRODUCTION

The objective of this chapter is to provide the truck technician with a general introduction to some of the repair techniques required of truck electronic management systems other than the engine and transmission. The subject of electronic engine management systems is handled in some detail in a companion to this textbook, and electronically managed transmissions and brake systems are dealt with in later chapters in this book. Any system that is managed by a computer will operate on the principles outlined in Chapter 6. Because an engine is complex in terms of the subsystems that enable it to operate, a more elaborate electronic management package is required to run it. By contrast, electric pulse wipers are also electronically managed but the circuit required to manage them is simple because the only input device is a multiposition switch. The output that results from its processing cycle depends on the selected switch position.

The electronics required to manage chassis subsystems such as the transmission and antilock braking system (ABS), although more complex than electric pulse wipers, tend to be simpler than engine management electronics because these systems are not as comprehensively monitored, and the range of outputs is limited. In all current trucks the various computers on the chassis (known as controllers) are required to network (talk) with each other. We use the term **electronic control unit (ECU)** to refer to each **controller** connected to the communication network. An ECU always has processing capability and can be classified as a computer.

The contents of this chapter continue the study of electronics that began in Chapter 6 and will be further developed in Chapter 12 and later chapters dealing with specific chassis systems such as transmissions and brakes.

UNDERSTAND THE PRINCIPLES FIRST

The first requirement of working on electronically managed chassis systems is to understand the fundamentals of the operating principles as outlined in Chapter 5 and Chapter 6. The next is to acquire the elementary skills required to operate the **electronic service tools (ESTs)** to perform circuit analysis. Since the trucking industry has entered the electronic age, the quality of original equipment manufacturer (OEM) technical service literature (on compact disc [CD] or online) has improved. However, the technician is required to accurately interpret the text in these service publications. A couple of generations ago, a diesel engine technician experienced in one OEM's product could probably successfully repair another OEM's equipment with little reference to service literature. This is simply not possible with today's technology. Each OEM publishes sequential troubleshooting procedures that must be accurately adhered to. Failure to exactly comply with each step or to scramble the troubleshooting sequence can make the outcome of the procedure meaningless.

SAE J-STANDARDS

Initially, the manufacturers of trucks in North America have cooperated to a much larger extent than automobile OEMs because of the fact that multiple manufacturers are involved in the chassis subsystem components. Since 1995, all highway electronic systems have been manufactured in compliance with Society of Automotive Engineers (SAE) standards known as J1587, J1708, and J1939. The first two SAE recommended practices, SAE J1587 and J1708, governed the electronic hardware compatibility and software protocols ("language" rules and regulations). The current SAE J1939 governs both electronic hardware and software protocols, the matter formerly covered by SAE J1587 and J1708. However, recently most manufacturers of truck components and subsystems have become more proprietary about granting access to their software and have charged increasingly high subscription fees. There are some notable exceptions such as Bendix, which has been the most open in educating the industry and enabling access to its electronic systems at no charge.

Complying with the hardware standards enables multiple onboard electronic systems to "share" hardware such as throttle position and vehicle speed sensors, avoiding unnecessary duplication of components. Compliance with software protocols means that multiple onboard electronic systems can "speak" to each other on the chassis data bus, allowing them to optimize vehicle performance. It also means that one manufacturer's software and EST can at least read another's using standardized SAE data bus protocols. Data bus navigation is fully explained later in this chapter. When two or more electronic systems are connected electronically, they can be described as being networked. Vehicle networking is known as multiplexing. This is fully explained in Chapter 12.

TROUBLESHOOTING TECHNIQUES

Troubleshooting electronic circuits is not only a lot easier than troubleshooting hydromechanical circuits; it produces faster, more accurate results. The guesswork that was always required with more complex hydromechanical systems has been almost eliminated. This is not to say that troubleshooting electronically managed circuits will not produce the odd problem that the sequential troubleshooting steps have not addressed. Every technician working on electronic circuits will have a nightmare tale about an intermittent failure that produced no codes or tattletales and defied every attempt in the book to solve it, but such instances are rare.

MULTIMETERS AND PERSONAL COMPUTERS

Two sets of skills are required to troubleshoot electronic circuits. The first is a knowledge of how to operate the ESTs recommended by the OEM. At minimum, a **digital multimeter (DMM)** and a microprocessor-based EST such as a **personal computer (PC)** with the OEM software is required. The term PC is used to describe any computer (desktop, laptop, or tablet) that uses a Windows operating environment. The next skill is simply the ability to read and accurately interpret wiring schematics and the OEM service literature. Armed with these skills, technicians are often able to repair a circuit malfunction without having truly understood the cause of the problem. Good technicians, however, should make it their business to fully understand the reasons for a circuit malfunction. This will enable the technician to build a set of diagnostic skills that, after some years, will be known as *experience*.

ADVANCED LEVEL SKILLS

Most OEMs offer training courses in the diagnosis and repair of their electronic systems. This specialized training becomes more essential as the complexity of the system increases. The technician who is fully familiar with one of the truck ABSs could probably diagnose and repair other truck ABSs given the appropriate EST and troubleshooting literature. This would not be smart when working on electronically managed engines and transmissions, when OEM specialized training is usually essential to avoid system damage and costly downtime. Electronically managed vehicle systems are not more complicated, but it is important to avoid guesswork and trial and error approaches to troubleshooting. Almost all current truck software is engineered to only function in a MS-Windows environment and as yet, little is available on **handheld (HH)** devices though this might change when **heavy-duty onboard diagnostics (HD-OBD)** eventually becomes mandatory across all model lines. All OEMs were required to manufacture one model line that was HD-OBD compliant beginning in model year (MY) 2010; the objective was to work out the bugs before it became mandatory across all model lines in 2013. This did not happen due to compliancy issues and at the time of writing, no firm date has been set for final implementation. Although HD-OBD only monitors emissions-related faults, many in the industry feel that it will make the bus more open to third-party ESTs and software.

SCOPE OF CHAPTER

The objective of this chapter is to cover some of the basic level electronic diagnostic practices. Many of these skills are taken for granted on the shop floor and practiced on a daily basis. At the end of the chapter, we take a brief look at truck supplement restraint systems (SRS). You might question why SRS appears in this chapter, but the fact is it is probably more appropriately located in this chapter than any other in the book.

11.1 TYPES OF EST

The term *electronic service tool* (EST) is generally used in the trucking industry to cover a range of electronic service instruments from onboard diagnostic/

malfunction lights to sophisticated computer-based communications equipment. The use of generic ESTs and procedures is reviewed in this section. Proprietary ESTs are designed to work with an OEM's specific electronics and are not discussed in any great detail in this text because they are system specific. All the OEMs provide courses that address their technology. The use of each proprietary EST is covered in these courses.

ESTs capable of reading electronic control unit (ECU) data are connected to the onboard electronics by means of a **data connector** generally known as a **diagnostic link connector (DLC)**. The common connector and protocols used by truck and heavy equipment OEMs usually allow the proprietary software of one manufacturer to at least read the parameters and conditions of their competitors. In short, this means that if a Navistar-powered truck has an electronic failure in a location where the only service dealer is Freightliner, some basic problem diagnosis can be undertaken.

CONNECTING TO THE BUS

To "talk" to the data bus on a heavy-duty vehicle, typically you require the following:

- Data link connector (DLC)
- Communications adapter (CA)
- Electronic service tool (EST)

DATA CONNECTORS

The data connector required to access a data bus depends on the generation of the communications bus on the vehicle. The physical connector is known as a **diagnostic link connector (DLC)** or **data link connector (DLC)** depending on the OEM. However, because both terms use the same acronym, and are interchangeable, DLC will be used throughout.

J1708 Data Connector

The connector used to access a J1587 data bus is a J1708 DLC. The J1708 DLC is a 6-pin **Deutsch connector** and its cavity assignments are shown in **Figure 11–1**.

J1939 Connector

Three types of J1939 connectors are in current usage. They are:

- J1939 black 9-pin
- J1939 green 9-pin for EPA MY 2013
- J1962 16-pin ALDL (Volvo-Mack in 2014)

The first generation J1939 DLC uses a 9-pin Deutsch connector with a key in the A-pin recess as shown in **Figure 11–2**. In Figure 11–2 the cavity assignments for a 9-pin data connector are shown along with the J1939 3-pin connector which is used to physically connect nodes (ECUs) to the bus within the chassis. A standard J1939 DLC incorporates J1587/1708, which may or may not be used depending on the OEM.

FIGURE 11–1 J1587/1708 connector cavity pin assignments

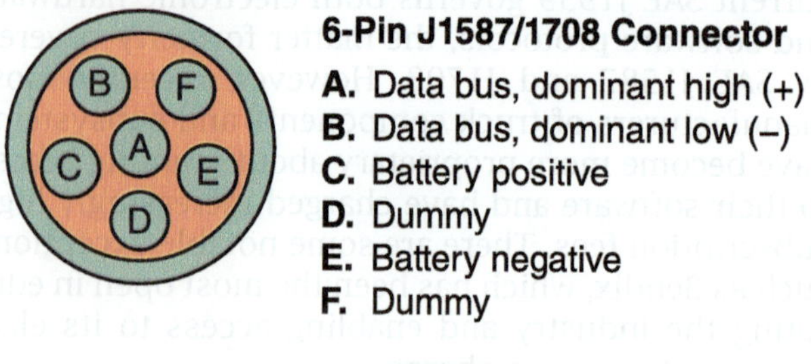

The second-generation J1939 data connector is almost identical in appearance to its predecessor except that it is green in color and is designed to block access by the older black colored version by using a reduced size F cavity in the receptacle. **Figure 11–3** shows a 2013 J1939 DLC.

It has been known for some time that the days of the 9-pin DLC are numbered but SAE and TMC have yet to agree on a universal standard. However, Volvo-Mack jumped the gun and elected to make the automotive assembly line diagnostic link (ALDL) or J1962 connector their data connector from MY 2014 on. Although using the J1962 connector is under discussion by the other OEMs no agreement had been made at the time of writing. **Figure 11–4** is an illustration of a J1962 connector, receptacle, and cables. The cavity pin assignments on a J1962 DLC are shown and discussed in Chapter 12.

Tech Tip

The post-2013 green colored J1939 9-pin data connector is backward compatible with pre-2013 receptacles. However, a smaller F-pin cavity on the green J1939 receptacle is designed to block access by a black pre-2013 plug.

COMMUNICATIONS ADAPTER

To connect most ESTs to a J1939 data bus a **communications adapter (CA)** is required. A CA is a serial link. CAs are supposed to be **RP 1210** compatible meaning that they should be universal. However, this is not always the case and some OEMs have been using blocking software with the result that their own CA is required to make the connection. **Figure 11–5** shows a generic CA, which is RP 1210 compliant and

FIGURE 11–2 J1939 connector cavity pin assignments

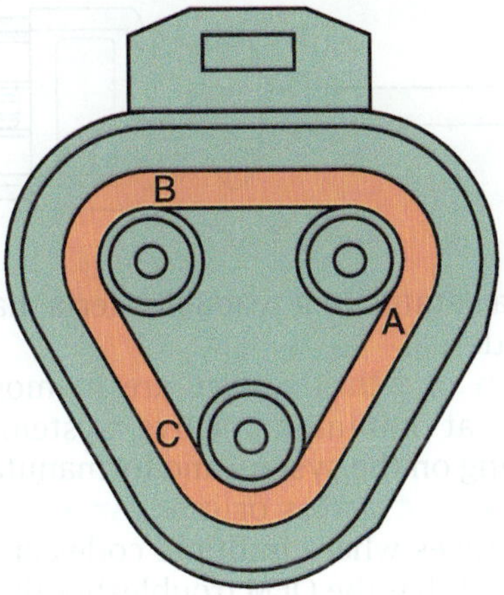

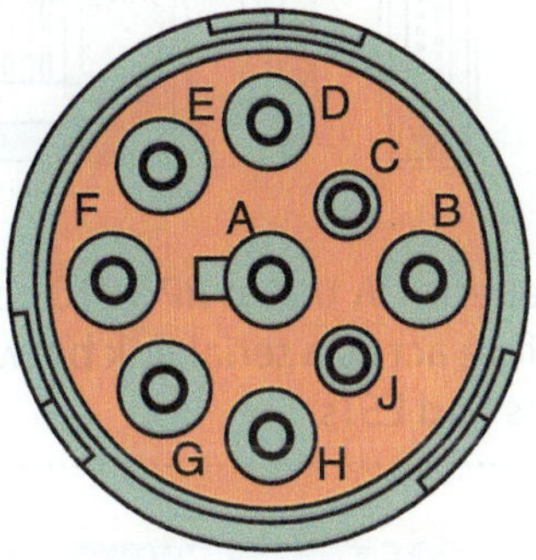

3-Pin J1939 Connector

A. CAN busline, dominant high (+)
B. CAN busline, dominant low (–)
C. CAN shield

9-Pin J1939 Connector

A. Battery negative
B. Battery positive
C. J1939 CAN busline, dominant high (+)
D. J1939 CAN busline, dominant low (–)
E. CAN shield
F. J1587 busline, dominant high (+)
G. J1587 busline, dominant low (–)

FIGURE 11–3 Green, post-2013 J1939 data connector: The cavity pin designations are unchanged but pin F on the new receptacle is smaller so it blocks access by the former black plug.

can be used as the serial link to connect most data link plugs to an EST.

ESTs

An EST can be a PC (desktop, laptop, or tablet) or any number of different types of HH devices. Most OEMs require the use of a PC operating in a Windows environment loaded with proprietary software. In most cases, the proprietary software is available only by subscription (and a fee) but there are some notable exceptions such as the Bendix example mentioned earlier. In addition to the OEM ESTs, there exists a wide range of generic ESTs and software packages. These generic diagnostic instruments and software are only as good as the OEM willingness to support the third party: some will do no more than read data fields off the bus.

DASH LIGHTS

All current trucks are equipped with dash warning lights to alert the operator of a problem and its severity. Typically those are:

- **MIL (malfunction indicator light).** Vehicle is safe to drive to a repair facility: may be an emissions-related problem.

FIGURE 11–4 A J1962 data connector used by Volvo-Mack in their MY 2014 chassis

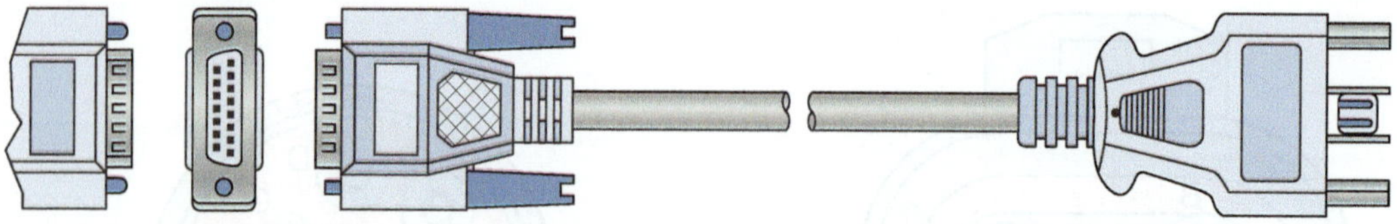

FIGURE 11–5 A generic CA that is RP 1210 compliant and can therefore act as a serial link between most data connectors and ESTs.

- **CEL (check engine light).** Vehicle (and the engine) should be checked for obvious items (oil level, coolant level, etc.).
- **SEL (stop engine light).** Stop vehicle immediately and do not attempt to restart.

Reading Codes Onboard

The simplest EST used for electronic circuit diagnosis is the system's own electronic warning light or lights and capacity for displaying data on the dash digital data display unit. When flash or blink codes are broadcast they can be difficult to interpret especially when multiple status alerts have to be displayed. The lights used to blink codes vary depending on the system. Usually they are mounted on the vehicle dashboard so the operator can be alerted, but in the case of some ABSs, they may be located on the chassis or system controller module. They may also share their function with other circuit warning lights. Electronic malfunction lights may be used initially as a driver alert but, in most cases, they have the ability to "blink" out numeric codes. These codes relate to a specific circuit malfunction that enables anyone to read and report the problem without specialized tools. In most current trucks, the dash digital display unit and its ability to display status in a readable alerts, has superseded blink codes.

Flash or **blink codes** are in most cases used to read, at minimum, active system fault codes. Depending on the system and its manufacturer, sometimes **historic codes** or **inactive codes** can also be read. In cases where multiple codes are displayed, it is essential that the OEM troubleshooting literature be consulted because certain types of circuit failure can trigger codes in functional circuits and components.

11.2 USING DIGITAL MULTIMETERS

A DMM is simply a tool for making electrical measurements. DMMs may have any number of special features, but essentially they measure electrical pressure or *volts*, electrical current or *amps*, and electrical resistance or *ohms*. A good-quality DMM with minimal features may be purchased for as little as $100. As the features, resolution, and display quality increase in sophistication, the price increases proportionally. A typical DMM and its features are shown in **Figure 11–6**, **Figure 11–7**, **Figure 11–8**, **Figure 11–9**, and **Figure 11–10**.

FIGURE 11–6 Typical DMM

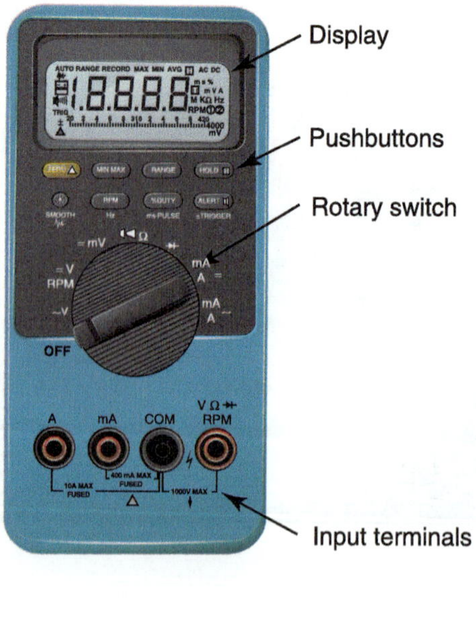

FIGURE 11–7 DMM input terminals

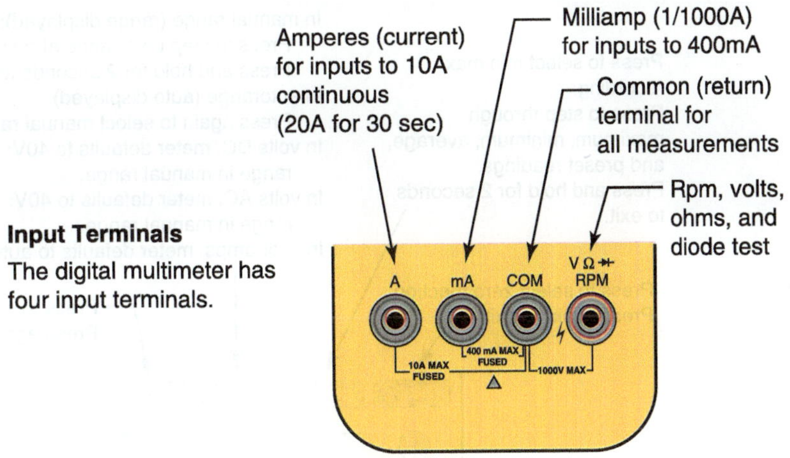

FIGURE 11–8 DMM rotary switch

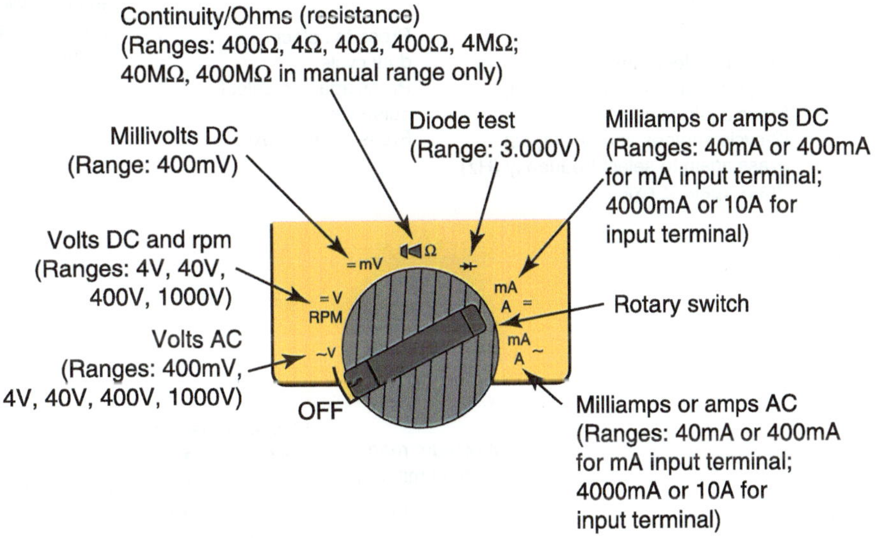

Because most electronic circuit testing *requires* the use of a DMM, this instrument should displace the analog multimeter and circuit testlight in the truck/bus technician's toolbox. Reliability, accuracy, and ease of use are all factors to consider when selecting a DMM for purchase. Some options the technician may wish to consider are a protective rubber holster (which will greatly extend the life of the instrument), analog bar graphs, and enhanced resolution. This section deals with the use of DMMs. A knowledge of basic electricity is assumed.

RESOLUTION

Resolution shows how fine a measurement can be made with the instrument. Digits and counts are used to describe the resolution capability of a DMM. A 3½-digit meter can display three full digits ranging from 0 to 9, and one half digit that displays either 1 or zero is left blank. A 3½-digit meter will therefore display 1,999 counts of resolution. A 4½-digit DMM can display as many as 19,999 counts of resolution.

Many DMMs have *enhanced resolution*, so the meter's reading power is usually expressed in counts rather than digits. For instance, a 3½-digit meter may have enhanced resolution of 4,000 counts. Basically, **meter resolution** is expressed in counts rather than digits. For example, a 3½-digit or 1,999-count meter will not measure down to 0.1V when measuring 200V or higher. However, a 3,200-count meter will display 0.1V up to 320V, giving it the same resolution as a 4½-digit, 19,999-count meter until the voltage exceeds 320V.

FIGURE 11–9 DMM pushbuttons

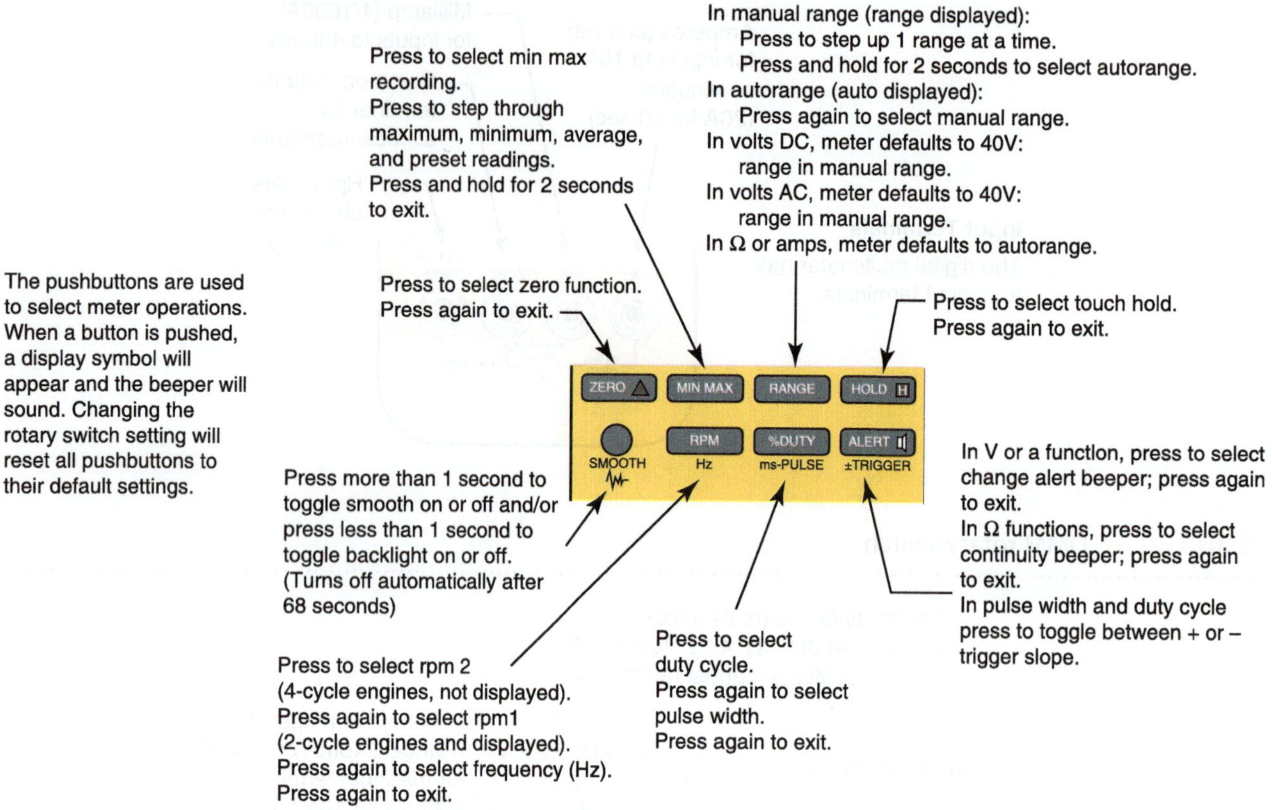

FIGURE 11–10 DMM data display

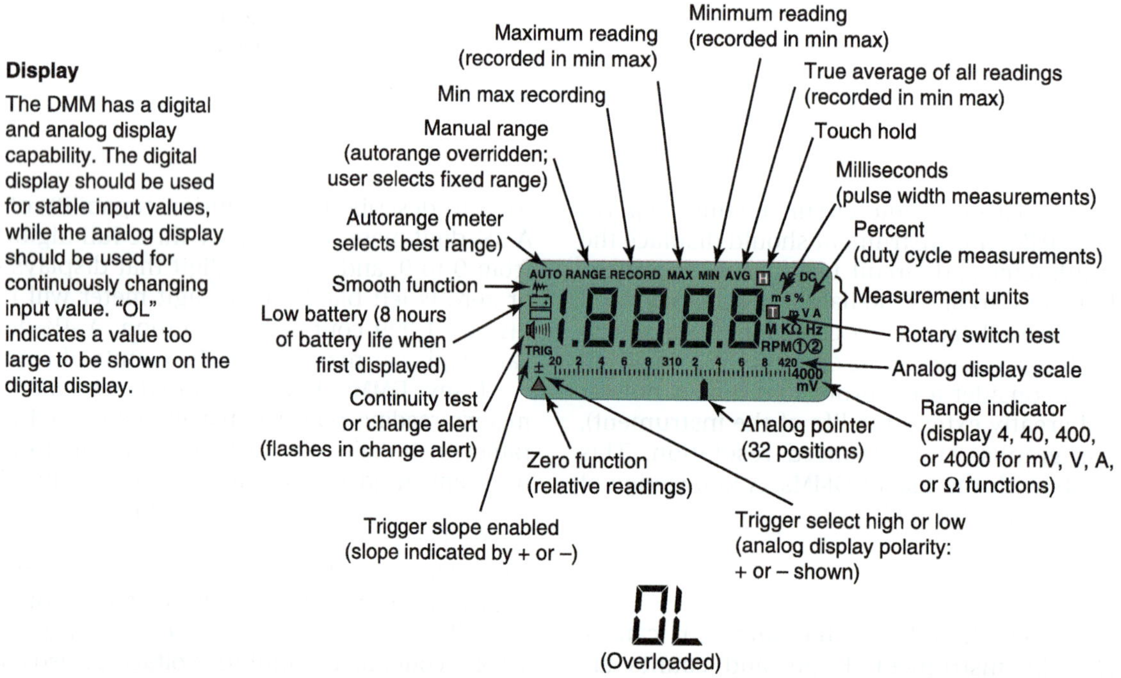

ACCURACY

Accuracy in a meter specification describes how close the displayed reading on the DMM is to the actual value of the measured signal. This is usually expressed as a percentage of the reading. An accuracy rating of ±1 percent in a DMM reading a voltage value of 10V means that the actual value could range between 9.9V and 10.1V. DMM accuracy can be extended by indicating how many counts the right display digit may vary. An accuracy rating of ±(1% + 2) would mean that a displayed voltage of 10V could have an actual value range between 9.88V and 10.12V.

Analog multimeters have accuracy ratings that vary between 2 and 3 percent of full scale. DMM accuracy ratings range between ±(.07% + 1) and ±(0.1% + 1) of the *reading*. If this sounds a little complicated, it is: but if you are going to invest $700 into a DMM rather than $100, you should understand what you are getting.

OPERATING PRINCIPLES

In any electrical circuit, voltage, current flow, and resistance can be calculated using *Ohm's Law*. A DMM makes use of Ohm's Law to measure and display values in an electrical circuit. A typical DMM will have the following selection options, chosen by rotating the selector:

Off	Shuts down DMM
V	Enables alternating current (AC) voltage readings
V-	Enables direct current (DC) voltage readings
mV-	Enables low-pressure DC voltage readings
Ω	Enables component or circuit resistance readings
α	Enables **continuity** (a circuit capable of being closed) testing; identifies an open/closed circuit
A	Checks current flow (amperage) in an AC circuit
A-	Checks current flow (amperage) in a DC circuit

CAUTION

Ensure that a DMM is used to make voltage measurements on all current trucks. Analog voltmeters can damage ECUs and sensitive electronic circuit components.

MEASURING VOLTAGE

Checking circuit supply voltage is usually one of the first steps in troubleshooting. This would be performed in a vehicle DC circuit by selecting the V-setting and checking for voltage present or high-/low-voltage values. Most electronic equipment is powered by DC. For example, home electronic apparatus such as computers, televisions, and stereos use rectifiers to convert household AC voltage to DC voltage.

The waveforms produced by AC voltages can be either sinusoidal (sine waves) or nonsinusoidal (sawtooth, square, ripple). A DMM will display the **root mean square (rms)** value of these voltage waveforms. The rms value is the effective or equivalent DC value of the AC voltage. Meters described as *average responding* give accurate rms readings only if the AC voltage signal is a pure sine wave. They will not accurately measure nonsinusoidal signals. DMMs described as *true-rms* will measure the correct rms value regardless of waveform and should be used for nonsinusoidal signals.

A DMM's capability to measure voltage can be limited by the signal frequency. The DMM specifications for AC voltage and current will identify the frequency range that the instrument can accurately measure (**Figure 11–11**). Voltage measurements determine:

- Source voltage
- Voltage drop
- Voltage imbalance
- Ripple voltage
- Sensor voltages

MEASURING RESISTANCE

Most DMMs will measure resistance values as low as 0.1 ohm. Some will measure high-resistance values up to 300MΩ (megohms). Infinite resistance or resistance greater than the instrument can measure is indicated as OL or flashing digits on the display. For instance, an **open circuit** (one in which there is no path for current flow) would read OL on the display.

Resistance and continuity measurements should be made on open circuits only. Using the resistance or continuity settings to check a circuit or component that is energized will damage the test instrument. Some DMMs are protected against such accidental abuse—the extent of damage will depend on the model. For accurate low-resistance measurement, test lead resistance, typically between 0.2 and 0.5 ohm depending on quality and length, must be subtracted from the display reading (**Figure 11–12**). Test lead resistance should *never exceed* 1 ohm. Some DMMs have a "zero" function that takes into account test lead resistance.

If a DMM supplies less than 0.3V DC test voltage for measuring resistance, it is capable of testing resistors isolated in a circuit by diodes or semiconductor junctions, meaning that they do not have to be removed from the circuit board. Resistance measurements determine:

- Resistance of a load
- Resistance of conductors
- Value of a resistor
- Operation of variable resistors

FIGURE 11–11 DMM set up for making voltage measurements

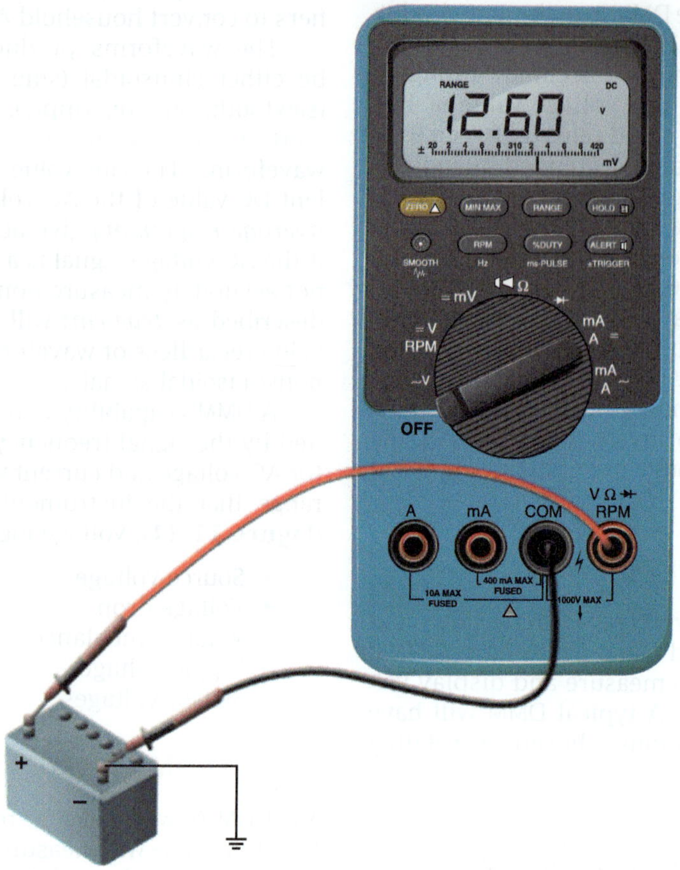

FIGURE 11–12 DMM set up for measuring resistance

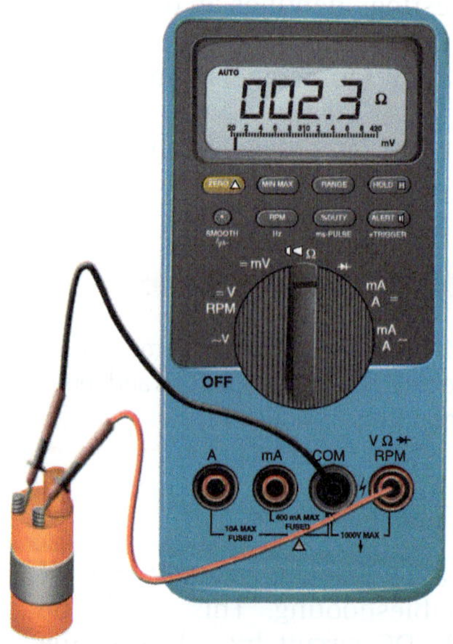

Continuity is a quick resistance check that distinguishes between an open and a closed circuit. Most DMMs have audible continuity beepers that beep when they detect a closed circuit, permitting the test to be performed without looking at the meter display. The actual level of resistance required to trigger the beeper varies from model to model.

Continuity tests determine:

- Fuse integrity
- Open or shorted conductors
- Switch operation
- Circuit paths

Diode Testing

A diode is an electronic switch that can conduct electricity in one direction while blocking current flow in the opposite direction. They are commonly enclosed in glass cylinders; a dark band identifies the cathode or blocking terminal. Current flows only when the anode is more positive than the cathode. Additionally, a diode will not conduct until the forward voltage pressure reaches a certain value, which is 0.3V in a silicon diode. Some meters have a diode test mode. When testing a diode with the DMM in this mode, 0.6V is delivered through the device to indicate continuity; reversing the test leads should indicate an open circuit in a properly functioning diode. If both readings indicate an open-circuit condition, the diode is open. If both readings indicate continuity, the diode is shorted.

MEASURING CURRENT

There are two ways to measure current flow through an electrical circuit using an ammeter:

- Direct reading
- Indirect reading

True current measurements should be made in series, unlike voltage and resistance readings, which are made in parallel. When using a DMM to make a direct-reading measurement of current flow, the test leads are plugged into a separate set of input jacks that routes the current to be measured through the meter. Current measurements determine:

- Circuit overloads
- Circuit operating current
- Current in different branches of a circuit

When the test leads are plugged into the current input jacks and are used to measure voltage, a direct short across the source voltage through a low-value resistor inside the DMM, called a *current shunt*, occurs. High current flows through the meter. If not adequately protected, both the meter and the circuit can be damaged (**Figure 11–13**).

A DMM should have current input fuse protection of high enough capacity for the circuit being tested.

FIGURE 11–13 DMM set up for measuring current flow

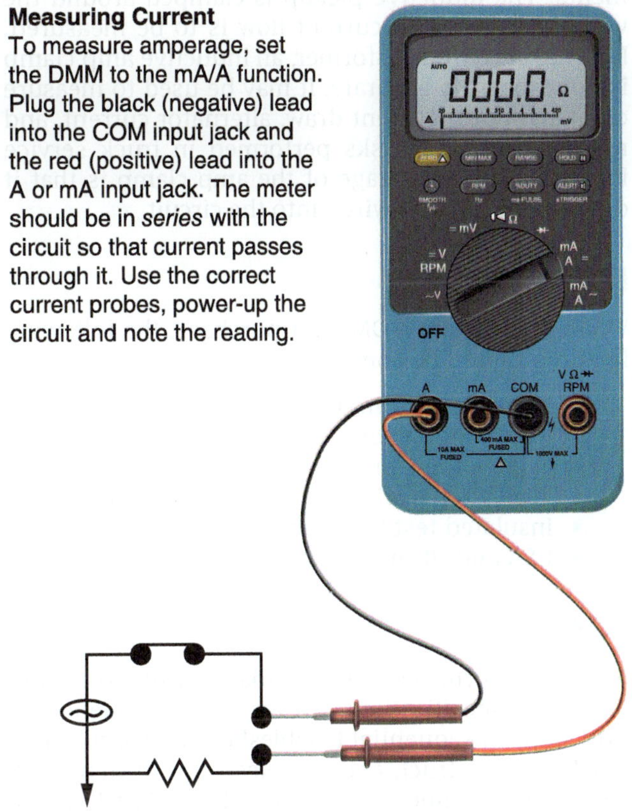

This is mainly important when working with high pressure (220V+) circuits.

Indirect Readings

When making current measurements that exceed the DMM's rated capacity, there are two ways to do this. It should be noted that indirect current measurement is not as accurate as direct measurement.

Current Transformers

A current transformer measures AC current only. The output of a current transformer is 1mA per 1A. This means that a current flow of 100A is reduced to 100mA, which can be handled by most DMMs. The test leads would be connected to the mA and common input jacks and the meter function switch set to mA AC. Current transformers provide a rather inaccurate circuit test that is used for ballpark reckoning only. There are no applications for this tool in truck electrical systems, but they can be used on the main electrical supply in service facilities.

Inductive Amp Clamp

An **inductive amp clamp** can be used to measure AC or DC. It is a transducer that produces an output of 1mV per ampere. The test leads are connected to the V and common input jacks. The DMM function switch should be set to the V or mV scale, selecting V-AC for AC current or V-DC for DC current measurements. The inductive pickup is clamped around the wire through which current flow is to be measured. Like the current transformer, an inductive amp clamp is not especially accurate. It may be used to measure cranking motor current draw, alternator current, and many other daily tasks performed in truck service facilities. The advantage of the amp clamp is that it does not have to be wired into the circuit.

DMM FEATURES

When considering a DMM for purchase, the following features should be considered:

- Fused current inputs
- Use of high-energy fuses (600V+)
- High-voltage protection in resistance mode
- Protection against high-voltage transients
- Insulated test lead handles
- CSA and UL approval

SOME TYPICAL DMM TESTS

Always perform tests in accordance with truck and bus OEM specifications. Never jump sequence or skip steps in sequential troubleshooting charts. Most DMM tests on truck and bus electronic systems will be used in conjunction with an EST or PC and proprietary software. The following tests assume the use of a Fluke 88 DMM but do not differ much when other DMMs are used.

1. Engine Position: Fuel Injection Pump Camshaft, Engine Camshaft, and Crank Position Sensors
   Hall effect sensors:
   (a) Cycle the ignition key and then switch off.
   (b) Switch the meter to measure V-DC/rpm.
   (c) Identify the ground and signal terminals at the Hall sensor. Connect the positive (+) test lead to the signal terminal and the negative (–) test lead to the ground terminal. Crank the engine.
   At cranking speeds, the analog bar graph should pulse. At idle speeds or faster, the pulses are too fast for bar graph readout.
   (d) Press the duty cycle button once. Duty cycle can indicate square wave quality (poor-quality signals have a low duty cycle). Functioning Hall sensors should have a duty cycle of about 50 percent, depending on the sensor. Check the specifications.
2. Potentiometer-Type Throttle Position Sensor (TPS)
   Resistance test:
   (a) Key off.
   (b) Disconnect the TPS.
   (c) Select Ω on DMM. Connect the test probes to the signal and ground terminals. Next, move the accelerator through its stroke, observing the DMM display.
   (d) The analog bar should move smoothly without jumps or steps. If it steps, there may be a bad spot in the sensor.
   Voltage test:
   (a) Key on, engine off.
   (b) Set the meter to read DC V. Connect the negative lead to ground.
   (c) With the positive lead, check the reference voltage value and compare to specs. Next, check the signal voltage (to the ECM) value through the accelerator pedal stroke. Check the values to specification. Also observe the analog pointer: As with the resistance test, this should move smoothly through the accelerator stroke.
3. Magnetic Sensors
   Magnetic or variable reluctance sensors function similarly to a magneto. The output is an AC voltage pulse, the voltage and frequency values of which rise proportionally with rotational speed increase. Voltage values range from 0.1V up to 5.0V, depending on the rotational speed and the type of sensor. Frequency varies with the number of teeth on the reluctor wheel and the rpm. Vehicle speed sensors (VSS), engine speed sensors (ESS), and ABS wheel speed sensors all use this method of determining rotational speed. Test using the V-AC switch setting and locating test leads across the appropriate terminals.
4. Min/Max Average Test for Lambda (λ), Exhaust Gas or O_2 Sensors
   The test procedure outlined here relates to a zirconium dioxide type O_2 sensor used on a natural gas engine. The procedure is similar when applied to a diesel particulate filter (DPF) NOx sensor but the values will differ.
   (a) With the key on, the engine running, and the DMM set at V-DC, select the correct voltage range.
   (b) Connect the negative test lead to a chassis ground and the positive test lead to the signal wire from the lambda sensor. Press the DMM min/max button.
   (c) Ensure that the engine is warm enough to be in closed loop mode (100 mV–900 mV O_2 sensor output). Run for several minutes to give the meter time to sample a scatter of readings.
   (d) Press the min/max button slowly three times while watching the DMM display. A maximum of 800 mV and a minimum of fewer than 200 mV should be observed. The average should typically be around 450 mV.

(e) Next, disconnect a large vacuum hose to create a lean burn condition. Repeat steps c and d to read the average voltage. Average voltage should be lower, indicating a lean condition.

(f) The same test can be performed using propane enrichment to produce a rich air-fuel ratio (AFR) condition and, therefore, higher voltage values.

(g) Lambda sensor tests can be performed while road testing the vehicle; 450 mV normally indicates stoichiometric fueling (15:1 AFR), but check to specifications.

5. Thermistors
   Most **thermistors** used in computerized engine systems have a negative temperature coefficient (NTC), meaning that as sensor temperature increases, their resistance decreases. They should be checked to specifications using the DMM ohmmeter function and an accurate temperature measurement instrument.

OEMs seldom suggest random testing of suspect components. The preceding tests are typical procedures. Circuit testing in today's computerized vehicle system management electronics is highly structured and part of a sequential troubleshooting procedure. It is important to perform each step in the sequence precisely as instructed. Never forget that skipping a step can invalidate every step that follows.

BREAKOUT BOXES AND BREAKOUT TS

The DMM is often used in conjunction with a **breakout box** or **breakout T**. Breakout devices are designed to be T'ed into an electrical circuit to enable circuit measurements to be made on both closed (active) and de-energized circuits. The idea is to access a circuit with a test instrument without interrupting the circuit. A breakout T normally describes a diagnostic device that is inserted into a simple two- or three-wire circuit such as that used to connect an individual sensor, whereas a breakout box accesses multiple wire circuits such as main harness connectors for diagnostic analyses of circuit conditions.

Connecting into the Circuit

Most of the electronic management system OEMs use a breakout box that is inserted into the interface connection between the engine electronics and chassis electronics harnesses. The face of the breakout box displays a number of coded sockets into which the probes of a DMM can be safely inserted to read circuit conditions. Electronic troubleshooting sequencing is often structured based on the data read by a DMM accessing a circuit. A primary advantage of breakout diagnostic devices is that they permit the reading of an active electronic circuit, for instance, while an engine is running.

CAUTION

When a troubleshooting sequence calls for the use of breakout devices, always use the recommended tool. Never puncture wiring or electrical harnesses to enable readings in active or open electronic circuits. The corrosion damage that results from damaging wiring conduit will create problems later on and the electrical damage potential can be extensive.

Connector Dummies and Test Lead Spoons

Diagnostic connector dummies are used to read a set of circuit conditions in a circuit that has been opened by separating a pair of connectors. Dummies and test lead spoons are a means of accessing the wiring and connector circuitry with a DMM without damaging the connector sockets and pins. Ensure that correct dummies/spoons are used.

CAUTION

The terminals in many connectors are especially vulnerable to the kind of damage that can be caused by attempting to insert DMM probes, paper clips, and other inappropriate devices. Even more important, remember that it is always possible to cause costly electrical damage by shorting and grounding circuits in a separated electrical connector.

SHOP TALK

When disassembling or performing a multiple step, electronic troubleshooting sequence on a large multiterminal connector, print or photocopy the coded face of the connector(s) from the service literature and use it as a template. The alphanumeric codes used on many connectors can be difficult to read, and using a template is a good method of orienting the test procedure. Just make sure you orientate the side of the connector you are working with correctly. OEMs orientate connector images to either or both sides.

TESTING SEMICONDUCTORS

Testing a diode is shown in **Figure 11–14**. The objective of these tests is to familiarize you with how diodes and transistors function and fail. To perform the tests you will need a DMM in ohmmeter mode: In these diagrams, an analog-type needle and rotary scale are used to indicate the difference between high and low readings. Use the diode test mode as well while doing the tests, and record the results.

FIGURE 11–14 Testing semiconductors

Testing a Diode

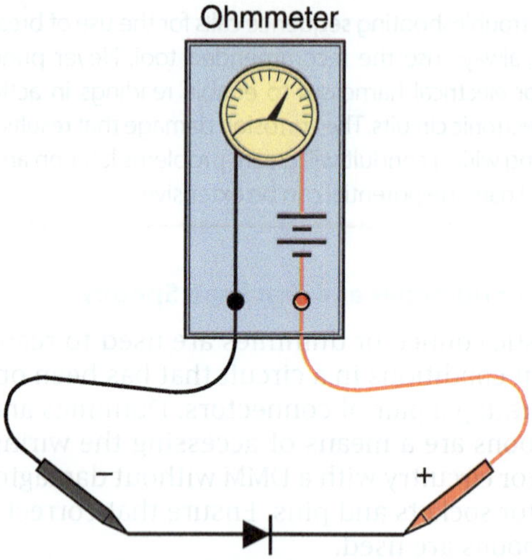

Step 1. Connect the ohmmeter leads to the diode. Notice if the meter indicates continuity through the diode or not.

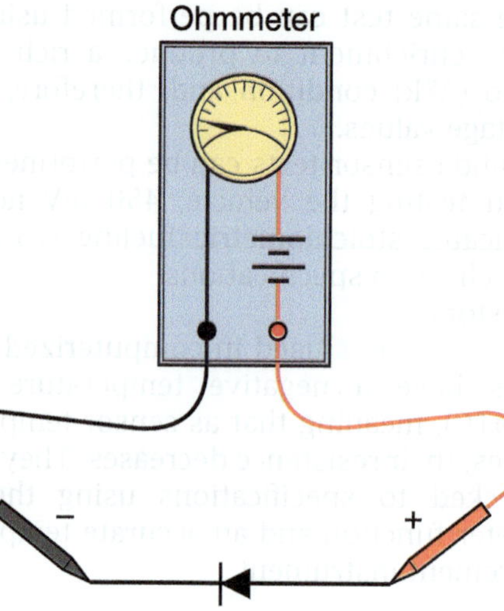

Step 2. Reverse the DMM leads to diode terminals and note if the meter indicates continuity. The ohmmeter should read continuity through the diode in only one direction. If continuity is not indicated in either direction, the diode is open. If continuity is indicated in both directions, the diode is shorted.

Testing a Transistor

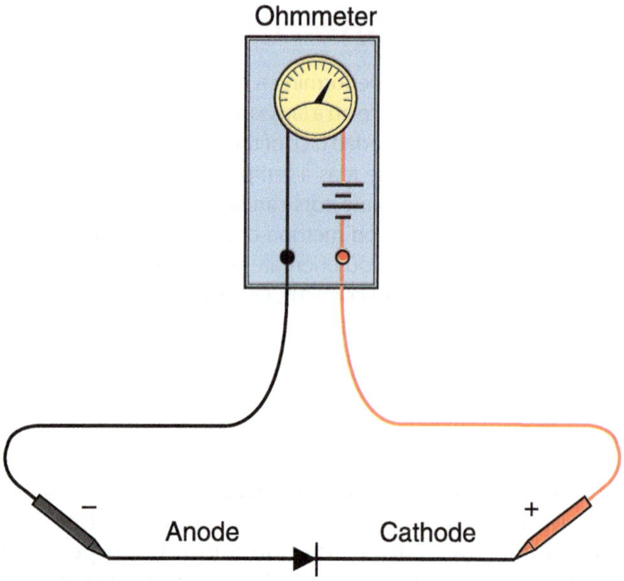

Step 1. Using a good diode, determine which ohmmeter lead is positive and which is negative. The ohmmeter indicates continuity through the diode only when the positive lead is connected to the anode of the diode and the negative lead is connected to the cathode.

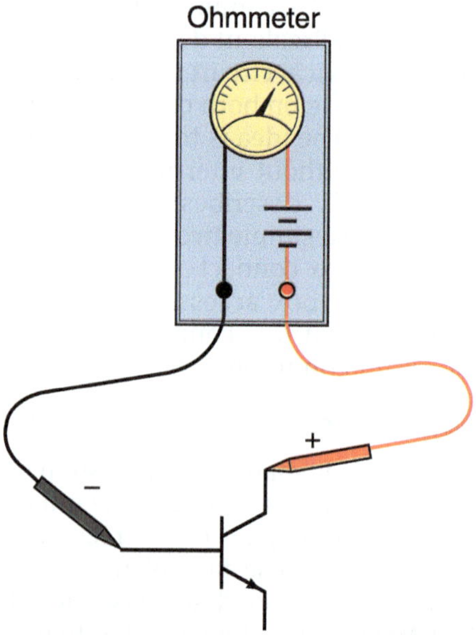

Step 2. Replace the diode with a transistor: if it is an NPN, connect the positive lead to the base and the negative lead to the collector. The ohmmeter should indicate continuity producing a similar reading to that obtained when the diode was tested.

FIGURE 11–14 (continued)

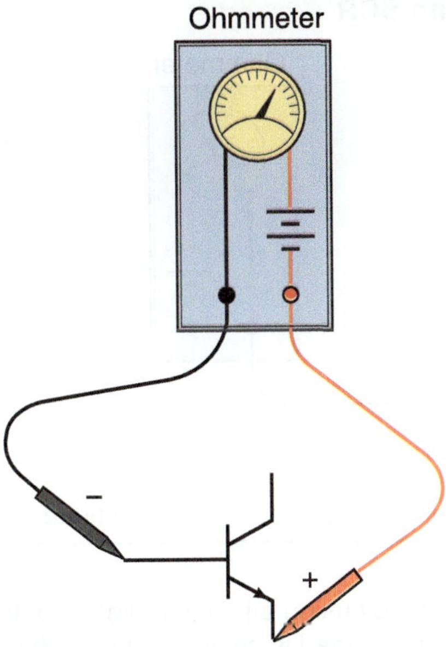

Step 3. With the positive lead still connected to the base of the transistor, connect the negative lead to the emitter. The ohmmeter should again indicate a forward diode junction. If the ohmmeter does not indicate continuity between the base-collector or the base-emitter, the transistor is open.

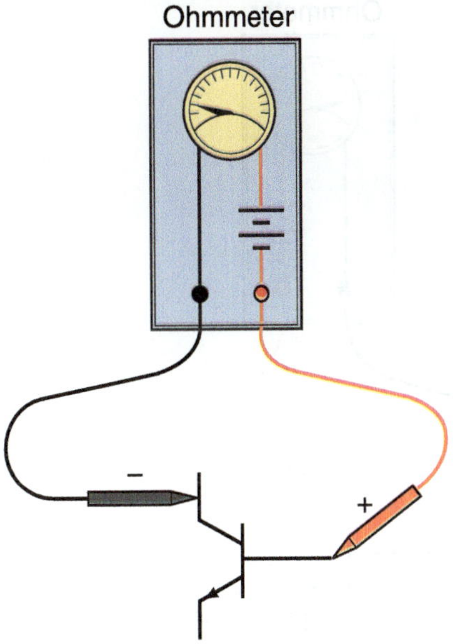

Step 4. Connect the negative lead to the base and the positive lead to the collector. The ohmmeter display should read infinity or no continuity.

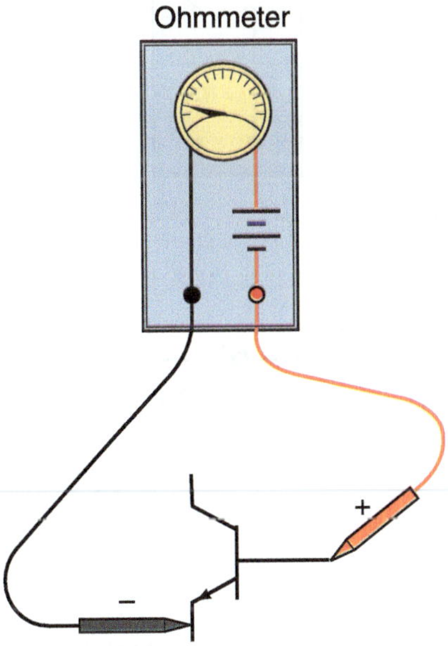

Step 5. With the negative lead connected to the base, reconnect the positive lead to the emitter. There should be no continuity. If a very high resistance reading is indicated, the transistor is "leaky" but may still function in the circuit. If a very low resistance reading is produced, the transistor is shorted.

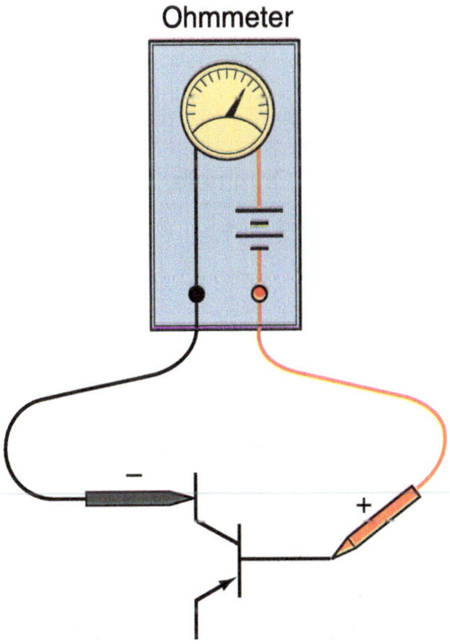

Step 6. To test a PNP transistor, reverse the polarity of the ohmmeter leads and repeat the preceding test. When the negative lead is connected to the base, a forward diode junction should be indicated when the positive lead is connected to the collector or emitter.

FIGURE 11–14 (continued)

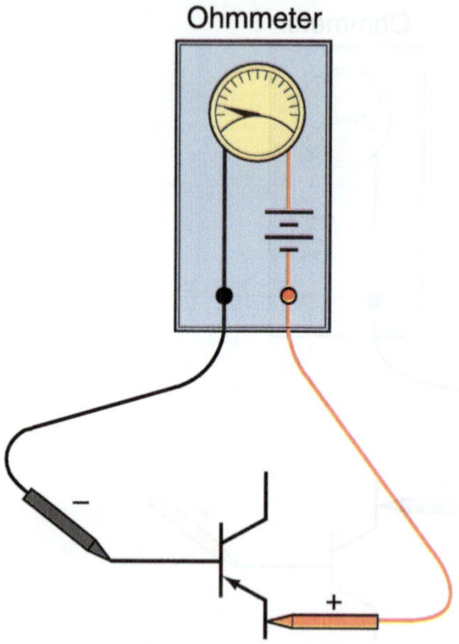

Step 7. When the positive ohmmeter lead is connected to the base of a PNP transistor, and the negative lead is connected to the collector or the emitter, no continuity should be indicated.

Testing an SCR

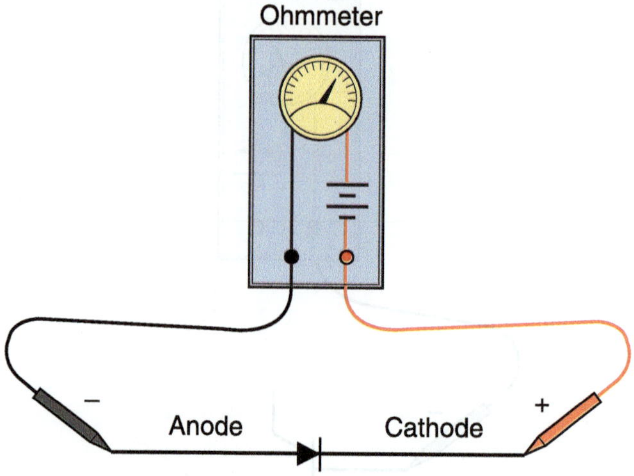

Step 1. Begin by using a junction diode, to determine which ohmmeter lead is positive and which is negative. The ohmmeter should indicate continuity only when the positive lead is connected to the anode of the diode and the negative lead is connected to the cathode.

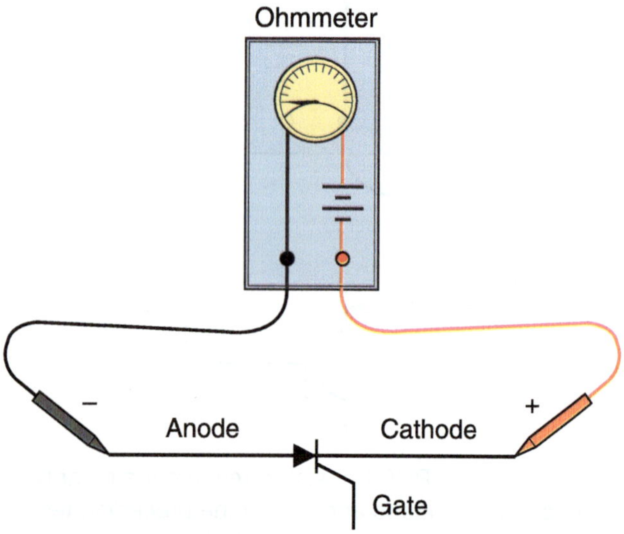

Step 2. Connect the positive ohmmeter lead to the anode of the SCR and the negative lead to the cathode. The ohmmeter should indicate no continuity.

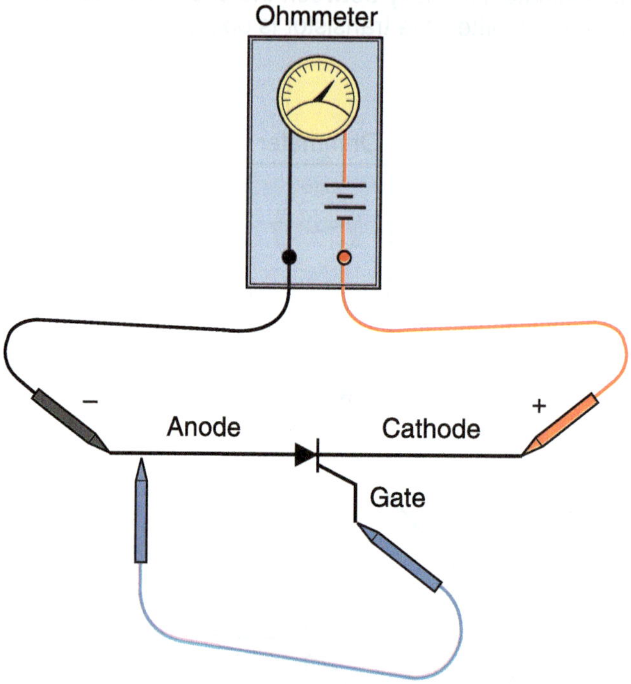

Step 3. Using a jumper lead, connect the gate of the SCR to the anode. The ohmmeter should indicate a forward diode junction when the connection is made. If the jumper is removed, the SCR may continue to connect or it may turn off depending on whether the ohmmeter can supply sufficient current to keep the SCR above its holding current.

FIGURE 11–14 (continued)

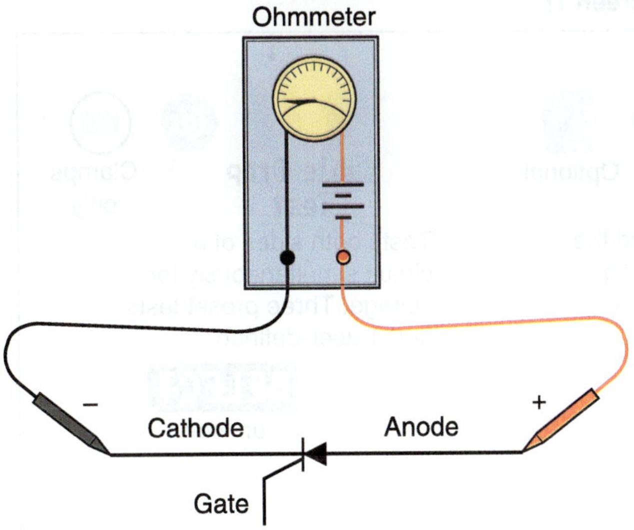

Step 4. Reconnect the ohmmeter leads to the SCR so that the cathode is connected to the positive lead and the anode is connected to the negative lead. The DMM reading should indicate no continuity.

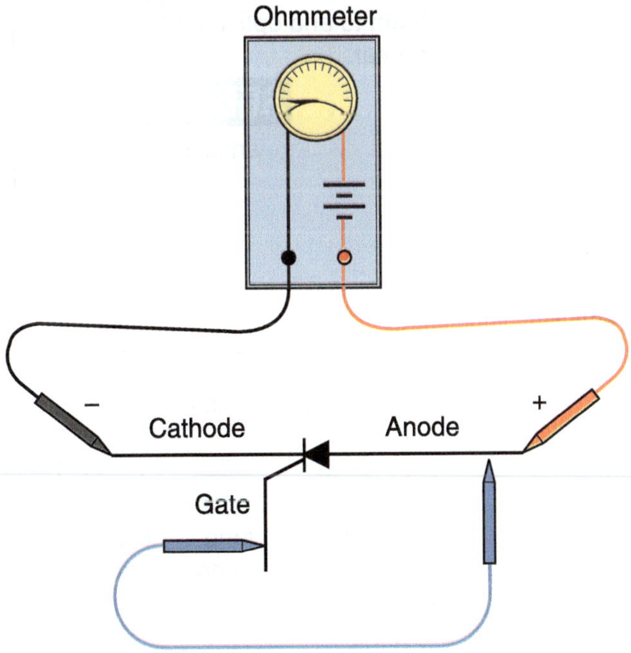

Step 5. If a jumper is used to connect the gate to the anode, the ohmmeter should indicate no continuity although SCRs designed to switch large currents (50 amperes or more) may indicate some leakage current.

11.3 MULTIFUNCTIONAL HANDHELD DIAGNOSTIC TOOLS

We first introduced the Midtronics EXP in Chapter 7 when we examined just the battery testing capabilities of this multifunctional, handheld diagnostic tool. This EST boasts a broad range of electrical circuit troubleshooting capabilities and there are other similar examples. Any technician who specializes in electrical diagnostics will find this type of multifunction tool invaluable. Its features include:

- Alphanumeric keypad, icons, and scroll navigation
- DMM with temperature evaluation and scope mode
- Battery conductance testing
- Cranking circuit testing
- Charging circuit testing
- Digital signal processing (DSP) to analyze both the amplitude and frequency of ripple patterns to identify open or shorted diodes and open-phase conditions
- Conductance cable testing for detailed analysis of voltage drop conditions
- Data read/write transfer via SD card for easy updates
- Print option
- Graphical display window with icon-driven menu options

At this moment in time the Midtronics EXP is recommended by two of the truck OEMs for chassis voltage electrical tests of batteries, cranking circuits, and charging circuits. Regardless of what brand of truck you are working on, technicians can increase diagnostic accuracy and reduce troubleshooting time using this type of tool. **Figure 11–15** shows the main menu display options on the Midtronics EXP.

11.4 HANDHELD DIAGNOSTIC INSTRUMENTS

The Pro-Link 9000 served the first couple of generations of truck chassis electronics but is now effectively obsolete. The Pro-Link family of ESTs can be described as generic **reader/programmer**. This type of simple, some might call them primitive, electronic data reader established itself as the industry standard EST for many years. Depending on how well the OEM supported NEXIQ (the manufacturer) with software (this ranged from very well to providing almost nothing), Pro-Link could read and program most systems with the correct

FIGURE 11–15 Midtronics EXP main menu options

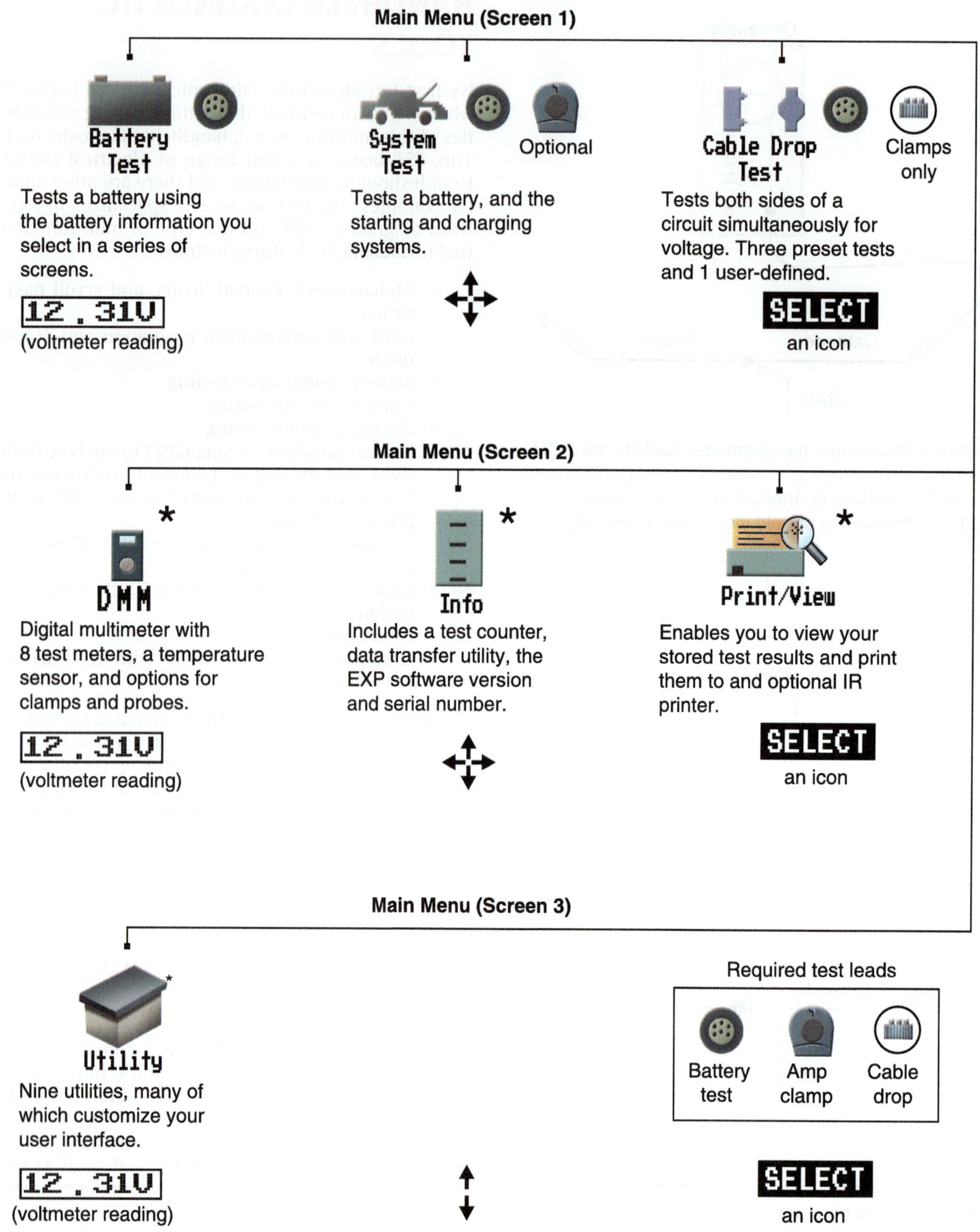

FIGURE 11–16 Typical data connector hardware

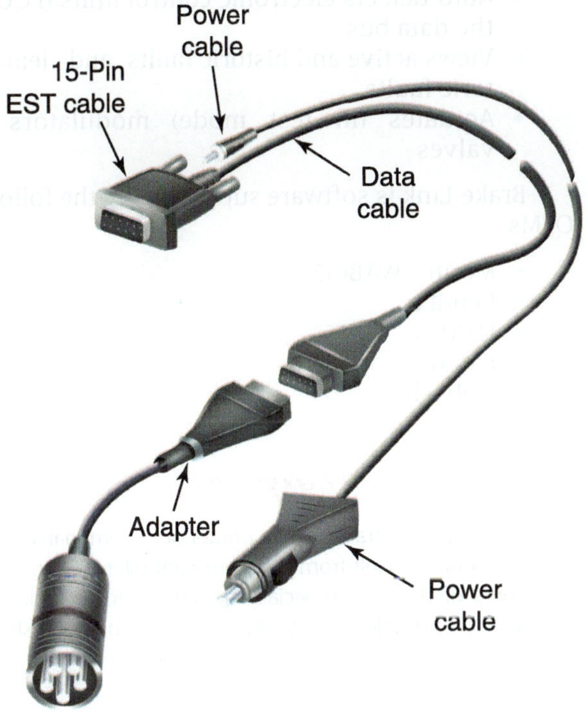

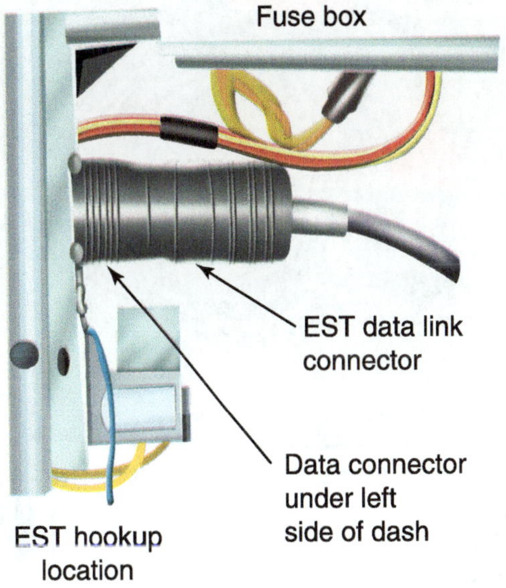

software cartridge or card. The last generation of the Pro-Link 9000 introduced the **multiprotocol cartridge (MPC)**. This enabled simpler updates. An MPC update required replacing the MPC card, which is a credit card-sized, data retention card. **Figure 11–16** shows the hardware required to connect a typical HH-EST to a chassis data bus.

CURRENT DIAGNOSTIC TOOLS

NEXIQ phased out Pro-Link 9000, updating it to a new generation of handheld diagnostic ESTs (HH-ESTs). These are equipped with better graphics, easier navigation, and more powerful diagnostic features. As each new generation of diagnostic instruments emerges, user-friendliness improves and today most HH-ESTs can be updated by connecting to the web and downloading new files. In most cases, technology-literate users can acquire some expertise with the instruments simply by exploring the menu fields while connected to an ECU. These are microprocessor-based test instruments that may do the following (depending on the system):

1. Access **active codes** and historic codes; erase historic (inactive) codes.
2. View system identification data.
3. View data on system operation with the key-on or the truck moving.
4. Run diagnostic tests on system subcomponents, such as solenoid testing.
5. Reprogram customer data parameters on engine and chassis systems.
6. Act as a serial link (communications adapter) to connect the vehicle electronics via the Internet to a centrally located mainframe for proprietary data programming (some systems only).
7. Snapshot system data parameters to assist intermittent fault-finding solutions.
8. Enable updates by CD, flash card, cartridge, or online.

PRO-LINK IQ

The **Pro-Link iQ** is a more powerful HH-EST. It is somewhat limited in its ability to network with some of the more recent engine management systems but can be useful when working with Bendix and WABCO brake electronics and Allison transmissions. Proprietary software can be downloaded to the Pro-Link iQ via the Internet (and a subscription fee), but it is important to check whether the diagnostics downloaded are compatible with the generation of onboard software you want to diagnose.

The Pro-Link iQ uses a large touch screen. The instrument is enabled for a heavy-duty standard read capability so it is useful as a first-level diagnostic tool, especially in general service facilities with

FIGURE 11–17 Pro-Link iQ diagnostic tool

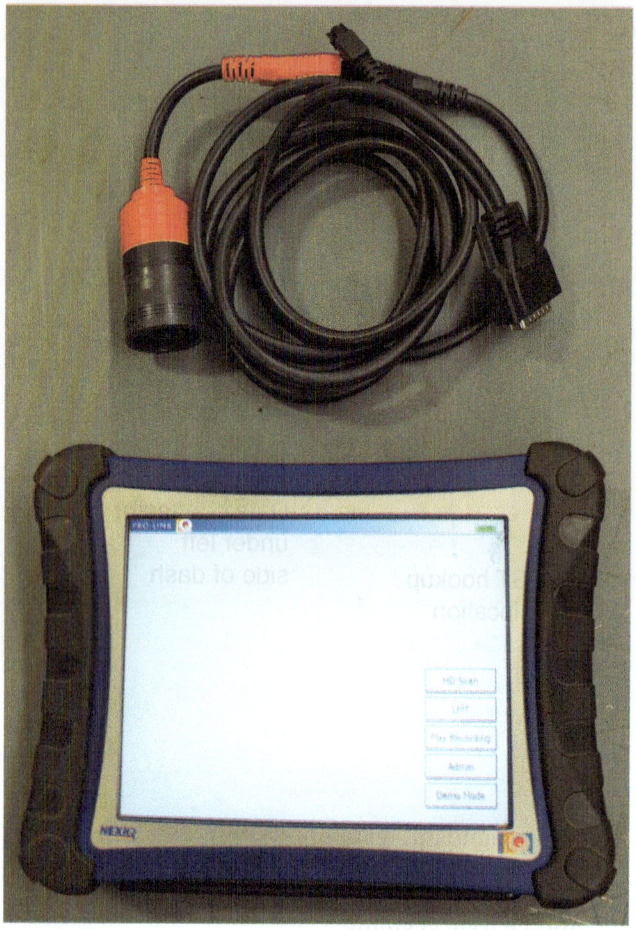

limited OEM software support. The touch screen is icon- and menu-driven, requiring a minimal level of technical skills to navigate those data bus systems it can access. The kit comes with 6- and 9-pin Deutsch connectors to connect to the chassis data bus, and in most cases, wireless connectivity. **Figure 11–17** is an image of a Pro-Link iQ HH diagnostic tool.

Pro-Link GRAPHIQ

NEXIQ's **Pro-Link GRAPHIQ** is a close relative of the Pro-Link 9000 it replaced. It boasts a few minor improvements over the earlier model, such as an eight-line color data display window and some basic graphing capability. However, this diagnostic tool is best suited to working on older truck electronics. It comes with heavy-duty standard reader capability.

Brake-Link

NEXIQ's **Brake-Link** HH-EST is designed to work exclusively with air brake electronic systems. Because its software is fully supported by the brake system OEMs, it actually has good practical application in truck shops. The EST's functions are simple:

- Tests trailer ABS with power line carrier (PLC) communications
- Auto-detects electronic control units (ECUs) on the data bus
- Views active and historic faults, and clears historic faults
- Actuates (in test mode) modulators and valves

Brake-Link is software supported by the following OEMs:

- Meritor WABCO
- Bendix
- Haldex
- Eaton
- Wabash

SHOP TALK

Most HH-ESTs can transmit to a printer or PC terminal any data that it can read itself from a system controller (ECU or ECM). Printouts of data are especially helpful when analyzing the causes of a condition or explaining a problem to a customer. Some HH-ESTs feature wireless connectivity.

Read-Only ESTs

If the term *scan tool* is going to be applied to any EST we use to access a truck data bus, it should be used to describe a new generation of read-only HH-ESTs such as the JDR 050 and JDR100 CANbus readers. This category of EST has been designed with simplicity in mind, measuring less than 3 × 3 inches, with a simple three-line data display window and no more than three command keys. The JDR 050 uses just two command keys and displays fault codes and system status for up ten ECUs on the J1939. The JDR 100 adds a third command key and boasts slightly enhanced fault code readout capability, including engine live engine status. Both ESTs are targeted for operations that primarily service rather than repair trucks.

CONNECTING THE EST

This section outlines the general procedure required to connect a handheld diagnostic tool to a chassis data bus. The guidelines here are general in nature and vary according to what extent the OEM supports the EST software.

Startup

This is an initializing process. It permits the chassis electronic system to communicate with the handheld EST. It requires connection of the data link (6-pin or 9-pin) to the chassis data bus.

- Turn the vehicle ignition on. Connect the HH-EST to the serial communication port. Ensure that the unit powers up. Depending on the chassis system, you may have to supply a dedicated power-up feed to the unit.

- The first information to be displayed by the EST is the system software identification. This identifies the chassis system you are working with.
- Select the Enter menu option. This initiates the loading sequence.
- When the application has finished loading, a screen confirms this.
- Pressing the Enter menu option produces a range of options such as:
  - Monitor parameters
  - Fault codes
  - Diagnostic tools
  - Customer programming
  - English/metric toggle
- The touch pad (if equipped) or up and down arrow keys can be used to scroll the five menu options and make a selection. In most cases a return key or function key permits the exiting of an application or series of applications.

English/Metric Toggle

Data can usually be displayed in either English or metric units by HH-ESTs. Again this is menu-driven and selection is made by using the touch screen or up-down keys.

Monitor Parameters

This option allows data to be downloaded from the system control module (ECM/ECU) to be displayed by the EST graphics. Depending on the specific system you are working with, a long list of parameters may be displayed. Use the menu to select those parameters you wish to monitor to customize a list. Alternatively, it is possible to scroll through the data list line by line. This feature is useful on ESTs with more limited display capability.

FAULT CODES

Use the main menu graphic display and select the fault codes field. This will then present you with three or four options depending on the MY of the vehicle:

- Active codes
- Inactive (historic) codes
- Pending codes (new in 2010)
- Clear codes

Active codes can be erased from some controller electronics when a vehicle is not being operated. If erased, the code will not be displayed as "active" until the vehicle is started.

Active Codes

An active code is any system code that is currently active. If there are no active codes logged into the system, the display window will read blank or Empty. The EST display window usually indicates an alpha explanation of any displayed active code. In addition, it will identify the SPN/PID and FMI and possibly proprietary system codes logged. SAs, SPNs, MIDs, SIDs, PIDs, DTCs, and FMIs are explained in a little more detail later in this chapter. The proprietary system code is the OEM code, usually the same code that would be flashed on the diagnostic lights. For an example of how a code is logged, if a TPS signal (used by both the engine and transmission management electronics) produces a voltage above normal (shorted high), the FMI code plus an explanation would be displayed. When more than one active code is logged into the system, depending on the EST, the up and down arrow keys may be required to scroll the list. The main menu option will get you out of this field.

Inactive Codes

Inactive codes are codes that have been previously logged into the system. Although no longer active they have not been cleared. Some OEMs use the term *historic codes* to describe inactive codes. The EST readout for an inactive code usually displays the following:

- The sensor, switch, or circuit that produced the fault
- A description of the fault
- The PID, FMI, and the OEM fault code
- The number of separate occurrences of the fault

Pending codes

Since 2010, the self-diagnostic capacity in some controllers has been able to identify pending codes. This makes use of a set of fault mode indicators (FMIs) that were introduced in 2010: the objective was to add some prognostic (see Chapter 4) capacity to self-diagnostics by identifying performance drop-offs that fell short of outright failure but suggested that failure might be imminent.

Clear Codes

Depending on the vehicle MY and the OEM, active, inactive, and pending codes can be cleared from a chassis electronic system. Technicians should refrain from clearing any codes until the root cause of a prob lem has been located and repaired. To clear codes, the engine should not be running and the ignition key should be on. It also is usually required that inactive codes be read before they are cleared.

Locate fault codes in the menu and press Enter. Next, use the menu options to select the clear faults option and press Enter. A prompt to confirm the clearing of codes will usually appear. This is followed by a confirmation that all inactive codes have been cleared.

CHANGING PASSWORDS

It is usually necessary to key in a password to perform certain functions, especially those that affect performance programming. Different passwords may be used for customer data, vehicle data, and fleet data programming access to the system or chassis management control module—or the same password may be used for all functions. Use the main menu to select the change password function. Locate program parameters in the menu and press Enter. It is necessary to enter the current password before programming a new password. If the wrong password is input, you will be instructed to press the Enter key and try again. Passwords can be numeric or alpha. In older handheld ESTs with numeric keypads, numeric passwords were used in most cases. A password reprogram sequence is concluded by an onscreen confirmation of the new password and statement that it was successfully programmed.

SHOP TALK

Always double-check that the password has been correctly input. After a password has been input to a system, no future access can be achieved without it. Read the number back to yourself to ensure that it is correctly input.

Customer Programming

Customer data programming is all programming that the owner of the vehicle has ownership of (as opposed to *proprietary* data programming, which the OEM has ownership of). This category of programming includes data that may have to be changed frequently, for example, daily or weekly, or because some critical hardware such as a differential carrier has been changed. The customer data programming on a vehicle management system may be subdivided into categories such as vehicle data, fleet data, change of ECM/ECU, and so on. However, the means of changing this programming is similar. You should exercise caution. Changing customer data can produce significant performance problems. Use the menu options to select customer programming. Press Enter. At this point, you will usually be presented with a menu of options such as:

- Programming history or log
- View programmed parameters
- Reprogram parameters

In most cases, only the reprogram parameters option is password protected. This means that you can enter and read the first two program options.

Reprogram Parameters

This function permits customer data to be altered. Remember that this can create significant vehicle performance problems. For instance, an incorrectly programmed differential carrier ratio will result in each ECM/ECU that uses vehicle road speed data receiving incorrect road speed signals. Use the menu selections to enter the Reprogram Parameters field.

When in the Reprogram Parameters field, you will be presented with a list of options. The number of options varies depending on the system. Use the menu graphics to scroll the options and press Enter to select. Always consult the OEM service literature before attempting to change any customer data. The OEM procedure should always be used. In all cases, a password is required to alter customer data programming. This may be the factory default password (if none has been programmed since the vehicle was manufactured) or any numeric or alpha password programmed.

CAUTION

Exercise extreme care when altering customer data. Significant performance problems and component damage may result from incorrectly programming data to an ECM or ECU.

11.5 PCs AND OEM SOFTWARE

Until recently, knowledge of PCs has been required of most technicians working with engine electronics but to a much lesser extent when working on other chassis electronics systems. This has changed. The PC used with the electronic system OEM software is being used more and more in the trucking industry as the primary EST. HH-ESTs serve their purpose as a first-level diagnostic tool but any in-depth diagnostics are likely to require a PC loaded with the appropriate OEM software.

A PC is a computer and, as such, most of what we said about vehicle computers in Chapter 6 also applies, so it would be a good idea to review that first. A brief description of a typical PC, its operating system, and how they are used in the trucking industry is provided here.

HARDWARE AND SOFTWARE

Computer hardware is the simplest aspect of computer technology to understand. Computer hardware consists of the things that you can touch and hold, such as system housings, monitors, a keyboard, a mouse, and a printer. Hardware also describes the guts of the system such as the motherboard, microprocessor, memory cards and chips, disk drives, and modems. Most people with a little technical inclination, the ability to read instructions, and a couple of hours to spare could assemble a current PC system without breaking a sweat. Serious problems with hardware in today's computers are not common but technicians need to know that a disc platter hard drive seldom lasts more than four years in a shop environment.

The computer problems that can make life working with computer systems difficult are software-related. You cannot touch software. It consists of program instructions and commands written in various different types of programming language. The computer system needs software from the moment we switch it on. When we wish to use a computer as a word processor to type a letter, we need software to tell the hardware in the computer system how to behave. Software makes the hardware in the computer system produce results such as the production of a letter or a connection to the Internet.

THE HARDWARE

Computer hardware is simple and the source of few real problems in a current PC system. It actually gets blamed for malfunctions that are usually the fault of the software in the system. When diagnosed as defective, computer hardware is seldom repaired—it is simply replaced. Does this sound familiar? That is exactly what we tend to do with many vehicle systems today. The key is accurate diagnosis. The following is a description of some PC hardware from the user's point of view (**Figure 11–18**)

SYSTEM HOUSING

This is the term we use to describe the flat (desktop) or vertical (tower) box into which most of the guts of the computer are located. From the outside, we

FIGURE 11–18 Common computer hardware components include a keyboard, mouse, microphone, system unit, disk drives, printer, monitor, digital camera, speakers, and modem.

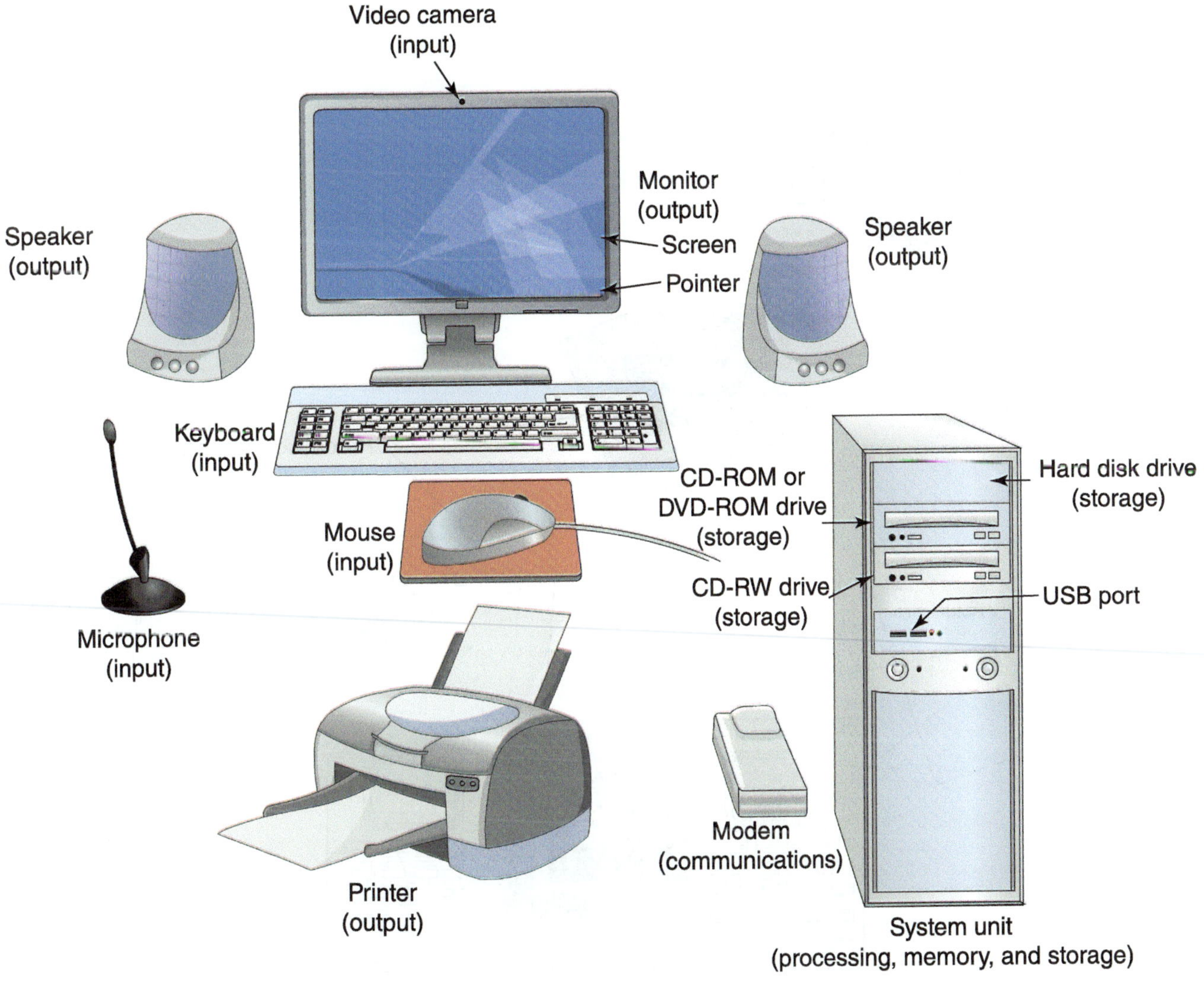

can observe an on/off switch, USB slots, CD and DVD drives, buttons that control disk drive systems, and perhaps a speaker volume control. Inside are components that make the system operate as a computer. In older systems, some of the hardware components listed next would have been located outside the system housing. In a typical PC today, we will find the following components:

- **Motherboard**. This is the main circuit board. Most of the other components are attached to this either directly or indirectly.
- **Processor**. This is usually called the CPU or central processing unit. The processor is the brain of the processing cycle, and the speed at which it operates plays a big role in describing computer performance. Processors are manufactured by companies such as Intel and AMD and their speed is specified in hertz. **Figure 11–19** shows some examples of processors.
- **Random access memory (RAM) chips**. RAM is main memory capability, so total RAM plays a big role in defining the computing power of the system. RAM is electronically retained and the CPU can only manipulate data in electronic format.
- **Memory chips**. Data can also be retained magnetically on chips that are attached to the motherboard. Figure 11–19 shows an example of a memory chip.
- **Hard drive**. The hard drive is a means of retaining information within the computer housing. There are two types of hard drive in current use:
  - Hard disk drive: consists of platters (many layers) of data retention disks that can be loaded with vast quantities of information.
  - Digital hard drive: digital hard drives eliminate moving parts and reduce (hopefully) drive failures. The present cost is high but prices are expected to rapidly drop.
- **USB flash drives**. Portable data retention media that use flash memory and, therefore, have no moving parts. They are fast, low cost, and reliable. **Figure 11–20** shows a typical USB flash drive.

FIGURE 11–19 Processors and memory chips

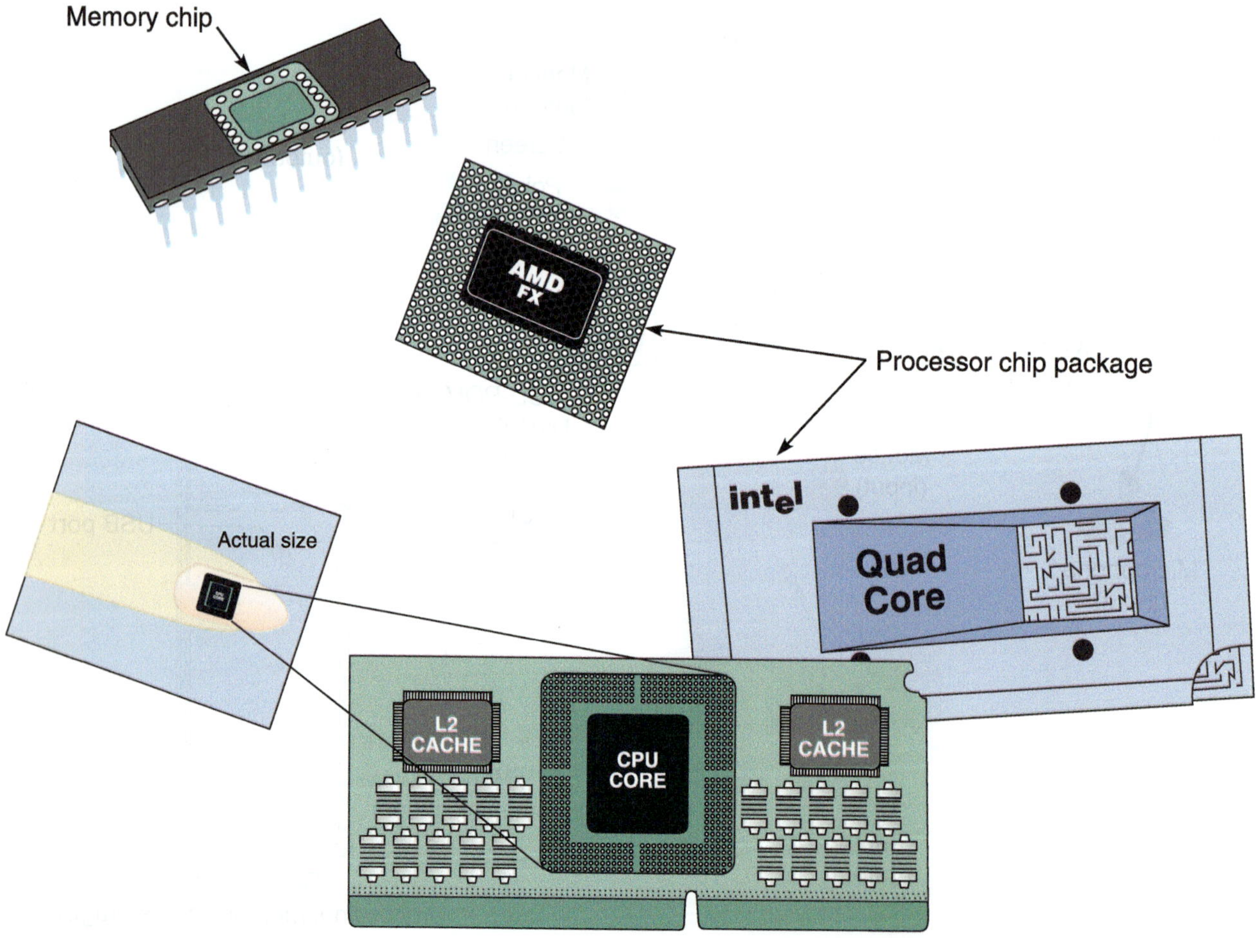

FIGURE 11–20 A typical USB flash-driven data retention medium

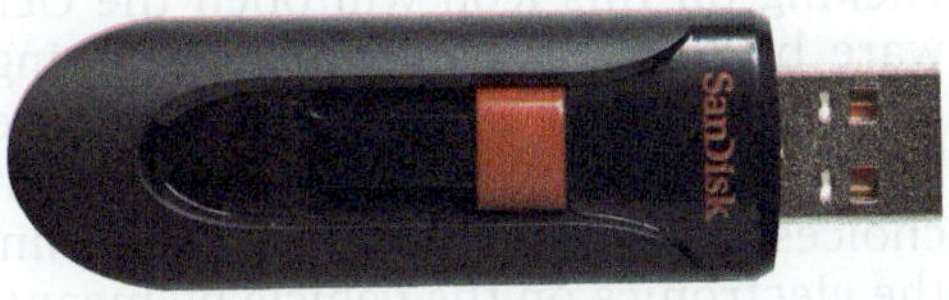

- **Optical disc reader/writer**. CD and DVD drives are a means of optically reading data using a laser. Most OEM software is provided to their dealers on CD and DVD today because they can hold very large quantities of data compared to diskettes. The disc is essentially a thin metal disc coated with plastic. Most computers today have disc "burners"; that is, they will write to CD or DVD format. At this moment in time, there are four categories of optical disc: CD-ROM (can only be "read"), CD-RW (can be recorded to), DVD-ROMs (digital video ROM), and DVD-R. **Figure 11–21** shows a typical optical disc drive unit.
- **Modem/network card**. The means of linking a PC electronically to other computer systems by means of the telecommunications systems.
- **Ports**. A means of linking the PC to a power supply, peripherals, network lines, and the data connectors of a typical truck electronic system.

Peripherals

Peripherals describe the hardware components we attach to the system housing. Some of them are used to input information into the PC processing cycle, whereas others perform specific tasks such as printing out documents. Some peripherals are essential (such as a keyboard and mouse), whereas others are optional (such as printers and scanners).

FIGURE 11–21 Optical disc tray from a PC

- **Monitor**. An LCD display that displays information on a TV-like screen. The monitor can be both an output (display) and input (touch screen and OR capable)
- **Mouse**. Moving a mouse on a flat pad positions a cursor (pointer) on the monitor screen. It contains command and function keys that make using computer systems much faster.
- **Keyboard**. The primary input device on a PC system. Consists of what appears to be a typewriter keyboard, but also has a number of command and function keys. These command and function keys may have different functions according to what software is being run.
- **Trackball, joystick, or touchpad**. Some PCs, especially portable ones described as notebook or laptop, use a trackball, joystick, or touchpad in place of a mouse. Used in conjunction with a couple of command and function keys, they do exactly what a mouse does.
- **Printer**. When on-paper documentation is required of onscreen or retained information, the printer is used to produce it. For instance, when using an "electronic service manual" from an OEM data hub, if the technician requires a printed copy of the procedure described electronically, the printer can generate it.
- **Scanner**. A device used to copy paper images or text into electronic format so it can be displayed, read, or electronically transferred using a PC system.
- **Modem**. This converts the digital signals of the computer processing cycle into the analog signals required to transmit them through the telecommunications system.

PC SOFTWARE

When any computer system is switched on, it goes through *boot-up*. During boot-up, the operating system (OS) is transferred into the processing cycle of the computer. The OS is an essential set of commands that instruct the processor on how to handle the computer hardware and how to interpret program instructions as they come in from other software. All computers have an OS, and the system that was used by most of industry up until 2007 was based on a specific Microsoft OS known as DOS. DOS was enhanced with a GUI known as MS Windows that made the computer much more user-friendly. Versions of MS Windows released after 2007 ceased to use the older DOS platform as the basis for the GUI. Current versions of Windows use integrated OS architecture, which enables smoother boot and easier troubleshooting. Recent versions of Windows have been designed as a common tablet and

PC OS: this has created some volatility when used on PCs with no touchscreen capability.

Just about every PC used in a truck dealership, fleet, or service facility today uses one of several generations of Windows. They all do essentially the same thing: make the computer easier to use and minimize the role of the keyboard in inputting information.

Instead, icons represent commonly required commands and drop-down menus are used to expand on them. The mouse is used to view and select these options.

- **DOS**. Today, it is not necessary to know much about DOS because it launches automatically into the PC processing cycle on versions of Windows that use it when booted. On boot-up, we go through DOS to the GUI or Windows display. Post-2007 Windows use an integrated OS that flows seamlessly from boot to GUI.
- **Windows**. Several generations exist; Windows is the GUI used by most of industry today. This Microsoft product has competitors, but they are not used much by the truck industry. After the boot-up procedure, we will see the Windows display on the screen. This uses icons to represent the software options already loaded onto the system hard drive. To make a selection, the mouse is used to move the cursor over the icon representing the desired software package, and the mouse is left-clicked. The left button of a mouse is a go-to command in Windows applications. The right button drops down menu options that vary according to the location of the cursor on the screen and the program.
- **Program software**. Program software is the instructions that get the computer processing cycle to process specific functions. For instance, if you wanted to type a letter and print it out, first you would need word processing software to turn the keypad into a typewriter. As keys are pressed on the keyboard, these are input into the processing cycle of the computer and displayed on the monitor as letters. The spelling of each word in the document can be checked by a function called spell check. When the user is happy with the appearance of the letter displayed onscreen, the system can be instructed to print it. The contents of the letter are digitally transferred to printer driver software, which then instructs the printer to print out a copy of the letter onto paper. Computers use hundreds of different types of program software to produce many different outcomes.
- **OEM software**. When a PC is used in conjunction with truck OEM software, it is generally loaded onto the hard drive of the computer, either from a CD or web download. This means that after the PC has booted up, an icon will be displayed on the Windows screen. Left-clicking on this icon will open the OEM software by loading it into the processing cycle of the computer. In most cases, a large number of choices will be available. Some of these choices may require that the PC be linked to the electronics on the vehicle by means of the data connector (J1708 or J1939) or the OEM data hub (by means of a modem and phone line handshake), or both. For instance, when performing a **handshake** connection between the vehicle electronics and the OEM data hub, in just about every case the chassis vehicle identification number (VIN) and some kind of password are required to complete the connection. Most OEMs have taken a lot of trouble to make their systems easy to use, so do not get put off the first time you attempt to use this technology. Being familiar with Internet protocols helps. A handshake connection with an OEM data hub that allows you to download data from it to a vehicle ECM is not much different from downloading a game file from an Internet site.
- **Networking and OEM LANs**. Almost all OEMs have their own subscription-accessed LANs or local area networks. These are usually driven by a data hub and are used to track such things as vehicle history, warranty claims, vehicle location, sales data, programming options, and so on. These systems are discussed a little later in this chapter. **Figure 11–22** shows the idea behind a network system linked by the telecommunications system.
- **Phone-based apps**. There are currently over 7,000 Google Android apps relating to transportation. At the time of writing none of these are of much consequence for technicians because they are directed at drivers and logistics but this will change. The change will happen when OEMs discover they cannot ignore the portability, low cost, and computing potential of the mobile phone. There is a good tire tracking and maintenance app but this is something a fleet manager is likely to find value in.

CONNECTING A PC TO A TRUCK DATA BUS

The following is the sequence of events required to access the data bus in a typical truck. We are going to use as our primary reference the Freightliner procedure to give the exercise a frame of reference, but it is not much different from most other OEM procedures. To optimize the performance of today's powerful

FIGURE 11–22 A network can be quite large and complex, connecting users all over the country.

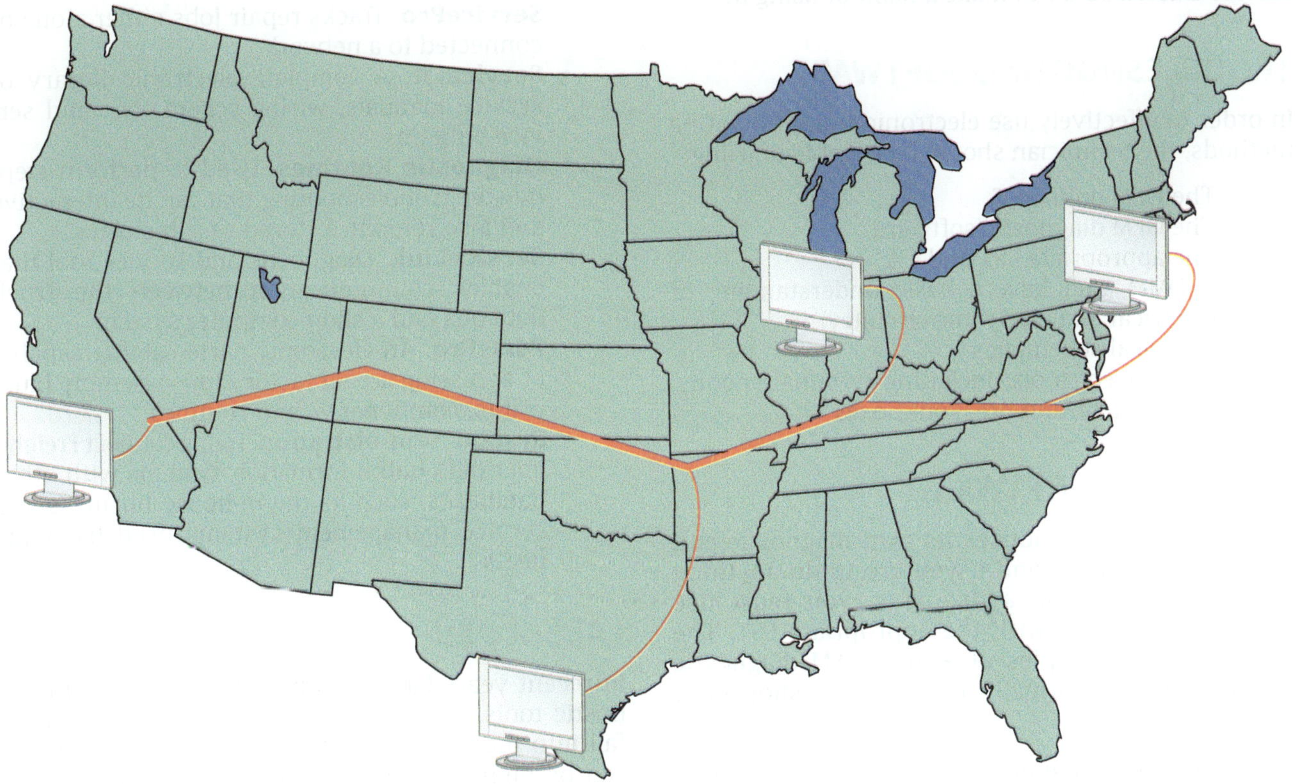

software diagnostics, the role of the PC or other type of EST is often to quarterback interactions between the vehicle data bus and the OEM data hub host using an online connection.

1. Power-up PC. This boots the computer and loads the operating system (Windows). When the GUI appears on the screen, select the appropriate icon to load the OEM software into main memory.
2. Once the OEM software has loaded into main memory, you will be prompted to make an online connection and connect the PC station to the chassis data bus via the data connector and communications adapter (CA). The CA LEDs will illuminate when a successful bus connection is made. It is possible to work with the vehicle electronics only if no online connection is possible.
3. Once connected to the vehicle data bus, you will be prompted to select the controller (denoted as source address [SA] or message identifier [MID]) you wish to work with. SAs and MIDs are explained in more detail in Chapter 12; for our purposes in this chapter, SAs or MIDs refer to any controller with an address on the bus such as the engine, transmission, ABS, and so on.
4. Having successfully accessed the SA or MID you want to work with, you can begin to route your way through the data within that system. This will include customer data programming, audit trails, active and historical codes, and so on.

SERVICE INFORMATION SYSTEMS

All current truck OEMs use online **service information systems (SIS)**. Most prefer that the first step in any troubleshooting procedure is to log on to their online SIS whether they are working on an electronic engine or one of an earlier vintage. Some examples of engine OEM SIS are:

- Caterpillar Service Information System (SIS)
- Detroit Diesel Power Service Literature (PSL)
- Freightliner ServiceLink (ServiceLit)
- John Deere StellarSupport
- Mack Trucks Electrical Information System (EIS)
- Navistar Service Information (NSI)
- Paccar ServiceNet
- Volvo IMPACT

Most OEM service information systems are updated daily. They tend to be easy to navigate. Technicians should learn how to use the Web site search engines. In most cases, information searches are by engine serial number and may include searches of the parts data. It is important to note that most online SIS support engines manufactured before the Web site service existed so try to make a habit of using it.

TROUBLESHOOTING SOFTWARE

In order to effectively use electronic troubleshooting methods, the technician should be capable of using:

- The OEM online SIS
- The OEM diagnostic software
- The appropriate EST and CA
- A DMM (and have a basic understanding of electrical and electronic circuitry)
- Electrical schematics
- The correct tools, including terminal spoons, breakout Ts, and breakout boxes

DIAGNOSTIC SOFTWARE

Each engine OEM produces its own diagnostic and calibration software and if you are to do anything more than just scan a data bus, you must use this software along with the appropriate EST. The following are examples of engine OEM diagnostic software; terms normally used on the shop floor appear in bold:

- Allison calibration configuration tool (ACCT)
- Allison **Diagnostic Optimized Connection** (DOC)
- Bendix **ACom** diagnostics
- Caterpillar **Electronic Technician** (ET)
- Freightliner **ServiceLink** and **EZ-wiring**
- John Deere **ServiceAdvisor** Mack Trucks **Premium Tech Tool** (PTT)
- Meritor **Toolbox** Software.
- Navistar **Master Diagnostics (MD)**
- Paccar **Electronic Service Analyst (ESA)**
- Volvo **V-Cads** Pro (Volvo computer-aided diagnostic system and programming)

You should note that some versions of the preceding software may have limited functions: Software for dealerships, body builders, and fleet applications may all be different depending on the level of programming required. For instance, a Volvo dealership uses V-Cads Pro (full function), whereas fleets would be restricted to V-Cads Elite, which cannot enable proprietary (Volvo) data programming.

UMBRELLA SOFTWARE PACKAGE

Freightliner uses a comprehensive umbrella software program known as ServicePro. This is really a package of six Freightliner software programs with live links to the data hubs of its major component suppliers. For ServicePro to function to its full capability, a connection to both the vehicle and Freightliner's data hub is required. The ServicePro umbrella package comprises:

- **ServicePro**. Tracks repair jobs either alone or connected to a network
- **ServiceLit**. A complete electronic library of service manuals, wiring schematics, and service bulletins
- **Diagnostic Routines**. Used to perform step-by-step troubleshooting that can be interactive and also reference stored case histories
- **ServiceLink**. Diagnostic and service tool that enables communication between the truck data bus and a shop computer system
- **PartsPro**. An electronic parts catalog capable of accessing actual factory line assembly blueprints, components, and routines
- **System Administration**. Helps connect Freightliner data hub information systems with other databases, such as the in-house business and service management systems used by many fleets

LABSCOPES

In recent years labscopes have become useful diagnostic tools. **Labscopes** or **scopemeter** can either fall into the category of HH-ESTs in which case they can be highly portable, or they can consist of software downloaded to a PC and the hardware necessary to make a connection between the vehicle electronics and the computer. Compact HH labscopes are limited in their usefulness by the size of the display screen; although accurate, it can make interpreting the data difficult. A PicoScope scopemeter consists of software (Pico 9000) downloaded onto a PC (for no charge) and the connection hardware required to bus into the vehicle components. **Figure 11–23** shows

FIGURE 11–23 A PicoScope kit designed to be used in conjunction with Pico 9000 automotive software on a computer.

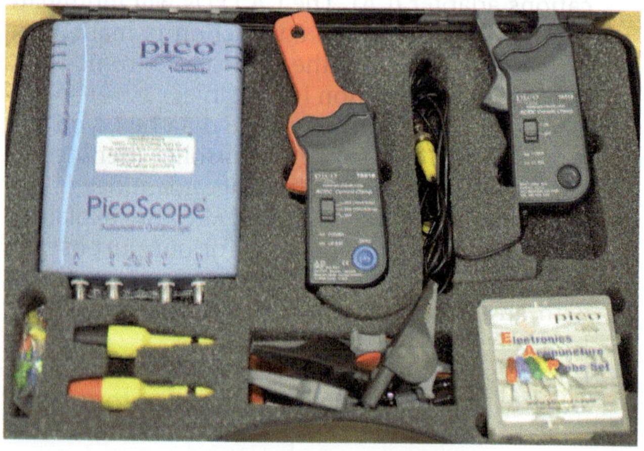

one type of Pico labscope kit. In Chapter 12 we will take a look at some tasks that can be performed by a labscope or scopemeter; in addition, using a labscope is a useful training tool because it helps with understanding how electronic and electrical circuits function.

ELECTRONIC TROUBLESHOOTING

Electronically managed systems are capable of malfunctioning hydromechanically, so the technician must not assume that every problem is electronically based. The major OEMs produce excellent sequential troubleshooting service literature and integrate it into their SIS. To use SIS and its related diagnostic software effectively, the technician should be capable of using a DMM, have a basic understanding of electrical and electronic circuitry, and be capable of interpreting schematics.

SELECTING AN EST

For purposes of field practice, we are going to divide ESTs capable of accessing truck, bus, and heavy equipment data buses into four general categories:

1. PC platform with OEM software and OEM data hub connectivity via limited access subscriptions usually only available to OEM dealers. This package represents the highest level of technician access. It is required to enable OEM software updates, OEM programming and reprogramming, interactive/guided troubleshooting trees, view parameters, and access to the OEM service literature and databases.
2. PC platform with aftermarket software (DG and J-Pro) and depending on the OEM, limited data hub connectivity. This type of EST option is best suited to independent service dealers that work with a range of different OEM products that do not require a full suite of access and reprogramming options. Depending on the OEM, the software may provide some guided troubleshooting and enable some customer data programming, but always have the capability to read the chassis data bus, view operational parameters, and ID active, historic and pending codes, and erase them.
3. Triage tools. Handheld (HH) or tablet based ESTs that can view operational parameters, read and erase **diagnostic trouble codes (DTCs)**. Often these ESTs are well supported by the third-party manufacturer (with updates/help line) and are used as a **triage tool** in OEM dealers for a fast analysis of a complaint before routing the vehicle to the appropriate repair channel in the shop. Included in this category of EST are system-specific HH-ESTs that are designed to function with one chassis system (such as brakes) in which case they usually offer an expanded suite of capabilities including guided diagnostics.
4. PM tools. These are most basic ESTs that can be used to access a truck chassis data bus. They can read generic bus language, ID serial numbers, pull data from a fleet truck, and usually read and erase DTCs. Typically such a tool is used in highway truck stop repair facilities and fleet refueling islands.

Table 11–1 shows the four categories of EST and how their application repair level.

WORKING ON THE DATA BUS

Some OEMs suggest that an EST be connected to the chassis data bus any time a truck enters a service facility. However, the following list covers some obvious reasons for connecting to the data bus:

- Dash warning light illuminated
- Driver complaint
- Truck in shop for maintenance
- Dash alert indicating an active code

EST Software Fields

There is some variance by OEM on exactly what will be displayed on an EST display after a connection has been made to the data bus. Some typical fields are:

- Client updates
- Repair and maintenance history
- Search previous work history
- Updates and technical bulletins
- Diagnostics
- Testing
- Calibrating
- Programming and downloads

TABLE 11–1 Four Categories of EST

Full Suite EST	Partial Suite 3rd Party EST	Triage EST	PM EST
PC Windows platform, RP 1210 CA, OEM data hub access	PC Windows platform, RP 1210 CA, 3rd party data hub access	HH EST or tablet app, RP 1210 CA, limited web access	HH EST; some have wireless connectivity to fleet management software hub

CAUTION

Some OEM software requires a technician logon every 90 days or less and often a change of user password every 90 days; exceeding this time period will lock the technician out. In event of lock out, the technician will require access renewal by the OEM service operations department.

SEQUENTIAL TROUBLESHOOTING

Sequential troubleshooting takes place in stages. It is critical that the instructions in each stage be precisely undertaken before proceeding to the next; subsequent stages will generally be rendered meaningless if a stage is skipped. In other words, technicians are required to approach every troubleshooting challenge as an entirely new one and avoid shortcuts that might have worked with a previous similar problem. Structured sequential troubleshooting software simplifies the sequencing of the troubleshooting stages and in the end saves a lot of time.

SHOP TALK

Diagnostic trees are often used by truck OEMs to troubleshoot malfunctions. The "root" of the tree is the problem. The "branches" are the various different paths that circuit testing will route the diagnostic technician. These tests are sometimes referred to as "leaves." Never skip tests or sections within a diagnostic tree. Some diagnostic trees are driven by OEM software and may have hundreds of steps.

Diagnostic Path

A typical diagnostic path to locate and repair a root cause of a problem varies by OEM and the complexity of the subsystem but it might look something like this:

- **Select symptom:** usually a range of options are provided and the technician must click on each.
- **View DTC information:** most software arranges this in order of priority so it is usual to start at the top of the list and work down.
- **Run diagnostics:** this will route the technician through troubleshooting trees which may require test routines.

USING THE CORRECT TOOLS

It cannot be emphasized enough how important it is to use the correct tools. Generally, test light circuit testers should not be used nor should analog test meters. When the use of a breakout box is mandated, make sure that it is used. Also, it should be remembered that electrostatic discharge can damage solid state components, so it is good practice to wear a **ground strap** when opening up any housing containing a microprocessor. When testing separated Deutsch, Weatherpac, and other sealed-type connectors with a DMM, ensure that the sockets have been identified before probing with the test leads. Use socket adaptors and tests lead spoons to avoid damaging terminal cavities. Never spike or strip back wires.

SNAPSHOT TESTS

Most systems will accommodate a **snapshot test** readout from a controller to facilitate troubleshooting intermittent problems. Such problems often either do not generate codes, or codes are logged with no obvious cause. Snapshot mode troubleshooting can be triggered by a variety of methods (codes, conditions, manually) and will record data frames before and after the trigger. These can be recorded to a handheld or PC (laptop) EST, while the vehicle is actually running. Because data frames can be snapshot both before and after the trigger, it is possible to analyze a wide spectrum of data that may have contributed to a problem. Each data frame can be examined individually after the event. The portability of handheld ESTs makes them ideal for this type of testing.

11.6 BUS CODES AND PROTOCOLS

We can now take a look at a partial listing of data bus codes that have been adopted by all the North American truck engine/electronics OEMs. The first generation heavy-duty data bus was **SAE J1587**, which covered the common software protocols on the bus, while **SAE J1708** covered the hardware protocols. The current **SAE J1939** covers both hardware and software protocols. The universal acceptance of these data bus protocols enables the interfacing of electronic systems manufactured by different OEMs on truck and bus chassis, along with allowing generic software to at least read other OEMs' electronic systems. The truck technician who works with multiple OEM systems, for instance in an independent service shop, may find it easier to work using these codes rather than use the proprietary codes.

SAs AND MIDs

The term **message identifier (MID)** is used to describe a major vehicle electronic system, usually with independent processing capability. The term was introduced with J1587 but it has persisted with J1939. On a J1939 data bus, the term **source address (SA)** replaced MID but you should note that many OEMs continue use MID in preference to SA in their training literature. Each major electronically managed chassis system is assigned a unique address on the data bus. Examples would be the engine, transmission, ABS, and collision warning systems—each of which are assigned an address we identify numerically as its SA or MID.

Within each SA or MID on the bus, there are subdirectories, which allow the technician to create a troubleshooting route to get to the root cause of a problem. **Diagnostic trouble codes (DTCs)** are generated when a system fault is detected. The composition of a DTC is analyzed a little later in this chapter.

PIDs and SPNs

A bus DTC always includes a **parameter identifier (PID)** on J1587 or **suspect parameter number (SPN)** on J1939. PIDs and SPNs are usually primary subsystems common to all different OEM types covered by the SA or MID. In both J1587 and J1939, the EST will (usually) display the SPN or PID in alpha so it is not necessary to remember the numeric codes.

SIDs

Subsystem identifiers (SIDs) are used by both J1587 and J1939 to describe subsystems that fall within an SA or MID address. The J1939 SID scheme is an improvement on the J1587 version in that there is a single look-up table, not differentiated by the MID if falls under: this simplifies troubleshooting.

FMIs

Failure mode indicators (FMIs) are indicated whenever an active or historic code is detected by the system controller (SA or MID). FMIs have been expanded in J1939: J1587 tended to identify outright failures whereas the expanded J1939 FMI code detection allows for imminent or predictive failures to be identified. This feature can help identify a "pending" failure before it cripples the vehicle and, combined with telematics, can used to mine data for condition based maintenance (CBM).

The failure code actually displayed on the EST be the OEM specific but *all* North American truck electronics use the same FMIs so system failures can at least be read and categorized by their competitor's diagnostic software. FMIs help make life easier for the truck technician because any electronically detected failure must be categorized into one of the codes outlined in **Table 11–2**, which shows current J1939 FMI.

SHOP TALK

Common SA, MID, SID, PGN, PID, and FMI codes are used by all the OEMs manufacturing trucks in North America and most imports certified as on-highway. This usually allows one OEM's diagnostic EST to at least read its competitor's diagnostic codes.

TABLE 11–2 J1939 FMIs

Failure Mode Indicators (FMI)	Description
00	Data valid but above normal operating range (most severe)
01	Data valid but below normal operating range
02	Data erratic, intermittent, or incorrect
03	Voltage above normal or shorted high
04	Voltage below normal or shorted low
05	Current below normal or open circuit
06	Current above normal or grounded circuit
07	Mechanical system not responding properly
08	Abnormal frequency, pulse width, or period
09	Abnormal update rate
10	Abnormal rate of change
11	Failure mode not identifiable
12	Bad intelligent device or component
13	Out of calibration
14	Special instructions
15	Data valid but above normal (least severe)
16	Data valid but above normal (moderate severity)
17	Data valid but below normal (least severe)
18	Data valid but below normal (moderate severity)
19	Received network data in error
20	Data drifted high
31	Not available

DTCs

A J1939 DTC includes the following:

- A SPN (a PID from J1587)
- A SID (same as J1587)
- An FMI (expanded from J1587)
- Occurrence count (same as J1587)

Advanced Diagnostics

The contents in this chapter focuses on the basics of troubleshooting bus-detected faults: this guides the technician through some elementary troubleshooting using proprietary or generic ESTs. Chapter 12 will take a closer look at multiplexing bus architecture, messages, and troubleshooting.

11.7 TROUBLESHOOTING CWS AND CMS

The VORAD **collision warning system (CWS)** and **Wingman collision mitigation system (CMS)** were studied in theory in Chapter 6. VORAD is non-interventionary so it can do no more than audibly and visually alert the driver when a potential collision is detected through its **driver interface unit (DIU)**. Wingman is known by Bendix as an **active cruise with braking (ACB)** system that is capable of "mitigating" a potential collision by broadcasting messages to the vehicle data bus. Both VORAD and Wingman ACB have three levels of driver alert:

- Stationary object alert (SOA)—the front antenna (radar) detects a stationary object in the path of travel that could cause an accident. A stalled vehicle would be an example.
- Following distance alert (FOA)—actuated when in cruise control (CC) and a vehicle detected by the front antenna detects a slower vehicle and a potentially dangerous closing speed. Audible and visual alert.
- Impact alert (IA)—the highest level of alert that will be broadcast whether the vehicle is in CC or not. An IA requires the driver to take immediate action.

In addition, Wingman can alert the driver by displaying a brake overuse warning (BOW) on the DIU. When a Wingman ACB equipped truck is in CC mode and traveling on an extended downhill grade, it is not uncommon for the service brakes to be excessively applied. To prevent this, a BOW alert is broadcast to the DIU (see **Figure 11–24**) alerting the driver to disengage Wingman and use gearing to help manage vehicle speed.

FIGURE 11–24 Wingman DIU

CMS MITIGATION LEVELS

Just like the transmission or brake systems, CWS and CMS have an address on the chassis data bus (see Chapter 12). While CWS could only receive messages off the bus, CMS can both receive and broadcast messages. This allows the system to send command requests to the engine and braking systems. Wingman command requests are broadcast on three different levels:

- Engine dethrottle
- Apply engine brake
- Apply chassis service brakes (up to 30% of potential)

Now we can take a look at some simple troubleshooting of the system. The CMS electronics have an address on the chassis data bus. This means that the fault codes can be displayed either directly from Wingman/VORAD electronics and codes or by connecting to the data bus connector and using the SAE codes. For any particular fault code, the occurrence count (OC) and duration can either be reset or cleared. To clear a code:

1. Select the fault code.
2. Position the display selection arrow on Clr Ct/Min display line.
3. Press Enter.

Displayed Fault Code	Code	SID	FMI	Description
RAM	11	254	12	CPU Internal scratch memory error
Flash	12	254	12	CPU Program memory error
FPGA	13	254	12	CPU gate array—will not program
DSP	14	254	12	CPU Digital signal processor—will not program
Battery	15	254	12	CPU internal battery low
NVRAM	16	254	12	CPU internal data memory corrupted
No DSP data	17	254	9	CPU processor not receiving antenna data
Frequency INJ	21	2	8	Antenna test signal bad
F/E data or CLK	22	2	2	Antenna serial link bad
Microwave	23	1	2	Antenna radar not transmitting

Displayed Fault Code	Code	SID	FMI	Description
Brake	31	3	5	Brake input not corrected
Speaker	41	4	5	Speaker not connected
	42			NO FAULTS FOUND
R. Steering	51	5	5	Right steering sensor not connected
L. Steering	52	5	5	Left steering sensor not connected
STR misaligned	53	3	13	Steering column sensor misaligned
Speedometer	61	6	5	Speedometer monitor
Right signal	71	7	5	Right turn signal input not connected
Right BSS	72	10	5	Right side sensor not connected
Left signal	73	8	5	Left turn signal input not connected
Left BSS	74	11	5	Left side sensor not connected
Volume	91	9	5	DIU volume control bad
Range	92	9	5	DIU range control bad
DRVR Disp	92	9	5	DIU not connected

WINGMAN/VORAD OPERATING PARAMETERS

The Wingman and VORAD radar antenna operates at a frequency of 24.725 GHz and can "see" through fog, snow, smoke, and heavy rain. The system uses low-power microwaves and is designed to not interfere with other vehicle collision warning systems (CWS), police radar, or other computer management systems. It can detect objects as small as a motorcycle but not animals or people.

The VORAD CWS cab display has three lights: red, orange, and yellow. The first warning light is the yellow light, which illuminates when a vehicle is within 3 seconds of the current speed of the host vehicle. The second warning light is orange, which is illuminated along with the yellow light when a vehicle is within 2 seconds of the host vehicle at the current speed. The red light is illuminated along with the orange and yellow when a vehicle is detected to be in a 1-second range. When the warning lights are illuminated, a distinct tone is delivered. There are five distinct tones, one for each of the following:

- 2-second alert
- 1-second alert
- Slow-moving vehicle alert
- Stationary vehicle alert
- Blind spot alert

The tones sound only once per incident, whereas the warning lights should stay illuminated throughout the incident. The idea behind both Wingman and VORAD is to alert the driver, not to annoy; but the key difference is that Wingman will intervene to mitigate a situation, its logic detects as being potentially dangerous by applying the service brakes when the chassis is operating in cruise control. Because Wingman and VORAD will default to err on the side of safety, false alerts/interventions are always possible.

11.8 ELECTRICAL WIRING, CONNECTOR, AND TERMINAL REPAIR

Most of the wiring and connectors used on electronically managed chassis systems in North America are produced by a couple of manufacturers, so although the procedures outlined in this section are specific to one OEM, they are representative of those required for all the major OEMs.

WEATHERPROOF CONNECTORS

The following sections sequence the disassembly, repair, and reassembly procedure required for sealed connectors used in truck electronic circuits. This procedure is similar to that required for the assembly and repair of most truck electronic wiring and connectors.

Installation of Weatherproof Square Connectors

Electrical connectors used in truck electrical and electronic circuits today require the use of the correct procedure and the correct tools. This means that connectors should be first identified, then use the appropriate SIS to determine the correct assembly procedure and what special tools are required. The instructions reproduced here outline typical procedures used for terminal installation and connector assembly:

1. Position the cable through the seal and correct cavity of the connector. See **Figure 11–25.**
2. Strip the end of the cable using wire strippers to leave 5.0±0.5 mm (0.2±0.02 in.) of bare conductor.
3. Squeeze the handles of the crimping tool together firmly to cause the jaws to automatically open.
4. Hold the "wire side" facing you.

FIGURE 11–25 Inserting wire in a connector

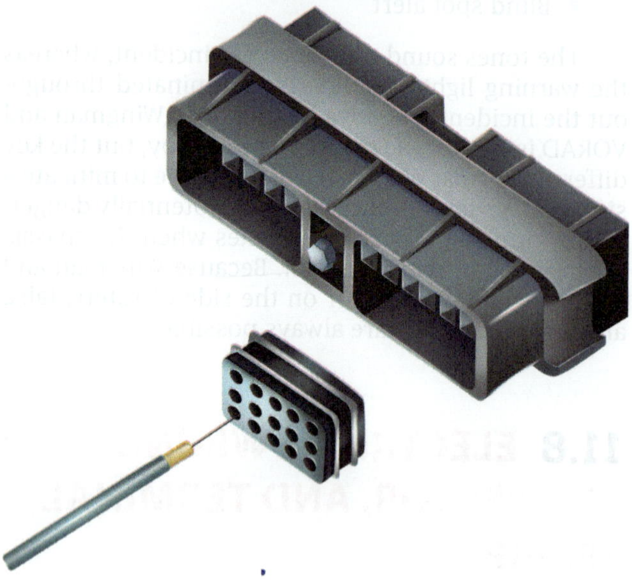

5. Push the terminal holder to the open position and insert the terminal until the wire attaching portion of the terminal rests on the 20–22 anvil. Ensure that the wire core wings and the insulation wings of the terminal are pointing toward the upper jaw of the crimping tool. See **Figure 11–26**.
6. Insert the cable into the terminal until the stripped portion is positioned in the wire core wings and the insulation portion ends just forward of the insulation wings. See **Figure 11–27**.
7. Squeeze the handles of the crimping tool until the ratchet automatically releases and the crimp is complete. Note: For faster, more efficient crimping operation, a bracket or bench rest may be used to cradle one handle of the tool. The operator can apply the terminals by grasping and actuating only one handle of the tool. See **Figure 11–28**.
8. Release the crimping tool with the lock lever located between the handles in case of jamming.
9. Align the **locking tang** of the terminal with the lettered side of the connector.
10. Pull the cable back through the connector until a click is heard. See **Figure 11–29**. Position the seal into the connector. Note: For ECM 30-pin connectors, put the locking tang opposite the lettered side.

FIGURE 11–26 Terminal and crimping tool position

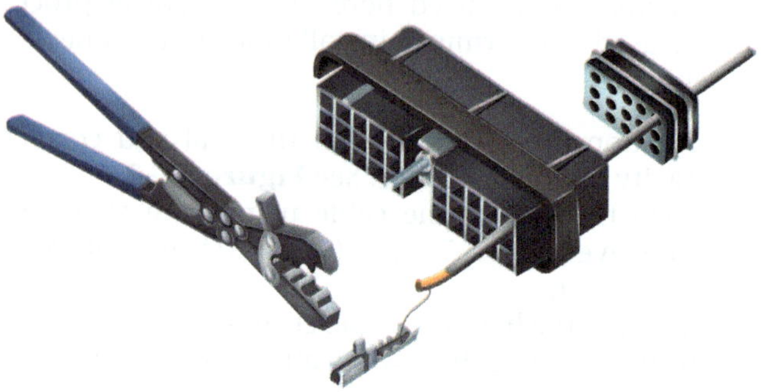

FIGURE 11–27 Cable to terminal alignment

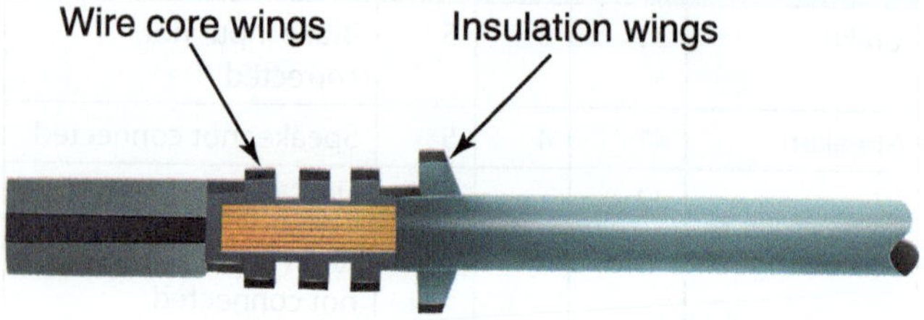

FIGURE 11–28 Crimping tool operation

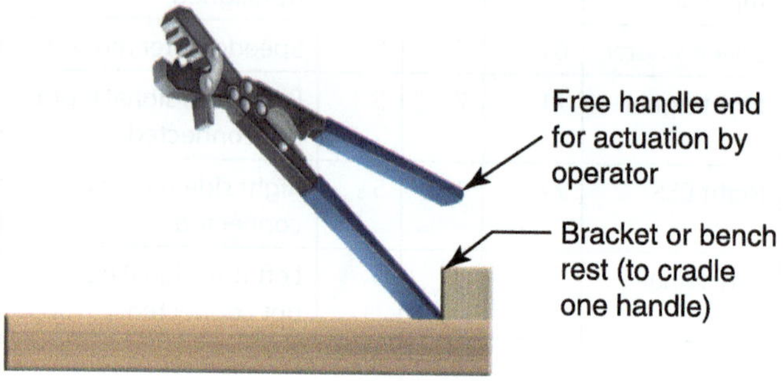

FIGURE 11–29 Pulling the terminal to seat

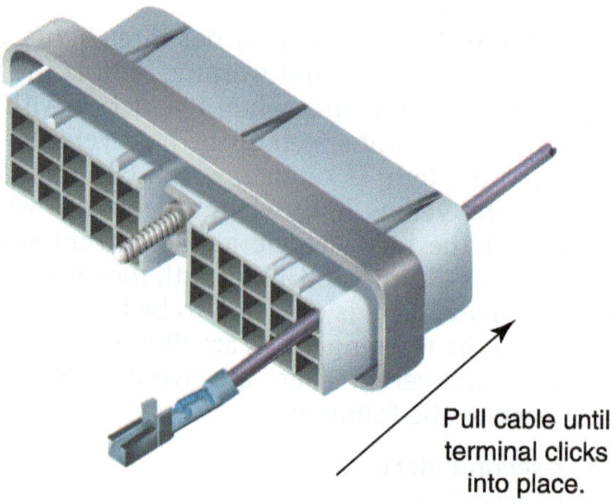

FIGURE 11–30 Terminal removal

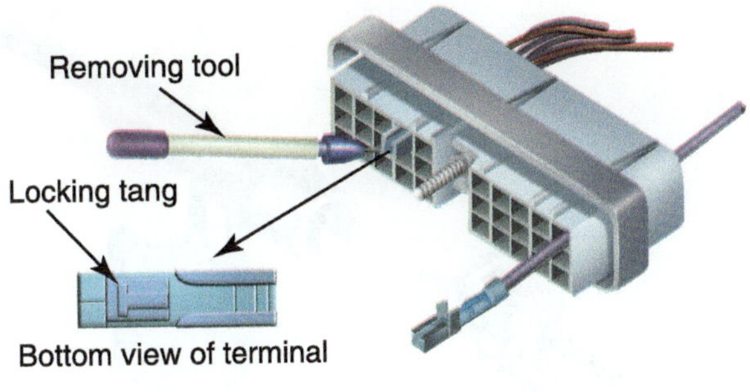

Tech Tip

When assembling pull-to-seat terminals in a connector assembly, the cable is pushed through the seal and cavity of the connector before crimping the terminal to the cable. It should be stripped of insulation *after* it is placed through the seal and connector body.

Removal and Repair

A tang on the terminal locks into a tab molded into the plastic connector to retain the cable assembly. Remove weatherproofed square terminals using the following guidelines:

1. Insert the **tang release tool** into the cavity of the connector, placing the tip of the tool between the locking tang of the terminal and the wall of the cavity. See **Figure 11–30**.
2. Depress the tang of the terminal to release it from the connector.
3. Push the cable forward through the terminal until the complete crimp is exposed.
4. Cut the cable immediately behind the damaged terminal to repair it.
5. Follow the installation instructions for crimping the terminal and inserting it into the connector.
6. Push the crimped terminal into the connector until it clicks into place. Gently tug on the cable to make sure it is secure. See Figure 11– 30.

Installation of Round Terminal Connectors

Use the following guidelines for terminal installation:

1. Insert the terminal into the locating hole of the crimping tool using the proper hole according to the gauge of the cable to be used. See **Figure 11–31**.
2. Insert the cable into the terminal until the stripped portion is positioned in the cable core wings and the seal and insulated portion of the cable are in the insulation wings. See **Figure 11–32**.

FIGURE 11–31 Terminal position

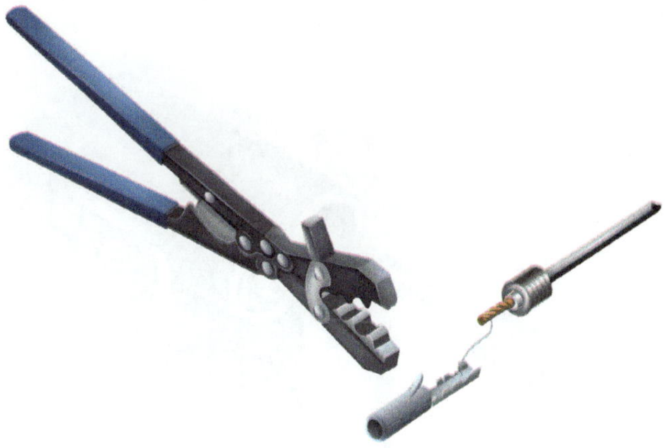

FIGURE 11–32 Cable and terminal position before and after crimping

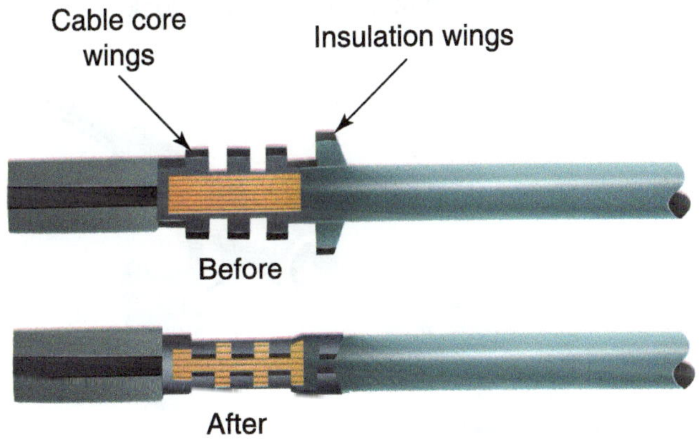

3. Compress the handles of the crimping tool until the ratchet automatically releases and the crimp is complete. A properly crimped terminal is shown in Figure 11–32.
4. Release the crimping tool with the lock lever located between the handles in case of jamming. Pull the cable until the terminal clicks into place. See **Figure 11–33**.

Removal and Repair

Two locking tangs are used on the terminals to secure them to the connector body. Use the following instructions for removing terminals from the connector body:

1. Disengage the locking tang, securing the connector bodies to each other. Grasp one-half of the connector in each hand and gently pull apart.

FIGURE 11–33 Inserting a terminal in a connector

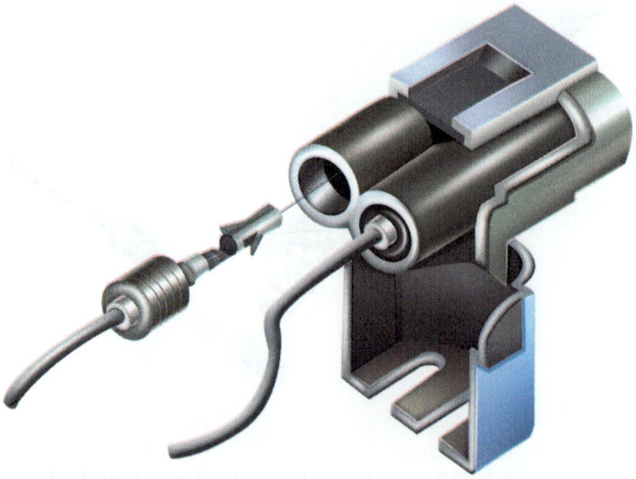

FIGURE 11–34 Unlatched secondary lock

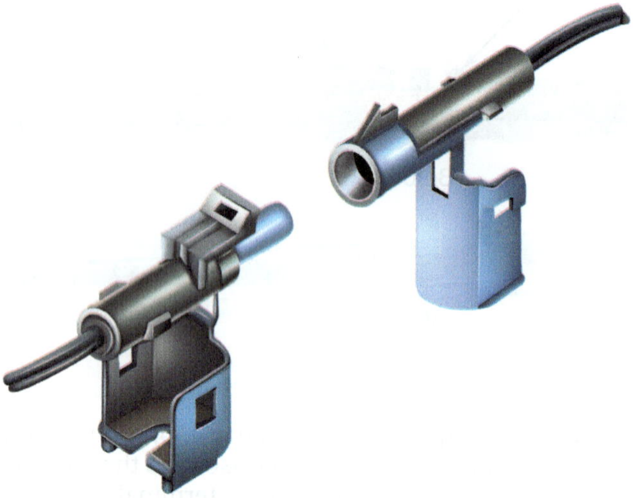

2. Unlatch and open the secondary lock on the connector. See **Figure 11–34**.
3. Grasp the cable to be removed and push the terminal to the forward position.
4. Insert the tang release tool straight into the front of the connector cavity until it rests on the cavity shoulder.
5. Grasp the cable and push it forward through the connector cavity into the tool while holding the tool securely in place. See **Figure 11–35**.
6. The tool will press the locking tangs on the terminal. Pull the cable rearward (back through the connector). Remove the tool from the connector cavity.

FIGURE 11–35 Removal tool procedure

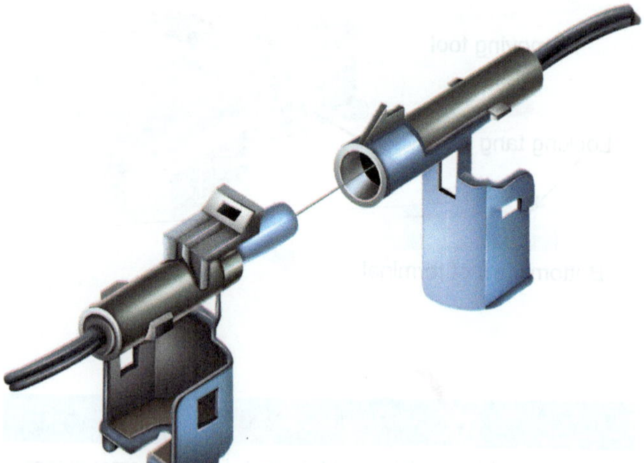

FIGURE 11–36 Proper cable seal position

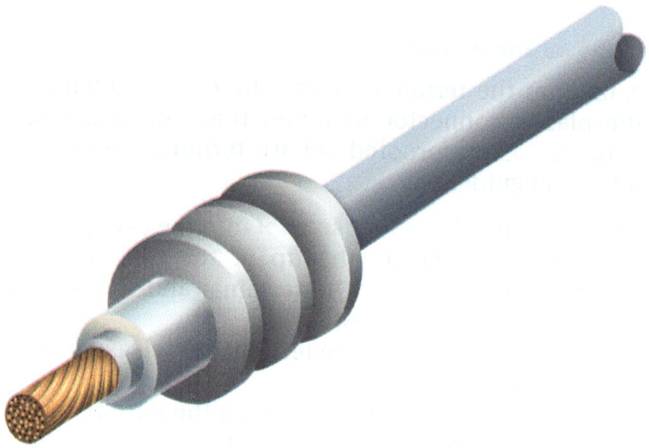

7. Cut the wire immediately behind the cable seat and slip the new cable seal onto the wire.
8. Strip the end of the cable strippers to leave 5.0 6 0.5 mm (0.2 6 0.02 in.) of bare conductor. Position the cable seal as shown. See **Figure 11–36**.
9. Crimp the new terminal onto the wire using the crimp tool. See **Figure 11–37**.

DEUTSCH CONNECTORS

At least one Deutsch connector was used by all the OEMs on chassis up to MY 2014 as a DLC; after MY 2014, Volvo-Mack chassis no longer use a Deutsch DLC. However, Deutsch connectors of a various different types are used within a typical truck data bus (including current Volvo-Mack), some of which can be seen in Chapter 12. Deutsch connectors have cable seals that are integrally molded into the connector.

FIGURE 11–37 Crimping procedure

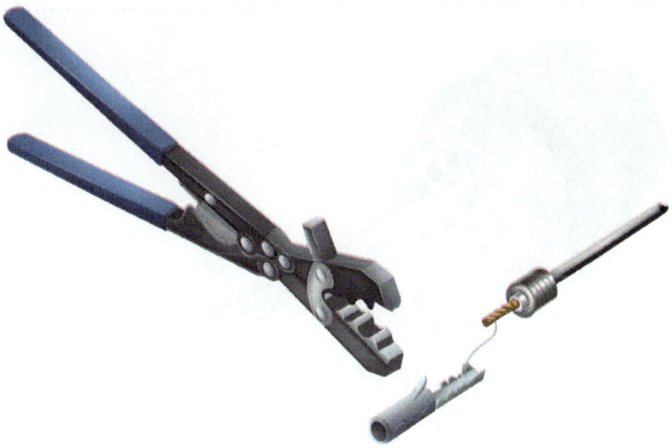

They are push-to-seat connectors with cylindrical terminals. The diagnostic terminal connectors are gold plated.

Installation of Deutsch Connectors

Use the following instructions for installation:

- Strip approximately ¼ inch (6 mm) of insulation from the cable.
- Remove the lock clip, raise the wire gauge selector, and rotate the knob to the number matching the gauge wire that is being used.
- Lower the selector and insert the lock clip.
- Position the contact so that the crimp barrel is 1⁄32 inch (0.8 mm) above the four indenters. See **Figure 11–38**. Crimp the cable.

FIGURE 11–38 Setting wire gauge selector and positioning the contact

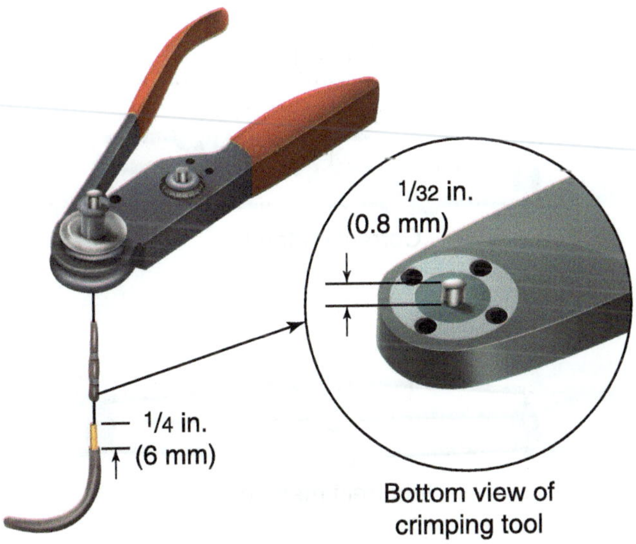

FIGURE 11–39 Pushing a contact into a grommet

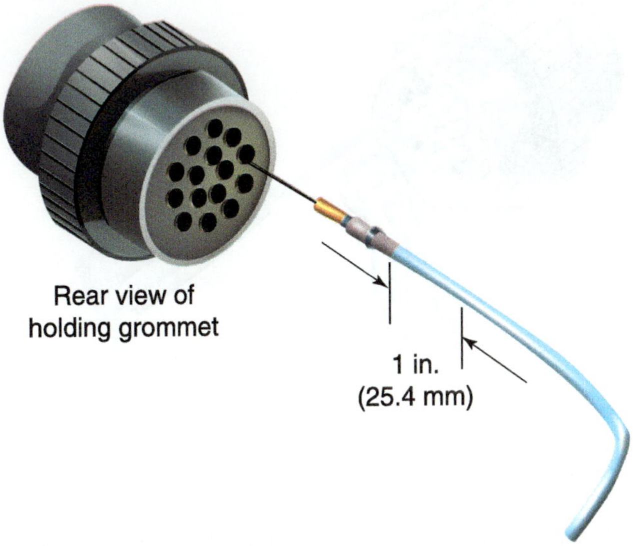

- Grasp the contact approximately 1 inch (25.4 mm) behind the contact crimp barrel.
- Hold the connector with the rear grommet facing you. See **Figure 11–39**.
- Push the contact into the grommet until a positive stop is felt. See **Figure 11–40**. A slight tug will confirm that it is properly locked into place.

FIGURE 11–40 Locking terminal into connector

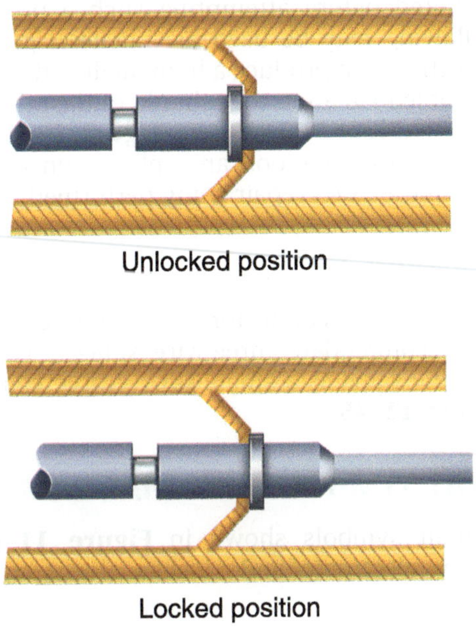

FIGURE 11–41 Removal tool position

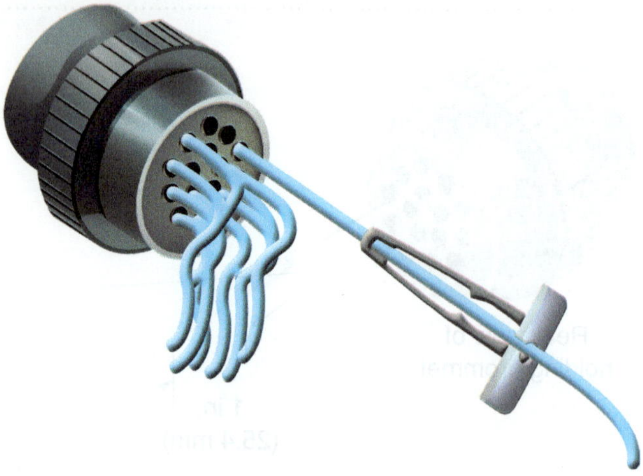

Removal

The appropriate size removal tool should be used when removing cables from connectors.

1. With the rear insert toward you, snap the appropriate size remover tool over the cable of contact to be removed. See **Figure 11–41**.
2. Slide the tool along the cable into the insert cavity until it engages and resistance is felt. Do not twist or insert tool at an angle. See **Figure 11–42**.
3. Pull the contact cable assembly out of the connector. Keep reverse tension on the cable and forward tension on the tool.

FIGURE 11–42 Terminal contact and wire removed

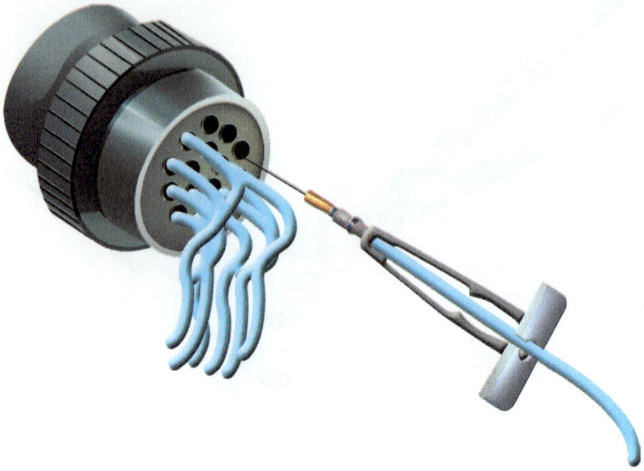

11.9 SPLICING GUIDELINES

Not all wiring in electronic circuits can be spliced, so in all cases, the OEM service literature should be consulted before attempting such a procedure. The following may be used as a general guideline. The objective is to produce a high-quality, tight splice with durable insulation that should outlast the life of the vehicle (**Figure 11–43** and **Figure 11–44**). The selection of crimping tools and splice connectors will depend on the exact repair being performed.

FIGURE 11–43 Spliced wire

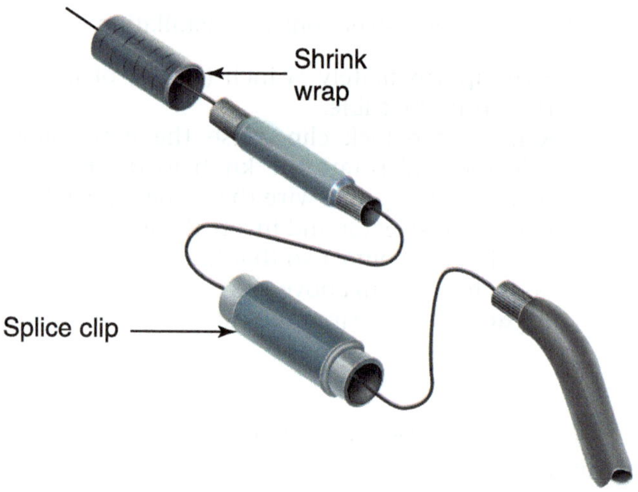

FIGURE 11–44 Multiple splices

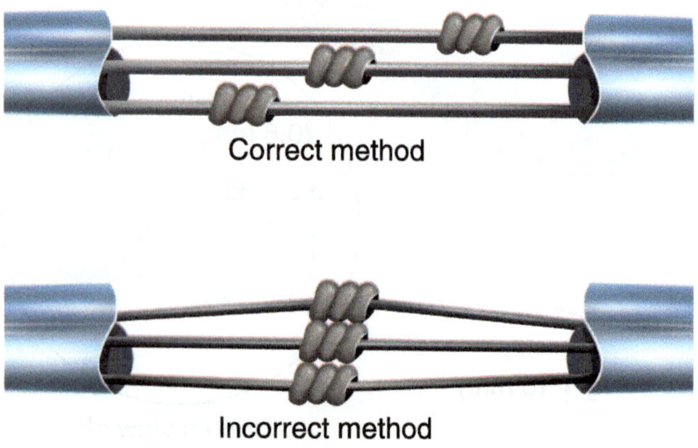

THREE-WIRE SPLICE

Three-way splice connectors are commercially available to accommodate three-wire splices. The technique is the same as a single **butt splice** connector. See **Figure 11–45**.

CIRCUIT SYMBOLS AND DIAGRAMS

The circuit symbols shown in **Figure 11–46** are typical in electronic and chassis electrical wiring

FIGURE 11–45 Three-way splice

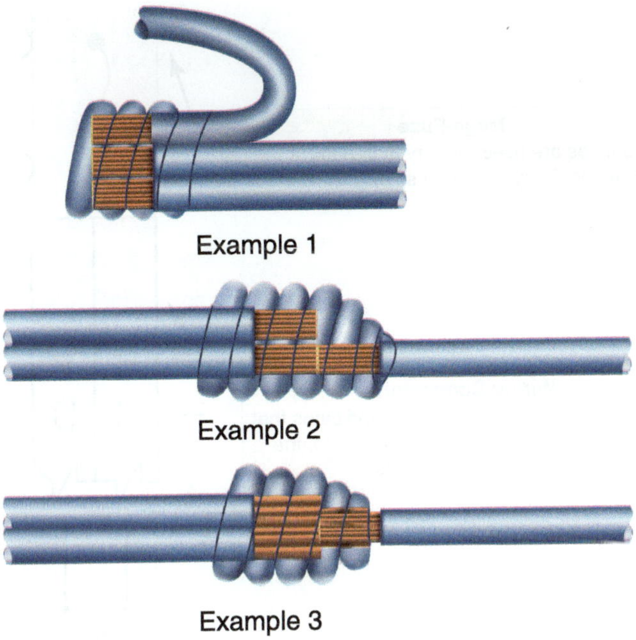

schematics, whereas those shown in **Figure 11–47** are typical of those used in multiplexed electronic wiring schematics. The technician should be able to easily identify most of the symbols displayed here because they are so commonly used. They do vary slightly depending on the OEM. Note that the twisted pair symbol used for the vehicle data bus is different in each schematic.

ISO WIRING CODES

Some manufacturers are using ISO wire color codes on their chassis electrical systems. **Table 11–3** shows these codes.

11.10 SUPPLEMENTAL RESTRAINT SYSTEM

Supplemental restraint systems (SRS) are becoming commonplace in trucks and although these systems are not mandatory as they are in automobiles, there is an increasing awareness that they can

FIGURE 11–46 Electrical circuit symbols

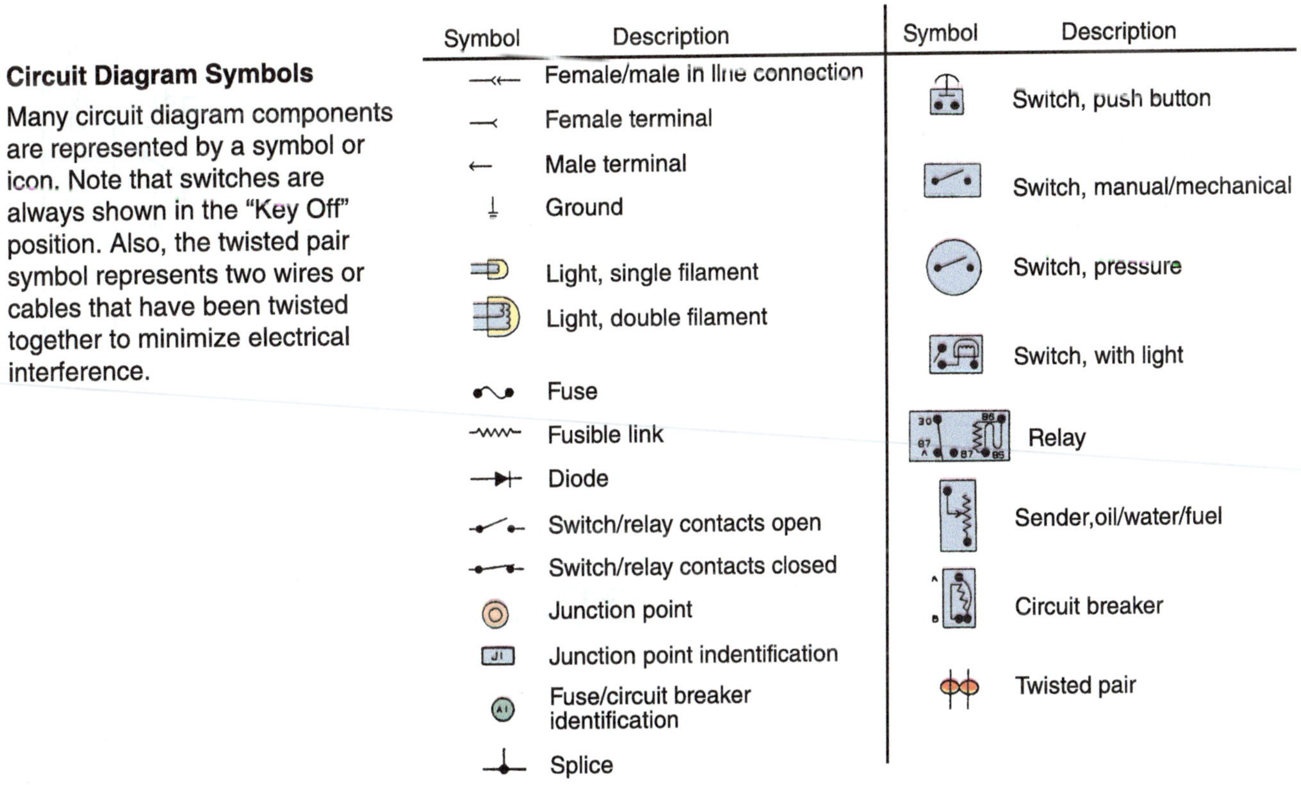

FIGURE 11–47 Circuit symbols used with electronic schematics

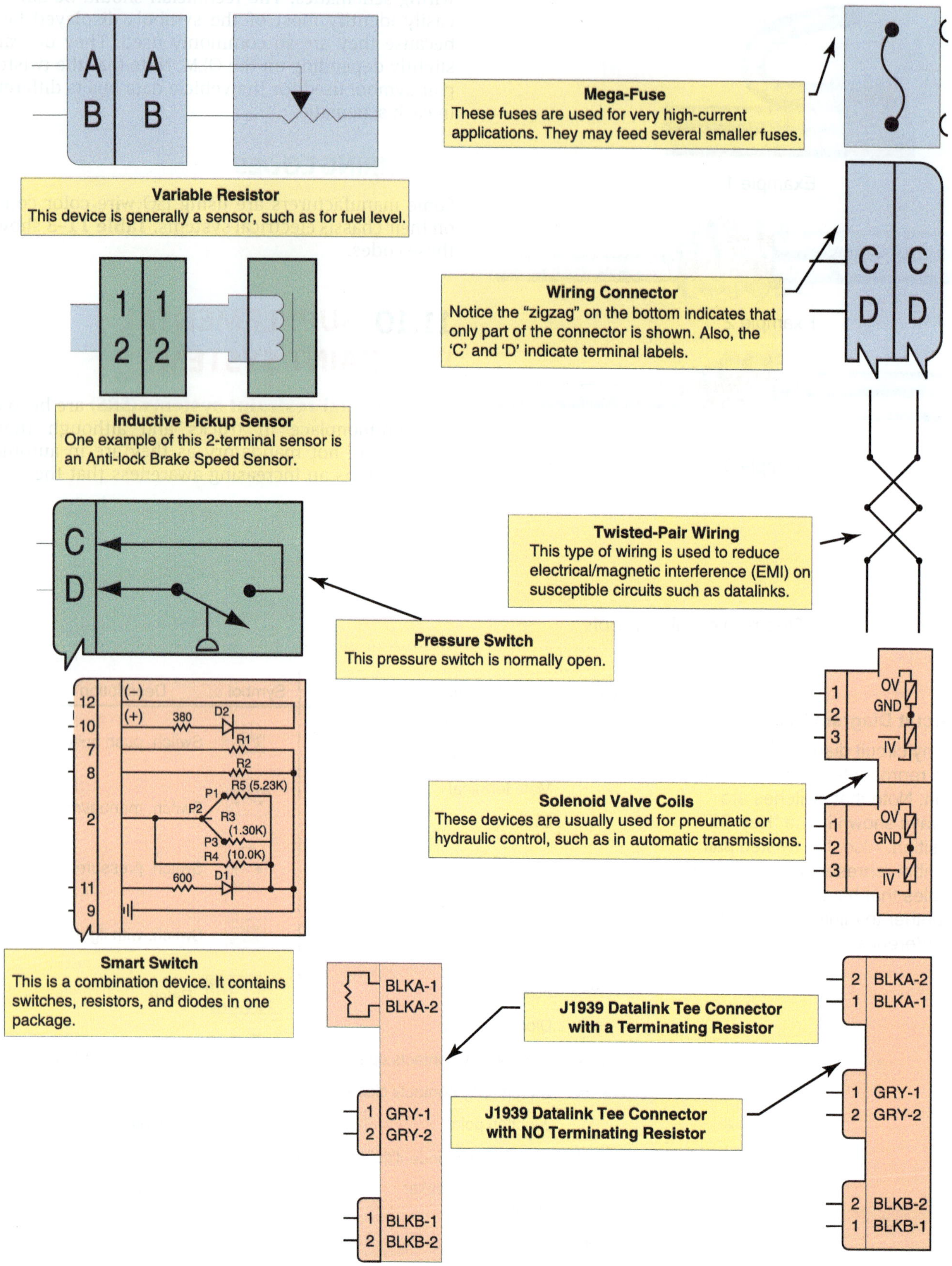

TABLE 11–3 ISO Wire Color Codes

Color	Abbr.	Standard Description
Black	Bk	Ground, general
Black-red	Bk-r	Power, ignition—run
Black-white	Bk-w	Ground, clean or isolated
Blue-dk	Dkbl	Back up, windshield wiper, trlr aux
Blue-lt	Ltbl	HVAC, circulation fans, 1922 pos.
Blue-lt-white	Ltbl-w	Water-oil gauges and indicators, engine and trans
Brown	Br	Tail, marker, panel lamps
Gray	Gy	Electronic engines, txl insulation
Green-dk	Dkg	R/h turn signal, drivers display, data record, 1587 POS—1939 NEG
Green dk-white	Dkg-w	Starting aids, fuel heaters, tail gate, material control, winch
Green-lt	Ltg	Headlamps, road lamps, drl
Green-lt-white	Ltg-w	Axle controls and indicators
Orange	O	Abs, ebs, 1587 neg.
Pink	Pk	Start control, ignition, charging, volt–ammeter, 1922 neg.
Pink-white	Pk-w	Fuel control and indicators, speed limiter–shutdown
Purple	Prp	Engine fan, pto, auto lube/oil
Purple-white	Prp-w	Utility, spot, interior, emergency lighting
Red	R	Power distribution, constant
Red-white	R-w	Pneumatic/hydraulic brakes, retarder, stop lamps
Tan	T	Mph/rpm signals, horn, flasher, pyrometer, turbo
White	W	Transmission, sxl insulation
Yellow	Y	L/h turn signal, 1939 pos.
Yellow-white	Y-w	Air bag

significantly protect drivers in the event of a serious accident. As with automobile SRS, the effectiveness of the system is dependent on the driver actually wearing the seat belt while driving. In this section, we describe the following types of SRS used in a Freightliner Cascadia chassis:

- Air bag system deployed from the steering wheel
- The RollTek cab rollover protection system

It should be noted that in chassis spec'd with both systems, they are designed to deploy interactively with each other. Because the operation of each system differs when the vehicle is equipped with only an air bag, for purposes of this description, we will assume that the vehicle is equipped with both a steering wheel air bag and rollover protection.

AIR BAG SRS

A steering column **air bag** SRS is only effective when a seat belt is worn. The electronic readiness of the system is indicated on the driver dash display during key-on self-test. Should the SRS indicator remain on after the self-test, there is a problem with the system. Air bag SRS is mastered by an **air bag control unit (ACU)**. The deployment algorithm used to trigger and inflate the air bag depends on other onboard electronics such as a road speed signaling, ABS reporting, a CWS, and rollover protection electronics. We will call this the trigger threshold and this is signaled by:

- **Crash sensor**: produces a signal corresponding to the rate of deceleration
- **Safing sensor**: senses only deceleration

FIGURE 11–48 Cascadia steering column located air bag showing the module connectors

The two sensors are in series so signals from both are required to trigger deployment. Deployment is managed by the ACU. When the trigger threshold has been met, the air bag gas propellant is ignited by an initiator. This event inflates a nylon fabric bag tailored to the cab and seat configuration. The objective is to dissipate the deceleration forces over as large an area of the driver's body as possible.

Air Bag Deployment Principle

The input circuit data that the ACU evaluates to trigger air bag deployment consists of data broadcast on the chassis data bus. When the deployment threshold is met, the deployment algorithm kicks in within a time frame of around 10 milliseconds. After this, the ACU outputs a signal to an igniter device. When hit with a voltage signal from a capacitor, the igniter explodes a solid combustible charge, which in turn generates (primarily) nitrogen gas that inflates the bag. Because SRS devices are used in accidents, and accidents can take out batteries and wiring, capacitors are relied on to trigger SRS igniters. During inflation events, some sodium hydroxide is generated. Though not classified as toxic, sodium hydroxide is known to cause allergic reactions in some persons.

Deployment Time

Using the example of a head-on crash, the entire sequence that produces full air bag inflation takes place within 100 milliseconds. Air bag deflation is designed to take place over a longer period, ranging

FIGURE 11–49 Air bag module retainer fastener location on a Freightliner Cascadia steering column

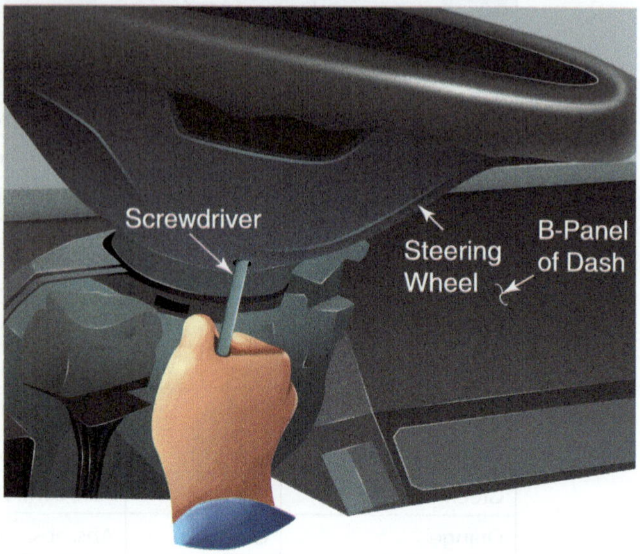

from 500 milliseconds to 1 second. During deflation, the nitrogen gas (and talcum powder used to cool and lubricate the bag during deployment) escapes through vents in the air bag fabric.

Deflation

At the completion of deployment, what remains is nitrogen, which immediately dissipates into the air; talcum powder residues; and sodium hydroxide. When accident victims suffer a reaction to an air bag deployment, the reaction is caused by the small amount of sodium hydroxide, not by the talcum powder. Air bags will also deploy in the event of a fire. When temperatures exceed 300°F (150°C) the air bag combustible charge autoignites without the assistance of the igniter. **Figure 11–48** shows an air bag module assembly as it appears on a Cascadia chassis.

Air Bag Removal

The following outlines some guidelines used by OEMs regarding the removal of an air bag module. Always consult OEM service literature before undertaking the task on an actual chassis.

1. Apply the parking brake system and chock the vehicle wheels.
2. Disconnect the battery banks. WAIT for 2 minutes. This allows the capacitors to leak off any accumulated charge.
3. Next you can loosen the fasteners on the underside of the steering wheel as shown in **Figure 11–49**.
4. Holding the air bag module away from your body, carefully lift it away from the steering wheel hub cavity. Disconnect the air bag module connector (see Figure 11–49) while doing this.

FIGURE 11–50 Air bag module connections

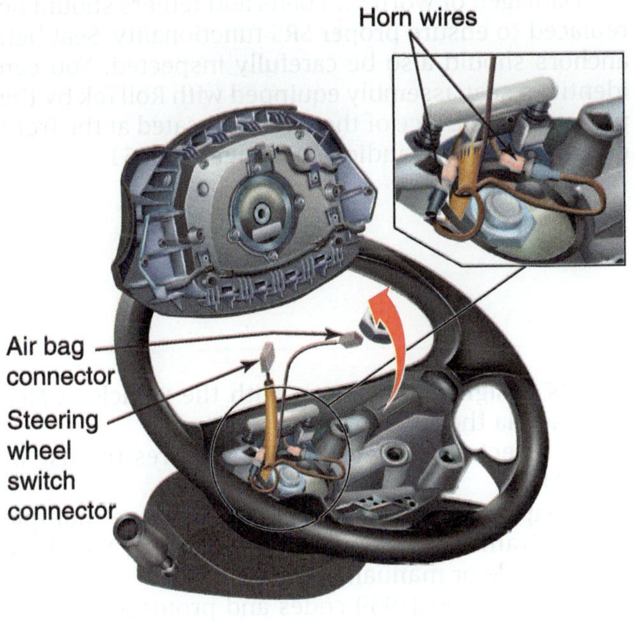

Air Bag Installation

A new air bag assembly must be considered to be armed and potentially dangerous. Once again, you should always consult the OEM installation guidelines before installing the assembly. If the air bag was previously deployed, you must carefully inspect the clockspring (see Figure 11–48) for signs of melting or other damage. If damage has occurred you will have to remove the steering wheel and replace the clockspring. To install the new air bag, use the following procedure:

1. Connect the steering wheel switch connector and the connector from the air bag module. Carefully position the air bag module into the module steering wheel cavity.
2. Install and torque the air bag module fasteners at the base of the steering wheel (**Figure 11–50**).
3. Reconnect the batteries.
4. Turn the ignition key-on.
5. Connect an EST with FTL ServiceLink to check for active codes.
6. If there are any chassis active codes, troubleshoot and deactivate them.
7. Use ServiceLink to erase any inactive/historic codes.
8. Cycle the ignition key. If the SRS indicator goes out after the self-check sequence and there are no active codes, the air bag module is armed and ready.

ROLLTEK ROLLOVER PROTECTION

Another SRS option on current trucks is a rollover protection system. The **RollTek** rollover protection system is designed to work in conjunction with a seat belt. It can be used on both the driver and passenger seats of a truck. When a seat is equipped with RollTek a sensor is mounted in the seat base. The sensor connects to the RollTek module that can activate two outputs:

FIGURE 11–51 RollTek protection system components

- A side-roll air bag (**Figure 11–51**)
- A seat pull-down device

When the RollTek module senses a rollover condition, it triggers gas cylinders located in the base of the seat. Once triggered, the gas cylinders activate power cinches that tighten the lap and shoulder belts against the occupant while at the same time lowering the seat. This tightening action moves the occupant down and away from the steering wheel and cab ceiling. As the seat is pulled into its lowest position, a side-roll air bag deploys on the outboard side of the seat.

System Readiness

As with the steering column air bag system, the operational readiness of RollTek SRS is indicated by an SRS indicator light on the dash. This light illuminates for approximately 10 seconds after the ignition key is turned on. In the event of the indicator light remaining on or failing to illuminate at key-on, an SRS problem is indicated. Always use OEM service literature when working with an armed SRS circuit. An SRS should be regarded as armed any time it has not been visibly deployed.

Damaged or worn seat belts and tethers should be replaced to ensure proper SRS functionality. Seat belt anchors should also be carefully inspected. You can identify a seat assembly equipped with RollTek by the presence of the face of the module located at the front of the seat base as indicated in Figure 11–51.

SUMMARY

- A continuity test is a quick resistance test that distinguishes between an open circuit and a closed circuit.
- A dark band identifies the cathode on a diode.
- Circuit resistance and voltage are measured with the test leads positioned parallel to the circuit.
- Direct measurement of current flow is performed with the test leads located in series with the circuit.
- A Hall effect probe can be used to approximate high-current flow through a DC circuit.
- Handheld ESTs such as the Pro-Link iQ can be used to access active and historic codes, read system identification data, perform diagnostic testing of electronic subcomponents, reprogram customer data, act as a serial link for mainframe linkage, and perform snapshot data analysis.
- The ESTs used to service, diagnose, and reprogram truck chassis management systems are onboard diagnostic lights, DMMs, scanners, generic reader/programmers, proprietary reader/programmers, and PCs.
- Flash codes are an onboard method of accessing diagnostic codes. Most systems will display active codes only. Some will display active and historic (inactive) codes.
- Most OEMs use the PC and proprietary software as their primary diagnostic and programming EST.
- ESTs designed to connect with the vehicle ECM(s) do so via the data connector.
- Most electronic circuit testing requires the use of a DMM.
- A snapshot test is performed to analyze multiple data frames before and after a trigger, usually a fault code or manually keyed.
- SAE J1587 and J1939 codes and protocols numerically code all onboard electronic systems, parameters, and failure modes.
- The PC is connected to the vehicle data bus by means of a 6-, 9, or 16-pin connector known as a data connector. Once connected, the OEM software accesses the appropriate SA or MID so that the system can be read and diagnosed.
- Technicians should remember that *any* system using J1587 or J1939 protocols can at least be read using simple general heavy-duty software.
- DTCs use SPN, PID, and SID routing paths to identify FMIs making circuit diagnosis easy.
- Sealed electronic circuit connectors must be assembled using the correct OEM tooling and components.
- SRS is becoming common in trucks and is scheduled to become mandatory in the near future. Current SRS has focused on steering column air bags, seat-located side bags, and rollover protection.

REVIEW QUESTIONS

1. What EST is used to read flash codes?
   a. dash diagnostic lights
   b. HH-EST
   c. diagnostic fork
   d. PC
2. The appropriate EST for performing a resistance test on a potentiometer-type TPS isolated from its circuit is a:
   a. DMM
   b. scanner
   c. generic HH-EST
   d. PC
3. Which of the following is the correct means of electronically coupling a HH-EST with a truck chassis ECM?
   a. modem
   b. chassis data connector
   c. any Deutsch connector
   d. parallel link connector

4. When resistance and continuity tests are made on electronic circuit components, the circuit being tested should be:
   a. open
   b. closed
   c. energized
5. DMM test lead resistance should never exceed:
   a. 0.2 ohm
   b. 1.0 ohm
   c. 10 ohms per inch
   d. 100 ohms per inch
6. The output (signal) of a pulse generator rotational speed sensor is measured in:
   a. V-DC
   b. ohms
   c. V-AC
   d. amperes
7. Which of the following cannot usually be performed by a generic HH-EST?
   a. erase historic fault codes
   b. customer data programming
   c. snapshot testing
   d. proprietary data programming
8. Which of the following procedures cannot be performed using a PC and the OEM software?
   a. erase historic fault codes
   b. customer data programming
   c. permanently erase active fault codes
   d. solenoid cycle tests
9. Which of the following methods is commonly used to upgrade the software on handheld ESTs?
   a. replace the docking head
   b. upgrade the cartridge PROM chips or cards
   c. reprogram the software cartridge from a mainframe
   d. replace the RAM chip
10. When reading an open TPS circuit, which FMI should be displayed?
    a. 0
    b. 3
    c. 5
    d. 9
11. How many counts of resolution can be displayed by a 4½-digit DMM?
    a. 1,999
    b. 19,999
    c. 9,995
    d. 99,995
12. The specification that indicates how fine a measurement can be made by a DMM is known as:
    a. root mean square (rms)
    b. percentage deviation
    c. resolution
    d. hysteresis
13. When a DMM is set to read AC, which is the term used to describe the averaging of potential difference to produce a meter reading?
    a. root mean square (rms)
    b. percentage averaging
    c. resolution
    d. saw toothing
14. Which port on an HH-EST data connector is used to connect the printer?
    a. SAE J1708 data connector
    b. SAE J1939 connector
    c. Parallel output port
    d. RS-232 port
15. Which of the following components are inserted into terminal sockets to enable testing by DMM probes and minimize physical damage?
    a. cycling breakers
    b. noncycling breakers
    c. terminal dummies
    d. diagnostic forks
16. Which electrical voltage value does a DMM output when in diode test mode?
    a. 0.3V
    b. 0.6V
    c. 0.9V
    d. 1.1V
17. Technician A states that the accuracy of an analog meter is usually no better than 3 percent. Technician B states that most DMMs have an accuracy factor that is within 0.1 percent of the reading. Who is correct?
    a. Technician A only
    b. Technician B only
    c. both A and B
    d. neither A nor B
18. The component that can be safely inserted into electronic circuits so they can be tested without interrupting them is called a:
    a. testlight
    b. diagnostic fork
    c. breakout box
    d. DMM test probe
19. Technician A states that using the snapshot test mode when using a handheld reader/programmer to troubleshoot an intermittently occurring fault code will help identify the conditions that produced the code. Technician B states that snapshot test mode can only analyze a historic code. Who is correct?
    a. Technician A only
    b. Technician B only
    c. both A and B
    d. neither A nor B
20. When verifying the *signal* produced by a potentiometer-type TPS, the DMM test mode selected should be:
    a. V-AC
    b. V-DC
    c. diode test
    d. resistance
21. Describe the difference between an active code and a historic code.
22. What is meant by the term *GUI*? Give some examples.
23. What is meant by the term *networking*?

24. Which of the following PC components would be usually described as a *peripheral?*
    a. system housing
    b. hard disk drive
    c. CPU
    d. printer
25. If a PC is to be networked to the Internet or an OEM data hub, which of the following hardware devices is usually required?
    a. a digital camera
    b. speakers
    c. a modem
    d. a CD drive
26. Which of the following J1587 identifiers is equivalent to a J1939 suspect parameter number (SPN)?
    a. MID
    b. PID
    c. FMI
    d. SA
27. Technician A says that a J1939 SA is a module with an address on the data bus. Technician B says that J1587 FMIs are identical in number to current J1939 FMIs. Who is correct?
    a. Technician A only
    b. Technician B only
    c. both A and B
    d. neither A nor B
28. Which of the following is the J1939 equivalent of the J1587 MID?
    a. SPN
    b. FMI
    c. PGN
    d. SA
29. What color is the 2013 J1939 data connector?
    a. black
    b. blue
    c. green
    d. red
30. Technician A says that the post-2013 J1939 data connector is backward compatible with pre-2013 data connectors. Technician B says the F-pin on a post-2013 J1939 data connector is smaller than that on a pre-2013 data connector. Who is correct?
    a. Technician A only
    b. Technician B only
    c. both A and B
    d. neither A nor B

12

Prerequisite: Chapter 11

MULTIPLEXING

OBJECTIVES

After reading this chapter, you should be able to:

- Describe a typical vehicle data bus.
- List the key data bus hardware components.
- Define the word *multiplexing*.
- Describe how multiplexing can make data exchange more efficient.
- Outline how a J1939/CAN 2.0 data bus functions.
- Define the terms "source address" and "suspect parameter number."
- Access J1587/1708 and J1939 data buses using a data connector.
- Explain how a "smart" ladder switch operates.
- Identify the fields that make up a data frame message packet on a truck data bus transaction.
- Explain how messages are prioritized on a serial data bus.
- Outline some of the 2014 changes to J1939.
- Identify the different generations of J1939 data bus connectors.
- Explain how FETs are used as relays to effect data bus outcomes.
- Access a controller source address on a truck chassis data bus with multiple networked electronic systems.
- Describe the composition of a diagnostic trouble code.
- Outline the procedure required to access a failure mode indicator (FMI) using electronic service tools.

KEY TERMS

algorithm
bandwidth
body control module (BCM)
bus arbitration
bus systems
bus topology
central gateway (CGW)
characteristic impedance
chassis management module (CMM)
client
communications adapter (CA)
controller area network (CAN)
controller
data bus
data bus communications
data connector
data framediagnostic/data link connector (DLC)
diagnostic trouble code (DTC)
electromagnetic interference (EMI)
failure mode indicator (FMI)
field effect transistor (FET)
gateway
global PGN
hexadecimal code
ISO 9141
ISO 11898
J1587/1708
J1939
ladder switch
mapping
message identifier (MID)
model year (MY)
neural network
occurrence rate (OC)
packet
parameter group (PG)
parameter group number (PGN)
power line carrier (PLC)
programmable logic controllers (PLCs)
protocols
RP1210
scopemeter
server
signal detect and actuation module (SAM)
smart switch
source address (SA)
specific PGN
subsystem identifiers (SIDs)
suspect parameter number (SPN)
terminating resistor
twisted-wire pair

INTRODUCTION

In the days of sailing ships, there existed a chain of command that extended from a captain through a first mate and lieutenants and ended up at the bottom of the chain with sailors, who would actually put into effect the instructions that began with a captain's whim. Communication was verbal. The captain delegated to subordinates, and the sailors at the end of the chain of command were entrusted with specific tasks such as raising a sail, turning the tiller, or dropping an anchor. Each task would probably be specialized to some extent and require some training. In this way, the critical functions of the ship were relayed and undertaken by word-of-mouth communications.

If we go back a generation and take a look at a highway truck of that era, most of the critical functions were managed directly by the driver using mechanical, pneumatic, hydraulic, and electrical signals. An accelerator pedal required a mechanical linkage to connect the driver's boot with the engine fuel control mechanism. Likewise, every electrical function on the chassis had to be hard-wire connected from a dash switch through a relay to whatever type of actuator was being controlled. Over the years, as chassis systems became more complex, more and more miles of wire were required to enable the electrical and mechanical outcomes required to manage the vehicle. That is until **data bus communications** were introduced.

INFORMATION INTERCHANGE

Data bus communications make managing a modern truck chassis seem similar to the way a large sailboat was managed a couple of hundred years ago. Instead of having a ship's captain bellow out some orders to be passed down a chain of command, a module broadcasts digital signals to a data backbone. These signals can either be read or responded to by other modules connected to the data backbone, or ignored if the message does not fall within that module's realm of responsibility.

Data bus communications make use of a technology called *multiplexing* that has been used in truck computer control systems since the first electronic control module (ECM) was introduced on trucks in the 1980s. An ECM or electronic control unit (ECU) is a system controller. In the early days, multiplexing was usually module to module and truck technicians did not need much of an understanding of what it was all about. For instance, one engine manufacturer used this type of communication to allow a chassis management module to "talk" to the fuel-injection control module: This was a dedicated bus connection that meant the technician could not troubleshoot bus malfunctions. Things have changed. Today, technicians regularly access the truck data bus to troubleshoot, reprogram, and read the systems within it.

THE NEURAL NETWORK

Simply put, multiplexing means data sharing between multiple system control modules. Our definition of a control module is one that has both processing and outcome-switching capability: it is commonly known as a system **controller**. If you set up a chassis computer system so that all of the control modules "speak" the same language and provide a common, shared communication path between them, then you have a multiplexed system. Today, most trucks use powertrain multiplexing systems that communicate using just a couple of universal "language" systems. But technicians should note that manufacturers may use proprietary buses for non-powertrain functions or distinct powertrain networks such as those found on a diesel-AC hybrid rock truck; proprietary buses tend to be only accessible using manufacturer-subscription software.

Some manufacturers use the term **neural network** to describe a comprehensively multiplexed truck chassis; this analogy makes the comparison of a multiplexed chassis with a human body's central nervous system. This is not really a fair comparison because a multiplexed truck chassis is much less complex and more logically organized than a human body.

MULTIPLEXING SIMPLIFIES HARDWARE

Although the term sounds complex, multiplexing actually simplifies truck electronics. It does this by giving electronic subsystems a common communication language, and, by using a **data bus** or information highway, it allows data signals to take the place of hard wire in the electronic input and output circuits. Current trucks network the electronic controllers in a chassis in a way that simplifies the hardware, eliminates miles of hard wiring, reduces the number of in-out (I/O) pins on modules, and optimizes vehicle operation by giving electronic subsystems an accountability that extends beyond the hardware they control making possible options such as antirollover and directional stability management. An example of how the volume of hard wire can be reduced by multiplexing is demonstrated by the International Trucks Diamond Logic system that has decreased the number of wires to the instrument gauge cluster from 67 to just 7.

Nice-to-Know versus Must-Know

So what do you really have to know about multiplexing today? If you have ever used a 6-pin or 9-pin **data connector** to connect a diagnostic reader or shop computer to read the electronic systems on a truck, you already have some experience in accessing a truck data bus. Not all OEMs use the full potential of multiplexing technology but in this chapter we will reference the Freightliner and International Navistar (Diamond Logic) systems, which have made the fullest use of the technology since its onset, incorporating full suites of "smart" switches, relays, and circuit protection devices. Regardless of OEM, the technician today must have a basic understanding of multiplexing to work on any chassis system that is managed by electronics.

12.1 MULTIPLEXING, CLIENTS, AND SERVERS

We have already said that multiplexing refers to transactions between a series of networked computers and other solid state devices. In the professional world, if you consult an accountant or a lawyer, you become their customer and are known as their **client**. In the computer world anyone or anything that wants something is referred to as a client. For a transaction to take place, there must be someone or something that fulfills that need. The fulfillment of a client need is provided by a **server**. So the data backbone serves as a path for connecting clients with servers. The messages used to make these connections are known as **packets**. There are rules about packets and how they are constructed—think of the difficulty and expense of sending an odd-shaped package by courier.

SPEED AND VOLUME

The roads used by trucks are known as highways. The roads used by telecommunications systems can be wires, fiber optic cables, light beams, or radio waves. The speed limit on a vehicle highway is rated in miles per hour. The speed limit on a telecommunications channel is measured in *baud* and *K-baud*. More K-baud is better and faster. Maximum volume on a vehicle highway is defined by the number of lanes. An interstate with more lanes is generally considered to be better and the number of lanes on a road can be likened to the telecommunications term **bandwidth**. They both represent the volumetric capacity of the road or transmission medium. One refers to the number of cars and trucks that can travel simultaneously, the other to the number of packets or data volume that can be pumped down a channel. On the highway, because there are other cars on the road, each containing people with their own specific objectives, you can closely compare the dynamics of road travel with data multiplexing. Multiple packets traveling a data highway heading from one place to another, using common transmission mediums, move at pretty much the same rate but usually have distinct destinations.

Multiplexing modernizes electronic communications by making electronic circuits a lot more like our highway system. In doing so, miles of unnecessary hard wire and replicated components can be eliminated. Multiplexing uses electrons modulated by some strict **protocols** (rules and regulations) to simplify subsystem-to-subsystem electronic transactions. Analog inputs are converted to digital signals by the receiving processor on the bus. These signals can then be broadcast digitally to other processors on the network, and then changed back to analog format to effect actions.

12.2 MULTIPLEXING BASICS

As we said before, if you have worked on trucks manufactured during the 1990s and later, you already have some experience with multiplexing, even though it may have been unknowingly. Beginning in 2002, a new generation of truck chassis began a second generation of truck multiplexing. Today most of the major commercial truck original equipment manufacturers (OEMs) use multiplexing, though some incorporate more advanced versions of the technology. All this replicates technology that has been used in city and highway buses since the late 1990s.

There is a good reason that buses moved ahead of trucks in their use of multiplexing. In a typical bus, the driver sits in the front and the engine, transmission, and most of the chassis electronics are located in the back. To hard-wire connect the driver with these complex electronic systems located in the back of the vehicle, miles of wire had to travel the length of the chassis. A data bus consisting of a **twisted-wire pair** provides a simple solution. Drivers generate command messages from their position at the front of the bus, and these are sent as message packets via an ECU down the communication backbone. The message packets can then be read by any module connected (networked) to the data bus. The modules process the messages and generate whatever outcomes are desired. More advanced multiplexing can eliminate electromechanical switches. By using switches that generate communication signals to the data bus rather than changes in electrical status, more miles of hard wire can be eliminated.

Our initial definition of *multiplexing* was to describe it as chassis information-sharing between the different on-board computer controlled systems to optimize vehicle performance. For instance, some of the key input sensor signals are required by more than one of the chassis control modules. The throttle position sensor (TPS) is a good example of such a signal. The TPS signal is a required input for the engine management, transmission management, antilock braking system/automatic traction control/antirollover electronics (ABS/ATC/ARE), and collision warning/mitigation systems. The TPS signal is hard-wired to one module (usually the engine ECM), and from there broadcast over the data bus so that any other networked controllers that require this signal can pick it up. Some electronic systems that are commonly multiplexed include the following:

- Engine control module (ECM)
- Fuel-injection control module (FICM)
- Body control module (BCM)
- Transmission control module (TCM)
- Instrument cluster unit (ICU)
- Dash display module (DDM)
- Signal detect/actuation module (SAM)
- Antilock braking system (ABS)
- Automatic traction control (ATC) system
- Antirollover electronics (ARE)
- Electronic immobilizer system (EIS)
- Collision warning system (CWS)
- Collision mitigation system (CMS)
- Global positioning system (GPS) (one-way)
- Satellite communications systems (SCS)
- Supplemental restraint systems (SRS)
- Lane departure warning (LDW) systems

Exchange of information between computer-controlled systems reduces the total number of sensors required, and it optimizes vehicle performance by making each system perform according to the requirements of the vehicle system rather than each of its subsystems.

STAR NETWORK STRUCTURE

First-generation data transmission in a vehicle required that every input and output signal had to be allocated an individual conductor; that meant a dedicated wire and terminals. This was necessary because binary signals can only be transmitted using binary "on" or "off" states. On/off ratios worked fine to transmit continually changing values such as the status of an accelerator pedal-travel sensor. As data exchange volumes between controllers increased over the years, it no longer made sense to handle communications by conventional hard wire and plug-in connectors such as that shown in **Figure 12–1**. Note that in the layout shown in Figure 12–1, each module shown requires a direct connection to any other module it has to communicate with. It is a modification on a star network. In more recent systems, the objective is to keep wiring-harness complexity down to a minimum to reduce both costs and technical malfunctions. For this a serial data bus is preferred and is described a little later in this section.

POWER LINE CARRIER

A more primitive method of multiplexing known as **power line carrier** has been in use in trucks since

FIGURE 12–1 Star network data transmission—no common multiplex bus

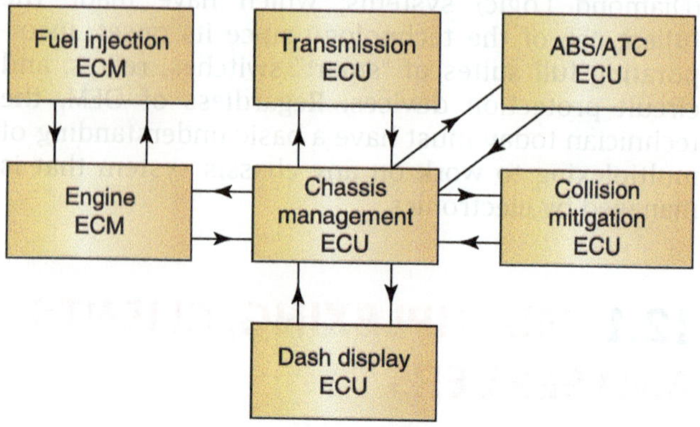

1995. Power line carrier enables communication transactions to take place on a non-dedicated communication wire. Power line carrier communications were introduced when ABS became an *option* on highway trailers; its use later became mandatory in 2001, 3 years after the introduction of *mandatory* ABS. Trailer ABS requires that a warning light be illuminated in the dash of the truck in the event of a trailer ABS malfunction. Because all the wires on a standard SAE J560, 7-pin connector between truck and trailer were already dedicated, power line carrier technology was used to convert a communication signal to a radio frequency signal and then superimpose it over the 12-volt auxiliary power wire. A transducer converts the radio frequency signal back to a signal that the receiver modules can use. These signals originally used the **J1587/1708** data bus protocols, but since 2010 they have been gatewayed into J1939. The term **gateway** is used to describe a module that allows two or more networks using different protocols to "speak" to each other; these may also be known as transducer modules.

TYPES OF LOGIC

Current engine and transmission management systems use mostly open-loop (fuzzy) processing as opposed to closed-loop logic so messages from any data bus controller can influence how the system is being managed at a given moment. This adds to volume of exchange data between the various controllers managing a chassis. To illustrate the difference between closed-loop and fuzzy logic, we will use an example you should be familiar with if you have worked on or driven a truck using cruise control.

Closed Loop

The older hard cruise, commonly used in most autos, requires that the driver set a road speed of, say, 60 mph. After the road speed has been selected, road speed sensor input drives the management of the powertrain to maintain the desired road speed regardless of conditions. This type of closed-loop operation puts the road speed sensor in command of the powertrain, and that can be a disadvantage. For instance, if the vehicle were traveling through mountainous terrain, attempting to maintain the input road speed at exactly 60 mph would waste fuel.

Open Loop

Most trucks today use "soft" or "smart" cruise. When the driver inputs a road speed cruise control command into a soft cruise system, the circuit "thinks" for itself. It uses multiple inputs and programmed instructions to manage fueling. For instance, if the soft road speed were set at 60 mph and the vehicle were traveling through undulating hills, the processing cycle would map an algorithm to "learn" from the terrain and in doing so, uses a variety of inputs to plot an actual road speed. From a fuel efficiency point of view, it makes sense to allow the truck to slow slightly through each uphill climb and accelerate above set speed on each downgrade to prepare for the next hill. In addition, when acceleration is required, the event can be softened to avoid abrupt fueling increases. Hard cruise is an example of closed-loop operation, whereas soft or smart cruise is an example of fuzzy logic.

SHOP TALK

The term algorithm is a software term used by most OEMs that describes a programmed sequence of operating events: the rules and conditions for producing a processing outcome from a chassis control module. This is also less commonly known as mapping.

SERIAL DATA TRANSMISSION

So you can see now that many of the problems of data transfer volume using conventional interfaces (hard wire) can be simplified by using **bus systems** or data highways. The basis of serial data transmission is to use a single communications channel that delivers "instructions" rather than electrical signals to the controller modules. The example adopted by most vehicles is **controller area network (CAN)**, a data bus system developed by Robert Bosch and Intel for vehicle applications. CAN is a serial data transmission network used for the following applications in a vehicle:

- ECM networking
- Comfort and convenience electronics
- Mobile on-board and external communications

CAN 2.0

CAN 2.0 (second generation) is the basis for SAE **J1939**, the high-speed network in standard use on trucks and buses in North America. J1939 was conceived as a powertrain bus, but today it can be used for non-powertrain functions depending on the OEM. The J1939 bus is designed to function from 125 K bits per second (Kb/s) up to 1 Mb/s. However, 500 Kb/s is a typical maximum for 2014, making it a class C bus. The current J1939 is the equivalent of the automobile CAN-C powertrain bus. Class A and B buses use slower speeds and no longer have applicability in trucks. In comparative terms, the current version of J1939 is about 50 times faster than the J1587/1708 (original) heavy-duty data backbone.

In any powertrain data bus, transfer rates must be high to ensure the required real time responses are met—so you can bet that transaction speeds will continue to increase in the coming years. That said, J1939 is not expected to endure beyond 2020 due to insufficient bandwidth.

Multiplexing Clock Speeds

Microprocessor clock speeds of at least 16 MHz are required for J1939 transactions. A clock speed of 16 MHz translates to an ability to make up to 16 million binary "decisions" per second; millions of these binary decisions are required to process a simple output command. Most current truck engine and transmission management processors have clock speeds of at least 16 MHz and are designed to broadcast on both the J1587/1708 and J1939 buses.

ECU Networking

CAN uses serial data transmission architecture. *Serial* means single track as opposed to multiple track, which is known as *parallel* transmission. Several system controllers such as the engine, fuel system, ABS/ATC, electronic transmission, dash display, and collision mitigation system (CMS) modules can be networked on the serial data bus. Each ECU is assigned equal priority and connected using a linear bus structure as shown in **Figure 12–2**. One advantage of this structure is the fact that should one of the stations (subscribers) fail, the remaining stations continue to have full access to the network. The probability of total failure is, therefore, much lower than with other logical arrangements such as loop or star (hub) structures, like that shown in Figure 12–1. With loop or star structures, failure of the central or command ECU necessarily results in total system failure.

ISO 9141 AND ISO 11898

Control module interconnectivity and data protocols fall under the International Standards Organization's **ISO 9141** and **ISO 11898**. Because American-engineered and -built equipment is marketed throughout the world, it makes sense that both the hardware and communication protocols be internationally common. Most OEMs manufacturing equipment on

FIGURE 12–2 J1939 linear data bus topology and nodes

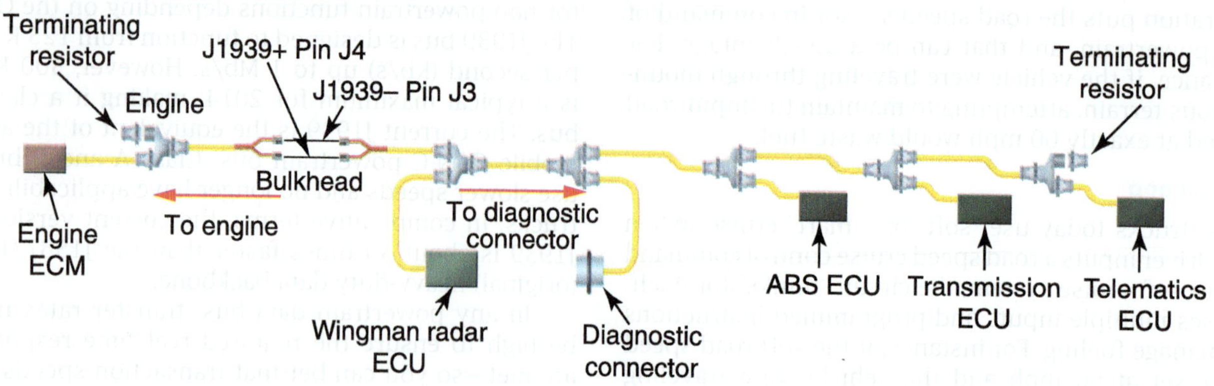

the J1939 multiplexing platform ensure that it is ISO 9141 and 11898 compliant. These ISO standards specify the requirements for data interchange among the various electronic control units (ECUs) on vehicles to facilitate readout, diagnosis, adjustments, and data exchange. It does not apply to system-specific equipment or proprietary CAN connections.

Bus Topology

The word *topology* means an organized structural system that remains unaffected when subjected to a sequence of events. The term **bus topology** is used in serial multiplexing to indicate that no single controller networked to the bus has more priority or status than another: No ECM/ECU is "in charge," in other words. For instance, if a data bus were to be cut in half, theoretically all of the controllers on each side of the break would be able to communicate with each other but not with the controllers on the other side of the break. Figure 12–2 shows the serial bus topology used in a truck data backbone; as you can see, it is linear.

Content-Based Addressing

CAN transactions are simplified. Instead of addressing individual stations in the network, an identifier label is assigned to every message pumped out onto the data highway. Each message is coded with a unique 11- or 29-bit identifier that identifies the message contents. For instance, engine-speed data is of significance to the engine, transmission, and CMS but probably has no importance to the climate-control module. Each station is designed to process only those messages whose identifiers are stored in its acceptance list. This means that all other messages are simply ignored.

Content-based addressing means that a signal can be sent to multiple stations. However, the sensor only needs to send its signal directly to one controller, from which it can then be broadcast on the bus network. Also, because it is easy to add more stations (addresses) to an existing CAN bus system, a large number of equipment variations are possible.

Assigning Priorities

In a multiplex system, controllers networked to the data bus can be labeled for data content and/or the priority of any message sent. In a vehicle management network, some signals are necessarily prioritized over others. An example of a high-priority signal would be the accelerator pedal angle or TPS signal. Any change in the TPS signal value must be responded to at high speed and is, therefore, allocated a higher priority than a signal that changes relatively slowly, such as that from the coolant-temperature sensor. A brake-request signal would have the highest priority status, and a critical signal such as a panic stop event can effectively shut down the data bus for a nanosecond (one-billionth of a second) or so.

Bus Arbitration

Handling traffic on the data highway requires rules and regulations. Just as every driver on an interstate is taught to cede to emergency vehicles, so must there be rules to handle data traffic. When the bus is free, any station can transmit a message. However, if several stations start to transmit simultaneously, **bus arbitration** awards first access to the message with the highest priority, with no loss of time or data bits. Lower-priority messages are shuffled by order of importance to automatically switch to receive and repeat their transmission attempt the moment the bus is freed up.

Message Tagging

In a busy airport with insufficient runways, there is an orderly lineup for takeoff at the beginning of each runway. The protocol (rule) on the airport runway is usually to sequence takeoffs in lineup order. The protocol in the truck data bus is necessarily different.

Each message is assigned a *tag*, a **data frame** of around 100 to 150 bits for transmission to the bus. The tag codes a message for sequencing transmission to the data bus and also limits the time it consumes on the bus. Tagging messages ensures that queue time until the next (possibly urgent!) data transmission is kept to a minimum. Each message or packet can be called a data frame. A data frame packet is made up of consecutive fields, as indicated in **Figure 12–3**, that show how a typical packet is constructed and the number of bits allocated per field.

Packet Construction

A typical J1939 message packet is constructed of the following fields:

- **Start of Frame/Message.** Start of frame announces the start of a packet and synchronizes all stations on the data bus.
- **Arbitration/Priority Field.** Data to announce the type of message and its priority. When this field is broadcast, the transmitter tags the transmission of each bit with a check to ensure that no higher-priority station is also transmitting.
- **Target Identifier.** The target may be global (broadcast to all controllers with addresses on the bus) or specific (directed to one controller on the bus). The target tag is known as a protocol data unit (PDU):
  - PDU 1 = destination specific
  - PDU 2 = global
- **Source Address.** The source address field is sized at 8 bits. It identifies the controller that originally broadcast the message.
- **Data Field.** The data field is sized between 0 and 56 bits. A message with data length 0 can be used for synchronizing distributed processes.
- **Cyclic Redundancy Check (CRC) Field.** The CRC proofs a message to identify possible transmission interference: It self-checks the output message contents.

FIGURE 12–3 Message bit encoding of a J1939 data packet

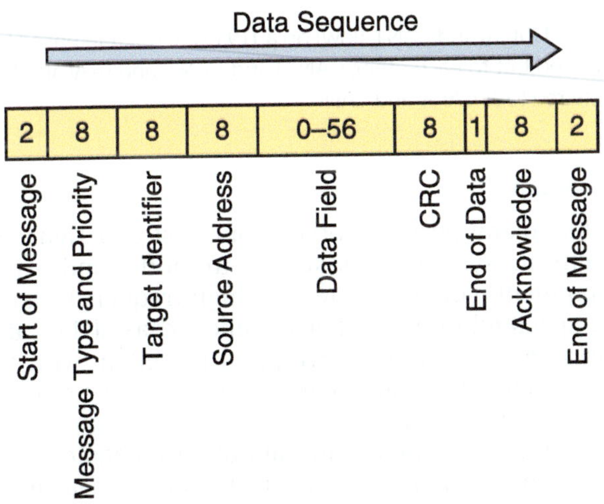

- **End of Data.** A 1-bit pulse to indicate that the packet data is complete.
- **Ack Field.** The ack or acknowledgment field contains acknowledgment signals by which the receiver stations on the bus indicate they have received a non-corrupted message.
- **End of Frame Field.** The end of frame (EOF) indicates the end of a message.

Integrated Diagnostics

A vehicle multiplexing bus system must be provided with monitoring capability so that transmission, transaction, and reception errors can be detected. This would include the check signal in the data frame and monitoring components in a transmission. It is a feature of a multiplex bus that each transmitter receives its own transmitted message again and by file comparison can detect any deviation or corruption. If a station identifies an error, it sends out an error flag that halts the current transmission. This prevents other stations from receiving a faulty message.

In the event of a major station defect, it is possible that all messages, both good and bad, might be terminated with an error flag. To protect against this, a multiplex bus system has a built-in function that enables it to distinguish between intermittent and permanent errors. Furthermore, this feature can sometimes localize station failures and route the data network to an appropriate failure strategy. The process of localizing a station-type failure is based on statistical evaluation of error factors, so it is not entirely foolproof.

Multiplexing Standardization

Multiplexing standardization for vehicles has been orchestrated by the International Organization for Standardization (ISO) internationally and by the SAE (Society of Automotive Engineers) in North America. In North America, CAN-C (light-duty vehicles) and SAE J1939 (heavy-duty, on and off-highway vehicles), define the hardware and software protocols of multiplexed components and data transfer. The international ISO 9141 and 11898 CAN 2.0 contains the core architecture on which the J1939 standards are constructed. The first generations of SAE J1939 accommodated the older J1587/1708 hardware and software protocols, but since 2010 some of the engine OEMs are no longer supporting the older bus protocol.

THE J1939 BACKBONE

A multiplexing data bus must be stable and must be protected against unwanted radiation interference that could corrupt the data signals. The data bus or backbone used in a J1939 multiplexed chassis is a pair of twisted wires color coded yellow and green. The reason for using a *twisted-wire pair* is to minimize the surface area on which external low-level radiation or **electromagnetic interference (EMI)** can act on the bus. The idea is to prevent the data backbone from acting like a radio antenna. Reducing EMI (in the form of electromagnetism, radar, microwaves) susceptibility is critical to protecting data from transmission corruption.

Minimizing EMI

The SAE standard for the J1939 data bus requires that the data bus wires twist through a full cycle once per centimeter—that is, 2½ times per inch. At either end of the data bus, a **terminating resistor** is used. The function of the terminating resistors is to:

- Prevent the twisted pair from acting like an antenna to attract signal interference
- Suppress data signals at the end of line to prevent data collisions

Terminating resistors test at 120 ohms. Their function is explained in a little more detail later in this section. Also, there cannot be any open T-connector cavities. At points where the backbone has to be spliced, special T-connectors with gold-plated contacts are used. Gold plating minimizes resistance because the nominal voltage on the data bus is low, usually around 3.5 volts or lower, and packet contents are read by voltage differentials of around 2.0 volts.

The J1939 data backbone uses stubs to make the physical connection between the backbone and each controller networked to it. The maximum length of a J1939 data backbone is 40 meters (130 ft.) and each stub should be no longer than 1 meter (40 in.). Exceeding these dimensions can cause problems with data transmission accuracy. It is not recommended to locate more than 10 nodes (controllers) on a J1939 bus.

Heavy Bus/Light Bus

Data buses can be divided into two types. They are differentiated by physical appearance and function. What each is called varies by manufacturer.

- Heavy or high bus (sometimes Hi bus)
- Light or low bus (sometimes Lo bus)

In this text, to avoid confusion with the terms used to describe each wire on the bus, we use the terms *heavy bus* and *light bus*. Each bus category is physically and functionally different.

Heavy Bus. A heavy-bus twisted-wire pair is wrapped with bare copper wire and foil shielded. Most OEMs suggest that it should not be repaired. The copper wire is known as the *drain*. The drain is required to be single-point grounded to battery negative. A heavy-bus twisted-wire pair transacts messages at the highest speeds typically at 500 Kb/s (but up to 800 Kb/s to 1 Mb/s) and is used for real-time data signaling. Real-time signaling is required by accelerator pedal status and electronically controlled transmission system-initiated input/output transactions.

Light Bus. A light-bus twisted-wire pair has no foil shielding. Light-bus signaling includes such things as audio system transactions, for instance, an increase in audio volume as road/engine speed increases. However, the use of light-bus systems is increasing on late-model vehicles for less critical multiplexing transactions.

Twisted-Wire Geometry

The high-speed and volume of data pumped through a truck chassis data bus requires wiring that protects against unwanted interference that could result in message corruption. As indicated earlier, current standards require that data bus wire pairs be twisted through a full cycle once per centimeter (2½ times per inch). There are two reasons for twisting the pair of data wires used in the bus:

- *To provide immunity to magnetic fields.* The voltage induced into any conductor by a magnetic field depends on distance, so by twisting the data wires, induced voltage effect will be about equal on both wires. In digital signaling using binary pulses, signal transceivers (any modules networked to the bus) are designed to look at the difference between CAN H (CAN+ or CAN1) and CAN L (CAN– or CAN2), not voltage levels referenced to ground.
- *To provide consistent capacitance values.* The distance between the two conductors on the bus alters the capacitance between them, with capacitance decreasing as the wires become further apart. The insulation material used to wrap the conductor wires also impacts on capacitance because it acts like the dielectric discussed in Chapter 5. On the data bus, the dielectric properties of insulation should be constant. Substituting ordinary wire for J1939 cable can corrupt communications on the bus by altering the impedance.

SHOP TALK

CAN H (high) and CAN L (low) on a J1939 are not referenced to ground but to each other. CAN H and CAN L and may also be referred to as CAN+ and CAN–.

Impedance

If, like many persons today, your first experience of electronic systems was setting up audio and visual equipment, you will know that high-frequency cables are specified in ohms. For instance, coaxial TV cable has a center conductor wrapped in Teflon, with the other conductor braided around it. This type of cable has what is known as a **characteristic impedance** of around 75 ohms. You cannot test characteristic impedance with an ohmmeter; it is just a measure of how the conductor appears (regardless of length) to a high-frequency signal.

A truck J1939 data bus must be limited to a characteristic impedance of 120 ohms. This is achieved by maintaining the dielectric constant (wire insulation) and keeping the physical spacing between the two conductor wires constant by having a consistent twist pattern.

Once again, because this is characteristic impedance, you cannot measure it with an ohmmeter. Do not confuse the 120-ohm characteristic impedance with the 120-ohm terminating resistors used at the end of data backbone, which *can* be measured with an ohmmeter.

Terminating Resistors

Terminating resistors are used to absorb signal energy, leaving no energy for reflections or echoes that result in electronic noise. They may be integrated into controllers or into bus stubs: **Figure 12–4** shows some examples of terminating resistors. Electronic noise causes scrambling of signals. It is a little like listening to an outdoor public address system where you are receiving audio input from several speakers at different distances.

Besides minimizing electronic noise, terminating resistors also provide a low-resistance path for current to flow between CAN H and CAN L. This permits capacitance to discharge and cancel rapidly. The length of time for a capacitor to discharge is proportional to the resistance of the conductor it is discharging through. If system capacitance cannot rapidly discharge when a device is trying to transmit at a low level, then the voltage differential between CAN H and CAN L remains high. This can shut down the data bus.

Missing Terminating Resistors

If only one terminating resistor is missing from a J1939 data backbone it will likely function problem free, though if you scoped the waveform it could be noticed due to longer capacitive discharge times and increased signal reflections. However, if both terminating resistors are missing, no communication can take place. When this happens, there is an increase in the length of time required for each signal pulse pumped down the bus to neutralize, and the resulting signal reflections scramble the data bus so no transactions are possible. **Figure 12–5** compares a pulsed scope signal on a J1939 data bus equipped with both terminating resistors with one that has had both terminating resistors removed.

Testing for Terminating Resistors

Figure 12–6 shows how to test for the presence or absence of terminating resistors at the **diagnostic link connector (DLC)** cavity pins. A missing terminating resistor will result in lower resistance. However, a more common problem is an additional terminating resistor on the bus, usually caused by the installation of an aftermarket controller on the bus that has an integral terminating resistor.

Protecting Bus Integrity

When adding controllers to the bus and making interbus connections, it is critically important to use OEM-approved hardware. **Figure 12–7** shows some approved J1939 bus connectors and **Figure 12–8** shows three Tee and splice connectors.

Loss of J1939

In most trucks, if the J1939 connection is lost, the driver will be alerted by dash display alert sometimes accompanied by a chassis signal such as flashing the hazard lights. Low-bus wires are unshielded and may be repaired under some circumstances. Strictly observe the OEM instructions when repairing twisted-wire pairs. This includes maintaining the existing twisting cycles and gauge size, and using the correct solder and interbus connection hardware.

FIGURE 12–4 Terminating resistors on a J1939 data bus

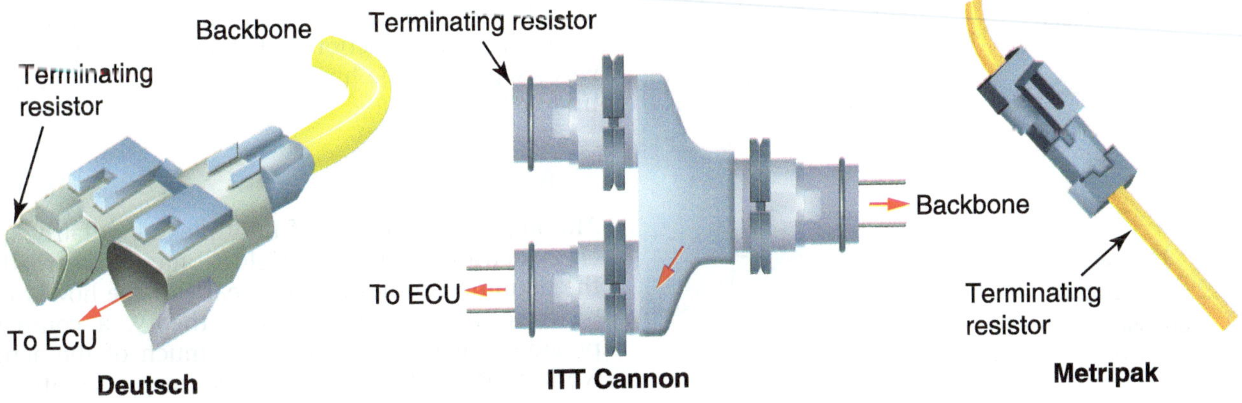

FIGURE 12–5 Comparison of J1939 scope patterns with and without 120-volt terminating resistors.

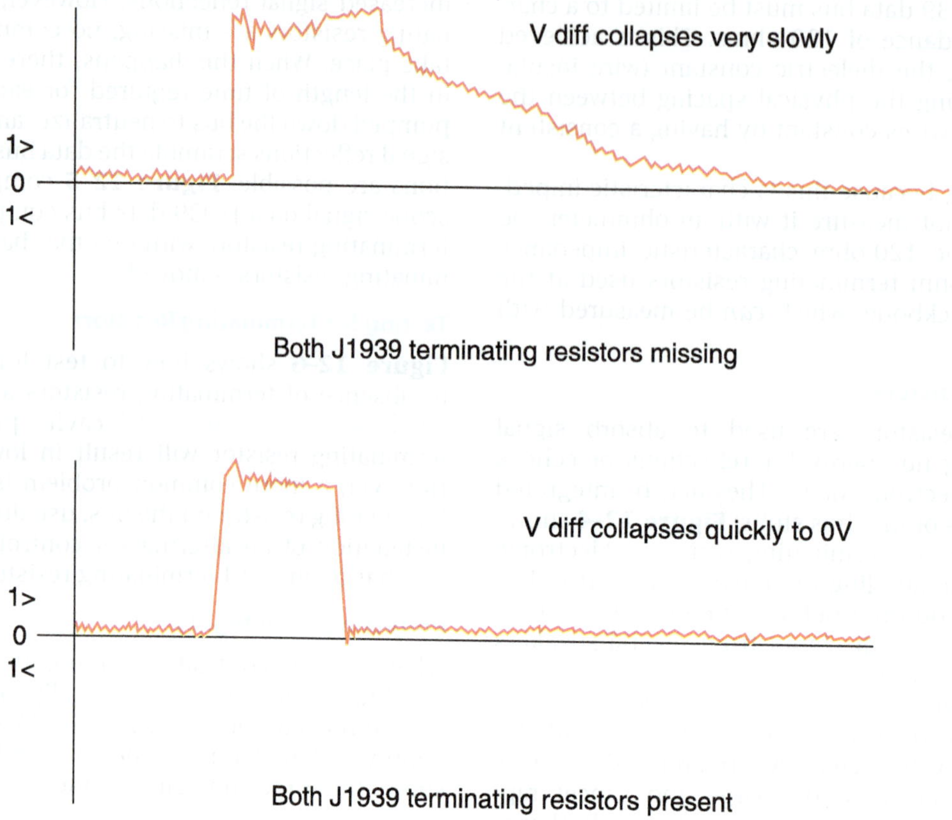

SHOP TALK

When repairing low-bus twisted wires, avoid twisting the wires together prior to soldering. Lay the wires you wish to solder so they contact each other, and then apply a small amount of tin solder. Twisting the wires together and applying a large blob of solder can create unacceptably high circuit resistance.

FIGURE 12–6 Testing a J1939 bus for the presence of terminating resistors

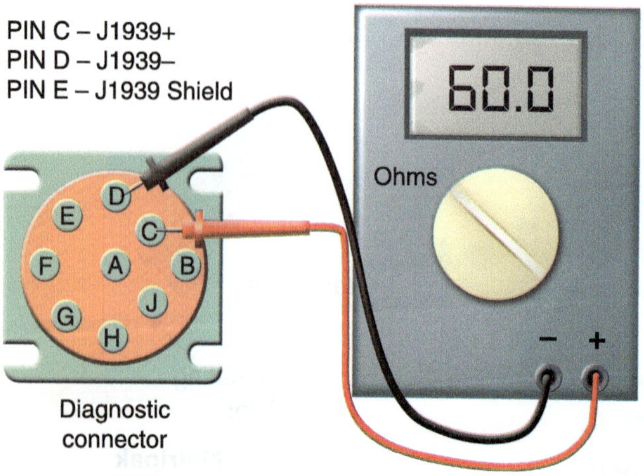

CAUTION

Most OEMs suggest that it is not good practice to attempt repairing physical damage to a data bus. Replace it according to OEM guidelines.

As truck chassis management electronics become increasingly dependent on the data bus, the consequences resulting from a data bus malfunction increase in severity. Physical damage to any data bus described as high bus (shielded) warrants replacement, not repair.

High-bus lines are shielded. A J1939 high bus consists of the twisted-wire pair (communication wires) already discussed, a shield wire, and often a couple of filler wires that function to keep the communication wires separate. The harness is usually wrapped with tin foil–like shielding.

SUMMARY OF MULTIPLEXING BASICS

Although the definition of multiplexing is fairly specific when applied to vehicle electronics, if we broaden the definition somewhat, it is easy to see how it can be effective. In organizational theory, a company president tends not to do very much of the actual work himself. He delegates those responsibilities to

FIGURE 12–7 Approved J1939 interbus connections

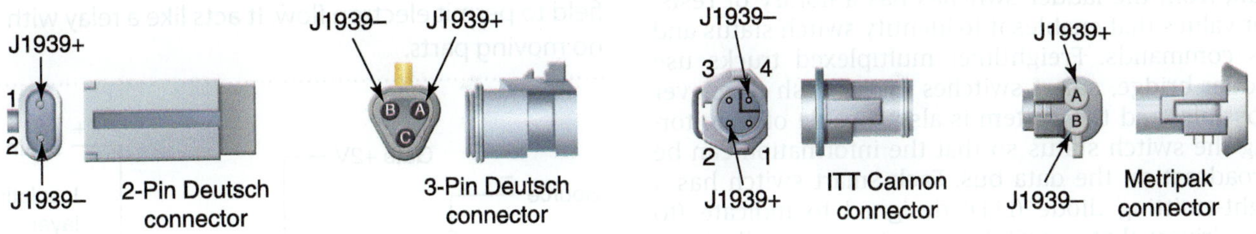

FIGURE 12–8 J1939 bus Tee and splice connections

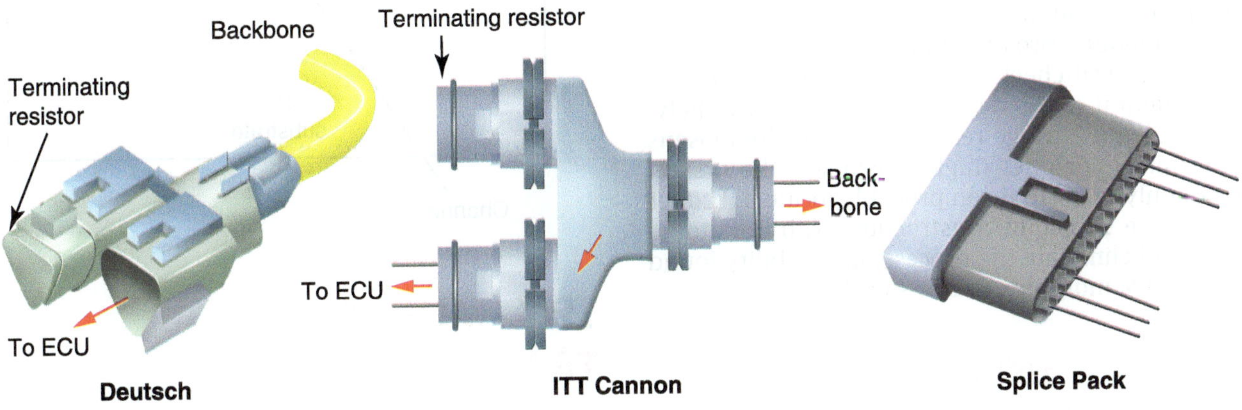

those who report to him. Communications within the company flow through a chain of command. In an effective company, reports also flow back up the chain of command.

Clearly, there are going to be occasional problems in the two-way flow of information, but in a good company there are checks and balances built into communication systems that quickly identify problems and rectify them. Using the example of a truck electrical system, rather than control a 20-amp blower motor directly through a switch that could overheat, we use a low-potential current through the control switch that acts to control a relay or virtual relay that carries the high-current load.

12.3 MULTIPLEX SWITCHING

In computer technology, for many years electronics have been used to simulate relays in devices known as **programmable logic controllers (PLCs)**. These small computers can simulate thousands of relays in a fraction of the space of a chip using elaborate stacking of transistors and silicone controlled rectifiers (SCRs). Now some of the principles used in PLCs are moving beyond the controller housing. A key factor that has moved us into a second generation of multiplexing is this "smart" switching and processing technology that we will take a closer look at in the next section of this chapter. In addition, PLCs have powerful processing capability that makes them more capable of processing the video signals used by more recent CMS and LDW systems.

STANDARD SWITCHES

Every truck on the road today is equipped with many different switches, and multiplexed trucks will not reduce this to any great extent. A standard switch operates on an electromechanical basis; moving the switch changes the electrical status of the wires connected to it. Some examples of typical standard switches would be cruise control on/off, cruise control set/resume, hazard warning switch, and ignition switch. Although a switch may be hard-wired, making it standard by classification, its status can also be broadcast on the data bus.

SMART SWITCHES

Smart switches are used to describe two distinct types of switches used in multiplex circuits. A smart switch may have some processing capability and use that to broadcast switch status onto the data bus. Alternatively, the term *smart switch* can be used to describe a **ladder switch**, so named because it contains a ladder of resistors, usually five per switch, known as a ladder bridge. The processor that receives data from the ladder switches has a library of resistor values that enables it to identify switch status and its commands. Freightliner multiplexed trucks use ladder bridge, smart switches in the dash whenever possible, and the system is also capable of monitoring the switch status so that the information can be broadcast on the data bus. Each smart switch has a light-emitting diode (LED) designed to indicate (to the driver) that a switch request has actually been effected. Sometimes the LED will flash while a particular action is in the process of being effected and stay on when completed. Smart switches can be toggle (two-position on/off), multiposition, or spring-loaded, momentary operation.

The module broadcasting to the data bus is designed to self-check smart switch operation and signal a fault if one is detected. It does this by analysis of the ladder bridge resistances. If a wire loosens or a terminal corrodes (high resistance), the system will not only know it, it can probably locate it. Ladder switches are simple in construction, with little to go wrong, and eliminate the processing capability found in some OEM multiplex circuit switches.

SHOP TALK

When troubleshooting, if you disconnect a smart switch, a code will be logged immediately. Always use the system self-diagnostics to locate problems.

FIELD EFFECT TRANSISTORS

Now it is time to take a look at what is happening at the other end of the processing circuit. First, we should examine the role of **field effect transistors (FETs)**. These are becoming commonplace in truck chassis electronics because they are inexpensive to manufacture and function reliably. FETs are electronic relay switches that perfectly complement circuits that are controlled by smart switches. In their simplest formats, there are two types of FETs: N channel and P channel. The channel behaves as a resistor that can conduct current from the source side to the drain side. The gate controls the resistance and, therefore, the operation of the semiconductor device by saying whether, or how much, current flows through the device. So, depending on the gate voltage, the FET can be used either as a straightforward switch (it is in most current truck applications) or as an amplifier. As a relay, the FET has some great advantages in that gate channel resistance is so high that, firstly, there is almost no current flow in the gate-current circuit, and, secondly, there is all but no effect on external components connected to the device. **Figure 12–9** shows a schematic of an FET.

FIGURE 12–9 FET operation: An FET uses a positive charge to the gate, which then creates a capacitive field to permit electron flow. It acts like a relay with no moving parts.

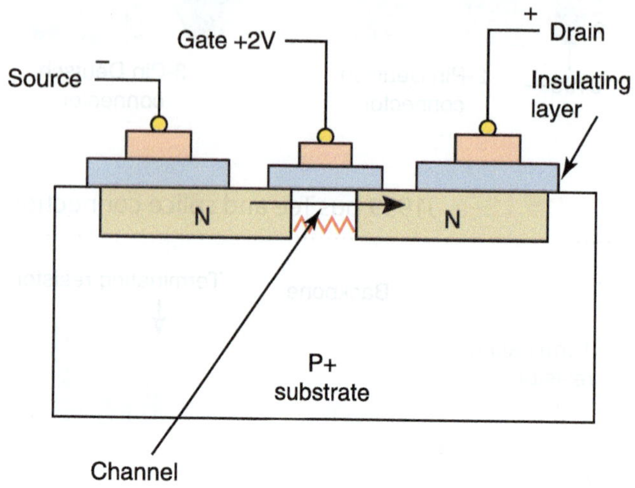

12.4 MULTIPLEX TRANSACTIONS

Next is a discussion on how some typical multiplexed communications take place.

WAKE UP

In the same way that the electronics in your automobile "wake up" (usually on door open/dome light), the multiplexing electronics in a truck are designed to reactivate following a period of inactivity. Wake-up signals are sourced from the ignition switch, dome light, brake lights, anti-theft circuit, and headlights.

SIMPLE MULTIPLEXING TRANSACTION

To show how a simple multiplexing transaction takes place, we are going to take a look at a simple switch operation from switch to signal to outcome. The suspension dump smart switch is used to illustrate the execution of a data bus–enabled command. The mechanical objective (the desired outcome) of this operation is to exhaust air from the chassis air suspension. The operator actuates the suspension dump ladder switch. This causes the dash suspension dump switch to start flashing and alters the resistance status on the ladder bridge so a *packet* is broadcast to the data bus. The data on this packet means nothing to any of the modules on the data bus until it is delivered to the module with the algorithm designed to process this input.

Using the example of a Freightliner multiplexed chassis, this would be the bulkhead module (BHM).

The BHM then broadcasts a command signal over the J1939 data bus to another module with some switching capability, the chassis control module (CHM). The CHM in turn generates an output signal to the FET that acts as a relay for the suspension dump solenoid. The signal puts the FET in forward bias, completing the solenoid ground circuit and energizing the coil. Source power for the solenoid is the power distribution module. At the point the action is completed—that is, when the suspension dump has been effected and air drains from the suspension air bags—the dash LED on the suspension dump switch ceases to flash and remains on. This is a simple data bus transaction that is becoming more commonplace as multiplexing technology develops.

SHOP TALK

When adding electrical loads to a multiplexed truck chassis, always consult the OEM service literature. Splicing into circuits can cause electrical problems that become very difficult to troubleshoot and can destroy controller electronics.

CAUTION

Never splice into existing fuses in chassis power distribution modules to source a V-Bat (battery voltage) requirement. In dealerships today, it is not uncommon to hear horror stories that result when a truck driver splices into a "hot" wire to power up his CB radio. Most OEMs provide non-dedicated terminals in their power distribution module that can be used for auxiliary electrical requirements such as CBs. Use the OEM literature and wiring schematics when connecting auxiliary electrical equipment.

DIAMOND LOGIC™

The International Trucks Diamond Logic multiplexing system functions a little differently from the Freightliner system. The components of the Diamond Logic multiplex system are the electronic system controller (ESC), switch packs, remote power module, and remote air solenoid module.

Electronic System Controller

The ESC is used to manage data bus activity on the International Trucks system. The logic and programming that is specific to the operation of the data bus is logged in the ESC module.

An ESC-managed data bus enables a range of optional features to be added to the truck by checking boxes on a computer screen or by dragging and dropping graphics on a virtual electronic wiring diagram displayed on the EST screen. After adding options to the chassis electronics, International Trucks requires that users access their data hub and upload the modified data files. As with any other type of modification to the vehicle files, the purpose is to make any new or modified files available to other service facilities and technicians.

International Trucks, in common with other OEMs, uses satellite communications to network their data hub with the chassis data bus electronics. This can mine data for condition-based maintenance (CBM) and prognostics profiles and may eventually eliminate the hard-wire requirement for upload and download transactions with the International data hub. The system would function similarly to the fleet satellite communications in current use.

Switch Packs

The switch packs contain long-life microswitches located behind what looks like a typical truck dash rocker switch. The ESC communicates the switch packs monitoring the switch status. Switch status is, therefore, monitored by the ESC and then broadcast over the data bus to be picked up by other modules networked to the data backbone such as engine, transmissions, brakes, and the remote air solenoid modules.

Each truck may have up to two 6-switch packs and one 12-switch pack. Some low-optioned, late-model trucks may also use a 3-switch pack that has the same physical dimensions as a 6-switch pack. The switch packs are designed to be daisy-chained together; this requires that the first switch pack connect directly to the cab harness and the remainder connect to the previous switch pack in series. The switch packs must be connected in correct order or addressing problems will result.

Remote Power Module

The remote power module has two connectors. one for body (chassis) switch inputs and one for power outputs. The connectors have six inputs and six outputs that provide the multiplex system with some versatility. Three power modules are available, which can provide up to eighteen 20-amp outputs.

A real advantage of the International Trucks system is that using its Diamond Logic software and an EST, electronic gauges and rocker switches can be relocated by dragging and dropping images on the EST display screen. This means that the physical location of switches is completely soft and can be arranged according to user preference. Aftermarket equipment can be installed and powered up by either a rocker switch in the cab or by running a single wire into a remote switch input. Connecting the selected remote switch input to battery voltage will instruct the module to enable 12-volt power to the selected power module output. Biasing the switch to ground opens the electrical supply to the added circuit.

Remote Air Solenoid Module

In much the same way as in the Freightliner system, the air modules are a means of controlling air-operated accessories, such as differential locks, suspensions, or transaxles, off the data bus. A system air supply is routed to one input on the air module. Modular air solenoids are supplied by this single air source. The on/off status of the solenoids is controlled by the ESC using the data bus.

VIRTUAL FUSING

Navistar FETs are designed to feed back the current values they conduct to the controller that actuates them. This allows a microprocessor to shut off an FET if it is conducting more current than it is specified for. For instance, if too many lights were installed on a circuit, causing an excess current draw, the controller could shut down that circuit. Virtual fusing is designed to have performance characteristics identical to circuit breakers. The difference is that processor logic is used to determine the manner in which virtual circuit breakers cycle. We classify them into type 1 (cycling) and type 2 (noncycling) in the same way electromechanical breakers are categorized.

Type 1 Virtual Fusing

Type 1 is used on headlamp and wiper circuits. In a current-overload condition on the circuit, the ESC first turns the circuit off; then after ½ second, it retries the circuit. If the current overload no longer exists, the circuit functions normally. However, if the overload condition continues, the ESC doubles the length of time before each retry, 1 second, 2 seconds, 4 seconds, up to 512 seconds. If the 512 recycle is achieved, the ESC shuts down the circuit until the ignition key is cycled.

Type 2 Virtual Fusing

Type 2 virtual circuit-breaking logic is used on all the remaining FETs. Should a circuit be subject to a current overload, the ESC shuts off the circuit, then retries after ½ second. However, using type 2 virtual breaker logic, if the short or overload still exists after the initial retry, the ESC shuts off the FET and does not initiate a retry until the ignition key is cycled. If three successive key cycles take place with an overloaded circuit in memory, the ESC will only briefly (100ms) permit the FET to be switched on after each key cycle just to test the overload condition.

BENEFITS OF NETWORKING

There are four primary benefits of networking a truck electronic system:

1. Greatly decreased hard-wire requirement, reducing the size of the wiring harnesses. This impacts on cost, weight, reliability, and serviceability.
2. Sensor data such as vehicle speed, engine temperature, and throttle position is shared, eliminating the need for redundant sensors.
3. Networking allows greater vehicle content flexibility because functions can be added by making simple software changes. Existing systems would require additional modules and additional I/O pins for each function added.
4. Many additional features can be added at little or no additional cost. For instance, driver preference data, once installed in memory, can be routed on the data bus to multiple processors to share such diverse information as seat preference, mirror positions, radio station presets, and engine governor type (LS or VS).

12.5 ACCESSING THE DATA BUS

Truck technicians are required to access the chassis data bus to read, troubleshoot, and reprogram the systems networked to it. The photo sequence in this section illustrates the procedure required. A number of different generic and proprietary tools may be used to make the connection, but commonly today a handheld, laptop, or portable computer system is required.

Whatever kind of computer is used, it must be loaded with either generic reader or OEM read/diagnose/reprogram software. The connection is made using one of two types of standardized data connector.

J1587/1708 CONNECTOR

A J1587 data bus was the original truck data backbone; it is accessed using a 6-pin *Deutsch connector.* J1587 is an SAE standard that governs the communications protocols used for data transactions. J1708 governs all the hardware standards used on a J1587 data bus. **Figure 12–10** shows the cavity pin assignments used on a J1708 connector.

J1939 CONNECTOR

The J1939 data bus is more recent and it covers both hardware and software standards. Because a truck

FIGURE 12–10 J1587/1708 connector cavity pin assignments

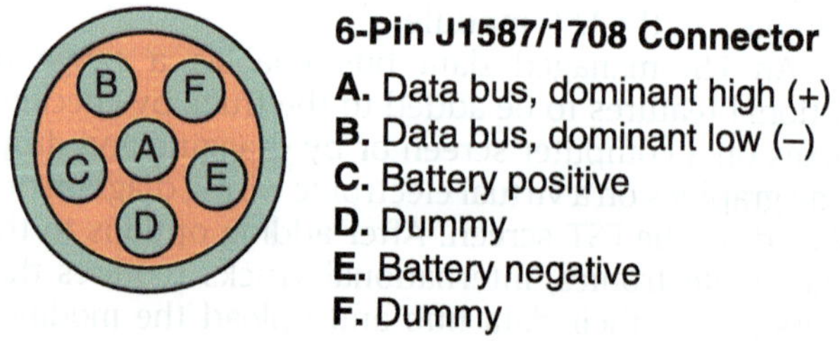

FIGURE 12–11 J1939 connector cavity pin assignments

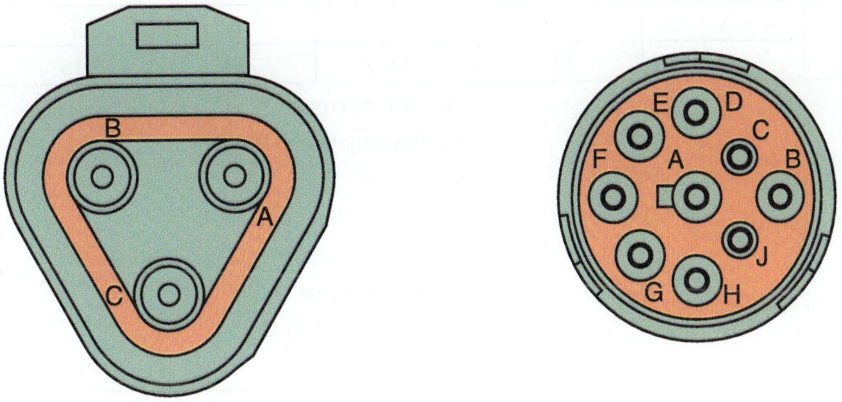

3-Pin J1939 Connector

A. CAN busline, dominant high (+)
B. CAN busline, dominant low (–)
C. CAN ground

9-Pin J1939 Connector

A. Battery negative
B. Battery positive
C. J1939 CAN busline, dominant high (+)
D. J1939 CAN busline, dominant low (–)
E. CAN shield
F. J1587 busline, dominant high (+)
G. J1587 busline, dominant low (–)

chassis equipped with a J1939 data bus may also equipped with the older J1587/1708 data bus, the data connector used has some extra pins. There are three types of J1939 currently in usage:

- Black 9-pin Deutsch (pre-2014)
- Green 9-pin Deutsch (2014 but back-compatible)
- 16-pin J1962 connector (2014 Volvo-Mack)

The last two DLCs were introduced by OEMs for **model year (MY)** 2014. The cavity pin assignments for a black J1939 DLC are shown in **Figure 12–11**. High-bus connections within the data bus backbone are made on a 3-pin connector with cavity pin assignments as shown in Figure 12–11, whereas low-bus connections use a 2-pin connector (see Figure 12–7).

Green 2014 J1939 Connector

The 2014 green DLC is also a Deutsch connector but is designed to be backward-compatible while blocking access to the earlier black version. The blocking feature is accomplished by using a smaller receptacle bore on the F-cavity assignment. **Figure 12–12** shows face on views of the 2014 J1939 DLC and **Figure 12–13** shows a side view of the same set of connectors.

J1962 Connector

There is consensus that the nine pins in a J1939 DLC are insufficient for the ever-increasing requirements of data bus access and for some time, alternatives have been discussed at SAE and TMC meetings: The automotive 16-pin J1969 connectors was one of the alternatives under discussion. Despite the fact that as of 2014 no general agreement had been finalized, Volvo-Mack elected to go ahead and adopt the automotive J1962 connector for 2014. For this reason, the cavity pin assignments for a J1962 connector are identified in Table 12–1.

FIGURE 12–12 Face on view of green 2014 J1939 DLCs and receptacles

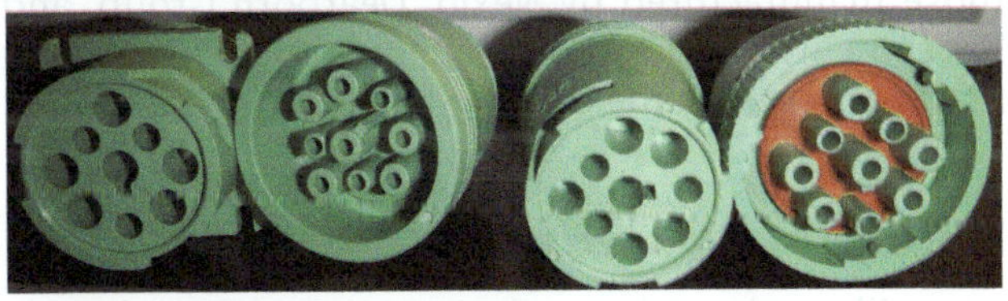

FIGURE 12–13 Side view of green 2014 J1939 DLCs and receptacles

TABLE 12–1 J1962 Cavity Pin Assignments

1	2	3	4	5	**6**	7	8
9	10	11	12	13	**14**	15	16

1. OEM discretion	9. OEM discretion
2. Bus positive + J1708	10. Bus negative – J1708
3. DLC+	11. DLC –
4. Chassis ground	12. OEM discretion
5. Signal ground	13. OEM discretion
6. **J1939 + high**	14. **J1939 – low**
7. K-line of ISO 9141	15. L-line of ISO 9141
8. OEM discretion	16. V-Bat

Communications Adapters

To connect all PCs and some HH-ESTs to a J1939 data bus, in addition to the appropriate DLC, a serial **communications adapter (CA)** is required. Although most CAs claim to be **RP1210** compliant, the truth is some OEM devices are not and will actually block access. CAs were discussed in Chapter 11 and they are mentioned here again because it is important not to condemn a CA without first verifying that this really is what's causing the problem. Generally, generic RP1210 compliant CAs are recommended; those manufactured by Nexiq, Dearborn Group, and Noregon will serve most general requirements but may not access some OEM systems.

MAKING THE CONNECTION

You will need an electronic service tool (EST) loaded with the OEM software plus the appropriate DLCs, CAs, and cabling. Most troubleshooting, customer programming, and communication transactions can be undertaken without network access. Examples of ESTs are:

- Handheld reader/programmer (scan tool) equipped with a heavy-duty reader cartridge
- Handheld reader/programmer (scan tool) equipped with OEM-specific software
- Laptop/notebook personal computer (PC) and OEM software
- PC station and OEM software

Identify SA or MID

Depending on the EST software you are using or if you are using general heavy-duty reader software, after connecting the data connector, you will have to identify the controller address you want to communicate with; this is assigned a **source address (SA)** in J1939 or **message identifier (MID)** in J1587. This means the actual electronic system on the data bus you want to talk to. The following are some examples of typical controller addresses used on truck data buses:

- Engine controller SA 00/MID 128
- Transmission controller SA 03/MID 130
- Shift console controller SA 05
- Brakes and traction control SA 11/MID 136
- Engine retarder SA 15
- Climate (cab) control SA 25
- Body controller SA 33
- Off-vehicle gateway SA 37
- Exhaust emissions controller SA 61
- Instrument cluster controller SA 23/MID 140
- Cab controller SA 49
- Vehicle (chassis) management system SA 71/ MID 142

All the SAs and MIDs connected to the chassis data will be displayed as menu choices on the EST you are using to read the system. If you wish to access the transmission electronics, you will scroll to the SA 03 (MID 130) menu option and select it. Once you have entered the transmission electronics, everything displayed on the EST display screen will only relate to the transmission. Menu choices will route you through the **subsystem identifiers (SIDs)** and parameter groups (PGs) groups/parameters (PIDs) specific to the transmission. Many current trucks use a **central gateway (CGW)** module to coordinate bus activity and permit communication between buses that use different communication protocols (languages). **Figure 12–14** shows the typical physical locations of controllers on a typical medium-duty truck chassis and **Figure 12–15** shows a firewall-mounted, **chassis management module (CMM)** which, depending on the OEM, can be known by names such as **body control module (BCM)**, **signal detect and actuation modules (SAM)**, and CGW modules.

FIGURE 12–14 Location of the modules on a typical medium-duty truck, data bus

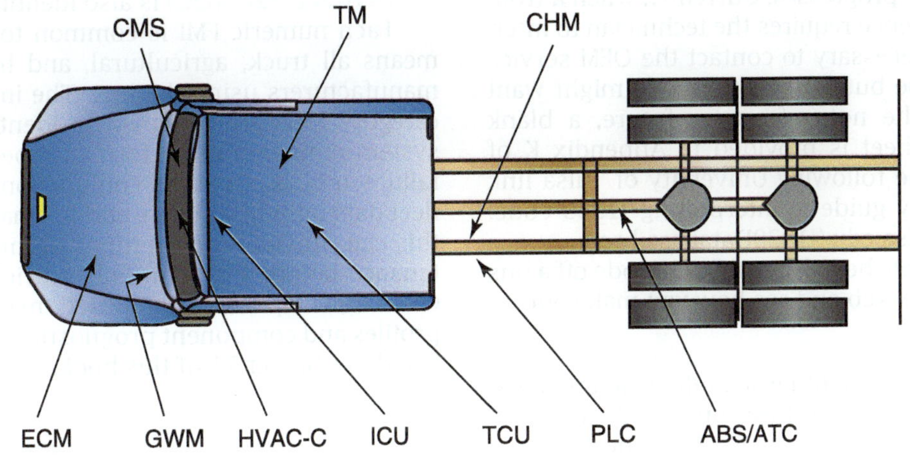

FIGURE 12–15 Location of a typical body controller module on the vehicle firewall

Parameter Groups

In J1939, a **parameter group (PG)** is a set of parameters belonging to the same topic within a controller; they also share the same transmission rate. Each PG has a message length that varies between 8 bytes (minimum) and 1,785 bytes (maximum). Each PG is assigned a unique **parameter group number (PGN)** that can be identified in J1939 message code. When PGNs are broadcast on the bus, the message can be directed to all of the controllers (SAs) with addresses on the bus, in which case it is described as a **global PGN**.

Alternatively, PGNs can be targeted to a single SA on the bus in which case it is described as a **specific PGN**. For purposes of troubleshooting, a **suspect parameter group (SPG)** is assigned to either a value/status condition or a component when an abnormal controller condition is identified. The source (SRC) is also identified in a message that broadcasts a **suspect parameter number (SPN)**. The function of an SPN is to:

- Translate PGNs
- Identify the DTC and related SIDs and FMIs

It is important to understand what these terms represent when undertaking more advanced level J1939 diagnostics.

Hexadecimal Code

Some of the more challenging problems that can arise when troubleshooting data bus problems may require the technician to interpret **hexadecimal code**. The hexadecimal system is useful because it can represent each byte as a pair of hexadecimal digits making it easier to read than binary numbers. Sixteen numeric digits are required: Digits 0 to 9 are represented by their decimal numeric equivalents while digits 10 to 16 are represented by the alpha letters A through F. To translate hexadecimal to binary, convert each hexadecimal digit into its 4-bit binary equivalent.

Deciphering J1939 hexadecimal code is beyond the scope of this textbook, but this is likely to change as the technology progresses. Currently, when a troubleshooting sequence requires the technician to interpret code, it is necessary to contact the OEM service operations hotline but for students who might want to prepare for the not-so-far-away future, a blank J1939 decoder sheet is provided in Appendix K of this book. Use the following University of Tulsa link for a user-friendly guide to interpreting J1939 code: http://tucrrc.utulsa.edu/J1939Database.html.

Download some hexadecimal J1939 code off a bus network or off the web and see how you make out.

Identify FMIs

If you were troubleshooting a fault in a transmission with a controller address on J1939, a source (SRC) message would be broadcast on the bus as a **diagnostic trouble code (DTC)**, which would route you to the problem controller (SA) and enable you to navigate through the appropriate PGNs/PIDs or SIDs to locate the failure mode indicator (FMI) that had triggered the fault code. A more detailed analysis of the structure of a DTC appears in Chapter 11. FMIs are designed to categorize failures and impending failures. Currently, there are 30 numeric FMIs though not all are used. Along with the DTC and the FMI, an **occurrence rate (OC)** is also identified.

Each numeric FMI is common to J1939 and that means all truck, agricultural, and heavy-equipment manufacturers using the bus. The impending failure category FMIs are designed to identify component/system reduction in performance before an outright failure occurs. Used in conjunction with OEM and fleet data mining software and telematics/downloads, this can provide opportunities of intervention maintenance before a system fails outright. It also helps OEMs develop condition-based maintenance (CBM) profiles and component prognostics. Current FMIs are listed in Chapter 11 of this book.

CONNECT TO CHASSIS DATA BUS

Connecting to the chassis data bus is easy, and to comprehensively read every SA/MID networked to the bus you need only to have generic heavy-duty reader software and an EST. **Photo Sequence 5** shows you where to look for the connector and how to make the connection.

Photo Sequence 5

ACCESSING A TRUCK DATA BUS WITH AN EST

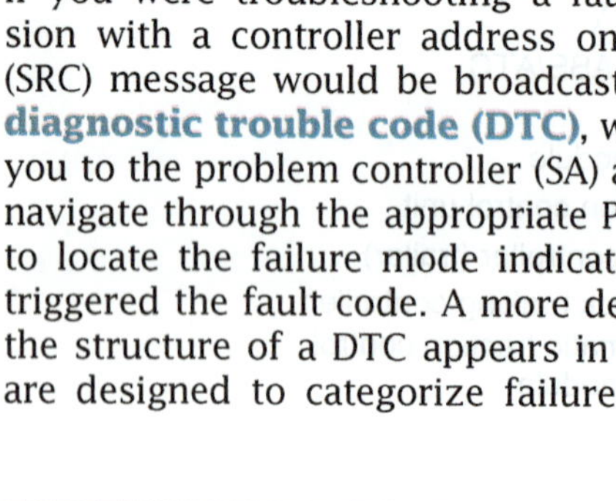

P 5–1 Locate and identify the chassis bus connector. The TMC recommends that this be located on the left side of the steering column, but this can be difficult to find, as on this older truck that uses a J1708 6-pin connector.

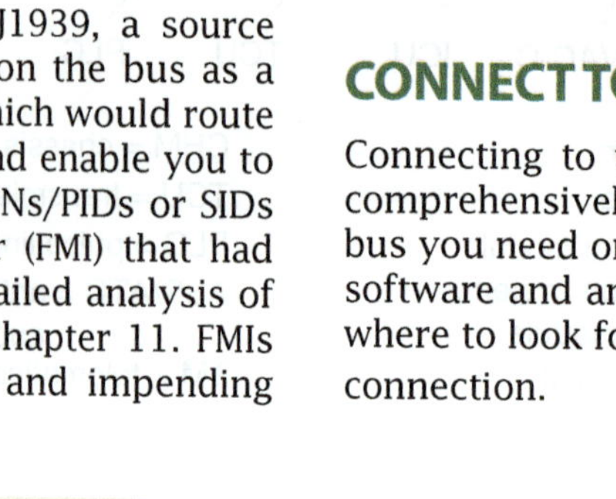

P 5–2 This 2013 Navistar Eagle positions a J1939 connector next to the ignition key making it easy to locate, and it stays protected from dirt and corrosion.

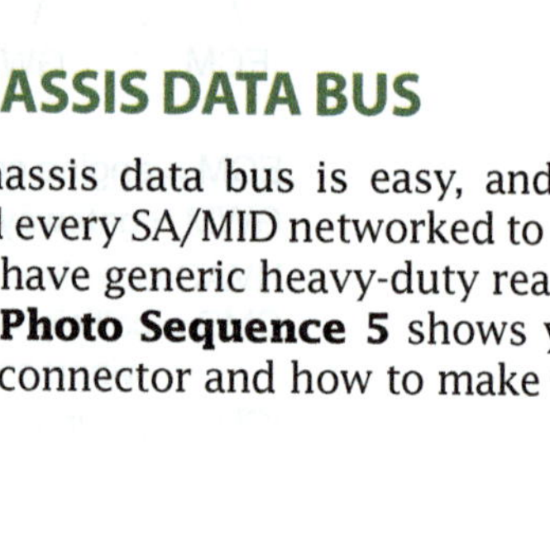

P 5–3 The green-colored J1939 9-pin connector shown in this image is a 2014 data bus connector: it is backward-compatible with earlier J1939 9-pin connectors, but the socket is designed to block pre-2014 black connectors.

P 5-4 To connect to this 2013 Navistar Eagle chassis data bus, a black J1939 connector is required. Make sure to orient the connector pins to the dash mounted socket.

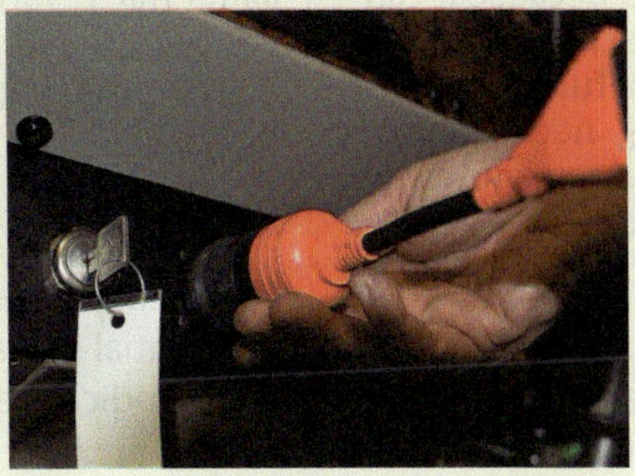

P 5-5 Insert the connector into the dash connector socket and lock it with a clockwise twist.

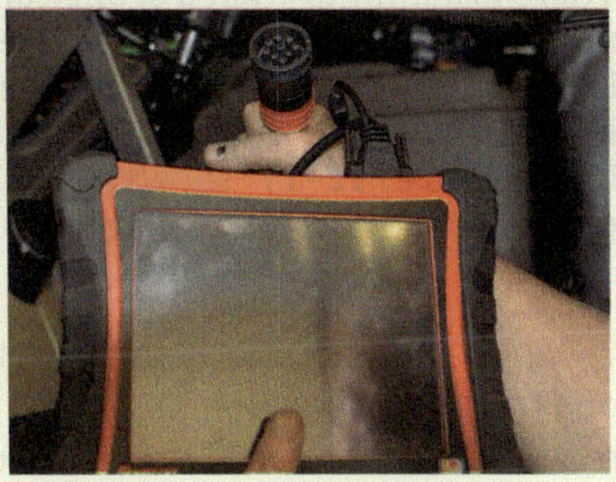

P 5-6 Most HH-ESTs, such as this Snap-On iQ, do not require a communications adapter (CA). So to start working with the chassis data bus, switch on the EST and switch on the vehicle ignition circuit. The EST onscreen menu will then take over with prompts and options.

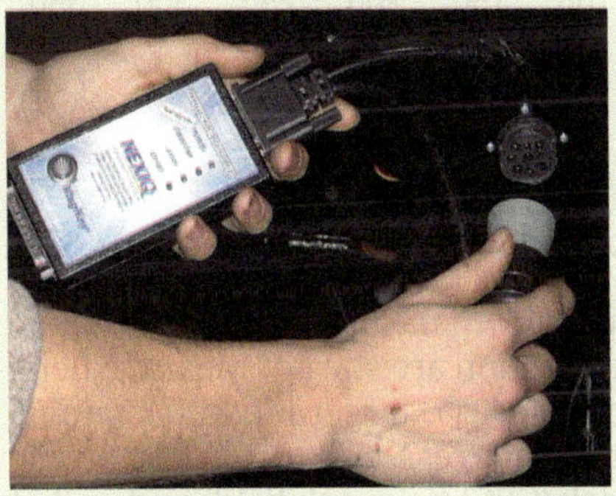

P 5-7 If the EST is a PC with proprietary software, a CA is required to make the connection. LEDs on the CA will indicate the connection protocol (J1708 or J1939) and the communication status.

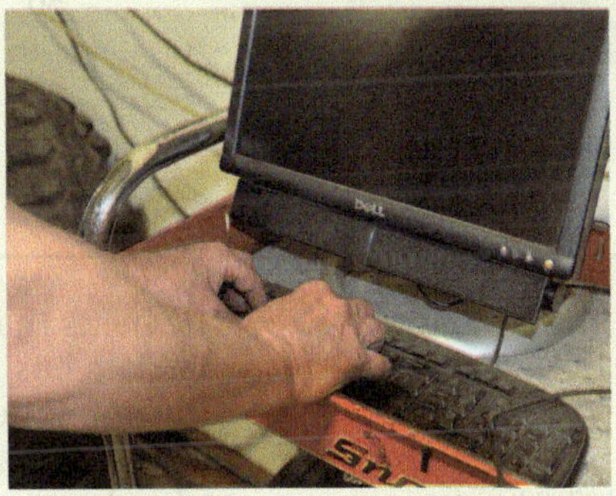

P 5-8 Shop computers can be laptops or dedicated desktops mounted on a shop cart, such as the unit shown in the above image.

P 5-9 Entry page to a typical OEM service information and diagnostic software. Almost all recent systems are required to be online, before the full capacity software suite is made available to the technician.

SCOPING THE BUS

The presence or absence of bus node activity can be verified by using a **scopemeter** (also known as a lab-scope) such as the Pico kit shown in **Figure 12–16**. There is a limit to what a scope can do because it cannot be used to interpret message packets, but as a first-level diagnostic tool it can be useful in identifying and isolating noncommunicating nodes on the bus. **Figure 12–17** shows a J1939 message scope pattern: note that the voltage differential of the core message remains within 2 volts; the 3-volt spikes indicate end of message stop bits. J1939 uses programming languages such as C, C++, and Python. In the binary messaging used in message packets, the 0 represented by the lower-voltage state is dominant while the 1, represented by the higher-voltage state, is recessive. In other words, on the bus, zeros rule: two successive 0s override a recessive 1.

FIGURE 12–16 A PicoScope meter that consists of a CA and connection hardware: the software required to use this is a free web download.

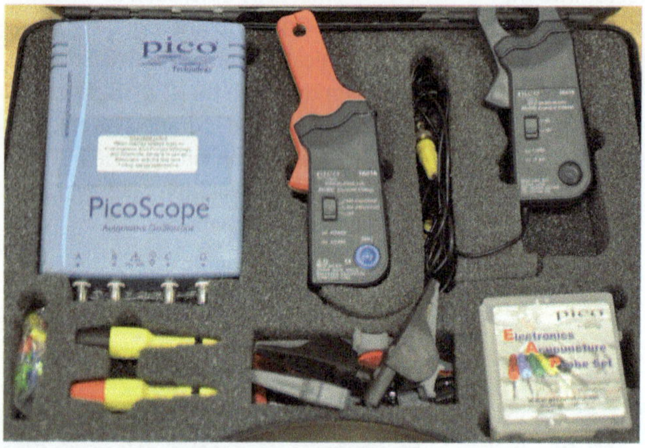

FIGURE 12–17 Message scope pattern of a J1939 bus packet. Note that the message core shows 2-volt differential spreads: the pair of 3-volt differential spikes indicate end-of-message stop bits.

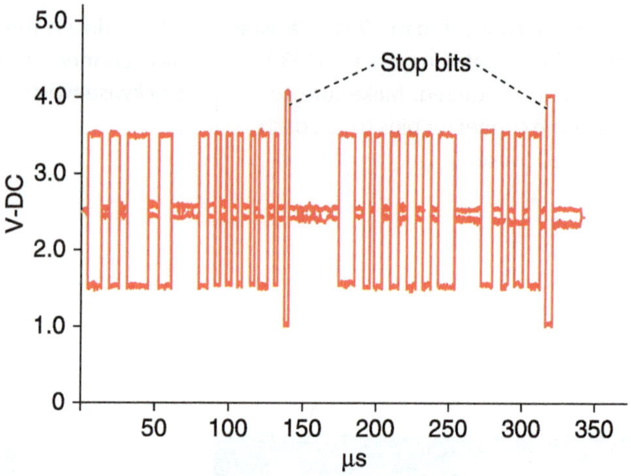

SUMMARY

- Multiplexing refers to the ability of electronic components to exchange information by means of a network known as a data bus.
- Multiplexing can eliminate miles of chassis harness wiring and duplication of hardware devices such as throttle position sensors by giving electronic subsystems a common communication language and, by using a data bus or information highway, allows data signals to take the place of hard wire in the electronic input and output circuits.
- The data bus acts as the "information highway" in a multiplexed electronic truck chassis.
- A "smart" ladder switch contains a ladder of resistors: The processor that receives a signal from the switch can interpret switch status data by comparing ladder resistances with a programmed library of resistance values that identify the switch, its status, and circuit integrity.
- The fields that usually make up a J1939 data frame of a data bus transaction are start of message field, arbitration field (priority), target identifier (global or specific), source address (SA), data field, cyclic redundancy check (CRC) field, ack field, and end of frame field.
- Access to a truck data bus is obtained by using one of four types of DLC. In addition to the first-generation 6-pin Deutsch connector (J1587/ 1708 data bus) and a black 9-pin Deutsch connector (pre-2014 J1939 data bus), a green 9-pin Deutsch connector and 16-pin J1962 were added for post-2014 J1939 data buses.
- The green J1939 DLC is backward compatible with the pre-2014 black DLC but a green J1939 receptacle is designed to block access by the earlier DLC.
- Trucks equipped with a J1939 data backbone manufactured before MY 2010 were also fitted with a J1587/1708 bus. Since MY 2010, some OEMs no longer support the J1587 protocol.

- When accessing a specific ECM/ECU on a truck data bus with multiple electronic systems you must first select the controller you want to work with (SA or MID) if using generic reader software (such as a Nexiq EST).
- A scopemeter is a one way of identifying node activity on a data bus but it cannot be used to interpret message packets.

REVIEW QUESTIONS

1. What is used to rate the speed of transmission on a data bus?
   a. bandwidth
   b. K-baud
   c. protocol
   d. CAN
2. Which of the following terms is used to rate the data volume that can be transmitted on a data bus?
   a. bandwidth
   b. K-baud
   c. frequency
   d. bus arbitration
3. When observing a scoped view of a J1939 message packet, what is the usual voltage differential spread of content data if the bus is functioning normally?
   a. 1.0 volt
   b. 1.5 volts
   c. 2.0 volts
   d. 5.0 volts
4. Which of the fields in a data packet contains the priority rating of the message?
   a. start of frame
   b. arbitration
   c. control
   d. cyclic redundancy
5. Which field in a data packet is used by receiver stations to indicate that they have received an uncorrupted message?
   a. arbitration
   b. control
   c. cyclic redundancy
   d. ack
6. What is used to indicate the end of a data packet message?
   a. CRC field
   b. ack field
   c. EOI
   d. EOF field
7. What current SAE J-standard is used as the backbone of a multiplex data bus in current medium- and heavy-duty highway trucks?
   a. J1850
   b. J1939
   c. J1962
   d. J1667
8. Which of the following best describes the role of an FET in a smart switch transaction?
   a. transformer
   b. electronic noise suppression
   c. relay or amplifier
   d. signal filter
9. What will result if a smart switch is disconnected at the connector?
   a. vehicle will stop
   b. data bus will shut down
   c. default mode operation
   d. a fault code will be logged
10. How many cavity pins are used in the J1939 Deutsch DLC?
    a. one
    b. four
    c. six
    d. nine
11. Technician A says that power line carrier (PLC) technology allows trailer ABS electronics to communicate with the tractor data bus. Technician B says that PLC signals are superimposed over the blue auxiliary power line in a standard 7-pin trailer connector. Who is correct?
    a. Technician A only
    b. Technician B only
    c. both A and B
    d. neither A nor B
12. Which of the following colors would be associated with a J1939 twisted-wire pair?
    a. white and black
    b. green and yellow
    c. red and green
    d. gray and white
13. Which of the following is true of a post-2014 J1939 DLC?
    a. black 6-pin Deutsch connector
    b. green 6-pin Deutsch connector
    c. black 9-pin Deutsch connector
    d. green 9-pin Deutsch connector
14. Which OEM uses a 16-pin J1962 DLC to access its post-2014 J1939 data bus?
    a. Freightliner
    b. Navistar
    c. Volvo-Mack
    d. Paccar
15. Technician A says that the F-cavity in a green J1939 9-pin receptacle is smaller than in the black pre-2014 equivalent. Technician B says that a green 9-pin J1939 DLC cannot connect to a black 9-pin receptacle. Who is correct?
    a. Technician A only
    b. Technician B only
    c. both A and B
    d. neither A nor B
16. What is the usual source address (SA) on a J1939 bus for an electronic transmission controller?
    a. 00
    b. 01
    c. 03
    d. 05

17. How many cavity pins are in a J1962 DLC?
 a. 6 c. 16
 b. 9 d. 24

18. Technician A says that a ladder switch is type of smart switch that uses internal resistors to signal switch status. Technician B says that some smart switches have processing capability. Who is right?
 a. Technician A only c. both A and B
 b. Technician B only d. neither A nor B

19. When a global PGN is broadcast on a J1939 bus, which controller receives the data?
 a. the engine controller only
 b. the brake controller only
 c. the body controller only
 d. any controller with an address on the bus

20. Which cavity pin on a 9-pin J1939 connector is used for the J1939 dominant low (–) transmission?
 a. A c. C
 b. B d. D

13

Prerequisites: Chapters 1, 2, and 3

HYDRAULICS

OBJECTIVES

After reading this chapter, you should be able to:

- Explain fundamental hydraulic principles.
- Apply the laws of hydraulics.
- Calculate force, pressure, and area.
- Describe the function of pumps, valves, actuators, and motors.
- Describe the construction of hydraulic conductors and couplers.
- Outline the properties of hydraulic fluids.
- Identify graphic symbols.
- Interpret a hydraulic schematic.
- Perform maintenance procedures on truck hydraulic systems.
- Identify safe practices when working with mobile hydraulics.
- Safely troubleshoot hydraulic leaks and understand the danger of hydraulic pinhole injection injuries and how to respond if you suspect such an injury.

KEY TERMS

accumulator
Bernoulli's Principle
cam ring
check valve
closed-center
closed-loop
coupler
dash size
directional control valve
ferrule
flow
flow control valve
fluid
force
head
head pressure
hydrodynamics
hydrostatics
kilopascals (kPa)
laminar flow
land
linear actuators
manometer
open-center
Pascal's Law
pilot-operated relief valve
pinhole injection
pressure
prime mover
quick-release coupler
ram
reservoir
rotary actuators
servo
spool valve
swashplate
torque
Torricelli's tube
vane
work

INTRODUCTION

Fluid power refers to the practice of using gases or liquids in confined circuits to transmit power or multiply human or machine effort. A **fluid** is any substance that has fluidity, that is, able to flow and yield to alter shape. For our purposes, any gas or liquid can be classified as a fluid. In a modern truck, we use both pneumatics and hydraulics in different ways. Here are some examples:

Pneumatic circuits:

- Air brakes
- Air suspensions
- Wiper motors
- HVAC controls
- Air starters
- Control circuits in standard transmissions
- Actuators in automated transmissions

Hydraulic circuits:

- Hydraulic brakes
- Power steering systems
- Automatic transmissions
- Fuel systems
- Wet-line kits for dump boxes
- Torque converters
- Lift gates

The term *hydraulics* is used to specifically describe fluid power circuits that use liquids—especially formulated oils—in confined circuits to transmit force or motion. The practice of hydraulics predates recorded history and is thought to have played a role in the construction of the Egyptian pyramids. In this chapter we focus on understanding basic hydraulics. In Chapters 28, 30, and 31, the focus is on pneumatics as they apply to truck brake systems.

The truck technician is required to have some knowledge of hydraulics to properly understand engine operation, fuel systems, and many types of transmissions, in addition to any auxiliary hydraulic apparatus on the vehicle.

Although hydraulics may have been used by the ancient Egyptians and Romans, it did not really become a science until the work of mathematician Blaise Pascal (France, 1623–1662). Pascal is so important to the understanding of modern hydraulics that most programs of study begin with what we know today as **Pascal's Law**, which states:

> Pressure applied to a confined liquid is transmitted undiminished in all directions and acts with equal force on all equal areas, at right angles to those areas.

13.1 FUNDAMENTALS

Although we do not often differentiate between them, hydraulics can be divided into two branches that we know as **hydrostatics** and **hydrodynamics**. *Hydrostatics* is the science of transmitting force by pushing on a confined liquid. In a hydrostatic system, transfer of energy takes place because a confined liquid is subject to pressure. A good example of a hydrostatic circuit would be the wet-line kit we use on trucks to actuate a dump box. *Hydrodynamics* is the science of moving liquids to transmit energy. A water wheel or turbine could be classified as a hydrodynamic device, but the best example on a truck chassis would be the torque converter used to impart drive to an automatic transmission. We can define hydrostatics and hydrodynamics as follows:

- *Hydrostatics*: low fluid movement with high system pressures.
- *Hydrodynamics*: high fluid velocity with lower system pressures.

PRESSURE

Pressure is force applied to a specific area. When a confined fluid is subject to pressure, the force applied to the area of confinement will be uniform throughout (Pascal's Law). When a confined liquid in a vessel is pushed on, the pressure that results acts evenly on all of the walls of the vessel. This characteristic makes it possible to transmit force or "push" through pipes. In hydraulics, we use liquids rather than gases as hydraulic fluids, because liquids are not easily compressed. Because of this incompressibility, we can relay action almost instantaneously so long as the circuit is full of liquid and contains no air. Pressure is usually expressed in pounds per square inch (psi) or **kilopascals (kPa)**.

Creating Pressure

Pressure is created when there is resistance to flow in a circuit. If we draw a column of water into a bicycle pump that has a plunger with a section area of 1 square inch and is equipped with a pressure gauge and then trap the water into the pump with a finger, we can say the following:

- If no force is applied to the pump plunger, the gauge will read zero.
- If 10 pounds of linear "push" is applied to the pump, the gauge will read 10 psi.
- If 15 pounds of linear "push" is applied to the pump, the gauge will read 15 psi.
- If the finger closing the pump outlet is removed, the water will be displaced as force is applied to the pump.

We can conclude from this that the pressure of a confined liquid is always in proportion to the applied force. In addition if the restriction (in our example here, the finger) is removed, gauge pressure drops to zero.

Atmospheric Pressure

The Earth's atmosphere extends upward for around 50 miles (80 km) and we know that air exerts a force because of its weight. A column of air measuring 1 square inch extending 50 miles into the sky would weigh 14.7 pounds at sea level as shown in **Figure 13–1**. We express atmospheric pressure at sea level as 14.7 pounds per square inch absolute (psia) or 101.3 kPa, its metric equivalent. If we stood on a

FIGURE 13–1 Atmospheric pressure

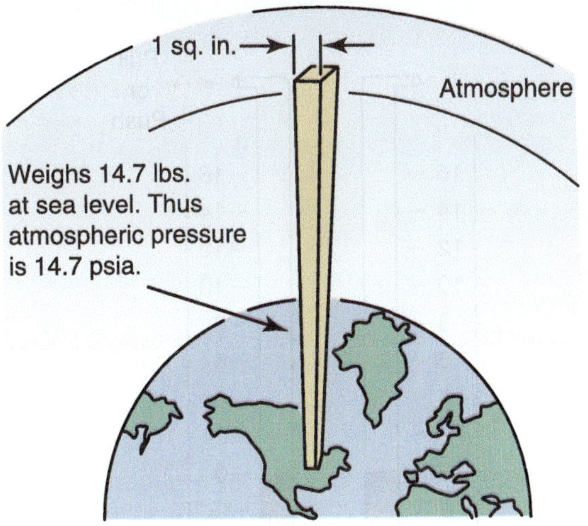

high mountain, the column of air would measure less than 50 miles (80 km) and the result would be a lower weight of air in the column. Similarly, if we were below sea level, in a mine, for instance, the weight of air would be greater in the column. In North America we sometimes use the term *atm* (short for *atmosphere*) to describe a unit of measurement of atmospheric pressure; the Europeans use the unit *bar* (short for *barometric* pressure). The relative equivalents of these values appear a little later in this chapter.

Force

Force is push or pull effort. The weight of one object placed on another exerts force on it proportional to its weight. If the objects were glued to each other and we lifted the upper one, a pull force would be exerted by the lower object proportional to its weight. Force does not necessarily result in any work done. If you were to push on the rear of a parked transport truck you could apply a lot of force, but that effort would be unlikely to result in any movement of the truck. The formula for force (F) is calculated by multiplying pressure (P) by the area (A) it acts on.

$$F = P \times A$$

Calculating Force

In hydraulics, force is the product of pressure multiplied by area. For instance if a fluid pressure of 100 psi acts on a piston sectional area of 50 square inches it means that 100 pounds of pressure acts on each square inch of the total sectional area of the piston. The linear force that results can be calculated as follows:

$$\text{Force} = 100 \text{ psi} \times 50 \text{ sq. in.} = 5{,}000 \text{ lb.}$$

Head Pressure

If you have ever drained fuel out of a tank through its drain valve, you will have observed that the fuel runs out of the tank faster when the tank is full and progressively slower as the tank empties. This occurs because the pressure that forces the fuel from the tank is defined by the weight of the fuel in the tank, and as the tank empties, less weight results in lower pressure. We express this as **head pressure**. By definition, **head** is the vertical distance between two levels in a fluid and it can be expressed using linear or pressure values. For instance, a 10-foot head of water is equivalent to 4.4 psi (30 kPa). The weight of a fuel or hydraulic oil is less than that of water so we would have to first reckon its specific gravity and then calculate the head required to move it.

Vacuum

A perfect vacuum is a volume that has been evacuated of all matter. That means that no air (mixture of nitrogen and oxygen molecules) can be present in a true vacuum. A true vacuum is difficult to achieve but we tend to use the word *vacuum* to describe any pressure condition lower than atmospheric pressure. Partial vacuums are used in hydraulics so that when a pump creates a low-pressure area on its inlet side, the static head pressure of our atmosphere acts on the reservoir oil to push it toward the pump. In this way, pumps are used to move fluid through circuits whether they are on a wet-line kit, steering gear circuit, or fuel subsystem.

Pressure Scales

There are a number of different pressure scales used today but all are based on atmospheric pressure. One unit of atmosphere is the equivalent of atmospheric pressure and it can be expressed in all these ways:

1 atm = 1 bar (European) = 14.7 psia = 29.92 in.Hg (inches of mercury) = 101.3 kPa (metric)

However, each of the above values is not precisely equivalent to the others:

1 atm = 1.0192 bar

1 bar = 29.53 in.Hg = 14.503 psia

1 in.Hg = 13.6 in.H_2O @ 60°F

An Italian named Evangelista Torricelli (1608–1647) discovered the concept of atmospheric pressure. He inverted a tube filled with mercury into a bowl of the liquid and then observed that the column of mercury in the tube fell until atmospheric pressure acting on the surface balanced against the vacuum created in the tube as shown in **Figure 13–2**.

At sea level, vacuum in the column in **Torricelli's tube** would support 29.92 inches of mercury. Mercury

FIGURE 13–2 Torricelli: Atmospheric pressure expressed in inches of mercury (Hg).

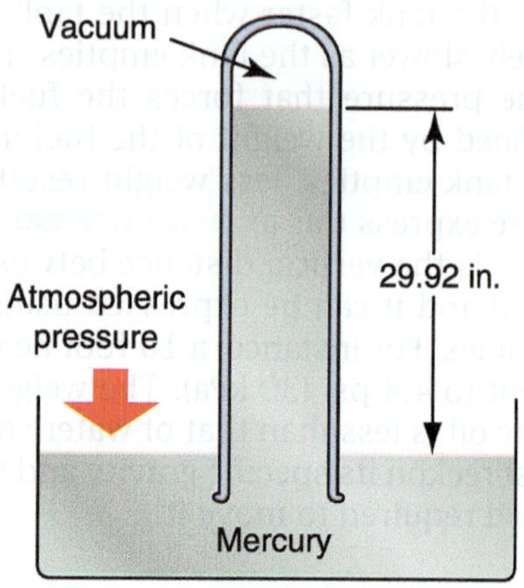

is usually abbreviated to its elemental symbol Hg. To this day, we use manometers to measure pressure values both below and above atmospheric. To convert 1 psi to its equivalent in inches of mercury, then, 1 psi is equal to 20 in.Hg (or more precisely, 2.0360 in.Hg).

Manometer

A manometer is a single tube arranged in a U-shape mounted to a linear measuring scale as shown in **Figure 13–3**. It may be filled to the zero on the calibration scale with either water (H_2O) or mercury (Hg), depending on the pressure range to be measured.

A **manometer** can measure either push or pull on the fluid within the column. Actual examples of how we use manometer readings are as follows:

- Vacuum restriction on a truck fuel subsystem primary circuit is specified in inches of mercury (in.Hg).
- Inlet restriction of a truck air filter system is specified in inches of water (in.H_2O).

Note that an inch of mercury is equivalent to 13.6 inches of water—so it is important that you do not mix the two scales. In practice, we tend to use pressure gauges that read values below atmospheric pressure rather than manometers. These pressure gauges display values in inches of mercury on a rotary scale.

Absolute Pressure

Absolute pressure uses a scale in which the zero point is a complete absence of pressure. Gauge pressure has as its zero point atmospheric pressure. A gauge, therefore, reads zero when exposed to the atmosphere. To avoid confusing absolute pressure with gauge pressure, absolute pressure is expressed as psia. Gauge pressure is usually expressed as psi or psig.

FIGURE 13–3 Manometer shown at atmospheric pressure: Medium in column can be either H_2O or Hg.

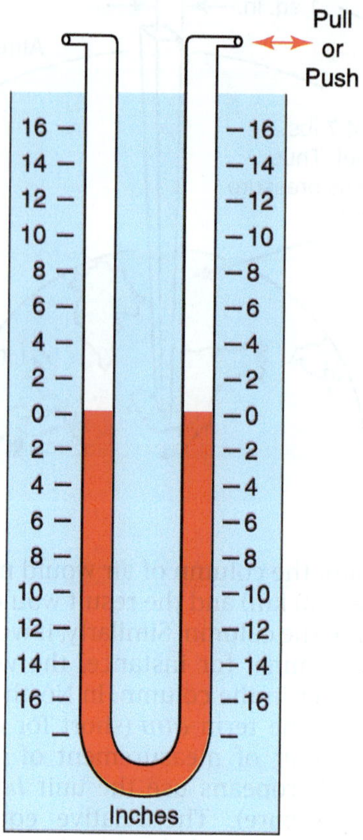

HYDRAULIC LEVERS

Hydraulic levers can be used to demonstrate Pascal's Law. Pressure equals force divided by the sectional area it acts on. Similarly, force equals pressure multiplied by area.

$$F = P \times A$$

So if we want to calculate the force exerted by a cylindrical ram we would use the following formula:

Force exerted = hydraulic pressure (psi) × cylindrical area (diameter^2 × 0.7854)

If we build a simple hydraulic circuit such as that shown in **Figure 13–4**, we have the building blocks of a hydraulic jack. A variation on the hydraulic jack principle can be seen in the type of hydraulic circuit shown in **Figure 13–5**, the basis of a hydraulic ram. This consists of a pair of unequally sized cylinders connected by a pipe. In the figure, one of the cylinders has a sectional area of 1 sq. in. and the other 50 sq. in.—so the following would be true:

- Applying a force of 2 lb. on the piston in the smaller cylinder lifts a weight of 100 lb. supported on a piston in the larger cylinder.

FIGURE 13–4 Simple hydraulic lever: Pressure multiplies force.

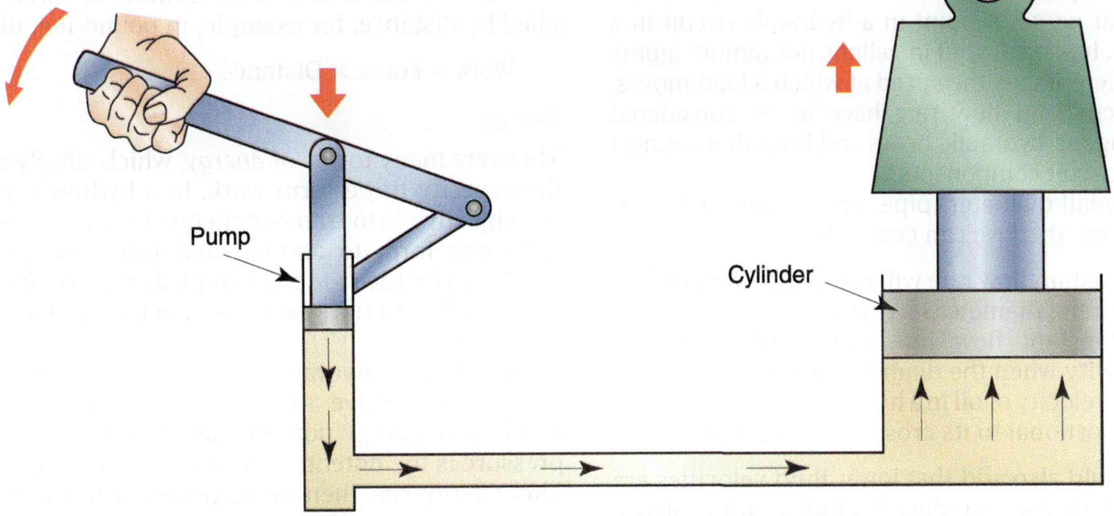

- Applying a force of 2 lb. on the piston in the smaller cylinder produces a circuit pressure of 2 psi because the force is applied to a sectional area of 1 sq. in.
- The circuit potential is 2 psi and because this acts on a sectional area of 50 sq. in. it can raise 100 lb.
- Accordingly, if a force of 10 lb. is applied to the smaller piston, the resulting circuit pressure would be 10 psi and the circuit would have the potential to raise a weight of 500 lb.

This principle is applied every time we use a hydraulic jack. A hydraulic jack is a simple hydraulic lever as is a cab lift ram or dump ram. There has to be a trade-off in this type of circuit just as there is with any mechanical lever. When 2 lb. of force is applied to the smaller cylinder to raise a 100-lb. weight 1 in. in the larger cylinder, the piston in the smaller piston must move through 50 in. as shown in **Figure 13–6**. This movement is accomplished using multiple strokes and a check valve in typical hydraulic circuits.

FIGURE 13–5 Pressure multiplies force.

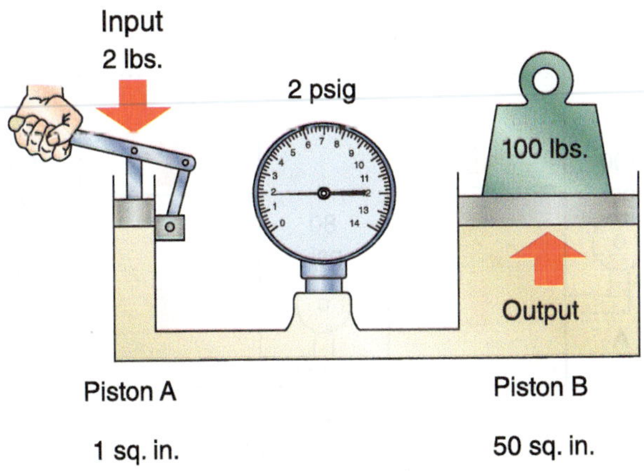

FLOW

Flow is the term we use to describe the movement of a hydraulic medium through a circuit. Flow occurs when there is a difference in pressure between two points. In a hydraulic circuit, flow is created by a device such as a pump. A pump exerts push effort on a fluid. Flow rate is the volume or mass of fluid passing through a conductor over a given unit of time; an example would be gallons per minute (gpm).

FIGURE 13–6 Applied force and distance in a hydraulic circuit

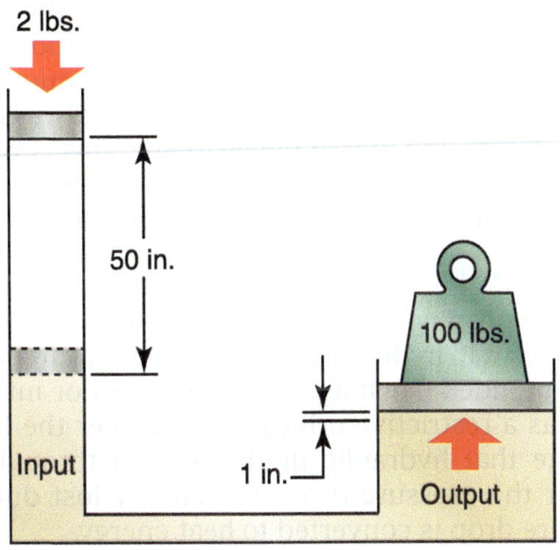

Measuring Flow

Flow can be measured in two ways: velocity and flow rate. The velocity of a fluid in a confined circuit is the speed at which the fluid moves through it. It is measured in feet per second (fps). Flow rate is the volume of fluid that passes a point in a hydraulic circuit in a given time. It is measured in gallons per minute (gpm). Flow rate determines the speed at which a load moves. Fluid velocity and flow rate have to be considered when sizing the hydraulic hoses and lines that connect hydraulic circuit components.

If a small-diameter pipe opens into a larger-diameter one, then we can conclude:

- A constant flow rate will result in lower velocity when the diameter increases.
- A constant flow rate will result in higher velocity when the diameter decreases.
- The velocity of oil in a hydraulic line is inversely proportional to its cross-sectional area.

We should also add that lower fluid velocities are generally desirable to reduce friction and turbulence in the fluid. The same works for hydraulic cylinders. Given an equal flow rate, a small cylinder will move faster than a larger cylinder. If the objective is to increase the speed at which a load moves, then:

- Decrease the size (sectional area) of the cylinder.
- Increase the flow to the cylinder (gpm).

The opposite would also be true, so if the objective were to slow the speed at which a load moves, then:

- Increase the size (sectional area) of the cylinder.
- Decrease the flow to the cylinder (gpm).

Therefore, the speed of a cylinder is proportional to the flow it is subject to and inversely proportional to the piston area.

Pressure Drop

In a confined hydraulic circuit, whenever there is flow, a pressure drop results. Again, the opposite applies. Whenever there is a difference in pressure, there must be flow. Should the pressure difference be too great to establish equilibrium there would be continuous flow. In a flowing hydraulic circuit, pressure is always highest upstream and lowest downstream. This is why we use the term *pressure drop*. A pressure drop always occurs downstream from a restriction in a circuit.

Flow Restrictions

Pressure drop will occur whenever there is a restriction to flow. In hydraulics, a restriction in a circuit may be unintended (such as a collapsed line) or intended (such as a restrictive orifice). The smaller the line or passage that hydraulic fluid is forced through, the greater the pressure drop. The energy lost due to a pressure drop is converted to heat energy.

Work

Work occurs when effort or force produces an observable result. In a hydraulic circuit, this means moving a load. To produce work in a hydraulic circuit, we must have flow. Work is measured in units of force multiplied by distance, for example, in pound-feet (lb-ft).

$$\text{Work} = \text{Force} \times \text{Distance}$$

Energy

There are many forms of *energy*, which simply means the capacity to perform work. In a hydraulic circuit, the objective is to transfer energy. We transfer energy from one form to another and from one point to another. The idea is to accomplish this as efficiently as possible and not waste too much by transforming it to heat.

In a typical hydraulic circuit, mechanical energy is required to drive a hydraulic pump to create flow and kinetic energy potential in the fluid. Fluid under pressure is the potential energy of a hydraulic circuit. This energy can then be reconverted to mechanical energy to move a load—that is, kinetic energy or the energy of motion. The prime source of the energy used in a hydraulic system may be the heat value of the fuel used in an engine, or the electrical energy in a battery.

The term **prime mover** is used to describe the machine that creates the mechanical energy required to power a hydraulic pump, though it is also sometimes used to describe the pump itself.

Bernoulli's Principle

Bernoulli's Principle states that if flow in a circuit is constant, then the sum of the pressure and kinetic energy must also be constant. As we said, when fluid is forced through areas of different diameters, the result is changes in fluid velocity. Fluid flow through a large pipe will be slow until the large pipe reduces to a smaller pipe; then the fluid velocity will increase. **Figure 13–7** shows a hydraulic circuit consisting of two cylinders connected by a pipe, with force being applied to one of the cylinders. Bernoulli proved that in such a circuit, when force is applied to the cylinder

FIGURE 13–7 Bernoulli's Principle

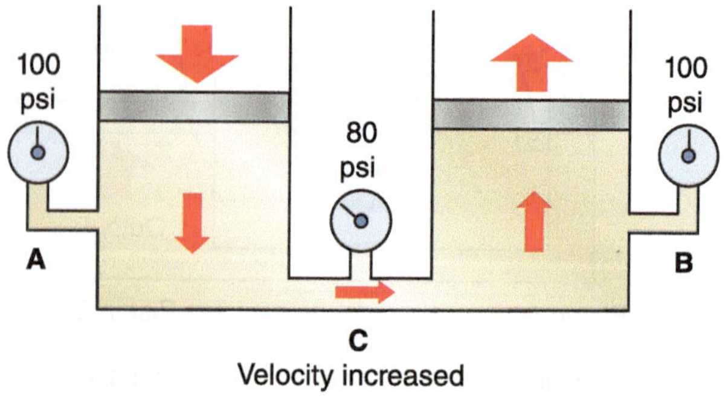

on the left, the pressure measured at the connecting pipe will be less than that measured at either cylinder. This happens because the increase in fluid velocity that occurs in the connecting pipe means that some of the pressure energy has been converted to kinetic energy. In the cylinder on the right, assuming that there have been no frictional losses, the pressure should return to the same value as the cylinder on the left because the kinetic energy has been converted back into pressure.

Laminar Flow

Flow of a hydraulic medium through a circuit should be as streamlined as possible. Streamlined flow is known as **laminar flow**. Laminar flow is required to minimize friction. Changes in section, sharp turns, and high-flow speeds can cause turbulence and cross-currents in a hydraulic circuit, resulting in friction losses and pressure drops.

TYPES OF HYDRAULIC SYSTEM

Hydraulic systems can be grouped into three main categories:

- Open-center systems
- Closed-center systems
- Closed-loop systems

The primary difference between the three systems is defined by what is done with the hydraulic oil in the circuit after it leaves the pump.

Open-Center Systems

In an **open-center** system, the pump runs constantly, which means that oil flows through the circuit continuously. A valve is used to manage the circuit and when this valve is in its "open" or neutral position, fluid is allowed to return to the reservoir. An example of an open-center hydraulic system on a truck is power-assisted steering. **Figure 13–8** shows an example of an open-center spool valve.

FIGURE 13–8 Open-center spool valve

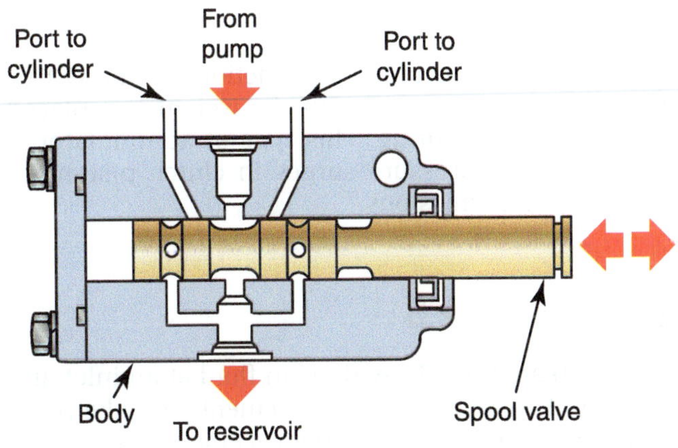

Closed-Center Systems

In a **closed-center** system, the pump can be "rested" during operation whenever flow is not required to operate an actuator. This means that the control valve blocks flow from the pump when it is in its "closed" or neutral position. A closed-center system requires the use of either a variable displacement pump or proportioning control valves. Closed-center systems have many uses on agricultural and industrial equipment, but on trucks, they would be used on garbage packers and front bucket forks.

Closed-Loop Systems

A **closed-loop** system is one in which oil is circulated from the pump to a motor and back to the pump without being returned to the reservoir. The most common use in trucks is to drive the mixing barrels in cement trucks. Also used in off-highway hydrostatic drive powertrains.

Summary of Basics

This concludes the introduction to some of the basic principles of hydraulics. It is important for you to understand some of the language of hydraulics before proceeding to the remainder of this chapter. You should also know that an elementary understanding of hydraulics is required to understand many truck systems such as transmissions, steering, brake, suspension, and fuel systems, to mention a few.

13.2 HYDRAULIC COMPONENTS

Most of the components in a hydraulic circuit are interrelated in much the same way as components in an electrical circuit. The components in the circuits described in this chapter are consistent with those found in mobile hydraulic systems.

RESERVOIRS

A **reservoir** in a hydraulic system has the following roles:

- Stores hydraulic oil
- Helps keep oil clean and free of air
- Acts as a heat exchanger to help cool the oil

A reservoir is typically equipped with:

- **Filler cap**. This is the means of replenishing oil in the hydraulic circuit. Depending on the system, the filler cap may be equipped with a vent or be held under pressure.
- **Oil-level gauge or dipstick**. This provides a means of verifying that there is sufficient oil in the reservoir.
- **Outlet and return lines**. These provide a means of conducting the oil out of and back into the reservoir. Although the lines can be

positioned on top of or at the sides of a reservoir, the ends of each should be located near the bottom within the reservoir. If an oil return line discharges above the fluid level in a reservoir, foaming can result.

- **Baffle(s)**. The function of a baffle or baffles in a reservoir is to separate return oil from that being drawn out by the pump. However, baffle(s) may be unnecessary on some systems, depending on how the lines and filters are arranged in the circuit.
- **Intake filter**. The intake filter is often just a screen or strainer designed to remove larger particles.
- **Drain plug**. Drain plugs provide a means of dumping the hydraulic system oil during system service. Some are magnetic to attract and hold small metal particles in the system.

HYDRAULIC JACKS

A hydraulic jack consists of a simple hand-actuated pump plunger, a reservoir, two check valves, and an output piston. Referencing Figure 13–4, when the pumping plunger is pulled up, a partial vacuum is created in the plunger chamber. This causes atmospheric pressure acting on the oil in the reservoir to push fluid into the plunger chamber through the inlet check valve. While the pumping plunger is being pulled upward through its stroke, load-induced pressure acting on the output piston holds the outlet check valve closed.

When the plunger chamber has been charged with oil (plunger at the top of its stroke), the pressure in the plunger chamber should be close to atmospheric pressure, as is the reservoir. When downward force is applied to the pumping plunger, the outlet check valve opens as soon as the pressure in the plunger chamber exceeds that in the output piston cylinder, permitting flow and pressure to charge the output cylinder.

Output flow of a simple hydraulic jack is determined by the volume displaced in a single pump cycle. The force developed is defined by the force potential applied to the pumping plunger. This simple hydromechanical device forms the basis of hydraulic floor jacks and the jacks used to raise cabs on cab-over-engine (COE) chassis.

ACCUMULATORS

A mechanical spring is a simple **accumulator**. When a spring is compressed, it "stores" energy and can also be used to dampen shock and buffer pressure rise. Accumulators used in hydraulic circuits perform the following roles:

- Store potential energy
- Dampen shocks and pressure surges
- Maintain a consistent pressure

In an actual hydraulic circuit, an accumulator is usually used for a specific function in one of the above-mentioned roles. An accumulator that stores potential energy is often used to back up a hydraulic pump in the event of pump failure or at system startup. For instance, in the event of oil supply failure in large off-highway dump trucks using hydraulically assisted brakes, an accumulator is used to feed several charges of oil into the circuit to effect an emergency stop. Accumulators can also be used to dampen shock loads and pressure surges and to facilitate smooth operation of a hydraulic circuit. Although it uses a self-contained circuit, any vehicle shock absorber is a type of accumulator. In terms of operating principles, accumulators generally fall into one of three categories:

- Gas-loaded
- Weight-loaded
- Spring-loaded

Gas-Loaded Accumulators

Although liquids tend not to compress, gases do, so they can be used to load a volume of oil under pressure. In this type of accumulator, the gas (usually a relatively inert gas such as nitrogen) and hydraulic oil occupy the same chamber but are separated by a piston, diaphragm, or bladder. When circuit pressure rises, incoming oil to the chamber compresses the gas. When circuit pressure drops off, the gas in the chamber expands, forcing oil out into the circuit. Most gas-loaded accumulators are precharged with the compressed gas that enables their operation. **Figure 13–9** shows a sectional view of a typical gas-loaded accumulator.

Weight-Loaded Accumulators

A weight-loaded accumulator uses a gravity piston and cylinder. Weight is loaded to the piston so that it acts on hydraulic oil in the accumulator chamber. When circuit pressure rises, oil is unloaded into the accumulator chamber, raising the piston and the weight it supports. When circuit pressure drops off, gravity forces the weight down, charging oil to the circuit.

Spring-Loaded Accumulators

A spring-loaded accumulator functions similarly to a weight-loaded version except the weight is replaced by a mechanical spring. This type of accumulator is used to buffer pressure surges in clutch pistons in automatic transmissions.

13.3 PUMPS

Pumps create flow. They draw in fluid at an inlet and displace it to an outlet. Displacement can take place in two ways. A nonpositive displacement pump picks

FIGURE 13–9 Hydraulic accumulator

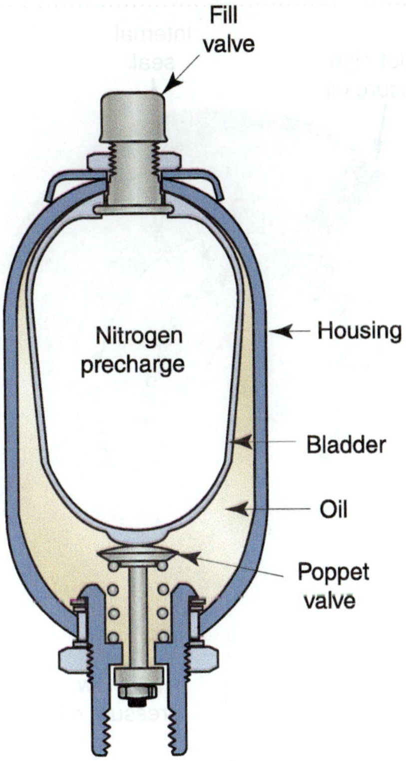

up fluid and moves it. Examples would be a water wheel or the coolant pump used on a typical diesel engine. **Figure 13–10** shows a comparison between positive and nonpositive pumping principles.

A positive displacement pump not only creates flow but backs it up. Fluid picked up by a positive displacement pump is sealed and held while the pump turns. Most hydraulic circuits only use positive displacement pumps and we can divide these into two general categories:

- Fixed-displacement pumps
- Variable-displacement pumps

FIGURE 13–10 Nonpositive and positive displacement pumping principles

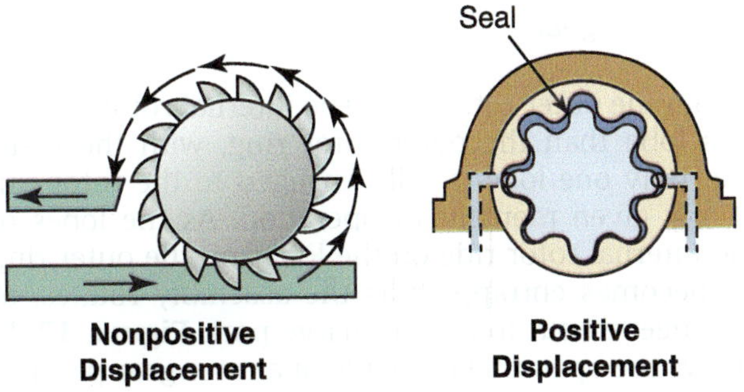

When studying pumps, you should remember that hydraulic pumps only create flow, not pressure. Pressure results from creating resistance to flow.

FIXED-DISPLACEMENT PUMPS

A fixed-displacement pump will move the same amount of oil per cycle with the result that the slug volume picked up by the pump at its inlet equals the slug volume discharged to its outlet per cycle. This means that the speed of the pump will determine how much hydraulic oil is moved. Fixed-displacement pumps are often used as an assist in a subsection of a hydraulic circuit and to move diesel fuel in truck fuel subsystems.

VARIABLE-DISPLACEMENT PUMPS

Variable-displacement pumps are positive displacement pumps designed to vary the volume of oil they move each cycle even when they are run at the same speed. They use an internal control mechanism to vary the output of oil—usually with the objective of maintaining a constant pressure value and reducing flow when demand for oil is minimal.

GEAR PUMPS

Gear pumps are widely used in mobile hydraulics because of their simplicity. They are also widely used to move fuel through diesel fuel subsystems and as engine lube oil pumps. Three types of gear pumps are used:

- External gear
- Internal gear
- Rotor gear

External-Gear Pumps

Two close-fit, intermeshing gears are driven within a housing. One of the gears is a drive shaft and this drives the second because it is in mesh with it. As the gears rotate, oil from the inlet is trapped between the teeth and the housing, and it is then moved around the outside of the gear teeth to the outlet of the pump housing. The close mesh of the gear teeth forms a seal that prevents oil from backing up to the inlet. Oil is discharged through the outlet side of the pump housing from which the circuit is charged. This principle is demonstrated in **Figure 13–11**, and **Figure 13–12** shows a cutaway of a typical gear pump.

Internal-Gear Pumps

The internal-gear pump also uses a pair of intermeshing gears, but now a spur gear rotates within an annular internal gear, meshing on one side of it. Both gears are divided on the other side by a crescent-shaped separator. The drive shaft turns the spur gear that meshes with the annular internal gear

FIGURE 13–11 External-gear pump

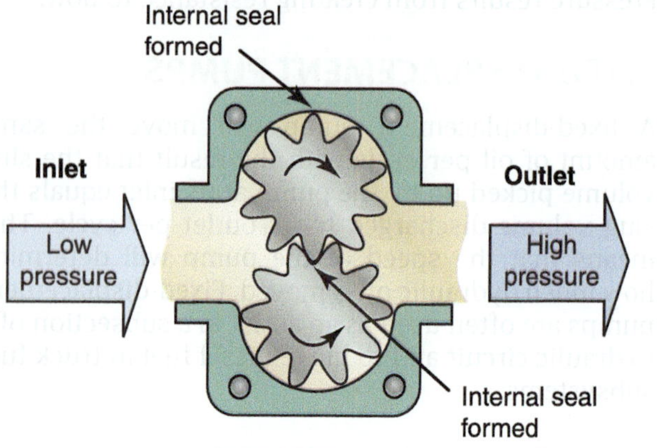

FIGURE 13–12 External tooth-gear pump cutaway

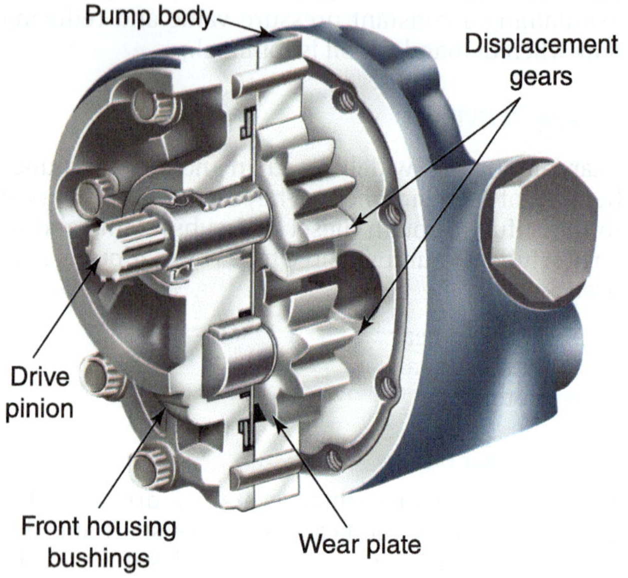

driving it—meaning that both gears will rotate in the same direction. As the gear teeth come out of mesh, oil from the inlet is trapped between their teeth and the separator and unloaded to the outlet. As with the external-gear pump, when the gear teeth mesh, a seal is formed, preventing backup. **Figure 13–13** shows the operating principle of an internal-gear pump, and **Figure 13–14** is a cutaway showing the internal components.

Geroter Pumps

A geroter or rotor-gear pump is a variation of the internal-gear pump. An internal rotor with external lobes rotates within an outer rotor ring with internal lobes. No separator is used. The internal rotor is driven

FIGURE 13–13 Operating principle of an internal gear pump

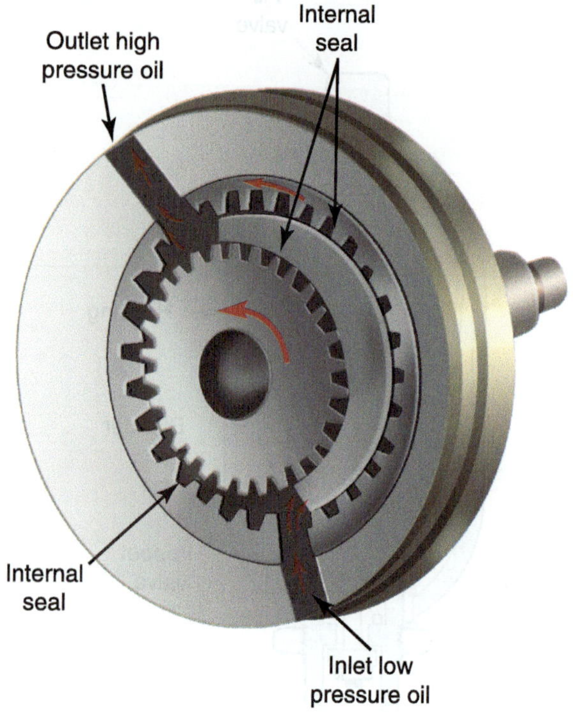

FIGURE 13–14 Internal-gear pump components.

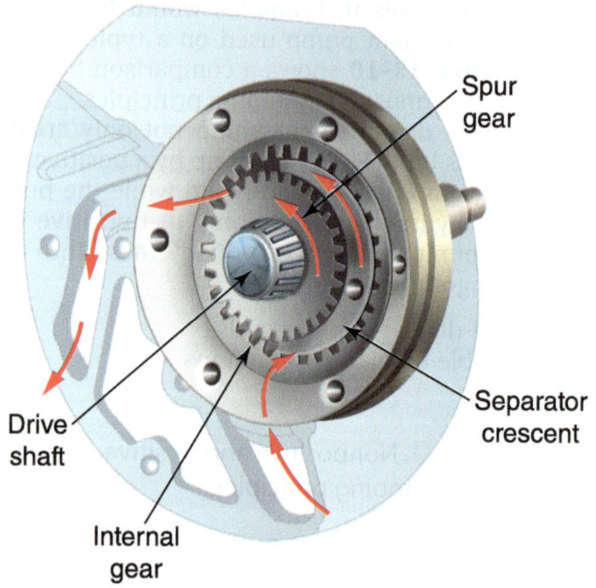

within the outer rotor ring. The internal rotor has one less lobe than the outer rotor ring, with the result that only one lobe is fully engaged to the rotor ring at any given moment of operation. As the lobes on the internal rotor ride on the lobes on the outer ring, oil becomes entrapped; as the assembly rotates, oil is squeezed out to the discharge port. **Figure 13–15** shows the operating principle of a rotor-gear pump.

FIGURE 13–15 Rotor-type pump

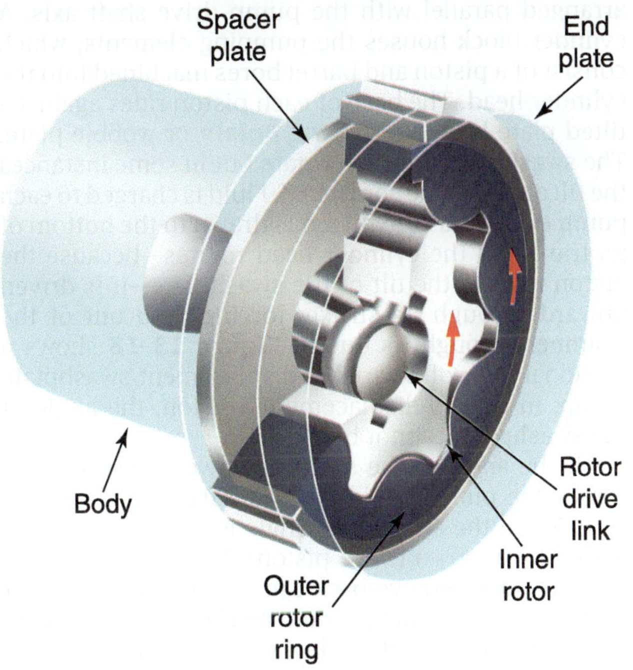

FIGURE 13–16 Balanced vane pump

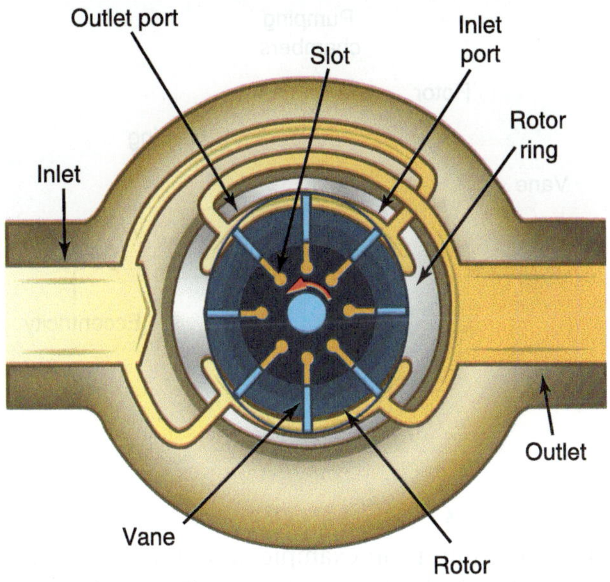

VANE PUMPS

Vane pumps are also used extensively in hydraulic circuits. Truck power-assisted steering systems commonly use vane pumps. Vane pumps move fluid by using a slotted rotor fitted with sliding vanes: The rotor turns within a stationary liner known as a **cam ring**. A major advantage of vane pumps is their ability to compensate for internal wear without experiencing any notable loss of efficiency. A worn vane will simply slide out further to contact the cam ring wall. There are two types:

- Balanced
- Unbalanced

Balanced Vane Pumps

In a balanced vane pump, the rotor turns within a stationary liner. The rotor is fitted with slots within which sliding vanes are inserted. When the rotor is turned, initially centrifugal force throws the vanes outward so they contact and seat against the liner walls. As the vanes follow the contour of the liner wall, fluid from the pump inlet is trapped between the crescent-shaped "chambers" formed between vanes; fluid pressure is then directed against the back of the vane that helps load it into the cam ring wall. These chambers are continually expanding and contracting as the rotor turns; oil from the inlet is trapped into the chamber formed by a pair of vanes. Turning the rotor causes the chamber to contract as it brings it into alignment with the outlet. This action repeats itself twice per revolution because there are a pair of inlet ports and a pair of discharge ports. **Figure 13–16** shows the operating principle of a balanced vane pump.

Unbalanced Vane Pumps

The unbalanced vane pump uses the same principle as the balanced version, with the exception that the operating cycle only occurs once per revolution. This pump locates the axis of the liner offset to that of the rotor and has only one inlet and one outlet port. In operation, inlet oil becomes entrapped between a pair of vanes when they are expanded and align with the inlet port. As the rotor turns, the chamber is contracted before the rotation of the pump brings the outlet port into alignment with the chamber. The disadvantage of the unbalanced vane pump is the radial load that is placed on one side of the rotor, as each chamber formed between the vanes is contracted between fill and discharge; this can cause the bearing to fail. No such force acts on the opposite side of the pump because the inlet oil is under little or no pressure. **Figure 13–17** shows the operating principle of an unbalanced vane pump.

PISTON PUMPS

There is a wide variety of piston pumps, beginning with the most simple and including some of the more complex pumps used in hydraulic circuits. There are three general types of piston pumps:

- Plunger pumps
- Axial piston
- Radial piston

Plunger-type pumps are seldom found on hydraulic circuits, but the latter two are used on systems that demand high-flow and high-pressure performance.

FIGURE 13–17 Unbalanced vane pump

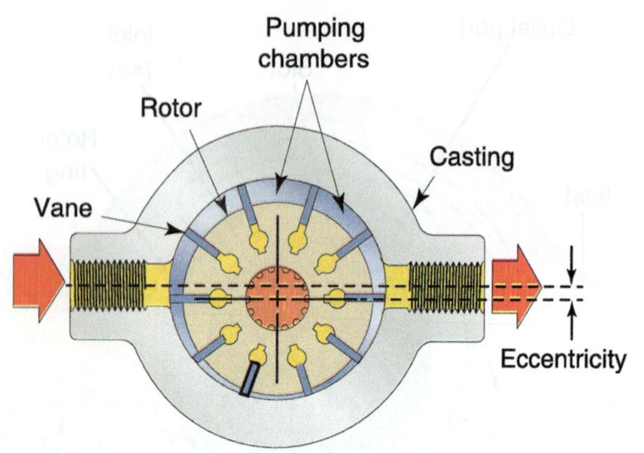

Plunger Pumps

A bicycle pump is an example of a plunger pump as are the fuel hand-priming pumps used on many diesel fuel subsystems. A plunger reciprocates within a stationary barrel. Fluid to be pumped is drawn into the pump chamber formed in the barrel on the outward stroke of the plunger. This fluid is then discharged on the inboard stroke of the plunger.

Axial Piston Pumps

In an axial piston pump, the pumping elements are arranged parallel with the pump drive shaft axis. A cylinder block houses the pumping elements, which consist of a piston and barrel bores machined into the cylinder head. The base of each piston rides against a tilted plate known as a **swashplate** or wobble plate. The swashplate does not rotate but in some instances the tilt angle can be controlled. Fluid is charged to each pump element as the piston is drawn to the bottom of its travel. As the cylinder head rotates—because the piston follows the tilt of the swashplate—it is driven upward through its stroke, forcing fluid out of the chamber through its outlet. **Figure 13–18** shows a sectional view of a variable-displacement swashplate pump; in a fixed-displacement version, the angle of the swashplate cannot be altered.

If the angle of the swashplate were fixed (it is in some), the pump would be of the fixed-displacement type. When the swashplate angle is movable and controlled by a **servo** (slave piston) the distance the pistons are able to move back can be either increased or decreased, so the pump would be classified as *variable displacement*. The farther the pistons travel, the greater the amount of oil that can be displaced per pump cycle. **Figure 13–19** shows how varying the swash angle can be used to alter the displacement of the pump.

FIGURE 13–18 Sectional view of a swashplate pump: Note the means of varying the swash angle.

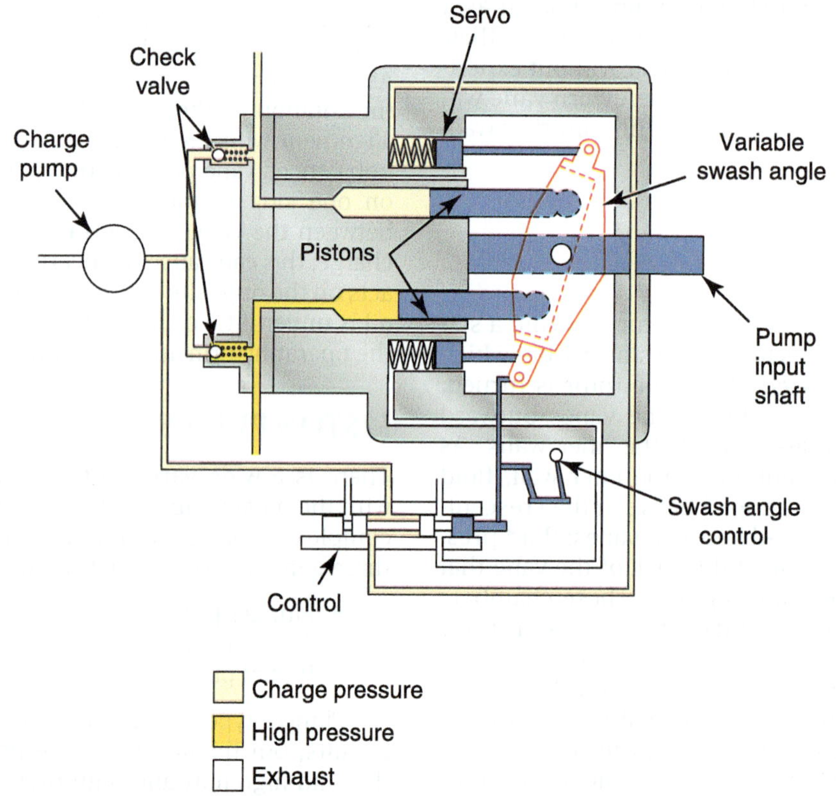

FIGURE 13–19 Effect of swash angle on displacement

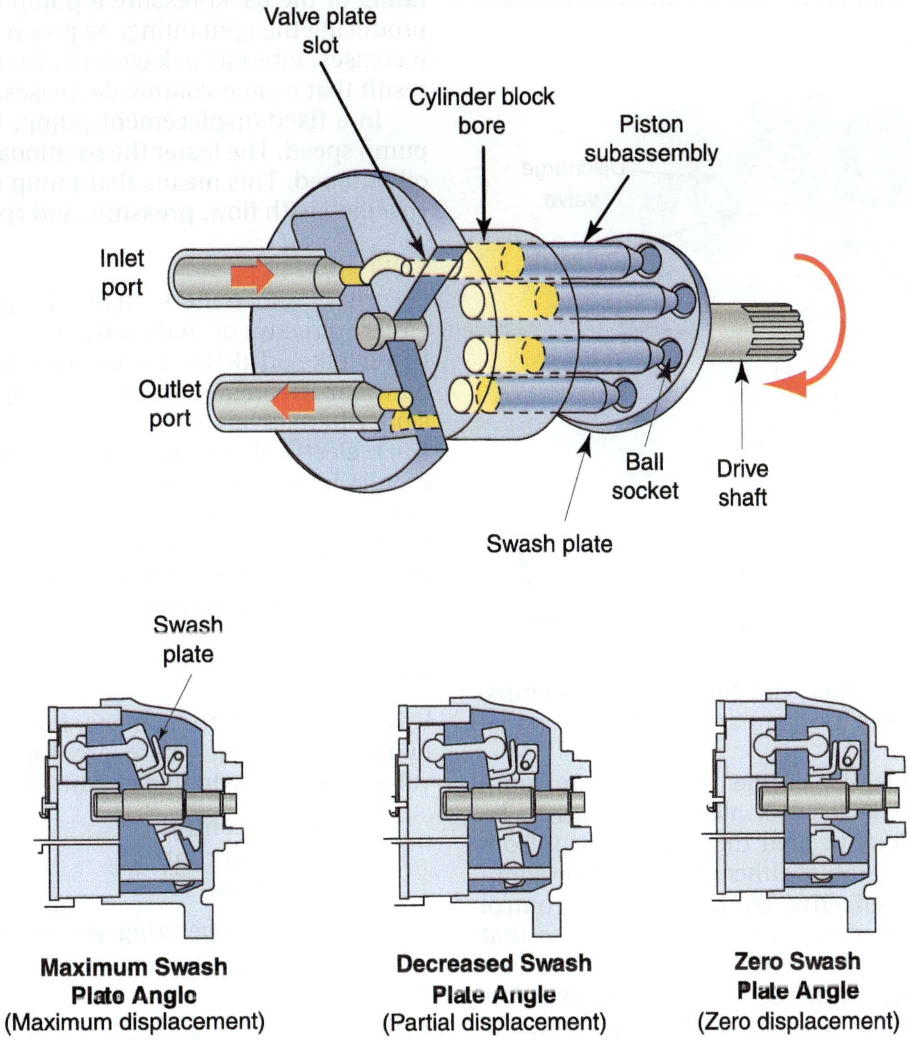

Radial Piston Pumps

Radial piston pumps are capable of high pressures, high speeds, high volumes, and variable displacement. However, they cannot reverse flow. Radial piston pumps operate in two ways:

- Rotating cam
- Rotating piston

Rotating Cam. In a rotating-cam pump, radially opposed pistons are located in a fixed pump body; a central drive shaft is machined with an eccentric cam. As the cam profile is rotated, it forces the pistons outward to effect a pumping stroke. Each piston is loaded by spring force to ride the cam profile. This means that the outward stroke of the piston is effected by cam profile and the inboard stroke by spring force. Oil inlet and outlet passages are located at either end of each pump element. Spring-loaded valves in the inlet and outlet ports permit oil to flow into and out of the piston bores.

During the inlet stroke, the piston is returned by the spring that loads it to ride the cam profile, creating low pressure in the piston bore. This low pressure in combination with oil pressure unseats the inlet valve, allowing the pump chamber to be charged with oil. When the pump chamber is filled with oil, the inlet valve is closed by its spring.

After the pump chamber has been charged, the cam rotates and acts on the piston, driving it outward through its pumping stroke; oil in the pump chamber is pressurized and unseats the discharge valve, unloading it into the outlet circuit. The pumping cycles of the pistons work sequentially as the cam rotates, producing a continuous flow of oil. Four-, six-, and eight-piston versions of the pump are common. Oil output depends on the rotational speed of the pump—assuming it is of the fixed-displacement type.

A common truck application of cam-actuated radial piston pumps is the high-pressure pump used

FIGURE 13–20 Rotating cam-type radial piston pump

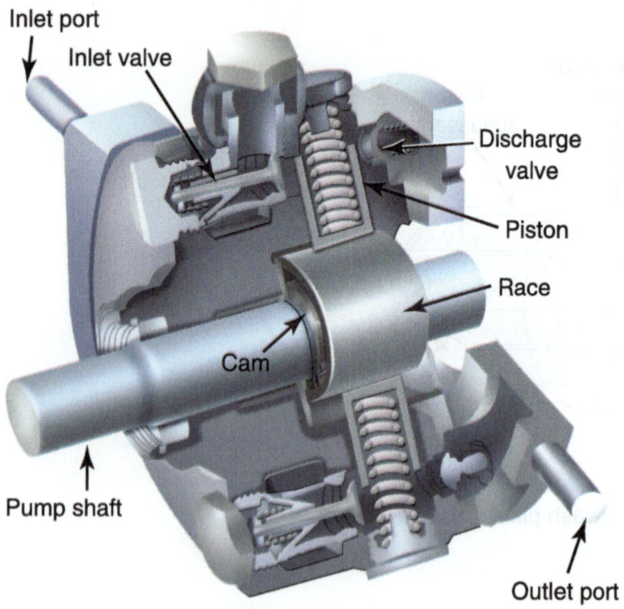

on some Bosch common rail fuel injection pumps. **Figure 13–20** shows the components used in a radial piston pump.

When a variable-displacement rotating-cam pump is required, a stroke control mechanism has to be used. This functions to hold the pistons away from the actuating-cam profile, either shortening or eliminating the piston-effective stroke. A stroke control valve admits oil into the center of the pump so that it acts on and holds off the pistons. The result is that while the camshaft continues to turn, pump chamber output can be held at near zero. When the hydraulic circuit requires pump outlet flow once again, hydraulic pressure at the center of the pump is dropped. This allows the piston springs to return the pistons to ride the cam profiles, and pumping action resumes.

Rotating-Piston Type. A rotating-piston-type pump functions similarly to an unbalanced vane pump. Radially reciprocating pistons are arranged within an eccentric cylinder. As the cylinder assembly rotates, the pistons are loaded by centrifugal force against the pumping housing wall. This permits the piston bores to be filled with oil from inlet ports and as the cylinder is rotated, the pistons are driven inward, displacing the oil to outlet ports. Variable displacement is achieved by adjusting the axis of the outer housing to that of the cylinder. This alters piston travel and, therefore, the volume of oil displaced per cycle.

PUMP PERFORMANCE

Hydraulic pumps are usually rated by delivery rate—in other words, the volume it can displace over a given time period. This is usually expressed in gallons per minute (gpm) at a specific rpm such as 1,500 or 1,200 rpm. This specification should be accompanied by a rating of the backpressure a pump can sustain while producing the gpm rating. As pump delivery pressure increases, internal leakage will also increase, with the result that usable volume decreases.

In a fixed-displacement pump, flow is related to pump speed. The faster the rotational speed, the more oil pumped. This means that pump ratings should be specified with flow, pressure, and rpm values.

Pump Drives

On trucks and trailers, hydraulic pumps are driven either directly or indirectly by the truck engine. Power take-off (PTO) devices may be driven directly by the truck engine or from gearing in the transmission. Alternatively, electrical power produced by the truck electrical system can be used to drive electric motors located anywhere on either the truck or trailer chassis. For instance, a two-tier car transporter trailer can be equipped with extensive hydraulic circuits that often use several electric motors to actuate a number of independent hydraulic circuits.

VALVES

Valves are used to manage flow and pressure in hydraulic circuits. There are three basic types of valves used in hydraulic circuits:

- Pressure control
- Directional control
- Flow control

The general operating principles of these three categories of valve are shown in **Figure 13–21**.

PRESSURE-CONTROL VALVES

Hydraulic systems are designed to operate at specific working pressures so pressure-control valves are used to regulate pressure in the system or in sections of a system. Many different types can be used. Examples would be pressure-reducing and safety relief valves. The objective is to separate two circuits working at different pressures. They can also act as safety relief valves to limit the maximum working pressure of a specific system or circuit.

Counterbalance Valves

Counterbalance valves are used to maintain backpressure in a circuit or component. They are designed to block a fluid flow path until pilot pressure (direct or remote) overcomes the spring tension setting of the valve and opens the spill circuit. These are often used in a double-acting lift cylinder to prevent sudden or uncontrolled dropping.

Relief Valves

Two general types of pressure-relief valves are used in hydraulic circuits.

FIGURE 13–21 Three types of hydraulic valves

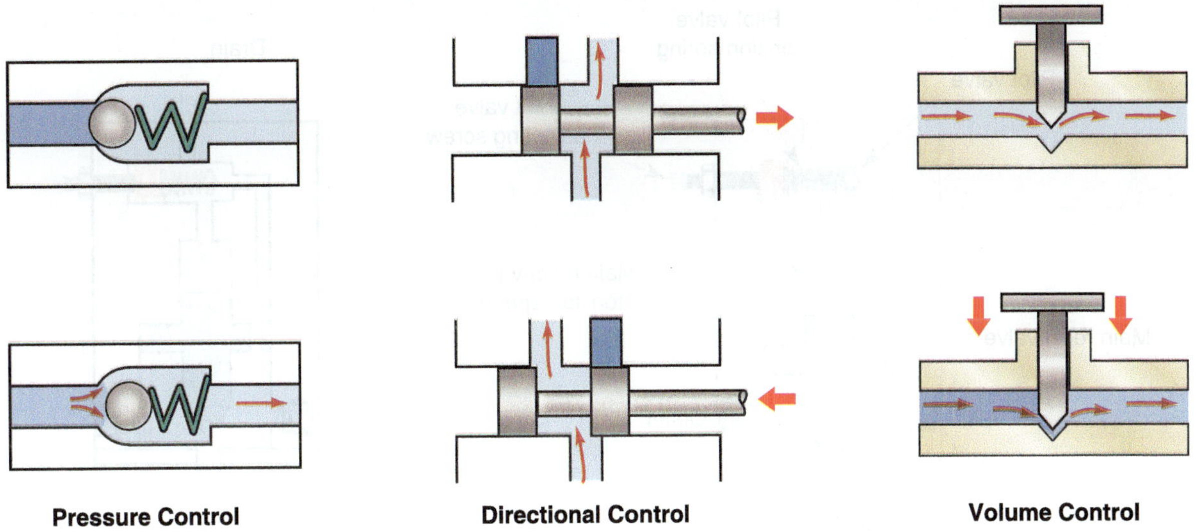

Directly Operated Valves. A directly operated relief valve consists of two operating ports: a ball poppet and a tension spring. The tension spring loads the ball poppet against the valve seat, giving the valve a normally closed operating status. When hydraulic pressure acting on the ball poppet is sufficient to overcome the spring pressure the ball unseats, permitting fluid to spill back to the reservoir, bypassing the circuit, and dropping circuit pressure. When system pressure returns to normal, spring pressure once again loads the poppet onto the valve seat. Directly operated relief valves may be adjustable: A screw located behind the tension spring permits adjustment of the valve opening pressure. **Figure 13–22** shows an illustration of a directly operated relief valve.

Pilot-Operated Valves. A **pilot-operated relief valve** is used to handle large volumes of fluid with little pressure differential. It consists of a main relief valve, through which the main pressure line passes, and a smaller relief valve that acts as a trigger to control the main relief valve. The valve body has a secondary port that directs flow back to the reservoir when the main relief valve opens. A passage in the valve body directs fluid flow to the pilot valve: As inlet pressure increases so does pressure in the pilot passage. When inlet pressure equals the pilot valve setting, it opens, releasing oil trapped behind the main relief valve and opening it. Excess oil then spills through the discharge port, preventing further pressure rise. At the moment inlet pressure drops below the valve setting pressure, the valves close. **Figure 13–23** shows an illustration of a pilot-operated relief valve.

FIGURE 13–22 Directly operated pressure-relief valve

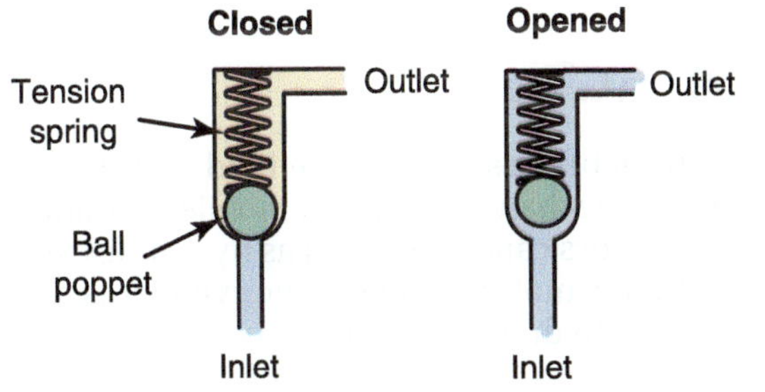

FLOW CONTROL VALVES

Flow control valves are used to regulate the speed of an actuator. These function by defining a flow area and can be classified as adjustable and nonadjustable. They are rated by operating pressure and capacity. They can be used in meter-in or meter-out applications. There are two general types of flow control valves:

- Bleed-off
- Fixed orifice

When meter-in is used to control flow, the valve is placed in series between the control valve and the actuator. This method of flow control is used in hydraulic circuits that continually resist pump flow such as those used to actuate a dump truck ram. A meter-out flow control valve is required when an actuator has the potential to run away. The meter-out flow control valve would be placed between the actuator and the reservoir to control flow away from the actuator.

A bleed-off flow control valve is placed between the pump outlet and the reservoir and it meters diverted flow as opposed to working flow. The result is that metered flow is directed back to the reservoir

FIGURE 13–23 Pilot-operated pressure-relief valve

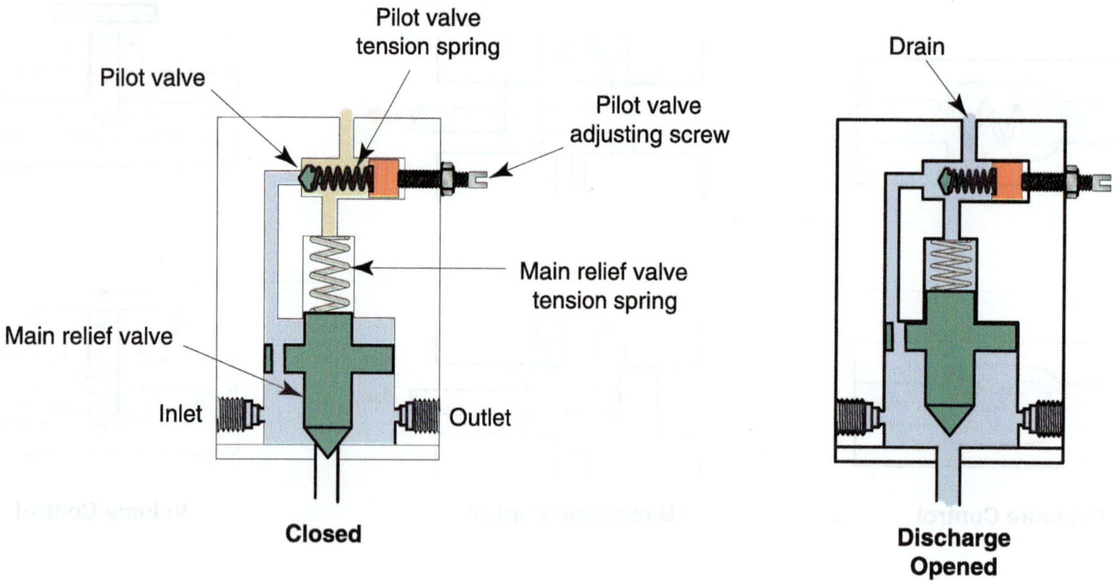

at the load pressure rather than the relief valve pressure. Bleed-off-type control valves are the least precise means of controlling flow. A simple fixed restriction in a hydraulic circuit can be used to control flow when it is located in series so that it controls actuator speed by either diverting or slowing flow.

DIRECTIONAL CONTROL VALVES

Directional control valves direct the flow of oil through a hydraulic circuit. They include:

- Check valves
- Rotary valves
- Spool valves
- Pilot valves

The center section of Figure 13–21 shows the operating principles of directional control valves.

Check Valves

A **check valve** uses a spring-loaded poppet that seats and unseats. It is designed to permit flow in one direction and to close to prevent flow in the other.

Rotary Valves

A rotary valve uses a rotary spool that turns to open and close oil passages and thereby option the routing of oil to different subcircuits. Rotary valves are commonly used as pilots for other valves in hydraulic systems with numerous subcircuits.

Spool Valves

Spool valves use a sliding spool within a valve body to open and close up hydraulic subcircuits. They are the primary type of directional control valve used in hydraulic circuits and function to start, actuate, and stop most activity in a typical circuit. A spool valve consists of a body and a cylindrical spool machined with annular recesses and lands. The annular recesses are the grooves machined into a spool, whereas the **lands** are the raised portions. Figure 13–8 shows a typical spool valve.

Though there is no limit to the number of lands that can be used in a spool valve, two-, four-, and six-land spools are most common. The spool recesses direct oil to and from sections of a hydraulic circuit, whereas the lands are used to seal off sections of a circuit. Spool valves may be controlled manually, hydraulically, pneumatically, or electrically. Spool valves are used extensively in hydraulic systems and automatic transmissions.

Pilot Valves

Pilot valves may be controlled mechanically, hydraulically, or electrically. The idea is to locate this type of control valve close to the function it controls, simplifying the hydraulic circuit. Pilot valves may use poppet, rotary, or spool operating principles and are used extensively in hydraulic circuits and automatic transmissions. Figure 13–23 shows a compound relief valve that uses a hydraulically actuated pilot trigger.

13.4 ACTUATORS

Hydraulic actuators convert the fluid power from the pump into mechanical work. In mobile hydraulic systems, actuators can be grouped as hydraulic cylinders and hydraulic motors. A hydraulic cylinder is a **linear actuator**. A hydraulic motor is a **rotary actuator**.

HYDRAULIC CYLINDERS

There are two types of hydraulic cylinders:

- Piston
- Vane

Although they are described briefly here, vane-type hydraulic cylinders are nearly obsolete.

Piston-Type Cylinders

In hydraulics, both single-acting and double-acting cylinders are used. In both types, a piston is moved linearly within the cylinder bore when it is subject to pressurized oil. In application, a dump truck ram would require only a single-acting cylinder, whereas COE cab-lift cylinders would have to be double acting.

Single-Acting Cylinders. In a single-acting cylinder, hydraulic pressure is applied to only one side of the piston. Single-acting cylinders may be either outward- or inward-actuated. When an outward-actuated cylinder has hydraulic pressure applied to it, the piston and rod are forced outward to lift the load. When the oil pressure is relieved, the weight of the load forces the piston and rod back into the cylinder. One side of a single-acting cylinder is dry. The dry side must be vented so that when oil pressure on the pressure side is relieved, air is allowed to enter, preventing a vacuum. In an outward-actuated single cylinder, the "dry side" or vent chamber is on the rod side of the piston.

When an inward-actuated cylinder has hydraulic pressure applied to it, the rod is pulled inward into the cylinder. This means that oil pressure acts on the rod side of the piston and the vent chamber is on the opposite side as shown in **Figure 13–24**.

A seal on the piston prevents oil leakage from the wet side to the dry side of the cylinder and a wiper seal is used to clean the rod as it extends and retracts. To prevent dirt from entering the dry side of the cylinder, a porous breather is used in the air vent.

A **ram** is a single-acting cylinder in which the rod serves as the piston. The rod is just slightly smaller than the cylinder bore. A shoulder on the end of the rod prevents the rod from being expelled from the cylinder. A ram-type cylinder eliminates the need for a dry side because oil fills the entire inner chamber when actuated.

Double-Acting Cylinders. Double-acting cylinders can provide force in both directions, which means that oil pressure can be applied to either side of the piston within the cylinder. When oil pressure is applied to one end of the cylinder to extend it, oil from the opposite side returns to the reservoir. This is reversed when the cylinder is retracted. Good examples of double-acting cylinders would be those required to raise and lower truck cabs on COE applications.

Double-acting cylinders may be balanced or unbalanced. In the balanced double-acting cylinder the piston rod extends through the piston head on both sides, giving an equal surface area on which hydraulic pressure can act. In an unbalanced double-acting cylinder a piston rod is located on one side of the piston. The result is that there is more surface area on which hydraulic pressure can act on the side without the rod. It should be noted that actual balance or unbalance of these cylinders depends on load. If the loads are not equal on either side, the balances will vary. **Figure 13–25** shows a sectional view of a typical double-acting cylinder.

Vane-Type Cylinders

Vane-type cylinders may be found in some much older hydraulic systems. A vane-type cylinder provides rotary motion. It consists of a cylindrical barrel, within which a shaft and vane rotate when pressurized oil enters. As the shaft vane rotates, it closes off an inlet passage in the top plate, leaving only a small orifice to discharge the oil; this slows the rotating vane as it comes to the end of its stroke. Most vane-type cylinders are double acting. Pressurized oil would be directed through one chamber to swing left, the other to swing right. Double-acting vane-type cylinders can be used in applications such as backhoes because they enable a boom and bucket to be swung rapidly from trench to pile. An alternative to a double-acting vane cylinder would be a pair of cylinders.

FIGURE 13–24 Single-acting cylinder

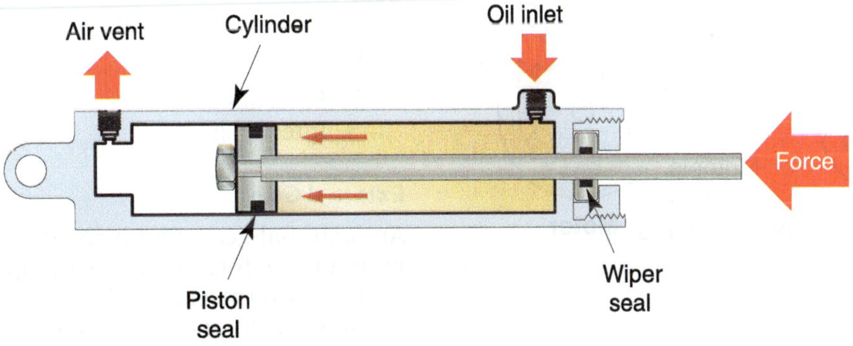

FIGURE 13–25 Double-acting cylinder

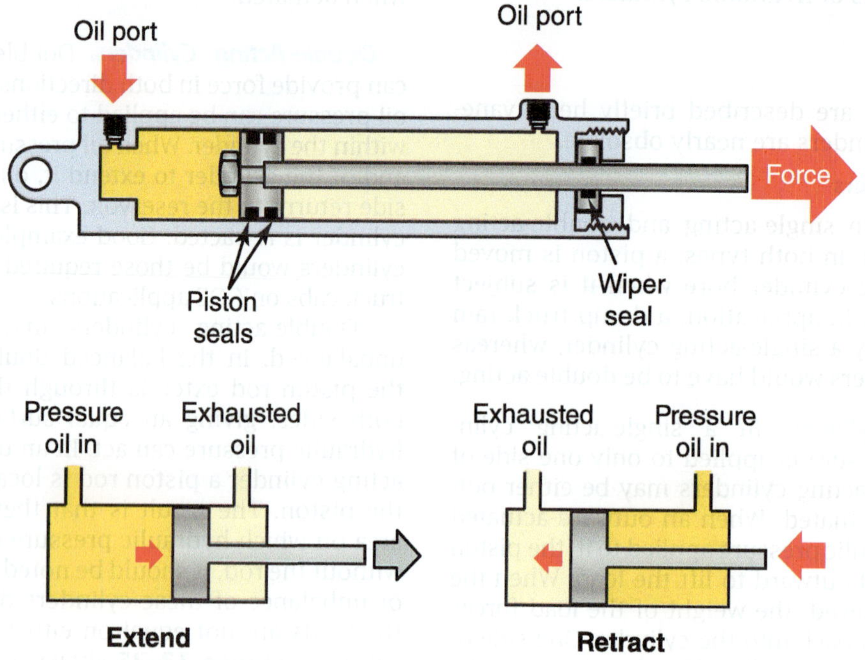

13.5 HYDRAULIC MOTORS

Hydraulic motors function oppositely from hydraulic pumps:

- **Pump**—draws in oil and displaces it, converting mechanical force into fluid force.
- **Motor**—oil under pressure is forced in and spilled out, converting fluid force into mechanical force.

A pump drives its fluid, whereas a motor is driven by fluid. **Figure 13–26** shows this principle.

FIGURE 13–26 Hydraulic pump versus motor principles

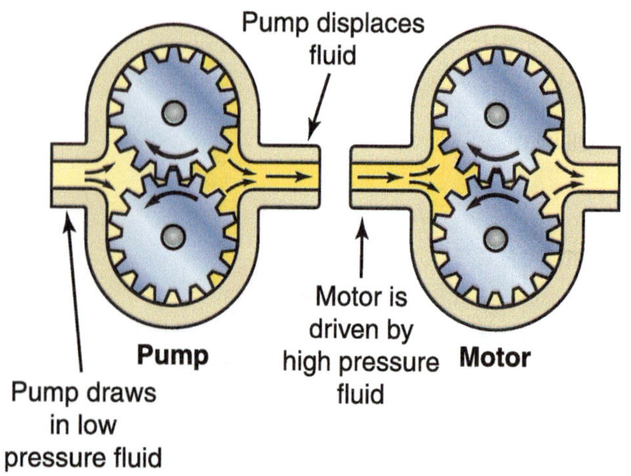

The work output of a motor is in the form of twisting force we describe as **torque**. Like pumps, motors can be of either fixed displacement or variable displacement. In a fixed-displacement motor, the speed is varied by managing the input flow of oil. This type of motor usually has a fixed torque rating. Variable-displacement motors can have both variable speeds and output torque. Input flow and pressure usually remain constant, so the speed and torque are managed by altering the displacement per cycle. There are three categories of hydraulic motor:

- Gear motors
- Vane motors
- Piston motors

All hydraulic motors rotate, driven by incoming hydraulic oil under pressure.

GEAR MOTORS

Gear motors are similar to gear pumps except that they function in reverse. There are two categories of gear motor:

- External gear
- Internal gear

External Gear

An external-gear motor is driven by pressurized hydraulic oil forced into the pump inlet, which acts on a pair of intermeshing gears, turning them away from the inlet, with the oil passing between the external

FIGURE 13–27 External-gear motor

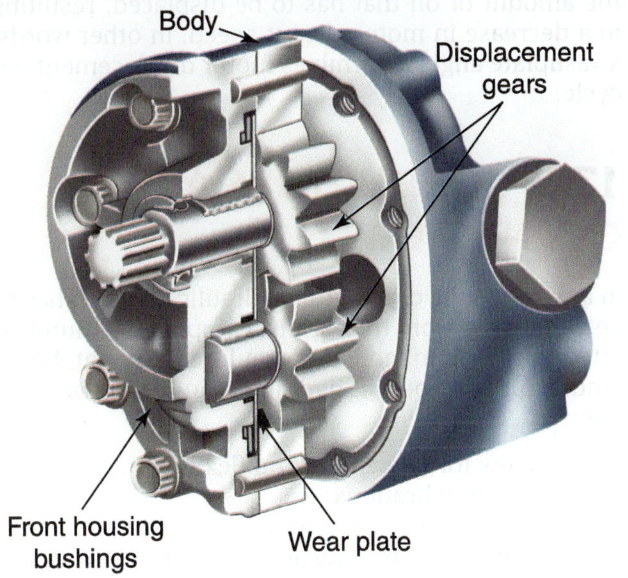

gear teeth and the pump housing. Hydraulic oil is then discharged at a lower pressure to the pump outlet. An output or drive shaft is integral with the axis of one of the two gears, allowing the motor to convert hydraulic potential to mechanical energy. **Figure 13–27** shows an external-gear hydraulic motor.

Internal Gear

An internal-gear motor is similar to an internal-gear pump in that an external-lobed rotor is driven within an internal-lobed rotor ring. The axis of the internal rotor is offset from that of the rotor ring. Pressurized hydraulic oil is delivered to the motor inlet, charging the cavity formed between the inner and outer rotor lobes, forcing both to rotate to discharge the oil at lower pressure through the outlet. The motor drive shaft is connected to the inner rotor. **Figure 13–28** shows an internal-gear hydraulic motor.

VANE MOTORS

Vane motors use both balanced and unbalanced operating principles and again function similarly to vane pumps operating in reverse. However, vane-type hydraulic motors are normally equipped with springs to load the vanes into the liner wall. This is necessary because incoming oil at the inlet is at a high enough pressure (especially at startup) to overcome any amount of centrifugal force developed by pump rotation and prevent the vanes from contacting the pump liner or rotor ring. Once in motion, the combination of centrifugal force and oil pressure is usually sufficient to load the vanes into the liner wall.

A balanced version of a vane-type motor (**Figure 13–29**) will locate a pair of diametrically opposed inlet and outlet ports to prevent offset loading of the output shaft connected to the rotor. Balanced vane motors are only capable of fixed displacement but they operate with higher efficiencies than gear motors do, and rotational direction can be reversed simply by switching the direction of fluid flow.

FIGURE 13–28 Internal-gear motor

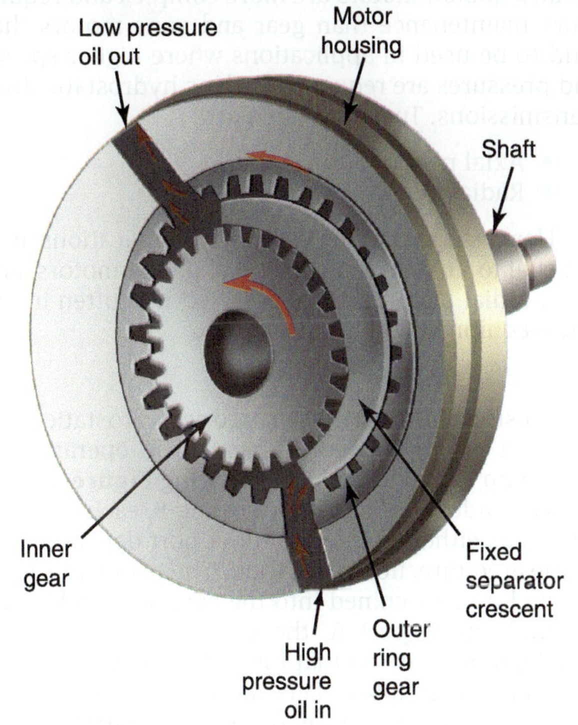

FIGURE 13–29 Balanced vane motor in operation

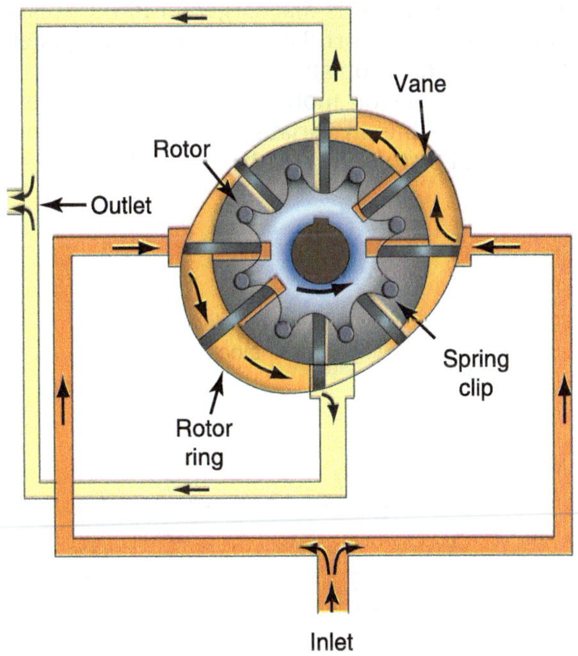

PISTON MOTORS

Because piston motors are more complex and require more maintenance than gear and vane motors they tend to be used in applications where higher speeds and pressures are required, such as hydrostatic drive transmissions. Two types are used:

- Axial piston
- Radial piston

However, in mobile hydraulic applications it is unusual to find anything but axial piston motors, and when radial piston motors are used it is often in off-shore equipment.

Axial Piston Motors

Axial piston motors are often used as hydrostatic drive units on off-highway equipment. The operation of this pump is described by referencing **Figure 13–30**.

The endcap contains ports A and B, each of which can act as either an inlet or outlet port depending on the desired direction of rotation. The pistons reciprocate in bores machined into the cylinder block rotor. As fluid enters port A, the piston bore is charged with high-pressure oil that forces the piston against a fixed-angle swashplate. This allows the piston to slide down the swashplate face, turning the cylinder block that in turn rotates the output shaft. As the cylinder block continues to turn, other piston bores align with the inlet port A, allowing each to be similarly actuated in sequence. Oil is discharged at lower pressure through the outlet port B as it comes into register at the end of each piston stroke.

Most axial piston motors are fixed displacement and not reversible. In variable-displacement applications of axial piston motors, the swashplate angle can be adjusted using an arm and lever assembly. The greater the swashplate angle to the shaft, the larger the amount of oil that has to be displaced, resulting in a decrease in motor output speed. In other words, swashplate angle determines motor displacement per cycle.

FIGURE 13–30 Axial piston motor

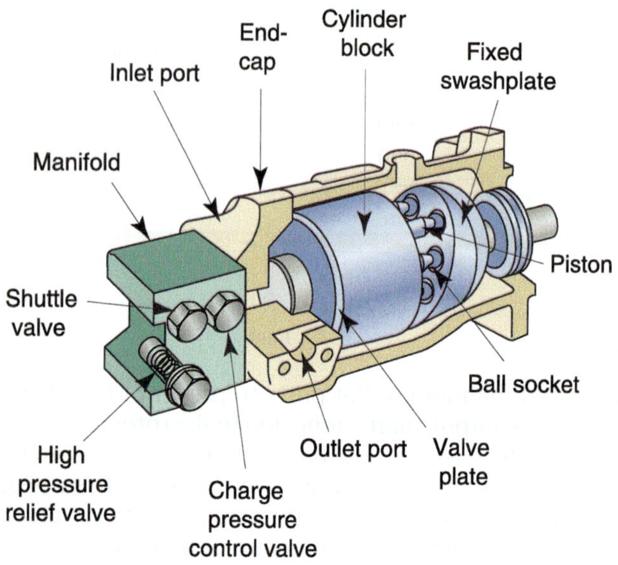

13.6 CONDUCTORS AND CONNECTORS

In any hydraulic circuit, the hydraulic medium has to be conducted from the various components plumbed into the circuit. In mobile hydraulic equipment, hoses tend to function best as hydraulic conductors because they:

- Allow for movement and flexing
- Absorb vibrations
- Sustain pressure spikes
- Enable easy routing and connection on chassis

HYDRAULIC HOSES

The selection of hoses in a hydraulic circuit is important because they play a major role in determining how the system performs. The flow requirements of a conductor determine what the internal diameter of a hose should be. For instance, a hose that can withstand the pressure specification but that has too small an internal diameter will restrict circuit flow, causing overheating and pressure losses. The size of any hydraulic hose is determined by its inside diameter. This is sometimes indicated as **dash size** in 1/16-inch increments: Because each dash number indicates 1/16 inch, a #4 dash hose would be equivalent to 4/16 inch or 1/4 inch.

Table 13–1 shows some common dash sizes and their nominal equivalents.

A large-diameter internal hose has to be stronger to sustain the working pressures of a hydraulic circuit. Generally, as the hose sectional inside diameter increases, its working pressure decreases. This can be observed by referencing **Table 13–2**. Another consideration for hose selection is that the hose must be compatible with the hydraulic fluid used in the

TABLE 13–1 Dash Size and Nominal Diameter

Dash Size	Nominal Diameter
# 4	1/4 inch
# 6	3/8 inch
# 8	1/2 inch
# 10	5/8 inch
# 12	3/4 inch

TABLE 13–2 Hose Working Pressures

Hose Size	Working Pressure: Single-Wire Braid	Working Pressure: Multiple-Wire Braid	Working Pressure: Multiple Spiral Wire Braid
¼ inch #4	3,000 psi	5,000 psi	NA
⅜ inch #6	2,250 psi	4,000 psi	5,000 psi
½ inch #8	2,000 psi	3,500 psi	4,000 psi
⅝ inch #10	1,750 psi	2,750 psi	3,500 psi
¾ inch #12	1,500 psi	2,250 psi	3,000 psi
1 inch #16	800 psi	1,875 psi	3,000 psi
1½ inch #24	500 psi	1,250 psi	3,000 psi
2 inch #32	350 psi	1,125 psi	2,500 psi

system. There are four general types of hoses used in hydraulic circuits:

- Fabric braid
- Single-wire braid
- Multiple-wire braid (up to 6 wire braid)
- Multiplespiral wire (up to 6 wire spirals)

These types of hydraulic hoses are shown in **Figure 13–31.**

Fabric Braid

Fabric braid hoses are used for petroleum base hydraulic and fuel oils at low pressures. There are various categories of fabric braid hose depending on application. They are often used on the low-pressure side of a fuel subsystem or the line that connects a hydraulic reservoir to a pump. They are also commonly used in return circuits. Typically, fabric braid hoses are constructed using a synthetic inner tube with fiber reinforcing and sometimes with an additional spiral wire to prevent collapse. The outer wrapping can be either synthetic rubber or fiber braid. If fabric braid hose is to be used at lower than atmospheric pressures, the maximum low-pressure specification must be observed. Ratings of up to 40 inches Hg inlet restriction are available but the specified rating drops as the inside diameter increases. Fabric braid line is not recommended for use as pressure line in mobile hydraulic circuits.

FIGURE 13–31 Four types of hydraulic hoses

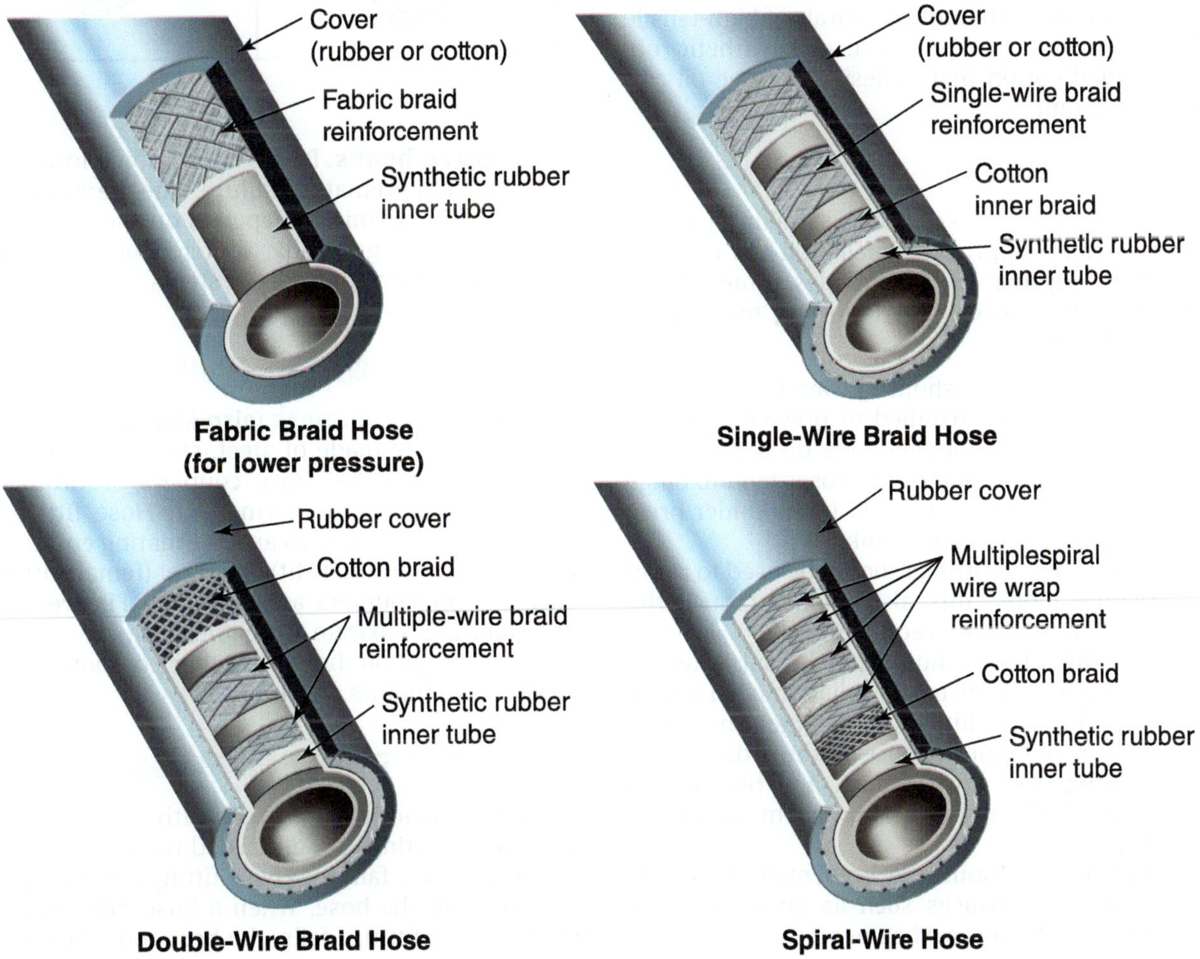

Single-Wire Braid Hose

Single-wire braid hose is commonly used as the conductor for the pressure side of diesel fuel subsystems and some hydraulic circuits. The inner tube is usually manufactured with synthetic rubber reinforced with a single braid of high-tensile steel wire. The outer wrap consists of a synthetic rubber or braided cotton that is designed to be oil and abrasion resistant. Single-wire braid is often used in truck, farm, and heavy equipment hydraulic circuits and in fuel subsystem applications.

Double-Wire Braid Hose

Double-wire braid hose is commonly used as the conductor for high-pressure hydraulic circuits. The inner tube is usually manufactured with synthetic rubber reinforced with at least two braids of high-tensile steel wire. The outer wrap again consists of a synthetic rubber or braided cotton that is designed to be oil and abrasion resistant. Double-wire braid is often used in truck, farm, and heavy equipment hydraulic circuits using higher pressure.

Spiral-Wire Hose

Spiral-wire braid hose is used in high-pressure hydraulic circuits and is the preferred option in systems that have to withstand high-pressure surges. The inner tube is usually manufactured with synthetic rubber reinforced with multiple spirals of high-tensile steel wire. The outer wrap consists of a synthetic rubber or braided cotton that is designed to be oil and abrasion resistant.

INSTALLING HOSES

Some care should be taken when installing hydraulic hoses because improper installation can result in rapid failure. **Figure 13–32** provides some tips on routing hydraulic hoses. The following precautions should be observed:

- **Taut hose**. There should always be some slack when a hose is installed to prevent straining even when the coupled ends do not move in relation to each other. Taut hoses tend to bulge and weaken in service and hoses under pressure can fractionally shrink.
- **Loops**. Angled fittings should be used to avoid creating loops; this also should reduce the total length of hose required.
- **Twisting**. Hoses should not be subjected to twisting either during installation or in operation. When installing hoses, tighten the fitting on the hose—not the hose on the fitting.
- **Rubbing**. Use hose clamps and brackets to ensure that hoses do not contact moving parts on the chassis.
- **Heat**. Hoses should be prevented from contacting hot surfaces such as engine components. Shield hoses when necessary.

FIGURE 13–32 Hydraulic hose routing options.

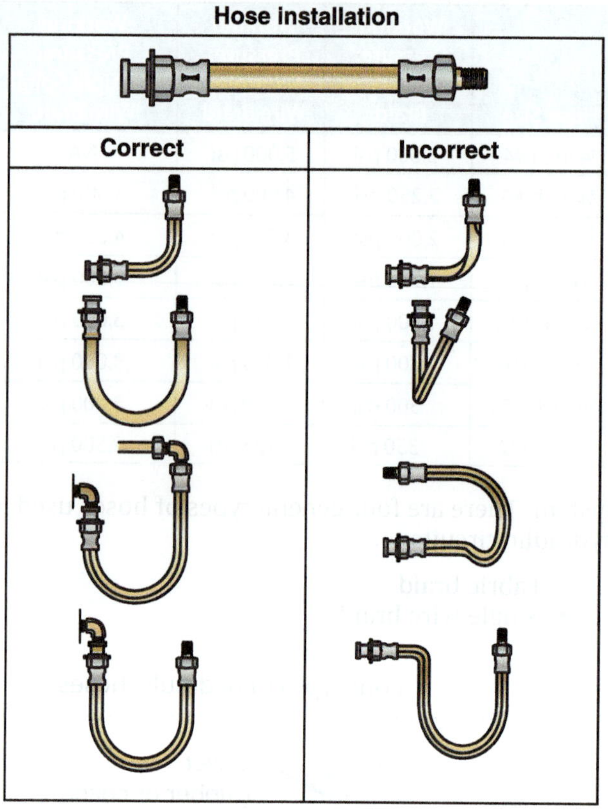

- **Sharp bends**. Bends choke downflow and can lead to premature failure. The maximum bend radius permitted depends on the type of hose used but generally tighter bends are permissible in low-pressure systems.

COUPLERS (CONNECTORS)

Hydraulic hose **couplers** (also known as *connectors* and *fittings*) are made of steel, stainless steel, brass, or fiber composites. Hose couplers or fittings can be either reusable or permanent. Hose fittings are installed at the hose ends and the mating end consists of either a nipple (male fit) or socket (female fit). Some examples of couplers are shown in **Figure 13–33**. Adapters are separate from the hose assembly and are used to couple hoses to other components such as valves, actuators, or pumps.

PERMANENT HOSE FITTINGS

When hydraulic hose is fitted with permanent hose fittings, the fittings are discarded with the hose in the event of a hose failure. These fittings are crimped or swaged onto the hose. When a hose fitted with permanent hose fittings fails, the hose must be replaced

FIGURE 13–33 Assorted couplers

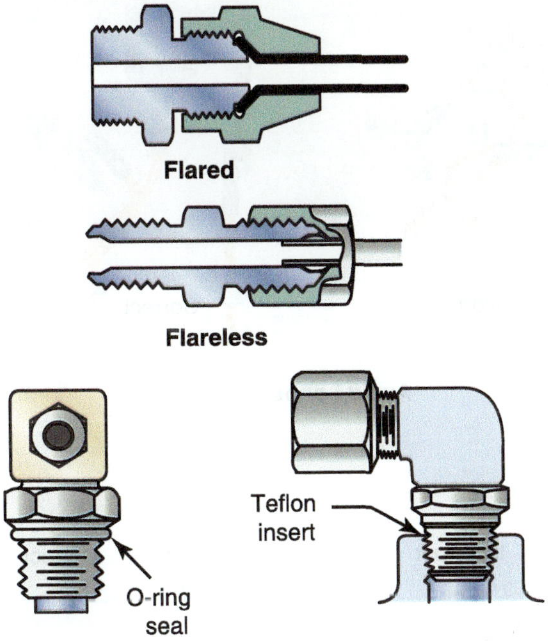

either as an assembly or one must be made up using stock hose cut to length fitted with either crimp-type or reusable fittings. **Figure 13–34** shows some standard permanent hose couplers.

REUSABLE HOSE FITTINGS

Reusable fittings are common in truck shops because a hose assembly station, some stock hose, and an assortment of reusable fittings can replace many of the hundreds of different types of hoses used on various original equipment manufacturer (OEM) truck chassis. When a hose with reusable fittings wears out, the fittings can be removed and assembled onto new stock hose. Reusable fittings are usually screwed onto hose, although some types of low-pressure hoses may use press fits.

ASSEMBLING HOSE FITTINGS

Figure 13–35 shows some hose-to-fitting assembly methods. Because of the high pressure used in many hydraulic circuits, the OEM assembly procedure must be precisely observed.

CAUTION

Separation of a fitting and hose at high pressure can be dangerous! Never reuse a suspect fitting and observe the manufacturer assembly procedure to the letter.

FIGURE 13–34 Permanent hose couplers

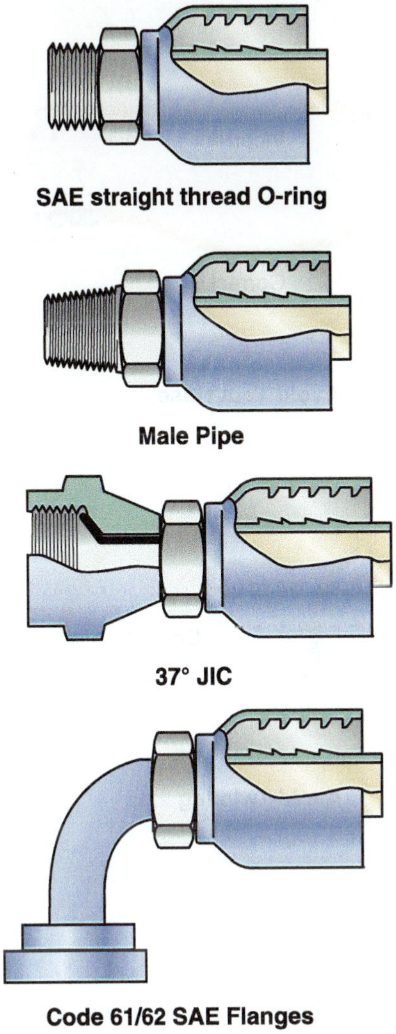

Sealing Fittings

Fittings can be sealed to couplers using the following:

- Tapered threads
- O-rings
- Nipple and seat (flare)

When making hydraulic connections, ensure that the coupler fittings are compatible with each other.

Adapters

Adapters are separate from the hose assembly. They have the following functions:

- To couple a hose fitting to a component
- To connect hydraulic lines in a circuit
- To act as a reducer in a circuit
- To connect a pair of hoses on either side of a bulkhead

FIGURE 13–35 Installing hydraulic hoses

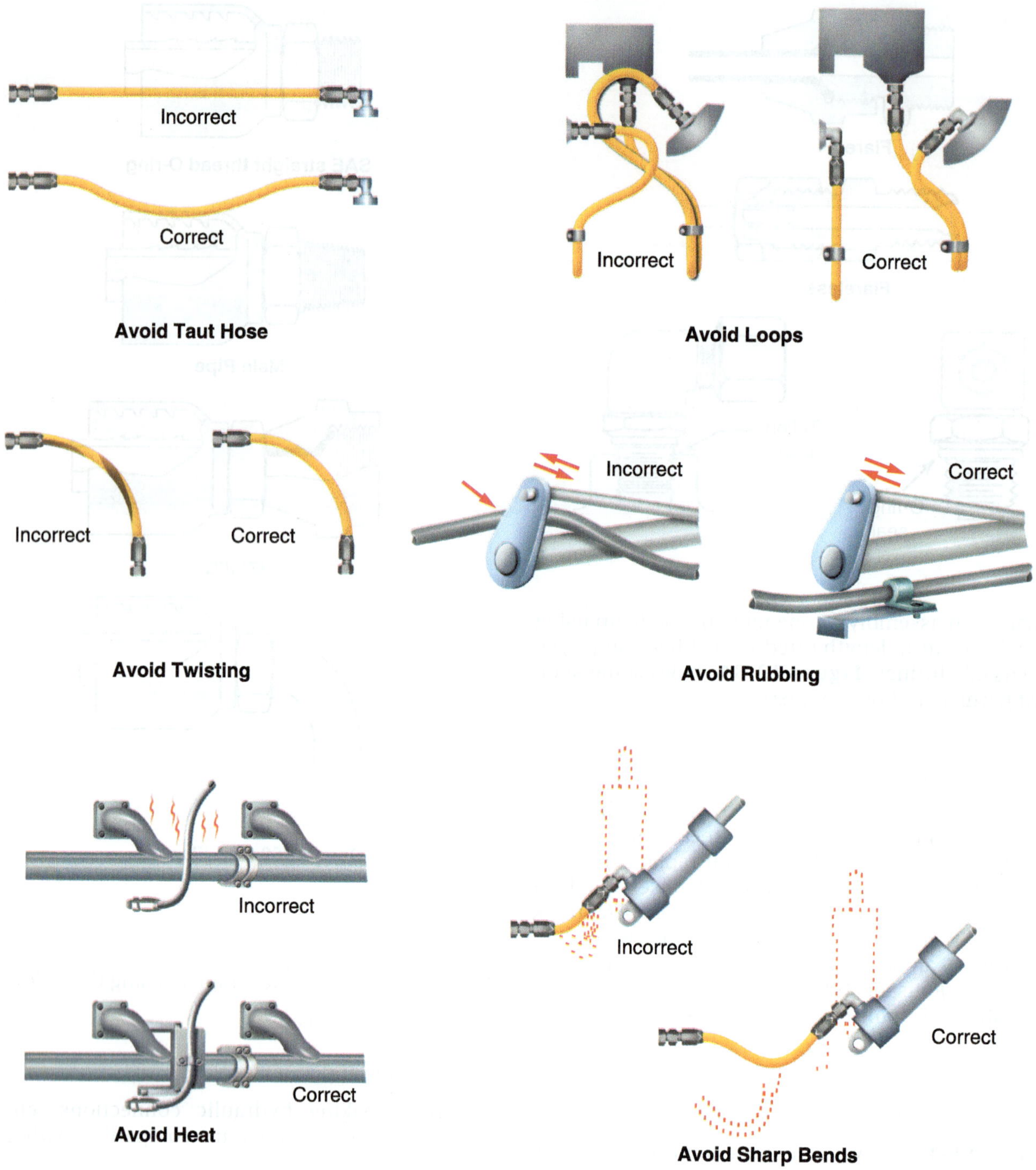

COUPLING GUIDELINES

When making hydraulic connections the following guidelines should be observed:

- Torque the fitting on the hose, not the hose on the fittings.
- Couple male ends before female ends.
- Ensure that the sealing method of each fitting to be coupled is the same.
- Use 45- and 90-degree elbows to improve hose routing.
- Use hydraulic pipe seal compound only on the male threads and on thread-seal unions.
- Use two wrenches when tightening unions to avoid twisting hose.
- Never overtorque hydraulic fittings.

SHOP TALK

When tightening the fittings on a pair of hydraulic couplers, always use two wrenches to avoid twisting hoses or damaging adapters.

PIPES AND TUBES

Pipes used in hydraulic circuits are generally made from cold-drawn, seamless mild steel. It should never be galvanized because the zinc can flake off and plug up hydraulic circuits. Tubing can also be used; it has the advantage of being able to sustain some flex—hence its use in vehicle brake systems. Tubing should be manufactured from cold-drawn steel if used in moderate-to-high-pressure circuits. When used in low-pressure circuits, copper or aluminum tubing may be used.

Pipe Couplers

Pipe joints are made by threading the outside diameter (od) and then turning it in to a tap-threaded bore. The standard threads used in vehicle applications are:

- National standard tapered thread (NPT)
- National standard dryseal taper pipe thread (NPTF)

An NPT coupling is sealed by the interference contact between the flanks of the mating threads, and as a consequence pipe sealing compound is required to ensure a seal. In an NPTF dryseal coupling, the roots and crests of the opposing threads contact before the flanks contact and a seal is created by crushing the thread crests. Dryseal couplers are intended for one-off usage and usually do not require thread seal compound.

SHOP TALK

When coupling pipe connectors, remove burrs from both the male and female threads, then apply sealing compound to the male threads only.

Tube Couplers

Tube couplers are designed so that the fitting can be tightened while the tube remains stationary. Two types of tube coupling seal are used: flared and flareless. Flared fittings seal by the metal-to-metal crush created when the flared mating surfaces contact between a pair of fittings. Flareless fittings seal by compression, either with or without a ferrule.

Flared Fittings. Flared fittings are used in thin wall tubing and the flare angle is usually 45 degrees (Society of Automotive Engineers [SAE] standard), although some fittings use 37 degrees (joint industry committee or JIC standard). The following types of flare fitting are used:

- **Inverted flare**. A 45-degree flare is rolled to the inside of the fitting body. It is commonly used in hydraulic brake circuits.
- **Two-piece flare**. A tapered nut aligns and seals the flared end of the tube.
- **Three-piece flare**. A three-piece flare fitting consists of a body, sleeve, and nut and fits over the tube. The sleeve free-floats, permitting clearance between the nut and tube and aligning the fitting. When tightened, the sleeve is locked without imparting a twist to the flared tube.
- **Self-flaring**. These fittings use a wedge-type sleeve that, as the sleeve is tightened, is forced into the tube end, spreading it into a flare. This type of fitting combines the advantages of high strength and fast assembly.

CAUTION

Never attempt to cross-couple SAE and JIC fittings: The result will be to damage both.

Flareless Fittings. Flareless fittings are commonly used in vehicle fluid power circuits, especially pneumatic circuits. They require no special tools for assembly and are reusable. There are three basic types:

- **Ferrule fittings**. Consist of a body, a compression nut, and a **ferrule**. A wedge-shaped ferrule is compressed into the fitting body by the compression nut, creating a seal between the tube and the body.
- **Compression fittings**. These are used with thin-walled tubing and are sealed by crimping the end of the tube to form a seal.
- **O-ring fittings**. O-ring fittings use a principle similar to ferrule-type fittings except that a compressible rubber compound O-ring replaces the ferrule. They differ from ORFS fittings in that they rely on the O-ring alone to achieve a seal. As the compression nut is torqued, the O-ring compresses, forming a seal between the tube and the fitting body. Several types of O-ring are used, including round section, square section, D-section, and steel-backed.
- **ORFS fittings**. ORFS (O-ring face seal) fittings have become popular in recent years because they provide a triple seal in separate mated contact faces. ORFS fittings meeting SAE J1453 standards as shown in **Figure 13–36** are often described as universal because the body and the nut of the fitting remain the same even when switching from standard- to metric-sized lines or pipes. They are also rated to withstand greater levels of vibration and wider ranges of temperature variance. **Figure 13–37** shows how an ORFS fitting achieves its seal on three mating plains.

SHOP TALK

When tightening tube fittings, torque them only until snug. Overtorquing any kind of line fitting is a direct cause of leaks. Use two wrenches when torquing fittings, one to hold the fitting body, the other to turn the compression nut.

FIGURE 13–36 ORFS fitting

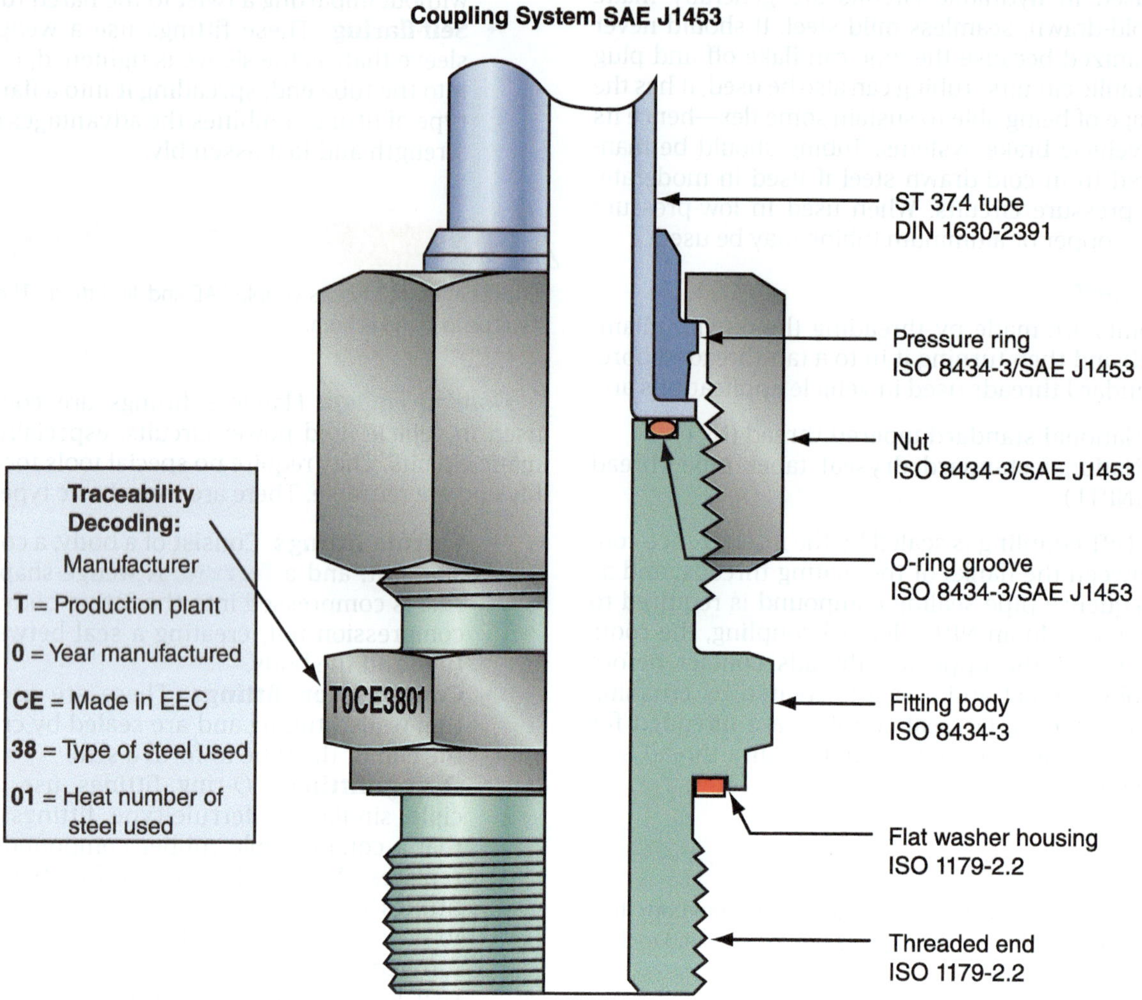

Quick-Release Couplers

When hydraulic lines have to be frequently connected and disconnected, quick-disconnect couplers are used. A typical application would be a wet-line kit on a dump trailer that has to be connected to multiple tractor units. A quick coupler is a self-sealing device that shuts off flow when disconnected. This eliminates the need to bleed a system each time a disconnect/connect takes place. **Quick-release couplers** consist of a male and female coupler. There are four types:

- **Double poppet**. A double-poppet quick connector has a self-sealing poppet in both the male and female halves. When coupled, each poppet is forced off its seat, allowing oil flow through the coupling. When disconnected, the poppets in both the male and female halves are closed by spring force, sealing both sides of the disconnected circuit. A connected double poppet coupler is locked into position by a ring of balls held into position by a spring-loaded sleeve. When the sleeve is retracted, spring pressure on the ring of balls is relieved, permitting the disconnect.
- **Sleeve and poppet**. Sleeve-and-poppet quick-release couplers have a self-sealing poppet in one half and a tubular valve and seal in the other half of the coupling. This type of quick coupler provides faster sealing on disconnect, minimizing the entry of air or oil leakage from the coupling.
- **Sliding seal**. Sliding seal couplers use a sliding gate that at disconnect covers a port on each coupling half. These tend to be prone to some leakage during connect and disconnect.
- **Double-rotating ball**. Double-rotating ball quick connectors couple by inserting the line plug into the coupler body while turning a locking valve lever. The action of rotating the locking lever forces back a pair of sealing

FIGURE 13–37 Sealing principle of an ORFS fitting

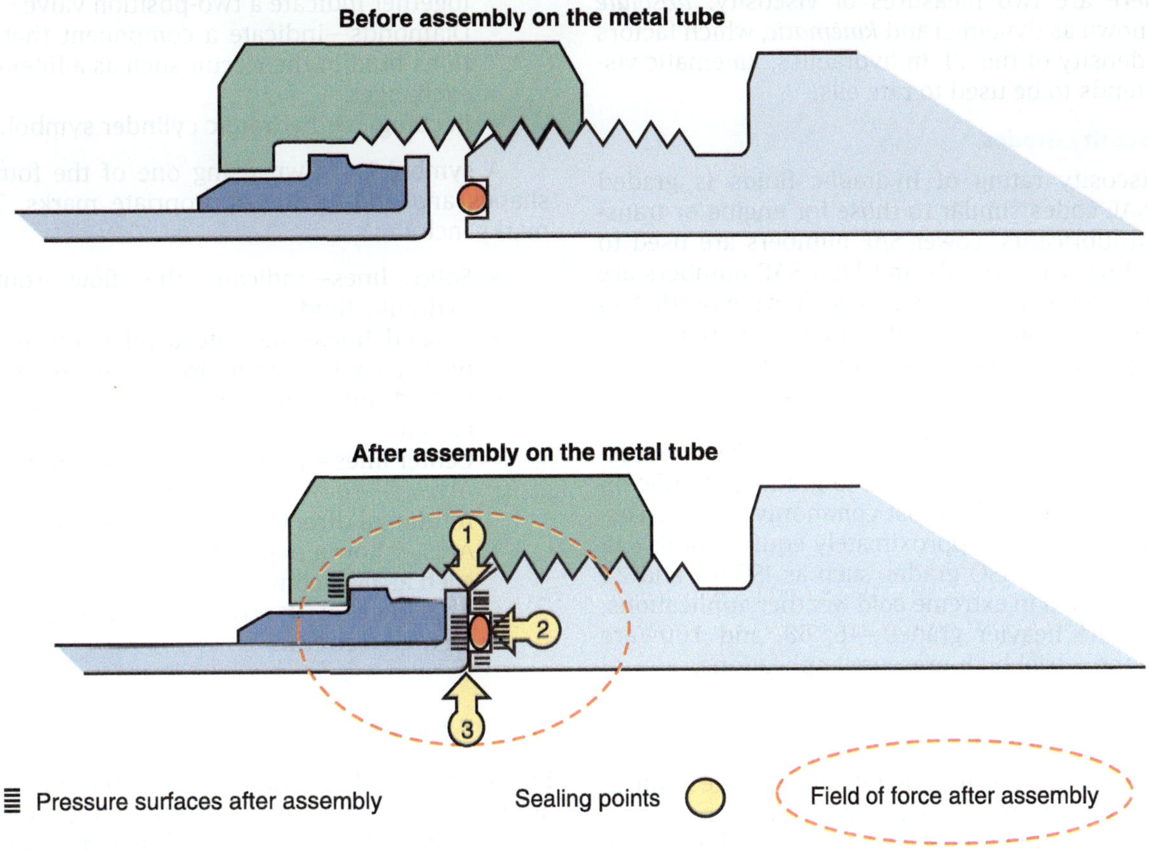

balls, permitting circuit flow. When disconnected, the locking lever is turned and spring pressure loads the sealing balls into recesses in the locking valve lever, sealing the circuit. Most double-rotating ball quick couplers have an automatic lock release that permits both immediate sealing and hose release in the event of an unintended hydraulic circuit separation. For instance, if such a union is made between a tow unit and a trailer, the separation would occur without damage to the hydraulics of either circuit.

13.7 HYDRAULIC FLUIDS

Hydraulic fluid is the medium used to transmit force through a hydraulic circuit. Most hydraulic fluids are refined from petroleum-base stocks and as such are seldom compatible with vehicle brake fluids. Hydraulic fluids used in truck hydraulic systems may be specialty hydraulic oils but engine and transmission oils are also commonly used.

Always check when adding to or replacing hydraulic oil. Synthetic hydraulic oils are commonly used in today's hydraulic circuits because they have wider temperature operating ranges and offer greater longevity. Hydraulic oils must:

- Act as hydraulic media to transmit force
- Lubricate the moving components in a hydraulic circuit
- Resist breakdown over long periods
- Protect circuit components against rust and corrosion
- Resist foaming
- Maintain a relatively constant viscosity over a wide temperature range
- Resist combining with contaminants such as air, water, and particulates
- Conduct heat

VISCOSITY

Viscosity is a rating of a liquid's resistance to fluid shear but is also a measure of its resistance to flow. Petroleum oils tend to thin as temperature increases and begin to gel at lower temperatures. In a hydraulic circuit, if the viscosity of hydraulic fluid is too low (fluid is thin), the chances of leakage past seals and poor surface lubrication increase. If viscosity is too high (fluid is thick), more pump force is required to push fluid through the circuit and the operation of actuators may be sluggish.

There are two measures of viscosity: *absolute* (also known as dynamic) and *kinematic*, which factors in the density of the oil. In hydraulics, kinematic viscosity tends to be used to rate oils.

SAE Viscosity Grades

The viscosity rating of hydraulic fluids is graded using SAE codes similar to those for engine or transmission lubricants. Lower SAE numbers are used to denote low-viscosity oils and high SAE numbers are used to denote high-viscosity oils. This means that an SAE 10W hydraulic oil will flow more readily than an SAE 30-graded oil at a given temperature.

ISO Viscosity Grades

International Standards Organization (ISO) viscosity grades are commonly used to grade hydraulic oil viscosity. ISO 32 is the most commonly used viscosity grade and this is approximately equivalent to SAE 10 grade. Lighter ISO grades such as ISO 15 and 22 tend to be used in extreme cold weather applications, whereas the heavier grades—46, 68, and 100—are used in industrial high-pressure applications.

VISCOSITY INDEX

Viscosity index (VI) is used to rate oil's change in thickness as temperature changes. If oil thickens at low temperatures and becomes very thin at high temperatures, it has low VI. When oil can retain a relatively consistent viscosity through a wide temperature range it can be described as having high VI. In a hydraulic circuit, the hydraulic medium is required to be thick enough to prevent leakage while also offering low-flow resistance through the circuit, so high VI is essential.

13.8 SYMBOLS AND SCHEMATICS

When working with vehicle hydraulic schematics, it is important to be able to interpret standard symbols. Symbols are two-dimensional figures that roughly approximate the components they represent in a hydraulic schematic. The symbols used in hydraulic schematics are standardized by the American National Standards Institute, or ANSI, or by the ISO. This helps makes life easier because most manufacturers use either ANSI or ISO symbols in their schematics. The graphics used in hydraulic schematics consist of shapes and marks.

SHAPES AND MARKS

The shapes used in hydraulic schematics are:

- Circles/semicircles—pump or motor symbol.
- Squares—(or envelope) used to represent one position or path through a valve. Two squares together indicate a two-position valve.
- Diamonds—indicate a component that conditions fluid in the circuit such as a filter or heat exchanger.
- Rectangles—hydraulic cylinder symbol.

A symbol is drawn using one of the four basic shapes and adding the appropriate marks. Typical marks include:

- Solid lines—indicate the flow route for hydraulic fluid.
- Dashed lines—indicate a pilot line that connects a control circuit to a slave circuit.
- Dotted lines—indicate a return or exhaust circuit.
- Center lines—used to enclose assemblies.
- Arrows—show the direction of fluid flow or rotational direction of pumps and motors.
- Arcs—show a point of restriction in the circuit such as flow control valve.
- Angled line with arrowhead—a point of adjustment in a circuit.

ANSI SYMBOLS

Figure 13–38 shows most of the ANSI symbols you are likely to come across. Use this to interpret some typical hydraulic circuits. Some truck manufacturers make their schematics a little more user friendly for technicians not accustomed to working exclusively on hydraulic circuits.

13.9 MAINTENANCE

As with all chassis systems, proper maintenance is a key to having a hydraulic system perform properly and minimize repair downtime. Maintenance problems result from:

- Low reservoir oil level
- Clogged oil filters
- Contaminated or improper oil in the system
- Dirt in the hydraulic circuit

Preventive maintenance schedules should be observed. Most routine maintenance on hydraulic circuits takes little time and can eliminate costly component repair and replacement. This section takes you through some typical maintenance procedures.

SAFE PRACTICE

Truck hydraulic circuits are designed to run at high pressures and support high loads. It is essential that you work safely around chassis hydraulic equipment. Some basic rules are:

- Never work under any device that is only supported by hydraulics. A raised dump box or chassis hoist must be mechanically supported before you work under it. Just as when using a floor jack, you must use some means of mechanically supporting any raised equipment or components.

FIGURE 13–38 ANSI schematic symbols

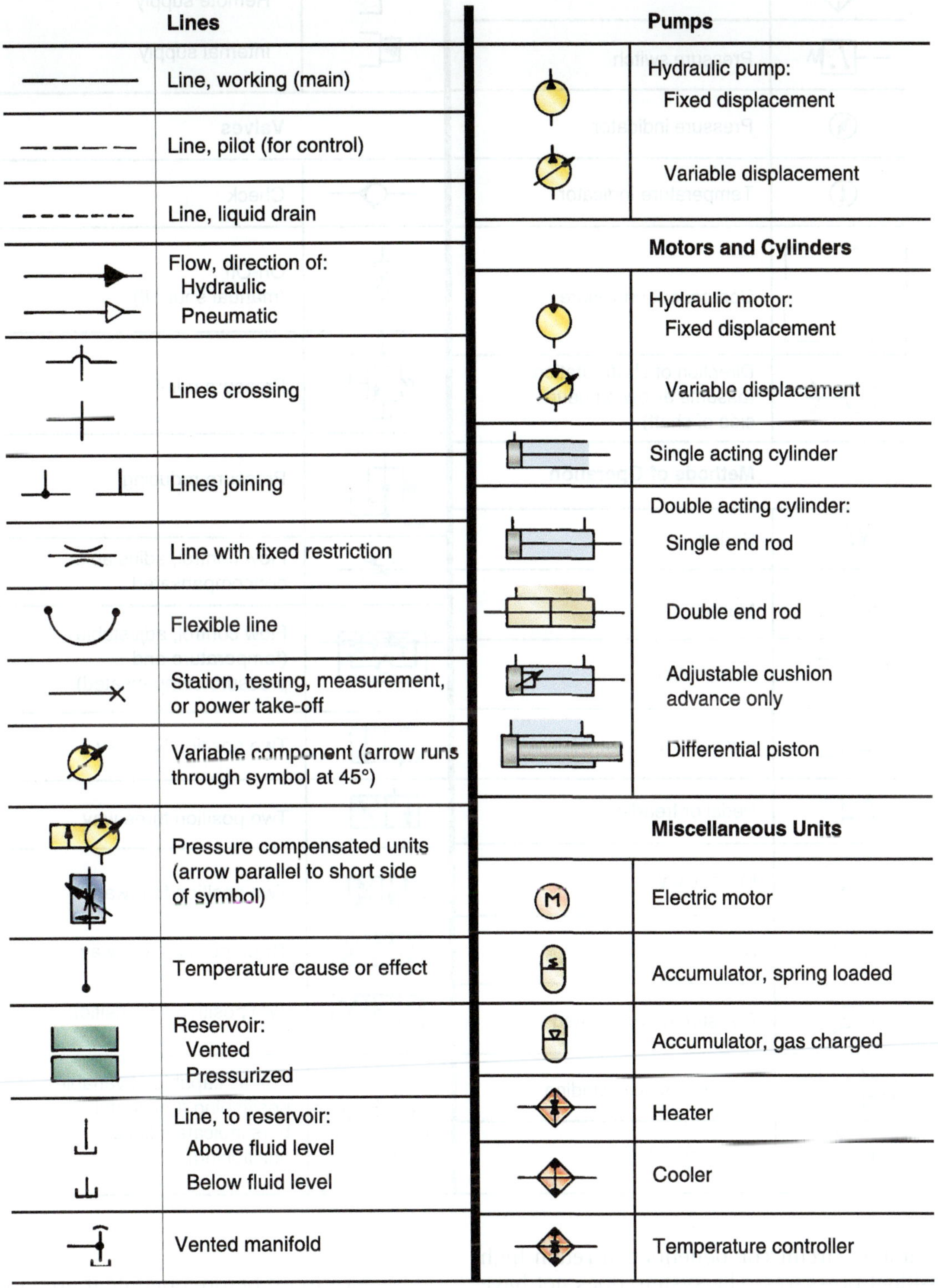

(continued)

FIGURE 13–38 (continued)

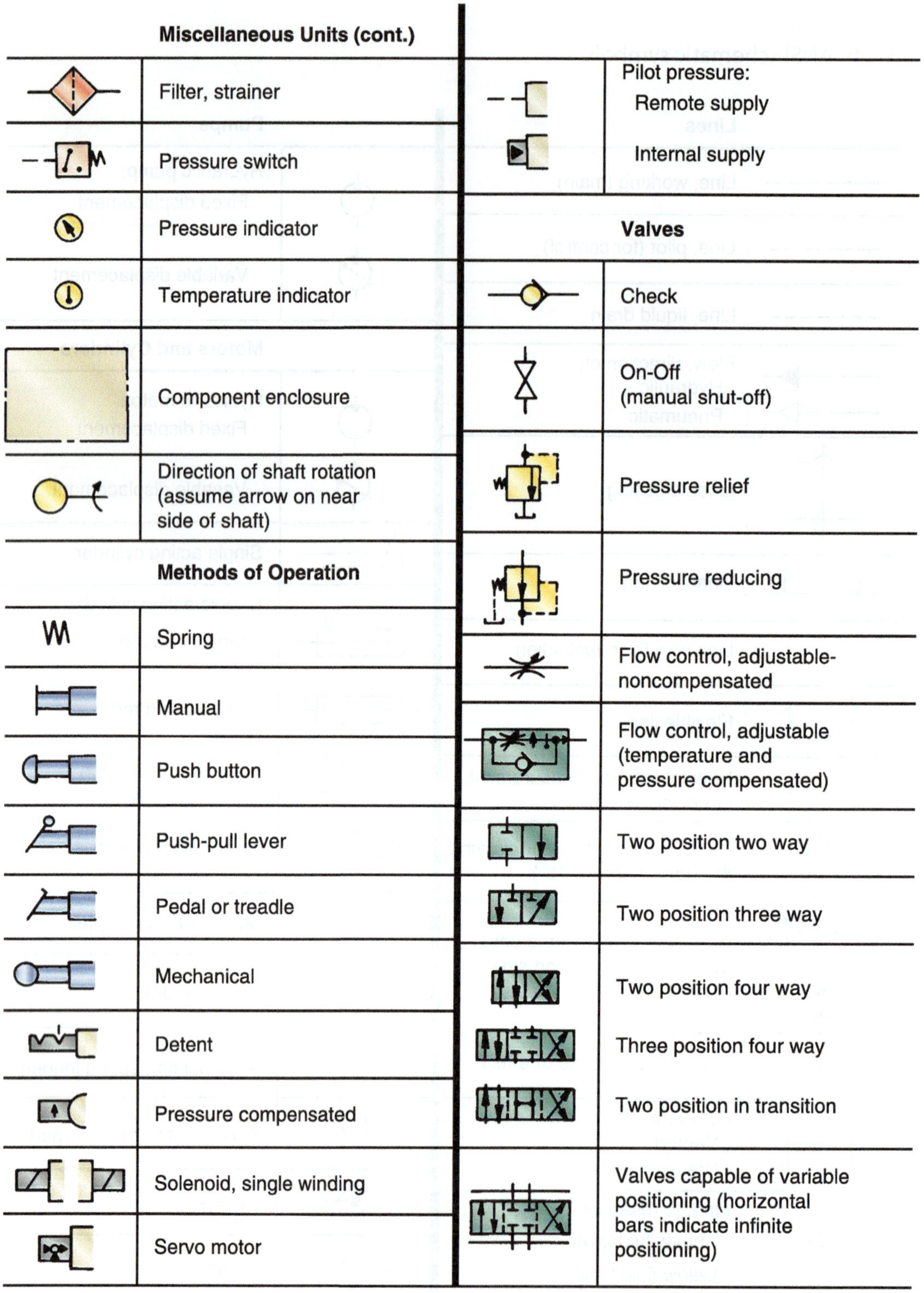

- Hydraulic circuit components can retain high residual pressures. The system does not have to be active for this to be a potential. Try to ensure that pressures are relieved throughout the circuit before opening it up. Crack hydraulic line nuts slowly and be sure to wear both safety glasses and gloves.

Pinhole Injection Injury

A major leak in a hydraulic circuit is rapidly identified because it makes a mess. A pinhole leak may be microscopic in size and a **pinhole injection** injury occurs when high-pressure droplets are injected into human flesh often without the knowledge of the injured person. The injection takes place in much the same manner as a hypodermic syringe injection. Pinhole droplet injection can occur when the injured person is wearing heavy leather gloves. Often no more than a minor stinging sensation is felt at the time of injury. It is not until several hours later when tissue damage has occurred that the injury becomes painful and this is often too late.

Make sure you are aware of the early symptoms of pinhole leak injection. If you suspect this type of injury seek immediate medical attention and be aware that some hospital emergency personnel may not be familiar with pinhole injection injuries; this will require you to do some explaining but do it anyway, it is preferable to losing a limb.

WARNING

Pinhole hydraulic leaks can cause maiming and death if not identified. A pinhole injection injury may occur unbeknownst to the injured person; this type of injury must be identified shortly after the injection intrusion into the flesh. Make sure you understand the section on pinhole hydraulic leaks that appears ahead of this warning.

CAUTION

Never use a bare or gloved hand when investigating hydraulic leaks especially in bundled clusters of hoses. Wear safety glasses and gloves and use a cardboard strip when attempting to source small leaks in blind clusters of hydraulic hose.

DRAIN, FLUSH, AND FILL

A draining, flushing, and refilling procedure is the only way to remove contaminants and degenerated hydraulic fluid from a hydraulic circuit. The frequency with which this is performed will depend on the severity of the working conditions, and in many cases simply draining and replacing the hydraulic oil will be sufficient. Sludging oil, solid deposits, and the appearance of varnishing on components are indicators that draining the system should be followed by system flushing before refilling with new hydraulic oil.

Drain System

Gravity-drain the system, clean any sediment from the reservoir, and replace the filter elements. Inspect the old oil. If it shows signs of severe degeneration, flushing the system should be considered. Indicators of severe degeneration of hydraulic oil are burned odor, excessive stickiness, discoloration, and lacquer flakes in oil.

Flush System

Most hydraulic system flushing should be performed with the specified service hydraulic oil. After draining, fill the system reservoir with the specified oil, bleed the hydraulic pump, and then operate the equipment to circulate it through the entire system. All the valves should be actuated to ensure that the flushing oil reaches every part of the circuit. Then drain and dispose of the flushing oil.

Fill the System

Remembering that the intrusion of dirt into hydraulic systems probably causes more problems than any other single problem, ensure that everything used in the refill procedure is clean. A new filter should be used. Fill the reservoir to the prescribed level with new hydraulic oil. Then run the equipment through its entire working cycle several times to ensure that any air is purged from the circuit. The equipment should run smoothly. Finally, recheck the oil level in the reservoir.

PREVENTING LEAKS

Leaks in hydraulic circuits can be internal or external. Internal leaks do not result in the loss of hydraulic oil, but system efficiency is reduced. External leaks are easier to identify because hydraulic oil is lost from the system and usually it is easy to source this.

Internal Leaks

A small amount of internal leakage in a hydraulic system is intended. Oil that seeps past pistons, valve spools, and other moving components acts to lubricate the system. Excessive internal leakage indicates wear of mating components. This can cause hydraulic cylinders to creep or drift. Internal leakage is increased when low-viscosity oils are used because of the thinning that results at higher temperatures. When system operation appears sluggish or shows signs of creeping, check the hydraulic oil first and then test the system to isolate the source of the internal leakage.

External Leaks

Although external leaks tend to be easier to source, they also have the potential to be more hazardous. Every connection point in a hydraulic system is a potential source of a leak. When reservoir oil level is low, inspect the system for possible leaks. Pinhole leaks can be hard to detect, but they can be dangerous, both because escaping high-pressure oil represents a fire hazard and can penetrate the skin. Air leaks in suction lines can also create problems and be difficult to locate.

CAUTION

Pinholes in hydraulic lines result in high-pressure oil escaping from the system in minute droplets, often so small that they cannot be observed by naked eye. High-pressure oil droplets can penetrate the skin, often with little external evidence, and the result can be blood poisoning. Use a piece of cardboard to search for leaks in hydraulic lines—never a bare (or even a gloved) hand! If you suspect hydraulic pinhole injection (possible indicators are redness and swelling) seek *immediate* medical help.

When an external leak appears to be sourced at hydraulic connectors, exercise extreme caution when tightening—especially if the circuit is under pressure. Torque the line nuts enough to stop the leak only, remembering that the threads on the fittings may be stripped. After repairing an external leak, operate the equipment to get the oil to operating temperature and then carefully reinspect the repair.

CHECKING SYSTEM PERFORMANCE

Before and after repairs, and as part of preventive maintenance, hydraulic systems should be checked. Some OEMs have system-specific checklists and these should be used when available. Typical checks should be performed in the following fields.

Check Reservoir

This check should be performed regularly because it can be performed in seconds and can identify small problems before they become major problems. Before opening the system filler cap, clean it and the surrounding area with a clean rag. When dipsticks are used, clean the dipstick with a lint-free cloth before checking the level. Look for:

- **Oil aeration**. Foaming oil or bubbles can indicate an air leak in the system, usually on the suction side of the circuit.
- **Change in oil level**. A reservoir oil level that requires constant topping up usually indicates an external leak in the system.
- **Milky oil**. This almost always indicates the presence of water in the system. Often this will require system flushing because hydraulic oil and water emulsify under pressure.

Check Oil Lines and Connections

Both low- and high-pressure sides should be checked for leaks and physical damage. Low-pressure side leaks can result in air being pulled into the system, whereas high-pressure side leaks will produce external leakage. Pinched lines can create restrictions that result in foaming, overheating, and reduction in hydraulic efficiency.

Check Circuit Components

A thorough visual inspection should be made of all of the hydraulic circuit components, including the oil cooler, valves, cylinders, pumps, and motors. The circuit should be run through its working cycle for leaks and to see whether the flow is sufficient to run the system. Check the oil cooler fins for plugging up and check the system valves for operation, remembering that wear can cause internal leakage. Motors should not be allowed to run hot. If this happens, check that the oil supply is adequate and that the oil cooler is functioning properly. Also check for leaks around the motor hose fittings, shaft, seals, and body mating surfaces.

FIGURE 13–39 Hydraulic system analyzer

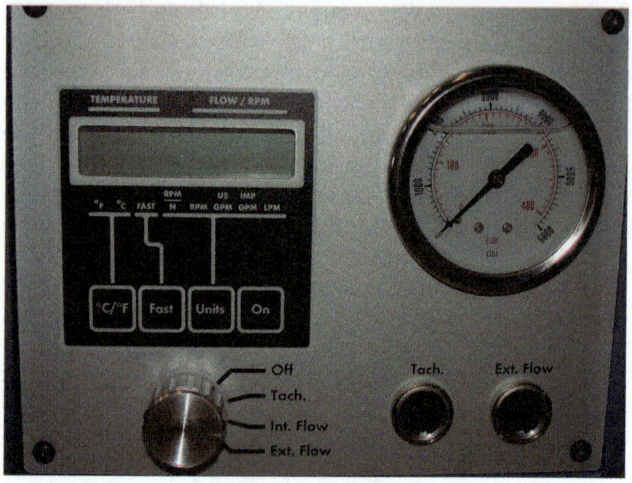

Flow and Pressure Testing

Various types of hydraulic analyzers are available for testing hydraulic circuits. **Figure 13–39** shows one type of hydraulic analyzer. **Figure 13–40** shows how the analyzer would be plumbed into a circuit to perform a pump output test. Typically, a hydraulic analyzer checks:

- **Flow**. Flow-testing determines whether the pump is producing its specified output.
- **Pressure**. Pressure-testing verifies the operation of the relief valve and, in closed-center systems, pump performance.
- **Temperature**. Hydraulic oil temperature should be measured because system pressure and flow specifications are calculated with the system at operating temperature.
- **Leakage**. Leakage-testing identifies leakage in a specific circuit component.

When you are required to test a vehicle hydraulic circuit, observe the OEM test cycle. A detailed test profile and photo sequence is provided in Chapter 25, using a truck power steering circuit. It is recommended that you record the results of each test step using an OEM chart such as that in **Figure 13–41**.

A typical test cycle consists of the following steps:

- Install the hydraulic tester into the circuit to be tested between the pump and the control valve.

FIGURE 13–40 Circuit diagram for pump test

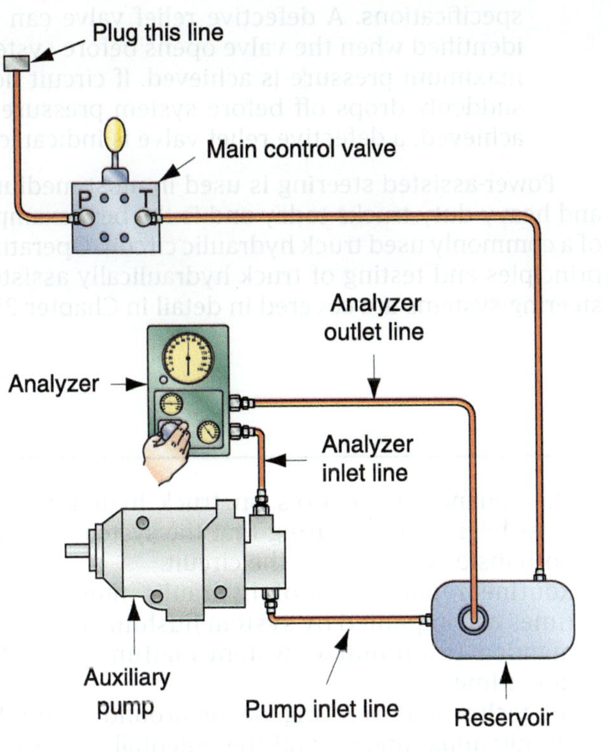

The pressure line should be connected to the inlet port of the tester.

- Connect the hydraulic tester outlet (return) to the system reservoir.
- After installation, check the system oil level.
- Start the engine and gradually close the hydraulic tester gate valve to load the circuit and bring the temperature up to normal operating value.
- Next, open the gate valve and record system flow at zero pressure.
- Move the system control valve into one of its power positions and gradually close the tester gate valve, recording flow in 250 or 500 psi increments from zero up to full system pressure.
- Repeat the above test in each of the system control valve power positions.

SHOP TALK

When performing test steps 6 and 7, ensure that the system oil temperature is the same. This may require that you allow the oil to circulate and cool between each test.

Diagnosing Test Results

You should make sure that you consult the OEM literature when diagnosing hydraulic circuit test results, but here are some typical guidelines used to test a simple hydraulic cylinder-type circuit:

- Pump flow at maximum pressure should be at least 75 percent of the pump flow measured at zero pressure. OEM specifications can require higher percentage values. A pump that fails to meet 75 percent of this value is probably worn.

FIGURE 13–41 Truck hydraulic circuit test chart

Truck Chassis OEM:				Truck Serial Number:		
Truck Owner/ID:				Truck Model Year:		
Engine:				Pump drive:		
Technician ID:				Date of Test:		
Pump Tests		Pump OEM: Relief valve psi:		Pump drive: Hydraulic fluid temperature:		
Pressure		**500 psi**	**1,000 psi**	**1,500 psi**	**2,000 psi**	**2,200 psi**
Flow		gpm	gpm	gpm	gpm	gpm
Circuit Test: Record flow data into chart below.						
Implement		**500 psi**	**1,000 psi**	**1,500 psi**	**2,000 psi**	**2,200 psi**
Chute lift	Raise	gpm	gpm	gpm	gpm	gpm
	Lower	gpm	gpm	gpm	gpm	gpm
Chute swing	Right	gpm	gpm	gpm	gpm	gpm
	Left	gpm	gpm	gpm	gpm	gpm
Tag axle	Raise	gpm	gpm	gpm	gpm	gpm
	Lower	gpm	gpm	gpm	gpm	gpm
Folding chutes	Fold	gpm	gpm	gpm	gpm	gpm
	Unfold	gpm	gpm	gpm	gpm	gpm

- When flow at each pressure value throughout testing is within 75 percent of pump flow at zero pressure, the hydraulic circuit components are probably functioning properly.
- When pressure drops off before full load, a leaking circuit is indicated. To determine whether the leakage is occurring in the control valve or the cylinder, disconnect the cylinder return hose and move the control valve to a power position. If oil leaks at the cylinder return, it is probably leaking internally and is defective. If no oil exits the cylinder return line, the control valve may be at fault.
- A relief valve should open within a window of its specified pressure so check the OEM specifications. A defective relief valve can be identified when the valve opens before system maximum pressure is achieved. If circuit flow suddenly drops off before system pressure is achieved, a defective relief valve is indicated.

Power-assisted steering is used in most medium- and heavy-duty trucks today and is the best example of a commonly used truck hydraulic circuit. Operating principles and testing of truck hydraulically assisted steering systems are covered in detail in Chapter 25.

SUMMARY

- Fundamental hydraulic principles include Pascal's Law; Bernoulli's Principle; and how force, pressure, and sectional area are used in hydraulic circuits to produce outcomes.
- A typical simple hydraulic circuit consists of a reservoir, pump, valves, actuators, conductors, and connectors.
- Hydraulic pumps convert mechanical energy into hydraulic potential.
- Valves manage flow and direction through a hydraulic circuit.
- Actuators such as hydraulic cylinders and motors convert hydraulic potential into mechanical movement.
- Hydraulic oil is used to store and transmit hydraulic energy through a hydraulic system.
- ANSI and ISO graphic symbols are used to represent hydraulic components and connectors in hydraulic schematics.
- Maintenance procedures on truck hydraulic systems begin with ensuring that the system is clean both inside and outside the circuit.
- Routine replacement of hydraulic fluid, sometimes accompanied by system flushing, is recommended to minimize system malfunctions and downtime.
- All technicians working on or around hydraulic circuits must understand the potential severity of pinhole injection injuries and what to do if one is suspected.
- Hydraulic circuit testers are used to analyze hydraulic circuit performance.
- A hydraulic tester consists of flow gauge, pressure gauge, temperature gauge, and gate valve.

REVIEW QUESTIONS

1. Which of the following best describes the operating principles of a truck wet-line kit?
   a. hydrostatics
   b. hydrodynamics
   c. hydrology
   d. hydromechanics
2. Which of the following is a correct specification of atmospheric pressure at sea level?
   a. 14.7 inches Hg
   b. 29.8 psi
   c. 101.3 kPa
   d. 29.8 inches H_2O
3. Convert a manometer reading of 12 inches Hg into pounds per square inch (psi).
   a. 6 psi
   b. 14.7 psi
   c. 24 psi
   d. 101.3 psi
4. When two hydraulic cylinders in the same circuit are subjected to equal flow, which of the following should be true?
   a. The larger cylinder will raise less weight.
   b. The smaller cylinder will move faster.
   c. The larger cylinder will require more pressure.
   d. The smaller cylinder will raise more weight.
5. When there is a fixed restriction to flow in a hydraulic circuit, what should happen to the pressure downstream from the restriction?
   a. remains unchanged
   b. reduces
   c. increases
   d. surges
6. In an open-center hydraulic circuit, which of the following should be true when describing pump operation?
   a. runs constantly
   b. produces higher operating pressures

c. manages to have rest cycles
d. flow is constant regardless of rpm

7. When a hydraulic pump discharges the same slug volume it picks up per cycle, it should be described as:
a. variable displacement
b. constant cycle
c. fixed displacement
d. nonpositive cycle

8. How is hydraulic oil routed through a typical external-gear pump?
a. through the center between the intermeshing gears
b. around the outside between gear teeth and pump liner body
c. through the inlet to the gear output shaft
d. around the idler gear shaft to the outlet port

9. A vane-type hydraulic pump has one inlet port and one outlet port; it would be classified as:
a. balanced
b. unbalanced
c. variable displacement
d. twin cycle

10. Which type of hydraulic pump would usually have higher operating efficiencies?
a. external gear
b. balanced vane
c. unbalanced vane
d. radial piston

11. Which of the following is not a function of a hydraulic accumulator?
a. dampen pressure surges
b. store potential energy
c. maintain consistent circuit pressure
d. charge hydraulic actuators

12. When applied to hydraulic flow in a circuit, what does the term *laminar flow* mean?
a. high turbulence
b. vortex flow
c. streamlined flow
d. flow through elbows

13. If a hydraulic pressure of 1,000 psi acts on a piston measuring 25 square inches, what amount of linear force would result?
a. 25 lb.
b. 250 lb.
c. 2,500 lb.
d. 25,000 lb.

14. Which type of gear pump uses a crescent-shaped separator located between an external spur gear and an annular internal gear?
a. internal gear
b. external gear
c. rotor gear
d. triple gear

15. Which of the following are used in an axial piston-type hydraulic pump?
a. rotating cam
b. swashplate
c. radial piston
d. compressor type

16. What specification(s) is normally used to rate hydraulic pump performance?
a. flow in gpm
b. pressure in psi
c. revolutions per minute (rpm)
d. a, b, and c

17. What is used to drive truck chassis hydraulic pumps?
a. electricity
b. engine
c. transmission
d. a, b, and c

18. Which of the following is the primary function of a flow-control valve?
a. control the speed of an actuator
b. control circuit pressure
c. manage circuit resistance
d. switch circuit outcomes

19. What type of actuators would be required in the cab-lift hydraulics used on a COE chassis?
a. single acting
b. double acting
c. separate lift and return rams
d. vane-type motor

20. What force is used to turn a hydraulic motor?
a. electrical
b. diesel engine
c. transmission
d. hydraulic

21. What type of hydraulic hose should be able to sustain the highest working pressures?
a. spiral wire
b. single-wire braid
c. double-wire braid
d. fabric braid

22. When comparing a section of ⅜-inch single-wire braid hydraulic hose to a ½-inch single-wire braid, which of the following should be true?
a. The ⅜-inch hose will have a higher specified flow rate.
b. The ⅜-inch hose will have a higher specified working pressure.
c. The ½-inch hose will have a lower specified flow rate.
d. The ½-inch hose will have a higher specified burst pressure.

23. Which of the following threads used in hydraulic fittings is described as dryseal?
a. UNF
b. UNC
c. NPT
d. NPTF

24. Which of the following should be true of a hydraulic fluid described as being of high viscosity?
a. It thickens as temperature increases.
b. More pump force is required to push it through the circuit.
c. It has better resistance to cold weather gelling.
d. It performs better through a wide temperature range.

25. Which of the following symbols would be used to represent a pump in a hydraulic schematic?
a. square
b. rectangle
c. diamond
d. circle

14

Prerequisite: Chapters 1, 2, 3, and 4

CLUTCHES

OBJECTIVES

After reading this chapter, you should be able to:

- Outline the operating principles of a clutch.
- Identify the components of a clutch assembly.
- Explain the differences between centrifugal, pull-type, and push-type clutches.
- Describe the procedure for adjusting manual and self-adjusting clutches.
- Explain how to properly adjust the external linkage of a clutch.
- Describe the function of a clutch brake.
- Outline the procedures required to work safely around heavy-duty clutches.
- Troubleshoot a clutch for wear and damage.
- Identify some typical clutch defects and explain how to repair them.
- Outline the procedure for removing and replacing a clutch.
- Explain the difference between a three-pedal and a two-pedal clutch system.

KEY TERMS

adjusting ring
antirattle springs
AutoClutch
centrifugal clutch
clutch
clutch actuator
clutch brake
clutch brake squeeze
clutch free pedal
clutch pack
dampened discs
DM (datalink mechanical) clutch
free pedal
free travel
friction discs
Kwik-Adjust
Kwik-Konnect
launch phase
lockstrap
pull-type clutch
push-type clutch
release bearing
rigid discs
self-adjusting clutch
three-pedal system
throwout bearing
torque-limiting clutch brake
touch point
two-pedal system
two-piece clutch brake
Urge-to-Move
wear compensator
wet clutch

INTRODUCTION

Because of the rapid emergence of automated manual transmissions (AMTs) in truck chassis, clutches have undergone some changes in recent years. Or better said, the way clutches are actuated have undergone these changes. This means that any study of clutches today must include those used in automated three- and two-pedal systems. In what is known as the two-pedal systems that are common today, the clutch must be actuated mechanically, but managed by an electronic control unit (ECU). For example, a term that Eaton use for ECU-managed, mechanical actuation is DM, an acronym for *datalink mechanical*. Regardless of application, the role a clutch plays does not change that much whether it is in a mechanical standard transmission or an AMT.

14.1 CLUTCH FUNCTION

The function of a **clutch** is to transfer torque from engine flywheel to the transmission. At the moment of clutch engagement, the transmission input shaft may either be stationary, as when the truck is stationary, or rotating at a different speed than the flywheel, as in the case of upshifting or downshifting. At the moment the clutch fully engages, however, the flywheel and the transmission input shaft must rotate at the same speed.

Torque transmission through a clutch is accomplished by bringing a rotating drive member connected to the engine flywheel into contact with one or more driven members splined to the transmission input shaft. Contact between the driving and driven members is established and maintained by both spring pressure and friction surfaces. Pressure exerted by springs on the driven members is unloaded by the driver by depressing the clutch pedal. This "releases" the clutch.

A clutch is equipped with one or two **friction discs**. These have friction surfaces known as facings (**Figure 14–1**). When the clutch pedal in the cab is depressed, the clutch pressure plate is drawn away from the flywheel, compressing the springs and freeing the friction disc(s) from contact with the flywheel friction surface. At this point, the clutch is disengaged and torque transfer from the engine to the transmission is interrupted.

As the clutch pedal is released, the pressure plate moves toward the flywheel, allowing the springs to clamp the disc(s) between the flywheel and the pressure plate. The discs are designed for moderate slippage as they are brought into contact with the rotating flywheel. This minimizes torsional (twisting) shock to the drivetrain components. As clutch-clamping pressure increases, the discs accept the full torque from the flywheel. At this point, engagement is complete, and engine torque is transferred to the transmission. Once engaged, a properly functioning clutch transmits engine torque to the transmission without slippage.

14–1 The clutch uses (A) one; or (B) two friction discs to couple the engine to the transmission.

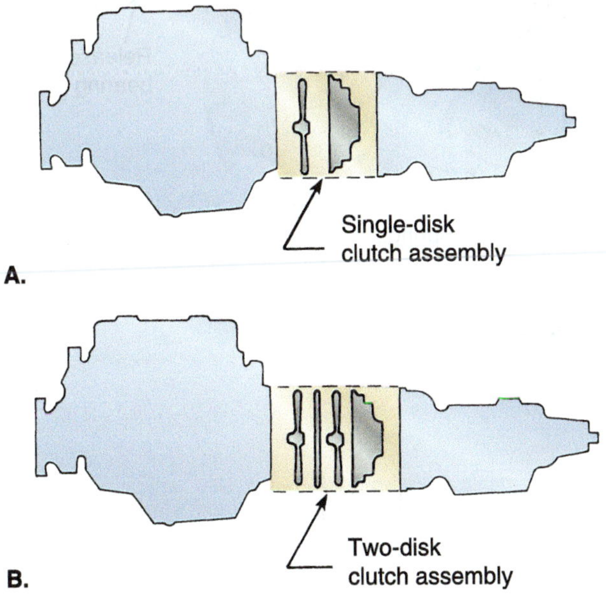

TWO- AND THREE-PEDAL SYSTEMS

The engaging and disengaging of a clutch may be an action controlled by the driver or one managed by the transmission management electronics. Automated transmissions (see Chapter 20) are widely used today. When a clutch is used in an automated transmission, its function does not change but control of it is moved from the driver to the shift electronics. In studying clutches in this chapter, when clutch packs used in automated transmissions are referenced, the following terms are used:

- **Three-pedal system**: automated transmission that uses a driver actuated clutch pedal. The driver is required to use the clutch pedal to break torque when stopping or starting or as an option.
- **Two-pedal system**: automated transmission system with no clutch pedal (the "pedals" refer to the accelerator and brake pedal). The clutch is actuated mechanically either by using a **clutch actuator** or by using a centrifugal clutch (both are discussed late in this chapter).

14.2 CLUTCH COMPONENTS

A clutch assembly is illustrated in **Figure 14–2**. Clutch components can be divided into two basic groups: driving and driven members.

FIGURE 14–2 An angle-spring clutch pack with two spring dampened discs

DRIVING MEMBERS

The driving members of a clutch are the cover assembly (contains the pressure plate) and, if the clutch has two friction discs, an intermediate plate. Although it is not part of the clutch pack, the engine flywheel drives the clutch assembly and should be inspected and serviced whenever work is performed on the clutch.

Flywheel

The flywheel may either be flat faced or use a pot design (**Figure 14–3**). The flywheel must be precisely perpendicular to the crankshaft with almost no runout (as little as 0.005 inch/0.127 mm). The surface of the flywheel that contacts the friction discs is machined smooth and should be dry of lube oil or grease.

Clutch Cover

The clutch cover assembly is constructed of cast steel, cast iron, or stamped steel (**Figure 14–4**). Cast cover assemblies provide maximum ventilation and heat dissipation and are often used on heavy-duty Class 8 trucks. Stamped steel clutch covers tend to be used on lighter-duty trucks.

FIGURE 14–3 (A) Some clutches are installed on flat-faced flywheels; and (B) others are installed in pot flywheels.

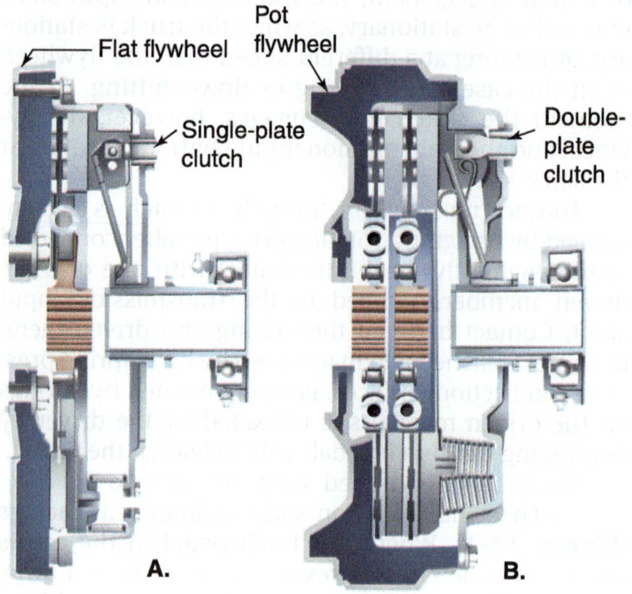

FIGURE 14–4 Exploded view of an angle-spring clutch assembly. The outside of the adjusting ring is threaded and turns in the clutch cover plate to adjust the clutch.

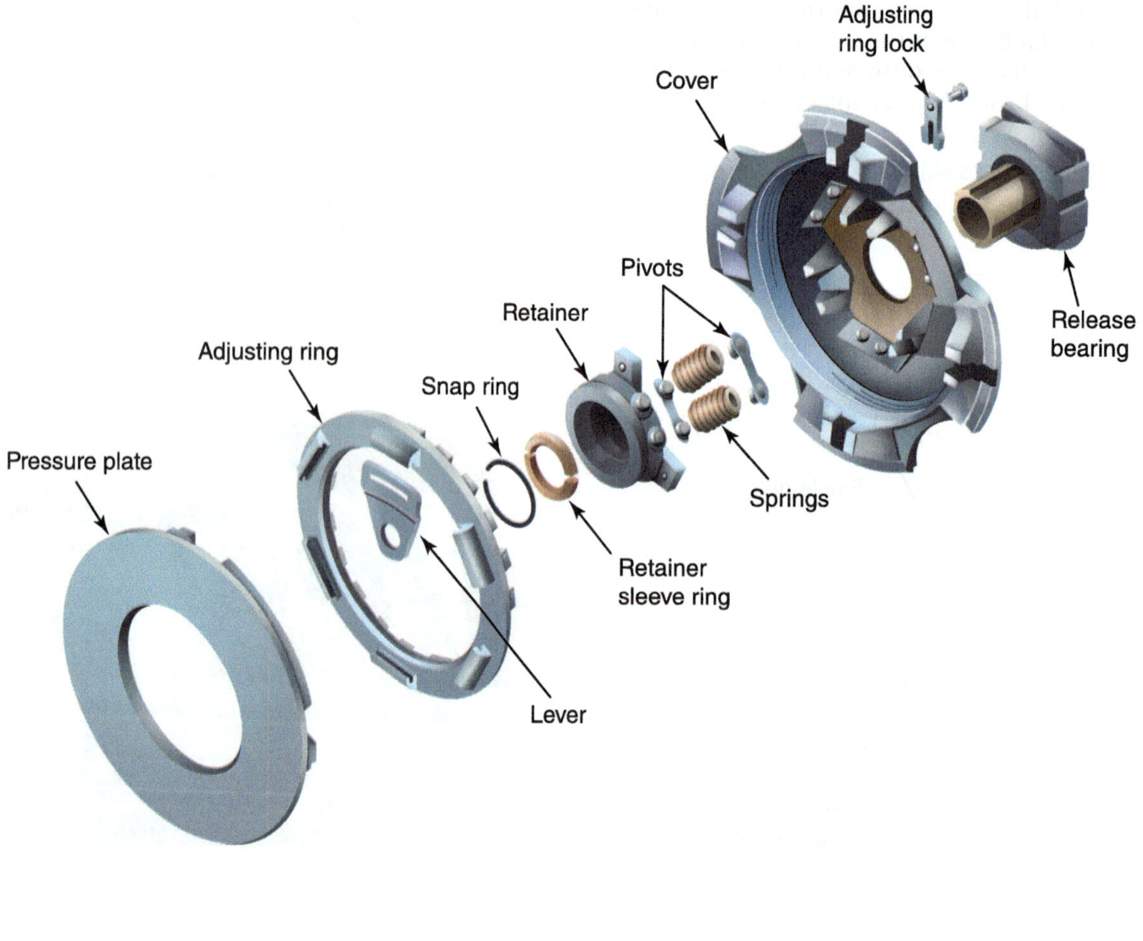

The clutch cover is bolted to the flywheel and rotates with it. The clutch cover contains the pressure plate, which is fitted to the cover with the pressure springs. Most truck clutch cover assemblies contain the levers that move the pressure plate back and forth, thereby making and breaking contact with the friction discs.

Pressure Plate

The clutch pressure plate is machined smooth on the side facing the driven disc (or discs). The pressure plate mounts on pins or lugs on the clutch cover and is free to slide back and forth on these pins or lugs. When the clutch pedal is released, spring pressure is applied to the pressure plate, clamping the friction discs to the flywheel with enough linear force to prevent slippage.

Pressure Springs and Levers

Pressure springs are located between the clutch cover and the pressure plate. Both coil spring and diaphragm spring designs are used.

Clutches with Coil Springs. Most clutch designs use multiple coil springs to force the pressure plate against the driven discs. In some clutches, coil springs are positioned perpendicular to the pressure plate and are equally spaced around the perimeter of the cover.

Other clutches use fewer coil springs and angle them between the cover and a retainer (**Figure 14–5**).

Angle-spring designs require 50 percent less clutch pedal effort. They also provide a constant plate load regardless of the thickness or wear on the friction facings of the driven discs.

The angle-spring clutch illustrated in Figure 14–5 operates under indirect pressure. Three pairs of coil springs are located away from the pressure plate rather than directly on it. Spring load is applied through a series of six levers. In this clutch design, the plate load and release bearing load are not directly proportional to the spring load.

The engaged position of a new angle-spring clutch is shown in **Figure 14–6A**. Pressure plate load is the result of the axial load of the springs multiplied by the lever ratio. The pressure springs are positioned at an angle to the center line of the clutch with each end attached to the flywheel ring (cover) and the release sleeve retainer. As the release sleeve retainer including the release sleeve and release bearing—is moved toward the flywheel, the springs pivot freely without bending or buckling. Connected to the retainer are levers that, when forced forward by the retainer, multiply the force of the springs. Pivot points on the levers press the pressure plate against the driven discs.

When the clutch pedal is released (**Figure 14–6B**), spring load increases but axial load decreases, resulting in reduced pedal effort. The plate load is defined by the axial spring force multiplied by the lever ratio.

FIGURE 14–5 The coil springs in this clutch are angled between the cover and a retainer. Levers multiply the spring force against the pressure plate.

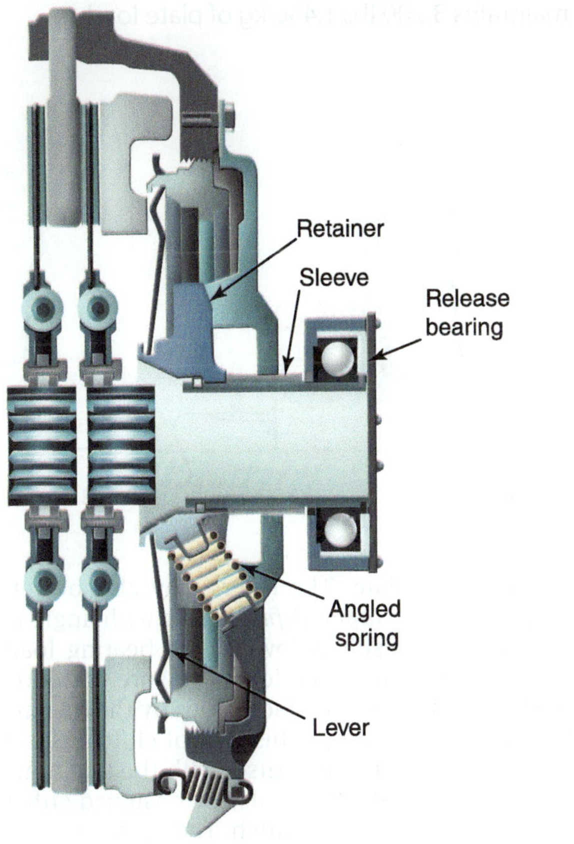

Axial spring force changes with release bearing movement but not in direct proportion to the spring load.

When friction facings wear (**Figure 14–6C**), spring load reduces but axial load remains constant. As the clutch wears, the release sleeve assembly moves toward the flywheel, and the pressure springs elongate, reducing their tension. The axial spring force, however, remains essentially constant. This means that pressure plate force remains constant throughout the life of the clutch. When the clutch is adjusted to compensate for friction facing wear, the pressure plate position does not move, but the rotating **adjusting ring** moves the levers toward the transmission, pushing the release sleeve and bearing assembly in that direction also. This reestablishes the internal spring position to the original setting for continued clutch use.

Clutches with Diaphragm Springs. Clutches equipped with a diaphragm or Belleville spring assemblies are often used in medium-duty trucks. Like coil spring-equipped clutches, diaphragm and Belleville spring clutches operate under indirect pressure using a retainer and lever arrangement to exert pressure on

FIGURE 14–6 Clutch operation: (A) In the engaged position, pressure plate load (3,200 lb./1,450 kg) is the result of the axial load (500 lb./225 kg) of the spring load multiplied by the lever ratio; (B) when the clutch is released, spring load increases but the release bearing load reduces to 420 lb./190 kg, which results in the reduced pedal effort; (C) when facings wear, spring load reduces but axial load remains at 500 lb./225 kg and maintains 3,200 lb./1,450 kg of plate load.

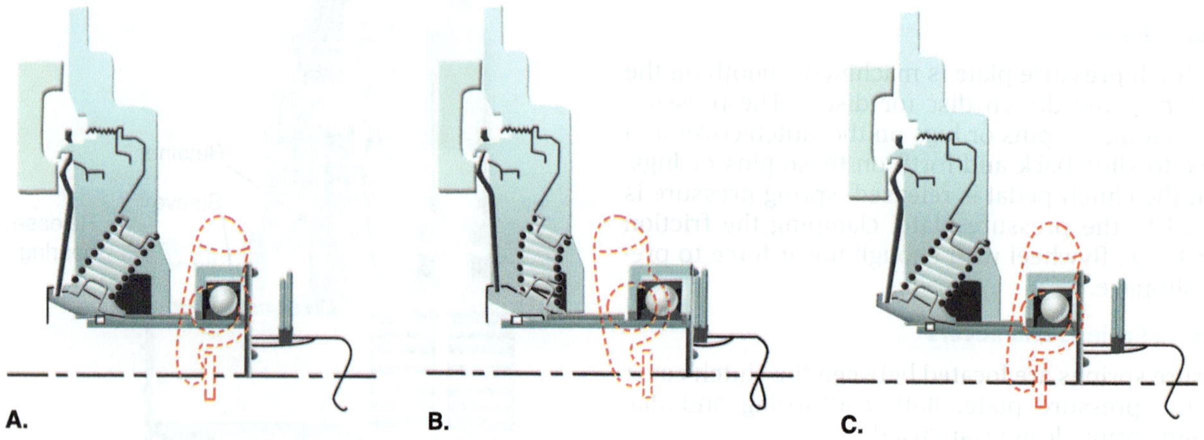

the pressure plate. The levers may also be referred to as *fingers* or *tapered fingers.* As with angle-spring clutches, the result is low release bearing load and constant pressure plate load. Pressure plate load on the friction material surface varies by the thickness of the diaphragm spring. This type of clutch uses either a single disc or multiple discs with the addition of an intermediate plate. It can also be designed either as a push-type or pull-type clutch. The actual design used is determined by the clutch linkage/vehicle design, space requirements inside the vehicle, and the torque load required from the clutch assembly.

In the diaphragm clutch assembly shown in **Figure 14–7**, the clutch cover assembly is bolted directly to the engine flywheel. The clutch cover assembly drives the pressure plate by means of drive straps.

In a push-type clutch, pressing the clutch pedal moves the release bearing toward the engine flywheel. As the clutch's diaphragm spring fingers are depressed, the pressure plate retracts from contact with the driven disc(s) and the clutch is disengaged.

When the clutch pedal is released, the release bearing and clutch diaphragm spring fingers move away from the engine flywheel. The diaphragm Belleville spring exerts pressure through its levers to the pressure plate. This results in the driven disc(s) being *locked up* between the friction surfaces of the pressure plate and the engine flywheel in the single disc type.

Torque flow through a multiple disc assembly is the same as with a coil spring or angle-spring type. The pressure plate diaphragm Belleville spring exerts pressure through the rear disc to the intermediate plate and to the forward disc onto the flywheel friction face. This locks the clutch assembly together.

On a conventional direct pressure clutch, as the friction facings wear, pressure plate load loss occurs as a result of spring elongation. A diaphragm clutch that uses a Belleville spring design maintains the

FIGURE 14–7 Diaphragm clutch operation: (A) new, engaged position; (B) released position; (C) worn, engaged position

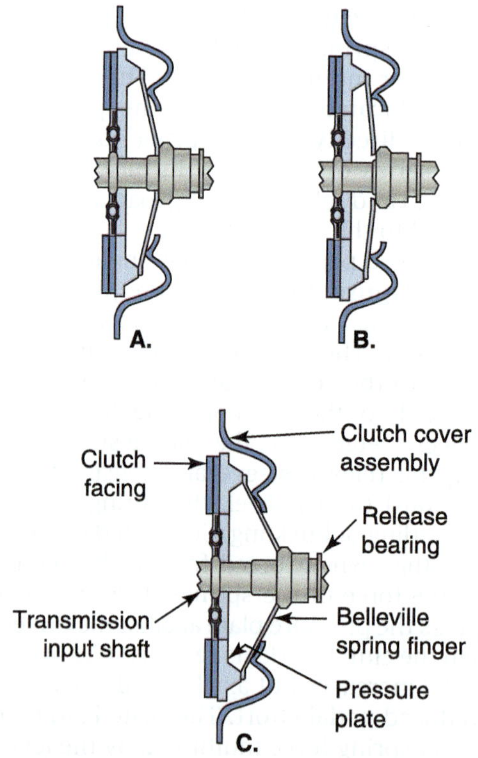

same plate load at the new (Figure 14–7A) and fully worn (Figure 14–7C) positions. Figure 14–7B shows the same clutch in its fully released position.

Intermediate Plate

In a clutch with two friction discs, an intermediate, or center, plate separates the discs. The plate is machined smooth on both sides because it is clamped between two friction surfaces. An intermediate plate increases the torque capacity of the clutch. It does this by doubling the contact friction area.

Some intermediate plates have drive slots machined in their outer edge. These slots fit over and are driven by hardened steel drive pins press-fit into holes in the flywheel rim (see Figure 14–2). Other intermediate plates have four or more drive lugs that fit to, and are driven by, slots in the clutch cover.

Clutches with heavy-duty intermediate plates may use antirattle springs to reduce wear between the intermediate plate and the drive pins and improve clutch release. Without the springs, the drive slots in the plate would wear excessively, resulting in poor clutch release. **Antirattle springs** are curved spring steel plates that are wedged between the edge of the intermediate plate and the inside wall of a pot type flywheel. These are spaced equal distances apart (three springs—120 degrees apart; four springs—90 degrees apart).

Separator Pins

Some two-plate clutches use separator pins. These are also known as intermediate plate pins. Their function is to maintain an equal distance between each friction disc and the intermediate plate when the clutch is released.

Adapter Ring

Some two-plate clutches use an adapter ring when the clutch is installed on a flat flywheel. The adapter ring is bolted between the clutch cover and the flywheel. It is sized to provide the needed depth to accommodate the second clutch disc and the intermediate plate.

Clutch Adjustment Mechanisms

As the friction facings wear, some method must be provided to adjust the clutch and compensate for friction material wear. Both manually adjusted and self-adjusting clutches are used.

Manually Adjusted Clutches. These clutches have a manual adjusting ring that permits the clutch to be adjusted to compensate for friction facing wear. The ring is positioned behind the pressure plate and is threaded into the clutch cover. A **lockstrap** or lock plate secures the ring so that it cannot turn during normal operation. When the lockstrap is removed, the adjusting ring can be rotated in the cover to adjust for wear. This forces the pivot points of the levers to advance, pushing the pressure plate forward and compensating for wear. A manual adjusting ring is shown in Figure 14–4. An improved method of manual adjustment is the **Kwik-Adjust** mechanism used on the Eaton family of Easy Pedal and Easy Pedal Plus clutches. Kwik-Adjust (**Figure 14–8**) permits a fast manual clutch adjustment using a hex wrench.

FIGURE 14–8 Kwik-Adjust mechanism

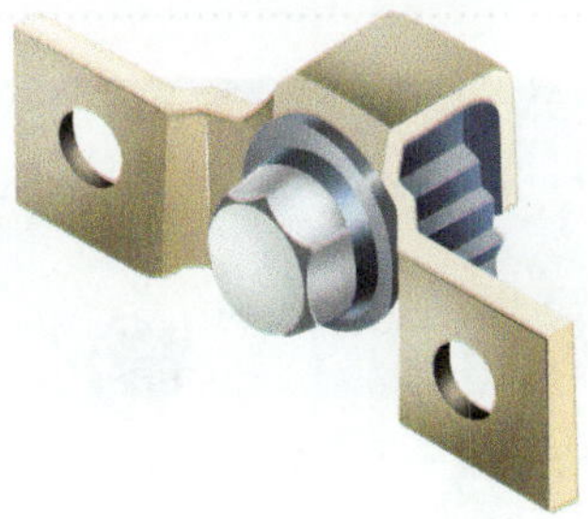

Self-Adjusting Clutches. **Self-adjusting clutches** automatically take up the slack between the pressure plate and clutch disc as wear occurs. Various types are used, some of which option but do not recommend manual adjustments. A clutch adjusting ring has teeth that mesh with a worm gear in a wear compensator. The **wear compensator** is mounted in the clutch cover and has an actuator arm that fits into a hole in the release sleeve retainer. As the retainer moves forward each time the clutch is engaged, the actuator arm rotates the worm gear in the wear compensator. Rotation of the worm gear is transferred to the adjusting ring in the clutch cover, removing slack between the pressure plate and the driven discs.

DRIVEN MEMBERS

The driven components of a clutch consist of the clutch friction discs. Clutch discs are lined on both sides with friction material and have an internally splined hub. The hub fits over the splined end of the transmission input shaft. When the discs are clamped between the pressure plate and the flywheel, torque is transmitted through the discs to the splined input shaft, rotating the shaft at engine speed.

Disc Design

There are two basic disc designs: rigid and dampened (**Figure 14–9**). **Rigid discs** are steel plates to which friction linings, or facings, are bonded or riveted. Rigid discs are most often used in two-plate clutches where the torque loads can be distributed over a greater surface area. A rigid disc cannot absorb torsional shock loads and its misapplication can damage the transmission and input shaft. Torsional vibration can also cause a rigid disc to crack.

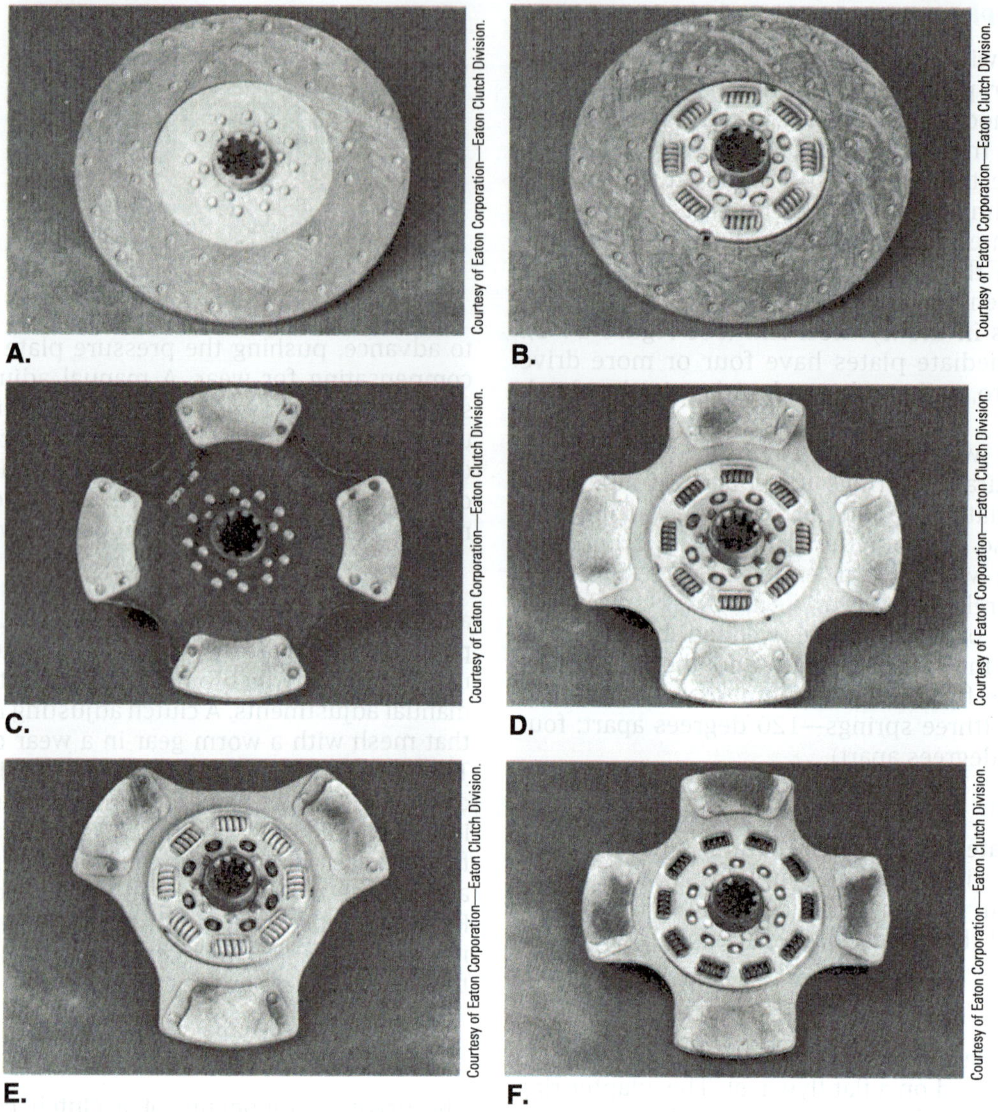

FIGURE 14–9 Types of friction discs: (A) rigid organic; (B) dampened organic; (C) rigid 4-button ceramic; (D) dampened 4-button ceramic; (E) dampened 3-button ceramic; and (F) super dampened 4-button ceramic

Most manufacturers specify the use of dampened discs in trucks equipped with high-torque-rise engines. **Dampened discs** have coaxial dampening springs incorporated into the disc hub. When engine or driveline torque is first transmitted to the disc, the plate rotates on the hub, compressing the springs. This action absorbs the shock and torsional vibration caused by low rpm, high-torque engines. The cushioning effect extends clutch life and protects other driveline components from torsional overloads.

Friction Facings

Friction facings are critical to clutch life and performance because they directly receive the torque of the engine any time the clutch is engaged. There are two types of friction facings:

- Organic: made from glass, mineral wools, and carbon fibers
- Metallo-ceramic: made from a mixture ceramics combined with copper and iron

Organic Facings. Organic friction facings (see Figure 14–9) are usually molded to the full surface of the disc and are, therefore, called "full faced." Grooves in the facing allow worn particles to be thrown off rather than accumulate on the face of the disc. Today, organic friction facings are not often used. When you see them it is usually on light-duty commercial vehicles.

Metallo-Ceramic Facings. Ceramic friction facings (see Figure 14–9) have a higher coefficient of friction, heat tolerance, and torque capacity than organic friction facings. Ceramic friction facings grab quicker with less slip. They also offer a longer service life, making them popular on pickup and delivery (P&D) vehicles, vocational trucks, and linehaul applications.

Ceramic friction facings consist of small pads or buttons riveted to a disc or isolator. The disc can be round with slots machined in it between each button or it might be a scalloped, paddle wheel configuration. Three, four, or more buttons can be installed on each face of the disc. Kevlar friction facings are used in some aftermarket applications. These are said to be able to sustain greater abuse.

RELEASE MECHANISMS

There are major differences in the way clutches are released or disengaged. All clutches are disengaged through the movement of a **release bearing**. Another commonly used term for a release bearing is **throwout bearing**. The release bearing is a unit within the clutch assembly that mounts on the transmission input shaft but does not rotate with it. The movement of the bearing is controlled by a fork attached to the clutch pedal linkage. As the release bearing moves, it forces the pressure plate away from the clutch disc. Depending on the design of the clutch, the release bearing will move in one of two directions when the clutch disengages. It will either be pushed toward the engine and flywheel, or it will be pulled toward the transmission input shaft.

Push-Type Clutches

In a **push-type clutch** (**Figure 14–10**), the release bearing is not attached to the clutch cover. To disengage the clutch, the release bearing is pushed toward the engine. When the pedal of a push-type clutch is initially depressed, there is some free pedal movement between the fork and the release bearing (normally about ⅛ inch/3 mm). After the initial movement, the clutch release fork contacts the bearing and forces it toward the engine.

As the release bearing moves toward the engine, it acts on release levers bolted to the clutch cover assembly. As the release levers pivot on a pivot point, they force the pressure plate (to which the opposite ends of the levers are attached) to move away from the clutch discs. This compresses the springs and disengages the discs from the flywheel, allowing the disc (or discs) to float freely between the pressure plate and flywheel, breaking the torque between the engine and transmission.

When the clutch pedal is released, spring pressure acting on the pressure plate forces the plate forward once again, clamping the plate, disc, and flywheel together and allowing the release bearing to return to its original position.

FIGURE 14–10 Cutaway of a 14-inch (356 mm) push-type clutch assembly on a pot (recessed) flywheel.

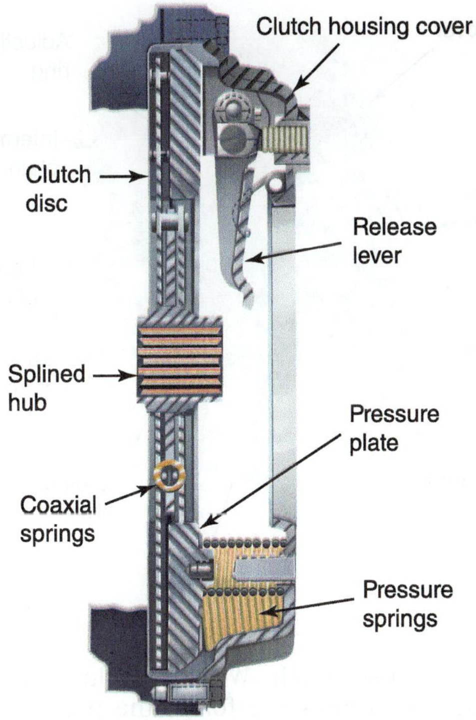

Push-type clutches are used predominantly in light and medium-duty truck applications in which a clutch brake is not required. This type of clutch has no provisions for internal adjustment. All adjustments normally are made externally via the linkage system.

Pull-Type Clutches

As the name implies, a **pull-type clutch** does not push the release bearing toward the engine; instead, it pulls the release bearing toward the transmission. In clutches with angled coil springs or a diaphragm spring, the release bearing is attached to the clutch cover by a sleeve and retainer assembly (**Figure 14–11**). When the clutch pedal is depressed, the bearing, sleeve, and retainer are pulled away from the flywheel. This compresses the springs and causes the pivot points on the levers to move away from the pressure plate, relieving pressure acting on the pressure plate. This action allows the driven disc or discs to float freely between the plate(s) and the flywheel.

On pull-type clutches with coil springs positioned perpendicular to the pressure plate, the release levers are connected on one end to the sleeve and retainer; on the other end they are connected to pivot points (**Figure 14–12**). The pressure plate is connected to the levers near the pivot points. Therefore, when the levers are pulled away from the flywheel, the pressure plate is also pulled away from the clutch discs,

FIGURE 14–11 Components of a pull-type clutch

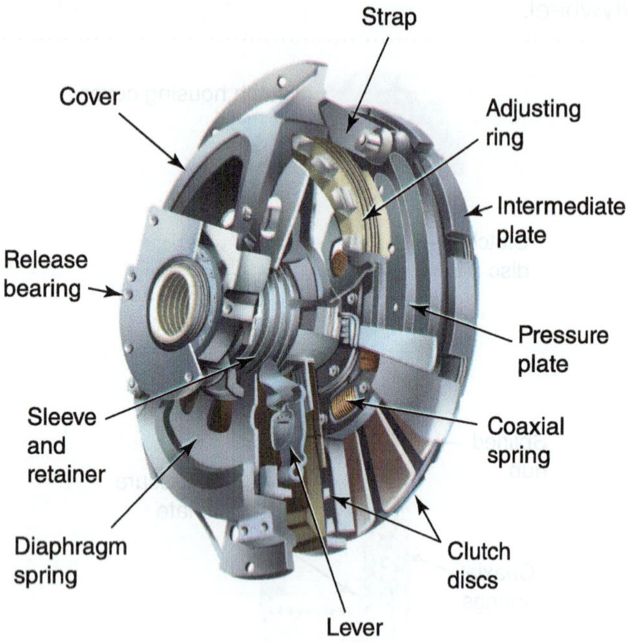

disengaging the clutch. When the clutch pedal is released, spring pressure forces the pressure plate forward against the clutch disc and the release bearing, sleeve, and retainer return to their original position. Pull-type clutches are used in both medium- and heavy-duty applications and are adjusted internally.

Centrifugal Clutches

Eaton introduced heavy-duty **centrifugal clutches** with their UltraShift AMT. **AutoClutch** is a 15½-inch, two-friction plate clutch pack usually faced with ceramic material. This UltraShift is a two-pedal system. Its centrifugal clutch uses centrifugal force resulting from an increase in engine revs to engage. It drops out of lock (disengages) when the engine spools down. AutoClutch uses a ball-and-ramp engagement hub that occupies no more space than a standard heavy-duty clutch pack. The assembly uses a standard bell housing. In addition, AutoClutch has a built-in inertia brake that speeds up automated upshifts by acting as a clutch brake, and the transmission electronics are designed to message the engine electronics to drop rpm to time downshifts.

CLUTCH BRAKES

Most pull-type clutches have a component not found on push-type clutches: a **clutch brake**. The clutch brake is a disc with friction surfaces on either side; it is mounted on the transmission input shaft splines between the release bearing and the transmission (**Figure 14–13A**). Its purpose is to slow or stop the

FIGURE 14–12 Cutaway of a Lipe single-plate, pull-type clutch

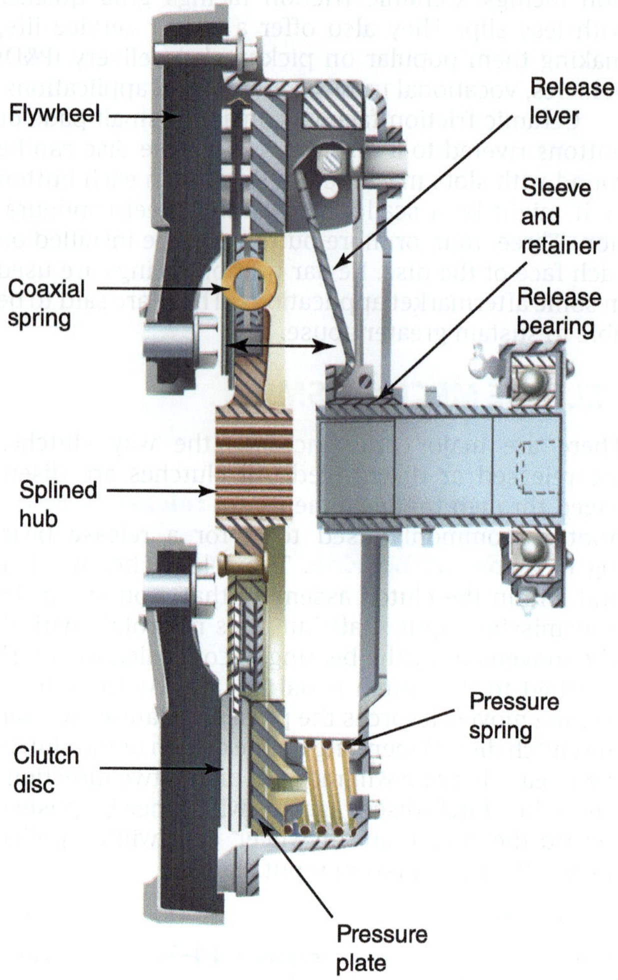

transmission input shaft from rotating to allow gears to be engaged without clashing (grinding) from a neutral-to-first or neutral-to-reverse shift. This can greatly extend transmission life. Clutch brakes are used only on nonsynchronized transmission systems.

Only 70 to 80 percent of clutch pedal travel is needed to fully disengage the clutch. The last ½ to 1 inch (12 to 25 mm) of pedal travel is used to engage the clutch brake. When the pedal is fully depressed, the fork squeezes the release bearing against the clutch brake, which forces the brake disc against the transmission input shaft bearing retainer (**Figure 14–13B**). The friction created by the clutch brake facing stops the rotation of the input shaft and countershaft. This allows the transmission gears to mesh without clashing. Eaton currently manufactures two types of clutch brake:

- Torque limiting
- Kwik-Konnect

FIGURE 14–13 Clutch brake: (A) clutch engaged, brake neutral; and (B) clutch disengaged, clutch brake engaged

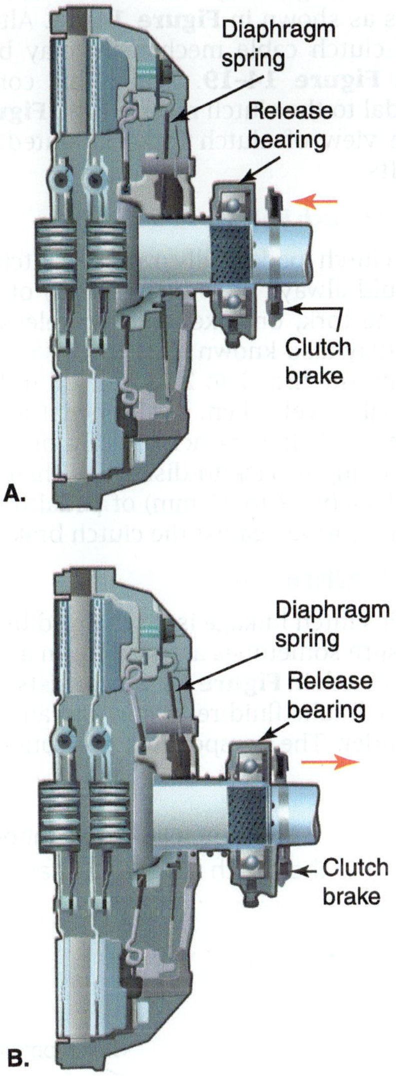

Conventional Clutch Brake

Conventional clutch brakes are designed to be used when shifting from neutral to first or reverse. A conventional clutch brake consists of a steel washer faced on both sides with friction material, or discs. The steel washer has two tangs that engage machined slots in the transmission input shaft. This mounting arrangement allows the brake to slide back and forth on the input shaft while turning with the shaft. This category of clutch brake is not currently used but may be found on older chassis.

TORQUE-LIMITING CLUTCH BRAKES

A **torque-limiting clutch brake** (**Figure 14–14**) is designed to slip when torque loads of 20 to 25 lb-ft.

FIGURE 14–14 Torque-limiting clutch brake

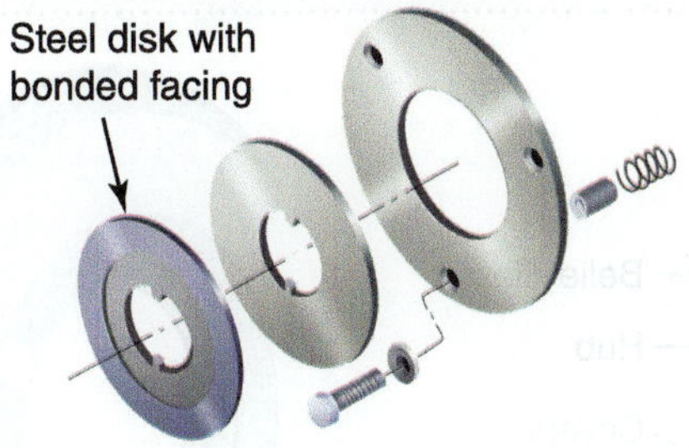

(27 to 34 N·m) are reached. This protects the brake from overloading and the resulting heat damage. Torque-limiting clutch brakes are commonly used in today's trucks. As shown in **Figure 14–15**, a clutch brake has a hub and lugs that engage to a groove on the input shaft. The hub and washers are designed to slip when a specified torque is exceeded. A clutch brake is primarily used only when shifting into first or reverse while the vehicle is stationary. Clutch brake nomenclature is shown in **Figure 14–16**.

Two-Piece Clutch Brakes

A **two-piece clutch brake** can be quickly installed without removing the transmission. Its objective is to replace a damaged clutch brake on an otherwise OK clutch assembly. The replacement can be performed without removing the transmission. The damaged clutch brake has to be cut off the transmission input shaft. Care should be taken not to damage the transmission input shaft when cutting out the defective clutch brake with an oxyacetylene torch. Eaton's brand name for its two-piece clutch brakes is **Kwik-Konnect** (**Figure 14–17**).

Torque-Limiting Clutch Brake

A torque-limiting clutch brake (seldom used) enables faster upshifts in addition to shifting into first and

FIGURE 14–15 Torque-limiting clutch brake

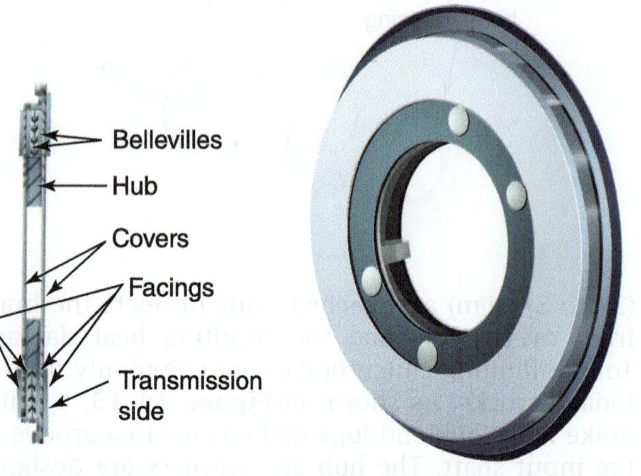

FIGURE 14–16 Nomenclature: torque-limiting clutch brake

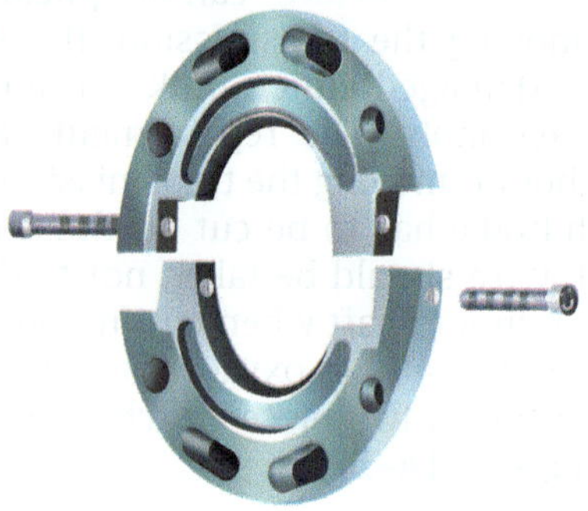

FIGURE 14–17 Two-piece clutch brake designed for installation without removing the transmission from the vehicle.

reverse. When the truck is moving and the clutch brake is engaged, it slows the transmission input shaft, allowing the speed of the transmission input shaft to synchronize quickly with the transmission countershafts. Quicker engagement results in faster shifts. This type of clutch brake provided some forgiveness when driver shifting was imprecise, but it is no longer manufactured.

CLUTCH LINKAGE

Clutches on heavy-duty trucks are controlled by mechanical, pneumatic, or hydraulic linkages. As indicated earlier, clutches may be entirely driver controlled, entirely electronically controlled, or a combination of the two, depending on the type of transmission used. In this section, we generally assume that a clutch is driver controlled unless otherwise indicated. In a driver-controlled clutch, the linkage connects the clutch pedal to the release bearing. This connection may be entirely mechanical or hydraulic.

Mechanical Linkages

Mechanical linkages may consist of rods, bell cranks, and levers as shown in **Figure 14–18**. Alternatively, a flexible clutch cable mechanism may be used as shown in **Figure 14–19**. Again, this connects the clutch pedal to the clutch release fork. **Figure 14–20** is a close view of clutch forks mounted on clutch cross-shafts.

Mechanical Clutch Release Action

With the clutch pedal fully raised (clutch applied), there should always be ⅛ inch (3 mm) of **free play** between the fork, or yoke, and the release bearing. This free play also known as **free travel**, should be taken up by the first 1 to 2 inches (25 to 50 mm) of clutch pedal travel. Then, as the pedal is depressed farther, the fork fingers act directly on the release bearing, pulling it back to disengage the clutch. The last ½ to 1 inch (12 to 25 mm) of pedal travel forces the release bearing against the clutch brake.

Hydraulic Clutch Linkage

A hydraulic clutch linkage is disengaged by hydraulic fluid pressure sometimes assisted by an air servo cylinder. The clutch in **Figure 14–21** consists of a master cylinder, hydraulic fluid reservoir, and an air-assisted servo cylinder. The components are connected both

FIGURE 14–18 The clutch linkage connects the clutch pedal to the clutch release lever and fork.

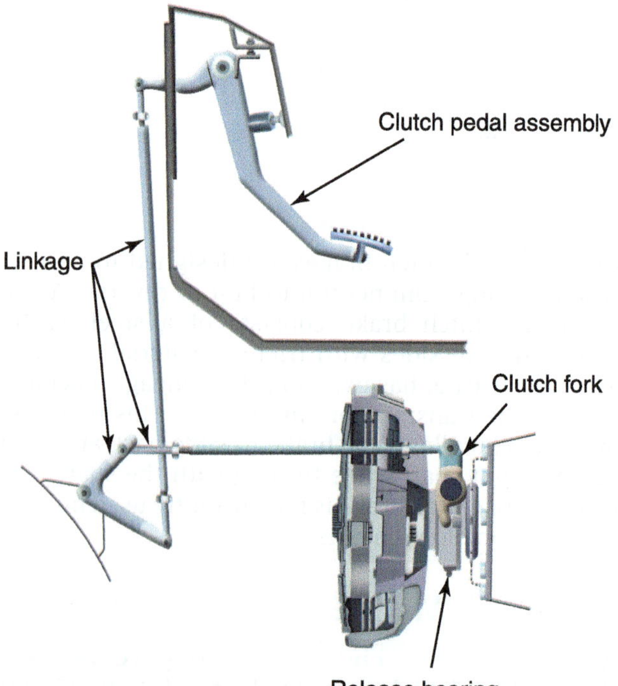

FIGURE 14–19 Mechanical cable clutch control linkage

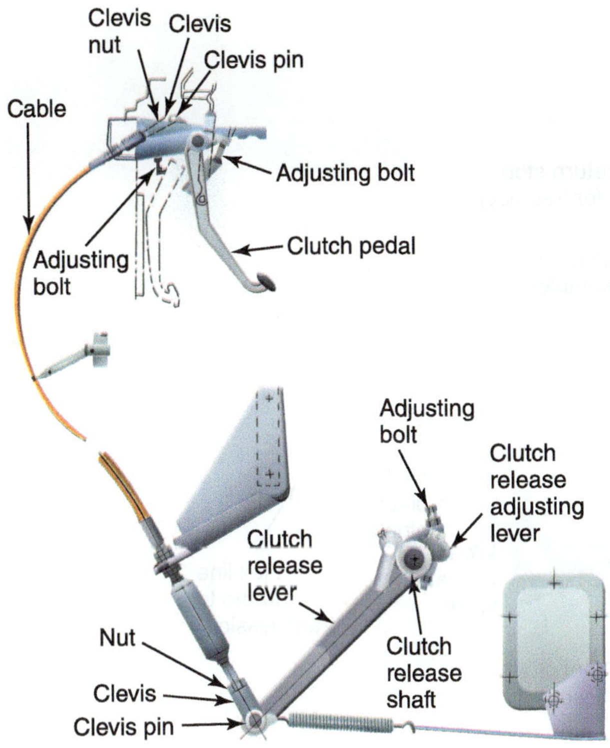

FIGURE 14–20 Clutch fork mounted between clutch cross-shafts: The forks engage with a pull-type clutch release (throw-out) bearing to disengage the clutch.

by rigid and flexible hydraulic lines. When a truck driver depresses the clutch pedal, an actuating plunger forces the piston in the master cylinder to move forward. This movement closes off the reservoir and forces hydraulic fluid through the circuit to a reaction plunger and pilot valve in the servo cylinder.

Hydraulic Servo-Assist Clutch Release Operation

Hydraulic pressure forces the reaction plunger to move forward to close off an exhaust port and to seat the pilot valve. When the plunger is moved farther, it unseats the pilot valve, which allows air to enter the servo cylinder, exerting pressure onto the rear side of the air piston. The movement of the air piston assists in clutch pedal application. As clutch pedal pressure increases, the air piston is moved farther forward and air pressure overcomes the hydraulic pressure in the reaction plunger. This causes the pilot valve to reseat, preventing any more air from reaching the air piston. The pilot valve and reaction plunger remain in this position until there is a change in the pressure.

When the hydraulic pressure decreases, the return spring returns the reaction plunger and the pilot valve seats itself, which in turn uncovers the exhaust port and allows the air to exhaust from the servo cylinder.

SOLO CLUTCHES

Adjustment-free clutches such as the Eaton Fuller Solo medium- and heavy-duty clutch family have become commonplace. These are capable of providing close to zero maintenance during the life of the clutch. A wear indicator is used to monitor remaining clutch life and provides an alert when replacement is required. Once the **clutch pack** assembly has been correctly set up, the release bearing position is maintained as the clutch wears, and free travel is maintained. The release bearing requires grease on a preventive maintenance (PM) schedule. You should know how to verify the adjustment of a Solo clutch and this procedure is covered at the end of this chapter; adjustment is verified only at inspection.

WET CLUTCHES

Eaton/Dana Corporation in the past has used fully automated **wet clutches** on some versions of its AutoShift transmission used in medium-duty applications. The AutoShift automated transmission was a first-generation two-pedal system. This wet clutch technology used a creep feature known as **Urge-to-Move**. Urge-to-Move loads a continuous 45 lb-ft. (61 N·m) of torque to the transmission through the wet clutch pack so that the driver senses engagement any time the shift lever is in the drive position. The Urge-to-Move torque feature gives a typical truck the ability to move ahead on a 6 percent grade. This system requires a cooler for the auto transmission fluid.

EATON DM CLUTCH

Eaton states that the acronym **DM** represents **datalink mechanical clutch** module and it uses this term to describe both its centrifugal clutches and

FIGURE 14–21 Components used in one type of hydraulic clutch circuit

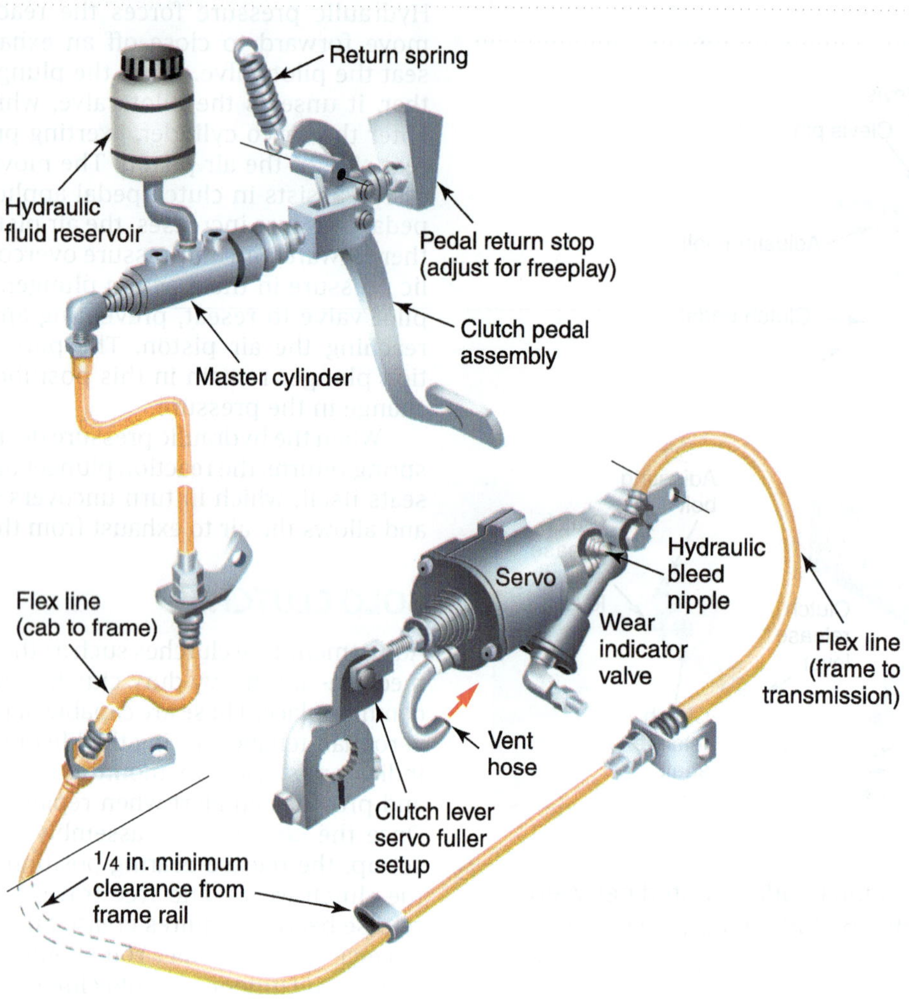

those that use clutch actuators to engage and disengage the unit. The DM clutch is one option that is used with the Eaton UltraShift transmission; it should not be confused with the AW wet clutch.

A DM clutch module used with an UltraShift transmission is a 15.5-inch (394 mm) twin plate mechanical clutch that is integrated with the UltraShift transmission. The UltraShift transmission is a two-pedal system so the DM clutch module requires no clutch pedal and linkage. Dry metallo-ceramic friction facings are typically used in DM clutches.

DM OPERATION

Because the UltraShift transmission is a two-pedal system, there is no driver-actuated clutch pedal. This eliminates the adjustment and maintenance associated with driver-actuated clutch release systems. The UltraShift DM is centrifugal clutch that relies on centrifugal force generated by its rotational speed to bring it into full engagement. During idle, the clutch is disengaged. As engine speed increases, a **touch point** is reached somewhere between 750 and 850 rpm. The touch point occurs when the clutch departure (gap) is taken up. Beyond the touch point rpm, clamping load begins to generate, meaning that the clutch engages and transfers engine torque to the transmission input shaft.

Engine speed is controlled by a combination of transmission and engine controller logic during launch of the truck to provide a smooth start. The clutch continues to generate additional clamping load as engine speed increases. Full clutch clamping force is achieved at approximately 1,350 rpm. The **launch phase** occurs between the touch point rpm (750–850 rpm) and the rpm (around 1,350 rpm) at which full clamp load is achieved. When the vehicle is brought to a stop, engine and transmission logic are used to modulate clutch friction disc clamping pressure until full disengagement occurs at approximately 800 rpm.

14.3 TROUBLESHOOTING

Clutches fail when they are abused. Minimize the potential for abuse and clutches last a lot longer. This is exactly what happens when clutch control becomes the responsibility of the drivetrain electronics rather than the driver. The automated transmission is the best thing to have happened to extend clutch life. In this section, we take a look at some of the factors that cause clutch failures and the strategies we use to troubleshoot them.

HEAT

A major cause of clutch failure is excess heat. Frictional heat generated between the flywheel, driven discs, intermediate plate, and pressure plate can cause surface melting and friction material failure. Wear potential is practically nonexistent when the clutch is fully engaged. However, considerable heat can be generated at the moment of engagement when the clutch is picking up the load. An improperly adjusted or slipping clutch will rapidly generate sufficient heat to self-destruct. **Table 14–1** provides some guidelines on troubleshooting clutch problems.

Where manually actuated clutches are used, training of both drivers and service technicians is important in extending clutch life. Critical points to cover in driver and service technician training are learning to start in the correct gear, recognizing clutch malfunctions, and identifying the need for clutch adjustment. Service personnel should recognize the effects of driver abuse and how it reduces clutch life.

Starting in the Correct Gear

An unloaded truck can be started in a higher transmission gear ratio than when partially or fully loaded. Drivers should be shown what ratios can be used for safe starts when the truck is empty or loaded.

If the truck is diesel powered, a general rule for both empty and loaded trucks is to select a gear ratio that allows the unit to start moving with an idling engine or just enough accelerator pedal travel to prevent stalling. After the clutch is fully engaged, the engine should be accelerated to near governed speed for the upshift into the next higher gear.

Gear Shifting Techniques

Many drivers upshift into the next gear, or skip-shift into a higher gear, before the vehicle has reached sufficient road speed. This shifting technique can be as damaging as starting off in a gear that is too high. This can space the engine speed and driveline speeds far apart, requiring the clutch to absorb the speed difference as friction, thereby generating heat.

TABLE 14–1 Troubleshooting Clutch Operation

Symptom	Probable Cause	Remedy
Clutch does not release or does not release completely	Clutch linkage and release bearing need adjustment	Adjust clutch linkage and release bearing.
	Worn or damaged linkage	Lubricate linkage. Make sure the linkage is not loose. If condition still exists, replace linkage.
	Worn or damaged release bearing	Replace release bearing.
	Worn or damaged splines on input shaft	Replace input shaft.
	Clutch housing loose	Tighten fasteners to specified torque. If necessary, replace fasteners.
	Worn or damaged pressure plate	Replace pressure plate and cover assembly.
	Worn or damaged center plate	Replace center plate.
	Center plate binding	14-inch clutch: Inspect drive pins in flywheel housing and slots in center plate.
		15½-inch clutch: Inspect tabs on center plate and slots in cover. Service as necessary.
	Damaged hub in clutch discs	Replace discs.
	Linings worn below specified dimension	Replace discs.
	Linings damaged	Replace discs.
	Oil or grease on linings	Clean linings. If oil or grease cannot be removed, replace discs.
	Linings not as specified for vehicle operation	Install correct type of linings.
	Damaged pilot bearing	Replace pilot bearing.

(continued)

TABLE 14–1 (continued)

Symptom	Probable Cause	Remedy
Clutch pedal hard to operate	Damaged bosses on release bearing assembly	Replace release bearing assembly.
		Make sure clutch is correctly adjusted.
	Worn or damaged clutch linkage	Lubricate linkage. If condition still exists, replace linkage.
	Worn or damaged components	Replace pressure plate and cover assembly.
Clutch slips on engagement	Driver keeps foot on clutch pedal	Use correct vehicle operating procedure.
	Clutch linkage and release bearing need adjustment	Adjust release bearing and clutch linkage.
	Worn or damaged components	Replace pressure plate and cover assembly.
	Worn or damaged linings	Replace discs.
	Oil or grease on linings	Clean linings. If oil or grease cannot be removed, replace discs.
	Linings not as specified for vehicle operation	Use correct linings.
	Worn or damaged flywheel	Service flywheel as necessary. See procedure of engine or vehicle manufacturer.
Noisy clutch	Clutch linkage and release bearing need adjustment	Adjust clutch linkage and release bearing.
	Worn or damaged linkage	Lubricate linkage. If condition still exists, replace linkage.
	Worn or damaged release bearing	Lubricate release bearing. If condition still exists, replace release bearing.
	Worn or damaged clutch housing	Replace clutch housing and pressure plate assembly.
	Clutch housing loose	Tighten fasteners to specified torque. If necessary, replace fasteners.
	Damaged hub or broken spring(s) in clutch discs	Replace clutch discs.
	Linings worn below specified dimension	Replace clutch discs.
	Linings damaged	Replace clutch discs.
	Oil or grease on linings	Clean linings. If oil or grease cannot be removed, replace linings.
	Damaged pilot bearing	Replace pilot bearing.
Vibrating clutch	Worn or damaged splines on input shaft	Replace input shaft.
	Pressure plate and cover assembly out of balance	Remove, check balance, and install pressure plate and cover assembly. If condition still exists, replace pressure plate and cover assembly.
	Worn or damaged splines in hub of clutch discs	Replace discs.
	Loose flywheel	Tighten fasteners to specified torque. If necessary, replace fasteners. Check flywheel mounting surface for damage. Replace if necessary.

Vehicle or Clutch Overloading

All clutches are designed and recommended for specific vehicle applications and loads. These limitations should not be exceeded. Overloading can damage both the clutch and drivetrain. If the total gear reduction in the powertrain is not sufficient to handle overloads, the clutch will wear because it will be forced to assume the load at a higher speed differential.

Riding the Clutch Pedal

Riding the clutch pedal means operating the vehicle with the clutch partially engaged. This is destructive to the clutch, because it permits slippage and generates excessive heat. Riding the clutch will also put constant thrust load on the release bearing, which can thin out the lubricant and cause excessive wear on the release pads. Release bearing failures are often the result of this type of driving practice. A similar problem occurs if a driver attempts to hold the vehicle on an incline with a slipping clutch. A slipping clutch generates more heat than it is capable of dissipating and damage occurs.

Coasting

Coasting with the clutch released and the transmission in gear can cause the driven discs to rotate at a very high rpm through the multiplication of ratios from the final drive and transmission. This can result in the clutch friction facings being thrown off the clutch discs by centrifugal force. Driven disc speeds as high as 10,000 rpm can occur coasting down an unloading ramp. Engaging the clutch while coasting can also result in a damaging shock load being placed on the clutch as well as the remainder of the drivetrain.

Driver/Service Technician Communication

Drivers should report erratic clutch operation as soon as possible. This allows service technicians to make the necessary inspection, internal clutch adjustment, linkage adjustments, and lubrication checks before a total clutch failure occurs.

The importance of **free pedal** cannot be overemphasized. This is the free play travel in the clutch pedal in the cab. A free pedal of 1½ to 2 inches (37 to 50 mm) is usually specified. Free pedal is directly related to free travel at the release fork. Free pedal should be a checklist component of the daily driver report form. Changes in free pedal measurement are an indicator to the service technician that problems may exist with internal clutch components, so make a habit of noting it whenever you operate a truck.

14.4 MAINTENANCE

Clutches should be checked periodically for proper adjustment and lubrication. Actual maintenance varies with the design of the clutch. Most clutches today are self-adjusting. This means that once the clutch has been installed and the initial adjustment made, no further adjustment to bearing free play should be necessary over the life of the clutch. In these cases, unnecessary manual adjustments can shorten clutch life. Other clutches are manually adjusted and must be adjusted periodically as the friction linings wear. All clutches require regular inspection.

LUBRICATION

Some clutches have sealed release bearings that never need lubrication. Other clutches have release bearings fitted with grease fittings (**Figure 14–22**), and these must be periodically lubricated. Some clutches are equipped with lubrication extension tube assemblies that allow the release bearing to be greased without removing the inspection cover of the bell housing. These grease-fitting extensions reduce maintenance and downtime.

A clutch release bearing should be lubricated according to schedule. Typically, this means whenever the clutch is inspected, once a month, every 6,000 to 10,000 miles (9,700 to 16,000 km), or whenever the chassis is lubricated, whichever comes first. Off-road or other severe service applications require more frequent service intervals. A good-quality extreme-pressure (EP) grease with a temperature performance range of −10°F to +325°F (−23° to 163°C) should be used to lube release bearings.

WARNING

Replacement release bearings are not prepacked with grease. They should be lubricated when the clutch is installed in the vehicle or premature failure will result.

FIGURE 14–22 Some release bearings have a grease fitting and must be periodically lubricated.

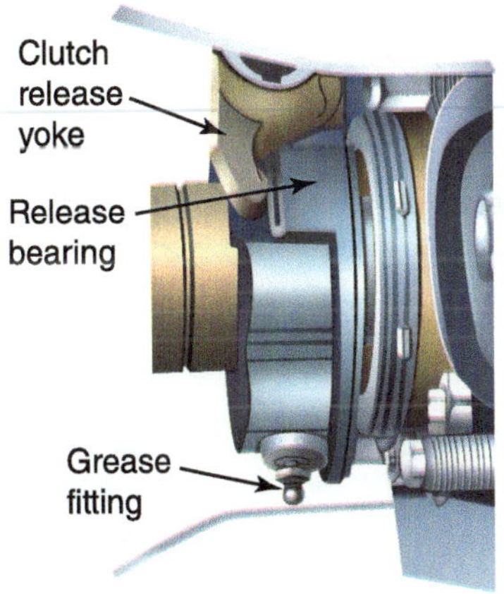

A small quantity of grease should be applied between the release bearing pads and the clutch release fork at normal service intervals. Eaton has recently made available an extended-life (EL) release fork equipped with rollers. Because this design significantly reduces friction, it should deliver on the promise of extended service life.

Many clutches have grease fittings in the clutch housing bosses (**Figure 14–23**) for the clutch cross-shaft assembly. These fittings should be greased whenever the release bearing is lubricated. Whenever the release bearing and other lubrication points on the clutch are serviced, all pivot points on the clutch linkage should also be lubricated (**Figure 14–24**).

FIGURE 14–23 Some transmission housings have grease fittings where the clutch release cross-shaft passes through the housing bosses.

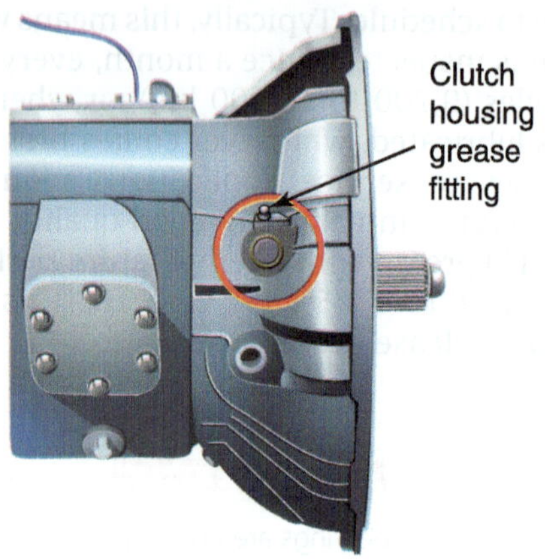

FIGURE 14–24 Clutch linkage lubrication points

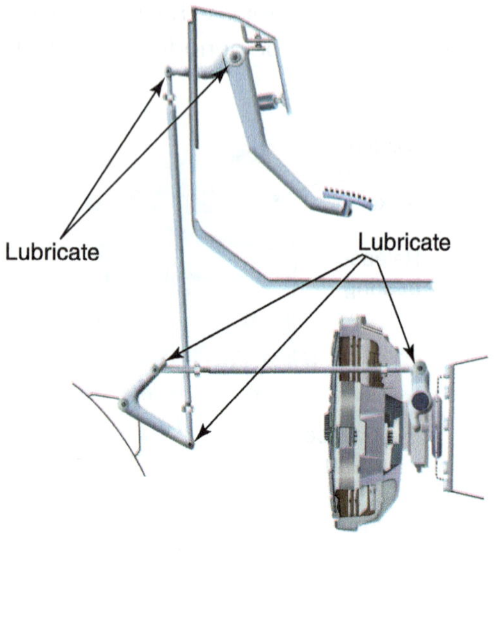

CLUTCH ADJUSTMENTS

Because many clutches used today never require adjustment, if you observe symptoms of an out-of-adjustment clutch, reference original equipment manufacturer (OEM) service literature before jumping in and possibly doing something that could cause the clutch to fail. In this section we take a look at some of the indicators that suggest the need for an adjustment—or other service intervention.

Free Pedal

As mentioned earlier, **clutch free pedal**, or the initial free travel of the clutch pedal, should be 1½ to 2 inches (37 to 50 mm) for both push-type and pull-type clutches. Free pedal is determined by placing your hand or foot on the clutch pedal and gently pushing it down until an increase in push effort is felt. Movement after this point will cause the release bearing to begin disengaging the clutch. In a push-type clutch, free pedal is set to 1½ to 2 inches (37 to 50 mm) to obtain the desired ⅛-inch (3 mm) free travel clearance between the clutch release bearing and clutch release levers (**Figure 14–25**), or diaphragm spring, whichever is used.

In a pull-type clutch, the ⅛-inch (3 mm) free travel clearance occurs between the release yoke fingers and the clutch release bearing pads (**Figure 14–26**).

FIGURE 14–25 In a push-type clutch, the desired free travel clearance is ⅛ inch. Total release bearing travel on a push-type clutch must be approximately ⅝ inch.

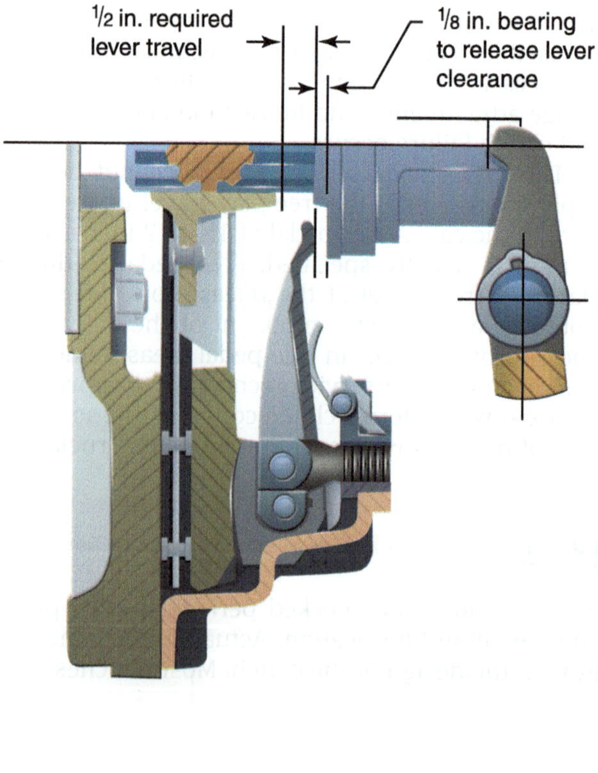

FIGURE 14–26 On a pull-type clutch, there should be ⅛ inch clearance between the release fork and the boss on the release bearing.

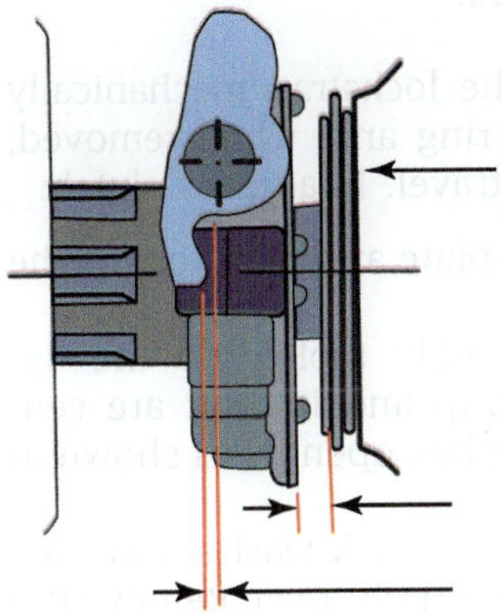

This ⅛-inch free travel at the release bearing should produce 1½ to 2 inches (37 to 50 mm) of free pedal.

Free pedal dimensions are greater than free travel at the release bearing specifications because as movement transfers through the linkage, it is amplified. Too much free pedal prevents complete disengagement of the clutch. Too little free pedal causes clutch slippage, heat damage, and shortened clutch life.

As the friction disc facings wear through normal operation, free pedal will gradually decrease. If inspection indicates clutch free pedal travel is less than ½ inch, adjustment of the clutch is required. DO NOT WAIT UNTIL NO FREE PEDAL EXISTS BEFORE MAKING THIS ADJUSTMENT. Remember that the method of setting free pedal and free travel is different between push-type clutches and pull-type clutches. Use the correct method for adjusting each type of clutch and refer to the service manual for specifications and setting procedures.

Push-Type Clutch Adjustment

In a push-type clutch, adjusting the external clutch linkage to obtain 1½ to 2 inches (37 to 50 mm) of free pedal should result in the specified ⅛-inch clearance between the release bearing and the clutch release forks. Before making a linkage adjustment, inspect the clutch linkage for wear and damaged components. If excessive free play is present in the clutch pedal linkage due to worn components, repair as necessary. Excessive wear of the release linkage can give a false impression of release bearing clearance.

Once the free pedal is set to between 1½ and 2 inches (37 to 50 mm), it is recommended that the free travel clearance be double-checked. To do this:

1. Set the parking brakes and chock the wheels.
2. Remove the clutch inspection cover from the transmission bell housing.
3. Measure the clearance between the release bearing and the clutch release fork. Clearance must be within specifications for the release bearing to release properly.
4. If this clearance is not present, adjust the linkage until the specified ⅛-inch clearance is obtained. Remember, the linkage must be in good condition to obtain accurate results.

Pull-Type Clutch Adjustment

If a pull-type clutch is correctly installed (with clutch brake squeeze properly set), the only adjustments that should be required during the life of the clutch assembly are free travel adjustments. Free travel adjustment is usually an internal adjustment; however, some clutch models are equipped with an external quick-adjust mechanism described a little later in this chapter. The following sections deal with how to assess and correct, if necessary, clutch adjustments.

Pull-Type Clutch Preadjustment Considerations

Before making adjustments to a pull-type clutch, review the following conditions to ensure optimum clutch performance:

1. **Clutch brake squeeze** (increased resistance) begins at the point the clutch brake is initially engaged. Optimum clutch brake squeeze begins 1 inch (25 mm) from the end of the pedal stroke or above the floor board (**Figure 14–27**). Adjustment is made by shortening or lengthening the external linkage rod.
2. Optimum free pedal is 1½ to 2 inches (37 to 50 mm) (**Figure 14–28**). This adjustment is made internally in the clutch, never with the linkage.

FIGURE 14–27 The last inch of clutch pedal travel should squeeze the clutch brake.

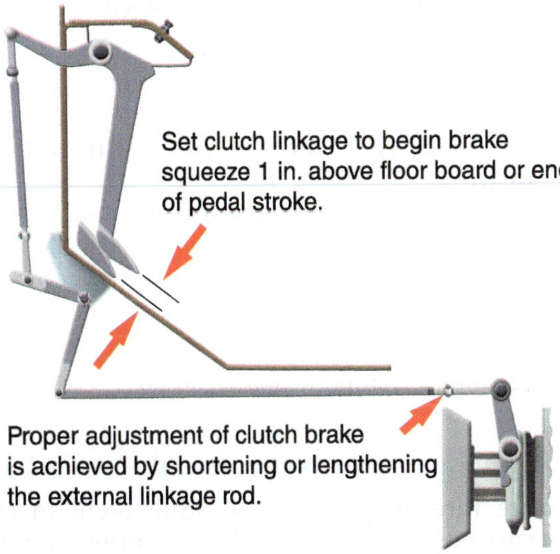

FIGURE 14–28 The first 1½-inches (37 mm) of pedal travel should take up the clearance between the fork and the release bearing.

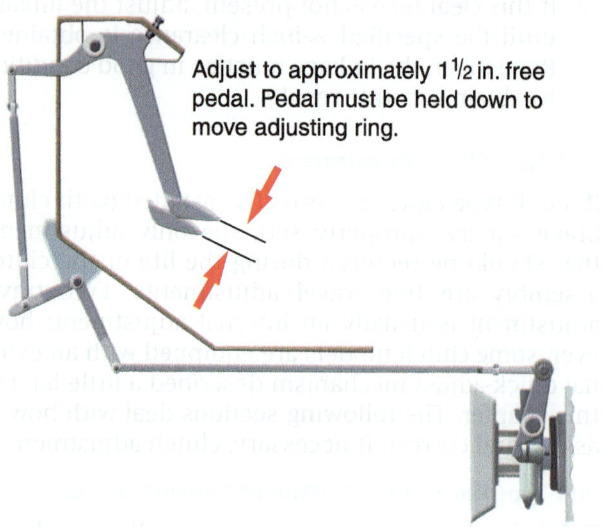

3. Release travel is the total distance the release bearing moves during a full clutch pedal stroke. A typical release travel distance of ½ to 9⁄16 inch (12 to 14 mm) is required to ensure that the release bearing releases sufficiently to allow the friction discs to turn freely with no clutch drag. Optimum free travel is ⅛ inch (3 mm) (see Figure 14–26). These adjustments are accomplished by using the adjusting ring.
4. Internal adjustment of the adjusting ring should be made before attempting linkage adjustments.
5. Internal clutch adjustments should be made with the clutch pedal down (clutch released position).
6. Turning the adjusting ring clockwise moves the release bearing toward the transmission. Turning the adjusting ring counterclockwise moves the release bearing toward the engine.
7. Linkage adjustment on a pull-type clutch should only be made:
    - At initial dealer preparation to set total pedal stroke and yoke throw.
    - To compensate for linkage or clutch brake wear.
    - When worn or damaged linkage components are replaced.

SHOP TALK

Linkage adjustments are not normally required.

Internal Adjustment Mechanisms: Angle-Spring Clutch

There are three basic types of adjustment mechanisms currently in use on angle coil spring clutches used in heavy-duty truck applications. Two are manual-adjusting mechanisms and the third category includes several types of self-adjusting mechanisms. **Photo Sequence 6** shows the procedure for adjusting various types of clutches.

Lockstrap Mechanism. The lockstrap mechanically locks the clutch adjusting ring and, when removed, permits adjustment of free travel. To adjust a clutch:

1. Remove the inspection plate at the bottom of the clutch housing.
2. Rotate the clutch assembly (bolted to the flywheel) until the lockstrap and its bolt are centered in the inspection plate opening as shown in (**Figure 14–29**).
3. Remove the cap screw and lock washer that fastens the lockstrap to the clutch cover. Remove the lockstrap.
4. Push the clutch pedal to the bottom of pedal travel. Use another person or a block of wood to hold the pedal at full travel. Hold the pedal in this position when the adjusting ring is moved. The pedal should be depressed when turning the adjusting ring and up when making a measurement.
5. Rotate the adjusting ring to obtain the specified clearance of the release bearing. Use a screwdriver or an adjusting tool as a lever against the notches on the ring to turn the adjusting ring (**Figure 14–30**). When the adjusting ring is moved one notch, the release bearing will move 0.023 inch (0.6 mm). Moving the ring three notches will move the release bearing approximately 1⁄16 inch (1.5 mm). Turning the adjusting ring clockwise moves the release bearing toward the transmission, increasing pedal free travel. Turning the adjusting ring counterclockwise moves the release bearing toward the engine, decreasing pedal free travel.

FIGURE 14–29 For adjustment, the lockstrap and bolt must be centered at the clutch inspection cover opening.

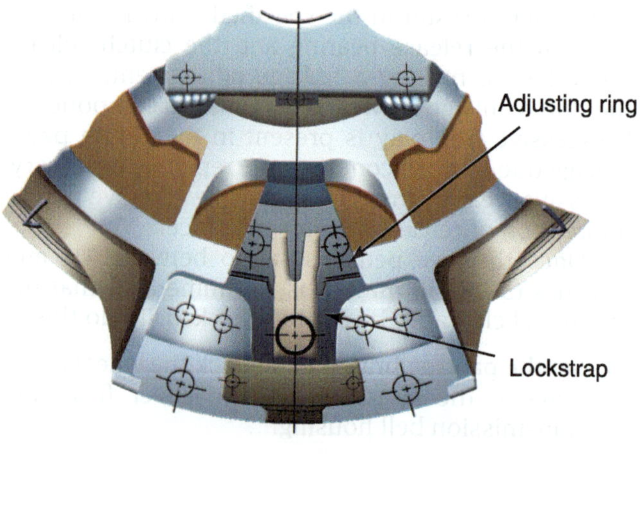

Photo Sequence 6

CLUTCH ADJUSTMENT

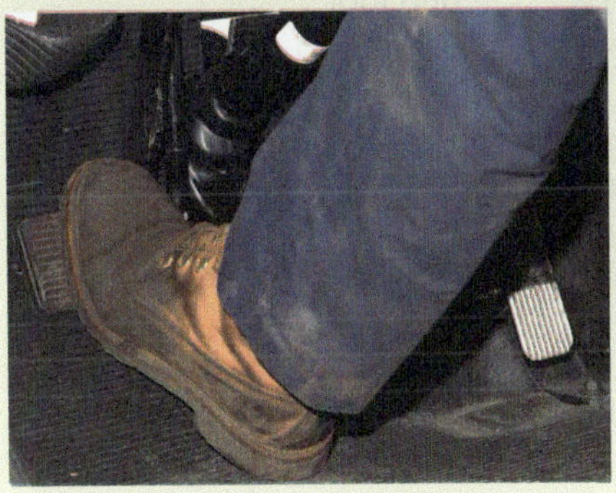

P 6–1 Assess the need for a clutch adjustment by first checking clutch brake squeeze. Clutch brake squeeze should begin 1 inch from the floor at the end of the pedal stroke. Clutch brake squeeze should be verified before attempting an internal adjustment of a clutch. In most cases, clutch brake squeeze will not require adjustment, but it should always be checked.

P 6–2 To adjust the clutch brake, either lengthen or shorten the external linkage by loosening the locknut and turning the adjusting rod either clockwise or counterclockwise.

P 6–3 Next check the clutch pedal free play. This is the pedal travel that results before the clutch disengagement occurs. Free travel is always adjusted internally: Attempting to correct this externally will result in incorrect clutch brake squeeze. Clutch pedal free play should be between $1\frac{1}{2}$ and 2 inches.

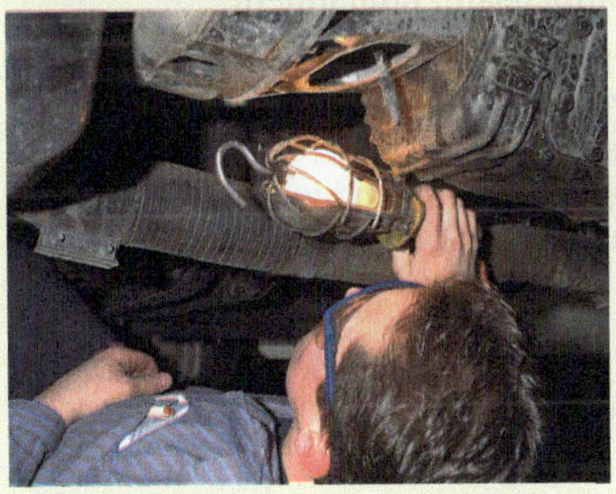

P 6–4 Remove the clutch inspection plate to check the condition of the clutch brake, release travel, and cross-shaft yoke free play. Use a trouble light to inspect the clutch brake. Release travel (the total travel of the release bearing through a full stroke of the clutch pedal) should be between $\frac{1}{2}$ inch and $\frac{9}{16}$ inch. Internal free play is the distance between the cross-shaft yoke and the release bearing when the clutch is fully engaged. It should measure $\frac{1}{8}$ inch.

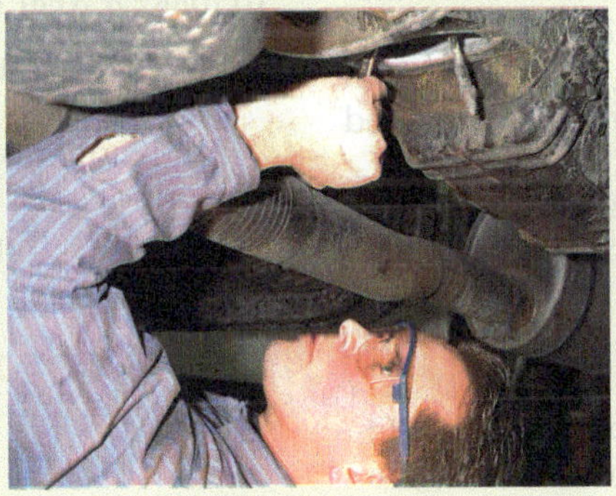

P 6–5 To make an adjustment, fit the adjusting tool to the clutch and have the clutch pedal fully applied by a second person. This releases the clutch, permitting the adjusting ring to be rotated by the adjusting tool. Three notches of CW travel will move the release bearing approximately $\frac{1}{16}$ inch.

P 6–6 Install the clutch inspection plate. Drive the truck a short distance to verify the clutch performance after the adjustment.

FIGURE 14–30 Special adjusting ring tools are available for adjusting the clutch.

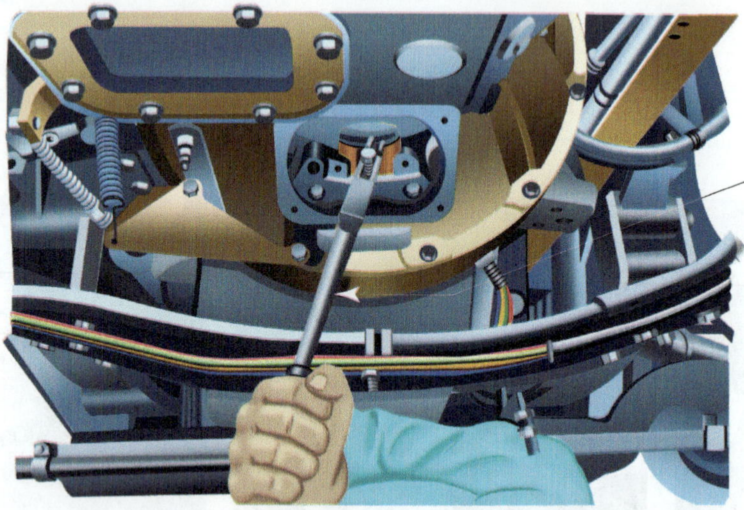

6. Release the clutch pedal, fully engaging the clutch to check release bearing travel. This should measure between ½ and 9⁄16 inch (12 and 14 mm). When this is set, the adjusting ring can be assumed to be correctly set.
7. Install the lockstrap. Insert and torque the lockstrap bolt.
8. Release the clutch pedal.
9. Check the clearance between the yoke and the wear pads. The clearance should be ⅛ inch (3 mm) (see Figure 14–26). If out of specification, adjust the clutch linkage according to the vehicle manufacturer's procedures. As indicated previously, adjustments at the clutch linkage should seldom be required if the clutch was initially installed correctly.
10. Install the inspection hole cover. Install and tighten the cap screws to the manufacturer's specifications.

SHOP TALK

The lockstrap must be located at bottom dead center (BDC) to adjust the clutch. Bar the engine to this position, using an appropriate tool.

Kwik-Adjust Mechanism. This manual adjust mechanism shown earlier in Figure 14–8 permits free travel adjustment without the use of special tools or removing bolts. The adjustment is made using a socket wrench to turn the adjusting bolt. When the adjusting bolt is turned, the clutch adjusting ring is rotated.

1. Using a ¾-inch (19 mm) or ⅝-inch (16 mm) socket (12 point) or box-end wrench, depress the adjusting nut and rotate to make the adjustment (**Figure 14–31**). The Kwik-Adjust will automatically reengage at increments.
2. Ensure that the adjusting nut is locked in position with the flats aligned to the bracket.

Wear Compensator. Wear compensators were used on clutches into the 1990s; they are not common today but you still occasionally see them, especially in vocational trucks. A wear compensator automatically adjusts for friction facing wear each time the clutch is actuated (**Figure 14–32**). When friction facing wear exceeds a predetermined amount, the adjusting ring is advanced and free pedal travel returned to original specifications. To make a wear compensator adjustment:

1. Manually turn the engine flywheel until the adjuster assembly is in line with the clutch inspection cover opening.
2. Remove the right side bolt and loosen the left bolt one turn (**Figure 14–33A**).

FIGURE 14–31 Performing the external manual adjustment with Kwick-Adjust

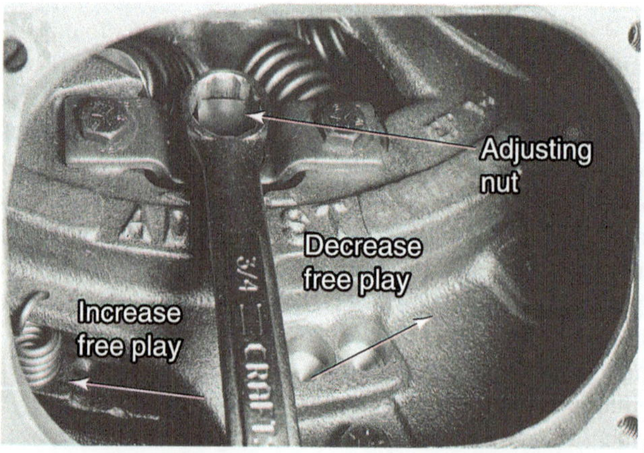

Courtesy of Eaton Corporation—Eaton Clutch Division.

FIGURE 14–32 Clutch wear compensator used on older, self-adjusting clutches

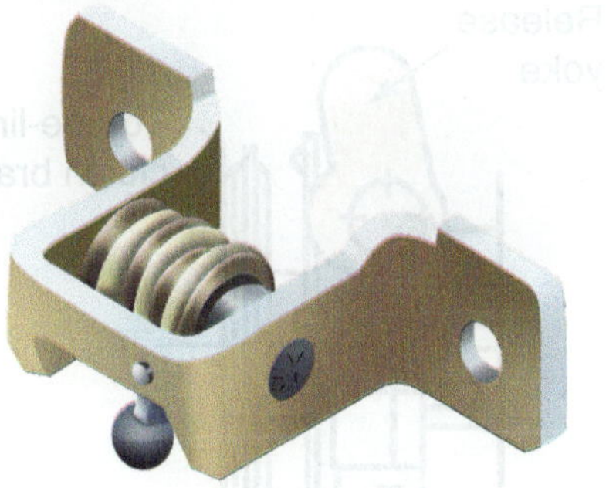

FIGURE 14–33 Steps in making a clutch adjustment on an older, self-adjusting clutch equipped with a wear compensator

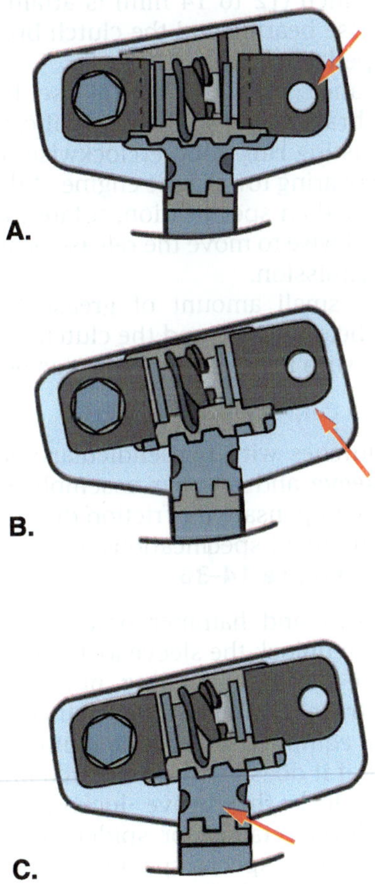

3. Rotate the wear compensator upward to disengage the worm gear from the adjusting ring (**Figure 14–33B**).
4. Advance the adjusting ring as necessary (**Figure 14–33C**).
5. Rotate the assembly downward to engage the worm gear with the adjusting ring. The adjusting ring may have to be rotated slightly to reengage the worm gear.
6. Install the right side bolt and torque both bolts to specifications.
7. Visually check that the actuator arm is inserted into the release sleeve retainer. If the assembly is properly installed, the spring will move back and forth as the pedal is full stroked. Note: The clutch will not compensate for wear if the actuator arm is not inserted into the release sleeve retainer or if release bearing travel is less than ½ inch (12 mm).

CAUTION

Do not pry on the inside gear teeth of the adjusting ring. Doing so can damage the teeth and prevent the clutch from self-adjusting.

SHOP TALK

Each lower gear tooth of the adjusting ring represents about 0.010 inch (0.25 mm) of release bearing movement. Therefore, six notches of rotation translates to 0.060 inch (1.5 mm) or about 1⁄16-inch (1.5 mm) release bearing movement.

When a self-adjusting clutch is out of adjustment, check for the following:

- Actuator arm incorrectly inserted into the release bearing sleeve retainer
- Bent adjuster arm
- Seized or damaged clutch components, such as the adjusting ring

After identifying and repairing or replacing the defective condition or component, readjust the bearing setting.

Internal Adjustment: Nonclutch Brake Clutches

The following steps outline the typical internal adjustment procedure for manual and self-adjusting angle-spring clutches not equipped with a clutch brake.

1. Remove the clutch inspection plate below the bell housing.
2. With the clutch engaged (pedal up), measure the clearance between the release bearing and clutch housing.
3. If clearance is not within specifications—typically 1⅞ inches (45 mm) for a single-plate clutch and ¾ inch (19 mm) for a two-plate clutch (**Figure 14–34**)—continue with steps 4 and 5 below; otherwise skip to step 6 below.

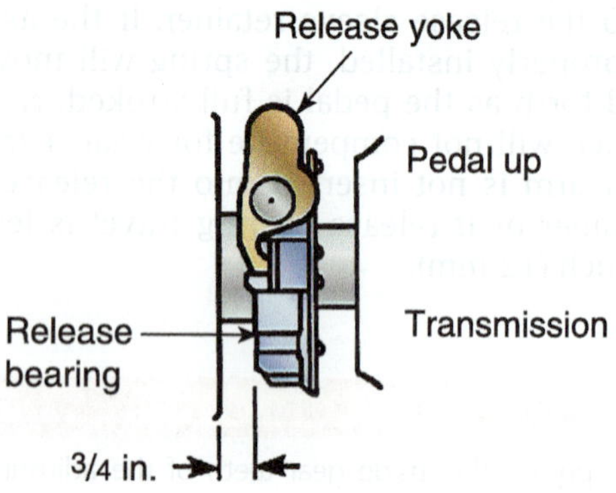

FIGURE 14–34 Two-plate clutch release bearing clearance on angle-spring clutches without a clutch brake. The ¾-inch specification is used on synchronized transmissions only. On a nonsynchronized transmission, the specification is ½ inch.

4. Release the clutch by fully depressing the clutch pedal.
5. Using the internal adjustment procedure outlined previously for lockstrap, Kwik-Adjust, and wear compensator adjustment mechanisms, advance the adjusting ring until the clearance specification is obtained. If the clearance between the release bearing and clutch housing is lower than specification, rotate the adjusting ring counterclockwise to move the release bearing toward the engine. If the clearance is higher than specification, rotate the adjusting ring clockwise to move the release bearing toward the transmission.
6. Apply a small amount of grease between the release bearing pads and the clutch release fork.
7. Check the linkage adjustment.

Internal Adjustment: Clutch Brake Clutches

The following steps outline the typical internal adjustment procedure for manual and self-adjusting angle-spring clutches equipped with a clutch brake.

1. Remove the inspection plate below the clutch housing.
2. With the clutch engaged (pedal up), measure the clearance between the release bearing and the clutch brake. This is the release travel. If clearance (**Figure 14–35**) is less than ½ inch (12 mm) or greater than 9⁄16 inch (19 mm) (typical), continue with steps 3 and 4 below; otherwise proceed to step 5.
3. Release the clutch by fully depressing the clutch pedal.
4. Using the internal adjustment procedures previously described for lockstrap, Kwik-Adjust, and wear compensator adjustment mechanisms, advance the adjusting ring until a distance of ½ to 9⁄16 inch (12 to 14 mm) is attained between the release bearing and the clutch brake with the clutch pedal released.
5. If clearance between the release bearing and clutch brake is less than specification, rotate the adjusting ring counterclockwise to move the release bearing toward the engine. If the clearance is greater than specification, rotate the adjusting ring clockwise to move the release bearing toward the transmission.
6. Apply a small amount of grease between the release bearing pads and the clutch release fork.
7. Proceed with linkage adjustment as needed.

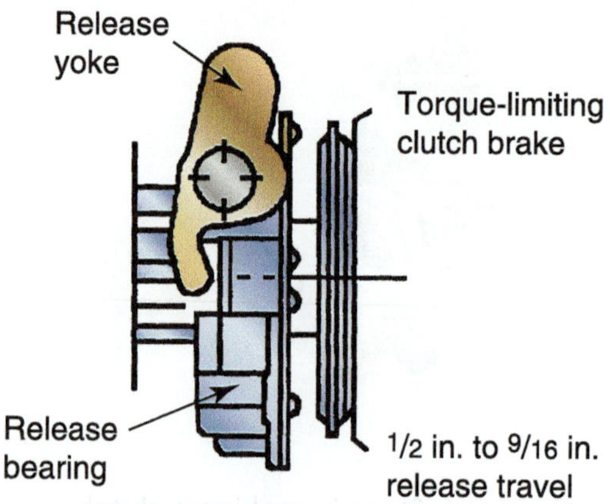

FIGURE 14–35 Clearance between release bearing and clutch housing on clutches with a clutch brake

Clutches with Perpendicular Springs

Pull-type clutches with perpendicular springs use a threaded sleeve and retainer assembly that can be adjusted to compensate for friction disc facing wear.

The adjustment specifications and procedure are illustrated in **Figure 14–36**.

1. Use a drift and hammer or a special spanner wrench to unlock the sleeve locknut.
2. Turn the slotted adjusting nut to obtain the release travel clearance of ½ inch (12 mm) if the clutch is equipped with a clutch brake and ¾ inch (19 mm) if it does not have a clutch brake.
3. Securely lock the sleeve locknut against the release lever retainer, or spider, using the drift and hammer or special spanner wrench.
4. Adjust the clutch linkage to obtain a yoke-to-bearing free travel clearance of ⅛ inch (3 mm) if the vehicle is equipped with a nonsynchronized transmission with a clutch brake. On vehicles equipped with synchronized transmissions (without a clutch brake) the yoke clearance should be ¼ inch (6 mm) and that should result in approximately 3 inches of free pedal.

FIGURE 14–36 Clutch-adjusting mechanism for a clutch with perpendicular springs

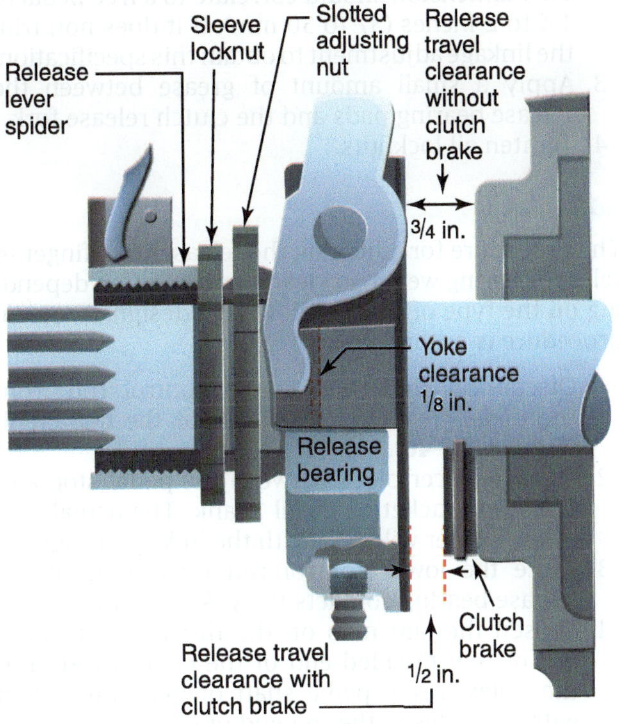

CLUTCH LINKAGE INSPECTION

The clutch linkage should be inspected carefully. The clutch will not operate properly if the linkage is worn or damaged. Inspect the linkage according to the following procedure.

1. Depress the clutch pedal and have another person check the release fork for movement. The smallest movement of the clutch pedal should result in movement at the release fork. If the release fork does not move when the clutch pedal moves, locate and correct the free play condition.
2. The linkage must move when the pedal is actuated. Make sure the linkage is not obstructed and every pivot point operates freely. Ensure that the linkage is not loose at any point and if it binds, locate and service the cause of the condition.
3. Check that the pedal, springs, brackets, bushings, shafts, clevis pins, levers, cables, and rods are not worn or damaged. In a hydraulic clutch, check for leaks and that the reservoir is filled to the specified level. Replace missing or damaged components. Do not attempt to straighten any damaged parts.
4. Lubricate every pivot point in the linkage. Use the lubricant specified by the manufacturer of the vehicle. An NLGI #2 multipurpose lithium grease is typically specified for bushings and other pivot points.

CLUTCH LINKAGE ADJUSTMENT

As outlined earlier, there are three types of linkage mechanisms currently used in truck applications: mechanical, hydraulic, and pneumatic. Adjustment methods vary between truck manufacturers, so always follow the procedure listed in the truck service literature. The following is general information on adjusting the clutch linkage.

Pedal Height. On some vehicles, the travel height of the clutch pedal is adjusted. The height is set by stop bolts. If the pedal height is not correct, the amount of free pedal will not be correct (**Figure 14–37**). Consult the manufacturer's service literature for pedal height specifications.

Total Pedal Travel. The *total pedal travel* is the complete distance the clutch pedal must move. It can be adjusted with bumpers and stop bolts in the cab or with stop bolts and pads on the linkage. Total travel makes sure that there is enough movement of the pedal to correctly engage and disengage the clutch (**Figure 14–38**). Consult the manufacturer's service literature for specifications.

Free Pedal Travel. As explained earlier, the free pedal is the travel of the pedal before the release bearing starts to move. It is typically 1½ to 2 inches (37 to 50 mm). If free pedal travel is more than 2 inches (50 mm), the clutch might not fully release. The clutch friction discs could be loaded to contact the flywheel continually, causing excessive wear. If free pedal travel is reduced to zero by wear, the clutch could slip and continually load the throwout bearing.

FIGURE 14–37 Pedal height adjustment point

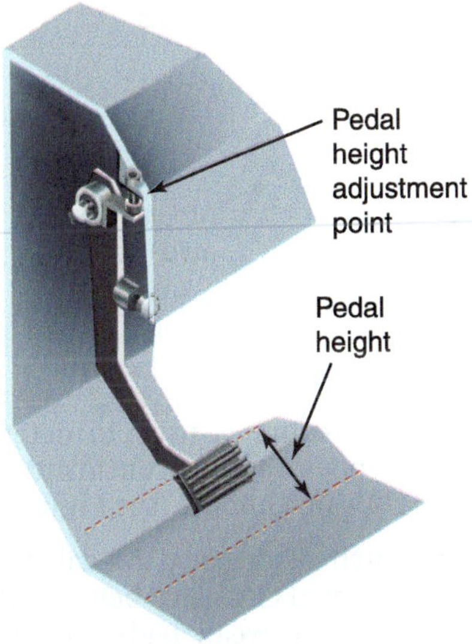

FIGURE 14–38 Total pedal travel adjustment point

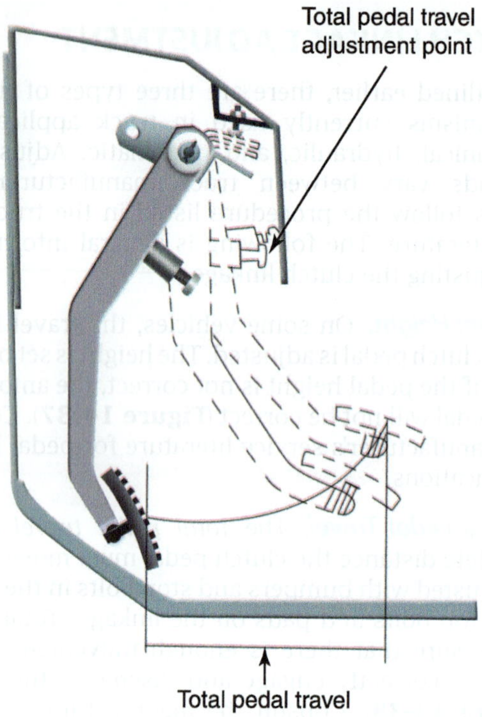

FIGURE 14–39 This linkage is adjusted by lengthening or shortening the control rod.

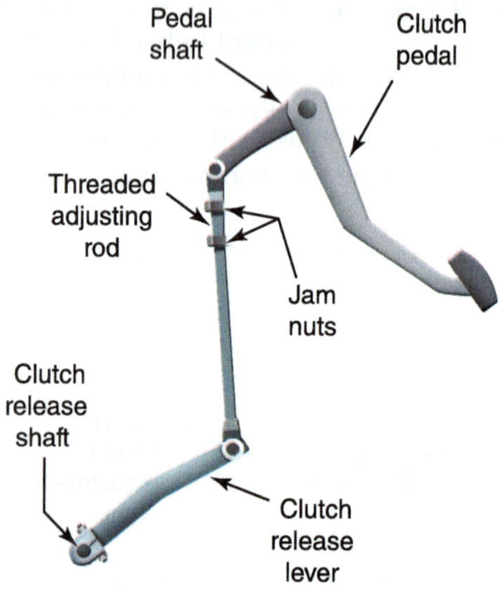

Checking Linkage Adjustment

The following procedure is used to determine if a linkage adjustment is required on manual and self-adjusting angle-spring clutches (see **Figure 14–39**).

1. With the clutch disengaged (pedal down), measure the free travel clearance between the release yoke fingers and wear pads on the release bearing. If clearance is either greater or less than ⅛ inch (3 mm), adjust the external linkage to obtain the ⅛-inch (3 mm) clearance (see Figure 14–26).
2. This dimension should correlate to a free pedal of 1½ to 2 inches (37 to 50 mm). If it does not, trim the linkage adjustment to obtain this specification.
3. Apply a small amount of grease between the release bearing pads and the clutch release fork.
4. Tighten all locknuts.

Adjustment

The procedure for adjusting the release yoke finger to release bearing wear pad clearance can differ, depending on the type of truck and linkage design. A typical procedure is outlined here:

1. Disconnect the lower clutch control rod from the pedal shaft (Figure 14–39) or the bell crank (**Figure 14–40**).
2. Place a spacer block between the pedal stop and the stop bracket or pedal shank. The actual size of the spacer will vary with the linkage design.
3. Force the lower control rod forward until the release bearing contacts the yoke fingers.
4. Loosen the jam nuts on the threaded adjusting rod of the threaded end of the control rod until the holes in the pedal shaft or bell crank align with the holes in the rod end or clevis.
5. Reconnect the linkage members and remove the spacer block from the pedal stop.
6. Operate the clutch pedal and recheck the yoke-to-bearing clearance. If the clearance is still insufficient, it might be necessary to adjust the pedal height or travel. By turning the pedal stop in, the pedal will return farther and yoke-to-bearing clearance will be increased.

FIGURE 14–40 Adjustments to this linkage are made by adjusting the clevis on the lower control rod.

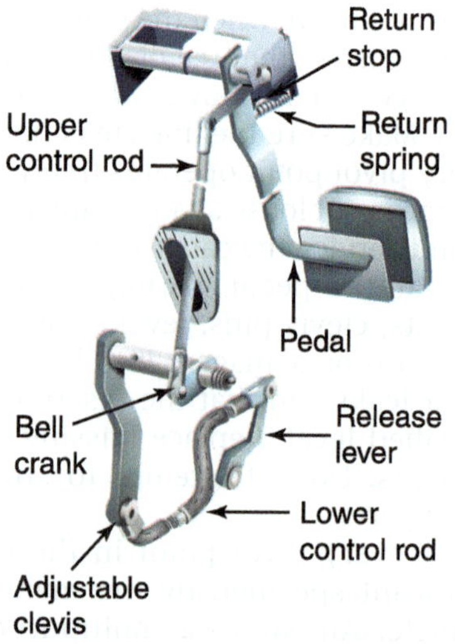

Clutch Brake Setting

To check the clutch brake setting, depress the clutch pedal in the cab and note the point at which the clutch brake engages by observation through the inspection cover. With the release travel and free travel settings, clutch brake squeeze should occur approximately 1 inch (25 mm) from the end of the pedal stroke (see Figure 14–28). To check this:

1. Insert a 0.010-inch (0.25 mm) thickness gauge between the release bearing and the clutch brake.
2. Depress the clutch pedal to squeeze the thickness gauge.
3. Let the pedal up slowly until the gauge can be pulled out and note the position of the pedal in the cab. It should be ½ to 1 inch (12.5 to 25 mm) from the end of the pedal stroke.
4. To adjust the clutch brake setting, shorten or lengthen the external linkage according to service literature procedure. If the specified adjustment cannot be obtained, check the linkage for excessive wear and pedal height.
5. Reinstall the inspection cover.

14.5 CLUTCH SERVICING

When it is determined that the clutch is not operating properly and is in need of servicing, the transmission and clutch cover assembly must be removed to access the clutch pack assembly. Parts that are worn or damaged must be replaced. Generally, components of the clutch pack assembly are not serviceable separately. As a rule, if the pressure plate, springs, release levers, and other parts of the component are damaged, the complete clutch assembly should be replaced. Today, clutch rebuilding is usually performed only by clutch specialty rebuilding shops.

The following sections explain how to remove a clutch, inspect clutch components for damage, and install a clutch.

SERVICING PRECAUTIONS

For many years, clutch facings were made with asbestos fibers. Asbestos has been proven to cause numerous health hazards when inhaled in sufficient quantities or over an extended period. The Occupational Safety and Health Administration (OSHA) has placed stringent regulations on working environments where asbestos brake linings or clutch facings must be handled. The least that technicians should do to protect themselves from exposure is to wear a particulate mask when working on a clutch.

Many manufacturers have developed nonasbestos facings using natural and man-made materials in place of asbestos. Current OSHA regulations do not cover nonasbestos fibers. Medical experts do not agree about the possible long-term risks of working with and breathing nonasbestos fibers. However, some experts believe that long-term exposure to some nonasbestos fibers could be harmful. Because of this, technicians should follow the precautions outlined here whenever working on any clutch.

1. Whenever possible, work on clutches in a separate area away from other operations.
2. Wear a particulate mask approved by the National Institute for Occupational Safety and Health (NIOSH) or OSHA during all clutch service procedures. Wear the mask from removal through installation.
3. Never use compressed air or dry brushing to clean clutch assemblies. OSHA recommends the use of sealed cylinders that enclose the clutch. These cylinders have pump-driven, high-efficiency vacuum filters and sealed arm sleeves. If such equipment is not available, wet with solvent and clean clutch parts in a well-ventilated area.
4. During disassembly, carefully place all parts on the floor to avoid getting dust into the air. Use an industrial vacuum cleaner with a high-efficiency filter to clean dust from the clutches. After using the vacuum, remove any remaining dust with a rag soaked in water or solvent and allow to air dry.
5. When cleaning the work area, never use compressed air or dry sweeping. Use an industrial vacuum with a high-efficiency filter and wet rags. Used rags should be disposed of with care to avoid getting dust into the air. Use a particulate mask when emptying vacuum cleaners and handling used rags.
6. Technicians should wash their hands before eating or drinking. Work clothes should not be worn home and should be laundered separately.

TRANSMISSION/CLUTCH REMOVAL

The following is a general procedure for removing the transmission and clutch. Always consult manufacturer service literature for specific instructions and specifications.

SHOP TALK

When preparing to remove a transmission from a truck chassis, attempt to raise the rear of the truck and place on jack stands so that the engine and transmission are on a level plane. There are two benefits: First, you will not be fighting gravity when the transmission is reinstalled and, second, you will have more room to work under the chassis.

Transmission Removal

When removing a transmission from a chassis, pay special attention to any auxiliary equipment that may have to be disconnected or removed. In chassis such as fire trucks and garbage packers that can be heavily optioned with auxiliary equipment, transmission removal and replacement times may be triple the OEM time estimate.

1. Remove the shift lever from the transmission. If necessary, remove the shift tower assembly from the transmission.
2. Mark the yoke or flange of the drive shaft and the output shaft of the transmission. The marks on the drive shaft and the output shaft should ensure that the drive shaft is correctly reinstalled.
3. Remove the drive shaft.
4. Disconnect the electrical connections from the transmission.
5. Disconnect the air lines from the transmission.
6. If used, remove the return spring from the clutch lever on the transmission. Mark and disconnect the clutch linkage from the clutch housing on the transmission.
7. Rotate the release yoke so it clears the release bearing when it is removed.
8. If a hydraulic system is used on the clutch, disconnect the pushrod and the spring from the release fork. Remove the hydraulic cylinder from the bracket on the transmission. Use wires to support the cylinder on the frame.
9. Use a sling or jack to maintain transmission alignment (**Figure 14–41**). Do not allow the rear of the transmission to drop or allow the transmission to hang unsupported in the splined hubs of the friction discs as this could distort them, resulting in poor clutch operation or clutch release problems.
10. Remove the fasteners that attach the transmission to the brackets on the frame.
11. Remove the bolts and washers that attach the bell housing to the engine. Pull the transmission and bell housing straight out from the engine. Remove the transmission from the vehicle (**Figure 14–42**).
12. If used, remove the clutch brake assembly from the input shaft of the transmission.

FIGURE 14–41 Maintain proper alignment angle when pulling the transmission away from an engine.

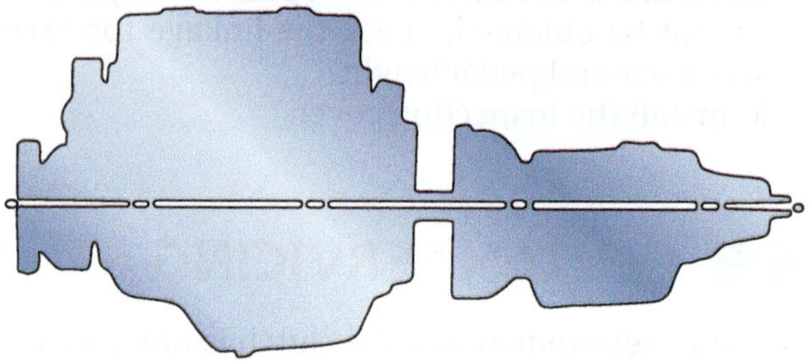

FIGURE 14–42 When removing the transmission, make sure that it is fully supported and pull it straight out.

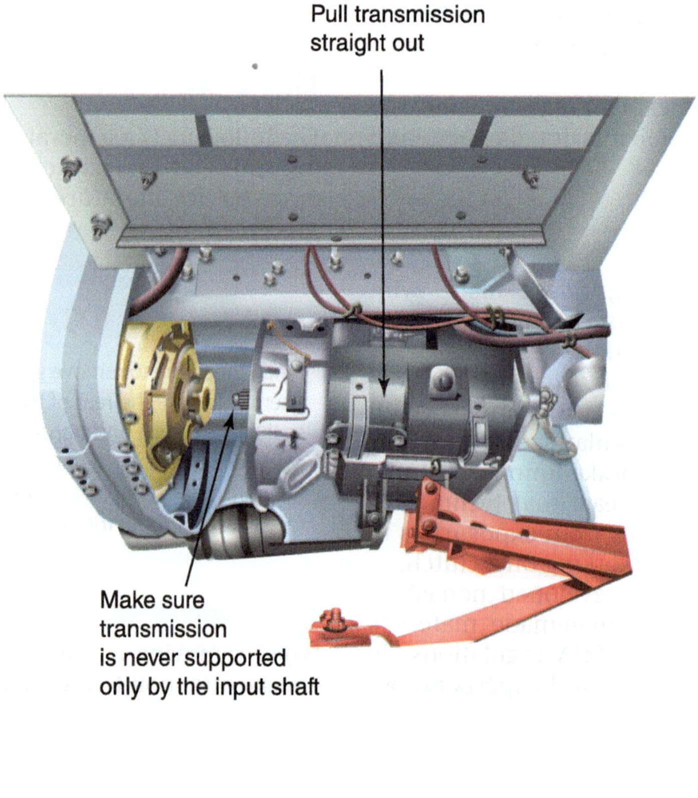

CAUTION

Make sure that the transmission does not hang by the input shaft in the pilot bearing bore in the flywheel. The clutch assembly and the pilot bearing can be damaged if the transmission is supported by the input shaft.

SHOP TALK

If the clutch is not being replaced, mark the cover of the clutch and the flywheel. The marks ensure that the clutch is correctly reinstalled in its original position on the flywheel.

CAUTION

Engine mounts are sometimes located on the transmission bell housing; if so, the engine must be securely supported before removing the transmission.

Removing Clutch from Flywheel

Ensure that you have a strategy for supporting the weight of a clutch pack before removing the fasteners. The weight of a heavy-duty clutch assembly can result in injuries if you are not prepared.

1. Install a clutch alignment tool (pilot shaft) through the release bearing and the friction discs and into the flywheel pilot bearing (**Figure 14–43**). The alignment tool supports the clutch assembly during removal. If an alignment tool is not available, use an appropriate size transmission input shaft.
2. Pull the bearing back using the release tool shown and insert two appropriately sized spacers (wooden shipping blocks will do) between the clutch cover and the release bearing (**Figure 14–44**). The spacers relieve the clutch spring force and enable easier reinstallation.
3. Loosen the mounting bolts around the flywheel in a progressive crisscross pattern, but do not remove them.
4. Completely remove the top two bolts that fasten the pressure plate and cover assembly to the flywheel. Install two ⅜-inch-diameter, 2½-inch-long guide studs in the holes (**Figure 14–45**).
5. Connect a lifting device to the pressure plate and cover assembly (**Figure 14–46**).
6. Remove the remaining bolts that fasten the pressure plate and cover assembly to the flywheel.
7. Lift the pressure plate and cover assembly over the alignment tool and off the flywheel.
8. Remove the rear friction disc.
9. Remove the center plate.
10. Remove the front friction disc.
11. Remove the alignment tool from the flywheel.

FIGURE 14–43 Install an alignment tool (pilot shaft) into the clutch before loosening the clutch cover.

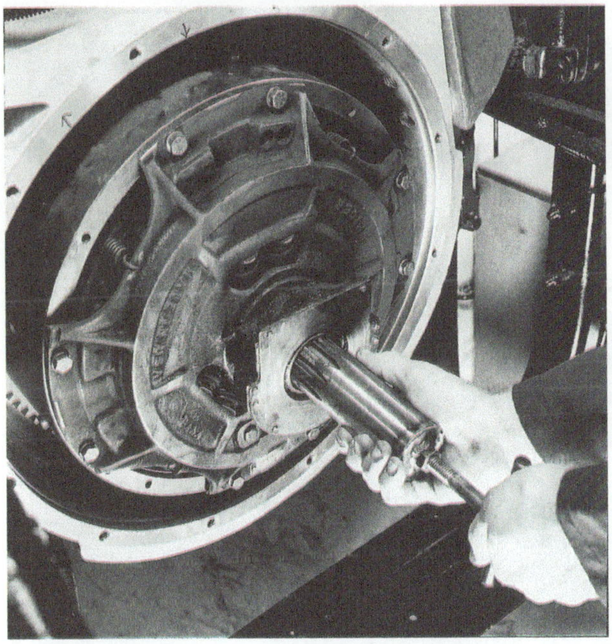

Courtesy of Eaton Corporation—Eaton Clutch Division.

FIGURE 14–44 Place a spacer between the release bearing and the clutch cover to unload the clutch spring force.

Courtesy of Eaton Corporation—Eaton Clutch Division.

SHOP TALK

When removing a 15½-inch clutch, the discs and the pressure plate can remain within the cover.

FIGURE 14–45 Install guide studs into the top of the cover to support the clutch and guide it aligned horizontally during removal.

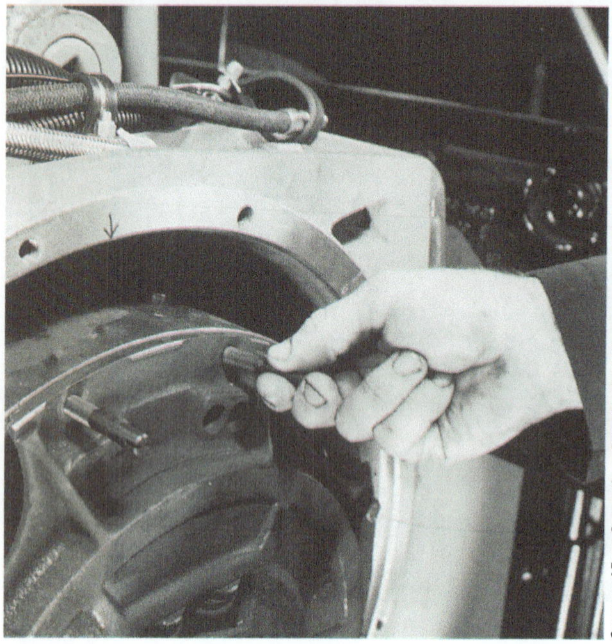

Courtesy of Eaton Corporation—Eaton Clutch Division.

FIGURE 14–46 Use a lifting device to support the heavier 15½-inch clutches.

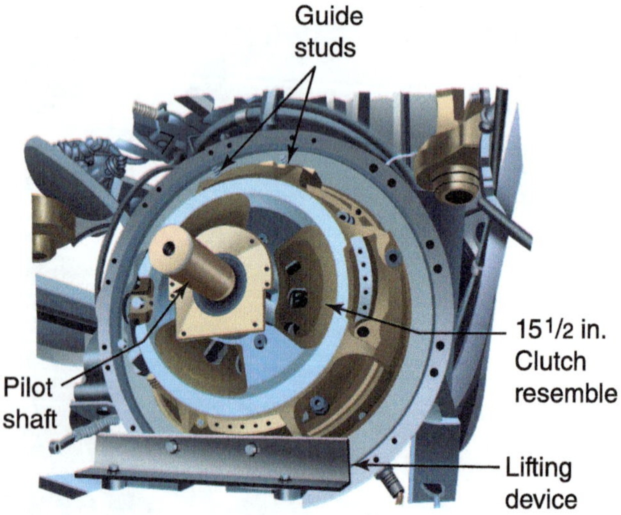

Pilot Bearing Replacement

Each time the clutch assembly is serviced or the engine is removed, the pilot bearing in the flywheel should be removed and replaced. Because a pilot bearing is often damaged during removal, it is important to inspect it before undertaking this procedure. A seized pilot bearing can damage the transmission input shaft, so this should also be inspected. Use an internal puller or a slide hammer to remove the pilot

FIGURE 14–47 Use a puller to remove the pilot bearing.

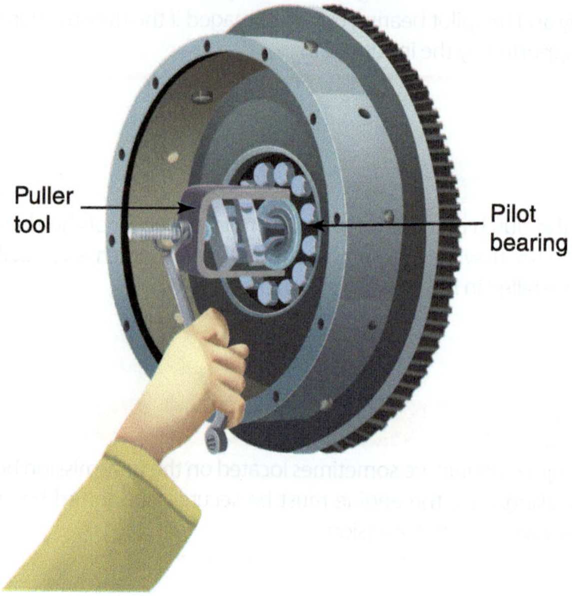

bearing (**Figure 14–47**) from the flywheel. Discard the used pilot bearing.

CLUTCH INSPECTION

After removing the clutch from the flywheel, inspect the clutch components for damage or signs of wear. It is important that the cause of wear be determined and the problem corrected.

Transmission Clutch Housing and Flywheel Surfaces

The engine and transmission should be in alignment. To verify this, perform the following inspection procedure. Surfaces being indicated must be clean for accurate measurements.

1. Inspect the surface face of the flywheel for wear or damage. Make sure that the flywheel is not cracked. Minor heat discoloration is a normal wear condition that can be removed with an emery cloth. Some wear or damage can be removed by machine-grinding the flywheel face. If wear or damage on the flywheel cannot be removed, it must be replaced.
2. Inspect the teeth of the ring gear on the outer surface of the flywheel. If the teeth are worn or damaged, replace the ring gear or the flywheel and inspect the starter drive teeth.
3. Inspect the mating surfaces of the transmission bell housing and the engine flywheel housing (**Figure 14–48A**). Wear on either housing flange will result in misalignment. Wear will usually

FIGURE 14–48 (A) Clutch (bell) housing inspection points; and (B) areas of maximum wear

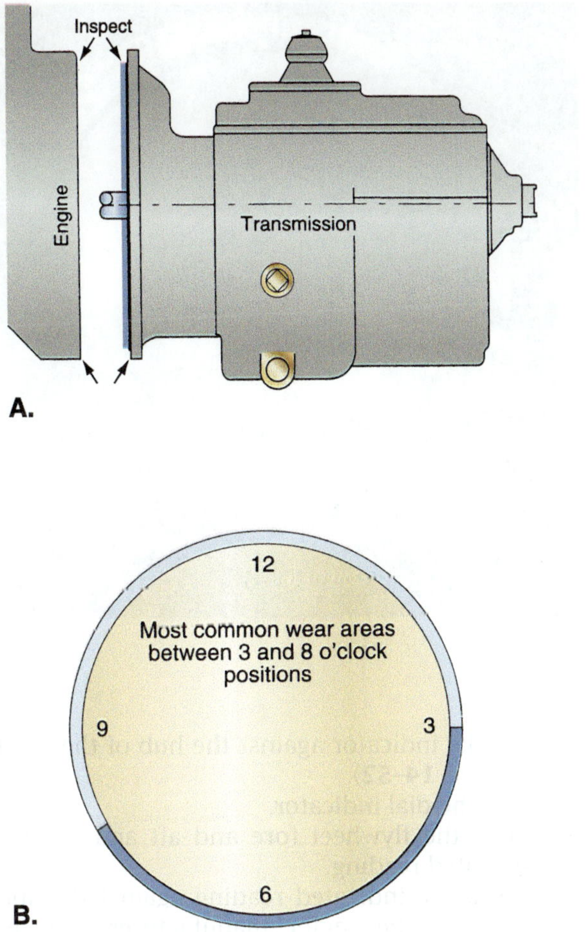

occur on the lower half of these surfaces, with the most wear occurring between the 3 o'clock and 8 o'clock positions (**Figure 14–48B**). Replace the clutch housing or flywheel housing, if worn.

SHOP TALK

When measuring runout dimensions on a flywheel, force the flywheel toward the engine so that crankshaft endplay is not a factor of the flywheel runout measurement.

Flywheel Outer Surface Runout. To make this check:

1. Secure a dial indicator to the flywheel housing with the gauge finger on the flywheel near the outer edge (**Figure 14–49**).
2. Zero the dial indicator.
3. Manually rotate the flywheel one revolution in the direction of engine rotation. On some engines the crankshaft can be rotated by putting a socket on the nut that holds the pulley on the front of the crankshaft or by a toothed ratchet that engages to the ring gear teeth. If access to the front of the crankshaft is difficult, use a spanner wrench on the teeth of the flywheel to rotate the crankshaft.
4. Record the reading on the dial indicator, marking the high and low points.
5. The acceptable runout on the outer surface of the flywheel is a specified amount multiplied by the diameter of the flywheel in inches. For example, maximum permissible runout may be specified as 0.0005 inch (0.0127 mm) per inch (25 mm) of flywheel diameter, making the maximum total indicated difference between the high and low points 0.007 inch (0.18 mm) or less for a 14-inch (356 mm) clutch, and 0.008 inch (0.20 mm) or less for a 15½-inch (394 mm) clutch. Check service specifications for exact tolerances.

FIGURE 14–49 Checking flywheel face runout

Courtesy of Eaton Corporation—Eaton Clutch Division.

Checking Flywheel Housing Pilot Runout. To make this check:

1. Secure the dial indicator to the crankshaft.
2. With the dial gauge plunger against the housing pilot, rotate the crankshaft one revolution in the direction of engine rotation (**Figure 14–50**).
3. Use a marker or soapstone to mark the high and low points and record dial indicator readings. Total difference between high and low points should not exceed manufacturer's specifications, which normally range between 0.006 and 0.015 inch (0.152 and 0.381 mm).
4. If runout exceeds specifications, service the flywheel as required.

FIGURE 14–50 Checking runout of the flywheel housing bore. This important specification is also known as flywheel housing concentricity.

Courtesy of Eaton Corporation—Eaton Clutch Division.

FIGURE 14–51 Checking runout on pilot bearing bore

Courtesy of Eaton Corporation—Eaton Clutch Division.

Checking Flywheel Housing Face Runout. Proceed as follows:

1. Position the dial gauge plunger to contact the face of the engine flywheel housing.
2. As described previously, rotate the crankshaft through one revolution, marking high and low points. Typically, the total difference between the high and low points should not exceed 0.008 inch.

Checking Runout on the Pilot Bearing Bore. Proceed as follows:

1. Install a dial indicator so that the base of the indicator is on the mounting surface of the flywheel housing. Ride the plunger of the dial indicator against the pilot bearing bore (**Figure 14–51**).
2. Zero the dial indicator.
3. Manually turn the crankshaft one revolution in the direction of the engine rotation.
4. Record the reading on the dial indicator. Maximum runout for the surface of the bore of the pilot bearing is typically 0.005 inch (0.127 mm). If runout exceeds specifications, service the crankshaft as required. If these limits are exceeded, the problem must be corrected or misalignment will cause premature wear to the drivetrain components.

Check Crankshaft Endplay. Proceed as follows:

1. Install a dial indicator so that the base of the indicator is on the flywheel housing. Set the plunger of the dial indicator against the hub of the flywheel (**Figure 14–52**).
2. Zero the dial indicator.
3. Force the flywheel fore and aft and record the indicated reading.
4. Check the indicated reading against the specification of the engine manufacturer. Service the crankshaft as required. Usually the crank thrust washers are at fault.

Flywheel Drive Pin Inspection/Removal

On 14-inch (356 mm) clutches, inspect the drive pins in the flywheel housing. The drive pins should not be filed or ground because this can result in unequal load pressure on the pins, causing them to break. Replace worn or damaged drive pins using the following procedure:

1. Remove the flywheel. Consult the manufacturer's specifications.
2. Remove the setscrew(s) that fasten each drive pin in the flywheel housing (**Figure 14–53**).
3. Use a hammer and punch to remove the drive pins from the flywheel housing.

WARNING

Wear eye protection when driving pins with a hammer and punch.

FIGURE 14–52 Checking crankshaft endplay

FIGURE 14–53 Remove the drive pin setscrew and force the pins out with a drift and hammer.

CAUTION

Do not attempt to file the drive pin slots on the intermediate plate to obtain correct clearances. Doing so can result in a clutch hanging up or shearing of the drive pins.

Clutch Release Fork and Shaft Inspection

Inspect the shaft and the release fork (**Figure 14–54**). Ensure that the release fork is straight and that the tips of the fork are not worn or damaged. Replace forks

FIGURE 14–54 Check the release shaft and fork for wear and smooth operation.

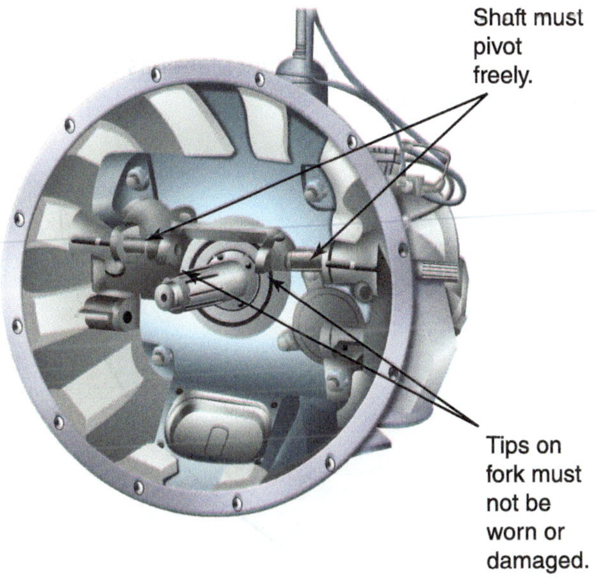

that are worn or damaged. The shaft should rotate freely in the clutch housing. It should not have any side-to-side movement in the housing. Inspect the clutch shaft bushings in the housing. Replace any components that are worn or damaged.

Input Shaft Inspection

Inspect the splines on the input shaft. Ensure the splines are not worn or damaged. Inspect the area of travel of the release bearing for damage. Use an emery cloth to remove small scratches on the input shaft. Replace input shafts that are worn or damaged. Wear or damage on the input shaft can cause the clutch to operate incorrectly.

Pressure Plate and Cover Assembly

Remove dirt and contamination from the pressure plate and cover assembly with nonpetroleum-based cleaning solvents. Inspect the cover for wear and damage. Make sure that the springs inside the cover are not broken. If a spring is broken, the clutch cover must be disassembled to replace the spring or the clutch must be replaced.

Inspect the pressure plate for wear or damage. Replace plates that are cracked (**Figure 14–55**). Minor heat discoloration is normal. Heat marks can be removed with an emery cloth. If heat discoloration cannot be removed, replace the center plate. Continue to check the center plate on each side of the plate according to the following procedure.

1. Place the pressure plate and cover assembly on a bench so that the plate faces toward you.
2. Use a thickness gauge and a straight edge to measure any scratches or scoring on the pressure plate. A caliper is used to measure thickness (**Figure 14–56**). If damage to the surface of the plate is more than the manufacturer's specifications (typically 0.015 inch/0.380 mm), replace the pressure plate.

FIGURE 14–55 This pressure plate has cracked due to extreme heat.

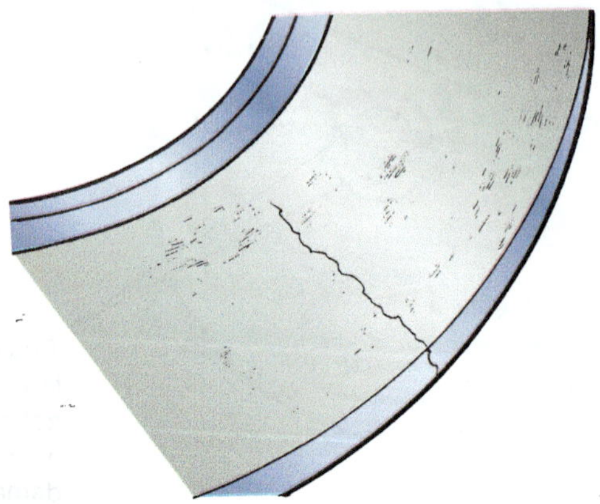

FIGURE 14–56 Measure the thickness of the pressure plate as well as the depth of any scratches.

3. Make sure that the surface of the pressure plate is not warped. Place a straightedge across the surface of the pressure plate. Attempt to insert a thickness gauge under any gap that appears between the straightedge and the pressure plate (**Figure 14–57**). Measure the pressure plate at four positions. If the gap exceeds the manufacturer's specifications, replace the pressure plate.
4. Measure the runout of the pressure plate to make sure the surface is parallel. Put the base of a dial indicator inside the center of the plate. Place the plunger of the dial indicator on the surface of the plate (**Figure 14–58**). Set the dial indicator at zero. Rotate the dial indicator one complete turn around the surface of the pressure plate. If the reading on the dial indicator is more than 0.002 inch (0.05 mm), replace the pressure plate.

Clutch Discs

Inspect the clutch friction discs for wear or damage (**Figure 14–59**). Make sure that the coaxial dampening springs are not loose in the hub (springs that rattle are not loose or broken). If the springs have rotary movement, they can be considered to be loose and in need of replacement. Make sure that

FIGURE 14–57 Use a straightedge and a thickness gauge to check the pressure plate for warpage.

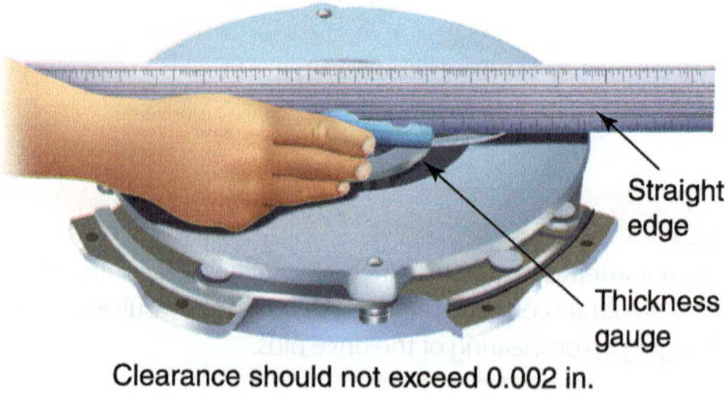

FIGURE 14–58 Check pressure plate runout.

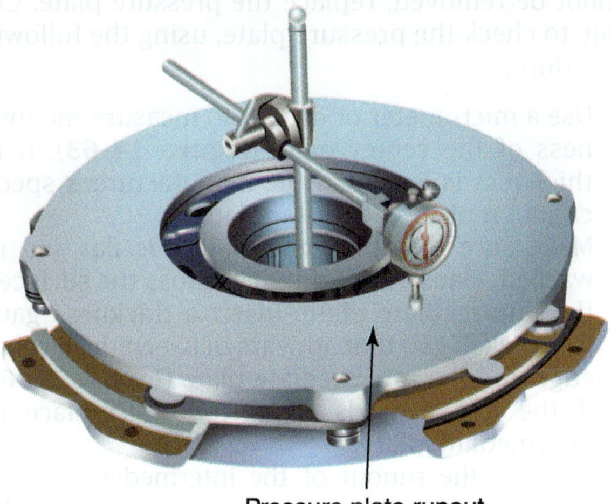

FIGURE 14–59 Check the friction discs for: (A) broken springs; and (B) cracked hub.

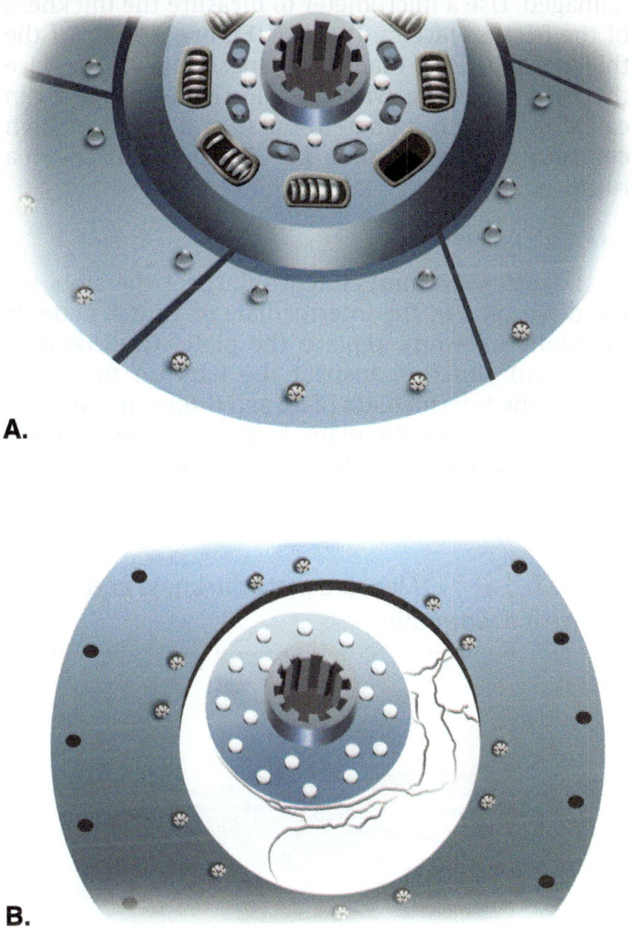

the splines in the hub are not damaged. Check that the hub has not separated from the disc. Replace discs that are worn or damaged. Use a cleaning solvent with a nonpetroleum base to remove grease and oil from the discs. If grease and oil cannot be removed, replace the disc.

Ceramic Lining

A ceramic facing is fastened to the disc with rivets. Replace the disc if the friction face is loose or damaged. Replace the disc if the friction face is worn flush with or below the rivets (**Figure 14–60**).

FIGURE 14–60 Check for excessive wear and heat damage.

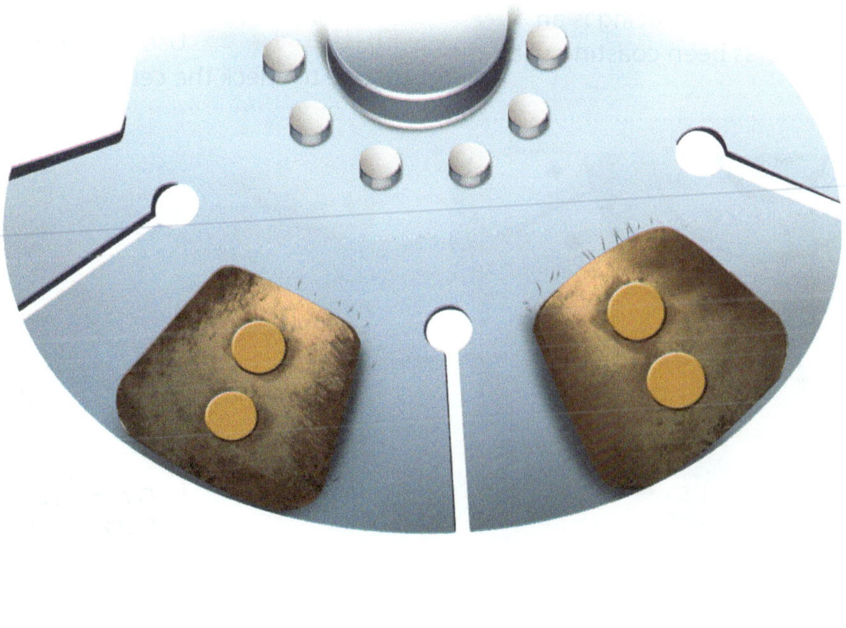

Molded Organic Lining

An organic friction face is integrally molded onto the disc. Replace the disc if the friction face is loose or damaged. Use a micrometer to measure the thickness of the friction face on the disc (**Figure 14–61**). If the thickness is less than OEM specifications, replace the disc. Coasting with the clutch disengaged can cause the clutch linings to burst as a result of high centrifugal forces. **Figure 14–62** shows a disc with a ruptured lining.

Intermediate Plate

On 14-inch (356 mm) clutches, inspect the slots for the drive pins in the intermediate or center plate. If the slots are worn, replace the plate. On 15½-inch (394 mm) clutches, inspect the tabs on the outer edge of the intermediate plate and if they are worn or damaged, replace the plate. Inspect the center plate for wear or damage. Make sure that the plate is not cracked. Heat discoloration is a normal condition. The heat discoloration can be cloth. If heat discoloration cannot be removed, replace the pressure plate. Continue to check the pressure plate, using the following procedure.

1. Use a micrometer or caliper to measure the thickness of the center plate (**Figure 14–63**). If the thickness is less than the manufacturer's specification, replace the center plate.
2. Make sure that the center plate is flat and not warped. Place a straightedge across the surface of the intermediate plate. Insert a thickness gauge under each gap that appears between the straightedge and the intermediate plate (**Figure 14–64**). If the gap exceeds specifications, replace the intermediate plate.
3. Measure the runout of the intermediate plate to make sure that the surface of the plate is parallel. Place the base of a dial indicator inside the center

FIGURE 14–61 Measure plate thickness and replace it if worn thin.

FIGURE 14–62 A burst friction facing is an indication that the truck has been coasting with the clutch disengaged.

FIGURE 14–63 Measure the thickness of the center plate and replace it if it is worn smaller than the manufacturer's specifications.

FIGURE 14–64 Use a straightedge and thickness gauge to check the center plate for warpage.

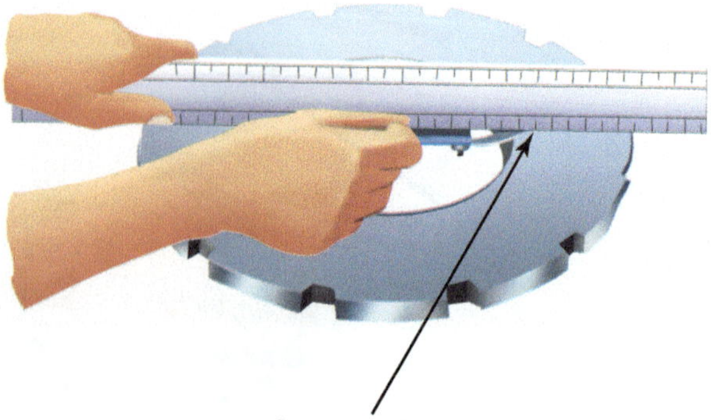

of the plate. Load the plunger of the dial indicator on the surface of the plate (**Figure 14–65**). Zero the dial indicator. Rotate the dial indicator one complete turn on the surface of the intermediate plate. If the reading on the dial indicator exceeds specifications, replace the intermediate plate.

4. If either runout or warpage of the plate exceeds specifications, machine the pressure plate. Do not grind more material from the plate than necessary. Do not machine the plate smaller than minimum specification. If the plate cannot be ground flat within the minimum thickness specification, replace it.

Pilot Bearing

Although the pilot bearings should be replaced when the clutch is removed, inspect the used pilot bearing for wear and damage. Determine and correct the cause of the wear and damage.

CLUTCH INSTALLATION

Installation procedures will vary according to clutch size and model. The following procedures for 14-inch (356 mm) diameter and 15½-inch (394 mm) clutches are general in nature. Follow service literature instructions.

Installing a 14-Inch (356 mm) Diameter Clutch

Begin the installation by installing new drive pins in the flywheel. The pins should be spaced out equally with their shanks press-fit into the flywheel. It is recommended that a drive pin aligning tool be used during this installation. Install new drive pins according to the following procedure:

1. Insert the drive pin in the flywheel housing. Use a machinist's square to make sure that the flat sides of the pin are at a 90-degree angle to the top of the housing (**Figure 14–66**).
2. Use a C-clamp to install the drive pin in the housing. Press the pin into the housing until the head of the pin touches the inner bore of the housing. *Make sure that the heads are square with the friction surface as misaligned drive pins may cause a clutch release problem.*
3. Put two small spacers on the friction surface face of the flywheel and carefully set the intermediate plate over the drive pins. The spacers should prevent getting your fingers pinched by the intermediate plate and give you a finger hold when removing it.
4. Turn the intermediate plate in one direction as far as it will go. Use a 0.006-inch (0.15 mm) thickness gauge to check the clearance between the drive pin and the drive slot (**Figure 14–67**). Check the same side of each pin. The minimum clearance between the drive pins and the drive slots is 0.006 inch (0.15 mm). If the specified clearance is not obtained, realign the drive pins and recheck the squareness.
5. Next, recheck the clearances. This check is necessary to ensure that the clutch will release properly when installed. *Do not file the drive pin slots on the intermediate plate to obtain correct clearances.* Doing so will cause an unequal load on the pins. This is a frequent cause of poor release or the clutch not releasing at all. It can also result in broken drive pins.

FIGURE 14–65 Use a dial indicator to check the center plate for runout.

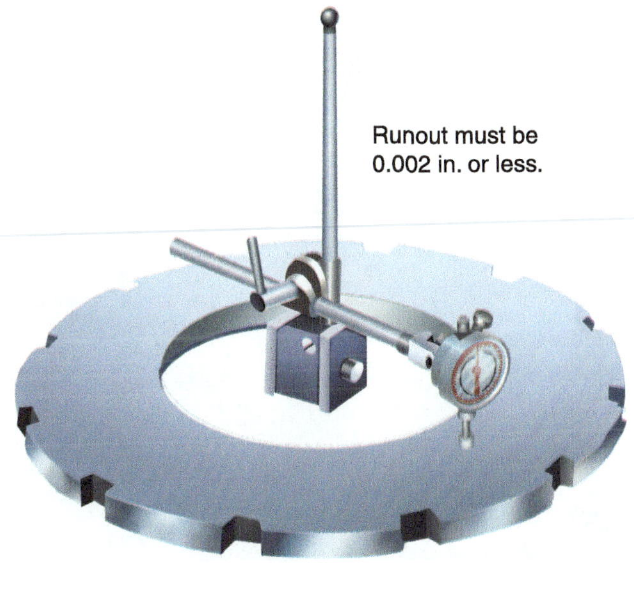

FIGURE 14–66 Use a machinist's square to ensure the drive pin is square with the surface face of the flywheel.

Courtesy of Eaton Corp.—Eaton Clutch Division.

FIGURE 14–67 Use a thickness gauge to check the pin-to-plate clearance.

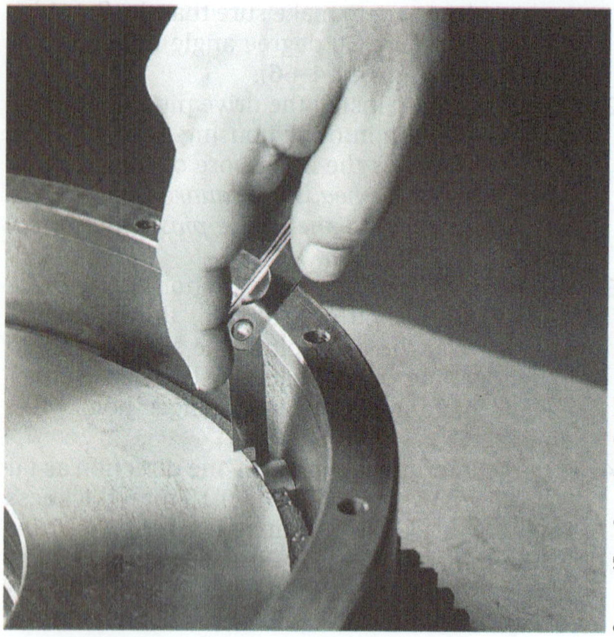

Courtesy of Eaton Corporation—Eaton Clutch Division.

6. If the alignment and clearance are correct, lock each of the six pins in place with two new, ⅜-inch by ⅜-inch setscrews (**Figure 14–68**).
7. Reinstall the flywheel to the engine crankshaft, making sure that the chalk marks are lined up. Refer to the engine specifications for torque specs on the flywheel mounting bolts.

FIGURE 14–68 Installing setscrews to hold pins in place

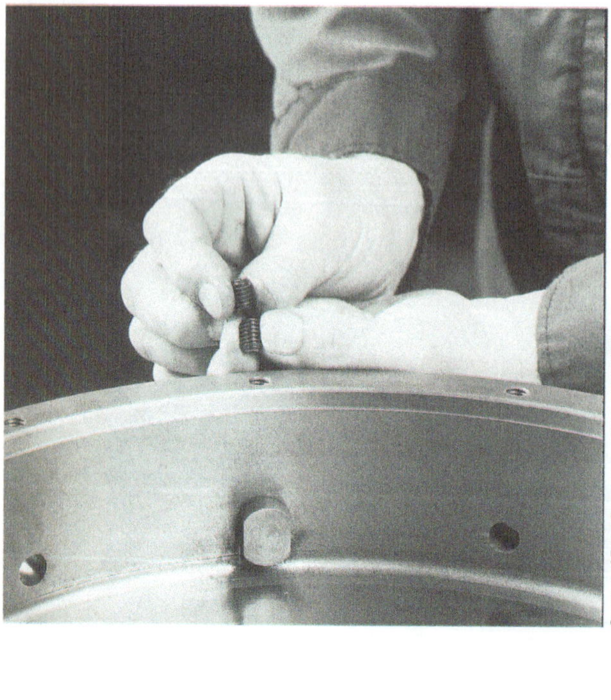

Courtesy of Eaton Corporation—Eaton Clutch Division.

FIGURE 14–69 Use a driver to install the pilot bearing.

8. Remove the pilot bearing and replace with a new bearing (**Figure 14–69**).
9. Place the front friction disc against the flywheel with the side stamped "flywheel" facing the engine (**Figure 14–70**). Remember: It is essential that the side stamped "flywheel" face the engine and the side stamped "pressure plate" face the transmission.
10. The relative position of the buttons on the front and rear driven discs is not important. If equipped, ensure that the Kwik-Adjust mechanism is aligned with the opening in the bell housing of the transmission.

FIGURE 14–70 Installing front friction disc

Courtesy of Eaton Corporation—Eaton Clutch Division.

FIGURE 14–71 Installing intermediate plate

Courtesy of Eaton Corporation—Eaton Clutch Division.

11. Install the intermediate plate by positioning the drive slots on the drive pins and remove the aligning tool (**Figure 14–71**).
12. If you are installing a clutch requiring antirattle springs make sure that they are properly installed. Space them equally between the drive pins with the rounded sections toward the flywheel face. For safety reasons, you should wear gloves when installing antirattle springs.
13. Insert the aligning tool through the hub of the rear friction disc with the side stamped "pressure plate" facing the transmission and install it after the intermediate plate (**Figure 14–72**). Note: It is essential that the side stamped "pressure plate" faces the transmission and the side stamped "flywheel" faces the engine.
14. Reinsert the aligning tool through the hub of the front friction disc and into the pilot bearing. The relative position of the buttons on the front and rear discs is not important.
15. Position the clutch cover over the guide studs installed at the top of the flywheel (**Figure 14–73**). Make sure that the adjustment mechanism is aligned with the opening in the bell housing of the transmission. Start six ⅜-inch × 1¼-inch grade 5 mounting bolts with lockwashers and tighten them finger tight.
16. *Lightly* tap the aligning tool to make sure it is centered and seated into the pilot bearing. Remove the guide studs and replace them with bolts and lockwashers.

FIGURE 14–72 Installing the rear friction disk and aligning tool

Courtesy of Eaton Corporation—Eaton Clutch Division.

FIGURE 14–73 Installing clutch cover over the aligning tool

Courtesy of Eaton Corporation—Eaton Clutch Division.

17. Tighten the bolts in a crisscross sequence to pull the clutch into alignment with the flywheel pilot (**Figure 14–74**). Start with the lower left-hand bolt. Failure to tighten the bolts in this manner can cause permanent damage to the clutch cover or create an out-of-balance condition.

FIGURE 14–74 Torque sequence for tightening bolts

Courtesy of Eaton Corp.—Eaton Clutch Division.

18. To achieve final torque, progressively tighten all bolts 35 to 40 lb-ft. As the bolts are tightened, the wooden spacers should fall out. If they do not fall free, be sure to remove them. You may have to lightly tap the aligning tool with a mallet to remove them.

CAUTION

Tap on the outer race of the pilot bearing only. Make sure it is seated properly in the bearing bore. This bearing must have a press fit within the pilot bearing bore. Never stake a loose pilot bearing, because it will eventually spin in the bore.

CAUTION

Some heavy-duty clutches have thicker intermediate plates and thinner Kevlar "super" buttons than standard clutches. Do not mix these components.

Installation of a Typical 15½-Inch Clutch

Begin the installation procedure by ensuring that the work area is clear and that all component mating surfaces are clean.

1. Insert two 7⁄16-inch (5 inches long) guide studs into the two upper mounting holes of the flywheel (**Figure 14–75A**).
2. Rotate the flywheel to level the guide studs (**Figure 14–75B**).
3. Insert the aligning tool through the release bearing sleeve into the new clutch (**Figure 14–75C**).
4. Guide the rear friction disc onto the aligning tool with the side stamped "pressure plate" facing the pressure plate (**Figure 14–75D**).
5. Place the intermediate plate in the clutch cover and align the driving lugs of the plate with the slots provided (**Figure 14–75E**). The separator pins in the intermediate plate must be flush with the lugs on the pressure plate side.
6. Install the front friction disc onto the aligning tool with the side stamped "flywheel" facing the engine (**Figure 14–75F**).
7. Remember: It is essential that the side stamped "flywheel" faces the engine and the side stamped "pressure plate" faces the transmission. The relative positions of the buttons on the front and rear driven discs are not important.
8. If equipped, make sure that the Kwik-Adjust mechanism is aligned with the opening in the bell housing of the transmission.
9. Position the clutch pack over the guide studs and slide it forward until contact is made with the flywheel surface (**Figure 14–76A**). Because a 15½-inch (394 mm) clutch assembly weighs over 150 pounds, a hoist may be required to lift it into place.
10. Start six 7⁄16-inch × 2¼-inch grade 5 mounting bolts with lock washers and tighten them finger tight (**Figure 14–76B**).
11. *Lightly* tap the aligning tool to make sure it is centered and seated into the pilot bearing (**Figure 14–76C**). Remove the guide studs and replace them with bolts and lock washers.
12. Tighten the bolts in the crisscross pattern sequence shown in Figure 14–74 to pull the clutch into position in the flywheel. You should start with the lower left-hand bolt. Failure to tighten the bolts in this manner can cause permanent damage to the clutch cover or create an out-of-balance condition. To achieve the final torque, progressively tighten all bolts to 45 to 50 lb-ft. (60 to 68 N·m). Note: As the bolts are tightened, the wooden spacers should fall out. If they do not fall free, remove them. This is important because the clutch will not function with the spacer blocks in place. After removing the spacer blocks, remove the aligning tool. You may have to *lightly* tap on the aligning tool with a mallet to remove it.
13. Using a ¼-inch (6 mm) diameter flat nose drift, lightly tap each of the four positive separator pins toward the flywheel (**Figure 14–76D**). After tapping, the pins should be flush against the flywheel.

FIGURE 14–75 Installation of 15½ -inch clutch: (A) Install two guide studs; (B) level flywheel guide slots; (C) insert aligning tool; (D) install rear driven disc; (E) install intermediate plate; and (F) install front disc.

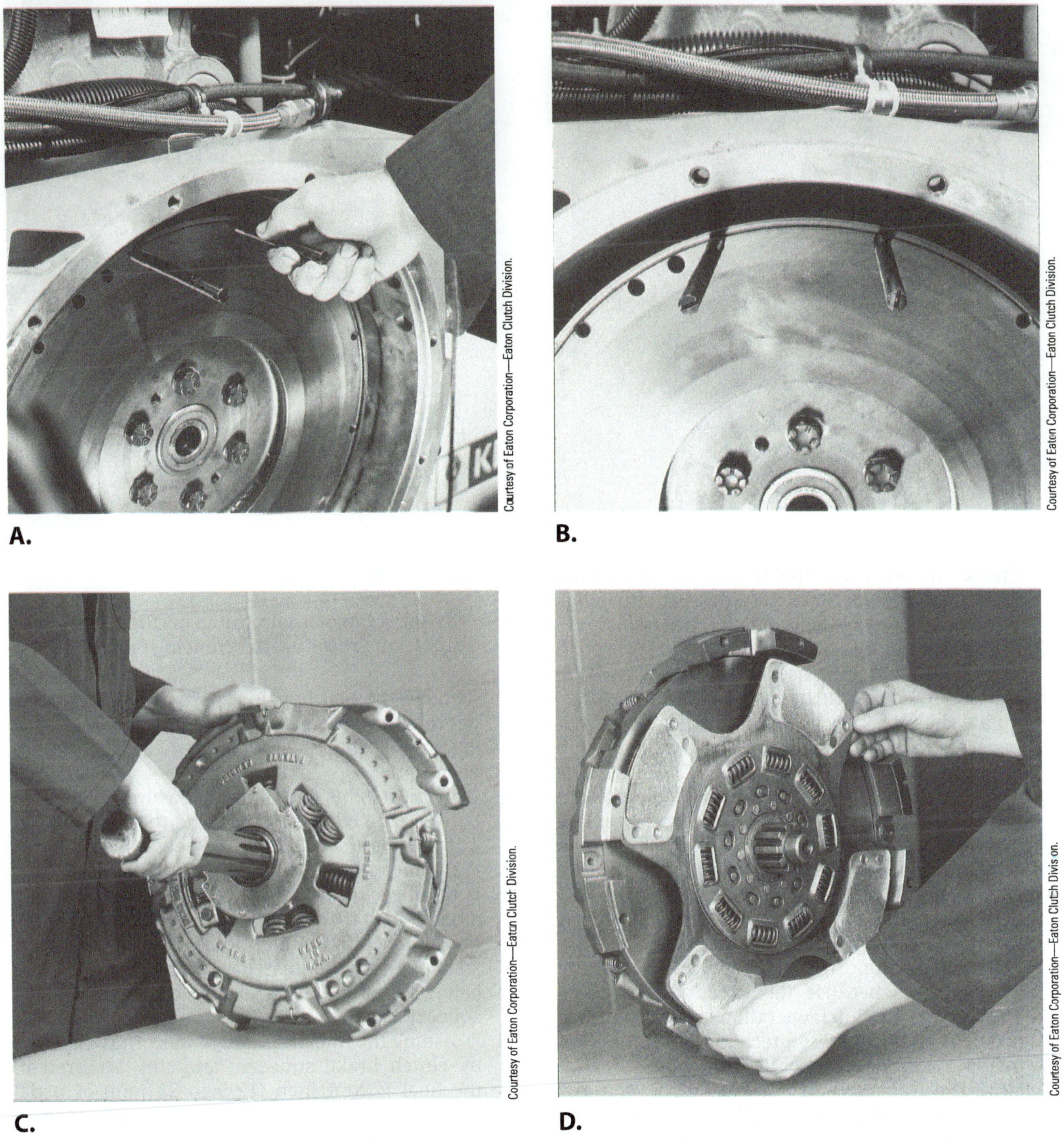

REINSTALLING THE TRANSMISSION

To reinstall the transmission:

1. Shift into gear.
2. Inspect the transmission input shaft for wear. If worn, replace.
3. Using a clean, dry cloth, wipe the shaft clean.
4. Check for wear on transmission bearing caps. See original equipment (OE) service literature if replacement is required.
5. If a clutch brake is used, be sure to install it on the input shaft of the transmission at this time (see Figure 14–15).

FIGURE 14–75 (continued)

E.

F.

6. Check for wear on the fingers of the clutch release yoke. Also, check the cross-shaft bushings (**Figure 14–77A**). Replace them if necessary.
7. Check that neither cross-shaft protrudes through the release fork because this could cause side loading of the release bearing (**Figure 14–77B**).
8. Rotate the release yoke so that it clears the wear pads of the release bearing as the transmission is moved forward (**Figure 14–77C**). *Be careful*: The release yoke fingers may not be elevated to the straight out position after clearing the release bearing because they could damage the clutch cover as the transmission is moved forward. Make sure that the transmission is properly aligned with the engine as it is moved into position.
9. Do not allow the rear of the transmission to drop and hang unsupported in the splined hubs of the friction discs because this could distort them and cause poor clutch operation or clutch release problems. Taking these precautions will prevent bending and distortion of the clutch discs.
10. Move the transmission forward but never force it into the flywheel housing. If it does not enter freely, investigate the cause and make the adjustments until it does.
11. Mate the transmission with the flywheel housing and install the mounting bolts.
12. Torque the bolts to specification.
13. Attach the clutch release linkage. Installation of the transmission is complete and adjustments can be made to the clutch and, if necessary, the linkage.

SOLO CLUTCH SETUP

To verify the setup of a Solo clutch perform the following verification procedure, referencing **Figure 14–78**.

1. Remove the bell housing inspection cover. Fully depress the clutch and then completely release to ensure that the release bearing is as far rearward as possible.
2. Using a set of inside calipers, measure the distance between the release bearing and the clutch brake. The measurement should be between 0.490 inch (12.4 mm) and 0.560 inch (14.2 mm).
3. If the measurement is within these values, hold a 0.010-inch (0.25 mm) thickness gauge between the clutch brake and release bearing and have someone fully depress the clutch. The thickness gauge should be firmly clamped by clutch brake squeeze.
4. Providing the thickness gauge was firmly clamped by clutch brake squeeze, and this occurred in the final ½ to 1 inch (12.5 to 25 mm) of pedal travel, use a ¼-inch (6 mm) flat nose punch to lightly tap the separator pins toward the flywheel (**Figure 14–78A**). You will have to rotate the clutch through a revolution if you are performing this through the clutch inspection cover. The objective is to be sure that each of the four pins is flush against the flywheel.
5. Now measure the distance between the release yoke and release bearing wear pad as shown in **Figure 14–78B**. If the free travel measures

FIGURE 14–76 Installation of 15½-inch clutch (continued): (A) Install clutch cover; (B) install mounting bolts; (C) remove aligning tool; and (D) bottom out positive separator pins after torquing mounting bolts.

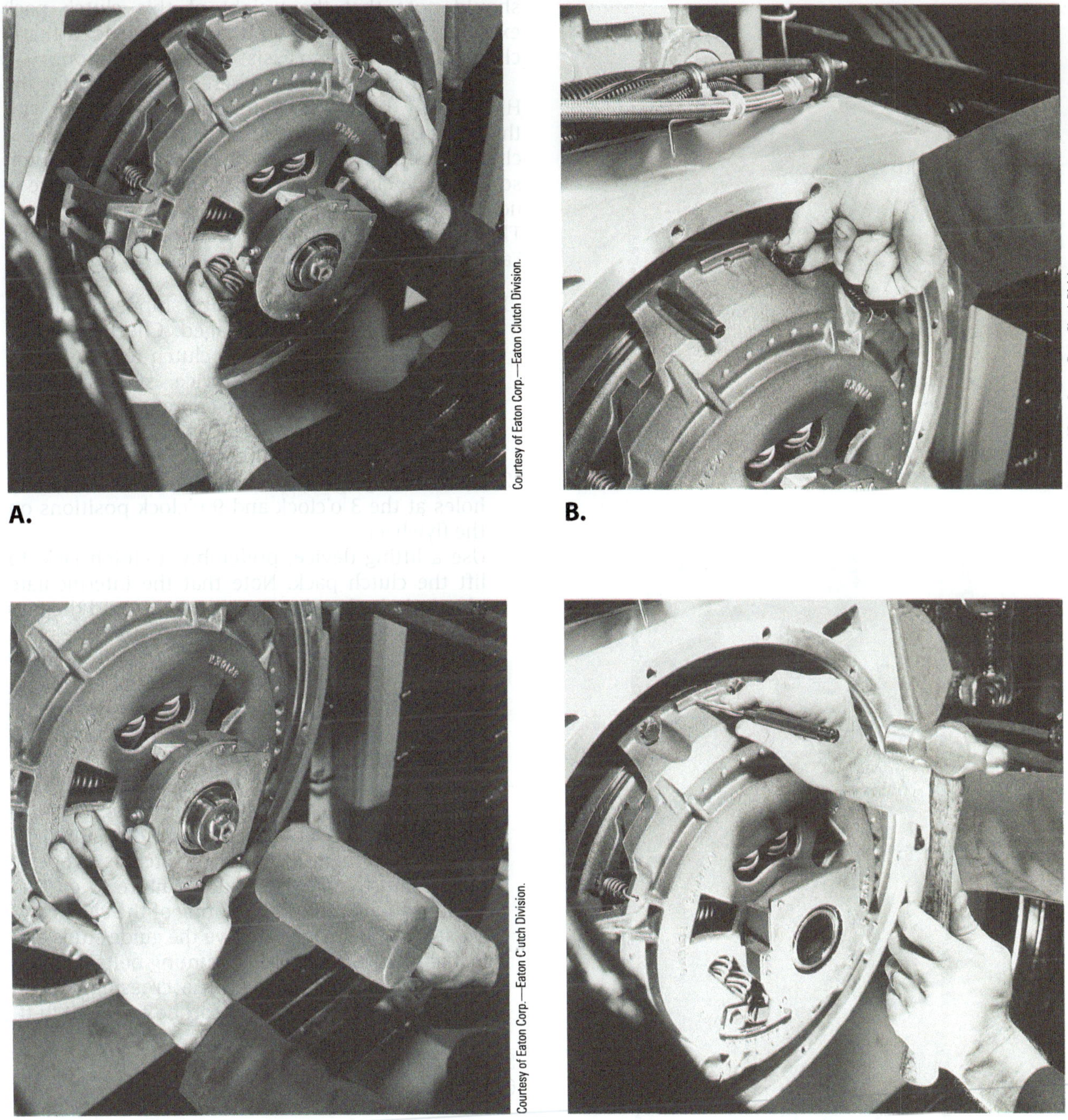

A. Courtesy of Eaton Corp.—Eaton Clutch Division.

B. Courtesy of Eaton Corp.—Eaton Clutch Division.

C. Courtesy of Eaton Corp.—Eaton Clutch Division.

D. Courtesy of Eaton Corp.—Eaton Clutch Division.

between 1/16 and 1/8 inch (1.5 and 3 mm), you can assume that the clutch is properly set up.

6. Make sure that the transmission is in neutral and start the engine. Attempt to shift into first gear. If the shift occurs without grinding, the clutch is properly adjusted.

Solo Clutch Problems

If the measurements detailed in steps 2, 3, 4, and 5 of the preceding verification procedure are out of specification, you will have to troubleshoot the cause. Use Eaton service literature to troubleshoot Solo clutches. The Eaton Heavy Duty Troubleshooting guide is a free download from http://www.roadranger.com.

FIGURE 14–77 Install the transmission, being careful to rotate the release yoke so it clears the wear pads of the release bearing.

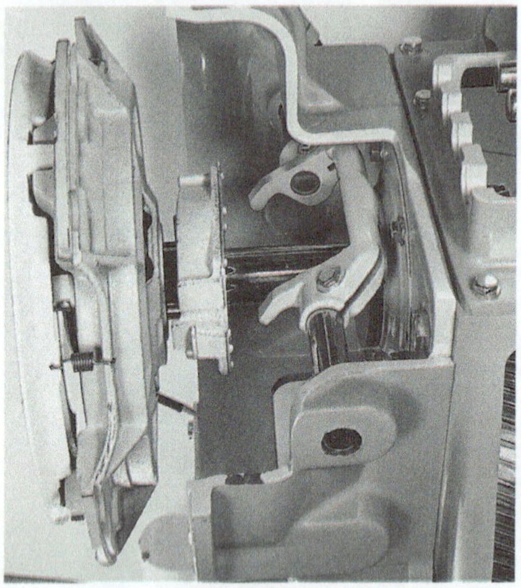

Courtesy of Eaton Corporation—Eaton Clutch Division.

A.

Courtesy of Eaton Corporation—Eaton Clutch Division.

B.

Courtesy of Eaton Corporation—Eaton Clutch Division.

C.

Eaton DM Clutch Installation

Figure 14–79 shows the Eaton procedure required to fit its centrifugal DM clutch to a flywheel. Technicians should note that the weight of this clutch pack exceeds 180 lb. (82 kg) so it is recommended that a clutch dolly be used to assist in the installation.

The procedure outlined here is that used for the HD version of the DM clutch used exclusively with the Eaton UltraShift transmission. The centrifugal DM clutch pack operates in a normally released condition so when the clutch is bolted to the flywheel there is no pressure plate pressure to hold the discs in place. The following procedure is that recommended by Eaton for installation.

1. The clutch pack is normally shipped with the clutch locking device engaged. Check that it is engaged before bolting the clutch to the engine. To engage the locking device, simultaneously push in on the locking device return spring and turn the lock approximately one-third of a turn in the counterclockwise direction.
2. Install two guide studs into the clutch mounting holes at the 3 o'clock and 9 o'clock positions on the flywheel.
3. Use a lifting device, preferably a clutch jack, to lift the clutch pack. Note that the intermediate plate is bolted to the cover assembly and the rear driven disc is in place between the pressure plate and intermediate plate. DO NOT unbolt the intermediate plate from the cover assembly. With a pilot shaft through the cover assembly and the rear disc, install the front driven disc over the splines on the pilot shaft.
4. Slide the clutch pack over the guide studs and start six of the clutch mounting bolts. Begin at the lower left when tightening the clutch mounting bolts to ensure that the clutch is properly pulled into the flywheel pilot. Failure to do this could result in improper piloting of the clutch and cause clutch damage. Remove the guide studs and install the two remaining mounting bolts. Tighten the clutch mounting bolts in a crossing pattern as on any other clutch and torque to 40–50 lb-ft. (54–68 N·m).
5. Remove the pilot shaft.
6. Install the transmission.

Drive Coupler for UltraShift AW3

The AW wet clutch should be serviced with the UltraShift transmission and not covered in this section. However, a separate bolt-on drive coupler is used with the UltraShift AW3 transmission and the procedure for its mounting is outlined here. The drive coupler mounts to the engine flywheel, which is the same as that used with a 14-inch (356 mm) Solo clutch. The following outlines Eaton procedure for installing the drive coupler to the flywheel.

FIGURE 14–78 Solo clutch verification procedure

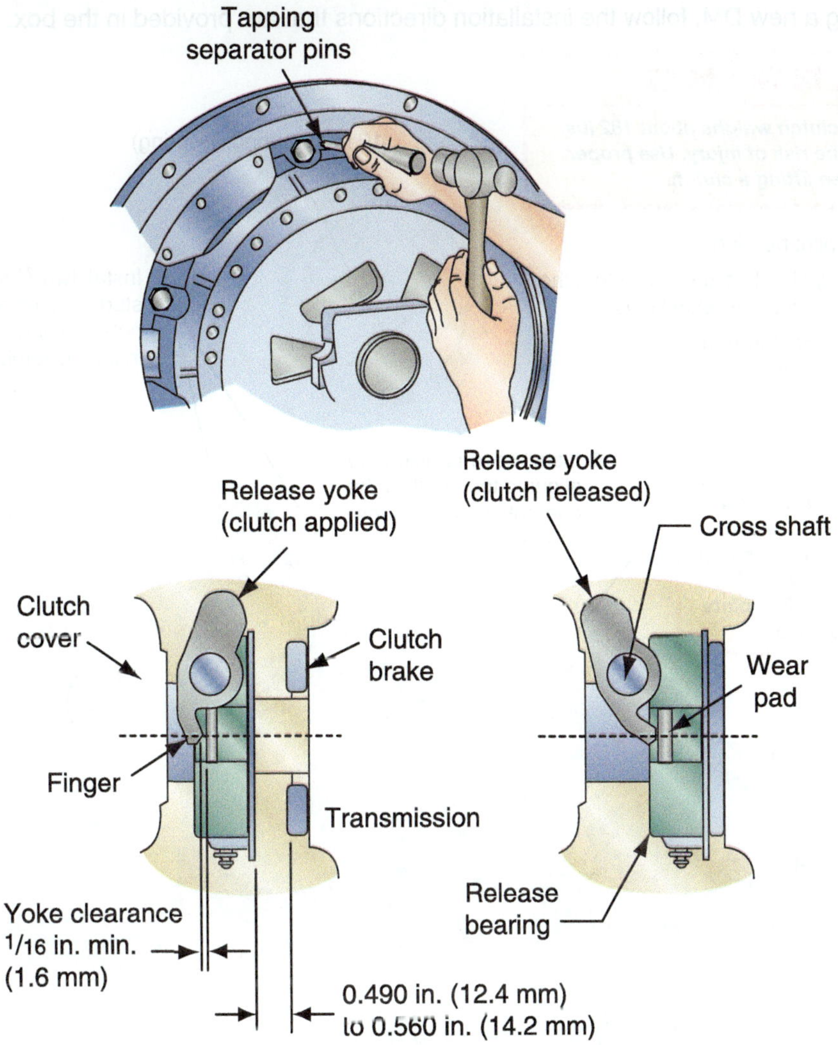

1. The drive coupler should be handled with care to avoid damage to the mating surfaces.
2. Check the flywheel concentricity: It must be within the OEM specifications.
3. Install the flywheel onto the engine crankshaft, using the engine OEM-recommended procedure.
4. Install the input shaft pilot bearing.
5. The Eaton drive coupler alignment tool must be used when installing the drive coupler.
6. Install the drive coupler using the specified mounting hardware and torque to specification currently 35–40 lb-ft. (47–55 N·m).

AW Clutch fluid

Eaton recommends two possible fluids for the wet clutches used on UltraShift AW AMTs:

- Synthetic Dexron III or IV
- Allison TranSynd

The clutch oil level should be checked at the dipstick with the engine running at idle rpm and the transmission in neutral. See Figure 20–2 to identify the dipstick/fill location.

CONCLUSION

Clutch friction discs are regarded in the industry as consumable. That means clutches tend to be routinely replaced and because the task requires a relatively low skill level, you can bet that it is one regularly assigned to entry-level techs. Often, apprentice techs are evaluated on their competence to perform clutch and brake jobs before being given more challenging tasks. **Figure 14–80** shows a clutch assembly mounted on an out-of-chassis transmission with the friction discs and intermediate plate separated so each can be seen. **Figure 14–81** shows a clutch pack with the friction discs and intermediate plate closed up.

FIGURE 14–79 Eaton DM clutch installation

Note: If installing a new DM, follow the installation directions that are provided in the box.

WARNING

An assembled clutch weighs about 182 lbs. (82 kg). Avoid the risk of injury. Use proper equipment when lifting a clutch.

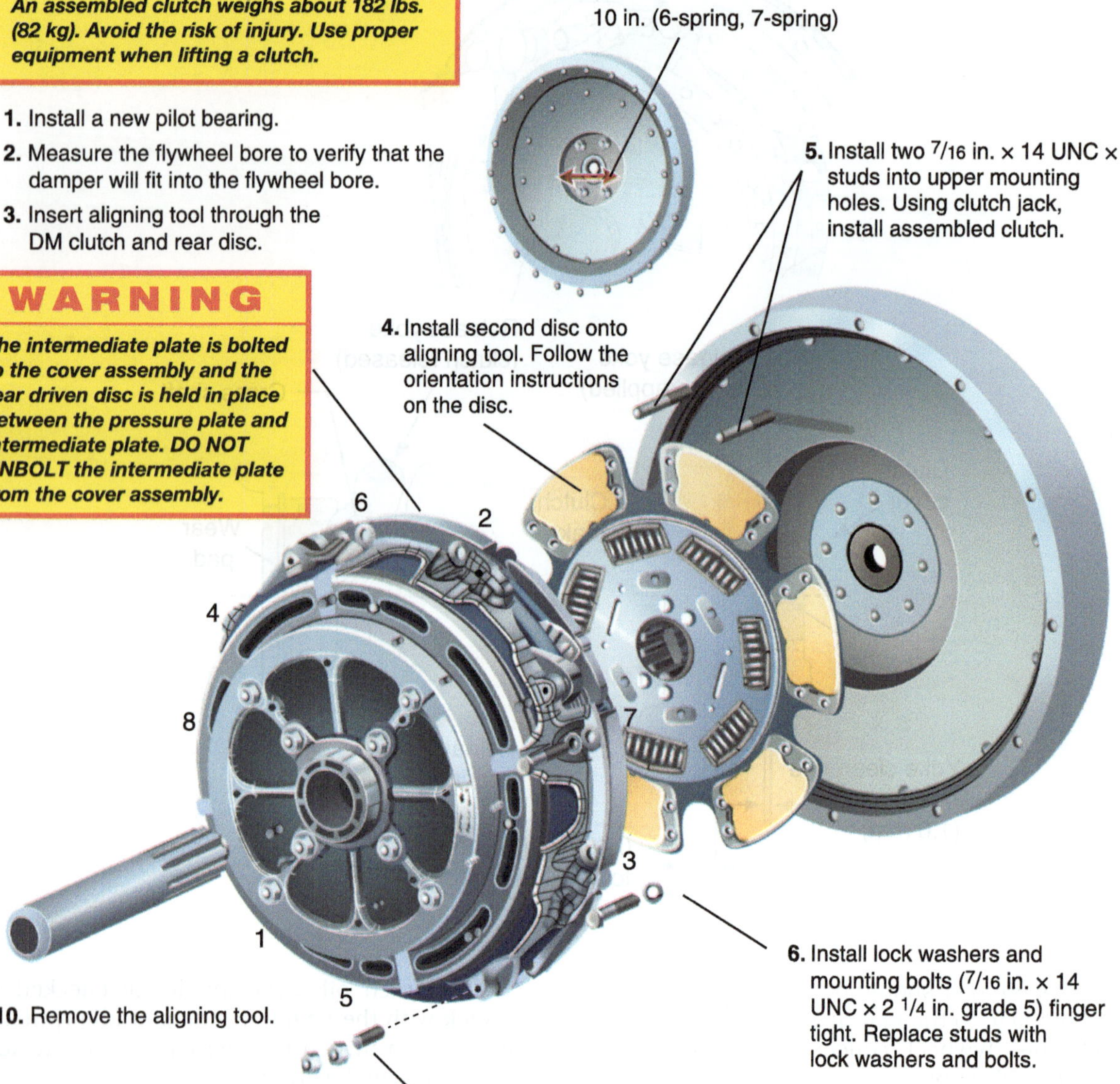

1. Install a new pilot bearing.
2. Measure the flywheel bore to verify that the damper will fit into the flywheel bore.
3. Insert aligning tool through the DM clutch and rear disc.

WARNING

The intermediate plate is bolted to the cover assembly and the rear driven disc is held in place between the pressure plate and intermediate plate. DO NOT UNBOLT the intermediate plate from the cover assembly.

4. Install second disc onto aligning tool. Follow the orientation instructions on the disc.
5. Install two 7/16 in. × 14 UNC × 5 in. studs into upper mounting holes. Using clutch jack, install assembled clutch.
6. Install lock washers and mounting bolts (7/16 in. × 14 UNC × 2 1/4 in. grade 5) finger tight. Replace studs with lock washers and bolts.
7. Tighten mounting bolts in a crisscross pattern starting with a lower left bolt (1, 2, 3, 4, 5, 6, 7, 8). Torque to 40–50 lb-ft. (54–68 N·m).
8. Using a section of 5/16 in. × 18 UNC × 3 in. threaded rod (jack screw), install two nuts on one end and lock them together. This will allow you to turn the jack screw in and out of the cover assembly.
9. Install the jack screw into one of the four holes located adjacent to the clutch mounting bolts. This forces the pressure plate forward, clamping the discs and holding them in place. The hole chosen should be at the 6 o'clock position to allow for removal after the transmission is installed.

CAUTION

Do not overtighten the jack screw. Over-torquing can cause permanent clutch damage.

10. Remove the aligning tool.

FIGURE 14–80 Clutch pack mounted on a transmission clutch shaft with the friction discs and intermediate plate spaced apart.

FIGURE 14–81 Clutch pack mounted on a transmission clutch shaft with the friction discs and intermediate plate closed up.

SUMMARY

- The function of a clutch is to transfer torque from the truck engine flywheel to the transmission.
- The components of a clutch assembly can be grouped into two basic categories: driving members and driven members. Driving members include the flywheel, clutch cover, pressure plate, pressure springs and levers, intermediate plate, adapter ring, and adjustment mechanisms. Driven members include friction discs and the transmission input shaft.
- A mechanical clutch can be released or disengaged by one of three methods: push-type clutch, pull-type clutch, or centrifugal clutch.
- Clutch brakes are found in some pull type clutches and can be grouped into four types: conventional, limited torque, torque limiting, and two-piece clutch brakes.
- The clutch linkage, which connects the clutch pedal to the release fork or yoke, can be one of three types: mechanical, hydraulic, and air control.
- The major cause of clutch failure is excess heat. Most heat-related damage is related to driver abuse. Heat damage results from starting in the incorrect gear, shifting or skip-shifting, riding the clutch pedal, and improper coasting.
- Free pedal, or pedal lash, is the amount of free play in the clutch pedal in the cab. This measurement is directly related to free travel, which is the clearance distance between the release yoke fingers and the clutch release bearing pads.
- Clutches should be checked periodically for proper adjustment and lubrication.
- The manual adjustment methods used on most angle coil spring clutches are lockstrap and Kwik-Adjust.
- Adjustment-free clutches such as the Solo clutch should automatically adjust for the life of the clutch pack. The only maintenance required is periodic lubrication of the release bearing mechanism.
- Pull-type clutches with perpendicular springs use a threaded sleeve and retainer assembly that can be adjusted to compensate for disc lining wear.
- Asbestos and nonasbestos fibers used as clutch friction facings may pose a health risk. Technicians should take the appropriate safety precautions when servicing clutches.
- The following should be checked in a clutch inspection: transmission bell housing, flywheel housing, flywheel drive pin, release fork and shaft, input shaft, pressure plate and cover assembly, clutch discs, friction facings, center plate, and pilot bearing.

REVIEW QUESTIONS

1. Which of the following clutch components is a driven member?
   a. clutch cover
   b. clutch friction disc
   c. intermediate plate
   d. pressure plate
2. Which of the following is not a part of the clutch cover assembly?
   a. pressure spring
   b. release lever
   c. pressure plate
   d. adapter ring
3. Which of the following clutch components is splined to the input shaft of the transmission?
   a. clutch friction disc
   b. pressure plate
   c. release bearing
   d. flywheel
4. What is the function of a *current* clutch brake used in a heavy-duty drivetrain?
   a. minimize gear clash during a neutral to first gear shift
   b. minimize gear clash during downshifts to all ratios
   c. minimize gear clash during upshifts to all ratios
   d. none of the above
5. Which of the following driving procedures will result in clutch or driveline damage?
   a. driving with a foot on the clutch pedal
   b. using the clutch as a brake to hold the truck on a hill and incline
   c. coasting downhill with the transmission in gear and the clutch disengaged
   d. all of the above
6. Which of the following could be a cause of poor clutch release?
   a. damaged drive pins
   b. worn clutch linkage
   c. worn release bearing
   d. all of the above
7. How often should a clutch be inspected and lubricated?
   a. every month
   b. every 6,000 to 10,000 miles (9700 to 16,000 km)
   c. any time the chassis is lubricated
   d. all of the above
8. When adjusting clutch linkage, Technician A says that pedal free travel should be about 1½ to 2 inches. Technician B says that free travel should be more than ½ inch measured at the release forks. Who is correct?
   a. Technician A only
   b. Technician B only
   c. both A and B
   d. neither A nor B
9. When servicing clutches, Technician A always wears an OSHA-approved particulate mask. Technician B uses compressed air to clean off clutch components before removing a clutch pack. Who is correct?
   a. Technician A only
   b. Technician B only
   c. both A and B
   d. neither A nor B
10. In the procedure required to perform a clutch replacement on a truck, which of the following steps should be performed first?
    a. Unbolt the clutch from the flywheel.
    b. Remove the clutch brake.
    c. Install a clutch alignment tool through the release bearing.
    d. Unbolt the transmission bell housing from the flywheel.
11. On which of the following types of flywheels would the center intermediate plate be driven by drive pins?
    a. flat flywheel
    b. pot flywheel
    c. torque converter
    d. all of the above
12. Failure analysis determines that the facing of a clutch disc has burst. Technician A says that the cause is the driver using the clutch as a brake. Technician B says that the problem might be the driver coasting downhill with the transmission in gear and the clutch disengaged. Who is correct?
    a. Technician A only
    b. Technician B only
    c. both A and B
    d. neither A nor B
13. When installing a 14-inch clutch, Technician A installs new drive pins in the flywheel. Technician B files the drive pin slots to ensure that proper clearances are met. Who is correct?
    a. Technician A only
    b. Technician B only
    c. both A and B
    d. neither A nor B
14. When regarding the operation of a typical 15½-inch clutch, which of the following components constantly rotates at engine rpm?
    a. rear clutch disc
    b. front clutch disc
    c. intermediate plate
    d. clutch brake

15. When inspecting the mating faces of a clutch bell housing, which of the following can be regarded as the critical wear area?
    a. at the 12 o'clock position
    b. at the 3 o'clock position
    c. between 2 o'clock and 10 o'clock counterclockwise
    d. between 3 o'clock and 8 o'clock clockwise

16. Which of the following adjustments is made on nonsynchronized transmissions only?
    a. pedal height
    b. total pedal travel
    c. clutch brake squeeze
    d. free travel

17. A driver complains of a hard clutch pedal. Technician A says that the problem may be damaged bosses on the release yoke assembly. Technician B says that the clutch linkage may be worn or damaged. Who is correct?
    a. Technician A only
    b. Technician B only
    c. both A and B
    d. neither A nor B

18. Technician A says that when the driver releases (disengages) the clutch while coasting downhill, the clutch may overheat. Technician B says that overheating may result when the driver rides the clutch pedal. Who is correct?
    a. Technician A only
    b. Technician B only
    c. both A and B
    d. neither A nor B

19. There are grooves worn in a clutch pressure plate. Technician A says that the clutch discs could be worn or damaged. Technician B says that the clutch pressure springs could be worn or damaged. Who is correct?
    a. Technician A only
    b. Technician B only
    c. both A and B
    d. neither A nor B

20. Which of the following should be used to remove minor heat discoloration from a clutch pressure plate?
    a. cleaning solvent with a nonpetroleum base
    b. ammonia
    c. emery cloth
    d. none of the above

15

Prerequisite: Chapter 14

STANDARD TRANSMISSIONS

OBJECTIVES

After reading this chapter, you should be able to:

- Identify the types of gears used in truck transmissions.
- Interpret the language used to describe geartrains and calculate gear pitch and gear ratios.
- Explain the relationship between speed and torque from input to output in different gear arrangements.
- Identify the major components in a typical transmission—including input and output shafts, main shaft and countershaft gears, and shift mechanisms.
- Describe the shift mechanisms used in heavy-duty manual truck transmissions.
- Outline the role of main and auxiliary (compound) gear sections in a typical transmission, and trace the powerflow from input to output in different ratios.
- Describe the operating principles of range shift and splitter shift air systems.
- Define the roles of transfer cases and PTOs in heavy-duty truck operation.

KEY TERMS

air filter/regulator assembly
automated manual transmission (AMT)
axis of rotation
backlash
bottoming
cab-over-engine (COE)
climbing
constant mesh
direct drive
drive gear
driven gear
drivetrain
gear pitch
main shaft
overdrive
powerflow
rotation
shift bar housing
shift fork or yoke
slave valve
transfer case

INTRODUCTION

When a truck is operated on a highway, engine torque is transmitted through the clutch to the input shaft of the transmission. Gears in the transmission are used to alter this input torque and speed to suit the driving requirements of the vehicle. The transmission output shaft turns the vehicle **drivetrain** downstream from the transmission, typically consisting of a propeller shaft, drive axle carrier, drive axle shafts, and drive wheel assemblies. The different gear ratios used in a truck transmission give the truck the ability to handle driving conditions that range from a full load stationary start to highway cruise speeds. A transmission also has to provide a reverse capability. Engines always rotate in one direction only, so the transmission must be capable of reversing input rotation at the output. If this were not possible, a vehicle would only be capable of moving in one direction.

ROLE OF TRANSMISSION

The role of the transmission is to manage the drivetrain. When the truck is being accelerated and driven down a highway, the role of the engine is to apply torque to the drivetrain components. However, this powerflow may be reversed during deceleration when the drive-axle wheels are driving the engine, providing what we call engine braking, because the engine role switches to become power absorbing rather than power producing. During deceleration, the role of the transmission components is also reversed: Output becomes input, and the transmission gear ratio will determine how aggressively the engine "brakes" the drivetrain.

STANDARD AND AUTOMATED TRANSMISSIONS

We use the terms *standard, manual,* or *mechanical* to describe transmissions that require manual shifting between gear ratios. Unlike automobiles, a vast majority of trucks on our highways still use what we refer to as *standard* or *manual transmissions.* However, many of these so-called standard transmissions are shifted electronically, that is, shifted by computer controls. These electronically managed transmissions are usually known as **automated manual transmissions (AMTs)**. The standard transmission platform introduced in this chapter is the basis of the AMTs studied in Chapter 20. The internal components and operating principles of the standard transmission change little when adapted for electronic controls.

MECHANICAL VERSUS AMTS

In the year 2013, the U.S. and Canadian market share of AMTs in Class 8 commercial trucks was 27 percent (truckinginfo.com). The projection is for this percentage to grow to 50 percent by the year 2017. The bottom line is that, at the moment of this writing, a huge percentage of trucks on our roads use 100 percent mechanical transmissions in which the only electronic component is the tailshaft road speed sensor. For that reason, it will continue to be important for technicians to understand manually shifted, standard transmissions operating behind standard mechanical clutch packs for a few more years.

RATIOS AND SHIFTING

Heavy-duty truck transmissions can use between five and twenty ratios. The actual number depends on application. Because of the size, expense, and complexity of automatic transmissions—all of which increase with the number of output gear ratios used—fully automatic transmissions have been a hard sell to truck linehaul fleets. However, AMTs are increasing their market share because they can reduce or eliminate driver manual shifts and, coincidentally, the driver skill level (and training) required to operate a truck. There is another factor. Stop for a coffee at a truck stop, get into conversation with some truck drivers and you will find a majority of them say they prefer manual shift transmissions.

STANDARD AND AMT COMPONENTS

AMTs often use identical components found in the fully mechanical standard transmission manufactured by the same original equipment manufacturer (OEM). What changes is the shifting hardware. It is, therefore, essential that you have a good understanding of this chapter and Chapter 16 before proceeding to study Chapter 20 on automated transmissions.

15.1 MAIN AND AUXILIARY GEARING

Truck transmissions may consist of a single housing (main box) or multiple housings. A single-housing transmission may use single or multiple countershafts. As the requirement for gear ratios increases, the single housing (main box or front section) can be "compounded" by adding an auxiliary section or sections to it. A compound with two gear ratios can convert a five-speed (ratio) main box into a transmission with ten output speeds or more.

COMPOUNDING RATIOS

A majority of light- and medium-duty trucks use a transmission that consists of a single main box, whereas most heavy-duty truck transmissions consist of a main box coupled to at least one compound or auxiliary housing. **Figure 15–1** shows a sectional view of a twin-countershaft transmission with a main and an auxiliary box. This transmission has thirteen output ratios. The auxiliary or compound housing is a separate unit coupled up to the main housing. Most compounded transmissions use this arrangement because it means that a single main box design can be used as a platform for a wide range of ratio options, which is achieved by coupling it to different auxiliary sections. Some transmissions are compounded within the main section, but this design is not common today.

The main box in a typical truck transmission contains gearing for five forward speeds and one reverse speed. An auxiliary housing typically contains gearing for two or three speeds. Two-speed auxiliary gearing is used to produce high- and low-speed ranges. Three-speed auxiliary gearing may add **overdrive** or deep-reduction gearing. Some transmissions use double compounding. A double-compounded transmission consists of a main box plus two auxiliary sections. A compound or auxiliary box always multiplies the gear ratios produced in the main box. Depending on the application and the OEM, compounding is used to produce seven, eight, nine, ten, thirteen, fifteen, eighteen, and twenty speed standard transmissions.

FIGURE 15–1 A thirteen-speed twin-countershaft transmission with main and auxiliary gearing contained in separate housings, bolted together.

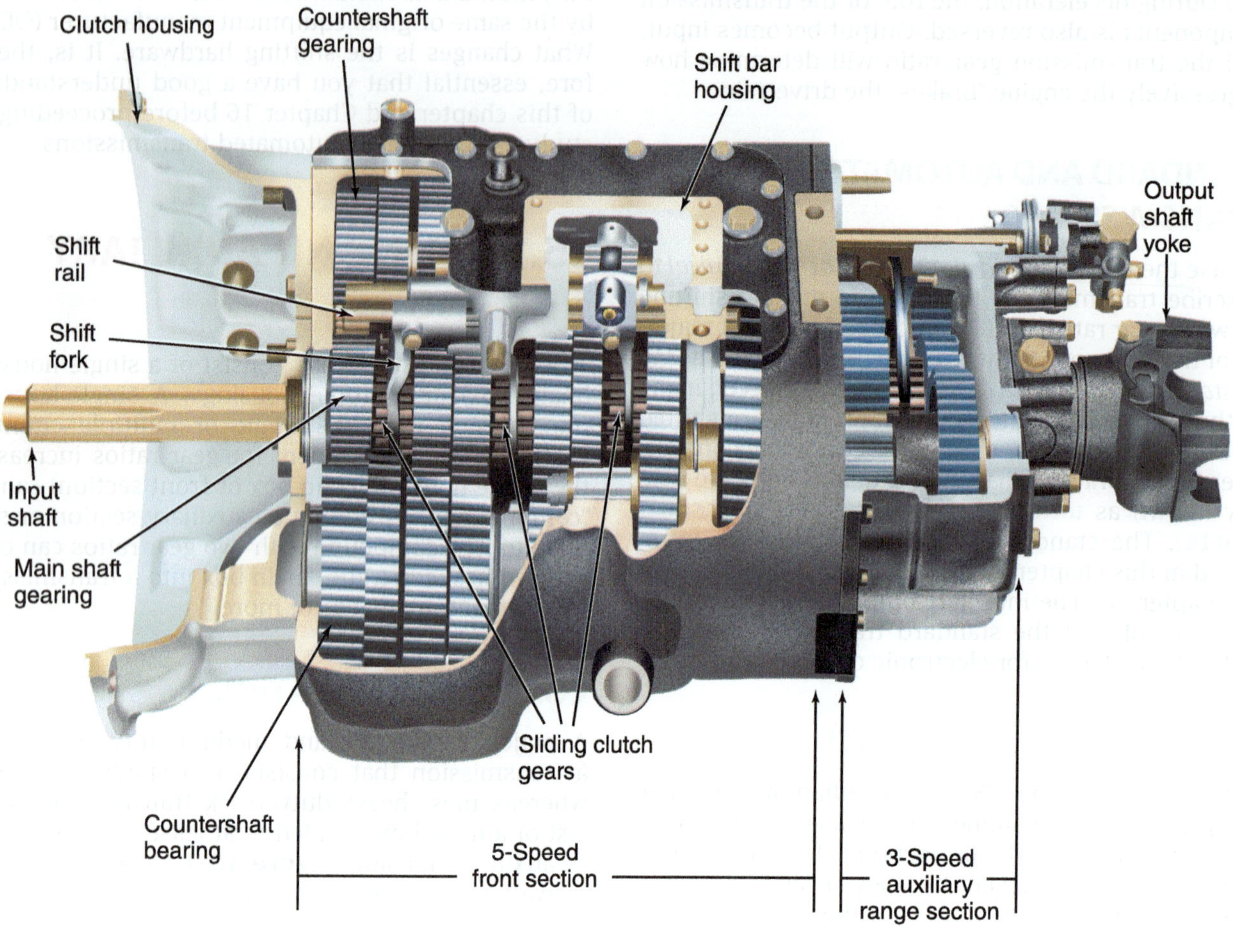

GEARS

Gears are used in a transmission to transmit rotational motion or torque. They are usually mounted on shafts. Gears positioned on parallel shafts with their teeth in mesh transmit rotating motion from one shaft to another.

A gear can be used in any of the following roles:

1. The shaft can drive the gear.
2. The gear can drive the shaft.
3. The gear can be left free to turn on the shaft (it idles).

Sets of gears can be arranged to do the following:

1. Multiply torque and decrease speed
2. Increase speed and decrease torque
3. Transfer torque and speed unchanged

Any study of transmissions requires some knowledge of gear terminology. Here are some of the common gear terms you should be familiar with:

- **Drive gear (input).** A **drive gear** or driving gear is one that drives another gear or causes another gear to turn.
- **Driven gear (output).** A **driven gear** is one that is driven by a drive gear or a shaft.
- **Rotation and direction of rotation.** **Rotation** is the term we use to describe anything that turns about on an axis. Direction of rotation is described by comparison to the rotation of the hands of a clock. As time progresses, the hands of a clock rotate in a right-hand direction when we face the front of the clock, a direction we know as clockwise (CW). Rotation in the opposite direction is known as counterclockwise (CCW) rotation or left-hand rotation. To determine the direction of rotation of gears in a transmission, you should face the front of the transmission—that is, the end that couples to the engine.
- **Speed of rotation.** The rotating frequency of gear shafts is measured in complete rotations per minute stated as *revolutions per minute,*

or *rpm*. Speed of rotation in a transmission is often described in terms of relative or comparative speed. These terms include faster, slower, increased speed, decreased speed, **direct drive** (same speed or one-to-one ratio), underdrive, and overdrive (output faster than input).

- **Axis of rotation**. The **axis of rotation** is the centerline around which a gear or shaft revolves.
- **Drive and coast**. The *drive side* of gear teeth is in contact with teeth of another gear while torque is being applied. Usually the drive side of gear teeth is subject to the most wear. The *coast side* of gear teeth is opposite to the drive side. The roles of the drive and coast sides of gear teeth are reversed when the driveline torque direction is reversed. An example would be during deceleration of the vehicle.
- **Constant mesh**. Any gears that are positioned to remain in permanent mesh with each other—that is, they are not engaged or disengaged from gears by shifting action—are said to be **constant mesh**.
- **Backlash**. The term **backlash** describes rotational free play between mating gears. All intermeshing gears should have some backlash to permit heat expansion of metal and proper lubrication. Out-of-specification backlash results in noise and rapid wear. Backlash increases as mating gears and shaft bearings wear.
- **Bottoming**. A **bottoming** condition occurs when the teeth of one gear contact the root of the mating gear. Bottoming seldom occurs in a two-gear drive combination but is more common in multiple-gear drive combinations. The mating action of a two-gear drive combination tends to thrust the gears apart, so bottoming is unlikely in this arrangement.
- **Climbing**. Excessive wear in gears, bearings, and shafts can create conditions in which mating gears are forced apart enough to cause a condition known as **climbing**. When the apex (high point) of the teeth on one gear climbs over the apex of the teeth on its mating gear, a climbing condition occurs. Climbing results in a loss of drive until other teeth are engaged and the condition is usually short-lived because it causes rapid destruction of the gears.

GEAR DESIGN

You should be familiar with some basic gear geometry because it is required when performing geartrain adjustments and failure analysis. **Figure 15–2** shows some of the terms used to describe gear tooth geometry; you should reference this in the following explanations. **Gear pitch** refers to the number of teeth per given unit of pitch diameter. You can determine gear pitch by dividing the total number of teeth in a gear by the gear's pitch diameter. Pitch diameter is the pitch line-to-pitch line diameter (reference Figure 15–2) between teeth across the face of a gear. For example, if a gear has 36 teeth and a 6-inch pitch diameter, it has a gear pitch of 6 (**Figure 15–3**). Gears must have the same pitch in order to operate together. A 5-pitch gear will mesh only with another 5-pitch gear, a 6-pitch only with a 6-pitch, and so on.

FIGURE 15–2 Basic gear tooth nomenclature

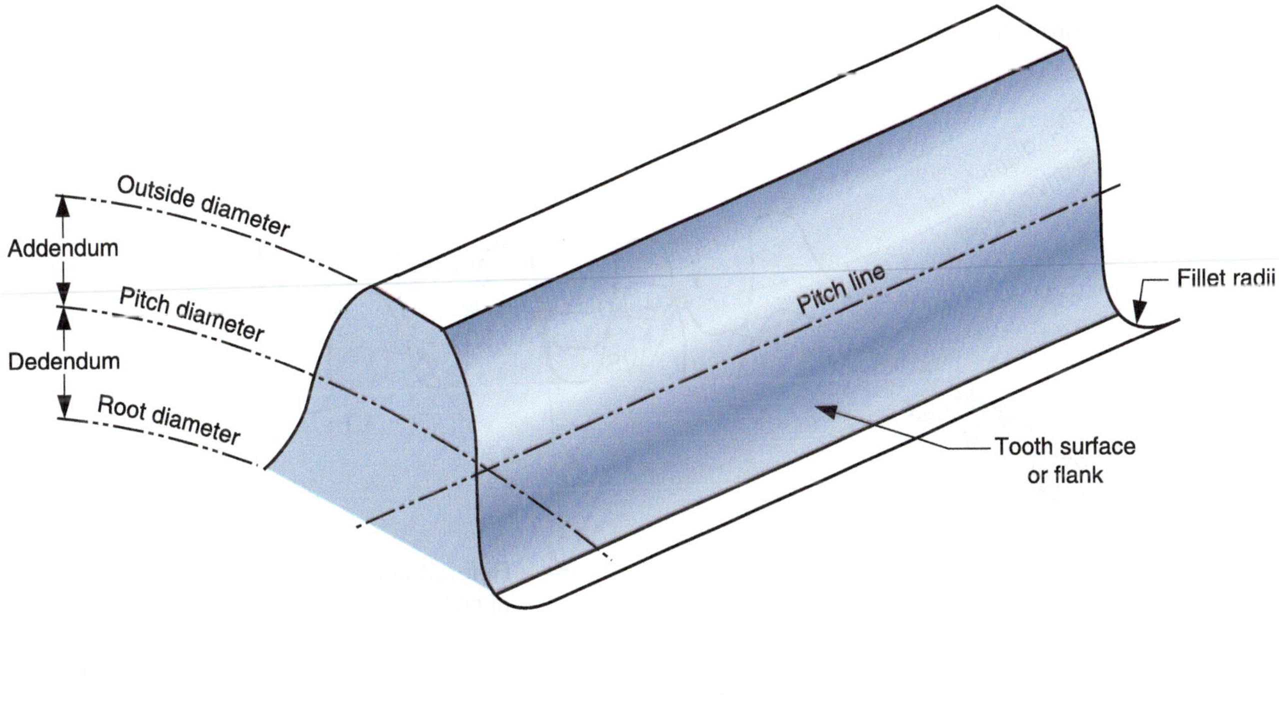

FIGURE 15–3 Determining gear pitch

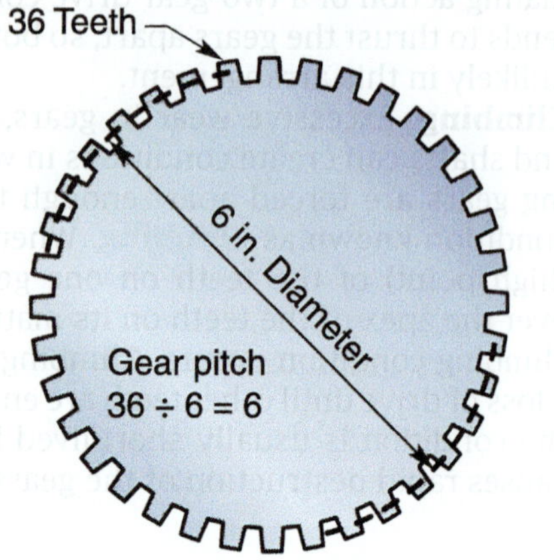

The teeth of a pair of mating gears, while in mesh, pass through the stages of contact (**Figure 15–4**) that we can describe as *coming into mesh, full mesh,* and *coming out of mesh.* Coming into mesh occurs at the initial contact of the dedendum (lower tooth section closer to the root) on the driving gear with the addendum (upper tooth section closer to the apex) of the driven gear. At this point of torque transfer, tooth loading (LT) is relatively light because most tooth loading is sustained by gear teeth that are in full mesh. As the gears rotate, contact between intermeshing teeth progresses in a sliding action. Sliding velocity decreases until it zeros when the contact points reach the intersection of their common pitch lines. At full mesh, when the two teeth meet at their common or operating pitch line, there is only a rolling motion; there is no sliding action. However, at this stage, tooth loading is at its greatest. Coming out of mesh, the two mating teeth again move in a sliding action, essentially the opposite of the initial contact stage.

FIGURE 15–5 Gear tooth profile tolerance zone

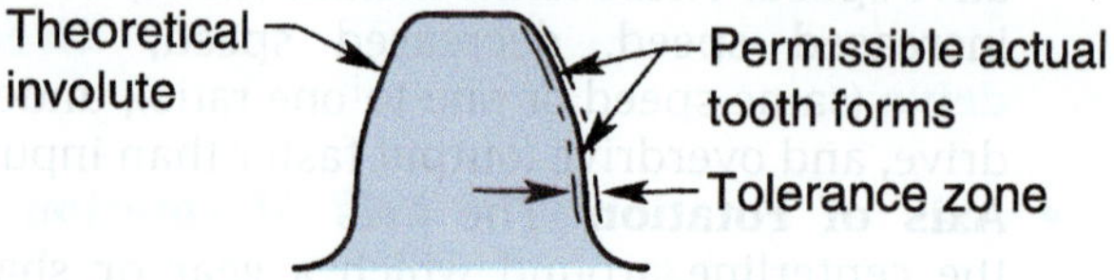

FIGURE 15–6 Gear tooth leads tolerance.

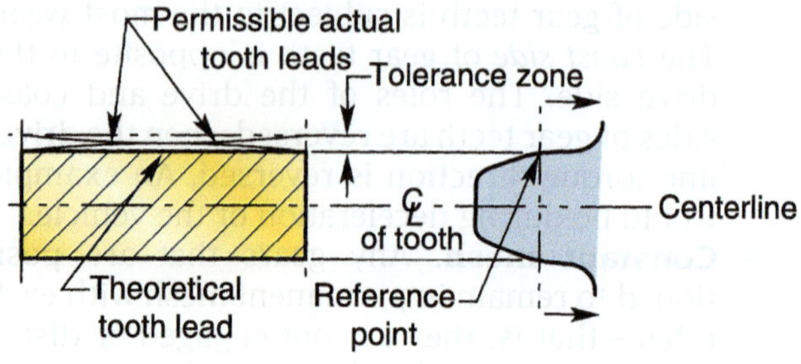

Each gear tooth profile has a theoretical perfect shape. Each manufacturer determines acceptable tolerances for the gears it uses in its assemblies, as shown in **Figure 15–5**. Actual tooth profile can fall anywhere within the tolerance window. A gear tooth must be parallel with the centerline of its shaft—again within a specified tolerance window. This tolerance window is illustrated in **Figure 15–6**. Having a tolerance zone makes it possible to mate gears with minor variations without causing operational failures.

Some manufactured transmission main shaft gearing is crowned, as shown in **Figure 15–7**. *Crowning*

FIGURE 15–4 Three stages of tooth gear contact

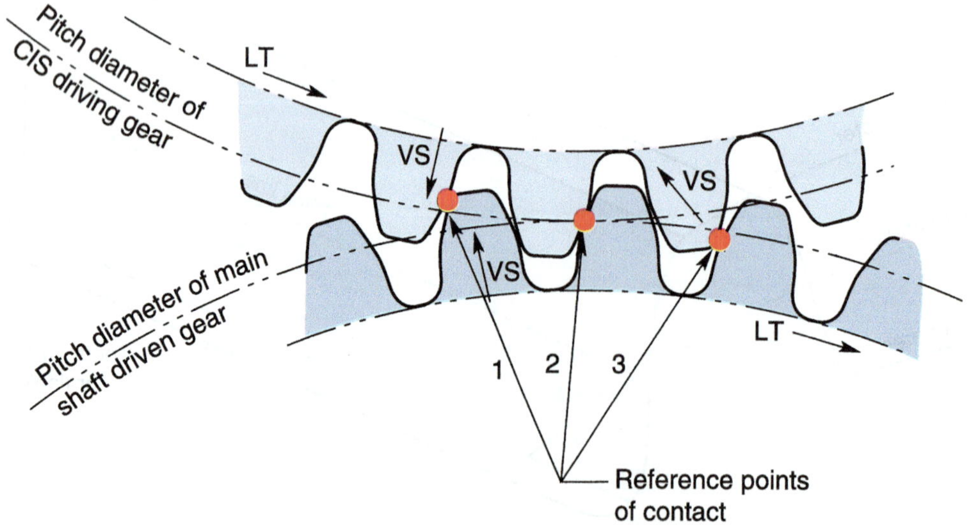

FIGURE 15–7 Gear tooth crowning

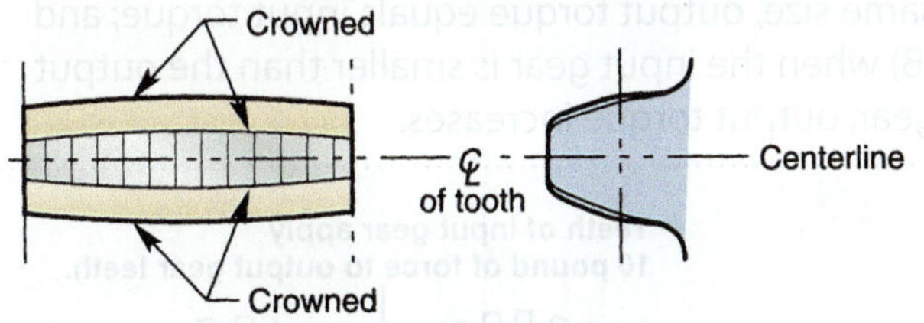

is a slight curvature machined from end to end on a main shaft gear tooth. Crowning prevents highly concentrated contact loads that could result in tooth surface damage.

GEAR RATIOS

As we stated earlier in this chapter, only gears with matching pitch can be meshed. When in mesh, the relative speed of the meshed gears depends on the ratio of the number of teeth on the driving (input) gear to the number of teeth on the driven (output) gear. In **Figure 15–8A**, the two gears are the same size and have the same number of teeth. Because both gears are the same size, the speed of both gears has to be equal so that a direct drive, reverse ratio results. If the objective was to produce direct drive but maintain the same direction of rotation, a third gear of the same size and number of teeth would be required.

In **Figure 15–8B**, a small drive (input) gear turns a larger driven (output) gear. Because the driving gear has fewer teeth than the gear it is driving, the speed of the driven (output) gear must be slower than that of the input gear. This drive arrangement produces a reduction ratio, because the output speed is reduced. When the driving gear has more teeth than the driven gear (as shown in **Figure 15–8C**), the output speed of the driven gear increases. Whenever output speed exceeds input speed, it is commonly referred to as *overdrive*.

The relative speed of two meshed gears can always be calculated by comparing the size and speed of each. Multiply the number of teeth of one gear by the speed of that gear, and divide that number by the number of teeth on the other gear. For example, if we wanted to calculate the speed of a driven gear with 75 teeth when the driving gear has 45 teeth and is rotating at 200 rpm we would use the following formula:

$$\frac{\text{teeth on driven gear}}{\text{teeth on driving gear}} = \frac{\text{rpm of driving gear}}{\text{rpm of driven gear}}$$

Blending the known data into this equation gives us:

$$\frac{75 \text{ teeth}}{45 \text{ teeth}} = \frac{200 \text{ rpm}}{X \text{ rpm}}$$

FIGURE 15–8 (A) When gears of equal size are meshed, output speed equals input speed; (B) a small gear driving a large gear reduces output speed; and (C) a large gear driving a small gear increases output speed (overdrive). In all cases, the output gear rotates in the opposite direction of the input gear.

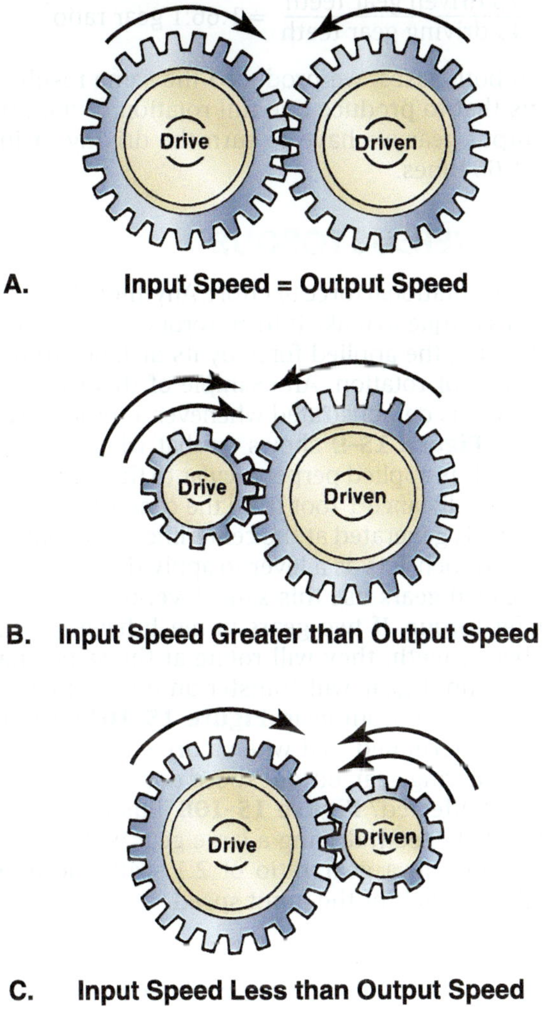

(Note that the values on top of each line are divided by those below the line.)

$$\text{So X rpm} = \frac{45 \times 200}{75} = \frac{9000}{75} = 120 \text{ rpm}$$

In this way, an input speed of 200 rpm is reduced to an output speed of 120 rpm. The relationship between input and output speeds is expressed as gear ratio. Gear ratio can be calculated in a number of ways depending on the known data. One method is to divide the speed of the driving (input) gear by the speed of the driven (output) gear or shaft. So using the preceding example we have:

$$\frac{200 \text{ rpm}}{120 \text{ rpm}} = 1.66{:}1 \text{ gear ratio}$$

Gear ratio can also be calculated by dividing the number of teeth on the driven gear by the number of teeth on the driving gear. Again, using the preceding data, the equation would be:

$$\frac{75 \text{ driven gear teeth}}{45 \text{ driving gear teeth}} = 1.66{:}1 \text{ gear ratio}$$

In both cases, we produced the same result, telling us that to produce one full rotation of the driven or output gear we have to turn the driving or input gear 1.66 times.

SPEED VERSUS TORQUE

Torque is rotational force or effort. Anything that rotates requires torque to make it turn. Torque is calculated by multiplying the applied force by its distance from the centerline of rotation. An example of this simple law of physics is demonstrated whenever a torque wrench is used. **Figure 15–9** shows us that when a force of 10 pounds is applied perpendicular to the centerline of a bolt at a distance 1 foot from the centerline, 10 lb-ft. of torque is generated at the centerline of rotation. The torque wrench acts as a lever to apply this force.

Meshed gears use this same leverage principle to transfer torque. If two gears in mesh have the same number of teeth, they will rotate at the same speed, and the input gear will transfer an equal amount of torque to the output gear (**Figure 15–10A**). We have already determined that when a driven gear is larger than its driving gear, output speed decreases. But how is torque affected? **Figure 15–10B** illustrates a small gear with 12 teeth driving a large gear with 24 teeth. This results in a gear ratio of 2:1, with the output speed being half of the input speed.

FIGURE 15–9 A torque wrench is a good example of how rotational force or torque is generated.

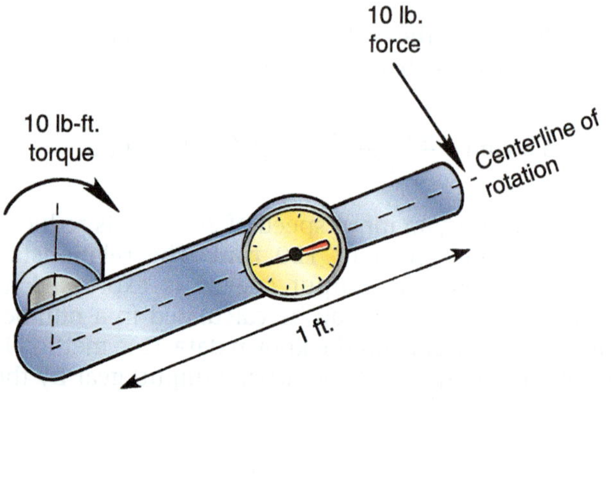

FIGURE 15–10 Gear teeth are actually a series of levers used to transfer torque. (A) If gears are the same size, output torque equals input torque; and (B) when the input gear is smaller than the output gear, output torque increases.

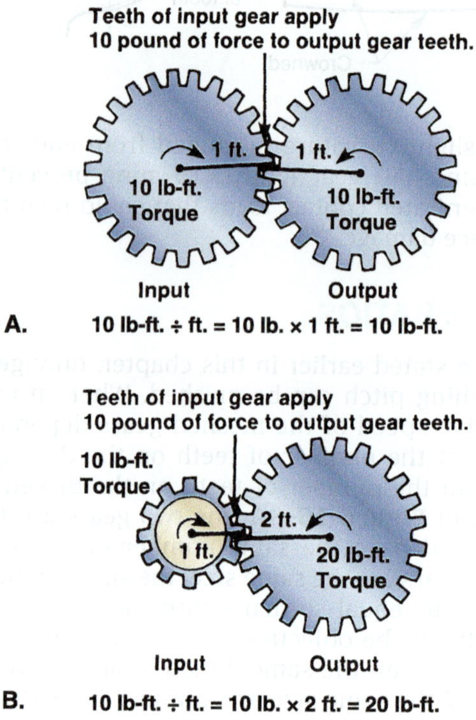

In this example, torque at the input shaft of the driving gear is 10 lb-ft. The distance from the centerline of this input shaft to the gear teeth is 1 foot. This means the driving gear transfers 10 pounds of force to the teeth of the larger driven gear. The distance between the teeth of the driven gear and the centerline of its output shaft is 2 feet. The result is that torque at the output shaft is:

$$10 \text{ pounds (force)} \times 2 \text{ feet (distance)} = 20 \text{ lb-ft.}$$

In this instance, torque has doubled. The torque increase from a driving gear to a driven gear is directly proportional to the speed decrease. When speed is halved, torque doubles.

If we go back and take a look at the results of the first equation, we calculated a gear ratio of 1.66:1, meaning that the output gear rotates 1.66 times slower than the input gear. But in reducing the output speed, output torque has increased by a ratio of 1.66 times. So if we say that torque at the input gear input is 100 lb-ft., then the gear reduction increases output torque by 1.66 times:

$$\text{Output torque} = 1.66 \times 100 \text{ lb-ft.} = 166 \text{ lb-ft.}$$

Most of the gearing in standard transmission gearing is speed reducing/torque increasing. In some transmissions, the top gear produces a 1:1 ratio where speed and torque are transferred directly from the input to the output shaft. In this case, engine flywheel and transmission output shaft rpm would be equal. In some transmissions, the top gear or gears are overdrive gear combinations, resulting in a speed increasing/torque-reducing relationship between input and output.

We express overdrive gear ratios using a decimal point, so if a gear ratio were said to be 0.85:1, to rotate the output shaft through one revolution the input shaft would only have to turn 0.85 times. In other words, to turn the output shaft through one full revolution, the input shaft would only have to rotate through 0.85 revolution. This would result in an output speed that was 1.176 times greater (1.00 divided by 0.85) than input speed. However, if you recall what we said about the relationship between speed and torque, we know that output torque is reduced to 0.85 of input torque. So if input speed is 100 rpm and input torque is 100 lb-ft., the output values can be calculated to be 117.6 rpm and 85 lb-ft. torque.

GEAR MOUNTING

Gears can be mounted or fixed to shafts in a number of ways. They can be internally splined to a shaft, in which case they can be allowed to slide longitudinally on the shaft, or they can be *keyed* to a shaft. In this case, the key positions the gear on the shaft. Gears can also be manufactured as an integral part of the shaft. Gears that are required to turn on a support shaft use bushings or bearings.

IDLER GEARS

An idler gear is often placed between a drive gear and a driven gear to act as an intermediary to transfer torque without changing the rotational direction or gear ratio of the gear set (**Figure 15–11**). This allows the driven gear to rotate in the same direction as the drive gear. Idler gears are used in reverse gearing.

FIGURE 15–11 An idler gear is used to transfer motion without changing rotational direction.

FIGURE 15–12 Spur gear and pinion arrangement

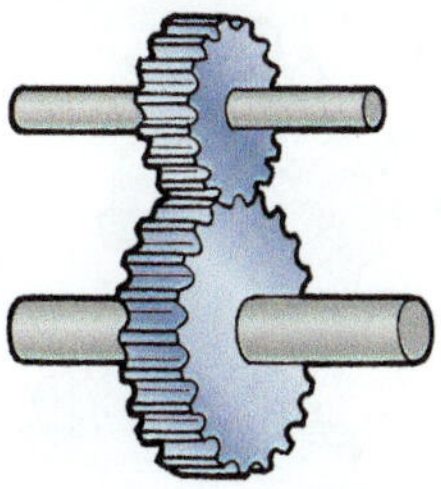

If two idler gears are used, the driven gear will rotate in the direction opposite to that of the drive gear.

Idler gears can also serve as a transfer medium between the drive and driven gears in place of a chain drive or belt drive. Idler gears are designed to not affect the relative speeds of either the drive or driven gears.

TYPES OF GEARS

Different types of gears are used in transmission assemblies. The following types are commonly used.

Spur Gears

The spur gear is the simplest gear design used in standard transmissions. As shown in **Figure 15–12**, spur gear teeth are cut straight, parallel to the gear shaft. During operation, enmeshed spur gears have only one tooth in *full contact* at any given moment with two and a half teeth in partial contact. Use of multimesh gearing can increase the partial contact zone to four teeth and thereby reduce peak loads applied to individual gear teeth. The straight-tooth design is an advantage of a spur gear. Straight teeth minimize the possibility of popping out of gear, a critical consideration during acceleration, deceleration, and reverse operation. For this reason, spur gears are often used in the reverse geartrain.

A disadvantage of spur gears is noise, specifically a clicking noise that occurs as teeth come into contact with each other. At higher turning speeds, the clicking noise becomes a constant whine. Helical gears are quieter and, because of this, are often used in preference to spur gears to eliminate gear whine.

SHOP TALK

When a small gear is meshed with a larger gear, as shown in Figure 15–12, the smaller gear is often called a pinion or pinion gear, regardless of its tooth design.

Helical Gears

A helical gear has teeth that are cut at an angle—that is, they are helical to the gear axis of rotation (**Figure 15–13**). This allows multiple teeth to be in

FIGURE 15–13 Helical gearing arrangement

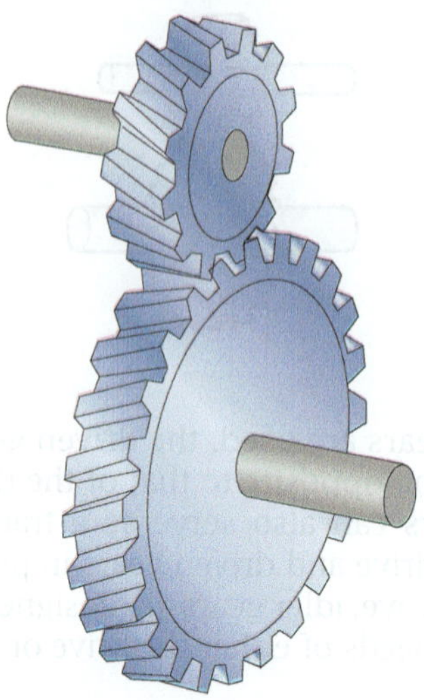

mesh at the same time. Helical gears permit the tooth load to be evenly distributed and are used in applications requiring high torque transfer loads. In addition, helical gears perform more quietly than spur gears because they create a wiping action as they come into and out of mesh with mating gears. The main disadvantage of helical gears is the axial thrust they create in operation. As torque transfers from one gear to its mating gear, fore or aft thrust is created on each shaft. This fore or aft force is known as *axial thrust force*; it has to be absorbed by thrust washers, taper roller bearings, other transmission gears, shafts, and, ultimately, the transmission housing.

Helical gears can have either right- or left-handed geometry, depending on the direction of the spiral when viewed face-on. It is possible to determine whether a helical gear is left- or right-handed using the hand-rule: Hold the gear alternately in each hand. The hand in which the helical teeth align with the thumb determines whether it is left or right cut. When designed to mesh, helical gears must be mounted on parallel shafts, with one gear using right-handed geometry and its mate using left-handed geometry. Two helical gears manufactured with helices in the same-direction spiral cannot mesh in a parallel-mounted arrangement. Both helical and spur gears are commonly used in heavy-duty truck transmission gearing.

15.2 GEARTRAIN CONFIGURATIONS

A truck standard transmission usually consists of a **main shaft** and one, two, or three countershafts. These shafts are mounted parallel to one another. Torque from the engine is transmitted from a transmission input shaft at the front of the transmission through gearing to the countershaft(s). This torque path travels through the countershaft(s) until it reaches the selected gearing. This gearing routes the torque path into the transmission main shaft and from there out to a compound (auxiliary gearing) and the driveline components.

Single-countershaft transmissions are used in a few heavy-duty applications but are more common in medium-duty trucks. Most heavy-duty trucks use a double- or twin-countershaft design (**Figure 15–14**). In a twin-countershaft transmission, torque delivered to the transmission input shaft is divided equally between two shafts. This means that each countershaft gear set carries only half the load of a similar gear set used on a single-countershaft design.

One design of truck transmissions features triple countershafts. **Figure 15–15** shows a sectional view of a triple-countershaft transmission. Note that the countershafts are equally spaced around the main shafts. This design evenly distributes the torque load around the three countershafts, minimizing deflection and containing gear tooth loading to a minimum. The principles are the same as in a twin-countershaft transmission. The difference is that torque is divided between three instead of two countershafts.

15.3 MECHANICAL SHIFT MECHANISMS

Standard transmissions are shifted manually, which means the driver makes gear selections on the basis of how the operating requirements of the vehicle are perceived in relation to engine load and road conditions. There are three general methods of shifting standard transmissions: sliding gear, collar shift, and synchronized shift.

SLIDING GEARSHIFT

A single countershaft transmission that uses a sliding gearshift mechanism has a main shaft and a countershaft supported parallel to each other. To change from one speed/torque range to another, gears on the main shaft are moved longitudinally until they engage the desired gear on the countershaft. To effect this movement, spur-cut sliding gears are used. These have a shift collar integral with the gear. When the driver makes a shift, a **shift fork** or **yoke** engages with a collar on the gear and slides it on the main shaft until it is brought into mesh with a mating gear on the countershaft.

Some older heavy-duty truck transmissions used this type of sliding gearshift arrangement for all speed ranges. However, this type of shifting is unsynchronized, and it was difficult for drivers to precisely match the speed of meshing gears. As a result, grinding and gear clash were a problem, especially with

FIGURE 15–14 A nine-speed, twin-countershaft transmission; note the use of an idler shaft and gear required to option reverse rotation.

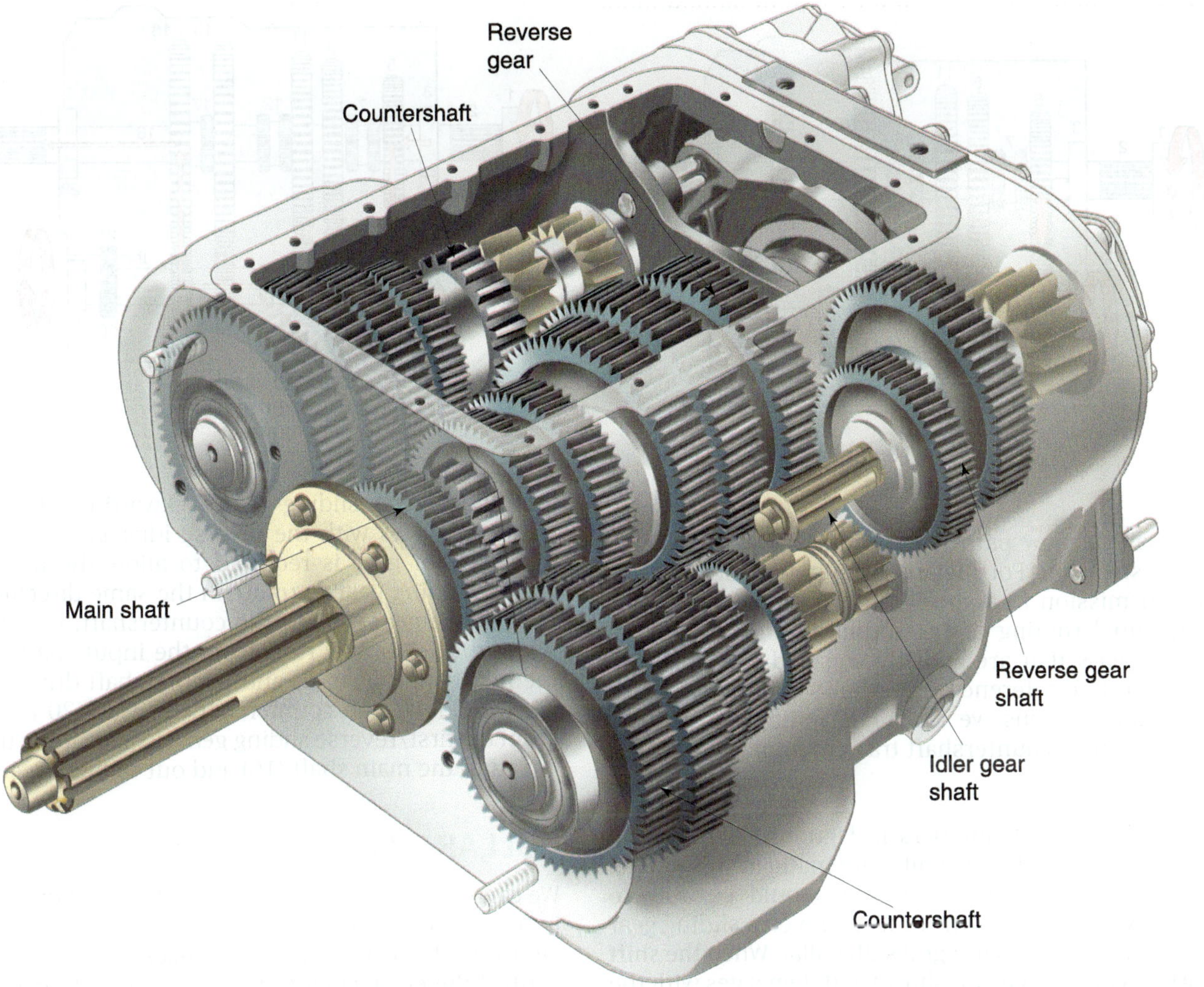

FIGURE 15–15 Layout of the triple-countershaft transmission

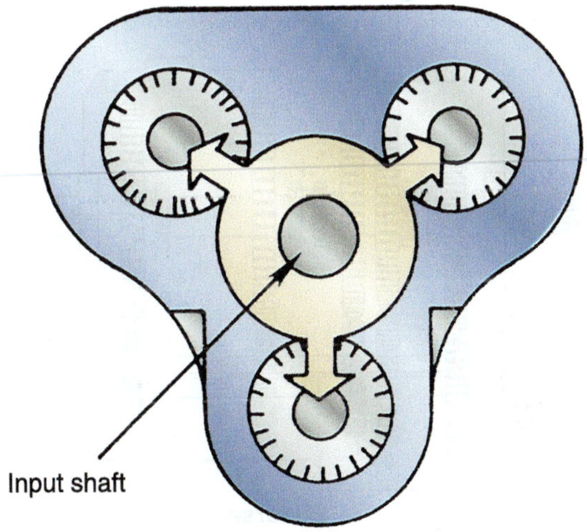

inexperienced drivers. Sliding gearshift mechanisms resulted in transmissions prone to rapid wear, gear teeth chipping, and fracture.

In modern standard, single-countershaft transmissions, the only gear ratios that generally use spur-type sliding gears are first and reverse. The other gears in these single-countershaft transmissions use some form of sliding clutch collar or synchronizer to engage each gear range. In addition, all nonsliding gears are in constant mesh with their mating countershaft gears. **Figure 15–16** illustrates a 5-speed single-countershaft truck transmission that uses a combination of sliding gear and collar shift mechanisms to change ratios.

POWERFLOW

To properly understand how a transmission functions, it is important to know something about **powerflow** paths. *Powerflow* is a general term that describes the

FIGURE 15–16 Typical components of a five-speed single-countershaft transmission; powerflow shown is neutral with the engine at idle.

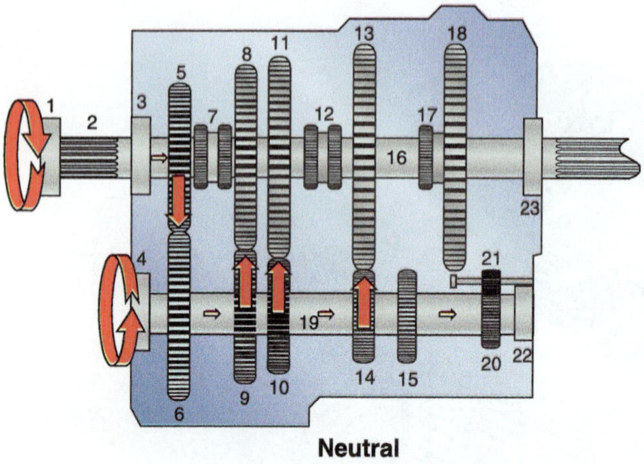

FIGURE 15–17 First-gear powerflow in a typical five-speed single-countershaft transmission

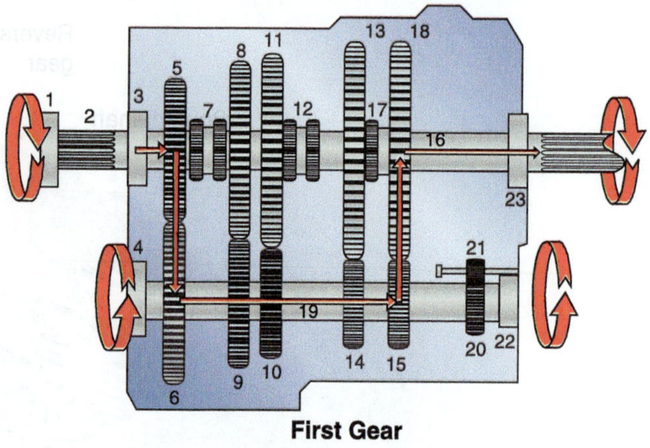

mechanical flow path that conducts torque through a transmission. For instance, the powerflow through a transmission in a specific gear ratio would be the sequential routing of the torque flow path through those gears that are actually engaged. We will from time to time reference powerflows through some of the transmissions we study, beginning here with a typical single-countershaft transmission.

First Gear

Using the callout numbers in Figure 15–16, we will walk you through how shifts are made in this single-countershaft transmission. You should be able to identify callout 18 as the first and reverse sliding gear and callout 17 is its integral shift collar. When the shift fork or yoke (this is not shown, but it engages with the shift collar 17) is moved by the gearshift lever in the cab, it slides the collar and gear either to the front or rear of the transmission housing. Sliding it forward (to the left) engages the first and reverse sliding gear on the main shaft with the first gear (callout 15) on the countershaft. This results in directing powerflow through the first gear as shown in **Figure 15–17**. The torque path is as follows: It flows from the engine flywheel through the clutch plate splines to the transmission input shaft, then (2) through the input shaft gear (5) to the countershaft-driven gear (6). Powerflow is then transmitted through the countershaft to the first gear (15) and up to the first and reverse sliding gear (18) on the main shaft (16). Because the first and reverse sliding gear is splined to the main shaft, powerflow is directed through the main shaft and out to the vehicle driveline.

Reverse Gear

When reverse is selected, the shift lever is moved from neutral to reverse. The shift fork forces the first and reverse sliding gear backward (to the right) until it engages with the reverse idler gear (21). The reverse idler gear is required to allow the first and reverse sliding gear to rotate in the same direction as the reverse gear (20) on the countershaft. Powerflow (**Figure 15–18**) now runs from the input shaft (2) to the input shaft gear (5) and countershaft-driven gear (6), then down the countershaft to gears 20 and 21. From the first/reverse sliding gear (18), torque is transferred to the main shaft (16) and out to the driveline.

COLLAR SHIFT

We can now take a closer look at collar-shift operation in the single-countershaft transmission first shown in Figure 15–16. This transmission uses collar shifting to control the engagement of second, third, fourth, and fifth gears. In a collar-shifting arrangement, all gears

FIGURE 15–18 Reverse-gear powerflow in a typical five-speed single-countershaft transmission

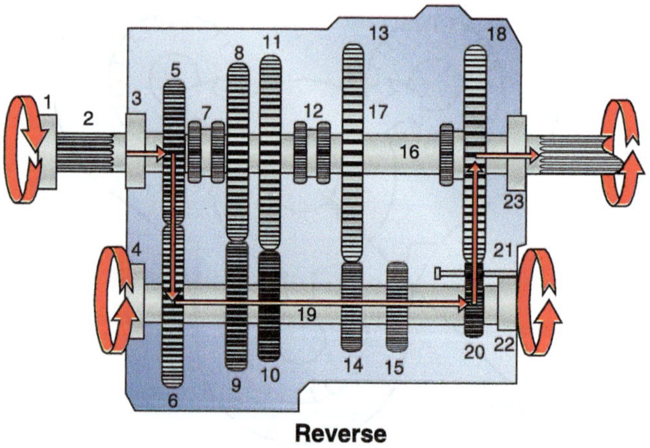

on the countershaft are fixed to the countershaft. On the other hand, the main shaft gears are free to *freewheel* (float) and do so around either a bearing or bushing. These bearings or bushings are positioned on a nonsplined section of the main shaft. The main shaft gears are in constant mesh with their mating countershaft gears. This means they turn at all times.

Study the neutral powerflow path of the five-speed transmission as shown in Figure 15–16. The input shaft (2) rotates at engine speed any time the clutch is engaged. The input shaft gear (5) is integral with the input (clutch) shaft, so it has to rotate with it. The input shaft gear meshes with the countershaft-driven gear (6), so it, the countershaft, and all the gears fixed to the countershaft also have to rotate.

The countershaft gears transfer torque to their mating gears on the main shaft. But main shaft gears 8, 11, and 13 all freewheel on the main shaft on bearings or bushings. Because they are freewheeling, they cannot transmit torque to the main shaft, so it does not turn. This results in no output to the driveline.

To enable torque transfer to the main shaft, one of the freewheeling main shaft gears must be locked to it. This can be accomplished in a number of ways and one method uses a shift collar and shift gear. The shift gear is internally splined to the main shaft at all times. The shift collar is splined to the shift gear. In **Figure 15–19**, you can see that the main gears have a short, toothed hub. The teeth on the main gear hub align with the teeth on the shift gear. The internal teeth of the shift collar mesh with the external teeth of the shift gear and hub.

When a given speed range is not engaged, the shift collar simply rides on the shift gear. When the driver shifts to engage that speed range, the shift fork moves the shift collar and slides it into mesh with the teeth of the main gear hub. At this moment, the shift collar rides on both the shift gear and main gear hub, locking them together. Power can flow from the main gear to the shift gear, then to the main shaft and out to the propeller shaft.

FIGURE 15–19 Collar shift mechanism using a shifter gear, shift collar, and splined, hub-driven gear

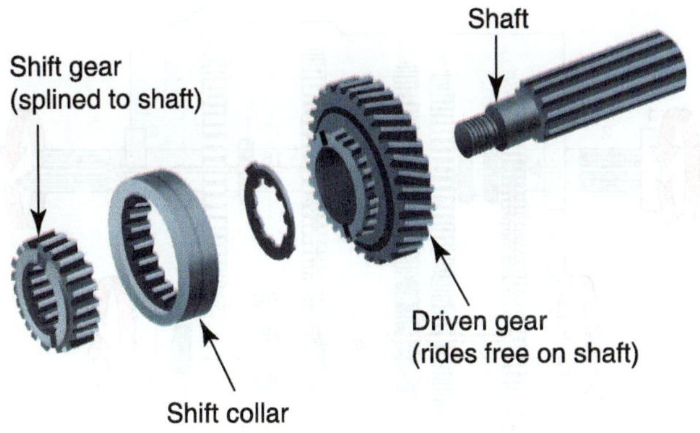

A second, more common method of locking main gears to the main shaft does not use a shift gear. Instead, the shift collar is splined directly to the main shaft. This shift collar, also called a *clutch collar* or a *sliding clutch*, is designed with external teeth. These external teeth mesh with internal teeth in the main gear hub or body when that speed range is engaged. Most shift collars or sliding clutches are positioned between two gears so they can control two-speed ranges depending on the direction in which they are moved by the shift fork (**Figure 15–20**).

FIGURE 15–20 Exploded view of a five-speed main shaft, showing the relative position of the sliding clutches and gears: One sliding clutch controls first and reverse shifts, another controls second and third shifts, and a third controls fourth and main gear engagement. Note: The main drive gear is not shown but remember that this is part of the input shaft/drive gear assembly.

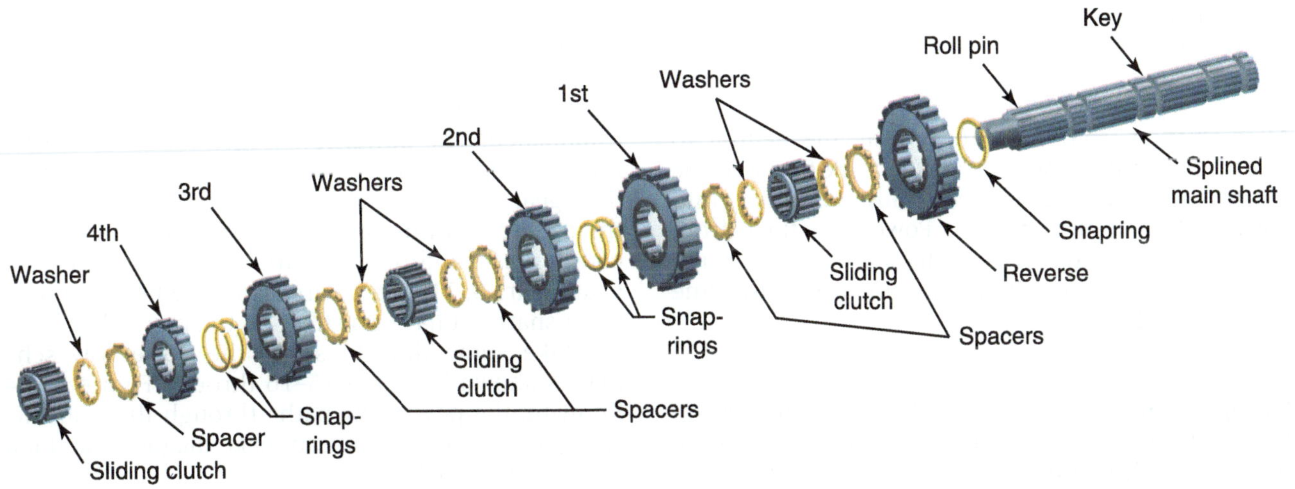

FIGURE 15–21 Second-gear powerflow in a typical five-speed single-countershaft transmission

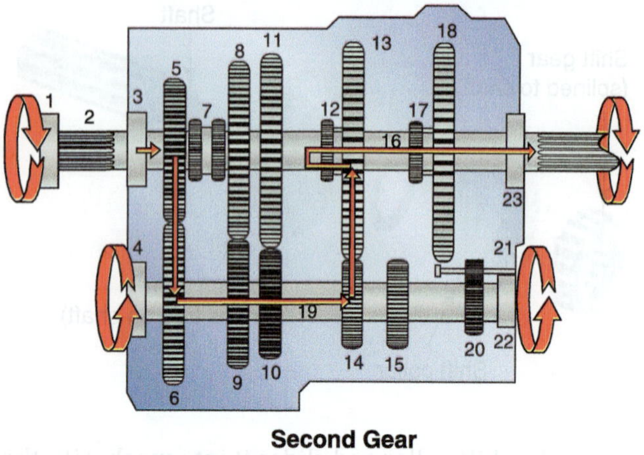

Second-Gear Powerflow

To shift from first to second gear, the operator pushes the clutch pedal down to disengage the clutch. The gear-select lever is then moved to neutral, disengaging the first and reverse gear (18). Moving the shift lever from neutral to the second-gear position causes the second and third shift collar (or sliding clutch) to move (12) back toward the second gear (13) on the main shaft. The shift collar engages second gear and locks it to the main shaft. Power flows from 2 to 5 and 6 and along the countershaft to 14. Gear 14 drives gear 13, which in turn drives main shaft output to the driveline (**Figure 15–21**).

Third-Gear Powerflow

Once again the driver returns to neutral before shifting to third gear. Moving from neutral to third gear moves the second and third shift collar (or sliding clutch) (12) forward toward the third gear (11), locking it to the main shaft. Power flows from 2 to 5 and 6, along the countershaft to 10, up to 11, through the shift collar (12) to the main shaft and out to the driveline (**Figure 15–22**).

Fourth-Gear Powerflow

After shifting from third to neutral, the neutral to fourth gearshift causes the shifter fork to move the fourth and fifth shift collar (or sliding clutch 7) into mesh with the fourth gear (8). Power now flows from 2 to 5 and 6, along the countershaft to 9, through 8 and 7 to the main shaft, and out to the driveline (**Figure 15–23**).

Fifth-Gear Powerflow

Shifting from fourth to neutral, then from neutral to the fifth gear, causes the shifter fork to move the fourth and fifth shift collar (or sliding clutch 7) into mesh with the input shaft gear (5). This locks the input shaft (2) directly to the main shaft (16). Input and output speeds are the same. So a 1:1 gear ratio is established. The powerflow (**Figure 15–24**) is from 2 to 5 through 7, then to the main shaft and out. In fifth gear the clutch shaft gear is still meshed with the countershaft-driven gear (6). This means that the countershaft and its gears are all turning. The main shaft gears 8, 11, and 13 are also freewheeling on the main shaft but have no effect on the powerflow.

FIGURE 15–22 Third-gear powerflow in a typical fivespeed single-countershaft transmission

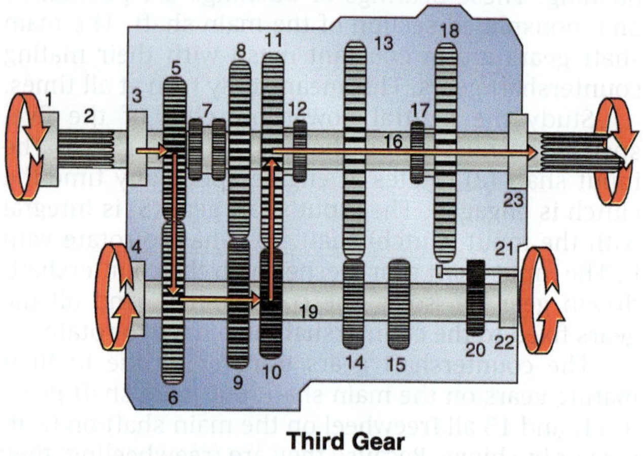

FIGURE 15–23 Fourth-gear powerflow in a typical five-speed single-countershaft transmission

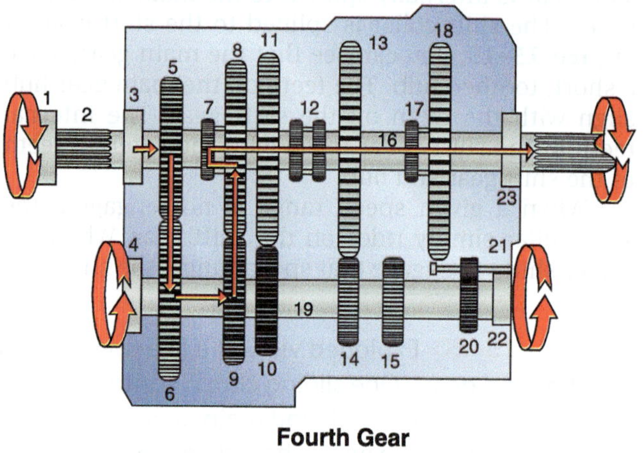

Take some time to study the powerflow schematics shown in Figure 15–16 through Figure 15-24. Knowing the powerflow paths through the transmission you are working on can make diagnosis and failure analysis much easier.

FIGURE 15–24 Fifth-gear powerflow in a typical five-speed single-countershaft transmission

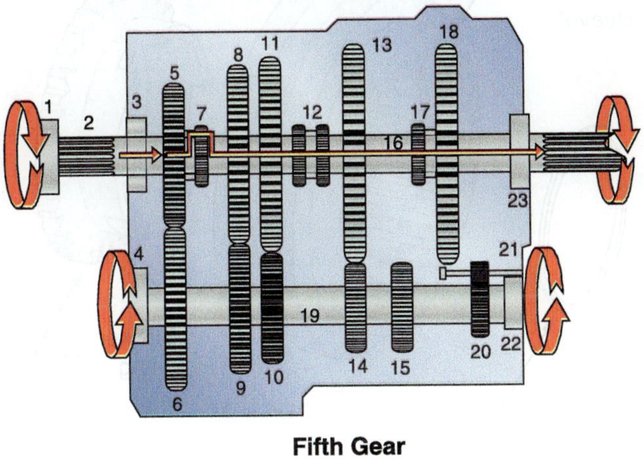

SHOP TALK

As explained in Chapter 14, a clutch brake is used to slow gear rotation in order to complete a shift into first or reverse from neutral when the vehicle is stationary. The clutch brake is actuated by depressing the clutch pedal completely to the floor. For normal upshifts and downshifts, only partial disengagement of the clutch is needed to break engine torque; in other words, it is not necessary to drive the clutch pedal to the floor to shift.

OPERATING A NONSYNCHRONIZED TRANSMISSION

The sliding gear and collar-shift mechanisms described in the preceding sections are used on nonsynchronized standard transmissions. Shifting nonsynchronized transmissions requires double-clutching between gears. We will outline this using the example of a seven-speed transmission coupled to an engine with a governed speed of 2,100 rpm.

Upshifting

To move away from a standing start with the engine at idle, depress the clutch and move the shift lever to the first-gear position. Release the clutch and accelerate the engine to governed speed. When the tach displays governed speed, depress the clutch, release the accelerator pedal, and shift to neutral. Next, engage the clutch and allow engine speed to drop approximately 750 rpm. Then, release the clutch and shift to second gear. Reengage the clutch and accelerate back up to governed speed. Continue shifting through the gear ratios until the seventh gear ratio is achieved.

SHOP TALK

The 750 rpm drop used in the description of shifting procedure varies according to engine-governed speed and torque rise profile.

Downshifting

To progressively downshift from seventh gear, allow engine speed to drop approximately 750 rpm before depressing the clutch and moving the shift lever to neutral. Now reengage the clutch and accelerate engine rpm to governed speed. Depress the clutch and move the shift lever into the sixth-gear position. Now reengage the clutch. Continue downshifting in this manner.

SYNCHRONIZED SHIFTING

A synchronizer performs two important functions. Its main function is to bring components that are rotating at different speeds to a common or synchronized speed. In a transmission, a synchronizer ensures that the main shaft and the main shaft gear about to be locked to it are both rotating at the same speed. The second function of the synchronizer is to actually lock these components together. If the driver and the synchronizer are both doing their jobs, the result is a clash-free shift. Most standard transmissions having five to seven forward speeds are completely synchronized except for first and reverse gears. The reason for not synchronizing first and reverse gears is that they are only selected when the vehicle is stationary. Synchronizing of engine speed and driveline speed is not required when the drive shaft is not turning, in fact, a synchronizer could cause hard shifting in these gear positions because a synchronizer needs gear rotation to do its job.

Many transmissions use both main and auxiliary gearing to provide a large number of forward ratios. In these transmissions synchronizers are normally used only in the two- or three-speed auxiliary gearing. Block or cone and pin synchronizers are the most common in heavy-duty truck applications. Both block or cone and pin synchronizers are often referred to as sliding clutches or dog clutches.

Block or Cone Synchronizers

Figure 15–25 illustrates a block or cone synchronizer. The synchronizer sleeve surrounds the synchronizer assembly and meshes with the external splines of the clutch hub. The clutch hub is splined to the transmission output (main) shaft and is held in position by two snaprings. The synchronizer sleeve has a small internal groove and a large external groove in which the shift fork rests. Three slots are equally spaced around the outside of the clutch hub. Inserts fit into these slots and are able to slide freely back and forth.

FIGURE 15–25 Block- or cone-type synchronizer components

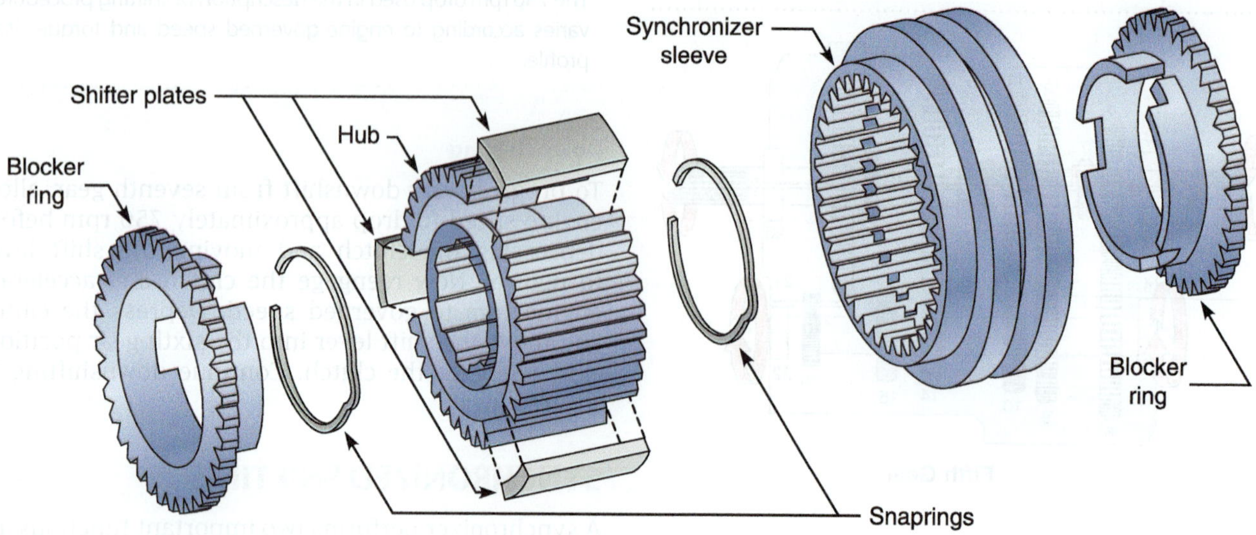

These inserts are designed with a ridge in their outer surface. Insert springs hold the ridge in contact with the synchronizer sleeve internal groove.

The synchronizer sleeve is precisely machined to slide on the clutch hub smoothly. The sleeve and hub sometimes have alignment marks to ensure proper indexing of their splines when assembling to maintain smooth operation. Brass or bronze synchronizing blocker rings are positioned at the front and rear of each synchronizer assembly. Each blocker ring has three notches equally spaced to correspond with the three inset notches of the hub. Around the outside of each blocker ring is a set of beveled dog teeth, which are used for alignment during the shift sequence. The inside of the blocker ring is shaped like a cone. This coned surface is lined with many sharp grooves.

The cone of the blocker ring makes up one-half of a cone clutch assembly. The second or mating half of the cone clutch is part of the gear to be synchronized. As you can see in **Figure 15–26**, the shoulder of the main gear is cone shaped to match the blocker ring. The shoulder also contains a ring of beveled dog teeth designed to align with dog teeth on the blocker ring.

Operation. In a typical five-speed transmission, two synchronizer assemblies are used. One controls second- and third-gear engagements, the other synchronizes fourth- and fifth-gear shifting. When the transmission is in first gear, the second/third and fourth/fifth synchronizers are in their neutral position (**Figure 15–27A**). All synchronizer components are now rotating with the main shaft at main shaft speed. Gears on the main shaft are meshed with their countershaft partners and are freewheeling around the main shaft at various speeds. To shift the transmission into second gear, the clutch is disengaged and the gearshift lever is moved to the second-gear

FIGURE 15–26 Gear shoulder and blocker ring mating surfaces

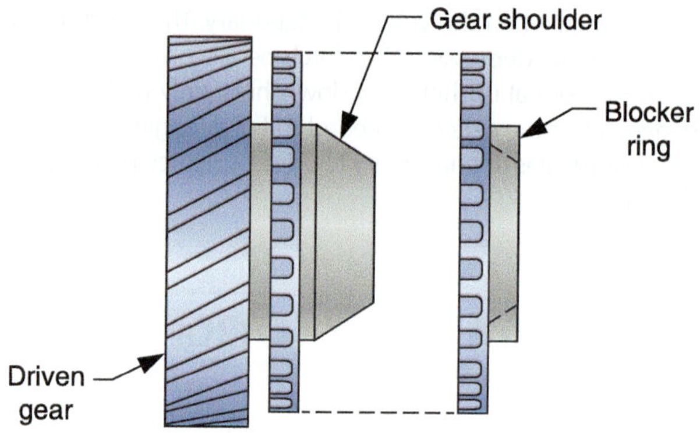

position. This forces the shift fork on the synchronizer sleeve backward toward the second gear on the main shaft. As the sleeve moves backward, the three inserts are also forced backward because the insert ridges lock the inserts to the internal groove of the sleeve.

The movement of the inserts forces the back blocker ring's coned friction surface against the coned surface of the second-gear shoulder. When the blocker ring and gear shoulder come into contact, the grooves on the blocker-ring cone clutch cut through the lubricant film on the second-speed gear shoulder and a metal-to-metal contact is made (**Figure 15–27B**). The contact generates substantial friction and heat. This is one reason that bronze or brass blocker rings are used. A nonferrous metal such as bronze or brass minimizes wear on the hardened steel gear shoulder. If similar metals were used, wear

FIGURE 15–27 (A) Block- or cone-type synchronizer in the neutral position; (B) initial contact of the blocker ring and gear shoulder; and (C) final engagement as the synchronizer sleeve locks the driven gear to the synchronizer hub and shaft.

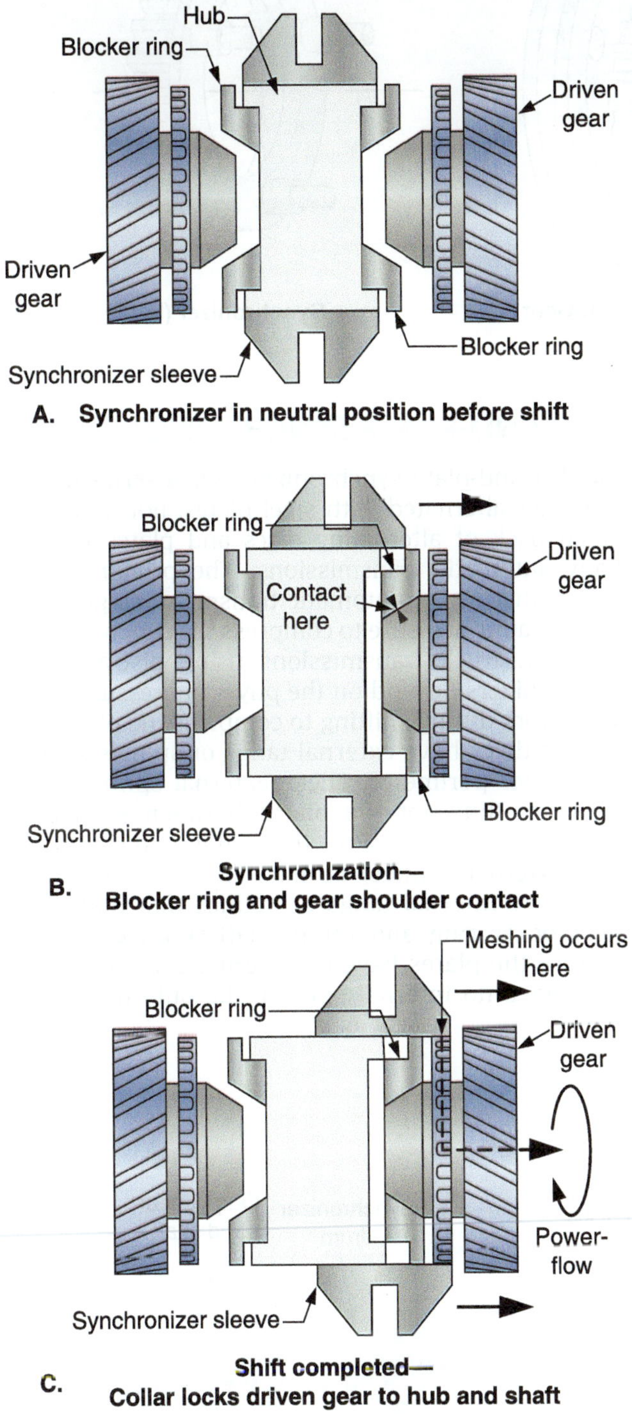

would be accelerated. This frictional coupling is not strong enough to transmit loads for long periods. But as the components reach the same speed, the synchronizer sleeve can now slide over the external dog teeth on the blocker ring and then over the dog teeth on the second-speed gear shoulder. This completes the engagement (**Figure 15–27C**).

Powerflow is now from the second-speed gear, to the synchronizer sleeve, to the synchronizer clutch hub, to the main output shaft, and out to the propeller shaft. To disengage the second-speed gear from the main shaft and shift into third-speed gear, the clutch must be disengaged as the shift fork is moved to the front of the transmission. This pulls the synchronizer sleeve forward, disengaging it from second gear. As the transmission is shifted into third gear, the inserts again lock into the internal groove of the sleeve. As the sleeve moves forward, the forward blocker ring is forced by the inserts against the coned friction surface on the third-speed gear shoulder. Once again, the grooves on the blocker ring cut through the lubricant on the gear shoulder to generate a frictional coupling that synchronizes the gear and shaft speeds. The shift fork can then continue to move the sleeve forward until it slides over the blocker ring and gear shoulder dog teeth, locking them together. Powerflow is now from the third-speed gear, to the synchronizer sleeve, to the clutch hub, and out through the main shaft.

Pin Synchronizers

Figure 15–28 shows the components of a typical pin-synchronizing unit. The clutch drive gear has both internal and external splines. It is splined to the main (output) shaft and held to it with two snaprings. The sliding clutch gear is loosely splined to the external splines of the clutch drive gear. The sliding clutch gear is made with a flange that has six holes with slightly tapered ends. The shift fork fits into an external groove cut into this flange.

Aluminum alloy blocker rings have coned friction surfaces designed with several radial grooves. Three pins are fastened to each blocker ring, and each pin is machined with two distinctly different diameters. The outer stop synchronizer rings are machined with an internal cone friction surface. The bores of the outer stops have internal teeth that spline to the external teeth on the shoulders of the gears to be synchronized. Finally, the smaller-diameter ends of the blocker-ring pins rest in the holes of the sliding clutch gear.

Operation. When the pin synchronizer is in a neutral (disengaged) position, the clutch drive gear, sliding clutch gear, and the blocker ring all rotate with the main (output) shaft. The gears on either side of the synchronizer, as well as the outer rings splined to the gear shoulders, are driven at varying speeds by the countershaft gears. To shift into a pin-synchronized gear, the clutch is disengaged and the shift lever is moved to the desired speed ratio. This moves the shift fork and the sliding clutch gear toward the gear to be synchronized. The three holes in the sliding clutch gear butt against the side of the

FIGURE 15-28 Pin-type synchronizer

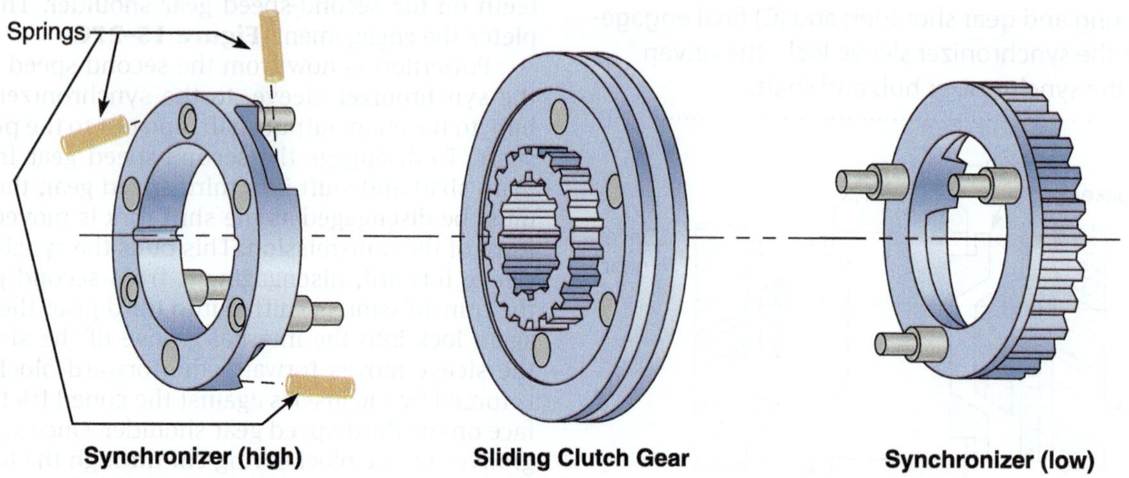

larger diameter ends of the three pins on the blocker ring. This forces the blocker ring into the adjoining outer stop ring. The two coned friction surfaces of the blocker and outer rings come into contact. The friction coupling equalizes the turning speed of these components. In addition, because the driven gear is splined to the outer stop ring, it is now turning at main shaft speed.

As with the block or cone synchronizer, this frictional coupling cannot be held for long. When all components are turning at the same speed, the holes in the sliding clutch gear are able to slide over the larger diameter portion of the pins. The internal teeth of the sliding clutch gear slide over the external teeth of the gear, completing the clutch engagement. To disengage the gear from the shaft, the clutch is disengaged and the shift lever is moved to neutral. This moves the shift fork and sliding clutch gear away from the main gear, disengaging the sliding clutch gear teeth from the main gear teeth.

DISC-AND-PLATE SYNCHRONIZERS

The disc-and-plate synchronizer uses a series of friction discs alternated with steel plates, not unlike the clutch pack of alternating discs and plates used in many automatic transmissions. The primary difference is that where automatic transmissions generally use hydraulic pressure to compress the clutch and disc assembly, truck transmissions using disc-and-plate synchronizers depend on the physical pressure of the shifter fork during shifting to compress the plates.

The discs have external tangs or splines around their outer perimeter. These external splines allow the discs (but not the plates between each disc) to engage with the slots the plates (which do not have external splines) do not. The plates do, however, have internal tangs or splines that mate with the blocker ring and input shaft (**Figure 15-29**). Because the plates have no external splines and the discs have no internal splines, the only movement

FIGURE 15-29 Disc-and-plate-type synchronizer

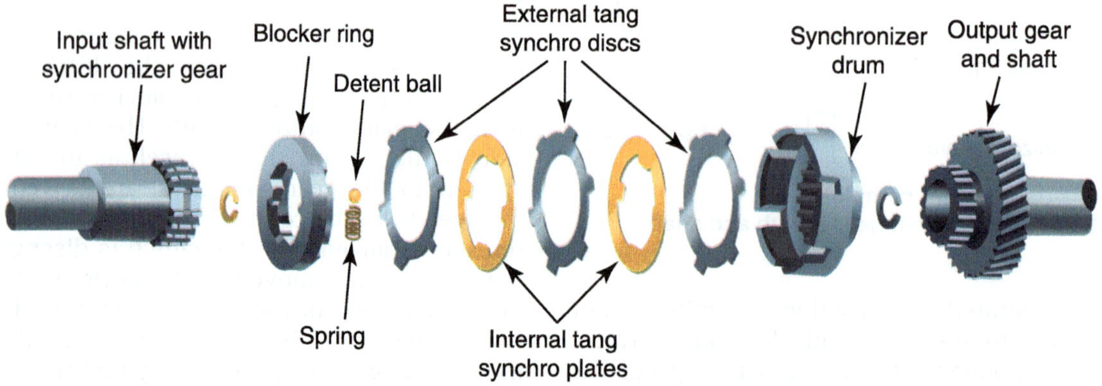

that can be transmitted between them comes from clamping friction between the two. If minimal pressure is applied to the disc-and-plate assembly, the amount of friction between the two will be small, and the amount of movement transferred will be correspondingly small. As more pressure is applied to the disc-and-plate assembly, friction increases and more movement is transferred.

The purpose of the disc-and-plate synchronizer is to match the speed of input gears with output gears during shifting. This occurs when the truck driver disengages the clutch and moves the gearshift lever, causing the shifter fork to move the synchronizer drum forward. This forward motion causes the discs and plates to rotate the blocker ring around the input shaft until the drive lugs on the blocker ring engage with the outer edge of the detent ball depression cut into the synchronizer gear.

Once the blocker ring has locked into the synchronizer gear, any additional forward motion of the shifter forks will compress the discs and plates together, causing them to rotate at the same speed. When the speeds of the input gear and blocker gear are synchronized to the speed of the output gear and both are turning at the same speed, the following should be true:

- Thrust force is released.
- The blocker disengages from its locked position.
- The drum can be moved forward to engage both gears.

PLAIN SYNCHRONIZERS

A plain synchronizer uses fewer working parts and is similar in design to a block synchronizer (**Figure 15–30**). As is typical with most synchronizers, the hub is internally splined to the main shaft. Mounted on the hub, and also connected to it, is a sliding sleeve that is controlled by the shift fork movement. In operation, the friction generated between the hub and the gear tends to synchronize their speeds. Pressure on the sliding sleeve prevents the sleeve from moving off the hub and also prevents the teeth on the sleeve from engaging the gear teeth until sufficient pressure has been put on both friction areas to synchronize the components. When in sync, any further movement of the sleeve will allow the sleeve teeth to engage the gear teeth.

DRIVING SYNCHRONIZED TRANSMISSIONS

A synchronizer simplifies shifting and helps the driver make clash-free shifts. In operation when a shift is required, the driver should disengage the clutch and move the shift lever machined inside the synchronizer drum that houses the discs and plates. This means that when the synchronizer drum turns, the clutch discs (with external tangs) turn with it but toward the desired gear position. When the synchronizer ring makes initial contact with the desired gear, the blockers automatically prevent the shift collar from completing the shift until the gear and main shaft speeds are matched. At that time, the blocker neutralizes automatically and a clash-free shift is the result. Steady pressure on the shift lever helps the synchronizer do its job quickly. When synchronized, the shift lever moves into gear smoothly and easily.

If the driver jabs at the synchronizer or teases it by applying and releasing pressure, the synchronizer

FIGURE 15–30 Plain-type synchronizer

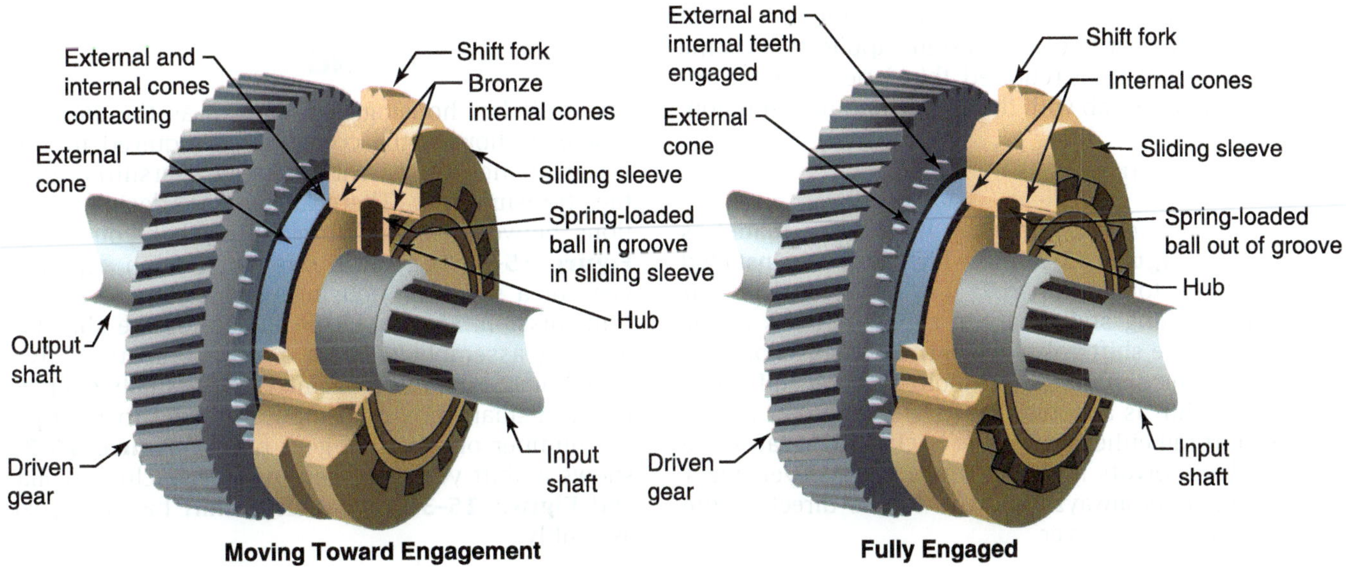

cannot do its job. It can take up to 1 or 2 seconds to match asynchronous speeds, and gentle but constant shift lever pressure produces faster synchronization action. It is possible to override a blocker if the lever is forced into gear. Forcing a transmission into gear defeats the purpose of the synchronizer and can cause gear clash and ultimately damage the transmission.

When the vehicle is stationary, fully depress the clutch, move the shift lever into first gear, and accelerate to an engine rpm that allows sufficient vehicle momentum to select the next higher gear. Drivers should be aware of the engine-torque-rise rpm values. These define the upper and lower rpm shift limits. Using a progressive shift technique and not running engine rpm up to governed speed on each shift saves fuel.

When downshifting, a similar procedure is used. Consider shifting from top (fifth) gear to fourth gear. As the shift point is approached (normally about 100 rpm over the shift point), disengage the clutch and move the shift lever with a steady, even pressure toward fourth gear. The synchronizer will pick up fourth gear, get it up to driveline speed, and allow a clash-free shift from fifth to fourth. After the shift, reengage the clutch and at the same time accelerate the engine to keep the vehicle moving at the desired speed. If further downshifts are required, continue in a similar manner.

Downshifting into first gear is not synchronized and will require a double-clutch operation to complete a clash-free shift. It is possible to double-clutch on all shifts if desired. This aids the synchronizer in its function by getting the engine speed and driveline matched.

15.4 SHIFT HOUSING ASSEMBLY

A shift housing assembly consists of a tower assembly and shift bar housing. In some applications some additional linkage is required; this depends on the orientation of the cab with the engine and transmission.

SHIFT TOWER

The shift tower assembly can be direct actuated as shown in **Figure 15–31**, in which case it is installed directly on top of the **shift bar housing**. The gear lever actuating stub floats in clevis recesses in the center of the shift yokes when the transmission is in neutral. The shift yokes are bolted to the shift rails. When a shift is affected, the gear lever finger moves the shift rail either forward or backward. Because the shift lever pivots in the tower, the gear lever finger movement is always opposite to the direction the driver moves the lever.

FIGURE 15–31 Typical direct shift lever and tower assembly

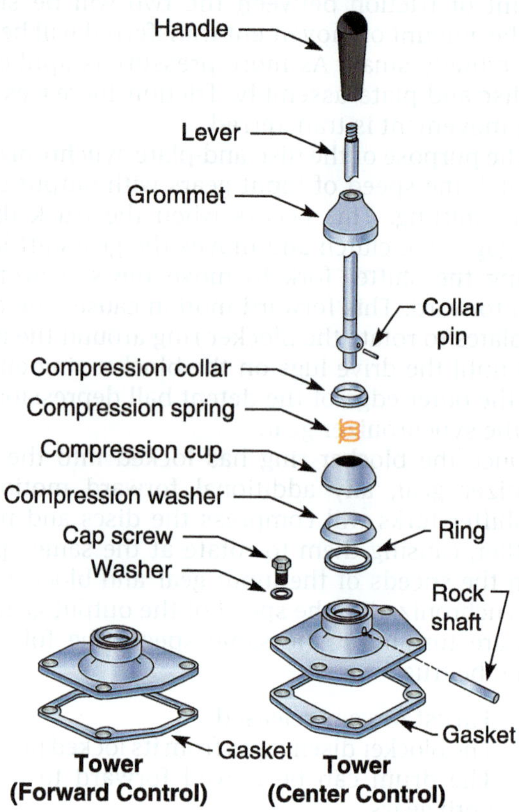

In a **cab-over-engine (COE)** truck, the gear lever is not usually located directly above the transmission because the driver position is further forward in the cab. In most COE trucks, a remote shift lever uses a linkage rod to connect the gear lever with the shift rail housing on top of the transmission (**Figure 15–32**).

SHIFT BAR HOUSING

The shift bar housing, also known as the *shift rail assembly*, houses the components required to convert gear stick movement into range gearshifts within the transmission. There are some design variations between manufacturers but what you see in **Figure 15–33** is typical; this figure is an exploded view of a shift bar housing used in a seven-speed transmission. **Figure 15–34** illustrates the shift bar housing assembly used in a commonly used five-speed main box; this transmission is usually coupled to an auxiliary box or compound (used to multiply the number of available gear ratios). **Figure 15–35** shows a shift yoke engaged into a dog clutch collar and **Figure 15–36** illustrates a shift bar and yoke assembly.

FIGURE 15–32 Remote-control assembly used to connect the shift tower with the shift bar housing in a COE application.

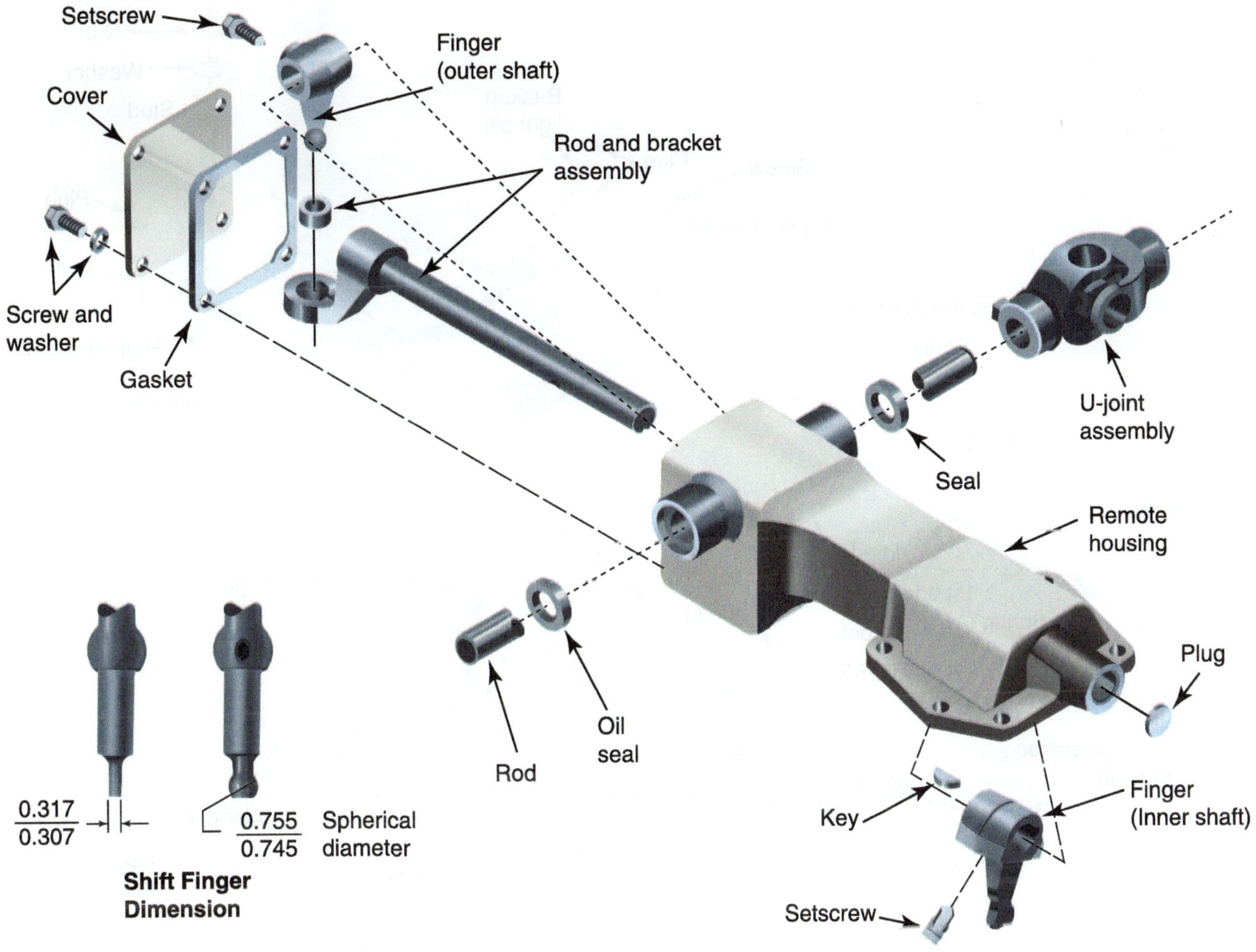

> **SHOP TALK**
>
> Auxiliary (compound) transmission gearing requires its own control system. The controls can be mechanical, air, or electronic-over-air operated. Air-operated controls are discussed later in this chapter.

OPERATION

The gearshift lever in a direct shift tower arrangement, or the remote-actuated shift finger in a COE application, fits into slots formed by the shift bracket or yoke blocks. If you take a look at Figure 15–33 and Figure 15–34, the shift brackets or blocks can either be clamped to a shift rail or integral with the shift fork or yoke. The shift bars or rails are designed to slide in the shift bar housing when actuated by the gear lever. The shift forks or yokes are clamped or bolted to the shift bars. When the gearshift lever is moved, it slides the shift bars and thereby moves the yokes. The shift yoke fits over a shift collar or synchronizer so an actual shift can be made.

After a shift has been effected, the shift bar must be held in position so that the selected gear ratio is maintained. To hold the selected gear and prevent unwanted rail movement, the shift bars are machined with recesses; these recesses align with a detent mechanism. The *detent mechanism* consists of a spring-loaded detent steel ball or poppet. The spring loads the steel ball into the recess in the shift bar. The detent ball holds the shift bar in position and prevents unwanted movement of the other bars. When drivers make a shift, they must overcome the spring pressure of the detent ball to move the shift bar.

FIGURE 15–33 Shift housing control components for a seven-speed transmission; shift brackets, rods, and forks are used.

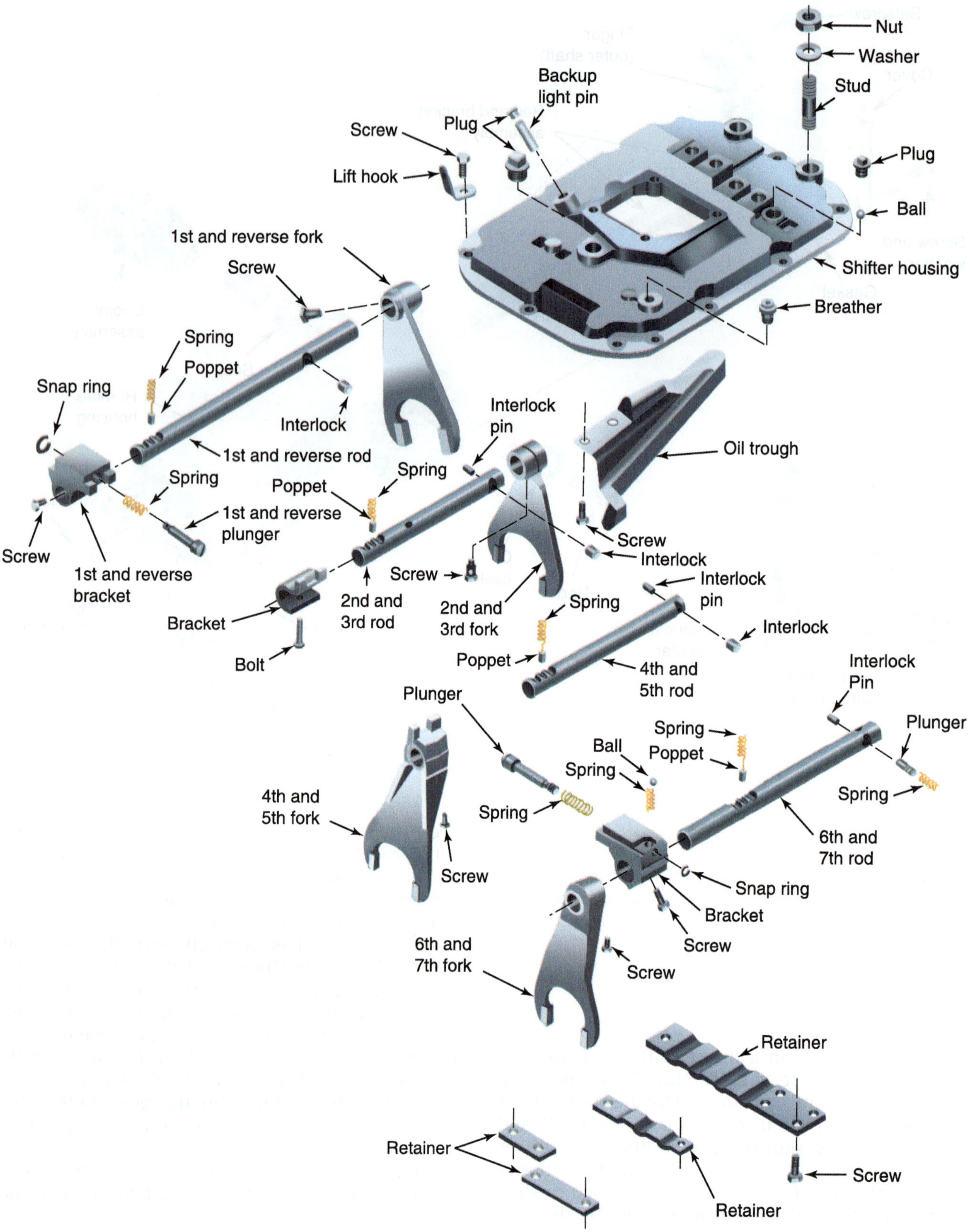

FIGURE 15–34 Shift housing control components for a five-speed main section of an eighteen-speed transmission; the shift blocks, yoke bar, and shift yokes correspond to the brackets, rods, and forks shown in Figure 15–33.

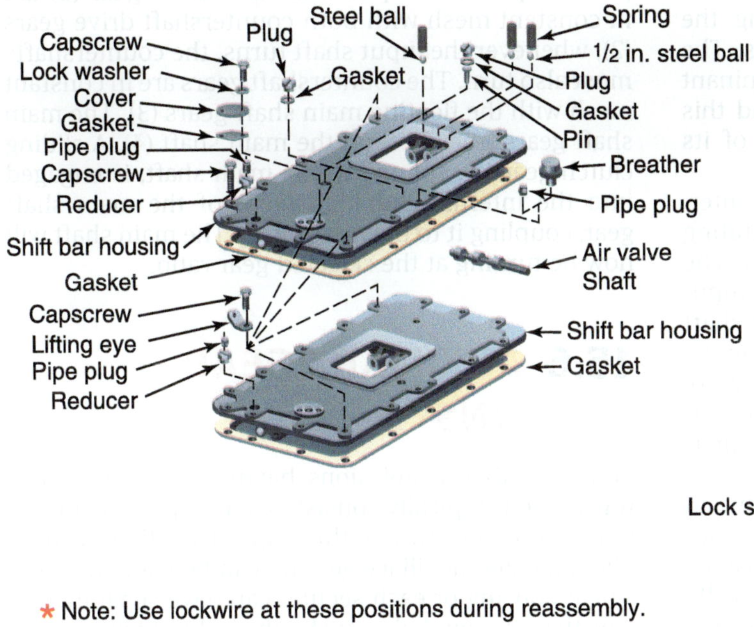

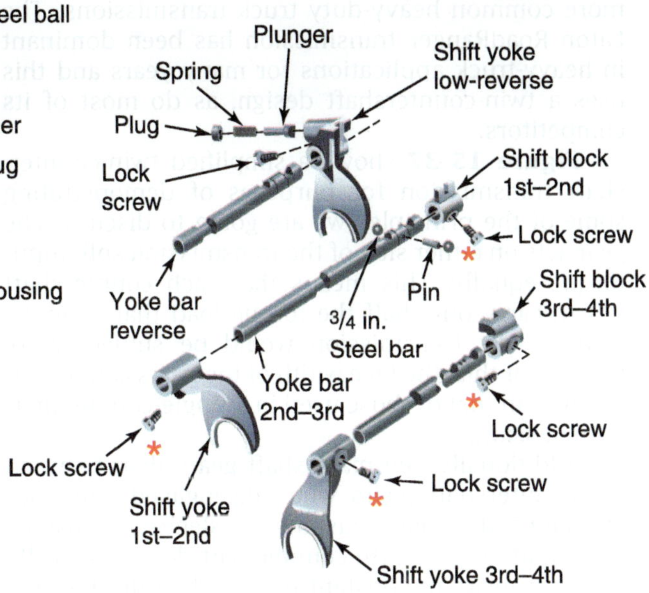

FIGURE 15–35 Shift fork engaged into a sliding clutch collar in a RoadRanger transmission

FIGURE 15–36 Shift rail and fork assembly in a RoadRanger transmission

SHOP TALK

In troubleshooting a transmission complaint of slipping out of gear, one of the first things you should check is the detent assemblies. Broken springs and seized detent balls can result in unwanted shift rail movement.

15.5 TWIN-COUNTERSHAFT TRANSMISSIONS

Now that the basics of gears, shift mechanisms, and powerflow have been discussed, a closer examination of countershaft transmissions equipped with both a main and auxiliary gear section should follow. Most heavy-duty truck standard transmissions are compounded, usually with a single auxiliary section. Some have a main box and two auxiliary sections.

Twin-countershaft transmissions having nine to eighteen forward speed ranges are among the more common heavy-duty truck transmissions. The Eaton RoadRanger transmission has been dominant in heavy-truck applications for many years and this uses a twin-countershaft design, as do most of its competitors.

Figure 15–37 shows a simplified twin-countershaft transmission for purposes of demonstrating some of the principles we are going to discuss. The gear sets on either side of the transmission split input torque equally. This means that each countershaft has to carry only half the torque load that a single-countershaft transmission would be subjected to. Because of this, the face width of the gears can be narrower than that of those used in a single-countershaft transmission.

Additionally, the main shaft gears float between the countershaft gears when disengaged, eliminating the need for gear bushings or sleeves. Transmissions that use the twin-countershaft design typically use spur gears in constant mesh. When disengaged, the main shaft gears freewheel around the main shaft because they are in constant mesh with the countershaft drive gears. The motion is not transferred to the actual shaft itself, however, until the sliding clutch gear is moved into engagement. The output shaft will then turn at the same speed as the main shaft gear. The sliding clutch gear that engages with the main shaft gear is typically splined to the main shaft.

Figure 15–38 shows a twin-countershaft transmission that looks more like something you would see in a truck; it illustrates the powerflow through the transmission when in low range. Note how the torque path is split. The input shaft and drive gear (1) are in constant mesh with both countershaft drive gears (2); whenever the input shaft turns, the countershafts must also turn. The countershaft gears are in constant mesh with the floating main shaft gears (3). The main shaft gears freewheel on the main shaft (4). A sliding clutch gear (5), splined to the main shaft, is engaged into the internal clutching teeth of the main shaft gear, coupling it to the main shaft. The main shaft will now be turning at the selected gear ratio.

15.6 AUXILIARY GEAR SECTIONS

Countershaft transmissions having ten or more forward speeds typically consist of a five-speed main section or box and a two- or three-speed auxiliary section. The main and auxiliary gearing can be contained in a single housing or each section can have separate casings that are coupled to each other. The auxiliary gearing is usually located behind the main box.

The main box shown in the ten-speed transmission shown in Figure 15–38 has a five-speed ratio; the auxiliary section that couples to it has two ranges or speeds. There is a direct (or high) range and a low range in this transmission. Shifts in the main section are effected mechanically by the driver using the shift stick. However, selection of the gears in the auxiliary

FIGURE 15–37 Simplified twin-countershaft gear train

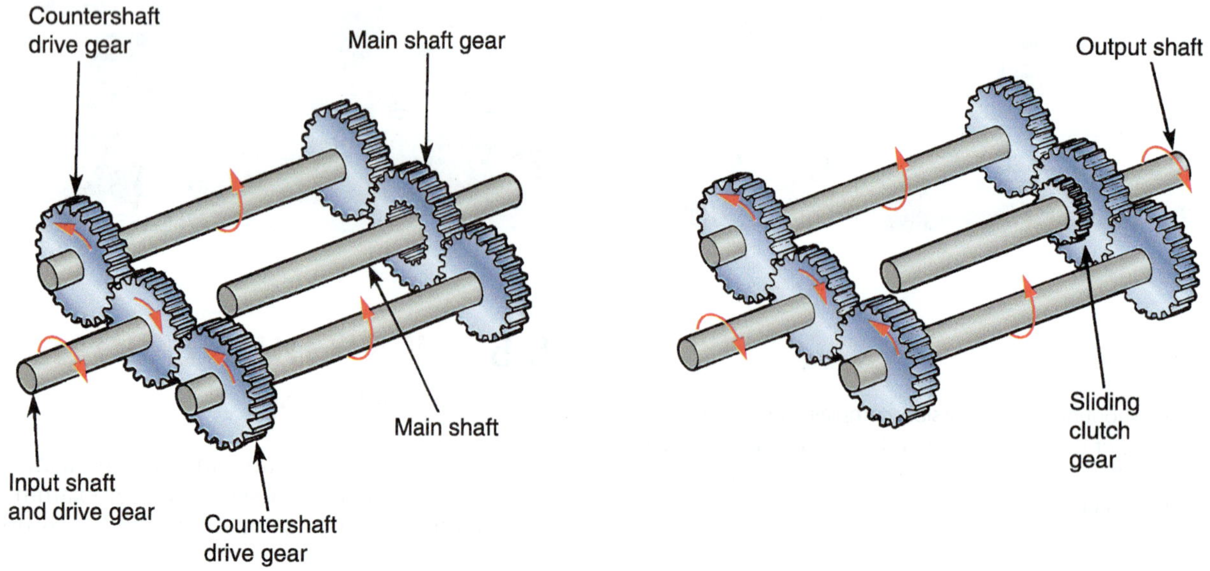

FIGURE 15–38 Low-range powerflow in a typical ten-speed transmission with a five-speed main section and a two-speed auxiliary section; the main section first-gear engagement is shown.

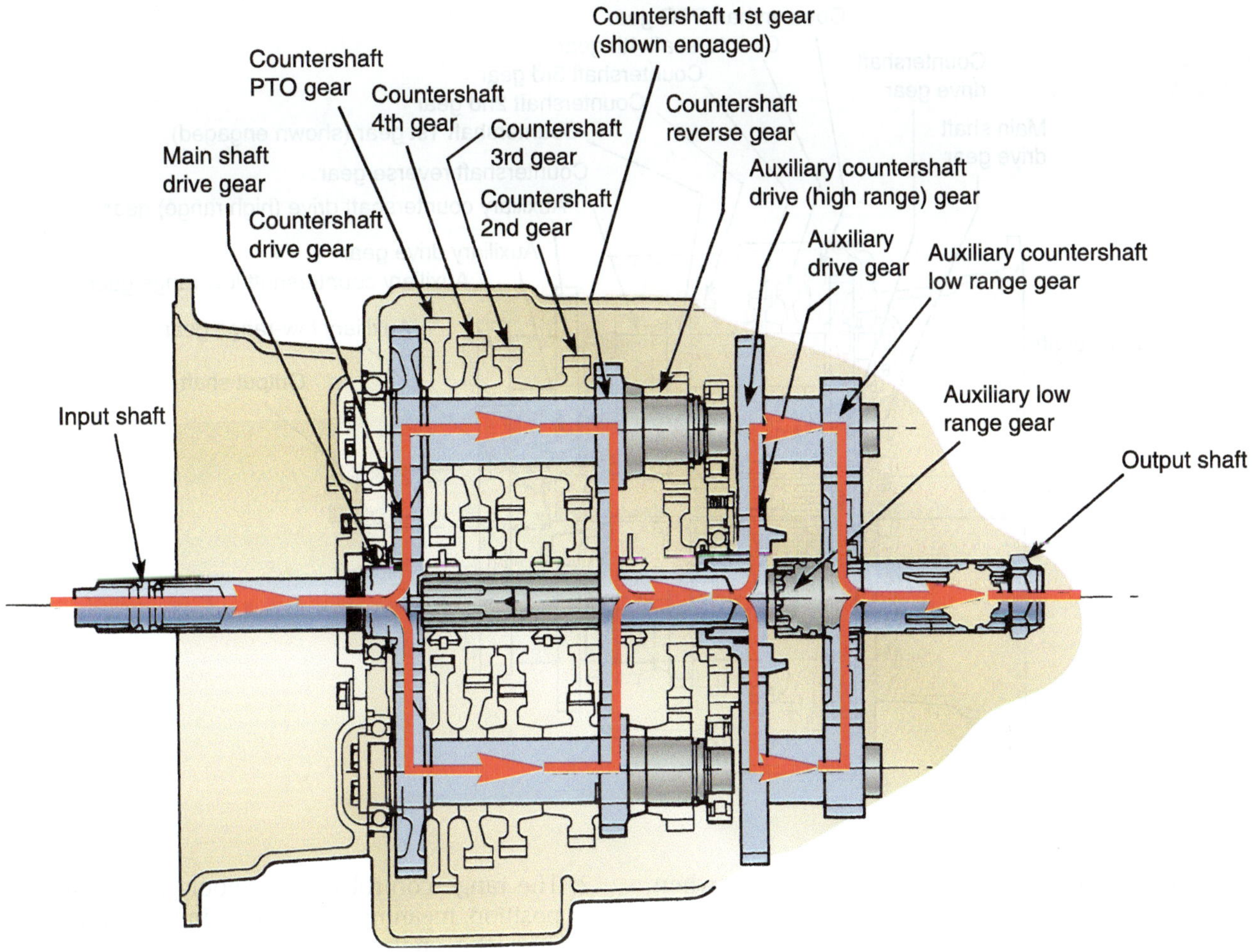

section is made by a driver-actuated, air-operated piston. The driver uses a pneumatic switch, usually located on the gear lever, that moves the auxiliary section into low- or high-range ratios. The driver controls this range selection mechanism through the use of a master control valve switch mounted on the gearshift tower in the operating cab. Follow the powerflow through the transmission in Figure 15–38; it is operating in low range. The following is true:

1. Torque from the engine flywheel is transmitted to the transmission input shaft.
2. Splines on the input shaft mesh with internal splines in the hub of the drive gear.
3. Torque is split between the two countershaft drive gears that mesh with the main shaft drive gear.
4. Torque is transferred by the two countershaft gears to whichever main shaft gear is engaged. Figure 15–38 shows first-gear engagement. Shifting and powerflows in the main section are similar to those in the five-speed transmission discussed earlier in this chapter under the heading Mechanical Shift Mechanisms. A notable difference is the use of a power take-off (PTO) gear located between the countershaft drive gears and fourth gears.
5. Internal splines in the hub of the main shaft gear transfer torque to the main shaft by means of a sliding clutch gear.
6. The main shaft transfers torque to the auxiliary drive gear through a self aligning coupling gear located in the hub of the auxiliary drive gear.
7. Torque is split between the two auxiliary countershaft drive gears.
8. Torque is delivered by the two countershaft low-range gears to the auxiliary low-range gear.
9. Torque is transferred to the output shaft through a self-aligning sliding clutch gear that locks the low-range gear to the output shaft.
10. The output shaft connects by means of a yoke to the driveline.

FIGURE 15–39 High-range powerflow in a typical ten-speed transmission with a five-speed main section and a two-speed auxiliary section; the main section first gear is engaged.

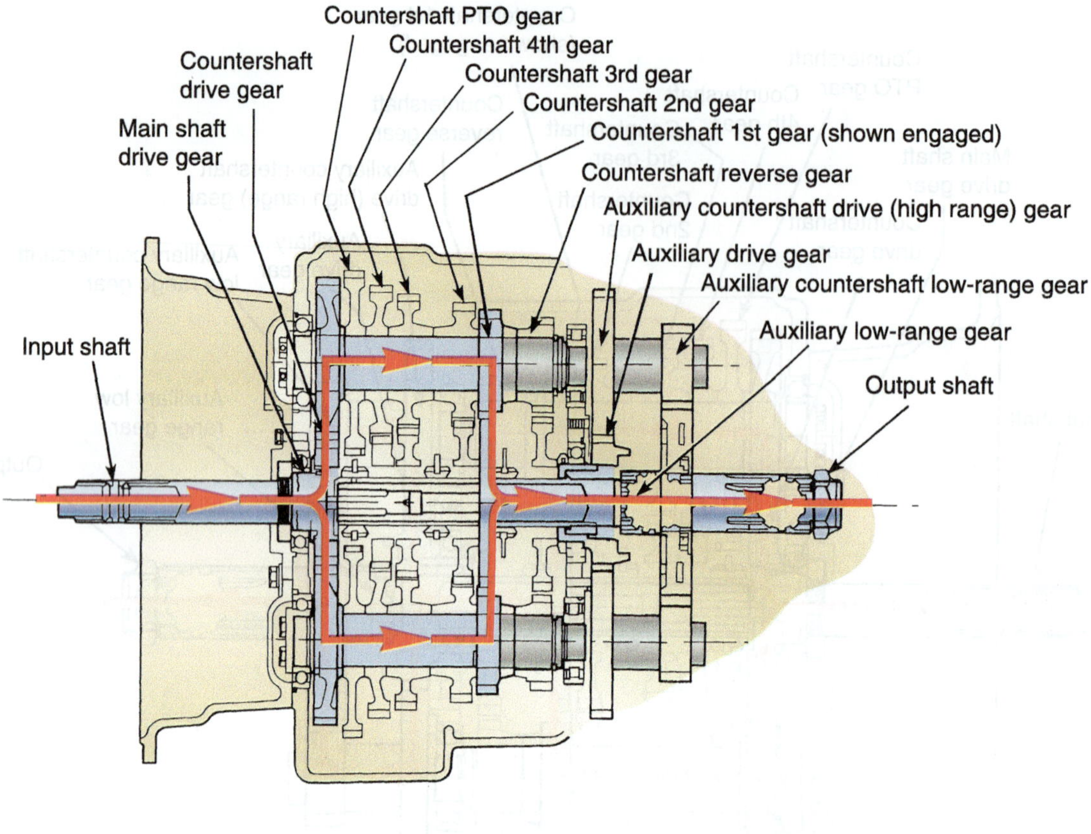

Figure 15–39 shows the same transmission when power is directed through the high-range (direct-drive) gearing of the auxiliary section. In this range, the sliding clutch gear locks the auxiliary drive gear to the output shaft. The low-range gear on the output shaft is now allowed to freewheel.

The ten forward speeds are selected by using a five-speed shifting pattern twice, the first time with the auxiliary section engaged in low gear or low range; the second time engaged in high gear or high range. By using the same shifting pattern twice, the shift lever position for sixth speed is the same as first, seventh the same as second, eighth the same as third, ninth the same as fourth, and tenth the same as fifth. **Figure 15–40** illustrates the gearshift lever pattern and range control button positions for this model of transmission. **Figure 15–41** shows what the shift knob looks like.

UPSHIFT SEQUENCE

From a standstill, accelerating to a highway cruise speed, you should perform the following sequence:

1. Start the engine and build air pressure to system pressure.
2. The range control button should be in the down position, meaning that you are in the low range.
3. Start the engine, fully depress the clutch, and shift into first gear. Accelerating, shift through second, third, fourth, and fifth gear ratios.
4. When in fifth and preparing to upshift, pull the range control button up, and then shift the gear lever to sixth speed. As the shift stick passes

FIGURE 15–40 An example of gearshift lever pattern and range control button positions

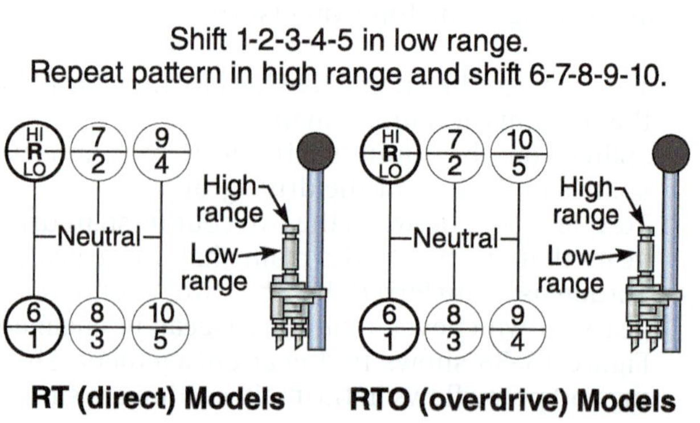

FIGURE 15–41 Shift knob on an Eaton Fuller ten-speed transmission

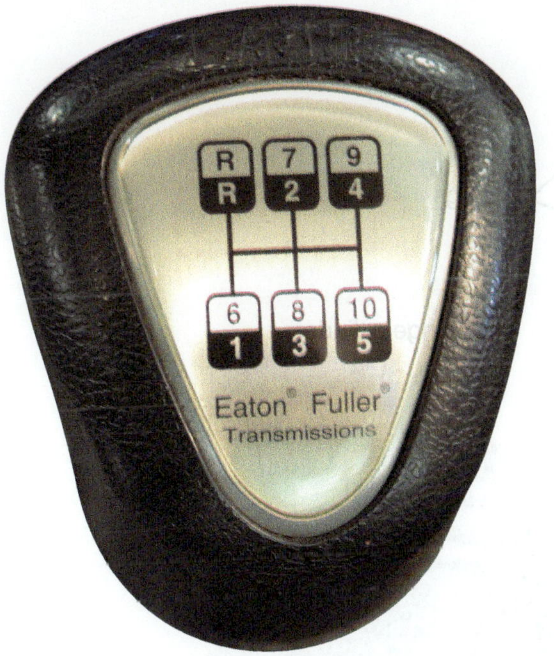

through the neutral position, the transmission will shift from low range to high range.
5. With the transmission in high range, shift progressively through seventh, eighth, and ninth to tenth.

DOWNSHIFT SEQUENCE

When progressively downshifting this transmission, you would use the following procedure:

1. Move the gear lever from tenth through each successive gear speed ratio until you are in sixth gear.
2. When in sixth and preparing to downshift, push the range control button downward, and then move the gear lever to fifth speed. Again, as the gear lever passes through the neutral position, the transmission pneumatics will make the shift from high to low range.
3. Now that the transmission is in low range, you can downshift through each of the four remaining gear ratios.

GENERAL OPERATING INSTRUCTIONS

When driving a truck with an auxiliary section, you should observe the following precautions:

- Use the range control button only as described.
- Do not attempt to shift from high range to low range at high speeds.
- Do not make range shifts with the truck moving in reverse gear.

Skip-shifting is possible on most standard transmissions. Depending on the load, a 10-speed transmission can most likely be shifted as a 6-, 7-, 8-, or 9-speed transmission by selectively bypassing certain gears in the shifting sequence. When skip-shifting, the range control button should be placed in the high range before the shift that passes fifth gear. When skip-shifting down, it is necessary to push the range control button down to low range before the shift that passes sixth.

HIGH-/LOW-RANGE SHIFT SYSTEMS

An air-operated auxiliary section gearshift system consists of the following:

- Air filter/regulator
- Slave valve
- Master control valve
- Range cylinder
- Fittings and connecting air lines

Figure 15–42 illustrates a typical air-operated gearshift control system used to engage high- and low-range gearing in the auxiliary section. Note the location of the range and splitter cylinders and how they connect with the control pneumatics.

Air Filter/Regulator

The **air filter/regulator assembly** (**Figure 15–43**) minimizes the possibility of moisture-laden air or impurities from entering the system. The regulator is used to reduce chassis system air supply pressure to the range valve and the **slave valve**. If you follow the air plumbing shown in **Figure 15–44**, you can see that air from the chassis air system is supplied to the air supply port on the air regulator. When filtered, the air is then routed to the air regulator. Transmission air pressure is typically regulated at between 57 and 62 psi. Next, the air passes through the ¼-inch supply air line and ⅛-inch overdrive (OD) range valve supply air line to the supply ports of the slave valve and range valve. Depending on the position of the gearshift-mounted range valve, air will pass through either the low-range air line or the high-range air line to the range shift cylinder.

Range shifts can be made only when the gearshift lever is in, or passing through, neutral. This means that the driver can preselect a range shift while in gear. As the gear lever is moved through neutral, the actuating plunger in the shift bar housing releases the slave valve, allowing it to move to the selected range position.

FIGURE 15–42 Components of a typical air-operated gearshift control system used for high- and low-range shifting

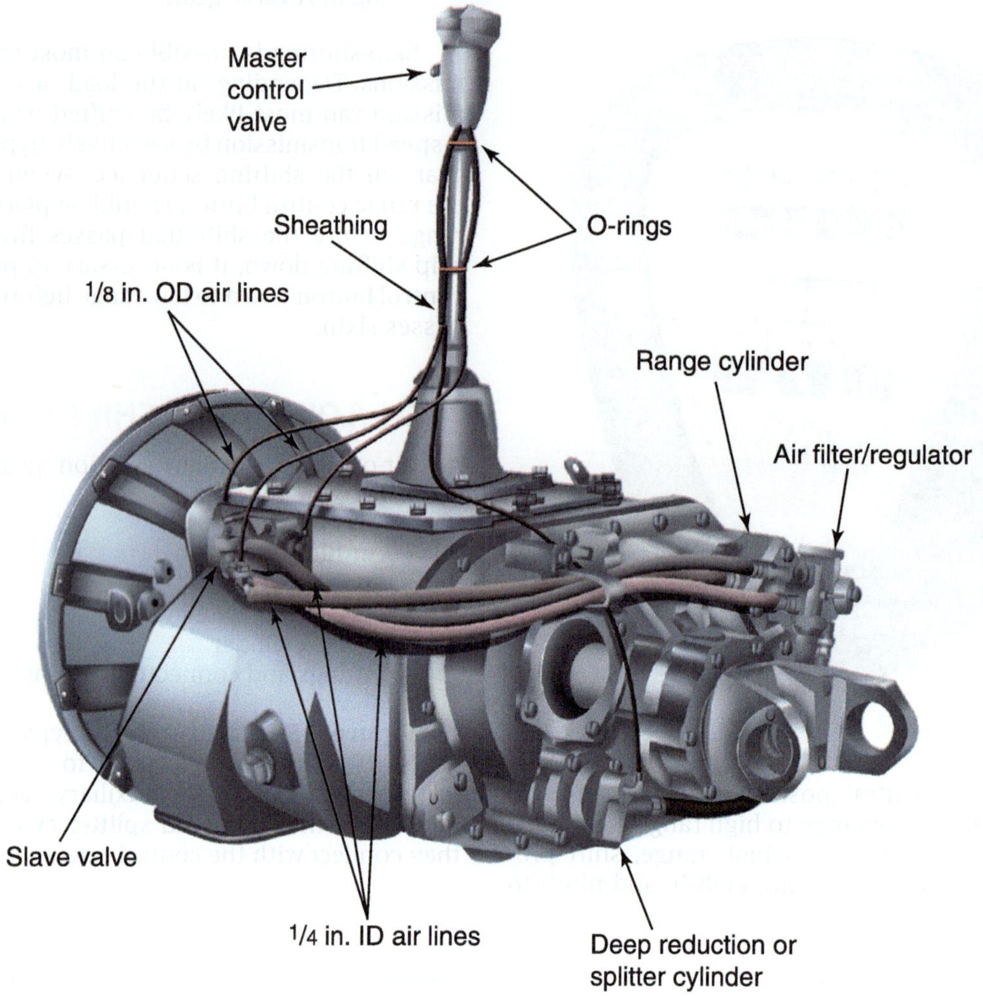

FIGURE 15–43 Air filter/regulator assembly

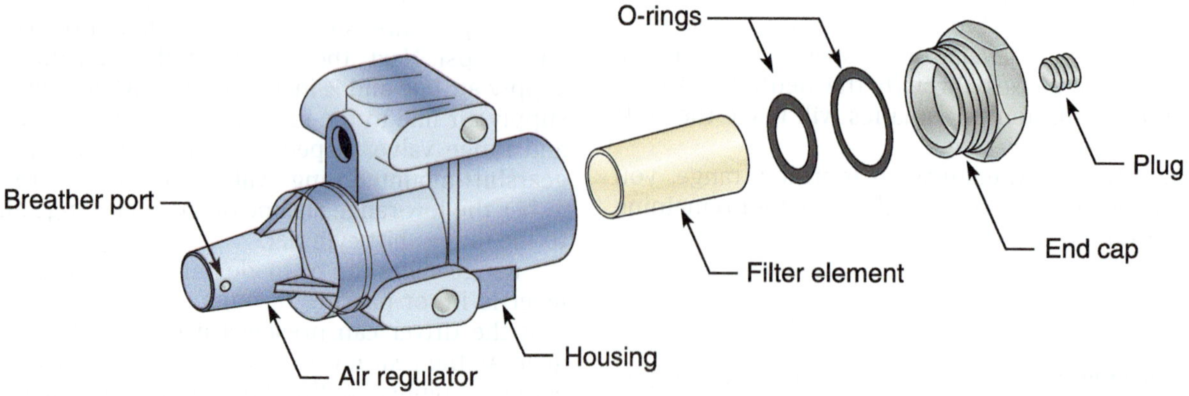

FIGURE 15–44 Range shift air system operation

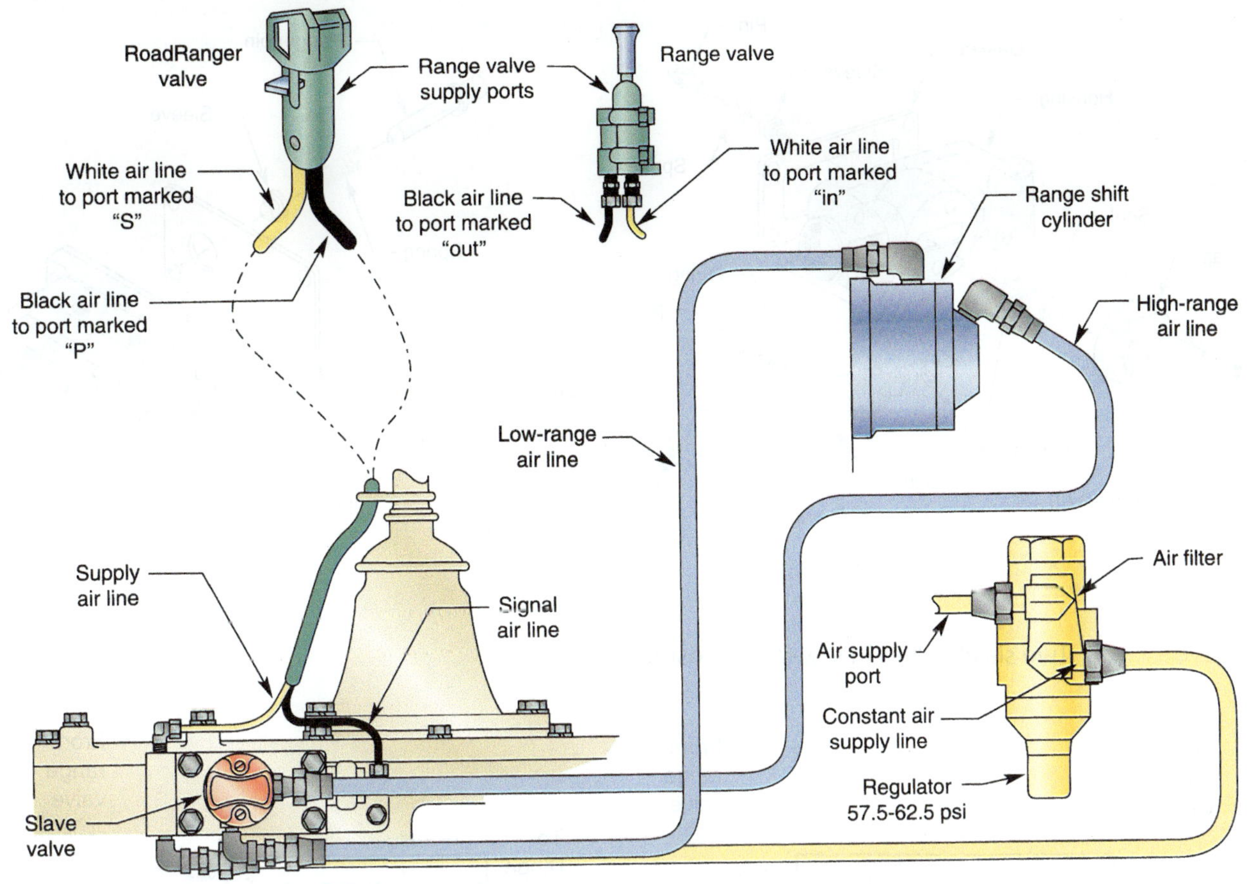

Slave Valve

The slave valve (**Figure 15–45**) can be of the piston or poppet type. The slave valve distributes inlet air pressure to both the low- and high-range air circuits that connect to the range cylinder. The piston (or plunger) controls when and where transmission air pressure is distributed. Slave valve operation (low range and high range) is illustrated in **Figure 15–46** and **Figure 15–47**.

An air valve shaft protruding from the shift bar housing prevents the actuating piston in the slave valve from moving while the gearshift lever is in any gear position. It releases the piston when the lever is moved to or through neutral (**Figure 15–48**).

Range Cylinder

The range cylinder is located at the auxiliary section at the rear of the transmission. In low range, air pressure is supplied from the slave valve to the low-range port or the range cylinder (**Figure 15–49**). This air pressure forces the range cylinder piston to the rear, engaging the low-range gearing. Air pressure behind the piston and in the high-range air line is exhausted from the slave valve. In high range, air pressure is supplied from the slave valve to the high-range port of the range cylinder cover. This air pressure forces the range cylinder piston forward, engaging the high-range gearing. The air pressure that acted on the front of the piston and in the lower range air line during low-range operation exhausts from the slave valve.

Range Valve

The master control valve is sometimes referred to as the range valve. Constant air pressure is supplied to the inlet port of the range valve (**Figure 15–50**) by the slave valve. In low range (control switch down), air routes through the valve slide and exits the outlet port. This air returns to the slave valve to the end port (or P-port), depending on the type of slave valve used.

In high range (control switch up), the valve slide prevents the air from passing through the range valve. Air pressure that was in the outlet line is now

FIGURE 15–45 Slave valves: (A) piston type; and (B) poppet type

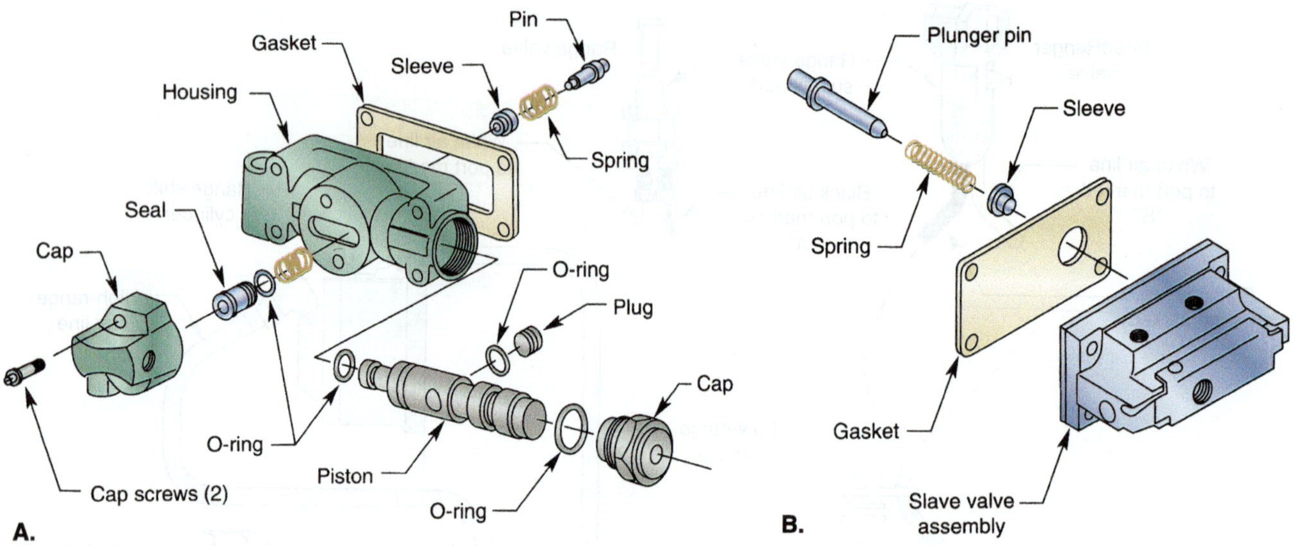

FIGURE 15–46 Two slave valves in low-range position

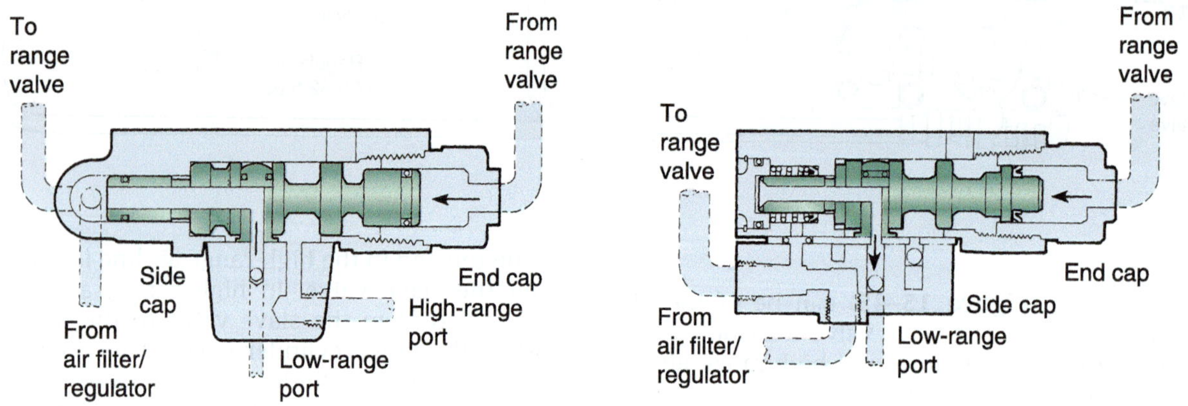

exhausted out the bottom of the range valve. This means that the transmission defaults to high range.

Additional Auxiliary Gearing

The auxiliary gear set can include a third-gear option in addition to the high- and low-range gears. This third gear can be used to provide overdrive or underdrive options.

- Overdrive option one: provides overdrive gear ratios in both low- and high-gear ranges. This design is used in eighteen-speed overdrive transmissions.
- Overdrive option two: provides additional overdrive gear ratios in the high-gear range only. This design is used in thirteen-speed overdrive transmissions.
- Underdrive option one: provides an underdrive gear ratio in the high-gear range only. This design is used in thirteen-speed underdrive transmissions.
- Underdrive option two: provides a deep reduction gear that produces a single additional low/low-gear ratio.

Split Shifting

The third gear of the auxiliary section is engaged or disengaged from the powerflow by means of an air-activated splitter shift system. The splitter control button is located on the gear lever.

FIGURE 15–47 Two slave valves in high-range position

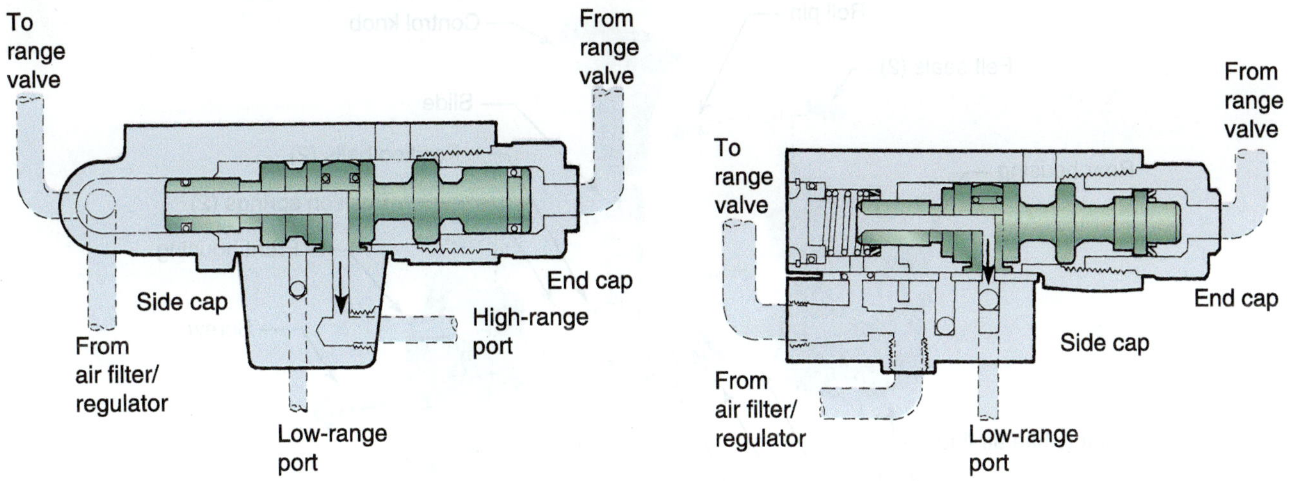

FIGURE 15–48 Components of the slave valve pre-selection system

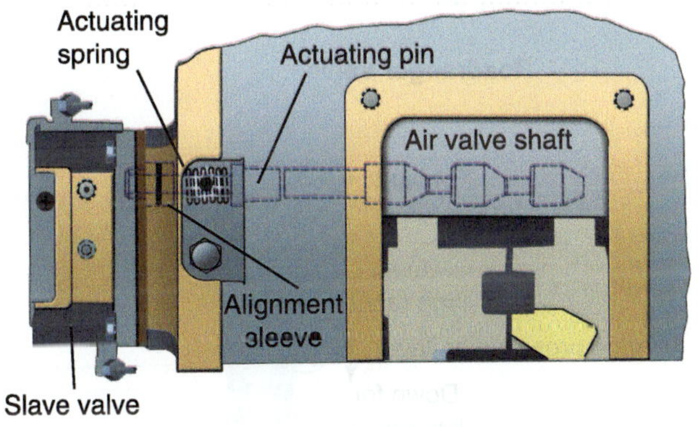

FIGURE 15–49 Range cylinder components that engage low gearing

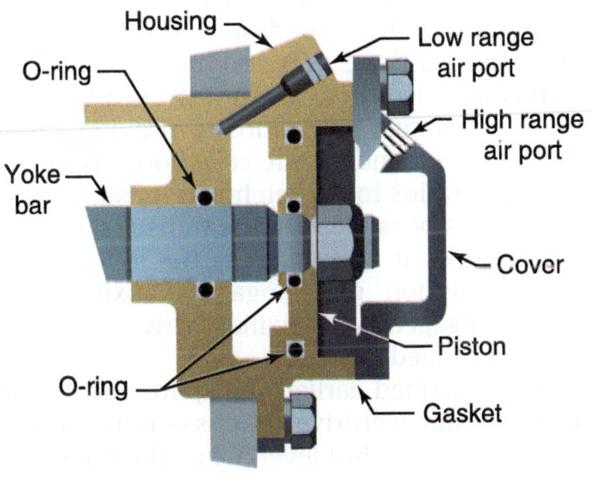

Figure 15–51 illustrates a typical splitter air system equipped with both the high-/low-range selector and splitter selector mounted on the gearshift lever. The system shown is used on typical thirteen-speed overdrive or underdrive transmissions. The splitter gear system in a thirteen-gear transmission is used only while in high range and splits the high-range gearing either into overdrive or underdrive ratios depending on the transmission model. Splitter systems used on eighteen-gear transmissions are used to split both high- and low-range gearing.

Splitter Cylinder Operation

In a basic splitter system, constant air from the air filter/regulator assembly is supplied to the splitter cylinder through a port on the cylinder cover (see Figure 15–42). This air supply acts on the front side of the shift piston in the cylinder. An insert valve installed in the cover provides the air required to move the splitter piston either forward or rearward. This movement is transferred to the yoke or shift fork bar and onto the shift fork (**Figure 15–52**). The movement of the yoke or shift fork engages a sliding clutch, which in turn locks the third auxiliary gear to the main output shaft.

When operating in either the high or low range, the air required to make the splitter selection is supplied to the gearshift lever valve (**Figure 15–53**) from the fitting at the S-port of the slave valve mounted on the transmission housing. When an overdrive (button forward) selection is made (**Figure 15–54A**), air is routed through the shift tower valve and is supplied to the left port of the cylinder cover of the splitter cylinder. When the splitter control button on the shift tower is in the direct or rear position (**Figure 15–54B**), the S-port of the shift tower valve

FIGURE 15–50 Master control valve (range valve) exploded view

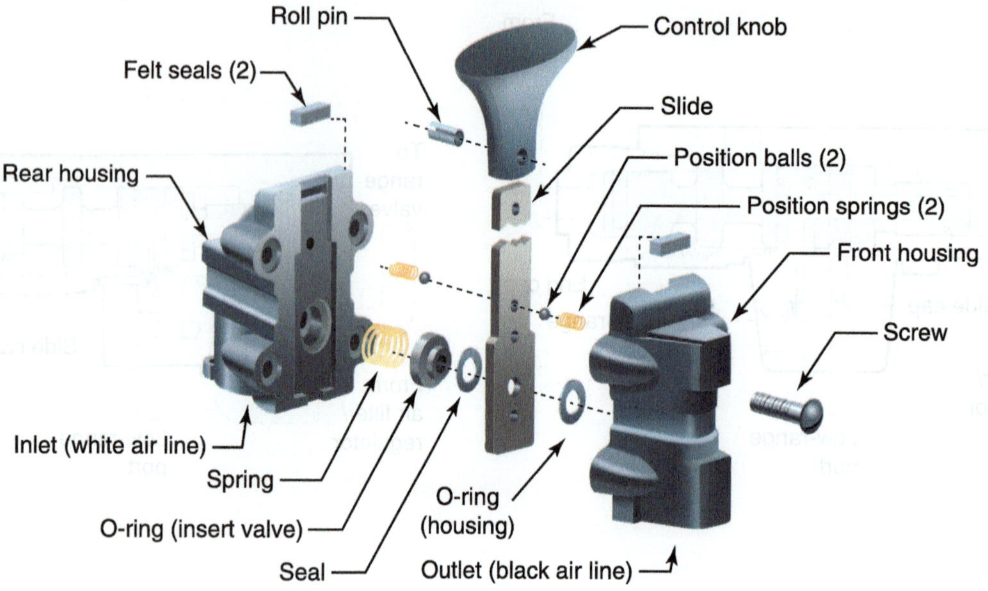

FIGURE 15–51 Shift tower using both air-activated high/low-range button and split shift button.

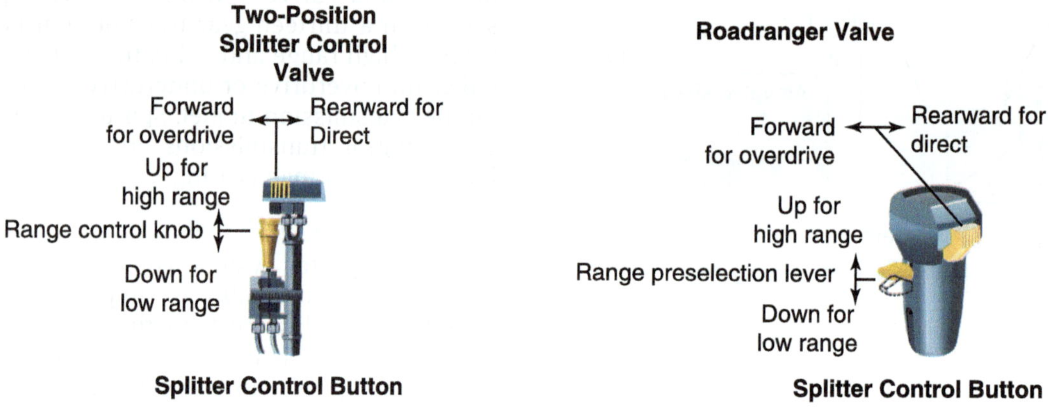

is closed and no air is supplied to the left port of the splitter cylinder cover.

EIGHTEEN-SPEED TRANSMISSIONS

Figure 15–55 illustrates a transmission with eighteen forward speeds. The main or front gearing is a five-speed transmission. In this particular transmission, main gears are referred to as low, first, second, third, and drive, rather than first, second, third, fourth, and drive. The auxiliary section contains an overdrive gear known as the front drive gear. The high-range gear in this particular auxiliary section is called the rear-drive gear. There is also a low-range gear.

Torque transferred through the front-drive gear produces an overdrive condition; torque transferred through the rear-drive gear produces a direct drive. The direct and overdrive gearing is used to split the high- and low-range gear ratios to provide a greater number of speed and torque selections. **Table 15–1** lists the gear ratios for the eighteen forward and four reverse speeds possible with the transmission shown in Figure 15–55. It also includes the following data: main transmission gear engaged, auxiliary range (high/low) engaged, and auxiliary drive gear (direct/overdrive) engaged.

As we described earlier, the splitter shift system for the direct and overdrive options is air-controlled by a button on the gearshift lever. When the transmission

FIGURE 15–52 Splitter piston components and operation

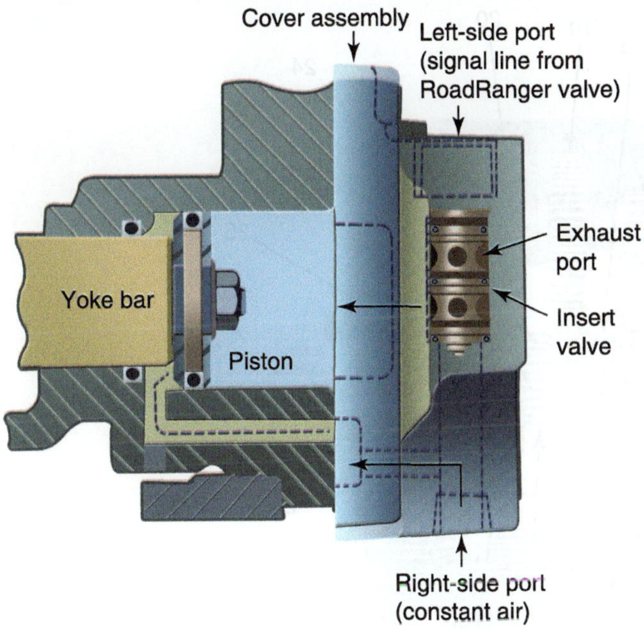

FIGURE 15–53 Exploded view of shift lever range selection and splitter valve assembly

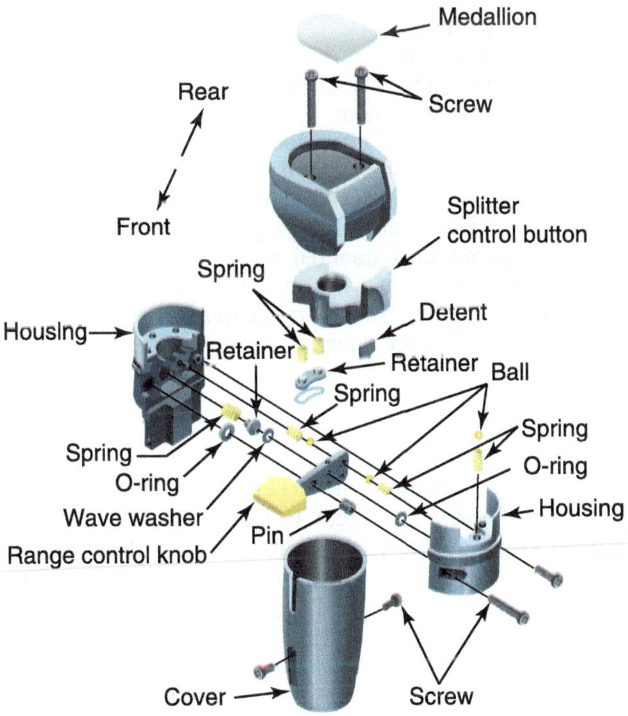

is in the high or low ranges, the splitter control button can be moved forward to shift the auxiliary transmission from a direct to overdrive condition.

FIGURE 15–54 Split shift operation

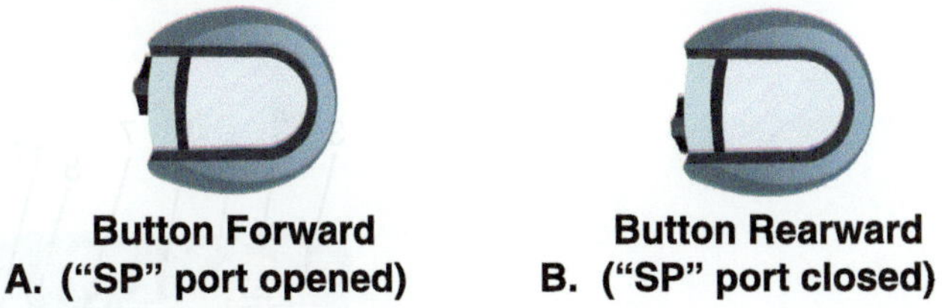

Low-Range, Direct Powerflow

In these gears (first, third, fifth, seventh, ninth, and #1 Reverse in Table 15–1), powerflow is through the rear auxiliary drive gear (direct) and auxiliary low-range gear (**Figure 15–56**). A complete powerflow description follows:

1. Torque from the engine is transferred to the transmission's input shaft.
2. Splines on the input shaft engage internal splines in the hub of the main drive gear.
3. Next, torque is split between the two countershaft drive gears.
4. Torque is delivered through the twin countershafts to the mating countershaft gears of the engaged main shaft gear.
5. Internal clutching teeth in the hub of the engaged main shaft gear transfer torque to the main shaft through the appropriate sliding clutch.
6. The main shaft transmits torque directly to the rear auxiliary drive gear by means of its engaged sliding clutch.
7. The rear auxiliary drive gear splits torque between the two auxiliary section countershaft gears.
8. Torque is delivered through both auxiliary section countershafts to the engaged reduction gear on the output shaft.
9. Torque is next transmitted to the output shaft by means of the sliding clutch.
10. The output shaft delivers torque to the driveline.

Low-Range, Overdrive Powerflow

In these gear selections (second, fourth, sixth, eighth, tenth, and #2 Reverse in Table 15–1), powerflow is through the front auxiliary gear (overdrive) and the auxiliary low-range gear (**Figure 15–57**). Powerflow through the main box is identical to that outlined in steps 1 through 5 under the previously described Low-Range, Direct Powerflow description. However, the sliding clutch between the front and rear auxiliary drive gears moves forward when the driver engages the splitter control button on the shift tower. The sliding clutch locks the auxiliary front-drive gear to the range main shaft. The powerflow through the auxiliary section of the transmission is as follows:

1. The front auxiliary drive gear splits torque between the two auxiliary countershaft drive gears.

FIGURE 15–55 Components of an eighteen-speed transmission with a five-speed main section and a three-speed auxiliary section

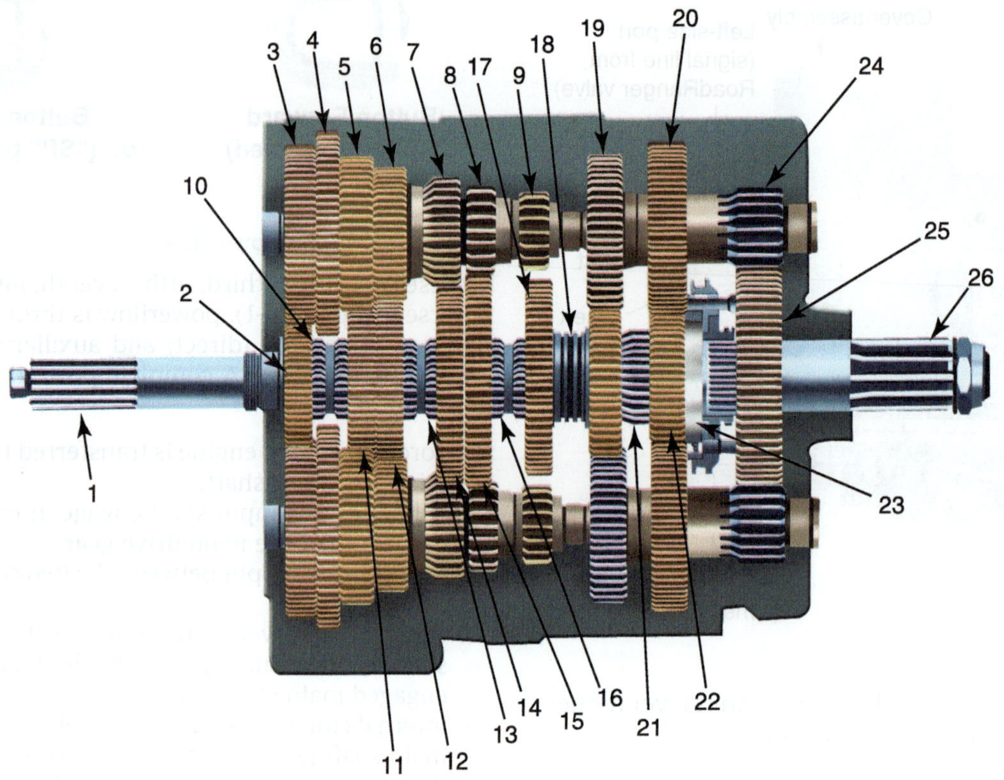

1. Input shaft
2. Main drive gear
3. Countershaft drive gear
4. Countershaft PTO gear
5. Countershaft 3rd gear
6. Countershaft 2nd gear
7. Countershaft 1st gear
8. Countershaft low gear
9. Countershaft reverse gear (idler gear not shown)
10. Main drive gear/3rd gear sliding clutch
11. Main shaft 3rd gear
12. Main shaft 2nd gear
13. 2nd gear/1st-gear sliding clutch
14. Main shaft 1st gear
15. Main shaft low gear
16. Low gear/reverse sliding clutch
17. Main shaft reverse gear
18. Front auxiliary drive gear (connected to main shaft)
19. Auxiliary countershaft front drive gear
20. Auxiliary countershaft rear-drive (HI range) gear
21. Auxiliary drive gear/rear-drive gear sliding clutch
22. Auxiliary rear-drive gear (HI range)
23. Hi-range/low-range synchronizer
24. Auxiliary countershaft low-range gear
25. Auxiliary low-range gear
26. Output shaft

2. Torque is delivered through both countershafts to the low-range gear on the range main shaft or output shaft. Once again, the high/low synchronizer is used to lock this low-range or reduction gear to the output shaft.
3. Torque is transferred to the output shaft through the sliding clutch of the synchronizer.
4. Torque is delivered to the driveline as low-range overdrive.

High-Range, Direct Powerflow

In these gear selections (eleventh, thirteenth, fifteenth, seventeenth, and #3 Reverse in Table 15–1), powerflow is through the rear auxiliary drive gear (**Figure 15–58**). This gear is locked to the auxiliary output shaft by the front/rear sliding clutch and the high side of the high-/low-range synchronizer. This locks the rear auxiliary drive gear directly to the output shaft. All other gears in the auxiliary section are freewheeling.

High-Range, Overdrive Powerflow

In these gear selections (twelfth, fourteenth, sixteenth, eighteenth, and #4 Reverse in Table 15–1), powerflow is through the front auxiliary drive gear, which is locked to the output shaft by the front/rear sliding

TABLE 15–1 Gear Ratios for an Eighteen-Speed Fuller Transmission

Gear Selected	Main Section Gear Engaged	Range High/Low	Direct or Overdrive	Gear Ratio ___:1
1	Low	Low	Direct	14.71
2	Low	Low	Overdrive	12.45
3	First	Low	Direct	10.20
4	First	Low	Overdrive	8.62
5	Second	Low	Direct	7.43
6	Second	Low	Overdrive	6.21
7	Third	Low	Direct	5.26
8	Third	Low	Overdrive	4.45
9	Fourth (main)	Low	Direct	3.78
10	Fourth (main)	Low	Overdrive	3.20
11	First	High	Direct	2.70
12	First	High	Overdrive	2.28
13	Second	High	Direct	1.94
14	Second	High	Overdrive	1.64
15	Third	High	Direct	1.39
16	Third	High	Overdrive	1.18
17	Fourth (main)	High	Direct	1.00
18	Fourth (main)	High	Overdrive	0.85
#1 Reverse	Reverse	Low	Direct	14.71
#2 Reverse	Reverse	Low	Overdrive	12.45
#3 Reverse	Reverse	High	Direct	3.89
#4 Reverse	Reverse	High	Overdrive	3.29

Note: The relative speed of the PTO gear to input rpm is 0.696 for both the right and bottom PTO gears on the countershaft.

FIGURE 15–56 Low-range direct powerflow in an eighteen- speed twin-countershaft transmission.

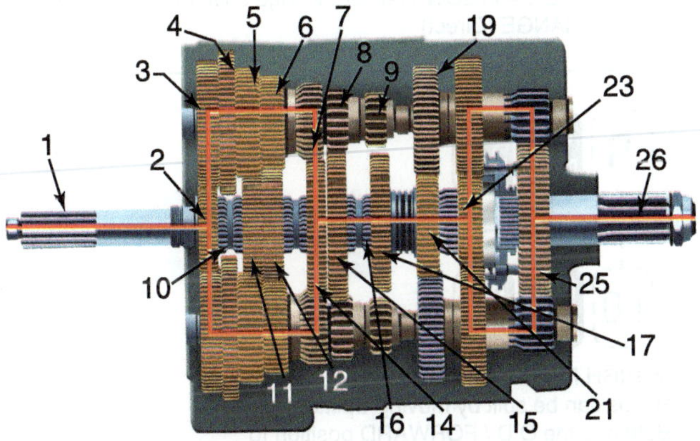

Note: See Figure 15–55 to interpret number codes.

FIGURE 15–57 Low-range overdrive powerflow in an eighteen-speed twin-countershaft transmission.

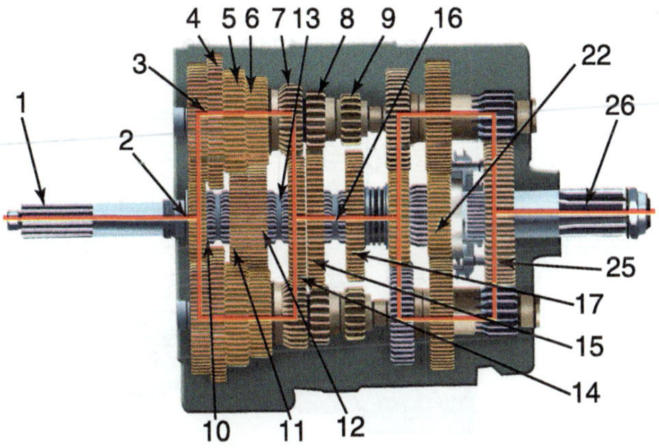

Note: See Figure 15–55 to interpret number codes.

FIGURE 15–58 High-range direct powerflow in an eighteen-speed twin-countershaft transmission

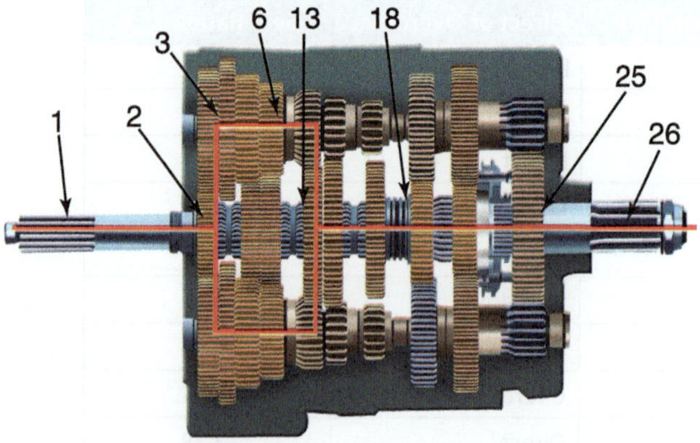

Note: See Figure 15–55 to interpret number codes.

clutch (**Figure 15–59**). This splits torque between the two auxiliary drive gears. Torque is then delivered through both auxiliary countershafts to the engaged rear auxiliary drive gear. This rear auxiliary drive gear is locked to the output shaft by the high side of the high-/ low-range synchronizer. Powerflow is through the rear gear to the synchronizer sliding clutch and out through the output shaft to the driveline.

THIRTEEN-SPEED TRANSMISSIONS

The thirteen-speed transmission illustrated in **Figure 15–60** is similar to the eighteen-speed transmission described previously in many ways. Once again, the main or front gearing is a five-speed transmission. The auxiliary section contains (moving from front to back) a high-range gear, a low-range gear, and an overdrive gear. In some models, this overdrive gear is replaced with an underdrive gear.

FIGURE 15–59 High-range overdrive powerflow in an eighteen-speed twin-countershaft transmission.

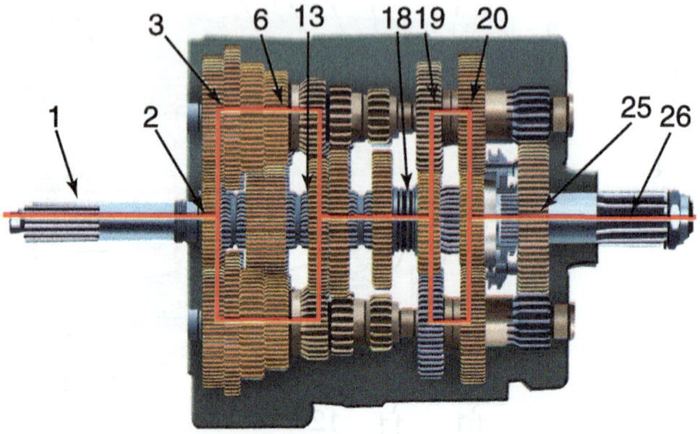

Note: See Figure 15–55 to interpret number codes.

FIGURE 15–60 Components of a thirteen-speed transmission with a five-speed main section and two-speed auxiliary section shown in low-range, direct drive.

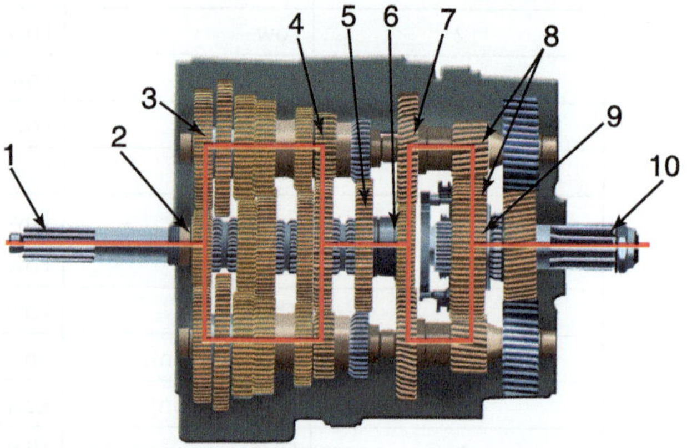

Note: See Figure 15–55 to interpret number codes.

As shown in **Figure 15–61**, the first five gear ratios (low, first, second, third, and fourth) are found with the splitter control button in the rearward (direct drive) position and the range selector in its low-range (down) position. Torque is delivered along both countershafts to the engaged low-range gear on the range

FIGURE 15–61 The shift pattern for the thirteen-speed transmission that is shown in Figure 15–60.

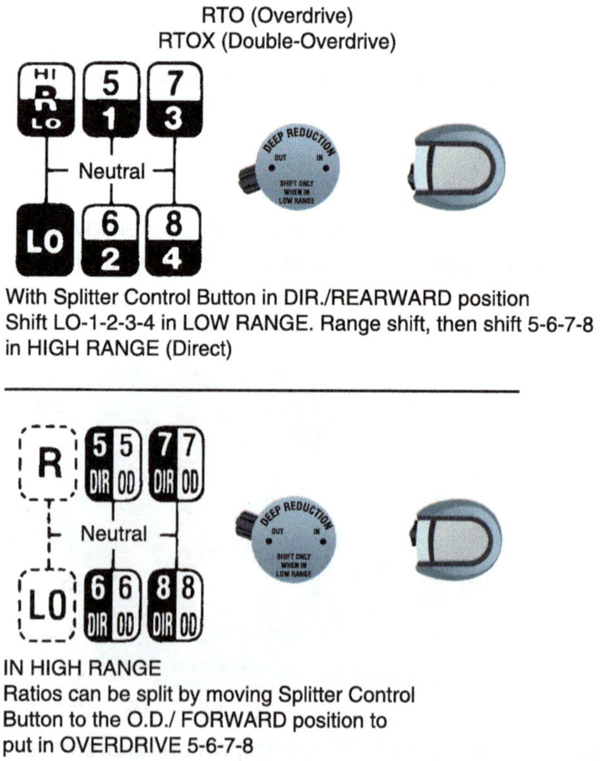

FIGURE 15–62 High-range direct powerflow in a thirteen- speed twin-countershaft transmission

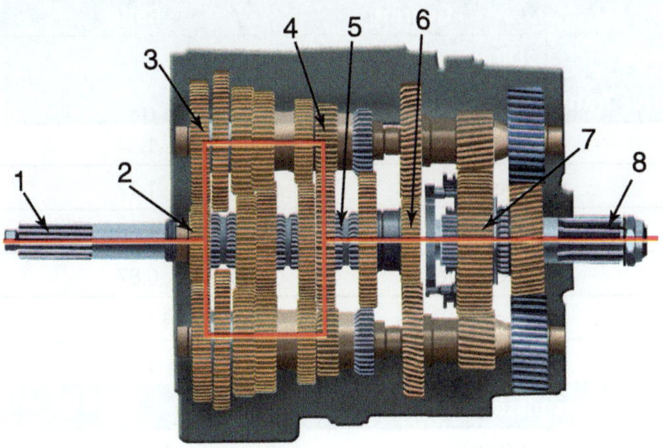

Note: See Figure 15–55 to interpret number codes.

FIGURE 15–63 High-range overdrive powerflow in a thirteen-speed twin-countershaft transmission

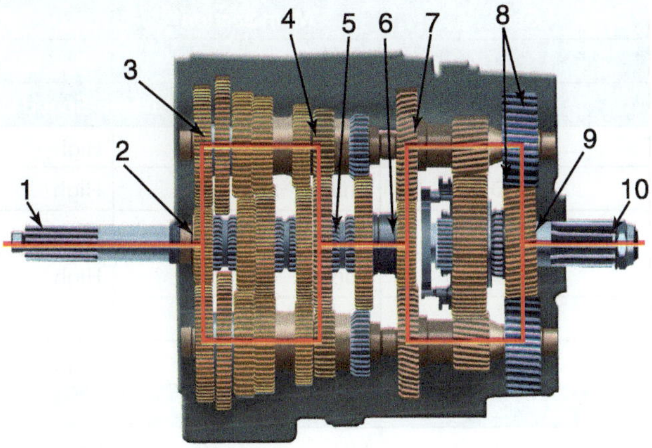

Note: See Figure 15–55 to interpret number codes.

mainshaft or output shaft. This creates low-range powerflows through the auxiliary gearing for each of the five speeds of the main section as shown in Figure 15–60.

The driver shifts to the high range by pulling up on the range selector. This action moves a sliding clutch that locks the auxiliary drive gear directly to the range main shaft or output shaft. Torque is delivered through the range main shaft and/or output shaft as high-range direct powerflows for the next four gear ratios—fifth, sixth, seventh, and eighth (**Figure 15–62**).

While in the high range only, the fifth, sixth, seventh, and eighth gear ratios can be "split" by moving the splitter control button on the cab shift lever to the OD or forward position. Activating the splitter shift system in this way moves a sliding clutch that locks the overdrive splitter gear in the auxiliary section to the output shaft. Torque is delivered along both auxiliary countershafts to the auxiliary overdrive gears to the output shaft overdrive gear and out through the output shaft (**Figure 15–63**). High-range overdrive powerflows are possible for four gear ratios: overdrive fifth, overdrive sixth, overdrive seventh, and overdrive eighth.

As you can see in **Table 15–2**, the fifth gear overdrive gear ratio falls between the fifth and sixth gear high-range direct ratios. So if the driver wished

TABLE 15–2 Gear Ratios for Thirteen-Speed and Deep Reduction Transmissions (Typical)

Gear	Main Section Gear Engaged	Range Low or High	Additional Auxiliary Gearing Direct or Special Gear	Ratio
Thirteen-Speed Overdrive Transmission				
LOW	Low	Low	Direct	12.56
1	First	Low	Direct	8.32
2	Second	Low	Direct	6.18
3	Third	Low	Direct	4.54
4	Fourth	Low	Direct	3.38
5	First	High	Direct	2.46
5 OD	First	High	Overdrive	2.15
6	Second	High	Direct	1.83
6 OD	Second	High	Overdrive	1.60

(continued)

TABLE 15–2 (continued)

Gear	Main Section Gear Engaged	Range Low or High	Additional Auxiliary Gearing Direct or Special Gear	Ratio
Thirteen-Speed Overdrive Transmission				
7	Third	High	Direct	1.34
7 OD	Third	High	Overdrive	1.17
8	Fourth	High	Direct	1.00
8 OD	Fourth	High	Overdrive	0.87
R1	Reverse	Low	Direct	13.13
R2	Reverse	High	Direct	3.89
Thirteen-Speed Underdrive Transmission				
LOW	Low	Low	Direct	12.56
1	First	Low	Direct	8.32
2	Second	Low	Direct	6.18
3	Third	Low	Direct	4.54
4	Fourth	Low	Direct	3.38
5 UD	First	High	Underdrive	2.86
5	First	High	Direct	2.46
6 UD	Second	High	Underdrive	2.12
6	Second	High	Direct	1.83
7 UD	Third	High	Underdrive	1.56
7	Third	High	Direct	1.34
8 UD	Fourth	High	Underdrive	1.16
8	Fourth	High	Direct	1.00
R1	Reverse	Low	Direct	13.13
R2	Reverse	High	Direct	3.89
Deep Reduction 8LL (8 1 2 Speed) Transmission				
LOW-LOW	Low	Low	Deep Reduction	14.05
LOW	Low	Low	Direct	9.41
1	First	Low	Direct	6.23
2	Second	Low	Direct	4.62
3	Third	Low	Direct	3.40
4	Fourth	Low	Direct	2.53
5	First	High	Direct	1.83
6	Second	High	Direct	1.36
7	Third	High	Direct	1.00
8	Fourth	High	Direct	0.74
R1	Reverse	Low	Deep Reduction	14.69
R2	Reverse	Low	Direct	9.83
R3	Reverse	High	Direct	2.89

to shift progressively through all gears, the correct sequence would be:

1. High-range direct fifth gear with the splitter button in the rear position (disengaged)
2. Double-clutch, engaging the splitter button into the forward position while in neutral and shifting into high-range fifth overdrive
3. Double-clutch, disengaging the splitter button into the rear position while in neutral and shifting into high-range direct sixth gear

Thirteen-Speed, Underdrive

Thirteen-speed underdrive transmissions operate on the same principles except that the underdrive gear ratio for a particular gear is selected before the direct drive ratio. Underdrive ratios for a typical thirteen-speed transmission are also listed in Table 15–2.

Deep-Reduction Transmissions

Figure 15–64 illustrates the shift pattern for a ten-speed transmission. The forward gear ratios are low-low, low, and first through eighth. Low-low is a special deep-reduction gear for maximum torque. It is used to produce maximum drivetrain torque for high-load, standing starts, using a deep-reduction gear in the auxiliary section. This low-low gear is engaged by activating a split shifter or dash-mounted deep-reduction valve. It can be operated in the low range only.

Constant air from the air filter/regulator assembly is supplied to the reduction cylinder at the port on the right side of the cylinder cover. With the deep-reduction lever in the "Out" position, the valve is opened and air is supplied to the center port of the cylinder cover, moving the reduction piston forward to disengage the deep-reduction gearing. When the lever is moved to the "In" position, the valve is closed and no air is supplied to the center port. Constant air from the air filter/regulator assembly moves the reduction piston rearward to engage the reduction gearing.

As shown in **Figure 15–65**, the powerflow is routed through both countershafts and countershaft deep-reduction gears to the output shaft deep-reduction gear, which is locked to the output shaft by

FIGURE 15–64 Shift pattern for a ten-speed transmission with a deep reduction gear in the auxiliary housing

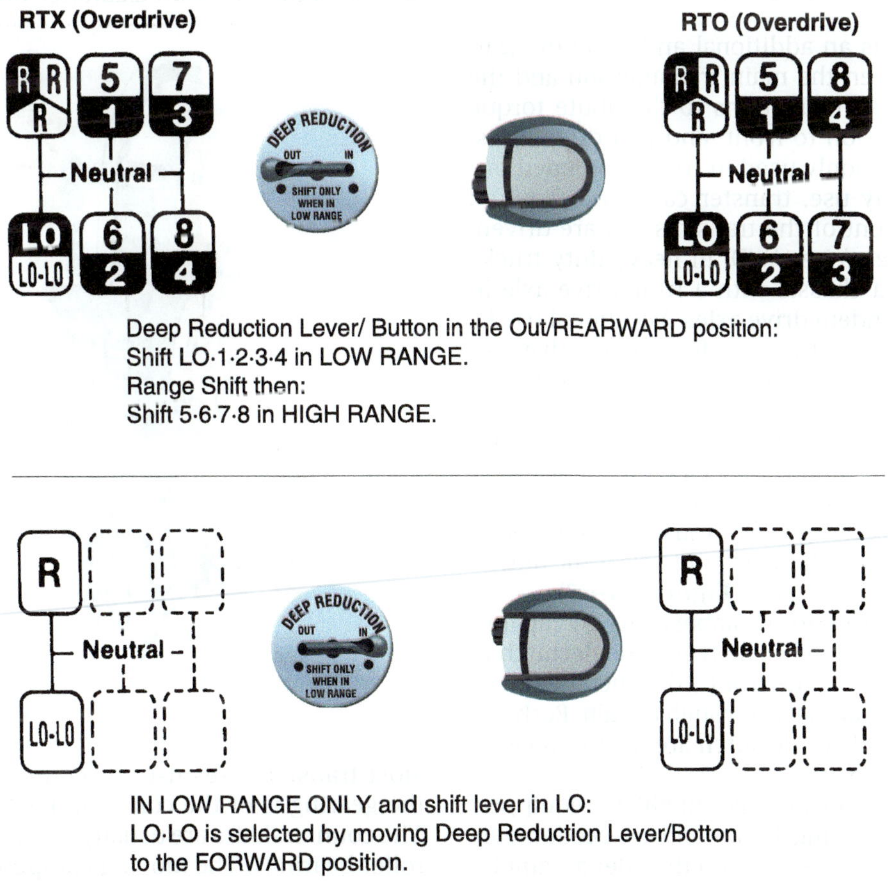

FIGURE 15–65 Deep reduction (low-low) powerflow in a ten-speed transmission

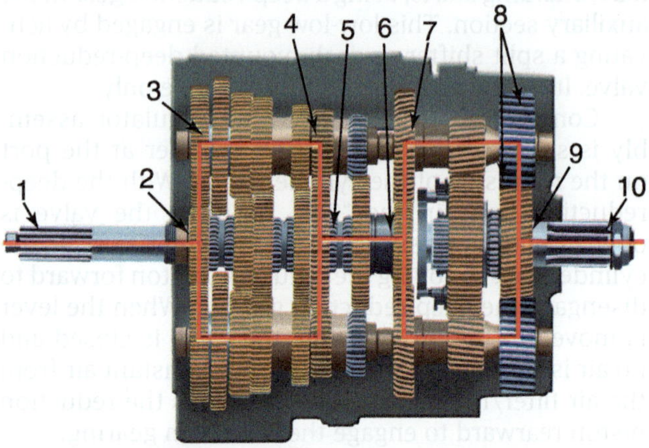

Note: See Figure 15–55 to interpret number codes.

the sliding clutch. In shifting from low-low to low, the driver double-clutches, releasing the split shifter and moving to low-range low. Low through fourth gears are low-range gear ratios. The driver then range-shifts into high range for gears five through eight.

15.7 TRANSFER CASES

A **transfer case** is an additional and separate gearbox located between the main transmission and the vehicle drive axles. It functions to distribute torque from the transmission to front- and rear-drive axles. Although not commonly used in trucks intended primarily for highway use, transfer cases are required when axle(s) in front of the transmission are driven. The term *all-wheel drive* (AWD) in heavy-duty trucks usually refers to a chassis with a front-drive axle in addition to rear tandem drive axles. Vocational trucks use these three-axle drive configurations that are essential in some on/off and off-highway applications. **Figure 15–66** shows a typical AWD configuration.

Transfer cases can transfer drive torque directly using a 1:1 gear ratio or can be used to provide low-gear reduction ratios additional to those in the transmission. The drop box design of a transfer case housing permits its front drive shaft output to clear the underside of the main transmission (**Figure 15–67**).

Most transfer cases are available with PTO capability and front axle declutch. The front axle declutch is used to option-drive to the front axle when negotiating steep grades or slippery or rough terrain. Both the PTO and front axle drive declutch are driver engaged by dedicated shift levers.

In addition, a transfer case might be equipped with an optional parking brake and a speedometer drive gear that can be installed on the idler assembly.

FIGURE 15–66 Location of a transfer case on a typical (A) 4 × 4 and (B) 6 × 6 drive arrangement

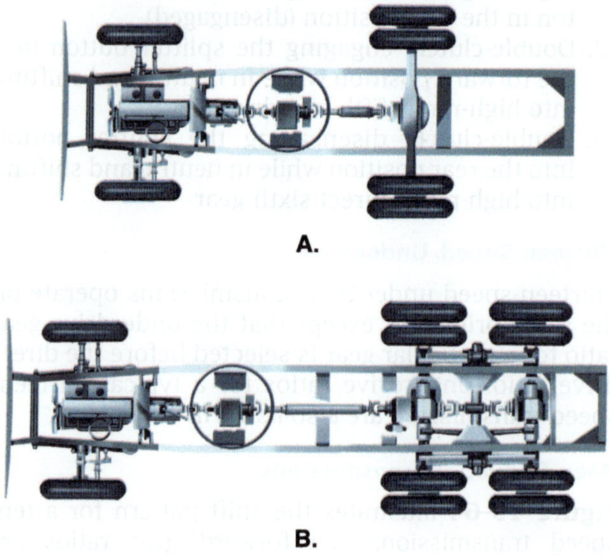

FIGURE 15–67 Single-speed, direct drive transfer case for front-wheel-drive 4 × 2 vehicles; drop box design allows for shaft clearance under the engine.

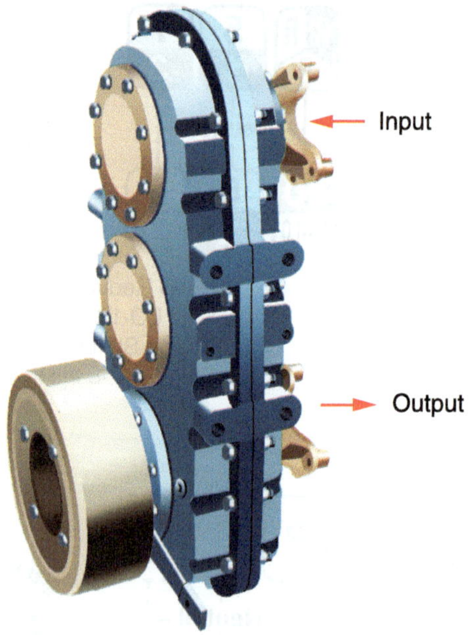

Most transfer cases use the countershaft design and the gearing is of the constant-mesh, helical-cut type. The countershafts are usually mounted in ball or taper roller bearings. The moving components of a transfer

FIGURE 15–68 Three-shaft design transfer case

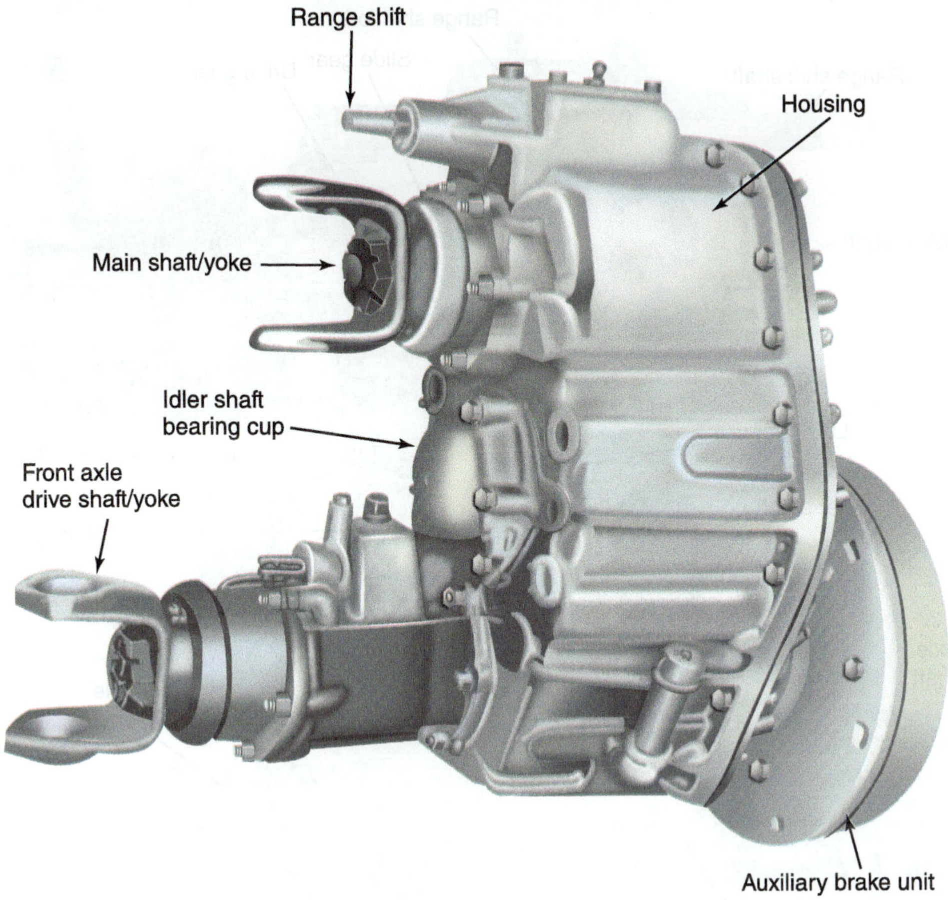

case are splash-lubricated by oil in the housing during operation. Some transfer cases use an auxiliary oil pump externally mounted to the transfer case.

Transfer cases may also be equipped with a driver- controlled, air-actuated differential lockout to improve traction under extreme conditions. A conventional transfer case is shown in **Figure 15–68**. This unit is a two-speed, three-shaft design. It is commonly adapted to incorporate a PTO and front axle declutch capability. A cutaway view of a fully equipped, three-shaft transfer case is shown in **Figure 15–69**. Note the means used to effect range shifts and the PTO output.

Another type of transfer case is the cloverleaf four-shaft design as shown in **Figure 15–70**. This two-speed, four-shaft design can also be adapted to incorporate a PTO and a mechanical-type auxiliary brake. Sectional views of a cloverleaf-type transfer case are shown in **Figure 15–71**. As can be seen in the sectional views, most transfer cases use simple gear designs; most have access panels to allow the technician to inspect and service the unit. The torque range window of transfer cases varies with application but you should note that this is specified to the transmission, not the engine, output torque range.

POWER TAKE-OFF UNITS

A variety of accessories on heavy-duty trucks require an auxiliary drive. Auxiliary drive can be sourced directly from the engine or by means of the transmission or transfer case. Auxiliary drive systems on trucks are usually known as PTOs, and that is the term we use in this text. When a PTO is coupled to the transmission or transfer case it is an assembly such as the one shown in **Figure 15–72**, and a typical installation is shown in **Figure 15–73**.

The PTO is simply a means of using the chassis engine to power accessories, eliminating the need for an additional auxiliary engine. There are six basic types of PTOs classified by their installation location or drive source:

- Side mount
- Split shaft
- Top mount

FIGURE 15–69 Cutaway view of a conventional-type transfer case (three-shaft design)

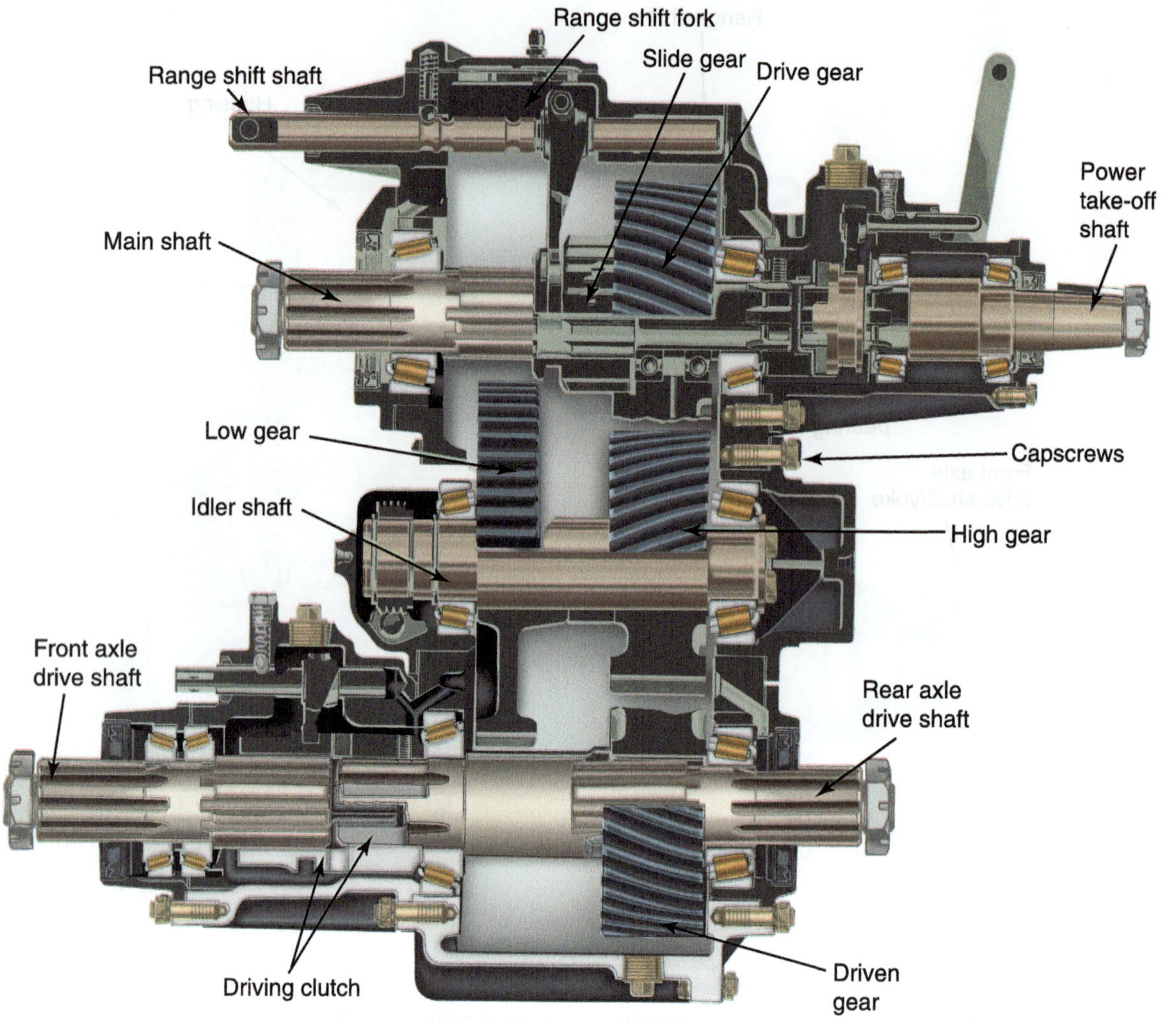

FIGURE 15–70 Cloverleaf design transfer case

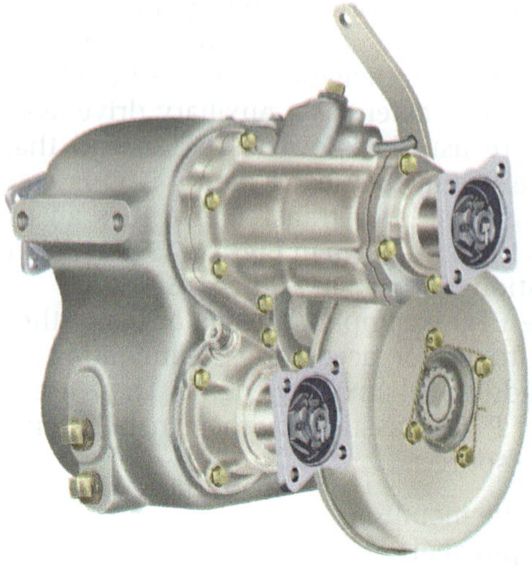

- Countershaft
- Forward crankshaft driven
- Rear crankshaft or flywheel driven

A side-mount PTO is bolted to the side of the main transmission and is the most common type found on trucks. A split-shaft PTO transmits torque from the chassis drive shaft and is located behind the transmission; split-shaft PTOs require special mounting to the chassis frame. A top-mount PTO is usually found mounted to the top of an auxiliary transmission. A countershaft PTO is mounted behind the transmissions and when used replaces the bearing cap located at the rear of one countershaft on a twin-countershaft transmission.

The forward crankshaft–driven PTO is driven by gearing at the front of the engine crankshaft. It is used in applications that require auxiliary power while the truck is in motion, such as cement mixers. Clutch-type

FIGURE 15–71 Sectional views of a cloverleaf design transfer case

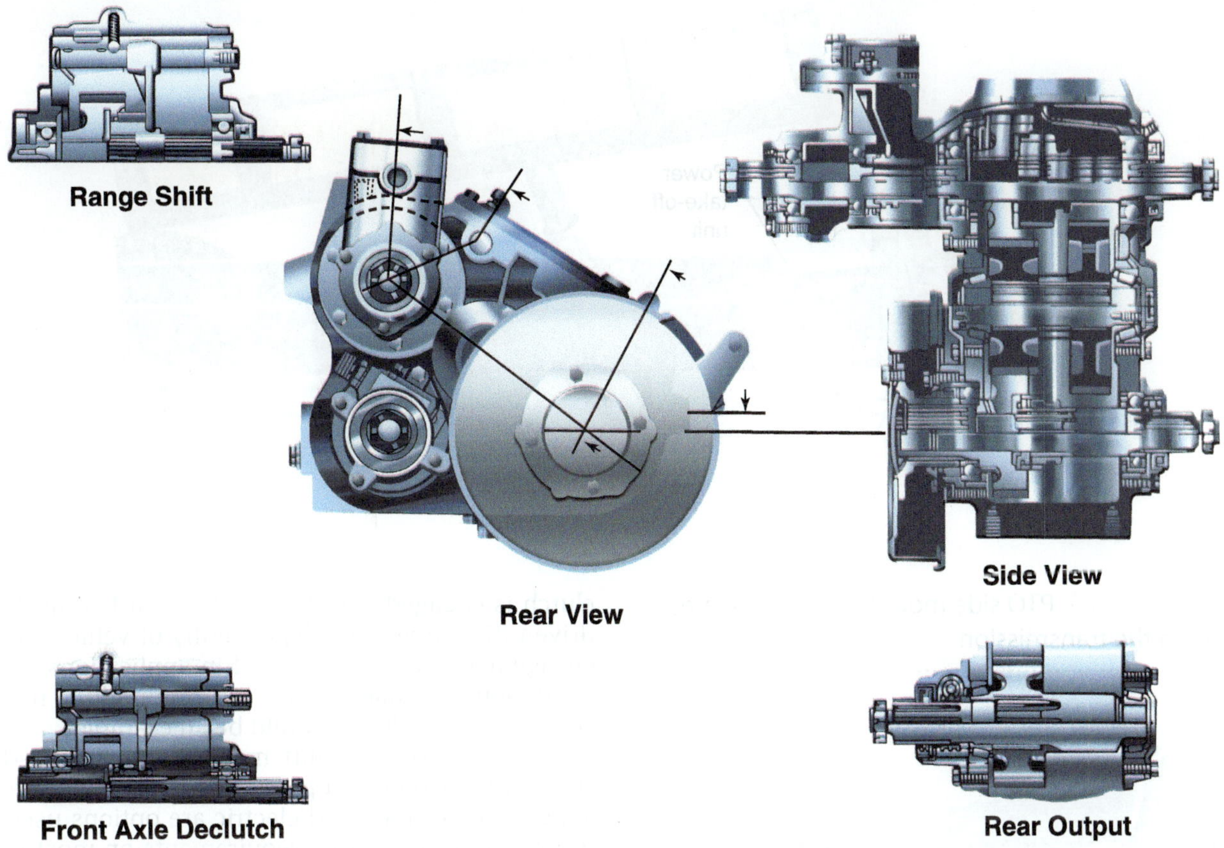

FIGURE 15–72 PTO unit essentially serves as an extension of the transmission.

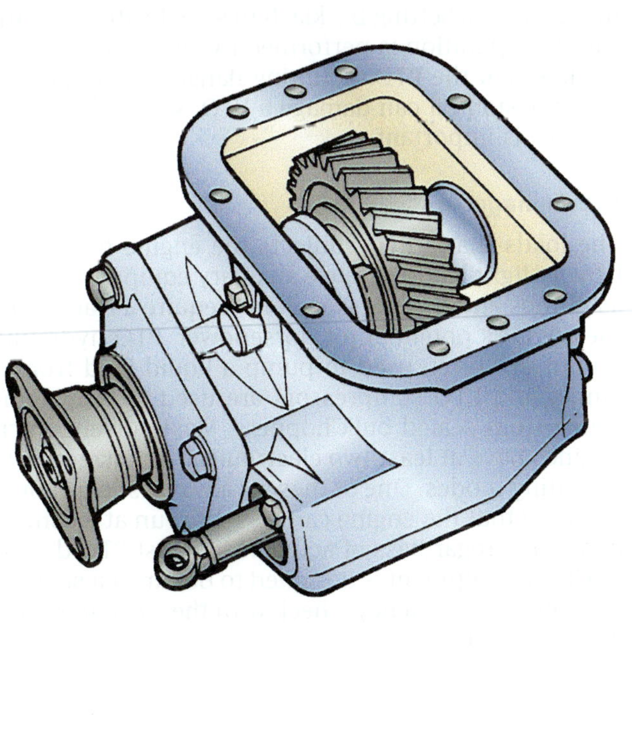

crankshaft-driven PTOs are used so that engagement/disengagement can take place while the engine is running. A flywheel PTO is sandwiched between the bell housing and the transmission. Like the forward crankshaft-driven PTO, a flywheel PTO permits continuous operation.

The objective of a PTO is to provide driving torque to auxiliary equipment such as pumps and compressors. The driven equipment can be mounted either directly to the PTO or indirectly using a small drive shaft. The PTO input gear is placed in constant mesh with a gear in the truck transmission (**Figure 15–74**). Also, refer to Figure 15–38 and Figure 15–55 earlier in this chapter. You can see that the PTO couples into these transmissions by means of a dedicated PTO gear on the transmission countershaft. Any rotation of the PTO countershaft gear drives the PTO input gear. The on/off capability required by a PTO is provided within the PTO assembly.

Gears in most PTO units can be either spur or helical. Establishing the correct mesh between the PTO drive gear and its partner in the transmission is critical. Too much or too little backlash can produce

FIGURE 15–73 Typical installation of a PTO unit

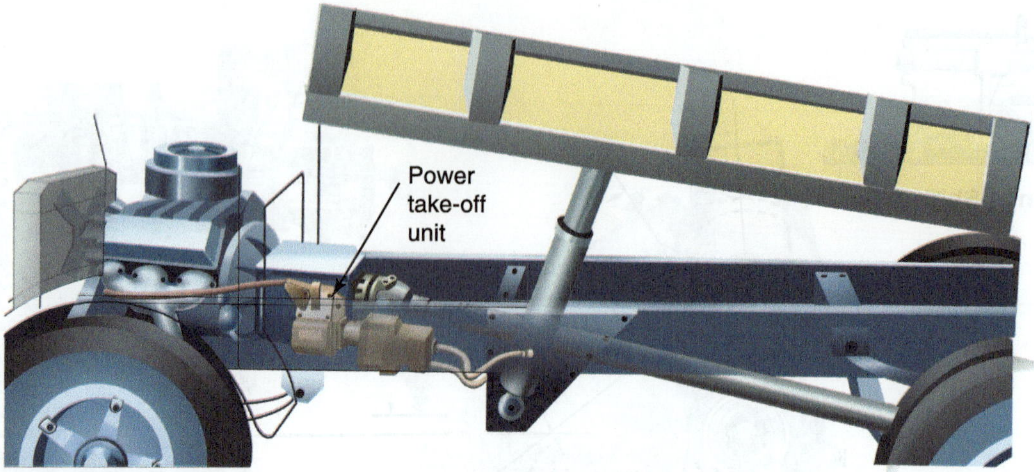

FIGURE 15–74 PTO side mount installation interface with the transmission

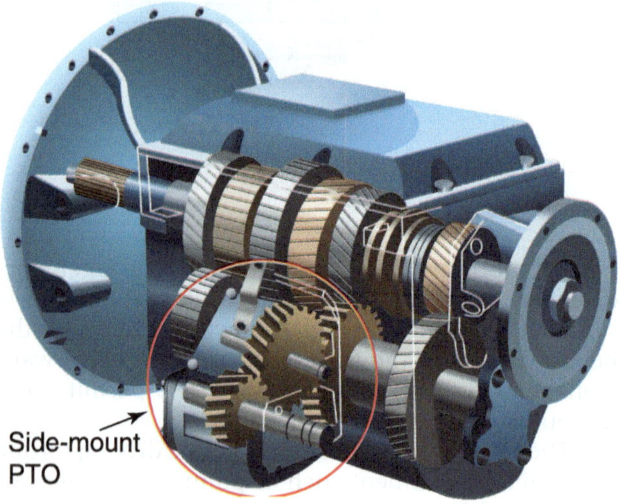

problems. Gear ratio is also critical in PTO operation. Gear ratio must be set to the torque capacity and operating speed required of the driven equipment. An exploded view of a typical PTO unit is shown in **Figure 15–75**.

As previously discussed, some PTOs function whenever the vehicle engine is running. These can operate at a fixed ratio of engine speed, whereas others vary depending on which gear the transmission is in. Other PTOs are designed to work only when the transmission has been shifted to the neutral position. All transmission-mounted PTOs are clutch dependent in that they only transmit torque when the vehicle clutch is engaged. Flywheel and forward crankshaft-driven PTOs operate independently of vehicle clutch engagement, so they are used in applications where continuous auxiliary power is required; a mobile cement mixer cylinder would be an example.

Several types of shift mechanisms are used to connect the PTO with operator controls. Cable, lever, air, electric-over-air, and electric are options used on trucks. Because the PTO requirements on most highway trucks are modest—usually a hydraulic pump or compressor—the cable-type control system is most commonly used. Some current units are equipped to use electronic-over-electric PTO controls.

Most PTOs feature simple designs and rugged construction. Care should be taken to properly mount a PTO. Setting backlash to specification is critical; this operation is performed by setting the gasket thickness at the PTO mounting flange. Contaminated transmission oil can damage a PTO, so regular maintenance is important.

PTO Management

The loads that a PTO subjects the engine to tend to be very light loads because auxiliary equipment, such as pumps and compressors, only require a fraction of the engine's potential for horsepower. PTO hydraulic pumps are often used to pump a liquid load from a tanker and PTO compressors are used to blow light solids from sealed bulk hoppers. Most current truck engines have at least two electronically managed PTO operating modes. One of these is always isochronous, meaning that the engine can be set to run at a continuous rpm, regardless of actual load. Most PTO-driven auxiliary equipment is designed to be run at a set rpm for optimum efficiency; check with the OEM literature before operating.

FIGURE 15–75 Exploded view of a typical PTO unit

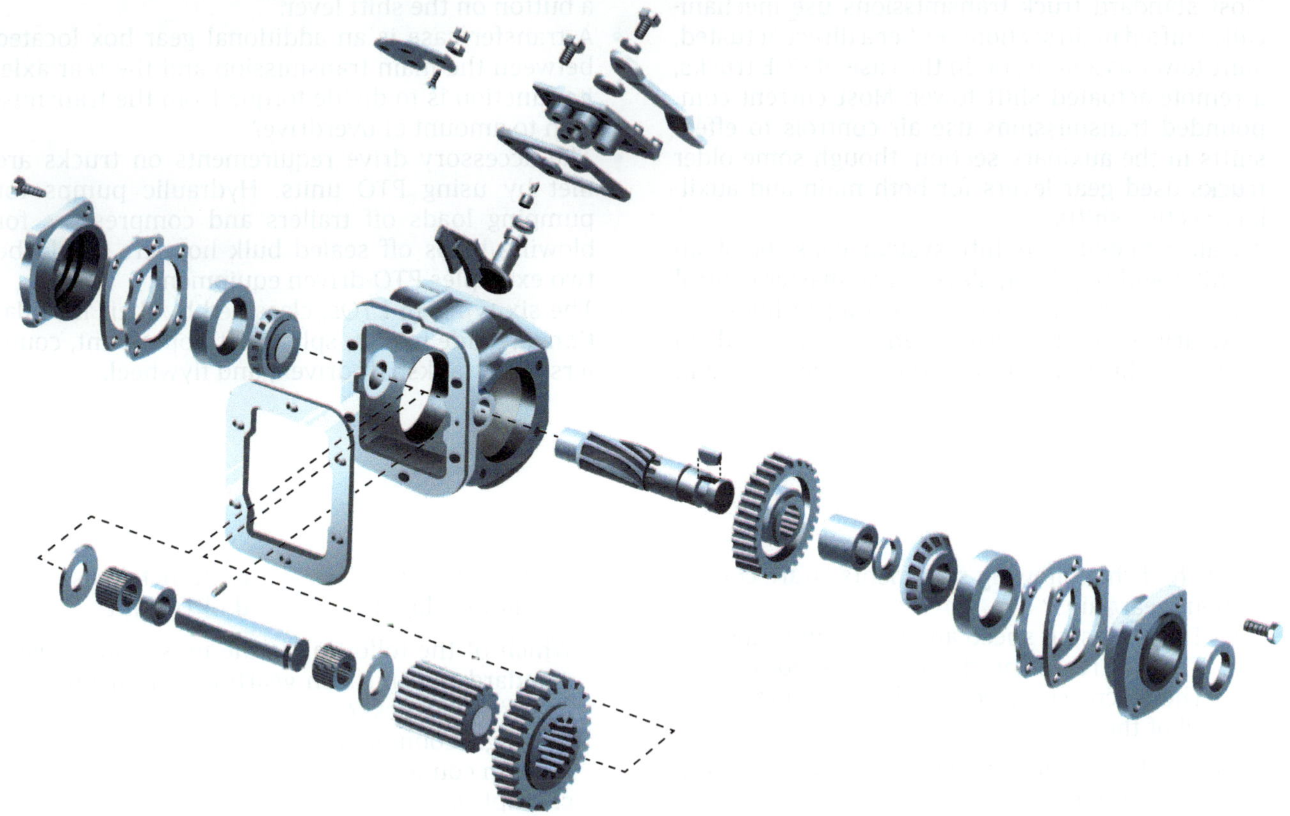

SUMMARY

- Engine torque is transferred through the clutch to the input shaft of the transmission, which drives the gears in the transmission. The transmission manages the drivetrain. It is the driver's means of managing drivetrain torque and speed ratios to suit chassis load and road conditions.
- A transmission enables the engine to function over a broad range of operating requirements that vary from a fully loaded standing start to cruising at highway speeds.
- Light-duty truck transmissions have a limited number of gear ratios and a single set of gears called main gearing, contained in a single housing. Most heavy-duty truck transmissions consist of two distinct sets of gearing: the main or front gearing and the auxiliary gearing located directly on the rear of the main gearing.
- Auxiliary gearing compounds the available ratios in a transmission. Most heavy-duty trucks use at least one compound, whereas some use two compounds.
- *Gear pitch* refers to the number of teeth per unit of pitch diameter.
- The three stages of contact through which the teeth of two gears pass while in operation are coming-into-mesh, full-mesh, and coming-out-of-mesh.
- The relationship of input to output speeds is expressed as *gear ratio*.
- Torque increase from a driving gear to a driven gear is directly proportional to speed decrease. Therefore, to increase output torque there is a resultant decrease in output speed and vice versa.
- The major types of gear tooth design used in modern transmissions and differentials are spur gears and helical gears.
- A heavy-duty standard transmission consists of a main shaft and one, two, or three countershafts.
- Standard transmissions can be generally classified by how they are shifted. Sliding gear, collar shift, and synchronized shift mechanisms are used to effect shifts in standard transmissions.
- Synchronizers have two primary functions. First, they bring two components rotating at different speeds to a single, synchronized speed; and second, they lock these components together. Block or

cone and pin synchronizers are the most common in heavy-duty transmissions.

- Most standard truck transmissions use mechanically shifted main sections: either a direct-actuated, shift tower assembly, or, in the case of COE trucks, a remote-actuated shift tower. Most current compounded transmissions use air controls to effect shifts in the auxiliary section, though some older trucks used gear levers for both main and auxiliary section shifts.
- An air-actuated gearshift system consists of an air filter and regulator, slave valve, master control valve, range cylinder, and connecting air lines.
- Auxiliary section gearing can be optioned to include a third gear in addition to the high- and low-range gears. This third gear is engaged or disengaged by a splitter shift system air activated by a button on the shift lever.
- A transfer case is an additional gear box located between the main transmission and the rear axle. Its function is to divide torque from the transmission to amount of overdrive?
- The accessory drive requirements on trucks are met by using PTO units. Hydraulic pumps for pumping loads off trailers and compressors for blowing loads off sealed bulk hoppers would be two examples PTO-driven equipment.
- The six types of PTOs, classified by their installation, are side mount, split shaft, top mount, countershaft, crankshaft driven, and flywheel.

REVIEW QUESTIONS

1. Which of the following statements is/are correct about geartrains?
   a. They transmit speed and torque unchanged.
   b. They decrease speed and increase torque.
   c. They increase speed and decrease torque.
   d. all of the above
2. Which of the following gear ratios indicates the most amount of overdrive?
   a. 2.15:1 c. 0.85:1
   b. 1:1 d. 0.64:1
3. Which type of gear is more likely to develop gear whine at higher speeds?
   a. a spur gear c. a worm gear
   b. a helical gear d. all of the above
4. When an idler gear is placed between the driving and driven gear, the driven gear:
   a. rotates in the same direction as the driving gear
   b. rotates in the opposite direction of the driving gear
   c. remains stationary
5. When sliding gearshift mechanisms are used, which gears do they control?
   a. high and low gears in the auxiliary section
   b. first and reverse gears in the main box
   c. direct and overdrive gears in the auxiliary section
   d. all of the above
6. Which type of transmission requires double-clutching between shifts?
   a. synchronized b. nonsynchronized
7. The component used to ensure that the main shaft and the main shaft gear about to be locked to it are rotating at the same speed is known as a:
   a. synchronizer c. transfer case
   b. sliding clutch d. PTO unit
8. Which of the following is the most widely used standard transmission geartrain arrangement in heavy-duty trucks?
   a. single countershaft
   b. twin countershaft
   c. triple countershaft
   d. double planetary
9. Which of the following is true of a twin-countershaft standard transmission?
   a. It divides input torque between two separate shafts.
   b. It allows the face width of each gear to be reduced.
   c. It allows the main shaft to float between the countershafts.
   d. all of the above
10. In a range shift air control system, which component distributes inlet air pressure to both the low- and high-range air hoses running to the range cylinder?
    a. a slave valve
    b. a splitter valve
    c. an air regulator
    d. an air compressor
11. To provide for overdrive gearing in the high- and low-auxiliary transmission ranges, which of the following components is added to the basic range-shift air control system?
    a. a splitter cylinder
    b. a button-actuated shift lever valve
    c. an additional slave valve
    d. both a and b

12. What component retains a shift rail in position following a shift?
    a. a detent ball
    b. a slave valve
    c. a range cylinder
    d. a dog clutch
13. An auxiliary mechanical drive for truck accessories driven by the transmission is known as a(n):
    a. transfer case
    b. power take-off (PTO) unit
    c. auxiliary transmission
    d. synchronizer
14. Which of the following statements is true regarding the function of a transfer case?
    a. It transfers torque from the transmission to a front drive axle.
    b. It transfers torque from the transmission to the rear drive axle(s).
    c. It may provide an additional gear reduction in the drivetrain.
    d. all of the above
15. Which of the following components in a typical heavy-duty, twin-countershaft transmission could be described as *floating*?
    a. countershafts
    b. the main shaft
    c. the input shaft
    d. the output shaft

16

Prerequisite: Chapter 15

STANDARD TRANSMISSION SERVICING

OBJECTIVES

After reading this chapter, you should be able to:

- Explain the importance of using the correct lubricant and maintaining the correct oil level in a transmission.
- List the preventive maintenance inspections that should be made periodically on a standard transmission.
- Explain how to replace a transmission rear seal.
- Describe the procedure for troubleshooting standard transmissions.
- Identify the causes of some typical transmission performance problems, such as unusual noises, leaks, vibrations, jumping out of gear, and hard shifting.
- Outline the procedure for overhauling a typical twin-countershaft transmission.
- Disassemble and reassemble a transmission auxiliary drive section.
- Disassemble and reassemble a transmission main case.
- Analyze the procedure for performing failure analysis on transmission components.
- Troubleshoot an air shift system.

KEY TERMS

brinelling
countershaft
fretting
jumpout
mainshaft
slipout
timing
yoke sleeve kit

INTRODUCTION

Properly operated and maintained, a standard transmission will give hundreds of thousands of miles of trouble-free operation. It is not uncommon for transmission manufacturers to offer limited warranties of up to 1,000,000 linehaul miles (1,600,000 km). Like all mechanical truck components, however, a transmission is subject to wear, fatigue, abuse, and manufacturing defects. Even when operated properly under ideal conditions, adjustments will eventually be necessary and parts will have to be replaced. This chapter explains how to maintain, troubleshoot, and service a heavy-duty standard truck transmission. Much of the service procedure outlined in this chapter can also be applied to automated manual transmissions (AMTs) because the fundamental mechanical platform of AMTs is based on standard transmissions.

16.1 LUBRICATION

Proper lubrication is one of the keys to a good preventive maintenance program. Maintaining the correct transmission oil level and ensuring that the oil is of the original equipment manufacturer

(OEM)-specified quality are both critical to ensure that a transmission achieves its expected service life.

Most standard transmissions are splash lubricated internally. Splash lubrication works by having some of the rotating components turn in a bath of oil in the sump. Some of this oil is picked up by gear teeth and used to splash lubricate those that are not in direct contact with the lube. Some transmissions also use a pump to help circulate oil to critical wear areas on the transmission that are not effectively lubricated by splash.

Standard transmission lubrication requirements will be properly met if the following procedures are observed:

1. Use the correct grade and type of oil.
2. Change the oil on schedule.
3. Maintain the correct oil level and inspect it regularly.

RECOMMENDED LUBRICANTS

Only the lubricants recommended by the transmission manufacturer should be used in the transmission. Most transmission manufacturers today prefer synthetic lubricants formulated for use in transmissions; they also suggest a specific grade and type of transmission oil. In the past, heavy-duty engine oils and straight mineral oils tended to be more commonly used in truck transmissions, but demand for extended transmission lube service intervals has driven the industry toward using synthetics almost exclusively.

Today, Eaton recommends synthetic E-500 lubricant (PS-164 R7 spec) for its transmissions. E-500 lube is designed to run 500,000 linehaul miles (80,000 km) with no initial drain interval required. E 250 lubricant is rated for 250,000 (400,000 km) linehaul miles before a change is required. Most synthetic transmission lubricants exceed the stated viscosity grade ratings, so they can be expected to perform effectively through various geographic and seasonal temperature conditions.

OIL CHANGES

Although most transmission manufacturers recommend the use of synthetic lubes in their transmissions, mineral oil gear lubes are still available and used by some operators. These mineral-based lubricants are slightly cheaper than synthetic lubricants to purchase, but because their projected service life is a fraction of synthetic lubricants', it really does not make economic sense to use them. The service requirements of both mineral-based and synthetic lubricants are covered here.

Mineral-Based Lubes

Most manufacturers recommend an early initial oil change after the transmission is placed in service when using mineral-based gear lubes. It is usually recommended that this first oil change be made any time after 3,000 miles (4,800 km) but before 5,000 miles (8,000 km) of linehaul road service. In off-highway use, the first transmission oil change should be made after 24 and before 100 hours of service. Although transmission oil is never required to be changed as frequently as engine oil, this first oil change is important because it will flush out of the transmission any cutting debris created by virgin gears meshing. However, modern manufacturing tolerances are of a much higher standard than those of a generation ago, so most OEMs no longer require a low-mileage oil change on new equipment.

A number of factors influence the appropriate scheduling of mineral-based transmission oil change periods. A key factor is the application: Linehaul operation tends to be kind to transmission oil, whereas operating on a construction site can reduce performance life. Depending on the transmission, manufacturer suggestions for mineral-based transmission oil change intervals vary between 50,000 and 100,000 miles (80,000–160,000 km) for linehaul applications. However, for purposes of this textbook, the Eaton recommended mineral oil change interval of 60,000 linehaul miles (96,00 km) is referenced. Off-highway operation usually requires oil change intervals ranging from 1,000 hours to a maximum of 2,000 hours of service. In each housing, the type of oil used can influence the oil change interval. Practice shows that using synthetic lubricants can greatly extend the performance life of the oil.

Synthetic-Based Lubes

Because transmission manufacturers almost exclusively recommend the use of synthetic gear lubes today, it follows that older, mineral-based lubes are not a realistic option. Eaton recommends E-500 lube for its transmissions. E-500 lubricant is designed to run 500,000 linehaul miles (80,000 km). Furthermore, when using E-500 lubricant, no initial drain interval is required for the transmissions. Modern production machining accuracy has all but eliminated the tendency of a gear set to produce break-in cuttings when new. As a result, the initial drain interval, once considered so important to maximize transmission service life, can be eliminated. E-250 lubricant is rated for 250,000 linehaul miles (40,000 km) before a change is required. Synthetic lubricants formulated for transmissions exceed the stated viscosity grade ratings, so they can be expected to perform effectively through various geographic and seasonal temperature conditions.

It is important that synthetic lubricants not be mixed with mineral-based lubes in transmissions. Although mixing of dissimilar lubricants may not produce an immediate failure, the service life of the lube can be greatly reduced. Mixing dissimilar lubricants can sometimes thicken oil and produce foaming.

CAUTION

Do not add transmission lubricant without first checking what lubricant the transmission is using. Mineral-based gear oils, mineral-based engine oils, and synthetic gear lubes are all approved for use in transmissions and none of them are particularly compatible. Mixing transmission oils causes accelerated lube breakdown, resulting in lubrication failures.

Checking Oil Level

The transmission oil level should be routinely checked at each A-type service, typically at intervals of 5,000 or 10,000 highway miles (8,000 and 16,000 km, respectively). When adding oil to transmissions, care should be taken to avoid mixing brands, weights, and types of oil. When topping up or filling the transmission oil level, it should be exactly even with the filler plug opening, as shown in **Figure 16–1**. More recent transmissions may have a sight glass such as that shown in **Figure 16–2**, which makes it easier to quickly check transmission oil level. Overfilling can cause oil aeration and underfilling results in oil starvation to critical components.

FIGURE 16–1 The oil level must be level with the bottom of the plug hole.

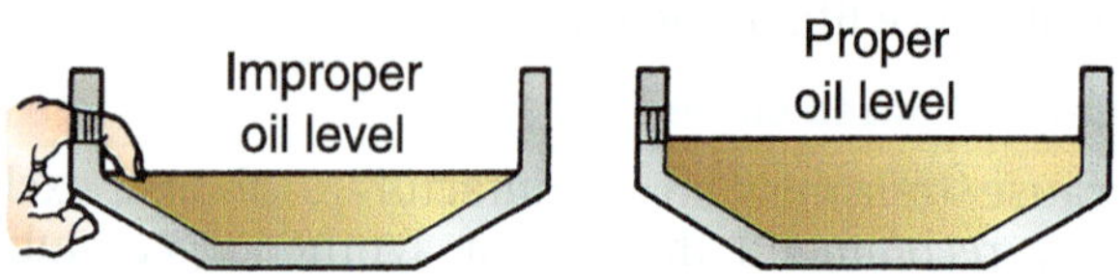

FIGURE 16–2 An oil level sight glass on an UltraShift AMT; the fill plug is located next to it.

SHOP TALK

Transmission oil should be exactly level with the filler plug opening. One finger joint equals approximately 1 gallon (4 liters) of transmission oil, so testing oil level by dipping your finger into the oil filler hole is not a good practice.

Draining Oil

Drain the transmission when the oil is warm. Remember that it can be hot enough to severely burn your hand. To drain the oil, remove the drain plug at the bottom of the housing. Allow it to drain for at least 10 minutes after removing the plug. Check the drain plug for cuttings and thoroughly clean it before reinstalling.

Oil Filter

Some transmissions are equipped with an oil filter. The oil filter is usually the key to the greatly extended oil change intervals on recent model transmissions. It should be changed at each transmission oil change or according to the OEM-recommended PM schedule; this may require a filter change when the oil is not required to be changed. It is not necessary to prime a new filter prior to installation (**Figure 16–3**).

Refill

Wipe away any dirt around the filler plug. Then remove the plug from the side of the transmission and refill with the appropriate grade of the new oil. Fill until the oil just begins to spill from the filler plug opening. If the transmission housing has two filler

FIGURE 16–3 The oil filter should be changed when the transmission oil is changed or according to the OEM PM schedule.

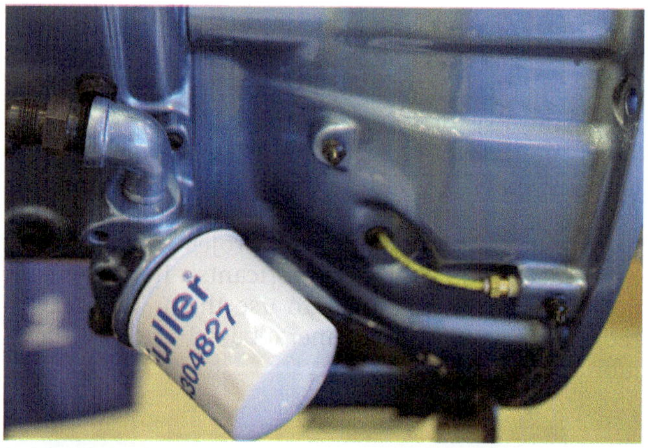

plugs, fill both until the oil is level with each filler plug hole.

CAUTION

Do not overfill the transmission. Overfilling usually results in oil breakdown due to aeration caused by the churning action of the gears. Premature breakdown of the oil will result in varnish and sludge deposits that plug up oil ports and build up on splines and bearings.

SHOP TALK

When draining transmission oil, check for metal particles in the oil. Metal particles may indicate excessive wear and may warn of an imminent failure. It is not unusual for a newly broken in transmission to have minute metal particles held to a magnetic drain plug.

16.2 PREVENTIVE MAINTENANCE INSPECTIONS

A good preventive maintenance (PM) program can help avoid failures, minimize vehicle downtime, and reduce the cost of repairs. Often, transmission failure can be traced directly or indirectly to poor maintenance. **Figure 16–4** identifies key areas of a transmission that should be routinely checked. We have outlined here some typical good maintenance practices, but you should remember that some operating conditions may either raise or lower maintenance standards. The numbers used reference callouts in Figure 16–4.

DAILY MAINTENANCE

Many of the practices listed here are part of the driver pretrip inspection.

- **Air tanks**. Drain air tanks to remove water or oil. To be sure of removing all liquid

FIGURE 16–4 Inspection points on a standard transmission

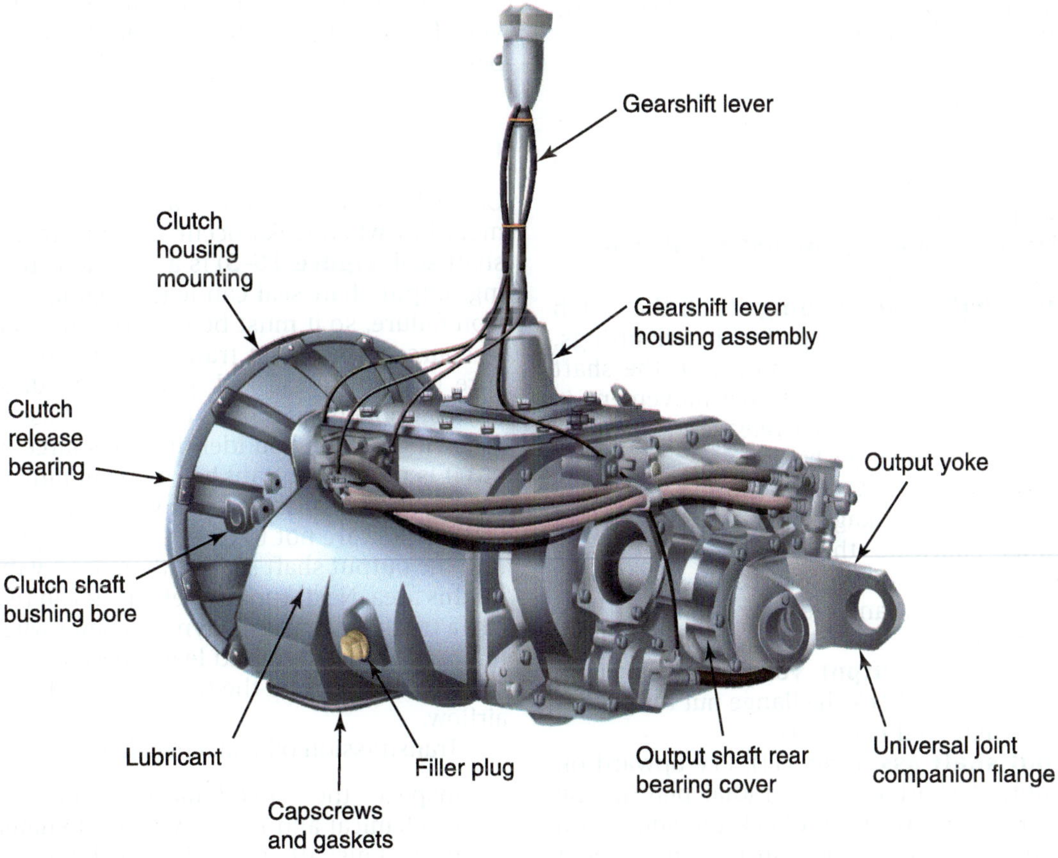

contaminants from an air tank, the drain cock must be fully opened and all air discharged. Most of the liquid will drain from the tank after the air has been bled.

- **Oil leaks**. Visually check for oil leaks around bearing covers, PTO covers, and other machined surfaces. Check for oil leakage on the ground before starting the truck each morning.
- **Shifting performance**. Report any shifting performance problems such as hard shift or jumping out of gear.

A INSPECTION PM

The following should be checked at each A or lube inspection, approximately every 10,000 miles (16,000 km):

- **Fluid level**. Remove filler plug(s) and check lubricant level. Top off if necessary and tighten plugs securely.
- **Fasteners and gaskets.** Use a wrench to check the torque on bolts and plugs, paying special attention to those on power take-off (PTO) covers/flanges and the rear bearing cover assembly. Look for oil leakage at all gasket mating surfaces.
- **Output yoke seal**. Check for leaks around the seal, especially if the transmission has recently been serviced or rebuilt.

B INSPECTION PM

- **Air control system** (1). Check for leaks, worn hoses and air lines, loose connections, and loose fasteners.
- **Bell/clutch housing mounting flange** (2). Check fastener torque.
- **Clutch shaft yoke bushings** (4). If the clutch shaft bushings are equipped with zerk fittings, grease them lightly. Pry upward on the shaft to check for wear. If excessive movement is found, remove the clutch release mechanism and check for worn bushings.
- **Cab-over-engine (COE) remote shift linkage**. Check the linkage U-joints for wear and binding. Lubricate the U-joints. Check any bushings in the linkage for wear.
- **Air filter**. Check and clean or replace the air filter element.
- **Transmission output yoke** (10). Uncouple the U-joint and check the flange nut for proper torque. Tighten if necessary.
- **Output shaft assembly** (11). Pry upward on the output shaft to check radial play in the mainshaft rear bearing. Check the splines on the output shaft (12) for wear from movement and chucking action of the U-joint yokes.

C INSPECTION PM

- **Check lubricant change interval**. This means checking the type of lubricant used in the transmission. Remember that many synthetic lubes are performance rated for oil change intervals up to 500,000 linehaul miles (800,000 km), so it is wasteful to change them more frequently. If an oil change is required, drain and refill the transmission with the specified oil. Transmission oil analysis can be used to establish more precise oil change intervals that are better suited to the actual operating condition of the truck.
- **Gearshift lever** (8). Check for bending and free play in the tower housing. A lever that is excessively loose indicates wear.
- **Shift tower assembly** (9). Remove the air lines at the slave valve and remove the shift tower from the transmission. Check the tension spring and washer for wear and loss of tension. Check the gearshift lever spade pin/shift finger for wear. Also take a look at the yokes and blocks in the shift bar housing, checking for wear at all critical contact points.

PM practices vary by manufacturer and by fleets, many of which have tailored PM to exactly suit their needs. **Table 16–1** outlines a preventive maintenance schedule recommended by Eaton Roadranger, but you should note that this assumes the use of mineral-based oil.

CHECKING FOR OIL LEAKS

Oil leakage in standard transmissions is not common, but when leaks occur the transmission output shaft seal (**Figure 16–5**) is a common culprit. A leaking output shaft seal can lead to complete transmission failure, so it must be repaired. It can be difficult to source oil leaks on transmissions, however, so be sure to identify the leak path before disassembling any components. You can use a checklist such as the following to help you identify the source of oil leaks, but it is most important before beginning to pressure wash the transmission, removing all evidence of oil leaks. Take care not to directly aim the pressure hose into the output shaft seal. Rather, use a wiper to clean up this area. Next, run the vehicle. It can really help to run a truck on a chassis dynamometer when attempting to locate hard-to-find leaks, because leaking oil is not smeared all over the transmission housing by the airflow.

Transmission oil leak checklist:

1. Inspect the speedometer connections. Both mechanical and inductive speedometers connect to the tailshaft assembly. Check the speedometer sleeve torque and ensure the O-ring or gasket has

TABLE 16–1 Preventive Maintenance Recommendations* (Mineral-Based Lube Schedule)

PM Operation	Vehicle Mileage											
	Daily	5,000	10,000	20,000	30,000	40,000	50,000	60,000	70,000	80,000	90,000	100,000
Bleed air tanks and listen for leaks	X											
Inspect for oil leaks	X											
Check oil level			X	X	X	X	X	X	X	X	X	X
Inspect air system connections				X		X		X		X		X
Check clutch housing capscrews for looseness				X		X		X		X		X
Lube clutch pedal shafts				X		X		X		X		X
Check remote control linkage				X		X		X		X		X
Check and clean or replace air filter element				X		X		X		X		X
Check output shaft for looseness				X		X		X		X		X
Check clutch operation and adjustment						X				X		
Change transmission oil		X**					X					X

* Repeat schedule after 100,000 miles.

**Initial fill on new units using mineral base lubes.

FIGURE 16–5 Rear seal and slinger assembly that seals the transmission output yoke.

not failed. Hydraulic thread sealant applied to the threads can sometimes cure this type of leak.

2. Inspect the rear bearing cover assembly bolts for tightness. These fasteners should be torqued to specification. Thread sealant is required to seal the fastener threads on the rear bearing cover assembly. Leakage will result if this sealant fails.
3. Check the rear bearing cover gasket and nylon collar. Verify that the collar and gasket are installed at the chamfered hole that is aligned near the mechanical speedometer opening. The fastener must be torqued to specification with thread sealant applied to the fastener threads. Use the same procedure to check an electronic tailshaft speed sensor.
4. Check the output yoke retaining nut for tightness. The retaining nut should be torqued to specification (typically 450 to 600 lb-ft./610 to 814 N·m). You will need a yoke holder to hold the yoke stationary. A torque wrench capable of handling 600 lb-ft. (814 N·m) or a lower rated torque wrench used with a 4 × 4 torque multiplier is required to accurately torque the nut.
5. Check the PTO manifolds and gaskets, making sure that the fasteners are torqued to specification.
6. Check the cover plates and manifolds bolted to the transmission housing for cracks or breaks, including the front bearing cover plate, front housing, shift bar housing, rear bearing cover, and the clutch housing. All of these components can be a source of oil leakage.
7. Check the front bearing cover plate. Oil return grooves that have been damaged by contact with the input shaft can cause oil leakage. If the grooves are damaged you must replace the front bearing cover. If the front bearing cover has an oil seal installed, inspect the oil seal. Damaged or worn seals must be replaced to prevent oil leakage; in the case of the front bearing cover seal, the transmission will have to be pulled.
8. An oil leak in the front bearing cover, in the auxiliary housing, or in some housings (such as the air

breather in the shift bar housing) can be caused by a damaged O-ring in the range air shift cylinder. The damaged O-ring bleeds air into the transmission, pressurizing it and resulting in oil leakage.

9. If the transmission is equipped with an oil cooler and oil filter, inspect the connectors, hoses, and filter element to make sure that they are tight and leak free.
10. Inspect the oil drain plug and oil fill plug for leakage. The oil drain plug and oil fill plug should be torqued to specification.

CAUTION

Overtorquing a transmission output yoke retaining nut can cause output shaft bearing failure. Using a 1-inch drive air gun to tighten yoke retaining nuts should be avoided, although it is okay to remove the nut using this tool.

CAUTION

Many housing bolt holes used to couple flanges and manifolds to the transmission are bored completely through the housing. These fasteners must have thread sealant applied to the threads. If oil leakage is observed at the PTO covers, replace the PTO cover gaskets and thoroughly clean the fasteners and their mating threads before reinstalling.

SHOP TALK

Always check for a plugged transmission breather when identifying the cause of a transmission oil leak. When transmission oil is raised from cold to operating temperatures, it expands to occupy a greater volume. If the breather is plugged, this can cause seal failure.

REAR SEAL REPLACEMENT

If your troubleshooting has determined that the output shaft oil seal is leaking, the end yoke has to be removed and the seal and slinger replaced. Check the yoke for signs of wear and damage (**Figure 16–6**). Pay special attention to the yoke seal surface. Any wear beyond normal polishing will require the yoke to be replaced or resleeved. There should be no rust, burrs, nicks, or grooves on the seal surface.

SHOP TALK

Do not attempt to repair a visibly worn or damaged yoke by polishing it with crocus cloth. Minor scratches can either track oil under the seal or draw in contaminants.

FIGURE 16–6 Check the yoke seal wear surface for wear and damage.

Yoke Sleeve Kits

Resleeving of yokes with a **yoke sleeve kit** is a common practice but not one that is recommended by Eaton. The Eaton recommended practice requires either replacement or turning the yoke down followed by the fitting of an Eaton sleeve kit. Current transmissions relocate the seal surface into the tone wheel for an easier repair. If an aftermarket yoke sleeve kit is to be used, make sure that the wear ring manufacturer's installation instructions are observed.

SHOP TALK

Wear rings that increase the original yoke diameter can cause the new seal to wear more rapidly due to increased seal lip pressure.

Replacement Procedure

1. Block the truck wheels and place the transmission in a low gear to facilitate removal.
2. Disconnect the drive shaft by removing the U-joint from the output yoke.
3. Remove the output yoke nut and use a puller to remove the output yoke.
4. Remove the oil seal from the rear bearing cover.
5. Wipe the bore of the bearing cover with a clean cloth to remove any contaminants.
6. Install the new seal and slinger with the proper driving tool. **Figure 16–7** shows how a slinger is installed. Make sure that both the seal and slinger are not cocked at installation. Damage to either can result in leakage.
7. Reinstall the output yoke over the transmission output shaft, prevent the yoke from rotating, and torque the nut to specification (**Figure 16–8**).

FIGURE 16–7 Install the slinger using the proper driver.

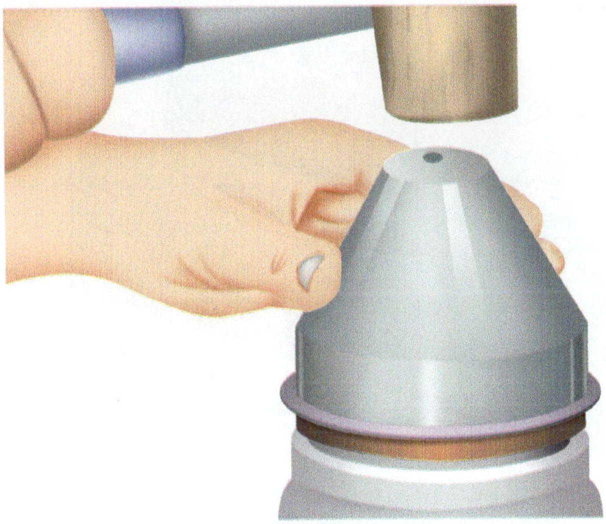

FIGURE 16–8 Torque the yoke nut to OEM specifications while holding the yoke stationary.

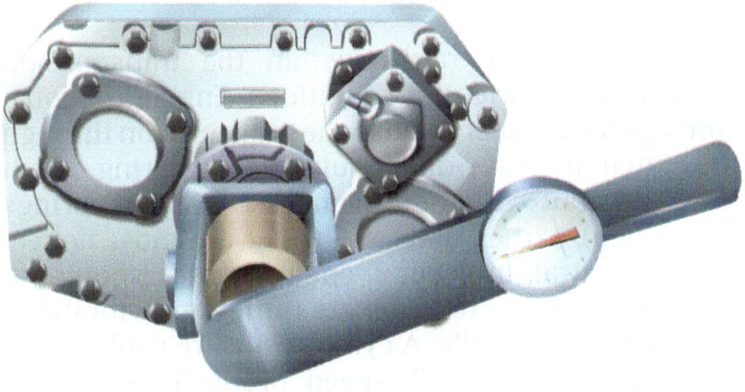

CAUTION

Do not overtorque the yoke nut, which can damage the bearing. Never perform this operation using a 1-inch drive air gun.

16.3 TROUBLESHOOTING

To isolate the root cause of any transmission problem, you have to follow a troubleshooting procedure. Begin by discussing the problem and its symptoms with the driver. Check any information you can get from the driver with the vehicle history and with audit trails off the data bus, if available. Even a non-electronic transmission creates tattletales on the vehicle data network. You can also check out both the truck and driver histories using terminal information tracking software, such as ProManager, GuardDog or equivalent, if available.

Visually inspect the transmission. Look for obvious problems such as broken transmission mounts, fittings, or brackets and signs of leakage. Check the air system for leaks. Road test the truck whenever possible. Technicians usually get second- or third-hand reports of trouble experienced with the unit, and these reports do not always accurately describe the actual conditions. Sometimes symptoms may point toward a transmission problem but the true problem is actually elsewhere in the drivetrain, such as in an axle, propeller shaft, U-joint, engine, or clutch. This is especially true of complaints of noise and vibration. Road testing the unit will provide the technician with firsthand evidence of a complaint and lessen the reliance on secondhand reports from drivers. Riding with a driver can also be informative, but if the technician is licensed to drive a heavy-duty truck, road testing can be more effective.

Troubleshooting should always begin with information gathering. Next, you will have to analyze the information and attempt to identify a root cause. Nothing is more helpful than experience with troubleshooting and knowledge of the product, but these skills take time to acquire.

SYMPTOMS

Next, we will take a look at some typical symptoms produced by transmission problems and give you some clues as to what might have caused them.

Noisy Operation

This complaint is common and can be caused by so many different things that you should always begin with a road test. Noise has a way of reverberating through the drivetrain, so let us begin with some non-transmission noises that can be mistaken for transmission noise, which includes:

- Engine fan out of balance or blades bent
- Defective vibration damper
- Crankshaft out of balance
- Flywheel out of balance
- Flywheel mounting bolts loose
- Engine rough at idle, producing a rattle in the geartrain
- Clutch assembly out of balance
- Engine mounts loose or broken
- PTO engaged
- U-joints worn out
- Drive shaft out of balance
- U-joints out of phase or at excessive working angle
- Hanger bearings in driveline dry or out of alignment
- Wheels out of balance

- Tire treads humming or vibrating at certain speeds
- Flywheel housing not concentric

It is your responsibility as a technician to determine the cause, so it is really important that you do not jump to conclusions based on a driver complaint. First determine the source of the noise. Next, if you are sure it is coming from the transmission, attempt to define the character of the noise. Be aware of the exact conditions that produce the noise. Ask questions such as: Does the noise occur only in one gear ratio? Does it occur at a specific road speed?

Growling. Low-frequency noises such as growling, humming, or a grinding sound are usually caused by worn, chipped, rough, or cracked gears (**Figure 16–9**). Once the symptoms are first noted, as gears continue to wear, a growling/grinding noise becomes more noticeable, usually in gear positions that throw the most load on the worn or damaged gears. Growling can also be caused by improper timing of the transmission during reassembly or improper timing due to a gear turning on the **countershaft**.

Thumping or Knocking. A thumping or knocking noise can begin as growling. As bearings wear and retainers start to break up, a thumping or knocking noise can be produced. This noise is heard at low shaft speeds in any gear position. Irregularities on gear teeth such as cracks, bumps, and swells can also cause thumping or knocking. Bumps or swells can be identified as highly polished spots on the face of the gear tooth (**Figure 16–10**) and these are often caused by rough handling during assembly. In most housings, the noise produced will be more prominent when the gear is loaded, making it easier to identify the problem gear. At high speeds a knocking or thumping noise can turn into a howl.

FIGURE 16–9 Cracked gears can cause such problems as whining, knocking, grinding, or thudding noises.

FIGURE 16–10 Bumps or swells on the gear teeth are caused by improper handling of gear before and during assembly.

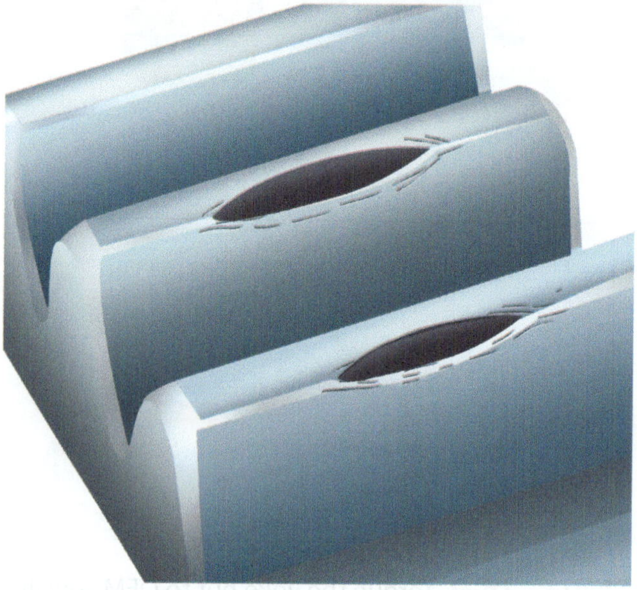

Rattles. Metallic rattles within the transmission result from a variety of conditions. Engine torsional vibrations are transmitted to the transmission through the clutch. In heavy-duty equipment featuring clutch discs without vibration dampers (coaxial springs in the hub), rattles may result that are most noticeable in neutral. In general, engine speeds should be 600 rpm or above to eliminate rattles and vibrations noticeable during idle. A cylinder misfire could cause an irregular or lower idle speed, producing a rattle in the transmission. Another cause of rattles is excessive backlash in the PTO unit mounting.

Vibration. Drivetrain vibrations can be very difficult to source. Never forget that they can be produced by any component in the drivetrain, so your troubleshooting must involve everything from the engine to the drive wheel assemblies. Most driveline vibrations are produced only at a specific road speed or in a specific gear; these clues can help you determine whether the cause is in the transmission or elsewhere.

If you have isolated the cause of the vibration to the transmission, first note the speeds at which it is most pronounced. Use what you know of the transmission to make your diagnosis. Remember that the input shaft rotates at engine speed when the clutch is engaged, and the output shaft rotates at drive shaft speed. Make note of the gear the vibration occurs at, and note whether it occurs only in high range or only in low. Make sure that you have some evidence that the cause is in the transmission before removing it from the chassis.

Whining. Gear whine is usually caused by insufficient backlash between mating gears, such as improper shimming of a PTO. A high-pitched whine or squeal can also be caused by mismatched gear sets. Such gear sets are identified by an uneven wear pattern on the face of the gear teeth. Pinched bearings with insufficient axial or radial clearance can also produce a whine.

Noise in Neutral. Possible causes of noise while the transmission is in neutral include:

- Misalignment of transmission
- Worn or seized flywheel pilot bearing
- Worn or scored countershaft bearings
- Worn or rough reverse idler gear
- Sprung or worn countershaft
- Excessive backlash in gears
- Worn mainshaft pilot bearing
- Scuffed gear tooth contact surface
- Insufficient lubrication
- Use of incorrect grade of lubricant

Noise in Gear. Possible causes of noise while the transmission is in gear include:

- Worn or damaged mainshaft rear bearing
- Rough, chipped, or tapered sliding gear teeth
- Noisy speedometer gears
- Excessive endplay of mainshaft gears
- Most of the conditions that can cause noise in neutral

Oil Leaks

Because the transmission in most trucks is located in the ground effect airflow, oil leaks can be difficult to source. Always clean the transmission before attempting to locate the source. Possible causes of oil leakage are:

- Oil level too high
- Defective front or rear seals
- Nonshielded bearing used as front or rear bearing cap (where applicable)
- Transmission breather plugged (pressurizing transmission housing, resulting in seal failure)
- Fasteners loose or missing from shift tower, shift bar housing, bearing caps, PTO, or covers
- Oil drain-back paths in bearing caps or housing plugged with varnish, dirt, or silicone, or covered with gasket material
- Failed gaskets, or gaskets shifted or squeezed out of position
- Cracks or porosity in castings
- Drain or fill plugs loose
- Oil leakage from engine
- Speedometer adapter or connections

Gear Slipout

When a sliding clutch is moved to engage with a **mainshaft** gear, the mating teeth must be parallel. Tapered or worn clutching teeth will attempt to *walk* apart as the gears rotate, causing the sliding clutch and gear to slip out of engagement. **Slipout** will generally occur when pulling with full power or decelerating with chassis momentum turning the drivetrain.

Any condition that prevents full engagement of a sliding clutch and gear can lead to a slipout condition. Begin by inspecting the shift tower assembly and ensuring that it moves freely and completely into position. If there are problems here, you may have to pull the shift rail housing. Visually check the floorboard opening, driver seat position, and, in housings using a remote shift linkage, other causes of binding. Some things to look for include:

- Remote shift linkage bent or out of adjustment
- Sloppy ball and socket joints
- Worn or loose engine mounts, or loose on frame
- Shift rail control clevis loose on rail, or control slot worn
- Taper wear or damage on gear clutch teeth
- Transmission and engine out of alignment, caused by out-of-specification concentricity at the flywheel housing or bell housing
- Weak or broken detent spring or wear on the detent notch of the yoke bar (**Figure 16–11**)

Vibrations set up by an improperly aligned driveline and lower than regulated air pressure can add to a slipout problem. **Jumpout** occurs when a fully engaged gear and sliding clutch are forced out of engagement. It generally occurs when a force sufficient to overcome the detent spring pressure is applied to the yoke bar, moving the sliding clutch to a neutral position. Conditions that result in jumpout include:

- Heavy and long gearshift levers that, over bumpy terrain, swing enough to overcome detent spring tension (**Figure 16–12**).

FIGURE 16–11 Gear slipout or jumpout can be caused by weak or broken detent springs or worn yoke bar notches.

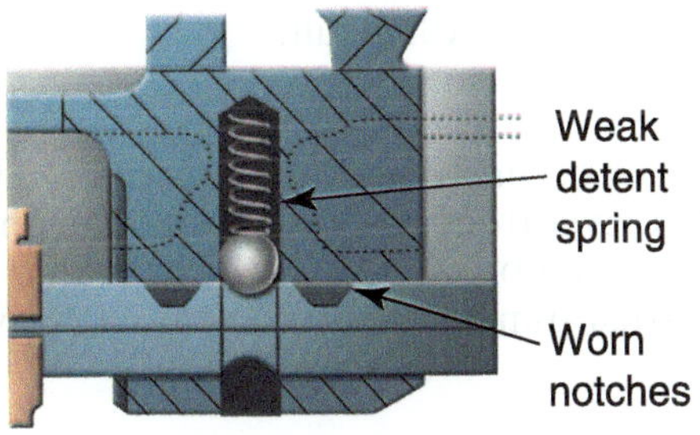

FIGURE 16–12 When passing over rough road surfaces or uneven terrain, the whipping action of a long heavy-shift lever can overcome the detent spring tension, causing the transmission to slip or jump out of gear.

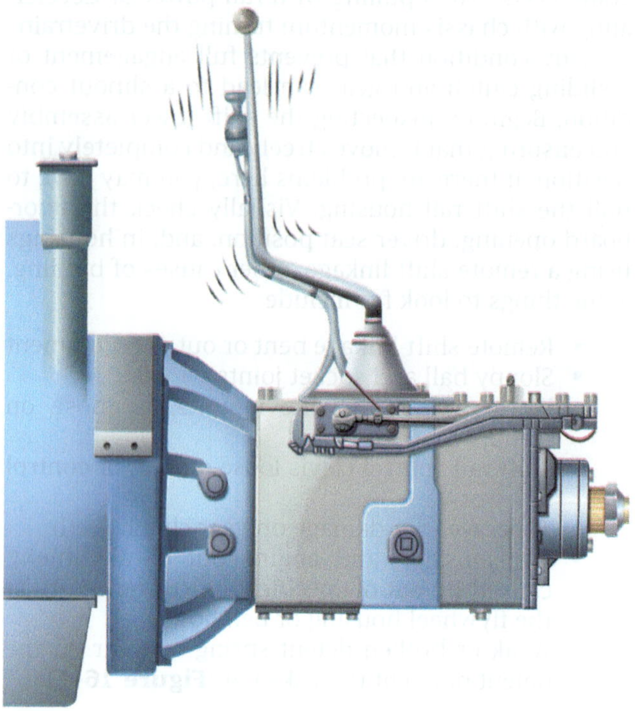

- Shift rail rod detent springs broken, or badly worn detent notches
- Shift rail bent
- Shift fork pads not square (twisted) with shift rail
- Excessive endplay in drive gear, mainshaft, or countershaft, caused by worn bearings or retainers
- Thrust washers worn excessively or missing

Jumpout in the auxiliary section usually occurs with the splitter gear set. If torque is not sufficiently broken during splitter shifts, the sliding clutch gear might not have enough time to complete the shift before torque is reapplied to the gears. As torque is reapplied, the partially engaged clutch gear jumps out of the splitter gear. Because the gears still have torque applied to them, damage to the clutching teeth of the mating gears will result.

Hard Shifting

Hard shifting occurs when the effort required to move a gearshift lever from one gear position to another is excessive. It can be caused by the gear shift linkage or by internal transmission problems. First, you will have to determine whether the problem is caused by the linkage or shift tower assembly, or whether it is a transmission problem. Remove the shift tower from the transmission. Next you can slide the shift blocks into each gear position using a prybar or screwdriver. If the yoke bars slide easily, the problem is probably caused by the linkage assembly. Here are some common causes of hard shifting:

- No lubricant in a remote linkage control unit used in a COE application or binding linkage U-joints
- Broken down or overheated transmission lubricant, causing varnishing and sludge deposits on splines of shaft and gears
- Badly worn or bent shift rods
- Sliding clutch gears binding on shaft splines
- Clutch teeth burred, chipped, or otherwise damaged
- Clutch pilot bearing seized, rough, or dragging
- Clutch brake engaging too soon when clutch pedal is depressed

Table 16–2 lists some of the more common transmission problems and their possible causes. This can be a useful prompt when investigating transmission problems.

TOWING PRECAUTIONS

Trucks must be towed properly to prevent damage to the transmission and other drivetrain components as well as the frame. When towing a truck with a disabled transmission or other problem, standard transmissions require rotation of the main box countershaft and mainshaft gears to provide adequate lubrication. These gears do not rotate when the vehicle is towed with the rear wheels on the ground and the drivetrain connected. Under these conditions the mainshaft can be driven at high speed by the rear wheels. Without lubrication, it will be rapidly destroyed. You should note that coasting with the transmission in neutral can produce the same damage. The procedure to observe when towing should be as follows:

1. Pull the axle shafts or disconnect the driveline.
2. The recommended method of towing a truck is to use a rigid tow bar and connect it to the front tow hook or pin (**Figure 16–13A**). This prevents vehicle and frame damage and can be used with a vehicle that is loaded or unloaded. If you are using a wrecker without a tow bar, you can use the alternate method shown in **Figure 16–13B**. Place the lift bar cradle under the front axle and raise the truck for towing.
3. Tow the truck with its drive wheels off the ground (**Figure 16–14**) if it is impractical to disconnect the axle shafts or driveline.

TABLE 16–2 Transmission Troubleshooting Guidelines

Possible Corrections

1. Instruct driver on proper driving techniques.
2. Replace parts (after trying other listed possible corrections).
3. Loosen lock screw and retighten to proper torque.
4. Look for resultant damage.
5. Smooth with emery paper.
6. Reset to proper specifications.
7. Install missing parts.
8. Check air lines or hoses.
9. Tighten part.
10. Correct the restriction.
11. Recheck timing.
12. Clean part.
13. Apply thin film silicone.
14. Apply sealant.

Possible Cause

Problem	Worn yoke pads	Bent yoke bar	Weak or missing detent spring	Burr on yoke bar	Interlock ball or pin missing	Too strong detent spring	Cracked shift bar housing	Breather hole plugged	Damaged insert valve	Defective regulator	Loose hose or fitting	Pinched air hose	Sticking slave valve piston	Damaged O-ring	Improperly mounted gasket	Air hose hooked to wrong place	Pinched air line or connector	Air cylinder piston nut loose	Air cylinder piston cracked	Gear twisted out of time on shaft	Cracked gear or burr on tooth	Excessive mainshaft gear tolerance	Twisted mainshaft	Tapered clutching teeth	Worn yoke slot in clutch gear	Broken key (front section)	Broken key (auxiliary)	Yoke installed backwards	Synchronizer spring broken	Failed synchronizer	Race left off auxiliary countershaft	Bearing failure
Slipout (splitter)	1 2								2	2	9						10			2 4				2	2 1						7 2	2
Slipout (range)										2	9						10			2 4				2								
Slipout or jumpout (front section)	2		2 7																	2 4				2	2	2						
Slow shift (splitter)									2	2	9	10		2	6		10															
Slow shift or won't shift (range)										2	9	10	12 13	12 13 2			10	9 12	2 13								2 4		2 7	1 2		
Hard shift or won't shift (front section)		2 3		5		2	2													2 4			2									
Able to shift front section into 2 gears at once					7																					2						
Grinding on initial lever engagement	2																			2 4												
Lever locks up or sticks in gear																							2									
Noise																				2 4	5 2	6									7 4	2
Gear rattle at idle																						6										
Vibration																																
Burned mainshaft washer																																
Input shaft splines worn or input shaft broken																																
Cracked clutch housing																																
Broken auxiliary housing																																
Burned synchronizer										2																		2 6	2 7			
Broken synchronizer																																
Heat																				2 4											2	2
Twisted mainshaft																																
Drive set damage																																2 4
Burned bearing																															2	
Oil leakage								10																								
Overlapping gear ratios																8 6																

(*continued*)

TABLE 16–2 (continued)

Possible Corrections

1. Instruct driver on proper driving techniques.
2. Replace parts (after trying other listed possible corrections).
3. Loosen lock screw and retighten to proper torque.
4. Look for resultant damage.
5. Smooth with emery paper.
6. Reset to proper specifications.
7. Install missing parts.
8. Check air lines or hoses.
9. Tighten part.
10. Correct the restriction.
11. Recheck timing.
12. Clean part.
13. Apply thin film silicone.
14. Apply sealant.

Possible Cause

Problem	Rough running engine	Preselecting splitter	Not using clutch	Starting in too high of gear	Shock load	C/s brake not working (push)	Clutch brake not adjusted (pull)	Clutch brake tangs broken (pull)	Not equipped with c/s or clutch brake	Linkage obstructions	Improper linkage adjustment	Bushings worn in control housings	Improper clutch adjustment	Broken engine mount	Clutch failure	Low oil level	High oil level	Poor quality oil	Too great operating angle	Infrequent oil changes	No silicone on "O" rings	Excessive silicone on "O" rings	Mixing oils or using additives	Pin hole in case	Damaged rear seal	Loose or missing capscrews	Improper towing of truck or coasting	Misalignment engine to transmission	Output shaft nut improperly torqued	Improper driveline setup	Worn suspension	Tires out of balance, loose lug nuts
Slipout (splitter)		1	1 4																													
Slipout (range)																																
Slipout or jumpout (front section)			1 4							10	6			2														2 6				
Slow shift (splitter)		1	1 4																		13	12 13										
Slow shift or won't shift (range)																					13	12 13										
Hard shift or won't shift (front section)			1 4							10			6	2 4																		
Able to shift front section into 2 gears at once																																
Grinding on initial lever engagement						2 1 8	6 1	2 6	7 1	10	6	2	6																			
Lever locks up or sticks in gear										10	6																					
Noise																2 4		2 4		2 4			2 4									
Gear rattle at idle	6																															
Vibration															2														6	6	2 6	2 6
Burned mainshaft washer																2 4 6											2 4 1					
Input shaft splines worn or input shaft broken				1 2	1 2									2 6		2 6												6 2		6		
Cracked clutch housing															2													2				
Broken auxiliary housing																														2 6		
Burned synchronizer																		2 6					2 6									
Broken synchronizer															2														6	6	2 6	2 6
Heat																6 4	6 4	2 6	2 6	2 6			2 6									
Twisted mainshaft				1 2	1 2																											
Drive set damage																2 4 6			2 6													
Burned bearing																2 4 6		2 4 6		2 4 6			2 4 6									
Oil leakage																	6							2 4	2 4	6,7 14						
Overlapping gear ratios																																

FIGURE 16–13 Towing methods: (A) recommended; and (B) alternate

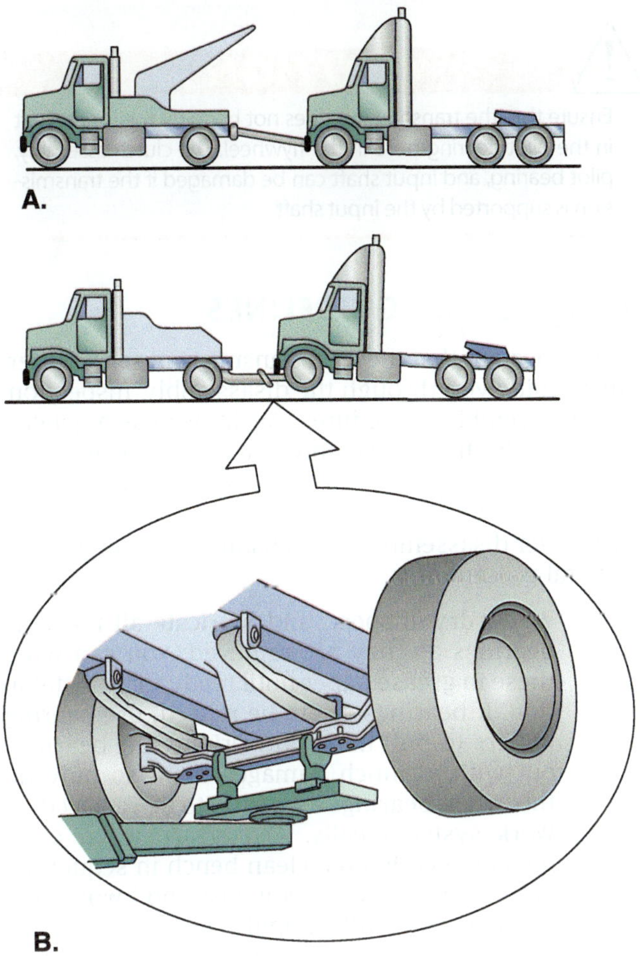

FIGURE 16–14 If the axle shafts or driveline cannot be pulled or disconnected, tow the truck with the drive wheels off the ground to avoid transmission damage.

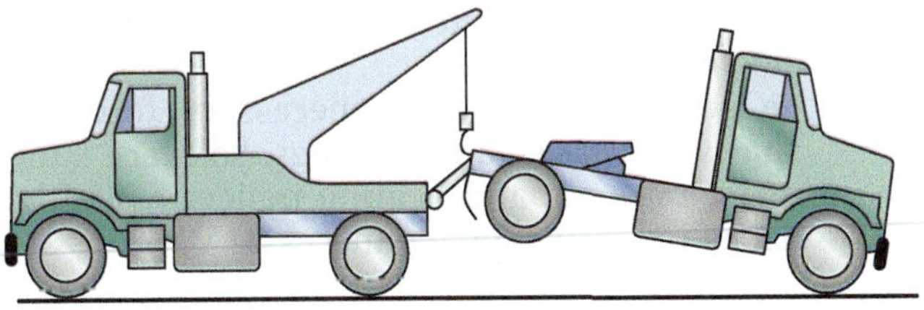

> **CAUTION**
>
> When the driveline or axle shafts are reinstalled, the axle nuts must be tightened to the correct torques, the axle shafts properly installed (RH and LH), and the drive shafts properly phased. Refer to the OEM service literature for the correct torque values.

16.4 TRANSMISSION OVERHAUL

Although troubleshooting may not have identified the exact cause of a transmission problem, once you have made the decision to pull the transmission from the truck, you would better be sure that you are working with a transmission problem and not one caused by another drivetrain component. Having made the decision to pull the transmission, you should examine each component as it is disassembled from the chassis and then do the same as you take apart the transmission itself.

> **CAUTION**
>
> Technicians are strongly advised to obtain the disassembly/reassembly procedure for the specific transmission they are working on. Some generational differences exist. The information in this chapter has been updated to cover current transmissions, but in all cases, the procedures outlined in this textbook should be regarded as general in nature. Eaton Roadranger service literature is available as a free download from http://www.roadranger.com.

REMOVING THE TRANSMISSION

Again, it is a good idea to begin the transmission removal procedure by referencing the OEM service literature. We have outlined here a general procedure that applies to a typical truck chassis but remember that each chassis is distinct. Pulling a transmission can be quick and easy in some COE chassis and long and complex in chassis such as garbage packers.

> **SHOP TALK**
>
> If possible, attempt to level the engine and transmission by raising the rear of the truck and supporting it on stands before beginning to remove the transmission. This has two advantages. First, it can greatly increase the amount of room under the truck. Second, it can make reinstallation much easier because you will not be fighting gravity while inserting the input shaft into the clutch assembly.

1. Remove the chassis battery ground cable.
2. Drain the lubricant from the transmission.
3. Remove the shift lever from the transmission. Remove the shift tower and/or shift rail housing from the transmission and seal the opening with a rag. See **Figure 16–15**.
4. Scribe a mark on the transmission output yoke, the drive shaft flange, and the output shaft stub on the transmission. Aligning the scribe marks on

FIGURE 16–15 Removing the shift rail assembly from a transmission

reinstallation will ensure that these components are correctly installed.

5. Separate the U-joint at the transmission and remove the drive shaft.
6. Disconnect any electrical connections from the transmission.
7. Disconnect all chassis-to-transmission air lines.
8. Remove the clutch lever return spring, then mark and disconnect the clutch linkage.
9. If a hydraulic clutch is used, disconnect the slave cylinder assembly and use mechanics wire to support it on the frame or cross-member. It should not be necessary to separate any hydraulic lines.
10. Support the transmission with either an overhead hoist or, from below, a transmission jack. The transmission should be securely supported.
11. Remove the fasteners that attach the transmission to any transverse frame brackets. Next take a look at the engine/transmission mounts. If the engine mounts are located on the engine flywheel housing, simply remove them. Some OEMs, however, place engine mounts on the transmission bell housing. If this is the case, you will have to support the full weight of the engine securely before uncoupling the bell housing.
12. Remove the bell housing to flywheel housing bolts. Gently separate the transmission bell housing from the flywheel housing by pulling rearward. Try to pull the transmission away from the engine while maintaining alignment.
13. Remove the transmission from the vehicle. You may have to use a frame scissor jack or a frame supported chain hoist to raise the truck chassis high enough to get the transmission out from under the frame. Never place any part of your body under any component supported on a hydraulic jack or chain hoist.
14. Use a cherry picker to raise the transmission from the transmission jack and mount it on a transmission overhaul fixture.

CAUTION

Ensure that the transmission does not hang by the input shaft in the pilot bearing bore in the flywheel. The clutch assembly, pilot bearing, and input shaft can be damaged if the transmission is supported by the input shaft.

DISASSEMBLY GUIDELINES

It is important that the OEM transmission service literature guide you through the disassembly, inspection, and reassembly procedures. Again, we use a general approach to the disassembly of a transmission.

General Precautions

Whenever disassembling a transmission, follow these general precautions:

- Wash, dry, inspect, and lubricate all reusable bearings as they are removed; you can wrap these in grease paper until ready for reinstallation. If bearings are to be reused, use bearing pullers to pull them out. Driving the bearings out with a punch damages them so only do this when bearings are going to be replaced.
- Work systematically. Arrange removed components neatly on a clean bench in sequential order of removal. This is just good workshop practice and will simplify reassembly. Placing removed parts on clean paper reduces contamination.
- Remove snaprings with snapring pliers. This allows you to reuse snaprings on reassembly. Using a screwdriver distorts and damages snaprings.
- Work cleanly. Do not risk dirt contamination when reassembling a transmission—it produces more field failures than any other single cause.
- Be gentle. Never use unnecessary force on shafts, housings, and bearings. Soft-face hammers, prybars, and bearing pullers make the use of excessive force unnecessary.

INSPECTION

Failure analysis, the process by which you determine what caused the failure, could be the most important step in the overhaul procedure. By the time the transmission is in pieces on the workbench, you should know exactly what caused it to fail. In addition, reusing a questionable component is a high-risk practice that could generate a *comeback* failure, with your employer assuming the cost of the overhaul. When assessing the reusability of transmission components, evaluate the age, application, and mileage of the vehicle. The following are some tips for inspecting transmission components during and after disassembly.

BEARINGS

The service life of a transmission is often determined by the life of the bearings. Transmission bearings fail due to vibration and dirt. Some of the more prominent reasons for bearing failures are:

- Abrasion wear due to dirt
- Fatigue of hard-surfacing on raceways or balls
- Wrong type or grade of lubricant, or broken-down lubricant
- Lack of lubricant caused by low lubricant levels or bad driving practices
- Vibration breakup of retainer, brinelling of races, fretting corrosion, and seized bearings
- Bearings adjusted tight or loose
- Improper fit of shafts or bore
- Acid etch of bearings due to water in lube
- Overloading of vehicle. Improperly spec'd transmission with the engine producing more torque than transmission can sustain
- Engine torsionals causing low-frequency vibration that pounds out bearing hardened surfaces

SHOP TALK

Engine torsionals refer to the frequency of torsional pulses delivered to the drivetrain as the force of each engine cylinder power stroke is unloaded into the crankshaft. Because of today's practice of managing engines at slower speeds and higher torque to produce better fuel economy, lower-frequency, high-torque torsionals are produced and this has resulted in some transmission failures.

Bearing Inspection

Clean and inspect a bearing as follows:

1. Wash all bearings in clean solvent. Allow the bearing to dry. Check balls, rollers, and raceways for pitting, discoloration, and spalling. Reject any bearings that are questionable.
2. Lightly lubricate the bearing and check axial and radial clearances with its race.
3. Reject bearings with excessive clearances. This is an indicator of wear and is usually accompanied by visual clues.
4. Check the bearing race and shaft fits. The bearing inner bore should be tight on the shaft to which it is fitted. Outer races are sometimes interference fit into the housing bore, so check this using the OEM procedure. If an interference fit bearing race turns in its bore, the housing has to be replaced or repaired.

Analysis of Bearing Failures

We now take a more detailed look at why transmission bearings fail.

Dirt. More than 90 percent of transmission bearing failures are caused by abrasive dirt. Dirt usually enters the bearings during assembly. Research by one major truck manufacturer concludes that this is why a reconditioned transmission unit has less than half the service life expectancy compared with a new one. Dirt can also enter through the seals or breather, or even from using dirty containers to add or change lubricant. Dirt in a transmission because of poor assembly practices is avoidable so make a practice of working cleanly.

Softer material such as grain chaff, smoke particulate, and light dust forms an abrasive paste within the bearings themselves. The rolling action of the bearing entraps the abrasives, which now work like lapping paste, eroding the hard-surfacing. As bearings wear they become noisy. The abrasive action tends to increase rapidly as ground steel particles from the bearing hard-surfacing add to the abrasives.

Hard, coarser materials can enter the transmission during assembly from hammers, drifts, or power chisels. These materials can also be produced from within the transmission due to raking teeth, causing minute cuttings. These cuttings mixed with the lube produce small indentations in the critical contact areas of bearings and gear teeth.

Fatigue. Bearing fatigue is characterized by flaking or spalling of the bearing race. It is a metallurgical failure of the bearing that can be caused by old age or a manufacturing problem. Spalling is granular weakening of the hardened surface steel that causes it to flake away from the race. Because of their rough surfaces, spalled bearings will run noisy and produce vibration.

Normal fatigue failure occurs when bearing service life is at an end and the bearing hard-surfacing appears to be evenly worn through into the base metal below it. All bearings will eventually succumb to fatigue failure.

Lubrication Failures. Bearing failures due to poor lubrication are characterized by discoloration, spalling, and sometimes bearing break-up. Failures can result from low oil level, contaminated lube, improper grade oil, and mixing of incompatible oil types. The use of additives falls into the last category. Lubrication failures are avoidable so ensure that PM practices are adhered to.

Brinelling. A **brinelling** condition (**Figure 16–16**) can be identified as tiny indentations high on the shoulder or in the valley of the bearing race. It can be caused by improper bearing installation or removal.

FIGURE 16–16 A brinelled race: Note that the indentations run across the race line.

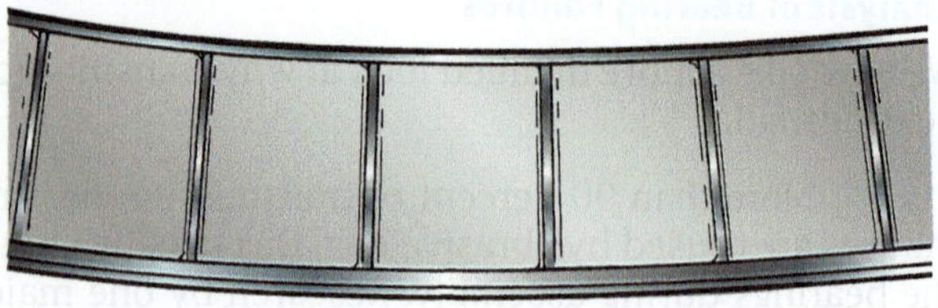

Driving or pressing unevenly on a bearing can fractionally twist it, meaning that any force absorbed by the bearing is not distributed evenly by the race. This type of damage can be avoided by using correct drivers or pullers.

Fretting. Sometimes the bearing outer race can pick up the machining pattern of the bearing bore as a result of vibration. This action is called **fretting** (**Figure 16–17**). A fretted bearing can be mistakenly diagnosed as one that has spun in the bore. Fretting in itself does not require replacement of the bearing.

Contamination. Scoring, scratching, and pitting of bearing contact surfaces identify contamination failures. Contamination is caused by very fine particles suspended in the lubricant or by the use of older extreme pressure (EP) oils. This type of lube contamination will result in abrasive action and polishing of the bearing contact surfaces, which can significantly shorten the life of the bearing.

SHOP TALK

Store new bearings in their shipping wrappers until ready for use. Used bearings should be cleaned in solvent, lubricated, and wrapped in greasy paper until ready for installation.

GEARS

Visually inspect gear teeth for frosting and pitting. Frosting of gear teeth is normal and not an indicator of transmission failure. Frosted gears often heal in operation. When the frosting appearance progresses to pitting that can be felt by stroking a fingernail over the surface, the gear will have to be replaced.

Check for gears with worn clutching teeth that should be replaced. Check for taper and for reduced length and chipping of the clutching teeth. Check the axial clearance of each gear. Where excessive clearance is identified, check the gear snapring, washer, spacer, and gear hub for wear. Check the OEM-recommended axial clearance between the mainshaft gears.

FIGURE 16–17 A fretted outer race: Note that the wear indentations run with the race line.

Synchronizers

Check the synchronizers for burrs, uneven and excessive wear at contact surfaces, and metal particles. Examine the blocker pins for wear or looseness. Replace any suspect parts. Remember to check the synchronizers for the auxiliary drive and low-range gears.

SPLINES

Check the shaft splines for wear and damage. If the sliding clutch gears, companion flange, or clutch hub have worn into the land walls of shaft splines, you will have to replace the shaft. Check the splines and the components that slide on them for binding.

Limit Washers

Visually inspect the contact surfaces of all limit washers. Limit washers define thrust tolerances so they should be replaced if they are scored or reduced in thickness.

Gray Iron Components

Check all gray iron parts for cracks and porosity. Replace or repair parts found to be damaged. Most OEMs recommend replacing any housing components found to be defective. Welding or brazing porosity or cracks is generally not recommended.

Clutch Yokes

Check the clutch release components. Replace yokes worn at cam surfaces. Replace the bearing carrier if worn at the contact pads.

Shift Bar Housing

Visually inspect the shift yokes and blocks at the pads and the lever slot. Check the rail action by sliding fore and aft, noting the detent performance. Replace any defective or worn components. Check the yokes for alignment and twisting; replace sprung yokes. Check lock screws in the yokes and blocks. Tighten and rewire those found loose.

If the housing has been disassembled, check the neutral and engage notches of the shift rails for wear from the detent balls. Each rail should slide smoothly and engage into each detent position with a snapping noise.

Shift Tower

Check the spring tension on the shift lever and replace the tension spring and washer if the lever flops around too freely. If the housing was disassembled, check the spade pin and its corresponding slot, replacing both if worn.

16.5 GENERAL OVERHAUL GUIDELINES

The procedure for reassembling most transmissions reverses the disassembly procedure with a few exceptions. Before beginning, the transmission and auxiliary section housings should be clean. The transmission overhaul kit should contain new gaskets and O-rings. The following are some preassembly tips:

- Gaskets can be held in place during assembly using a high-tack spray adhesive. As a rule you should avoid using silicone unless instructed by the assembly literature.
- Use the O-ring assembly lubricant provided in the overhaul kit. This lubricant has a silicone base but does not set.
- Thread sealant should be used on most of the fasteners used in assembling a transmission. In addition to clamping components together, transmission bolts more often than not seal the threaded holes in which they are installed.
- Torque. As you reassemble the transmission, refer to the manufacturer's specifications for proper installation and torque of all fasteners, setscrews, locknuts, and plugs.

Timing Transmissions

Later in this chapter the **timing** of a specific ten-speed transmission is outlined, so in this section we will look at some general guidelines to timing gears in the main box. Before reassembling the front section of the transmission, the drive gears of the mainshaft and the countershafts must be marked in the following way:

1. Before placing each countershaft assembly into the housing, mark the tooth located directly over the keyway of the countershaft drive gear, as shown in **Figure 16–18A**. In many transmissions, you might find that this tooth is stamped with an "O" to identify it.
2. Mark any two adjacent teeth on the main drive gear.
3. Mark the two adjacent teeth located directly opposite the first set marked on the main drive gear. As shown in **Figure 16–18B**, there should be an equal number of unmarked gear teeth on each side between the marked teeth.

FIGURE 16–18 (A) On the countershaft, mark the tooth opposite the keyway; (B) on the main drive gear, mark two teeth on opposite sides of the gear.

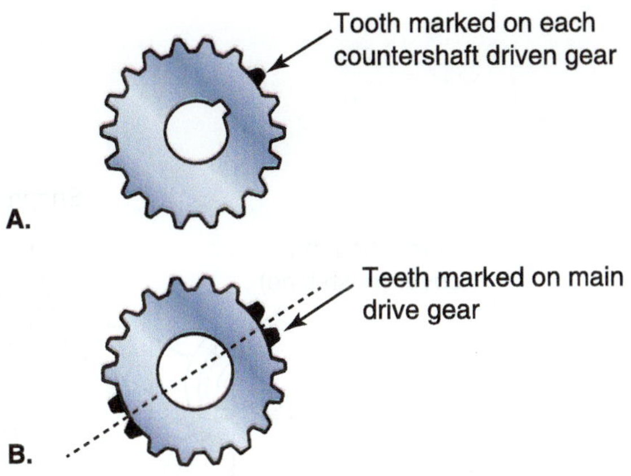

FIGURE 16–19 When properly timed, the marked teeth on the mainshaft and countershaft drive gears will mesh.

4. After the mainshaft and countershaft are installed into the housing, the marked teeth should be meshed exactly as shown in **Figure 16–19**.

16.6 STEP-BY-STEP OVERHAUL

This section divides the transmission into subunits and addresses the strip-down and reassembly procedure for each subunit. Sometimes general information that appears earlier in the chapter is repeated.

CHANGING AN INPUT SHAFT

The usual reason for having to replace the input shaft of a transmission is excessive clutch wear on the splines. The input shaft can usually be removed

FIGURE 16–20 Components of a typical input shaft assembly

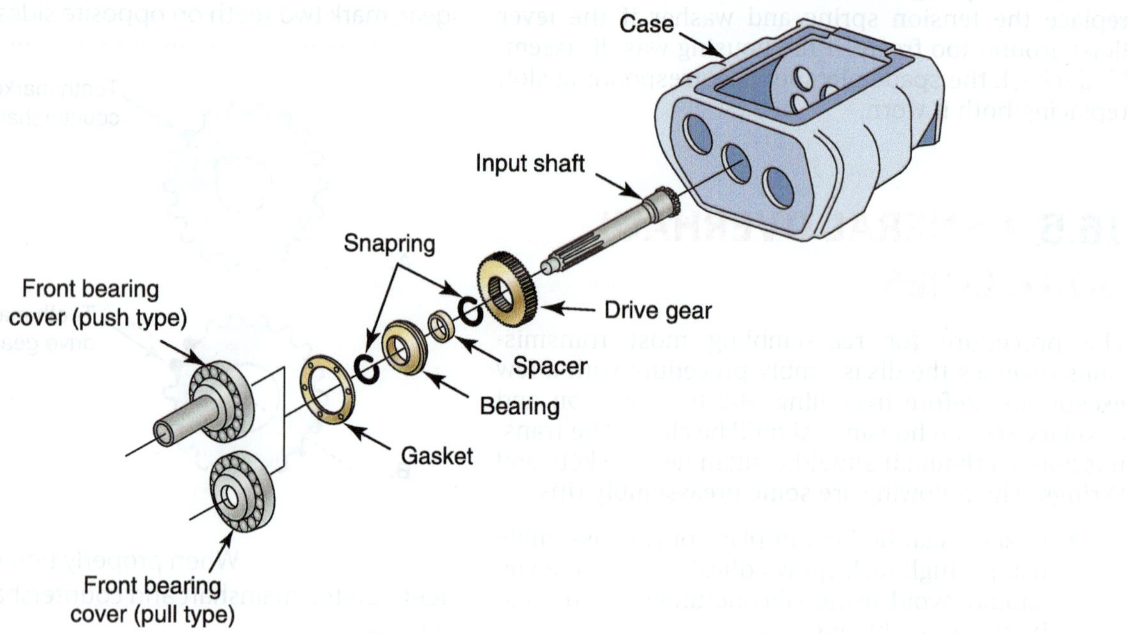

without too much disassembly of the transmission (**Figure 16–20**), but if you want to also change the main drive gear, the main section will have to be disassembled.

Input Shaft Removal

You can remove the input shaft without disassembling the main case, using the following procedure.

1. Remove the six fasteners from the front bearing cover.
2. Remove the front bearing cover and gasket. Clean any gasket material from the case and cover and discard the rubber shipping seal if it is there.
3. Remove the bearing retaining snapring.
4. Use a mild steel drift and drive the input shaft toward the rear case as far as possible. Then pull the input shaft forward sufficiently to fit a bearing puller over the bearing.
5. Draw the input shaft bearing from the housing using the bearing puller.
6. Remove the drive gear internal snapring.
7. Pull the input shaft forward and out of the drive gear and housing.
8. Inspect the bushing in the front input shaft pocket and replace if damaged.

Reinstallation of an Input Shaft

To reinstall an input shaft without disassembling the main box, you can use the following procedure:

1. If necessary, install the bushing in the pocket of the input shaft.
2. Install the input shaft so that it engages with the splined teeth of the main drive gear. The current design input shaft spline teeth have noticeable clearance to the main drive gear internal spline teeth; this is normal.
3. With the bearing external snapring to the outside, position the bearing on the input shaft.
4. Drive the input shaft bearing over the input shaft using a flanged end bearing driver that contacts both the bearing inner and outer races. Drive the bearing until it seats in the front case.
5. Reinstall the input shaft retaining snapring.
6. Install the front bearing cover along with a new gasket. Torque the six fasteners to specification.

SHIFT RAIL HOUSING

The procedure for removing and then reassembling a shift tower and rail housing is covered next. You can reference Figure 15–32, Figure 15–33, and Figure 15–34 if you have trouble identifying the components of a typical shift tower and shift rail housing assembly. Basic disassembly is as follows:

Shift Rail Housing Disassembly

1. Remove the screws securing the gearshift lever housing to the top of the transmission, jar the lever slightly to break the gasket seal, and then remove the shift lever from the shift bar housing. You remove a remote control shift hub in the same manner.
2. Remove the fasteners securing the shift rail housing to the transmission, jar the housing to break

the gasket seal, and lift the assembly away from the transmission housing.

3. Tilt the shift rail assembly and remove the detent tension springs and balls from their bores in the housing.
4. Secure the shift rail housing in a vise. Slide the third fourth speed shift bar to the rear, and remove the yoke (fork) and shift block from the rail. You may have to use sidecutters to cut the yoke bolt lock wires.
5. Slide the first second speed shift bar to the rear; remove the yoke and shift block from the rail. As the neutral notch in the bar clears the rear boss, remove the small interlock pin from the bore.
6. Remove the actuating plunger from the center boss bore.
7. Slide the low reverse shift rail to the rear; remove the yoke from the rail. As the rail is pulled from the housing, two interlock balls will drop from the rear boss bottom bore.
8. If required, remove the plug, spring, and plunger from the low reverse speed shift yoke bore.

Shift Rail Housing Assembly

A shift rail assembly should proceed as follows:

1. Install the reverse stop plunger in the low/ reverse shift yoke. Install the spring in the bore on the plunger shank.
2. Install the plug and tighten it to compress the spring. Back the plug out one and a half turns and stake the plug through the small hole in the yoke.
3. With the shift rail housing in a vise, hold the notched end of the low reverse shaft and install the bar in the lower bore of the shift rail housing bosses. Slide the bar through the low reverse yoke and install the yoke lock screw. Tighten the lock screw and then secure with a lock wire.
4. Install the actuating plunger in the center boss bore.
5. Install one interlock ball in the rear boss top bore so that the ball rides between the low reverse and the first second shift rails.
6. Install the first second shift rail in the middle bore of the housing bosses. Position the shift block on the rail between the center and rear bosses, and the first second yoke on the rail between the front and center bosses. Just prior to inserting the notched end of the rear boss bar, install the small interlock pin vertically in the neutral notch bore. Install the block and yoke lock screws. Install the lock wire securely.
7. Install the second interlock ball in the rear boss top bore so it rides between the first/second and third/fourth shift bars.
8. Install the third/fourth speed shift rail in the housing boss upper bore with the third/fourth yoke positioned between the front and center bosses and the shift block between the center and rear bosses. Install the block and yoke lock screws and lock wire securely. Reinstall the oil trough, if so equipped.
9. Remove the assembly from the vise and install the three detent balls, one in each bore, on the housing top. Install the three tension springs, one over each tension ball, in the housing bores.
10. Reinstall the shift rail housing and replace the gasket.
11. Reinstall the shift tower assembly and replace the gasket.

OUTPUT YOKE

1. Prior to removing the output yoke, lock the transmission by engaging the two mainshaft gears with their mainshaft sliding clutches.
2. Use a large breaker bar to turn the output shaft nut from the output shaft.
3. Pull the yoke straight to the rear and off the output shaft (**Figure 16–21**).

AUXILIARY SECTION REMOVAL

To remove the auxiliary section from the transmission:

1. Hold the transmission output stationary using a yoke spanner, allowing it to lock against the frame. Use a 1-inch air gun or breaker bar to remove the yoke nut from the output shaft.
2. Pull the yoke from the output shaft splines.
3. Remove the speedo drive gear or replacement spacer from the hub of the yoke. If not already removed, remove the shift bar housing from the main box. Remove the fasteners that bolt the auxiliary section to the main box.
4. Insert the puller screws in the puller threads in the auxiliary housing flange. Depending on the

FIGURE 16–21 Removing the output yoke from the output shaft

FIGURE 16–22 Removing the auxiliary section using a chain hoist

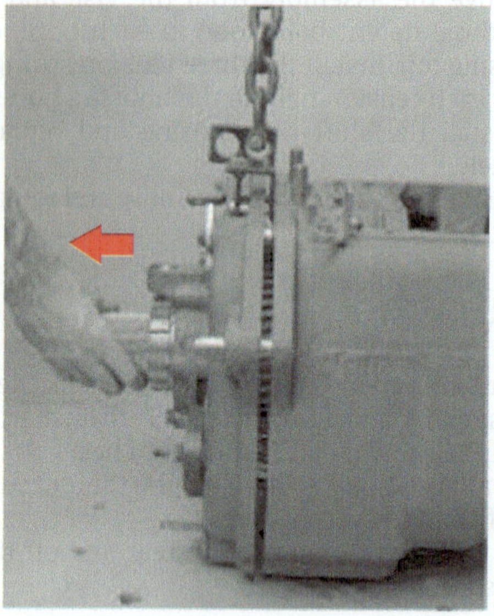

generation either two or three puller threads are located to ensure that the housing is separated evenly. You will probably have to clean the threads using a M10 × 1.5 thread tap. Oil the holes.

5. Run in the puller screws evenly to break the auxiliary box housing away from the front section. Stop just before the auxiliary section clears the alignment dowels. Remove the pusher screws.
6. Attach an auxiliary section hanger bracket (**Figure 16–22**) to support the weight. In most housings you will have to jiggle the auxiliary section before it moves free of the main box.

BELL HOUSING REMOVAL

To remove the bell (clutch) housing and front bearing cover from the transmission main box, use the following procedure:

1. Remove the clutch release mechanism by separating the yoke bolts from the clutch release shafts.
2. Unscrew the nuts from the studs and turn out the bolts that clamp attach the bell housing to the transmission housing.
3. Tap on the clutch housing with a rubber mallet to break the gasket seal and separate the bell housing.

AUXILIARY SECTION

Although this procedure is not so commonly performed today, it is not unusual to overhaul the auxiliary section of a truck transmission in isolation from the main section. This section addresses this procedure, beginning with disassembly. We use the example of a current ten-speed Roadranger transmission, but as always, technicians are advised to consult the service literature (reference http://www.roadranger.com) for the specific transmission they are working on. **Figure 16–23** shows an exploded view of the auxiliary section components.

DISASSEMBLY

Clamp the auxiliary section upright into a soft jaw vise (**Figure 16–24**) or mount to a stand.

1. If the auxiliary countershafts, bearing assemblies, shims, auxiliary housing, and main case are to be reused, remove one of the countershaft bearing covers and mark the parts as upper or lower countershaft as shown in **Figure 16–25**. Providing the countershafts are reassembled with all their parts in their original positions, reshimming will not be required.
2. Mark the output bearing cover and the rear of the housing so that they can be reinstalled in an identical way on reassembly. (It can be installed in two opposite positions.)
3. Remove the speedometer rotor/seal sleeve and separate the O-ring from the output shaft.
4. Use a mild steel drift to drive the output shaft far enough forward so that the auxiliary countershafts can be removed.
5. Separate the range synchronizer assembly and range yoke assembly from the output shaft.
6. Continue to move the output shaft forward until it can be separated from the housing.
7. Remove the four fasteners from each auxiliary housing countershaft rear bearing cover.
8. Remove the auxiliary countershaft bearing races.
9. From the back of the auxiliary housing, remove the capscrews securing the lock cover. Then remove the lock cover and O-ring.
10. From the rear of the auxiliary housing cover, remove the output bearing cover plate fasteners. Drive the seal out using a mild steel drift and then remove the output bearing and its cup.

WARNING

As the output shaft is moved forward, the auxiliary countershafts may drop from the housing, causing personal injury or damage to the countershafts.

Disassemble the Range Yoke

1. Remove the range bar O-ring and range bar pin.
2. Remove one of the range yoke snaprings, taking care not to overstretch the snapring.
3. Finally, remove the yoke bar from the range yoke bore.

FIGURE 16–23 Exploded view of the auxiliary section housing components

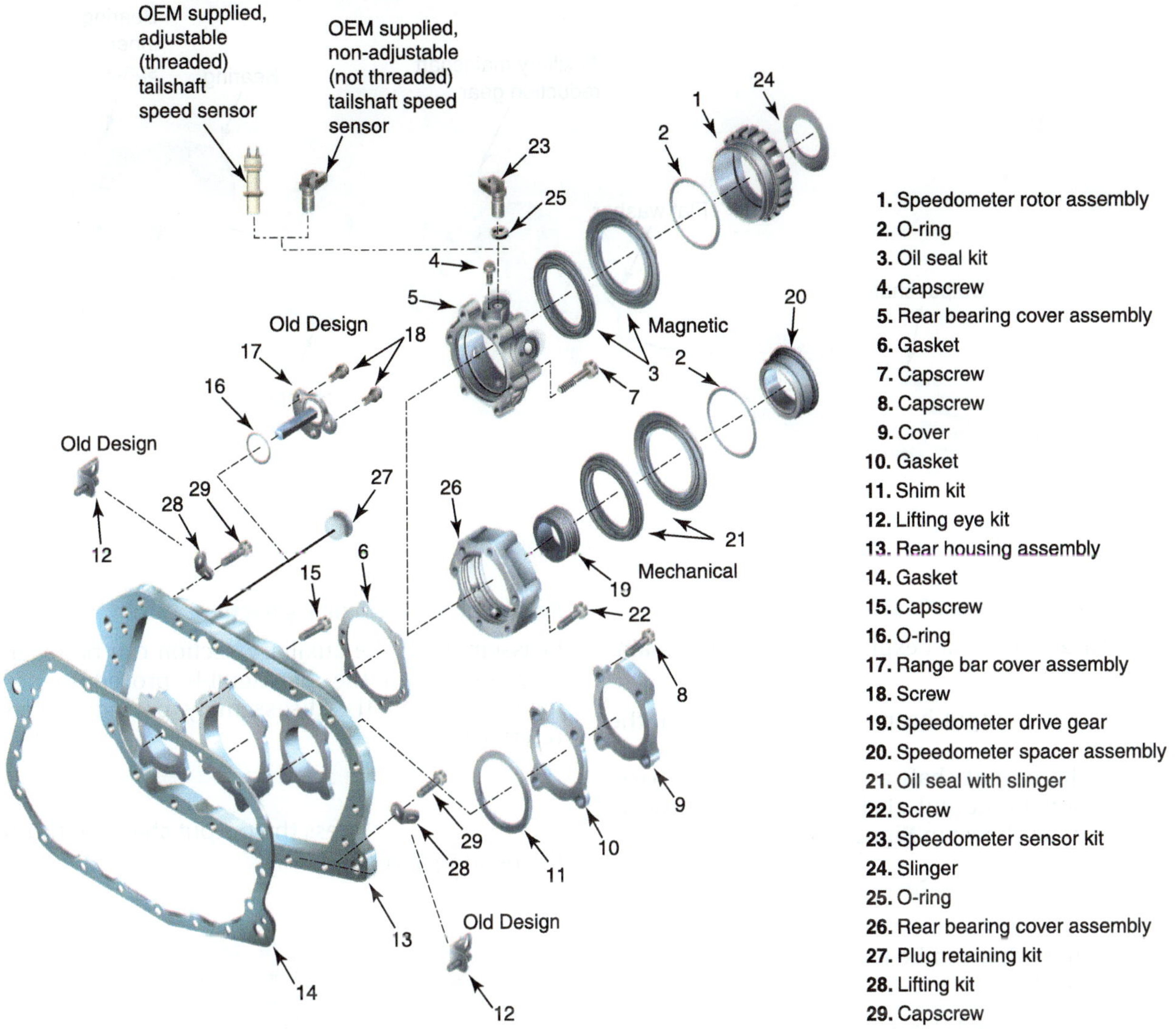

FIGURE 16–24 Auxiliary section in a vise prior to disassembly

Courtesy of Eaton Corporation

FIGURE 16–25 Marking the upper countershaft parts prior to disassembly so that reshimming is not required on reassembly.

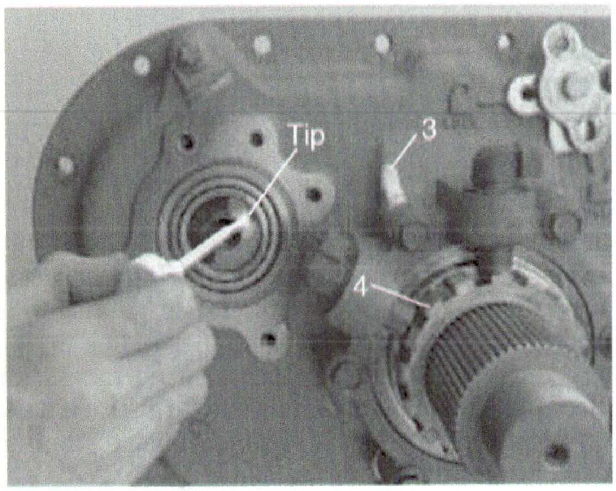

Courtesy of Eaton Corporation

FIGURE 16–26 Exploded view of the output shaft assembly

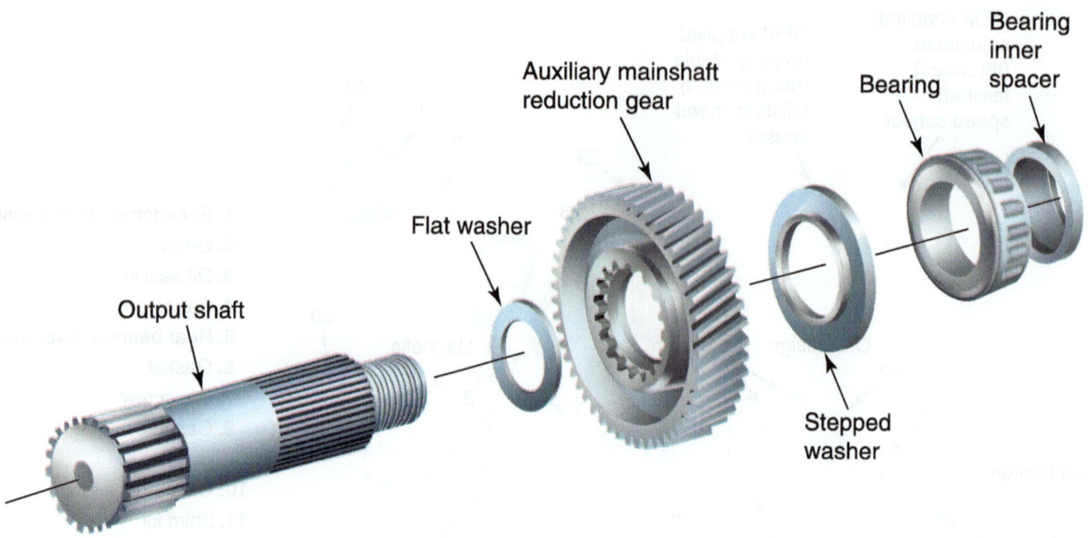

Disassemble the Output Shaft

Figure 16–26 shows an exploded view of the mainshaft assembly.

1. Remove the output bearing inner cover from the output shaft.
2. Using the output shaft assembly gear as a base, press out the output shaft through the bearing and gear as shown in **Figure 16–27**.

Disassemble the Synchronizer

Figure 16–28 shows an exploded view of a synchronizer assembly that can be used as a guide throughout this procedure.

1. Place the larger LO range synchronizer on the bench and cover it with a rag to contain the springs when they are released.
2. Hold both sides of the HI range synchronizer and pull to separate.
3. Remove the sliding clutches from the synchronizer ring LO range pins.

Finish Auxiliary Section Disassembly

To complete the disassembly of the auxiliary section a bearing puller and slide hammer will be required. Use the following procedure:

1. To disassemble the auxiliary countershaft, fit a bearing puller over the bearing. It may be necessary to cut and remove the steel cage and roller assembly in order to get the puller in position.
2. To remove the countershaft bearing races from the auxiliary housing, attach a bearing race puller to the cone. Connect a slide hammer to the race puller and drive from the housing.

Reassemble the Auxiliary Section

Reassembly of the auxiliary section can be regarded as a reversal of the disassembly procedure, so the figures used in the disassembly procedure can be referenced.

FIGURE 16–27 Press the output shaft out through the bearing and gear.

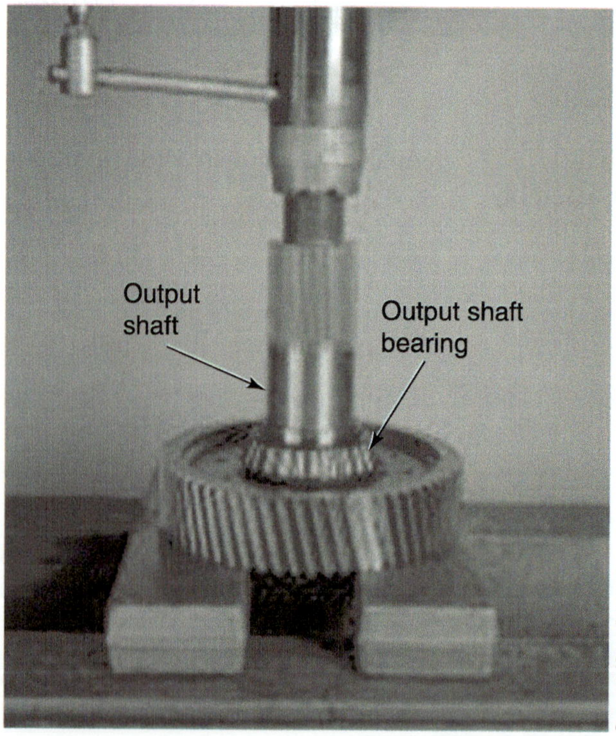

FIGURE 16–28 Exploded view of a synchronizer assembly

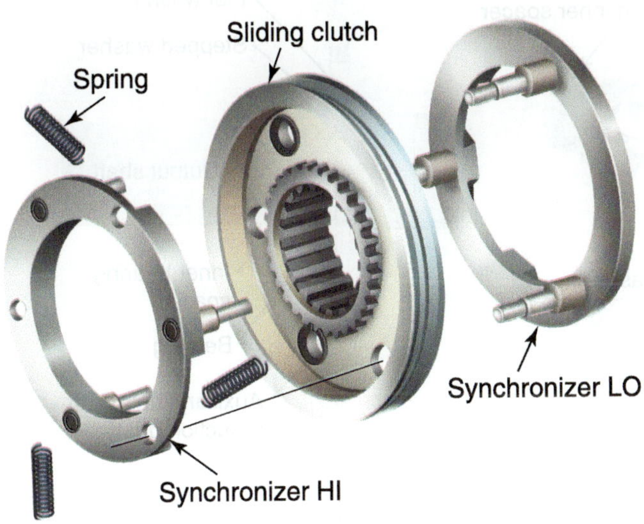

Assemble the Synchronizer

1. Place the larger LO range synchronizer on a bench with the pins facing upward.
2. Holding the sliding clutch with the recessed side up, position it onto the LO range synchronizer pins.
3. Install the three springs into the HI range synchronizer bores.
4. Lower the HI range synchronizer ring onto the LO range synchronizer ring as shown in **Figure 16–29**. Rotate the HI range synchronizer until the springs seat against the pins.

FIGURE 16–29 Seating the synchronizer springs

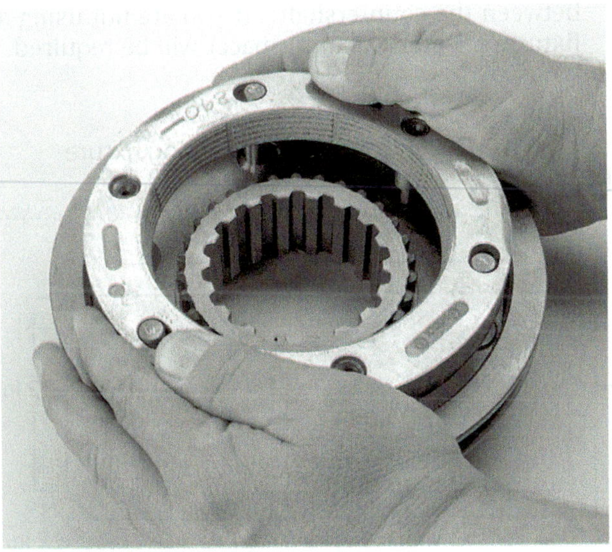

Courtesy of Eaton Corporation

Assemble the Output Shaft

Reference Figure 16–26 for an exploded view of the output shaft assembly.

1. Use toolmaker's dye to mark one tooth on the auxiliary mainshaft reduction gear. Do the same on the tooth diametrically opposite the first one marked.
2. Stand the output shaft upright on a bench with the threaded end up.
3. Referencing the detail in **Figure 16–30**, assemble the output shaft components, beginning with the flat washer and working backward.
4. With the auxiliary mainshaft reduction gear clutching teeth facing down, slide the gear over the output shaft.
5. Ensuring that the smooth side is facing up, slide the stepped washer onto the output shaft and against the reduction gear.
6. Either heat or press the output bearing cone (with the long rollers) taper side up tightly against the stepped washer. Place the bearing inner spacer onto the output shaft.

CAUTION

Output bearings are matched sets: Do NOT mix bearings from other assemblies. The result will be damaged bearings.

CAUTION

If using heat to install bearings, ensure that the temperature never exceeds 275°F (135°C). Excessively high temperature damages the bearings and results in premature failure.

Assemble the Range Yoke

Use **Figure 16–31** to facilitate the assembly of the range yoke using the following procedure:

1. Install the range yoke onto the range yoke bar. The flat side of the range yoke should face toward the notched end of the range yoke bar.
2. Install the range yoke snaprings in the grooves on each side of the range yoke. Ensure that they fit tightly and are not overstretched. If they are loose replace them.
3. Install the round pin in the range yoke bar.
4. Slide the range yoke bar O-ring into place to hold the pin in place.

Assemble the Auxiliary Countershaft

Reference **Figure 16–32** when assembling the auxiliary countershaft and use the following procedure:

1. Install a washer on the rear of the countershaft.
2. Using a bearing driver, press a new bearing onto the rear of the countershaft. Ensure that the

FIGURE 16–30 Output shaft assembly sequence

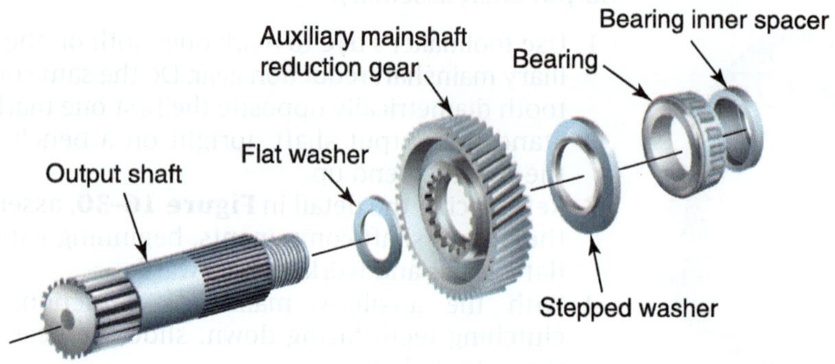

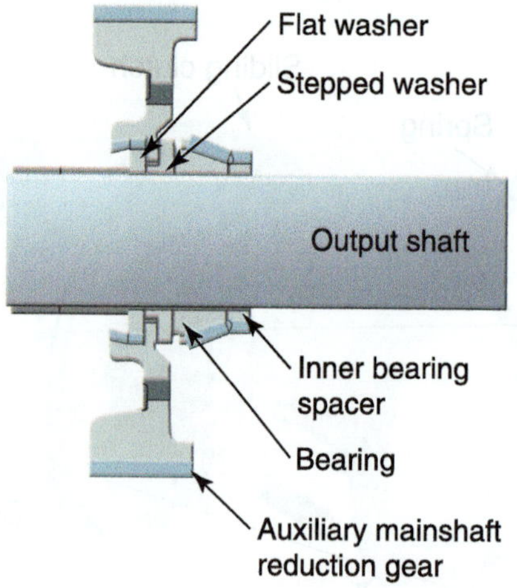

FIGURE 16–31 Exploded view of the range yoke

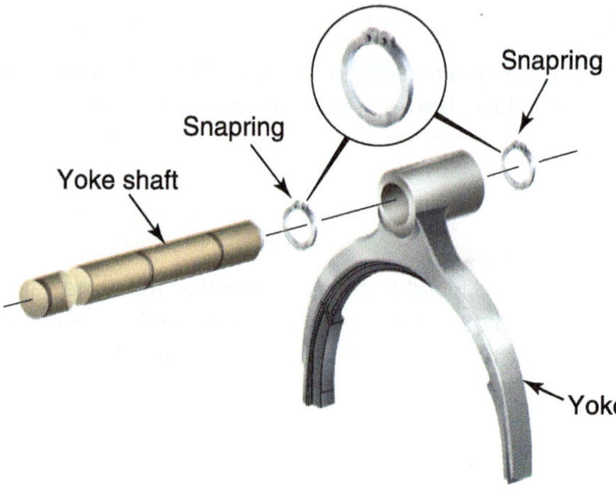

FIGURE 16–32 Exploded view of the auxiliary countershaft assembly

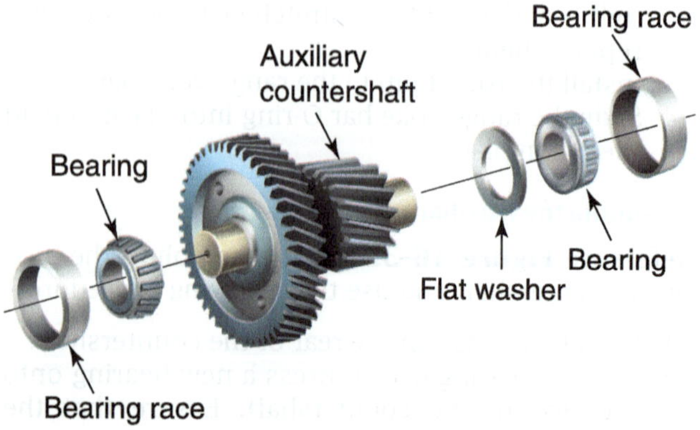

driver used contacts the inner race of the bearing only.

3. Using a bearing driver, press a new bearing onto the front of the countershaft.

Final Assembly

To assist in final assembly of the auxiliary section, Eaton recommends that the tool drawn in **Figure 16–33** be manufactured out of a 2-inch (50 mm) section of wood. This will hold the auxiliary shafts in position during assembly.

1. In order to time the countershaft, identify and mark the two teeth on each countershaft marked with a "0" with toolmaker's dye.
2. Place the countershafts on a fixture or clean flat surface with the marked teeth turned down to the middle of the fixture.
3. Position a 2-inch (50 mm) spacer (a socket will do) between the countershafts. If you are not using a fixture, a 3-inch (75 mm) spacer will be required.

FIGURE 16–33 Countershaft retaining fixture blueprint

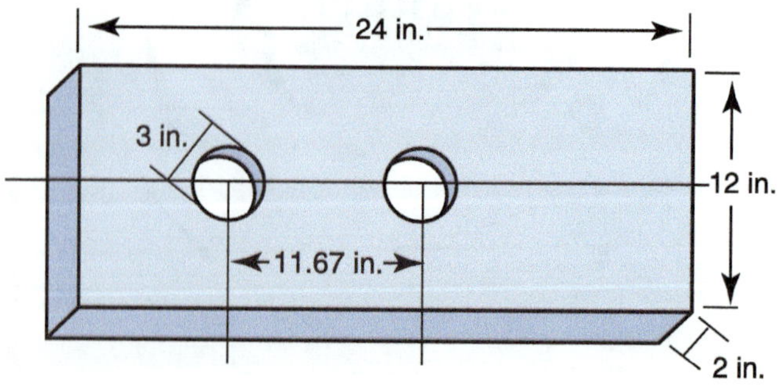

4. Install the synchronizer assembly onto the output shaft. The low-range synchronizer ring (side with external friction material) must face the auxiliary mainshaft reduction gear.
5. Match the timing marks on the auxiliary mainshaft reduction gear with the timing marks on the countershafts (**Figure 16–34**). The output shaft will rest on the spacer (see step 3).
6. With the round hole in the range yoke bar facing up, slide the range yoke assembly into the groove on the range synchronizer.
7. Lightly oil the bearing bores on the auxiliary housing and place it over the countershaft assemblies and output shaft assembly (**Figure 16–35**).
8. Install the output shaft bearing race in the housing bore.
9. Ensure that the output shaft bearing inner spacer is on the output shaft; this should have been installed when the output shaft assembly was put together.
10. Heat the rear output shaft bearing cone (or use an appropriate driver) and install the outer bearing cone, tapered side down onto the shaft.
11. Install a new gasket.
12. Using the orientation marks made during disassembly as a guide, install the rear bearing cover into its original position.
13. Apply Eaton or equivalent thread sealant to the six rear bearing cover retaining capscrews and torque to specification.
14. Insert the countershaft bearing races in their bores. If the original parts are being reused in their original locations, the shims can also be installed at this time.

FIGURE 16–34 Matching the timing marks on the auxiliary mainshaft reduction gear with those on the countershafts

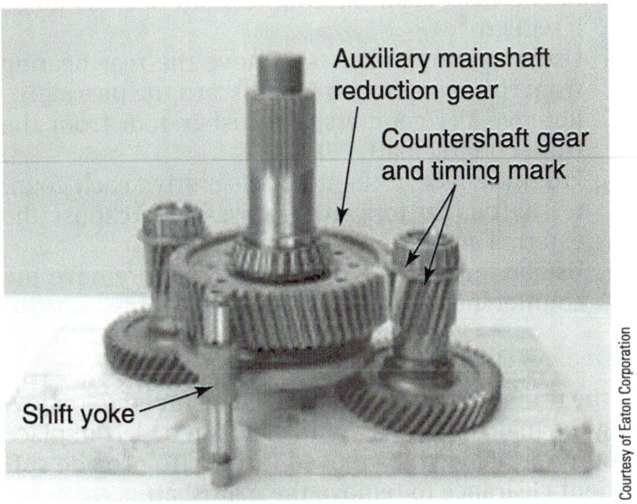

FIGURE 16–35 Assembling the auxiliary housing over the countershafts and output shaft assembly

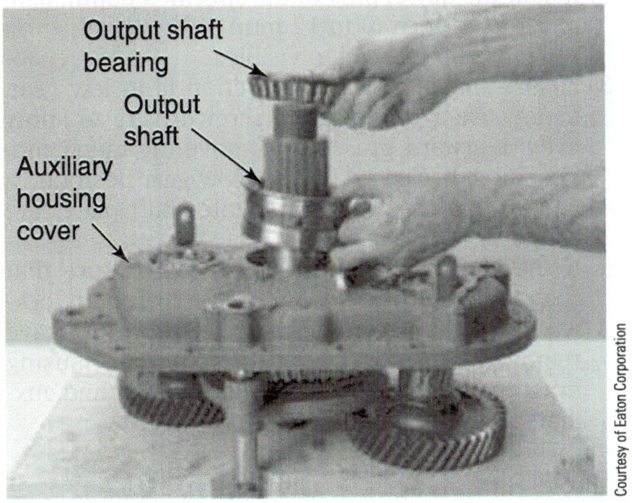

15. If shimming is required (new parts), install the auxiliary countershaft shim and support tools. Final installation and shimming of the countershafts has to be done after the auxiliary section is installed on the front section. If shimming is not required, the auxiliary countershaft bearing covers can be installed; the shim tools will help prevent the countershafts from dropping excessively when the auxiliary section is installed.
16. Mount the auxiliary section into a soft-jawed vise.
17. On the back of the auxiliary housing, install the range alignment lock cover and O-ring. Rotate the cover to the unlocked position and then secure the cover in place with a capscrew. Newer units do not contain a lock cover, so the shift bar housing has to remain off until the auxiliary section and range cylinder are assembled.
18. Install the O-ring over the output shaft and insert it against the output bearing.
19. Install the speedometer rotor/seal sleeve.
20. Install the output yoke over the output shaft. The yoke should slide on and stop before contacting the speedometer rotor. As the output shaft is installed, the output yoke will contact the speedometer rotor.
21. Inspect the output nut for damage and wear. If the nylon locking material is visibly damaged or worn, a new output nut should be used.
22. Lightly oil the output shaft and output nut threads. Install the output nut and torque to specification.

Install the Auxiliary Section

1. If not previously installed, install the output yoke or flange and the output nut. The yoke should

slide on and stop before contacting the speedometer rotor. As the output shaft nut is installed, the output yoke will contact the speedometer rotor. Torque the output nut to specification.

2. Install the dowel pins to the specified depth; usually ⅜ to ½ inch (9 to 12 mm) of the shoulder on the pins should remain visible.
3. Ensure that the auxiliary section is in low gear. If not, use a pair of large screwdrivers to apply even, rearward pressure to move the range sliding clutch backward into the low-gear position.
4. Ensure that the auxiliary countershaft plates are installed.
5. If the shiftbar housing is installed, ensure that the range bar lock cover is rotated counterclockwise into the unlock position. Newer units have removed the lock cover so the shiftbar housing must remain off until the range cylinder and auxiliary sections are assembled.

Main Case Disassembly

The steps throughout this procedure adhere to the sequence recommended by Eaton Roadranger. Technicians are advised to download the procedure for the specific transmission they are working on from http://www.roadranger.com. Reference **Figure 16–36** while working through this step.

Disassemble the Upper and Lower Reverse Idlers

Newer model reverse idler shafts use a snapring (**Figure 16–37**) on the front in place of the capscrew used in older models.

1. Remove the capscrew (or nut) on the front of the reverse idler shaft. If the idler shaft spins, use a large screwdriver between reverse idler bosses to jam the shaft.
2. Remove the snapring.

FIGURE 16–36 Exploded view of the upper and lower reverse idler gear assemblies

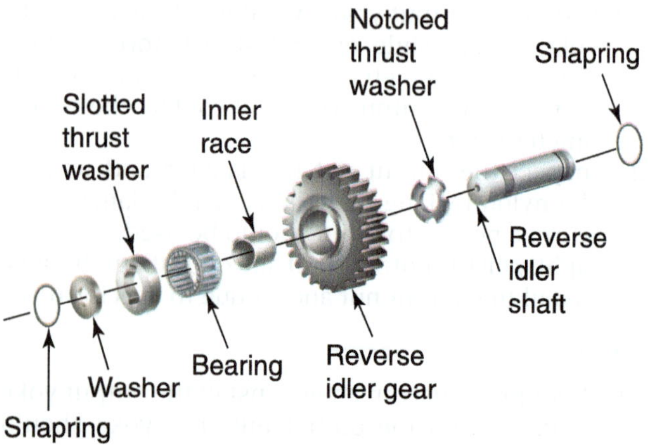

FIGURE 16–37 Pull the rear bearing from the countershaft.

3. Thread a slide hammer puller into the rear of the reverse idler shaft. The threads are M12 × 1.75.
4. Using the slide hammer, pull out the reverse idler shaft. Note: If desired, the upper reverse idler gear can be removed without removing the mainshaft. To do so, remove the snapring from the mainshaft reverse gear, and move the mainshaft reverse gear forward.

Remove the Upper and Lower Countershaft Bearings

To remove the mainshaft assembly, only the upper countershaft bearings need to be removed. This procedure will damage the bearings and should not be performed unless bearing replacement is planned.

1. To prevent the mainshaft pilot from falling out of the input shaft pocket, temporarily install the auxiliary drive gear on the mainshaft.
2. Remove the snapring from each countershaft rear groove.
3. Remove the bearing retainer bolts on the front countershaft. Install the countershaft pusher tool (Eaton special tool T8) to the studs on the front of the case. Push the countershaft back as far as it will go.
4. Use a bearing puller to remove the rear bearing from the countershaft and discard the bearing.
5. Remove the countershaft pusher tool from the front countershaft.
6. From the rear of the main case, drive each countershaft as far forward as possible to expose the external snapring.
7. Use an appropriate bearing puller to remove the countershaft front bearings.

Remove the Mainshaft Assembly

If the transmission overhaul is to be limited to removing the mainshaft, the input shaft can be left in place but it does have to be moved forward to provide sufficient clearance to remove the mainshaft.

FIGURE 16–38 Main case and input bearing cover

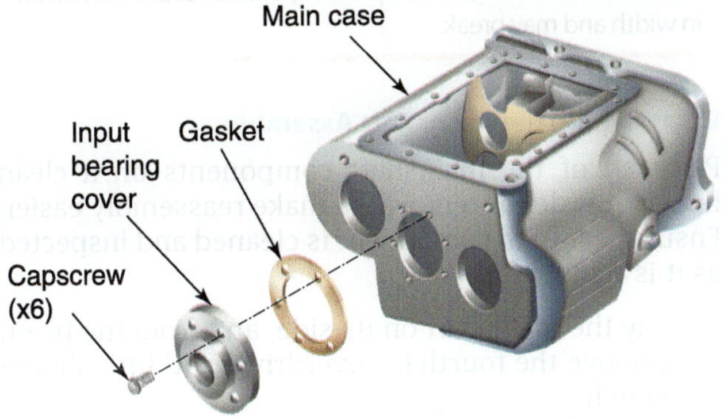

1. Remove the six fasteners from the input bearing cover and remove the cover as shown in **Figure 16–38**.
2. Pull the input shaft forward until the main drive gear contacts the case wall.
3. Remove the reverse gear retaining snapring from the mainshaft.
4. Rotate the mainshaft reverse gear to engage the sliding clutch, and then move the gear forward over the sliding clutch to a position flush with the next gear on the mainshaft.
5. Remove the tanged washer from the mainshaft.
6. Lift out and remove the upper reverse idler gear.
7. Remove the countershaft front and rear bearings by first grasping the mainshaft and pulling rearward until the reverse gear contacts the case wall and then removing the snaprings from the rear of the countershaft. Install the bearing pusher tool on the front countershaft bearing (Eaton special tool #T8) in the main case as shown. Use the tool to push the countershaft rearward through the front bearing. Move the countershaft slowly and stop when resistance is felt. Remove the lower tool T8 from the front countershaft and attach a bearing puller (Eaton special tool #T9) to the upper countershaft rear bearing. Pull the bearing. If the front of the countershaft remains in the front bearing, drive the countershaft forward to expose the snapring. Then use the bearing puller (Eaton special tool #T9) to remove the bearing. If required, repeat this procedure for the lower countershaft bearings.
8. Insert a large screwdriver between the main drive gear and the upper countershaft driven gear to lever the countershaft to the side and away from the mainshaft.
9. Ensure that the mainshaft reverse gear is moved forward and held against the first speed gear.
10. Position the mainshaft hook tool (Eaton special tool #T20) or rope around the mainshaft as shown in **Figure 16–39**. Note: Keep the upper countershaft forward against the case front wall.

FIGURE 16–39 Pulling the mainshaft assembly

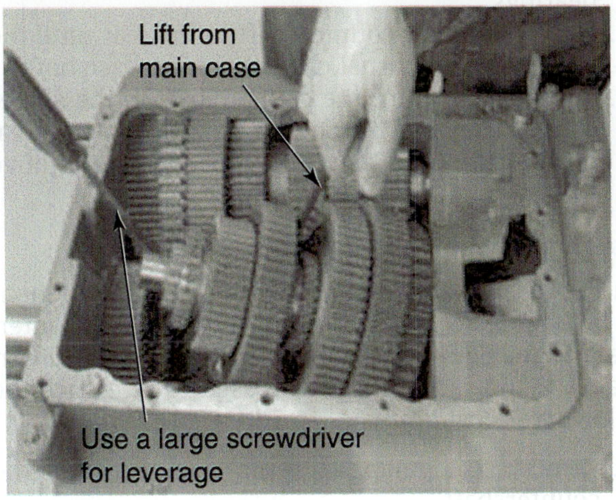

11. Pull the mainshaft to the rear to free the pilot from the input shaft pocket.
12. Tilt the mainshaft front up and lift the assembly from the case.

CAUTION

Take care when removing the mainshaft assembly. The sliding clutch on the front and the reverse gear on the rear of the shaft can slip off the shaft.

Remove the Countershaft Assemblies

Other than the PTO gears, the upper and lower countershaft assemblies are the same, but they should not be mixed. Mark the countershafts as UPPER or LOWER as you remove them. The mainshaft and main drive gear have to be removed before removing the countershaft assemblies.

1. To prevent the mainshaft pilot from falling out of the input shaft pocket, temporarily install the auxiliary drive gear on the mainshaft.
2. Remove the snapring from each countershaft rear groove.
3. Remove the bearing retainer bolts on the front countershaft and install the countershaft pusher assembly (Eaton special tool #T8) to the studs on the front of the main case. Force the countershaft back as far as it will go.
4. Use a bearing puller to remove the rear bearing from the countershaft. Discard the bearing.
5. Remove the pusher assembly (Eaton special tool #T20) from the front countershaft.
6. Working from the rear of the case, drive each countershaft forward as far as possible to expose the external snapring.

7. Use an appropriate bearing puller (Eaton special tool #T9) to remove the countershaft front bearings.
8. Move the upper countershaft to the rear until the front bearing journal clears the front case bore.
9. Swing the front of the countershaft to the center of the case, and lift out the countershaft assembly. You can use the same procedure to remove the lower countershaft.

Disassemble the Countershaft Assemblies

During disassembly, mark each removed component as belonging to either the upper or lower countershaft. **Figure 16–40** is an exploded view of a countershaft assembly that should be used to reference the following procedure.

1. Remove the snapring from the front of the countershaft.
2. Place the countershaft assembly in a press. Using the rear face of the fourth or overdrive gear as a base, press this gear, the drive gear, and the PTO gear from the countershaft.
3. Reposition the countershaft in the press. Using the rear face of second speed gear as a base, press the third speed gear and the second speed gear from each countershaft.
4. Inspect the key. Remove and replace it if it is damaged. NOTE: Now use the procedure outlined earlier in this chapter to remove the input shaft.

CAUTION

Do not use the PTO gear as a pressing base because it is narrow in width and may break.

Disassemble the Mainshaft Assembly

Place all of the mainshaft components on a clean bench in order of removal to make reassembly easier. Ensure that each component is cleaned and inspected as it is removed.

1. Lay the mainshaft on its side, and from the front, remove the fourth (or overdrive) and fifth sliding clutch.
2. Place the mainshaft in a vertical position, pilot end down.
3. Remove the snapring (if applicable).
4. Remove the key, washer, and reverse gear. Mark the keyway location. When removing the washer and reverse gear, note their orientation. Keep the washer with the reverse gear.
5. Remove the first and reverse sliding clutch.
6. Remove the washer and first gear.
7. Remove the washer and second gear.
8. Remove the washer and sliding clutch.
9. Remove the washer and third gear.
10. Remove the washer.
11. Remove the fourth or overdrive gear.
12. Remove the washer.

FIGURE 16–40 Exploded view of a countershaft assembly

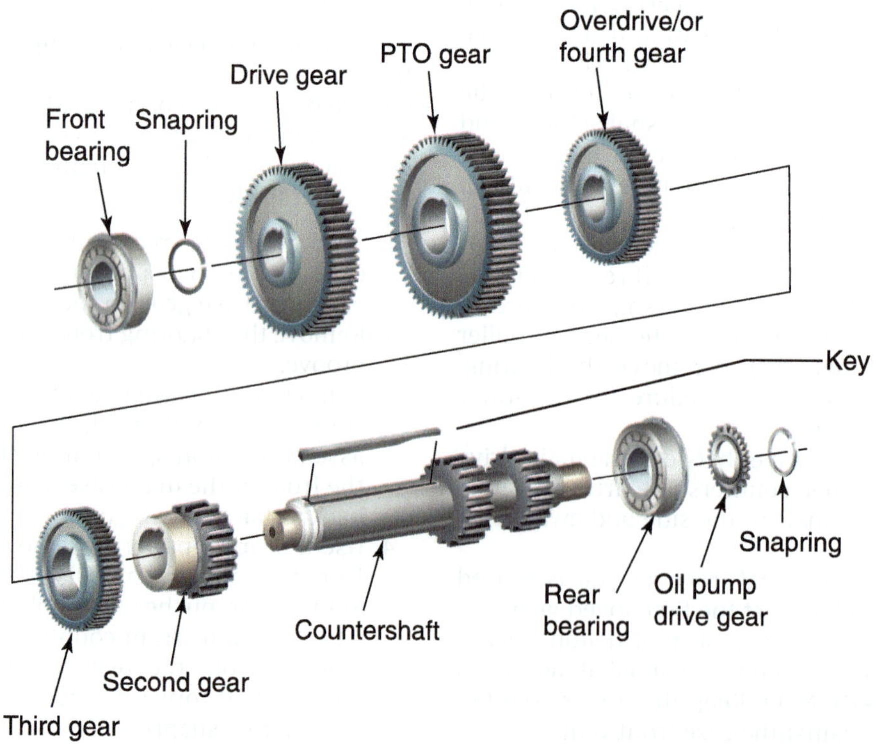

Main Case Reassembly

Ensure that the three magnetic discs are firmly attached to the bottom of the main case. If not, apply 3M Scotch Grip or an equivalent adhesive to the discs and attach them to the main case. Inspect for damaged or worn clutch housing or rear support studs and repair or replace if required. Apply Eaton®Fuller® thread sealant #71205 or an equivalent to any replacement studs before installing them.

CAUTION

Make sure that the rear support stud or capscrew does not protrude past the case housing into the transmission. If it does, it may contact the range yoke and interfere with low- and high-range shifts.

Assemble the Mainshaft Assembly

This section outlines the procedure to assemble a mainshaft with nonselective (not adjustable) tolerance washers. The correct mainshaft key must be used with the specified mainshaft washers. Refer to the parts database to confirm the proper parts. Previous design levels can be updated to the current design, and note that any keyway on the mainshaft can be used for mainshaft key location. Use **Figure 16–41** as a guide for the reassembly sequence.

1. With the mainshaft pilot end down, secure the mainshaft in a vise equipped with brass jaws or wood blocks. Mark one keyway for reference to aid in assembly as shown in **Figure 16–42**.
2. With the mainshaft pilot end down, install the washer. Rotate the washer until the washer splines and mainshaft splines align.

FIGURE 16–42 Mark one mainshaft keyway as a reference for reassembly and install a plastic line to hold the washer in place.

Courtesy of Eaton Corporation

3. Starting at the bottom of the mainshaft, insert a plastic line into the marked keyway to lock the washer in place.
4. With the clutching teeth facing downward, position the direct/overdrive gear on the mainshaft.
5. Install a washer. Rotate it until the washer splines and mainshaft splines align. Push the plastic line up to lock the washer on the mainshaft.
6. With the clutching teeth facing up and against the spacer, install the third gear.
7. Position the washer against the gear. Rotate the washer until the washer splines and mainshaft

FIGURE 16–41 Exploded view of the mainshaft

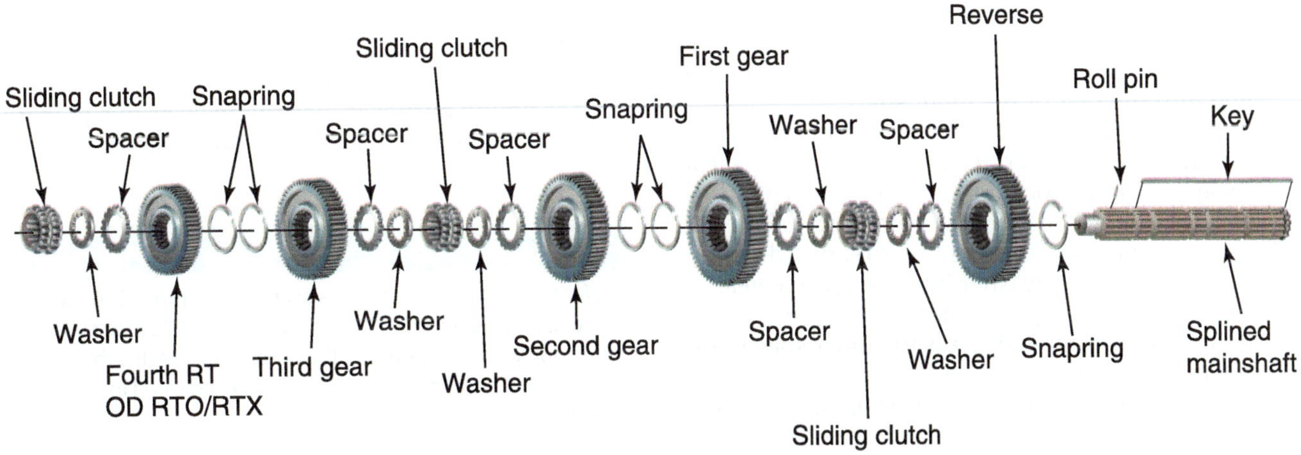

splines align. Push the plastic line up to lock the washer in place.

8. With the missing internal splines aligned with the plastic line, install the sliding clutch.
9. Position the next washer in the next available groove. Rotate the washer until the washer splines and mainshaft splines align.
10. With clutching teeth facing downward, position the second gear on the mainshaft.
11. Position the washer against the gear. Rotate the washer until the washer splines and mainshaft splines align. Again, push the plastic line up to lock the washer on the mainshaft.
12. With the clutching teeth facing up, install the first gear onto the shaft against the previously installed gear.
13. Position the washer against the gear. Rotate the washer until the washer splines and mainshaft splines align. Push the plastic line up to lock the washer on the mainshaft.
14. With the missing internal splines aligned with the plastic line, install the first/reverse sliding clutch.
15. From the rear, install the reverse gear over the sliding clutch teeth with the snapring groove facing up.
16. Install the reverse gear washer in the final groove.
17. From the mainshaft rear, install the mainshaft key with the raised end pointed up and facing out into the spline with the plastic line. Pull the plastic line from the spline while installing the key.
18. Remove the mainshaft from the vise and lay it on its side.
19. Install the front sliding clutch into position.

Install the Mainshaft Assembly

1. Move the countershaft forward as far as possible and block it to the side by inserting a large screwdriver or prybar between the main drive gear and the countershaft. Now make sure that the input shaft is pulled forward before positioning the mainshaft.
2. Hold the reverse gear against the next gear, and lower the mainshaft assembly into position while guiding the shaft rear through the case bore. A mainshaft hook (Eaton special tool #T20) or piece of rope can be used to support the mainshaft.
3. Move the mainshaft forward to position the pilot end into the rear of the input shaft.
4. Install the reverse idler bearing and inner race into the reverse idler gear. Now place the gear in its approximate location. Rest it against the case wall, but do not install the reverse idler shaft yet.
5. Slide the mainshaft reverse gear rearward.
6. Install the tanged washer into the reverse gear.
7. Install the snapring into the reverse gear.
8. If previously removed, press the inner spacer into the mainshaft rear bearing.
9. Slide the mainshaft rear bearing and splined spacer over the mainshaft and seat the bearing with a soft hammer or maul. You may need to lift the rear of the mainshaft slightly to be able to fit the bearing into the case bore.

Assemble the Countershafts

Identify the upper or lower countershafts, and mark them if necessary. The lower countershaft has a larger 47-tooth PTO gear, and if the transmission has an oil pump, the rear of the shaft will be extended to support the oil pump drive gear. You may wish to reference Figure 16–40 while performing this procedure.

1. If previously removed, install the key in the countershaft keyway.
2. Lubricate the shaft and inside diameter of each gear.
3. Install the second speed gear (smallest gear). To install it, align the gear keyway with the countershaft key and press the gear on the countershaft.
4. Install the third speed gear (next smallest gear). To install it, align the gear keyway with the countershaft key and press the gear on the countershaft.
5. Perform one of the two following steps:
   a. For OD transmissions, select the largest diameter gear, place it over the shaft, and press it into place.
   b. For direct drive transmissions, install the fourth speed gear (next smallest gear). To install it, align the gear keyway with the countershaft key, long hub to the countershaft front, and press the gear on the countershaft.
6. Install the PTO gear. This gear is identified by the rounded teeth on one side. To install the PTO gear, align the gear keyway with the countershaft key, rounded teeth facing up (shaft rear), and press the gear on the countershaft.
7. Install the countershaft drive gear. This gear is identified by the tooth marked with an "0." To install the drive gear, align the gear keyway with the countershaft key, "0" to the front, and press the gear on the countershaft.
8. Install the countershaft drive gear retaining snapring in the groove on the shaft front. If the snapring groove is not fully exposed, either the gears are not oriented properly or they are not seated. The countershaft may have to be partly disassembled to correct the problem.
9. For timing purposes, use toolmaker's dye to mark a "0" on the tooth lined up with the keyway.

Tech Tip

Refer to the countershaft cross section in **Figure 16–43** (cross section for an overdrive [OD] transmission) while reassembling the countershaft and gears. If assembled correctly, the countershaft will have the same shape as that indicated in the cross section.

FIGURE 16–43 Countershaft cross section of an OD model transmission.

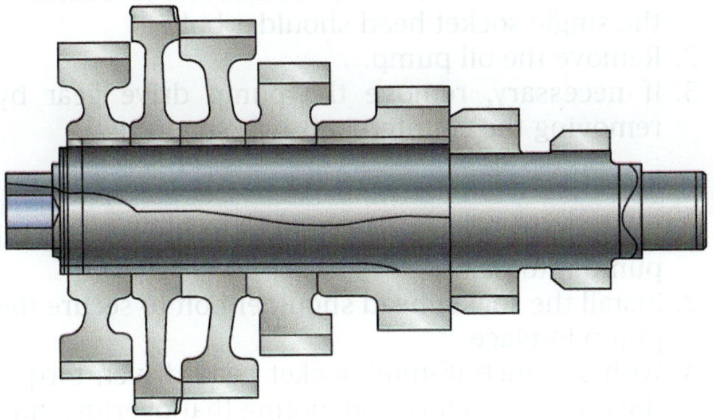

CAUTION

Do not attempt to install the countershaft in the transmission if you cannot install the snapring.

Assemble the Lower Reverse Idler Gear

Reference Figure 16–36 when undertaking the following procedure:

1. Position the reverse idler bearing and inner race into the reverse idler gear.
2. Hold the reverse idler gear and flat thrust washer in place in the main case, and insert the idler shaft, threaded end to the front, through the rear case bore and into the gear. Do not insert the reverse idler shaft into the main case support boss yet.
3. Position the slotted thrust washer in front of the gear with a slot facing up. Note the word FRONT on the slotted washer, and make sure that it faces the front of the transmission. Continue to feed the idler shaft forward through this washer and into the hole in the case boss.
4. Slide the reverse idler shaft into the support boss bore, and with a soft bar and driver, drive the reverse idler shaft fully into position.
5. Secure the shaft in position according to one of the three procedures that follow:
   - Nut Fastener Design: Inspect the nylon locking insert in the nut, and replace it if it is worn. Install the nut and washer on the front of the shaft and torque to specification.
   - Capscrew Fastener Design: Apply the specified thread sealant to the capscrew threads, and install the capscrew and washer. Torque the capscrew to specification.
   - Snapring Design: Install the washer and snapring on the idler shaft using a socket. Place the snapring over the front of the reverse idler shaft seal using a large socket. Tap lightly into place.

Install the Countershafts

Ensure that the countershaft assemblies have been marked for proper position, 47-tooth PTO gear in the lower position. Before installing the countershafts, the lower reverse idler gear assembly should be installed. If the transmission is equipped with an internal oil pump, this should also be installed.

1. On the drive gear of each countershaft assembly, mark the tooth aligned with the gear keyway and stamp with an "O" for easy identification. Toolmaker's dye is recommended for making timing marks.
2. Insert the lower, 47-tooth PTO gear countershaft into the main case with the shaft seated in the lower countershaft case bores. A hook or piece of rope can be used to support the countershaft.
3. Place the upper countershaft (with the 45-tooth PTO gear) into the main case with the shaft seated in the upper countershaft case bores.

Install the Lower Countershaft Bearings

1. Place a countershaft support (Eaton special tool #T11) under the lower countershaft in the front case bore.
2. Insert a large screwdriver or prybar at the front of the countershaft between the case wall and countershaft to space the countershaft rearward approximately ¼ inch (6 mm).
3. Position the bearing over the rear of the countershaft. It does not matter which bearing you choose. The front and rear bearings are identical.
4. Place a flanged-end bearing driver (Eaton special tool #T10) over the bearing, and drive the bearing over the countershaft and into the case bore. When the bearing is fully seated, no gap should be visible between the bearing and the shaft. If the bearing is not fully seated, the snapring to be installed in step 6 will not fit. The bearing driver must bear on both the inner and outer races of the bearing. If the bearing driver contacts only one race, the bearing will be damaged.
5. If used, install the oil pump drive gear.
6. In the countershaft rear groove, install the rear snapring.
7. Remove the screwdriver or prybar.
8. If necessary, tap the countershaft forward to seat the rear bearing snapring in its bore.
9. Temporarily install the mainshaft rear bearing retainer using two capscrews. The cover prevents the countershaft from moving rearward when the front bearing is installed.
10. Remove the countershaft support tool and place the front bearing on the front of the countershaft.
11. Place a flanged-end bearing driver (Eaton special tool #T10) over the bearing, and drive the bearing until the inner race is flush with the front of the countershaft.
12. Remove the mainshaft rear bearing retainer.

Install the Upper Countershaft Bearings

1. Ensure that the lower countershaft and main drive gear timing marks are aligned.
2. Mesh the marked tooth of the upper countershaft with the two remaining marked teeth of the main drive gear.
3. Support and center the front of the countershaft by placing a support tool (Eaton special tool #T11) in the front bore.
4. Place a large screwdriver or prybar at the front of the countershaft between the case wall and countershaft to space the countershaft rearward approximately ¼ inch (6 mm).
5. Position a bearing over the rear of the countershaft. The front and rear bearings are identical, but the bearing driver must bear on both the inner and outer races of the bearing or it will be damaged.
6. Place a flanged-end bearing driver (Eaton special tool #T10) over the countershaft rear bearing, and drive the bearing into the case bore. The countershaft may need to be lifted slightly to start the bearing into the bore. When the bearing is fully seated, no gap should be visible between the bearing and the shaft. If the bearing is not fully seated, the snapring to be installed in step 7 will not fit.
7. In the countershaft rear groove, install the rear snapring.
8. Remove the screwdriver or prybar.
9. If necessary, tap the countershaft forward to seat the rear bearing in its bore.
10. Install the mainshaft rear bearing retainer. To install the bearing retainer, apply the specified thread sealant to the capscrew threads and install them in the bearing retainer. Torque the screws to specification.
11. Remove the timing block and place the remaining bearing on the front of the countershaft.
12. Position a flanged-end bearing driver (Eaton special tool #T10) over the bearing, and drive the bearing over the countershaft until the inner race is flush with the end of the countershaft.

Install the Upper Reverse Idler Gear

The installation procedure is the same as that for the lower reverse idler shaft outlined a little earlier in this chapter, so it is not repeated here. The new reverse idler shaft uses a snapring in place of the capscrew.

Install the Auxiliary Drive Gear

1. Install the auxiliary drive gear over the mainshaft splines.
2. Install the auxiliary drive gear retaining snapring in the mainshaft snapring groove. If necessary, slide the mainshaft rearward to fully expose the groove.

Oil Pump Removal and Installation

Remove the oil pump as follows:

1. Use a ¼-inch (6-mm) Allen socket driver to remove the single socket head shoulder bolt.
2. Remove the oil pump.
3. If necessary, remove the pump drive gear by removing the countershaft rear snapring.

Install the oil pump as follows:

1. From the rear of the transmission, position the oil pump into place.
2. Install the socket head shoulder bolt to secure the pump in place.
3. With a ¼-inch (6-mm) socket head driver, torque the bolt to specification, noting that overtorquing can bind up the pump.

Reinstall Transmission

Mount the transmission on a transmission jack and ensure it is secure; use a chain if necessary. If you ensured that transmission and engine were level prior to removing the transmission, reinstallation should present few problems. Install the clutch using a pilot shaft; remove the release chocks from the clutch pressure plate, then remove the pilot shaft. Level the transmission on the jack (see **Figure 16–44**); slide the clutch shaft through the release bearing ensuring that the release forks clear it. You may have to rotate the transmission clutch shaft before it engages and slides through the friction disc splines. Use the cross shaft and fork to help pull the transmission home so that the clutch shaft pilot stub is fully engaged in the

FIGURE 16–44 Reinstalling the transmission: This task is made much easier if you ensure that both the engine and transmission are perfectly aligned with each other.

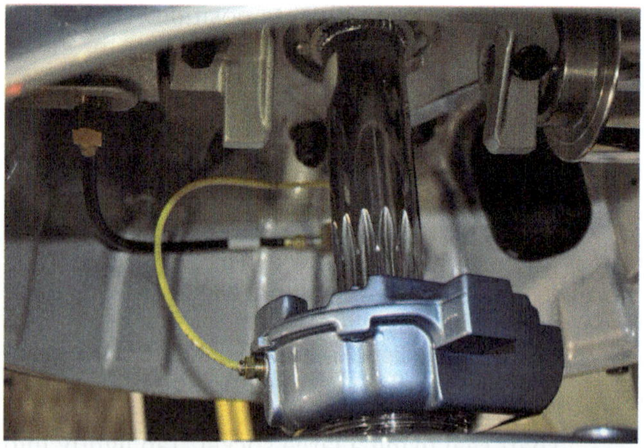

flywheel pilot bearing and the flywheel housing and bell housing flanges abut. Never attempt to pull a transmission home using bolts.

16.7 TROUBLESHOOTING AIR SHIFT SYSTEMS

Air shift systems can cause shifting problems in compounded transmissions that usually noticed because of either failing to make a range shift or shifting slowly. Troubleshooting consists of checking for leaks, testing the regulated air pressure, and ensuring that the airflow is correctly routed. Connecting the air lines properly after a transmission overhaul is obviously important, as is the practice of tagging the lines with identification labels on disassembly. Nevertheless, you should understand how to interpret air system schematics for those occasions when you have to reassemble the lines after someone else has failed to tag them. A summary of air system troubleshooting guidelines is provided in **Table 16–3**.

To troubleshoot an air system, chassis air pressure should be at normal system pressure but the engine should be off. The following are some typical air system problems with their possible causes:

INCORRECT AIR LINE CONNECTIONS

This problem is not as uncommon as you might think—air lines are often disconnected from a transmission without being tagged and end up being improperly connected. You should first refer to the OEM air system schematics to correctly identify and route the air lines. Figure 15–42 and Figure 15–44 in the previous chapter show the routing layout required by our typical system.

Begin troubleshooting by moving the gearshift lever into neutral. Then move the range selection control button up and down:

- If the air lines are crossed between the control valve and the slave valve, there should be constant air flowing from the exhaust port of the control valve while in high range.
- If the air lines are crossed between the slave valve and the range cylinder, the transmission gearing will be the reverse of the range selection. A low-range selection will result in a high-range engagement and vice versa.

AIR LEAKS

Begin troubleshooting by moving the gearshift lever into neutral. Then use a soap and water solution in a spray bottle to wet down the air lines and fittings in the transmission air circuit. You should cycle the range options while doing this, looking for bubbles sourced by the air leak.

- Steady leakage from the exhaust port of the control valve indicates a defect in the control valve, usually in the O-rings.
- Leakage from the slave valve breather indicates either a defective O-ring in the slave valve or defective O-rings in the range cylinder piston.
- Failure to shift into low-range or sluggish-range shift indicates a problem with the range cylinder, which is covered later in this chapter.

CAUTION

The use of improper sealants and lubricants during reassembly of air system components can cause improper operation of the pneumatic components and result in mechanical damage to the transmission. Use the OEM-recommended silicone-based lubricants and sealants. Avoid using silicone sealants to seal the splitter cylinder or range cylinder cover plates because such sealants can block air passages and cause erratic operation of air system components.

AIR REGULATOR PROBLEMS

With the gearshift lever in neutral, check the breather of the air filter/regulator assembly (see Figure 15–43). There should be no air leaking from this port. The complete assembly should be replaced if there is evidence of leakage.

Cut off the vehicle air supply to the air filter/regulator assembly, disconnect the air line at the fitting in the supply outlet, and install an air pressure gauge in the opened port. Bring the vehicle air pressure to normal. Regulated air pressure should be within the OEM specification. If not, replace the air regulator.

SHOP TALK

Do not adjust the lower screw on the regulator to correct an out-of-specification transmission circuit pressure reading. Regulators are not designed to be adjusted. If replacement or cleaning of the filter element does nothing to correct the air pressure readings, replace the complete assembly.

SLAVE VALVE PROBLEMS

The major problem encountered with slave valves is failed O-rings. Refer to Figure 15–45 for the location of O-rings in a typical slave valve. If any of the O-rings is defective, air leaks out of the exhaust on the slave

TABLE 16–3 Troubleshooting Air System Problems

Problem: Slipout (Splitter)	
Possible Cause	**Possible Correction**
Worn yoke pads	Instruct driver on proper driving technique. Replace damaged parts.
Damaged insert valve	Replace.
Defective regulator	Replace.
Loose hose or fitting	Tighten.
Pinched air line or connector	Correct restriction.
Gear twisted out of time on countershaft	Check for damage and replace components.
Tapered clutching teeth	Replace damaged parts.
Worn yoke slot in clutch gear	Instruct driver on proper driving technique. Replace damaged parts.
Inner race left off front of auxiliary C/S	Install missing part.
Bearing failure	Replace damaged parts.
Improper preselection of splitter	Instruct driver on proper driving technique.
Not using clutch	Instruct driver on proper driving technique. Inspect for resulting damage and repair as needed.
Problem: Slipout (Range)	
Possible Cause	**Possible Correction**
Defective regulator	Replace.
Loose hose or fitting	Tighten.
Pinched air line or fitting	Correct restriction.
Gear twisted out of time on countershaft	Check for damage and replace components.
Tapered clutching teeth	Replace damaged parts.
Improper range shifting technique	Instruct driver on proper driving technique. Inspect for resulting damage and repair as needed.
Problem: Slow Shift (Splitter)	
Possible Cause	**Possible Correction**
Worn yoke pads	Instruct driver on proper driving technique. Replace damaged parts.
Defective regulator	Replace.
Loose hose or fitting	Tighten.
Pinched air hose	Correct restriction
Pinched air line or connector	Correct restriction.
Damaged O-ring in slave valve or in splitter cylinder assembly	Replace damaged part using thin film of silicon during reassembly.
Improperly mounted gasket	Replace part and set to proper specifications.
Improper preselection of splitter	Instruct driver on proper driving technique.
Not using clutch	Instruct driver on proper driving technique. Inspect for resulting damage and repair as needed.
Excessive silicone on O-rings	Clean part and apply proper amount.

TABLE 16–3 (continued)

Problem: Slipout (Splitter)	
Possible Cause	**Possible Correction**
Mixing oils or using additives	Check for damage and install proper grade oil.
Damaged O-ring	Replace.
Problem: Slow Shift or No Shift (Range)	
Possible Cause	**Possible Correction**
Defective regulator	Replace.
Loose hose or fitting	Tighten part.
Pinched air hose	Correct restriction.
Sticking slave valve piston	Clean part and apply thin film of silicone.
Damaged O-ring in range cylinder or slave valve	Replace, applying thin film of silicone during reassembly.
Improperly mounted gasket	Replace part and set to proper specifications.
Improper air line hookup	Check air lines and hoses.
Pinched air line or connector	Correct restriction
Range cylinder piston nut loose	Tighten.
Range cylinder piston cracked	Replace part applying a thin film of silicone during reassembly.
Synchronizer spring broken	Replace broken or missing parts.
Failed synchronizer	Replace.
No silicone on O-rings	Apply thin film of silicone.
Excess silicone on O-rings	Clean part and apply thin film of silicone.
Damaged O-ring	Replace.
Restricted slave valve breather	Clean and/or replace part.
Improper range shifting techniques	Instruct driver on proper driving technique. Inspect for resulting damage and repair as needed.
Problem: Burned Synchronizer	
Possible Cause	**Possible Correction**
Defective regulator	Replace part.
Yoke installed backwards	Replace part and set to proper specifications.
Synchronizer spring broken	Replace broken or missing parts.
Mixing oils or using additives	Check for damage and install proper grade oil.
Improper range shifting techniques	Instruct driver on proper driving technique. Inspect for resulting damage and repair as needed.
Problem: Overlapping Gear Ratios	
Possible Cause	**Possible Correction**
Improper air line hookup	Check air lines and hoses and set to proper specifications.
Problem: Air Leak Out of Range Valve in High Range Only	
Possible Cause	**Possible Correction**
Improper air line hookup	Check air lines and hoses and set to proper specifications.

(continued)

TABLE 16–3 (continued)

Possible Cause	Possible Correction
Problem: Constant Air Leak Out of Air Range Valve	
Damaged O-ring in range valve	Replace part applying a thin film of silicone during installation.
Problem: Air Leak Out of Slave Valve Breather	
Damaged O-ring in slave valve or range cylinder	Replace part applying a thin film of silicone during installation.
Pinched air line or connector	Correct restriction.
Range cylinder piston nut loose	Tighten nut.
Range cylinder piston cracked	Replace part applying a thin film of silicone during reassembly.
Problem: Transmission Case Pressurized—(Low Range Only)	
Damaged O-ring in range cylinder	Replace part applying a thin film of silicone during installation.
Problem: Transmission Case Pressurized—(All Range Positions)	
Damaged O-rings in dash-mounted deep reduction valve or splitter cylinder assembly.	Replace part, applying a thin film of silicone during installation.
Problem: Constant Air Leak Out of Insert Valve Exhaust Port	
Damaged insert valve	Replace part, applying a thin film of silicone during installation.
Damaged O-ring in splitter cylinder assembly	Replace part, applying a thin film of silicone during installation.
Problem: Constant Air Leak Out of Selector Valve Exhaust Port in "Direct" Position	
Damaged insert valve	Replace part, applying a thin film of silicone during installation.
Damaged O-ring in splitter cylinder assembly	Replace part, applying a thin film of silicone during installation.

valve. In normal operation, exhaust will occur only for an instant as each range shift is made.

In the event of continuous leakage, disconnect the air lines to the slave valve and remove the slave valve from the transmission. Disassemble the slave valve. Replace all the O-rings and lubricate them with the OEM-recommended silicone lubricant before reassembly. Carefully observe the recommended torque specifications during reassembly. If overtorqued, the piston may bind in the valve bore.

Poppet-Type Slave Valves

Defective poppet-type slave valves are usually not serviced but rather replaced. Before replacing a poppet-type slave valve, check for and correct any of the following conditions:

1. Restrictions in the air lines between the slave valve and the range valve.
2. Proper torque of the slave valve mounting fasteners (typically 8 lb-ft./11 N·m). If retorquing the

fasteners eliminates the leak, remove the valve from the transmission and retorque the base plate screws on the back of the valve to the OEM-recommended torque (typically 45 lb-ft./60 N·m). Now, reinstall the slave valve, substituting flat mounting washers for the original star washers. Flat washers will not dig into and damage the valve housing.

RANGE CYLINDER PROBLEMS

If any of the seals in the range cylinder assembly (see Figure 15–49) is defective, range shifts will be affected. The amount of air being lost will affect the severity of the problem, which can range from slow shifting to a complete failure to shift.

- An air leak at either O-ring "A" will result in complete failure to make a range shift and will cause a steady flow of air from the breather of the slave valve in both ranges.
- A leak at gasket "B" will result in a steady escape of air to atmosphere when in high range.
- A leak at O-ring "C" will result in a slow shift to low range and will also pressurize the transmission housing.
- A gasket leak will result in a slow shift to the direct range.
- Also, a cracked piston caused by improper installation will result in a failure to shift in either low or high range. It will also cause a steady leak out of the breather in the slave valve.

If any of the seals are defective, the range cylinder should be disassembled and new O-rings and gaskets installed. The following is the disassembly procedure for the range cylinder illustrated in Figure 15–49.

1. Remove the four fasteners and separate the range shift cylinder cover.
2. Remove the stop nut from the yoke bar.
3. Apply air to the low-range air line to pop the piston from the cylinder housing.
4. Remove the O-rings from the piston.
5. Install new O-rings, lubricate with recommended silicone lubricant, and reassemble the cylinder. Install a new gasket, but do not use gasket sealant.

WARNING

Do not direct the piston toward yourself or others during this step.

RANGE CONTROL VALVE PROBLEMS

Troubleshooting the range valve (see Figure 15–49) will vary slightly between valve models. With the gearshift lever in neutral and the high range selected, the ⅛-inch (3 mm) outside diameter air line is disconnected from either:

1. The outlet or "P" port of the range control valve (some models); or
2. The slave valve (some models).

When low range is selected, a steady stream of air should flow from the opened P port or the disconnected line. Select high range to shut off the airflow. This indicates the control valve is operating properly. Reconnect the air line.

If the control valve does not operate properly, check for restrictions and air leaks. Leaks indicate defective or worn O-rings. If the O-rings or parts in the range valve are defective, there will be constant leakage out of the exhaust located on the bottom of the valve. A defective insert valve O-ring will result in a constant leak through the exhaust in both ranges, and the valve will not make range shifts. A defective housing O-ring will result in a constant, low-volume leak through the exhaust in the low range only.

If the slide is assembled backwards, or if the air lines between the slave valve and the range valve are connected in reverse, there will be constant leakage through the exhaust in the high range. When installing the slide in the range valve, make sure that the slot in the slide faces the outlet port. When assembling a range valve, use a very small amount of silicone lubricant to avoid clogging the ports. A small amount of grease on the position springs and balls will help hold them in place during reassembly.

High-Range Operation

With the gearshift lever in neutral, select low range and disconnect the ¼-inch (6-mm) ID line at the port of the range cylinder cover. Make sure that this line leads from the high range or "H" port of the slave valve. When high range is selected, a steady stream of air should flow from the disconnected line. Select low range to shut off the airflow.

Now move the shift lever to a gear position and select the high range. No air should flow from the disconnected line. Return the gearshift lever to the neutral position. There should now be a steady flow of air from the disconnected line. Select the low range to shut off airflow and reconnect the air line. If the air system does not operate accordingly, the slave valve or actuating components of the shift bar housing assembly are defective.

Low-Range Operation

With the gearshift lever in neutral, select high range and disconnect the ¼-inch (6 mm) inside diameter air line at the fitting on the range cylinder housing. Make sure that this line leads from the low range or

"L" port of the slave valve. When low range is selected, a steady stream of air should flow from the disconnected line. Select high range to shut off airflow.

Move the gearshift lever to a gear position and select low range. No air should flow from the disconnected line. Return the gearshift lever to the neutral position. There should now be a steady flow of air from the disconnected line. Select high range to shut off the airflow and reconnect the air line. If the air system does not operate accordingly, the slave valve or actuating components of the shift bar housing assembly are defective.

SPLITTER SHIFT PROBLEMS

If the transmission fails to shift or shifts too slowly to or from the split position, the fault might be in the splitter shift air system or in related components of the range shift system. Perform the following checks with normal vehicle air pressure supplied to the system. The engine should be off, the transmission in neutral, parking brake set, and wheels chocked. The following checks should be made after proper air pressure has been verified at the air filter/regulator and all lines have checked out leak free.

Air Supply Test

With the gearshift lever in neutral, select the high or low range and loosen the connection at the "S" port of the valve mounted on the gearshift lever until you can determine that air is being supplied to the valve. If there is air to the valve, reconnect the fitting. If there is no air supplied to this valve, check for air line restrictions between this valve and the slave valve. Make sure the line is properly connected to the slave valve.

Shift Tower Valve Test

With the gearshift lever in neutral, disconnect the air line running from the gearstick valve to the splitter cylinder cover on the transmission at the splitter cylinder port. While in high or low range, move the splitter control button on the shift lever forward. Air should flow from the disconnected line at the splitter cylinder cover. When the splitter control button is moved rearward, airflow should stop. If the preceding conditions do not exist, the splitter control valve on the shift lever is either defective or the air line is restricted.

Splitter Cylinder

If any of the seals in the splitter cylinder assembly are defective, splitter shift operation will be affected, resulting in partial or total loss of shift capability (depending on the severity of the air leakage). See Figure 15–52, which shows the location of each splitter cylinder O-ring. Leakage at O-ring "A" results in a slow shift to engage the rear auxiliary drive gear, pressurizing of the transmission housing, or disengagement of the auxiliary gearing.

A leak at O-ring "B" results in slow shifting or complete failure to engage and disengage the front or rear auxiliary drive gearing and a steady flow of air from the exhaust port of the shift tower valve and/or cylinder cover when the splitter control button is in the rearward position.

A leak at gasket "C" results in a slow shift to disengage the rear auxiliary drive gear and a discharge of air to the atmosphere. Any continuous air discharge from the exhaust port of the cylinder cover usually indicates a defective insert valve. Exhaust should occur only briefly when the splitter control button is moved rearward while in low or high range. A defective insert valve, or leaking at the valve O-rings or inner seals, results in continuous air leakage and shift failure.

Two indications of defective O-rings or seals are:

- Constant air discharge from the exhaust port of the cylinder cover
- Constant air discharge from the exhaust port of the control valve while the splitter control button is rearward or forward (providing the control valve is operating properly)

The three O-rings on the valve OD can be replaced. If an inner seal is damaged, however, the complete assembly must be replaced.

AIR CONTROL DISASSEMBLY

Figure 15–42 and Figure 15–44 show the location of the air lines and components in a typical air control system. Disassembly of the air control system should be undertaken as follows:

1. Disconnect the two air lines running from the gear stick valve to the slave valve on the transmission at the "S" port and "P" port on the slave valve.
2. Remove the air line at the splitter cylinder cover (if equipped).
3. Remove the valve cover from the shift tower and disconnect the three air lines.
4. Loosen the jam nut and turn the shift tower valve and nut from the gearshift lever. Remove the valve cover, air lines, sheathing, and O-rings from the lever.
5. Disconnect and remove the air line between the splitter cylinder and air filter/regulator.
6. Disconnect and remove the air line between the slave valve and the air filter/regulator.
7. Remove the screws securing the air filter/regulator to the transmission and remove the air filter/regulator.

8. Disconnect the air hose between the slave valve and the range cylinder cover.
9. Remove the slave valve from the transmission housing.
10. Remove the spring and plunger pin from the bore in the transmission housing. Remove the slave valve gasket (see Figure 15–45B).

SUMMARY

- Scheduled lubrication services are key to a good transmission maintenance program.
- Most standard transmissions depend on splash lubrication, meaning that some of the rotating components contact, pick up, and circulate oil from the oil in the sump. Maintaining the correct oil level is critical for splash lubrication to be effective.
- Only lubricants recommended by the transmission OEM should be used in a transmission. These could be gear, engine, or synthetic oils.
- When mineral oils are used, the first oil change should be performed shortly after the transmission enters service, often between 3,000 and 5,000 miles (4,800–8,000 km) of operation in a linehaul application, or less in vocational service. This is not required with synthetics.
- In general linehaul application, it is good practice to schedule a transmission oil change from 50,000 to 100,000 miles (40,000–80,000 km) of service. Oil change intervals can be extended up to 500,000 (800,000 km) miles when synthetics are used.
- Good preventive maintenance (PM) lowers the incidence of breakdowns, minimizes downtime, and reduces the cost of repairs.
- Leakage in transmission rear seals is a relatively common problem in truck transmissions but one that is easily repaired.
- When diagnosing transmission complaints, it is important that you confirm that the transmission is the actual cause of the problem before removing it for repair.
- When disassembling a transmission, each component should be carefully inspected for abnormal wear and damage. Components that are not reusable should be ordered after disassembly, not discovered during reassembly.
- Ensure that a cause of failure is determined before reassembly, or the failure is likely to recur within a short period.
- Bearing failures occur because of dirt contamination and poor lubrication. More than 90 percent of bearing failures are caused by dirt. Cleanliness is critical when repairing and servicing standard transmissions.

REVIEW QUESTIONS

1. Which of the following OEM-approved lubricants would produce the longest service life in a standard transmission?
   a. synthetic E-500 gear lube
   b. SAE 50 grade heavy-duty engine oil
   c. multipurpose or EP gear lube
   d. SAE 50 grade transmission fluid
2. After breaking in a new transmission filled with E-500 synthetic lube on a truck used in a linehaul application, when should the first oil change take place?
   a. between 3,000 and 5,000 miles (4,800–8,000 km) of service
   b. after 25,000 miles (4,000 km) of vocational service
   c. after 100,000 miles (160,000 km) of linehaul service
   d. after 500,000 miles (800,000 km) of linehaul service
3. Which of the following would indicate that transmission oil in a truck standard transmission is at the correct level?
   a. Oil is visible through the filler hole.
   b. Oil is reachable with the first finger joint through the filler hole.
   c. Oil is level with the filler hole.
   d. Oil is at the high level on the dipstick.
4. Oil is discovered to be leaking from the rear seal area of a standard transmission. Technician A says that the problem could be a damaged speedometer sleeve O-ring or gasket. Technician B says that the problem could be caused by a cracked bearing cover. Who is correct?
   a. Technician A only
   b. Technician B only
   c. both A and B
   d. neither A nor B

5. Technician A says that a damaged transmission output yoke can be repaired by installing a yoke sleeve kit. Technician B says that a damaged transmission output yoke should always be replaced. Who is correct?
   a. Technician A only
   b. Technician B only
   c. both A and B
   d. neither A nor B
6. Which of the following statements is true?
   a. A transmission must be disassembled to replace an input shaft.
   b. An auxiliary section can be removed and overhauled without removing the transmission from the chassis.
   c. A shift rail assembly cannot be removed and repaired without first removing the transmission from the vehicle.
   d. A gearshift tower cannot be separated from a shift rail assembly.
7. A driver complains of driveline vibration. Which of the following would be the least likely cause?
   a. clutch failure
   b. tires out of balance
   c. output shaft nut improperly torqued
   d. failed synchronizer
8. A driver complains that the vehicle jumps out of gear when driven over a bumpy road surface. Technician A says that the problem could be worn detent notches in the shift rail. Technician B says that the problem could be worn gear clutch teeth. Who is correct?
   a. Technician A only
   b. Technician B only
   c. both A and B
   d. neither A nor B
9. Which of the following would most likely result in hard shifting?
   a. sliding clutch gears worn to a taper
   b. lack of lubricant in remote shift linkage
   c. weak detent springs
   d. worn shift fork pads
10. Technician A says that lack of lubrication damage can be caused in a standard transmission when coasting with the clutch depressed. Technician B says that when towing a truck from the front, the axle shafts should be pulled. Who is correct?
    a. Technician A only
    b. Technician B only
    c. both A and B
    d. neither A nor B
11. Which of the following is said by transmission OEMs to cause most transmission bearing failures?
    a. lack of lubrication
    b. dirt
    c. improper shifting
    d. operation on rough highways
12. When removing a transmission from a truck, Technician A states that it is usually necessary to uncouple the clutch from the flywheel before separating the transmission from the flywheel housing. Technician B says that when a transmission is being removed, it should be held in alignment with the engine to prevent the flywheel pilot bearing from being damaged. Who is correct?
    a. Technician A only
    b. Technician B only
    c. both A and B
    d. neither A nor B
13. Technician A says that an advantage of using synthetic transmission lubes is that they are compatible with all other gear lubes. Technician B says that high-gear lube level in a standard transmission can cause aeration and result in lubrication failures. Who is correct?
    a. Technician A only
    b. Technician B only
    c. both A and B
    d. neither A nor B
14. Which of the following transmission components are splined to the mainshaft in a standard, twin-countershaft transmission?
    a. first speed gear
    b. sliding clutch
    c. countershaft gears
    d. idler gear
15. Technician A says that when disassembling a standard, twin countershaft transmission, the mainshaft must be removed before removal of the countershaft bearings. Technician B says that before timing the auxiliary countershafts, the countershaft bearings must be fully seated. Who is correct?
    a. Technician A only
    b. Technician B only
    c. both A and B
    d. neither A nor B

17

Prerequisite: Chapter 13

TORQUE CONVERTERS

OBJECTIVES

After reading this chapter, you should be able to:

- Outline the function of the torque converter in a vehicle equipped with an automatic transmission.
- Explain how the torque converter is coupled between the crankshaft and the transmission.
- Identify the three main elements of a torque converter torus and describe their roles.
- Define torque multiplication and explain how it is generated in the torque converter.
- Define both rotary and vortex fluid flow and explain how each affects torque converter operation.
- Describe the overrunning clutch, lockup clutch, and variable pitch stators.
- Outline torque converter service and maintenance checks.
- Remove, disassemble, inspect, and reassemble torque converter components.

KEY TERMS

coupling point
flexplate
lockup torque converter
overrunning clutch
roller clutch
rotary flow
stall torque ratio
stator
stator assembly
torque multiplication
torus
turbine
variable pitch stator
vortex flow

INTRODUCTION

In Chapter 14 we looked at how a solid friction clutch is used to mechanically engage and disengage a standard (manual) truck transmission to the engine. A friction clutch has to be mechanically engaged and disengaged to break torque between the engine crankshaft and the transmission so that a gear change can take place. This type of mechanical clutch is controlled either directly by the driver or by the transmission electronics. The coupling that results is mechanical.

Fully automatic truck (and passenger car) transmissions use a type of fluid coupling known as a torque converter to transmit engine torque to the transmission. The coupling is fluid. Because it is a fluid coupling, some element of forgiveness between input and output speeds is provided. This would not be true in a mechanical coupling. Although we classify a torque converter as a *type* of fluid coupling, it should not be confused with the standard fluid couplings used commonly in off-highway equipment.

OPERATING PRINCIPLE

A torque converter transmits twisting force hydraulically using automatic transmission fluid, often simply called *transmission oil.* When the engine is running at idle, the torque converter is designed to behave like a disengaged mechanical clutch. When the engine rpm increases, the role of a torque converter is to transmit or multiply the twisting force or torque delivered to it by the engine crankshaft. It then directs torque through to the transmission to provide a number of ratios suitable for the particular load and speed of the vehicle.

Because a torque converter is a hydrodynamic device (see Chapter 13), the amount of torque transferred from the engine to the transmission by the torque converter is directly related to engine rpm. When the engine is turning at low idle speed, less twisting force is transferred through the torque converter because there is insufficient oil flow for power transfer to occur. As engine speed increases, however, there is a corresponding increase in fluid flow. This increased flow creates enough force to transfer a greater amount of engine power through the torque converter and into the transmission.

17.1 TORQUE CONVERTER CONSTRUCTION

Truck torque converters can be serviceable or nonserviceable. Many torque converters are welded together in such a way that prevents disassembly, so they can be regarded as nonserviceable. The only types of torque converters (T/Cs) that are readily serviceable are the types that are bolted together, such as that shown coupled to a fully automatic truck transmission in **Figure 17–1**.

SHOP TALK

T/Cs can be confused with fluid couplings because both use similar operating principles. The most fundamental difference is that torque converters use curved blades, whereas fluid couplings and fluid flywheels use straight pitch blades. T/Cs also use stators and have the ability to multiply torque, neither of which is characteristic of fluid couplings.

A typical T/C assembly is driven by a flexplate, which is not considered to be part of the T/C assembly. A T/C assembly is known as a **torus**. The torus consists of three moving components:

- Impeller
- Stator
- Turbine

FLEXPLATE

A **flexplate**, sometimes called a *flex disc*, is used to mount the T/C to the crankshaft (**Figure 17–2**). The flexplate is positioned between the engine crankshaft and the T/C. The purpose of the flexplate is to transfer crankshaft rotational force to the shell of the T/C assembly. The flexplate bolts to a flange machined on the rear of the crankshaft and to mounting pads located on the front of the T/C shell.

The flexplate also carries the starter motor ring gear. Often when a T/C is used, no flywheel is required because the combined mass of the T/C and flex disc acts like a flywheel to smooth out the power pulses produced by the engine. Eliminating the flywheel is more common with small-bore engines; some larger diesel automatic transmissions use both a flexplate and a flywheel. The flexplate also allows for a slight alignment tolerance between the engine and T/C assembly, so exact alignment during installation is

FIGURE 17–1 The torque converter mounts to the front end of an automatic transmission. It transfers engine torque from the crankshaft to the transmission gears.

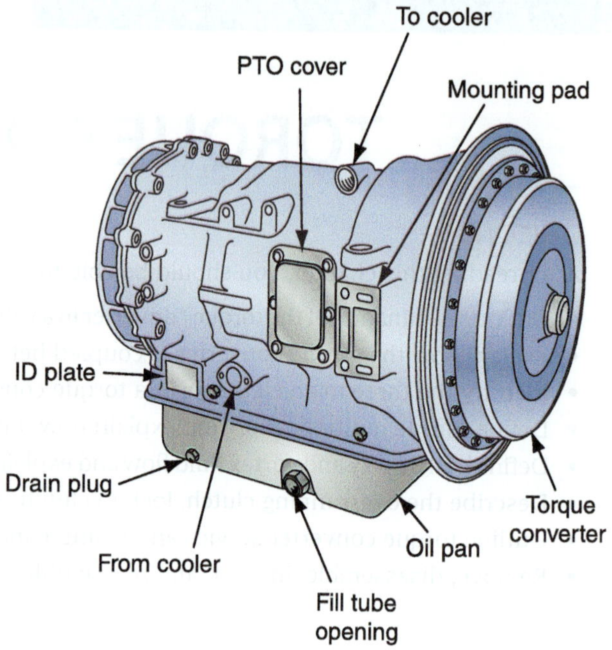

FIGURE 17–2 A flex plate located between the crankshaft and torque converter housing transfers crankshaft rotation to the torque converter.

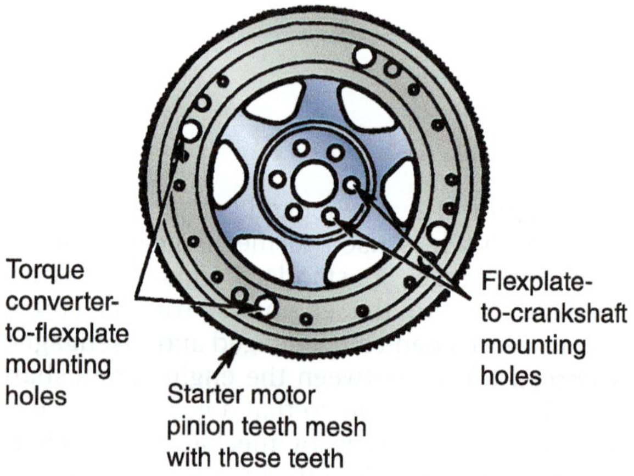

not critical to T/C operation. Some applications use multiple flexplates.

TORQUE CONVERTER COMPONENTS

Figure 17–3 shows a front view of a torque converter mounted within a typical automatic transmission. The T/C provides a fluid coupling between the engine and the transmission gearing that can multiply torque. The T/C has three main components (shown in **Figure 17–4**) and an optional lockup clutch.

Impeller or Pump

Directly coupled to the engine flywheel, the impeller rotates at engine rpm. The converter is full of fluid at all times. The impeller forms the rear half of the T/C assembly and is designed with a series of internal curved blades or vanes. One-half of the split guide ring is attached to the blades. As the impeller rotates, fluid enters from around the pump hub. Centrifugal force throws this fluid outward through the blades, around the split-guided ring, and toward the turbine. This acts to propel the turbine. The driving principle can be compared to the similar action of a boat propeller in water. The impeller is mechanically driven by the engine.

FIGURE 17–3 Front view of a torque converter assembly

FIGURE 17–4 The three main elements of a torque converter: the turbine, the stator, and the pump (or impeller)

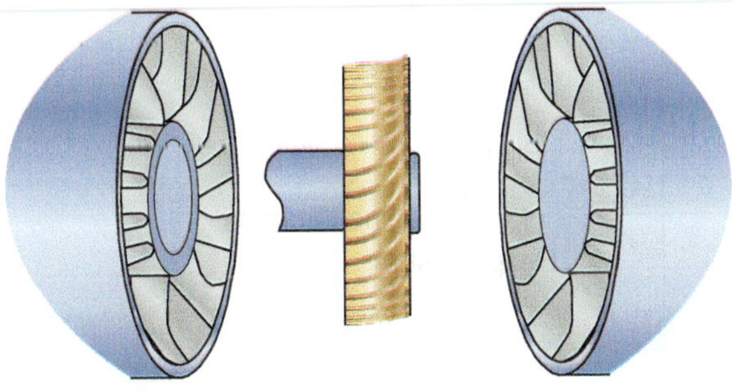

Turbine

The **turbine** is coupled (*splined*) to the transmission turbine shaft and forms the front half of the T/C assembly. The other half of the split guide ring is attached to the turbine vanes. Fluid from the impeller strikes the turbine vanes and flows around the split guide ring. In this way, the energy of the fluid from the impeller (an increase in impeller rpm increases the fluid velocity delivered to the turbine) is imparted to the turbine. As fluid exits the inner edge of the turbine vanes, it is directed back toward the impeller.

When the turbine is rotated by this propelling action of the T/C impeller, the turbine shaft is rotated and provides rotational input to the transmission gearing. No mechanical connection exists between the turbine and the impeller. Fluid acting on the turbine vanes exits near its hub, from which it is directed to the stator. The turbine is hydraulically driven by the impeller. **Figure 17–5** shows a torque converter with the impeller and turbine cutaway.

FIGURE 17–5 Cutaway view of a torque converter: Identify the impeller and the turbine.

Stator

The stator is the component that makes the T/C functionally different from a fluid coupling. It is located between the impeller and the turbine and is splined to the outer race of a one-way roller clutch mounted on the stator support. Fluid leaving the turbine does so in an opposite rotational direction to that of the turbine and impeller. This fluid acts on the stator vanes, locking it to the turbine shaft on a one-way clutch, and then is directed back to the impeller at an accelerated rate, increasing torque. This is known as **vortex flow**, and its effect is to multiply torque. So **torque multiplication** occurs during the vortex flow phase of T/C operation.

As a turbine increases speed, the fluid it unloads into the stator is directed on the reverse side of the stator vanes, causing it to *freewheel*. At this point, the stator ceases to multiply torque. As turbine rotational speed increases, flow through the stator smooths out and eventually stops. This is known as **rotary flow**. Torque multiplication, therefore, occurs only when pump and turbine rotational speeds are significantly different. It is, therefore, at a maximum at full stall and at a minimum at the coupling phase. **Figure 17–6** shows a cutaway view of a torque converter highlighting the stator.

Lockup Clutch

In any type of fluid coupling, a percentage of *slippage* will always exist. This slippage means that turbine speed will never equal the impeller rotational speed. When rotary flow has been achieved and other speed and range criteria are met, the converter is *locked up*, meaning that the impeller and turbine speeds will be equal. Lockup clutch components include a backing plate located in front of the turbine and bolted to the converter front cover; a lockup clutch plate splined to the turbine shaft; and a lockup clutch piston splined to the converter front cover and, therefore, always rotated at engine speed.

FIGURE 17–6 Torque converter cutaway highlighting the stator

Whenever fluid is charged between the front cover and the lockup clutch piston, the clutch disc, which is splined to the turbine shaft, is *sandwiched* and forced to rotate at cover rpm (engine rpm). In lockup mode, the hydraulic slippage mode is eliminated and the T/C acts as a solid mechanical transfer device.

T/C Exterior

The exterior of the T/C shell is shaped like two bowls facing each other and welded or bolted together (**Figure 17–7**). To support the weight of the T/C, a short, stubby shaft, located in a socket provided at the rear of the crankshaft, projects forward from the front of the T/C shell. This forms the frontal support for the T/C assembly.

At the rear of the T/C shell is a hollow shaft with notches or flats at one end, ground 180 degrees apart. This shaft is called the *pump drive hub*; the notches or flats drive the transmission pump assembly. At the front of the transmission within the pump housing is a pump bushing supporting the pump drive hub and providing rear support for the T/C assembly.

FIGURE 17–7 Torque converter input and output principle

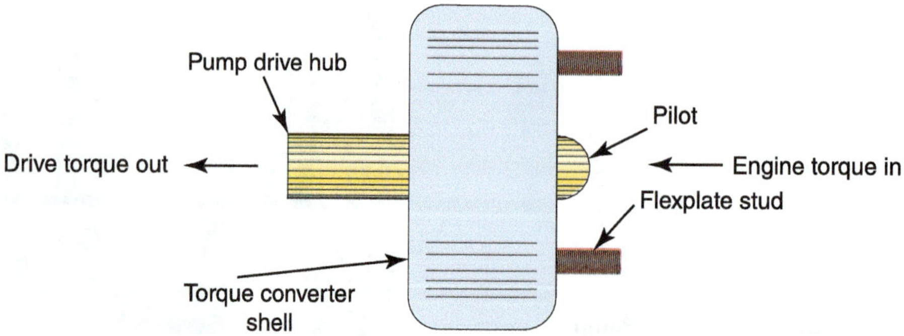

Operation

The pump or impeller forms one section of the T/C shell (**Figure 17–8**). The impeller has numerous curved blades that rotate as a unit with the shell. The size, shape, and number of blades in the impeller play a major role in determining the extent to which the T/C can multiply torque. The impeller turns at engine speed, acting like a pump to start the transmission oil circulating within the T/C shell.

The impeller and turbine face each other. Whereas the impeller is positioned with its back to the transmission housing, the turbine is positioned with its back to the engine. The curved blades of the turbine face the impeller assembly. The hub of the turbine is splined so that the turbine can connect to and drive the turbine shaft. The turbine shaft transfers engine torque (its powerflow path) to the transmission gearing.

The turbine blades are designed to have a greater curve than the impeller blades. This helps reduce oil turbulence between the turbine and impeller blades, turbulence that would slow impeller speed and reduce the converter's efficiency. **Figure 17–9** illustrates how turbulence leads to a loss of energy and operating efficiency. For instance, when water strikes the flat surface of a teacup, it changes direction and moves back toward the oncoming flow. It splashes away from the flat surface in many directions; its energy is scattered. When the teacup is turned right side up, as in **Figure 17–10**, the water flow strikes the curved inner surface of the cup, which is similar to the curved surface of the turbine blades.

FIGURE 17–8 An interior view of a typical torque converter showing the pump/impeller, turbine, and stator

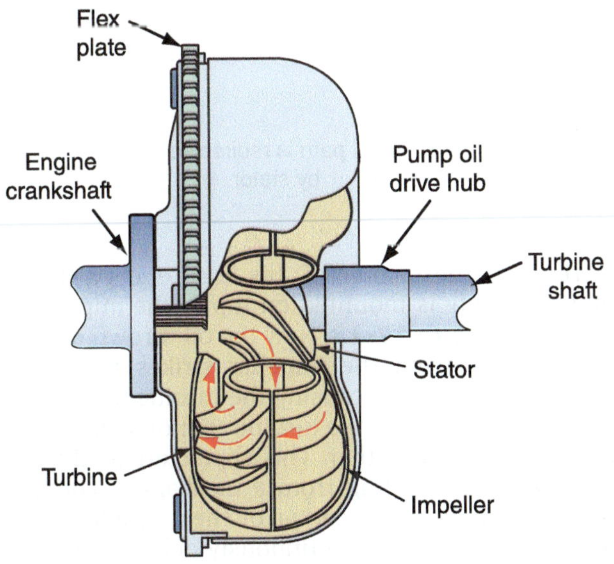

FIGURE 17–9 Contacting the flat back of the cup forces the water to splash away, scattering its energy.

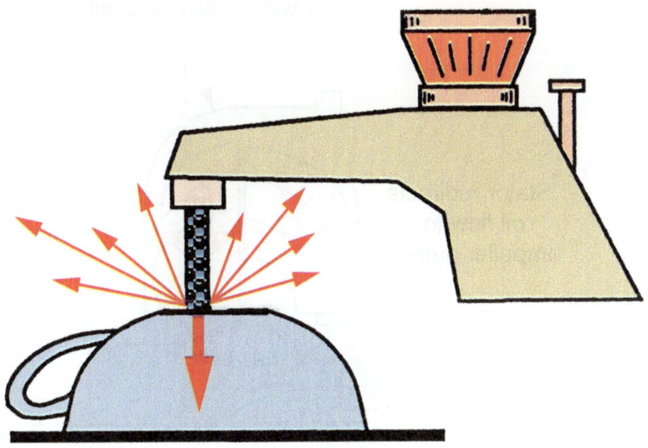

FIGURE 17–10 Contacting the curved surface of the cup concentrates and redirects the water's energy.

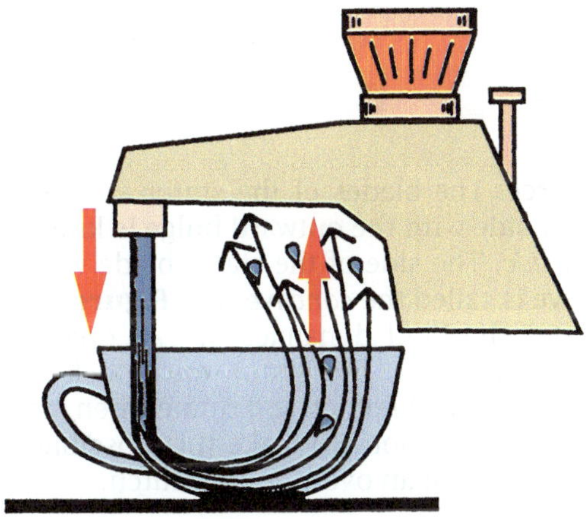

In the teacup example, the curved surface helps direct and concentrate the water's energy. As the water flows around the curve of the cup body, it pushes against the surface and moves away from the cup. The push of the water makes the cup a reaction member, just as the turbine is the reaction member in the T/C.

A fundamental law of hydraulics states, "The more the moving stream of fluid is diverted (changed), the greater the force it places on the curved reaction surface." Thus, oil in the T/C moves around the turbine blades, pushing against the blades as it moves away and placing additional force on the turbine blade that drives the turbine shaft. The stator is located between the pump/impeller and the turbine. It redirects the oil flow from the turbine back into the impeller in the direction of impeller rotation with minimal loss of

FIGURE 17–11 Converter operation

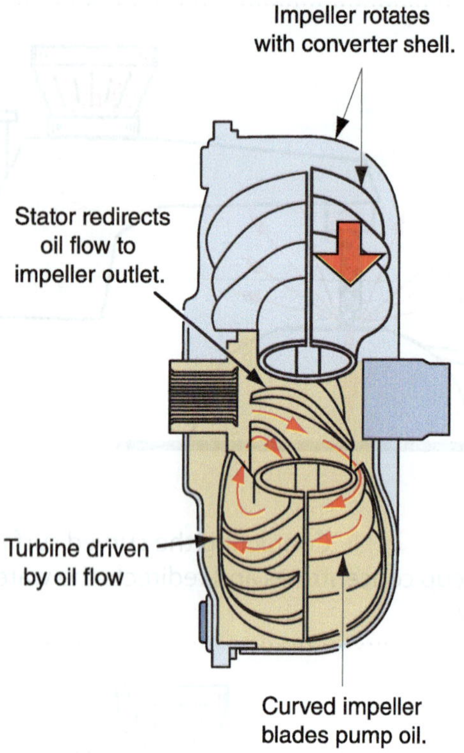

FIGURE 17–12 (A) Transmission fluid (oil) at rest in the T/C impeller/pump. The fluid is level and the pump is not spinning; (B) the pump is spinning, turning oil up and outward.

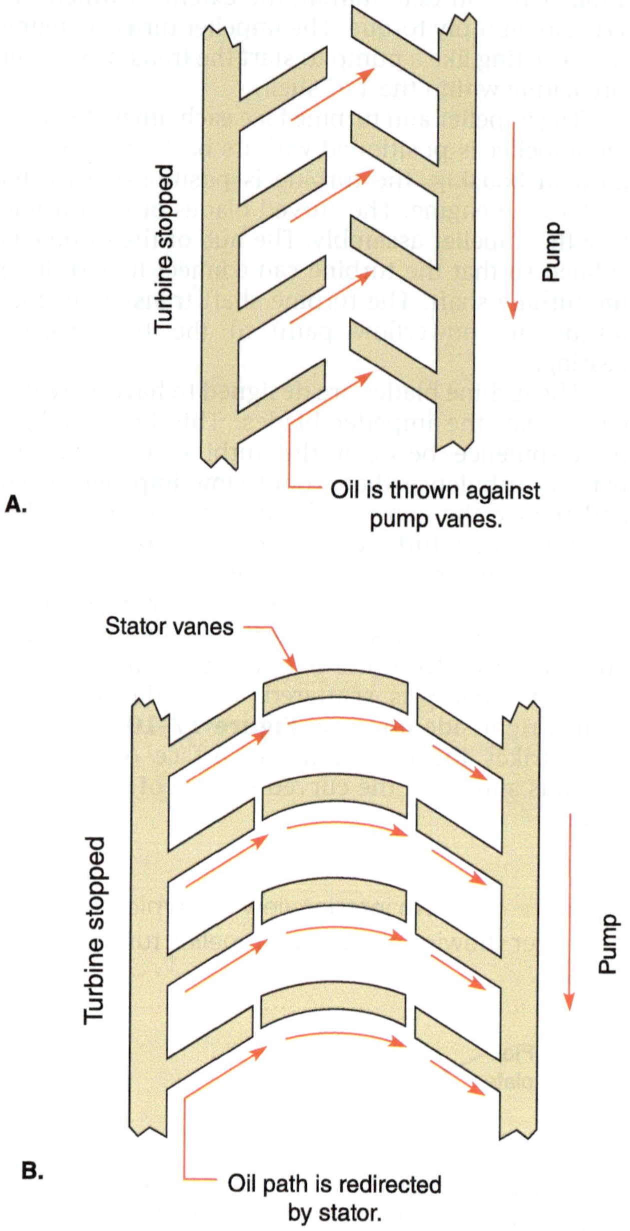

speed or force. The blades of the stator are curved, so the stator side with the outward bulge is known as the *convex side*. The side of the stator blade with the inward curve is called the *concave side*. **Figure 17–11** demonstrates the fluid dynamics of T/C operation. During T/C operation, the stator must lock when forced to turn in one direction and rotate when moved in the opposite direction. To make this possible, the stator is mounted on an overrunning clutch.

17.2 PRINCIPLES OF OPERATION

As we stated at the beginning of this chapter, transmission oil is used as the medium to transfer energy from the engine to the transmission when a torque converter is used. **Figure 17–12A** illustrates what happens to the oil when the T/C impeller is at rest, and **Figure 17–12B** shows how the oil behaves when it is being driven. As the pump impeller rotates, centrifugal force throws the oil outward and upward. This is due to the curved shape of the impeller housing. (Remember the previous example of the teacup and water.)

The faster the impeller rotates, the higher the centrifugal force. In Figure 17–12B, the oil is allowed to exit the housing without producing any measurable work. To harness some of this "lost" energy, the turbine assembly is mounted on top of the impeller, as shown in **Figure 17–13**. Now the oil thrown outward and upward from the impeller strikes the curved vanes of the turbine, causing the turbine to rotate.

The impeller and the turbine are not mechanically connected to each other. This characteristic differentiates a fluid coupling from a mechanical coupling. A hollow shaft at the center of the T/C allows fluid under pressure to be continuously delivered from an

FIGURE 17–13 As the T/C rotates, oil is thrown up and out of the pump (impeller) onto the turbine, and then returned to the pump. This action is called vortex flow.

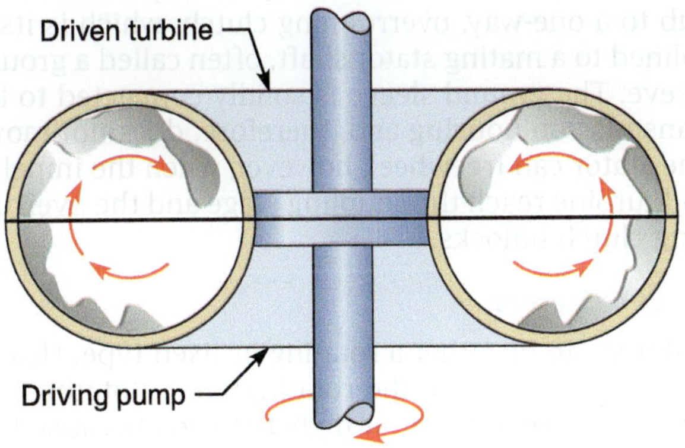

oil pump. It is important to note that the oil pump delivering the fluid is driven by the engine. A seal or combination of seals prevents fluid from being lost from the system.

SHOP TALK

There can be a mechanical connection between the impeller and turbine by the use of a lockup clutch. A lockup clutch eliminates slippage between the impeller and turbine under certain operating conditions. Under converter lockup, the engine, T/C impeller, T/C turbine, and transmission input shaft all rotate at the same speed. This helps to reduce the heat generated in the fluid and improves fuel mileage.

The turbine shaft is located within the hollow shaft at the center of the T/C. It is splined to the turbine and transfers torque from the T/C to the transmission main drive shaft or in the case of the Allison TC-10, the transmission turbine shaft. Oil leaving the T/C turbine is directed out of the converter assembly to an external oil cooler and then to the transmission oil sump.

When the transmission is in neutral, torque cannot be transferred from the engine to the transmission output because the powerflow between these two points has been mechanically disconnected. This does not mean, however, that the driver must physically place the transmission in neutral each time the vehicle is stopped. With the transmission in gear and the engine at idle, the truck can be held stationary by applying the service brakes. At idle, engine speed is slow. Because the impeller is driven by engine speed, it too turns slowly, creating less centrifugal force within the T/C. The small amount of driveline torque generated can easily be held using the service brakes.

When the accelerator is depressed, engine speed increases. Because the impeller is driven directly by the engine, its speed increases with engine speed. This increase in speed increases the centrifugal force of the transmission oil, and the oil is directed against the turbine blades. The force of the oil against the blades transfers torque to the turbine shaft and transmission proportional to engine speed. The faster the engine turns, the more force is applied.

T/C efficiency is always highest when input speed (engine speed) is greatest.

TYPES OF OIL FLOW

Understanding fluid flow within the T/C during its operation is essential. The two types of oil flow that occur within the T/C are *rotary* and *vortex* (**Figure 17–14**). Rotary flow describes the centrifugal force applied to the fluid as the converter rotates on its axis. Vortex flow is the circular flow that occurs as the oil is forced from the impeller to the turbine and then back to the impeller. If a toy pinwheel were held at arm's length and swung in a large circle, air movement at the outer circle would produce rotary flow, while the small circles cut by the pinwheel's propeller vanes would produce vortex flow (**Figure 17–15**).

As the speed of the turbine increases, it approaches the speed of the impeller that is driving it. The point at which the turbine is turning at close to the same speed as the impeller is referred to as the **coupling point**. Due to some fluid slippage that occurs between the turbine and the impeller, however, a true 1:1 coupling ratio is unobtainable unless some type of mechanical means is used to couple the two components. A lockup clutch is used to achieve this.

SPLIT GUIDE RINGS

Power can flow through the converter only when the impeller is turning faster than the turbine. When the impeller is turning at speeds much greater than the turbine, a great deal of oil turbulence occurs.

FIGURE 17–14 Rotary and vortex oil flow in the torque converter.

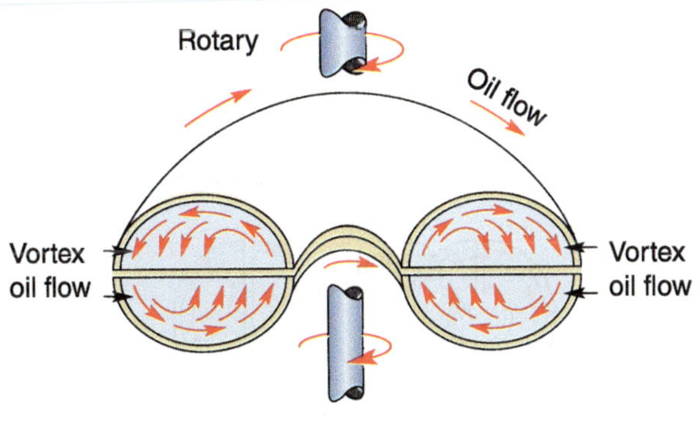

FIGURE 17–15 Rotary flow and vortex flow produced in the air by a pinwheel.

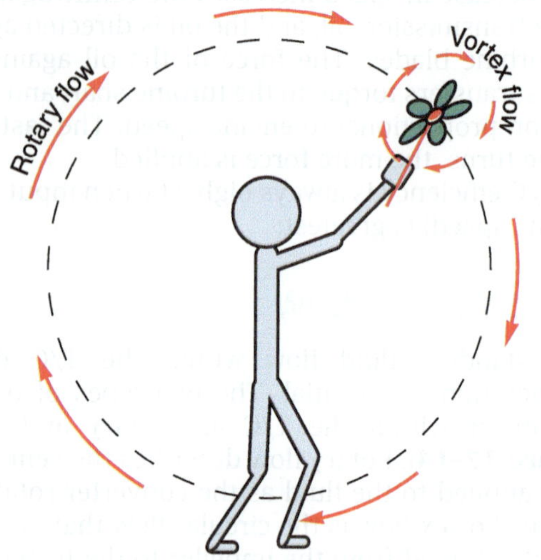

Fast moving oil exits the impeller blades, striking the turbine blades with considerable force. It then has a tendency to be thrust back toward the center of both impeller and turbine.

To control this fluid thrust and the turbulence that results, a split guide ring is located in both the impeller and turbine sections of the T/C (**Figure 17–16**). The guide ring suppresses turbulence, allowing more efficient operation.

STATOR

The **stator** is a small wheel positioned between the pump/impeller and the turbine (see Figure 17–4). The stator has no mechanical connection to either the impeller or turbine but fits between the outlet of the turbine and the inlet of the impeller so that all the oil returning from the turbine to the impeller must pass through the stator.

The stator mounts through its splined center hub to a one-way, overrunning clutch, which is itself splined to a mating stator shaft, often called a ground sleeve. The ground sleeve is solidly connected to the transmission housing and, therefore, does not move. The stator can freewheel, however, when the impeller and turbine reach the coupling stage and the overrunning clutch unlocks.

FIGURE 17–16 (A) The vanes are cut away to accommodate the split ring channel; (B) the vortex flow is guided and smoothed by the split ring; (C) without a split ring, fluid turbulence intensifies at the center.

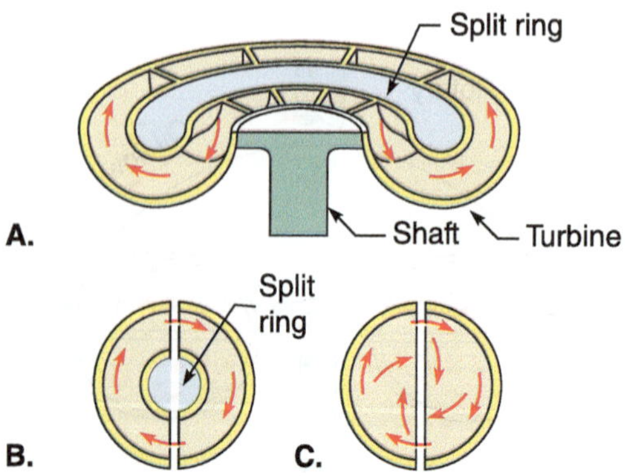

Types of Stators

A stator can be either a rotating or fixed type. Heavy-duty truck T/Cs use the rotating type. Rotating stators are more efficient at higher speeds because less slippage occurs when the impeller and turbine reach the coupling stage.

Torque Multiplication

The stator is the key to torque multiplication. It redirects the oil leaving the turbine back to the impeller, which helps the impeller rotate more efficiently (**Figure 17–17**). Torque multiplication can occur only when the impeller is rotating faster than the turbine. This results from the velocity or kinetic energy transferred to the transmission oil by impeller rotation, plus the velocity of the oil that is directed back to the impeller by stator action. During operation, the oil gives up part of its kinetic energy as it strikes the turbine vanes. The stator then redirects the fluid flow so that the oil reenters the impeller, moving in the same direction that the impeller is turning. This allows the kinetic energy remaining in the oil to help rotate the impeller even faster, multiplying the torque produced by the converter.

Overrunning Clutch

An **overrunning clutch** keeps the **stator assembly** from rotating when driven in one direction and permits overrunning (rotation) when turned in the opposite direction. Rotating stators generally use a roller-type overrunning clutch that allows the stator to freewheel (rotate) when the speed of the turbine and impeller reach the coupling point. The **roller clutch** (**Figure 17–18**) is designed with an inner race, rollers, accordion (apply) springs, and outer race. Around the inside diameter of the outer race are several cam-shaped pockets. The clutch assembly rollers and accordion springs are located in these pockets.

As the vehicle begins to move, the stator stays in its stationary or locked position because of the wide difference between the impeller and turbine speeds. This locking mode takes place when the outer race attempts to rotate counterclockwise. The accordion springs force the rollers down the ramps of the cam

FIGURE 17–17 (A) Without a stator, fluid leaving the turbine works against the direction the impeller or pump is rotating; (B) with a stator in its locked (non-coupling) mode, fluid is directed to help push the impeller in its rotating direction.

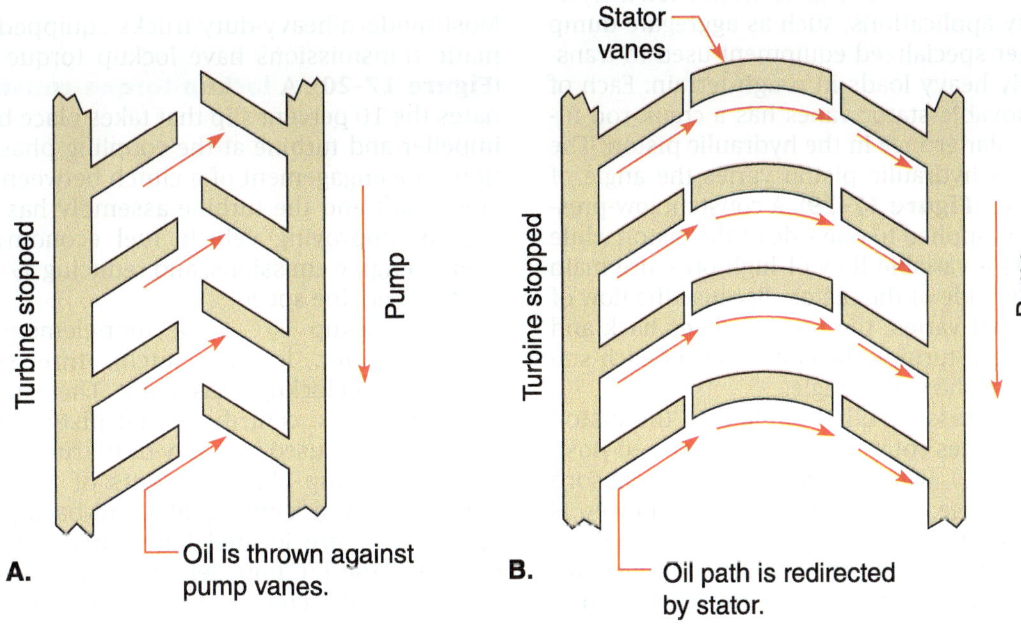

FIGURE 17–18 Roller-type overrunning clutch

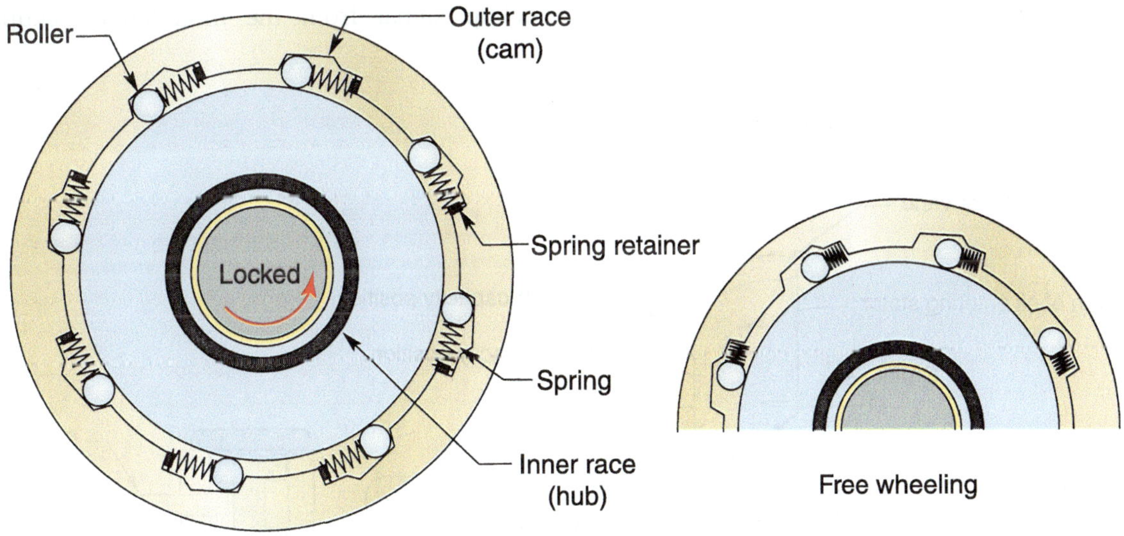

pockets into a wedging contact with the inner and outer races. When the overrunning clutch is locked, the stator will be held stationary.

As vehicle road speed increases, so does turbine speed until it approaches impeller speed. Oil exiting the turbine vanes strikes the back face of the stator, causing the stator to rotate in the same direction as the turbine and impeller. At this higher speed, clearance exists between the inner stator race and hub. The rollers in each slot of the stator are pulled around the stator hub. The stator freewheels or turns as a unit.

If the vehicle slows, engine speed also slows along with the turbine speed. This decrease in turbine speed allows the oil flow to change direction. It now strikes the front face of the stator vanes, halting the turning stator and attempting to rotate it in the opposite direction. As this happens, the rollers jam between the inner race and hub, locking the stator in position. In a stationary position, the stator now redirects the oil exiting the turbine so that torque is again multiplied.

Variable Pitch Stator

A **variable pitch stator** design is often used in T/Cs in off-highway applications, such as aggregate dump trucks or other specialized equipment used to transport unusually heavy loads in rough terrain. Each of a series of movable stator vanes has a crank rod fitted into a circular groove in the hydraulic piston. The movement of a hydraulic piston varies the angle of the stator vanes (**Figure 17–19**). A constant low-pressure oil force is applied to one side of the piston while a valve directs a variable flow of high-pressure main oil on the other side of the piston. Because the flow of high-pressure oil varies, the piston moves back and forth accordingly, turning the crank rods of each stator, which varies the stator angle.

When high-pressure oil is applied to the piston assembly, the vanes rotate to a partially closed position (low capacity), which causes the T/C to absorb less force. Because less of the available power is being absorbed by the T/C, more of it is available to the power take-off gearing for operation of auxiliary equipment. When the oil pressure decreases to the point that the pressure-regulating valve reduces the amount of high-pressure oil being delivered to the piston, the piston moves back to its normal position, allowing the stators to rotate to their usual position (high capacity, fully open). This directs the available fluid power so that its energy can be most efficiently used.

LOCKUP CLUTCHES

Most modern heavy-duty trucks equipped with automatic transmissions have lockup torque converters (**Figure 17–20**). A **lockup torque converter** eliminates the 10 percent slip that takes place between the impeller and turbine at the coupling phase of operation. The engagement of a clutch between the engine crankshaft and the turbine assembly has the advantage of improving vehicle fuel economy, lowering overall engine emissions, and reducing T/C operating heat and engine speed.

The lockup T/C is a four-element (impeller, turbine, stator, lockup clutch), three-stage (stall, coupling, and lockup phase) unit. There are two types of lockup T/Cs: centrifugal and piston. Piston lockups tend to be used in heavy-duty truck applications. A piston lockup clutch consists of three main elements: a piston, clutch plate, and back plate. These components are located between the T/C impeller cover and the T/C impeller. The piston and back plate rotate with the converter impeller. The clutch plate is located between the piston and the back plate and splined to the converter turbine.

The engagement of the lockup clutch is controlled by the lockup relay valve that receives its lockup signal from the modulated lockup valve.

FIGURE 17–19 Variable pitch stator assembly

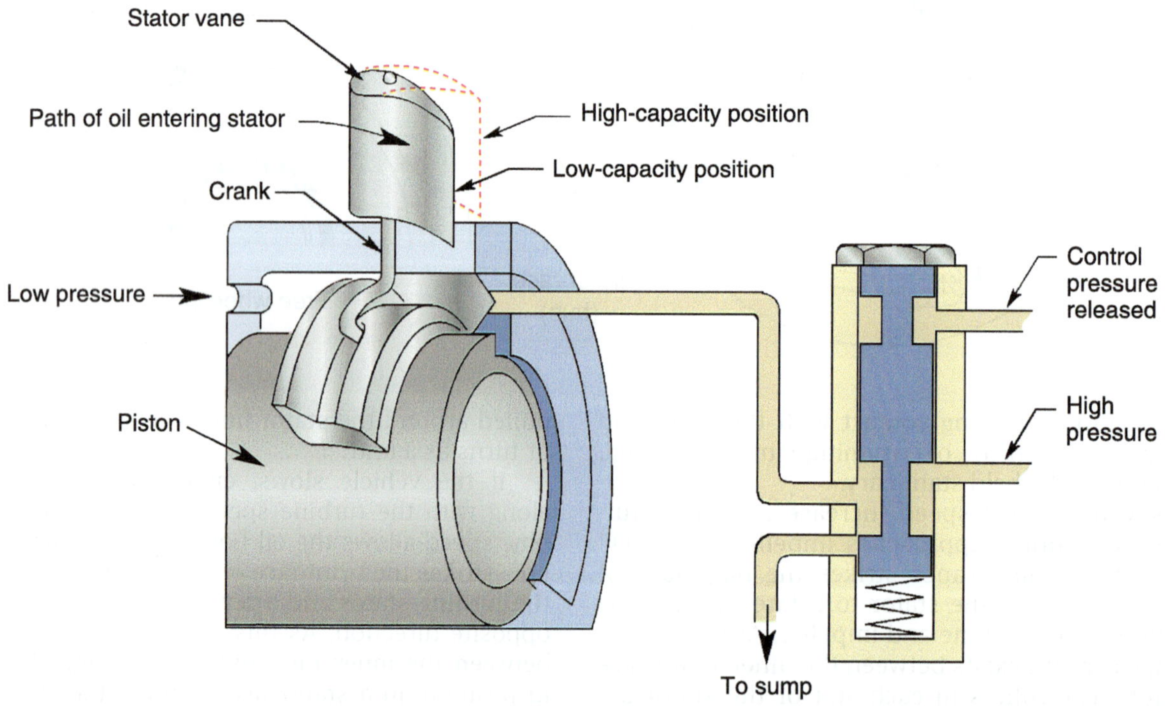

FIGURE 17–20 Components of a typical lockup torque converter

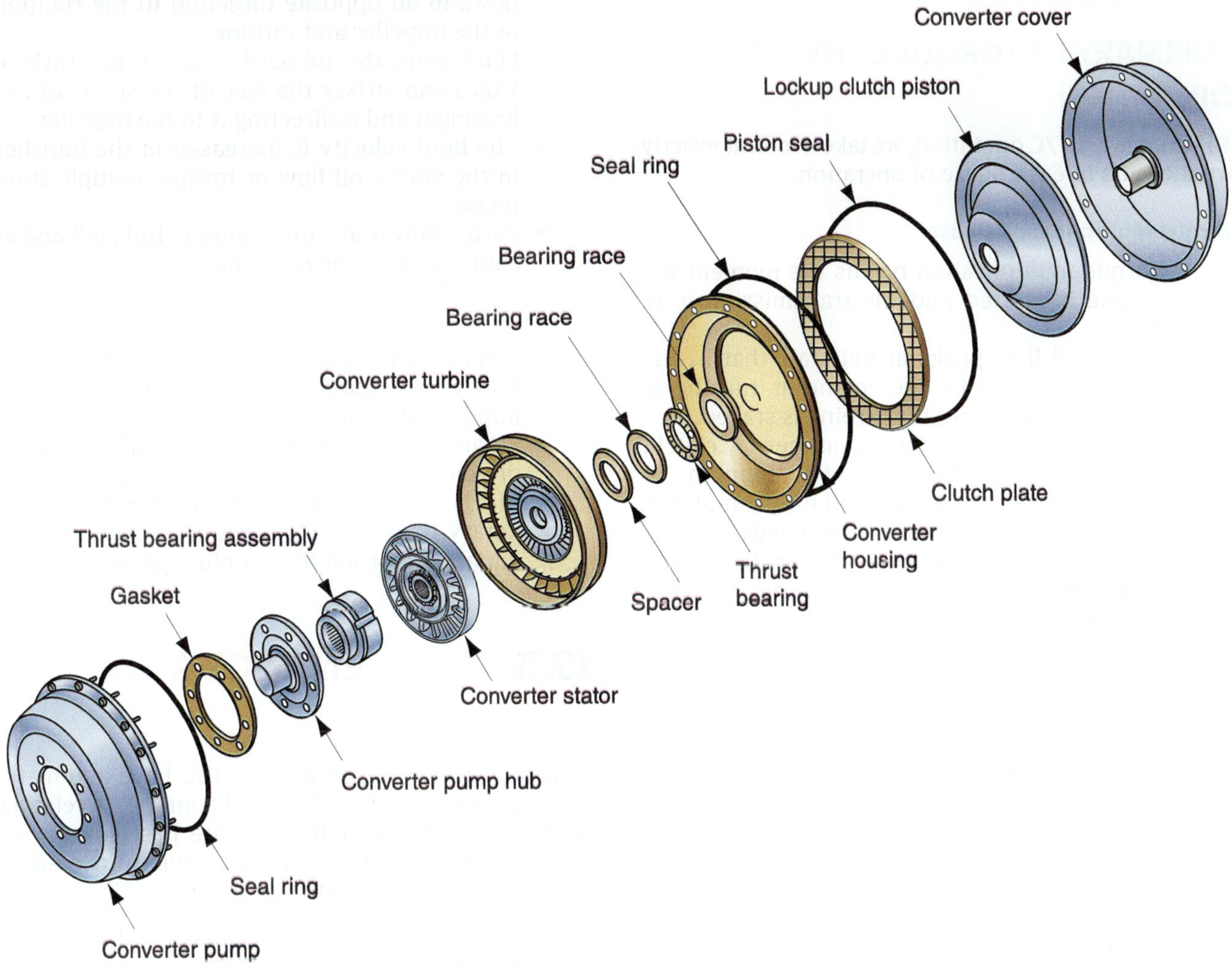

Clutch-apply pressure compresses the lockup clutch plate between the piston and the back plate, locking all three together. When the converter impeller and the turbine are locked together, they provide a direct drive one-to-one from the engine to the transmission gearing. The result is top vehicle speed performance and improved fuel mileage. As rotational speed of the output shaft decreases, the relay valve automatically releases the lockup clutch and standard T/C operation resumes. **Figure 17–21** shows how a torque converter connects to the automatic transmission it drives.

STALL TORQUE RATIO

The **stall torque ratio** identifies a torque converter's ability to multiply torque. It is always expressed as a ratio. For instance, a torque converter with a stall torque ratio of 2:1 is capable of multiplying engine torque by two times. The math is not usually that

FIGURE 17–21 Cutaway view of a torque converter showing how the turbine shaft connects with the automatic transmission.

simple: for instance, the T/C used on an Allison TC-10 has a stall torque ratio of 1.84:1.

SUMMARY OF TORQUE CONVERTER OPERATION

To summarize T/C operation, we take a look at exactly what occurs in each phase of operation.

Torque Multiplication Phase

- Torque multiplication begins the moment the engine is started and the transmission is in gear.
- Vortex oil flow peaks at full stall; that is, the impeller (driven by the engine) is turning at maximum rpm, and the turbine is stalled.
- Vortex oil flow locks the stator one-way clutch, causing rotary oil flow to be at a minimum.
- The rate of vortex oil flow and torque multiplication decreases as the turbine speed increases and begins to approximate the impeller rpm.
- As turbine speed approaches 90 percent of the impeller speed, vortex oil flow drops to a minimum, and the converter enters the coupling phase.

Coupling Phase

- The converter enters the coupling phase when vortex oil flow is at a minimum and rotary oil flow is at a maximum.
- In the coupling phase, turbine rpm is within 10 percent of impeller rpm.
- Rotary flow causes the stator one-way clutch to unlock, enabling it to freewheel.
- In the coupling phase, a slippage factor is always present, meaning that the turbine will always rotate at a slightly lower speed than the impeller.
- Torque converters equipped with a lockup clutch may apply it in the coupling phase.

Lockup Phase

- A lockup clutch locks the turbine to the front cover.
- Lockup phase means that the impeller and turbine rotate at the same speed.
- Lockup increases the 90 percent turbine to impeller drive efficiency of the coupling phase to 100 percent drive efficiency.

Vortex Flow

- Oil enters the impeller, passes around the split guide rings, and exits at the outer edge of the impeller vanes.
- The oil is next directed to act on the outer edge of the turbine vanes, to pass around the other half of the split guide ring, and to exit at the inside edges of the turbine vanes. This fluid flows in an opposite direction to the rotation of the impeller and turbine.
- Fluid exits the inboard edge of the turbine vanes and strikes the face of the stator vanes, locking it and redirecting it to the impeller.
- The fluid velocity is increased in the impeller in the vortex oil flow or torque multiplication phase.
- Vortex flow is at a maximum at full stall and at a minimum in the coupling phase.

Rotary Flow

- Rotary flow occurs when the direction of oil flow in the stator is the same as that of the impeller and turbine.
- Rotary flow occurs when the turbine is being driven within 10 percent of the impeller rpm; that is, with 90 percent efficiency or better.
- Rotary flow is at a minimum at stall and at a maximum during the coupling phase.

17.3 MAINTENANCE AND SERVICE

Because trucks fitted with T/Cs also have automatic transmissions, most of the checks and tests relating to the maintenance of the T/C are best covered in detail under general transmission service procedures. These checks and precautions include:

- Using the manufacturer's specified transmission oil
- Changing transmission oil at proper intervals and keeping the oil free of contamination
- Using the proper oil changing procedure and regularly checking oil levels
- Maintaining the T/C or transmission breather
- Maintaining the proper oil operating temperature

Chapter 19 fully covers the service and maintenance required on T/Cs. In this chapter we have isolated a few general checks and tests for T/Cs. As usual, the information presented here is not intended to replace OEM service literature.

REMOVING THE TRANSMISSION/ TORQUE CONVERTER

Drain the oil from the transmission before removal from the vehicle. For better drainage, the transmission should be warm. Make sure all linkages, controls, cooler lines, modulator actuator cables, temperature connections, input and output couplings, oil filler tubes, parking brake linkages, and mounting bolts are disconnected before the transmission is removed. Oil lines should be carefully placed out of the way and openings covered to keep out dirt. Position a jack or hoist sling relative to the transmission's center of gravity so that the weight of the unit will be balanced as it is lifted.

CAUTION

The T/C is free to move forward when the transmission is disconnected from the engine. To ensure that the T/C does not separate from the transmission while the transmission is being removed from the vehicle, install a retaining strap to hold the T/C in position.

For transmission overhaul work, including removal and servicing of the T/C, the transmission should be mounted to an overhaul fixture, as shown in **Figure 17-22**. The transmission power take-off (PTO) cover is removed and replaced with a holding plate that is attached to a fixture mounted to the stand. The fixture allows the transmission to be fully rotated so that all parts are easily accessible.

After the transmission has been mounted to the overhaul fixture, position the transmission front upward and remove the retaining strap used to hold the T/C in position. Attach a lifting tool and raise the T/C assembly from the transmission, as shown in **Figure 17-23**. You should clean the housing to inspect it—use mineral spirits or a power washer. Clean internally splined clutch friction plates with transmission fluid. Blow-dry the wet components with compressed air. Lightly oil any components that might rust. Make sure that water or cleaning solvent does not enter sealed, welded units.

FIGURE 17-22 Setup for mounting a transmission and torque converter to an overhaul stand

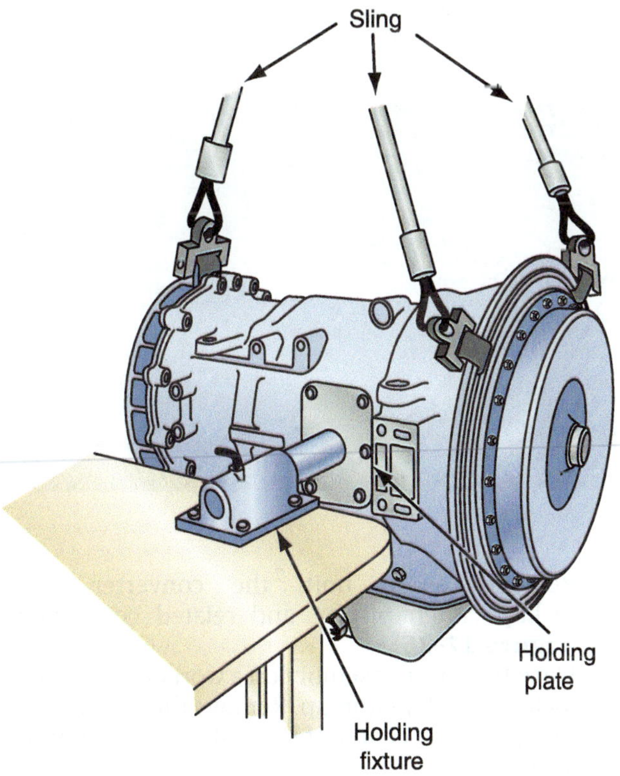

FIGURE 17-23 Lifting the torque converter from the transmission body

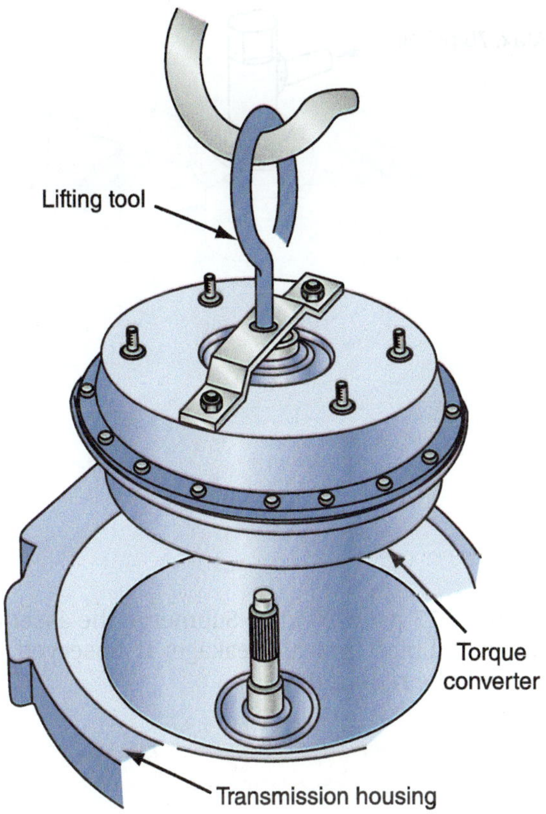

INSPECTION AND TESTING

Inspect the T/C housing for damaged slots, scoring, cracks at welded seams, missing weights, dents, missing lugs, or damaged threads. The unit should be replaced if any significant damage is observed.

Pressure Testing (Sealed/Welded Units)

This is a leakage test. Before performing this test on sealed T/C units, drain all of the oil from the T/C. The procedure is to pressurize the T/C housing with regulated compressed air and submerse it in water. **Figure 17-24** illustrates the basic setup used for this type of leak testing. Pressurize the housing using the original equipment manufacturer (OEM)-recommended test pressure, usually a maximum air

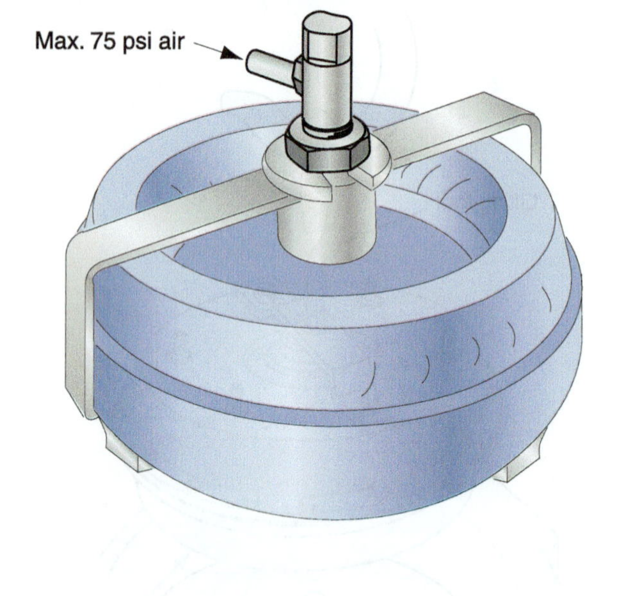

FIGURE 17–24 Tools and setup for leak testing sealed weld torque converters

pressure of 75 psi (520 kPa). Submerge the assembly in water. Bubbles indicate leakage; if observed, the T/C should be replaced.

> **CAUTION**
>
> Exhaust the air pressure at the test valve stem before loosening the nut holding the test fixture in place.

Check Endplay

Check endplay using a dial indicator fixture mounted as shown in **Figure 17–25**. An out-of-specification endplay measurement indicates unacceptable wear between the internal components. Zero the dial indicator and pry upward on the center screw to full travel. Record the gauge reading.

If the play exceeds the OEM specification on a sealed, welded T/C, the unit should be replaced. On a bolted shell unit, the converter can be disassembled and rebuilt. Standard servicing of bolted T/Cs requires the disassembly and inspection of internal components, even if the endplay is within specification.

TORQUE CONVERTER DISASSEMBLY

A typical disassembly and inspection procedure is outlined for a truck T/C. The procedure references the components identified in Figure 17–21.

1. Remove the six rubber ID retainers (and spacers, if used) from the converter cover assembly.
2. Remove the twenty-four nuts from the cover.

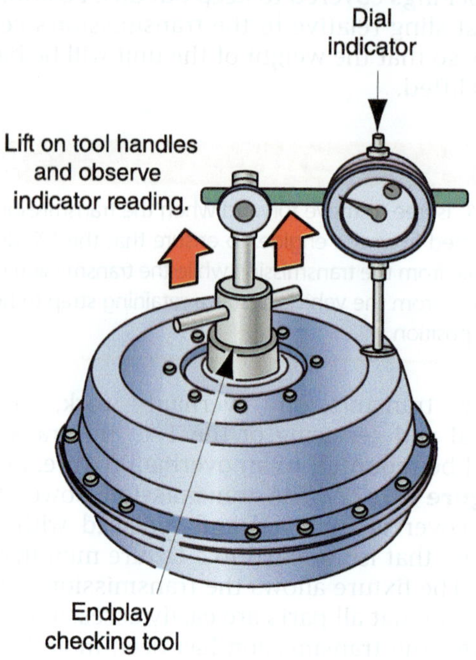

FIGURE 17–25 Tools and setup for torque converter endplay inspection and testing

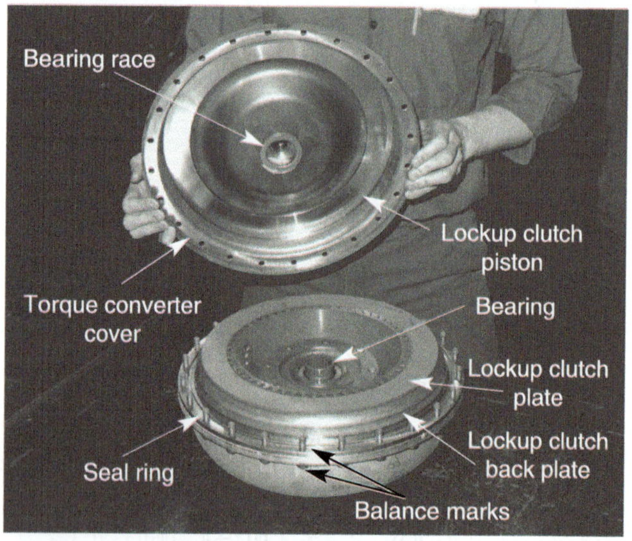

FIGURE 17–26 Removing (or installing) a converter cover assembly

3. Remove, as a unit, the converter cover, lockup clutch piston, and related components (**Figure 17–26**).
4. Place the cover assembly on a workbench with the lockup clutch piston up. Remove the bearing race. Compress the center of the piston and remove the snapring.

5. Turn the cover assembly over (piston down) and bump the cover sharply on a wood surface to remove the piston. Remove the seal ring retainer and seal ring from the cover. Remove the seal ring from the piston.
6. Remove the bushing only if replacement is necessary.
7. Remove the lockup clutch plate.
8. Remove the lockup clutch back plate from the T/C pump. Remove the seal ring from the plate.
9. Remove the roller bearing assembly, bearing race, and spacer from the hub of the turbine.
10. Remove the converter turbine assembly (**Figure 17–27**).
11. Grasp the stator and the roller race, as shown in **Figure 17–28**, and remove as a unit.
12. Position the stator assembly on the worktable so that the freewheel roller race is upward. Remove the roller race by rotating it clockwise while lifting it out of the converter stator.
13. Remove the ten rollers and ten springs from the stator assembly.
14. Check the needle bearing assembly. Wash and flush the needle bearing assembly thoroughly with dry-cleaning solvent or mineral spirits. Dry and lubricate it with transmission oil. Replace the freewheel race only and rotate the bearing while pressing on the freewheel race. If there is no roughness or binding, the needle bearing assembly can be left in the stator and cam assembly and can be reused. Do not mistake dirt or grit for a damaged needle bearing. Clean and oil the needle bearing if dirt is suspected. Check the needle bearing end of the freewheel race for a smooth finish. Replace the freewheel race if the bearing end is scratched or contains chatter marks.

FIGURE 17–27 Removing (or installing) a converter turbine assembly

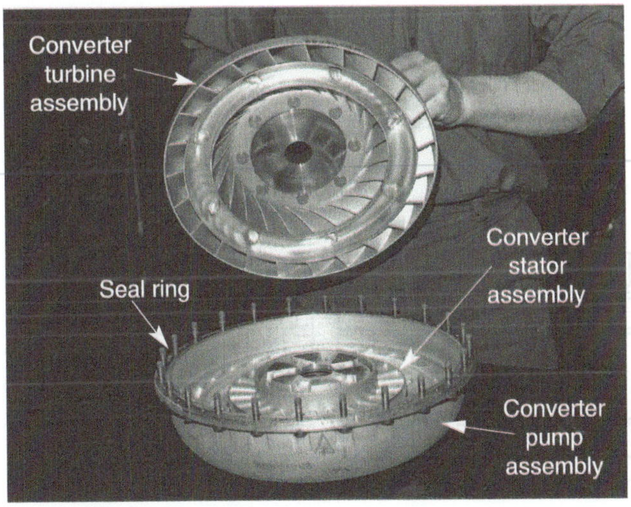

FIGURE 17–28 Removing (or installing) a converter stator and race assembly

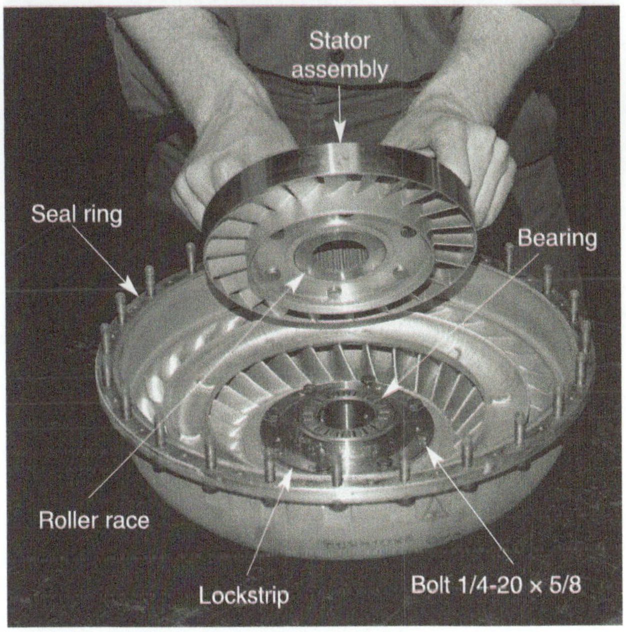

SHOP TALK

If it is necessary to rebuild the stator assembly, remove the needle bearing assembly before starting the stator rebuild. Always follow the T/C manufacturer's rebuild procedures for all T/C subassemblies. Do not disassemble the stator assembly unless parts replacement is needed.

15. If only the needle bearing assembly requires replacement, remove it carefully to avoid nicking the aluminum bore in which it is held.
16. Place the new needle bearing assembly, thrust race first, into the aluminum bore or the stator. Use a bearing installer and handle to install the thrust bearing (**Figure 17–29**).

CAUTION

Apply load only to the outer shell of the bearing during installation.

17. Drive the bearing assembly into the stator until the top of the outer shell is 0.025 to 0.035 inch (0.64 to 0.89 mm) above the shoulder in the side plate (**Figure 17–30**). The installing tool will seat on the stator area surrounding the bearing when the bearing is properly installed.
18. On some earlier units remove the bearing and bearing race from the converter pump hub.

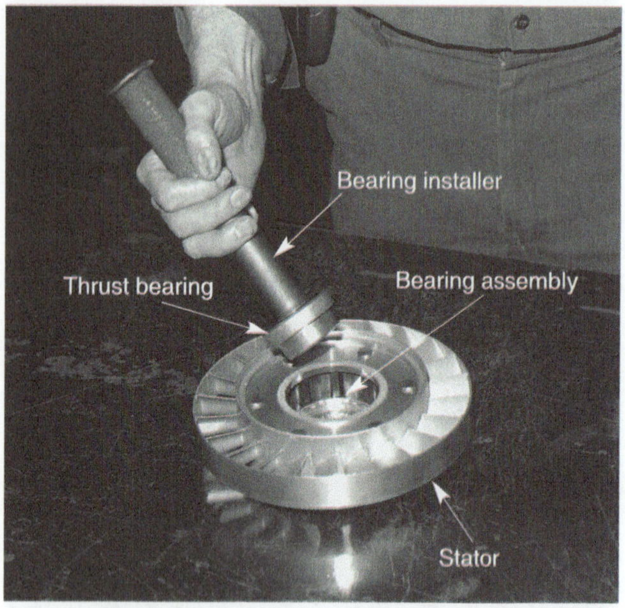

FIGURE 17–29 Installing the stator thrust bearing

On later model units, remove the needle bearing and bearing race, and remove the roller bearing from the converter pump hub. Remove the seal ring.

19. Flatten the corners of the lock strips and remove eight bolts and four lock strips from the converter pump hub.
20. Remove the hub and gasket from the pump. Remove the seal ring.

INSPECTION AND ANALYSIS

When performing a T/C overhaul, the following inspections and checks should be made on the converter and its related components.

Flywheel Assembly (If Used)

- Inspect the flywheel (**Figure 17–31**) for cracks, impact damage, and signs of overheating.
- Inspect the flexplate mounting holes for pulled, stripped, or crossed threads. Damaged bolt threads can be repaired. Any signs of cracks or severe impact damage will require component replacement.
- Inspect the flywheel crank adapter pilot for burrs, heat damage, or signs of misalignment. Light honing is permitted to remove slight irregularities.
- On flywheel assemblies with welded or pressed lockup pins, inspect for cracked welds or evidence of fluid leakage around the pin. Cracked welds can be repaired if the lockup pin is not damaged excessively on the backside of the flywheel. Leakage of the pressed pins can be repaired by welding the pin diameter to the flywheel face. A heli-arc or tungsten inert gas (TIG) welding process should be used.
- The seal ring bore should be free of scratches, nicks, scoring, and signs of overheat. Slight irregularities can be corrected with a crocus cloth. Inspect the seal ring bore for any signs of excessive grooving from the rotating seal ring. Any grooving or excessive wear noted

FIGURE 17–30 Typical stator thrust bearing

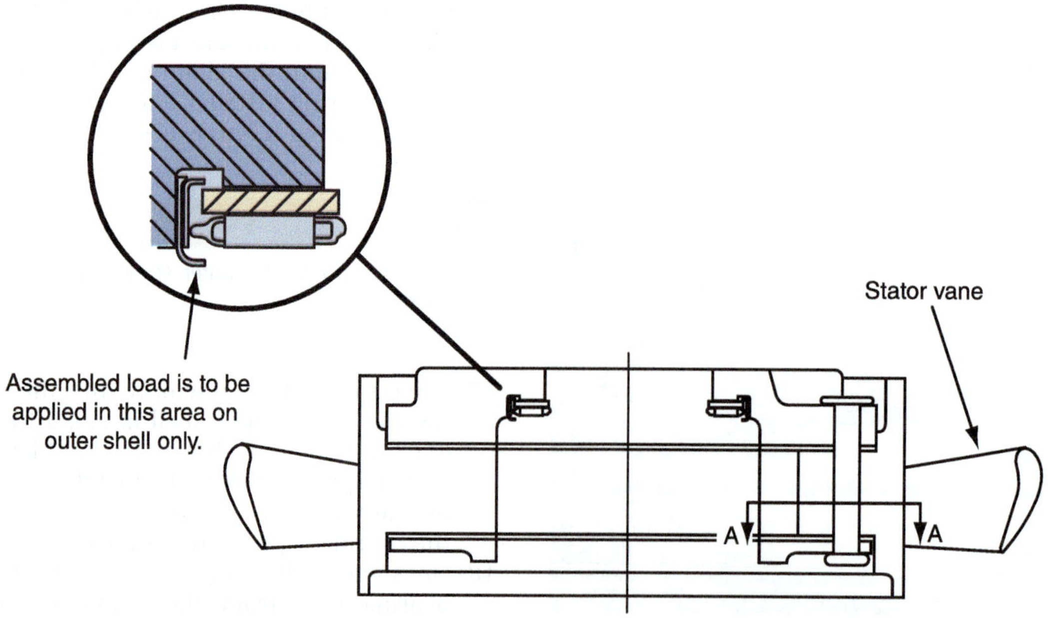

FIGURE 17–31 Flywheel inspection points: (A) front; and (B) back

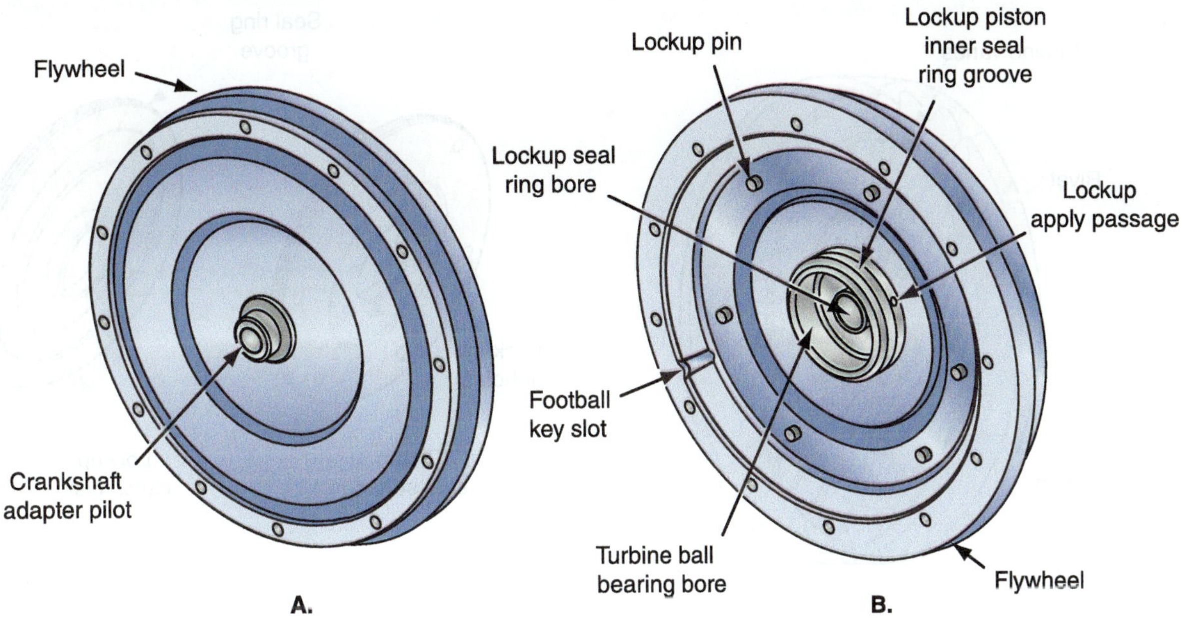

can be repaired by reworking with the available service bushing.

- Inspect the flywheel assembly for cracked, battered, or broken lockup pins. Any damage noted will require component replacement. Lockup apply passages should be open and free of debris.
- Inspect the ball bearing bore for scratches, grooves, nicks, scoring, and signs of overheating. Light honing is permitted to remove slight irregularities.
- Inspect the seal ring groove for cracked edges, nicks, burrs, and sharp edges. Light honing is permitted to remove slight irregularities. Any cracked or chipped edges will require component replacement.
- The football key slot should show no evidence of wear from movement of the lockup backing plate and football key. Excessive wear or elongation of the football key slot will require component replacement.
- Inspect the flywheel assembly for pulled, stripped, or crossed threads. Inspect the gasket surface for nicks, scratches, or burrs that could cause fluid leakage between the flywheel assembly and pump assembly. Light honing is permitted to remove slight irregularities. Damaged bolt threads can be repaired.

Lockup Clutch Back Plate

Inspect the lockup clutch back plate surface that contacts the friction plate for wear, scoring, scratches, and signs of overheating (**Figure 17–32**). The back plate should be flat to within 0.006 inch (0.15 mm). Light honing is permitted to remove slight irregularities.

Inspect the football key slot for evidence of wear from movement of the lockup back plate and football key. Excessive wear or elongation of the football key slot will require component replacement. Inspect the back plate for cracks, from the key slot to the ID of plate. Any damage noted will require part replacement.

Turbine Assembly

Inspect the turbine assembly (**Figure 17–33**) for cracked or broken vanes and signs of overheating. If the turbine assembly is taken apart, inspect the rivet holes for signs of wear or elongation. Any damage noted usually requires component replacement.

The turbine assembly hub should be checked for stripped, twisted, or broken splines. Any damage observed will require component replacement.

FIGURE 17–32 Lockup clutch back plate inspection

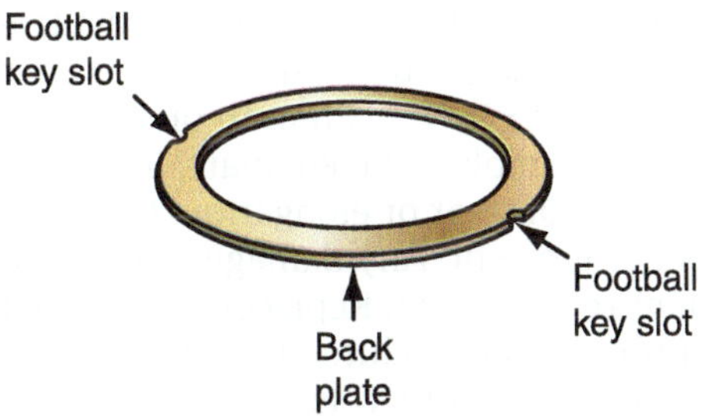

FIGURE 17–33 Torque converter turbine assembly inspection points

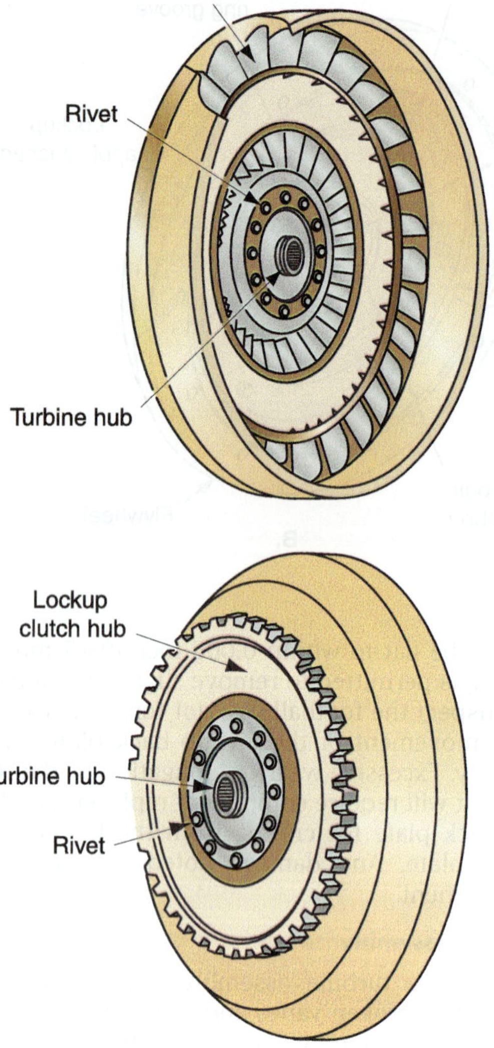

FIGURE 17–34 Lockup clutch piston inspection

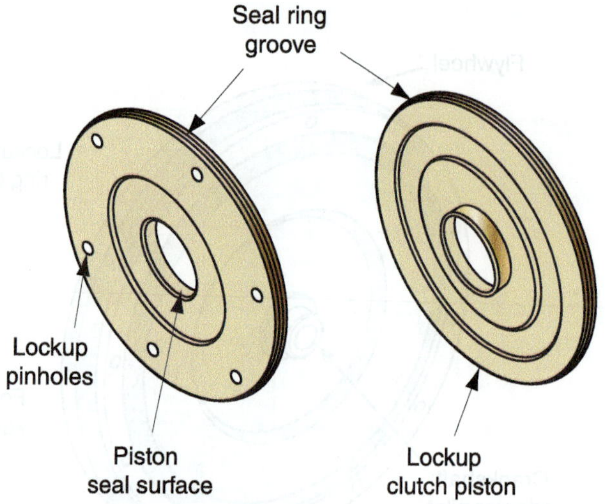

Inspect the roller bearing journal for scoring, pitting, scratches, metal transfer, and signs of overheating. Slight irregularities can be removed with a crocus cloth. Inspect the selective turbine thrust bearing spacer surface for scores, scratches, burrs, and signs of overheating. Light honing is permitted to remove slight irregularities.

Inspect the spline condition of the lockup clutch hub for notching due to lockup clutch plate movement. Replace lockup clutch hub if splines are notched deeper than 0.015 inch (0.38 mm) on any one side. If the turbine assembly is taken apart, inspect the rivet holes for signs of wear or elongation. Check the rivets for cracks or loose fit. Any damage noted will require component replacement. Replacement of the turbine, turbine hub, or lockup clutch hub (Figure 17–33) will require replacement of rivets.

Lockup Clutch Piston. Inspect the seal ring groove of the lockup clutch piston (**Figure 17–34**) for cracked edges, nicks, burrs, and sharp edges; remove any slight irregularities with light honing. Pistons with cracked or chipped groove edges must be replaced.

Check the lockup pinholes for signs of battering and elongation wear. Excessive elongation wear of the holes could indicate a wear condition of the lockup pins in the flywheel assembly. Excessive wear will require component replacement. Inspect the piston for cracks, warpage, and signs of overheating. Inspect the surface in contact with the friction plate for wear, scoring, scratches, signs of overheating, and flatness. The surface in contact with the friction plate should be flat to within 0.003 inch (0.08 mm) total indicated runout (TIR). Any damage noted will require component replacement. Inspect the seal surface for scratches, burrs, and nicks. Slight irregularities can be corrected with a crocus cloth.

Stator

Inspect the stator assembly (**Figure 17–35**) for cracked or broken vanes and signs of overheat. Any damage noted will require replacement of the stator assembly. Inspect the rivets for cracks or loose fit. Any damage noted will require component replacement. Replacement of the stator, stator thrust washer, cam, side plate washer, or stator cam washers will require replacement of rivets. Check the stator cam roller pockets for pitting, scoring, signs of overheating, and wear. Emery cloth can be used to clean up slight irregularities.

Inspect the stator thrust bearing race surface for scoring, scratches, nicks, and grooves. Light honing is permitted to remove slight irregularities. Signs of

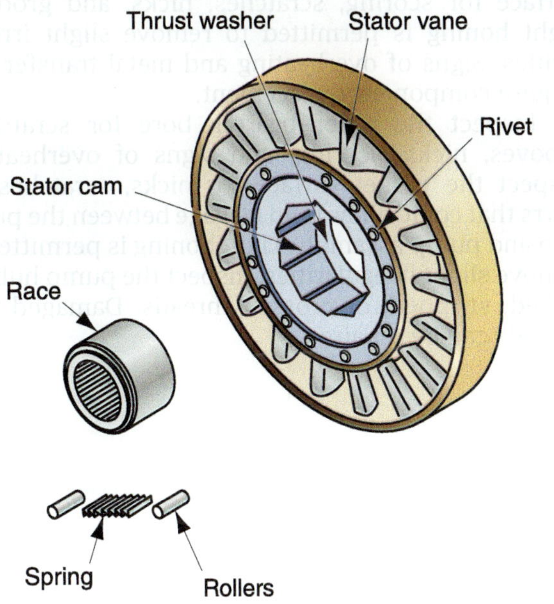

FIGURE 17-35 Stator assembly inspection points

overheating and metal transfer will require component replacement.

Check the stator freewheel roller surface, thrust bearing surface, and roller bearing surface for scoring, scratches, nicks, and grooves. Light honing is permitted to remove slight irregularities. Signs of overheating and metal transfer will require component replacement. Inspect the stator freewheel rollers for scoring, scratches, grooves, and signs of overheat at bend. The freewheel roller springs should show no signs of cracks, distortion, overheating, or breakage. Any damage noted will require component replacement.

Pump/Impeller

Inspect the pump (**Figure 17-36**) for cracked or broken vanes and indications of overheating. Any damage noted will require component replacement. Next, inspect the pump-to-flywheel surface for nicks, scratches, or burrs that could cause fluid leakage between the flywheel assembly and pump assembly. Inspect the pump-to-pump hub gasket surface for nicks, scratches, or burrs that could allow fluid leakage when assembled. Light honing is permitted to remove slight irregularities.

Pump (Impeller) Hub

Check the seal ring grooves on the pump hub (**Figure 17-37**) for cracked edges, nicks, burrs, and sharp edges. Light honing is permitted to remove slight irregularities. Any cracked or chipped edges noted will require component replacement.

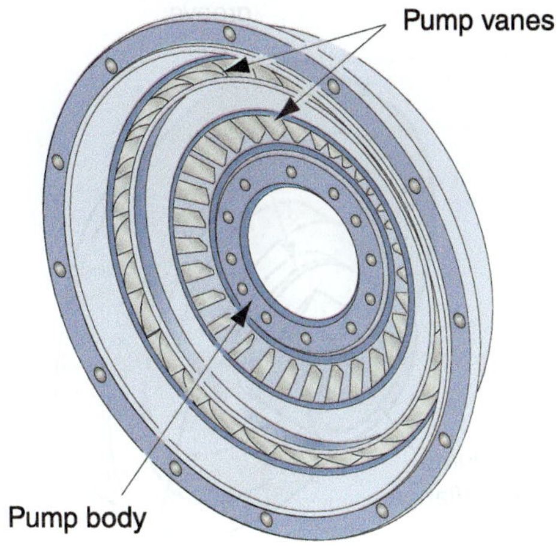

FIGURE 17-36 Pump/impeller assembly inspection points

Inspect the front seal surface for scoring, scratches, nicks, and grooves. No rework of any irregularities noted is allowed on the seal surface. The use of a crocus cloth or light honing on this surface could promote leakage past the front seal. Any irregularities observed will require component replacement.

There should be no signs of cracks, scoring, metal transfer, or heat damage on the pump drive flange. You can remove any slight irregularities with light honing. Inspect the snapring groove for burrs, cracks,

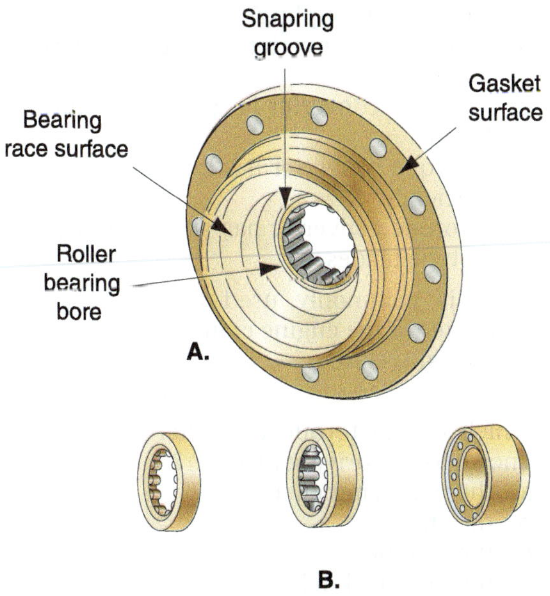

FIGURE 17-37 (A) Pump hub inspection points; (B) pump hub components

FIGURE 17–38 Pump hub inspection points

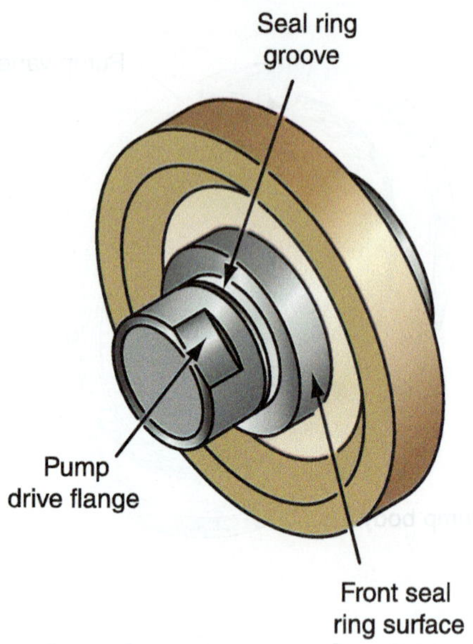

and nicks (**Figure 17–38**). The snapring must be able to lock securely in its groove. Check the bearing race surface for scoring, scratches, nicks, and grooves. Light honing is permitted to remove slight irregularities. Signs of overheating and metal transfer will require component replacement.

Inspect the roller bearing bore for scratches, grooves, nicks, scoring, and signs of overheating. Inspect the gasket surface for nicks, scratches, or burrs that could allow fluid leakage between the pump hub and pump assembly. Light honing is permitted to remove slight irregularities. Inspect the pump hub for pulled, stripped, or crossed threads. Damaged bolt threads can be repaired.

SUMMARY

- Automatic truck (and passenger car) transmissions use a type of fluid coupling known as a torque converter to transfer engine torque from the engine to the transmission.
- A flexplate, sometimes called a flex disc, is used to connect the torque converter to the crankshaft.
- Transmission oil is used as the medium to transfer energy from the engine-driven impeller to the turbine, which in turn drives the transmission.
- Two types of oil flow take place inside the torque converter: rotary flow and vortex flow.
- A converter lockup clutch enables a mechanical coupling of the engine and transmission.

REVIEW QUESTIONS

1. Which one of the following statements is true of a torque converter when comparing it with a standard fluid coupling?
   a. A torque converter produces more fluid slippage.
   b. A torque converter is capable of multiplying torque.
   c. A torque converter has lower torque transfer efficiency.
   d. A torque converter has straight impeller and turbine blades.
2. What is commonly used to mount a torque converter to the engine crankshaft?
   a. a lockup clutch
   b. a stator
   c. a flexplate
   d. a turbine
3. Which of the torque converter components is known as the driving member?
   a. the pump/impeller
   b. the stator
   c. the turbine
   d. the lockup clutch
4. Which torque converter component is splined to a shaft that connects to the forward clutch of the transmission?
   a. the pump/impeller
   b. the stator
   c. the turbine
   d. the flexplate
5. Which of the torque converter components is responsible for torque multiplication?
   a. the pump/impeller
   b. the stator
   c. the turbine
   d. the lockup clutch
6. What medium is used to transfer energy through a torque converter when it is not in lockup mode?
   a. engine oil
   b. transmission fluid
   c. friction discs
   d. brake fluid

7. Where is the stator located in a torque converter assembly?
   a. between the flexplate and pump/impeller housing
   b. between the pump/impeller and turbine
   c. between the turbine and the first clutch of the transmission
   d. on the crankshaft
8. When is rotary fluid flow at a maximum during torque converter operation?
   a. at full stall
   b. at initial pickup
   c. at coupling phase
   d. at torque multiplication phase
9. In a torque converter, which type of fluid flow has the effect of multiplying torque?
   a. rotary flow
   b. vortex flow
   c. primary flow
   d. secondary flow
10. In which of the following phases would the drive efficiency of a torque converter be at 100 percent?
    a. coupling phase
    b. torque multiplication
    c. secondary phase
    d. lockup
11. What is the function of split guide rings in a torque converter?
    a. to reduce turbulence
    b. to maximize turbulence
    c. to cool transmission fluid
    d. to multiply torque
12. In order for torque multiplication to take place in a torque converter, which of the following should be true?
    a. The pump/impeller must turn faster than the turbine.
    b. The turbine must turn faster than the pump/impeller.
    c. The stator must freewheel.
    d. The lockup clutch must be engaged.
13. Which of the following is true during the torque converter coupling phase?
    a. Turbine speed exactly equals impeller speed.
    b. Turbine speed is just slightly slower than impeller speed.
    c. Impeller speed is slightly slower than turbine speed.
    d. Torque multiplication is at maximum.
14. At torque converter coupling phase:
    a. vortex flow is at a minimum
    b. rotary flow is at a minimum
    c. the lockup clutch disengages
    d. slippage ratio is at a maximum
15. What is the most common lockup clutch design used in heavy-duty torque converters?
    a. split ring
    b. centrifugal
    c. piston
    d. overrunning
16. Technician A says that a torque converter (T/C) with a stall torque ratio of 1.5:1 can transmit 75 percent of engine torque to the transmission it is coupled to. Technician B says that when the engine rpm, T/C impeller rpm, and T/C turbine rpm are all identical, the T/C is operating in rotary flow mode. Who is correct?
    a. Technician A only
    b. Technician B only
    c. both A and B
    d. neither A nor B
17. If engine output torque to a torque converter with a stall torque ratio of 2:1 is 1,000 lb-ft. (1,356 N·m), what would be the maximum converter output torque specification?
    a. 500 lb-ft. (678 N·m)
    b. 1,000 lb-ft. (1,356 N·m)
    c. 2,000 lb-ft. (2,712 N·m)
    d. 3,000 lb-ft. (4,067 N·m)
18. Technician A says that engine and the T/C impeller should both turn at the same rpm. Technician B says that T/C turbine and transmission input shaft should both turn at the same rpm. Who is correct?
    a. Technician A only
    b. Technician B only
    c. both A and B
    d. neither A nor B
19. Technician A says that bolted torque converters can usually be serviced and repaired. Technician B says that welded torque converter housings can usually be serviced and repaired. Who is correct?
    a. Technician A only
    b. Technician B only
    c. both A and B
    d. neither A nor B
20. Technician A says that when a truck is stationary and the engine is running at idle, the torque converter behaves like a disengaged clutch. Technician B says that whenever the torque converter impeller is rotating, there is output to the transmission. Who is correct?
    a. Technician A only
    b. Technician B only
    c. both A and B
    d. neither A nor B

18

Prerequisite: Chapter 17

AUTOMATIC TRANSMISSIONS

OBJECTIVES

After reading this chapter, you should be able to:

- Identify the components of a simple planetary gearset.
- Explain the operating principles of a planetary geartrain.
- Define a compound planetary gearset and explain how its outputs are managed.
- Describe a multiple-disc hydraulic clutch and explain its role in the operation of an automatic transmission.
- Outline torque path powerflow through typical four- and five-speed automatic transmissions.
- Describe the hydraulic circuits and flows used to control automatic transmission operation.
- List the two types of hydraulic retarders used in Allison automatic transmissions and explain their differences.

KEY TERMS

annulus	epicyclic gearset	output retarder	priority valve
carrier	input retarder	planetary gearset	ring gear
clutch pack	multiple-disc clutch	planetary pinion gears	sun gear

INTRODUCTION

This chapter introduces hydromechanical automatic transmissions. The focus is primarily on transmissions manufactured by Allison. Allison has dominated the North American market for automatic transmissions in on-highway trucks for several decades. Allison's competitors for the on-highway, hydromechanical automatic transmission market are ZF and Voith. Many of the principles outlined for Allison transmissions in this chapter also apply to the products of other manufacturers of automatic transmissions.

Hydromechanical automatic transmissions are becoming a thing of the past due to the emergence of electronically controlled transmissions. However, they continue to be a factor in service repair facilities because of the type of vehicle application they are spec'd in. Buses and certain kinds of vocational vehicles such as garbage haulers commonly use automatic transmissions, and these types of vehicles have much greater longevity than linehaul fleet trucks. The result is that repair facilities will continue to see hydromechanical automatic transmissions for a few years yet.

DEFINITIONS

Because of the widespread use of *automated manual transmissions* (AMTs) (see Chapter 20) it is important that the following definitions be understood:

- Hydromechanical automatic transmission: a fully automatic transmission that uses no control electronics. Engine torque is transmitted to it by a torque converter and output ratios are optioned using planetary gearsets. Upshifts and downshifts take place with no direct assistance from the driver. This type of transmission is studied in this chapter.

- Electronic automatic transmission: fully automatic transmission that is controlled by computer. Engine torque is transmitted to it by a torque converter and output ratios are optioned using planetary gearsets. Upshifts and downshifts are managed electronically with no direct assistance from the driver. This type of transmission is studied in Chapter 21.
- Automated standard transmission: standard transmission that is electronically controlled. Engine torque is transmitted to it using a mechanical clutch. This type of transmission is studied in Chapter 20.

Most of the operating principles of the hydromechanical transmissions described in this chapter relate closely to those used in their more recent electronically controlled counterparts. The first generation of Allison electronically controlled automatics was known as Consolidated Electronic Controls (CEC), which essentially adapted the transmissions described in this chapter to some basic electronic management. In addition, most of the basic principles of more recent electronically controlled transmissions up until the most recent Allison TC-10 closely relate to those described here.

HOW AN AUTOMATIC TRANSMISSION WORKS

Fully automatic transmissions rely on planetary gearsets to transfer power and manage torque from the engine to the driveline. Planetary gearsets differ from the cluster gear arrangements used in standard manual shift transmissions. In hydromechanical automatic transmissions, the shifting of planetary gears is actuated by the use of hydraulic force. A complex system of valves is used to control and direct pressurized fluid in the closed system. The force generated by this fluid is used to apply and release the various clutches and brakes that control planetary gear operation. Factors such as road speed, throttle position, and governed engine speed are used to control and trigger shifts.

Planetary Gearsets

The **planetary gearset** is an old invention. Its first use was actually in early manually shifted transmissions. Although it was replaced by the cluster gear arrangements we looked at in Chapters 15 and 16, transmission designers did not forget the many advantages that planetary gearsets offered. **Figure 18–1** shows a simple planetary gearset. Automatic transmissions tend to use two or more planetary gearsets that work with each other to option a range of output ratios.

FIGURE 18–1 Typical simple planetary gearset

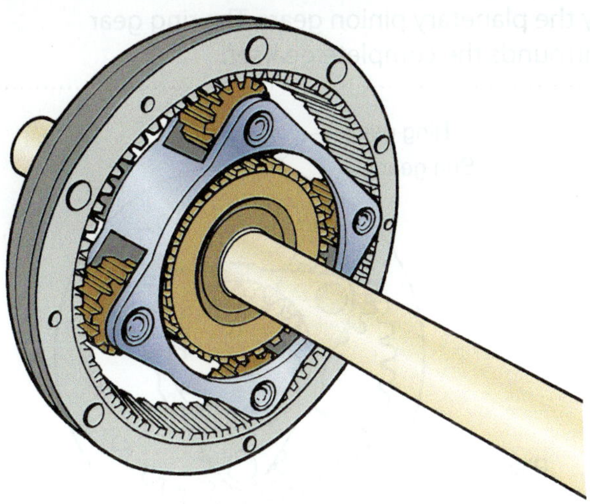

Simpson Geartrain

A Ford Motor Company engineer named Howard W. Simpson spent his retirement exploring the use of planetary gearing in automatic transmissions. Between 1948 and 1955, Simpson was granted close to two dozen patents relating to planetary gearsets. Simpson geartrains remain the standard in heavy-duty truck transmissions and in many passenger car automatic transmissions.

A Simpson geartrain combines a pair of simple planetary gearsets in tandem. This arrangement allows the torque load to be spread over a greater number of gear teeth for strength and also provides a large number of output gear ratios. We begin this chapter by looking at a simple planetary gearset and then we take a look at what happens when we add a second—that is, when we compound the geartrain. Planetary gearsets are the basic building blocks of both hydraulic automatic and electronic automatic transmissions.

You should note that this chapter is a prerequisite to understanding electronic automatic transmissions. Hydromechanical and electronic automatic transmissions share many common principles of operation. Planetary gearsets are also the foundation for the transmissions used in parallel hybrid drivetrain systems. These permit infinitely progressive input torque from two sources.

18.1 PLANETARY GEARSET COMPONENTS

A simple planetary or **epicyclic gearset** can provide overdrive, reverse, forward reduction, neutral, and direct drive ratios. It can also supply fast and slow

FIGURE 18–2 Planetary gear configuration is similar to the solar system, with the sun gear surrounded by the planetary pinion gears. The ring gear surrounds the complete gearset.

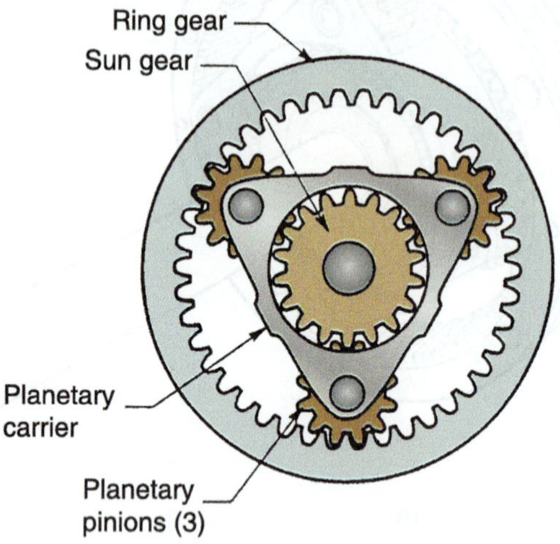

speeds for each operating range, with the exception of neutral and direct drive. There are three main components in a simple planetary gearset:

- **Sun gear** with external gear teeth
- **Carrier** with **planetary pinion gears** mounted to it
- An internally toothed **ring gear** or **annulus**

The sun gear is located in the center of the assembly (**Figure 18–2**). It is the smallest gear in the assembly and it is located at the center of the axis. The sun gear can be either a spur or helical gear design. It meshes with the teeth of the planetary pinion gears. Planetary pinion gears are small gears fitted into a framework called the planetary carrier; each turns on its own axis. The planetary carrier can be made of cast iron, aluminum, or steel plate and is designed with a shaft for each of the planetary pinion gears. For simplicity, we use the term *planetary pinions* in place of *planetary pinion gears.* Each planetary pinion is in constant mesh with both the sun gear and the ring gear of the assembly.

Planetary pinions rotate on needle bearings positioned between the planetary carrier shaft and the planetary pinions. The number of planetary pinions in a carrier depends on the load the gearset is required to carry. Passenger vehicle automatic transmissions might use only three planetary pinions, whereas heavy-duty automatics could use as many as five planetary pinions in a planetary carrier. The carrier and its pinions are considered as a single unit.

The planetary pinions rotate in mesh with the sun gear and are enclosed by the ring gear. The ring gear surrounds the components that make up a simple planetary gearset. In some ways, the ring gear acts like a band to hold the entire gearset together. The planetary pinions mechanically connect the sun gear with the ring gear. This means that when the planetary carrier is held stationary, the ring, the planetary pinions continue to rotate on their axes; however, input rotation is reversed by whichever of the other two members acts as output.

To help remember the design of a simple planetary gearset, use the solar system as an example. The sun is the center of the solar system with the planets rotating around it, hence the name *planetary gearset.* When considering the relative *gear* sizes of the three members that make up a simple planetary gearset, to determine ratios, the following is true:

- Sun gear: small gear
- Ring gear: midsize gear
- Planetary carrier assembly: large gear

ADVANTAGES OF PLANETARY GEARSETS

When compared with other gearsets, some notable advantages of a simple planetary gearset include:

- Constantly meshed gears. With the gears constantly in mesh there is little chance of damage to the teeth. Grinding or missed shifts are not a wear or damage factor.
- Gear forces are divided equally. Planetary gearsets are also very compact.
- Versatility. Seven combinations of speed and direction can be obtained from a single set of planetary gears.
- Additional variations of both speed and direction can be added through the use of compound planetary gears.

HOW PLANETARY GEARS WORK

Each member of a planetary gearset—the sun gear, planetary gear carrier, and ring gear—can revolve or be held at rest. Power transmission through a planetary gearset is possible only when one of the three component members is held stationary, or if two of the members are locked together with each other to act as drive members. In this case, the input-to-output ratio would be 1:1.

Any one of the three members—sun gear, pinion gear carrier, or ring gear—can be used as the driving or input member. At the same time, another member might be kept from rotating and thus becomes the held or stationary member. The third member then becomes the driven or output member. Depending on which member is the driver, which is held, and which is driven, either a torque increase or a speed increase is produced by the planetary gearset. Output

TABLE 18–1 Laws of Simple Planetary Gear Operation

Sun Gear	Carrier	Ring Gear	Speed	Torque	Direction
1. Input	Output	Held	Maximum reduction	Increase	Same as input
2. Held	Output	Input	Minimum reduction	Increase	Same as input
3. Output	Input	Held	Maximum increase	Reduction	Same as input
4. Held	Input	Output	Minimum increase	Reduction	Same as input
5. Input	Held	Output	Reduction	Increase	Reverse of input
6. Output	Held	Input	Increase	Reduction	Reverse of input
7. When any two members are driven, direct 1:1 output results.					
8. When no member is held or locked together, output cannot occur. The result is a neutral condition.					

direction can also be reversed through various combinations. You can take a look at the ratio and directional capabilities of a single planetary gearset by studying **Table 18–1**.

Table 18–1 summarizes the input and output rules of a simple planetary gearset. It shows relative output speed, torque, and direction of the various combinations available. It is also helpful to remember the following two points with regard to direction of rotation:

1. When an external-to-external gear tooth set is in mesh, there will be a change in the direction of rotation at the output (**Figure 18–3A**).
2. When an external gear tooth is in mesh with an internal gear, the output rotation for both gears will be the same (**Figure 18–3B**).

Maximum Forward Reduction

With the ring gear stationary and the sun gear turning clockwise, the external sun gear teeth rotate the planetary pinions counterclockwise. This option is shown as combination 1 in Table 18–1 and in **Figure 18–4**. The inside diameter of each planetary pinion pushes against its shaft, moving the planetary carrier clockwise. The small sun gear (driving) has to rotate several times to turn the larger planetary carrier through one complete revolution, resulting in an underdrive. This combination represents the most gear reduction or the maximum torque multiplication that can be achieved in one planetary gearset. Input speed is high, but output speed will be low.

Minimum Forward Reduction

In this combination, the sun gear is stationary and the ring gear rotates clockwise. This option is shown as combination 2 in Table 18–1 and in **Figure 18–5**. The ring gear drives the planetary pinions clockwise and walks around the stationary sun gear. The planetary

FIGURE 18–3 (A) With external teeth in mesh, there is a change in direction at the output; (B) when an external gear is meshed with an internal gear, both turn in the same direction.

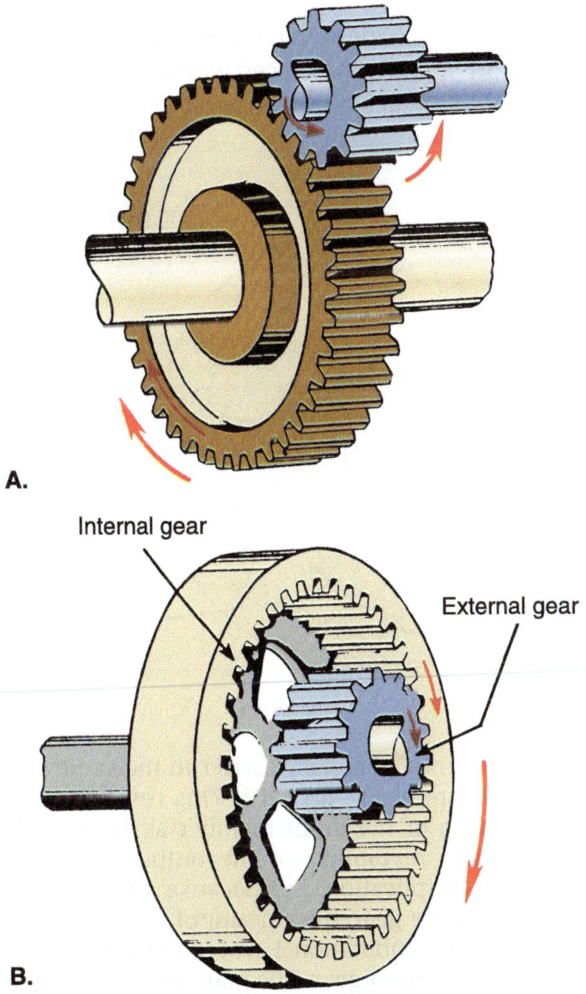

FIGURE 18–4 Maximum forward reduction (greatest torque, lowest speed) is produced with the sun gear as input, the ring gear stationary, and the carrier as the output.

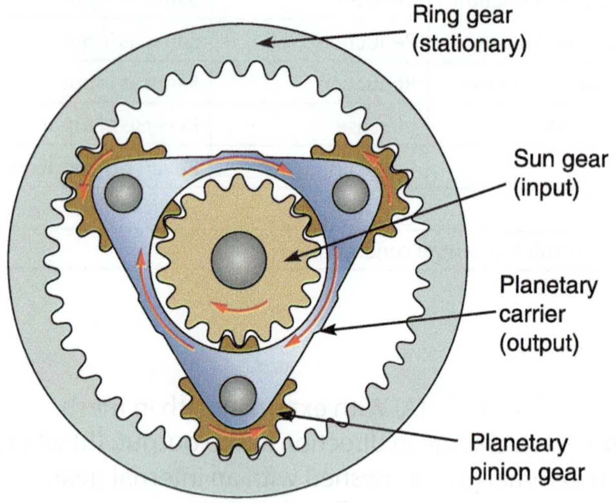

FIGURE 18–5 Minimum forward reduction (good torque, low speed) is produced with the ring gear as input, the sun gear stationary, and the carrier as the output.

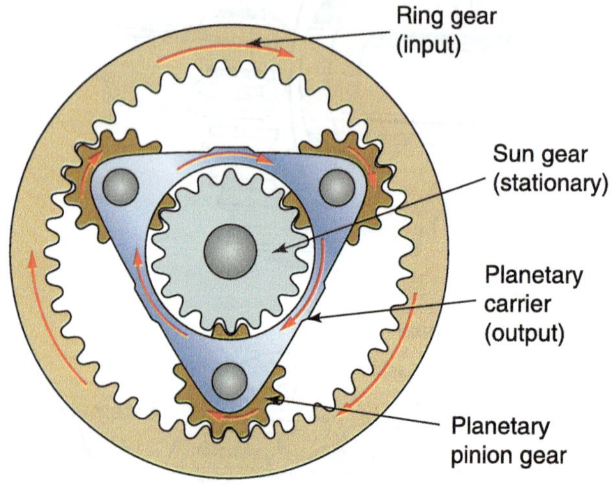

pinions drive the planetary carrier in the same direction as the ring gear—forward. This results in more than one turn of the input member as compared to one complete revolution of the output. The result is torque multiplication. But because a midsize gear is driving a large gear, the amount of reduction is not as great as in combination 1 where the smallest gear is driving the largest gear. The planetary gearset operates in a forward reduction mode with the midsize

FIGURE 18–6 Maximum overdrive (lowest torque, greatest speed) is produced with the carrier as input, the ring gear stationary, and the sun gear as output.

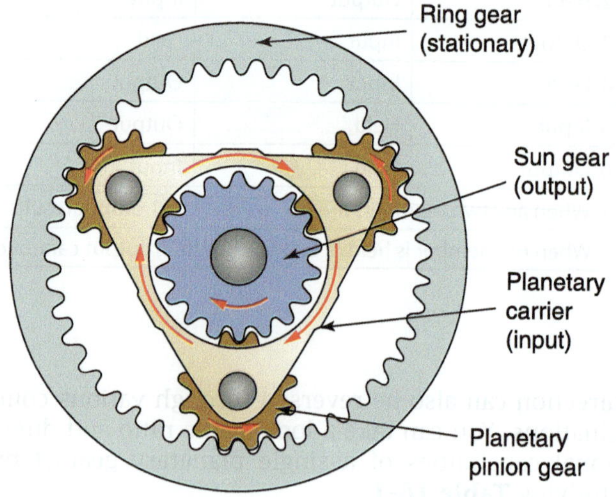

ring gear driving the larger planetary carrier. Therefore, the combination produces minimum forward reduction.

Maximum Overdrive

In this combination, the ring gear is stationary and the planetary carrier rotates clockwise. This option is shown as combination 3 in Table 18–1 and in **Figure 18–6**. The three planetary pinion shafts push against the inside diameter of the planetary pinions. The pinions are forced to walk around the inside of the ring gear, driving the sun gear clockwise. The carrier is rotating less than one turn input compared to one turn output, resulting in an overdrive condition. In this combination, the planetary carrier the large gear in the assembly is rotating less than one turn and driving the small sun gear at a speed greater than the input speed. The result is a fast overdrive with maximum speed increase.

Slow Overdrive

In this combination, the sun gear is stationary and the carrier rotates clockwise. This option is shown as combination 4 in Table 18–1 and in **Figure 18–7**. As the carrier rotates, the pinion shafts push against the inside diameter of the pinions and they are forced to walk around the held sun gear. This drives the ring gear faster and the speed increases. The planetary carrier turning less than one turn causes the pinions to drive the (midsize) ring gear one complete revolution in the same direction as the planetary carrier. As in combination 3, an overdrive condition exists, but the large-size carrier is now driving the midsize ring gear so a slow overdrive results.

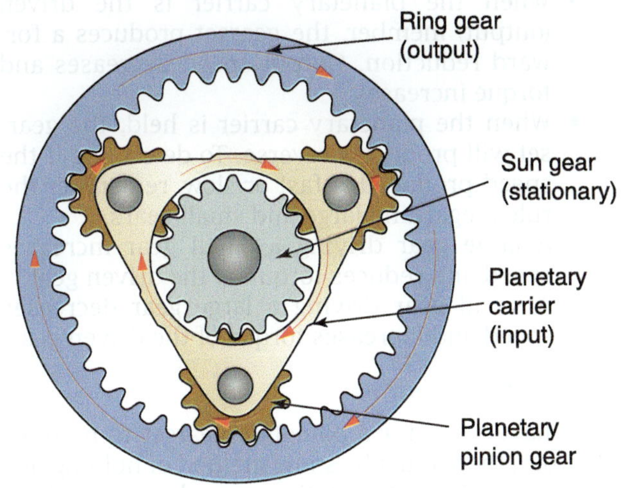

FIGURE 18–7 Slow overdrive (lower torque, greater speed) is produced with the carrier as input, the sun gear stationary, and the ring gear as output.

Slow Reverse

Here, the sun gear is driving the ring gear while the planetary carrier is held stationary. This option is shown as combination 5 in Table 18–1 and in **Figure 18–8**. The planetary pinions, driven by the external sun gear, rotate counterclockwise on their shafts. The external teeth of the planetary pinions drive the internal teeth of the ring gear in the same direction. While the sun gear is driving, the planetary pinions are used as idler gears to drive the ring gear counterclockwise. This means the input and output shafts are operating in opposite directions to provide a reverse powerflow. Because the driving sun gear is small and the driven ring gear is midsize, the result is slow reverse.

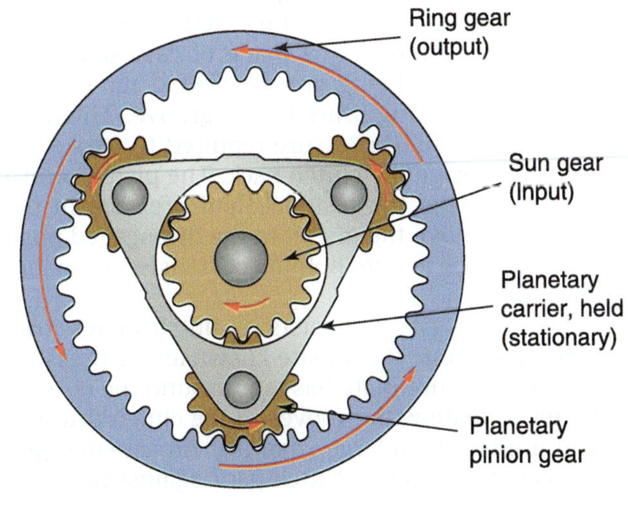

FIGURE 18–8 Slow reverse (opposite output direction, greatest torque, lowest speed) is produced with the sun gear as input, the planetary carrier stationary, and the ring gear as output.

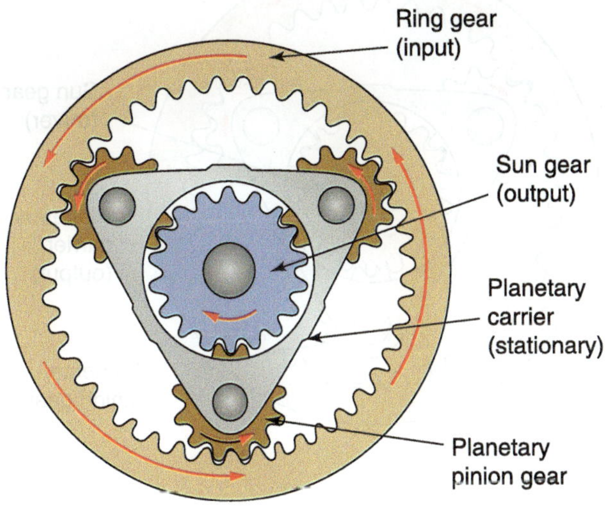

FIGURE 18–9 Fast reverse (opposite output direction, greater torque, low speed) is produced with the ring gear as input, the carrier stationary, and the sun gear as output.

Fast Reverse

For fast reverse, the carrier is still held as in slow reverse, but the sun gear and ring gear reverse their roles, with the midsize ring gear now being the driving member and the small sun gear driven. This option is shown as combination 6 in Table 18–1 and in **Figure 18–9**. As the ring gear rotates counterclockwise, the pinions rotate counterclockwise as well, while the sun gear turns clockwise. In this combination, the input ring gear uses the planetary pinions to drive the output sun gear. The sun gear rotates in reverse direction to the input ring gear. In this operational combination, the midsize ring gear rotating counterclockwise drives the small sun gear clockwise, providing fast reverse.

Direct Drive

In the direct drive combination, both the ring gear and the sun gear are input members. This option is shown as combination 7 in Table 18–1 and in **Figure 18–10**. They turn clockwise at the same speed. The internal teeth of the clockwise turning ring gear will try to rotate the planetary pinions clockwise as well. But the sun gear, an external gear rotating clockwise, will try to drive the planetary pinions counterclockwise. These opposing forces lock the planetary pinions against rotation so that the entire planetary gearset

FIGURE 18–10 Direct drive is produced if any two gearset members are locked together.

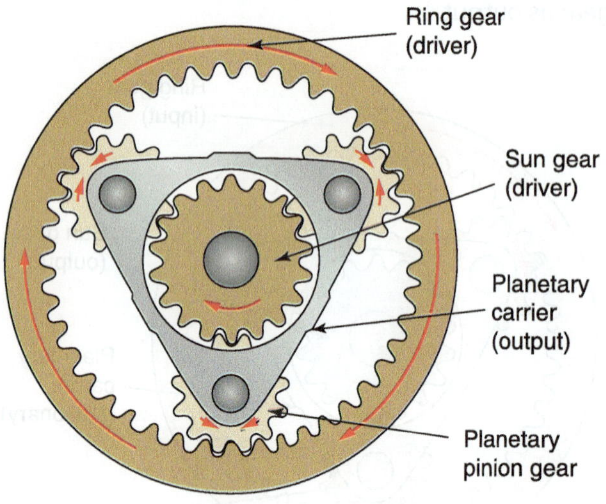

rotates as one complete unit. This ties together the input and output members and provides direct drive. For direct drive, both input members must rotate at the same speed.

Neutral Operation

Combinations 1 through 7 (see Table 18–1) all produce an output drive of various speeds, torques, and direction. In each case, one member of the planetary gearset is held or two members are locked for the output to take place. When no member is held stationary or locked, there will be input into the gearset but no output. The result is a neutral condition (**Figure 18–11**).

FIGURE 18–11 With no held or stationary member, there can be no output. The result is a neutral condition.

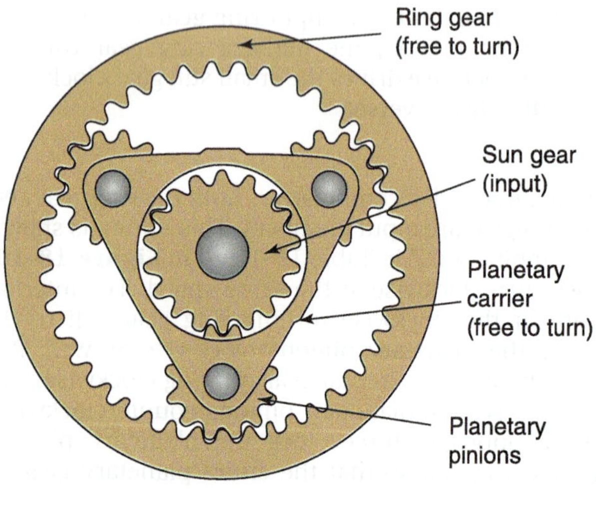

Summary of Simple Planetary Gearset Operation

- When the planetary carrier is the drive (input) member, the gearset produces an overdrive condition. Output speed increases and torque decreases.
- When the planetary carrier is the driven (output) member, the gearset produces a forward reduction. Output speed decreases and torque increases.
- When the planetary carrier is held, the gearset will produce a reverse. To determine if the speed produced is fast or slow, remember the rules regarding large and small gears.
- A large gear driving a small gear increases speed and reduces torque of the driven gear.
- A small gear driving a large gear decreases speed and increases torque of the driven gear.

Planetary Gear Controls

As we have seen, for a planetary gearset to produce an output, there must be some method of holding one planetary gear member stationary and applying drive torque to it. In most truck transmissions, this job is performed by **multiple-disc clutches**, which serve as both braking and power transfer devices.

A multiple-disc clutch uses a series of hollow friction discs to transmit torque or apply braking force. The discs have internal teeth that are sized and shaped to mesh with splines on the clutch assembly hub. In turn, this hub is connected to a planetary geartrain component so that gearset members receive the desired braking or transfer force when the clutch is applied or released.

Design. Multiple-disc clutches have a large drum-shaped housing that can be either a separate casting or part of the transmission housing (**Figure 18–12**). This drum housing holds all other clutch components: the cylinder, hub, piston, piston return springs, seals, pressure plate, clutch pack (including friction plates), and snaprings.

The cylinder in a multiple-disc clutch is relatively shallow. The cylinder bore acts as a guide for piston travel. The piston is made of cast aluminum or steel with a seal ring groove around the outside diameter. A seal ring seats in the groove. This rubber seal retains fluid pressure required to stroke the piston and engage the clutch pack. The piston return springs overcome the residual fluid pressure in the cylinder and move the piston to the disengaged position when clutch action (holding or transfer) is no longer needed.

The **clutch pack** consists of clutch plates, friction discs, and one thick plate known as the pressure plate. The pressure plate has tabs around the outside diameter to mate with grooves in the clutch drum. It is held in place by a large snapring. An actuated piston forces the engaging clutch pack against the fixed

FIGURE 18–12 Components of a typical heavy-duty transmission multiple disc clutch

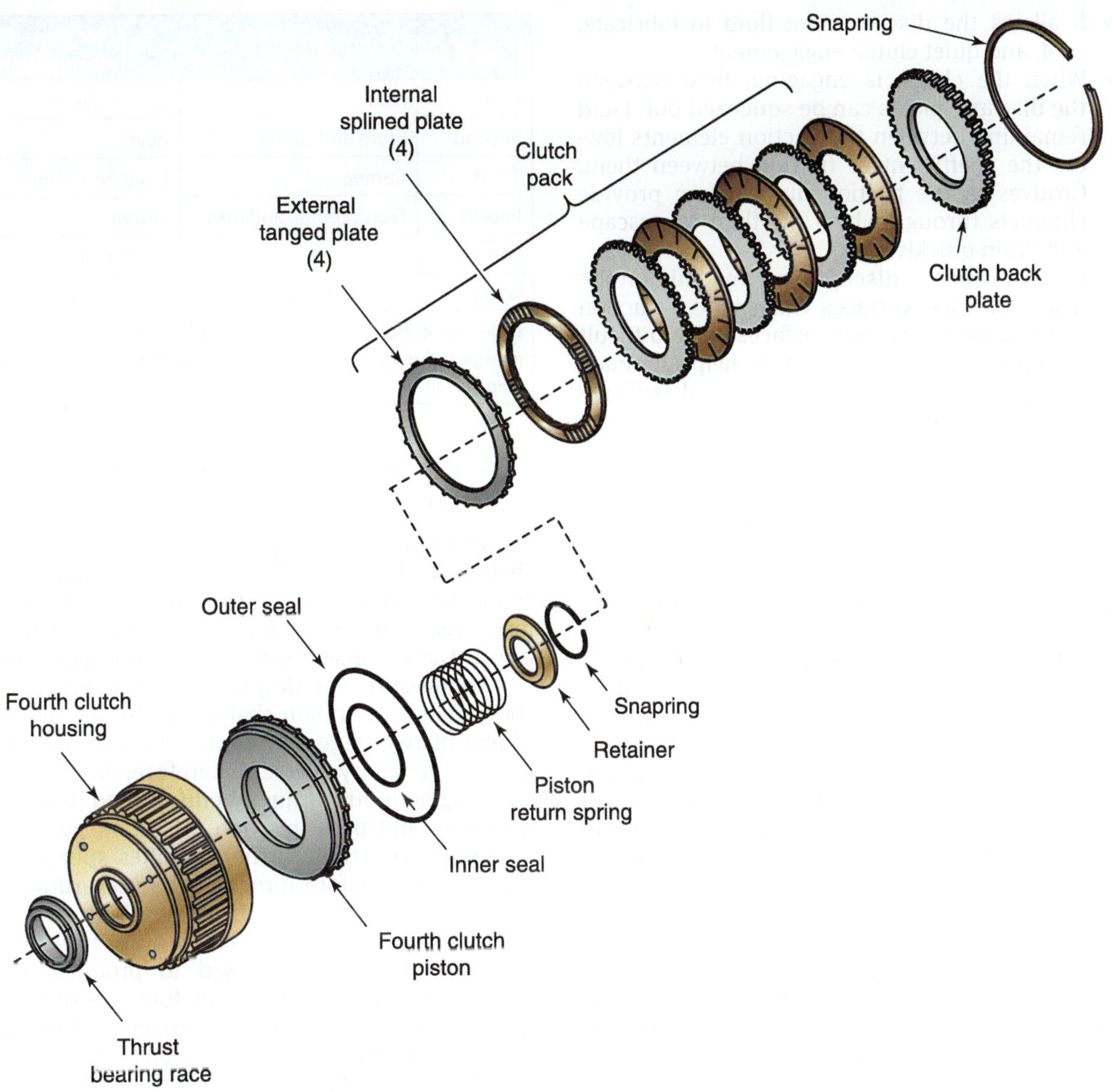

pressure plate. Because the pressure plate cannot move or deflect, it provides reaction to the engaging clutch pack.

The number of clutch plates and friction discs contained in the pack depends on the transmission size and application. The clutch plates are made from carbon steel and have several tabs in their outside diameter. These tabs are located in grooves machined into the inside diameter of the clutch drum housing. Clutch plates must be perfectly flat. Although the surface of the plate might appear smooth, it is machined to produce a friction surface to transmit torque.

The friction discs are sandwiched between the clutch plates and the pressure plate. Friction discs are designed with a steel core plate center with friction material bonded to either side. Cellular paper fibers, graphites, and ceramics are used as friction facings.

Cooling. During clutch engagement, friction takes place and develops heat between the clutch plates and friction discs. The transmission fluid absorbed by the paper-based friction material transfers heat from the discs to the plates. The plates then transfer the heat to the drum housing, where it can be cooled by the surrounding transmission fluid. Operating temperatures can reach 1,100°F (593°C), so clutch disc cooling is an important design consideration in transmission design.

To assist cooling and provide other performance advantages, friction disc surfaces are normally grooved. Grooving is generally thought to provide the following advantages:

- It allows the disc to store fluid to lubricate, cool, and quiet clutch engagement.
- When the clutch is engaging, fluid between the disc and plates can be squeezed out. Fluid remaining between the friction elements lowers the coefficient of friction between them. Grooves in the friction disc surface provide channels through which the fluid can escape and drain quickly.
- When a clutch is disengaging, any fluid on the disc and plate surfaces can create a suction effect, making the two surfaces more difficult to separate. Grooved surfaces help eliminate this problem and allow for clean disengagement at high speeds.

18.2 COMPOUND PLANETARY GEARSETS

Automatic transmissions for both passenger car and heavy-duty truck applications use more than one simple planetary gearset coupled to generate the required range of output gear ratios and direction. The number of planetary gearsets that can be combined is limited by practical weight, space, and cost considerations. The four-speed heavy-duty transmissions discussed in this chapter have three simple gearsets that we call front, center, and rear. Five-speed transmissions can add a low planetary gearset to the three-gear configuration for a total of four gearsets in this compound arrangement.

Any one of the three members of a simple planetary gearset can serve as the input or drive member. In addition, any of the three members could be held stationary. In a working transmission, this is not practical. There is a single path for input torque from the engine to supply drive torque. This input torque is directed to one of the planetary sun gears through the use of clutches, which are then used to option multiple torque path flows through the planetaries.

Similarly, having a clutch (brake) to hold each member of each planetary gearset is not practical. A four-speed, three-gearset transmission uses only five clutches, excluding the torque converter lockup clutch. A five-speed, four-gearset transmission uses six clutches. They function as follows:

- Forward clutch locks the converter turbine to the main input drive shaft.
- First clutch anchors the rear planetary ring gear.
- Second clutch anchors the front planetary carrier.
- Third clutch anchors the sun gear shaft.
- Fourth clutch locks the main input drive shaft to the center sun gear shaft.

TABLE 18–2 Planetary Gearset Combinations

Gear	Four-Speed Transmissions	Five-Speed Transmissions
First	Rear	Low and rear
Second	Front and center	Rear
Third	Center	Front and center
Fourth	Front, center, and rear	Center
Fifth	—	Front, center, and rear
Reverse	Center and rear	Center and rear

Note: In CR (close ratio) five-speed transmissions, all powerflows must pass through the low planetary gearset before reaching the output shaft.

- Low clutch (five speeds only) anchors the low planetary carrier.

The three planetaries are connected by a sun gear shaft assembly, main shaft assembly, and a planetary connecting drum. This interconnection of planetary input, reaction, and output components produces the forward and reverse speeds and torque ranges required.

Because the simple planetary gearsets are interconnected by common shafts and drums, the output of one planetary gearset can be used as the input of another planetary gearset. Therefore, one gearset can take engine input torque; modify speed, torque, and direction; and pass this onto a second gearset. The second gearset then further modifies speed, torque, and direction before directing the powerflow to the output shaft and vehicle drivetrain. With two simple planetaries working together, torque input is compounded. The gearsets used to produce forward and reverse ranges in typical four-speed and five-speed transmissions are summarized in **Table 18–2**. **Table 18–3** summarizes the clutches applied to produce the required powerflow paths.

TABLE 18–3 Clutch Application in Various Gear Ranges

Gear	Four-Speed Transmissions	Five-Speed Transmissions
Neutral	First	First
First	First and forward	Low and forward
Second	Second and forward	First and forward
Third	Third and forward	Second and forward
Fourth	Fourth and forward	Third and forward
Fifth	—	Fourth and forward
Reverse	First and fourth	First and fourth

GEAR RANGE POWERFLOWS

In all automatic transmissions, input torque is directed from the torque converter through the transmission planetary gearing and to the output shaft. Torque converter operation is covered in Chapter 17. In some earlier generation models, Allison converted their four-speed units into five-speed units by simply adding an extra "low" planetary gearset.

Oil Pump

An oil pump is used to circulate transmission fluid within the transmission for both lubrication and the hydraulic application of clutches. This pump is driven from the torque converter and is located between the torque converter and the transmission gearing and clutches (**Figure 18–13**). When the torque converter rotates, its rear hub drives the pump drive gear, which is in mesh with the driven gear. As the gears rotate, they draw oil into the pump and move the oil throughout the hydraulic system.

Because the pump is located ahead of the transmission gearing and clutches, it will not be driven when the vehicle is towed or pushed. Therefore, any time the vehicle must be towed or pushed more than a short distance (½ mile maximum), the driveline must be disconnected or the driving wheels raised. Failure to do this will result in premature wear and damage to internal components.

FOUR-SPEED TRANSMISSION POWERFLOW

The following powerflow paths are used in a typical four-speed transmission. Four-speed transmissions use front, center, and rear gearsets to generate four forward gears and one reverse gear. Five clutches are used. A sun gear shaft assembly and connecting drum provide connecting links between the planetaries.

Neutral

Figure 18–14 shows a cross section of a four-speed transmission. The arrows indicate the neutral gear powerflow. When the engine starts, the torque converter rotates, driving the oil pump in the transmission.

FIGURE 18–13 Typical transmission oil pump components

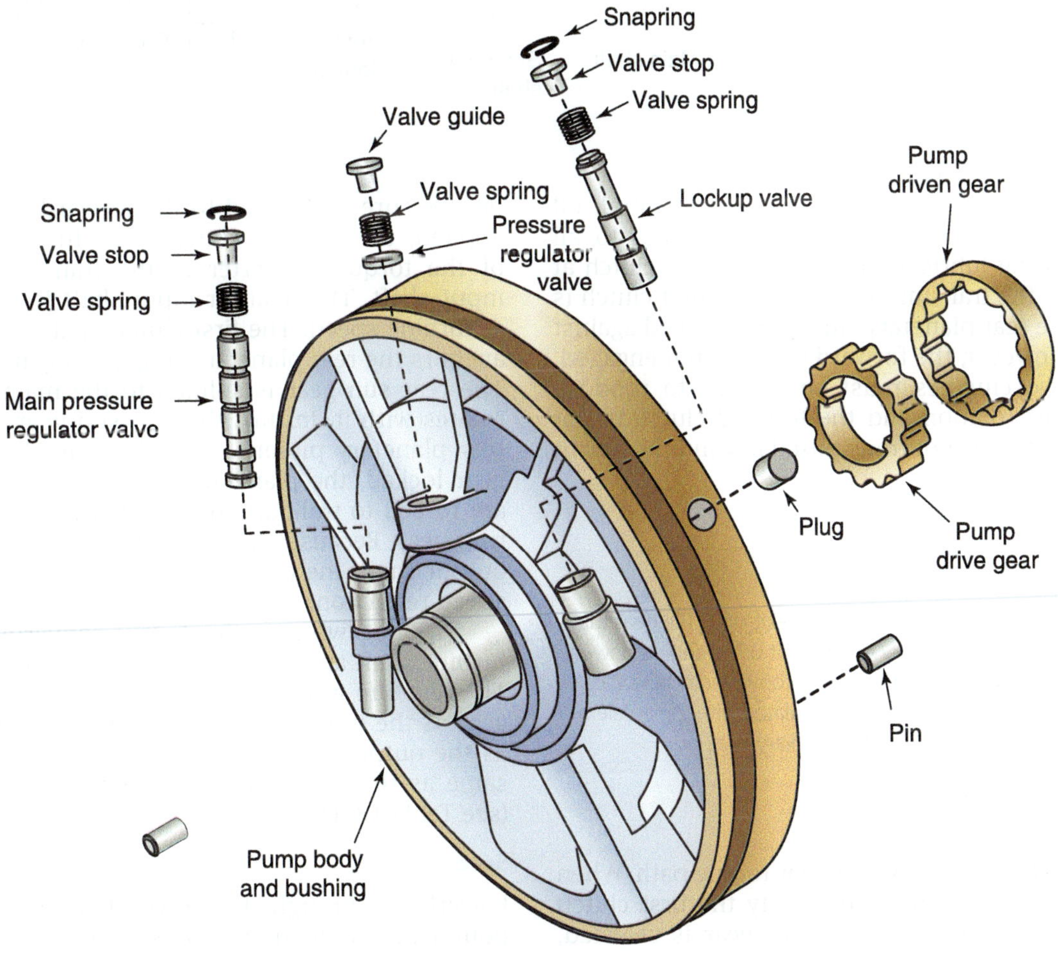

FIGURE 18–14 Typical four-speed neutral torque path powerflow

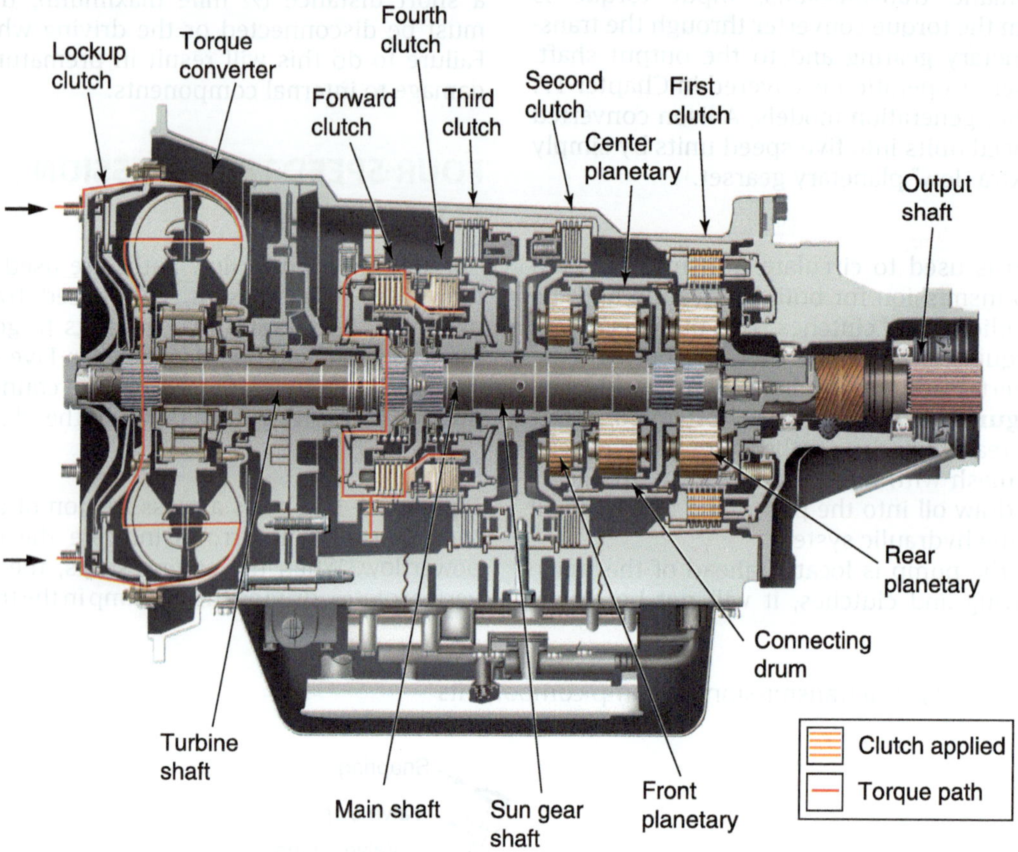

With the range selector (gear shifter) in neutral, oil pressure is routed to the torque converter, to all points needing lubrication, and to the first clutch at the back of the transmission. When the first clutch is applied, the rear planetary ring gear is locked against rotation. However, the forward clutch is not engaged. Because two clutches must be applied to produce output shaft rotation and the forward clutch is not engaged, torque is not transmitted beyond the fourth clutch hub.

SHOP TALK

Transmissions are fitted with a power take-off (PTO) gear that can be used to drive auxiliary equipment using engine power. The PTO gear is driven from the torque converter turbine. It can be engaged in neutral; its speed is regulated with the accelerator pedal and can be based on the auxiliary load.

First Gear

Figure 18–15 illustrates first-gear torque path. When the transmission is in neutral, only the first clutch is applied. But when any forward gear is selected, oil pressure is directed to the forward clutch. When applied, the forward clutch locks the turbine shaft of the torque converter to the transmission main input shaft. The main or input shaft is now turning at turbine speed. The first clutch is also applied and anchors the rear planetary ring gear against rotation. The rear sun gear is splined to the main shaft and rotates with it. In turn the rear sun gear rotates the rear planetary pinions. With the rear planetary ring gear locked, the planetary pinions and their carrier are forced to walk around internally within the held ring gear. Because the rear carrier is splined to the output shaft, the output shaft rotates at a maximum speed reduction.

The clockwise rotating torque converter turns the turbine shaft and main shaft in a clockwise direction. The rear sun gear also turns clockwise. With the sun gear as the input, the ring gear held, and the carrier as the output, the output rotational direction is the same as the input, and reduction is at a maximum (see Table 18–1).

Second Gear

Powerflow through the second gear is a compound gear reduction because it runs through two

FIGURE 18–15 Four-speed first-gear torque path powerflow

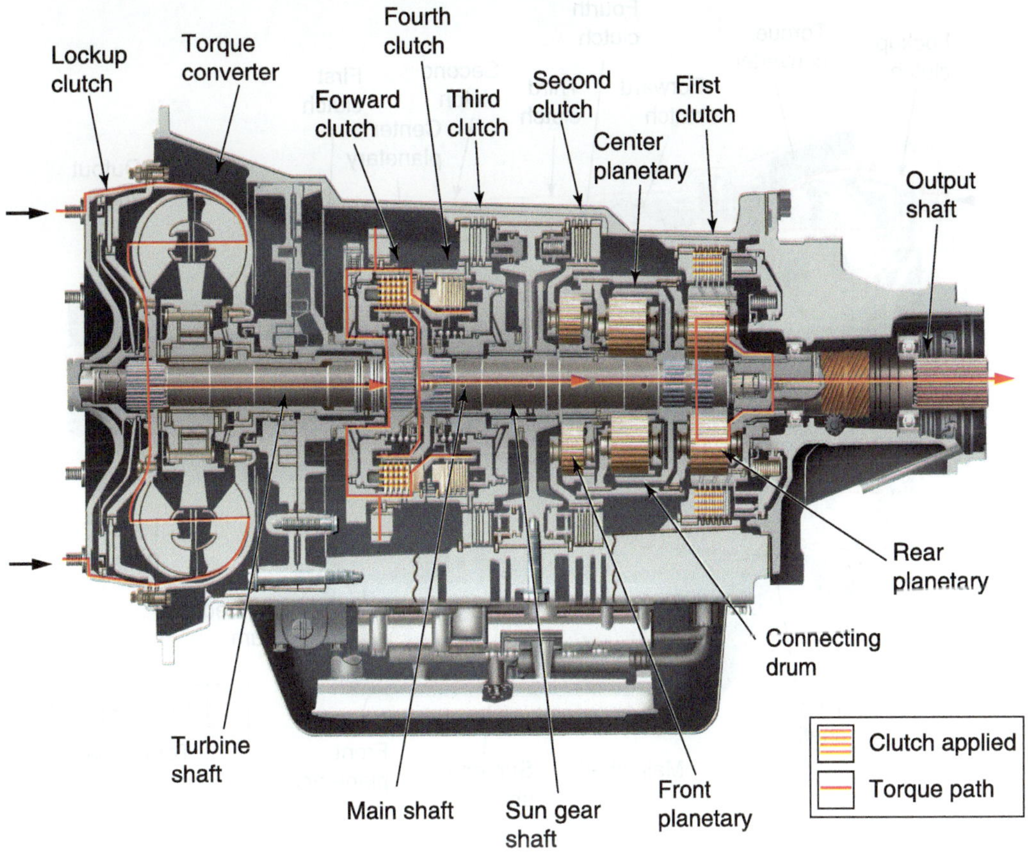

planetary gearsets before reaching the output shaft (**Figure 18–16**). The forward clutch and the second clutch are both applied. The forward clutch locks the turbine and main shafts together so that they rotate clockwise. The first clutch is released in second gear. The rear planetary assembly thereby freewheels and has no effect on second-gear operation.

Torque from the main input shaft is transmitted to the rear sun gear and the center ring gear immediately in front of the sun gear. This center ring gear is splined to the main shaft and becomes the driving member, rotating clockwise. It drives the center planetary pinions in a clockwise direction. In turn the pinions are in mesh with the center sun gear, turning it in a clockwise direction.

The center sun gear is splined to the sun gear shaft. A forward sun gear is splined to the sun gear shaft immediately in front of the center sun gear. This forward or front sun gear drives the front planetary pinions whose carrier is being held by the applied second clutch. This causes the pinions to rotate, driving the front ring gear. Remember, the input force from the center sun gear and sun gear shaft is in a counterclockwise direction. In the front planetary, the sun gear is the input, the carrier is held, and the ring gear is the output. This corresponds to combination 5 in Table 18–1. Notice that this combination produces a reverse in the output direction. This means that the front ring gear is now turning in a clockwise direction. The front ring gear is splined to the planetary connecting drum, which connects through the rear carrier to the output shaft. The end result is the output shaft rotating clockwise at a medium reduction.

Third Gear

Figure 18–17 shows the third-gear torque path. In third gear, the forward and third clutches are applied. The third clutch anchors the sun gear shaft against rotation. This prevents the integral center sun gear from rotating. The forward clutch again locks the turbine and main shafts together so that they can rotate clockwise as a unit. The rear sun gear is splined to both the main shaft and the center ring gear and rotates at input speed. With the center sun gear held stationary and the center ring gear rotating, the ring gear drives the center planetary pinions. This action rotates the center carrier at a minimum speed reduction. Both the center and rear carriers are splined to the planetary connecting drum and rotate as a unit. Because the output shaft is splined to the rear carrier,

FIGURE 18–16 Four-speed second-gear torque powerflow

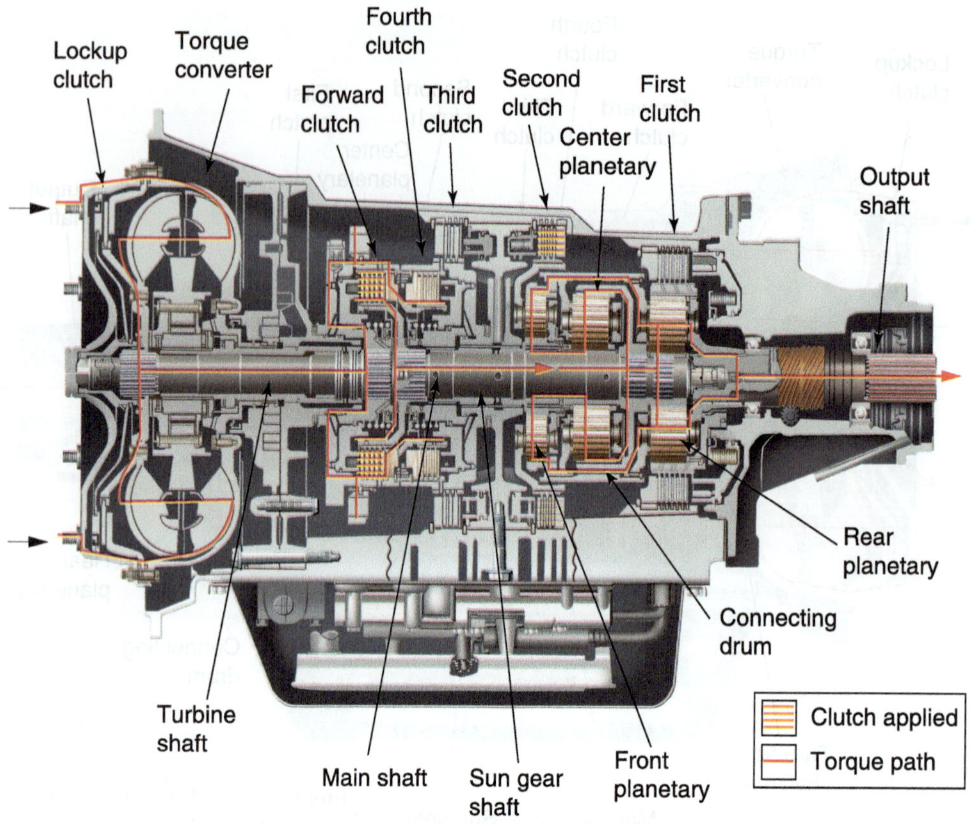

FIGURE 18–17 Four-speed third-gear torque powerflow

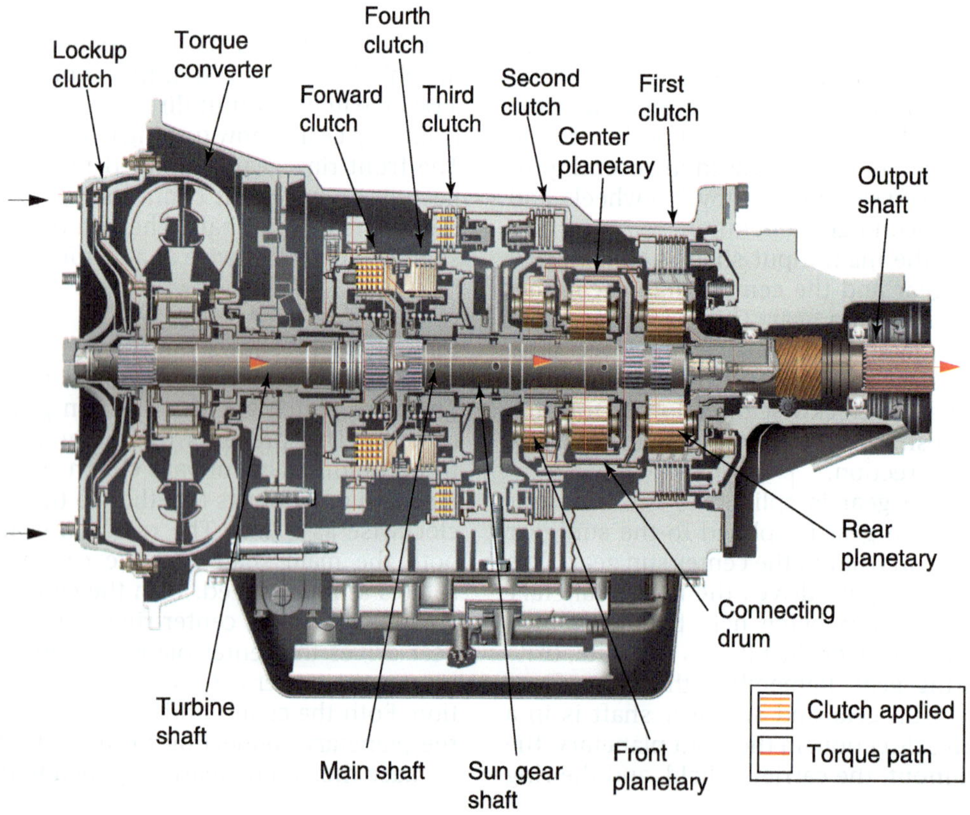

FIGURE 18–18 Four-speed fourth-gear torque powerflow

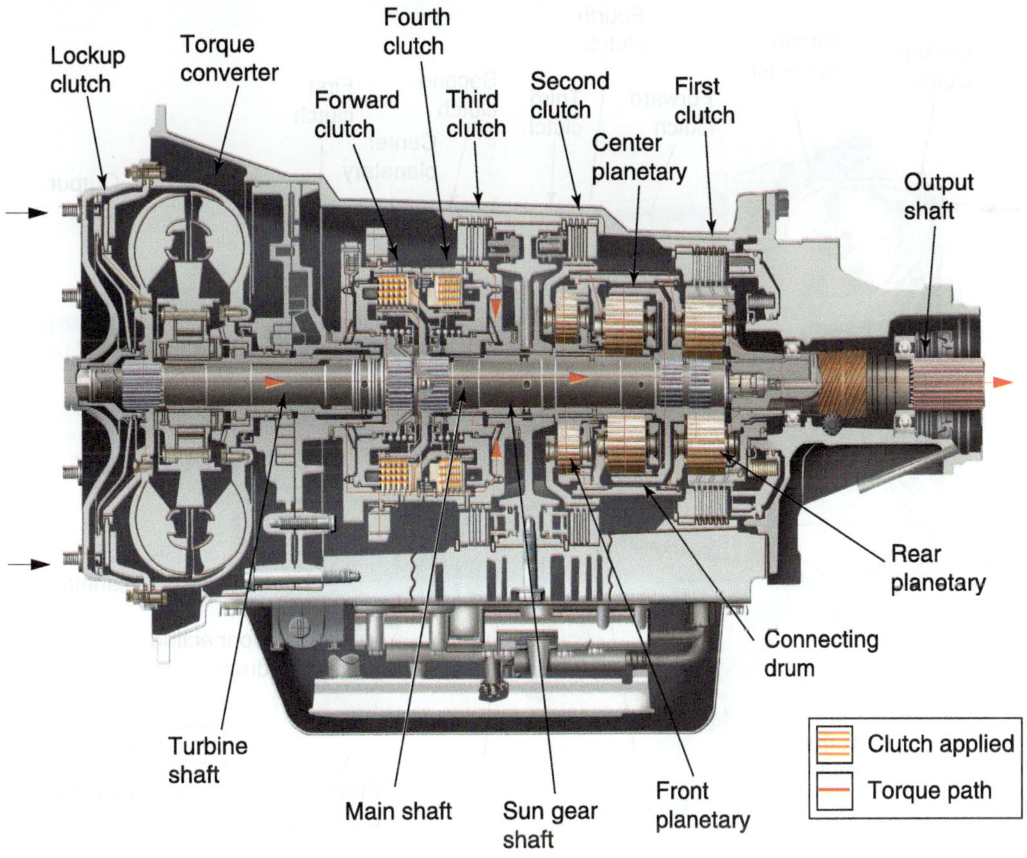

it rotates at the same speed as the center planetary carrier. In this gear, the center ring gear is the input, the sun gear is held, and the carrier is the output. The result is combination 2 in Table 18–1.

Fourth Gear

In this range (**Figure 18–18**), the forward and fourth clutches are applied. The transmission main shaft and center sun gear shaft are thus locked together and rotate as a unit at input speed. When this happens, both the center and rear sun gears rotate at the same speed, also rotating their respective carriers, which are both splined to the planetary connecting drum and output shaft.

With any two members of a planetary gearset rotating at the same speed, the result is direct 1:1 drive. In this case, although the front sun gear is being driven, no front planetary gearset member is being held. Therefore, the front planetary freewheels, with the power transfer being split to the center and rear planetary sets.

Reverse Gear

Reverse gear is the only gear in which the forward clutch is not applied (**Figure 18–19**). In reverse, the fourth clutch is applied and this rotates, at input speed, the center sun gear shaft with the front sun gear splined to it. The first clutch is applied also. It anchors the rear ring gear against rotation. The center sun gear rotates the center carrier pinions, which, in turn, rotate the center ring gear in the opposite direction. The center carrier is splined to the planetary connecting drum that connects to the rear carrier. The reverse rotation of the center ring gear rotates the rear sun gear. This causes the rear planetary pinions to drive the rear carrier in a reverse direction around the internal gears of the locked rear ring gear. Thus the output shaft that is splined to the rear carrier rotates in a reverse direction at a significant speed reduction. **Table 18–4** summarizes four-speed clutch application outcomes.

FIVE-SPEED TRANSMISSION POWERFLOW

It is important to remember that the only difference between five-speed and four-speed transmissions is that an additional low clutch and low planetary gearset are added to the back of the transmission.

FIGURE 18–19 Four-speed reverse gear torque powerflow

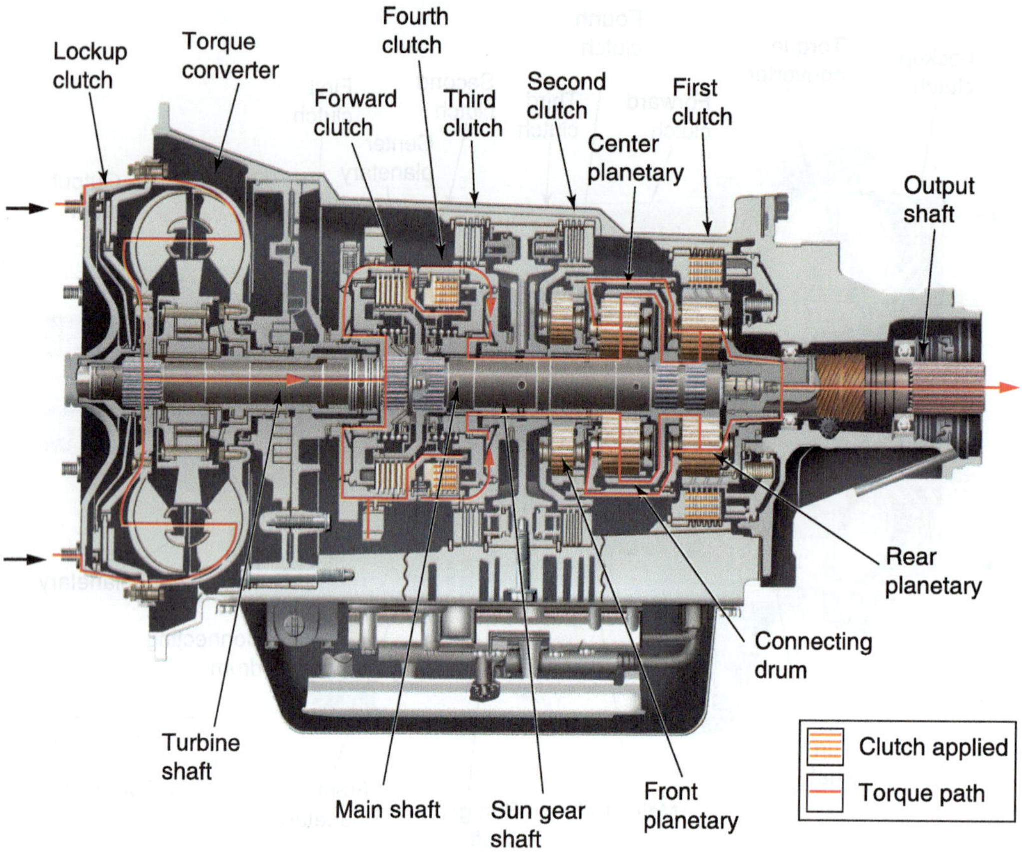

TABLE 18–4 Summary of Four-Speed Clutch Application

Gear Ratio	1st Clutch	2nd Clutch	3rd Clutch	4th Clutch	Forward Clutch
1st gear	ON				ON
2nd gear		ON			ON
3rd gear			ON		ON
4th gear				ON	ON
Reverse	ON			ON	

In deep ratio (DR) models, only first-gear powerflow is through the low planetary gearset. On close ratio (CR) models, the output shaft is splined to the low planetary carrier, so power must flow through the low gearset in all forward gears. We take a look first at DR then at CR powerflows.

DR Neutral

Figure 18–20 shows a DR transmission. Note the position of the low clutch and low planetary gearset at the rear of the transmission. All other components shown in Figure 18–20 through Figure 18–25 are identical to the four-speed transmissions. The neutral torque path is indicated in Figure 18–20. It is identical to the neutral torque path of four-speed transmissions discussed earlier. The forward clutch is not engaged. The first clutch is engaged, but power transfer is not possible. As before, the PTO gear can be engaged and operated with accelerator action.

DR Low (First) Gear

In low or first gear, both the forward and low clutches are applied (**Figure 18–21**). The forward clutch locks the turbine shaft and main shaft together so that they

FIGURE 18–20 Five-speed neutral torque powerflow in a deep ratio, five-speed transmission

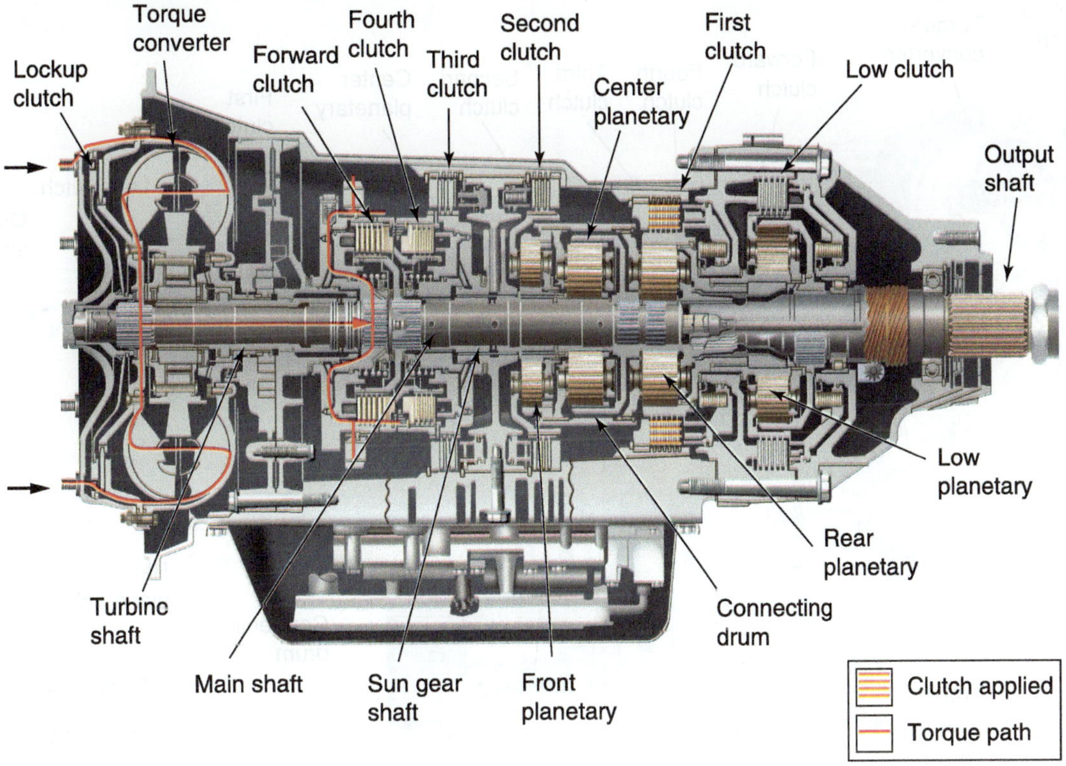

FIGURE 18–21 Five-speed low-(first) gear torque path powerflow in a deep ratio, five-speed transmission

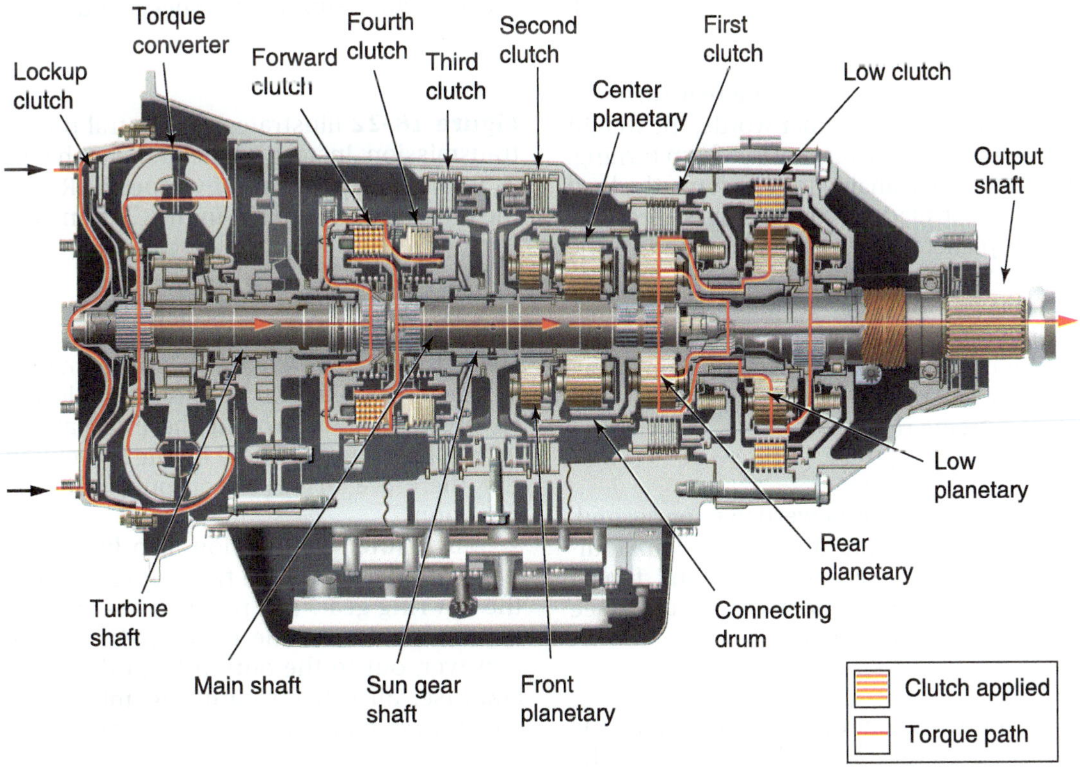

FIGURE 18–22 Neutral torque path powerflow in a close ratio, five-speed transmission

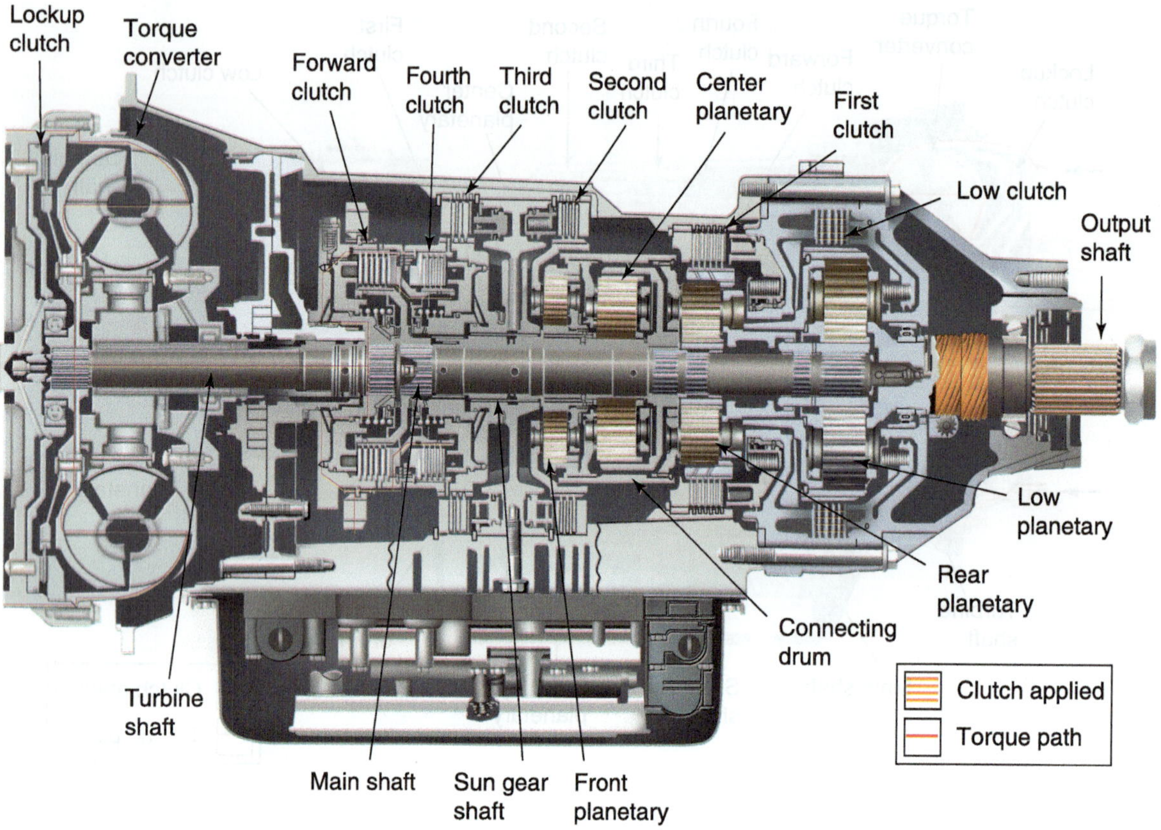

rotate as a unit. The low clutch holds the low planetary carrier stationary. The rear sun gear splined to the main shaft rotates the rear pinions. The rear carrier resists pinion rotation because the rear carrier is splined to the output shaft. In other words, the weight of the vehicle load keeps the rear carrier from turning.

With the rear carrier unable to overcome the load placed on it, the rear planetary ring gear is forced to rotate. The low planetary sun gear is splined to the rear ring gear and also rotates with it. The low sun gear rotates the low pinions. Because the low carrier is anchored against rotation by the applied low clutch, the pinions rotate the low planetary ring gear. The low ring gear is splined to the output shaft and rotates it at an extreme reduction.

DR Second through Fifth Gears

Five-speed, DR transmission powerflows for second, third, fourth, and fifth gears are identical to four-speed transmission first, second, third, and fourth gears, respectively. The low planetary is not used (see Figure 18–15 through Figure 18–18).

DR Reverse Gear

Five-speed, DR transmission reverse gear operation is identical to four-speed reverse (see Figure 18–19). The low planetary supply freewheels and does not affect output.

CR Neutral

Figure 18–22 illustrates the neutral position in a CR transmission. Instead of the first clutch being applied as in the four-speed and five-speed DR neutral flows, the low clutch is applied in the CR transmissions.

CR Low (First) Gear

The forward and low clutches are applied (**Figure 18–23**). Torque flows from the turbine shaft, forward clutch and housing, main shaft, low sun gear, low planetary pinions, and low carrier and is forced to walk around the internal gears of the locked low ring gear held by the low clutch. The carrier, therefore, drives the output shaft in a clockwise direction.

Low clutch application also locks the rear carrier stationary because the rear carrier is splined to the low ring gear. The rear sun gear rotates because it is connected to the main shaft. No drive exists, however, out to the output from this planetary set. Likewise, all other rotating members in the front and center planetary gearsets add nothing to the drive.

FIGURE 18–23 Five-speed low-(first) gear torque path powerflow in a close ratio, five-speed transmission

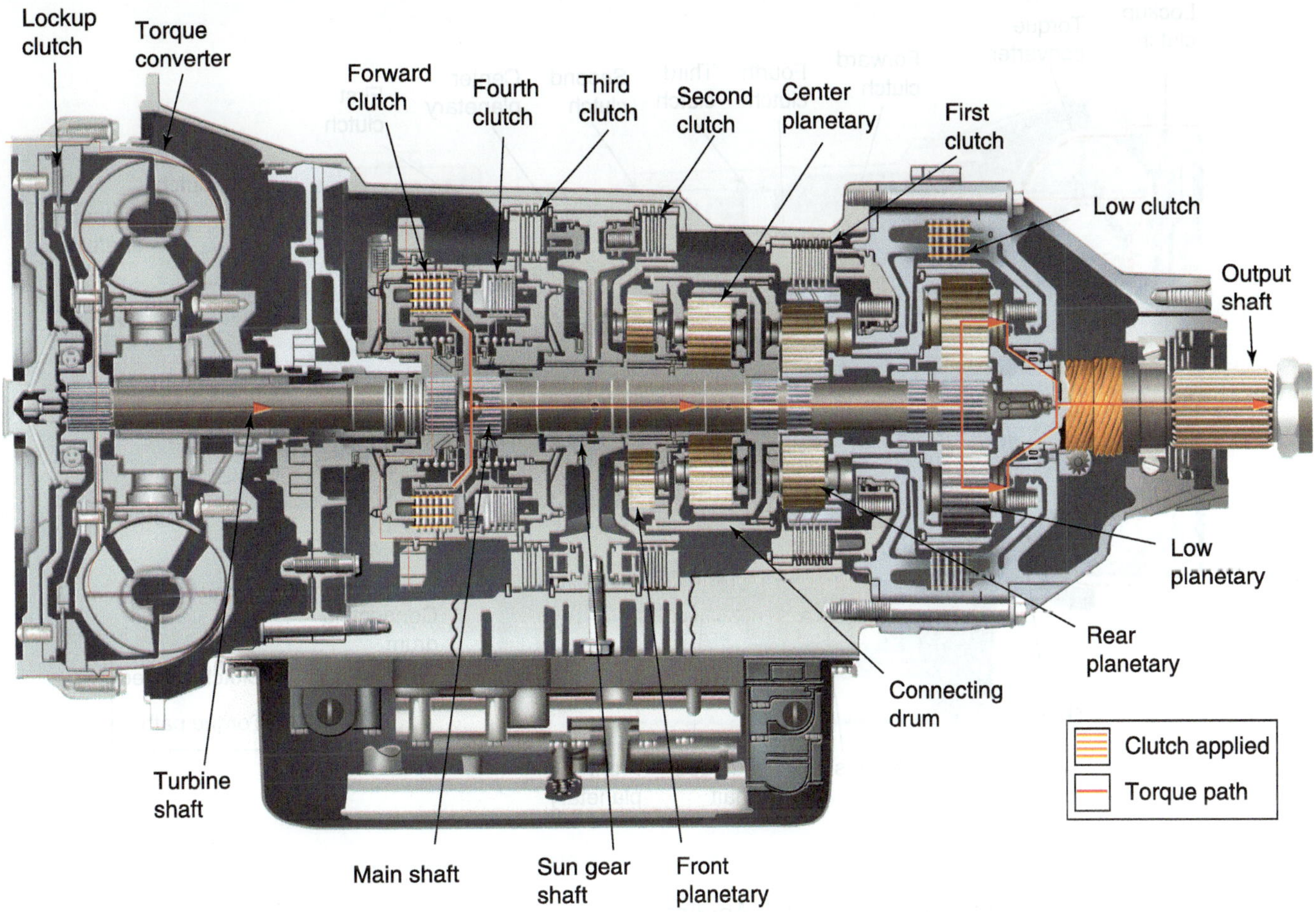

CR Second through Fifth Gears

The powerflow path for CR second, third, fourth, and fifth gears is the same as a four-speed transmission first, second, third, and fourth gears, respectively, except that the flow must pass through the low planetary gearset carrier to reach the output shaft.

CR Reverse Gear

In CR reverse (**Figure 18–24**), both the fourth and low clutches are applied. Torque travels from the turbine shaft, forward clutch housing (clutch released) to the fourth clutch (which locks onto the rear of the forward clutch housing) to the sun gear shaft (which becomes the driving member turning clockwise) and to the center sun gear, center pinions, and center ring gear (which is now turning counterclockwise). The rear sun gear is splined to the center ring gear. The rear planetary gears simply rotate around their pinions because the rear carrier is held by the applied low clutch.

The drive to the output shaft is through the center ring gear to the main shaft and low sun gear to the low planetary pinions. The low carrier is forced to walk around within the held low ring gear in a counterclockwise direction. The carrier is splined to the output shaft. **Table 18–5** summarizes five-speed clutch application outcomes.

18.3 TRANSMISSION HYDRAULIC SYSTEMS

The hydraulic circuits used to generate, direct, and control the pressure and flow of transmission oil within most automatic transmissions are similar in design and operation. As with clutches and planetary gearing, modifications are made to the basic system to incorporate a fifth gear or special requirements such as DR or CR applications.

Transmission oil is the torque-transmitting medium in the torque converter. Its velocity drives the converter turbine and its flow cools and lubricates the transmission. The pressure it is under is used to operate the various control valves and to apply the

FIGURE 18–24 Five-speed reverse gear torque path powerflow in a five-speed, close ratio transmission

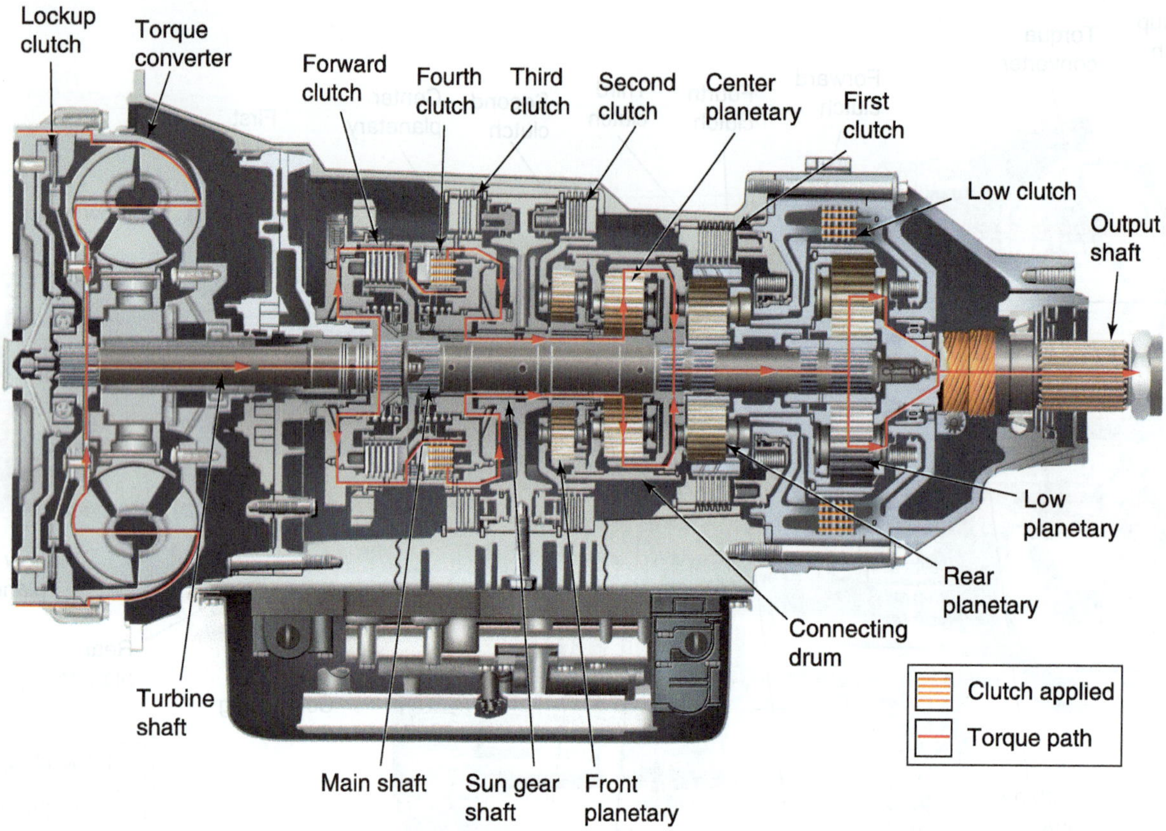

TABLE 18–5 Summary of Five-Speed Clutch Application

Gear Ratio	1st Clutch	2nd Clutch	3rd Clutch	4th Clutch	Forward Clutch	Low Clutch
1st gear					ON	ON
2nd gear	ON				ON	
3rd gear		ON			ON	
4th gear			ON		ON	
5th gear				ON		
Reverse	ON			ON		

clutches. **Figure 18–25** and **Figure 18–26** show flow-coded schematic diagrams of four-speed and five-speed automatic transmissions. These diagrams show the systems as they would function in neutral with the engine idling, and they are referenced throughout this section.

OIL FILTER AND PUMP CIRCUIT

Oil is drawn from the sump (transmission oil pan) through a filter by the torque converter-driven oil pump. The oil discharge by the pump flows into the bore of the main pressure regulator valve.

MAIN PRESSURE CIRCUIT

Main pressure is regulated by the main pressure regulator valve. Oil from the pump flows into the valve bore and through an internal passage in the valve to the upper end of the valve. Pressure at the upper end of the valve forces the valve downward against the spring until oil flows into the converter-in circuit.

FIGURE 18–25 Hydraulic control circuit of four-speed transmission

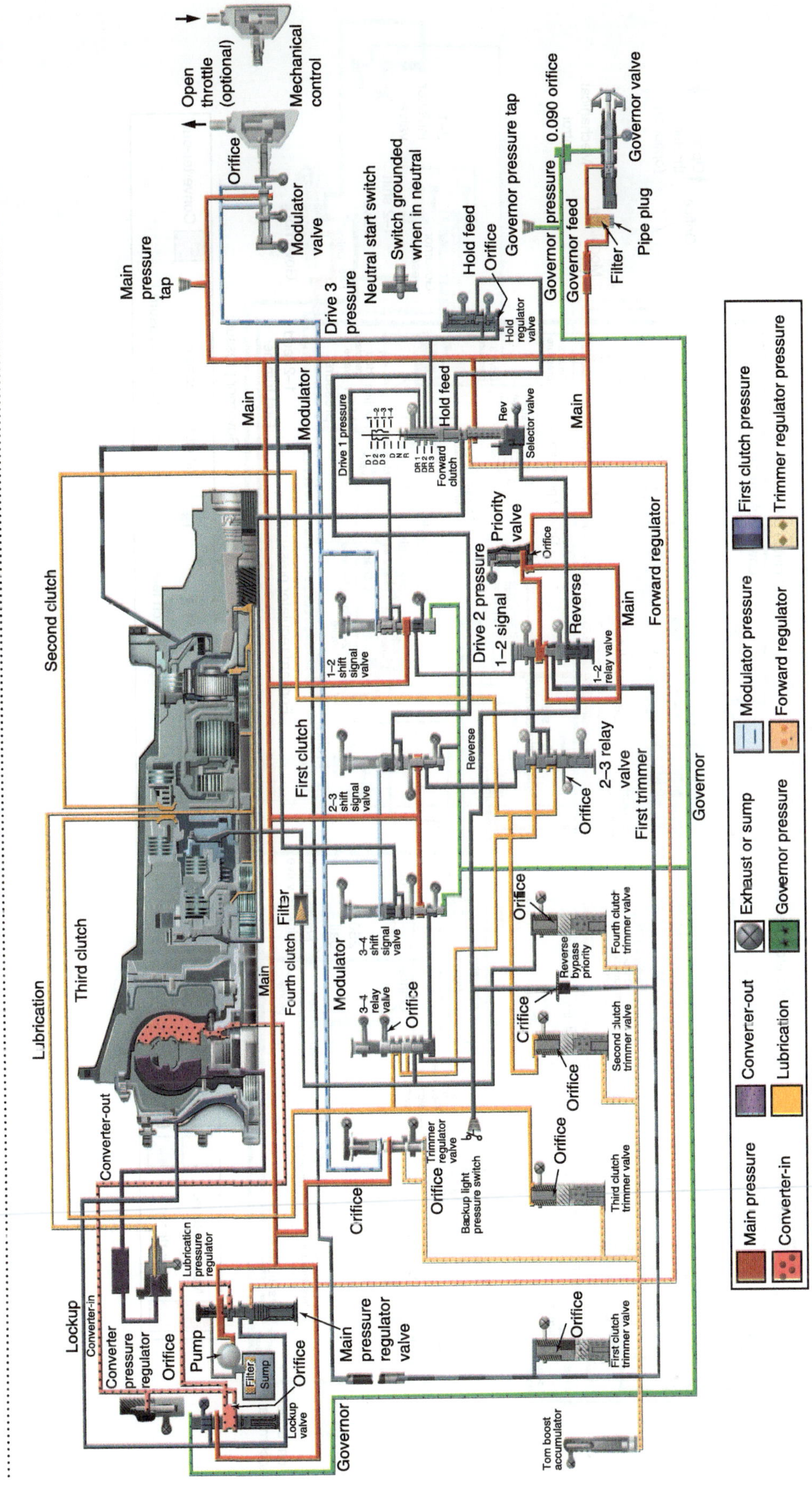

FIGURE 18–26 Hydraulic control circuit of five-speed transmission

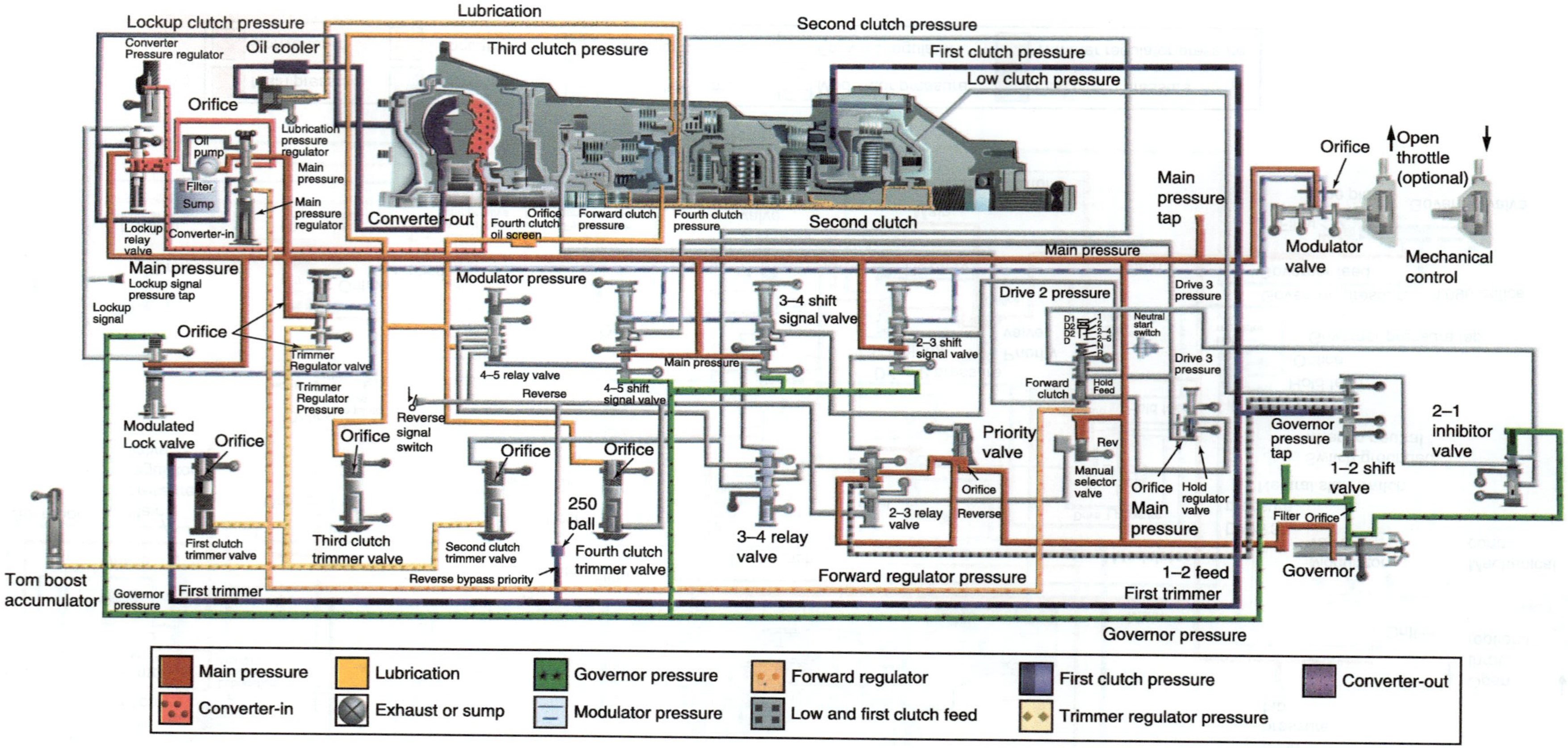

When flow from the pump exceeds the circuit requirement, the converter regulator valve opens and allows the excess to exhaust to sump.

Although main pressure is controlled primarily by the spring at the bottom of the valve, it is also affected by forward regulator pressure and lockup pressure. These pressures are not present at the regulator valve during reverse operation and main pressure is regulated at a higher value.

CONVERTER COOLER CIRCUIT

This circuit originates at the main pressure regulator valve. Converter-in oil flows to the torque converter. Oil must flow through the converter continuously to keep it filled and to remove heat generated by the converter. The converter pressure regulator valve regulates converter-in pressure by exhausting excessive oil to sump.

Converter-out oil, exiting the torque converter, flows to an external cooler (a vehicle original equipment manufacturer [OEM]-supplied component). Ram air or engine coolant is used to remove heat from the transmission oil at the cooler heat exchanger. From there, lubrication oil is directed through the transmission to those components requiring continuous lubrication and cooling. Oil in excess of that required by the lubrication circuit drains to the sump through the lubrication pressure regulator.

SELECTOR VALVE (FOUR-SPEED)

The selector valve masters control of the transmission hydraulic circuits. It is manually shifted by the driver to select the operating range desired. It can be shifted to any one of six positions. These are: neutral (N), reverse (R), drive (D), drive 3 (D3), drive 2 (D2), and drive 1 (D1). At each of these positions the selector valve establishes the hydraulic circuit for operation in the condition indicated. D, D3, D2, and D1 are forward ranges.

Forward Regulator Valve (Four-Speed)

The lowest gear attainable is first gear. In each of these positions, the transmission will start in first gear (except on second-gear start). Shifting within any range is automatic, depending on vehicle speed and throttle position. The forward regulator circuit originates at the manual selector valve and terminates at the main pressure regulator valve. In every forward range the circuit is pressurized. This pressure helps reduce main pressure to a lower value. In reverse, forward regulator pressure is not present, allowing for higher main pressure to control the higher torque produced in reverse operation.

Forward Regulator Valve (Five-Speed)

The selector valve is manually shifted to select the operating range desired. It can be shifted to any one of seven positions. These are: neutral (N), reverse (R), drive (D), drive 4 (D4), drive 3 (D3), drive 2 (D2), and drive 1 (D1). At each of these positions the selector valve establishes the hydraulic circuit to be activated for the range selected.

D, D4, D3, D2, and D1 are forward ranges. The lowest gear is first gear. This is available in D1 range only. Selecting the D2, D3, D4, or D positions will start the transmission in second gear and automatically shift to range, depending on vehicle speed and throttle position.

Some models do not upshift in D2 position under normal operating conditions. Again, the forward regulator circuit originates at the manual selector valve and terminates at the main pressure regulator valve. The circuit is pressurized in every forward range. This pressure aids in reducing main pressure to a lower value. In reverse, the forward regulator pressure is not present. This allows for the higher main pressure required for the higher torque produced in reverse range.

GOVERNOR CIRCUIT

Main pressure is directed to the governor valve. The speed of the transmission output controls the position of the governor valve. The position of the governor valve determines the actual pressure in the governor circuit. When the transmission output is not rotating, governor pressure is approximately 2 psi. Governor pressure increases with the speed of output rotation.

In a typical four-speed, governor pressure is directed to the 1–2, 2–3, and 3–4 shift signal valves, and the modulated lockup valve. In earlier models that do not include the modulated lockup valve, governor pressure is directed to the top of the lockup valve. In the five-speed, governor pressure is directed to the 2–3, 3–4, and 4–5 shift signal valves, 2–1 inhibitor valve, and the modulated lockup valve. In models that do not include the modulated lockup valve, governor pressure is directed to the top of the lockup valve.

At the 2–1 inhibitor valve, governor pressure acts upon the top of the valve. When governor pressure is sufficient to push the valve downward, road speed is too high to permit a downshift to first gear. So a 2–1 shift cannot occur until governor pressure (and road speed) is reduced. DR1 pressure is blocked from reaching the top of the 1–2 shift valve until road speed (and governor pressure) is decreased to a value that will allow the inhibitor valve to move upward.

In transmissions designed to start in second gear, the presence of governor pressure at the bottom of the 1–2 shift valve and forward regulator pressure on the 1–2 modulator valve, plus the assistance of spring force under the 1–2 shift valve, causes the shift valve to move upward against its stop. This allows main pressure to pass through the 1–2 shift signal valve to the 1–2 relay valve. The 1–2 relay valve is forced downward in its bore, allowing apply pressure to be directed to the second clutch, permitting second-gear start.

MODULATOR PRESSURE CIRCUIT

The modulator valve assists the governor pressure in moving any of the shift signal valves during an upshift. It can also delay a downshift. The modulator valve controls a regulated pressure derived from main pressure. The modulator valve receives main pressure and regulates it as a direct result of the accelerator position.

The modulator actuator varies modulator pressure as the accelerator pedal moves. It can be controlled in a number of ways. On gasoline engines, vacuum pressure is used. On diesel engines, air pressure or a mechanical linkage can be used to operate the modulator.

The mechanical linkage is most common. Modulator pressure is highest when the engine is running at idle and decreases in proportion to accelerator pedal travel. In a mechanical linkage system, the modulator is controlled by a push-or-pull cable connected to the throttle.

When there is no throttle pedal pressure, the valve is moved to the right by a spring at the left end of the valve. As the driver depresses the throttle pedal, a wedge-shaped actuator forces the modulator valve to the left, decreasing modulator pressure. When spring force at the left of the valve is in balance with the spring force and modulator pressure at the right of the valve, modulator pressure is regulated.

When the accelerator is depressed, the movement of the actuator cam forces the modulator valve to the left. This leftward movement reduces modulator pressure. When the pedal travel is reduced, the downward movement of the actuator cam allows the spring at the left to push the valve to another regulating position. Because the throttle pedal angle varies with load and engine speed, modulator pressure also varies. This varying pressure is directed to the 1–2, 2–3, and 3–4 shift signal valves (four-speed); the 2–3, 3–4, and 4–5 shift signal valves (five-speed); the trimmer regulator valve; and to the modulated lockup valve. Earlier models do not include the modulated lockup valve.

At the shift signal valves, modulator pressure assists governor pressure to overcome shift valve spring force. The surface areas and spring tension of each one of the shift valves are so calibrated that the valves will shift in the proper sequence at the right time. A decrease in modulator pressure will cause a downshift if governor pressure is not sufficient to oppose the shift valve spring force. The trimmer boost and trimmer regulator valve are controlled by modulator pressure.

At the modulated lockup valve, modulator pressure assists governor pressure in moving (or holding) the valve downward. The variation in modulator pressure will vary the timing of lockup clutch engagement and release. The primary purpose of lockup modulation is to prolong the engagement of the lockup clutch (at closed throttle) in lower ranges to provide greater engine braking.

All modulated lockup valves are adjustable, but not all transmissions have modulated lockup. Models with adjustable lockup can be adjusted by rotating the adjusting ring at the bottom of the valve until the desired shift point is attained (see Chapter 19).

TRIMMER REGULATOR VALVE

Late-generation Allison hydromechanical transmissions are equipped with trimmer regulator valves. The trimmer regulator valve reduces main pressure to a regulated pressure. The regulated pressure is raised or lowered by changes in modulator pressure.

Trimmer regulator pressure is directed to the lower sides of the first and second trimmer plugs on some models or to the lower sides of the first, second, third, and fourth trimmer plugs on models equipped with smooth shift. This varies the clutch apply pressure pattern of the trimmer valves. A higher modulator pressure at zero accelerator pedal travel will reduce trimmer regulator pressure. This results in a lower initial clutch-apply pressure. Conversely, a lower modulator pressure (open throttle) results in higher regulator pressure and a higher initial clutch apply pressure.

TRIMMER VALVES

The purpose of the trimmer valves is to avoid shift shock. Trimmer valves reduce pressure to the clutch apply circuit during initial application, and then gradually ramp up the pressure to operating maximum. This applies the clutch gently, and harsh shifts are prevented.

Although each trimmer valve is calibrated for the clutch it serves, all four trimmers function in the same manner. Each trimmer includes (top to bottom) an orificed trimmer valve, trimmer valve plug, one or two trimmer springs, and a stop pin.

When the clutch (except forward) is applied, apply pressure is sent to the upper end of the valve. Initially, the plug and valve are forced downward against the spring until oil is routed to exhaust. The escape of oil, as long as it continues, reduces clutch apply pressure. Oil flows through an orifice in the trimmer valve to the cavity between the trimmer valve and the trimmer valve plug. Pressure in this cavity forces the plug down to the stop. Because it is acting on a greater sectional area than at the upper end, the pressure below the trimmer valve pushes the trimmer valve to the upper end of the valve bore. This stops the routing of oil to the exhaust. When the escape of oil ceases, clutch pressure is at the maximum. The plug remains downward until the clutch is released.

On the release of the clutch, the spring pushes the trimmer components to the top of the valve bore. In this position, the trimmer is reset and ready to repeat the action at the next application of that clutch.

LOCKUP CIRCUIT

Lockup clutch engagement and release are controlled by the modulated lockup valve and the lockup relay valve. The modulated lockup shift valve is actuated by governor pressure or governor plug modulator pressure. At full throttle, governor pressure will move the valve downward. At part throttle, governor and modulator pressures will move the valve downward. The purpose of the modulated lockup valve is to prolong lockup clutch engagement while vehicle speed decreases (closed throttle). This feature provides engine-braking action at speeds lower than the normal lockup disengagement point.

In its downward position, the modulated lockup valve directs main pressure to the top of the lockup valve. Main pressure pushes the lockup valve downward. In its downward position, the valve directs main pressure to the lockup clutch. This engages the clutch.

The position of the lockup valve affects the volume of oil flowing to the torque converter. When the valve is upward (lockup clutch released), there is maximum flow to the converter. When the valve is downward (clutch engaged), converter-in flow is restricted by an orifice and flow is reduced.

When the lockup clutch is engaged, a signal pressure is sent to the main pressure regulator valve. This signal pressure reduces main pressure to a lower pressure schedule. On most models with adjustable lockup, modulator assistance at the modulated lockup valve is not present. By setting the adjusting ring to the shift point specifications, with the assistance of governor pressure, lockup engagement is initiated. Models that do not have modulated lockup do not include the modulated lockup valve. Governor pressure is directed to the top of the lockup valve.

PRIORITY VALVE

The **priority valve** ensures that the control system upstream from the valve will have sufficient pressure during shifts to perform its automatic functions. In many models, first clutch pressure can bypass the priority valve when the transmission is in reverse. Without the priority valve, the filling of the clutch might require a greater volume of oil (momentarily) than the pump could supply and still maintain pressure for the necessary control functions.

INHIBITOR VALVE

In five-speed transmissions, a 2–1 inhibitor valve prevents a downshift from second gear to first gear when road speed is too high. It will also protect the engine from overspeeding during downgrade operation in first gear by making a 1–2 upshift if road speed exceeds that which is safe for first-gear operation.

When road speed is great enough to produce a governor pressure that forces the 2–1 inhibitor valve downward, D1 pressure to the 1–2 shift valve is blocked. Thus, if a manual shift to drive 1 (first gear) is made, D1 pressure cannot reach the 1–2 shift valve. The transmission will remain in second gear until road speed (governor pressure) decreases to a value that permits the 2–1 inhibitor valve to move upward. In its upward position, D1 pressure can reach the 1–2 shift valve and move it downward to the first-gear position.

While operating in first gear, the 2–1 inhibitor valve will move downward if road speed (governor pressure) exceeds that which is safe for first-gear operation. This will block D1 pressure and exhaust the cavity above the 1–2 shift valve. The shift valve will move upward to give second-gear operation.

CLUTCH CIRCUITS

There are five separate clutches in four-speed transmissions and six clutches in five-speed models. As covered earlier, various clutch combinations are used in various drive ranges (see Table 18–3).

Each clutch has its own circuit. The first, second, third, and fourth clutches are each connected to a relay valve and to a trimmer valve. The forward clutch is connected directly to the selector valve and does not connect to a trimmer valve. Because the vehicle is not moving, the application of a trimmer valve is not required.

The low clutch is not trimmed because there is no automatic upshifting from, or downshifting to, this clutch. The low clutch circuit (five-speeds) connects the clutch to the 1–2 shift valve. The 1–2 shift valve also receives 1–2 (low and first) feed from the 2–3 relay valve. In neutral, the 1–2 shift valve is held upward by spring pressure. It cannot move downward unless drive 1 line is charged. In the upward position, 1–2 (low and first) feed pressure is sent to the first clutch. The first clutch circuit connects the clutch to the first trimmer valve and to either the 1–2 relay valve (four-speeds) or to the 1–2 shift valve (five-speeds).

In neutral the 1–2 shift valve is held upward by spring pressure. The 1–2 shift valve cannot move downward unless the 2–1 signal line is charged. This will not occur in neutral (vehicle standing) because there is no governor pressure to shift the 2–1 inhibitor valve. Only the first clutch is applied, so the transmission output cannot rotate. A bypass and check ball are provided between the reverse and first trimmer passages to ensure a rapid and more positive shift from first gear to reverse.

The first clutch is also applied in first-gear operation (four-speeds), second-gear operation (five-speeds), and reverse gear operation. Shifting the selector from neutral to D, D3, D2, or D1 charges the forward clutch circuit and applies the forward clutch. Shifting from neutral to reverse (R) charges the fourth clutch, while the first clutch remains charged. In reverse, fourth clutch (reverse signal) pressure is also directed to the bottom of 1–2 relay valve (four-speeds) or 2–3 relay valve (five-speeds). The pressure at this location prevents either relay valve from moving downward.

When the circuits are charged and the selector valve is at D, four-speed transmissions will automatically shift from first to second, second to third, and third to fourth. Five-speed models will automatically shift from second to third, third to fourth, and fourth to fifth. These shifts occur as a result of governor pressure and (if the throttle is not fully open) modulator pressure.

The position of the selector valve determines the highest gear that can be reached automatically while the vehicle is moving. Four-speed models that do not incorporate a second-gear start will automatically shift 1–2, 2–3, and 3–4 when the selector valve is in drive 4 position. In drive 3, automatic 1–2 and 2–3 shifts occur. In drive 2, an automatic 1–2 shift occurs. Drive 1 does not permit an upshift to occur unless an overspeed of the transmission output occurs.

The position of the selector valve in a second-gear start is the same but transmission performance will be altered. The transmission always starts in second gear when any forward range other than drive 1 is selected. When drive 4 is selected, the transmission starts in second gear and automatically shifts 2–3 and through to 3–4. In drive 3, an automatic 2–3 shift occurs. Drive 2 does not permit an upshift to occur unless an overspeed of the transmission output occurs. Drive 1 is the only position in which first gear is available and upshifting is not permitted unless an overspeed of the transmission output occurs.

In five-speed transmissions drive (D) automatic 2–3, 3–4, and 4–5 shifts can occur. In drive 3 (D3), automatic 2–3 and 3–4 shifts can occur. With some early transmissions an automatic 2–3 shift can occur in drive 2 (D2). In recent models, no shift from 2 can occur unless overspeed occurs. In drive 1 (D1), no shift from 1 can occur unless overspeed occurs. The various drive ranges limit the highest gear attainable by introducing a pressure that prevents governor pressure from upshifting the signal valves (unless governor pressure is well above that normally attained). This pressure is a regulated, reduced pressure sourced from the main pressure circuit at the hold regulator valve. Main pressure is directed to the hold regulator valve through the hold feed line when the selector lever is in drive 1, drive 2, or drive 3 position.

The pressure produced in the hold regulator valve is directed to the 3–4 shift valve (four-speeds) or the 4–5 shift valve (five-speeds) when the selector is in drive 3. The hold pressure is directed to the 2–3 and 3–4 (four-speeds) or the 3–4 and 4–5 (five-speeds) when the selector is in drive 2 position.

In some later generation five-speed models, hold pressure is directed to the 2–3, 3–4, and 4–5 shift valves. In other four- and five-speeds, pressure can be directed to all shift signal valves when the selector is in drive 1. Hold regulator pressure at each shift signal valve will push the modulator valve upward and exert downward force on the shift valve. So, when hold pressure is present, upshifts are inhibited except in extreme overspeed conditions.

AUTOMATIC UPSHIFTS

When the transmission is operating in first gear (four-speed) or second gear (five-speed), with the selector valve at drive (D), a combination of governor pressure and modulator pressure, or governor pressure alone, upshifts the transmission to second or third gear, respectively. At closed or part throttle, modulator pressure exists and will assist governor pressure. At full throttle, there is no modulator pressure, meaning that upshifts occur at higher road speeds. When the throttle is closed, shifts occur at lower wheel speeds.

Governor pressure is dependent on the rotational speed of the transmission output. The greater the output speed (vehicle speed), the greater the governor pressure. When governor pressure is sufficient, the first upshift will occur. With a further increase in governor pressure (and vehicle speed), the second upshift will occur. A still further increase will cause a third upshift. Modulator pressure, when present, will assist governing pressure in overcoming shift valve spring force.

In any automatic upshift, the shift signal valve acts first. This directs a shift pressure to the relay valve. The relay valve shifts, exhausting the applied clutch and applying a clutch for a higher gear.

AUTOMATIC DOWNSHIFTS

Automatic downshifts, like upshifts, are controlled by governor and modulator pressures. Downshifts occur in sequence as governor pressure and/or modulator pressure decrease. Low modulator pressure (full accelerator pedal travel) will hasten the downshift; high modulator pressure (zero accelerator pedal travel) will delay downshifts.

In any automatic downshift, the shift signal valve acts first. This exhausts the shift signal holding the relay valve downward. The relay valve then moves upward, exhausting the applied clutch and applying the clutch for the new lower gear.

Downshift and Reverse Inhibiting

As a result of valve sectional areas and spring forces, the system is designed to prevent downshifts at too rapid a rate or to prevent a shift to reverse while moving forward. For example, if the vehicle is traveling at a high speed in fourth gear (4) and the selector valve is inadvertently moved to D1, the transmission will not immediately shift to first gear. Instead it will shift 4–3–2–1 as speed decreases. (It will remain in fourth gear if speed is not decreased sufficiently to require an automatic downshift.)

The progressive downshift occurs because the hold regulator pressure is calibrated, along with the valve sectional areas, to shift the signal valves downward against governor pressure only when governor pressure decreases to a value corresponding to a safe downshift speed. So if speed is too great, governor pressure is sufficient to hold the shift signal valve upward against drive 3, drive 2, or drive 1 pressure, all of which are the regulated holding pressures originating in the hold regulator valve. As governor pressure decreases, all shift signal valves move downward in sequence.

18.4 HYDRAULIC RETARDERS

Hydraulic retarders are one of several types of auxiliary braking systems used on heavy-duty trucks. Some retarder systems are designed as an integral component of the automatic transmission. They provide some vehicle braking on severe downgrades and help maximize service brake life by providing braking action at the driveline. The retarder is applied by the driver as needed.

Integral retarders are usually available in two automatic transmission configurations: the input retarder and the output retarder. In one European import design used in some buses, the retarder is located in the center of the transmission, but it uses principles of operation identical to those discussed in this section.

INPUT RETARDERS

The **input retarder** is located between the torque converter housing and the main housing. It is designed primarily for highway applications. The unit uses a paddle wheel design with a vaned rotor mounted between stator vanes in the retarder housing.

Retardation or brake mode occurs when transmission oil is directed into the retarder housing. The oil causes resistance to the rotation of a vaned rotor. The retarding force is transmitted through the transmission down the vehicle driveline, slowing (braking) the vehicle drive wheels. Retarding capacity can be enhanced by downshifting.

Variable control is achieved by moving a hand lever or activating a separate foot pedal to regulate oil flow to the retarder. With the controls in the "off" position, the oil is evacuated from the retarder, leaving no drag on the rotor. Heat generated by the absorption of horsepower is dissipated by circulating the oil through a high-capacity transmission oil cooler.

OUTPUT RETARDERS

Most **output retarders** are mounted on the rear of the transmission without adding additional length to it. They apply retarding force directly to the driveline. The output retarder uses a two-stage principle. The first stage consists of a rotor/stator hydraulic design. The second stage can be a powerful, oil-cooled, friction clutch pack.

On activation of the output retarder, quick initial response is provided by a momentary application of the friction clutch pack while the hydraulic section is being charged with oil. The hydraulic section provides better retardation at higher speeds. As the vehicle slows, the second stage friction clutch pack phases in, providing low-speed retardation force capable of slowing the vehicle to a virtual stop.

Because it is located at the rear of the transmission and transmits retarding force directly to the driveline for greater effectiveness, the output retarder functions independently of engine speed or transmission ratio. Heat generated by retardation is dissipated through the transmission cooling system.

Activation of the output retarder can be by a hand lever or separate foot pedal for highway downhill speed control. Retarder control can also be incorporated into the service brake pedal for stop-and-go traffic operations. This method provides a smooth transition from retarder to service brakes, depending on braking effort needed. For flexibility in varied driving conditions both systems can be used together.

SUMMARY

- Automatic transmissions may be hydromechanical or electronic. All automatic transmissions upshift and downshift with no direct assistance from the driver.
- Factors such as road speed, throttle position, and governed engine speed control shifting between gears.
- Fully automatic transmissions use compounded planetary gearsets. The simple planetary gearset

consists of three main components: a sun gear, a carrier with planetary pinions mounted to it, and an internally toothed ring gear or annulus.

- Advantages of the planetary gearset are as follows:
  - Constantly meshed gears
  - Gear torque loads are divided equally
  - Compact and versatile
  - Additional ratio and direction variations added by compounding
- In a planetary gearset, any one of the three main members—sun gear, pinion gear carrier, or ring gear—can be used as the driving or input member.
- Depending on which member of the planetary gearset is the driver, which is held, and which is driven, torque or speed increases are produced.
- To produce an output, one of the planetary gear members must be held stationary. In heavy-duty truck transmissions this is achieved with multiple-disc clutches, which can serve as both braking and power transfer devices.
- Compound planetary combinations are several planetary gearsets coupled to produce the required gear ratios and direction. Four-speed, heavy-duty transmissions use three simple planetary gearsets—front, center, and rear. Five-speed transmissions add an additional low planetary gearset for a total of four gearsets.
- In all automatic transmissions, power flows from the torque converter through the transmission planetary gearing and out to the output shaft.
- Transmission oil is drawn from the sump through a filter by the input-driven oil pump.
- Oil pressurized by the pump flows into the bore of the main pressure regulator valve.
- The converter/cooler/lubrication circuit originates at the main pressure regulator valve.
- The selector valve/forward regulator circuit is manually shifted to select the operating range desired.
- Main pressure is directed to the governor valve. The speed of the transmission output shaft controls the position of the governor valve, which determines the amount of pressure in the governor circuit.
- The modulator actuator varies modulator pressure as the accelerator pedal moves.
- Lockup clutch engagement and release are controlled by the modulated lockup valve and the lockup relay valve.
- There are five separate clutches in four-speed transmissions and six clutches in five-speed models, each clutch having its own circuit.
- Hydraulic retarders are one of several types of auxiliary braking systems used on heavy-duty trucks; they are applied by the driver as needed.

REVIEW QUESTIONS

1. Which of the following can be considered the largest *gear* in a simple planetary gearset?
   a. the sun gear
   b. the ring gear
   c. the carrier assembly
   d. the pinions
2. What results when any two members of a simple planetary gearset are locked together to act as the input?
   a. reverse output
   b. direct 1:1 drive
   c. overdrive
   d. underdrive
3. When the planetary carrier of a simple planetary gearset is the held member:
   a. there is a reverse in rotation direction at the output member
   b. direct 1:1 drive occurs
   c. an overdrive condition results
   d. an underdrive condition results
4. What results if a large external tooth gear is meshed with and driving a smaller external tooth gear?
   a. Output speed increases; output direction is the same as input.
   b. Output speed decreases; output direction is the reverse of input.
   c. Output speed increases; output direction is the reverse of input.
   d. Output speed decreases; output direction is the same as input.
5. What is the function of multiple-disc clutches in a truck hydromechanical automatic transmission?
   a. to hold components stationary
   b. to lock components together to transfer torque
   c. to open and close hydraulic valves
   d. both a and b
6. In a close ratio transmission, all powerflows must pass through the ________ before reaching the output shaft.
   a. front planetary gearset
   b. rear planetary gearset
   c. low planetary gearset
   d. center sun gear shaft
7. What assists governor pressure in moving any shift signal shafts during an upshift?
   a. the priority valve
   b. the modulator valve
   c. the trimmer valve
   d. the inhibitor valve

8. What component helps avoid shift shock by reducing pressure to the clutch apply circuits during initial application and then gradually increasing the pressure?
   a. the priority valve
   b. the modulator valve
   c. the trimmer valve
   d. the inhibitor valve
9. Which type of hydraulic retarder is located between the torque converter and the transmission gearing?
   a. an input retarder
   b. an output retarder
   c. a torque retarder
   d. none of the above
10. Which type of hydraulic retarder applies a friction clutch while the unit is charging with hydraulic fluid?
    a. an input retarder
    b. an output retarder
    c. a torque retarder
    d. all of the above
11. Which of the following does the most to determine the position of the governor valve?
    a. input shaft speed
    b. output speed
    c. main pressure
    d. accelerator position
12. Which of the following does most to determine modulator pressure?
    a. engine speed
    b. torque converter turbine speed
    c. range shift position
    d. accelerator pedal position
13. Which of the following devices has an operating principle similar to that of a hydraulic retarder?
    a. a torque converter
    b. a planetary gearset
    c. a hydraulic clutch
    d. an S-cam brake
14. How many planetary gearsets are used in an Allison five-speed automatic transmission?
    a. two
    b. three
    c. four
    d. five
15. What regulates main pressure in an Allison automatic transmission?
    a. transmission input speed
    b. the main pressure regulator
    c. the trimmer valve
    d. the governor valve

19

Prerequisite: Chapter 18

AUTOMATIC TRANSMISSION MAINTENANCE

OBJECTIVES

After reading this chapter, you should be able to:

- Perform hot and cold transmission oil level checks.
- Identify the types of hydraulic fluid used in truck automatic transmissions.
- Change automatic transmission oil and filters.
- Inspect transmission oil for signs of contamination.
- Adjust the manual gear selector linkage, mechanical modulator control linkage, and air modulator control on a truck automatic transmission.
- Perform a transmission stall test.
- Perform engine speed and vehicle speed shift point tests.
- Describe basic transmission test stand procedure.
- Test the transmission valve body.
- Summarize some basic inspection and troubleshooting procedures for automatic transmissions.

KEY TERMS

adjusting ring
auxiliary filter
breather
engine stall point
stall test
TranSynd
valve body and governor test fixture
valve ring adjusting tool

INTRODUCTION

This chapter covers the periodic maintenance procedures required to keep hydromechanical automatic transmissions operating smoothly. As in Chapter 18, the emphasis is primarily on Allison transmissions. Although the focus in this chapter is on hydromechanical automatic transmissions, many of the procedures outlined also apply to electronically controlled automatic transmissions. This chapter is retained in this sixth edition because of the large number of hydromechanical automatic transmissions that remain in service and the fact that technician certification testing still requires technicians to understand its contents. Electronically controlled automatic transmissions are covered in Chapter 21.

SCOPE

The chapter includes some guidelines for inspection, care of the oil system and breather, checking temperatures and pressures, procedures for storage, linkage adjustment, and care of external coolers and piping. Procedures for adjusting shift points are included, along with some general considerations for overhaul work.

Specific disassembly, overhaul, and assembly instructions are beyond the scope of this textbook. Advanced troubleshooting and rebuilding of Allison automatic transmissions is a specialized skill. It is not that Allison and other automatic transmissions are that difficult to work on, but they require special tooling, support literature, and specialty training to repair them properly. These transmissions are expensive and the costs of misdiagnosis and comeback repairs can be considerable.

Students of truck technology programs are required to understand not just how automatic transmissions function but also how to perform some typical repairs, remove and reinstall reconditioned units, and make routine adjustments. Remember, however, that to work effectively on these transmissions, specialty training is recommended and it is essential to have the special tools. The procedures outlined in this chapter are those required for most hydromechanical Allison four- and five-speed transmissions.

19.1 INSPECTION AND MAINTENANCE

The transmission should be kept clean to make inspection and servicing easier. Clean with a shop pressure washer (being careful to direct the stream away from the breather), but avoid using solvents that could damage the aluminum housing. Inspect the transmission for loose bolts, loose or leaking oil lines, oil leakage, and the condition of the control linkage and cables. The transmission oil level should be checked regularly.

TRANSMISSION OIL CHECKS

Maintaining the proper oil level in an automatic transmission is very important. As explained in Chapter 18, transmission fluid or oil plays the following roles in an automatic transmission:

- It acts as the drive medium (transfers torque inside the torque converter).
- It acts as hydraulic media to apply clutches.
- It lubricates the transmission components.
- It cools the transmission components.

If the oil level is too low or too high, a number of problems can occur.

LOW OIL LEVEL

When the transmission oil level is low, oil will not completely cover the oil filter. This draws air into the pump inlet along with oil, which is then routed to the clutches and converter. The result of air in the hydraulic system is known as aeration. Air is compressible, so it compromises the operation of hydraulic circuits. The result is converter aeration, irregular shifting, overheating, and poor lubrication. Aeration of transmission oil alters its viscosity and changes its appearance to that of a thin, frothy liquid.

High Oil Level

At normal oil levels (FULL mark on dipstick with oil at operating temperature), the sump oil level should be slightly *below* the planetary gearsets (**Figure 19–1**). When the oil level is maintained above the FULL mark on the dipstick, the oil level in the sump rises so that the planetary gears run in oil, a condition that can cause foaming and aeration. Once again, aerated transmission fluid results in converter aeration,

FIGURE 19–1 Oil levels on a 4.3-inch (11 cm) transmission oil pan and a 5.1-inch (13 cm) oil pan

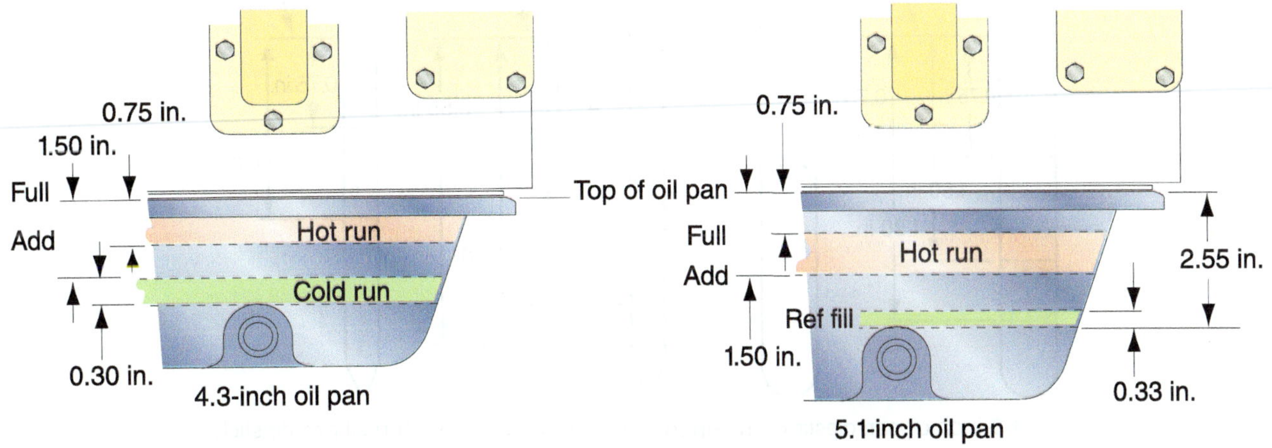

irregular shifting, overheating, and poor lubrication. If accidental overfilling occurs during servicing, the excess oil should be drained.

SHOP TALK

It should be noted that a defective oil filler tube seal ring will allow the oil pump to draw air into the oil from the sump, which will result in aeration of the oil.

Interpreting Oil Level Readings

Transmission input speed and oil temperature significantly affect the oil level. An increase in input speed will lower the oil level, whereas an increase in oil temperature will raise it. For these reasons, always check the oil level with the engine at its specified idle speed with the transmission in neutral. Both cold and hot level checks should be taken.

A cold level check is required to ensure that there is sufficient oil in the transmission until normal operating temperature is reached. The hot check is made when the transmission oil reaches normal operating temperature (160°F–200°F [71°C–93°C]) and is the more reliable of the two checks.

Dirt must not enter the filler tube when you are checking or adding oil to the transmission. Clean around the filler tube opening before removing it.

When testing transmission oil level, park the vehicle on a level surface and apply the parking brakes.

CAUTION

You should check the transmission oil level at least two times to ensure that an accurate reading is made. If the dipstick readings are inconsistent (some high, some low), check for proper venting of the transmission breather or oil filler tube. A clogged breather can force oil up into the filler tube and cause inaccurate readings. If the filler tube is unvented, the vacuum produced will cause the dipstick to draw oil up into the tube as it is pulled from the tube. Again, the result will be an inaccurate reading.

Cold Check

Operate the engine for approximately 1 minute to purge air from the system. Shift the transmission into drive, then to reverse, and then back into neutral. This charges the clutch cavities and hydraulic circuits with oil. The oil temperature should be between 60°F and 120°F (16°C and 49°C) for an accurate cold check. Wipe the dipstick clean and take the reading. If the oil level registers in the REF FILL (COLD RUN) band of the dipstick, the oil level is sufficient to run the transmission until a hot check can be made. Familiarize yourself with the appearance of the Allison dipstick (shown in **Figure 19–2**).

FIGURE 19–2 Oil pan dipstick calibration: 4.3-inch (11 cm) oil pan dipstick; 5.1-inch (13 cm) oil pan dipstick; and 7.0-inch (18 cm) oil pan dipstick

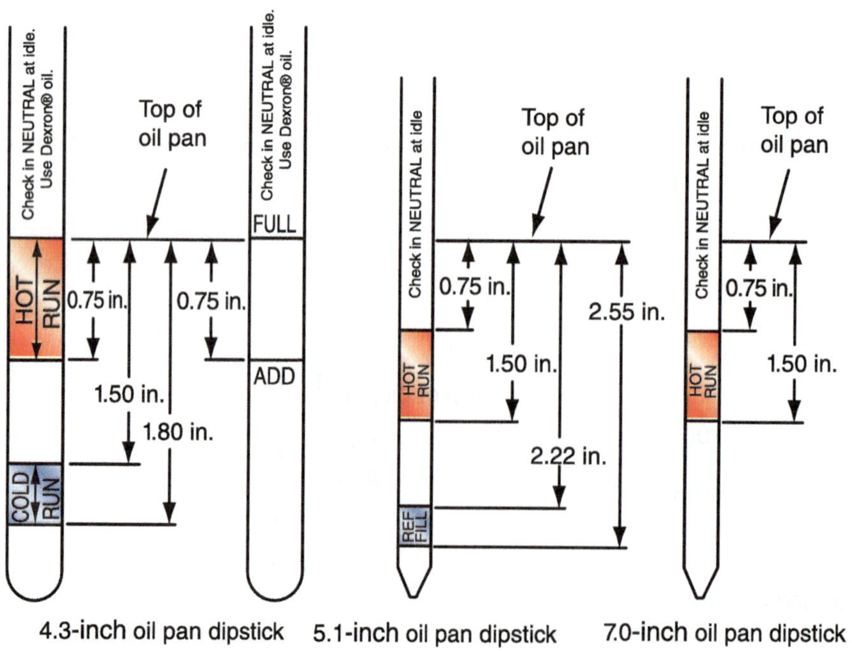

SHOP TALK

The REF FILL (COLD RUN) level is an approximate level and can vary with specific transmissions. To ensure proper operating levels, a hot oil level check must be performed.

If the oil level registers at or below the lower line of the REF FILL (COLD RUN) band, add oil to bring the level within the REF FILL (COLD RUN) band. Do not fill above the upper line of this band. If the oil level is above the upper line of the REF FILL (COLD RUN) band, drain some oil to correct the level. Next, you can operate the vehicle until the hot check temperature is reached. Then continue with the hot level check.

Hot Check

The oil temperature should be between 160°F and 200°F (71°C and 93°C) to make this test. With the engine at idle and the transmission in neutral, wipe the dipstick clean and check the oil level. If the oil level registers in the HOT RUN band (between ADD and FULL), the oil level is correct. If the oil level registers on or below the bottom line of the HOT RUN band or the ADD line, add oil to bring the level to the middle of the band. Note that 1 quart (liter) of oil will raise the level from the bottom of the band to the top of the band in most transmissions (from the ADD line to the FULL line).

Hydraulic Fluid Recommendations

Observe the original equipment manufacturer (OEM) hydraulic fluid specifications. For example, several automatic transmission manufacturers recommend Dexron, Dexron II, Dexron III, and type C-4 (ATD approved SAE 10W or SAE 30) oils for their automatic transmissions. Type C-4 fluids are the only fluids usually approved for use in off-highway applications. Type C-4 SAE 30 is specified for all applications in which the ambient temperature is consistently above 86°F (30°C). Some, but not all, Dexron II fluids also qualify as type C-4 fluids. If type C-4 fluids are to be used, check that the materials used in auxiliary equipment such as tubes, hoses, external filters, and seals are C-4 compatible.

Allison currently recommends the use of **TranSynd** synthetic oil in all their transmissions. TranSynd, formulated jointly by Allison and Castrol, can extend oil drain intervals by three times.

Cold Startup

The transmission should not be operated in forward or reverse gears if the transmission oil falls below a certain temperature. Minimum operating temperatures for recommended fluids are as follows:

TranSynd	–10°F (–23°C)
Dexron (I, II or III)	–10°F (–23°C)
Type C-4 SAE 10W	10°F (–12°C)
Type C-4 SAE 30	32°F (0°C)

When the ambient temperature is below the minimum fluid temperatures listed and the transmission is cold, preheat is required either by using auxiliary heating equipment or by running the engine in neutral at idle for at least 20 minutes to reach the minimum operating temperature. Failure to observe the minimum temperature limit can result in transmission malfunction or reduced transmission life.

OIL AND FILTER CHANGES

Change intervals are specified by Allison but must take into account Allison recommendations and the chassis application. Typical filter and oil change intervals are 25,000 miles (40,000 km) or 12 months for on-highway trucks and 1,000 hours or 12 months for off-highway applications. When the Allison-recommended TranSynd synthetic oil is used, oil change intervals can be extended by approximately 300 percent, depending on application.

Some operating conditions may require shorter oil and filter change intervals. Oil should be changed when there is visible evidence of dirt or high-temperature breakdown in the oil. High-temperature breakdown is indicated by discoloration and burned odor and can be corroborated by oil analysis. Severe operating conditions might require more frequent service intervals.

Oil Change Procedure

The key to oil change procedures is cleanliness. Contaminants should be prevented from getting into the transmission by handling the oil in clean containers and using clean fillers and funnels.

CAUTION

Containers or transfer devices that have been used for engine coolant solutions must not be used for transmission fluid. Antifreeze contains ethylene or propylene glycol, which, if introduced into a transmission, can cause the clutch plates to fail.

Place the dipstick on a clean surface when filling the transmission and keep replacement filters and seals in their packaging until ready for installation. Before draining the transmission, it should be at an operating temperature of 160° to 200°F (71°C to 93°C). Shift the gear selector to neutral before draining the oil from the sump pan.

Allison transmissions are equipped with either a standard or a heavy-duty oil pan (**Figure 19–3**). When changing oil and filters on a standard pan, the entire oil pan is removed from the base flange of the transmission to access the filter. The heavy-duty oil pan does not have to be removed, because the filter is accessed through an opening in the side of the oil pan.

FIGURE 19–3 Typical transmission housing with a standard and a heavy-duty oil filter and oil pan configuration

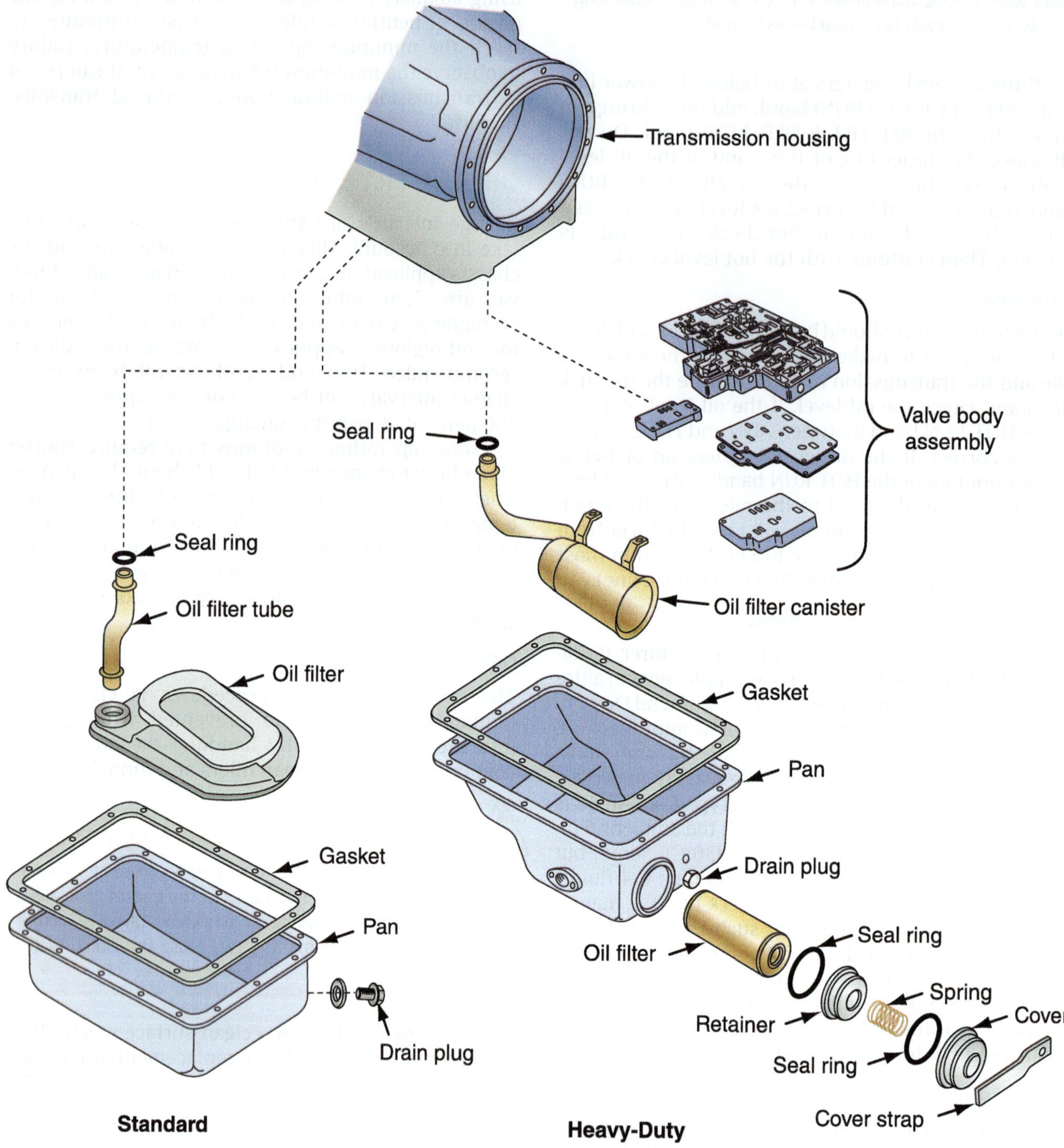

Standard Pan. The oil and filter change on an Allison transmission with a standard oil pan should be performed as follows:

1. Remove the oil drain plug and gasket from the right side of the oil pan. Allow the oil to drain.
2. Remove the oil pan, gasket, and oil filler tube from the transmission. Discard the gasket. Thoroughly clean the oil pan.
3. Remove the oil filter retaining screw. Remove the old filter and discard it. The oil filter intake

pipe is not part of the filter. It should be kept for reinstallation.

4. Install the filter tube into the new filter assembly. Install a new seal ring onto the filter tube. Lubricate the seal ring with transmission oil. Install the new oil filter, inserting the filter tube into the hole in the bottom of the transmission. Secure the filter with the screw torqued to 10- to 15 lb-ft. (14–20 N·m).

CAUTION

Do not use gasket-type sealing compounds anywhere inside the transmission. If grease is used for internal assembly of transmission components, use only greases approved by Allison.

5. Place the oil pan gasket onto the oil pan. Sealant should only be applied to the area of the oil pan flange outside the raised bead of the flange.
6. Install the oil pan and gasket, carefully guiding them into position. Ensure that no dirt or other material enters the pan during installation. Turn in the oil pan retaining screws by hand and then torque them sequentially to specification.
7. Install the filler tube at the side of the pan and tighten the tube fitting to specification.
8. Install the drain plug and gasket and tighten the plug.
9. Fill the transmission with oil to the proper level. Follow the OEM specifications for oil capacity. Recheck the oil level as described earlier in this chapter.

Heavy-Duty Oil Pan. The oil and filter change on an Allison transmission with a heavy-duty, high-capacity oil pan should be performed as follows:

1. After getting the transmission to operating temperature, remove the oil drain plug from the rear or side of the pan. Allow the oil to drain.
2. Remove the bolt and nut securing the filter-retaining strap. Remove the filter cover, spring, and retainer.
3. Remove the filter. Discard the filter retainer and cover O-ring gaskets.
4. Install the seal ring on the filter retainer, and then install the filter and retainer into the external access canister.
5. Install the cover seal ring onto the lip of the oil pan.
6. Install the spring into the cover and place it over the filter retainer.
7. Install the strap, bolt, and nut to the pan and torque to specification.
8. Install the drain plug and tighten.
9. Fill the transmission with oil to the proper level. Follow the OEM specification for oil capacity. Recheck the oil level.

GOVERNOR FILTER CHANGE

Allison recommends that the governor be inspected or replaced at every oil/filter change. A pipe plug can be used to retain the governor oil screen in older model Allison transmissions, as shown in **Figure 19–4**. Remove the pipe plug and inspect the filter. If it is undamaged, clean it in mineral spirits and reinstall it. If it is damaged, replace it. Install the filter open end first into the transmission cover and reinstall the pipe plug.

In later-generation hydromechanical transmissions, a hexagon plug and O-ring seal are used to retain the governor oil filter in the transmission rear cover, as shown in **Figure 19–5**. On others, the hexagon plug and filter location is on the output retarder. In both cases, the O-ring and filter should be changed at every oil change. Remember to torque the plug to specifications.

OIL CONTAMINATION

At each oil change, examine the oil drained from the transmission for evidence of dirt or water. Trace condensation is normal; this will emulsify in the oil during operation. However, if there is visible evidence of water in the oil, check the transmission cooler for internal leakage between the coolant and oil circuits. Oil in the coolant circuit of the transmission cooler is another indication of leakage.

FIGURE 19–4 Location of the governor filter on early model transmissions

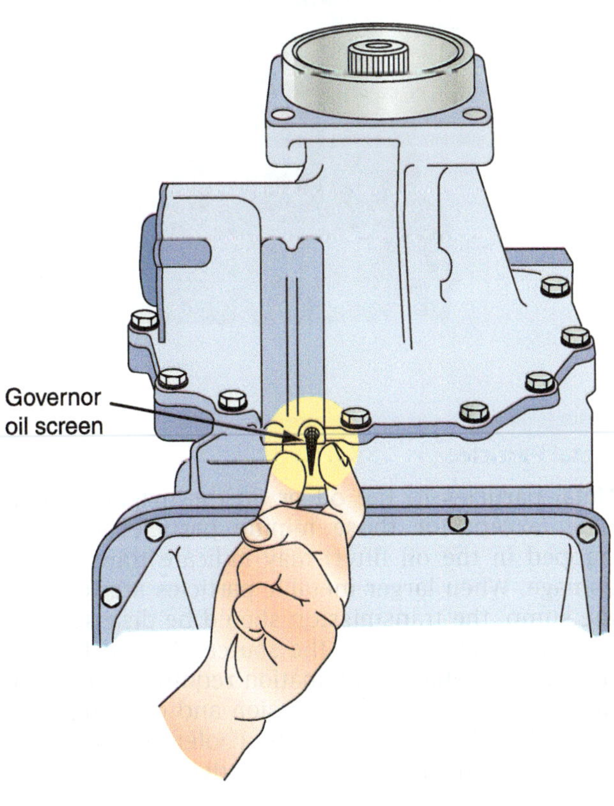

FIGURE 19–5 Location of the governor filter on later model transmissions

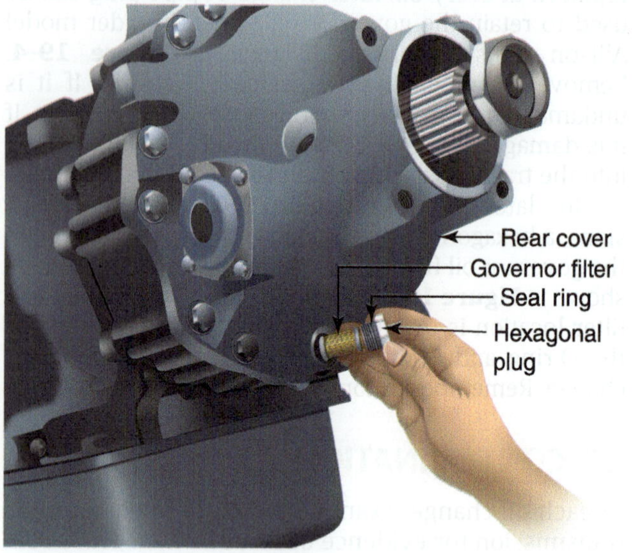

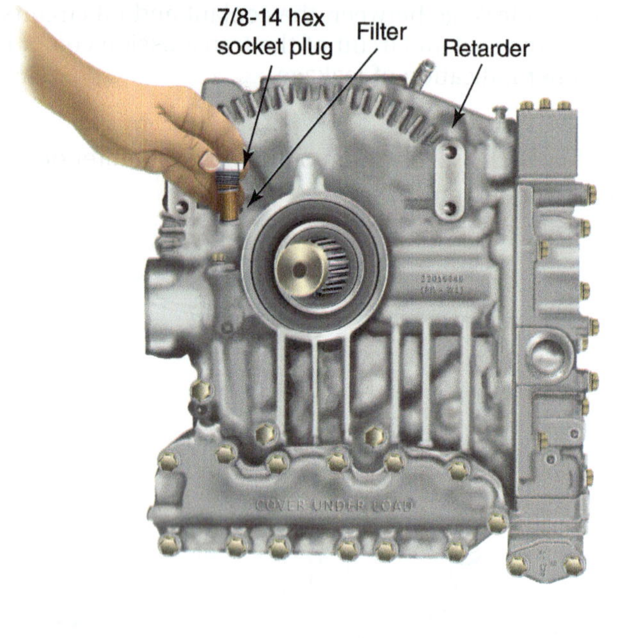

Metal Particles

Metal particles in the oil or on the magnetic drain plug (except for those minute particles normally trapped in the oil filter) may indicate transmission damage. When larger metallic particles are found in the sump, the transmission should be disassembled and inspected to locate the source. Beyond locating the cause, metal contamination requires a complete disassembly of the transmission and cleaning of all internal and external circuits, cooler, and all other areas where the particles could lodge.

CAUTION

Metal contamination normally requires the replacement of all the transmission bearings.

Coolant Leakage

If engine coolant leaks into the transmission oil circuit, action should be taken to prevent malfunction and complete failure. The transmission should be disassembled, inspected, and cleaned. All traces of the coolant and varnish deposits that result from coolant contamination should be removed.

Test kits can be used to detect traces of glycol in the transmission oil. You should note, however, that certain additives in some transmission oil can produce a positive reading. In the event of questionable test results, therefore, use a lab analysis of the oil.

Breathers

The **breather** is located at the top of the transmission housing, as shown in **Figure 19–6**. It prevents pressure buildup within the transmission and should be kept clean and unplugged. Exposure to dust and dirt will determine the frequency at which the breather requires cleaning. Use care when cleaning the transmission. Spraying steam, water, or cleaning solution directly on the breather can force water and cleaning solution into the transmission.

Auxiliary Filters

Some vehicles are equipped with an **auxiliary filter**, which is installed in the oil return line between the oil cooler and the transmission. This auxiliary filter is normally installed after debris or dirt has been introduced into the oil system because of failure. When such a failure or accident occurs, complete cleanup of the system cannot be ensured, even with repeated system flushing. The auxiliary filter, which should be installed before the vehicle is placed back in service, prevents debris from being circulated into the transmission and causing a repeat failure. This applies whether the transmission is overhauled or replaced with a new or rebuilt unit.

Most auxiliary oil filters are changed 5,000 miles (8,000 km) after their initial installation and at regular oil change intervals thereafter. High-efficiency filters are available for older transmissions and although these have no mileage limitations, they are not recommended for World Transmissions (WT). When used on older transmissions, high-efficiency filters should be changed when they become clogged or at 3-year intervals, whichever occurs first. High-efficiency filters have a differential pressure switch that monitors pressure drop across the filter. It triggers a dash-mounted warning light if pressure exceeds a certain limit, indicating that the filter is clogged.

FIGURE 19–6 Breather location on top of a transmission housing; also note the position of the manual selector shaft, neutral start switch, modulator control opening, pressure taps, plugs, and other key components.

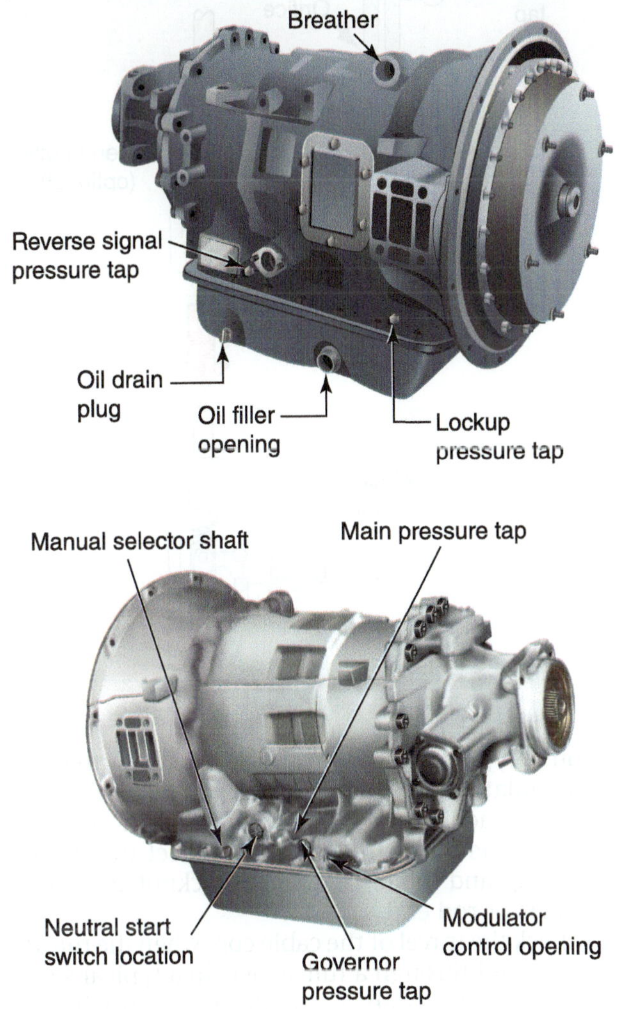

CAUTION

Allison does not recommend the use of extended service high-efficiency filters on any WT.

External Lines and Oil Cooler Inspection

Inspect all lines for loose or leaking connections, worn or damaged hoses or tubing, and loose fasteners. Examine the radiator coolant for traces of transmission oil. This condition may indicate a defective heat exchanger.

Extended operation at high operating temperatures can cause clogging of the oil cooler and can lead to transmission failure. The oil cooler system should be thoroughly cleaned after any rebuild work is performed on the transmission.

Transmission Oil Summary

We finish this section on transmission oils by outlining three myths highlighted in a recent Allison newsletter:

- **Myth 1: Oils are interchangeable.** Additive packages in lubricants made by different companies vary. Two oils with the same viscosity and basic properties can have additive packages that can conflict and result in breakdown. Some synthetic oil additives are incompatible with mineral-based stocks; therefore, check with Allison if you are swapping from one to the other.
- **Myth 2: Oil never wears out.** Oxidation limits the life of transmission oils, and the extent to which the oil is oxidized depends on running temperatures. Oil begins to oxidize at 150°F (65°C) and for each 25°F (10°C) increment rise above that temperature oil life is reduced by half.
- **Myth 3: Snake oil additives improve lubrication.** Major OEMs have extensive research and engineering capabilities when proofing lubricants and precisely balance the additive package to maximize operational life. Tampering with the additive package by dumping in additives of unknown chemistry unbalances the oil and can result in premature failure.

MANUAL SELECTOR LINKAGE INSPECTION

Proper adjustment of the manual gear range selector valve linkage is important. The shift tower detents must correspond exactly to those in the transmission. Failure to obtain proper detent in drive, neutral, or reverse gears can affect the supply of transmission oil at the forward or fourth (reverse) clutch adversely. The resulting low-apply pressure can cause clutch slippage and decreased transmission life.

Periodic inspections of the linkage should be made for bent or worn components, loose fasteners, and the accumulation of dirt and grease. Clean and lubricate all moving joints in the linkage. The manual selector lever should move easily and provide a firm detent in each position. The linkage should be adjusted so the stops in the shift tower match and are positioned by the detents in the transmission. When the linkage is correctly adjusted, the pin that engages the shift lever linkage at the transmission can be moved freely in each range.

Adjusting the Shift Linkage

The procedure for checking and adjusting the shift linkage in an automatic transmission is as follows:

1. Place the range gear selector control in the neutral position (**Figure 19–7**). Next, move the selector lever on the transmission to the neutral position.

FIGURE 19–7 Manual gear selector range positions and ratios optioned in each position.

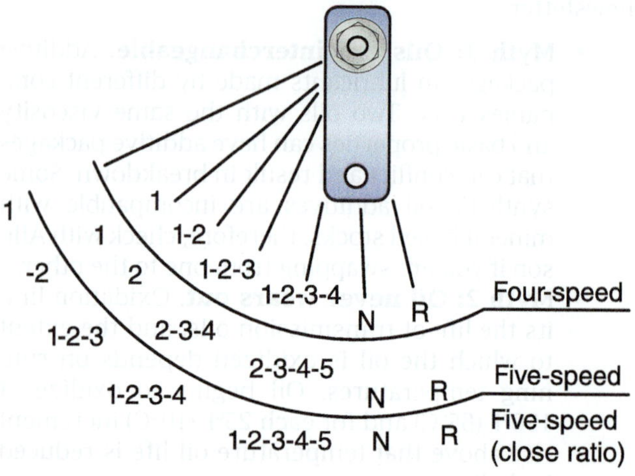

Adjust the linkage so that it matches the selector lever on the transmission. Connect the linkage to the selector lever.

2. Now shift through all selector positions, checking at each to ensure that the valve body detent positions correspond to the respective selector positions.
3. Check the neutral-start switch (see Figure 19–6 for location). The starter should not operate except when the range selector is at neutral.

MECHANICAL MODULATOR CONTROL

Most diesel-powered trucks use a mechanical modulator control linkage. Its function is to interpret a request for acceleration power and modulate shifting accordingly. To install and adjust a mechanical modulator linkage, you should use the following procedure:

1. Connect the throttle pedal end of the modulator cable housing to its mounting as shown in **Figure 19–8**. (Check Figure 19–6 for the modulator control location on the transmission.)
2. Fully depress the accelerator pedal and check to see whether the throttle linkage pushes or pulls the cable core while the linkage moves toward full travel. If it pushes the cable core, push the cable core farther until it reaches the end of its travel. If movement of the throttle linkage toward full-throttle position pulls the cable, pull the cable to the end of its travel.
3. Adjust the clevis or rod end on the cable core until it registers with the hole in the throttle linkage lever and the connecting pin can be freely inserted. With the pin removed, rotate the clevis one additional turn counterclockwise (viewing cable core from its end) for the pull-type arrangement, or one additional turn clockwise for the push-type arrangement. This ensures that the modulator does not prevent the throttle lever from reaching its fully actuated position. Install the clevis pin or rod end to connect the throttle linkage and cable. Tighten the locknut against the clevis or rod end.
4. Check the travel of the cable core when the throttle is moved through a full stroke. In a typical setup, the travel distance should be approximately 11.8 inches (30 cm) minimum and 19.16 inches (49 cm) maximum.
5. Some mechanical controls include a support bracket secured by the retainer bolt.
6. Check the cable routing. Bends must not be tighter than a 1-inch (25 mm) radius. The cable should be kept at least 6 inches (150 mm) away from the engine exhaust system components. The cable should follow throttle linkage movement. It is sometimes necessary to add a spring to ensure that this movement occurs smoothly through the pedal stroke.

FIGURE 19–8 Mechanical modulator control linkage operation

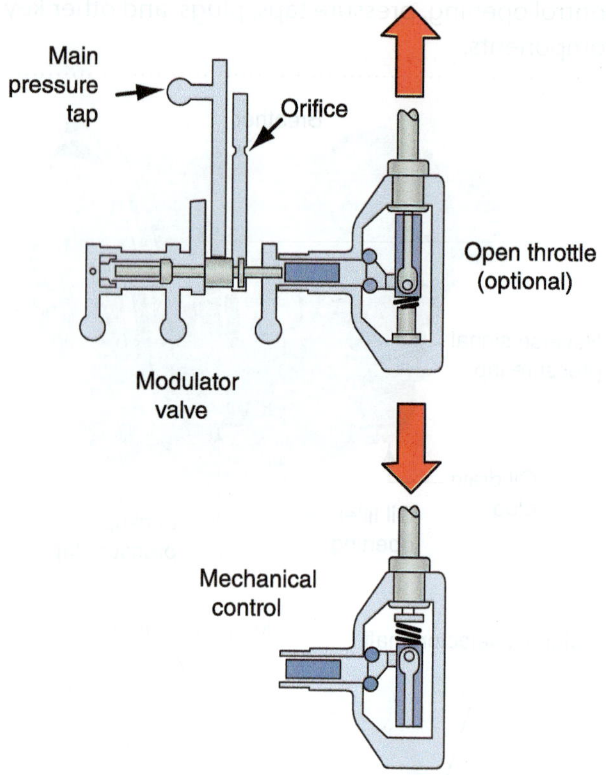

STALL TESTING

A **stall test** is used for troubleshooting some types of transmission problems. It is used to determine whether the engine or transmission is at fault when there appears to be a powertrain problem. The objective of a stall test is to drive the torque converter pump at its maximum or a specified speed while the turbine is held stationary, that is, fully stalled. Maximum vortex flow occurs in the torque converter during a stall test. Whenever a stall test is performed, the vehicle must be prevented from moving.

WARNING

To perform a transmission stall test, both the vehicle parking and service brakes should be properly adjusted and fully applied. The wheels over at least two axles must be chocked as an extra precaution. It also makes sense to chain the vehicle to floor anchors. No one should stand either in front of or behind the vehicle during the test.

You will have to locate the engine OEM stall test specifications. The **engine stall point**, expressed in rpm, should be used to analyze the test data. Make sure you consult this, especially when an electronically controlled engine is coupled to a hydromechanical automatic transmission.

Temperature readings at the converter-out (to cooler) line should be monitored during the test and through a 2-minute cool-down period. This helps verify the performance of the transmission cooling system.

CAUTION

Stall tests are usually specified for 15 seconds. Full stall should never be run for a period exceeding 30 seconds at any one time because of the rapid rise in oil temperature. Converter-out temperatures should not exceed 300°F (150°C). If the stall test has to be repeated, allow for a cool-down period. Monitor the engine temperature as well.

Test Procedure

The following steps outline a typical stall test used on a heavy-duty automatic transmission (make sure you consult the OEM service literature before undertaking the test):

1. Connect a tachometer of known accuracy to the engine.
2. Install a temperature probe into the converter-out (to cooler) line.
3. Start the engine and bring the transmission oil up to its normal operating range temperature (160°–200°F [71°C–93°C]).
4. Apply both the parking and service brakes. Block the wheels on two axles, including one drive axle, and if floor anchors are available, chain the vehicle to the floor. Ensure that all personnel are out of the way to prevent injury in the event of brake or restraint failure.
5. With the vehicle secured against movement, shift the selector control to any forward range with the exception of D1 or DR in deep ratio models. The high torque in this low gear may damage the transmission or the vehicle driveline. The stall test also can be performed in reverse if needed.
6. Accelerate the engine to the specified stall test speed. This may be engine-rated speed or some other engine speed in torque rise. The converter-out temperature should begin to climb rapidly. When the temperature reaches a minimum of 255°F (124°C), record the engine rpm when at rated speed. Release the throttle and shift to neutral.

WARNING

Never allow the converter-out oil temperature to exceed 300°F (150°C).

7. Immediately after the engine rpm has been recorded, note the converter-out temperature.
8. With the transmission remaining in neutral, run the engine at 1,200 to 1,500 rpm for 2 minutes and record the converter-out temperature.

Stall Test Results

Problems are identified by comparing stall test rpm readings with the engine and transmission OEM specifications. When interpreting test results, remember that environmental conditions such as ambient temperature, altitude, and engine parasitic loads all have an impact on engine performance and torque converter efficiencies. Because of these variables, stall speed deviations of 150 rpm from specification can be accepted as falling within normal range. Test result indicators can include the following:

1. Engine stall speed is more than 150 rpm below the OEM specification: indicates an engine problem. Continue by troubleshooting the engine.
2. Engine stall speed is more than 150 rpm above the OEM specification: indicates a transmission-based problem. This could include a range of problems, for example, slipping clutches, cavitation, or torque converter failure.
3. An extremely low stall speed, such as 30 percent of the specified engine stall rpm, with no engine tattletales such as smoking, could indicate a freewheeling torque converter stator.
4. If the engine stall speed conforms to specification but the transmission oil overheats, reference the cool-down check. If the oil does not cool down during the 2-minute cool-down check, a stuck torque converter stator could be indicated.

19.2 SHIFT POINT ADJUSTMENTS

In Chapter 18, we looked at how automatic transmissions upshift and downshift within a predetermined input (engine rpm) window and vehicle road speed ranges. Two key adjustments determine when these automatic shifts occur:

- Shift signal valve spring pressure rate
- Modulator spring force adjustment

To adjust either of these springs requires dropping the oil pan and removing the valve body.

The spring force of each individual gear range shift signal valve can be increased or decreased using a special tool. The modulator valve spring is adjusted in a similar way.

You can use these three methods of checking the transmission shift points:

- Engine rpm
- Speedometer readings
- Test stand tailshaft speed (transmission output speed)

You can use any of these methods to troubleshoot shift point irregularities and also use them to make adjustments. The engine rpm and speedometer reading test methods require using a chassis dynamometer or taking a test drive. Record the road speed and engine rpm. You can then adjust the shift valve tension to bring any out-of-range shift points into the required specification window.

The test stand method is the most accurate. This test is performed using a valve body and governor test stand, a specialized piece of test equipment. The valve body of the transmission is removed from the vehicle and mounted into the test stand. The test stand is designed to replicate a full range of vehicle running conditions so that the valve body can be precisely tested and calibrated.

The transmission shift points cannot be properly adjusted if the wrong governor is installed. Cross-check the part number on the governor with the parts listing for the transmission. Whatever method is used to test the shift points, you will have to perform the following steps:

1. Warm the transmission or test stand setup to normal operating temperature (160°–200°F [71°C–93°C]).
2. Check the engine no-load governor setting and adjust, if required, to conform to the transmission engine speed requirements.
3. Check engine performance before making any shift point adjustments.
4. Check the accelerator pedal linkage that controls the modulator valve mechanical actuator on diesel engines.
5. Check the shift selector linkage for proper range selection.
6. Test the accuracy of the dash-mounted tachometer using a test tachometer. Adjust dash readings to account for any error. Check the accuracy of the speedometer against another vehicle.

ENGINE SPEED (RPM) METHOD

To achieve the most accurate results, the vehicle should be loaded to its loaded operating weight. During the initial part of the test drive, note and record the full-load governed speed (rpm) of the engine. You may have to disable progressive shifting programming with some electronic engines. Governed or rated speed is the usual reference speed from which checks and adjustments are made. We use the term *rated speed* in the description that follows.

Shifts should occur at a specific engine rpm below rated speed. For example, on a typical four-speed automatic transmission, the first-to-second gear shift may be calibrated to occur at 400 rpm less than rated speed. The second-to-third shift may be set to occur at 300 rpm less than rated speed, and the third-to-fourth shift may be set to occur at 200 rpm less than rated speed. Consult the OEM service literature for the specific shift points calibrated for the model you are testing. **Table 19–1** shows some typical shift points relative to engine rated speed.

On close ratio (CR) models, the 1–2 upshift should occur at approximately 2 mph (3.2 km/h) below the lockup engagement speed. To establish this lockup engagement point, install a pressure gauge at the transmission lockup pressure tap (see Figure 19–6). With the selector in D1 (hold), note when the gauge needle moves and record the engine rpm. This is the engagement point for the 1–2 shift.

The first step in adjusting shift points is to determine the rated speed of the engine. In older hydromechanical engines, this is a hard value set by the engine governor. It is often displayed on the engine identification plate. Many hydromechanical Allison transmissions are coupled to electronically controlled engines. On these, you should use an electronic service tool (EST) to determine what the programmed rated speed is and whether any kind of progressive shifting or torque limiting programming is written into the software.

When you have established the engine rated speed, run the engine with the vehicle stationary and place the selector in drive. Begin the test drive and accelerate the engine to proceed through the shifts. Note the engine speed (at full throttle) at which the various shifts occur. Compare these shift points to the desired shift points listed in the transmission service literature.

Adjusting Shift Points

Adjustments to shift points can be made based on the engine rpm readings recorded during the test drive. Shift speeds can be altered by changing the positions

TABLE 19–1 Shift Point Engine Speeds for Select Transmissions*

(All speeds listed are subtracted from rated or full-load engine governed speed.)				
Shift	AT Series	MT Series (4-Speed)	MT Series (5-speed)	HT Series (all)
1–2	Less 400 rpm	Less 600 rpm	**	
2–3	Less 300 rpm	Less 200 rpm	Less 600 rpm	Less 100 rpm
3–4	Less 200 rpm	Less 200 rpm	Less 200 rpm	Less 100 rpm
4–5	Less 200 rpm	Less 100 rpm		

*Selected models and examples only. Check appropriate service literature to confirm speeds.

**On close ratio models, the 1–2 upshift should occur at approximately 2 mph (3.2 km/h) below the lockup engagement speed. To establish lockup engagement point, install pressure gauge at transmission lockup pressure tap. With the selector in D1 (hold), note the instant the gauge needle moves and record engine rpm. This is the engagement point.

of the adjusting rings that determine the application force of valve springs in the valve body. An exploded view of the valve body components is shown in **Figure 19–9**. Adjustment of the valves is performed using a **valve ring adjusting tool**. This tool has two small pins mounted parallel to each other. The pins are placed against the adjusting ring and pressure is applied inward to depress the spring and ring.

The **adjusting ring** is held in the shift signal valve bore by a pin that is press-fit through the valve body housing. When the ring is depressed by the adjusting tool, the slots on the ring that engage the pin are released. The adjusting ring can then be turned to adjust spring pressure. Clockwise rotation increases spring pressure; counterclockwise rotation decreases spring pressure. The slots on the adjusting ring that engage the pin are on a sloping ramp around the circumference of the adjusting ring. This alters the spring tension as the slots reengage the pin in the new ring position. A sectional view of this operation is shown in **Figure 19–10**.

Each notch in the adjustment ring will alter the shift point by an incremental rpm value. For example, one notch might be equal to 25, 35, 40, or 50 rpm of engine speed, based on the particular model transmission. Confirm the notch values by checking the service literature. If engine speed fails to reach the specified rpm before an upshift occurs, the shift point can be raised by increasing the spring force on the required shift signal valve (1–2, 2–3, 3–4, 4–5). If the upshift occurs at a higher than specified rpm or does not occur at all, the spring force will have to be reduced.

If more than one shift signal valve spring requires adjustment in the same direction, it might be necessary to adjust the spring force on the modulator valve in the same direction. If the modulator valve is not adjusted, closed throttle downshifts can be abnormally high or low, depending on the direction in which the shift signal adjusting rings were rotated. If all full-throttle upshift points are too low by approximately the same amount, check the adjustment on the modulator external linkage.

SHOP TALK

Only adjust the spring tension on those valves that fail to upshift at the appropriate speeds. You mainly should check the performance of the valve body.

SPEEDOMETER METHOD

When an accurate tachometer is not available for checking shift points, you can use the vehicle speedometer to check and adjust shift points. Begin the test by checking the top speed the vehicle can achieve in each gear before a shift occurs, remembering that some electronically managed engines may have programming that makes these values soft. On four-speed transmissions, check top speed in first, second, and third gears. On five-speed transmissions, check top speed in second, third, and fourth gears. (On some five-speeds, there might not be a third-gear selector position.)

When the top vehicle speed has been recorded for each gear, accelerate the vehicle at full throttle from a standing start and note the speed at which each upshift occurs. Compare the upshift rpm with the selected shift speeds recorded in the first part of the test drive. On four-speed transmissions, the 2–3 upshift should occur at approximately 2 mph (3.2 km/h) below the top speed for second gear. The 3–4 upshift should occur at about 2 mph (3.2 km/h) below the third-gear top speed. The 3–4 and 4–5 upshifts for five-speed transmission should occur about 2 mph (3.2 km/h) below third- and fourth-gear top speeds, respectively.

The 1–2 shift in four-speeds and the 2–3 shift in five-speeds are not adjusted relative to the top speeds. Instead, the 2–1 downshift (four-speeds) and 3–2 downshift (five-speeds) should occur, with closed throttle, at 3 to 5 mph (5 to 8 km/h). Speedometer or vehicle speed is a less direct indicator of engine speed than a tachometer reading. Although no exact relationship between vehicle speed and notch increments exists, the same general adjustment principles apply as in rpm adjustment. If the upshift vehicle

FIGURE 19–9 Low shift valve body assembly

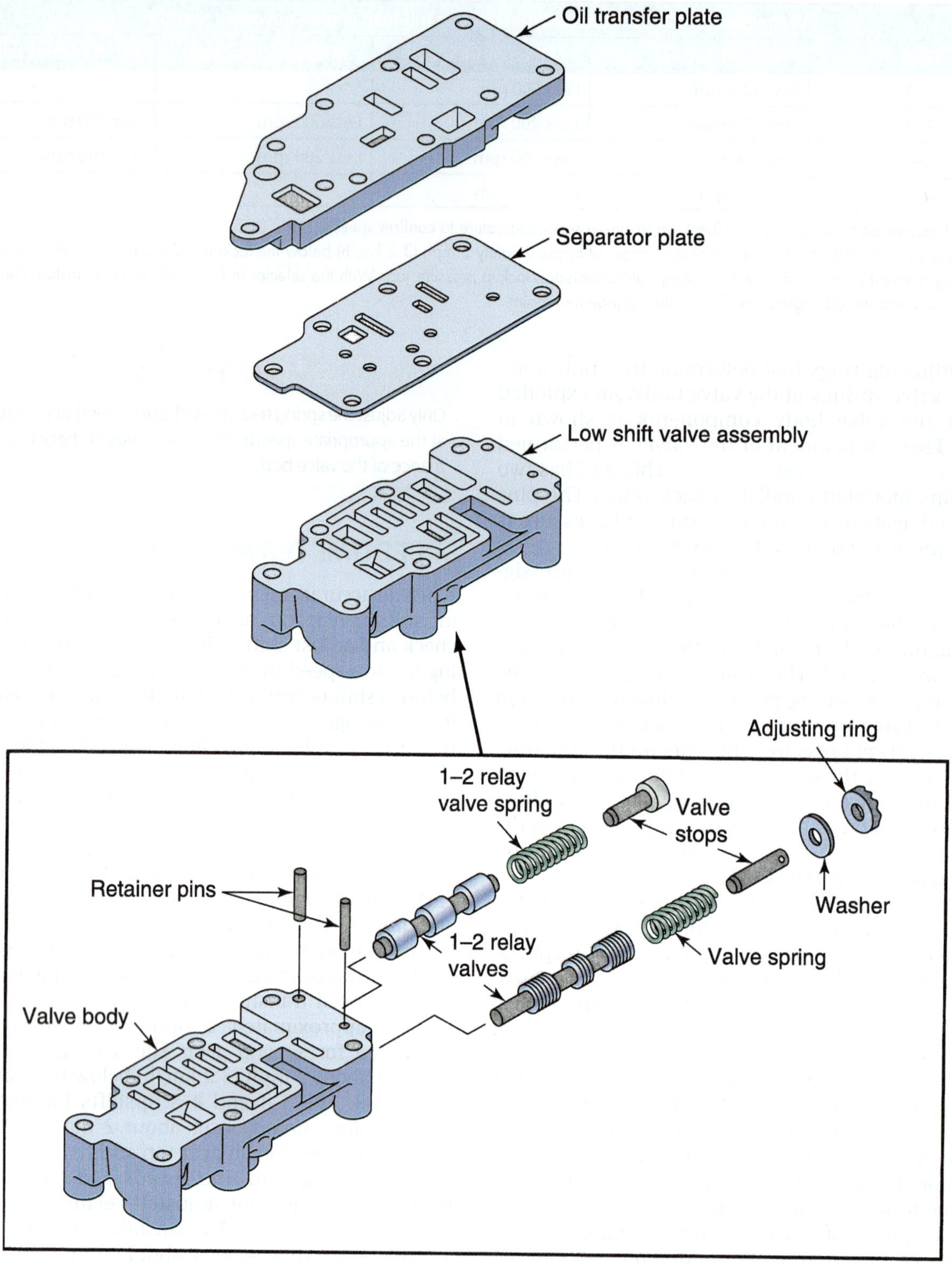

FIGURE 19–10 Adjusting spring retainer position using an Allison adjustment tool

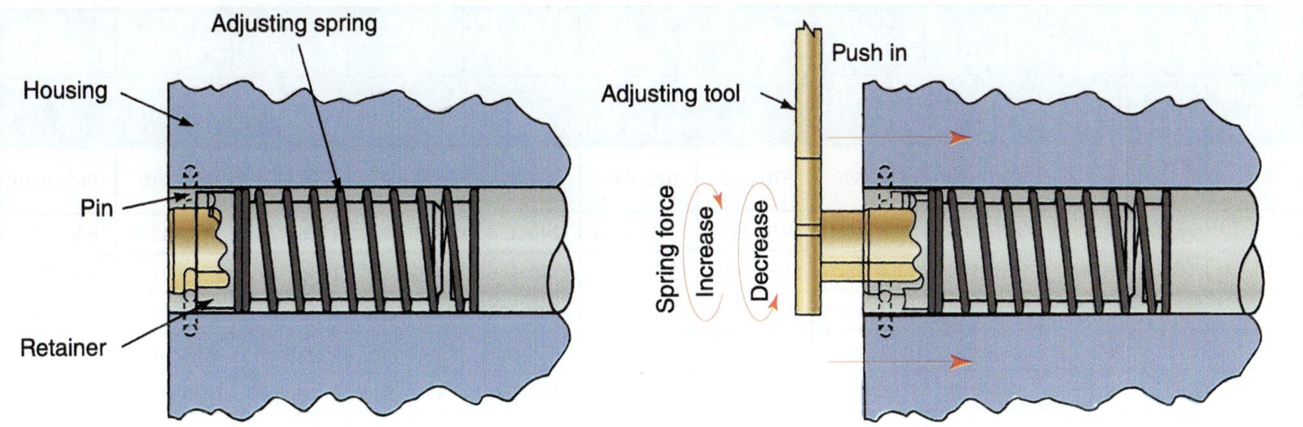

speed is below specification, the shift point can be raised by increasing spring tension on the appropriate shift valve. If the vehicle fails to shift or shifts at the maximum rpm for that gear, the shift point should be lowered by decreasing spring tension on the appropriate shift valve.

TEST STAND CALIBRATION

This is the preferred and most accurate method of checking shift points. **Figure 19–11** illustrates the **valve body and governor test fixture** used to check the operation of some transmission valve bodies. The valve body is bolted onto a manifold that resembles the lower portion of the transmission housing. This manifold is drilled and tapped to mate to all the ports and hoses required to route oil properly through the valve body for testing.

The manifold should be mounted to the test stand and all lines connected to it. The test stand can now be used to check five principal transmission valve body functions:

- Governor pressure
- Modulator pressure
- Hold regulator pressure
- Shift points (up-down/inhibit)
- Trimmer regulator operation

To test the operation of the valve body components, the test stand replicates transmission output shaft speed. The governor is gear driven from the test stand. A variable resistor control is used to increase or decrease output shaft rpm, which is displayed on a digital tachometer.

The applicable service literature provides tables listing the desired output shaft rpm levels for each shift. We provide an example of one of these specification tables used for transmissions driven by engines with rated speeds from 2,200 to 4,000 rpm in **Table 19–2**.

Output shaft speed is controlled on the test fixture using the variable resistor control. This indicates the rpm at which shifts occur should be recorded. Adjustments are then made to these shift valve springs, if required. The adjustment procedure is the same as described in the "Adjusting Shift Points" section earlier in this chapter.

During testing, main pressure to the clutch shift signal and relay valves in the valve body is controlled by the test stand main pressure regulator. An electric heater in the test bench is used to raise and maintain the fluid to normal operating temperature. With the main pressure at 150 psi (10 bar) and oil temperature at 100°F (38°C), governor output pressure can be checked at four output shaft speeds: 440, 800, 1,500, and 2,200 rpm, as shown in **Table 19–3**. No adjustment is possible for the governor assembly internal valve, but it can be removed and cleaned or replaced to ensure specified performance.

FIGURE 19–11 Automatic transmission test stand used for testing valve bodies and governors.

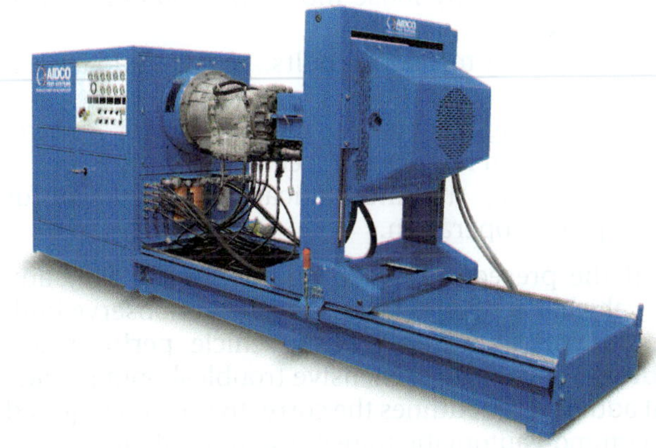

Power Test, Inc.

TABLE 19–2 Shift Points without Modulated or Adjustable Lockup

Engine Full Load Governed Speed (rpm)			2200	2400	2600	2800	3000	3200	3400	3600	4000
Throttle Position	Range	Shift	Transmission Output Speed (rpm) at Start of Shift								
Full	DR4 or DR2	1–2	530–590	530–590	570–630	625–685	685–740	740–795	795–855	850–910	965–1020
	DR4 or DR3	2–3	875–1015	875–1015	970–1110	1110–1250	1240–1380	1255–1400	1420–1565	1445–1590	1640–1780
		3C–3LU	1050–1400	1190–1600	1190–1600	1360–1615	1310–1600	1580–1870	1580–1870	1745–2155	2160–2450
	DR4	3–4	1475–1625	1580–1725	1725–1870	1870–2010	2085–2230	2160–2300	2300–2445	2445–2590	2735–2880
	DR3	4–3	1560 min	1850 min	1910 min	2020 min	2370 min	2440 min	2550 min	2720 min	3090 min
	DR2	3–2	945–1300	1190–1470	1280–1580	1340–1580	1460–1900	1530–1930	1620–2000	1730–2120	2080–2460
	DR1	2–1	660–760	770–960	740–1080	775–1025	910–1100	930–1130	960–1200	1040–1280	1200–1590
Closed	DR4	4–3*	150–390	150–390	250–675	20–675	225–640	425–780	555–855	20–590	570–950
		3–2*	370–470	370–470	390–615	355–640	445–700	510–780	550–840	350–700	600–815
		2–1	20–300	20–300	20–470	20–535	205–530	285–550	365–550	20–535	345–550

*4–2 Downshift is acceptable.

TABLE 19–3 Pressure Schedule for Select Governor Models

	Governor Output Pressure (psi)			
Governor Assembly	440 rpm	800 rpm	1500 rpm	2200 rpm
#1	3.5–6.5	17.5–20.5	42.8–48.4	58.6–64.8
#2	3.5–7.5	16.0–20.0	39.1–44.2	59.0–65.0
#3	8.5–11.5	27.5–30.5	54.3–60.8	82.6–91.0
#4	9.0–13.0	38.0–42.0	63.8–70.2	103.0–112.0
#5	8.5–11.5	27.5–30.5	54.3–60.8	82.6–91.0
#6	3.5–7.5	16.0–20.0	39.1–44.2	59.0–65.0
#7	9.0–13.0	38.0–42.0	63.8–70.2	103.0–112.0
#8	8.5–11.5	27.5–30.5	54.3–60.8	82.6–91.0

Note: This table provides an overview of general pressure ranges. Consult the proper manufacturer's parts catalog and service literature to determine exact governor assembly being tested and proper pressure schedule.

TROUBLESHOOTING

Troubleshooting is the systematic search for and locating of malfunctions in the engine or transmission. During initial troubleshooting, the engine and transmission should be regarded with equal attention, because a problem in the engine can fool you into thinking that it is in the transmission, and vice versa. Begin troubleshooting by:

- Visually inspecting for oil leakage and checking the oil level.
- Visually inspecting split lines, plugs, and the hose and tube connections at the transmission and cooler. (Oil leakage at split lines can be caused by loose mounting bolts or broken gaskets.)
- Checking torque on bolts, plugs, and connections where leakage is found.
- Checking to ensure the modulator control cable and linkage are free.
- Checking the mechanical shift linkage for proper operation.

If the preceding inspection checks are okay and the vehicle is operational, test drive and observe both the transmission and general vehicle performance. **Table 19–4** is a comprehensive troubleshooting guide that additionally outlines the corrective action required for common automatic transmission problems.

TABLE 19–4 Automatic Transmission Troubleshooting

Symptom	Probable Cause	Remedy
Harsh downshifts	1. Improper oil level 2. Engine idle speed too high 3. Manual selector linkage out of adjustment 4. Mechanical modulator out of adjustment 5. Valve sticking in governor 6. Downshift points too high 7. Valves sticking in control valve body	1. Adjust oil level. 2. Set engine idle speed to engine manufacturer's spec for an automatic transmission (500 rpm min). 3. Adjust linkage. 4. Adjust mechanical modulator. 5. Clean governor and governor screen. (Replace only if necessary.) 6. Adjust valve body. (Refer to service literature.) 7. Clean valve body. (Refer to service literature.)
Shift cycling	1. Improper oil level 2. Engine governed speed too low for transmission shift calibration 3. Governor screen clogged 4. Valve sticking in governor 5. Improper shift valve adjustment 6. Valves sticking in control valve body	1. Adjust oil level. 2. Set engine no-load governed speed to the engine to the engine manufacturer's spec or an automatic transmission (2,200 rpm min). 3. Remove governor screen and clean or replace. 4. Clean governor and screen. (Replace only if necessary.) 5. Adjust valve body. (Refer to service literature.) 6. Clean valve body. (Refer to service literature.)
Harsh upshifts	1. Improper oil level 2. Manual selector linkage out of adjustment 3. Mechanical modulator out of adjustment 4. Valve sticking in governor 5. Improper shift valve adjustment 6. Valves sticking in control valve body	1. Adjust oil level. 2. Adjust linkage. 3. Adjust mechanical modulator. 4. Clean governor and screen. (Replace only if necessary.) 5. Adjust valve body. (Refer to service literature.) 6. Clean valve body. (Refer to service literature.)
Downshifts occur at wide open throttle	1. Bent or loose modulator can 2. If used, check mechanical modulator for proper adjustment 3. Shift points set too high 4. Flanges out of parallel 5. Output flange retaining bolt improperly torqued 6. Slip joint splines seized 7. Drive line yoke needle bearings worn 8. Drive line out of balance or bent	1. Replace modulator and seal ring. 2. Adjust mechanical modulator. 3. Adjust valve body. (Refer to service literature.) 4. Flanges should be parallel within 1 degree. 5. Install flange bolt and lock tab washer correctly. 6. Lubricate slip joints properly. See vehicle manufacturer's specification. 7. Replace. 8. Rebalance.
Excessive retarder	1. Sticking low-speed valve 2. Sticking high-speed valve 3. Sticking priority valve 4. Leaking clutch piston seals 5. Leaking accumulator piston seals	1. Clean valve. (Refer to service literature.) 2. Clean valve. (Refer to service literature.) 3. Clean valve. (Refer to service literature.) 4. Replace seals. (Refer to service literature.) 5. Replace seals. (Refer to service literature.)
No retarder operation at	1. Sticking charging valve 2. Sticking clutch valve 3. Sticking regulator charging valve 4. Leaking clutch piston seals 5. Failed retarder clutch	1. Clean valve. (Refer to service literature.) 2. Clean valve. (Refer to service literature.) 3. Clean or replace valve. (Refer to service literature.) 4. Replace seals. (Refer to service literature.) 5. Replace clutch. (Refer to service literature.)
No retarder operation at high speed	1. Low air supply 2. Ruptured air line to retarder control valve 3. Sticking air control 4. Sticking retarder control valve	1. Check air supply. 2. Repair air line. 3. Repair or replace control. (Refer to service literature.) 4. Clean or replace valve. (Refer to service literature.)
Low main and/or retarder pressure	1. Sticking charging valve 2. Sticking clutch valve 3. Leaking clutch piston seals 4. Leaking accumulator piston seals 5. Leaking rotor hub seals	1. Clean or replace valve. (Refer to service literature.) 2. Clean or replace valve. (Refer to service literature.) 3. Replace seals. (Refer to service literature.) 4. Replace seals. (Refer to service literature.) 5. Replace seals. (Refer to service literature.)

19.3 POWER TAKE-OFFS (PTOs)

Automatic transmissions have special PTO application considerations because they transmit power differently from their standard transmission counterparts. Engine torque enters the automatic transmission through the torque converter. Because the torque converter has the ability to multiply engine torque, it provides an infinitely variable ratio capability that can get a vehicle moving quickly free from lugging and stalling. The advantage of this type of fluid coupling is the cushioning that it provides to the vehicle powertrain.

However, because of the infinitely variable ratio characteristic of the torque converter, the speed of the PTO output shaft changes with load and engine speed, not a desirable characteristic of most auxiliary equipment. When spec'ing PTO-driven equipment for automatic transmissions, it is essential to check the requirements of the equipment that has to be driven before fitting it to the transmission.

SUMMARY

- Automatic transmissions should be cleaned with a power washer to make inspection and servicing easier. Special care should be taken to avoid forcing water through the transmission breather.
- Inspect the transmission for loose bolts, loose or leaking oil lines, oil leakage, and the condition of the control linkage and cables.
- Maintaining the specified oil level in an automatic transmission is important, because either low or high oil levels can cause aeration of the transmission fluid.
- At each oil change, examine the oil that is drained for evidence of dirt or water.
- Metal particles in the oil or on the magnetic drain plug (except for the minute particles normally trapped in the oil filter) indicate that damage has occurred in the transmission.
- If engine coolant leaks into the transmission oil system, immediate action should be taken to prevent serious damage. Antifreeze destroys the clutch materials used in automatic transmissions.
- There are three methods of testing shift points in automatic transmissions. Adjusting shift points is part of routine preventive maintenance.
- Transmission disassembly, testing, and reassembly require the use of OEM service literature, special service tools, and specialized training.
- Truck technicians should be able to undertake first-level troubleshooting of Allison transmissions using diagnostic charts. During initial troubleshooting, the engine and transmission should be regarded with equal attention, because engine problems can often be misinterpreted as transmission problems, and vice versa.

REVIEW QUESTIONS

1. Which of the following conditions can cause aerated oil in automatic transmissions?
   a. low oil level
   b. high oil level
   c. a clogged breather
   d. both a and b
2. On most Allison transmissions, when the oil level reads at the FULL mark on the dipstick with the transmission at operating temperature, which of the following indicates the actual level of oil in the transmission?
   a. if it is slightly below the planetary gearsets
   b. if it is slightly above the top of the oil pan
   c. if it is slightly above the ring gears
   d. if it is at the axis of the sun gears
3. Which of the following could result in an inaccurate dipstick reading?
   a. failure to wipe the dipstick clean before taking the reading
   b. a clogged breather
   c. an unvented oil filler tube
   d. all of the above
4. When performing an automatic transmission cold check level, at what temperature should the oil be?
   a. between 0°F and 60°F (−17°C and 16°C)
   b. between 60°F and 120°F (16°C and 49°C)
   c. between 120°F and 160°F (49°C and 71°C)
   d. between 160°F and 200°F (71°C and 93°C)
5. When the transmission input speed is increased, which of the following should happen to the oil level in the sump?
   a. The oil level lowers.
   b. The oil level rises.
   c. There is no effect on the oil level.
   d. The oil aerates.
6. What effect will an increase in oil temperature have on the transmission oil level?
   a. to lower the oil level
   b. to raise the oil level
   c. no effect
   d. to aerate the oil

7. The ___ is located on top of the transmission and its function is to prevent pressure buildup within the transmission housing.
   a. auxiliary filter
   b. governor
   c. breather
   d. release valve
8. Technician A suggests that it is good practice to install an auxiliary filter after a transmission failure that resulted in debris being introduced into the hydraulic circuit. Technician B changes the auxiliary filter every 5,000 miles (8,000 km). Who is correct?
   a. Technician A only
   b. Technician B only
   c. both A and B
   d. neither A nor B
9. What procedure should be used to identify whether the engine or transmission is at fault after a drivetrain complaint in a chassis equipped with an automatic transmission?
   a. a stall test
   b. shift point adjustment
   c. an oil analysis
   d. all of the above
10. An automatic transmission upshifts prematurely. Technician A says that the mechanical modulator might have to be adjusted. Technician B says that the shift valve might have to be adjusted. Who is correct?
    a. Technician A only
    b. Technician B only
    c. both A and B
    d. neither A or B
11. If the engine stall speed is more than 150 rpm above the stall speed specified by the engine manufacturer, which of the following is more likely?
    a. an internal transmission problem
    b. an engine problem
    c. a torque converter problem
    d. a drive axle carrier problem
12. A stall speed test meets specification on a transmission with an oil overheat problem. Which of the following is a more likely cause?
    a. an engine problem
    b. a stuck torque converter stator
    c. a freewheeling torque converter stator
    d. either a or c
13. A stall speed that tests 35 percent below the OEM engine specification would most likely indicate:
    a. an engine problem.
    b. a stuck torque converter stator.
    c. a freewheeling torque converter stator.
    d. either b or c.
14. Which of the following would be the most accurate method of checking transmission shift points?
    a. engine speed (rpm) method
    b. speedometer method
    c. test stand method
    d. road test method
15. When a valve body is mounted to a test stand, which of the following components is driven by the test equipment?
    a. the modulator
    b. the transmission input shaft
    c. the governor
    d. the torque converter impeller

20

Prerequisite: Chapters 6, 15, and 16

AUTOMATED MANUAL AND HYBRID TRANSMISSIONS

OBJECTIVES

After reading this chapter, you should be able to:

- Explain how a standard mechanical transmission is adapted for automated shifting in three-pedal and two-pedal systems.
- Identify different OEM automated transmissions and interpret some of the serial number codes used.
- Describe the hardware changes that differentiate a standard Roadranger twin-countershaft transmission from its electronically automated version.
- Outline the electronic circuit components that are used to manage AutoShift and UltraShift transmissions.
- Outline the main box and auxiliary section actuator components required for AutoShift and UltraShift electronically automated manual transmissions.
- Describe how the electronic circuit components work together to perform the system functions and outline the role played by the transmission ECU.
- Explain exactly how a shift takes place in an automated manual transmission.
- Perform some basic diagnostic troubleshooting on automated manual transmission electronics.
- Identify the DT-12, Mercedes-Benz AGS, ZF Meritor SureShift, and FreedomLine transmissions.
- Identify hybrid-electric and hybrid-hydraulic transmissions.

KEY TERMS

automated manual transmission (AMT)
Automatic Gear Shift (AGS)
AutoShift
clutch actuator
electronic control unit (ECU)
FreedomLine
gear select motor
Hall effect principle
hybrid electric drive
hybrid hydraulic drive
hydraulic launch assist (HLA)
inertia brake
J1939 data bus
launch phase
logic power
rail select motor
SAE J1939
speed sensors
three-pedal system
touch point
two-pedal system
UltraShift
Urge-to-Move
wet clutch

INTRODUCTION

It takes 60 pounds of linear force from a driver's boot to depress a truck mechanical clutch, and a typical driving day for some truck drivers can require over 500 shifts. A higher degree of driver skill and training is required to operate the manually shifted, compounded gearing used in conventional truck transmissions. Surveys indicate that constant shifting of gears adds considerably to driver fatigue. This has vastly increased the market share of automated standard transmissions over recent years. Automated transmissions are fundamentally standard mechanical transmissions that use computer controls to eliminate some or all of the clutch work required of the truck driver. They achieve this using computer logic and shift-by-wire. This category of transmissions is most often described as **automated manual transmissions (AMTs)**, which is the term we use in this chapter.

SHIFTING MYTHS

Over a coffee in a truck stop, you might find it difficult to find a truck driver who admits to using the clutch to shift a standard transmission except for the moments of stopping and starting. It would be a fine world if every professional truck driver had such a feel for the drivetrain of the equipment operated that the clutch would become almost unnecessary. The reality is that few do. Those who do not and feel disinclined to use the clutch anyway should be appreciated because they create plenty of work for truck technicians by taking out clutches and transmissions on a regular basis. But there are some seasoned truck drivers who have developed an intuitive sense of timing engine rpm and clutch shaft speed that enables them to smoothly shift seldom using the clutch. A computer can do the same thing to perfection by simply monitoring shaft speeds and using mapped logic to shift gears. Fleets are discovering that adopting more costly AMTs is producing almost immediate dividends in reduced repair costs and downtime.

THREE- AND TWO-PEDAL SYSTEMS

Some AMT systems, especially the first-generation versions, allow those drivers who have a taste for manual shifting to continue to do so. Given the choice, however, most drivers happily relinquish shifting control to the computer that manages an AMT. Current AMTs can be divided into **three-pedal systems** and **two-pedal systems**. A three-pedal system is one that retains a clutch pedal (the *third* pedal) that is used for certain running conditions such as driveline stall, whereas a two-pedal system has a clutch but no clutch pedal. In two-pedal systems, clutch operation is usually controlled electronically and actuated by air or hydraulics.

AUTOMATED TRANSMISSION MANUFACTURERS

Eaton Fuller Roadranger has been the dominant supplier of transmissions to the heavy-duty truck market for many years. It was the first to introduce AMTs in the 1990s. More recently, the Fuller Roadranger family of AMTs has had competition from ZF Meritor, Volvo-Mack, Detroit, and Mercedes-Benz. Early versions of AMTs that appeared during the 1990s were the Eaton Fuller AutoSelect system and the ZF Meritor Engine SynchroShift (ESS) system, but these sold in low numbers and tended to be problematic. Current versions are the Fuller Roadranger **AutoShift** (three pedal) and **UltraShift** (two pedal), the ZF **FreedomLine** (two pedal), Detroit DT-12 (two pedal), and Mercedes-Benz **Automatic Gear Shift (AGS)** (two pedal).

ADVANTAGES OF AUTOMATED TRANSMISSIONS

Because of a prolonged industry-wide shortage of truck drivers, truck fleet managers increasingly have been looking at the benefits provided by automated transmissions. First, AMTs allow the driver to focus primarily on driving the vehicle rather than on managing shifting. This means less upfront driver training is required. Second, despite higher initial cost, the overall cost-effectiveness of AMTs is major factor in increasing sales.

There is no doubt that a computer can manage shifting to produce better fuel economy and reduce drivetrain damage (especially clutch abuse) better than all but the most experienced truck drivers. The Eaton **UltraShift** uses a modular design and theoretically requires zero servicing during its life. This transmission has been tested to produce substantially improved fuel economy over equivalent manual shift transmissions but then so do most AMTs because they optimize both upshift and downshift points.

For the technician with a basic understanding of standard transmissions, the operating principles of AMTs will not present any problems. In engine technology, the term *partial authority* is used to reference a hydromechanical system that has been adapted for computer controls, and that is the idea behind automated transmission systems. An automated transmission adapts a standard mechanical transmission platform for electronic management so that some original equipment manufacturers (OEMs) are describing AMTs as partial authority.

PARTIAL AUTHORITY ELECTRONICS

Automated transmissions are mastered by, and multiplexed to, other chassis electronics by the transmission controller module. We use the term **electronic control unit (ECU)** to describe this module. Depending on the manufacturer and the generation of the system, either one or two module systems are used. The advantage of multiplexing is to network the transmission electronics with other chassis electronic systems, allowing them to "talk" to each other and permit the sharing of input hardware and information.

MULTIPLEXING

In truth, there is no better argument for applying multiplexing technology to truck chassis than the relationship between the engine and transmission. Current automated transmissions are networked to the **J1939 data bus**, meaning that they both speak the same network language so no transducer modules are required. With high-speed multiplexing, apart from eliminating the duplication of common input hardware such as the throttle position sensor (TPS), exact synchronization of engine and transmission shift rpm is achievable without the need to rely on a truck driver's driving instincts.

SCOPE

In this chapter, we primarily reference the Eaton Fuller Roadranger UltraShift transmission because it is the current market leader. There are not a lot of fundamental differences between AMTs, and some limited references are made to the FreedomLine, DD12, and MB AGS systems when required. This chapter addresses only the electronic and electrical characteristics of AMTs; the assumption will be made that the operation of a truck standard mechanical transmission is fully understood. Automated transmission technology has emerged rapidly. In a short period these transmissions have undergone significant changes to the shift management system. The approach here is to reference more recent versions but also not to overlook some of the earlier ones.

20.1 TRANSMISSION IDENTIFICATION

For some reason, there is no standard serial number code sequence in transmissions even when manufactured by the same company, as you will observe in the following examples. AutoShift, UltraShift, DT-12, and ZF FreedomLine use somewhat similar identification nomenclature; the Mercedes-Benz AGS is a little different. This means that some familiarity with each OEM code sequence is required. Connecting a generic electronic service tool (EST) to the chassis data bus will identify that a module is reporting at the SA 03 or MID 130 address on the bus.

AUTOSHIFT/ULTRASHIFT

AutoShift and UltraShift are primarily differentiated by the clutch used in the transmission. **Figure 20–1** shows overhead and side views of the heavy-duty versions of AutoShift and Ultrashift AMTs. The product coding used by these transmissions is more or less identical to that used in other Roadranger twin-countershaft transmissions.

AutoShift and UltraShift product coding is interpreted as follows:

Serial number: RTO – 14 9 10 B - DM/AW/AS 3

R	Roadranger
T	Twin countershaft
O	Overdrive
14	Torque capacity × 100 in lb-ft.
9	Design level (Gear geometry)
10	Forward speeds
B	Ratio category
DM	UltraShift with DM clutch
AW	UltraShift with wet clutch
AS	AutoShift
3	Automation level/Gen III electronics

Figure 20–2 shows a medium-duty Eaton Fuller Roadranger UltraShift transmission with the wet-clutch option (AW) denoting the dipstick and fill location.

FREEDOMLINE

Figure 20–3 shows the ZF Meritor FreedomLine transmission. FreedomLine product coding is interpreted as follows:

Serial Number: M O – 16 Z 10 C - A 18

M	ZF Meritor
O	Overdrive (no letter = direct drive)
16	Input torque rating × 100 in lb-ft.
Z	Design platform codes:
	Z = FreedomLine
	F = 9 speed
	G = 10 speed
10	Number of forward gears
C	Ratio
A	Automated (S = shift by wire, M = manual)
18	Highest torque in top 2 gears = 1,800 lb-ft.

MERCEDES-BENZ AGS

Figure 20–4 shows the Mercedes-Benz AGS transmission. AGS product coding is interpreted as follows:

Serial Number: MBT – 560 S - 6 O

MBT	Mercedes-Benz Transmission
560	Input torque rating 560 lb-ft.
6	Number of forward ratios
O	Overdrive (D = direct drive)

20.2 ULTRASHIFT COMPONENTS

The Eaton Fuller Roadranger UltraShift AMT is networked to the J1939 data bus. The UltraShift ECU is a single-module system that manages input data,

FIGURE 20-1 An (A) overhead and (B) left side views of Eaton HD versions of AutoShift and UltraShift transmissions

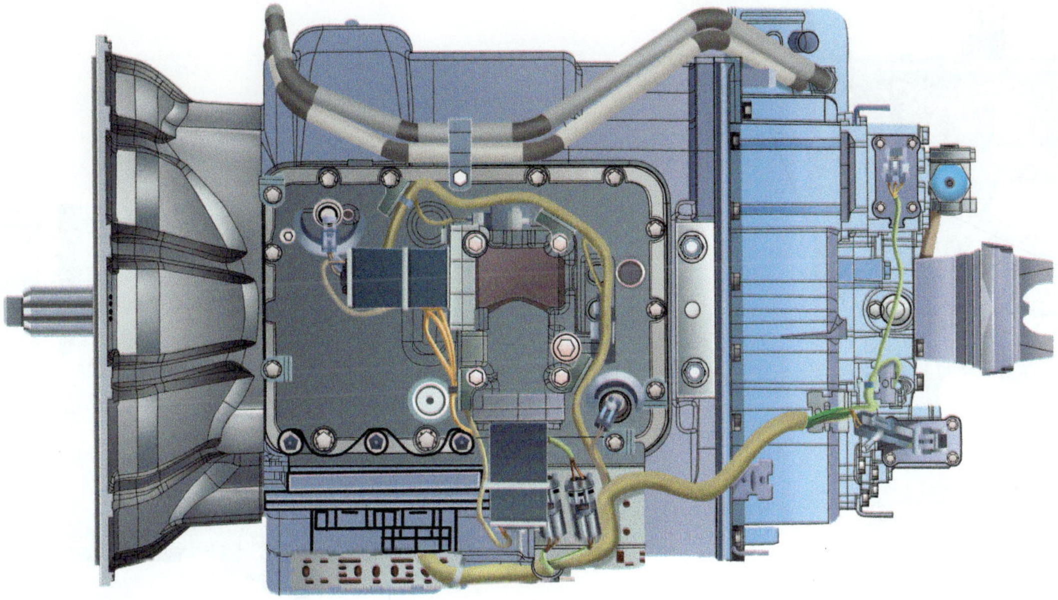

A.

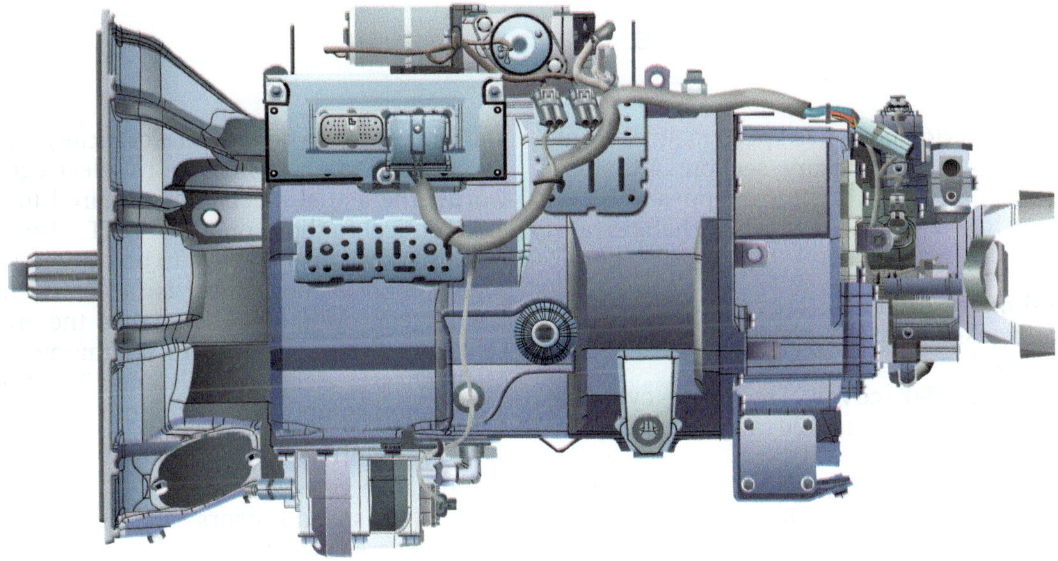

B.

processes and switches outputs, and communicates with the chassis data bus. The ECU is networked to the SA 03 (MID 130) address on the chassis data bus. UltraShift transmission components can be grouped as follows:

- Base standard gearbox
- Transmission automation components
- Vehicle automation components

For study purposes, the components in each subsection are examined separately.

BASE TRANSMISSION

The UltraShift transmission is based on a Roadranger twin-countershaft platform. A transmission with ten forward speeds is used as an example. A Fuller

FIGURE 20–2 An (A) overhead and (B) right side views of Medium-duty UltraShift transmission with the AW (wet-clutch) option showing the dipstick and fill location

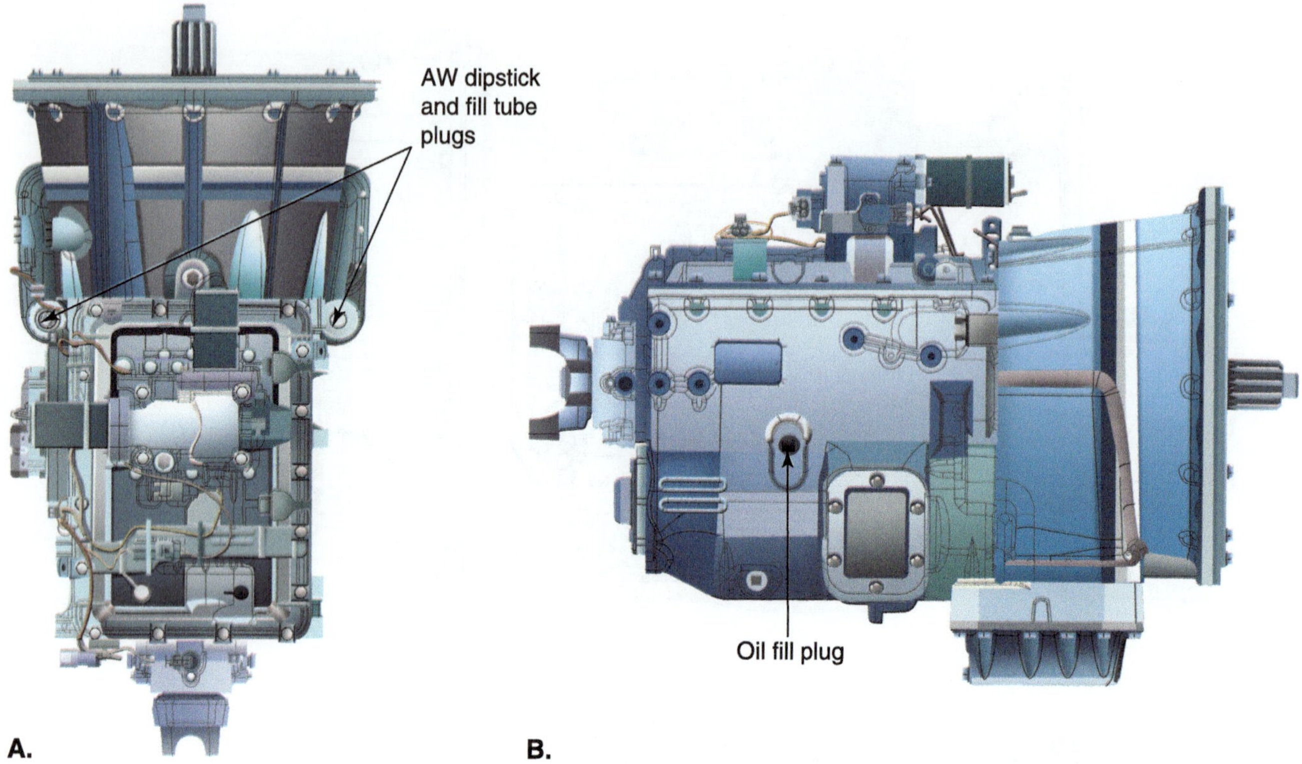

ten-speed transmission has a five-speed main or front case, with a two-speed auxiliary section. Although the base transmission has a similar appearance to nonautomated versions, the following changes are required:

- **Main box**. Screw bosses are added to the case so the transmission ECU can be mounted to the side.
- **Range cylinder cover**. The range cylinder cover is altered to house the range shift control mechanism.
- **Shift bar housing**. The shift bar housing has been machined to accept the **speed sensors**.
- **Shift yokes**. The shift yokes have been hardened at the fork pads to improve durability during shifts.

TRANSMISSION AUTOMATION COMPONENTS

Several components have to be added to the base transmission in order to automate main box and auxiliary section shifting. The UltraShift transmission has to perform electronically over electrically what the standard version of the transmission accomplishes mechanically during a shift. That means the UltraShift shift finger has to be moved electrically both fore and aft, and right and left in order to select a gear. In other words, the actual mechanics required to shift gears are identical to those on a non-AMT. Also, the engine, road, and gear speeds have to be computed to ensure that a nonclashing shift is accomplished. Shifts are further optimized by giving the transmission electronics the capability to communicate with the other chassis electronic systems. The transmission automation components include:

- X–Y shifter
- Range valve
- Shaft speed sensors
- Reverse switch
- Power module
- ECU
- Transmission harness
- Vehicle harness

Electric Shifter Assembly

The electric shifter assembly replaces the manual gearshift lever on the standard transmission. Its function is to effect shifts in the transmission main box by using a shift finger. The shift finger is moved to select the correct rail and gear position using the normal standard shift bar housing. The electric shifter components are the shift finger, gear select yokes, ball

FIGURE 20–3 FreedomLine two-pedal AMT

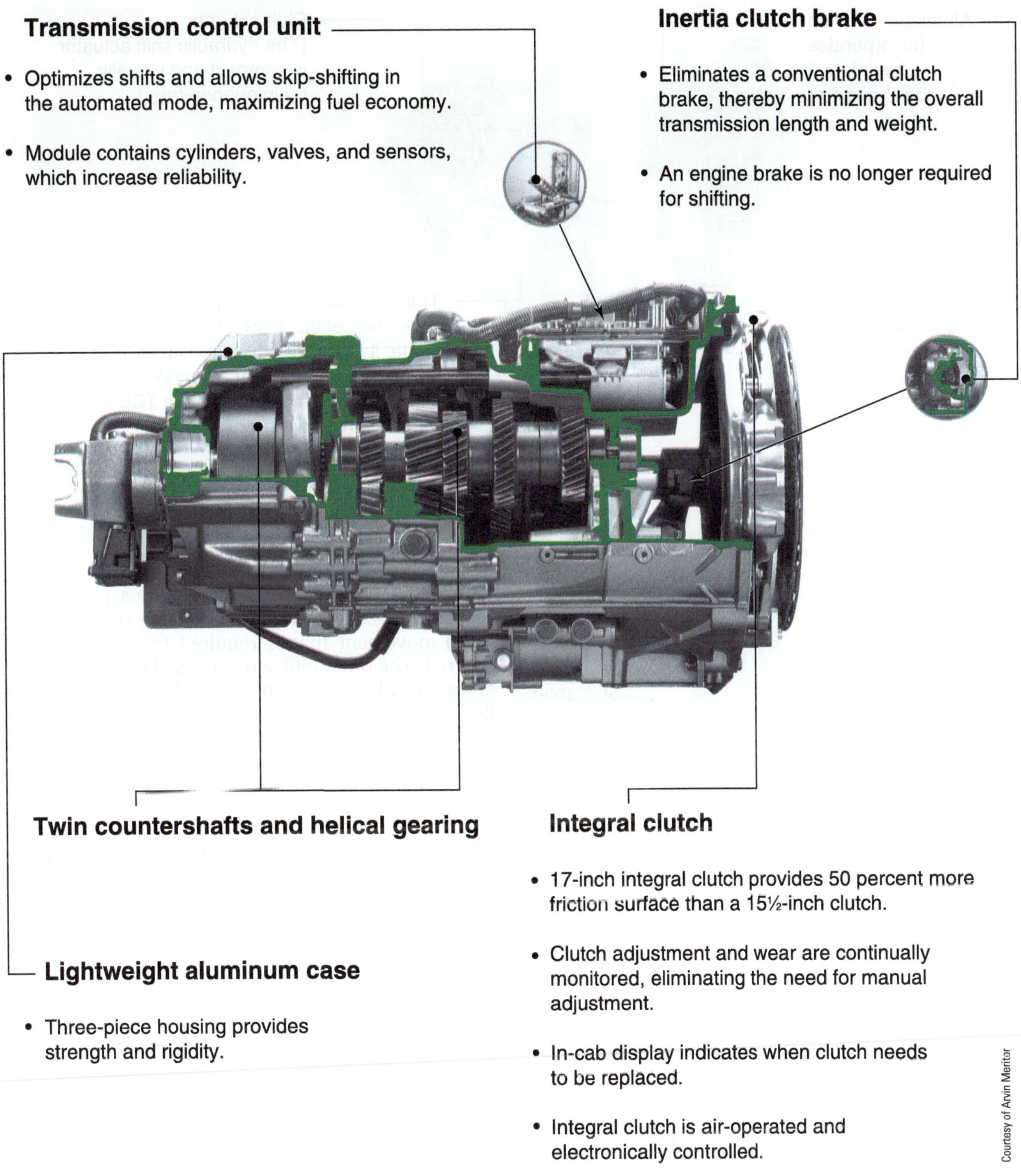

Courtesy of Arvin Meritor

screw assembly, electric motors, and position sensors. These components are shown in **Figure 20–5**.

The shift finger is the component that actually selects the commanded rail and gear. Its appearance is similar to that of the spade at the end of a manual Roadranger gearshift lever. The shift finger engages with the shift block recesses in the shift rails. When a shift is to be made, the shift finger is first positioned

FIGURE 20–4 Mercedes-Benz AGS two-pedal AMT

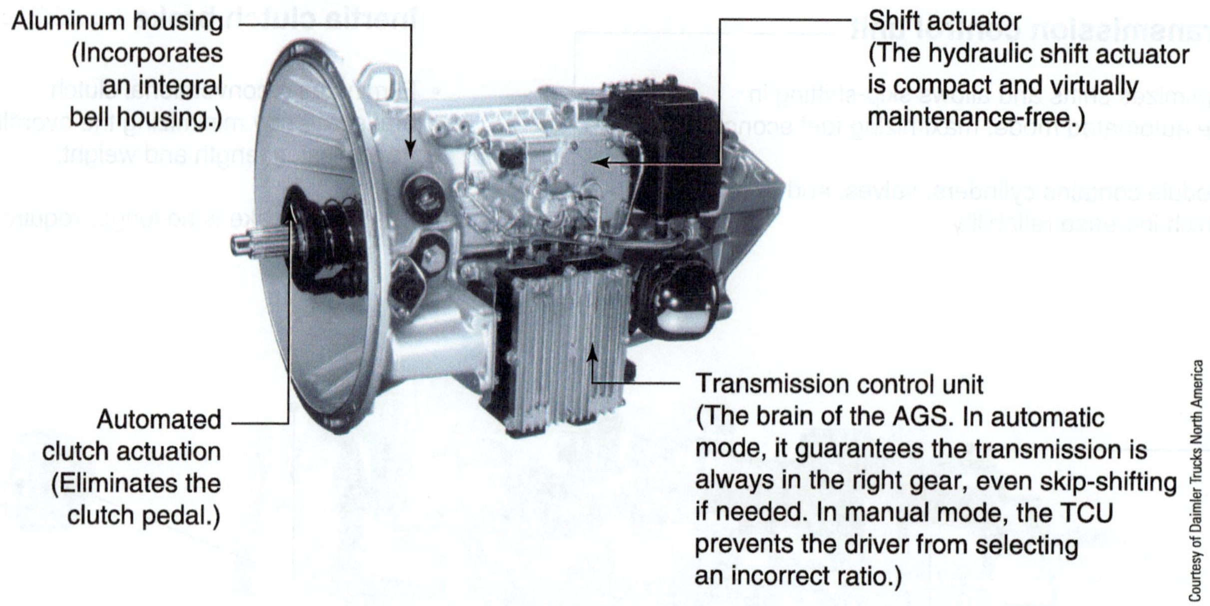

FIGURE 20–5 Electric shifter assembly

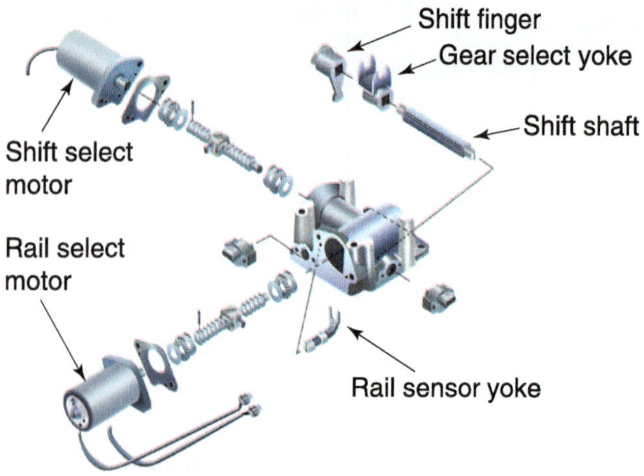

in the shift block of the selected rail, after which it can be moved either forward or backward to select the commanded gear.

The shift shaft is a square section shaft (see Figure 20–5) within the electric shifter housing. The shift finger rides on the shift shaft. This permits the shift finger to be moved along the rail and gear select directions. The shift shaft has a machined area on the end to enable it to rotate the position sensor. The gear select yoke also rides on the shift shaft. The yoke is driven by the gear select ball screw, which rotates the shift shaft for gear selection. The gear select sensor provides position data to the transmission ECU.

BALL SCREW ASSEMBLIES

The shift finger position is located by a pair of ball screws. One ball screw is used for the lateral (left to right) movement that is required to select the correct rail. The other ball screw is used for the forward or backward positioning required to select a gear. Remember that the actual means of selecting a gear is identical to that in a standard transmission: Shift finger movement first selects one of the three shift rails and then slides it fore or aft.

Ball Screw Operation

The ball screw assembly is made up of a worm gear, a ball screw nut, and ball bearings. Its operation resembles that of a recirculating ball steering gear. When the worm gear rotates, the ball bearings ride in the grooves of the worm threads. When a ball reaches the end of a ball screw block, it is forced into a tube to be returned to the beginning of the worm threads in the ball screw block (**Figure 20–6**).

Gear Select Ball Screw

The gear select ball screw nut has a pair of tabs, one on either side. These tabs drive the gear select yoke, which is located on the shift shaft with the shift finger. When gear selection is commanded, the ball screw block moves the gear select yoke, which in turn moves the shift shaft. This is shown in **Figure 20–7**.

Rail Select Ball Screw

The rail select ball screw has a slot machined into it. The shift finger has a half moon protrusion designed

FIGURE 20–6 Ball screw assembly

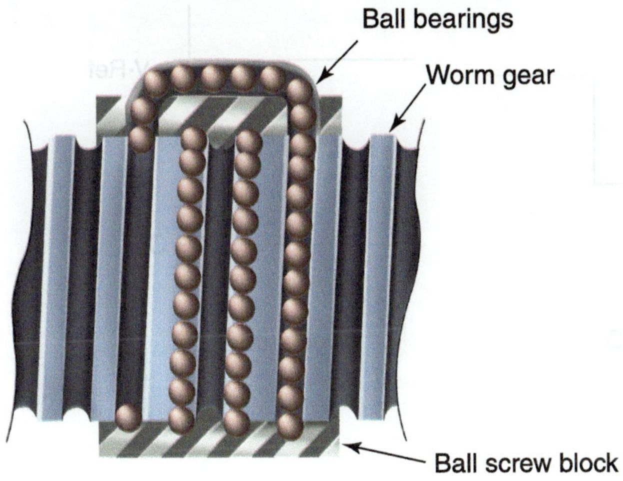

FIGURE 20–7 Gear select and rail select ball screw assemblies

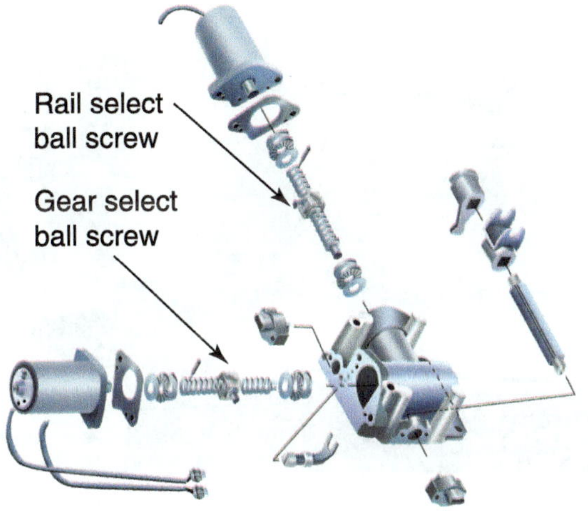

to interface with the slot cut into the rail select ball screw. When a rail selection is made, the rail ball screw drives the shift finger that slides up and down the shift shaft. The rail select and gear select ball screws are not interchangeable, despite their similarity in appearance. You can see both the gear and rail select screws in Figure 20–7. Review the operation of each using this figure.

ELECTRIC MOTORS

Each of the ball screw assemblies is driven by its own electric motor. The motors use a permanent magnet operating principle and are capable of forward or reverse rotation, depending on how current flow is directed by the ECU. The polarity of the current that flows through the motor can be changed to drive the

FIGURE 20–8 Gear select and rail select motors

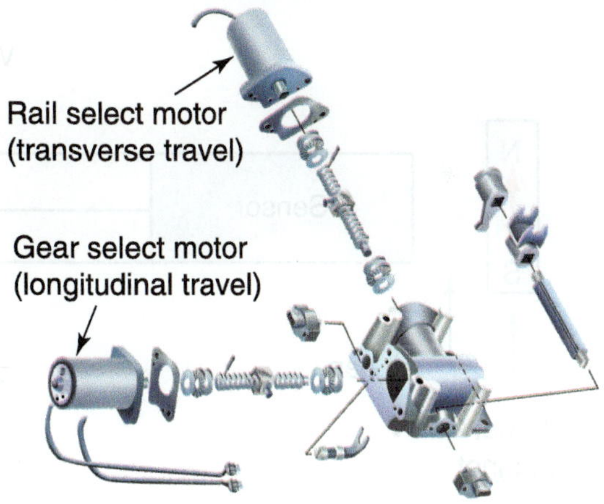

ball screw assemblies to a specific position to effect gear shifts. The **rail select motor** is pinned directly to the rail select ball screw and the **gear select motor** is pinned directly to the gear select ball screw. The electric motors used in each ball screw assembly are not interchangeable. **Figure 20–8** shows the location of the gear and rail select motors used in an UltraShift transmission.

Rail Select Motor

The rail select motor is a reversing direct current (DC) motor managed by the ECU. Its function is to move the shift finger transversely (side to side) to select one of the three rails in the Roadranger shift bar housing.

Gear Select Motor

The gear select motor is a reversing DC motor managed by the ECU. Its function is to move the shift finger forward and backward. This movement is responsible for actually moving the shift rail into and out of a selected gear position.

POSITION SENSORS

UltraShift and all current versions of AutoShift use Hall effect position sensors but note that the first generation of AutoShift used potentiometer position sensors.

A complete description of a potentiometer and Hall effect sensors is provided in Chapter 6. Regardless of electrical principle, both types of position sensor are supplied with V-Ref at 5V. The linear Hall effect principle used in position sensors is shown in **Figure 20–9**.

The shifter assembly incorporates two position sensors:

- The gear select sensor monitors shift finger position during gear selection. The signal returned to the transmission ECU confirms whether a shift has been successfully effected.

FIGURE 20–9 Linear Hall effect position sensor principle

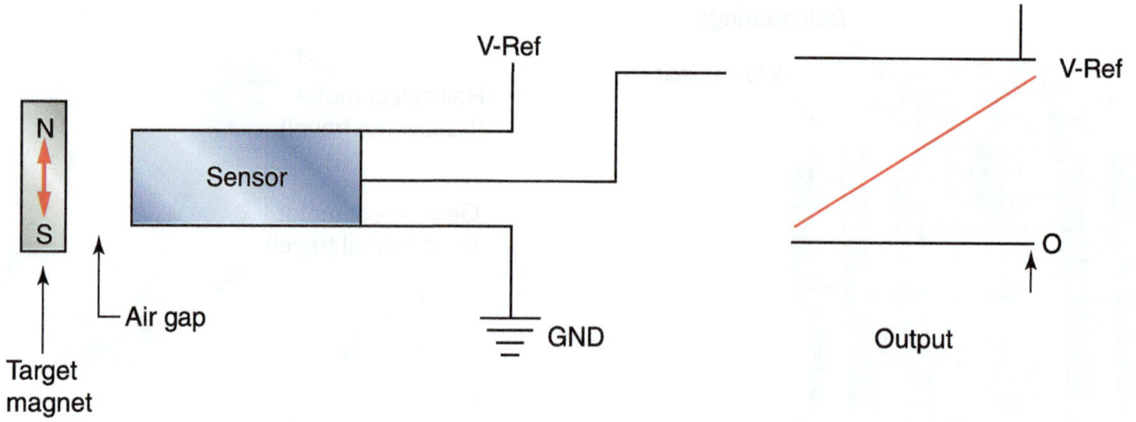

FIGURE 20–10 Location of position sensors

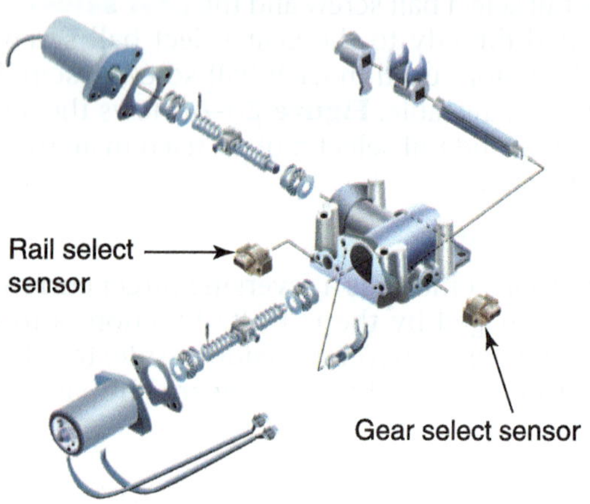

- The rail select sensor is connected to the rail select ball screw. The position sensor yoke arm fits over a pin on the side of the rail select screw block (**Figure 20–10**). Its signal allows the ECU to determine whether the shift finger is located over the correct rail before effecting a shift.

Electric Shifter Overview

The electric shifter is the means by which the ECU shifts gear ratios. A pair of motors is used to effect the shifts, whereas a pair of position sensors is used to verify that a shift has actually taken place. The shift motor tower on an UltraShift is shown in **Figure 20–11**. For a shift to occur, the following sequence of events takes place:

1. The rail select motor is energized.
2. The rail select motor turns the ball screw, which in turn moves the ball screw block.

FIGURE 20–11 Gear and rail select motors on a current UltraShift transmission

3. The ball screw block moves the shift finger into the commanded rail select position; that is, one of the three rails is selected.
4. The rail select sensor signals the transmission ECU that the shift finger is correctly located over the selected rail.
5. The gear motor is energized.
6. The gear motor turns its ball screw that, in turn, moves the ball screw block.
7. The ball screw block moves the gear select yoke on the shift shaft, moving the shift shaft.
8. When the shift shaft moves, the shift finger is moved forward or backward for gear selection.
9. The gear select sensor confirms that the desired shift has been effected by signaling the transmission ECU.
10. The shift sequence is completed.

SHOP TALK

Current versions of the three-pedal AutoShift are designed to break driveline torque when a predetermined driveline (road) speed is sensed. This prevents driveline component stall damage to the engine.

RANGE VALVE

On an AutoShift transmission, the range valve takes the place of the HI-LO flipper switch mounted on the gear lever assembly in a standard Roadranger transmission. The HI-LO flipper switch was a master/slave air valve that controlled auxiliary section shifts on the standard transmission. Similarly, the AutoShift range valve controls range cylinder air for auxiliary section shifts. The range valve contains a pair of solenoids, one for HI range and one for LO range. The range valve is shown in **Figure 20–12**.

Range Valve Subcomponents

The range valve consists of the following components:

- **Plunger**. The plunger located at the center of the solenoid has a sealing surface at either end and two grooves cut in its side to allow for air travel.
- **Air ports**. The solenoid contains three air ports:
  - A regulated air port is located at the bottom of the solenoid. It receives air from the air filter regulator. It has a seat to seal regulated air.
  - An exhaust port is located at the top of the solenoid. It is equipped with a seat used by the plunger to seal off the exhaust.
  - A range cylinder port is located at the side of the solenoid. It supplies air to the range cylinder.
- **Bias spring**. The bias spring is located inside the solenoid cartridge. It forces the solenoid plunger downward, sealing off the regulated air port.
- **Coil**. The transmission ECU uses the coil to control air at the range cylinder. The bias spring helps control the air. Switching at the ECU energizes the coil and the resulting magnetic field pulls the plunger into the center of the coil. When the coil is de-energized, the bias spring returns the plunger to its original position.

FIGURE 20–12 Range valve components

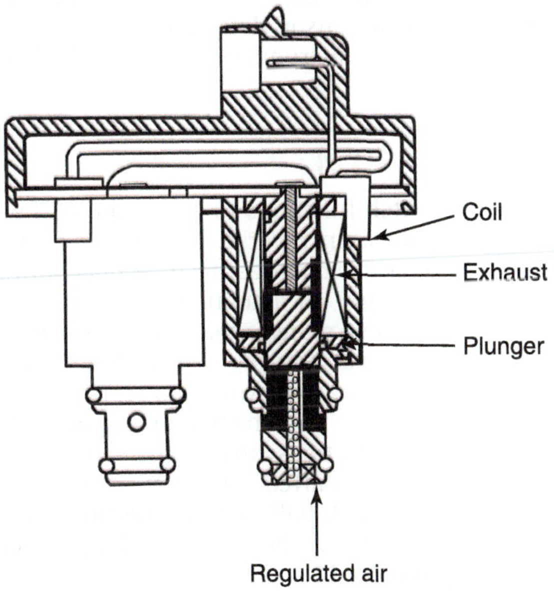

Range Valve Operation

Range valve operation is easily understood if you reference the diagrams in **Figure 20–13** through **Figure 20–15**. These figures demonstrate range valve operation through each phase.

FIGURE 20–13 Range valve: all-off position

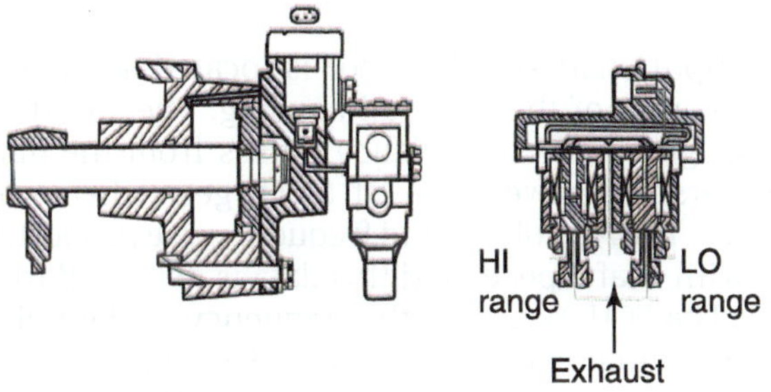

FIGURE 20–14 Range valve: LO range operation

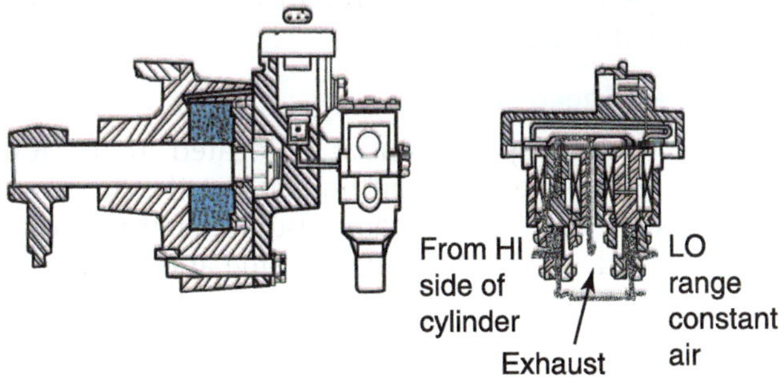

FIGURE 20–15 Range valve: HI range operation

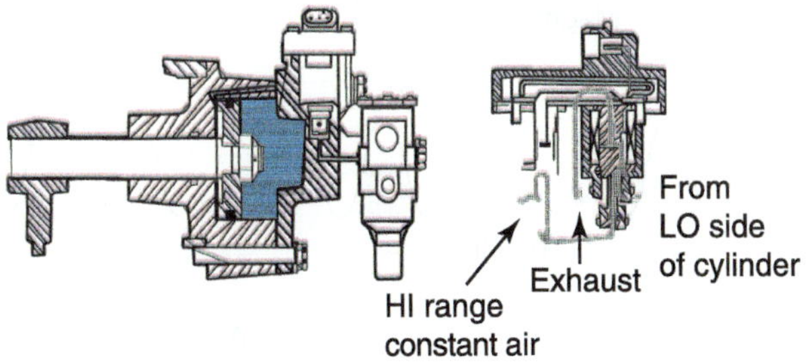

SPEED SENSORS

Speed sensors are part of the input circuit. Their function is to signal shaft speed data to the transmission ECU. Three-speed sensors are used in the transmission. They are responsible for signaling input shaft speed, main shaft speed, and output shaft speed data. All three sensors use an inductive pulse generator principle. A ferrous metal pulse wheel (chopper or tone wheel) with external teeth cuts through a magnetic field at a sensor coil. When the teeth of the pulse wheel cut through the magnetic field in the sensor, an alternating current (AC) voltage is induced in pulses. This principle is shown in **Figure 20–16** and explained in some detail in Chapter 6. In addition, a cutaway view of a tailshaft speed sensor in an UltraShift is shown in Figure 6–42.

The AC voltage value and frequency produced will rise in proportion to the increase in rotational speed. In this way, the transmission ECU can determine shaft speed by correlating the frequency and voltage value with rpm. These sensors can be tested by measuring voltage output.

Input Shaft Speed Sensor

The input shaft speed sensor is located at the right front corner of the shift rail housing. The speed sensor magnet and coil produce pulses from the upper countershaft power take-off (PTO) gear whenever it rotates. The AC voltage and frequency rise proportionally with shaft speed, and this data is delivered to the ECU. The ECU measures the frequency of the pulses, not the actual voltage generated. The upper countershaft is direct driven by the input shaft through the head set gears, so the input shaft speed is calculated by multiplying countershaft speed by the head set ratio. **Figure 20–17** shows the location of the input shaft speed sensor.

Main Shaft Speed Sensor

The main shaft speed sensor is located at the left rear corner of the shift rail housing at the auxiliary countershaft drive gear. Again, a pulse generator principle is used, so the AC voltage value and frequency fed to the ECU increase proportionally with shaft rotational speed. The auxiliary countershaft drive gear is meshed to the main shaft through the auxiliary drive gear and driven at the same speed. Figure 20–16 shows the location of the main shaft speed sensor.

FIGURE 20–16 AC pulse generator principle

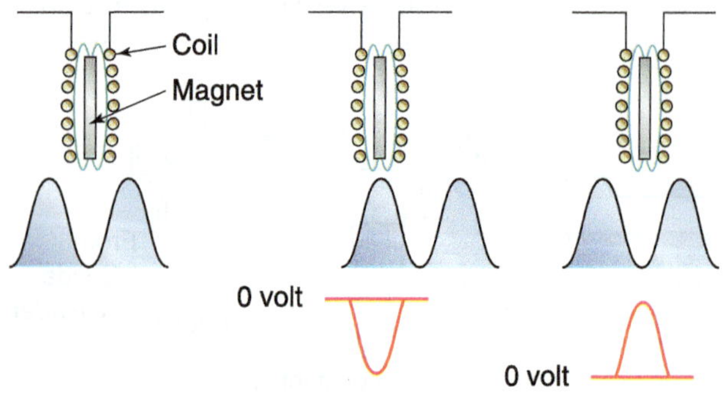

FIGURE 20–17 Location of speed sensors

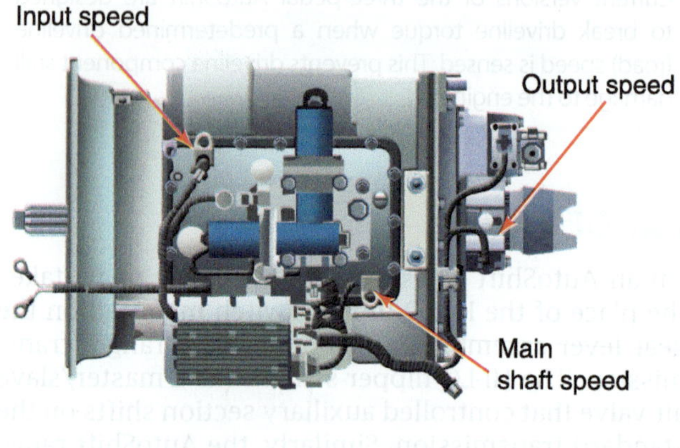

Tailshaft Speed Sensor

The output shaft speed sensor picks up pulses produced by a tone ring located on the output shaft. This tailshaft-located sensor inputs road speed data to the AutoShift ECU and to other chassis electronic systems. Road speed is determined by factoring the output shaft speed with the drive axle carrier ratio and the rolling radius dimension of the tires. A standard 16-tooth pulse wheel is used. Figure 20–17 shows the location of the tailshaft speed sensor.

Function of the Speed Sensors

Input shaft speed and main shaft speed data is used by the transmission ECU for synchronizing shifts in the transmission main box. The data generated by the main shaft and output shaft speed sensors is used by the ECU to verify range engagement. The output or tailshaft speed sensor data may also be shared with other chassis electronic systems as vehicle road speed data.

REVERSE SWITCH

The reverse switch functions to signal the ECU when the transmission is in reverse. It consists of a pair of normally open terminal contacts, a bias spring, a plunger contact, and an actuating ball. The reverse ball switch rides on a plunger that follows the reverse rail in the shift bar housing. When the transmission is not in reverse, the switch pin rests down in a groove on the reverse rail. The bias spring acts on the plunger contacts and the switch remains open. When the transmission is shifted into reverse, the switch pin acts on the switch ball, forcing the plunger contacts onto the terminals. Reverse switch operation is shown in **Figure 20–18**.

FIGURE 20–18 Reverse switch operation

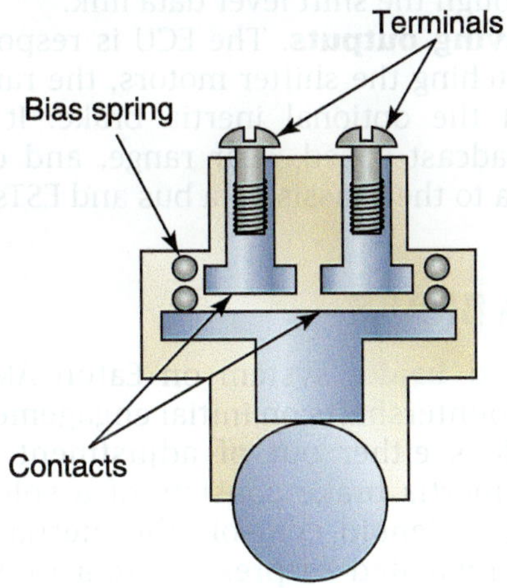

One of the reverse switch terminals is connected to the battery voltage. The other is connected to an ECU input. When the switch is closed (reverse gear selected), battery voltage is input both to the ECU (as a signal) and to the vehicle backup lights.

POWER MANAGEMENT

The way in which electrical power is managed by AMT electronics has changed with each generation. In recent versions, electrical power management is the responsibility of the transmission ECU. The first generation of AutoShift used a separate power module to connect the vehicle electrical system with the transmission electronics. The power management electronics are designed to protect the transmission electrical system from current overloads, voltage spikes, and reverse voltage problems (backfeeds). The power management electronics of AMTs incorporate the following components:

- Cycling circuit breaker (electromechanical or virtual) protects against circuit current overloads and is rated by amperage.
- Reverse voltage diode protects the circuit against accidental current backfeeds caused by reversing the polarity of connections.
- Reverse voltage zener diode acts like a voltage regulator to protect the circuit against voltage spikes; permits current flow but does not allow voltage to exceed a preset value. **Figure 20–19** shows what a zener diode looks like in a circuit diagram.

These circuit protection components are integrated into the ECU of the most recent versions AMT electronics and the electromechanical circuit breakers have been upgraded to virtual breakers.

FIGURE 20–19 Reverse voltage zener diode schematic

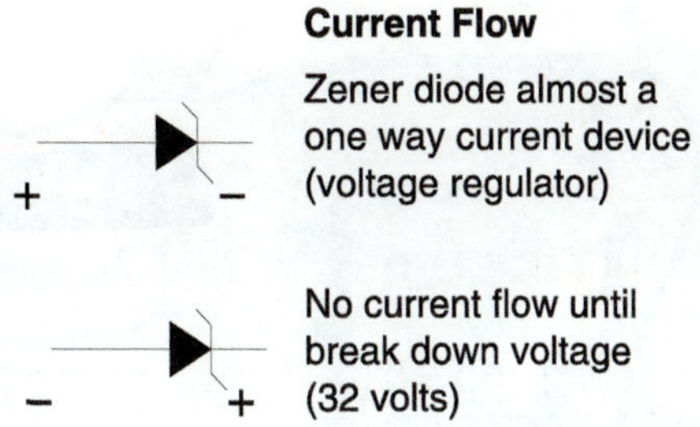

20.3 TRANSMISSION ECU

The transmission control electronics may use either two modules or a single module depending on the generation. As mentioned a little earlier, the two-module system loosely divided the controller functions into processing and switching. The two modules were networked by a proprietary bus connection. In the description here we reference the single-module system and refer to that module as the transmission ECU. The transmission ECU is multiplexed to the chassis J1939 bus so it can be regarded as being in real-time communication with the engine ECM.

The transmission ECU is located on the left side of the transmission. It is positioned by dowels and fastened by bolts threaded into the transmission case. It interfaces with the electronic circuit by means of a dedicated harness known as the transmission harness. All of the input and output circuits are connected to the ECU by means of this harness. The transmission ECU and its location on a current UltraShift transmission are shown in **Figure 20–20**

FIGURE 20–20 Left side view of an UltraShift AMT showing the location of the transmission ECU

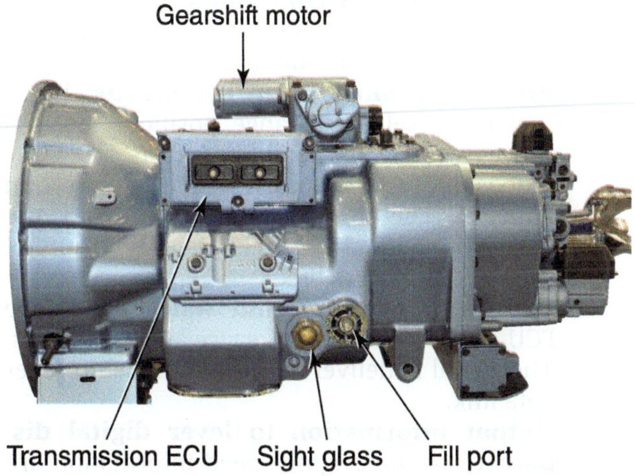

FIGURE 20–21 Location of the transmission ECU on a Detroit DT-12 AMT

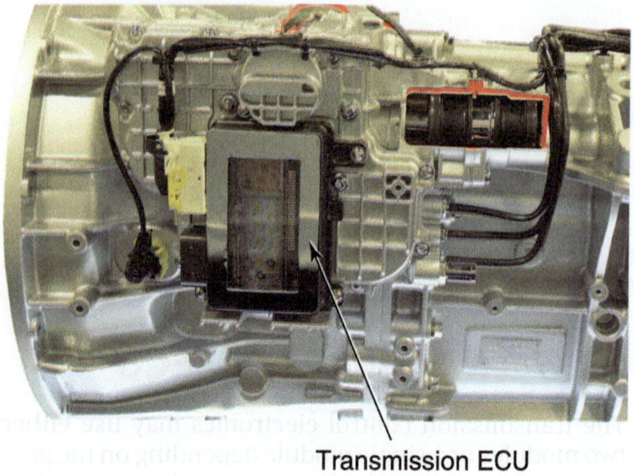

and **Figure 20–21** show the location of the ECU on a Detroit DT-12.

TRANSMISSION ECU FUNCTIONS

The transmission ECU performs the following functions:

- **Controlling logic power**. The ECU is responsible for switching system logic power on and off.
- **Managing shifting**. All processing functions required of the AutoShift transmission take place in the ECU. It processes desired outcomes based on programmed shift algorithms (logic maps) in its data (memory) banks and converts these commands to the ECU output drivers. The processing logic is based on programmed maps in system memory, data broadcast from the chassis data bus, and input data from the command and monitoring circuits.
- **Processing fault conditions**. The transmission ECU receives and can act on diagnostic fault codes when required.
- **Processing shift lever commands**. When the driver selects a range position, the ECU processes the command and effects providing it does not result in transmission or drivetrain damage. The ECU communicates with the driver-actuated shift mechanism by means of the shift lever data link.
- **Controlling service and wait lights**. The ECU can illuminate the service and wait lights. The signal is delivered through the shift lever data link.
- **Output information to lever digital display**. Gear and engine speed information are provided by the ECU to the digital gear status display. This status information is also sent through the shift lever data link.
- **Driving outputs**. The ECU is responsible for switching the shifter motors, the range valve, and the optional inertia brake. It can also broadcast speed, gear range, and diagnostic data to the chassis data bus and ESTs.

INERTIA BRAKE

The **inertia brake** system on Eaton AMTs helps stop the countershafts on initial engagements when the clutch is either out of adjustment or defective. The inertia brake consists of a solenoid and brake. The solenoid controls the inertia brake by switching regulated air pressure to act on the inertia brake piston and load the friction and reaction discs, which brake the countershaft. The solenoid is energized at 12V. The inertia brake is mounted at the 6- or 8-bolt PTO. When air pressure is applied, the PTO gear is connected with the grounded shaft, which in turn connects the countershaft to the transmission case, stopping it. When the inertia brake needs to be released, the transmission ECU ceases to energize the solenoid and the air supply acting on the inertia brake piston is exhausted. This frees up the friction and reaction discs, permitting the countershafts to turn freely. A sectional view of the inertia brake assembly is shown in **Figure 20–22** and its location on the transmission is shown in **Figure 20–23**.

CAUTION

When removing the inertia brake solenoid, ensure that system air pressure is completely relieved and the air/filter regulator is removed.

FIGURE 20–22 Sectional view of the inertia brake

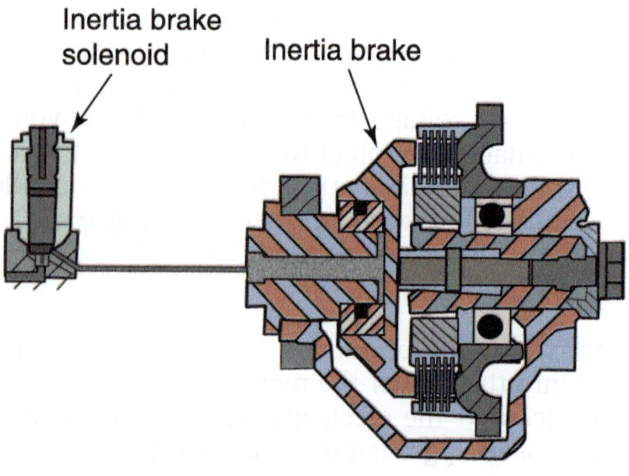

FIGURE 20–23 Location of inertia brake on transmission

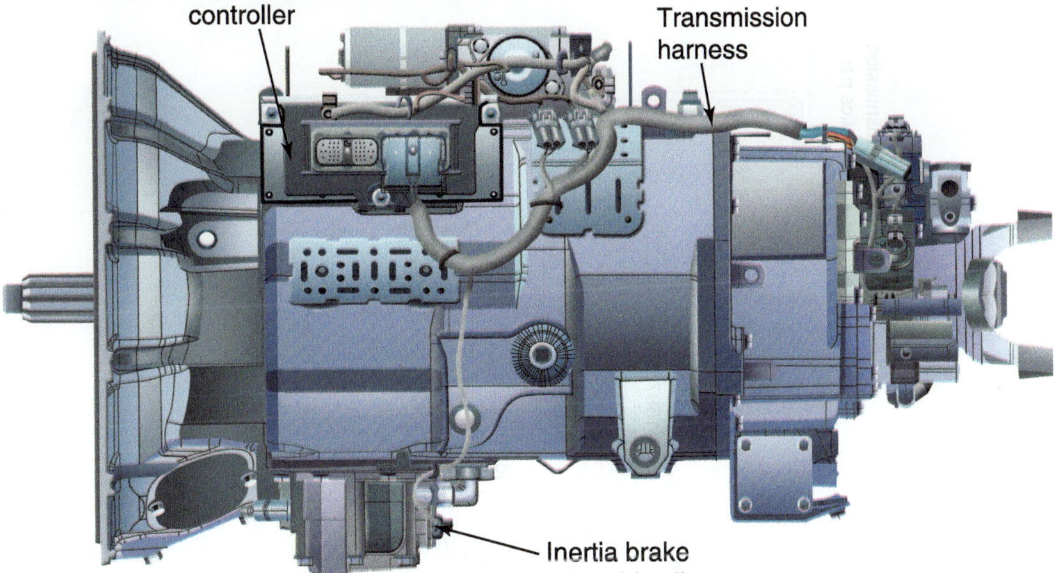

VEHICLE AUTOMATION COMPONENTS

The vehicle automation components connect the transmission electrical and electronic circuits with those of the vehicle. This group of components includes:

- Transmission ECU
- Gear display
- Shift lever
- Power connect relay
- Start enable relay
- Vehicle harnesses

CAUTION

The shifter module must be calibrated before a vehicle is placed in operation.

ELECTRIC SHIFTER CALIBRATION

The transmission ECU calibrates the electric shifter each time the ignition key circuit is shut down. The calibration procedure accounts for the clicking noise heard each time the key is turned off. When the calibration is complete, the ECU stores the shifter calibrations in its memory.

The calibration procedure requires that the neutral position be located by performing a rail sweep. The rail select and rail position motors move the shift finger, and the ECU records the point at which the sensors stop moving at the extremity of each sweep.

Data Retention

The ECU stores fault code data in memory before switching off the power connect relay during each shutdown.

SHIFT LEVER

The shift lever signals driver commands to the transmission ECU. It also displays data to the driver by means of service and wait lights. The chassis OEM determines the location of the shift lever assembly but, typically, it is located within comfortable reach of the driver beside the driver seat. Most AMTs opt to use the manufacturer shift lever assembly or a chassis OEM version. The shift lever used can be a touch keypad, steering column paddle, or shift lever and console. **Figure 20–24** is a complete wiring schematic of an UltraShift Gen3 DM3 AMT when an Eaton shift console is used. Note how the transmission electronic network into the chassis (J1939) data bus.

Shift Lever Operation

The shift lever is moved mechanically by the operator. **Hall effect principle** switches are used to signal the range request position of the shift lever to the ECU. A magnet is located at the bottom of the shift lever below the display console. Its magnetic field will have the effect of closing the normally open Hall effect switch, sending a current signal to the ECU. In this way, a range request command is made to the ECU.

FIGURE 20–24 Wiring schematic of an Eaton UltraShift Gen3 DM3 AMT showing all the electronic components; this diagram shows an Eaton shift console. Note the relationship between the transmission electronics and the J1939 data bus.

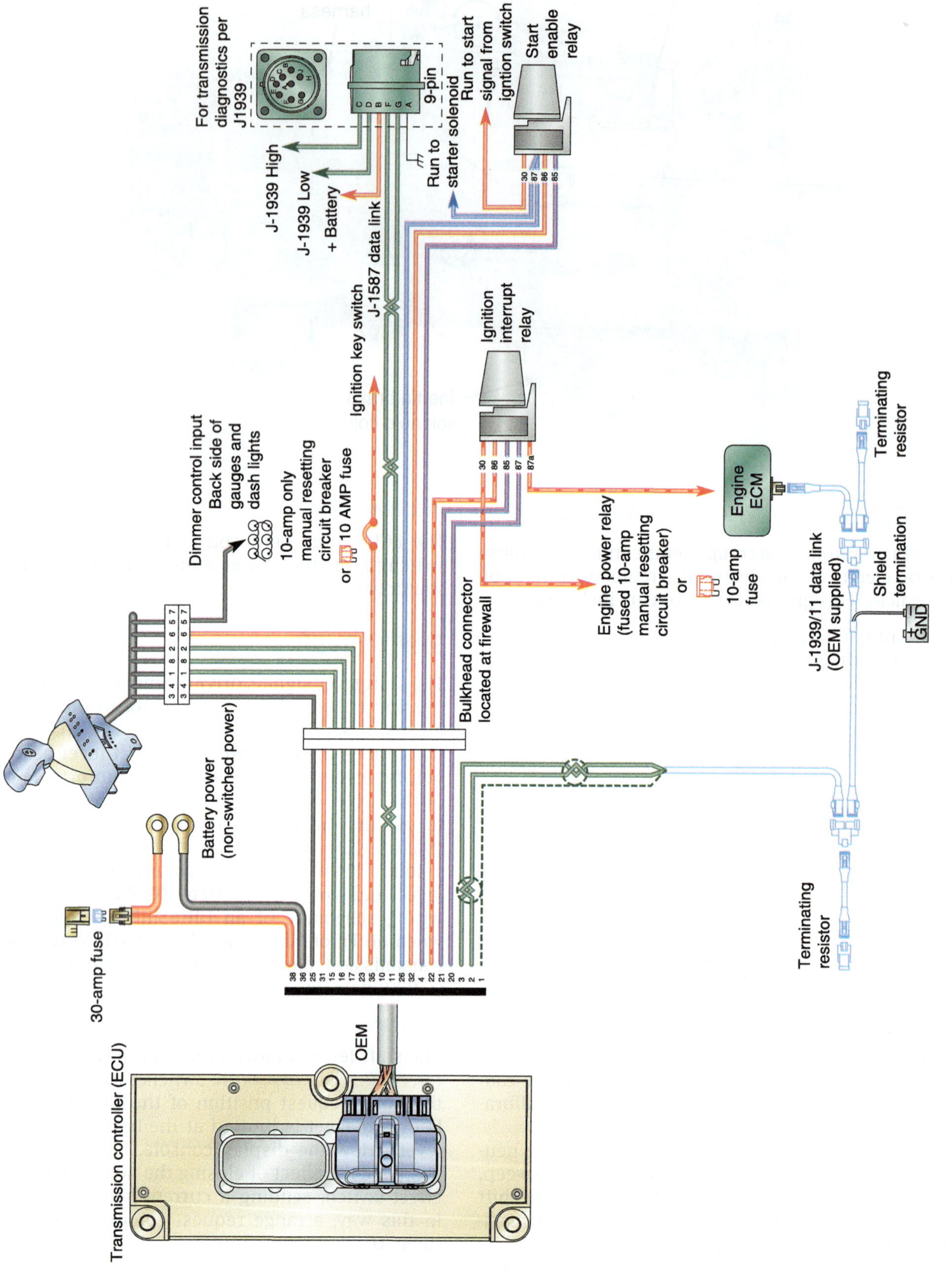

Service Light

The service light displays fault information to the driver or technician. A flashing or continuous illumination of the service light indicates that a fault code has been logged. Logged fault codes can then be pulse displayed by flashing numeric codes on the service light.

Detents

The detent mechanism locates the transmission shift lever in the correct location in each range select position. It provides an element of gearshift feel to the driver when shifting from one gear to another.

GEAR DISPLAY

The gear ratio status can be displayed in a number of ways. The display signals are output from the transmission ECU. The information displayed tells the driver the shift status in range and the target gears during shifting. When a shift is initiated, the target gear is displayed continuously on the gear display light-emitting diodes (LEDs) and, depending on the chassis dash electronics, on the driver digital dash electronics. When the transmission shifts to neutral, the target gear begins to flash. When a shift sequence is completed, the display ceases to flash and the new range status is displayed.

VEHICLE HARNESSES

Vehicle harnesses connect an AMT electrically and electronically with the truck chassis as can be seen in Figure 20–24. Typically up to three vehicle harnesses may be used:

- *The transmission interface harness* was used to connect the transmission harness to the tower harness and ECU.
- *The tower harness* connects the transmission ECU to the shift lever and power connect relay and provides a service port for voltage checks. It connects all the components in the shift tower.
- *The vehicle interface/dash harness* connects the transmission ECU to the chassis electrical and electronic circuits.

20.4 FAULT DETECTION AND DATA LINKS

All AMT transmission electronics are designed to detect faults within the transmission electrical/electronic circuits as well as faults that occur in other vehicle systems on which it is dependent. Connecting to the transmission electronics may be performed directly by means of a proprietary diagnostic connector or by means of the chassis J1939 data link connector (DLC). **SAE J1939**–compatible electronics permit AMT electronics to interact with other on-board vehicle electronic systems.

FALLBACK MODES

When a fault code is logged, the AMT electronics are designed to assess all remaining functionality and then default to a limited operational mode. In AutoShift and UltraShift, four fallback modes may be used. Each is classified by the number of mechanical shifts that can be performed.

Five-Speed Fallback

In a five-speed fallback, the detected fault has occurred in the auxiliary section. The main section is fully operational. Shifting is therefore limited to the five speeds in the main box, in whichever range gear in the auxiliary box that can be achieved. For instance, when only LO range is available, the transmission will shift from first through fifth gear ratios. When only HI range is available, shifting is limited to sixth through tenth gear ratios.

One-Speed Fallback

One-speed fallback indicates a transmission fault where all shifting is inhibited. The transmission remains locked in a single gear ratio, usually the one in which the fault was logged.

In-Place Fallback

An in-place fallback results in the transmission remaining in whatever state it is at the time of the occurrence. This could be in gear or in neutral.

Downshift Fallback

The transmission will downshift when appropriate until the vehicle is brought to a stop and the transmission is in neutral.

Fault Codes

The procedure for accessing Eaton AutoShift is simple and is outlined in **Figure 20–25**. Note that the codes have changed in the current generation of AutoShift electronics due to the adoption of a single-module system.

FAULT CODE STATUS

An active code indicates a problem active at the time of reading the system. An inactive code (historic code) indicates the occurrence of a fault that has either been logged in the past or is not detected at the moment of reading the system. The presence of an inactive code may indicate that the vehicle has not been driven in the proper manner. AMT system troubleshooting should be performed using the manufacturer service literature that contains step-by-step troubleshooting guides.

FIGURE 20–25 Accessing, interpreting, and clearing UltraShift fault codes

Fault Code Retrieval/Clearing AutoShift Gen3

Retrieving Fault Codes

1. Place the shift lever in neutral.
2. Set the parking brake.
3. Turn the ignition key ON but do not start the engine. If the engine is already running, you may still retrieve codes, however, do not engage the starter if engine stalls.

4. **To Retrieve Active Codes:** Start with the key in the ON position. Turn the key OFF and ON two times within 5 seconds ending with the key in the ON position. After 5 seconds, the gear display begins flashing two-digit fault codes. If there are no active codes the gear display will show code 25, followed by PD (Product Diagnostic mode). The vehicle will not start in PD mode. You must turn vehicle key OFF and allow the system to power down to exit PD mode before restarting. **To start engine turn key off and wait for the transmission to power down.**
5. **To Retrieve Inactive Codes:** Start with the key in the ON position. Turn the key OFF and ON four times within 5 seconds ending with the key in the ON position. After 5 seconds, the gear display begins flashing two-digit fault codes. If no codes are active, the gear display will flash code 25 (no codes).

6. Two digit fault codes will be displayed in the gear display. Record the codes. A 1- or 2- second pause separates each stored code, and the sequence automatically repeats after all codes have been flashed.

Clearing Fault Codes

1. Start with the key in the ON position. Turn the key OFF and ON six times within 5 seconds, ending with the key in the ON position.

   Note: If the codes have been successfully cleared, the service lamp will come on and stay on for 5 seconds.

2. Turn the key OFF and allow system to power down.

Fault Code Index AutoShift Gen3

Fault Codes	Description
11.	No ECU. Operation test
12.	Improper ECU configuration
13.	J1939 control device
14.	Invalid shift lever voltage
16.	High integrity link
17.	Start enable relay
25.	NO CODES
26.	Clutch slip
27.	Clutch disengagement
28.	Clutch system
31.	Momentary engine ignition interrupt relay
32.	Loss of switch ignition power test
33.	Low voltage supply
34.	Weak battery voltage supply
35.	J1939 communication link
36.	J1939 engine message test
37.	Power supply
41.	Range failed to engage
42.	Splitter failed to engage
43.	Range solenoid valve
44.	Inertia brake solenoid coil
45.	Inertia brake performance
46.	Splitter solenoid valve
51.	Rail position sensor
52.	Gear position sensor
56.	Input shaft sensor
57.	Main shaft sensor
58.	Output shaft sensor
61.	Rail select motor
63.	Gear select motor
71.	Unable to disengage gear
72.	Failed to select rail
73.	Failed to select gear
74.	Engine failed to respond
75*.	Power down in gear
83.	Shift lever missing
84.	Shift controller device not configured
85.	Shift control device incompatible

An 88 may appear in the gear display which indicates self-testing.
*This code only set inactive.

Pretests

Transmission pretests should be performed before progressing to specific fault code isolation. Prechecks require that electrical inputs and pneumatic supply to the transmission be functioning properly. The pretests eliminate obvious high-level problems 1 before progressing to a detailed, sequentially stepped diagnostic procedure.

Symptom-Driven Diagnostics

A transmission is, above all, a mechanical device. Though system nonelectronic components are comprehensively monitored, it is possible that problems can occur in which active or inactive electronic codes are not logged. Never assume that every fault must be electronically monitored.

Fault Isolation Procedure

The manufacturer's troubleshooting service literature is required for most AMT troubleshooting. In addition, but not instead of, you can reference **Figure 20–26** as a simple guide. In the case of Eaton AMTs, the troubleshooting guide is a free download from http://www.Roadranger.com. The objective of any diagnostic procedure is to isolate the failure from all its possible causes. In all AMTs, there are the mechanical components of the main gearbox and the auxiliary box; the pneumatic control system; the electrical components, such as shift motors; and a comprehensive electronic management system. It is critically important to follow the troubleshooting guide so that the diagnostic approach is systematic.

Transmission Lubrication

Eaton recommends synthetic E-500 lubricant (PS-164 R7 spec) for AutoShift transmissions. E-500 lube is designed to run 500,000 linehaul miles (800,000 km) with no initial drain interval required. The alternative is to use a straight 40W mineral oil meeting military spec Mil 2104H, but this reduces the oil drain interval

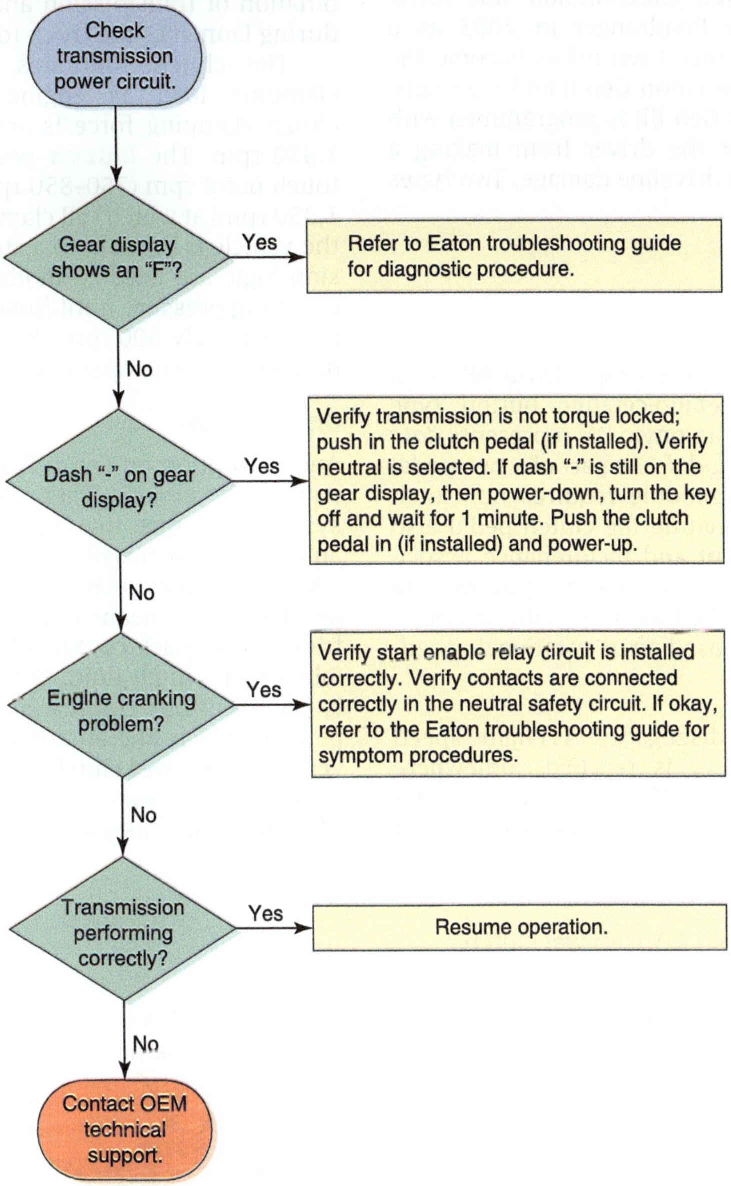

FIGURE 20–26 Eaton AMT troubleshooting tree

to a meager 60,000 miles (96,000 km or 1 year of operation). Most synthetic transmission lubricants exceed the stated viscosity grade ratings, so they can be expected to perform effectively through various geographic and seasonal temperature conditions.

Clutch Fluid

Eaton recommends two possible fluids for the wet clutches used on UltraShift AW AMTs:

- Synthetic Dexron III or IV
- Allison TranSynd

The clutch oil level should be checked at the dipstick with the engine running at idle rpm and the transmission in neutral.

20.5 AMT CLUTCH ACTUATORS

AMTs are designed to remove the driver's responsibility for managing gear ratios by mechanically shifting the transmission. AMT shift management becomes a function of the transmission controller; however, all AMTs use a mechanical clutch so a means of actuating it is required. This requires a **clutch actuator**, which may be any of the following:

- Centrifugal
- Hydraulic
- Pneumatic

In this section, we will take a look at a couple of the more common methods AMTs use to actuate clutches.

ULTRASHIFT

The UltraShift automated transmission was introduced by Eaton Fuller Roadranger in 2003 as a two-pedal system and since then it has become the market-leading AMT. The Eaton Gen II and later software (current version is Gen III) is programmed with algorithms that prevent the driver from making a shift that could result in driveline damage. Two types of clutch are available:

- UltraShift DM
- UltraShift AW

UltraShift DM

The DM clutch currently used with UltraShift is a 15½-inch, two-plate, metallo-ceramic button type with coaxial springs. The acronym DM represents *data link mechanical, clutch module*. As part of a two-pedal system, the DM clutch module requires no clutch pedal and linkage. Eliminating the clutch pedal also eliminates the adjustment and maintenance associated with driver-actuated clutch release systems. The DM clutch can be described as a centrifugal clutch because engagement is based on its rotational speed.

DM Clutch Operation

During idle, the clutch is disengaged. As engine speed increases, a **touch point** is reached somewhere between 750 and 850 rpm. The touch point occurs when the clutch departure (gap) is taken up. Beyond the touch point rpm, clutch clamping load begins to be applied. Engine speed is controlled by a combination of transmission and engine controller logic during launch of the truck to provide a smooth start.

The clutch continues to generate additional clamping load as engine speed increases. Full clutch clamping force is achieved at approximately 1,350 rpm. The **launch phase** occurs between the touch point rpm (750–850 rpm) and the rpm (around 1,350 rpm) at which full clamp load is achieved. When the vehicle is brought to a stop, engine and transmission logic are used to modulate clutch friction disc clamping pressure until full disengagement occurs at approximately 800 rpm. A parking pawl is integrated into more recent versions of UltraShift.

Ultrashift AW

Another option on Eaton Fuller AMTs is a fully automated **wet clutch**. This converts a two-pedal AMT system into one that handles similarly to a fully automatic transmission with a torque converter. The AW system flywheel connects to the transmission input shaft by means of a engageable/disengageable friction disc pack contained is a housing cooled and lubed with clutch fluid. Engagement/disengagement is via J1939 bus command. To make the wet clutch perform much like an automatic with a torque converter, Eaton's wet clutch uses a creep feature known as the **Urge-to-Move**. **Figure 20–27** shows an UltraShift transmission with a wet clutch.

FIGURE 20–27 UltraShift AW with wet clutch

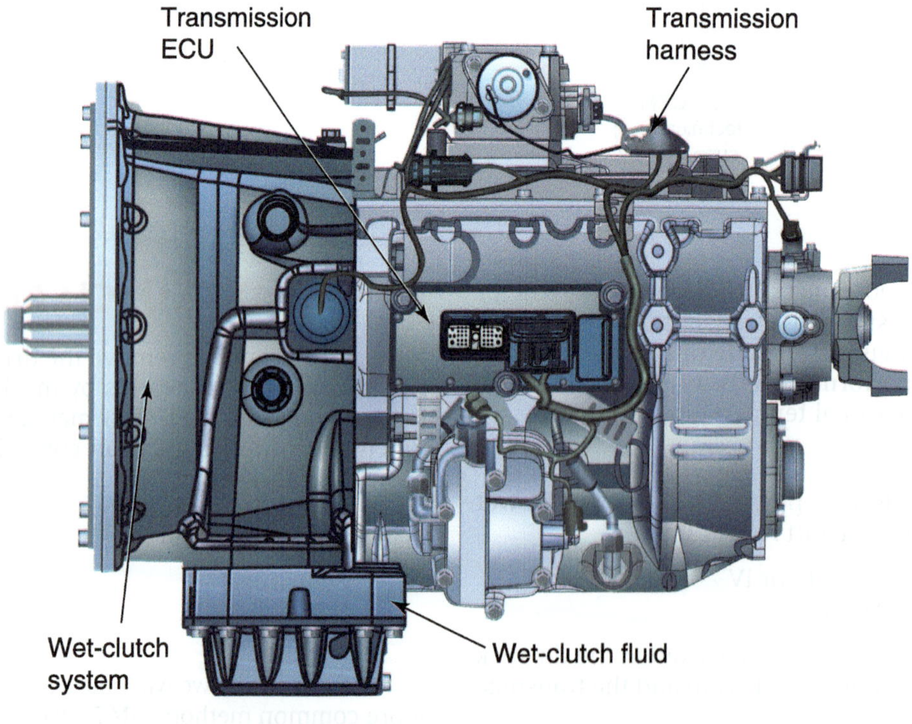

FIGURE 20–28 ECU-controlled pneumatic clutch actuator used a Detroit DT-12 AMT.

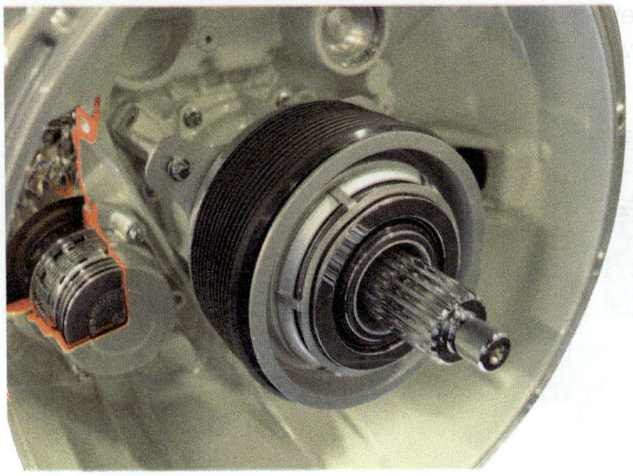

Urge-to-Move loads a continuous 45 lb-ft. (40 N·m) of torque to the transmission through the wet-clutch pack so that the driver senses engagement at any time the shift lever is in the drive position. The Urge-to-Move torque feature gives a typical truck the ability to move ahead on a 6 percent grade. This system requires a cooler for the transmission fluid and synthetic fluid (Dexron III or IV, or Allison TransSynd). The oil drain interval is specified at 150,000 miles (250,000 km) or 3 years, whichever comes the soonest.

Pneumatic Actuator

Another option for managing clutch release is to use a pneumatic actuator. The transmission ECU computes the appropriate shift points and electrically switches a pneumatic actuator to release the clutch. The Detroit DT-12 uses an ECU-controlled pneumatic clutch actuator shown in **Figure 20–28**.

20.6 HYBRID TRANSMISSIONS

Hybrid transmissions can be classified by the type of drivetrain they are located in. This can be:

- Hybrid electric
- Hybrid hydraulic

The purpose of a hybrid transmission is to option more than a single source of driveline torque and recapture driveline energy that would otherwise be lost. In this section, we will take a brief look at electric hybrid and hydraulic hybrid transmissions, both of which are required to play a role in "recycling" driveline energy that can—under the right circumstances—improve fuel efficiency. Hybrid electrical drive has become common in city transit buses as has hybrid hydraulic drive trains in medium-duty pick-up and delivery vehicles. **Figure 20–29** shows a cutaway of an Eaton hybrid transmission, which as you will see is a close relative of their AMTs.

FIGURE 20–29 Cutaway view of an Eaton hybrid-electric transmsion. Note the similarity between this and an AMT.

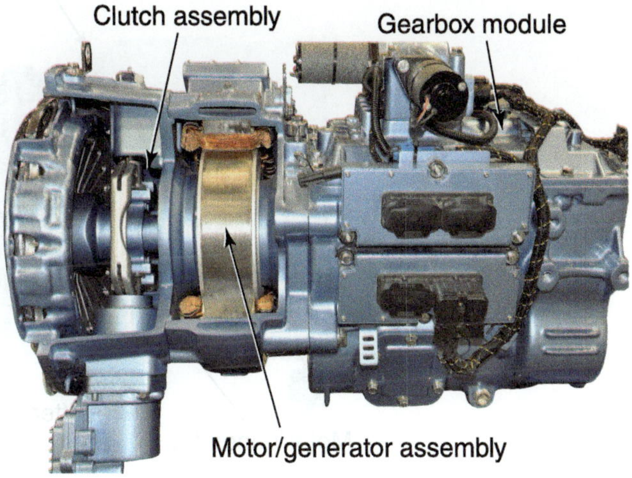

HYBRID ELECTRIC DRIVE

Hybrid electric drive has become common in many intracity commercial vehicles especially transit buses. Electric hybrid drive can produce improved fuel economy in start-stop type applications and reduce noise. The primary power source of the vehicle is typically a diesel or natural gas engine but there is a motor-generator unit within the hybrid transmission. When functioning as a generator, the motor-generator unit converts driveline mechanical energy into electrical energy to be stored in battery or capacitor banks. It then reverses its role to use the stored electrical energy to provide driveline torque. **Figure 20–30** shows an Eaton hybrid transmission.

Working on electric hybrid drivetrains requires specialized training. However, because of the abundance of these vehicles in intracity transit operations, it is critically important that technicians understand that potentially lethal voltages are stored in these systems. Class O rubber gloves should be worn when working anywhere in the vicinity of the chassis electrical systems and technicians should never take any risks when working around high-voltage equipment.

CAUTION

Hybrid electrical drivetrains produce and store lethal voltages. Make sure you have received specialized training before working on hybrid electrical vehicles and wear Class O rubber gloves when working close to the electrical systems.

FIGURE 20–30 An Eaton hybrid electric drive transmission: overhead and side views.

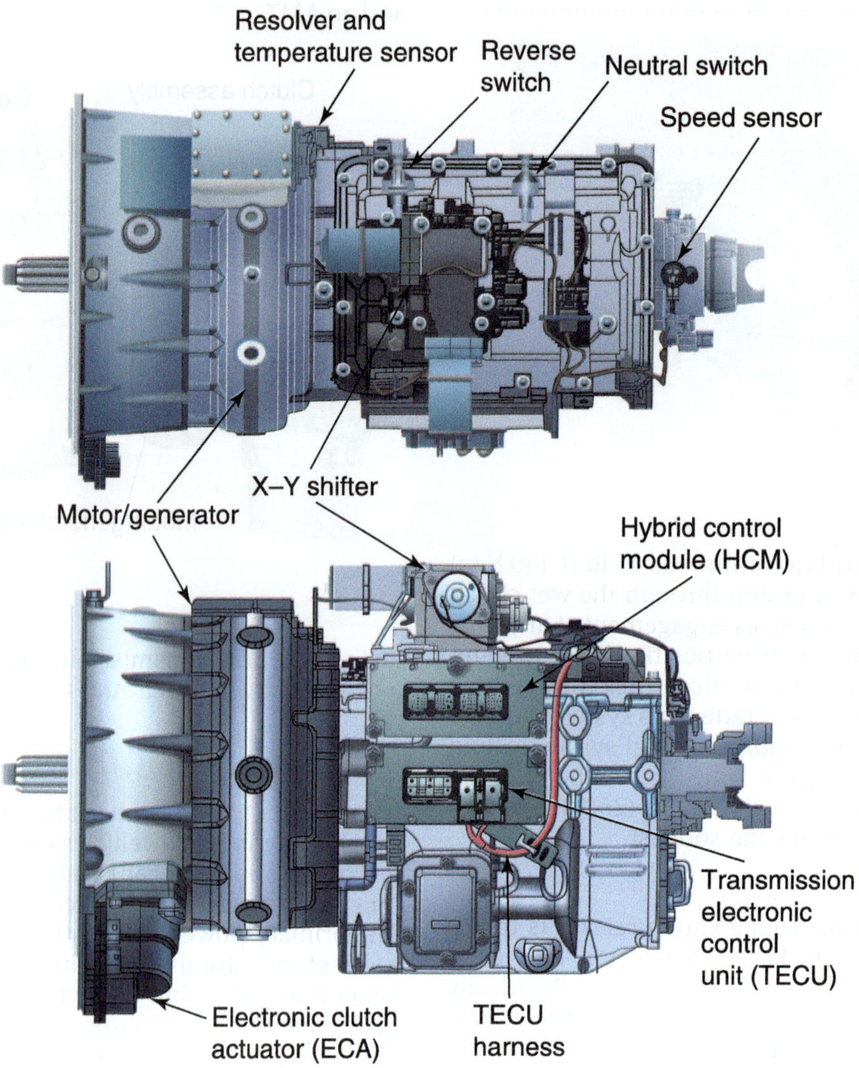

Rectifiers and Inverters

The motor-generator in hybrid electrical systems generates alternating current (AC) and, in turn, is driven by AC. However, the electrical energy it produces has to be stored in battery and capacitor banks as direct current (DC); this requires a cycle of rectification (AC to DC) and inversion (DC to AC). The generating principle is identical to the alternators studied in Chapter 8. The solid-state inverters seldom fail but these produce heat when operating, another source of potential danger. In summary, technicians are reminded that they must be trained before working on these systems, and caution must be exercized when working around the electrical systems on hybrid vehicles. **Figure 20–31** shows a cutaway of an Eaton hybrid electric transmission highlighting the motor-generator.

FIGURE 20–31 Cutaway view of an Eaton hybrid-electric transmission highlighting the motor-generator unit

HYBRID HYDRAULIC DRIVE

Hybrid hydraulic drive is being used in many intra-city commercial vehicles, especially in start-stop type applications. The system described briefly in this section is the Eaton **hydraulic launch assist (HLA)**. As with hybrid-electric systems the primary power source of the HLA vehicle is typically a diesel or natural gas engine but integrated into the drivetrain is a transfer case which can generate and receive accumulated hydraulic pressure. The energy storage device in the HLA system is a nitrogen gas-loaded accumulator (see Chapter 13) as shown in **Figure 20–32**.

The accumulator is charged by a hydraulic pump/motor unit driven by driveline torque. The role of the hydraulic pump/motor unit can then be reversed to

FIGURE 20–32 Eaton hydraulic launch assist (HLA); system layout on vehicle

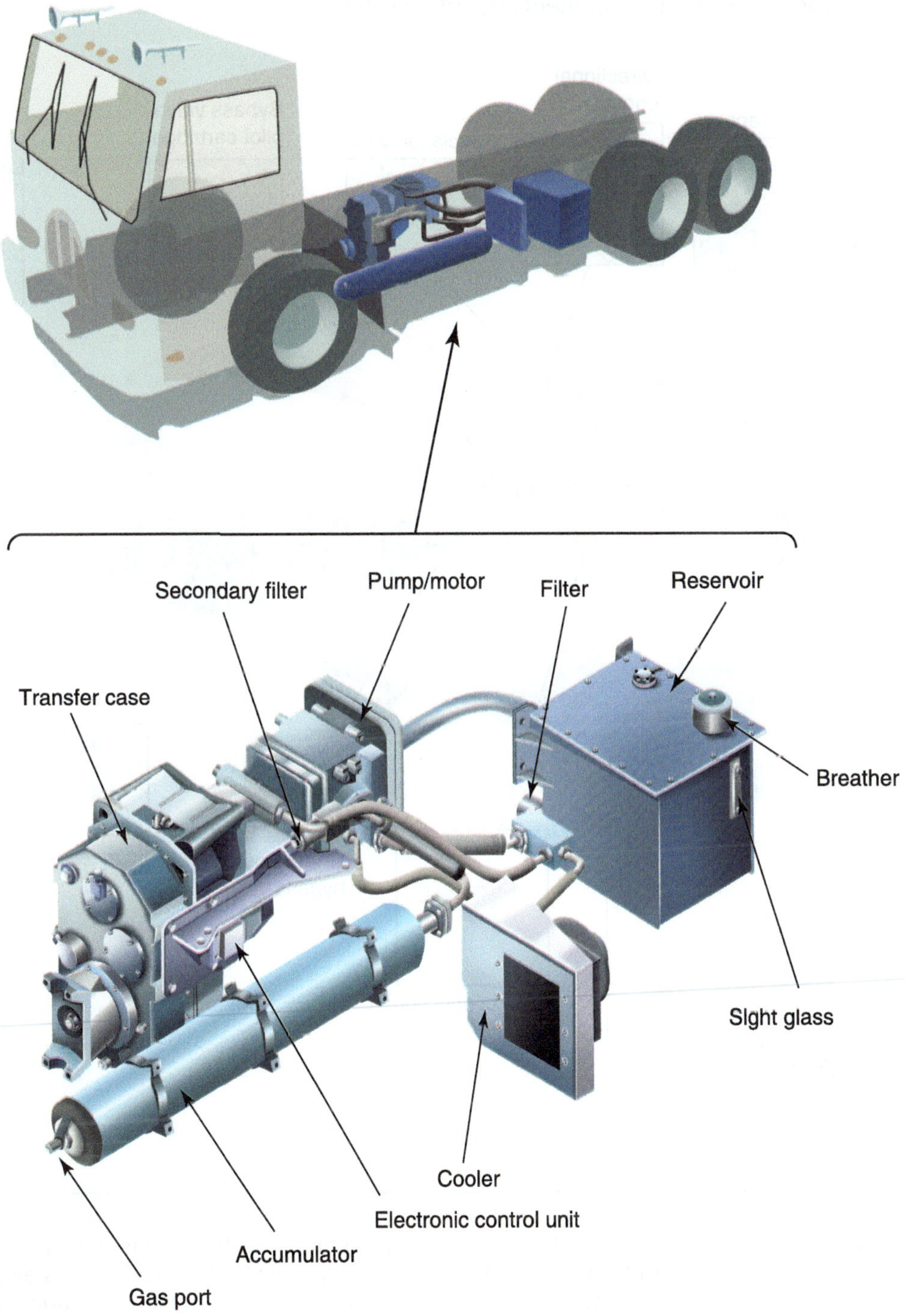

allow the accumulator to discharge stored hydraulic energy to the pump/motor unit to help produce drive torque. **Figure 20–33** shows the key components in an HLA system.

Working on hybrid hydraulic drivetrains requires specialized training. Technicians must be aware that hydraulic pressures of up to 2,000 psi can be stored in the system when the engine is not running. In addition, when the accumulated hydraulic pressure in the system is relieved, the nitrogen accumulator is under lethal pressure. If you work in a fleet that has hydraulic hybrid vehicles, ensure that you understand some of the hazards that are highlighted in **Figure 20–34**.

CAUTION

Hybrid hydraulic drivetrains are loaded under potentially lethal gas and hydraulic pressures. Make sure you have received specialized training before working on hybrid hydraulic vehicles.

FIGURE 20–33 Eaton HLA system components layout on vehicle

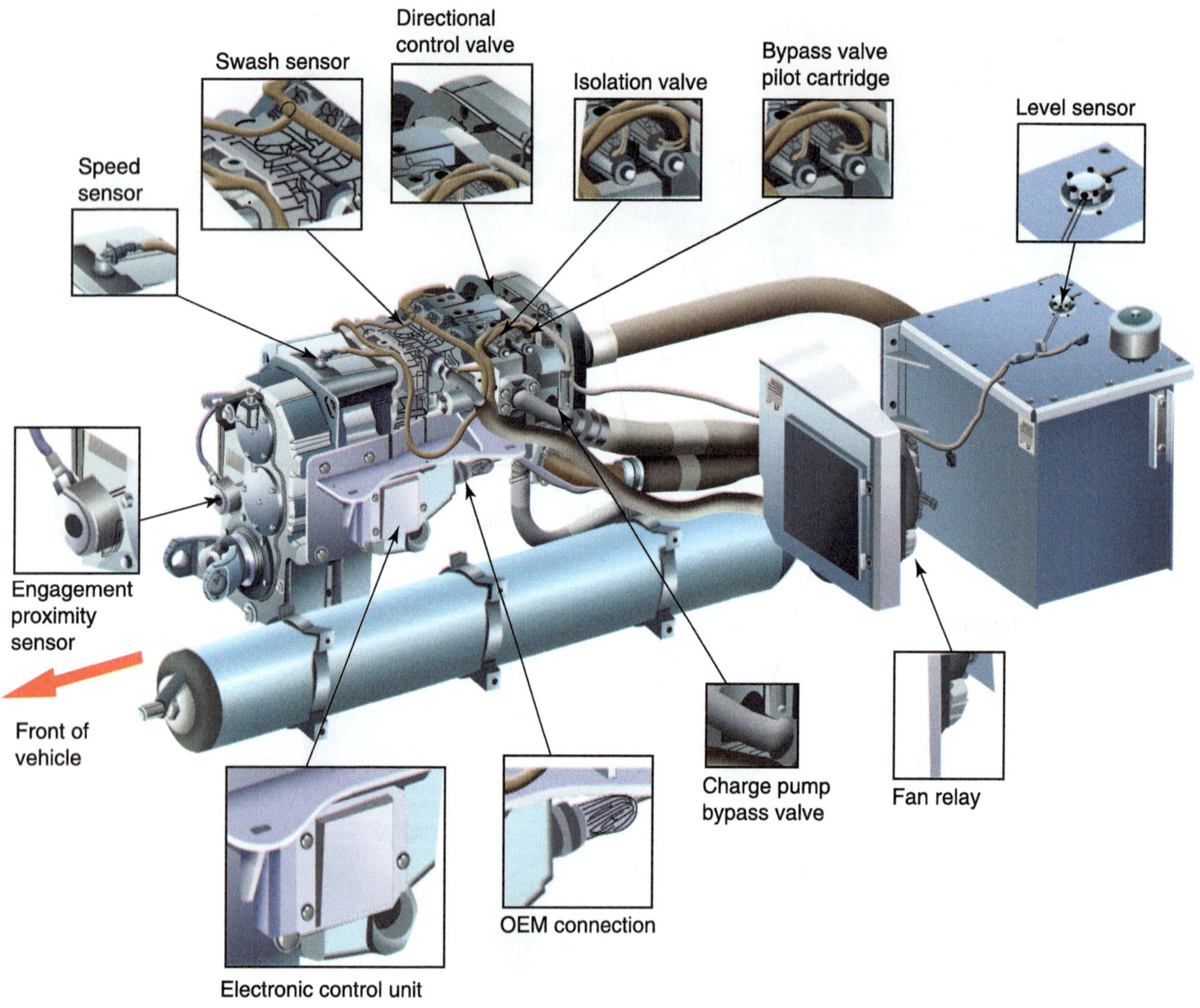

SUMMARY

- Automated standard transmissions (AMTs) use electronic controls to adapt a standard mechanical transmission for either semiautomated or fully automated shifting. This technology allows the driver to keep both hands on the steering wheel during shifting and significantly reduces driver fatigue.

FIGURE 20–34 Eaton HLA cautions

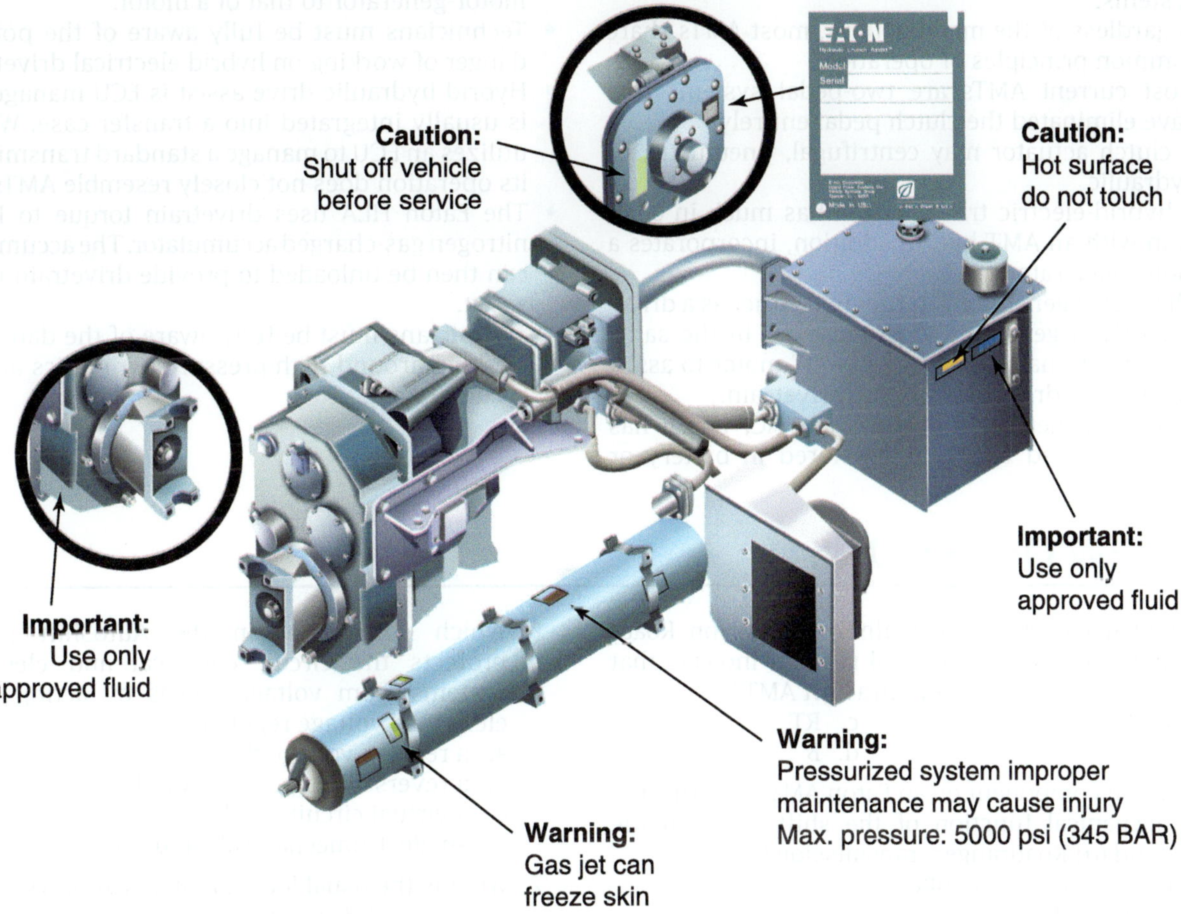

- AMTs can be divided into three-pedal systems that use a clutch pedal, and two-pedal systems that eliminate the clutch pedal. All AMTs use a clutch, but the clutch is controlled by the system electronic control unit (ECU).
- Eaton Fuller AMTs are based on a standard Roadranger twin countershaft mechanical transmission platform.
- Although some early generations of AMTs used a two-module controller system, all current generations of AMTs use a single controller. This is usually known as a transmission ECU. The transmission ECU performs the logic processing and output the switching operations required by the system.
- The transmission ECU manages shifting based on command signals received from the input circuit components and processes them using stored maps in its memory and data received by its J1939 multiplex connection. The ECU also logs an audit trail and fault codes.
- Shifts are effected by an electric shifter assembly, which can respond to manual and ECU commands. The electric shifter assembly replaces the gearshift lever in a standard Roadranger transmission. A pair of reversing electric motors actuated by the ECU precisely locates a shift finger in the shift block recesses in the shift rails to shift gear ratios.
- The rail select motor is used to move the shift finger transversely (side to side) with the objective of locating it precisely in the shift recess of one of the three shift rails.
- The gear select motor moves forward and backward. After the shift finger has been positioned over the correct shift rail, the gear select motor is energized to move the rail into gear.
- The means used by AMTs to select gears is identical to that in a standard transmission: Shift finger movement first selects one of the three shift rails and then slides that rail fore or aft.
- Fault codes can be displayed by the AMT diagnostic (service) lights or be broadcast to a digital driver information display on the dash. They can also be read by a handheld diagnostic EST or a personal computer (PC) with the appropriate software.

- The electronics are SAE J1939 compatible, which permits multiplexing with other chassis electronics systems.
- Regardless of the manufacturer, most AMTs share common principles of operation.
- Most current AMTs are two-pedal systems that have eliminated the clutch pedal entirely.
- A clutch actuator may centrifugal, pneumatic, or hydraulic.
- A hybrid-electric transmission has much in common with an AMT but, in addition, incorporates a motor-generator.
- The motor-generator can function either as a driveline-driven generator to produce AC in the same way an alternator does or as an AC motor to assist in applying drive toque to the drivetrain.
- A motor-generator unit generates AC, which has to be rectified to DC to be stored in battery or capacitor banks. When required, stored DC can then be inverted to AC to reverse the role of the motor-generator to that of a motor.
- Technicians must be fully aware of the potential danger of working on hybrid electrical drivetrains.
- Hybrid hydraulic drive assist is ECU managed but is usually integrated into a transfer case. While it utilizes an ECU to manage a standard transmission, its operation does not closely resemble AMTs.
- The Eaton HLA uses drivetrain torque to load a nitrogen gas-charged accumulator. The accumulator can then be unloaded to provide drivetrain torque assist.
- Technicians must be fully aware of the dangers of working around high-pressure hydraulics and gas accumulators.

REVIEW QUESTIONS

1. Which letters in the suffix of the Eaton Roadranger transmission serial number indicate that the transmission is an UltraShift AMT?
   a. AS
   b. DM
   c. RT
   d. B
2. Which component on an Eaton AMT performs the mechanical function of the shift lever on the standard Roadranger transmission?
   a. the gear select yoke
   b. the shift shaft
   c. the worm gear
   d. the shift finger
3. Which of the following correctly describes the gear select motor used on an AutoShift transmission?
   a. a reversing DC motor
   b. an unidirectional DC motor
   c. a reversing AC motor
   d. an unidirectional AC motor
4. What type of sensor is used to signal shift finger position to the transmission ECU in a current Eaton AMT?
   a. a piezoelectric sensor
   b. a photoelectric sensor
   c. a Hall effect sensor
   d. an inductive pulse
5. Which of the following components is responsible for effecting shifts in the transmission auxiliary box?
   a. the range valve solenoid
   b. the rail select motor
   c. the power module
   d. the reverse switch
6. Which component in the AutoShift system protects the circuit electrical and electronic systems from voltage spikes by acting as an electronic voltage regulator?
   a. a reverse voltage diode
   b. a reverse voltage zener diode
   c. a virtual circuit breaker
   d. an electromechanical circuit breaker
7. What is the usual location of an Eaton AMT ECU?
   a. on the shift tower
   b. on the left side of the transmission
   c. on top of the transmission
   d. under the cab dash
8. What is the function of the inertia brake used in the AutoShift transmission?
   a. It slows the input shaft during driveline torque release.
   b. It slows the countershaft during initial engagement.
   c. It is used to synchronize all shifts.
   d. It effects driveline retarding.
9. To which Eaton AMT component is the driver shift lever signal command directed?
   a. the dash display ECU
   b. the gear select motor
   c. the transmission ECU
   d. the power module
10. What component does a three-pedal AMT have that is *not* used on a two-pedal system?
    a. a clutch
    b. a clutch pedal
    c. an auxiliary box
    d. an ECU

11. When a five-speed fallback fault is displayed by Eaton AMT electronics, where is the source of the fault?
    a. fifth gear
    b. first gear
    c. fourth and fifth gears
    d. auxiliary section
12. What would be the performance outcome if an Eaton AMT defaults to one-speed fallback mode?
    a. All shifting is inhibited.
    b. First to second shifts are inhibited.
    c. Second to first downshifts are inhibited.
    d. Shifting is inhibited in one gear only.
13. Technician A says that when an AMT is described as a two-pedal system, the clutch is eliminated. Technician B says that when an AMT is described as three-pedal system, the clutch pedal is eliminated. Who is correct?
    a. Technician A only
    b. Technician B only
    c. both A and B
    d. neither A nor B
14. An Eaton AMT will not shift out of second gear and Technician A says that an in-place (logged) fallback failure strategy could be the cause. Technician B says that when the AMT is in downshift fallback, only downshifts can be made until the vehicle is brought to a standstill. Who is correct?
    a. Technician A only
    b. Technician B only
    c. both A and B
    d. neither A nor B
15. Technician A says that AMT faults can only be read by the OEM software. Technician B says that AMT fault codes can be accessed by connecting to the chassis data bus. Who is correct?
    a. Technician A only
    b. Technician B only
    c. both A and B
    d. neither A nor B
16. Which of the following AMT identification letters indicate that an UltraShift transmission uses a wet clutch?
    a. AW
    b. DM
    c. RT
    d. WC
17. Which of the following AMT identification letters indicate that an UltraShift is equipped with a centrifugal clutch?
    a. CC
    b. DM
    c. RT
    d. AW
18. Technician A says that the motor-generator used in an Eaton hybrid electric transmission generates AC. Technician B says that the motor-generator used in an Eaton hybrid-electric transmission can be driven by AC. Who is correct?
    a. Technician A only
    b. Technician B only
    c. both A and B
    d. neither A nor B
19. Technician A says that the UltraShift AMT uses a pneumatic clutch actuator. Technician B says that the Detroit DT-12 AMT uses a pneumatic clutch actuator. Who is correct?
    a. Technician A only
    b. Technician B only
    c. both A and B
    d. neither A nor B
20. What is the energy storage device used in an Eaton HLA transmission?
    a. capacitor bank
    b. battery bank
    c. nitrogen gas-loaded accumulator
    d. mechanical spring loaded accumulator

21

Prerequisites: Chapters 6, 11, 12, 17, 18, and 19

ELECTRONICALLY CONTROLLED AUTOMATIC TRANSMISSIONS (ECATs)

OBJECTIVES

After reading this chapter, you should be able to:

- Identify the Allison and Caterpillar families of electronically controlled, automatic transmissions (ECATs).
- Describe the modular designs used by Allison WT and TC10 transmissions.
- Outline how Allison WT, Allison TC10, and Caterpillar CX transmissions use full authority management electronics to manage shifting and communicate with other vehicle electronic systems.
- Identify some of the service and repair advantages of the modular construction of Allison WT and TC10 transmissions.
- Identify the components used in a Caterpillar CX and note the differences between six- and eight-speed models.
- Describe how the electronic control modules on both Allison and Caterpillar ECATs master the operation of the transmission.
- Define the terms *pulse width modulation*, *primary modulation*, and *secondary modulation*.
- Describe how base versions of Allison WT and Caterpillar ECATs use interconnected planetary gearsets to stage gearing to provide six forward ranges, reverse, and neutral.
- Outline the differences between Caterpillar CX six- and eight-speed transmissions.
- Describe the integral driveline retarder components and operating principles used in electronic automatic transmissions.
- Outline the function of the dropbox option in one Allison WT model.
- Describe the role of the electrohydraulic controls used in Allison and Caterpillar ECATs.
- Identify the electronic components used in ECATs and classify them as input circuit, processing, and output circuit components.
- Describe how ECAT electronics interface with the J1939 data bus to optimize vehicle performance.
- Describe how the electrohydraulic clutches are controlled.
- Outline the torque paths through WT and TC10 transmissions in each range selected.
- Describe how diagnostic codes are logged within ECATs and the manner in which they are displayed.
- Outline some maintenance practices used on ECATs.
- Perform some basic diagnostic troubleshooting on Allison and Caterpillar ECATs.
- Calibrate a Caterpillar CX transmission.
- Outline the operation of a Voith transmission.
- Identify the modules and components of an Allison TC10.

KEY TERMS

accumulator
adaptive logic
ADEM software
backfill pressure
control main pressure
counter-rotating torque converter
data bus viewer (DBV)
diagnostic data reader (DDR)
diagnostic optimized connection (DOC)
diagnostic trouble code (DTC)
digital display (DD)
dropbox
Drop-7
electrohydraulic control module
electrohydraulic valve body
electronic clutch pressure control (ECPC)
electronic control unit (ECU)
Electronic Technician (ET)
electronically controlled automatic transmission (ECAT)
fast adaptive mode
ISO 9141
J1939
main pressure
main pressure regulator
modules
planetary gearset
powershift
primary modulation
prognostics
pulse width modulation (PWM)
retarder
scavenging pump
secondary modulation
serial communications interface (SCI)
slow adaptive mode
source address (SA)
submodulation
TC10 Tractor Series (TS)
throttle position sensor (TPS)
transmission control module (TCM)
TranSynd
vehicle interface module (VIM)
World Transmission (WT)
WTEC controller

INTRODUCTION

This is one of the most complex chapters in this book and students are reminded that they should be familiar with the contents of the prerequisite chapters. A number of **electronically controlled, automatic transmissions (ECATs)** are available in North America. Many of these are found in buses, but they are becoming more common in vocational applications such as fire trucks and garbage packers and, more recently, on Class 8 highway trucks. They are frequently used in off-highway construction trucks. For over 30 years, Allison has dominated the heavy-duty highway automatic transmission market with the only real threat to that market share coming from offshore, specifically from the German companies ZF and Voith. In 2006, Caterpillar introduced its CX family of electronically controlled fully automatic transmissions for use in Class 8 vocational trucks. In 2013, Allison introduced its TC10 transmission, specifically designed for linehaul Class 8 trucks. This chapter focuses on:

- Allison **World Transmission (WT)**
- Allison TC10 **Tractor Series (TS)**
- Caterpillar CX

HISTORY

Allison began optioning computerized management of its automatic transmissions in 1982. The first versions of electronically controlled Allison transmissions were known as Allison Electronic Controls (ATEC); this evolved into Commercial Electronic Control (CEC). Both ATEC and CEC transmissions adapted Allison hydromechanical automatics for control by computerized logic. The degree of control by CEC is best described as *partial authority*. It should also be noted that ATEC and CEC are seldom seen today because they were replaced by Allison transmissions with full authority controls a generation ago. For that reason they are no longer covered in this textbook; previous editions may be referenced for information on the first generation of ECATs.

SHOP TALK

The term full authority is borrowed from the way we categorize engines. In the early days of engine management electronics, the term partial authority was used to describe a hydromechanical system that was adapted for electronic management. This description fits the first generation of electronic automatics (Allison CEC). A full authority engine was one that was designed to be managed by electronics and we also use the term to describe the ECATs introduced in this chapter.

CURRENT ALLISON FAMILY

The Allison WT and TC10 transmissions both use modular platforms and full authority electronic controls, but this is the only thing they have in common. The gearing arrangements in WT transmissions are driven entirely by complementary planetary gearsets while the newly introduced TC10 uses a revolutionary design that incorporates a torque converter, twin-countershaft main gearbox, and planetary gear output module. There are three general families of WT transmissions:

- 1000/2000 Series (light-duty on-highway)
- 3000/4000 Series (heavy-duty on-highway)
- 5000/6000 Series (heavy-duty off-road)

Although WT 3000 and 4000 transmissions can be used in linehaul trucks, the reality is that they are used primarily in vocational trucks and buses. The objective of the recently introduced Allison TC10 is to directly address the requirements of linehaul Class 8 trucks with a powershift ten-speed transmission. **Figure 21–1** shows a cutaway image of an Allison WT 4000.

CATERPILLAR FAMILY

Caterpillar offers three transmissions in its CX family: CX28, CX31, and CX35. These transmissions use a more traditional arrangement of complementary planetary gearsets but they significantly differ from their Allison competitors in the manner in which the electrohydraulic controls are managed. A variation of this transmission family has been used in off-road trucks and equipment for a number of years and is gaining popularity in its on-road version. CX transmissions are repair-friendly and are designed to be reconditioned during their projected service life.

ADVANTAGES OF ECATs

When comparing the performance of fully automatic electronically controlled transmissions with automated manual transmissions (AMTs), the most notable difference is that the automatics can be powershifted with little or no spin losses between ratio shifts. Our definition of the term **powershift** (in the context of automatic transmissions) is the ability to shift under full throttle demand without interrupting driveline torque. Another major advantage of fully automatic transmissions is the fact that a lower level of driver skill is required to handle range shifts, although this is also a feature of the AMTs that have become popular in recent years. Any transmission that is driven through a fluid coupling or torque converter, provides some forgiveness to many forms of driveline shock load, whether caused by bad driving or road conditions. When electronic controls are added to an automatic transmission package, range shift selection becomes more precise and in sync with engine rpm and load conditions. The result is that the drivetrain is better protected from driver abuse. All of the ECATs studied in this chapter are multiplexed on the **J1939** data bus to ensure that shifting is synchronized with the driver commands to the engine.

FIGURE 21–1 Cutaway view of an Allison WT 4000 transmission

Disadvantages

There are two disadvantages to the electronically controlled automatic transmissions. The first is upfront cost, which can be significantly higher than the cost of a comparable AMT. The second disadvantage is that they are more technically complex, so a higher level of skills is required to diagnose and recondition them.

21.1 ECAT MANAGEMENT

ECATs are managed by controllers known as **electronic control units (ECUs)**. ECU processing logic switches electromechanical devices to produce the desired shift ratio outcomes. The controller input data is combined with programmed maps in ECU memory to process an outcome. Input data refers to commands such as throttle position sensor (TPS), shift console requests, and monitoring sensor information such as shaft speeds, fluid pressure, and temperature.

After ECU logic processes a desired outcome, drivers are used to convert the results into action. ECU drivers receive low potential (voltage) signals resulting from the processing cycle and amplify these into higher potential electrical values required to switch the electrohydraulic devices in the transmission.

Switched output signals from ECU drivers are then relayed to actuate the electrohydraulic circuits within the transmission. These controlled electrohydraulic circuits ultimately determine how the transmission routes input torque from the engine to its output shaft. Within an ECAT, the electronics *manage* the hydraulic circuit. The loop that results is as follows:

- Logic processing (the system electronics and controller) drive
- Electrical devices (solenoids) that actuate
- Hydraulic devices (valves) that control
- Mechanical clutches that select ratios

ECAT HYDRAULICS

The hydraulic valves used in ECATs function similarly to their hydromechanical counterparts and it is important to underline the following:

- Hydraulic pressure works against spring pressure moving the valve.
- Higher hydraulic pressure versus lower hydraulic pressure: When the valve lands are of equal sectional area, higher pressure overcomes lower pressure either moving the valve or holding it in place.
- Valve lands of different sectional area: When equal hydraulic pressure acts on valve lands of different sectional area, pressure acting on the land with the larger sectional area will overcome that acting on the smaller land.

An **electrohydraulic valve body** is used to house the solenoids and actuating valves. Depending on the transmission, this component may be able to be serviced independently from the transmission.

Lube Pressure Management

In addition to being used as hydraulic media, transmission fluid used in automatic transmissions is used to lubricate and cool the components. The fluid is used to charge the torque converter during non-lockup operation. A lube relief valve is designed to trip open if the pressure exceeds a threshold value. In addition, most torque converters are equipped with a converter relief valve located in the pump assembly that trips open to allow fluid to return directly to the supply circuit when a threshold value is exceeded.

Main Pressure

Main pressure is initiated by a main pressure pump. For example, the main pressure pump in the Allison TC10 is a variable vane pump that provides fluid flow to the main pressure circuit. Main pressure is used for clutch applications and is managed by the main pressure regulator, the main modulation (mod) valve, and the main pressure relief valve.

Main Pressure Regulator

The main pressure regulator valve converts main pump fluid flow into main pressure. The pump regulator uses hydraulic pressure versus spring force and when the former overcomes the latter, the valve spills to the return circuit thereby reducing main pressure.

Main Modulator's Solenoid

The main modulator (mod) solenoid is used in some ECATs to control the pressure that acts on the main pressure regulator valve. When the solenoid is de-energized, the pressure acts with the main pressure mechanical spring force to increase main pressure; this means that main pressure is highest when the mod solenoid is not energized. When the mod solenoid is energized, the hydraulic pressure acting with spring pressure in the main pressure regulator is exhausted reducing main pressure. As current flow to the mod solenoid increases, more fluid is spilled to return and main pressure reduces accordingly.

Main Relief Valve

The main relief valve is located within the pump assembly in most ECATs. Its function is to protect the hydraulic circuit from excessive pressure under extreme cold start conditions. It is simple hydraulic pressure versus spring relief valve that spills to the sump when tripped.

Control Main Regulator

Control main pressure is a product of the transmission main pressure; it incorporates a regulator to remove pressure spikes and fluctuations, as well as providing smooth, consistent valve actuation. The control main pressure regulator incorporates a mechanical spring to absorb hydraulic pressure fluctuations. The main control regulator is normally located within the shift valve body assembly.

Exhaust Backfill Regulator and Relief Valves

The exhaust backfill regulator and a relief valve are designed to maintain a low exhaust **backfill pressure** within the clutch supply passages. A simple spring loaded relief valve controls backfill maximum pressure. Backfill is designed to maintain a slight hydraulic pressure within the circuit in order to eliminate air within the clutch and feed passages of the circuit. This helps to improve shift quality. The backfill regulator is usually located within the solenoid valve body assembly while the backfill regulator can be located anywhere in the backfill circuit.

CLUTCHES

Clutches are hydraulically actuated by the solenoid valve assemblies. They are therefore hydraulically actuated assemblies that consist of fiber friction discs and steel reaction plates that rotate independently when not actuated. Clutches can be designated as:

- Rotating: Rotating clutches connect components together or supply rotational input.
- Stationary: Stationary clutches are fixed position and when actuated do such things as hold planetary members.

Accumulators

Accumulators are used variously on ECAT hydraulic circuits to absorb pressure spikes and ensure precise shifts. For example, it is common to locate accumulators in each pressure control solenoid (PCS) and pressure control valve (PCV).

ECAT ELECTRICS AND ELECTRONICS

Controller logic is discussed briefly earlier in this chapter and in more detail in Chapters 6 and 11 so the objective is to underline some of the key electronic components and to describe how ECATs are networked to the chassis data bus.

Harnesses

ECATs commonly incorporate between one and three wiring harnesses. Typically, a chassis wiring harness is used to connect the transmission electronics with the chassis electrical system. Some systems use a separate harness to connect the shift console with the transmission ECU; an alternative is to use the J1939 bus to connect the shift console with the ECU. Most ECATs use an internal harness to connect the ECU with its electrohydraulic circuit. Almost all ECATs today are networked to the chassis data backbone; this is addressed later in this section.

Throttle Position Sensor

The **throttle position sensor (TPS)** is a key input command. It is based directly on accelerator pedal angle determined by the operator's boot. In all contemporary ECATs (networked to a data backbone), TPS command data is broadcast over the chassis data bus. TPS operation is studied in Chapter 6, either a potentiometer or Hall effect type TPS may be used.

Pressure Switches

Pressure switches are used in ECATs to provide valve position feedback to the transmission ECU. A range shift (upward or downward) would be aborted if the ECU did not recognize the anticipated change of pressure from the pressure switch within a specified amount of time. Pressure switches are used to confirm individual clutch actuation events, direction status (forward or reverse), and general transmission fluid pressure status. They can theoretically be located anywhere in the circuit in which the pressure has to be verified but in the example of an Allison TC10, a pair of pressure switch manifolds is used.

Shift status pressure switches are usually of the normally open (NO) type. They electrically close when the specified hydraulic pressure is present: Closing the pressure switch sends a signal to the ECU. This type of switch most often uses a single wire design and electrically grounds into the transmission housing to close the signal circuit.

"Monitoring" types of pressure switches are designed to signal an actual pressure value to the transmission ECU. Depending on the transmission, they may be located anywhere in the circuit and sometimes incorporate a temperature sensor (thermistor).

Solenoids

Solenoids are commonly located in ECAT electrohydraulic valve bodies. They are switched by the transmission ECU output drivers and their function is to direct hydraulic pressure into specific circuits and passages or exhaust the circuit. A variety of different types of solenoids may be used but some common categories are as follows:

- Nonlatching: Nonlatching solenoids have a plunger that is spring loaded to the "off" position. This plunger loads a check ball to block flow through the solenoid passage; in its "off" position, the solenoid exhaust port is open. When the plunger is energized, it retracts, closing the exhaust port and unseating the check ball, permitting flow through the solenoid passage (**Figure 21–2**). The moment the solenoid is de-energized, the spring closes the plunger. This type of solenoid usually permits hydraulic flow only when it is electrically energized...but the reverse may also be true.
- Latching: Latching solenoids must be momentarily energized in order to be switched, after which the plunger remains in position until energized again with reverse polarity. When a latching circuit solenoid is in the "open" position, it will permit hydraulic flow through its circuit until it receives an ECU switch command to close it.
- Variable bleed: Variable bleed solenoids (VBS) are proportioning solenoids. These are used to control multi-positional valves. When a proportioning solenoid is actuated, the mechanical position of the armature (to which the valve connects) depends on the current flowed through its coil. They are usually **pulse width modulation (PWM)** controlled.

In addition to being classified by operational principle, solenoids used in ECATs may also be classified by function. The following terms are used:

- On-off solenoids: These are three-port, two-position solenoids. They are used to option hydraulic fluid into a pair of circuits.

FIGURE 21–2 Solenoid operation

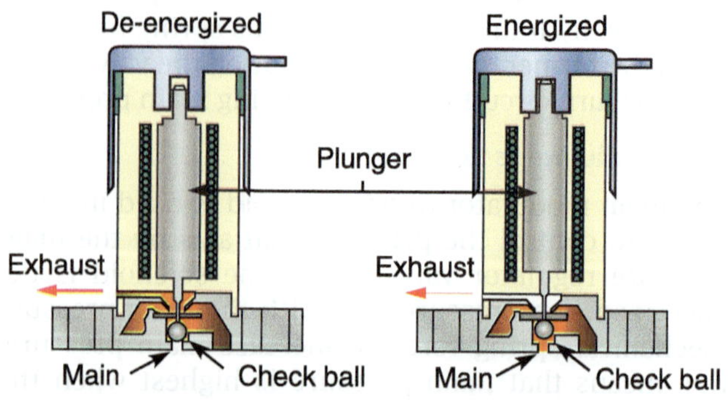

- Normal High (NH) solenoids: Used to describe a solenoid that outputs maximum control pressure when de-energized.
- Normal Low (NL) solenoids: Used to describe a solenoid that outputs minimum control pressure when de-energized.

ECAT J1939

The **source address (SA)** for an ECAT on the J1939 data bus is SA 03. The 3-pin connector terminal stub that connects an ECAT to J1939 has the following pin assignments:

- A terminal = CAN high yellow
- B terminal = CAN low green
- C terminal = Shield

In order to troubleshoot J1939 communication problems within an ECAT, the appropriate software for Allison transmissions should be used: Allison **data bus viewer (DBV)**, and on Caterpillar transmissions, **Electronic Technician (ET)**. Some basic bus diagnostics are introduced within this section, so that the concepts introduced in Chapter 12 can be applied to an actual vehicle system.

BUS TROUBLESHOOTING

The first level of diagnosing bus communication problems requires no more than an understanding of basic bus topology and a digital multimeter (DMM).

Bus Test

Ensure that the bus is powered down (ignition key off), and both termination resistors are in place: Use a DMM in resistance mode placing the probes across:

- 3-pin stub connector: Terminals A (CAN high) and B (CAN low)
- 9-pin diagnostic connector: Terminals C (CAN high) and D (CAN low)

Figure 21–3 shows the test points in a J1939 connector.

Test Results:

- **60 ohms:** Good
- **120 ohms:** One terminating resistor missing
- **0 ohms:** Short between CAN high and CAN low
- **40 ohms:** an additional resistor has been inadvertently been added
- **OL:** Open circuit, resistance exceeds the DMM's ability to perform a measurement

Bus Opens and Shorts

Bus backbone open circuits can affect *one* or *multiple controllers* on the bus but will most likely produce communication problems with the controller that is located the closest to the open circuit.

FIGURE 21–3 Powered-down resistance test points in on a J1939 diagnostic connector

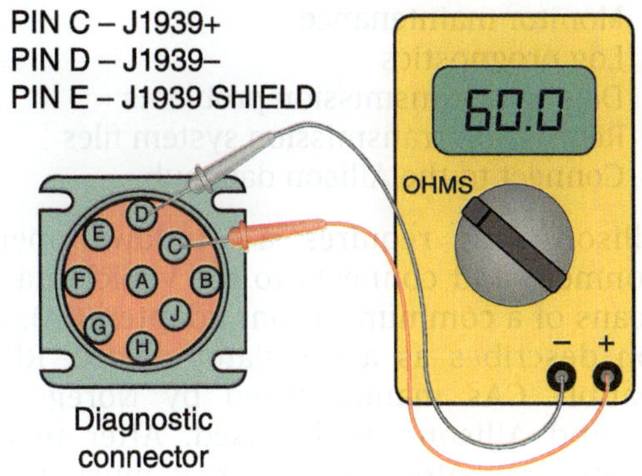

Shorts will usually affect *all J1939 communication* messages. A quick method of locating short circuits is by sequentially connecting and disconnecting controllers, triggering bus activity at each disconnection in between a sequence. Although simple open and short circuits can sometimes be identified by using these continuity checks, in most cases the proprietary software mentioned earlier will be required for accurate diagnosis.

Inductive Noise

Electronic noise can corrupt bus messages and you will need a bus viewer (Allison DBV or Cat ET) in order to identify this condition. This is mentioned at this point because it is a common occurrence, especially in applications where an original equipment manufacturer (OEM) chassis has been finished by an aftermarket body builder, or upfitter, prior to delivery to the customer. Examples would be garbage packers, dump trucks, fire trucks, and PTO installations. If pauses, errors, and dropped messages are detected while viewing a datalink, these can be due to induced electronic noise. The main causes are:

- Physical placement of the bus backbone close to high current wiring, motors or solenoids;
- Damage to the bus shield (check for proper grounding);
- Use of unshielded (J1939-15) cable stubs in place of specified shielded (J1939 11) cable stub sections.

21.2 ALLISON DOC

To do anything more than the most basic troubleshooting on Allison transmissions, proprietary software and factory training are required. Because two-thirds of this chapter is devoted to Allison transmissions, their software is described in this chapter. Allison **diagnostic optimized connection (DOC)** is the Allison software tool used to:

- Read transmission status
- Monitor maintenance
- Log prognostics
- Diagnose transmission problems
- Reprogram transmission system files
- Connect to the Allison data hub

Allison DOC requires a Windows operating environment and connects to the vehicle data bus by means of a communications adapter (CA), which Allison describes as a translator device. RP1210-compatible CAs manufactured by Noregon, DG, Nexiq, and Allison can be used. After installing DOC software onto a computer, it must be registered and activated; thereafter, it must be continued to be activated every 60 days to remain active. A registered licensee is supplied with a digital key required to receive system updates. DOC communicates with the Allison data hub when connected to the Internet.

DOC NAVIGATION

Anyone with a basic understanding of how to navigate within a Windows environment can navigate DOC. That said, Allison emphasizes that the importance of undertaking their in-house training programs, which are accessible with dealer clearance at http://allisonelearn.com. Some introductory guidelines are provided in this section.

CONNECTING DOC TO TRANSMISSION

Launch DOC, and select the "SmartConnect" option. This allows DOC to automatically choose the correct option and device protocol in order to connect to the transmission electronics. It is also possible to manually connect by unchecking the SmartConnect field and then selecting "connection devices and protocols." At the moment the connection is made, the general information and **diagnostic trouble code (DTC)** screen is displayed. In the event of a failure to connect, a help window appears with a connection troubleshooting checklist. An F4 connect icon enables the technician working with DOC to:

Access Pull-Down Menus

This divides menu pull-down options into:

- Files: Drive functions such as initiate TCM reflash and chassis data bus communication.
- Reports: Displays maintenance, prognostics, and diagnostic status.

Some of the options and tasks displayed by DOC software are briefly described in this section.

Help Fields

DOC help fields can:

- Access DOC demos
- Display service updates
- Access troubleshooting PDFs
- Run training videos
- Display hydraulics schematics
- Display Allison version information

TransHealth Monitor

This is only available for transmissions using more recent software. TransHealth monitor reports general status conditions on transmission status based on prognostics monitoring, logged codes, and maintenance logs. Some fields monitored are:

- Transmission fluid level and condition,
- TCM battery input voltage and ignition voltage,
- Check on failure logs to display the prevailing conditions (temperature/pressure data) when a DTC is set.

DTC AND DIAGNOSTIC INFORMATION

DOC must be connected to the chassis data bus and therefore the TCM to display the following diagnostic data fields:

- Data monitor
- Graphical views of data
- System calibration information
- Prognostics
- Custom data display options
- Shift inhibit block status

Performance Complaints Field

The performance complaints field can be used when no DTCs have been set in the TCM. It launches a troubleshooting tree, which is indexed to general troubleshooting guidelines; this can help isolate and source transmission problems that have not triggered a DTC.

Failure Records

Failure records are the Allison equivalent of an audit trail. This allows the technician to view the operating conditions that existed at the moment a DTC had been logged. In failure records, each DTC is listed with corresponding failure records; clicking the **+** sign next to the DTC displays the corresponding failure records.

DTC Tests

DOC DTC tests provide an overview of transmission component status. Data that can be displayed in this menu option include:

- Codes set since last DTCs had been cleared.
- Codes that were set since last ignition cycle.
- Every code checked by the system and with its corresponding result.

Graphical View

In common with most OEM software today, DOC can graphically display transmission data status using virtual gauges, thermometers, indicator lights, and graphs. This really helps with comparative analysis of data displayed during troubleshooting. Graphical view fields that can be displayed are those relating to:

- Temperature
- Speed
- Voltage
- Range selected—range attained
- Active and inactive DTCs
- StripChart data: includes turbine, input, output rpm
- Pressure switch data

Snapshot

From the earliest generation of ECATs, snapshot captures have been essential to the troubleshooting process. When in snapshot mode, system data is saved as a file that can be either played back or shared with other DOC-equipped computers. The file consists of a data stream of conditions that prevailed when the snapshot trigger had been initiated. In snapshot mode, the following options can be used to launch a data stream capture:

- Logging a specific DTC
- Logging any DTC
- On command using the DOC F5 trigger

In addition, bookmarks can be set with the F5 button. Snapshot data can also be exported to a spreadsheet for analysis.

Prognostics

Prognostics data can be monitored using Allison fourth-generation electronics and later software. **Prognostics** relate to predictive- and condition-based maintenance reporting and interventions that were described in Chapter 4; the objective is to identify a transmission problem before it results in a breakdown. Allison uses prognostics to monitor status conditions within the transmission that results first by alerting the operator, and (if ignored) subsequently inhibiting transmission performance and launching DTCs. Prognostics fields include:

- Service indicator status
- Oil monitor status and history: The % range on current Allison electronics runs between +100% and –100% life with a warning light illuminated at +2% of recommended oil life.
- Filter monitor status and history: The % range on current Allison electronics runs between +100% and –100% life with a warning light illuminated after 100 hours beyond 0%.
- Transmission health indicator light: Monitors clutch life and will produce an alert indication, when a service intervention is required.

After servicing or repair, the prognostics can be reset using the Action Request dropdown in DOC.

21.3 WT TRANSMISSIONS

The full authority Allison WT family of transmissions has been available for a generation. The family can be subdivided as follows:

- 1000/2000 series: light-duty truck
- 3000/4000 series: medium- and heavy-duty truck
- 5000/6000 series: heavy-duty off-highway

The focus in this chapter is generally on the WT 3000/4000 family of transmissions rather than on the lighter-duty 1000/2000 or off-highway 5000/6000 families, but as much as possible we generally avoid referring to specific transmissions. The electronics used to manage WTs are known as World Transmission Electronic Controls (WTEC). The ECU is also known as the **WTEC controller** or by the currently preferred Allison term, **transmission control module (TCM)**. The software generation used in a specific WT is identified by a Roman numeral, for example, Allison V (fifth) generation controls. Within this text, we try not to reference specific software generations because this kind of detail is not usually required until you take Allison technician training courses.

MODULAR CONSTRUCTION

WT series transmissions are made up of major components and subsystems called **modules**. Each module (**Figure 21–4**) is an integral subassembly that may be removed for service, repair, or replacement as a separate unit, providing a significant servicing advantage. The modules plus the external management electronics combine to form the complete transmission package. These modules may be grouped into the following four sections:

1. **Input modules**
   - Torque converter module
   - Converter housing module
   - Front support/charging pump module
2. **Gearbox modules**
   - Rotating clutch module
   - Converter housing module
   - P1 planetary module
   - P2 planetary module
   - Main shaft module
3. **Output module**
   - Rear cover module, P3 module, and output **retarder** module, or transfer gear module (depending upon the model)

FIGURE 21–4 Cross-sectional view of a typical WT transmission showing its modular construction

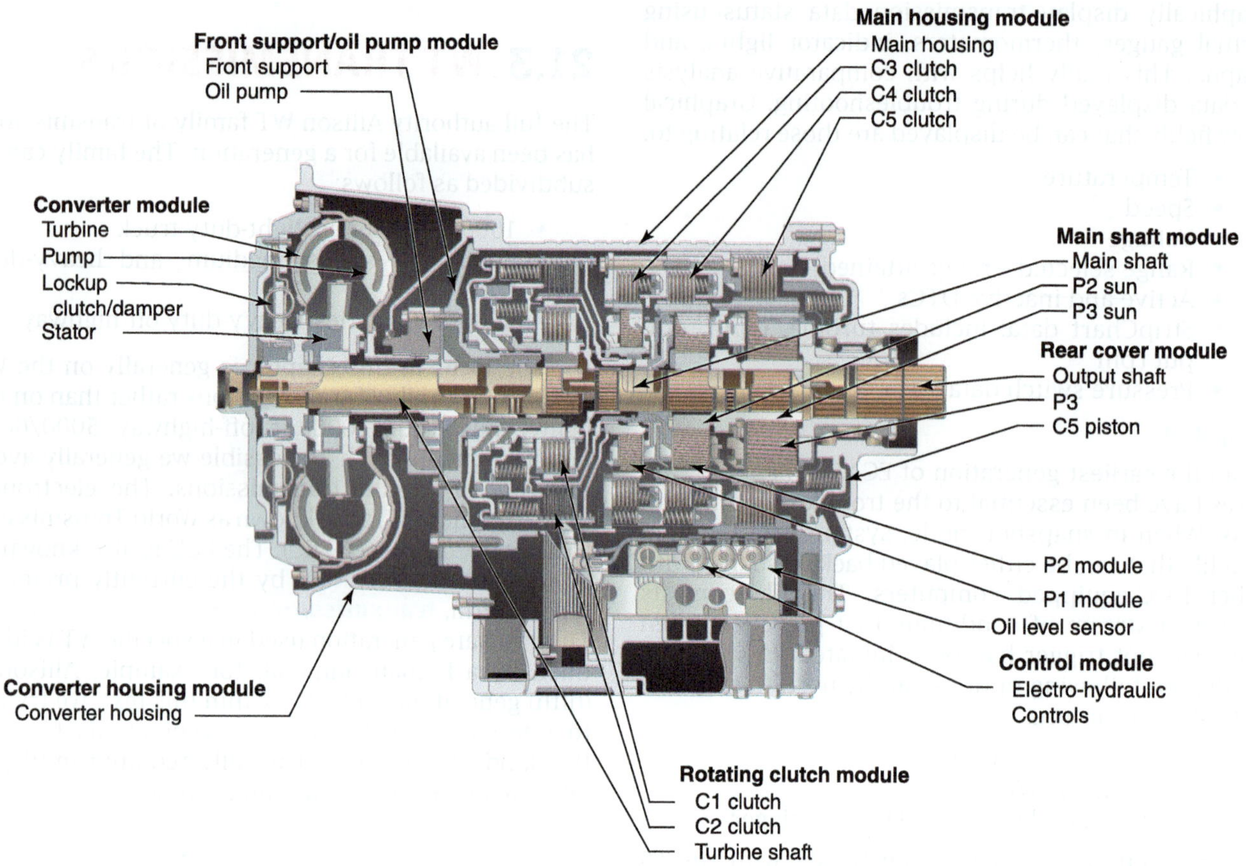

4. Transmission control module

- The transmission control module (TCM) is located outside of the main transmission housing. This unit is responsible for receiving input signals, performing logic processing, and producing the switching signals to effect the outcomes of processing. The electronic control module can also be referred to as the ECU and the WTEC controller.

Figure 21–5 shows the modular construction of a typical WT. Identify each of the modules.

WT IDENTIFICATION

A transmission identification plate is located on the right rear side of the transmission (**Figure 21–6**). This plate displays the transmission serial number, part number (assembly number), and model number. All three sets of numbers are required when ordering parts or identifying the transmission. The model data can be decoded as follows:

MD, HD, or B Automatic Truck Transmission-Automatic Bus/Coach Transmission

3, 4, or 5	Indicates transmission series
0	Close ratio
5	Wide ratio
6 or 7	Number of forward speed ratios (not present in B Series)
0 or 6	Reserved for major model changes
P	Power takeoff (PTO)
R	Output retarder
T	Transfer case (dropbox)

Figure 21–7 shows a cross-sectional view of a WT. You should reference this figure as we study each of the modules in the system.

WT INPUT MODULES

WT input modules are located between the engine and the transmission gearing. WTs are designed to couple (bolt) directly to the engine by means of flex-plates and hub adapter hardware. Note the location of the input modules in **Figure 21–8**.

FIGURE 21–5 Cross-sectional view of an Allison 4000 series transmission

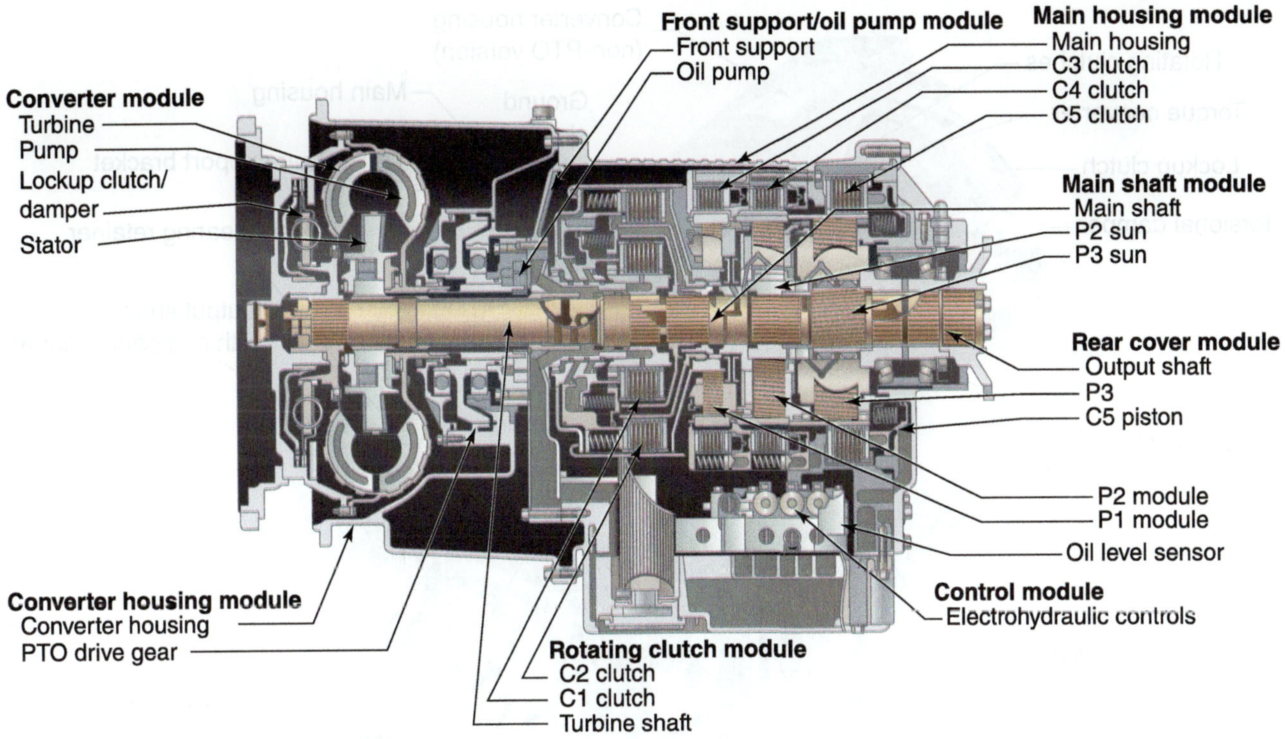

FIGURE 21–6 World transmission identification plate

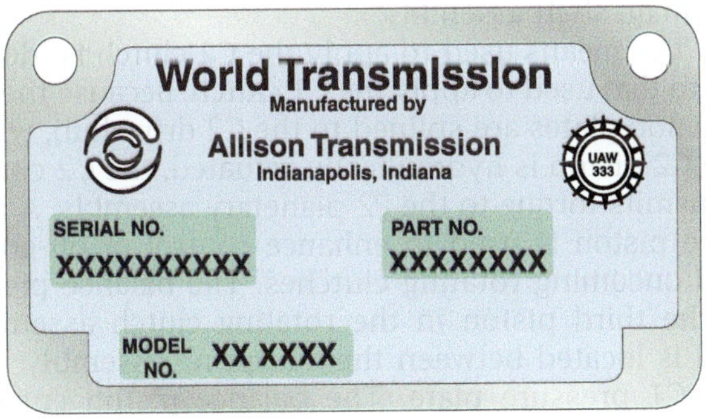

Torque Converter Module

The function of the torque converter is to couple the engine flywheel to the transmission and transmit engine torque. The operating principles are described in detail in Chapter 17. The WT torque converter is equipped with a lockup clutch. The torque converter lockup clutch disc has an integral torsional damper that is only active during lockup mode.

Converter Housing Module

The converter housing module physically connects the engine to the gearbox modules. The module may be specified for a PTO mount. In this case, the converter housing is approximately 4 inches longer than a non-PTO optioned housing. The turbine shaft passes through the converter housing to the transmission gearing. The helical PTO drive gear is contained in the converter housing and is driven at engine rpm by the torque converter pump drive hub. The PTO gear hub drives the charging pump.

Front Support/Charging Pump Module

The front support is bolted directly to the transmission main housing and incorporates the converter ground sleeve, through which the turbine shaft passes to the rotating clutch assembly. The charging pump functions to pressurize the transmission fluid. It is direct driven by the engine and, therefore, charges the transmission hydraulic circuits any time the engine is running. This supply of pressurized hydraulic fluid is required for transmission operation.

GEARBOX MODULES

The gearbox modules combined contain five clutch assemblies and three **planetary gearsets**. All WTs have the hardware for six forward ratios, neutral, and reverse. The MD and HD Series transmissions are available with either close-ratio or wide-ratio gearing. Chassis application will determine which is used. All B Series transmissions are built with

FIGURE 21–7 Offset cross-section of a typical Allison WT transmission

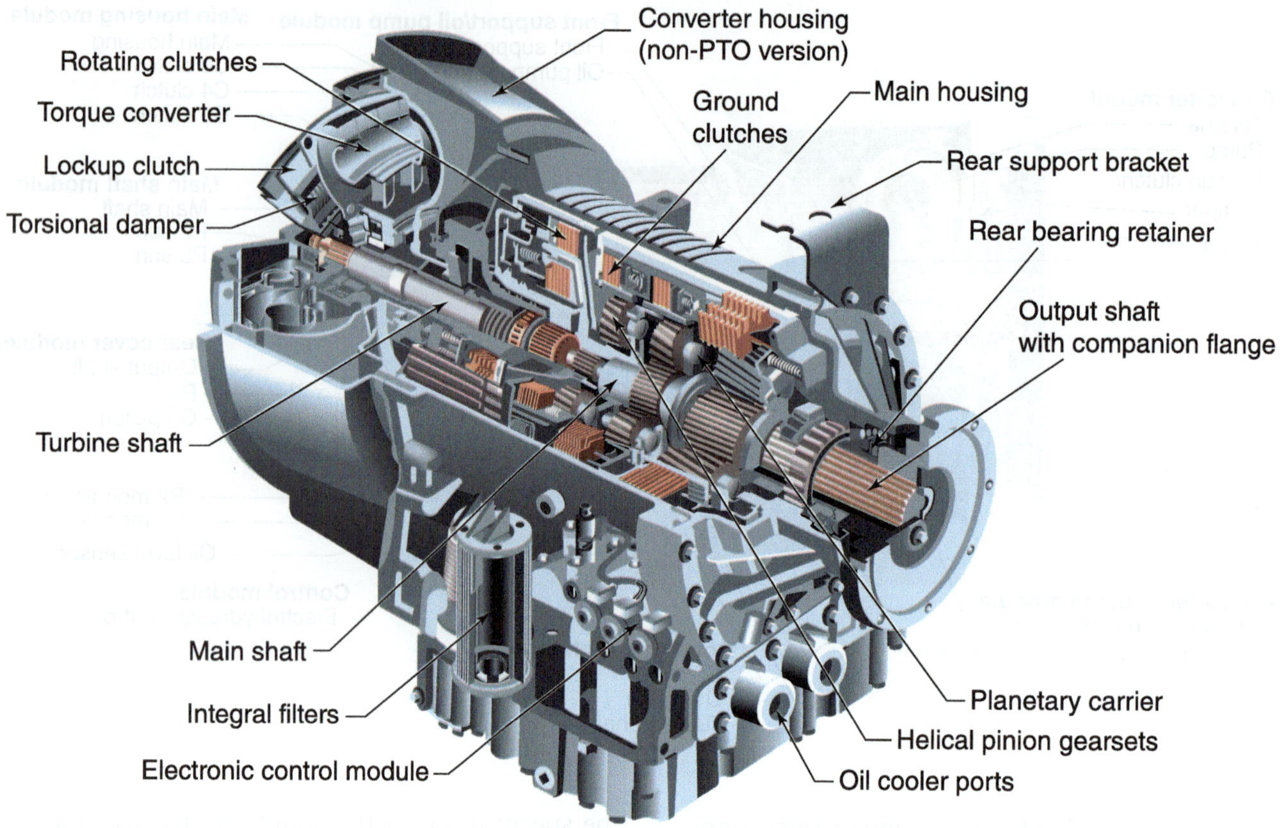

close-ratio gearing. Close- or wide-ratio gearing is determined by the physical characteristics of the planetary gearsets used in the assembly of the transmission. The upper two ranges of WT model transmissions have overdrive gear ratios. You can identify the gearbox modules used in a WT by referencing Figure 21–8.

Rotating Clutch Module

The rotating clutch module is splined to the turbine shaft and, therefore, rotates with the turbine. This module contains the turbine shaft assembly, the rotating clutch hub assembly, the C1 and C2 clutch assemblies, the rotating clutch drum, and the sun gear drive hub assembly for the P1 planetary gearset. The P1 sun gear supplies constant rotational input to the P1 planetary gearset. This means that whenever the turbine shaft rotates, so does the P1 sun gear.

The C1 and C2 clutches consist of pistons, return spring assemblies, a drive hub, and clutch reaction and friction plates. The reaction plates in both C1 and C2 clutches are splined to rotate with the rotating clutch hub assembly. The C1 clutch is applied by hydraulic fluid acting on the C1 piston. This action forces the piston and the clutch reaction plates against the clutch friction plates. Because the clutch friction plates are splined to the C1 drive hub, this will rotate, providing turning force or torque at turbine speed to the main shaft assembly.

The means used to apply the C2 clutch is identical to that used to apply the C1 clutch. Because the C2 reaction plates are splined to the C2 drive hub, when the C2 piston is hydraulically actuated, the C2 clutch transmits torque to the P2 planetary assembly. A balance piston is used to enhance control of off-going and oncoming rotating clutches. The balance piston is the third piston in the rotating clutch assembly and is located between the C1 spring assembly and the C1 pressure plate. The balance piston entraps fluid (lubrication pressure) between itself and the C1 piston, thereby balancing piston movement against exhaust backfill pressure behind the C1 piston.

Main Housing Module

The main housing module consists of the C3, C4, and C5 clutches, the ring gear component of the P1 planetary gearset, and the transmission main housing. All three clutches in the main housing module are stationary and, therefore, do not rotate when hydraulically applied, as in the case of the C1 and C2 clutches. The three clutches are applied and released by TCM output signals.

The purpose of the C3 clutch when actuated is to hold the P1 ring gear. The C3 clutch friction plates

FIGURE 21–8 WT transmission cross-section showing the modules and subcomponents

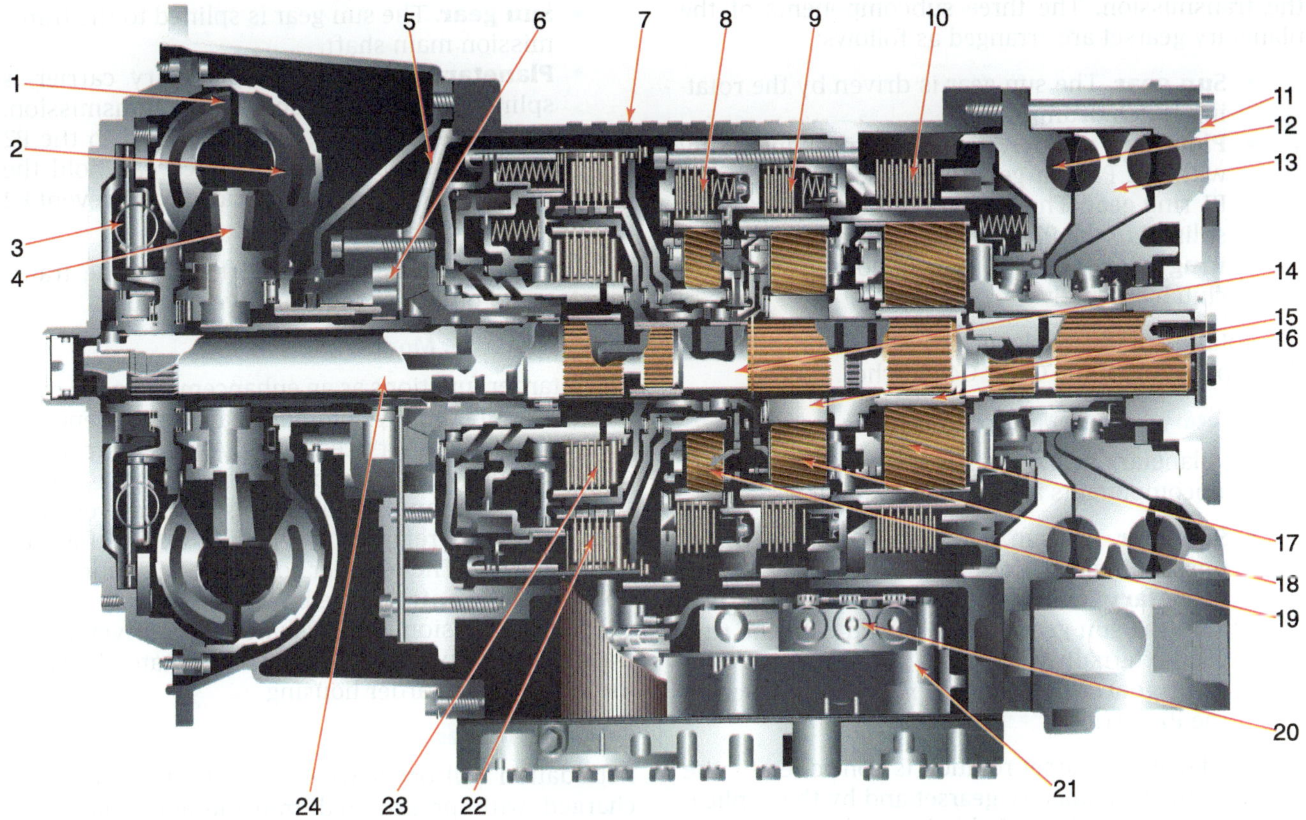

Torque converter module
1. Torque converter turbine
2. Torque converter pump
3. Lockup clutch damper
4. Torque converter stator

Front support/oil pump module
5. Front support
6. Oil pump

Main housing module
7. Main housing
8. C3 clutch
9. C4 clutch
10. C5 clutch

Retarder module
11. Retarder housing
12. Retarder stator
13. Rotor

Mainshaft module
14. Mainshaft
15. P2 sun gear
16. P3 sun gear

P3 module
17. P3 planetary gearset

P2 module
18. P2 planetary gearset

P1 module
19. P1 planetary gearset

Control module
20. Electro-hydraulic controls
21. Oil level sensor

Rotating clutch module
22. C2 clutch
23. C1 clutch
24. Turbine shaft

mesh with the P1 ring gear. These friction plates rotate with the P1 ring gear when it is not engaged. The C3 reaction plates are splined to the C3 clutch housing so that when the C3 clutch is applied the P1 ring gear is held stationary.

The function of the C4 clutch when applied is to hold the P2 ring gear. The P2 planetary ring gear is spline coupled to the P1 carrier assembly. The C4 clutch friction plates mesh with the P2 ring gear and will rotate with the P1 planetary carrier when the clutch is not applied. The C4 reaction plates are splined to the C4 clutch housing and, therefore, are held stationary. When the C4 clutch is applied, the P2 ring gear will be held stationary. Holding the P2 ring gear will prevent P1 planetary carrier rotation.

The function of the C5 clutch is to hold the P3 ring gear. The P3 planetary ring gear is spline coupled to the P2 planetary carrier assembly. The C5 friction plates mesh with the P3 ring gear and, therefore, rotate with it whenever the C5 clutch is not applied. The C5 reaction plates are splined with the main housing and are stationary. This means that when the C5 clutch is applied, the P3 ring gear is held. Preventing the P3 ring gear from rotating results in the P2 planetary carrier being held.

P1 Planetary Gearset Module

The P1 is the first of the three planetary gearsets in the transmission. The three subcomponents of the planetary gearset are arranged as follows:

- **Sun gear**. The sun gear is driven by the rotating clutch drum.
- **Planetary carrier**. Its pinions mesh internally with the P1 sun gear and externally with the P1 ring gear. The ring gear of the P2 carrier is splined to the P1 planetary carrier.
- **Ring gear**. The ring gear is housed in the C3 clutch assembly.

Rotation of the P1 planetary gears is controlled by the application of the C3 or C4 clutches.

P2 Planetary Gearset Module

The P2 planetary gearset is the second in the WT. Its three subcomponents are arranged as follows:

- **Sun gear**. The sun gear is splined to the transmission main shaft.
- **Planetary carrier**. Its pinions are meshed with the sun gear internally and with the ring gear externally.
- **Ring gear**. The ring gear is spline coupled to the P1 carrier.

P2 planetary gearset rotation is controlled by the action of the P1 planetary gearset and by the application of either the C4 or C5 clutches. Input rotation to the P2 planetary carrier can be provided either by the P2 ring gear or by the P2 sun gear, which is splined to the transmission main shaft.

Main Shaft Module

The main shaft passes through the center of the transmission and provides the primary path for transmitting torque through the transmission. The P2 and P3 sun gears are integral with the main shaft, enabling it to transmit torque to both the P2 and P3 planetary gearsets. The main shaft is also splined to the rotating clutch module. Input torque can be provided by the rotating clutch module via either C1 or C2.

OUTPUT MODULES

The output modules contain the components responsible for transmitting output torque to the vehicle driveline. There are some differences in the output module arrangements in WTs, depending on the specific model.

Rear Cover Module

The rear cover module houses the planetary carrier component of the P3 planetary gearset, the output shaft assembly, the C5 clutch piston, the speed sensor, and the output flange. When the parking brake option is installed, this is also located in the rear cover module. The P3 planetary gearset is arranged as follows:

- **Sun gear**. The sun gear is splined to the transmission main shaft.
- **Planetary carrier**. The planetary carrier is splined to the output shaft of the transmission.
- **Ring gear**. The ring gear is splined to the P2 carrier assembly. The C5 clutch can hold the ring gear stationary, which also will prevent P2 planetary carrier rotation.

Engine torque is transferred through the transmission and output at the rear cover module.

Output Retarder Module

The retarder functions as an enhancement to the overall vehicle braking system and may help extend the life of the vehicle service brakes. A large percentage of vehicle brake applications involve relatively light service application pressures, so skillful driver management of the retarder can greatly extend foundation brake life. The retarder consists of a stator, rotor, housing, and control valve body. The rotor is splined to the transmission output shaft and is driven at the transmission output shaft speed. It rotates between the stator and retarder housing.

Retarding Modes

Retardation will occur when the retarder housing is charged with pressurized transmission fluid. This will slow the rotation of the rotor and output shaft and, therefore, the vehicle driveline. The action of the retarder can be compared to that of a torque converter operating in reverse.

When the retarder is not in use, transmission fluid is drained from the retarder housing to an external **accumulator**. When the retarder is activated, it is charged with transmission fluid stored in the external accumulator. The control valve body is mounted to the right side of the transmission. Three types of retarder capacities are available in Allison WT: low-, medium-, and high-capacity configurations, which are determined by spring tension. For the retarder to be applied, the following is required:

- The vehicle dash-mounted Retarder Enable switch must be on.
- The vehicle must be moving.
- The TPS signal must indicate that accelerator travel is close to zero.
- The antilock brake system (ABS) must not be activated.

Retarder Controls

WT retarders are fitted with a variety of apply devices that vary with vehicle application. Some of the controls used on current transmissions are listed here:

- **Hand lever**. The driver selects either the off position or one of six levels of retardation.

Each successive level will increase the voltage return signal to the TCM with maximum retardation (braking) occurring at level 6.

- **Foot pedal**. A dedicated foot pedal that uses a potentiometer supplied with reference voltage returns a voltage signal to the TCM proportional with pedal travel.
- **Pressure switch**. This option integrates the operation of the WT retarder with the vehicle service brake system. The pressure switch is a normally open switch plumbed into the service brake application circuit that closes at a set trigger value. One model of pressure switch can be used to provide 100% retardation at the preselected service brake application pressure (trigger value). In order to accommodate various vocational applications, pressure switches are manufactured with 2, 4, 7, and 10 psi trigger values. Pressure switches are often used in multiples, of usually two or three, to provide increased retardation, which would be proportional with service brake application pressures.
- **Auto apply—auto full-on**. When this option is specified, the TCM can switch to full retardation when certain conditions are met (the Retarder Enable switch usually is "on," throttle position is close to "zero" travel, and the transmission output speed is *above a preset value*).
- **Auto apply—auto percent-on**. This option permits the TCM to vary the percentage of retarder application "on" time when certain preprogrammed conditions are met. (Note: 33% and 50% are typical.)
- **Combination**. Allison states that almost any of the above control options can be used together in a vehicle.

Retarder Electronic Controls

The retarder electronic controls consist of the following:

- **The TCM**. The TCM receives and interprets speed, resistance module, and temperature signals and then computes the appropriate retarder application response.
- **One or more resistance modules**. A resistance module is required for all retarder apply control units, so when multiple control units are used, each must have its resistance module.
- **Two pulse width modulated solenoids**. These solenoids control the regulation of retarder cavity pressure and the application of vehicle air pressure to the retarder accumulator. Both solenoids are normally closed, which means they must be electrically energized to perform their function.
- **A retarder temperature sensor**. This sensor inputs retarder temperature data to the TCM, which, when processed, could result in reducing retarder capacity in an overtemperature condition.

Retarder Operation

The retarder control valve solenoid is switched by the transmission TCM at a calibrated rate; when it is actuated, it permits main pressure to act on top of the retarder control valve. This action results in the following:

1. Main pressure enters a passage to charge the retarder charging circuit.
2. Vehicle system air pressure contacts the accumulator piston, discharging the stored transmission fluid into the retarder housing.

Fluid charged to the retarder housing by the accumulator is continuously supplemented by retarder charging pressure and effects the retarding action of the rotor. The transmission fluid circulates through the retarder housing during a retarding cycle and exits to be routed to the oil cooler before being rerouted to the retarder. During the retarding cycle, the torque converter-out fluid is routed directly into the transmission lubrication circuit.

The TCM manages the retarder effective-cycle. When output shaft speed is within the calibrated range, retarder capacity is at its maximum allowable value. When output shaft speed is either above or below the calibrated value, the TCM will modulate the retarder charging pressure at a lower value. The TCM manages the deactivation of the retarder at a low output shaft speed (typically 350 rpm) to smooth the transition of retarding effort to the vehicle service brakes.

When the retarder is deactivated, the apply process is reversed. The accumulator solenoid is deenergized, which exhausts the vehicle air pressure acting on the accumulator piston, and the retarder control valve solenoid is deactivated at a fixed rate. This controlled deactivation allows pressurized converter-out fluid to recharge the accumulator before it is rerouted to the oil cooler.

TRANSFER GEAR OR DROPBOX MODULE

One member of the WT family uses a transfer gear module. This transmission is also known as a **Drop-7** transmission. It consists of an output adapter housing, transmission output shaft adapter, transfer case, transfer case charging oil pump, C6 clutch assembly, and transfer case **scavenging pump** (auxiliary pump) mounted at the right side PTO provision. The addition of the **dropbox** (**Figure 21–9**) adds a forward gear to the WT and provides for full-time, all-wheel drive.

The transfer case is mounted to the adapter housing and is underslung below the main transmission to provide for forward and rear drive yokes in line with the transmission. The transmission output

FIGURE 21–9 WT transmission with dropbox

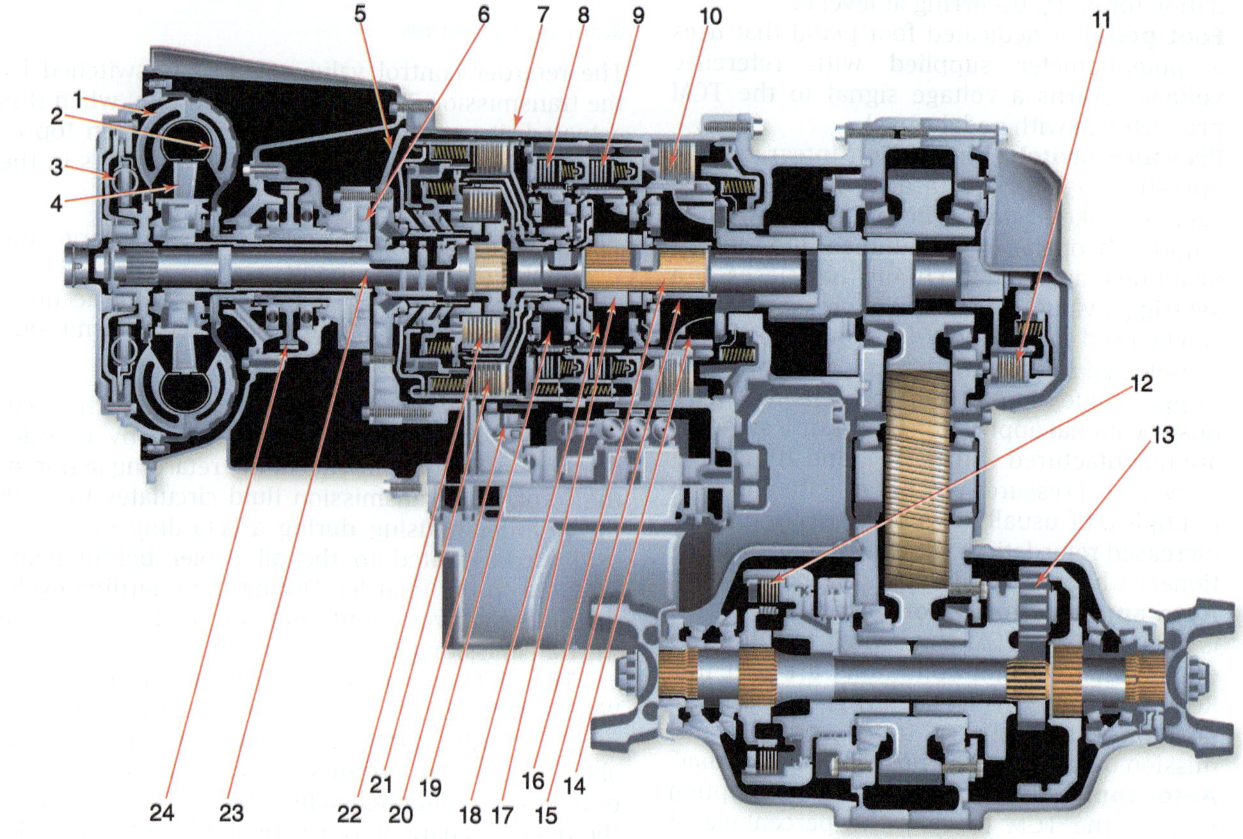

Torque converter module
1. Torque converter turbine
2. Torque converter pump
3. Lockup clutch damper
4. Torque converter stator

Front support/oil pump module
5. Front support
6. Oil pump

Main housing module
7. Main housing
8. C3 clutch
9. C4 clutch
10. C5 clutch

Dropbox module
11. C6 clutch
12. C7 clutch
13. Torque proportioning differential

P3 module
14. P3 planetary gearset

Mainshaft module
15. P3 sun gear
16. Mainshaft
17. P2 sun gear

P2 module
18. P2 planetary gearset

P1 module
19. P1 planetary gearset

Control module
20. Electro-hydraulic controls

Rotating clutch module
21. C2 clutch
22. C1 clutch
23. Turbine shaft
24. PTO drive gear

shaft adapter splines to the P3 carrier hub. The transfer case charging pump and transfer case drive gear are mounted on, and driven by, the output shaft adapter. The function of the drive gear is to direct torque downward into the transfer case. When the stationary C6 clutch is actuated, the main shaft is held to provide the dropbox with its extra low-range capability.

The transfer case idler gear is installed between the transfer case drive gear and driven gear and acts to transfer torque from one to the other. The driven gear transmits torque to the P4 planetary gearset, which functions to drive the forward and rear drive shafts. When the C7 clutch is engaged, differential action is locked out, meaning that the front and rear outputs are locked. When the C7 clutch is not engaged, dropbox output torque is split 30/70 with 30 percent to the front and 70 percent to the rear drivelines.

The transmission hydraulic circuit provides the pressure to apply the C6 and C7 clutches and to lubricate the dropbox components. The lubricating oil is returned from the dropbox sump to the main transmission housing by a scavenging pump. The dropbox oil pump functions to lubricate the dropbox components in the event that the vehicle is towed with the wheels on the ground.

ELECTROHYDRAULIC CONTROL MODULE

The WT **electrohydraulic control module** is mounted under the main transmission housing and, in fact, forms its bottom enclosure. This module houses the solenoids, sensors, valves, and regulators that manage the pressure and flow of transmission fluid to the clutches, the torque converter, and the lubrication/cooling circuits. The transmission filters are also located in this module.

WTEC ELECTRONICS

Electronic management of the WT is known as WTEC. Most notable is its ability to multiplex to the chassis data bus and better programmability to suit specific vehicle applications. First-generation versions of WT (WTEC I and II) could be networked to J1587. Current WTEC transmissions are designed to network to the J1939 bus on which they are assigned with a 03 source address. WTEC can broadcast range-inhibit, check transmission light data, and send range status to the data bus for display onto the appropriate dash displays. WTEC can also receive commands from the data bus to assist in such things as accident avoidance, stability control, and smart cruise electronics. The transmission electronic circuit may be divided into input circuit components, a microcomputer, and output circuit components. A wiring harness connects the various components in the system.

Input Circuit Components

- Shift selector(s)
- TPS
- Engine speed sensor
- Turbine speed sensor
- Output speed sensor
- C3 pressure switch
- Sump temperature sensor
- Coolant temperature
- Vehicle interface module (VIM)

Input Signals. The following is a partial list of the input signals that the WT TCM may be programmed to process:

- Secondary shift schedule
- PTO enable
- Shift selector transition
- Engine brake and preselect request
- Fire truck pump mode
- Automatic neutral
- Automatic neutral for PTO
- Automatic neutral for refuse packer
- Two-speed axle enable
- Manual lockup
- Antilock brake response
- Retarder enable
- Service brake status
- Differential clutch request
- Kickdown

WT OUTPUT CIRCUIT COMPONENTS

The following are WT TCM switched output circuit components:

- A–E clutch solenoids
- Lockup clutch solenoid F
- Forward latch solenoid G
- Vehicle interface module (VIM)
- PTO functions

Output Signals

The following is a partial list of the output signals that WT electronics may be programmed with to broadcast on the data bus to other chassis electronic management systems:

- Engine brake enable
- Sump/retarder temperature indicator
- Range indicator
- Output speed indicator
- PTO enable
- Engine overspeed indicator
- Two-speed axle enable
- Lockup indicator
- Service indicator
- Shift-in-progress indicator
- Retarder indicator
- Neutral indicator for PTO

Input Circuit

The WTEC input circuit is responsible for supplying the TCM with monitoring and command data it needs to compute the shift logic. Input from the operator is sent to the TCM by means of the shift selector. Other input signals come from sensors located in the transmission itself and from the **vehicle interface module (VIM)**. The VIM exchanges data between WT electronics and other vehicle electronic management systems using SAE J1939 communication protocols and hardware. All input signals feed data into the processing cycle of the TCM. Most of this data can be classified as monitoring data. Temperature sensors, shaft speed sensors, and pressure switches all monitor conditions within the transmission that are required to compute an outcome. Command input signals are usually operator generated and often require the TCM to make a change in outcome. The TPS, range shift selector, retarder controller, and PTO switches are all examples of command input signals.

Shift Selectors. A lever and two types of push-button shift selectors are available on WTs. **Figure 21–10** shows some examples of WT gear selectors. The strip-type push-button selector is not full function.

FIGURE 21–10 WT shift selectors

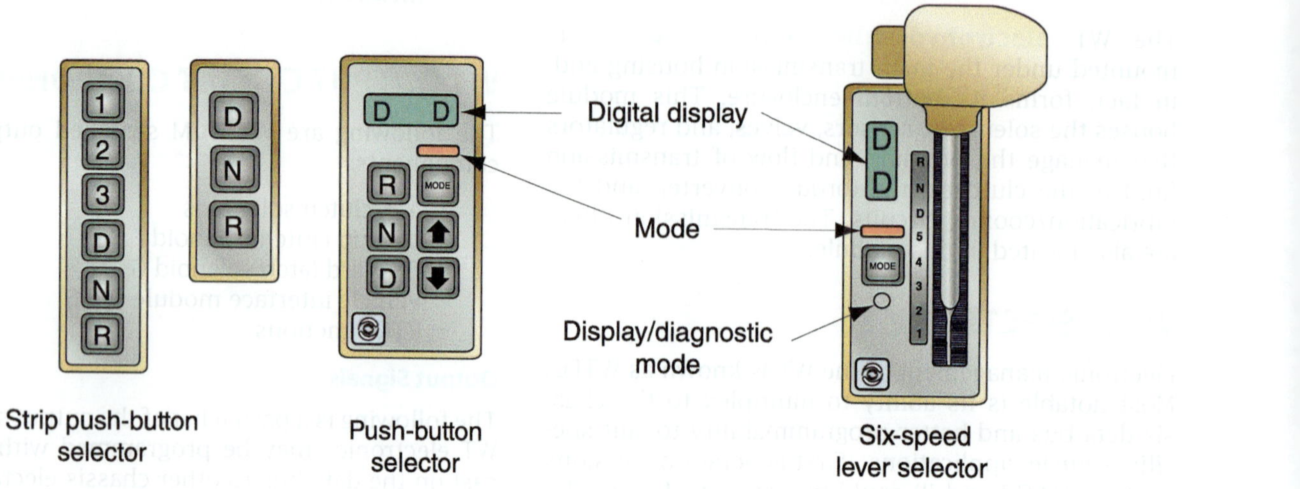

It allows the operator the least amount of input as to range selection and it does not have an illuminated **digital display (DD)**. It is not often used.

The full function push-button (non-strip-type) gear range selector allows the operator to select any available gear range programmed to WTEC by selecting the drive button (D), and using arrowed buttons (≤ ≥) in order to downshift or upshift. Using these buttons does not override the automatic shifting function of the transmission. If a higher or lower range is selected by the operator, the transmission will shift through the ranges to the selected range. This shifter has a digital display, which is explained later in this section.

A number of different types of lever selectors are used with WTs, the most common being the six-position lever selector shown in Figure 21–10. The lever selector is an electromechanical device that inputs command data to the TCM. It will not permit the operator to select a shift schedule that could result in damage to either the transmission or drivetrain. When the D range position is selected, WTEC will manage shift schedules using all the ranges programmed to WTEC. Selection of any of the numeric ranges will confine shifting within the ranges below the selected range. For instance, if the numeric position 3 were selected, only ranges 1–3 would be used by WTEC.

Vehicle Interface Module (VIM). The vehicle interface module (VIM) is a splashproof (as opposed to watertight) housing that provides the necessary relays, fuses, and connection points to connect the WT TCM with the vehicle electrical system. The VIM contains two 10-amp fuses and somewhere between two and six relays. WT electronics will operate on either 12- or 24-volt chassis electrical systems. The relays used must match the ignition voltage.

One of the 10-amp fuses protects the main power connection to the TCM. The other protects the feed from the ignition circuit to the TCM. The VIM will always have at least two relays. One is used to output a signal to the reverse warning circuit and the other is used for the neutral start circuit. As many as four other relays may be used for special function outputs. **Figure 21–11** shows a VIM unit.

Throttle Position Sensor (TPS). A dedicated TPS is only required when a WT is used with a hydromechanical engine and no chassis data bus. Because most current chassis are multiplexed, WT electronics rely on the TPS signal being broadcast from the engine controller over J1939. Allison refers to the data bus connection as a **serial communications interface (SCI)**. SCI is covered under "WT Output Circuit" later in this chapter.

Speed Sensors. Shaft speed sensors are required to input rotational speed data to the WT TCM. They use an induction pulse generator principle. Shaft speed

FIGURE 21–11 Vehicle interface module (VIM)

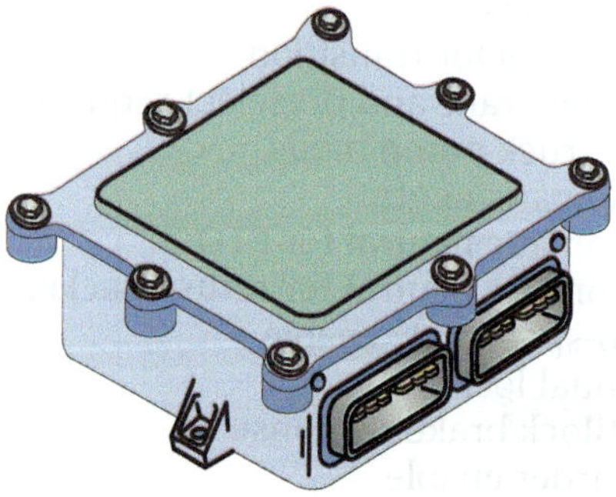

sensors are covered in Chapter 6, and Figure 6–39 shows their operating principles. In a typical shaft speed sensor, a toothed rotor cuts through a magnetic field, producing an AC voltage that is signaled to the TCM as frequency. Digital Hall effect sensors may also be used. Some typical WT speed sensors are shown in **Figure 21–12** and **Figure 21–13** highlight the 16-tooth reluctor wheel (standard) on a WT output shaft.

FIGURE 21–12 WT speed sensors

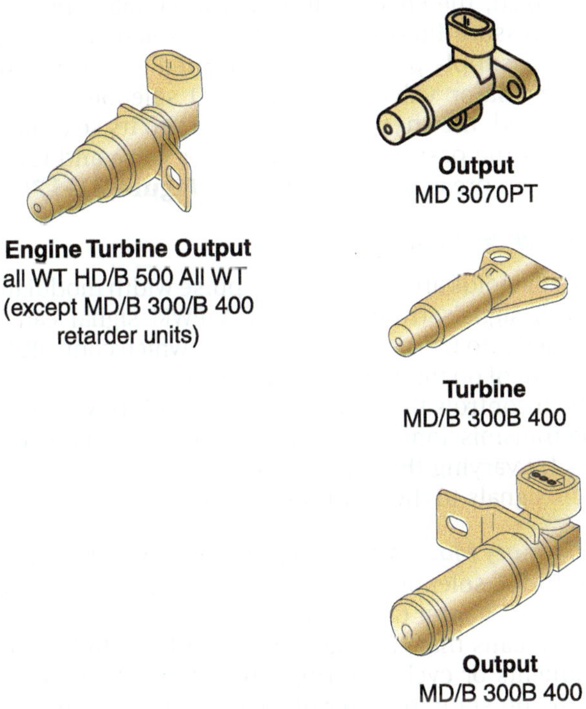

FIGURE 21–13 Output shaft of a WT transmission showing the speed sensor reluctor wheel at the rear of the assembly

WT TCM

A detailed description of microprocessing cycles is provided in Chapter 6. The WTEC TCM is responsible for processing the command, monitoring input data, plotting shift sequences and timing, and effecting the results of its computations into outcomes by switching clutches. The WT TCM has an electronically erasable PROM (EEPROM) chip, which means that both Allison and customer preference or application data can be reprogrammed.

Adaptive Logic. The significant improvement in WT shift logic in comparison with CEC is accounted for by replacing hard parameter processing outcomes with soft logic processing, better called **adaptive logic**. WT electronics continually compare actual transmission shift characteristics with the optimum shift calibration stored in TCM memory. When these actual shift characteristics fall outside the optimum parameters, the TCM alters solenoid shift modulation to correct the shift character.

This means that WT electronics can produce close to optimum shift quality by constantly adapting to changes in operating conditions. In simple terms, WT electronics constantly "learn" from actual performance conditions, factoring in variables such as grade, load, and engine power. After every shift is completed, the TCM will compare the shift to what ideally should have occurred during the shift and the next time a shift is made under similar conditions, the adjustments are effected.

Fast Adaptive Mode. Adaptive logic is, therefore, the establishing of the initial conditions for shifts. It can be categorized into two types: fast adaptive logic and slow adaptive logic. When the TCM is programmed to operate in **fast adaptive mode**, it makes large changes in initial shift conditions to adjust for major system tolerances. Major system tolerances are solenoid-to-solenoid, main pressure, and clutch-to-clutch variations. Any new or newly rebuilt transmission should be programmed to operate in fast adaptive mode.

Slow Adaptive Mode. **Slow adaptive mode** programming makes small changes in initial shift conditions in response to minor system changes. Minor system changes are conditions that occur over time and are often associated with wear. As internal transmission components wear and solenoids electrically degrade or drift, WT electronics will change from fast adaptive to slow adaptive mode to compensate.

Autodetect. The Allison WTEC TCM also has an autodetect feature. Autodetect works a little like Windows "plug-and-play." The autodetect mode is active in the system for the first 30 seconds of the first 24 to 49 engine starts (dependent on the component or sensor being detected). This means that WTEC searches for the following components or data inputs:

- Retarder: Present/Not Present
- Oil Level Sensor: Present/Not Present
- TPS: Analog (potentiometer type)/data bus (J1939 or J1587) compatible
- Engine Coolant: Analog/data bus (J1939 or J1587) temperature compatible

SHOP TALK

TPSs that output a pulse width modulation (PWM) signal (variable capacitance type, current Caterpillar, or Hall effect noncontact) are not autodetected and must be programmed to WTEC using Allison DOC software or the **diagnostic data reader (DDR)**.

After autodetect has been completed, it can be reset at any time using the DDR. During the first 24 starts, autodetect does not retain what was detected on any previous start. For example, if a retarder were detected on every engine startup to number 23, H solenoid diagnostics would be performed by the TCM until the system was shut down each time. If the H solenoid failed at startup number 24, autodetect would not recognize that a retarder was present and no diagnostics would be logged for the H solenoid. **Figure 21–14** shows some WT electronic circuit components.

The WTEC TCM is also programmed with Allison data designed to protect the transmission and drivetrain components from abuses. This type of programming would include inhibiting action when set to a full-throttle, neutral-to-range shift command or a high-speed direction change. The WTEC TCM logs diagnostic codes and can retain them with either *active* or *inactive* status. Codes can be accessed by the operator or service technician.

WT OUTPUT CIRCUIT

As in any vehicle electronic management system, the output circuit is responsible for effecting the outcomes of the processing cycle of the controller. In the Allison WT, the results of TCM processing are converted into actions by an electrohydraulic control module. This is mounted at the base of the transmission, and the channel plate forms its bottom enclosure. **Figure 21–15** shows a WT control module.

Solenoids

Two types of solenoids are used in the WT electrohydraulic control module. They are somewhat similar in construction but differ in their switching characteristics. Both types receive control main pressure from a supply port and may either route that pressure to exit through the solenoid regulator valve or to an exhaust port, depending on the switch status. The two types are:

1. NC solenoids: remain closed until energized by TCM electrical signal. When closed, main pressure fluid is blocked by the seated check ball, routing it directly to the exhaust passage. When energized, the check ball unseats, blocking the exhaust port and permitting main pressure fluid to exit to the solenoid regulator valve as shown in **Figure 21–16**.
2. NO solenoids: remain open until energized by TCM electrical signal. When the solenoid is not energized, the check ball is unseated, allowing main pressure fluid to flow into the solenoid through the supply passage and be routed to the solenoid regulator valve. When the NO solenoid is energized, the check ball is forced onto its seat blocking the passage to the regulator valve and routes main pressure to the exhaust (**Figure 21–17**).

Pulse Width Modulation (PWM)

WT solenoids are controlled by pulse width modulated (PWM) signals from the TCM. A PWM signal can be divided into **primary modulation**, which controls the amount of on time, and secondary or **submodulation**, which controls the current flow. The WT TCM controls the transmission clutch apply-and-release characteristics by varying the primary and **secondary modulation** signals to the control solenoids.

Primary Modulation. The primary modulation signal to WT solenoids is delivered at a frequency of 63 Hz. (See Chapter 6 for a full explanation of frequency.) This means that each second is divided into 63 equal segments or cycles, during which the voltage will be in the on state for a period. This period is expressed as a percentage. The percentage of time that a voltage is present within each $\frac{1}{63}$ of a second is referred to as the solenoid duty cycle. A 100 percent duty cycle would represent the maximum signal that could be delivered to the solenoid, whereas a 0 percent duty cycle would indicate no signal. As the pulse width (PW) or duty cycle is increased, the percentage of solenoid on time is increased. This permits the solenoid regulator valve to apply or release a clutch with optimum shift quality. **Figure 21–18** shows a PWM waveform in graphic form, expressing it as percentage of cycle.

Submodulation. Secondary or submodulation supplies a constant optimum average current to the solenoid at a high frequency (7,812 Hz). This permits the solenoid ball to be first pulled in and then held. The result is that the heat in the solenoid coil can be minimized. Secondary modulation, by managing current to a variable optimum requirement, permits consistent solenoid performance when variables such as system battery voltage and sump temperature are factors.

FIGURE 21–14 WTEC electronic control components

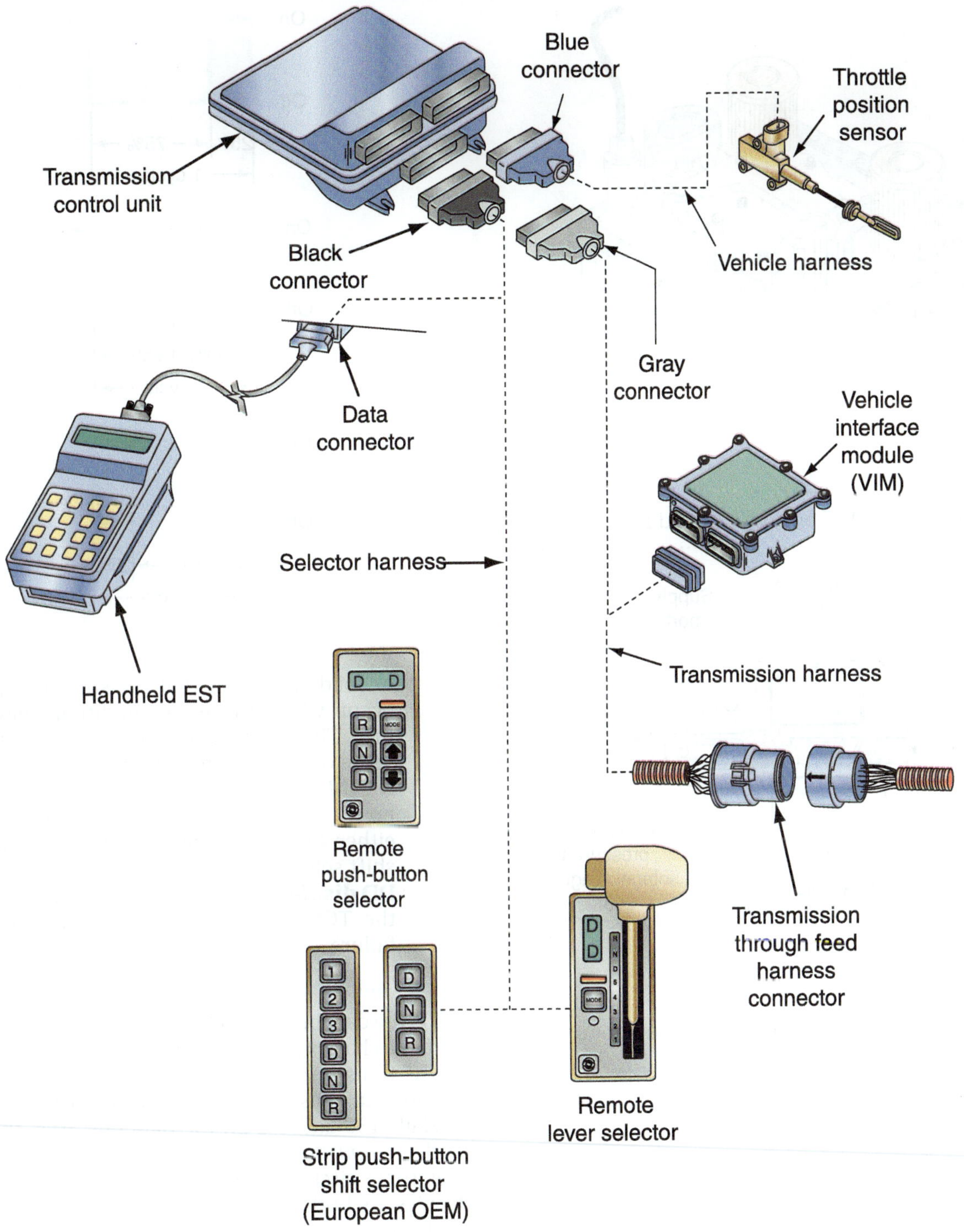

WT Digital Display (DD)

The data that appears on the digital DD (keypad, shift lever, dash display) is controlled by the TCM. In other words, it does not display data directly as a result of mechanically pushing buttons or moving the shift lever. The following information can be displayed:

- **Selector position**. The display is the acknowledgment by the TCM of the selected shift position.
- **DD flashing**. Flashing indicates that the requested shift is inhibited because of a condition such as inappropriate actual engine

FIGURE 21–15 WT control module

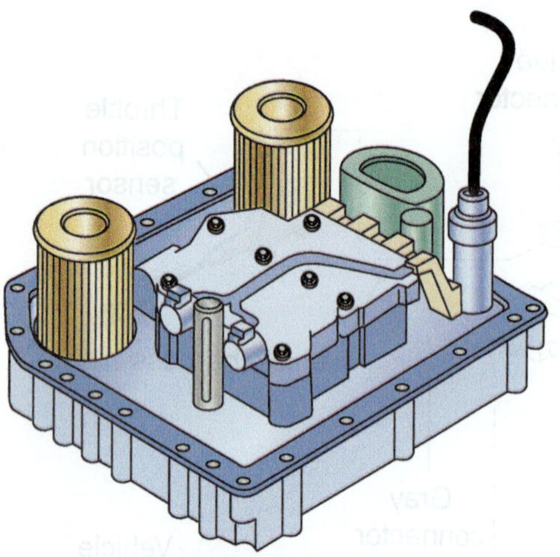

FIGURE 21–16 WT normally closed solenoid

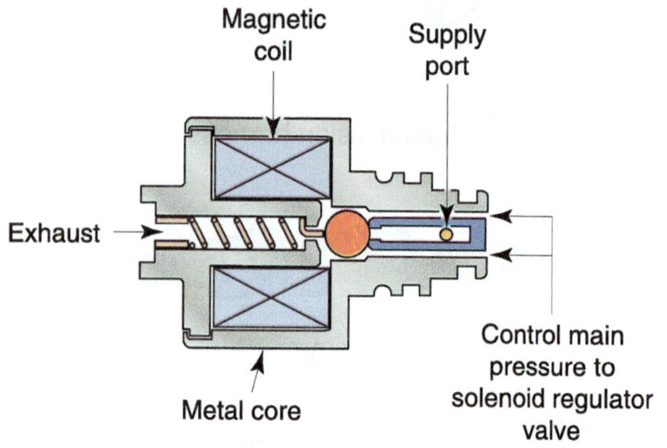

FIGURE 21–17 WT normally open solenoid

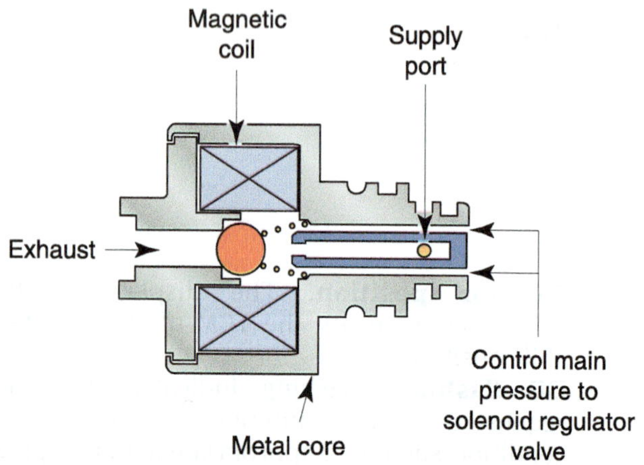

FIGURE 21–18 Pulse width modulation waveforms

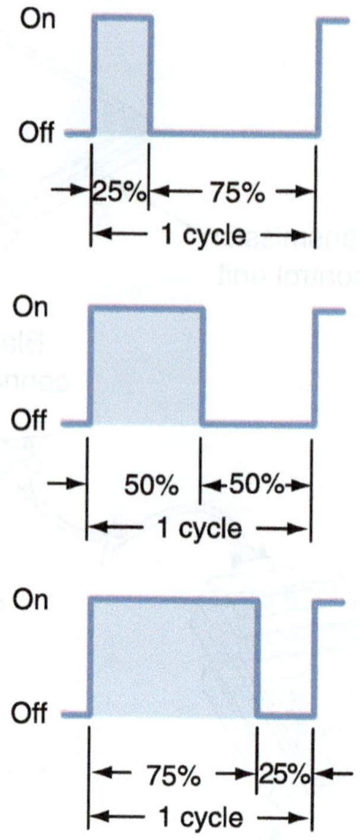

or road speed but that the shift may possibly occur when those preconditions have been met. The flashing display shows the numeric value of the current shift range.

- **DD blank**. A blank display indicates that a problem exists within the electrical circuit, either in the message communication to the shift selector or that the data link has failed.
- **DD displays "cat's-eye**." Communication with the TCM has been lost, indicating data link failure.

Serial Communications Interface (SCI)

SCI is the data bus portal that permits WTEC to interact with the chassis data bus. This optimizes the operation of the transmission and coordinates it with the powertrain and chassis systems. The SCI occupies the SA 03/MID 130 address on the data bus. It is, therefore, both an *input* and an *output* mechanism within the WT electronic circuit. Current Allison transmission electronics are both J1939 and **ISO 9141** compliant. Data that can be input through the SCI include:

- Throttle position
- Percentage load
- Cruise control status
- Road speed limit status
- Engine coolant temperature
- ABS active
- Kickdown signal
- Antislip logic

Data that can be output through the SCI include:

- Transmission sump fluid temperature
- Retarder temperature
- Transmission output speed
- Engine rpm (input rpm)
- Range position selected
- Actual range
- Lockup clutch status
- Retarder active status

TCM-Managed PTO Mode

WT provides for three main types of PTO management modes. The management modes are:

- Constant drive
- Clutch drive, operator activated
- Clutch drive, TCM activated

Constant-drive PTOs provide continuous PTO power whenever the transmission input is rotated. Clutch-drive PTOs provide PTO power when activated by the operator. Activating the PTO causes a solenoid in the PTO to apply main pressure to the rotating PTO clutch, which then transmits torque to the PTO output. Clutch-drive PTOs also may be controlled by the TCM. The TCM can, therefore, be programmed to *permit* or *inhibit* PTO operation when specific transmission or chassis operating conditions exist. An example would be an interlock/shift status or parking brake status during operation. This feature allows PTO operation to be integrated safely with other chassis functions.

SHOP TALK

In a general sense, the WT electronic management circuit is no different from systems that are used to manage most of today's engines. It consists of input, processing, and output circuits. It must also interact with other chassis electronic systems to optimize vehicle performance and therefore can both broadcast and receive data off the J1939 network.

WT HYDRAULIC CIRCUIT

The Allison WT hydraulic system is responsible for:

- Storing hydraulic fluid
- Generating hydraulic pressure
- Acting as the drive medium in the torque converter
- Controlling system pressures
- Directing flow through the various subcircuits
- Cooling and lubricating the transmission

The primary components of the hydraulic system are:

- Transmission fluid
- Charging pump
- Three integral filters
- Electrohydraulic control module
- Breather
- C3 pressure switch

There are seven subcircuits that make up the hydraulic circuit:

- Main pressure circuit
- Control main circuit
- Torque converter circuit
- Cooler/lubrication circuit
- Clutch apply circuit
- Exhaust circuit
- Exhaust backfill circuit

Torque converter operation is covered in detail in Chapter 17. Fluid pressure and flow through the hydraulic circuit is provided by the engine-driven charging pump. The charging pump draws fluid through a primary filter (under "suction") from the transmission sump. This fluid is then pumped through the main filter to supply the hydraulic circuit.

Solenoids and valves located in the electrohydraulic control module and managed by the TCM control the flow and pressure of the fluid, directing it to specific clutches to achieve range shifts. Fluid for the cooler and lubrication circuit flow through a third filter known as a lubrication filter.

The C3 pressure switch is located in the C3 clutch circuit. Its function is to monitor circuit pressure. An accumulator is located in the C3 pressure switch cavity. Its function is to dampen hydraulic pulsing in the clutch apply circuit.

Main Pressure Circuit

The main pressure circuit is responsible for supplying the hydraulic pressure required to manage the hydraulic circuit of the transmission. Main pressure is managed by the **main pressure regulator** valve. The primary function of the main pressure regulator valve is to modulate charging pressure values to main pressure values. Actual main pressure value is modulated based on input from the overdrive knockdown valve, the converter flow valve, and the control main pressure.

The main pressure regulator valve is a spool-type valve held in an upward position by spring force. Main pressure acts on the top of the main pressure regulator valve and forces it downward against spring pressure. When pressure is high enough to drive the valve downward enough, main pressure is spilled to the exhaust, thereby reducing main pressure.

Regulated main pressure is routed to seven areas within the hydraulic circuit. Each of the six solenoid regulator valves plus the control main regulator valve receives regulated main pressure. Pressure at the outlet of each solenoid regulator valve is clutch feed pressure. Pressure at the outlet of the control main regulator valve is control main pressure.

Solenoids and Solenoid Regulator Valves. The WT TCM switches the solenoids in the electrohydraulic control module. The solenoids direct control main pressure to the top of solenoid regulator valves, causing each to move inboard working against spring pressure. This blocks the exhaust backfill circuit and allows main pressure to move through the valve passage into the clutch feed circuit, and thereby applying the clutch. When control main pressure ceases to act on the top of the solenoid regulator valve, the spring pressure forces the valve back to the top of its travel, exhausting the clutch apply circuit. This releases the clutch.

Clutch Application. In order to obtain any forward or reverse range, two clutches must be applied. **Table 21–1** shows which clutches are applied in each transmission range, the corresponding energized solenoids, the C1 and C2 latch valve positions, and the converter flow valve position.

HYDRAULIC CIRCUIT OPERATION DURING AN ELECTRICAL FAILURE

The WT is managed electrically. In the event of a chassis electrical failure, the TCM and its switching actuators and solenoids will cease to function. Such an interruption of electrical power will result in the solenoid regulator valves locking in their NO or NC states. Should an electrical failure occur, the WT defaults to totally hydraulic operation. The C1 and C2 latch valves are used to accomplish this default mode of operation.

When a clutch is applied, clutch feed pressure from the solenoid regulator valve is routed through the latch valves to the clutch piston to actuate the clutch. Clutch apply pressure acting against the lands of the latch valves holds them in position or, in the case of an NC valve, permits the fluid to pass through the valve. When an electrical failure occurs, the latch valves cause the transmission to engage specific clutches based on the range the transmission was in when the failure occurred.

The latch valves are activated by the NC solenoid G. When G solenoid is energized, control main pressure flows to the top of the C1 and C2 latch valves. This pressure forces the valves downward to supply the flow passages required for clutch engagement. In case of an electrical failure, the latch valves and the two NO solenoids, A and B, will enable the transmission to operate in this default or "limp home" mode by reverting to total hydraulic operation.

WT TORQUE PATHS AND RANGE MANAGEMENT

The general operating principles and powerflow characteristics of torque converters and planetary gearsets are covered extensively in previous chapters and are not covered again in this section. The torque path routing in each gear range is described, referencing schematics. Toward the conclusion of this chapter, the range powerflows are shown in diagram form. In each range, the gearing is staged through the transmission. The stages are sequenced numerically, but the stages of gearing do not necessarily correlate with the numbering of the planetary gearsets. For instance, first range gearing is single stage and takes place in the P3 planetary gearset. As many as three stages of gearing may be used, depending on the range selected.

TABLE 21–1 Clutch Application Chart

Range	Clutches Applied	Solenoids Energized	C1 Latch Position	C2 Latch Position	Converter Flow Valve Position
Neutral	C5	A, B, E	Up	Up	Down
Reverse	C3, C5	A, B, C, E	Up	Up	Down
First	C1, C5	B, E, G	Down	Down	Down
Second	C1, C4, lockup	B, D, F, G	Down	Down	Up
Third	C1, C3, lockup	B, C, F, G	Down	Down	Up
Fourth	C1, C2, lockup	F, G	Down	Down	Up
Fifth	C2, C3, lockup	A, C, G, F	Down	Down	Up
Sixth	C2, C4, lockup	A, D, F	Up	Down	Up

Notes:

1. A and B solenoids are normally open when not energized.
2. C, D, E, F, and G solenoids are normally closed when not energized.
3. G solenoid is on in first and second ranges but is switched off for the duration of the 1–2 upshift and the 2–1 downshift.

> **SHOP TALK**
>
> Learning hydraulic circuit and torque powerflows helps the student/technician to visualize transmission operation so they are a key factor in achieving a proper understanding of Allison transmission operation. For the Allison specialist tech, knowing the hydraulic circuit factors and torque powerflows are essential to diagnose performance problems.

HYDRAULIC CIRCUITS

Students and technicians typically require some initial guidance to understand the complex hydraulic circuits used in ECATs. By following hydraulic circuit schematics and Allison service literature, it does not take long to understand the hydraulics that determine the powerflow through the transmission.

Neutral Operation

The transmission is in neutral when torque from the torque converter is not transmitted beyond the rotating clutch module and the P1 planetary sun gear (**Figure 21–19**). Both the C1 and C2 clutches are released. The C5 clutch is applied, but no output movement occurs because a driving clutch is not applied to move the planetary gears. The result is that no output torque occurs. If the output shaft is rotating while in neutral, the stationary clutch applied will depend on the shaft rotational speed. This varying application corresponds to the neutral ranges identified as N1, N2, N3, and N4. These ranges are dependent on the application of the clutches C5, C4, C3, and C2, respectively. The application of a specific clutch during neutral operation means that one of the two clutches required to ensure a range selection when the transmission is shifted to D is already engaged.

FIGURE 21–19 WT hydraulic schematic—neutral

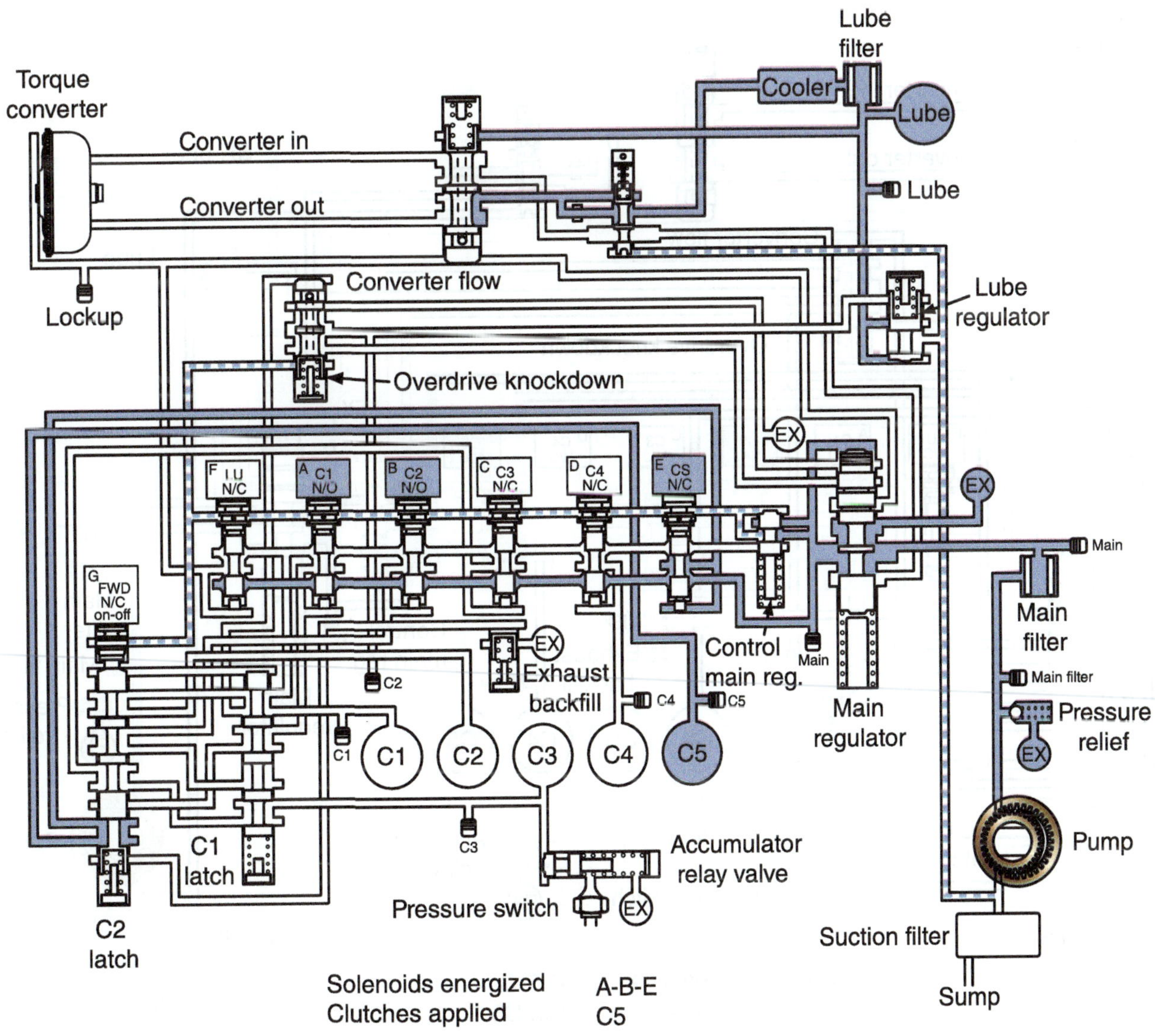

First Range Operation

Clutches C1 and C5 are applied (**Figure 21–20**). Torque is applied by the torque converter to the turbine shaft. Because the C1 is applied, the rotating clutch assembly acts to lock the turbine shaft and main shaft, causing them to rotate as a unit. Torque is transmitted through the main shaft to the P3 sun gear, which is splined to the main shaft. The P3 sun gear acts to rotate the P3 pinion carrier, which is splined to B the output shaft. The C5 clutch being engaged prevents the P3 ring gear from being rotated.

Torque route is as follows: converter turbine shaft, main shaft, P3 sun gear, P3 pinion carrier, and output shaft.

First Range Operation in WT with Dropbox/Transfer Case. Clutches C3 and C6 are applied. The C6 clutch holds the main shaft stationary. A WT with a transfer case has a main shaft that extends from the main housing through the P3 planetary gear assembly and output adapter to the C6 clutch. When the C3 clutch is applied, the P1 ring gear is held. Because the main shaft cannot rotate, torque input is through the P1 sun gear. Because the P1 ring gear is held, output torque is transferred through the pinion carrier.

The P1 planetary carrier is splined to the P2 ring gear, which acts on the P2 pinion carrier. The P2 sun gear is integral with the main shaft, which is being held by the C6 clutch. Torque is routed from the P2

FIGURE 21–20 WT hydraulic schematic—1st range

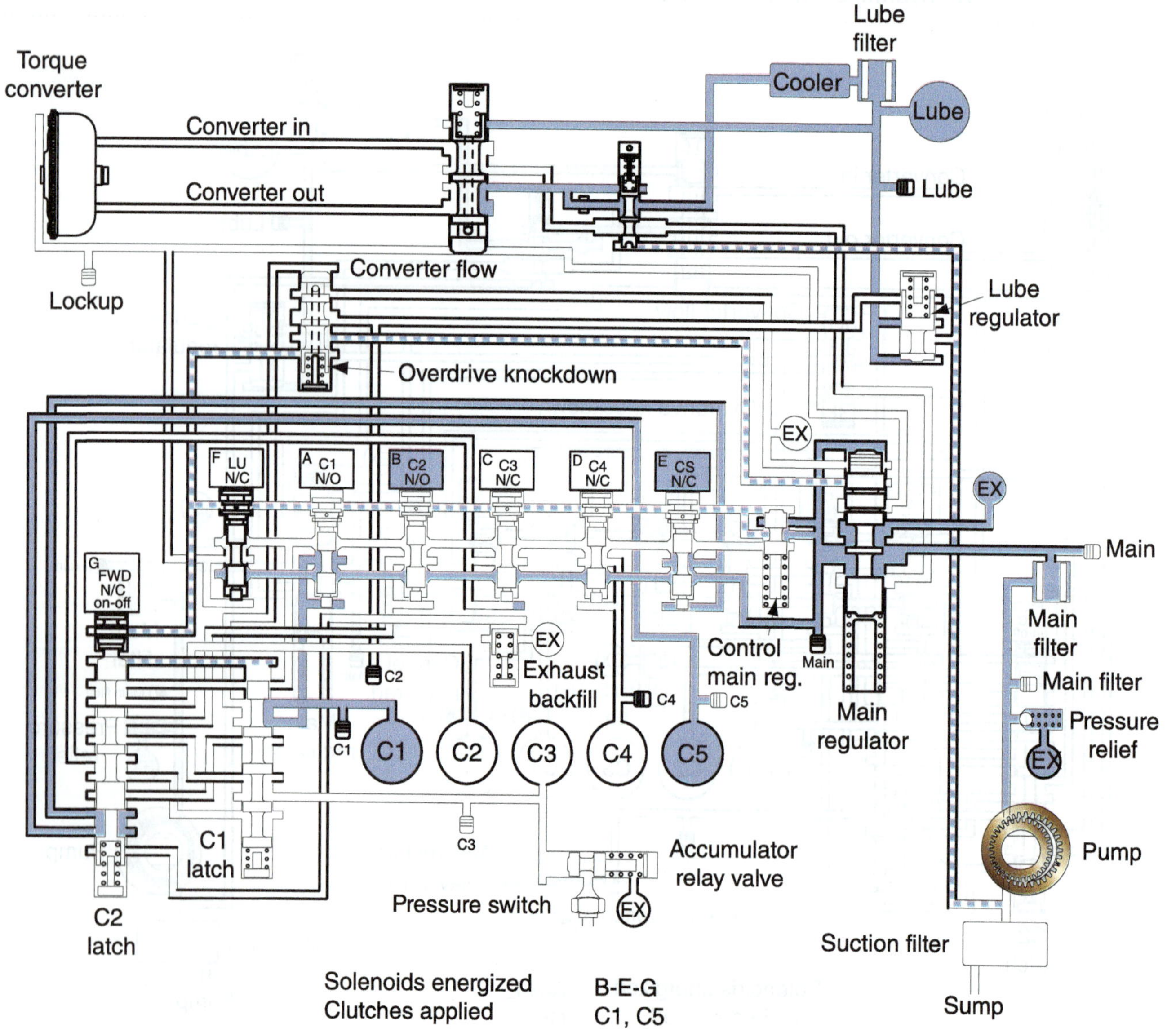

pinion carrier to the P3 ring gear, which is splined to it. Because the P3 sun gear is integral with the main shaft, which is being held by the C6 clutch, torque is output from the P3 pinion carrier, which is splined to the output shaft adapter. The output shaft adapter applies torque to the transfer drive gear.

Torque is transmitted to the transfer case by means of the transfer drive gear. This drives an idler gear, reversing the direction of rotation. The driven gear is splined to the P4 (differential) pinion carrier. The P4 sun gear is integral with the transfer case output shaft. Torque is, therefore, transferred from the P4 pinion carrier to the P4 sun gear, rotating the transfer case output shaft. When the C7 (differential lock) clutch is disengaged, transfer case output torque is split, 70 percent of which is directed at the rear output shaft and 30 percent at the front output shaft. When the C7 clutch is engaged, output torque is split evenly between the front and rear output shafts. Torque route is as follows: converter turbine shaft, P1 sun gear, P1 planetary carrier, P2 ring gear, P2 pinion carrier, P3 ring gear, P3 pinion carrier, output shaft adapter, transfer drive gear, P4 pinion carrier, P4 sun gear, and transfer case output shafts.

Second Range Operation

Clutches C1 and C4 are applied (**Figure 21–21**). The applied C1 clutch locks the turbine shaft and main shaft so they rotate as a unit. Torque is applied to the P2 sun gear (integral with the main shaft), rotating the P2 pinion carrier. The applied C4 clutch holds the C2 ring gear. Torque applied to the P3 planetary is now split, being transferred from the P2 pinion carrier to the C3 ring gear and by the C3 sun gear

FIGURE 21–21 WT hydraulic schematic—2nd range

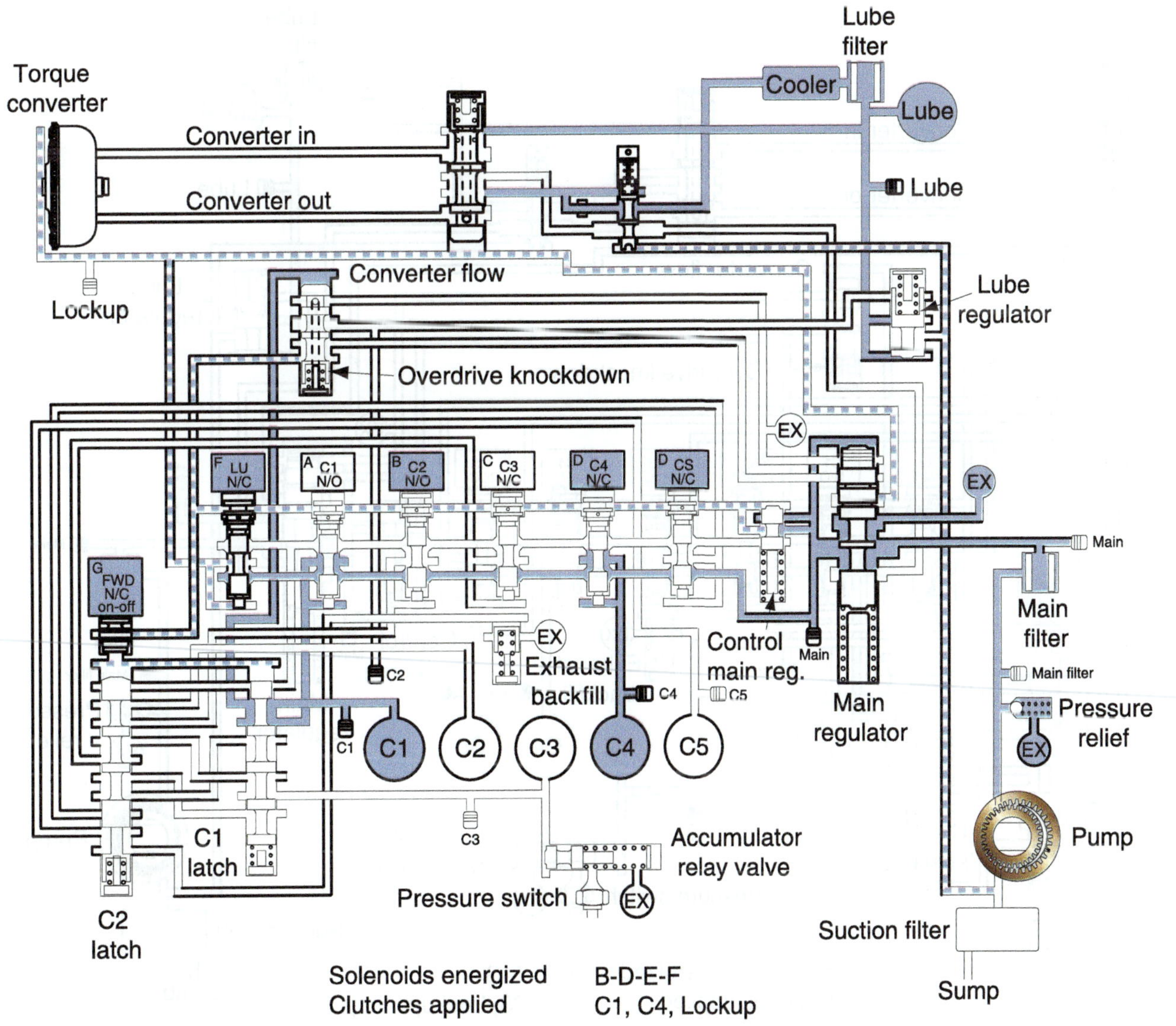

that is splined to and rotates with the main shaft. The P3 ring gear rotates at a slower speed than the P3 sun gear. This results in the P3 ring gear action resembling that of a held member. The P3 sun gear becomes the main input torque path for the P3 planetary gearset. This torque is applied to the P3 pinion carrier, which provides output torque to the transmission output shaft.

Torque route is as follows: converter turbine shaft, main shaft, P2 sun gear, P2 pinion carrier, P3 ring gear/main shaft, P3 sun gear, P3 pinion carrier, and output shaft.

Third Range Operation

Clutches C1 and C3 are applied (**Figure 21–22**). The C1 clutch locks the turbine shaft and main shaft so they rotate as a unit. The P1 planetary gearset provides first stage gearing with torque passing from the P1 sun gear (part of the rotating clutch) to the P1 pinion carrier. The C3 clutch holds the P1 ring gear. The P1 carrier is splined to the P2 ring gear. The P2 planetary gearset, therefore, is subject to input from the torque applied to both its ring gear and its sun gear, which is splined to and rotates with the main shaft.

The P2 ring gear will rotate at a lower rate than its sun gear, so the action of the P2 ring gear will resemble that of a held member in the planetary gearset. The P2 sun gear is, therefore, the main path for torque transmitted to the P2 pinion carrier. The P2 pinion carrier is the output from the P2 planetary in the third gear range. The P2 pinion carrier is splined to the P3 ring gear and provides one of two inputs to the P3 planetary gearset. The other input torque path is the P3 sun gear, which is splined to and, therefore,

FIGURE 21–22 WT hydraulic schematic—3rd range

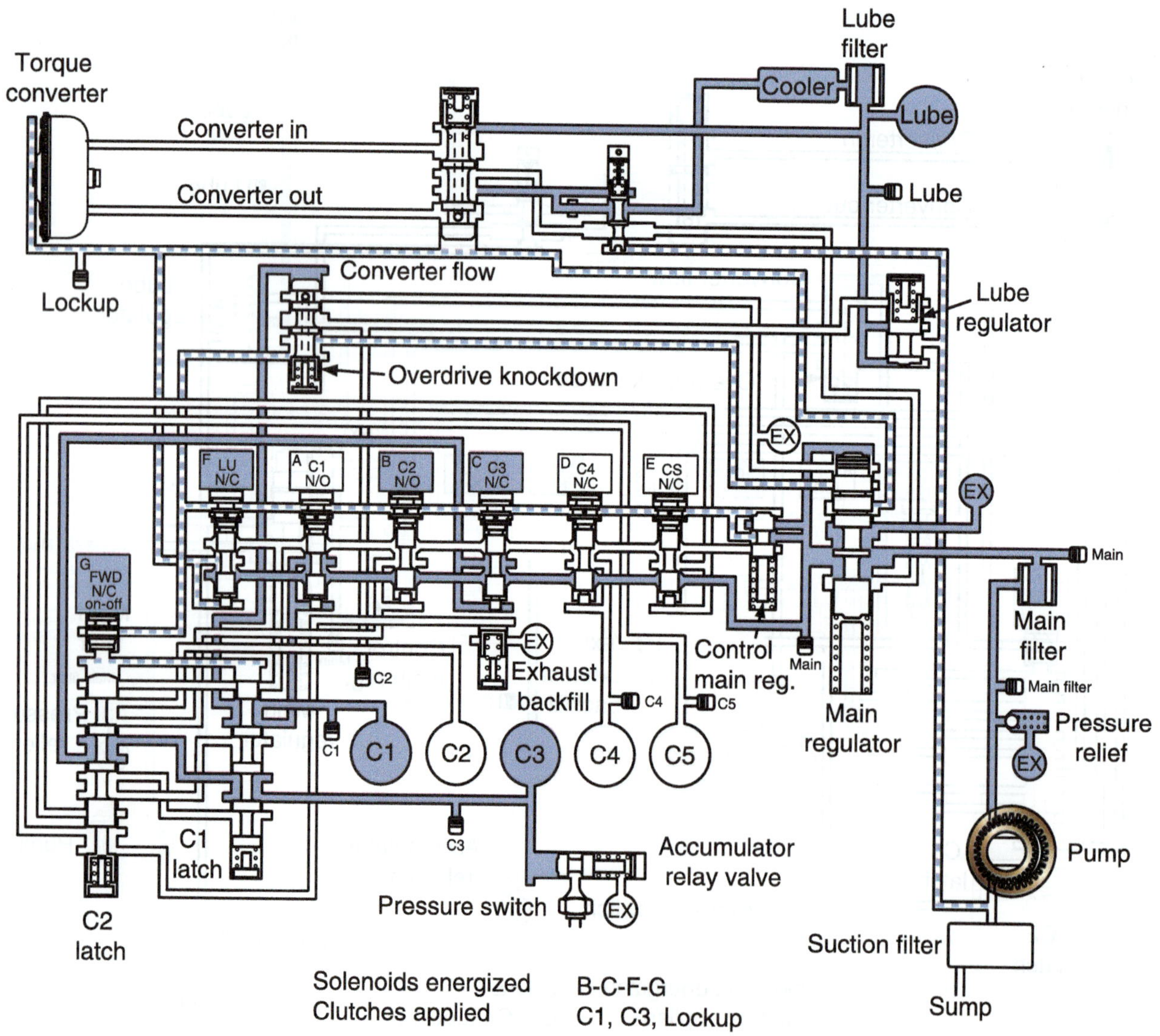

rotates with the main shaft. The P3 ring gear rotates at a lower speed than the P3 sun gear, so P3 ring gear action will resemble that of a held member. The P3 sun gear becomes the main torque path for torque transmitted to the P3 planetary gearset. This torque is transmitted to the P3 pinion carrier, which is splined to the output shaft of the transmission.

Torque path is as follows: converter turbine shaft, main shaft, P1 ring gear and P1 sun gear, P1 pinion carrier, P2 sun gear and P2 pinion carrier, P3 ring gear and P3 sun gear, P3 pinion carrier, and output shaft.

Fourth Range Operation

Clutches C1 and C2 are applied (**Figure 21–23**). The C1 and C2 clutches rotate with the turbine shaft. The C1 drive hub is splined to the main shaft so that when it is actuated, the main shaft will rotate at turbine shaft (input) speed. The C2 drive hub is splined to the P3 ring gear (through the P2 pinion carrier), which results in the P3 ring gear rotating at turbine speed. Because the C3, C4, and C5 stationary clutches are all disengaged, the remaining pinion carrier and ring gears rotate at the same speed and direction as the turbine shaft. The P3 ring gear drives the P3 pinion carrier, which is splined to the transmission output shaft. The result is direct drive.

Torque route is as follows: converter turbine shaft, main shaft, P3 sun gear and P3 ring gear, P3 pinion carrier, and output shaft.

Fifth Range Operation

Clutches C2 and C5 are applied (**Figure 21–24**). The P1 planetary gearset provides first stage gearing in fifth range with input torque routed from the

FIGURE 21–23 WT hydraulic schematic—4th range

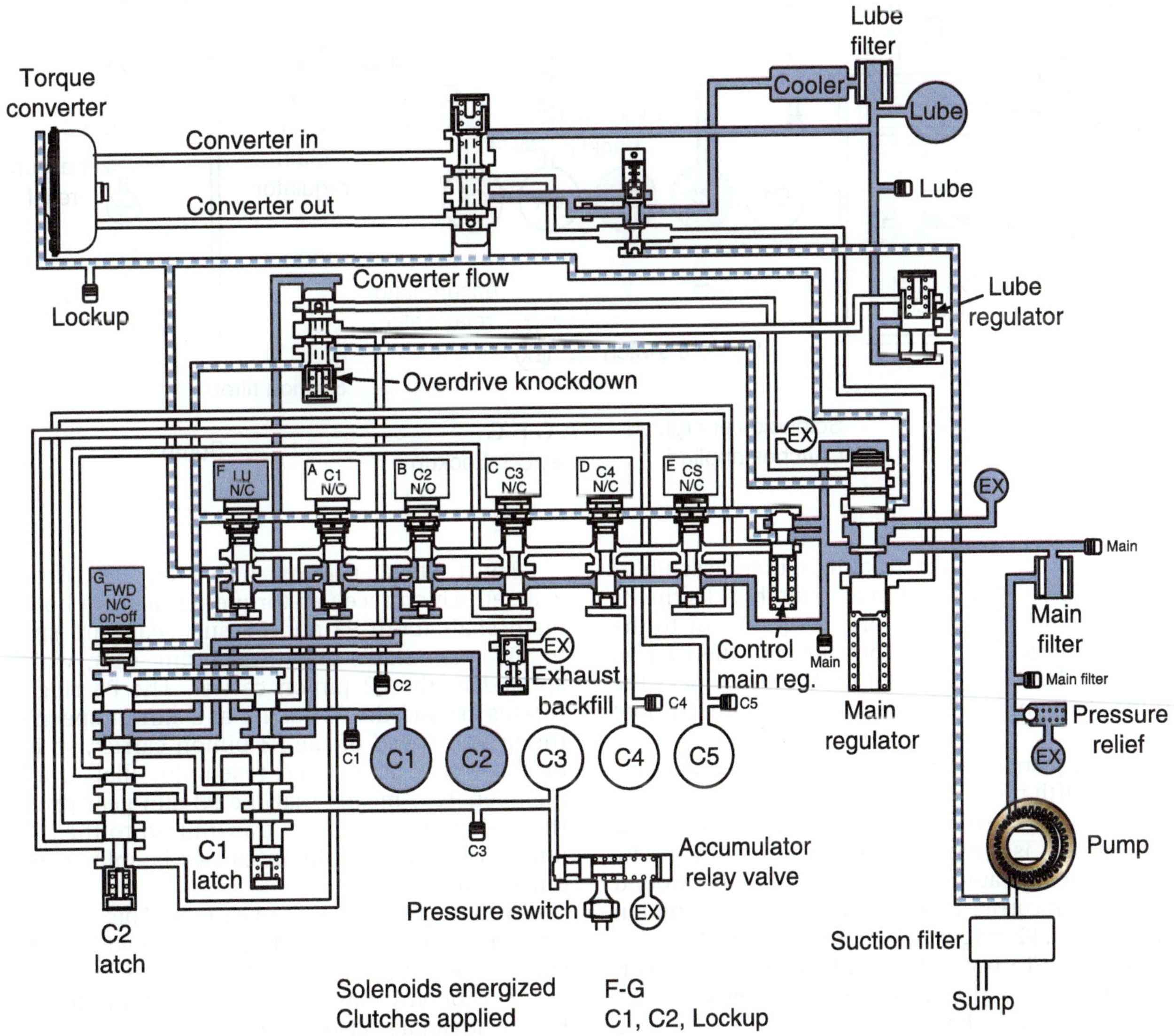

FIGURE 21–24 WT hydraulic schematic—5th range

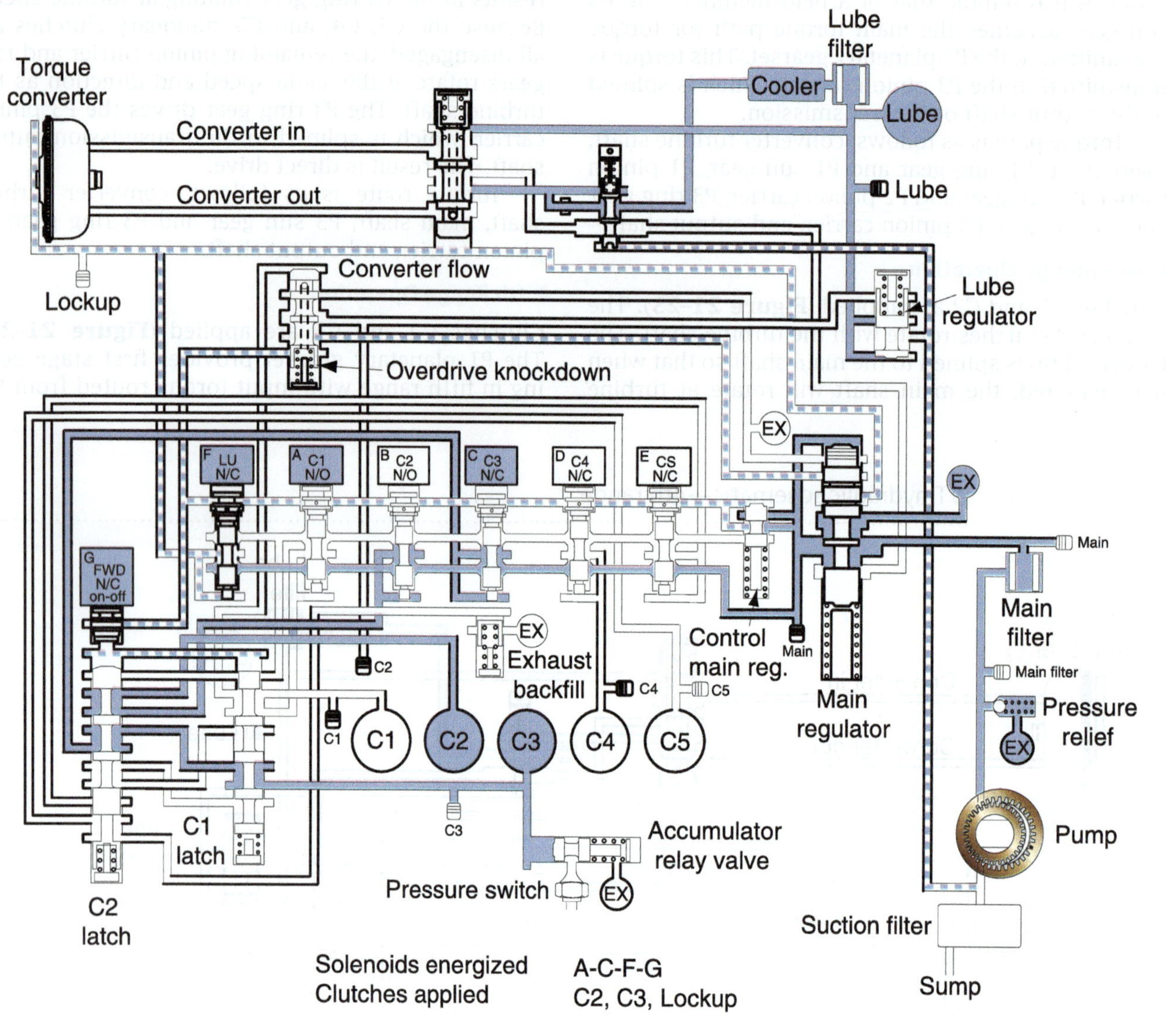

P1 sun gear (part of the rotating clutch assembly/turbine shaft). The C3 stationary clutch locks the P1 ring gear, holding it stationary. Torque input to the P1 planetary gearset is output through the P1 pinion carrier. The application of the C2 clutch locks the turbine shaft and the P2 pinion carrier so they rotate together.

The P2 planetary gearset provides second stage gearing in fifth range. The P2 pinion carrier is subject to torque applied from the C1 drive hub and the P2 ring gear that is splined to and rotates with the P1 pinion carrier. Because the P2 ring gear rotates around the P2 pinion carrier at a slower rate than the pinion carrier speed, P2 ring gear action resembles that of a held member. Torque is transmitted from the P2 carrier to the P2 sun gear, which provides second stage output to the main shaft.

Third stage gearing uses the P3 planetary gearset, which receives torque input from both the main shaft (both P2 and P3 sun gears are splined to the main shaft) and the P3 ring gear, which is splined to the P2 pinion carrier. The P3 ring gear rotates around the P3 carrier at a slower speed than the driven speed of the P3 pinion carrier, so it will produce gear action that resembles that of a held member. This means that the P3 sun gear provides the main path for torque transmitted to the P3 pinion carrier. The P3 pinion carrier is splined to the output shaft.

Fifth range is an overdrive range. This means that transmission output shaft speed exceeds the input speed. When overdrive gearing is selected, an increase in fuel economy may be achieved, but a reduction in torque advantage also occurs.

Torque route is as follows: converter turbine shaft, P1 sun gear, P1 pinion carrier, P2 ring gear and P2 planetary carrier, P2 sun gear, main shaft, P3 ring gear, P3 sun gear, P3 pinion carrier, and output shaft.

Sixth Range Operation

Clutches C2 and C4 are applied (**Figure 21–25**). The application of the C2 clutch locks the turbine shaft and the P2 pinion carrier, causing them to rotate as a unit. The P2 planetary gearset provides the first stage gearing in sixth range. The C4 clutch holds the P2 ring gear. With the P2 ring gear held and torque input to the P2 carrier, torque is transmitted to the P2 sun gear, which is splined to the main shaft.

Second stage gearing in sixth range takes place in the P3 planetary gearset. Torque is input to the P3 planetary gearset by the P3 sun gear (splined to the main shaft) and the P3 ring gear that is splined to the P2 pinion carrier. The P3 ring gear rotates around the P3 pinion carrier at a slower rate than the rotational speed of the P3 sun gear, so its action resembles that of a held member. The result is that the sun gear becomes the main path for delivering torque to the P3 pinion carrier. The P3 pinion carrier transmits torque to the transmission output shaft. Sixth range is also an overdrive range. The input shaft rotational speed is exceeded by the output shaft to a greater extent than when in the fifth range.

Torque route is as follows: converter turbine shaft, P2 pinion carrier, P2 sun gear, main shaft, P3 ring gear/main shaft, C3 sun gear, C3 pinion carrier, and output shaft.

Reverse Operation

Clutches C3 and C5 are applied (**Figure 21–26**). The C3 clutch holds the P1 ring gear. The C5 clutch holds the

FIGURE 21–25 WT hydraulic schematic—6th range

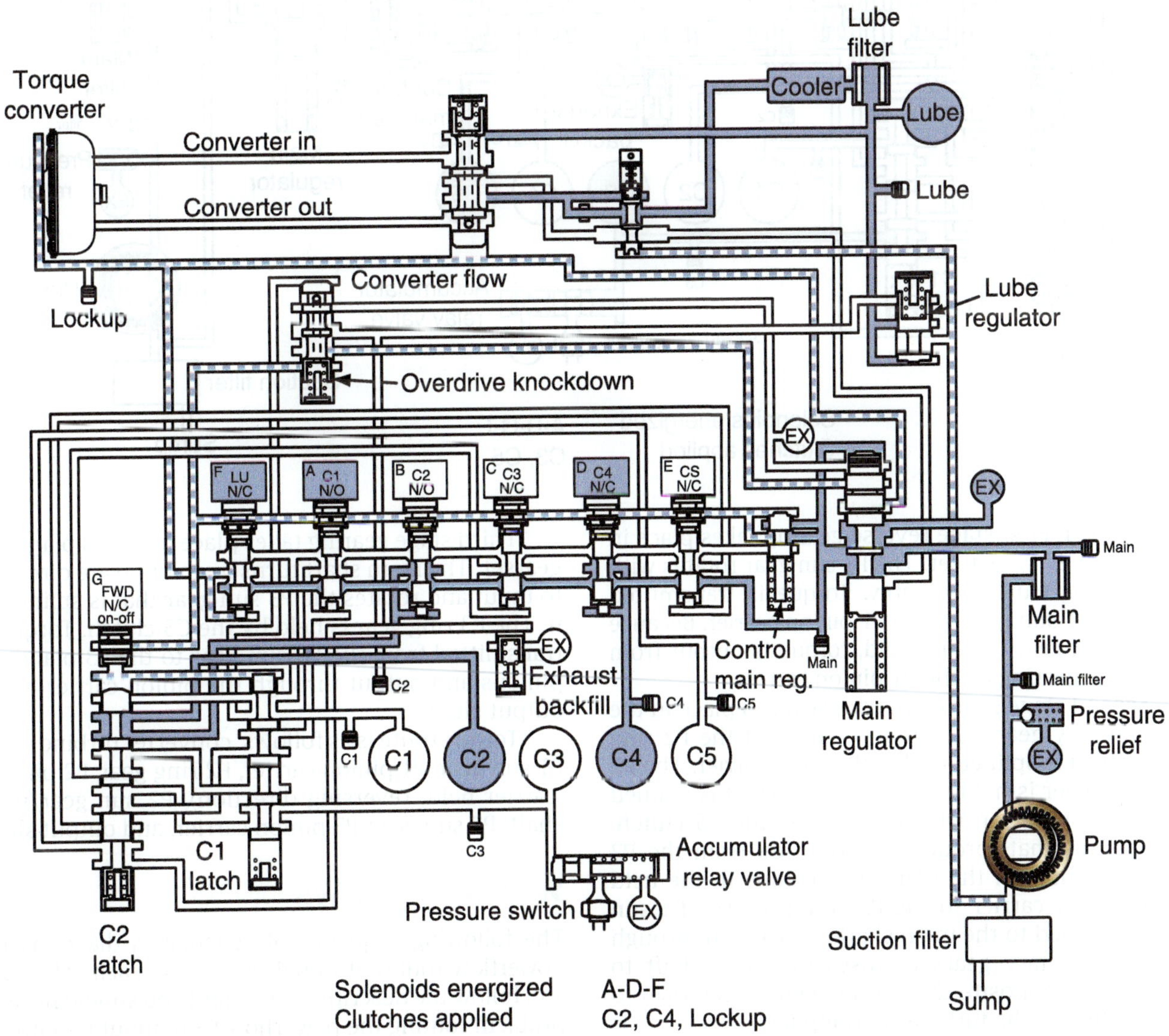

FIGURE 21–26 WT hydraulic schematic—reverse

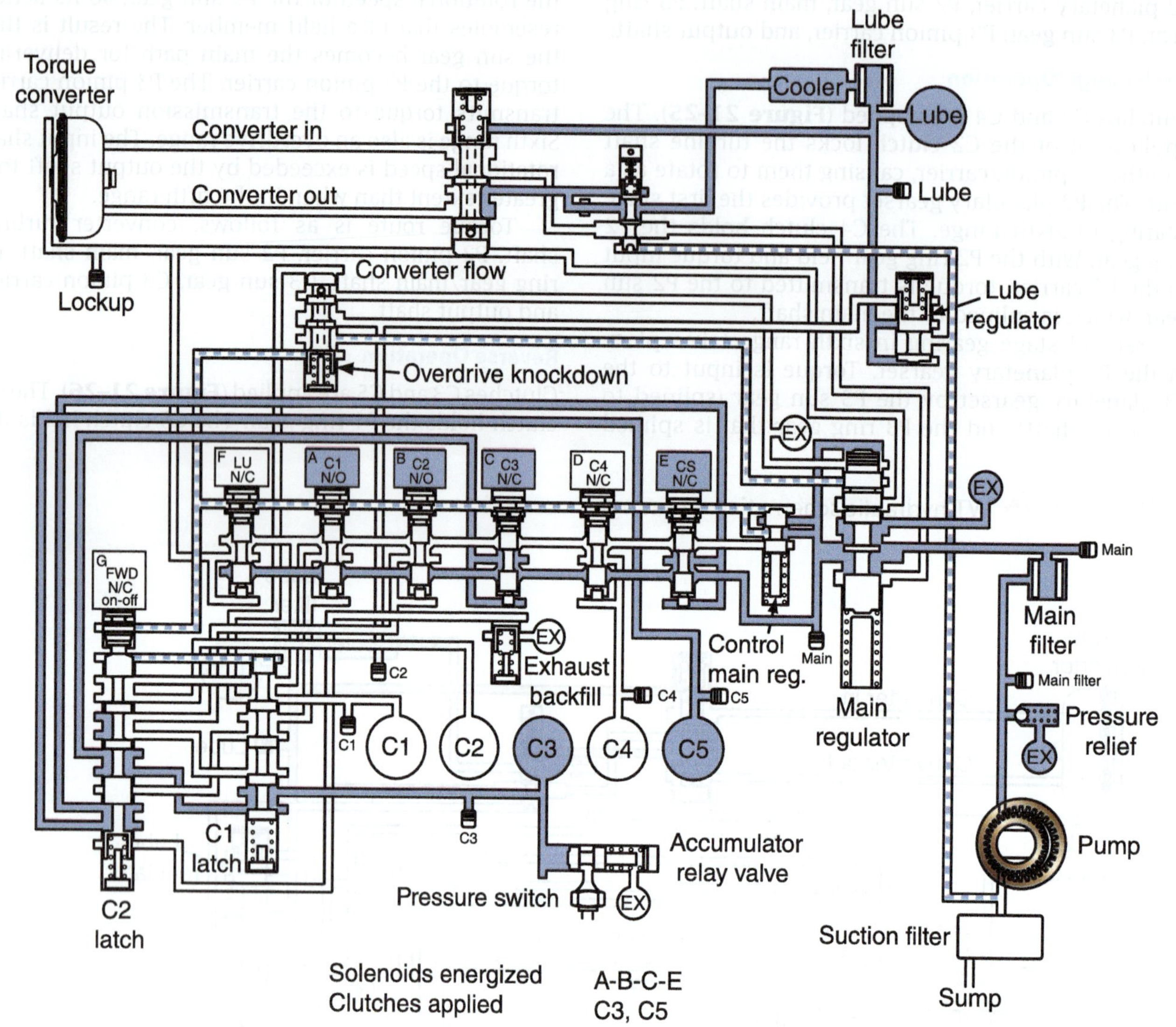

P3 ring gear. First stage reverse gearing takes place in the P1 planetary gearset. The P1 sun gear rotates with the rotating clutch assembly. Torque is transmitted from the P1 sun gear to the P1 pinion carrier. Because the P1 ring gear is being held, torque is output from first stage gearing by the P1 pinion carrier.

Second stage gearing in reverse takes place in the P2 planetary gearset. Torque is input at the P2 ring gear, which is splined to the P1 pinion carrier. The P2 pinion carrier is held stationary because it is splined to the P3 ring gear, which is held by the C5 clutch. This means that torque is transmitted from the P2 ring gear through the planetary pinions in the held P2 planetary carrier to the P2 sun gear. The P2 sun gear is splined to the main shaft. This action through the P2 planetary gearset causes the main shaft to rotate in an opposite direction (counterclockwise) to the rotation of the turbine shaft input.

Third stage gearing takes place in the P3 planetary gearset. The main shaft turns in an opposite direction to input and rotates the P3 sun gear that is splined to it. The P3 ring gear is held by the C5 clutch. Torque is transmitted from the P3 sun gear to the P3 planetary pinions and output through the pinion carrier to the output shaft.

Torque route is as follows: converter turbine shaft, P1 sun gear, P1 pinion carrier, P2 ring gear, P2 pinions (carrier held, reversing direction), P2 sun gear/main shaft, P3 sun gear, P3 pinion carrier, and output shaft.

WT POWERFLOWS

The following sequence of schematics diagrams the powerflow routing through a WT. Use each schematic and follow the descriptive text, and you should have no problem routing the powerflow from input to output.

Neutral Powerflow

Only the C5 clutch is applied (**Figure 21–27**).

- Torque from the turbine shaft is not transmitted beyond the rotating clutch module and the P1 sun gear.
- If the vehicle is rolling when in neutral, different clutches may be applied. This controls the speed of rotating components.

First Range Powerflow

In first range, the C1 and C5 clutches are applied (**Figure 21–28**).

- C1 locks the turbine shaft and main shaft together. The P3 sun gear is part of the main shaft module, so it becomes input for the P3 planetary gearset.

FIGURE 21–27 Neutral powerflow

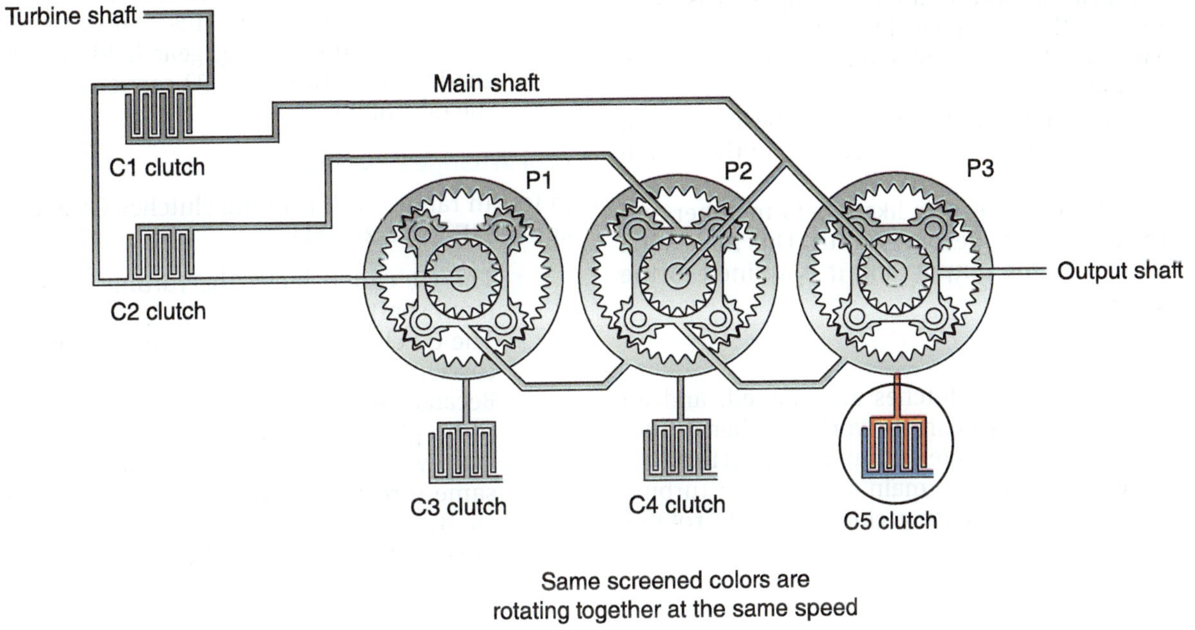

FIGURE 21–28 1st range powerflow

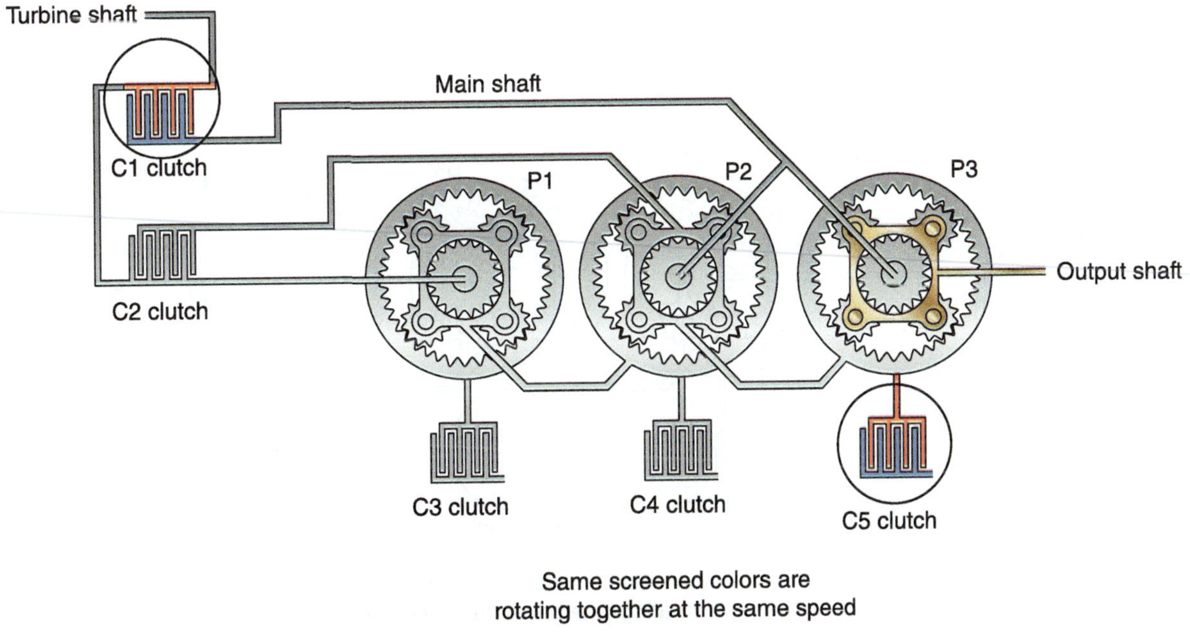

- C5 holds the P3 ring gear. The P3 carrier is splined directly to the output shaft. The P3 sun gear is input and the P3 ring gear is held, so the P3 carrier becomes the output.

Second Range Powerflow

In second range, the C1 and C4 clutches are applied, making the P2 and P3 planetary gearsets work together to provide the output (**Figure 21–29**).

- The C1 clutch locks the turbine shaft and main shaft together and this drives the P2 sun gear.
- The C4 clutch holds the P2 ring gear. The P2 sun gear is input and the P2 ring gear is held, making the P2 carrier the output.
- The P3 sun gear is splined to the main shaft so it rotates.
- The P3 ring gear is splined to the P2 carrier, but it rotates at a slower speed than the P3 sun gear.
- The P3 ring gear acts like a held member and the P3 sun gear becomes input. This makes the P3 carrier the output, and it is splined to the output shaft.

Third Range Powerflow

- The C1 and C3 clutches are applied, and all three planetary gearsets work together to provide the appropriate output (**Figure 21–30**).
- C1 clutch locks the main shaft to the turbine shaft and the rotating clutch module rotates the P1 sun gear.
- The C3 clutch holds the P1 ring gear. Because the P1 sun gear is input and the P1 ring gear is held, the P1 carrier becomes the output.
- The P2 sun gear rotates with the main shaft.
- The P2 ring gear is splined to the P1 carrier, so it rotates. Because the P2 ring gear rotates more slowly than the P2 sun gear, it acts like a held member. This makes the P2 sun gear the input, the P2 ring gear "held," and the P2 carrier the output.
- The P3 sun gear rotates with the main shaft.
- The P3 ring gear is splined to the P2 carrier, so it rotates with the P2 carrier at a speed slower than the P3 sun gear. This makes the P3 sun gear input, the P3 ring gear held, and the P3 carrier the output. The P3 carrier is splined to the output shaft.

Fourth Range Powerflow

In fourth range, both rotating clutches C1 and C2 are applied (**Figure 21–31**).

- The C1 clutch locks the turbine shaft to the main shaft.
- The C2 clutch locks the turbine shaft to the P2 carrier.
- Because no stationary clutches are applied, all three planetary sun gears, carriers, and ring gear will rotate at the same speed and in the same direction as the input from the turbine shaft.
- This results in a direct drive or 1:1 ratio.

FIGURE 21–29 2nd range powerflow

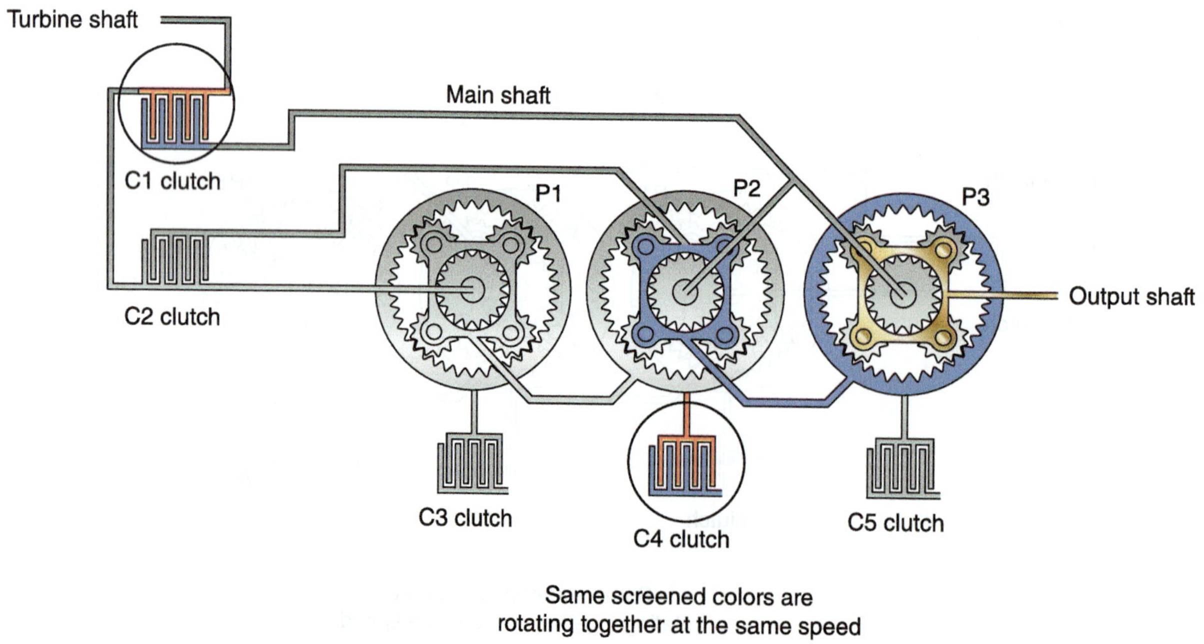

FIGURE 21–30 3rd range powerflow

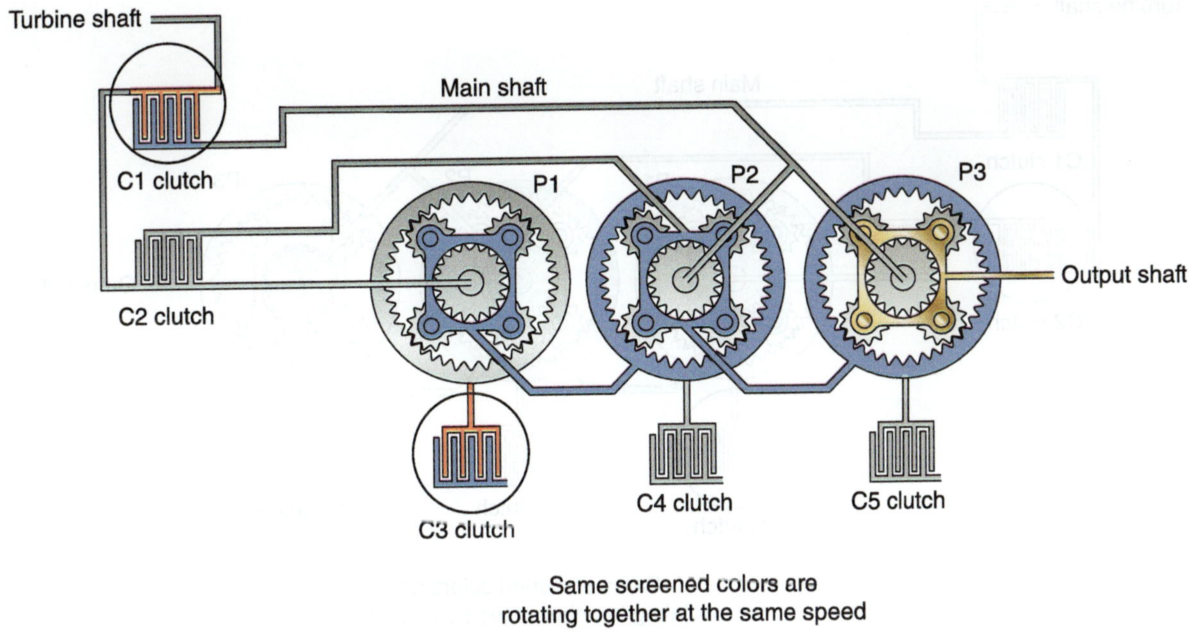

FIGURE 21–31 4th range powerflow

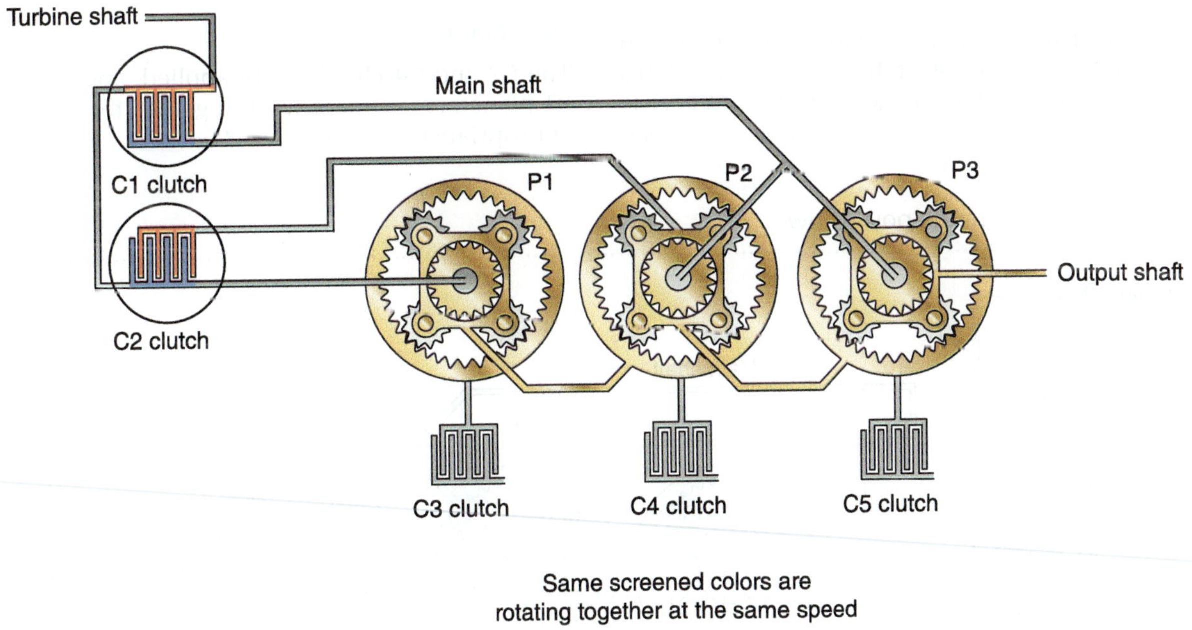

Fifth Range Powerflow

In fifth range, the C2 and C3 clutches are applied and all three planetary gearsets work together to provide the appropriate output (**Figure 21–32**).

- The P1 sun gear is rotating with the rotating clutch module.
- The C3 clutch is holding the P1 ring gear stationary, making the P1 carrier the output.
- The P2 carrier is rotating at turbine speed because the C2 clutch locks the turbine to the P2 carrier.
- The P2 ring gear is splined to and rotating with the P1 carrier. The P2 carrier is rotating faster than the P2 ring gear, so it is the input. The ring

FIGURE 21–32 5th range powerflow

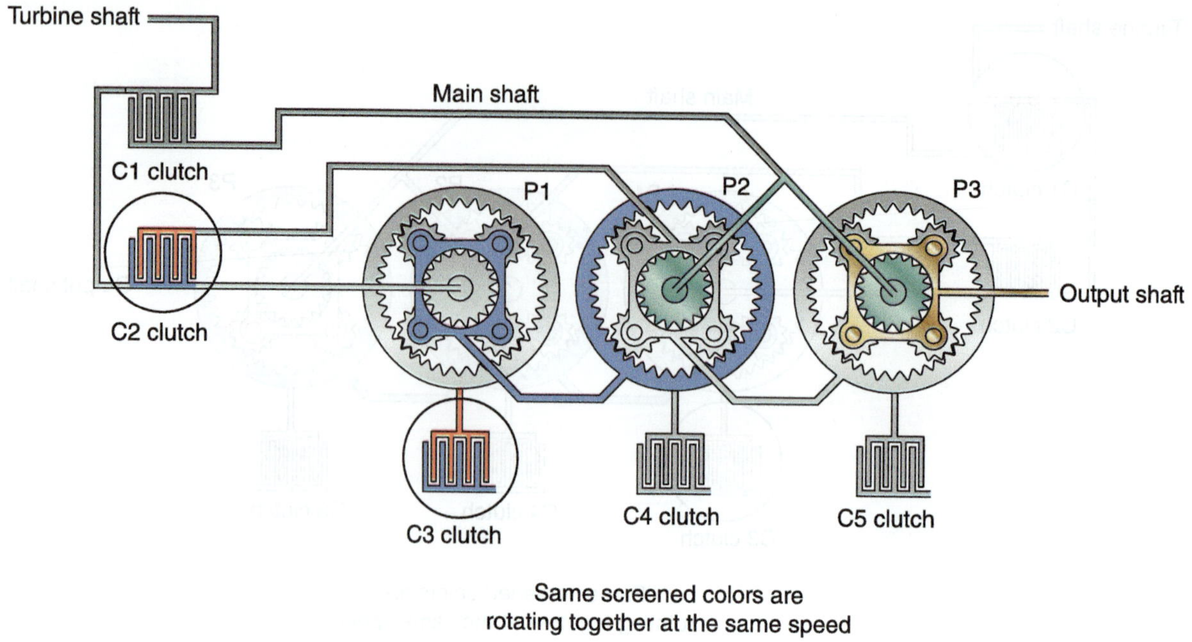

gear acts like a held member. This makes the P2 sun gear the output.

- The P2 sun gear rotates the main shaft and the P3 sun gear.
- The P3 ring gear is splined to and rotating with the P2 carrier. However, the P3 sun gear rotates faster than the P3 ring gear, so the sun gear is input and the ring gear acts as a held member. This makes the P3 carrier the output, and it is splined to the output shaft.
- This gear range produces an overdrive.

Sixth Range Powerflow

The C2 and C4 clutches are applied, and the P2 and P3 planetary gearsets work together to produce the appropriate output (**Figure 21–33**).

FIGURE 21–33 6th range powerflow

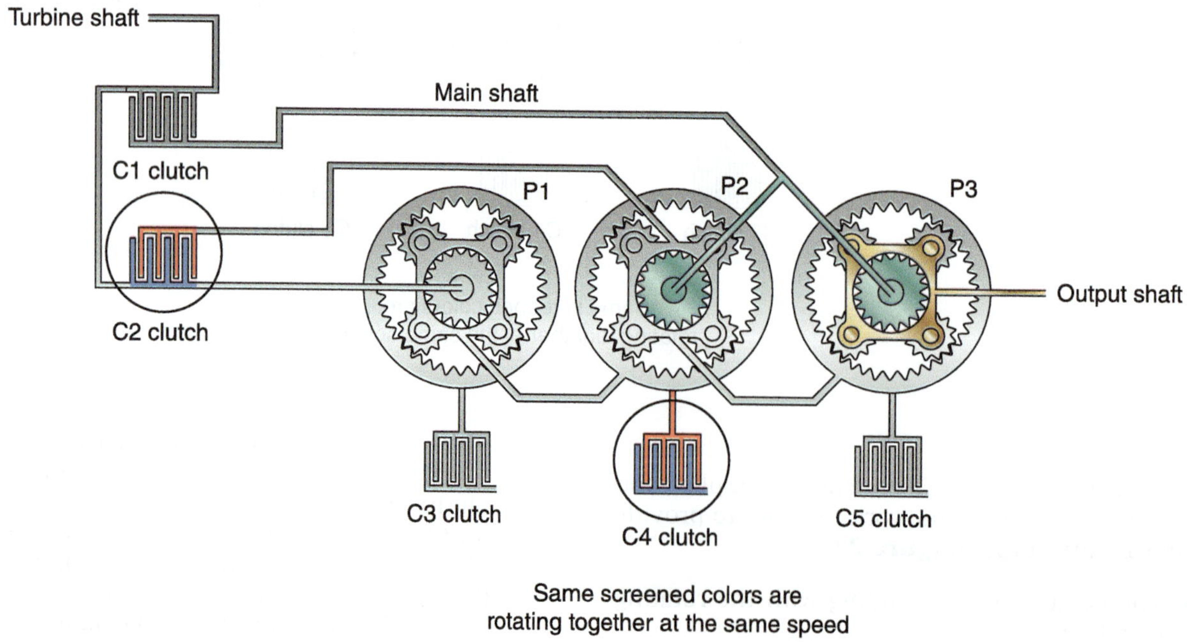

- The C2 clutch locks the P2 carrier to the turbine shaft.
- The C4 clutch holds the P2 ring gear. The P2 carrier is the input (from the turbine shaft) and the P2 ring gear is held so the P2 sun gear becomes the output.
- The P2 sun gear rotates the main shaft and the P3 sun gear. The P3 sun gear is input and the P3 ring gear acts like a held member. This makes the P3 carrier the output, and it is splined to the output shaft.
- This gearing produces an overdrive.

Reverse Range Powerflow

The C3 and C5 clutches are applied, and all three planetary gearsets work together to produce the appropriate output (**Figure 21–34**).

- The P1 sun gear rotates with the rotating clutch module.
- The P1 ring gear is held by the C3 clutch, making the P1 carrier the output.
- The P1 carrier is splined to the P2 ring gear. The P2 ring gear becomes the input for the P2 planetary set.
- The C5 clutch holds the P3 ring gear, which is splined to the P2 carrier. Because the P2 ring gear is the input and the P2 carrier is held, the P2 sun gear becomes the output. Because the carrier is the held member, the P2 sun gear rotates in an opposite direction (counterclockwise) to the input direction of rotation.
- The P2 sun gear rotates the main shaft in an opposite direction. Because the P3 sun gear rotates with the main shaft, it also turns in an opposite direction.
- The P3 sun gear becomes reverse input for the P3 planetary gearset.
- The P3 ring gear is held by the C5 clutch. The P3 carrier becomes reverse output, and it is splined to the output shaft.

SERVICING AND TROUBLESHOOTING

WT electronics are equipped with an extensive self-diagnostic capability. Fault codes are used to isolate the nature, source, and severity of a problem in the electronic circuit. To effectively troubleshoot WTs, the Allison DOC should be used. The following section outlines some of the diagnostic techniques recommended by Allison.

Lubrication

Allison currently recommends the use of **TranSynd** synthetic oil in all of their transmissions. TranSynd is formulated jointly by Allison and Castrol. If this oil is used, the oil drain intervals can be extended by as much as four times depending on the specific application. There are other approved lubricants, but note that Allison suggests that use of any fluid other than TranSynd results in greatly reduced service life. Chapter 19 outlines the methods of checking and changing of the transmission lubricant. It is a good practice to refer to either Allison or chassis OEM

FIGURE 21–34 Reverse range powerflow

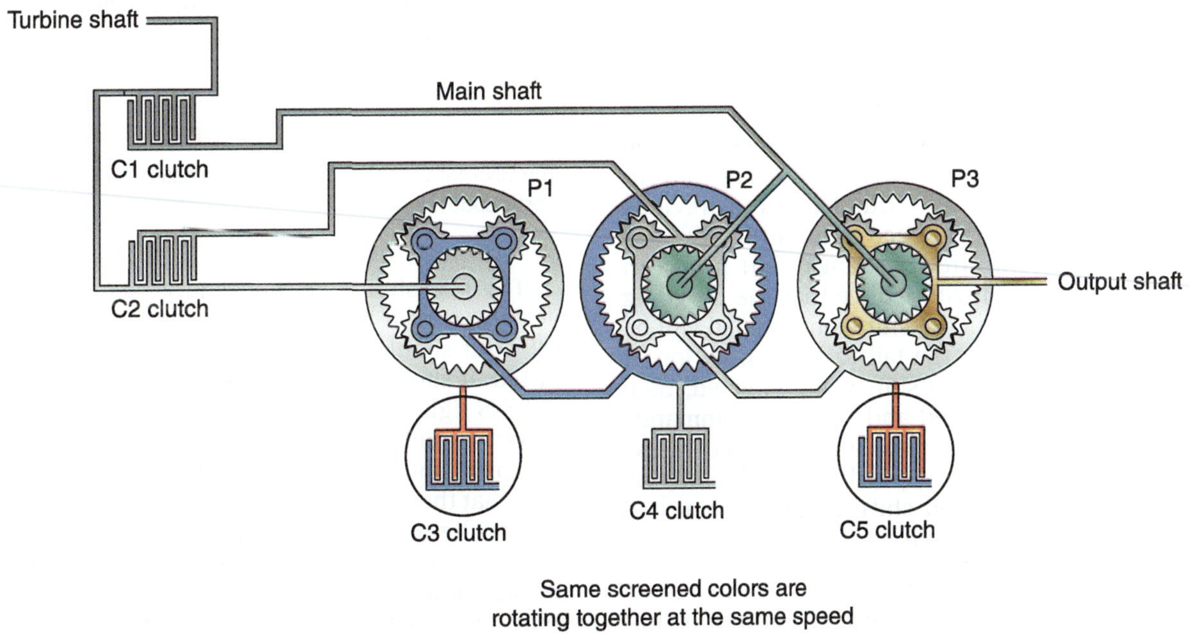

FIGURE 21–35 Cutaway view of a WT fluid filter

service literature when servicing automatic transmissions. A key to maximizing the operational life of a WT transmission is in performing oil and filter changes at the recommended intervals. **Figure 21–35** shows a cutaway view of a WT filter assembly.

Range/Shift Tests

Each time the WT electronics effect a shift, the shift logic is verified and the transmission range selection is checked. The following sections describe the verification tests that should be used.

Range Verification. WT continuously verifies the transmission range, whether or not a shift is in progress. Essentially, this involves verifying that the current range is the range commanded by the TCM. The test involves checking the current gear ratio by comparing the turbine and output shaft speeds. This ratio is then compared with the ratio logged in the TCM memory for the commanded range. If the two ratios do not match, a diagnostic code is logged.

Off-Going Ratio Test. The off-going ratio test is performed when a shift is in progress. Within a set time after the TCM has commanded a range shift, the TCM calculates the ratio between input and output speeds, comparing this ratio with that of the previous range. For instance, if the speed ratio of the previous range is still present after the range shift has been commanded, the TCM will deduce that the off-going clutch failed to release. The shift commanded will be repeated twice to verify the fail to shift condition. If this does not result in the shift being effected, then a diagnostic code is logged and the TCM commands the transmission into the previously selected range. The off-going ratio test is applied during the interval between the turbine speed shift initiation point and the pull-down detected point.

Oncoming Ratio Test. The oncoming ratio test is performed by WT electronics near the end of a shift in progress sequence. The oncoming ratio test checks turbine speed and output shaft speed to determine whether the transmission is in the new range commanded by the TCM. When the ratios fail to match, the TCM assumes that the oncoming clutch did not actuate and will log a fault code.

SHOP TALK

The off-going and oncoming ratio test can detect clutch slippage almost immediately. This provides a considerable advantage to the technician when troubleshooting WTs. WT electronics can command a hold in the previous gear range *in order* to protect the transmission when indications of severe clutch slippage warrants this failure strategy. In most instances, by the time the driver complained of a slippage problem, significant transmission damage had already occurred. This was a major issue with the first generation of Allison ECATs.

C3 Pressure Switch Tests

The C3 pressure switch is an important diagnostic indicator. The C3 pressure switch signals the TCM as to whether pressure is present or not present in the C3 clutch. For instance, when C3 pressure is not present and it should be, a diagnostic code is then logged and a DNS condition is immediately put into effect, with the result that the transmission remains within the current range.

If the opposite occurs, that is, C3 pressure is signaled in a range when it should not be, then the result is the same. A diagnostic code is logged and a DNS condition results.

DNS Light. Whenever WT electronics sense a condition that could damage the transmission, transmission electronics, or the vehicle drivetrain, shifting is restricted and the DNS light is illuminated.

SHOP TALK

The DNS light is not necessarily illuminated every time a diagnostic code is logged.

The TCM should be checked for fault codes whenever there is a transmission-related problem. The DNS light should illuminate every time the engine is started. It should remain lit for a few seconds only. If the DNS light fails to illuminate at startup, then the system should be checked immediately.

Continuous illumination of the DNS light indicates that the TCM has logged a diagnostic code. When the DNS light is first illuminated, 8 seconds of short beeps are emitted from the shift selector. The beeps are an audible alert to the operator that shifts are being restricted. Additionally, the SHIFT digit on the shift selector display will remain blank and the TCM may not respond to requests from the shift selector.

Limp Home Mode. When the DNS light is illuminated, the intent is to inform the driver that the transmission is not operating properly and requires repair. However, the transmission can usually be operated for a short time in the selected range in "limp home" mode, provided the ignition key is not switched off.

When the DNS light illuminates when the ignition key is cycled from off to on, the transmission will remain in neutral range until the fault code responsible for the DNS condition is cleared. The converter lockup clutch is always disengaged whenever transmission shifting is restricted or during any critical transmission malfunction.

Accessing Diagnostic Codes

DOC is the preferred method of accessing diagnostic codes logged within the WT TCM. All WT diagnostic codes are numeric and consist of a two-digit main code and a two-digit subcode (**Figure 21–36**). The WT TCM logs these codes and produces them for readout by sequencing either the most severe code or the most recent code first. This is followed by the order in which they were logged, beginning with the most recent and working backward. A maximum of five codes may be logged. As codes are added, the least current nonactive code is then dropped from the list. Should all the logged codes be active, the codes are listed in order of severity. When the number exceeds five, the first logged code in the sequence is dropped. Codes may be accessed either by DOC, or by some handheld ESTs, or by the shift selector mechanism. The DOC process for navigating DTCs is outlined earlier in this chapter.

CAUTION

Always use the appropriate Allison troubleshooting procedure and never make any assumptions based on a previous WT problem.

Figure 21–37 is a detailed cutaway view of a WT transmission showing how the clutches interact with the planetary gearsets.

21.4 ALLISON TC10

The Allison TC10 TS transmission is an ECAT that has been specifically designed for use in Class 8 linehaul trucks. It uses a torque converter to drive a gearbox module in which a turbine shaft options input torque to twin countershafts: the design of the gearbox can be compared to that used in a manually shifted transmission rather than any other existing fully automatic transmission. The gearbox module can output four forward drive ratios and two reverse ratios to the output module. However, the turbine shaft passes through the gearbox module and can impart drive torque directly to the output module bypassing the gearbox module, thus providing an additional two forward ratios, for a total of ten. The output module contains a single planetary gearset that options a high-low range feature. It is important to emphasize that there is not a division of torque via the dual countershafts, as would be in a typical manually shifted twin-countershaft transmission. **Figure 21–38** shows an overhead view of a TC10 with callouts indicating the components and modules in the assembly.

In common with the advantages of the multiplication factor that is provided in most other torque converter-driven fully automatic transmissions, the TC10 has excellent vehicle launch characteristics from converter "stall" and is powershifted meaning that there is low spin loss (rpm) between ratios compared with AMTs. Its low spin loss is aided by the fact that within the gearbox module almost all the components are rotated in the same direction. The transmission can be coupled to engines outputting up to 600 hp (447 kW) and up to 1,700 lb-ft. (2305 N·m) gross torque. **Figure 21–39** shows a photographic overhead view of the TC10 transmission.

TRANSMISSION CONFIGURATION

The TC10 base transmission is divided into three modules:

- Input module: comprises a two-pass torque converter and lockup clutch.
- Gearbox module: comprises a turbine shaft, twin countershafts, and four clutches.
- Output module: comprises a planetary gearset and three clutches.

All input drive to the gearbox module is via the turbine shaft, which is direct driven by the torque converter turbine. Each of the countershafts has an output gear: the output gears on the countershafts provide connect to a range gear that can deliver torque to the output module.

INPUT MODULE

The input module consists of torque converter. The operating principles of torque converters are covered in Chapter 17. The torque converter cover that is used in the TC10 is a welded unit and not field-serviceable. It uses a two-pass fluid flow design: when in-out oil flow is reversed, the torque converter clutch (TCC) engages and locks up the assembly putting it in direct-drive mode. In direct drive, torque converter input rpm equals its output rpm. Output from the torque converter is by means of the transmission turbine

FIGURE 21–36 WT diagnostic codes: if the problem has not been resolved at the completion of each set of diagnostic routines, contact the Allison dealership.

If codes read: Main Code	Sub Code	Circuit	Recommended procedures
13	12	ECU Input voltage low	1. Check: a. Battery ground and power connections. b. Battery state of charge. c. Vehicle charging system is not over- or under-charging. d. VIM fuse is good. e. VIM connections. f. OEM supplied wiring is correct. g. ECU connectors. 2. If all points test OK, contact Allison.
13	13	ECU Input voltage low (medium)	
13	23	ECU Input voltage high	
14	12, 23	Oil level sensor	1. Check: a. Oil level sensor. b. Check operation of engine speed sensor, output speed sensor, and temperature sensor. c. Wiring harness has no opens, shorts to ground, or shorts to battery. 2. If all points are checked, call distributor.
21	12, 23	Throttle position sensor	1. Check: a. TPS connector plug and receptacle. b. End of TPS cable is pulled out properly. c. TPS angle is in idle position. d. TPS provides specified stroke on TPS cable. e. Wiring harness to TPS has no opens, shorts between wires, or shorts to ground.
22	14, 15, 16	Speed sensors	1. Check: a. Speed sensor connectors are tight, clean, and undamaged. b. Speed sensor mounting bolts are properly torqued (24–29 N·m [18–21 lb-ft.]). c. Wiring harness to sensors has no opens, or shorts.
23	12, 13, 14, 15	Shift selectors	1. Check: a. ECU connectors are tight, clean, and undamaged. b. Remote shift selector connector is tight and jumper wire is cut. c. Wiring harness on remote has no opens, or shorts.
24	12	Sump oil temperature cold	1. Check: a. Air temperature is below −32°C (−25°F). (1) If yes, this is a correct response for temperature. (2) If no, check that main transmission connector is tight, clean, and undamaged. b. ECU connectors are tight, clean, and undamaged.
24	23	Sump oil temperature hot	1. Idle engine with parking brake applied, wheels chocked, and vehicle level. Check: a. Correct dipstick is installed. b. Fluid level is correct. (1) If fluid level is incorrect—correct fluid level. (2) If fluid level is correct—check for engine system overheating, causing transmission to overheat. 2. Check if ECU and transmission connectors are tight, clean, and undamaged.
25	00, 11, 22, 33, 44, 55, 66, 77	Output speed sensor	1. Check: a. Speed sensor connector is tight, clean, and undamaged. b. ECU connectors are tight, clean, and undamaged. c. Fluid level is correct. d. Sensor mounting bolt torque is correct (24–29 N·m [18–21 lb-ft.]). e. Wiring harness to sensor has no opens, shorts bet ween wires, or shorts to ground.
32	00, 33, 55, 77	C3 pressure switch open	1. Idle engine with parking brake applied, wheels chocked, and vehicle level. Check: a. Correct dipstick is installed. b. Fluid level is correct. 2. Check: a. Main transmission connector is tight, clean, and undamaged. b. ECU connectors are tight, clean, and undamaged. c. Wiring harness has no opens, shorts between wires, or shorts to ground.
33	12, 23	Sump oil sensor failure	1. Check: a. Main transmission connector is tight, clean, and undamaged. b. ECU connectors are tight, clean, and undamaged. c. Wiring harness has no opens, shorts between wires, or shorts to ground.
34	12, 13, 14, 15, 16	EEPROM	1. Recalibrate ECU; if this cannot be done, replace ECU.
35	00, 16	Power interruption EEPROM or write interruption	1. Check: a. ECU connectors are tight, clean, and undamaged. b. VIM connectors are tight, clean, and undamaged. c. OEM supplied wiring has correct power and ground connections. d. Power connections are battery direct. e. Ground connections are battery direct. f. Ignition switch connections are correct.
36	00	Hardware/ Software Not Compatible	1. Reprogram ECU. 2. If ECU cannot be reprogrammed or replaced, contact Allison.
41	12, 13, 14, 15, 16, 21, 22, 23, 24, 25, 26	Open or short to ground in solenoid circuit	1. Check: a. Main transmission connector is tight, clean, and undamaged. b. ECU connectors are tight, clean, and undamaged. c. Wiring harness is not pulled too tight, and there is no damage, chafing, or screws through harness. d. Wiring harness has no opens, shorts between wires, or shorts to ground. 2. Change wiring harness (optional).
42	12, 13, 14, 15, 16, 21, 22, 23, 24, 25, 26	Short to battery in solenoid circuit	1. Check: a. Main transmission connector is tight, clean, and undamaged. b. ECU connectors are tight, clean, and undamaged. c. Wiring harness is not pulled too tight, and there is no damage, chafing, or screws through harness. d. Wiring harness has no opens, shorts between wires, or shorts to ground. e. Unauthorized repairs have not been made. 2. Change harness (optional).
43	21, 25, 26	Solenoid low side circuit, open driver, or wire shorted to ground	1. Check: a. Transmission connector is tight, clean, and undamaged. b. ECU connectors are tight, clean, and undamaged. c. Wiring harness has no opens, shorts between wires, or shorts to ground. 2. Replace ECU. 3. If all points are checked, call distributor.
44	12, 13, 14, 15, 16, 21, 22, 23, 24, 25, 26	Solenoid circuit short to ground	1. Check: a. Transmission connector is tight, clean, and undamaged. b. ECU connectors are tight, clean, and undamaged. c. Wiring harness has no opens, shorts between wires, or shorts to ground.
45	12, 13, 14, 15, 16, 21, 22, 23, 24, 25, 26	Solenoid circuit open	1. Check: a. Transmission connector is tight, clean, and undamaged. b. ECU connectors are tight, clean, and undamaged. c. Wiring harness has no opens.

FIGURE 21–36 (continued)

If codes read			Recommended procedures
51	01, 10 ,12, 21, 23, 24, 35, 42, 43, 45, 46, 53, 64, 65	Offgoing ratio test (during shift)	1. Check: a. Output and turbine speed sensor connectors are tight, clean, and undamaged. b. Speed sensor wiring harness has no opens, shorts between wires, or shorts to ground. 2. Idle engine with parking brake applied, wheels chocked, and vehicle level. Check: a. Correct dipstick is installed. b. Fluid level is correct. 3. If all points check, call distributor.
52	01, 08, 32, 34, 54, 56, 71,72,78 79, 99	Offgoing C3 pressure switch (during shift)	1. Check: a. Output and turbine speed sensor connectors are tight, clean, and undamaged. b. Speed sensor wiring harness has no opens, shorts between wires, or shorts to ground. c. Main wiring harness to transmission has no shorts between wires or shorts to ground. 2. Idle engine with parking brake applied, wheels chocked, and vehicle level. Check: a. Correct dipstick is installed. b. Fluid level is correct. 3. If all points check, call distributor.
53	08, 18, 28, 29, 38, 39, 48, 49, 58, 59, 68, 69, 78, 99	Offgoing speed test (during shift)	1. Check: a. Turbine and engine speed sensor connectors are tight, clean, and undamaged. b. Speed sensor wiring harness has no opens, shorts between wires, or shorts to ground. 2. Idle engine with parking brake applied, wheels chocked, and vehicle level. Check: a. Correct dipstick is installed. b. Fluid level is correct. 3. If all points check, call distributor.
54	01, 07, 10, 12, 17, 21, 23, 24, 27, 32, 34, 35, 42, 43, 45, 46, 53, 54, 56, 64, 65, 70, 71, 72, 80, 81, 82, 83, 85, 86, 92, 93, 95, 96, 97	Oncoming ratio test (after shift)	1. Check: a. Turbine and output speed sensor connectors are tight, clean, and undamaged. b. Speed sensor wiring harness has no opens, shorts between wires, or shorts to ground. 2. Idle engine with parking brake applied, wheels chocked, and vehicle level. Check: a. Correct dipstick is installed. b. Fluid level is correct. 3. EEPROM calibration is correct for the transmission. 4. If all points check, call distributor.
55	17, 27, 87, 97	Oncoming C3 pressure switch (after shift)	1. Idle engine with parking brake applied, wheels checked, and vehicle level. Check: a. Correct dipstick is installed. b. Fluid level is correct. 2. Check: a. Output and turbine speed sensor connectors are tight, clean, and undamaged. b. Speed sensor wiring harness has no opens, shorts between wires, or shorts to ground. 3. Check: a. Transmission connector is tight, clean, and undamaged. b. ECU connectors are tight, clean, and undamaged. c. C3 pressure switch wiring has no opens, shorts between wires, or shorts to ground. 4. If all points check, call distributor.
56	00, 11, 22, 33, 44, 55, 66, 77	Range verification test (between shifts)	1. Check: a. Turbine and output speed sensor connectors are tight, clean, and undamaged. b. Speed sensor wiring harness has no opens, shorts between wires, or shorts to ground. c. Transmission connector is tight, clean, and undamaged. d. ECU connectors are tight, clean, and undamaged. 2. Idle engine with parking brake applied, wheels checked, and vehicle level.
			Check: a. Correct dipstick is installed. b. Fluid level is correct 3. If all points check, call distributor.
57	11, 22, 44, 66, 88, 99	Range verification test (C3)	1. Idle engine with parking brake applied, wheels chocked, and vehicle level. Check: a. Correct dipstick is installed. b. Fluid level is correct. 2. Check: a. Output and turbine speed sensor connectors are tight, clean, and undamaged. b. Speed sensor wiring harness has no shorts between wires or shorts to ground. 3. If all points check, call distributor. a. Transmission connector is tight, clean, and undamaged. b. ECU connectors are tight, clean, and undamaged. c. C3 pressure switch wiring has no opens, shorts between wires, or shorts to ground. 4. If all points check, call distributor.
61	00	Retarder over temperature	1. Check: a. Fluid level is correct. b. Retarder apply system is not allowing retarder and throttle to be applied at the same time. c. Fluid cooler is adequately sized for load. 2. If all points check, call distributor.
62	12, 23	Retarder temperature sensor	1. Check: a. Retarder temperature measured with DDR is consistent with code; determine if code is active using shift selector. b. Ambient temperature is above 178°C (352°F) or below –45°C (–49°F). c. Sensor connector is tight, clean, and undamaged. d. ECU connectors are tight, clean, and undamaged. e. Temperature sensor circuit has no opens, shorts, or shorts to ground. 2. If all points check, call distributor.
63	00, 26, 40	Input electrical fault	1. Check input wiring, switches, and connectors to determine why input states are different. 2. If all points check, call distributor.
64	12, 23	Retarder modulation request device fault	1. Use DDR to read retarder counts and identify problem wires. Check wiring for short to battery, ground wire open, or short to ground. 2. If all points check, call distributor.
65	00	Engine rating high (computer-controlled engines only)	1. Check if engine rating or governor speed is too high for transmission. 2. If all points check, call distributor.
66	00	Serial communications interface fault	1. Check: a. Serial connection to engine computer is tight, clean, and undamaged. b. SCI wiring harness has no open, shorts, or short to ground. c. If DDR is not available, also be sure that transmission ECU connections are tight, clean, and undamaged. 2. If all points check, call distributor.
69	12, 13, 14, 15, 16, 21, 22, 23, 24, 25, 26, 32, 33, 34, 35, 36, 41	ECU malfunction	1. Clear diagnostic code and retry vehicle start. 2. If code recurs, reprogram ECU. 3. If able, replace ECU. 4. If code continues to recur, call distributor.

FIGURE 21–37 Detailed cutaway view of a WT transmission highlighting clutch packs and planetary gearsets

shaft. The transmission turbine shaft connects to the torque converter turbine. **Figure 21–40** shows a sectional view of the TC10 input module.

FIGURE 21–39 Overhead, sectional view of an Allison TC10 transmission

GEARBOX MODULE

The gearbox module uses twin countershafts. The module receives input torque from the transmission turbine shaft, which is driven by the torque converter turbine. It houses four rotating clutches, which are engaged and disengaged to determine powerflow

FIGURE 21–38 An overhead view of an Allison TC10 showing modules and major components

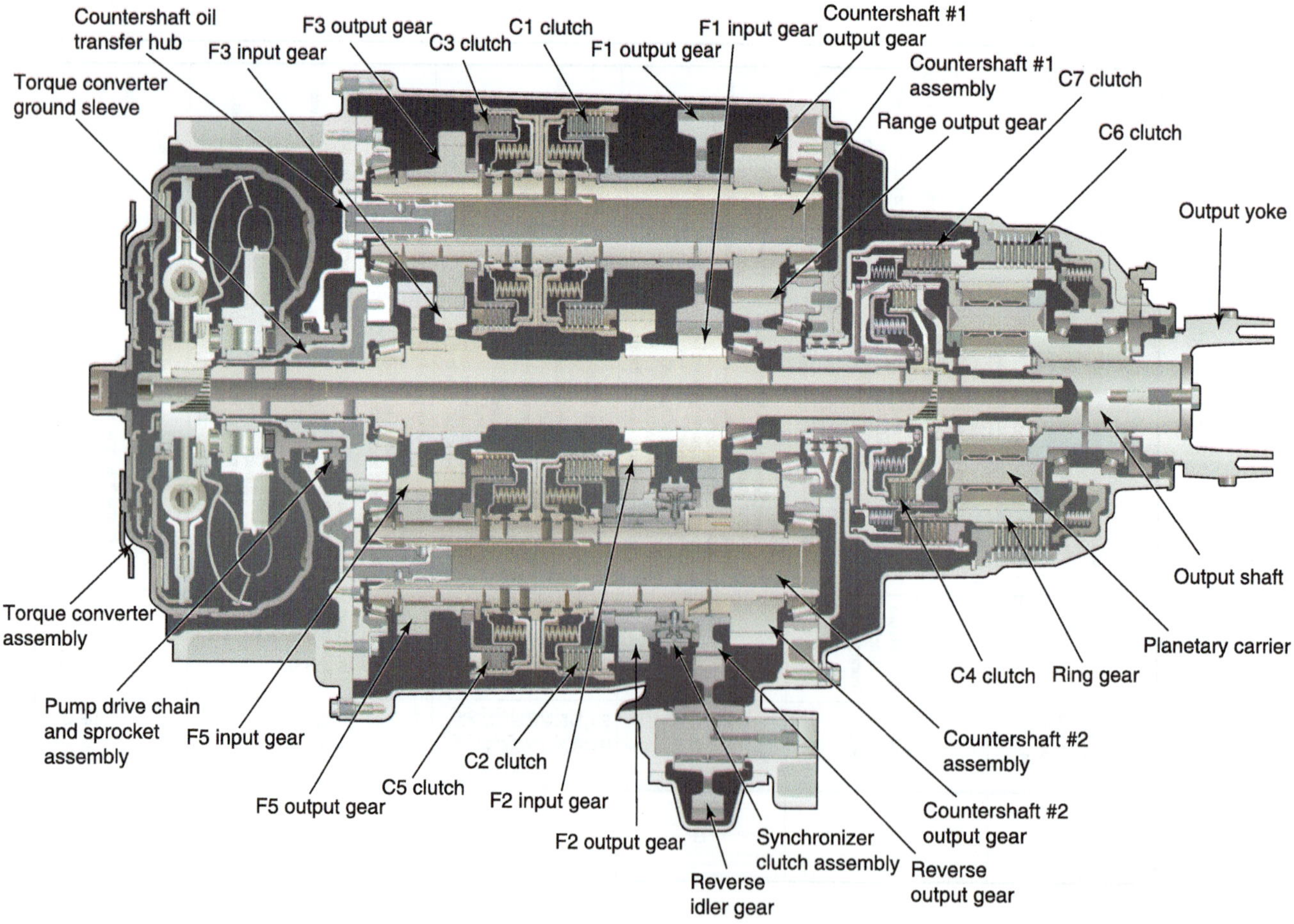

FIGURE 21–40 Sectional view of the TC10 front module

through the transmission. Gearbox module output to the output module is by means of the range output gear, except when the fourth and ninth ratios are selected when torque passes through the gearbox module directly to the output module. **Figure 21–41** shows a sectional view of the TC10 gearbox module.

Twin Countershafts

All of the gears on both countershafts with the exception of the two output gears rotate with the turbine shaft and freewheel on the countershafts at all times. Each of the two output gears rotates with the countershafts. Application of clutches in the gearbox module locks the freewheeling gears to the corresponding countershaft and provides a means of transmitting turbine shaft torque to the countershaft output gear. Torque is not divided between the countershafts in any gear ratio.

Range Output Gear

The range output gear is responsible for transmitting gearbox module torque in all gears except fourth and ninth. It is in mesh with the countershafts on either side of it.

Turbine Shaft

The turbine gear shaft is common to all three transmission modules because it extends through all three. It is driven by the torque converter turbine and connects with the planetary assembly in the output module.

Gearbox Module Ratios

The gearbox module provides for five ratio ranges. This means that the clutch action and powerflow for ratio ranges 1 through 5 is repeated for gear ratio ranges 6 through 10.

The planetary gearset in the output module provides high-low range option that dictates whether the transmission is in first gear or sixth gear. The status of the planetary gearset is determined by the actuation of the C6 clutch.

OUTPUT MODULE

The output module houses the planetary gearset, two rotating clutches, and one stationary clutch. The transmission output shaft is connected to the planetary carrier. **Figure 21–42** shows a sectional view of the TC10 output module.

FIGURE 21–41 Sectional view of the TC10 gearbox module

FIGURE 21–42 Sectional view of the TC10 output module

Output Module Ratios

There are two status options for the planetary gearset within the output module. These will determine whether the transmission is in low range (first to fifth) or high range (sixth to tenth). To achieve forward ratios 1 through 5, the planetary gearset in the output module operates in an underdrive mode. The sun drive is the input, the ring gear is held, and the carrier provides the output, because it is splined/connected to the output shaft. To produce forward ratios 6 through 10, the planetary operates in direct-drive mode with the sun gear and carrier driven together. A more detailed discussion of the powerflows will be reviewed later in this section. A listing of available ratios are documented in **Table 21–2**.

CLUTCHES

TC10 clutches are hydraulically actuated assemblies consisting of fiber friction discs and steel reaction plates that rotate independently when not actuated. The following clutches are used in the TC10:

- Gearbox module rotating clutches: These connect components together or supply rotational input. Clutches C1, C2, C3, and C5 are rotating clutches located in the gearbox module; when applied, these clutches engage countershaft gears with the countershaft so that they rotate with it.
- Output module rotating clutches: The C4 clutch imparts rotational input from the turbine shaft to the planetary sun gear, and the C7 clutch imparts rotational input from the range output to the planetary carrier.
- Stationary clutch: The C6 clutch is located in the output module and its function when actuated is to hold the planetary ring gear.

TABLE 21–2 Allison TC10 Ratios

Range	Ratio	Step
1st	7.40	
2nd	5.44	1.36
3rd	4.25	1.28
4th	3.43	1.24
5th	2.94	1.17
6th	2.16	1.36
7th	1.59	1.36
8th	1.24	1.28
9th	1.00 (DD)	1.24
10th	0.86 (OD)	1.17
R1	6.71	
R2	1.96	

Synchronizer Assembly

The synchronizer assembly is used to engage and sync either:

- F2 output gear, or
- Reverse output gear

A shift fork is used to position the synchronizer into either the forward (F2) or the reverse position. It engages the F2 output gear to select forward ratios. In neutral or either of the two reverse ratios, the synchronizer engages with the reverse output gear. **Table 21–3** shows which clutches and synchronizers have to be actuated in order to attain each gear ratio.

HYDRAULIC SCHEMATICS

In order to gain an understanding of how powerflows are routed through the TC10, we have reproduced three sets of hydraulic schematics; use Table 21–3 and the powerflow schematics that follow this section in order to understand TC10 powerflow. **Figure 21–43** demonstrates the hydraulic circuit in neutral, while **Figure 21–44** and **Figure 21–45** do this in first and fourth ranges respectively.

TC POWERFLOWS

The powerflow paths in the TC10 transmission are relatively easy to understand. We will introduce these by gear ratio in this section and it helps to cross-reference Table 21–3 for each range image.

Neutral

Gearbox module: Torque input to the transmission is via the turbine shaft that rotates in the gearbox module. In the gearbox module:

- Neither countershaft output gear is turning because all of the countershaft gears are freewheeling.
- Rotation stops at the C4 clutch hub.

Output module:

- The C6 clutch holds the planetary ring gear.
- Planetary components are not being driven, as a result there is no output shaft rotation.
- Application of the C6 clutch in neutral ensures a rapid response when a moving range is selected.

First Gear

Gearbox module: Torque input to the transmission is via the turbine shaft that rotates in the gearbox module.

- C1 and C6 clutches are applied.
- Shift fork is in the forward position.

TABLE 21–3 Clutch and Synchro Actuation by Gear Ratio: Note that SF Denotes Synchro in Forward Engagement and SR Reverse Engagement.

Range	C1	C2	C3	C4	C5	C6	C7	SF	SR
N						X			X
1	X					X		X	
2		X				X		X	
3			X			X		X	
4				X		X		X	
5					X	X		X	
6	X						X	X	
7		X					X	X	
8			X				X	X	
9				X			X	X	
10					X		X	X	
R1	X					X			X
R2	X						X		X

- C1 clutch locks F1 input gear to the #1 countershaft.
- F1 gear transfers torque from the turbine shaft to #1 countershaft driving its output gear.
- Countershaft #1 output gear drives the range output gear transmitting torque to the output module.

Output module:

- Range output gear drives the planetary sun gear via the C4/C7 clutch module.
- Applied C6 clutch holds the planetary ring gear meaning that it functions in underdrive mode.
- Planetary carrier becomes the output and drives the transmission output shaft which is splined to it.

Figure 21–46 shows the powerflow path through the TC10 in first range.

Second Gear

Gearbox module: Torque input to the transmission is via the turbine shaft that rotates in the gearbox module.

- C2 and C6 clutches are applied.
- Shift fork is in the forward position.
- C2 clutch locks F2 input gear to the #2 countershaft.
- F2 gear transfers torque from the turbine shaft to #2 countershaft driving its output gear.
- Countershaft #2 output gear drives the range output gear transmitting torque to the output module.

Output module:

- Range output gear drives the planetary sun gear via the C4/C7 clutch module.
- Applied C6 clutch holds the planetary ring gear meaning that it functions in underdrive mode.
- Planetary carrier becomes the output member and drives the transmission output shaft which is connected (splined) to it.

Figure 21–47 shows the powerflow path through the TC10 in second range.

Third Gear

Gearbox module: Torque input to the transmission is via the turbine shaft that rotates within the gearbox module.

- C3 and C6 clutches are applied.
- Shift fork is in the forward position.
- C3 clutch locks F3 input gear to the #1 countershaft.
- F3 gear transfers torque from the turbine shaft to #1 countershaft driving its output gear.
- Countershaft #1 output gear drives the range output gear transmitting torque to the output module.

Output module:

- Range output gear drives the planetary sun gear via the C4/C7 clutch module.
- Applied C6 clutch holds the planetary ring gear meaning that it functions in underdrive mode.

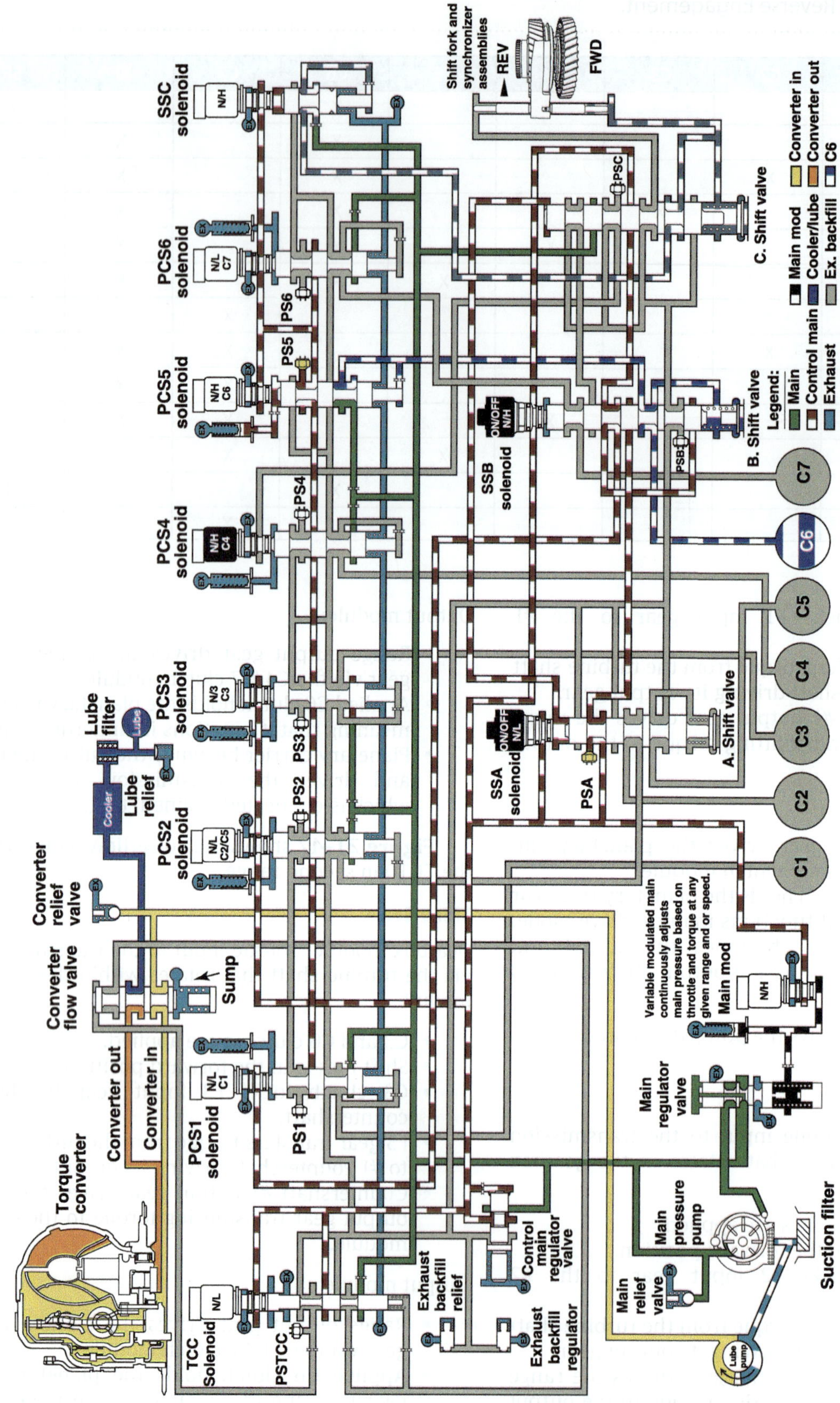

FIGURE 21–43 TC10 hydraulic schematic showing circuit status in neutral range

FIGURE 21–44 TC10 hydraulic schematic showing circuit status in 1st range

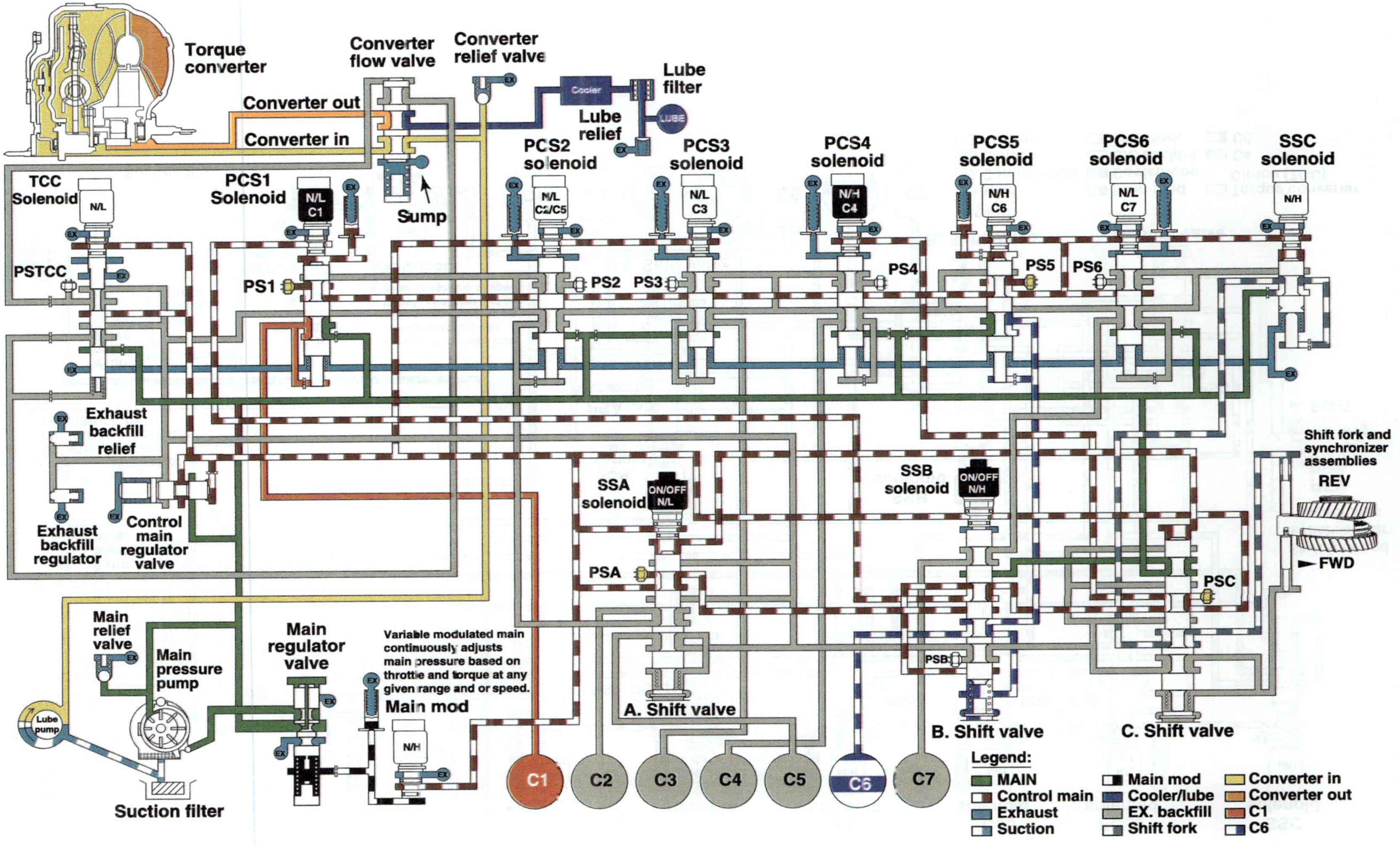

FIGURE 21–45 TC10 hydraulic schematic showing circuit status in 4th range

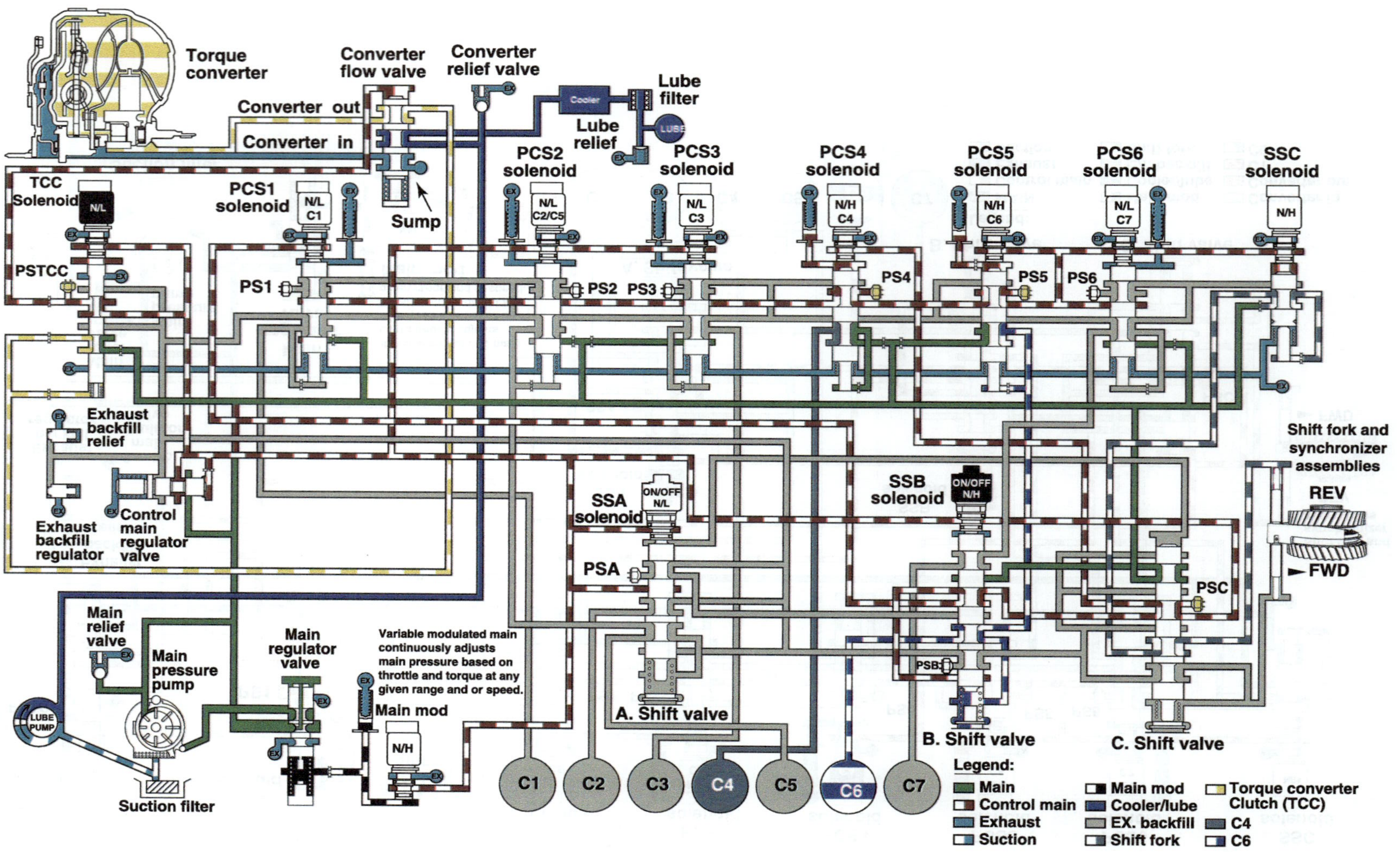

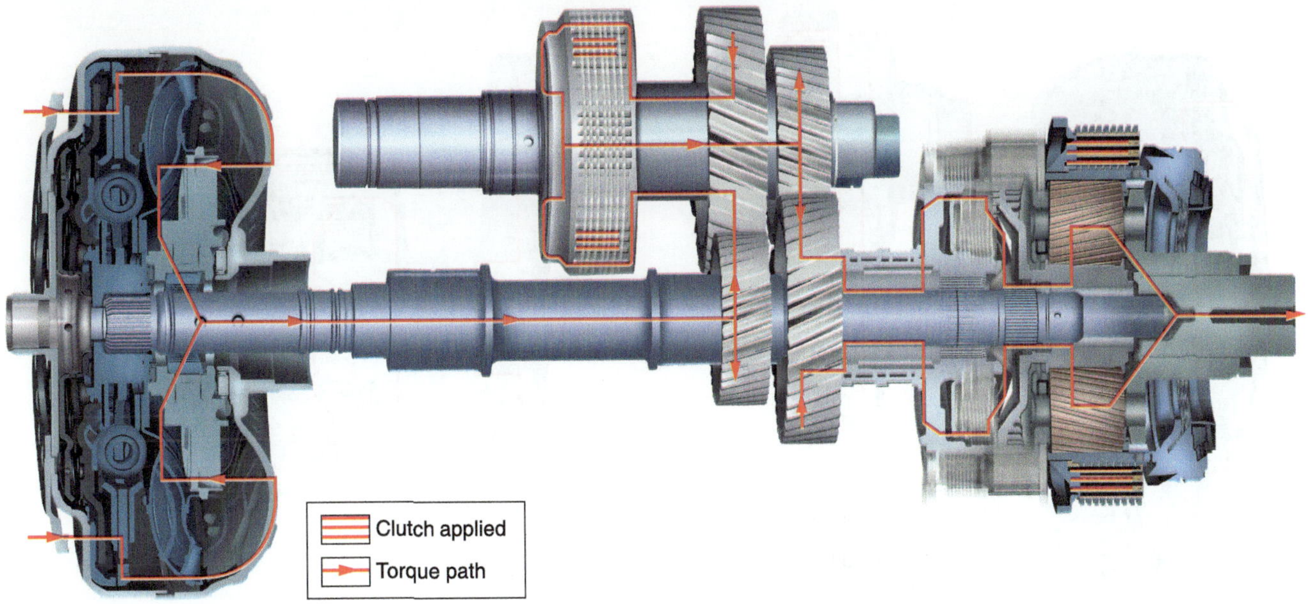

FIGURE 21–46 TC10 powerflow in 1st range

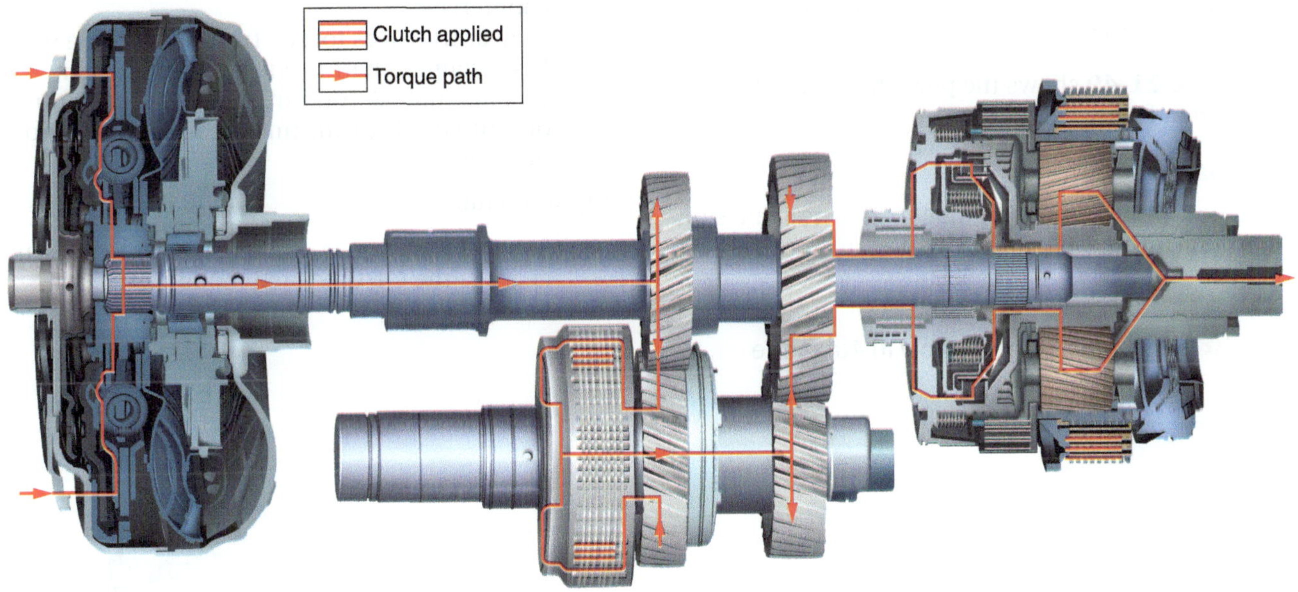

FIGURE 21–47 TC10 powerflow in 2nd range

- Planetary carrier becomes the output and drives the transmission output shaft which is splined to it.

Figure 21–48 shows the powerflow path through the TC10 in third range.

Fourth Gear

This is the direct-drive gear that bypasses the gearbox module.

Gearbox module: Torque input to the transmission is via the turbine shaft that rotates in the gearbox module and connects with the C4 clutch in the output module.

- No gearbox module clutches are applied.

Output module:

- Torque is transferred from the turbine shaft to the C4/C7 clutch module that drives the planetary sun gear.

FIGURE 21–48 TC10 powerflow in 3rd range

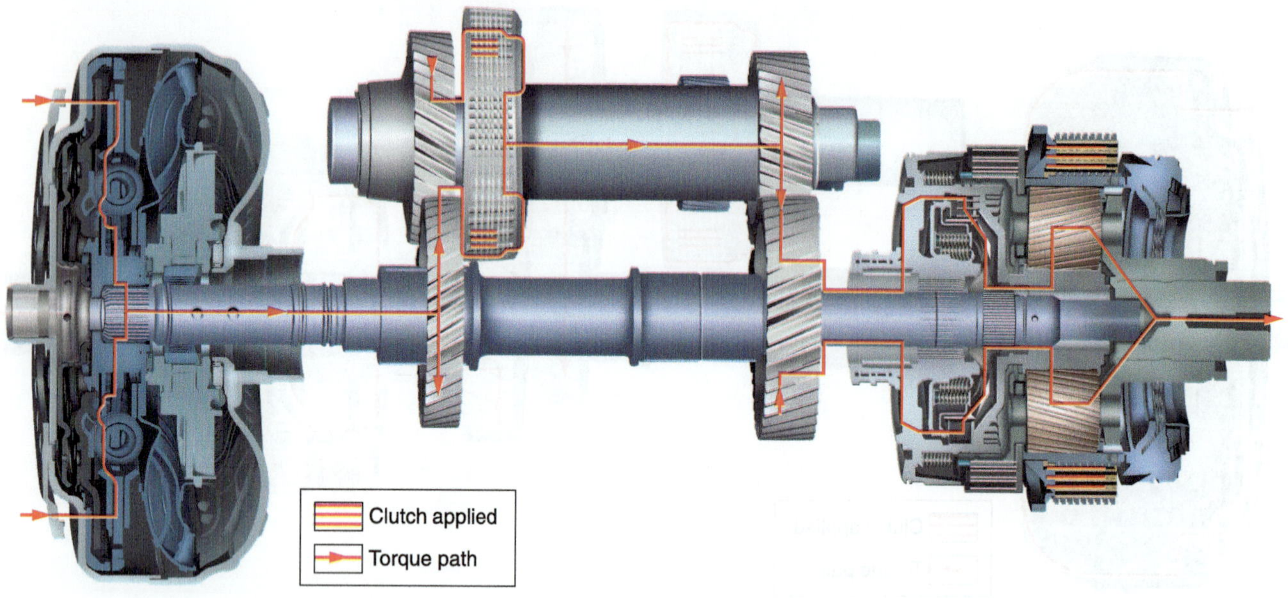

- Applied C6 clutch holds the planetary ring gear meaning that it functions in underdrive mode.
- Planetary carrier becomes the output and drives the transmission output shaft which is splined to it.

Figure 21–49 shows the powerflow path through the TC10 in fourth range.

Fifth Gear

Gearbox module: Torque input to the transmission is via the turbine shaft that rotates in the gearbox module.

- C5 and C6 clutches are applied.
- Shift fork is in the forward position.
- C5 clutch locks F5 input gear to the #2 countershaft.
- F5 gear transfers torque from the turbine shaft to #2 countershaft driving its output gear.
- Countershaft #2 output gear drives the range output gear transmitting torque to the output module.

Output module:

- Range output gear drives the planetary sun gear via the C4/C7 clutch module.

FIGURE 21–49 TC10 powerflow in 4th range

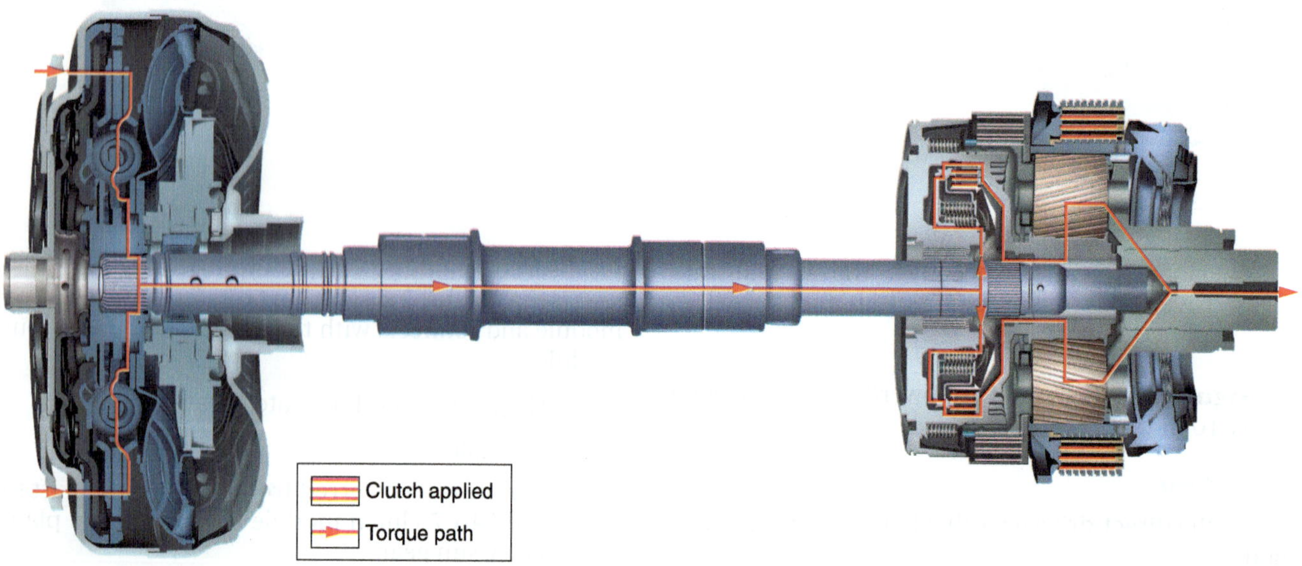

- Applied C6 clutch holds the planetary ring gear meaning that it functions in underdrive mode.
- Planetary carrier becomes the output and drives the transmission output shaft which is splined to it.

Figure 21–50 shows the powerflow path through the TC10 in fifth range.

Sixth Gear

Gearbox module: Torque input to the transmission is via the turbine shaft that rotates in the gearbox module.

- C1 and C6 clutches are applied.
- Shift fork is in the forward position.
- C1 clutch locks F1 input gear to the #1 countershaft.
- F1 gear transfers torque from the turbine shaft to #1 countershaft driving its output gear.
- Countershaft #1 output gear drives the range output gear transmitting torque to the output module.

Output module:

- Range output gear drives the planetary sun gear via the C4/C7 clutch module.
- Applied C7 clutch holds the planetary carrier meaning that the planetary gearset functions in direct–drive mode.
- With the planetary gearset in direct-drive mode, the carrier drives the transmission output shaft to which it is splined.

Figure 21–51 shows the powerflow path through the TC10 in sixth range.

Seventh Gear

Gearbox module: Torque input to the transmission is via the turbine shaft that rotates in the gearbox module.

- C2 and C6 clutches are applied.
- Shift fork is in the forward position.
- C2 clutch locks F2 input gear to the #2 countershaft.
- F2 gear transfers torque from the turbine shaft to #2 countershaft driving its output gear
- Countershaft #2 output gear drives the range output gear transmitting torque to the output module.

Output module:

- Range output gear drives the planetary sun gear via the C4/C7 clutch module.
- Applied C7 clutch holds the planetary carrier meaning that the planetary gearset functions in direct drive mode.
- With the planetary gearset in direct drive mode, the carrier drives the transmission output shaft to which it is splined.

Figure 21–52 shows the powerflow path through the TC10 in seventh range.

Eighth Gear

Gearbox module: Torque input to the transmission is via the turbine shaft that rotates in the gearbox module.

- C3 and C6 clutches are applied.
- Shift fork is in the forward position.

FIGURE 21–50 TC10 powerflow in 5th range

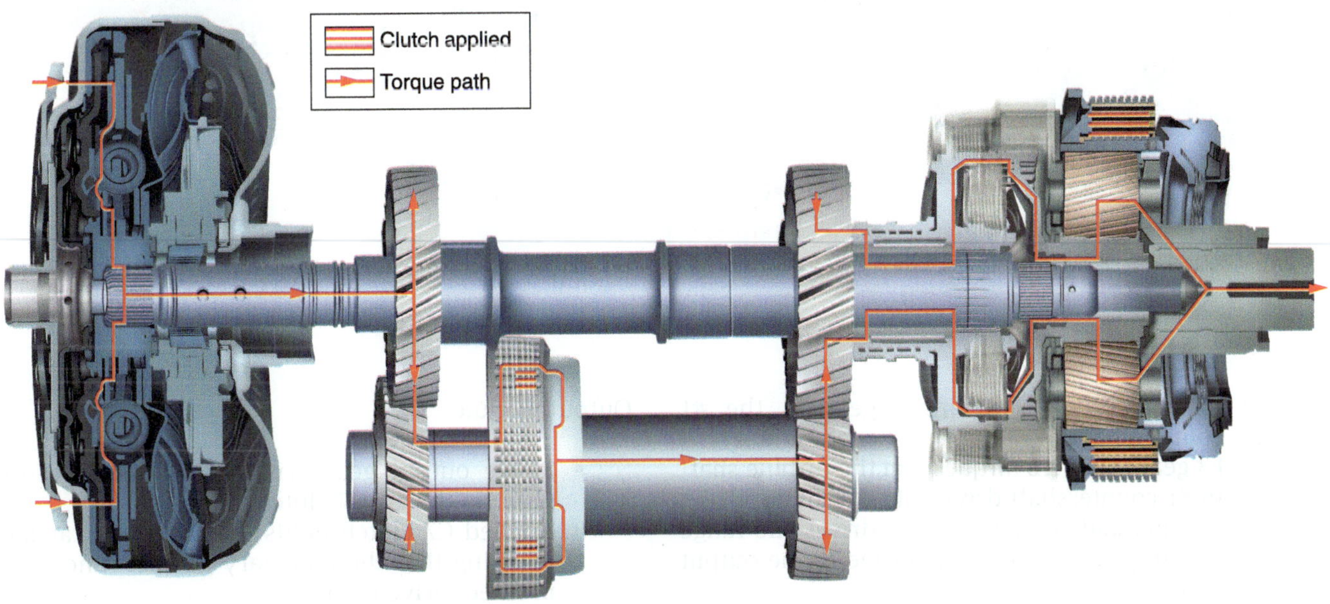

FIGURE 21–51 TC10 powerflow in 6th range

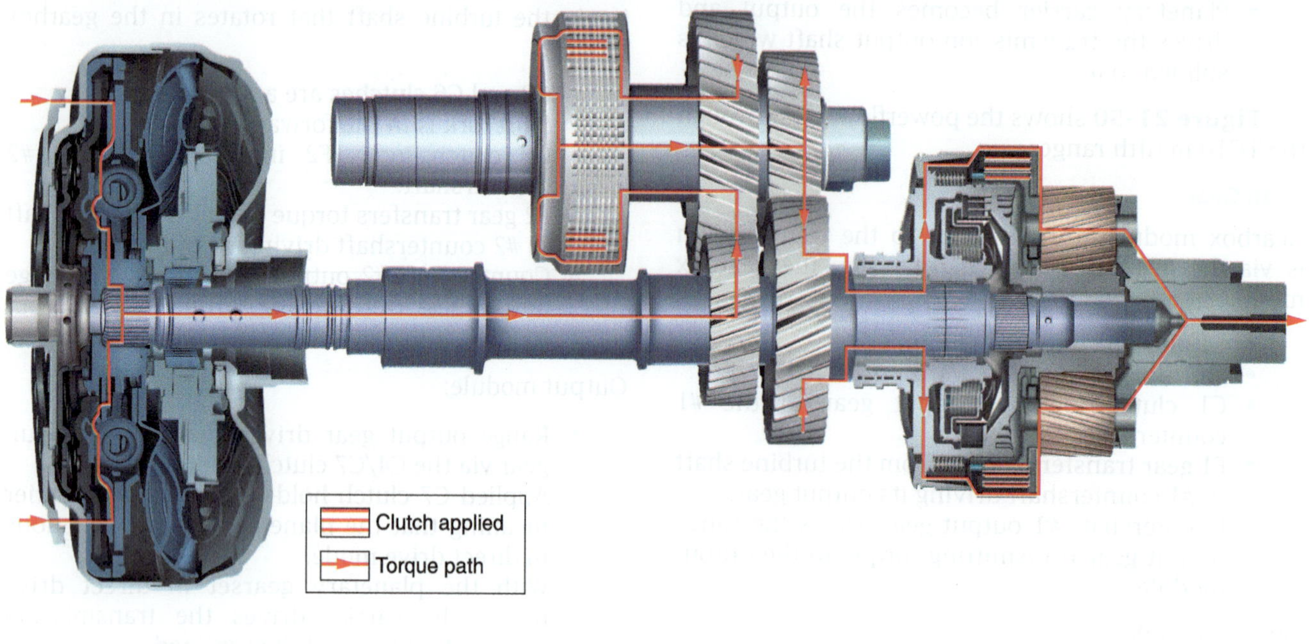

FIGURE 21–52 TC10 powerflow in 7th range

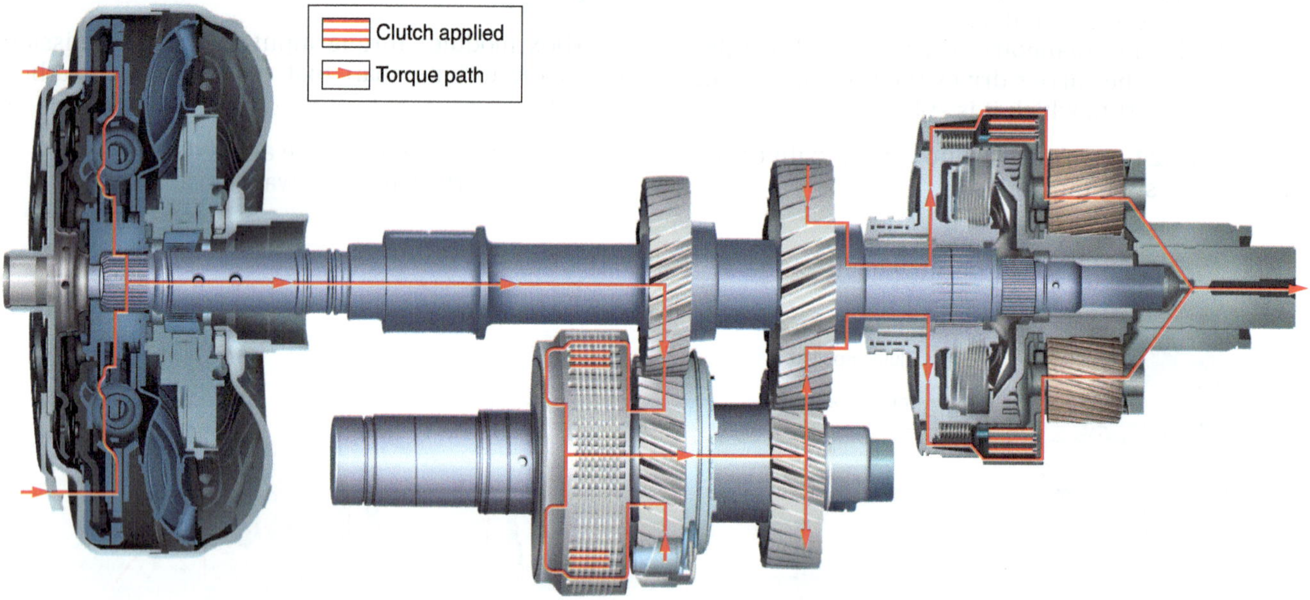

- C3 clutch locks F3 input gear to the #1 countershaft.
- F3 gear transfers torque from the turbine shaft to #1 countershaft driving its output gear.
- Countershaft #1 output gear drives the range output gear transmitting torque to the output module.

Output module:

- Range output gear drives the planetary sun gear via the C4/C7 clutch module.
- Applied C7 clutch holds the planetary carrier meaning that the planetary gearset functions in direct-drive mode.

- With the planetary gearset in direct-drive mode, the carrier drives the transmission output shaft to which it is splined.

Figure 21–53 shows the powerflow path through the TC10 in eighth range.

Ninth Gear

This is the direct-drive gear that bypasses the gearbox module.

Input module: Under most conditions the torque converter clutch is in lockup mode.

Gearbox module: Torque input to the transmission is via the turbine shaft which rotates within the gearbox module and connects with the C4 clutch in the output module.

- No gearbox module clutches are applied.

Output module:

- Torque is transferred from the turbine shaft to the C4/C7 clutch module that drives the planetary sun gear.
- Applied C6 clutch holds the planetary ring gear meaning that it functions in underdrive mode.
- Planetary carrier becomes the output and drives the transmission output shaft.

Figure 21–54 shows the powerflow path through the TC10 in ninth range.

Tenth Gear

Input module: Under most conditions the torque converter clutch is in lockup mode.

Gearbox module: Torque input to the transmission is via the turbine shaft that rotates in the gearbox module.

- C5 and C6 clutches are applied.
- Shift fork is in the forward position.
- C5 clutch locks F5 input gear to the #2 countershaft.
- F5 gear transfers torque from the turbine shaft to #2 countershaft driving its output gear.
- Countershaft #2 output gear drives the range output gear transmitting torque to the output module.

Output module:

- Range output gear drives the planetary sun gear via the C4/C7 clutch module.
- Applied C7 clutch holds the planetary carrier meaning that the planetary gearset functions in direct-drive mode.
- With the planetary gearset in direct-drive mode, the carrier drives the transmission output shaft to which it is splined.

Figure 21–55 shows the powerflow path through the TC10 in tenth range.

FIGURE 21–53 TC10 powerflow in 8th range

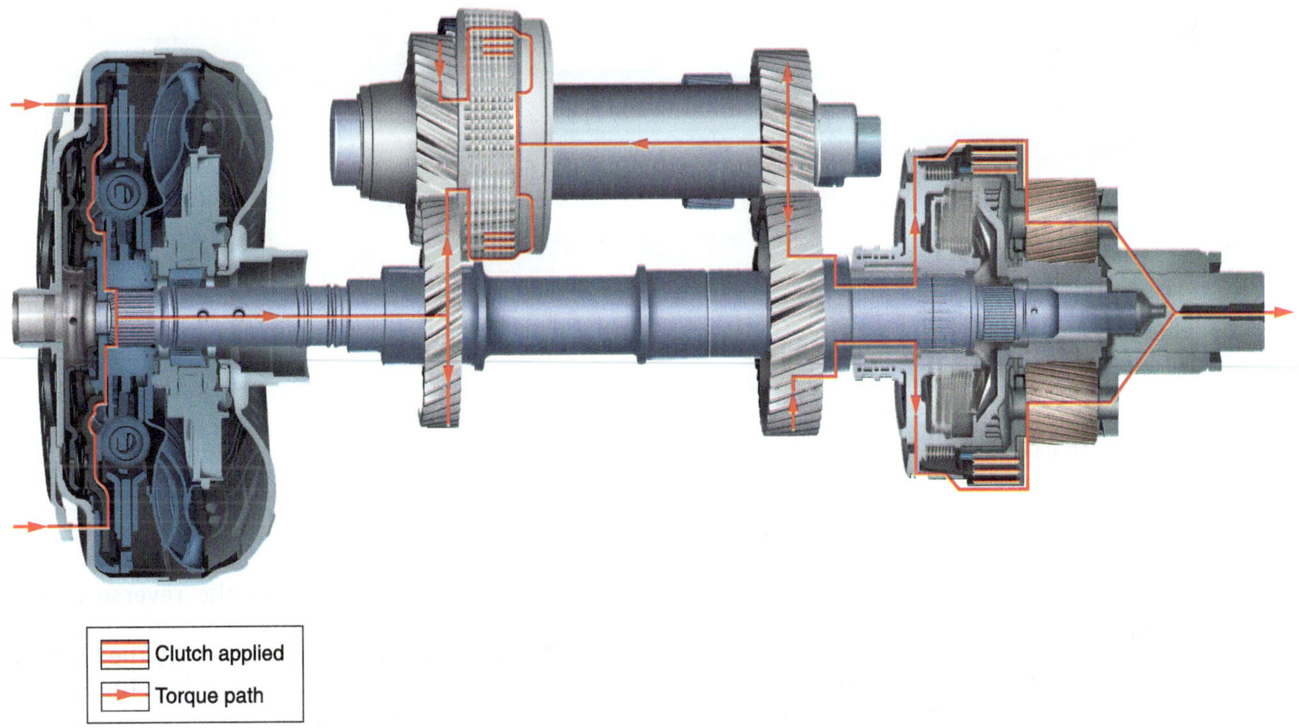

FIGURE 21–54 TC10 powerflow in 9th range

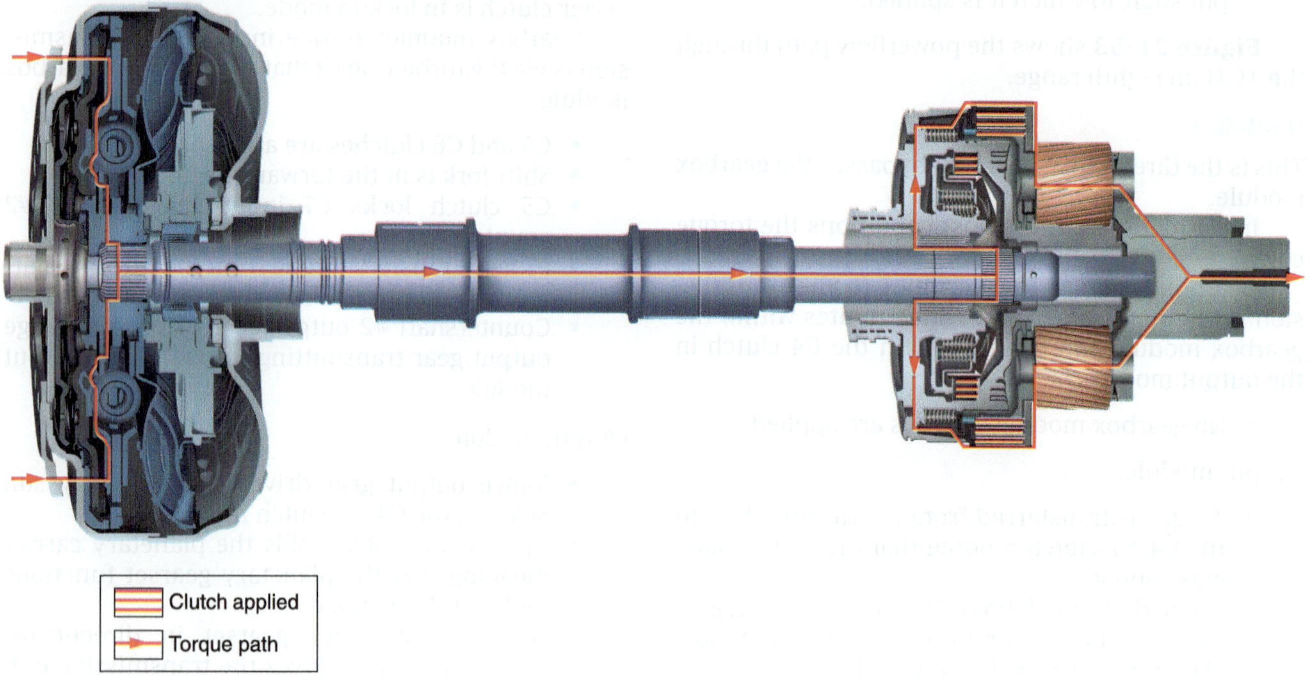

FIGURE 21–55 TC10 powerflow in 10th range

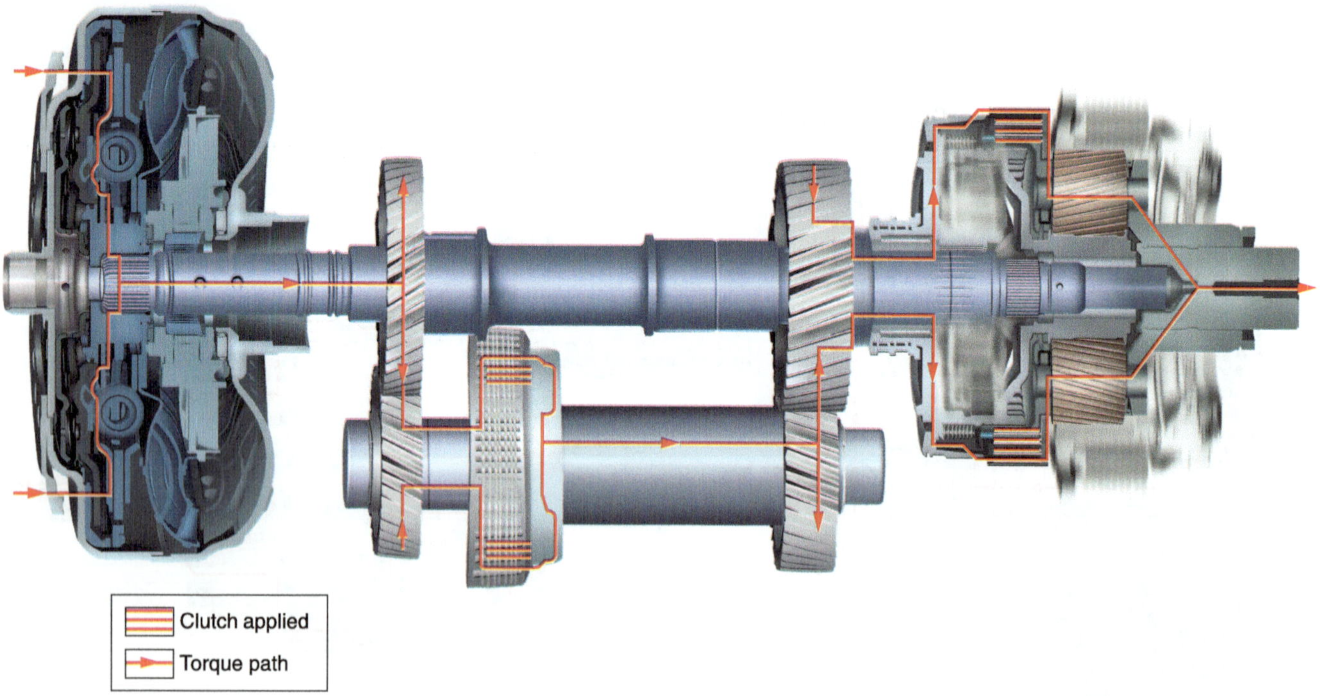

Reverse 1 gear

Gearbox module:

- Synchronizer is in the reverse position locking the reverse output gear to the C2 clutch hub.
- The C2 clutch hub locks the reverse output gear to countershaft #2.
- The reverse idler gear is driven by the F1 input gear on the turbine shaft, which in turn drives the reverse output gear.

- This rotates countershaft #2 in the same direction as the turbine shaft.
- Countershaft #2 output gear drives the range output gear to transfer torque to the output module.

Output module:

- Applied C6 clutch holds the planetary ring gear stationary.
- The range output gear drives the planetary sun gear via the C4/C7 clutch module.
- The planetary carrier drives the output shaft to which it is splined in reverse direction.

Figure 21–56 shows the powerflow path through the TC10 in R1 range.

Reverse 2 gear

Gearbox module:

- Synchronizer is in the reverse position locking the reverse output gear to the C2 clutch hub.
- The C2 clutch hub locks the reverse output gear to countershaft #2.
- The reverse idler gear is driven by the F1 input gear on the turbine shaft, which in turn drives the reverse output gear.
- This rotates countershaft #2 in the same direction as the turbine shaft.
- Countershaft #2 output gear drives the range output gear to transfer torque to the output module.

Output module:

- Applied C7 clutch holds the planetary carrier stationary meaning that it operates in direct-drive mode.
- The range output gear drives the planetary sun gear via the C4/C7 clutch module.
- The planetary carrier drives the output shaft to which it is splined in reverse direction.

Figure 21–57 shows the powerflow path through the TC10 in R2 range.

TC10 MAINTENANCE

This section covers only the most basic service maintenance procedures required on the TC10 transmission. Allison corporate training and access to Allison DOC is required to safely service TC10 transmissions.

Drain Fluid

To drain the transmission fluid the following steps are required:

1. Engine must be warmed to normal operating temperature.
2. Install a catch basin under the transmission and remove the drain plug from the oil pan.
3. Inspect the draining fluid for contaminants.
4. Install a new seal onto the drain plug, then install and torque it to specification.

FIGURE 21–56 TC10 powerflow in R1 range

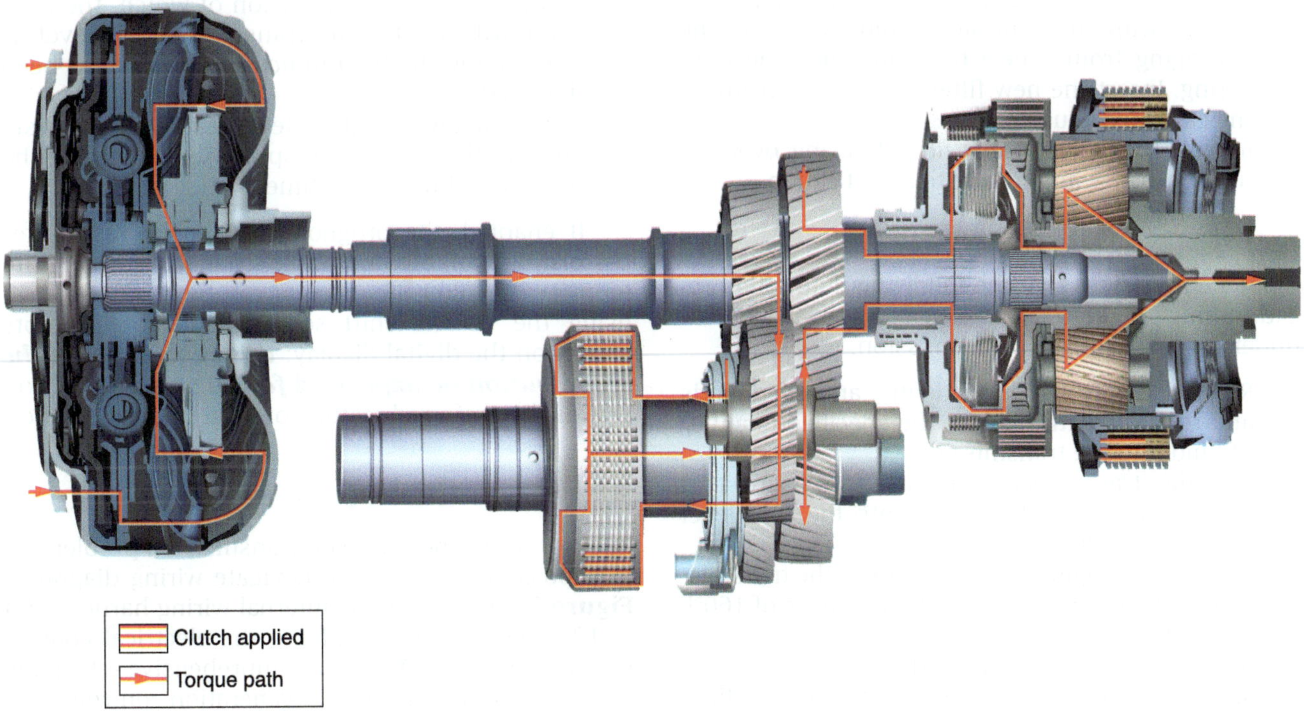

FIGURE 21–57 TC10 powerflow in R2 range

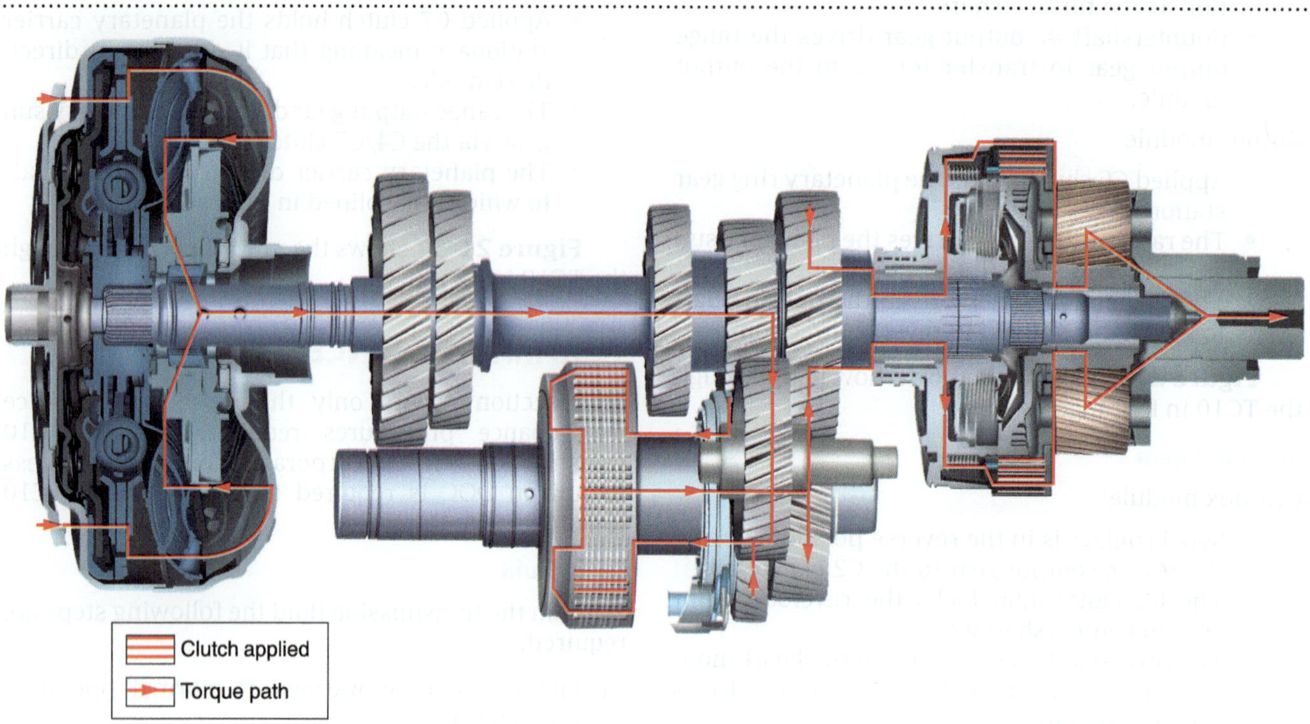

Replace Lube Filter

Use the following procedure to replace the lube filter:

1. Install a catch basin under the cooler return filter housing, then remove it to expose the filter cartridge.
2. Pull the filter cartridge from the housing.
3. Remove and replace the cover O-ring. Lube the O-ring with transmission fluid. Remove the packaging from a new filter and lube the filter O-ring. Insert the new filter over the pilot inside the cartridge housing.
4. Install the cooler return filter housing over the filter cartridge and torque to specification.

Fluid Refill

Allison emphasizes that the following fluid refill procedure must be adhered to exactly in order to avoid under- or over-filling the transmission.

1. Reference Allison DOC to identify approved fluids and refill capacities. Pump the specified quantity of fluid into the transmission.
2. Use the transmission dipstick to ensure that transmission fluid level falls into the cold level range on the dipstick band.
3. Operate the engine in order to warm the transmission to its normal operating temperature of 160°F to 200°F (71°C to 93°C).
4. Allison recommends that the oil level sensor and the shift lever display be used for the final checks of the oil level; this can be done using several different methods, one of which is described here. With the vehicle parking brake on, engine at operating temperature and running at base idle rpm, and the transmission placed in neutral range, depress the shift pad diagnostic button one time. This will initiate a 2-minute countdown at the conclusion of which, the indicator will display the transmission oil level as normal, low, high, or indicate that fault has been detected.
5. Correct a low or high oil level reading and retest.
6. To exit the oil level display field, depress the diagnostic button two times.

If enabled, the prognostics must be reset after performing an oil and filter change. The oil life monitor and filter life monitor can be manually reset using the vehicle shift selector and the prompts shown on the digital display; ***this requires that the mode button be depressed for 10 seconds***. Allison DOC is the preferred method of resetting prognostics.

ALLISON WIRING SCHEMATICS

In order to diagnose Allison transmission problems it is important to be able to navigate wiring diagrams. **Figure 21–58** shows the internal wiring harness of a TC10 transmission identifying the electrical components and **Figure 21–59** is a comprehensive schematic of Allison Transmission fifth-generation controls.

FIGURE 21–58 Internal wiring harness used in a TC10 showing the electrical component layout

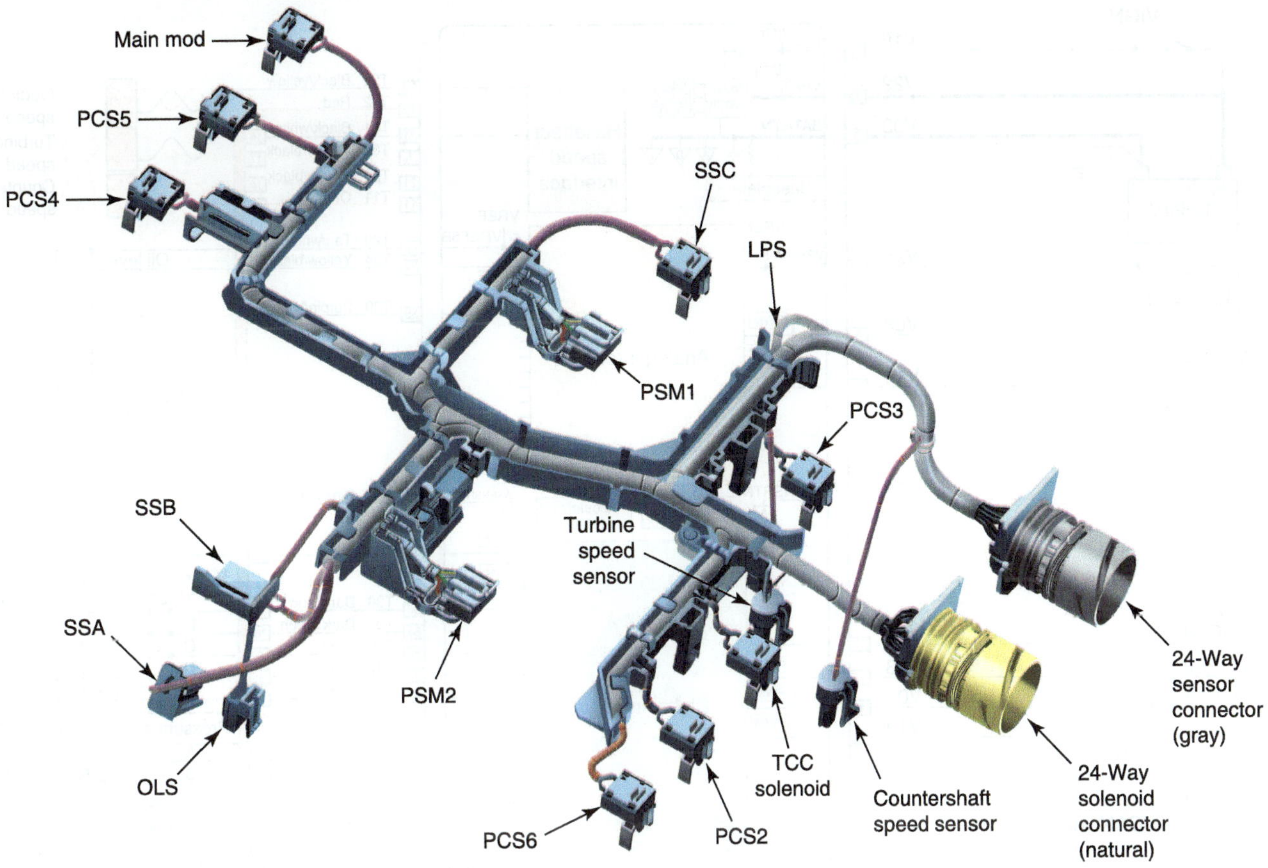

21.5 CATERPILLAR CX TRUCK TRANSMISSIONS

Caterpillar introduced its CX family of transmissions in 2006. This full authority ECAT is based on a design used for a number of years in Caterpillar off-highway 725-series articulated trucks. Currently, three versions of the on-highway version of this transmission exist:

- CX28 medium-duty six speed: rated for up to 400 BHP (300 kW) and 1,250 lb-ft. (1,700 N·m) torque input.
- CX31 medium- to heavy-duty six speed: rated for up to 525 BHP (390 kW) and 1,770 lb-ft. (2,400 N·m) torque input.
- CX35 heavy-duty, eight speed: rated for up to 625 BHP (466 kW) and 2,150 lb-ft. (2,915 N·m) torque input.

STANDARD EQUIPMENT

The CX family of electronic automatic transmissions feature integral torque converters, spin-on oil filters, oil level sensors, and up to three PTO mounting locations. Some of the electronic features of this series of transmissions include:

- J1939 electronic networking
- Compatibility with traction control, ABS, and OEM dash display electronics
- Extensive customer reprogrammability options
- Engine retarder option

The current versions of this transmission use an **ADEM software** ECU identical in external appearance to that used on many Caterpillar engines. Caterpillar refers to their transmission controller as an ECM. However, for consistency this will be referred to as an ECU in this text. The ECU is mounted on the left side of the transmission, a feature that helps reduce the length of electrical harnesses. Caterpillar SIS and ET are required to reprogram, calibrate, and troubleshoot the transmission electronics. Troubleshooting hydraulic malfunctions is made easier by a series of quick-couple external taps located at key points within the hydraulic circuit.

CX CONSTRUCTION

The CX transmission is built for rugged vocational applications and looks like it. It is slightly heavier in

FIGURE 21–59 Allison fifth-generation controls wiring schematic

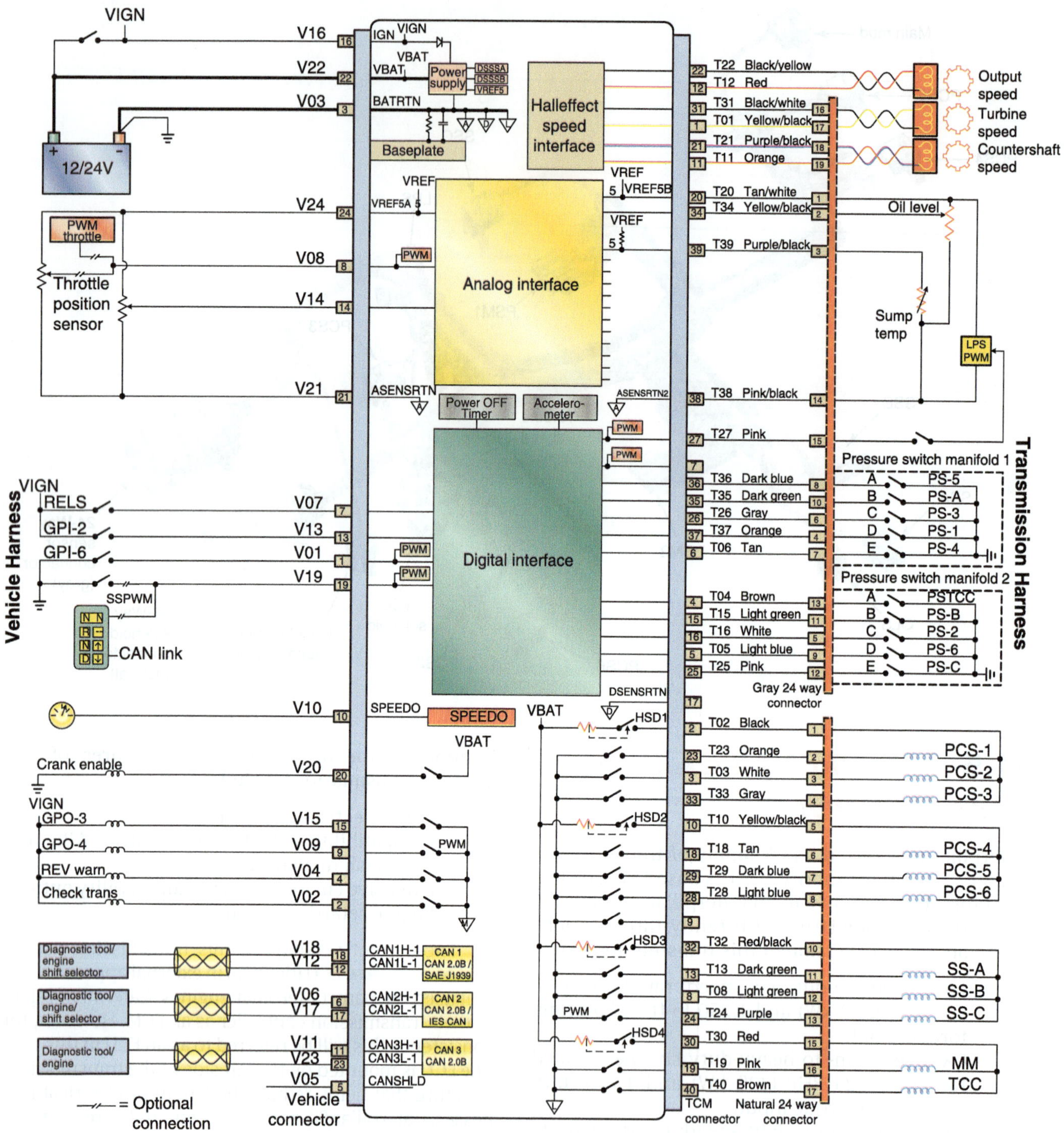

weight (around 25 lb [11 kg]) than an equivalent Allison. The transmission uses an integral torque converter with electronically controlled lockup known as lockup converter (LUC). The torque converter can be easily rebuilt using standard shop equipment. The CX28 and CX31 use three complementary planetary gearsets to produce six forward and one reverse speed. The heavier duty CX35 uses four complementary gearsets in order to produce eight forward speeds and one reverse gear range. Clutches are applied and released using Caterpillar's **electronic clutch pressure control (ECPC)** system. **Figure 21–60** shows a cutaway view of the six-speed versions of CX transmissions; the numeric callouts refer to the clutches used.

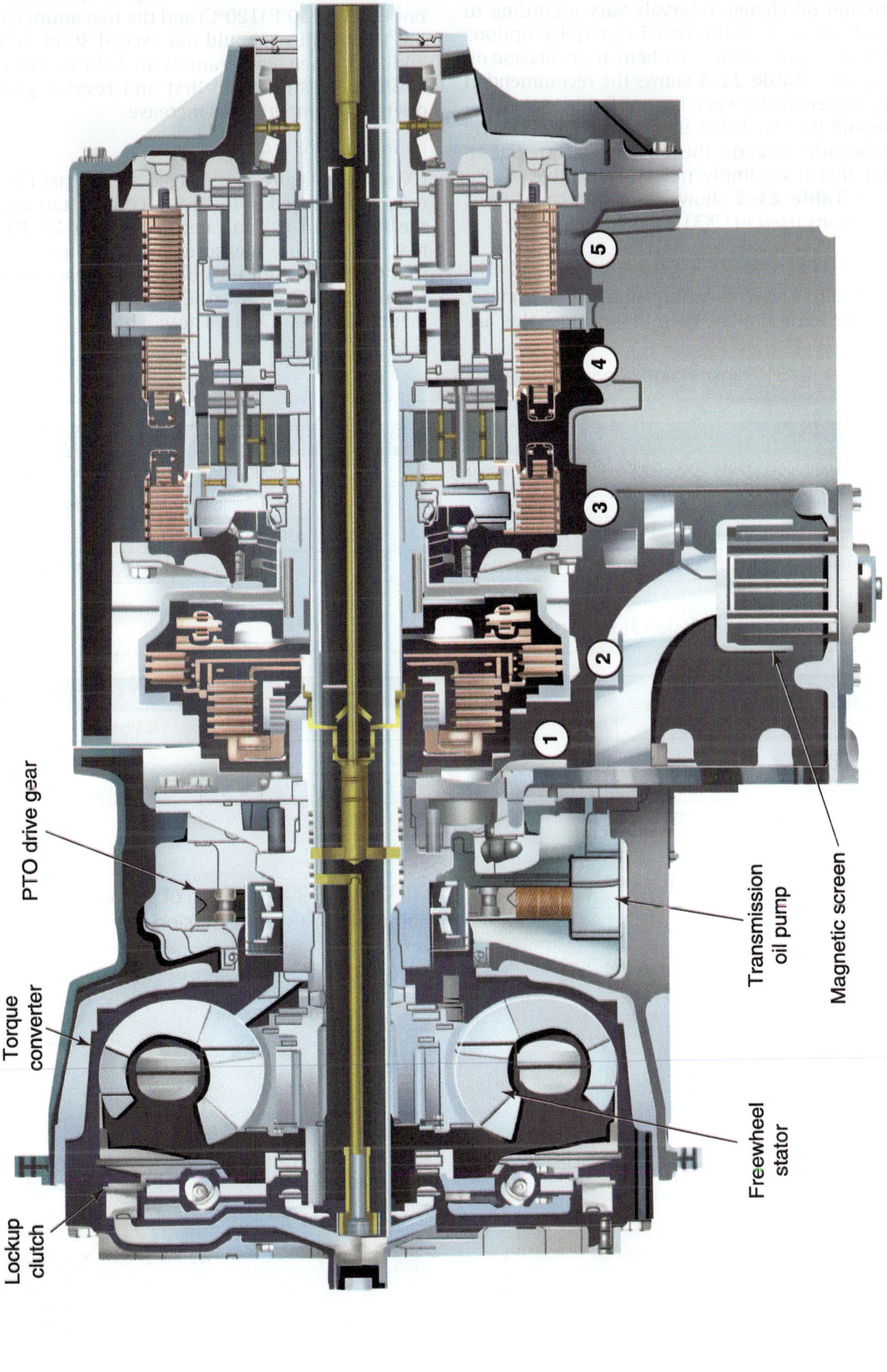

FIGURE 21–60 Cutaway view of the Caterpillar six-speed CX transmission showing the location of the clutches by number

MAINTENANCE

Transmission oil change intervals vary according to the type of transmission fluid used. Caterpillar options Dexron III or its proprietary synthetic transmission oil known as AT-1. **Table 21–4** shows the recommended oil change intervals for each type of fluid.

It should be noted that using Cat AT-1 synthetic oil significantly extends the oil change interval to the point that it is simply not cost effective to use Dexron III. **Table 21–5** shows the sump capacity of the pan options used in CX31 transmissions.

Operational Temperature Limits

CX transmissions have maximum and minimum temperature operating limits. When these thresholds are exceeded, fault codes are logged and shifts may be inhibited. The maximum sump temperature should not exceed 250°F (120°C) and the maximum converter out temperature should not exceed 300°F (150°C). In addition, when temperatures are below −10°F (−23°C), shifts are restricted to first and reverse gears until operating temperatures increase.

Main Pressure and Tap Locations

CX main pressure is managed at 350 psi (24 bar). It can be measured at the quick coupler tap located on the filter pad as indicated in **Figure 21–61**. When making pressure measurements, it is recommended that Caterpillar fluid-filled hydraulic gauges be used. **Figure 21–62** shows the location of valve adjustments on the CX31. These should only be used

TABLE 21–4 CX31 Transmission Service Intervals

Fluid Type	Service Interval: Engine Hours	Service Interval: Vocational Operation	
Dexron III general applications	1,000 hours	25,000 miles	40,000 kilometers
Dexron III severe applications	500 hours	12,000 miles	20,000 kilometers
Cat AT-1 synthetic, general applications	4,000 hours	150,000 miles	240,000 kilometers
Cat AT-1 synthetic, severe applications	3,000 hours	75,000 miles	120,000 kilometers

TABLE 21–5 CX31 Transmission Fluid Capacities

Transmission Type	Estimated Install Fill	Estimated Service Refill
Shallow sump	10 gallons (38 L)	4.5 gallons (17 L)
Standard sump	12 gallons (45 L)	6.5 gallons (24 L)
With retarder option	13 gallons (49 L)	7 gallons (26 L)

Note: Caterpillar requires the use of AT-1 when a retarder is used.

FIGURE 21–61 Location of quick-couple taps to measure system main pressure

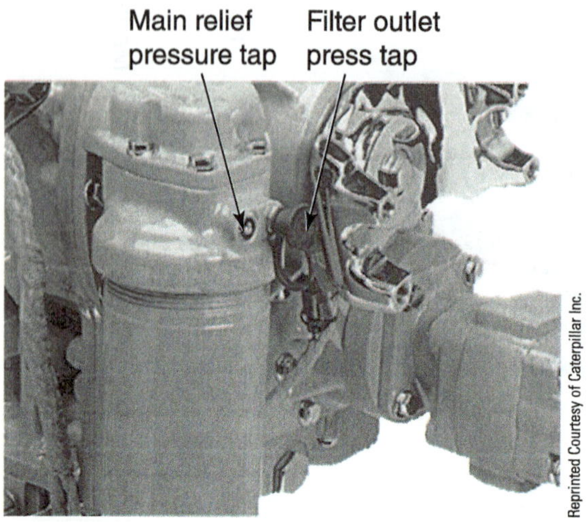

FIGURE 21–62 Location of pressure adjustments points on a CX31 transmission

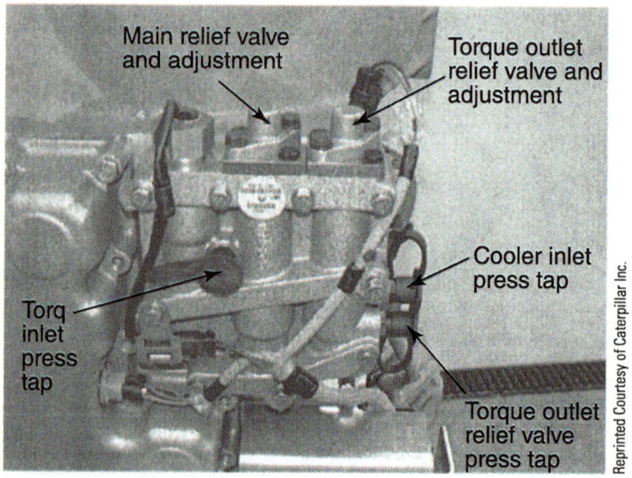

by technicians that have completed the required Caterpillar training and possess access to ET and SIS.

SHIFT PAD

The shift pad used on CX transmissions is a Caterpillar supplied component. It functions similarly to other transmission shift pads in that it can display shift status and provides the driver with the option of overriding a desired gear position. The shift pad can also be used to display the oil level (OL) in the sump. When the UP and DOWN buttons are depressed simultaneously, a 24-second countdown is initiated after which one of the following is displayed on the digital readout:

- OL ..... OK
- OL ..... Low ... (numeral indicating quarts/liters)
- OL ..... Hi ... (numeral indicating quarts/liters)

The CX transmission uses a Hall effect fluid level sensor described previously in Chapter 6 and shown in Figure 6–39. Caterpillar suggests that the numeral indicating the quarts/liters low or high is approximate and that the dipstick should be used for a more accurate reckoning. NOTE: When the sump is significantly overfilled, the display tends not to be accurate. **Figure 21–63** shows a CX shift keypad and **Figure 21–64** is an image of the Hall effect fluid level sensor used in a CX sump.

CX ELECTRONICS

Figure 21–65 is a schematic showing the layout of the input and output circuits used in a CX transmission. The key outputs are the ECPCs shown on the upper right of the schematic. The additional ECPC is used only on the CX35, which uses a fourth planetary gearset to obtain the eight gears.

FIGURE 21–63 CX transmission shift keypad

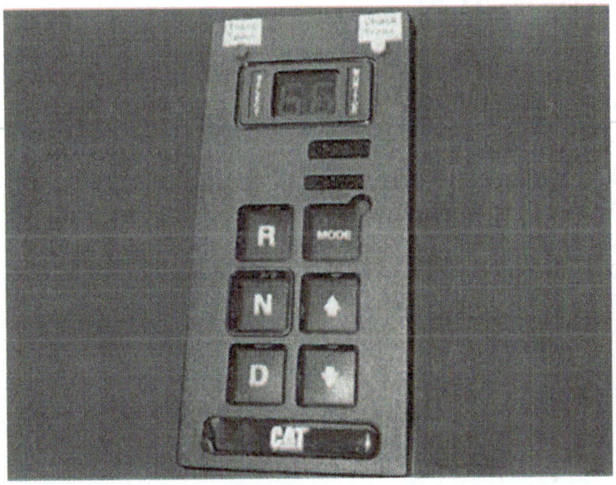

FIGURE 21–64 CX transmission Hall effect fluid level sensor

CX HYDRAULIC CIRCUIT

The transmission oil pump (see Figure 21-56) provides the fluid power required by CX transmissions. Main pressure is regulated at 350 psi (2,400 kPa). The CX hydraulic circuit actuates the ECPCs when they are PWM switched by the ECU. The ECPCs are externally mounted for easy access. They are solenoid-actuated hydraulic clutches that function to hold one of the planetary components stationary. To achieve output from the planetary gearsets, two clutches must be actuated so that they hold two planetary members. The specific ECPCs that are actuated determine which gear ratio is then output by the transmission. **Table 21–6** shows the ECPCs required in order to produce the six gear ratios in CX28 and CX31 transmissions.

ECPCs

The ECPCs are ECU-switched solenoids. Each ECPC is identical (same part numbers) and any unit can be substituted for a suspect solenoid when performing hydraulic troubleshooting. When PWM actuated, the ECPCs route main pressure (350 psi [2,400 kPa]) to engage clutches that hold planetary members. There are two phases to ECPC activation:

1. Clutch fill: Restrictive ports within the ECPC hydraulic circuit are designed to modulate the application and release of clutches. You can observe the activation sequence in **Figure 21–66**, **Figure 21–67**, and **Figure 21–68**.
2. Clutch engagement: This occurs when the clutch engagement circuit becomes fully charged with main pressure.

FIGURE 21–65 CX transmission electronic circuit schematics

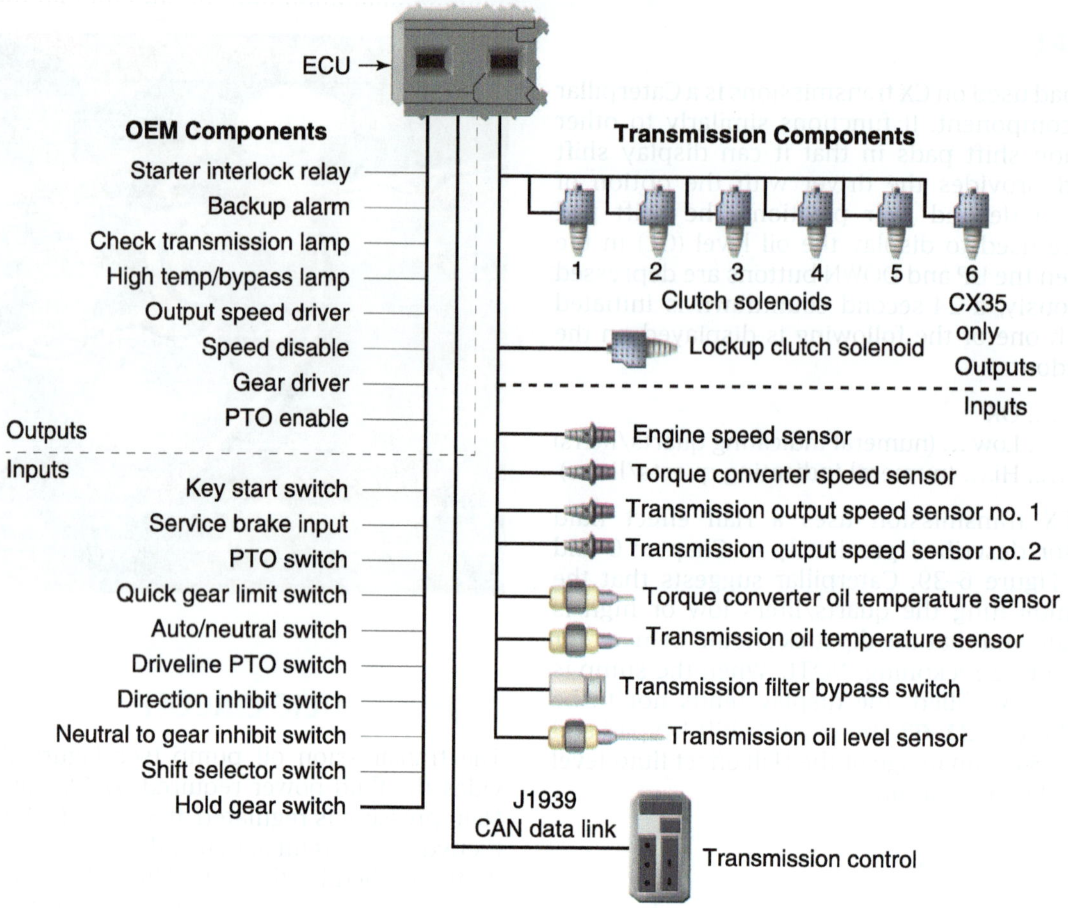

TABLE 21–6 Gear Ratios and ECPCs Actuated

Gear Ratio	ECPCs Actuated	Optional
First gear	1 and 5	
Second gear	1 and 4	LUC
Third gear	1 and 3	LUC
Fourth gear	1 and 2	LUC
Fifth gear	2 and 3	LUC
Sixth gear	2 and 4	LUC

The current load to ECPCs is spiked on actuation (Figure 21–63), and when the solenoid spool valve has completed its mechanical movement (Figure 21–64) current flow through the solenoid coil is reduced.

Hydraulics Troubleshooting

The task of troubleshooting CX transmission hydraulics is made much easier by the fact that every critical circuit can be accessed by external taps on the exterior of the transmission case. **Figure 21–69** shows the complete CX28/CX31 hydraulic circuit and how the ECPCs are located within it.

Table 21–7 can be used to record temperature and tap pressure data on a CX31 transmission. The tap locations on a CX31 are shown in **Figure 21–70** and **Figure 21–71**.

CX CALIBRATION

Shift quality in CX transmissions is unusually smooth. This is due to precise powershift trim that optimizes shift quality. This precise trim is "learned" through initial vocational type calibration and following this it is continually updated using adaptive logic. Calibration should be undertaken:

- After installing a new transmission
- After replacing, exchanging, or servicing the ECU
- After replacing or servicing clutches, seals, pumps, or ECPCs

FIGURE 21–66 CX transmission ECPC modulating valve: no ECU switched signal

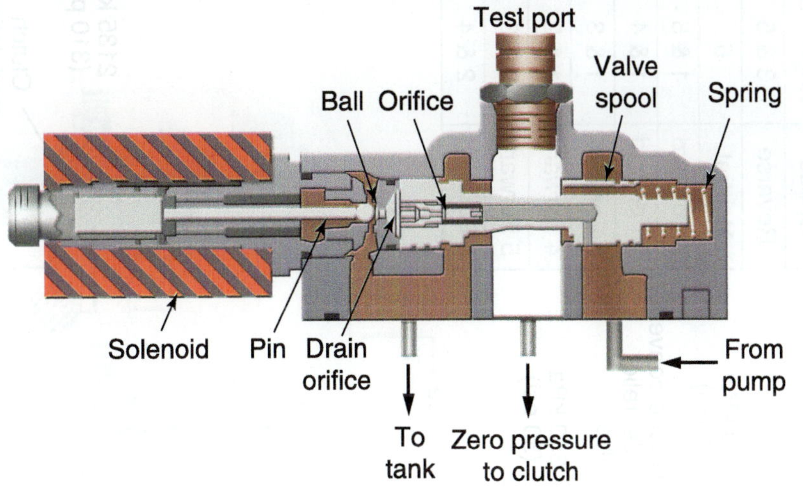

FIGURE 21–67 CX transmission ECPC modulating valve with ECU commanded signal at maximum

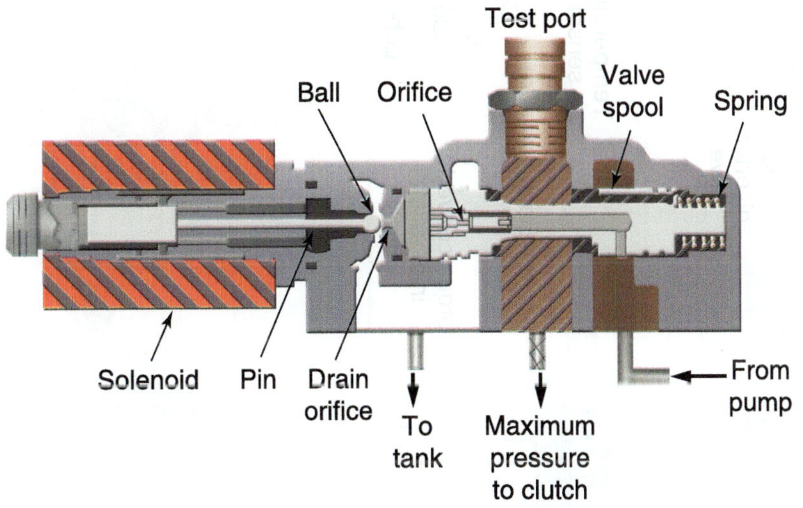

FIGURE 21–68 CX transmission ECPC modulating valve with ECU commanded signal below maximum

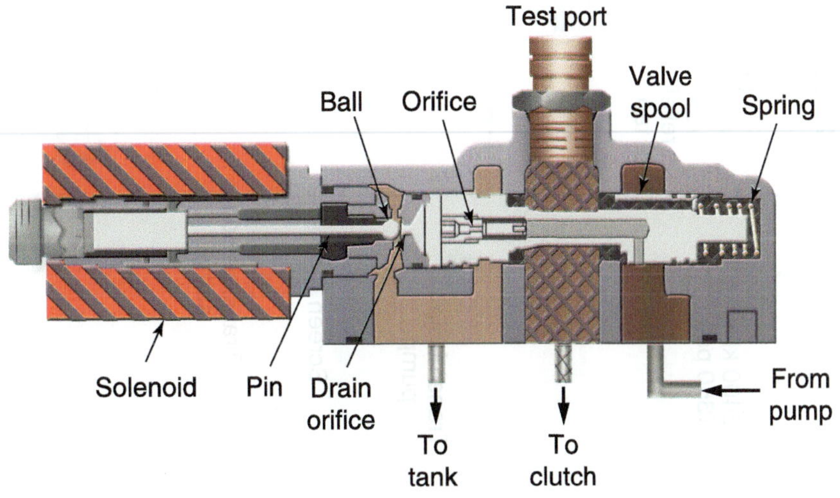

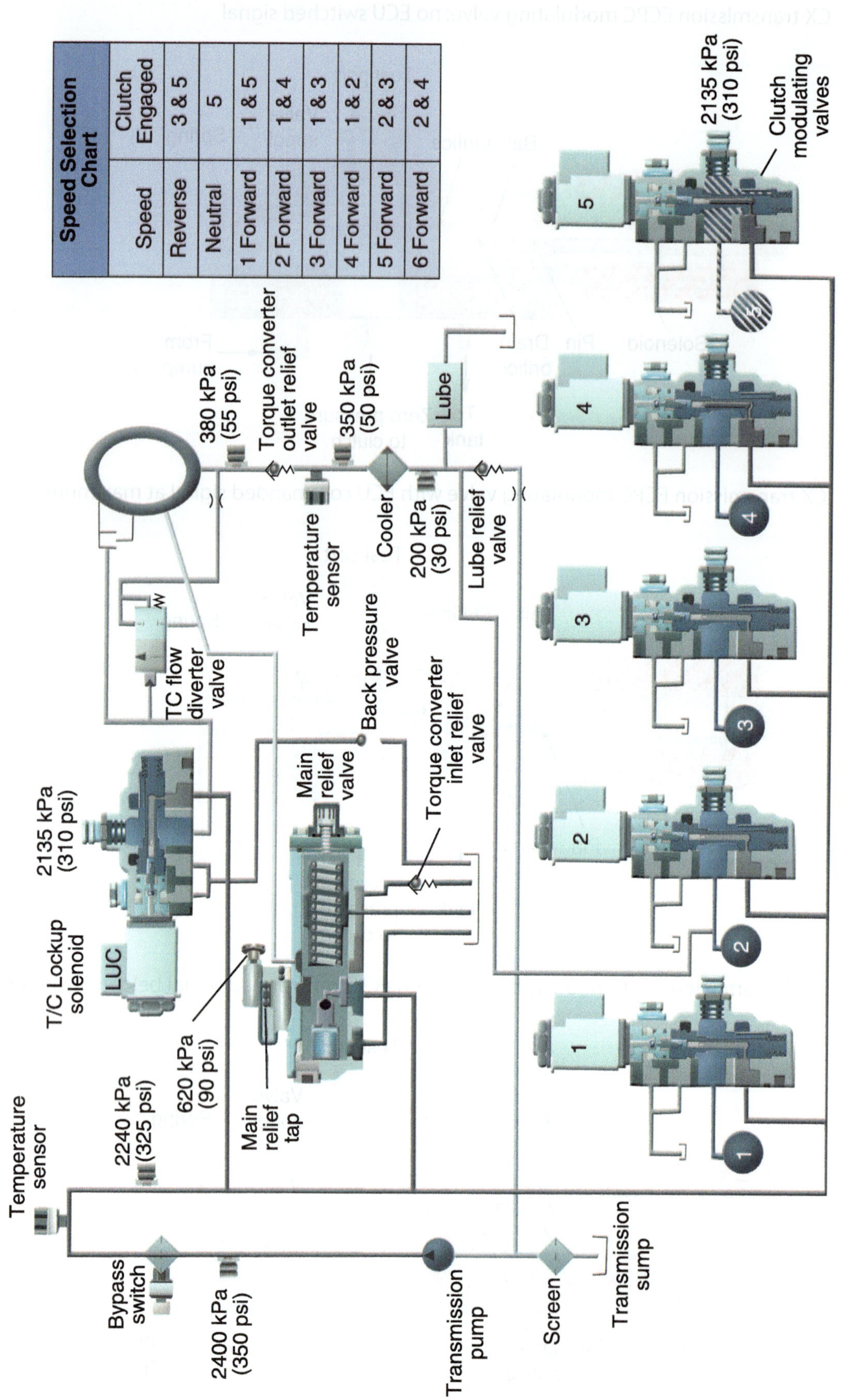

Speed Selection Chart	
Speed	Clutch Engaged
Reverse	3 & 5
Neutral	5
1 Forward	1 & 5
2 Forward	1 & 4
3 Forward	1 & 3
4 Forward	1 & 2
5 Forward	2 & 3
6 Forward	2 & 4

FIGURE 21–69 Complete CX28/CX31 transmission hydraulic circuit

TABLE 21–7 CX Transmission Test Port Data Chart

	50°C (122°F) Sump			80°C (176°F) Sump			90°C (194°F) + 121°C (250°F+) Sump		
Port Location	600 rpm Actual psi	1,400 rpm Actual psi	2,100 rpm Actual psi	600 rpm Actual psi	1,400 rpm Actual psi	2,100 rpm Actual psi	600 rpm Actual psi	1,400 rpm Actual psi	2,100 rpm Actual psi
1. TC outlet temp									
2. Filter inlet psi									
3. Main relief psi									
4. TC inlet psi									
5. TC outlet psi									
6. Cooler inlet psi									
7. Lube psi tap									
8. Lockup clutch									
9. ECPC 1									
10. ECPC 2									
11. ECPC 3									
12. ECPC 4									
13. ECPC 5									

FIGURE 21–70 Tap locations on underside of a CX31 transmission. The callout numbers correspond to the data test fields shown in Table 21–7.

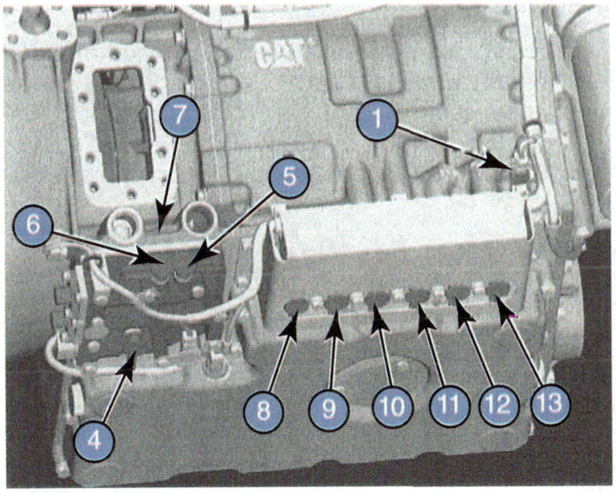

FIGURE 21–71 Tap locations on left side of a CX31transmission. The callout numbers correspond to the data test fields shown in Table 21–7.

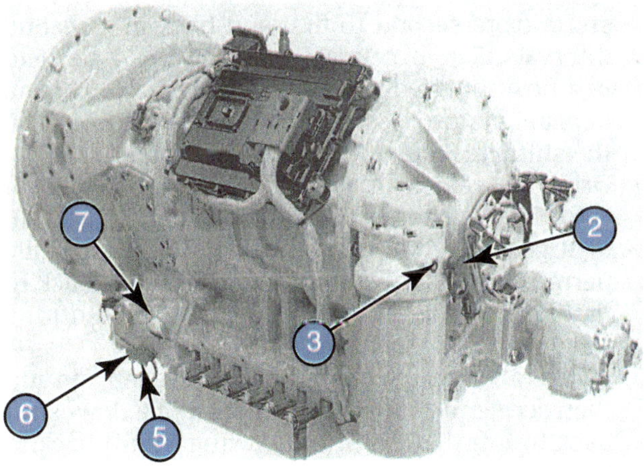

Calibration Procedure

To calibrate a CX transmission, the vehicle must be driven through an extensive road test while the technician responds to prompts from the software on a portable EST. The following steps map out the sequence required for the CX28 and CX31 but it should not be attempted without the appropriate Caterpillar training and software:

1. Check and correct the transmission oil level.
2. Connect an EST to J1939 and launch the ET.
3. Select <configuration> from the menu options, then <vehicle configuration parameters> followed by <transmission calibration configuration>. Program the calibration to <all clutches>.
4. Cycle power to the ECU in this sequence: key-on, key-off, key-on.
5. Drive the unit until the oil temperature is at normal operating temperature. This means above 60°C (140°F) and below 100°C (212°F).
6. In an area located away from a public highway, release the parking brakes and apply

FIGURE 21–72 Location of external components on a CX31 transmission: PTOs, ECU, and ECPC valves

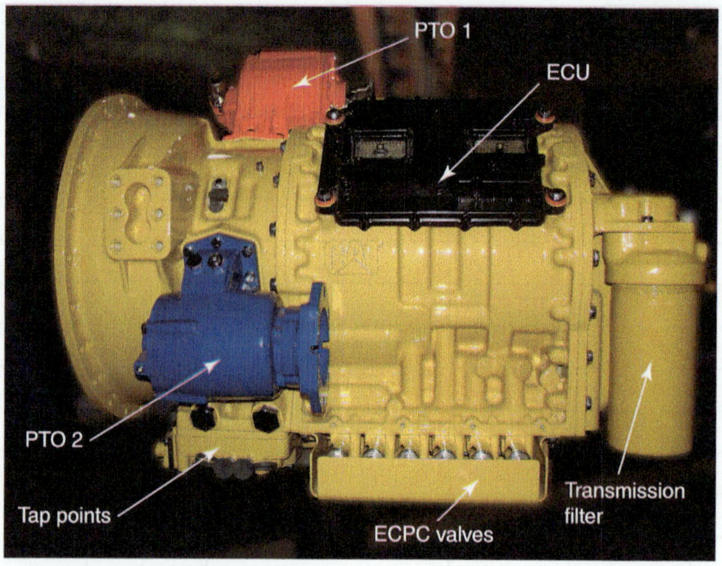

the service brakes; then shift from N to D then through N to R. Repeat this sequence *at least five times* or until each shift change feels similar.

7. In a safe place, accelerate the vehicle so that it shifts to second gear while maintaining an engine speed of 1,200 rpm. This will prompt a series of shifts from second to first and back in 4-second intervals. The accelerator pedal should be held in a position, to operate the engine at 1,200 rpm. Repeat this process at least three times or until the shifts feel smooth.
8. Drive the vehicle at approximately 50 percent of throttle to generate upshifts from first gear to fifth gear. Then allow the vehicle to coast while permitting the transmission to downshift back to first gear. Moderate braking may be used during the downshift sequence.
9. Accelerate through upshifts to fifth gear again and then accelerate sufficiently to prompt a downshift back to fourth gear. After allowing a shift back to fifth gear again, release the accelerator pedal and allow the vehicle to coast and return to fourth gear. The range for the shift point is 1,200 to 1,400 rpm on an engine programmed to achieve peak torque at 1,200 rpm. Depending on the engine peak torque rpm, the shift point will rise proportionally with peak torque rpm. This shift sequence should be repeated for a minimum of five times or until there is no noticeable change in shift quality.
10. Road test the vehicle, ensuring that the transmission upshifts and downshifts through all six speeds smoothly. If shifting continues to be harsh, troubleshoot the transmission using the ET.

EXTERNAL COMPONENT LOCATION

CX transmissions option three possible locations for PTOs:

- Left side
- Upper right side
- Rear

The rear PTO mounting location is recommended when higher torque is required. **Figure 21–72** shows a CX transmission with a pair of side mount PTOs. This figure also shows how the ECU and ECPC valves are positioned on the left side of the transmission.

21.6 VOITH DIWA TRANSMISSION

A short description is provided here of the Voith DIWA electronically controlled, automatic transmission. This transmission originates from Germany and has found some acceptance in transit bus applications. DIWA appears to be a German acronym that Voith chose not to interpret for its North American customers. The transmission uses a pair of complementary planetary gearsets located in the rear of the housing to provide four speeds.

COUNTER-ROTATING TORQUE CONVERTER

Unique to the Voith DIWA transmission is a **counter-rotating torque converter** located in the heart of the transmission housing (**Figure 21–73**). The

FIGURE 21–73 Voith DIWA transmission

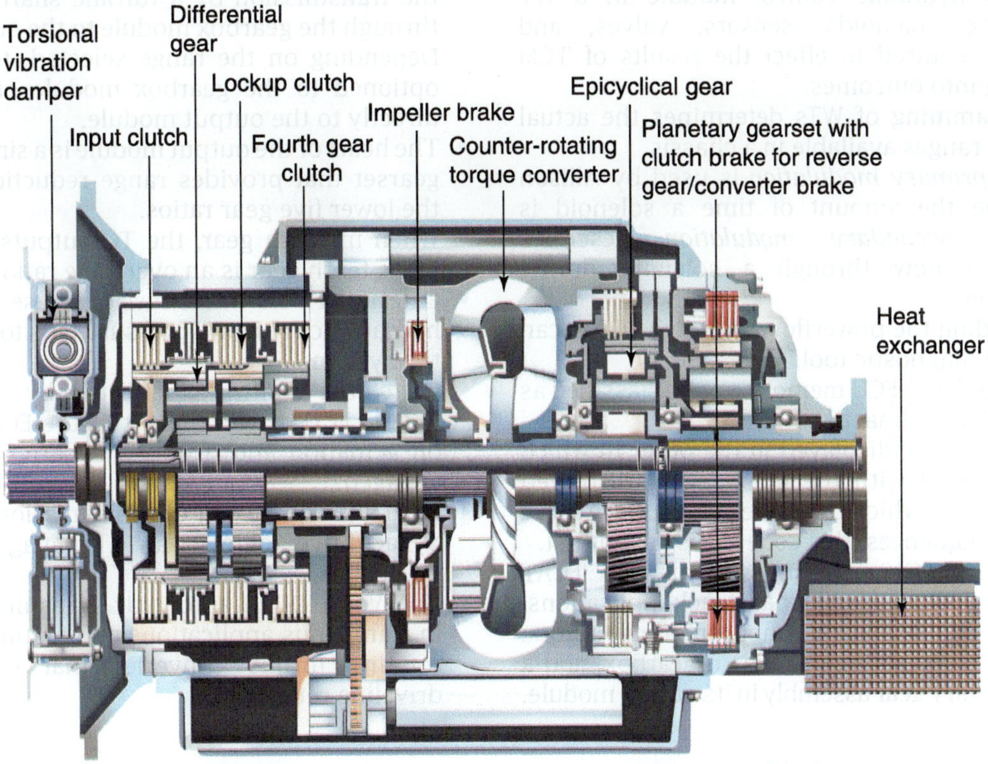

counter-rotating torque converter acts as a conventional torque converter when it receives drive torque from the engine via the transmission input shaft. However, the unit doubles as a driveline retarder. In retarding mode, driveline torque is absorbed by the converter retarder unit.

SUMMARY

- ECATs use an electronic control unit (ECU) that receives electronic monitoring and command inputs, processes the information using program instructions in memory, and outputs electrical signals. Allison currently uses the term transmission control module (TCM) in preference to ECU.
- The electrical signals output from the transmission ECU drive electrohydraulic components located within the ECAT housing.
- The electrohydraulics used in ECATs consist of solenoids and clutches: The status of clutches determine what happens to transmission input torque from the engine.
- All current truck ECATs are networked to the J1939 chassis data bus.
- The main advantages of ECATs over AMTs are the ability to powershift, close to zero spin losses at range change, and better forgiveness for driveline shock loads.
- The ECATs focused on in this chapter are the Allison WT family, the Allison TC10, and the Caterpillar CX family.
- The WT families are divided into light-duty 1000/2000 Series, medium-heavy-duty on-highway 3000/4000 Series, and the off-highway heavy-duty 5000/6000 Series.
- The WT uses a modular design and full authority management electronics to effect shifting and communicate with the chassis data bus.
- WT modules may be removed from the transmission and be serviced or replaced as separate units.
- WT modules may be grouped into input, gearbox, and output categories.
- The WT gearbox modules are the rotating clutch module, the main housing module, the P1 planetary module, the P2 planetary module, and the main shaft module.
- The WT output modules are the rear cover module, output retarder module, or the transfer gear/dropbox module (depending on the model).
- Both WT and CX transmissions option integral driveline retarders. These can supplement vehicle service brake applications. These function similarly to a torque converter driven in reverse.

- One WT model is available with a dropbox or transfer gear for four-wheel-drive vehicles.
- The electrohydraulic control module in a WT houses the solenoids, sensors, valves, and regulators required to effect the results of TCM processing into outcomes.
- The programming of WTs determines the actual number of ranges available in a chassis.
- The term *primary modulation* is used by Allison to describe the amount of time a solenoid is energized. *Secondary modulation* describes the current flow through a solenoid during energization.
- Understanding the powerflows through ECATs can be a useful diagnostic tool.
- Codes logged into ECU memory can be classified as active or historic (inactive).
- Codes are usually displayed in the order in which they are logged with the most recent displayed first. In cases in which one code can produce more severe consequences, this code is displayed first.
- The Allison TC10 transmission is an ECAT specifically designed for linehaul truck applications.
- The TC10 consists of an input module (torque converter), twin-countershaft main gearbox, and a single planetary gear assembly in its output module.
- The TC10 provides ten forward ratios and two reverse ratios. Engine torque is transferred to the transmission by a turbine shaft that extends through the gearbox module to the output module. Depending on the range selected, torque can be optioned to the gearbox module or transmitted directly to the output module.
- The heart of the output module is a single planetary gearset that provides range reduction gearing in the lower five gear ratios.
- When in ninth gear, the TC outputs direct drive, while tenth gear is an overdrive ratio.
- Caterpillar CX transmissions use an external hydraulic circuit and pressure taps to troubleshoot the system.
- CX clutch control is effected by PWM-actuated ECPCs. ECU driver current to the ECPCs is spiked on actuation and dropped off after the clutch is engaged.
- Final shift quality in CX transmissions is trimmed by adaptive logic in an ET-managed, road test procedure.
- The Voith DIWA automatic transmission is used in transit bus applications and features a counter-rotating torque converter that doubles as a driveline retarder.

REVIEW QUESTIONS

1. Technician A says that most ECATs will not shift when first started up under extreme cold conditions. Technician B says that extremely cold transmission fluid can create high enough transmission oil pressure to trip the main pressure relief valve. Who is correct?
   a. Technician A only
   b. Technician B only
   c. both A and B
   d. neither A nor B
2. What is the J1939 source address of the transmission ECU?
   a. 00
   b. 01
   c. 02
   d. 03
3. When performing a resistance key-off test on J1939 bus accessing a 3 pin bus connector, which terminal socket designation indicates CAN Low?
   a. A
   b. B
   c. C
   d. shield
4. How many reluctor teeth are on a standard tailshaft vehicle speed sensor (VSS) used to signal road speed?
   a. 8
   b. 16
   c. 32
   d. 64
5. Which of the following ECATs uses a single planetary gearset?
   a. WT 4000
   b. Voith DIWA
   c. CX31
   d. TC10
6. When using a DMM to perform a key-off resistance test across terminal designations D and C on a J1939 diagnostic connector, a reading of 120 ohms is produced. What does this suggest?
   a. The bus is functioning normally.
   b. An additional terminating resistor on the bus.
   c. A missing terminating resistor.
   d. A short between CAN high and the shield.
7. Which letter(s) in the transmission identification code would indicate that an Allison WT was equipped with a dropbox?
   a. P
   b. HD
   c. DB
   d. T
8. How many clutch assemblies are used in the basic WT gearbox?
   a. 1
   b. 3
   c. 5
   d. 7
9. How many planetary gearsets are used in the basic WT gearbox?
   a. 1
   b. 3
   c. 5
   d. 7
10. Which of the following is responsible for converting a command electrical signal into mechanical movement in Allison electronic transmissions?
    a. the actuator transistor
    b. the clutch piston
    c. the solenoid
    d. the main pressure

11. What drives the sun gear in the Allison WT P1 planetary gearset?
    a. P1 pinion carrier
    b. main shaft
    c. rotating clutch drum
    d. P2 pinion carrier
12. What force is used to effect the mechanical movement of clutch pistons in a WT?
    a. PWM signal
    b. main pressure
    c. electrical pressure
    d. clutch feed pressure
13. In which of the planetary gearsets in the WT does the reversal of input rotation take place?
    a. P1 c. P1 and P2
    b. P2 d. P2 and P3
14. Which of the following components is splined to the WT output shaft?
    a. the P3 pinion carrier
    b. the P3 sun gear
    c. the P3 ring gear
    d. the main shaft
15. Which of the following is required to connect the WT to the chassis electrical system?
    a. the VIM
    b. the transmission harness
    c. the DDR connector
    d. the TPS
16. When first put into service, for how many starts is the WT AutoDetect enabled?
    a. 1
    b. 12
    c. 24
    d. It is always enabled.
17. In which of the following applications would WT electronics require a dedicated TPS signal?
    a. WT is the only chassis electronic system.
    b. WT is used with J1939 electronics.
    c. WT is used with a Caterpillar engine.
    d. WT is used with J1587/1708 electronics.
18. Technician A says that WT electronics will always display the most recently logged trouble code first. Technician B says that WT electronics are capable of logging up to five active codes. Who is correct?
    a. Technician A only c. both A and B
    b. Technician B only d. neither A nor B
19. Technician A says that WT trouble codes can only be accessed by using a DDR with the appropriate software card or cartridge. Technician B says that the shift selector mechanism may be used to read active codes. Who is correct?
    a. Technician A only c. both A and B
    b. Technician B only d. neither A nor B
20. Technician A says that to put a WT into reverse, clutches C3 and C5 must be applied. Technician B says that in reverse range, the P3 pinion carrier becomes the reverse output. Who is correct?
    a. Technician A only c. both A and B
    b. Technician B only d. neither A nor B
21. How many gear speed ratios are used in a Caterpillar CX31 transmission?
    a. 4 c. 6
    b. 5 d. 8
22. How many gear speed ratios are used in a Caterpillar CX35 transmission?
    a. 4 c. 6
    b. 5 d. 8
23. What is used to hold planetary members in CX transmissions?
    a. ECPCs c. centrifugal force
    b. C-clutches d. solenoid pressure
24. What is the main pressure value used in Caterpillar CX transmissions?
    a. 55 psi (3.8 bar) c. 350 psi (24 bar)
    b. 250 psi (17 bar) d. 580 psi (40 bar)
25. Which of the following best describes the counter-rotating torque converter used in a Voith DIWA transmission?
    a. torque converter only
    b. driveline retarder only
    c. torque converter and driveline retarder
    d. planetary gearset
26. Technician A says that the planetary gearset in the output module provides an overdrive option. Technician B says that the TC10 twin countershafts always rotate when the turbine shaft rotates. Who is correct?
    a. Technician A only c. both A and B
    b. Technician B only d. neither A nor B
27. How many rotating clutches are there in a TC10 transmission?
    a. 1 c. 5
    b. 3 d. 6
28. Which of the following must occur to engage the torque converter clutch (TCC) in a TC10 transmission?
    a. Converter in-out oil flow is reversed.
    b. Main pressure must be supplied to the C1 clutch.
    c. The front planetary carrier and ring gear lock together.
    d. Oil is evacuated from the converter housing.
29. When the TCC is in fourth range, which of the following clutches must be engaged?
    a. C1 and C4 c. C1 and C7
    b. C4 and C6 d. C4 and C5
30. Which of the following check methods is recommended by Allison to ensure the oil level is correct at operating temperature after a fluid change?
    a. remove oil fill plug and finger check
    b. diagnostic sight glass when cold
    c. dipstick
    d. electronically on digital display

22

Prerequisites: Chapters 14, 15, and 16

DRIVESHAFT ASSEMBLIES

OBJECTIVES

After reading this chapter, you should be able to:

- Identify the components in a truck driveline.
- Explain the procedures for inspecting, lubricating, and replacing a universal joint.
- Describe the various types of universal joint wear.
- Outline the procedure for sourcing chassis driveline vibration.
- Remove and replace a driveshaft universal joint.
- Define and explain the importance of driveshaft phasing.
- Explain the importance of driveline working angles and how to calculate them.
- Phase a heavy-duty driveshaft assembly.
- Describe the procedure for balancing a driveshaft.

KEY TERMS

axle shims
brinnelling
broken-back driveline
center support bearing
compound angle
driveshaft
false brinnelling
galling
Glidecoat
hanger bearing
inclinometer
in-phase
journal cross
nonparallel driveline
out-of-phase
ovality
parallel-joint driveline
phase angle
pitting
propeller shaft
runout
slip spline
spalling
spider
trunnion
U-joint
universal joint
yoke
zerk fitting

INTRODUCTION

Driveshafts, or **propeller shafts**, as they are sometimes called, have a simple function: to transmit drive torque from one driveline component to another. This should be accomplished in a smooth, vibration-free manner. In a heavy-duty truck, that means transmitting engine torque from the output shaft of the transmission to a rear axle or to an auxiliary transmission. Driveshafts also are used to connect forward and rear axles on 6 by 4 tractors. **Figure 22–1** shows the driveshaft arrangement used in a tandem drive tractor equipped with an auxiliary transmission. Although this arrangement is not commonly used today, it is shown here as an example of a complex driveshaft assembly.

DRIVESHAFT FUNCTIONS

In most cases, a driveshaft is required to transfer torque at an angle to the centerlines of the driveline components it connects to. Because the rear drive axle is part of the suspension and not connected to the rigid frame rails of the truck, the driveshaft must be capable of constantly changing angles as the rear suspension reacts to road profile and load effect. In addition to being able to sustain constantly changing angles, a driveshaft must also be able to change in length when

FIGURE 22–1 Driveshafts transmit torque between driveline components.

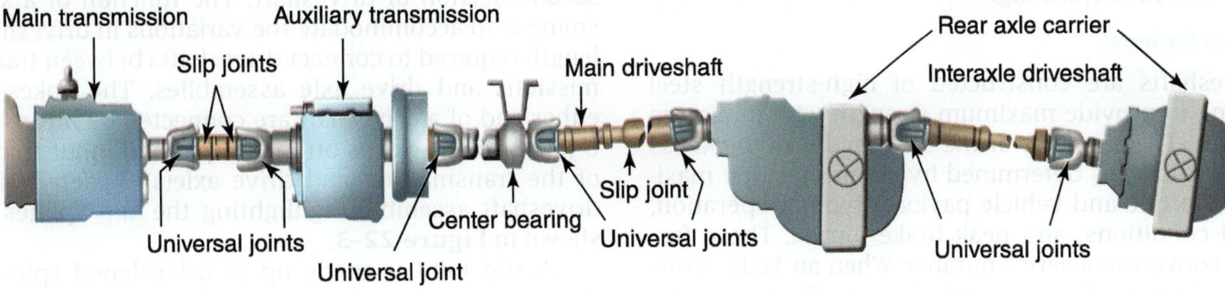

transmitting torque. When the rear axle reacts to road surface changes, torque reactions, and braking forces, it pivots both forward or backward, requiring a corresponding change in the length of the driveshaft.

22.1 DRIVESHAFT CONSTRUCTION

A driveshaft assembly is made up of U-joints (universal joints), **yokes**, **slip splines**, and propeller shafts. The propeller shafts have a tubular construction designed to sustain high-torque loads and be light in weight. **Figure 22–2** shows the components of a typical heavy-duty truck driveshaft. In this section, we will study the role that each of these components plays in the driveshaft assembly.

DRIVESHAFTS

To transmit engine torque from the transmission to the rear drive axles, driveshafts have to be durable and strong. If an engine produces 1,000 pound-feet of torque, when this is multiplied by a 12:1 gear ratio in the transmission, it produces 12,000 pound-feet of

FIGURE 22–2 Exploded view of a heavy-duty truck driveshaft

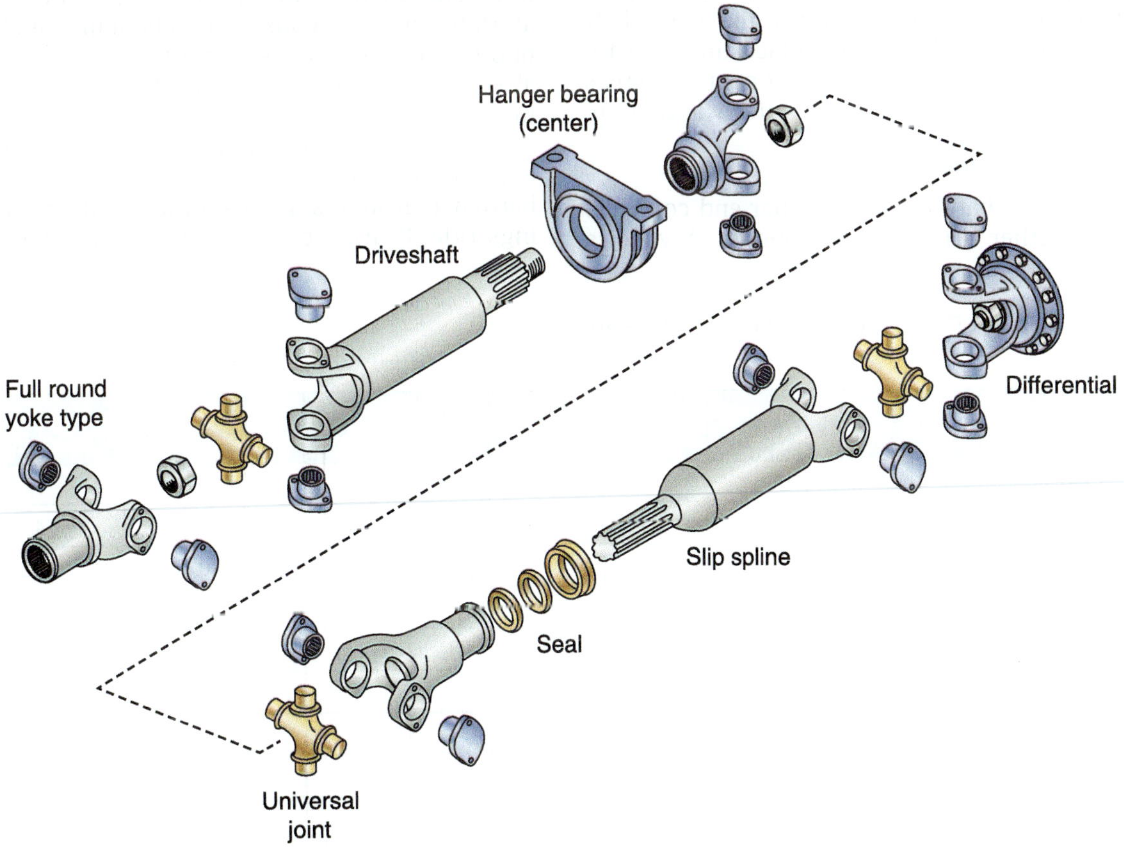

torque at the driveshaft. The shaft has to be tough enough to deliver this twisting force to a fully loaded axle without deforming.

Series Ratings

Driveshafts are constructed of high-strength steel tubing to provide maximum strength with minimum weight. The diameter of the shaft and wall thickness of the tubing is determined by several factors: maximum torque and vehicle payload, type of operation, road conditions, and peak brake torque. These factors convert to a series number. When an 1800-series driveshaft is referred to, this is a specification of the maximum torque, maximum shaft horsepower, and maximum rpm that the driveshaft can handle. The most common driveshaft rating series used on line-haul highway tractors are the four variations of the 1700-series. **Table 22–1** shows the specifications of some common driveshaft series used in truck applications; this table is for reference, there is no point in remembering its contents.

CONFIGURATIONS

One-, two-, and three-piece driveshaft assemblies are used, depending on the length of the driveline. The truck original equipment manufacturer (OEM)-recommended driveshaft rating should be adhered to when replacing a driveshaft. Oversized, undersized, overweight, and out-of-balance driveline components result in repeated driveline failures that can extend to powertrain components beyond the driveshaft assembly. Oversized driveshafts place unnecessary loads on transmission and axle components, whereas undersized driveshafts tend to fail quickly and often.

Driveshaft Components

In a simple driveshaft, a yoke at either end connects the shaft to other driveline components. A yoke is usually welded to the shaft tube at one end, whereas at the other, a slip spline connects to a slip yoke or a second section of driveshaft. The function of a slip spline is to accommodate the variations in driveshaft length required to connect driveshafts between transmissions and drive axle assemblies. The yokes at either end of a driveshaft are connected by means of U-joints to end yokes on the output and input shafts of the transmission and drive axle(s). A detail of a driveshaft assembly highlighting the slip splines is shown in **Figure 22–3**.

A slip joint is made up of a hardened splined shaft stub welded into the end of the driveshaft tube. The male end of the slip yoke with external splines is inserted into a female slip yoke with mating internal splines. The splines allow the driveshaft to change in length at the same time as it transmits full torque loads. The slip splines are lubricated with grease and, additionally, are polymer nylon coated. The polymer nylon coating used by Spicer is known as **Glidecoat** and it permits slippage while significantly reducing wear. A threaded dust cap contains a washer and seal (usually synthetic rubber or felt) that excludes contaminants and contains the grease.

U-JOINTS

A **universal joint** is commonly known as a **U-joint**. The purpose of a U-joint is to connect the driveline components while permitting them to operate at different and constantly changing angles. The hub of a U-joint consists of a forged **journal cross** also known as a **spider**. The forged cross may be machined for **zerk fittings** (grease nipples) and four sets of needle bearings. The ends of the universal cross are called **trunnions**. The trunnions are case-hardened ground surfaces on which the needle bearings ride. Bearing caps contain the needle bearings

TABLE 22–1 Truck Driveshaft Specifications by Series

Driveshaft Series	Shaft Torque Rating in N·m	Shaft Torque Rating in lb-ft.	Max Power Rating in kW	Max Power Rating in hp	Max Safe Shaft Operating rpm
1550	3,100	2,280	125	170	5,000
1610	4,670	3,450	180	240	4,500
1710	6,200	4,570	245	330	4,500
1710 HD	8,400	6,210	245	330	4,500
1760	6,200	4,570	270	360	4,500
1760 HD	8,400	6,210	270	360	4,500
1810	7,900	5,850	320	430	4,500
1810 HD	10,760	7,940	320	430	4,500
1880	14,050	10,330	375	500	3,000

FIGURE 22–3 A driveshaft with a slip joint

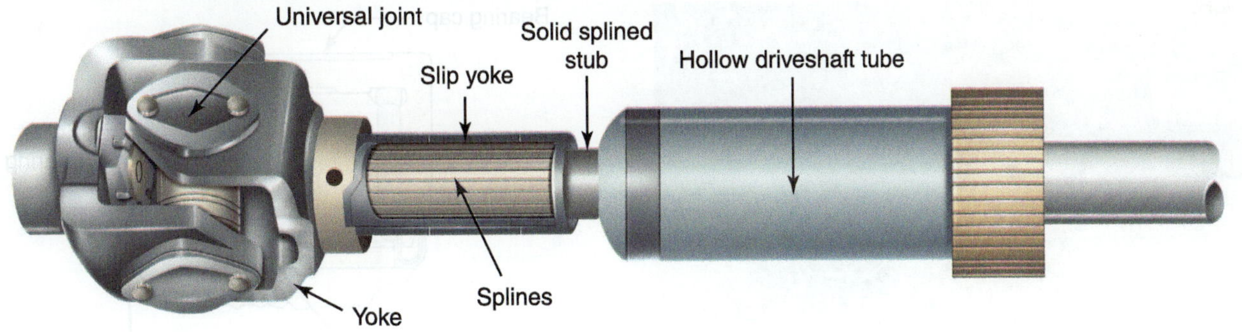

and these fit tightly to yoke bores to retain the U-joint within a pair of yokes offset with each other. The needle bearings and trunnions are lubricated with chassis grease and this is covered in detailed later in this chapter. A complete U-joint assembly with one bearing cap removed from the trunnion is shown in **Figure 22–4**.

In the center of each trunnion on some U-joints is a stand pipe (a type of check valve), which prevents reverse flow of the hot liquid lubricant generated during operation. When the U-joint is stationary, one or more of the trunnions has to be upright (**Figure 22–5**). Without the stand pipe, lubricant would flow out of the upper grease passage and trunnion, resulting in a partially dry startup that could cause wear. The stand pipe helps ensure adequate lubrication of the trunnions and needle bearings at each startup. Other U-joints have rubber check valves in each cross that perform the same function.

FIGURE 22–4 Universal joint assembly

Bearing Assemblies

Each cross has four bearing assemblies, one for each trunnion race. Each assembly consists of needle bearings in a bearing cup and a rubber seal around the open end of the cup. Needle bearings are used because of their strength and durability. They can sustain the high loads that result from the oscillating action of rotating driveshafts.

Needle bearings also require less radial space, which reduces the overall diameter of a U-joint. Most U-joints are factory lubricated with just enough lubricant to hold the rollers in place during assembly into the driveshaft. Additional grease should be applied after the U-joint has been installed. Needle bearings are contained in cups in the bearing cap assembly (**Figure 22–6**). The bearing cups fit over the trunnions on the U-joint and retain the cap assemblies within the yokes. Most bearing cups are circular and fit snugly into the bore of a yoke. Other U-joints have wing bearing cups rather than the common round cups. The wing bearing cup assembly bolts onto shoulders of the yoke, as shown in **Figure 22–7**.

FIGURE 22–5 A stand pipe at the end of each trunnion ensures constant lubrication and prevents dry startup.

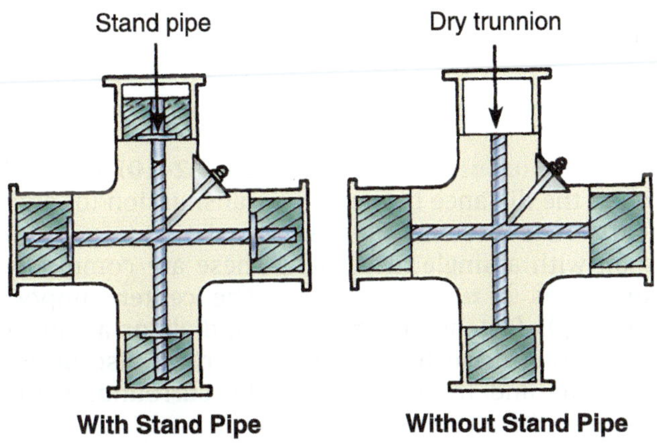

FIGURE 22–6 Bearing cup and needle rollers

FIGURE 22–7 Wing-type U-joint and yoke

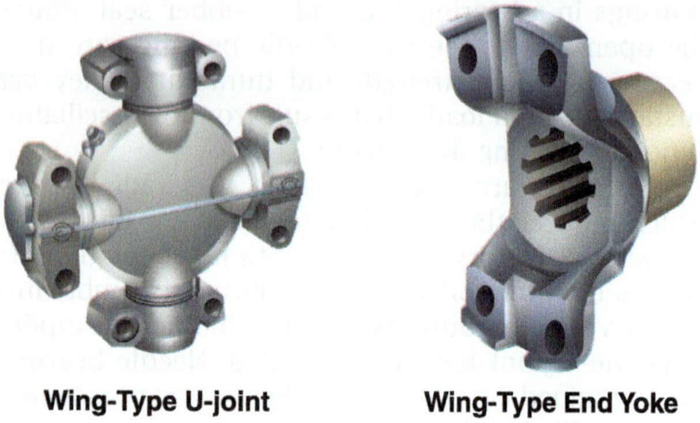

Each bearing cup is lubricated and has a seal designed to keep the bearings free of dirt and other contaminants. **Figure 22–8** shows a common seal design.

Mounting Hardware

U-joints can be connected to yokes in different ways. **Figure 22–9** shows four common methods of securing U-joints to yokes. Half round end yokes are clamped to the U-joints using bearing straps or U-bolts. Full-round end yokes use snaprings or bearing lock plates to secure the joint in the yoke bore.

FIGURE 22–8 Typical U-joint trunnion seal

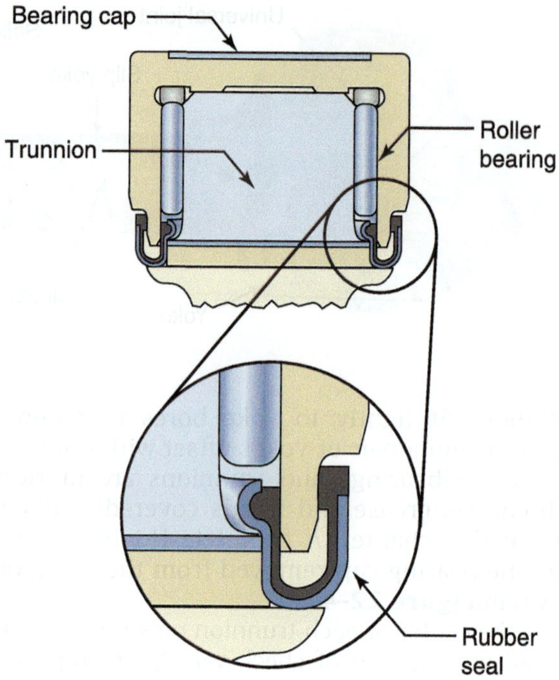

FIGURE 22–9 Four methods of fastening a universal joint in a yoke: (A) bearing plate; (B) snapring; (C) strap; and (D) U-bolts.

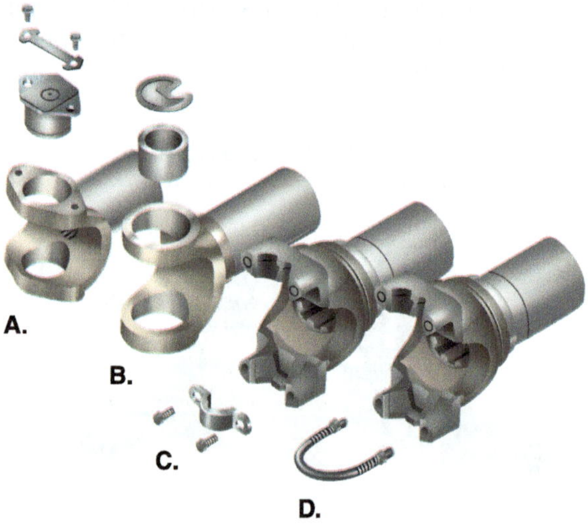

CENTER SUPPORT BEARINGS

Center support bearings (**Figure 22–10**) are used when the distance between the transmission (or auxiliary transmission) and the rear axle is too great to span with a single driveshaft. These are commonly known as **hanger bearings**. The center support bearing is fastened to the frame and aligns a pair of connecting driveshafts. Hanger bearings also buffer driveline and frame vibrations by surrounding the bearing with a rubber insulator.

A center support bearing is housed in a stamped steel bracket that aligns and fastens the bearing assembly to the frame cross-member. The rubber insulator supports the bearing with a small margin of forgiveness. Most hanger bearings are sealed so no maintenance lubrication is required for the life of the bearing. A center support bearing is usually required on driveshafts that exceed 70 inches (180 cm)

FIGURE 22–10 Hanger or center bearing

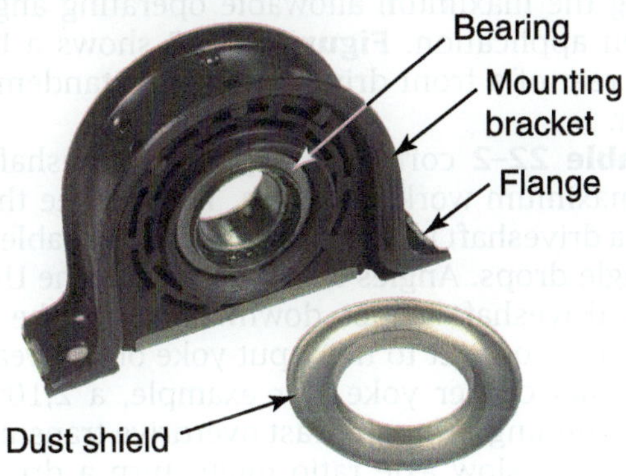

in length. **Figure 22–11** shows a section view of a hanger bearing and **Figure 22–12** presents an actual view of a heavy-duty hanger bearing and U-joint.

DRIVELINE GEOMETRY

Two types of driveline configurations are used to transmit drive torque to the drive wheels: the **parallel-joint driveline** and the **nonparallel driveline** or **broken-back driveline**. In the parallel-joint type, all the companion flanges or yokes in the driveline are on a parallel plane to each other with the working angles of the joints (see A and B dimensions in **Figure 22–13**) of a given shaft being equal and opposite. For instance, if the transmission output shaft centerline angles down 5 degrees from a true horizontal plane, the centerline at the front of the auxiliary main shaft or rear axle pinion shaft must be angled 5 degrees up.

FIGURE 22–11 Cross-section of a typical center bearing

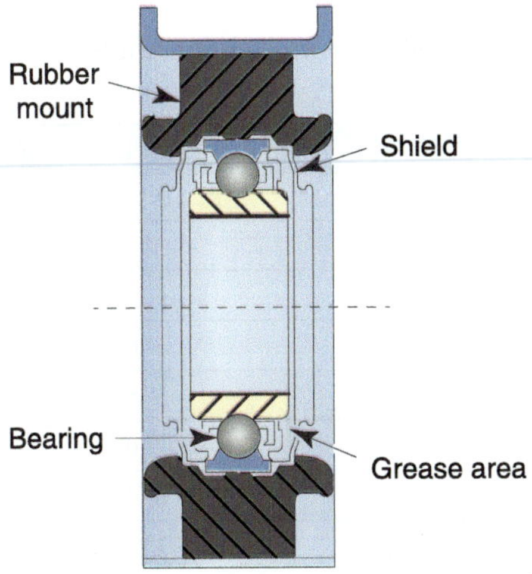

FIGURE 22–12 Hanger bearing and U-joint

With the nonparallel or broken-back installation, the working angles of the U-joints (A and B, in **Figure 22–14**) of the driveshaft are equal, but the companion flanges and/or yokes are not parallel. For example, the transmission yoke is angled 3 degrees down from a horizontal plane, whereas the rear axle pinion flange is angled up at 12 degrees. So long as the U-joint working angles of this propeller shaft remain equal, the shaft will run vibration free.

U-JOINT WORKING ANGLES

Proper U-joint working angles are required for vibration-free and long-lasting driveline operation. Most drivelines are angled on a vertical plane, as

FIGURE 22–13 Parallel joint driveline arrangement

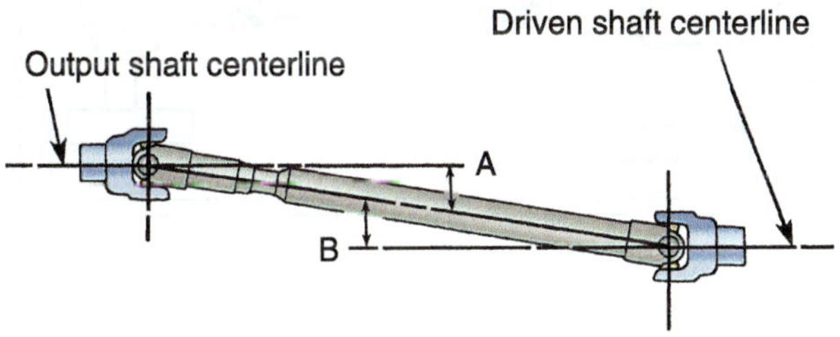

FIGURE 22–14 "Broken back"–type driveshaft; angles A and B are equal.

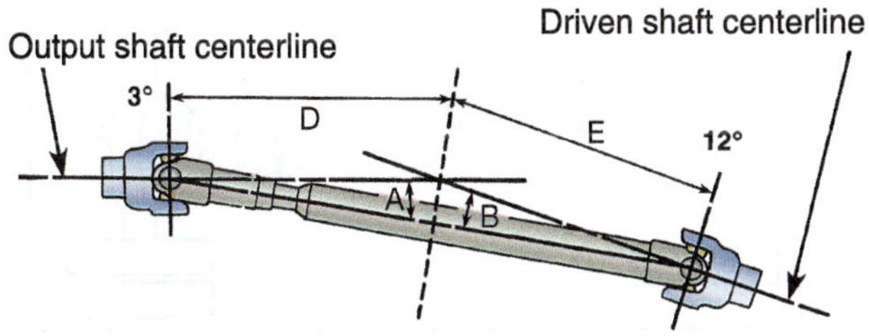

shown in **Figure 22–15A**, but on some trucks, the drivelines are also horizontally offset (angled), as shown in **Figure 22–15B**. When a driveshaft is angled on both the vertical and horizontal planes, a **compound angle** exists.

All U-joints have a maximum working angle at which they can smoothly transmit torque. This working angle depends in part on the U-joint size and design. Exceeding the maximum recommended working angles of U-joints can destroy a U-joint rapidly and also damage interconnected driveline components.

High working angles combined with high rpm tend to result in a reduced U-joint life. Unequal U-joint working angles can cause vibrations and contribute to U-joint, transmission, and differential problems. Ideally, the operating angles on each end of a driveshaft should be equal to or be within 1 degree of each other and have a 3-degree-maximum operating angle. Driveshaft rpm is the primary factor in determining the maximum allowable operating angles in a given application. **Figure 22–16** shows a U-joint at the rear of a front drive carrier on a tandem drive tractor.

Table 22–2 correlates expected driveshaft rpm with maximum working angles. You can see that the faster a driveshaft turns, the lower the allowable working angle drops. Angles are calculated at the U-joints as the driveshaft angles downward from the transmission to connect to the input yoke on the rear axle differential carrier yoke. For example, a 2,100 rpm engine running through a fast overdrive transmission and a fairly slow axle ratio might turn a driveshaft faster than 3,000 rpm. This would limit the maximum permissible angle to 5 degrees, which is measured as the truck sits on level pavement.

With equal working angles, the rear U-joint will slow down by the same amount that the forward joint

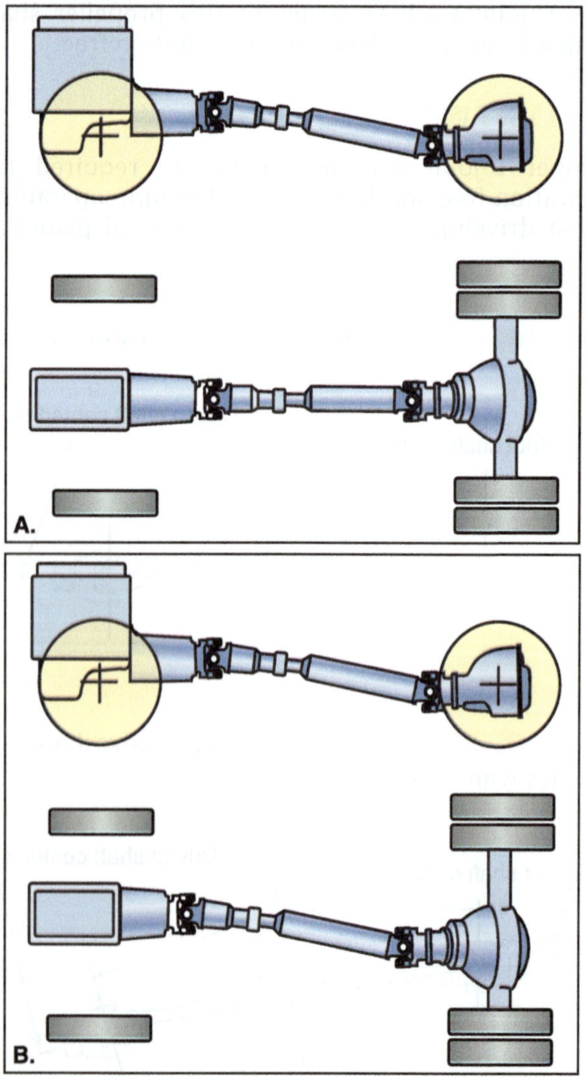

FIGURE 22–15 (A) One-plane-angle driveshaft; and (B) two-plane-angle driveshaft

FIGURE 22–16 U-joint at the output of a forward carrier assembly

TABLE 22–2 Maximum U-Joint Angles

Driveshaft rpm	Maximum Normal Operating Angles
5,000	3° 15'
4,500	3° 40'
4,000	4° 15'
3,500	5° 0'
3,000	5° 50'
2,500	7° 0'
2,000	8° 40'
1,500	11° 30'

speeds up during a rotation, resulting in U-joint cancelation. The driving and driven shafts will turn at constant and identical speeds. If the working angles of two opposed U-joints vary by more than 1 degree, the driveshaft will not rotate smoothly because it accelerates and decelerates during a cycle. The result is vibration and ultimate U-joint failure.

U-joint working angles become greater when the vehicle suspension flexes over uneven road surfaces. If this occurs at slow speeds, it should not be a major problem unless driveline angles are high to begin with. Driveline angles tend to present fewer problems these days than back when tractors with a short wheelbase were required to keep vehicle overall length within legal limitations. Relaxation of those legal restrictions has led to the use of longer wheelbases that allow driveshafts to be longer and eliminate aggressive U-joint working angles.

SHOP TALK

A U-joint, operating at an angle, causes a cyclical speed variation in the driveshaft so that it speeds up and slows down twice every revolution. The nonuniform output speed produces driveline pulsations any time it is operating at an angle. Working the same way but in reverse is the U-joint at the rear of the driveline that connects to the drive axle input shaft. In order to avoid harmful torsional vibrations that result from the nonuniformity, the rear U-joint needs to cancel out the nonuniform rotation of the forward U-joint. This can be accomplished by having the transmission output shaft and drive axle input shaft set at relatively similar angles, producing equal U-joint working angles. This is known as driveshaft U-joint angle cancellation. When U-joint angles at either end of the driveshaft are kept within at least 1 degree of each other, uncanceled, nonuniform rotation will be minimized. Gear clatter, gear seizures, synchronizer failures, and torsional vibration, which are caused by driveshaft vibration, will be eliminated.

The longer the driveshaft, the greater the weight and, therefore, the greater the radial forces, especially as driveshaft rpm increases. At high speeds, balance becomes more critical. This is why manufacturers limit tube length. For example, at 3,000 rpm the length of any single driveshaft section, measured between the centerline of the U-joints at either end, should not usually exceed 70 inches (180 cm).

DRIVELINE PHASING

Heavy-duty U-joints have a unique characteristic. Because they are always operating on an angle, they do not transmit constant torque or turn at a uniform speed during their 360-degree rotation (**Figure 22–17**).

With the drive yoke turning at a constant rpm, a driveshaft increases and decreases speed twice per revolution. To counterbalance this fluctuation above and below mean (average) driveshaft speed, the U-joints must be positioned **in-phase** with each other as shown in **Figure 22–18**. An **out-of-phase** condition produces an effect somewhat like that of a rope being snapped on one side when two kids turn a skipping rope; by contrast, if both kids at either end of the rope snap it at the same time, the resulting waves through the rope would cancel each other, and neither child would feel the reaction of the wave the snap causes.

PHASE ANGLE

When in phase, the slip yoke lugs (ears) and welded yoke at the opposite end should be perfectly in line as shown in **Figure 22–19**. If a driveshaft is assembled one spline out, the driveshaft is out of phase and the result can be a significant vibration. There should be opposing alignment arrows stamped on the slip yoke and on the tube shaft so that you can assemble a driveshaft in phase. If a driveshaft is not marked with timing arrows, do it yourself. Make a light score mark using a steel scriber. The term **phase angle** is used to refer to the relative rotational position of each yoke on a driveshaft assembly. Get it wrong, even if it is just a single spline out, and you will create a significant vibration.

FIGURE 22–17 A driveshaft increases and decreases speed twice per each revolution.

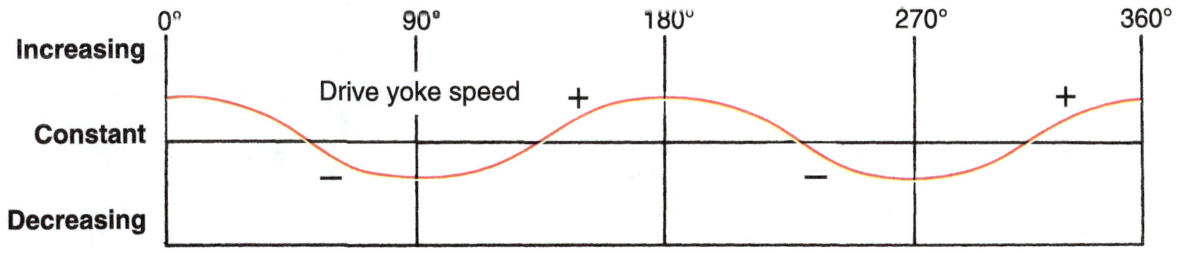

FIGURE 22–18 Two universal joints in-phase will cancel out the speed fluctuations in the driveshaft.

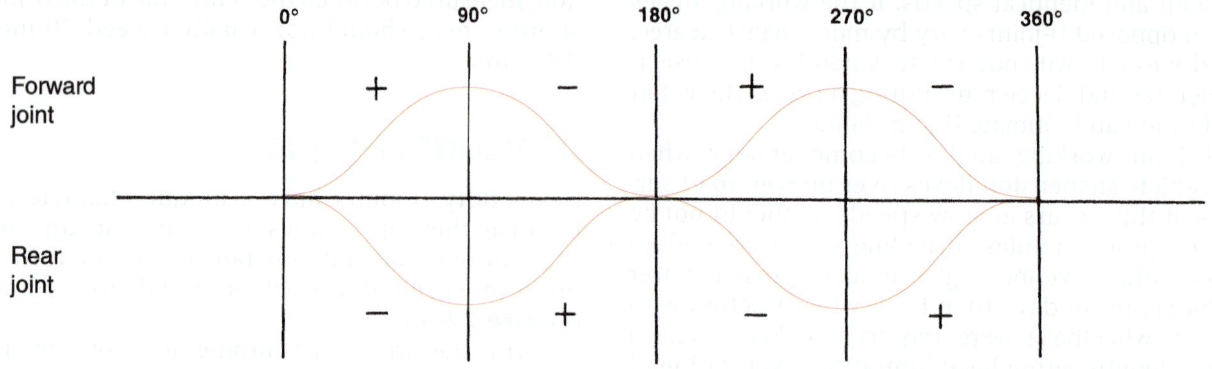

FIGURE 22–19 A driveshaft is in-phase when the lugs on the tube yoke and slip yoke line up with each other.

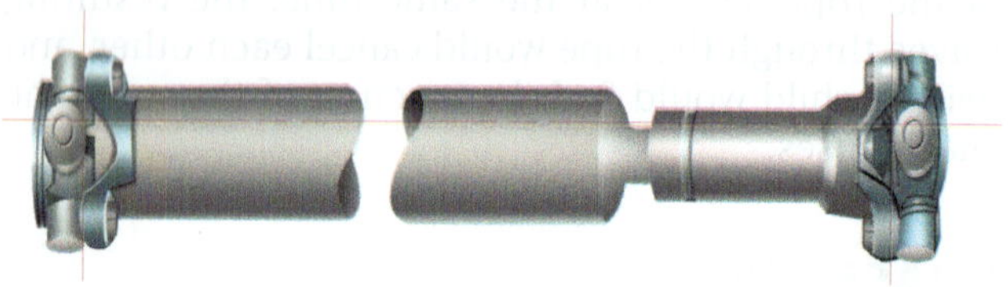

22.2 DRIVESHAFT INSPECTION

Whenever a truck is moving, its driveshafts are working. Driveshafts should be routinely inspected and lubricated. Driveline vibrations, U-joint failures, and hanger bearing problems are caused by such things as loose end yokes, excessive radial play (side to side), slip spline play, bent driveshaft tubing, and missing lube plugs in slip joint assemblies.

You can help limit driveshaft performance problems by performing some simple inspection procedures at each time the driveshaft is lubed.

1. Check the yokes on both the transmission and drive axle(s) for looseness as shown in **Figure 22–20**. If loose, disconnect the driveshaft and retorque the end yoke retaining nut to the OEM specification. If this does not correct the problem, yoke replacement may be necessary. If you do have to replace a yoke, check for the OEM recommendation regarding replacement frequency of the end yoke retaining nut.
2. If the end yokes are tight, check for excessive radial looseness of the transmission output shaft and drive axle input and output shafts in their respective bearings, as shown in **Figure 22–21**. Consult transmission and axle OEM specifications

FIGURE 22–20 Check U-joint yokes for radial play.

FIGURE 22–21 Check the transmission output yoke and the rear axle input yoke for excessive radial play.

for acceptable radial looseness limits and method of checking. If the radial play exceeds the specifications, the bearing should be replaced.

3. Check for looseness across the U-joint bearing caps and trunnions (**Figure 22–22**). This looseness should not exceed the OEM specification (typically as little as 0.006 inch [0.152 mm]).
4. Check the slip splines for radial movement. Radial looseness between the slip yoke and the driveshaft stub should not exceed the OEM specification (typically as little as 0.007 inch [0.178 mm]).
5. Inspect the driveshaft for damage, bent tubing (**Figure 22–23**), or missing balance weights. Ensure that there is no buildup of foreign material on the driveshaft, such as asphalt or concrete. Anything adhering to a driveshaft has the potential to unbalance it.
6. Check the hanger bearing visually, making sure it is mounted securely. Check for leaking lubricant from the bearing, damaged seals, or rubber insulator failure. Replace the hanger bearing if there is evidence of damage. Do not attempt to repair or lubricate it.

FIGURE 22–22 Check for excessive play in the U-joint bearings.

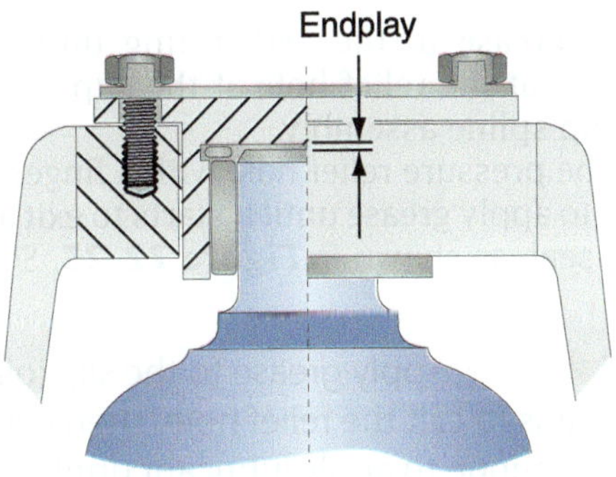

FIGURE 22–23 Points to inspect when checking a driveshaft tube for twisting, dents, missing weights, and buildup of foreign material

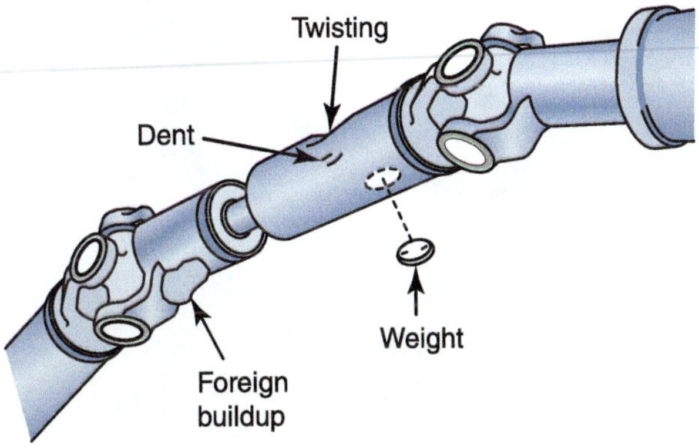

22.3 LUBRICATION

One of the most common causes of U-joint and slip joint problems is the lack of proper lubrication. Some U-joints are lubed for life; these may have significant longevity providing the driveshafts are not frequently removed. When U-joints are properly lubricated at recommended service intervals, they will meet or exceed their projected operating life span. Regular lubing ensures that the bearings have adequate grease and, additionally, that the trunnion races are flushed, which removes contaminants from the critical surface contact areas.

SHOP TALK

Clean grease fittings thoroughly to remove accumulated grease and abrasives before use. Any contaminants on the zerk nozzle can be forced through the nipple into the bearing during lubing.

U-JOINT LUBRICANTS

Heavy-duty driveshafts typically use lithium soap-based, extreme pressure (EP) grease meeting National Lubricating Grease Institute (NLGI) classification grades 1 or 2 specifications. Grades 3 and 4 are not recommended because of their greater thickness, meaning that they function less effectively when cold. Most lubed-for-life bearings use synthetic greases.

Lubrication Schedules

When you purchase a replacement U-joint kit, note that it contains only sufficient grease to protect the needle bearings during shipping and storage. Every replacement U-joint should be properly lubricated after installation with the recommended chassis grease, using the procedure outlined in this section.

After initial lubrication, U-joints should be lubed at each chassis lubrication service. Lubrication cycles obviously will vary, depending on the service requirements and operating conditions of the vehicle. A typical recommended lube cycle for U-joints is shown in **Table 22–3**.

TABLE 22–3 Lubrication Schedule

Type of Service	Miles	Time
City	5,000/8,000	3 months
On-highway	10,000/15,000	1 month
On-/off-highway	5,000/8,000	3 months
Extended (linehaul)	50,000	3 months
Severe usage off-highway	2,000/3,000	1 month

SHOP TALK

On-highway operation is generally defined as an application that operates the vehicle running less than 10 percent of total operating time on gravel, dirt, or unimproved roads. Vehicles running more than 10 percent operating time on poor road surfaces are classified as off-highway in terms of preventive maintenance.

LUBRICATING A U-JOINT

You should carefully follow the foregoing procedure for lubricating U-joints.

1. Apply chassis grease through either one of the two zerk fittings on the U-joint cross. Fit the grease nozzle over the zerk nipple as shown in **Figure 22–24**. Pump grease slowly into the fitting until each of the four trunnion seals pops. Allow a small quantity of the old grease to be forced out of the bearings. This flushes contaminants out of the bearings and helps ensure that all four have taken grease. The U-joint is properly greased when evidence of purged grease is seen at all four bearing trunnion seals.
2. If grease does not exit from the U-joint trunnion seals, try forcing the driveshaft from side to side when applying gun pressure. This allows greater clearance on the thrust end of the bearing assembly that is not purging. If the U-joint has two grease fittings, try greasing from the opposite fitting. If this does not work, proceed to the next step.
3. Back off the bearing cap bolts on the trunnion that is not taking grease and pop the cap out about ⅛ inch (3 mm). Apply grease. This usually will cure the problem, but if it does not you should remove the U-joint to investigate the cause.
4. After working on a U-joint that fails to take grease, make sure you torque the bearing cap bolts to the OEM specification.

CAUTION

If grease is only seen to exit from three of the four trunnion seals, the bearing is not properly lubed and will almost certainly fail. You must take the appropriate corrective action when U-joints fail to take grease. If a driveshaft separates because of a U-joint failure, it can take out air tanks and fuel tanks and generally cause damage that greatly exceeds the cost of replacing a single U-joint.

CAUTION

Half-round end yoke self-locking retaining bolts should not be reused more than five times. If in doubt as to how many times bolts have been removed, replace with new bolts.

LUBRICATING SLIP SPLINES

The chassis lubricant used for U-joints is satisfactory for use on slip splines. An EP grease meeting NLGI B grade 1 or 2 specifications is required. Slip splines should be lubed on the same service schedule as the U-joints on the intervals outlined in Table 22–3. Use the following procedure:

1. Apply grease to the zerk fitting until lubricant appears at the relief hole at the slip yoke end of the slip spline assembly.
2. Seal the pressure relief hole with a finger and continue to apply grease until it starts to exit at the slip yoke seal, as shown in **Figure 22–25**. Sometimes

FIGURE 22–24 Apply grease until it exits all four trunnion seals.

FIGURE 22–25 Apply grease to the slip joint until grease begins to exit the relief hole. Then cover the hole and continue to apply lubricant until grease begins to ooze out around the seal.

it is easier to purge the slip yoke by removing the dust cap and reinstalling it after grease appears.

CAUTION

In cold temperatures, you should drive the vehicle immediately after lubricating driveshafts. This activates the slip spline assembly and removes excessive lubricant. Excess lubricant in slip splines can freeze in cold weather to a wax constituency and force the breather plug out. This would expose the slip joint to contaminants and eventually result in wear and seizure.

LUBRICATING HANGER BEARINGS

Hanger bearings are usually lubed for life by the manufacturer and are not serviceable. However, when replacing a support bearing assembly, fill the entire cavity around the bearing with chassis grease to shield the bearing from water, salt, and other contaminants. You should pump in enough grease to fill the cavity to the edge of the slinger surrounding the bearing.

SHOP TALK

When replacing a hanger bearing, make sure you look for and do not lose track of the shim pack that is usually located between the bearing mount and cross-member. The shims set the driveshaft angles and omitting them will result in a driveline vibration.

22.4 U-JOINT REPLACEMENT

Replacing U-joints is a routine shop task and usually requires no special tools. However, an arbor press or U-joint puller can make the procedure easier and reduce the risk of damaging the yoke. In most truck applications, you do not have to raise the vehicle to perform a U-joint replacement. You should begin by removing the grease fittings because these are easily sheared during the removal process.

CAUTION

When removing a driveshaft with half-round- or flange-type yokes, support the weight of the driveshaft with a sling before separating the U-joints.

SHOP TALK

Before removing a driveshaft, mark the slip yoke assembly and tube shaft with a paint stick to ensure the correct phasing alignment on reassembly. If the shaft assembly is to be cleaned before reassembly, use a steel scribe to indent alignment marks.

REMOVING U-JOINTS

The key here is to use the least amount of force possible. Although U-joints are relatively inexpensive and commonly replaced, the driveshaft, yokes, and slip spline assembly are expensive and designed to last the life of the vehicle.

Bearing Plate U-Joints

The removal procedure is as follows:

1. If the U-joint is connected to full-round yokes with bearing cups, use a hammer and chisel to bend back the lock tabs away from the bolt heads.
2. Remove the four bolts on the two end caps on the yoke you are separating.
3. There are several ways to remove the bearing assemblies from the end yoke bores. A hydraulic jack can be used to apply pressure under the driveshaft, raising the bearing cup up out of the yoke bore, as shown in **Figure 22–26A**. If you use this separation method, you use the weight of the truck, so you can expect it to give suddenly. Gently tap on the outside of the yoke with a light hammer if the bearing cup does not loosen when initial pressure is applied. After the bearing cup has been exposed above the bore, it can be usually pulled out by hand, as shown in **Figure 22–26B**. Rotate the driveshaft 180 degrees and repeat the procedure to remove the bearing from the opposite side of the yoke. After separating the U-joint cups, ensure that the bearing cups stay matched to the trunnion from which they were removed. Do not reinstall the bearing caps on any trunnion other than on their original ones.
4. A safer method of removing the bearing cup assembly is with the use of a two-jaw puller designed for separating U-joints. Bolt the puller to the bearing assembly and apply torque to separate each bearing cap.
5. Free the trunnion from its yoke by tilting the U-joint cross until the trunnions clear the yoke bore.
6. Force the driveshaft inboard on the slip joint and then lower it to the ground.

Removing Other Types of U-Joints

The main problem when replacing U-joints is the potential to damage the driveshaft and slip yoke assembly. It is important to exert only enough force to separate the U-joint. Remember that using heat to free a seized bearing cap almost always destroys the U-joint, so this is not acceptable when you are removing the driveshaft for reasons other than replacing the U-joint. Many shops use a 50-pound slide hammer to separate U-joints. This can result in damaging the driveshaft tube when the U-joint cups seize in the yoke bores.

FIGURE 22–26 (A) Use a jack to force the bearing cup out of the yoke; (B) when the top of the cup is exposed, it can be removed by hand.

Courtesy of Chicago Rawhide

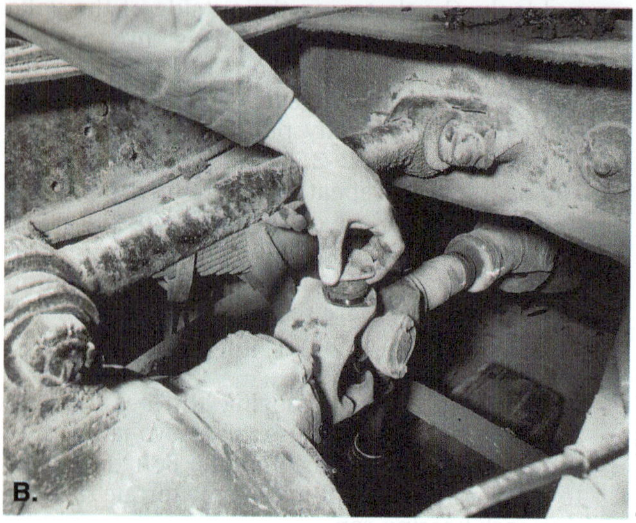

Courtesy of Chicago Rawhide

Snapring U-Joint. If the U-joints are secured in the yoke by snaprings, remove the snaprings and raise the driveshaft using either the jack or puller methods described earlier to separate the cups.

Half-Round Yoke Assemblies. If the driveshaft has half-round end yokes, remove the strap retaining bolts or the U-bolts. Then collapse the driveshaft by moving it inboard to separate the bearing cups from the yoke. An advantage of this design is the ease it lends to removing driveshafts.

Flange-Type Yokes. If the driveshaft has flange-type yokes, loosen and remove the fasteners securing the flange yoke to the transmission or drive axle carrier flange. Hold the shaft firmly when tapping the flanges free. When separated, compress the driveshaft by forcing it inboard and lower the assembly to the floor.

CAUTION

Never use a sledge hammer directly on a yoke to separate a U-joint. The result will almost certainly be a damaged driveshaft.

SHOP TALK

If only one end of the driveshaft requires service, disconnect that end, unscrew the slip shaft seal (dust cap) from the slip yoke assembly, and then separate the driveshaft at the slip spline joint. When removing the entire driveshaft, disassemble one end at a time, laying the disconnected end on the floor carefully. When reassembling, be sure that the arrows or marks on the shaft and slip joint are in line to keep the driveshaft yokes in phase.

General U-Joint Removal Precautions. Do not distort the driveshaft tube by applying excessive clamping force. Using an appropriate puller, such as that shown in **Figure 22–27**, is the best way to remove plate-type bearing caps. If a puller is not available or if the U-joint is not equipped with bearing plates, you can use an arbor press or hammer and soft round drift to remove the bearings. Another way to remove U-joints from yokes is shown in **Figure 22–28**, although this is the least preferred method because of its potential to damage either the yoke or the slip splines assembly. Support a pair of cross trunnions on the jaws of a vise and then tap on the yoke with hammer blows. Use a minimum amount of aggression. When the bearing cup can be pulled out by hand,

FIGURE 22–27 Plate-type bearings can be removed with a puller.

Courtesy of Dana Corporation

FIGURE 22–28 Support the cross on vise jaws, then tap the yoke to drive the bearing cup upward.

Courtesy of Chicago Rawhide

reverse the yoke and U-joint and repeat the procedure to remove the opposite bearing cap.

U-JOINT FAILURE ANALYSIS

Inspect the U-joints and bearing cups for signs of wear and damage. **Figure 22–29** shows some examples of U-joint failures.

- Cracks are stress lines caused by metal fatigue. Severe and numerous cracks will weaken the metal until it breaks.
- **Galling** occurs when metal is cropped off or displaced because of friction between surfaces. Galling is commonly found on trunnion ends.
- **Spalling** (surface fatigue) occurs when chips, scales, or flakes of metal break off due to fatigue rather than wear. Spalling is usually found on splines and U-joint trunnion races.

FIGURE 22–29 Types of universal joint damage: (A) cracks; (B) galling; (C) spalling; and (D) brinnelling

Courtesy of Arvin Meritor

Courtesy of Arvin Meritor

Courtesy of Arvin Meritor

Courtesy of Arvin Meritor

- **Pitting**—small pits or craters in metal surfaces—is caused by corrosion and can lead to surface wear and eventual failure.
- **Brinnelling** is a type of surface failure evidenced by grooves worn in the surface. Brinnelling is often caused by improper installation of the U-joints. Do not confuse the polishing of a surface known as **false brinnelling**, where no structural damage occurs, with actual brinnelling. False brinnelling is a manufacturing characteristic that can surface-polish bearing races without actually creating any damage. Technicians should learn to recognize this to avoid replacing bearings that are functionally sound.

YOKE INSPECTION

After removing the U-joint cross and bearing cups, inspect the yoke bores for damage or burrs. Minor bore irregularities can be removed with a rat tail or half-round file, followed by finishing with emery cloth (**Figure 22–30**).

Check the yoke bores for wear, using a go-no-go wear gauge. Use an alignment bar (a bar with approximately the same diameter as the yoke bore) to inspect for misalignment of the yoke lugs. Slide the bar through both yoke bores simultaneously, as shown in **Figure 22–31**. If the alignment bar does not pass through both yoke bores simultaneously, the yoke has been distorted either by disassembly malpractice or excessive torque and should be replaced. Next, clean and inspect the mating yoke with an alignment bar gauge. Do not risk reusing a defective yoke.

FIGURE 22–30 Clean the yoke bearing bore with emery cloth.

Courtesy of Chicago Rawhide

FIGURE 22–31 To check for misalignment, slide an alignment bar through both yoke bearing bores.

Courtesy of Dana Corporation

U-JOINT REASSEMBLY

Use the following procedure to assemble a driveshaft, installing new U-joints:

1. Place the slip yoke end of the driveshaft assembly in a bench vise so that you can see the phase mark made on disassembly. Locate the phase mark on the main section of the driveshaft and assemble by mating the slip joint. Double-check the phasing. An out-of-phase driveshaft will cause an immediate driveline vibration.
2. Remove the new U-joint from its packing and separate the bearing caps from the cross trunnions. Visually inspect the cross, ensuring that the one-way check valve in each trunnion lube hole is present. Then position the cross into the driveshaft yoke, aligning the lube zerk fitting as close as possible to the slip spline lube fitting. The zerk fitting should be directed toward the inboard side.
3. Paste some antiseize compound to the outside diameter of four bearing assemblies. This facilitates installation and removal next time around. Angle the cross and insert into the yoke, as demonstrated in **Figure 22–32A**. Fit the first bearing cup by inserting a trunnion into the needle bearings (**Figure 22–32B**), then push the bearing cap into yoke bore (**Figure 22–32C**). When the trunnion is inserted into the first bearing cap assembly, it is aligned. Insert the opposing bearing cup into the yoke bore and press home. You can now insert the lock tabs and fasteners, as shown in **Figure 22–32D**, but do not torque yet.
4. Move the cross back and forth to check the cross for binding. It should pivot freely with zero drag. You can now torque the fasteners, but do not bend up the lock tabs in case you have to loosen the bearing caps for lubing.

FIGURE 22–32 Align the bearing with the trunnion cross when installing a universal joint. Clamp close to the yoke and take care not to apply too much clamping force from the vice.

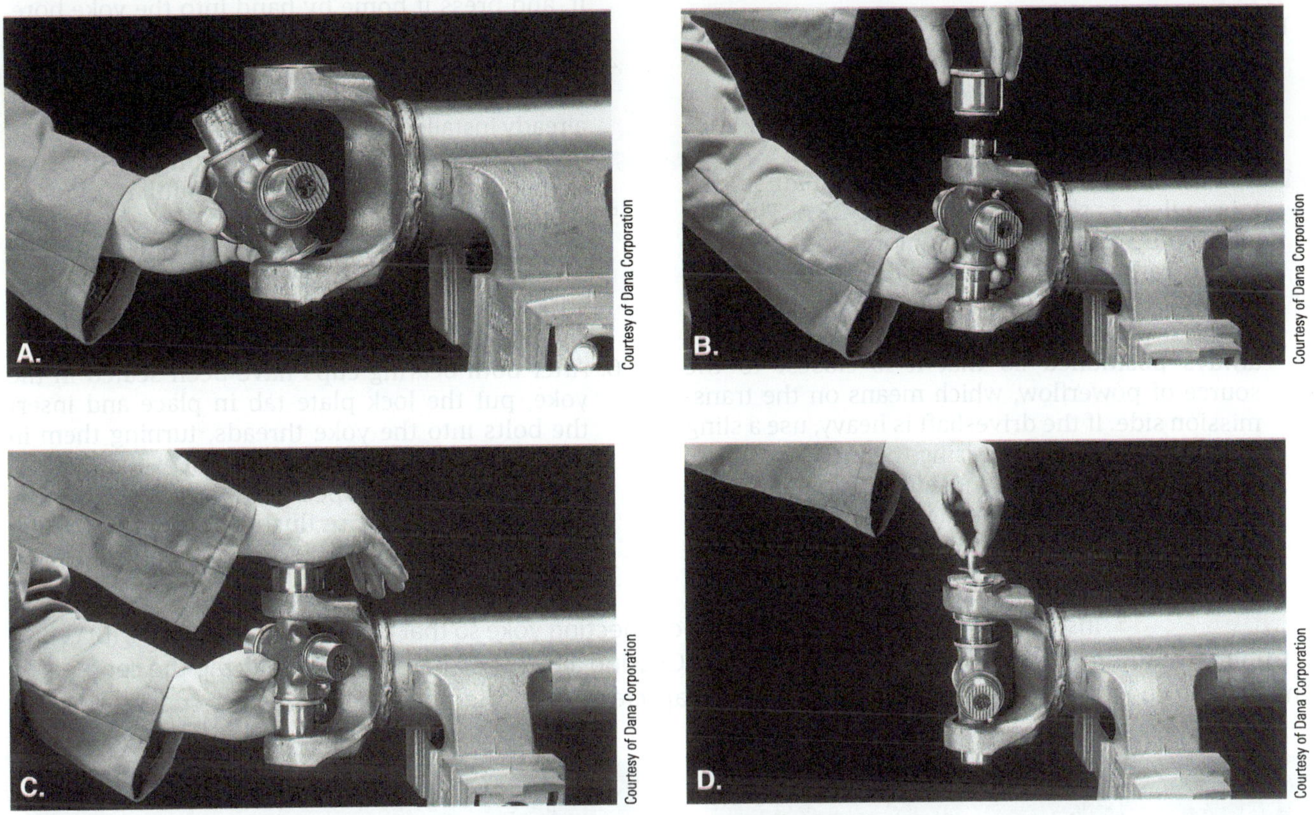

5. Now repeat the foregoing process to install the U-joint at the opposite end of the driveshaft. Again, position the cross in the yoke so that the lube zerk fitting aligns with the lube fitting at its opposite end.
6. For flange yoke applications, you can install the flange yoke, bearing assemblies, and fasteners at this time.

CAUTION

If the bearing cap binds in the yoke bore, gently tap with a ball peen hammer in the center of the bearing cap. Do not tap the outer edges of the bearing cap because this could damage either the bearing cup or the yoke.

SHOP TALK

Using guide bolts when installing U-joint caps can help minimize alignment problems. Two ⅜-inch fine thread bolts of the appropriate length should be used.

DRIVESHAFT INSTALLATION

Continuing from this procedure, you now have a U-joint installed at each end of the driveshaft assembly. Make a final inspection of the driveshaft, checking the following:

- Damage or dents on the driveshaft tubing.
- Spline movement with minimal drag.
- Both crosses should turn in their bearing cups without bind. A slight drag is acceptable (but no binding) but no looseness should be detectable.
- Yoke flanges should be free from burrs, paint, and contamination, any of which could prevent the bearing caps from sealing properly.

CAUTION

Once in use, bearing caps and their trunnions should remain matched. Also, never take assembly shortcuts by installing only the new bearing caps on a used trunnion because this will usually result in a rapid failure. Regard a U-joint cross, its four bearing assemblies, and mounting hardware as a unit, and replace as such.

Full-Round End Yoke

To continue installing a driveshaft with a full-round end yoke, use the following procedure:

1. First put the transmission in neutral and jack one drive axle wheel off the ground. This will allow you to rotate both the transmission output yoke and the differential carrier yoke. Turn each yoke so that the yoke bores are horizontal.
2. Swivel the U-joint cross on the driveshaft so that it is angled, as shown in **Figure 22–33A**. This will allow you to insert the trunnions into the yoke on the transmission and when both trunnions are located in the yoke bores, you can rest the weight of the driveshaft. Remember that a slip joint is always positioned so that it is closest to the source of powerflow, which means on the transmission side. If the driveshaft is heavy, use a sling to aid in handling and raising the driveshaft.
3. Apply a thin coating of antiseize compound to the outside of the two bearing cups. Now raise driveshaft slightly and move to one side of the yoke so that a trunnion protrudes through it as shown in **Figure 22–33B**. In this position, you can place a bearing cup over the protruding trunnion, align it, and press it home by hand into the yoke bore, as shown in **Figure 22–33C**.
4. Next slide the U-joint over to the opposite side so that the trunnion exits the bearing cup you have already installed by about half its length. Now align the second bearing cup over the trunnion first, then in the yoke bore, and press home until it is flush.
5. If a bearing cap assembly binds in the yoke bore during installation, you can tap it lightly with a hammer in the center of the cap plate. Never strike the bearing cup on the outer edges of the cap plate.
6. After both bearing cups have been seated in the yoke, put the lock plate tab in place and insert the bolts into the yoke threads, turning them in first by hand (see **Figure 22–33D**), then with a wrench. Torque to specification. Check that the bearing cup plates are flush with the yoke. Then back the fasteners off slightly.

FIGURE 22–33 Installing a driveshaft: (A) rotate the connecting yoke so that the lugs are horizontal; (B) center a trunnion in the yoke bore and install a bearing cup; (C) using a ball peen hammer, carefully tap the center of the bearing cup to seat it in the yoke bore; (D) install capscrews, lubricate, and torque screws to specs.

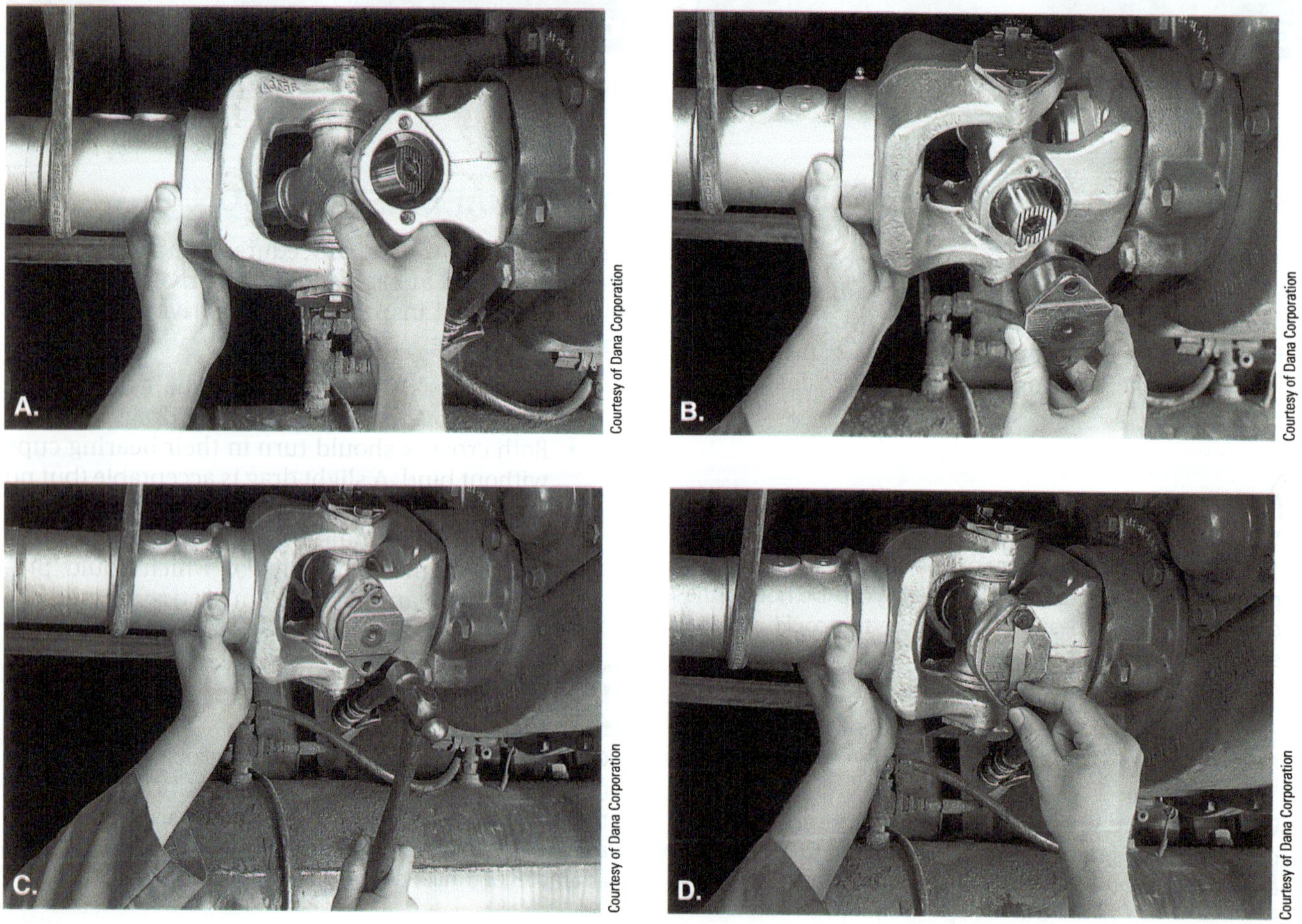

7. Install the zerk fitting(s) if you removed them and lubricate the U-joint assembly until the grease appears to exit at all four trunnion seals.
8. Torque the bearing cap bolts to specification. Next, bend the lock plate tabs up against the flat of the capscrew hex heads to lock them in place.
9. Repeat the process at the opposite end of the driveshaft. Do not forget to lube the slip joint when you have finished.

CAUTION

It makes sense to remove the grease fittings when installing a driveshaft. If they are knocked against the yoke, they tend to shear. You can easily reinstall them after the driveshaft has been installed.

Half-Round End Yoke

For driveshafts using half-round end yokes, first support the driveshaft more or less in position using slings. Then install the bearing cups onto the cross trunnions. Install the bearing cups into the end yoke shoulders and place the retaining straps over the bearing assemblies. Thread the self-locking capscrews into the threaded holes and torque the bolts to specification. Lubricate the cross and bearing assemblies.

Flange Yoke

Using slings to support the driveshaft, align the (permanent end) flange pilots of the driveshaft, flange yoke, and drive axle companion flange with each other. Align the bolt holes and then install bolts, lockwashers, and nuts to temporarily secure the driveshaft to the axle. Compress the slip joint assembly to position the opposite end of the driveshaft to the transmission companion flange. Align bolt holes and install bolts, lockwashers, and nuts. Torque the fasteners to OEM specifications.

22.5 CHASSIS VIBRATION DIAGNOSIS

Driveline vibrations are the source of many work orders written daily in truck shops and some of them can be very difficult to isolate. When investigating the source of a driveline vibration, you always should be aware that the cause can originate in an area other than in the driveline. It is probably not good practice to rely solely on a driver's report of a driveline vibration, so make a habit of road testing the vehicle.

The first challenge is to determine the source of the complaint. The cause of a vibration can be in the steering or suspension systems, in the engine or transmission systems or mounts, or in the wheels or tires. Road test the vehicle loaded and unloaded, if possible, while recording engine rpm and road speed. Note any irregularities and at what engine or road speeds they occur. If the problem is noticeable only when pulling a trailer, try coupling to a different trailer to see if the problem disappears. A diagnostic flowchart (**Figure 22–34**) can help locate the source of a driveline vibration.

DRIVELINE BALANCING

An unbalanced driveline causes transverse vibrations or bending movements in a driveshaft. This type of vibration is directly related to driveshaft rpm and usually is most noticeable at a specific driveshaft speed. The cause is imbalance in the driveshaft and, without using specialized equipment, it can be difficult to pinpoint.

Attempting to balance a driveshaft without spinning it up is so much of a hit-and-miss process that it is unrealistic to attempt it, given the cost of shop door rates today. Although some OEMs outline methods of achieving this in their service literature, it can take hours and produce no improvement. We outline here a simple method of dynamically balancing a driveshaft assembly using a balance sensor and strobe light.

Dynamic Balancing

You need a balance sensor and strobe light, preferably from a kit designed to dynamically balance driveshafts. The truck should be raised off the ground and placed on stands, ensuring that:

1. The frame is level.
2. The transmission and differential carrier housing(s) angles are the same when raised as when on the ground.
3. The chassis is secure enough on the stands so that the engine can safely drive the drivetrain to normal road speeds.

WARNING

Before running a stationary vehicle on stands at "road speed," ensure that any debris and rocks embedded in the tire treads are removed: Centrifugal force can loosen and propel rocks at deadly speeds from rotating tires. Ensure that all persons remain in front of the drive wheels while performing this test in a shop.

Fit the balance sensor to the chassis to be tested. The test procedure is similar to that used when balancing tire and wheel assemblies. Using the strobe light and white machinist's crayon, mark the driveshaft and spin it up. If out of balance on this initial test, begin by removing any counterweights tacked onto the driveshaft with a cold chisel. Remove any weld by lightly grinding with an angle grinder.

FIGURE 22–34 Chassis vibration diagnosis flowchart

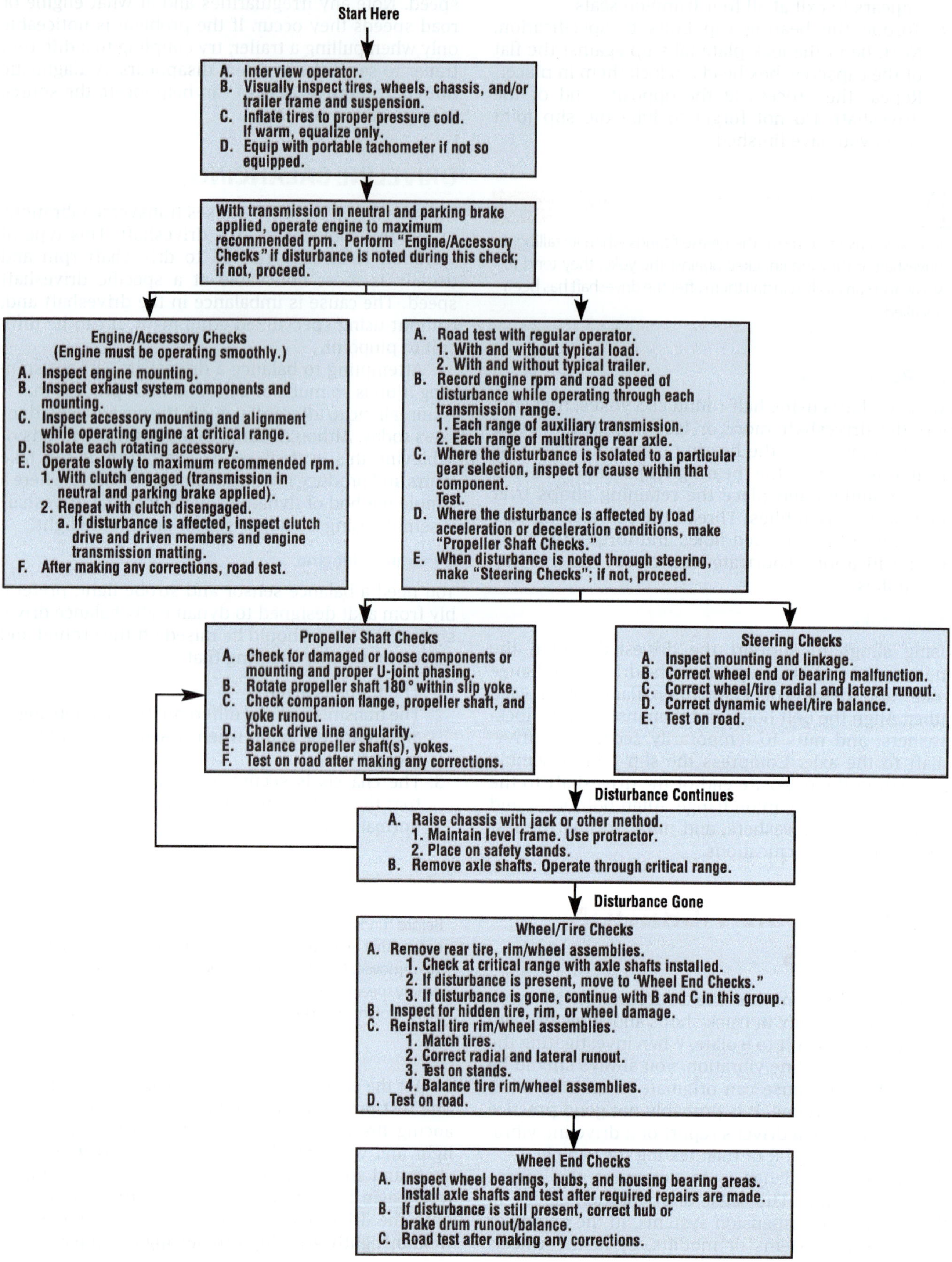

Run the vehicle at those equivalent road speeds noted on the road test, when the vibration was most noticeable. Use hose clamps to clamp counterweights (steel washers or block weights) into position with the clamp worm screw located over the weight washer, as shown in **Figure 22–35**. When the imbalance has been neutralized, tack weld the weight or weights into position. The weight of the weld tacks should be approximately equivalent to that of the hose clamp worm, so make a practice of always locating the worm over the balance weight.

CAUTION

Use small tack welds to attach weights. Larger welds can create another imbalance or, worse, distort driveshaft tubing.

Check Operating Angles

To determine whether vibrations are caused by improper driveline angles, run through the following routine:

1. Inflate all tires to the pressure at which they are normally operated. Park the vehicle on a surface that is level both from front to rear and from side to side. Do not attempt to level the vehicle by jacking up the front or rear axles. Shift the transmission into neutral and block the front tires. Jack up a rear drive wheel.
2. Rotate the wheel by hand until the output yoke on the transmission is vertical and lower the wheel you raised back to the shop floor. This simplifies measurement later. Check driveshaft angles in the same loaded or unloaded condition as when the vibration was noted. It is good practice to try to check driveline angles with the chassis loaded and unloaded.
3. To measure driveline angles, you can use a magnetic base protractor or an **inclinometer**, but the inclinometer is definitely preferred because it measures relative angles. The Spicer Anglemaster is an electronic inclinometer designed specifically to measure driveline angles; it is user-friendly and produces a digital readout. To use the magnetic base protractor or an electronic driveline inclinometer (**Figure 22–36**), place it on the component to be measured. A display window will show the angle and in which direction it slopes. Track the measured angles, working from the front of the drivetrain to the rear. A component slopes downward if it is lower at the rear than at the front. A component slopes upward when it is higher at the rear than it is in front.
4. Check and record the angle on the main transmission. This reading can be taken on the end yoke lug, with the U-joint cap removed, or on a flat surface of the main transmission parallel to the output yoke lug plane. Record your readings on a sketch of the driveline, for example, on one that looks like the diagram in **Figure 22–37**.
5. Now check the driveshaft angle between the transmission and drive axle carrier. On short tube length driveshafts, measure the angle of the driveshaft on either the tube or slip yoke lug with the bearing cap removed. On long tube length driveshafts, measure the angle on the tube at least 3 inches (75 mm) away from the yoke circumferential weld and at least 1 inch (25 mm) away from any balance weights, as shown in Figure 22–36. Make sure you remove any rust, scale, or flaking from the driveshaft tube so that you can obtain an accurate measurement.
6. Check the forward drive axle carrier input yoke angle by removing a bearing cap and measuring the angle of the yoke lug, or, alternatively, locate a flat surface on the axle housing parallel. Some differential carrier housings have a machined surface designed precisely for this purpose.
7. In tandem drive arrangements, measure the output yoke angle of the forward drive axle, the angle of the

FIGURE 22–35 Balancing a propeller shaft

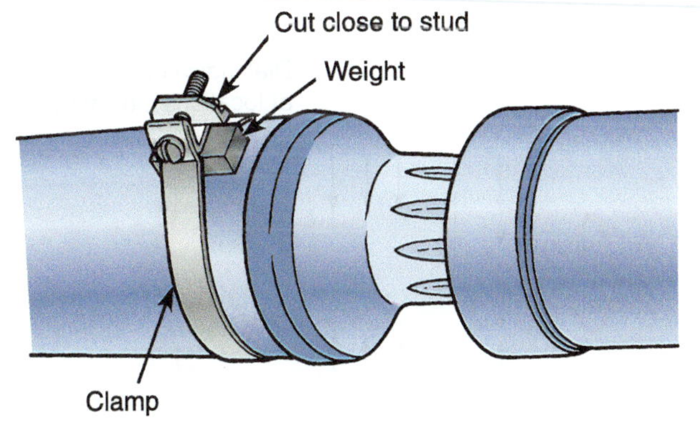

FIGURE 22–36 Checking the driveshaft angle—3 inches (75 mm) from the tube yoke weld

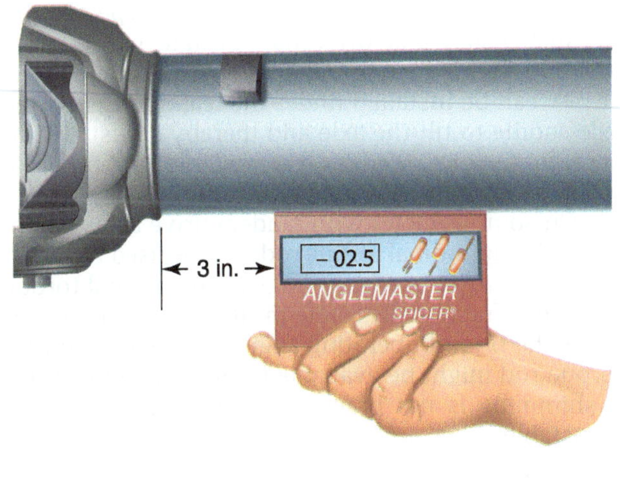

FIGURE 22–37 Method for calculating the driveshaft operating angles between the transmission and axle differential carriers

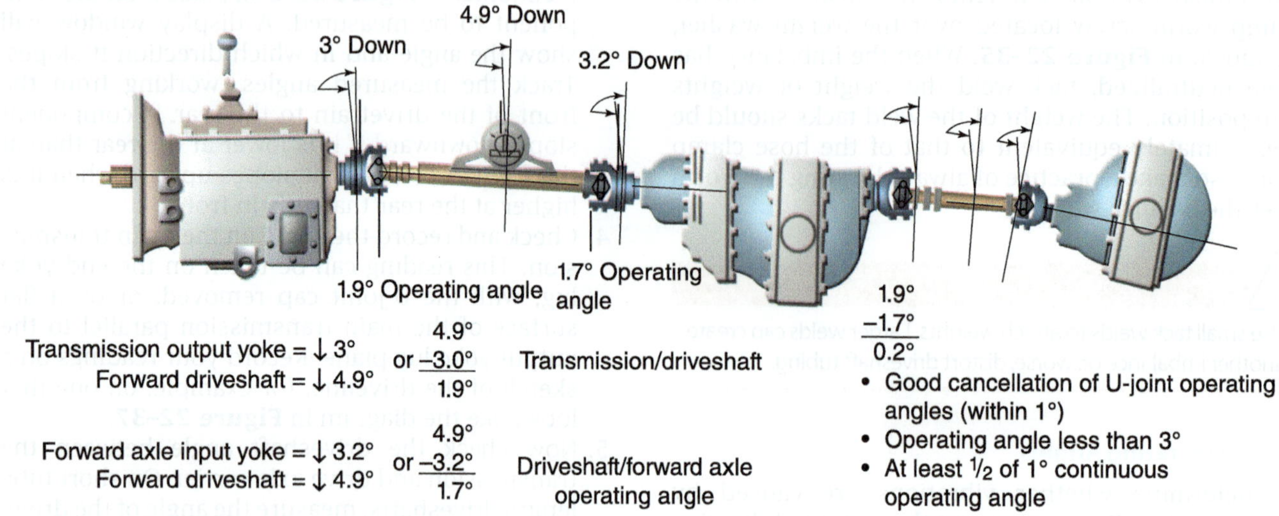

tandem driveshaft that connects the forward rear drive axle carriers, and the input yoke angle of the rear drive axle carrier. When you have measured all of the angles, record the data onto a drawing similar to that shown in Figure 22–37. Up to this point, you have not recorded any U-joint operating angles in the drawing, just the inclination of the components.

8. To determine U-joint operating angles, simply find the difference in inclination of the components. When the inclination occurs in the same direction on two connected components, subtract the smaller number from the larger to find the U-joint operating angle. When the inclination occurs in the opposite direction on two connected components, add the measurements to find the U-joint operating angle. Now compare the U-joint operating angles on the drawing to the guidelines provided earlier in Table 22–1.

Correcting U-Joint Operating Angles

The recommended method for correcting severe U-joint operating angles depends on the type of vehicle suspension and driveshaft design. On vehicles with a leaf spring suspension, thin wedges called **axle shims** can be installed under the leaf springs on the axle saddle to tilt the axle and thereby adjust U-joint operating angles. Wedges are available in a range of sizes to alter pinion angles.

On some vehicles with tandem drive axles, shimming of longitudinal torque rods can be used to adjust the drive carrier operating angle. Longitudinal torque rod shims fractionally rotate the drive axle pinion that alters the U-joint operating angle. A longer or shorter torque rod might be available from the manufacturer if shimming is not practical. Some torque rods are adjustable.

As a rule, the addition or removal of a ¼-inch (6 mm) shim from the rear torque arm will alter the axle angle approximately ¾ degree. A ¾-degree change in the pinion angle will typically change a U-joint operating angle by about ¼ degree. Factors that can cause the U-joint operating angle to change are:

- Suspension changes caused by worn bushings in the spring hangers, worn bushings in the torque rods, or incorrect air bag height
- Driveshaft adjustments
- Stretching or shortening the chassis
- Addition of an auxiliary transmission or transfer case into the driveline
- Worn engine or transmission mounts

Checking Driveshaft Runout

Any **runout** in a driveshaft can result in driveline vibrations. Driveshaft runout can be checked by using a dial indicator. **Figure 22–38** shows the locations used to measure driveshaft runout and the tolerance limits for

FIGURE 22–38 Allowable limits for total shaft runout

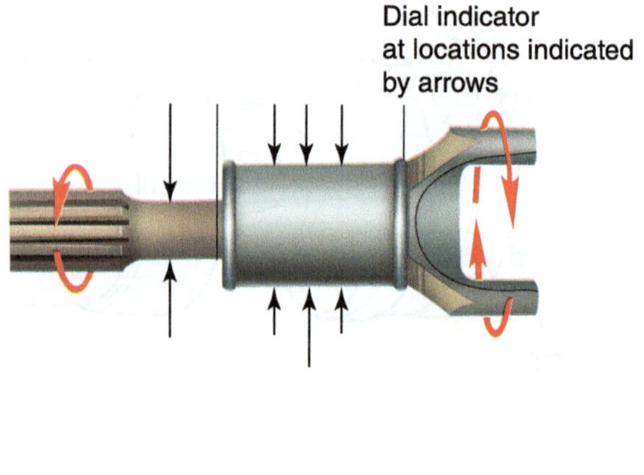

total indicated runout (TIR). Note that these are fine measurements. Before measuring, clean the surfaces from which you will take the dial indicator readings.

Runout versus Ovality

When checking for runout, it is important to distinguish between runout and ovality. Runout occurs when a driveshaft tube is slightly bent but still maintains its circularity throughout the tube. When you measure runout, the indicator needle will deflect once per full rotation. **Ovality** (**Figure 22-39**) occurs when a driveshaft tube is no longer circular but oval in shape. When indicating an oval driveshaft, the dial indicator needle will deflect twice per full revolution.

Even though a tube may be linearly straight, ovality will make it appear bent. A driveshaft tube with a small degree of ovality can be used. Check the OEM specifications, but usually up to a 0.010-inch (0.25 mm) TIR reading is acceptable. If this limit is exceeded, the driveshaft should be replaced.

YOKE RUNOUT AND ALIGNMENT

Not only must the driveshaft be straight but also the yokes attached to the transmission, auxiliary transmission, and drive axles. The assembly must be true and straight and in alignment with the shafts to which they are attached.

Checking Yoke Runout

Mount a dial indicator to the transmission or axle housing as shown in **Figure 22–40** and **Figure 22–41**. If the yoke is the half-round type, place the tip of the indicator measuring plunger onto the machined surface of the yoke shoulder. If the yoke is the full-round

FIGURE 22–39 Ovality should be kept to a minimum. Round and straight tubes are easier to balance and the least amount of correction weight will be required. (A) Bent driveshaft tube and (B) ovalled driveshaft tube.

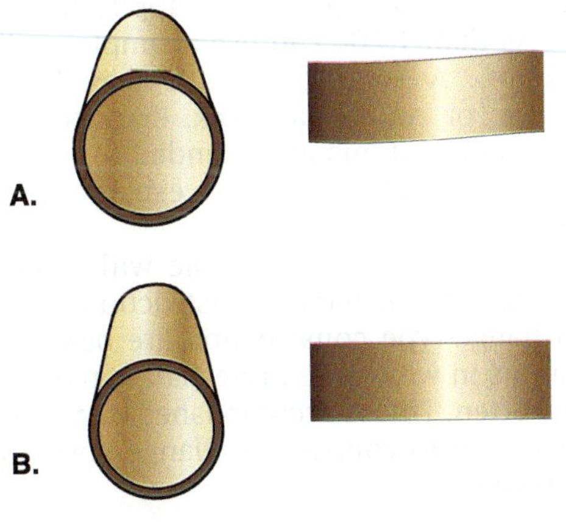

FIGURE 22–40 When checking half-round end yoke for runout, place the indicator plunger on smooth surface of shoulder.

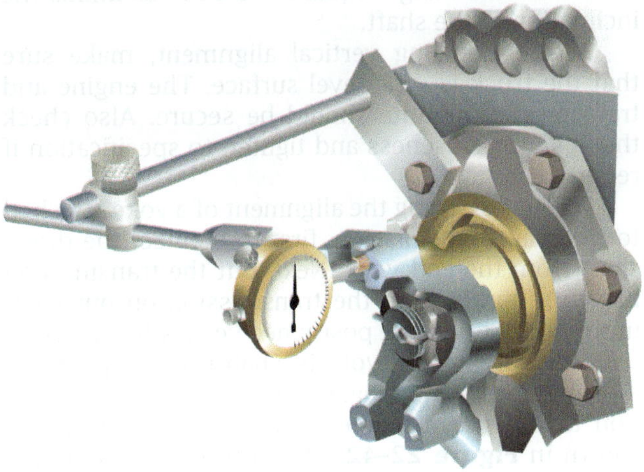

FIGURE 22–41 When checking a full-round end yoke for runout, be sure to leave enough surface for the indicator plunger.

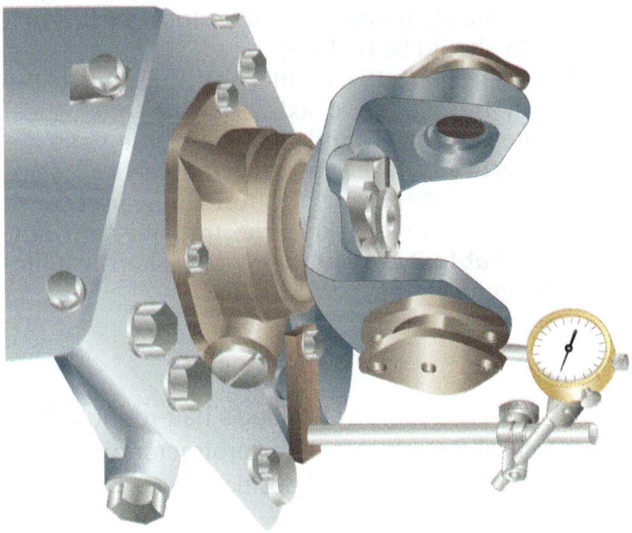

type, pop a bearing cup out just enough to insert the indicator plunger onto the cup as shown in Figure 22–41. Take the measurement and then rotate the yoke 180 degrees and measure again. If runout exceeds 0.005 inch (0.127 mm), the yoke should be replaced. Note that the reason for a yoke damaged in this manner is usually technician abuse caused by aggressive separation of a U-joint.

Checking Yoke Vertical Alignment

The machined shoulder of a half-round yoke or the yoke bores in a full-round yoke should be exactly

90 degrees to the centerline of the shaft to which the yoke is attached. If the shaft is horizontal (0 degrees), the yoke should be vertical (90 degrees); if the shaft angles away from a true horizontal plane, the yoke should be at an angle equal to 90 degrees minus the inclination of the shaft.

Before checking vertical alignment, make sure that the truck is on a level surface. The engine and transmission mounts should be secure. Also check the yoke for looseness and tighten to specification if required.

When measuring the alignment of a yoke attached to the main transmission, first disconnect the driveshaft from the end yoke. Next, put the transmission in neutral and rotate the transmission output shaft until the yoke lugs are positioned vertically one above the other. Then if the yoke is a half-round type, place a protractor with a magnetic base in a vertical position across the machined surfaces of the yoke, as shown in **Figure 22–42**. The protractor dial should read 90 degrees minus the inclination of the engine/transmission plane.

If the yoke is the full-round type, place the protractor base on the outside surface of the machined shoulder as shown in **Figure 22–43**. The protractor should read the angle of the engine/transmission inclination. If the output yoke is more than ½ degree out of vertical alignment, it is either loose or it is distorted and should be replaced.

When checking the input yoke attached to the pinion shaft of a drive axle, jack up one wheel and rotate the wheel until the lugs of the yoke are aligned vertically. Then measure the vertical alignment of the yoke, using the foregoing procedure just described for checking the transmission output yoke. **Table 22–4** outlines a driveshaft troubleshooting guide.

FIGURE 22–42 When checking a half-round yoke for vertical alignment, place the protractor on the smooth surface in a vertical position.

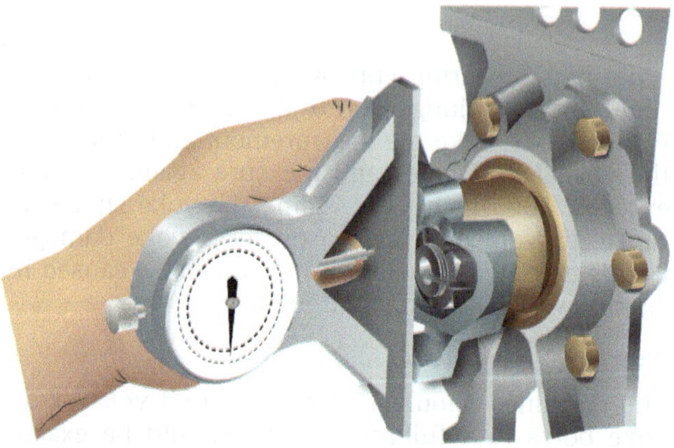

FIGURE 22–43 When checking full-round end yoke bore for vertical alignment, place the protractor on the outside of the shoulder.

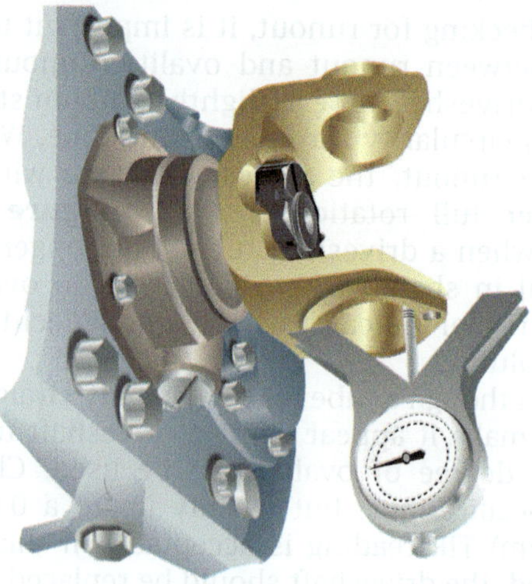

POWER TAKEOFF (PTO) DRIVESHAFTS

In some applications, an auxiliary power unit, such as a pump, can be directly mounted to the power takeoff (PTO) assembly, as shown in **Figure 22–44A**. However, it is more common to locate PTO-driven units remotely and drive them using a driveshaft as shown in **Figure 22–44B**. A PTO driveshaft transmits torque from the PTO output to the input flange or yoke of the driven accessory. The driveshaft should be tough enough to handle the torque and rpm required by the accessory with a margin of safety for any shock loads that might result.

PTO driveshafts are similar to a vehicle powertrain driveshaft. U-joints and slip joints allow the working angles between the PTO and the driven accessory to change because of movements in the powertrain from torque reactions and chassis deflections. PTO driveshafts designed for high-torque applications operating at more than 1,200 rpm are made of tubular steel.

Driveshafts designed for applications where rpm is less than 1,200 can use a simpler, solid shaft construction. A solid driveshaft may be hexagonal, square, or round bar-stock. Because most PTO driveshaft applications are intended for intermittent service, precision-balanced driveshafts are seldom required.

PTO driveshafts are available with coupling fittings using yoke or flange construction. Driveshafts using a flange-type construction are best suited for applications in which the driveshaft is required to be removed often. This is accomplished by removing the fasteners used to connect the flange yoke and companion flange.

TABLE 22–4 Driveshaft Troubleshooting Guide

Symptom	Probable Cause	Remedy
Vibration —low gear shudder —at certain speeds under full drive or full coast —under light loaded conditions	1. Secondary couple load reaction at shaft support bearing 2. Improper phasing 3. Incompatible driveshaft 4. Driveshaft weight not compatible with engine transmission mounting or ` driveshaft too long for speed 5. Loose outside diameter fit on slip spline 6. Excessively loose U-joint for speed 7. Driveshaft out of balance; not straight 8. Unequal U-joint angles 9. U-joint angle too large for continuous running 10. Worn U-joint 11. Inadequate torque on bearing plate capscrews 12. Torsional or inertial excitation	1. Reduce U-joint continuous running angle. Check with transmission or axle manufacturer—replace shaft bearing 2. Check driveshaft for correct yoke phasing 3. Use larger diameter tube 4. Install two-piece driveshaft with shaft support bearing 5. Replace spline or yoke 6. Replace U-joint 7. Straighten and balance shaft 8. Shim drivetrain components to equalize U-joint angles 9. Reduce U-joint continuous running angle 10. Replace U-joint 11. Inspect U-joint flex effort for looseness; torque to specification 12. Straighten and balance shaft. Check driveshaft for correct yoke phasing
Premature wear —low mileage U-joint wear —repeat U-joint wear —end galling of cross trunnion and bearing assembly —needle rollers brinnelled into bearing cup and cross trunnion —broken cross and bearing assemblies	1. End yoke cross hole misalignment 2. Excessive angularity 3. Improper lubrication 4. Excessive U-bolts torque on retaining nuts 5. Excessive continuous running load 6. Continuous operation at high angle/high speed 7. Contamination and abrasion; worn or damaged seals 8. Excessive torque load (shock loading) for U-joint and driveshaft size	1. Use alignment bar to check for end yoke cross hole misalignment; replace end yoke if misaligned 2. Check U-joint operating angles with an electronic driveline inclinometer; reduce excessive U-joint operating angles 3. Lubricate according to specifications 4. Replace U-joint kit 5. Reduce U-joint continuous running angle 6. Replace with higher capacity U-joint and driveshaft 7. Check U-joint flex effort; replace joint or yoke if necessary; clean and relubricate U-joint 8. Realign to proper running angle minimum 1 degree; torque bearing retention method to specification
Slip spline wear —seizure —galling	1. Improper lubrication	1. Lubricate slip spline according to specifications

FIGURE 22–44 (A) Some accessories are mounted directly to the PTO; (B) others are remote-mounted and driveshaft driven.

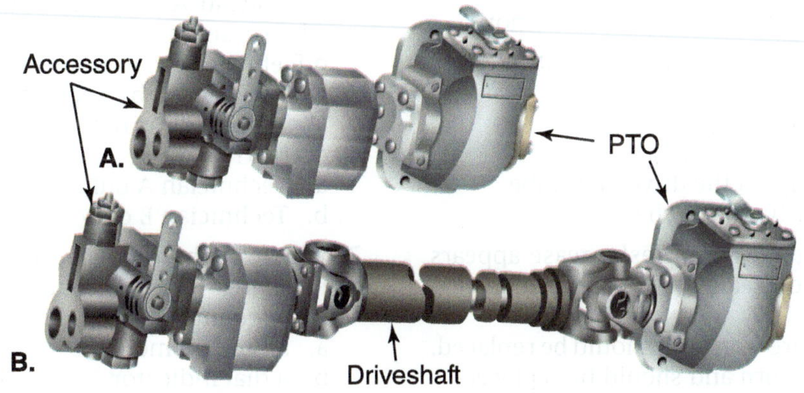

SUMMARY

- The components of a driveshaft assembly include the following:
  - Yokes
  - Universal joints
  - Slip spline
  - Hanger bearing
  - Propeller shaft(s)
- The purpose of U-joints is to connect drivetrain components while allowing the driveshaft(s) to operate at changing angles.
- Two types of driveline arrangements are used to transmit torque to the drive axle carriers: the parallel-joint type and the nonparallel or broken-back driveshaft type.
- Because of U-joint action, when a drive yoke turns at a constant rpm, the driveshaft speed increases and decreases twice per revolution. This fluctuation in driveshaft speed requires the U-joints to be in phase.
- When the length of a single section of driveshaft tubing exceeds 70 inches (180 cm) a hanger bearing should be used.
- Driveline vibrations are caused by loose yokes, excessive driveshaft radial runout, slip spline joint radial play, bent driveshaft tubing, and failing U-joints.
- A U-joint puller should be used to separate a U-joint from a yoke assembly. Avoid using high-impact tools such as sledge hammers because driveshaft damage can result.
- After removing the U-joint cross and bearing cups from a yoke, inspect the yoke bores for bore alignment, damage, and metal burrs. Burrs can be removed with a rat tail or half-round file and emery cloth.
- Measure the yoke bores for wear using a go-no-go wear gauge and check yoke bore alignment using a machined alignment bar.
- When diagnosing driveline vibrations, it should be noted that they often originate elsewhere in the chassis.
- An electronic inclinometer is the preferred instrument for measuring driveline component angles. A protractor also may be used.
- Driveline problems are common reasons for opening a work repair order. They are also the source of many questions in certification testing. Become familiar with the driveline troubleshooting guide at the end of this chapter.

REVIEW QUESTIONS

1. U-joint failures, hanger bearing failures, and driveline vibrations are caused by:
   a. loose yokes
   b. bent driveshafts
   c. missing balance weights
   d. all of the above
2. A truck driver complains of a driveline vibration. Technician A says that the problem might be a loose driveshaft yoke. Technician B says that the problem could be worn splines in the slip joint. Who is correct?
   a. Technician A only
   b. Technician B only
   c. both A and B
   d. neither A nor B
3. Which of the following would be least likely to affect driveshaft balance?
   a. missing balance weights
   b. missing U-joint zerk fitting
   c. asphalt buildup on the driveshaft tube
   d. dents in the driveshaft tube
4. When lubricating a U-joint, fresh grease appears at all four trunnion seals. Which of the following is probably true?
   a. The bearings are worn and should be replaced.
   b. The seals are worn and should be replaced.
   c. The trunnions are worn and the U-joint should be replaced.
   d. The U-joint has been properly lubricated.
5. The OEM recommendation for U-joint lubrication interval scheduling on a truck used for an on-highway operation should be:
   a. every 10,000 to 15,000 miles (16,000 to 24,000 km)
   b. once a year
   c. 3,000 to 5,000 miles (4,800 to 8,000 km)
   d. every 50,000 miles (80,000 km)
6. Technician A says that a driveshaft is in phase when assembled so that the yokes are 90 degrees offset. Technician B says that driveshaft phasing will be altered by separating a driveshaft at the slip joint and turning by one spline tooth. Who is correct?
   a. Technician A only
   b. Technician B only
   c. both A and B
   d. neither A nor B
7. Which of the following tools should be used to measure the inclination of a driveline component?
   a. an inclinometer
   b. a dial indicator
   c. a go-no-go gauge
   d. a slide hammer

8. Technician A says that when lubricating a U-joint and grease does not exit from one of the four trunnion seals, the U-joint must be immediately replaced. Technician B says that backing off the bearing cap on a U-joint sometimes helps a trunnion to take grease. Who is correct?
   a. Technician A only
   b. Technician B only
   c. both A and B
   d. neither A nor B
9. A driveshaft is in phase when:
   a. the driveshaft yoke is aligned on the same plane as the transmission yoke.
   b. the slip yoke and driveshaft tube yoke lugs are perfectly aligned.
   c. the driveshaft is perfectly balanced.
   d. all of the above.
10. The U-joint working angles on most truck driveshaft assemblies should generally not exceed:
   a. 3 degrees
   b. 5 degrees
   c. 8 degrees
   d. 10 degrees
11. Which of the following tools is preferred when separating a U-joint from a driveshaft?
   a. a sledge hammer
   b. a slide hammer
   c. a hydraulic jack
   d. a puller
12. Technician A says that driveshaft working angles can be adjusted by adding wedge shims to the spring seats of a leaf spring suspension system. Technician B says that loose engine or transmission mountings can cause excessive driveline angularity resulting in vibration. Who is correct?
   a. Technician A only
   b. Technician B only
   c. both A and B
   d. neither A nor B
13. Grooves worn into the trunnions by the bearings are called:
   a. brinnelling
   b. galling
   c. spalling
   d. pitting
14. Technician A says that a driveline vibration can be caused by worn clutch components. Technician B says that a truck should be road tested both loaded and unloaded to help determine the source of a driveline vibration. Who is correct?
   a. Technician A only
   b. Technician B only
   c. both A and B
   d. neither A nor B
15. What is the function of a driveshaft slip spline joint?
   a. It allows the driveshaft to change length while transmitting torque.
   b. It accommodates variations in driveline angles.
   c. It multiplies drive torque.
   d. All of the above.
16. What is the most common driveshaft series rating used on linehaul highway tractors?
   a. 1500 series
   b. 1600 series
   c. 1700 series
   d. 1800 series
17. For which of the following is an Anglemaster used?
   a. Checking shaft runout
   b. Checking driveshaft angles
   c. Measuring phase angle
   d. all of the above
18. Which of the following correctly describes an Anglemaster instrument?
   a. electronic inclinometer
   b. electronic caliper
   c. dynamic balance tool
   d. linear measurement tool
19. Technician A says that new stub shafts are shipped coated with blue plastic that should be completely removed prior to assembly. Technician B says that male slip spline stubs are coated with Glidecoat because it is wear resistant and lasts the life of the driveshaft. Who is correct?
   a. Technician A only
   b. Technician B only
   c. both A and B
   d. neither A nor B
20. What is the minimum total number of driveshaft U-joints that would have to be used on a tandem drive axle truck equipped with a hanger bearing on the front driveshaft?
   a. two
   b. three
   c. five
   d. six

23

Prerequisite: Chapter 22

HEAVY-DUTY TRUCK AXLES

OBJECTIVES

After reading this chapter, you should be able to:

- Identify the types of axles used on trucks and trailers.
- Define the terms dead axle, live axle, pusher axle, and tag axle.
- Outline the construction of a drive axle carrier assembly.
- Explain how a pinion and crown gear set changes the direction of powerflow.
- Describe differential action and list the reasons it is required.
- Identify the components required to create differential action.
- Describe the operation of the various drive axle configurations.
- Identify the components used in an interaxle differential or power divider.
- Explain how an interaxle differential lock functions.
- Define the term spinout and explain how it is caused.
- Trace the powerflow path through different types of differential carriers.

KEY TERMS

amboid gear set
axle range interlock
axle shaft
banjo housing
controlled traction
crown gear
dead axles
differential
differential carrier
differential lockout
double-reduction axle
final drive carrier
full floating
hypoid gear set
interaxle differential
interaxle differential lock
live axle
pinion
planetary double-reduction axle
power divider
pusher axle
single-reduction axle
spigot
spinout
spiral bevel gear
tag axle
tandem drive axle
torque proportional differential

INTRODUCTION

Take a simple vehicle, such as the wagon you had as a kid, and you can observe that both axles play a role in supporting any load you want to haul. You also will see that the forward axle has an additional role because it is used to steer. You may have graduated to a go-kart as you got older; because a go-kart is self-propelled, one of the axles is given the role of driving the wheels.

All axles support wheel assemblies but we categorize them by the other roles they play. The axles used on transport trucks are a little more sophisticated than those on children's toy vehicles, but they can still be grouped into three general categories:

- Axles that support vehicle load only. We see these on trailers, and lift and tag axles also fall into this category.
- Steering axles responsible for supporting vehicle load and steering. There are

different types of steering axles, including single steer axles used to steer most highway vehicles, double steer axles used on some vocational trucks, and trailing steer axles used in some trailer applications. Steer axles may also be drive axles on vocational trucks.

- Drive axles responsible for supporting vehicle load and drive wheels to propel the vehicle. Drive axles house drive gearing that delivers drive torque to wheels. Drive axles used on vocational trucks may also be steer axles.

TYPES OF AXLES

The role of drive axles is necessarily a little more complex because some gearing is required; first, to change the drive direction and, second, to reduce the speed ratio suitable for turning the drive wheels. Some refer to drive axles as **live axles** for this reason. A typical drive axle contains **axle shafts** and a **differential carrier** assembly. Drive torque from the carrier is transmitted to drive wheels by means of the axle shafts, also known as *half shafts* and *drive shafts*. You can see a couple of different types in **Figure 23–1**. A **differential** is required in any rigidly mounted drive axle to accommodate the different wheel speeds that result when a vehicle corners, causing the outboard wheel to "travel" further than the inboard wheel. Drive axles are usually the rear axle assemblies on trucks.

Steering and trailer axles are used to support the weight of the vehicle and its load. Sometimes these are referred to as dead axles. All axles on modern highway vehicles are suspension mounted. The suspension acts as an intermediary between the vehicle frame and the axle assembly. It provides some forgiveness between the road surface and the ride of the vehicle, improving comfort and handling. Steering axles are typically located in the front of the vehicle and, in most highway trucks, one steering axle is sufficient to support the forward weight and steer. An example of a truck steering axle is shown in **Figure 23–2**.

Commercial highway vehicles use a variety of axles that simply support load. These include truck rear dead axles, lift axles, and the range of axles used on commercial trailers both fixed and lift. Any axle used on a vehicle must be capable of supporting the load applied to it and in most cases will do this for the life of the vehicle. **Figure 23–3** shows an example of a typical trailer axle.

In this chapter, we will take a look at the operating principles of all three categories of axles, beginning with drive axles. Troubleshooting and repair of drive axles are handled separately in Chapter 24 and wheel end procedure is dealt with in Chapter 27.

FIGURE 23–1 Examples of heavy-duty axle assemblies: (A) rear driving; and (B) front driving

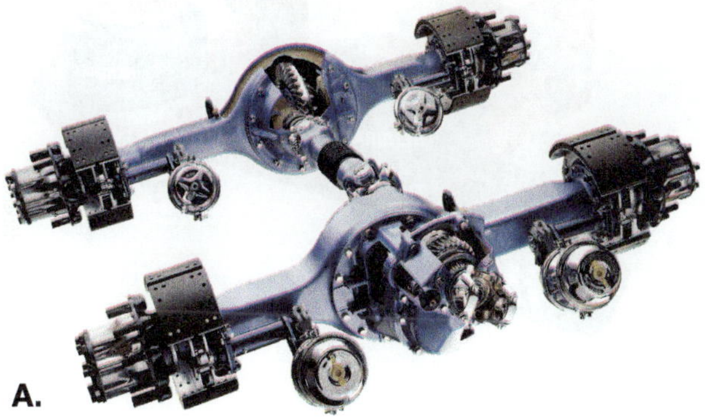

A.

Courtesy of Dana Corporation

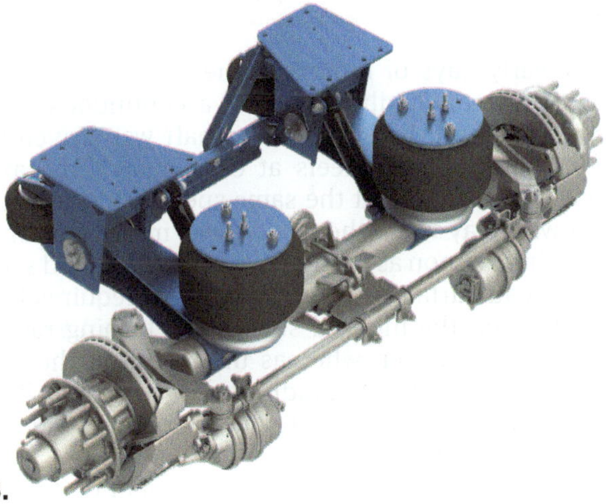

B.

Courtesy of Ingersoll Axles

FIGURE 23–2 Typical nondriving steering axle

Courtesy of Hendrickson

FIGURE 23–3 Tube beam trailer axle

23.1 DRIVE AXLES

In the early days of motoring, the driving wheels of a vehicle were rigidly geared to a continuous, one-piece axle shaft. When the axle shaft was driven by engine torque, the wheels at each end of the axle turned together and at the same speed. This arrangement was okay when the vehicle was moving straight ahead, but as soon as it turned, the wheels would scuff on the road surface. A vehicle in a turn requires that the wheel on the inboard side of the turning radius travel a little slower, whereas the wheel on the outboard side of the turning radius needs to travel a little faster. A better system was required and the need was met by differential gearing.

Differential gearing splits driveline torque between the two drive wheels on an axle. Each drive wheel separately connects to differential gearing by means of a half shaft. Drive torque is delivered to the drive axle by means of a driveshaft. The driveshaft turns a **pinion** in mesh with a **crown gear** (the term *ring gear* is also widely used) that changes the direction of the powerflow by 90 degrees. A differential is so called because it differentiates between the actual turning speed of each wheel with a bias to the wheel that offers the least resistance to turning effort.

The powertrain in a truck begins at the engine and ends at the drive wheels. Engine torque is transmitted through a transmission and driveshaft(s) to the **final drive carrier**(s) at the axles. The final drive carrier is another way of saying differential drive carrier. We use both terms in this chapter. At the final drive carrier, input torque is at right angles to the direction that the wheels turn in. This requires that the powerflow be turned 90 degrees so that the drive axle shafts turn in the same direction as the wheels. The gear set in the final drive axle carrier is designed to achieve this. You can see a typical drivetrain powerflow in **Figure 23–4**.

FIGURE 23–4 The drivetrain powerflow of a truck equipped with a single rear drive axle.

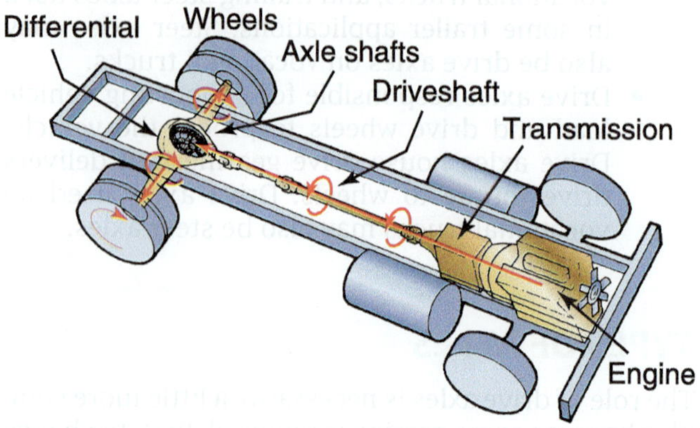

DIFFERENTIAL CARRIER CONSTRUCTION

A differential carrier consists of a number of major subcomponents, as shown in **Figure 23–5**. These include:

- Input shaft and pinion gear
- Crown or ring gear
- Differential spider
- Axle shafts or half shafts

FIGURE 23–5 Cutaway view of a differential carrier assembly showing the major components and the path of power flow through the unit

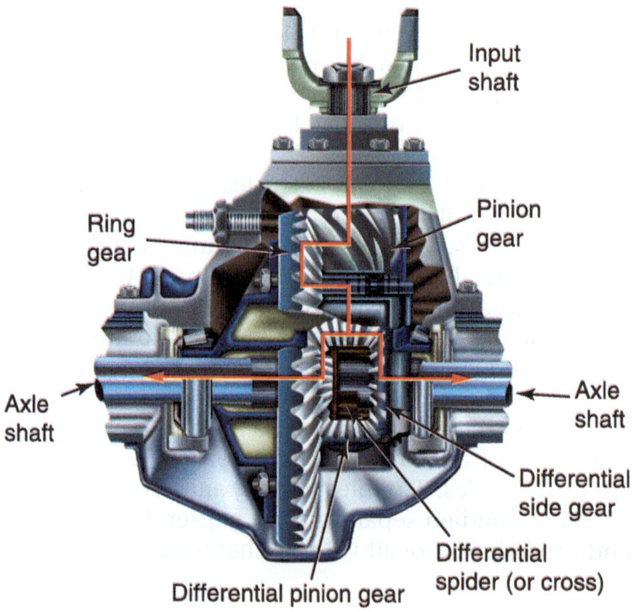

> **SHOP TALK**
>
> Each original equipment manufacturer (OEM) tends to use different words to describe the same components, but in this text we try to use the most common term. A crown gear is also a ring gear, but we usually use the first term. An axle shaft also can be known as a half shaft or drive shaft; once again, we usually use the first term. A drive shaft connects a drive wheel to a differential, whereas a driveshaft is a propeller shaft that delivers input torque to the final drive carrier. Confused? You bet but, nevertheless, you should try to become familiar with all the OEM terms so that you can accurately interpret service literature.

You can see in Figure 23–5 that the input yoke is splined to the input shaft. The pinion gear meshes with the crown gear and because they engage at right angles, the direction of torque flow is also changed by 90 degrees. The differential assembly is driven by the crown gear and the two axle shafts are turned by the differential gearing. **Figure 23–6** is an exploded view of a differential assembly showing the differential pinions and side gears. The differential side gears are machined with internal splines to which the axle shafts mesh. All of the gear components in a typical differential carrier are shown in exploded format in **Figure 23–7**.

FIGURE 23–6 Components of a typical differential

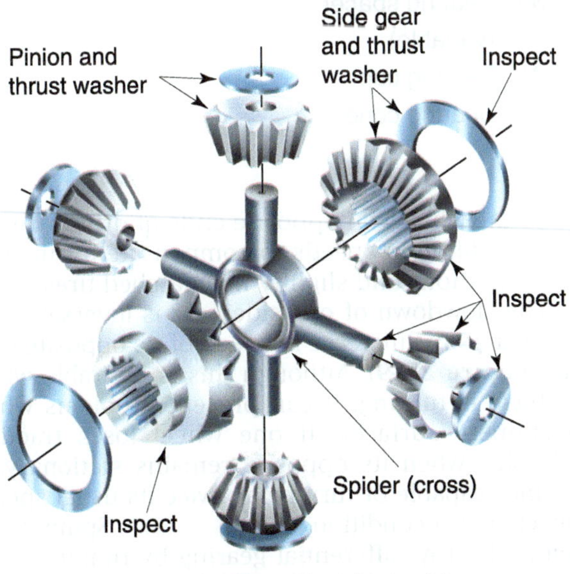

The gear components are held in position by a number of bearings and thrust washers. The final drive carrier assembly is mounted in the differential carrier (see number 29 in Figure 23–7). The differential carrier is itself flange mounted to the **banjo housing** in the drive axle assembly. The banjo housing is named because of its shape. After a differential carrier has been installed into the banjo housing, the axle shafts are then inserted into either side of the drive axle housing to engage with differential gearing.

DIFFERENTIAL OPERATION

When driveshaft torque is applied to the input shaft and drive pinion, they rotate in a direction perpendicular to the truck drive wheels. The drive pinion is beveled and tooth meshes to the crown gear. The crown gear is also beveled at a complementary angle to the pinion. The result of the pinion and crown arrangement is to change the torque flow path at right angles, enabling a longitudinal driveshaft to drive transverse axle shafts.

When the crown gear rotates, the differential case rotates with it because they are bolted together. The spider (or *cross*) mounted in the differential case halves also rotates with the crown gear. Differential pinions are mounted on protruding shafts on the differential spider. The pinions revolve on the spider shafts. The four spider pinions mesh to side gears. The side gears have internal splines to which the axle shafts mesh. As the side gears rotate, they transfer torque to the axle shafts, which in turn transfer driving torque to the drive wheels.

When the vehicle is moving in a straight-ahead direction, the ring gear, differential gears, and drive axles rotate at the same speed. In other words, the pinions rotate with the spider, acting as a sort of bridge between the spider and side gears. You can see this in **Figure 23–8A**.

Differential Action During Turns

When a truck moves forward in a straight-ahead direction, all the wheels should turn at the same speed, assuming they are the same size. But in a turn the outer wheel has to track through a larger turning radius than the inside wheel, meaning that it has to turn a little faster. Nondriven wheels rotate independently of one another on their own wheel ends, so they present no problems in turns, but driven wheels are in geared-mesh with the driveline. So, in a turn,

FIGURE 23–7 Components of a differential carrier assembly

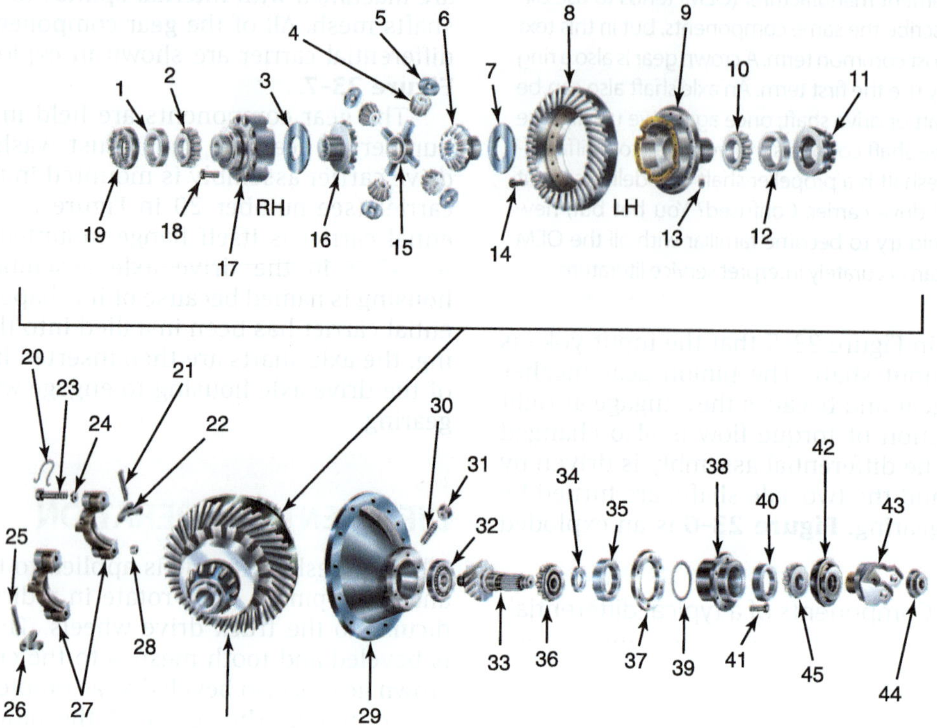

1. Bearing cup
2. Bearing cone
3. Thrust washer
4. Side pinion
5. Thrust washer
6. Side gear
7. Thrust washer
8. Ring gear
9. Flanged half differential case
10. Bearing cone
11. Bearing adjuster
12. Bearing cup
13. Nut
14. Bolt
15. Spider
16. Side gear
17. Plain half differential case
18. Cap screw
19. Bearing adjuster
20. Lock wire
21. Cotter pin
22. Adjuster lock
23. Cap screw
24. Flat washer
25. Cotter pin
26. Adjuster lock
27. Bearing cap
28. Dowel bushing
29. Differential carrier
30. Thrust screw
31. Locknut
32. Pilot bearing
33. Drive pinion
34. Bearing spacer (variable)
35. Bearing cup
36. Bearing cone
37. Bearing cage shim
38. Bearing cage
39. O-ring
40. Bearing cup
41. Cap screw
42. Oil seal
43. Yoke
44. Pinion nut
45. Bearing cone

the gearing has to be able to accommodate the wheel speed differential that results.

As a vehicle turns into a corner, the differential side gear on the inside of the turn slows. This causes the pinion gears to rotate or walk on the spider shafts. When the spider pinions start to rotate, this adds speed to the opposite side gear. The reduction in speed of one side gear is exactly equal to the increase in speed of its opposite. You can see this in **Figure 23–8B**. During this differential action, equal driving force is transmitted to each wheel. Differential action makes the vehicle easier to control and prevents tire scuffing during turns. A differential also allows the wheels on opposite ends of drive axles to rotate at different speeds to compensate for uneven road conditions and slightly mismatched tires.

The slowdown of one side gear is inversely proportional to the increase in speed of its opposite side gear (**Figure 23–9**). Although this is desirable when a vehicle is turning, it can present problems when on slippery surfaces. If one wheel loses traction and spins when its opposite remains stationary, it becomes capable of turning at twice its usual speed. This creates a condition called **spinout**. Spinout can rapidly destroy differential gearing by running it up at speeds at which it was not designed to operate.

FIGURE 23-8 (A) When traveling in a straight line, the pinions rotate with the spider, acting as a bridge between the spider and side gears. (B) When turning a corner, the side gear on the inside slows down, causing the pinions to rotate on the spider legs, which adds speed to the opposite side gear.

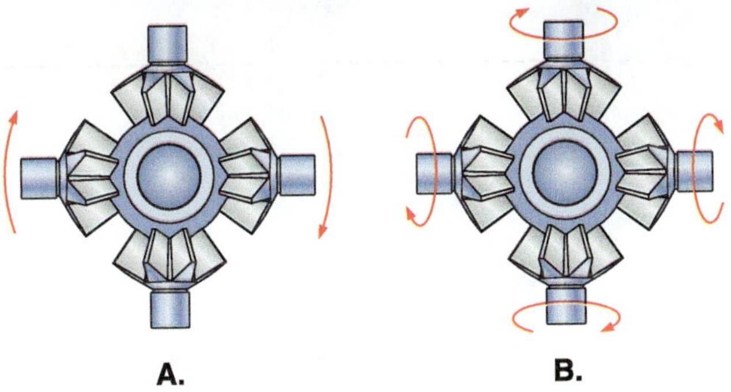

FIGURE 23-9 Turning a corner differential action

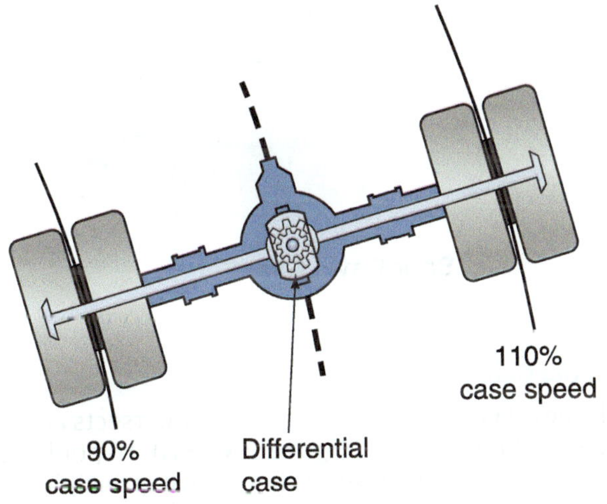

FINAL DRIVE AXLE CONFIGURATIONS

Medium-duty trucks are typically equipped with a single rear drive axle, whereas most heavy-duty highway trucks and tractors use tandem drive systems.

Tandem drive systems require two drive axles. They are used to accommodate the high gross vehicle weight (GVW) of larger trucks. Whether a single drive axle or tandem drive axle configuration is used, both function in a similar manner.

The next design factor in drive axles is the number of gear reductions built into the differential carrier gearing. Some of the more common drive axle configurations and gearing is as follows:

- A **single-reduction axle** is a differential carrier that uses only one gear reduction. The

FIGURE 23-10 Gear reduction is accomplished with a pinion gear and ring gear arrangement.

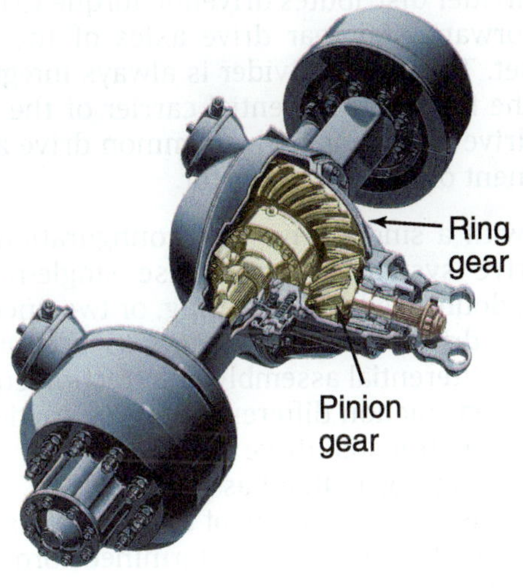

gear reduction is a function of the pinion gear and crown gear ratio. **Figure 23-10** shows a cutaway view of a commonly used single-reduction truck axle.

- A **planetary double-reduction axle** uses two separate gearing reductions. One gearing reduction is achieved by the pinion and crown gear ratio, whereas the other gear reduction is achieved by a planetary gearset.
- A two-speed axle assembly (often referred to as a dual range unit) has two different output ratios from the differential. The driver selects the ratios from controls located in the cab of the truck.
- A **tandem drive axle** (**Figure 23-11**) combines two single axle assemblies through the

FIGURE 23-11 Tandem drive axle assembly

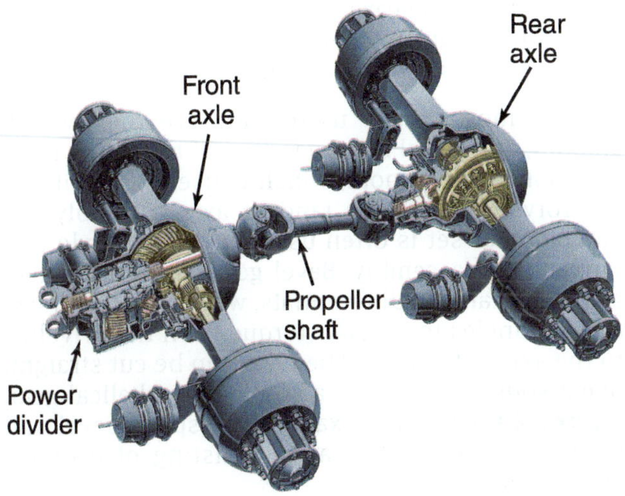

use of an **interaxle differential** or **power divider** and an interconnecting driveshaft that connects the two drive axles. The power divider distributes driveline torque to both the forward and rear drive axles of the tandem set. The power divider is always integral with the forward differential carrier of the tandem drive unit. This is the common drive arrangement of a highway tractor.

As with a single drive axle configuration, a tandem drive system unit may use single-reduction gearing, double-reduction gearing, or two-speed gearing. Some drive axle arrangements use a **controlled traction** differential assembly or a traction equalizer. A controlled traction differential uses a friction plate assembly to transfer drive torque from a slipping wheel to its opposite that has good traction. The friction plate assembly consists of a multiple disc clutch designed to slip when a predetermined torque value is reached.

Differential lockout assemblies control the operation of an interaxle differential used on a tandem drive axle. The objective is to temporarily lock out differential action to improve traction. Operation of the lockout system is remote controlled by the truck driver from the cab. Activation is usually by air pressure but vacuum and electrically activated systems exist. On two-speed differential carrier assemblies, the ratio is remote driver controlled. Two types of axle ratio shift systems are used: air shift and electric shift.

FIGURE 23–12 Bevel gear sets

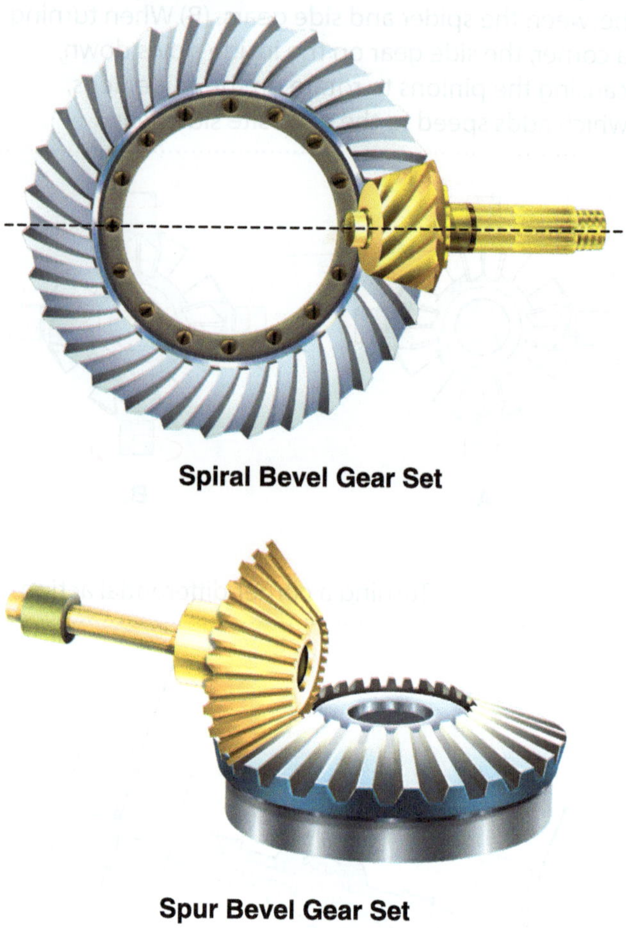

DIFFERENTIAL GEARING TYPES

A variety of gear types and arrangements are used in truck differential carriers. The common gear types include the following:

- Hypoid gearing
- Amboid gearing
- Spiral bevel gearing
- Spur bevel gearing
- Underslung worm drive gearing
- Overhead worm drive gearing

The most widely used gears are the hypoid, amboid, and spiral bevel gearing. Of these, the most common is the hypoid, which is used primarily on the forward axle of a tandem axle assembly. An amboid gear set is often used on the rear axle of a tandem axle assembly. Bevel gears are used extensively in rear axle differentials, where they intersect at right angles to transfer torque from the driveline to the axle half shafts. The teeth can be cut straight, like a spur gear, or with a spiral, like a helical gear. **Figure 23–12** shows examples of spiral bevel and spur bevel gear sets, each consisting of a crown and pinion.

Hypoid Gear Sets

Like bevel gears, a **hypoid gear set** intersects at right angles when meshed (**Figure 23–13A**). Hypoid gearing uses a modified spiral bevel gear principle that allows several gear teeth to absorb the driving force and allows the gears to run quietly. A hypoid gear set is typically found at the drive pinion gear and crown gear interface. This gear arrangement is designed so that the axis of the pinion gear does not intersect the axis of the crown gear but slightly below it. In addition, the teeth on the drive pinion gear and crown gear are cut spirally. Because of the curvature of the gear teeth, a sliding wiper action results between the two gear interfaces.

In a hypoid gear set, the pinion gear axis sits below the centerline of the crown gear axis, allowing the driveshaft to be mounted in a lower position. Because of this, hypoid differential carriers have a greater torque capacity than equivalent spiral bevel gear set carriers.

The hypoid pinion gear is typically larger in diameter and has larger teeth when compared with the construction of a spiral bevel gear. Hypoid pinion gears also have a larger tooth contact area. Their design means that, in operation, more teeth remain in contact longer with the crown gear teeth as the gears turn.

Hypoid gear sets drive forward on the convex side of the ring gear teeth. These design characteristics provide strength and quieter operation. Hypoid gears require the use of extreme pressure, or EP-rated, gear oils because of the extensive sliding action between the gear teeth. Synthetics tend to be recommended by most OEMs. Proper adjustment of backlash and tooth contact is critical to gear life.

FIGURE 23–13 (A) Hypoid gearing; (B) amboid gearing; and (C) spiral bevel gearing arrangements

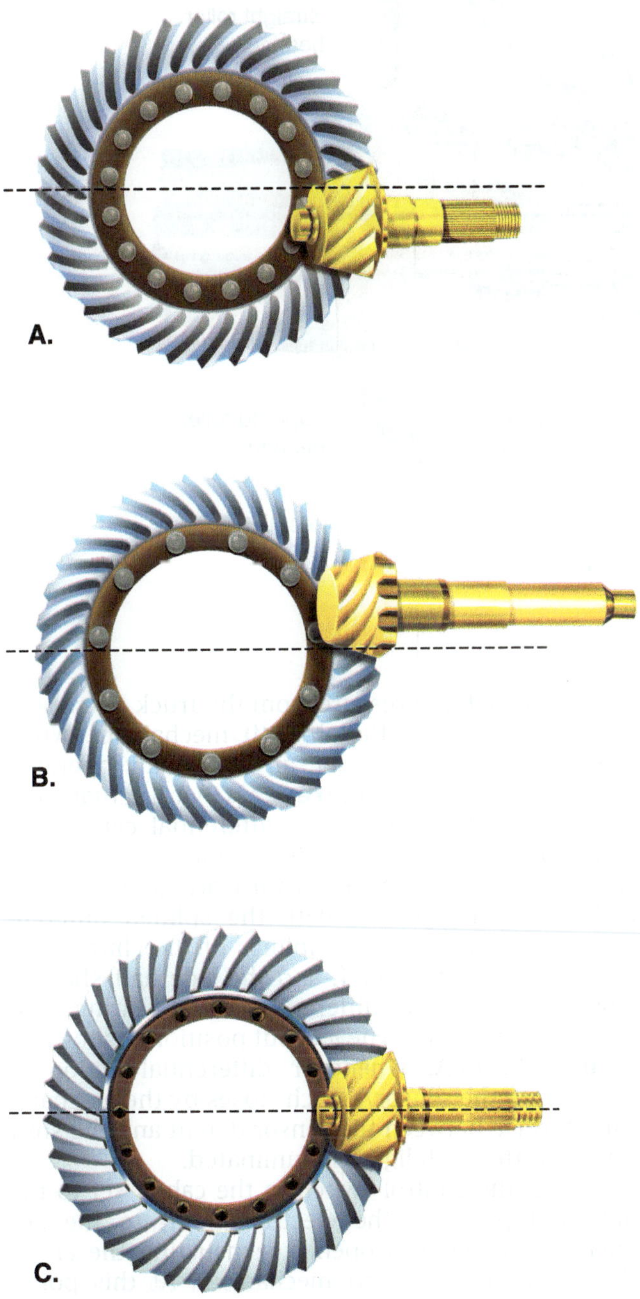

Amboid Gearing

An **amboid gear set** (**Figure 23–13B**) is similar to the hypoid gear set except that the axis of the drive pinion is located above the centerline of the crown gear. This requires the driveshaft to be mounted in a higher position than with the hypoid gearing arrangement. Amboid gear sets drive forward on the concave side of the ring gear teeth. This gearing arrangement typically is used on the rear axle of a tandem drive axle assembly.

Spiral Bevel Gearing

The **spiral bevel gear** set (**Figure 23–13C**) has a drive pinion gear that meshes with the ring gear at the centerline axis of the crown gear. This gearing provides strength and allows for quiet operation. These characteristics are achieved because of the overlapping tooth contact; for example, before one tooth rolls out of contact with another, a new gear tooth comes into contact.

SINGLE-REDUCTION AXLE

The single-reduction axle is the most common and basic type of gearing arrangement used in most drive axles on heavy-duty trucks. The axle assembly has only one gear reduction achieved by the crown and pinion ratio. The single-reduction axle uses hypoid, amboid, or spiral bevel, pinion, and crown gears. A single-reduction axle can be used as an individual assembly on a single drive axle chassis, as the rear axle on a tandem axle assembly, or as a front drive steering axle.

Operation

Figure 23–14 shows a single reduction axle, differential carrier mounted into the axle banjo housing. This example has a hypoid crown and pinion gear set and uses bevel gears in the differential. The drive pinion gear is straddle mounted on two tapered roller bearings and a straight roller-type pilot or **spigot** bearing. This straight roller bearing is mounted on the head of the drive pinion. The differential is mounted on two tapered roller bearings. Powerflow through this type of drive axle is through the drive pinion and crown gear, through the four differential pinions, to the two side gears, and from there to the axle shafts and wheels. On this type of drive axle, full-time differential action occurs during operation.

No-Spin Locking Differential

Some trucks equipped with single-reduction axles use a driver-controlled main differential lock, shown in **Figure 23–15**. This type of drive axle assembly has greater flexibility over the standard type of

FIGURE 23–14 Standard single-reduction carrier

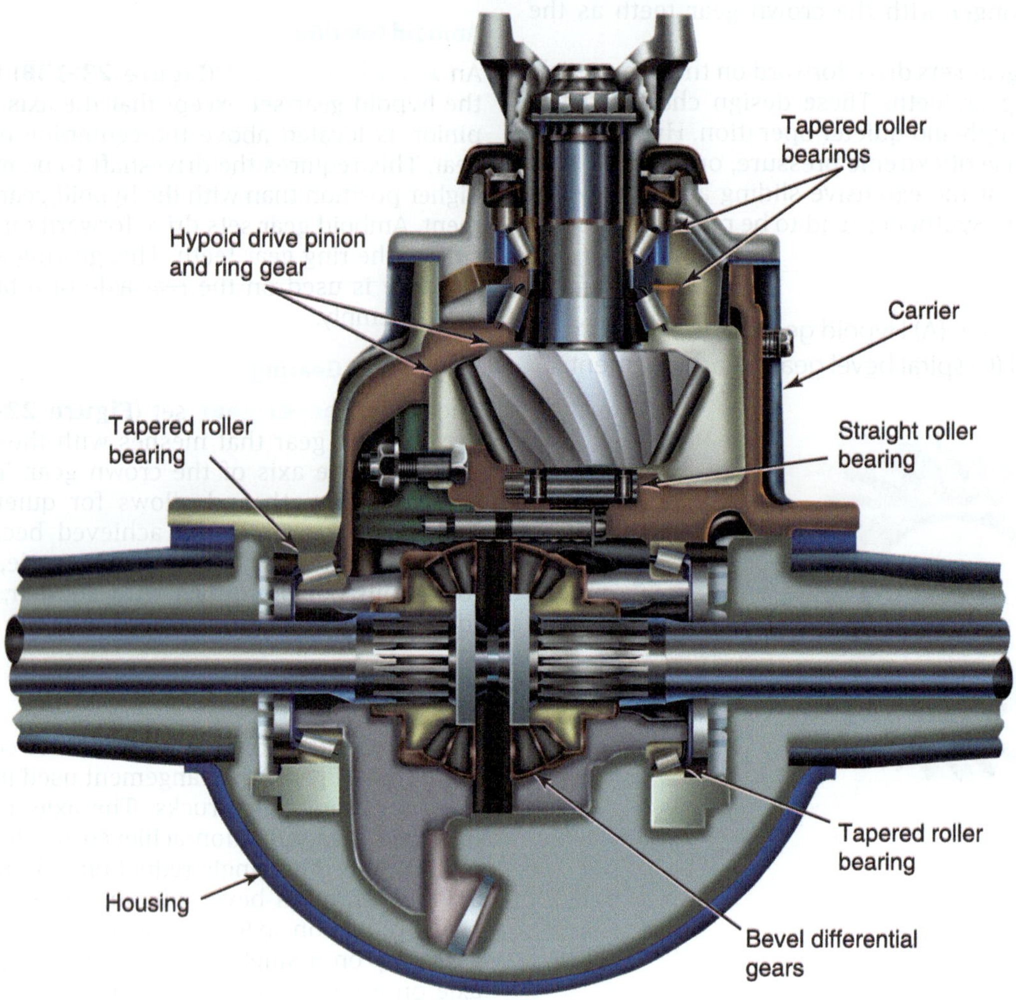

single-reduction axle because it can option equal drive torque to each drive wheel at the flick of a switch. The no-spin option is used when operating in mud, snow, ice, or wet road conditions because it locks out differential action and prevents wheel spinout.

The typical single-reduction carrier with differential lock is similar in most respects to standard differential carriers, as you can see by comparing Figure 23–14 and Figure 23–15. The differential lockout is operated by an air-actuated shift assembly mounted on the carrier and driver controlled from the cab. A single-reduction drive axle carrier equipped with a differential lock consists of the following (see Figure 23–15):

- Sensor switch
- Splined differential case
- Shift fork
- Shift collar
- Air shift mechanism

Air at system pressure from the truck air supply circuit is ported to the air shift mechanism. When the differential lock is set, an electrically operated air solenoid opens (**Figure 23–16**) to actuate the air shift mechanism on the differential carrier. On actuation, the air shift mechanism acts on the shift fork, causing it to move forward bringing the shift collar into engagement with the splined differential case. This causes the splines of the shift collar, the differential case half, and the axle half shaft to engage, locking the differential case, gearing, and axle shafts together. The lockout position is shown in **Figure 23–17A**. Whenever differential action is locked out, the sensor/switch closes by the mechanical interface between the sensor detent and the shift fork, and the dash light is illuminated.

When the control switch in the cab is set to the unlocked position, the electric circuit to the air-operated solenoid is opened, exhausting the air in the differential air shift mechanism. At this point,

FIGURE 23–15 Standard single-reduction carrier with differential lock

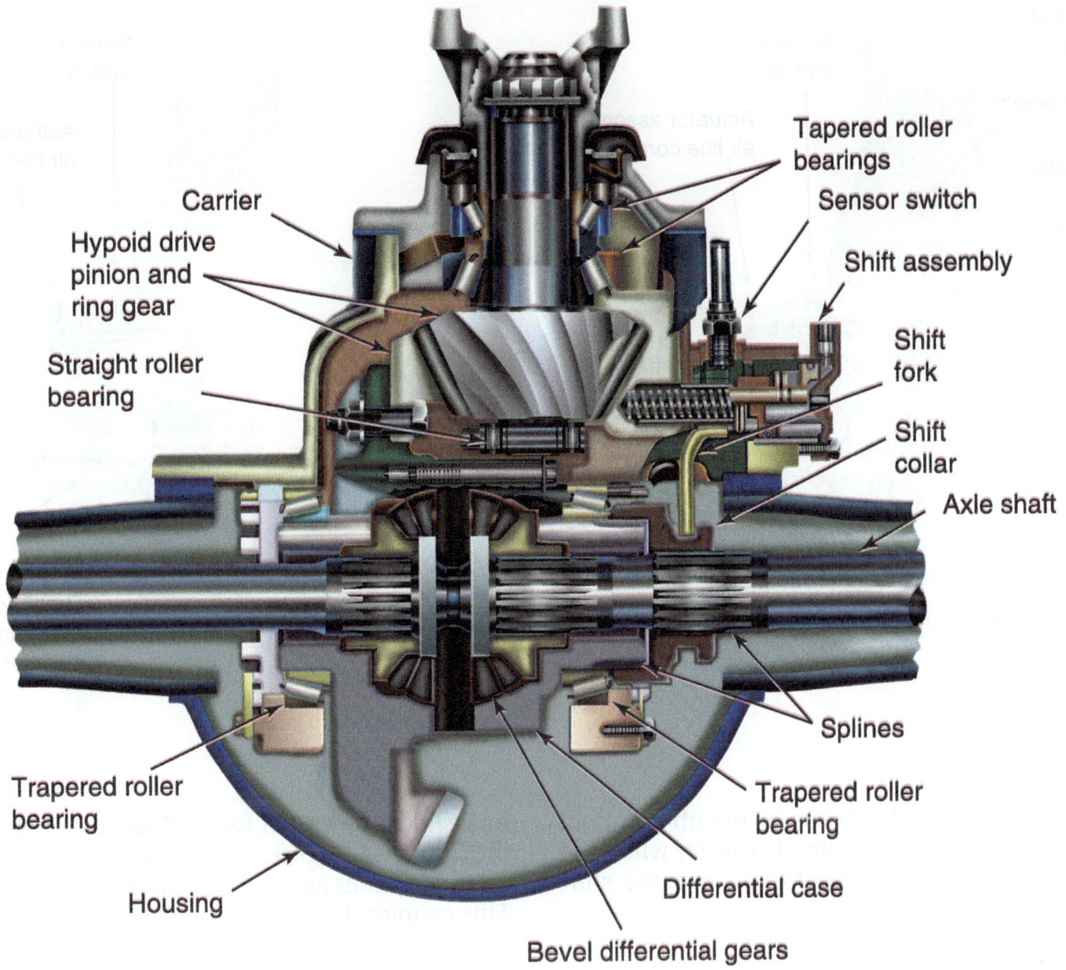

FIGURE 23–16 Electrical and pneumatic schematic for the differential lock

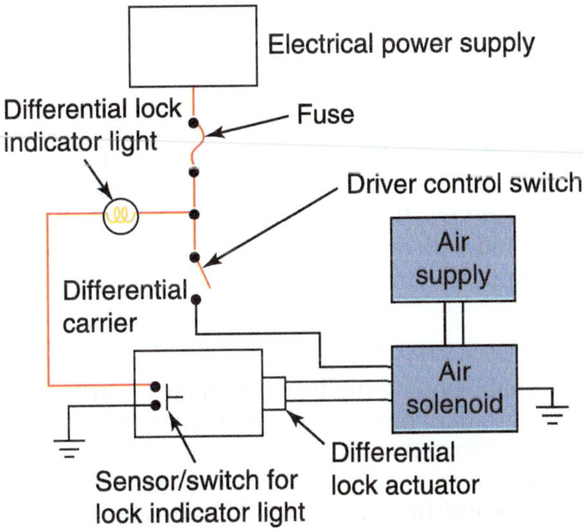

a spring inside the mechanism withdraws the shift fork and collar out of engagement to unlock the differential (**Figure 23–17B**). Normal differential action between both wheels then resumes.

Controlled Traction Differential

Another type of limited-slip differential is the controlled traction differential shown in **Figure 23–18**. A controlled traction differential is designed to inhibit normal differential action during uneven or slippery road conditions. It does this with a set of clutch discs and plates, known as a clutch pack.

The clutch pack shown in **Figure 23–19** consists of a large number of friction plates; for example, there are twenty-one in this particular limited slip differential. Ten of the friction discs are splined to a driver, and the remaining eleven have a pair of tangs each that fit into slots in the gear support case. Coil springs keep the clutch plates clamped together. A sliding clutch is actuated by a shift fork that is controlled by a cab-mounted air valve.

FIGURE 23–17 Differential lock assembly: (A) locked; and (B) unlocked

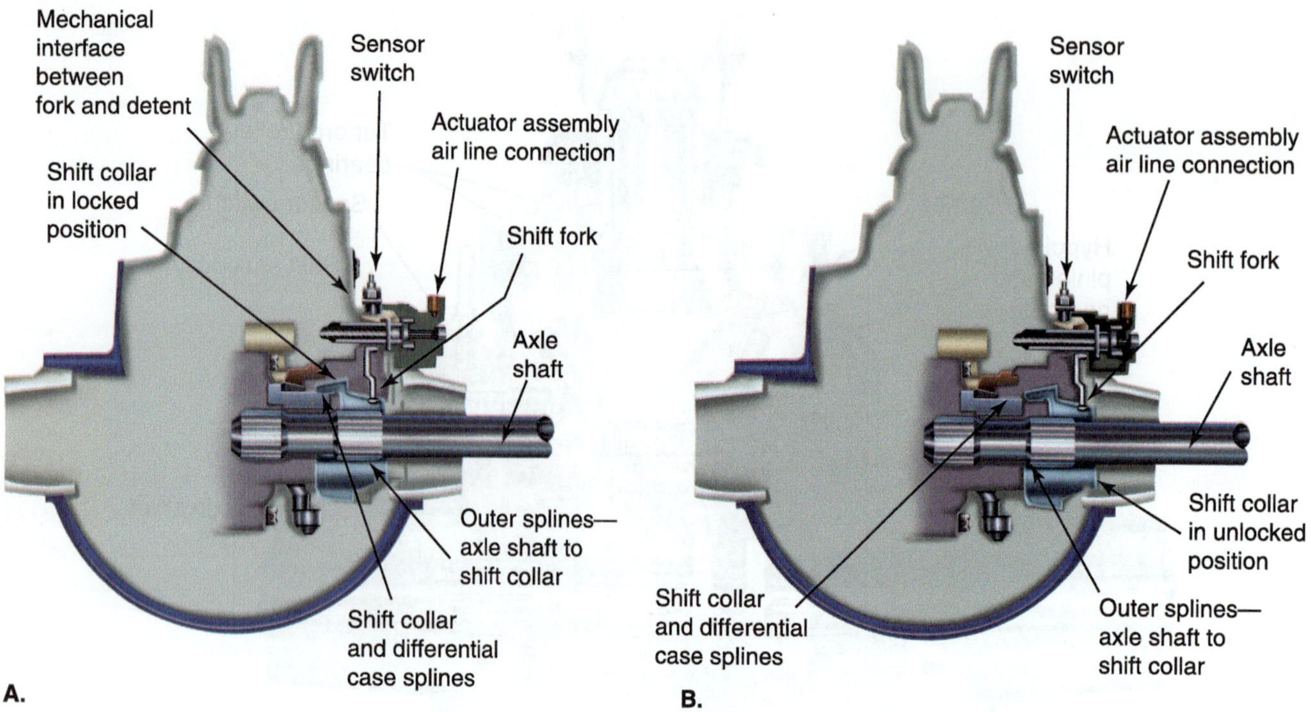

When the air control valve is in the off position, normal differential action occurs. However, when the driver actuates the traction control switch, the shift fork slides the clutch forward so that it moves inside the driver (see Figure 23–19) and engages with both its internal splines and splines on the end of the side gear. This couples the side gear to the gear support case and crown gear through the clutch pack. When the clutch plates are engaged, the differential will remain locked and both axle shafts rotate at the same speed.

FIGURE 23–18 Heavy-duty controlled traction differential

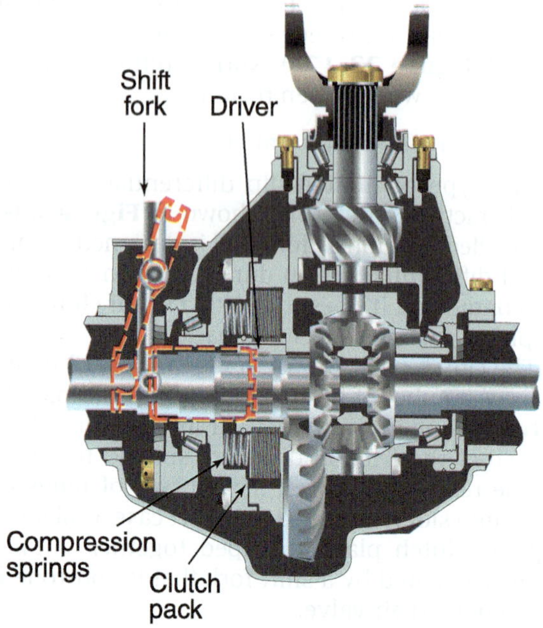

The clutch pack is designed to slip when a predetermined torque value is reached. This slipping action enables the truck to negotiate turns in a normal manner. The high torque created by the outside wheels during a turn causes the plates in the clutch pack to slip, allowing the side gear speed to increase and initiating the pinion walking action that occurs during normal differential operation.

Controlled traction differentials are available with a driver-controlled shift unit (as just described), and with a permanent driver that permits full-time limited slip operation. If the axle is equipped with a permanent control traction differential, the regular driver is substituted with a driver designed to engage with the side gear at all times.

Double-Reduction Axle

Vocational on-/off-highway vehicles and trucks required to haul heavy loads are sometimes equipped with **double-reduction axles**. A double-reduction axle uses two gear sets for a greater overall gear reduction and higher torque. This design is favored

FIGURE 23–19 Exploded view of a heavy-duty controlled traction differential

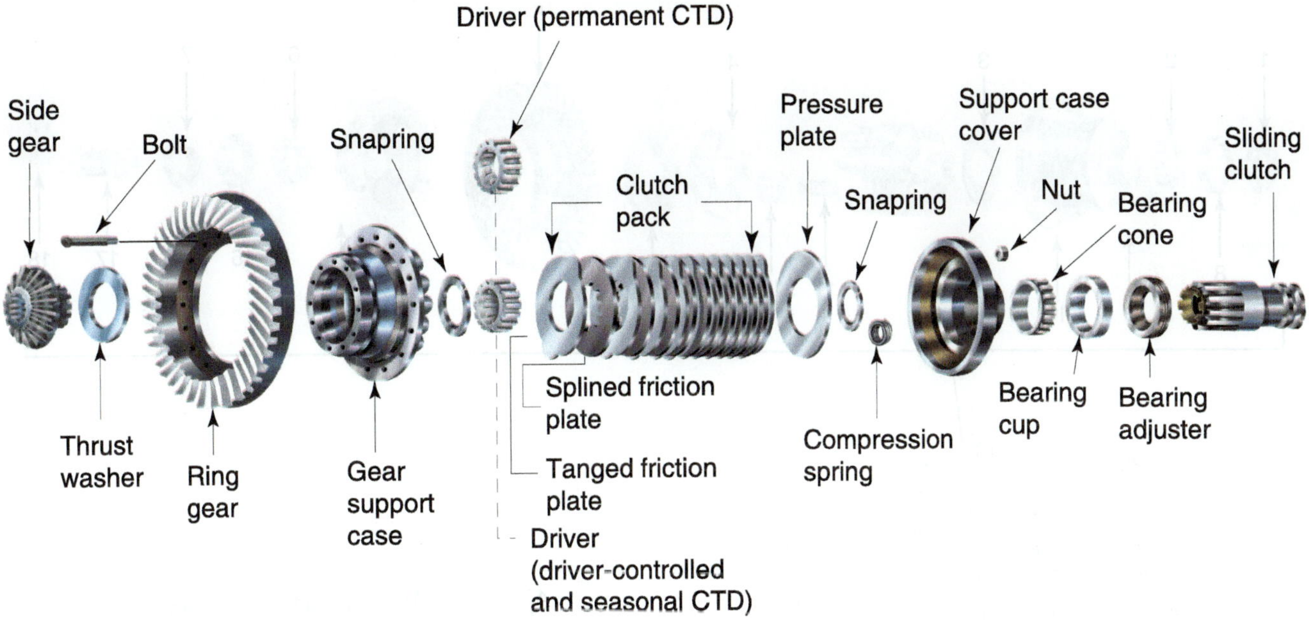

for severe service applications, such as dump trucks, cement mixers, and other heavy haulers.

The double-reduction axle shown in **Figure 23–20** uses a heavy-duty spiral bevel or hypoid pinion and crown gear combination for the first reduction. The second reduction is accomplished with a wide-faced helical spur pinion and gear set. The drive pinion and crown gear function just as in a single-reduction axle. However, the differential case is not bolted to the ring gear. Instead, the spur pinion is keyed to and driven by the ring gear. The spur pinion is, in turn, constantly meshed with the helical spur gear to which the differential case is bolted.

Another double-reduction gear arrangement is shown in **Figure 23–21**. This design uses planetary gears (**Figure 23–22**) to achieve the second reduction. In an axle with planetary double reduction, the ring gear has internal gear teeth. These teeth mesh with planetary pinion gears (idler pinions), which are mounted on pins fastened to the differential case. A third gear, the sun gear, is splined to the left-hand bearing adjuster and is held in position by a retainer.

In this design, the differential casing is not bolted to the crown gear. The differential casing is driven by the planetary pinions. The drive pinion drives the crown gear. Internal teeth on the crown gear form the ring gear of the reduction planetary gearset. As the crown gear/planetary ring gear rotates, it drives the planetary pinions, which rotate around the stationary sun gear. The planetary pinions drive the differential gear set. This arrangement results in a gear reduction and torque increase.

FIGURE 23–20 Cutaway view of a hypoid double-reduction differential carrier

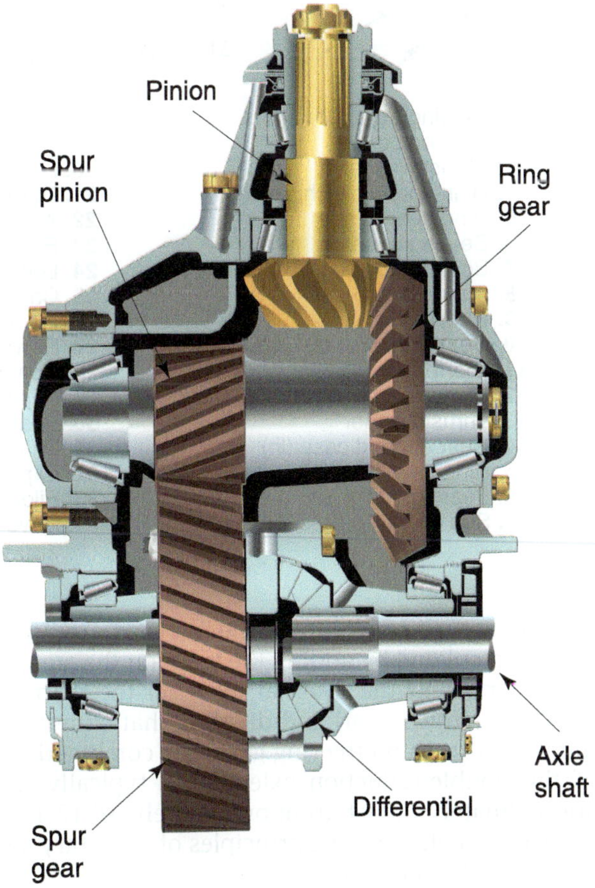

FIGURE 23–21 Exploded view of a planetary double-reduction axle

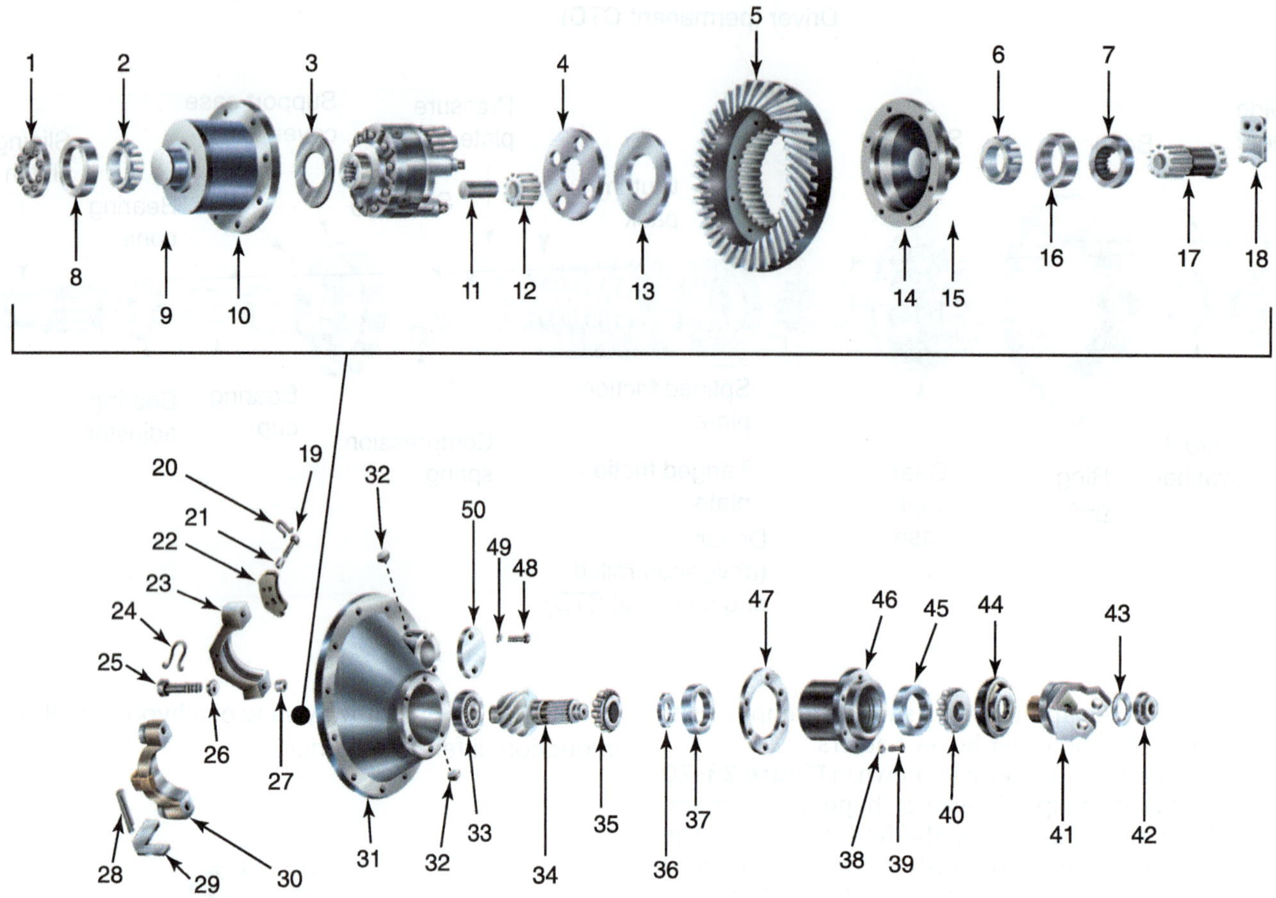

1. Bearing adjuster
2. Bearing cone
3. Thrust washer
4. High-speed clutch plate
5. Ring gear
6. Bearing cone
7. Bearing adjuster
8. Bearing cup
9. Ring gear bolt
10. Gear support case (rh)
11. Idler pinion pin
12. Idler pinion
13. Thrust washer
14. Gear support case (lh)
15. Ring gear nut
16. Bearing cup
17. Sun gear
18. Retainer
19. Cap screw
20. Lock wire
21. Dowel bushing
22. Adjuster lock
23. Bearing cap
24. Lock wire
25. Cap screw
26. Flat washer
27. Dowel bushing
28. Cotter pin
29. Adjuster lock
30. Bearing cap
31. Differential carrier
32. Expansion plug
33. Pilot bearing
34. Drive pinion
35. Bearing cone
36. Bearing spacer (variable)
37. Bearing cup
38. Lock washer
39. Cap screw
40. Bearing cone
41. Yoke
42. Pinion nut
43. Flat washer
44. Oil seal
45. Bearing cup
46. Bearing cage
47. Bearing cage shim
48. Cap screw
49. Lock washer
50. Shift fork cover

The actual gear ratio of the planetary reduction can be calculated by adding the teeth on the sun gear (stationary) and the internal teeth on the planetary ring gear (crown gear) and dividing that sum by the number of teeth on the ring gear. The combined ratio of the double-reduction axle ranges typically from approximately 5.04:1 to approximately 10.13:1. For more information on the principles of planetary gearing, refer to Chapter 18.

23.2 POWER DIVIDERS

If you take a quick glance at our highways, you can see that most heavy-duty trucks are equipped with two rear drive axles. This is known as a tandem drive arrangement (**Figure 23–23**). Tandem drive axle trucks require a gearing to split driving torque from the transmission to both the forward and rearward, rear drive axles. This gearing also must be capable

FIGURE 23–22 Planetary double-reduction gearing

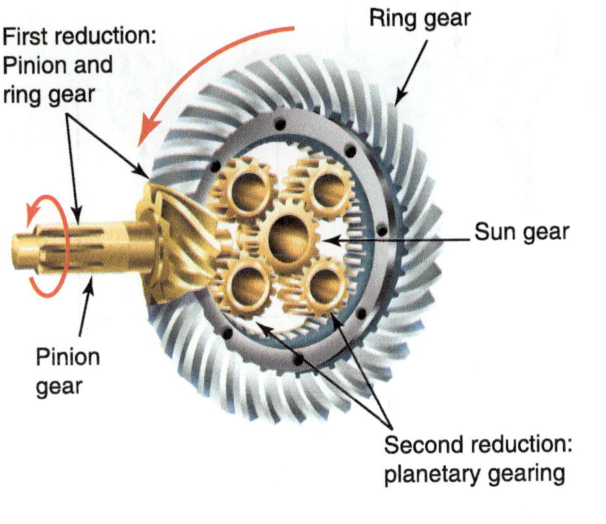

of allowing speed differences between the axles. The gear set used to accomplish this is known both as a power divider and an interaxle differential. We use both terms in this chapter.

INTERAXLE DIFFERENTIALS

A power divider has two functions:

- To divide driving torque between the two drive axles on a tandem drive chassis
- To provide differential action between the two drive axles

FIGURE 23–23 Typical tandem drive axles

An interaxle differential is an integral part of the forward rear axle differential carrier in a tandem drive truck. You can see a sectional view of a forward rear axle differential carrier in **Figure 23–24** and an exploded view of the components in **Figure 23–25**.

Most of the components of an interaxle differential are similar to those in a regular differential. These include a spider (or cross), differential pinion gears, a case, washers, and the side gears (**Figure 23–26**).

POWER DIVIDER OPERATION

During normal interaxle operation, torque from the vehicle driveline is transmitted to the carrier input shaft that is itself splined to the interaxle differential spider. At this point, the interaxle differential distributes torque equally to both axles, routed as shown in **Figure 23–27**. We can take a look at how this input torque is handled by each of the drive axles.

FIGURE 23–24 Components of a tandem forward rear drive axle

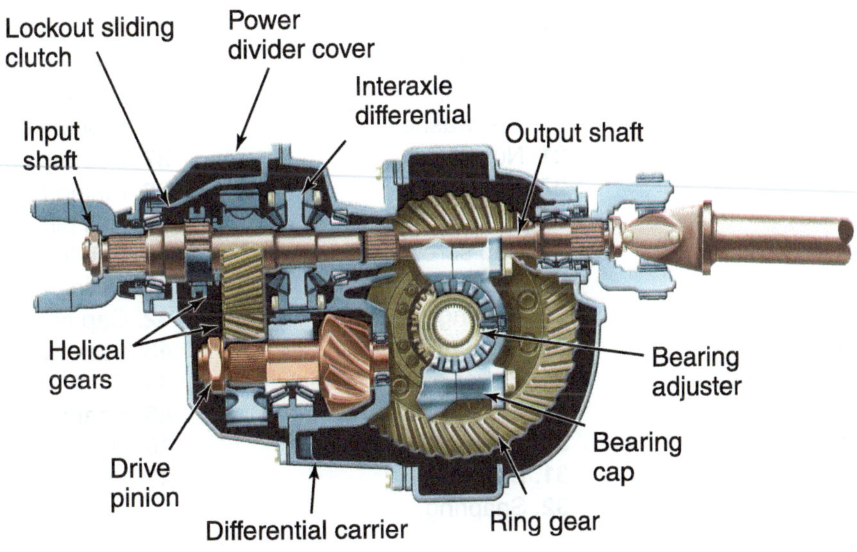

FIGURE 23–25 Exploded view of a power divider

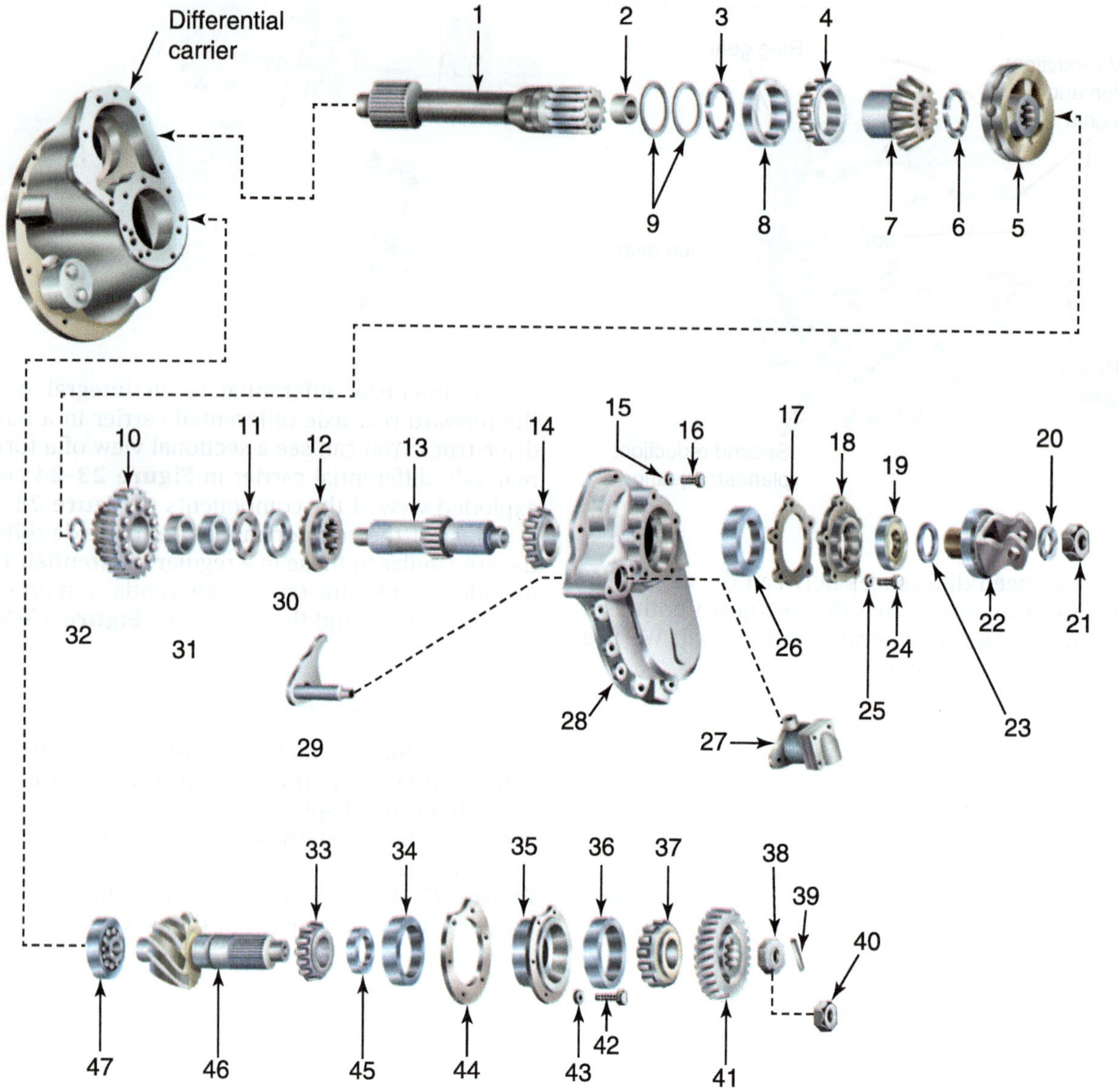

1. Output shaft
2. Bushing
3. Snap ring
4. Bearing cone
5. Interaxle differential assembly
6. Snapring
7. Side gear
8. Bearing cup
9. O-ring
10. Helical side gear
11. Thrust washer
12. Lockout sliding clutch
13. Input shaft
14. Bearing cone
15. Lock washer
16. Cap screw
17. Shim
18. Bearing cover
19. Oil seal
20. Flat washer
21. Nut
22. Yoke
23. Bearing spacer
24. Cap screw
25. Lock washer
26. Bearing cup
27. Lockout unit
28. Power divider cover
29. Shift fork and pushrod
30. D-washer
31. Bushings
32. Snapring
33. Bearing cone
34. Bearing cup
35. Bearing cage
36. Bearing cup
37. Bearing cone
38. Pinion-slotted nut
39. Roll pin
40. Self-locking nut
41. Helical gear
42. Cap screw
43. Lock washer
44. Shim
45. Bearing spacer (variable)
46. Drive pinion
47. Pilot bearing

FIGURE 23–26 Components of an interaxle differential

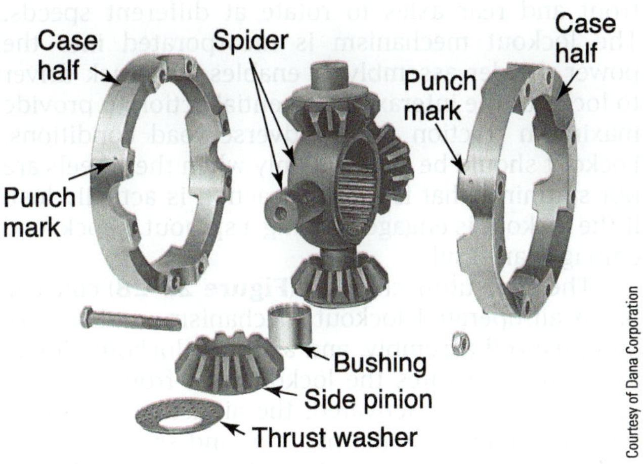

FIGURE 23–27 Torque distribution (powerflow) in a tandem rear drive axle when the interaxle differential is operating (lockup disengaged).

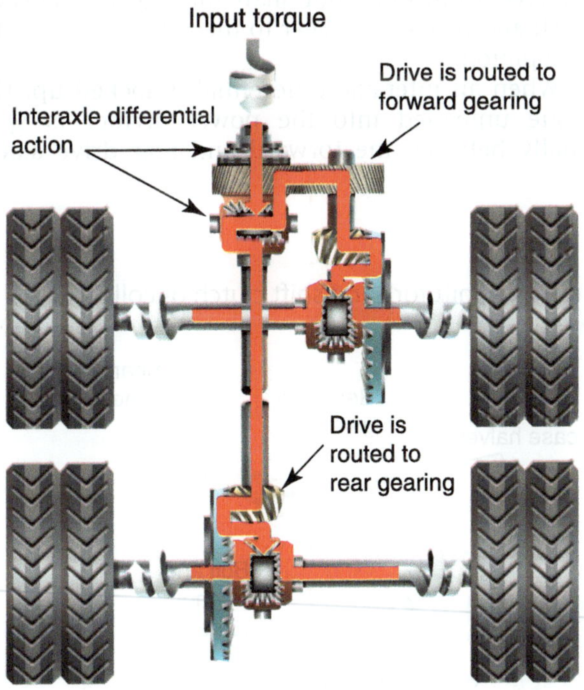

Forward Drive Axle

A helical gear is combined with the front interaxle differential side gear. The side gear is pressed onto a bushing that floats on the input shaft. As the input shaft rotates and drives the interaxle differential, the helical side gear is driven by the differential pinion gears. The helical side gear, in turn, drives the second helical gear fastened to the drive pinion. The drive pinion delivers torque to the front rear axle crown gear. This crown gear delivers torque to the front axle carrier differential. This conventional differential then delivers drive torque to the wheels on the forward drive axle, as discussed earlier in this chapter.

Rear Drive Axle

Within the power divider assembly (located in the forward differential carrier), the rear interaxle differential side gear is splined to the output shaft. The output shaft connects to the front differential carrier output yoke and the rear driveshaft. Torque required to drive the rear drive axle is transmitted from the output shaft side gear, through the output shaft, through the interaxle driveshaft (which connects the forward and rear drive axle carriers), to the drive pinion in the rear axle differential carrier.

The rear axle differential carrier functions exactly as described earlier. Driving torque is delivered to the drive pinion, which in turn drives the rear axle crown gear, which then delivers drive torque to the rear wheel differential and wheels.

Axle Speed Variations

The interaxle differential compensates for axle speed variations in the same way that the carrier differential functions between the two wheels of a single drive axle. As previously shown in Figure 23–26, the interaxle differential has a spider and four pinions held between two case halves. The side gears connected to the helical gear and to the end of the output shaft are meshed with these pinion gears.

One axle might want to turn faster than the other in two situations. First, when tires of the two axles are mismatched so that there is a difference in diameter between them, differential action between the axles is desirable. Second, when there is a reduction in traction for one axle, such as on a gravel road, some differential action is again desirable.

Because the power divider pinions drive both side gears (forward axle and rear axle) when turning with the spider, they can also walk on the spider shafts just as in a carrier differential. This allows the side gears to operate at different speeds when differential action is required. Although this enables the interaxle differential to compensate for mismatched tire problems, it also can permit the condition known as spinout, in which one axle spins on a poor traction surface when the other is immobilized on good pavement. Spinout can cause serious damage to the differential in seconds.

Spinout

The downside of differential gearing is that a reduction in speed of one side gear is exactly equal to the increase in speed of its opposite. This characteristic can allow for very high wheel speeds when one wheel has no traction on a slippery surface. Spinout can always be attributed to driver abuse and typically has three causes:

- Aggressive pull-away acceleration on a slippery or split coefficient surface
- Aggressive backing up under a trailer on a poor traction surface
- Loss of traction when mobile on a slippery surface combined with aggressive accelerator response

In every case of spinout, it is important to remember that the interaxle differential is more susceptible to spinout than the axle differential because it rotates at much higher speeds (i.e., driveshaft rpm), which can throw lubrication away from the cross and pinion bearing surfaces and generate intense heat. Drivers should be trained to avoid spinout conditions by engaging the interaxle lockout when conditions warrant it. Spinout damages power dividers for two reasons:

- Lack of lubrication: Oil is thrown off gears by centrifugal force.
- Shock load: Spinout can generate individual wheel speeds up to four times the normal speed. The centrifugal force generated throws lubricant away from the bearings and when the spinning wheel finds some traction, the shock load can damage the interaxle differential and other powertrain components.

Interaxle Differential Lockout

Most power dividers have a lockout mechanism that prevents the interaxle differential from allowing the front and rear axles to rotate at different speeds. The lockout mechanism is incorporated into the power divider assembly. It enables the truck driver to lock out the interaxle differential action to provide maximum traction under adverse road conditions. Lockout should be engaged only when the wheels are not spinning, that is, before traction is actually lost. If the lockout is engaged during a spinout, shock load damage can result.

The lockout mechanism (**Figure 23–28**) consists of an air-operated lockout mechanism, a shift fork and pushrod assembly, and a sliding lockout clutch. The driver activates the lockout unit from a switch in the cab. When activated, the air-actuated lockout mechanism extends the pushrod and shift fork. The shift fork is engaged with the sliding clutch, and when extended, it forces the clutch against the helical side gear. Because the clutch is splined to the input shaft and the input shaft is splined to the differential, moving the clutch into engagement with the helical side gear locks the side gear to the differential, preventing any differential action from occurring. Torque is delivered from the input shaft through the meshed clutch and helical side gear to the helical gear on the drive pinion.

When an interaxle differential is locked up, the torque unloaded into the power divider is split equally between the forward and rear drive axles.

FIGURE 23–28 Components of a power divider showing lockout shift fork and shift clutch or collar

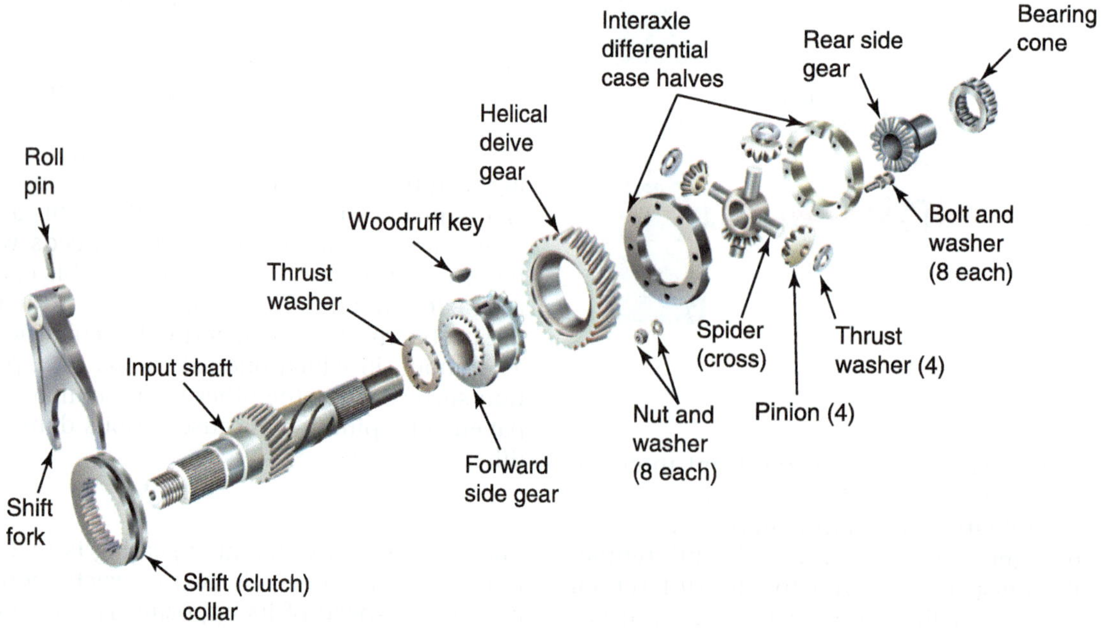

FIGURE 23–29 Torque distribution (powerflow) in a tandem rear drive axle when the interaxle differential is not operating (lockup engaged).

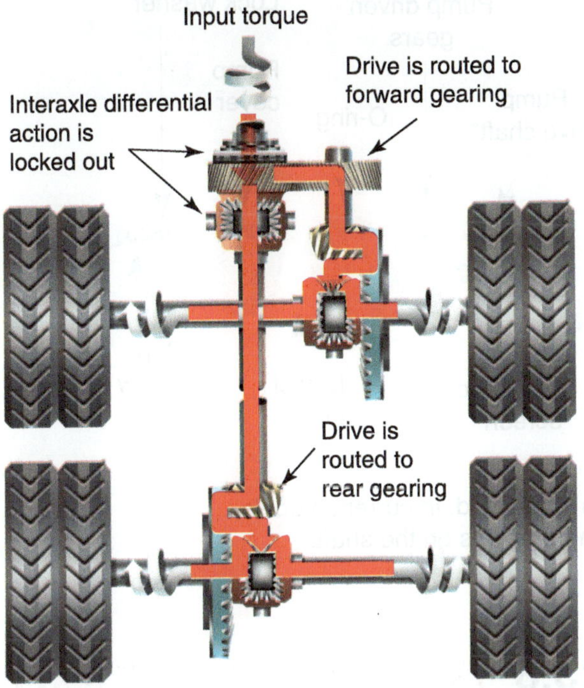

Figure 23–29 shows the torque flow path when the **interaxle differential lock** is engaged in a tandem drive axle truck and **Figure 23–30** shows a cutaway view of an interaxle differential on a forward carrier.

DRIVE AXLE GEARING LUBRICATION

Most single-reduction axles are splash lubricated by the churning of gears running in oil as the differential components rotate. Directional baffles and galleries are designed into the differential carrier to direct lubricant to critical friction points.

Some drive axle carriers, especially those with a through-shaft type power divider, use an oil pump to route lubricant to bearings, gears, and other lubrication points. Lubing the interaxle differential gearing is critical, and it is difficult to ensure that lubricant gets into the power divider gears without the assistance of a lube pump. **Figure 23–31** shows a cutaway view of a splash lubricated differential carrier with a lube pump.

A differential carrier lube pump typically uses an external gear pumping principle. The lube pump shown in **Figure 23–32** is driven by a drive gear that meshes with the input shaft splines. When the truck is moving in a forward direction, an external gear-type lube pump delivers pressurized oil to the interaxle differential gearing. A magnetic strainer screen keeps the system clean by trapping particles and a magnet holds any ferrous metal cuttings.

The drive axle carrier shown in **Figure 23–33** has an externally mounted oil pump, that is, one directly driven by the drive pinion shaft. Oil is drawn from the bottom of the carrier through an oil line to the pump assembly. The filtered oil is pumped

FIGURE 23–30 Cutaway view of a power divider in a front differential carrier

FIGURE 23–31 Splash lubricated differential carrier

FIGURE 23–32 Differential lube system

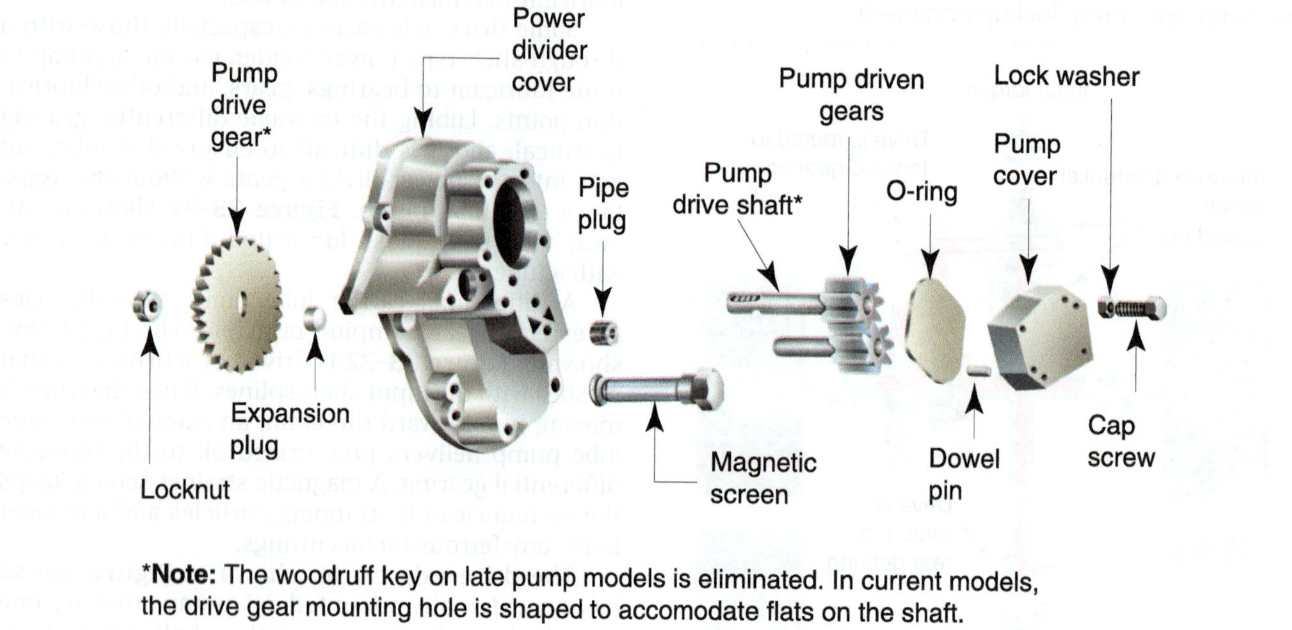

FIGURE 23–33 Power divider with an externally mounted oil pump

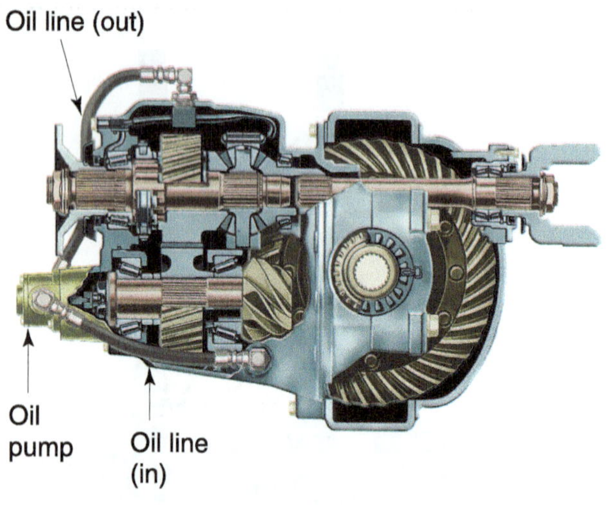

through another external oil line to the top of the carrier, where it is routed to the forward bearing and to the interaxle differential gearing. A ball check valve in the outlet line traps oil in the system when the vehicle is stationary.

Some drive axle carriers have a rotor-type pump instead of the external gear type. External spin-on oil filters are used with some systems, and these provide the advantage of entirely removing entrapped contaminants from the system.

23.3 TWO-SPEED DRIVE AXLES

Some drive axle carriers are equipped with a shift mechanism that allows the final drive to function either as a single-reduction axle or a double-reduction axle. This type of axle is known as a two-speed or dual range axle. It is similar to the double-reduction axle discussed earlier with one major difference. A double-reduction axle can operate only in double-reduction mode. The two-speed axle can be operated in either the single-reduction or double-reduction mode.

The double-reduction and the two-speed drive axle carriers share similar design principles in that they both use a planetary gearset to achieve the second reduction ratio. The major difference between the two is that the planetary sun gear on the two-speed carrier can be moved into two positions (high and low). The sun gear on a double-reduction carrier is always held in one position (low).

A typical two-speed drive axle carrier uses a hypoid pinion and crown gear set to achieve the first reduction ratio and a four-pinion planetary gearset to achieve the second reduction ratio. An example of a two-speed drive axle is shown in **Figure 23–34**. Bevel gears are used in the main differential.

The hypoid pinion is supported by three bearings in the carrier. Two tapered roller bearings are mounted on the pinion shaft in front of the gear head. The third bearing is a roller or spigot bearing that is mounted on the nose of the pinion. The main differential and the planetary gearset are mounted in differential case halves, which are contained in the

FIGURE 23–34 Hypoid planetary two-speed drive unit

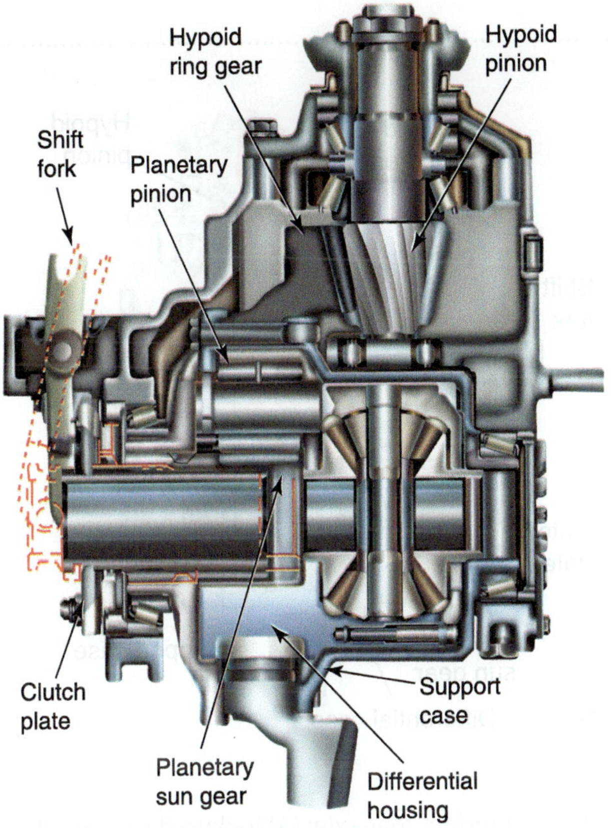

support case. The support case is mounted in the carrier on two tapered roller bearings.

An air or electric shift unit is mounted on the carrier, and it is actuated by a switch in the cab of the vehicle. The switch can be actuated by the driver to shift the drive unit into high- or low-gear range. In high-gear range, the sun gear is engaged with the support case and the planetary pinions are locked in place. When the planetary pinions are locked, there is no second reduction, and the torque is delivered directly from the hypoid gear set to the main differential (**Figure 23–35A**). This hypoid single-reduction drive provides a lower numerical ratio for higher road speeds but lower torque.

In low-gear range, the sun gear is shifted to engage with the clutch plate. The four planetary pinions are then free to rotate and become the second reduction gearset (**Figure 23–35B**). The hypoid gear set ratio is now multiplied by the planetary gearset ratio. This provides the drive axle carrier with a higher ratio that increases torque but reduces the top road speed. In some two-speed drive axles, the sun gear is referred to as a sliding clutch gear. Two-speed differential carriers require a speedometer input signal from the control circuit to correctly display vehicle road speed.

TWO-SPEED TANDEM AXLES

Tandem drive axles also can be equipped with two-speed differential gearing on both the front and rear drive axles. In the high-range mode, the powerflow is through the pinion and crown gear only on both axles. The second reduction planetary gearing is bypassed, as shown in **Figure 23–36**.

However, when the low-range mode is engaged, the planetary gearing is no longer bypassed and now generates the second reduction. This results in increased torque and reduced output speed. The resulting powerflow through the drivetrain is shown in **Figure 23–37**.

A two-speed drive axle carrier effectively doubles the number of drive gears that the truck can be operated in. A typical progressive shift pattern might be first gear, low range; first gear, high range; second gear, low range; second gear, high range; and so on. The two-speed axle also provides the truck with the high-reduction ratios required for low-speed hauling of heavy loads, along with low-reduction ratios best for highway cruising speeds.

THREE-SPEED DIFFERENTIAL

A three-speed differential is basically a tandem two-speed axle arrangement with the capability of operating the two drive axles in different speed ranges at the same time. The third speed is actually an intermediate speed between the high and low range. The three speeds are obtained as follows:

1. High speed: Both axles are shifted into high range (single reduction).
2. Low speed: Both axles are shifted into low range (double reduction).
3. Intermediate speed: The forward axle is shifted into high range (single reduction) and the rear axle is shifted into low range (double reduction). The interaxle differential equalizes the differences in rpm between the two axles to produce a speed midway between the high and low ranges.

DUAL RANGE AXLE SHIFT SYSTEMS

Although some vehicles are equipped with electrical shift units, most axles are equipped with pneumatic shift systems. There are two air-actuated shift system designs used to select the ratio of a dual range tandem axle or to engage a differential lockout:

1. **Standard System.** For range selection, a cab-mounted air shift valve operates two air shift units mounted on the drive axle carriers. The interaxle differential lockout is of the fully pneumatic type that uses air pressure to engage the lockout and spring pressure to release it.

FIGURE 23–35 Hypoid planetary two-speed drive unit powerflows: (A) In high range the planetary pinions are locked, there is no second reduction, and power flows directly from the hypoid gearset to the main differential. (B) In low range the sun gear engages the clutch plate, and powerflow is through the planetary pinions (second reduction) to the main differential.

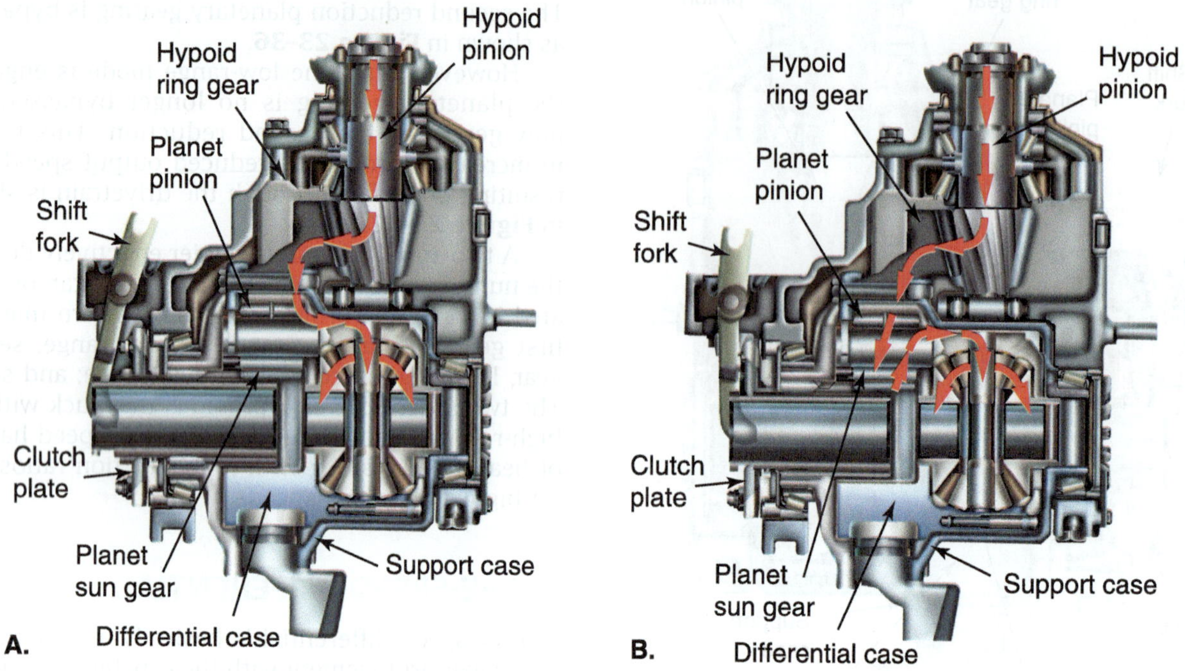

FIGURE 23–36 Torque distribution in the high-range mode of a tandem rear axle: (A) lockup disengaged; and (B) lockup engaged.

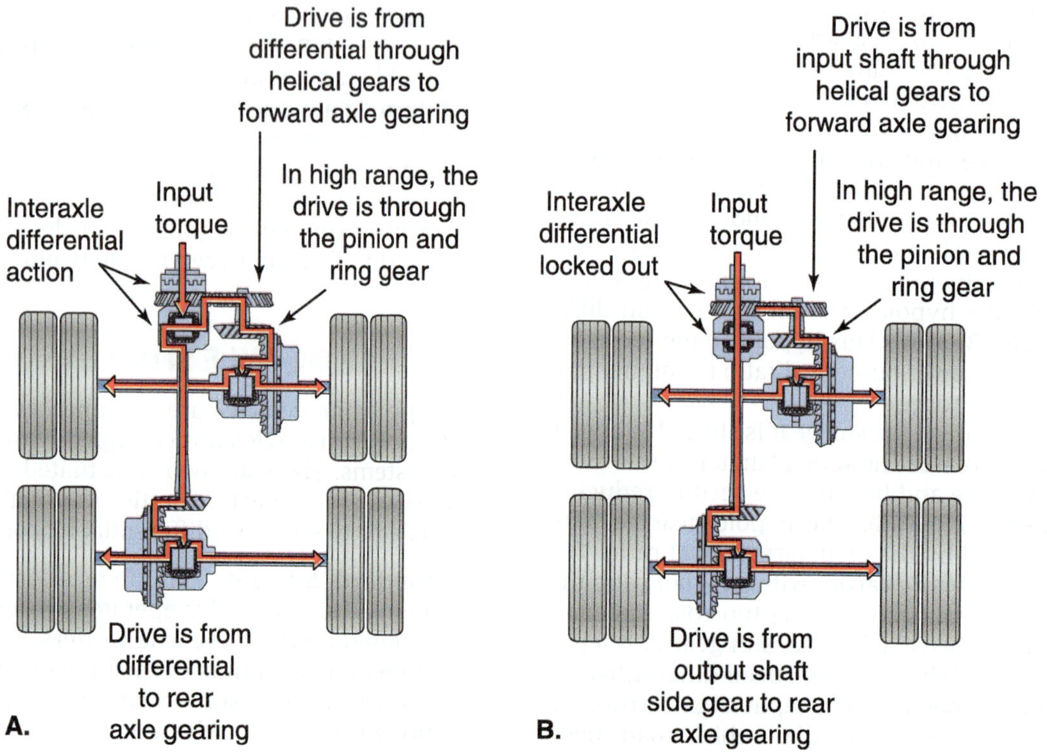

FIGURE 23–37 Torque distribution in the low-range mode of a tandem rear axle: (A) lockup disengaged; and (B) lockup engaged.

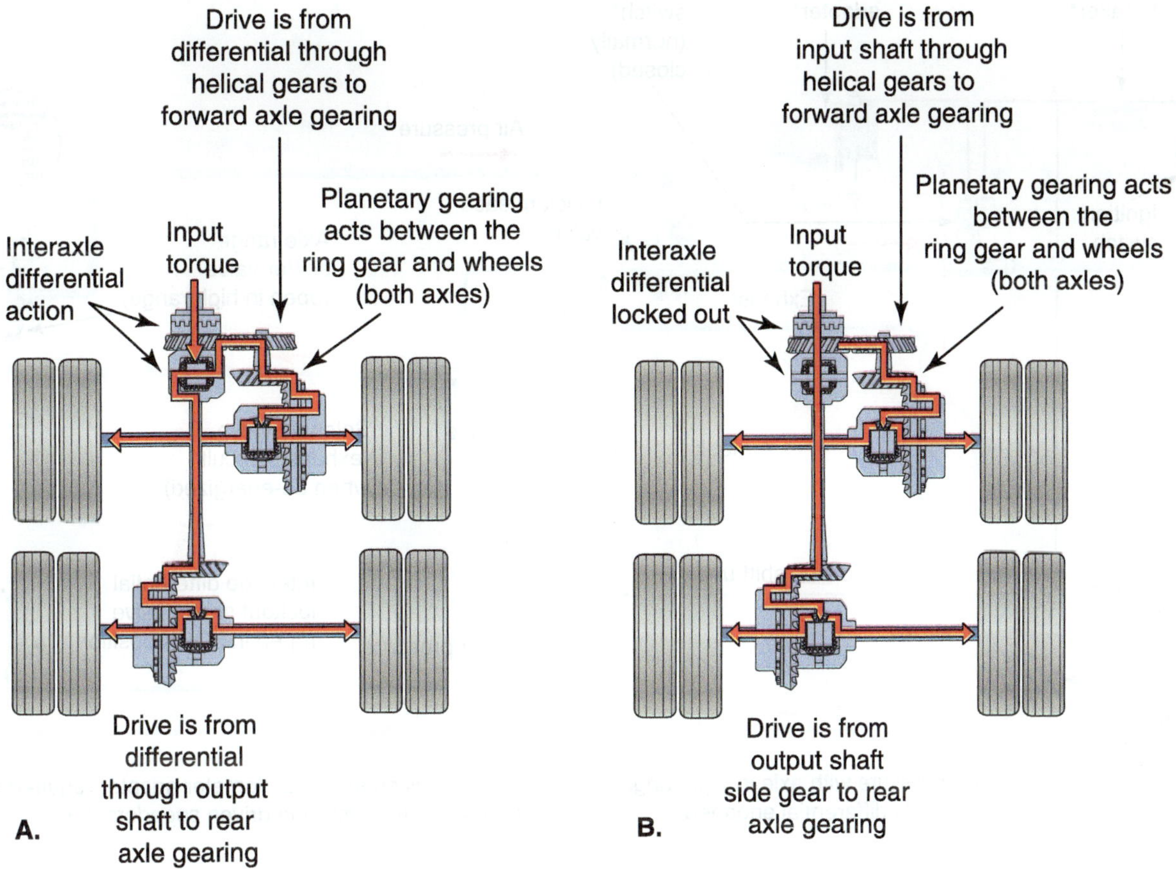

2. **Axle Range Interlock System.** An **axle range interlock** system has an added feature to prevent axle reduction shifting when the interaxle differential is locked out. The axle range air shift valve for this system includes an interlock pin assembly to provide the interlock feature. The interaxle differential lockout is of the fully pneumatic type. It is equipped with a control valve, which sends air pressure to the shift valve interlock pin. Another lockout design is the reverse-air-type interaxle differential lockout. It is spring actuated to lock the differential carrier and requires air pressure to unlock the differential.

Standard System

The standard dual range shift system is shown in **Figure 23–38**, and consists of:

- A manually operated air shift valve to change axle ratio.
- A quick-release valve that provides for fast release of air pressure from the axle shift units.
- Two air shift units mounted on the axles. These shift units are mechanically connected to the axle shift forks and sliding clutches, which, in turn, shift axles into low or high range.

For vehicles not equipped with automatic safety brakes, an ignition circuit-controlled solenoid valve exhausts the system and downshifts the axles when the ignition switch is turned off. The electrical circuit is protected by a circuit breaker. For vehicles equipped with transmission-driven mechanical speedometers, the system must include a speedometer adapter that compensates speedometer readings when the axle is in low range. The adapter is operated by an electrical switch mounted on or near the quick-release valve. The switch is normally closed (NC) and is opened by air pressure. For vehicles using transmission tailshaft road speed sensors, road speed data automatically adjusts between ranges.

With the axle ratio in low range, the switch is closed and the adapter is energized. The adapter operates with a ratio compatible with the axle low

FIGURE 23–38 Standard system for axle range selection and interaxle differential lockout

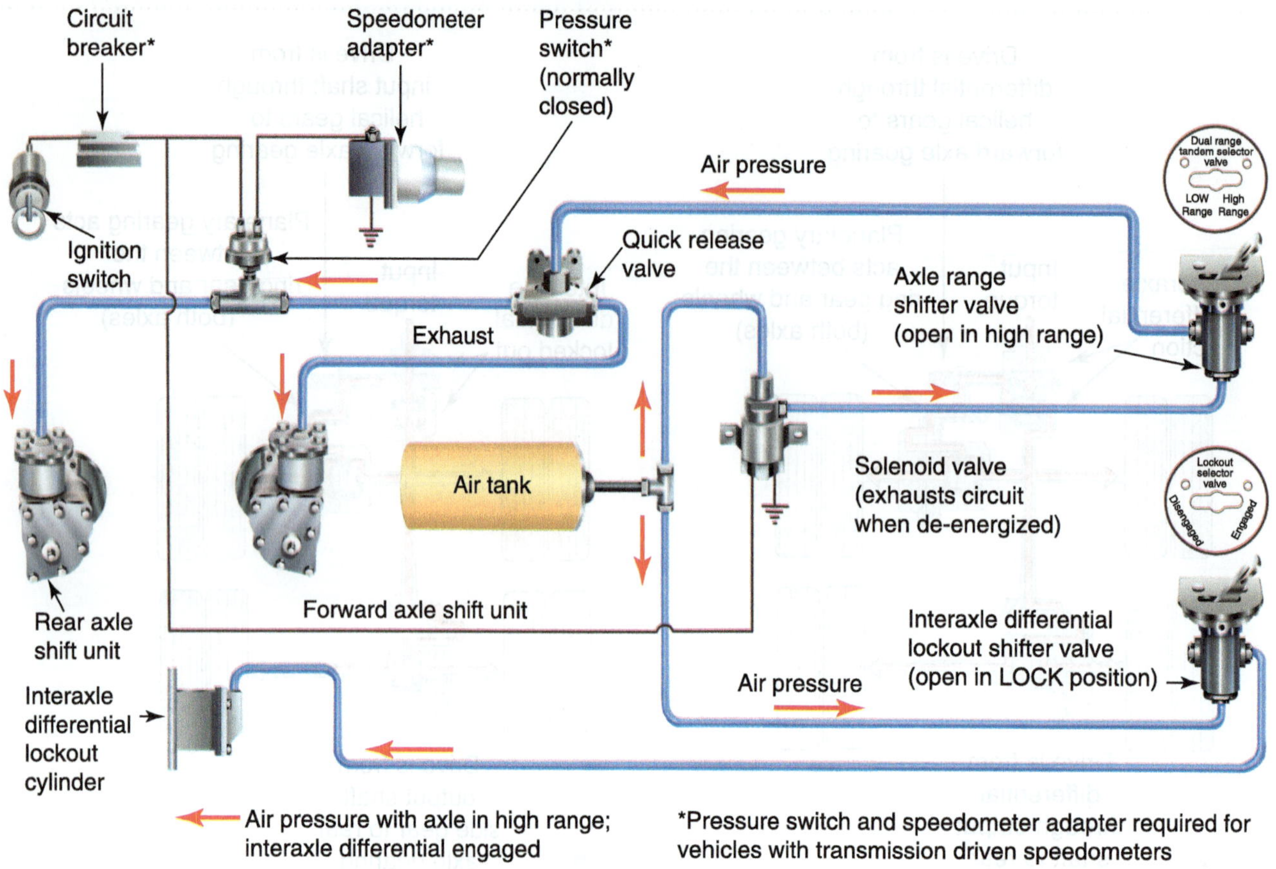

range for correct road speed readings. With the axle in high range, the air lines are pressurized and the pressure switch is open. The adapter now operates with a 1:1 ratio for the correct road speed readings in high range. You can identify this switch in the schematic in Figure 23–38.

Axle Shift Operation

When the shift valve is moved to the high position, the valve opens and air pressure is distributed through the quick release valve to both axle shift units. When driveline torque is next interrupted, the shift units will shift both axles to the high range. When the shift valve is moved to the low position, the valve closes. Air pressure in the shift units is exhausted through the quick release valve. When driveline torque is interrupted, both axles are shifted into the low range and held in this position by the shift unit return springs.

Interaxle Differential Lockout System

The lockout air shift system (see Figure 23–38) consists of:

- A manually operated air shift valve that controls engagement or disengagement of the interaxle differential.
- A lockout cylinder that operates under air pressure. This cylinder is mechanically connected to a shift fork and sliding clutch. The clutch engages or disengages a differential helical side gear to lock or unlock the differential.

The standard lockout unit is air operated to engage lockout and spring-released to disengage lockout. The air piston is mechanically connected to the shift fork and sliding clutch. The clutch engages or disengages the helical side gear to lock or unlock the interaxle differential.

When the air shift valve lever is moved to the unlock position, the valve closes and air pressure in the cylinder is exhausted. Air pressure at the piston is then discharged. Spring pressure moves the piston, shift fork, and sliding clutch. At this time, the clutch is disengaged from the helical side gear and the interaxle differential is unlocked and functions normally.

When the air shift valve lever is moved to the lock position, the valve is opened and supplies air to the lockout cylinder (shown in **Figure 23–39**). Air pressure enters the cover and moves the piston, shift fork, and sliding clutch. The clutch engages the helical side gear and the interaxle differential is locked out, preventing differential action.

Axle Range Interlock System

The axle range interlock feature in the system shown in **Figure 23–40** is designed to prevent axle shifting when the interaxle differential is locked out. The basic shift system operates in the same way as the standard shift system to shift the axle and engage or disengage the lockout. However, it varies by adding an interlock pin assembly to the axle range shift valve and an interlock control valve to the lockout cylinder. These two components are interconnected with air lines.

Figure 23–41 illustrates the operation of the axle range lockout. When air is supplied to the lockout cylinder from the lockout shifter valve in the cab, the lockout piston forces the shaft and fork forward so that the sliding clutch engages with the helical side gear (Figure 23–41A). Movement of the piston allows the interlock control valve to open. Air pressure is

FIGURE 23–39 Air shift valve operation

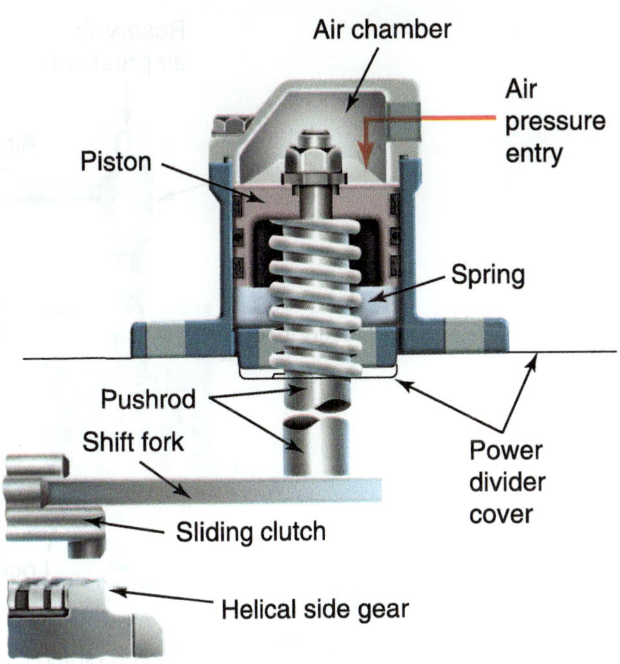

FIGURE 23–40 Axle range interlock system and interaxle differential lockout

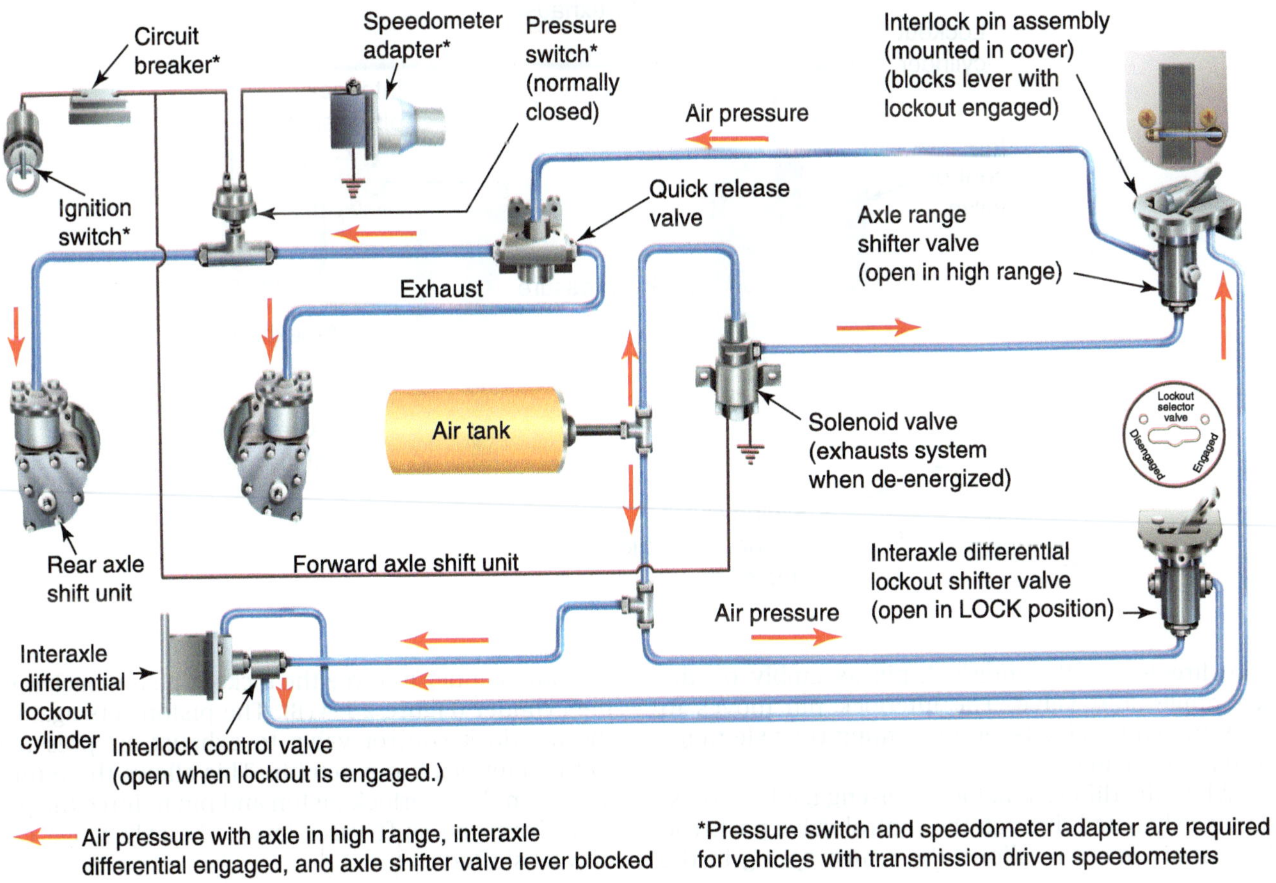

FIGURE 23–41 Axle range interlock operation: (A) lockout engaged; and (B) lockout disengaged

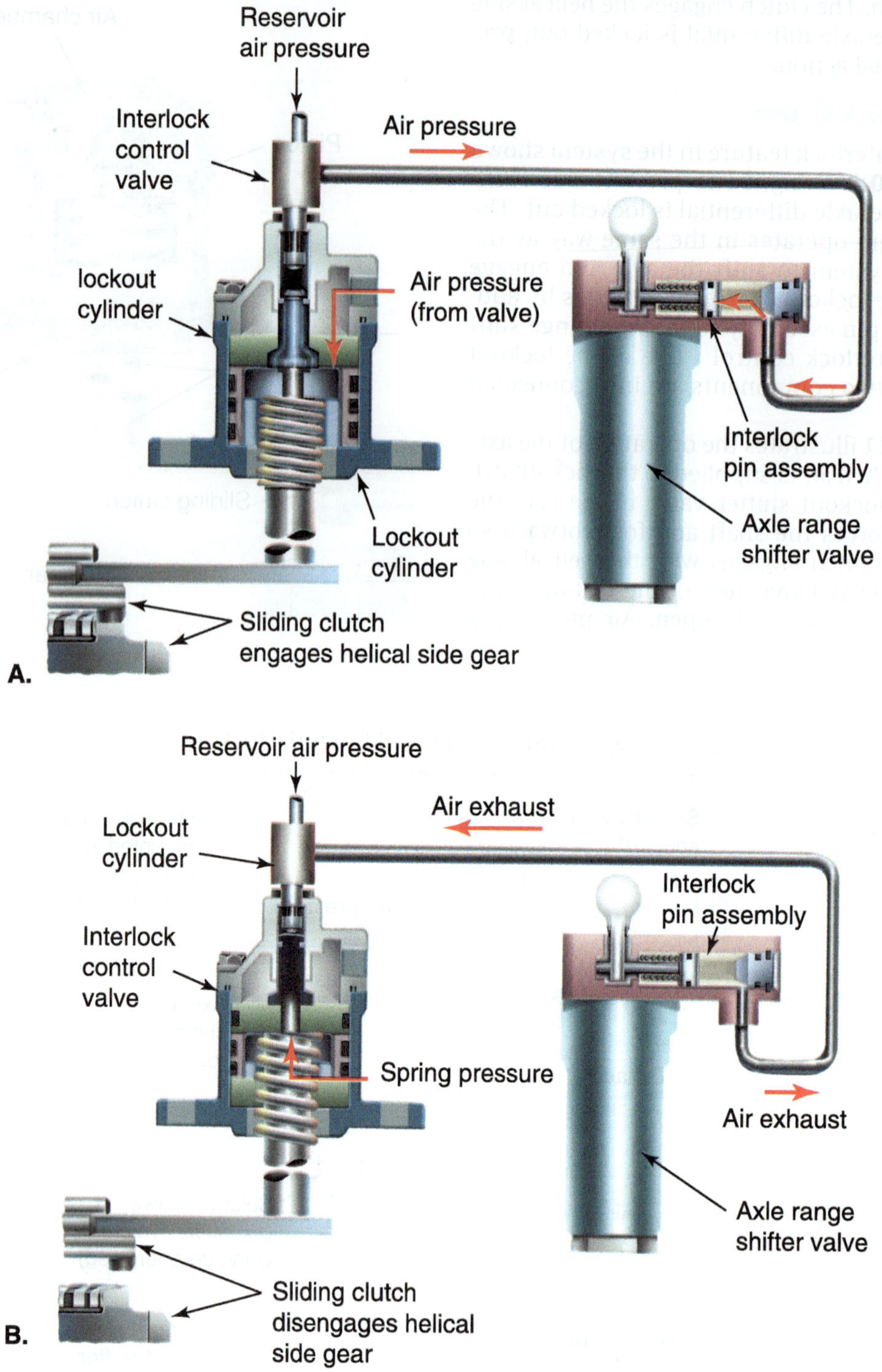

then directed to the interlock pin assembly on the axle range shift valve. The interlock pin moves to block the shift valve lever, preventing the axle range from being changed.

When the differential lock is disengaged, air pressure is exhausted through the control valve port. The drop in air pressure allows the return spring in the lockout cylinder to force the lockout piston back into the cylinder (Figure 23–41B). The piston acts against the interlock control valve to exhaust air pressure to the interlock pin assembly. This allows the return spring on the interlock piston and pin to force the pin away from the shift valve lever. The axle range can now be shifted.

TORQUE PROPORTIONAL DIFFERENTIALS

Torque proportional differentials are used in off-highway and a small number of vocational truck applications, most notably those manufactured by Mack Trucks. The objective of a torque proportional differential is to vary the distribution of torque between a pair of drive axles. When one or both wheels on an axle lose traction, a torque proportional differential can direct up to 75 percent of available transmission output torque to the carrier with the better traction.

Design

The power divider lockout used in a torque proportioning differential is bolted to the interaxle power divider housing. The power divider driving cage is supported within the lockout housing by double rows of ball bearings. Inserted in two staggered rows of slotted bores in the driving cage are radial plungers, known as peanuts or wedges. The radial plungers are free to slide either inward or outward. The companion flange is itself splined to the splined end of the driving cage.

Operation

When the vehicle is moving in either direction and tire traction on the forward and rear drive axles is equal, torque drive to the inner and outer cams is also equal. This results in the radial plungers wedging against the contour of the inner and outer cams. When the radial plungers are wedged in this manner, the inner and outer cams lock and rotate as a unit. This produces a powerflow in which drive torque delivered to the forward power divider is divided equally between the four drive wheels.

However, if either bevel pinion (and, consequently, the inner and outer power divider cams) is forced to rotate faster than the driving cage, either the inner or outer cam overruns the driving cage, forcing half the radial plungers to slide down and half to slide up the inner cam. The cam that overruns the driving cage forces the other cam (by means of the radial plungers) to reduce its speed proportionally. In this manner, torque to the wheels is modulated so that it protects each individual drive carrier in the tandem drive set and will not permit more that 75 percent of available drive torque to be transmitted through one axle. This mitigates the potential for spinout failures.

23.4 DRIVE AXLE SHAFT CONFIGURATIONS

Two drive axle shaft configurations are used to provide support between the axle hub and the vehicle wheels:

- Semifloating axle shaft
- **Full floating** axle shaft

SEMIFLOATING AXLES

In the semifloating axle shaft, drive torque from the differential carrier is delivered by each axle half shaft directly to the drive wheels. A single bearing assembly, located at the outer end of the axle, is used to support the axle half shaft. The part of the axle extending beyond the bearing assembly is either splined or tapered to a wheel hub and brake drum assembly. The main disadvantage of this type of axle is that the outer end of each axle shaft is required to support the weight of the truck over the wheels. If an axle half shaft were to break, the wheel would fall off. This design is simple but seldom used in current trucks.

FULL FLOATING AXLES

In the full floating axle shaft, the axle half shafts transmit only driving torque to the wheels. Typically, the wheels and hubs are mounted on a pair of opposed tapered roller bearing assemblies located in the wheel assembly. With this design, the truck weight is supported by the wheel bearings, relieving the axle shafts from any role in supporting the vehicle weight. This design also has the advantage of allowing the axle half shafts to be removed without jacking up the truck's wheels. If a fully floating axle shaft should break, there is no danger of the wheel falling off, because the axle shaft's only role is to transmit drive torque to the drive wheels.

23.5 NONDRIVING AXLES

There are three categories of nondriving or **dead axles**. Because they are covered elsewhere in this book, they simply are introduced here. The three categories are:

- Steering axles
- Lift and tag axles
- Trailer axles

STEERING AXLES

Steering axles are used on the front of most heavy-duty trucks, most of which are steered by a single axle. Some vocational trucks use a pair of steering axles, one of which is a driven axle.

The most common type of steering axle uses a forged I-beam design, shown in **Figure 23–42A**. The front suspension is directly mounted to flat machined pads on each end of the I-beam. The steering control mechanism is linked to a steering control arm, steering knuckles, and kingpins. Truck steering systems

FIGURE 23–42 Various steering axle designs: (A) standard I-beam axle; (B) tubular axle; and (C) Center-Point™ steering axle

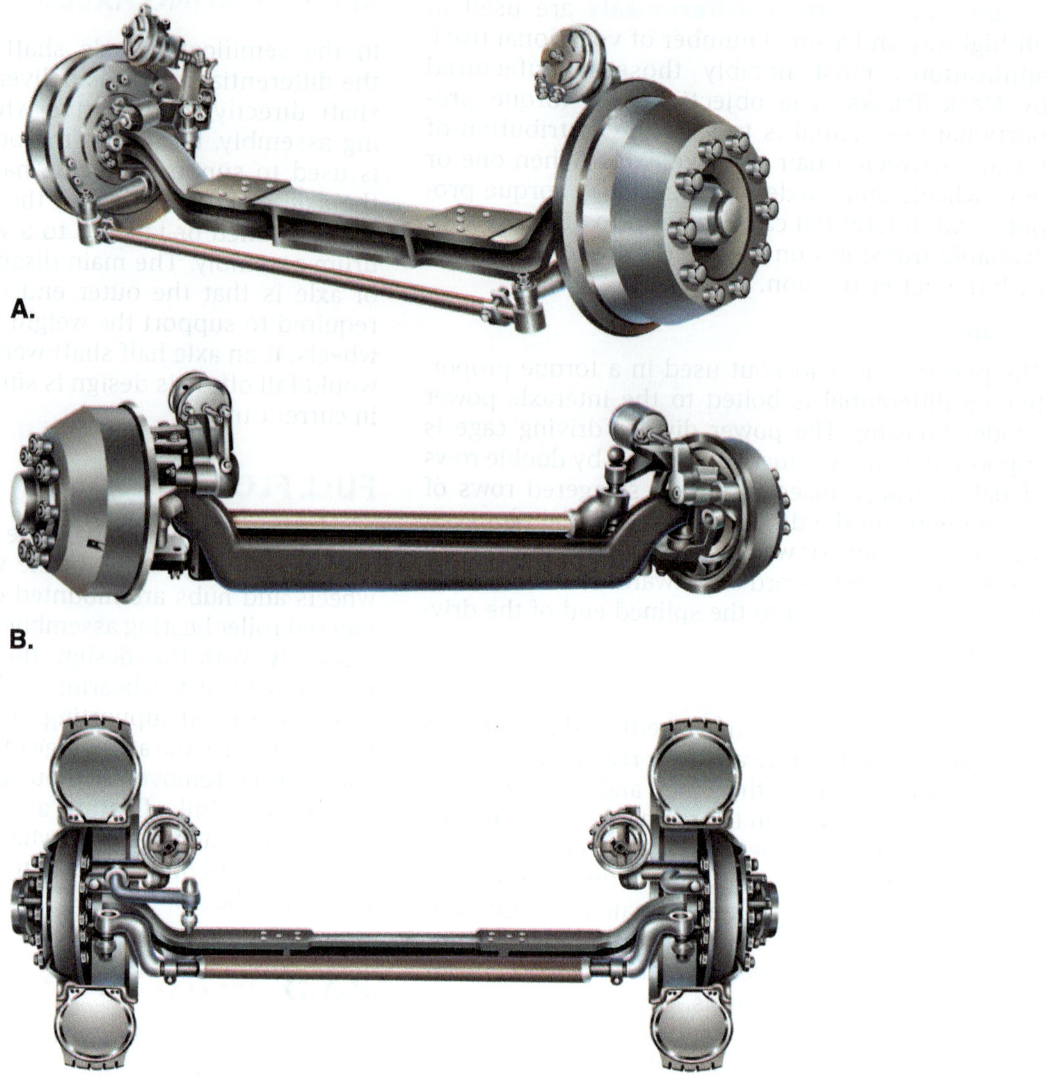

are covered in detail in Chapter 25. Two other steering axle designs are illustrated in **Figure 23–42B** and **Figure 23–42C**.

The steering knuckle or kingpin can be either of the tapered or straight type. The tapered kingpin can be drawn into the axle center by tightening a nut at the upper pin end. The straight kingpin type, shown in **Figure 23–43**, uses either one or two flats and is held in position by a tapered draw key. The steering knuckle that supports the kingpin contains bushings at the top and bottom and machined grooves designed to retain grease. Each steering knuckle is connected to the other by means of a cross tube or tie-rod, threaded at each end (for adjustment purposes) and held in position by clamp bolts. The cross tube has right-hand and left-hand threads at opposite ends to allow for toe (tracking) adjustments.

TRUCK DEAD AXLES

Some tractors also are equipped with rear dead axles. These axles provide additional load-carrying capacity, which may be required in jurisdictions with strict weight over axle regulations. If a dead axle is mounted ahead of a drive axle, it is referred to as a **pusher axle**. When a dead axle is mounted behind a drive axle, it is referred to as a **tag axle**.

FIGURE 23–43 Permanently sealed straight knuckle pin design

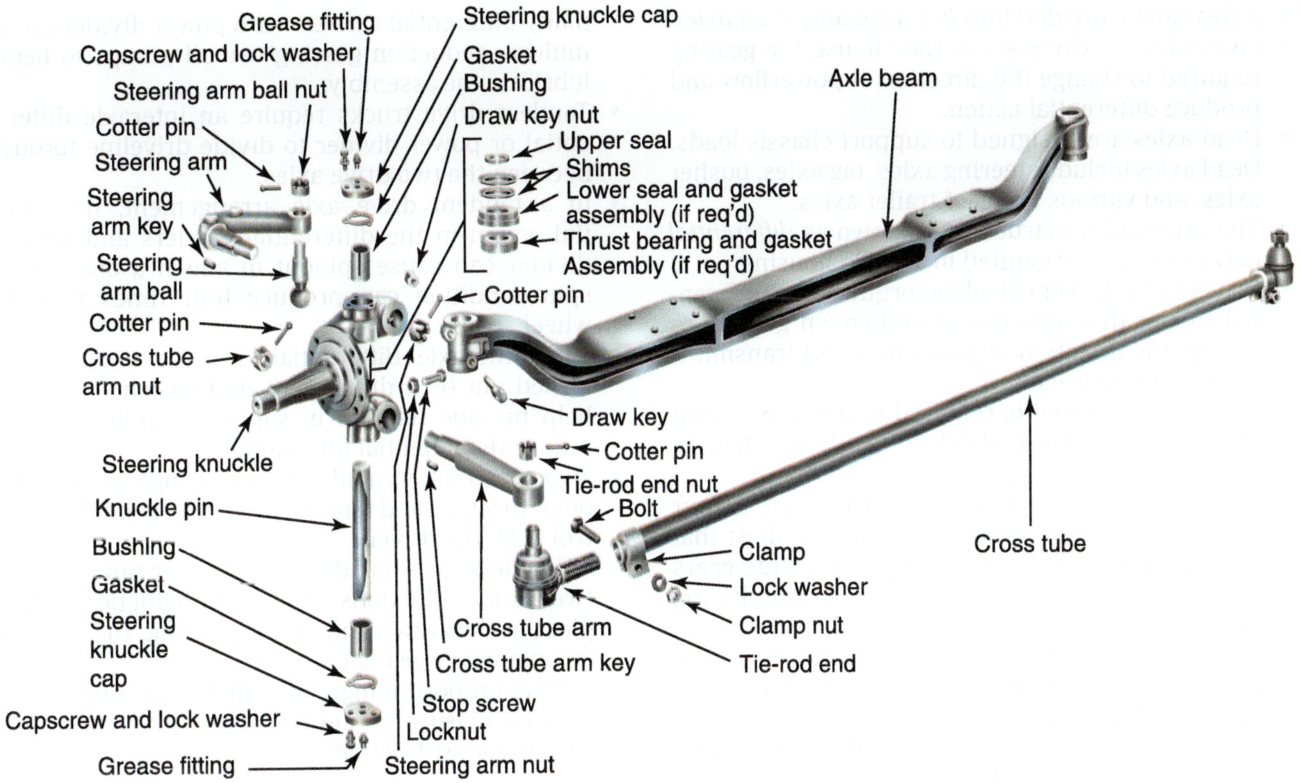

Dead axles are often lift axles, meaning that they can be raised and lowered by a cab dash-located control switch. A lift axle is usually supported on air bags when in the lowered position. The air pressure charged to the air bags is usually controllable and determines what percentage of the vehicle load is supported by the axle. An advantage of a lift axle is that it can be raised when the vehicle is travelling unloaded or turning and lowered when required to help support the chassis load.

TRAILER AXLES

Trailer axles are used to support the trailer and provide a means to mount the trailer suspension system and foundation braking components. Onhighway trailer axles are manufactured with capacities for single axles rated from 17,000 to 34,000 pounds (7,700 to 15,400 kg) but these weights may be substantially higher for offroad trailers. Axles are available with round, box, or rectangular beams; with straight and drop center designs; and with cam, wedge, or disc brakes in many sizes (**Figure 23–44**). Most brake spiders are welded to the axle beam because they have to sustain high braking forces. More information on trailer brake systems and axle configurations is provided in Chapters 31 and 33.

FIGURE 23–44 Trailer axle configurations

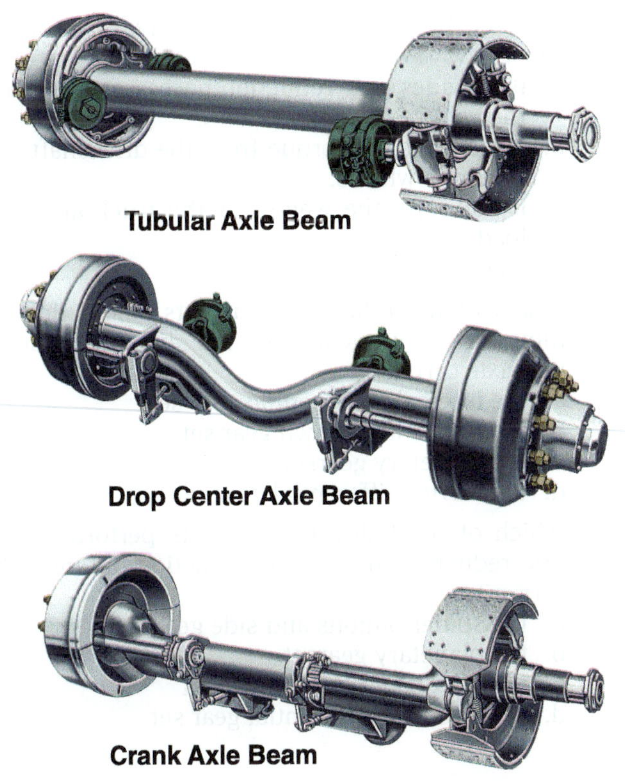

SUMMARY

- Axles can be divided into *live axles* and *dead axles.*
- Live axles are drive axles; they house the gearing required to change the direction of powerflow and produce differential action.
- Dead axles are designed to support chassis loads. Dead axles include steering axles, tag axles, pusher axles, and various types of trailer axles.
- The drive axles in trucks are known as *differential carriers* and are mounted in a banjo housing.
- Driveshafts deliver driveline torque to the differential carrier that uses pinion and crown gearing to change the direction of powerflow and transmit it to the drive wheels.
- Differential gearing is required in drive axles and enables an outboard wheel to turn faster than an inboard wheel through a turn.
- The critical differential components are a four shaft spider with pinion gears on each shaft that mesh to side gears on either side. The side gears are splined to the axle shaft responsible for driving the wheels.
- During differential action, the slowdown of one side gear is inversely proportional to the increase in speed of its opposite side gear.
- Most single-reduction axles are splash lubricated by gear oil as the geared components rotate, but many differential carriers with power dividers and multiple-reduction gearing use oil pumps to help lubricate the assembly.
- Tandem drive trucks require an interaxle differential or power divider to divide driveline torque between the two drive axles.
- In a tandem drive axle arrangement, differential action in the differential carriers and power divider can cause spinout in which a one-wheel spin condition can produce four times normal wheel speed.
- Most interaxle differentials can be temporarily locked out by a driver-activated switch. This can help provide traction in slippery conditions and reduce the potential for spinout.
- Two-speed differential carrier gearing can be used on both single and tandem drive axle vehicles; control is by the driver.
- A double-reduction differential carrier can produce two ratio reductions. The first reduction takes place at the crown and pinion gearing, the second at a planetary gearset.
- Torque proportioning power dividers reduce spinouts by biasing torque transfer to the drive axle with the most traction.

REVIEW QUESTIONS

1. What is the function of a rear drive axle on a truck?
   a. It provides gear reduction necessary to drive the vehicle.
   b. It redirects the torque from the driveshaft to the drive wheels.
   c. It supports the weight of the truck and its load.
   d. all of the above
2. Which of the following gear sets would determine the output gear ratio in a single-reduction rear axle carrier?
   a. the interaxle differential gear set
   b. the pinion and crown gear set
   c. the planetary gearset
   d. the carrier differential gear set
3. Which of the following gear sets performs the first reduction in a double-reduction differential carrier?
   a. the spider pinions and side gear set
   b. the planetary gearset
   c. the pinion and crown gear set
   d. the interaxle differential gear set
4. Which of the following gear combinations accommodates a speed difference between the two drive wheels on a drive axle?
   a. the crown and pinion
   b. the differential pinions and side gear
   c. the idler pinions and sun gear
   d. the ring gear and sun gear
5. When a truck is traveling in a straight-ahead direction, which of the following should be true?
   a. Differential pinions do not turn on the spider shafts.
   b. Differential pinions walk on the spider shafts.
   c. Differential side gears do not turn.
   d. Differential action is required.
6. Which pinion and crown gear arrangement positions the pinion gear below the centerline of the crown gear?
   a. amboid
   b. hypoid
   c. spiral bevel
   d. spur bevel

7. Which pinion and crown gear arrangement positions the pinion gear above the centerline of the crown gear?
   a. amboid
   b. hypoid
   c. spiral bevel
   d. spur bevel
8. Which of the following restricts normal differential action except during cornering?
   a. no-spin locking differentials
   b. two-speed axles
   c. controlled traction differential
   d. double-reduction differential
9. What is the usual way of supporting a differential carrier input shaft in front of the pinion?
   a. bushings
   b. single tapered roller bearing
   c. opposed tapered roller bearings
   d. ball bearings
10. In which of the following differentials would you find friction plates alternately splined to a driver and tanged to the gear support case?
    a. controlled traction differentials
    b. no-spin locking differentials
    c. two-speed tandem axles
    d. all of the above
11. In which of the following axles would you be likely to find a planetary gearset?
    a. single-reduction differential carrier
    b. trailer axles
    c. double-reduction differential carrier
    d. lift axles
12. In a tandem axle power divider, which component drives the helical gear set?
    a. the input shaft
    b. the drive pinion
    c. the ring gear
    d. the forward side gear
13. What type of bearing is used as a pilot/spigot for supporting a differential carrier input shaft behind the pinion?
    a. bushings
    b. straight roller bearing
    c. opposed tapered roller bearings
    d. ball bearings
14. Which of the following would be a factor in defining the reduction ratio in differential carrier gearing?
    a. the number of teeth on crown gear
    b. the number of splines on side gears
    c. the number of teeth on differential pinions
    d. the number of teeth on side gears
15. Technician A says that during differential action on a drive axle in a turn, the speed increase on the outboard wheel has to be inversely proportional to the speed decrease on the inboard wheel. Technician B says that pinion spiders walk at any time there is differential action. Who is correct?
    a. Technician A only
    b. Technician B only
    c. both A and B
    d. neither A nor B
16. Technician A says that a *tag axle* on a highway tractor is located in front of a drive axle or axles. Technician B says that a *pusher axle* is located behind a drive axle. Who is correct?
    a. Technician A only
    b. Technician B only
    c. both A and B
    d. neither A nor B
17. Which of the following best describes the two axles on standard tandem trailer?
    a. steer axles
    b. lift axles
    c. dead axles
    d. drive axles
18. What is the most common method of managing an interaxle differential lockout device?
    a. pneumatically engaged, spring-released
    b. spring-engaged, pneumatically released
    c. hydraulically engaged, spring-released
    d. spring-engaged, electrically released
19. Technician A says that a single wheel spinout on a tandem drive tractor can destroy a carrier assembly by shocking the gearing when the spinning wheel finds traction. Technician B says that a spinout condition can cause a lubrication failure of a final drive carrier. Who is correct?
    a. Technician A only
    b. Technician B only
    c. both A and B
    d. neither A nor B
20. Which of the following pavement surfaces would fit the description of a split coefficient surface?
    a. wheels on both sides of tractor on high friction coefficient surface
    b. wheels on both sides of tractor on low friction coefficient surface
    c. driving the tractor in snow
    d. wheels on right side in gravel, wheels on driver side on pavement

24

Prerequisite: Chapter 23

HEAVY-DUTY TRUCK AXLE SERVICE AND REPAIR

OBJECTIVES

After reading this chapter, you should be able to:

- Describe the lubrication requirements of truck and trailer dead axles.
- Outline the importance of not mixing synthetic- and mineral-based gear lubes.
- Outline the lubrication service procedures required for truck drive axle assemblies.
- Perform some basic level troubleshooting on differential carrier gearing.
- Outline the procedure required to disassemble a differential carrier.
- Disassemble a power divider unit.
- Perform failure analysis on power divider and differential carrier components.
- Reassemble power divider and differential carrier assemblies.

KEY TERMS

banjo housing
fatigue failures
inclinometer
protractor
shock failure
spinout failure
tooth contact pattern
torsional failures

INTRODUCTION

Because of the low cost of reconditioned differential carriers, they tend not to be rebuilt very frequently in the field today. This means that the role of the technician often is to diagnose a malfunction and to remove the differential carrier assembly and replace it with a rebuilt/exchange unit if a major repair is required. That said, most truck technician college programs usually include a differential carrier disassembly/reassembly shop task because it helps students develop diagnostic skills. For that reason, we outline the procedure in this chapter.

SERVICING AXLES AND DIFFERENTIAL CARRIERS

Servicing of heavy-duty truck axles consists of lubrication, inspection, diagnosis of malfunctions, and, when required, disassembly and reassembly. It is essential for truck technicians to understand the lubrication requirements of differential carrier assemblies, steering, and trailer axles. A large percentage of all axle failures have their origin in the lubricant and the maintenance practices used in an operation. Safety is important. Axles support the vehicle and its load, so an axle failure can produce fatal consequences. For instance, the lubrication of the wheel bearings is directly dependent on the oil in the axle differential carrier, so low oil level can result in a vehicle that is dangerous to operate.

SCOPE

This chapter focuses on the lubrication and servicing requirements of heavy-duty truck axles, which are skills that are required by any truck technician. We will also study the overhaul procedure of final drive units with the objective of developing some troubleshooting skills.

It should be studied along with Chapters 25, 26, and 27. Chapter 27 deals specifically with the critical adjustments of wheel end procedure, so wheel bearing lubrication and adjustment is mentioned only in passing in this chapter.

24.1 DRIVE AXLE LUBRICATION

The efficiency and life of mechanical equipment is as dependent on proper lubrication as it is on sound engineering design. Mechanical components rely on lubrication to:

- Provide a lubrication film between the moving parts to reduce friction
- Help cool components subject to friction
- Keep dirt and wear particles away from mating components

Proper lubrication depends on using the right type of lubricant at the proper intervals and maintaining the specified capacities. The recommended lubrication practices and specifications covered in this section are general in nature and typical of manufacturers' procedures. However, it is advisable to refer to the original equipment manufacturer (OEM) service literature for specific instructions.

AXLE LUBE SERVICE

With the recent widespread adoption of synthetic lubricants, manufacturers have significantly revised their lubricant drain intervals. Synthetic oils provide many advantages and most manufacturers endorse their use. The upfront cost of purchasing synthetic lubricants is greater when compared to mineral-based lubes, but if the drain interval is increased by three times, the synthetic becomes the more economical option. Generally, synthetic oils can be said to possess:

- Higher lubricity
- Much higher boil points than mineral-based oils (600°F vs. 350°F [315°C vs. 177°C])
- Better filming properties (boundary lubrication)
- Better cold weather performance

Eaton Roadranger CD-50 is one example of a synthetic lubricant that meets the current API-GL-5 specification and has extended drain intervals. In fact its advertising suggests that drain intervals can be extended by five to seven times over that of conventional gear lubes. However, because of the range of OEM recommendations that apply to using synthetic- and mineral-based lubes, we generally avoid using specific time periods or mileage values when discussing service intervals in this text. Some OEMs suggest that the initial drain and flush of the factory-fill axle lubricant on a new drive axle is unnecessary when using synthetic lubricants. Extending drain intervals reduces labor costs and the result is that most major fleets have embraced synthetic lubricants.

Up until the early 2000s, OEMs recommended an initial draining of rear axle oil in a new or reconditioned unit after the first 1,000 miles (1,600 km) and before 3,000 miles (4,800 km). The objective was to remove fine wear particles generated during break-in. These metallic particles can cause accelerated wear on gears and bearings if not removed. However, modern close-tolerance manufacturing techniques have nearly eliminated break-in wear and synthetics have superior ability to suspend contaminants. It should be noted that when synthetic gear oils are used, they should not be contaminated with mineral-based lubricants.

MIXING LUBRICANTS

When a vehicle is subject to service in multiple truck shops when on the road, the inevitable result is mixing of the rear axle lubricant. Despite the fact that some manufacturers of synthetic oils state that their product is compatible in a mix with other gear lubes, most service experts in the trucking industry suggest that mixing rear axle lubricants accelerates breakdown. Every effort should be made to avoid mixing different gear lubes in the axle housing because some of the additives will conflict. In some cases, when petroleum-based stock and synthetic lubes are mixed, thickening and foaming can result, producing premature failures.

APPROVED LUBRICANTS

All lubricants used in differential carrier assemblies must meet the American Petroleum Institute (API) and Society of Automotive Engineers (SAE) GL standards. Although it is difficult to find a gear lube on the market that is not approved, there are wide variations in what you can expect from a gear lube in terms of service and drain intervals. The best practice is to observe the OEM recommendations. Large fleets will sometimes set their own drain interval standards based on the lubricant they are using and the type of use their vehicles are subjected to. Gear lubricants classified as GL-1, GL-2, GL-3, GL-4, and GL-6 are either obsolete or no longer approved for use. GL-5 is an extreme pressure gear lubricant suitable for use in hypoid gear arrangements. It is consistent with the U.S. military specification known as MIL-L-2105-D.

Although differential gear lubes run at around 225°F (107°C) once at operating temperature, most severe wear occurs while getting the oil up to that temperature. For this reason it is important to use a gear lube of the correct viscosity. API-GL-5 rated

TABLE 24–1 Axle Gear Lube Viscosity

Ambient Temperature Range	Proper Grade
−40°F to −15°F (−40°C to −26°C)	75W
−15°F to 100°F (−26°C to −38°C)	80W-90
−15°F and above (−26°C and above)	80W-140
10°F and above (−12°C and above)	85W-140

petroleum and synthetic lubes are available in several viscosities suitable for a range of low operating temperatures. **Table 24–1** shows appropriate gear lube grades to use for the operating temperatures.

All the OEMs currently approve of the use of synthetic lubricants meeting the GL-5 performance classification. Synthetics adapt better to operating temperature extremes and can extend service intervals. They have superior centrifugal throw-off resistance, so they also have the potential to reduce the incidence of **spinout failures**.

OEMs do not generally approve the use of additives in differential carrier gear lube. Additives are most often applied to gear lube to suppress the symptoms of a mechanical condition, such as gear whine. Although an additive can work to suppress noise, the severity of the mechanical damage is likely to be greater when the repairs are eventually undertaken.

LUBE CHANGE INTERVALS

If mineral-based gear lube is used, an initial lube change should be made at 1,000 to 3,000 miles (1,600 to 4,800 km) with subsequent lube changes at 100,000-mile (160,000 km) intervals for linehaul operation; that is, terminal-to-terminal highway operation. For other types of operations, the lube should be changed more frequently. If the truck does not accumulate enough mileage to require a lube change on the basis of mileage completed, it is good practice to change the lubricant once yearly. When using synthetics go with the OEM recommendation or fleet practice. In most, but not all, cases, use of a synthetic gear lube greatly extends service intervals.

If the lube level falls below its proper level between changes, it should be replenished as needed. If loss is excessive, troubleshoot the problem. Use an API-GL-5 gear lube. To maintain proper viscosity levels, however, do not mix lube grades when adding to an existing supply.

Checking Lube Level

Remove the fill hole plug located in either the **banjo housing** or differential carrier housing using the guidelines introduced earlier in this chapter. The lube should be level with the bottom of the fill hole, as shown in **Figure 24–1**. To be seen or touched is not sufficient; it must be exactly level with the fill hole. Do not overfill the housing because this can cause problems with aerating the oil. When checking the lube level, also check and clean the housing breather. The breather is usually located on the banjo housing offset from the banjo. If this plugs, the wheel seals can be blown out.

FIGURE 24–1 Correct and incorrect ways of checking the lube level

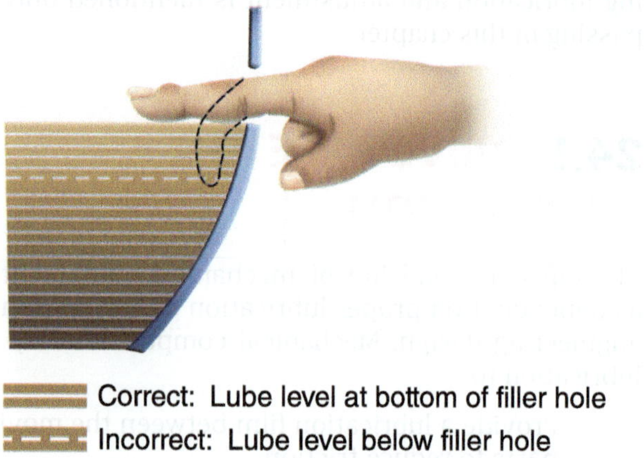

Draining Axle Lube

Drain the lubricant when still warm from both the differential carrier housing and the interaxle differential. Remember that rear axle lubricant runs at high temperatures, so you should take precautions not to get burned. The location of the drain plug is shown in **Figure 24–2**. Check the drain plug for metal cuttings. Metal debris on the drain plug can be an indication of

FIGURE 24–2 Removing the drain plug from the axle housing

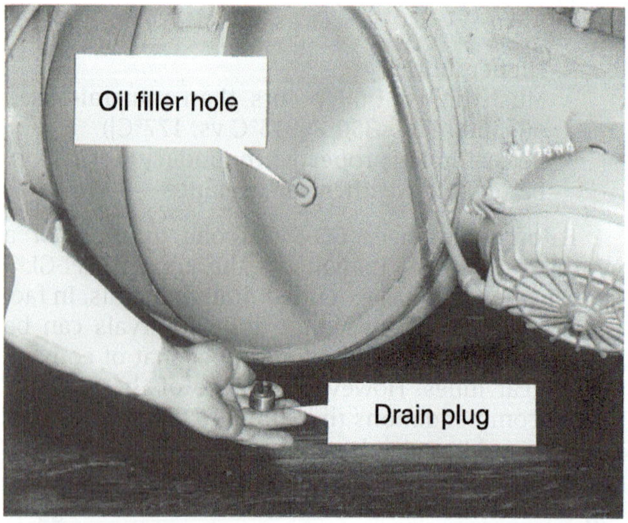

Courtesy of Dana Corporation

damage or imminent failure, so attempt to identify it. A few small particles should not be a problem. Clean the drain plug and replace it after the lube has been completely drained. The oil should be disposed of by a recycling refiner.

SHOP TALK

Draining lubricants when warm ensures that contaminants are still suspended and also reduces drain time.

Refill the Rear Axle

Make sure you use the specified lubricant. It makes sense to record the lubricant used on the vehicle data file using an electronic service tool (EST). It takes only a couple of minutes and it can help avoid mixing chassis lubricants. Pump oil into the fill hole. This may be located either in the differential carrier or in the banjo housing. With the truck parked on a level surface, the oil level should be exactly equal to the bottom of the fill.

If the drive axle is equipped with an interaxle differential, this also should be filled. Some OEMs recommend that this be filled first with a couple of quarts of lube, followed by filling the axle housing to the correct level. The interaxle differential is filled using the top filler plug hole, as shown in **Figure 24–3**.

The angle of the differential carrier pinion usually determines which oil fill port should be used to fill and set the oil level in a rear axle. Measure the differential carrier pinion angle using a **protractor** or **inclinometer**, as shown in **Figure 24–4**. If the angle is less than 7 degrees (above horizontal), use the fill hole located in the side of the carrier. If the angle is more than 7 degrees (above horizontal), use the hole located in the banjo housing.

FIGURE 24–3 Location of the power divider oil fill and oil drain plugs on a forward rear tandem axle carrier

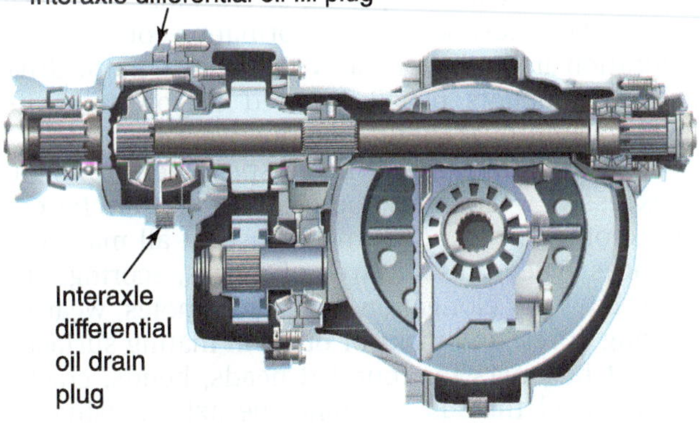

FIGURE 24–4 Proper lubricant levels

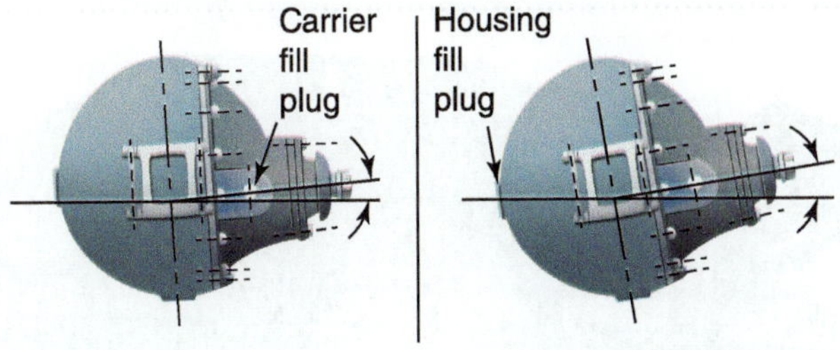

Some rear drive axles have only one lube fill hole, which is located in the banjo housing. When this is the case, you have no choice but to use this lube fill hole regardless of differential carrier pinion. Rear differential carrier axles sometimes have a smaller plug located nearby, usually just below the lubricant level plug. If there is a plug in this hole, it is for a lubricant temperature sensor that is not installed. It should not be used as a fill hole or for determining the correct axle oil level.

Assuming that you have completely drained and refilled the rear axle housing lubricant, drive the truck unloaded for a couple of miles to circulate the lubricant throughout the axle and differential carrier. Leave for 5 minutes to settle, then remove the fill plug and check the oil level. It should be exactly level with the bottom of the fill hole, as shown in Figure 24–1.

Drive Axles with Lube Pumps

To drain axles equipped with a lube pump, remove the magnetic strainer from the power divider cover and inspect for wear material in the same manner as the drain plug. The location of the strainer is shown in **Figure 24–5**. Wash the magnetic strainer in solvent and blow dry with compressed air to remove oil and metal particles.

WHEEL BEARING LUBRICATION

The lube in the differential carrier housing is responsible for lubricating the axle wheel end assemblies. One reason that the oil level in the differential carrier is critical is that the wheel bearings are lubricated by the rear axle lube. There is no external visual means of checking lubricant level in the wheel end, so the importance of making sure that the drive axle lubricant level is correct cannot be overemphasized. Wheel bearings are entirely lubricated by the gear oil in the banjo housing. There should always be some lube in the hub cavity of each wheel end (**Figure 24–6**).

FIGURE 24–5 Location of magnetic strainer for axles with lube pump

Courtesy of Dana Corporation

FIGURE 24–6 Location of wheel-end hub lube cavity

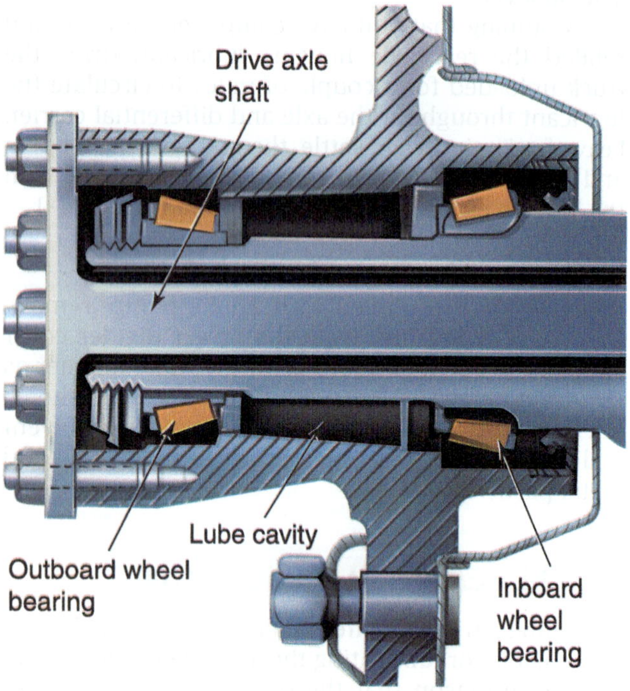

CAUTION

On most drive axles, there is no external visual means of checking lubricant level in the wheel end, so the importance of making sure that the drive axle lubricant level is correct cannot be overemphasized. Raising each side of an axle with a jack ensures that oil fills the wheel end hub cavity. Make a final check of the differential carrier oil level after tilting the axle from both sides.

When drive wheels are installed, the hub cavities should be prelubed with the same lubricant used in the differential carrier. Otherwise, they can be severely damaged before gravity and the normal action of the differential gearing and axle shafts can distribute lube to the wheel hubs.

Procedure

If the wheel assemblies have not been pulled, you should use this procedure to ensure that the wheel ends have a supply of lubricant.

1. Fill the drive axle with lube to the correct level using the fill hole in either the differential carrier or banjo housing.
2. Jack up the left side of the axle. Maintain this position for 1 minute to allow lube to drain to the wheel end on the right side.
3. Jack up the right side of the axle. Maintain this position for 1 minute to allow lube to drain to the wheel end on the left side.
4. Lower the axle to a level position and then check the lube level at the differential carrier fill hole. Add lube if necessary.

24.2 TRUCK AXLE CLEANING AND INSPECTION

The exterior of a differential carrier assembly can be pressure washed or steam cleaned when mounted in the housing, providing the breather has been plugged. Once removed from the banjo housing, the differential carrier should be disassembled before any attempt is made to clean it. When disassembled, the drive axle subcomponents should be washed in shop solvent and air-dried. The housing casting can be washed with solvent and a brush and then soaked in a hot tank using mild alkaline solution. Dry with compressed air and lightly spray with lubricant to prevent surface rusting before reassembly.

DIFFERENTIAL CARRIER INSPECTION

Once disassembled, inspect the components for damage and wear. Look for pitting or cracking along gear contact surfaces. Scuffing, deformation, or heat discoloration are indicators of excessive heat in the drive axle components usually related to low lubricant levels or improper lubrication practices.

Before reusing a gear set, inspect teeth for signs of excessive wear. Check tooth contact pattern for evidence of incorrect adjustment. Inspect all machined surfaces. They must be free of cracks, scoring, and wear. Look for elongation of fastener holes, wear on bearing bores, and nicks or burrs in mating surfaces. Inspect fasteners for rounded heads, bends, cracks, or damaged threads. Examine the axle housing for cracks or leaks. You can check out a leak using a dye penetrant kit. Also check for loose studs or cross-threaded holes. Inspect machined surfaces for nicks and burrs.

24.3 CARRIER, AXLE, AND GEARING IDENTIFICATION

Each OEM has similar methods of coding its identification data. Identification of the differential carrier by the manufacturer is either stamped on the carrier itself or on a metal tag attached to the differential carrier. An example of a carrier identification tag is shown in **Figure 24–7**.

The complete drive axle assembly is usually identified by the specification number stamped onto the banjo housing (**Figure 24–8**). This number identifies all the component parts of the axle, including specific OEM requirements such as yoke or flange. In addition, some axles might have a metal identification tag riveted to it. This usually falls off after a couple of years' service.

The crown and pinion matched set must be replaced together. Check the appropriate OEM axle parts listings for part numbers. To identify gear sets, both parts of the gear set are stamped with such information as the number of pinion and ring gear teeth, individual part number, and matched set number, as shown in **Figure 24–9**.

FIGURE 24–9 Crown gear and pinion identification

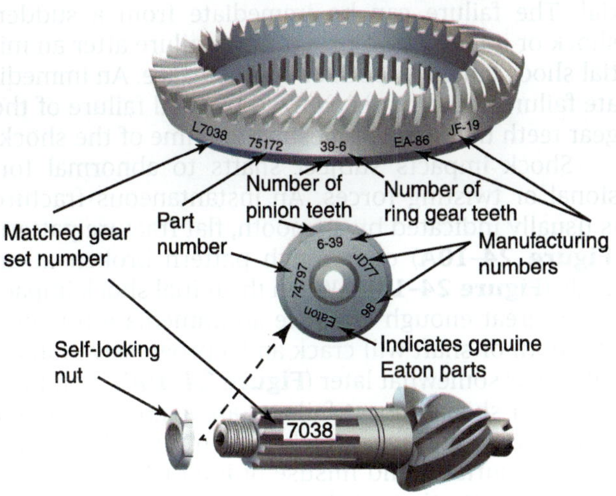

FIGURE 24–7 Differential carrier identification

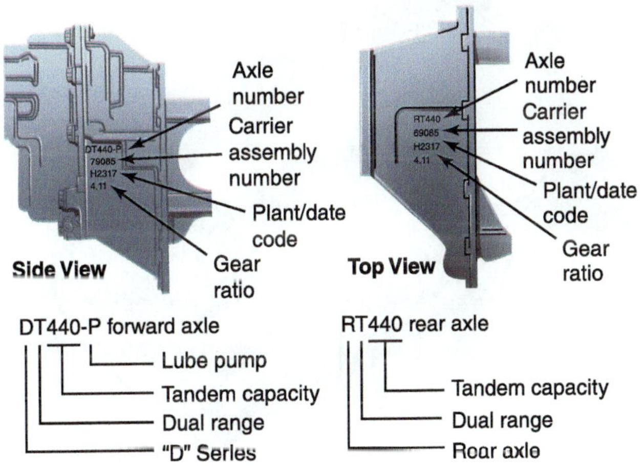

FIGURE 24–8 Axle specification number identification

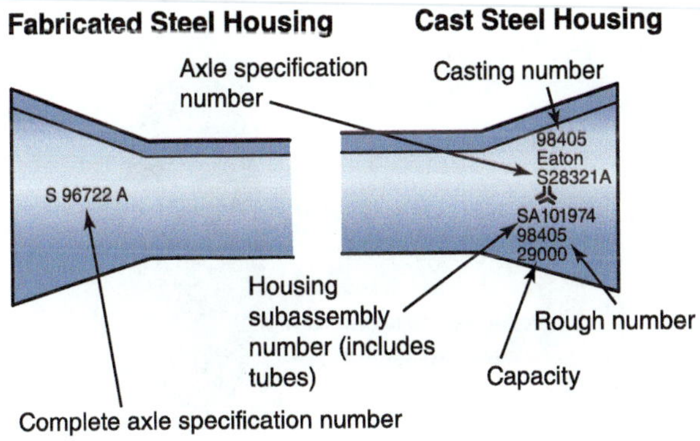

24.4 FAILURE ANALYSIS

Differential carriers are more subject to driver-abuse failures than any other component on a truck except perhaps the clutch. When a drive axle fails, failure analysis should answer two key questions: What happened? How can a recurrence be avoided? Sometimes failure analysis is complicated and it takes lab equipment to determine the cause. But most failures can be identified in the shop. In this section, we take a look at some of the skills required to identify and determine the root cause of a failure so recurrences can be avoided.

TYPES OF FAILURE

Differential carriers usually fail because of one or more of the following reasons:

- Shock load
- Fatigue
- Spinout
- Faulty lubrication
- Normal wear

Shock Failures

Shock failures occur when the gear teeth or shaft have been stressed beyond the strength of the material. The failure can be immediate from a sudden shock or it could be a progressive failure after an initial shock cracks a tooth or shaft surface. An immediate failure could be recognized by total failure of the gear teeth that were in mesh at the time of the shock.

Shock impacts subject shafts to abnormal torsional or twisting forces. An instantaneous fracture is usually indicated by a smooth, flat fracture pattern (**Figure 24–10A**) or a rough pattern broken at an angle (**Figure 24–10B**). When the initial shock impact is not great enough to cause an immediate fracture, the tooth or shaft will crack and a progressive failure will occur somewhat later (**Figure 24–10C**). Common causes of shock impact failures are aggressive trailer hookups, a spinning wheel suddenly grabbing onto the road surface, and misuse of interaxle differential lockouts. Note that all these are failures that can be attributed to driver abuse.

CAUTION

Most driver-abuse generated failures do not cause an instantaneous equipment failure. The equipment failure can take place some time after the driving incident that generated it. This is important to remember when attempting to attribute blame in fleets that do not assign drivers dedicated trucks.

Fatigue Failures

A **fatigue failure** is usually defined as a progressive destruction of shaft or gear teeth material. This type of failure can usually be attributed to overloading of the drivetrain. The contact surfaces of gears and shafts are case hardened (chemically hardened, usually by nitriding) for wear resistance. When under the case-hardened surface, a softer, ductile core is retained to provide some yield to stresses. Fatigue failures can occur both in the surface or core of shafts and gears. They can be classified in three ways:

- Surface failures. Surface failures are identified by cracked, pitted, or spalled tooth surfaces. The cause is usually geartrain overloading. As the failure progresses, the surface material (i.e., the case-hardened skin) flakes away. In some instances, an elliptical design (or "beach" marks) can be created on the surface pattern. **Figure 24–11** illustrates typical surface failures.
- Torsional (or Twisting) failures. **Torsional failures** in shafts are usually indicated by a star-shaped fracture pattern; a classic example of this is shown in **Figure 24–12**. An extremely high shaft load can initiate the crack, which then progresses to the center of the core to complete the fracture.

FIGURE 24–10 (A) Smooth and (B) rough fracture patterns; and (C) a fractured tooth as a result of instantaneous shock

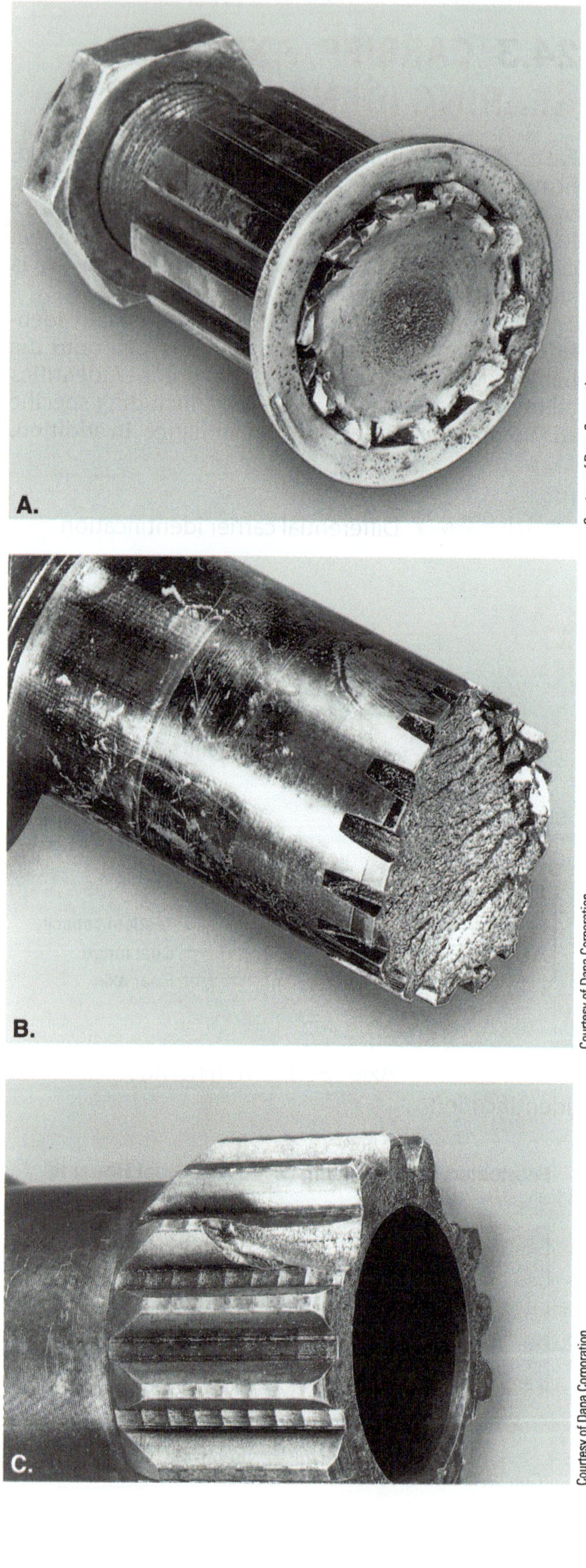

Courtesy of Dana Corporation
Courtesy of Dana Corporation
Courtesy of Dana Corporation

FIGURE 24–11 Surface failure patterns as a result of fatigue: (A) cracks and spalling; and (B) pitting and spalling

Courtesy of Dana Corporation

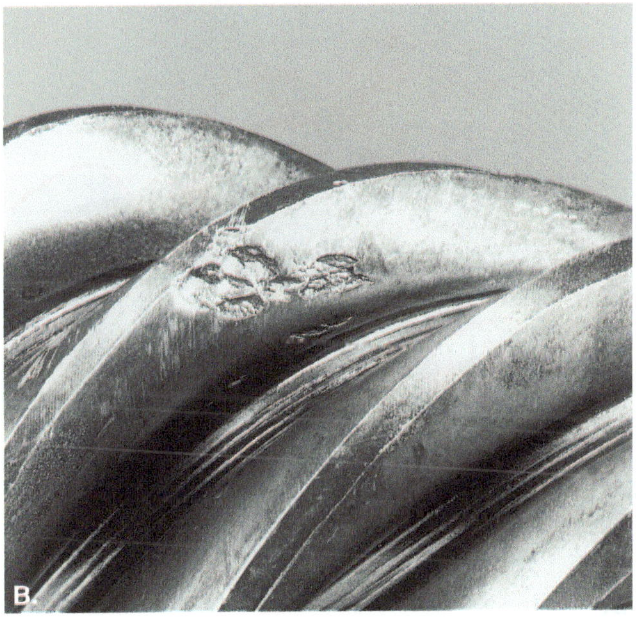

Courtesy of Dana Corporation

- Bending failures. In shafts, rotating and bending forces will usually cause a scalloped (or spiral) type of fracture pattern, as shown in **Figure 24–13**. Overloads progressively crack the shaft until it weakens sufficiently for a complete fracture. In gears, bending forces initially crack the tooth surface. As the crack continues to grow, a "beach" mark design is formed in the fracture pattern (Figure 24–13A). The tooth is finally weakened to a point where it breaks. Bending failures are generally caused by overloading the truck beyond the rated capacity or by abusive handling of the drivetrain over rough terrain.

FIGURE 24–12 Star-shaped torsional failure

Courtesy of Dana Corporation

Spinout Failures

Spinout is the result of differential action that spins one wheel up to abnormal speeds when on a slippery surface (spinout is fully explained in Chapter 23). Spinout is a far too common cause of differential carrier failures and is always attributable to driver abuse.

On a single rear drive axle, spinout occurs when differential action allows one wheel to remain stationary, as shown in **Figure 24–14A**. On a tandem rear drive axle system, spinout occurs in the interaxle differential when one of the axles remains stationary and the other axle spins out on a slippery surface (**Figure 24–14B**). In differential action, the increase of speed at one wheel (or axle) is inversely proportional to the decrease of its opposite. When a wheel or axle is held stationary, the speed of its opposite is twice what it should be relative to drive shaft speed. This high-speed rotation throws off the lube film and metal-to-metal contact occurs, creating friction, heat, and a subsequent failure. Spinout can produce both immediate and delayed component failures.

Spinouts can initiate other types of failure. If a spinning wheel suddenly grabs to the road surface, a shock load will result. If shock is severe enough, this can result in broken gear teeth or shaft fracture. Spinout failures can be prevented by proper handling of the truck when traction loss is encountered. As soon as a spinout condition develops, brake the truck and engage the interaxle differential lockout. If this does

FIGURE 24–13 Bending failure patterns as a result of fatigue: (A) beach mark; and (B) spiral pattern

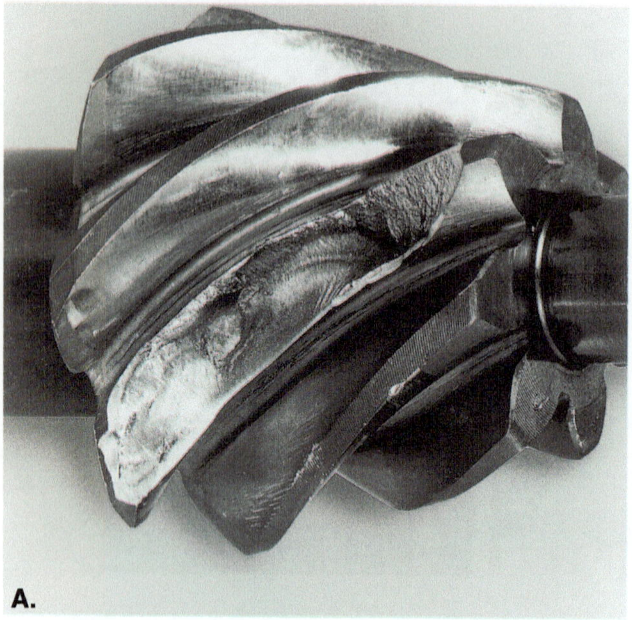

FIGURE 24–14 (A) Spinout on single rear drive axles; and (B) spinout on tandem rear drive axles

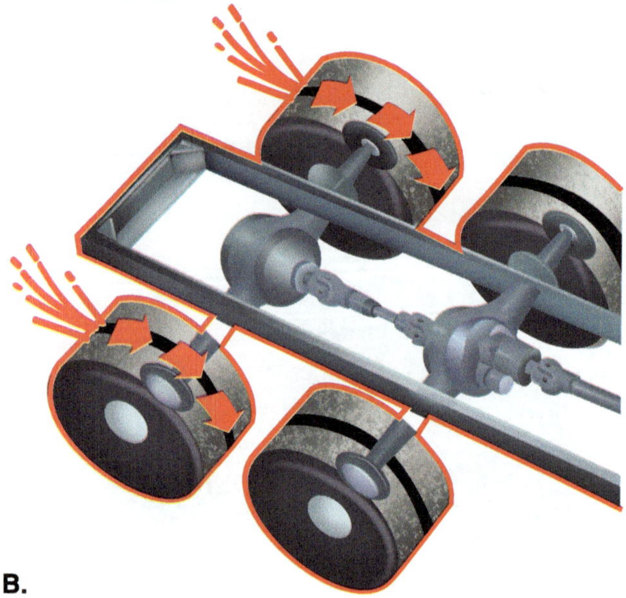

not eliminate spinout, then sand, salt, chains, or traction rugs might help. The bottom line is that even if the vehicle has to be towed out of the problem, the cost will be significantly less than destroying drivetrain components.

Lubrication Failures

Lubrication deficiencies can greatly affect the life of bearings, gears, and thrust washers in differential carriers. Here are some typical reasons:

- *Contaminated lube* from moisture, use of dirty oil, or break-in particles can cause etching, scoring, or pitting of the contact surfaces. Foreign material in the lube acts as an abrasive and shortens service life.
- *Inadequate lubrication* is a common problem, usually resulting from underfilled rear axle housings. The resulting metal-to-metal contact creates friction that causes overheating and further breakdown of the protective film. The end result is seizure and localized welding of mating components and is recognizable by severe scoring or galling and actual melting of surface materials. Components are usually black and discolored from tempering.

- *Improper lube* does not provide the lubricating film that is required to prevent metal-to-metal contact. Included in the improper lube category is lubricant that is never changed but rather just topped up over the years with whatever oil is handy at the moment of the service. This results in a brew of lubricants with conflicting properties. This causes surface etching and lubricant breakdown, creating metal-to-metal contact and surface originated failures.
- *Incorrect viscosity lubricant* for the operating temperatures could result in breakdown of the lube film. Mixing incompatible lubricants also can stratify the lubricant into differing viscosities, damaging internal components. Using heavy gear lubes in extremely cold conditions can result in almost no lubrication during cold startup.

Normal Wear

Drive axles are engineered and manufactured to achieve a reasonable useful life. As any OEM will tell you, all manufactured components will eventually wear under normal use. During break-in, some wear will occur in any mechanical assembly. In many instances, this type of wear is beneficial. Mating components improve their mesh or surface contact. Preventive maintenance is important to minimize wear.

Some marks and patterns on surface finishes are the result of manufacturing processes. These should be recognized and not classified as a failure condition. It is important to recognize normal wear to eliminate unnecessary component replacement. Technicians should become acquainted with and be able to identify normal wear patterns of components. Washers and seals should, however, always be replaced when the axle is being repaired.

IDENTIFY ROOT CAUSE

Problems associated with different models of axles and types of gearing can be specific to one model only. It is important to identify the root cause of a failure to avoid repeat failures. In all troubleshooting of final drive problems, technicians should learn to see the drivetrain as a set of components working together and to realize that a failure in one component could have its source in another. Universal joints, drive shafts, transmissions, tires, and the final drive gearing all play roles in delivering driving torque to the drive wheels, a fact that is important to remember when making a diagnosis. Some typical troubleshooting symptoms and their probable causes are listed in **Table 24–2**.

DIFFERENTIAL CARRIER OVERHAUL

Overhaul of the differential carrier varies somewhat between different makes and models of the same units. However, the general procedure will follow a similar pattern. The overhaul procedure presented here is based on a single-reduction differential carrier. Refer to the OEM service literature for the procedure to overhaul specific differential carrier.

TABLE 24–2 Drive Axle Troubleshooting

Symptom	Probable Cause	Remedy
Noisy on turns only	Differential pinion gears tight on spider Side gears tight in differential case Differential or side gears defective Excessive backlash between side gears and pinion	Overhaul drive axle and make necessary adjustments
Intermittent noise	Crown gear not running true Loose or broken differential bearings	Overhaul axle and replace defective crown gear or differential bearings
Constant noise	1. Lubricant incorrect 2. Lube level low 3. Crown gear teeth chipped or worn; loose or worn bearings 4. Crown gear and pinion not in adjustment for correct tooth contact 5. Too much or too little pinion-to-gear backlash or overlap of wear pattern	1. Verify type and class of lubricant used 2. Check lube level and fill if needed 3. Overhaul axle and replace defective crown gear, pinion, or bearings 4. Adjust crown gear and pinion for correct tooth contact 5. Adjust gear backlash
Rear wheels do not drive (driveline rotating)	1. Broken axle shaft 2. Crown gear teeth stripped 3. Differential pinion or side gear broken 4. Differential spider broken	1. Replace broken axle 2. Overhaul axle and replace defective crown gear, pinion, or spider

REMOVING A DIFFERENTIAL CARRIER

This procedure outlines the steps required to remove a differential carrier from the banjo housing of a truck drive axle.

1. If the final drive is a dual range axle, shift the axle to the low range.
2. Use a hydraulic jack to raise the truck axle housing so that the wheels just clear the floor.
3. Place axle stands under each side of the axle housing to hold the truck in the raised position, as shown in **Figure 24–15**. Try to make sure that the axle housing is horizontally close to level.
4. Remove the drain plug from the bottom of the axle housing and drain the lubricant from the assembly. Dispose of the lube properly.
5. Disconnect the driveline U-joint from the pinion yoke or flange on the carrier.
6. Back off the capscrews or stud nuts from the flanges of both axle shafts at each wheel. Do *not* completely remove the fasteners.
7. Next, loosen the tapered spring dowels from the axle shaft flanges. Use a brass drift and a large hammer to loosen the dowels, as shown in **Figure 24–16**, remembering that when they loosen, they all pop at once and can fly off like bullets if the axle retaining nuts have been removed.
8. When the axle shaft has popped free, you can remove the nuts, washers, and tapered spring dowels. Pull both axle shafts from the axle housing.
9. On dual range drive axles, disconnect shift unit air lines. Remove shift unit, catching any oil that escapes from reservoir.

FIGURE 24–15 Supporting truck with axle stands under each side of the axle housing

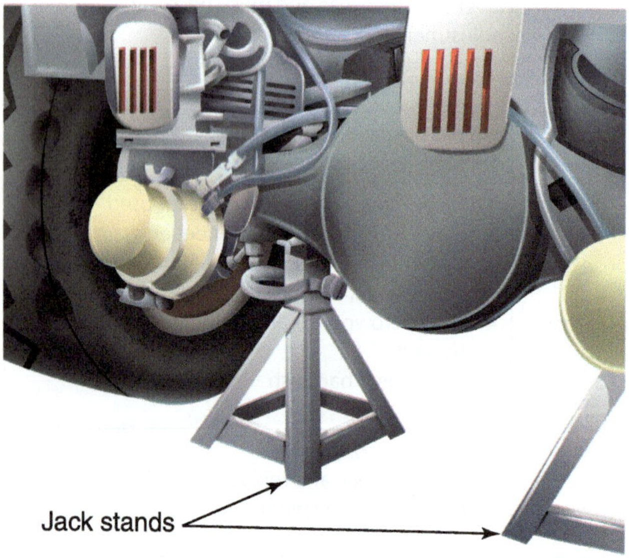

FIGURE 24–16 Loosen the tapered dowels with a brass drift and large hammer.

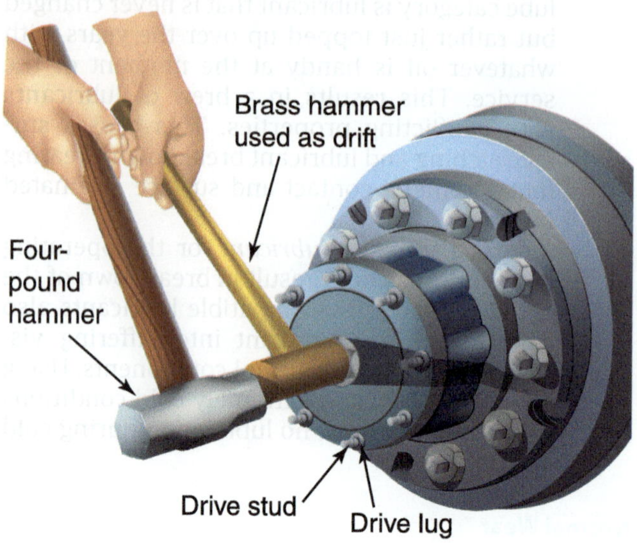

10. Place a differential jack, a transmission jack with differential cradle, or a hydraulic roller jack (**Figure 24–17**) under the differential carrier to support the assembly. The differential carrier should be properly fixed to the jack supporting it, remembering that most of the weight is on the opposite side of its mounting flange.
11. Remove all but the top two carrier flange-to-housing capscrews or stud nuts and washers.
12. Back off the top two carrier-to-housing fasteners, leaving them attached to the assembly. The fasteners will hold the carrier in the housing.
13. Loosen the differential carrier from the axle housing. Use some thrust force applied to the nose of the differential carrier to achieve this.

FIGURE 24–17 Support the differential carrier with a hydraulic jack.

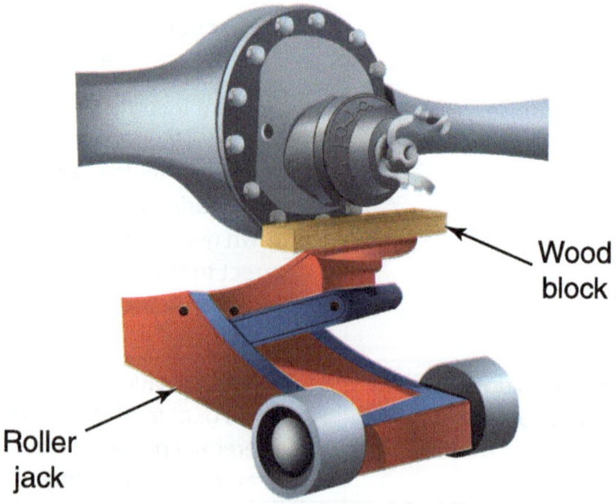

14. Carefully remove the carrier assembly from the banjo housing by wriggling the jack from side to side while pulling away. If you have to use a prybar, make sure that you do not damage the flange faces or gearing.
15. Remove and discard the carrier flange-to-banjo housing gasket.
16. Move the differential carrier from under the truck, raise it by its yoke, and mount in a repair stand, as shown in **Figure 24–18**. You can use a cherry picker or chain hoist for this procedure.

SHOP TALK

You sometimes have to use more force to pop axle shafts than can be delivered using a drift and 4-pound hammer, as shown in Figure 24–16. When this method does not work, use a 16-pound sledgehammer directly on center of the axle shaft flange; use a ¼ swing of the sledgehammer, allowing the weight of the hammer do all of the work.

WARNING

The use of a full face shield is recommended when using a sledgehammer rather than relying on safety glasses alone.

CAUTION

Most of the weight of a differential carrier assembly is on the inboard side of its mounting flange. Ensure that the assembly is properly fastened to the jacking device and that your body is never positioned under the carrier.

FIGURE 24–18 Place the removed differential carrier in a repair stand.

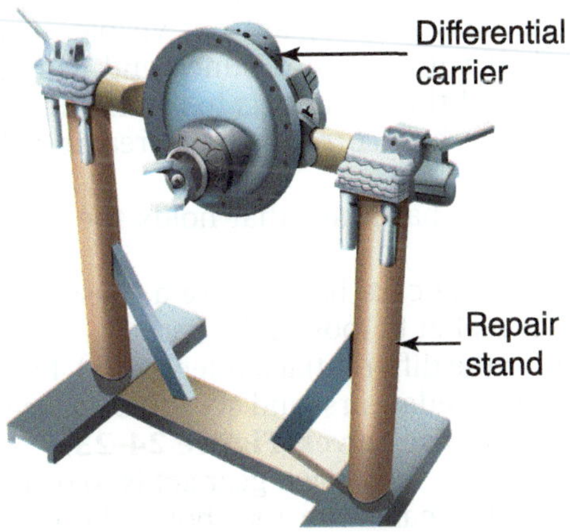

DIFFERENTIAL CARRIER DISASSEMBLY

Before disassembling the carrier, visually inspect the hypoid gear set for damage. If no damage is apparent, you should be able to reuse the gear set. Measure the backlash of the gear set and record the dimension (see the section on adjustment in this chapter). On reassembly, the crown and pinion backlash should be adjusted to the same dimension. The best overhaul results are obtained when used gearing is adjusted to run in established wear patterns. Omit this procedure if the gear set is to be replaced. To remove the differential and ring gear from the carrier, use the following procedure:

1. Loosen the jam nut on the thrust screw, if applicable (some carriers do not have a thrust screw).
2. Remove the thrust screw and jam nut from the differential carrier (**Figure 24–19**).
3. Rotate the differential carrier assembly in the repair stand until the ring gear is at the top of the assembly.
4. Mark one carrier leg and bearing cap for the purpose of correctly matching the parts when the carrier is reassembled. A center punch and hammer can be used to mark these components (**Figure 24–20**). Avoid using paint because it may contaminate the lubricant when placed back in service.
5. Remove the cotter keys, pins, or lock plates that hold the two bearing adjusting rings in position. Use a small drift and hammer to remove the pins. Each lock plate is held in position by two capscrews (**Figure 24–21**).

FIGURE 24–19 Thrust screw, jam nut, and thrust block

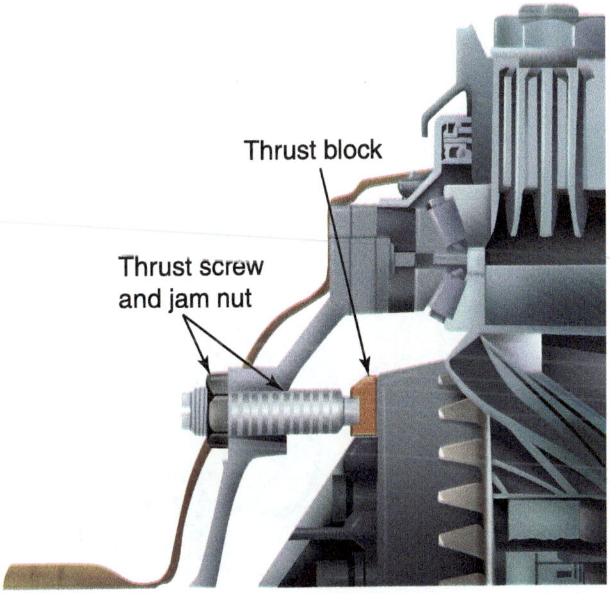

FIGURE 24–20 Marking the carrier components for reassembly

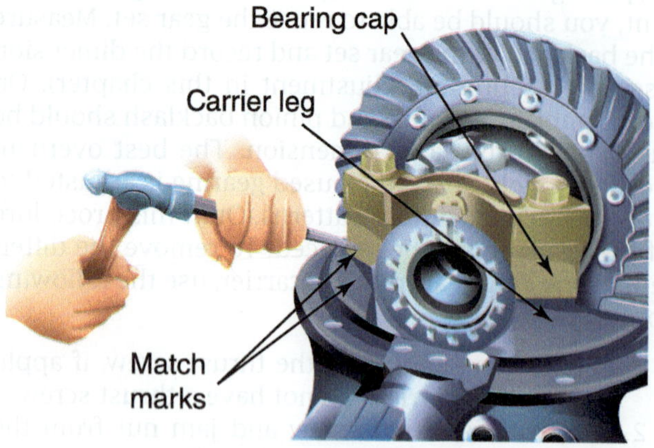

FIGURE 24–21 Removing the (A) cotter keys or (B) lock plate from adjusting rings

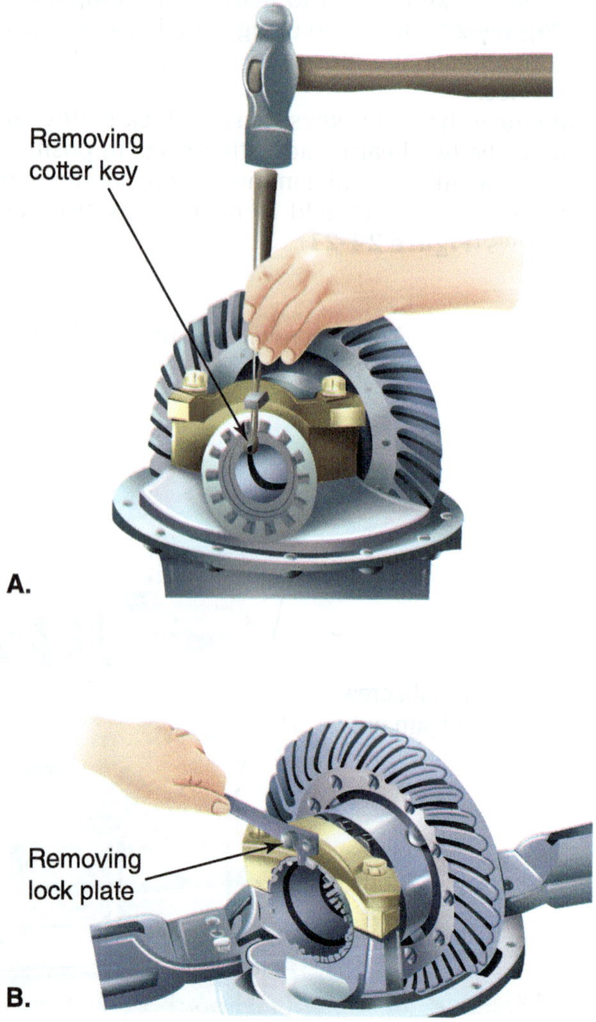

FIGURE 24–22 Removal of the bearing cap and adjusting ring

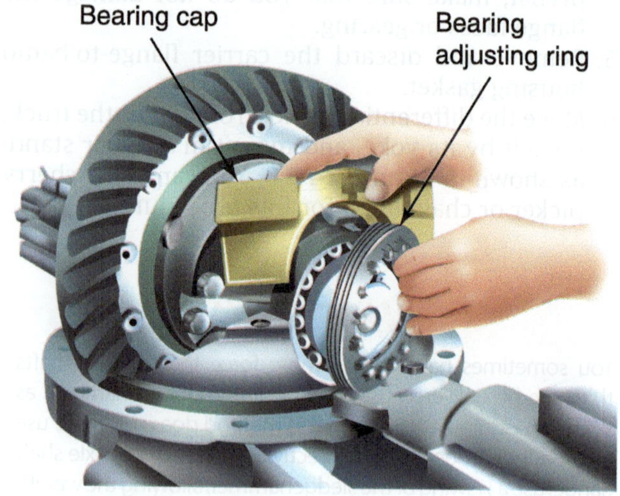

6. Remove the capscrews and washers that hold the two bearing caps on the carrier. Each cap is held in position by a pair of capscrews and washers. When reusing the crown and pinion gear set, remove the left-hand bearing cap, adjuster, and lock as a unit. This will help return the gear set to its original adjustment during reassembly.
7. Remove the bearing caps and bearing adjusting rings from the carrier (**Figure 24–22**).
8. Safely lift the differential and crown/ring gear assembly from the carrier and place it on a workbench.
9. Remove the thrust block (if provided) from inside the carrier. (The thrust block will fall into the carrier when the thrust screw is removed.)

Differential and Crown Gear Disassembly

To disassemble the differential and crown/ring gear assembly, you should use the following procedure:

1. If the matching marks on the case halves of the differential assembly are not visible, mark each case half with a center punch and hammer. The purpose of these marks is to match the plain half and flange half correctly when reassembling the carrier.
2. Remove all hardware that holds the case halves together.
3. Separate the case halves. Use a brass, plastic, or leather mallet to loosen the parts.
4. Remove the differential spider (cross), four pinion gears, two sidegears, and six thrust washers from inside the case halves (**Figure 24–23**).
5. If the pinion and crown gear set is to be replaced, remove the hardware that holds the gear to the

FIGURE 24–23 Disassembling the differential and crown gear

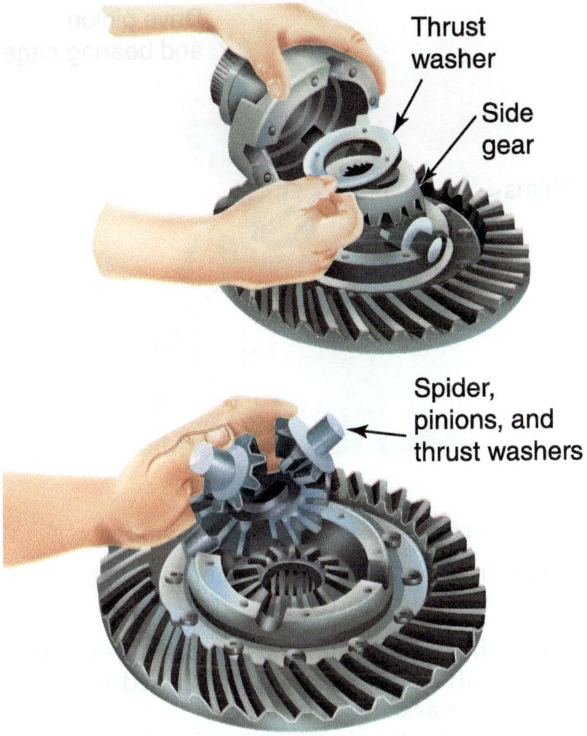

flange case half. If rivets hold the ring gear to the flange case half, remove them as follows:

- Carefully center punch each rivet head in the center on the crown gear side of the assembly.
- Drill each rivet head on the ring gear side of the assembly to a depth equal to the thickness of one rivet head. Use a drill bit that is 1⁄32 inch (0.8 mm) smaller than the body diameter of the rivets, as shown in **Figure 24–24**.
- Drive the rivets through the holes in the ring gear and flange case half. Press from the drilled rivet head.
- Separate the case half and ring gear using a press. Support the assembly under the ring gear with metal or wooden blocks and press the case half through the gear (**Figure 24–25**).
- If the differential bearings need to be replaced, remove the bearing cones from the case halves. Use a bearing puller or press to remove them.

CAUTION

Do not remove the rivet heads or rivets with a chisel and hammer because this can damage the flange case half or enlarge the rivet holes, resulting in loose rivets.

FIGURE 24–24 (A) Drill and punch out rivets. (B) Never use a hammer and chisel to remove rivets.

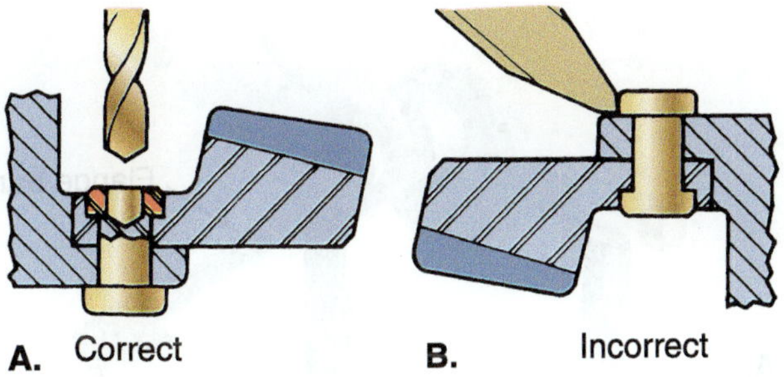

FIGURE 24–25 Pressing the flange case half out of the crown gear

Removing Drive Pinion and Bearing Cage

To remove the drive pinion and its bearing cage from the carrier, use the following procedure:

1. Fasten a yoke bar or flange bar to the input yoke or flange to prevent if from turning while you remove the nut, as shown in **Figure 24–26**.
2. Remove the retaining nut from the drive pinion using an air gun. Then remove the yoke or flange bar.
3. Remove the yoke or flange from the drive pinion. If the yoke or flange is tight on the pinion, use a puller to remove it.
4. Remove the fasteners that retain the bearing cage in the carrier (**Figure 24–27**).
5. Remove the cover and seal assembly and the gasket from the bearing cage. If the cover is tight on the bearing cage, use a brass drift and hammer for removal.

FIGURE 24–26 Removing the input (A) flange or (B) yoke

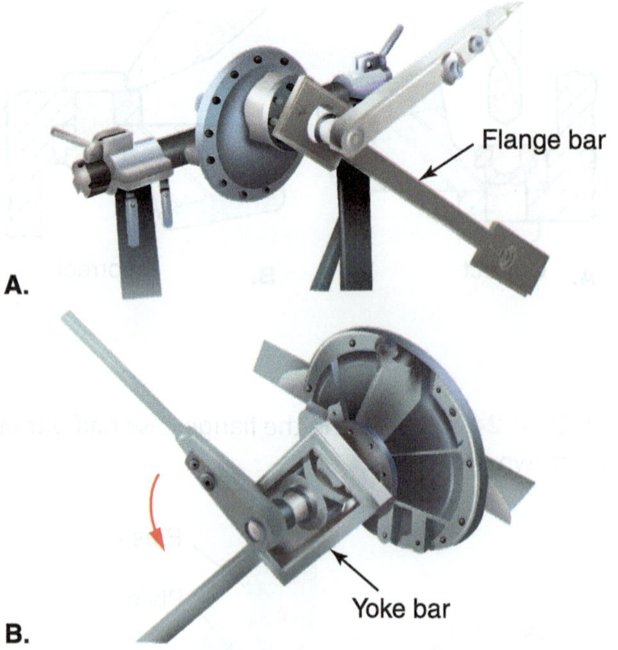

FIGURE 24–27 Bearing cage removal

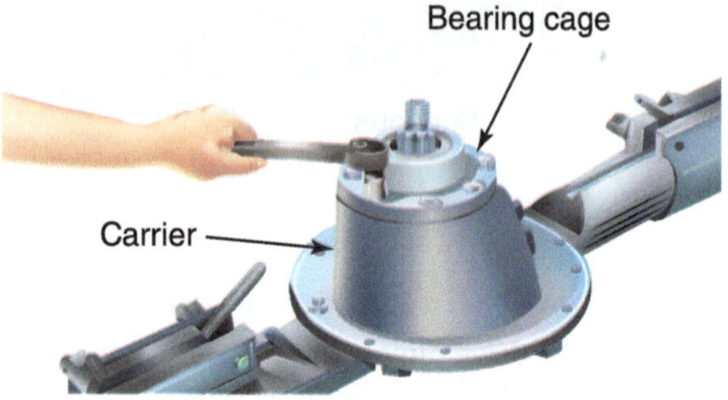

6. If the pinion seal is damaged, remove the seal from the cover. Use a press and sleeve or seal driver. If a press is not available, use a screwdriver or small prybar for removal. Discard the pinion seal.
7. Remove the drive pinion and bearing cage, plus the shims from the carrier (**Figure 24–28**).
8. If the shims are in good condition, keep them and use them for reassembly of the carrier.
9. If the shims are to be discarded because of damage, measure the shim pack total thickness with a micrometer. Record the dimension. It will be needed to calculate the depth of the drive pinion in the carrier when the gear set is installed.

FIGURE 24–28 Removing the drive pinion and bearing cage

Drive Pinion and Bearing Cage Disassembly

To disassemble the drive pinion and bearing cage, use the following procedure:

1. Place the drive pinion and bearing cage in a press. The shaft end should be positioned toward the top of the assembly.
2. Support the bearing cage under the flange area with metal or wooden blocks.
3. Press the drive pinion through the bearing cage (**Figure 24–29**). Do not allow the pinion to fall to the floor from the press when the bearing is free.
4. If the pinion oil seal is mounted directly in the outer bore of the bearing cage, remove the seal at this time. Be careful that the mounting surfaces of the bearing cage are not damaged.

FIGURE 24–29 Pressing the drive pinion from the bearing cage

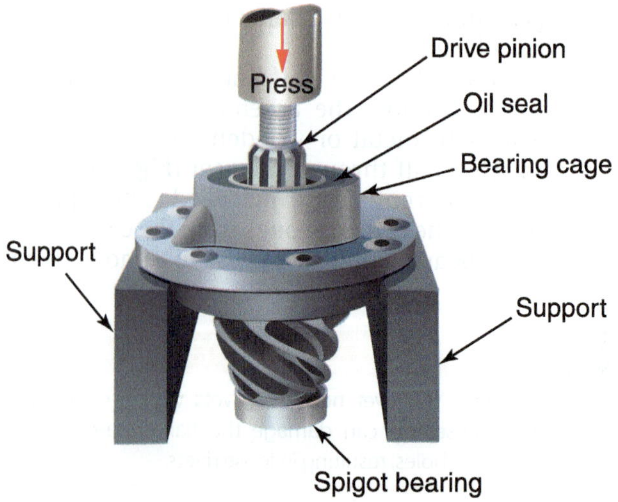

5. If the seal is a one-piece design (without mounting flange), discard the seal. If the seal is a triple-lid design (with flange), inspect the seal for damage. If the surfaces of the seal and yoke or flange *are smooth and not worn or damaged*, it is possible to reuse the seal again during reassembly of the unit. However, replacement is recommended to prevent the cost associated with premature seal failure in the rebuilt unit.
6. If the pinion bearings need to be replaced, remove the inner and outer bearing cups from the inside of the cage. Use a press and sleeve, a bearing puller, or a small drift and hammer.
7. If the pinion bearings need to be replaced, remove the inner bearing cone from the drive pinion with a press or bearing puller. The puller must fit under the inner race of the cone to remove the cone correctly without damage (**Figure 24–30**).
8. If the spigot bearing needs to be replaced, put the drive pinion in a vise. Install a soft metal cover over each vise jaw to protect the drive pinion.
9. Remove the snapring from the end of the drive pinion (**Figure 24–31**).
10. Remove the spigot bearing from the drive pinion with a bearing puller.

DIFFERENTIAL CARRIER REASSEMBLY

The reassembly procedure is essentially a reversal of the disassembly procedure. The procedure outlined here is general and students are urged to consult the service literature for the specific units being worked on.

FIGURE 24–30 Pinion bearing removal

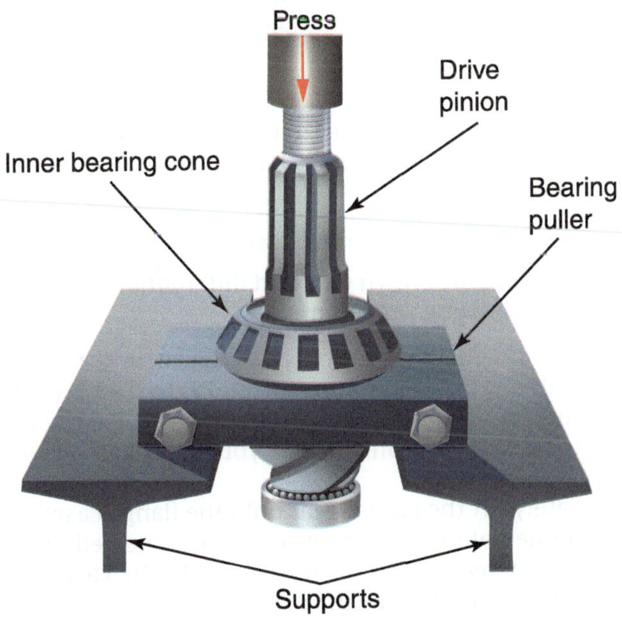

FIGURE 24–31 A snapring secures the spigot bearing to the pinion shaft.

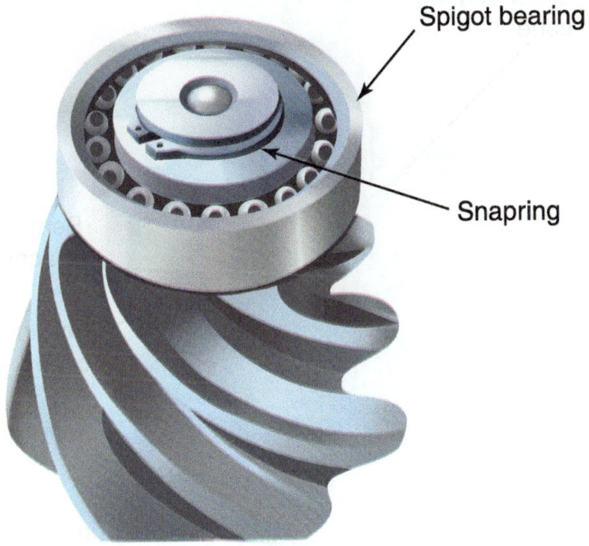

Drive Pinion and Bearing Cage Reassembly

To reassemble the drive pinion and bearing cage, use the following procedure:

1. Place the bearing cage in a press. Support the bearing cage between metal or wooden blocks.
2. Press the bearing cup into the bore of the bearing cage until the cup is flat against the bottom of the bore. Use a sleeve of the correct size to install the bearing cup.
3. Put the drive pinion in a press with the gear head (teeth) toward the bottom.
4. Press the inner bearing cone onto the shaft of the drive pinion until the cone is flat against the gear head. Use a sleeve against the bearing inner race if necessary.
5. To install one-piece spigot bearing assemblies, put the drive pinion in a press with the gear head teeth toward the top. Press the spigot bearing onto the end of the drive pinion until the bearing is flat against the gear head. Install the snapring to secure bearing. Be sure that the snapring is securely seated.
6. Apply axle lubricant on the bearing cups in the cage and bearing cones.
7. Install the drive pinion into the bearing cage.
8. Install the bearing spacer(s) onto the pinion shaft against the inner bearing cone (**Figure 24–32**).
9. Install the outer bearing cone onto the pinion shaft against the spacer(s).
10. At this time, adjust the pinion bearing preload. (Refer to the section on axle adjustments and checks later in this chapter.)
11. Once the preload has been set and adjusted, adjust the thickness of the shim pack in the

FIGURE 24–32 Drive pinion assembly

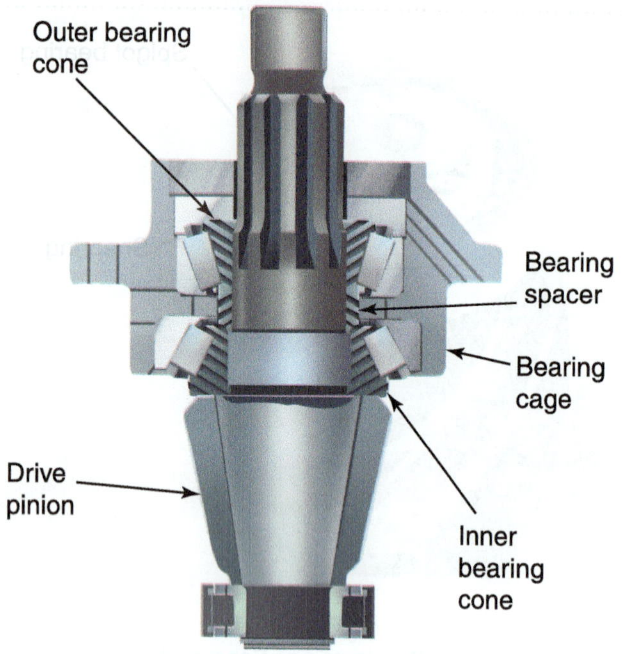

FIGURE 24–33 Shim pack installation

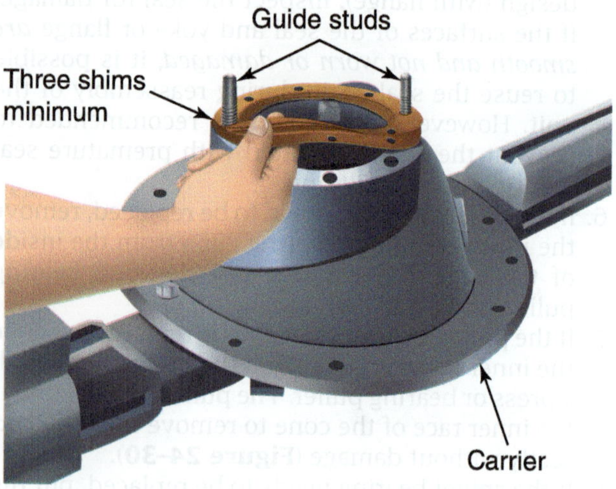

FIGURE 24–34 The cover and seal assembly and gasket should be mounted to the bearing cage.

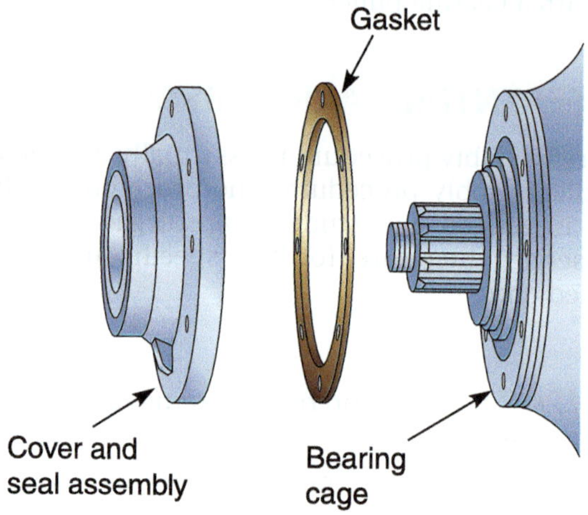

pinion cage. (Refer to the section on axle adjustments and checks later in this chapter for the adjustment procedure.)

Install the Drive Pinion and Bearing Cage

When the drive pinion and bearing cage are assembled and adjusted, they are ready to be installed into the carrier. Install the assembly into the carrier as follows:

1. If a new drive pinion and crown gear set is to be installed, or if the depth of the drive pinion has to be adjusted, calculate the thickness of the shim pack. (Refer to the section on axle adjustments and checks later in this chapter for the adjustment procedure.)
2. Install the correct shim pack between the bearing cage and carrier.
3. Align the oil slots in the shims with the oil slots in the bearing cage and carrier. The use of guide studs will help align the shims (**Figure 24–33**).
4. Install the drive pinion and bearing cage into the carrier. If necessary, use a rubber, plastic, or leather mallet to tap the assembly into position.
5. If used, install the cover and seal assembly and gasket over the bearing cage (**Figure 24–34**).
6. Align the oil slots in the cover and gasket with the oil slot in the bearing cage.
7. Install the bearing cage to the carrier and tighten all hardware to the correct torque value.
8. Install the input yoke or flange onto the drive pinion. The yoke or flange must be against the outer bearing for proper installation.
9. Tighten the pinion nut to the correct torque value. Use the yoke or flange bar during the torquing operation.

Crown Gear and Differential Assembly

To assemble the crown gear and differential assembly, use the following procedure:

1. Expand the ring gear by heating it in a tank of water to a temperature of 160°F to 180°F (71°C to 82°C) for 10 to 15 minutes. Never use a torch for this operation because you could damage the hardening.
2. Safely lift the crown gear onto the flange case half immediately after the gear has been heated. If the crown gear does not fit easily onto the case half, reheat it and try again.

3. Align the fastener holes of the crown gear and the flange case half. Rotate the crown gear as needed.
4. Install the fasteners that clamp the crown gear to the flange case half. Install the bolts from the gear side of the assembly. Bolt heads must be against the crown gear (**Figure 24–35**).
5. Tighten the bolts to the correct torque value.
6. If rivets are used to hold the crown gear to the flange case half, install the correct size rivets in pairs opposite each other from the case half side of the assembly. The rivet heads must be against the flange case half (**Figure 24–36A**). Press the rivets into position from the crown/ring gear side of the assembly using a riveter machine and the manufacturer's specified pressure. Hold riveting pressure for at least 1 minute and check for gaps between the back surface of the ring gear and the case flange using a 0.003-inch (0.076 mm) thickness gauge (**Figure 24–36B**).
7. Install the bearing cones on both of the case halves (**Figure 24–37**). Use a press and sleeve if necessary to install the cones.
8. Apply axle lubricant on the inside surfaces of both case halves, spider, thrust washers, side gears, and differential pinions.
9. Place the flange case half on a bench with the crown gear teeth facing upward.
10. Install one thrust washer and side gear into the flange case half (**Figure 24–38**).
11. Install the spider, differential pinions, and thrust washers into the flange case half (**Figure 24–39**).
12. Install the second side gear and thrust washer over the spider and differential pinions.
13. Put the plain half of the differential case over the flange half and gears. Rotate the plain half as needed to align the match marks.
14. Install hardware into the case halves. The distance between fasteners must be equal.

FIGURE 24–35 Crown/ring gear installation to flange case half

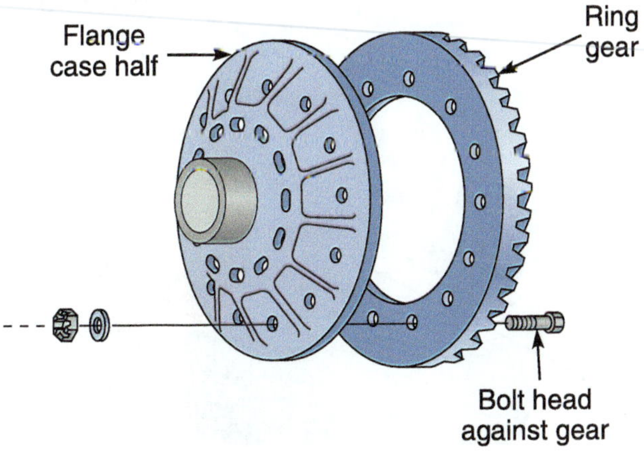

FIGURE 24–36 (A) Install rivets in pairs; (B) check for gaps between the rear gear and case flange.

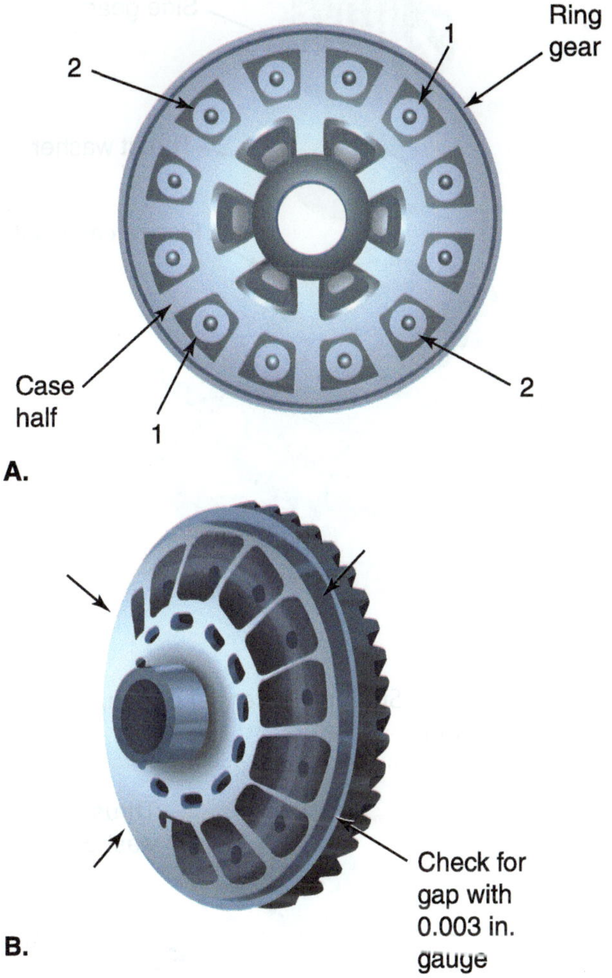

FIGURE 24–37 Bearing cone installation in case half

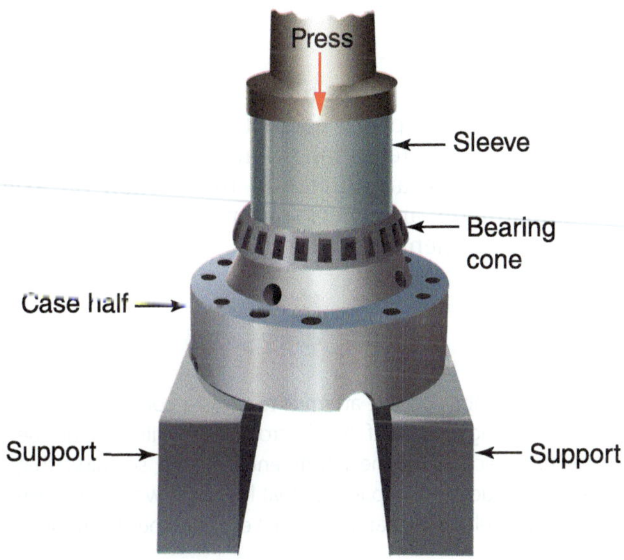

FIGURE 24–38 Side gear installation

FIGURE 24–39 Spider, differential pinion, and side gear installation

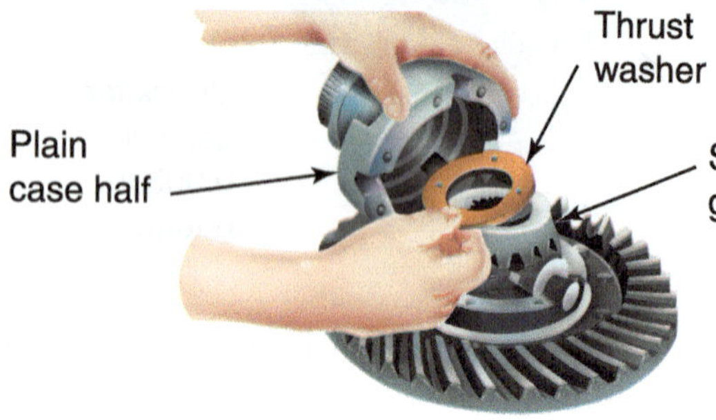

15. Tighten the fasteners to the correct torque value in a pattern opposite each other.
16. Check the differential rolling resistance of the differential gears. (Refer to the section on axle adjustments and checks later in this chapter for the adjustment procedure.)

SHOP TALK

During assembly, do not attempt to press a cold crown gear onto the flange case half. A cold crown gear will damage the case half because of the interference fit. The scraping that results produces metal particles that lodge between the components, resulting in gear runout that exceeds specifications.

Installing the Crown Gear Assembly into the Carrier

To install the assembled crown gear and differential assembly into the carrier, use the following procedure:

1. Clean and dry the bearing cups and bores of the carrier legs and bearing caps.
2. Apply axle lubricant on the inner diameter of the bearing cups and onto both bearing cones that are assembled on the case halves.
3. Apply a suitable adhesive in the bearing bores of the carrier legs and bearing caps.
4. Install the bearing cups over the bearing cones that are assembled on the case halves.
5. Safely lift the differential and ring gear assembly and install into the carrier.
6. Install both of the bearing adjusting rings into position between the carrier legs, as shown in **Figure 24–40**. Turn each adjusting ring hand tight against the bearing cup.
7. Install the bearing caps over the bearings and adjusting rings in the correct location as marked before removal (**Figure 24–41**).
8. Tap each bearing cap into position with a light leather, plastic, or rubber mallet. The caps must fit easily against the bearings, adjusting rings, and carrier.
9. Install the hardware that holds the bearing caps to the carrier. Tighten all hardware by hand first and then torque to the correct values.
10. Do not install the cotter keys or lock plates that hold the bearing adjusting rings in position.

Continue the overhaul procedure by performing the following checks or adjustments:

1. Adjust preload of differential bearing.
2. Check runout of ring gear.
3. Adjust backlash of ring gear.

FIGURE 24–40 Installation of adjusting rings

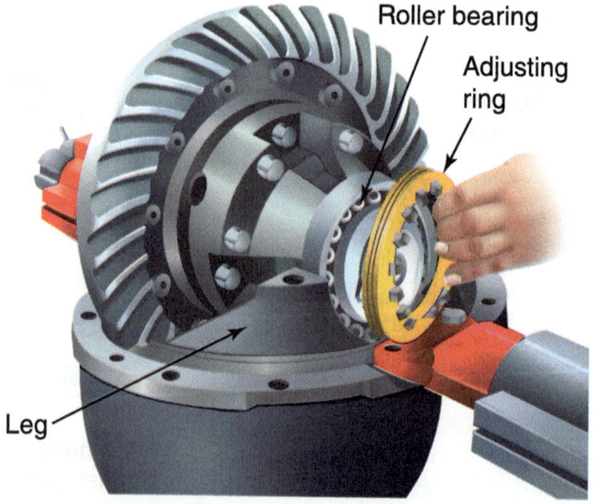

FIGURE 24–41 Installation of bearing caps using match marks

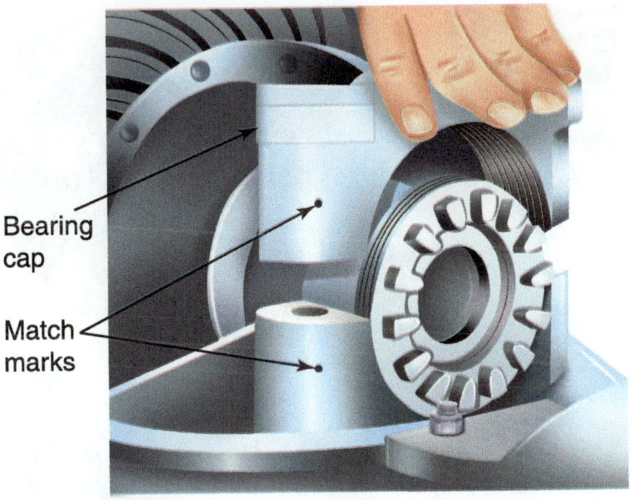

4. Check and adjust tooth contact pattern.
5. Adjust the thrust screw.

These checks and adjustments are described in the section on axle adjustments and checks later in this chapter.

Installing the Differential Carrier into the Banjo Housing

Now that the differential carrier has been assembled, you can install it into the axle banjo housing using the following procedure:

1. Clean the inside of the axle housing and the carrier mounting flange. Use a cleaning solvent and clean shop cloths to remove dirt and foreign matter. Blow dry the cleaned areas with air. Wear appropriate eye protection.
2. Inspect the axle housing for damage. Repair or replace if necessary.
3. Check for loose studs in the mounting surface of the housing where the carrier fastens. Remove and clean the studs that are loose.
4. Apply a liquid sealing adhesive such as Loctite™ to the threaded holes and install the studs into the axle housing. Tighten the studs to the correct torque value.
5. Apply silicone or a gasket (depending on OEM recommendation) to the mounting flange of the banjo housing (**Figure 24–42**).
6. Install hardware in the four corner locations around the carrier and axle housing. Hand-tighten the fasteners.
7. Carefully push the carrier into position. Tighten the four fasteners two or three turns each in a pattern opposite each other.
8. Repeat step 7 until the four fasteners are tightened to the correct torque value.

FIGURE 24–42 Application of silicone gasket material to the mounting surface of the axle housing

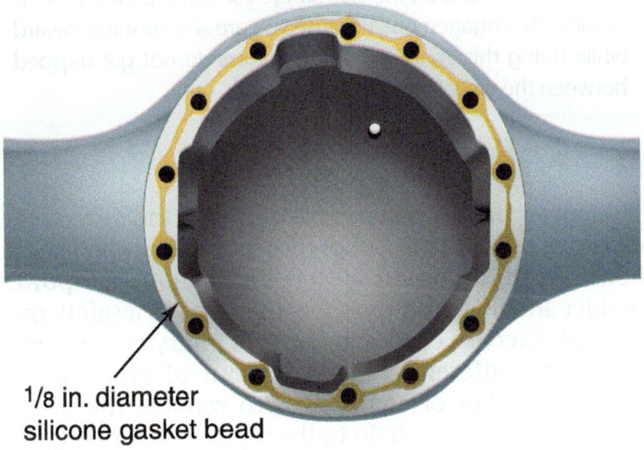

FIGURE 24–43 Installing the gaskets and axle shafts into the axle housing and carrier

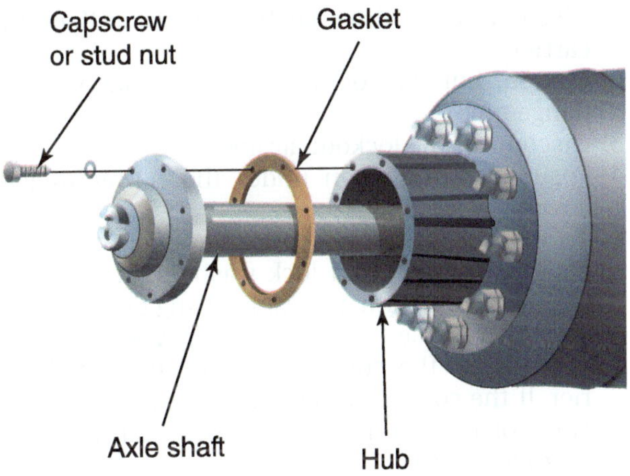

9. Install the other fasteners that hold the carrier in the axle housing. Tighten the fasteners to the correct torque value.
10. Connect the driveline universal joint to the pinion input yoke or flange on the carrier.
11. Install the gaskets and axle shafts into the axle housing and carrier (**Figure 24–43**). The gasket and flange of the axle shafts must fit flat against the wheel hub.
12. Install the hardware that fastens the axle shafts to the wheel hubs. Tighten to the correct torque value.
13. If the wheel hubs have studs, install the tapered dowels at each stud and into the flange of the axle shaft. Use a punch or drift and hammer if needed. Install the hardware on the studs and tighten to the correct torque value.

SHOP TALK

When inserting axle shafts into the axle assembly, pushing radially downward on the axle flange can help guide the shaft splines into engagement. Because you are also pushing inward while doing this, make sure your fingers do not get trapped between the wheel hub and axle flange.

Power Divider Replacement

The power divider can be replaced with the differential carrier assembly in or out of the truck chassis and banjo housing. During removal and installation, the power divider should be supported with a sling as a safety precaution. Exercise caution to prevent injury or damage.

If the differential carrier is out of chassis, use a cherry picker or chain hoist to remove the power divider. Fasten the chain to the input yoke or flange. If the differential carrier is in the chassis, use a transmission jack or a cherry picker and sling to remove the power divider. Wrap the sling strap around the power divider and attach it to the hoist hook (**Figure 24–44**).

To replace a power divider, use the following procedure:

1. Disconnect the drive shaft from the differential carrier.
2. Loosen, but do not remove, the input yoke or flange nut.
3. Disconnect the lockout air line.
4. Position a drain pan under the power divider cover.
5. To remove the power divider assembly, remove cover capscrews and lock washers. Support the power divider with a suitable lifting device.
6. Tap the reverse face of the input yoke or flange to separate the cover from the differential carrier. If the cover does not separate easily, tap the sides of the cover near the dowel pin locations (**Figure 24–45**).
7. Drain the lube from the differential.
8. Pull the power divider assembly forward until it is free of the carrier.
9. With the power divider removed, lift the differential gearing off the output shaft side gear.
10. If necessary, remove the output shaft by disconnecting the interaxle driveline. Remove the nut and output shaft yoke. Pull the output shaft assembly out of the carrier.
11. After repairing or replacing all defective components, lubricate them before replacing the power divider into the carrier housing.
12. If the output shaft was removed, lubricate the O-rings and then install the shaft assembly into the differential carrier and housing cover (**Figure 24–46**). Be sure to lubricate the seal lip.
13. Make sure that the yoke or flange is clean and dry. Then install the yoke or flange self-locking nut. Torque the nut to the correct torque value.

FIGURE 24–44 Attaching power divider to a chain hoist and sling

A.
Courtesy of Dana Corporation

B.
Courtesy of Dana Corporation

14. Install the interaxle differential on the output shaft side gear (with the nuts facing away from the side gear) (**Figure 24–47**).
15. Use a silicone rubber gasket compound on the differential carrier mating surfaces.

FIGURE 24–45 Power divider cover dowel pin locations

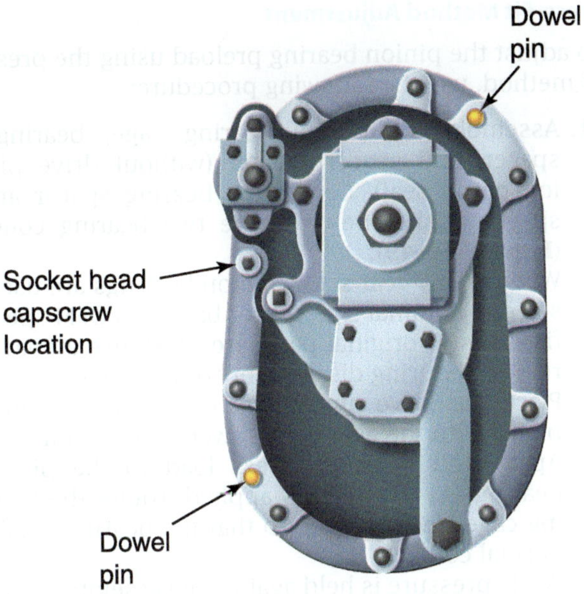

FIGURE 24–46 Installation of the shaft assembly into the differential carrier

16. Make certain that the dowel pins are installed in the carrier. Then install the power divider to the carrier. Use a transmission jack or hoist and sling to support the power divider.
17. During installation, rotate the input shaft to engage the input shaft splines with the interaxle differential. After installation, again rotate the input shaft

FIGURE 24–47 Installation of the interaxle differential

to check for correct assembly. The output shaft should turn when the input shaft is rotated.
18. Install the power divider cover capscrews and lock washers, torquing capscrews to the correct torque value.
19. Check and adjust the input shaft endplay. With the power divider assembled to the differential carrier, check endplay with a dial indicator. Endplay typically should be 0.003 to 0.007 inch (0.076 to 0.178 mm) but check the service specifications. If necessary, adjust the endplay.
20. After the input shaft endplay is within specifications, complete the assembly procedure by connecting the drive shaft to the differential yoke.
21. Connect the lockout air lines.
22. Fill the axle to the proper lube level.

Endplay Adjustment

Input shaft endplay requirements will vary with operating conditions, mileage, and rebuild procedures. To measure and adjust endplay, do the following:

1. With the power divider assembled to the differential carrier, measure endplay with a dial indicator positioned at the yoke end of the input shaft (**Figure 24–48**).
2. Move the input shaft axially and measure the endplay. A new power divider or a used unit rebuilt with all new components should produce an endplay reading of 0.003 to 0.007 inch (0.076 to 0.178 mm). A rebuilt unit with reused components will typically read from 0.013 to 0.017 inch (0.33 to 0.43 mm).

FIGURE 24–48 Measuring endplay with a dial indicator

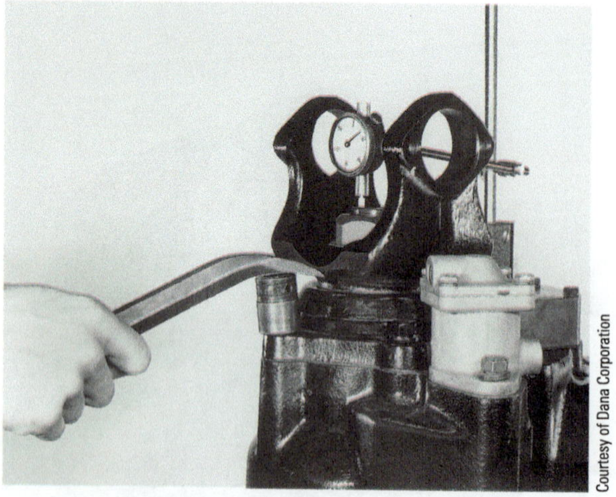

3. If the endplay reading is not correct, remove the input shaft nut, flat washer, and yoke. Remove the bearing cover capscrews and lock washers. Then remove the cover and shim pack.
4. To increase endplay, add shims to the shim pack. To decrease endplay, remove shims from the shim pack.
5. To reassemble the input shaft, install the adjusted shim pack and bearing cover. Install capscrews and lock washers and torque to correct value.
6. Install the yoke, flat washer, and nut. Tighten the nut snugly. Tap the end of the input shaft lightly to seat the bearings.
7. Measure the input shaft endplay again with the dial indicator. If endplay is still incorrect, repeat steps 3 through 7.
8. With the endplay correct, seal the shim pack to prevent lube leakage. Then torque the input shaft nut and cover capscrews to the correct value.

24.5 AXLE ADJUSTMENTS AND CHECKS

This section introduces the differential carrier adjustments, checks, and tests that the truck technician must be capable of performing; some have been referred to previously in the text. For the most part, the procedures described here are general in nature. The truck technician should refer to OEM service literature for specific procedures.

PINION BEARING PRELOAD

Most differential carriers are provided with a press-fit outer bearing on the drive pinion gear. Some older rear drive axles use an outer bearing, which slips over the drive pinion. The procedures for adjusting both types follow.

Press-Fit Method Adjustment

To adjust the pinion bearing preload using the press-fit method, use the following procedure:

1. Assemble the pinion bearing cage, bearings, spacer, and spacer washer (without drive pinion or oil seal). Center the bearing spacer and spacer washer between the two bearing cones (**Figure 24–49**).
2. When a new gear set or pinion bearings are used, select a nominal size spacer based on OEM specifications. If original parts are used, use a spacer removed during disassembly of the drive.
3. Place the drive pinion and cage assembly in a press, with the gear teeth toward the bottom.
4. Apply and hold the press load to the pinion bearing. As pressure is applied, rotate the bearing cage several times so that the bearings make normal contact.
5. While pressure is held against the assembly, wind a cord around the bearing cage several times.
6. Attach a spring scale to the end of the cord (**Figure 24–50**). Pull the cord with the scale on a horizontal line.
7. As the bearing cage rotates, read the value indicated on the scale.
8. Preload normally is specified as torque required to rotate the pinion bearing cage, so take a reading only when the cage is rotating. Starting torque will give a false reading.
9. To calculate the preload torque, measure the diameter of the bearing cage where the cord was wound. Divide this dimension in half to get the radius.
10. Use the following procedure to calculate the bearing preload torque:

FIGURE 24–49 Assembly of the pinion bearing cage

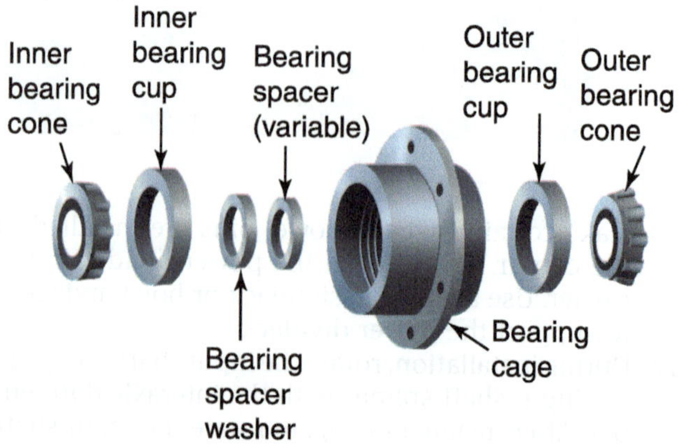

FIGURE 24–50 Cage in press to check bearing preload

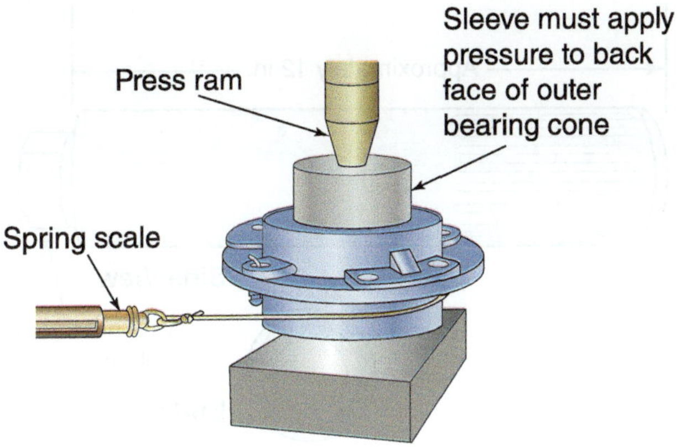

Standard.

Pull (lb) × radius (inches) = preload (lb-in.)

or

Preload (lb-in.) × 0.113 (a conversion constant) = preload (N·m)

Metric.

Pull (kg) × radius (cm) = preload (kg-cm)

or

Preload (kg-cm) × 0.098 (a conversion constant) = preload (N·m)

Examples.

We can convert the foregoing equations into examples by applying some data to them:

Standard

7.5 lb × 3.31 in. = 24.8 lb-in. (preload)

or

24.8 lb-in. × 0.113 = 2.8 N·m (preload)

Metric

3.4 kg × 8.4 cm = 28.6 kg-cm (preload)

or

28.6 kg-cm × 0.098 = 2.8 N·m (preload)

11. If necessary, adjust the pinion bearing preload by changing the pinion bearing spacer. A thicker spacer will decrease preload, whereas a thinner spacer will increase the preload.
12. Once the correct bearing preload has been established, note the spacer size used. Select a spacer 0.001 inch (0.025 mm) larger for use in the final pinion bearing cage assembly procedures. The larger spacer compensates for slight expansion of the bearing, which occurs when pressed on the pinion shank. The trial spacer pack should result in correct pinion bearing preload in three times out of four cases.

Yoke Method of Adjustment

To adjust the pinion bearing preload using the yoke or flange method, proceed as follows:

1. Assemble the complete pinion bearing cage as recommended in the press-fit method.
2. A forward axle pinion is equipped with a helical gear. For easier disassembly during bearing adjustment procedures, use a dummy yoke (if available) in place of the helical gear.
3. Install the input yoke or flange, nut, and washer on the drive pinion. The yoke or flange must be against the outer bearing. If the fit between the yoke or flange splines and drive pinion splines is tight, use a press to install the yoke or flange (**Figure 24–51**).
4. Temporarily install the drive pinion and cage assembly in the carrier (**Figure 24–52**). Do not install shims under the bearing cage.
5. Install the bearing cage to the carrier capscrews. Washers are not required at this time. Hand-tighten the capscrews.
6. Fasten a yoke or flange bar to the yoke or flange (**Figure 24–53**). The bar will hold the drive pinion in position when the nut is tightened.
7. Tighten the nut on the drive pinion to specification, typically 400 to 700 lb-ft. (542 to 950 N·m).
8. Remove the yoke or flange bar.

FIGURE 24–51 Using a press to install the yoke or flange to the drive pinion

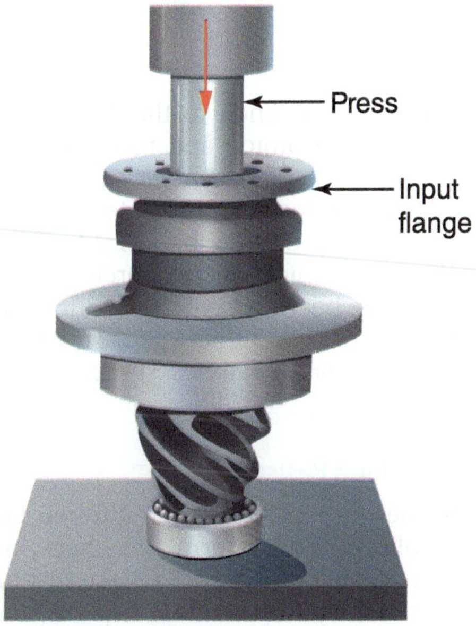

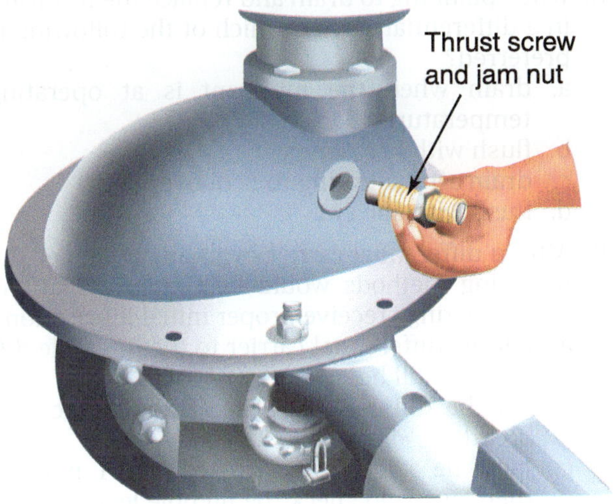

FIGURE 24–74 Installing the jam nut on the thrust screw

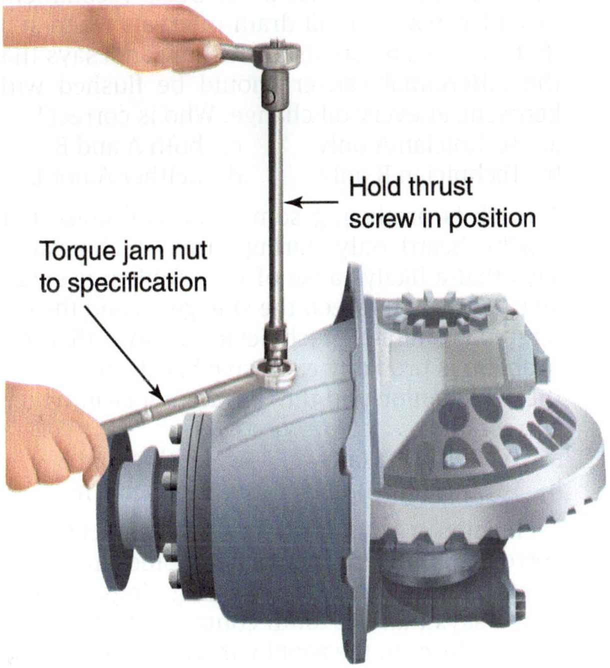

FIGURE 24–75 Tighten the jam nut to the correct torque value.

SUMMARY

- Adhering to OEM-recommended lubrication schedules is the key to ensuring the longest service life from both drive and dead axles.
- Knowing the correct procedure to check lubricant level is essential. The level is correct when lubricant is exactly level with the bottom of the fill hole.
- Because most OEMs approve of the use of synthetic lubricants in final drive carriers, lubrication drain schedules have been greatly increased in recent years. Drain schedules are determined by the actual lubricant used and the type of application to which the vehicle is subjected.
- When topping up lubricant during servicing, ensure you do not mix synthetic- and mineral-based lubricants because the result can be rapid degradation.
- Servicing of axles on heavy-duty trucks consists of routine inspection, lubrication, cleaning, and, when required, troubleshooting and component overhaul.
- Failure analysis is required to prevent recurrent failures.
- Drive axle carrier components usually fail for one of the following reasons:
  - Shock load
  - Fatigue
  - Spinout
  - Lubrication problems
  - Normal wear
- Most differential carriers are replaced as rebuilt/exchange units, so the role of the technician is, more often than not, to diagnose the problem and then, if necessary, to replace the defective assembly as a unit.
- The technician who has disassembled and reassembled differential carriers should find troubleshooting procedures easier to follow.
- Follow the OEM procedure when disassembling differential carriers. Taking a few moments to measure shim packs and gear tooth contact patterns on disassembly can save considerable time when reassembling the carrier.
- A crown and pinion gear set often can be reused when rebuilding a differential carrier. Make sure that you inspect it properly on disassembly.
- Crown and pinion gear sets are always replaced as a matched pair during a rebuild.
- When setting crown and pinion backlash, it is increased by moving the crown gear away from the drive pinion and decreased by moving the crown gear toward the drive pinion.

REVIEW QUESTIONS

1. Technician A says that most OEMs recommend extended rear axle oil drain intervals when synthetic lubricants are used. Technician B says that the differential carrier should be flushed with kerosene at every oil change. Who is correct?
   a. Technician A only
   b. Technician B only
   c. both A and B
   d. neither A nor B
2. A truck is producing some rear axle noise that can be heard only during turns. Technician A says that a likely cause of the problem is excessive backlash between the side gears and the differential pinions. Technician B says that the problem is probably excessive backlash between the drive pinion and the crown/ring gear. Who is correct?
   a. Technician A only
   b. Technician B only
   c. both A and B
   d. neither A nor B
3. Technician A says that crown gear backlash is correctly adjusted by alternately loosening and tightening the bearing adjusting rings. Technician B says that the tooth contact pattern can be moved closer to the root by increasing the crown gear backlash. Who is correct?
   a. Technician A only
   b. Technician B only
   c. both A and B
   d. neither A nor B
4. Which of the following would indicate that the level was correct when checking lube oil level in a differential carrier housing with the fill plug removed?
   a. oil spilling from the fill hole
   b. oil level exactly even with the bottom of the fill hole
   c. oil level within one finger joint of the fill hole
   d. oil level within 1 inch (2.5 cm) of the fill hole
5. Technician A says that spinout failures are often caused by a defect in the interaxle differential gearing. Technician B says that spinout failures are unavoidable during winter operation. Who is correct?
   a. Technician A only
   b. Technician B only
   c. both A and B
   d. neither A nor B
6. Which gear lube viscosity rating is usually recommended for use in a differential carrier for a truck operating in temperatures of −30°F (−34°C)?
   a. 75W
   b. 80W
   c. 80W-90
   d. 80W-140
7. Which type of lubricant is recommended when hypoid gearing is used in a differential carrier?
   a. petroleum-based lubes only
   b. API GL-4
   c. any EP-rated lubricant
   d. any API-rated lubricant
8. When planning to drain and replace the lubricant in a differential carrier, which of the following is preferred?
   a. drain when the lubricant is at operating temperature
   b. flush with kerosene
   c. drain after allowing to cool overnight
   d. flush with solvent
9. After a differential carrier overhaul, which of the following methods would best ensure that the wheel bearings receive proper initial lubrication?
   a. Fill the differential carrier to the interaxle differential fill hole.
   b. Fill the differential carrier through the axle breather.
   c. Jack the axle up at each side for 1 minute, then level and check the fill hole level.
   d. Remove each axle shaft rubber fill plug and fill to the indicated oil level.
10. Which of the following measuring tools should be used to check crown gear runout?
    a. a fish scale
    b. a torque wrench
    c. thickness gauges
    d. a dial indicator
11. Which type of failure is usually indicated when a shaft fails, producing a star-shaped fracture pattern?
    a. fatigue failure
    b. surface failure
    c. single shock load
    d. torsional failure
12. When a gear tooth separates, leaving a smooth, flat fracture pattern, what is the usual cause?
    a. fatigue failure
    b. surface failure
    c. single shock load
    d. progressive torsional failure
13. What is the correct method for removing rivets from a crown gear assembly?
    a. Use a chisel and punch.
    b. Use a drill and punch.
    c. Use an oxyacetylene torch.
    d. Use a hammer and punch.
14. What is the correct method for fitting a crown gear to a flange case half?
    a. Use an oxyacetylene torch.
    b. Place in an oven.
    c. Heat in a water tank to 170°F (77°C).
    d. Cold press into position.
15. The rear wheels on a tandem drive axle truck will not turn but the drive shaft does. Technician A says that the first step is to pull the forward differential carrier. Technician B says to pull both axle shafts. Who is correct?
    a. Technician A only
    b. Technician B only
    c. both A and B
    d. neither A nor B

25

Prerequisites: Chapters 1, 2, 3 and 4.

STEERING AND ALIGNMENT

OBJECTIVES

After reading this chapter, you should be able to:

- Identify the components of the steering system of a heavy-duty truck.
- Describe the procedure for inspecting front axle components for wear.
- Explain how toe, camber, caster, axle inclination, turning radius, and axle alignment affect tire wear, directional stability, and handling.
- Describe the components and operation of a worm and sector shaft and a recirculating ball-type steering gear.
- Explain how to check and adjust a manual steering gear preload and backlash.
- Identify the components of a power steering gear and pump and explain the operation of a power steering system.
- Outline the operating principles of a truck rack and pinion steering gear.
- Describe the components and operation of an electronically variable power steering system.
- Describe the components and operation of a load-sensing power-assist steering system.

KEY TERMS

Ackerman geometry
backdrive
camber
caster
cross tube
dog-tracking
drag link
electronically variable steering (EVS)
kingpin
kingpin inclination (KPI)
laser alignment
light beam alignment
oversteer
photo alignment
Pitman arm
power steering analyzer
rack and pinion steering gear
steering control arm
steering gear
tie-rod
toe
toe-in
toe-out
understeer
unitized axle
WinAlign
yaw

INTRODUCTION

The steering system in a heavy-duty truck is expected to deliver precise directional control of the chassis at both gross and unloaded vehicle weight. It has to be able to minimize driver effort while retaining some road feel. Truck steering systems can be either manual or power-assisted. Current power-assist units use a hydraulic-assist circuit to reduce steering effort. Some hydraulic power-assisted systems are available with electronic or load-sensing controls. For any truck certified to operate on a highway, power steering must have full mechanical redundancy. This means that in the event of a loss of hydraulic assist, the vehicle can be manually steered.

In addition to its role in managing vehicle directional control, the steering system is integral with the front suspension, steer axle, and wheel/tire components. **Figure 25–1** shows a typical manual steering gear system from steering wheel to steering knuckle. Although it is not common, manual steering gear is still used in trucks today and a couple of national linehaul fleets use manual steering almost exclusively.

FIGURE 25–1 Manual steering gear components

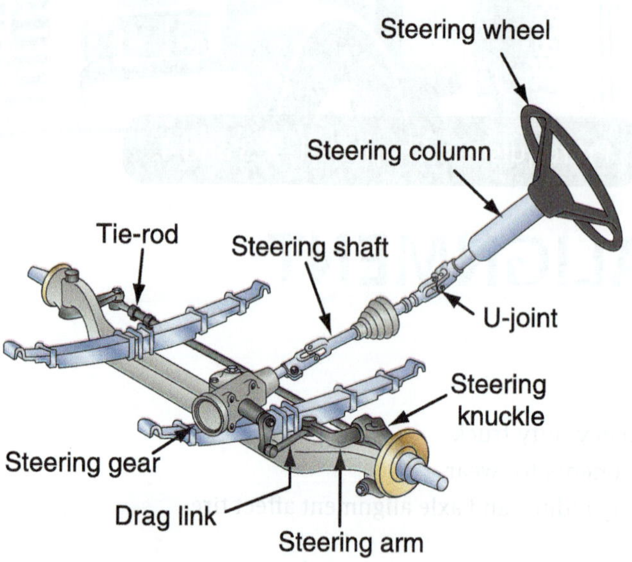

Most truck steering gear found on trucks today is power-assisted. Power-assisted steering on highway vehicles is required to function with full mechanical redundancy in the event of a loss of the power-assist system. Do not confuse power-assisted steering with the full power steering systems found on some off-highway equipment. This equipment often uses hydrostatic steering; in the event of a loss of hydraulic power, this will result in no steering capability.

In this chapter, we take a look at truck steering systems. We begin by learning steering system components, then studying the theory of steering geometry and alignment. Truck technicians routinely check and repair steering alignment, so it is important to understand how to diagnose steering problems, which requires some knowledge of steering gear operation. Finally, we take a look at some recent innovations in truck steering systems, including rack and pinion systems.

25.1 STEERING SYSTEM COMPONENTS

A truck steering system is made up of a steering wheel, steering column, steering gear, and steering linkages that actually move the steering tires. All of the components are introduced first, but some of these components are complex and are dealt with in greater detail later in the chapter.

STEERING WHEEL

The steering wheel is the means used by the driver to control the directional tracking of the vehicle. It is, therefore, the primary input to the steering system. The steering wheel used on a truck is made of a tempered steel rod shaped to a circular rim and supported by spokes. The spokes extend from the wheel hub, which turns on a bushing or bearing at the top of the steering column. A steering wheel is usually covered with a hard plastic compound to make it easy to handle.

Driver effort applied to the steering wheel at the rim becomes torque in the steering column. It, therefore, makes sense that the larger the steering wheel diameter, the more torque is generated from the same amount of driver effort. Steering wheels on heavy-duty trucks are typically 22 to 24 inches (56 to 61 cm) in diameter. This large diameter can help a driver control a vehicle equipped with power-assisted steering when the power-assist fails. Many vehicles are equipped with an electric horn at the center hub of the steering wheel.

WARNING

Always consult original equipment manufacturer (OEM) service literature before undertaking any work on a steering wheel. An armed air bag can be extremely dangerous if triggered. Supplement restraint systems (SRS) are covered in Chapter 11 of this book.

STEERING COLUMN

The steering column connects the steering wheel to the steering gear. **Figure 25–2** shows a complete steering column assembly. The major components of the steering column assembly are a jacket (tube), bearing assemblies (staked in place in the top and bottom of the jacket), a steering column shaft, and wiring and contact assemblies for the electric horn. At the upper end of the steering column shaft are threads machined for a nut that holds the steering wheel in place. Straight external splines on the column shaft match internal splines on the steering wheel hub. This allows the two components to be coupled with zero axial play. The lower end of the column shaft has external splines that mate to internal splines on the column yoke.

The steering column assembly is mounted to the dash steering column bracket by support brackets located under the cover housing. Coupled to the steering column upper shaft by a pair of yokes and the U-joint assembly is the lower shaft assembly. The U-joint permits some angular deviation between the upper and lower column shafts. The lower column shaft assembly connects to the steering gear.

The steering column assembly usually supports some other switch and control components because the driver's hands should be close by when the vehicle is moving. A multifunction stalk switch, which contains the turn signal switch at a minimum, is usually

FIGURE 25–2 Steering column for a conventional cab

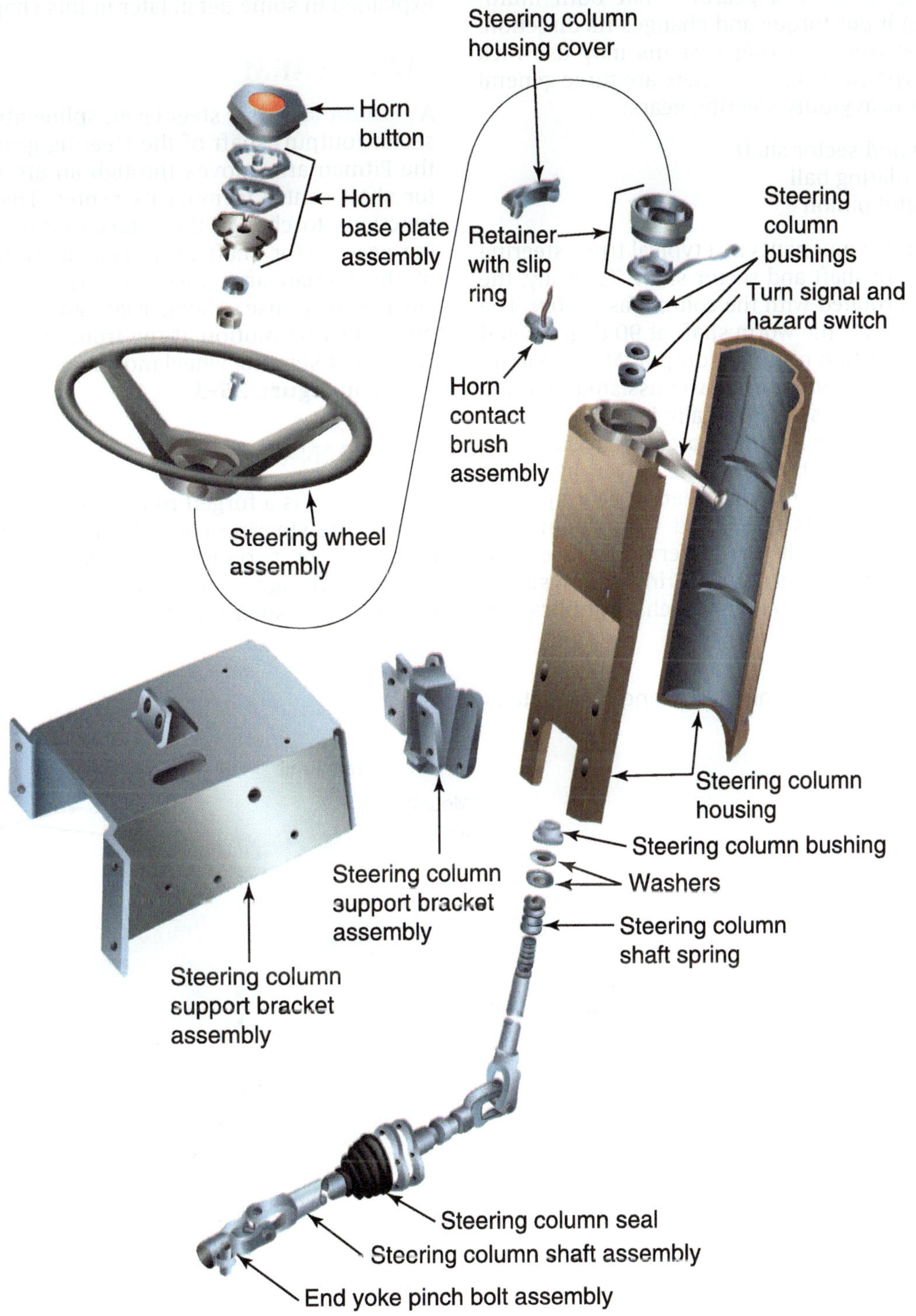

mounted on the left side of the steering column. If the vehicle is equipped with a trailer service brake control valve, it is clamped onto the right side of the steering column.

As long as the steering column components are routinely lubricated, they seldom fail. Because of the U-joints, steering shafts are phased; this phasing must be retained, or the result will be binding in the assembly. The U-joints in a steering shaft should be lubed in the same manner as those in a drive shaft assembly. When greased, lube should exit all four trunnion journals. Try to make sure it does. Many steering column related problems are caused by U-joint failures.

STEERING GEAR

The **steering gear** is a gearbox that both multiplies steering input torque and changes its direction. Recent model truck steering systems may use rack and pinion-type steering gear. There are three general categories of heavy-duty steering gears:

- Worm and sector shaft
- Recirculating ball
- Rack and pinion

The critical components in a typical truck steering gear are a worm shaft and sector shaft. Typically, the worm shaft is rotated with the column assembly. The sector shaft bisects the worm shaft at 90 degrees and changes the direction of torque output. Most contemporary trucks use hydraulic power-assisted steering gear. The hydraulic assist is usually integral with the steering gear, with the hydraulic pressure required by the system produced by an engine driven hydraulic pump. However, some manual steering gear systems are still used by fleets concerned with minimizing equipment costs. Some drivers prefer the precise road-feel that is a feature of manual steering gear systems. For a truck that is operated in a linehaul application, the requirement for power-assist is really limited to parking and yard operation. Steering gear systems are explained in some detail later in this chapter.

PITMAN ARM

A **Pitman arm** is a steel lever, spline attached to the sector (output) shaft of the steering gear. The end of the Pitman arm moves through an arc with the sector shaft center forming its center. The Pitman arm functions to change the rotary motion of the steering gear sector shaft into linear motion. The length of the Pitman arm affects leverage and, therefore, steering response. A longer Pitman arm will generate more steering motion at the front wheels for a given amount of steering wheel movement. A Pitman arm is shown in **Figure 25–3**.

DRAG LINK

A **drag link** is a forged rod that connects the Pitman arm to the steering control arm. The drag link can be a one- or two-piece component. The length of two-piece design is adjustable, which makes it easy to center the steering gear with the wheels straight

FIGURE 25–3 Typical manual steering gear installation

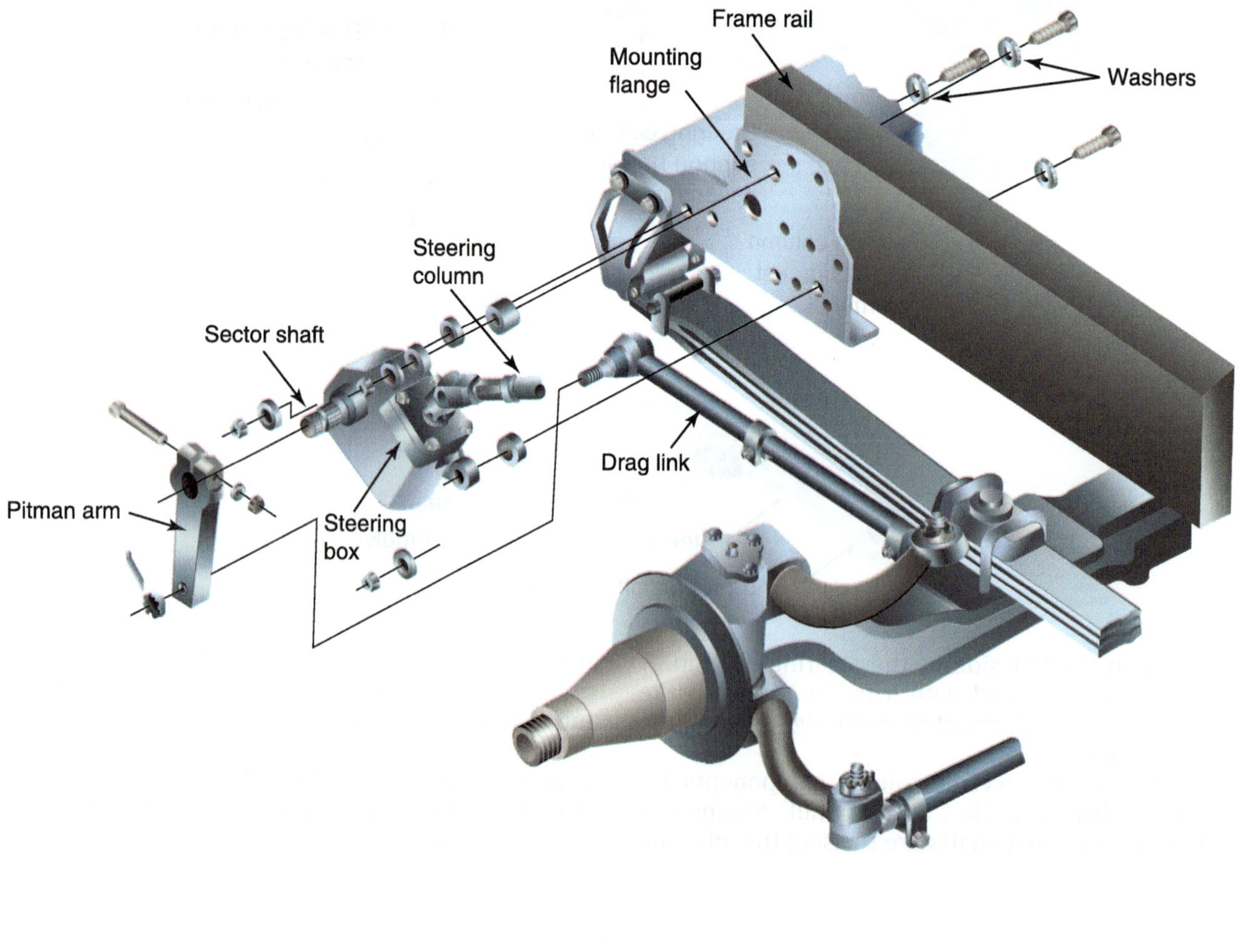

ahead. One-piece drag links are used in systems with very close tolerances. Other components are used to make adjustments to the system when a one-piece drag link is used. The drag link is connected at each end by ball and socket joints. Ball and socket joints help isolate the steering gear and Pitman arm from axle motion. Figure 25–3 shows an adjustable-type drag link.

STEERING CONTROL ARM

The **steering control arm** connects the drag link to the steering knuckle on the driver side of the vehicle. A steering control arm is also known as a steering arm, a control arm, and a steering lever. It is usually a drop-forged, tempered steel component. When the drag link is moved in a linear direction, the steering control arm moves the steering knuckle, which changes the angle of the steering knuckle spindle.

STEERING KNUCKLE

Steering knuckles (**Figure 25–4**) mount to the rigid front axle beam by means of steel pins known as **kingpins** or knuckle pins. They provide the ability for the pivoting action required to steer the vehicle. The steering knuckle incorporates the spindle onto which wheel bearings and wheel hubs are mounted, plus a flange to which the brake spider is bolted. A steering control arm is attached to the upper portion of the left side steering knuckle and Ackerman (tie-rod) arms are attached to both left and right steering knuckles.

A kingpin can be either tapered (**Figure 25–5A**) or straight (**Figure 25–5B**). Tapered pins are drawn into the axle center and secured by tightening a nut at the upper pin end. Straight kingpins are secured to the axle with tapered draw keys that bear against flats on the pin. Tapered pins are usually sealed and may

FIGURE 25–4 Components of a steering axle

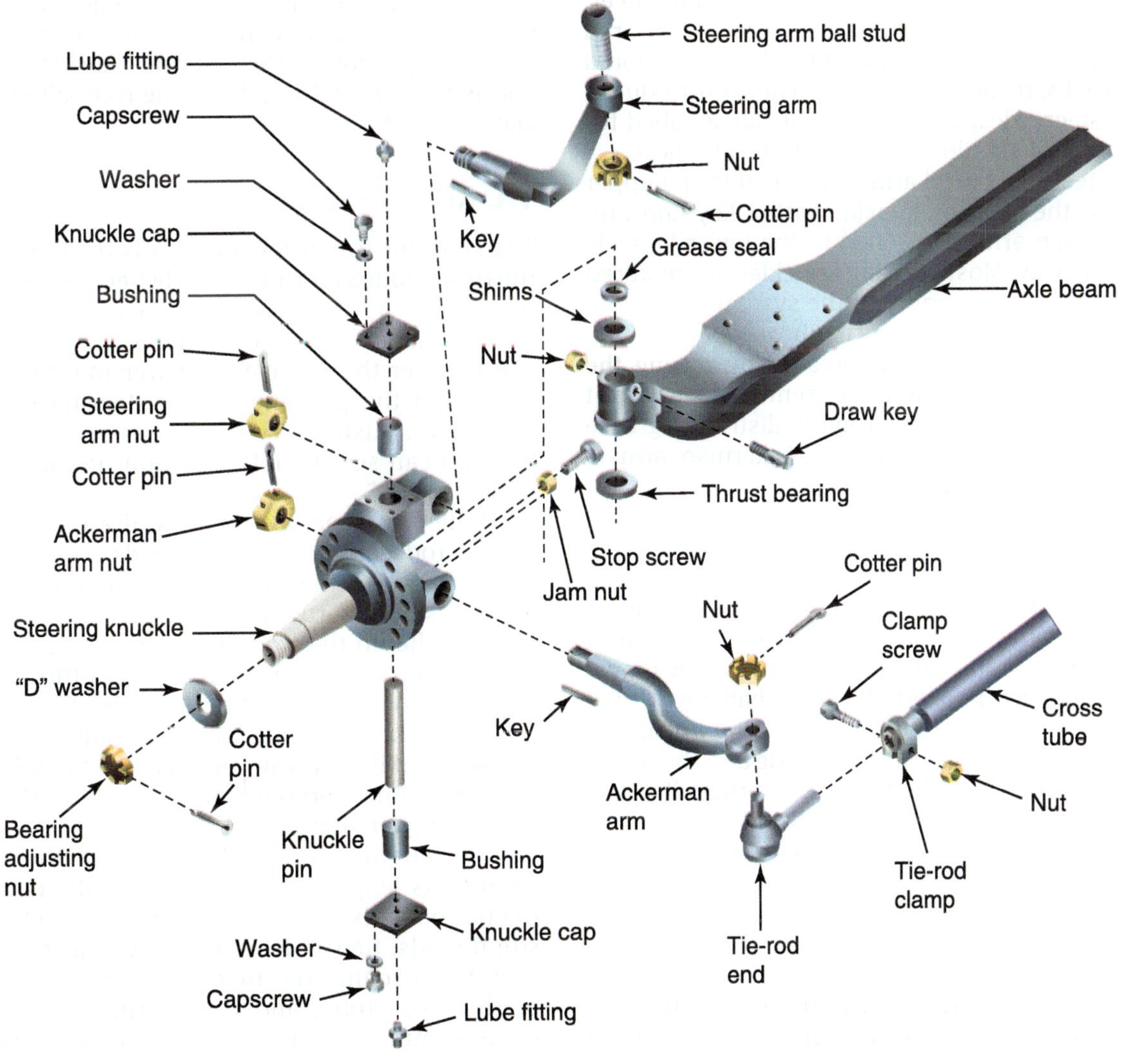

FIGURE 25–5 (A) Tapered knuckle pin; and (B) straight knuckle pin

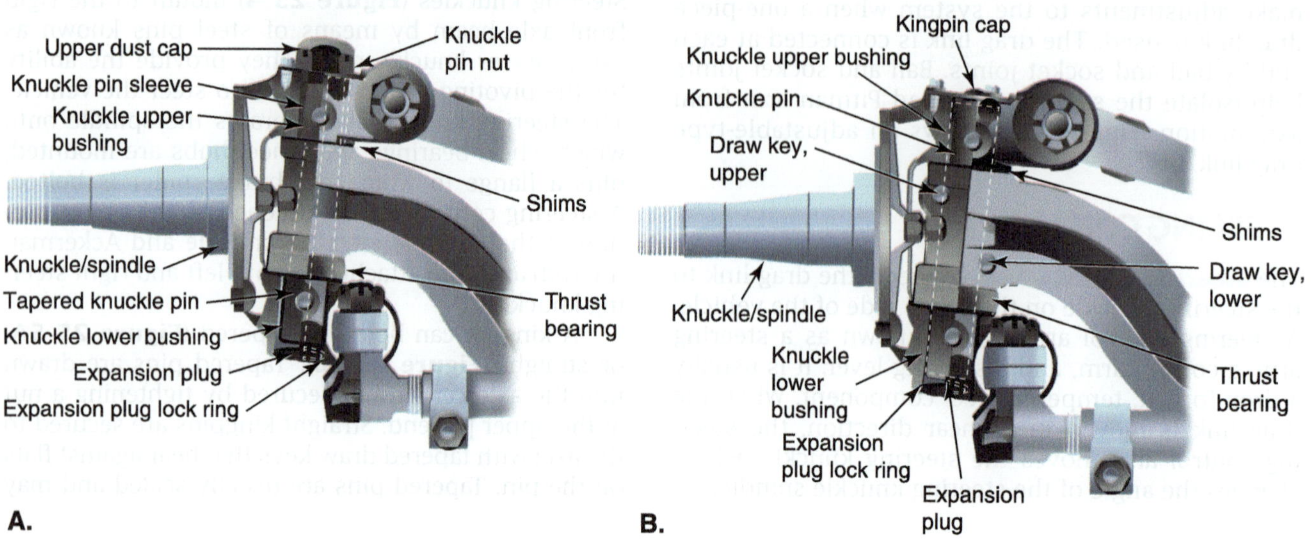

not require periodic lubricating. Straight pins have a cap on either end (top and bottom) to retain grease. Zerk-type grease fittings are used to lubricate steering knuckles. Good-quality chassis grease is required to lubricate most steering knuckles, but you should check the OEM recommendation. On newly introduced unitized steer axles, the steering knuckle is lubed for life; no attempt should be made to lubricate these.

The steering control arm is fastened to the upper knuckle of the steering knuckle with a key and nut. The Ackerman arm is fastened to the lower knuckle in the same way. Most steering knuckles are required to have regular lube service. If the system is not connected to an automatic-lube circuit or lubed for life, steering knuckles should be greased by raising the steering axle off the ground. This removes the weight from the wheels and helps properly distribute grease through the kingpin. A left-side Ackerman arm is shown in Figure 25–5.

BALL AND SOCKET JOINTS

A ball joint and socket joint are manufactured using a forged steel ball with a threaded stud attached to it. A socket shell grips the ball. The idea is to allow movement in any direction off the fixed ball pivot. Ball and socket joints are used in the steering linkage to accommodate motion between the axle and the frame rails that results when the front axle springs flex and do so without affecting steering. Ball and socket joints are used in the end of a steering control arm, idler arms, and tie-rod ends.

ACKERMAN ARM

An Ackerman or tie-rod arm is the means used to transfer and synchronize steering action on both steer wheels on a steering axle. Ackerman arms are drop-forged, tempered steel levers that are angled to define the steering geometry required on turns. One end of an Ackerman arm is keyed and bolted to the lower portion of the steering knuckle. The other end is taper bored to allow a tie-rod ball stud to be clamped to it.

TIE-ROD ASSEMBLY

The steering arm or lever controls the movement of the driver side steering knuckle because it connects directly to it. The steering knuckles are required to be connected to each other so they act in unison to steer the vehicle. Transferring this steering motion to the passenger side steering knuckle is achieved by using a **tie-rod** or **cross-tube** assembly. You can see the left side of a tie-rod assembly in Figure 25–5.

A tie-rod assembly consists of a cross tube and two tie-rod ends. The tie-rod ends are ball sockets that connect to the Ackerman arms on each steering knuckle. A tie-rod or cross tube is a steel rod that runs parallel to the front axle and extends from one side to the other. Each tie-rod end is threaded onto either end of the cross tube to connect it to tapered sockets in the Ackerman arms. One side has left-hand threads, and its opposite side has right-hand threads. This allows the tie-rod length to be adjusted by either lengthening or shortening.

The length of a tie-rod assembly defines the toe setting. Its length is determined by the position of the tie-rod ends. Because each tie-rod end has opposite threads, tie-rod length can be adjusted in position by loosening the tie-end clamps at either end of the cross tube, and then turning it. When the adjustment is set, the tie-rod end clamps should be tightened on the cross tube to hold the adjustment setting. This is the method used to set steering toe, and it is described in detail later in this chapter. Although it is not best practice, tie rod adjustment is often checked and adjusted on trucks using a simple mechanical tram gauge method in the absence of any more sophisticated front-end alignment equipment.

25.2 FRONT-END ALIGNMENT

The components of a truck front axle, steering system, and suspension must be precisely aligned to ensure the vehicle steers and tracks true. When properly aligned, the wheels of both a loaded and unloaded vehicle should contact the road properly, allowing each wheel to rotate without scuffing, dragging, or slipping on the road. A front end that is properly aligned will result in:

- Easier steering
- Longer tire life
- Directional stability
- Less wear on front-end components
- Better fuel economy
- Increased safety

The primary alignment angles are toe, caster, and camber. Kingpin inclination, turning angle, Ackerman geometry, and axle alignment will also have some effect on tire wear and steering characteristics. Front-end alignment should be checked at regular intervals, particularly after the front suspension has been subjected to severe impact loads. Before checking and adjusting alignment, components such as wheel bearings, tie-rods, steering gear, shock absorbers, and tire inflation should be inspected and corrected when necessary.

TOE

Toe is the tracking angle of the tires from a true straight-ahead track. If a line were to be drawn through the centerline of each tire and compared to the centerline of the vehicle (the true straight-ahead position), the amount of deviation between the two lines indicates the toe angle of the tire. When the tire centerline is exactly parallel with the vehicle centerline, the toe angle is zero. When the front end of the tire points inward toward the vehicle, the tire has **toe-in**. When the front of the tire points outward from the vehicle, the tire has **toe-out**. You can see toe-in and toe-out by looking at **Figure 25–6**, in which you are observing the tires from an overhead viewpoint.

The ideal toe angle when a vehicle is running loaded down a highway is zero. We set toe angles statically. The objective of setting toe at a specified angle when aligning the front end is to have zero toe at highway speeds. Incorrect toe angles not only accelerate tire wear but also can have an adverse effect on directional stability of the vehicle. In fact, incorrect toe angles have the potential to cause more front tire wear than any other incorrect alignment angle. Too much toe-in produces scuffing, or a featheredge, along the inner edges of the tires. Excessive toe-out produces a similar wear pattern along the outer edge of the tires. In extreme cases of toe-in or toe-out, feathered edges develop on the tread across the entire width of the tread face of both radial and bias-ply tires.

Because the objective is to have a zero toe angle under running conditions, most truck specifications require a zero or a slight toe-in setting. When a fully loaded vehicle is moving at highway speeds, there is a slight tendency of steering tires to toe-out. Any looseness in the steering linkage and tie-rod assembly also will contribute to the toe-out tendency. Today, most radial steering tires are set with zero toe angle and bias-ply tires are set with a fractional toe-in.

FIGURE 25–6 Toe-in (left) and toe-out (right)

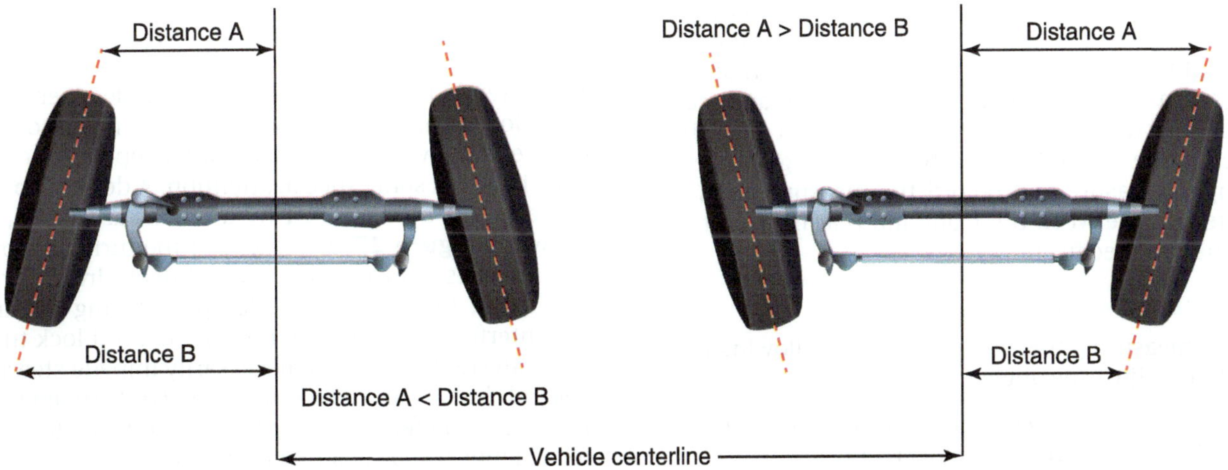

CAMBER

Steering tires also are designed to use a positive **camber** angle setting. We take a closer look at camber later in this chapter but, because it influences the toe setting, we define it here. Camber is a measure of the angle a wheel leans away or toward the frame. Positive camber means that the tires lean away from the truck frame at the top. A positive camber setting is used to help compensate for that slight tendency of steering tires to toe-out when the vehicle is moving.

Measuring Toe

Toe is specified in either degrees or fractions of inches depending on the type of alignment instruments you are using. When a toe-in setting is required, settings of 1⁄16 inch ± 1⁄32 inch (1.6 ± 0.8 mm) are usually specified. The toe specification usually depends on whether radial or bias-ply tires are used. Adjustment of toe angle or dimension requires lengthening or shortening the tie-rod dimension. This is achieved by loosening the tie-rod end clamp bolts and then rotating the cross tube.

To determine the causes of excessive tire wear that you suspect to be related to toe angle, you should first check kingpin inclination, camber, and caster. Correct, if necessary, in this order. These are covered later in this chapter. You should not make an adjustment to toe angle until the other factors of front-wheel alignment are known to be within specifications.

When measuring toe angle, the front suspension should be neutralized. To neutralize the suspension, roll the vehicle back and forth about a half vehicle length. This relaxes the front suspension and steering linkages. Neutralizing the front suspension is important before making front-end adjustments, especially if the vehicle has been jacked up on either side to scribe the tires. This operation causes the front wheels to angle as each is returned to the floor. When possible, use a scissor jack that cradles both sides of the front axle when working on front ends.

If you began by neutralizing the front end, both front wheels should at this point be in the exact straight-ahead position. Toe-in measurements should be taken across the hub centerline, front, and rear on each tire (**Figure 25–7**). Make sure that the wheels are on the ground and fully supporting the vehicle weight. Measure and record the measurements. The difference between the front and rear measurements is the toe specification.

Adjusting Toe Angle

After measuring toe, you can use the following procedure to adjust toe angle:

1. Use the steering wheel to turn the steering gear to the exact midposition (overcenter).

FIGURE 25–7 To determine toe, measure between the centerlines of the tires, both front and back position. The differences between the two measurements is the toe.

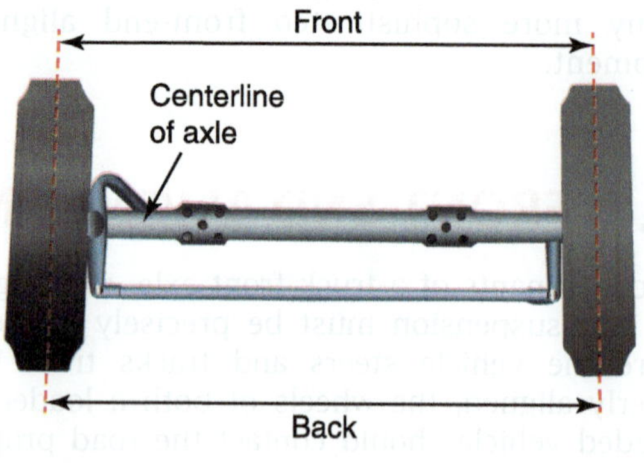

2. Loosen the clamping bolts on both tie-rod ends. Raise the weight off the axle.
3. Turn the tie-rod cross tube in the direction necessary to bring toe to the specified setting; turning the cross tube either lengthens or shortens it. After making the adjustment, lower the vehicle to the floor. Measure the toe setting again.
4. When the toe angle is set to specification, torque the tie-rod end clamp bolts to specification.

CAUTION

Recheck the toe setting after any change in caster or camber angle.

SHOP TALK

When tie-rod ends are replaced, they must be threaded into the tie-rod cross tube sufficiently so that clamp pressure is applied directly over the threads under the clamp. The threaded end should extend past the slot in the cross tube.

Deep Drop Tie-Rod Ends

Toe-in tends to increase slightly on some steer axles when loaded and stationary. It is, therefore, important that vehicles with setback axles and deep drop tie-rod ends be given some special attention. A deep drop tie-rod end can be identified by a downward curve, as shown in **Figure 25–8**. Because of the turning geometry that results from setback and deep drop tie-rod end axles, the tie-rod end clamps if facing forward can interfere with the axle I-beam on a full lock turn. This interference will not necessarily impede steering control, but it can reduce the effective turn angle of the vehicle. Refer to the appropriate service literature for diagnosis and repair procedures.

FIGURE 25–8 A typical deep drop tie-rod end. Note top-mounted curve.

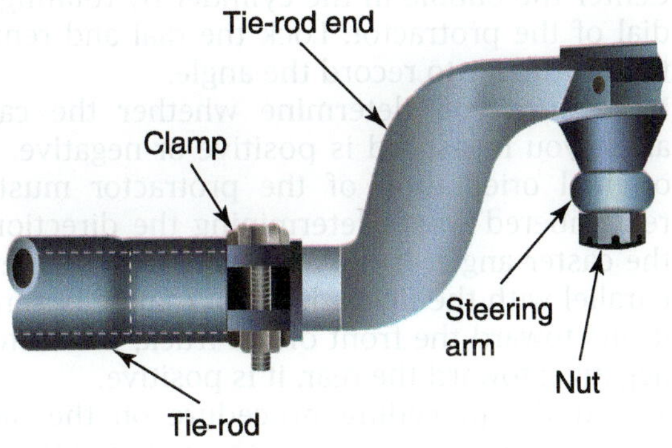

CASTER

Caster is the forward or rearward tilt of the kingpin centerline when viewed from the side of the vehicle. Zero caster occurs when the centerline of the kingpin is exactly vertical. Positive caster indicates that the kingpin is tilted rearward, as shown in **Figure 25–9**. Negative caster indicates that the kingpin is tilted forward.

Caster is a directional stability angle only. Incorrect caster by itself will not affect tire wear. Most heavy-duty trucks are designed with some degree of positive caster. Positive caster creates a trailing force in the front wheels, which tends to keep them tracking straight ahead. Positive caster tends to make the steering axle wheels want to return to a straight-ahead position after a turn has been made. The principle is exactly the same as that used in tilting the front fork of a bicycle, which makes it possible to ride the bicycle without holding the handlebars. It is the opposite of the wobble-wheel effect of a supermarket pushcart when mobile.

FIGURE 25–9 Caster is the forward or rearward (shown) tilt of the kingpin when viewed from the side of the vehicle.

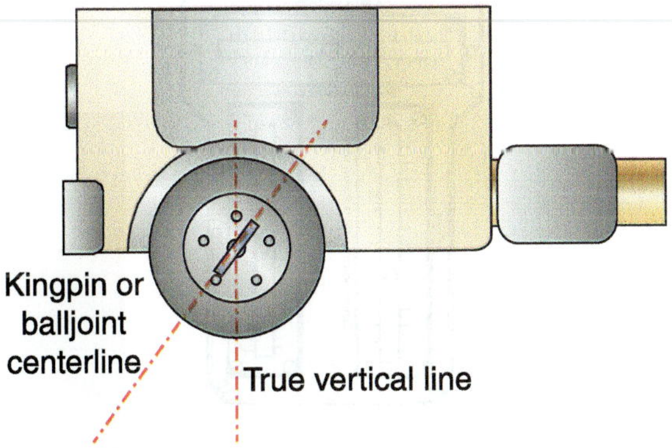

Positive caster also means that when the front wheels of the truck are turned, one side of the vehicle raises slightly when the other side is lowered. When the steering wheel is released, the weight of the vehicle forces the lifted side downward, resulting in the wheels returning to a straight-ahead steering track. In other words, positive caster produces a condition in which the right wheel resists a right turn and the left wheel resists a left turn. The further into a turn, the more this resistance increases. Caster settings generally affect steering performance in the following ways:

- Too little caster can cause wheel instability, wandering, and poor wheel recovery.
- Too much caster can result in hard steering, darting, oversteer, and low-speed shimmy.

Recommended Caster Settings

Vehicle manufacturers specify varying caster settings, depending on the truck model, what type of load it is to carry, whether it is a tandem or single drive axle, and whether it has manual or power steering. The objective is to provide just enough positive caster to ensure wheel stability and good wheel recovery. The following settings are some examples used for vehicles with manual steering gear, measured weight down on a level floor:

- Tandem drive axle: ½–1½ degrees positive
- Single drive: 1½–2½ degrees positive
- No more than ½ degree difference between the left and right wheels
- Positive caster on the left wheel should not be greater than on the right.

Caster specifications are based on vehicle design load (no payload), which will usually result in a level frame. If the frame is not level when alignment checks are made, this must be factored in the caster measurement. With the vehicle on a smooth, level surface, measure the frame angle with a protractor or inclinometer placed on a frame rail, as shown in **Figure 25–10**. Positive frame angle is defined as forward tilt (front end down) and negative angle as tilt to rear (front end high).

FIGURE 25–10 Measuring frame angle

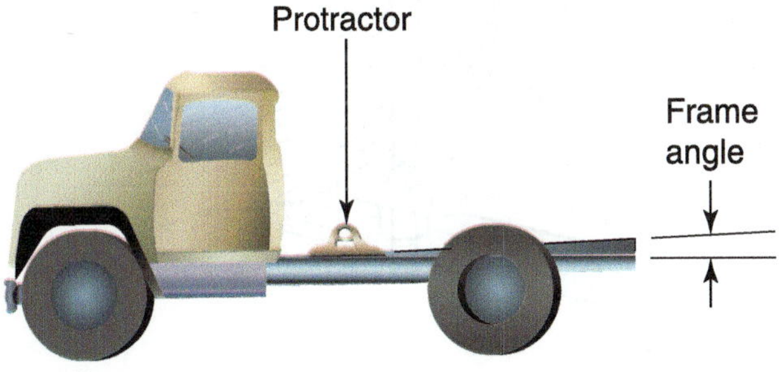

The frame tilt angle should be added or subtracted, as required, from the caster specification in the service literature, noting the following:

1. Positive frame angle should be subtracted from the caster specification.
2. Negative frame angle should be added to the caster specification.

For instance, if the specified caster setting is a positive 1 degree and you measure the vehicle frame angle to be positive 1 degree, the measured caster is ± ½ degree. This calculates to the desired 1 degree ± ½ degree caster angle when the chassis settles to level the frame under load.

Caster should be measured with the vehicle on a level floor. If the vehicle is equipped with a manually controlled air-lift axle, adjust it so that the frame is at normal operating height. Replacement of worn suspension, frame, and axle components should be done before caster is measured.

Measuring Caster

Caster can be measured in a number of ways. One way is to use a protractor. Another is to use a radius gauge and caster gauge. Modern computerized alignment equipment is the fastest, most accurate method of measuring caster (and other alignment angles) and, increasingly, shops are using this.

Protractor. A protractor measures the angle between its base and true horizontal, as determined by a bubble cylinder on one side of the dial (similar to that used in a carpenter's level). To measure caster angle:

1. Place the protractor on the machined surface of the front axle pad as shown in **Figure 25–11**. The axle pad must be free of dirt and debris. Make sure that the protractor contacts both U-bolts so that it is parallel with the truck frame.
2. Center the bubble in the cylinder by rotating the dial of the protractor. Lock the dial and remove the protractor to record the angle.
3. Make sure you determine whether the caster angle you measured is positive or negative. The original orientation of the protractor must be remembered when determining the direction of the caster angle. If the axle pad surface is level/parallel with the floor, the caster angle is zero. If it tilts toward the front of the truck, the caster is negative; toward the rear, it is positive.
4. Repeat the preceding procedure on the opposite side of the vehicle. If the caster difference on each side exceeds ½ degree, the axle is probably twisted. Also, remember that the right side should not have less positive caster than the left. If the right side is less, it will tend to lead the vehicle to the right, particularly on crowned roads.

FIGURE 25–11 To measure caster, place the protractor on a clean, flat surface of the axle pad. Be sure that the axle pad is clean and the protractor touches the U-bolts.

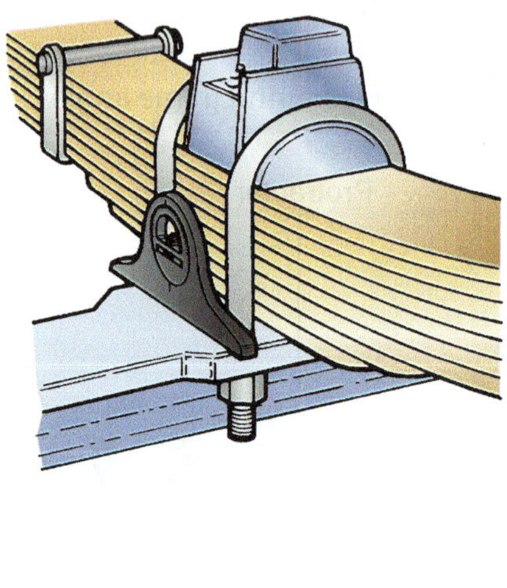

Camber/Caster Gauges. **Figure 25–12** shows a camber/caster gauge. The following steps must be performed before making both camber (explained later in this chapter) and caster measurements. Wheel runout should be checked using a tram bar or dial indicator to make an accurate camber/caster reading.

1. Drive the truck onto a level surface.
2. Raise the front end and position a radius gauge (**Figure 25–13**), with the lock pins installed, under each front wheel. The degree scale should be pointing outward.
3. Turn the steering wheel so that the front wheels face straight ahead.

FIGURE 25–12 One type of camber/caster gauge

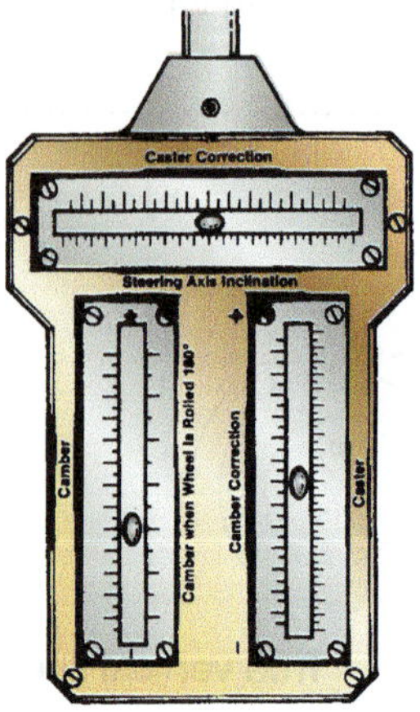

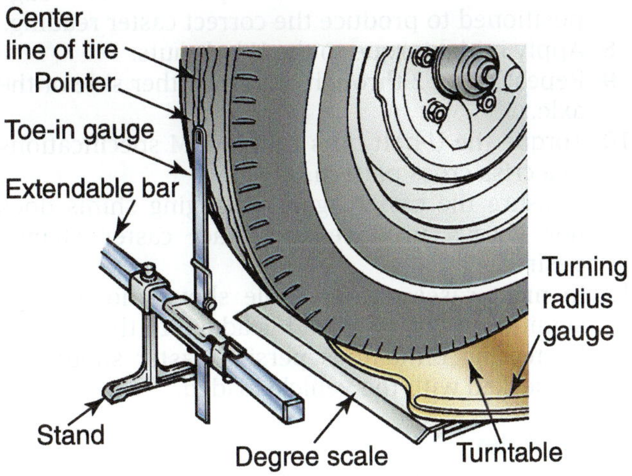

FIGURE 25–13 Simple wheel alignment equipment, including a turning radius gauge and toe-in gauge

4. Lower the tires onto the radius gauges. Be sure that the tires contact the center of the turntables, as shown in Figure 25–13.
5. Remove the lock pins that hold the turntables in a fixed position.
6. Zero both radius scales.
7. Apply the brakes when using a full-floating turning gauge, but if a semifloating type is used, leave them released.
8. Install the wheel clamp to the wheel rim, as shown in **Figure 25–14**, and the camber/caster gauge to the wheel clamp. On smaller wheels the camber/caster gauge is usually fitted directly to the wheel hub instead of the wheel rim. Follow the equipment OEM's instructions.
9. To check and adjust wheel runout, first make sure that the wheel clamp fingers rest firmly against the rim and that the gauge rests firmly against its seat, as shown in Figure 25–14.
10. Raise the front wheel off the turning radius gauge.
11. Note the reading on the camber scale.
12. When turning the wheel, note the point of the greatest wheel runout and mark the tire in that location. Next, position this location in the forward or rearward direction. In this way, the true wheel centerline is brought into the vertical position, splitting the runout.
13. Lower the front end so that the tires sit again on the turning radius gauges.

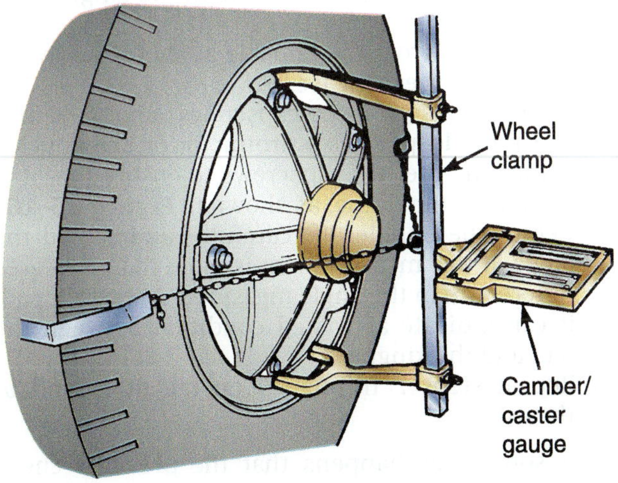

FIGURE 25–14 The camber/caster gauge attached to the wheel clamp.

SHOP TALK

Some camber/caster gauges have a runout compensation adjustment. Follow the OEM procedure when using this equipment.

Measure Caster. To measure caster using radius gauges, use the following procedure:

1. Set both turning radius scales at zero.
2. Make a right-hand turn with the steering wheel so that the left front wheel is turned to 20 degrees.
3. Turn the adjusting screw to center the left wheel caster gauge bubble.
4. Now turn the steering wheel to the left until the left front wheel is at the 20-degree left-hand turn position.
5. Rotate the gauge to level it.
6. Visually align the center of the bubble with the left- or right-hand caster graduation and read the degree of caster. The caster is positive if the bubble is toward the positive sign and negative if the bubble is toward the negative sign.
7. Repeat this procedure on the opposite side of the truck.

If the caster difference from side to side exceeds ½ degree, the axle is probably twisted. Also remember that the right side should not have less positive caster than the left. If the right side has less, it will tend to lead the vehicle to the right, particularly on crowned roads.

Caster Shims. Caster angle is changed by using caster shims, which are angled metal wedges inserted between the spring seat and the axle pad, as shown in **Figure 25–15**. Shims are typically manufactured in ½-degree increments, from ½ degree to about 4 degrees, and in a choice of 3-, 3½-, or 4-inch (7.7-, 9-, or 10-cm) widths to accommodate typical spring pack widths.

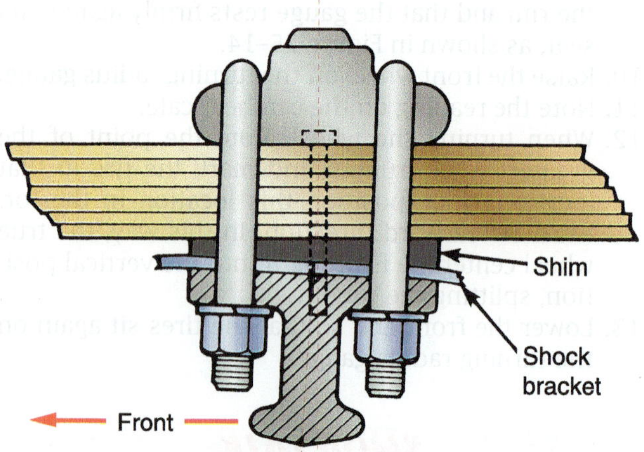

FIGURE 25–15 Caster angle is changed by inserting metal shims between the spring and axle pad.

When inserting caster shims under spring seats, you should observe the following:

- Use only one shim on each side. Do not stack them.
- The width of the shim should equal that of the spring.
- Do not use shims of more than 1-degree difference from side to side.

Changing Caster. After determining the caster desired on each side of the axle, use the following procedure to install the caster shims:

CAUTION

When changing caster, be sure to block the wheels so that the truck will not roll. Also, use safety stands to support the springs when changing the shims.

1. Loosen all spring U-bolts on each side just enough to relieve clamping tension (see Figure 25–15).
2. On one side, fully back off the U-bolt nuts without completely removing them from the U-bolt threads.
3. With a frame jack, raise the springs from the axle seat. Do not lift the wheel from the floor.
4. If a caster shim has been previously installed, remove it and make sure that the angle is factored in the calculation of desired caster angle. In the interest of safety, use a tool to remove the shim. Never insert your fingers between the axle and the springs.
5. Remove dirt and debris from the exposed axle pad.
6. Ensure that the spring center bolt cap protrudes sufficiently to pass through the caster shim and locate to the axle alignment recess.
7. Insert the shim, aligning the hole with the center bolt, and slowly lower the spring onto the axle, making sure that the center bolt seats the axle alignment recess. The caster shim must be installed with the thick part of the wedge positioned to produce the correct caster reading.
8. Apply initial torque to the U-bolt nuts.
9. Repeat steps 2 through 8 on the other side of the axle.
10. Torque the U-bolt nuts to the OEM specifications in a criss-cross pattern.
11. Measure the caster again. Changing shims does not always produce the exact caster change desired.
12. In most cases, caster angle should not change significantly when a load is added to the vehicle. If steering difficulties persist, caster should be rechecked with the vehicle loaded.

Twisted Axles

In a few cases, steering problems can be traced to a twist in the front axle. A twisted steer axle can be caused by an actual structural twist or one induced by improper use of caster shims. A twisted axle should be suspected whenever:

- The difference in caster angle exceeds ½ degree from side to side
- The caster shims in place differ by 1 degree or more
- A low-speed shimmy exists, and there is no evidence of looseness elsewhere in the steering system
- A leading or darting condition persists

Checking Steer Axle Twist. To detect a twisted axle, you can use the following procedure:

1. Roll the vehicle onto a level floor, leaving the steering wheel free for at least the last 10 feet to permit the tires to seek a true straight-ahead position.
2. Remove the capscrews that hold the kingpin cap in place.
3. Lift the cap from the knuckle assembly and wipe away any excess grease.
4. Partially insert two capscrews about ¼-inch (6.3 mm) deep. Place a protractor on the knuckle contacting the two capscrews to ensure that it is parallel to the frame, as shown in **Figure 25–16**.
5. Center the bubble, lift carefully, and record the angle. If this measurement is more than ½ degree different than the angle measured at the axle pad, it can indicate a twisted condition between the pad and the kingpin.
6. Repeat steps 2 through 5 on the other end of the axle.

It sometimes happens that the previous check will not indicate a twisted axle at either end, but there

FIGURE 25–16 Checking for a twisted axle using a machinist's protractor on the steering knuckle

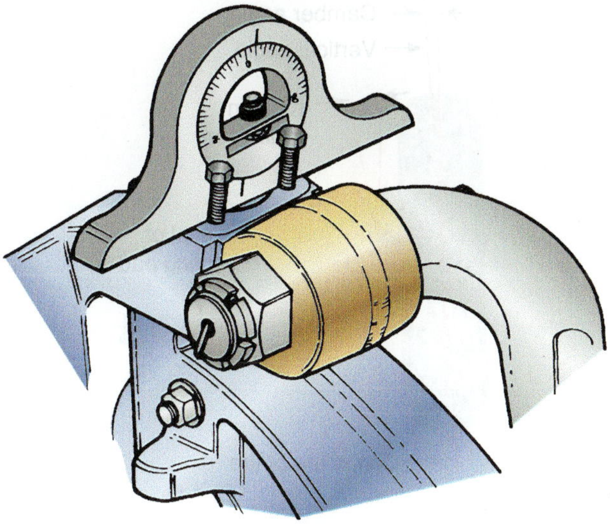

FIGURE 25–17 Camber is the inward or outward tilt of the tire when viewed from the front of the vehicle.

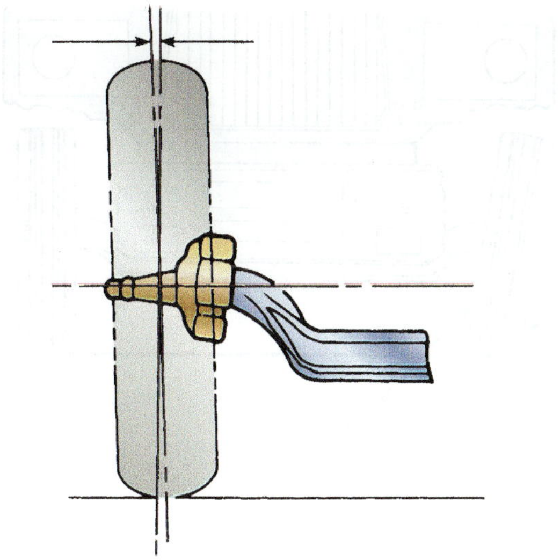

remains a side-to-side caster difference exceeding ½ degree or the steering problems previously mentioned still exist. This can be because of a twist in the axle that can be detected only in its natural or unloaded state. This can be determined by the following check:

1. Loosen all U-bolts on the front axle enough to relieve tension when keeping center bolts in position in the axle recesses.
2. Using a frame jack, lift the vehicle weight from the axle.
3. Measure the angle at both kingpins and axle pads. Any differences in these readings should identify a twist in the axle.

CAMBER

Camber is the inward or outward tilt of the top of the wheels when viewed from the front of the vehicle, as shown in **Figure 25–17**. If the centerline of the tire is exactly vertical, the tire has a camber setting of zero degrees. If the top of the tire tilts outward, it has positive camber, and if the top of the tire tilts inward, it has negative camber. Positive and negative camber is illustrated in **Figure 25–18**.

Camber settings of 1¼ degree for the curb side and 1¾ degree for the driver side wheels are typical. The amount of camber used depends on the **kingpin inclination (KPI)** setting. Kingpin inclination is the inclination of the steering axis to a vertical plane that is the equivalent of steering axis inclination (SAI) in automotive front ends.

Incorrect camber angle results in abnormal wear on the side of a tire tread. Excessive positive camber causes the tire to wear on its outside shoulder. Likewise, excessive negative camber causes the tire to wear on its inside shoulder. Unequal camber in the front wheels also can cause the steering to lead to the right or left. The truck will lead to the side that has the most positive camber.

Camber adjustments are not often made on heavy-duty truck axles. Some service facilities with specialized equipment adjust camber by cold bending axle beams, but, generally, axle manufacturers do not approve of this practice. Bending an axle can damage the structural integrity of the beam, increase the risk of metal fatigue, and lead to the axle breaking. Heating and bending the axle should never be attempted because it removes the heat tempering.

Measuring Camber

The preparatory procedure that has to be performed before measuring camber is detailed earlier in the section on caster. Perform those steps first. Then stabilize the front end. Next, visually align the center of the bubble with the left- or right-hand camber graduation in order to read the camber from the camber scale. Camber is positive if the bubble gravitates toward the positive sign, and it is negative if it gravitates toward the negative sign. It should be emphasized that camber should be checked but seldom has to be adjusted in truck front ends.

Kingpin Inclination

KPI is the amount in degrees that the top of the kingpin inclines away from the vertical, viewed from the front of the truck, as shown in **Figure 25–19**. KPI, in conjunction with camber angle, places the approximate center of the tire tread footprint in contact with

FIGURE 25–18 (A) Positive camber results in wear on the outer edge of the tire; (B) negative camber results in tire wear on the inner edge.

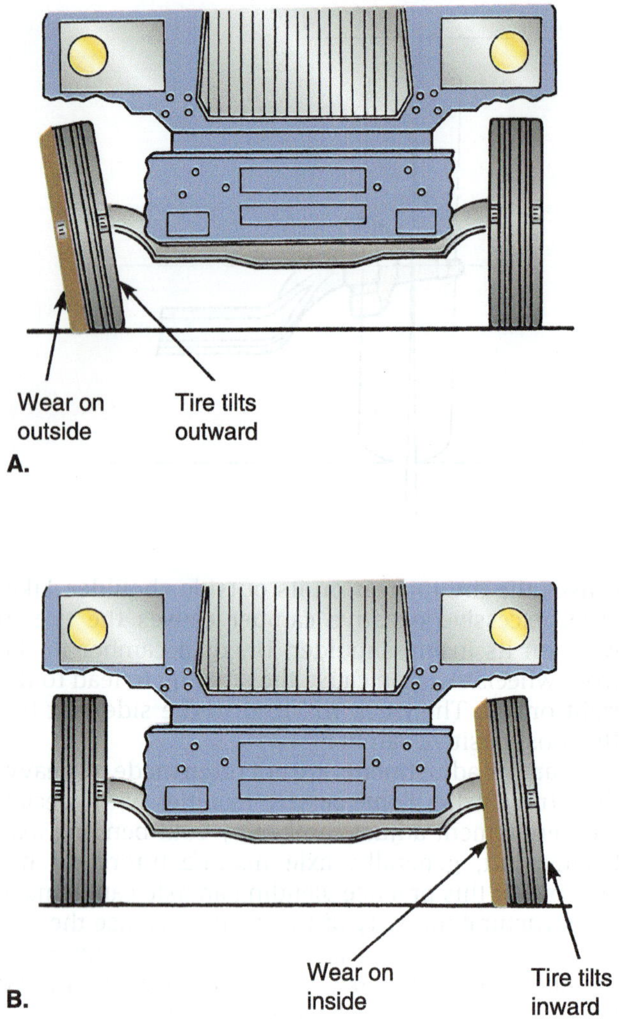

FIGURE 25–19 Kingpin inclination and camber angle as viewed from the front of the truck

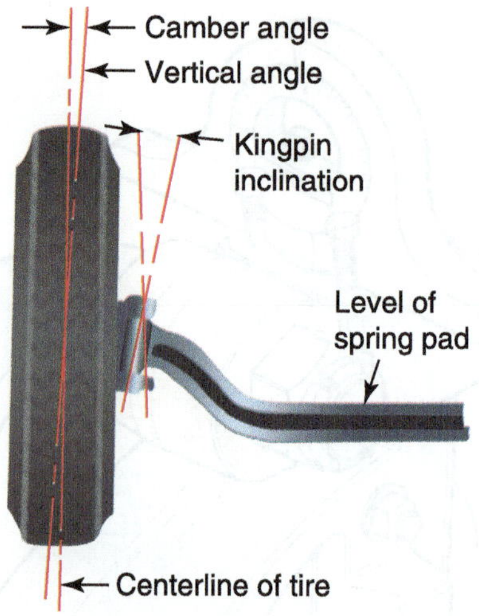

If turning angle does not meet specifications, it should be corrected. To make an adjustment to the turning radius, use the following procedure:

1. Block the rear wheels and raise the truck off the ground. Position support stands under the front axle.
2. Turn the wheels to the extreme right turn until the steering gear bottoms or tire-to-chassis contact is made.
3. Back off the steering wheel one-quarter turn or until ½- to 1-inch (12.5 mm to 2.5 cm) clearance is obtained between the tire and chassis.
4. Next check both front tires for clearance at full steering lock. When the specified clearance is set, back the wheel stop screw out until the screw contacts the axle stop (see Figure 25–20) and tighten the jam nut.

the road. KPI reduces steering effort and improves the directional stability. KPI cannot be adjusted in trucks. Once set, KPI should not change unless the front axle has been bent. Corrections or changes to this angle are accomplished by replacement of broken, bent, or worn parts.

Turning Angle or Radius

Turning angle or radius is the degree of movement from a straight-ahead position of the front wheels to either an extreme right or left position. Two factors limit the turning angle: tire interference with the chassis and steering gear travel. To avoid tire interference or bottoming of the steering gear, adjustable stop screws are located on the steering knuckles (**Figure 25–20**). Turning radius or angle should be checked using the radius gauge described earlier.

FIGURE 25–20 The maximum turning angle of a wheel is determined by a stop screw.

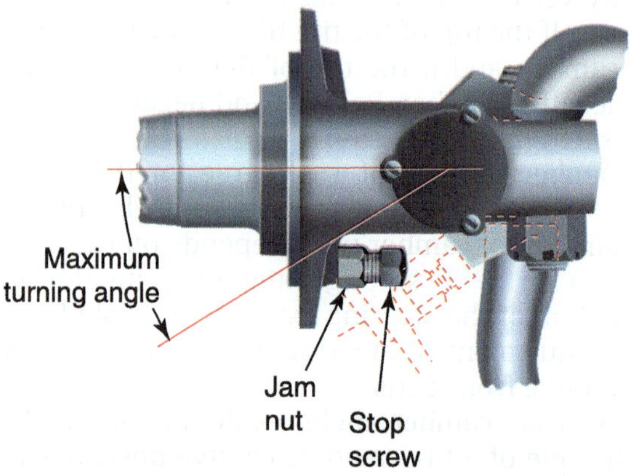

5. Repeat the same procedure on the left extreme turn and set the left steering knuckle stop screw.

SHOP TALK

The power steering gear pressure relief valve should open just before the steering stop screw contacts the axle stop. You may have to adjust the power steering gear so that power-assist stops approximately 1 degree before the steering stops contact. Failure to do this will result in slamming of the steering stops on full lock turns.

Driving Front Axles

A stop screw is located on each end of the axle housing for the purpose of limiting the amount of the turning angle of the wheels. These screws are not adjusted in accordance with the frame and tire interference, as in conventional steer axles. The stop screws in front drive axles limit the operating angle of the universal joints in the drive axle shafts. You should reference the OEM service literature before making adjustments.

ACKERMAN GEOMETRY

Ackerman geometry is the means used to steer a vehicle so that the tires track freely during a turn. During a turn, the inboard wheel on a steer axle has to track a tighter circle than the outer wheel. In other words, the outer tire has to travel just a little further because of the wider arc it has to follow. Ackerman geometry is also known as toe-out during turns. It allows the inner and outer wheel to turn at different angles so that both wheels can negotiate the turn without scrubbing. This is shown in **Figure 25–21**.

To achieve the required Ackerman angle, the steering linkage must be arranged so the axes of the steering wheels intersect on the axis of the rear axle (see Figure 25–21). On single drive axles, the centerline of the drive axle is the rear axis. On tandem drive axles, the intersecting axis is centered between the two drive axles (**Figure 25–22**).

The Ackerman angle built into the steering geometry actually provides perfect rolling angles for both front wheels at only one turning angle. Anything other than this turning angle introduces an error, the size of which depends on the length and inclination of the tie-rod arms. As a result, the lengths of the cross tube and tie-rod arms have to be selected so as to minimize the error throughout the full range of steering angles possible on a particular front end. In trucks, the Ackerman steering error has to be smallest at low turning angles where the most steering corrections are made and at the highest road speeds.

Toe-out on turns is accomplished by having the ends of lower steering arms (those that connect to the tie-rods) closer together than the kingpins, as shown in **Figure 25–23**. Actual toe-out during a turn depends on the length and angle of the steering control arms and the length of the cross tube. Even if the toe-in setting with the wheels in a straight-ahead position is correctly adjusted, a bent steering arm can cause the toe-out on a turn to be incorrect, causing tire scuffing.

FIGURE 25–21 Ackerman geometry requires that an ideal turning radius is one where the turning angles of the front wheels intersect the centerline of the drive axles at the same point.

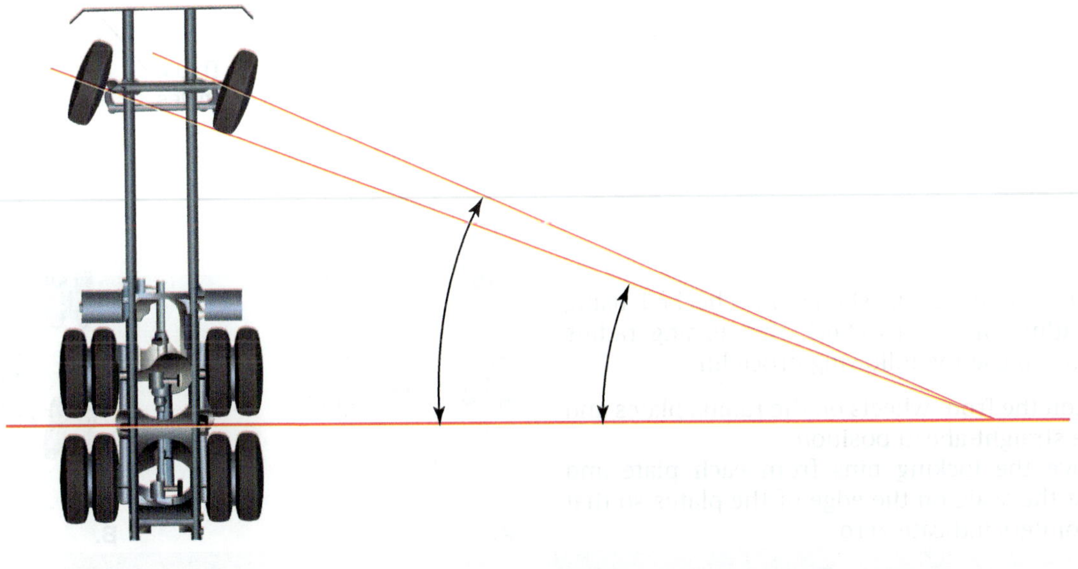

FIGURE 25–22 On single drive axles, wheelbase is measured from the centerline of the front axle to the centerline of the rear axle. On tandem drive axles, the wheelbase is measured from the centerline of the front axle to the center point between the two drive axles.

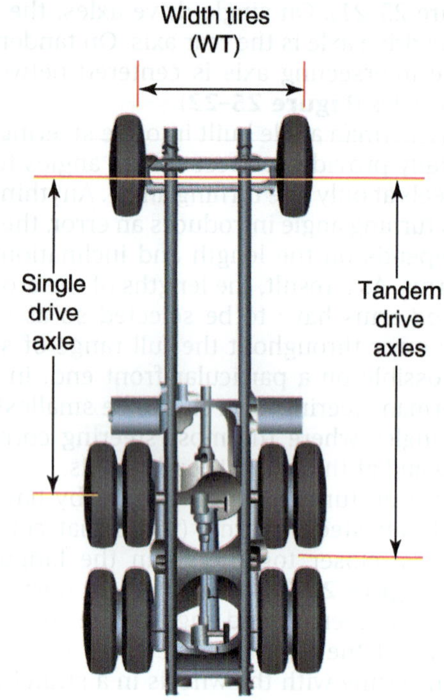

FIGURE 25–23 In a toe-out condition on turns, the inside wheel turns at a greater angle than the outside wheel.

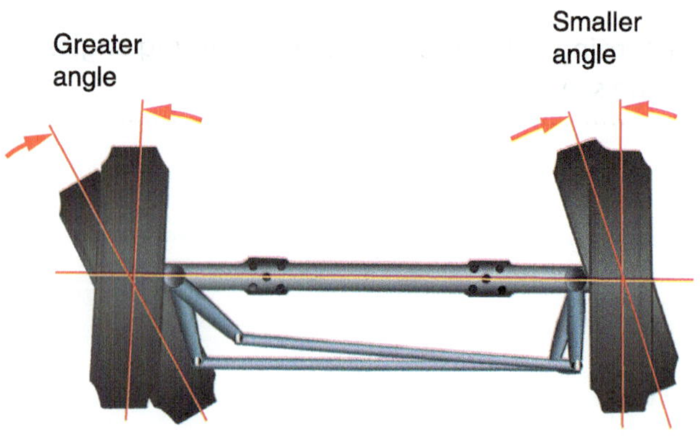

Turning radius angle should be checked using turning radius plates. To check the turning radius angle, you can use the following procedure:

1. Position the front wheels on the radius plates and in the straight-ahead position.
2. Remove the locking pins from each plate and adjust the scale on the edge of the plates so that the pointers indicate zero.
3. Next, turn the wheels to the right until the indicator gauge at the left wheel reads 20 degrees. Then read and record the angle of the right wheel.
4. The right wheel should then be turned to an angle of 20 degrees. The left wheel should read the same turn angle as the right wheel did in the previous step. If the angle is different, check the steering arms for damage and replace any that are bent. Do not attempt to straighten them.

AXLE ALIGNMENT

When a vehicle is running on the highway, all of the axles should track perpendicular to the vehicle centerline; that is, the rear wheels should track directly behind the front wheels when the vehicle is moving straight ahead. When this happens, the thrustline created by the rear wheels is parallel to the vehicle centerline, as shown in **Figure 25–24A**.

However, if the axles are not running perpendicular to the vehicle centerline, the rear wheels will not track directly behind the front wheels, and the thrustline of the rear wheels deviates from the centerline of the vehicle, as shown in **Figure 25–24B**. The steering controls fight the vehicle thrustline, resulting in an uncentered steering wheel and accelerated front tire wear, and which causes the vehicle to **oversteer** when turning in one direction and **understeer** when turning in the other

FIGURE 25–24 (A) Correct thrust line; and (B) incorrect thrust line

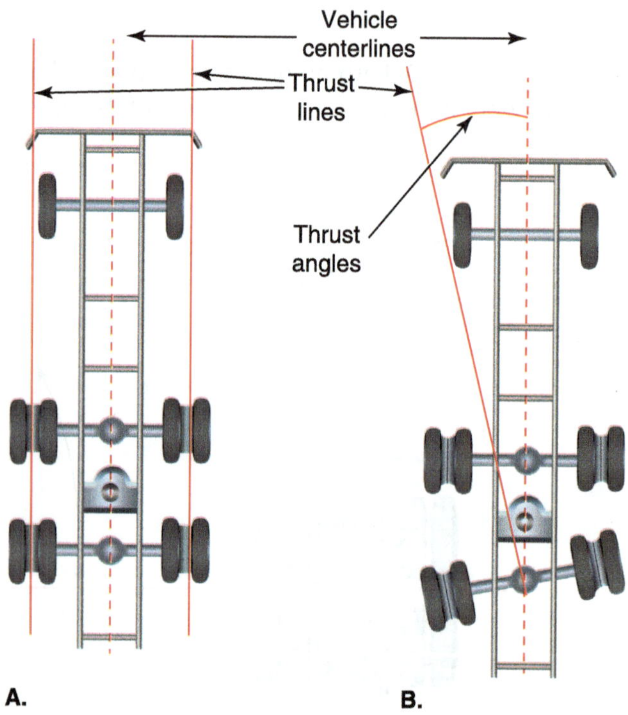

direction. Oversteer is an overresponse to steering input, which results in vehicle **yaw** or lateral tracking off the intended turning radius. Understeer is an underresponse to steering input, most often causing steering tire slip at high speeds. Incorrect directional tracking can occur on single-axle vehicles, tandem-axle vehicles, and trailers. On a single-axle vehicle, the rear-axle thrustline can be off if the entire axle is offset or if only one wheel has an improper toe angle. On a tandem axle, there are a number of different combinations that can cause incorrect tracking. **Figure 25–25** illustrates some of those combinations.

One method of checking a single axle for misalignment is to clamp a straightedge across the frame so that it is square with the frame rails on each side. Then measure from the straightedge to the center of the hub. The distances on each side should be within ⅛ inch (3 mm) of each other. If not, the axle must be aligned.

Trailer Tracking

It is also possible for the trailer axles to be out of alignment and cause a tracking problem. Depending on the severity of the trailer misalignment, it might be possible to see the effects of the misalignment as the trailer travels down the road. Usually, the trailer will travel at an angle to the tractor. Misalignment also makes it very hard to back up the trailer. This is commonly called **dog-tracking**, a condition that is illustrated in **Figure 25–26**.

FIGURE 25–26 A dog-tracking trailer

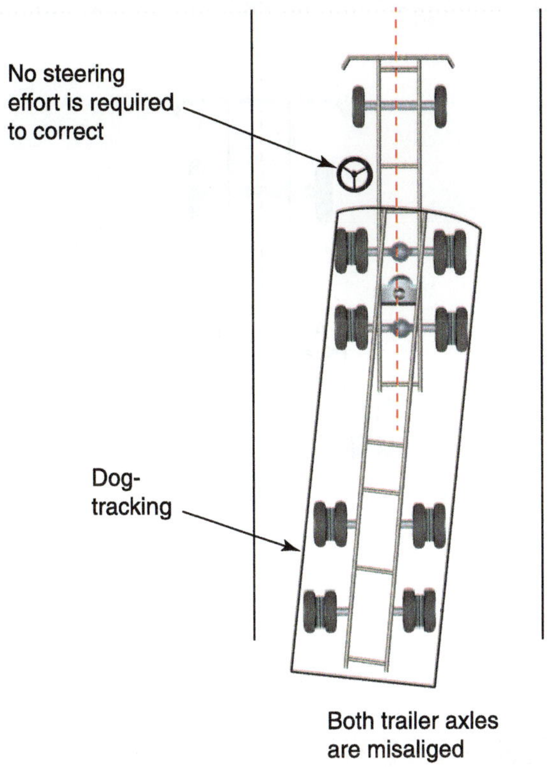

FIGURE 25–25 Typical tracking problems of tandem-axle vehicles

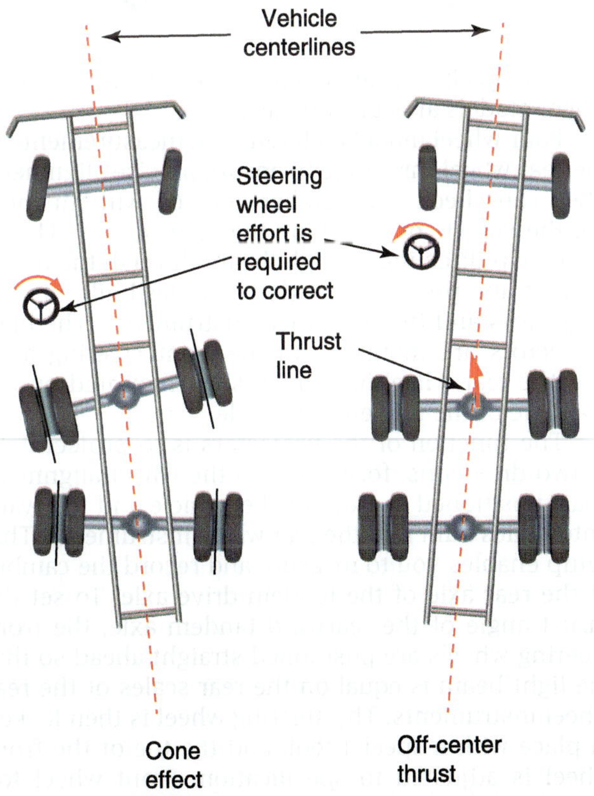

Measurement

For a tandem axle, the setup, measurement, and specifications for the forward-rear axle are the same as a single drive axle. The forward-rear axle should be square within ⅛ inch (3 mm). It is also necessary to check and make sure that the rear-rear axle is parallel with the forward-rear axle. A trammel bar or tape measure can be used to measure from the hub center to the hub center on each side of the vehicle. These measurements should be within ¼ inch (6 mm) of each other. If either axle or both axles are misaligned, steering and tire wear problems will result. Additional information on aligning tandem axles is given in Chapter 26.

Axle Offset

Another problem is an axle that is not centered with the centerline of the vehicle, as shown in **Figure 25–27**. When an axle is offset and the vehicle is driven straight down a highway, the steering wheel should be centered and the vehicle will not dog-track but, as soon as it is cornered, it will oversteer in one direction and understeer in the other.

FIGURE 25–27 Axle offset: dimension A should equal dimension B.

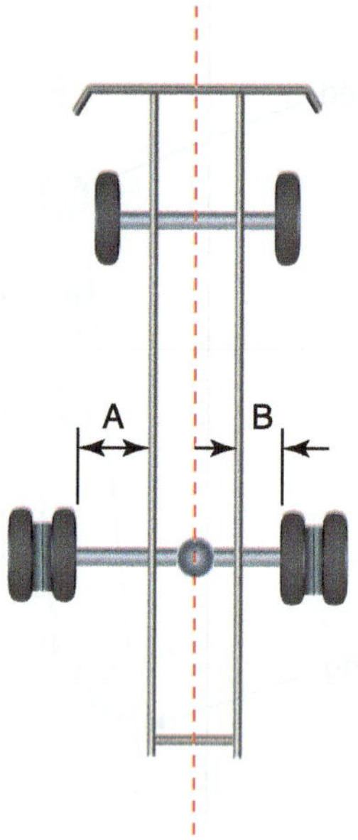

FIGURE 25–28 Hunter shop floor computer station loaded with WinAlign HD Software.

25.3 ELECTRONIC ALIGNMENT SYSTEMS

As mentioned earlier in this chapter, wheel and axle alignment measurements can be taken in a number of ways. A number of electronic alignment systems exist. Any electronic alignment system uses computers to analyze and display alignment conditions. The electronic methods include:

- Light beam alignment
- Laser alignment
- High-resolution camera

Figure 25–28 shows a Hunter alignment shop floor computer console used with their **WinAlign** HD software. This Windows platform software uses wireless interface with wheel-mounted sensors.

ALIGNMENT SYSTEMS

Light beam, **laser alignment**, and **photo alignment** systems use wheel-mounted instruments to project light beams onto charts, scales, and sensors to measure toe, caster, and camber and to note the results of alignment adjustments. The high-resolution camera photo alignments system is at present available only for light-duty applications (the Hunter system is known as Hawkie Elite) but a truck version is expected to be on the market in the near future.

In all three systems, wheel-mounted projectors can be focused to handle any length wheelbase. The system checks all alignment angles. Two- and four-wheel models are currently available.

Four-wheel models reference all measurements to the rear wheel thrustline to ensure precise alignment. They can check rear camber and toe without moving the vehicle or switching the instruments. This is done by redirecting the light beam through the use of mirrors and precisely positioned scales built into the projectors and the rear wheel instruments. The light projectors are mounted on the front steering axle, and the rear wheel instruments are mounted on the rear axle of the tandem drive axle.

The function of the projectors is to project light in two directions: forward onto the wheel alignment chart positioned in front of the vehicle and rearward onto scales built into the rear wheel instruments. This setup enables you to measure and record the camber of the rear axle of the tandem drive axle. To set the thrust angle of the rearward tandem axle, the front steering wheels are positioned straight ahead so that the light beam is equal on the rear scales of the rear wheel instruments. The steering wheel is then locked in place with a special tool, and the toe of the front wheel is adjusted to specification. Front wheel toe readings are taken off the chart positioned in front of the vehicle.

To adjust the rear axle of the tandem drive axle, the rear scales of the rear wheel instruments are lifted to expose mirrors. These mirrors redirect the light beam from the projector back onto scales mounted in the rear toe boxes of the projectors. The rearward tandem axle can then be aligned, based on the readings shown on the rear toe box scales.

To align the front steering axle, the front wheels are set straight ahead and camber and caster measurements are taken off the chart positioned in front of the vehicle. Adjustments are made as needed. With the wheels still in the straight-ahead position, toe can be measured and adjusted as needed.

To align the forward drive axle of a tandem drive truck, the rear scales of the rear wheel instruments are lifted to expose the mirrors. The beams from the projectors are now directed back into the rear toe boxes of the projectors. Adjustments are then made based on the reading on the rear toe box scales. **Figure 25–29** shows a technician fitting a self-centering wheel adaptor on a highway tractor alignment system.

FIGURE 25–29 Technician is fitting a self-centering wheel adaptor to a highway tractor front wheel. The wheel adaptor uses a digital signal processor and wireless technology to interface with WinAlign software.

Courtesy of Hunter Engineering Company

Trailer Axles

A light beam alignment system also can be used to align trailer axles. A gauge bar attached to the trailer kingpin works with the system to set trailer wheels straight ahead as referenced to the trailer centerline. The light projectors are easily adjusted to compensate for wheel distortion, runout, and mounting discrepancies. During operation, light beams are projected onto the trailer gauge scales to show the position of the wheels. The wheels are adjusted to obtain equal measurement side to side on the scales.

An air control assembly allows the operator to control the trailer brakes during the alignment operation. The kingpin flag assembly is mounted on the kingpin approximately perpendicular (90 degrees) to the trailer geometric centerline, as shown in **Figure 25–30A**. The projectors or wheel adaptor sensors are mounted on the rearmost wheels of the trailer with the rear light beams directed toward the kingpin flag. The forward wheels are chocked, the air valve assembly is connected, and the trailer brakes are released. The projectors are then adjusted by jacking up the rearmost axle and focusing the projector light beams on a special compensating panel positioned between the projector and the kingpin flag. This adjustment compensates for any wheel distortion, runout, or mounting discrepancies.

When this adjustment is made, the rearmost axle of the trailer can be adjusted until the light beam readings are equal on both kingpin flags. The rear wheel instruments are then installed on the front axle wheel with the mirrors facing the projectors, as shown in **Figure 25–30B**. The rearmost wheels are chocked. The front axle is then raised, and the projectors are adjusted to compensate for irregularities. The mirrors of the rear wheel instruments are exposed to redirect the light beams back into the rear toe boxes of the projectors. The front axle can then be adjusted until the readings on the rear toe box scales are equal. This procedure can equalize the thrustline of all axles to the geometric centerline. Individual toe and total toe cannot be determined.

SENSOR AND COMPUTER ALIGNMENT SYSTEMS

The most advanced alignment systems in use today are controlled by microprocessors. They use sensors mounted at each wheel for fast, precise alignment. A five-step alignment analysis procedure is shown in **Figure 25–31**. Alignment readings, specifications, and step-by-step instructions are displayed on a video display monitor. Keyboard-entered specifications are automatically compared against the actual angles of the vehicle with the results displayed on the display screen. Specifications can be retained in computer

FIGURE 25–30 Typical wheel adaptor alignment setup for trailer axle alignment: (A) setup for rearmost axle alignment; and (B) setup for forward axle alignment

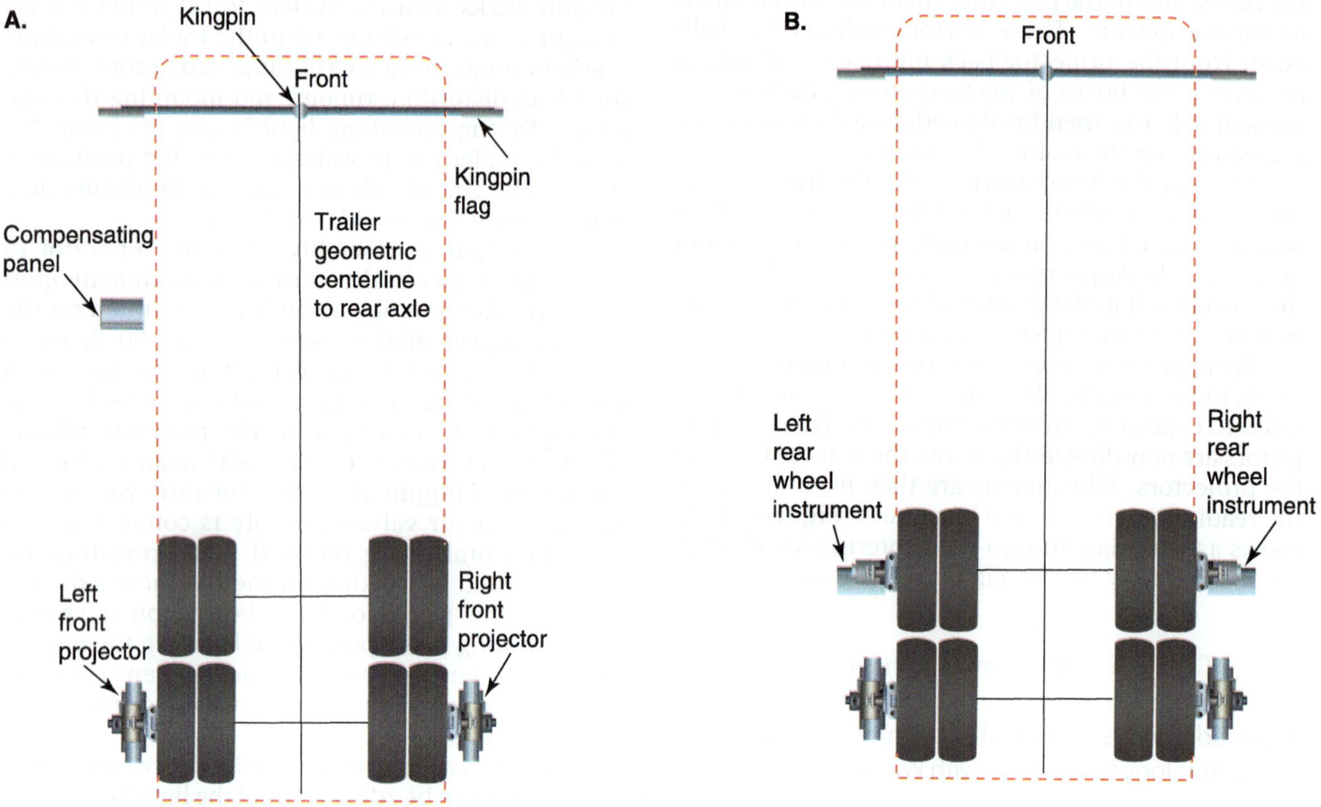

memory for future use and correlated with the OEM specifications referenced online or on compact disc (CD).

Most computer-controlled alignment equipment is user-friendly and provides step-by-step instructions and graphics on the display monitor. As adjustments are made on the truck, these are automatically displayed on the monitor, enabling a high degree of precision.

A typical display screen is shown in **Figure 25–32**. This shows an initial analysis made by Hunter WinAlign software as displayed by a Windows-driven PC.

WinAlign automatically calculates the correction for the technician. As the adjustment is made, the arrow moves across the bar graph target, guiding the technician. When the adjustment comes within spec,

FIGURE 25–31 Five-step alignment analysis sequence used on a tandem-axle highway tractor.

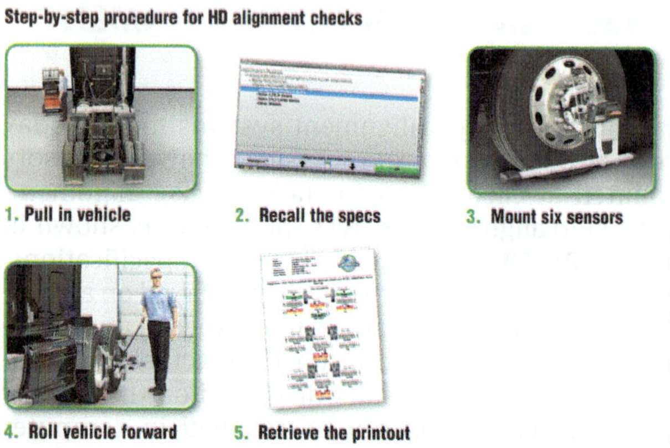

Courtesy of Hunter Engineering Company

FIGURE 25–32 A Hunter WinAlign display screen showing an initial analysis of alignment data

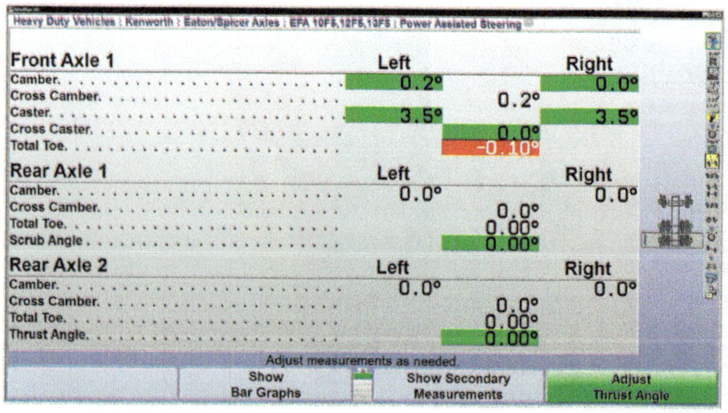

Courtesy of Hunter Engineering Company

FIGURE 25–33 WinAlign automatically calculates the correction for the technician. As an adjustment is made, the arrow moves across the bar graph target until it comes within spec.

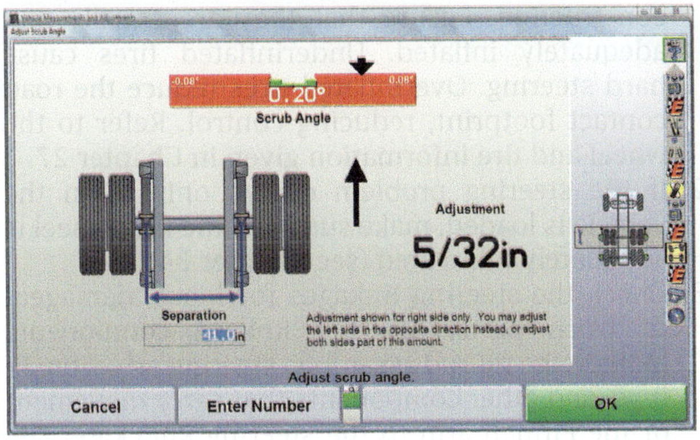

Courtesy of Hunter Engineering Company

the bar graph changes color from red to green, as shown in **Figure 25–33**.

A typical system provides for four-wheel alignment with four sensors (two axles simultaneously) or for two-wheel alignment with two sensors (one axle at a time). All wheels are aligned to a common centerline for precise alignment. By moving instruments, the system can also check both rear axles of a tandem drive axle as well as the front steering axle. This means all axles can be aligned relative to a common centerline.

Today's computer software makes alignment adjustments easy. After the truck specifications and vehicle identification number (VIN) are keyed or downloaded into the system computer, the alignment can begin. By pressing the appropriate key, the thrust angle of the rear drive axle is computed and displayed on the display monitor. Adjustments to this angle can then be made until it is within specifications. The caster and toe are then measured and adjusted on the front steering axle. Once again, a few simple keystrokes compute the data and track alignment progress.

After the front axle is aligned, the front sensors are moved to the forward axle of the tandem pair. The thrust angle of this axle is then measured by the sensors and adjusted as needed. Like the light beam system explained earlier, this computer-controlled system can be adapted to align trailer axles by using a gauge bar attached to the trailer kingpin, offering a much greater degree of precision than equivalent bazooka alignment equipment.

Sensor-driven, computerized alignment systems make truck alignments an exact science. They will also measure and display frame offset angles. This allows technicians to true truck and trailer chassis and suspensions. A WinAlign display screen showing frame offset is shown in **Figure 25–34**.

FIGURE 25–34 Frame offset screen display using WinAlign software

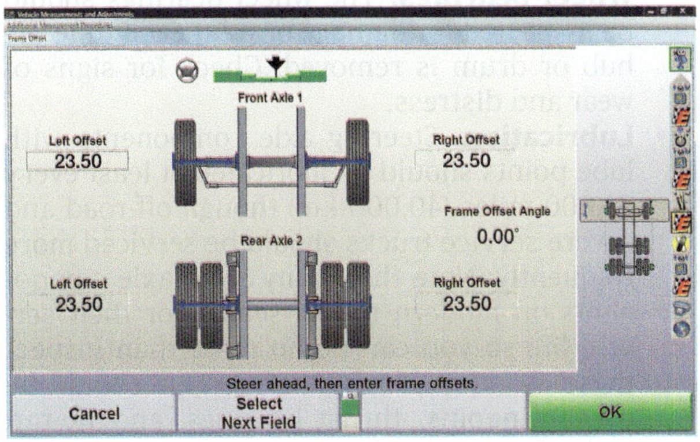

Courtesy of Hunter Engineering Company

25.4 STEERING AXLE INSPECTION

Specific service procedures and maintenance intervals can be found in OEM service literature. The following procedures are general guidelines for periodic steering axle inspection:

- **General Inspection.** A thorough visual inspection for proper assembly, broken parts, and looseness should be performed each time the vehicle is lubricated. In addition, ensure that the spring-to-axle mounting nuts and the steering connection fasteners are secure.
- **Wheel Alignment.** Front steering wheel alignment should be checked periodically according to OEM-recommended service intervals (TMC RP 642 recommends every 80,000 to 100,000 miles (130,000 to 160,000 km). If excessive steering effort, vehicle wander, or uneven and excessive tire wear is evident, the wheel alignment should be checked immediately.
- **Steering Axle Stops.** Although steering axle stops should be checked periodically, they seldom need adjustment. However, if the steering turning radius is insufficient or excessive (resulting in tire or wheel contact with the frame), the stops should be adjusted.
- **Tie-Rod Ends.** Tie-rod ends should be inspected each time the axle is lubricated. Check for torn or cracked seals and boots, worn ball sockets, or loose fasteners.
- **Steering Knuckle Thrust Bearings.** The knuckle thrust bearings should be checked each time the hub/drum is removed. Knuckle vertical play should be adjusted each time the knuckle pin is removed for service, at each axle overhaul, or whenever excessive knuckle vertical movement is noted.

- **Kingpins.** The kingpin and its bushings should be inspected whenever the pin is removed for service, at axle overhaul, or looseness is noted.
- **Wheel Bearings.** The wheel bearings should be inspected for damage or wear each time the hub or drum is removed. Check for signs of wear and distress.
- **Lubrication.** Steering axle components with lube points should be lubricated at least every 25,000 miles (40,000 km) though off-road and severe-service trucks should be serviced more frequently. Note that many steer axle components on modern trucks sealed for their service life so you can do no more than inspect these. Good-quality chassis grease should be used. Kingpins, thrust bearings, and tie-rod ends should be lubricated at each preventive maintenance (PM) service interval. Those without grease fittings are usually permanently lubricated. Consult the OEM service literature for the recommended steering gear lubricant. This may be gear lube, engine oil, or hydraulic transmission fluids. Because most steering gear have a number of lubricant options, you should try to top up with the same lubricant that is already in the gear or reservoir.

WARNING

When a vehicle is operated at temperatures below 30°F (−1°C) with SAE 90 weight oil in a manual steering gear, it can turn to a grease-like consistency, resulting in stiff, sluggish steering. This can compromise accident avoidance maneuvers by slowing steering response. When operating in temperatures continuously below 30°F (−1°C), install a lighter oil in manual steering gear, such as SAE 75 weight oil.

INSPECTION PROCEDURE

When a steering problem is reported, systematically inspect the vehicle steering system, front and rear suspensions, and trailer suspension. In most cases, a road test will be required, but never take a truck out onto a road until you are sure it is roadworthy. If a reported problem occurs only when the vehicle is loaded, you should test drive the vehicle loaded. In addition when you are checking out steering systems, remember that other chassis systems can cause steering problems.

WARNING

All steering mechanisms are critical safety items. A vehicle should be deadlined (out-of-service [OOS] report) when a defect is reported. It is essential that instructions in the service literature are adhered to. Failure to observe these procedures may cause loss of steering with life-threatening results.

Perform the following sequence of checks when symptoms such as rapid tire wear, hard steering, or erratic steering indicate a problem in the steering system.

1. Check that the front tires are the same size and construction. Ensure that they are equally and adequately inflated. Underinflated tires cause hard steering. Overinflated tires reduce the road contact footprint, reducing control. Refer to the wheel and tire information given in Chapter 27.
2. If the steering problem occurs only when the vehicle is loaded, make sure that the fifth wheel is adequately lubricated (see Chapter 34).
3. Check the steering linkages for loose, damaged, or worn parts. Steering linkage components include the tie-rod assembly, steering arms, bushings, and other components that carry movement of the Pitman arm to the steering knuckles. The wheels should turn smoothly from stop to stop through a full turn cycle.
4. Inspect the drag link, steering drive shaft(s), and upper steering column for worn or damaged parts.
5. Ensure that the steering column components, especially the U-joints, are adequately lubricated.
6. Check the front axle wheel alignment, including wheel bearing adjustment, caster and camber angles, and toe-in.
7. Check the rear axle alignment. Rear axle misalignment can cause hard or erratic steering. If needed, align the rear axle(s).
8. Inspect the front axle suspension for worn or damaged parts.
9. With the front wheels straight ahead, turn the steering wheel until motion is seen at the wheels. Align a reference mark on the steering wheel with a mark on a ruler and then slowly turn the steering wheel in the opposite direction until motion is again observed at the wheels. Measure the lash (freeplay) at the steering wheel. Too much lash exists if the steering wheel movement exceeds:
   - 3 inches (7.6 cm) for a 24-inch (61 cm) steering wheel
   - 2¾ inches (7 cm) for a 22-inch (56 cm) steering wheel
   - 2¼ inches (5.6 cm) for a 20-inch (51 cm) steering wheel
10. Turn the steering wheel through a full right and full left turn. If the front wheels cannot be turned to the right and left axle steering stops without binding or noticeable interference, one or more of the following may be the cause:
    - The worm shaft bearings or the sector shaft mesh are adjusted too tightly.
    - The worm shaft or worm shaft bearings are worn or damaged.

- The sector shaft is worn or damaged. Instructions on troubleshooting and repair of steering gear problems can be found later in this chapter.

11. Secure the steering wheel in the straight-ahead driving position. Move the front wheels from side to side. Any play in the steering gear bearings will be felt in the drag link ball stud at the Pitman arm. If any bearing play exists, make adjustments in the following order:
    - Adjust the worm shaft bearings.
    - Adjust the sector shaft total mesh preload.
12. Remove the safety stands, lower the vehicle, and remove the chocks.
13. Check the tractor fifth wheel. Lack of lubrication between the fifth wheel plate and trailer upper coupler can cause serious steering problems. This is most common in tractor trailer combinations that seldom uncouple. With many fifth wheels, the only way to ensure adequate lubrication between lower and upper coupler plates is to separate the tractor and apply grease directly to the fifth wheel plate.

SHOP TALK

Before performing the remaining checks, apply the parking brakes and chock the rear tires. Raise the vehicle until the front tires leave the pavement and then place safety stands under the frame rails. Be sure that the stands will support the weight of the vehicle.

WARNING

Do not drive a vehicle with too much lash in the steering gear. Excessive lash is a sign of an improperly adjusted steering gear or worn or otherwise damaged steering gear components. Driving the vehicle in this condition could result in a loss of steering control.

STEERING AXLE PROBLEM DIAGNOSIS

If general troubleshooting and inspecting of a steering system suggest a problem in the steering axle, you can use the following procedure to identify worn or out-of-adjustment components.

CAUTION

Before performing any servicing procedure on the front suspension, set the parking brake and block the drive wheels to prevent the vehicle from moving. After jacking the truck up until the steering axle wheels are raised off the ground, support the chassis with safety stands. Never work under a vehicle supported only by a jack.

FIGURE 25–35 Inspecting knuckle vertical play

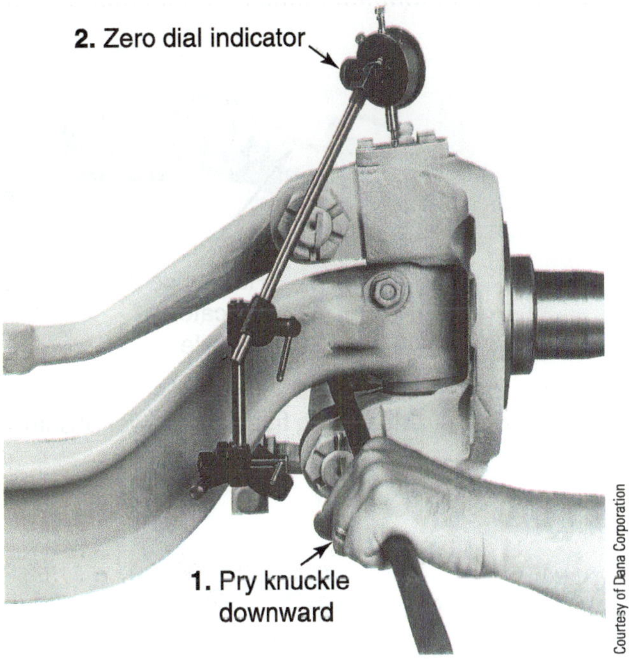

Steering Knuckle Inspections

Steering knuckles account for a number of steering problems and many of these problems originate from lack of lubrication. The following tests should be performed to check out steering knuckles and kingpins.

Steering Knuckle Vertical Play.

1. Mount a dial indicator on the axle beam.
2. Position the indicator plunger on the knuckle cap (**Figure 25–35**).
3. Pry the steering knuckle downward.
4. Zero the dial indicator.
5. Lower the front axle to obtain the dial indicator reading. If the reading exceeds the OEM specification (typically 0.04 inch [1 mm]), inspect the thrust bearings. Replace them if necessary.

Kingpin Upper Bushing FreePlay.

1. Mount a dial indicator on the axle, as shown in **Figure 25–36**. Place the indicator plunger on the upper part of the knuckle, as shown.
2. Move the top of the wheel in and out with a push/pull motion. Have someone monitor the dial indicator readings. Readings that exceed the OEM specifications (typically 0.015 inch [0.38 mm]) indicate the need for bushing replacement.

Kingpin Lower Bushing FreePlay.

1. Mount the dial indicator on the axle. Reference the plunger on the lower tie-rod end socket of the steering knuckle, as shown in **Figure 25–37**.

FIGURE 25-36 Inspecting freeplay in upper bushing

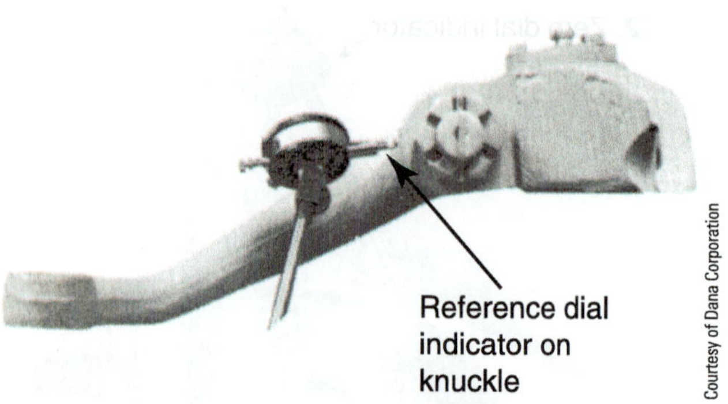

FIGURE 25-37 Measuring freeplay in lower bushing

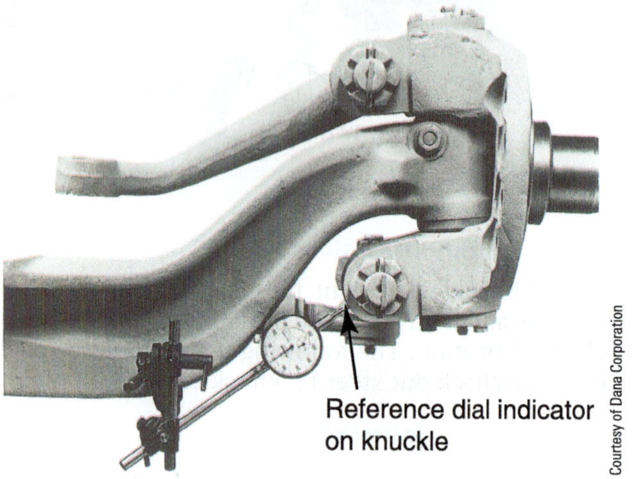

2. Move the bottom of the wheel in and out with a push/pull motion. Have an assistant read the dial indicator.
3. A dial indicator reading that exceeds the OEM specifications (typically 0.015 inch [0.38 mm]) indicates that the lower bushing should be replaced.

Kingpin Upper Bushing Torque Deflection.

1. Mount the dial indicator to the axle, referencing the upper knuckle steering arm socket area (**Figure 25-38**).
2. Have someone apply the foot brake. Try to roll the wheel forward and backward and note the deflection.
3. Readings in excess of the manufacturer's specifications (typically 0.015 inch [0.38 mm]) indicate that the top bushing should be replaced.

Lower Bushing Torque Deflection Test.

1. Mount the dial indicator on the axle and the plunger on the lower bushing area.

FIGURE 25-38 Measuring upper bushing torque deflection

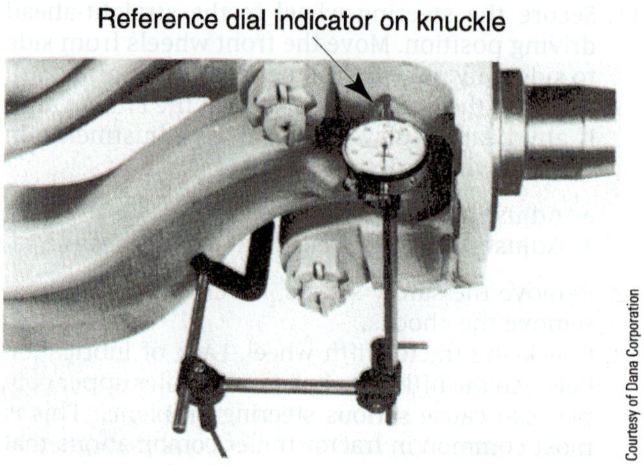

2. Have someone apply the foot brake. Try to roll the wheel forward and backward and note the deflection.
3. Readings that exceed the OEM specifications (typically 0.015 inch [0.38 mm]) indicate that the lower bushing should be replaced.

Tie-Rod Inspection

1. Shake the cross tube. Movement, or looseness, between the tapered shaft of the ball and the cross-tube socket members indicate that the tie-rod end assembly should be replaced. If no obvious play is evident, a dial indicator can be used to check and verify that any play is within OEM specifications.
2. The threaded portion of both tie-rod ends must be inserted completely beyond the tie-rod split recess, as shown in **Figure 25-39**. This is

FIGURE 25-39 Correct (A) and incorrect (B) tie-rod installation

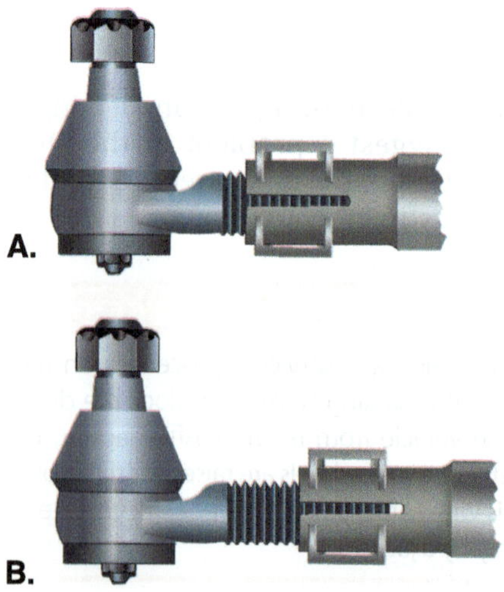

essential for adequate clamping. If this is not so, components will have to be replaced. Check to see if the cross-tube or tie-rod ends are at fault.

3. Ball and socket torque (exclusive of boot resistance) should be 5 lb-in. (0.04 N·m). or more on disconnected tie-rod end assemblies. Replace assemblies that test less than 5 lb-in. (0.04 N·m). Loose assemblies will adversely affect the steering system performance and might prevent adjustment of the steering assembly to the vehicle OEM alignment specifications.
4. If the tapered shank-to-tie-rod arm connection is loose or the cotter pin is missing, disconnect and inspect these components for worn contact surfaces. If either one is worn, replace it.

Wheel Bearing Inspection

The following procedure can be used on steer axles that do not use preset wheel bearing assemblies:

1. Remove the wheel bearings from the wheel hub.
2. Thoroughly clean the bearings, spindle, hubcap, and hub cavity. Parts can be washed in a suitable commercial solvent. The parts should be free of moisture or other contaminants.
3. Clean hands and tools to be used before assembly.
4. Closely examine the bearing cup, rollers, and cage checking for grooving, etching, staining, and scoring, as shown in **Figure 25–40** and **Figure 25–41**.
5. Replace any damaged or distressed bearings with mated bearing assemblies.
6. Lubricate bearings after making certain that they are free of moisture or other contaminants. Refer to Chapter 27 for additional information on installing bearings and the procedure used for preset hubs.

FIGURE 25–40 Inspect bearings for damage, nicks, and gouges.

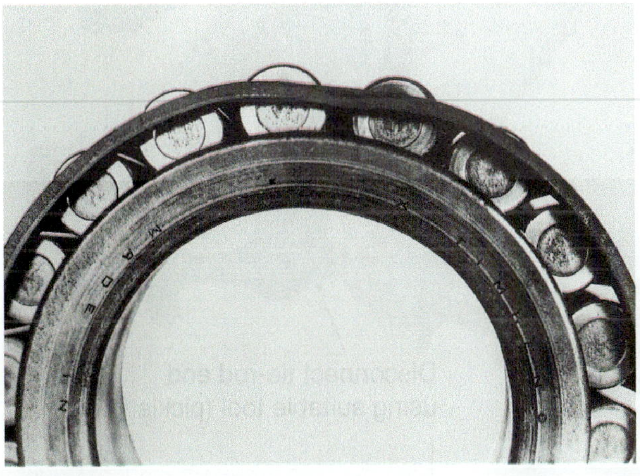

Courtesy of Dana Corporation

FIGURE 25–41 Inspect bearing races for grooving, etching, staining, and scoring.

Courtesy of Dana Corporation

Adjusting Knuckle Vertical Play

If the vertical play in the knuckle exceeds the OEM specifications, adjustments can be made by removing or installing shims between the axle beam and the upper part of the knuckle. To do so, follow these steps:

1. Remove the capscrews, lockwashers, and grease caps from the steering knuckle.
2. Remove the nut and draw key.
3. Partially drive out the kingpin, using a hammer and brass drift. Drive from the top down. Remove the kingpin far enough to facilitate removal and replacement of the shims.
4. Adjust the shim pack (**Figure 25–42**) to obtain the manufacturer's specified vertical play (typically 0.005 to 0.025 inch [0.127 mm to 0.635 mm]).
5. With the correct shims installed, drive the kingpin back into the knuckle assembly.
6. Recheck the vertical play and readjust if necessary.
7. With the adjustment complete, ensure that the kingpin flat is aligned with the draw key so it can be secured.
8. Install the draw key and nut. Torque the nut to the OEM specification.
9. Install the knuckle grease cap gasket (if applicable) or use a silicone compound on the grease cap mounting surface. Install the grease caps.
10. Install the capscrews and the lockwashers. Torque the capscrews to the OEM specification.

FIGURE 25–42 With the knuckle pin partially removed, remove or install shims to adjust vertical endplay.

Photo Sequence 7 outlines the procedure for measuring steering kingpin wear and vertical endplay.

STEER AXLE REPAIRS

When worn or bent components require steering axle work, you can use the following shop practices as guidelines.

Preliminary Steps

Before working on a truck front end, you should perform the following:

1. Set the parking brake and block the drive wheels to prevent vehicle movement.
2. Jack the vehicle until steering axle wheels are raised from the ground. The jack of choice is a scissor jack with mechanical locks that can be used in place of safety stands. If using other types of jacks, support the raised vehicle with safety stands.

WARNING

Never work under a vehicle supported only by a jack.

Steering Knuckle Disassembly

1. Back off the slack adjuster to ensure that the brake shoes are in the released position and well clear of the drum.
2. Remove the hubcap from the wheel assembly, followed by the cotter pin, nut, washer, and outer bearing cone assembly on traditional wheel ends. On preset wheel end assemblies, use the OEM instructions to pull the wheel.
3. Remove the wheel and hub assembly.
4. Disconnect the air line from the brake assembly.
5. Remove the foundation brake assembly from the axle spider.
6. To disconnect the tie-rod end from the Ackerman arm, remove the cotter pin and slotted nut, as shown in **Figure 25–43**.
7. Disconnect the tie-rod end using a suitable tool, such as a pickle fork (**Figure 25–44**).
8. Disconnect the drag link from the steering arm by removing the cotter pin and slotted nut.

FIGURE 25–43 To remove a tie-rod, first remove the cotter pin and castellated nut.

FIGURE 25–44 Use a pickle fork to separate the tie-rod end from the tie-rod arm.

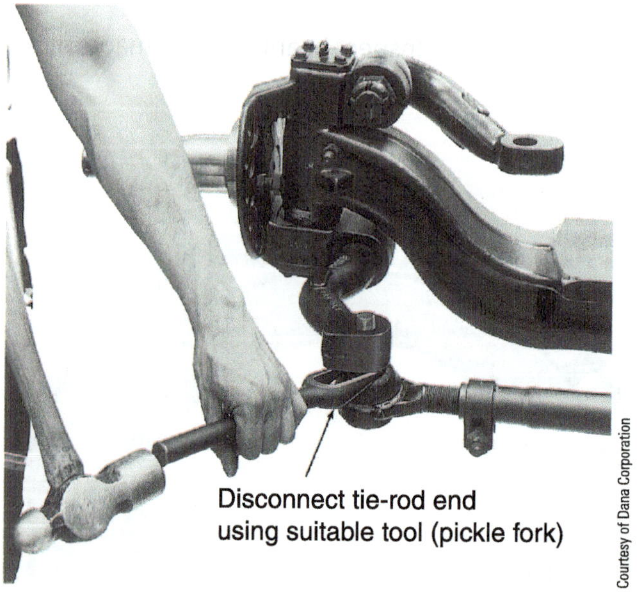

Photo Sequence 7

MEASURING KINGPIN WEAR AND VERTICAL PLAY

P 7–1 Apply the vehicle parking brakes and block the rear wheels.

P 7–2 Jack the steer axle and lower onto axle stands so that the tires on either side are approximately 1 inch (25 mm) off the shop floor.

P 7–3 Turn the steering wheel to full lock so that you can easily access the steering knuckle as shown.

P 7–4 Mount a dial indicator on the steer axle I-beam, positioning the measuring plunger so that it contacts inside of the steering knuckle. A magnetic base dial indicator works best to make this measurement.

P 7–5 Correct dial indicator mounting position to measure kingpin wear. After the dial indicator is properly mounted, zero the dial.

P 7–6 Use a helper to apply inboard and outboard thrust to the top of the tire as shown, while you observe the indicator dial.

(continued)

Photo Sequence 7 (Continued)

P 7–7 Make a note of the reading, and compare it to the OEM maximum permissible specification; this will vary somewhat according to the specific kingpin design. Excess wear will require replacement of the kingpin bushings.

P 7–8 To check kingpin vertical play, mount a dial indicator onto a bracket integral with the steer axle, and load the plunger onto the top of the upper knuckle joint cap. Zero the indicator dial.

P 7–9 Use a large prybar to apply upward (as shown) and downward thrust force on the wheel. Check the indicator readings to OEM specifications. An out-of-spec reading will require steering knuckle disassembly and possible replacement of the thrust bearing.

CAUTION

Never use heat on any steering system components, because it removes the temper and destroys them.

Kingpin Removal

1. Remove the capscrews and lockwashers to remove the knuckle caps.
2. Remove the nut from the draw key; then, drive the key out using a hammer and brass drift (**Figure 25–45A**).

FIGURE 25–45 To remove the kingpin, (A) drive out the draw key and (B) drive out the knuckle pin.

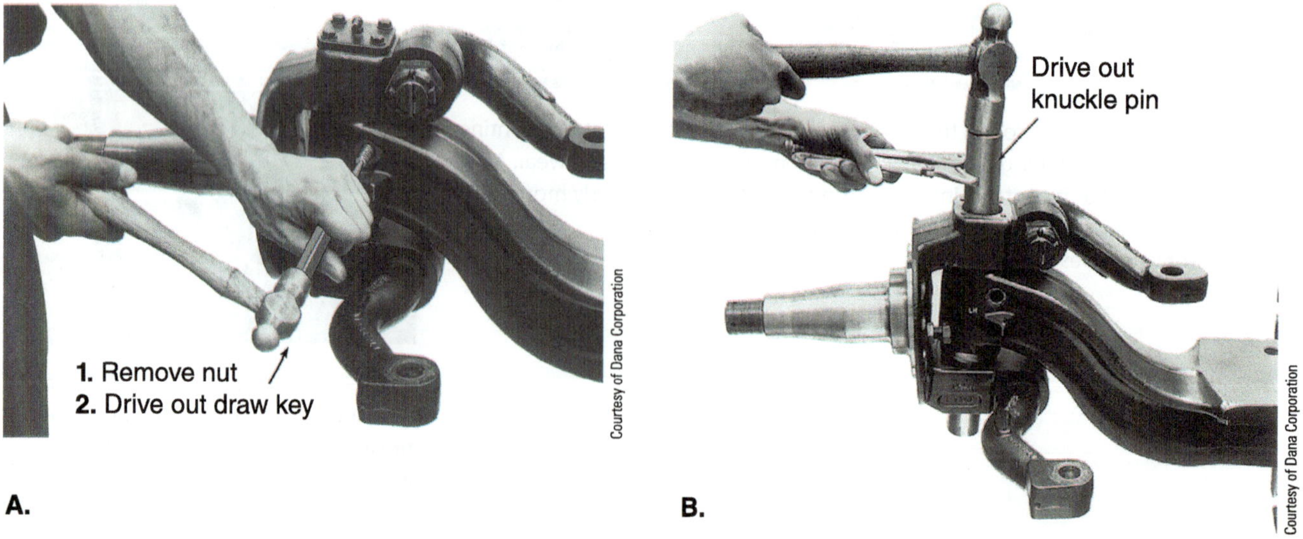

3. Drive the kingpin out with a hammer and brass drift, as shown in **Figure 25–45B**.
4. Remove the steering knuckle from the axle beams and discard the thrust bearing.

Cleaning

After disassembly and before attempting inspection, clean the components as follows:

1. Wash steel components that have ground or polished surfaces in suitable cleaning solvent. Dry parts with clean lintless rags.
2. Wire brush or pressure wash castings, forgings, and other rough-surface parts that are susceptible to accumulation of mud, road dirt, or salt.

Kingpin Seal Replacement

1. Remove the grease seal from the knuckle upper arm.
2. Install a new seal, using a suitable pilot drift or mandrel that will not damage the seal as it is installed.

SHOP TALK

When installing the grease seal, be sure the lip is pointing toward the axle side of the knuckle (**Figure 25–46**). This is essential for correct seal operation.

Ackerman Arm Replacement

1. Disconnect the tie-rod end from the Ackerman arm (or drag link from the steering arm).
2. Remove the cotter pin and nut from the arm at the knuckle end.
3. Drive the arm out of the knuckle using a suitable brass drift and hammer.
4. Install a new key in the slot provided in the Ackerman arm.
5. Install the nut and tighten it to the correct torque. Install the cotter pin.

FIGURE 25–46 The seal lips must face the axle when installed in the knuckle.

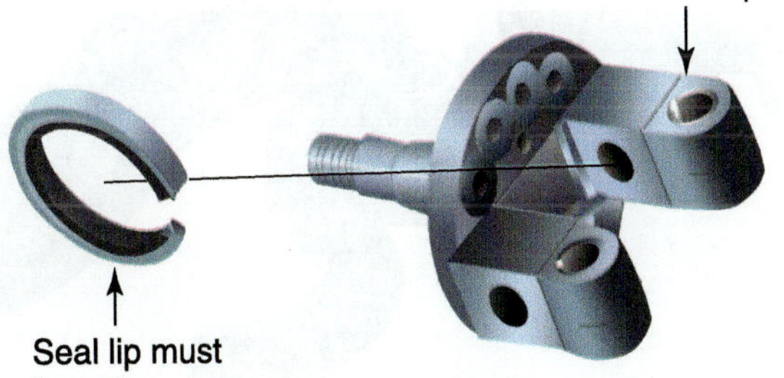

6. Reconnect the tie-rod ends. Install the nut and tighten it to the correct torque. Install the cotter pin.

CAUTION

Never use heat on any steering components.

SHOP TALK

Tighten the nut to the minimum torque first and then continue tightening to align the cotter pin hole. Verify that the torque does not exceed the maximum specification. If it does, replace the nut.

Tie-Rod End Replacement

1. Disconnect the tie-rod end using a pickle fork.
2. Loosen the clamp nut, apply penetrating the threads, and unscrew the tie-rod end.
3. Install the new tie-rod end. Thread the end into the cross tube past the tube split (see Figure 25–39).
4. Attach the tie-rod end stud to the tapered bore in the Ackerman arm. Tighten to the correct torque. Install the cotter pin.
5. Adjust toe-in using the procedure described earlier. Tighten the cross-tube clamp screw to correct the torque.

Tech Tip

If the tie-rod end slips in the tapered bore while attempting to torque it, apply upward pressure on the cross-tube end by hand (or by jack and wood block if this does not work) until the stud bites and begins to seat.

Kingpin Bushing Replacement

1. After removing the kingpin, drive the upper bushing out of the steering knuckle using a suitable piloted drift or mandrel.
2. Drive the lower bushing out in the same manner.
3. Remove all foreign material from the steering knuckle and axle beam, using a wire brush on the machined surfaces. Clean the knuckle bushing bores.
4. Lightly lubricate the outside diameter of the bushing to assist installation.
5. Align the bushing seam as shown in **Figure 25–47**.
6. To install the bushings, first hand-start the bushing in the bore. Then, using a suitable piloted drift or bushing mandrel, drive the bushings home, as shown in Figure 25–47. The bushings are usually recessed slightly below the surface of the knuckle. On some models, the lower bushing might be recessed until it is flush with the inside surface of the knuckle.

FIGURE 25–47 Bushing and seal installation

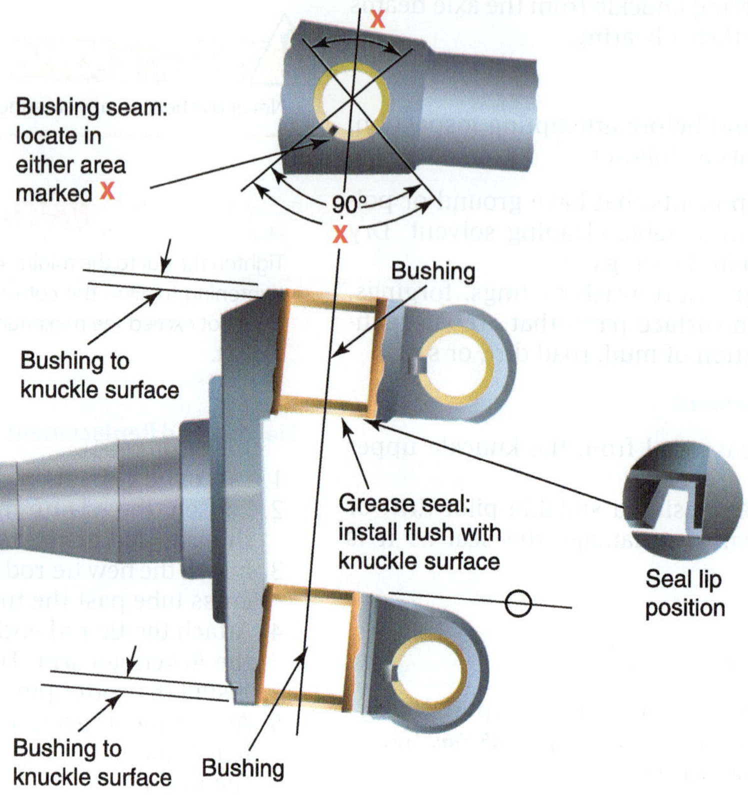

Steering Knuckle Assembly

Use the following procedure to assemble a steering knuckle assembly:

1. Lightly lubricate the thrust bearing areas of the steering knuckle, axle end, and kingpin bore of the axle beam (**Figure 25–48**).
2. Position the thrust bearing on the knuckle with the seal facing up, as shown in **Figure 25–49**.
3. Mount the steering knuckle onto the axle beam.
4. Wedge the steering knuckle upward and fit shims to fill the gap at the knuckle upper end. This step is a preadjustment of knuckle vertical play.
5. Position the kingpin flat, aligned with the draw key hole, and start installing it by pushing it in by hand.
6. Install the kingpin into the knuckle and axle beam from the top. If necessary, tap the pin home using a hammer and brass drift.

FIGURE 25–48 Lubricate the steering knuckle and axle before assembling the two.

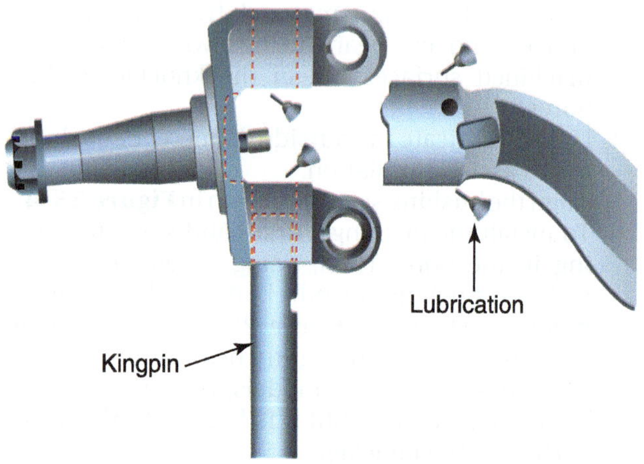

FIGURE 25–49 Position the thrust bearing with the seal facing up.

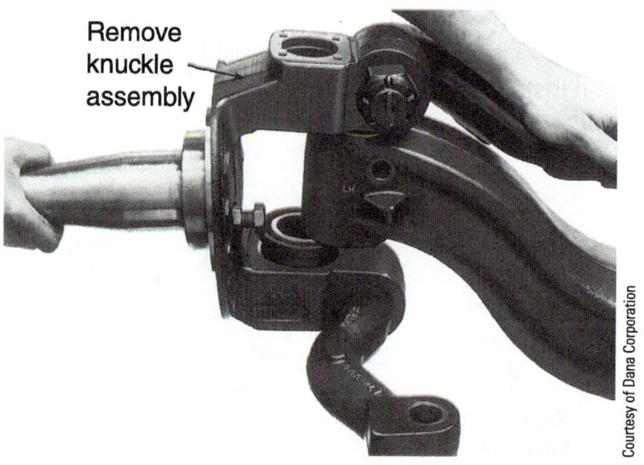

FIGURE 25–50 To check knuckle vertical play, (A) reference the dial indicator at the top of the knuckle pin, and (B) simulate axle loading with a jack.

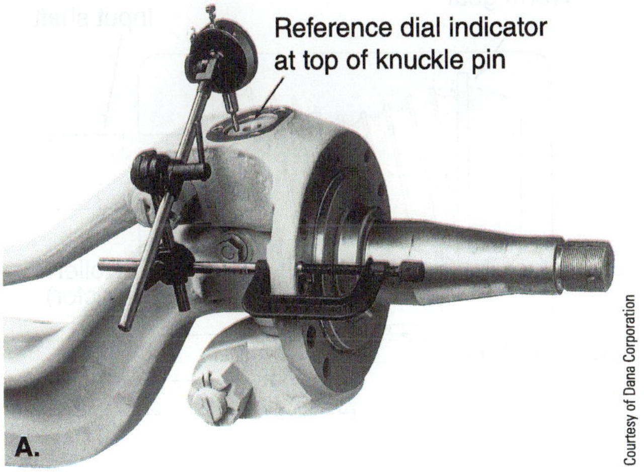

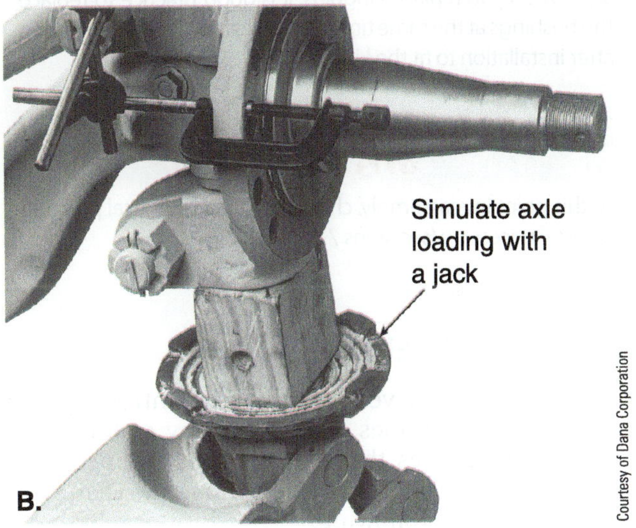

7. To check knuckle vertical play, center the steering.
8. Mount the dial indicator to the flange and set the plunger to reference the top of the kingpin, as shown in **Figure 25–50A**. Zero the dial indicator.
9. Simulate axle loading with a jack (**Figure 25–50B**) and note the dial indicator reading.
10. If the vertical play is outside of OEM specifications, either add or remove shims to set the specified endplay.
11. When the adjustment is complete, check the draw key opening and kingpin flat alignment, as shown in **Figure 25–51**.
12. Install a new draw key, using a soft face hammer.
13. Install the draw key nut and tighten it to the correct torque.

FIGURE 25–51 Align the draw key hole with the knuckle pin flat.

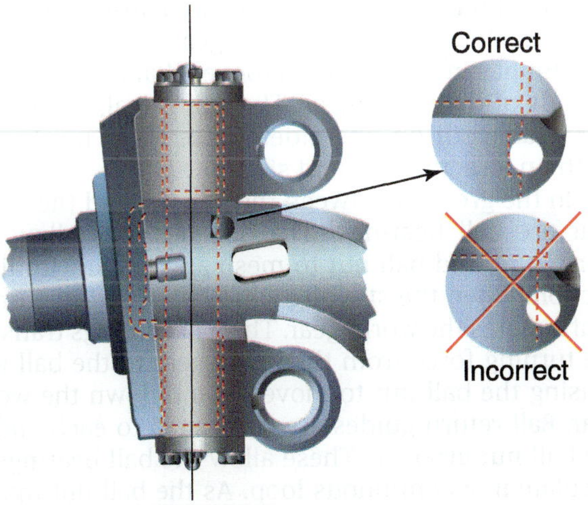

14. Apply silicone compound (**Figure 25–52**) to the kingpin grease cap mounting surface, or install a gasket if applicable.
15. Install capscrews and lockwashers to secure the grease cap.
16. Tighten the capscrews to the OEM specification.
17. Attach the drag link to the steering arm. Install and torque the nut to the OEM specification. Install the cotter pin.
18. Attach the tie-rod end to the Ackerman arm. Install and torque. Install the cotter pin.
19. Reinstall the brake and wheel assemblies, following the OEM service literature procedure.

FIGURE 25–52 Apply silicone rubber gasket compound to the knuckle flat before installing the cap.

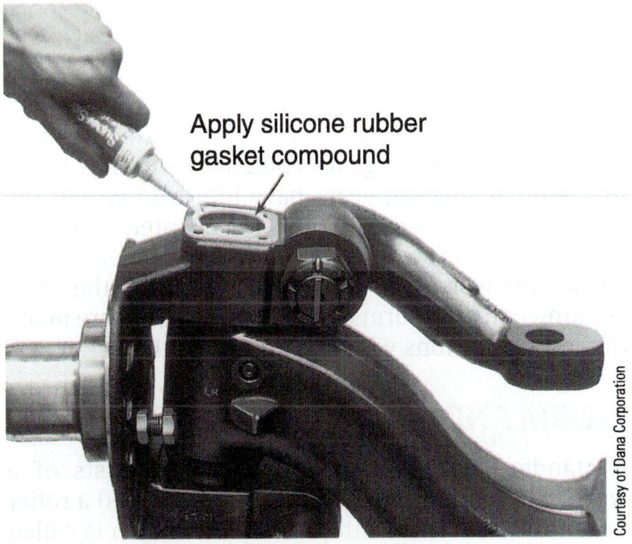

SHOP TALK

Always replace the kingpin if there is evidence of surface damage. When you replace kingpins, it is good practice to replace the bushings at the same time. Some bushings must be reamed after installation to fit the kingpin.

SHOP TALK

At this point in reassembly, check the knuckle vertical play and adjust if necessary (see steps 7 through 10).

UNITIZED STEER AXLES

Unitized axles have recently been introduced to reduce and sometimes eliminate many of the maintenance procedures that we have just outlined. One example is the Meritor Easy Steer Plus axle. Easy Steer Plus features the following:

- Unitized hub, bearing, and seal assembly. This eliminates most of the critical wheel end adjustment procedures, including bearing adjustment.
- Bearing steel hub assembly. Using high-strength steel allows greater strength and lighter weight.
- Integrated seals and permanently lubed. The seals run on machined bearing steel and are designed to last as long as the bearings. Synthetic grease lubricates the moving components for the life of the axle.
- Integral brake spider. A lightweight casting that eliminates the tie-rod arm joint and spider to steering knuckle joint.
- Other features. Easy Steer Plus also features a bolt-on steering control arm and threaded kingpin cap that improves serviceability.

25.5 MANUAL STEERING GEARS

A steering gear is often referred to as a steering box. It is an assembly of gears contained within a housing. Manual steering gears have lubricant within the housing. As we said earlier in this chapter, two basic types of manual steering gears are used in heavy-duty trucks: the worm and sector shaft type and the recirculating ball and worm type. However, there are many different variations within these categories.

WORM AND SECTOR STEERING GEARS

A standard worm gear arrangement consists of a worm or driving gear and a driven gear called a roller or sector gear. The tooth pattern on a worm is called a thread, and the number of teeth it has determines whether it is a single-thread or double-thread worm. The worm gearing used in steering systems is slightly modified. The driving worm gear has more of an hourglass shape (**Figure 25–53**) and is attached to the input shaft, itself linked to the steering shaft. The threads of the driving worm gear are meshed with the threads of a sector gear, sometimes called a roller. The driven sector gear transfers the rotary motion of the worm gear input shaft to the output shaft. The output shaft is splined to the Pitman arm, which translates the rotary motion of the shaft to linear motion. Movement in the Pitman arm is transmitted through a drag link to the steering arm and knuckle and thus to the wheels.

FIGURE 25–53 An hourglass-shaped worm gear and sector roller manual steering gear are shown turned in an off-center position.

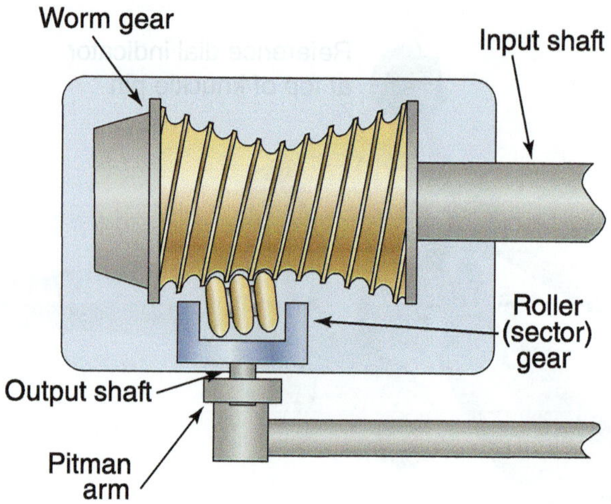

RECIRCULATING BALL GEARS

The input shaft of this type of steering gear is also connected to a worm gear, but the worm gear used is straight (**Figure 25–54**) in a recirculating ball steering gear. Mounted on the worm gear is a ball nut. The ball nut has internal spiral grooves that mate with the threads of the worm gear. The ball nut also has exterior gear teeth on one side. These teeth mesh with teeth on a sector gear and shaft.

In the grooves between the ball nut and the worm gear are ball bearings. The ball bearings allow the worm gear and ball nut to mesh and move with little friction. When the steering wheel is turned, the input shaft rotates the worm gear. The ball bearings transmit the turning force from the worm gear to the ball nut, causing the ball nut to move up and down the worm gear. Ball return guides are connected to each end of the ball nut grooves. These allow the ball bearings to circulate in a continuous loop. As the ball nut moves

FIGURE 25–54 A recirculating ball type steering gear

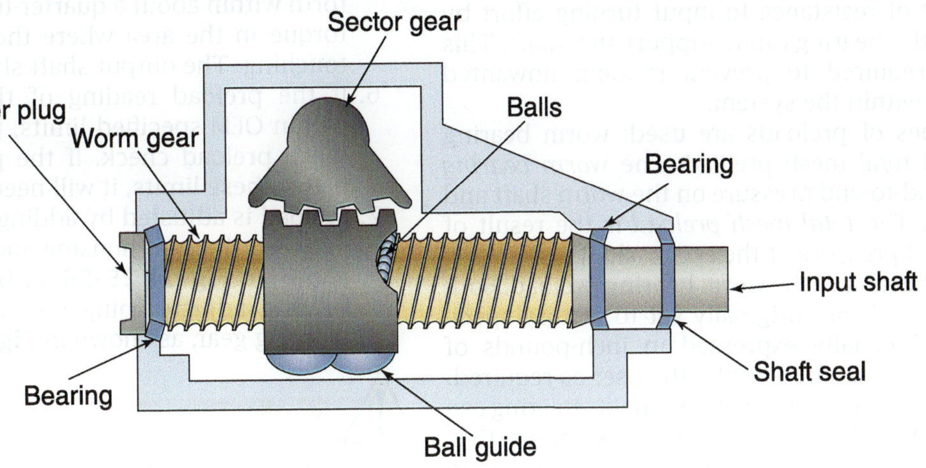

up or down on the worm gear, it causes the sector gear to rotate, which, in turn, causes the Pitman arm to swivel back and forth. This motion is transferred to the steering arm and knuckle to turn the wheel.

MANUAL STEERING GEAR TROUBLESHOOTING

Manual steering gears tend to be reliable, usually requiring only routine maintenance through their service life. The housing is filled with lubricant but has few moving parts. Depending on service application, some adjustments will be required throughout the service life of a steering gear, but when these are specified to trucks today, it is almost always in highway applications.

Because of the increased physical effort required to steer a manual steering gear it makes little sense ever to spec them to non-linehaul applications. Some typical adjustments required in manual steering gear are described here.

Worm and Sector Gear Set

When the vehicle is driving in straight-ahead direction, the steering gear should be in the center of its travel. This means that the sector gear (or roller) should be positioned at the center of the worm gear. Mesh between the worm and sector is at maximum in this center position, which is illustrated in **Figure 25–55A**. The minimum clearance in the center position produces tight response to steering inputs and good road feel at highway speeds.

When the steering wheel is turned, the worm gear rotates, moving the sector gear to one side away from the center. It also reduces the extent of mesh, as shown in **Figure 25–55B**. This results in more clearance between the sector shaft gear and worm gear teeth. The increased clearance that results is known as backlash. The backlash travel area of a manual steering gear begins to be felt around a half-turn each way off the overcenter position. Steering gear backlash can be felt by wiggling the Pitman arm.

The reason for the hourglass shape of the worm gear shown in Figure 25–55 is to increase the ratio as the steering cuts into a turn. In this figure, you can see that the ratio increases as the turn becomes tighter. Because most driving is done at or near the straight-ahead position, it follows that most wear occurs in the center of a worm gear. In order to maximize the service life of the steering gear, there is increased groove depth in the off-center of the worm, permitting overcenter adjustments where the worm gear is most likely to wear.

FIGURE 25–55 (A) Overcenter position; and (B) backlash area

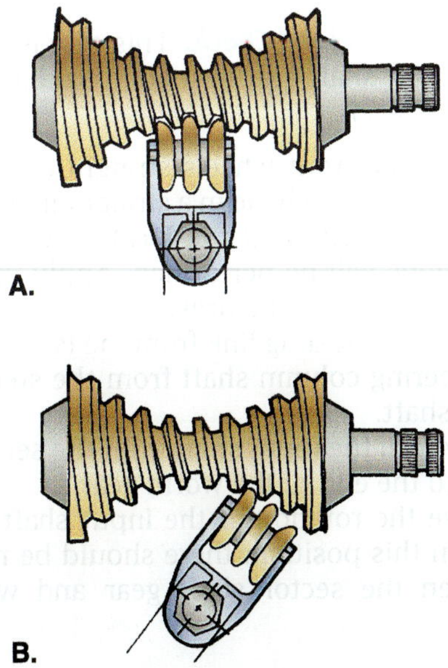

Worm Shaft Preloads

Steering gear worm shafts are assembled with a specified amount of resistance to input turning effort by preloading the bearings that support the shaft. This preload is required to prevent random unwanted movements within the system.

Two types of preloads are used: worm bearing preload and total mesh preload. The *worm bearing preload* is end-to-end pressure on the worm shaft and its bearings. The *total mesh preload* is the result of the combined pressure of the sector shaft gear acting on the worm gear and the worm bearing preload. Both types of preloads are originally set to factory specifications and usually expressed in inch-pounds of torque. Preloads can be reset by the user as required.

When preloads are out of adjustment, steering can be significantly affected. From a driver's perspective, this will mean constant corrections are required to drive a straight line. Insufficient preload permits lost motion within the gear set, meaning that the driver must turn the steering wheel farther than normal to get a response from the front wheels. On the other hand, excessive preloads result in hard steering, darting, and oversteer.

Measuring and Adjusting Preloads

Although incorrect preloads can cause the steering problems we have described, they are not the only cause of these conditions. Before checking steering gear preload, check the steering linkages for worn or binding areas first. If the steering linkage components check out okay, then proceed with the preload check.

SHOP TALK

There are two distinct preload specifications, called high point adjustment and full mesh preload. Do not attempt shortcutting the process by making just one of the adjustments.

Worm Shaft Preload Check. This test is performed with the sector shaft gear located in the backlash area (see Figure 25–55).

1. Position the front wheels straight ahead. If possible, drive the vehicle in a straight line for a short distance, stopping at the location where service operations will be performed. Apply the parking brakes and chock the tires.
2. Disconnect the drag link from the Pitman arm and the steering column shaft from the steering gear input shaft.
3. Slowly turn the input shaft until the sector gear is close to the end of the worm gear.
4. Reverse the rotation of the input shaft a quarter turn. In this position, there should be no contact between the sector shaft gear and worm gear teeth.
5. Place an inch-pound torque wrench on the end of the input shaft. Turn the torque wrench back and forth within about a quarter-turn arc to record the torque in the area where the gear teeth are not touching. The output shaft should not turn.
6. If the preload reading of the moving shaft is within OEM-specified limits, proceed to the total mesh preload check. If the preload is above or below these limits, it will need to be adjusted.
7. Preload is adjusted by adding or removing shims between the gear housing and the cover plate, as shown in **Figure 25–56**, or by turning the worm bearing adjuster plug on the recirculating ball steering gear, as shown in **Figure 25–57**.

CAUTION

Do not force the sector shaft gear all the way to the end of the worm, or damage can result.

CAUTION

Steering gear adjustments should never be taken lightly, because the consequences can be fatal. Always use the OEM service literature when making any adjustments to steering boxes.

SHOP TALK

When adjusting worm shaft preload, removal of shims increases preload; adding shims decreases preload.

Adjusting Worm Shaft Preload. If you have already tested worm shaft preload and it is out of specification, you should use the following procedure to correct it:

1. Clean the outside of the steering gear.
2. Remove the fill plug and the drain plug from the steering gear. Drain the lubricant and dispose of it in an environmentally safe manner.
3. Remove the worm shaft cover screws and the worm shaft cover.
4. If worm shaft preload tested lower than specified (requires less torque to turn) shims will have to be removed. Removal of shims increases preload; adding shims decreases preload. Shims are available in a variety of sizes. Typical sizes include 0.002, 0.003, 0.005, 0.0075, and 0.010 inch (0.05, 0.08, 0.13, 0.2, 0.3 mm). Remove only one shim at a time, as necessary, using a knife blade. The size of the shim to be removed or added will depend on the amount of lash. Use a micrometer to identify the thickness of the shims.
5. Assemble the worm shaft cover and shims.

FIGURE 25–56 Exploded view of a typical manual steering gear

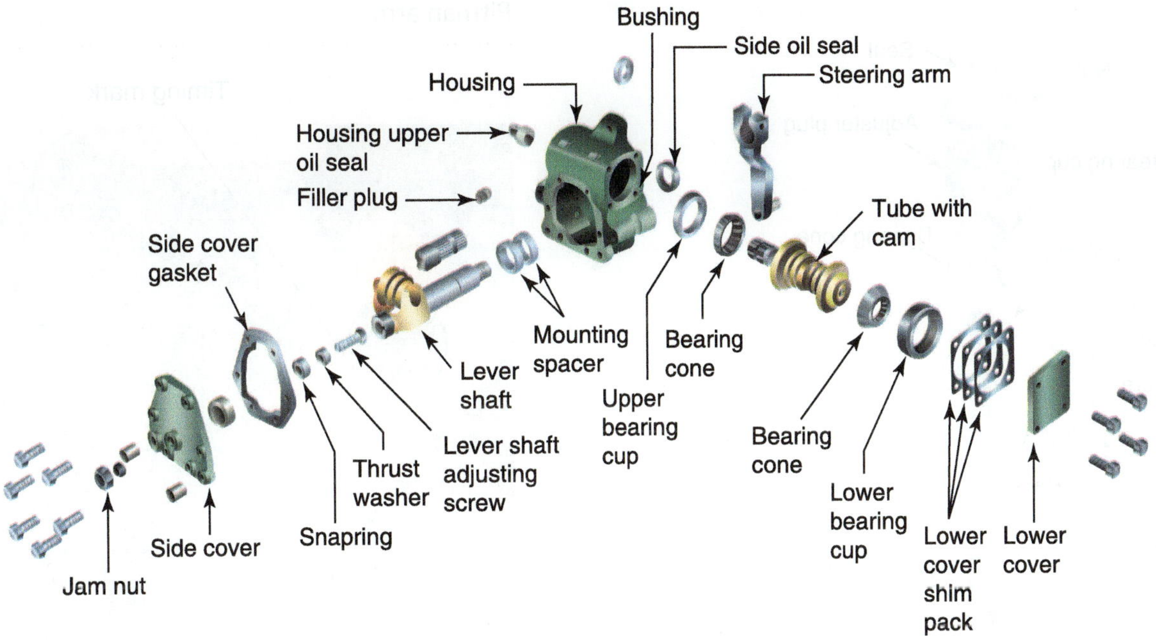

6. Install the worm shaft cover screws. Make sure you have a light preload before final torquing the fasteners.
7. Using a criss-cross pattern or sequence, incrementally tighten the worm shaft cover screws to the OEM torque specification.
8. Recheck the worm bearing preload. If necessary, repeat the adjustment procedure modifying shim pack thickness as appropriate.

SHOP TALK

Do not final torque the screws immediately. Rotate the worm shaft assembly with one hand while slowly tightening the worm cover screws with the other hand. This prevents damaging the bearings if the initial shim pack thickness is inadequate.

Full Mesh Preload Check. Do not check the sector shaft full mesh preload unless the worm shaft bearing preload has been previously checked to be within the specifications. Checking sector shaft full mesh preload when the worm shaft bearing preload is out of adjustment will produce inaccurate measurements.

1. After checking the worm shaft bearing preload to be within specs, center the steering gear by turning the worm shaft from stop to stop, counting the number of turns required. Starting at the end of travel, turn back exactly half that number of full turns. This will place the sector shaft gear teeth in the approximate center of the worm gear.
2. Check the alignment mark on the sector shaft output stub. If the alignment mark is at a right angle to the worm shaft, the gear is centered as shown in **Figure 25–58**. If not, turn the worm shaft using the input shaft until the sector shaft alignment mark is at a right angle to the worm shaft.
3. Rotate the worm shaft with a torque wrench approximately 180 degrees through the center position. Compare the torque reading with the manufacturer's specification. If the sector shaft total mesh preload measurement is not within specifications, adjust the preload.

Full Mesh Preload Adjustment.

1. If an adjustment is necessary, loosen the adjustment screw locknut, as shown in **Figure 25–59**.
2. Center the steering gear.
3. Turn the slotted adjustment screw clockwise to increase the preload or counterclockwise to decrease the preload and then check the adjustment. Repeat until the preload is within specifications.
4. Hold the adjustment screw in place with a screwdriver while tightening the adjustment screw locknut to specifications.
5. Recheck the preload.
6. Fill the steering gear with lubricant, as follows:

FIGURE 25–57 Exploded view of a typical recirculating ball type manual steering gear

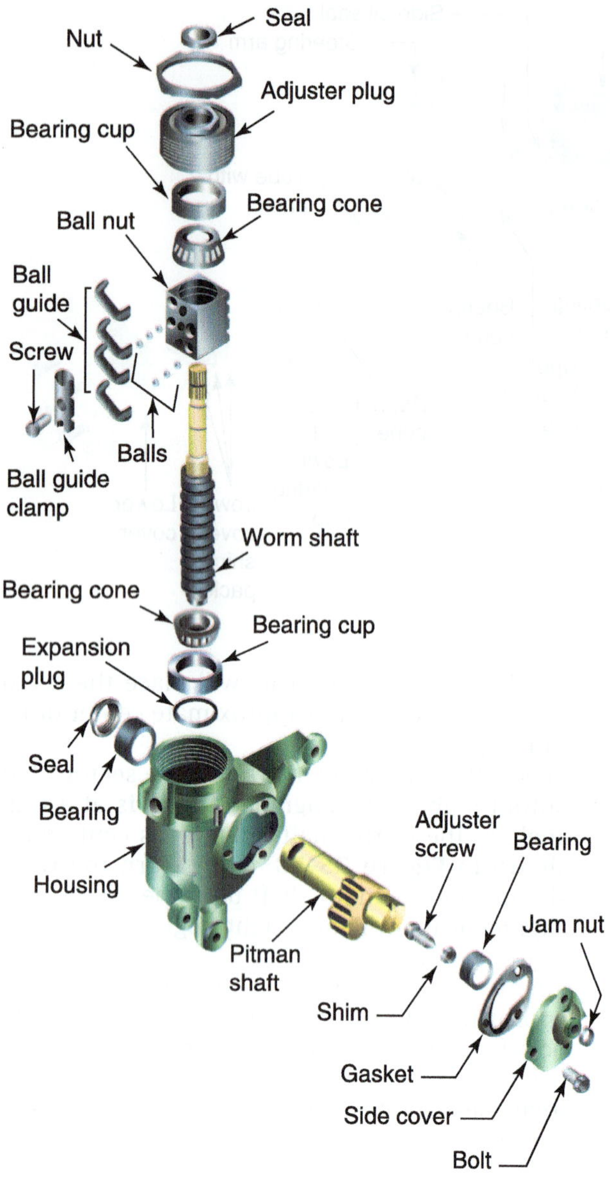

a. If not already done, remove the filler plug (and pressure relief fitting) from the steering gear housing. Thoroughly clean the plug.
b. Add the recommended lubricant until it is within ½ inch (12.5 mm) of the filler hole.
c. Install the filler plug (and pressure relief fitting) and tighten it to OEM specifications.

7. Center the steering wheel and reconnect the steering driveline and linkage.

Backdrive Check. The steering gear should turn smoothly, whether operated by the steering wheel by means of the input shaft, or in reverse, by the front wheels acting on the steering linkage through to the sector shaft in the steering gear. Reversing steering by making the usual output path the input path is called **backdrive**. Backdrive is an important factor in proper wheel recovery.

FIGURE 25–58 When the gear is centered, the sector shaft alignment mark should be timed to the Pitman arm.

FIGURE 25–59 To adjust the total mesh preload, first loosen the adjustment screw locknut. Then, turn the adjustment screw as necessary and observe torque reading.

To check steering system backdrive, turn the steering gear through full travel on either side, being careful not to bottom out the gear. Return the gear to center by rotating the Pitman arm by hand. This will take some physical effort, but the average technician should be able to handle this. If any roughness or binding is noted during this test, you will have to investigate the cause. Consult the OEM service literature.

Centering the Steering System

Driving a truck with the steering gear off-center is not uncommon. Off-center steering gear operation occurs when the front wheels point straight ahead but the sector shaft gear is not centered on the worm gear. This causes both erratic steering and unnecessary wear on the steering gear.

When the front wheels are in a straight-ahead position, the steering control arm can be located in only one point. Similarly, when the sector shaft gear is positioned exactly centered in the mesh or non backlash area on the worm gear, the Pitman arm also can only be located in one position. The distance between these two points is, therefore, a precise and measurable value; this concept is shown in **Figure 25–60**. The mechanism that connects these two points is the drag link. This is represented by dimension A in Figure 25–60. If a drag link is not precisely adjusted to this required length, the steering gear will be forced to operate off-center.

SHOP TALK

If caster adjustment is required, it should be done before centering the steering system because it will change the settings.

FIGURE 25–60 With the steering gear centered and the front wheels straight ahead, the drag link length must equal dimension A. If not, the gear could be forced off-center.

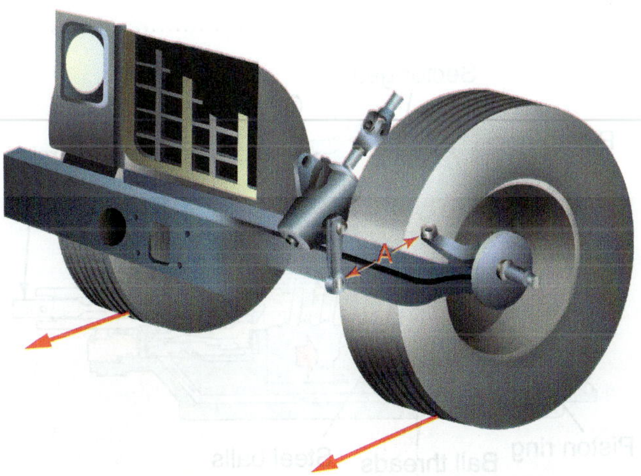

Centering Procedure. Use the following procedure to center the steering system:

1. Point the front wheels straight ahead. The preferred method is to pull the vehicle onto a level floor, leaving the steering wheel free for at least the last 10 feet (3 m) so that the front wheels will seek their natural straight-ahead position. Other methods include sighting the front wheels with the rear wheels and equalizing the distances from the front wheel (not the tire) to the spring at identical points on each side of the vehicle. When you have finished, you must be able to assume that the vehicle wheels are pointing in a true straight-ahead position.
2. Place chocks around the wheels to prevent the vehicle from moving.
3. Remove the drag link and disconnect the steering shaft from the gear. Starting with the steering gear in either backlash area, turn the input shaft toward the center until the Pitman arm will no longer wiggle. This is one edge of the mesh area and should be identified temporarily with a mark on the housing in line with a mark on the input shaft (**Figure 25–61**).
4. Locate the other edge of the mesh area and put another temporary mark on the housing in line with the mark previously made on the input shaft.
5. The midway point between the two marks on the housing is the center of the steering gear. (If the mesh area is more than one full turn of the input shaft, be sure to consider this when locating the midway point.) Mark this point on the housing. This mark should line up with the mark on the

FIGURE 25–61 The midway point between the ends of the mesh area is the center of the steering gear. It should be noted with a permanent mark on the gear housing.

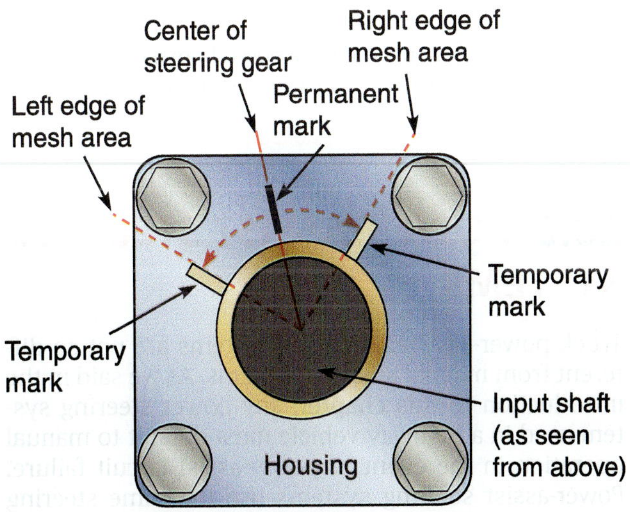

input shaft when the vehicle is moving straight ahead on a level road.

6. Double-check the center mark by measuring the amount of input shaft turn on either side of the mark within the mesh area. These arcs should be equal with a minimum of a one-third turn on each side of center.
7. Permanently mark the center position with paint on the housing and on the input shaft. Also make a permanent mark on the Pitman arm in line with a mark on the housing. These two sets of marks will serve as reference points to identify the center of the gear.
8. At this point you should have centered both the wheels (steering linkage assembly) and the steering gear. Now you can adjust the drag link to the exact length needed to connect it with the Pitman arm and the steering control arm. Make sure that you do not disturb either setting you have made when doing this. Next, install the drag link in accordance with the OEM service literature.
9. With the wheels and steering gear still straight and centered, reconnect the steering column shaft to the steering gear input shaft. Align the steering wheel king spoke to this straight-ahead position by removing and repositioning the wheel on the steering column shaft splines if necessary. (Do not attempt to align the steering wheel by changing the length of the drag link.)
10. As a final check, drive the vehicle on a level surface and bring it to a stop with the steering wheel free. The wheels should be in their natural straight-ahead position, with the steering wheel centered, and the center marks on the steering gear sector shaft should be in alignment. If not, readjust the length of the drag link. If the center marks are in alignment but the kingspoke of the steering wheel is not, remove and reposition the wheel on the steering column.

CAUTION

The Pitman arm should not be removed from the sector shaft to center the steering. This can dangerously affect the steering geometry.

25.6 POWER-ASSIST STEERING SYSTEMS

Truck power-assisted steering systems are not so different from manual steering systems. As we said in the introduction to this chapter, any power steering system used in a highway vehicle must default to manual operation in the event of power-assist circuit failure. Power-assist steering systems use the same steering axle and steering linkage as manual systems and both have similar steering column assemblies. The real difference is in the addition of a hydraulic-assist circuit.

Most trucks today use hydraulic power-assist. There are some older air-assist systems still around but these tend to be aftermarket units installed to trucks with a conventional manual steering gear system. Truck hydraulic power-assisted steering systems contain a dedicated pump that delivers hydraulic fluid under pressure to a power-assist steering gear. A typical power-assist steering gear is shown in **Figure 25–62**. In a gear such as this, the hydraulic control circuit is located within the steering gear so it is known as an integral system. There are also some nonintegral systems, but these can be regarded as obsolete.

POWER STEERING PUMP

The power steering pump is used to produce the hydraulic circuit flow required for the power-assist to the steering gear. They can be either gear- or belt driven, depending on the engine application, but the majority are gear driven by an engine accessory drive—heavy-duty trucks exclusively use gear-driven power steering pumps. Power steering pumps are usually mounted near the front of the engine. Most power steering pumps are similar in construction.

Two basic types are most commonly used on heavy-duty trucks:

- Roller
- Vane (**Figure 25–63**)

Occasionally you will come across the gear or rotary pump on trucks but these are more common in large, off-highway applications. All pumps used in power steering circuits use a positive displacement pumping principle. The hydraulic principles used in truck power-assisted steering circuits are studied in detail in Chapter 13.

FIGURE 25–62 Cutaway view of a typical power steering gear

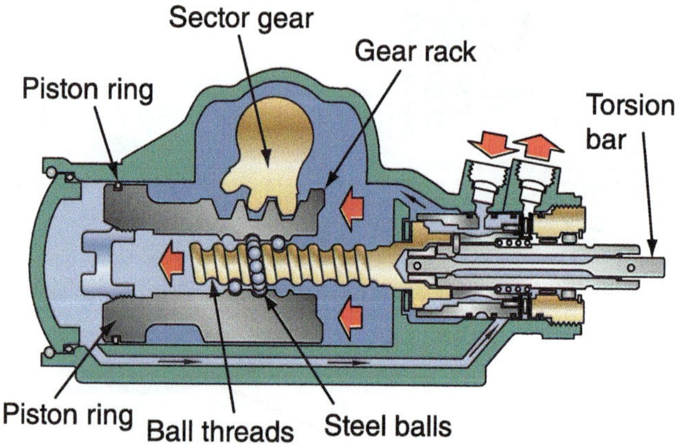

FIGURE 25–63 A vane type pump

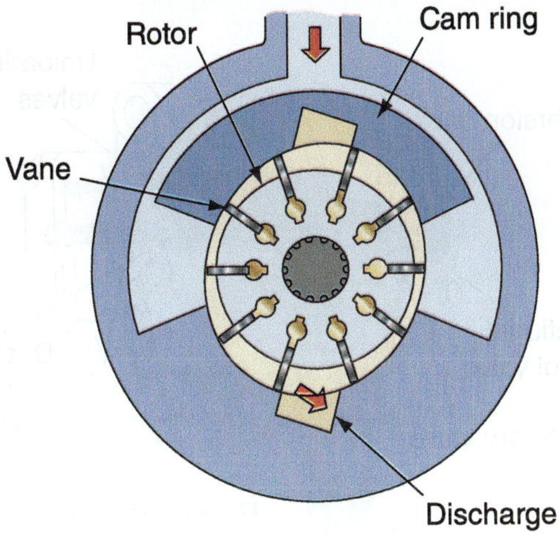

Pump Operation

The pump shaft is driven by the engine and turns the pump rotor. Vanes or rollers turn with the rotor and a combination of centrifugal force and hydraulic pressure loads them outward against a cam ring or liner. The inside of the cam ring is oval. When the rotor turns in the liner, the volume of the space changes between the rotor, the cam ring chamber, and two of the vanes. When two vanes move past the inlet side of the cam ring, the volume between rollers increases, creating a low-pressure area. As the rotor turns, the volume between the vanes and the seal they create in the cam ring is reduced, creating pressure rise, after which the hydraulic oil is discharged through the outlet. A typical pump has two inlet and two outlet ports. This helps smooth pump operation and increase pump displacement by providing two delivery pulses per revolution.

Hydraulic fluid used in the pump is stored in a reservoir that might be remotely mounted or might be an integral part of the pump assembly. Fluid is routed between the pump and the reservoir and between the pump and the steering gear via hoses and lines. Peak pressure is controlled by a relief valve. The fluid used in the hydraulic circuit can be engine oil, automatic transmission fluid (ATF), hydraulic fluid, and synthetic power steering fluid depending on OEM specification and user preference.

Flow Control and Pressure Relief Valves

The flow control valve in the pump regulates pressure and flow output from the pump (**Figure 25–64**). This valve is necessary because of variations in engine rpm and the requirement for consistent power-assist when the engine is operated at low speeds. It is positioned in a chamber subject to pump outlet pressure at one end and supply hose pressure at the other. A spring is used at the supply pressure end to help maintain a balance.

As the fluid is discharged from the pump rotor, it passes the end of the flow control valve and is forced through an orifice that causes a slight drop in pressure. This reduced pressure, aided by the springs, holds the flow control valve in the closed position. All pump flow is directed to the steering gear.

When the engine speed increases, the pump can deliver more flow than is required to meet system

FIGURE 25–64 Control valve operation

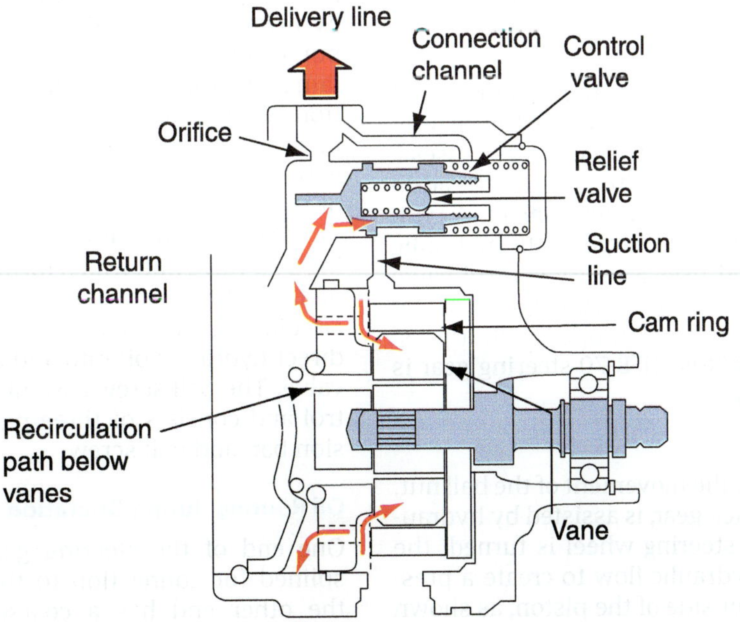

requirements. Because the outlet orifice restricts the amount of fluid leaving the pump, the difference in pressure at the two ends of the valve becomes greater until pump outlet pressure overcomes the combined force of supply line pressure and spring force. The valve is pushed down against the spring, opening a passage that returns the excess flow back to the inlet side of the pump.

A spring and ball contained inside the flow control valve form a pressure relief valve and are used to relieve pump outlet pressure. This is done to protect the system from damage due to excessive pressure when the steering wheel is full lock held against the stops. Because flow in the system is severely restricted in this position, the pump would continue to build pressure until a hose ruptured or the pump destroyed itself.

When outlet pressure reaches a preset level, the pressure relief ball is forced off its seat, creating a greater pressure differential at the two ends of the flow control valve. This allows the flow control valve to open wider, permitting more pump pressure to flow back to the pump inlet, and pressure is held at a safe level.

Power Steering Gears

A couple of different types of power-assist steering gearboxes are used in most medium- and heavy-duty trucks, notably those manufactured by TRW/Ross, Saginaw, and Sheppard. Some of the more popular models of steering gear are:

- Sheppard M90 axle weight ratings 8,000 to 12,000 pounds (3,600 to 5,400 kg)
- Sheppard M90 axle weight ratings 10,000 to 14,000 pounds (4,500 to 6,350 kg) linehaul
- Sheppard M90 axle weight ratings 16,000 to 20,000 pounds (7,200 to 9,000 kg) vocational
- TRW/Ross HFB-64 medium-duty linehaul
- TRW/Ross HFB-70 heavy-duty linehaul

TRW/Ross Steering Gear

For purposes of understanding the operation of power-assist steering gear, we focus on the market leading TRW/Ross HF series. This steering gear uses a recirculating ball steering gear principle with the addition of a hydraulic-assist control mechanism. A typical power steering gearbox is charged with hydraulic fluid under pressure and uses a rotary control valve to control the flow of fluid into, through, and out of the gear. Some gears are also equipped with a pressure relief valve. A TRW/Ross HFB-70 steering gear is shown in **Figure 25–65**.

Operation

In a power steering gear, the movement of the ball nut, also called a piston or rack gear, is assisted by hydraulic pressure. When the steering wheel is turned, the rotary valve changes hydraulic flow to create a pressure differential on either side of the piston, as shown in **Figure 25–66**. The unequal pressure causes the piston to move toward the lower pressure, reducing the effort required to turn the wheels.

FIGURE 25–65 Cutaway view of a TRW HFB-70 power steering gear

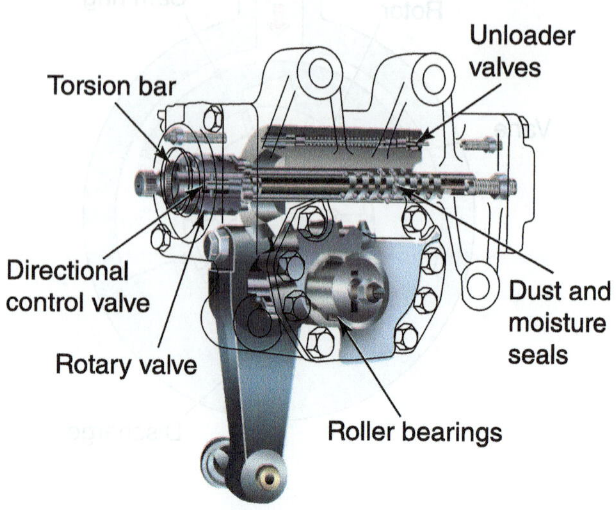

The integral-type power steering has the rotary control valve and a power piston integrated with the gearbox. The rotary valve directs the oil pressure to the left or right chamber to steer the vehicle. The spool valve is actuated by a lever or a small torsion bar located in the worm gear.

HYDRAULIC OPERATION

The driven end of the worm gear (called a ball screw in power steering gears) rotates on a ball bearing contained in the valve body. Hydraulic oil under pressure enters and exits the power steering gear by means of hydraulic lines connected to threaded ports in the valve body. A pressure relief valve contained in the valve body prevents overpressurization of the power steering gear. Hydraulic pressure in excess of the setting of the relief valve causes the valve to open a channel to the reservoir return side of the gear.

The ball screw assembly is held in the valve housing by a valve nut, which forms the outermost element of the rotary control valve. The valve nut contains circular channels and radial passages, which serve to direct hydraulic oil into and out of the rotary control valve. The ball screw assembly forms the rotary control and consists of three parts: the input shaft, torsion bar, and ball screw.

Oil Routing during Operation

One end of the steering gear input shaft is finely splined for connection to the steering column, and the other end has a coarse spline that fits more

FIGURE 25–66 Oil flow in a TRW HFB-70 during (A) right cut; (B) neutral; and (C) left cut

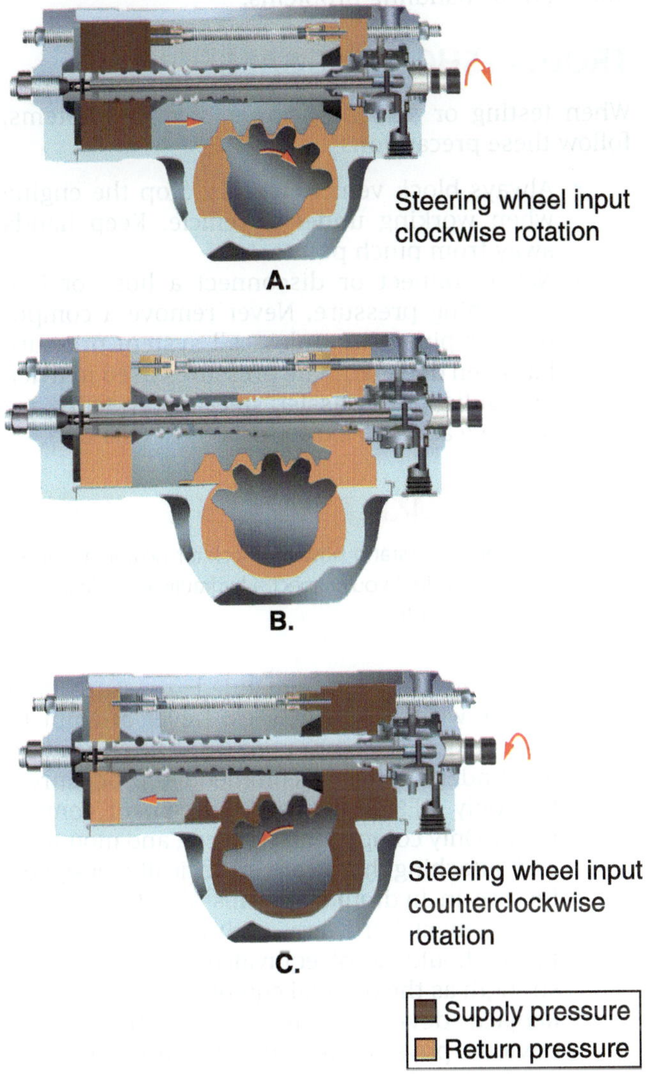

loosely with a similar spline inside the worm screw. The coarse splines form mechanical stops that limit the amount or relative rotation between the ball screw and input shaft. A torsion bar connects the input shaft to the ball screw. Six evenly distributed longitudinal grooves are machined into the outer surface of the input shaft and correspond to six grooves machined into the bore of the ball screw. Holes extend from the outside surface of the ball screw into the six grooves in the bore. These holes allow pressurized oil to enter and exit the two inner elements of the rotary control valve.

The six grooves in the bore of the ball screw are connected alternately to each side of the piston through three pairs of the drilled holes. The other three holes admit pressurized oil directly to three of the six grooves in the input shaft. The other three grooves in the output shaft carry oil to the return line connection. The length of the six pairs of grooves cut into the ball screw and input shaft allows large pressure changes to be achieved with a small rotational displacement of the valve elements.

Rotary Control Valve

The rotary control valve is an open-center type that allows a continuous flow of oil (through the longitudinal grooves in the input shaft and bore of the ball screw) when held in the neutral position by the torsion bar. When steering effort is applied, the input shaft and ball screw tend to turn in unison; however, the spring action of the torsion bar results in the input shaft rotating slightly in advance of the ball screw. The six pairs of grooves that form the rotary control valve are displaced from their neutral flow position.

As steering effort increases, so does the amount of displacement. Depending on the direction steered, the groove displacement of the input shaft directs hydraulic oil through the appropriate drilled passages in the ball screw to one side or the other of the piston. Hydraulic pressure acting on the piston surface eliminates much of the piston's resistance to movement. Spring force exerted by the torsion bar causes the ball screw to rotate as piston resistance is removed. As the ball screw rotates, the relative groove displacement is eliminated, and the rotary valve returns to a neutral position.

Steering Feel

Moderate effort at the steering wheel produces smaller valve displacements and lower power assist and in this way provides good steering feel. Good steering feel is a critical element in any power-assist system. At increased displacements, the pressure rises more rapidly, giving increased power assistance and faster response. Maximum pressure is developed after approximately 3 degrees displacement, providing a direct feel to the steering.

Groove displacement is limited by the freeplay of the stop spline mesh between the input shaft and ball screw. The splines take up the steering movement while allowing the torsion bar to hold the groove displacement. The torsion bar and stop splines form two parallel means of transmitting the steering torque. When no steering torque is applied, the torsion bar returns the valve grooves to a neutral position, allowing the pressurized oil to flow to the return line as shown in Figure 25–66.

Full Cut Pressure Relief

Power-assisted movement of the valve nut within its bore is limited by poppet valves installed in both piston faces. When the piston approaches its extreme travel in either direction, the stem of the limiting poppet valve makes contact at the end of the piston bore. As piston travel continues, the limiting poppet is unseated and some hydraulic power-assist is removed as pressurized oil is diverted to the return line. As more and more power-assist is removed by the action of the limiting poppet valves, steering effort increases. The piston can travel to the extreme ends of its bore; however, the maximum steering assistance available is reduced to protect the steering components in the axle.

The bypass valve is located in the valve body and permits oil to flow from one side of the piston to the other when it is necessary to steer the vehicle without the hydraulic pump in operation. Oil displaced from one side of the piston is essentially transferred to the other side, which prevents reservoir flooding and cavitation in the pressure line. The pressure relief valve is located in the valve body and limits internal hydraulic pressure to a preset maximum. The pressure relief valve can be set to various pressures; however, its setting is typically 150 psi (10 bar) lower than the power steering pump relief valve setting.

25.7 TROUBLESHOOTING POWER STEERING SYSTEMS

When there is a complaint about a hydraulic power steering system, as a first step try to talk to the driver. Get a detailed explanation of the problem, including the operating conditions under which it occurs. Inquire about any recent repairs, modifications, or prior adjustments. Next, visually check the entire steering system, looking for obvious problems. If the truck appears roadworthy, test drive it or ride with the driver with the objective of reproducing the problem.

SHOP TALK

Power steering gear is almost never field-rebuilt in today's truck shops. The reason is both the critical role steering gear plays in vehicle safety and the need to have specialized equipment to properly bench test steering gear. The function of the technician is to diagnose system malfunctions and replace a power steering gear with a rebuilt exchange unit when necessary.

Just as with manual steering systems, the steering gear is usually the least likely cause of a steering problem. Go through the process of visually inspecting the tires, front end, suspension, steering linkages, and axles for damage or misalignment. Tires account for a large percentage of steering problems, so check them for mismatch, inflation, and obvious defects.

If the rear axles are out of line, the truck will be difficult to steer. Also check for proper location and lubrication of the fifth wheel. Take a close look at the steering gear mounting, making sure that the fasteners are tight and it has not shifted. Note that a slight movement of the steering gear on its mounts can produce serious handling problems.

TROUBLESHOOTING SAFETY

When testing or servicing power steering systems, follow these precautions:

- Always block vehicle wheels. Stop the engine when working under a vehicle. Keep hands away from pinch points.
- Never connect or disconnect a hose or line containing pressure. Never remove a component or pipe plug unless all system pressure has been depleted. The pressures used in truck hydraulic power-assist steering systems are massive and can exceed 1,500 psi (103 bar).

WARNING

Make sure you understand how to check for power steering leaks and what to do if you suspect a hydraulic pinhole injection injury (Chapter 13).

- Never exceed recommended pressure and always wear safety glasses. Never attempt to disassemble a component until after reading and understanding recommended procedures.
- Use only OEM-suggested replacement components. Only components, devices, and mounting and attaching hardware specifically designed for use in hydraulic systems should be used. Replacement hardware, such as hoses and fittings, should be of equivalent size, type, and strength as the original equipment.
- Replace devices with stripped threads or obvious physical damage. Repairs requiring machining should not be attempted.

CAUTION

As with manual steering gear, the Pitman arm should NEVER be removed from the sector shaft to center the steering. In some instances, the repair could require disassembly of the steering gear.

HYDRAULIC TESTS

If a visual inspection of the vehicle suggests that the power steering system is at fault, check the hydraulic supply circuit first (see **Photo Sequence 8**). Power steering gear performance is dependent on an adequate supply of oil pressure and oil flow through the circuit. When diagnosing power steering problems, oil pressure and oil flow must meet the OEM

Photo Sequence 8

FLOW AND PRESSURE TEST A HYDRAULIC ASSIST STEERING GEAR SYSTEM

P 8–1 Use a high-pressure washer or steam cleaner to thoroughly clean the area around the steering gear.

P 8–2 Park the vehicle in a stall and ensure that the front wheels are tracking in the straight-ahead position. Fit a shop exhaust pipe to the truck exhaust stack.

P 8–3 Place a drain pan under the steering gear to catch power steering fluid spilled during this procedure.

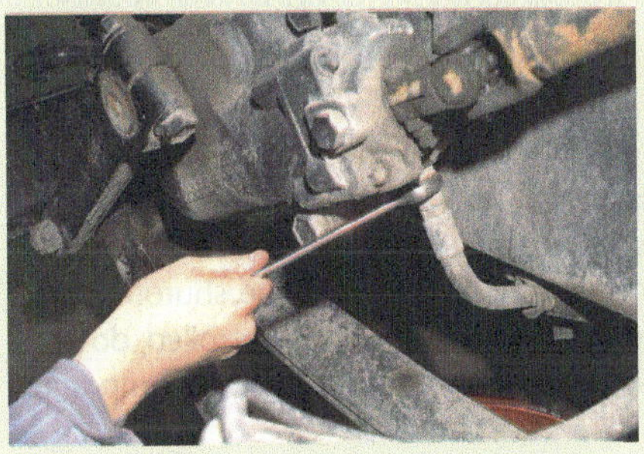

P 8–4 Separate the high-pressure supply hose from the steering gear. Be aware that the fluid in this line may have some residual high pressure. Some fluid will drain from the system, so the drain pan should be placed directly under the steering gear.

P 8–5 Connect the power steering analyzer in series with the high-pressure supply hose and the nipple on the inlet side of the steering gear. Now the system will function as normal except that the hydraulic fluid will be routed through the power steering analyzer.

P 8–6 Start the engine. Run it at idle speed and record the pressure and flow readings on the power steering analyzer. Check to specifications.

P 8–7 Next close down the flow valve until the pressure increases to the test value specified in the OEM service literature. This may be between 500 and over 1,000 psi (34 and 69 bar), depending on the system. Record the flow in gpm at this pressure value. If the flow is lower than specified, the power steering pump may be defective and require replacement.

P 8–8 Test the pump relief valve and flow control pressures by closing down the flow valve until the flow gauge reads zero. Record the peak pressure reading on the gauge. Perform this two or three times but for no longer than 5 seconds separated by a 30-second interval. Check the gauge pressure.

P 8–9 Compare all test results to OEM specifications and perform repairs as required. Ensure that after the power steering analyzer is removed, the engine is run, and the steering turned from lock to lock and back to center before attempting to remove the truck. This will ensure that no air remains in the high-pressure circuit.

specifications. Pressure and flow specifications vary considerably, so follow the vehicle manufacturer's recommendations.

High system oil temperatures reduce the efficiency of the power steering pump and the steering gear. High temperatures are caused by restriction to flow or inadequate system oil capacity to allow for heat dissipation during normal operation. A supply pump that constantly operates at maximum pressure relief will also generate more heat than can be dissipated.

Various types of pressure gauges and flow meters are available and can be used to diagnose power steering problems. You can purchase a specialty **power steering analyzer** or use more crude, shop-made test components built into a manifold gauge set. The advantage of specialized power steering analysis equipment is speed and accuracy. A pressure gauge that reads at least 3,000 psi (207 bar) and a flow meter with a capacity up to 10 gpm should be used to check pressures and oil flow. A shutoff valve downstream from the pressure gauge makes it possible to isolate the steering pump from the steering gear and by closing the valve, maximum pump relief pressure can be read. A simple thermometer in the reservoir will indicate system oil temperatures. **Figure 25–67** illustrates a typical installation of test equipment in the hydraulic system.

SHOP TALK

A power steering analyzer is the preferred method of assessing the performance of a hydraulic power-assist steering circuit. The power steering analyzer consists of hoses, quick-release couplers, a flow meter, pressure gauge, and flow control valve.

FIGURE 25–67 Equipment for testing hydraulic supply circuit

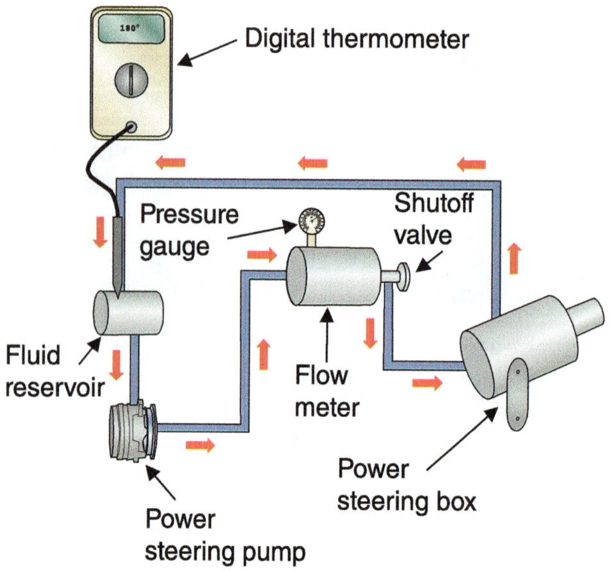

Test Procedure

The following procedure is general in nature, so make sure you consult the vehicle OEM service literature before actually performing it.

1. Make necessary gauge/meter connections.
2. Start the engine and check the system oil level, ensuring that oil flow is in the proper direction through the flow meter.
3. Place the thermometer in the reservoir.
4. Run the engine at the correct idle speed and steer from lock to lock several times to allow the system to warm up (140°F–160°F [60°C–70°C]).
5. With the engine running at a specified idle speed, slowly turn the shutoff valve until closed and read the pressure at which the pressure relief valve opens. (Open the shutoff valve as quickly as possible to avoid heat buildup or possible damage to the steering pump.) This pressure reading should equal the maximum pump pressure specified by the chassis OEM. Check your specifications.

CAUTION

A malfunctioning pressure relief valve might not relieve pump pressure, and closing the shutoff valve could result in severe pump damage or rupture of high-pressure hoses. Constantly observe the pressure gauge when closing the shutoff valve. If pressure rises rapidly or appears to be uncontrolled, do not completely close the valve before inspecting the pump and pressure relief valve.

6. With the engine running at a specified idle speed, the vehicle stationary on the shop floor, and with a normal load on the front axle, steer the wheels from a full right to a full left turn and observe the flow meter. The flow should not fall below the minimum gpm flow specification.
7. Increase engine speed to approximately 1,500 rpm and note the flow rate with the steering wheel stationary. Check this reading against the maximum flow rate specifications. Excessive oil flow can cause high operating temperatures and sluggish or heavy steering response.

SHOP TALK

It is important that the flow be checked at a normal operating temperature and with a load on the front axle. Inadequate flow will cause binding and uneven or intermittently hard steering.

8. If the supply pump is performing to specification, install a ½-inch (12.5 mm) spacer between the axle stops on one side, as shown in **Figure 25–68** and turn the steering wheel to full lock in the direction necessary to pinch the spacer block. Record the maximum pressure reading. The maximum

FIGURE 25–68 Install a spacer between the axle stop and the axle when making a leakage test.

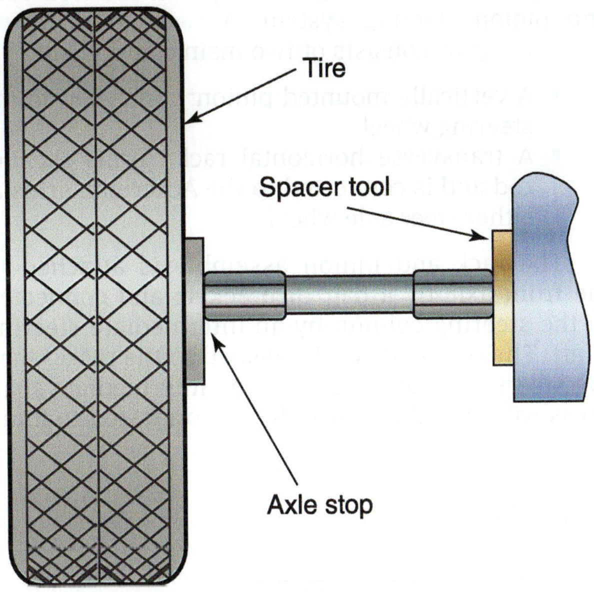

pressure reading should be within 100 psi (689 kPa) of that recorded in step 5 for pump relief pressure when the shutoff valve was closed. Remove the spacer and repeat the test in the opposite direction. Record the pressure. If the pressure does not meet the recorded maximum pressure reading, the steering gear is worn internally and should be repaired or replaced.

9. Normal system back-pressure should be 50 to 75 psi (345 to 482 kPa) with the engine idling and the steering wheel stationary. Back-pressure is checked with the system at a normal operating temperature.
10. Steering system oil temperature is best checked after 2 hours of normal operation. The ideal operating temperature should range between 140°F and 160°F (60°C and 70°C). Normal operation in this range will allow for intermittently higher temperatures that will be encountered during periods of heavy steering usage.
11. Visually check for the presence of air mixed with the oil in the steering system. The oil should be clear. Any signs of frothing indicate air entry, and steering performance will be affected. Carefully check for leakage on the suction side of the steering pump. Drain and refill the system and bleed for air.

SHOP TALK

Before any steering gear repairs are attempted, the preceding hydraulic supply tests should be undertaken and corrections made as required. Many times, steering gears have been repaired or replaced needlessly because a hydraulic supply system had not been tested thoroughly.

P/S TROUBLESHOOTING TERMS

Technicians should be familiar with the terms used to describe common power steering problems experienced with power steering (P/S) gear systems. Here are a few:

- Hard steering. Unit steers but gets little or no power steer assist.
- Wheel cut problems. Excessive or restricted wheel cut, not enough turns to lock.
- Steering kick back. Steering track reacts to bump in road or change in surface camber. Also known as backlash or kick-steer.
- Binding. Change in steering effort that is inconsistent—usually caused by a mechanical and not a P/S problem.
- Darting and oversteer. Caused by erratic power-assist response to input turning effort.
- Directional pull. Steering pulls to right or left usually when off-center but although less common, can be when directionally centered. Requires constant driver effort to maintain tracking.
- Road wander. Excessive steering lash that requires continual driver correction to tracking direction.
- Nonrecovery. Steer wheels do not readily return to straight-ahead direction.
- Shimmy. Steering wander (requires constant driver correction) sometimes accompanied by wheel oscillation (shake).
- Noise. Clicking or clunking when steering wheel is turned.

25.8 RACK AND PINION STEERING

Rack and pinion steering has been used on automobiles for many years and few would disagree that in small vehicles, it provides tighter steering in which the driver has better control. It was not used on medium- and heavy-duty trucks because rack and pinion steering was thought of as appropriate only in light-duty applications until Freightliner introduced ThyssenKrupp rack and pinion steering on its Cascadia chassis in 2007.

Truck conventional steering gear systems are robust but even the most recent systems have shortcomings in terms of responsiveness and precision. In addition, there is often a "dead" spot overcenter, the most used, straight-ahead position. This results in overcenter play and the need for adjustments. Despite the great improvements that have taken place in truck conventional steering systems since the 1990s, it is likely that the trucking industry will embrace rack and pinion steering because drivers will come to prefer it. **Figure 25–69** shows an external view of a Freightliner rack and pinion steering system on a 2008 Cascadia chassis.

FIGURE 25–69 Rack and pinion steering gear on a Freightliner Cascadia

RACK AND PINION PRINCIPLE

You should reference **Figure 25–70** to follow the description of the Freightliner power-assisted rack and pinion steering system. A **rack and pinion steering gear** consists of two main components:

- A vertically mounted pinion: connected to the steering wheel
- A transverse horizontal rack: forms the tie-rod and is connected to the Ackerman arms at either steer axle wheel

The rack and pinion assembly is attached to the front axle by a pair of brackets and connected to the steering column by an intermediate steering shaft. The pinion is tooth-meshed to the rack. When the steering wheel is turned, the intermediate shaft turns with it and because this is connected to input

FIGURE 25–70 Components used in a Freightliner rack and pinion system

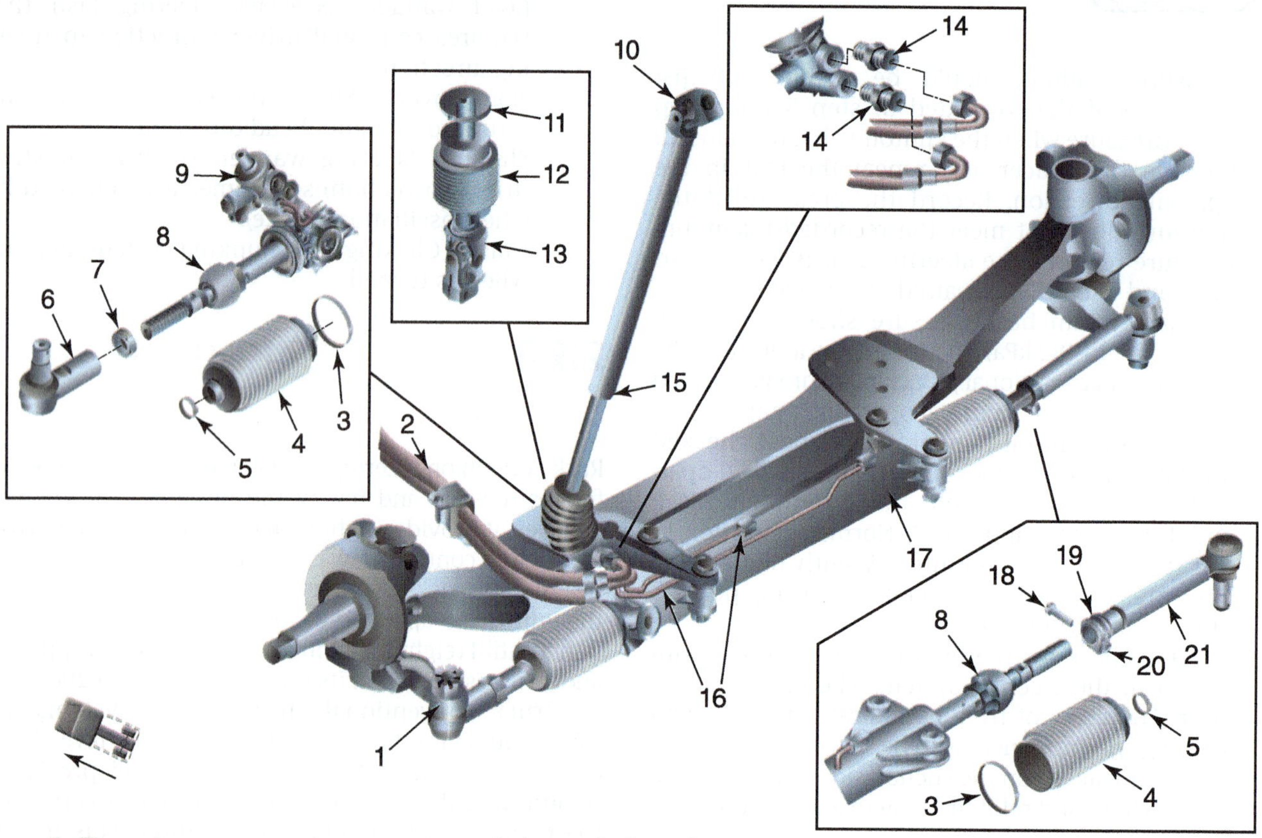

1. Tie-rod arm
2. Lines to reservoir and pump
3. Large bellows clamp
4. Bellows
5. Small bellows clamp
6. Left outer tie-rod
7. Left tie-rod jam nut
8. Inner tie-rod ball joint
9. Input shaft seal cover
10. I-shaft upper yoke
11. Lower yoke boot clamp
12. Lower yoke boot
13. Lower end yoke
14. Power steering fluid line fittings
15. I-shaft
16. Hard lines for power-assist
17. Rack and pinion gear
18. Pinch bolt
19. Right tie-rod clamp
20. Pinch-bolt nut
21. Right outer tie-rod

shaft, torque is applied to the torsion bar and pinion. Twisting the torsion bar applies power-assist via the rack-mounted piston. This linear motion actuates the tie-rods that are in turn connected to tie-rod arms.

The linear motion of the rack can be compared to the linear movement of a tie-rod in a conventional steering gear system. The rack is connected to the tie-rod/Ackerman arms on each wheel on either side of the steer axle. When the rack is moved either to the left or to the right, the steer axle wheels turn. Figure 25–70 shows how the components used in a truck power-assisted, rack and pinion steering system interact with each other.

ADVANTAGES

Rack and pinion steering is simple. There are fewer components, fewer pivot points, and fewer potential wear points; all this results in lower maintenance costs. The simplicity of the system also translates into a 30 percent reduction in total steering system weight. Because of the rack and pinion system layout, the steering column no longer restricts engine compartment access. In a conventional cab configuration, the rack and pinion steering intermediate shaft descends almost vertically to the axle rack assembly. But probably the biggest advantage is better road feel. The road feel factor is difficult to describe beyond saying such things as it makes driving more pleasurable; but the bottom line is that better road feel translates into reduced driver fatigue.

25.9 ELECTRONIC AND LOAD-SENSING STEERING

With a few exceptions, power steering assist has become standard on highway trucks. It does not make economic sense to spec manual steering gear into heavy highway vehicles in an era when it is difficult to both hire and retain drivers. Manual steering has become yesterday's technology. Computer-controlled, speed-proportional steering will be tomorrow's technology. Using electronics can provide driver-selectable controls and automatic integrated adaptive/active vehicle steering, suspension, braking, and traction control systems.

Two systems that provide reduced parking effort combined with improved high-speed road feel are covered in the following sections. Both the electronic variable and load-sensing systems discussed operate with a conventional integral steering gear. Steering as we have studied it so far in this chapter is a hydromechnanical system. In the trucking industry, there has always been some skepticism when predominantly hydromechanical systems move to adapt to computer controls, but actual experience has shown that computer-controlled systems outperform their hydromechanical predecessors in every case.

Computer-enhanced steering is only available as an optional add-on to an existing system, but that will change. The next time you take a plane trip, remind yourself that every directional control system on the aircraft is computer controlled, as is the ability to land blind. These are features that should make us all more comfortable in accepting electronic steering in a truck.

ELECTRONICALLY VARIABLE STEERING

To better understand the advantages offered by **electronically variable steering (EVS)**, the limitations of conventional power steering must be understood.

In a conventional power steering system, the demand for power assistance is opposite to its availability. Designers are forced to compromise power steering pump optimization pressures. This results in a parasitic power loss that can amount to 40 to 50 percent of the power consumed by the steering system. It also results in light steering effort at highway speeds because the flow control setting is often high, as dictated by dry park conditions.

Optimized power steering control has the following features:

- Improved steering wheel road feel
- Directional stability enhanced by a steering effort proportional to driving speed
- Reduced dry park effort at the steering wheel
- Reduced system power consumption

Because conventional power-assisted steering systems do not sense or respond to these variables, the electronically controlled version can provide a correction toward the ideal. Ideally, such a system changes the driver road feel and effort as a function of road speed and lateral acceleration. The electronic system shown in **Figure 25–71** offers single-switch, operator-selectable steering modes. The system consists of a modified, integral power steering gear that includes a revised high-gain rotary control valve; a pulse width modulated (PWM), solenoid-controlled, variable-flow control pump; a speed sensor input usually sourced from the chassis data bus; and an electronic control module for closed-loop solenoid current control.

Variable assist is provided by changing the integral-pump control flow supplied to the steering gear. The steering gear provides the driver a variable torque input based on the variable flow control. Steering gear features include freeplay control, a high-efficiency manual section, low mechanical ratio, and a direct rotary valve. Power-assist is controlled by adaptive features built into the steering gear, such as torsion bar spring-rate and valve "feel" characteristics.

FIGURE 25–71 Electronic variable power steering system that uses a dash-mounted switch to allow drivers to operate in any one of three modes for maximum road feel and control.

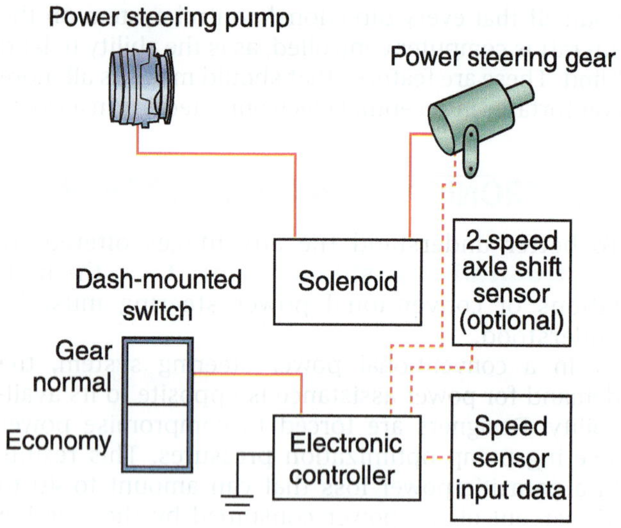

Because the system modulates pump flow control to the steering gear, the steering gear rotary control valve helps achieve appropriate effort or "feel" changes. The potential to bypass pump flow internally, combined with the low power steering gear total loop drop at reduced flow, is the source of power consumption improvements. From the driver's perspective, it is a little like having manual steering on the highway when running at highway speeds and enhanced power-assisted steering when it comes to parking maneuvers.

EVS Pump Design

The EVS steering system uses a solenoid-controlled variable-flow control concept. The pump assembly includes a high-efficiency vane pump, flow control, and proportional solenoid-type throttle valve. The system is designed with fail-safe features. When the solenoid is not energized, flow is low (at around 60 percent of its normal capacity), similar to conventional power steering systems.

EVS Electronic Control Unit (ECU)

The purpose of the electronic controller is to change the PWM actuation signal to the pump solenoid with respect to the road-speed sensor signal. The controller incorporates switches that allow for three potential operating modes with a light-emitting diode (LED) display to assist in troubleshooting.

In economy mode, flow control is at maximum below 15 mph (24 km/h); at its minimum above 50 mph (80 km/h); and it decreases proportionally from 15 to 50 mph (24 to 80 km/h). The normal mode has high flow at low speeds and decreases as speed increases (similar to economy mode), except the minimum flow is set at 3.6 gpm/13.6 L/min and is reached at about 40 mph (64 km/h). In this mode, the steering system acts much like a conventional power steering system with 3.6 gpm/13.6 L/min flow control. If the vehicle is stationary for longer than 10 minutes, the system switches into a standby mode. This mode reduces flow to minimum power-assist as long as the speedometer indicates zero. When road speed is detected, the system automatically switches to its original mode.

The PWM actuation circuit operates the solenoid valve by a fixed-frequency PWM signal with closed-loop duty cycle control that uses the solenoid current as feedback (**Figure 25–72**). This ensures that the solenoid positioning is relatively constant with changes in vehicle battery voltage. It also allows the controller to provide current to the solenoid directly from the battery without need for high-current voltage regulators.

LOAD-SENSING POWER STEERING

The nonelectronic hydraulically controlled load-sensing system shown in **Figure 25–73** has a demand-type flow control that uses steering load pressure to control its variable-flow power-assist. This system provides for speed proportioning of flow to the steering gear from a flow-controlled pump. However, it does not provide driver-selectable operational modes or programmable flow settings. Instead, the system relies on a control strategy in which the pressure requirement for power steering is a function of vehicle speed.

The load-sensing system is similar to a conventional hydraulic power-assist circuit, except that a modification has been made to the flow-control

FIGURE 25–72 Speed sensor input and battery power is fed to the PWM control unit. The control unit output signal drives the control valve.

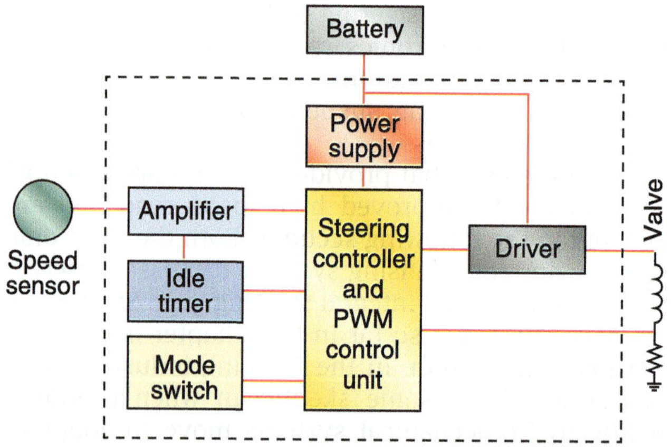

FIGURE 25–73 This hydraulically controlled load-sensing steering pump incorporates a pressure-sensing variable flow control feature that varies power-assist according to demand.

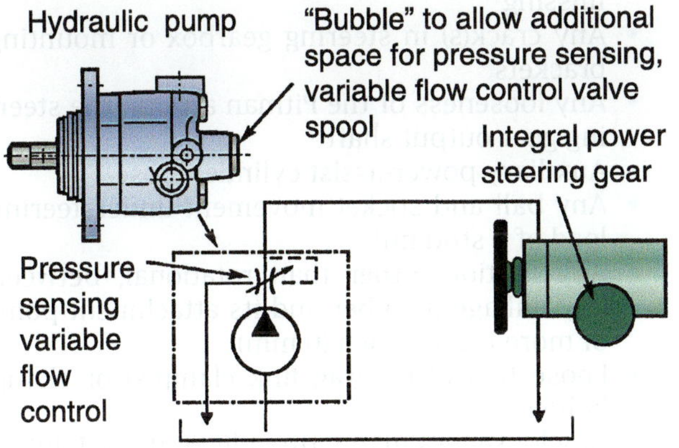

section of the pump as shown. Compared to the electronic version, the load-sensing system provides increased flow only to the steering gear "on demand" in response to steering pressure increase designated as "load sensing."

The load-sensing pump components include a high-efficiency, balanced vane pumping cartridge and a combination flow/relief/load-sense valve. The valve construction and operation within the circuit are shown in **Figure 25–74**. The valve spring has two functions. The first is to restrict the movement of the variable-orifice spool valve according to the steering load. The second is to load the relief poppet valve at the relief pressure setting.

The small variable-orifice valve is exposed to return pressure on the opposite end of the spool. As pump outlet pressure increases, a progressively larger opening is produced by the moving spool. This larger opening allows a greater volume of fluid to flow into the spring-chamber end of the large flow control valve. Because the fluid must pass through a fixed orifice to escape this spring chamber, more fluid flow produces a higher pressure. The higher pressure adds to the load imparted to the flow control spool by the spring and increases the restriction of flow to the return circuit. As a result, more flow is forced through the primary control orifice and delivered to the steering gear. A conventional pressure-compensated flow control valve maintains a constant flow rate regardless of pressure, whereas the pump used on this system increases flow as pressure is increased.

25.10 COMMERCIAL VEHICLE SAFETY ALLIANCE (CVSA) OOS STEERING CRITERIA

CVSA OOS are not maintenance standards. They define the point at which highway equipment becomes dangerous to operate. OOS are used to issue citations during safety inspections. The following OOS apply to truck and tractor steering:

FIGURE 25–74 Combination flow/relief/load-sensing valve using a dual function spring

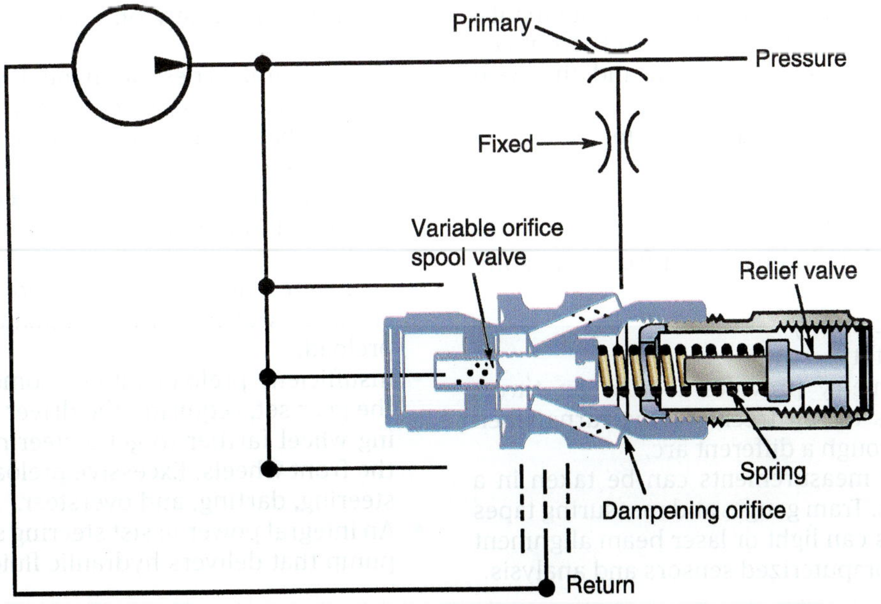

TABLE 25–1 Maximum Steering Wheel Freeplay

Steering Wheel Diameter	Manual Steering	Power Steering
16 inch.	2 inch.	4½ inch.
18 inch.	2¼ inch.	4¾ inch.
20 inch.	2½ inch.	5¼ inch.
22 inch.	2¾ inch.	5¾ inch.
24 inch.	3 inch.	6 inch.

Steering wheel freeplay: (on vehicles equipped with power steering the engine should be running when tested). Check OOS specification in **Table 25–1**.

Steering column assembly:

- Any absence or looseness of U-bolt(s) or positioning clamp(s)
- Worn, defective, or repair welded universal joint(s)
- Steering wheel not properly secured

Front axle beam and all steering components other than steering column:

- Any visible crack(s)
- Any obvious welded repair(s)
- Any steering gear mounting bolt(s) loose or missing
- Any crack(s) in steering gearbox or mounting brackets
- Any looseness of the Pitman arm on the steering gear output shaft
- Auxiliary power-assist cylinder loose
- Any ball and socket movement understeering load of a stud nut
- Any motion, other than rotational, between any linkage member and its attachment point of more than ¼ inch (6 mm)
- Loose tie-rod or drag link clamp(s) or clamp bolt(s)
- Any looseness in any threaded linkage joint
- Loose or missing nuts on tie-rods Pitman arm, drag link, steering arm, or tie-rod arm
- Any steering system modification or other condition that interferes with free movement of any steering component

SUMMARY

- Steering systems used in trucks must deliver precise directional control of the vehicle and its load, in both loaded and unloaded conditions, and at highway and park/stall speeds.
- Truck steering systems are either manual or power-assisted. Power-assist systems are required to default to manual operation in the event of a loss in the power-assist circuit. Power-assist systems can use either hydraulic or air-assist circuits.
- Improper steering adjustments and front-end alignment can lead to suspension and tire wear problems.
- A properly aligned front end results in:
  - Easier steering
  - Increased tire life
  - Directional stability
  - Less wear and maintenance on front-end components
  - Better fuel economy
  - Increased safety
- Ackerman geometry provides toe-out on turns, permitting tires to roll freely during turns when each travels through a different arc.
- Axle alignment measurements can be taken in a number of ways. Tram gauges and measuring tapes may be used, as can light or laser beam alignment systems with computerized sensors and analysis.
- The most accurate and easiest to use alignment systems in use today are computer-controlled and feature in-memory specifications, step-by-step instructions, and user-friendly displays. Keyboard-entered specifications are automatically correlated to the actual angles measured on a vehicle with the results displayed on the monitor screen.
- Steering axle components should be inspected and lubricated routinely on a preventive maintenance schedule.
- Two general types of manual steering gear are used in heavy-duty trucks. They are the worm and sector shaft type and the recirculating ball and worm type.
- Some worm shaft preload is necessary to prevent unwanted random movement within a steering system.
- Two types of steering gear preloads have to be checked: worm bearing preload and total mesh preload.
- Insufficient preload allows some lost motion in the gear set, requiring the driver to turn the steering wheel farther to get a steering response from the front wheels. Excessive preloads result in hard steering, darting, and oversteer.
- An integral power-assist steering system contains a pump that delivers hydraulic fluid under pressure

to a power steering gear. The steering gear has an integral hydraulic control valve that measures steering effort and responds to provide the appropriate amount of assist in turning the wheels.

- A typical power-assist steering gear is fundamentally similar to a manual recirculating ball steering gearbox with the addition of a hydraulic assist: It must default to manual operation in the event of a hydraulic circuit failure.
- Power-assisted steering is standard on trucks on our highways, but integral power steering gear is seldom field repaired. Technicians diagnose steering problems and, if necessary, replace a defective steering gear with a reconditioned unit.
- A disadvantage of conventional hydraulic power-assisted steering gear is that the availability of power-assist is inversely proportional to the need for it. Dry park maneuvers are negotiated when engine speeds are lowest.
- Truck rack and pinion steering systems were introduced in 2007. They provide better road feel, resulting in reduced driver fatigue, over conventional steering gear systems.
- A truck rack and pinion steering gear consists of vertically mounted pinion connected to the steering wheel and a transverse horizontal rack connected to the Ackerman arms at either steer axle wheel.
- Electronically variable steering or EVS is a computer-controlled enhancement to a conventional integral power-assist steering gear and offers better directional stability, steering effort proportional to driving speed, and greatly reduced system power consumption.

REVIEW QUESTIONS

1. Which of the following assemblies uses U-joints?
   a. the steering wheel
   b. the steering column
   c. the tie-rod
   d. the drag link
2. Which of the following components contains ball and socket joints?
   a. the steering gear
   b. the Pitman arm
   c. the steering knuckle
   d. the tie-rod end
3. Which of the following components connects the Pitman arm to the steering control arm?
   a. the drag link
   b. the tie-rod
   c. the cross tube
   d. the steering column
4. How are straight kingpins secured to the steering knuckle?
   a. with a nut and bolt
   b. with a cotter pin
   c. with a draw key
   d. with tapered bearings
5. What is the likely result of an out-of-alignment front end on a truck?
   a. directional instability
   b. steer tire wear
   c. hard steering
   d. all of the above
6. Which of the following alignment angles has the potential to cause the most tire wear?
   a. toe
   b. camber
   c. caster
   d. Ackerman angle
7. When performing a front-end alignment, Technician A checks and corrects the toe setting first. Technician B says that to measure toe using a tape and tram gauge, it is necessary to make measurements between the tires at both the front and back at hub height. Who is correct?
   a. Technician A only
   b. Technician B only
   c. both A and B
   d. neither A nor B
8. Which of the following front-end angles is defined as the forward or rearward tilt of the kingpin centerline when viewed from the side of the vehicle?
   a. camber
   b. caster
   c. toe
   d. kingpin inclination
9. Which of the following can be defined as the inward or outward tilt of the top of the tires when viewed from the front of the vehicle?
   a. turning radius
   b. Ackerman angle
   c. camber
   d. toe
10. Which of the following alignment angles can be corrected using shims on a leaf spring front end?
    a. toe
    b. caster
    c. camber
    d. turning radius
11. Which of the following truck alignment angles can usually only be adjusted by bending the front axle?
    a. caster
    b. camber
    c. kingpin inclination
    d. toe-in

12. When inspecting a steering axle, Technician A uses a dial indicator to check knuckle vertical play. When checking wheel alignment, Technician B uses a protractor to check caster setting. Who is correct?
    a. Technician A only
    b. Technician B only
    c. both A and B
    d. neither A nor B
13. Which of the following best describes a power-assist steering gear system used on most current trucks?
    a. external air-assist
    b. external hydraulic assist
    c. hydrostatic
    d. integral hydraulic
14. When setting worm shaft preload on a manual worm and sector-type steering gear, what would happen if you reduce the size of the shim pack?
    a. The pump leaks.
    b. The preload increases.
    c. The preload reduces.
    d. The sector shaft seizes.
15. Technician A states that, in a conventional power steering system, demand for power assistance is opposite to its availability. Technician B says that newer load-sensing hydraulically controlled power steering systems provide increased flow from the steering pump in response to an increase in steering pressure. Who is correct?
    a. Technician A only
    b. Technician B only
    c. both A and B
    d. neither A nor B
16. What type of hydraulic pump is used on most highway truck power-assist steering gear systems?
    a. gear
    b. vane
    c. plunger
    d. centrifugal
17. After properly connecting a power steering analyzer into a steering circuit with the engine running, what should happen when the flow control valve on the analyzer is closed?
    a. The system pressure goes up.
    b. The system pressure goes down.
    c. The wheels turn off-center.
    d. The gpm gauge reading increases.
18. When adjusting drag link length before installing it, which of the following should be true?
    a. The steering gear should be centered.
    b. The front wheel should be in a true straight-ahead direction.
    c. The steering wheels should be a full lock on one side.
    d. both a and b
19. When diagnosing a steering system using a power steering analyzer, which of the following should be true to obtain accurate results?
    a. The engine oil should be at operating temperature.
    b. The steering system hydraulic oil should be cold.
    c. New steering system hydraulic oil must be used.
    d. The steering system hydraulic oil should be above 140°F (60°C).
20. In a truck rack and pinion steering system, which of the following components is directly connected to the Ackerman arms?
    a. the pinion
    b. the rack
    c. the power cylinder
    d. the Pitman arm

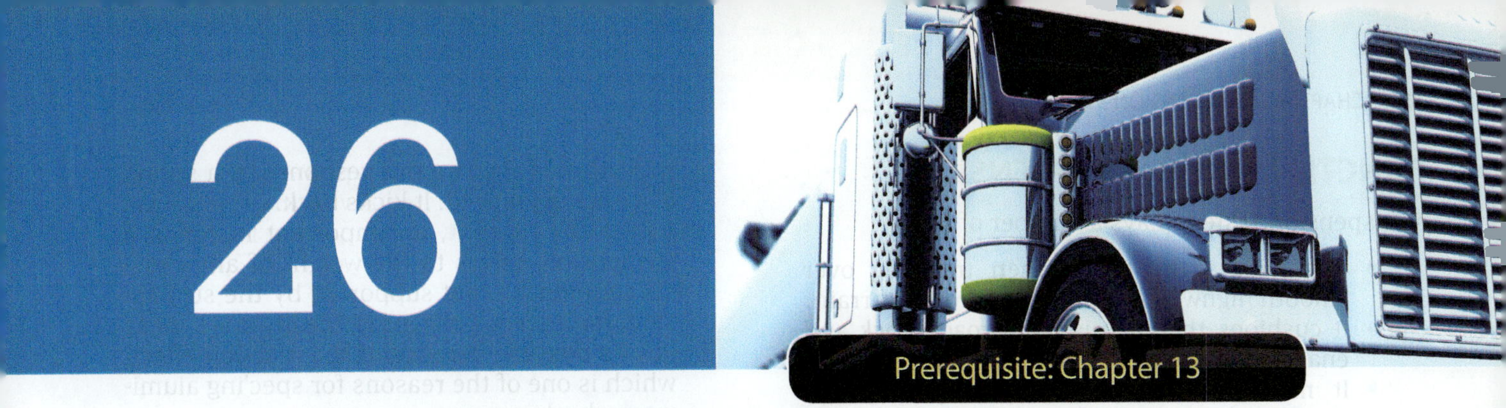

SUSPENSION SYSTEMS

OBJECTIVES

After reading this chapter, you should be able to:

- Identify the suspension systems used on current trucks.
- Describe the components used on mechanical leaf and multileaf spring suspension systems and explain how they work.
- Describe a fiber composite spring.
- Identify equalizing beam suspension system components and explain how they function.
- Describe how rubber spring and torsion bar suspensions function.
- Identify air spring suspension system components and explain how they function.
- Outline the advantages of combining different spring types in a truck suspension system.
- Troubleshoot suspensions and locate defective suspension system components.
- Outline suspension system repair and replacement procedures.
- Explain the relationship between axle alignment and suspension system alignment.
- Troubleshoot common suspension problems.
- Outline some common suspension repair procedures.
- Perform full chassis suspension system alignments.
- Describe the operation of the cab air suspension system.

KEY TERMS

adaptive suspension
air spring
bogie
combination air/leaf spring suspension
combination air/torsion spring suspension
combination rubber/torsion spring suspension
constant rate spring
equalizing beam suspension
height control valve
interleaf friction
jounce
jounce blocks
leaf spring
leveling valve
oscillation
rebound
roll stability control (RSC)
shock absorbers
solid rubber suspension
spring pack
torsion bar suspension
unsprung weight
variable rate spring
walking beam

INTRODUCTION

A suspension system supports the frame on a vehicle. It acts as an intermediary between the axles and the frame. The axles are subject to whatever forces they encounter when running down the highway. With no suspension, these forces would be transferred directly to the truck frame. Think of how rough it was to ride in one of those wagons you had as a kid. Given this kind of pounding and punishment, without the forgiveness of a suspension, the truck frame, its load, and the driver would not last too long.

FUNCTIONS OF THE SUSPENSION

A suspension system plays a number of roles:

- It stabilizes the truck when traveling over smooth highway as well as over rough terrain.
- It cushions the chassis from road shock and enables the driver to steer the truck.
- It maintains the proper axle spacing and alignment.
- It provides a smooth ride both when loaded and unloaded.

Because trucks are specified by the heaviest load they are expected to haul, suspensions first must be capable of supporting that maximum load. However, a heavy-duty suspension that performs comfortably when fully loaded can be harsh and unforgiving when not loaded down. A really good suspension should perform well, both loaded and unloaded, at off-highway and highway speeds.

CATEGORIES OF SUSPENSIONS

For study purposes, we have divided truck suspension systems into four general categories:

- Leaf spring
- Equalizer beam: leaf spring and solid rubber spring
- Rubber block and torsion bar
- Air spring: pneumatic-only and combination air/leaf spring

Each type of suspension is studied separately in this chapter, including its servicing and repair requirements. As truck designers continuously improve suspension dynamics, suspension systems that use a combination of spring types are becoming more common. For instance, there are many suspensions using a combination of air and leaf springs currently on our roads. Less common are suspensions using a combination of solid rubber and torsion bars. Kenworth has a unique combination air and torsion bar suspension. By using combination spring suspensions, original equipment manufacturers (OEMs) can take advantage of two or three different types of springs while eliminating the disadvantages.

COMMON SUSPENSION TERMS

Before we begin looking at the different suspension categories used on trucks, here are some suspension terms that we will be using:

- **Jounce** literally means "bump": In suspension terminology, it is the most compressed condition of a spring. For instance, many suspensions use **jounce blocks** to prevent frame-to-axle contact known as suspension slam.
- **Rebound** is the reactive response of a spring after being jounced: It kicks back.
- **Unsprung weight**, an important factor in a suspension, refers to the weight of any chassis components not supported by the suspension, for instance, the axles. Ideally it is kept as low as possible because of the reaction effect, which is one of the reasons for spec'ing aluminum wheels.
- **Oscillation** is either rhythmic or irregular vibrations or movements in a suspension. For instance, a good suspension will minimize jounce/rebound oscillations by using dampening devices such as shock absorbers and multileaf spring packs.

There are more terms to learn, but that will do for a start. Knowing something about suspensions is important for the novice truck technician because you get to work on them a lot. Suspension work can be heavy and dirty but also interesting because of the way in which they have advanced over the years. A generation ago, heavy-duty vocational trucks were rough riding when loaded and were difficult to steer and control when unloaded. That has changed. Today, careful spec'ing of a truck suspension can provide a comfortable ride loaded and unloaded and steer through rough off-highway terrain as well as on a smooth highway.

26.1 LEAF SPRING SUSPENSIONS

A **leaf spring** is a steel plate or stack of clamped steel plates. They have been used since the first vehicles hit our roads; for instance, they were key comfort components in horse-drawn buggies. Most leaf springs used in trucks today are manufactured from spring steel. Spring steel is middle-alloy steel that has been tempered, that is, heat treated. The result is to provide a leaf spring plate with considerable ability to flex without permanently deforming. Leaf springs may consist of a single leaf or a series of leaves clamped together, known as a **spring pack**.

SPRING PACK PRINCIPLES

A spring pack consists of a stack of tempered steel leaves clamped together by a centerbolt. The stack forms a semielliptical shape. When an ellipse is supported on either side and loaded in the center, the bending (deflection) that takes place is proportional to the load applied. If you take a look at the spring pack shown in **Figure 26–1**, you can see that it is a semielliptical shape.

Most multileaf spring packs used in trucks and trailers are semielliptical.

FIGURE 26–1 Semielliptical spring pack—constant rate spring

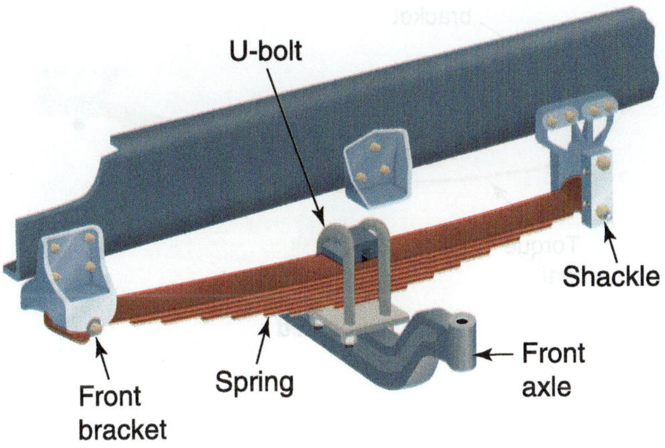

SPRING PACK SELF-DAMPENING

The front axle spring pack shown in Figure 26–1 is commonly used in truck applications. The reason for using multiple leaves clamped together rather than a single piece of metal cut to the same shape has to do with what happens when a load is applied to the spring. As a load is applied to a spring pack, it begins to deflect. This deflection causes bending of each individual leaf in the pack. Because the leaves are clamped to each other with some force, any movement that takes place has to first overcome friction between the leaves. We call this **interleaf friction**. Interleaf friction provides a self-dampening characteristic to the spring pack. Dampening is required in most springs to limit oscillations after a spring compression.

Spring Pack Clamping Force

Two factors ensure that a spring pack retains its self-dampening characteristics. First, when a spring pack is assembled, the individual leaves must never be lubricated or painted. This would reduce interleaf contact friction. Second, the function of the centerbolt that clamps the leaves is critical. The tension it loads the leaves under helps define the self-dampening ability of the spring assembly. In the event of a broken centerbolt, much of the self-dampening properties of a spring pack are lost.

SHOP TALK

When assembling multileaf spring packs, never paint or lubricate the contact surfaces of the individual leaves. The result would severely limit the self-dampening characteristics of the spring.

Most springs have to be dampened after sustaining a compression load. A notable television commercial made by a manufacturer of shock absorbers shows an automobile rhythmically bouncing up and down a highway, overreacting to every little bump, with the message that a new set of shocks will cure the problem. Shock absorbers dampen suspension oscillation. The advantage of the multileaf spring pack is that often they can so effectively dampen suspension oscillations that shock absorbers can be eliminated.

TYPES OF LEAF SPRING ASSEMBLIES

Until relatively recently, leaf spring assemblies were used on a majority of truck and trailer suspensions. Although heavy-duty air suspensions have been around for over 50 years, recently they have become more common in transport applications. Many current truck suspensions use **combination air/leaf spring suspensions**. A combination suspension exploits the advantages of each type of spring to provide good unloaded and fully loaded suspension performance. For our purposes, we have divided leaf springs into two general types: constant rate and progressive or variable rate auxiliary.

CONSTANT RATE

Constant rate springs are leaf-type spring assemblies that have a constant rate of deflection. For example, if 500 pounds deflect the spring assembly 1 inch, 1,000 pounds deflect the same spring assembly 2 inches. A typical constant rate spring is mounted to an axle saddle with U-bolts. The front end of the spring is pinned to a stationary bracket. The rear end of the spring is pin mounted to a flex bracket known as a spring shackle. The shackle allows for variations in spring length between jounce and rebound.

In the constant rate multileaf spring pack shown in Figure 26–1, individual spring leaves are clamped by a centerbolt and the spring pack is secured to the axle housing by U-bolts. Some spring packs use spring clips to ensure that the alignment between the leaves in the stack is maintained. You can see an example of a spring using spring clips (also known as alignment clips) in **Figure 26–2**. In this figure, the constant rate main spring is supplemented by an auxiliary spring pack that only becomes a factor once the main spring has been deflected some distance. This type of auxiliary spring is sometimes known as a helper spring because it is used only when the chassis is fully loaded or when the main spring is jounced. Constant rate spring packs are used in both front and rear axle applications on trucks and trailers.

A variation on the multileaf spring packs just described is the tapered leaf spring design. Tapered leaf spring packs are constructed with leaves that are thicker in the center than at the ends, as shown in **Figure 26–3**. This type of design is a little more

FIGURE 26–2 Rear axle application—constant rate spring pack with auxiliary pack

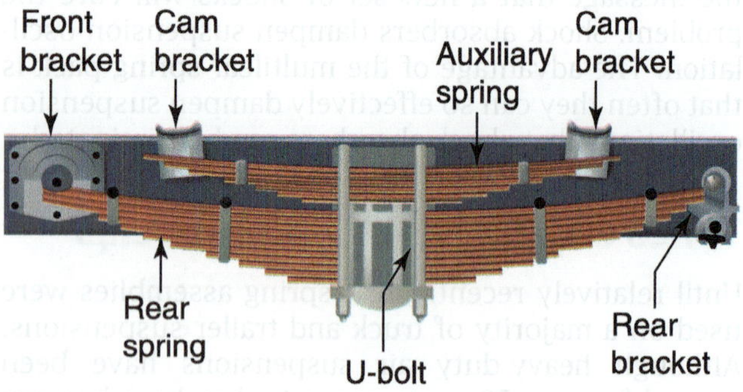

FIGURE 26–3 Taper leaf spring assembly

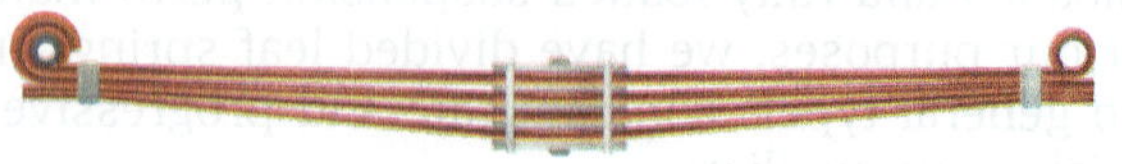

FIGURE 26–4 Progressive spring operation

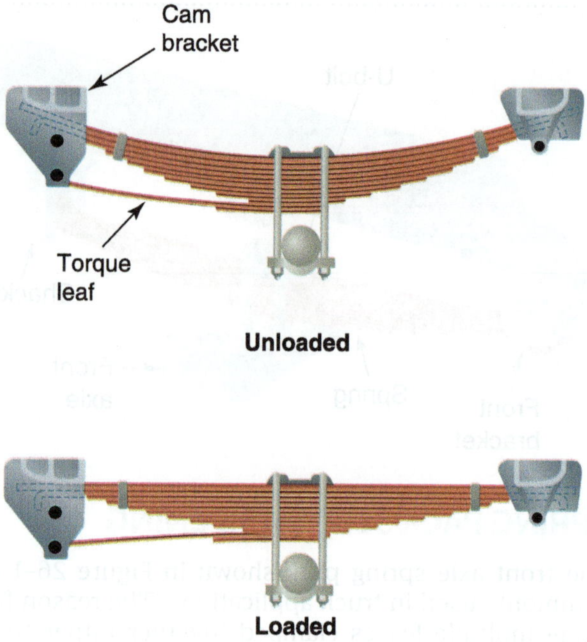

expensive but results in fewer individual leaves, providing lighter weight. Tapered leaf springs are similar to conventional springs in usage and installation.

VARIABLE RATE

Variable rate springs (also known as progressive or vari-rate springs) are leaf-type spring assemblies with a variable deflection rate obtained by varying the effective length of the spring assembly. This is accomplished by using a cam bracket. As the spring assembly deflects, the point of contact on the bracket moves toward the center of the spring assembly, shortening the effective length (**Figure 26–4**). Variable rate spring assemblies also incorporate a progressive feature in that the lower spring leaves are separated at the ends. As the spring assembly deflects under load, these leaves come in contact, providing increased capacity and stiffness.

The variable rate spring shown in Figure 26–4 is commonly used in truck applications. It features a torque leaf that helps prevent axle windup. Axle windup is axial rotation of the axle under aggressive braking or acceleration. A torque spring also plays a role in dampening suspension oscillations. Another example of a variable or progressive rate spring would be the single tapered leaf type. In a tapered leaf spring, when subjected to light load forces, flexing takes place in the ends where the leaf is thinner. However, when subjected to a high load force, spring deflection would encounter progressively more resistance as the thicker midsection of the spring becomes a factor.

APPLICATIONS

Spring-type suspensions are used as both front and rear suspensions on tractor and trucks and on trailer suspension systems. They are spec'd to handle the load requirements for which the vehicle is rated and, if these are exceeded, the life of the suspension components will be reduced.

A good suspension attempts to maintain good tire-to-road contact throughout operation, which is not easy to achieve in trucks designed with rigid suspensions in order to sustain heavy loads. One method of achieving this is by using a **walking beam** on tandem axle arrangements. A walking beam is a pivoting beam that maintains axle spacing and alignment when pivoting to maximize tire-to-road contact. The beams are usually manufactured from heat-treated alloy steels in cast I-beam sections.

FRONT SUSPENSION

The front suspension plays a special role in the suspension because it directly affects driver comfort. Until recently, steel spring suspensions were used exclusively on the front of trucks, but air springs have now become an option. Many types of front steel spring assemblies are used. Some common types include *taper-leaf eye and slipper, multileaf eye and slipper, taper-leaf shackle,* and *multileaf shackle.* An exploded view of a commonly used front suspension, multileaf shackle spring, is shown in **Figure 26–5**.

In all of these front suspension arrangements, the spring assemblies are mounted to the axle by means

FIGURE 26–5 Front suspension, multileaf shackle spring

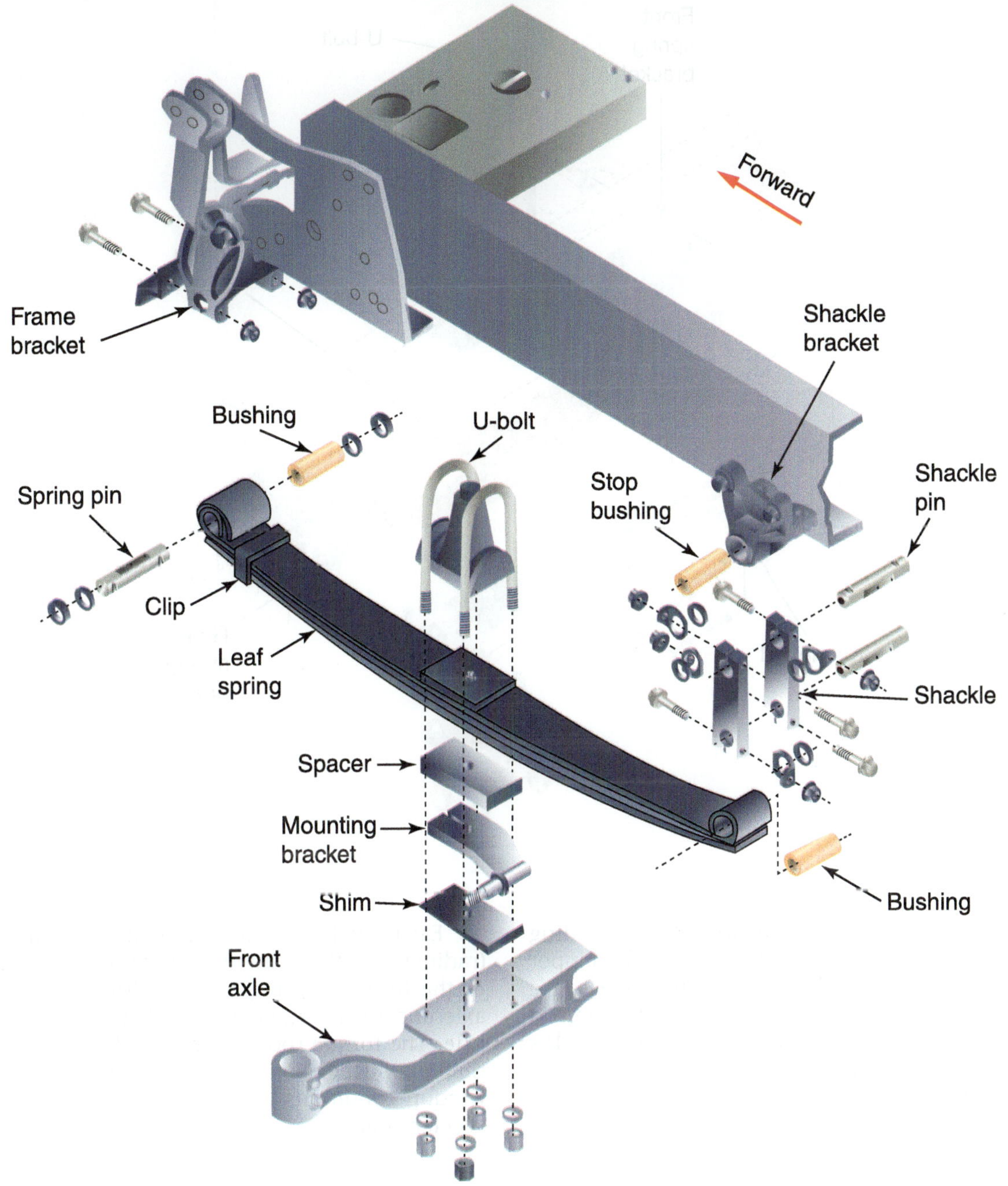

of U-bolts, with the front-end eye of the spring pin (permitting pivot action) to a stationary frame bracket. The rear of each spring is also pinned but permitted some longitudinal movement by spring shackles, which are suspended from a frame-mounted bracket. Spring shackles allow for variations in spring length during spring flexing.

As in any spring pack, the individual leaves are clamped under some tension by a centerbolt, which also acts as a locating dowel to position the spring on the axle. Such springs sometimes have spring clips, also known as alignment clips. These limit sideways spread and vertical separation of the individual leaves.

Rear Suspension—Single Axle

Single rear axle spring assemblies generally are one of two types: *eye and slipper spring* or *semielliptical radius rod*. **Figure 26–6** illustrates an example of semielliptical radius rod spring assemblies. This spring is

FIGURE 26–6 Single drive axle spring suspension

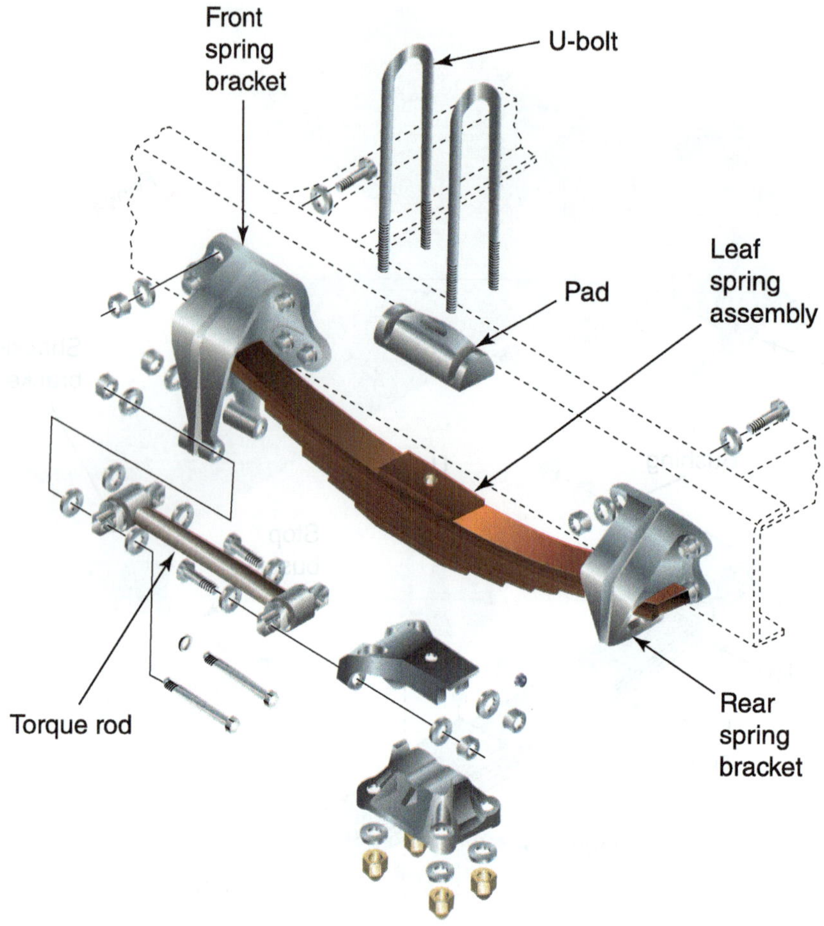

clamped to the axle with U-bolts, with the ends riding in slipper brackets mounted to the frame rails. Slipper brackets do not control longitudinal forces because they do not hold the spring. A radius or torque rod attached to the axle seat forward spring bracket holds the axle in alignment and controls longitudinal forces applied to the suspension.

Rear Suspension—Tandem Axle

The tandem axle rear spring suspension shown in **Figure 26–7** uses a six-point equalizing leaf spring design to compensate for axle articulation, both side to side and front to rear. Four spring packs are attached to the axles with U-bolts. On both sides of the vehicle, the forward end of the forward spring and the rear end of the rear spring ride in brackets mounted on the frame rails. Steel wear shoes are cast into each bracket. At the center, between the forward and rear springs, the springs ride on an equalizer, which pivots on a sleeve in the equalizer bracket. The equalizer used in this suspension is short, but it is effective in helping balance suspension forces between the axles.

Each axle is held in alignment by a pair of torque (radius) rods that extend forward from the axle seats to the forward spring brackets for the forward rear axle and to the equalizer brackets for the rearmost axle. The torque rods control the longitudinal forces that the suspension is subjected to as well as axle windup conditions caused by aggressive braking or acceleration.

Bogie Suspensions

Bogie spring assemblies include the taper-leaf, the camelback, solid rubber spring, and the walking beam. A **bogie** is a general term used to describe a pair of axles arranged with common suspension members designed to act and react together. The ends of both camelback spring assemblies and walking beams are mounted in rubber shock insulators. One type of taper leaf spring bogie has springs with eyes at both ends to accommodate pin connections at the axle housings. Metal-backed rubber insulators are installed in the spring ends to minimize road shock.

FIGURE 26–7 Tandem axle, equalizer spring suspension

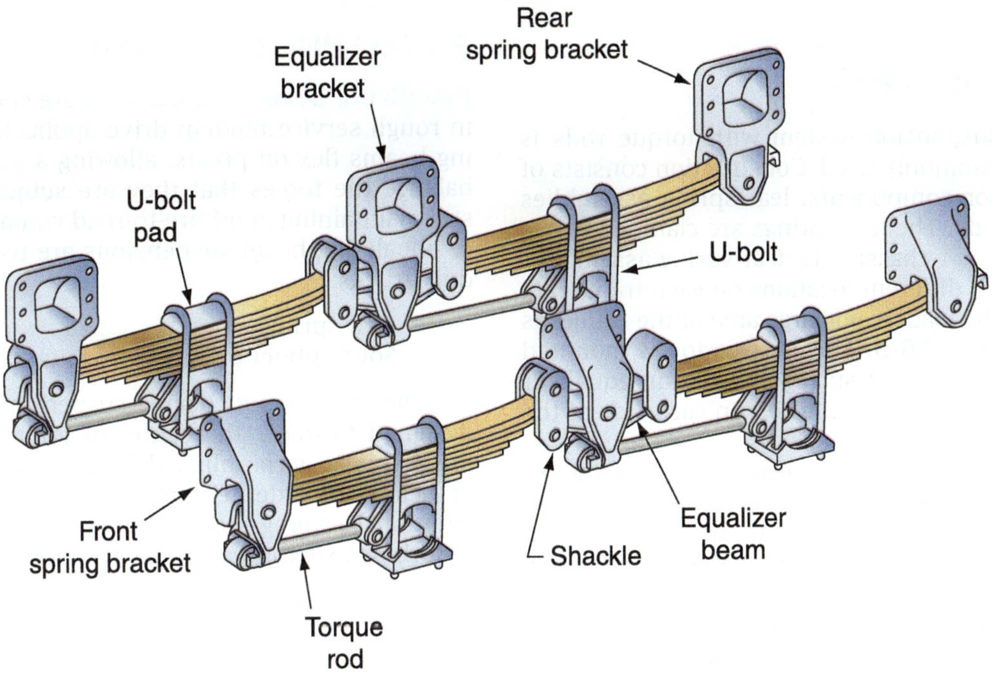

SPRING SUSPENSION WITH SHOCK ABSORBERS

Although multileaf spring packs have good self-dampening characteristics, most front spring suspension systems use **shock absorbers** (**Figure 26–8**). Shock absorbers dampen spring oscillation. They control body sway and eliminate excessive tire wear, front wheel shimmy, and spring breakage. They also improve the ride qualities of the truck and are

FIGURE 26–8 Semielliptical, steel spring suspension with shock absorbers

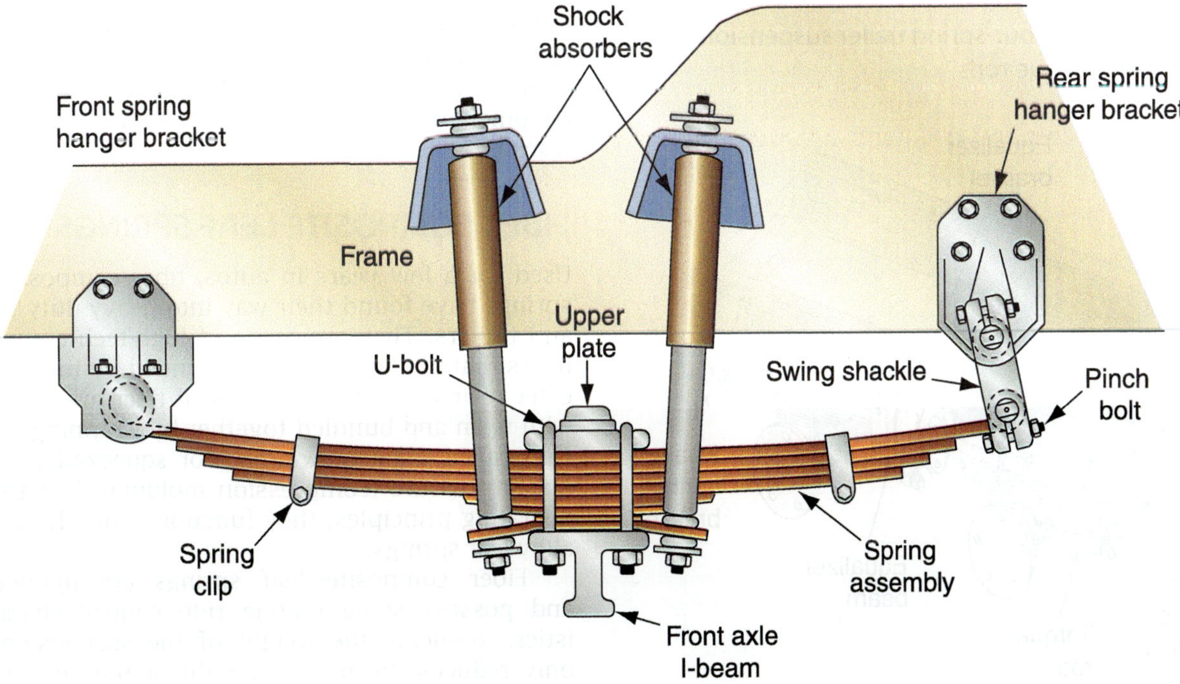

especially useful when the truck is empty or only partially loaded.

SPRING SUSPENSION WITH TORQUE RODS

The spring suspension system with torque rods is simple and commonly used. Construction consists of just two major components: leaf spring assemblies and torque rods. The leaf springs are clamped to the axles. On a 6 × 4 chassis, the leaf spring assemblies mount at three different locations on each frame rail, distributing the load over a large area of the frame, as shown in **Figure 26–9**. This suspension is mounted to the frame by a front spring bracket, an equalizer bracket, and a rear spring bracket on each side of the frame.

On a 4 × 2 chassis, the leaf spring assemblies mount at two locations. The spring pack brackets and frame cross-members are bolted to the frame rail with the same hardware, resulting in the suspension being an integral part of the chassis. Torque rods provide stability by minimizing axle windup caused by driveline torque and load transfer during braking. Suspension alignment is accomplished by inserting shims at the torque rod end mount.

Some four-spring suspension systems use six torque rods instead of four. Four longitudinal torque rods are used between the axle and forward spring bracket, and a pair of transverse torque rods connects the center of each axle banjo housing to a frame cross-member on the truck. Transverse torque rods permit higher driveline torque forces while also providing additional stability, especially when trucks are operated on rough terrain.

FIGURE 26–9 Four-spring trailer suspension system with torque rods

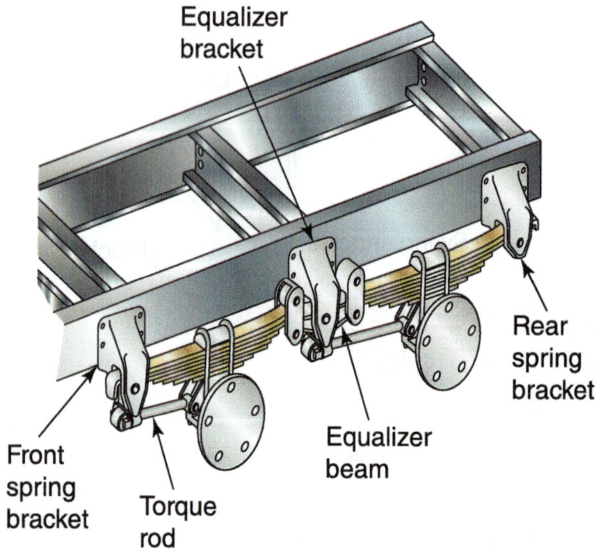

EQUALIZING BEAM SUSPENSIONS

Equalizing beam suspensions are commonly used in rough service tandem drive applications. Equalizing beams flex on pivots, allowing a pair of axles to balance the forces that they are subjected to while still maintaining good tire-to-road contact. Two types of equalizing beam suspensions are used on heavy-duty trucks:

- Leaf spring type
- Solid rubber or rubber cushion type

The equalizing beam suspension system is designed to lower the center of gravity of the axle load. This is accomplished by placing the beam on or below the axle centerline. This design provides additional leverage to work with the torque rods in absorbing axle torque and road shock.

LEAF SPRING EQUALIZING BEAM

The leaf spring-type suspension (**Figure 26–10**) uses a leaf spring pack on each equalizer beam. The springs are mounted on saddle assemblies above the equalizing beams and pivot at the front on spring pins and brackets. The rear of each spring pack has no rigid attachment to the spring brackets but is free to move forward and backward to compensate for spring deflection.

Equalizing beams use a lever principle to distribute the load equally between axles and to reduce the effect of road irregularities. The torque rods permit absorption of windup torque, which is the tendency of the axles to turn backward or forward on their axis due to starting or stopping forces. The cross tube connecting the equalizing beams ensures correct alignment of the tandem and prevents damaging load transfer.

FIBER COMPOSITE LEAF SPRINGS

Used for a few years in autos, fiber composite leaf springs have found their way into heavy-duty trucks and trailers. They are made of high-tech composite fibers that are laminated and bonded by tough polyester resins. The long strands of fiber are saturated with resin and bundled together by wrapping (a process called filament winding) or squeezed together under pressure (compression molding). In terms of operating principles, they function similarly to steel alloy leaf springs.

Fiber composite leaf springs are lightweight and possess some unique ride control characteristics. Reducing the weight of the suspension not only reduces the overall weight of the vehicle but

FIGURE 26–10 Equalizing beam suspension system—leaf spring type

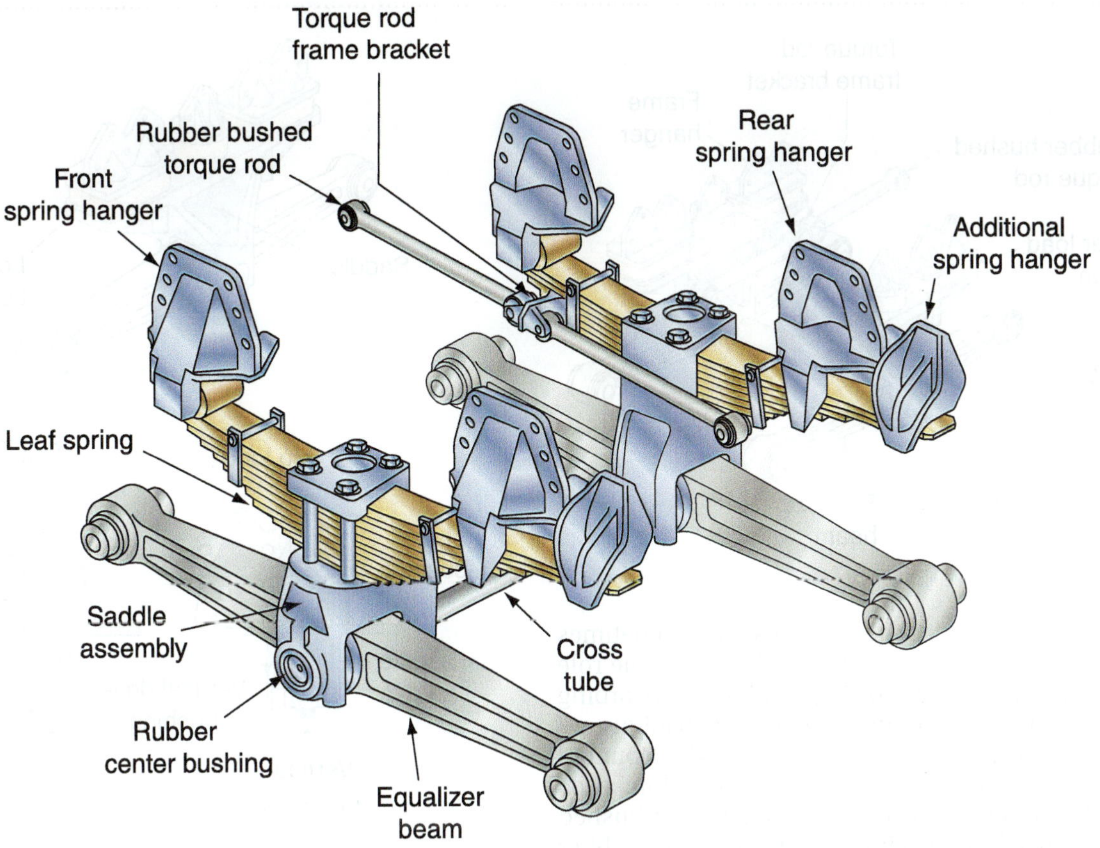

also decreases the unsprung mass of the suspension itself. This reduces the spring effort and amount of shock control required to keep the wheels in contact with the road. The result is a smoother-riding, better-dampening, and faster-responding suspension. Other advantages of using fiber composite springs include the following:

- **Quieter ride.** A fiber composite spring does not resonate or transmit sound like a steel spring. In fact, it actually dampens noise.
- **No spring sag.** All steel springs sag with age, whether they are leaves or coils. Spring sag affects ride height, which in turn alters wheel alignment, handling, steering, and braking. A weak spring can load the suspension unevenly, allowing the wheel under the weak spring to lose traction when accelerating or braking. However, according to the manufacturers of fiber composite springs, there is no sag with age.
- **Less body roll.** In applications in which the leaf springs are mounted transversely, the spring also acts like a sway bar to limit body sway and roll when cornering. This load transfer characteristic also permits softer than normal spring rates to be used for a smoother ride.

With fiber composite leaf springs, there is little or no danger of the spring suddenly snapping. The laminated layers create a built-in fail-safe mechanism for keeping the spring intact should a problem arise.

26.2 CUSHION AND TORSION SUSPENSIONS

This category of suspension is used in specialty vocational applications. There are a wide range of types and manufacturers competing in this field. They may be more costly to purchase but often have significant advantages such as much lower maintenance costs and greater versatility over the types of terrain they can handle. The *cushion* used in this type of suspension can be either a solid rubber or an air spring but it is more common to use solid rubber.

SOLID RUBBER SPRING EQUALIZING BEAM

This type of suspension system (**Figure 26–11**) uses solid rubber springs and is known as a **solid rubber**

FIGURE 26–11 Equalizing beam suspension system—solid rubber spring type

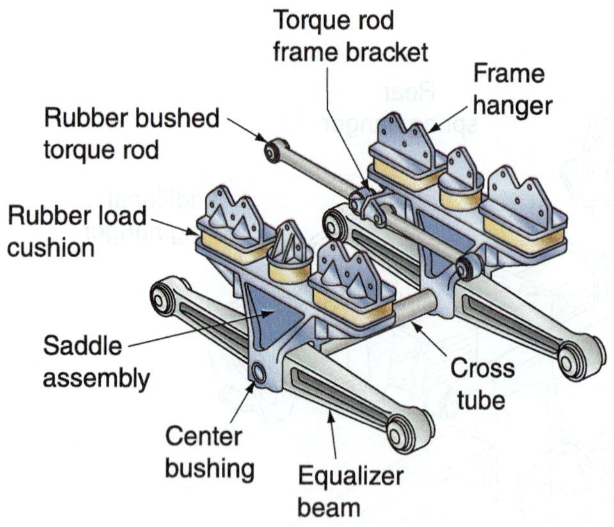

FIGURE 26–12 Sectional view of the rubber cushion-type equalizing beam suspension

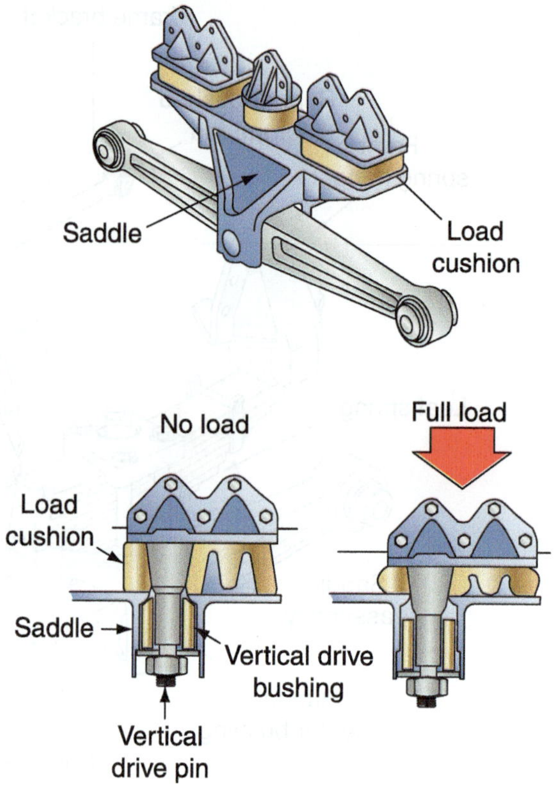

suspension. Solid rubber springs are sometimes known as rubber cushions. They fulfill the same role as leaf springs in supporting load and absorbing road shock. On these units, rubber load cushions are mounted on a saddle assembly at each side. Mounted between frame brackets and the suspension, each rubber block unit is secured by four rubber-bushed drive pins, each of which passes through the rubber cushion. All driving, braking, and cornering forces are transmitted through these pins.

Unloaded, the spring rides on the outer edge of the spring load cushions. As the load increases, the crossbars of the cushions are progressively brought into contact to absorb the additional load. Cushioning and alignment are accomplished by the four drive pins encased in rubber bushings. The bushings permit the drive pins to move up and down in direct relation to movement of the load cushions, as shown in **Figure 26–12**. Solid rubber spring suspensions handle loaded conditions well but tend to be a little unforgiving when run in an unloaded condition.

RUBBER BLOCK AND TORSION BAR SUSPENSIONS

A rubber block and **torsion bar suspension** system uses a combination of both types of springs to maximize the advantages of each. They are known as **combination rubber/torsion spring suspensions** and have had a recent resurgence of popularity in both truck and trailer applications because of their versatility. This category of suspensions has some special advantages for vehicles required to negotiate rough, off-highway terrain along with normal highway operation.

CONTROL LINK, RUBBER SPRING, AND EQUALIZING BEAM

The Chalmers suspension shown in **Figure 26–13** is described in this section. A Chalmers suspension is an eight-link, rubber spring, and equalizing beam suspension. This type of suspension uses a combination of control link and an equalizing beam coupled to a variable rate rubber spring. The objective is to maximize the advantages of each. The basic principle of the Chalmers suspension is to separate the roles of:

- Supporting the vehicle load
- Guiding the axles

Operating Principle

A Chalmers suspension uses molded, cylindrical rubber springs. The rubber springs can be described as variable rate, self-damping. On the tandem drive (shown in Figure 26–13) or tandem dead axle configuration, one variable rate rubber spring is used on each side. The variable rate of deflection of each spring minimizes side roll and self-adjusts to load and road force loads. When the vehicle sways, body roll is reduced because the deflection rate of the outer spring (in relation to the sway angle) increases while

FIGURE 26–13 Chalmers rubber cushion and torsion spring suspension system

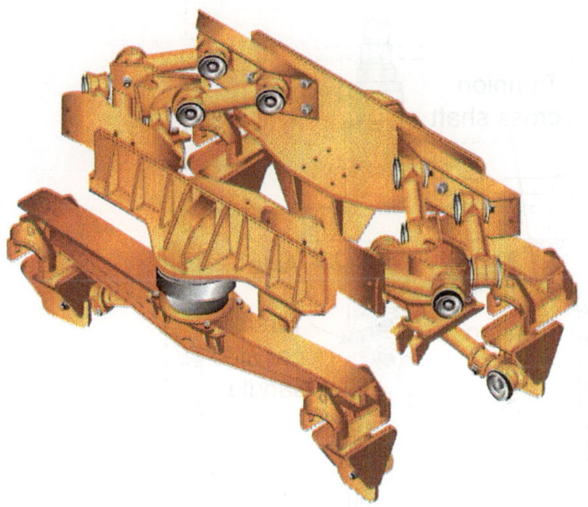

that of the inner spring decreases. This provides good cornering stability.

Each rubber spring is located at the center of a walking beam, the end of which rests on each of the tandem axles. As with any other walking beam system on a tandem-axle application, bump intensity is reduced by half on receiving impact, after which the rubber springs perform final cushioning. The beams support the load and can be regarded as floating, allowing almost unlimited articulation.

Axle Guidance

Each axle is located longitudinally, transversely, and torsionally by four rubber-bushed control links. Two longitudinal control links are attached below the axle center and do most of the work in holding the axles parallel to each other. Another pair of upper links completes the parallelogram. The upper links attempt to maintain a constant planing angle for each axle and limit windup torque. Because they are angle mounted, they locate the axles transversely. This crossover configuration allows for wider splaying of the upper links; they are attached to the two chassis rails dividing forces between them. These twin diagonal links help hold the suspension symmetrical. The roll-center of the suspension is formed by the central tower connections of the upper links.

Advantages

This rubber spring and eight-link suspension permits almost unlimited free articulation to take place. This wide range of articulation does the following:

- Allows each of the four wheels to maintain its share of the load at all times
- Maximizes traction
- Minimizes wheel hop
- Reduces tire wear
- Minimizes maintenance

The suspension is significantly lighter compared to equivalent air spring and leaf spring walking beam suspensions. It is especially suited to vehicles that are required to navigate rough off-highway conditions. The hollow molded rubber springs are said to be virtually indestructible and the only other maintenance item is periodic replacement of the rubber bushings in the control links.

AIR AND TORSION ROD SUSPENSION

In the late 1990s, Kenworth introduced **combination air/torsion spring suspension**. This suspension combines the soft ride of an air spring with the terrain versatility of a torsion bar suspension. Sway bars set axle tracking. The suspension is designed to provide improved articulation over a standard air suspension and the result is to minimize axle road hop under a wide range of loads. **Figure 26–14** shows the principal components of this unique Kenworth suspension.

26.3 AIR SPRING SUSPENSIONS

An increasing number of heavy-duty trucks and trailers are equipped with **air spring** suspension systems. These suspensions may be fully pneumatic (all air springs) or combination air/leaf spring suspension discussed earlier and shown in **Figure 26–15** and **Figure 26–16**. The air bag or air spring suspension system provides a smooth shock- and vibration-free ride with a preset constant frame height. Using it in combination with steel leaf springs helps it to overcome some of its disadvantages.

Air suspensions have been used as the rear suspension on highway tractors and on trailers for many years, but more recently they have gained acceptance on the front steer axle of trucks. The key to making air springs possible on front axles is to dampen the spring oscillations effectively so that steering control is not compromised. This is achieved on current systems by combining the air spring with steel or composite leaf springs and shock absorbers. An example is the Hendrickson Airtek suspension used in the unitized Steertek front axle assembly.

ADAPTIVE SUSPENSION

The air springs on the suspension system can either take the place of mechanical leaf-type springs or be used in conjunction with them. Air springs minimize road shock transfer to the truck frame, the cargo, and the driver/operator. The air spring

FIGURE 26–14 Kenworth air and torsion rod suspension

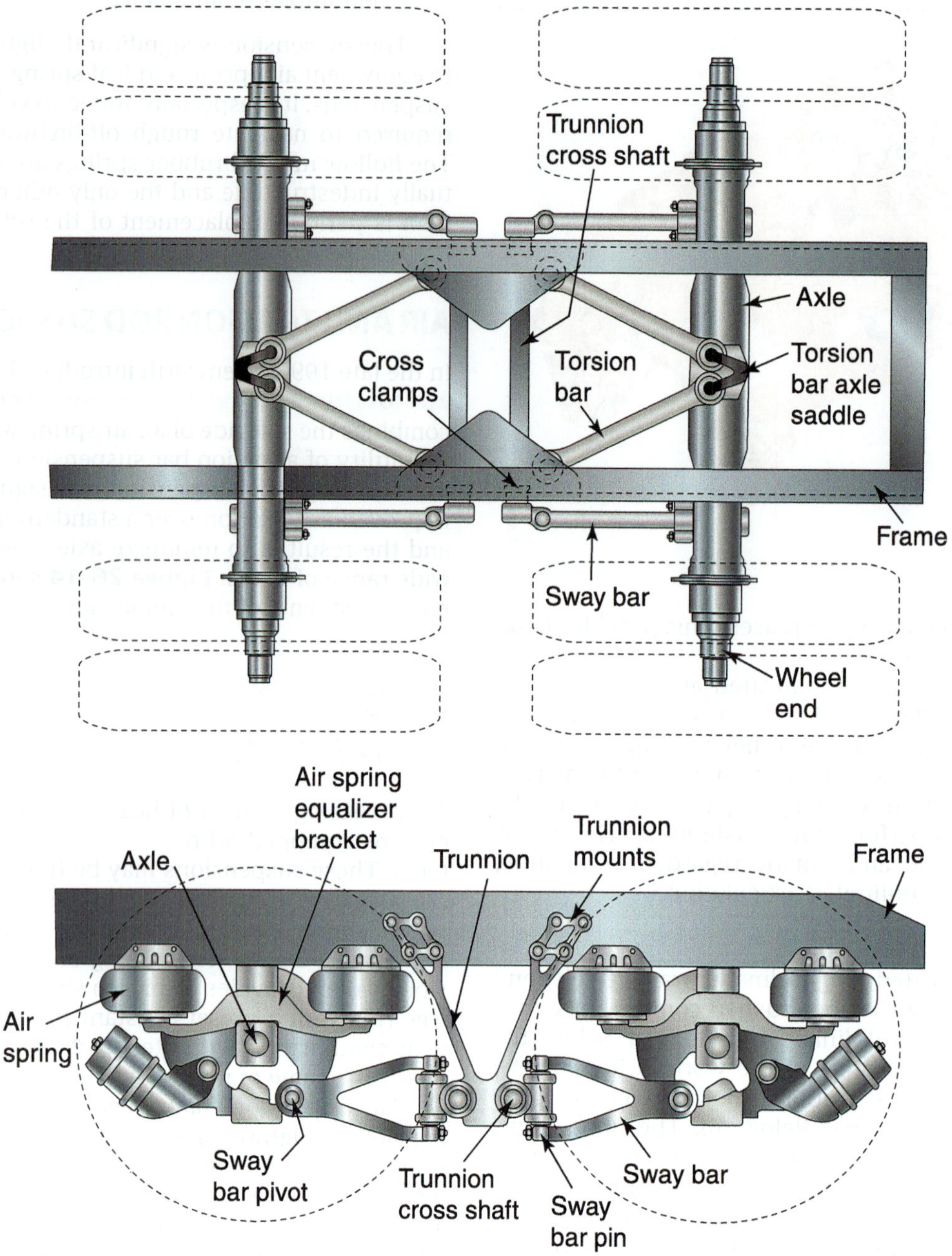

suspension system adjusts to load conditions automatically, providing a low rate suspension with light or no loads and a high rate suspension with heavier loads. The term **adaptive suspension** is often used to describe this feature. The primary disadvantage of the air springs is a zero ability to dampen suspension oscillations. For this reason, they use auxiliary dampening mechanisms such as shock absorbers. In this section, we will take a look at both fully pneumatic and combination air/leaf spring suspensions. Some features of air spring suspensions include the following:

- Front pivotal bushing controls chassis roll and axle alignment while permitting vertical up or down spring travel.
- Suspension brackets attach to the frame by welding or bolting.
- Trailing arms are steel box beams.
- The axle connection is a seat assembly, which is welded as part of the beam and U-bolted.
- The air suspension circuit incorporates a pressure protection valve so the leaks in the suspension system do not siphon the chassis air system.

FIGURE 26–15 Combination leaf and air spring suspension system

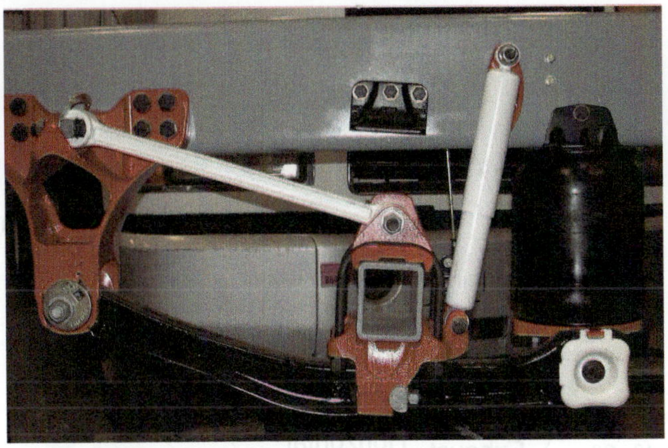

AIR SUSPENSION COMPONENTS

An advantage of air spring suspensions is their simplicity combined with the ability to adapt to load and road conditions. The key to the adaptive capability of an air suspension is the height control valve. The major components of an air suspension system are:

- Height control valve
- Pressure regulator
- Air lines
- Air bags or springs
- Shock absorbers

Height Control Valve

An air spring suspension system is managed by a simple lever-actuated valve known as a **height control valve** (also known as a **leveling valve**), shown in **Figure 26–17**. The height control valve automatically maintains chassis ride height. The height control valve is usually mounted at the rear of the truck frame, as shown in **Figure 26–18**, but other arrangements also are used. The valve has a lever rigidly connected to the rear axle assembly by means of a linkage rod. This height control lever controls air into and out of the air springs. When the axle moves upward, as when passing over a road bump, the lever is forced upward. The reverse would occur if the axle dropped into a pothole. Moving the height control lever charges and discharges air to the air springs. To prevent rapid cycling, a delay mechanism is built into most height control valves.

Figure 26–19 illustrates a height control valve with the valve lever horizontal, that is, in its neutral

FIGURE 26–16 Combination air/steel spring suspension installation

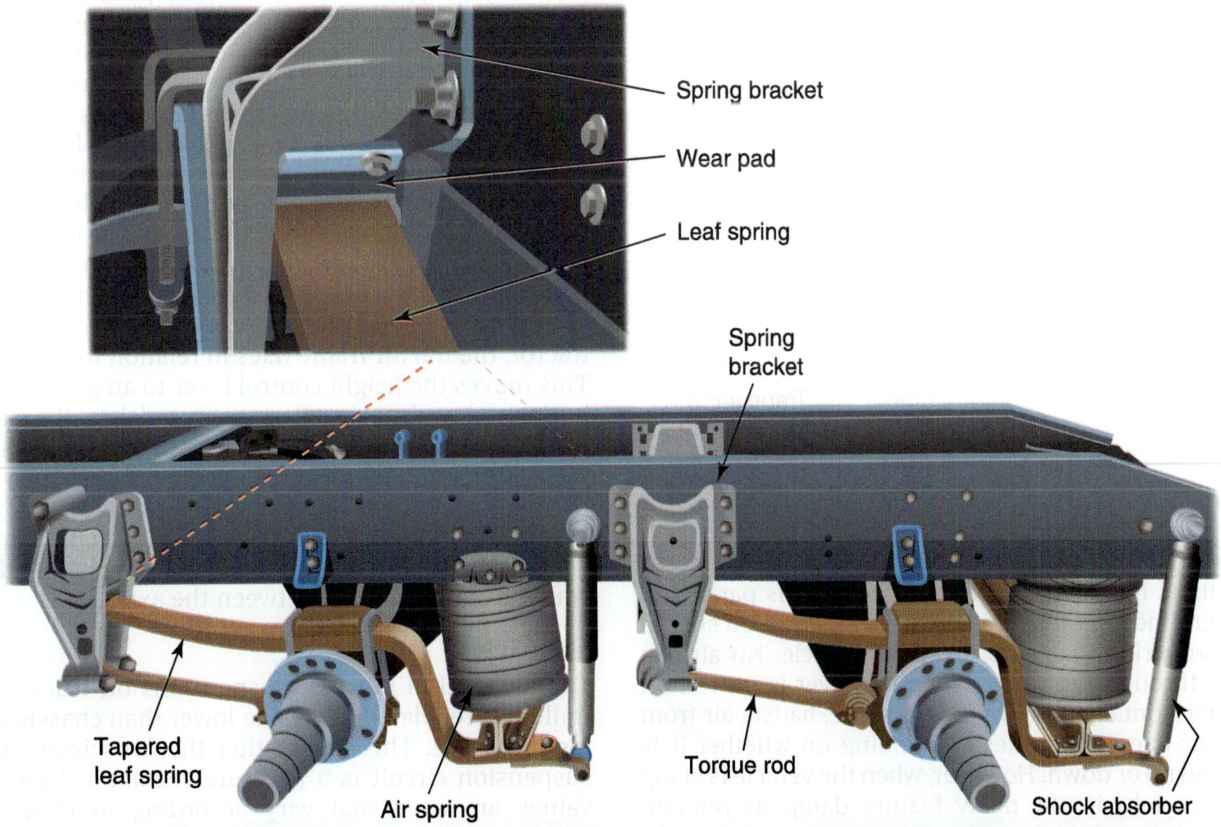

FIGURE 26–17 Height control valve

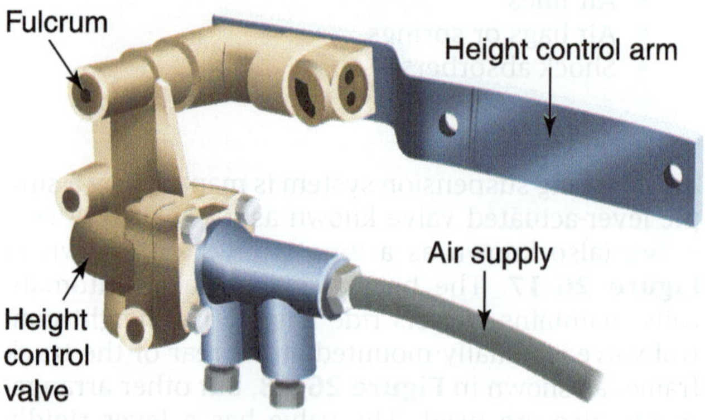

FIGURE 26–18 Location of the height control valve and components

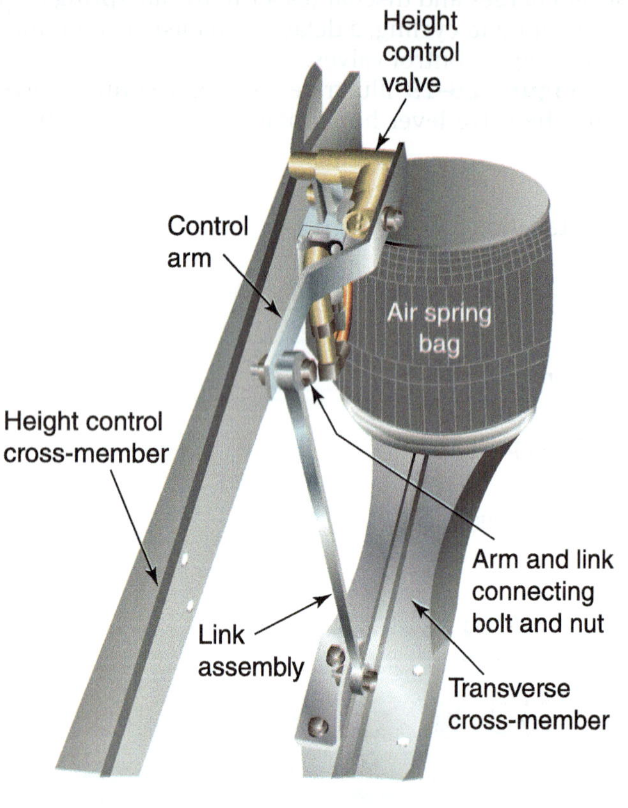

position. In a neutral position, the lever is parallel to the ground, and that means the air suspension should be properly inflated to level the vehicle. No air can pass through the valve. When the lever is moved off its horizontal position, it meters or exhausts air from the air bags it supplies, depending on whether it is moved up or down. However, when the vehicle is being driven, a hydraulic delay feature dampens random

FIGURE 26–19 Height control valve operation

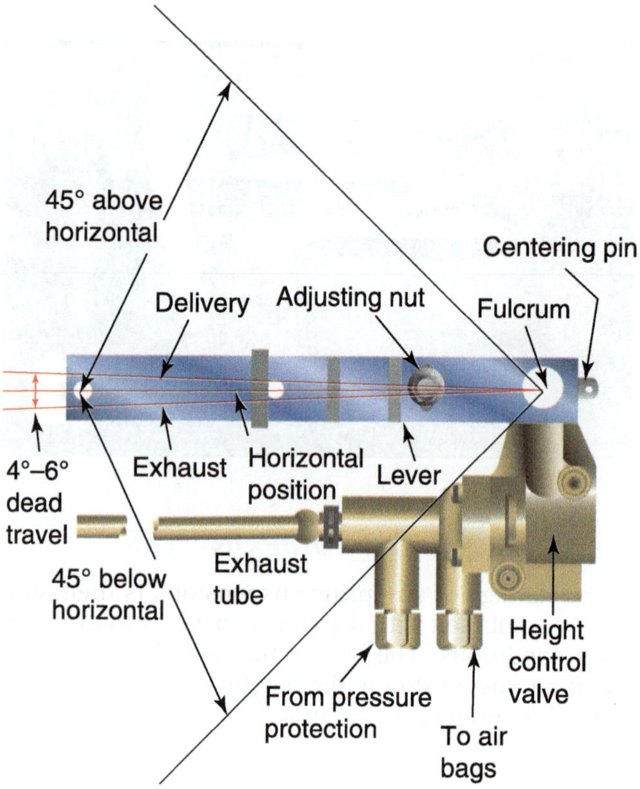

inputs from typical road shock (bumps and potholes) to keep the valve from rapid cycling between opening and closing. This is known as a delay feature.

When a load is applied to the chassis, such as coupling a trailer to the tractor, the frame lowers in relation to the axles. This moves the height control lever to an angle above horizontal, and the valve opens to meter air to the air springs after the built-in time delay. As air passes through the height control valve to charge the air bags and inflate the suspension, the frame rises to resume its set ride height and neutralizes the valve.

Similarly, when the trailer is uncoupled from the tractor, the tractor frame rises in relation to the axles. This moves the height control lever to an angle below horizontal and again, after a time delay, the valve exhausts air until the frame once again resumes its set ride height. Air lines, which are located between the front and rear axle air springs, transfer air pressure between the axles. In this way, the height control valve maintains a consistent frame rail height and an equal load distribution between the axles.

Regulator

The air used in most air suspensions on trucks and trailers is regulated to a value lower than chassis system pressure. This means that the first device in a suspension circuit is a pressure regulator. Different values are used that vary according to OEM and application, but 90 psi (620 kPa) is typical. Most regulators are combined with a pressure protection check valve used to prevent the suspension system from siphoning chassis air in the event of a serious leak.

Air Springs

The air springs or air bags used in air suspension systems are of either the reversible sleeve type or the convoluted type. Examples of both are shown in **Figure 26–20**. The most common type of air spring in use today on trucks and trailers is the reversible sleeve type, shown in Figure 26–20A. The convoluted air spring is also available in three styles: single, double, and triple convolutions. A double convoluted air spring is shown in Figure 26–20B.

The major components of air springs are:

- **Stud.** Used to attach the spring to the suspension; usually manufactured as a part of the bead assembly.
- **Combo Stud.** Serves the dual purpose of mounting the spring to the suspension and providing a threaded air port to which an air line can be coupled.
- **Bead Plate.** Usually crimped onto the bellows during manufacture; it enables pressure testing following assembly.
- **Bellows or Bag.** The rubber inflatable spring member that is charged with compressed air. Usually manufactured from at least four plies (layers) of material: an inner layer, two or four plies of cord-reinforced fabric, and an outer cover. The rubber construction withstands temperatures down to −65°F (−54°C).
- **Bumper or Jounce Block.** Prevents pedestal to frame plate contact in the event of sudden air pressure loss or severe jounce. Supports the frame when the air springs are fully exhausted.
- **Pedestal or Piston.** Used only on the reversible sleeve air springs; usually made of aluminum, but steel, fiberglass, or hard rubber also can be used. The pedestal provides a mount for the air spring in the form of tapped holes or studs.
- **Piston Bolt.** Attaches the piston to the bellows assembly. In some cases, it may be extended to serve as a means of attaching the spring to the suspension.
- **Blind Nut.** A permanent part of the bead plate assembly; it provides an alternate mounting system to the stud.
- **Air Fitting.** Provides an air port to charge the spring.
- **Girdle Hoop.** A ring between the convolutions of the convoluted type spring; used to provide lateral stability.

FIGURE 26–20 Components of (A) reversible sleeve and (B) convoluted air springs

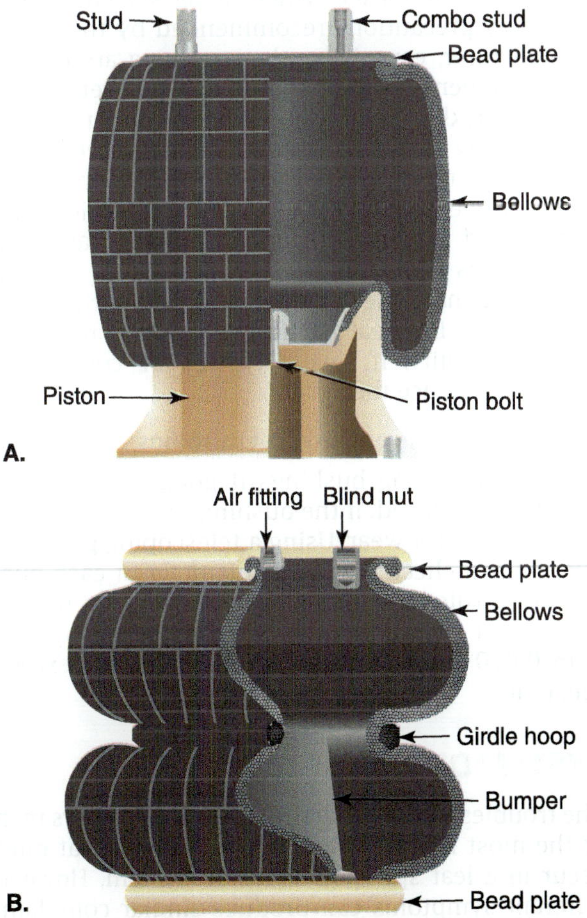

Longitudinal and Transverse Forces

Because of their adaptive ability, air springs are good at supporting vertical suspension forces but cannot sustain longitudinal and transverse suspension forces. For this reason, other suspension mechanisms must be used to sustain chassis suspension forces other than those applied on a vertical plane. Pivoting beams, tapered leaf springs, and torque rod combinations are used in combination with air springs to absorb nonvertical plane road shock, tractive, and braking forces. Lateral and longitudinal torque rods are used to minimize axle windup caused by driveline torque and load transfer during braking.

Shock Absorbers

Air springs have no self-dampening ability so they usually require combinations of shock absorbers (see Figure 26–16) to suppress oscillation cycles. The shock absorbers must be spec'd to the specific air suspension because the travel sweep of an air spring suspension can be considerable, and on no account can shock absorbers be used for jounce control or limit rebound travel. Properly spec'd shock absorbers can function for years with little attention. However, an out-of-balance wheel end produces aggressive high frequency oscillations that can result in premature failures. An external failure of a shock absorber is easy to spot due to visible damage or evidence of leakage, but the shock absorber hydraulics can fail internally, rendering it useless.

26.4 SERVICING MECHANICAL SUSPENSIONS

Servicing a spring suspension system consists of periodic checks and inspections, troubleshooting to identify defective components, and component replacement. Suspension alignment is discussed separately at the end of this chapter.

PERIODIC MAINTENANCE CHECKS

Steel spring suspensions tend to be low maintenance over the lifetime of the vehicle. The U-bolts should be tightened on a preventive maintenance schedule, along with lubricating spring pin bushings and visually inspecting the spring clearance. On a new truck or trailer, it is good practice to retorque U-bolt nuts after the first month or 1,000 miles (1,600 km) of operation. Thereafter, tightening of U-bolts should be built into a preventive maintenance schedule C-type service occurring typically between 35,000 and 50,000 linehaul miles (56,000 and 80,000 km), or every 6 months, depending on the application.

If spring pins fail to take grease, the first thing you should do is remove the weight from the spring and try again. If this does not work, try using a hand grease gun with the weight off the spring—a hand grease gun develops more pressure than an air-actuated grease gun. If this fails you can try heat, using a rosebud (make sure the bracket/shackle is not made of aluminum alloy). But when you spend extra time properly lubing a spring pin, remind yourself that the pin/bushing will fail if not lubricated. It is only a question of when the failure will occur.

Inspect spring ends to ensure that they have not shifted and come into contact with the sides of the equalizer or hanger brackets, as shown in **Figure 26–21**. Spring end contact indicates that the spring assemblies are not seated on the axle housing or there is a need for suspension alignment. Check U-bolt seats, U-bolts, and spring assemblies for integrity.

CAUTION

When checking U-bolts, torque to the original specifications. Rusty U-bolts should be disassembled, cleaned, and lubricated to ensure that the clamping pressure achieved by torquing is accurate.

FIGURE 26–21 Spring end clearance

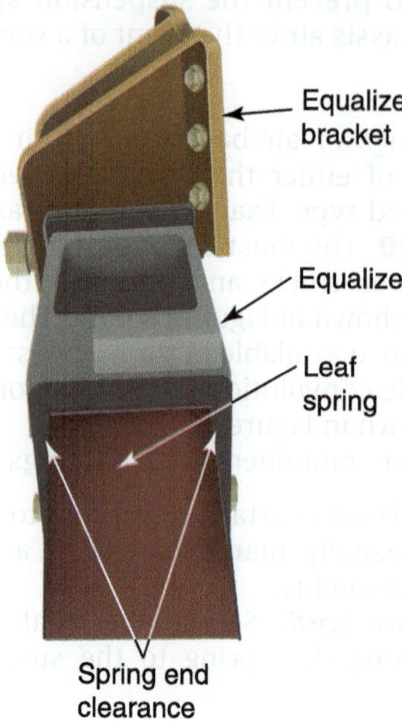

A wire brush and solvent or pressure washer can be used to clean the suspension system and remove dirt, grease, and scale. When using solvents, follow the safety precautions recommended by the solvent manufacturer, especially when working around rubber components. Inspect the shackle brackets and the spring shackles for cracks, wear, and other damage. Replace any damaged components. Inspect spring leaves for cracks and corrosion. If any cracked or broken leaves are observed, most suspension OEMs recommend that spring pack be replaced as a unit. An option is to rebuild the spring pack by replacing only the damaged leaf. Never paint leaf springs. This would lower interleaf friction and reduce the dampening capability. If severe rusting or corrosion is observed, replace the spring.

Inspect the Spring and Bracket Bushings

Replace the spring bushings if gouged, cracked, or otherwise damaged. If the bushings are not damaged, inspect them for wear. Using a telescoping gauge and micrometer, check the inside diameter of each bushing. As a rule, the inside diameter of any bushing should not exceed the diameter of its pin by more than 0.020 inch. Replace any bushing that exceeds this limit.

SYSTEM DIAGNOSIS

The troubleshooting chart in **Table 26–1** covers many of the most frequent causes of problems that might occur in a leaf spring suspension system. However, different symptoms can produce similar complaints

TABLE 26–1 Leaf Spring Suspension Troubleshooting Guide

Symptom	Probable Cause	Remedy
Vehicle leans to one side	1. One or more broken spring leaves	1. Replace the spring assembly.
	2. Weak or fatigued spring assembly	2. Replace the spring assembly.
	3. Unmatched spring design/load capacity spring assemblies	3 Install the correct spring assemblies as originally specified for the vehicle.
	4. Bent or twisted frame rail	4. Check the frame rails for bends and twists; correct as needed.
Vehicle wanders	1. One or more broken spring leaves	1. Replace the spring assembly.
	2. Wheels out of alignment	2. Adjust the wheel alignment.
	3. Caster incorrect	3. Install correct caster shims.
	4. Steering gear not centered	4. Adjust steering.
	5. Drive axles out of alignment	5. Align the drive axles.
Vehicle bottoms out	1. Excessive weight on the vehicle causing an overload	1. Reduce the loaded vehicle weight to the maximum spring capacities.
	2. One or more broken spring leaves	2. Replace the spring assembly.
	3. Weak or fatigued spring assembly	3. Replace the spring assembly.
Frequent spring breakage	1. Vehicle overloaded or operated under severe conditions	1. Reduce the loaded vehicle weight to the maximum spring capacities; caution the driver on improper vehicle handling.
	2. Insufficient torque on the U-bolt nuts	2. Torque the U-bolt nuts to the manufacturer's value.
	3. Loose centerbolt allowing the spring leaves to slip	3. Check the spring leaves for damage; if damaged, replace the spring assembly; if not, tighten the centerbolt nut to the manufacturer's value.
	4. Worn or damaged spring pin bushings allowing spring endplay	4. Replace the spring pin and bushing.
Noisy spring	1. Loose U-bolt nut or centerbolt allowing spring leaf slippage	1. Inspect the components for damage; replace damaged components as necessary; torque all fasteners to the manufacturer's values.
	2. A loose, bent, or broken spring shackle or front spring bracket impairing the spring flex	2. Inspect the shackle and bracket for damage; replace damaged components as necessary; torque the fasteners to the manufacturer's values.
	3. Worn or damaged spring pins allowing spring endplay	3. Replace any worn or damaged spring pins.
Rough ride	Refer to applicable subject in **Table 26–2**.	

relating to axle and wheel alignment problems, so be careful about jumping to conclusions.

Rough Ride Diagnosis

There are two ways of describing rough ride conditions: harmonic and erratic. Harmonic ride problems are those that occur rhythmically, for instance, a once-per-revolution energy input from such things as bent or imbalanced wheels. The rhythmic vibration that results continues as long as the critical road speed is maintained. Harmonic ride problems are more likely to be noticed on smooth roads. The sources of harmonic ride problems are typically wheels, rims, and brake drums or hubs.

Erratic ride problems are those in which the suspension transfers, rather than absorbs, momentary energy inputs produced when the tires hit bumps or holes in the road. Wavy asphalt, or a series of bumps, might cause repetition of the harsh, jarring motion in the cab, but the motion ceases after the tires pass over the bumps. Harsh ride problems tend to occur on rough roads. This section contains general information on correcting rough ride problems and is not intended to replace OEM service literature.

When checking a vehicle with a rough ride complaint, first visually check for signs of damaged or missing suspension components. Repair or replace the components as required. Then test drive the truck. During the test drive, try to duplicate the conditions that initiated the complaint. Raise the truck until the tires are off the ground and the weight is removed from the leaf springs; use a frame jack if one is available. Make sure to mechanically support the vehicle. Then, using a guide such as that provided in Table 26–2, attempt to locate the source of the problem.

LEAF SPRING REPLACEMENT

The exact replacement procedures will differ somewhat for each type of leaf spring assembly, but the basics remain similar. The following is a general outline of spring pack replacement.

Remove the Spring Pack

1. Chock the tires, place a floor jack under the truck frame, and raise the truck sufficiently to relieve the weight from the spring to be removed. Then place safety stands under the frame.
2. Remove shock absorbers where used.
3. Remove U-bolts, spring bumper, and retainer or U-bolt seat.
4. Remove the lubricators or zerk fittings (not used on springs equipped with rubber bushings).
5. Remove the nuts from the spring shackle pins or bracket pins.
6. Slide the spring off the bracket pin and shackle pin and remove from chassis.

TABLE 26–2 Rough Ride Diagnosis

Symptom	Probable Cause	Remedy
Harsh ride, tires off the ground	Seized front spring shackle pins not allowing the springs to flex	Replace seized shackle pins.
Harsh ride, tires on the ground	1. Tires improperly inflated.	1. Adjust the tire pressure.
	2. The frame is bottoming out against the suspension.	2. Check the suspension for weak or damaged springs or components; inspect the springs for "gull-winging" when the vehicle is loaded; replace the spring assembly as necessary; reduce the overall loaded weight on each axle to conform with the maximum spring load capacities on the vehicle specification sheet; do not exceed the maximum spring load capacities.
	3. The vehicle's normal loaded weight is markedly below the spring load capacity.	3. Contact the manufacturer for the correct application of a lower rated spring. Replace the spring assembly.
	4. When the vehicle is loaded, the front axle spring shackle angle is not within the rearward 3- to 18-degree angle.	4. Contact the manufacturer for shackle angle corrective measures.
	5. The weight on the tractor fifth wheel is causing overloading on the front axle springs.	5. If possible, move the fifth wheel toward the rear of the vehicle; otherwise, change the loading pattern on the trailer.
	6. There is a loaded weight differential between the rear axles greater than 800 pounds.	6. Contact the manufacturer for corrective measures.
	7. Forces from the trailer suspension are acting on the tractor fifth wheel, causing a rough ride condition.	7. Review the ride problems in this table that might apply to the trailer suspension; perform the corrections, as necessary.

Install a Spring Pack

Although the actual installation procedures will vary with each spring type, the pivot end of the spring is usually fastened to the frame bracket first. The shackle or slipper end then can be fastened by aligning it to the other frame bracket. When installing the U-bolts or spring clips to secure the axles, do not final torque until the springs have been placed under normal load. This requires lowering the chassis weight onto the spring.

Maintenance

U-bolts that loosen in service place undue strain on the spring centerbolt. OEMs state that centerbolts should be retorqued at specified intervals but this is seldom done. If you are torquing centerbolts on a spring not clamped by U-bolts, remember that they are under considerable tension should the centerbolt fail.

Loose spring clips or U-bolts can result in axle misalignment, which adversely affects tire wear. When torquing U-bolt nuts, follow the recommended tightening sequence after lubricating the U-bolt threads with a little lubricant. The lubricant helps ensure the specified clamping pressure of the spring pack.

To correctly torque U-bolt nuts, first tighten all the nuts until they are snug. Next, tighten the nuts until approximately one-third of the recommended torque is achieved. Repeat tightening of the nuts, using the same sequence and gradually increasing the torque through a second, third, and fourth stage, until the recommended final torque is attained.

Shock Absorber Inspection

OEMs suggest checking shock absorbers at intervals of 12,000 miles (20,000 km) but you can assume that if they are regularly failing at low mileage, another suspension-related problem is the cause. To check shock absorbers on a truck, you can use the following sequence:

1. To check for noise, make sure that it is not coming from some other part of the vehicle. Then check shock absorber brackets and mounting stud nuts to ensure that they are tight and that the shock absorber is not striking or rubbing on the frame or some other part of the chassis.
2. If a truck has just been operated on the highway (a short test drive is sufficient), shock absorbers can be checked using an infrared temperature gun or by manual contact. The shock absorber cylinder should feel warmer than the frame to which it is attached. If this is not the case, suspect a hydraulic problem and disconnect the shock absorber at one end to test (see #5).
3. Check the rubber mounting bushings and replace if worn.
4. Disconnect the bottom mount and work the shock absorber by hand to check that the outer tube is not contacting or rubbing against the fluid reservoir tube.
5. When the shock absorber is disconnected, check the piston movement by pulling and pushing the absorber down and up slowly through its travel stroke, making sure that the piston does not bind in the pressure tube. Typically, there should be considerable resistance when extending the absorber but only slight resistance when collapsing it. Also note the rate of effort for distance of travel. This rate should not vary.
6. If it is determined that the shock absorber is defective, replace it.

CAUTION

Do not operate a vehicle with a shock absorber removed or defective because this places undue stress on other suspension components.

SHOP TALK

Shock absorbers are pretty good at suppressing the normal oscillations produced by an air suspension but when they routinely fail at low mileage, the cause is often an out-of-balance axle end. This can pound out shocks, drastically reducing their service life. See Chapter 27 for a detailed explanation of the consequences of out-of-balance axle end conditions.

Shock Absorber Replacement

1. Remove the nuts from the top and bottom bolts/studs from the eyes of the shock absorber.
2. Remove the shock absorber from the vehicle.
3. Insert new rubber bushings into the eyes of the new shock absorber.
4. Install the top bolt/stud through the top mounting bracket on the frame rail.
5. Place a flat washer over the bolt and thread the bolt into the upper rubber bushings of the new shock absorber.
6. Place a second flat washer over the bolt and then turn the nut finger-tight.
7. Extend the lower part of the shock absorber until the mounting holes line up in the lower bracket and shock absorber eye.
8. Install the bottom bolt/stud through the lower mounting bracket on the axle.
9. Place a flat washer over the bolt and thread the bolt into the lower rubber bushing of the new shock absorber.
10. Place a second flat washer over the bolt and turn the nut finger-tight.
11. Check to make sure that the rubber bushings are seated properly and tighten the nuts to the specified torque.

SHOP TALK

Some shock absorber mount brackets have a stud welded to the bracket, rather than a nut and bolt. This does not alter the installation procedure.

Radius Rod Replacement

Before replacing a radius rod on a leaf spring suspension system, secure the truck parking brakes and chock the wheels. Make sure that the replacement radius rod is the correct length because they come in a range of sizes. Some radius rods are adjustable. Then replace the radius rod by observing the following procedure:

1. Note the number of axle alignment washers (refer to **Figure 26–22**, item No. 4) at the forward end of the radius rod that is being removed.
2. Remove the fasteners that attach the radius rod to the forward spring bracket and to the axle seat.
3. Remove the radius rod and any axle alignment washers. Retain the axle alignment washers for use when installing the new radius rod.
4. At all points where steel components (including bolts, washers, and nuts) contact an aluminum forward spring bracket, apply Alumilastic™ compound, or an equivalent, on the mating surfaces. Failure to apply Alumilastic compound, or an equivalent, to areas where aluminum and steel components contact each other can result in electrolytic corrosion of the metals, resulting in damage to the suspension and severe seizure of the components, making removal difficult.
5. Position the replacement radius rod so that the radius rod pins are between the rear side of the forward spring bracket and the front side of the axle seat.
6. Install the hex-head bolts, hardened washers, and locknuts in the axle seat and the radius rod rear pin.
7. Install any previously removed axle alignment washers between the radius rod front pin and the forward spring bracket. Install the hex-head bolts, hardened washers, and locknuts in the radius rod front pin and the forward spring bracket.
8. Tighten the radius locknuts to the torque value specified by the manufacturer.
9. Check the axle alignment. If necessary, adjust the rear axle alignment.

FIGURE 26–22 Radius rod attachment (top view)

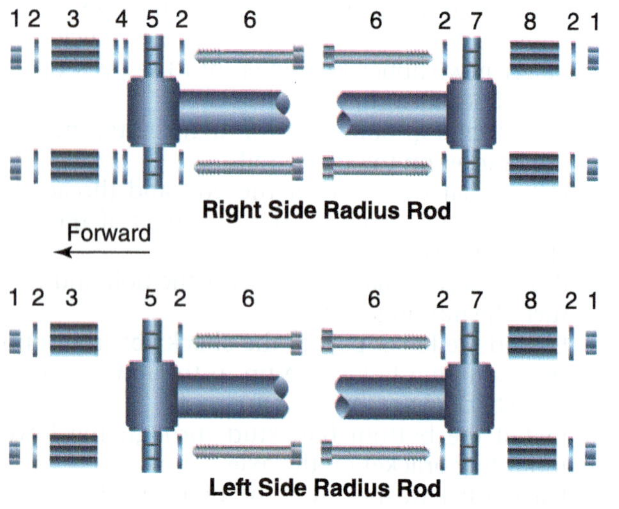

1. Hex locknut
2. Hardened washer
3. Forward spring bracket
4. Alignment washers (install only on one side, right-side installation shown)
5. Radius rod front pin
6. Hexhead bolt
7. Radius rod rear pin
8. Axle seat

Spring Bracket Replacement

Failure to replace a worn, cracked, or damaged spring bracket can result in breakage of the bracket itself, which in turn could cause a loss of vehicle control. Before replacing a spring bracket, chock the front tires. Then raise the truck and block the wheels of the axle(s) you are not going to be working on. Raise the truck frame so that all the weight is removed from the leaf springs. Then support the frame with safety stands. Then use the following procedure:

1. If removing the forward spring, note the number of axle alignment washers, if any, between the bracket and the radius rod front pin. Remove the fasteners that attach the radius rod to the bracket and remove any axle alignment washers.
2. Remove the fasteners that attach the spring bracket to the frame rail and remove the spring bracket. Note the fastener type and grade. If Huck™ fasteners are used, you may have to cut them out with an oxyacetylene torch.
3. Place the new spring bracket on the frame rail. Align the mounting holes and install the spring bracket fasteners, hardened washers, and locknuts.
4. If replacing the forward spring bracket, install the nuts for the top two bolts on the outboard side of the frame rail and install the nuts for the bottom four bolts on the inboard side of the frame rail (**Figure 26–23A**).
5. If replacing the rear spring bracket, install the nuts for the top two bolts on the outboard side of the frame rail and install the nuts for the bottom two bolts on the inboard side of the frame rail (**Figure 26–23B**).
6. Tighten the locknuts to the torque value specified by the manufacturer.
7. Check the axle alignment. If necessary, adjust the rear axle alignment.

FIGURE 26–23 Spring frame brackets

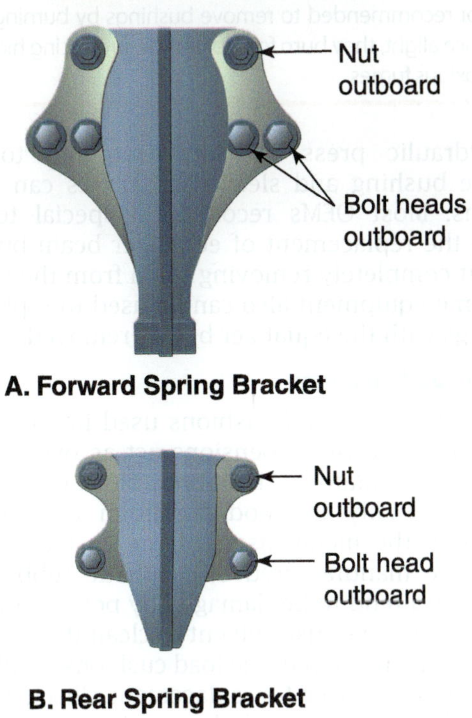

> **SHOP TALK**
>
> It is common practice to use SAE grade 8 fasteners in suspension systems but not universal. Grade 5 bolts flex more than grade 8 bolts and that is desirable in some applications so replacing them with grade 8 fasteners is not appropriate. The body bound bolts with an interference-fit shank used by some OEMs are always grade 5. When replacing suspension Huck™ fasteners with bolts, it is generally safe to use grade 8 bolts.

> **CAUTION**
>
> Failure to properly torque suspension fasteners can result in abnormal tire wear and damage to the springs, spring brackets, and frame rail.

EQUALIZER REPLACEMENT

Before replacing suspension equalizers, make sure that the front wheels of the truck are chocked. Then raise the rear of the truck and support the axles with safety stands. Raise the truck frame so that all the weight is removed from the leaf springs; then support the frame with safety stands. A frame scissor jack equipped with mechanical support stops is the best means of hoisting the vehicle. To the equalizer bracket or beam, remove the wheel assemblies on that side of the vehicle. Then refer to **Figure 26–24** and replace the equalizer as follows:

1. If removing an equalizer from a truck with tandem drive axles, remove the cotter pin from the outboard end of each spring retainer pin; then pull the retainer pins.
2. Remove the equalizer cap and tube assembly locknut, inboard bearing washer, bolt, and outboard bearing washer.
3. Insert a bar between the bottom of the equalizer and the equalizer bracket. Gently lever the weight of the equalizer off the equalizer cap and tube assembly. Insert a piece of bar stock through the inboard equalizer cap and tube assembly bolt hole and lightly tap the cap and tube assembly out of the equalizer.
4. Thoroughly clean the equalizer bushings and inspect them for damage or defects. Replace the bushings if necessary.
5. Apply multipurpose chassis grease to the inside of the equalizer bushing and then install the bushings in the equalizer.

FIGURE 26–24 Equalizer assembly components

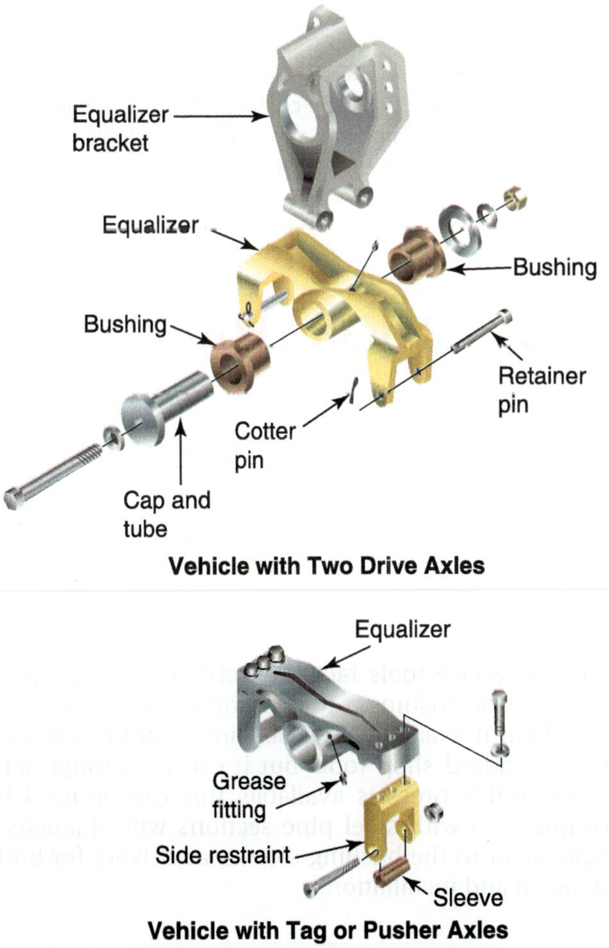

6. Install the replacement equalizer in the equalizer bracket.
7. Apply multipurpose chassis grease to the equalizer cap and tube assembly. Start the cap and tube assembly into the equalizer, through the equalizer bracket.
8. Push the equalizer cap and tube assembly partway through the equalizer and then place the inboard wear washer(s) between the inboard equalizer bushing and the equalizer bracket. Push the cap and tube assembly the rest of the way into the equalizer bracket.
9. Place the outboard bearing washer on the equalizer cap and tube assembly bolt and install the bolt in the cap and tube assembly.
10. Install the inboard bearing washer and locknut on the cap and tube assembly bolt. Tighten the locknut to the torque value specified by the manufacturer.
11. Install the wheel assemblies. Remove the safety stands from under the frame and axle and lower the truck.
12. At any time radius rods are loosened or an equalizer bracket has been replaced, you must check the rear axle alignment, adjusting axle alignment if necessary.

CAUTION

You must apply Alumilastic compound, or an equivalent, to areas where aluminum and steel contact each other or the result will be metal corrosion and severely seized components.

26.5 SERVICING EQUALIZING BEAMS

Power wash the suspension and inspect the components for cracks or damage. Inspect rubber bushings for damage or deterioration and plan to replace them if they show any indications of fatigue.

EQUALIZING BEAM BUSHINGS

Special service tools facilitate replacement of equalizing beam bushings. It is recommended that you use them if available. The bushings can be removed with standard shop tools but it can be a tough job. If a portable press is available, this can be used in conjunction with steel pipe sections with diameters equivalent to the bushing sleeves as drivers for both removal and installation.

CAUTION

It is not recommended to remove bushings by burning them out; once alight, they burn for a long time, producing high heat and noxious fumes.

Hydraulic press pressures required to drive out the bushing and sleeve assemblies can exceed 50 tons. Most OEMs recommend special tools to enable the replacement of equalizer beam bushings without completely removing them from the chassis. The same equipment also can be used to replace the bushings with the equalizer beams removed.

Inspect Load Cushions

Jounce blocks or load cushions used in some types of equalizer beam suspensions act as overload and shock load insulators. The jounce block must appear to be externally in good condition and securely bonded to the metal insert of the cushion. Jounce blocks are manufactured of synthetic rubber compounds and tend to be damaged by petroleum-based products, so never use solvent to clean them.

Visual inspection of the load cushions is sufficient to determine the need for replacement. Check them for cracks, deterioration, and abnormal wear by removing the insulator cap (if so equipped). Also check the height of cushion against the specified height. They do get pounded out in time and a height reduction exceeding ¼ inch (6.3 mm) suggests that they should be replaced. An equalizer beam axle should not be operated with defective load cushions because they can result in premature failure of other suspension and frame components, including the spring packs.

EQUALIZING BEAM OVERHAUL

Make a thorough inspection before you start disassembling the suspension to ensure that you have replacement parts in place before undertaking the work. The following procedure describes a general approach to reconditioning an equalizing beam suspension.

Preparation for Overhaul Repair

1. Block the wheels on both axles of the truck.
2. Remove the drive axle shafts from both rear axles.
3. Remove the four saddle caps that lock the center bushing of the equalizer beams to the saddle assemblies.
4. Disconnect the torque rods from the axle housing by loosening the locknuts and driving the shaft from the bushing and axle housing bracket.
5. Using an overhead hoist (A-frame works best), raise the truck frame sufficiently so that the lower part of the saddle clears the top of the axle housings. Support the frame securely. Block the axles to prevent them from pivoting on the wheels.

6. Roll axles, with the beams attached, out from under the truck.

Overhaul Procedure

There are three main types of equalizer beam mounting configurations:

- Bolt-type beam end mounting (**Figure 26–25**)
- Tube-type beam end mounting (**Figure 26–26**)
- Ball and socket beam end mounting (**Figure 26–27**)

FIGURE 26–25 Three-piece adapter bolt-type beam end mounting

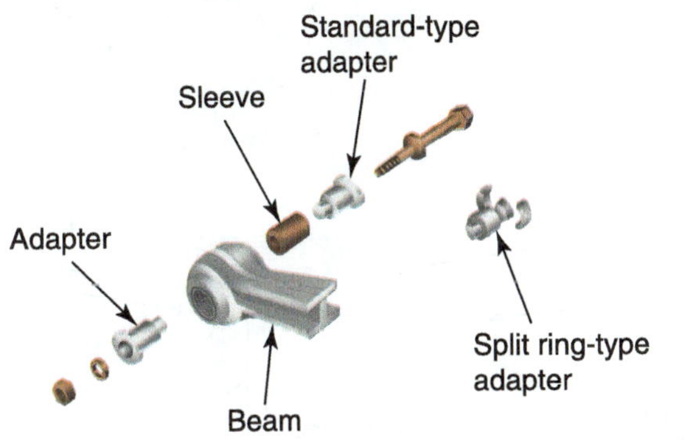

FIGURE 26–26 Tube-type beam end mounting

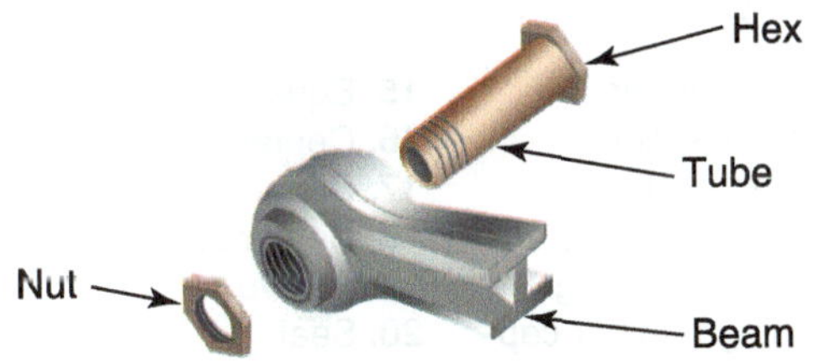

The replacement procedure outlined here primarily references bolt-type beam ends.

1. Disconnect the equalizer beam from the axle housing, referencing Figure 26–24.
   a. Remove the equalizer beam end bolt.
   b. Drive the bushing adapters and sleeve (if so equipped) out of the bushing and axle housing brackets. Adapters are notched to help keep the chisel in place between the axle bracket and adapter. Drive the chisel first on one side and then on the other side of the adapter to wedge out the adapter.
   c. After the first adapter is removed, the opposite adapter can be driven out with an impact hammer or heavy bar and hammer.
2. Separate the equalizer beams from the cross tube.
3. After both equalizer beams have been separated from the axles, separate the beams from the cross tube by pulling them apart by hand. The cross tube floats in the inner sleeve of the center bushings. The slide float is 3 inches plus, which will polish the cross tube where it enters the center bushing on each side. This visible polished area on the cross tube is normal.
4. Remove the saddle assemblies, springs, or cushions. Reference the *leaf spring-type* equalizer beam (shown in **Figure 26–28**).
   a. Support the spring and saddle assembly on a floor jack and safety stand. Loosen the spring-aligning setscrew. Remove the spring top pad saddle bolts and nuts. Remove the top pad. Lower the floor jack and remove the saddle.
   b. Remove the spring assembly by repositioning the floor jack under the spring. Loosen the locknut on the spring pin draw key (item 3 in **Figure 26–29**).

FIGURE 26–27 Ball and socket beam end mounting

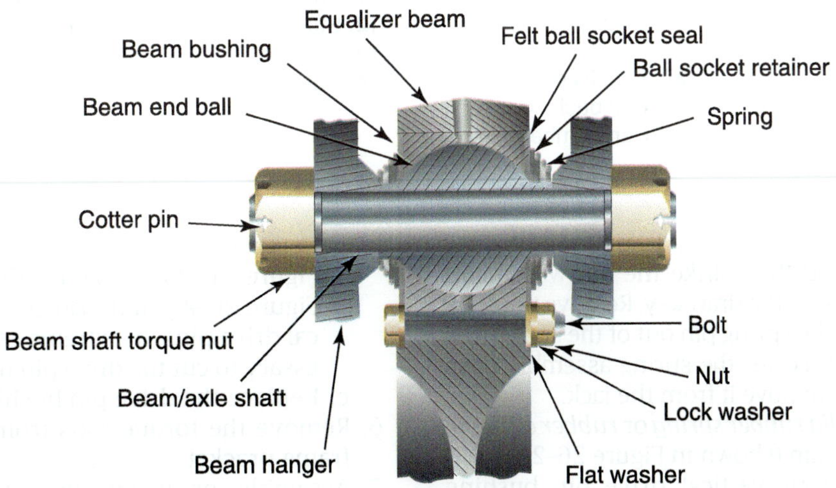

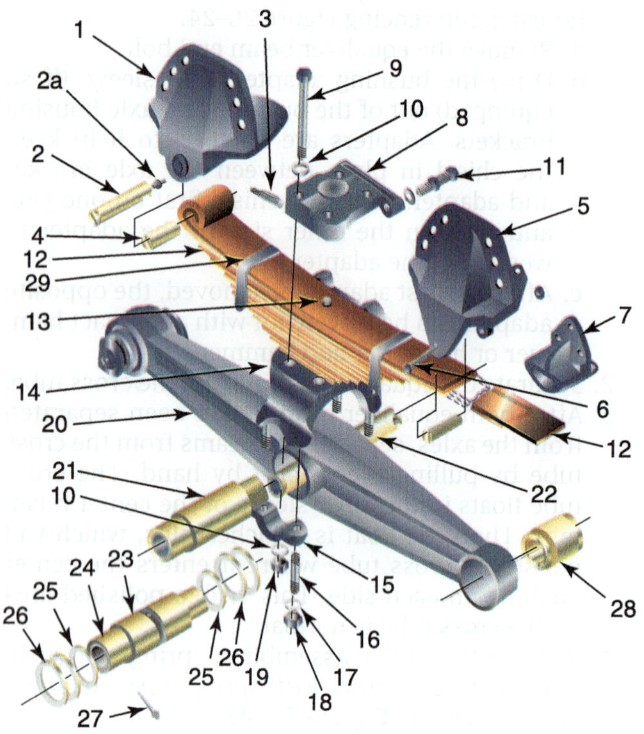

FIGURE 26–28 Leaf spring-type, equalizer beam suspension

1. Spring bracket
2. Spring pin
2a. Zerk fitting
3. Draw key
4. Bushing
5. Spring bracket
6. Rebound bolt with spacer
7. Spring bracket (extended leaf only)
8. Top pad
9. Top pad bolt
10. Washer
11. Setscrew
12. Leaf spring assembly
13. Center bolt
14. Saddle assembly
15. Saddle cap
16. Stud
17. Washer
18. Nut
19. Nut
20. Equalizer beam
21. Center bushing
22. Cross tube
23. Center bushing bronze
24. Sleeve
25. Seals
26. Washer
27. Zerk fitting
28. End bushing
29. Part number

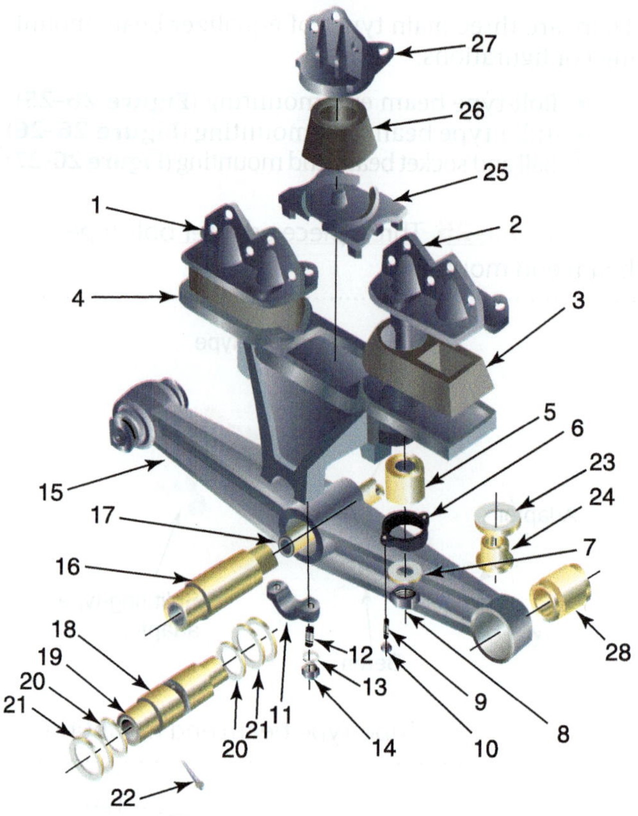

FIGURE 26–29 Solid rubber spring, equalizer beam suspension system

1. Frame hanger
2. Frame hanger
3. Rubber cushion
4. Saddle assembly
5. Drive pin bushing
6. Bushing retainer cap
7. Rebound washer
8. Drive pin nut
9. Stud
10. Nut
11. Saddle cap
12. Stud
13. Washer
14. Nut
15. Equalizer beam
16. Center bushing
17. Cross shaft
18. Center bushing bronze
19. Sleeve
20. Seal
21. Thrust washer
22. Zerk fitting
23. Rebound bumper
24. Stop nut
25. Plate
26. Rubber cushion
27. Frame hanger
28. End bushing

c. Back off the nut sufficiently to protect the draw key threads and then strike the nut with a soft hammer to loosen the draw key. Remove the draw key and drive the spring pin out of the spring and spring bracket. Lower the spring assembly from the frame and remove it from the jack.

5. Reference the *solid rubber spring* or *rubber cushion-type* equalizer beam (shown in Figure 26–29).
   a. Remove the four vertical drive pin bushing retainer caps (item 6 in Figure 26–29).
   b. Remove the drive pin nuts (item 8 in Figure 26–29). Separate the saddle (item 4 in Figure 26–29) and load cushions from the vertical drive pins. In extreme cases, it might be necessary to cut the drive pin nuts to remove them.
   c. Replace the drive pin bushings if required.
6. Remove the torque rods from the cross-member frame bracket.
7. Assemble or install new spring, cushion, or insulator and saddle assemblies. Assemble

the *leaf spring-type* equalizer beam (shown in Figure 26–28).

a. Seat the spring in the spring saddle with the head of the spring center bolt positioned in the hole provided in the saddle.
b. Position the spring top pad over the cup on the main spring leaf.
c. Install the saddle-to-top pad bolts and nuts. Run the nuts up snug but do not tighten completely at this time.
d. To properly position the spring in the saddle, tighten the spring aligning setscrews to the OEM-specified torque. Tighten the aligning screw locknuts.
e. Tighten saddle-to-top pad bolt nuts to the OEM-specified torque.
f. Using a roller jack, position the spring (with saddle) in the front and rear spring mounting brackets.
g. Align the spring eye with the spring pin bore in the front bracket. Install the spring pin, aligning the draw key slot in the pin with the draw key bore in the bracket.
h. Install the draw key, lockwasher, and nut. Tighten the nut to the OEM-specified torque.
i. Install the zerk fitting in the spring pin and grease.

8. Assemble a *solid rubber spring* equalizer beam (shown in Figure 26–29).
   a. With the drive pin bushings assembled to the saddle and the bearing caps installed, assemble or install the rubber load cushion on the saddle.
   b. Apply a light coat of multipurpose grease to the drive pins on the frame brackets.
   c. Assemble the frame brackets to the saddle by inserting the drive pins through the load cushions and the drive pin bushings.
   d. Install drive pin nuts and washers. Washers are installed with the flange down. Tighten the nuts snug. If the frame hangers were removed, tighten the nuts to the specified torque after the hangers have been assembled to the chassis frame.
   e. Assemble the preassembled saddles, load cushions, and frame hangers to the chassis frame. Install the mounting bolts, lockwashers, and nuts.
   f. Tighten the mounting bolt nuts to the specified torque.
   g. Tighten the drive pin nuts to the specified torque.

Cross Shafts

To install equalizer beam cross shafts or tubes (see Figure 26–28 and Figure 26–29), you should use the following assembly/installation procedure. When assembling the suspension, the equalizer beam, mounting nuts, and the equalizer beam saddle cap, mounting nuts should have final torque applied with the wheels of the vehicle on the ground.

a. Align the beam end bushings with the hanger brackets and insert bushing adapters. Lubricate the adapters with a light coat of multipurpose grease. The flat surface on adapters should be in the vertical position.
b. Tap the adapters into the bushing until the adapter flanges contact the outer faces of the hanger brackets.
c. Install the bolt and locknut. Do not tighten the nut at this time.

CAUTION

When reinstalling rubber-bushed equalizing beams to the axles and saddles, do not tighten the nuts until the torque rods are installed and the equalizing beams are level with the frame. This ensures that both the suspension and rubber bushing are in a neutral state, reducing the potential for creating unwanted torsional windup, which could affect ride quality. Then tighten all nuts to recommended torque.

Final Assembly

1. Place one end of the cross tube into the center sleeve of the installed equalizer beam. Ensure that the tube seats into the sleeve.
2. Assemble the other equalizer beam to the cross tube and position the beam in the spring saddle.
3. Roll the axle assemblies with the equalizer beams under the center of the saddle. Ensure that the outer bushings line up with the center of the saddle legs.
4. Lower the vehicle frame centering saddles on the beam end bushings.
5. Install the saddle cap and nuts. Do not tighten until the torque rods are installed and the equalizer beams are level with the frame. Torque the nuts to the recommended torque.
6. The cross center shaft is free to rotate or float in the bushing assemblies at each side. The function of the cross shaft or tube is to ensure alignment between the left and right equalizer beams.
7. The saddle caps secure the sleeve with the bushing. The clamping of the caps against the sleeve and bushing has no effect on the movement of the cross shaft.
8. Install the torque rods to the axle brackets and the frame brackets. When tightening, rap the bracket with the hammer to drive the taper of the torque rod stud into the bracket. If the ends of torque rods have straddle mount ends (two holes), as shown in **Figure 26–30**, use spacers between the bracket and cross-member for proper axle setting. Do not exceed ½ inch (12.7 mm) when adjusting. Tighten the shaft nuts to specified torque.

FIGURE 26–30 Torque rod disassembly

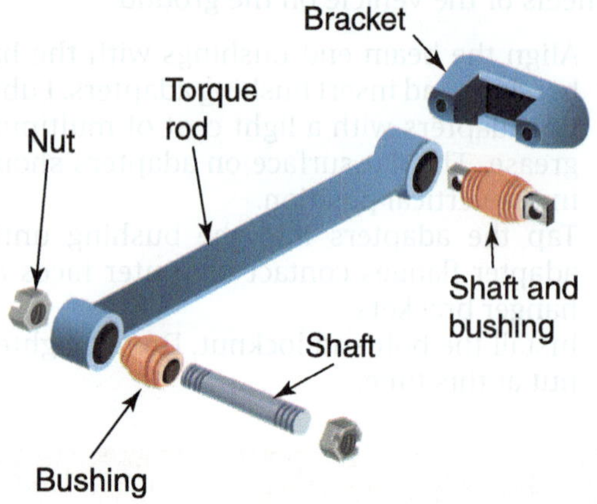

9. Tighten the equalizer beam end fittings to the OEM-specified torque.
10. Install the axle shafts. Connect the propeller shafts, brake lines, and interaxle lockout air lines.
11. On vehicles equipped with air brakes, check the brake piping for air leaks.

Summary of Equalizing Beam Assembly

The materials used in equalizer beam manufacture are aluminum, cast steel, and nodular cast iron, and each has different service characteristics. Aluminum and cast steel equalizer beams are manufactured in such a way that they can be stressed in either direction.

Nodular cast iron equalizer beams are manufactured in such a way that they can be stressed in one direction only. It is, therefore, critical that nodular iron equalizer beams be installed with the correct side up. To identify nodular cast iron beams, an arrow and the word "UP" is usually cast on the side of the beam. Also, reinforcing gate pads are usually designed into each end and middle of the top side of nodular iron beams.

26.6 SERVICING AIR SUSPENSIONS

Air suspension systems tend to require a little more attention in inspection and preventive maintenance schedules than mechanical suspensions. Some of the commonly required service procedures are outlined in this section, including repairs and height control valve adjustment.

AIR SUSPENSION INSPECTION

A daily check of the air suspension should be incorporated into the driver pretrip inspection. The truck or trailer should be observed to be level and riding at the correct height in both loaded and unloaded conditions. Any broken or loose components should be noted on the checklist.

Monthly or B schedule service should include inspections of clearances around the air springs, tires (these provide good evidence of suspension problems), shock absorbers, and other moving parts in the suspension system. When air spring bellows make contact with other chassis components, the problem almost always does not originate in the air bag assembly.

At each quarterly or C schedule service, all welds should be inspected, and the frame attachment joints and fasteners, cross-members, and pivot connections visually inspected for wear, cracks, and corrosion. U-bolt locknuts should be retorqued every 100,000 miles (160,000 km).

Air springs should last almost indefinitely in most applications. However, they fail rapidly if rubbed or scuffed and are easily punctured by contact with sharp objects. When an air spring fails, the chassis weight settles onto the pedestal jounce blocks, which have almost no ability to buffer road forces when used in this way. In the event of air spring failure, drivers should be instructed to proceed to the closest service facility, exercising some caution. Try to determine what caused the air spring to fail.

Air springs can be pulled apart if their stroke is not limited because they are not designed to be stretched in this way. Some shock absorbers used on air suspensions are designed to limit rebound travel, so, when replacing them, do not lift the vehicle without the shock absorbers in place.

Many types of air controls are available with air suspension systems. The most common system automatically regulates frame height by controlling air supplied to the air springs. When used in conjunction with other types of suspensions, such as leaf spring suspension, an operator-controlled pressure regulator is often used. It is important to drain or purge the air suspension system periodically to prevent the buildup of moisture and contaminants.

CHECKING A RIDE HEIGHT CONTROL VALVE

Ride height control valves (also known as leveling valves) are too often improperly diagnosed as defective and unnecessarily replaced. A simple diagnostic routine can determine whether the valve is defective or whether proper adjustment has not been performed. The diagnostic routine outlined here references one type of height control valve but the procedure is similar to that used on equivalent types. To verify the performance of a height control valve, perform the following service checks:

1. Remove the bolt that attaches the height control link to the height control valve lever.

2. Ensure that air reservoir pressure is above 100 psi (690 kPa).
3. Raise the height control valve lever 45 degrees above horizontal. Air pressure at the air springs should begin to increase within 15 seconds, which will cause the truck frame to rise.

CAUTION

Most ride height control valves have a reaction delay that can be as long as 15 seconds. This is used to prevent continuous correction cycling. Remember this when diagnosing height control valve problems.

4. Lower the height control valve lever 45 degrees below horizontal. Air pressure at the air springs should begin to decrease within 15 seconds. Let air escape until air spring height is approximately 10 inches (25 cm).
5. Raise the valve lever to 45 degrees above horizontal again, until air spring height is approximately 12.5 inches (32 cm). Release valve lever.
6. Check the valve body, all tubing connections, and air springs for leaks with soapy water.
7. Check air spring height 15 minutes after performing step 5.
8. If no leaks are found, the valve functions as just described, and air spring height has not changed in 15 minutes, the valve is functioning properly.
9. Install the bolt that attaches the height control lever to the height control link. Torque nut to specification.

RIDE HEIGHT ADJUSTMENT

Some versions of height control valves have a centering pin and bosses. The pin is positioned in the bosses after setting the height. To adjust the height control valve, do the following:

1. Remove the plastic center pin from the bosses, as indicated in **Figure 26–31**.
2. Ensure that at least 100 psi (690 kPa) air pressure is indicated on the dash air pressure gauges of the truck.
3. Disconnect the vertical link from the height control valve lever.
4. The height adjustment is checked at the rearmost axle. Place a straightedge on the centerline of the top metal plate on the air springs, as shown in **Figure 26–32**. Do not position the straightedge on the frame rails. Measure the distance from the bottom edge of the straightedge to the top edge of the transverse cross-member. Reference the OEM service literature for the correct specification distance.
5. If the distance is less than specified, raise the valve control lever and hold it up until the distance is correct. Release the lever. Loosen the lever adjusting nut, which is located 1⅞ inch (4.75 cm) from the lever pivot point, so that the steel portion of the lever moves but the plastic portion does not. Position the lever so that the link-to-lever bolt can be inserted through the lever and link.
6. If the distance is more than specified, lower the valve lever and hold it down until the distance is correct. Release the lever. Loosen the lever adjusting nut. Position the lever and link so that the lever bolt can be inserted through the lever and link.
7. Install the lever link bolt nut. Tighten the nut 115 to 130 inch-pounds (13 to 15 N·m).
8. Center the plastic portion of the valve lever and insert the alignment pin through the boss on the valve body and plastic arm.
9. Tighten the lever adjusting nut to lock the steel and plastic portions of the lever together; then tighten to specification.
10. Remove the centering pin (if equipped) and discard.

The procedure for adjusting trailer ride height is covered in **Photo Sequence 9**.

FIGURE 26–31 Height control valve centering pin

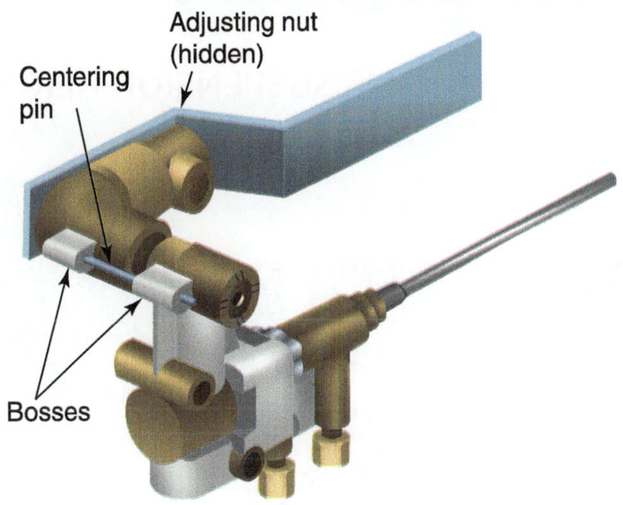

FIGURE 26–32 Ride height adjustment

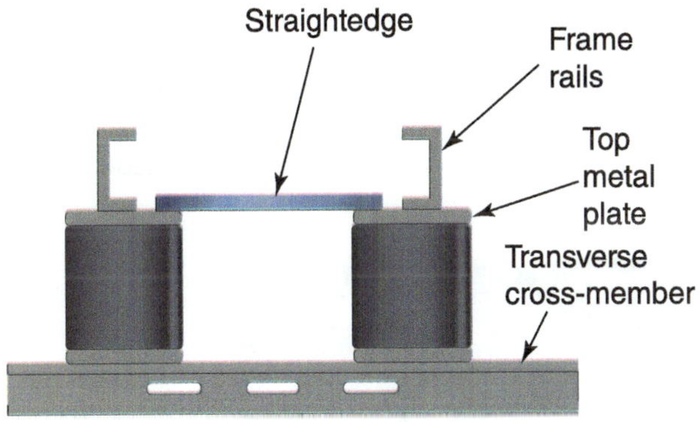

Photo Sequence 9

AIR SUSPENSION HEIGHT CONTROL ADJUSTMENT

P 9–1 Begin by blocking a set of wheels on each axle of the vehicle as shown. A tractor (used in this procedure) should be uncoupled and a truck or trailer, unloaded, to perform this check.

P 9–2 Throughout the suspension check procedure outlined here, system air pressure must be above 90 psi (620 kPa); this truck must be run to raise system pressure to its cutout value before performing the suspension check procedure.

P 9–3 A tractor air suspension showing the height control valve and linkage. NOTE: Modern air ride height control valves seldom require adjustment during their operational life so this procedure focuses on checking ride height.

P 9–4 Referencing the OEM orientation points, use a tape measure to check the ride height dimension. Some systems require the use of a tram gauge or wood blocks to make this measurement.

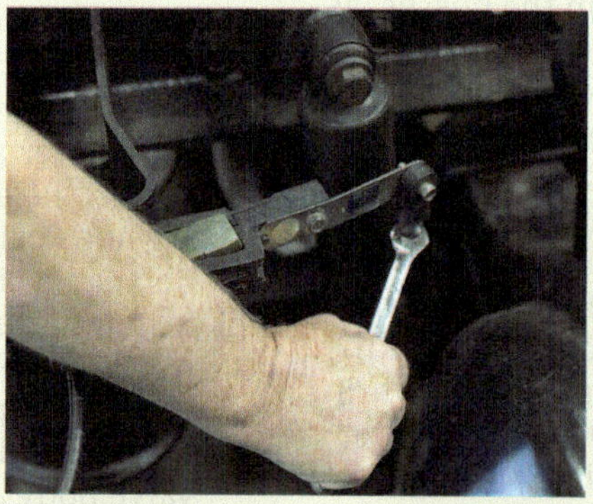

P 9–5 To test ride height valve operation, separate the actuating arm from the control linkage; doing this will allow you to move the actuating arm up or down without interfering with the ride height set point.

P 9–6 Moving the actuating arm of the height control valve downward exhausts the air from the system. Do this until all air is discharged by the air spring(s) supplied by the valve; note that on some systems there is a response lag.

P 9–7 Ensure that chassis system pressure is above 90 psi (620 kPa), then raise the ride height control valve actuating lever. This will charge the air springs to raise the frame above specified height; note that on some systems there is a response lag.

P 9–8 Level the actuating arm and re-connect the linkage. The ride height should reset to specification. Should any adjustment be required, consult the OEM service literature; the adjustment location on this height control valve is show in the above image.

AIR SPRING REPLACEMENT

The procedure for replacing air springs differs slightly depending on whether the bag is of the convoluted type or reversible bellows type. The description provided here references the more common reversible bellows-type air spring, as shown in **Figure 26–33**.

1. Chock the wheels and exhaust the air pressure from the air spring.
2. Raise the truck frame to remove the load from the suspension. Support the frame with safety stands.
3. Remove the locknuts and washers that connect the air spring to the air spring mounting bracket (**Figure 26–34**).
4. Remove the air lines connected to the air spring.
5. Remove the brass air fittings from the air spring.
6. Remove the locknuts and washers that connect the air spring to the frame hanger and then those that attach the pedestal to the saddle. Remove the air spring from its upper and lower mountings.
7. Install the replacement air spring to the frame hanger by inserting the studs on the air spring into the appropriate holes on the hanger bracket.
8. Slide the air spring pedestal onto the spring mounting saddle. You can inflate the bag using shop air to extend the spring if required.
9. Install the locknuts and washers to secure the air spring to the bracket. If the pedestal is aluminum and uses bolt fasteners, lightly paste some Aluminastic™ compound or equivalent on the threads to prevent reaction. Tighten and torque the locknuts.

FIGURE 26–33 Reversible sleeve air spring upper mounting

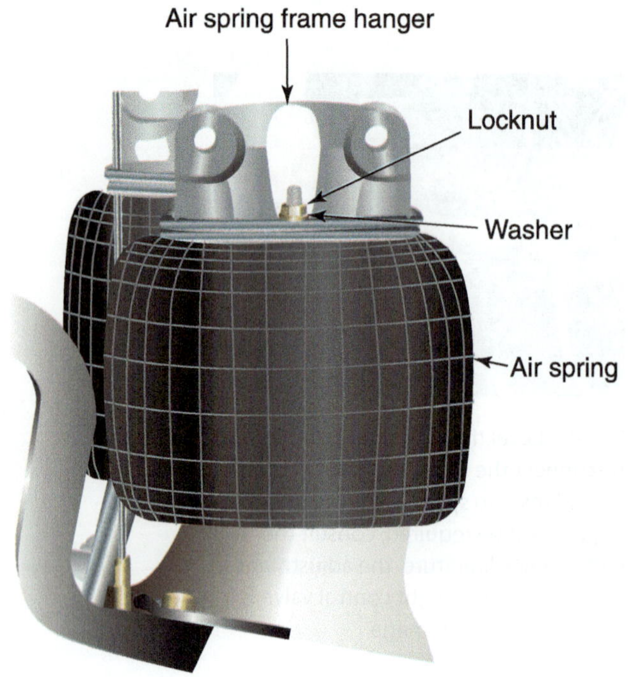

10. Install the washers and locknuts to secure the air spring to the frame hanger. Tighten and torque the locknuts.
11. Install the brass air fitting to the air spring using a thread sealant.

FIGURE 26–34 Air spring replacement

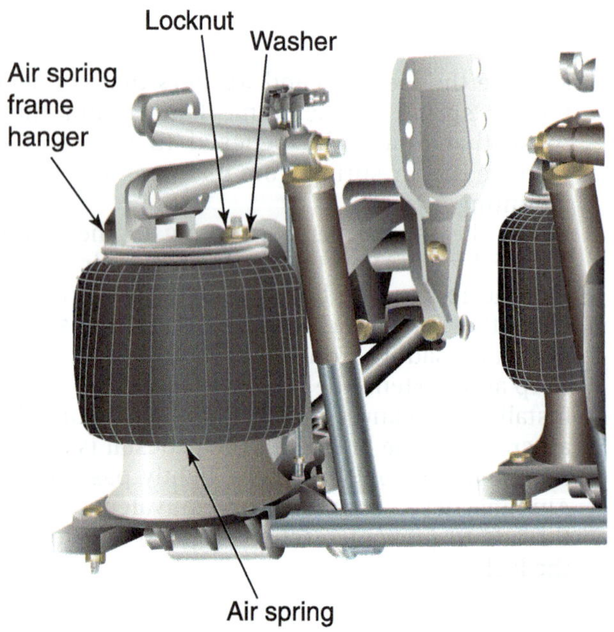

12. Install the air lines to the brass fitting.
13. Supply air pressure to the air spring. Adjust the height of the truck or trailer to specification.

SHOCK ABSORBER REPLACEMENT

Shock absorbers probably require more attention than any other components on an air suspension. They are worked extensively because air springs have limited self-dampening capability but it is easy to exceed their performance capacity if the axle end is out of balance. Make sure that the bushings are replaced along with the shocks and remember that defective bushings can cause premature shock absorber failure. To replace a defective shock absorber, refer to **Figure 26–35** and observe the following procedure:

1. Remove the locknut and washer that connect the shock absorber to the shock absorber frame hanger. Then remove the washer and locknut that connect the shock absorber to the air spring mounting bracket.
2. Pull the shock absorber mounting eye bushings and separate it from its mounting studs.
3. Check the removed shock absorber stroke dimensions and ensure that the replacement shock is compatible. Install the replacement shock absorber along with a new set of bushings to the studs on the hanger and the bracket.
4. Install the washers and locknuts to the studs. Tighten and torque all locknuts.

FIGURE 26–35 Shock absorber location on combination air/steel spring suspension

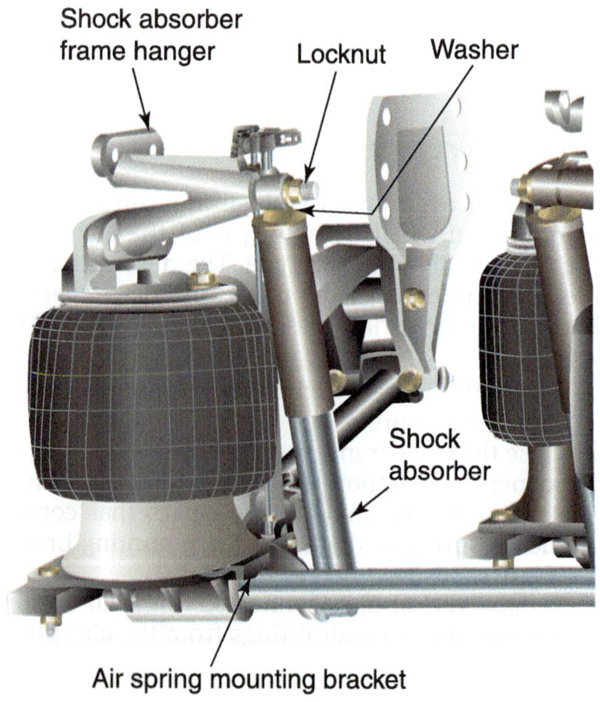

REPLACING A HEIGHT CONTROL VALVE

Make sure that you use the procedure outlined previously in this chapter to performance check the height control valve before replacing it. One OEM reports that 50 percent of height control valves submitted for warranty function to specification. Reference **Figure 26–36** and use the following procedure to replace a defective leveling valve:

1. Exhaust air pressure from the truck chassis air system. Then disconnect the air lines from the height control valve.
2. Remove the height control valve threaded extension link from the valve and axle by removing the fasteners.
3. Remove the mounting fasteners that connect the height control valve to the truck frame.
4. Disconnect the brass air fittings from the height control leveling valve and install them in the replacement valve.
5. Secure the new height control valve to the truck frame, torquing the fasteners to specification. Reattach the air lines to the valve, making sure that each connects to the correct port.
6. Install the threaded extension rod (see Figure 26–36) to the height control valve using the appropriate fasteners.
7. Build up chassis system air pressure to cut out and then adjust the height control valve for ride height using the method described earlier.

FIGURE 26–36 Leveling valve fitting and actuation components

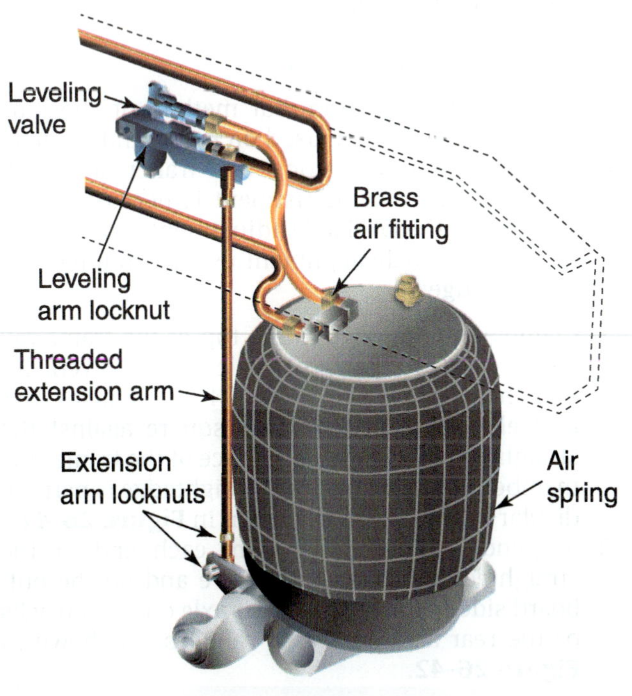

AIR CIRCUIT MAINTENANCE

Air lines route air pressure to and from the air springs, height control valve, and pressure protection/regulator valve. They should be routinely inspected and repaired when chaffed or leaking. **Figure 26–37** illustrates a typical air plumbing installation for an air suspension system and **Figure 26–38** shows a less common type of heavy-duty air-suspension system.

26.7 SUSPENSION ALIGNMENT

Alignment of a truck suspension system is closely allied to axle alignment. For a suspension to operate properly, the axles must be aligned with the vehicle tracking a true thrust line. This section should be studied in conjunction with Chapter 25.

AXLE ALIGNMENT

Several types of equipment and methods can be used to check axle alignment. These methods include light and laser beam alignment equipment and computer-controlled sensor systems. A straightedge and trammel can also be used to check axle alignment. This last method is the most difficult and inaccurate method of checking axle alignment and should be used only when no other alignment equipment is available.

Before performing axle alignment checks or making torque rod adjustments, remove road debris from the suspension system and perform the following preliminary procedures:

1. Park the vehicle on a level surface, having first neutralized the suspension by moving backward and forward in a straight line. Block at least a couple of wheels.
2. Release the parking brake.
3. Check the rear wheels for runout and correct if necessary.

Laser Alignment

Laser alignment equipment is the preferred method of checking axle alignment on both tractors and trailers because of its greater accuracy. **Figure 26–39** shows a typical laser alignment setup connected to a tandem-axle tractor. Check both the equipment and truck OEM service literature for the recommended operating procedures.

Laser aligners use a soft laser light projected from the center of the rear frame gauge target to the front frame gauge target. The self-centering frame gauge targets are used to locate the chassis centerline. The rear axle is in precise alignment when the beam of light is centered between the front and rear frame gauge targets. If the beam of light is not centered, the rear axle has to be adjusted. When the rear axle is properly aligned, the position of the forward rear

FIGURE 26–37 Air plumbing diagram

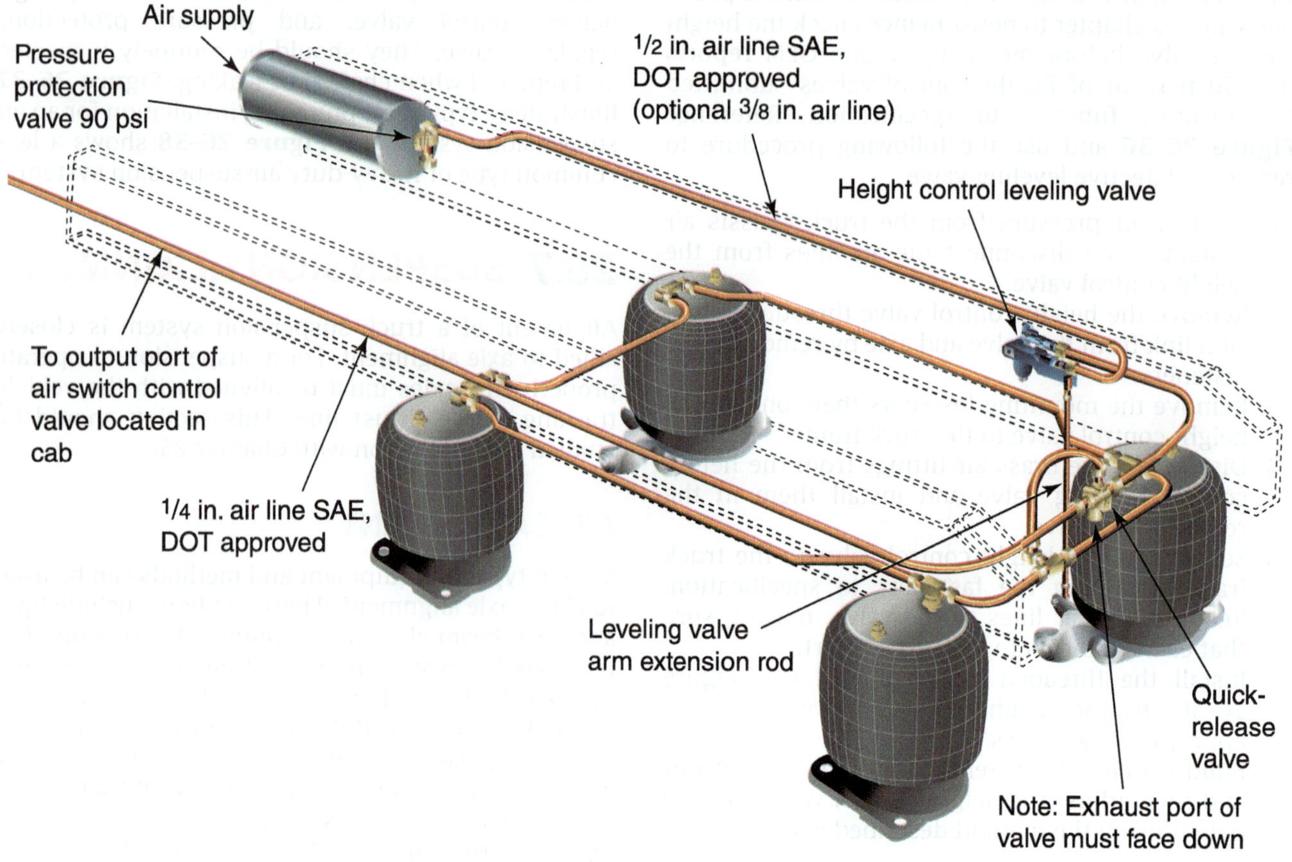

axle can be checked using a tram bar, as shown in **Figure 26–40**.

The forward rear axle can then be adjusted until it is exactly parallel with the rear axle. The alignment procedures are the same for a trailer, except that an extender target is clamped to the kingpin. The extender target ensures that the rear axle is aligned with the center of the kingpin.

FIGURE 26–38 A heavy-duty air spring suspension

Straightedge and Tram Method

The straightedge and tram bar method of checking and aligning axles can be used on most tandem drive suspensions, air suspensions, and trailer bogies. In the event that an axle adjustment is necessary, you should refer to the OEM axle adjustment procedure.

To check out axle alignment using a trammel bar and straightedge:

1. Clamp a straightedge to the top of the frame rail ahead of the forward rear axle on 6 × 4 vehicles and ahead of the rear axle on 4 × 2 vehicles. Use a steel right-angled framing square against the straightedge and outside surface of the frame side member to ensure that the straightedge is perpendicular to the frame, as shown in **Figure 26–41**.
2. Suspend a plumb bob from each end of the straightedge in front of the tire and on the outboard side of the forward rear axle on 6 × 4 trucks or the rear axle on 4 × 2 vehicles, as shown in **Figure 26–42**.

FIGURE 26–39 Typical laser aligner setup

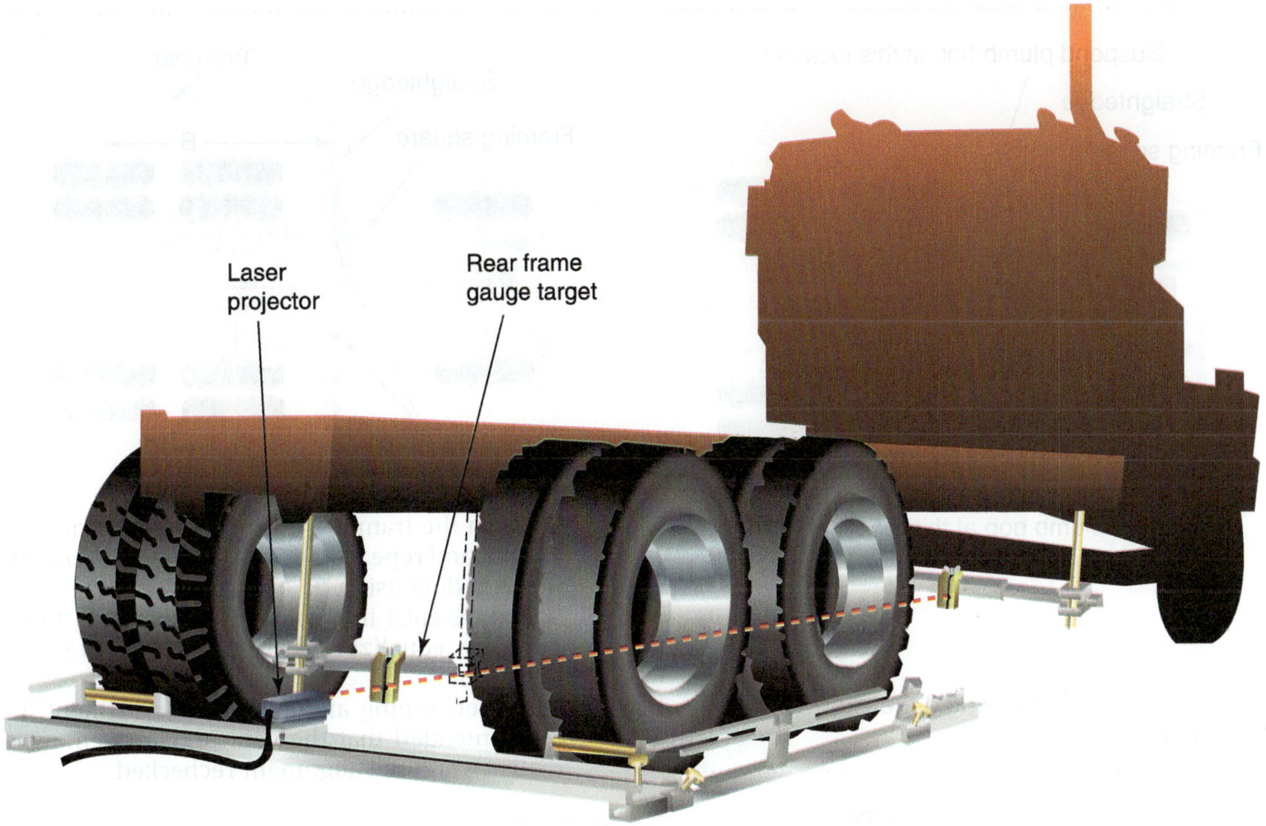

FIGURE 26–40 Tram bar installation to measure axle spread

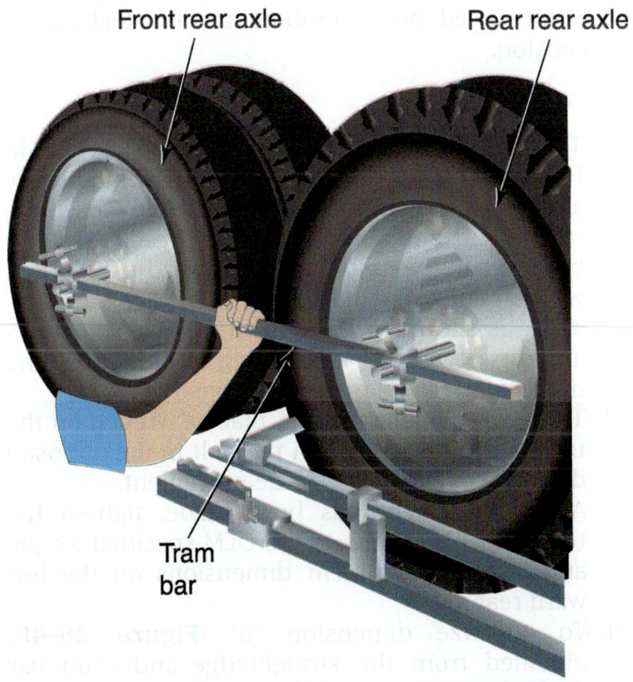

FIGURE 26–41 Straightedge location on the frame

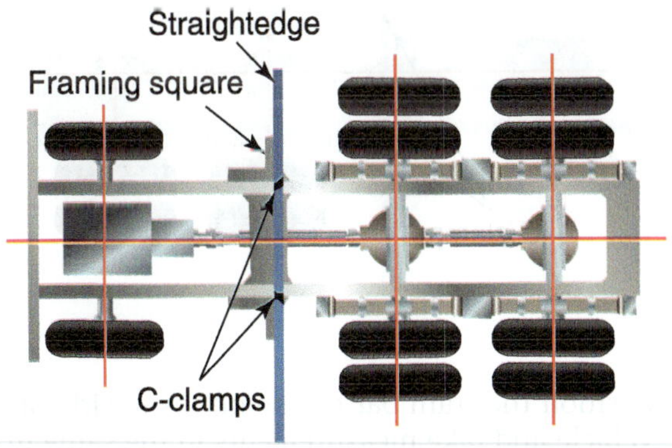

3. Position an adjustable tram bar so that the two pointers engage to the center holes in the drive axle flanges of both rear axles, as shown in **Figure 26–43**.
4. Measure the distance between the cord of the plumb bob and the pointer on the forward rear axle of 6 × 4 vehicles or the rear axle on 4 × 2 vehicles. Record this measurement as dimension "A," as indicated in Figure 26–43.

FIGURE 26–42 Installation and location of plumb bob

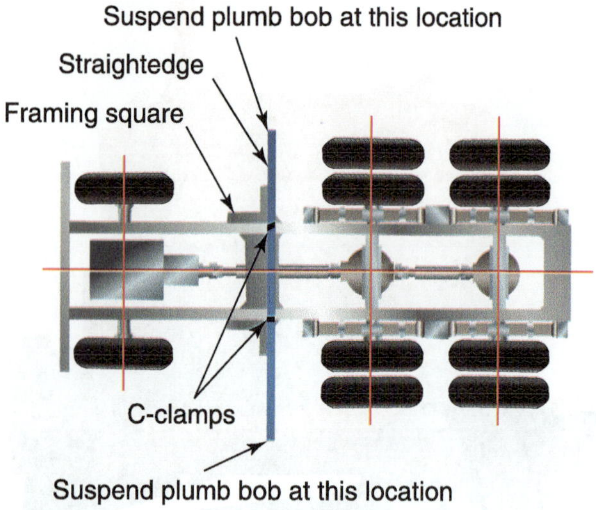

FIGURE 26–43 Measuring dimension "A" axle alignment

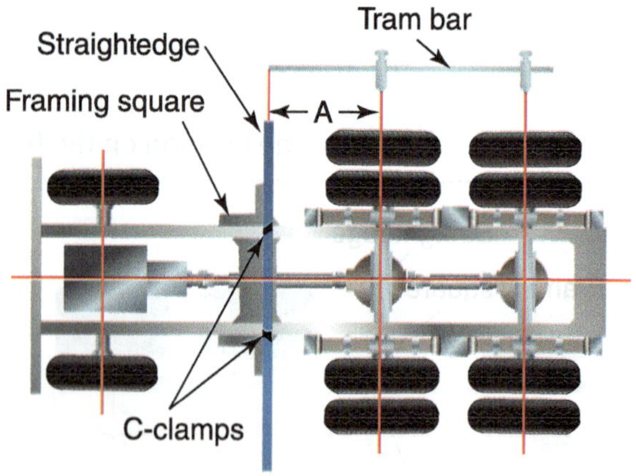

5. Position the tram bar on the opposite side of the vehicle and take measurements in the same manner as indicated in step 4. Record this measurement as dimension "A" also.
6. Any difference in dimensions from side to side in excess of 0.125 inch (3.175 mm) must be equalized.
7. Again, position the adjustable tram bar so that the two pointers engage to the center holes in the drive axle flanges of both rear axles, as shown in **Figure 26–44**. Measure the distance between the plumb bob and the pointer on the rear axle and record this measurement as dimension "B."

FIGURE 26–44 Measuring dimension "B" axle alignment

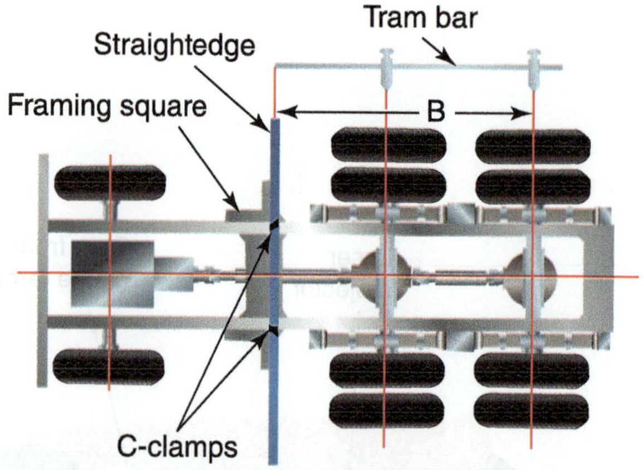

8. Position the tram bar on the opposite side of the vehicle and repeat step 7, recording this measurement as dimension "B" also.
9. Any difference in dimensions from side to side must be equalized if in excess of 0.125 inch (3.175 mm).
10. After performing an alignment adjustment, it is recommended that the vehicle be driven a short distance and the alignment rechecked.

Axle Adjustments

Adjustment of the axle location for alignment purposes varies with the type of suspension used on a truck or trailer. For purposes of this description, we will use the example of a suspension that uses eccentric bushings located at the torque leaf mounting eyes to make an adjustment, followed by the adjustment procedure used on a combination air/steel spring suspension.

Eccentric Bushing.

1. To equalize dimension "A" (as shown in Figure 26–43), measure using the straightedge and tram bar alignment procedure and loosen the torque leaf bolt nuts enough to free the rubber bushings in the castings. Mark the adjusting bolt, washer, and front hanger with chalk or a paintstick, as shown in **Figure 26–45**. This provides a means of visually identifying eccentric bushing movement as you make the adjustment.
2. To adjust the axle setting, place a wrench on the torque leaf bolt and turn the bolt in the opposite direction of the desired axle movement.
3. After adjustment has been made, tighten the torque leaf bolt nuts to the OEM-specified torque and recheck alignment dimensions on the forward rear axle.
4. To equalize dimension "B" (**Figure 26–46**) obtained from the straightedge and tram bar

alignment procedure, loosen the torque leaf bolt nut at the equalizer bracket enough to free the rubber bushings in the castings. Again, mark the adjusting bolt and equalizer bracket with chalk or a paintstick so you can visually identify the eccentric bushing movement while making the adjustment.

FIGURE 26–45 Identifying eccentric bushing movement for dimension A

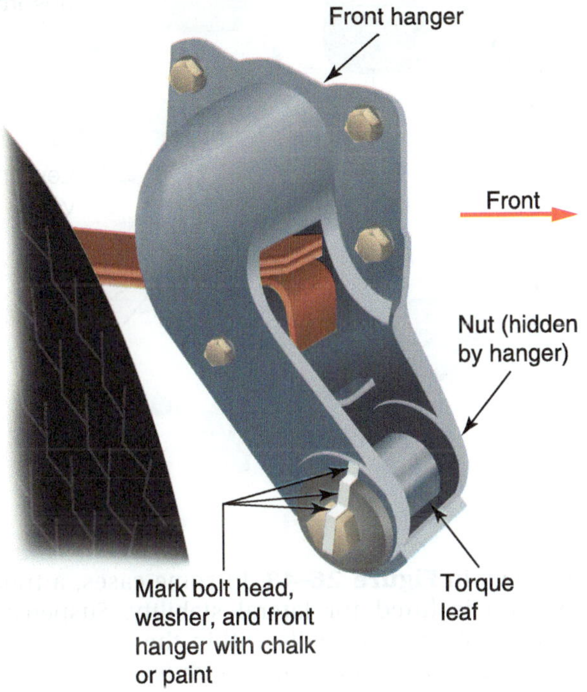

FIGURE 26–46 Identifying eccentric bushing movement for dimension B

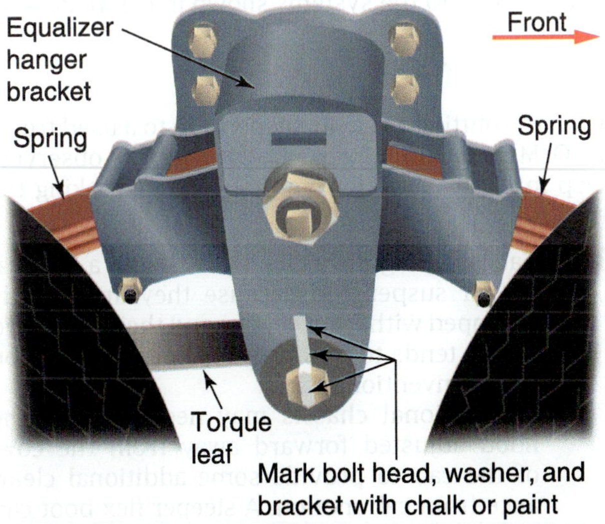

5. To adjust the axle, place a wrench on the torque leaf bolt and turn the bolt in the opposite direction of the desired axle movement.
6. After the adjustment has been made, tighten the torque leaf bolt nuts to the OEM-specified torque and recheck the alignment dimension on the rear axle.

Torque Rods on Combination Air/Steel Spring. Adjustment of the axle alignment in this common combination air/steel spring suspension is accomplished by means of shims installed at the torque rod end mounting or threaded rod ends.

1. To equalize dimension "A" using the straightedge and tram bar alignment procedure shown in Figure 26–42, install or remove shims at the forward rear axle torque rod end mount, as shown in **Figure 26–47**. Installation of shims at the torque rod forward end mount moves the axle forward. Installation of shims at the torque rod axle end mounting moves the axle rearward. Tighten the torque rod mounting nuts to the specified torque and check alignment dimensions on the forward rear axle. Make sure that the suspension is in a relaxed state before checking the measurement after each adjustment.
2. Equalize dimension "B" (6 × 4 vehicles only) using the straightedge and tram bar alignment procedure (see Figure 26–44 by installing or removing shims at the rear axle torque rod end mounting (see Figure 26–47). Installation of shims at the forward end mounting moves the axle forward. Installation of shims at the torque rod axle end mounting moves the axle rearward. Tighten torque rod mounting nuts to specified torque and check alignment dimensions on the rear axle.
3. Following the installation of the alignment shims, inspect the torque rod mounting nuts. A minimum of two threads of the mounting bolt must extend through the nut to allow the mounting nut locking feature to function properly.

FIGURE 26–47 Location of alignment shims

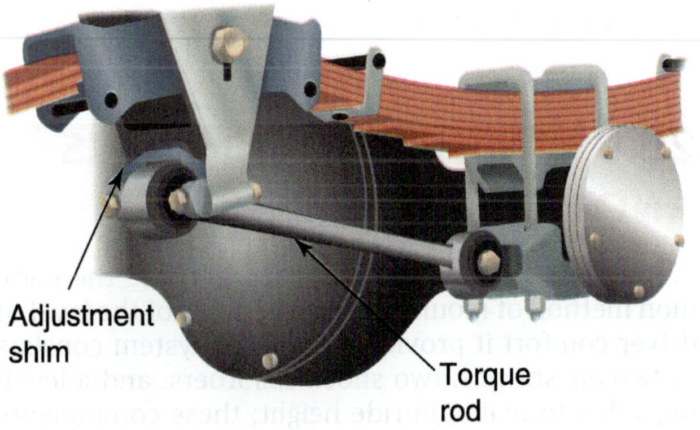

4. When performing axle alignment, it is preferred to have alignment shims installed at the torque rod forward mounting. However, if axle alignment cannot be obtained by installing shims at the forward mounting, it is permissible to install additional shims at the axle end mounting.

26.8 ROLL STABILITY CONTROL

Roll stability control (RSC) is a simple add-on to an antilock brake/automatic traction control (ABS/ATC) system, requiring the addition of one simple pilot valve. The system is designed to aid vehicle stability and limit rollover conditions. It uses controller electronics that are either integral with an ABS/ATC control unit or multiplexed directly to the chassis data bus. This electronic control unit (ECU) can both pick up and broadcast data over the bus. The system is available on straight trucks, tractors, and semitrailers.

The RSC controller monitors the vehicle center of gravity, monitors wheel speed, and continuously plots the lateral acceleration limit (rollover threshold). A common feature is virtual gyroscope logic such as that used in a phone. When a likely rollover condition is sensed, the system can broadcast commands over the chassis data bus that can reduce engine torque, engage engine/driveline retarders, apply pressure to the drive axle brakes, and modulate pressure to the trailer brakes.

RSC electronics make accident avoidance maneuvers much safer in trucks. Accident avoidance maneuvering requires the steering and suspension system to react almost instantly to a series of high-speed inputs without any loss of control, which is a tough task in a heavily loaded transport truck and trailer. The RSC system is not unlike the directional stability controls used in many automobiles today. When wheel speeds on both sides of an axle are monitored, chassis electronics can detect even fractional degrees of yaw and then broadcast corrective action to the other electronic subsystems connected to the data bus. This system is becoming used extensively in the trucking industry, especially in applications with a high center of gravity.

26.9 CAB AIR SUSPENSIONS AND DRIVER SEATS

Cab air suspension systems have become the common method of mounting a cab because of the level of driver comfort it provides. A typical system consists of two air springs, two shock absorbers, and a leveling valve to maintain ride height; these components are shown in **Figure 26–48**. In some cases, a transverse rod is fitted for lateral stability. Suspension travel is not great—usually within the range of 1 to 3 inches (25 to 76 mm)—but the difference to driver comfort is considerable, especially on vehicles spec'd with unforgiving, heavy-duty suspensions.

FIGURE 26–48 Exploded view of two-point cab air suspension

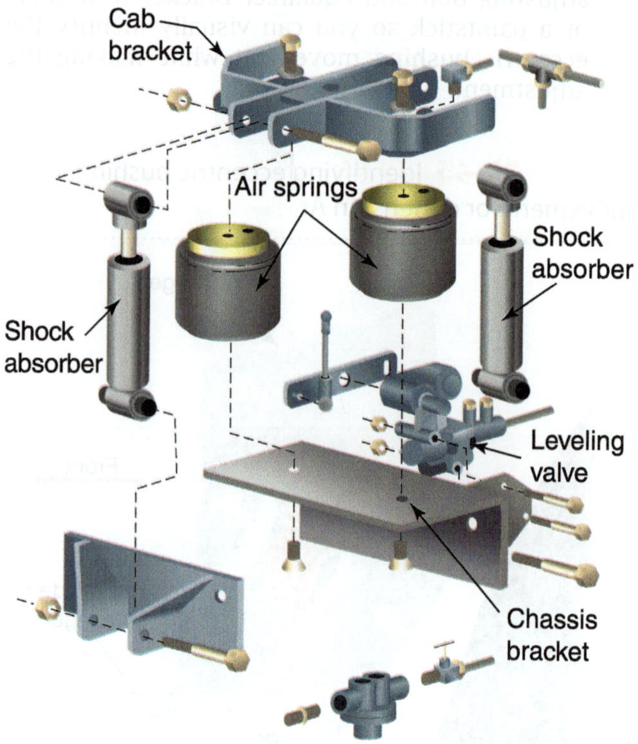

Because no two truck cabs are alike, there are many variations in size and mounting brackets. Some systems use single-point mounts, featuring a single air bag and shock absorber, but the majority are dual-point, similar to the systems shown in Figure 26–48.

RETROFITTING

When retrofitting a cab air suspension to a used truck, the OEM installation instructions should be observed. Keep in mind the following points when making the installation:

- Cab-over-engine (COE) chassis are a natural for air suspension because they are already equipped with front pivots, and their increased height tends to produce more pitching motion than conventional cabs.
- Conventional chassis may need to have the hood adjusted forward away from the cowl of the cab to provide some additional clearance from cab motion. A sleeper flex boot can

FIGURE 26–49 Air-suspended driver's seat

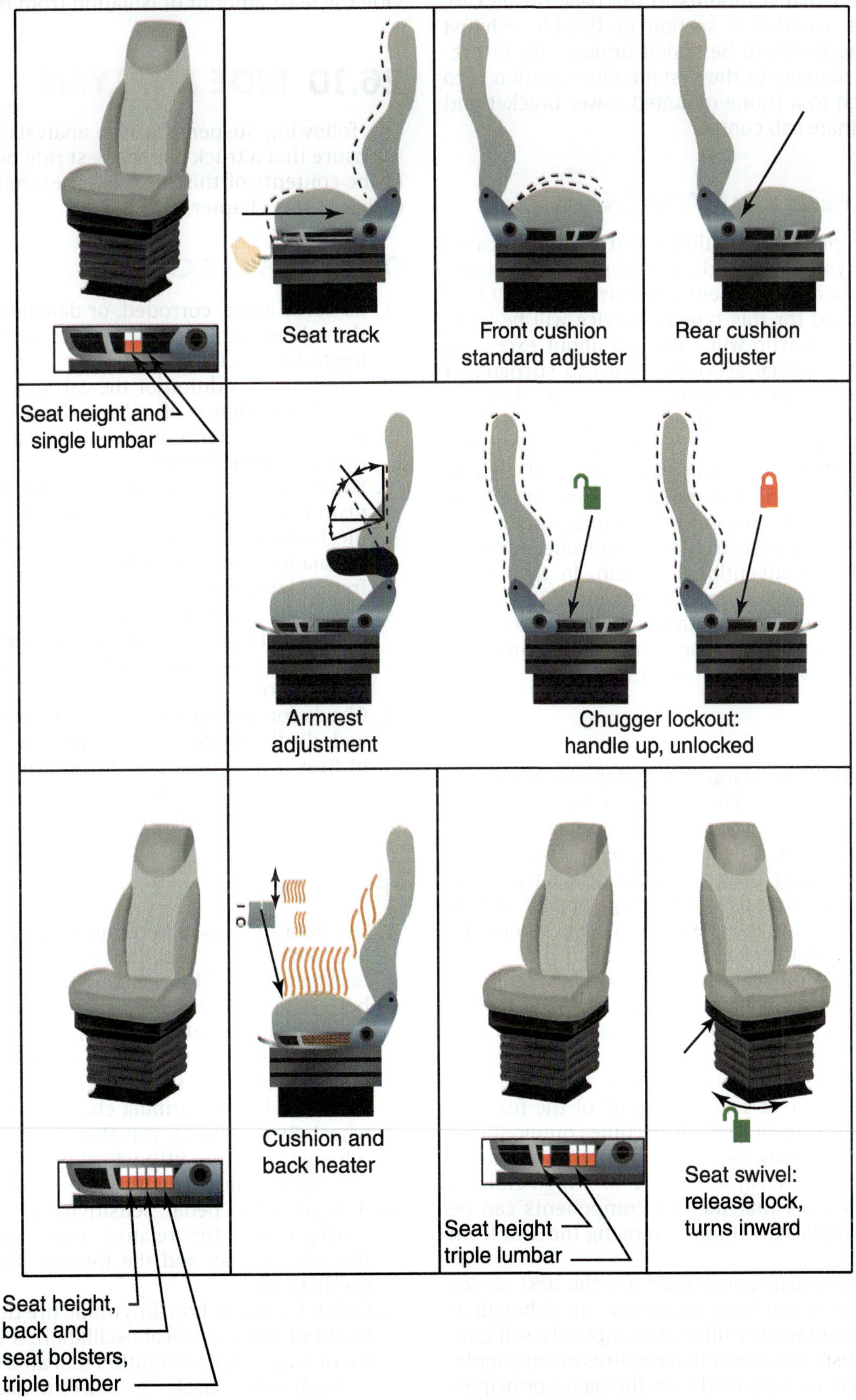

restrict motion somewhat but should have enough flexibility not to tear.

- If the exhaust mounts to the back of the cab, check whether a section of flexible exhaust piping needs to be added underneath to prevent damage to the system. One solution is to switch to a frame-mounted tower bracket and eliminate cab contact.

CAB SUSPENSION MAINTENANCE

Moving components should be routinely lubricated and the air system checked for oil admission. Periodic shock absorber replacement is required on most systems. The need for this type of service will be identified by drivers who will complain about excessive bouncing. The actual service life of cab suspension shock absorber varies with the application. They can outlast the vehicle in a linehaul application.

Electronically managed active cab suspensions that are capable of reading vibrations and activating solenoids to cancel or minimize them have been introduced by cab suspension manufacturers. It is also possible to use electronically controlled air valving on a more conventional system. In these new technology systems, a cab suspension ECU can be networked to the chassis data bus and coordinate cab ride improvement depending on the terrain and conditions that the vehicle encounters.

DRIVER SEATS

Driver comfort is to a large extent dependent on having a comfortable seat. There are three types: mechanically suspended, pneumatically (air) suspended, and solid mounted. A fleet using the latter may have problems retaining drivers. Mechanical suspension seats use a mechanical, free-moving support system to somewhat isolate the driver from suspension jolts transferred from the chassis to the cab.

Air suspension seats do a better job of isolating the driver from suspension and frame jolts. They consist of an air bag, jounce cushion, and shock absorber assembly. The air bag is charged with compressed air from the vehicle pneumatic circuit. It shows a typical air suspended driver seat with some of the features available. This type of seat has become commonplace on trucks used today, especially by fleets wanting to retain their drivers. An advantage of a high-end, air suspended seat is that the key components can be repaired or replaced without removing the seat from the truck.

Mechanical suspension seats are the best choice if the truck does not have an air system, although at least one manufacturer offers a completely self-contained gas suspension seat that requires no air supply. A gas suspended seat works on the same principles used in many office chairs. Solid-mounted seats are the most economical, and, as the name implies, provide the least amount of isolation from the road.

26.10 RIDE ANALYSIS

The following suspension ride analysis can be used to ensure that a truck has the best ride possible. Most of the contents of this section are examined in detail earlier in this chapter.

QUICK VEHICLE CHECKS

1. Look for loose, corroded, or damaged threads on the cab mounting bolts, which may be contributing to fore-and-aft pitching.
2. Check the condition of the cab mounting insulators for evidence of collapsing or checking due to weather deterioration. Replace any insulators found to be defective.
3. Where applicable, check the cab latch mechanism for sufficient tension. Freeplay in the latch mechanism can result or contribute to chassis resonance. Also check for elongated latches or a fretted latch bar.
4. Check the condition of the driver seat. If it is in poor condition, recommend an air suspension seat replacement with vertical as well as fore-and-aft isolation.
5. Check for proper tire inflation, making sure to include the trailer tires. Under- or overinflation of tires is a common contributor to a harsh ride. After the tires have been properly inflated, check them for lateral runout.

TEST DRIVE

1. Use your mirrors to determine whether bogie hop (also known as wheel hop) is occurring. Ideally, a test drive should be made bobtailing, then with both a loaded and unloaded trailer. If the fifth wheel is adjustable, put it in the optimum position with relation to the load during the road test and then continue the test drive. You may have to consult a bridge formula chart to locate the fifth wheel. By moving it, you should be able to determine whether the fifth wheel position was either causing or influencing the ride complaint.
2. If accelerator pedal sensitivity is a suspected cause, check for relative movement between the cab, engine, and the throttle linkage if it is mechanical.
3. Check for wheel hop. Rhythmic tire deflection is a result of wheel vertical oscillation. It is an indicator of bogie hop, a condition that always produces a rough ride. Check tire/wheel dynamic balance

and obvious wheel imbalance indicators such as a fractured brake drum.
4. If you suspect a front axle shimmy, align and balance the steer axle wheels.

Because a rough ride may be the result of more than one circumstance, test drive the vehicle after each repair or alteration has been made to make certain that all problems have been corrected.

ROUGH RIDE CHECKLIST

1. Check first for any loose, worn, or damaged components.
2. Check wheel balance.
3. Check the shock absorbers throughout the chassis.
4. Check wheel and tire runout—laterally and radially. This also includes hub and drum runout.
5. Check that the rear spring shackles have a rearward slant to it and are not vertically positioned. If they are vertical, this can cause a hop problem because the spring can option moving forward or backward under load.
6. If the springs are inadequate, this can cause the vehicle to bounce. This can cause axle roll, causing steering problems. Check the vehicle for overloading or underloading. Axles and suspensions are quite often spec'd heavier than required in a given application and this can result in a rough ride that is expensive to cure.
7. Check for unmatched tires, especially on the duals. Also check for proper tire pressure.
8. Check the front-end suspension alignment, including camber and toe settings.
9. Check the cross-members. Loose cross-members can cause frame vibrations.
10. Check the driveline angle, U-joint phasing, and any looseness in hanger bearings or U-joints. Check the drive shafts between the drive axle carriers. Also, check that the drive axle short shafts are bottoming in the slip joint generating a hop when the drive axle passes over a bump. Check for worn or twisted splines on the shaft or slip yoke.
11. Check the torque rods throughout the chassis, making sure that they are secure and that the bushings are not worn out. Also, check the angle of these torque arms. If the angle is too acute, they can cause an axle housing to rotate forward and backward as the tandem oscillates, causing binding U-joints and a hop problem.
12. On the rear axle housings, check the mounting of the wheels and matching of the tire, brake drum, and hub balance.
13. Check the rear suspension bushings.
14. Check rear axle alignment and tracking.
15. On walking beam tandem suspensions, check the center shaft (cross shaft) for a bend that could cause the rear suspension to bind. Check the mounting of the saddles to the frame. Fractional movement of the saddles on either side can cause binding in the center pivots, resulting in rear suspension vibration.
16. The frame can cause a problem by flexing too much because of either inadequate cross-members or lack of strength. If a frame has too much flexibility, it can flex, causing a rhythmic vibration, which can resonate throughout the frame, making it feel like it is coming from the front wheels. If the frame has been extended and sleeved to accommodate a long dump body (grain type or gravel type), the frame itself might have enough strength with the added channel to hold the load, but because of the added distance hop problems are caused. Another problem that can be encountered when stretching out the frame and moving rear axles to the end of the frame is overloading the front axle.
17. A sliding fifth wheel positioned too far forward or too far back and can cause hops and vibrations.
18. Do not overlook the trailer, especially if the complaint is that the hop occurs only when the trailer is hitched to the tractor. A trailer with the rear axle badly out of alignment, mismatched tires, out-of-round tires, worn-out suspension systems, or broken springs can transfer vibrations to the tractor.
19. Frame misalignment can also cause a binding effect throughout the drivetrain and contribute to a rough ride condition.
20. If cast spoke wheel front axles are heavily loaded, flex of the rim between the spokes can cause an out-of-round effect. A disadvantage of overloading cast spoke wheels is that when overloaded, they tend to permanently deform.

CVSA OOS SUSPENSION CRITERIA

Commercial Vehicle Safety Alliance (CVSA) out of service (OOS) conditions are not maintenance standards. They define the point at which highway equipment becomes dangerous to operate. OOS are used to issue citations during safety inspections. The following OOS apply to truck, tractor, and trailer suspensions:

- Any suspension U-bolt(s), spring hanger(s), or other axle positioning part(s) that are cracked, broken, loose, or missing, resulting in shifting of an axle from its normal position
- Any leaves in a leaf spring assembly that are broken or missing
- Any broken main leaf in a leaf spring assembly (includes assembly with more than one main spring)
- Missing rubber spring
- One or more leaves displaced in a manner that could result in contact with a tire, rim, brake drum, or frame

- Broken torsion bar spring in a torsion bar suspension
- Deflated air suspension, that is, system failure, leak, and so on
- Any part of a torque, radius, or tracking component assembly that is cracked, loose, broken, or missing (does not apply to loose bushings in torque or track rods)

SUMMARY

- A suspension system supports the frame on a vehicle and acts as an intermediary between the axles and the frame. With no suspension, road forces would be transferred directly to the truck frame.
- A suspension system plays a number of roles:
  - It supports the chassis and its load.
  - It stabilizes the vehicle over both smooth and rough terrain.
  - It cushions the chassis from road shock, enabling the driver to steer.
  - It maintains the proper axle spacing and alignment.
- There are four general categories of suspension used on trucks:
  - Leaf spring
  - Equalizer beam: leaf spring and solid rubber spring
  - Rubber or air cushion cushion and torsion bar
  - Air spring: pneumatic-only and combination air/leaf spring
- *Jounce* describes a spring in its most compressed state, whereas *rebound* is a spring when it extends after reacting to jounce.
- Unsprung weight is the vehicle weight not supported by the suspension; it includes the wheel and axle assemblies. Because unsprung weight reacts directly through the suspension to the frame, it is kept as light as possible.
- Constant rate and progressive or variable rate springs are two types of leaf spring suspension used on trucks. Auxiliary springs are helper springs and only become a factor when a vehicle is fully loaded.
- Steel springs are often assembled into semielliptical spring packs consisting of a stack of sprung steel plates clamped by a centerbolt. Spring packs are self-dampening because of interleaf friction.
- Shock absorbers are used in suspension systems to dampen suspension oscillations. Shock absorbers reduce tire wear, front wheel shimmy, and spring breakage.
- Leaf spring and rubber cushion are both equalizing beam types of suspension used on heavy-duty trucks.
- A majority of today's trucks are equipped with air-only or combination air/steel spring suspension systems. Air suspensions use truck system air pressure to keep the air springs charged with compressed air.
- Air bags can be either the reversible sleeve type or the convoluted type.
- A reversible sleeve air bag consists of an inflatable rubber compound bag mounted on a pedestal assembly.
- Convoluted air springs can be single, double, or triple convoluted.
- A major advantage of truck air suspensions is that they are adaptive, having the capability to adapt to changing load and road surface conditions.
- Ride height is managed by a height control valve in air suspensions. Most ride height control valves have a built-in reaction delay that should be recognized when troubleshooting the system.
- Air springs have no self-dampening capability so they almost always use shock absorbers.
- Equalizer beam suspensions are used in tandem drive and bogie arrangements to effectively balance suspension stresses and maximize tire-to-road contact.
- Axle alignment is a key to a properly functioning suspension system. Axles can be aligned using laser beam alignment equipment and cruder shop equipments such as a plumb bob, a straightedge, and tram bars.
- The air suspension cab system is the most common method of mounting a cab on a truck chassis because of the driver comfort it provides.
- Driver seats may be solid mount, mechanically suspended, or pneumatically suspended. Air driver seats are the most common in current trucks because of the comfort they provide the driver.

REVIEW QUESTIONS

1. What is used to load the leaves in a spring pack under tension to provide its self-dampening properties?
   a. spring clips
   b. U-bolts
   c. centerbolt
   d. spring pins
2. In an air suspension, what happens to the air in the air bags when a height control valve lever is raised upward off horizontal?
   a. The air is exhausted.
   b. The bag is further inflated.
   c. The suspension air circuit is dumped.
   d. The chassis system pressure is reduced.

3. A spring pack has a single cracked leaf. Technician A says that the spring pack should be replaced as a unit. Technician B says that the fractured leaf can be repair welded providing it is not the main leaf. Who is correct?
   a. Technician A only
   b. Technician B only
   c. both A and B
   d. neither A nor B
4. Axle alignment on a tandem rear leaf spring suspension is adjusted by:
   a. adjusting torque rods length.
   b. rotating the eccentric adjustment bolts.
   c. both a and b.
   d. neither a nor b.
5. A truck recently had both front spring packs replaced. The driver complains that it is difficult to steer the truck straight ahead. Technician A says that this could be caused by installing the caster shims backward. Technician B says that this condition could be caused by seized spring pins. Who is correct?
   a. Technician A only
   b. Technician B
   c. both A and B
   d. neither A nor B
6. Which of the following would be the most likely road handling result when a spring pack center-bolt breaks on a front suspension?
   a. hard steering
   b. bouncing front end
   c. U-bolt failure
   d. leaf breakage
7. What is the function of torque arms on a suspension system?
   a. to retain axle alignment
   b. to control axle torque
   c. to adjust alignment
   d. all of the above
8. The most compressed condition of a spring is known as:
   a. rebound.
   b. jounce.
   c. oscillation.
   d. failed.
9. A leaf-type spring suspension that has a variable deflection rate obtained by varying the effective length of the spring assembly is known as:
   a. progressive spring suspension.
   b. constant-spring suspension.
   c. an auxiliary spring suspension.
   d. none of the above.
10. What type of spring is usually mounted on top of heavy-duty truck rear spring assemblies and is used only when a truck is under a heavy load?
    a. progressive spring
    b. constant spring
    c. auxiliary spring
    d. air spring
11. What device is used primarily to absorb energy and dampen suspension oscillation?
    a. equalizer bracket
    b. torque rod
    c. springs
    d. shock absorber
12. What device is used to control the air in and out of the air springs in an air suspension system?
    a. height control valve
    b. air pressure regulator
    c. pressure protection valve
    d. quick-release valve
13. Which type of air spring is mounted on a pedestal?
    a. single convoluted
    b. double convoluted
    c. reversible sleeve
    d. jounce block
14. A truck with an air suspension system is being unloaded. After unloading the truck, the driver observes that one side rises. Technician A says that this could be caused by a plugged quick-release valve. Technician B says that a defective leveling valve could cause this problem. Who is correct?
    a. Technician A only
    b. Technician B only
    c. both A and B
    d. neither A nor B
15. When lubing a spring pack assembly, neither the fixed bracket spring pin nor the shackle pin will take grease. What should you do *first*?
    a. Change the zerk fitting.
    b. Remove the weight from the spring.
    c. Replace the spring pins.
    d. Replace the spring pack.

27

Prerequisite: Chapter 26

WHEELS AND TIRES

OBJECTIVES

After reading this chapter, you should be able to:

- Identify the wheel configurations used on heavy-duty trucks.
- Explain the difference between standard and wide-base wheel systems and stud- and hub-piloted mountings.
- Identify the common types of tire-to-rim hardware and describe their functions.
- Explain the importance of proper matching and assembly of tire and rim hardware.
- Outline the safety procedure for handling and servicing wheels and tires.
- Identify the different means of balancing tire and wheel assemblies.
- Describe brake drum mounting configurations.
- Perform wheel runout checks and adjustments.
- Properly match tires in dual and tandem mountings.
- List the major components of both grease- and oil-lubricated wheel hubs.
- Perform bearing and seal service on grease-lubricated front and rear wheel hubs.
- Perform bearing and seal service on oil-lubricated front and rear wheel hubs.
- Understand the concept behind axle-end dynamic balance and the consequences of an out-of-balance condition.
- Perform front and rear bearing adjustment.
- Describe TMC wheel end procedure.
- Outline the procedure for installing preset bearing wheels.

KEY TERMS

aspect ratio
bias ply
ConMet PreSet hub
counteract balancing beads (CBB)
footprint
hertz
internal balancing agent (IBA)
jounce
lip-type seal
Meritor tire inflation system (MTIS)
nitrogen-filled tires
oscillation
preset hubs
radial tire
rebound
retread
rim
section height
section width
sipes
Technology and Maintenance Council (TMC)
tire pressure control system (TPCS)
tire pressure inflation system (TPIS)
tire pressure monitoring system (TPMS)
Unitized Hub System (UHS)
unitized seal
unsprung weight
wheel balance
wheel dolly
wheel end
wheel hop
wide-base singles
wide-base tires
wide-base wheels

INTRODUCTION

Often considered part of the vehicle's suspension system, wheels and tires play a vital role in the safe operation of all heavy-duty trucks and trailers. They carry all the weight of the vehicle and have to operate on a wide range of road conditions. After fuel costs, tires represent one of the biggest cost factors of operating a truck.

Wheels and tires must be properly maintained and serviced. Improperly mounted, matched, aligned, or inflated tires can create a dangerous on-road situation. Poorly maintained tires will also wear unevenly and at a faster rate. Chances of a blowout or other major failure are also increased. Wheel bearings, lubricant seals, and other components in the wheel hub that keep the wheel and tire turning smoothly on the axle spindle also require regular maintenance and service.

WHEEL CONFIGURATIONS

For many years, there was little difference in the wheel configurations used on heavy-duty trucks. Today, advances in technology and changing customer needs have increased the number of wheel configurations available. It is now possible to operate on cast spoke or disc wheel systems in steel or aluminum. **Rims** can be removable or part of the wheel body. They can be single or multipiece.

Wheels can be clamp mounted, stud piloted, or hub piloted. Dual-wheel or wide-base single wheels are available. Wheel bearings can be either grease or oil lubricated. Several wheel seal designs can be used. The tires mounted to the wheels can be bias ply or radial, tube or tubeless, standard or low profile. Brake drums can be inboard or outboard mounted. The wheel system chosen will affect payload, fuel efficiency, tire mileage, and hardware requirements. It will also have a major effect on service and maintenance procedures.

WEIGHT FACTORS

Regardless of the system used, truck wheel and tire assemblies are heavy. Truck designers attempt to keep this weight as light as possible because it represents unsprung weight (see Chapter 26 for an explanation). Another consideration is wheel balance. In truck wheel assemblies, much of the mass (weight) of rotating wheel components is in the hub and drum assemblies. For this reason, to maximize the operational life of the tires and suspension, the axle end should be in dynamic balance.

SAFETY FACTORS

Inflation pressures can also exceed 100 psi. This combination of weight and pressure can create dangerous work situations if proper safety and working procedures are not followed. A simple but key skill required by truck technicians today is **wheel end** procedure. Wheel end procedure refers to the installation of wheel assemblies to axle spindles. In some types of wheel end, the technician is required to set bearing endplay. The procedure used to achieve this is based on an industry standard that every technician should adhere to. More recently, **preset hubs** have been introduced and are now widely used. In a preset hub, the seal has been factory installed and bearing endplay set. The technician simply places a self-piloting hub on the axle spindle and torques it to specification.

27.1 WHEELS AND RIMS

The term *wheel* may be used in several ways. First, it can refer to the entire rotating assembly on the end of an axle and, therefore, include tires mounted on rims, a rim spacer, a hub assembly, and brake drum. The term also can be used to describe the hub assembly to which the rims and brake drum (or rotor) are attached. When we talk about the type of wheel system used on a truck, we are using the second definition. On trucks today, two general types of wheel configurations are used along with several methods of attaching tire assemblies to the wheels. The two general types of wheels used are:

- Cast spoke wheels
- Disc wheels

Spoke wheels are gradually becoming obsolete but there are still plenty on our roads. On a spoke wheel, the rim and wheel are separate components. On a disc wheel, the rim is a distinct section of a wheel assembly. Disc wheels can be subdivided into two types by the means used to mount them:

- Stud piloted
- Hub piloted

We begin by getting an understanding of each type of wheel by category.

CAST SPOKE WHEELS

Cast spoke wheels were used almost universally but this has changed as operators have come to understand the advantages of disc wheels. That said, cast spoke wheels are not yet obsolete. Though they have the disadvantage of being heavy, they are tough and can withstand more punishment than disc wheels, making them a more likely choice on dump, construction, refuse, and some leased trucks and trailers.

A spoke wheel consists of a one-piece casting that includes an integral hub and spokes. **Figure 27–1** shows an exploded view of a typical cast spoke wheel. They are usually manufactured of cast steel, and

FIGURE 27–1 Components of a cast spoke wheel

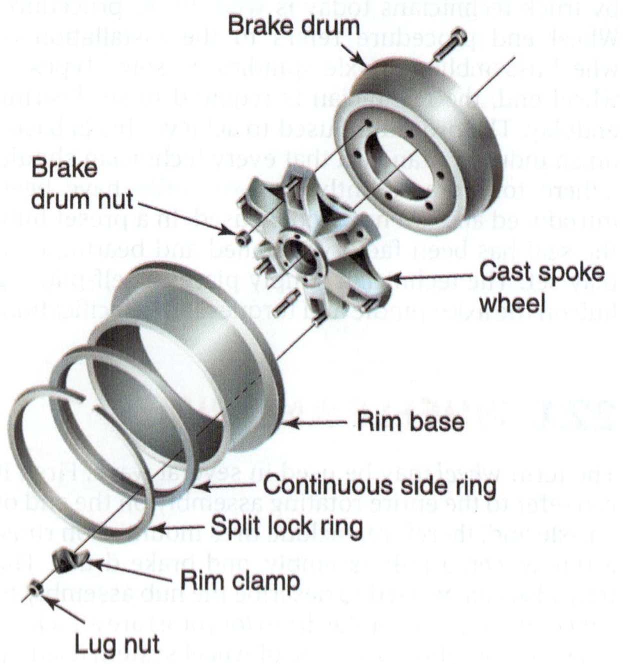

the hub and wheel mounting surfaces are machined after casting. Tires are mounted on a separate rim or rims that are clamped onto the spokes. In dual-wheel applications (two tire/rim assemblies mounted on a wheel), a spacer band is positioned between the inner and outer rims in the manner shown in **Figure 27–2A**. This spacer band is clamped between the two tire rims and provides exact spacing between them (**Figure 27–2B**).

Three, Five, and Six Spokes

Spoke wheels are usually manufactured in three-, five-, and six-spoke configurations. Six-spoke designs are often used on heavily loaded front axles. Five or six spokes are used on drive axle duals, but six-spoke designs are often preferred because of the added wheel clamping force on the rim, which reduces the chance of rim slippage. Three-spoke wheels have wider spokes, using two wheel clamps per spoke. Trailers are their most popular application.

Axial Runout

As you can see in Figure 27–2, spoke wheels use multipiece rims that clamp to the spokes with wheel

FIGURE 27–2 (A) Position of the spoke wheel dual mounting spacer band; and (B) cross-section view of mounted dual wheels

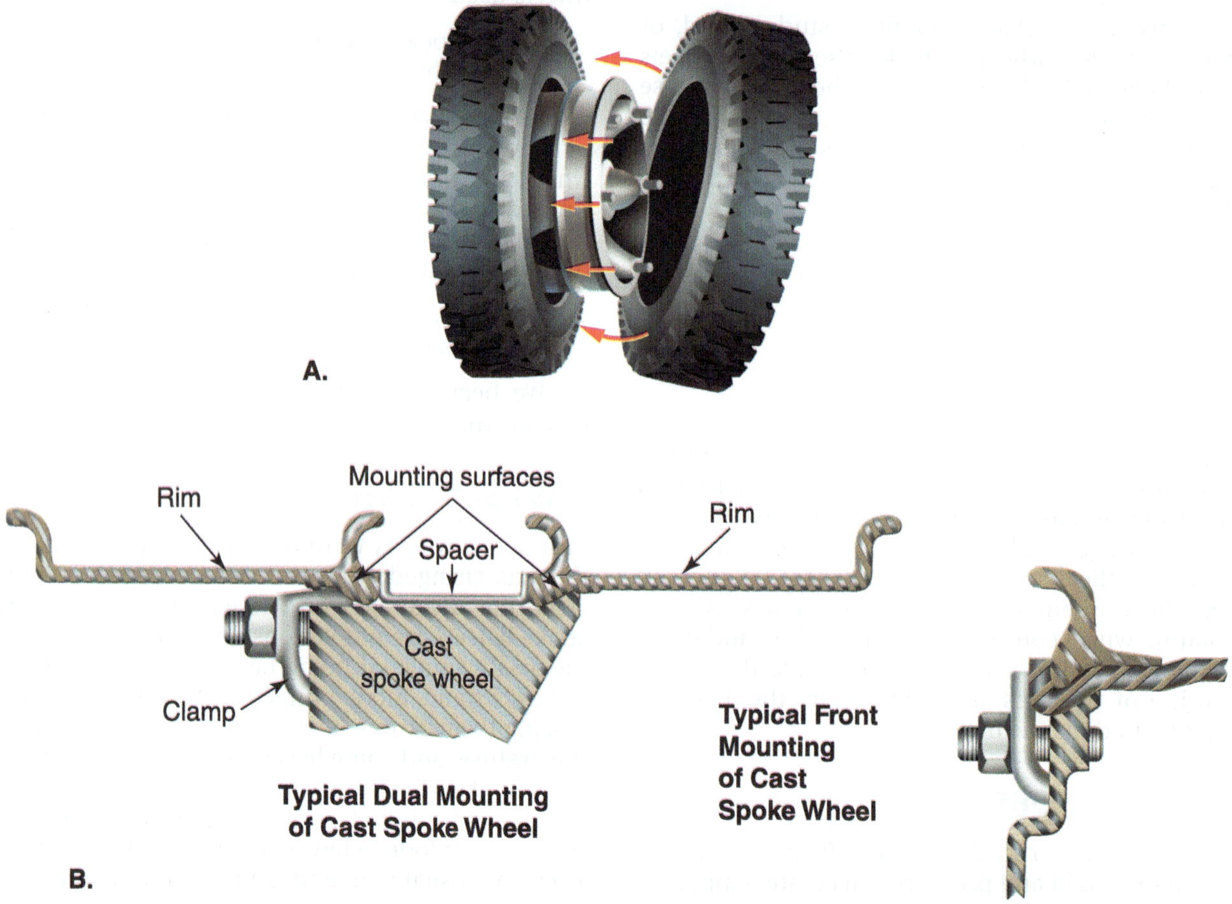

clamps. If the clamps are not installed correctly, the tire/rim might produce a wobble condition known as axial runout. Axial tire runout is more of a problem in cast spoke wheels than disc wheels, but it can be minimized by adhering to the proper installation and torquing sequence and by checking runout after assembly. Although cast spoke wheels experience greater alignment and balance challenges because of their extra weight and rim mounting, they usually encounter fewer outright failure problems and, notably, fewer "wheel-off" incidents than disc wheel assemblies.

DISC WHEELS

In disc wheels, the tire rim forms the wheel and this mounts to the hub assembly (as shown in **Figure 27–3**). Stud holes in the center of disc wheels allow them to be mounted to the hub studs with nuts. These hub studs pass through both drum and disc wheels.

A real advantage of disc wheels is that they run true and produce few alignment problems. Disc wheels use one-piece forged steel or aluminum construction, which makes them lighter, reducing vibration (and tire running temperatures) and extending tire life. Because of their ability to extend tire life, disc wheels are popular with cost-conscious fleet managers. By improving driver comfort and vehicle handling, they have also become a driver preference, although some drivers might tell you that they just look better than cast spoke wheels.

Because wheel assemblies are part of the unsprung weight of a vehicle, there is always a handling advantage to reducing their weight. Also, a weight reduction can mean increased payload, so aluminum disc wheels have steadily gained in popularity. Aluminum is not only lighter than steel but also dissipates heat faster, resulting in tires running cooler. Disc wheels can be used in single and dual configurations. There are also two different mounting systems: stud piloted and hub piloted.

Stud-Piloted Wheel

A stud-piloted wheel uses the stud and a ball seat to center and secure the wheel. **Figure 27–4A** illustrates a stud-piloted mounting for a single disc wheel. The wheel simply mounts onto studs on the hub and is secured using single cap nuts. **Figure 27–4B** shows how a dual disc wheel configuration is mounted.

Inner cap nuts screw onto the hub studs. The flange on the nut caps rests in the seat created between the inner and outer wheels. This helps center the two wheels and clamp them together. Finally, outer cap nuts screw onto the threaded ends of the inner cap nuts, thereby securing the entire assembly

FIGURE 27–4 Aluminum disc wheel stud-piloted mounting configurations: (A) single wheel and (B) dual wheel

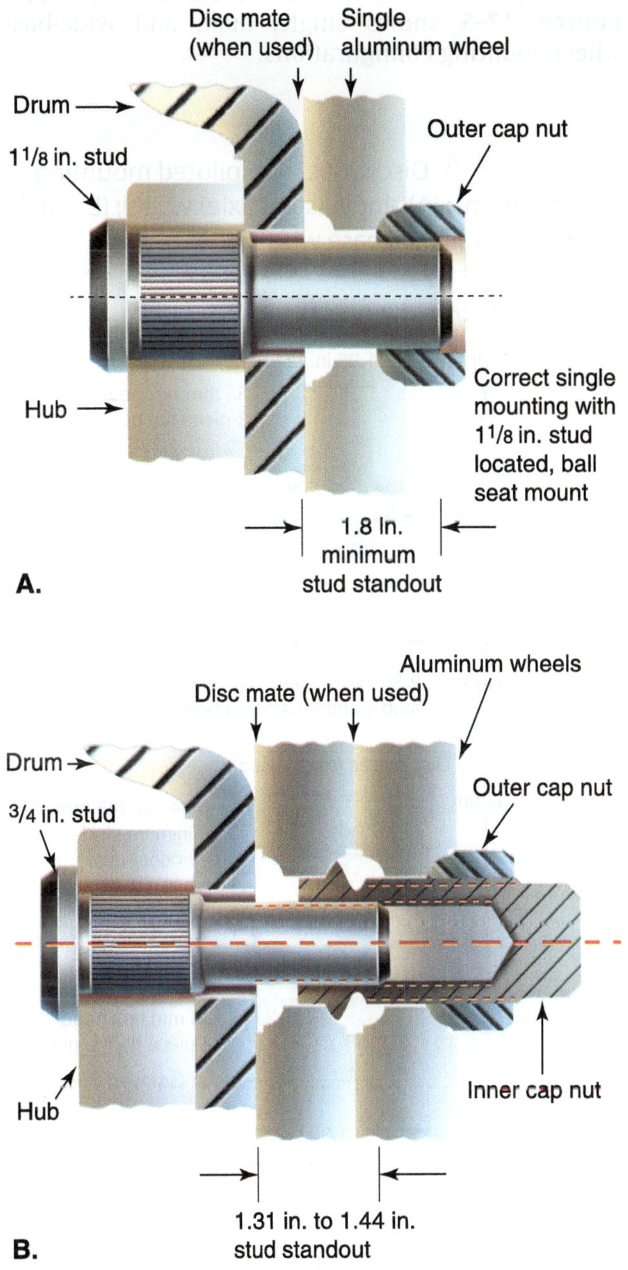

FIGURE 27–3 Components of a typical disc wheel

to the hub. The torquing procedure for stud-piloted wheels is covered later in this chapter. Stud-piloted disc wheels have been used less following the introduction of hub-piloted wheels, which use a simpler mounting procedure that better centers the wheel. **Figure 27–5** shows single, dual, and wide-base wheelmounting configurations.

FIGURE 27–5 Disc-wheel hub piloted mounting configurations: (A) single (steer axle) wheel; (B) dual wheel; and (C) wide-base wheel

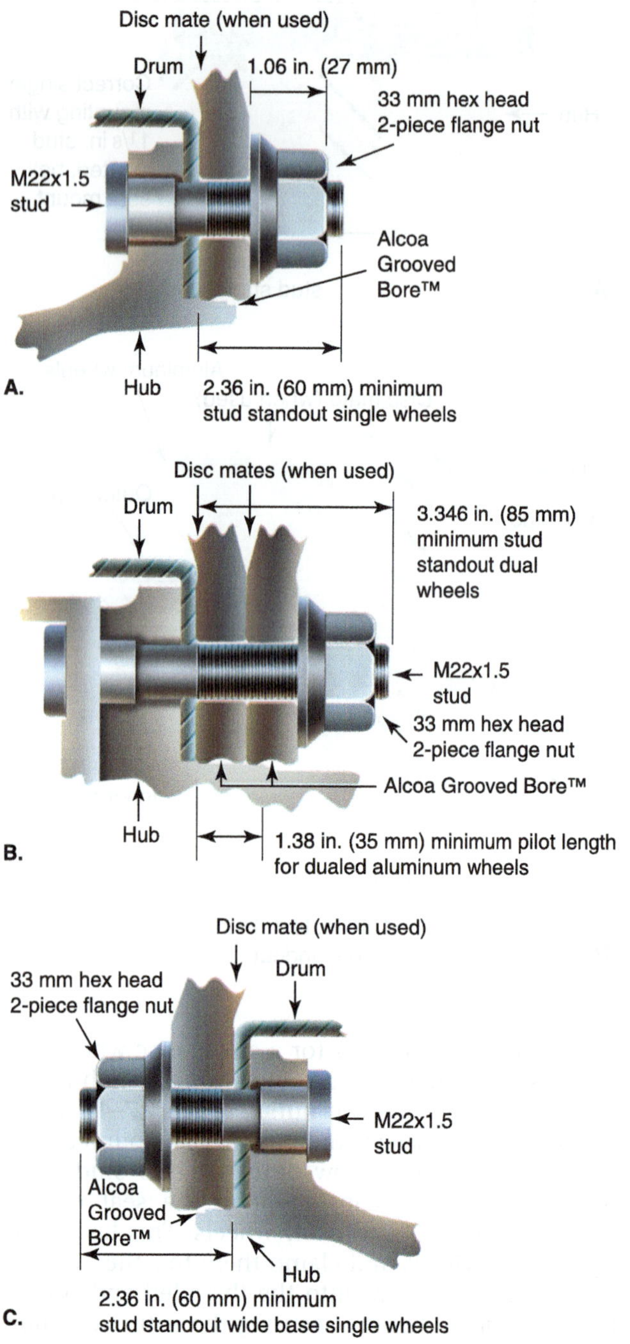

Hub-Piloted Wheel

Hub-piloted wheels use the hub and two-piece flange nuts to center and secure the wheel. The hub-piloted system simplifies centering and clamping wheels to hubs. The nuts and studs do no more than provide clamping force. You can see what the arrangement looks like in Figure 27–5.

A hub-piloted wheel uses one cone locknut per stud, eliminating the need for inner cap nuts. This significantly reduces the amount of wheel fastening hardware compared to stud-piloted wheels. Over- or undertorquing of stud-piloted wheels can cause broken studs and cracked or loose wheels.

The single flange nuts of hub-piloted wheels are less susceptible to this problem. In stud-piloted systems, a loose inner nut can easily go undetected, eventually pounding out the nut ball seat. With hub-piloted systems, both the inner nut and its ball seat are eliminated.

SHOP TALK

Learn how to identify stud- and hub-piloted disc wheels: Improper torquing procedure and sequencing of stud-piloted wheels is a major cause of wheel failure. With the cone locknut design, a flat washer is seated directly against the wheel face. The nonrotating washer prevents galling of the wheel surface.

WIDE-BASE WHEELS

Today's **wide-base wheels** (Figure 27–5C) are usually known as **wide-base singles** but can also be referred to as *high-flotation*, *super single*, *wide-body*, *duplex*, or *jumbo wheels*. The technology has rapidly evolved. Today, one wide-base wheel and tire replaces a traditional set of duals without any reduction of payload specification and a 2 to 5 percent improvement in freight efficiency. You can see some of the available configurations in **Figure 27–6**.

If wide-base wheels were to be used for a tractor/trailer combination, instead of a total of eighteen tires, only ten would be required: A pair of traditional steering tires would be used at the steering axle, and eight wide-base wheels would be used everywhere else on the rig, replacing the duals. Compared to steel dual wheels and tires, aluminum wide-base wheels and tires are significantly lighter in weight, meaning increased payloads and fuel efficiency.

Wide-base wheels have been available for a couple of decades now but only recently have they begun to catch on. In 2013, 5 percent of linehaul commercial rigs were using wide-base wheels. The reason for renewed interest in wide-base wheels has everything to do with the potential for operators to lower fuel costs and increase gross payloads. The downside is that the cost of one wide-base tire is currently greater

FIGURE 27–6 Wide-base wheel mounting compared to dual configurations

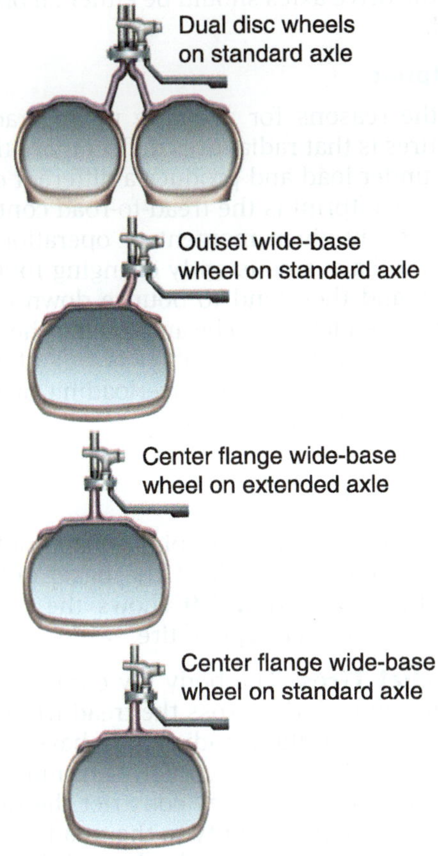

than that of a pair of duals and the infrastructure for replacement and repair needs to be improved. Later in this chapter there is a description of the tires used on wide-base wheels.

TIRE-TO-RIM HARDWARE

A tire can be mounted to a rim in a number of ways. The simplest is the single-piece rim. A fixed flange formed into the edge of the rim supports both sides of the tire, as shown in **Figure 27–7**. Single-piece rims are used in combination with tubeless tires.

Tube-type tires are mounted on the rim using various side ring or lock ring combinations. Side ring and lock ring designs vary from manufacturer to manufacturer, so it is important to always use properly matched components. Two types of side rings are used:

- **Split Side Rings.** In two-piece assemblies, the side ring retains the rim on one side of the rim. The fixed flange supports the other side (**Figure 27–8A**). The split side ring is designed so that it acts as a self-contained lock ring as well as a flange.
- **Flange or Continuous Side Rings.** In three-piece assemblies, the flange, or continuous side ring, supports the tire on one side of the rim. The continuous side ring is, in turn, held in place by a separate split lock ring, as shown in **Figure 27–8B**.

FIGURE 27–7 Sectional view of a single-piece demountable rim used with tubeless tires

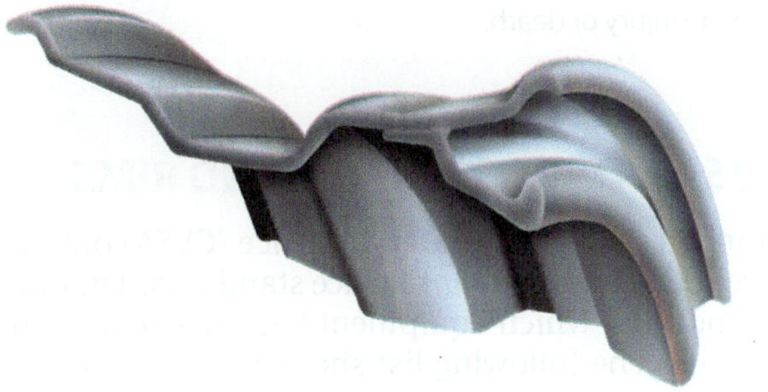

FIGURE 27–8 Side ring configurations: (A) split side ring; and (B) continuous side ring with separate split lock ring to secure it to the rim

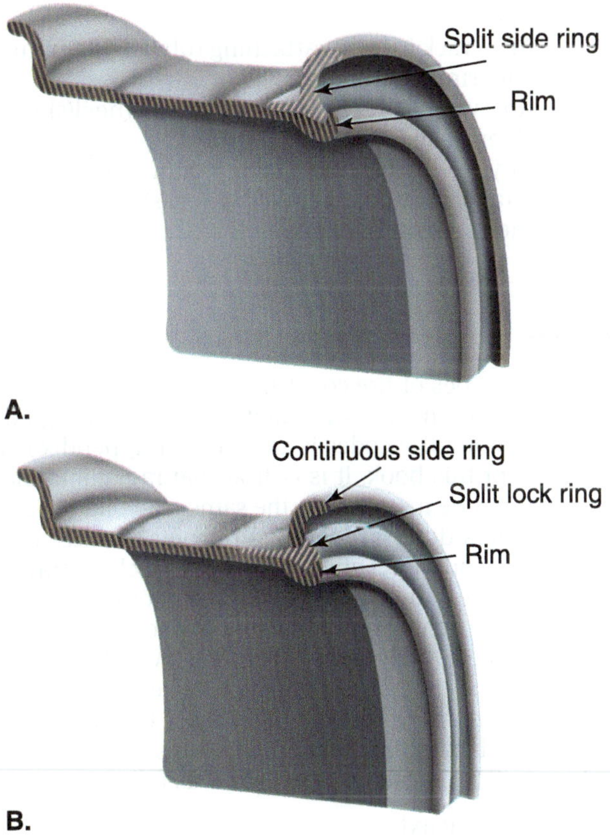

WARNING

A tire safety cage should be used any time a tire is inflated off the truck; a split rim can separate during inflation and cause severe injury or death.

CVSA OOS FOR WHEELS AND RIMS

Commercial Vehicle Safety Alliance (CVSA) out of service (OOS) are not maintenance standards. They define the point at which equipment becomes dangerous to operate. The following list shows some examples:

- Lock or side ring that is bent, broken, cracked, improperly seated, sprung, or mismatched
- Wheels and rims that are cracked or broken or have elongated bolt holes
- Fasteners on spoke or disc wheels that are loose, missing, broken, cracked, stripped, or otherwise ineffective fasteners
- Any cracks in welds attaching disc wheel disc to rim
- Any crack in welds attaching tubeless demountable rim to adapter
- Any welded repair on aluminum wheel(s) on a steering axle
- Any welded repair other than disc to rim attachment on steel disc wheel(s) mounted on the steering axle

27.2 TIRES

Two basic types of tire construction are used in heavy-duty applications: **bias ply** and **radial tires**. A guide to interpreting tire codes is provided in Appendix L at the back of this book. It is critical that radial and bias ply tires not be installed on the same axle . Radial and bias ply tires differ in their tread profile, surface contact, and handling characteristics. As truck operators have become more aware of fuel economy, the role tires play in achieving better fuel economy has been studied. Fleets today target tires with low rolling resistance and the tire manufacturers are likely to focus on the fuel economy potential of tires more than longevity and retreadability. Because of this, tires have become axle specific and can be categorized as follows:

- Steering axle
- Drive axle(s)
- Trailing axles

DUALS

Dual wheels should not have mismatched tires. The tires on an axle should be of the same construction, tread pattern, and nominal size. Mismatched tires on opposite sides of the same axle can cause drive axle failure by continually working the differential, so if the left duals are radial, the right duals also should be radials. If the vehicle has two or more drive axles, the tires on the drive axles should be either all bias ply or all radial.

Tire Footprint

One of the reasons for avoiding mixing radial and bias ply tires is that radial tires deflect more than bias ply tires under load and produce a different dynamic **footprint**. Footprint is the tread-to-road contact area of a tire at any given moment of operation. Radial tires tend to have a constantly changing footprint in operation, and they tend to bounce down the road. This bounce tendency can be amplified if the axle end (tire/wheel/drum/hub assembly) is out of balance. The action can cause cyclic overloading of bias ply tires if mixed with radial tires.

TIRE CONSTRUCTION

The easiest way to learn how bias ply and radial tires are constructed is to look at the basic differences between them. **Figure 27–9** shows the basic construction used in each type of tire.

- **Radial Tires.** The body ply cords are placed perpendicularly across the tread from bead to bead. In addition, radial tires have belt plies that run circumferentially around the tire under the tread. They constrict the radial ply cords and give rigidity to the tread.
- **Bias Ply Tires.** The body ply cords lie in a diagonal direction from bead to bead. The tires may also have narrow plies under the tread, called breakers, with cords strung in approximately the same direction as the body ply cords.

Body Ply, Breaker, and Belt Materials

Tire body plies, breakers, and belts can be made of polyester, rayon, nylon, fiberglass, and steel braids.

FIGURE 27–9 Construction of bias ply and radial tires

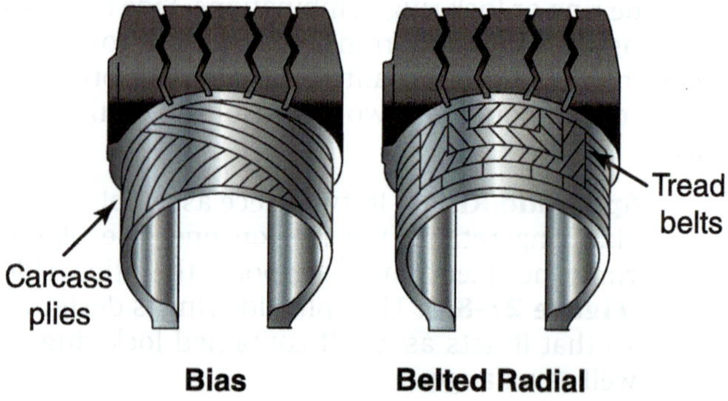

FIGURE 27–10 Three typical tire tread designs

In radial ply tires, these materials can be used in various combinations, such as steel body steel belt, polyester body fiberglass belt, or nylon body steel belt.

Tread Design

The basic categories of highway truck tread designs are:

- Rib (**Figure 27–10A**)
- Lug (**Figure 27–10B**)
- High-traction lugs (**Figure 27–10C**).

Rib-Type Tread

Tires with rib-type tread are all position tires. They can be used on all wheel positions at highway speeds. These tires are usually recommended for front steering axles on both tractors and straight trucks used for highway applications along with trailer axles. The open groove design provides optimum steering control and good skid resistance.

Lug-Type Treads

Cross lug or cross rib and rib lug-type tires are designed for drive wheel service and are suitable for most highway operations. They are manufactured with blocks and grooves (see Figure 27–10C) that cut across the tread pattern, which add traction and aggressiveness. They may also have circumferential grooves, usually located close to each shoulder. This tread design provides maximum resistance to wear and better traction. The tires are suitable for some off-road operations but do not provide maximum off-road traction.

High-Traction Lug Treads

High-traction lugs can also be known as mud and snow lugs. High-traction lugs are designed to provide good traction on drive axle wheels for both on- and off-road service. They are not great when it comes to getting the best fuel mileage and they tend to wear faster than standard lugs, so they should be used only when maximum traction in mud or snow is required. They are most commonly used on cement, refuse, dump, and construction vehicles.

Unidirectional Tread Designs

Unidirectional tread patterns are designed to rotate in one direction only. They are spec'd into applications in which high linehaul mileage is required. When unidirectional tires are installed backward, they will wear more quickly though no damage is likely to the casing.

Sipes

The tiny notches along the edges of grooves within the tread blocks and ribs are known as **sipes**. Sipes help relieve rolling stress that initiates river erosion wear during wet pavement operation.

RETREADS

Retreads have acquired a bad rap and are falsely blamed for 100 percent of the "alligators" we see littered over our roads. This thinking goes back to the days where retread technology was in a primitive phase of development. Today's tires are engineered to be retreaded several times during their service life, making retreads a fact of life in any trucking operation. The bottom line is that the cost of a retread is about 50 percent of a new tire and it can be expected to last as long as the new tire before the next retread is required.

WIDE-BASE TIRES

Earlier in this chapter we outlined the reasons many operators were considering adopting wide-base wheels. A wide-base wheel replaces a set of duals. **Wide-base tires** can be used on tractor drive and trailer axles. Using wide-base wheels and the single tires that replace the dual set does the following:

- Reduces overall wheel end weight
- Reduces tire rolling resistance
- Increases payload potential
- Spreads the axle permitting lower center of gravity and greater stability

Wide body tires have been more popular in Japan and Europe where more costly fuel hastened their introduction. As fuel prices continue to be volatile in North America, the fuel economy benefits have gradually increased the acceptance of wide-base wheels. The disadvantage is the lack of a good supply infrastructure; many operators and fleets do not want to downtime a vehicle because of a single failed tire. **Figure 27–11** compares wide-base wheels with a traditional dual-wheel configuration. Note how the center of gravity can be lowered, which is a real advantage in tanker or hopper body trailers.

Weight Saving

Wide-base single tires such as the Michelin X1 family of tires can be retrofitted without affecting driveline gear ratios or top-speed capability. Compared to steel dual wheels and tires, aluminum wide-base wheels and tires are significantly lighter in weight, meaning increased payloads and fuel efficiency. The weight saving per axle fitted with wide-base singles ranges from 170 to 190 pounds (77 to 86 kg). The weight saving for a highway semicombination could be 800 pounds (363 kg).

Reduced Rolling Resistance

Today, tire manufacturers claim that wide-base singles can replace a set of duals in any application. In addition, they are said to provide equal or better traction than duals along with actually increasing the axle footprint width; this marginally increases stability. However, the main reasons for renewed interest in wide-base singles can be mostly attributed to the fact that operators can achieve better fuel economy and increase payloads. When a rig is run with just some of the axles equipped with wide-base singles, there is some significant potential for an improvement in fuel economy. In a typical semicombination rig equipped with duals, rolling resistance can be divided as follows:

- Steer axle: 16 percent
- Drive axles: 44 percent
- Trailer axles: 40 percent

FIGURE 27–11 How wide-base singles enable reduced weight, lower rolling resistance, and increased payloads.

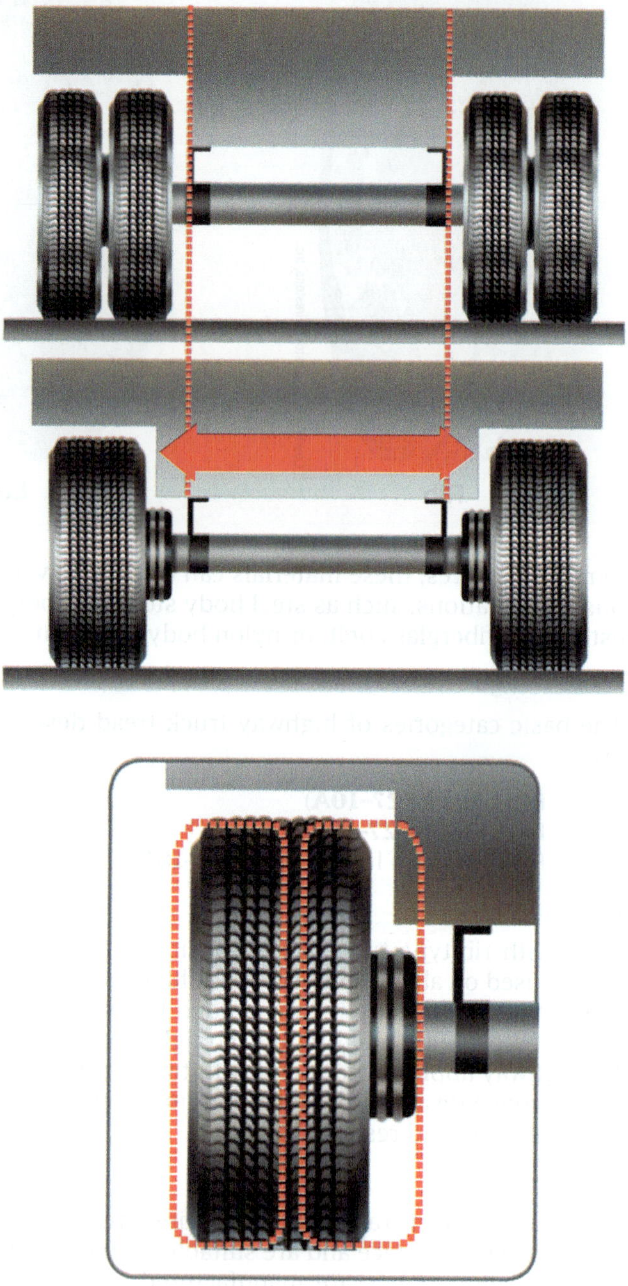

Wide-base singles provide a potential for reducing rolling drag by around 5 percent. What this translates to in fuel savings depends on cartage weights, but the savings can be significant throughout a fleet.

Wide-Base Single Construction

Michelin X1 wide-base singles use what it calls Infinicoil technology. Infinicoil is a ¼-mile (400 m) steel cable that is wrapped underneath the tire tread in the direction of its rotation. Its objective is to reduce some of the flexing cycling that are a characteristic of radial tires. Reducing the flexing provides a more consistent dynamic footprint said to be rectangular (similar to a set of duals) rather than oval as it was in the first generation of super singles.

Repair and Retread

The infrastructure for replacement and repair of wide-base tires has improved in recent years. In addition, the current generation of these tires is retreadable, though in some cases a proprietary process has to be used. Today's wide-base tires are designed to be retreaded several times over their lifecycle. They are notably more sensitive to overinflation, which can greatly shorten their service life.

TIRE SIZE

Tire size is determined by the type of application intended for the truck. The nominal 22.5- and 24.5-inch (57 and 62 cm) wheel/tire size are commonly used on Class 8 transport trucks. Nominal size 19.5-inch (50 cm) tires are gaining a growing share of the high-cube truck/trailer market because they are better suited for applications in which low trailer floor and fifth wheel heights are required to get them under 110- to 120-inch (2.8 to 3.0 m) door openings. Size 17.5-inch (45 cm) tires are used on some high-cube applications, but these smaller wheel/tire combinations can accommodate only a 12$\frac{1}{16}$-inch (31 cm) brake drum, whereas a 19.5-inch (50 cm) wheel can take only a 15-inch (38 cm) brake drum.

Nominal tire size and actual revolutions per mile do not necessarily correlate, which is something that should be remembered when new tires are installed. Two manufacturers of tires or the same manufacturer with different tread profiles, all of the same nominal size, may have very different rolling radii. Revolutions per mile data on the drive axle tires must be programmed to the appropriate controller on the chassis data bus, so whenever new drive axle tires are installed on a truck, check the revolutions per mile data, then check or reprogram this data.

RIB DESIGNS

Tire rib designs have grooves that run circumferentially around the tire. In addition to the three main categories of tire tread design shown earlier in Figure 27–10 there are many other designs. Rig grooves can be of a continuous straight groove such as in the low rolling resistance (LRR) tire shown in **Figure 27–12**. Continuous straight grooves roll in a straight line with minimal rolling resistance and low noise, so they are the tire of choice in high-mileage linehaul applications where fuel economy is important. Some tires are manufactured with "decoupling" grooves on the shoulders of the tire, which act as barriers to reduce shoulder wear. Decoupling grooves work well in linehaul applications but they tend to fail prematurely in urban use because they are prone to stone retention and tearing as a result of curbing.

FIGURE 27–12 A low rolling resistance straight groove rib design

If better grip is necessary, tires with more aggressive rib designs are required. Examples are the zig-zag geometry shown in **Figure 27–13** and the super-grip lug design shown in **Figure 27–14**.

FIGURE 27–13 A high-traction zig-zag rib groove design

FIGURE 27–14 Rough terrain lug tire grooves

FIGURE 27–15 Department of Transportation tire coding identification for (A) a new tire; and (B) a retread

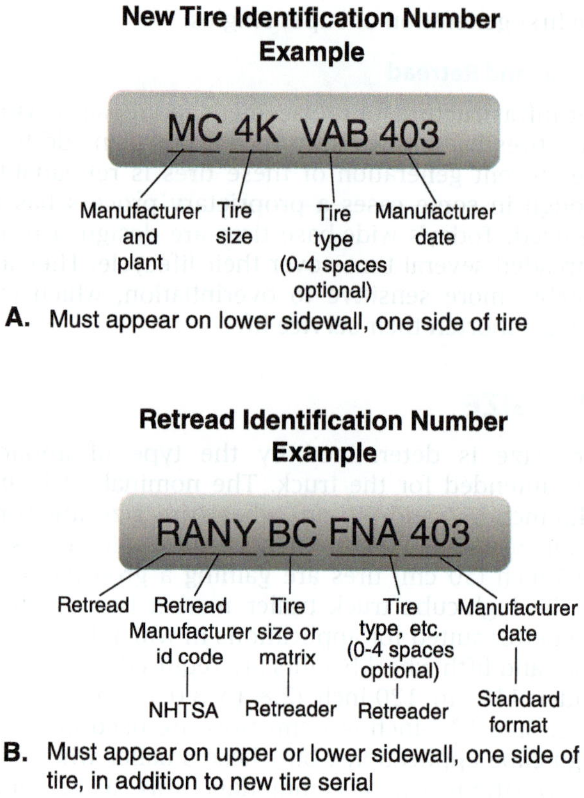

DOT CODES

The type of tire tread design used is determined by vehicle application and in some cases by personal preference. The rib-type tread is most common for on-highway operations and is almost always used on steering (front) axles, regardless of what is used on the rest of the rig. All tires (new and retread) sold in the United States are required to have a Department of Transportation (DOT) number cured into the lower sidewall on one side of the tire. **Figure 27–15** shows the standard format designated by the U.S. federal government Department of Transportation for both new and retread truck tires.

SHOP TALK

Revolutions per mile data on the drive axle tires must be correctly programmed to the appropriate chassis data bus source address (SA). Whenever tires are replaced or swapped on a vehicle, ensure that you check and reprogram tire revolutions per mile to the appropriate controller on the chassis data bus. This data is used to calculate and broadcast road speed data to the instrument cluster, engine, transmission, collision warning, and other controllers networked to the bus.

LOW-PROFILE TIRES

Low-profile radial truck tires are used to improve fuel economy. The low profile is achieved by reducing the tire **aspect ratio**. The aspect ratio of a tire is calculated by dividing the tire **section height** (tread center-to-bead plane) by its **section width** (sidewall to sidewall).

Low-profile tires are exactly what the name suggests: squatter in height than conventional radials. Advantages offered by low-profile radials include lighter weight (up to 10 percent less than standard radials), lower rolling resistance (again about 10 percent less), greater vehicle stability due to a lower center of gravity, a better footprint as a result of improved pressure distribution, good retreadability, improved fuel economy, better traction, and increased tread life. **Figure 27–16** shows a sectional view of a low aspect ratio tire compared with a conventional radial of the same nominal size.

Tire Terminology

Tires represent a significant percentage of the overall cost of maintaining a fleet and, as a consequence, fleet operators usually assign specialist tire technicians to

FIGURE 27–16 Dividing height by width equals the profile.

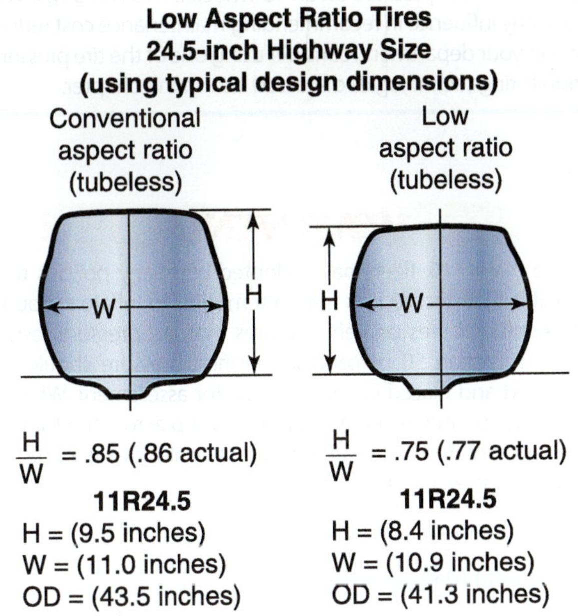

the management of tires. Consequently, most fleet truck technicians seldom have much to do with tires and may be expressly prohibited from working on them. However, the tire is part of the wheel assembly and can act as a tattletale for a range of abnormal running conditions. The first step to understanding tire inspections, maintenance procedure, and failure analysis is interpreting the language of tires. **Figure 27–17** shows the terms used to describe a radial tire and **Figure 27–18** is a cross section of a tubeless tire.

FIGURE 27–17 Terms used to describe a typical radial tire

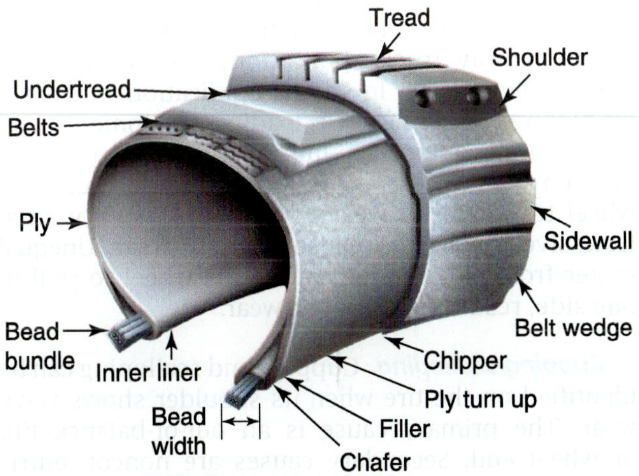

FIGURE 27–18 Cross section of a tubeless tire

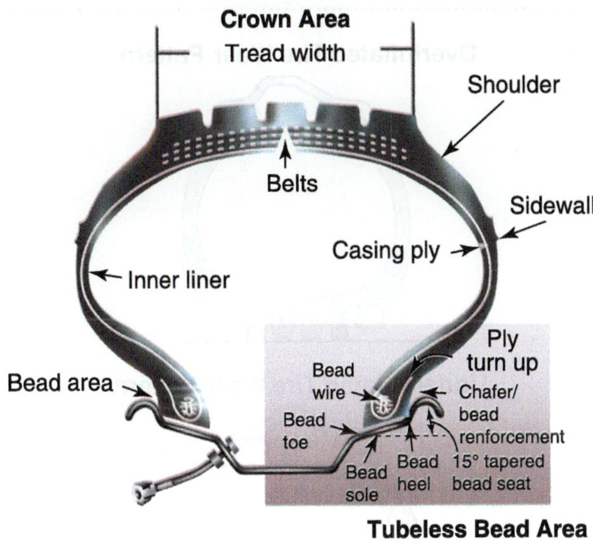

27.3 TIRE MAINTENANCE

Proper tire care and maintenance comes second only to increasing fuel mileage in its ability to reduce overall cost per mile of truck operation. However, most tires wear fast or fail prematurely because of neglect. One problem is getting drivers and service personnel to maintain the correct tire pressure—tires operated at lower than recommended pressure wear quickly. To maximize the life of tires, the rotating mass of the axle end must be in balance. in this regard, it is important to point out that a tire represents only a percentage of the rotational mass of an axle end. A perfectly balanced tire operating on an out-of-balance wheel end will have a reduced operational life, because it is the tire that takes the brunt of the resulting punishment. Careless driving habits also can result in tire damage and contribute to shortening the life of a tire.

TIRE INSPECTION

Regular inspection of tires is the first step in increasing fuel economy and tire longevity. Regular inspection will help to identify underinflation, overinflation, and misalignment before they can cause damage. Minor damage that can be detected and repaired during an inspection can sometimes save a tire that would otherwise blow out.

Underinflation

Underinflation causes abnormal wear at the sides of the tread because the outer edges of the tire carry the load, whereas the center tends to flex up, away from the road. This also causes the tire to run

FIGURE 27–19 Wear effects of over- and underinflated tires

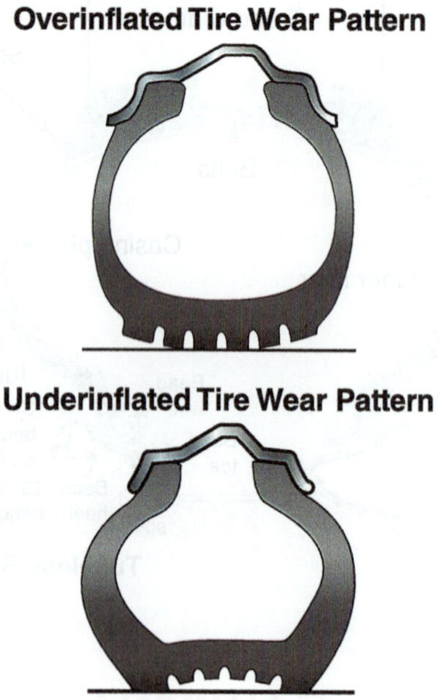

hotter. Any tire that is determined to be underinflated should be inflated to the specified pressure. Driving on an underinflated tire, even for a short distance, can cause severe cord damage and belt separation. **Figure 27–19** shows how underinflation and overinflation produces tire wear.

SHOP TALK

Contrary to the beliefs of some truck drivers, tire pressure cannot be checked with a hammer. The only condition a hammer can identify is a completely deflated tire. Tire pressures must be checked with a pressure gauge. In fleets where this is not a driver's responsibility, it must be done by a technician.

Overinflation

Overinflation causes abnormal wear in the center of the tread because it bulges into a smaller footprint that has to carry more than its share of the load, as shown in Figure 27–19. Overinflation is not so common a problem as underinflation, but it does shorten the life of the tire.

Check tire pressure when the tires are cool. When a tire is in use it heats up because the air in the tire expands, raising the air pressure. The pressure rise from cold to operating temperature is normally about 10 percent, so if the specified inflation pressure is 100 psi, the operating pressure at running speeds and loads will typically be around 110 psi.

Tech Tip

It typically takes around 30 minutes for a technician to check and correct tire pressure on an 18-wheeler. This is costly. If you have any influence in recommending maintenance cost reduction in your department, consider using one of the tire pressure monitoring technologies discussed later in this chapter.

WARNING

Some major U.S. fleets have adopted company policies that prohibit their technicians from attempting to inflate seriously underinflated tires on vehicles. Tires that are pressure tested to be less than 50 percent of specified pressure should be removed and placed in a safety cage for assessment. When a tire blows during inflation, it tends to unzip across the thinner and less reinforced sidewall; the force of air alone exiting the tire can damage eyesight.

Mechanical Irregularities

Tires in various positions on the truck will wear differently. Because of alignment factors and the fact that they are used to steer the vehicle, the front tires wear at a faster rate than the rear tires. The way in which tires wear is often a clue to the cause. When toe-in is excessive, tire wear appears as feathered edges on the inside edge of the tread design, as shown in **Figure 27–20**. A toe-out condition will show feathered edges on the outer edge of the tire tread design, as shown in **Figure 27–21**.

Camber. Camber is the inward or outward tilt from the frame at the top of the wheel. Too much positive camber (wheel tilts outward) causes the outside edge of the tire to wear prematurely. Too much negative camber (wheel tilts inward) causes the inside edge of the tire to wear first. The effects of out-of-specification camber are shown in **Figure 27–22**.

Caster. Positive caster is the backward tilt of the axle, or inclination of the kingpin (**Figure 27–23**). In on-highway truck applications, caster problems almost never occur. In special application vehicles—off-highway trucks, dock loading equipment, and agricultural equipment—improper caster can cause problems. Too little positive caster will cause the wheel to wander or weave, resulting in spotty wear. Excessive caster may cause shimmy wear. Unequal caster from side to side causes the wheel to pull to one side, resulting in uneven wear.

Cupping/scalloping. Cupping and scalloping can be identified on the tire when its shoulder shows wavy wear. The primary cause is an out-of-balance tire or wheel end. Secondary causes are nonconcentric

FIGURE 27–20 Tire wear from excessive toe-in

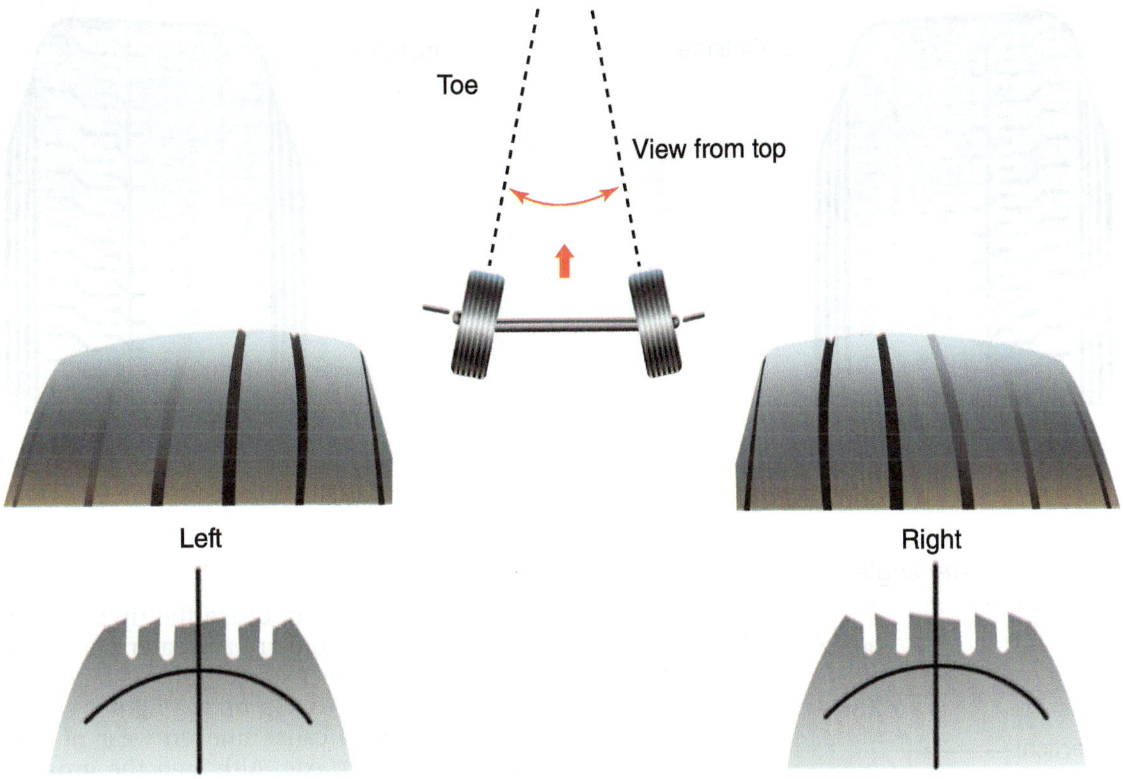

FIGURE 27–21 Tire wear from excessive toe-out

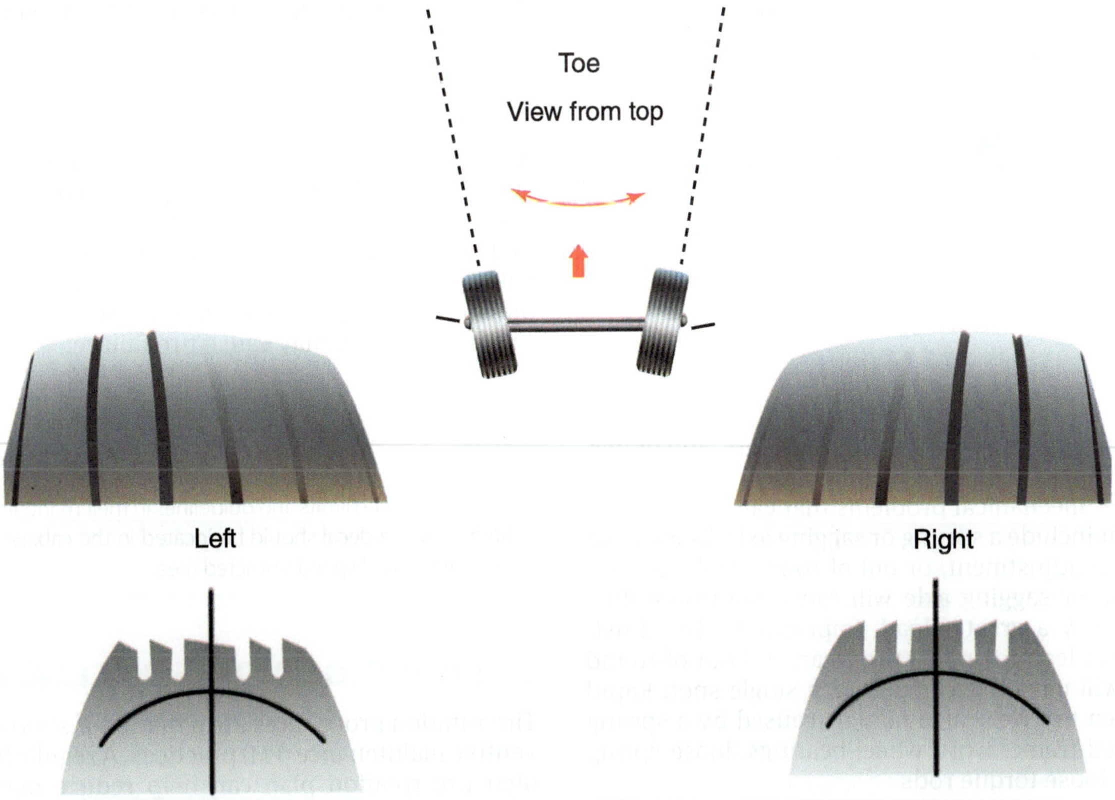

FIGURE 27–22 Tire wear from excessive positive camber and excessive negative camber

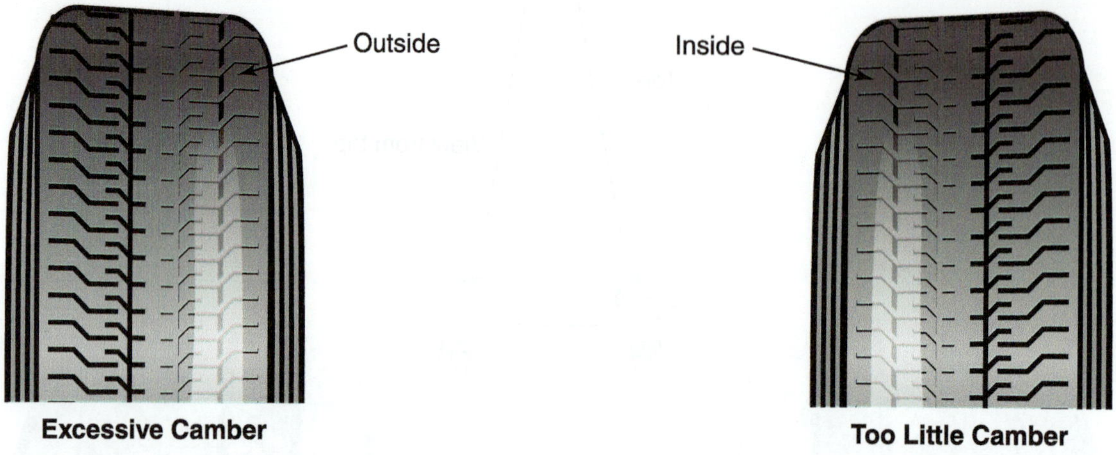

FIGURE 27–23 Caster angle

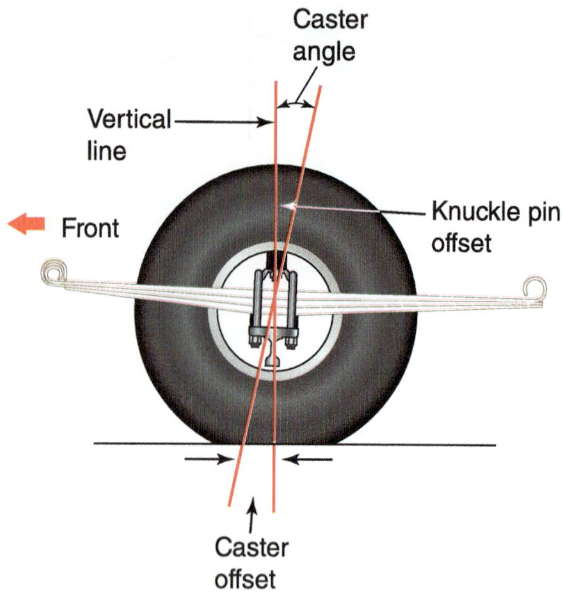

mounting (on rim), failed shock absorbers, and underinflation.

Other mechanical problems that can cause excessive wear include a sprung or sagging axle, brakes that are out of adjustment, or out-of-round brake drums. A sprung or sagging axle will cause the inside dual tire to carry a greater load. Improper brake adjustments will lead to spotty tire wear, and out-of-round drums will usually wear tires in a single spot. Rapid or uneven tire wear also may be caused by a sprung or twisted frame, worn wheel bearings, loose spring clips, or loose torque rods.

Load Distribution

Improper loading can cause the tires on one side of a truck or trailer to carry a greater percentage of the load. This can affect starting from standstill, causing the drive wheels to slip on the lighter side and wear faster, as well as cause uneven wear of the tires on either side of an axle. Although the gross load may not be excessive, one wheel, one axle, or one side of the truck may be overloaded because of improper distribution of the load. On semitrailers, distribute the load so that each axle and the fifth wheel are carrying their share of the load. **Figure 27–24** shows some incorrect (left side) and correct (right side) methods of distributing loads on vehicles.

Other Causes of Tire Wear

All tires are speed rated. Exceeding rated speed creates heat. Excessive heat produced by running a vehicle at higher speeds will shorten tire life. At higher speeds, the tire can become distorted. Higher speeds can cause small cuts to enlarge, causing a blowout. High-speed distortion exerts a strain on the tire fabric, which may cause tire failure.

SHOP TALK

Manufacturers of speed restricted tires caution operators not to exceed the speed limits and guidelines in their manuals. To alert the operator, a decal should be located in the cab, warning of the limitations of speed restricted tires.

TIRE ROTATION AND REPLACEMENT

Tire rotation procedures are a matter of shop and preventive maintenance (PM) practices. A regularly scheduled tire rotation plan can help reduce overall tire

FIGURE 27–24 Load distribution do's and don'ts

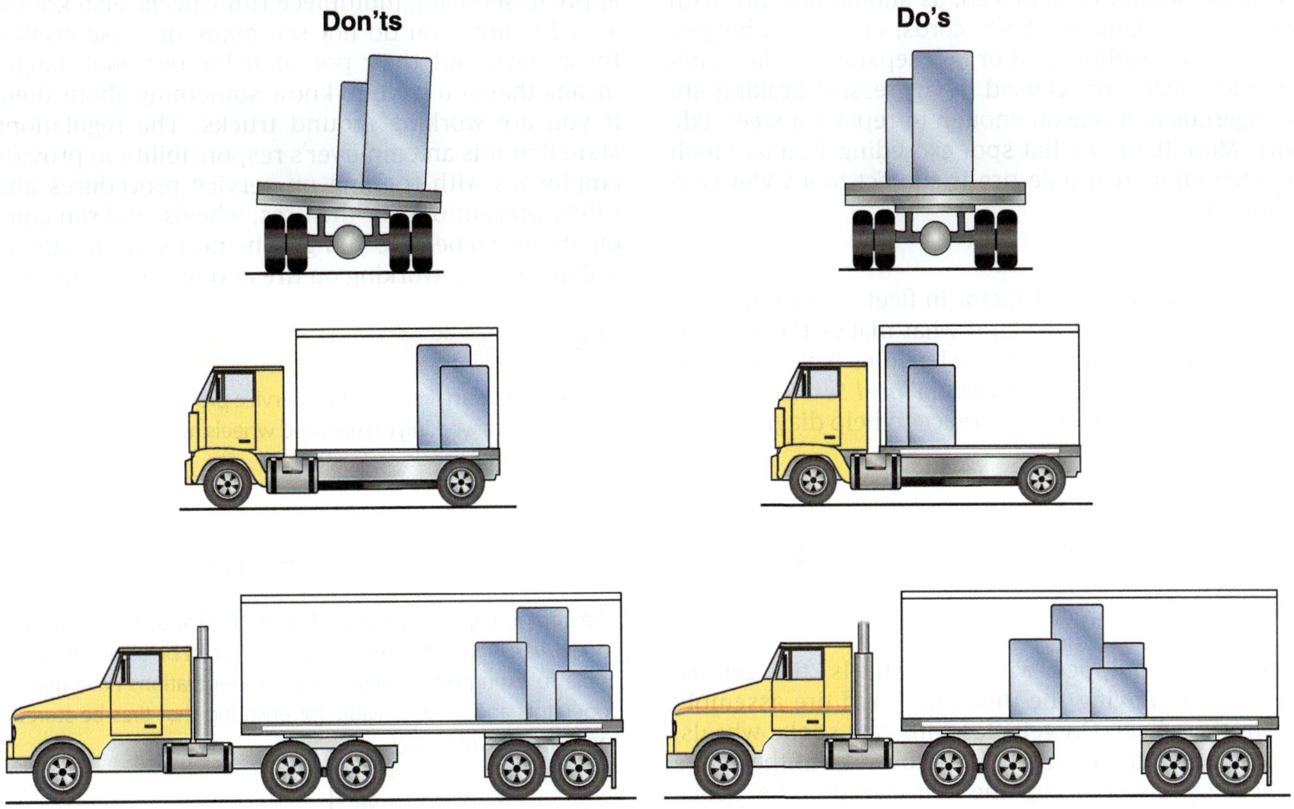

costs. Relocation of tires from the front-to-rear wheel positions depends on the type of truck being operated and the size and type of tires (**Figure 27–25**). Some fleets have their own regulations about what tires are placed on steering axles. For instance, they may follow the practice of never putting a used tire of any sort on a steer axle.

FIGURE 27–25 Suggested tire rotation patterns for (A) two-axle truck; (B) tractor and semitrailer; (C) three-axle truck; and (D) tractor with tandem-axle drive and semitrailer with tandem axle

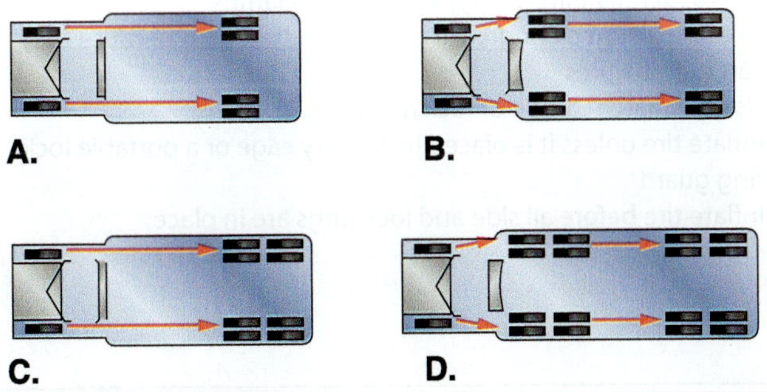

Tread Depth

Federal Motor Carrier Safety Regulations require a vehicle to have at least 4/32 inch (3 mm) of tread depth on its steer front tires. It is not recommended that retreaded tires be installed on steer axle wheels but it is not actually illegal in most jurisdictions. Tires that fail the standard for steer axles may be relegated to drive axle or trailer axles and used until the legal minimum of 2/32 inch (1.5 mm) of tread remains. Any tire with less than 2/32 inch (1.5 mm) of tread (measured within a groove and not over wear bars) must be taken out of service. Tires measuring:

- Less than 4/32 inch (3 mm) thread depth on a steer axle
- Less than 2/32 inch (1.5 mm) thread depth on any other axle

are subject to CVSA OOS citation. Most operators understand the importance of steering axle tires. Because of liability factors, most will not take risks with tires used to steer rigs.

Regrooving

Tires with the word *regroovable* molded to the sidewall may be regrooved. These tires, along with recapped and retreaded tires, should never be used as front steer tires. Tires with visible fabric breakage, or tires that have been repaired with a blowout patch or boot, should be replaced, as should any tire with exposed or damaged body cords, or bumps, bulges, or knots caused by cord or belt separation. Flat spots on steer axle tires caused by aggressive braking are dangerous and reason enough to replace a steer axle tire. More than one flat spot exceeding 1 square inch in size on a steer axle tire is subject to a CVSA OOS citation.

Tire Reference Manual

Tires are a major cost factor in fleet operations so it makes sense to understand what makes them fail. It is recommended that service facilities have a copy of the TMCs *Radial Tire and Disc Wheel Service Manual* (TMC item T0122) as a reference to help diagnose tire conditions.

27.4 TIRE, RIM, AND WHEEL SERVICE

On trucks equipped with disc wheels, tire service involves removing the disc wheel and tire assembly from the wheel. On vehicles with cast spoke wheels, the rim and tire assembly is removed from the spoke wheel by dismounting the rim clamps. The spoke wheel/hub is not disturbed.

The range of wheel systems and wheel/rim combinations makes a detailed illustration of tire-to-rim mounting and dismounting beyond the scope of a textbook. Detailed tire-to-wheel/rim changing procedures are available from all major tire and wheel manufacturers in their service literature.

The Occupational Safety and Health Administration (OSHA) has established rules and regulations that apply to servicing multipiece rim wheels, also known as split rims. You do not see many of these around these days, but their potential for personal danger means that you have to know something about them if you are working around trucks. The regulations state that it is an employer's responsibility to provide employees with training on service procedures and safety precautions for the tires, wheels, and rim combinations, so be sure you get the necessary hands-on training before working on tire and wheel assemblies.

CAUTION

You cannot learn tire and wheel service procedure from a book alone. Before working on tires and wheels make sure you have received some hands-on training.

SHOP TALK

Tires are a major fire hazard when trashed and stored in large quantities. Once a tire fire has ignited, it can be almost impossible to extinguish. Federal and state regulations have made operators and garages liable for ensuring that tires be stored and disposed of legally.

TIRE AND RIM SAFETY

Anyone working with tires should become familiar with some basic tire and rim safety handling before attempting to demount and mount tires. Some of these basic rules are outlined as Do's and Don'ts in **Table 27–1**.

TABLE 27–1 Tire and Rim Safety

Do's	Don'ts
• Instruct all tire and rim-handling personnel how to mount tires safely • Remove valve core to completely deflate tire before disassembling tire from rim • Use proper tools both to mount and demount tires • Use approved rust-retarding compounds to keep rims clean and free from rust and corrosion • Use correct size rim for specified tire • Avoid rim damage when changing tires • Examine inside of tire before mounting and dry thoroughly if any moisture is found • Use proper tubes and flaps with radial tires • Install side or lock ring split directly opposite (180 degrees) from valve stem slot • Use correct tire lubricant, but sparingly, to minimize the possibility of fluid entering the tire, especially in radial tire applications	• Overinflate tires • Overload rims • Remove tire from rim before completely deflating • Attempt to correct seating of side or lock rings by hammering while tire is inflated; *always* remove air pressure first • Use corroded, damaged, or distorted rims, rings, or rim parts • Fail to identify different types of wheel clamps • Use petroleum oil or grease on tire beads or rims because they can ruin the tires • Use mismatched side or lock rings • Inflate tire unless it is placed in a safety cage or a portable lock ring guard • Inflate tire before all side and lock rings are in place

When working on tires, technicians should ensure that they remain out of the trajectory of any of the tire or wheel components as they are broken down. Under some circumstances, the trajectory can deviate beyond the indicated danger zone. You should never attempt to seat split rings when a tire is totally or partially inflated and you should be aware that air pressure in an inflated truck tire mounted on a rim/wheel contains an almost explosive energy potential. This can cause tire/rim components to burst, which can result in personal injury or death.

Tire Service Accidents

Working around truck tires is not dangerous if all safety precautions are properly observed. However, tire accidents are more common than they should be because so many shortcuts are taken. It does not hurt to be aware of the top three causes of tire service related injuries to technicians (data from *Transportation Topics 9.13*).

1. **Lifting accident.** A jacked axle unexpectedly drops. Ensure that mechanical stands are placed under any raised axle end.
2. **Blowout during inflation.** Before inflating any tire on the vehicle, inspect it for damage. Do not attempt to inflate any tire on the vehicle when pressure tested is at 50 percent or less below specified. Remove it and inflate in a tire service cage.
3. **Rollover accident.** Driver attempts to move vehicle during tire service injuring the technician. Always use lockout tags when servicing vehicles.

REMOVAL FROM A VEHICLE

Before removing any tire and wheel assembly or tire and rim assembly from the vehicle, set the spring brakes and chock the wheel assemblies not being removed. Jack up the vehicle using a heavy-duty jack.

Raise the axle where the wheel and tire assembly is being removed and support it on adequate capacity jack stands. Mechanically cage the parking brake chamber when working on any wheels equipped with spring brakes. Apply the parking brake when working on front brakes.

When removing the wheel nuts from spoke wheels, loosen (do not remove) the nuts ¼ to ½ inch (6 to 12 mm) and then rap on the clamps (lugs) with a hammer to free them from the wheel. It is important to free the clamps from the wheel before removing the wheel nuts. Failure to free the clamps may cause them to spring from the wheel under extreme pressure, which could cause serious personal injury.

SHOP TALK

Disc wheel nuts for right-side wheel generally have right-hand threads, and wheel nuts for left-side wheels usually have left-hand threads.

It is important when demounting aluminum wheels to ensure that the end of the wheel wrench or socket is smooth. Burrs on the end of the wrench or socket can tear grooves in the wheel around the cap nuts. This not only ruins their appearance; it may, in time, cause the disc to crack.

CAUTION

Never raise a vehicle with a jack placed under a leaf spring. When the wheel has been raised, use heavy-duty axle stands placed under the axle and do not rely on a hydraulic jack alone.

CAUTION

When handling tire assemblies, remember to lift properly, using your legs rather than your back, as shown in **Photo Sequence 10**. Make use of a wheel dolly whenever possible.

DEFLATION AND DISASSEMBLY

Examine the tire/rim assembly for proper component seating before removing it from the vehicle. Some tire manufacturers suggest deflating tires completely before removing them from the vehicle for any type of service. When deflating a tire, first reduce tire pressure by pushing the Schrader valve core plunger; then remove the complete valve core, as shown in **Figure 27–26**. You should be wearing safety glasses and avoid positioning yourself in front of the air discharge.

SHOP TALK

The valves used on truck tires are known as Schrader valves, identical to those used on cars and bicycles. It is probably the only common component found on all vehicles, ranging from a bicycle to the heaviest off-highway earth moving equipment.

Avoid using steel hammers to disassemble or assemble rim components. Hammering rings or rims with steel hammers may cause fracturing of steel from the hammer and tire assembly components. Use a composite plastic or rubber-covered, steel-headed hammer. The procedure for manually breaking down (dismounting) tube and tubeless tires from rims is illustrated in **Figure 27–27**. Make sure that the tire is completely deflated before beginning.

RIM INSPECTION

Rims and rings must be matched by size and type. These components cannot be interchanged except

Photo Sequence 10

MOUNT A SET OF DUALS TO A CAST SPOKE WHEEL

P 10-1 Before disassembling the wheel, check for cracks in the cast spokes, bent and damaged studs, and evidence of rim rotation on the wheel.

P 10-2 Use a wire brush to remove rust from the exposed part of the studs and nuts. Badly rusted studs must be replaced.

P 10-3 With an air wrench, back off the nuts about halfway on all the studs. Important: Do not completely remove the wheel nuts.

P 10-4 Strike the inside edge of the wedge clamps, with nuts loosened only, with an 8-lb hammer until the wedge clamps release; when they release, the tire rim should drop forward. When this happens, the nuts and clamps can be removed, followed by the rims and spacer band.

P 10-5 Before reassembling the wheel, check that the 28-degree bevel mounting surfaces on the cast spokes are clean and free of lubricant. The studs should be clean and rust free. Ensure that the wedge clamps are those specified for the wheel.

P 10-6 Lift the inside tire and rim assembly onto the wheel and properly position it. Ensure that the valve is centered between two spokes.

P 10-7 Now fit the spacer band over the wheel and tap it into position with a rubber mallet.

P 10-8 Now fit the outside tire and rim to the wheel. Again, ensure that the valve is centered between two spokes.

P 10-9 Install the wedges and nuts over the studs. Use a hammer (tire hammer or 8-lb sledge) gently on the rim and rotate the wheel to set runout.

P 10–10 Initially torque the nuts in the sequence to 50 lb-ft. Important: Do not use an air impact wrench in place of a torque wrench. Now check wheel runout by rotating the wheel and using an upright hammer head to eyeball the runout. Correct runout if necessary.

P 10–11 Torque the wheel nuts in sequence to 80 percent of final torque. Then torque to the OEM final torque specification.

FIGURE 27–26 Removing tire valve core (Schrader valve)

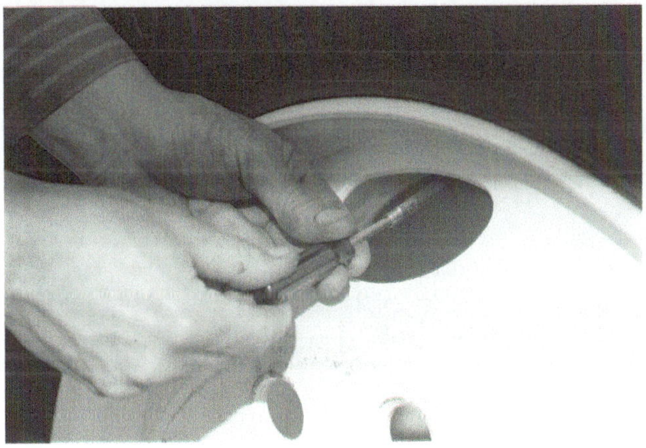

as provided for in a Multipiece Rim/Wheel Matching Chart. Select the proper tire size and construction to match the original equipment manufacturer (OEM) rim or wheel rating and size. The diameter of the tire must match the diameter of the rim. Never use any rim or wheel component you cannot positively identify.

Thoroughly inspect all metal surfaces when the tires are being checked, including the contact areas between duals and the inboard side of wheels. Examine for the following:

- Excessive rust or corrosion buildup
- Cracks in metal
- Bent flanges
- Deep tool abrasions on rings or in gutter areas
- Damaged or missing rim drive plates
- Matched rim parts

It is not a bad practice to mark defective parts for destruction to ensure that they will no longer be used, given the potential danger. Remember that a leak in a tubeless tire assembly can be caused by a cracked rim. Do not put a tube in a tubeless assembly to correct this problem. Cracked rims should be destroyed (use a torch) to avoid accidental reuse. Never attempt to weld or otherwise repair cracked, bent, or out-of-shape components; it is illegal as well as dangerous.

MOUNT TIRE TO RIM

Remove rust, corrosion, dirt, and other foreign material from all metal surfaces using a wire brush or wheel, as shown in **Figure 27–28**. This is especially important in the rim gutter and bead seat areas. Check the mating surfaces of side/lock rings in multipiece assemblies.

Paint the rim with fast-dry primer to help prevent rust from forming, as shown in **Figure 27–29**. If you do this, always allow paint to dry before assembling components. Apply lubricant to bead seat area, tire bead, tire flap, and rim mating surfaces just prior to mounting the tire. Use only those lubricants recommended by the rim and tire manufacturer and never use petroleum, silicon, or water-based lubricants. These can damage the rubber, cause rust buildup, or produce tire-to-rim slipping. Install the tire onto the rim either manually or using a bench tire/rim installer.

AUTOMATED TIRE SERVICE STATION

A much better and safer method of servicing truck tires on rims is to use an automated tire service station such as the Hunter equipment shown in **Figure 27–30**. Using this equipment effectively

FIGURE 27–27 Removing a tire from a disc wheel: (A) separating tire bead from wheel; (B) lubricating the tire bead; (C) prying the bead over the wheel; and (D) removing the second bead

FIGURE 27–28 Clean wheel/rim components before assembly.

FIGURE 27–29 Painting the rim

FIGURE 27–30 Features of the Hunter TCX 640 HD

The Hunter TCX640HD is an efficient electrohydraulic-powered tire changer that combines a range of unique features and capabilities to make wheel service for trucks, buses, tractors and other specialized machinery faster, easier, and more profitable.

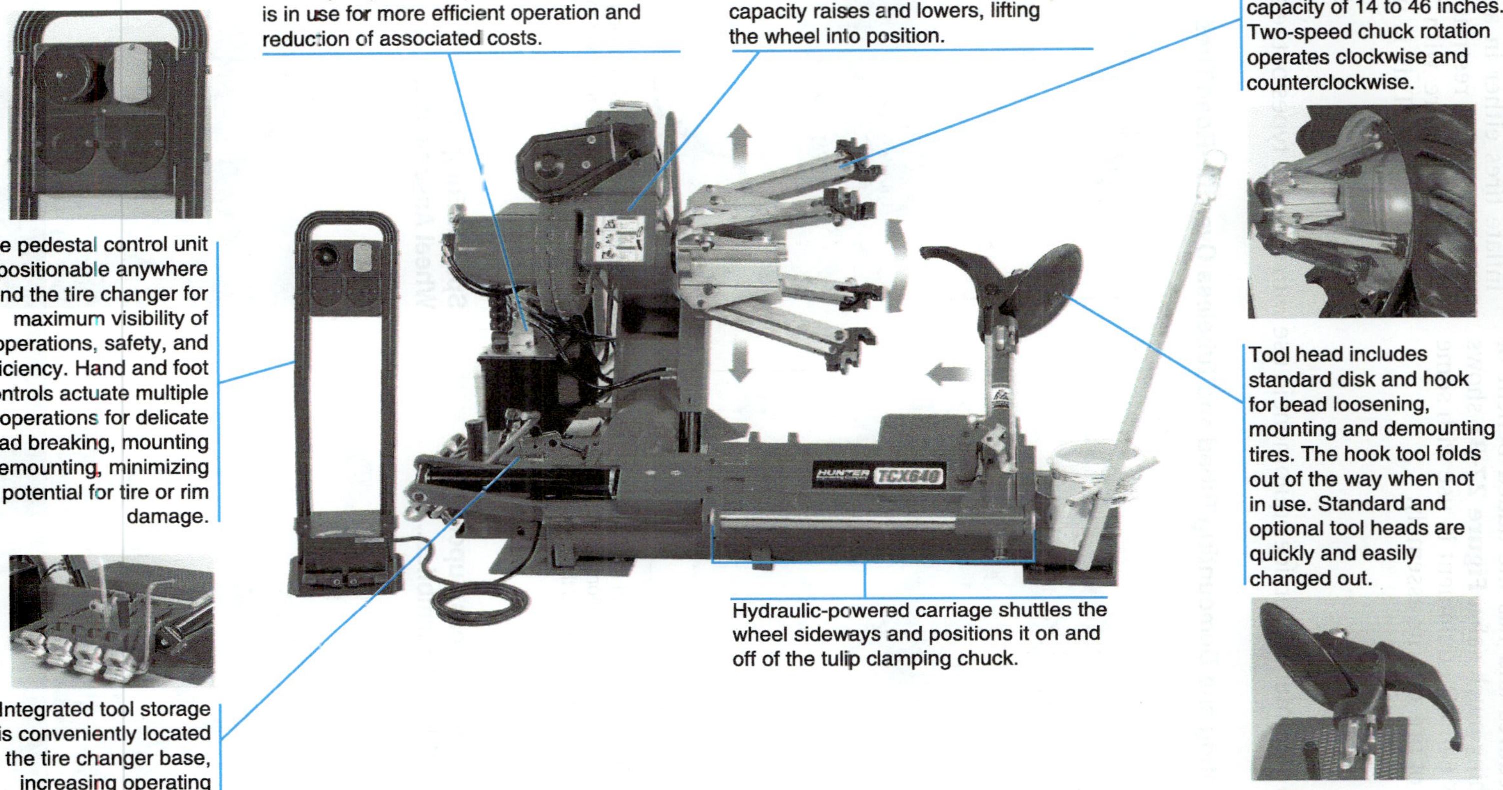

requires less training than manual tire breakdown/assembly procedures and has the added benefit of making everything much safer. **Figure 27–31** shows the Hunter TCX 640 HD equipment in use with some different types of tire and rim assembly.

Inflation

Inflate tires either in a safety cage (**Figure 27–32**) or in a portable restraining device. Seat the bead by charging the tire with air with the Schrader valve core removed. This is usually sufficient to seat the

FIGURE 27–31 Hunter tire service equipment in use with different types of tire and rim assembly

Mounting and Demounting Tubed and Tubeless Over-the-Road Tires

Position the wheel pliers and roll the first bead onto the rim. Reposition the pliers and roll the second bead on. It's that easy!

Bead loosening and demounting can be as easy as rolling the tire off of the rim in a single step.

Tubed and Tubeless Super Single Wheel Assemblies...

The standard hook tool is used to mount and demount tubed and super single tires or more difficult-to-handle wheel assemblies.

Special Equipment Wheel Assemblies...

Tractor, implement and other specialty machine wheel assemblies are serviced using the same tools and techniques.

Courtesy of Hunter Engineering Company, Inc.

FIGURE 27–32 Inflating a tire in a safety cage

FIGURE 27–33 Using a clip-on chuck with a remote in-line valve and gauge when inflating

beads if you have applied the appropriate lubricant to the bead seat. When you are sure that the bead has seated, install the Schrader valve core and inflate the tire to its specified pressure. Use a clip-on air chuck with a remote in-line valve and gauge (**Figure 27–33**) that enables you to stand clear of the tire as it inflates.

WARNING

Igniting quick start (ether) inside a tire is a far too common dangerous practice used to seat tire beads. The explosion that results depends on the proportions of air and ether combined inside the tire and the consequences can be deadly.

SPOKE WHEEL INSTALLATION

When mounting the tire/rim assembly onto the wheel, check that you have the correct nuts, studs, and lugs/clamps. Spoke wheels use rim studs. Rim studs are threaded on both ends with a nonthreaded section in the middle of the stud. The studs are coated with an anaerobic locking compound that essentially glues them into the wheel spoke. If a stud has been struck during removal (not unusual), you will have to pull it and install a new one using a stud tool.

Rim lug nuts should be kept tight and checked on a regular basis. Checking alignment of the rim/wheel installation is important because the rims can be drawn out of alignment when improperly tightened. Photosequence 10 outlines some of the key steps when mounting a set of duals to a cast spoke wheel.

The following are general installation instructions for installing a dual set of tire/rims to cast spoke wheels:

1. Slide the inner tire/rim over the cast spoke wheel and push it back into position against the tapered mount stops. Make sure that the valve stem faces out and is centered between the two spokes. It should clear the disc brake calipers, if the wheel is equipped with disc brakes.
2. Slide the spacer ring over the wheel, tapping it into position against the inner tire with a tire hammer. Check the spacer ring for concentricity by rotating the spacer ring around the cast spoke wheel.
3. Lift and slide the outside rear tire and rim assembly on the wheel, making sure that the valve stem faces inboard and is located 180° across from the inner valve stem.

CAUTION

Watch your fingers and back when installing tire/rim assemblies onto cast spoke wheels. It is good practice to wear gloves, and lifting with the tire/rim behind you is easier on your back.

4. Assemble all the wheel lugs and nuts. Turn the nuts on their studs until each nut is flush with the end of each stud. Tap the outside tire/rim with a tire hammer to keep it concentric while you are doing this.
5. Turn the top nut until it is snug.
6. Rotate the wheel and rim until the opposite or near-opposite nut is at the top position and snug the nut.

7. Rotate the wheel and use a criss-cross pattern to snug the remainder of the nuts.
8. Ensure that you rotate the wheel and rim and torque each nut when it is in the top rotational position: Because the entire weight of the tire and rim assembly is on the top spoke, this criss-cross sequence will help ensure an even application of force at all points on the rim, keeping the rim in alignment.
9. Repeat the sequence of tightening the nuts incrementally to the OEM torque.
10. After operating the vehicle approximately 50 miles (80 km), check the lug nuts for tightness in the same sequence. Once each week inspect and retorque the wheel stud nuts.

Spoke Wheel Runout

Any time a wheel hub or tire has been reinstalled, the wheel and tire runout should be checked after the wheel stud nuts have been torqued to specifications. To check runout, position a wooden block (or tire hammer) on the floor approximately ½ inch (12 mm) from the tire, as shown in **Figure 27–34**. Rotate the wheel and watch the variation in gapping between the tire and block. If the wheel runout exceeds ⅛ inch (3 mm), it should be corrected.

Position a piece of chalk on the wood block as shown and rotate the wheel through one rotation so that the chalk marks the tire high spots. The high and low (unmarked) areas indicate which lug nuts have to be loosened and which have to be tightened to correct the condition. Slightly loosen the lug nuts 180 degrees opposite to the chalk marks and tighten those on the chalk-marked side of the tire. Do not overtorque the nuts. Recheck runout and repeat until runout is within ⅛ inch (3 mm). If runout cannot be corrected in this way, inspect for part damage or dirt between the mating parts.

FIGURE 27–34 Checking tire runout using a hammer head and chalk

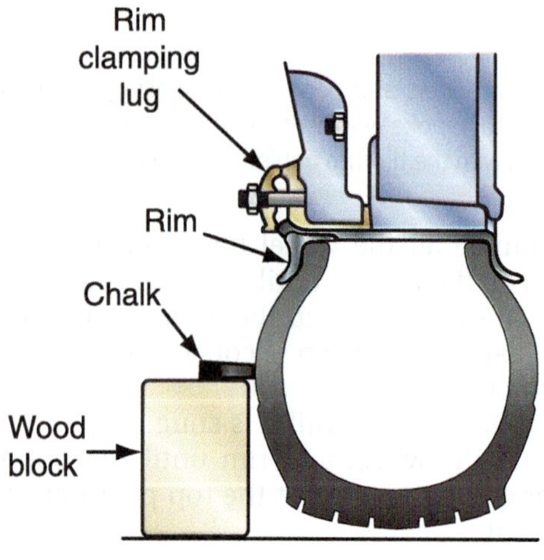

DISC WHEEL INSTALLATION

Disc wheels are mounted to wheel hubs with either threaded studs and nuts or with headed wheel studs. A headed wheel stud either has serrations on the stud body or a single flat machined into the head of the stud to prevent it from turning in the wheel hub, as shown in **Figure 27–35**. In some disc wheel systems, the end of the stud that faces away from the hub is stamped with an "L" or "R," indicating that left- or right-hand threads are used. Studs marked with an "L" are installed on the driver side of the vehicle, and right-hand threads are installed on the passenger side. This helps ensure that wheel rotation does not loosen the fasteners. Other systems use right-hand threads only. Whatever type of fastener is used, ensure that all hardware is in good condition.

Stud-Piloted Disc Wheels

The following are general installation instructions for installing stud-piloted dual disc wheels that use conical nuts.

1. Slide the inner/rear or front tire and wheel in position over the studs and push back as far as possible. Use care to avoid damage to threads on the studs and inspect the valve stem to caliper for clearance.
2. Install the outer wheel nut on front wheels and the inner wheel nut on rear dual wheels. Run the nuts on studs until they come into contact with the wheel. Rotate the wheel a half turn to allow parts to seat naturally.
3. Draw up the stud nuts alternately, following the sequence (criss-cross pattern) as illustrated in **Figure 27–36**. Do not fully tighten the nuts at this time. This procedure should allow a uniform seating of nuts and ensure an even face-to-face contact of wheel and hub.

FIGURE 27–35 Headed wheel stud shanks for disc wheels.

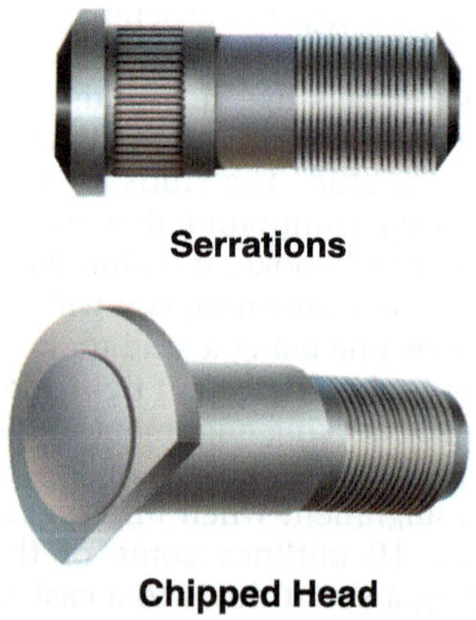

FIGURE 27–36 Disc wheel torquing sequence: Arrows indicate half turn to ensure even seating of the wheel to the hub.

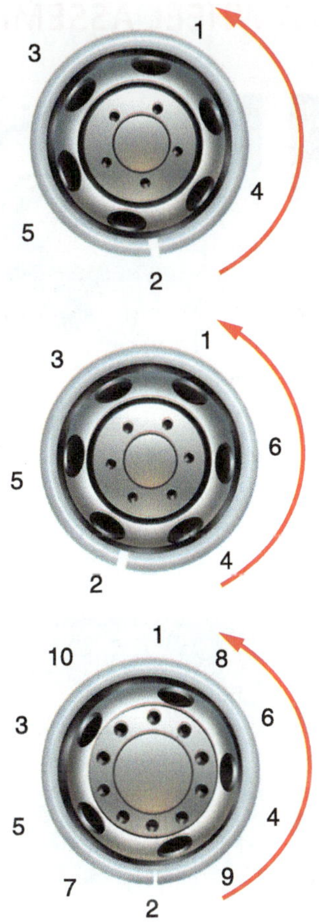

4. Continue tightening the nuts to torque specifications using the same alternating method.
5. Install the outer rear wheel and repeat the preceding method. Be sure that both inner and outer tire valve stems are accessible.
6. After operating the vehicle approximately 50 miles (80 km), check the stud nuts for tightness. Some break-in seating may be encountered, and the nuts will loosen. Retighten all nuts to specified torque.

CAUTION

To check and tighten the inner wheel torque, first loosen the outer wheel nuts several turns and tighten the inner nuts and then retighten the outer nuts. To avoid losing the seating of the outer wheel when checking the inner wheel torque, loosen alternate outer nuts, tighten the inner nuts, and retighten the outer nuts. Then loosen the remaining outer nuts, tighten inner nuts, and retighten the outer nuts. OEMs suggest that disc wheels require weekly inspections and torque checks.

Hub-Piloted Disc Wheels

Figure 27–37 illustrates a cross-section view of a disc wheel flange nut installation, and **Photo Sequence 11** runs through the installation of a set of dual hub-piloted wheels to an axle. The procedure used to assemble hub-piloted disc wheels is as follows:

1. Slide the inner rear or front tire and wheel in position over the studs and push back as far as possible. Use care so that the threads on studs are not damaged.
2. Position the outer rear tire and wheel in place over the studs and push back as far as possible. Again, use care so that the threads on the studs are not damaged.
3. Run the nuts on the studs until the nuts contact the wheel(s). Rotate the wheel assembly a half turn to permit parts to seat.
4. Draw up the nuts alternately following the (crisscross) sequence illustrated in Figure 27–36. Do not fully tighten the nuts at this time. This allows uniform seating of the nuts and ensures even face-to-face contact of wheel and hub (see Figure 27–37).
5. Continue tightening the nuts to torque specifications using the same alternating sequence.

FIGURE 27–37 Cross section of disc wheel with flange

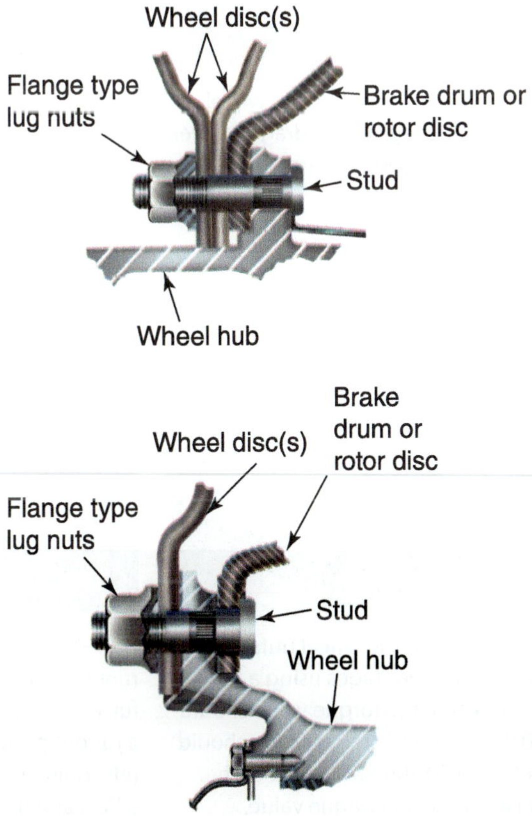

Photo Sequence 11

INSTALL A SET OF HUB-PILOTED DUALS TO A WHEEL ASSEMBLY

P 11-1 Before beginning the disassembly of the wheel, perform a visual inspection. Check that the wheel nuts are property engaged and look for damaged studs.

P 11-2 Remove rust and road dirt from the wheel studs with a wire brush.

P 11-3 Use an impact wrench to remove the wheel nuts.

P 11-4 Carefully remove the wheels, ensuring that they are not dragged over the studs damaging the threads.

P 11-5 Before reassembling the wheel, clean the hub, wheel, and hub/brake drum mounting faces of rust, dirt, and loose paint. Visually inspect all the studs. Do not paint or apply any other substance on the mounting faces.

P 11-6 If an outboard mounted drum is used, make sure that the brakes are released and lift drum into position, ensuring that it is not sitting on the pilot ledge. Lubricate the fasteners. (Hub-piloted studs must be lightly lubricated to ensure that the correct amount of clamping force is achieved.) Mount the wheels to the hub.

P 11-7 Tighten the wheel nuts in sequence in three stages using a torque wrench. First-stage torque value should be 50 lb-ft. Second-stage torque should be between 50 and 80 percent of the OEM-specified final torque value.

P 11-8 As a final step, check the wheel runout by rotating the wheel through a full revolution. Do not attempt to correct a runout problem by loosening and retorquing the nuts. Disassemble the wheel assembly and locate the problem.

TUBE-TYPE RADIAL TIRES

Use only rims approved for radial tire use by the rim OEM. Clean the rim components, removing all rust and other foreign material. Make certain that rim parts match and are not sprung or broken. Check the inside of the tire before mounting and dry thoroughly if any moisture is found. Sparingly lubricate the tire beads and the tube/flat and rim surface with an approved lubricant to minimize the possibility of fluid entering the tire. Use tubes and flaps that are compatible.

Radial tubes have a permanent red band on the valve stem, below the cap threads, or have either the word "Radial" or the letter "R" molded or stamped on either the valve stem, sleeve, or ferrule affixed to the valve stem.

Mounting Procedure

Position the tire assembly in a safety cage and inflate to the recommended operating pressure, deflate completely, and then reinflate to the correct pressure. This will allow the tube, flap, and tire to properly seat. Visually check the slot and side ring gap (on a two-piece rim) to make sure that the bead is seated. A further check should be made by laying the tire flat and measuring the space between the rim flange and one of the lower sidewall rim line rings. Take measurements around the circumference of the rim flange. If the spacing is uneven, deflate the tire completely and then disassemble, remount, and reinflate.

TUBELESS-TYPE RADIAL TIRES

Again make sure to use only rims approved for radial tire usage by the OEM. Clean the rim, removing all rust and other foreign material. Lubricate the tire beads and rim bead seats with an approved rubber lubricant.

Mounting Procedure

Position the tire assembly in a safety cage and inflate to the recommended operating pressure. Because of radial truck tire construction, it may be necessary to use an inflation aid to help seat the beads of tubeless radial tires. There are three types of inflation aids:

- Metal ring inflation aid: ring rim and uses standard shop compressed air.
- Rubber ring inflation aid: more common; basically it reduces leak path during inflation. It helps to have a large supply of compressed air to help this type to seal.
- Rapid air discharge: consists of a 5-gallon (20 L) air tank and 2-inch (50 mm) discharge hose and dispersal nozzle. The TSI Cheetah™ system is the best known of the air-actuated bead seating shop tools.

Check the bead seating by laying the tire flat and measuring the space between the rim flange and one of the three lower sidewall rim line rings. Take the measurements around the circumference of the rim flange. If the spacing is uneven, deflate the tire completely and then demount the tire, remount, and reinflate.

SHOP TALK

When you have problems seating a tire bead into a rim, use the tire inflation aids described in this chapter to help seat beads.

DUAL TIRE MATCHING

Matching of dual tires is important for several reasons:

- Tire life greatly increases when tires are properly matched in tread pattern, diameter, and circumference.
- Improperly matched tires cause costly mechanical problems due to differential carrier failure resulting from constant differential action.

Improper traction also results from mismatched tires and can cause failure of both tires in a short operating time. The term *mating tires* basically refers to matching tires to the same nominal size. Matching the tread patterns also must be considered in mating. Both duals should be of the same tread design.

When pairing tires in a dual assembly, the tire diameters should not differ by more than ¼ inch (6 mm) or the tire circumference by more than ¾ inch (19 mm). The total tire circumference of one driving rear axle must match, as nearly as possible, the total tire circumference of the other driving rear axle.

CAUTION

The larger the diameter of the tire, the more likely it will be to overdeflect and overheat. Smaller diameter tires may have insufficient road contact and, as a result, wear faster and unevenly. Tread or ply separation, tire body breaks, and blowouts can occur from mismatched duals.

There should be sufficient space between dual tires for air to flow and cool the tires, plus prevent them from rubbing against each other. Rims and wheels of the same size, but of different makes and types, can have different offsets, which would affect dual spacing. If there is sidewall contact between tires or between the inside tire and the chassis, refer to the tire OEM catalog to determine the minimum dual spacing. Refer to the rim or wheel OEM catalog to determine the correct offset.

The installation of a new tire next to a used or worn tire is considered mismatching. It is also critical to use tires of the same construction (bias or radial) on the drive axle, as we have already outlined in this chapter. If you cannot avoid a diameter variation between dual tires, place the larger tire on the outside. If two tires are of equal size but one is slightly more worn than the other, place the less-worn tire on the outside.

Duals must be checked for adequate spacing. Make sure that they are not kissing (contacting each other), especially at the 6 o'clock position. When performing this test, the tires should be fully inflated. There are several methods of measuring dual sizes:

- **Square Method.** Using a square is the standard method of checking dual diameter matching on the vehicle (**Figure 27-38**). The square leg must be placed parallel to the floor to avoid the tire "bulge." Measure the distance (if any) between the tire tread and the square arm with a ruler. It should not exceed ¼ inch (6 mm).
- **Straightedge.** A straightedge can be placed across the four tires on a dual axle to compare tire diameters. Measurements should be taken from the straightedge to the tire tread where gaps show (**Figure 27-39**). The measurement should be doubled to obtain the diameter difference.
- A **taut string** can be used in place of the straightedge.
- **Tire Meter.** A tire meter checks a single tire for size when it is not mounted to the vehicle (**Figure 27-40**).
- **Tape Measure.** A flexible tape measure can be used to check the circumference of an unmounted tire (**Figure 27-41**). Make sure that the tape runs around the tread center line of the tire. A difference of ¾ inch (19 mm) in circumference is normally acceptable in mated tires.

FIGURE 27-38 Using a square to check dual tire matching

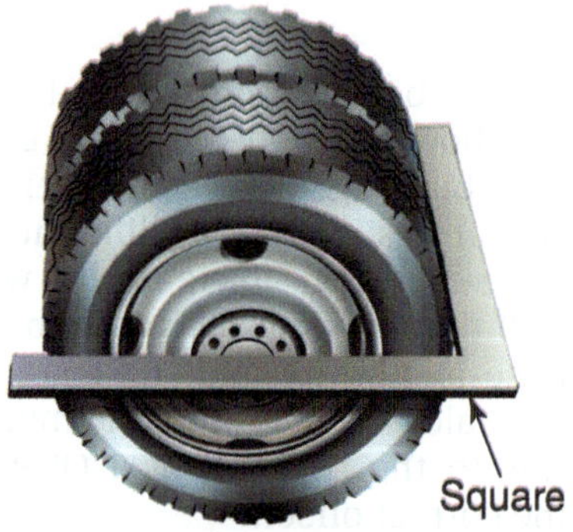

FIGURE 27-39 A straightedge positioned across the tires will detect difference in tire size.

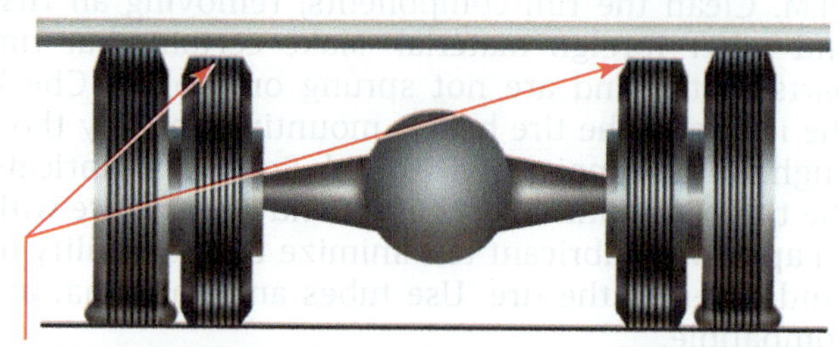

FIGURE 27-40 Measuring the size of an unmounted tire using a tire meter

FIGURE 27-41 Measuring tire circumference with a flexible tape measure

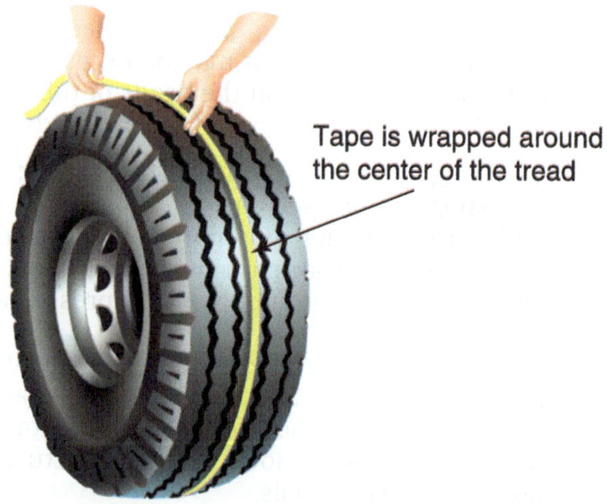

FIGURE 27–42 Dual spacing with wheel assemblies

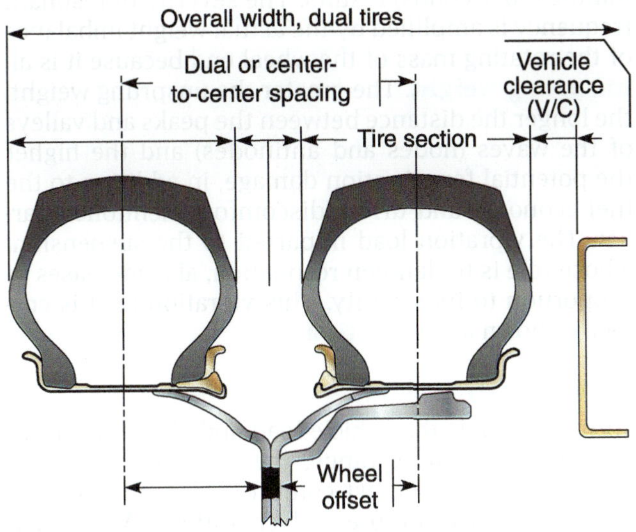

FIGURE 27–43 Dual spacing with rim assemblies

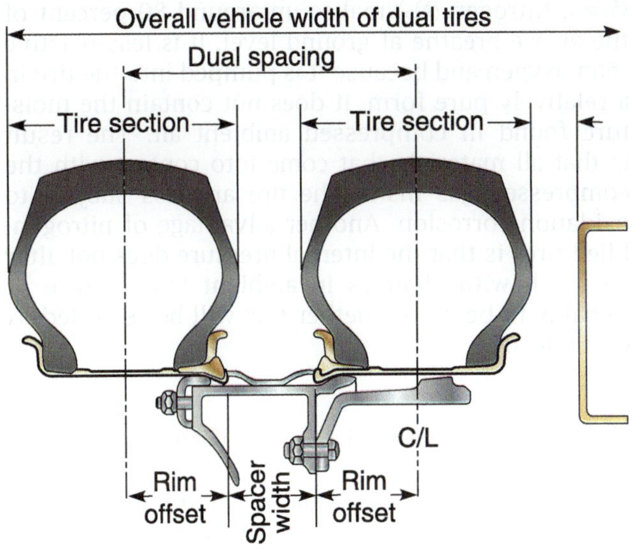

Dual Spacing

Dual spacing for vehicles using disc wheels is determined by the sum of the offset of both wheels used (**Figure 27–42**). Dual spacing for cast spoke wheels with rims is determined both by the offset of the demountable rims used and by the width of the spacer band (**Figure 27–43** and **Figure 27–44**).

Three types of spacer bands are available—corrugated, channel, and corrugated-channel. Inspect spacer bands for concentricity to ensure that the bands have not been distorted or bent. If a spacer band must be replaced because of distortion, misalignment, or corrosion, make sure that the replacement band is of equal size. An improperly sized band can alter the overall vehicle width and may not allow the rims to seat properly on the wheel.

Checking Tire Pressures

An accurate pressure gauge is required to check tire pressure. Discourage drivers from checking tires by rapping them with a mallet; this test can do no more than identify a complete flat. Any pressure gauge reads with the most accuracy in center of its sweep. This means that truck tire gauges should be purchased with a 200 psi (13.8 bar) reading capability. When checking tire pressures on trucks, the check should be done with the tire cold; that means not driven for 3 hours. It also means that no part of the tire should be in direct sunlight. Actual ambient temperature does not matter. Even when a vehicle is equipped with automatic tire inflation system, tire pressures should be checked once per month with a master pressure gauge.

FIGURE 27–44 Sectional view of mounted duals on a cast spoke wheel

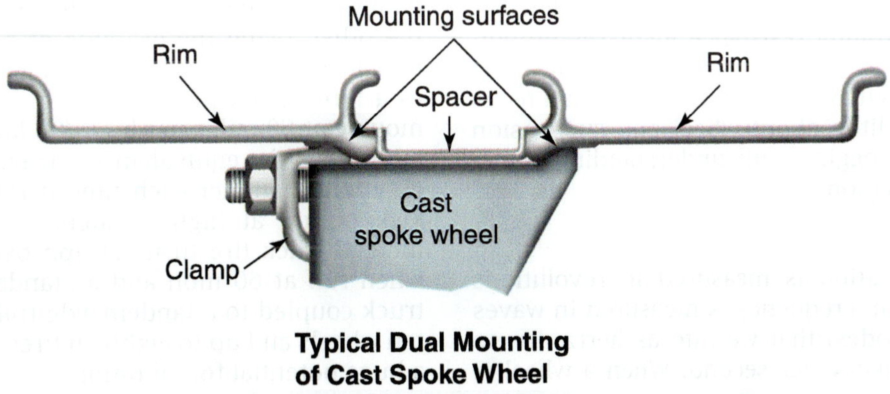

Nitrogen Filled

Some fleets are experimenting with **nitrogen-filled tires**. Nitrogen (N_2) makes up around 80 percent of the air we breathe at ground level. It is less reactive than oxygen and because it is pumped into the tire in a relatively pure form, it does not contain the moisture found in compressed ambient air. The result is that all materials that come into contact with the compressed gas inside the tire are less subject to oxidation corrosion. Another advantage of nitrogen-filled tires is that the internal pressure does not alter so much with changes in ambient temperature. It remains to be seen whether this will be regarded as cost effective.

27.5 WHEEL BALANCING

Wheel balance is the equal distribution of weight in a wheel assembly with the tire mounted. This is an important factor that affects tire wear and vehicle control. In this section we will take a look at the general principles of wheel imbalance conditions. The procedure for performing tire and axle end balancing will be outlined at the end of the chapter.

AXLE-END VIBRATION

The causes and consequences of axle-end imbalance are complicated but it is worth understanding them because it costs the trucking industry millions of dollars annually. Because some manufacturers of on-highway truck tires state that their product does not require balancing on installation, we wrongly assume the entire axle end to be in balance. It should be noted that the tire represents only a percentage of the dynamic rotating mass of a truck or trailer axle-end assembly. When an axle end is dynamically balanced, tire life can be doubled because it is the tire (even one with perfect static balance) that assumes most of the punishment.

In an axle-end out-of-balance condition, the resonant effect is reflected through the suspension, the frame, and the cab right through to the driver's seat. This fact produces a number of other consequences that affect vehicle ride, handling, and suspension life. In addition, tire rolling resistance increases proportionally with the extent of out of balance, meaning that there can be a major hit on fuel economy. It helps to understand a little about wheel and suspension dynamics, which begins with understanding something about resonation.

Wheel Dynamics

The speed of rotation is measured in revolutions per minute or rpm. Frequency is measured in waves (nodes and antinodes) that we rate as hertz. **Hertz** is the wave frequency per second. When a wheel is out of balance it has a tendency to hop so if **wheel hop** frequency is measured at ten times per second (10 hertz), then a resonant frequency of 600 times per minute (60 seconds) results. The severity of resonant frequency is amplified by the actual weight imbalance of the rotating mass of the wheel end because it is all **unsprung weight**. The greater the unsprung weight, the longer the distance between the peaks and valleys of the waves (nodes and antinodes) and the higher the potential for vibration damage, in addition to the fuel economy and driver discomfort mentioned earlier. The vibration load imparted to the suspension, whose role is to dampen resonation, also increases in proportion to its severity. This vibration load is correctly known as **oscillation**.

Suspension Dynamics

Springs support the vehicle load and all springs have a working operating range, the limits of which are known as **jounce** (most compressed condition) and **rebound** (least compressed condition). As we discussed in Chapter 26, some springs have a built-in damping action such as multileaf spring packs, which use interleaf friction to suppress oscillation. Because of this self-damping action, multileaf spring packs on truck suspensions seldom require shock absorbers.

However, the most common truck suspension is the air suspension. Air springs have no self-damping capacity so they use shock absorbers. Shock absorbers dampen axle oscillation by using simple hydraulics to smooth the resonant nodes and antinodes and they are pretty good at doing this provided a balanced axle end and normal road conditions. However, when the severity of the oscillation exceeds the shock absorber's ability to suppress the resulting resonation, the vibration conducts to the chassis, and coincidentally ends up destroying the shock absorber.

Root Cause

The heavy spot in a wheel end is subject to centrifugal force for most of its rotation and comes into contact with the road surface just once per revolution. This is the basis of rotational reverberation frequency. You can get a better understanding of this effect by striking a truck tire with a 4-pound shop hammer on one side while feeling the reaction on the other. Using the example of a real world truck tire, just 6 ounces of static imbalance can multiply to a force of 60 pounds (27 kg) when the wheel is moving at 60 miles per hour (96 km per hour) down a highway, the equivalent of striking the tire with a 60-pound hammer each time it rotates. This radial kick occurs at high frequency. An average commercial truck tire turns at approximately 600 rpm when run at 60 mph and a standard tandem-drive truck coupled to a tandem-axle trailer has a total of ten wheels and up to eighteen tires. That amounts to a huge potential for vibration.

Importance of Axle-End Balance

The true cost of out-of-balance axle ends is only just becoming understood. Driver comfort, fuel economy, and premature equipment failure (tires, suspensions, and shocks) are all affected by axle-end imbalance so ensuring balanced wheel ends has the potential to save the trucking industry millions of dollars. Although an out-of-balance steer axle-end condition is usually going to be identified by the driver before it becomes too severe, it is important to recognize that it is a factor of every wheel that contacts the road surface.

ASSESSING WHEEL IMBALANCE

Radial truck tires tend to be more sensitive to balance and alignment problems than bias ply tires. A typical tattletale of an out-of-balance axle end is irregular tire tread cupping. A tire/wheel assembly that is out of balance or not rolling true causes uneven tread wear and vibration. In such cases, check:

- Inflation pressure. This is always the first check.
- Proper bead seating. A twisted bead seating can cause vibration, especially on front axles.
- Balance. If the tire/wheel assembly is out of balance, it should be balanced either statically or dynamically. Most OEMs recommend dynamic wheel balancing, which can be performed on the vehicle or by using an internal balancing agent.
- Alignment. Different service conditions may require different settings.

There are different procedures and methods of balancing axle-end assemblies and obviously some work better than others. The following are some methods used to balance tires/axle ends on trucks:

1. Correct lateral or radial runout. Runout must be corrected to within ⅛ inch (3 mm) before attempting the balancing operation. In some cases it may be necessary to remount the tire onto the rim/wheel assembly.
2. Check for loose wheel bearings or kingpin play. Either, or both, must be corrected before attempting to balance the wheel.
3. Check for wheel weight distribution. Check for equal amounts of weights on each side of the rim/wheel when a wheel is known to have been static balanced. For example, where a 16-ounce (454 g) weight is required, 8-ounce (227 g) weights should be installed on each side of the rim/wheel assembly directly opposite to each other.
4. Check preparation of rim/wheel. If you are going to balance a wheel, make sure that all old balance weights, mud, dirt, and foreign material are removed from the rotating assembly before attempting the balance operation.
5. Be sure that tire pressure is correct.
6. Inspect rim/wheel. Inspect side ring/lock ring openings on the rim assembly. The gap should not be less than 3⁄32 inch (2 mm) or greater than 5⁄16 inch (8 mm). Anything other than this could indicate an improperly seated lock ring assembly.
7. Balance the wheel.

BALANCING WHEELS

A large percentage of out-of-balance axle-end complaints on truck wheels focus on the front steering axle because driver control (and comfort) is most affected. Maximum front tire service life is achieved when the front axle ends are balanced, tire inflation pressures are maintained, and front axle alignment is properly set. However, today there is an increasing awareness of the high cost of axle end out-of-balance problems on every wheel on a rig.

Static Balance

This is accomplished using wheel weights (see **Figure 27–45**). When static balancing a tire/wheel assembly a maximum of 18 ounces (510 g) of wheel weights may be used to balance any one tire. If more weight is required, it is suggested that the tire be removed from the rim/wheel assembly, rotated 180 degrees, and remounted. This will, in many cases, bring the assembly within the acceptable limits but it should be noted that this procedure can be time consuming and does not balance the brake and hub assembly. Also note that the use of lead weights is now illegal in many jurisdictions.

Dynamic Balance

There are five methods of dynamically balancing tire/wheel assemblies:

1. Spin balancing. This method is most commonly used on truck front steer axle wheels. The truck tire is raised off the shop floor and spun up, driven by an electric tire balancer. A strobe light is used to identify the balance points to attach weights. The advantage of this method is that the entire wheel, brake drum/disc included, is balanced but it is time consuming. Because it is often used, the procedure for spin-balancing wheels on the vehicle is outlined later in this section.
2. Off-the-vehicle dynamic balancing. This method uses the same principles as the previous method except that the wheel assembly is removed from the chassis and spun up using a specialized wheel balancer.
3. Internal balancing agent (IBA). This is probably the best means of balancing axle ends because **internal balancing agents (IBAs)** dynamically adjust to correct an out-of-balance location in a wheel assembly. **Counteract balancing beads (CBB)** are a proven IBA technology that provide

FIGURE 27–45 Weights attached to the rim flange and brake drum are used to balance the wheel assembly.

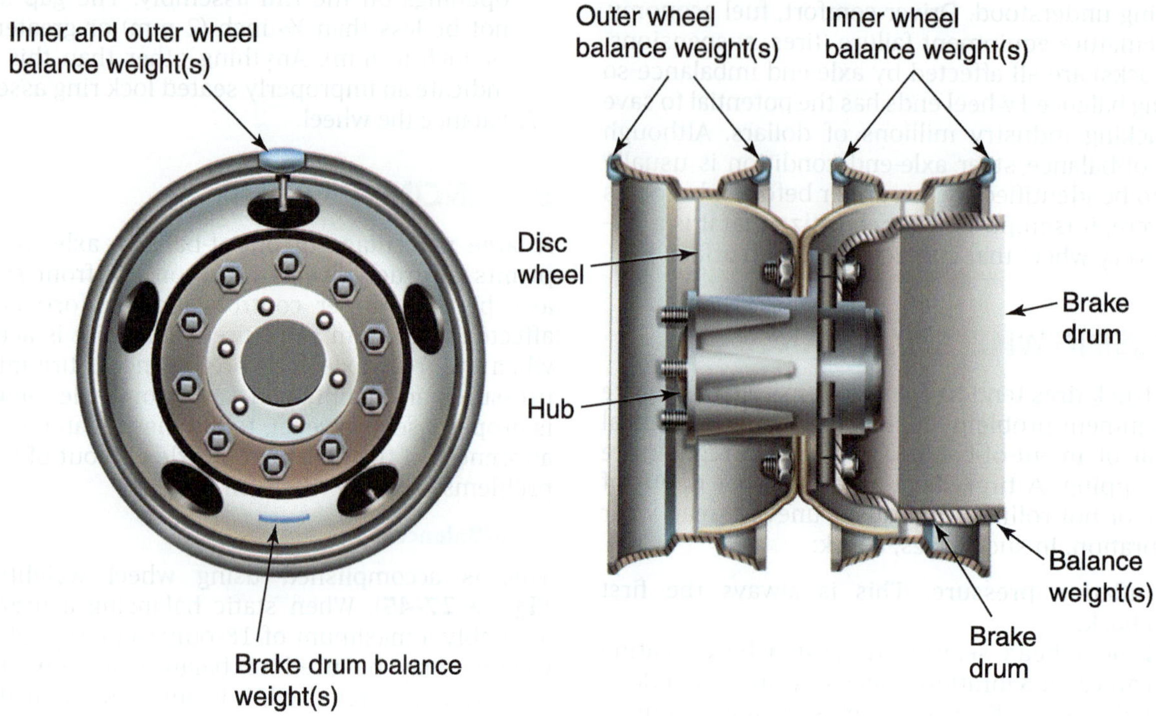

a simple solution to a known out-of-balance axle end and are cheap enough to use as a preventive measure in all wheel ends. CBBs are Teflon-coated tempered glass beads that are injected into the tire through the Schrader stem or are placed directly into a disassembled tire. They are commonly used on wide-base singles (the effects of imbalance tend to be more pronounced) but they can be used on all tires to correct axle-end balance problems. CBBs are conducive to electrostatics, which provides them with a memory, but they will respond to correct either a sudden or progressive change in a state of balance condition. **Figure 27–46** shows axle-end out-of-balance effect and a demonstration of how an electrostatic balancing agent can remedy the problem.

4. Balancing fluids. As with counteract balancing beads, balancing fluid is injected into the mounted tire assembly. These fluids are available commercially and are best used as a stopgap temporary repair.
5. External balancing rings. Products such as Balance Masters make counterbalance products for anything that rotates including truck wheels. In a truck wheel application, these use mercury within a tubular ring attached to the brake drum; centrifugal force generated by the rotating wheel locates the mercury in the counterweight position to balance the assembly. A significant disadvantage is upfront cost.

WHEEL BALANCE PROCEDURE

A couple of commonly used methods are outlined here. The dynamic spin method requires the use of an electric spinner and strobe light shown in **Figure 27–47**. The spinner should be power rated at around 8 horsepower for truck wheels. The wheel to be balanced is first raised off the ground and then the drive roller of the electric spinner is loaded to contact the tire. This accelerates the wheel assembly to road speed. The wheel assembly includes the tire, wheel, and brake drum.

Prebalance Inspection

Perform the following before running the spin test:

1. Measure and correct air pressure.
2. Measure and adjust wheel runout.
3. Check wheel bearing endplay and correct if necessary.
4. Inspect the front end for loose or defective steering components.
5. Remove any rocks lodged in tire grooves.

WARNING

Never stand in front or behind a tire that is spun up on a dynamic wheel spinner. A small rock could be thrown out by centrifugal force with the effect of a bullet.

FIGURE 27–46 Out-of-balance axle end and CBBs

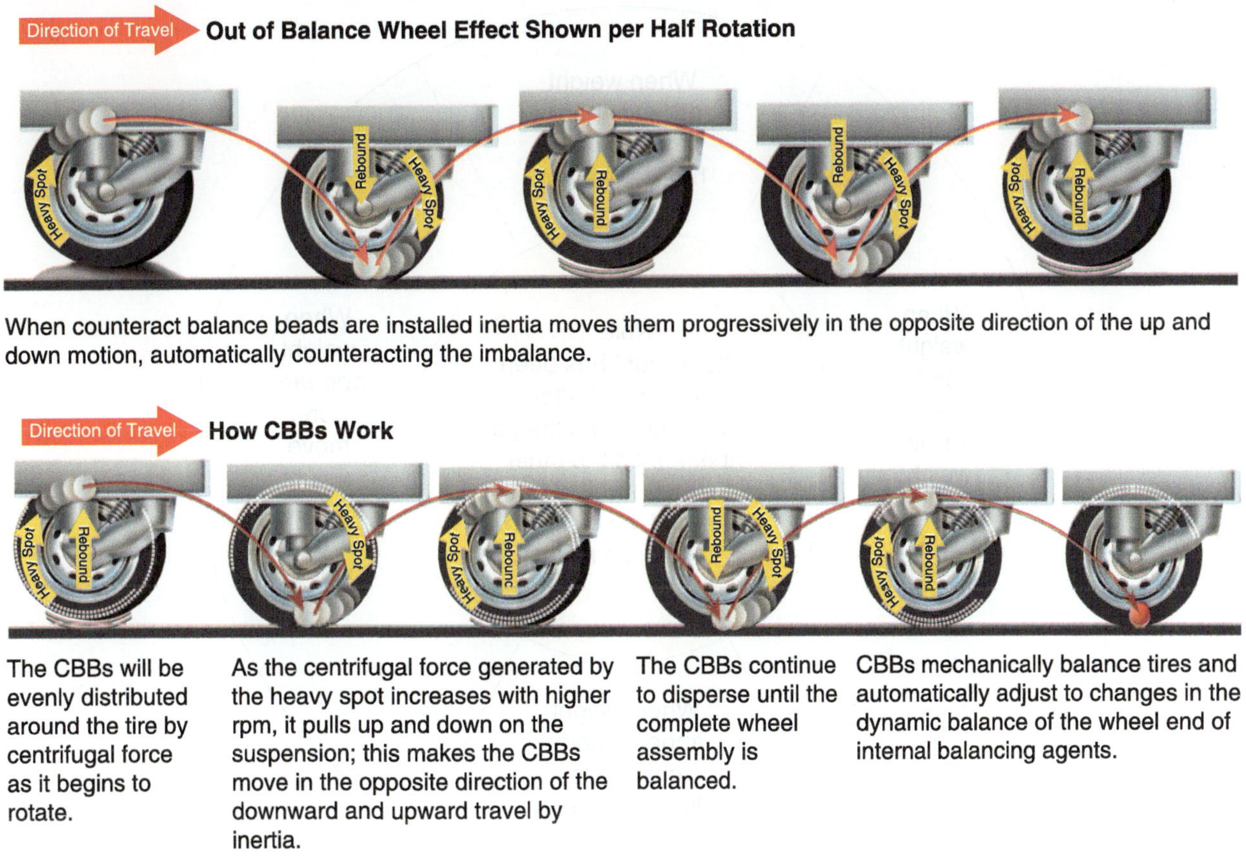

FIGURE 27–47 Dynamic wheel balance equipment: a technician using a spin balancer and strobe light to balance a steer axle end

Courtesy of Hunter Engineering Company, Inc.

CAUTION

When spinning up drive axle wheels, make sure that you remove the axle shafts before running up the wheels. Failure to do so can damage driveline components.

Kinetic Test Procedure

1. Run up the wheel fast enough to determine peak and break speeds on the strobe indicator.
2. Use **Figure 27–48** to apply and adjust balance weights.

CAUTION

Lead balance weights are illegal in some jurisdictions. Nonlead balance weights or internal (inside the tire) balance compensating media are required in these instances.

Counterbalance Beads

A simple, cost-effective alternative to the spin method of balancing axle ends is to install the CBBs described earlier in this chapter in every tire on a truck rig. This may be done by injection through the Schrader valve core, or installed directly into the tire when in broken down state. A typical tire treatment consists of depositing a 10-ounce (283 g) charge of CBBs into each tire on a rig.

FIGURE 27–48 Dynamic balance weight chart

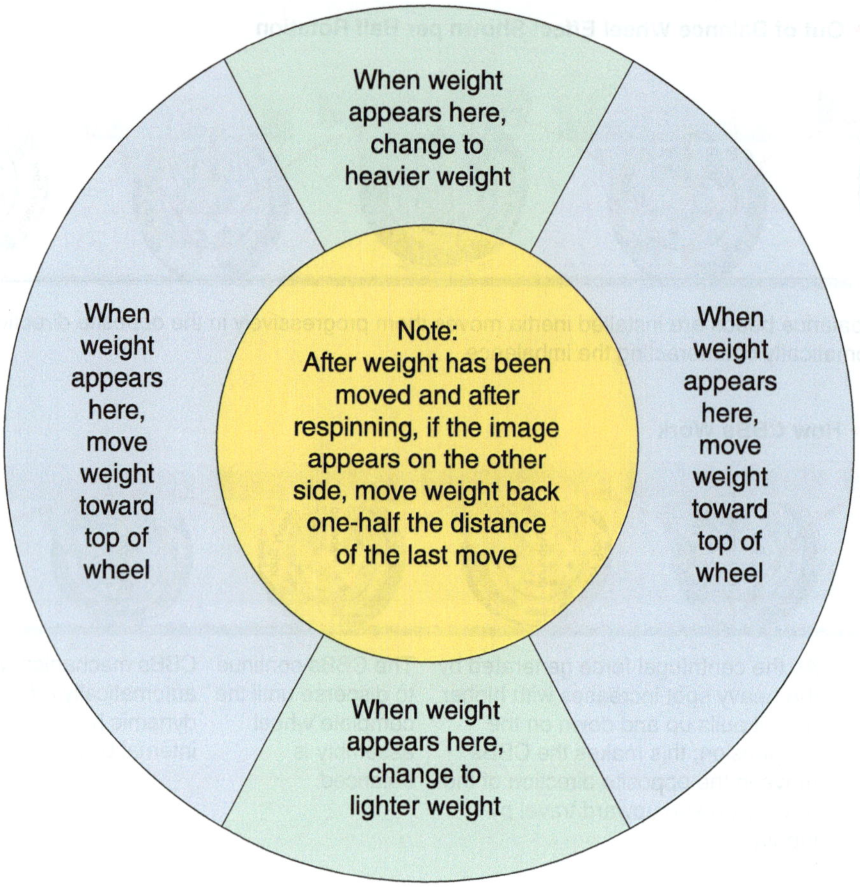

27.6 CHASSIS TIRE PRESSURE MANAGEMENT

Tire pressure monitoring and management systems have become commonplace in the trucking industry over the past few years. They have been widely accepted because of the payback in increased tire longevity. All cars made in 2008 on must be equipped with a tire pressure monitoring system, and it is only a matter of time before trucks are legislated to have at least a pressure monitoring system.

TIRE PRESSURE MONITORING SYSTEM

Some fleet operators claim that a wireless **tire pressure monitoring system (TPMS)** can pay for itself within a single year in reduced per unit tire expenses. The TPMS has been widely used on automobiles for a generation. A TPMS is an entry-level approach to managing vehicle tire costs. All systems monitor tire pressure, some also monitor temperature. If the pressure falls short of a threshold value, the operator is alerted to the condition. After the alert has been broadcast, intervention is required; that is, the driver or a technician must inflate the tire(s) to the specified value.

The earliest types were equalization systems that distributed air from one tire to another. Today active systems known as **tire pressure inflation system (TPIS)** such as the Spicer Dana **tire pressure control system (TPCS)**, **Meritor tire inflation system (MTIS)**, and Hendrickson are widely used. TPIS monitors tire pressure throughout the chassis and uses chassis system air pressure to inflate tires when mobile. We are going to use the Meritor system as the basis for the description here but most others can be closely compared.

MTIS OPERATION

Air from the chassis air supply is routed to the MTIS controller. The location of the tire inflation system components on a typical trailer is shown in **Figure 27–49**. The MTIS controller then routes air to the individual wheels by means of air supply lines that pass through the axle tubes, exit at the axle spindle hub caps, and connect to the valves on the duals. A rotary union is located at the hub cap; this is the only moving

FIGURE 27–49 (A) External view of a TPIS and (B) TPIS diagnostic tools.

component in the system. It allows the outer air supply hoses connected to the tire valves to rotate with the wheels' rotary union bearings. Air is supplied to each tire connected into the circuit on an as-needed basis. System troubleshooting for air leaks can be performed with a soap solution and spray bottle.

MTIS uses antilock brake system (ABS) sourced air. A pressure protection valve set at 80 psi (550 kPa) means that the system will only receive air from the pneumatic circuit when system pressure is above 80 psi (550 kPa) so that braking pressures are not compromised. A system regulator sets the pressure. Adjusting the pressure is simple with each ¼ turn of the regulator being equivalent to 5 psi (30 kPa). An indicator light is used to alert the driver of the vehicle in the event of severe pressure loss, providing some advance warning of a tire or system failure. **Figure 27–50** is a detail of a TPIS.

FIGURE 27–50 TPIS hub cutaway

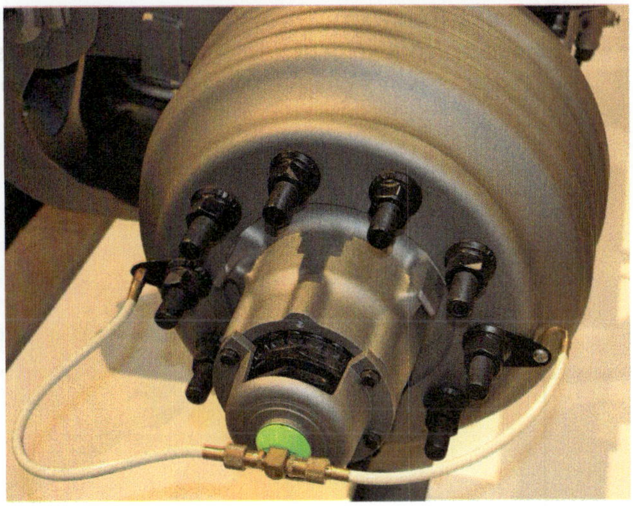

MTIS Safety

Safety features designed into the MTIS include a flow switch. When leakage exceeds 4 cubic feet (112 cubic liters) per minute, the flow switch isolates tire circuit, preventing further air bleed-down. An additional safety feature is thermalert, designed to signal a wheel end overheat condition. The thermalert device is an axle plug designed to melt at 280°F (138°C). When the thermalert melts, a very loud whistling noise is produced by air leakage. This is designed to alert drivers to an imminent danger condition, allowing them to pull over to the side of the road.

MTIS Retrofit

The MTIS system can be installed new or retrofitted to vehicles. Because there is only one moving component in the system, almost no maintenance is required. Field feedback suggests that fitting the system to trailers pays for itself in a short period because of tire maintenance savings and increased longevity. A typical installation time for a retrofit to a trailer is around 4 hours.

Alternate TPIS

There are a number of options to MTIS including the Hendrickson Tiremaax Pro system shown in **Figure 27–51**.

Tech Tip

Meritor recommends that you rely on MTIS to manage final tire inflation pressures. On installation, tires should be inflated to 80 percent of specified pressure. Next, the system should be allowed to build pressure to the specified tire running pressure. This is said to be more accurate than inflating manually.

FIGURE 27–51 Hendrickson Tiremax automatic inflation system

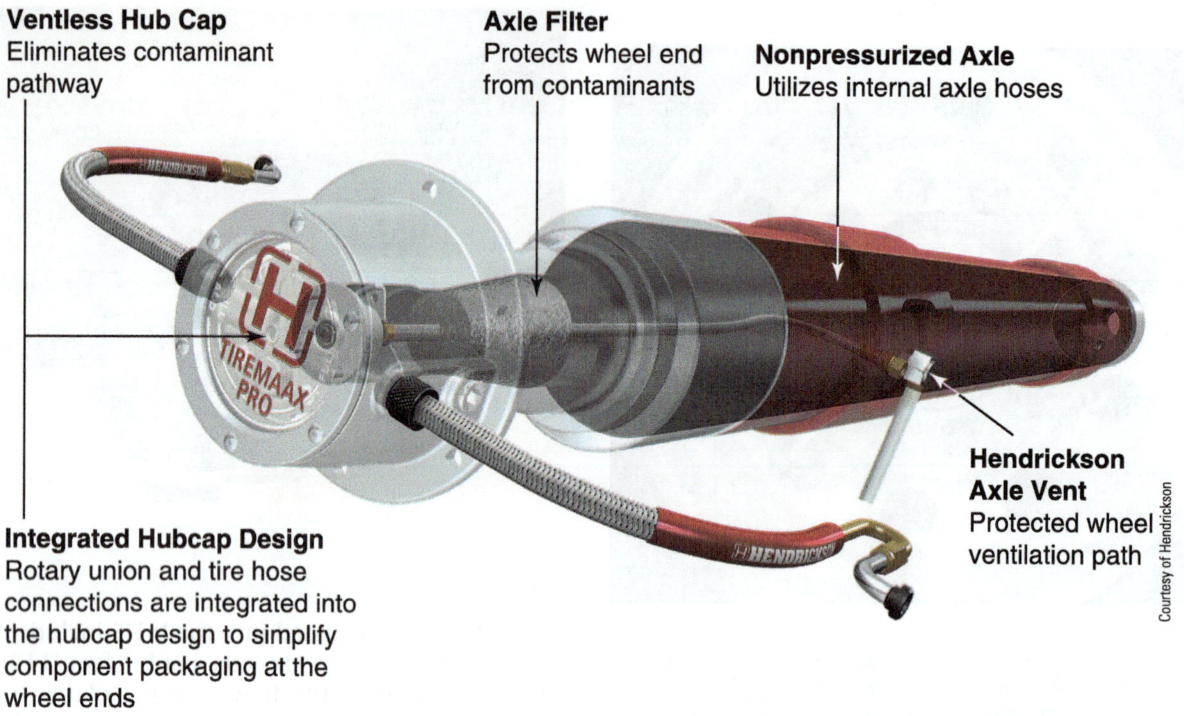

CVSA OOS STANDARDS FOR TIRES

This section contains some typical CVSA OOS standards for tires. Once again, it should be said that an OOS standard is not a maintenance standard; it defines the point at which a condition becomes dangerous.

OOS Any Tire on the Steering Axle(s) of a Power Unit

- With less than 4/32-inch (3 mm) tread measured at any point on a major tread groove
- With body ply or belt material exposed through the tread or sidewall
- With any visible tread or sidewall separation
- Has a cut where the ply or belt material is exposed
- Labeled "Not for Highway Use"
- A tube-type radial tire without radial tube stem markings
- Mixing bias and radial tires on the same axle
- Tire flap protrudes through the valve slot in the rim and contacts stem
- Boot, blowout patch, or other ply repair
- Weight exceeds tire load limit
- Tire is flat or has noticeable (can be heard or felt) leak
- Any bus equipped with recapped or retreaded tire(s)
- Mounted tire that contacts any part of the vehicle

OOS Tires Used on Other Than the Steering Axle of a Power Unit

- Weight exceeds tire load limit
- Tire is flat or has noticeable (can be heard or felt) leak
- Has exposed body ply or belt material
- Has any tread or sidewall separation
- Has a cut where ply or belt material is exposed
- Mounted tire that contacts any part of the vehicle or its mate (duals)
- Labeled "Not for Highway Use"
- With less than 2/32 inch (1.5 mm) tread when measured at any point on a major tread groove

27.7 WHEEL HUBS, BEARINGS, AND SEALS

For anyone employed in the trucking industry, having both a theoretical and hands-on knowledge of wheel end procedure is mandatory. In this section we will take a close look at wheel end components and finish outlining the procedures used to set up and adjust them. A wheel hub assembly has tire rims and brake drums that are mounted to it and contains the bearings and mounting hardware that attach the rotating components of the wheel to an axle spindle. **Figure 27–52** is a sectional view of the wheel end

FIGURE 27–52 Cross-section view of spoke and disc wheel assemblies (single-wheel configurations) showing the axle mounting and wheel bearings

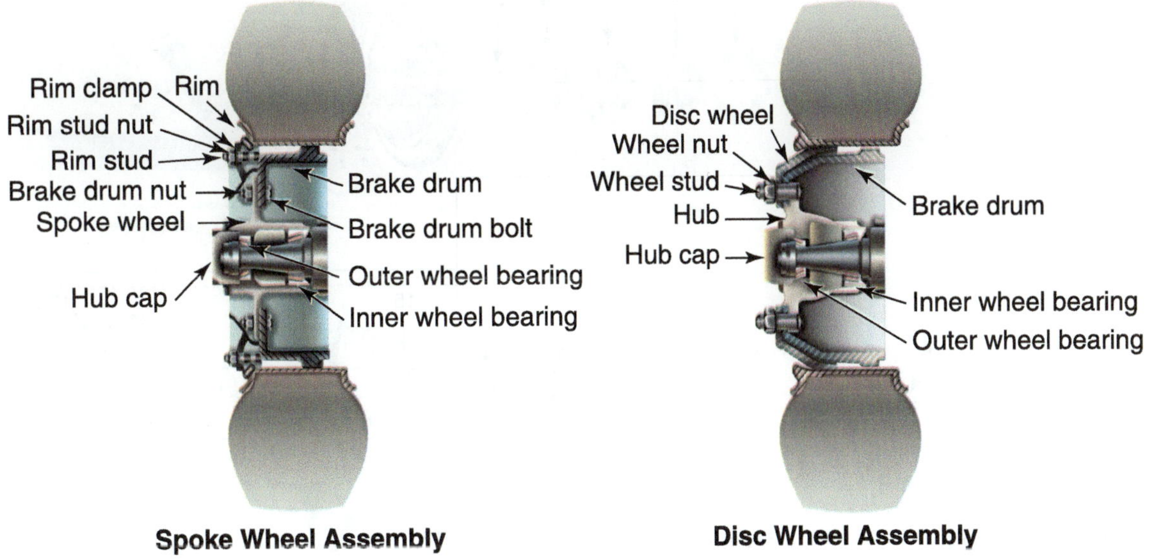

components on both cast spoke and disc wheels. **Figure 27–53** is an exploded view of the wheel end components.

TAPERED ROLLER WHEEL BEARINGS

A tapered roller wheel bearing assembly consists of a cone, tapered rollers, a roller cage, and a separate cup or race press-fit into the hub. All the bearing components carry the load the axle is subjected to with the exception of the cage, which contains and spaces the rollers around the cone. A tapered roller bearing should always be regarded as an assembly—the cup and cone should remain matched during their service life and be replaced together when necessary. Each axle hub has one inner and one outer tapered roller bearing assemblies and they are configured to support the entire weight of the vehicle.

Typically, the bearings are locked into position on the axle spindle by an adjusting nut and lock nut (jam). However, the means of adjusting and locking wheel bearings into position change. In adjustable wheel bearing systems, split forged nuts or castellated nuts and cotter pins could also be used to secure the hub to an axle. In a preset axle hub, no bearing adjustment is required to be made by the technician; the hub assembly is simply torqued onto the axle spindle to a predetermined torque specification.

Bearing adjustment and wheel hub installation are dealt with later in this chapter. Cast spoke wheels integrate the wheel and hub into a single unit. On disc wheels, the hub is a separate component in the assembly. On drive axles, the hub is also the interconnecting point for transmitting drive torque from the drive axle carrier to the wheels. Truck drive axle shafts are full floating and their only function is to transmit drive torque from the carrier to the drive wheels.

Roller Bearing Operating Principle

Tapered roller bearings are suited to applications where major radial and thrust loads have to be sustained. When used as wheel bearings they are used in opposed pairs; that is, with the rollers facing each other. The conical-shaped bearings, therefore, have a common intersection point on the shaft onto which they are mounted, allowing them to accommodate a range of loads while performing effectively. The shaft in the case of a wheel bearing is the axle spindle. Tapered roller bearings when used as wheel bearings are set with a fractional endplay; in other applications, they may be set with a fractional preload.

WHEEL SEALS

Wheel bearings are lubricated using either grease or oil. A wheel bearing lubricated with axle grease is known as a dry bearing; these are not common today. A wheel bearing lubricated with oil, usually gear oil, is known as a wet bearing. Most wheel ends use a wet bearing assembly. Whether a wet or dry bearing is used, the lubricant has to be contained in the hub with a seal. A wheel seal is a dynamic seal. It is installed into the axle hub, usually with a slight interference fit. A couple of different principles are used to seal the stationary (nonrotating) axle spindle and the rotating hub cavity in which the lubricant is contained.

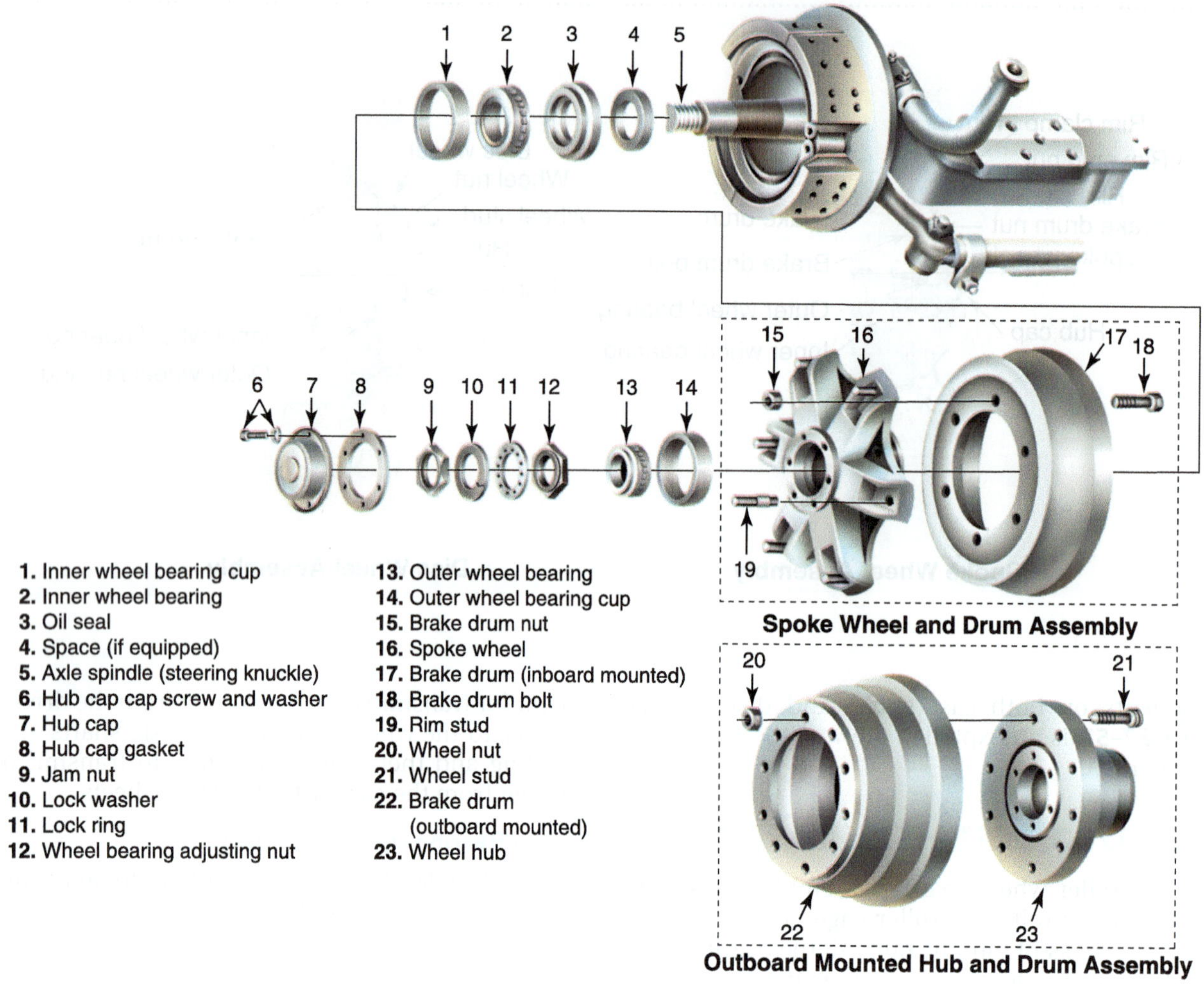

FIGURE 27–53 Typical front axle components (oil lubricated)

Unitized Seals

In a **unitized seal**, a static seal is made at both the axle hub installation bore and the axle spindle. The dynamic portion of the seal is self-contained within the seal itself. In other words, it turns within itself. A unitized seal is shown in **Figure 27–54**. The outer shell fits tight to the seal bore in the axle hub and the rubber inner shell fits tight to the axle spindle. The dynamic seal is created at the sealing lip assisted by spring pressure.

Lip-Type Seals

In a **lip-type seal**, the dynamic seal is created by having the seal medium, usually leather or synthetic rubbers, spring-loaded to ride on a wiper ring or wiper race. You can see a sectional view of lip-type seals in **Figure 27–55**. On the left side of this figure a lip seal using a standard wiper is illustrated, and on the right side, one with a grit guard wiper race. The grit guard type offers better protection when operated in dusty construction or agricultural environments.

Brake Drums

On cast spoke wheels, the brake drum is mounted to the inboard side of the wheel/hub and is held in place with nuts. Servicing inboard brake drums on spoke wheels requires removing the wheel/hub and drum as a single assembly. Because this involves removing the hub by separating the bearing assembly, bearing and seal service are required on reassembly.

On disc wheels, the brake drum is usually mounted on the outboard side of the disc hub. The drum fits over the wheel studs and is secured between the wheel(s) and hub. This enables the wheel and drum to be dismounted without disturbing the hub nut, so the bearings and seal remain untouched. The brake drum location on both cast spoke and disc wheel assemblies is shown in Figure 27–53.

FRONT AXLE WHEEL END SERVICE

This procedure makes use of a **wheel dolly**, a heavy-duty single- or dual-wheel hoist on rollers. The device

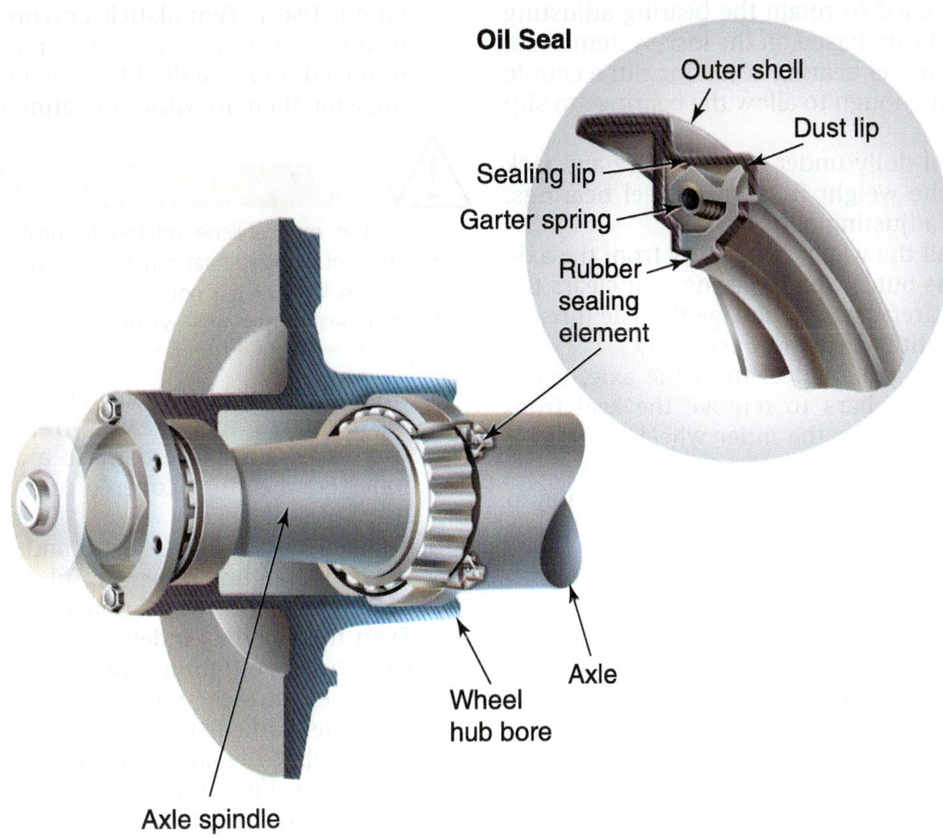

FIGURE 27–54 Example of a unitized oil seal

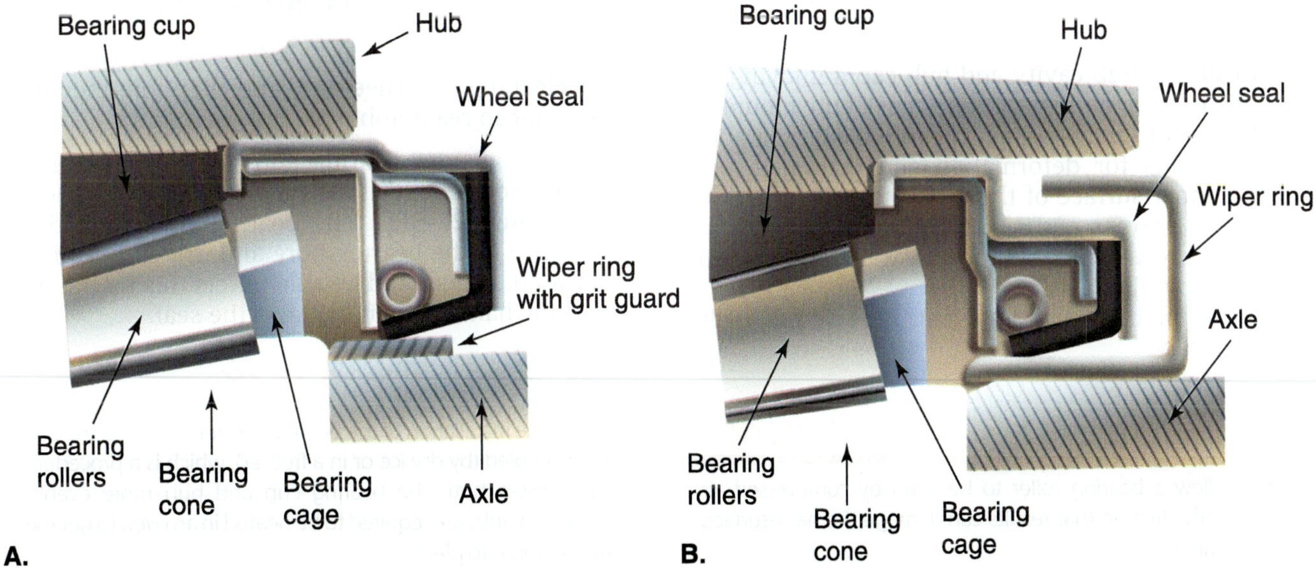

FIGURE 27–55 Typical metal-encased lip-type seals used for wet (oil-lubricated) bearings: (A) oil-lubricated wheel seal and (B) wiper ring with grit guard

can be rolled under a wheel, which is supported by a pair of forks when raised. To remove a front wheel from a truck steer axle without separating the tire/rim on a cast spoke wheel, proceed as follows:

1. Chock the rear tires and set the parking brakes.
2. Raise the front axle until the tires clear the shop floor. Place safety stands under the axle.
3. Back off the slack adjuster to release the brakes.

4. Remove the hub cap, hub cap gasket, outer bearing jam or locknut, washer(s), and whatever other hardware is used to retain the bearing adjusting nut. This will vary based on the lock system used.
5. Back off the wheel bearing adjusting nut a couple of turns, just enough to allow the bearings to slip their races.
6. Place a wheel dolly under the front tire and jack to remove the weight from the wheel bearings. Remove the adjusting nut.
7. Carefully pull the wheel dolly away from the axle and catch the outer wheel bearing as it clears the spindle. Be careful not to damage the axle threads with the weight of the wheel assembly.
8. After separating the wheel from the axle, use a couple of pinch bars to remove the seal from the hub and remove the inner wheel bearing. If the axle spindle has a seal wear ring, use a ball peen hammer in one location on the wear ring to enlarge it. A couple of taps are usually sufficient to pop it off the spindle.

SHOP TALK

If pulling more than one wheel, be sure to keep all of the components of each wheel together and separate from the other wheels.

On disc wheels, the procedure is simple. First remove the disc wheel and then follow the preceding procedure to pull the hub drum assembly.

Cleaning and Inspection

Next you will have to inspect the hub cavity, cap, and bearings.

1. Clean the hub cavity and hub cap, removing the old lubricant. Inspect the wheel hub mounting flange for wear, warpage, or rough edges. Inspect the studs for deformation and cracks. Inspect the inner surface of the wheel or hub for cracks, dents, or wear, and replace if required.
2. Use solvent to clean the brake drum. Wipe the axle spindle clean.
3. Wash the bearings in solvent. Allow them to dry naturally.

CAUTION

Never allow a bearing roller to be spun by compressed air because the friction that results can damage the hard-surface contact areas.

4. Visually inspect the bearing cones and cups. Replace them if they are worn, cracked, pitted, or otherwise damaged. Wheel bearing cups use an interference fit to the hub. On some aluminum hub bores, the hub has to be heated in an oven to expand the bore for both removal and installation. Use a Tempilstick crayon to avoid overheating. Wheel bearing cups on ferrous hubs are removed and installed by driving them out and pressing them in without heating the hub.

CAUTION

Never use oxyacetylene torches to heat aluminum hubs. This type of localized heat can weaken and often destroy the aluminum. If an oven is not available, you should replace the hub and bearing assembly as a unit.

5. Dip cleaned wet bearings in lubricant or pack dry bearings with axle grease. Wrap the bearings in wax paper and place them in a clean box or carton. Keep bearings covered until ready to install the new seal.
6. Inspect the spindle bearing and seal surface for burrs or roughness. Be careful not to scratch the sealing surfaces when polishing out roughness. Even the smallest indentations can permit lubricant to seep out under the sealing lip.
7. Remove surface nicks, burrs, grooves, and machine marks from the axle spindle.
8. Ensure that the shaft and bore diameters match those specified for the seal selected—do not attempt to reuse old seals. Make sure that the new seal faces in the same direction as the original. In general, the lip faces the grease or lubricant to be retained.

The procedure for removing bearing cups is shown in the first part of **Figure 27–56.**

Reassembly

After cleaning the wheel hub and prepping the bearings, you can reassemble the hub as follows:

1. To install new bearing cups, use a mandrel or sleeve driver to drive them into the hub, as shown in Figure 27–56D. The bearing cups should be square in the bearing bore and driven until fully seated. You should be able to feel this by the ring of the hammer as it contacts the seal.

SHOP TALK

To ease the installation of the bearing cup into the hub, the cup can be cooled (by dry ice or in a freezer), which is a procedure that stresses both the bearing cup and hub more evenly. Aluminum hubs are required to be heated in an oven to get the cup to drop into place.

2. If the hub uses wet bearings, apply some fresh oil to the bearings immediately prior to installation. In the case of dry bearings, pack the hub cavity between the two bearing cups with an approved

wheel bearing grease up to the level of the cup inside diameter.

3. Pack dry (grease-type) bearing cones, using a pressure packer. If a pressure packer is not available, force the grease into the cavities between and around the rollers and cage by hand.

CAUTION

Never pack wet bearings with grease. Grease-coated wheel bearings inhibit the ability of gear lube to properly lubricate the bearing assembly.

SHOP TALK

The reason dry bearings are seldom used on current equipment is that grease does not lubricate as effectively as gear oil. Gear oil has a much wider temperature operating range than grease.

4. With the hub lying flat, inboard side up, install the inboard bearing into its race, making sure that both the cup and cone are properly lubricated. Prelube the seal by wiping the lip with the lubricant to be retained. Place the prelubed seal into the hub with the lip facing toward the bearing cone.

5. Seat the seal using the correct installation tool. Seal drivers should have an outside diameter approximately 0.010 inch (0.254 mm) smaller than the bore size. The center of the tool should be open so that pressure is applied only at the outer edge of the seal, as shown in Figure 27–56E or by press fitting shown in **Figure 27–57**. A section devoted entirely to wheel seal installation appears a little later in this chapter.

CAUTION

Never hammer or use a punch directly on any part of a seal. Force must be applied evenly around the outer edge to avoid cocking the seal. Wheel seals are expensive. Failed wheel seals are more expensive because of the labor required to replace them!

6. If a spacer is used, position it on the spindle using a light film of lubricant. Align the hole and pin. If a wear ring is used on the spindle, select the correct driver and drive it home; it should be flush with the spindle race flange. Wear rings have an interference fit, so using a fluid-type gasket dope on its inside diameter can facilitate installation (by acting as a lubricant) and seal minor imperfections.

7. Use a wheel dolly to lift the wheel hub/assembly, then align and center it on the spindle. Taking

FIGURE 27–56 Wheel bearing removal, cup replacement, seal and wheel installation

A. Removal of the wheel assembly should always be done with a wheel dolly.

B. Grease or oil seals should be removed with a special seal removal tool to avoid damage to the hub. New seals must always be used in replacement.

C. Bearings and seals should be inspected for wear or signs of probable failure.

Courtesy of SKF USA Inc.

(continued)

FIGURE 27–56 (continued)

D. Bearing cups should be installed using a special bearing installation tool. A hammer should never be used directly to drive bearing cups into position!

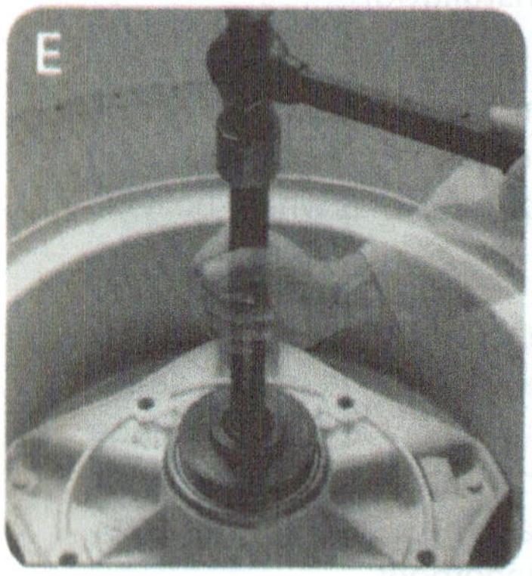

E. Scotseals should be seated in the bore using a special installation tool with centering plug. Never use direct hammer blows on the seal—it will destroy the seal's ability to contain lubricants and protect the bearings.

F. Remounting of the wheel assembly on the spindle should always be done using a wheel dolly. Be careful in moving the assembly onto the spindle because it is necessary to avoid damaging the seal.

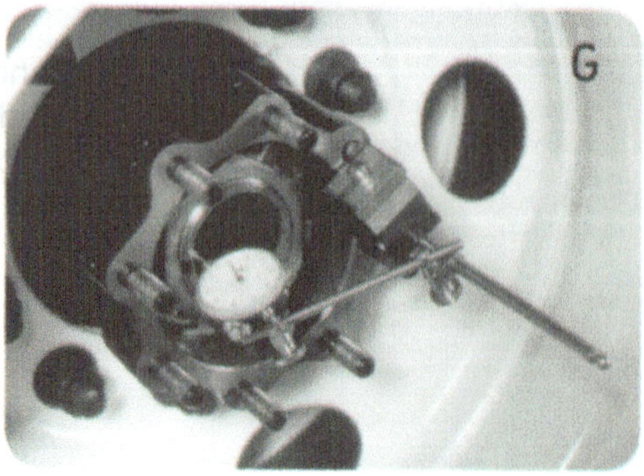

G. Spindle nut torque adjustment should follow the manufacturer's specification on TMC RP618. End-play adjustment should be verified using a dial indicator.

Courtesy of SKF USA Inc.

FIGURE 27–57 Press-fitting a grease retaining seal into position

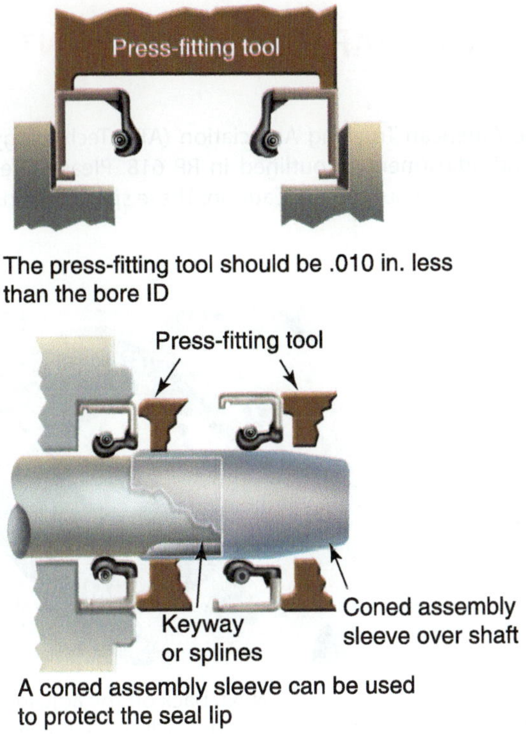

FIGURE 27–58 Cutaway view of an axle end highlighting the wheel bearings

care to maintain the alignment (misalignment can damage the seal), push the wheel/hub assembly onto the spindle until the seal is in contact with its wear ring on the bearing spacer or spindle. Install the outer bearing cone, washer, and adjusting nut in reverse order of removal. A cutaway of a typical axle-end assembly is shown in **Figure 27–58**.

8. Adjust the wheel bearing according to the TMC procedure covered in the next section. After bearing adjustment, secure the locknut and locking device. Install the hub cap and fill with oil to the correct level.

27.8 WHEEL BEARING ADJUSTMENT

Because of an unacceptable number of heavy truck wheel-off incidents in the United States and Canada, some of which have been the result of bearing maladjustments, all the manufacturers of wheel end hardware have approved a single method of wheel bearing adjustment. This method was agreed by consensus at meetings of the **Technology and Maintenance Council (TMC)** committee of the American Trucking Association, and the trucking industry has embraced this single standard throughout the continent. Wheel bearing adjustment is a simple but highly critical procedure. It is recommended that technicians learn the procedure, follow it precisely, and use no other method of adjusting wheel bearings. The TMC-recommended procedure (RP 618) is reprinted here, word for word.

SHOP TALK

The TMC adjustment procedure outlined here does not apply to hubs with preset bearing and seal assemblies.

TMC WHEEL BEARING ADJUSTMENT

This procedure was developed by the TMC's Wheel End Task Force, and it is important to remember that it represents the combined input of manufacturers of wheel end components. **Photo Sequence 12** runs through the procedure outlined next.

Step 1. **Bearing Lubrication.** Lubricate the wheel bearing with clean lubricant of the same type used in the axle sump or hub assembly.

Step 2. **Initial Adjusting Nut Torque.** Tighten the adjusting nut to a torque of 200 lb-ft. while rotating the wheel.

Step 3. **Initial Back-Off.** Back the adjusting nut off one full turn.

Step 4. **Final Adjusting Nut Torque.** Tighten the adjusting nut to a final torque of 50 lb-ft. while rotating the wheel.

Step 5. **Final Back-Off.** Use **Table 27–2**.

Step 6. **Jam Nut Torque.** Use **Table 27–3**.

Step 7. **Acceptable Endplay.** The dial indicator should be attached to the hub or brake drum with its magnetic base. Adjust the dial indicator so that its plunger is against the end of the

Photo Sequence 12

WHEEL END PROCEDURE: TMC METHOD OF BEARING ADJUSTMENT

The following photo sequence outlines some of the key steps in the American Trucking Association (ATA) Technology and Maintenance Council (TMC) Recommended Practice (RP) for wheel end adjustment as outlined in RP 618. Please reference Table 27–2 for the final back-off specification and Table 27–3 for the specified jam nut torque. Caution: These specifications vary by axle location, spindle TPI, and jam nut dimension.

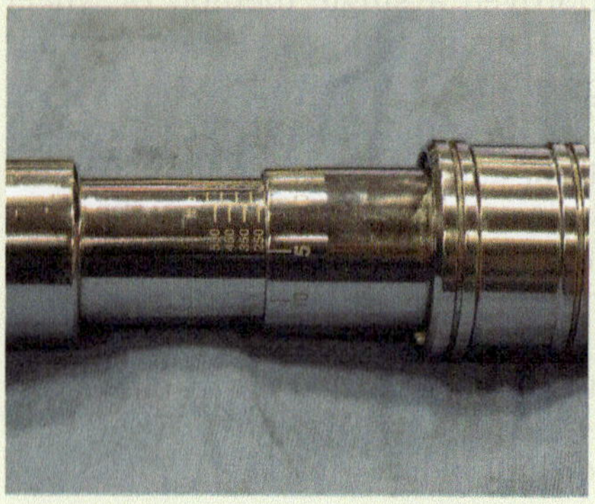

P 12–1 After lubing the bearing and mounting the wheel, install the spacer and adjusting nut. Select a suitably sized, click-type torque wrench, set the torque to 200 lb-ft. (270 N·m).

P 12–2 While slowly rotating the wheel, torque the adjusting nut until it clicks at the set value (200 lb-ft. [270 N·m]).

P 12–3 Next, back off the adjusting nut one full turn. This will leave the wheel assembly loose.

P 12–4 Reset the torque wrench to 50 lb-ft (68 N·m) and while rotating the wheel, torque the adjusting nut until the torque wrench clicks. Final back off: consult Table 27–2 and back off adjusting nut through the specified rotation. Next, on a double nut system, install the locking plate and jam nut. Torque the jam nut to the torque specified in Table 27–3.

P 12–5 Install a dial indictor onto the wheel assembly so that the measuring plunger tip contacts the axle stem.

P 12–6 Properly installed dial indicator on a wheel assembly, set to measure wheel end endplay.

CAUTION

The consequences of not observing RP 618 are wheel-off incidents that may cause fatalities.

P 12–7 Zero the dial indicator.

P 12–8 Apply inboard and outboard thrust to the wheel assembly while observing the indicator readings. If the needle fails to deflect from the zero setting, remove the jam nut, loosen the adjusting nut, and repeat the entire procedure.

P 12–9 Inboard and outboard thrust on the wheel assembly should produce an indicator reading of between 0.001 and 0.005 inch (0.025 and 0.127 mm) to indicate that the wheel bearing has fractional endplay. If the reading exceeds 0.005 inch (0.127 mm), remove the jam nut, loosen the adjusting nut, and repeat the entire procedure.

TABLE 27–2 Final Back-Off

Axle Type	Threads per Inch	Final Back Off
Steer (single nut)	12 18	⅙ turn* ¼ turn*
Steer (double nut)	14 18	½ turn ½ turn
Drive	12 16	¼ turn ¼ turn
Trailer	12 16	¼ turn ¼ turn

*Install a cotter pin to lock the axle nut in position.

TABLE 27–3 Jam Nut Torque

Axle Type	Nut Size	Torque Specifications
Steer (double nut)	Less than 2⅝″ 2⅝″ and over	200–300 lb-ft. 300–400 lb-ft.
Drive	Dowel-type washer Tang-type washer	300–400 lb-ft. 200–275 lb-ft.
Trailer	Less than 2⅝″ 2⅝″ and over	200–300 lb-ft. 300–400 lb-ft.

spindle with its line of action approximately parallel to the axis of the spindle. Grasp the wheel or hub assembly at the 3 o'clock and 9 o'clock positions. Push and pull the wheel end assembly in and out while oscillating the wheel approximately 45 degrees. Stop oscillating the hub so that the dial indicator tip is in the same position as it was before oscillation began. Read the bearing endplay as the total indicator movement. Acceptable endplay is 0.001 to 0.005 inch (0.025 to 0.127 mm).

This is a simple hands-on procedure that you must become familiar with before working on the shop floor. Observing the TMC RP 618 wheel end procedure takes only a little longer than the numerous shortcut methods that result in hit-or-miss adjustments that make our roads unsafe.

SHOP TALK

A truck wheel-off incident can have fatal consequences. When such an incident occurs at highway speeds, any other vehicle on the road is no match for the weight of a bouncing wheel assembly. Wheels have been falling off trucks since trucks first traveled our highways. However, our roads are more congested today and the consequences tend to be more severe. The media have rightly placed some focus on such incidents and forced the trucking industry to take action. The first thing that can be said about wheel-off incidents is that they are almost always related to a service practice error. In few cases is the cause sourced to equipment failure. For this reason, anybody working in the trucking industry must become aware of what is required to safely work on wheel ends.

Wheel End Preload

Wheel end bearing endplay has always been a contentious issue. When wheels are assembled in a truck factory assembly line, the wheel bearings are preloaded using sophisticated equipment to precisely set a fractional preload, because taper roller bearings last longer this way. The problem is there is no currently approved method of setting bearing preload on the shop floor. At the time of writing there is a TMC task force examining shop floor instruments that can set wheel bearings with the kind of precision required to ensure they do not fail. Until this type of tooling is validated by the TMC, continue to use the TMC RP 618A method of wheel end adjustment, or risk litigation if you are the cause of wheel-off incident.

PRESET HUB ASSEMBLIES

In response to wheel-off incidents, most OEMs have made preset (bearing) and unitized hub systems available. A preset hub is a preadjusted but serviceable assembly, whereas a unitized hub is preadjusted and nonserviceable. A couple of examples are the **ConMet PreSet hub** assembly and the Dana Spicer **Unitized Hub System (UHS)**. These types of hub systems are projected to become adopted almost universally over the coming years, if for no other reason than they limit the liability that fleets and service facilities have assumed over wheel-off incidents. Another variation on the UHS is the unitized axle, such as the Meritor Easy Steer Plus described in Chapter 25. In this unitized steer axle, the hub, brake spider, and steering knuckle are all integrated in a single assembly.

The ConMet PreSet hub uses a precision-dimensioned spacer between the taper roller bearings in the axle hub that allows the assembly to be torqued to 300 lb-ft. (406 N·m) on installation, eliminating bearing adjustment by the technician. The ConMet PreSet hub is shown in **Figure 27–59**, with the key components called out in **Figure 27–60** They are available for truck steer axle, drive axle, and trailer axle hubs.

The Spicer Dana Unitized Hub System more recently introduced represents an improvement on the low maintenance system (LMS) for hubs that was similar to the ConMet prehub. It unitizes the key hub components and lubricates them with synthetic grease for life. This eliminates the need for bearing adjustment, seal replacements, and even periodic lubrication top-ups. The UHS hub is shown in **Figure 27–61** and an Eaton preset hub is shown in **Figure 27–62**.

FIGURE 27–59 ConMet PreSet hub

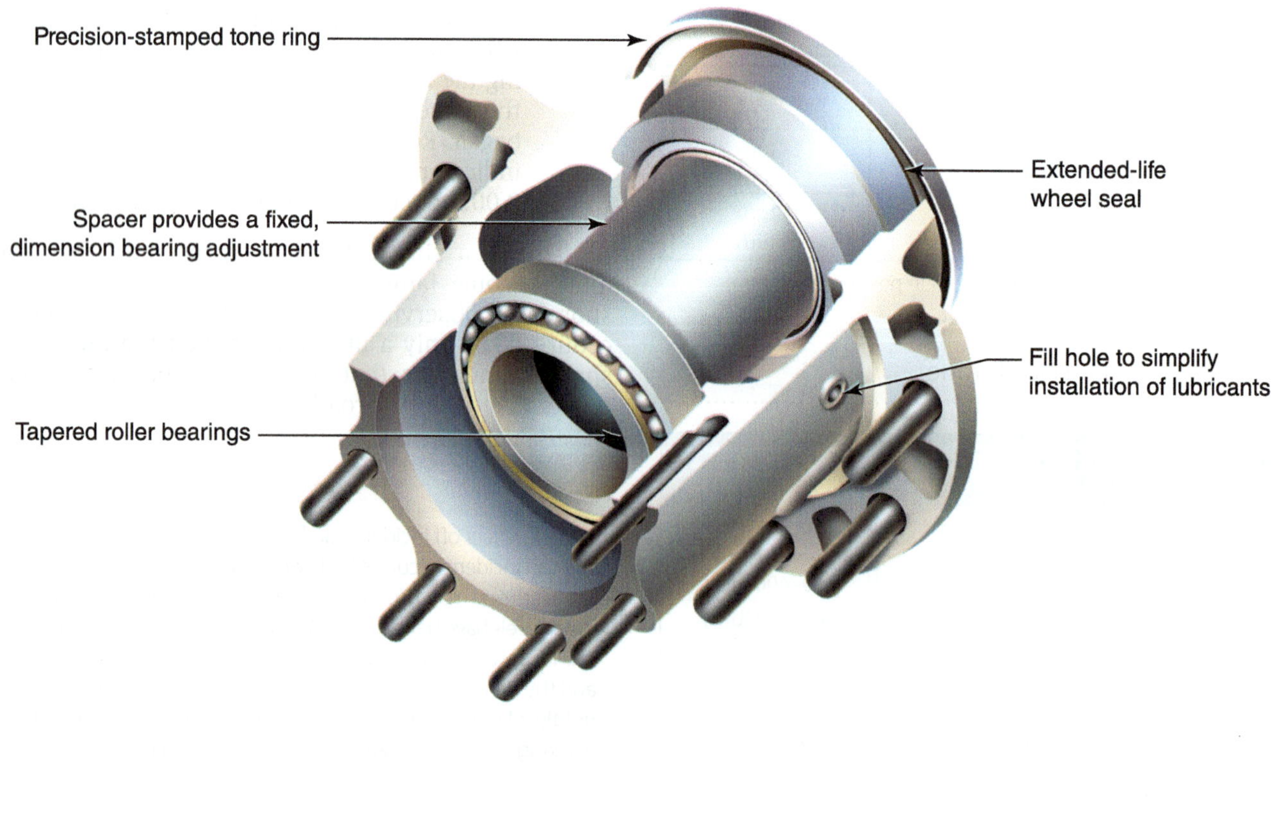

FIGURE 27–60 Cutaway photograph of a ConMet PreSet hub

FIGURE 27–61 A unitized hub

FIGURE 27–62 An Eaton preset hub locked into position.

Figure 27–63 shows some examples of wheel end bearing setting mechanisms.

WHEEL END SUMMARY

Make sure that you know exactly what you are working with when servicing wheel ends. If you are not sure, ask questions. Learn how to identify wheel ends that require bearing adjustment and the newer preset hubs. After having made a wheel bearing adjustment, make sure that you lock and set the adjusting nut with the appropriate locking/jam mechanism.

27.9 WHEEL SEAL SERVICE

We have already looked at the installation of wheel seals as part of other service procedures, but we will finish by focusing on the seals themselves and their removal and installation. Wet seals are used on most of the trucks you see on the road today. This means that the bearings are lubricated by liquid gear oil or specialty synthetics. Dry lubed bearings are almost never seen unless you are working with much older equipment. What you will see are the unitized hub assemblies that are lubed with synthetic oils for the life of the hub.

The operating principles of the two general types of wheel seals, lip type and unitized, are described earlier in the chapter. Here we will take a look at seal removal and installation as if it were a stand-alone procedure.

SEAL REMOVAL AND INSPECTION

Wheel seals should be removed with a pair of pry-bars. By levering on either of its side, the seal can be worked free of the hub without risk of damaging the bearing. Failing this, you can drive out an inboard seal using a mild steel drift and hammer. Position the drift through the outer opening of the hub and against the bearing. Tap the bearing and seal out through the brake drum side of the hub; you will need a helper to catch the bearing.

FIGURE 27–63 Bearing setting hardware: (A) jam nut and D-shaped lock ring; (B) Pro-Torq® nut; (C) Eaton axle with adjusting nut and jam nut; and (D) Rockwell axle with adjusting nut and jam nut

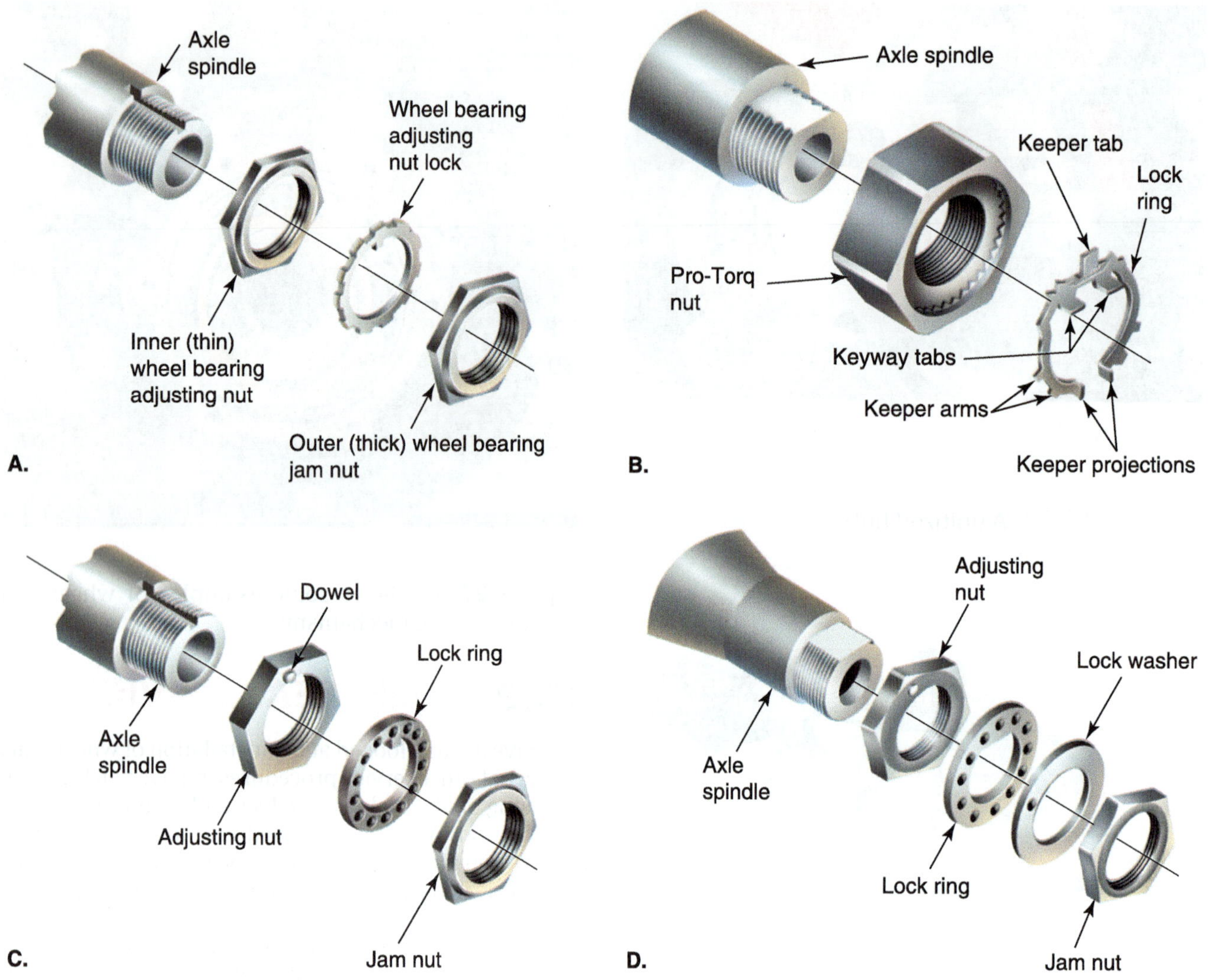

CAUTION

Avoid using brass drifts to drive out wheel seals if you plan on reusing the bearing—brass particulate is difficult to remove from a roller bearing assembly.

When using the drift method of removing a seal, take care not to damage either the bearing cone or the seal bore. The wiper or wear ring that is used with oil-lubed, lip-type wheel seals should be removed using a ball peen hammer and tapping on the ring in one location to expand it. Never use a chisel to cut the ring because it could damage the machined surface on the axle spindle.

Clean and inspect the bearings and hub cavity as explained earlier in this chapter. With wet lube systems, inspect for porous or cracked hubs that could be a source of oil leakage. Visually inspect the hub and remove any minor nicks or burrs with an emery cloth.

Wheel seal installation requires special tools to locate, drive, and seat them. Attempting to do this without these special tools risks damaging these expensive assemblies. Remember that there are minor differences in installation, depending on the OEM and seal design.

Lip-Type Seal Installation

Apply a thin coat of gasket dope (use a thin base-type with some lubricity) to the wear ring shoulder of the axle spindle. Using the specified installation driver (a typical example is shown in **Figure 27–64**), install the wiper ring on the axle spindle shoulder. Hit the

FIGURE 27–64 Installation driver used to install wear rings to the axle spindle when a lip-type seal is installed.

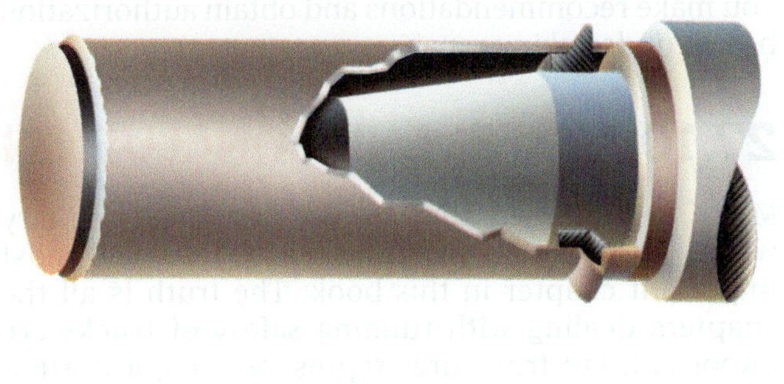

driver with a 4-pound (2 kg) mallet until it contacts the shoulder. Check the position of the wear ring.

SHOP TALK

The care with which a wear or grit ring is installed will affect seal performance. A damaged ring will significantly shorten seal life.

Next select the specified seal driver that fits to the wheel bearing cone bore. Most lip-type seals are coated with a sealant (it is dry and looks like paint) and do not require additional sealant. However, a light coating of gasket dope or silicone around the outside of a lip-type seal can seal any minor imperfections in the seal bore. Wet the lip of the seal with a light film of axle lube. Then drive the seal into the hub until it bottoms into the hub bore (as shown in **Figure 27–65**). You should know when the seal bottoms by the feel and sound of the hammer on the driver. Do not continue to drive the seal after it is seated because damage to the seal can result.

FIGURE 27–65 Driving the seal into the hub

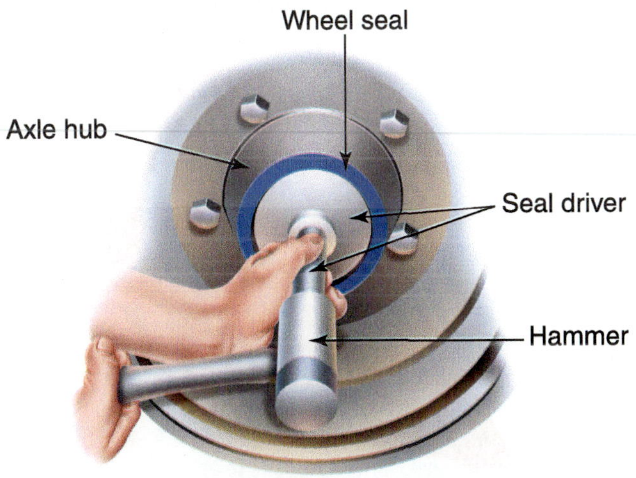

UNITIZED SEAL INSTALLATION

If a seal is being converted to a unitized seal from a lip-type seal, make sure that you remove the wear rings from the axle spindle using a ball peen hammer, as described earlier. To install a unitized seal in the wheel hub, consult the OEM literature. Some OEMs stress that nothing be applied to the outside of the seal, especially to those that use a rubber compound exterior. Lubricate the dynamic part of the unitized seal by applying oil to the integral seal and rotating it.

Next, seat the seal in the recess of the installation tool adapter, as shown in **Figure 27–66**. Then insert the centering plug of the tool in the bore of the inner bearing cone, as shown in **Figure 27–67**. The center plug prevents cocking of the seal in the bore.

SHOP TALK

Wheel seal replacements are routine service facility activities. Because of the high cost of seals, trainee technicians do not commonly practice seal replacement in a training environment. The first couple of times you replace wheel seals in a real-world setting, read the instructions and ask questions. The bottom line is that if you experience comeback failures after replacing wheel seals, your days as a truck technician will not last long.

Hold the installation driver handle firmly and strike the head squarely with a 4-pound (2 kg) mallet until the seal is seated, as shown in **Figure 27–68**. If you used sealant, remove any excess with a wiper. Next, check that the integral seal turns properly and shows some in-and-out movement. Now you are ready to install the wheel assembly; use the procedure described earlier in the chapter.

FIGURE 27–66 Seating a unitized seal in the recess of the installation tool

FIGURE 27–67 Centering the plug of the tool in the hub bore

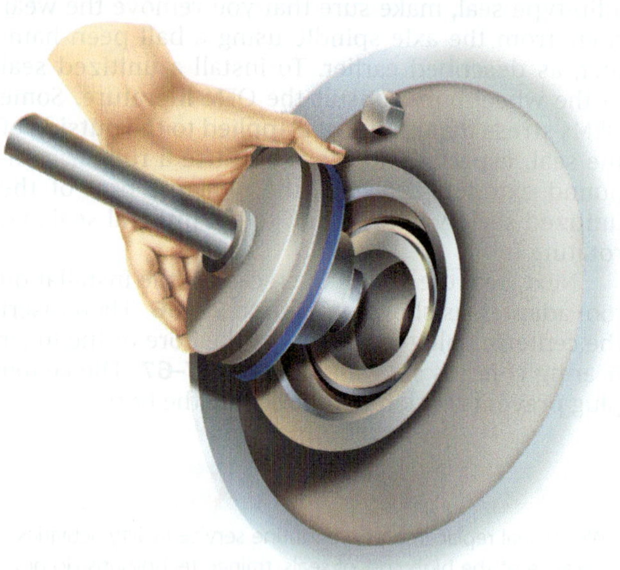

FIGURE 27–68 Seating the seal with a hammer

WHEEL SEAL FAILURE CONSEQUENCES

When a wheel seal fails, oil spills into the foundation brake assembly. This fouls the brake drum and brake linings. In most cases this requires the replacement of the brake linings and possibly the brake drum because gear oil saturates the brake linings, bakes into them, and glazes them to a mirror finish. For the sake of brake balance, you also should replace the linings on the opposite side of the axle.

Shop procedure requires a technician to *recommend* good practice. When you perform a simple repair that becomes more complex because of coincidental damage, as described earlier, make sure that you make recommendations and obtain authorization before undertaking repairs.

27.10 WHEEL END CONCLUSION

Many persons working in the trucking industry would not hesitate to claim that this is the most important chapter in this book. The truth is all the chapters dealing with running safety of trucks are important. But truck tires represent a major portion of the cost of running any kind of trucking operation, and service deficiencies regarding wheel end procedure have been a consistent headache over a number of years.

Although tires wear out, truck wheels seldom do. If they do, it is caused by overloading, by operation over severe road surfaces, or by poor PM practices. If every truck technician working on wheel ends could remain constantly aware of the on-road dangers created by service malpractice and the consequences of imbalanced axle ends, few problems would ever occur. Taking shortcuts is not only expensive but helps create a negative public profile of the trucking industry and those employed in it.

LOOSE LUG NUT INDICATORS

Loose lug nut indicators are a low cost means to enable drivers and technicians to immediately identify a wheel lug nut that is loosening. The indicator is a brightly colored and/or fluorescent tag that is fixed to each wheel lug nut; when the pointer tag shifts it can be easily visually identified. There are a number of these on the market and most have a unit cost of well under a dollar. **Figure 27–69** shows an example made by Torque Tight.

FIGURE 27–69 Loose wheel lug nut indicator

SUMMARY

- Wheels and tires must be properly inspected during daily driver inspections and on preventive maintenance schedules.
- Improperly mounted, matched, aligned, or inflated tires create potentially dangerous on-road situations.
- Wheel bearings and wheel seals are key to keeping the wheel assemblies turning smoothly and safely.
- The rim supports the tire. Three general categories of tread design are used on trucks: rib tread, general duty lugs, and high-traction lugs. These thread designs are used on bias ply and radial tires.
- One wide-base wheel and tire can replace a traditional dual-wheel assembly. Wide-base wheels are categorized as high flotation, super single, wide body, duplex, or jumbo wheels. They have considerable weight saving and fuel economy advantages.
- Compared to steel dual-wheel assemblies, aluminum wide-base wheels and tires are significantly lighter in weight.
- Three basic types of wheels are used in truck applications: cast spoke, stud-mounted disc, and hub-mounted disc.
- Tires should not be mismatched. Mismatching includes the mixing of nominal tire sizes and tread designs on a chassis. It is especially important that a set of duals never be mismatched.
- The tire body and belt material can be constructed of rayon, nylon, polyester, fiberglass, steel, or the newest synthetic rubber compounds.
- All tires (new and retread) sold in the United States and Canada must have a DOT number cured into the lower sidewall on one side of the tire.
- Proper tire care and maintenance is second only to fuel mileage in overall cost per mile of truck operation.
- Improper loading can cause the tires on one side of the truck or trailer to carry a greater percentage of the load than those on the other side.
- Tires represent only a portion of the dynamic rotating mass of an axle end. Even if a tire is statically balance, the axle end may be significantly out of balance, which can substantially reduce the service life of the tire and suspension components.
- Excessive heat produced by running a vehicle at higher-than-rated speeds shortens tire life.
- Wheel balance is the equal distribution of weight in a wheel with the tire mounted.
- Out-of-balance axle ends drastically shorten tire life, compromise fuel economy, and damage suspension components. Key indicators of an out-of-balance wheel end are premature tire by irregular cupping and rapid failure of shock absorbers.
- Improper wheel bearing adjustment can result in bearing runout, steering problems, or serious wheel-off incidents.
- Technicians should learn and always use the TMC RP 618 method of adjusting bearings when working on wheel ends with adjustable bearings.
- Preset axle hubs such as the ConMet PreSet, the Spicer Dana UHS, and others eliminate the need for the technician to adjust wheel bearings. These wheel hubs are installed on the axle and torqued to a specified value.
- Two general categories of wheel seals are used on trucks today: lip-type seals and unitized seal assemblies.
- Wheel seals should always be installed using OEM seal drivers to avoid damaging them.
- Dynamic wheel balancing performed on vehicle is preferred over static tire balancing. The routine installation of an internal balancing agent (IBA) is a cost-effective alternative.
- Loose lug nut indicators identify a wheel lug nut that is loosening; they consists of tags that are fitted to each wheel lug nut and when the pointer shifts it can be visually identified.

REVIEW QUESTIONS

1. Which of the following wheel types uses tires mounted to rims that mount to the wheel using lugs?
   a. disc wheels
   b. cast spoke wheels
   c. hub piloted
   d. stud piloted
2. If a set of dual tires were properly matched, which of the following would be true?
   a. Both tires would have the same nominal size.
   b. Both tires would have the same tread design.
   c. Both tires would have equal wear.
   d. All of the above.
3. Which of the following methods should be used to completely deflate a truck tire?
   a. The Schrader valve pin should be depressed.
   b. The Schrader valve core should be removed.
   c. The split rim should be separated.
   d. A hydraulic bead buster should be used.
4. Which type of tire tread is recommended for use on steering axles for on-highway trucks?
   a. rib
   b. radials only
   c. lugs
   d. bias ply only

5. What should be done to match multipiece rim components properly?
   a. measure the component size
   b. locate the part numbers and cross-check matching
   c. use components from a single OEM
   d. all of the above
6. Technician A says that he seats radial tire beads by using a Cheetah rapid discharge air tank. Technician B says that he seats radial tire beads using a metal ring inflation aid. Who is correct?
   a. Technician A only
   b. Technician B only
   c. both A and B
   d. neither A nor B
7. What is the maximum allowable wheel runout permitted on cast spoke wheels?
   a. 1⁄16 inch (0.75 mm)
   b. 1⁄8 inch (1.5 mm)
   c. 3⁄16 inch (2 mm)
   d. 1⁄4 inch (3 mm)
8. Where is the brake drum mounted on a cast spoke wheel assembly?
   a. on the inboard side of the wheel hub
   b. on the outboard side of the wheel hub
   c. the drum is mounted separately from the hub
   d. none of the above is correct
9. Which type of wheel assembly presents the *least* maintenance and alignment problems?
   a. stud-piloted disc wheels
   b. cast spoke wheels
   c. hub-piloted wheels
   d. any wheels using radial lugs
10. Which type of hub requires heating in an oven to remove the bearing cups?
    a. ferrous
    b. aluminum
    c. cast spoke hubs
    d. any hub using an aluminum hub cap
11. Oil-lubricated drive axle wheel bearings receive oil from:
    a. the banjo housing
    b. a cavity built into the hub cap
    c. an on-chassis auto-lubricator
    d. all of the above
12. Which type of wheel seal fits tightly to both the hub and axle spindle and moves within itself?
    a. wiper ring
    b. unitized
    c. lip type
    d. all of the above
13. Which of the following methods is used to lock the bearing setting on an axle or spindle?
    a. jam nuts
    b. castellated nuts with cotter pins
    c. split forging nuts
    d. all of the above
14. Which of the following is the preferred method of repairing a wheel seal leak that has glazed the brake shoe linings?
    a. Replace the wheel seal only.
    b. Replace the wheel seal and bearings.
    c. Replace the wheel seal and damaged brake shoes.
    d. Replace the wheel seal and brake shoes on both sides of the axle.
15. Technician A says that an advantage of disc wheels over cast spoke is that they produce fewer wheel alignment problems. Technician B says that in stud-piloted disc wheel systems, a loose inner nut can easily go undetected, eventually pounding out the ball seat. Who is correct?
    a. Technician A only
    b. Technician B only
    c. both A and B
    d. neither A nor B
16. Which of the following is said to be an advantage of low-profile radial tires?
    a. better traction
    b. increased tread life
    c. greater vehicle stability
    d. all of the above
17. Technician A states that any tire that tests 50 percent or less of its specified pressure should be removed from the truck and checked out in a safety cage. Technician B uses a clip-on air chuck with a remote in-line valve and gauge to inflate tires to 80 percent of specified pressure on a TPIS equipped vehicle. Who is correct?
    a. Technician A only
    b. Technician B only
    c. both A and B
    d. neither A nor B
18. Technician A says that wheel axial runout on a cast spoke wheel can be eliminated only by removing the wheel rims and breaking down the tires. Technician B adjusts cast spoke wheel radial runout by chalk-marking high points, then loosening the lug nuts on the chalk-marked side and tightening those 180 degrees opposite. Who is correct?
    a. Technician A only
    b. Technician B only
    c. both A and B
    d. neither A nor B
19. What must be done to install a ConMet PreSet hub to an axle?
    a. Set bearings using the TMC method.
    b. Torque to specification.
    c. Install new bearing races.
    d. All of the above.
20. When using the TMC method of adjusting wheel bearings, endplay must be within what range?
    a. 0.001–0.005 inch (0.025–0.127 mm)
    b. 0.001–0.010 inch (0.025–0.254 mm)
    c. 0.005–0.010 inch (0.127–0.254 mm)
    d. 0.005–0.050 inch (0.127–1.27 mm)

28

Prerequisite: Chapter 13

TRUCK BRAKE SYSTEMS

OBJECTIVES

After reading this chapter, you should be able to:

- Identify the components of a truck air brake system.
- Explain the operation of a dual-circuit air brake system.
- Understand what is meant by pneumatic and torque imbalance.
- Describe the role played by the Federal Motor Vehicle Safety Standard No. 121 (FMVSS No. 121) on present-day air brake systems.
- Identify the major components and systems of an air compressor.
- Outline the operating principles of the valves and controls used in an air brake system.
- Explain the operation of an air brake chamber.
- Outline the functions of the hold-off and service circuits in truck and trailer brake systems.
- Describe the operation of S-cam and wedge-actuated drum brakes.
- Describe the operating principles of slack adjusters.
- List the components and describe the operating principles of an air disc brake system.
- Outline the factors that determine brake balance in truck rigs.

KEY TERMS

aftercooler
air dryer
bobtail
bobtail proportioning valve
brake chamber
brake fade
brake pad
brake shoe
check valve
coefficient of friction
compressor
dash control valve
desiccant
disc brakes
double check valve
drum brakes
dryer reservoir module (DRM)
dual-circuit application valve
Federal Motor Vehicle Safety Standard No. 121 (FMVSS No. 121)
foot valve
gladhands
gross vehicle weight (GVW)
inversion valve
lightly loaded vehicle weight (LLVW)
modulating valve
National Highway Transportation Safety Administration (NHTSA)
park-on-air
pop-off valve
pressure protection valve
pressure reducing valve
purge tank
quick-release valve
ratio valve
redundancy
relay valve
safety pressure relief valve
service brake priority
spring brake priority
supply tank
system pressure
tractor protection valve
trailer application valve
treadle valve
two-way check valve
wet tank

Note: Because the trucking industry in North America continues to use Standard system weights and measurements in compliance specifications, not all the values used in this chapter have been converted to their equivalent metric units.

INTRODUCTION

The brakes on a truck are probably its most vital safety component. It takes a truck with a 450-horsepower (336 kW) engine and an 80,000 lb (36,287 kg) load around 90 seconds to accelerate to 50 mph (80 kph). From that speed, the same truck must be brought safely to a stop in less than 5 seconds. If brakes were rated in horsepower, trucks would be required to have brakes rated between 10 and 20 times the horsepower rating of the engine.

EVOLUTION OF BRAKES

Brakes have evolved since the earliest wagon brake. A wagon brake consisted of a hand-actuated lever with a friction block of leather or wood on one end. When the wagon brake lever was pulled, the friction block was forced against the rotating outer rim of the wheel. This type of brake became obsolete when the pneumatic tire was introduced. Today, truck brakes are either hydraulically actuated or air actuated. Hydraulic brake systems tend to be used on light-duty trucks that usually do not have to be coupled to trailers. Medium- and heavy-duty trucks use air-actuated brakes almost exclusively.

HYDRAULIC BRAKES

Hydraulic brake systems use liquid confined in a circuit to transfer mechanical force from the driver's foot to the **brake shoes**. Depressing the brake pedal creates a mechanical force that is transmitted through a pushrod to a piston in the master cylinder. This piston forces the brake fluid through the brake lines to pistons in calipers or wheel cylinders located at each wheel assembly. The hydraulic pressure actuating these pistons forces **brake pads** or shoes against a disc or drum that rotates with the wheel assembly.

Brake pads and shoes are designed with friction facings. When these friction facings are forced into a rotating disc or drum, the energy of motion (kinetic energy) is converted into heat energy. The friction generated by the contact of brake friction faces with the rotating disc/drum retards or slows its motion and can eventually stop it. When hydraulic brake systems are used on light-duty and a small number of medium-duty trucks, they can be either four-wheel disc brakes (more common) or disc brakes on the front wheels and **drum brakes** on the rear wheels.

The potential energy of a hydraulic brake system is the mechanical force applied to the brake pedal. In other words, the greater the amount of foot pressure applied to the brake pedal, the greater the amount of retarding or braking effort delivered to the wheels by a hydraulic brake system. Even when hydraulic brakes are power assisted (and they mostly are in trucks), brake force at the wheel is directly proportional to the mechanical application pressure at the foot valve. A detailed account of hydraulic brakes is provided in Chapter 29.

WHY AIR BRAKES ARE USED IN TRUCKS

Air brake operation is similar in some ways to hydraulic brake operation, except that compressed air is used to actuate the brakes. The potential energy of an air brake system is in the compressed air itself. The retarding effort delivered to the wheels has nothing to do with mechanical force applied to the brake pedal. The source of the potential energy of an air brake system is the vehicle engine, which drives an air **compressor**. The compressor discharges compressed air to the vehicle air tanks.

The brake pedal, actuated by the driver, meters (precisely measures) some of this compressed air out to the **brake chambers**. Brake chambers are used to convert air pressure into mechanical force to actuate foundation brakes. Foundation brakes consist of shoes and drums or rotors and calipers that convert the mechanical force applied to them into friction. In other words, in an air brake system, braking effort at the wheel is achieved in the same way as in a hydraulic brake system. The force that actuates the foundation brakes is converted to friction, which retards movement. The friction is converted to heat energy, which then must be dissipated (transferred) to the atmosphere.

Air brakes are used almost exclusively in heavy-duty trucks and trailers. They have some very definite advantages over hydraulic brakes in heavy-duty highway vehicles, including:

- Air is limitless in supply. It must be cleaned, compressed, stored, and distributed properly in an air brake system. Minor leaks do not result in brake failures.
- The air brake circuit can be expanded easily so that trailers can be coupled and uncoupled from the tractor circuit by a person with no mechanical knowledge.
- In addition to providing the energy required to brake the vehicle, compressed air also is used as the control medium. That is, it is used to determine when and with how much force the brakes should apply in any situation.

- Air brakes are effective even when leakage has reduced their capacity. This means that, in the event of air leakage, the system can be designed with sufficient fail-safe devices to bring the vehicle safely to a standstill.

FMVSS 121

Legislation introduced in 1975 called **Federal Motor Vehicle Safety Standard No. 121** (known as FMVSS 121 in the United States, CMVSS 121 in Canada) governs all air brake system requirements. It has been modified in small ways to keep up to date with technology, but from the beginning it required all highway vehicles using air brakes to use a dual-circuit application circuit. This means that almost every air brake–equipped vehicle on the highways today has dual-circuit brakes that are FMVSS 121–compliant. The dual circuits are called primary and secondary circuits. It also is now a requirement of FMVSS 121 that all highway tractors manufactured after March 1, 1997, and trailers manufactured after March 1, 1998, be equipped with antilock braking systems (ABS).

ABS is explained in Chapter 30. However, it is important to note here that ABS does not significantly alter the basic brake system that is introduced in this chapter because the brake system is required to have full **redundancy** performance. The redundancy requirement means that the legally required stopping distances must be met in the event of an electrical or electronic failure of the ABS. **Figure 28–1** and **Figure 28–2** show the layout of a current ABS air brake system on a tractor and trailer. It continues to remain important to understand a non ABS air brake systems because over 10 percent of the commercial vehicles on our roads have no ABS. You can see how a current ABS air brake system looks when it is integrated into the heavily optioned pneumatic circuit schematic shown in Figure 28–1. As you study this chapter, reference the components discussed in the schematics provided.

DEFINING STOPPING DISTANCE

FMVSS 121 specifies that a tractor/trailer combination must be brought to a complete standstill within 250 feet (76 m) at its gross vehicle rate. This stopping distance has to be achieved with the rig loaded, and from any speed. In addition, the **National Highway Transportation Safety Administration (NHTSA)** requires the same vehicle at its **lightly loaded vehicle weight (LLVW)** to stop in 230 feet (70 m). This change represented a 30 percent reduction in stopping distances over the pre-2010 requirement and the result is that drum brakes have become larger and disc brakes have become common on commercial vehicles.

28.1 AIR BRAKES: A DRIVER'S POINT OF VIEW

If a driver climbs into a parked tractor/trailer combination with no pressure in its air tanks, with the engine stopped the air pressure gauges indicating system pressure in the primary and secondary circuits will both read zero. When the ignition key is turned on, a low-pressure warning light and buzzer will alert the driver that the air pressure is low. After the engine is started, the compressor is driven and the air it compresses is discharged and contained in a **supply tank**, which in turn feeds primary and secondary tanks. In humid and cold climates, an air dryer is usually located between the compressor and the supply tank. As pressure builds in the system, pressure gauges indicate the primary and secondary circuit pressure. At 60 psi (4 bar) (sometimes a little higher), the low-pressure warning light extinguishes and the buzzer ceases. Pressure will continue to build until governor cut-out pressure is achieved.

Governor cut-out pressure in most truck applications will vary between 120 (8 bar) and 130 psi (9 bar), but we will use 125 psi (8.5 bar) for this description. At governor cut-out, the compressor is "unloaded"; that is, it continues to rotate but no longer compresses air. At the moment of governor cut-out, governed air pressure is achieved. Both the primary and secondary dash pressure gauges should indicate 125 psi (8.5 bar) in the air tanks.

CHARGING THE TRAILER CIRCUIT

If the driver then depresses the red octagonal Trailer Supply button (older tractors may require that the yellow diamond-shaped System Park button be actuated before the Trailer Supply button), air will be charged to the trailer air tanks. When both the trailer air system is charged and the tractor brakes are released, the vehicle can be moved. The source of the air available to the **dash control valves** is the primary tank, but this is backed up by the secondary circuit and two-way **check valve**. A check valve is one that only permits airflow in one direction; a **two-way check valve** is a shuttle valve that is supplied with air from two sources and has a single outlet. The valve is designed to shuttle bias airflow from the higher of the two sources to the outlet.

MOVING THE RIG

To move the vehicle, both the System Park and Trailer Supply dash buttons must be depressed. This releases the parking brakes by charging air to the spring brake hold-off circuits on both the tractor and trailer. Any time an air brake–equipped vehicle with properly functioning brakes is moved, air must be

FIGURE 28–1 Highway tractor air brake system with ABS: This schematic shows the pneumatic circuit of a heavily optioned highway tractor.

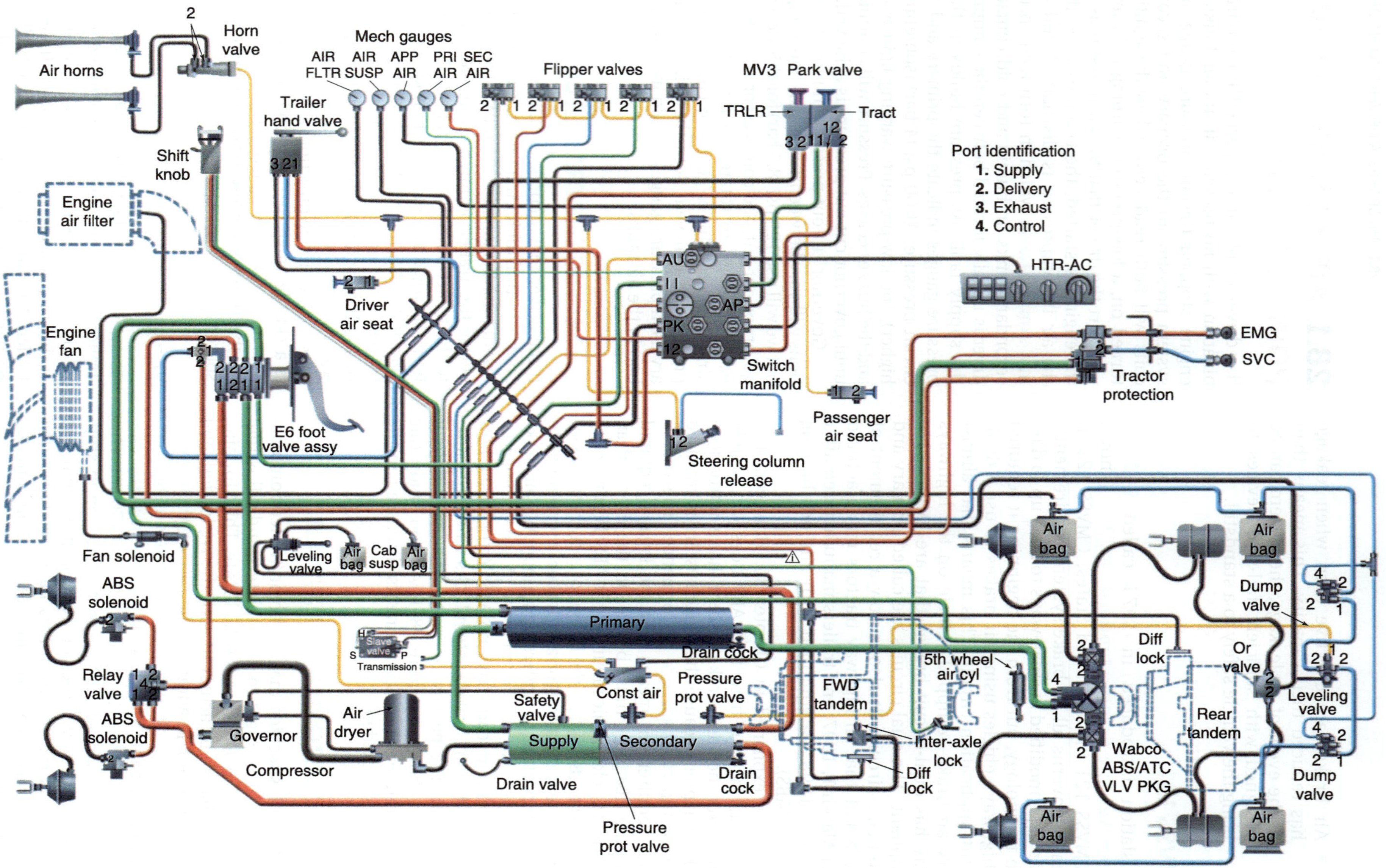

FIGURE 28–2 Trailer air brake system with ABS

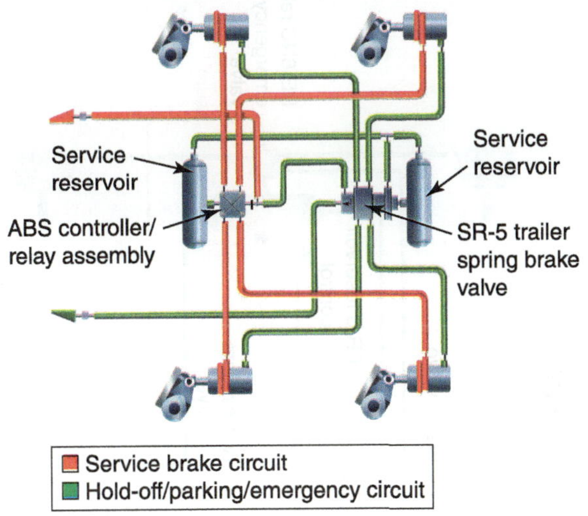

in the hold-off circuit, which puts the parking brakes into release mode.

SERVICE BRAKE APPLICATION

When the vehicle is being moved, if the driver applies the foot pedal, a service application of the brakes occurs. When a service application is made, air is delivered from the primary and secondary reservoirs to act on the service chambers. In the commonly used spring brake chambers, which can be used on all but the steering axle brakes, the service chamber is the one closest to the slack adjuster. The air pressure that acts on the service chambers is modulated and varies according to the amount of braking required. Under light-duty braking, there will be low pressure in the service chambers. Under panic braking, system pressure might be delivered to the service chambers.

During service braking, air is present in both the service and hold-off chambers in a spring brake unit. Some trucks have a service application gauge. This tells the driver how much pressure is being delivered to the service chambers during service braking. The majority of routine service brake applications involve service application pressures of 20 psi (1.4 bar) or less. Service brakes perform the braking requirements of a moving vehicle. The braking requirements of a moving vehicle have to be variable. The **foot valve** (treadle valve) is the means of precisely varying service application pressures.

MAINTAINING SYSTEM AIR

As air is used by the system, pressure will drop. This can be observed by watching the gauges that indicate pressure in the primary and secondary circuits. When the pressure drops to between 20 and 25 psi (1.4 and 1.7 bar) below governor cut-out point, the air pressure in the system has to be returned to its governed pressure. At this value, known as cut-in pressure, an air governor signals the compressor to return to its effective pumping cycle. The compressor will remain in its effective cycle until system pressure has returned to 125 psi (8.5 bar), at which point the governor will signal the compressor to switch to cut-out mode.

PARKING THE RIG

Most air brake–equipped vehicles on the road today are required to have a means of mechanical parking. The means used to park air brake–equipped vehicles is usually with spring brake assemblies. A driver brings a vehicle to a standstill using the service brakes that is, using the foot valve. When the vehicle must be held parked, the driver will exhaust the hold-off air by pulling out the dash system park valve. This discharges the hold-off air in both the tractor and the trailer circuits. The full force of the springs in the spring brake assemblies is then applied to the slack adjusters, and from there it is applied to the foundation brakes in both the tractor and the trailer. The term "dynamiting" often is used by truck drivers to describe exhausting the hold-off air. The result is the mechanical locking up of all the spring brakes on the rig.

This chapter attempts to explain why and how everything described here happens in an air brake system. The approach is to explain everything by system first, and afterward to take a look at exactly how the different components and valves in the system operate. This chapter deals with the operating principles of truck air brake systems and should be followed by studying Chapters 30 and 31, which address ABS and servicing.

28.2 AIR BRAKE SUBSYSTEMS

Truck air brake systems can be divided into a number of subsystems. You might find it helpful when beginning to learn about air brake systems to reference an earlier system that is not equipped with ABS. **Figure 28–3** represents pre-ABS on a tractor, trailer, and converter dolly.

FMVSS 121 legislation requires a truck air brake system to have separate service and parking brake systems. It also requires all trucks equipped with air brakes to have a dual service application circuit. The term *service brakes* describes the running brakes on the vehicle applied by the foot valve. Toward the end of the 1990s, federal legislation required that the trucking industry manufacture all new trucks and trailers with electronically controlled ABS. The conversion to ABS actually began well ahead of the legislation.

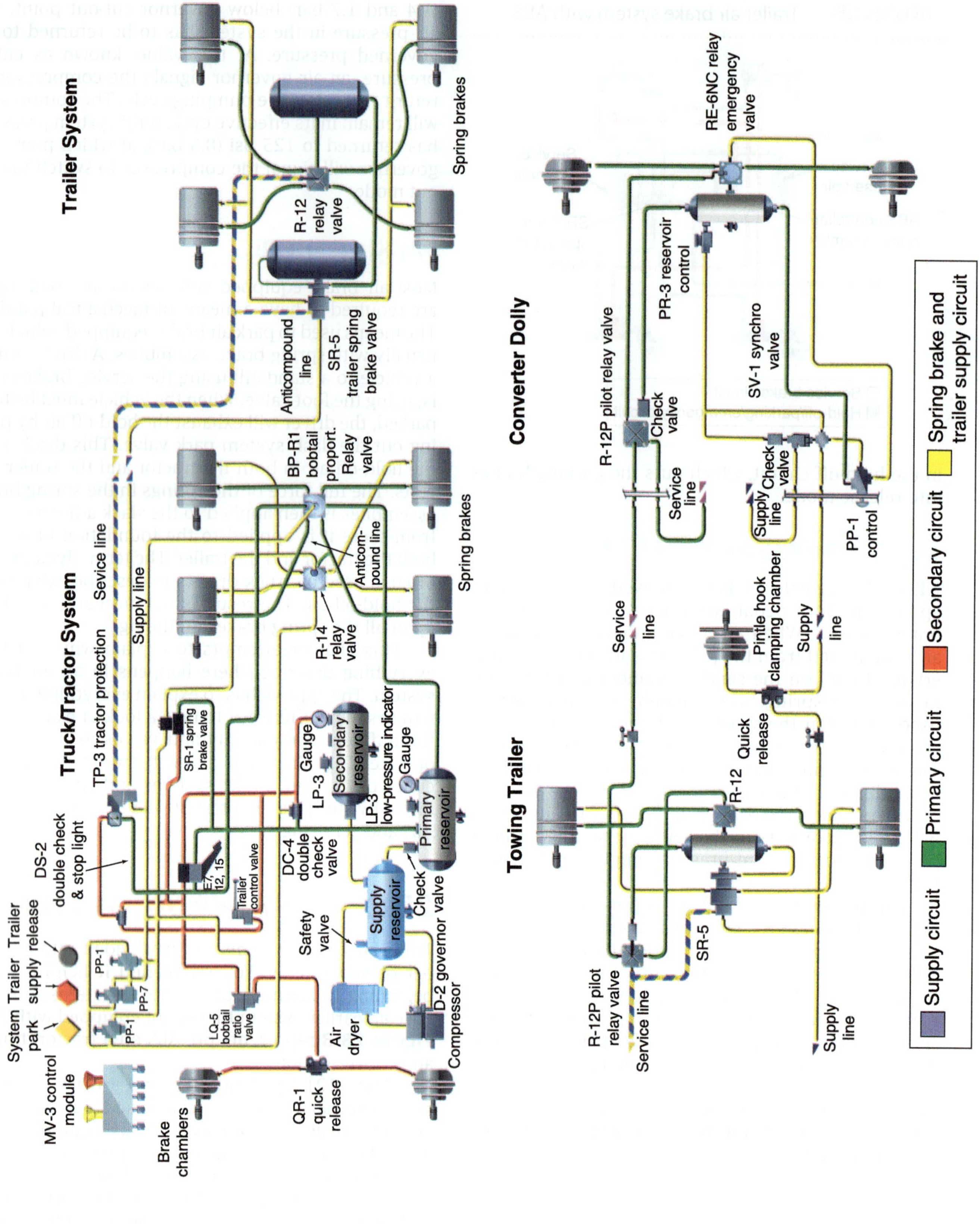

FIGURE 28–3 A pre-ABS tractor, trailer, and converter dolly air brake system

If you take a look at Figure 28–3 and compare it with Figure 28–1 you will observe that ABS changes very little in a typical brake circuit. This is because a truck ABS must have a built-in redundancy to non-electronic operation in the event of an ABS circuit failure.

A complete air brake system on a tractor/trailer combination is made up of the following subsystems:

- The air supply circuit. The air supply circuit is responsible for charging the air tanks with clean, dry, filtered air. It consists of an air compressor, governor, air dryer/aftercooler, supply tank, low-pressure switch, and system safety (pop-off) valve.
- Primary circuit. In a tractor/trailer combination, the primary circuit usually is responsible for actuating the tractor rear wheel brakes and actuating the trailer service brakes when applied from the foot valve. The foot valve is also known as a **treadle valve** or **dual-circuit application valve**. The primary circuit components consists of a primary tank, the foot valve, pressure gauge, quick-release or rear axle relay valves, and ABS solenoids if the truck is ABS equipped. This circuit is sometimes called the *rear axle circuit.*
- Secondary circuit. In a tractor/trailer combination, the secondary circuit is usually responsible for actuating the front axle service brakes and the trailer brakes when actuated by the trailer hand control valve on the tractor. It consists of a secondary tank, foot valve, pressure gauge, quick-release or relay valves, and ABS solenoid valves. This circuit is sometimes called the *front axle circuit.*
- Dash control valves and the parking/emergency circuit. The control circuit consists of two or three dash-mounted valves, which manage the parking and emergency circuits. The parking/emergency circuit consists of spring brake assemblies, **tractor protection valve**, and the interconnecting plumbing.
- Trailer circuit. The trailer service and parking/emergency brakes are controlled by the tractor. The tractor and trailer air systems are connected by a pair of hoses and couplers known as **gladhands**. The trailer brake circuit consists of gladhands, air hoses, air tanks, pressure protection valves, and relay valves.
- Foundation brakes. The function of foundation brakes is to convert the air pressure supplied by the application circuits (primary and secondary) into the mechanical force required to stop a vehicle. The parking and emergency brakes also use the foundation brakes to effect mechanical application of the brakes. Foundation brakes consist of brake chambers, slack adjusters, S-cams or wedges, shoes, pads, linings, drums, and discs. Vehicles equipped with ABS additionally will have wheel speed sensors mounted on the wheel assemblies.

Note the location of the dual-circuit air brake system components in the figures used in this book. Our approach to the study of air brake systems begins by introducing and explaining the operation of each of the subcircuits, starting with the air supply circuit and working through to the foundation brakes. Next, a more detailed description of the critical air brake components and valves is provided.

The student who requires more detail on how a specific particular component functions when studying one of the systems can refer ahead in the chapter. For instance, the foot valve plays a role in all service braking, whether sourced from the primary or secondary circuits, so in order to understand all of its functions, the sections on the primary circuit, secondary circuit, and the description of the valve itself should be studied. Chapter 31 introduces what is required to maintain, adjust, and repair air brake systems and components.

28.3 THE AIR SUPPLY CIRCUIT

The air supply circuit on a truck is responsible for charging the air tanks with clean, moisture-free air to be used by the brake system and supply the remaining air requirements of the vehicle. The potential energy of an air brake system is compressed air. The air is compressed by the supply circuit and then distributed and stored in reservoirs (tanks) at pressures typically around 125 psi (8.5 bar). Valves controlled by the driver can then deliver compressed air to the brake system components. Trucks today are required to have their parking/emergency brakes fully applied by mechanical force in the event that there is no system air pressure. So air is required first to release the parking/emergency brake circuit and then to effect service braking when needed. **Figure 28–4** shows the location of the components in a typical truck air supply circuit.

AIR COMPRESSORS

The air compressor is driven by the engine. Compressors used on current trucks may be either single or multiple cylinder. An air compressor drive usually comes directly from an engine accessory drive gear. In some cases, a compressor may be controlled by a clutch, and, in light-duty and some older trucks, it may be belt driven. **Figure 28–5** shows some typical flange-mounted, gear-driven air compressors, and **Figure 28–6** is an illustration of a less common, cradle-mounted, belt-driven unit. These are usually only found as retrofits or on gasoline-fueled engines. In this textbook, we make the assumption that all compressors are gear driven by the engine.

FIGURE 28–4 Truck air supply circuit components: This system shows a three-reservoir system but the supply (wet), primary, and secondary reservoirs can be integrated into a single tank divided into three chambers.

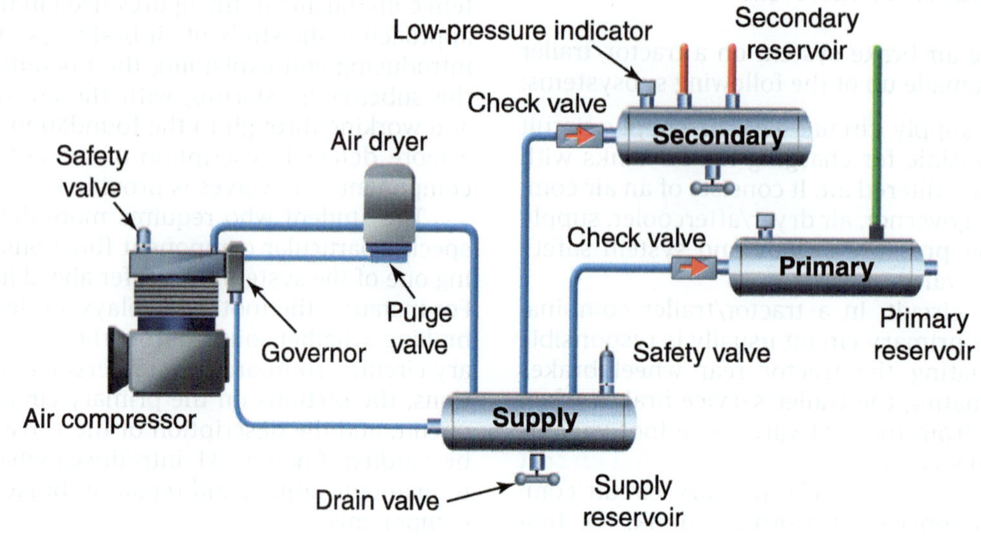

FIGURE 28–5 Some typical single and two cylinder air compressors. These units are designed to be flange-mounted and driven by an engine accessory drive.

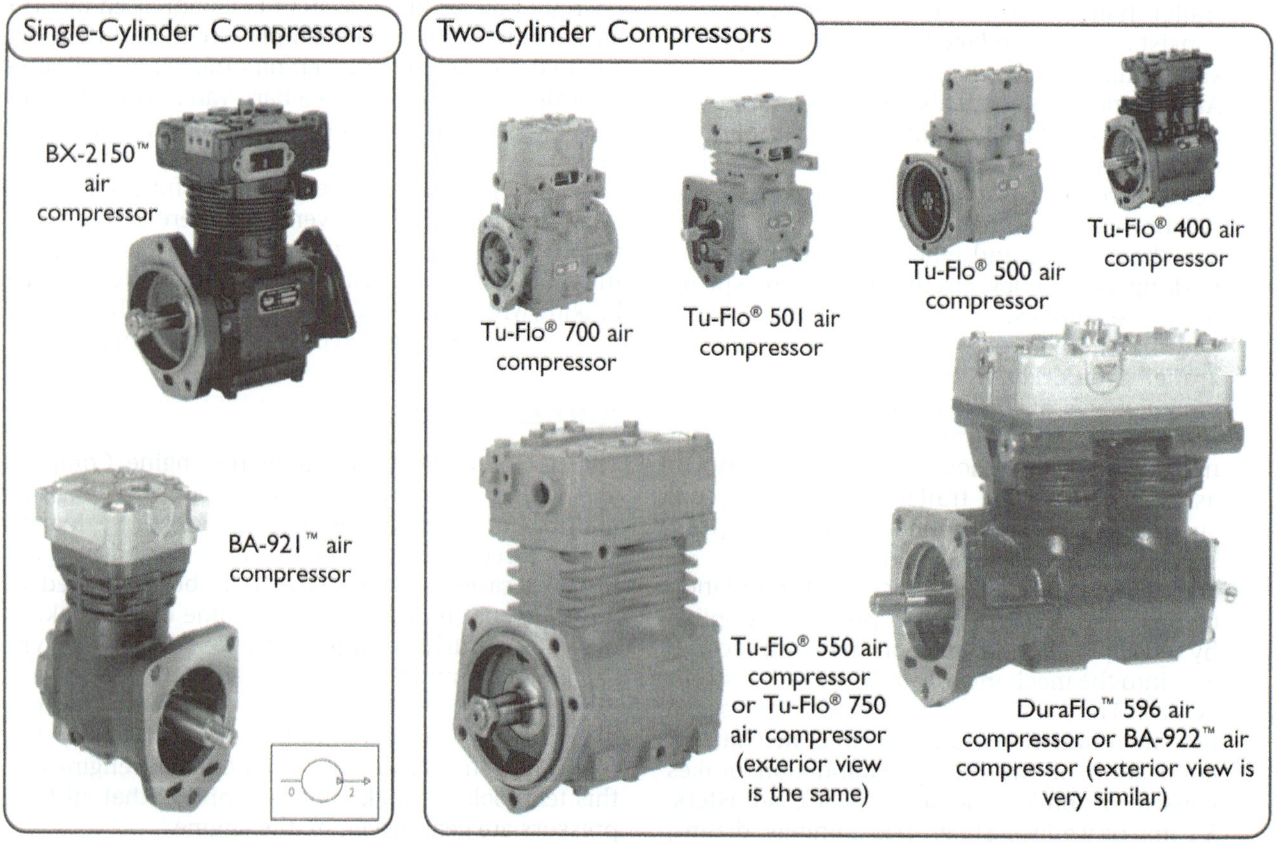

FIGURE 28–6 Exploded view of a belt-driven air compressor

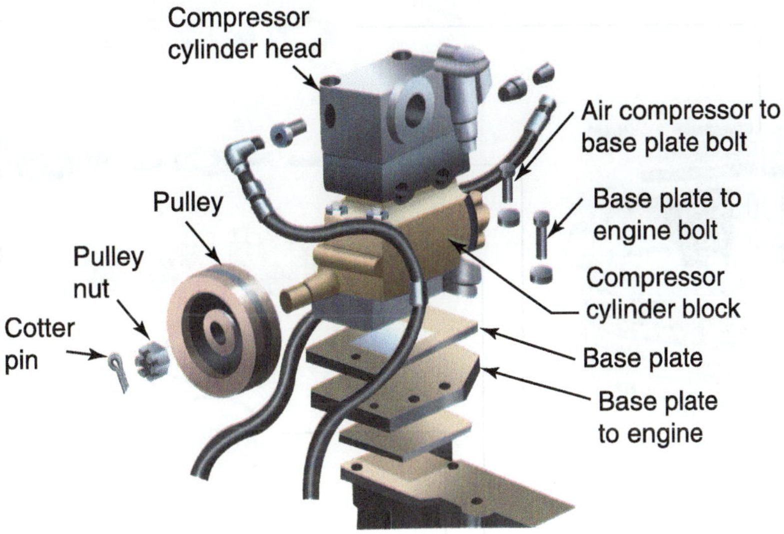

The components of an air compressor are much like those of an engine. The compressor crankshaft is supported by main bearings, which are lubricated with engine oil. Connecting rods are attached to the crank throws and support pistons with rings to seal them in the cylinder bore. When the crankshaft is rotated by the engine accessory drive, the pistons reciprocate in bores machined in the compressor housing. The connecting rod bearings and cylinder walls also are lubricated by engine oil. The engine oil is pumped directly into the compressor crankshaft.

The spill and throw-off oil from the bearings drains to a sump. The oil level in the sump is determined by the location of the oil return line. Because the oil pumped into the compressor is allowed to drain back from its sump to the engine crankcase, the lubricating oil in the compressor is constantly changing.

The cylinder head in the compressor contains valves that are designed to open in one direction only. An inlet valve is designed to open when low pressure is created in the cylinder. A discharge valve is designed to open only when cylinder pressure is greater than the pressure in the discharge line.

Compressor Operation

When the engine is run, it rotates the compressor crankshaft, causing the pistons to reciprocate in the cylinder bores. The internal components of the compressor are shown in an exploded view in **Figure 28–7**. On the downward stroke of the piston, low pressure is created in the cylinder. This low pressure unseats the inlet valve, which allows filtered air to be pulled into the cylinder. When the piston reaches the bottom of its stroke, it stops and then reverses. As the piston is driven upward, it pressurizes the air charge, closing the inlet valve. When the pressure in the cylinder exceeds the pressure on the outlet side of the discharge valve, the discharge valve opens. When the discharge valve is open, the compressed air in the cylinder is unloaded into the discharge line toward the system supply tank. This operating cycle is shown in **Figure 28–8**.

Compressor Cycles

Most compressors are direct driven by the vehicle engine. The compressor is continuously rotated, and it may be in either a loaded cycle or unloaded cycle. These cycles are controlled by an air governor. When the compressor is in a loaded cycle, it compresses the air and delivers it to the discharge circuit. The governor senses pressure in the supply tank by means of a small gauge signal line. When the governor signals the compressor into unloaded cycle, an air signal is delivered from the governor unloader port to unloader pistons located in the compressor cylinder head. When the unloader pistons are actuated, they hold the inlet valves open. The piston then pulls in air on its downstroke and immediately blows it back out of the inlet valve on the upstroke. In other words, no compression takes place. The unloaded cycle means that the compressor is being rotated but is not performing the work of compressing air. The compressor goes into loaded cycle at governor cut-in.

Governor cut-in occurs when the system pressure has dropped below a specified value and requires recharging. The compressor goes into unloaded cycle at governor cut-out. This is the specified maximum system pressure.

FIGURE 28–7 Air compressor components

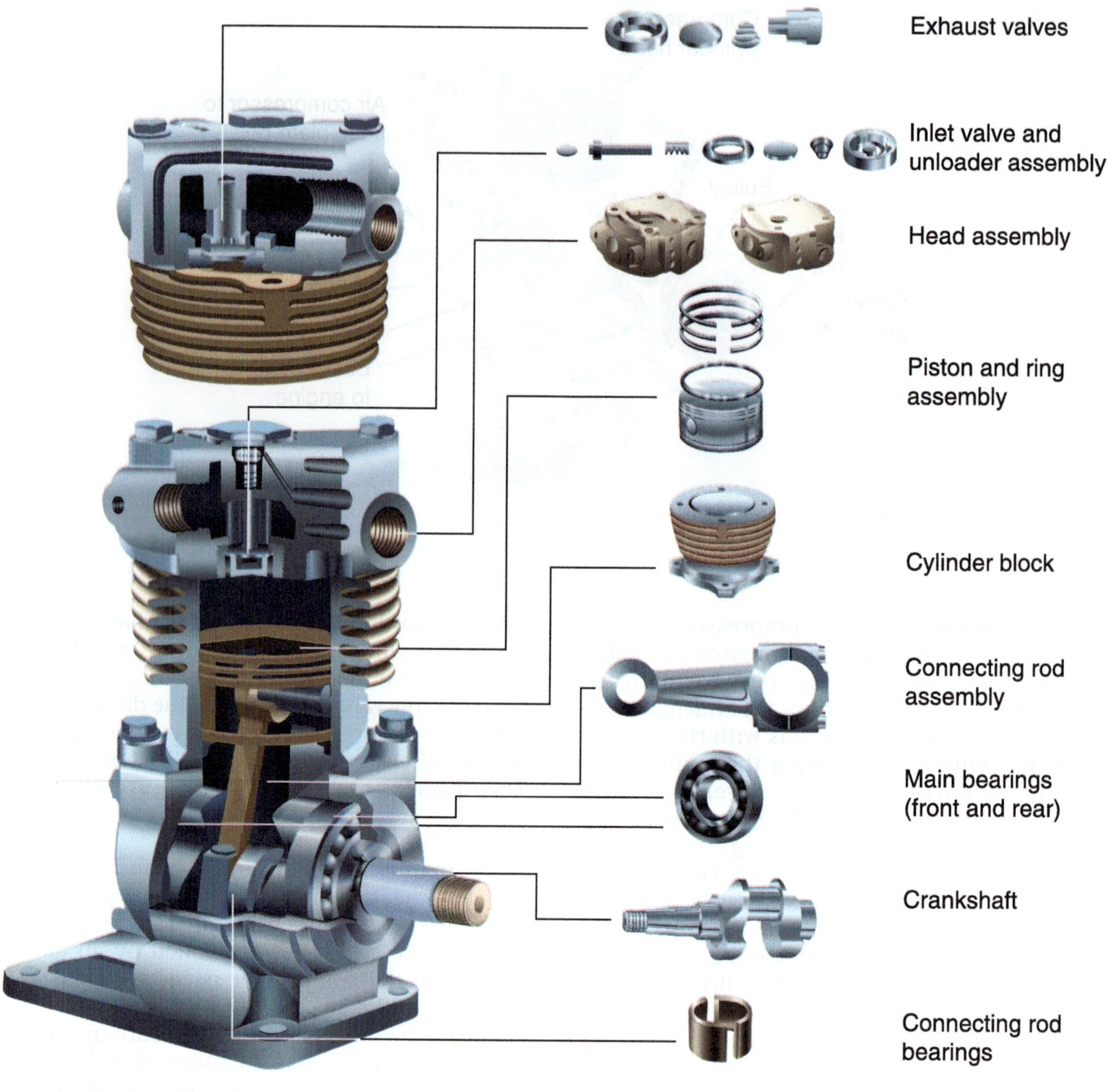

GOVERNORS

The function of the air governor is to monitor system air pressure and manage the loaded and unloaded cycles of the compressor. The governor assembly can be mounted directly to the air compressor (not generally recommended because of higher temperatures close to the compressor) or remotely, often on the firewall of the engine compartment. **Figure 28–9** shows a typical mounting location of an air governor.

Air pressure in a truck air system must be carefully managed at what is called system pressure. System pressure is a range of pressures between governed pressure (maximum system pressure) and compressor cut-in pressure (minimum system pressure). Governed pressure is usually set between 120 and 130 psi (8 and 9 bar) on trucks, but you should note the following:

- FMVSS 121 specifies that setting governor cut-out pressure at any value between 115 and 135 psi (7.9 and 9.3 bar) is acceptable.
- The Technology and Maintenance Council (TMC) of the American Trucking Association (ATA) recommends that governor pressure be set at 125 psi (8.5 bar).

FIGURE 28–8 Air compressor operation

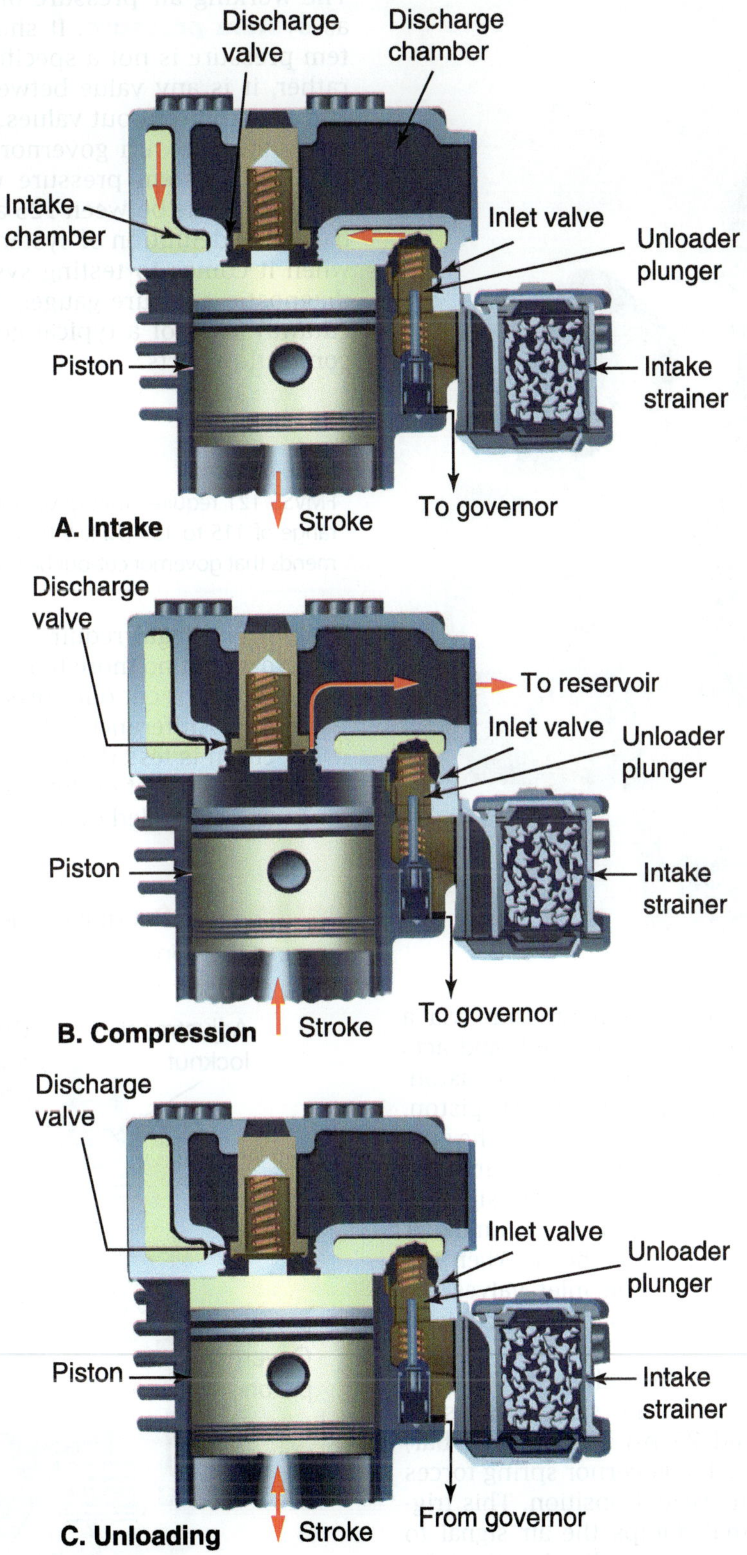

FIGURE 28–9 Compressor-mounted air governor

The governor measures system pressure via a signal line that runs from the supply tank and acts on the lower sectional area of the governor piston. The governor spring acts in opposition to this piston. Governor spring tension is adjusted to define the governor cut-out pressure value. When the system pressure achieves the specified maximum, an air signal is sent from the governor unloader port to the unloader pistons in the compressor cylinder head. When the unloader pistons are actuated, the inlet valves are held open and the compressor is no longer capable of compressing air.

As system air is used by the vehicle air circuits, system pressure drops. When the system pressure drops to between 20 and 25 psi (1.4 and 1.7 bar) below governed pressure, the governor spring forces the governor piston to its seated position. This triggers an exhaust valve that dumps the air signal to the compressor unloader pistons, allowing the inlet valves to seat once again. Governor cut-in occurs at the moment the unloader signal to the compressor is exhausted by the governor and the compressor returns to its effective cycle.

SYSTEM PRESSURE

The working air pressure on a truck is referred to as **system pressure**. It should be noted that system pressure is not a specific value on any system; rather, it is any value between the governor cut-in and governor cut-out values. For instance, on an air brake system with governor cut-out set at 125 psi (8.5 bar), system pressure would indicate any air pressure value between 125 and 100 psi (8.5 and 6.9 bar). This definition of system pressure is important when it comes to testing system performance with diagnostic pressure gauges. **Figure 28–10** shows a cutaway view of a typical governor, identifying the connection ports.

SHOP TALK

FMVSS 121 requires that governor cut-out be set within the range of 115 to 135 psi (7.9 to 9.3 bar). The ATA TMC recommends that governor cut-out be set at 125 psi (8.5 bar).

- It is a legal requirement that governor cut-in occur at no more than 25 psi (1.7 bar) less than governor cut-out pressure.
- If the difference between governor cut-out and cut-in is less than 20 psi (1.4 bar), the result is too frequent cycling of the compressor loaded and unloaded cycles.

FIGURE 28–10 Cutaway view of a governor showing the connection ports

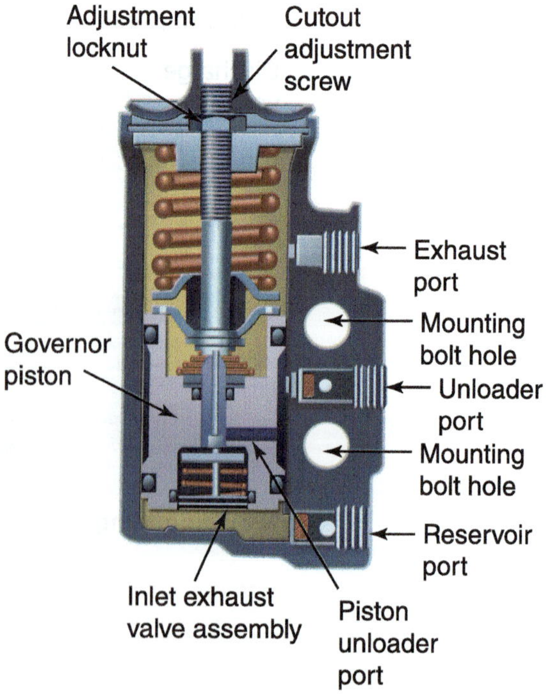

AIR DRYER

An **air dryer** is a common option on an air brake system. Most trucks on our roads today are equipped with an air dryer, especially those operating in humid or cold conditions. **Figure 28–11** shows the location of an air dryer in the supply circuit. The air compressed by the compressor is either atmospheric or ported boost air from the turbo, both of which contain some percentage of vaporized moisture. This moisture can harm air system valves if it is not removed.

The air dryer is located in series between the compressor discharge and the supply tank. More recent air brake–equipped trucks may use a dryer reservoir module, which combines several supply side subcomponents into a single module; however, this changes none of the operating principles of the subcomponents. We will take a closer look at dryer reservoir modules a little later in this section.

Compressed air is heated to temperatures up to 300°F (150°C) or higher when using a boost air source. Air dryers can use two methods of separating moisture from the compressed air:

- Aftercooling
- Desiccant absorption

Some air dryers use only one of these principles to remove moisture from air, whereas others use both.

Aftercooling

The air dryer functions first as a heat exchanger, that is, an **aftercooler**, to reduce the air temperature, dropping it to a value cool enough to condense any vaporized moisture it receives. Actually, first-stage cooling of the hot compressed air occurs in the discharge line that connects the compressor with the dryer. In some older applications, copper lines were used for this purpose. Current steel braided discharge lines still play a role in cooling compressed air. They do not achieve this as efficiently as copper line, but steel outlasts copper and is more flexible. In addition, air dryers often use external fins to increase cooling efficiency. When compressed air enters the dryer, it first passes through an oil filter that removes trace quantities of oil that have been discharged by the compressor.

Desiccant

Next, the air is passed through a desiccant pack designed to absorb water. A **desiccant** is a moisture removal agent; most air dryers contain desiccant media that may have to be replaced when it becomes depleted or saturated with oil. The desiccant pack may be contained in a cylinder located inside the dryer assembly or be a replaceable cartridge. The desiccant pack absorbs vaporized moisture in the air and condenses it. The liquid moisture then drains to a sump so that it can be discharged by the dryer purge valve. Dry air exits the dryer and is routed to the supply tank. Some dryers use a twin cartridge system; this allows one desiccant pack to regenerate while the other continues to dry. Twin cartridge dryers are effective in air systems where the compressor is required to cycle continuously such as with trucks with central tire inflation systems and numerous air accessories. **Figure 28–12** shows a cutaway view of a typical air dryer.

FIGURE 28–11 Typical air dryer installation

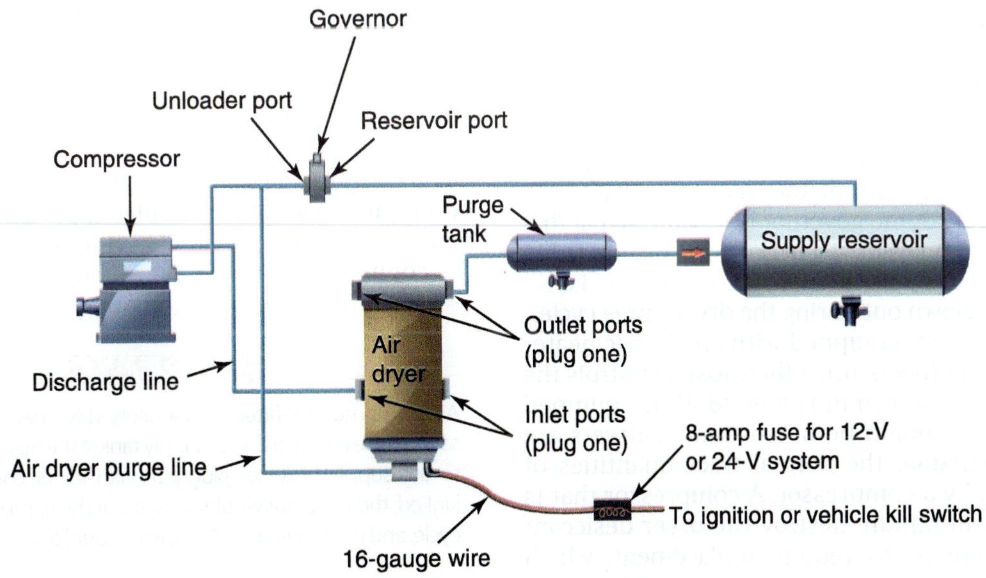

FIGURE 28–12 Cutaway view of an air dryer

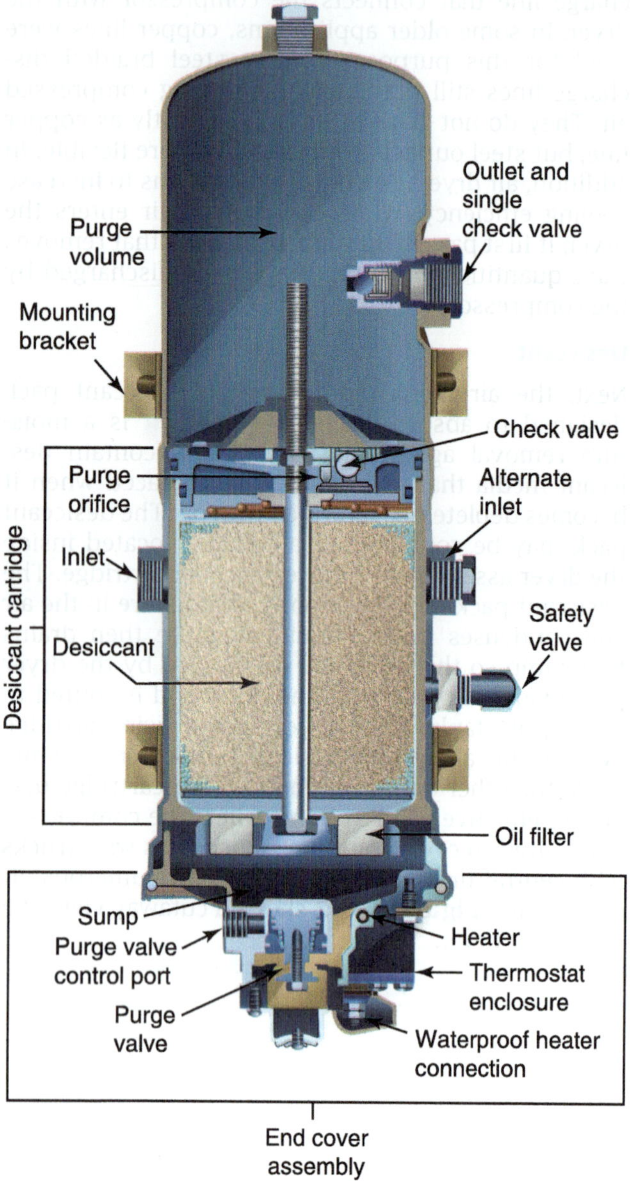

Sump and Purge Valve

Water and any oil that drains into the sump must be discharged or purged from the dryer. A signal from the unloader port of the governor (the same signal that "unloads" the compressor at governor cut-out) is used to trigger a purge valve in the air dryer sump. Any liquid in the sump is blown out during the dryer purge cycle.

Some dryers are equipped with an electric heater to prevent winter freeze-up. A thermostat controls the heater. The dryer shown in Figure 28–12 is equipped with a heater. It should be noted that air dryers are designed to separate the normal trace quantities of oil discharged by a compressor. A compressor that is pumping oil will rapidly destroy the dryer desiccant packs. Desiccant packs require replacement, which

FIGURE 28–13 Cutaway of an air dryer with a spin-on replaceable cartridge

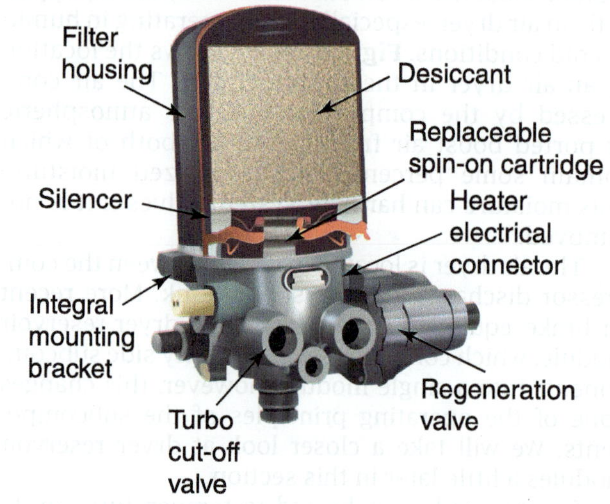

is largely dependent on how much contaminant they have been exposed to. **Figure 28–13** shows a cutaway of an air dryer with a spin-on replaceable cartridge.

Dryer Reservoir Module

A **dryer reservoir module (DRM)** integrates a number of the components outlined earlier in this chapter. **Figure 28–14** and **Figure 28–15** show two views of a Bendix DRM. A DRM consists of:

- The secondary reservoir
- Integrated dryer
- Governor
- Four-valve pressure protection pack

The components that make up the DRM assembly all function as they do in a nonintegrated assembly. They are commonly used in current trucks. **Figure 28–16** is a schematic that shows how the components of a DRM interrelate.

Trailer Air Dryers

Trailer air dryers are used to protect trailer air systems when they are hauled by tractors without air dryers and when the trailer air circuits have to be charged when uncoupled to a tractor. Dry-bulk hoppers would be a typical example of a trailer application that would benefit from an air dryer circuit. **Figure 28–17** shows the components and circuit of a trailer dryer system.

SHOP TALK

Air dryers must be fitted with a safety valve. Because the system safety valve is located in the supply tank, if the hose from the dryer to the supply tank got plugged (with ice or contaminants) or kinked, the governor would not cut out the compressor effective cycle, and the resulting high pressure could explode the air dryer.

FIGURE 28–14 Bendix DRM

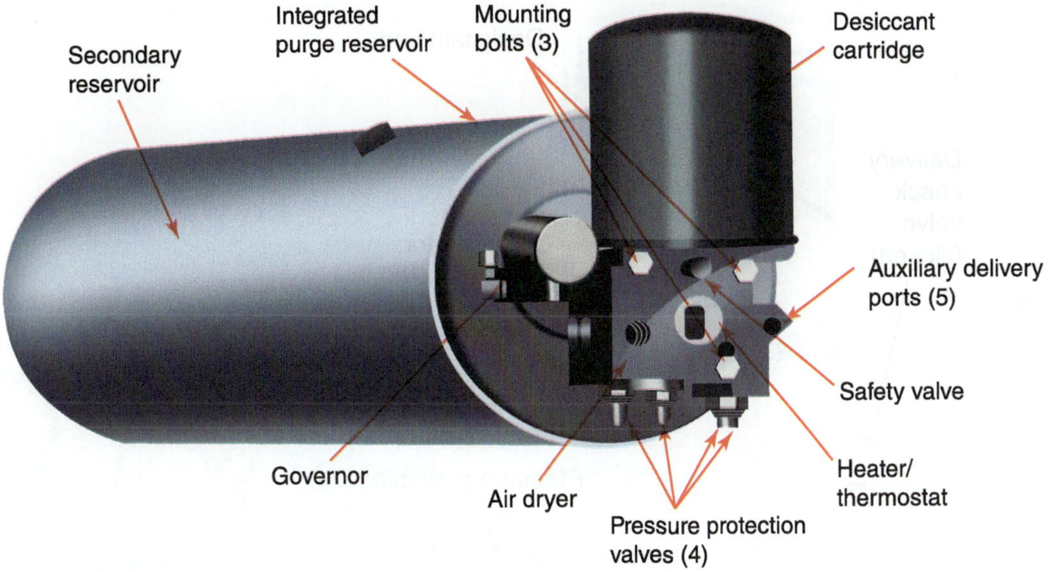

ALCOHOL EVAPORATORS

Not all of the moisture in the air can be removed. Alcohol evaporators are used in some systems to help prevent freeze-ups. Alcohol vapor is introduced into the system by the alcohol evaporator. This forms a solution with water and lowers its freezing point. Alcohol evaporators should be located downstream from the air dryer. It is important only to use alcohols intended for use in air brake systems. Brake valves contain moving parts and most are lubricated on assembly to last for the life of the component. Pure alcohol (methyl hydrate or methyl alcohol) can dry out lubricants in brake valves, causing premature failure. Alcohols sold specifically for use in brake systems contain brake valve lubricant.

FIGURE 28–15 A Bendix DRM assembly shown removed from the truck.

CAUTION

Alcohol evaporators MUST be located downstream from the air dryer in the supply circuit. Alcohol will turn air dryer desiccant into mush if pumped through the system.

SUPPLY TANK

The supply tank (or *reservoir*) is the first tank to receive air from the compressor. Therefore, in a typical system, it would be located downstream from the compressor. The supply tank is sometimes referred to as the **wet tank** because in a system with no air dryer, this is where the air cools and condensing moisture collects. If a compressor discharged oil, this also would collect in the wet tank.

Some systems also include a **purge tank**, which is located upstream from the supply tank. This acts to further reduce the chance of pumping contaminants through the system. The function of the supply tank is to supply all the other tanks in the truck chassis air system. In the brake system, this includes the primary and secondary tanks.

In common with all other tanks on board a truck chassis, the internal walls of the supply tank are coated with an anticorrosive liner. The anticorrosive liner protects the steel walls of the air tank from corrosion by moisture. The liner itself, however, is susceptible to the carbon coking that can be caused by a compressor pumping oil. Reservoirs must be equipped with draining devices, usually known as dump valves. A draining device can be as simple as a hand-actuated draincock or an electrically heated, automatic drain valve. Many reservoirs are pressure

FIGURE 28–16 Schematic showing operating principle and components of a Bendix DRM

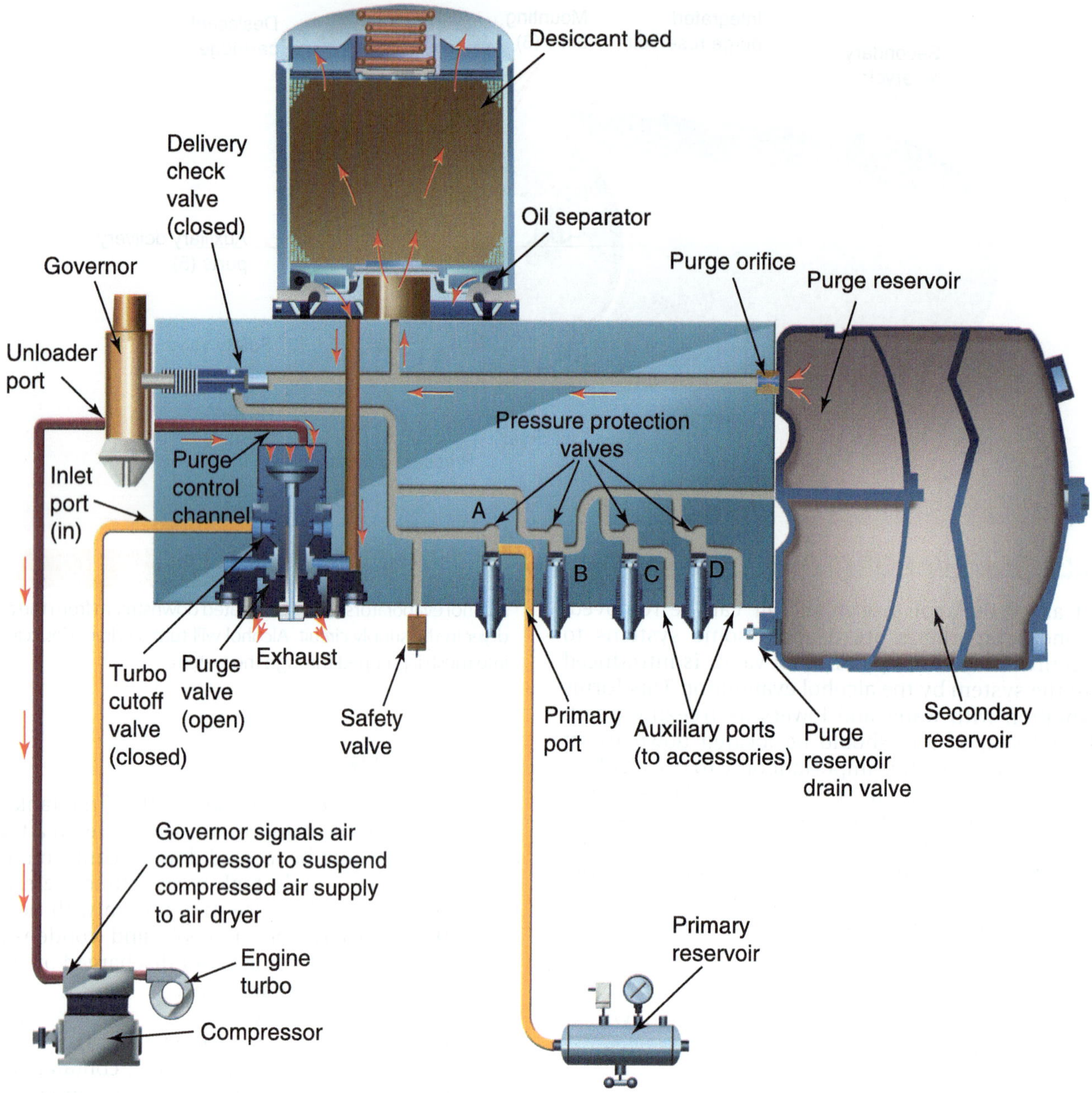

protected by single check valves. A single check valve is a one-way device that will allow a tank inlet to be charged with air but will not discharge through the inlet. **Figure 28–18** shows some typical air tanks, dump valves, single check valves, and low-pressure indicators.

The maximum working pressure of the tank is usually 150 psi (10.3 bar). To ensure that this pressure is not exceeded, a **safety pressure relief valve** is fitted. This is often known as a **pop-off valve**. Should something such as a debris-restricted signal hose cause the governor to malfunction and not signal compressor cut-out, the safety valve would trip at 150 psi (10.3 bar) to prevent damage to the chassis air circuits. **Figure 28–19** shows a typical safety pressure relief valve.

Moisture Ejectors

Automatic moisture ejectors (see Figure 28–18) respond to changes in tank pressure and expel any moisture that has collected in the tank. They are sometimes known as spitter valves or automatic dump valves. All moisture ejectors must be able to manually open. As mentioned earlier, some moisture ejectors are equipped with thermostatically actuated heaters that prevent winter freeze-up.

FIGURE 28–17 Trailer air dryer system

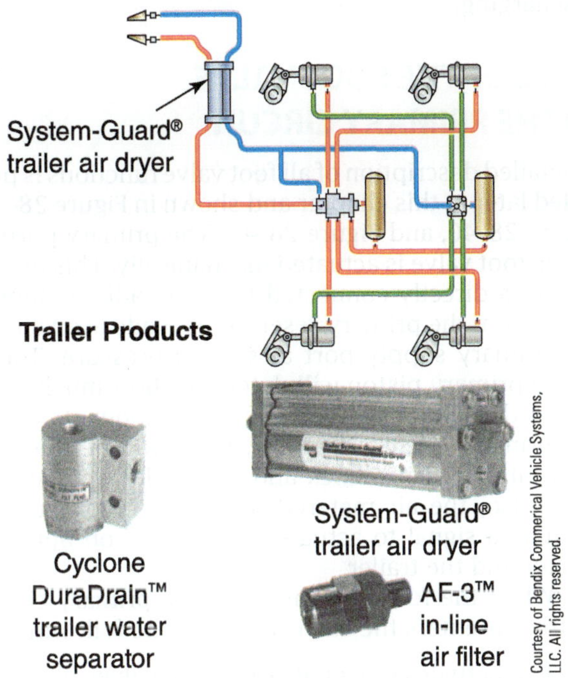

FIGURE 28–18 Typical air reservoirs, dump valves, check valves, and low-pressure indicators

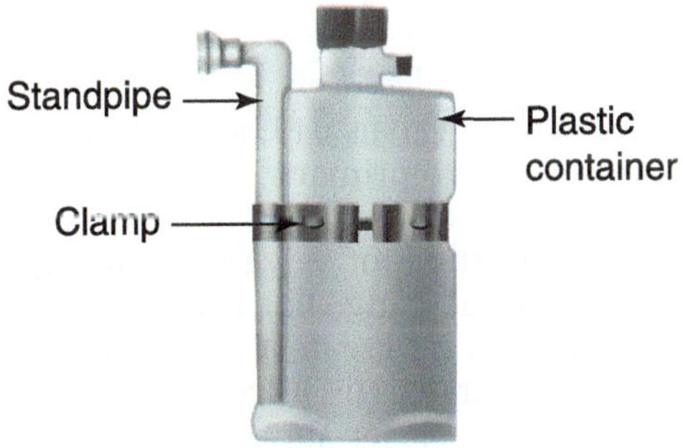

FIGURE 28–19 Safety pressure relief valve, also known as a pop-off valve

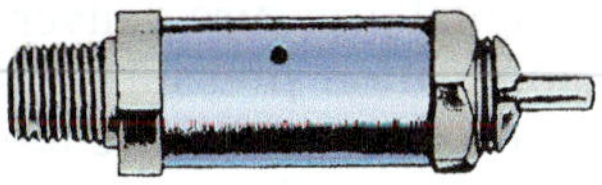

Other Supply Circuit Valves

Other types of valves can be used in the supply circuit. Note that the same valves may be used elsewhere in the circuit. Their functions are briefly described here:

- **Double check valve**: used when a component or circuit must receive or be controlled by the higher of two possible sources of air pressure.
- **Pressure protection valve**: a normally closed, pressure-sensitive control valve. Some are externally adjustable. It can be used to delay the charging of a circuit or reservoir, or to isolate a circuit or reservoir in the event of a pressure drop-off.
- **Pressure reducing valve**: used in any part of a circuit that requires a constant set pressure lower than system pressure.

SHOP TALK

Air tanks are pressure vessels. All vehicle air tanks are required to be hydrostatically tested after assembly. They should never be repair-welded. Even if the weld was sound, the heat causes interior anticorrosive liner to melt and peel away from the tank wall. Failed air tanks always should be replaced.

LOW-PRESSURE INDICATOR

In an air brake system, if the system air pressure drops too low, the brakes will cease to function properly. A low-pressure indicator is a pressure-sensing electrical switch that closes at any time the pressure drops below its preset value. The low-pressure indicator should be specified to close at 50 percent governor cut-out pressure. In a typical system, that would be at approximately 60 psi (4 bar). When the switch closes, a light and a buzzer are actuated.

Air Tank Volume

FMVSS 121 requires that the total volume of compressed air stored on board the tractor dedicated to the brake system be a minimum of 12 times the total brake chamber volume. In dual-circuit brake systems, at least three air tanks are required. Most have more.

It should be pointed out that what at first sight appears to be a single air tank may in fact be more; some air tanks are divided by internal bulkheads.

One common design uses a single cylindrical tank divided into supply and secondary chambers by a bulkhead. The supply tank charges the secondary chamber by means of an external line, one-way check valve, and a separate primary tank, which also is pressure protected by a one-way check valve. To determine the number of separate chambers there are in an air tank, count the draincocks and spitter valves. Every air tank is required to have its own draincock or spitter valve.

28.4 PRIMARY CIRCUIT

On a typical tractor/trailer combination, the primary service circuit is responsible for actuating the brakes over the tractor drive-axle tandem and foot-actuated applications of the trailer service brakes. This is why the primary circuit sometimes is referred to as the rear brake circuit. The secondary circuit is responsible for actuating the tractor front axle brakes and hand valve-actuated applications of the trailer service brakes. This explains why it is sometimes known as the front brake system. **Figure 28–20** is a schematic of a modern ABS-equipped tractor showing disc brakes on the right side and drum brakes on the left; the purpose of the schematic is to show that pneumatically there are no differences between the two types of foundation brakes.

PRIMARY TANK

The primary tank receives air directly from the supply tank. If the primary and secondary tanks are supplied by the supply tank in series, the air must pass through the primary tank first. The supply air enters the primary tank by means of a one-way check valve.

This means that compressed air can only travel one way from the supply tank. For instance, if the supply tank pressure were to drop below that of the primary tank, air could not drain back. The one-way check valve helps isolate the primary circuit from the remainder of the chassis air system. Should a severe leak take place elsewhere in the system, the primary circuit air supply would be pressure protected. Like all air tanks, the primary tank must be equipped with a draincock or an automatic moisture ejector.

The primary tank has two main functions. The first is to make a supply of air at system pressure available to the relay valves for direct application of the service brakes. The second is to make a supply of air at system pressure available to the foot valve. The foot valve and primary circuit relay valves are plumbed directly into the primary tank.

Primary circuit pressure must be monitored by a dash gauge. A signal line supplies this gauge either directly from the tank or indirectly through the foot valve manifold. The gauge used in a specific truck chassis may be a single unit with separate, color-coded needles to indicate primary and secondary circuit pressure, or it may be two separate gauges, each identified as primary and secondary. The primary circuit is usually coded green, so, in a single-unit, two-needle gauge, the green needle would indicate primary circuit pressure.

The primary tank also supplies air to the dash control valve assembly, the push-pull valves, or modular dash control valve assembly. This supply is ensured by a two-way check valve, so that in the event of an upstream primary circuit failure, air would be routed to the dash control valves from the secondary tank by shuttling a two-way check valve. In older systems that used a two-way check valve supplied from the primary and secondary tanks, if a line ruptured downstream from the two-way check valve, the air in both circuits would be lost. In the current modular dash control valve assembly, only one circuit is capable of discharging.

ROLE OF THE FOOT VALVE IN THE PRIMARY CIRCUIT

A detailed description of all foot valve functions is provided later in this chapter and shown in Figure 28–39, Figure 28–40, and Figure 28–41. The primary portion of the foot valve is actuated mechanically. This means that it is directly connected to the treadle. A supply of air from the primary reservoir is made available to the primary supply port at system pressure. Travel of the primary piston will determine how much air is metered out to actuate whatever components are in the primary circuit. Typically, these will be the tractor's tandem drive axles and the trailer's brake system. Also, the air metered out by the foot valve is used as a signal to actuate relay valves on both the tractor and the trailer.

When the foot valve is actuated, primary circuit source air enters the valve and does the following:

- It actuates or signals the brake valves plumbed into the primary circuit.
- It actuates the relay piston in the foot valve; that is, it actuates the secondary circuit.
- It reacts against the applied foot pressure to provide "brake feel."

Brake Valves in the Primary Circuit

In most cases, relay valves will be used to actuate the tractor brakes on the primary circuit. These usually will be the rear tractor brakes. The relay valve signal is delivered from the foot valve using a small-gauge signal line. The small-gauge line is used to reduce lag. Signal lag tends to be greatest when the volume of compressed air is greatest. The relay valve is also connected directly to a closer supply of compressed air, either by a large-gauge air hose or sometimes directly coupled to the tank itself.

A **relay valve** is actuated by the signal pressure and is designed to deliver a much larger volume from the local supply of air to the service brake chambers it supplies. The signal pressure delivered to the relay valve and the application pressure it delivers to the brake chambers are designed to be identical in most brake systems. The low volume of signal air (in relation to the application air) speeds up application and release times.

Some systems, mostly much older systems, used quick-release valves in place of the relay valves to route air to the tractor rear service brakes. A **quick-release valve** acts like a T fitting when air is applied to it from a source, but it exhausts air at the valve when source air is removed; the objective is to speed release times. When a quick-release valve is used to

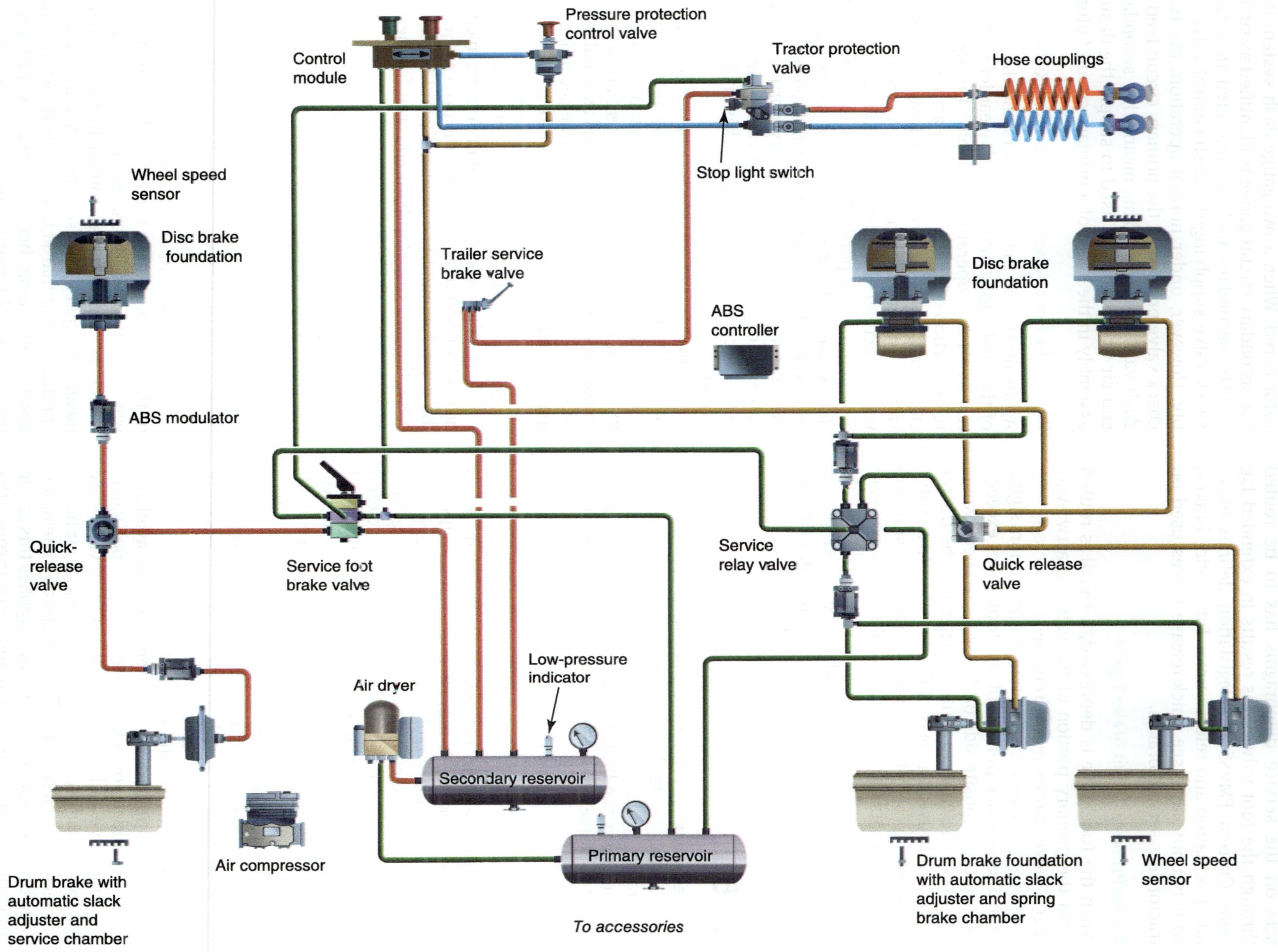

FIGURE 28–20 Typical tractor ABS brake circuit and components identifying key valves and both disc and S-cam actuated brake options

actuate the rear tractor service brakes, the air that acts on the service diaphragms has to be routed through the foot valve. This results in extended lag time. Current FMVSS 121 legislation defines maximum actuation and release times for service brakes, so it is unusual to see quick-release valves used on tractor rear service brakes.

Trailer Primary Circuit Brake Signal

When the foot valve is depressed, a signal is routed from the primary portion of the valve to a two-way check valve downstream from the **trailer application valve**. Whenever the trailer application signal from the primary portion of the foot valve exceeds that on the opposite side of the two-way check valve, it shuttles the valve to route the signal to the tractor protection valve. The bias of the two-way check valve is designed to select the highest source pressure and route that to its outlet. Whenever the signal pressure from the foot valve is greater than the signal pressure from the trailer application valve, the trailer service application signal is provided by the primary circuit.

28.5 SECONDARY CIRCUIT

The secondary service circuit is usually responsible for actuating the brakes over the tractor steering axle and trailer application valve-sourced applications of the trailer service brakes. This is why the secondary circuit is sometimes referred to as the front brake circuit.

SECONDARY TANK

The secondary tank can receive air directly from the supply tank, or air may be routed to it indirectly through the primary tank. Air in the secondary tank is pressure protected by a one-way check valve. This one-way check valve isolates the secondary circuit from the rest of the chassis air system, meaning that loss of compressed air elsewhere in the circuit will not affect the secondary circuit air supply. Like all air tanks, the secondary tank must be equipped with a draincock or automatic moisture ejector.

The secondary tank has two main functions. The first is to make a supply of air at system pressure available to the relay valves or quick-release valves for direct application of the service brakes. The second is to make a supply of air at system pressure available to the secondary or relay portion of the foot valve. The foot valve and secondary circuit relay valves are plumbed to the secondary tank by means of air hoses.

Secondary circuit pressure is required to be monitored by a dash gauge. A signal line supplies this gauge either directly from the tank or indirectly through the foot valve manifold. If a single-unit gauge with separate needles to indicate primary and secondary circuit pressures is used, the secondary circuit needle indicator is red. When a two-gauge dash system is used, the secondary circuit gauge is identified as such.

The secondary tank is connected to a two-way check valve supplying the dash control valve assembly. However, under normal operation, the two-way check valve prioritizes the primary circuit feed to the dash valve assembly. This means that secondary circuit air would only be used to supply the dash valve assembly in the event of a primary circuit failure.

Role of the Foot Valve in the Secondary Circuit

Again, make sure that you understand the description of the dual-application foot valve earlier in this chapter. When it is operating normally, the secondary or relay portion of the foot valve is actuated pneumatically, that is, by air pressure. Whatever air pressure value is metered by the primary portion of the foot valve to actuate the primary circuit is also metered to actuate the relay portion of the valve. The relay portion of the foot valve is designed to act exactly as a relay valve. This means that it will replicate the signal pressure value from the primary portion of the foot valve, sending out exactly that value using secondary tank source air to actuate the brakes on the secondary circuit.

In the event of total failure of the primary circuit air supply, the foot valve primary piston will travel downward until it mechanically contacts the relay piston. Because the relay piston meters air according to the distance it travels, it functions as usual, metering secondary source air to actuate the secondary circuit brakes. In the event of a total failure of the secondary circuit air, no air will be metered out to actuate the secondary circuit when the relay piston is displaced.

Remember, whenever a failure occurs in either the primary or secondary circuits, the driver is alerted to the condition audibly and visually and is required to bring the unit to a standstill and wait for mechanical assistance.

Brake Valves in the Secondary Circuit

Relay valves or quick-release valves may be used to actuate the tractor brakes on the secondary circuit. In most cases, the secondary circuit is responsible for the front service brakes. In many older systems, because the front brakes were located close to the foot valve, they were actuated directly by the relay portion of the foot valve. In other words, the air directed out of the relay portion of the foot valve was the air that was actually used to charge the front service brake chambers. This air was routed to the chambers by means of a quick-release valve. This provided a short application lag that actually helped ensure balanced application timing, because the air had to travel further to reach the rear brakes. Because a quick-release valve was used, release timing lag was exactly what it would be if a relay valve was used.

Front Brake Proportioning

Most front axle braking is pneumatically proportioned in non-ABS systems, whether quick-release or relay valves are used. How the application pressures are proportioned depends on how the truck is being operated. Although proportioning is theoretically unnecessary in ABS, in fact it is often still used because of the redundancy requirement of FMVSS 121. The two common proportioning valves used to manage steer axle braking are the ratio valve and the bobtail proportioning valve. Both ratio and bobtail proportioning valves are described in some detail later in this chapter.

Manually switched front axle limiting valves have not been used for a number of years, but they may sometimes be seen on older vehicles. These operated simply by limiting the application pressure to the front axle brakes by 50 percent whenever the *slippery road* setting was selected. When air directed to the front axle service brakes passes through an automatic ratio valve, it is proportioned to limit application pressure under conditions of less severe braking but to permit full front wheel braking under severe braking conditions.

Bobtailing

Research into the braking dynamics of **bobtail** tractors (a tractor running without a trailer) and unloaded and loaded tractor/trailer combinations on both dry and slippery pavement has shown that drivers maintain better control under panic and severe braking when full application pressures are delivered to the front axle brakes. For this reason, it has been mandatory since 1984 for all North American highway tractors to be equipped with front axle brakes.

Bobtail proportioning valves alter the brake application proportioning when a tractor is being bobtailed. The brakes over the dual drive axles on a typical tractor have much greater capacity than the front brakes because they are designed to handle the vehicle braking requirements when a loaded trailer is coupled to the tractor. When a tractor is bobtailed, the rear brakes become much too powerful for a vehicle that has most of its weight located close to the front axle. In this instance, it is desirable that the steer axle performs a much larger percentage of the vehicle braking.

The braking dynamics of a bobtail tractor are similar to those of a car. When a bobtail tractor is braked, load transfer, suspension dive, and the low percentage of weight over the drive axles throw upward of 60 percent of the braking onto the steer axle. A bobtail condition is sensed when there is no air in the trailer air supply line. Bobtail proportioning applies full application air pressure to the front axle for all brake application pressures and proportions the application pressure to the rear axles. That is, it makes the braking of the rear axles less aggressive except under severe braking, when the rear axles also receive full application pressure.

Trailer Secondary Circuit Brake Signal

The trailer application signal from either a dash or steering column-mounted trailer application valve (also known as the spike, trolley valve or *broker brake* [slang]) is sourced from the secondary circuit. A supply of secondary circuit air at system pressure is available at the trailer application valve. When the lever on the valve is actuated, air is metered to the trailer service application signal line. Downstream from this application valve is a two-way check valve.

The two source supplies to the two-way check valve are the trailer application valve and the foot valve. The bias of the two-way check valve is designed to select the highest source pressure and route that to its outlet. Whenever the signal pressure from the trailer application valve is greater than the signal pressure from the foot valve, the trailer service application signal is provided by the secondary circuit.

DASH CONTROL AND THE PARKING/EMERGENCY CIRCUIT

The parking and emergency circuits in a truck air brake system consist of dash-mounted hand control valves, spring brake chambers, and self-applying fail-safe features. FMVSS 121 legislation requires that all highway vehicles equipped with air brakes have a means of mechanically applying the parking brakes. Also, because air brakes are vulnerable to loss of braking capability when air leakage occurs, a fail-safe mechanically actuated emergency brake system is required. Figure 28–20 shows a tractor emergency brake schematic.

SPRING BRAKES

Spring brake assemblies, dual-chamber brake actuators designed to perform both service and the parking/emergency brake function, are used by most trucks and trailers today. Service applications are performed by air pressure, whereas the parking/emergency function is performed mechanically, that is, by powerful springs. The two chambers are known as service and hold-off (emergency) chambers. Each chamber receives its own air supply that is directed to act on a diaphragm. When the service chamber receives air pressure, it performs service braking.

Hold-Off Circuit

The hold-off diaphragm must have air supplied to it in order to release—that is, cage—the main spring. Without air in the hold-off chambers, a truck with a properly functioning brake system will not move. When the chassis system air pressure is zero, there is no air acting on the hold-off diaphragms on the truck, so the

FIGURE 28–21 Pre-ABS spring brake valve circuit and connections

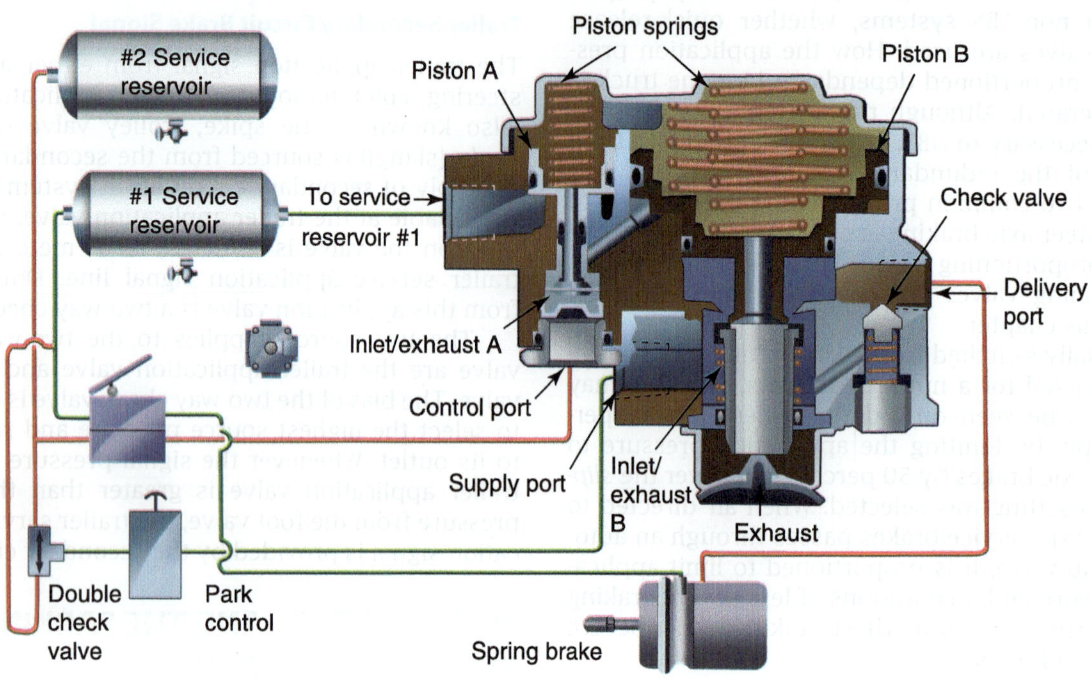

main springs are fully applied to the spring chamber pushrod. This means the system is in park mode.

When air pressure builds in the system and the parking brakes are released, air is directed to the hold-off chambers and the force of the main springs is "held off" the brake chamber pushrod. At least 60 psi (4 bar) must be present in all the spring brake hold-off chambers before the vehicle parking brakes can be considered to be released. **Figure 28–21** shows a pre-ABS spring brake valve circuit and its connections.

The air pressure that acts to hold off the spring brakes is controlled by valves in the tractor dash. The valves that manage the hold-off air pressure usually limit the hold-off pressure to a value below system pressure. This has a way of extending the service life of hold-off diaphragms, and in a true emergency it would result in faster application of the spring brakes. It should be underlined that the effectiveness of parking brakes is always dependent on the brake stroke adjustment at the slack adjusters. In other words, out-of-adjustment service brakes also mean that the parking and emergency brakes are out of adjustment. The following summarizes the operation of spring brakes in a tractor/trailer combination:

- When no air is present in the hold-off circuit, the mechanical force of the spring is fully applied to the chamber pushrod and the brake is in park mode.
- To release the parking/emergency brakes, air pressure greater than 60 psi (4 bar) must be present in the hold-off circuit.
- In the event of severe leakage in the hold-off circuit, the spring brakes will begin to apply the moment pressure drops below 60 psi (4 bar).
- The effectiveness of the parking/emergency brakes depends on the brake stroke adjustment at the slack adjusters in the same way that service brakes do.
- Because 60 psi (4 bar) air pressure is required to hold off a spring brake, when they are fully mechanically applied they exert about the same amount of force on the foundation brakes as a 60 psi (4 bar) service application.
- Spring brake chambers may not be used on steering axle brakes.
- The spring pressure available in the spring brake assembly can be used for emergency service brake applications when a **modulating valve** or **inversion valve** is used. Application signal pressure is used to proportion hold-off air pressure (that is, bleed it down) for a one-time emergency stop.

DASH CONTROL VALVES

Dash control valves are push-pull air valves that enable the driver to release and apply the vehicle parking brakes and supply the trailer with air pressure. Most of these valves are pressure sensitive, so they will automatically move from the applied to exhaust position if the supply pressure drops below a certain minimum.

The dash valve assembly receives its air supply from the primary tank, but it is backed up with air from the secondary circuit by means of a two-way check valve. The two-way check valve receives air from two sources, in this case the primary and secondary reservoirs, and it is designed to shuttle to output whichever source arrives first or is at the higher pressure.

Under normal conditions, primary circuit air is output by the two-way check valve, but, in the event of a primary circuit failure, secondary circuit air is routed to the dash control valve assembly. The more recently introduced modular control valve assembly is equipped with a two-way check valve combined with pressure protection for each air pressure source. This means that if an air line failed downstream from the modular control valve, only one of the source air circuits would bleed down.

DASH VALVE CONFIGURATIONS

Two types of dash valve configurations are used: two-valve and three-valve systems. The function, color, and shape of each of the three valves are determined in FMVSS 121 legislation, so they function exactly the same regardless of which original equipment manufacturer (OEM) chassis they are in. The shape is important because it means the valves can be identified by touch. The shank size of each valve is different, so it is not possible to install an incorrect knob on a valve. Every tractor system is required to have the first two valves: the system park and trailer supply valves. The third is optional and, in fact, is not often seen in current systems. It is known as the tractor park valve. All three are described later and can be seen in several of the air system figures earlier in this chapter.

System Park

The control knob for the system park valve must be diamond shaped and yellow. It masters the trailer supply valve and, if used, the tractor park valves. In older systems, neither of these valves could be actuated until the system park valve had been actuated. When the system park valve is pushed inward, it directs air to the spring brake hold-off chambers in the tractor brake system. The tractor hold-off chambers are charged with air that is routed from the system park valve and directed through quick-release valves before being delivered to the hold-off chambers.

When the system park valve is pulled out (exhausted), both the tractor and trailer parking brake systems are put into parking mode—that is, the spring brakes are applied by exhausting all hold-off air. In applications using the current modular control valve assembly, the trailer (trailer supply valve pushed in) may be supplied with air with the tractor brake spring brakes applied, but, in these, trailer supply air is immediately cut when the system park valve is exhausted.

When the system park valve is pulled outward, the air supply to the trailer supply valve and the tractor hold-off circuit is exhausted. The trailer immediately goes into park mode. The air in the hold-off circuit up to the quick-release valves is exhausted at the system park dash valve and the air in the hold-off chambers is dumped at the quick-release valve exhaust. Because the system park masters the trailer supply valve and, when used, the tractor park valves, the whole tractor/trailer combination is immediately put into park mode.

Trailer Supply

The trailer supply valve must be octagonal in shape and red. Unless a modular control valve is used, the trailer supply valve can only be actuated after the system park valve has been actuated. When this happens, air is directed from the valve to the trailer supply port of the tractor protection valve, and, from there, directed by means of a hose and gladhands to the trailer air circuit. In more recent applications using a modular control valve assembly, the trailer may be charged with air before the system park valve is depressed; however, when both valves are depressed, pulling the system park valve out also will cut the trailer air supply. The immediate result is that the trailer supply valve pops out and the trailer parking brakes are applied.

Whenever the trailer supply valve is actuated (pushed in) and the system pressure is above 60 psi (4 bar), air is directed into the trailer spring brake hold-off chambers, releasing the parking brakes. When the system park valve is pulled out, the trailer supply air is cut, applying the trailer parking brakes. **Figure 28–22** shows a cutaway view of a dash trailer supply valve and **Figure 28–23** shows some examples of dash circuit control valves.

FIGURE 28–22 Trailer supply valve

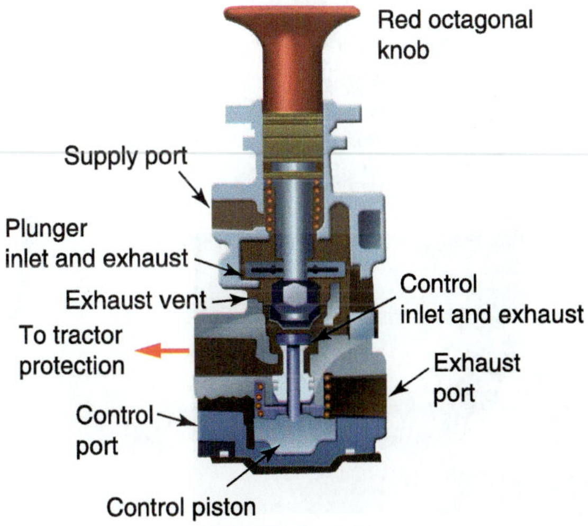

FIGURE 28–23A Color- and shape-coded dash control valves

FIGURE 28–23B Dash-mounted trailer service brake valve

Tractor Park

The tractor park valve, when used, must be round and blue. It is used in dash three-valve control systems and is mastered by the system park valve. In a dash three-valve system, when the system park valve is depressed, air is made available to the trailer supply and tractor park valves. Both these valves must be depressed before the rig is taken out of park mode. However, to apply the parking brakes in the whole rig, only the system park valve has to be pulled. You will come across these valves in some older truck air brake systems.

SHOP TALK

Make sure that you understand the function of each dash control valve in the system. Block the wheels and get into a truck with a fully charged air system. Use the controls and observe the effect.

28.6 TRAILER CIRCUIT

The trailer service and parking/emergency brake circuits are controlled entirely by the tractor brake system. Although many technicians are employed to work exclusively on trailers, in order to effectively work on trailer brakes, the operation of the tractor brake system must be understood first. **Figure 28–24** shows the trailer brake circuits, but also see Figure 28–3 and Figure 28–4, which show the tractor air brake circuits.

TRAILER AIR SUPPLY

The compressed air required by the trailer air circuits is supplied by the tractor. Two hoses couple the tractor air and brake circuits to the trailer. The source of the trailer air supply is the tractor dash control manifold. To send air to the trailer from the tractor, the system park dash valve must first be actuated, followed by the trailer supply valve. This sends air to the tractor protection valve and shuttles it to its on position. The trailer supply air then passes through the tractor to trailer hoses that are coupled to the trailer by means of gladhands. The air in the trailer is stored in reservoirs or tanks. FMVSS 121 requires that the volume of air dedicated to the brake system stored on the trailer be at least eight times the total brake chamber volume.

TRAILER SERVICE SIGNAL

The trailer service brake signal can be sourced from either the tractor foot valve or the trailer application valve. This signal is delivered to the tractor protection valve even when the tractor is being bobtailed. However, the trailer service signal air is prevented from exiting the tractor protection valve when the valve is in its off position. It will be off any time the trailer

FIGURE 28–24 Tandem-axle trailer service and hold-off circuits

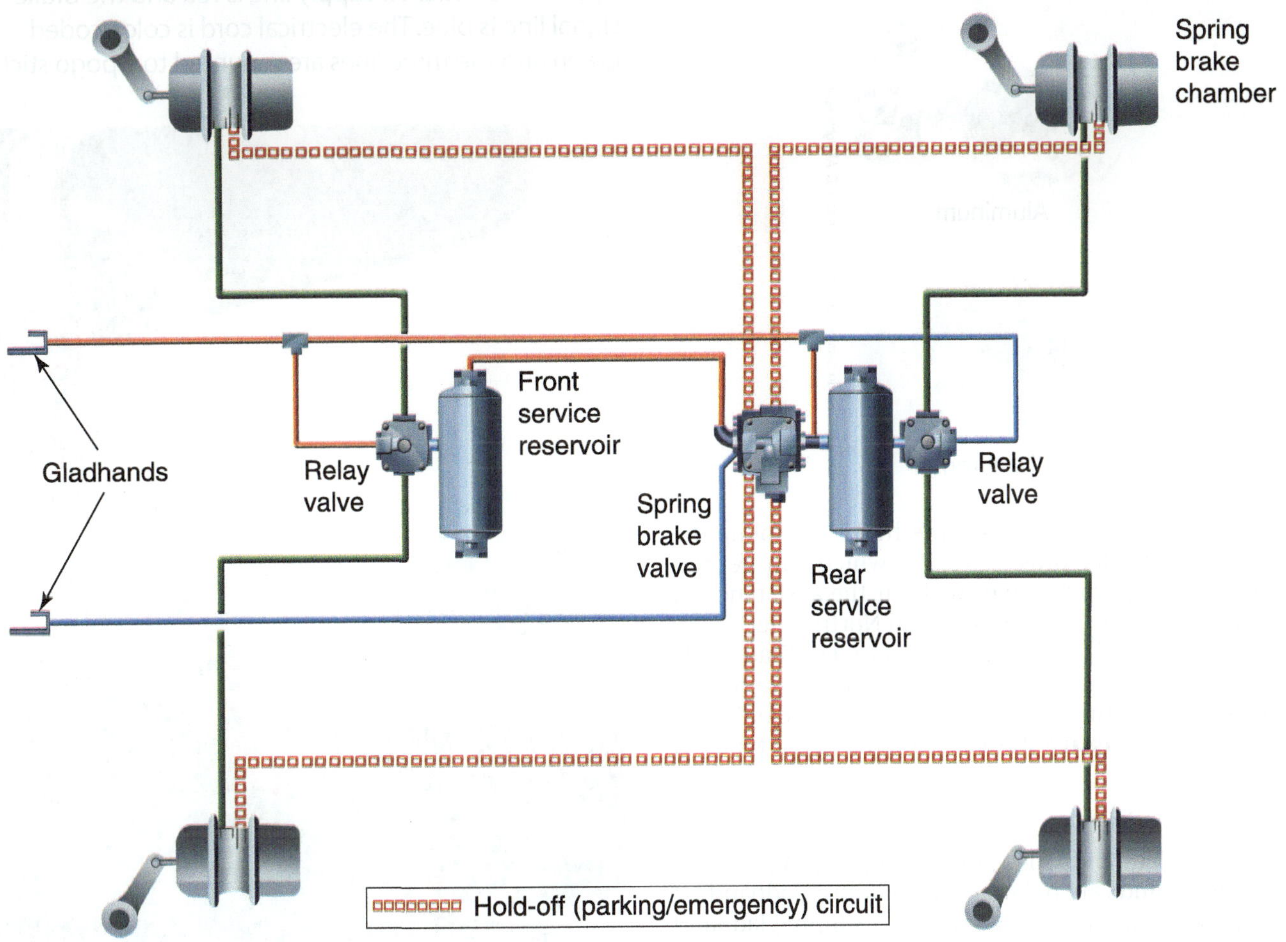

supply dash control valve is not actuated. As soon as the tractor protection valve is shuttled to the on position, the trailer service brake signal can be transmitted through the tractor protection valve.

HOSES

Two hoses are required to couple the air and brake system of the tractor with that of the trailer. Flexible hoses are required because the tractor and trailer articulate at the fifth wheel. The two hoses are usually connected directly to the tractor protection valve. One hose is known as the trailer supply hose, and the other is known as the service brake hose.

Trailer Supply Hose

The trailer supply hose acts as a flexible coupling between the tractor protection valve and the trailer. This hose is normally red. It has the following functions:

- To conduct air from the tractor air system to charge the trailer air tanks
- To master the hold-off (emergency) trailer brake system

Whenever air (at over 60 psi [4 bar]) is present in the trailer supply hose, the trailer air supply will be charged. This air is routed to the trailer air reservoirs and to activate the trailer hold-off (emergency) circuit. Whenever the hold-off circuit is activated, the trailer spring brakes will release.

Trailer Service Signal Hose

The trailer service signal hose has a single function and that is to route a tractor service signal to the relay valves on the trailer. It is normally blue. Whenever the tractor protection valve is open, the trailer service brake signal is routed through this hose directly to the trailer relay valve(s). When this occurs, the trailer service brakes are actuated, using stored air from the trailer reservoirs.

GLADHANDS

At the end of each hose are gladhands. They are called gladhands because, when two are coupled, they have the appearance of a pair of hands in a handshake. The gladhands on the tractor hoses connect with a pair of gladhands on the trailer.

FIGURE 28–25 Gladhands

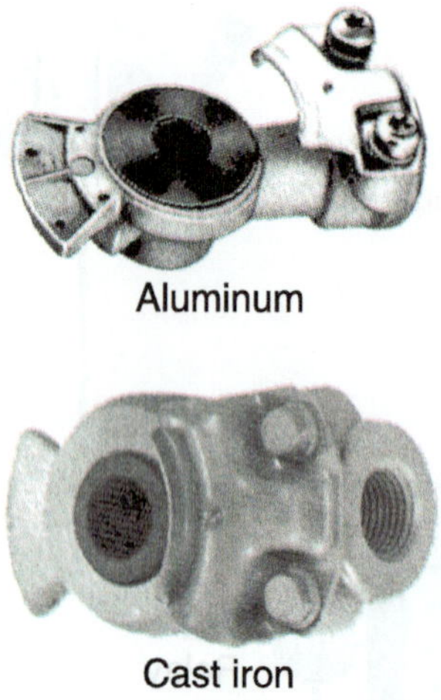

Gladhands provide a fast method of coupling the tractor and trailer brake circuits, which can be performed by any truck driver without the assistance of a truck technician. All gladhands in North America are universally sized, but some are polarized. Gladhands are sealed by rubber grommets. These deteriorate and require routine replacement—this is the only service required. **Figure 28–25** shows different types of gladhands.

Polarized Gladhands

Polarized gladhands make cross-coupling impossible. When gladhands are polarized, the supply gladhand on the tractor will only engage with the supply gladhand on the trailer and likewise with the service gladhands.

Quick-Release Gladhands

Quick-release gladhands are sometimes used in the service supply signal to speed up brake release times. When a quick-release gladhand is used in the service signal line to the trailer, it exhausts at the gladhand rather than at the signal source. This can speed up the release of service brakes after an application and reduce drag at the foundation brakes. Brake lining wear, drum life, and release drag are all improved. Quick-release service signal dump may also be a function of a tractor protection valve.

Dummy Gladhands

These are located on the tractor. They allow the live gladhands to be connected when the tractor is being bobtailed. Dummy gladhands dead-end the hoses and keep the trailer supply and service lines safely out of the way of rotating drive shafts. They should be vented to allow for proper operation of some types of bobtail proportioning valve.

Color-Coded Supply Hoses

Figure 28–26 shows a set of color-coded trailer air lines. The trailer supply hose is normally red in color while the trailer (brake) signal line is blue. In the image, the electrical connection is shown as green in color; the mounting shaft is known as a pogo stick.

FIGURE 28–26 Color-coded trailer air hoses on a tractor. The trailer air supply line is red and the brake signal line is blue. The electrical cord is color-coded green and the three lines are mounted to a pogo stick.

TRAILER BRAKE CONFIGURATIONS

Today most trailer parking/emergency brake systems use one of two pneumatic systems:

- Service brake priority
- Spring brake priority

Service Brake Priority

Until the mid-1990s, most trailers were manufactured with **service brake priority**. A service brake priority system ensured that the service brake reservoir on the trailer would fill with air before the hold-off chambers of the spring brakes were charged to release the parking brakes. This resulted in a delay when hooking up to a trailer whose air tanks were empty or near empty. Though not common, these systems are still permitted today.

In the event of trailer system air leakage in a service brake priority system, air pressure bleeds down in the service and hold-off chambers simultaneously. This results in spring pressure (from the hold-off chamber) gradually taking over as both the service application and hold-off pressures drained down. Most of these service brake priority, park-on-air systems modulate the hold-off pressure at around 90 psi (6 bar). This results in faster parking/emergency response whenever air was exhausted from the trailer supply line because less compressed air has to be exhausted from the hold-off circuit before the brakes are applied with spring pressure.

Spring Brake Priority

Most current trailers meeting FMVSS 121 standards use park-on-spring brakes; we know these as **spring brake priority** systems. The trailer combination valves used to achieve spring brake priority incorporate a quick-release valve to direct air to the hold-off diaphragms in the spring chambers. When the trailer supply air is activated (from the tractor), air is first directed to the spring brake hold-off chambers, causing them to release almost immediately. When the air pressure exceeds 60 psi (4 bar) in the hold-off circuit, air from the trailer supply line can be directed into the service brake tanks to be made available for service braking. In most cases, the hold-off pressure is regulated below system pressure, usually at a pressure value around 90 psi (6 bar).

Because a spring brake priority system first directs air to the hold-off chambers of the trailer spring brakes, a trailer can be moved shortly after the air lines have been coupled. This speeds trailer shunting operations and, in the event of a breakdown, allows a trailer to be moved in the minimum amount of time. With a spring brake priority system, it is possible to couple to a trailer and move it onto the highway before there is any air in the tanks available for service braking. Most trailers today use spring brake priority but FMVSS 121 continues to permit the use of service brake priority systems.

Relay-Emergency Valves

Trailers manufactured before 1975 and those that do not have to comply with FMVSS 121 may be equipped with relay-emergency valves. In such cases, the trailer may not be equipped with spring brake chambers because there was no legal requirement for a mechanical park mode. The brakes on these older trailers were single-chamber service brakes. The relay-emergency valve functioned as a relay valve to effect service braking on the trailer. When the trailer supply air line is charged with air (from the tractor) and a service application signal is received by the relay-emergency valve, it uses air stored in the trailer air tanks to apply the brake chambers. However, when the trailer air supply is cut (trailer supply valve out/trailer breakaway), the trailer brakes go into a **park-on-air** mode.

In park-on-air mode, the relay-emergency valve applies system air pressure to the service chambers. When a trailer is parked for a period of time (or there is an air leak), causing system pressure to bleed down, the result is a parked trailer with no brakes. This means that when coupling a tractor to a trailer, the trailer must be charged with air from the tractor so the brakes can be applied before the fifth wheel and upper coupler are engaged.

CAUTION

When coupling a tractor to a trailer not equipped with spring brake chambers, always connect the gladhands and charge the trailer with air before attempting to couple the fifth wheel.

TRAILER BRAKE VALVES

Most trailers use a multifunctional brake valve in a single housing, capable of managing both service and parking/emergency brake requirements, pressure protecting the trailer system, and modulating hold-off pressure. The brake system may not be the only pneumatic system on the trailer. Many trailers use air suspensions and air-actuated lift axles. The air supply also can be used for other application-specific functions. In all trailer air systems, the trailer supply line is the means of releasing the parking brakes. This means that in any (current) uncoupled trailer, the parking brakes should be fully applied. However, as soon as the trailer supply is activated, the trailer parking brakes are released.

Generally, when coupling a tractor to a trailer, the tractor should first be mechanically coupled to the trailer at the fifth wheel before applying air to the trailer supply circuit. There are some exceptions to this. Most jurisdictions in the United States and Canada require drivers to observe a specific procedure when they are tested, which requires air lines to be attached first. See Photosequence 15 in Chapter 34 for a full explanation of what should be done when coupling a tractor to a trailer.

Because trailer parking/emergency brakes are only released when air is present in the trailer supply air line, in the event of a trailer breakaway (trailer separation from the tractor at the fifth wheel, which is a very unusual occurrence), the trailer supply line would sever, causing immediate application of the trailer parking/emergency brakes.

TRAILER SERVICE BRAKES

Trailer service brakes are controlled by the tractor. The service brake signal to the trailer may be sourced from either the tractor foot valve or the trailer application valve (dash or steering column mounted). You should understand exactly how these two application valves operate to manage the trailer service signal; review the descriptions that appear earlier in this chapter. The service signal is directed to the tractor protection valve. If this is shuttled to its on position, it is then passed through to the trailer service gladhand. Next, the service signal is directed to a trailer relay valve.

The relay valve is designed to replicate the service signal pressure but uses a supply of air stored in the trailer service tanks. This air is sent to the trailer service chambers. This means that if the pressure of the service signal to the relay valve is 20 psi (1.4 bar), it will deliver a 20 psi (1.4 bar) service application using air from the trailer service tanks. When the service signal pressure drops, the air acting on the trailer spring brake chambers is exhausted at the relay valve. Service signal air is exhausted either at the tractor or, in the event that a quick-release gladhand is used, at the gladhand coupler.

28.7 FOUNDATION BRAKES

So far, we have dealt with the components that manage the vehicle braking. The foundation brakes are the components that perform the braking at each wheel on the end the axles. They consist of a means of actuating the brake, a mounting assembly, friction material, and a drum or rotor onto which braking force can be applied. Three types of foundation brakes are found on today's highway trucks:

- S-cam
- Disc
- Wedge

Figure 28–27 shows all three types.

FIGURE 28–27 Foundation brake types: (A) S-cam brake; (B) Wedge brake; (C) Disc brake

A.

B.

C.

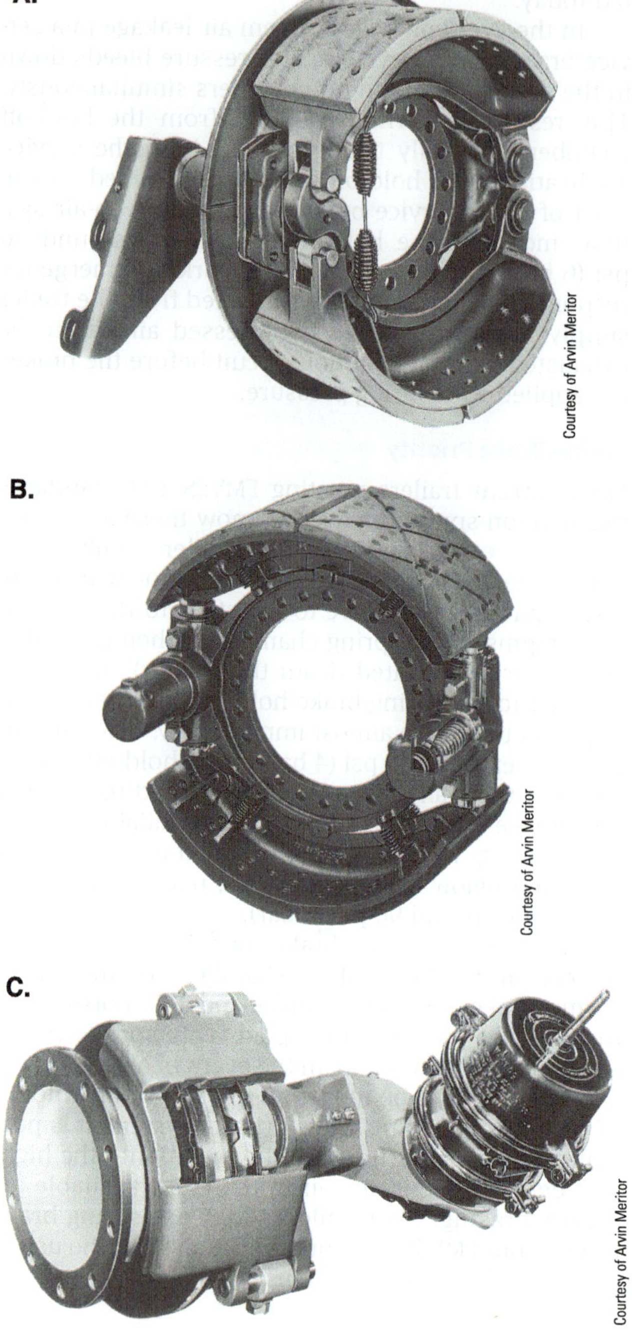

Courtesy of Arvin Meritor

Courtesy of Arvin Meritor

Courtesy of Arvin Meritor

S-CAM FOUNDATION BRAKES

In an S-cam brake system air pressure is used to manage braking, which requires the use of mechanical actuators to convert air pressure into mechanical action. The actuator in an S-cam brake system is a brake chamber. Extending from the brake chamber is a pushrod. A clevis is threaded onto the pushrod, and a clevis pin both connects the pushrod to a slack adjuster and allows it to pivot. The slack adjuster is splined to the S-camshaft, which is mounted in fixed brackets with bushing-type bearings.

When force from the brake chamber pushrod is applied to the slack adjuster, the S-camshaft is rotated. The S-camshaft extends into the foundation assembly in the wheel and its S-shaped cams are located between two arc-shaped brake shoes. The foundation assembly is mounted to the axle spider, which is bolted to a flange at the axle end. The brake shoes may be either mounted on the spider with a closed anchor pin that acts as a pivot, or open-anchor mounted, meaning that they pivot on rollers onto which they are clamped by springs. Opposite the anchor end of the shoe is the actuating roller that rides on the S-cam. The shoes are lined with friction facings on their outer surface. Attached to the wheel assembly and rotating around the brake shoes is a drum. When brake torque is applied to the S-camshaft, the S-shaped cams are driven into actuating rollers on the brake shoes, forcing the shoes into the drum. This action is shown in **Figure 28–28**.

When the shoe is forced into the drum, the friction that retards the movement of the drum is created. In mechanical terms, the energy of motion is converted into friction, which must be dissipated by the drum as heat. Some cam-actuated brakes use flat cam geometry. This means that the shape of the actuating cam is flat instead of being S-shaped. Both types are shown in **Figure 28–29**.

S-Camshafts

The function of the S-camshaft is to convert the brake torque applied to it by the slack adjuster into the linear force that spreads the brake shoes and drives them into the brake drum. The slack adjuster is splined to the S-camshaft and retained by washers and snaprings. Two different spline configurations, known as coarse spline and fine spline, are used. They are also oriented left or right. They must never be interchanged.

S-camshafts are supported on the axle housing and on the brake spider by nylon bushings that permit the shaft to rotate to transmit brake torque to the shoes. The bushings should be lubricated with chassis grease on a routine basis.

FIGURE 28–28 S-cam brake operation: rotation of the S-cam is clockwise to actuate the brake in this foundation brake.

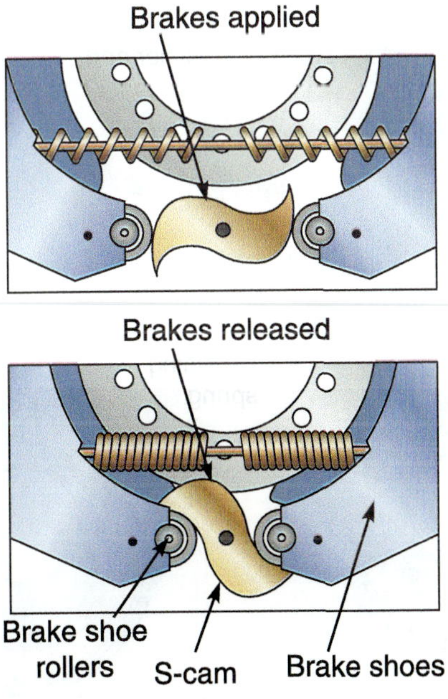

FIGURE 28–29 Cam geometry: (A) S-cam and rollers; (B) flat cam and roller

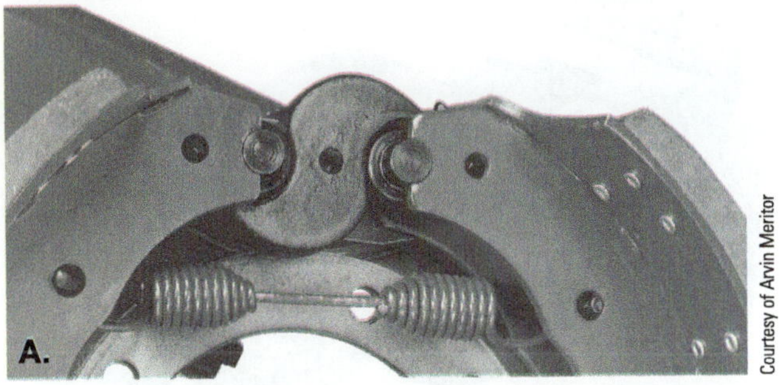

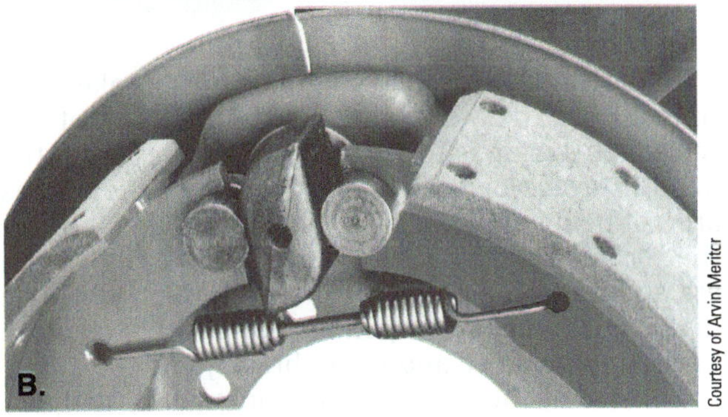

CAUTION

S-cams are oriented with left and right cam profiles and must never be interchanged.

Spiders

The brake spider is bolted or welded to the axle end and provides a mounting for the foundation brake components. It must be tough enough to sustain both the mechanical forces and heat to which it is subjected. The brake shoes are mounted to the spider by means of either open or closed anchors. The anchor is the pivot end of the shoe.

Brake Shoes

Brake shoes are arc-shaped to fit to the inside diameter of the drum that encloses them. They are mounted to the axle spider assembly. Each shoe pivots on its own anchor pin retained in a bore in the brake spider. Today, most shoes are of an open-anchor design. This means that the anchor end of the shoe does not enclose the anchor pin that forms the pivot. In most cases, open-anchor shoes are retained at the anchor end by a pair of springs that clamp the shoes to each other, forcing them against their individual anchors. On the actuation end of the brake shoes, a roller is retained by a clip spring or pin, as shown in **Figure 28–30**.

FIGURE 28–30 Brake shoe rollers

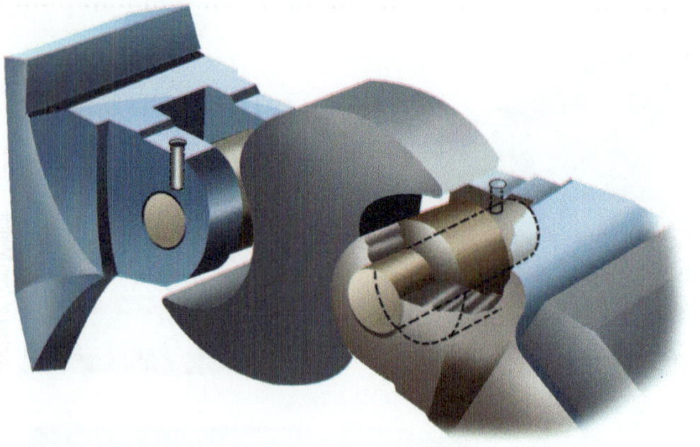

A single retraction spring loads the brake shoe rollers onto the S-cam profile. The function of the retraction spring is to hold the shoes away from the inside surface of the brake drum when the brakes are not being applied. When the brakes are applied, the S-cam rotates, acting on the brake shoe rollers and forcing the brake shoes into the drum. The major advantages of open-anchor brake shoes are lower maintenance and faster brake jobs.

Closed-anchor shoes operate identically to open-anchor shoes. However, on the anchor end of the shoe, a pair of eyes encloses the spider-mounted anchor pin on either side. The disadvantage of the closed-anchor brake shoe design was more maintenance time required to perform brake jobs, mainly because the anchor pins tend to seize in their spider bores.

Friction faces are attached to the shoes. These are graded by their **coefficient of friction** rating, which is given a letter code. Coefficient of friction describes the aggressiveness of any friction surface. For instance, sheet ice has lower friction resistance than an asphalt surface, so it can be said to have a lower coefficient of friction. However, the letter codes used on truck shoe friction linings are too general to use as a reliable guide for replacement. The common width of a brake shoe in 2009 was 7 inches (17.78 cm) but since 2010 FMVSS 121 stopping requirements, 8-inch (20.3 cm) brake shoes have become more common. Severe-duty and off-road applications may use larger sizes.

Ensure that the OEM recommended friction linings are used when relining shoes. It also should be noted that coefficient of friction alters with temperature, generally decreasing as temperature increases. The friction facing is either riveted or bolted to the shoe. The general practice today is to replace the whole shoe with one with new friction lining attached at each brake job. This reduces the overall time required for the brake job. **Figure 28–31** shows how a typical S-cam foundation brake is assembled. **Figure 28–32** shows an S-cam foundation brake assembly.

WARNING

All truck technicians should be aware of the hazards of inhaling brake dust. Brake dust may contain asbestos, a known carcinogen, and other microparticulates that are potentially harmful. Wear a dust particulate mask, eye protection, and never clean up foundation brakes using compressed air. If a specialty brake solvent is used, be sure not to inhale the fumes nor expose it to heat.

FIGURE 28–31 Q-brake, S-cam assembly: (A) removing return spring; (B) removing retaining springs to separate shoes. This drops the shoes in seconds.

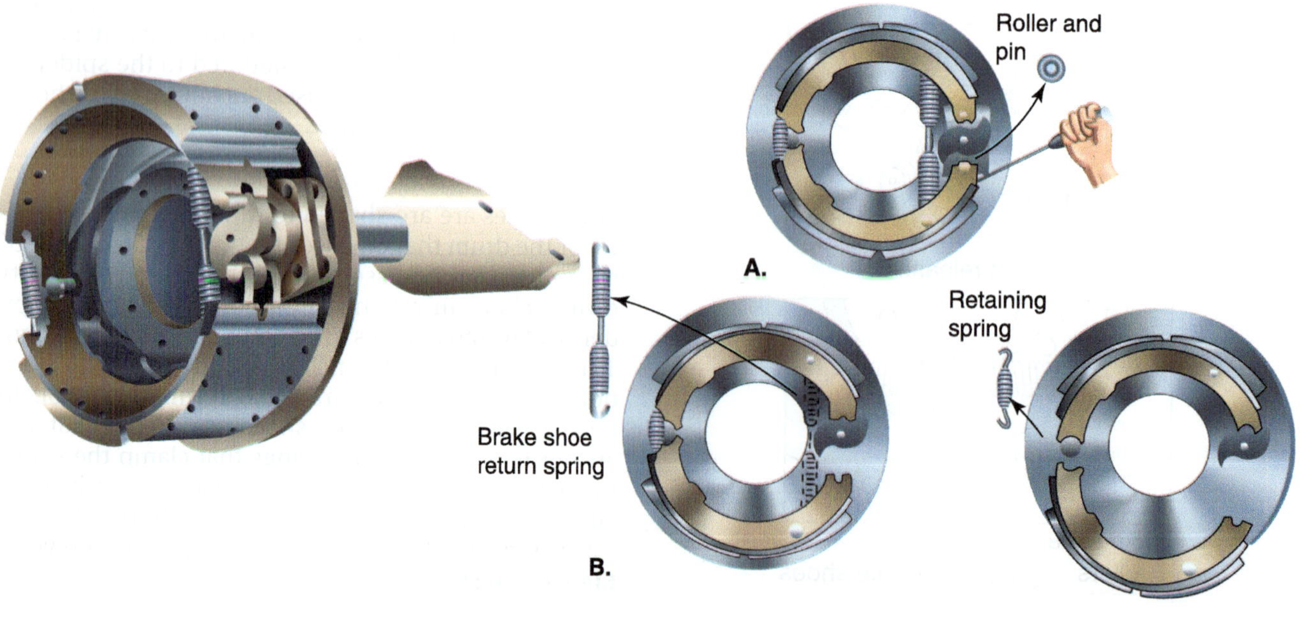

FIGURE 28–32 An S-cam foundation brake assembly

Edge Codes

Letters imprinted on the edge of heavy-duty friction facings are used to rate the coefficient of friction (**Table 28–1**). The first letter indicates the cold coefficient of friction, the second the hot coefficient of friction. The further down in alphabetical sequence, the higher the coefficient of friction—that is, the more aggressive the lining. So a G rating indicates a higher coefficient of friction than an F rating.

Remember, because of the wide range of coefficient of friction covered by each letter rating, most industry opinion is that this is an inadequate rating of actual brake performance. In fact, the TMC of the ATA has stated that up to a 40 percent variance is possible between shoes with the same nominal ratings. Two letters are always used to rate any lining. For instance, a lining code rated FF would have both cold and hot coefficient of friction ratings within the range 351 to 450.

Brake Drums

Brake drums are manufactured of cast alloy steels or fabricated steel. The two have different coefficients of friction, so they should never be mixed. Brake drums are manufactured in several sizes, but 16.5 inches (42 cm) is by far the most common. The friction facings used in brake linings used today, which do not use asbestos, tend to be harder. This means that the friction facings on the shoes last longer but also can be much tougher on brake drums. It has become routine in some applications for the brake drums to be changed at each brake job along with the shoes. Brake drums tend to harden in service. After they show heat checks (small cracks) and heat discoloration (blue shiny areas), they should be replaced rather than machined to oversize.

Because drums are relatively low in cost, they are rarely machined after they have been in service. However, because they are prone to out-of-round distortion when they are stored for any length of time, it is good practice to machine drums when new, after fitting them to the wheel assembly, to ensure they are concentric.

When drums are replaced, the correct size and weight class should be observed to ensure that the brakes are balanced. Drums of different weights will run at different temperatures. When drums are machined, ensure that the maximum machine and service dimensions are observed. For a 16.5-inch (42 cm) drum, the maximum permitted service dimension is 0.120 inch (3 mm) and the maximum machining dimension is 0.090 inch.(2.3 mm). **Figure 28–33** shows a heavy-duty brake drum used with S-cam actuated brakes.

TABLE 28–1 Letter Grading of Coefficient of Friction

Letter	Coefficient of Friction Rating
D	0.150–0.250
E	0.251–0.350
F	0.351–0.450
G	0.451–0.550
H	Over 0.551

Brake Fade

Drum brakes are more susceptible to brake fade. **Brake fade** has a number of causes mostly relating to reduction in the coefficient of friction of the brake contact materials. However, thermal expansion of a brake drum under prolonged braking enlarges its diameter, meaning that the actuation mechanism such as a slack adjuster is no longer functioning at its best mechanical efficiency. For this reason, for trucks operating with heavy loads in mountainous regions, disc brakes are better option than either type of drum brake.

FIGURE 28–33 A heavy-duty brake drum used with an S-cam foundation brake assembly.

S-Cam Foundation Brake Hardware

The S-cam foundation brake hardware includes all the springs, anchor pins, bushings, and rollers on the spider assembly. It is good practice to replace everything at each brake overhaul. The springs used in the foundation brake assemblies must sustain high temperatures, which causes them to lose their tension over time, so it is critically important that these be changed.

Cam rollers should be lubricated on the roller pocket race and never on the cam contact face. Lubricating the cam roller pocket race helps avoid flat spots on the cam contact face. The cam contact faces should never be lubricated because the roller is designed to rotate with the S-cam when brake torque is applied to it.

WEDGE BRAKE SYSTEMS

Wedge foundation brakes are used in some applications, especially on the front axles of vocational trucks. In this brake design, the slack adjuster and S-camshaft are replaced by a wedge-and-roller mechanism that spreads the shoe against the drum. They may be single- or double-actuated.

In a single-actuated wedge brake system, the shoes are mounted on anchors. A single wedge acts to spread the shoes, forcing them into the drum. Dual-actuated wedge brakes tend to be used in heavy-duty applications. The dual-actuated wedge brake assembly has an actuating wedge located at each end of the shoe. The actuating principle of a wedge brake assembly is shown in **Figure 28–34**.

The brake chambers in a wedge foundation brake system are attached directly to the axle spider. The wedge-and-roller actuation mechanism is enclosed within the actuator and chamber tube. When air is delivered to the service chamber, the actuating wedge is driven between rollers that in turn act on plungers that force the shoes apart. When the service application air is exhausted from the service chambers, retraction springs pull the brake shoes away from the drum. The actuating wedges usually are machined at a 14-degree pitch, but they also may be machined at 12 or 16 degrees, so they must be carefully checked on replacement.

FIGURE 28–34 Wedge-actuated brake system

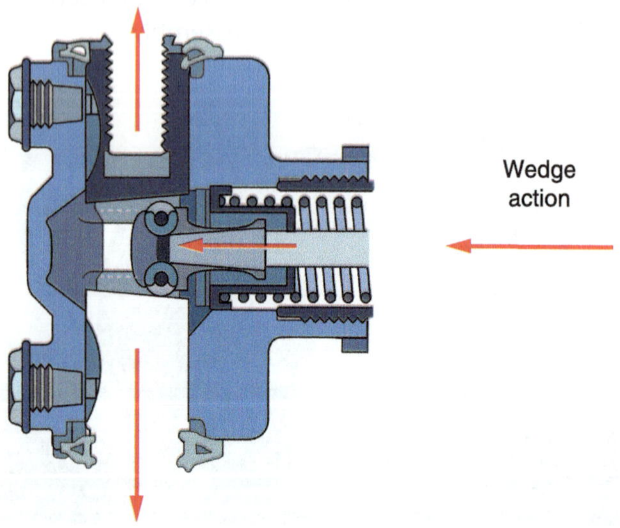

Self-Adjusting Mechanism

A self-adjusting mechanism is standard on all wedge brakes, and this is contained in the wedge brake actuator cylinder. It consists of a serrated pawl and spring that engages to helical grooves in the actuator plungers. A ratcheting action prevents excessive lining-to-drum clearance occurring as the friction linings wear. Despite the fact that dual-actuated wedge brakes have higher braking efficiencies than S-cam brakes, they tend to be rarely used today. They are seen by the trucking industry as having a much higher overall maintenance cost. Of note is the tendency of the automatic adjusting mechanism and plungers to seize when units are not frequently used. This often necessitates the disassembly of the foundation brake assembly to repair the problem.

AIR DISC BRAKES

Air **disc brakes** have been around for a couple of decades, but it is only recently that they have become widely used in the trucking industry. This usage has become more widespread since 2010 when FMVSS 121 reduced stopping distances by 30 percent. Disc brakes have been proven to possess better stopping efficiency than drum brakes.

An example of a disc brake is the bicycle brake. Disc brakes have become the automobile brake of choice because they provide better control and higher braking efficiencies than drum brakes. A disadvantage of air disc brakes in trucks is higher initial and maintenance costs, but as they become more common prices are becoming more competitive. **Figure 28–35** shows a cutaway view and operating principle of air disc brakes. Although the operating principles of air disc brakes remain similar, the actuating mechanisms used differ somewhat between manufacturers and have changed over the years.

Disc Brake Operation

The air disc brake assembly consists of a caliper assembly, a rotor, and an actuator mechanism as shown in **Figure 28–36**. The actuator mechanism is a brake chamber, usually acting on an eccentric lever to amplify its force and convert the linear force produced by the chamber into brake torque to a pair of threaded tubes that act on the pads. The objective of an air disc brake system is to apply clamping force to a caliper assembly

FIGURE 28–35 Air disc brake assembly cutaway. As the actuating beam moves the inner brake pad into contact with the rotor, the caliper slides away from the rotor pulling the outer brake pad into the rotor. This creates two-way, balanced clamping force on the rotor.

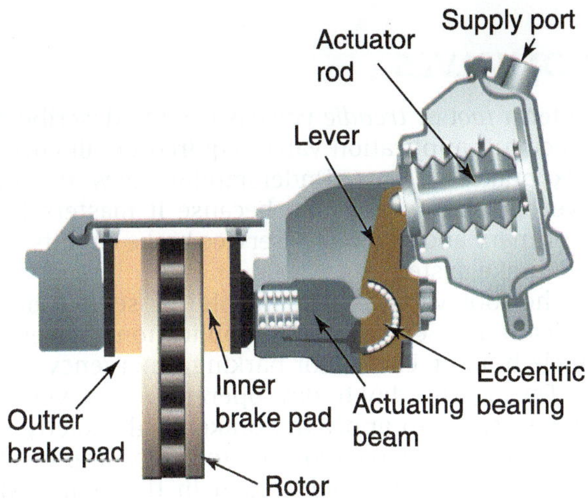

FIGURE 28–36 Air disc brake foundation assembly

within which the pads are mounted. Running between the friction pads of the caliper is a rotor that is bolted to and, therefore, rotates with the wheel assembly. When the power beam is actuated, the threaded tubes force the brake pads into the rotor. **Figure 28–37** shows the air chamber actuator used with a typical disc brake system. Because the caliper is floating, when the inboard-located pad is forced into the rotor (see **Figure 28–38**), the outboard friction pads are also forced into the outboard side of the rotor with an equal amount of force. There are also some older air-actuated disc brakes that use brake chambers that act on a slack adjuster to actuate the caliper assembly. Figure 28–38 highlights the calipers and friction facings on a Meritor steer axle disc brake assembly.

FIGURE 28–37 A typical air disc brake actuator

FIGURE 28–38 Calipers and friction facings used on a steer axle air disc brake system.

Air disc brakes provide the advantage of better brake control and higher braking efficiency than S-cam actuated brakes. They also are almost immune to brake fade and boast a slight overall weight advantage over S-cam brake assemblies. Brake fade is caused by a number of factors but in disc brakes fade relates almost exclusively to high temperature reduction in pad friction coefficient; thermal expansion of the rotor does not impact on brake efficiency. The weight advantage of disc brakes though slight has made their use more popular on trucks operating in jurisdictions with tough weight-over-axle regulations.

Advantages Air Disc Brakes

Loaded trucks have enormous braking requirements. Because the principle of braking a vehicle requires that the energy of motion (kinetic energy) be converted to heat energy, the important factor in determining how long brakes last between service intervals is how well the foundation brakes sustain and dissipate heat. This has influenced the development and design of air disc brakes as they have evolved over the past few years and has resulted in the current generation of air disc brakes that are substantially beefed up when compared with those of a decade ago. No longer do the friction pads outlast rotors, which was a real problem with the first generation of air disc brakes.

Air disc brakes have a lower total sectional area of the friction faces than equivalent brake shoes, which means that more force must be applied to them in order to make them run hotter and less efficiently under braking. However, a well-designed rotor located in the air flow is able to dissipate the heat it absorbs, especially when engineered to maximize cooling.

The primary advantage of air disc brakes is that they have superior brake performance characteristics. With the introduction of reduced tractor-trailer stopping distances by FMVSS 121 in 2010, many manufacturers were forced to adopt air disc brakes to meet the new requirements. Because air disc brake became more common after 2010, this immediately improved the parts availability infrastructure.

28.8 AIR BRAKE SYSTEM COMPONENTS

This section describes how different components and valves used in an air brake system function. Most have been introduced previously. For instance, a foot valve plays a role in all service braking, so it has a role in both primary and secondary circuit application pressure management. The description of the foot valve provided in this section outlines all its functions, so some information is repeated from earlier sections.

FOOT VALVES

The term *foot* or *treadle valve* is used to describe the dual-circuit application valve required on all current truck brake systems. Understanding how the foot valve operates is essential because it masters both the primary and secondary service brake circuits on a truck brake system.

The foot valve is responsible for service brake applications. Service brake applications cover all vehicle braking except for parking/emergency braking. During the day-to-day operation of a vehicle, it is necessary to apply the brakes with exactly the right amount of force required to safely stop a vehicle. Too little force could result in the vehicle not stopping in time, whereas too much force could lock the wheels.

The foot valve is the driver's means of managing the service brakes in an air brake system. It plays the same role as the brake pedal in your car. **Figure 28–39** shows some typical foot valves.

FIGURE 28–39 Types of dual-circuit brake foot pedal control valve. They also are called service application or treadle valves.

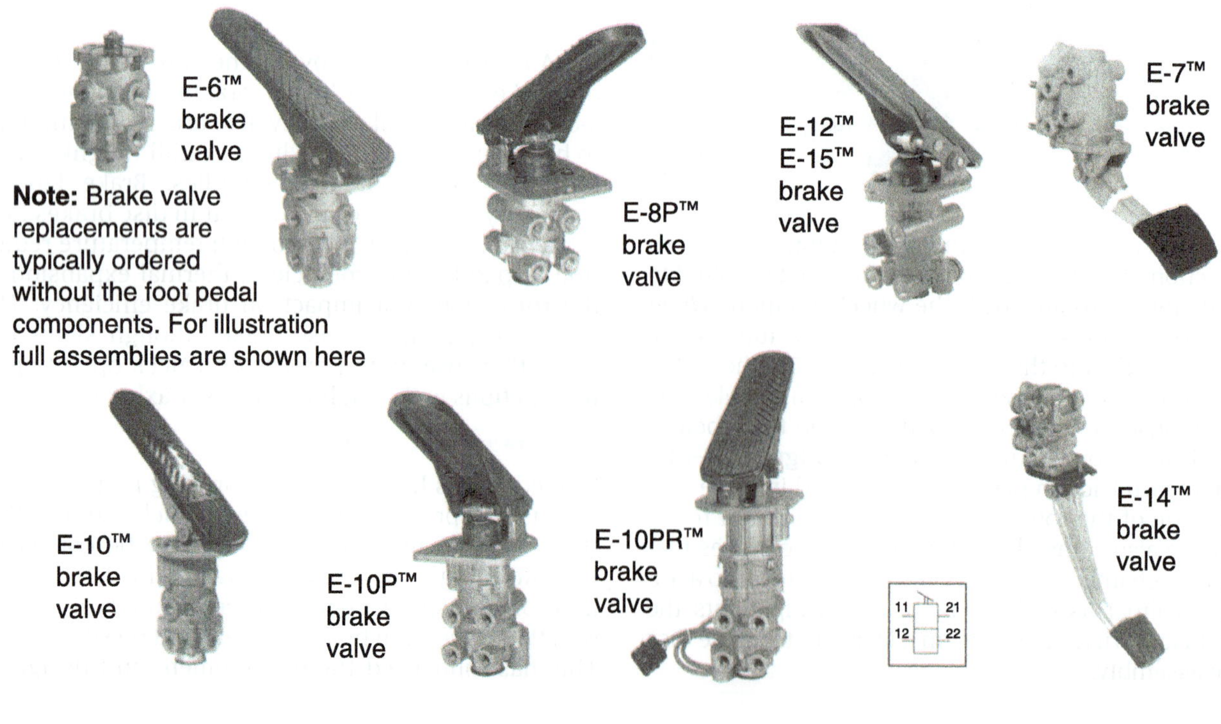

Brake Feel

In a hydraulic brake system, the application pressure delivered from the brake pedal is always associated with the amount of mechanical force applied by the foot, even in a power-assist system. Because of this relationship with the amount of pressure applied and the actual braking force, the driver is provided with what is known as "brake feel." The potential energy of an air brake system is compressed air. This means that the mechanical force applied to the brake pedal has no direct relationship with the resulting brake force. Instead, a different means of providing the driver with brake feel is required. In other words, the foot valve in an air brake system must be made to at least *feel* like a brake pedal in a hydraulic brake system. That is, the pedal should produce increased resistance as braking force increases. The main function of the foot valve is to provide the driver with a means of precisely managing the correct amount of brake force for each stopping requirement.

Brake feel is provided by directing whatever air pressure value has been metered out to the primary circuit to act *against* the applied foot pressure. This means that the further the treadle travels, the more pedal resistance is felt because the air pressure below it is greater. The idea is to provide air brakes with the *feel* of hydraulic brakes. In hydraulic brakes, whether power assisted or not, actual braking force is always proportional to the mechanical force applied to the brake pedal.

Foot Valve Construction

The foot valve is divided into two sections (**Figure 28–40** and **Figure 28–41**). The section closest to the treadle (foot pedal) is known as the primary section and is responsible for primary circuit braking. The section further from the treadle is known as the relay section and is responsible for secondary circuit braking.

The primary section of the foot valve receives a supply of air from the primary tank at system air pressure. The relay section of the foot valve receives a supply of air directly from the secondary tank at system air pressure. When the driver depresses the brake pedal, the supply air from each section of the foot valve is metered out to apply the brakes on each circuit, primary and secondary. The air from each circuit never comes into direct contact with the other when in the foot valve.

Foot Valve Operation

The primary section of the foot valve is actuated mechanically; that is, by foot pressure. This causes the plunger under the treadle to move the primary piston. Any primary piston travel off its seat results in the closing of the exhaust valve and then, depending on primary piston travel, air being metered from the primary supply port out to the primary circuit service

FIGURE 28–40 Cutaway view of a typical dual-circuit service brake valve

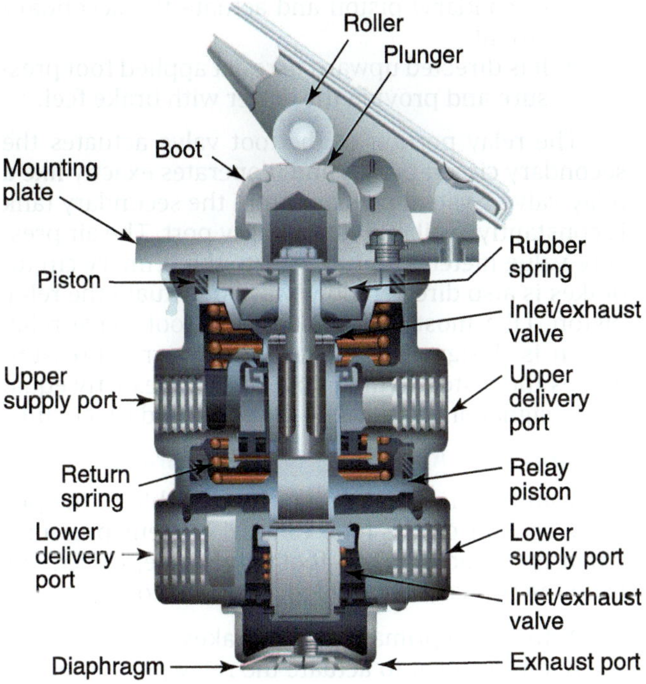

FIGURE 28–41 Suspended pedal dual-circuit service brake valve

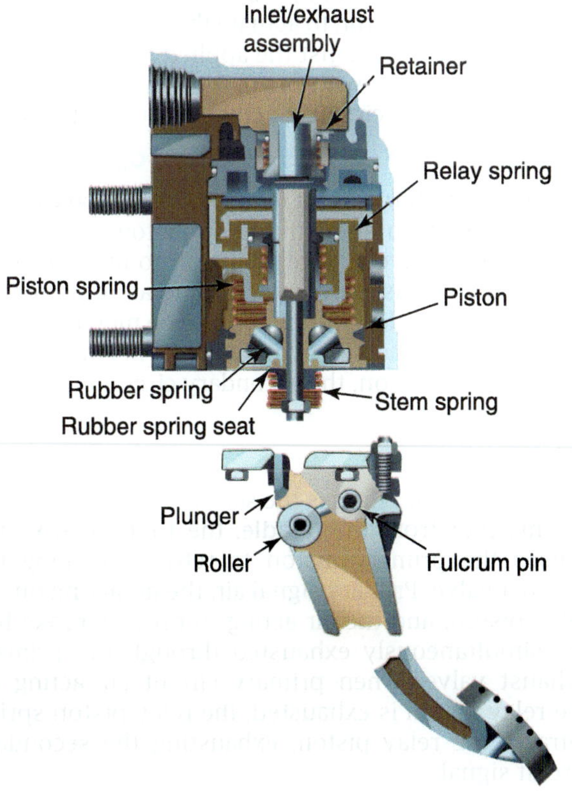

brake components. This metered air performs two other essential functions:

- It is directed downward to act on the relay (secondary) piston and actuate the secondary circuit.
- It is directed upward to resist applied foot pressure and provide the driver with brake feel.

The relay portion of the foot valve actuates the secondary circuit brakes and it operates exactly like a relay valve. System pressure from the secondary tank is constantly available at its supply port. The air pressure value metered out to actuate the primary circuit brakes is also directed downward to actuate the relay piston. Like most relay valves, the foot valve relay piston is designed so the actuating signal pressure moves the piston to meter out pressure exactly equal to the signal pressure value to the secondary circuit.

Primary Circuit Application: Normal Operation

Air from the primary reservoir is available to the primary section of the foot valve at system pressure. When the treadle on the foot valve is depressed, air from the primary circuit is metered out to:

1. Actuate the primary circuit brakes
2. Act as a signal to actuate the relay portion of the foot valve
3. Act against the applied foot pressure to provide brake feel

If the treadle on the foot valve is depressed so that 20 psi (1.4 bar) of air is metered out to actuate the primary circuit brakes, 20 psi (1.4 bar) also will act on the relay piston to actuate the secondary circuit brakes and to act against the applied foot pressure to provide the driver with brake feel. In normal operation, the primary circuit is actuated mechanically.

Secondary Circuit Application: Normal Operation

When the primary circuit is actuated, primary circuit air is directed to enter the relay piston cavity. This air pressure moves the relay piston to first close the secondary exhaust valve and, next, to allow air to flow out of the secondary supply port to be metered out to actuate the secondary circuit brakes. This means that, in normal operation, the secondary circuit is actuated by air pressure.

Releasing Brakes

When the driver releases the service brakes by removing his foot from the treadle, the foot valve spring causes the primary piston to retract, opening the exhaust valve. Primary signal air, the air acting on the relay piston, and the air acting to provide brake feel are simultaneously exhausted through the primary exhaust valve. When primary circuit air acting on the relay piston is exhausted, the relay piston spring retracts the relay piston, exhausting the secondary circuit signal.

CIRCUIT FAILURES

The main reason for having a dual-circuit service application system is to provide a fail-safe in the event of one circuit failure. This is one form of redundancy offered by current truck brake systems. When one circuit fails, the companion circuit is designed to bring the vehicle to one safe stop. Total failures of one of the two circuits are rare but not unknown. Physical damage to either of the circuit air tanks would be an example of such a failure.

In the event of a failure of either circuit, the driver is trained to bring the vehicle to a standstill and await mechanical assistance. When such a failure occurs, the driver is alerted to the condition by gauge readings, warning lights, and a buzzer. As a fail-safe for both the primary and secondary circuits, the spring brakes would bring the vehicle to a standstill.

Primary Circuit Failure

In the event of a primary circuit failure, the foot valve treadle will depress the primary piston, which will not meter any air because none will be available at the primary supply port. This means that no air will be metered out to actuate the primary circuit brakes, to actuate the relay piston, or to provide brake feel. The driver's foot on the treadle will now travel further with little resistance until the primary piston physically contacts the relay piston. The relay piston will now be actuated mechanically and, according to the distance it travels, meter out air to actuate the secondary circuit brakes. The secondary circuit will operate normally, except that instead of being actuated by air from the primary circuit, it will be actuated mechanically.

Secondary Circuit Failure

In the event of a secondary circuit failure, the primary circuit would operate normally with no loss of brake feel. The air metered into the relay piston cavity would move the relay piston, but no secondary circuit air would be available at the supply port, so none would be metered out to the secondary circuit service brakes.

SHOP TALK

One OEM survey of warranty returns indicates that over half of the foot valves submitted function perfectly. This misdiagnosis results from a failure by the technician to understand how the valve operates normally. Take some time to understand exactly how this important valve masters the operation of the service brakes.

TRAILER APPLICATION VALVE

The trailer application valve is often known by such names as *broker brake* (slang), *spike*, or *trolley valve*. The valve isolates the trailer service brake function

FIGURE 28–42 Trailer control brake valves

from that of the tractor. It is a hand-operated graduated control valve that is usually mounted on the steering column of the truck. The name *broker brake* came about because a tractor owner-operator (broker) could use the company-owned trailer brakes to perform a large percentage of the rig's braking, thereby saving the tractor's brakes. Typical trailer application valves are shown in **Figure 28–42**.

In a typical system, the trailer application valve receives a supply of air from the secondary circuit at system pressure. When the application valve is applied, this air is metered out to the tractor protection valve to be used as the trailer service brake signal. This signal also could be delivered from the foot valve, in which case the source air would be the primary circuit. For this reason, a two-way check valve is required. The operation of a two-way check valve is described in some detail later in this section.

Essentially, both possible sources of the trailer service brake signal are routed to the two-way check valve located before the tractor protection valve. The two-way check valve is designed to shuttle and direct the higher source inlet pressure to the outlet. This means that in the event of a service signal from both the foot valve and trailer application valve, the higher pressure signal would be the one routed through the tractor protection valve.

RELAY VALVES

A relay valve is a simple remote-controlled brake valve. It can be described as a proportioning pilot valve. Some examples are shown in **Figure 28–43**. The relay valve is designed to have a constant supply of air at system pressure close by. When it receives a remote signal, it uses the local supply to actuate brake chambers.

Most relay valves are designed to output the same pressure to the brake chambers as the signal pressure. This means that a 25 psi (1.7 bar) signal delivered to a relay valve would result in a 25 psi (1.7 bar) output. Most current truck air brake systems use relay valves because there is some distance between the application valves (foot valve and trailer spike) and the foundation brakes on a vehicle. This distance varies from the service brakes on the tractor unit to those most far away on the trailer or trailers in the rig. If the air sent to the brake chambers were sourced from the application valve, a considerable "lag" time would result. To speed up the application of brakes, air from the application valve is often used only as a signal. The air used to actuate the vehicle service brakes is contained

FIGURE 28–43 Relay valves

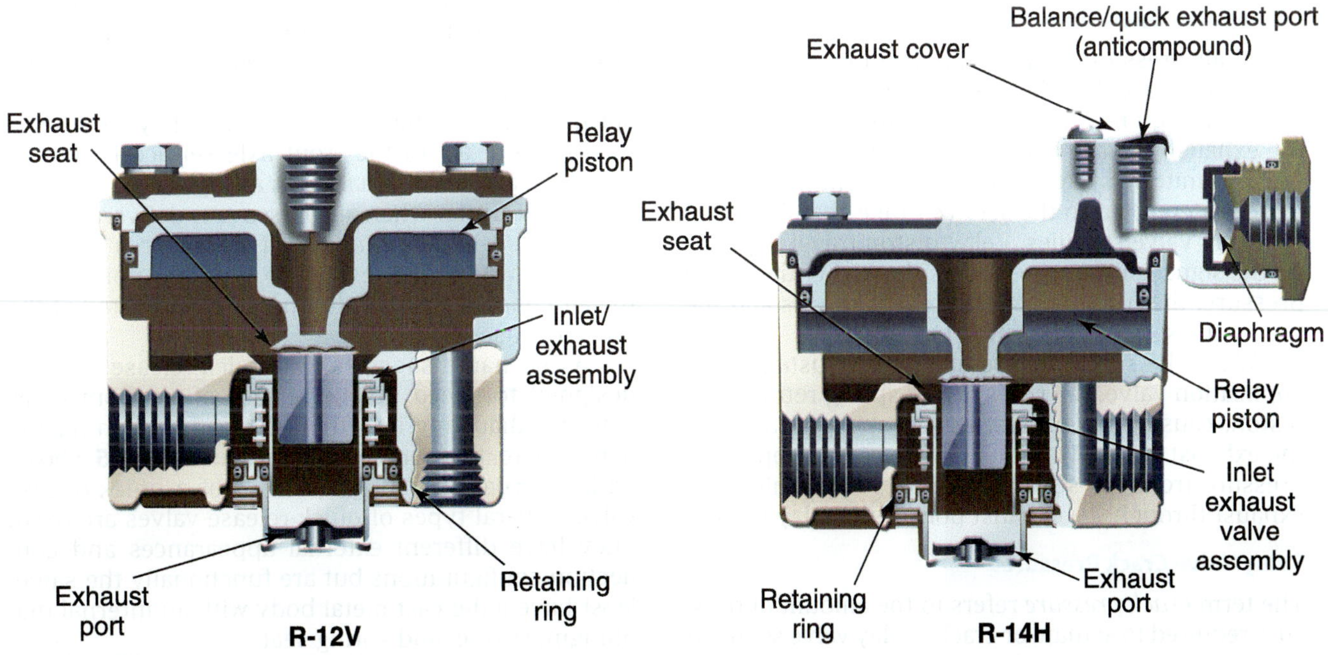

FIGURE 28–44 Primary circuit connection to a relay valve

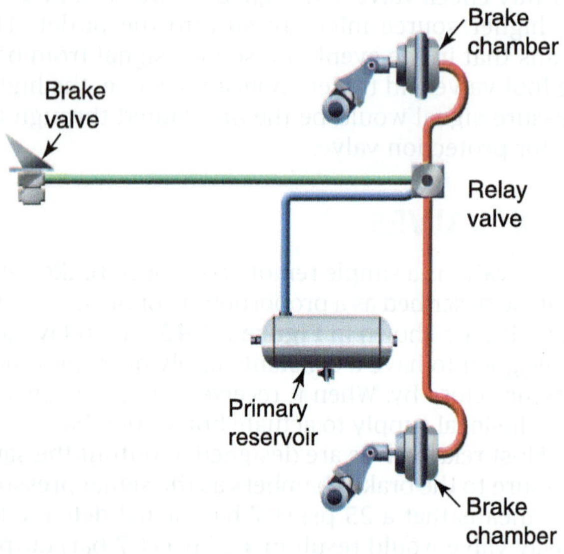

in a tank mounted as close as possible to the foundation brakes (**Figure 28–44**). Relay valves are used to reduce brake application and release times.

A relay valve receives an air signal directly from the application valve. The signal line from the application valve is of a smaller gauge to minimize the volume of signal air. The relay valve is designed to meter a locally contained supply of air at system pressure, delivering it to the service brake chambers. The signal volume of air is small compared to the volume of air required to actuate a service brake chamber. Because this larger volume of air is stored close to the service brake it supplies, the actuation time is reduced.

Relay Valve Operation

Signal air pressure from the application valve enters the cavity above the relay piston and moves the piston down through its stroke. The exhaust valve seals off the exhaust passage and the inlet valve moves off its seat, permitting system pressure air from the air tank to be metered out to the service brake chambers that the relay valve supplies. Relay piston travel depends on the signal air pressure value. The higher the signal pressure, the further the relay piston travels and the higher the service application pressure.

When the signal air pressure is exhausted at the application valve, the relay piston spring retracts the piston, causing its exhaust seat to move away from the exhaust valve. With the exhaust passage open, air pressure from the service chambers is permitted to exhaust through the exhaust port of the relay valve.

Relay Valve Crack Pressures

The term *crack pressure* refers to the amount of pressure required to actuate or crack a relay valve so that it can begin to meter air. Air is compressible. The larger the volume of confined compressed air, the greater its compressibility. There is a slight delay between the moment an air signal leaves the application valve and the moment it acts on the relay piston. This delay increases as the distance from the application valve to the relay valve increases.

The service brakes in a double trailer combination are actuated by the foot valve in the tractor. The air signal from the tractor foot valve has to travel a considerable distance to actuate the service brakes on the rearmost trailer axle. If all the relay valves on the rig had identical crack pressures, those service brakes closest to the foot valve would be applied before those furthest away, creating a potential danger. In order to balance and properly time the service brakes on a tractor/trailer combination, relay valves are manufactured with different crack pressures.

Generally, those relay valves designed to function furthest away from the tractor foot valve would have the lowest crack pressures. Those on the tractor itself would have slightly higher crack pressures. When the service brakes are applied, the brakes are applied evenly through the whole rig.

Valve crack pressures are defined by the strength of the relay piston return spring. Crack pressures range from zero (not common) to values around 4 to 7 psi (0.275 to 0.482 bar). It should be noted that approximately 6 psi (0.41 bar) of air pressure is required to rotate the S-cams sufficiently to contact the brake shoes against the drum. Properly timed crack pressures on a tractor/trailer combination can help prevent trailer jackknife and other brake balance problems.

QUICK-RELEASE VALVES

Quick-release valves are designed to speed the exhausting of air from an air circuit. They are used in brake and other chassis air systems. They commonly are used to distribute air to the hold-off chambers in spring brake assemblies. In pre-ABS, they were used to distribute air to the front axle service brakes. A quick-release valve usually has a single inlet port and two outlet ports. When air pressure is delivered to its inlet port, it is distributed as if it were in a T-union. However, when air pressure to the inlet port drops, the air charged to the outlet ports is exhausted at the quick-release valve.

As its name suggests, the quick-release valve is designed to speed up the release of air from a circuit by exhausting it at the valve rather than having it return to the source valve. **Figure 28–45** shows both external and cutaway views of a quick-release valve. Several types of quick-release valves are used. They have different external appearances and connection configurations but are functionally the same. Most have a die-cast metal body with an internal diaphragm, spring, and spring seat.

FIGURE 28–45 Quick-release valve: (A) external view; (B) sectional view

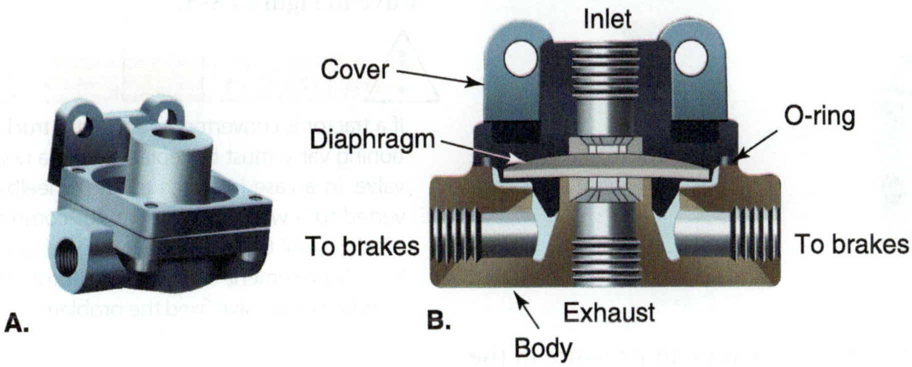

RATIO VALVES

Some years ago, a front-wheel limiting valve (also known as a "wet-dry" switch) was used in many air brake systems. Although not currently legal, there are still a few examples of front-wheel limiting valves around. This valve was switched from the cab and reduced the application pressure to the front axle service brakes to 50 percent of application pressure when toggled to the *wet road* setting. Trucks meeting the FMVSS 121 standard today cannot use a manually switched limiting valve, so when the front axle brakes are required to be ratio modulated, an automatic **ratio valve** must be used. **Figure 28–46** shows both external and sectional views of ratio valves.

Automatic ratio valves replaced the front-wheel limiting valve. An automatic ratio valve limits front axle braking under light and medium-duty applications. Typically, the valve functions as follows. The ratio valve will not crack until the application pressure exceeds 10 psi (0.69 bar). With application pressures between 10 and 40 psi (0.69 and 2.76 bar), the front axle service brakes receive 50 percent of the application pressure. Between 40 and 60 psi (2.76 and 4 bar), the ratio valve graduates the ratio of air delivered to the front axle service brakes to 100 percent. With any application pressure greater than 60 psi (4 bar), the front brakes receive 100 percent of the application pressure. This means that, under severe braking, the front axle brakes receive 100 percent application pressure.

Automatic ratio valves are considered safer than front-wheel limiting valves because of the importance of the front axle brakes under any severe braking condition. All research and analysis of panic brake performance on heavy-duty rigs has shown that the driver can maintain better control when the front axle service brakes are provided with 100 percent application pressure. Mandatory ABS makes this even truer. Theoretically, steer axle ratio valves are not required on ABS. However, because ABSs are required to meet performance standards in the event of an ABS failure, they still appear on some chassis.

FIGURE 28–46 Ratio valves: (A) typical ratio valves; (B) sectional view of a ratio valve

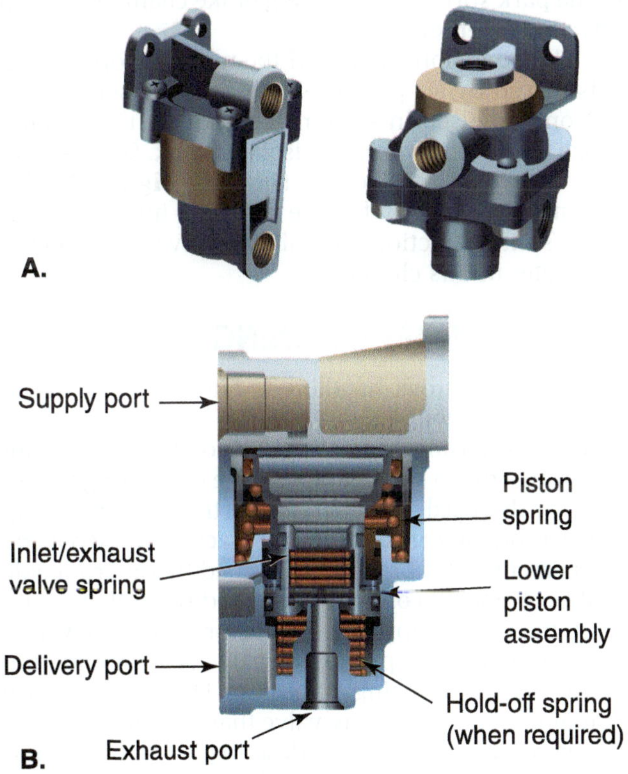

INVERSION VALVES

Inversion valves are used on some brake systems. They are used to improve braking under emergency conditions that have produced a complete system failure. They function by modulating the parking brake hold-off air to provide service braking in a failure

FIGURE 28–47 Inversion valves

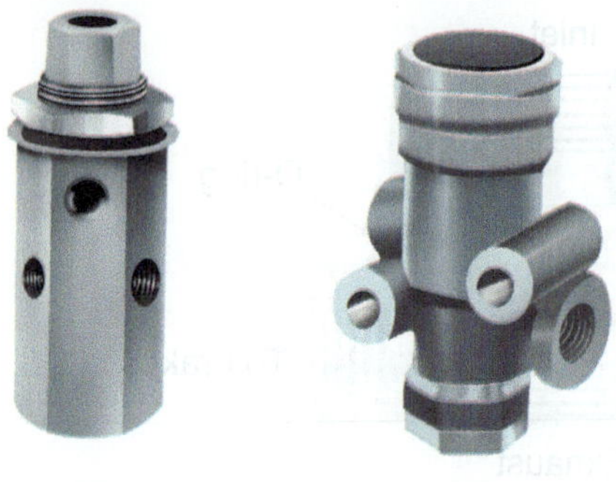

condition, resulting in total loss of air pressure in the application circuit source air.

When an inversion valve is in failure mode, it permits the application signal from the foot valve to proportionally relieve the spring brake hold-off pressure (by exhausting air from it) to assist in stopping the vehicle. This takes advantage of the mechanical force of the park spring in the spring brake chamber to help effect an emergency stop.

The inversion valve is of the normally open type. It is closed by air pressure from another source and can only be used in systems that source the parking brake (hold-off) air from an isolated tank. Some examples of inversion valves are shown in **Figure 28–47**. An inversion valve feature is built into many of today's multifunction, modulating valves described a little later in this chapter.

BOBTAIL PROPORTIONING VALVES

A highway tractor's handling is quite different when the unit is not coupled to a trailer. A bobtail tractor has a large proportion of its weight loaded over the front axle with a much smaller percentage of the weight over the rear drive axles. Bobtail proportioning valves help balance braking in a bobtailed tractor by increasing the percentage of braking performed by the front axle brakes. A bobtail condition is sensed when there is no air in the trailer supply to the tractor protection valve.

The bobtail proportioning valve is a combination of two valves in a single housing. The lower portion contains a standard relay valve that functions exactly as the relay valves already described. The upper portion houses a proportioning valve designed to reduce normal service brake application pressure any time the tractor is not coupled to a trailer. The control port on the valve is connected to the trailer supply air circuit. During bobtail mode, the front brakes receive full application pressures, whereas the service braking to the rear brakes is proportioned. This condition occurs until severe braking is required. When application pressures are at 90 percent, the rear axle brakes receive full application pressure.

The less aggressive braking to the rear axles by a bobtail proportioning valve greatly improves bobtail vehicle handling, especially under slick road conditions. Note the location of the bobtail proportioning valve in Figure 28–3.

CAUTION

If a tractor is converted to a straight truck, the bobtail proportioning valve must be replaced with a nonproportioning relay valve. In a case in which a long-wheelbase tractor was converted to a wrecker, the operator complained of lazy service braking until the treadle was floored when the brakes would lock. Replacement of the bobtail proportioning valve with a standard relay valve fixed the problem.

SPRING BRAKE CONTROL VALVE

A spring brake control valve is a combination valve that is used to manage both service and parking/emergency brake functions to spring brake chambers. This valve also may incorporate the functions of an inversion valve, regulate spring brake chamber hold-off pressure, and have anticompounding features. **Figure 28–48** shows some examples of spring brake control valves. **Figure 28–49** shows a current multifunction, modulating valve. It performs four functions:

- Limits hold-off pressure to the spring brakes
- Provides quick release of air from the hold-off chamber, providing faster actuation of spring brakes
- Prevents compounding by having an anticompounding feature
- Provides inversion valve braking in the event of full system pneumatic failure

SHOP TALK

Current trailer spring brake control valves usually prioritize air to the hold-off circuit over air to the air tanks. These are known as spring brake priority systems. When air builds up in the air system on a spring brake priority system, it is first delivered to the hold-off circuit, enabling the release of the parking brakes. This means that a vehicle with a disabled air supply system can be moved much more quickly because a much smaller volume of air has to be transferred to the vehicle before release of the parking brakes becomes possible.

CAUTION

In spring brake priority systems, because air is delivered to the trailer hold-off circuit before the trailer tanks are charged, the trailer can be moved before any service braking is possible. You should charge trailer tanks before attempting to move the trailer.

FIGURE 28–48 Spring brake control valves and hold-off circuit

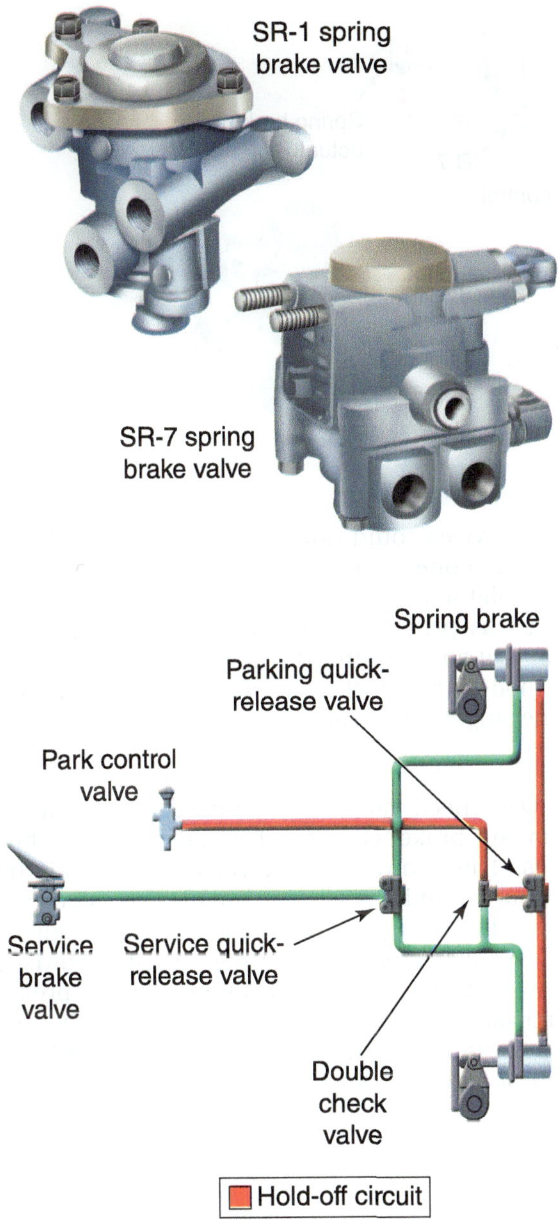

TRACTOR PROTECTION VALVES

The tractor protection (TP) valve has two main functions. First, it acts as the means of routing and switching the trailer air supply and trailer service signal from the tractor. Second, it functions to protect the tractor air supply in the event of a trailer breakaway or total air loss. When a service brake signal is made from either the foot or trailer application valve, it is routed to the TP valve. This signal will only be transmitted to the trailer if the TP valve is switched on. The TP valve is switched on only when the trailer supply valve in the dash is activated. This prevents loss of air from the trailer service signal line when the tractor is not connected to a trailer. **Figure 28–50** shows a cutaway view of a two-line TP valve. **Figure 28–51** shows how a typical TP valve is connected into the tractor air brake circuit.

Most current TP valves are known as two-line valves. They have two inlet lines, one from the trailer supply dash control valve or valve assembly and the other from the trailer service signal circuit. The two outlet ports of the TP valve are connected to the two air hoses that enable the tractor air circuit to be coupled to that of the trailer. When no air is present in the trailer supply inlet to the TP valve, it is in an off state. This means that whenever a tractor service application is made, the signal is delivered to its service port, but it is prevented from passing beyond.

When the tractor is coupled to a trailer and the tractor hoses are connected to the trailer gladhands, the TP valve is switched to its on state. Trailer supply air passes from the tractor to charge the trailer air circuit. When the TP valve is in its on state, its spool valve shuttles to permit service signal air to be applied from the tractor application circuit to the trailer service signal circuit.

For a TP valve to be in the on state, the air pressure must be above its trigger value. This trigger value is usually 45 psi (3 bar). In some cases it may be less (FMVSS 121 requires it to be between 20 and 45 psi [1.4 and 3 bar]). It should be noted that when system pressure has dropped to this value, some spring force from the spring brake assemblies is already being applied to the foundation brakes. In the event of a trailer breakaway or massive air loss in the trailer, the TP spool valve will shuttle to the off position after the air pressure drops below the trigger value. After the TP valve shuttles to an off position, the tractor air circuit becomes isolated from the trailer air circuit.

SHOP TALK

FMVSS 121 requires that the TP valve isolate the tractor air supply from that of the trailer when pressure drops to between 20 and 45 psi (1.4 and 3 bar).

Most current TP valves are two-line systems. Two-line systems are usually controlled by a two-dash valve manifold. The TP valve trailer supply air is sourced directly from the dash control valve assembly.

In a three-line TP valve, the control system may use a two- or three-dash valve control system. The three-line TP valve operates much as the two-line version, with the exception that when the valve is switched to its on state, the trailer air supply is sourced directly from one of the tractor air tanks. Current TP valves may have secondary features and incorporate a quick-release valve (speeds service signal dump), brake light sensing port, and a two-way check valve, as shown in Figure 28–50.

FIGURE 28–49 Multifunctional spring brake modulating valve: incorporates hold-off regulator, quick-release, anticompounding and inversion valve features.

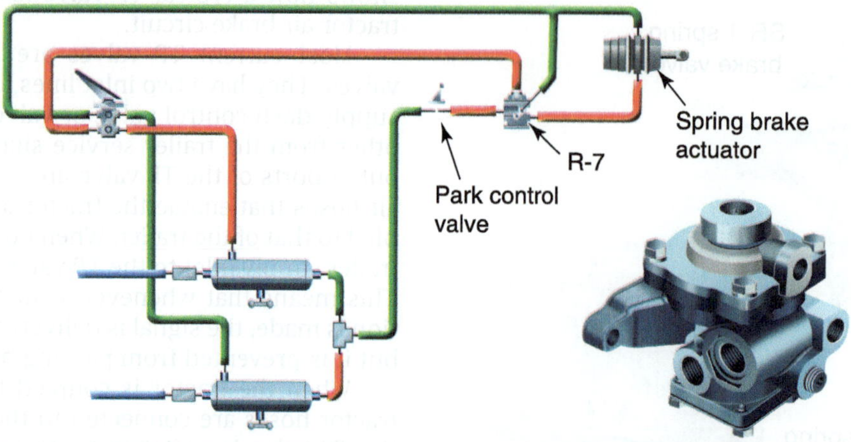

FIGURE 28–50 Two-line tractor protection valve

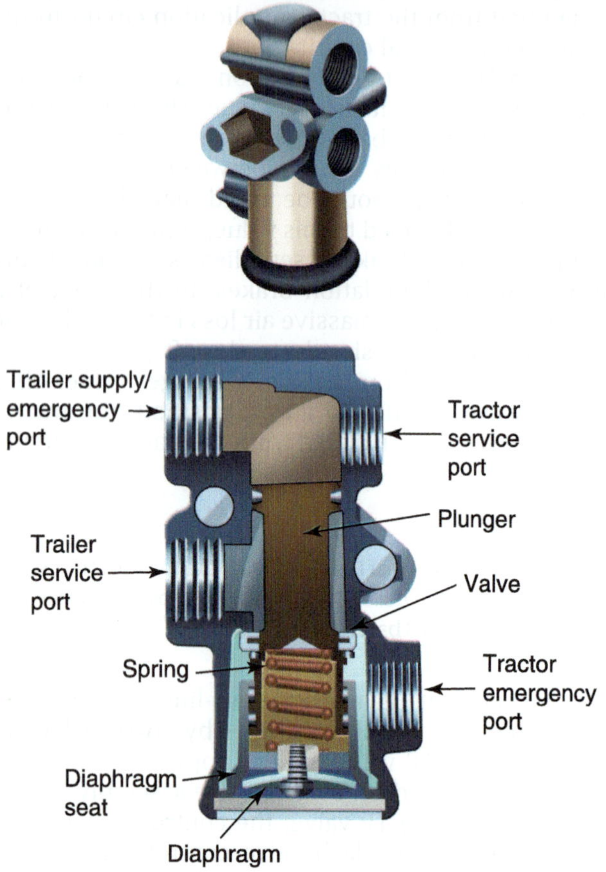

CHECK VALVES

A check valve routes airflow in one direction and, in most cases, prevents any backflow. A couple of different types of check valves are used in air brake systems to perform such tasks as pressure protection and priority routing.

One-Way Check Valve

These devices could not be simpler. They permit air to travel in one direction only. They are often located at the inlet to a tank or subcircuit, in which case they will act as pressure protection devices. This would mean that air can enter the tank or subcircuit but not exit though the check valve. **Figure 28–52** shows an in-line single check valve.

Two-Way Check Valves

Two-way check valves can be used in several areas of an air brake circuit. They are disc or shuttle valves. The two-way check valve receives air from two sources and has a single outlet. The valve core is designed to shuttle so the higher-pressure source air is always delivered to the outlet. In other words, the two-way check valve will transmit the higher-pressure source air to the outlet. It is recommended that these valves be mounted horizontally. **Figure 28–53** shows a typical double check valve.

STOP LAMP SWITCHES

The stop lamp switch is designed to close an electrical circuit when line pressure is sensed. Its function is to illuminate the brake lights on the vehicle. It is rated by the amount of air pressure required to close the electrical circuit. It can be located at various points in the service application circuit.

BRAKE CHAMBERS

Brake chambers convert the compressed air back into the mechanical force required to stop a vehicle. A simple service brake chamber consists of a two-piece housing within which is a diaphragm, a pushrod, and a retraction spring. The two sections of the housing are clamped together so that the diaphragm forms an

FIGURE 28–51 Current tractor protection valve schematic showing plumbing: This circuit uses a newer, multifunction TP valve.

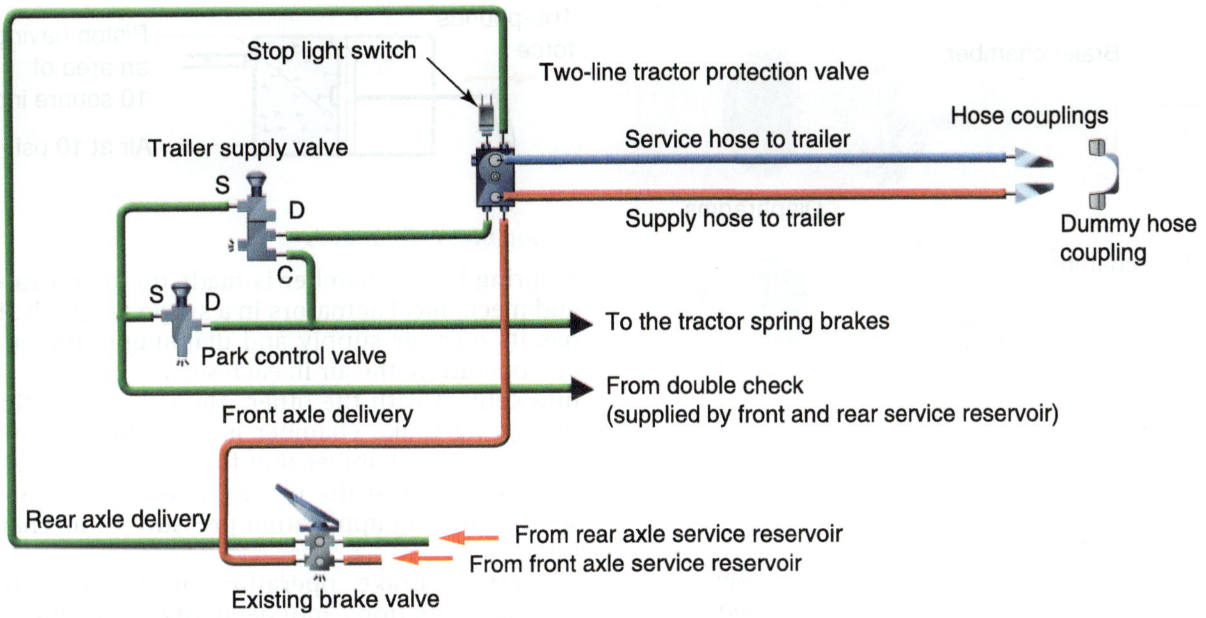

FIGURE 28–52 In-line single check valve

FIGURE 28–53 Double check valve

airtight seal. The pushrod is linked to the foundation brakes by means of a slack adjuster.

When compressed air acts on the flexible rubber diaphragm, force is applied to the pushrod plate. The amount of linear force is proportional to the amount of air regulated by the driver using the service brake application valve. **Figure 28–54** shows a typical assortment of air brake chambers. A single chamber is shown at the top of the figure. The spring in this assembly is relatively light duty and functions simply to return (retract) the pressure plate after a brake application. This service chamber can be integrated into different types of double chambers, also shown in Figure 28–54.

The amount of linear force delivered by a brake chamber can be calculated by knowing the sectional area of the diaphragm and the amount of air pressure applied to it (**Figure 28–55**). A common size of diaphragm is 30 square inches. If a 100 psi (6.9 bar) service application was applied to a 30-square-inch brake chamber, the linear force developed could be calculated as follows:

$$\text{Linear force} = 30 \text{ sq. in.} \times 100 \text{ psi} = 3{,}000 \text{ lb}$$

Or using the metric system in which the diaphragm is 200 square centimeters and the pressure acting on it is 101 psi (7 bar):

$$\text{Linear force} = 200 \text{ sq. cm.} \times 7 \text{ bar} = 1{,}400 \text{ kg}$$

Brake Chamber Stroke

The distance between the pressure and nonpressure plates of the brake chamber defines the limits of its stroke. After the chamber pushrod plate is bottomed or stroked out, no more linear force can be applied to the foundation brake regardless of the air pressure. In fact, a stroked-out brake chamber may not apply the brakes at all, resulting in a complete loss of braking.

FIGURE 28–54 Examples of single and double air brake chambers. The double units combine service and park/emergency brake functions.

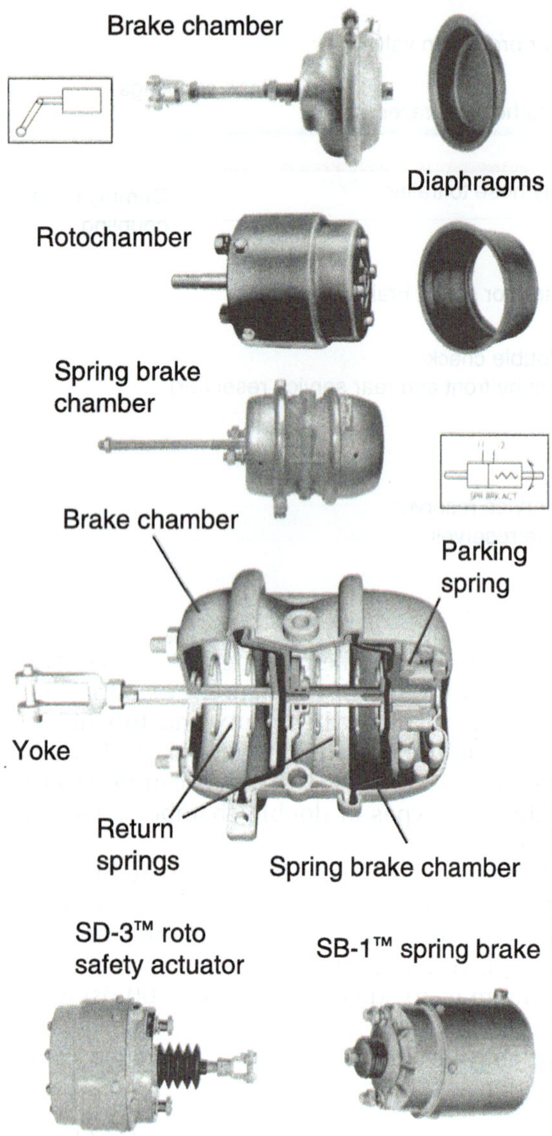

Correct chamber pushrod travel is maintained by properly adjusting the brakes as they wear by means of the slack adjuster, either automatically or manually. **Figure 28–56** shows a sectioned view of a single-pot brake chamber. Note the terms used to describe the different subcomponents.

Brake Chamber Size

Brake chambers are sized by the effective sectional area of the diaphragm. As of now, this is always measured in square inches. For instance, a brake chamber with a 30-square-inch diaphragm is known as a series 30. Most tractor drive and trailer axles use either series 30 or series 24 diaphragms. Tractor steering axles and air disc brakes are more often fitted with series 20, 16, or smaller brake chambers.

FIGURE 28–55 Compressed air used as potential energy.

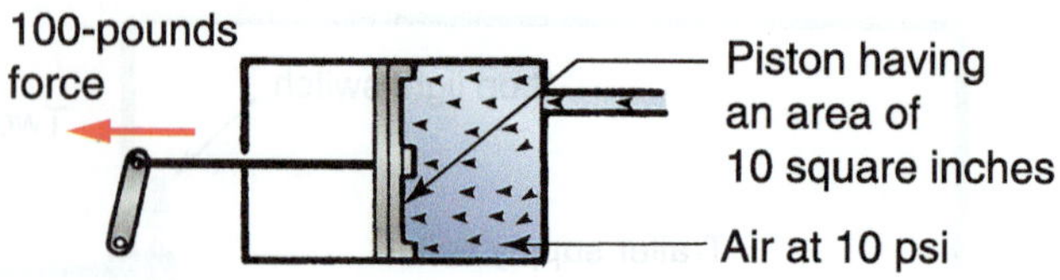

Spring Brake Chambers

A spring brake chamber is made up of separate air and mechanical actuators in a single assembly. Each has its own air supply and diaphragm. Under normal operation, the air in each section does not come into contact with the other. The forward portion of the spring brake chamber is a simple air-actuated service brake chamber that functions exactly as the unit described in the previous section. It converts service braking application pressure to mechanical force.

Service brake operation of a spring brake chamber functions independently from the parking/emergency function. Figure 28–55 shows how the brake chamber converts the potential energy of compressed air into mechanical force, and **Figure 28–57** shows how either air pressure or spring force can be used to effect mechanical movement. **Figure 28–58** demonstrates how brake chamber pushrod movement actuates the foundation brakes. Spring brake chambers are required on both drum and disc foundation brake assemblies. **Figure 28–59** shows a spring brake chamber designed for use on either an S-cam or air disc brake foundation assembly.

Hold-Off Chamber

The rear section of the spring brake chamber houses a large compressed spring capable of exerting a force of up to 1,800 lb (816 kg). This spring is sometimes known as a *maxi spring*, named for the manufacturer of the first spring brakes. The spring brake section of the spring brake chamber has a diaphragm that functions to "hold off" the spring. An air pressure of 60 psi (4 bar) is required to hold off the spring. When hold-off pressure is applied to the hold-off chamber from the park release circuit, the hold-off diaphragm (sometimes known as an "emergency" diaphragm) acts against a cage plate that prevents the spring from loading the intermediate pushrod and service chamber pushrod with spring force. Because the function of this spring is to mechanically apply the brake for parking, hold-off air must be applied to the spring brake chambers on an air brake-equipped vehicle before it can be moved.

Whenever no air acts on the hold-off diaphragm, the mechanical force of the spring acts on the intermediate pushrod, meaning that the chamber is in

FIGURE 28–56 Cutaway view of a typical air brake chamber

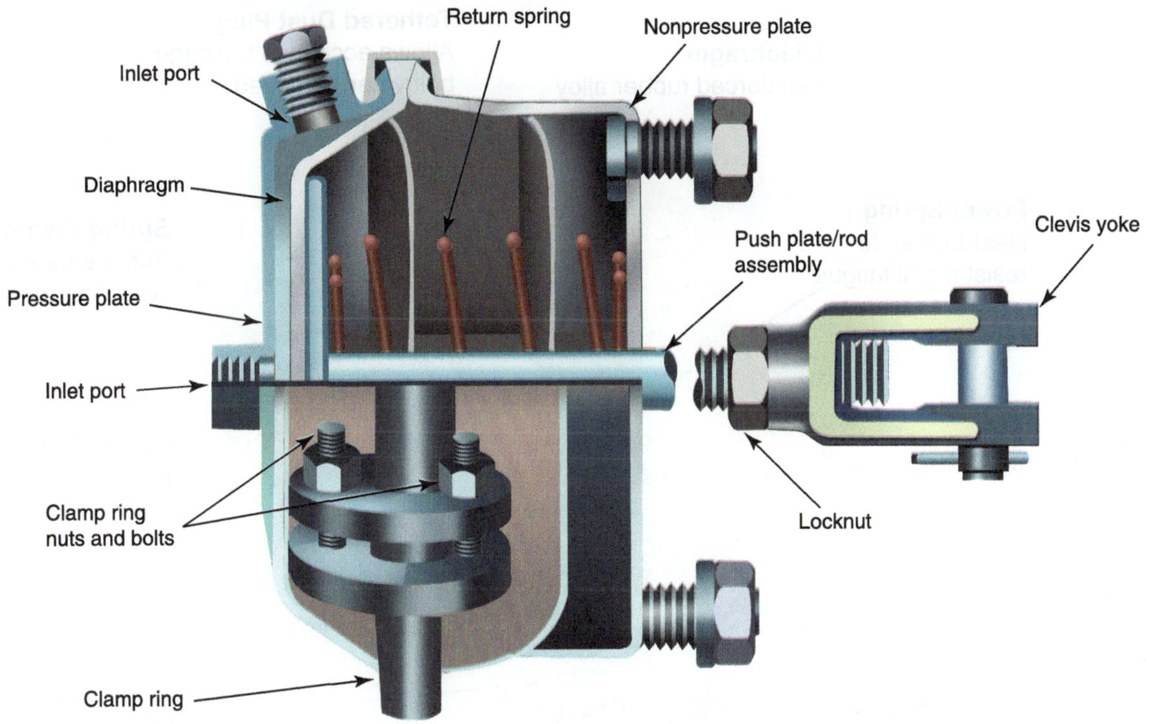

park mode. When there is no air pressure in the hold-off chamber, the force of the spring acts on the intermediate pushrod and is transmitted to the service chamber pushrod and plate. This mechanically applies the brake. **Figure 28–60** shows cutaway views of the different types of spring brake chamber used with S-cam and wedge foundation brake assemblies. The operating principle of a spring brake chamber is shown in **Figure 28–61**. Be sure to understand the operation of a spring brake chamber.

FIGURE 28–57 Potential energy of a compressed spring compared with compressed air.

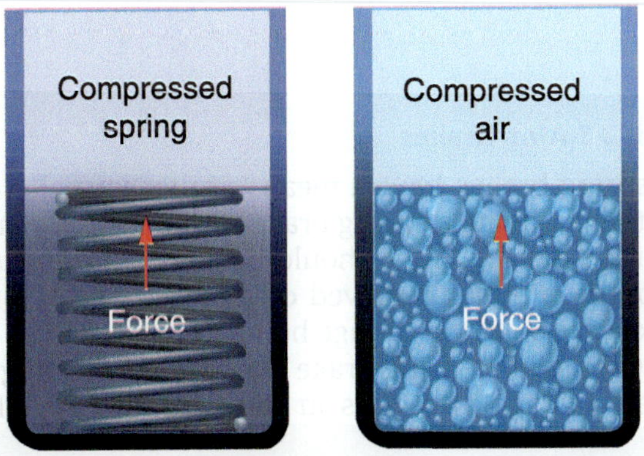

WARNING

It takes 1,725 lb (782 kg) of force to cage the power spring in a spring brake chamber. This spring force can kill. Never take any risks with spring brake chambers.

FIGURE 28–58 Converting the potential energy of compressed air into stopping power

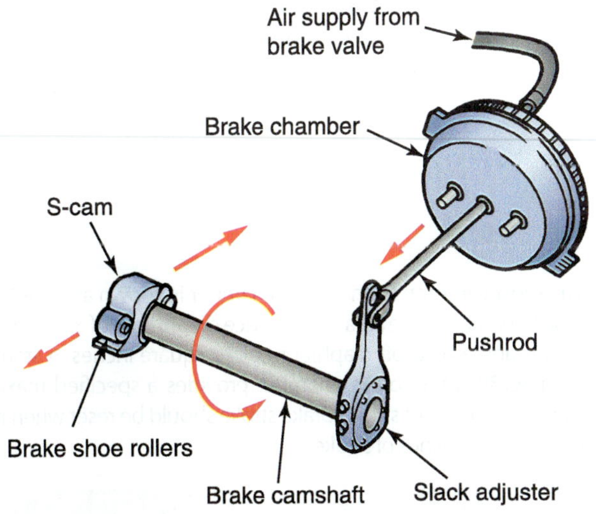

FIGURE 28–59 Actuator for air disc brake

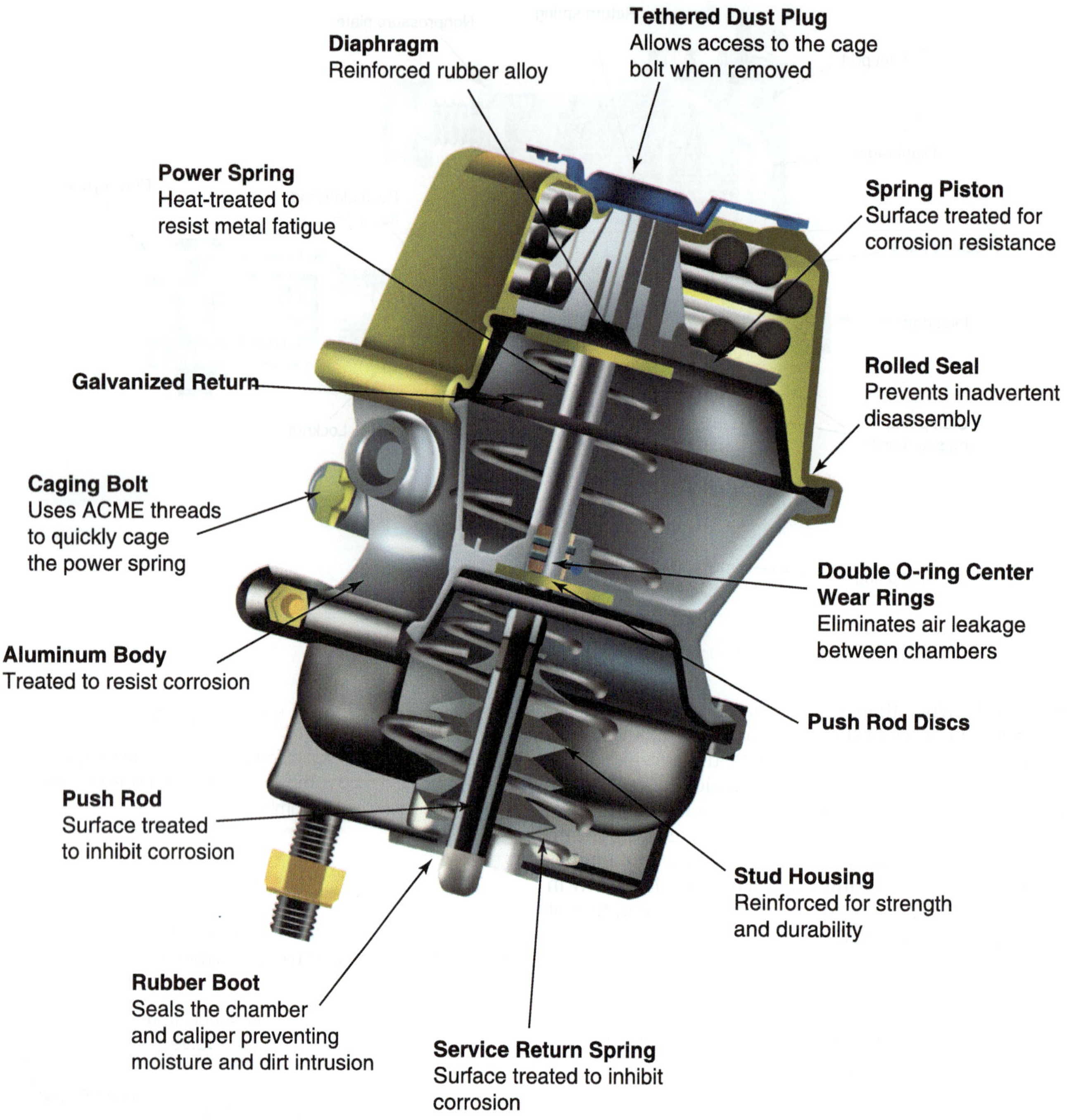

SHOP TALK

The most common spring brake chamber is known as a 30/30. A 30/30 spring brake area has a service diaphragm of 30 square inches and a hold-off diaphragm of 30 square inches. A standard 30/30 spring brake chamber provides a specified maximum of 2 inches of stroke. Brake stroke should be reset when it exceeds 1.75 inches of stroke.

Caging Spring Brakes

All spring brakes have a means of mechanically caging them. Caging a spring brake compresses the main spring. A spring brake should be mechanically caged before it is either removed or replaced. Most spring brakes are sold with a cage bolt assembly.

Caging of a spring brake consists of inserting the cage bolt lugs into slots in the internal cage plate,

FIGURE 28–60 Spring brake chamber types used with different foundation brakes. Note the difference in size between the power spring and return springs.

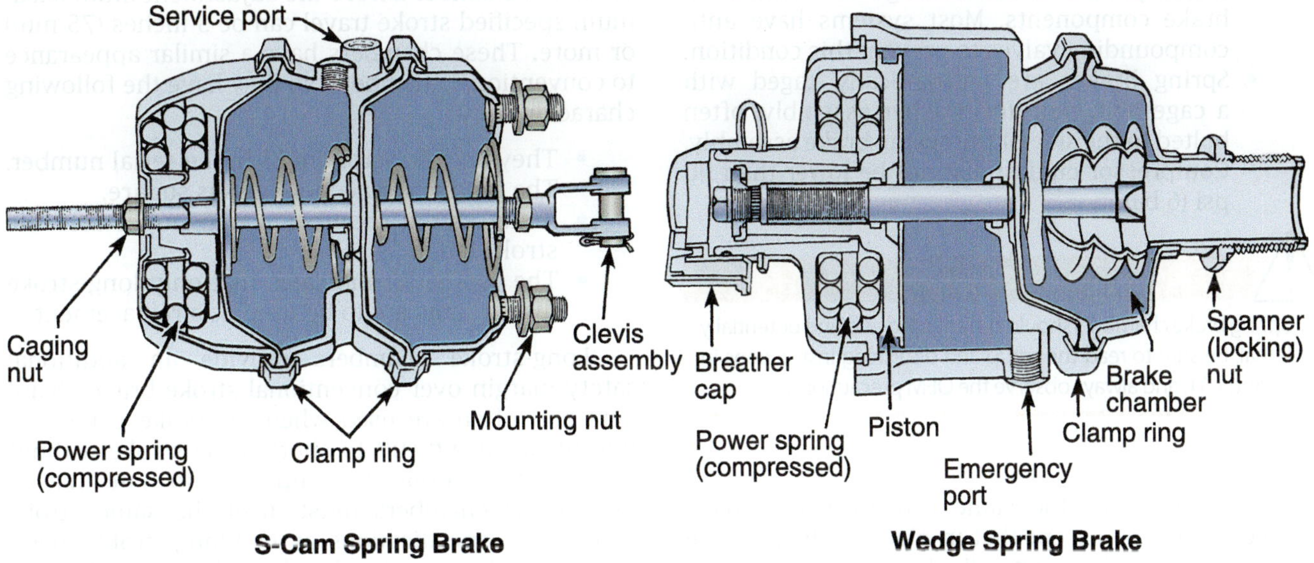

FIGURE 28–61 Spring brake chamber operation: Observe the role of the power spring.

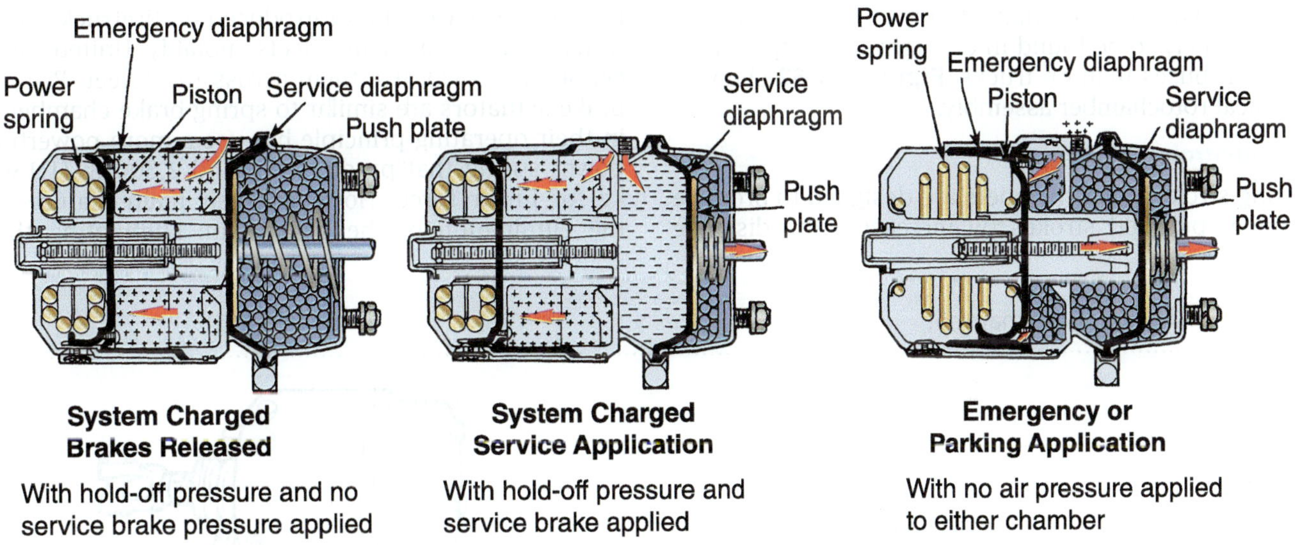

rotating the bolt, and then compressing the spring using the cage nut and washer. A caged spring brake should be regarded as armed and dangerous. The procedure for disarming spring brake assemblies is covered in Chapter 31 and is demonstrated in the video shorts that accompany this book.

Summary of Spring Brake Operation

- No air pressure in either the hold-off or service circuits: The brake is mechanically applied by the full force of the main spring in the spring brake assembly. This condition is used to mechanically "park" the brake or to effect emergency braking in the event of total loss of air.
- Air pressure in the hold-off circuit, none in the service chamber: The parking brake is released and the vehicle is capable of being moved. To release the parking brakes and move the vehicle, air pressure greater than 60 psi (4 bar) must be present in the spring brake hold-off chambers.
- Air pressure in the hold-off circuit, service application pressure in the service chamber: Service brakes are applied at the service brake application pressure value. This condition exists at any time the vehicle is mobile and the service brakes are applied.
- No air in the hold-off circuit, service application pressure in the service chamber: Braking

is compounded by having spring and air pressure applied to the foundation brake assembly simultaneously. Compounding of brake application pressures can damage the foundation brake components. Most systems have anticompounding valves to prevent this condition.
- Spring brakes are mechanically caged with a cage bolt, nut, and washer assembly, often bolted to the side of the spring brake assembly.
- Compressor cut-in must not be lower than 90 psi (6 bar).

CAUTION

Spring brake chambers should be regarded as being potentially lethal. Be sure to read the setup and disarming instructions in Chapter 31, and always observe the OEM precautions.

Rotochambers

Rotochambers use the same principles as typical brake chambers, except that in place of the pancake diaphragm, a rolling-type diaphragm is used. This produces a constant output force throughout the pushrod stroke and provides much longer stroke.

Rotochambers are available in a number of sizes and provide a wide range of output forces. They are more likely to be found in off-highway equipment or transit buses than on trucks. **Figure 28–62** shows a typical rotochamber assembly.

Long-Stroke Chambers

Long-stroke brake chambers are designed to produce longer pushrod strokes by increasing the distance between the pressure and nonpressure plates of the chamber. This will typically produce an extra ½ to 1 inch (12.5 to 25 mm) of stroke travel over the conventional chamber before the adjustment limit. Maximum specified stroke travel can be 3 inches (75 mm) or more. These chambers have a similar appearance to conventional chambers, so they have the following characteristics:

- They use the acronym LS in the serial number.
- The air line connection boss is square.
- A tag at the chamber clamp bolt states the stroke length.
- The cast letter indicates that only long-stroke diaphragms are to be used as a replacement.

Long-stroke chambers provide an additional safety margin over conventional stroke brake chambers and often are used when the brake configuration makes it difficult to shorten the stroke without creating brake drag. It is important to note that all long-stroke chambers must meet the same stroke adjustment specifications as non-long-stroke chambers. Using long-stroke chambers does not eliminate the need for routine brake stroke adjustment.

Piston Brakes

Piston brakes were first introduced on firetrucks and transit buses but some fleets, notably United Parcel Service (UPS), use them across their fleet. Piston brake actuators are similar to spring brake chambers in their operating principle but use a more powerful power spring that produces around 2,600 lb (1,180 kg) of linear force. However, their main feature is the elimination of the diaphragm. Eliminating the

FIGURE 28–62 Rotochamber

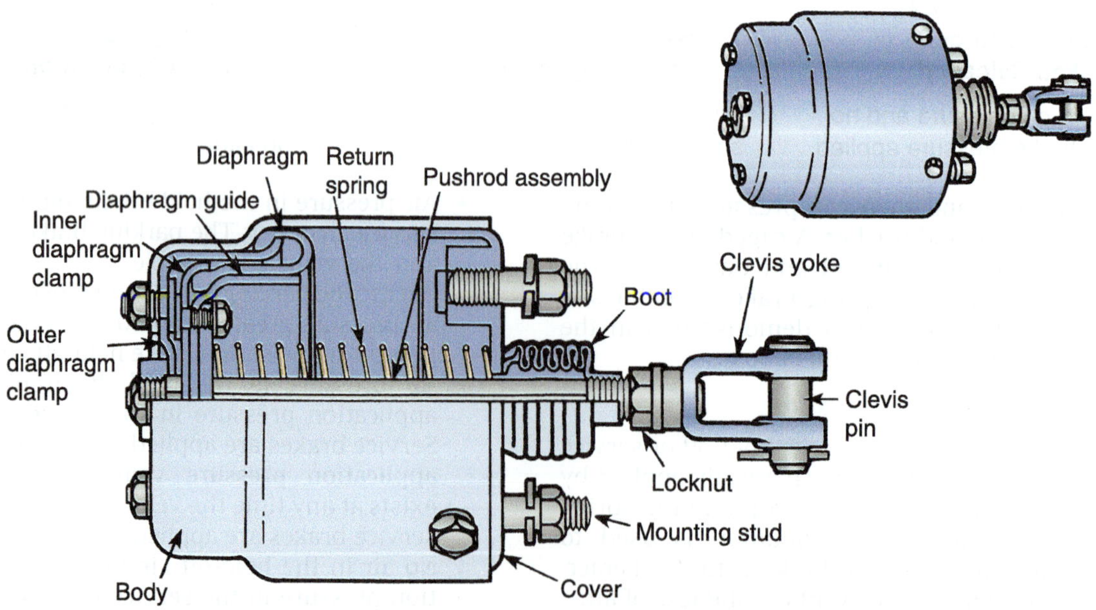

diaphragm for a linear-moving piston results in less air contamination and a 3-inch (75 mm) stroke travel. In addition, they seldom fail. They have an appearance similar to a diaphragm spring brake assembly with a slightly extended power spring hub. Piston brakes are designed to work with S-cam and air disc brakes. They are currently manufactured by Haldex (check them out at www.haldex.com), which markets them as MaxiBrake chambers.

DD3 Safety Actuator

The double diaphragm (DD) 3 safety actuator, like a truck spring brake unit, has three functions:

- Service braking
- Emergency braking
- Park braking

The DD-3 features a mechanical roller locking mechanism that locks the chamber pushrod for parking. Because of this locking roller mechanism, the DD-3 requires the use of special control valves. They are no longer often used on highway trucks and their use on transit buses is expected to diminish over the next few years. DD-3 safety actuators are available with diaphragm effective surface areas of 24 and 30 square inches, respectively.

SLACK ADJUSTERS

The linear force produced by the brake chamber must be converted into brake torque (twisting force). Rotation of the S-cam forces the shoes into the drum to effect braking. Slack adjusters are used to connect the brake chamber with the S-cam shaft. Figure 28–58 shows the roles of the slack adjuster in the actuation of the brakes. First, the slack adjuster is the mechanical link between the brake chamber and the foundation brake assembly. It is attached to the brake chamber pushrod by means of a clevis and pin and to the S-camshaft by means of internal splines on the slack adjuster and external splines on the S-camshaft. The slack adjuster is also a lever. As a lever, it amplifies the force applied to it by the brake chamber. The amount of leverage potential depends on the distance between the clevis pin that connects it to the chamber pushrod and the fulcrum formed at the centerline of the S-cam. This is known as the leverage factor. For instance, if the leverage factor of a slack adjuster was 6 inches (0.5 foot), and a linear force of 3,000 pounds (1361 kg) was applied to it by the brake chamber, then the brake torque could be calculated as follows:

Brake torque = Force × Distance
Brake torque = 3,000 lb × 0.5 ft. = 1,500 lb-ft. torque

Figure 28–63 shows a manual and an automatic slack adjuster. The leverage factor of a slack adjuster is critical to the correct operation of a brake system.

FIGURE 28–63 (A) Manual and (B) automatic slack adjusters

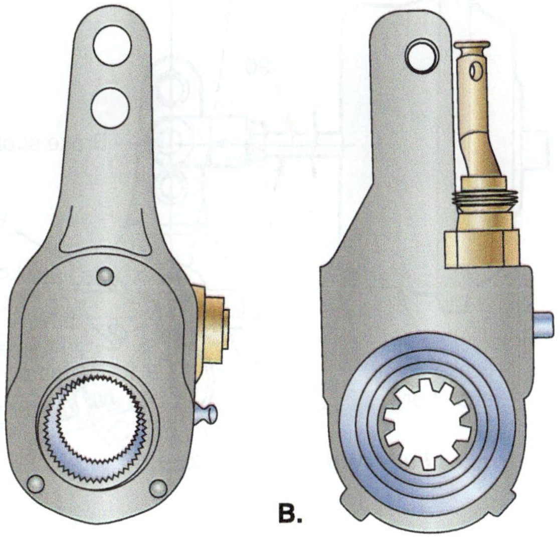

For this reason, it is important whenever a slack adjuster is changed that the clevis pin be located in the correct slack adjuster hole. A difference of a ½ inch (12.5 mm) can drastically alter the brake torque applied to the foundation brake. This will result in unbalanced braking.

Finally, the slack adjuster is used as the means of removing excess free play that occurs as the brake friction facings wear. All current slack adjusters are required by FMVSS 121 to be capable of automatic adjustment. Two general types of automatic slack adjusters are available. The truck technician should be aware of how a manual slack adjuster operates because some remain in use today on truck and trailer units manufactured before 1995.

Manual Slack Adjusters

A manual slack adjuster is located on the S-camshaft by means of internal splines on a gear. External teeth on the gear mesh with a worm gear on the adjusting wormshaft. The wormshaft is locked to position by a locking collar that engages to a hex adjusting nut.

This locking collar prevents rotation of the wormshaft and sets the adjustment. As friction face wear occurs, slack results; that is, the chamber-actuated pushrod will have to travel farther to apply the foundation brake. This slack is removed by adjusting the brakes. The brake is adjusted by depressing the locking collar and engaging a closed wrench (usually 1/16 inch [1.59 mm]) over the hex on the wormshaft. The wrench can now rotate the wormshaft to remove the slack.

After the brake has been properly adjusted to obtain the correct amount of pushrod free play, the adjustment is set by reengaging the locking collar. **Figure 28–64** shows the critical components of a typical manual slack adjuster.

FIGURE 28–64 Manual slack adjuster

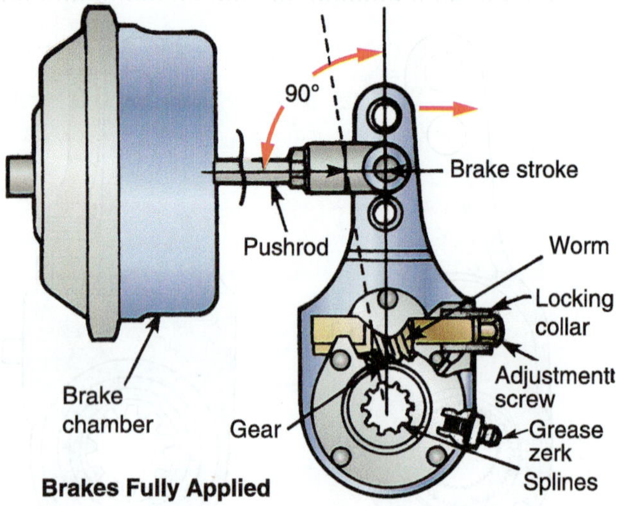

Stroke-Sensing Automatic Slack Adjusters

The stroke-sensing automatic slack adjuster responds to increases in pushrod stroke to actuate an internal cone clutch and rotate an adjusting wormshaft. This wormshaft meshes with the external teeth of the gear engaged to the splines of the S-camshaft. This type of slack adjuster operates on the assumption that if the applied stroke dimension is correctly maintained, the drum-to-lining clearance will be correct. **Figure 28–65** shows a cutaway view of an automatic slack adjuster and **Figure 28–66** shows a Meritor style slack adjuster.

Clearance-Sensing Automatic Slack Adjusters

Clearance-sensing automatic slack adjusters respond only to changes in the return stroke dimension, so they function almost opposite to stroke-sensing models. They "adjust" only when the return stroke dimension increases, so if the applied stroke dimension became excessive, no adjustment would occur. Clearance-sensing automatic slack adjusters work on the assumption that if the shoe-to-drum clearance is normal, the stroke should be within the required

FIGURE 28–65 Stroke-sensing automatic slack adjuster

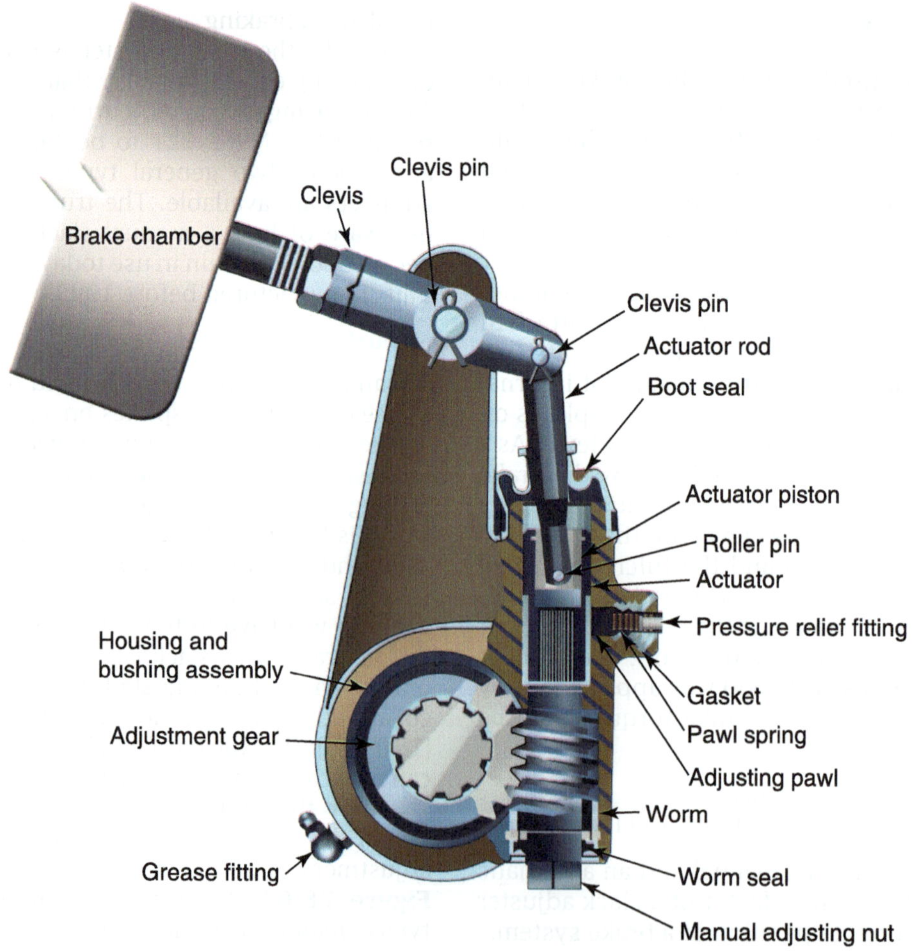

FIGURE 28–66 Automatic slack adjuster

adjustment limits. **Table 28–2** lists types of automatic slack adjusters.

SHOP TALK

Slack adjusters without a grease fitting should not be assumed to be functional. Slack adjusters should be lubricated with a low-pressure grease gun.

CAUTION

The TMC warns against performing routine manual adjustments of automatic slack adjusters. Automatic slack adjusters that require frequent adjustments to maintain brake stroke within specification are an indicator of a failed slack adjuster or serious brake problems. More discussion on this can be found in Chapter 31.

GAUGES

Air pressure gauges in air brake circuits are used to monitor air pressure in the primary and secondary circuits and to give the driver an indication of what the application pressure is. Either individual gauges or a two-in-one gauge can be used to display the system pressure in the primary and secondary circuits.

When a two-in-one gauge is used, a green needle indicates primary circuit pressure and a red needle indicates secondary circuit pressure. If a loss of pressure occurs in either of the circuits, the driver is alerted to the condition after the pressure drops below 60 psi (4 bar). The alert is usually visible (gauges/warning lights) and audible (buzzer).

Some systems also have an application pressure gauge. This gauge tells the driver the exact value of a service application. Most application pressure gauges read the primary circuit pressure only, so when the trailer application valve is used, there may be no indicated reading.

28.9 BRAKE SYSTEM BALANCE

Two factors determine the amount of braking force applied to each wheel in a truck air brake system. Because the control and application circuits are managed by compressed air, the first factor of a balanced brake system is pneumatic balance. For a service application to be balanced, the application valve, for example, the foot valve, must deliver a signal pressure to relay valves. The relay valves are designed to replicate the signal they receive by sending a local supply of air to service brake chambers they supply. The idea is that, during a service brake application,

TABLE 28–2 Types of Automatic Slack Adjusters

Manufacturer	Type	Features
Bendix	Clearance-sensing/apply stroke	Infinitely variable adjustment; antireverse lock; left- and right-hand versions
Gunnite	Clearance-sensing/apply stroke	Incremental adjustment; limited correction for overadjustment
Haldex	Clearance-sensing/release stroke	Incremental adjustment; limited correction for overadjustment
Rockwell	Stroke-sensing/release stroke	Incremental adjustment; antireverse feature; front and rear axle versions are different

each of the brake chambers on the vehicle is charged with air at exactly the same pressure value.

You could argue that mandatory ABS has compensated for many brake balance shortcomings that can arise on tractor/trailer combinations. Numerous tests have proven that rookies driving ABS-equipped rigs can safely outbrake seasoned truck drivers driving the same rig with the ABS electronically defeated. However, there is a limit to the extent ABS can compensate for fundamental brake problems, the most notable of which is that the applied brake stroke on all the wheel ends must be within limit. In this section, we will look at some causes of brake imbalance.

PROPORTIONING AND TIMING APPLICATION PRESSURES

Most brake systems without ABS have some ability to proportion service application pressures to ensure that the vehicle can be safely stopped with maximum driver control. A good example would be the ratio valve discussed earlier in this chapter. This proportions braking to the front axle brakes according to the service application pressure. Timing is another factor. Because the signal pressure must travel much further to reach the brakes at the rear of the trailer, unless some accommodation is made, the front brakes on the tractor unit apply well before those on the rear of the trailer. Pneumatic timing means that the brakes on a tractor/trailer combination are applied simultaneously.

BRAKE TORQUE BALANCE

Brake torque balance means that the mechanical braking force applied to each wheel assembly should be equal. If, for instance, the brakes were out of adjustment on one side of an axle and properly adjusted on the other, when the service brakes were applied it would be possible to lock up the brakes on one side (assuming no ABS) and have no braking force applied to the other. Many factors other than maladjusted brakes contribute to brake torque imbalance. Mismatched friction facings, glazed linings on one wheel (usually caused by a leaking wheel seal), improper brake chamber pushrod setting, mismatched slack adjuster lengths, glazed drums, and irregularly worn drums could all result in brake torque imbalance.

In the operation of a tractor/trailer combination, each axle and each wheel assembly must bear exactly the proportion of total vehicle load it is designed to carry. Each foundation brake on each wheel must be properly adjusted so it performs its share of the required vehicle braking. Also, the brake applications have to be timed so that each application and release of the brakes is practically simultaneous.

FACTORS AFFECTING BRAKE BALANCE

Anything that changes on the brake or wheel assembly can affect the brake balance. Tire treads and their coefficient of friction (aggressiveness) with the road surface both impact the vehicle braking efficiency. The fitting of low-profile tires also can produce overbraking because of the lower radii of the tires.

Although system air pressure is available to the driver for braking the vehicle, it is only under emergency service applications that this amount of pressure would be required. The majority of service brake applications are made at an application pressure of 20 psi (1.4 bar) or less. This means that a variance of 4 or 5 psi (0.276 to 0.345 bar) in the normal pressure required to move the shoes from a retracted position to drum contact (typically 6 psi) (0.41 bar) on different axles that might be relatively unimportant at a 60 psi (4 bar) application pressure becomes critical at low application pressures of 15 to 20 psi (1 to 1.4 bar), where most of the braking action takes place.

Note: Around 95 percent of braking on a loaded tractor/trailer combination involves brake application pressures of 25 psi (1.72 bar)or less; 85 percent of braking involves brake application pressures of 15 psi (1 bar) or less.

The key to brake balance is proper maintenance of the brake system. Balanced braking means that the braking performance of the vehicle is functioning properly. Vehicle braking requirements are always specified for the maximum load that the vehicle is designed to haul. No truck or tractor/trailer combination run at partial or empty loads will brake at optimum efficiency; in most cases, they will be overbraked. In fact, it should be noted that most unloaded trucks are more difficult to handle than when they are loaded because the specifications are developed for vehicle combinations at their GVW. Three major factors affect brake balance and, thus, braking performance:

- Brake torque balance between the tractor and trailer. To have balanced brakes between the tractor and trailer, the brakes on each axle of each unit must produce brake torque in proportion to the load applied up to the capacity of each axle. Brakes are designed and rated for the maximum axle rating such as a 23,000-pound (10,500 kg) single axle or a 34,000-pound (15,420 kg) tandem. Even with ABS an empty or partially loaded rig will tend to overbrake, which is why LLVW is reckoned into current brake legislation. Lightly loaded axles are more likely to lock up. This is often what causes the black tire streaks on the pavement near intersections. Overbraking causes wheel lockup. The result is flat-spot tires.

- Air timing balance between the tractor and trailer. Air timing is defined as the time required for signal air to be converted into a brake application. A braking event begins the instant the driver's foot moves the brake pedal. When a trailer and tractor are coupled in a combination, it is important that the trailer brakes be applied simultaneously with those on the tractor.
- Rubber on the road. All brake force must be transmitted by means of the tires that make contact with the road pavement. The importance of ensuring that all the tires on a truck rig are properly matched, balanced, and in good condition cannot be overemphasized. Loss of control during panic braking on rigs is often caused by the inability of ABS to compensate for defective tires.

Crack Pressures

The brakes on each axle of the entire rig should receive air at the same time and at the same pressure value. To ensure this, relay valves are often specified with different crack pressures. The crack pressure of a relay valve is the air pressure value required to overcome the relay valve piston return spring. Setting higher crack pressures on the tractor relay valves that are located close to foot valve and at a lower value on those on the rearmost trailer axles helps achieve balanced brake applications.

For this reason, when replacing relay valves, always properly match the serial numbers. Two valves of identical appearance may actually perform differently. Sequenced crack pressures are most critical in tractor/train combinations such as A and B train doubles. In this type of application, the lowest relay valve crack pressures would be on those valves on the pup trailer and the highest over the tractor axles. Similarly, on the 53-foot (16 m) doubles that are becoming a common sight on the highways, the rear trailer brakes would be spec'd with the lowest crack pressures.

Pressure Differential between Tractor and Trailer

There is usually a difference between tractor and trailer air pressure when these units are coupled. Pressure differentials between axles are created by the various valves used in the brake actuation system. Each valve used in a tractor or trailer brake system creates additional potential for pressure drop between input and output pressure. The pressure to the front axle chamber of a tractor normally will have close to the same pressure as the output line of a foot valve. The pressure to the rearmost axle chamber in a multitrailer rig will have the greatest differential. The differential value is always greatest when service brake application pressures are lowest. Because of the number of valves and length of plumbing required between the foot valve and the rearmost trailer axles, it is difficult to achieve identical application pressures on all axles of a combination.

COMPATIBILITY ISSUES

Balanced brakes extend brake service life and provide drivers with the tools they need to safely stop a truck during emergency braking. A tractor/trailer jackknife condition often is caused by unbalanced brakes. The condition occurs as often as it does because of the following:

- There is a tendency in the trucking industry to maintain tractor brakes to higher overall standards than trailer brakes.
- Compatibility of the tractor brake system with that of the trailer: When the tractor brakes more aggressively than a trailer on a slippery pavement, the chances of jackknife increase.

Properly functioning brake systems on a tractor and a trailer can produce compatibility problems when they are coupled. The effects of brake imbalance can produce one or more of the following conditions:

- Uneven brake lining wear
- Reduced stopping capability
- Brake fade
- Brake drag
- Jackknife
- Wheel hop: suspension damage
- Trailer surge: tractor brakes apply before trailer
- Reduced brake drum life
- Tire wear
- Tire flat-spotting

BRAKE IMBALANCE TERMS

This is a summary of some of the terms used to describe brake imbalance in trucks. As you go through this list, think about how many of these problems can be eliminated when a tractor/trailer rig is equipped with ABS.

- **Brake torque imbalance.** Brake torque imbalance occurs when the foundation brakes have variable aggressiveness. Ideally, the brakes on a vehicle should be applied with the same amount of mechanical force and the critical friction surfaces should have the same coefficients of friction.
- **Jackknife.** Jackknife occurs when the rear wheels of the tractor lock, causing the tractor to spin. When this condition occurs, the forward momentum of the trailer during the tractor spin makes it difficult to steer out of this condition.
- **Trailer swing.** Trailer swing occurs when the trailer wheels lock under braking, causing the trailer to spin around while the tractor remains in a straight-ahead position. An experienced

truck driver often can steer out of a trailer swing condition.

- **Plow-out.** The tractor steering wheels lock under braking, making the tractor difficult to steer. Most plow-out is driver error, and experienced drivers can avoid it.
- **Pneumatic timing.** Pneumatic timing describes the timing actuation of each brake chamber on the rig. Ideally, all the foundation brakes should be pneumatically actuated at exactly the same moment, whether they are on the tractor or the rearmost trailer of a truck train.
- **Pneumatic balance.** Pneumatic balance is the ability of the air system balance on a rig to manage the correct air pressure at every brake chamber on the rig during braking and release.

Consequences of Brake Imbalance

The reason or reasons for brake torque or timing imbalance may not be easy to detect. The causes of pneumatic imbalance include the use of oversized air lines on trailers, air leaks, sticking relay valves, and the use of incompatible air valves. Torque imbalance occurs when some brakes are stronger or weaker than others. There can be a number of causes such as drums machined beyond tolerance, nonuniform adjustment, oil-soaked linings, polished drums and glazed linings, nonuniform drum-to-lining contact, linings with varying friction ratings, incorrect slack adjuster length, nonuniform chamber sizes, or an incorrectly specified axle rating.

Acceptable pneumatic balance usually requires that the air delivered to each axle does not vary more than 2 psi (0.14 bar) during a 10- to 20-psi (0.7 to 1.4 bar) application. Although ABS can help, pneumatic timing is most critical in a multiple trailer combination, because the signal air pressure has to travel much further before it arrives at the pup axles.

Maintenance and Wear Conditions That Cause Brake Imbalance

- Scored, glazed, or "bell-mouthed" brake drums. Often, brand-new drums deform to an eccentric shape in storage. It is no longer regarded as good practice to machine used drums, but it is good practice to machine new brake drums immediately prior to installation.
- Linings contaminated by leaking oil seals or excessive grease pumped into the inboard S-cam bushing. This contamination gets baked into the friction material, and the result is lining glazing.
- Brake shoe return springs that stretch and lose their temper caused by excessive heat. These springs should be replaced at every brake overhaul.
- Rollers that have become flat spotted. S-cam rollers should be replaced at each brake overhaul.
- Retrofitting of aftermarket or remanufactured brake valves that can destroy pneumatic balance. Relay valve crack pressures should be matched when replacing valves.
- Defective valves. Valves can slow down or fail when their moving parts become gummed up (water and oil sludge), corroded (water contamination), and carboned due to oil vapor and heat.
- S-cam profiles that wear significantly. Although S-cams may not have to be replaced at each brake job, they should be carefully inspected.
- S-camshaft slop of any kind. Worn camshaft bushing bearings, bushings, and splines degrade performance. Bushings should be inspected and replaced if necessary at every brake overhaul.
- Bent spiders. Bent spiders reduce lining-to-drum contact and the effective leverage of the shoe. Brake spiders tend to bend when seized anchor pins are removed. Use a proper anchor pin puller rather than a sledge hammer and punch to remove seized anchor pins.

MECHANICS OF S-CAM, DISC, AND WEDGE BRAKES

The performance of S-cam, disc, and wedge brakes varies significantly, and this can produce problems in a tractor/trailer combination in which the foundation brakes are mixed. For instance, a tractor equipped with S-cam brakes coupled to a trailer equipped with wedge brakes will often cause trailer overbrake/tractor underbrake because in moderate to severe braking, double-actuated wedge brakes have superior brake mechanics. A tractor equipped with air disc brakes hauling a trailer with S-cam brakes can cause the opposite problem because of the higher mechanical efficiency of disc brakes over an S-cam actuated drum. **Figure 28–67** is a demo axle set up to show the critical differences between air disc brake and S-cam brakes.

FIGURE 28–67 A demo axle set up to show the differences between air disc and S-cam foundation brakes.

FMVSS NO. 121 REQUIREMENTS

FMVSS No. 121 governs the requirements of air brake systems used on North American highways. Here is a sampling of FMVSS requirements:

- Maximum application time (signal to chamber) for each service chamber must not exceed 0.45 second on a truck or tractor and 0.60 second on a trailer. Application time is defined as the time required to increase the pressure from atmospheric to 60 psi (4 bar).
- Maximum release time for trucks and tractor units is 0.55 second.
- Parking brakes must be able to hold the vehicle loaded to its gross weight stationary on a 20 percent grade.
- Unapplied pressure loss in the supply (wet) tank must not exceed 2 psi (0.14 bar) for single vehicles and 3 psi (0.21 bar) for a tractor/trailer combination in 1 minute.
- A tractor/trailer combination manufactured after 2010 must be brought to a complete standstill within 250 feet (76 m) with the rig loaded, from any legal speed. The stopping distance required of tractor/trailer combinations built before 2010 is 355 feet (108 m) and that of a LLVW rig, 230 feet (70 m).

28.10 AIR START SYSTEMS

Air cranking circuits do not ideally fit into either the electrical cranking circuit or air supply circuit chapters. However, although not commonly used, there are enough air starters used on trucks to justify a short explanation of them somewhere in this book, so we will deal with them here.

WHY USE AIR CRANKING MOTORS?

Air starters enjoyed some popularity in the 1980s. Some engine OEMs recommended them because of the high cranking speeds they produced. Electric starter motors crank at between 200 and 250 rpm, whereas air starters crank at around double that speed, typically 450 rpm. The problem with an air starter, however, is that they will only crank when there is a supply of compressed air to drive the pneumatic motor. Although air start systems incorporate a dedicated large-volume reservoir, at best, 15 seconds cranking was attainable. Having used up the onboard supply of compressed air, the reservoir would have to be recharged with compressed air from an outside source. Although it is easy enough to start a truck diesel engine within 15 seconds in the middle of summer, this is not so in cold weather.

SYSTEM COMPONENTS

The primary system components of a typical air starter system are:

- Dedicated air tank
- Pneumatic relay (pilot) valve
- System air pressure gauge
- Starter control valve
- Air starter motor

Most air starter motors are lubricated with diesel fuel oil. A generic air starter system is illustrated in **Figure 28–68**.

FIGURE 28–68 Generic air starter system

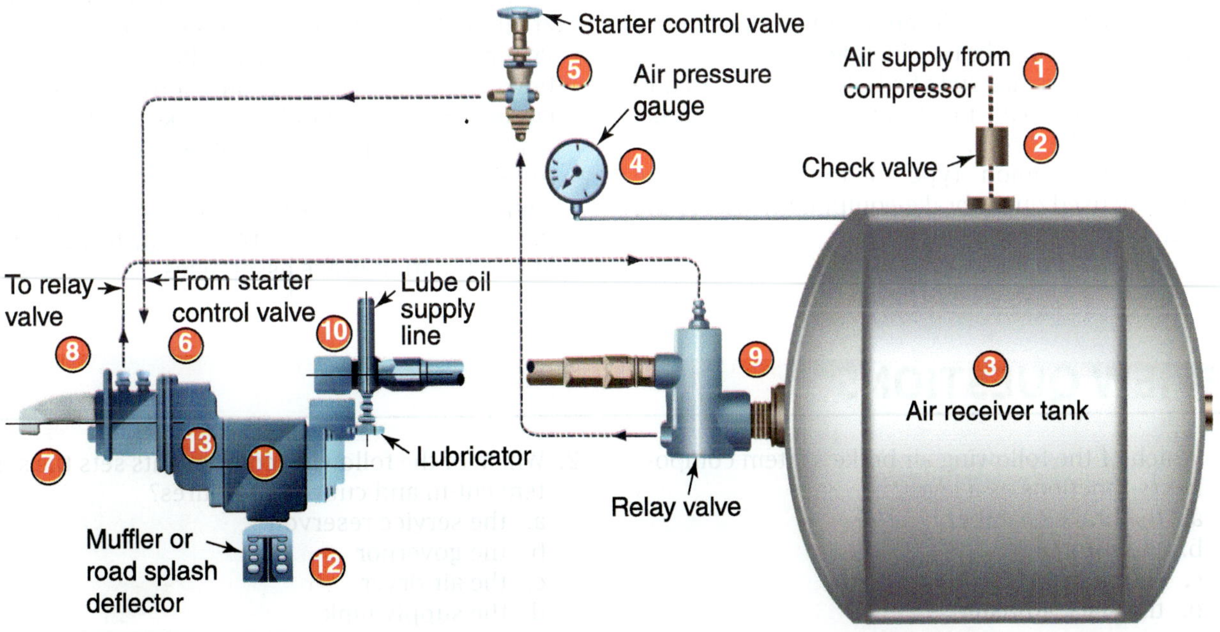

System Operation

Using the callout numbers in Figure 28–68, the following sequence explains air cranking circuit operation:

1. Air is supplied to the air cranking circuit by the truck air supply circuit.
2. Air from the supply circuit passes through a one-way check valve that isolates the air cranking circuit from the supply circuit.
3. Compressed air is stored in a dedicated reservoir. This large-volume reservoir is something of a problem (because of its size) in itself and it can be positioned on the chassis or in place of a passenger seat in the cab. It is charged either by the truck supply circuit or optionally by an outside source (i.e., shop air supply/truck gladhand).
4. An air pressure in the cab indicates the system pressure to the driver.
5. The starter control valve is located in the cab and is actuated by the driver to initiate cranking. This can be electric-over-air or a direct pneumatic pilot valve.
6. Air from the starter motor supply hose actuates a spring-loaded pinion into mesh with the flywheel ring gear.
7. The ring gear to starter motor pinion is preengaged, meaning that they are in mesh before the starter motor turns.
8. Control air leaves the pinion drive housing and is routed to the relay valve.
9. Air opens the relay valve, which opens the air supply from the starter tank to the cranking motor.
10. During cranking, the starter motor is lubricated by diesel fuel.
11. Compressed air from the air tank acts on the starter motor turbine vanes to rotate them at a speed of 6,500 rpm, which is gear reduced to crank the engine at around 450 rpm.
12. A muffler dampens the noise of air leaving the starter motor vanes.
13. The air starter motor drives reduction gearing that connects to the drive pinion for cranking the engine.

SUMMARY

- Air compressors are single-stage, reciprocating piston air pumps that are either gear- or belt-driven.
- Air dryers are used to help eliminate moisture and contaminants from the truck's air system.
- Dual-circuit application or foot valves, trailer application valves, bobtail proportioning valves, ratio valves, quick-release valves, relay valves, TP valves, dash control valves, double check valves, and check valves are some of the critical valves of an air brake system.
- The potential energy of compressed air is changed into mechanical force in an air brake system by slack adjusters and brake chambers.
- Brake chambers are specified by sectional area in square inches. The most common spring brake chamber is the 30/30.
- The most common type of foundation brake assembly used on air brake-equipped trucks is the S-cam type.
- Slack adjusters multiply the force applied to them by the brake chamber into brake torque.
- Brake torque applied to the S-camshafts results in the shoes being forced against the drum.
- Air discs operate by using an air-actuated caliper to squeeze brake pads against both sides of a rotor.
- Wedge brakes use a drum, a pair of shoes, and air-actuated wedges are used to force the shoes against the drum.
- Brake torque balance refers to the ability of a brake system to apply balanced mechanical brake force at all the foundation brake assemblies.
- Pneumatic timing refers to the ability of an air brake system to time the air control and actuation circuits so all the foundation brake components are applied at exactly the same moment.
- An air dual-circuit brake system is composed of a supply circuit, primary circuit, secondary circuit, parking/emergency control circuit, trailer circuit, and foundation brake assemblies.

REVIEW QUESTIONS

1. Which of the following air brake system components functions as a lever?
   a. the brake chamber
   b. the rotor
   c. the caliper
   d. the slack adjuster
2. Which of the following components sets the system cut-in and cut-out pressures?
   a. the service reservoir
   b. the governor
   c. the air dryer
   d. the supply tank

3. What is the function of a system safety pop-off valve?
   a. It releases excess pressure from the compressor.
   b. It pressure protects air in the supply tank.
   c. It alerts the driver when air pressure has dropped below system pressure.
   d. It relieves pressure from the supply tank in the event of governor cut-out failure.
4. The function of the foot valve or dual-circuit application valve is to:
   a. increase the flow of air to the service reservoirs.
   b. reduce the flow of air from the service reservoirs.
   c. meter air to the service brake circuit.
   d. meter air to the brake hold-off chambers.
5. The potential energy of an air brake system is:
   a. compressed air.
   b. foot pressure.
   c. spring pressure.
   d. vacuum over atmospheric.
6. Which of the following components is a requirement on every air tank?
   a. an air dryer
   b. a safety valve
   c. a draincock
   d. a pressure protection valve
7. What is the minimum total storage volume of air dedicated to brake system requirements of an air brake-equipped trailer?
   a. equal to the total volume of all the brake chambers
   b. two times total chamber volume
   c. four times total chamber volume
   d. eight times total chamber volume
8. The function of an air brake chamber is to:
   a. convert the energy of compressed air into mechanical force.
   b. multiply the mechanical force applied to the slack adjuster.
   c. reduce the air pressure for a service application.
   d. increase the air pressure for parking brakes.
9. An air brake–equipped truck is traveling on a highway at 60 mph (100 kph) and the driver makes a service brake application. Which of the following should be true in all of the spring brake chambers on the vehicle?
   a. Only the hold-off chambers are charged with air.
   b. Only the service chambers are charged with air.
   c. Both the service and hold-off chambers are charged with air.
   d. Neither the service nor the hold-off chambers are charged with air.
10. Technician A says that automatic slack adjusters can improve brake balance and reduce downtime and maintenance costs. Technician B says that automatic slack adjusters should not be adjusted after installation. Who is correct?
    a. Technician A only
    b. Technician B only
    c. both A and B
    d. neither A nor B
11. When an inversion valve becomes functional during a brake failure, what force is used to brake the vehicle?
    a. air pressure
    b. spring pressure
    c. hydraulic pressure
    d. vacuum
12. Which of the following determines the amount of leverage gained by a slack adjuster?
    a. the slack adjuster length
    b. the pushrod stroke length
    c. the S-cam geometry
    d. the number of internal splines
13. Which of the following is an advantage of an open anchor brake shoe?
    a. servo action
    b. the anchor pins never have to be changed
    c. faster brake jobs
    d. the retraction springs last longer
14. The brake shoes, retraction, and fastening hardware are mounted to:
    a. a spider
    b. a backing plate
    c. the drum
    d. the axle spindle
15. In order for a service signal to pass through a TP valve, which of the following must be true?
    a. The service gladhand must be connected to the trailer.
    b. The trailer supply dash control valve must be on.
    c. There must be air in the primary circuit.
    d. There must be air in the secondary circuit.
16. What actuates the relay piston in a foot valve on an air brake system when the valve is operating normally?
    a. foot pressure
    b. air pressure
    c. spring pressure
    d. mechanical pressure
17. Which of the following does the most to ensure that the driver has brake feel at the foot valve in a truck air brake system?
    a. spring pressure
    b. metered primary circuit service pressure
    c. metered secondary circuit service pressure
    d. system cut-out pressure

18. Which of the following best describes the term coefficient of friction?
    a. the heat rating of friction materials
    b. the aggressiveness of a friction material
    c. the longevity of friction materials
    d. the melting point of friction materials
19. Technician A says that in a tractor/trailer B-train combination, the crack pressures of the relay valves over the pup axles would be the highest on the rig. Technician B says that quick-release valves on the service gladhands can speed up service brake release times on multitrailer combinations. Who is correct?
    a. Technician A only
    b. Technician B only
    c. both A and B
    d. neither A nor B
20. A trailer air brake system compliant with a spring brake priority system has no air in its on-board air tanks. When coupled to a tractor air supply, where is the air directed to first?
    a. the service reservoirs
    b. the service brake chambers
    c. the hold-off chambers
    d. the main air reservoir
21. What is the usual color of the tractor-to-trailer service hose?
    a. green
    b. blue
    c. red
    d. white
22. What is the usual color of the tractor-to-trailer supply hose?
    a. green
    b. blue
    c. red
    d. white
23. When polarized gladhands are used, which of the following should be impossible?
    a. a full circuit brake failure
    b. cross-connecting the trailer supply and service hoses
    c. compounding service and parking brakes
    d. wheel lock-up
24. Technician A says that air disc brakes usually produce higher mechanical braking efficiencies than equivalent S-cam brakes. Technician B says that the reason wedge brakes produce higher mechanical braking efficiencies than equivalent S-cam brakes is that they are usually double-actuated. Who is correct?
    a. Technician A only
    b. Technician B only
    c. both A and B
    d. neither A nor B
25. In the event of a total failure of the primary circuit in an FMVSS 121 compliant vehicle, how is the relay/secondary circuit actuated in the foot valve?
    a. pneumatically
    b. hydraulically
    c. mechanically
    d. cannot be actuated

29

Prerequisite: Chapter 13

HYDRAULIC BRAKES AND AIR-OVER-HYDRAULIC BRAKE SYSTEMS

OBJECTIVES

After reading this chapter, you should be able to:

- Describe the principles of operation of a hydraulic brake system.
- Identify the major components in a truck hydraulic brake system.
- Describe the operation of drum brakes in a hydraulic braking system.
- Describe the operation of wheel cylinders and calipers.
- Identify the different types of disc brake calipers.
- List the major components of a master cylinder.
- Identify the hydraulic valves and controls used in hydraulic brake systems.
- Explain the operation of a hydraulic power booster.
- List the major components of an air-over-hydraulic braking system.
- Outline some typical maintenance and service procedures performed on hydraulic and air-over-hydraulic brake systems.
- Bench bleed a master cylinder.
- Identify the methods used to bleed brakes.
- Describe the operation of a typical hydraulic ABS.

KEY TERMS

brake spring pliers
calipers
combination valve
Federal Motor Vehicle Safety Standard No. 105 (FMVSS 105)
fixed caliper
floating caliper
hydraulic control unit (HCU)
hydraulic power booster
hygroscopic
load proportioning valves (LPVs)
master cylinder
metering valve
pressure differential valve
primary circuit
proportioning valves
residual pressure valve
secondary circuit
servo brake
slave cylinders
sliding caliper
wheel cylinders

INTRODUCTION

Hydraulic brakes are used on light- and medium-duty trucks up to Class 7 weight category. Hydraulic brakes work on the principle that if mechanical force is applied to a liquid within a closed circuit, it can be used to transmit motion or multiply and apply force (see Pascal's Law in Chapter 13). A basic hydraulic brake system (see Figure 29–3) consists of a **master cylinder** actuated mechanically (by foot pressure) hydraulic lines and flexible hoses that connect the master cylinder and **wheel cylinders** with the foundation brakes. Foundation brakes convert the hydraulic pressure into mechanical braking force.

Depressing the brake pedal moves a piston within a master cylinder. This piston forces fluid under pressure through brake lines and hoses to wheel cylinders/calipers. The force applied at the master cylinder can be amplified at the wheel cylinders/calipers if the pistons are of larger diameter than the master cylinder. This concept is shown in **Figure 29–1** and **Figure 29–2** and explained in greater detail in Chapter 13.

The amount of braking force delivered at the wheels in a hydraulic brake system is always proportional to the applied force (foot pressure) at the master cylinder, even when power-assist devices are used. Most hydraulic brakes use some form of assist even in the lightest duty applications. Hydraulic brake standards must comply with **Federal Motor Vehicle Safety Standard No. 105 (FMVSS 105)**. This federal legislation performs the same role in hydraulic brake standards as FMVSS 121 does with air brakes. FMVSS 105 requires that all highway-use hydraulic brake systems be dual circuit and have the ability to perform a mechanical emergency stop and park operation.

Although hydraulic brakes are seldom used in large highway trucks, they are used extensively in all sizes of off-highway heavy equipment and in the air-over-hydraulic systems used on medium-duty trucks. In this chapter, we make the assumption that the reader has some knowledge of basic hydraulics, as outlined in Chapter 13.

FIGURE 29–1 Force and movement are transmitted equally.

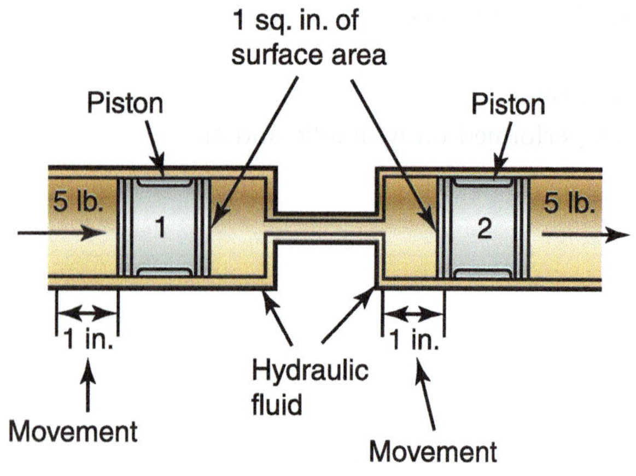

FIGURE 29–2 Force increases—movement decreases.

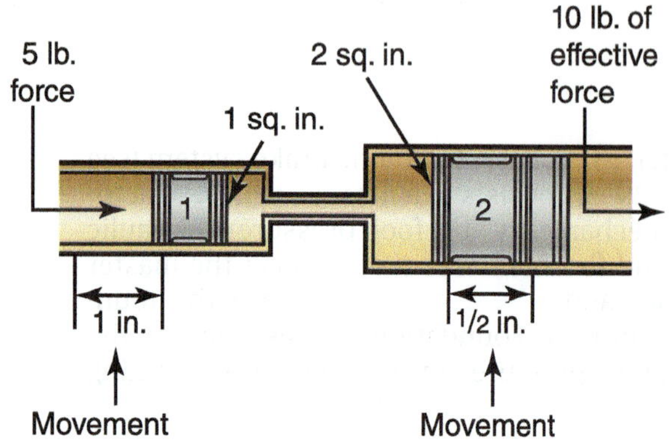

29.1 HYDRAULIC BRAKE SYSTEM COMPONENTS

This section examines the components in a hydraulic brake system, beginning with hydraulic fluids, continuing with the hydraulic circuit components, and concluding with the foundation brakes. **Figure 29–3** shows a simple hydraulic brake circuit in a light-duty four-wheel disc brake system. **Figure 29–4** shows a schematic of a disc brake system used in light- and medium-duty trucks. Both systems are dual circuit, which is mandatory for on highway applications.

HYDRAULIC BRAKE FLUID

Brake fluid standards are set by the Society of Automotive Engineers (SAE) and Department of Transportation (DOT). There are four categories of brake fluid in current use known as DOT 3, DOT 4, DOT 5, and DOT 5.1. DOT 3, 4, and 5.1 use a polyglycol base. These are **hygroscopic**, meaning that they are designed to absorb moisture that enters the system. This is an advantage for a brake fluid because it prevents the formation of water droplets that could cause localized corrosion, freezing, or overheating, resulting in a loss of brake application force at the wheel ends. DOT 5 is a silicone-based synthetic brake fluid that is not hygroscopic. It has higher compressibility than DOT 3 and 4 fluids and produces a spongy feel on initial application and is not recommended for use in an antilock brake system (ABS). **Table 29–1** shows the constituency and boil points for the brake fluids in current use.

SHOP TALK

Hygroscopic brake fluid can absorb airborne moisture rapidly. If DOT 3, 4, and 5.1 brake fluids are left uncovered in a shop environment, they can be ruined in as little as 1 hour.

Identifying the brake fluid used in a system is especially important when servicing hydraulic brakes. Service considerations for hydraulic fluids are covered later in this chapter.

FIGURE 29–3 Four-wheel hydraulic disc brake system with power assist: typical of many school bus systems

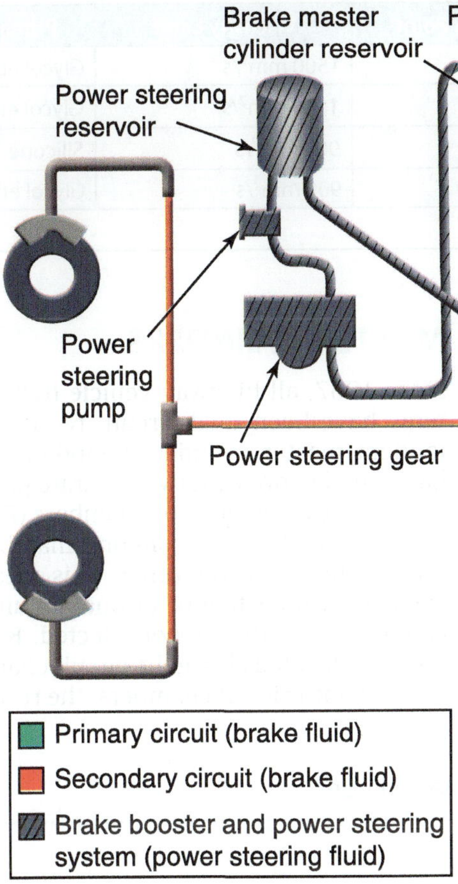

FIGURE 29–4 Schematic of a dual-circuit, split disc hydraulic brake system

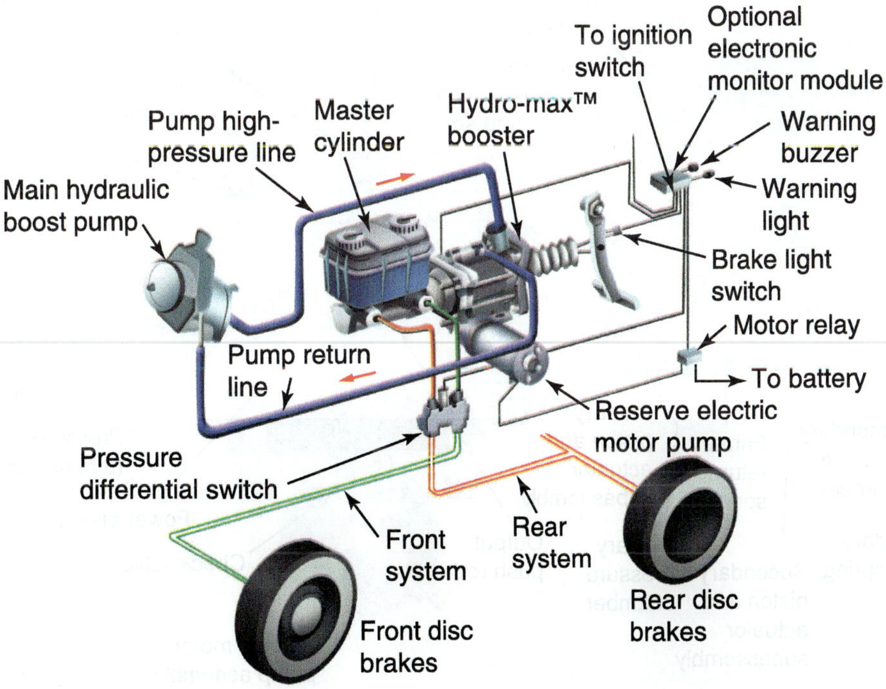

TABLE 29–1 Boil Points and Constituency of Brake Fluids

Brake Fluid Category	Dry Boiling Point	Wet Boiling Point	Viscosity Limit	Primary Constituent
DOT 3	205°C (401°F)	140°C (284°F)	1500 mm^2/s	Glycol ether
DOT 4	230°C (446°F)	155°C (311°F)	1800 mm^2/s	Glycol ether/borate ester
DOT 5	260°C (500°F)	180°C (356°F)	900 mm^2/s	Silicone
DOT 5.1	260°C (500°F)	180°C (356°F)	900 mm^2/s	Glycol ether/borate ester
Wet boiling point defined as 3.7% water by volume.				

WARNING

Brake fluid splashed into eyes can cause extreme irritation and potentially damage eyesight. Always wear safety glasses when working on brake systems.

CAUTION

Brake fluids contain compounds that can break down paint. Handle brake fluid carefully and if it comes into contact with paint, immediately remove it using alcohol followed by water.

MASTER CYLINDERS

Since 1967, all highway vehicle hydraulic brake systems have been dual circuit, requiring the use of a tandem master cylinder. A tandem master cylinder has a single bore with two separate pistons and chambers. The two sections or chambers (**Figure 29–5**) are designed so the fluid from one chamber cannot come into contact with the other. This means that, in the event of a total failure in one circuit, the hydraulic action of the other is not affected. Brake fluid reservoirs are located above the circuit chambers and gravity feed both circuit chambers; the reservoir is usually

FIGURE 29–5 Operating principle of a dual-circuit master cylinder

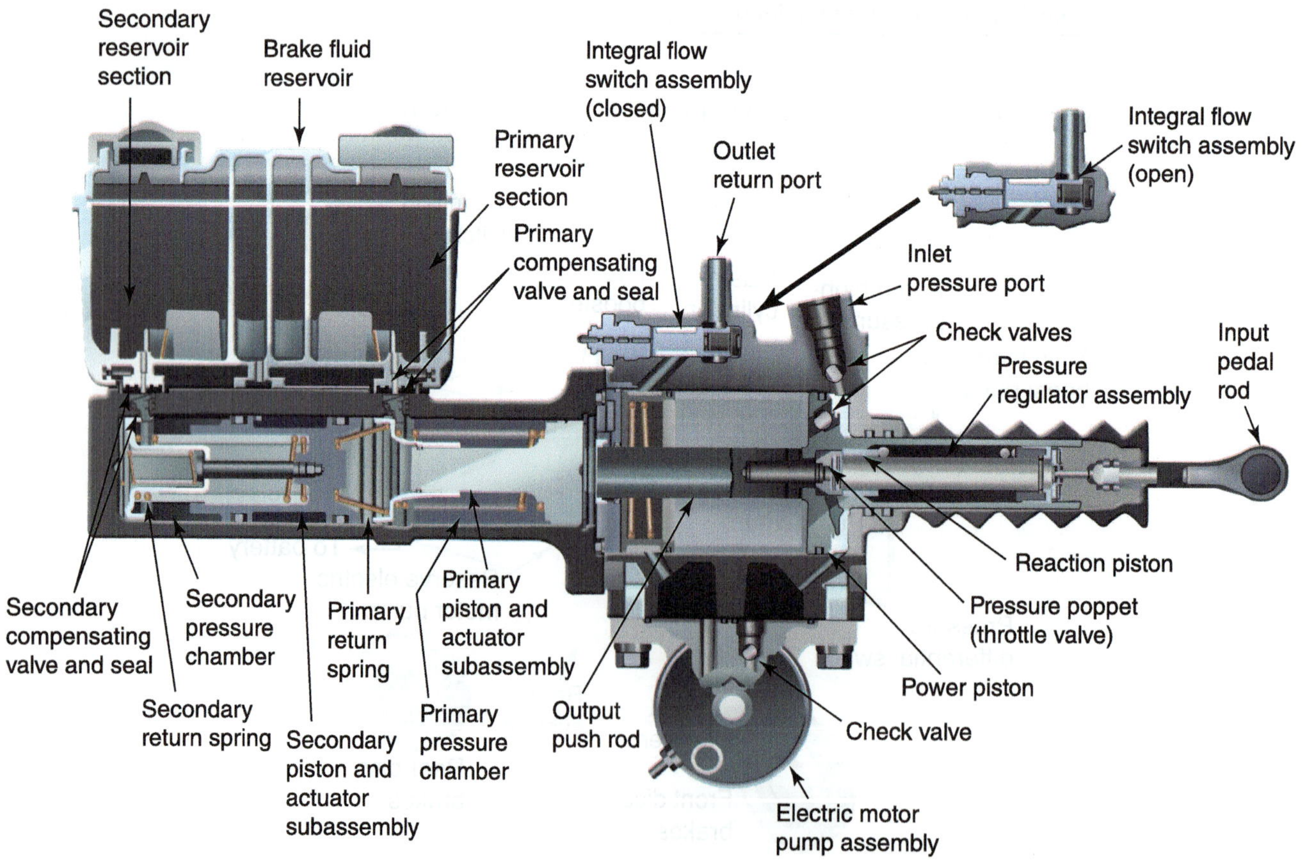

divided at the 50 percent fill point so that in the event of a serious leak, one circuit cannot draw all of the fluid from the reservoir. A master cylinder is so called because it hydraulically actuates one or more **slave cylinders**. The force produced by a wheel cylinder or caliper is always proportional to the force delivered to the master cylinder.

Dual-Circuit Master Cylinder Construction

Each chamber of a tandem master cylinder bore is connected to the reservoir by means of inlet ports. A dual-circuit master cylinder has two internal pistons, one in each of its chambers. The pistons identified as primary and secondary pistons are held in their retracted position by a spring (Figure 29–5). When the master cylinder is actuated by foot pressure, the primary piston is pushed past the primary inlet and replenishment ports, sealing the primary chamber, allowing it to be developed in the primary chamber.

This hydraulic pressure actuates the brakes on the **primary circuit** and, simultaneously, acts upon the secondary piston. This moves the secondary piston past its inlet and replenishment ports, applying pressure to the brakes on the **secondary circuit**. When the actuating pressure on the master cylinder is released, compensating ports help vent the brake fluid in the chambers back to the reservoirs. Figure 29–5 shows the operating principle of a heavy-duty master cylinder.

In the event of a secondary circuit hydraulic failure, the primary circuit will function as normal, but the fluid it sends to act on the secondary piston will not produce any braking effect. A brake warning light would alert the driver that half of the hydraulic circuit had failed.

If a failure occurs in the primary circuit, the brake pedal would have to travel much further until a stub on the primary piston mechanically contacts the secondary piston. Although the secondary piston is actuated mechanically in this failure mode, the brakes on the secondary circuit should otherwise apply normally.

The hydraulic brake plumbing from the dual-circuit master cylinder can be longitudinally split or diagonally split. The longitudinally split system operates the front and rear brakes from each of the tandem circuits (see Figure 29–3). The diagonally split system operates a right front wheel and left rear wheel from one circuit and the left front wheel and right rear wheel from the other. They are popular on front-wheel-drive vehicles that have a much greater load transfer to the front axle under severe braking.

Residual Pressure Valves

Some brake drum brake systems are equipped with a **residual pressure valve** (see **Figure 29–6**) and this is usually (but not always) located in the master cylinder. The valve is designed to retain a predetermined amount of pressure in brake lines when the brakes are released typically set at values between 7 and 14 psi (0.5 to 1.0 bar), but 12 psi (0.8 bar) is typical in truck applications. The shoes on drum brakes are fitted with a heavy retraction spring so the actual residual pressure is insufficient to overcome this spring pressure and load the shoes into the drum but residual line pressure does speed brake applications. For example, gravity acting on brake fluid in a master cylinder chamber typically exerts no more than 1 or 2 psi (0.07 to 0.14 bar).

Hydraulic Power Boosters

In hydraulic brake systems, the use of a vacuum or **hydraulic power booster** to assist the master cylinder in applying the brakes is common. Figure 29–6 shows an exploded view of a hydro-max booster assembly. The use of a booster reduces the pedal effort and the pedal travel required to apply the brakes as compared to a non-power-assisted system. **Figure 29–7** shows the circuit layout of a Meritor Wabco hydraulic power boost (HPB) circuit highlighting the role of the **hydraulic control unit (HCU)** which is bussed to the ECU. Because of federal regulations requiring that vehicles stop in a shorter distances, vacuum boosters are not often seen on medium-duty trucks. In other words, trucks with hydraulic brakes require hydraulic power assist because otherwise they would be unable to stop within the legislated distance.

Figure 29–8 is a graph that shows typical system performance of a hydro-max system. Note how the applied pedal pressure correlates to braking performance.

A hydraulic booster unit is comprised of an unrestricted open-center valve or an accumulator and reaction feedback mechanism, a large-diameter boost power piston, a reserve electric motor pump, and an integral flow switch. A hydraulic booster unit is powered either by the truck power steering pump (**Figure 29–9**) or by an electric over-hydraulic pump dedicated to the booster. The reserve electric motor pump provides backup assist in the event the power steering portion of the hydraulic booster fails. The backup pump is activated by the integral flow switch.

A hydraulic pressure accumulator booster is used in some light- and medium-duty vehicles. A nitrogen gas accumulator is used as backup to a hydraulic booster and in some applications it is charged by a driveline driven pump to provide some regenerative brake capacity.

Booster Master Cylinder

The master cylinder used with the hydraulic booster is a dual-split system using separate brake fluid reservoirs, pistons, and output ports for the front and rear brake circuits. In some applications, the master

FIGURE 29–6 Exploded view of a hydro-max booster assembly

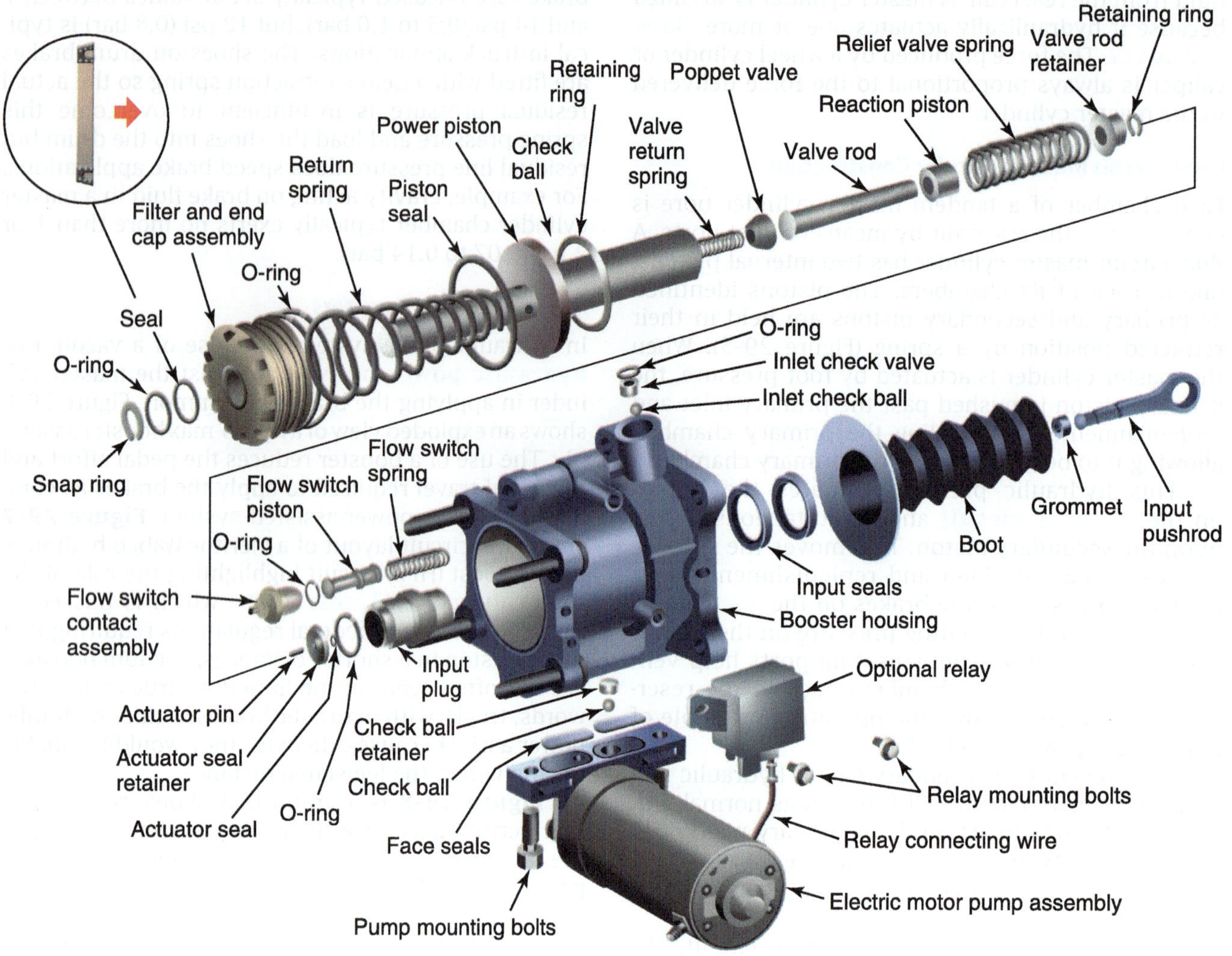

cylinder is internally divided into three compartments: primary, front, and rear. The booster can be located in the vehicle hydraulic circuit in one of two ways. First, it can be integrated with the power steering hydraulic pump and steering gear, operating in series with the power steering gear. In this case, the system flow and pressure demands during simultaneous steering and braking are considerable. The fluid flow path is designed to prioritize the brake circuit over the power steering circuit. Second, the unit can be powered by a dedicated hydraulic power source, that could be an additional power steering pump.

A flexible or rigid pressure line runs from the hydraulic power source to the booster inlet pressure port. Both maximum inlet port pressure and general principles of operation are similar to steering pump-driven systems.

A typical hydraulic booster system can incorporate the following mechanical and electrical components (see **Figure 29–10**):

- A pressure warning light
- A controller module
- Drum/drum, disc/drum, or disc/disc foundation brakes
- A relay/integral flow switch
- The vehicle battery(s)
- Electrical connectors and wiring
- A buzzer
- A warning light for the reserve electrical motor

System Operation

During normal system operation, fluid flow from the hydraulic power source enters the inlet pressure port of the booster, flows through the throttle valve and power piston—which acts on the flow switch—and exits via the outlet (return) port, unrestricted. When force applied to the brake pedal by the vehicle operator (multiplied by the lever ratio of the pedal mechanism), the movement activates the throttle

FIGURE 29–7 Circuit layout of a hydraulic power booster brake system highlighting the role of the HCU

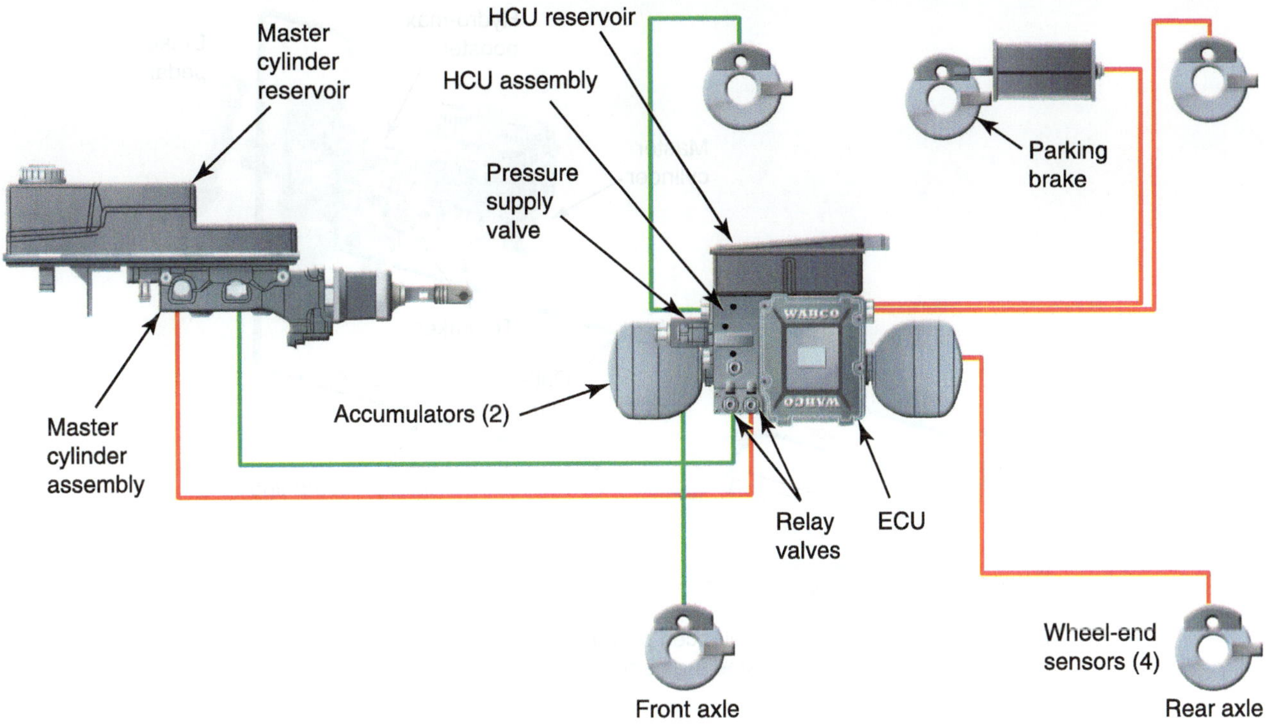

valve, restricting flow through the booster, applying boosted force to the master cylinder primary piston (see Figure 29–8). A reverse-reaction piston inside the power piston subassembly provides the driver with brake feel during an application of the brake pedal. Fluid pressure through the flow switch opens (interrupts) the reserve motor pump electrical circuit during normal operation. A separate check valve in the motor pump prevents backflow through the motor pump during normal power applications.

FIGURE 29–8 Pedal pressure correlated to performance in a hydro-max brake system. Circuit layout of a hydraulic power booster brake system.

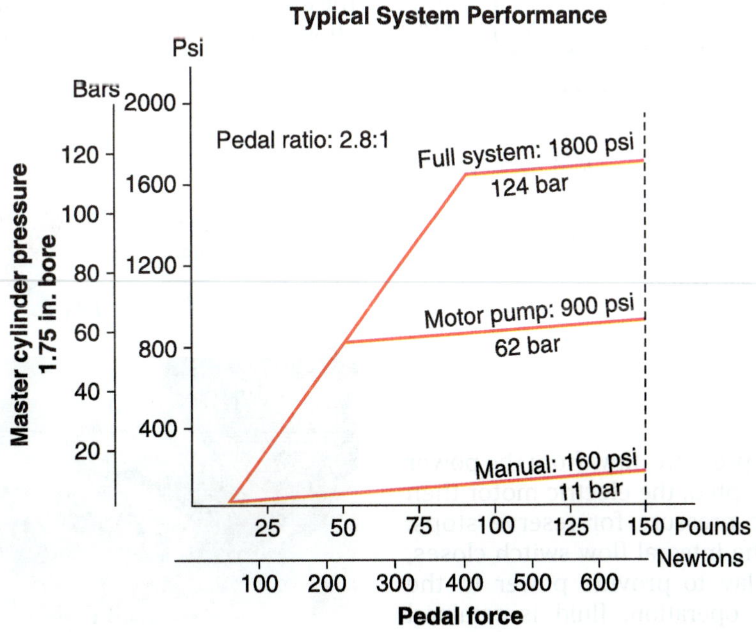

FIGURE 29–9 Hydraulic booster charged by the power steering pump.

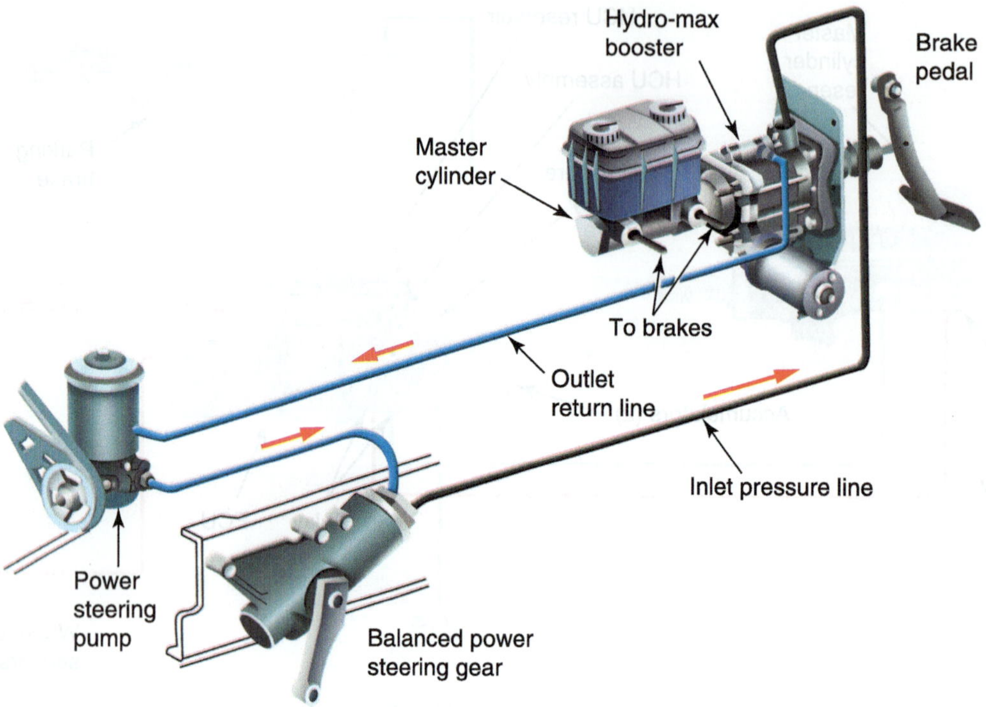

FIGURE 29–10 Operation and warning system used on a hydro-max brakes system

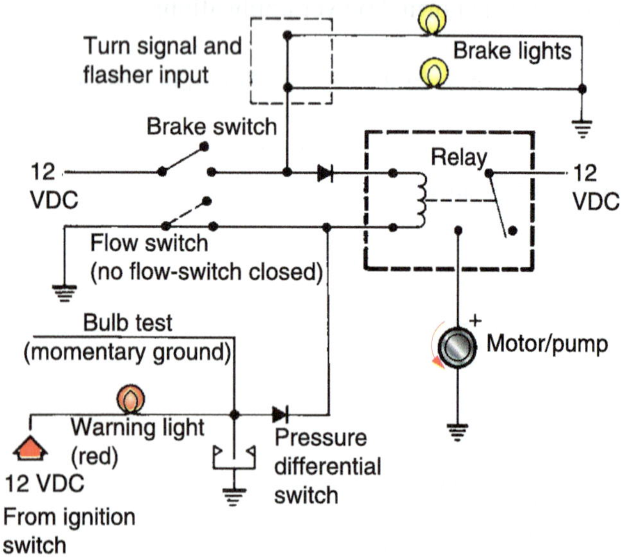

If normal operating pressure/flow from the power steering pump is interrupted, the electric motor then provides the hydraulic pressure for reserve stops. On flow interruption, the integral flow switch closes, energizing a power relay to provide power to the motor. During reserve operation, fluid is retained within the booster by the inlet port check valve. The motor pump recirculates fluid within the booster assembly with pressure supplied on demand via the throttle valve. In this case, the number of applications is limited by the:

- Electrical capacity of the vehicle
- Hydraulic accumulator pressure

Unassisted (unboosted) braking is also possible in the event that power-assist systems fail. **Figure 29–11** shows a medium-duty hydro-max brake assembly.

FIGURE 29–11 Medium-duty hydro-boost system

Pressure Differential Valve

All dual-circuit hydraulic systems have a **pressure differential valve** and brake warning light system, operated by a mechanically or hydraulically actuated electric switch. Its function is to indicate to the driver when one-half of the system is not functioning. It typically consists of a cylinder inside of which is a spool valve (Figure 29–4). Each end of the spool valve is subject to hydraulic pressure from either side of the circuits within the brake system (primary and secondary circuits). When a brake application is made in a dual-circuit hydraulic system, the pressure in the primary and secondary circuits should be equal. If this is the case, the spool valve remains in a balanced position midpoint in the cylinder, and the brake warning light circuit remains open (not grounded). However, if pressure in one of the two circuits is lost, the metal spool valve will be moved off-center, contacting the warning lamp ground terminal. Closing the brake warning light circuit illuminates the dash brake warning light. On some brake systems, the parking brake also will illuminate the brake warning light circuit when applied. **Figure 29–12** shows an optional Bendix controller module schematic, which is widely used.

Metering Valves

Metering valves are incorporated into some hydraulic brake systems that use disc brakes on the front axle and drum brakes on the rear axle. Disc brakes have much lower lag times than drum brakes. Lag time means the time between the first movement of the brake pedal and the moment that the friction material makes contact with the rotating surface at the foundation brakes. Metering valves delay the application of the front brakes until pressure has been established in the rear brake circuit. This helps balance the application timing because disc brake pads are typically within 0.001 inch (0.025 mm) of the rotor when unapplied, whereas brake shoes may have to move up to 0.250 inch (6 mm) before contacting the drum. This also helps reduce the amount of light-duty braking performed by the front-wheel brakes; in other words, it balances front to rear braking.

Proportioning Valves

Proportioning valves also are used in systems using front disc and rear drum brakes. During hard braking, the vehicle weight is transferred forward onto the front axle. This is known as load transfer, and the more severe the braking, the greater the load

FIGURE 29–12 Widely used optional Bendix controller module brake circuit

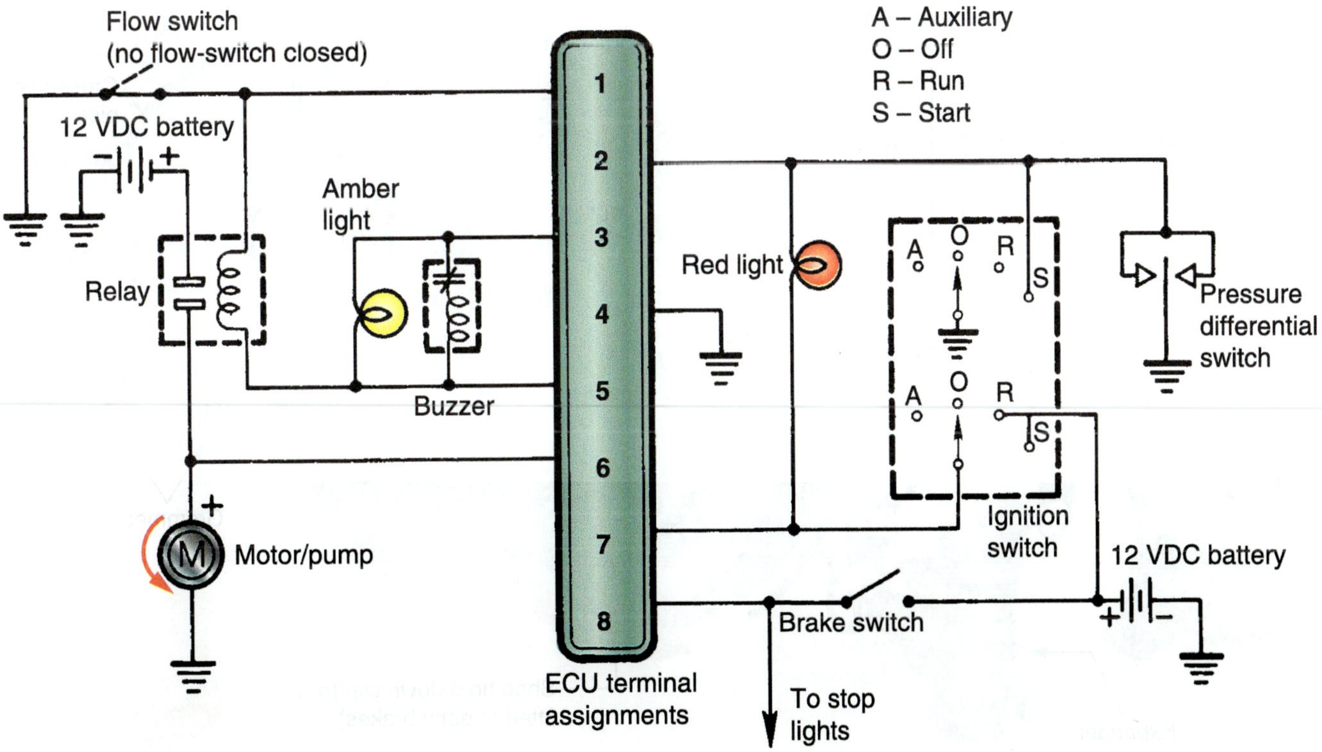

transfer onto the front axle. Load transfer can result in rear wheel brake lockup. Proportioning valves prevent pressure delivered to the rear wheel brakes from exceeding a predetermined pressure value under hard braking conditions.

Load Proportioning Valves

Some trucks use load-sensitive, rear-wheel proportioning valves called **load proportioning valves (LPVs)**. Pressure is applied to the rear wheels according to how much weight is on the rear of the vehicle. The vehicle weight (height) is sensed by mounting the valve on a frame cross-member and using a linkage and lever system attached to the rear axle housing. In this way, as the load increases over the rear axle and the distance between the frame and axles diminishes, the rear brakes can assume a greater proportion of the required vehicle braking. The result is that when the vehicle load is light, the rear wheels are less likely to be overbraked. The LPV is a hydraulic restriction valve. An example is referenced later in Figure 29–21.

Combination Valves

The pressure differential switch, metering, and proportioning valves can either be housed in a single valve body called a **combination valve** or integrated into the master cylinder assembly. When a combination valve is separate from the master cylinder assembly it is normally mounted close to it.

HYDRAULIC DRUM BRAKES

In a typical hydraulic drum brake system, the brakes are actuated by wheel cylinders. The manner in which the brake shoes are mounted will determine whether the system is self-energizing or servo-assisted. **Figure 29–13** shows the components of an automatic adjusting, two leading shoe, drum brake assembly. Note the layout of the components.

Wheel Cylinders

Hydraulic drum brakes are actuated by a wheel cylinder. The wheel cylinder responds to hydraulic pressure delivered to it by the master cylinder. Wheel cylinders usually contain a pair of pistons, which, when actuated hydraulically, force brake shoes apart and into the drum.

Some wheel cylinders are single actuating. They use a single piston to actuate an anchor pin pivoted brake shoe. Cups seal the pistons in the wheel cylinder bore. When the master cylinder relieves the pressure to the wheel cylinder, the brake shoes retraction

FIGURE 29–13 Components of an automatic adjusting, two leading shoe drum brake

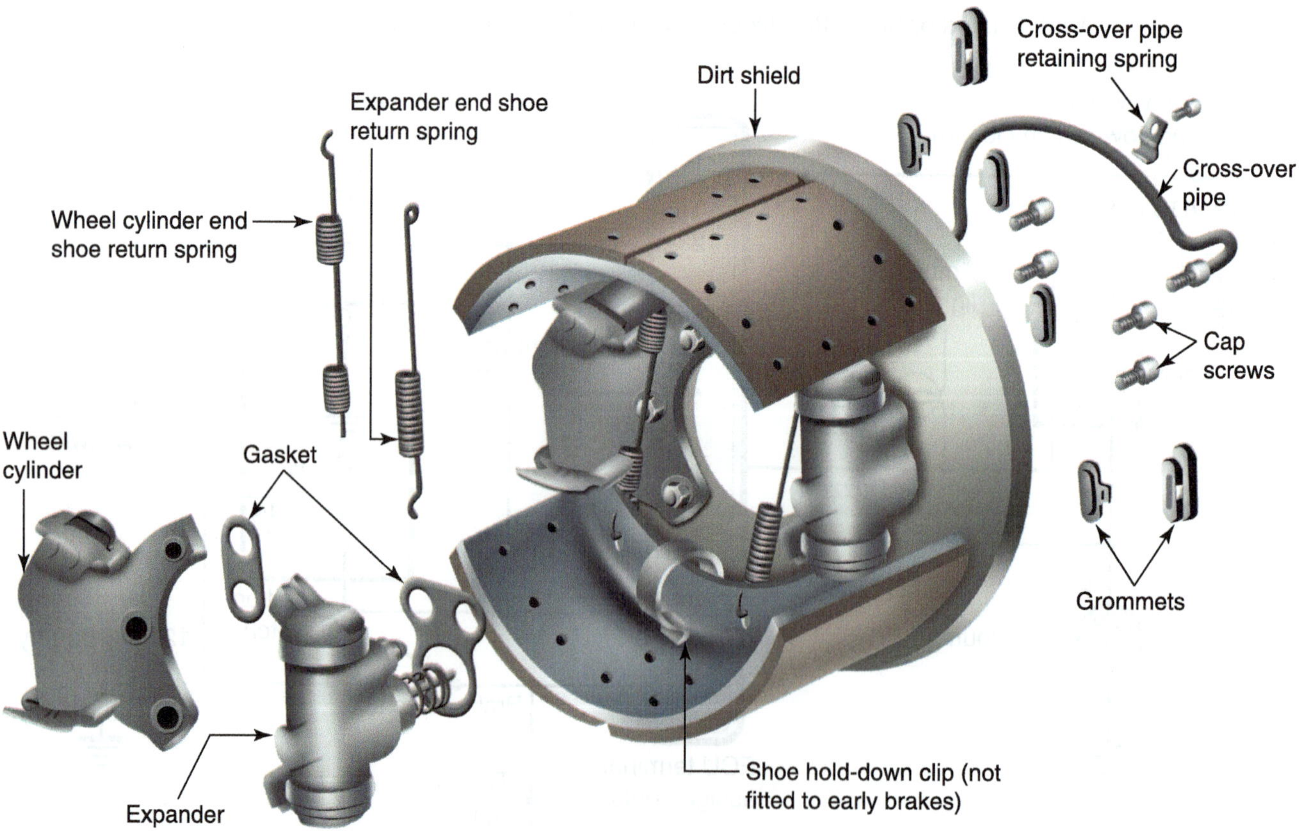

springs pull the shoes away from the drum and force the pistons back into their unapplied position in the wheel cylinder. Most wheel cylinders on current hydraulic brake systems are double acting, which means that each wheel cylinder has a pair of pistons each applied to a different shoe. **Figure 29–14** shows a double-acting wheel cylinder with the components identified. **Figure 29–15** shows both a single-acting front-wheel cylinder and a double-acting rear-wheel cylinder.

Drum Brakes

As with the air brake, a hydraulic drum brake assembly consists of a cast-iron drum that is bolted to, and rotates with, the vehicle wheel, and a backing plate that is attached to the axle. The shoes, wheel cylinders, automatic adjusters, and linkages are mounted to the fixed backing plate. Additional hardware is required for the parking brakes. The shoes are lined with friction facings, which contact the inside of the drum when the brakes are applied. The shoes are forced outward by the wheel cylinder pistons actuated by hydraulic pressure from the master cylinder.

When the shoes are forced into the brake drum, the kinetic energy (energy of movement) of the rotating drum is transformed by means of friction to heat energy and then dissipated to atmosphere. As a brake shoe is forced against the drum, the frictional drag acting on its circumference tends to draw the shoe toward the drum. When the rotation of the drum corresponds to the outward rotation of the shoe, the frictional drag will act to pull the shoe tighter against the inside of the drum. This is known as self-energizing.

FIGURE 29–14 Exploded view of a double-acting wheel cylinder

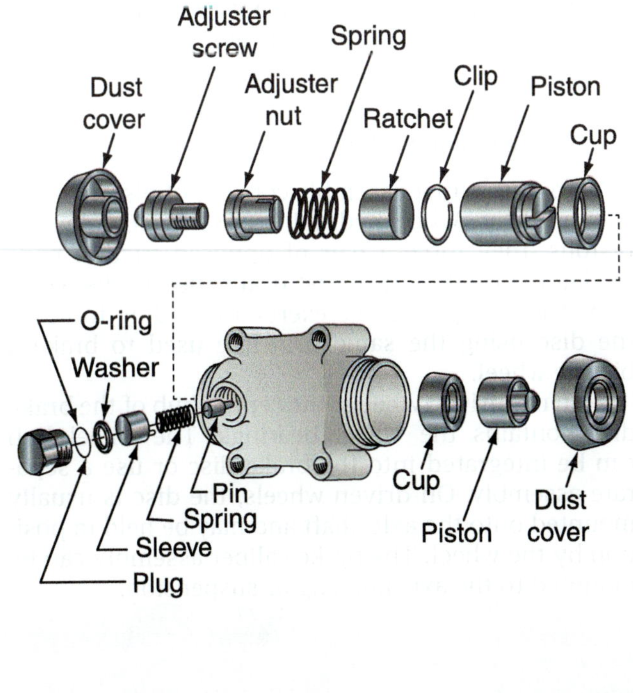

FIGURE 29–15 (A) Exploded view of a single-acting, front-wheel cylinder; (B) exploded view of a double-acting, rear-wheel cylinder

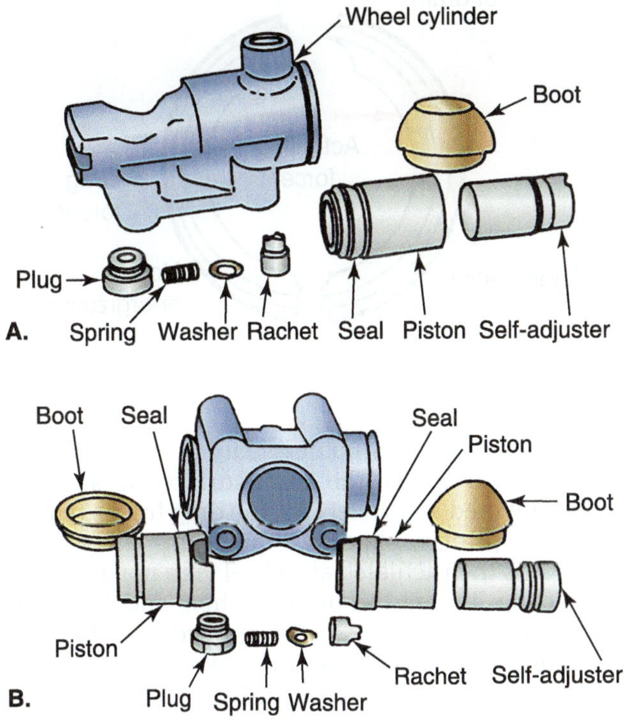

Non-Servo Drum Brakes

In non-servo drum brakes, each shoe is separately anchored at its heel end. Actuating force is applied to the opposite end, the toe. When the brakes are applied with the vehicle in forward motion, the forward or leading shoe is energized. The trailing shoe receives no energization unless the brakes are applied when the truck is in reverse. This reverses the roles of the leading and trailing shoes. For obvious reasons, the forward or leading shoes will tend to wear faster, because most of the vehicle braking will be performed when the vehicle is in forward motion.

Figure 29–16 demonstrates the action of a non-servo brake. The brakes are anchored on a pivot and only one of the shoes is energized. Note that their roles are reversed as the direction of rotation of the drum is reversed.

Servo-Type Drum Brakes

In many hydraulic brake designs, the self-energizing effect is supplemented by a servo action. **Servo brake** action occurs when the action of one shoe is governed by an input from the other. When hydraulic force is applied to the toe ends of the brake shoes, the primary shoe is designed to react first because it has a weaker return spring (and possibly different coefficient of friction/color). The shoe is drawn off the

FIGURE 29–16 Non-servo brake action

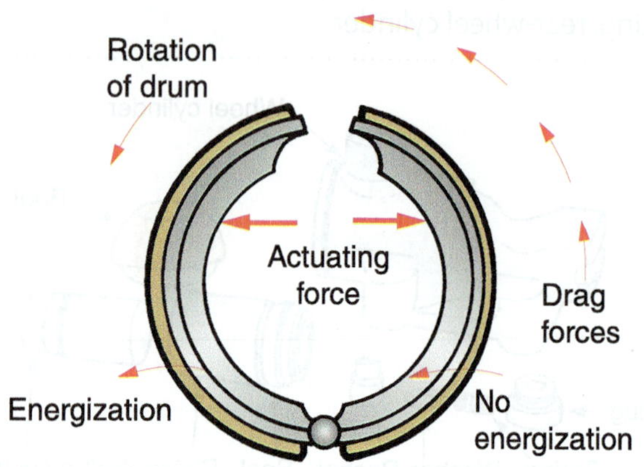

anchor (which at this point is only acting as a stop) and is forced against the drum surface. Because the shoe is pivoted at its heel end, the drag forces tend to rotate the primary shoe with the drum. The secondary shoe is then actuated at its heel via a nonanchored connecting link by the primary shoe, and it, too, is forced into the drum. In the applied position, the servo brake makes both shoes behave as if they were one continuous shoe, increasing the total surface area contact. The actuating force pushes on one end, and the other is held fast by the anchor. Drag forces cause the whole assembly to be energized. This concept is demonstrated in **Figure 29–17** and **Figure 29–18**. There are several different configurations of servo brake. Some hydraulic piston pressure acts against the secondary shoe, but this is incidental.

FIGURE 29–17 Servo brake action

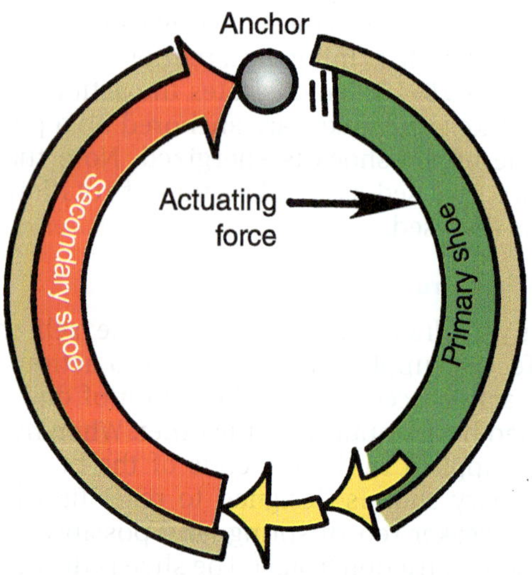

FIGURE 29–18 How a shoe becomes self-energizing

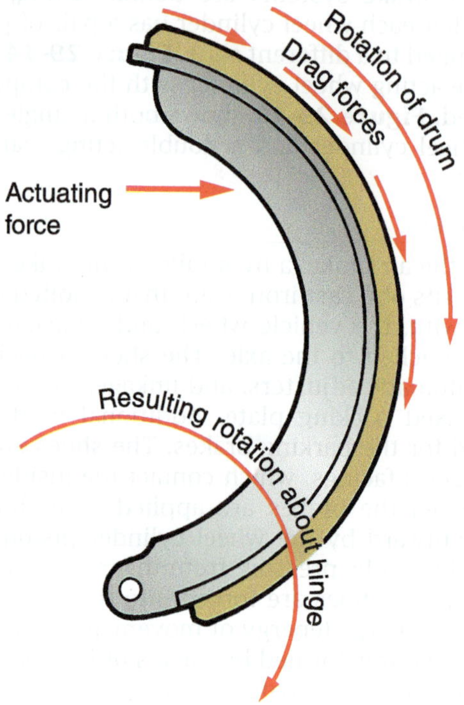

HYDRAULIC DISC BRAKES

These may be used on all four wheels of a vehicle, or combined with disc brakes on the front wheels and drum brakes on the rear. The actuation circuit works similarly to that used on drum brakes but higher pressures are required. Hydraulic pressure from the actuation circuit is transmitted via brake lines to one or more pistons on each disc brake foundation assembly (see **Figure 29–19**). The principal components of a disc brake foundation assembly are:

- Piston(s)
- Caliper
- Wheel mounted disc

A brake caliper assembly consists of piston(s) and friction discs. When hydraulic pressure actuates the pistons, they force a pair of opposed friction pads to clamp a rotating disc that is attached to the wheel hub. The clamping force exerts a retarding force on the disc using the same principle used to brake a bicycle wheel.

On nondriven wheels, the central hub of the brake disc contains the wheel bearings. The wheel hub can be integrated into the brake disc or use a separate assembly. On driven wheels, the disc is usually mounted onto the axle shaft and may be held in position by the wheel. The brake caliper assembly can be mounted to the axle housing or suspension.

FIGURE 29–19 Key components of a disc brake assembly

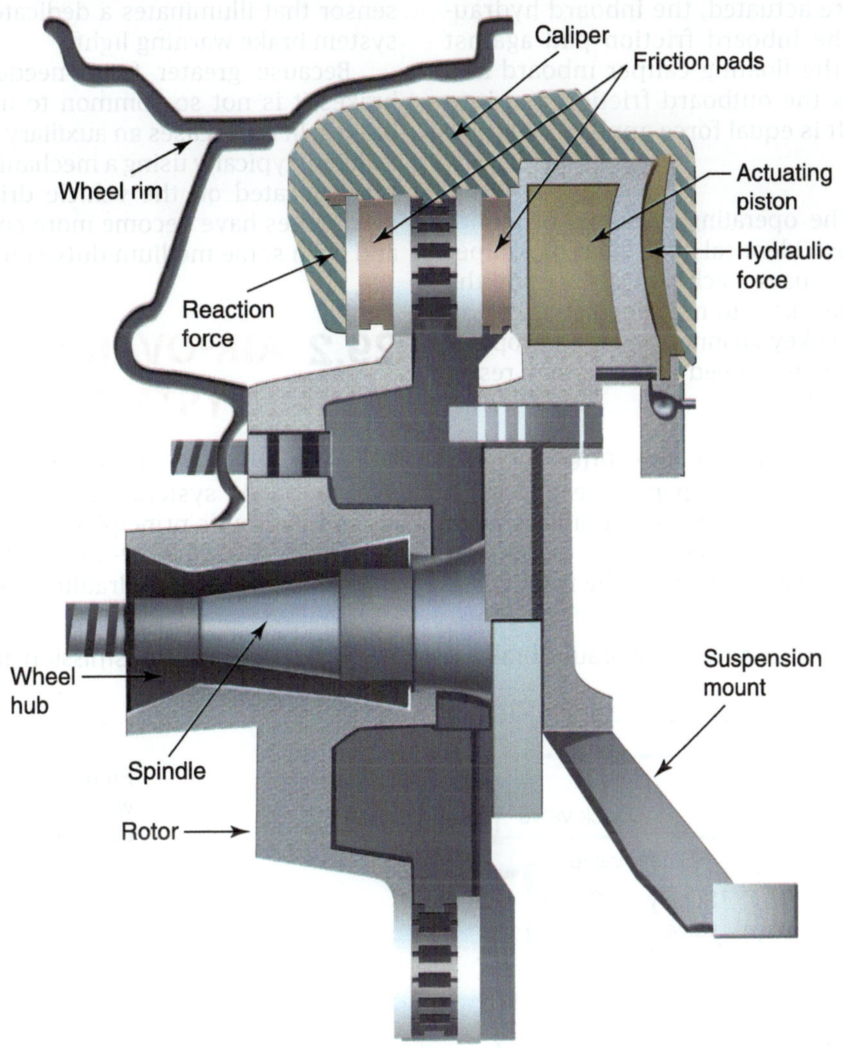

Disc brakes require greater application force and a substantial amount of heat that has to be dissipated mostly through the disc, so they are designed to maximize air flow through the assembly. Power boost circuits are required in commercial vehicle applications to ensure there is sufficient hydraulic assist to pedal effort. Although they require greater hydraulic pressure to actuate them, disc brakes have higher mechanical efficiencies than drum brakes and brake fade is rare. Brake disc rotors tend to throw off water by the centrifugal force they generate, so brake performance is only compromised momentarily under wet conditions.

Types of Caliper Assembly

The function of a disc brake caliper is to convert hydraulic energy into mechanical force—the "squeeze" force that is exerted on the rotor that turns with the wheel. A caliper assembly may contain from one to four hydraulic pistons; when the pistons are actuated, they exert uniform pressure on the friction pads. There are three types of caliper assembly:

- Fixed caliper
- Floating caliper
- Sliding caliper

Fixed Caliper. A fixed caliper is bolted in place and does not move when the brakes are applied; pistons on both sides of the rotor sandwich the rotating disc to apply the brakes.

Floating Caliper. A floating caliper is a one-piece casting that uses single-piston actuation and because it is hydraulically actuated on one side only acts like a clamp. The caliper is attached to an anchor plate on mounted to the axle spindle on threaded locating pins that allow the caliper to slide back and forth. When the brakes are actuated, the inboard hydraulic piston forces the inboard friction pad against the rotor, moving the floating caliper inboard and in doing so clamps the outboard friction pad into the rotor. The result is equal force applied on either side of the rotor.

Sliding Caliper. The operating principle of a sliding caliper is almost identical to a floating caliper with exception that it uses machined surfaces on the caliper assembly that "key" to machined surfaces on the anchor plate. The key channels must be properly lubed and wear in the machined surfaces may result in tapered friction pad wear.

Wear Sensors. Most calipers are fitted with a mechanical or electronic friction pad wear alert. A mechanical wear alert produces an audible squeal when friction pads are worn enough to permit an alloyed steel tab to make contact with the rotor when the brakes are applied. An electronic wear sensor does the same thing when brake pad wear grounds a sensor that illuminates a dedicated dash light or the system brake warning light.

Because greater force needed to apply a disc brakes it is not so common to use them as parking brakes. In most cases an auxiliary park brake circuit is utilized, typically using a mechanically actuated drum brake located on the vehicle drive shaft. Hydraulic disc brakes have become more common in light-duty and even some medium-duty commercial trucks.

29.2 AIR-OVER-HYDRAULIC BRAKE SYSTEMS

Some Class 6 and 7 trucks are equipped with an air-over-hydraulic system. Such systems combine both air and hydraulic principles to braking. The schematics shown in Figure 29–19 and **Figure 29–20** show two typical air-over-hydraulic brake systems.

FIGURE 29–20 A typical air-over-hydraulic brake system (A/T—automatic transmission and M/T—manual transmission)

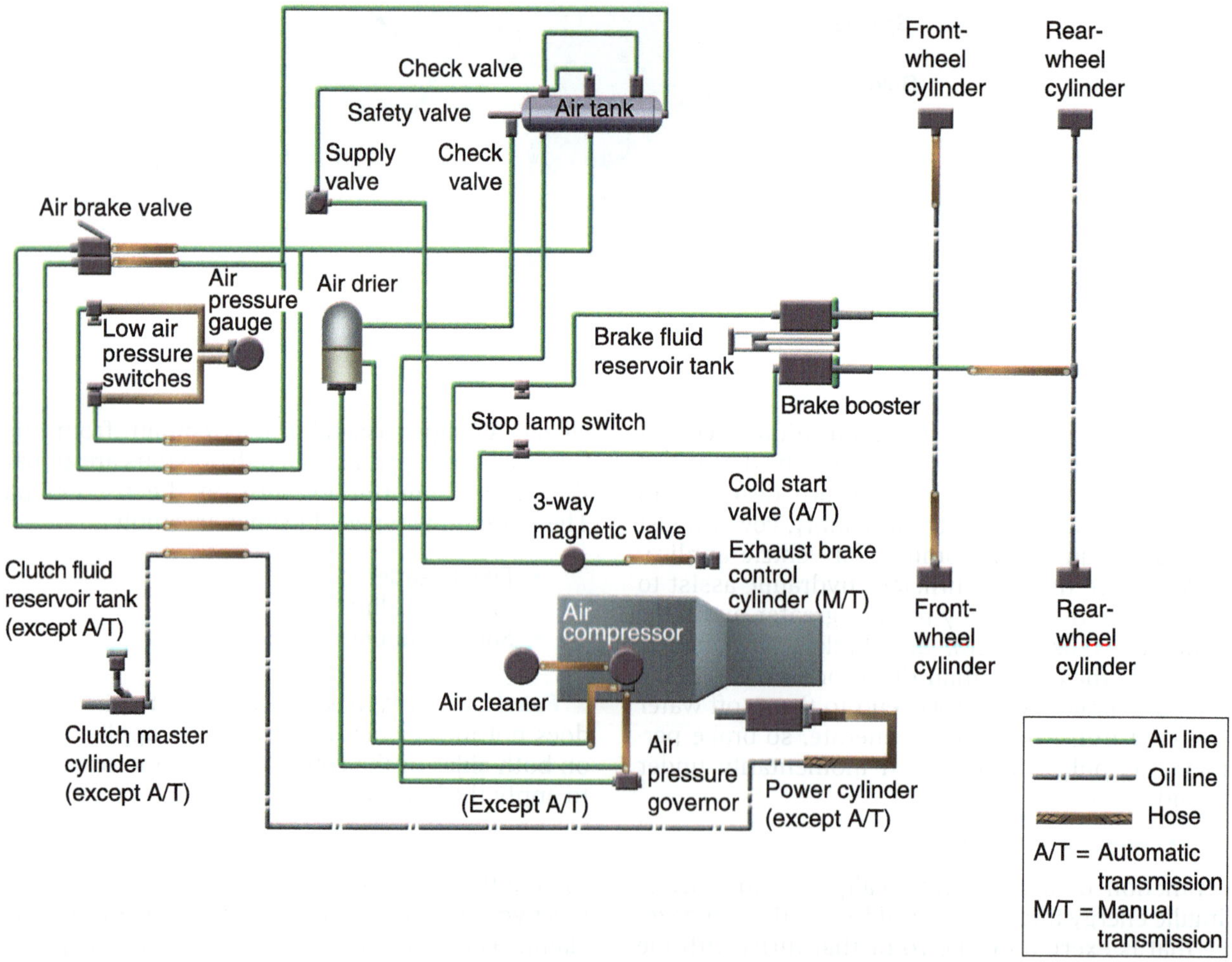

There are several variations of air-over-hydraulic brakes used on trucks. In general, the control circuit is managed by air pressure and the application circuit by hydraulics. Depressing the brake pedal (treadle valve) moves the application valve plunger. This allows modulated air to flow through quick-release valves and to be directed to the booster assemblies. Separate booster assemblies are used for front and rear brakes. Air pressure applied to the booster moves an actuating piston, which increases hydraulic pressure.

The actual amount of hydraulic pressure increase depends on the air pressure applied to the diaphragm and its sectional area. This hydraulic pressure is routed to the wheel cylinders, which effects vehicle braking in the same manner as in a hydraulic brake system.

Hydraulic pressure forces the pistons of each wheel cylinder outward. Outward movement of the pistons applies brake pressure on brake drum or disc **calipers**. A brake lining wear switch is mounted in each booster to indicate excessive travel of the air piston. A front-wheel limiting valve is often used on front axle brakes without antilock braking systems (ABS). This limiting valve proportions fluid pressure during light braking, both to balance the braking and reduce brake lining wear. LPVs also can be used to modulate hydraulic pressure to front and rear brakes based on vehicle loads. The general setup of the air circuit is identical to that required on a vehicle equipped with air brakes. **Figure 29–21** shows one type of LPV.

FIGURE 29–21 Location of one type of load proportioning valve. The torsion bar deflects as load increases.

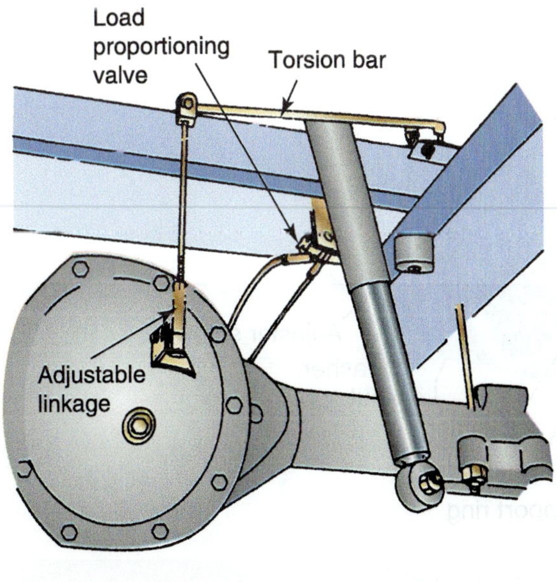

SYSTEM REDUNDANCY

All current air-over-hydraulic systems are required to be dual-circuit systems. As with full air and hydraulic systems, a sudden leak in one portion of a system should not result in a total loss of braking. Because the power for the boost force is sourced from an engine-driven compressor, air-over-hydraulic brakes for medium-duty trucks are required to have fail-safe backup (known as *redundancy*) from an electric boost compressor. Other fail-safe features include buzzers and lights that warn of low air pressure in the system in much the same manner as an air brake-equipped truck. **Figure 29–22** shows the location of air-over-hydraulic brake components on a truck.

PARKING BRAKES

The most common type of parking brake on air-over-hydraulic systems is the spring brake canister type but electric parking brakes are finding some traction in the market place. Its parking brake action is obtained by forcing a wedge assembly between the plungers in the wheel cylinder on each rear wheel. Engagement of the wedge assembly by the spring brake expands the plunger in the wheel cylinder, forcing the brake shoes into the brake drum, as shown in **Figure 29–23**.

Cable-actuated parking brakes are also found in some air-over-hydraulic systems. The mechanically actuated brake drum unit is mounted at the rear of the transmission so that it acts on the rear wheels through the drive shaft. For a transmission-located parking brake to function both rear wheels must be firmly on the ground. Another method is to incorporate a parking brake in each rear wheel, which is cable controlled from the cab area.

Electric parking brakes on current trucks often function through an automated manual transmission (AMT) where the electronic control unit (ECU) electrically engages an internal parking brake when the foot brake is fully depressed and the truck is stationary. Electrically controlled driveline brakes are also used and operate using a spring-apply, electric-release system.

CONFIGURATIONS

Air-over-hydraulic systems can be used with either drum brakes or disc brakes on all four wheels or mixed disc and drum systems. There are other variations, however. Some systems use hydraulic discs on the front and full air S-cam drums in the rear. A delay valve keeps the hydraulic front brakes from coming on before the slower-acting air brakes. Some medium-duty trucks mix hydraulic and air brake

FIGURE 29–22 Location of air-over-hydraulic brake components

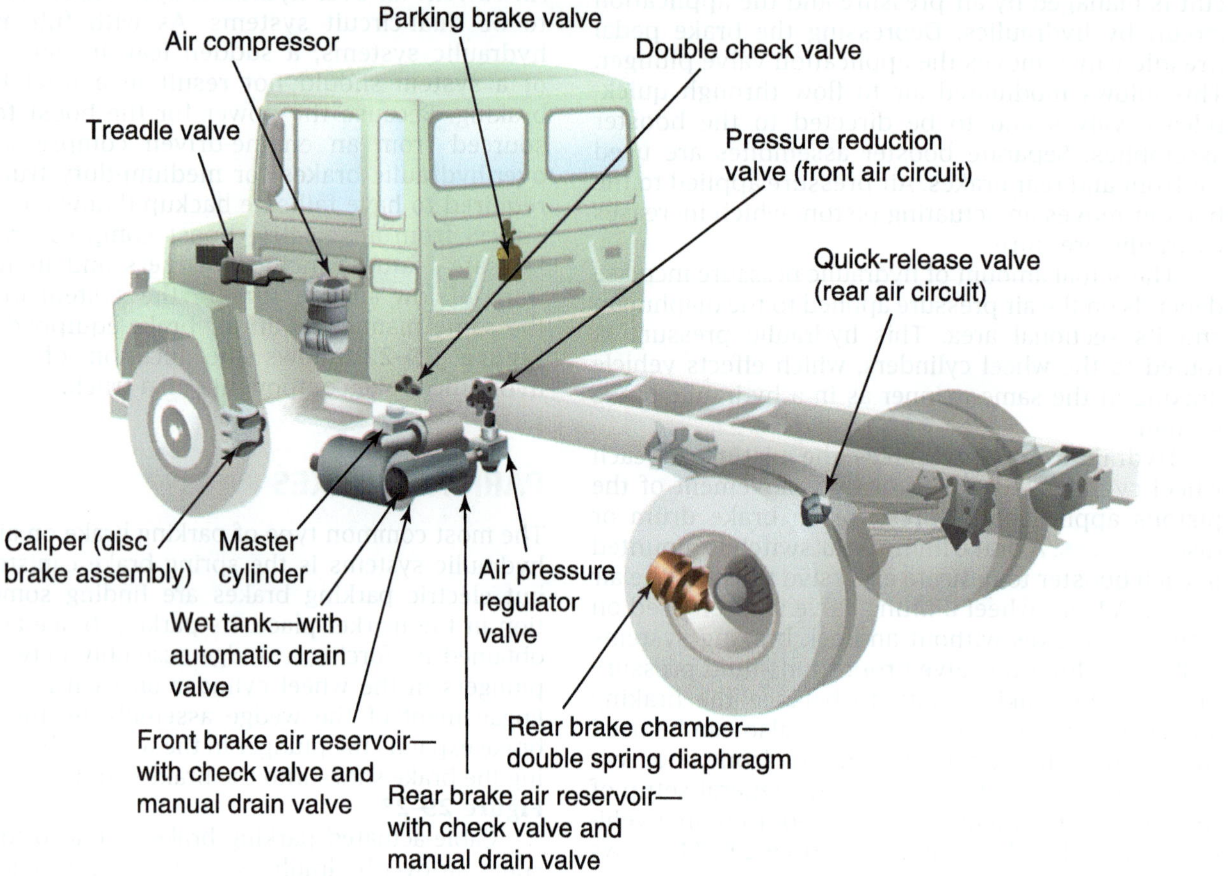

FIGURE 29–23 Spring brake canister components

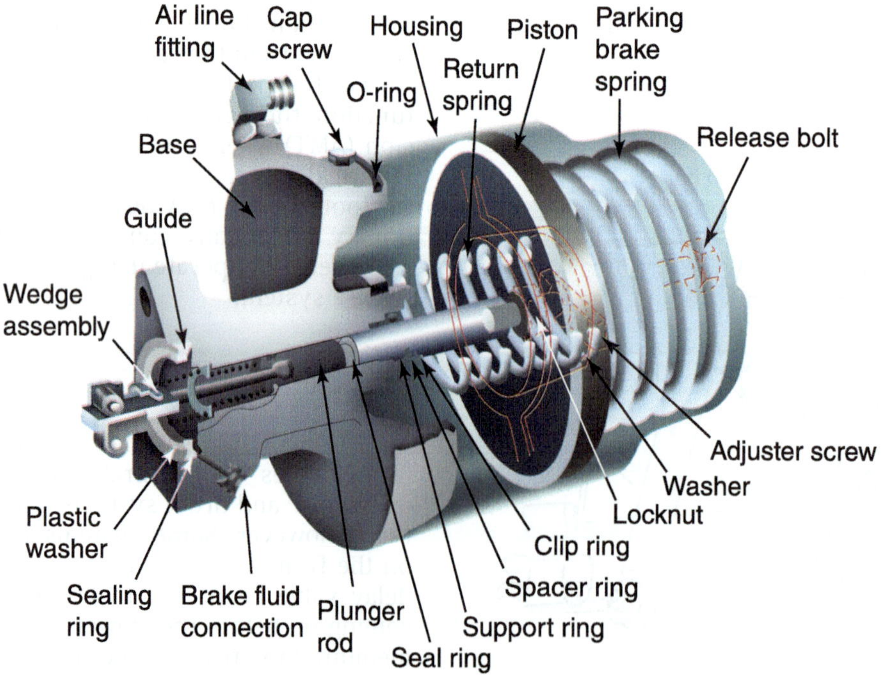

FIGURE 29–24 System with hydraulic discs in front and air-actuated, S-cam brakes in the rear

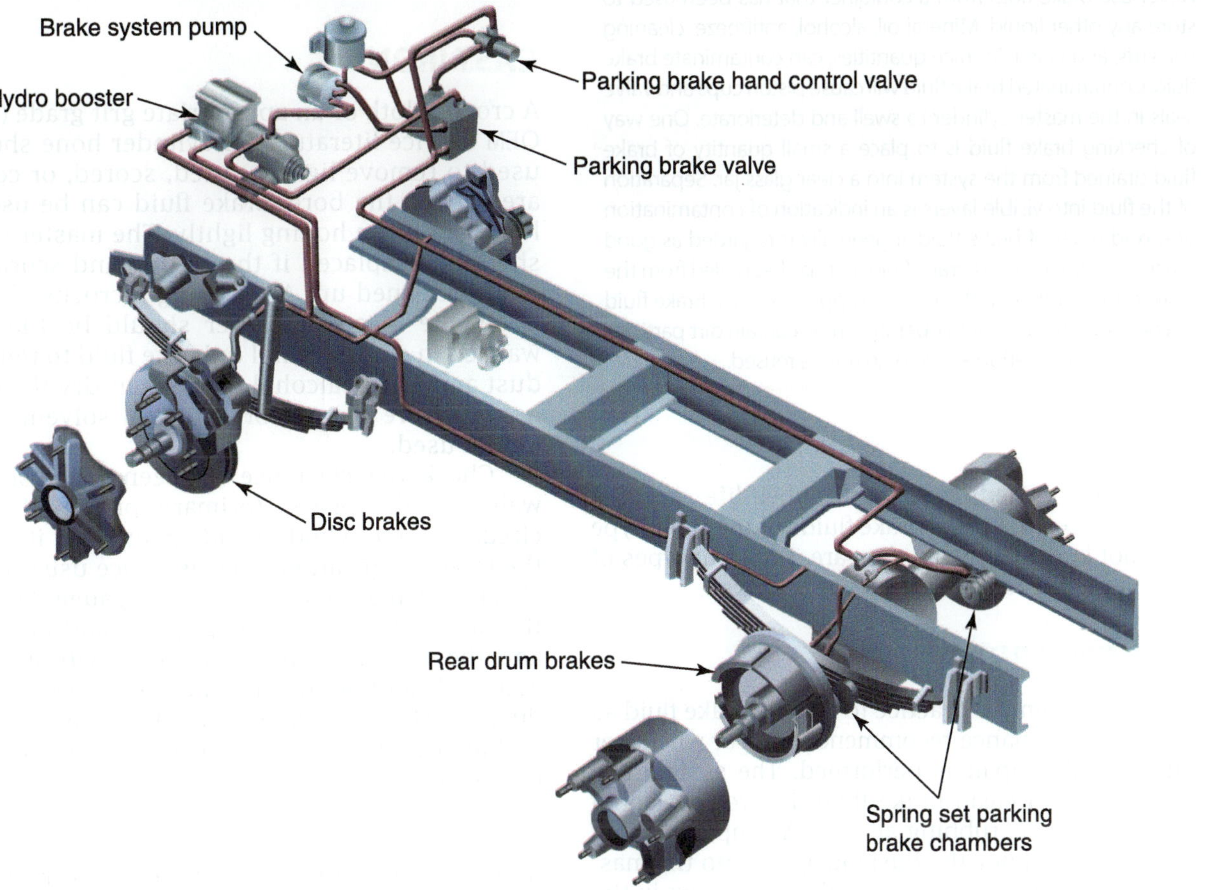

systems. **Figure 29–24** shows a schematic of a truck with hydraulic discs on the front axle and air-actuated S-cam brakes in the rear.

29.3 HYDRAULIC BRAKE SERVICE PROCEDURES

The service procedures recommended by the manufacturers of hydraulic brake systems should be observed. There will be differences among original equipment manufacturer (OEM) system maintenance and repair practices. For this reason, it is always good practice to consult the manufacturer service literature when performing repairs. This section describes some general maintenance and repairs commonly performed on hydraulic brake systems.

HEAVY-DUTY BRAKE FLUID

It is important to use the manufacturer's specified brake fluid to ensure proper system operation. Currently used DOT brake fluids and their performance properties were listed earlier in Table 29–1. An approved heavy-duty brake fluid is designed to retain a proper consistency at all operating temperatures. It is also designed not to damage any of the internal rubber components and to protect the metal parts of the brake system against corrosion and failure.

Some hydraulic brake systems use a non-petroleum-based hydraulic brake fluid such as SAE J1703 or SAE J17021. Other hydraulic systems use petroleum-based brake fluids (mineral oil). It is important to ensure that the correct brake fluid is used in the vehicle brake system and incompatible fluids are not mixed. Use of the wrong brake fluid can damage the cup seals of the wheel cylinders and pistons and can result in brake failure.

Avoid leaving the top off a brake fluid container for any longer than is necessary to pour out fluid and seal the containers tightly. Because of the hygroscopic characteristics of some brake fluids, they can rapidly absorb moisture from air and become saturated within an hour. Water content will have the effect of reducing the fluid boil point to that of water. All brake

CAUTION

Never use brake fluid from a container that has been used to store any other liquid. Mineral oil, alcohol, antifreeze, cleaning solvents, and water, in trace quantities, can contaminate brake fluid. Contaminated brake fluid will cause piston cups and valve seals in the master cylinder to swell and deteriorate. One way of checking brake fluid is to place a small quantity of brake fluid drained from the system into a clear glass jar. Separation of the fluid into visible layers is an indication of contamination or mixed types of brake fluid. It generally is regarded as good practice to discard used brake fluid that has been bled from the system. Contaminated fluid usually appears darker. Brake fluid drained from the bleeding operation may contain dirt particles or other contamination and should not be reused.

fluids, once opened, have a short shelf-life, and it is good practice not to mix brake fluids of the same type if it is not known how old they are. Different types of brake fluid, of course, must never be mixed.

CHANGING BRAKE FLUID

It is a recommended practice to change brake fluid as per OEM maintenance recommendations or whenever a major brake repair is performed. The system can be flushed with clean brake fluid, isopropyl alcohol usually known as rubbing alcohol. A simple flushing technique is to pour the flushing agent into the master cylinder reservoir and open all bleed screws in the system. The brake pedal is then pumped to force the flushing agent through the system. Ensure that containers are located at each bleed screw nipple to catch the fluid. Replace all the rubber components (seals, cups, hoses, and so on) in the system before adding fresh brake fluid to a flushed, cleaned system.

SERVICING A MASTER CYLINDER

A master cylinder should be removed from a truck using the following procedure:

1. Disconnect the negative battery cable.
2. Disconnect the pressure differential (brake light warning) switch if equipped.
3. Disconnect and cap the hydraulic lines at the master cylinder to prevent dirt from entering. (On some trucks using remote reservoirs, it may be necessary to disconnect the lines connecting the reservoirs with the master cylinder.)
4. Remove the master cylinder from the brake booster assembly.

Clean the master cylinder and any other parts to be reused in clean alcohol. Petroleum-based products should not be used to clean the components. Inspect all the disassembled components to determine whether they are in serviceable condition. Two types of master cylinder are used in truck applications.

CAST IRON

A crocus cloth or an appropriate grit grade (consult OEM service literature) of cylinder hone should be used to remove lightly pitted, scored, or corroded areas from the bore. Brake fluid can be used as a lubricant when honing lightly. The master cylinder should be replaced if the pitting and scoring cannot be cleaned up. After using a crocus cloth or a hone, the master cylinder should be thoroughly washed in clean alcohol or brake fluid to remove all dust and grit. If alcohol is used, air-dry the components before reassembly. Mineral solvents should not be used.

Check the clearance between the inside bore wall and the piston (primary piston of a dual-circuit master cylinder) and ensure that it is within the OEM specification. The clearance usually can be checked using a feeler/thickness gauge. Measuring the bore with a telescoping gauge and micrometer and the piston's outside diameter with a micrometer and subtracting the difference also will produce the clearance dimension. If the clearance is not within specification, the master cylinder should be replaced.

Aluminum Bore

Aluminum is more susceptible to corrosion and physical damage than cast iron. Inspect the bore for scoring, corrosion, physical damage, taper, and pitting. If the bore is scored or badly pitted and corroded, the assembly should be replaced. An aluminum master cylinder bore can never be cleaned with an abrasive material because this will remove the anodized surface. Anodizing provides a measure of wear and corrosion protection. If this is removed, the result will be rapid failure. Clean the bore with a clean piece of cloth wrapped around a wooden dowel and wash thoroughly with alcohol. Some bore discoloration and staining is normal and does not indicate corrosion.

Overhauling a Master Cylinder

Rebuilding aluminum master cylinders may require resleeving using the usual machine shop processes. If it is determined that a master cylinder requires reconditioning without resleeving, the following general procedure can be used:

1. Remove the snapring, thrust washer, primary piston, rubber sleeve, and spring.
2. Remove the connection fitting for the rear brake circuit.
3. Remove the sealing ring, valve seat, valve, valve spring, and disc.

4. Remove the stop bolt.
5. Remove the secondary piston with the rubber sleeves and the spring.
6. Remove the connection fitting for the front brake circuit.
7. Remove the sealing ring, valve seat, valve, and valve spring.
8. Clean all parts in brake fluid or denatured alcohol.
9. Before reassembly, lubricate the cylinder bore, pistons, and sleeves with brake fluid.
10. When the master cylinder has been reassembled, reinstall and bleed the brakes.

SHOP TALK

Use a rebuild kit and assembly fluid or brake fluid to reassemble the master cylinder. Take special care to ensure that new rubber components are not damaged, crimped, or pinched during reassembly. Note how the components fit together when replacing the seals, especially the direction the rubber seals face on the pistons. Some rebuild kits provide a primary piston assembly. The assembly procedure simply reverses the disassembly sequence.

METERING VALVE SERVICE

On systems with front disc and rear drum brakes, inspect the metering valve whenever the brakes are serviced. A trace amount of fluid inside the protection boot does not indicate a defective metering valve, but evidence of a larger amount of fluid indicates wear and the need to replace it. Ensure that the brake lines are correctly connected to the metering valve ports when replacing the valve. Crossed lines will result in the metering valve proportioning oppositely, producing front brake drag.

SHOP TALK

A pressure bleeder ball and OEM software is usually required to bleed ABS brakes (shown later in Figure 29–34). Use regulated shop air pressure, at a MAXIMUM of 15 psi (103 kPa), to the bleeder ball. When a pressure bleeder is used to bleed a system that includes a metering valve, the valve stem must either be pushed in or pulled out to open the valve. Manual bleeding using the brake pedal develops sufficient pressure to overcome the metering valve, and the stem does not have to be held open. Always follow the OEM recommended procedure for bleeding ABS brakes.

PRESSURE DIFFERENTIAL VALVES

The pressure differential valve should recenter automatically on the first application of the brakes after repair work. However, some pressure differential valves may require manual resetting. After repairs have been completed, open a bleeder screw in a portion of the hydraulic circuit that was not worked on. Turn on the ignition to illuminate the dash warning light and then slowly depress the brake pedal. This should bias the valve piston, causing the brake warning light to go out.

If too much pressure is applied during this procedure, the valve piston will go over center, causing the light to flicker out and then come back on. If this occurs, reverse piston movement starts by closing the open bleed screw and opening a bleed screw in the other hydraulic circuit. Again, slowly depress the pedal to center the valve piston.

PROPORTIONING VALVE

The proportioning valve also should be inspected whenever the brakes are serviced. To check valve operation, install a pair of hydraulic gauges upstream and downstream of the proportioning valve and ensure that the rear brake pressure is proportioned to specification. If this is not the case or the valve is leaking, it must be replaced. Make sure that the valve port marked R (*rear*) is connected to the rear brake lines.

WHEEL CYLINDERS

Wheel cylinders should be at least externally inspected during any routine brake job. Any evidence of leakage should be a reason to recondition the unit.

Removing and Installing a Wheel Cylinder

A wheel cylinder should be removed, observing the following general procedure:

1. Remove the wheel, drum, and brake shoes.
2. Remove the wheel cylinder mechanical linkages.
3. Disconnect the brake line from the wheel cylinder.
4. Remove the brake cylinder retaining bolts and lock washers and separate the cylinder from the backing plate.
5. Install the wheel cylinder by reversing the removal sequence.

Disassembly and Reassembly of a Wheel Cylinder

Wheel cylinders are disassembled by removing the dust boot and then the piston or pistons and cup or cups, depending on the style of cylinder. Next, remove the piston spring. Ensure that the bore is inspected and measured to specification. Cast-iron bores may be resurfaced using a hone, but the piston-to-bore clearance must be measured to specification using the procedure described earlier for master cylinders. Aluminum wheel cylinders may be resleeved with a steel insert using the appropriate machine shop processes.

To reassemble the wheel cylinder, lubricate the cylinder bore, pistons, and cups in brake fluid and observe the following general procedure:

1. Install the spring in the wheel cylinder bore.
2. Install a new cup in each end of the cylinder bore with the cup lips facing toward the spring.
3. Insert a piston in either end with the flat side contacting the cup.
4. Install a new rubber boot on each end of the cylinder. Ensure that the boot lip is seated in its groove on the end of the cylinder.
5. Install the bleeder screw in the cylinder body.
6. Mount the wheel cylinder to the backing plate and torque the fasteners.
7. Connect the hydraulic fluid inlet line.
8. Bleed the brakes using the procedure described later in this chapter.

CAUTION

Leaking at the wheel cylinder seals after a brake job may occur resulting from the repositioning of cups onto dirt or sludge.

Removing Brake Drums and Shoes

The procedure for disassembling hydraulic foundation brakes does vary by vehicle and weight class but generally the following procedure may be used:

1. Begin by backing off the brakes using a brake spoon.
2. After pulling the brake drum away from the brake assembly, disconnect the retraction spring(s) and remove the brake shoes from the backing plate.

To reassemble the wheel-end assembly, reverse the disassembly procedure. Use of a pair of **brake spring pliers** will assist in properly connecting the retraction spring(s).

SERVICING BRAKE DRUMS

Whenever a brake drum is removed to inspect the linings or wheel cylinder, make a habit of evaluating the other components in the braking system. Before inspecting drum brakes, release the parking brake. Evidence of grease or oil at the center of the brake assembly indicates that the axle seal needs replacing. If brake fluid drips out of the wheel cylinders when the rubber dust boots are removed, the wheel cylinder needs reconditioning or replacing. Contaminated linings should be replaced. Check for cracked or worn hoses and bent or corroded tubes.

Drum Removal and Inspection

Some brake drums can be replaced without removing the hubs from the vehicle. This is possible on models in which the brake drum is mounted outboard from the hub. The drum is fastened to the hub by the wheel studs and nuts. The drum can be removed after removing the tire and wheel assembly. On other models, the hub and drum must be removed from the vehicle.

Drum replacement requires the removal of nuts and bolts securing the hub and drum. The hub and drum can then be separated. Before installing a replacement drum, wash the drum thoroughly with denatured alcohol to remove grease, oil, and other residue. To inspect a drum, proceed as follows:

1. Visually check the drum for cracks, scoring, pitting, or grooves.
2. Check the edge of the drum for chipping and fractures.
3. Drums that are blue, glazed, or heat-checked have been overheated and should not be reused. Severe heat-checking in drums will cause rapid lining wear and, even when machined, the drums will fail prematurely.

CAUTION

Severe overheating is caused by drums that are machined too thin, improper lining-to-drum contact, incorrect lining friction ratings, or vehicle overloading. A probable driver complaint would be brake fade.

4. If the brake linings appear to be unevenly worn, check the drum for a barrel-shaped or taper condition (**Figure 29–25**). Consult the OEM specifications for the maximum drum taper specifications and the procedure required to measure brake shoes. Also, check the shoe heel-to-toe dimension to ensure that the arc is not spread.

Tech Tip

Servo brake replacement shoes usually have different lining thicknesses. In most cases, the shoe with the thicker lining should be installed in the secondary shoe position facing the rear of the vehicle.

5. Check the brake drum inside diameter (ID) to make sure it is within specification. The maximum or "discard" spec is often cast into the drum.
6. Check for a drum out-of-round condition using an inside micrometer or drum gauge. Measure the drum at inboard and outboard locations on the machined surface at four locations. Compare to the OEM specifications. An out-of-round drum condition produces a pulsating brake pedal condition.
7. Clean the inside of the drum with a water-dampened cloth. New drums may be corrosion protected with a light layer of grease. This must be

FIGURE 29–25 Drum wear conditions

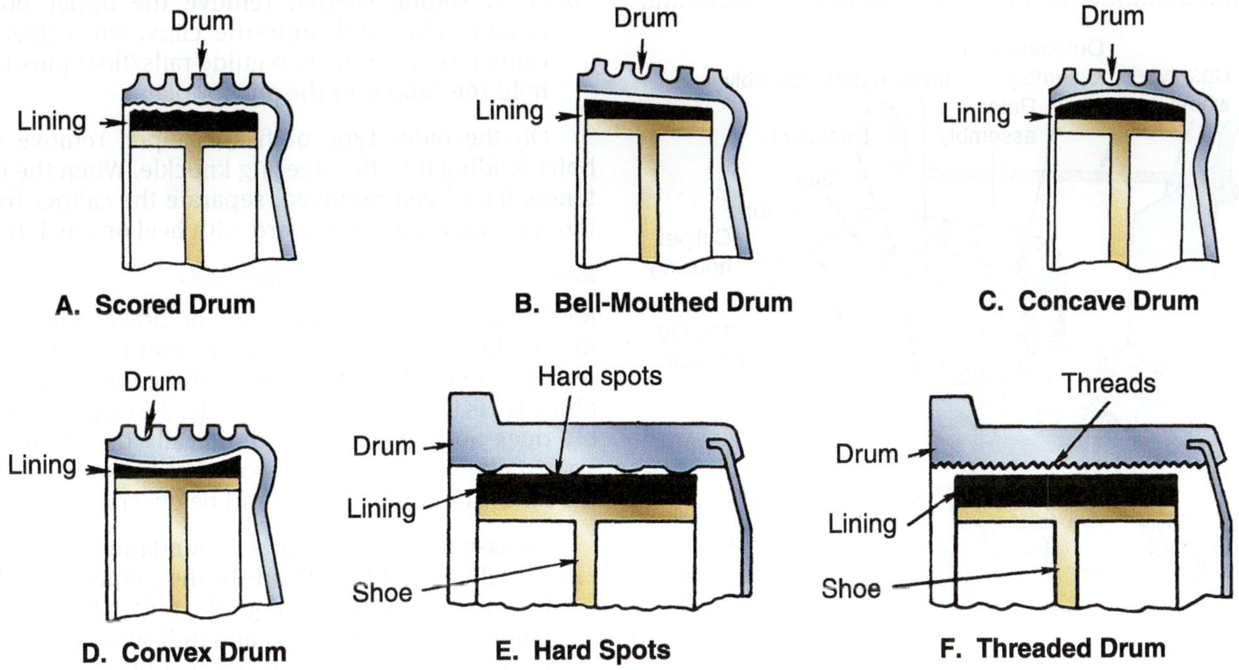

removed with solvent from the friction surface before installing the drum.

SHOP TALK

Never paint the outside of the drum. Paint will act as an insulator and slow heat dissipation. The heat absorbed by the drum during braking must be dissipated to the atmosphere around the drum. Overheated brakes result in brake fade!

Brake drums are installed by reversing the order of removal. If reusing brake drums, it is good practice to install them in their original location. Torque the fasteners to the specification recommended in the service manual.

DISC BRAKE REMOVAL

First, remove the wheel assembly. Next, the caliper assembly must be removed from the brake assembly (consult **Figure 29–26**). This procedure will depend on the caliper design, but the general procedure is accomplished as follows:

1. Take care when removing the caliper assembly not to damage the bleed screw fitting.
2. Mark the location of all calipers when removing them, for instance, as right side and left side, so they can be positioned correctly during installation. **Figure 29–27** shows the disc brake assembly components.
3. On a sliding or floating caliper, install a C-clamp on the caliper with the solid end of the clamp on the caliper housing and the worm end on the outboard brake pad bracket. Crack the bleed screw on the caliper so fluid is not backed up into the

FIGURE 29–26 Hydraulic disc brake assembly

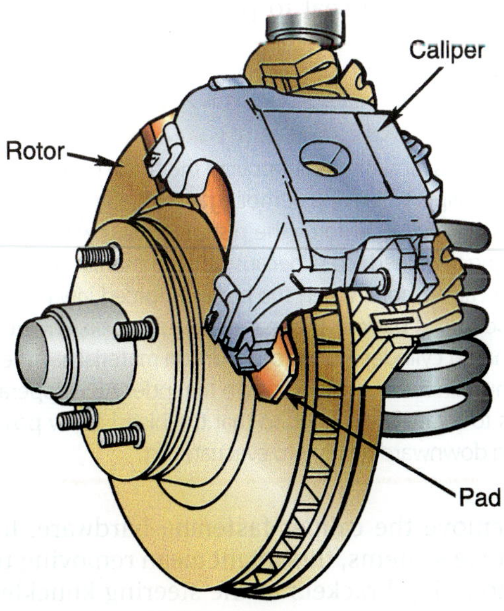

FIGURE 29–27 Cutaway view of a hydraulic disc brake assembly

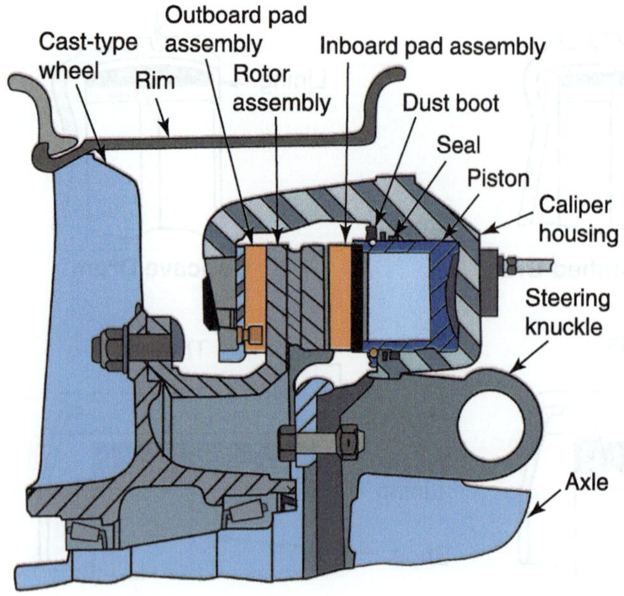

master cylinder. Tighten the clamp until it bottoms the piston in the caliper bore, collect the brake fluid in a container, and then remove the clamp. Bottoming the piston will create enough clearance for the brake pads to slide over the ridge of rust that accumulates on the edge of the rotor.

4. Disconnect the brake hose from the caliper and cap the brake line and nipple to prevent dirt from entering the hose or caliper assembly. If the object of the service procedure is only to replace the brake pads, the brake hose may not have to be disconnected. If the brake hose is disconnected, cap it with a seal to prevent contaminants from entering.

CAUTION

Many OEMs recommend removing the brake hose or supporting the caliper when disassembling a disc brake assembly, especially on ABS. Never force the piston inboard without at least opening the bleed screw because contaminants tend to collect in the caliper bore. When the piston is forced inward, this dirt can be forced back to the master cylinder. Also, any sludge in the master cylinder is disturbed. Foreign material half the width of a human hair has been known to render ABS inoperative. It helps to rotate the caliper, so that the bleed screw position is facing downward helping to evacuate dirt.

5. Remove the caliper fastening hardware. In front brake systems, this might mean removing the two mounting brackets to the steering knuckle bolts. Support the caliper when removing the second bolt to prevent the caliper from falling.
6. On a sliding caliper, remove the upper bolts, retainer clip, and antirattle clips. On a floating caliper, remove the two guide rails/float pins that hold the caliper to the anchor plate.

On the older type of fixed caliper, remove the bolts holding it to the steering knuckle. When the fasteners have been removed, separate the caliper from the rotor by prying it upward with heel or pinch bars.

Removing and Installing Brake Pads

Removing and installing a set of brake pads to a disc brake assembly first requires that the caliper be removed using the procedure outlined earlier. If the objective is to replace the pads only, the caliper assembly does not have to be hydraulically disconnected. To install a new set of pads to a disc brake assembly, the following procedure should be observed:

1. Disassemble the disc brake foundation.
2. Visually inspect the inside of the caliper for leaks. Clean the caliper and guide rails. Inspect the guide rails for alignment and indications of corrosion or scoring. Remove any rust buildup using a fine file. Replace the rails if there is any doubt as to their condition.
3. Place a block of wood or metal bar across the face of the caliper pistons. Force the pistons back into the caliper bores. This provides the clearance required for the caliper assembly to slide off the rotor.
4. Measure the friction pad dimension. This should be a minimum of 1⁄32 -inch (0.8 mm) above the head of riveted pads and 2⁄32 -inch (1.6 mm) total thickness on bonded pads. It is not reommended that any pad be returned to service when it is anywhere close to its minimum specification.

CAUTION

In cases where only the brake pads are to be replaced the caliper does not need to be hydraulically disconnected. Make sure that when the caliper assembly is separated, it is not allowed to hang by the brake hoses that can internally damage the hose.

5. Lubricate the guide rails/float pins at both ends of the caliper using a silicone-based or other approved grease.
6. Install the new pads in the caliper. In some cases, the pads are installed through slots in the lining abutment rails. Install the inboard pad first with the metal side against the pistons. Install the outboard pad with the lining side facing the inboard pad.
7. Slide the caliper assembly over the rotor and onto the guide or support rails. Install the support key for the inboard side and lock it in place with the retaining bolt or screw.

8. Install the lining retaining spring. Typically, the ends are seated in grooves at the caliper rails and the sides extend over the top of the lining backing plates. **Figure 29–28** shows a heavy-duty dual-piston caliper assembly in an exploded view.

SHOP TALK

When axial or radial runout specifications are provided by OEMs, they are expressed in total indicated runout (TIR). TIR is calculated by adding the most positive to the most negative reading through one complete revolution when using a dial indicator. For instance, if, after zeroing the dial indicator, a maximum positive reading of 0.001 inch (0.0254 mm) and maximum negative reading of 0.002 inch (0.0508 mm) occurs through one rotation, perform the following calculation to obtain the TIR:

-0.001 inch $+ 0.002$ inch $= 0.003$ inch (0.0762 mm) TIR

INSPECTING ROTORS

To inspect the rotors, it may or may not be necessary to remove the hub and rotor assembly. This will depend on the configuration used by the OEM. It also may be necessary to separate the rotor from the hub. The following checks are required:

1. Visually inspect the rotor surface for scores, cracks, and heat-checks. This check is critical and will determine whether those steps that follow are required. Reject any rotor with evidence of heat-checking.
2. Next, check the rotor or hub and rotor assembly for lateral runout and thickness variation. To check lateral runout:
   - Mount a dial indicator on the steering arm or anchor plate with the indicator plunger contacting the rotor 1 inch (25 mm) from the edge of the rotor.
   - Adjust the wheel bearings so they are just loose enough for the wheel to turn; that is, set them with a little preload.
   - Check the lateral runout on both sides of the rotor. Zero the indicator and measure the TIR.
   - The lateral runout TIR is always a tight dimension, seldom exceeding 0.0015 inch (0.0381 mm).
3. If the lateral runout exceeds the specification, the rotor will have to be either machined or replaced.
4. Thickness variation also should be measured at 12 equidistant points with a micrometer at about 1 inch (25 mm) from the edge of the rotor.
5. If the thickness measurements vary by more than the specified maximum, usually around 0.002 inch (0.0512 mm), the rotor should be machined or replaced. This measurement can be made using two dial indicators or a micrometer. If using dial indicators, position them opposite each other and zero the dials. Rotate the rotor and observe the indicator readings for the proper tolerance. Some light scoring or wear is acceptable.

FIGURE 29–28 Components of a hydraulically actuated, two-piston caliper assembly

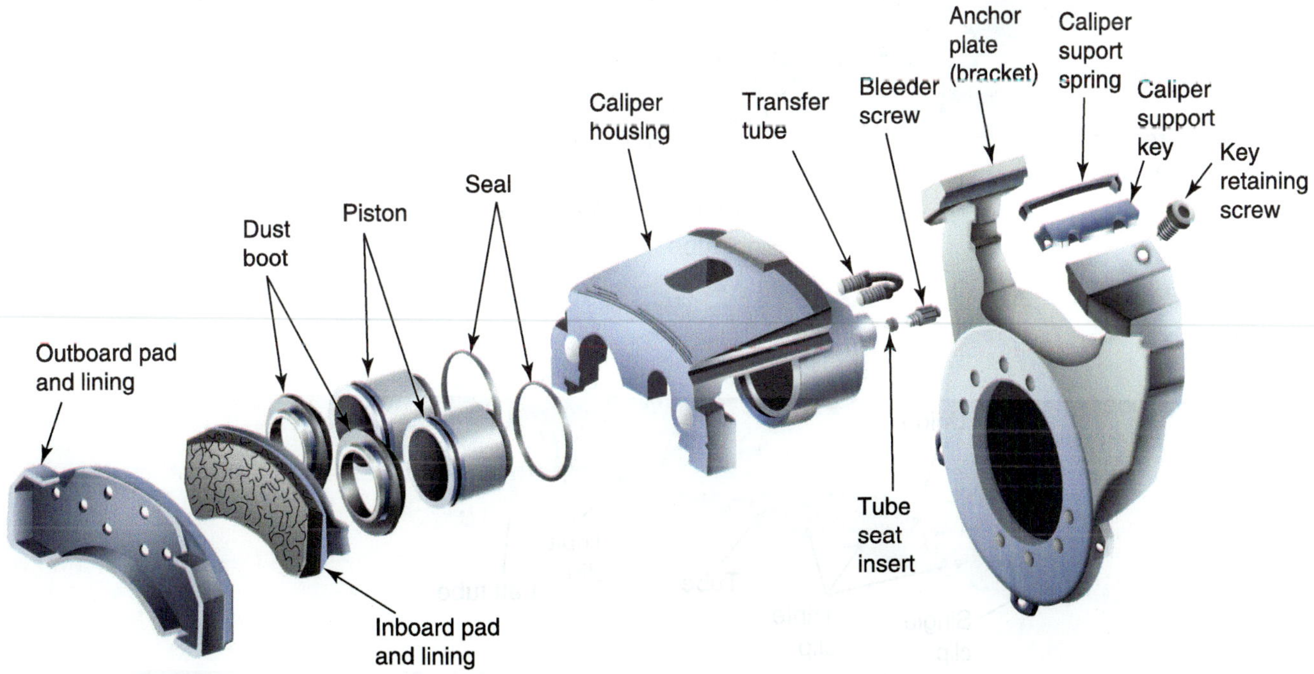

If cracks are evident, the hub and rotor assembly should be replaced.

6. With the wheel bearings adjusted to zero end-play, check the radial runout. Rotor radial runout should not exceed the OEM TIR specification, usually around 0.030 inch (0.762 mm). Set a dial indicator on the outer edge of the rotor and turn through 360 degrees.
7. Next, inspect the wheel bearings and adjust them using the OEM-specified procedure.
8. Wear ridges on rotors can cause temporary improper lining contact if the ridges are not removed, so it makes sense to skim rotors at each brake job.

SHOP TALK

Excessive rotor runout or wobble increases pedal travel because the condition opens up the caliper piston and can cause pedal pulsation and chatter when the brakes are applied.

Most rotors fail because of excessive heat. Dishing and warpage are caused by extreme heat. These conditions almost always require the replacement of the rotor. Excessive rotor radial runout can cause noise from caliper housing-to-rotor contact.

Brake rotors have a minimum allowable thickness dimension cast on an unmachined surface of the rotor. This is the replacement thickness. It is illegal to machine a rotor that will not meet the minimum thickness specifications after refinishing. In other words, if a rotor will not "true up" during skimming before the minimum allowable thickness is obtained, it must be replaced.

BRAKE LINES AND HOSES

Brake lines and hoses must be repaired using the OEM-specified components. **Figure 29–29** shows the brake line system as it would appear on a typical truck. Note that the fasteners that hold the lines are clamped to the vehicle frame.

CAUTION

Never clamp brake hoses off with Vise-Grip™ or "locking pliers." The result will be internal damage of the hose and premature failure.

FIGURE 29–29 Typical truck hydraulic brake line circuit routing

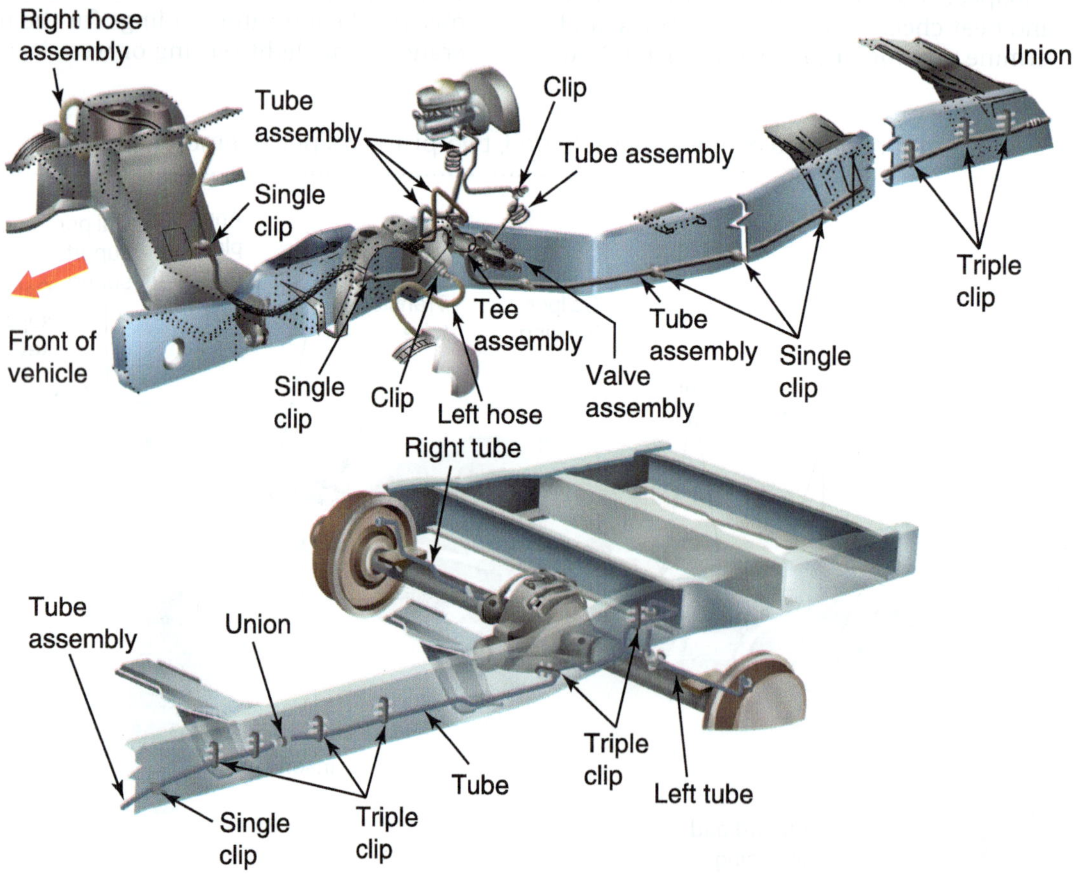

When brake hoses have to be disconnected, ensure that the lock clips are removed and line wrenches used to separate the hose nuts, as shown in **Figure 29–30**. Steel brake lines are usually double flare or ISO flare, as shown in **Figure 29–31**. Note the different seats and the fact that they cannot be interchanged. **Figure 29–32** shows how the double flare is made using a double-flaring anvil and cone.

BLEEDING MASTER CYLINDERS

Most OEMs recommend that a master cylinder be bench bled prior to installation. In most cases a new/recon master cylinder comes with a priming kit, which contains fittings for the brake line holes and a set of plastic tubes. The OEM procedure should be referenced but the following ten steps are typical:

1. If the master cylinder is in the vehicle, remove the lid from the brake fluid reservoir and use a large syringe to pull all of the fluid from the reservoir. Remove lines, the unit from the vehicle making sure there is a drip pan underneath to catch brake fluid.
2. Mount the master cylinder in a vise so that the jaws clamp the mounting flange. Make sure the master cylinder is level and the brake hose attachments face upward. Never tighten the vise jaws on the master cylinder body because it is easily damaged.
3. Attach the hoses from the fluid reservoir to the fittings on the master cylinder and install the brake light switch on the new master cylinder.
4. Connect the plastic fittings supplied in the bench bleeding kit into the brake line ports. Use something to hold the brake fluid reservoir in place above the master cylinder.
5. Fill the fluid reservoir with new brake fluid. Leave the cap off the reservoir.
6. Fluid should begin to drip from the two plastic fittings: depending on the specific unit this may take around 15 minutes. Avoid pumping the piston while waiting because this will result in mixing air with the brake fluid inside the master cylinder.
7. Once fluid is dripping from both plastic fittings, attach a section of clear plastic tubing about 18 inches (0.5 m) long to each fitting and run the other end of each tube into a jar about one-third full of brake fluid. The jar should be positioned so that the ends of the tubes are higher than the master cylinder.

FIGURE 29–31 Two common types of line flares and their seats

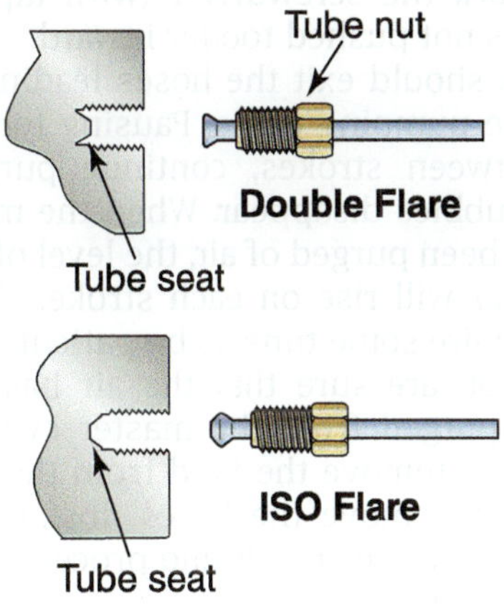

FIGURE 29–32 (A) Anvil folds tubing; and (B) cone performs second fold and doubles seat thickness.

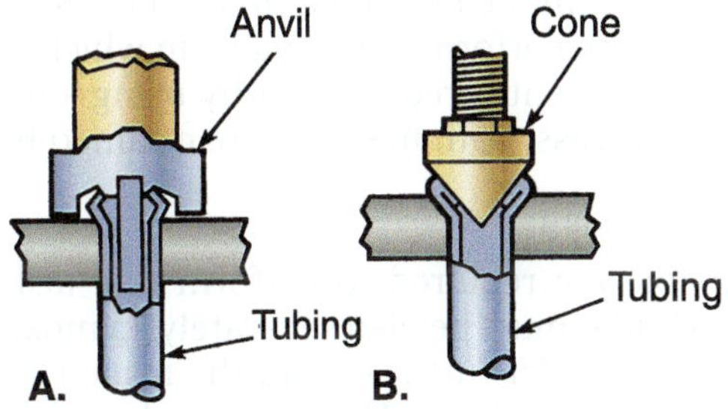

FIGURE 29–30 Brake hose replacement. (A) Remove the lock clip from the chassis bracket; and (B) disconnect the female end first.

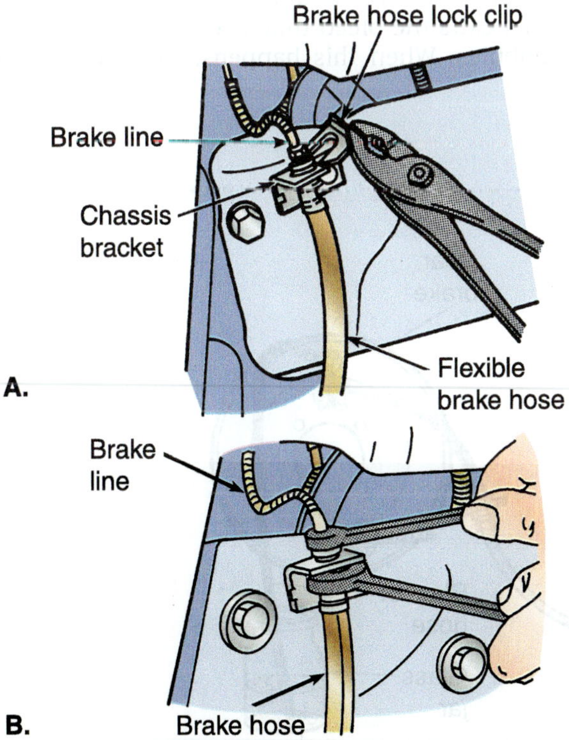

8. With a large screwdriver, gently force the master cylinder piston into its bore, typically about 1 inch (25 mm) but reference the OEM specification. Mark the screwdriver (with tape) to make sure it is not pushed too far inward.
9. Bubbles should exit the hoses leading to the jar after the pumping pulse. Pausing for short periods between strokes, continue pumping until these bubbles disappear. When the master cylinder has been purged of air, the level of brake fluid in the jar will rise on each stroke. This process may require some time so be patient.
10. When you are sure that the air has been completely purged from the master cylinder, use a syringe to remove the fluid from the master cylinder and remove the hoses. Install the master cylinder into the vehicle and proceed to bleed the brake system.

BLEEDING BRAKES

Bleeding of the brakes is an essential component of servicing hydraulic brakes. The process involves purging air from the hydraulic circuit. Bleeding of the brakes may be performed manually, in which case no special equipment is required, or by using a pressure bleeding process. Both methods are outlined here.

Manual Bleeding

Two people are required to perform this task. Each wheel cylinder must be bled separately. In most cases, the wheel cylinder furthest from the master cylinder is bled first but the OEM-recommended procedure should be referenced. In addition, some current vehicles may have to be connected to the chassis data bus to properly bleed the brakes.

Assuming that the procedure begins with the wheel cylinder furthest from the master cylinder, this means that the right rear-wheel cylinder would be first and then the left rear. After bleeding the rear brakes, the right front brake and then the left front should follow. Ensure that the master cylinder reservoir is topped with brake fluid during the bleeding procedure. The following procedure is used but you should note that if the master cylinder has been bench bled, the first four steps can be eliminated:

1. Loosen the master cylinder line nut to the rear brake circuit one complete turn.
2. Push the brake pedal down slowly by hand to the floor of the cab. This will force air trapped in the master cylinder to escape at the fitting.
3. Hold the pedal down and tighten the fitting. Release the brake pedal.
4. Repeat this procedure until bubble-free fluid exits at the fitting, and then tighten the line fitting.
5. To bleed a wheel cylinder, loosen the bleed fitting on the wheel cylinder or the caliper with a 6-point box wrench. Attach a clear plastic or rubber drain hose to the bleed fitting, ensuring that the end of the tube fits snugly around the fitting.
6. The person at the wheel should submerge the free end of the tube in a container partially filled with brake fluid, as shown in **Figure 29–33**. Loosen the bleed fitting three-quarters of a turn.
7. The person in the cab should pump then push the brake pedal all the way down and hold.
8. The person at the wheel should then close the bleed fitting and the person in the cab can then return the pedal to the fully released position.
9. Steps 6, 7, and 8 should be repeated until the fluid that exits the bleed tube is completely free of air bubbles. When this happens, the person at the

FIGURE 29–33 Manual brake bleeding technique

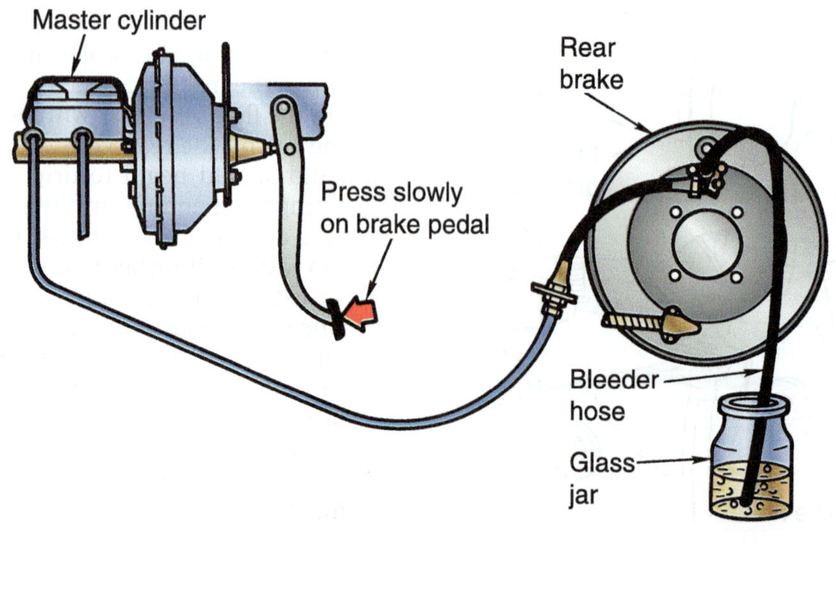

wheel can close the bleeder fitting and remove the tube.

10. Repeat this procedure at the wheel cylinder or caliper on all the calipers and wheel cylinders beginning at the one furthest from the master cylinder and working toward it.
11. Keep the master cylinder reservoir topped up throughout the bleeding process.
12. When the bleeding operation is complete, top up the master cylinder with brake fluid to the appropriate level in the reservoir.

Pressure Bleeding Brakes

Pressure bleeding equipment must be of the diaphragm type to prevent air, moisture, oil, and other contaminants from entering the hydraulic system. The adapter must limit pressure to the reservoir to 35 psi (240 kPa). When pressure bleeding current hydraulic ABS, OEM software (such as Meritor ToolBoxTM) is usually required to actuate the modulator valves to enable the removal of all traces of air from the system. This requires an EST connection to chassis data bus.

1. Clean all dirt from the top of the master cylinder and remove the reservoir caps.
2. Install an appropriate master cylinder bleeder adapter. A typical pressure bleeder is shown in **Figure 29–34**. Connect the hose from the bleeder equipment to the bleeder adapter and open the release valve on the bleeder equipment.
3. Using the correct size box wrench over the bleeder screw, attach a clear plastic or rubber hose over the screw nipple. Insert the other end of the hose in a glass jar containing enough fluid to cover the end of the hose.

FIGURE 29–34 Pressure bleeder

4. Bleed the wheel cylinder or caliper furthest from the master cylinder first, and then the next nearest, and so on until all the valves have been bled.
5. Open the bleeder screw and observe the fluid flow at the end of the hose.
6. Close the bleeder screw as soon as the bubbles stop exiting the hose and the fluid flows in a solid stream.
7. Remove the wrench and hose from the bleeder screw.
8. Fill the master cylinder to the correct level in the reservoir.

Bleeding an Air-over-Hydraulic System

To bleed a typical air-over-hydraulic system, follow this procedure:

1. Adjust the air pressure in the air reservoir to 28 to 43 psi (193 to 296 kPa). If the air pressure drops below this range during the bleeding operation, start the engine.
2. When the air pressure is 28 to 43 psi (193 to 296 kPa), stop the engine.
3. Bleed the air from each air bleeder screw in the following order:
    - Front brakes and air booster hydraulic cylinder plug
    - Right front-wheel air bleeder screw
    - Left front-wheel air bleeder screw
    - Rear brakes and air booster hydraulic cylinder plug
    - Right rear wheel air bleeder screw
    - Left rear wheel air bleeder screw
4. Fill the brake reservoir with fluid. Refill frequently during bleeding so no air enters the brake lines.
5. Attach a vinyl tube to the bleeder screw. Insert the other end of the tube into a bottle partially filled with brake fluid.
6. Depress the brake pedal and at the same time turn the bleeder screw to the left.
7. Tighten the bleeder screw when the pedal reaches the full stroke. Release the pedal.
8. Repeat until no bubbles are observed.

29.4 HYDRAULIC ABS

Full treatment of ABS on air brake systems is provided in Chapter 30. Brief coverage is provided here, focusing on the differences in the system components when ABS is used on a hydraulic brake system. The objective of any ABS is to provide the maximum amount of braking force to each wheel without causing it to lock up. This means that the brake application pressures are modulated (regulated) or pulsed at high speed. The pulsing that results in hydraulic ABS may be felt at the brake pedal.

ABS CONFIGURATIONS

When hydraulic ABS is used in truck applications, it is almost always on light-duty, two-axle units, so we will examine a typical two-axle application. ABS can be configured in three ways:

- **Single-Channel Rear Wheel.** Only the rear wheels are controlled by ABS. Each rear axle wheel is equipped with a sensor, but hydraulic pressure is modulated to both rear axle wheels.
- **Three-Channel.** All four wheels are equipped with wheel speed sensors but each front wheel is modulated individually (two channels) and the rear wheels are modulated as a pair (one channel).
- **Four-Channel.** All four wheels are equipped with wheel speed sensors and each wheel on the vehicle is modulated individually. **Figure 29–35** shows a four-channel (4S/4M) system used on a school bus application.

Today's ABS are self-diagnostic and are designed to revert to normal hydraulic brake operation when either a system or subcircuit failure is detected. The vehicle operator is alerted when a malfunction occurs by a warning light.

Integrated and Nonintegrated Systems

An integrated ABS incorporates all the critical system hydraulic actuator components in a single unit at the master cylinder assembly. Integrated systems combine a booster/master cylinder with a valve block. A pump and motor are used to provide pressurized brake fluid to an accumulator. The pump is typically electrically driven and used to maintain pressure in the accumulator at a prescribed value, usually above 2,000 psi (138 bar).

In a nonintegrated system, the actuator components, hydraulic boost, master cylinder, and valve block are separate components. Nonintegrated ABS operation is mostly the same in principle as integrated systems. Because these systems vary from manufacturer to manufacturer, always consult the specific OEM service literature before attempting a repair.

ABS COMPONENTS

Hydraulic ABS is a computer-managed system and, as such, requires inputs, control modules, and an output circuit to effect the results of processing. Because ABS is designed to default to normal hydraulic braking in the event of a failure, all of the usual hydraulic proportioning built into a normal ABS is still used. This is known as system redundancy and is required by federal law in all ABS.

Wheel Speed Sensors

The function of a wheel sensor is to provide the ABS electronic control unit with wheel speed data. A pulse generator is used. A stationary permanent magnet is wound with coiled wire. A rotating toothed ring cuts through the magnetic field of the coil, generating an alternating current (AC) voltage. Both the voltage and frequency increase proportionally with rotational speed. The voltage/frequency is input to the ECU, and it correlates frequency to wheel speed. The wheel speed sensor is located in the wheel hub assembly. Wires connect it to the ECU. **Figure 29–36** shows a couple of different types of wheel speed sensors.

Other ABS Inputs

The ABS ECU receives inputs from other chassis switches and sensors, including the following:

- Ignition switch: turns system on and off.
- Pressure transducer: inputs primary and secondary hydraulic pressure values.

FIGURE 29–35 Location of ABS components on a 4S/4M small truck or school bus chassis

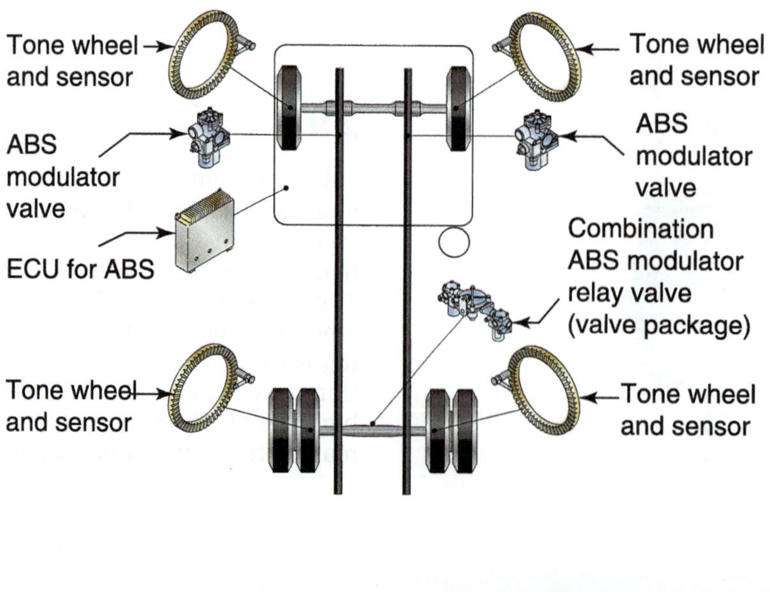

FIGURE 29–36 Typical ABS wheel sensors: (A) front and (B) rear.

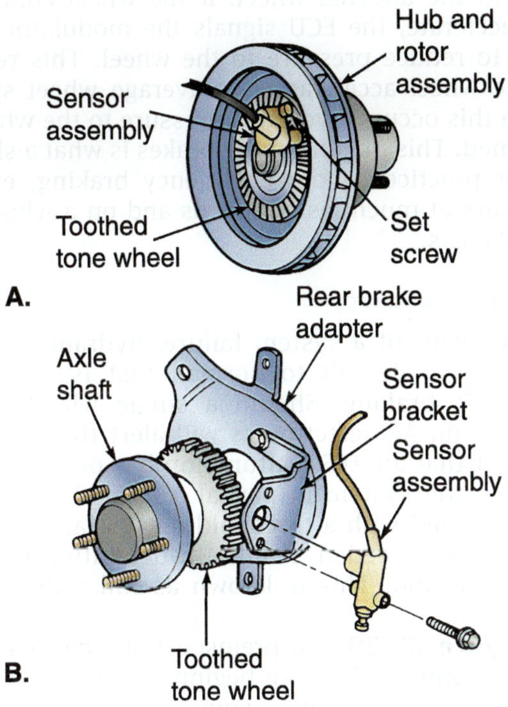

- Differential pressure switch: detects primary and secondary pressure differentials.
- Low fluid level switch: activates brake warning light and disables ABS function.
- Brake pedal sensor: a potentiometer-type sensor whose signal is used to compare pedal travel with braking force required.
- Lateral accelerometer: inputs cornering speed data. ABS can be managed with respect to load transfer effect.

Electronic Control Unit (ECU)

Sensor input is delivered to the ECU. Sensor input is received from the wheel speed sensors and the brake pedal sensor. The ECU processes the input information and determines the required outcomes. Essentially, the ECU is programmed to prevent wheel lockup during hard braking. It manages the braking pressure delivered to the wheels by switching solenoids in a control valve assembly. This energizing of solenoids modulates, that is, alternately relieves (when lockup is sensed) and resumes, the hydraulic braking force to the wheel. **Figure 29–37** shows the role of the control unit in ABS mode.

FIGURE 29–37 ABS operation: potential brake lock condition

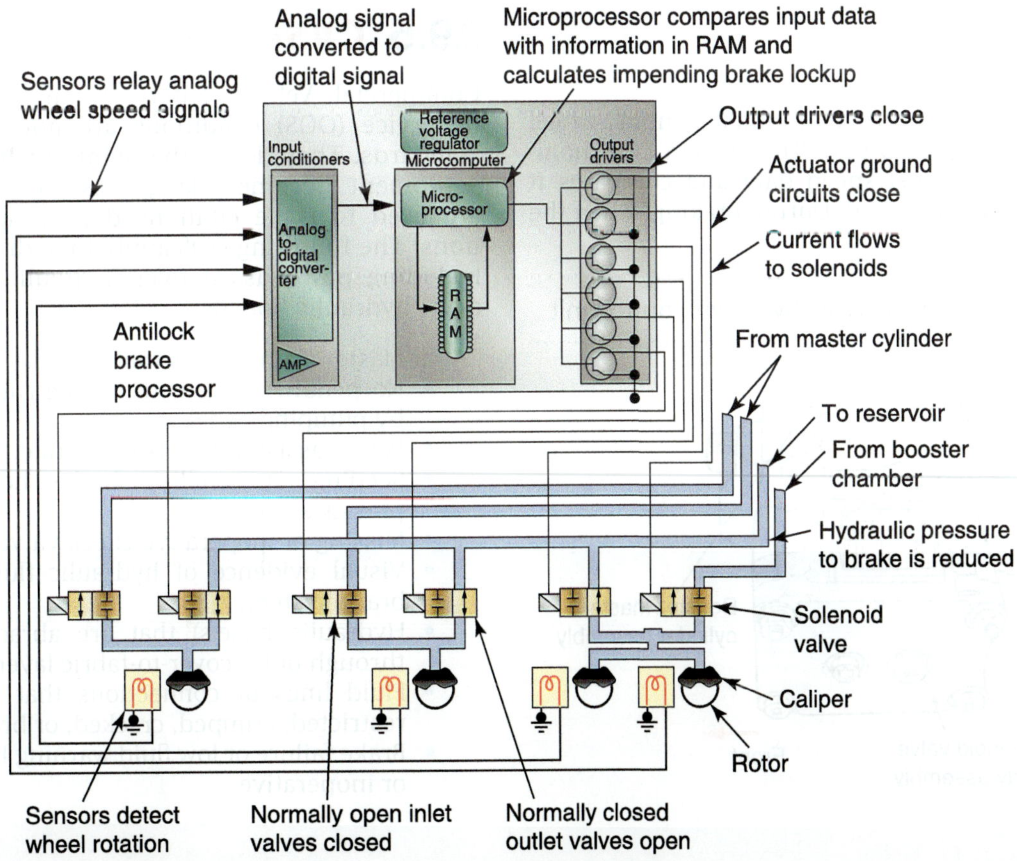

Modulator Assembly

This unit also is known as a hydraulic control or modulator/valve assembly. Its function is to effect the results of ECU processing into action. The unit consists of ECU-controlled solenoids and valves.

When the ECU receives the same wheel speed data for all wheels on the vehicle, braking is under normal mode; that is, no modulation of pressure to the wheels takes place. However, when the vehicle is being braked and one wheel is slowed proportionally more than the others, hydraulic pressure to the wheel is briefly relieved. The solenoids operate the modulator valves that act to relieve hydraulic pressure to the wheels. The solenoids are controlled or switched by the ECU.

The modulation speed on hydraulic ABS is faster than on air brake systems, mainly because of the incompressibility of hydraulic brake fluid. Modulation frequency can be switched up to 15 times per second on most truck hydraulic ABS. This is known as modulation frequency cycling. **Figure 29–38** shows a typical ABS hydraulic control unit.

Other ABS Outputs

- Brake warning lamp can be illuminated when a hydraulic or ABS malfunction occurs.
- Antilock warning light is illuminated when an ABS malfunction occurs.
- Diagnostic display is used to flash or blink out system codes to aid in troubleshooting the system.

ABS OPERATION

Wheel speed sensors continuously input wheel speed data to the ABS ECU. The system ECU monitors individual wheel speed data and compares it with average wheel speed. During braking, when the ECU senses that a high rate of wheel deceleration is occurring in a wheel, the modulator solenoid for the wheel first functions not to increase hydraulic pressure to the affected wheel. If the wheel continues to decelerate, the ECU signals the modulator solenoid to reduce pressure to the wheel. This results in the wheel accelerating to average wheel speed. When this occurs, hydraulic pressure to the wheel is resumed. This pulsing of the brakes is what a skilled driver practices under emergency braking, except it occurs at much faster speeds and on a wheel-by-wheel basis.

FIGURE 29–38 Typical ABS hydraulic control unit

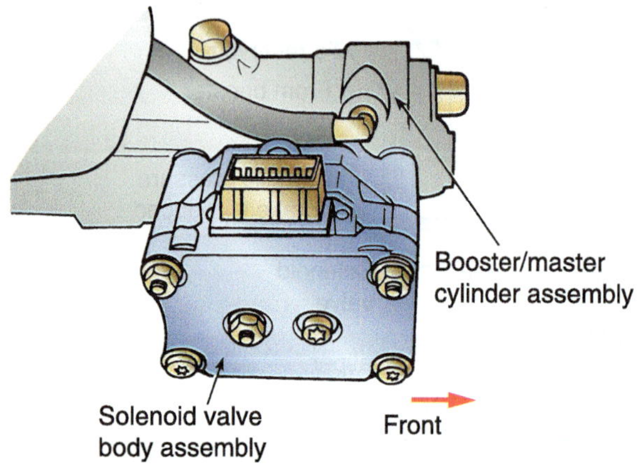

System Redundancy

In the event of a system failure, hydraulic ABS is required to default to normal, that is, non-ABS, hydraulic braking. Should a single wheel speed sensor fail, ABS electronics will alert the driver to the fact that an ABS malfunction has occurred and operate the system in partial ABS mode. Systems are designed with a threshold of ABS failure fields that, once exceeded, results in defaulting to non-ABS operation. This is known as full redundancy operation.

Figure 29–39 is a preinspection checklist that can be completed before beginning a brake service procedure on a vehicle equipped with hydraulic brakes. This can be useful for technicians who are more familiar with working on air brake systems.

29.5 CVSA OOS

Commercial Vehicle Safety Alliance (CVSA) out-of-service (OOS) conditions are not maintenance standards. They define the point at which highway equipment becomes dangerous to operate. OOS are used to issue citations during safety inspections. The following OOS apply to hydraulic brakes, including power-assist over hydraulic and engine drive hydraulic booster:

- Master cylinder less than ¼ full.
- No pedal reserve with engine running except by pumping pedal.
- Power-assist unit fails to operate.
- Seeping or swelling brake hose(s) under pressure.
- Missing or inoperative check valve.
- Visual evidence of hydraulic fluid from the brake system.
- Hydraulic hose(s) that are abraded (chafed) through outer cover-to-fabric layer.
- Fluid lines or connections that are leaking, restricted, crimped, cracked, or broken.
- Brake failure or low fluid warning light on and/or inoperative.

FIGURE 29–39 Preinspection brake service checklist

PREBRAKE JOB INSPECTION CHECKLIST

Owner ____________ (Last) ____________ (First) Phone ____________ Date ___/___/___

Address ______________________________ Licence No. ____________

Make __________ Model __________ Mileage __________ Serial No. __________ Year __________

Special Key for Hubcaps/Wheels Location ______________ Owner Use Parking Brake Yes ☐ Yes ☐

4 Drum ☐ 4 Disc ☐ Disc/Drum ☐ P/B No ☐ Yes ☐ Vacuum ☐ Hydro ☐ ABS ☐

Owner Comments ______________________________

1. CHECKS BEFORE ROAD TEST

	Safe	Unsafe
Stoplight Operation		
Brake Warning Light Operation		
Master Cylinder Checks		
Fluid Level		
Fluid Contamination		
Under Hood Fluid Leaks		
Under Dash Fluid Leaks (No Power)		
Bypassing		
BRAKE PEDAL HEIGHT AND FEEL		

Check One		Check One	
Low		Spongy	
Med		Firm	
High			

Power Brake Unit Checks

VACUUM	Safe	Unsafe	HYDRO	Safe	Unsafe
Vacuum Unit			Hydro Unit		
Engine Vacuum			P/S Fluid		
Vacuum Hose			P/S Belt Tension		
Unit Check Valve			P/S Belt Condition		
Reserve Braking			P/S Fluid Leaks		
			Reaerve Braking		

Tire Pressure Specs

	Front	Rear
Record Pressure Found		

RF ______ LF ______ RR ______ LR ______

Tire Condition

RF ______ LF ______ RR ______ LR ______

2. ROAD TEST

	Yes	No	RF	LF	RR	LR
Brake Pull						
Brake Clunk						
Brake Scraping						
Brake Squeal						
Brake Grabs						
Brakes Lock Prematurely						
Wheel Bearing Noise						
Vehicle Vibrates						

STEERING WHEEL MOVEMENT WHEN STOPPING FROM 2–3 MPH YES/NO/RGT/LFT

Does ABS Work	YES	NO
Pedal Pulsation when Braking	YES	NO
Steering Wheel Oscillation when Braking	YES	NO
No Stopping Power	YES	NO
Warning Light illuminates when Braking	YES	NO
Difference in Pedal Height after Cornering	YES	NO
Suspension Dive	YES	NO

3. In Shop Checks On Hoist

	Yes	No	RF	LF	RR	LR
Brake Drag						
Intermittent Brake Drag						
Brake Pedal Linkage Binding						
Wheel Bearing Looseness						
Missing or Broken Wheel Fasteners						
Supension Looseness						

	Front					Rear				
	Right			Left		Right			Left	
	Spec	Safe	Unsafe	Safe	Unsafe	Spec	Safe	Unsafe	Safe	Unsafe
Mark Wheels and Remove										
Caliper/Piston Stuck RF LF RR LR										
Mark Drums and Remove										
Measure Rotor Thickness or Drum Diameter										
Measure Rotor Thickness Variation										
Measure Rotor Runout										
Lining Thickness										
Tubes and Hoses										
Fluid Leaks										
Broken Bleeders										
Leaky Seals										
Self-Adjuster Operation										

	Safe	Unsafe
Parking Brake Cables and Linkage		

- Insufficient vacuum reserve to permit one full brake application after engine is shut off.
- Vacuum hose(s) or line(s) that are restricted, abraded (chafed) through outer cover to cord ply, or collapse when vacuum is applied.
- Low vacuum warning device inoperative.

SUMMARY

- Hydraulic brakes tend to be used only on light- and medium-duty highway trucks and a wide range of off-highway applications.
- Hydraulic brakes are based on a principle that a liquid does not compress. Pressure applied to one portion of the circuit is transmitted equally to all parts of the circuit.
- In hydraulic braking systems, the mechanical force of the driver stepping on the brake pedal is converted to hydraulic pressure usually assisted by some type of power boost.
- At the wheel, the hydraulic force is once again changed back into the mechanical force required to brake the truck.
- Drum brakes can be servo, in which the action of one shoe is governed by input from the other, or non-servo, in which the shoes are separately anchored.
- Hydraulic disc brakes can be of the fixed caliper or floating/sliding caliper type.
- When the brake pedal is depressed, the master cylinder forces brake fluid to the calipers or wheel cylinders, changing mechanical force into hydraulic pressure; the wheel cylinders and calipers change hydraulic pressure back into mechanical force, braking the vehicle.
- Pressure differential valves, metering valves, proportioning valves, combination valves, and load proportioning valves are all operational components of a hydraulic brake system.
- In hydraulic brake systems, a hydraulic power booster is used to assist the master cylinder in applying the brakes. This means that the mechanical force applied by the driver's boot is amplified hydraulically.
- Some Class 6 and Class 7 trucks are equipped with air-over-hydraulic brake systems. This combines some of the advantages of air and hydraulic brake systems.
- A hydraulic ABS is designed to modulate hydraulic application pressures to the wheel cylinders to permit maximum braking force without locking the wheels.
- A typical hydraulic ABS consists of wheel speed sensors, an electronic control module, and a modulator assembly.
- Hydraulic ABS is capable of modulation cycles at up to 15 times per second.
- All hydraulic ABS are required to default to full redundancy mode in the event of a complete system failure. Some systems permit operation in partial ABS in the event of subsection failure.

REVIEW QUESTIONS

1. In hydraulic disc brakes, the pads are forced against the rotors by:
   a. the shoes.
   b. the pistons.
   c. the braked anchors.
   d. the return springs.
2. As the pressure of fluid received by the wheel cylinder of a drum brake system increases, which of the following happens?
   a. The wheel cylinder cups and pistons are forced apart.
   b. The brake shoe retracting springs push the pistons back together.
   c. The pads are pushed against the rotor.
   d. The brake drum locks to the wheel cylinder.
3. Which of the following safety devices delays application of front disc brakes in a disc/drum system until pressure is built up in the hydraulic system?
   a. a pressure differential valve
   b. a metering valve
   c. a proportioning valve
   d. a load proportioning valve (LPV)
4. In a typical air-over-hydraulic brake system, depressing the treadle causes which of the following?
   a. Hydraulic fluid flows through quick-release valves.
   b. Air pressure forces the pistons of each wheel cylinder outward.
   c. An LPV redistributes air pressure to the front and rear brakes.
   d. Air pressure is sent to the booster assemblies.

5. Which of the following correctly describes the function of a load proportioning valve (LPV) in an air-over-hydraulic system?
   a. It increases the percentage of braking performed by the rear axle(s) as vehicle load increases.
   b. It decreases the percentage of braking performed by the rear axle(s) as vehicle load increases.
   c. It changes the frame-to-axle height dimension.
   d. It compensates for front-wheel lockup.
6. In a hydraulic brake system, brake fluid transmits any pressure applied:
   a. equally to all parts of the enclosed circuit.
   b. to any surface that will yield.
   c. only to slave cylinders in the circuit.
   d. to any surface that will not yield.
7. In a hydraulic brake system, which of the following best describes what happens when the driver's foot is applied to the brake pedal?
   a. Hydraulic force is converted to mechanical force.
   b. Mechanical force is converted to hydraulic force.
   c. Hydraulic force is converted to mechanical force then back to hydraulic force.
   d. Mechanical force is converted to hydraulic force then back to mechanical force.
8. What component in a hydraulic brake system helps fluid in the master cylinder return to the reservoir after a brake application?
   a. a compensating port
   b. a metering valve
   c. a pressure differential valve
   d. a check valve
9. Technician A says that in non-servo drum brakes, the forward or leading shoe is energized in normal braking when the vehicle is in forward motion. Technician B says that in non-servo drum brakes, the leading and trailing shoes exchange roles when the vehicle reverses motion. Who is correct?
   a. Technician A only
   b. Technician B only
   c. both A and B
   d. neither A nor B
10. In hydraulic drum brakes, the shoes are mechanically forced against the drums by:
    a. a master cylinder
    b. the wheel cylinders
    c. braked anchors
    d. return springs
11. A typical TIR specification for disc rotor lateral runout in a hydraulic brake system would more likely be which of the following?
    a. 0.0015 inch (0.0381 mm)
    b. 0.0050 inch (0.127 mm)
    c. 0.0100 inch (0.254 mm)
    d. 0.0150 inch (0.381 mm)
12. In a hydraulic drum brake system, what force pulls the shoes away from the drum when the brakes are released?
    a. hydraulic force at the master cylinder
    b. retraction spring force at the brake shoe assembly
    c. hydraulic force at the wheel cylinder
    d. brake pedal spring force
13. Which of the following devices would play a role in illuminating the brake warning light in the event of a complete hydraulic failure of the front brake circuit?
    a. pressure differential valve
    b. metering valve
    c. proportioning valve
    d. load proportioning valve (LPV)
14. A typical maximum rotor thickness variation specification required by an OEM would likely be:
    a. 0.002 inch (0.0508 mm).
    b. 0.005 inch (0.127 mm).
    c. 0.020 inch (0.508 mm).
    d. 0.050 inch (1.270 mm).
15. Technician A says that when replacing disc brake pads, most OEMs recommend that fluid be removed from the master cylinder before compressing the caliper piston with a C-clamp. Technician B says that the caliper bleed screw should be opened before compressing a caliper piston with a C-clamp. Who is correct?
    a. Technician A only
    b. Technician B only
    c. both A and B
    d. neither A nor B
16. Which of the following best describes the electrical operation of a typical ABS wheel speed sensor?
    a. Hall effect sensor
    b. thermistor
    c. potentiometer
    d. pulse generator
17. Which of the following components would best be described as the "brain" of a hydraulic ABS?
    a. master cylinder
    b. wheel speed sensor
    c. ECU
    d. modulator valve
18. Which of the following would represent the typical maximum modulation cycle speed of a hydraulic ABS?
    a. 2 times per second
    b. 5 times per second
    c. 15 times per second
    d. 60 times per second

19. Technician A says that an ABS ECU reads the frequency produced by a wheel speed sensor and interprets it as wheel speed. Technician B says that the basis of hydraulic ABS is that hydraulic pressure to the brakes over one wheel is relieved when a lockup condition is sensed. Who is correct?
    a. Technician A only
    b. Technician B only
    c. both A and B
    d. neither A nor B

20. Technician A says that when hydraulic ABS fails electronically, the system defaults to normal hydraulic braking. Technician B says that a failure of one ABS wheel sensor would result in a vehicle that should not be driven on the highway. Who is correct?
    a. Technician A only
    b. Technician B only
    c. both A and B
    d. neither A nor B

30

Prerequisites: Chapters 6, 28, and 29

ABS AND EBS

OBJECTIVES

After reading this chapter, you should be able to:

- Describe how an antilock brake system (ABS) works to prevent wheel lockup during braking.
- List the major components of a truck ABS.
- Describe the operation of ABS input circuit components.
- Outline the role of the ABS module when managing antiskid mode.
- Explain how the ABS module controls the service modulator valves.
- Explain what is meant by the number of channels of an ABS.
- Describe how trailer ABS is managed.
- Outline the procedure for diagnosing ABS faults.
- Describe the procedure required to set up and adjust a wheel speed sensor.
- Explain how an electronic brake system (EBS) manages service brake applications.
- Describe some ABS add-ons such as stability control electronics.
- Outline the reasons why an EBS has to meet current FMVSS 121 requirements.

KEY TERMS

ACom software
antilock braking system (ABS)
antilock braking system (ABS) module
chopper wheel
controller
electronic brake system (EBS)
electronic control unit (ECU)
electropneumatic
exciter ring
modulator
NEXIQ Brake-Link
power line carrier (PLC)
Pro-Link iQ
pulse generator
pulse wheel
redundancy
reluctor wheel
remote diagnostic unit (RDU)
stability control electronics (SCE)
tone wheel
wheel speed sensors
yaw

INTRODUCTION

An **antilock braking system (ABS)** is an electronically managed brake system that monitors and controls wheel speed during braking. ABS is designed to work with standard air brake systems and it has been mandatory on all commercial vehicle brake systems since the late 1990s. It is a requirement of ABS that it has full system **redundancy** to a Federal Motor Vehicle Safety Standard No. 121 (FMVSS 121) pneumatic control circuit in the event of an electronic failure. Put in other words, *system redundancy* means that should any part of the ABS fail, the pneumatic brake control circuit functions exactly as it would in any non-ABS truck rig.

A fully electronic braking system (EBS) is a computerized air brake system in which the control circuit is managed electronically and in which the brake pedal is simply an input to an EBS control module. Although full EBS is used in Europe, when it is used in North America it must be in conjunction with an FMVSS 121–compliant pneumatic control system. This makes it an expensive option, so it is not regularly used. However, the major truck and bus original equipment manufacturers (OEMs) are testing EBS extensively, and efforts continue to be underway to modify FMVSS 121 to accommodate brake-by-wire. There is a big advantage to using EBS because not only can it speed service brake applications, it can also increase the effectiveness of accident avoidance technologies, resulting in safer overall rig operation.

Most of this chapter addresses the current ABS that has been mandatory on both tractors and trailers for a generation. A short introduction to EBS also is provided. Because ABS is managed by computer, it will help you understand this chapter better if you read Chapter 6 first. We also take a brief look at truck stability enhancement systems. Stability enhancement systems in trucks address antirollover (roller avoidance/roll stability control) and directional tracking controls by interacting with other vehicle electronics using multiplexing.

It is assumed that students reading this chapter fully understand the contents of Chapters 28 and 29, along with the fundamentals of electronic controls covered in Chapter 6. It is important to stress that at this moment in time, ABS does not substantially change the control, actuation, and foundation brakes of non-ABS due to the system redundancy requirement.

ABS OBJECTIVES

The idea behind ABS is simple. The system monitors wheel speeds at all times and controls brake application pressures when wheel lock is imminent. The main benefits of ABS are better vehicle stability and control during severe braking, which is another way to describe a *panic stop*. The ABS **electronic control unit (ECU)** (the system controller) receives and processes signals from the wheel speed sensors (which signal wheel speed to the ECU). When the ECU detects a wheel lockup, the unit activates the appropriate modulator valve (system output, an electrically controlled slave valve), and service air pressure is pulsed. ABS permits a less experienced truck driver to maintain much better control during panic stops. **Figure 30–1** is a schematic of a typical ABS with the ABS components highlighted.

Because of the requirement for system redundancy, ABS does not require any significant changes to the dual-circuit air brake system. Both the control and actuator circuits of the dual-circuit air brake system function pretty much as in a system without ABS. They use compressed air and mostly the same components. In the event of a typical ABS failure, the affected wheel or axle is ABS disabled and reverts to normal braking. The other wheels retain ABS function.

30.1 BASICS OF ABS OPERATION

When compared to the electronic systems required to manage a transmission or an engine, an ABS is simple. An input circuit is used so that ABS can monitor individual wheel speeds under braking. The wheel sensors use a pulse generator principle. This consists of a toothed ring located at the axle ends (on the wheel assembly) or within the axle. The analog alternating current (AC) voltage output from these sensors is transmitted to an ECU. The ECU ignores the voltage value and, instead, uses the AC frequency to determine wheel speed.

The function of the ECU is to receive input information, process it, and then generate an output by switching a **modulator**. The "brain" of the ECU is a microcomputer. The ECU monitors the input information—in this case, wheel speed data—and, based on its programmed software, generates an outcome. Although the process can be called logic processing, the role of ABS is simple; that is, to prevent wheel lockup. Because the ECU is supplied with information on exactly how fast each wheel is rotating, it can calculate the point at which a wheel is about to lock up, and outputs a signal to reduce braking force to the affected wheel. On an air brake system, this is achieved by controlling air through a modulator. The modulator is a solenoid-controlled valve capable of dumping air being delivered to the brake chamber. This exhausting of air to the brake chamber is momentary and is immediately followed by recharging the brake chamber. The cycling of on/off air to the brake chamber can occur at high speed.

SHOP TALK

An ABS is a simple computer-controlled system. As such, it requires system inputs, processing capability, and outputs. The steps required to produce an outcome are less complex than in an electronically managed transmission, but electronically the process is similar.

SPLIT-COEFFICIENT STOPS

Because in many systems the brakes on each side of the vehicle are individually controlled, ABS permits maximum efficiency stops even when the vehicle is run on a split-coefficient surface. A split-coefficient surface might occur when one side of the truck is running on ice and the other on bare pavement. This type of braking situation is hazardous even for experienced drivers. The **antilock braking system (ABS) module** is capable of cycling the brakes on the ice side of the vehicle at a different rate from that on the dry side. This enables it to obtain maximum road traction through the stop.

FIGURE 30–1 Typical ABS schematic of a highway straight truck highlighting the electronic components

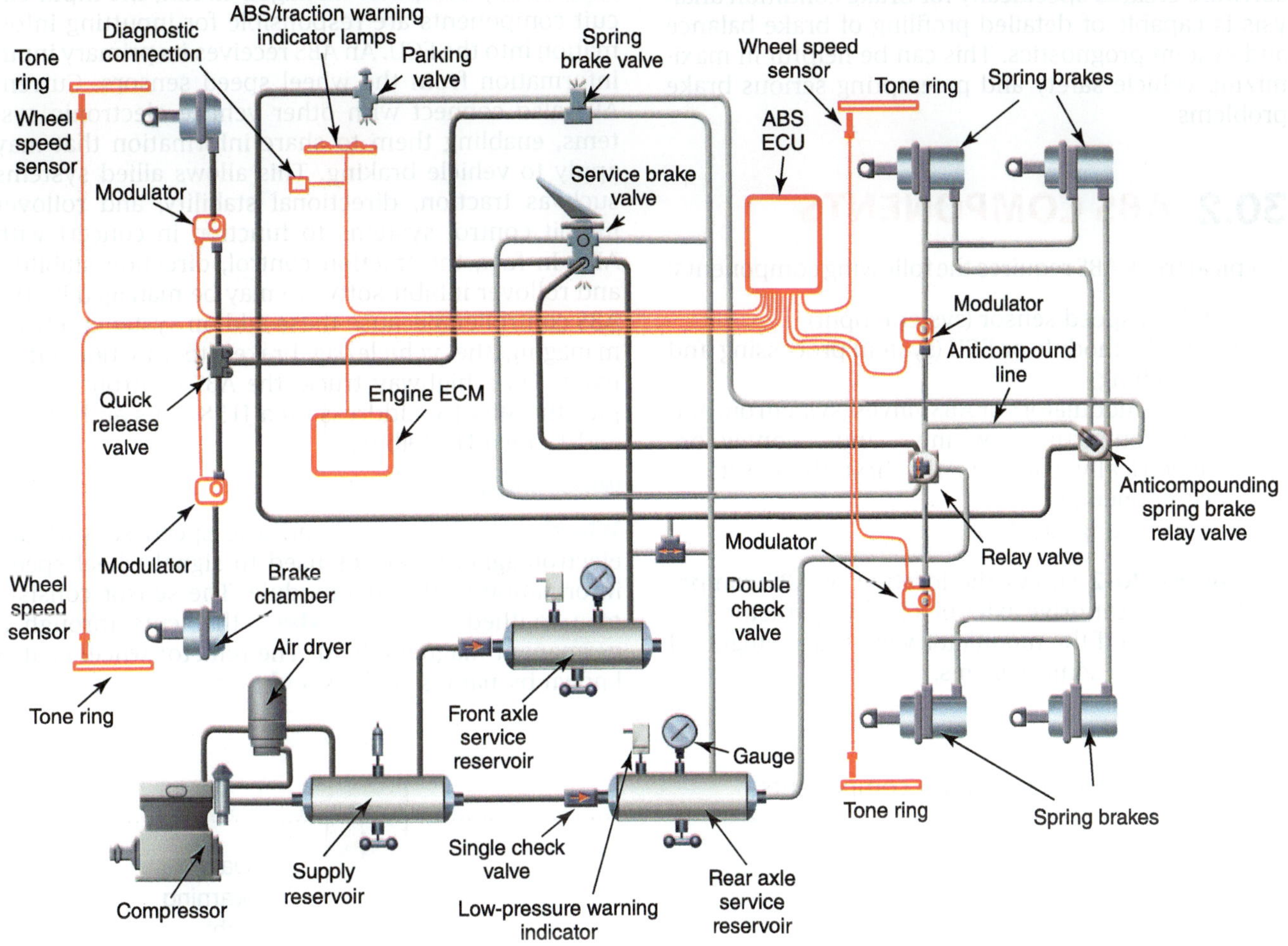

CAUTION

When driving an ABS-equipped truck, drivers have a tendency to expect too much. Always exercise care when braking, especially under bobtail conditions, even when the vehicle ABS is fully functional.

ABS FAILURES AND DIAGNOSIS

When the truck is operating, the ABS module continually monitors its own circuitry and performance. If a fault is detected, ABS will alert the driver and shut down either the entire system or a portion of the system. In the event of any type of ABS failure, a warning lamp alerts the driver as to the status of the system. This warning lamp may also be used to display blink code diagnostics. Because the ABS module is capable of self-diagnosis, the technician can then troubleshoot the system using the diagnostics written to the system software.

TYPES OF ABS

Although all truck ABS operate on more or less the same principles, each manufacturer uses different hardware, and most of it is specific to the system. This means that most components cannot be interchanged. To make things slightly more complicated, ABS is a rapidly developing technology, which has resulted in each OEM producing different generations of the system and different methods of diagnosing system faults.

ABS TROUBLESHOOTING

Troubleshooting on some early systems was limited to a light-emitting diode (LED) display on the ECU to signal fault codes, or flash codes displayed through a dash warning light in the cab. This has progressed to electronic service tool (EST)–based diagnostic software that accesses the ABS electronics either directly or via the chassis data bus. A new generation of specialty handheld ESTs along with PC-based OEM software are used to troubleshoot truck ABS. The OEM software created specifically for brake condition analysis is capable of detailed profiling of brake balance and system prognostics. This can be helpful in maximizing vehicle safety and preempting serious brake problems.

30.2 ABS COMPONENTS

A typical truck ABS requires the following components:

- Wheel speed sensor (system input)
- An ABS module or ECU (system processing and switching)
- Brake modulators or ABS valves (system output)
- Interconnecting wiring and connectors (electrically connects the first three sets of components)
- Networking capability

Figure 30–2 shows the location of ABS components on the rear drive axles of a highway tractor. The ABS module and the modulator valves are integrated into one unit in some systems.

INPUT CIRCUIT COMPONENTS

As with any computer-managed circuit, the input circuit components are responsible for inputting information into the ECU. An ABS receives its primary input information from the wheel speed sensors. Current ABS also connect with other vehicle electronic systems, enabling them to share information that may apply to vehicle braking. This allows allied systems such as traction, directional stability, and rollover inhibit control systems to function in concert with ABS. In fact, the traction control, direction stability, and rollover inhibit software may be managed by the ABS controller, because these add-on systems rely on managing the vehicle/rig brakes to function effectively. On a highway truck, the ABS controller occupies the MID 136 address on a J1587 bus and SAs 11 and 12 on a J1939 bus.

Wheel and Axle Speed Sensors

Wheel speed sensors and axle speed sensors are electromagnetic devices used to signal wheel speed information to the ABS module. The sensor consists of a toothed **reluctor wheel** that cuts through a permanent magnetic field. The reluctor wheel is also known by names such as a **chopper wheel**, **exciter**

FIGURE 30–2 Location of ABS components

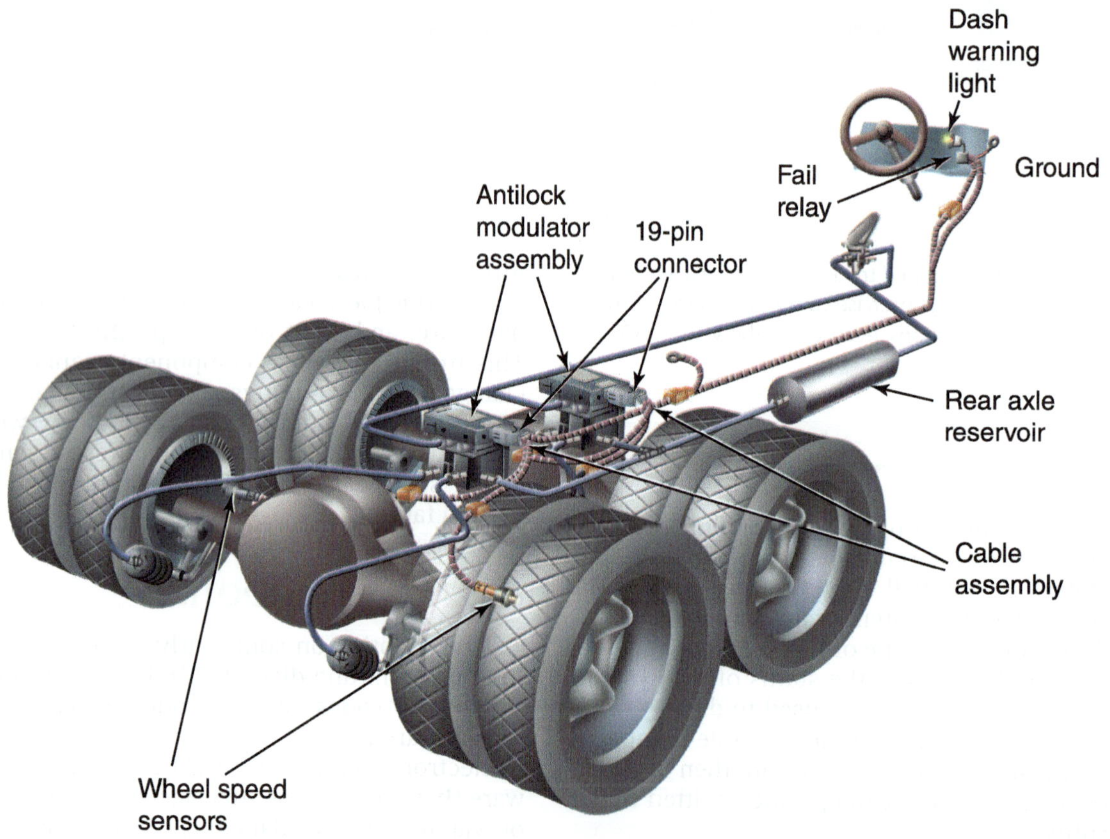

FIGURE 30–3 Major components of a typical ABS system: reluctor wheel, wheel speed sensor, modulator valve, and ECU or ABS module

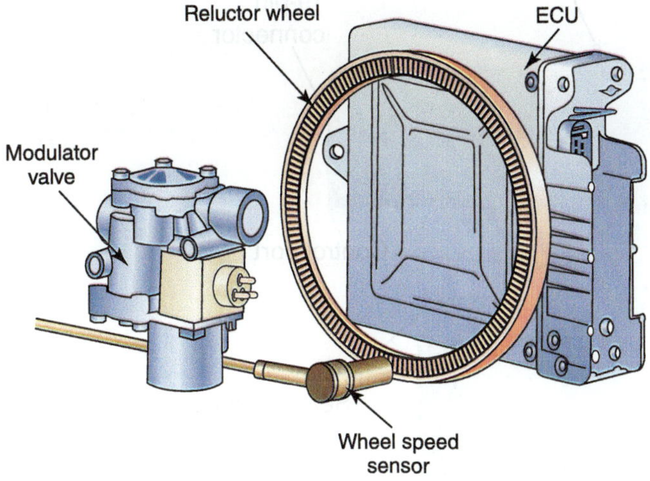

ring, or **tone wheel**, depending on the OEM. However, for the sake of consistency, it is referred to as a *reluctor wheel* in this text. The complete sensor unit that includes the reluctor wheel and permanent magnet is referred to as a **pulse generator**. A typical pulse generator assembly can be seen in the center of **Figure 30–3**. A photo view of a reluctor wheel can be observed in the axle-end cutaway in Figure 27–59.

Reluctor Operation

When the reluctor wheel rotates, it cuts through a permanent magnetic field located in the stationary sensor. This makes it act like a generator, producing a simple AC signal. It shares operating principles with other shaft speed sensors on the chassis, including the road speed (transmission tailshaft) sensor and engine rpm sensor. A reluctor or exciter ring can be integral with the drum or disc or isolated.

Signal Processing

Both the voltage and frequency of this AC signal will rise in exact proportion to wheel speed. This analog voltage signal is sent to the ECU, which converts the frequency pulse to an actual wheel speed. On a typical ABS, the frequency varies from 0 Hz (wheel locked or stationary) to around 400 Hz. A graphic representation of the frequency signals produced by a wheel speed sensor is shown in **Figure 30–4**. The ABS module must be capable of rapidly processing the wheel speed data. This data must be converted into a digital format to enable it to be processed by the system ECU.

SHOP TALK

Systems that use a 100-tooth wheel sensor do so for a reason. When divided into 360 degrees of one full tire revolution, the 100 teeth on the **pulse wheel** equal 3.6 degrees each. When a radial tire on a loaded-to-specification truck is run down the highway at 60 mph (97 kmh), its dynamic footprint is about 3.6 degrees. One hundred-tooth ABS reluctor wheels have become the industry standard.

ELECTRONIC CONTROL UNIT

Depending on who manufactures the system, the ABS module can be called the system **controller** or an ECU, the preferred term used in this text. The ECU is a simple computer. It receives input signals from the wheel speed sensors. Because these input signals are in analog form, they must be converted to a digital format before they can be processed. A microprocessor (the brain of the ECU) manages the processing cycle. It is programmed with data that contains the rules and regulations that allow it to produce outcomes. The outcomes required of the ECU in an ABS are electrical control signals to solenoids in the modulator.

The ABS ECU continuously monitors input wheel speed information from the sensors in the circuit. Its function is calculation and comparison to information and instructions programmed into its memory. When an imminent wheel lockup condition is detected, the ECU outputs an electrical signal to the solenoids that control the air routed to the brake chambers. This function of the ECU is known as switching.

Speed

The speed at which any computer "thinks" is known as its clock speed. This speed in most ABS modules is around 200 times a second. Although this might sound slow compared to a desktop computer, it is fast enough to manage braking in a truck air brake system. Truck air brake systems are capable of modulating air pressures three to seven times per second,

FIGURE 30–4 AC frequency signals produced by a wheel speed sensor at low and high speeds.

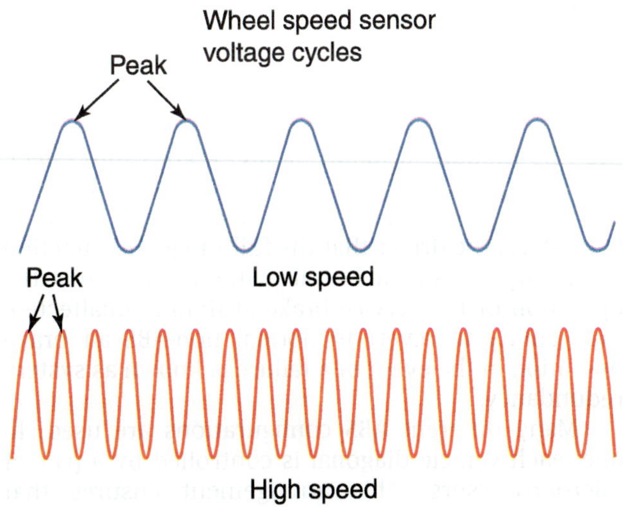

FIGURE 30–5 An older style modulator controller assembly with a diagnostic LED window

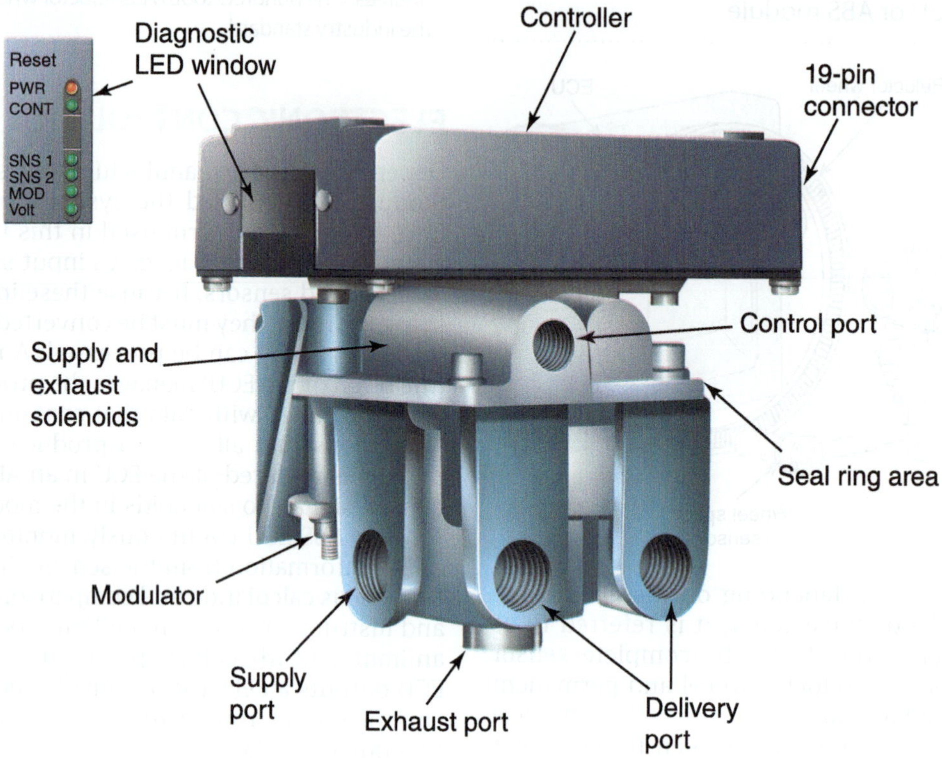

depending on the system. Some ABS modules have two microcomputer circuits, one to act as a primary processor and the second to act as a backup. This provides for a little additional safety.

Some ABS ECUs contain a diagnostic window and manual reset switch, and they may be integral with the modulator valve assembly (**Figure 30–5**). Most ABS modules are insulated and protected from road debris. When located in the chassis, they function in a harsh environment and have to be protected from physical damage and electromagnetic radiation. In some cases, the ECU is physically located on the modulator assembly and electrically bussed to it by means of sealed pin connectors and a harness.

Failure Strategy

When a system or subsystem failure is detected by the ABS module, a failure strategy goes into effect. This can mean that the entire system or a portion of it defaults to normal non-ABS operation. A warning light alerts the driver that the full-range ABS function is no longer available but also should mean that the operation of the service brake system is unaffected. This ability to revert to normal non-ABS air brake operation in the event of a failure is known as system redundancy.

Many different ABS configurations are used. In one, each vehicle diagonal is controlled by a pair of microprocessors. This arrangement ensures that faults are reliably detected within the microprocessors. Because both microprocessors are fed the same input information when the system is functioning properly, they should produce the same results. Each microprocessor additionally runs constant self-tests and monitors the program execution time. These two monitoring levels function independently of each other and ensure reliable and fast detection of faults.

The performance of both the input and output circuit devices is monitored. For example, wheel speed sensors are monitored electrically, enabling open circuits, short circuits, and shunts to be detected even when the vehicle is standing. The air gap between the sensor and the reluctor wheel is also checked through software monitoring. These measures identify not only electronic defects but malfunctions resulting from mechanical vibration in the axle area and electrical interference. In the case of pressure modulating valves, voltage and current are measured in actuated condition and are compared with a permissible range of values. All components are activated and checked each time the ignition is switched on. The ABS warning light is switched off if the system is operational.

ABS Control Logic

We know that the brain of an ECU processing cycle is the central processing unit (CPU). It performs such tasks as data comparison, fetching and carrying data from memory, and arithmetic and logic calculations at very high speeds. We group all of these functions together and collectively call them logic processing.

It is much easier to describe logic processing in a simple electronic management system such as ABS than in the more complex management systems required on engines or transmissions. The ABS module monitors wheel speed data on a continuous basis. This data is logged in the random access memory (RAM; main memory) of the processing cycle. When the vehicle is braked, the rate of deceleration of each monitored wheel is compared with a profile programmed into system memory. This essentially tells the CPU how fast a wheel can be decelerated without locking up.

Under braking, if the *rate of deceleration* identified in the computer's programmed memory is exceeded, the antilock strategy kicks in and the CPU signals the modulator solenoids to momentarily dump service application pressure to the wheel about to lock up. This immediately changes the rate of deceleration so that revised wheel speed data is now input to the computer processing cycle. Air pressure to the service chamber is now resumed.

The ABS ECU can not only manage modulator release/apply cycling at high speeds, but it can accomplish this on a wheel-by-wheel basis. In air brake systems, the maximum cycling frequency is slower than in a hydraulic brake circuit, usually up to a maximum of about seven times per second. The reason for the slower cycling of air brake systems is due to the compressibility of air when compared to hydraulic brake circuits that use "incompressible" hydraulic fluid. The action of ABS occurs in a pulsating manner. It simulates pumping of the brakes by the driver except, of course, that the "pumping" takes place at higher speeds and is wheel-by-wheel specific. Regardless of what any experienced driver may tell you, ABS substantially reduces rig stopping distances every time it has been tested.

OUTPUT CIRCUIT COMPONENTS

The output circuit is switched by the ABS module. Outputs are the devices that effect the results of the ABS module processing cycle. The main output device in an ABS is the brake modulator assembly. A brake modulator assembly combined with an ABS control module is shown in Figure 30–5.

Brake Modulators or ABS Valves

The brake modulator or ABS valve assembly is controlled by the ECU. Again, the term used to describe this valve body varies by manufacturer. In this text, the term *modulator* generally is used. The typical brake modulator assembly contains from one to four solenoid valves. The actual number of solenoid valves used depends on the number of brake circuits that have to be regulated from the unit. Each

FIGURE 30–6 Modulator valve in balanced position during a normal service application

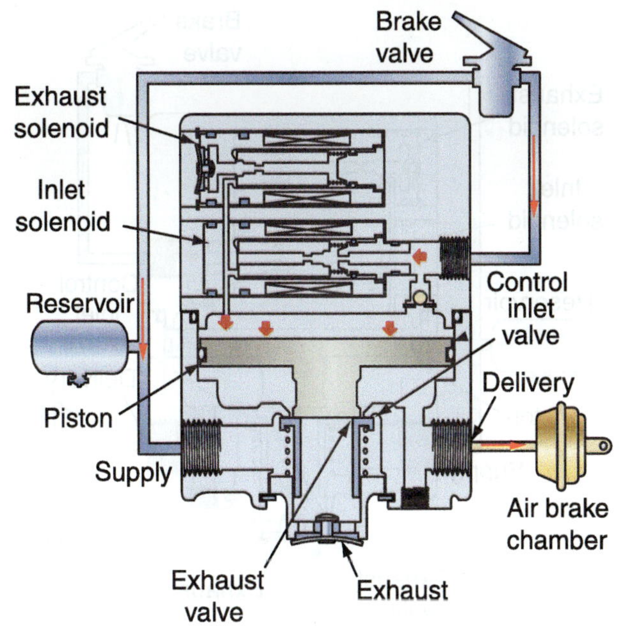

solenoid is a switched output from the ECU. They are capable of response times as fast as 3 milliseconds.

Each modulator valve consists of a solenoid, a piston, and a body housing. During normal (nonantilock) operation, the inlet solenoid is open and the exhaust solenoid remains closed (**Figure 30–6**). This permits service application air to pass through and be directed to the service chambers. The modulator valve functions as a pilot-controlled relay valve and, when energized, is used to divert pressure to the brake chambers to exhaust. The air-in port is always to the electrical side of the connector.

The modulator valve is located downstream from the relay valve. In practice, it may be integral with the relay valve, but it still functions "downstream" from it. When the system is not managing an antilock stop, the modulator simply routes the air from the relay valve to the service chamber. In this instance, control pressure from the foot valve directs an air signal through to the relay valve, and the relay valve uses reservoir air and delivers it to the service brake chamber, enabling a service brake application in exact proportion to the control pressure. However, when the ABS module determines that a wheel lockup condition is about to occur based on the input signal from the wheel speed sensor, it switches, or energizes, the solenoids in the modulator to exhaust air pressure routed to the service brake chambers (**Figure 30–7**). This momentarily relieves the service application pressure on the brake chamber(s).

When the modulator solenoid and piston are energized in ABS mode, control pressure is exhausted

FIGURE 30–7 Modulator valve exhausting during a normal service application

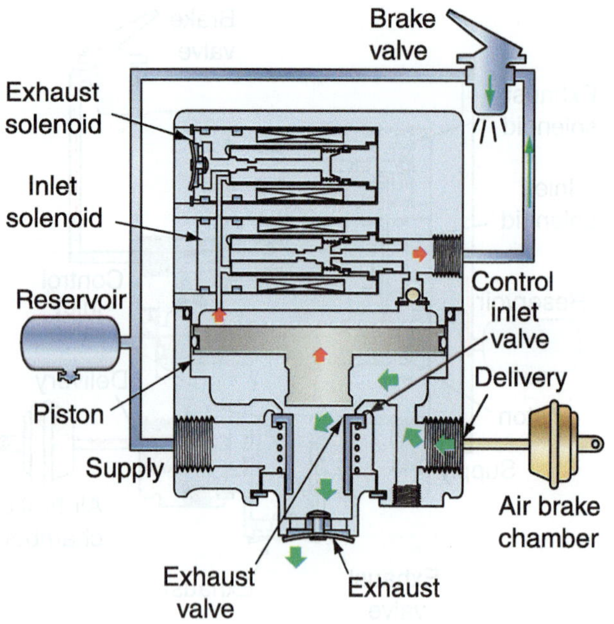

both through check valves located in the solenoid housing and out the exhaust port of the brake valve (**Figure 30–8**). As the modulator piston moves, the exhaust is opened, allowing the air from the underside of the piston to discharge through the exhaust port of the modulator. Study Figure 30–6, Figure 30–7,

FIGURE 30–8 Modulator valve in ABS dump mode during a service application

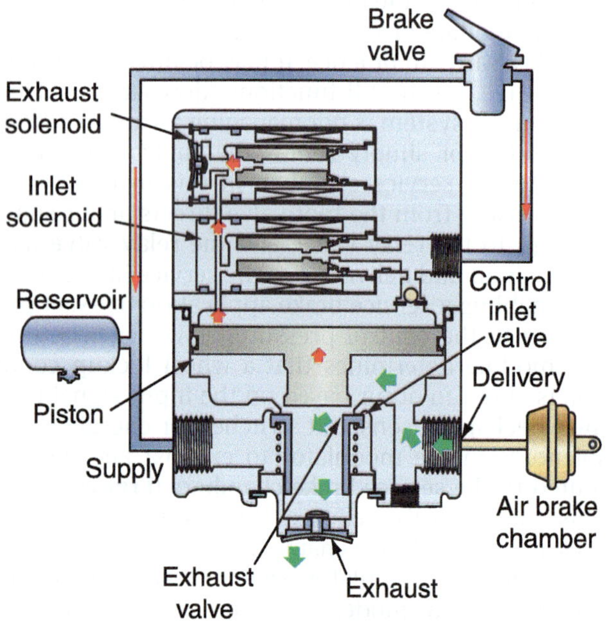

FIGURE 30–9 ABS modulator valve operation

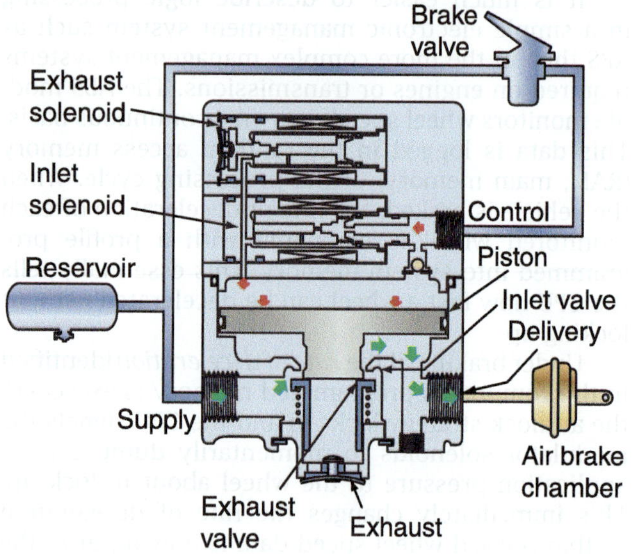

and Figure 30–8, and then look at **Figure 30–9**, which captures the content of all three figures.

Fail Relay Coil

A fail relay coil (**Figure 30–10**) is installed between the chassis power source and the ABS electronic system. The fail light power flows through the normally closed relay contact. Should a failure occur in the system, the fail relay coil will not energize, illuminating the failure indicator light in the dash, alerting the driver. In figure 30–10, the fail relay is shown in a no power position.

Alternate Designs

Some modulator valve assemblies, instead of using solenoids, use motor-driven stepped pistons to regulate braking pressure. They perform the same function as solenoids; they dump service application air when switched to do so by the ABS module. To improve severe braking performance, some modulators are designed to provide feedback to the ECU with the objective of smoothing braking on split coefficient road surfaces.

As each new generation of ABS arrives on the marketplace, the response times are improved. The faster the modulator regulates service braking pressure when switched by the ECU, the better the system performance. It is important to remember that the cycling performance of an air brake ABS is somewhat slower than a hydraulic ABS because of the compressibility of air.

Dash Warning Indicator Light

Most ABS have a warning light. The warning light may be mounted on the dash or above the windshield.

FIGURE 30–10 Electrical layout of a typical ABS system is shown with no power from fail relay circuit.

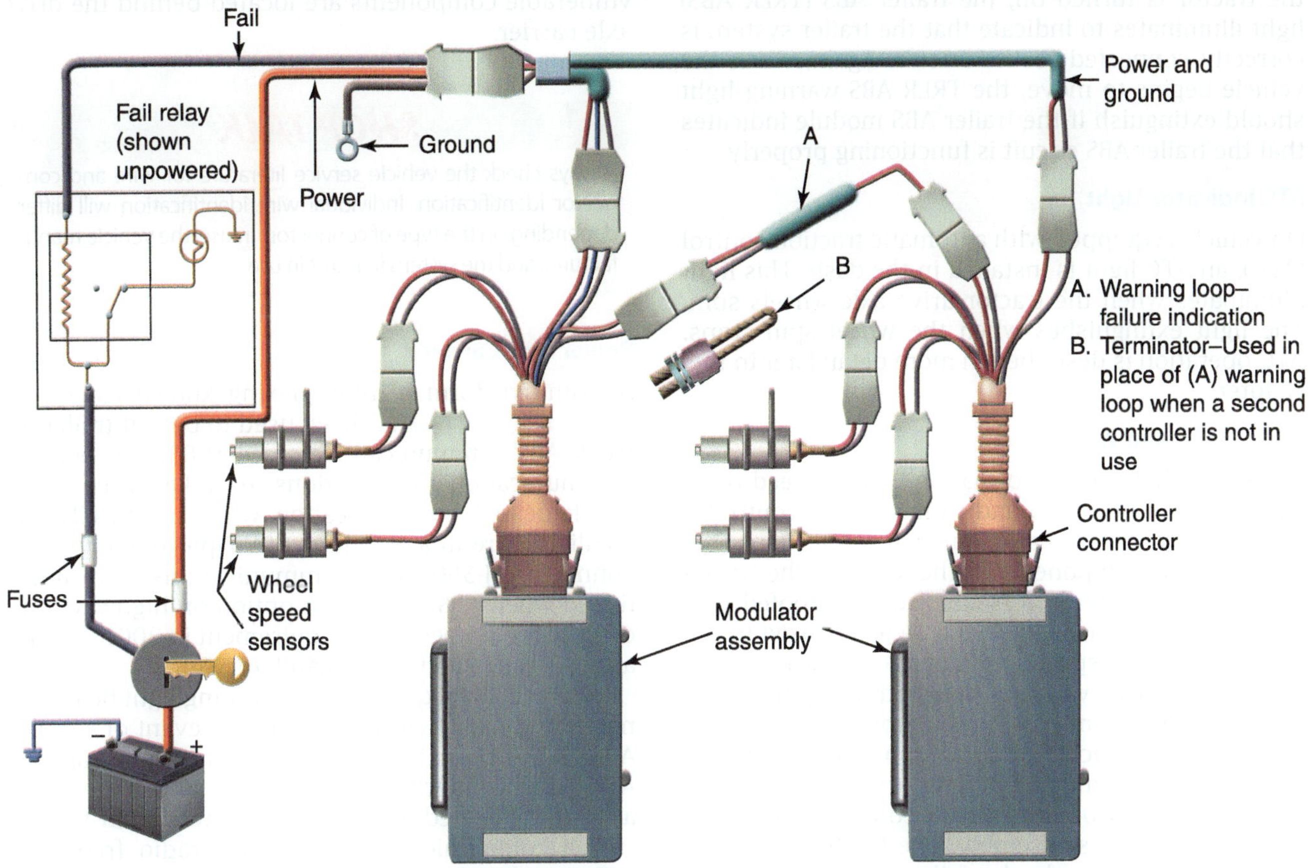

This light is used to indicate to the driver that the system is properly operating on startup, to alert the driver in the event of a malfunction, and to signal diagnostic codes to the technician.

The ABS tractor warning light (TRAC ABS) will illuminate when the ignition switch is turned on. A TRAC ABS light is shown prominently located between the tachometer and speedometer in **Figure 30–11**. The driver would find it difficult not to notice it. Depending on the OEM and the generation of the ABS, this light will go off when the vehicle starts to move. If the light stays on, a failure in the ABS is indicated. The light is also designed to illuminate when a malfunction occurs during a trip. Whenever an ABS malfunction occurs, standard full-capability air braking is still available. Most current systems will close down ABS only on the portion of the ABS circuit in which the malfunction has been detected.

Trailer ABS Warning Light

In a tractor/trailer combination in which both vehicles are equipped with ABS, there is either a second dash light (all systems manufactured after 2001) or a left forward trailer bulkhead light that can be seen

FIGURE 30–11 ABS tractor (TRAC ABS) warning light prominently located in driver's field of vision.

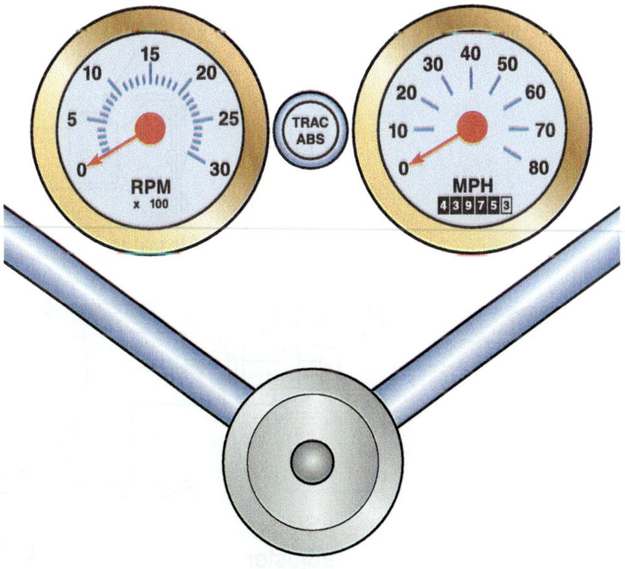

in the driver side mirror that indicates the operating status of the trailer ABS. When the ignition circuit in the tractor is turned on, the trailer ABS (TRLR ABS) light illuminates to indicate that the trailer system is correctly connected to the tractor. Again, when the vehicle begins to move, the TRLR ABS warning light should extinguish if the trailer ABS module indicates that the trailer ABS circuit is functioning properly.

ATC Indicator Light

On vehicles equipped with automatic traction control (ATC), an ATC light is installed in the dash. This light illuminates when the tractor drive axle wheels spin. The light extinguishes when the wheel spin stops. ATC operation is described in more detail later in this chapter.

Connectors and Wiring

The wires that carry information (wheel speed data) and electrical power to and from the ABS controller module are harnessed and joined at connections that plug into the components in the system. The wiring harnesses and connectors are weatherproofed and sealed to the connectors. The wire gauge used in the wire harnesses is specific to the task performed.

The antilock system is powered up by the chassis electrical system. The power supply is fuse or circuit breaker protected. In a tandem-axle system, the electrical supply may be directly to each modulator assembly or can be joined in a common connector and wiring harness assembly. The fail line from the fail relay coil grounds the fail relay whenever the modulator assembly is powered up and in a nonfailure mode. An electrical schematic for a typical ABS is illustrated in Figure 30–10. An axle-mounted ABS is illustrated in **Figure 30–12**. Note that the more vulnerable components are located behind the drive axle carrier.

SHOP TALK

Always check the vehicle service literature for wire and connector identification. Individual wire identification will differ depending on the type of connectors in use, the vehicle manufacturer, and the system features in use.

Power Line Carrier

A primitive form of multiplexing known as **power line carrier (PLC)** has been used to permit trailer-to-truck ABS communications since 1985. PLC enables communications transactions to take place on a nondedicated communication wire, specifically the auxiliary wire in a standard ATA 7-pin tractor/trailer connector (J-560). PLC communications were introduced when ABS became an option on highway trailers, but its use became a requirement in 2001, 3 years after the introduction of mandatory ABS.

Trailer ABS requires that a warning light be illuminated in the dash of the truck in the event of a trailer ABS malfunction. Because all the wires on a standard ATA 7-pin connector between truck and trailer were already dedicated, PLC technology was used to convert a communication signal to a radio frequency (RF) signal and then superimpose it the 12-volt auxiliary power wire. A transducer converts the RF signal back to a signal that any networked controllers on

FIGURE 30–12 Axle-mounted ABS components

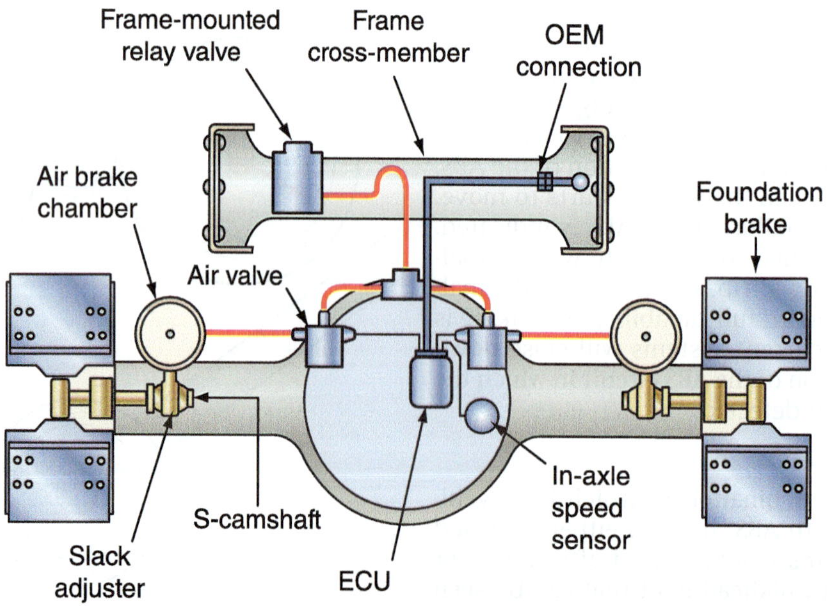

the data bus can use. These signals use J1587/J1708 or J1939 hardware and protocols and are diagnosed accordingly.

30.3 ABS CONFIGURATIONS

There are a number of types of ABS on trucks today. Just as in automobiles, there is some difference between high-performance and basic systems. The type of ABS that will function best on a truck depends on number of axles, axle configuration, axle load, brake circuit distribution, and brake force distribution. Another factor is whether the rig is spec'd with stability enhancement, traction control, and rollover avoidance electronics. For most applications, the steering axle wheels should be included in the ABS loop because the more severe the braking, the greater the role played by the front axle brakes.

DEFINING CHANNELS

The term *channel* refers to either an input or an output path to the ECU. However, in describing ABS, each manufacturer uses the word to mean slightly different configurations. For instance, a system described as a six-channel system fitted to a tandem drive straight truck (two steering wheels/four drive wheels) would normally consist of six inputs, that is, wheel speed data inputs to the ECU and six outputs or modulators. This means that braking on each wheel of the truck would be managed individually. A six-in/six-out channel system would represent a top-of-the-line system and is required on trucks equipped with stability enhancement features such as antirollover and directional stability electronics.

A more modest system used on truck chassis would be a four-in/four-out system (known as 4S/4M), but a six-in/four-out system (known as 6S/4M) can also be used. In the six-in/four-out configuration, there would be a wheel speed sensor on each of the six wheels on the vehicle but only four modulators. This means that each modulator would manage a pair of service brake chambers.

Channel Arrangements

Regardless of the number of ABS channels, on most tractors or straight trucks, there is only one ABS module managing the system. Truck ABS, whether air or hydraulic, can operate as either split or diagonally arranged systems. A split system means that the modulator valve manages a pair of service chambers located on one side of the vehicle. This arrangement has a certain advantage when a vehicle is run on a split-coefficient surface, for example, pavement on one side and gravel shoulder on the other. Diagonal arrangements also are used, often on straight trucks, in which case the following is true:

- Diagonal 1 controls the right front wheel and the left rear wheels.
- Diagonal 2 controls the left front wheel and the right rear wheels.

If a system fault occurs in one diagonal, the other diagonal will continue to provide the ABS function. Diagonal configurations of ABS do not perform as well on split-coefficient surfaces.

SIX-CHANNEL SYSTEM

The six-channel system is often described as a 6S/6M system. The "S" stands for *sensor* and the "M" stands for *modulator*. The six-channel ABS can be used on straight trucks or tandem drive tractors. A single ECU receives inputs from a wheel speed sensor located on each wheel of the chassis. An ABS modulator for each wheel is located downstream from the relay valve responsible for actuating the service chambers. In this way, the wheel speed of each wheel is monitored and air pressure to the service chambers can be modulated based on the exact conditions at each wheel.

Six-channel ABS have become the most common system on newer trucks, which tend to be spec'd with stability enhancement electronics. If the only consideration was a stand-alone ABS on an air-braked truck, a 4S/4M system has a proven track record of effectiveness. For this reason, there are currently more 4S/4M systems on our roads but this will change, and 6S/6M systems will become commonplace as operators seek to maximize rig safety. **Figure 30–13** shows a full six-channel (six-in/six-out) system that includes a traction control option, that is explained in the next section.

SIX/FOUR CHANNEL SYSTEM

The 6S/4M ABS locates a wheel speed sensor on each of the six wheels of the vehicle, so the speed of each wheel is monitored. However, the system produces only four outputs, meaning that four of the six wheels are managed in pairs. If a wheel lockup condition is sensed by the ABS module in either of the wheels grouped in a pair, the service application pressure will be modulated to both wheels managed by the modulator.

FOUR-CHANNEL SYSTEM

The four-channel system is usually referred to as a 4S/4M system. A wheel speed sensor is located on each of the steering axle wheels and on two of the four rear wheels. The rear brake wheel speed sensors can be installed on either the forward or rear axle of the tandem, and the actual location usually depends on the type of suspension. One sensor must be located

FIGURE 30–13 Six-channel ABS configuration with ATC

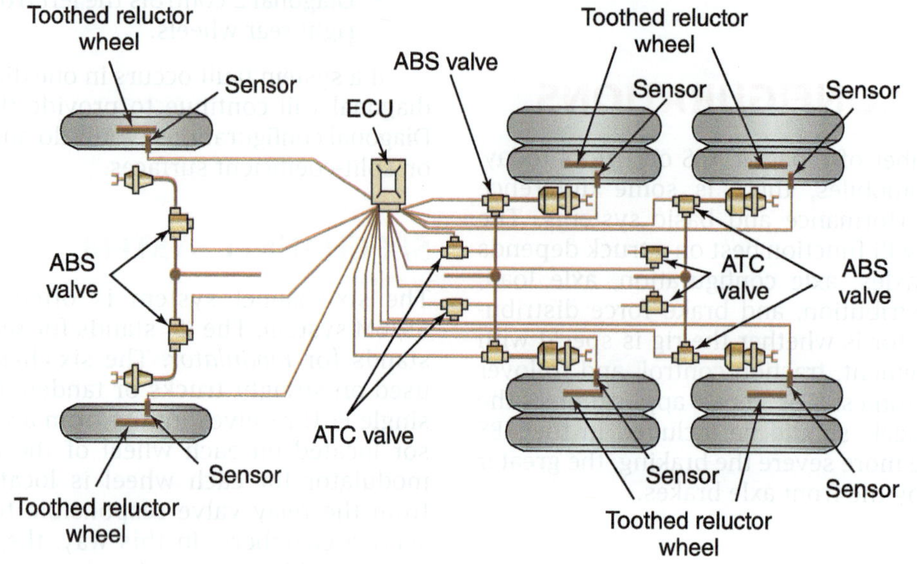

FIGURE 30–14 Some typical four-channel ABS configurations

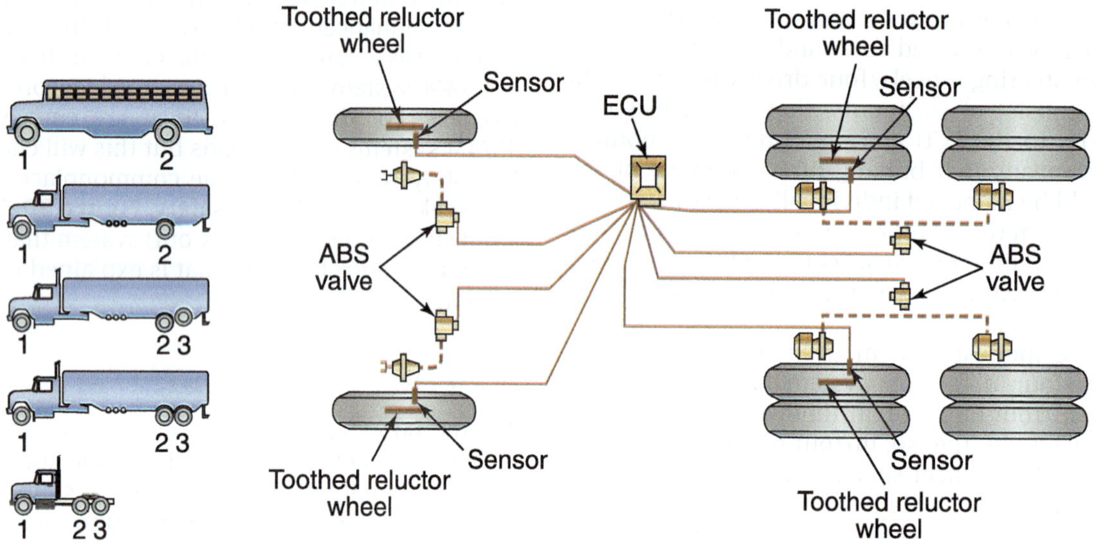

on either side, so both sensors have to be on one axle. **Figure 30–14** shows some typical four-channel ABS configurations.

TWO-CHANNEL SYSTEM

Today, a two-channel system can be found on trailers, buses, vans, and straight trucks. There are limited benefits to an ABS that only functions over the drive axles on a vehicle. However, it is not totally without merit on a tractor because drive axle wheel lock is a major cause of trailer jackknife. Also, when a tractor is bobtailed, ABS on the drive axles can reduce fishtail, especially where no bobtail proportioning valves are used. **Figure 30–15** shows a typical two-channel ABS on a trailer application that meets the current minimum legal requirements.

CAUTION

ABS effectiveness is fully dependent on the foundation brake adjustment. Use stroke indicators to check brake adjustment status.

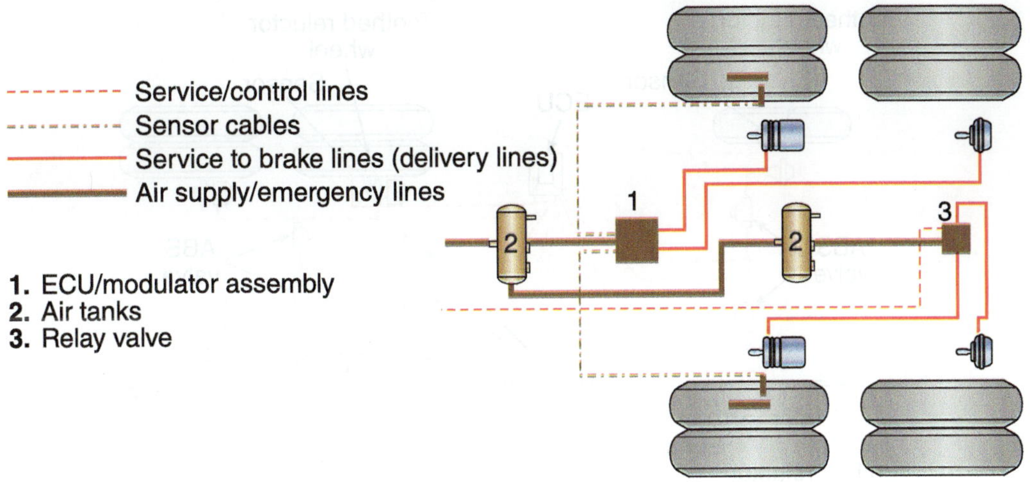

FIGURE 30–15 Two-channel ABS configuration on a highway trailer

30.4 AUTOMATIC TRACTION CONTROL (ATC) SYSTEMS

An electronic ABS also can be used to control a wheel spinout condition if the system is set up with traction control. In fact, relatively little extra hardware or expense is required, so ATC has become a popular option. The module that manages the system is now called an ABS/ATC module.

ATC OPERATION

Wheel rotational speed is monitored in an ABS, so when a drive axle wheel begins to spin out, brake pressure is applied to the brakes on that wheel. This transfers driveline torque to the wheels with better traction and arrests the spinout condition. In most current systems, an ABS/ATC module is multiplexed with an electronically managed engine. This means that when a spinout condition is detected by the ABS/ATC module, it can signal the engine controller to reduce drive torque to the drive axles as part of the traction control strategy.

ATC systems are an option with most four-channel or six-channel ABS. Just as wheel deceleration can be individually controlled to avoid a skid, acceleration can also be checked to avoid spinning a wheel. When an ATC-equipped vehicle accelerates on a split-coefficient surface, the brakes can be managed to intermittently apply to any drive wheel that is sensed to exceed a preprogrammed acceleration threshold, in other words, is about to spin out.

ATC uses two pneumatic solenoid valves and a pair of double check valves. The ATC valve assembly can either be a separate unit or integral with the ABS modulator valve package. Most wheel spinout occurs at relatively low vehicle speeds, usually less than 20 mph (32 kmh). The ATC systems, therefore, are designed to operate only at vehicle speeds below a certain value. Above 20 mph (32 kmh), an engine derate may take place when a spinout condition is detected. **Figure 30–16** shows a four-channel ABS with the ATC option.

30.5 TRAILER ABS

All current trailers (anything built after March 1, 1998) require ABS. The minimum requirement is quite basic, however, and a 2S/1M system on a four-axle trailer meets the legal requirement for ABS. Any ABS monitors wheel speed and controls the wheel rotational speed under braking. Trailer ABS operates on the same fundamentals already introduced in discussing tractor and straight truck ABS. Again, the various system configurations are described by the number of inputs (wheel speed sensors) and outputs (modulators) to the ABS module. Some typical configurations used are:

2S/1M = two wheel speed sensors and a single modulator valve
2S/2M = two wheel speed sensors and two modulator valves
4S/2M = four wheel speed sensors and two modulator valves
4S/3M = four wheel speed sensors and three modulator valves
6S/3M = six wheel speed sensors and three modulator valves

TRAILER ABS OPERATION

The operating principles are identical to those on the tractor. Signals from the wheel speed sensors are

FIGURE 30–16 Four-channel ABS with an ATC option

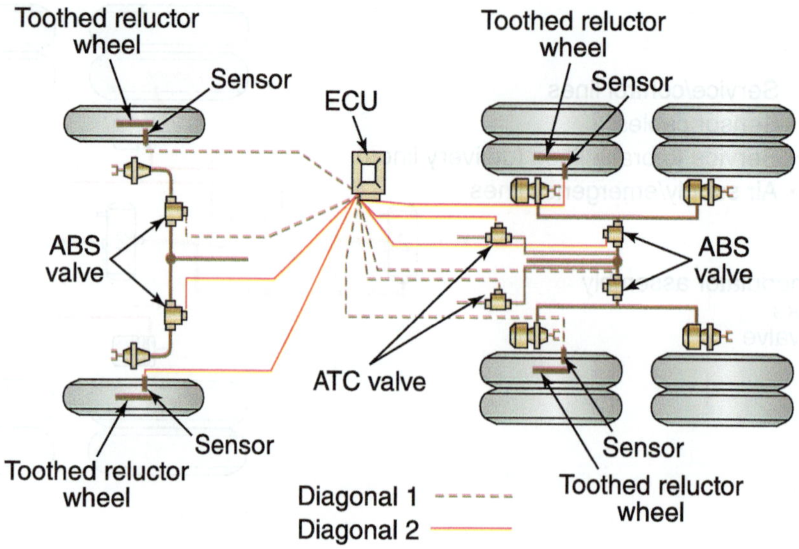

received and processed by the ECU. When no lockup condition is sensed, the ABS module allows the service brakes to operate exactly as in non-ABS mode. When imminent wheel lockup is sensed, the ECU begins to manage braking in ABS mode. This means that the ABS relay valve or modulator can cycle the apply/release pressure to the service chamber pressure to prevent a lockup condition. **Figure 30–17** shows some of the more commonly used trailer ABS options.

Trailer ABS has its own ECU that manages antilock braking independently from that of the tractor. Current trailer ABS modules are powered up by the auxiliary pin (pin number 7) on an ATA/SAE 7-pin trailer plug (SAE J560 plug). **Figure 30–18** shows a trailer plug and the location of the number 7 auxiliary pin. This pin is currently allocated for PLC RF signals that are superimposed over the electrical current.

Some pre-2007 trailer ABS are powered up off the brake light circuit, meaning that the system was only electrically active under braking and only when there was no problem with the brake light circuit. A major disadvantage of this system was that drivers had no way of knowing if there was an ABS malfunction until they applied the brakes. The processing functions of a trailer ABS module are almost identical to those on a tractor ABS:

- If an ABS relay valve malfunctions, the wheels controlled by the valve return to standard service braking.
- If a sensor malfunctions in a two-sensor system, control of that affected side of the trailer returns to standard service braking.
- If one sensor malfunctions in a system that has four sensors on a tandem axle, the ABS will continue to function. In this event, the ECU manages the system using the wheel speed data on the functioning sensor on the same side of the tandem.
- If a problem completely disables the ABS, control returns to standard braking.
- All current trailer ABS use a 100-tooth reluctor wheel.
- Many different manufacturer ABS components are interchangeable, but be sure that components can be cross-referenced before actually doing this. For instance, some Haldex solenoids interchange with Arvin Meritor.

TRAILER ABS WARNING LIGHT

Trailer ABS should have a failure warning light mounted on the trailer. Auxiliary-powered systems also should have a second light on the tractor dash. The warning lights indicate the status of the system. The system automatically performs a self-check each time the ECU is activated. The warning light stays on until the self-check is completed. A self-check includes checking the output of the wheel speed sensors. Therefore, the vehicle must be moving at least 4 mph (6 kmh) to complete the self-check. Then, if the system is functioning correctly, the warning light will go out. If a fault exists, the light will stay on. The warning light will always go out when the ECU is deactivated.

SHOP TALK

Bulb Test: Depress the ABS switch for 16 seconds. The ABS light should illuminate even when no codes are present.

FIGURE 30–17 Typical trailer ABS configurations

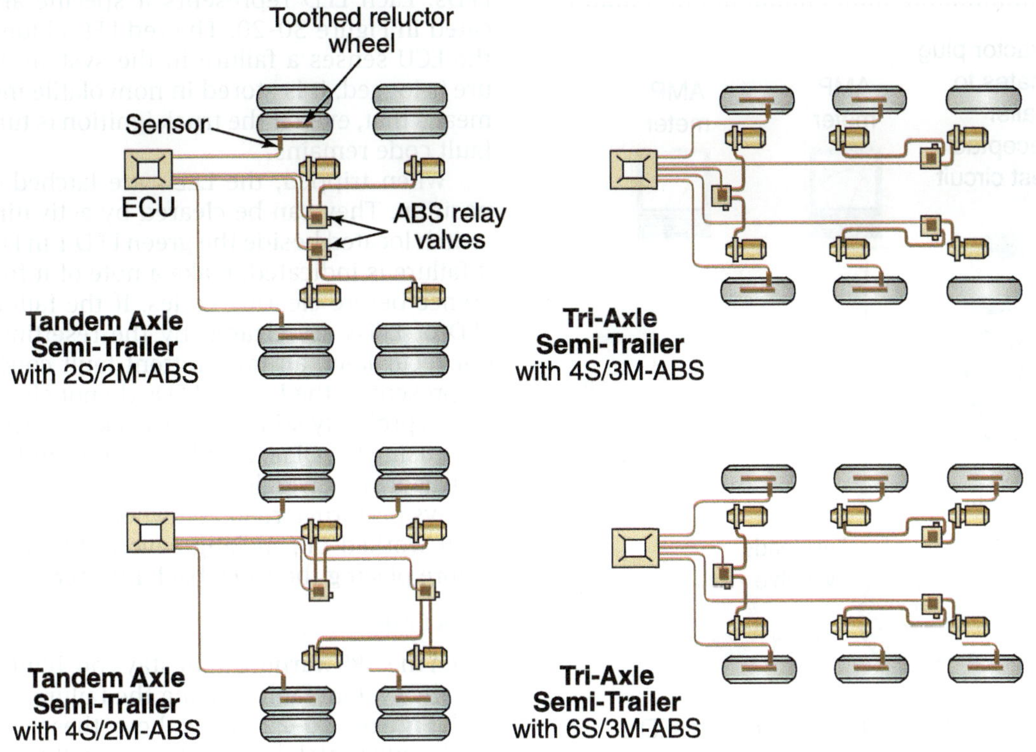

FIGURE 30–18 Trailer electrical socket power test schematic

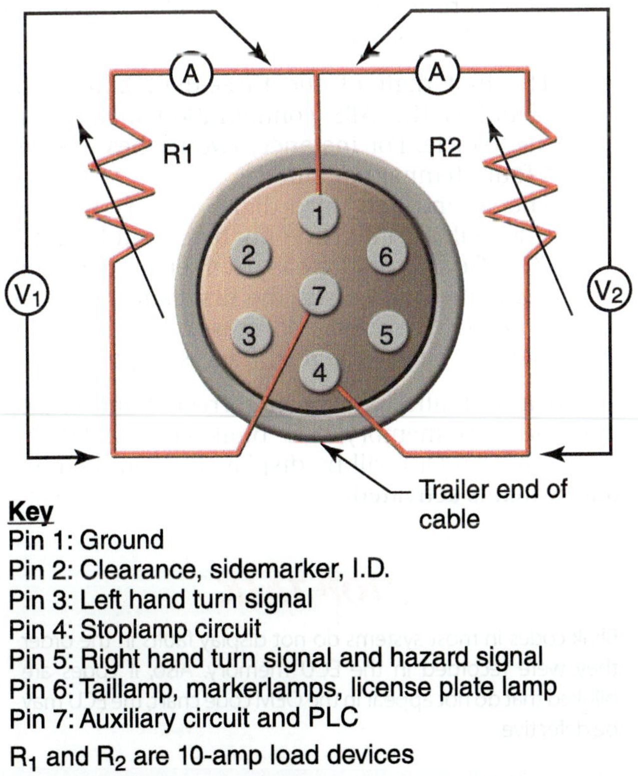

When the ECU is activated depends on how the system is powered. A brake light–powered ECU is activated each time the brakes are applied and is deactivated each time the brakes are released. An auxiliary-powered ECU is activated when the ignition is turned on and remains activated until the ignition is turned off. Therefore, how the warning light operates depends on the power source. **Figure 30–19** shows the tractor ammeter power test circuits in an ATA seven-wire trailer receptacle but check the OEM service literature before using this type of light load to test the circuits. Some will advise that only the ammeters integral with digital multimeters (DMMs) be used.

30.6 ABS DIAGNOSTICS

The ECU or the modulator controller usually provides a quick and easy method of checking ABS failures or faults, as they are called when stored in ECU memory. There are two onboard diagnostic systems in common use: LED indicators (shown in Figure 30–5 and **Figure 30–20**) and blinking light codes (shown in **Figure 30–21**). On all current systems, the ECU diagnostics can be accessed off the chassis data bus. The following account of ABS roughly follows Arvin Meritor systems, but most other systems are similar to this model.

FIGURE 30–19 Tractor power test circuits using an ammeter

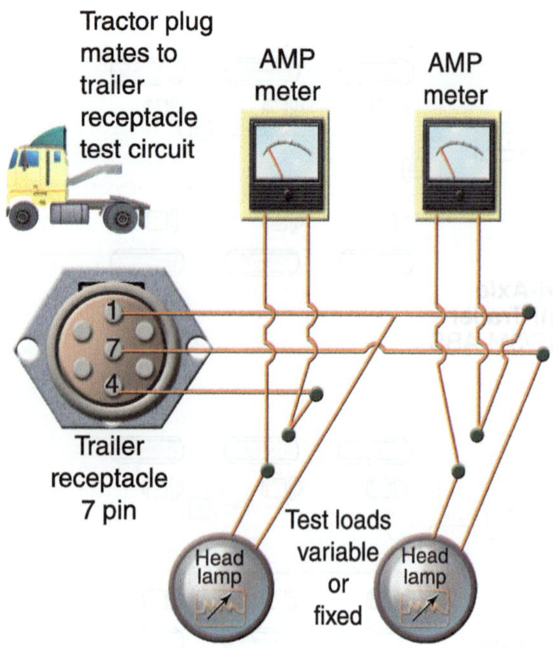

FIGURE 30–20 Typical ECU diagnostic LED window located on a first-generation controller assembly.

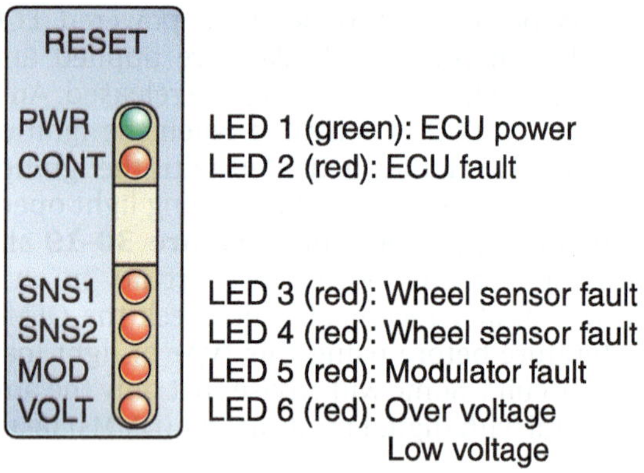

WARNING

The ABS is an electrical system. When working on ABS, the same precautions that must be taken with any other electrical system should be taken (see Chapters 2, 10, and 11). There is a danger of electrical shock or sparks that can ignite flammable substances. Always disconnect the battery ground cable before working on the electrical system. This is of special concern when working with flammable or explosive truck or trailer cargos.

LED Indicators

In this system, if the dash warning light remains on after the truck reaches a speed of 4 to 7 mph (6 to 11 kmh), with or without blinking on the initial startup of the truck, inspect the ECU for illuminated LEDs. Each LED represents a specific area, as indicated in Figure 30–20. The red LEDs illuminate when the ECU senses a failure in the system. When a failure is logged, it is stored in nonvolatile memory. This means that, even if the truck ignition is turned off, the fault code remains.

When tripped, the LEDs are latched into the on position. They can be cleared by activating the reset switch located beside the green LED 1 in Figure 30–20. If failure is indicated, make a note of it for future reference before clearing codes. If the failure indicator LED or LEDs are cleared by the resetting procedure but recur later, an intermittent fault condition could be present. If the LED or LEDs do not clear, the condition is probably still active. Check all wiring and hardware using the OEM troubleshooting guide. Remember to perform a "wiggle wire" test at each connection. Always perform a road test whenever the brakes have been worked on, making several brake applications before placing the truck back into service.

Blink Codes

When the ABS warning light stays on, blink codes identify the system fault causing the failure. Figure 30–21 and **Figure 30–22** show how these operate. The blink code switch and light are usually located on the relay/fuse panel. In an Arvin Meritor system, when the blink code switch is activated, the light flashes are usually either 0.5 second or 2.5 seconds. The pauses between the flashes also can be either 0.5 second or 2.5 seconds. The cycle starts with a 2.5-second flash followed by a three-digit blink code. Blink codes can be interpreted as follows:

- The first digit of the three-digit blink code identifies the ABS configuration installed in the vehicle. For instance, two flashes indicate a four-channel system.
- The second and third digits identify the specific fault. A fault can only be erased from the ECU after the failure has been repaired. If more than one fault exists, the first fault displayed must be corrected and erased from the ECU memory before the next fault is displayed.

When all faults have been corrected and erased from the ECU memory, the blink code 2-0-0 (No Faults—System OK) will be displayed when the blink code switch is activated.

SHOP TALK

Blink codes in most systems do not display faults in the order they were recorded in the ECU memory. Also, if codes are blinked that do not appear in the OEM code chart, the ECU may be defective.

FIGURE 30–21 Example of a typical blink code, the only means of reading some first-generation systems. The diagnostic chart required to interpret the codes is located in the service literature.

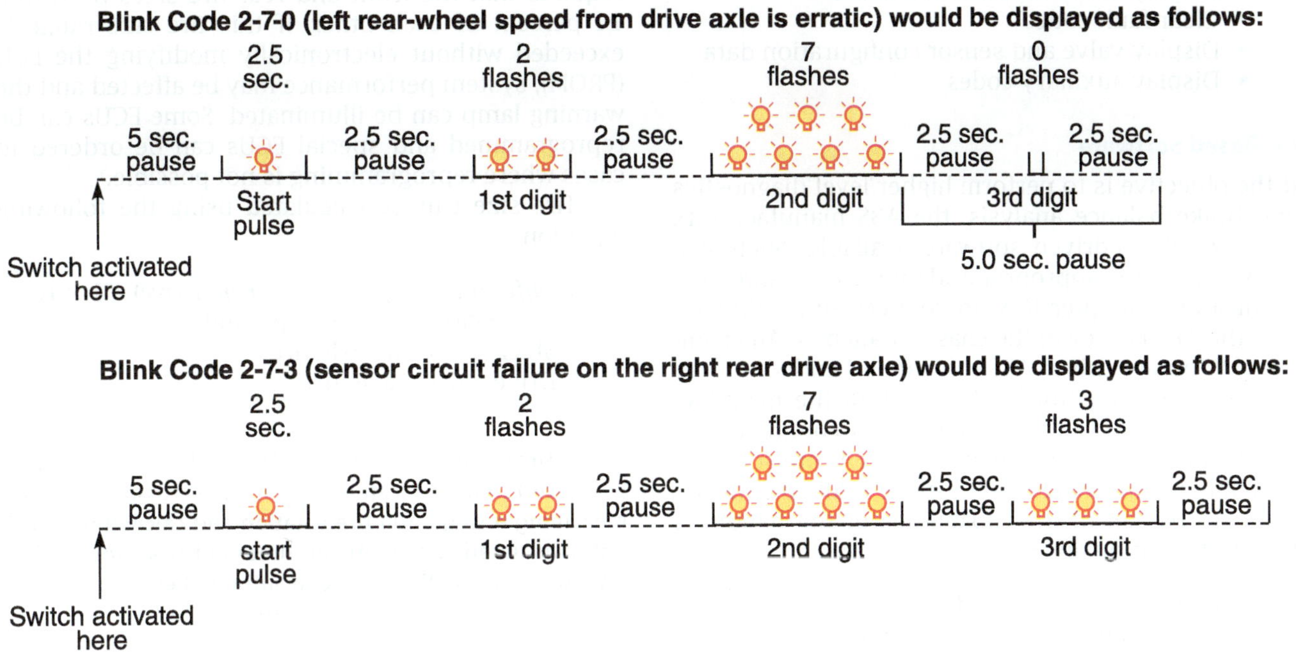

FIGURE 30–22 A typical "no-fault" blink code

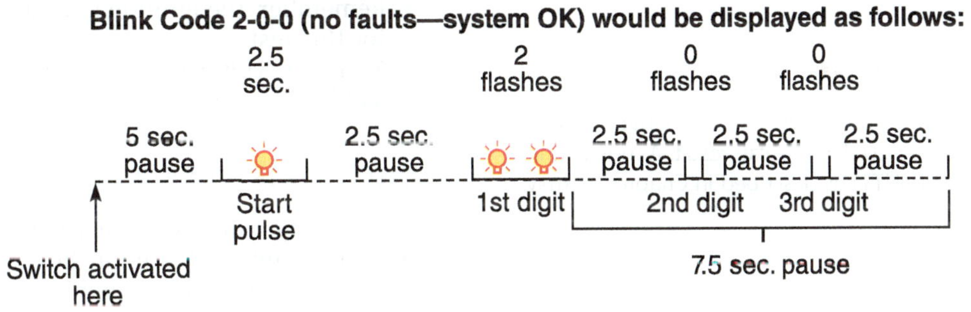

Remember, when servicing ABS components that they can be delicate and must be handled with care. If a fault indicates that a component is not functioning properly, it should be removed and replaced as directed in service literature. Some typical service procedures are covered in the next section.

SELF-DIAGNOSTICS

Self-diagnostics are written into the software of all the current systems. Most ABS are designed to perform a self-test on startup. When the ABS circuit is first energized, a clicking noise can be heard from the modulator valve solenoids. There may be some read and display capability on the ABS module, but in most cases it is accepted that one type of EST will be used to navigate the system electronics. Within this section, we will reference Haldex-Midland ABS in describing the basics of trailer ABS diagnosis.

ESTs

ABS manufacturers make available handheld ESTs and software to troubleshoot their systems. NEXIQ also manufactures generic full-function handheld ESTs such as the **Pro-Link iQ** that are ABS user friendly. The **NEXIQ Brake-Link** is a specialty generic brake system EST designed to diagnose all the OEM ABS. A Bendix **remote diagnostic unit (RDU)** or Haldex digital diagnostic unit (DDU) are two handheld ESTs that come equipped with the appropriate wiring and connectors to bus directly to the ABS. Most of these ESTs perform the following functions:

- Display active and inactive codes
- Clear fault codes
- Display valve and sensor configuration data
- Display auxiliary codes

PC-Based Software

If the objective is to perform higher level diagnostics and brake balance analysis, the ABS manufacturers make Windows-driven software available. Such systems require the appropriate adapter cables and communications adapter (CA) to connect either directly into the ABS ECU or to the chassis data bus. An example of Windows-driven ABS software is Bendix's **ACom software** and Meritor ToolBox but all the manufacturers have equivalent versions. It is commendable that Bendix continues to make Bendix ACom software available for no charge. Full feature brake diagnostic software has all of the capabilities of the handheld ESTs but additionally will:

- Display odometer readings
- Display an audit trail (archived faults)
- Accept tire data programming
- Provide service reminder data
- Analyze brake performance and display percentage of braking by each wheel
- Archive data trails

SHOP TALK

Never use a battery charger to perform a dynamic check of a trailer ABS because ECU damage can result. A mobile universal trailer tester (MUTT) as described in Chapter 33 may be used.

30.7 ABS SERVICE PROCEDURE

Technicians are reminded that there is no substitute for the OEM service literature and software-guided troubleshooting when maintenance and service procedures are undertaken on ABS. More recent ABS tend to rely more on PC software, and specialty handheld diagnostic tools, whereas some of the early systems used single-system scan tools and flash codes. Regardless of which EST is used to access the ABS electronics, the fundamental procedures are similar. It should be noted that PCs loaded with the appropriate software are more user-friendly than the blink/flash code systems because interpretation errors are minimized. We will reference Arvin Meritor systems in this section, but most of these procedures are similar in other OEM systems.

TIRE SIZE RANGE

Proper ABS/ATC operation with a standard ECU requires that the front and rear tire sizes be within 14 percent of each other. If this tire size range is exceeded without electronically modifying the ECU (PROM), system performance may be affected and the warning lamp can be illuminated. Some ECUs can be reprogrammed and special ECUs can be ordered in cases where reprogramming is not possible.

Tire size can be calculated using the following equation:

$$\%\ \textit{Difference} = (\textit{rpm Steer} \div \textit{rpm Drive}) - 1 \times 100$$

rpm = tire revolutions per mile
Steer = steering axle tire
Drive = drive axle tire

Note: Tire rolling radii expressed in revolutions per mile (or kilometer) should be referenced in a tire crossover manual. Measuring a single full tire rotation and calculating this is not considered sufficiently accurate in trucks and trailers, although the practice is used in automobiles.

ABS VOLTAGE CHECK

- Voltage should measure between 11 and 14 volts on a 12-volt system (18 and 30 volts for a 24-volt system).
- Remember, the ignition key must be turned on for this test.
- Measure voltage at the pins indicated in **Table 30–1**.

LOCATION OF SENSORS

On the steering axles of most trucks, the sensor is located on the inboard side of the steering knuckle. On drive axles, the drum assembly sometimes has to be removed to access the sensor.

TABLE 30–1 Voltage Check Pins

ECU	Connector	Pins
Cab-mounted	18-Pin	7 and 10
		8 and 11
		9 and 12
Basic	15-Pin	7 and 4
		8 and 9
Frame-mounted	X1-Gray	1 and 12
		2 and 11

TABLE 30–2 ArvinMeritor Codes

Blink Code Identification	
First Digit (Type of Fault)	Second Digit Specific (Specific Location of Fault)
1 No faults	1 No faults
2 ABS modulator valve 3 Too much sensor gap 4 Sensor short or open 5 Sensor signal erratic 6 Tooth wheel	1 Right front steer axle (curb side) 2 Left front steer axle (driver's side) 3 Right rear drive axle (curb side) 4 Left rear drive axle (driver's side) 5 Right rear/additional axle (curb side)* 6 Left rear/additional axle (driver's side)*
7 System function	1 J1922 or J1939 datalink 2 ATC valve 3 Retarder relay (third brake) 4 ABS warning lamp 5 ATC configuration 6 Reserved for future use
8 ECU	1 Low power supply 2 High power supply 3 Internal fault 4 System configuration error 5 Ground

*Tandem, lift, tag, or pusher axle, depending upon the type of suspension.

CODE INTERPRETATION

Troubleshooting of any ABS must be performed using the OEM service literature. **Table 30–2** may be used to interpret Meritor WABCO and Haldex-Midland codes.

HALDEX CODES

Haldex codes also are grouped into categories. The following description uses the codes as they would be read by the Haldex EST or Haldex Windows software.

- **Codes 00, 07, and 8.8.** No faults detected. 00 is dynamic no-fault (moving vehicle) and 07 is static (vehicle parked) no-fault. 8.8 indicates that the system is undergoing a self-test sequence, which includes testing the display LEDs. DB indicates that an electrical connection has been made between the chassis and the ECU.
- **Codes 03, 04, 05, and 06.** Indicate a short or open in a wheel speed sensor. The resistance in the indicated sensor should be tested with a DMM. The resistance value should read 980 to 2,350 ohms.
- **Codes 13, 14, 15, and 16.** Indicate that the output of a moving wheel sensor is insufficient, usually caused by excessive gap between the sensor and the reluctor/exciter wheel. Measure the AC voltage produced when rotating the wheel at one revolution every 2 seconds. It should produce a reading of between 50 and 750 millivolts (0.050 – 0.750 V-AC) depending on the OEM, but 200 millivolts (0.2 V-AC) is typical. If the sensor gap is excessive (sensor dropped out of socket), no fault will be logged because the ECU will assume that the vehicle is parked.
- **Codes 23, 24, 25, and 26.** Indicate intermittent loss of wheel sensor input when the vehicle is moving down the road. Causes can be difficult to diagnose, but some possibilities are a broken sensor retaining clip, damaged reluctor wheel, or excessive wheel bearing endplay.
- **Codes 42 and 43.** Occur when the unit is moving and indicate that the wheel speed is slow to recover when ABS releases the brake during an ABS event. Possible causes are a dragging foundation brake, a kinked or otherwise damaged brake hose, or a defective modulator valve.
- **Codes 62, 63, 68, and 69.** Indicate that one of the solenoids or its cable is open or shorted. Disconnect the indicated solenoid and check the resistance across the connectors. Readings across the bottom 2 pins should be 7 to 9 ohms. Between either bottom pin and the top pin the reading should be between 3.5 and 4.5 ohms. Remember to check for corrosion at the pins and connectors before condemning a solenoid.
- **Codes 72, 73, 78, and 79.** Indicate a short to ground in a modulator solenoid or its cable. The most likely cause is a damaged cable or solenoid. Disconnect the solenoid connector and check for continuity between each solenoid terminal and chassis ground. Perform tests as outlined in the Haldex service literature. This code also can be caused by a defective ECU.
- **Codes 82, 83, 88, and 89.** Indicate a short to battery positive in a modulator solenoid or its cable. The most likely cause is a damaged cable or solenoid. Disconnect the solenoid connector and check for continuity between each solenoid terminal and chassis positive. Perform tests as outlined in the Haldex service literature. This code also can be caused by a defective ECU.
- **Code 90.** Indicates ABS voltage is below 8.5 volts. Most likely caused by damaged or corroded terminals or wires. Use voltage drop testing between the trailer connector and the ABS power circuit to locate fault. This fault also can be caused by undersized wiring.

If performing the test using a battery-powered ABS tester, check the battery voltage.

- **Code 92.** High voltage caused by ABS power supply voltage above 16 volts. Check tractor voltage regulator or battery-powered tester voltage.
- **Code 80, 93, 99, E0–E9, or EA–EF.** Defective ECU. Replace ECU and retest system.
- **Code CA.** Invitation to clear all fault codes. Dynamic faults will not be erased until the vehicle is moved at a speed exceeding 6 mph (10 kmh).
- **Code CC.** A CC code is displayed the third consecutive time that a CA is attempted. It is an invitation to clear configuration, so it should be avoided. If a CC is displayed, first power down the system and then power up again.

Sensor Adjustment

1. Push the sensor in until it contacts the tooth wheel.
2. Do not pry or push sensors with sharp objects.
3. Sensors will self-adjust during wheel rotation.

Sensor Signal (Output) Voltage Test

Voltage should typically be 0.200 volt AC at 30 rpm.

1. Turn the ignition off.
2. Disconnect the ECU (see the wiring diagram).
3. Raise the wheel from the ground and release the brakes.
4. Use a DMM on AC volts mode range set to read up to 1 V-AC.
5. Rotate the wheel by hand at 30 rpm (one revolution per 2 seconds).
6. Measure voltage at the pins indicated in **Table 30–3**.

Photo Sequence 15 shows the procedure for using this method of checking wheel speed sensor output voltage.

Sensor Resistance

The sensor circuit resistance should be between 300 to 2,200 ohms. Resistance can be measured at the sensor connector or at the pins on the ECU connector using a DMM. To measure resistance:

1. Turn the ignition off.
2. Set the DMM to read ohms/resistance.
3. To measure resistance at the sensor connector, disconnect the ECU connector from the ECU.
4. Measure resistance at the pins indicated and check to specification.

Dynamometer Testing Vehicles with ATC

Vehicles equipped with ATC and stability enhancement electronics must have the systems electronically disabled before running the vehicle on a chassis dynamometer. Although the preferred means of achieving the disabling is electronic, this is not always possible with older systems. Most current truck OEMs provide dynamometer operation mode programming on the chassis or ABS electronics; it can be selected by going into customer data programming. Another typical method of disabling ATC is to press and hold the blink code switch for at least 3 seconds. When the system configuration code appears, ATC has been disabled. The ATC lamp illuminates and stays on while the system is disabled. Some systems require that the ATC fuse or breaker be removed. Check with OEM service literature.

TABLE 30–3 Sensor Check Pins

ECU	Sensor	Connector	Pins
Cab-mounted	LF	6-Pin	4 and 5
	RF	9-Pin	4 and 5
	LR	15-Pin	5 and 6
	RR	15-Pin	8 and 9
	LR (third axle)	12-Pin	5 and 6
	RR (third axle)	12-Pin	8 and 9
Basic	LF	18-Pin	12 and 15
	RF	18-Pin	10 and 13
	LR	18-Pin	11 and 14
	RR	18-Pin	17 and 18
Frame-mounted	LF	X2—Black	7 and 8
	RF	X2—Black	5 and 6
	LR	X3—Green	1 and 2
	RR	X3—Green	3 and 4
	LR (third axle)	X4—Brown	3 and 4
	RR (third axle)	X4—Brown	5 and 6

Testing an ABS Modulator Valve

Measure the resistance across each valve solenoid coil terminal and ground on the ABS valve to ensure 4 to 8 ohms for a 12-volt system (8 to 16 ohms for a 24-volt system).

- If the resistance is greater than 8 ohms for a 12-volt system (16 ohms for a 24-volt system), clean the electrical contacts in the solenoid. Check the resistance again.
- To check the cable and the ABS valve as one unit, measure resistance across the pins on the ECU connector of the harness. Check the diagram of the system you are testing for pin numbers.

Photo Sequence 13

WHEEL SPEED SENSOR TESTING

P 13–1 Because the parking brakes must be released to perform this check on any wheels other than those on the steer axle, ensure that a pair of wheels on two separate axles are blocked as shown.

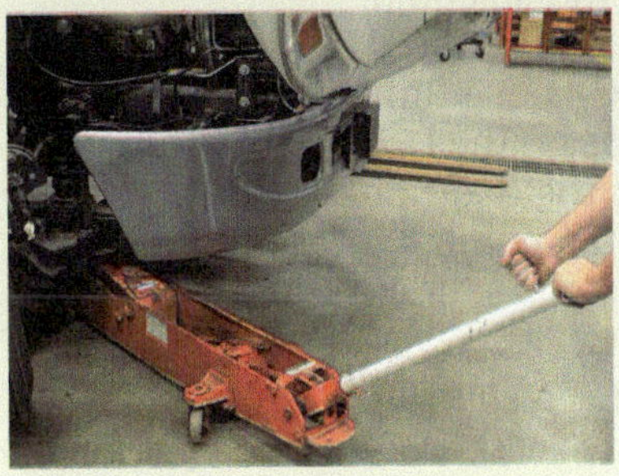

P 13–2 For ease of demonstration, we are going to check the wheel speed sensor on the front axle of this truck. Jack the truck up as shown.

P 13–3 Lower the steer axle I-beam onto a pair of heavy-duty safety stands as shown.

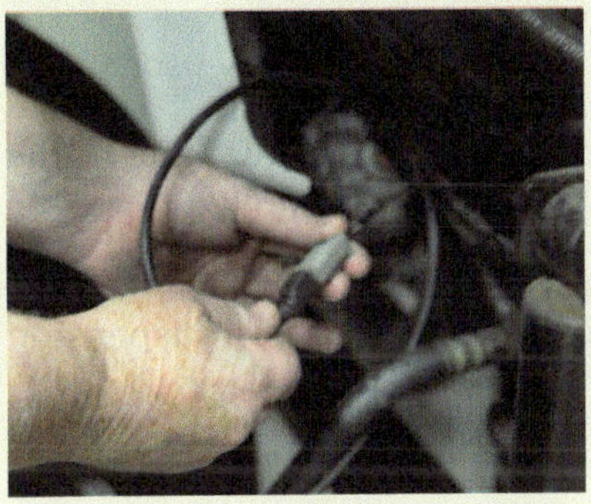

P 13–4 Locate the harness connector to the wheel speed sensor: It may be necessary to cut nylon ties or disconnect clamps to do this.

P 13–5 Separated wheel speed sensor connector.

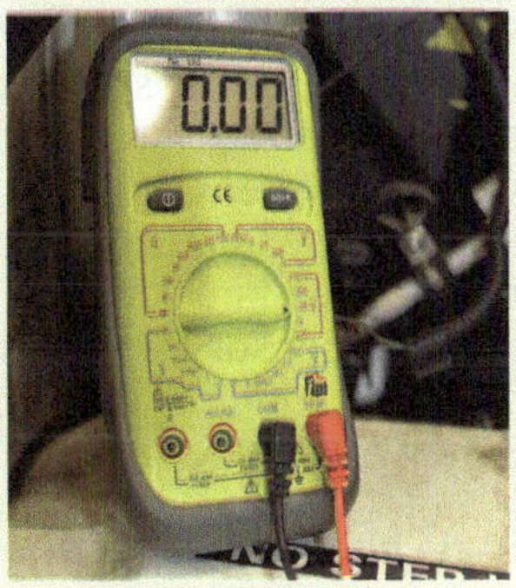

P 13–6 Set a DMM to AC test mode as shown: If the DMM is not autoranging, ensure that it is set to accurately display readings of 1 V-AC or less.

P 13–7 Connect a pair of jumper wires (alligator clips can help here) to the wheel speed sensor connector terminals.

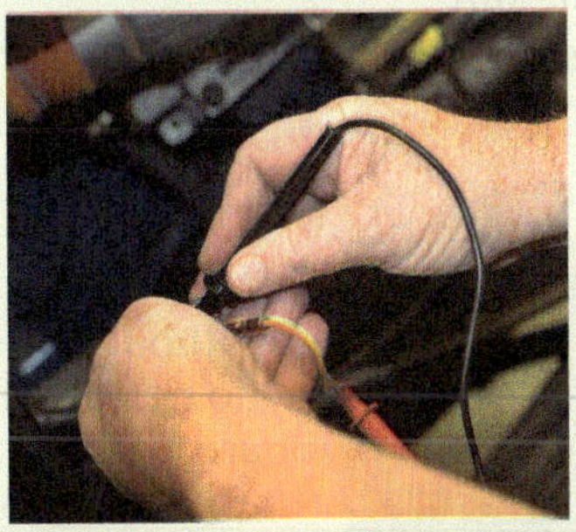

P 13–8 Connect the DMM test leads to the jumper wires.

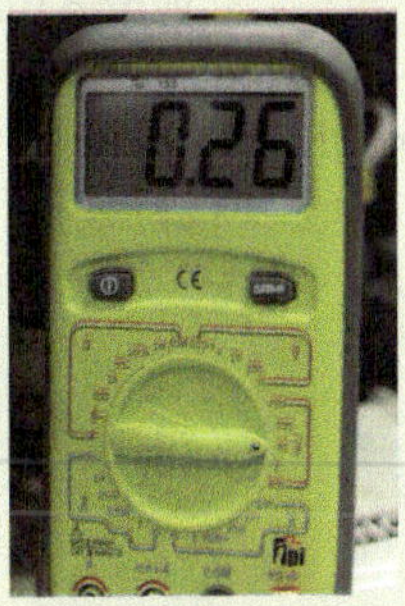

P 13–9 Spin the wheel up to around 30 rpm. This should produce a reading of between 0.050 and 0.750 V-AC. Check to OEM specification. If the OEM spec is only specified as resistance, change the DMM to read ohms and do not rotate wheel. The specification should read between 300 and 2200 ohms depending on the type; check to OEM specifications.

ATC Valve

Measure the resistance across the two electrical terminals on the ATC valve to ensure 8 to 14 ohms for a 12-volt system (16 to 28 ohms for a 24-volt system).

- If the resistance is greater than 14 ohms for a 12-volt system (28 ohms for a 24-volt system), clean the electrical contacts on the solenoid. Check the resistance again.
- To check the cable and ATC valve as one unit, measure resistance across the pins on the ECU connector of the harness. Check the diagram of the system you are testing for pin numbers.

Front Axle Wheel Speed Sensor Installation

When working on wheels and axles, the technician should be careful of ABS components such as sensors because they are easily damaged (**Figure 30–23**). To replace the sensor in a front axle:

1. Connect the sensor cable to the chassis harness.
2. Install the fasteners used to hold the sensor cable in place.
3. Apply an OEM-recommended lubricant to the sensor spring clip and sensor.
4. Install the sensor spring clip. Make sure that the spring clip tabs are on the inboard side of the vehicle.
5. Push the sensor spring clip into the bushing in the steering knuckle until the clip stops.
6. Push the sensor completely into the sensor spring clip until it contacts the tooth wheel.
7. Remove the blocks and safety stands.

Rear Axle Wheel Speed Sensor Installation

1. Apply an OEM-recommended lubricant to the sensor spring clip and sensor.

FIGURE 30–23 When working on wheels and axles, the technician should be careful of ABS components such as sensors.

FIGURE 30–24 ABS operation

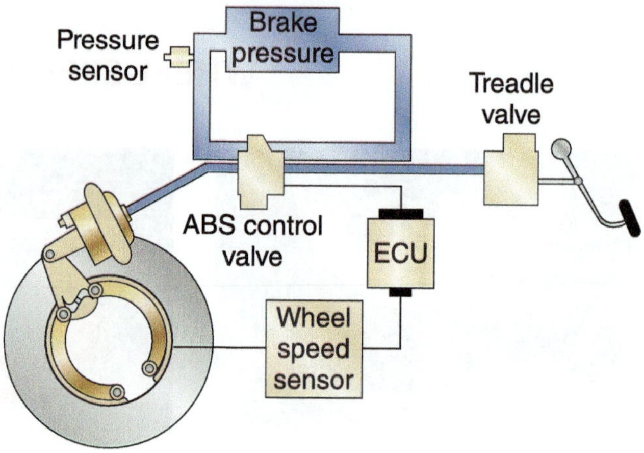

2. Install the sensor spring clip. Make sure that the spring clip tabs are on the inboard side of the vehicle.
3. Push the sensor spring clip into the mounting block until it stops.
4. Push the sensor completely into the sensor spring clip until it contacts the tooth wheel.
5. Insert the sensor cable through the hole in the spider and axle housing flange. Route the cable to the frame rail. Be sure to route the cable in a way that will prevent pinching or chafing and will allow sufficient movement for suspension travel.
6. Connect the sensor cable to the chassis harness.
7. Install the fasteners that hold the sensor cable in place.
8. Install the brake drum on the wheel hub.
9. Complete the installation per the vehicle manufacturer's manual.

Figure 30–24 shows an operational schematic of an air-actuated ABS.

30.8 ABS ADD-ONS

ABS lends itself to some add-on enhancements. Although these are currently optional, because they do not significantly add to the system hardware, they are commonly found on today's vehicles. There is a good possibility that these will be legislated as a required safety feature on highway tractor trailers in the near future. When ABS is networked to the chassis data bus, the ABS ECU can both broadcast and receive data from the chassis data backbone and enable these add-ons that are mostly logic driven. Logic driven means that little additional hardware is required to enable the system to function.

Add-ons to ABS can be generally grouped under the heading of **stability control electronics**

(SCE) though some OEMs prefer the term "stability *enhancement* electronics" so that the capabilities of the system are not overstated. We have already taken a look at the most common stability enhancement system, ATC, so this section will examine the others. Stability enhancement systems manage yaw control, roll stability, and directional tracking by using both the vehicle service brake and broadcasting messages on the chassis data bus. Some of these systems have also been touched on in other chapters because they can fall under another chassis system. An example is roll stability control (RSC), which is covered in Chapter 26.

YAW AND TRACKING

Yaw is the tendency of a vehicle moving in a forward direction to turn about its vertical axis. You see an example of yaw when you observe an aircraft landing, countering a strong crosswind. To counter the effect of the crosswind, the airplane is required to angle sideways in order to move in a straight-ahead direction. This is considered a good thing if the pilot wants the aircraft to end up on the runway.

On a highway, yaw can always be considered undesirable. When the friction at the tires of a truck is not sufficient to oppose the side forces it counters, one or more of its tires can slide, resulting in spinout. As the wheelbase on a vehicle becomes shorter, the result is lower dynamic yaw resistance. Factors that influence yaw stability are:

- Length of wheelbase
- Steering geometry
- Suspension type
- Weight distribution
- Tracking geometry of the vehicle

Additional Hardware

Yaw management electronics share the ABS wheel speed sensor circuit as well as using some additional inputs that include:

- Lateral accelerometer
- Steering angle sensors
- Yaw rate sensors

When any threshold values are breached, the system electronics intervene to assist the driver in controlling the condition. Intervention includes multiplex command broadcasts to the chassis data bus; this can result in an immediate reduction in engine output torque to help correct the condition.

ROLLOVER AND JACKKNIFE

Other running conditions that can be managed by SCE are potential rollover and jackknife conditions. A rollover can result when a vehicle with a high center of gravity negotiates a turning maneuver on high friction road surface. Jackknife is another somewhat common condition that is caused when the driving speed exceeds a threshold, causing lateral thrust and resulting in vehicle slide on a road surface with a lower coefficient of friction, usually a wet surface.

It should be noted that jackknife is a common result of unbalanced braking, and ABS when installed on both the tractor and trailer has greatly reduced jackknife incidents. A possible negative outcome of mandatory ABS when it was first introduced was evidenced in the increase in the number of trailer swing incidents. Trailer swing occurs when the trailer brakes lock. This can happen when a tractor is ABS equipped (post-1997 chassis) and the trailer is not (pre-1998 trailer). The result can be a tractor with perfectly balanced brakes and overbraked trailer, making panic braking on the rig a tricky operation.

STABILITY MANAGEMENT

Stability management electronics require multiplexing. High-speed communication between the stability management electronics and the brake and vehicle drivetrain are system requirements. The objective is to help the driver avoid a potential accident. The emphasis here is on help; the driver is still key to any accident avoidance strategy.

Steering Angle Sensors

A steering angle sensor enables the stability management electronics to capture driver steering input and intervene if a yaw condition is indicated. This sensor also can provide an indication of an increase in lateral acceleration that may result in a potential rollover event.

Brake Demand Sensors

A brake demand sensor is designed to input an indication of the amount of braking requested by a driver by measuring the application pressure (requested at foot pedal) and comparing it with the application pressure calculated by the SCE to effect a stop. For instance, in a braking maneuver, if the system electronics calculates that a 40 psi (276 kPa) application pressure is required and the driver is requesting only 20 psi (138 kPa), automatic compensation can result. A sudden change of application pressure status by the driver will override the temporary change made by the system.

OPERATING STRATEGIES

SCE constantly compare the actual vehicle running dynamics with performance models logged in system memory. When an operational condition is identified

FIGURE 30–25 Rollover stability control response to a potential rollover situation

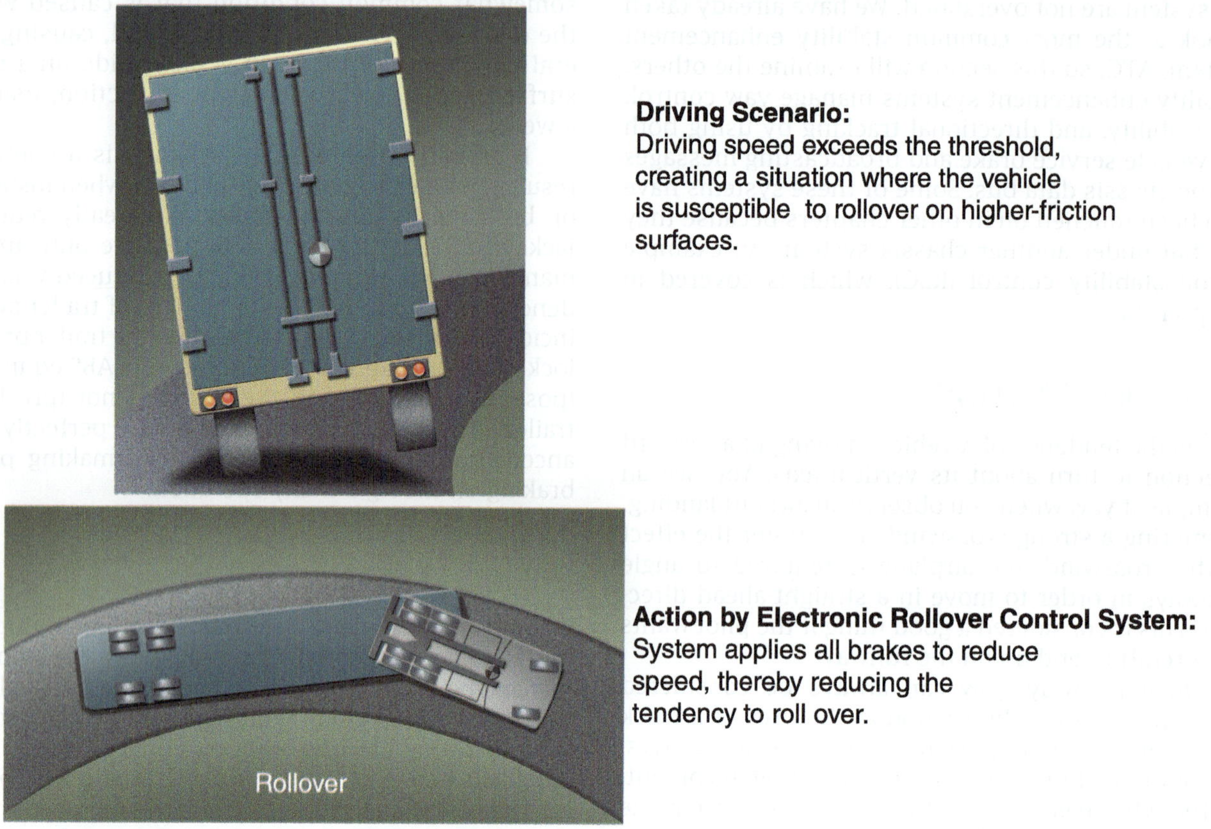

that might cause the vehicle to leave the *desired* travel path, the electronics intervene.

Rollover Avoidance

Tractor/trailer rollovers can occur during a rapid lane change, slippery road surfaces, or encountering a split coefficient surface. Stability electronics use lateral accelerometers, steer angle, and yaw rate sensors to detect and respond to a potential rollover condition faster than a driver can react to the situation. The system can instantly reduce drivetrain torque and while simultaneously applying braking effort to selected chassis brakes to attempt to correct the condition. **Figure 30–25** shows one brake system manufacturer's rollover control response strategy.

Oversteer and Understeer

Although road steering feel is much better in today's trucks, especially those with rack and pinion steering, than those of a generation ago, over- and understeer in trucks is more of a problem in trucks than in automobiles due to loading variables. In an oversteer condition, the system electronics can apply the outside front brake, whereas in an understeer, the inside front brake would be applied.

Jackknife

Jackknifing events have been greatly reduced on our highways following the adoption of mandatory ABS on trucks and trailers. However, further jackknifing protection can be achieved by SCE, as shown in **Figure 30–26**. Jackknifing control response can involve biasing braking effort to the rearmost brakes on the vehicle or rig.

System Options

If you have driven an automobile equipped with SCE on an icy road, you will have a good idea of how much safer the vehicle is to handle. Due to its much longer wheelbase, a truck is more difficult to put into yaw than an automobile, but the instant the directional tracking is lost in a truck, it becomes much more difficult to recover. For this reason, directional stability electronics have become a desirable option in current trucks. **Figure 30–27** shows one manufacturer's description of its directional stability options and the running maneuvers each attempts to provide an additional measure of safety.

FIGURE 30–26 Directional stability control response to a potential jackknife situation

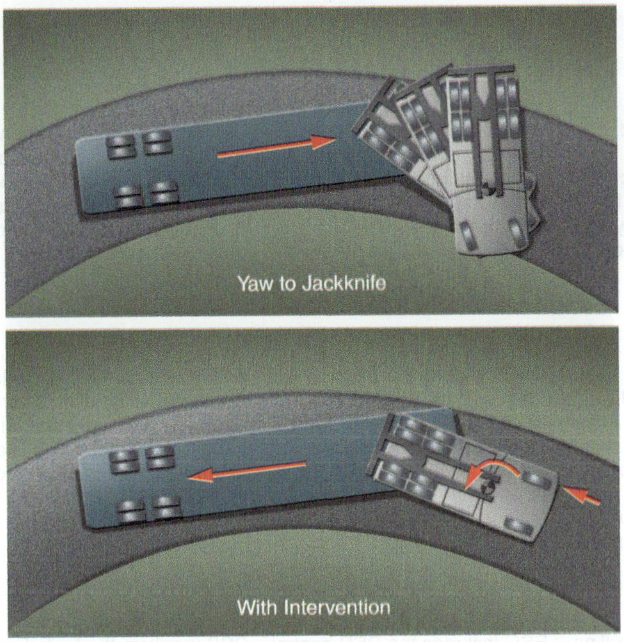

Driving Scenario:

Driving speed exceeds the threshold, and the resulting lateral force causes the vehicle to slide or jackknife on lower-friction surfaces.

Action by Electronic Stability Control:

System applies the appropriate brakes to reduce speed and properly align the vehicle, thereby reducing the tendency to slide or jackknife.

30.9 ELECTRONIC BRAKING SYSTEMS (EBS)

Electronic brake systems (EBS) are currently used on European trucks, and U.S. companies Bendix and Arvin Meritor have established themselves as major players in the systems used there. EBS is an option in the United States, but the system can only be fitted to one with full FMVSS 121–compliant air control circuit as a backup. The problem is the current FMVSS 121 standards, which assume that any use of EBS would be in addition to the required dual-circuit air brake system with pneumatic controls.

Until FMVSS 121 addresses EBS issues, any EBS introduced in North America can only be in addition to a FMVSS 121-compliant dual-circuit air brake system. The European EBS cost more than non-EBS, so they have to be sold on the advantages they offer. These advantages include better stopping performance, faster service braking response, driver stability control, and better antirollover protection (when EBS is designed to interface with antirollover electronics). When EBS is networked to a chassis data bus, all of the rig safety electronics are improved due to faster intervention and response times. The manufacturers of EBS have proven that their systems lower the percentage of brake-related accidents when used in Europe, but so far this has not yet influenced the FMVSS 121 requirements. North American trucks using EBS also must have a completely functional pneumatic control circuit redundancy (i.e., failsafe backup), making EBS a costly option. However, it is used, although not frequently.

EBS Advantage

The control circuit in a truck FMVSS 121 air brake system is pneumatic. By control circuit, we mean all the air signals used to "pilot" or control slave valves such as relay valves in the system. Under ideal circumstances, air pressure signals travel at the speed of sound. Usually the conditions of a pneumatic control circuit are far from ideal, so lag times reduce the efficiency of the control circuit during both application and release.

Electrical signals travel at closer to the speed of light. In a true EBS, the pneumatic control circuit would be eliminated. It would be replaced by a "brake-by-wire" electrical control circuit, which would respond much faster to changes in the command signals from the driver's boot. The actual braking force is still provided by the potential energy of compressed air metered to brake chambers by electrically controlled modulators. **Figure 30–28** shows a circuit schematic of a typical EBS system.

EBS OPERATION

The real advantages of EBS are with control and monitoring. The EBS in testing and in the European market have essentially adapted an air brake system to electronic control and monitoring. This means the

FIGURE 30–27 Running maneuvers and directional stability control options in a Meritor WABCO system

Road Conditions	Common Maneuvers	Roll Stability Control (RSC)	Electronic Stability Control (ESC)	Roll Stability Support (RSS)
High-Friction Surfaces (Dry Pavement)	On Ramp	When critical lateral acceleration is measured while entering a ramp, RSC intervenes by reducing engine torque, engaging the engine retarder and applying drive axle and trailer brakes.	When accelerating while entering a ramp, ESC includes the RSC control functions; the intervention would be similar to RSC and include steer axle braking.	When accelerating while entering a ramp, RSS monitors the trailer's lateral acceleration. If critical lateral acceleration is detected, full trailer brake application is applied to the outside wheels to reduce vehicle speed.
High-Friction Surfaces (Dry Pavement)	Exit Ramp	When critical lateral acceleration is measured while exiting a ramp with excess speed or with a reducing radius, RSC will reduce speed through engine torque reduction, engaging the engine retarder and applying drive axle and trailer brakes.	ESC includes the RSC control functions; the intervention would be similar to RSC and include steer axle braking	RSS monitors the trailer's lateral acceleration. If critical lateral acceleration is detected, full trailer brake application is applied to the outside wheels to reduce vehicle speed.
High-Friction Surfaces (Dry Pavement)	Avoidance	Assists the driver in controlling severe avoidance maneuvers that may exceed stability limits. RSC intervention: engine torque reduction, engaging the engine retarder and applying drive and trailer axle brakes.	ESC includes the RSC control functions; the intervention would be similar to RSC and include steer axle braking	RSS monitors the trailer's lateral acceleration. If critical lateral acceleration is detected, full trailer brake application is applied to the outside wheels to reduce vehicle speed.
Low-Friction Surfaces (Rain/Snow/Ice)	On Ramp	With low friction on ramps the ATC function of RSC reduces the spinning of drive axles to increase traction during acceleration and to maintain vehicle lateral stability and directional control while driving through curves.	With low friction on ramps, ESC has the ATC function of RSC and additional directional control by selectively applying brakes based on the inputs from steer and yaw angle sensing.	While low-friction surfaces do not typically generate lateral acceleration conditions that can cause rollover, the risk of rollover increases when the vehicle suddenly encounters a surface with a higher friction. The ABS function of RSS intervenes to reduce potential swing out or jackknifing.
Low-Friction Surfaces (Rain/Snow/Ice)	Exit Ramp	On low-friction exit ramps, ATC will reduce engine torque when the rear axle speeds exceed the steer axle speeds to reduce jackknife potential.	On low-friction exit ramps, ESC has the ATC function of RSC and the additional directional control by selectively applying brakes based on the inputs from steer and yaw angle sensing.	While low-friction surfaces do not typically generate lateral acceleration conditions that can cause rollover, the risk of rollover increases when the vehicle suddenly encounters a surface with a higher friction. The ABS function of RSS intervenes to reduce potential swing out or jackknifing.
Low-Friction Surfaces (Rain/Snow/Ice)	Avoidance	Low-friction avoidance maneuvers generate little lateral force. ATC assists in maintaining control of the vehicle by reducing engine torque.	During low-friction avoidance maneuvers, ESC provides directional control by selectively applying brakes based on the inputs from steer and yaw angle sensing.	While low-friction surfaces do not typically generate lateral acceleration conditions that can cause rollover, the risk of rollover increases when the vehicle then suddenly encounters a surface with a higher friction. The ABS function of RSS intervenes to reduce potential swing out or jackknifing.

FIGURE 30–28 Typical EBS circuit

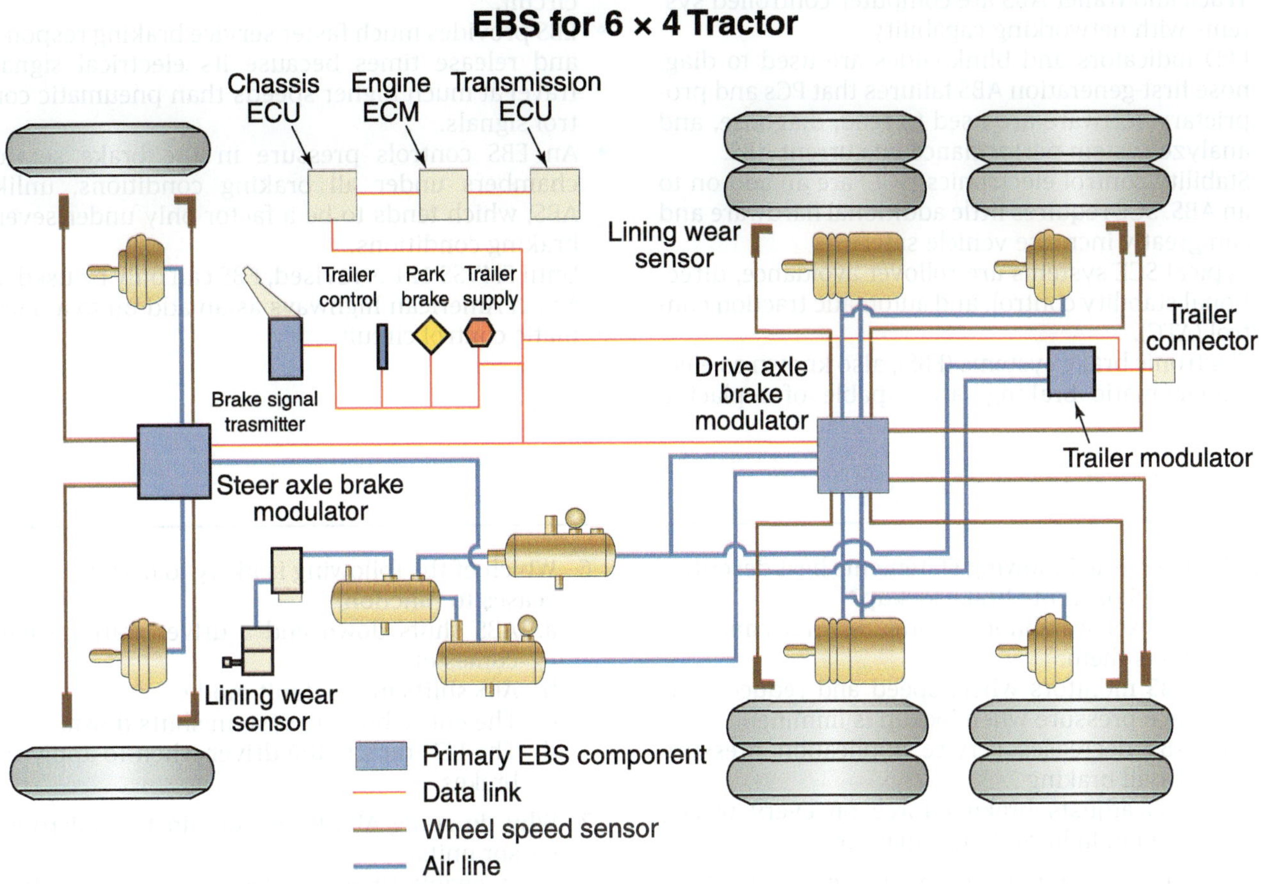

replacement of the dual-circuit pneumatic foot valve with an electrical foot control valve, known as a brake signal transmitter. The electrical foot control valve signals the EBS control module with a driver-initiated brake request. The resulting electrical signal is processed by the EBS ECU, which processes it to switch **electropneumatic** valves, known as brake modulators. Brake modulators are located over each axle or axle grouping with brakes and convert the electrical signals they receive from the EBS ECU into an air pressure value, which is delivered to brake chambers.

The European experience of EBS has produced benefits in addition to faster response. EBS can improve air brake system performance in the following areas:

- Balance
- Proportioning
- Antilock

Under normal braking conditions, brake pressures can be balanced to provide the best lining wear. When subjected to heavier braking, the system can be programmed to go into proportioning mode, applying each foundation brake with respect to the load sensed and the wheel rotational speed. Brake-by-wire electronic heavy-duty truck air brakes will be introduced in North America and become the standard means of controlling truck brake systems—it is only a question of time.

SUMMARY

- Antilock brake systems (ABS) are designed to help prevent wheel lockup during severe braking.
- An ABS uses microcomputer technology to sense and reduce braking force on wheels that are beginning to lock up under braking.
- The components of a typical ABS are wheel speed sensors, control modules, and brake modulators.
- Several ABS configurations are used on tractors and trailers, depending on the number of axles,

axle configuration, axle load, brake circuit distribution, and brake force distribution.
- Truck and trailer ABS are computer-controlled systems with networking capability.
- LED indicators and blink codes are used to diagnose first-generation ABS failures that PCs and proprietary software are used to read, diagnose, and analyze system performance on current ABS.
- Stability control electronics (SCE) are an add-on to an ABS. SCE requires little additional hardware and can greatly increase vehicle safety.
- Typical SCE systems are rollover avoidance, directional stability control, and automatic traction control (ATC).
- Electronic brake systems (EBS), also known as electropneumatic braking, are capable of replacing the current FMVSS 121–compliant pneumatic control circuit with an electronically managed control circuit.
- EBS provides much faster service braking response and release times because its electrical signals travel at much higher speeds than pneumatic control signals.
- An EBS controls pressure in the brake service chambers under all braking conditions, unlike ABS, which tends to be a factor only under severe braking conditions.
- Until FMVSS 121 is revised, EBS can only be used on North American highways as an add-on to a pneumatic control circuit.

REVIEW QUESTIONS

1. Which of the following statements best describes how ABS prevents brake lockup?
   a. ABS senses abnormal braking forces and corrects them.
   b. ABS monitors wheel speed and reduces service pressure when lockup is imminent.
   c. ABS decreases service application pressure for all braking.
   d. ABS adjusts braking force on every wheel, even in light braking situations.
2. Which of the following provides the most critical input signal to the ABS module?
   a. slack adjuster travel
   b. wheel speed sensor
   c. modulator valve
   d. engine ECM serial bus
3. Which of the following best describes the type of sensor used to signal wheel speed data to an ABS module?
   a. pulse generator
   b. transducer
   c. Hall effect sensor
   d. Thermistor
4. Which of the following is the main output device on a truck ABS?
   a. Wheatstone bridge
   b. wheel speed sensor
   c. power supply
   d. modulator assembly
5. Which of the following is the control component in an ABS modulator valve?
   a. engine ECM serial bus
   b. relay valve
   c. solenoid or motor
   d. data display window
6. Which of the following is likely to occur if an ABS ceases to function?
   a. ABS shuts down and a driver warning light comes on.
   b. ABS shifts into default mode.
   c. The entire braking system shuts down.
   d. The LED directs the driver when to apply the brakes.
7. Why do some ABS ECUs contain two microprocessor units?
   a. to enable them to process large amounts of information
   b. to enable them to provide continuous diagnostic information
   c. one microprocessor acts as a fail-safe backup to the other
   d. to double the processing speed
8. Which of the following is true of the ABS tractor warning light?
   a. It illuminates when the ignition key is turned on.
   b. It illuminates when the vehicle is braked.
   c. It goes off when there is an ABS failure.
   d. It goes off only when the vehicle is stationary.
9. When does an ATC light illuminate?
   a. when the interaxle differential engages
   b. whenever a spinout condition occurs
   c. whenever the ABS is cycling
   d. when a wheel speed sensor fails
10. Which of the following ABS channel configurations is required in a truck equipped with stability control electronics (SCE)?
    a. 2S/2M
    b. 4S/4M
    c. 6S/4M
    d. 6S/6M

11. What is the maximum modulation speed of a typical air brake ABS?
    a. 2 times per second
    b. 7 times per second
    c. 20 times per second
    d. 200 times per second
12. Technician A says that air brake ABS can be modulated at much higher speeds than hydraulic ABS because of the compressibility of air. Technician B says that a truck air brake ABS control module can "think" (clock speed) at a frequency of 200 times per second. Who is correct?
    a. Technician A only
    b. Technician B only
    c. both A and B
    d. neither A nor B
13. Before testing a dual-drive axle truck equipped with ATC on a chassis dynamometer, what should be done?
    a. ATC must be enabled.
    b. ATC must be disabled.
    c. The power divider must be locked out.
    d. The drive shafts on the rear drive should be pulled.
14. An ABS light displays fault codes that are not interpreted in the OEM service literature. Which of the following is the likely problem?
    a. defective wheel speed sensor
    b. defective modulator valve
    c. defective ECM
    d. defective data connector
15. Technician A says that in a true EBS, the brake pedal does no more than produce an electrical signal to the EBS control module. Technician B says that current FMVSS 121 legislation requires that when EBS is used on North American highways, it is in addition to pneumatic controls. Who is correct?
    a. Technician A only
    b. Technician B only
    c. both A and B
    d. neither A nor B
16. When is a tractor ABS self-test performed?
    a. each time the ignition key is switched off
    b. every time the brakes are released
    c. when a fault code is logged
    d. each time the ECU is activated

31

Prerequisite: Chapter 28

AIR BRAKE SERVICING

OBJECTIVES

After reading this chapter, you should be able to:

- Understand the safety requirements of working on an air brake system.
- Perform basic maintenance on an air brake system.
- Diagnose common compressor problems.
- Describe the procedure required to service an air dryer.
- Performance test an air dryer.
- Check out the service brakes on a truck.
- Test the emergency and parking brake systems.
- Verify the operation of the trailer brakes.
- Understand the OOS criteria used by safety inspection officers.
- Diagnose some brake valve failures.
- Describe the procedure required to overhaul foundation brakes.
- Determine brake freestroke and identify when an adjustment is required.
- Recognize different types of brake stroke indicator and identify an overstroke condition.
- Outline some common service procedures used on air disc brake systems.
- Outline some of the changes that have taken place in commercial tractor/trailer air brake systems since the introduction in 2010 of FMVSS 121 revised stopping distances.

KEY TERMS

applied stroke
automatic slack adjuster (ASA)
brake torque balance
cage bolt
Commercial Vehicle Safety Alliance (CVSA)
decelerometer
Federal Motor Vehicle Safety Standard No. 121 (FMVSS 121)
freestroke
governor
hold-off circuit
hold-off diaphragm
National Highway Traffic Safety Administration (NHTSA)
out-of-service (OOS)
piggyback assembly
plate-type brake testers
pneumatic timing
service diaphragm
sideslip

Note: The trucking industry in North America continues to use Standard system weights and measurements in compliance standards. It is important not to confuse the equivalent metric standards (usually averaged when converted) provided in this chapter with the standard values that continue to be used by law enforcement agencies. In North America, new truck certification standards are governed by FMVSS and on-the-road compliance standards are managed by Commercial Vehicle Safety Alliance (CVSA). CVSA out-of-service standards are sometimes quoted in this text but it is important to note that these are subject to frequent change. Australian new truck compliance standards are covered by Australian Design Rule legislation, which are notably similar to FMVSS. Enforcement standards vary and are the responsibility of each jurisdiction.

INTRODUCTION

This section covers some of the most often performed service and repair procedures. As much as possible, describing original equipment manufacturer (OEM)-specific procedures is avoided, so the approach attempts to be generic. When performing service and repair procedures, whether it be in a training center or service garage, technicians should use the OEM service literature specific to the equipment on which they are working. The air brake system is one of the most important safety systems on a truck. Maintenance of the system is simple and can be performed by any trained technician with ordinary shop tools.

IMPORTANCE OF WORKING SAFELY

Although proper maintenance of the brake system is essential to the operation of any highway vehicle, brake work is often the first experience that truck technicians have with vehicle systems. Brake work represents a large percentage of the operating maintenance and repairs required on a vehicle and is regarded as requiring fewer skills than other repairs, such as engine work. It is true that brake work can be performed with less training than that required for other chassis systems, but it must always be performed with precision. Quite simply, the consequence of a failed engine repair job is a truck on the side of the road, but the consequence of a failed brake repair job can be loss of life. Never regard brake service and repair work as something that is less important than that required on the other vehicle systems. Use the service literature when performing repairs and make a habit of asking questions when you do not understand something.

31.1 MAINTENANCE AND SAFETY

The approach in this chapter follows the air brake system through, circuit by circuit, and identifies critical Commercial Vehicle Safety Alliance (CVSA) **out-of-service (OOS)** criteria. Technicians must never forget that OOS standards are not maintenance standards. They identify a potentially dangerous defect. In addition, we will take a look at common troubleshooting strategies, and some repair practices are also examined. Repair practices required on hydraulic and antilock brake systems (ABS) are covered in earlier chapters dedicated to those subjects.

AIR BRAKE SAFETY CONSIDERATIONS

Before beginning any repairs, it is important to realize that compressed air can be dangerous if not handled with some respect. The potential energy of compressed air can be compared to the potential energy in a compressed coil spring. When this force is released from a storage tank or brake hose, it can cause serious injury.

The wheels of the vehicle should always be blocked by some means other than the air brake system. Wheel chocks or wedges are used, and these should be placed on both sides of a wheel, not on a steering axle. For many of the tasks required to maintain or service an air brake system, the air will have to be drained from one or more tanks. When draining an air system, wear eye protection and point the release jet of air away from people. Do not disconnect a chassis hose under pressure from its coupler because the hose may whip around as air escapes and cause an injury.

COMPRESSED AIR

Follow the OEM-recommended procedures when working on any air device so as to avoid injury or damage from components that, when released, might be subject to mechanical (spring) or pneumatic propulsion. As system pressure is drained and the emergency brakes are applied, exercise care in the vicinity of brake chambers, pushrods, and the foundation brake assembly. Air tanks closest to the compressor should contain a safety valve specified to trip at the required pop-off pressure, usually 150 psi (10.3 bar).

When air tanks have to be replaced, do not select old, used reservoirs as replacements. Air tanks should never be repair-welded, even as a temporary measure in an emergency, because of the internal coating. This plasticized coating crinkles and separates from the tank wall when heated. The operation of a spring brake chamber must be fully understood (see Chapter 28), and every truck technician should know that the force exerted by a spring brake can produce fatal consequences.

When these spring brake units have to be removed from a vehicle, they should be caged. Caging requires connecting an external screw known as a **cage bolt** to the spring brake chamber cage plate and mechanically compressing the main spring. Note the warning tags on such units and the fact that it is illegal to dispose of a spring brake chamber in a garbage or scrap metal bucket unless it has been disarmed.

GENERAL MAINTENANCE

Maintenance of an air brake system can be broken down into the major areas of pneumatic factors, brake torque balance, and safety considerations.

Pneumatic Factors

A properly functioning or balanced braking system is one in which the braking pressure reaches each actuator at the same time and at the same pressure value.

Brake balance is discussed in some detail in Chapter 28, but some of the factors of brake balance are also addressed in this section. Two factors that affect brake balance are application and release times. This is known as **pneumatic timing**.

In meeting **Federal Motor Vehicle Safety Standard 121 (FMVSS 121)** standards for application and release times, vehicle manufacturers carefully select tubing and hose sizes, the valve and fitting flow areas, and valve crack pressures. Air application and release performance is also dependent on the size and volume of the brake chambers and the distance the air must travel to get to them.

Original equipment (OE) performance is engineered into the vehicle brake system by the manufacturer. The role of the vehicle owner or truck technician is to preserve braking performance to the original standards as the vehicle ages. For example, when replacing tubing or hose, the replacement plumbing must match the original size and grade. When replacing valves in the air brake system, ensure that the function and ratings of the replacement valve are comparable with the valve being replaced. For ratings cover factors such as crack pressure, regulated hold-off pressure values, and other terms relating to brake balance (see Chapter 28). Most valves can be crossed over to a different brand on replacement, providing the function and ratings exactly match.

SHOP TALK

The brake plumbing can affect pneumatic timing on a truck brake system. When replacing brake hose and fittings, be sure to use the same sizes, or the flow dynamics (movement of air through the system) will be altered. One air brake OEM has determined that replacing a straight pipe nipple with a 90-degree elbow is equivalent to adding 7 feet (2.13 m) of brake hose into the circuit. Adding a 45-degree elbow is equivalent to adding 3 feet (0.91 m) of hose into the circuit. Either can disrupt pneumatic timing.

Figure 31–1 and **Figure 31–2** show test instrument locations for performing pressure and torque balance tests on air brake–equipped tractor/trailer combinations. The test instruments used are described a little later in this chapter.

FIGURE 31–1 Brake torque and pressure tests on the service circuit of a tractor/semitrailer combination

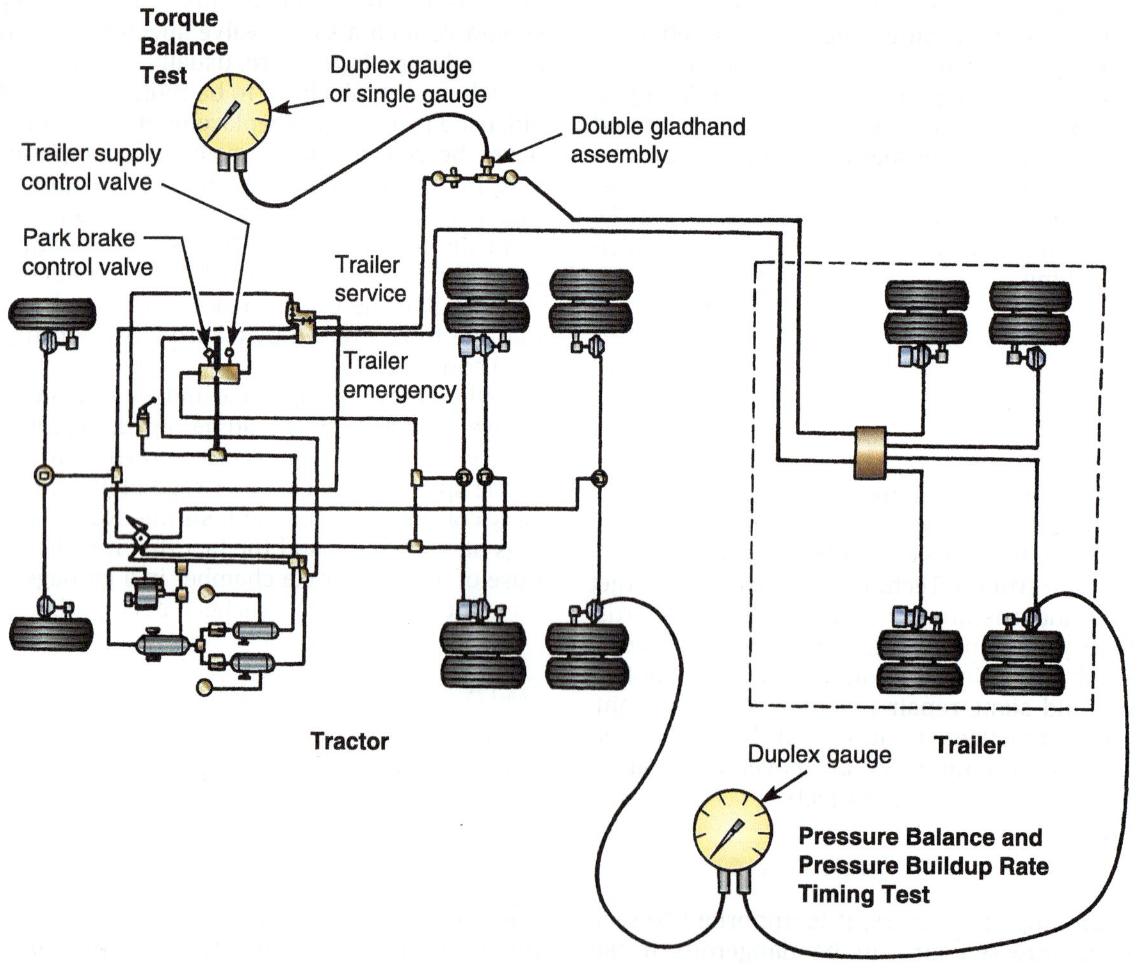

FIGURE 31–2 Multiple trailer configuration test setup: Valve crack pressures should be highest on tractor and lowest on the second trailer.

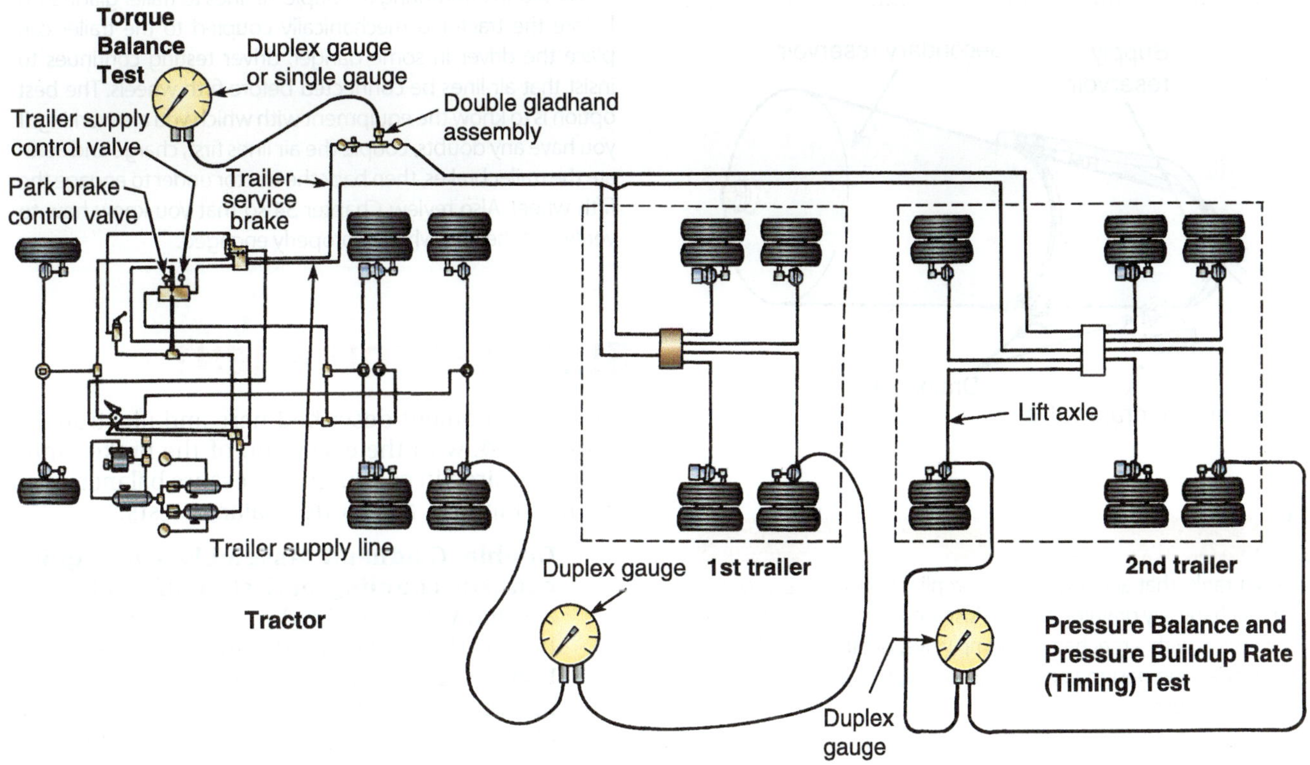

Brake Torque Balance

Brake torque balance is when the amount of mechanical braking effort applied to each wheel under braking is balanced. The mechanical components of the typical brake system consist of a number of levers that must work together to ensure that one wheel is not overbraked and another underbraked.

A ½-inch (12.5 mm) difference in slack adjuster length or a shoe that has had its arc spread can make a large difference in the amount of braking force applied to each wheel assembly. For a brake system to meet the required standards, it must be maintained on a regular basis throughout the life of the vehicle. Because brake failures can have lethal consequences, brakes are a crucial element of any preventive maintenance program.

Draining Air Tanks

The contaminants that collect in air brake reservoirs consist of water condensed from the air and a small amount of oil from the compressor. This water and oil normally pass into the air tank in the form of vapor because of the heat generated during compression. There is probably no more simple, yet more important, maintenance than reservoir draining. If water is allowed to collect in any air tank, the storage capacity of the tank is reduced. An air tank with insufficient reserve capacity means that the air pressure drops off too rapidly when air demand on the system is highest. The result can be inadequate braking in high-demand situations.

Reservoir Drain Valves

Truck drivers are instructed to drain air tanks daily. The common type of drain valve is a cable-release, spring-loaded drain valve. This enables the driver to pull on the cable for a couple of seconds and to dump the small amount of moisture sitting on the drain valve. Any significant amount of moisture in the tank will exit only after all the compressed air has been dumped.

Options to the cable-release dump valve are the automatic drain or spitter valve and the manual draincock. Least popular with drivers is the manual draincock because this often requires that the driver climb under the chassis to open it; however, it is the most effective valve for ensuring that the system is completely drained. Automatic drain valves are most popular among drivers because they do not require daily draining. All automatic drain valves (spitter valves) should be checked periodically for proper operation. **Figure 31–3** shows a typical supply, secondary air tank equipped with an automatic drain valve (supply tank) and a manual draincock (secondary tank).

FIGURE 31–3 Supply, secondary air tank equipped with an automatic drain valve and manual drain cocks.

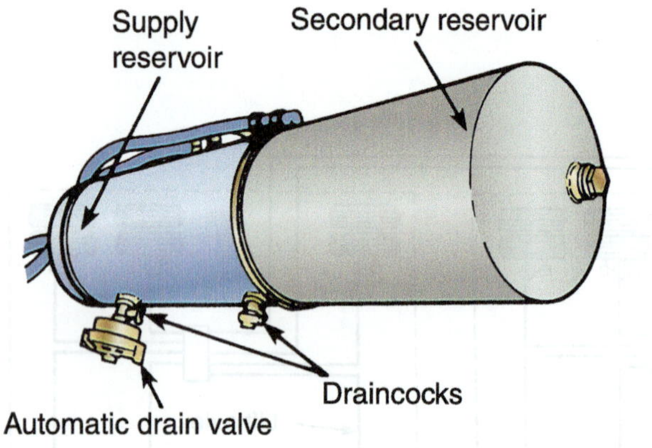

CAUTION

All air tanks that are FMVSS 121 compliant must have a means of mechanically draining them. Most automatic drain valves are equipped with a mechanical dump valve, but it can be difficult to locate. One design uses a Schrader-type plunger that is difficult to access.

System Contamination

Many current air brake valves contain small orifices and passages that make them more susceptible to contaminants. The prevention of freeze-ups in the system is equally important. The use of air dryers is critical in maintaining a supply of clean, dry air, especially in geographic areas of high humidity. Although alcohol evaporators do nothing to eliminate moisture from an air brake system, they can prevent freeze-ups.

SHOP TALK

Trailer hookups. In just about all jurisdictions in the United States and Canada, driver license testing requires the air lines to be coupled to the trailer before backing the tractor under the trailer to couple the fifth wheel. Any observer of day-to-day truck yard coupling procedures would quickly realize that this practice is seldom observed by truck drivers on a daily basis—except when undergoing mandatory state testing. In pre-FMVSS 121 system trailers (pre-1975) and in a small number of current special application trailers, an air-actuated, nonmechanical park mode is used. The problem with older park-on-air trailers was that, in the event of an air leak, the air used to brake the trailer bled off, leaving the trailer with no brakes. This meant that air had to be supplied to the trailer to ensure that it was braked before coupling a tractor. The reason drivers can safely engage the fifth wheel to a trailer before the air lines is that most trailers with spring brakes (in compliance with the FMVSS 121 requirement for a mechanical park mode) can be assumed to be in park mode and, therefore, the trailer is going to hold—providing the brakes are in adjustment.

Although attempting to couple air lines to trailer gladhands before the tractor is mechanically coupled to the trailer can place the driver in some danger, driver testing continues to insist that air lines be connected before fifth wheels. The best option is to know the equipment with which you are working. If you have any doubts, couple the air lines first, charge and actuate the trailer brakes, then back the tractor under to engage the fifth wheel. Also review Chapter 34 so that you know how to verify that the fifth wheel is properly engaged.

31.2 TEST EQUIPMENT

The test equipment described here and illustrated in **Figure 31–4**, with the exception of the brake-timing unit, is required in conducting pressure balance, pressure buildup rate, and torque balance tests.

- **Double Gladhand Assembly (with quick-connect coupling and shut-off cock).** This assembly is used in the service line connection to the trailer for the torque balance test. **Figure 31–5** shows the components that make up this test equipment, which can be assembled from odds and ends found around any truck shop. This equipment is designed to use shop air to simulate trailer supply and service applications in place of a tractor. **Figure 31–6** shows how the unit would be coupled into the circuit for performing tests.
- **Duplex Air Gauges.** Two sets are required. Each set includes one duplex gauge and two 25-foot (7.5 m) air hoses. **Figure 31–7** shows the readings on a duplex air gauge during a pressure balance test. Duplex gauges may be used for monitoring application pressures and balance pressures in different parts of the air brake circuit.
- **Additional Air Hoses for Long Single-Trailer and Multiple-Trailer Tests.** Two hoses of sufficient length and with quick connection couplings are required. These hoses are used for making the test connections between the service brake chambers on the various axles of the tractor and trailer(s) and the air gauge(s).
- **Decelerometer.** The **decelerometer** measures what is called load transfer dynamic pitch: It provides some indication of brake torque balance. A decelerometer measures the change in angle of a chassis when it is subjected to aggressive braking. This will be dependent on the chassis configuration (bobtail, straight truck, and tractor/trailer combination), the suspension type, payload (loaded or empty), and type of load (static = fixed solid load, or dynamic = liquid or livestock). The placement

FIGURE 31–4 Air brake test equipment

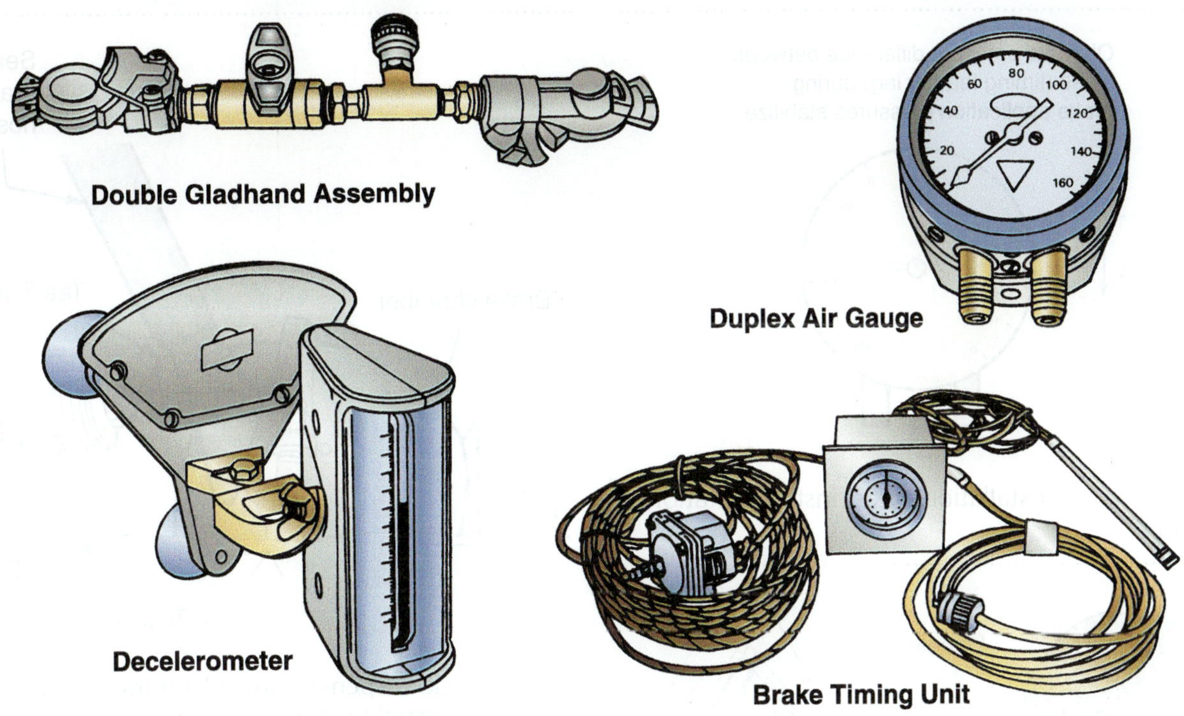

FIGURE 31–5 Double gladhand assembly components

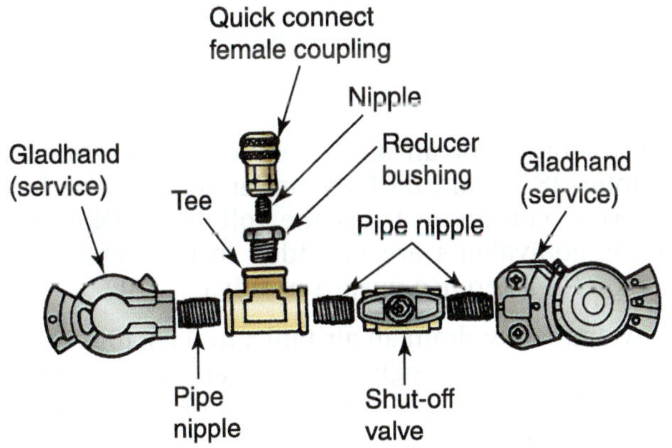

FIGURE 31–6 Installation of a double gladhand assembly

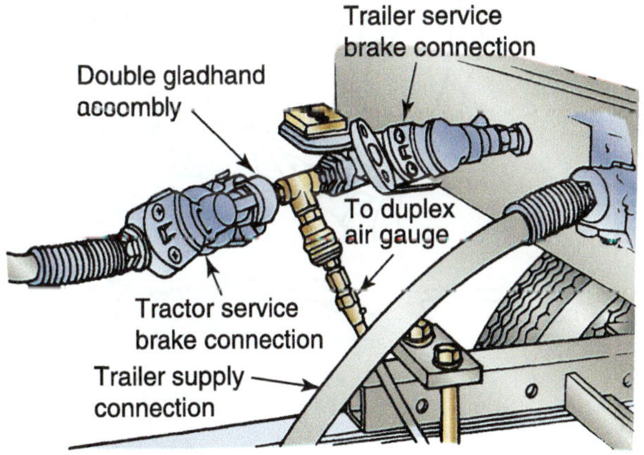

of a decelerometer in its test location in a vehicle cab is shown in **Figure 31–8**.

- **Brake Timing Unit.** This is a sophisticated test instrument that can accurately record brake application and release times. These test units are available from the OEM air brake manufacturers. **Figure 31–9** shows a test line connection on a service brake hose.
- **PC Software.** The read and diagnostic software used by most ABS manufacturers will display detailed brake performance data such as percentage braking by wheel, load transfer effect, adjustment, and fade.

31.3 ASSESSING BRAKE SYSTEM PERFORMANCE

Five tests can be used to verify that the system is functioning properly. These tests are static leakage checks only and do not replace functional, dynamic

FIGURE 31–7 Gauge readings for a pressure balance check

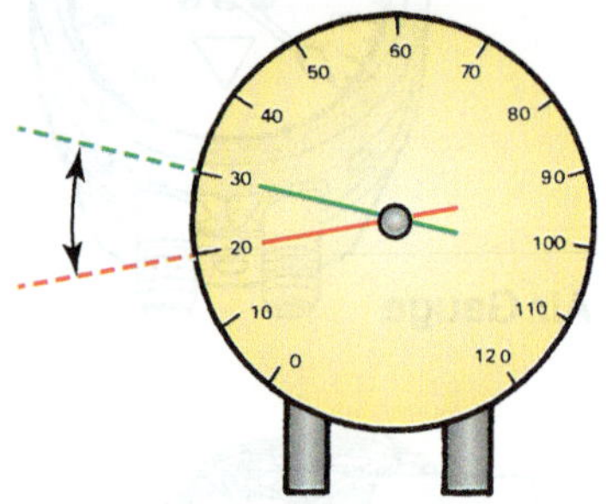

FIGURE 31–8 Installation of test instrumentation in a cab

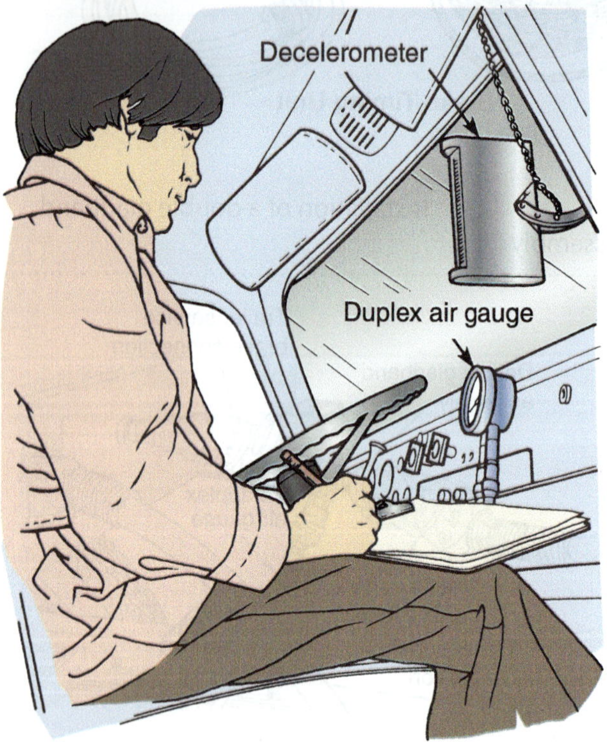

control and performance tests required to fully check out the air brake system. In addition, the Bendix 121 brake test procedure (which includes a check sheet) is outlined in the Student Guide that accompanies this textbook. Before beginning the testing, perform the following checks:

- Examine all tubing for kinks or dents.
- Examine all hoses for signs of wear, drying out, or overheating.

FIGURE 31–9 Connecting a test gauge hose at a service brake chamber

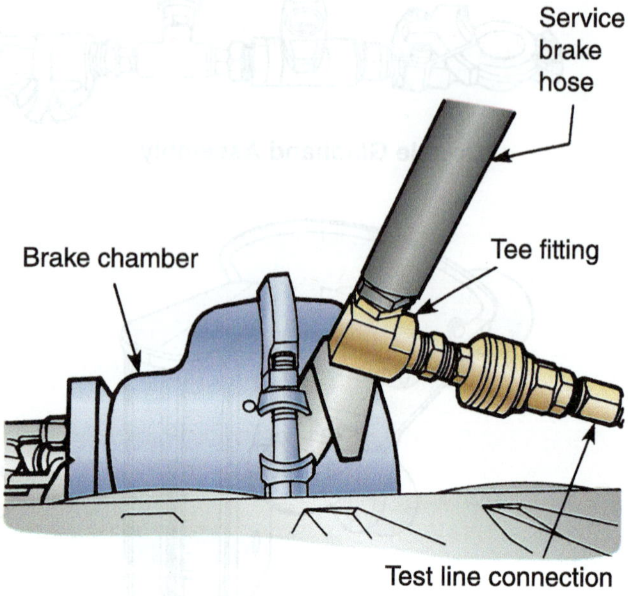

- Check suspension of all tubing. It should be supported to eliminate vibration.
- Check suspension of all hoses. Position so that the hose will not abrade or be subject to excessive heat.

TEST 1. GOVERNOR CUT-OUT, LOW-PRESSURE WARNING, AND PRESSURE BUILDUP TEST

The operation of an air **governor** as described in Chapter 28 should be fully understood. To perform governor cut-out test, do the following: (Note that metric equivalents are provided in brackets in this section for reference purposes only.)

1. Completely drain all air tanks to 0 psi (0 kPa).
2. Start the truck engine and run at a 1,500 rpm (no load). The low-pressure warning light in the cab should be on. On some trucks equipped with an ABS, the warning light will also come on momentarily when the ignition is first turned on.
3. The dash warning light should extinguish at or slightly above 60 psi (4 bar).
4. Note and observe the pressure buildup time. Pressure should build from 85 to 100 psi (6 to 7 bar) within 25 seconds. This is an FMVSS 121 requirement, and the test is often used by enforcement officers to verify the performance of the truck air supply circuit because it can be performed quickly. The actual citation standard varies by jurisdiction and may be up to 2 minutes.

5. Record the governor cut-out pressure value (the air compressor unloads) and check it to specification. This value typically will be around 120 to 130 psi (8.3 to 9.0 bar).
6. Discharge air from the system by pumping the foot valve and note the governor cut-in value (air compressor effective cycle resumes). The difference between cut-in and cut-out pressures is required not to exceed 25 psi (1.7 bar). The difference should not be smaller than 20 psi (1.4 bar) to avoid frequent compressor cycling.
7. The minimum air compressor governor cut-in pressure for trucks is 100 psi (7 bar). Minimum cut-in for buses is 85 psi (6 bar).

If the air brake system tested does not meet the required specifications, the following troubleshooting sequence can be used to correct the problem:

1. If the low-pressure warning light does not come on:
   - Check the wiring.
   - Check the bulb in the warning light.
   - Repair or replace the bulb or low-pressure warning switch(es).
2. If the governor cut-out value is higher or lower than specified by the truck manufacturer:
   - Adjust the governor using an air pressure gauge of known accuracy; some newer governors are not adjustable and may have to be replaced. Use **Figure 31–10** and locate the adjusting screw. Back off the locknut and either increase or decrease the governor spring tension to alter the governor cut-out. **Photo Sequence 14** shows the procedure required to adjust an air system governor.
   - Check the governor unloader signal.
   - Check the compressor unloader mechanism.
   - Reset or replace the governor.

FIGURE 31–10 Sectional view of a governor assembly. Note the location of the adjusting screw.

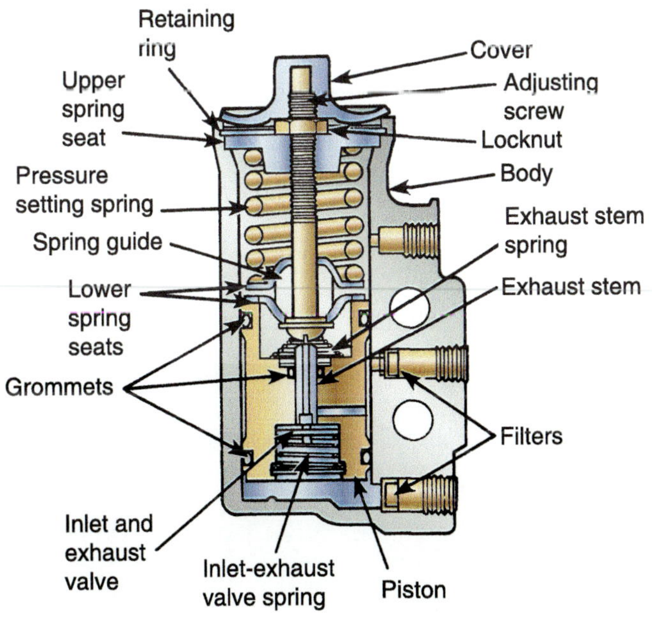

3. If the low-pressure warning circuit does not activate when the pressure is below 60 psi (4 bar):
   - Check the electrical circuit/sender unit with a digital multimeter (DMM).
   - Repair or replace the faulty low-pressure sender.
4. If buildup time from 85 to 100 psi (6 to 7 bar) exceeds 25 seconds:
   - Examine the compressor air supply filter and clean or replace as necessary.
   - Check the compressor discharge port and line for excessive carbon buildup. Clean or replace as necessary and then check the air dryer.
   - With the air system charged and the compressor in the unloaded mode, listen at the compressor inlet for leakage. If leakage can be heard, apply a small amount of oil around the unloader pistons to verify. If no leakage is indicated at the inlet, leakage could be occurring through the compressor discharger valves.
   - Check the compressor drive for slippage.

TEST 2. RESERVOIR AIR SUPPLY LEAKAGE TEST

To perform the air tank supply leakage test, run the truck engine until the system is at system cut-out pressure and then shut the engine off. Then do the following:

1. Allow pressure to stabilize for at least 1 minute.
2. Observe the dash gauge pressures for 2 minutes and note any pressure drop.

- Pressure drop for a tractor or straight truck: 2 psi (14 kPa) drop within 2 minutes is the maximum allowable for *either* circuit service tank (primary and secondary).
- Total maximum pressure drop permitted in a tractor or straight truck air system is 3 psi (21 kPa) in 5 minutes.
- Pressure drop for a tractor/trailer combination: 6 psi (42 kPa) drop within 2 minutes is the maximum allowable in all the service tanks.
- Pressure drop for a tractor/trailer train (multiple trailers): no more than 8 psi (55 kPa) drop within 2 minutes is allowable in the trailer service tanks.

Photo Sequence 14

GOVERNOR ADJUSTMENT

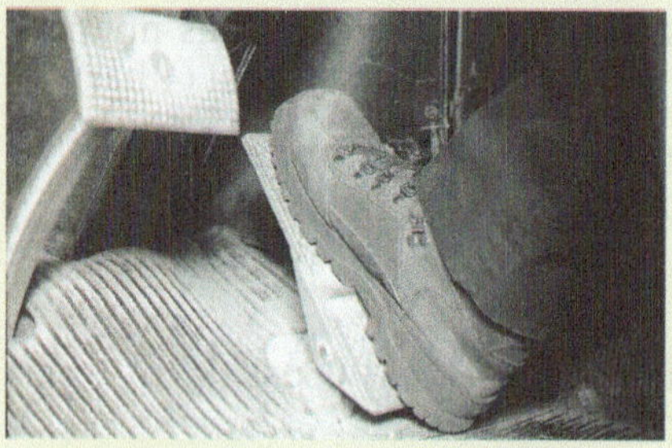

P 14–1 With the engine running, apply the brake several times to drop system air pressure until the governor cuts in.

P 14–2 Now run the engine at 1,500 rpm, building system air pressure to cut out. When the governor cuts out, observe the pressure on the dash gauge.

P 14–3 Determine the specified governor cut-out pressure in the OEM service literature. If the pressure on the dash gauge does not equal the specified governor pressure, the governor must be adjusted.

P 14–4 Shut off the engine and remove the dust boot on the governor.

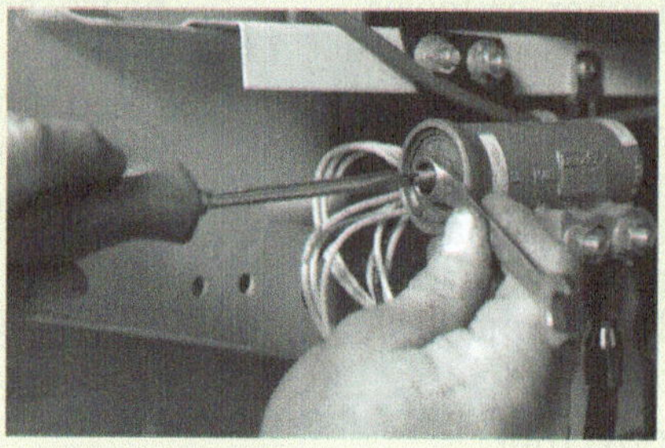

P 14–5 Back off the locknut with a $\frac{7}{16}$-inch (11 mm) wrench. Turn the governor adjusting screw counterclockwise to increase the cut-out pressure or clockwise to decrease the pressure. When the governor adjusting screw is turned $\frac{1}{4}$ turn, the governor cut-out pressure is changed approximately 4 psi (28 kPa).

P 14–6 Start the engine and repeat steps 1 and 2 to verify that the governor cuts out at the specified pressure.

P 14–7 With the engine running and the system reading governor cut-out pressure dump air from the system using the treadle valve and note the pressure when the governor cuts in. Cut-in pressure should not be more than 25 psi (1.7 bar) below governor cut-out pressure. Cut-in pressure is not adjustable.

SHOP TALK

Maximum allowable leakage/drop-off rates are often defined by local jurisdictions (state and provincial governments). These regulations may differ from the test values used here. Contact your local transportation enforcement office to identify the specifications used in your area.

To locate a leak in the supply circuit of a truck air system, use a leak detector or soapy solution. The cause is likely to come from one or more of the following:

- Supply lines and fittings (tighten)
- Supply tank
- Safety (pop-off) valve in supply reservoir
- Governor
- Compressor discharge valves
- Air dryer and its fittings
- Alcohol injection components

TEST 3. MANUAL PARKING/EMERGENCY SYSTEM TEST

Before performing an emergency system test, run the engine and idle in the range of 600 to 900 rpm and make sure that the system pressure is correct (between cut-in and cut-out). Then do the following:

Straight Trucks and Tractors

Manually operate the park control valve, checking that the parking brakes apply, and release promptly as you pull out and push in the control valve button.

Tractor/Trailer Combinations

- Manually operate the Trailer Supply valve (usually red and octagonal) and check that the trailer brakes apply and release promptly as you pull out and push in the control button.
- Manually operate the system park control button (usually yellow and diamond or square shaped) and check that all parking brakes (tractor and trailer) apply and release promptly.

If performance is sluggish in either test, check for:

- Dented or kinked lines
- Improperly installed hose fitting
- A defective trailer combination relay emergency valve
- Defective ABS modulator(s)

If the trailer parking brakes do not release and apply when the Trailer Supply valve is cycled, check:

- Tractor protection control
- Trailer spring brake valve/combination relay/emergency valve

Figure 31–11 shows a typical dash control module assembly used on a current two-dash valve system; note the shape and color coding of the valve knobs. **Figure 31–12** shows a sectional view of the internal components of a typical push-pull dash valve.

TEST 4. AUTOMATIC EMERGENCY SYSTEM TEST

This test should be performed with the engine stopped and the air brake system at full system or cut-out pressure with the wheels chocked and the parking brakes released.

1. Drain the primary circuit (rear axle) tank in the tractor until the pressure reads 0 psi (0 kPa).
   - Secondary circuit tank should not lose pressure. (This can be read on dash gauge.)
   - On tractor/trailer combination vehicles, the trailer air system should remain fully charged.
   - Neither the tractor nor trailer parking brakes should apply.
2. With no air pressure in the primary tank, make a brake application.
   - Front axle brakes should apply and release.
   - On tractor/trailer combinations, the trailer brakes also should apply and release.
   - The stoplights should illuminate.
3. Slowly bleed down the secondary (front axle) tank pressure.
   - The dash System Park push-pull valve should pop out when the secondary circuit pressure drops to between 35 and 45 psi (2.4 and 3.1 bar).
   - The tractor protection valve (TPV) should close between 45 and 20 psi (3.1 and 1.4 bar), and the trailer supply hose should be exhausted.
   - Trailer parking brakes should be applied immediately after the TPV closes.
4. Close the draincocks, recharge the system, and this time drain the secondary (front axle) tank to 0 psi (0 kPa).
   - The primary (rear axle) tank should not lose pressure.
   - On tractor/trailer combinations, the trailer air system should remain fully charged.
5. With no air pressure in the secondary (rear axle) reservoir, make a brake application.
   - The rear axle brakes should apply and release.
   - On tractor/trailer combination vehicles, the trailer brakes also should apply and release.

If the vehicle fails to pass the tests outlined, use an OEM brake schematic to check routing and valve

FIGURE 31–11 Tractor dash control two-valve module

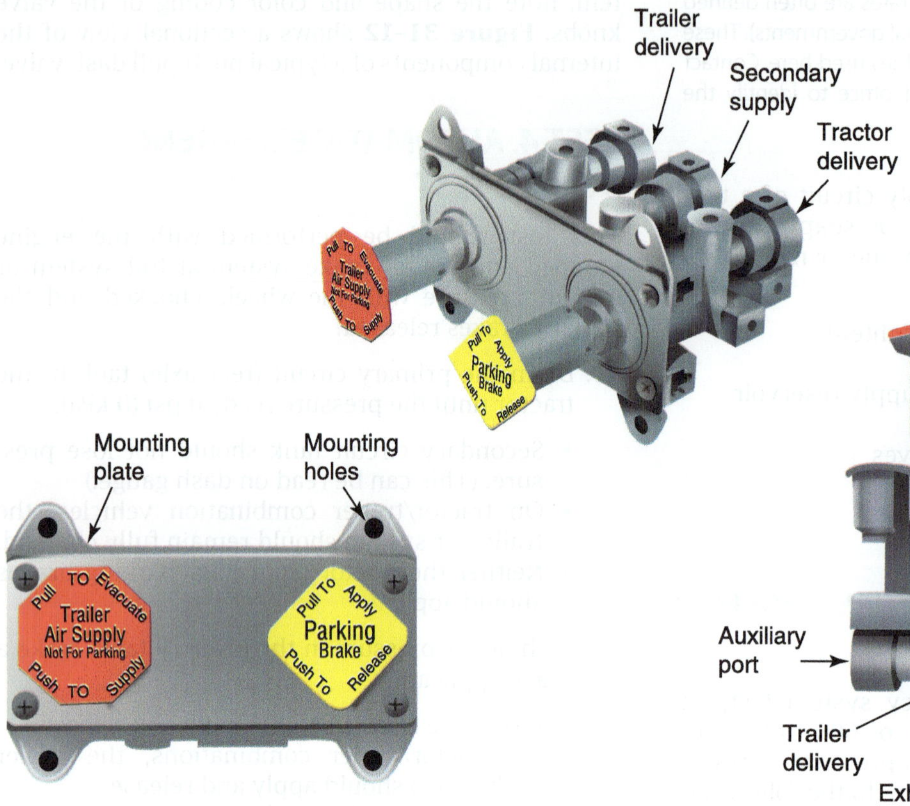

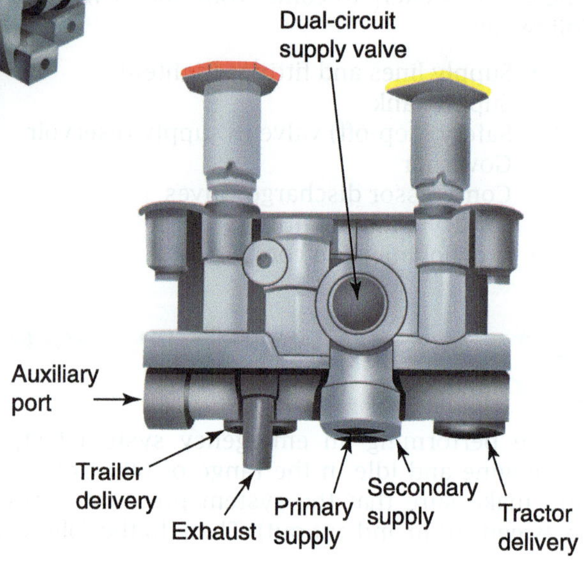

FIGURE 31–12 Sectional view of a nonmodular push-pull type control valve

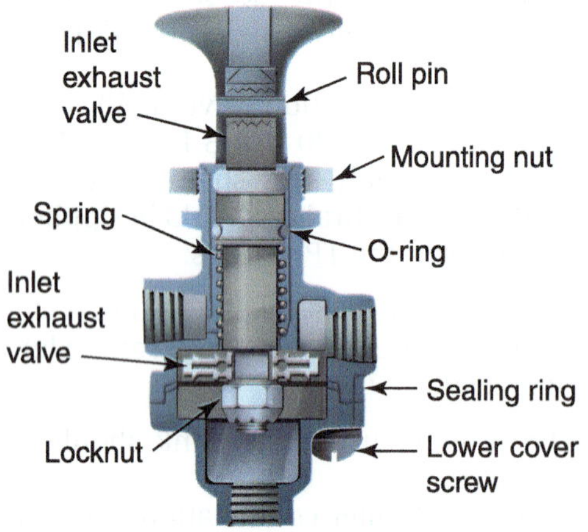

location and test the following components for leakage and proper operation:

- Fittings
- Kinked hose or tubing
- Single check valves
- Double check valves
- Tractor protection valve
- Tractor protection control valve
- Parking control valve
- Antilock modulators
- Trailer spring brake control valve
- Inverting relay spring brake control valve

TEST 5. AIR BRAKE SYSTEM OPERATIONAL CHECKS

This test outlines a more detailed operational checklist of truck air brake system performance. Perform the following in sequence:

1. Block a set of wheels on the tractor (not the steering axle) and the trailer to prevent the unit from moving when the brakes are released.
2. Inspect the primary and secondary reservoir inlet check valves for correct operation by doing the following:
   - Build air pressure up to system pressure.
   - With the ignition key on, open the draincock at the supply air reservoir and completely drain the reservoir.
   - Pressure in both the primary and secondary reservoirs should remain at system air pressure. Use the dash gauges to read this. If loss of air is evident in either system, the one-way inlet check valve could be defective.

3. With the System Park brake dash valve (diamond shaped and yellow) and the Trailer Supply control valve (octagonal and red) in their released positions (pushed in), open the draincocks in both the primary and secondary tanks.
4. On a coupled tractor/trailer combination, the following should occur when the draincocks are opened in the primary and secondary tanks:
   - The Trailer Supply dash valve (octagonal and red) should pop out (trailer park brakes applied) when the circuit tank (primary or secondary) with the highest pressure reaches 40 ± 6 psi (2.75 ± 0.4 bar). This dash valve could pop immediately if air is depleted rapidly at the trailer supply line.
   - When air pressure in the tank with the higher pressure reaches 30 ± 5 psi (2.0 ± 0.34 bar), the System Park (diamond-shaped and yellow) dash valve may pop out. When the air pressure drops to 25 ± 5 psi (1.7 ± 0.34 bar), this dash valve must pop out.
5. Close all reservoir draincocks.
6. Build up air supply in the chassis system to approximately system cut-out pressure.
7. With the Trailer Supply dash valve pushed in (trailer parking brakes released), disconnect the trailer emergency gladhands from each other. The Trailer Supply valve should pop out instantly, cutting the supply of air exiting the hose at the gladhand.
8. If the Trailer Supply knob pops out, reconnect the trailer supply hose to the trailer gladhand. Push the System Park in to the released position (in) and pull the Trailer Supply knob out. This will apply the trailer parking brakes and release the tractor brakes.
9. Check the air pressure for leakage by observing air gauges on the instrument panel. Leakage should not be greater than 2 psi (14 kPa) in 1 minute.
10. Open the draincock in the secondary air tank. The draincock must be opened all the way for quick loss of air. Loss of air can be monitored by observing the dash air gauges. Only the secondary circuit should show a loss of air. With the ignition key on, the low-pressure indicator buzzer in the cab should sound at between 60 and 70 psi (4.0 and 4.8 bar).
11. Apply the service brakes and observe the slack adjusters and service brake chamber pushrods. The following should occur:
    - The low air warning buzzer should come on.
    - The secondary circuit gauge should indicate zero.
    - All the service brakes throughout the vehicle should apply.
    - Check for special application valves if this test does not confirm that the primary circuit is properly functioning.
12. Close the draincock in the secondary air reservoir.
13. Build the air supply in both the primary and secondary circuits to system pressure.
14. Fully open the draincock in the primary reservoir. Observe the secondary circuit air pressure gauge for loss of air; none should occur.
15. Apply the service brakes. What happens here will depend on how the brake system has been optioned.
    - The low air pressure buzzer and warning light should come on.
    - The primary circuit dash gauge should read zero.
    - At minimum, the front service chambers on the tractor and all the service brakes on the trailer should apply. On some systems, all the service chambers will apply.
    - If the brake chamber pushrods move as described, the secondary circuit is functioning properly.
16. Close all draincocks, build the air pressure to system pressure, and return the vehicle to service.

31.4 SUPPLY CIRCUIT SERVICE

The supply circuit of an air brake system includes all the components that compress the air, manage the cycling of the compressor, regulate the system air pressures, remove contaminants from the compressed air, and store it in a supply tank. For a full description of the operating principles of the supply circuit components, refer to Chapter 28.

AIR COMPRESSOR SERVICE

Air compressor service should be included in a scheduled maintenance plan to ensure its proper operation and to extend its service life. The OEM maintenance manual should be referred to for specific maintenance instructions. The following would be typical:

- Every 5,000 miles (8,000 km) or monthly. Service, clean, and replace (as necessary) the air cleaner filter elements. A typical air induction system for a compressor is shown in **Figure 31–13**.
- Every 25,000 miles (40,000 km) or every 3 months. Perform the supply circuit operational test outlined earlier in this chapter to verify that the system meets FMVSS 121 requirements.
- Every 50,000 miles (80,000 km) or every 6 months.
  - Inspect the compressor air discharge port, inlet and discharge lines for restrictions, oil, and carboning.

FIGURE 31–13 Typical air induction system for an air compressor

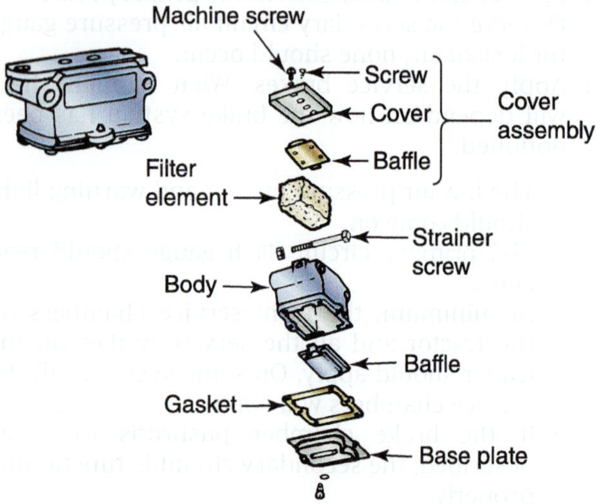

- Check external oil supply (if fitted) and return lines for kinks and flow restrictions.
- Check for noisy compressor operation.
- Check pulley and belt (if equipped) alignment and tension.
- Check compressor mounting bolts and retorque if necessary.

Compressor Air Supply

A compressor must have an unrestricted supply of clean filtered air. Part of the preventive maintenance program will be to check the compressor induction system. More frequent maintenance will be required when the vehicle is operated in dusty or dirty environments. Some common methods of providing the compressor with clean filtered air are:

- **Polyurethane Sponge Strainer**. This type of strainer element should be cleaned or replaced. Cleaning requires that it be in a commercial solvent or a detergent and water solution. The element should be saturated in clean engine oil and then squeezed dry before replacing it in the strainer. Replace the gasket whenever the air strainer is removed from the compressor intake header.
- **Dry Element/Pleated Paper Air Filter**. Remove the spring clips from either side of the mounting baffle, separate the cover, and discard the pleated paper filter. Replace the pleated paper filter and remount the cleaned cover, ensuring that the filter is properly positioned. Again, replace the air strainer gasket whenever the air strainer is removed from the compressor intake header.
- **Intake Adapter**. Many compressors are fitted with intake adapters that permit the compressor intake to be connected to the engine air intake system. This means that the air delivered to the compressor is filtered by the engine air cleaner system. The delivery of air to the compressor can be upstream or downstream from the turbocharger compressor. When the compressor intake air is ported off downstream from the turbocharger compressor, boosted air is delivered to the compressor. This reduces some of the work that has to be performed by the compressor. When the engine air filter is changed, the compressor intake adapter should be checked. If loose, remove the intake adapter, clean the strainer plate, and replace the intake adapter gasket. Check line connections both at the compressor intake adapter and at the engine or engine air cleaner. Inspect the connecting line for kinks and ruptures and replace it if necessary.

When replacing a compressor assembly, ensure that the mating flange is properly configured to the engine to which it is to be installed. **Figure 31–14** shows some different types of compressor mounting flanges required to mount the compressor to different OEM engines.

FIGURE 31–14 Various compressor mountings to meet engine OEM requirements

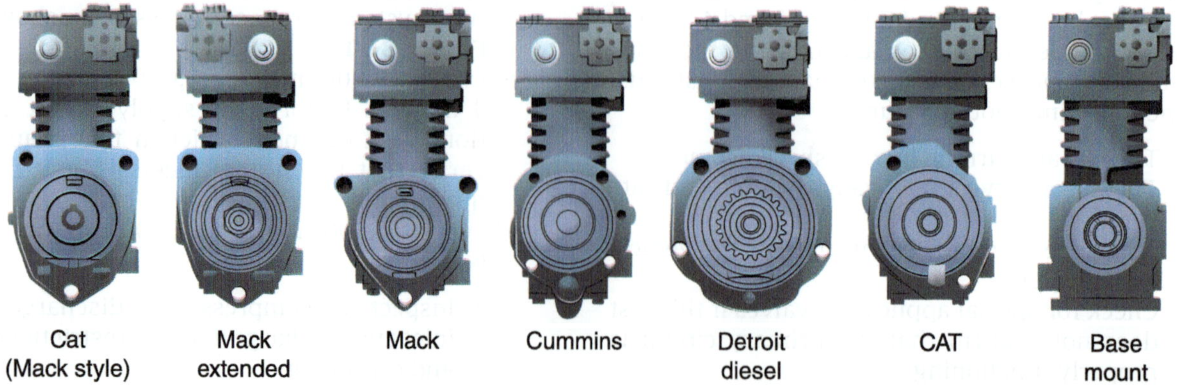

Compressor Troubleshooting

Today, compressors are almost never repaired in the field, because it usually is cheaper for a service facility to stock rebuilt exchange units than the dozens of subcomponents required to rebuild a compressor. Labor and warranty are also factors. Some exceptions are made—for example, when a persistently occurring malfunction with an OEM product is identified, such as failing valves, it may be cost effective simply to replace the valves.

However, disassembling and reassembling an air compressor during training is a good exercise that will improve the technician's ability to diagnose compressor complaints. **Figure 31–15** shows an exploded view of an air compressor. The technician is usually required to diagnose the cause of failure and, unless the problem is external, to replace the compressor with a rebuilt exchange unit (**Table 31–1**). Remember, all air compressors will pump trace quantities of the engine oil used to lubricate them. Replacing a compressor diagnosed with trace oil in the discharge is unnecessary. The key is to inspect the air dryer for excessive oil contamination.

CAUTION

All compressors pass minute quantities of oil into the air circuit. Avoid replacing a compressor that is passing trace oil because warranty will usually be denied by the OEM. The warranty also will be denied if the drain-back ports are plugged with silicone.

AIR DRYER SERVICE

No two vehicles operate under identical conditions, so maintenance and service intervals will vary. Experience is a valuable guide in determining the best maintenance interval for an air dryer. The usual OEM-suggested interval will be somewhere around 25,000 miles (40,000 km) linehaul or every 3 months. Service for a typical air dryer, shown in **Figure 31–16**, consists of the following checks:

1. Check for moisture in the air brake system by opening reservoirs, draincocks, or valves and checking for water. If moisture is present, the desiccant may require replacement. However, the following conditions also can cause water accumulation and should be considered before replacing the desiccant:
   - An outside air source has been used to charge the system. This air may not have been passed through the air dryer.
   - Air usage is higher than normal for a typical highway vehicle. This can be caused by accessory air demands or some unusual air requirement that does not allow the compressor to spend sufficient time in its unloaded cycle. Also check for air system leakage.
   - The air dryer is newly installed in a system that previously had no air dryer. The entire air system may be saturated with moisture, and several weeks of operation may be required to dry it out.
   - The air dryer is too close to the air compressor, resulting in the air being too hot to condense the moisture.
   - In areas in which a minimum 30-degree range of temperatures occurs in 1 day, small amounts of water can accumulate in the air brake system because of condensation. Under these conditions, the presence of small amounts of moisture is normal and should not be considered as an indication that the dryer is not performing properly.

Note: A small amount of oil in the system may be normal and should not in itself be considered a reason to replace the desiccant. See the foregoing caution on replacing compressors that pass trace oil. Desiccant gummed up with oil should not be confused with some staining evident in the desiccant bed. Oil-stained desiccant can function adequately.

2. Check mounting bolts for tightness. Retorque to specifications. A major service event should be conducted every 300,000 miles (483,000 km) or 36 months. This requires the air dryer to be disassembled and rebuilt. The overhaul procedure would usually include replacing the desiccant cartridge.

Note: The desiccant change interval may vary from vehicle to vehicle. Although typical desiccant cartridge life is 3 years, many will perform adequately for a longer period. In order to take maximum advantage of desiccant life and ensure that replacement occurs only when necessary, perform an operation and leakage test to determine the need for service.

Air Dryer Operation and Leakage Tests

- Test the outlet port check valve assembly by building the air system pressure to governor cut-out and observing a test air gauge installed in the supply tank. Rapid loss of pressure could indicate a failed outlet check valve.
- Locate a failed outlet check valve by bleeding the system down, removing the check valve assembly from the end cover, subjecting air pressure to the unit, and applying a soap solution to the check valve side. Leakage should not exceed a 1-inch (25 mm) bubble in 1 second.

FIGURE 31–15 Exploded view of a typical compressor

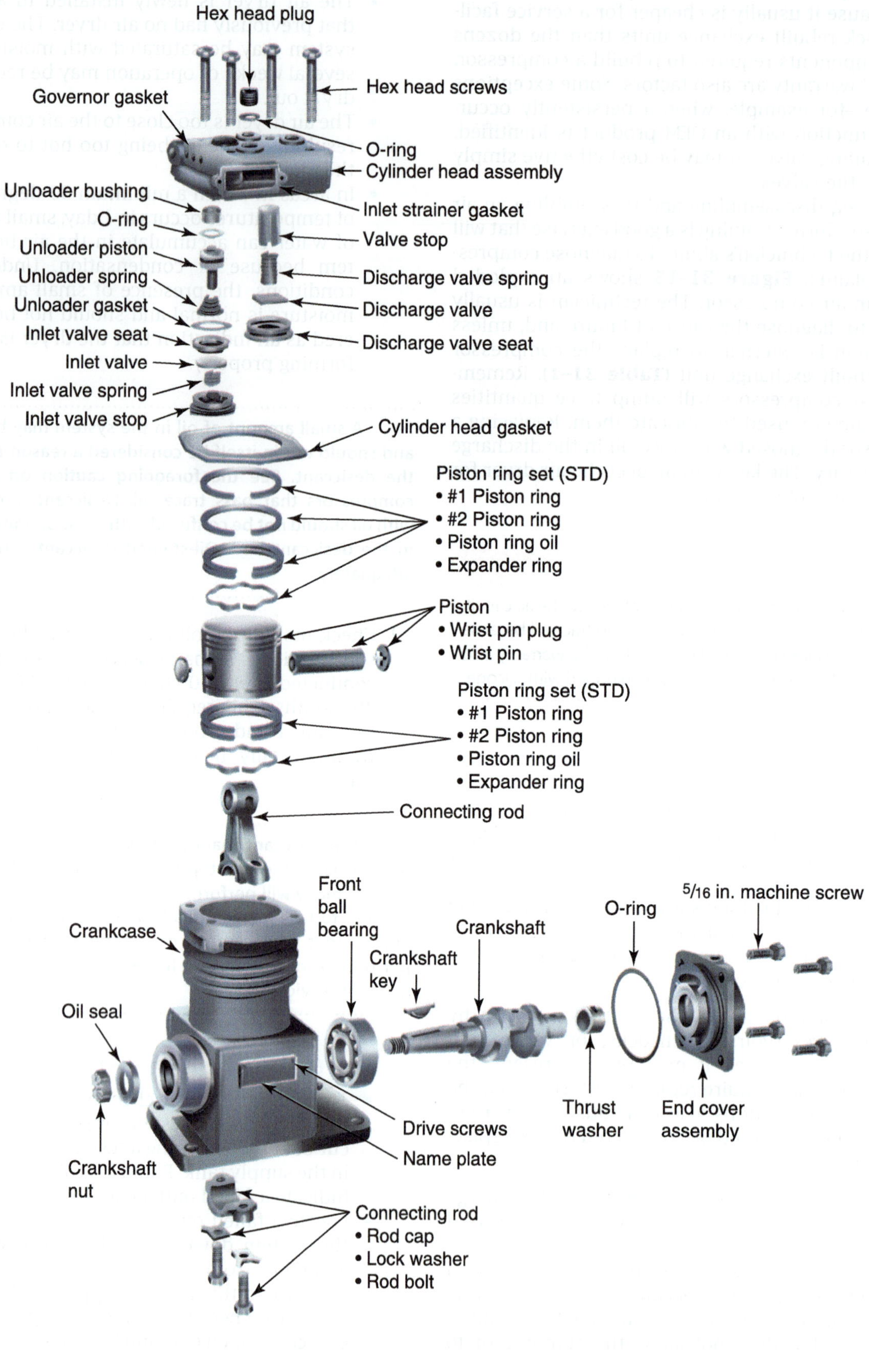

TABLE 31–1 Symptoms of Compressor Problems

Symptom	Possible Cause	Remedy
Compressor fails to maintain sufficient pressure or adequate air supply.	1. Restricted intake cleaner	1. Clean/replace element
	2. Restricted intake tract	2. Clean/repair compressor
	3. Failure of intake or exhaust valves to seal	3. Repair intake and exhaust valves or replace compressor
	4. Compressor drive slippage	4. Adjust or replace the compressor drive mechanism
	5. Failed or maladjusted governor	5. Adjust or replace governor
	6. Defective cab gauge	6. Replace cab gauge
	7. Excessive system leakage	7. Locate and repair leaks
	8. Worn-out compressor	8. Replace compressor
Compressor operates with excessive noise.	1. Loose drive mechanism	1. Tighten, repair, or replace pulley as necessary
	2. Restricted cylinder head intake/discharge tracts	2. Repair or replace as necessary
	3. Worn-or burned-out bearings	3. Replace bearings as necessary
	4. Improper lubrication	4. Service lubrication system as necessary
	5. Excessive wear	5. Overhaul or replace compressor as necessary
Compressor not unloading (excessive pressure).	1. Defective unloader pins or seals	1. Replace pins/seals or compressor as necessary
	2. Defective governor	2. Replace governor
	3. Restricted reservoir line to governor	3. Repair or replace line as necessary
	4. Stuck or binding unloader mechanism	4. Repair unloader mechanism or compressor
	5. Defective gauge	5. Replace gauge
Compressor pumps excessive oil.	1. Excessive wear	1. Overhaul or replace compressor
	2. Plugged air cleaner	2. Clean or replace element
	3. High inlet vacuum (obstructed intake)	3. Service intake circuit
	4. Restricted oil return line flooding compressor	4. Repair or replace return line
	5. Excessive oil pressure	5. Service lubrication system
	6. Defective compressor rear main seal	6. Replace rear main seal or the compressor
	7. Failed or improperly installed compressor piston rings	7. Remove and reinstall rings or replace compressor
	8. Back pressure from engine crankcase	8. Check engine crankcase ventilation system/blowby

FIGURE 31–16 Exploded view of a typical air dryer

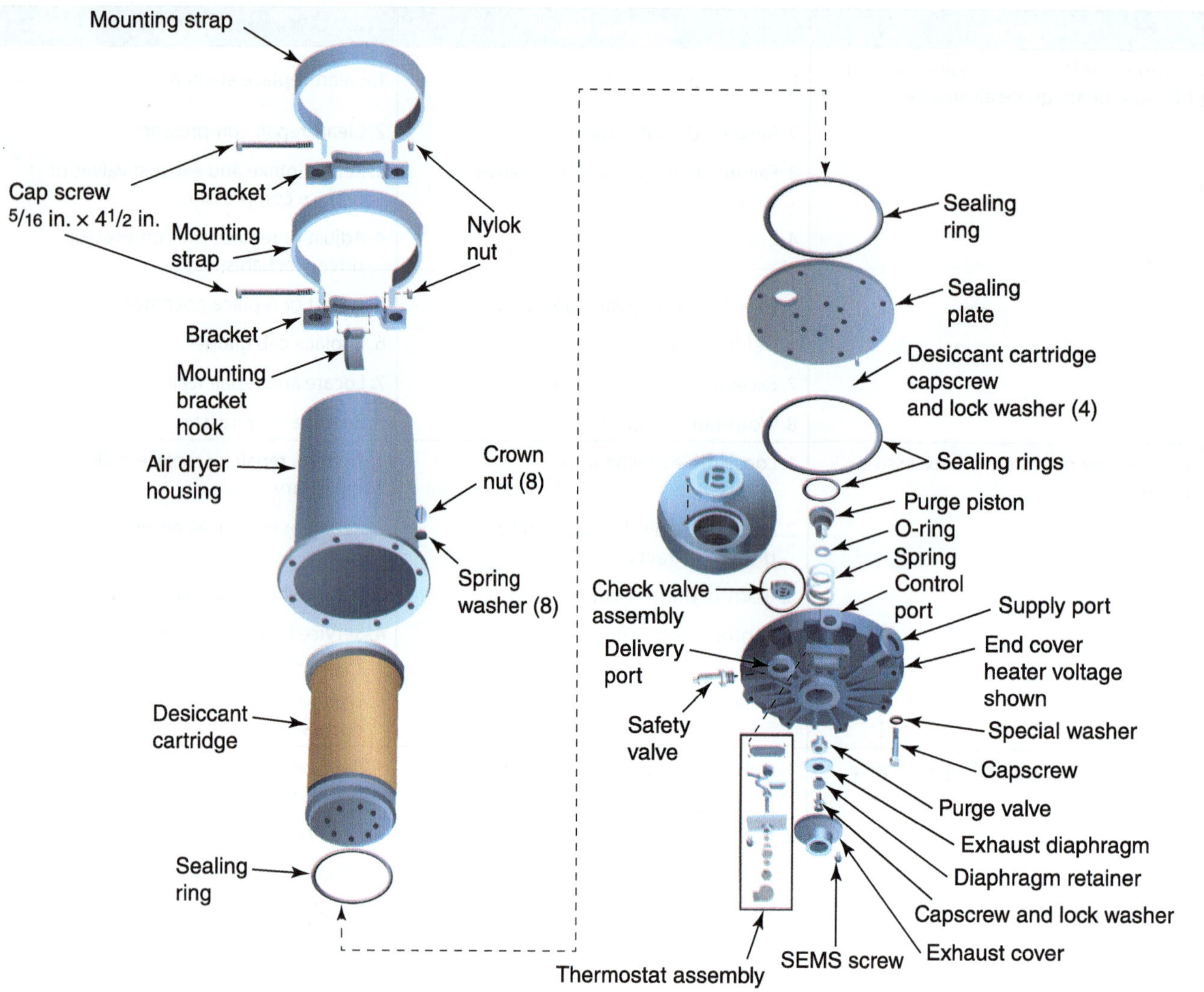

- Check for excessive leakage around the dryer purge valve. With the compressor in effective cycle (loaded mode), apply a soap solution to the purge valve housing assembly exhaust port and check that leakage does not exceed a 1-inch (25 mm) bubble in 1 second. If it does, service the purge valve housing assembly.
- Close all reservoir draincocks. Build up system pressure to governor cut-out and check that the purge valve dumps an audible volume of air. Use the foot valve to pump down the service brakes to reduce system air pressure to governor cut-in. Note that the system once again builds to cut out pressure and is followed by an audible purge at the dryer.
- Check the operation of the safety valve by pulling the exposed stem when the compressor is loaded (compressing air). Air should exhaust when the stem is held, and the valve should reseat when the stem is released.
- Check all lines and fittings leading to and from the air dryer for leakage and integrity.
- Check the operation of the end cover heater and thermostat (**Figure 31–17**) assembly during cold weather operation as follows:
  - **Electrical supply to the dryer**. With the ignition key on, use a DMM or testlight to check for voltage to the heater and thermostat assembly. Unplug the electrical connector at the air dryer and place the DMM test leads on each of the pins of the male connector. If there is no voltage, look for a blown fuse, broken wires, or corrosion in the vehicle wiring harness. Check to see whether a good ground path exists.
  - **Thermostat and Heater Operation**. Turn off the ignition switch and cool the end cover assembly to a temperature below 40°F (44°C). Using an ohmmeter, check the

FIGURE 31–17 Air dryer thermostat assembly

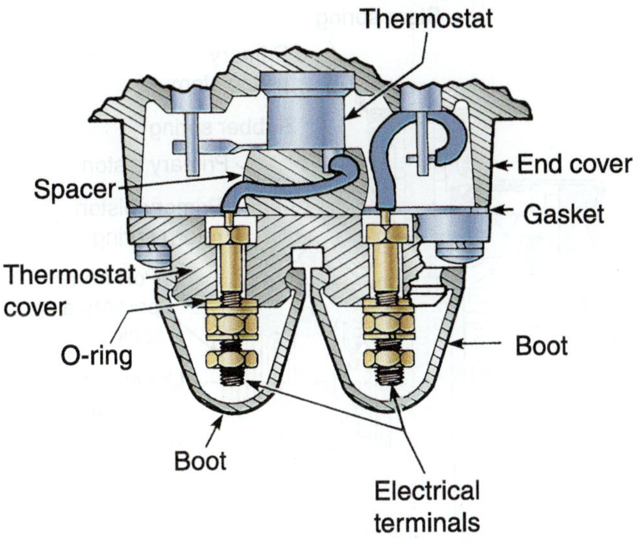

resistance between the electrical pins in the female connector. Compare the readings with the values specified by the OEM. If the resistance is higher than the maximum specified, replace the purge valve housing assembly, which includes the heater and thermostat. Now heat the end cover assembly to a temperature greater than 90°F (32°C) and check the resistance. The resistance should exceed 1,000 ohms. If the resistance values obtained are within the specified limits, the thermostat and heater assembly are operating properly. If the resistance values obtained are outside specifications, replace the purge valve housing assembly.

Some current air dryers are integral with the supply tank and also may have the governor built into the assembly. Each type of air dryer requires a different rebuild procedure. Although this is not especially difficult, you should always use the manufacturer service literature to guide you through the procedure.

Compliance Issues Related to the Supply System

Because maintaining a reliable source and supply of compressed air is critical to ensuring that a vehicle can stop as required, a number of regulations addressing potential deficiencies are enforced by most jurisdictions. The regulations place certain brake conditions into more severe categories than other conditions.

Out-of-Service Conditions of the Air Supply System

Out-of-service (OOS) conditions are the most serious vehicle safety defects. The presence of these conditions normally means that a vehicle is detained until repairs are made. OOS standards are not maintenance standards. They define a vehicle that is considered to be dangerous to operate. The following OOS criteria are defined by the **Commercial Vehicle Safety Alliance (CVSA)** and are used by vehicle safety enforcement officers across North America:

- Low-pressure warning device is missing, inoperative, or does not activate at 55 psi (3.8 bar) and below or half of the governor cut-out pressure, whichever is less. (Note that the low-air warning may be audible or an indicator light, or both. At least one must function normally.)
- An air reservoir is separated from its original attachment points.
- A compressor has loose mounting bolts; a cracked, broken, or loose pulley; or cracked or broken mounting brackets, braces, or adapters.

Other Compliance Requirements

When vehicles undergo periodic maintenance inspections or when inspections are carried out at roadside, additional criteria may apply, based on regulations for a particular jurisdiction. These criteria are as follows:

- Compressor buildup time must be less than 3 minutes to get from 50 psi to 90 psi (3.4 to 6.2 bar) with the engine run at 1,200 rpm.
- Governor cut-out pressure must be between 115 and 135 psi (7.9 and 9.3 bar).
- Governor cut-in must be above 100 psi (7 bar) on trucks and above 85 psi (6 bar) on buses.
- Sufficient compressed air reserve must be available so that a full brake application does not drop reservoir pressure by more than 18 psi (1.24 bar).
- The one-way check valves must be present and functional.

31.5 SERVICE APPLICATION CIRCUIT

This section deals with service application circuit maintenance and repair requirements.

AIR BRAKE FOOT VALVE PM

When you are checking air brake foot valve operation, remember that OEMs also refer to this as a treadle or dual-circuit application valve. Make sure that you know how it functions normally. Refer to Chapter 28 if you are not sure. **Figure 31–18** shows a typical foot valve and a sectional view of its internal components. Every 3 months or 25,000 miles (40,000 km):

- Visually check for physical damage to the brake valve, such as broken air lines and broken or missing components.

FIGURE 31–18 Typical foot valve: (A) treadle external components; and (B) cutaway of foot valve

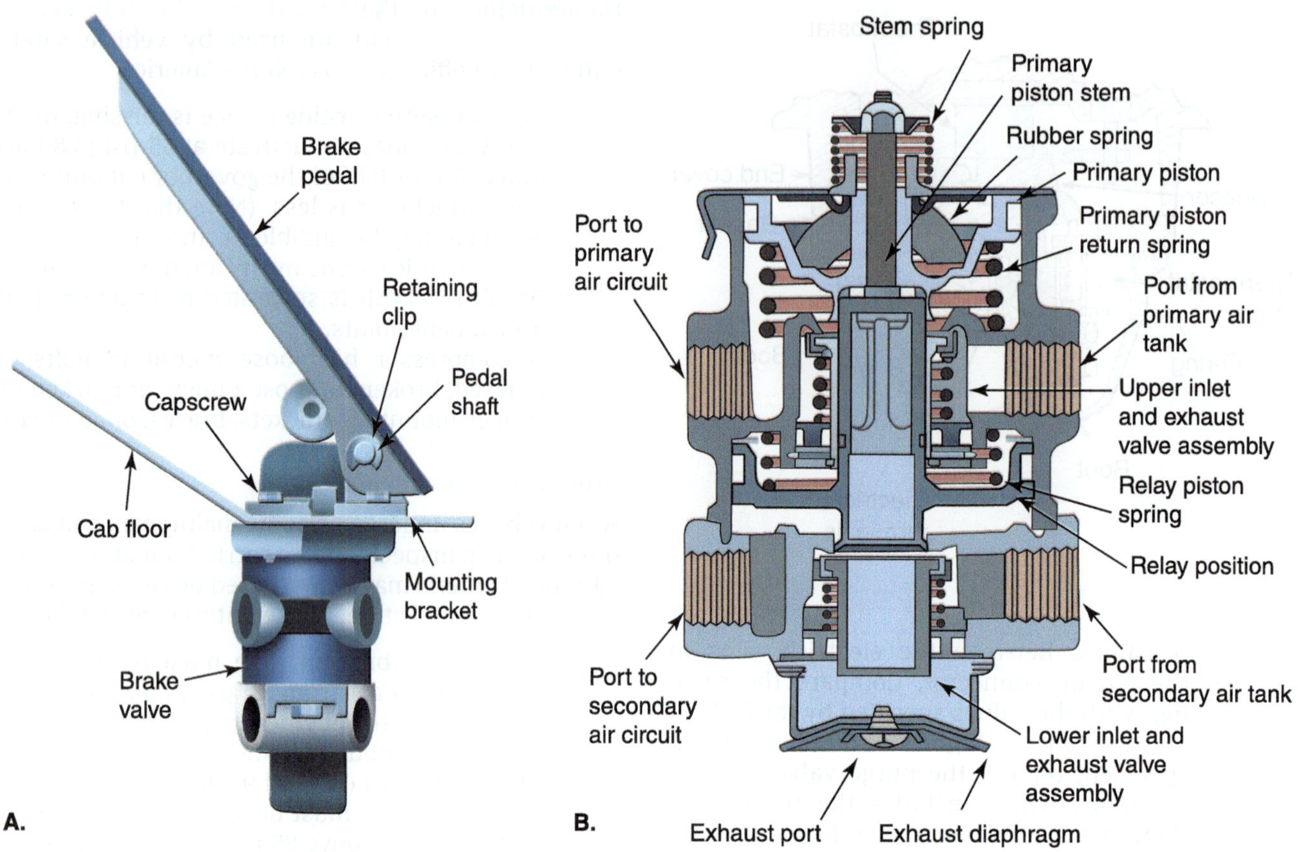

- Clean any accumulated dirt, gravel, or foreign material away from the heel of the treadle, plunger boot, and mounting plate.
- Using light oil, lubricate the treadle roller, roller pin, and hinge pin.
- Check the rubber plunger boot for cracks, holes, or deterioration and replace if necessary. Also, check mounting plate and treadle for integrity.
- Apply two to four drops of oil between the plunger and mounting plate. Never overlubricate!

Every 12 months or 100,000 miles (160,000 km):

- Disassemble the treadle assembly and clean the external components with mineral spirits; replace all rubber components or any worn or damaged part. Check foot valve operation before returning the vehicle to service.

PERFORMANCE CHECKS FOR FOOT VALVES

The foot valve is the key component in the service brake system of a vehicle. A good preventive maintenance (PM) program will incorporate the following tests.

Function Check

- Test the application pressure of both primary and secondary circuits using accurate test gauges. Depress the treadle to several positions between the fully released and fully applied positions. The application pressure on each test gauge should be equal in each test position and in correct proportion to the movement of the brake pedal (**Figure 31–19**).
- Following the release of a full application stroke of the pedal, the reading on the test gauges should drop to zero promptly.

Note that the primary circuit application pressure should normally be about 2 psi (14 kPa) greater than the secondary circuit application pressure when both supply tanks are at the same pressure.

Leakage Test

- Make and hold a high-pressure (over 80 psi [5.5 bar]) service application. You can use dash gauges (if equipped) or install a master gauge.
- Coat the exhaust port and body of the brake valve with a soap solution.
- Leakage permitted is a 1-inch (25 mm) bubble in 3 seconds. If the brake valve does not

FIGURE 31–19 Brake test application pressures

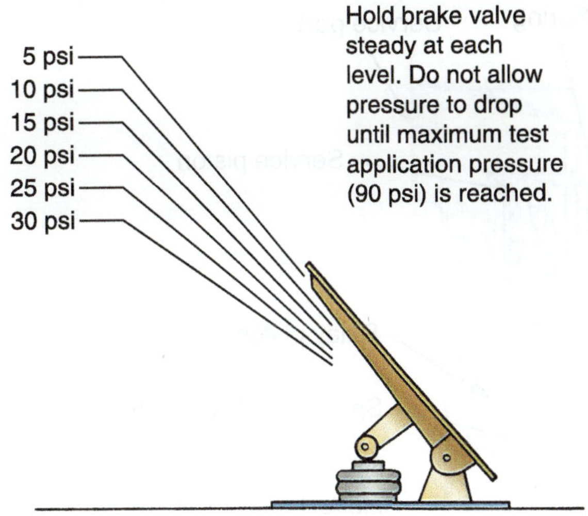

function as described or if leakage is excessive, it is recommended that it be replaced with a new or remanufactured unit.

Removal of a Foot Valve

The removal of a foot valve on some chassis can be labor-intensive. Ensure that the valve really requires replacement either because of a performance defect or leakage. It is not smart to make trial-and-error replacements of foot valves.

- Block the vehicle wheels or park the vehicle by mechanical means. Drain all air system reservoirs.
- Identify and disconnect all supply and delivery lines at the brake valve.
- Remove the brake valve and treadle assembly from the vehicle by removing the three capscrews on the outer bolt flange of the mounting plate. The basic brake valve alone can be removed by removing the three capscrews on the inner bolt flange.

SHOP TALK

Carefully label and code every line as it is removed from the foot valve assembly. Failure to do this can considerably lengthen the time required to install the replacement valve.

PREVENTIVE MAINTENANCE OF RELAY VALVES

No two vehicles operate under identical conditions, so once again, the maintenance practices and intervals will vary. Experience is a valuable guide in determining the best maintenance interval for any one particular operation. Typical recommended maintenance practices are as follows:

- Every 3 months or 25,000 miles (40,000 km), check each relay valve for proper operation.
- Every 12 months or 100,000 miles (160,000 km), each relay valve should be performance tested to ensure that the signal pressure and output application pressures are correct. One OEM recommends disassembling the valve and cleaning the components with mineral spirits. This is a labor-intensive practice that is seldom observed, but when it is, replace all rubber components and other components that are visibly worn or damaged. Check for proper operation before placing the vehicle in service.

Operational and Leakage Test

- Block the wheels, fully charge the air brake system, and adjust the brakes.
- Make several service brake applications and check for prompt application and release at each wheel.
- Check for inlet valve and O-ring leakage. Make this check with the service brakes applied when the relay valve is used to control the service brakes. Make this check with the spring brakes released when the relay valve is used to control the spring brakes. Coat the exhaust port and the area around the retaining ring with a soap solution; a 1-inch (25 mm) bubble in 3 seconds is permitted.
- Check for exhaust valve leakage. Make this check with the service brakes fully applied if the valve controls the service brakes.
- Coat the outside of the valve where the cover joins the body to check for seal ring leakage; no leakage is permitted.
- If the valve is used to control the spring brakes, place the park control in the released position and surround the balance port with a soap solution to check the diaphragm and its seat. Leakage equivalent to a 1-inch (25 mm) bubble in 3 seconds is permitted. Note: If the anticompound feature is in use, the line attached to the balance port must be disconnected to perform this test.

If the valves do not function as described, or if leakage is excessive, replace the valves with new or remanufactured units. With current ABS, each manufacturer has specific methods for diagnosing its components and the technician should always refer to these. Although some components and subcomponents can be cross-referenced between manufacturers, never assume this to be the case.

FIGURE 31–20 Sectional view of a typical antilock relay valve

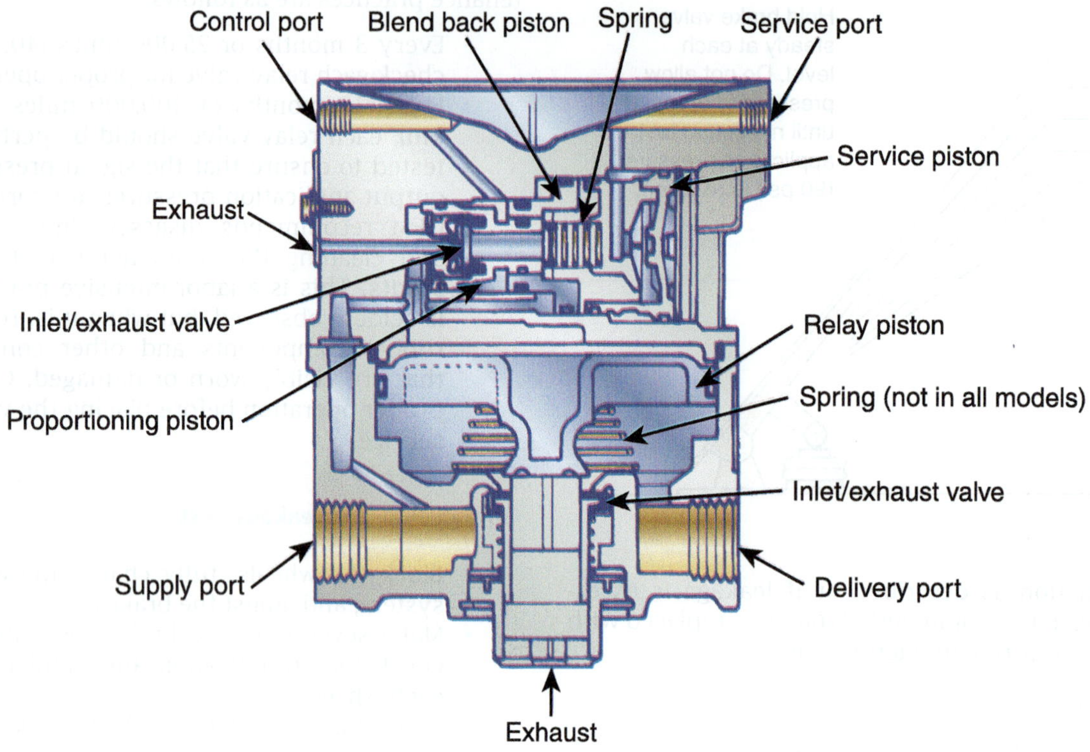

Figure 31–20 shows a sectional view of a typical antilock relay valve.

Removal of a Relay Valve

- Block and hold the vehicle by means other than air brakes.
- Drain the air brake system reservoirs.
- If the valve is to be removed, label the air lines before disconnecting them to facilitate installation.
- Disconnect the air lines from the valve.
- Remove the valve from the reservoir, or if remotely mounted, remove the mounting bolts and then the valve.

CAUTION

Drain all reservoirs before attempting to remove the inlet exhaust valve.

TROUBLESHOOTING RELAY VALVES/ SERVICE APPLICATIONS

Relay valves can cause a number of brake performance problems. They are vulnerable to external damage and freeze-up because of their location on the vehicle axles.

Complaint: Leaks

Determine when the leakage occurs and try to locate the source.

For static leakage:

1. Check for air brake system contamination.
2. Check for back-feeds. Back-feed occurs when air from one circuit forces its way into another; for instance, when hold-off air bleeds into the service circuit. Back-feeds can result from the following failures:
   - Spring brakes—emergency diaphragm or piston rod seal
   - Anticompound—double check or diaphragm in spring brake relay
3. Check for hold-off circuit bleed-down. Bleed-down can be caused by:
   - Reduction of signal pressure (park relay)
   - Parking system leakage
   - Park control valve leakage

SHOP TALK

Before replacing a park-relay combination valve that leaks from its exhaust port, make sure that the root of the problem is not caused by a defective center seal in one of the spring brake chambers it supplies.

For dynamic leakage:

1. Check for air brake system contamination.
2. Check for reduction of signal pressure (service relay).
   - Total system leakage
   - Brake valve leakage
3. Check for back-feed.
   - Spring brake relay anticompound diaphragm (only occurs with less than full service brake application)

Complaint: Brakes Will Not Apply/Release

1. Check the foundation brakes.
   - Worn, damaged, improperly installed or maintained components
2. Check the brake adjustment.
   - Automatic slack malfunction—over/under adjustment
   - Improper manual adjustment—over /under-adjustment
3. Make sure that the signal pressure is reaching the relay valve.
4. Check for freeze-ups.
   - Relay piston seized in bore—causing no application pressure or constant application pressure
5. Check for system contamination.
   - Inlet/exhaust sticking in relay
   - Air pressure not reaching relay
6. Check for line restrictions.
   - Pinched, kinked air lines
   - Frozen moisture in lines
7. Make sure that the chamber/slack adjusters are not binding.
   - Incorrect chamber/slack adjuster angle
8. Application valve sticking. Tractor treadle (foot) valves are notorious for sticking because they have the potential to be exposed to so much dirt especially in vocational and construction applications.
   - Foot valve plunger sticking
   - Pedal movement restricted (often by dirt trapped under the treadle)

Complaint: Slow Application/Release

1. Check the brake adjustment.
   - Underadjusted
   - Automatic slack adjusters
2. Check for air brake system contamination.
   - Internal relay valve components sticking
   - See "line restrictions"
3. Check for freeze-out.
   - Moisture in air system
   - Mechanical components
4. Check for line restrictions.
   - Pinched, kinked
   - Air system contamination
5. Check the foundation brakes.
   - Worn, defective, or improperly installed components
   - Rust corrosion
6. Check for chambers/slack adjusters binding.
   - Chamber/slack adjuster angle incorrect
   - Bent spider

Note: The procedure for replacing a service diaphragm in a spring brake unit is covered later in this chapter.

Trailer Application Valves

These valves are responsible for service applications of the trailer brakes independently from the tractor brakes. They meter service signal air to the trailer service application circuit in proportion to the mechanical travel of the valve. **Figure 31–21** shows a sectional view of a trailer hand application valve.

Quick-Release Valves

Quick-release valves are used in both the service application circuit and the hold-off circuits. They distribute as a tee fitting when supplied with air but exhaust at the valve when the source air ceases. Because they are so simple, diagnosing a malfunction is fairly easy. A common failure is discharge leakage to exhaust when actuated. **Figure 31–22** shows a cutaway view of a quick-release valve.

Tractor Protection Valve (TPV)

The TPV plays an important role in routing service application signal air to the trailer. The full operation of a TPV is explained in Chapter 28. **Figure 31–23** shows a cutaway view of a three-line TPV. When checking out current TPVs, remember that some incorporate a quick-release function to accelerate service signal dump. This should not be confused with unintended leakage.

COMPLIANCE ISSUES RELATED TO THE SERVICE BRAKE SYSTEM

Any defect in the service brake application circuit directly affects the ability of the driver to stop the vehicle in the FMVSS 121–required stopping time. CVSA OOS standards should not vary by jurisdiction, but the thoroughness with which they are enforced does despite the efforts of the CVSA to maintain

FIGURE 31–21 Sectional view of a trailer hand control brake valve

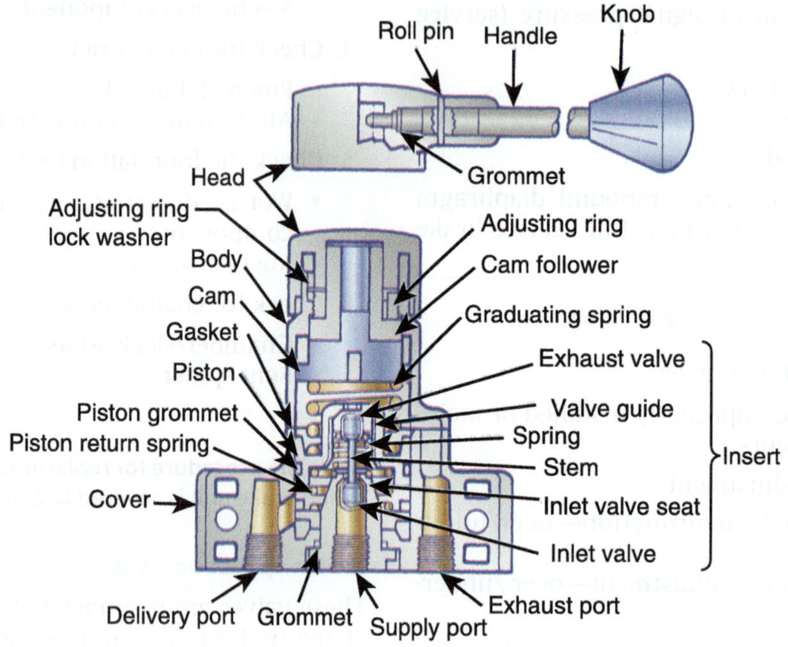

FIGURE 31–22 Cutaway view of a quick-release valve

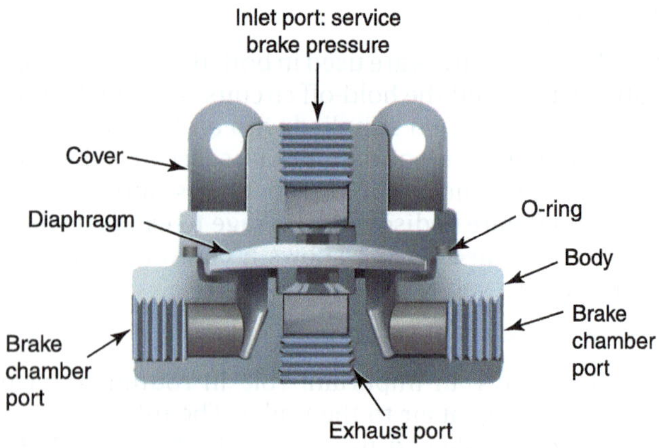

consistency. OOS regulations grade some service brake defects into more severe categories than others.

OOS Conditions of the Service Brake System

OOS conditions are the most serious vehicle safety defects. The presence of these conditions normally means that a vehicle is detained until repairs are made. A fine may be levied for an OOS infraction. The following OOS test criteria defined by the CVSA are used by vehicle safety enforcement officers across North America.

With the governor cut-in, reservoir between 80 and 90 psi (5.5 and 6.2 bar), engine idling, and service

FIGURE 31–23 Cutaway view of a three-line tractor protection valve

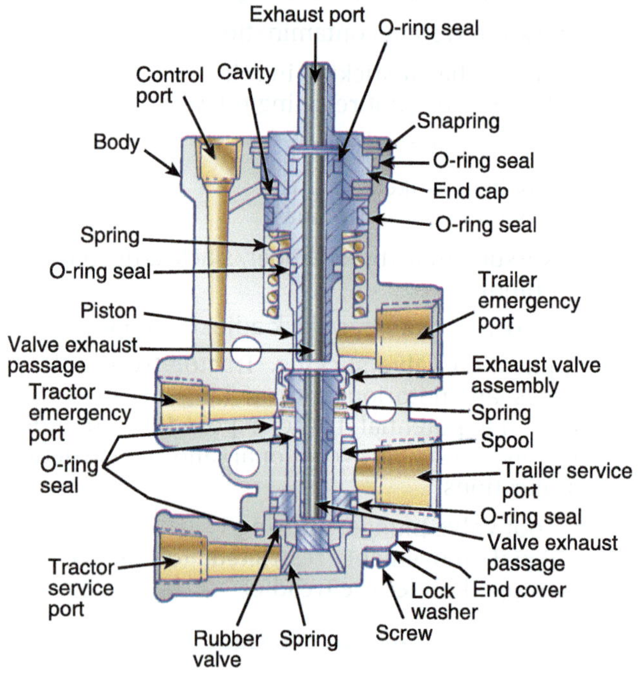

brakes fully applied, reservoir pressure must be maintained. Reject the vehicle if:

1. Brake hose bulges or swells under pressure.
2. Audible leak(s) can be heard from any brake hose.
3. Improperly joined or spliced brake hose is discovered.

4. Brake hose is visibly cracked, broken, or crimped.
5. Brake tubing has an audible leak at other than a proper connection.
6. Brake tubing is cracked, damaged by heat, broken, or crimped.
7. Audible air leak is detected at any brake chamber.
8. Mismatched chamber sizes or slack adjuster lengths occur on a steer axle.
9. Any nonmanufactured hole(s), missing cage bolt hole plug, or crack(s) exist in a spring brake assembly.

SHOP TALK

OOS standards are NOT maintenance standards. They indicate that the brakes have become a safety hazard and that the vehicle should not be operated on public roads. Operators who identify OOS standards as their maintenance threshold standards are courting danger. Many non-OOS faults can cause serious brake system underperformance problems.

31.6 PARKING/EMERGENCY CIRCUIT

FMVSS 121 requires that all air brake–equipped vehicles have a means of mechanically applying the parking brakes. In most truck and trailer air brake systems, the parking and emergency circuits are essentially one circuit. This means that they have in common the same actuation hardware and control pneumatics. It cannot be emphasized enough that proper parking/emergency brake performance still requires that the brake adjustment be correct. **Figure 31–24** shows an exploded view of a current two-valve dash control module; these are not normally field repaired. In most cases, they are diagnosed as defective and then replaced as a unit.

COMPLIANCE ISSUES RELATED TO THE PARKING/EMERGENCY BRAKE SYSTEM

The regulations are based on FMVSS 121 performance requirements and place some brake deficiency conditions into more severe categories than others. Again, those classified as OOS conditions are considered to be the most severe.

OOS Conditions of the Parking/ Emergency Brake System

The following OOS criteria are defined by the CVSA and used by vehicle safety enforcement officers across North America:

1. The parking brakes are applied on a tractor/straight truck and an attempt is made to move the vehicle under engine power. The brakes must prohibit vehicle movement.

FIGURE 31–24 Exploded view of a two-valve dash control module

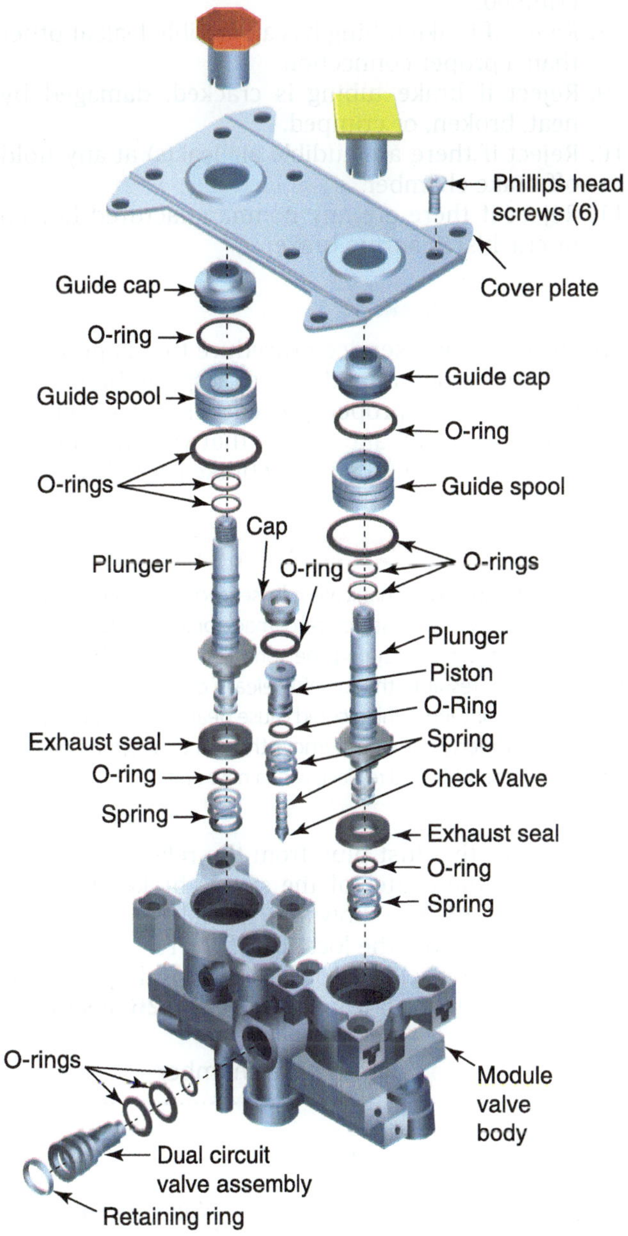

2. On a tractor/trailer combination, with the trailer parking brakes applied, the parking brakes should remain locked when an attempt to move the unit under tractor power is made.
3. A tractor/trailer combination must have a parking brake grade holding capacity that allows it to remain stationary when facing either uphill or downhill on a 20 percent grade when loaded to its gross vehicle weight rating (GVWR).
4. Reject if any brake hose bulges or swells under pressure.
5. Reject if there are audible leak(s) at any brake hose.

6. Reject if there are improperly joined or spliced brake hose.
7. Reject if a brake hose is cracked, broken, or crimped.
8. Reject if brake tubing has an audible leak at other than a proper connection.
9. Reject if brake tubing is cracked, damaged by heat, broken, or crimped.
10. Reject if there are audible air leak(s) at any hold off brake chamber.
11. Reject if there are any nonmanufactured hole(s) or crack(s) in spring brake.

Caging a Spring Brake

This is a common service procedure that is perfectly safe when the potential dangers of spring brakes are realized and the proper procedures are observed. Consult Chapter 28 and ensure that you fully understand the operation of a spring brake assembly.

WARNING

Do not attempt to mechanically release (cage) the spring when the spring brake shows structural damage or when safety hooks have been removed. Caging the spring or disassembling the chamber can result in the forceful release of the spring chamber and its contents, which could cause death, severe personal injury, or property damage. Remove the complete spring brake chamber assembly and replace with a new unit.

1. Remove the dust plug from the release tool keyhole in the center of the spring brake chamber. **Figure 31–25** shows a typical spring brake assembly. Note the location of the release stud.
2. Remove the release tool from the side pocket of the adapter as shown in **Figure 31–26**. Insert the release tool through the chamber keyhole into the pressure plate as shown in **Figure 31–27**.
3. Turn the release tool ¼ turn clockwise.
4. Pull on the release tool to seat the release stud cross pin in the cross pin area of the pressure plate.
5. Assemble the release tool washer and nut onto the release tool finger-tight.
6. Tighten the release tool nut with a hand wrench (do not use an impact wrench), as shown in **Figure 31–28**, and make certain that the pushrod is retracting.
7. Do not overtorque the release tool nut. Typical torque values are as follows. Refer to the applicable OEM service literature for specific torque values.
   - S-cam type: 35 lb-ft. (47 N·m)
   - Wedge type: 27 lb-ft. (37 N·m)
8. To ensure that the compression spring is fully caged, the release tool length beyond the nut should typically measure:
   - 30^2-inch (762^2 mm) chamber: 2.875 inches (73 mm) minimum

FIGURE 31–25 Spring brake chamber

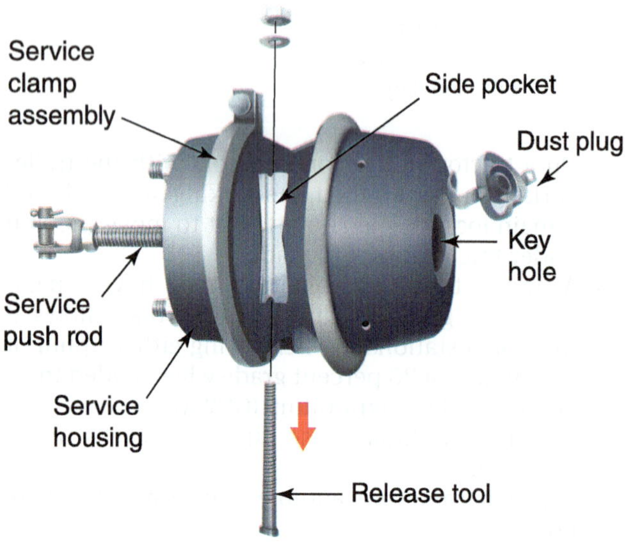

FIGURE 31–26 Release tool removal from spring brake: (A) Insert release tool into cage key hole; (B) rotate the release tool clockwise to lock the lugs into the cage plate.

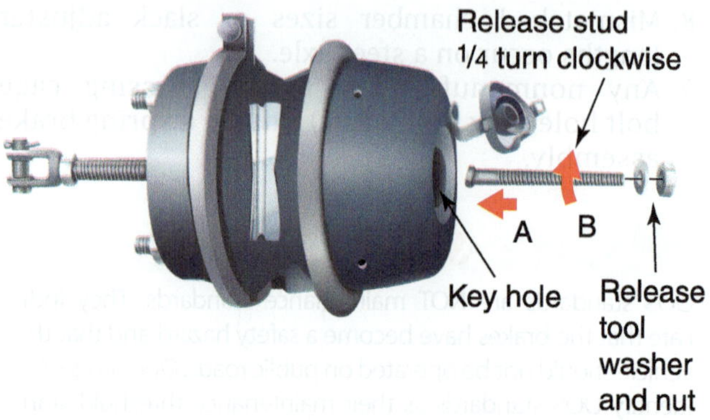

FIGURE 31–27 Insert the release tool lugs into the pressure plate.

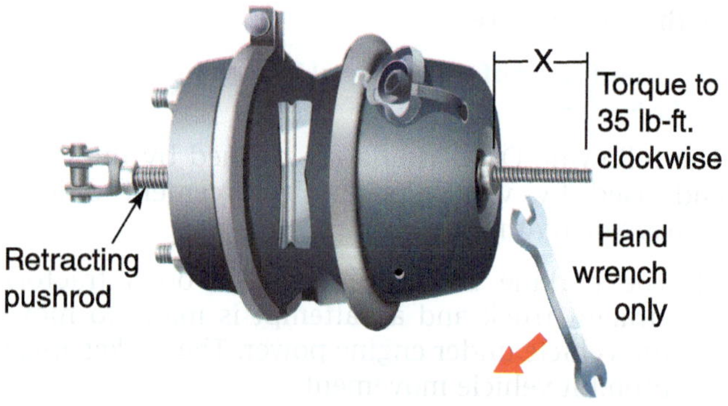

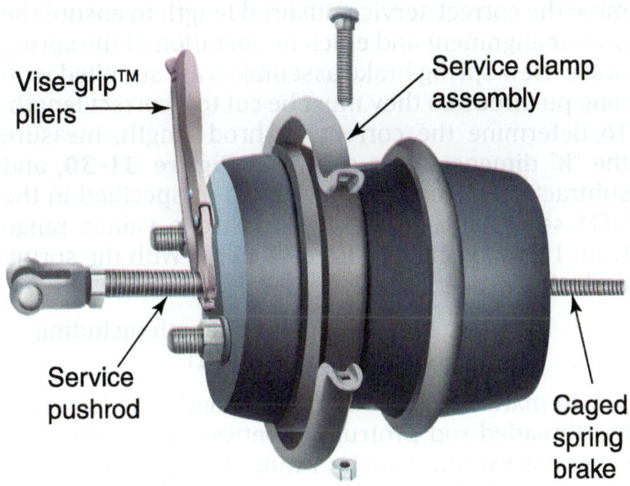

FIGURE 31–28 Caging the main spring

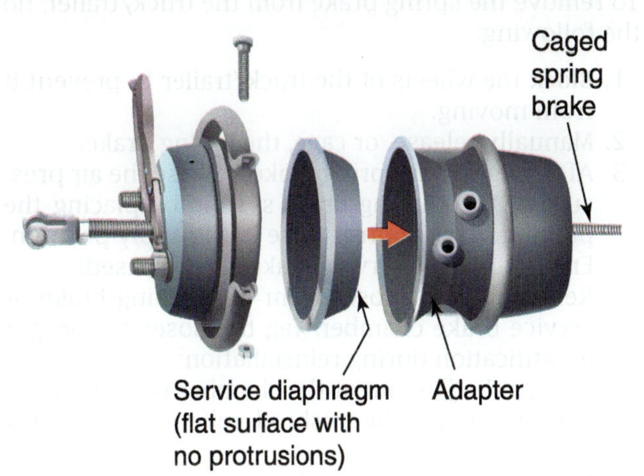

FIGURE 31–29 Service diaphragm replacement

- 24²-inch (610² mm) chamber: 2.915 inches (74 mm) minimum

9. Refer to the applicable OEM service literature for specific tool lengths.

Service Diaphragm Replacement

Current practice in the trucking industry is to discourage replacement of the service diaphragm in a spring brake assembly. In other words, it is recommended that failure of either the hold-off or service diaphragm warrants spring brake assembly replacement. Having said that, it remains common practice to replace the service diaphragm on a spring brake chamber. The following procedure is required:

1. Manually cage the spring brake chamber as outlined earlier.
2. To prevent the sudden release of the **piggyback assembly** (self-contained spring brake and service chamber flange fitted to the existing service chamber) or service pushrod assembly and to facilitate the installation of the new diaphragm, prevent the service pushrod from retracting by clamping it in place with Vise-Grip™ pliers.
3. Remove the service clamp assembly as shown in **Figure 31–29** and discard the old diaphragm.
4. Inspect the service clamp assembly, adapter wall and lip, housing, service return spring, and service pushrod. If any structural damage is noted, replace with new parts.
5. Wipe the surface of the service pushrod plate clean of any oil, grease, or dirt. Check to see that weep holes in the housing are not plugged.
6. Place the new service diaphragm in the adapter and center the housing over the diaphragm and adapter.
7. Make sure that the diaphragm is properly seated between the adapter and housing lip and reassemble the service clamp assembly. Torque the carriage nuts to the manufacturer's specification (typically 18–25 lb-ft. [24–34 N·m]). Check the carriage bolts and clamp assembly for proper seating around the adapter and housing lip and remove the Vise-Grip™ pliers from the service pushrod.
8. Apply a maximum of 120 psi (8.3 bar) air pressure to the service port and check the diaphragm seal for leakage by applying a soap-and-water solution to the service clamp area. No leakage is allowed.
9. Uncage the compression spring and reassemble the release tool in the side pocket of the adapter.
10. Replace the dust plug in the release tool keyhole in the center of the chamber.

WARNING

Always cage the compression spring with the release tool. Never rely on air pressure alone to keep the spring compressed.

Changing Position of the Mounting Bolts, Clamps, and Air Ports

To change the position of the chamber mounting bolts, clamps, or air ports, do the following:

1. Manually cage the spring brake chamber.
2. If air pressure was used to aid the caging process, exhaust the air pressure.
3. To prevent the sudden release of the housing and to facilitate rotation, prevent the pushrod from retracting by clamping the pushrod with Vise-Grip™ pliers.
4. Loosen the carriage bolts on the clamp and rotate the central housing to locate air ports in the desired position.

Spring Brake Removal

To remove the spring brake from the truck/trailer, do the following:

1. Block the wheels of the truck/trailer to prevent it from moving.
2. Manually release, or cage, the spring brake.
3. After caging the spring brake, release the air pressure in the parking brake system by placing the parking brake control valve in the apply position. Ensure that the service brakes are released.
4. Remove the air hoses from the parking brake or service brake chamber. Tag the hoses for proper identification during reinstallation.
5. Loosen the jam nut at the chamber pushrod yoke and disconnect the pushrod yoke from the slack adjuster.
6. Remove the mounting stud nuts and remove the complete spring assembly from the mounting bracket.

CAUTION

The loaded main or power spring in a spring brake assembly exerts a linear force of up to 1,800 lb (816 kg) and is potentially lethal. Mechanical caging devices must be used whenever the spring chamber assembly is removed and replaced from the foundation brake mounting plate. Most current spring brake units are sealed with a band clamp that has to be destroyed to separate the spring chamber to prevent the unit from being disassembled.

Disarming Spring Brakes

Spring brakes should never be disposed of in scrap metal containers unless they have first been disabled. Disarming a spring brake requires that it is installed in a disarmament chamber because the power spring tension can propel it to a velocity exceeding 700 mph (1,126 kmh). The disarmament chamber is a heavy steel cylindrical chamber within which the spring brake is placed. After the chamber door has been closed, the chamber has a pair of openings on either side large enough to allow the head of an oxyacetylene cutting torch to enter.

The spring brake chamber band should be cut on either side. In some cases, this will cause the spring chamber to separate (Exercise caution: This may occur with explosive force!) and release the spring. Other times, a hammer and long chisel will be required to separate the chamber housing. The procedure for disarming a spring brake chamber is shown in the video series that accompanies this book.

CAUTION

Never open a disarmament chamber until the main spring has been observed through the inspection windows to be separated from the housing and is not under tension.

Spring Brake Installation

When installing a new spring brake chamber, determine the correct service pushrod length to ensure the proper alignment and efficient operation of the spring brake. New spring brake assemblies are supplied with long pushrods so they must be cut to a correct length. To determine the correct pushrod length, measure the "B" dimension, as shown in **Figure 31–30**, and subtract the setup stroke as listed or specified in the OEM service literature (typical stroke values range from 1⅜ to 1¾ inches [36 to 45 mm]). With the spring brake fully caged:

B – setup stroke = pushrod length including clevis (**Figure 31–31**)

To mark the pushrod cutoff length, the length of the threaded rod protruding between the clevis legs must not exceed ⅛ inch (3 mm); this prevents interference with the operation of the slack adjuster, as shown in **Figure 31–32**.

When the proper pushrod length has been marked off, the pushrod can be cut to length with the spring brake fully caged, as shown in **Figure 31–33**.

Install the spring brake chamber by performing the following procedures:

1. Install the clevis and jam nut on the pushrod, and install the spring brake mounting studs to the mounting bracket. (Refer to the OEM service literature for the stud installation torque, which is typically 135 to 155 lb-ft. [185 to 210 N·m]) The clevis must be adjusted so that it has full thread engagement on the pushrod (from flush to 3⁄16 inch [5 mm]) protrusion).

FIGURE 31–30 Determining required pushrod length

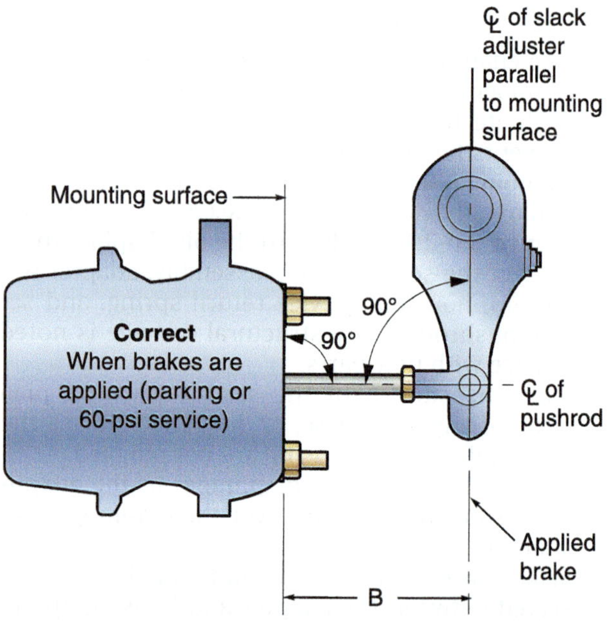

FIGURE 31–31 B dimension − setup stroke = pushrod/clevis length.

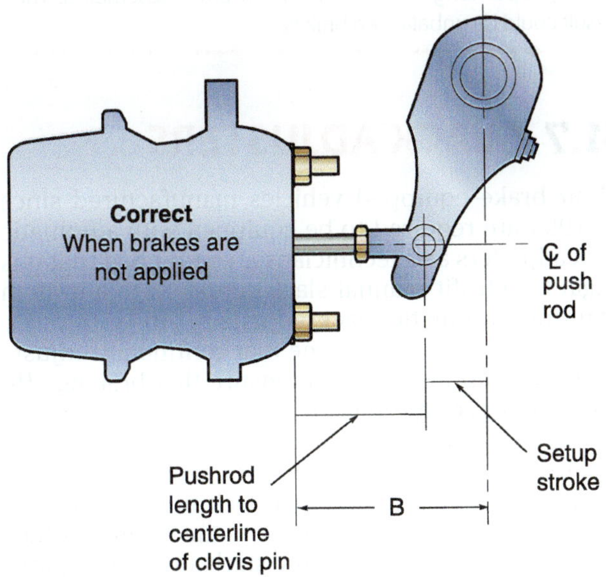

FIGURE 31–32 Installation of clevis onto pushrod

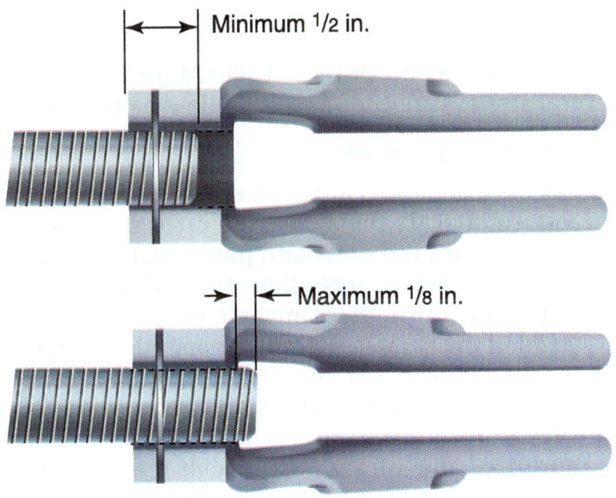

2. Connect the service and emergency air lines to the proper air ports, and connect the clevis to the slack adjuster.
3. Uncage the compression spring by loosening and removing the release tool nut and washer.
4. Reassemble the release tool and install it into the side pocket of the adapter.
5. Install the dust plug into the release tool keyhole in the center of the chamber.
6. Adjust the slack adjuster to the specified setup stroke. Using a full service brake application, the following conditions must occur:

FIGURE 31–33 Cut the pushrod to the correct length with the spring brake fully caged.

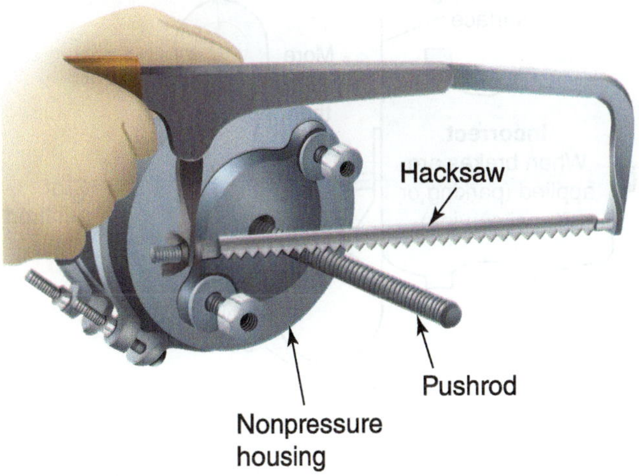

- Pushrod should be 90 degrees to the centerline of slack adjuster.
- Pushrod should be 90 degrees to the mounting face of the spring brake.

7. If the setup results in a condition shown in **Figure 31–34** or **Figure 31–35**, the spring brake is misaligned and must be corrected by the following:
   - Shorten the pushrod and align the spring brake on the mounting bracket (see Figure 31–34).
   - Align the spring brake on the mounting bracket or if required, replace the spring brake assembly (see Figure 31–35).
8. If misalignment cannot be corrected, consult with the foundation brake manufacturer to verify the correct mounting bracket position.

FIGURE 31–34 Pushrod too long

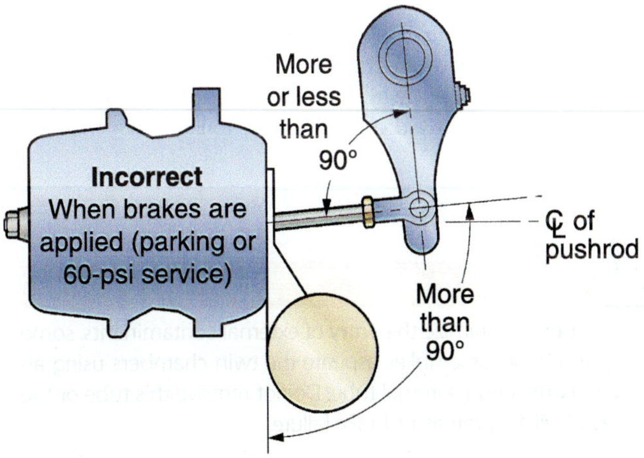

FIGURE 31–35 Pushrod too short

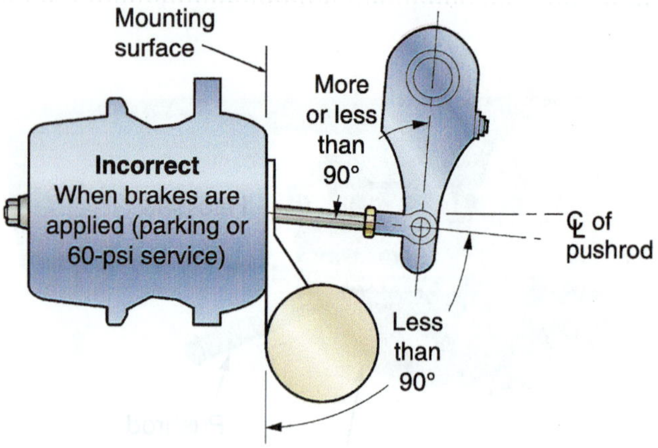

9. When the spring brake and pushrod are set as shown in Figure 31–30, release the brakes and readjust the slack adjusters to the shortest possible stroke without the brakes dragging.
10. Ensure that the release tool is inside the pocket of the adapter and that the dust plug is installed in the release tool keyhole in the center of the chamber.

CAUTION

Setting the 90-degree angle as outlined in step 6 ensures the best mechanical performance of the actuation circuit. However, note that Department of Transportation (DOT) inspectors check **applied stroke**, not the angle at full travel. All air brake–equipped commercial trucks and trailers manufactured after model year (MY) 2007 must be equipped with stroke indicators on the slack adjusters.

Tech Tips

1. If a brake chamber actuation rod kicks to one side, it indicates a broken spring.
2. Measure the actuation rod length during replacement when it is caged, not when extended.
3. OEMs are getting away from using threaded actuation rods and moving toward chassis-specific spring brake chamber assemblies.

CAUTION

In order to minimize the entry of external contaminants, some spring brake assemblies aspirate the twin chambers using an interconnecting external tube. Do not remove this tube or the result will be premature brake failure.

CAUTION

When replacing brake actuator units do not mix standard, long stroke, super long stroke, and piston brake assemblies. The result could be unbalanced brakes.

31.7 SLACK ADJUSTERS

All air brake–equipped vehicles manufactured since MY 1994 are required to be equipped with automatic slack adjusters and technicians are reminded that it is illegal to retrofit manual slack adjusters to a system requiring automatic slack adjusters. An **automatic slack adjuster (ASA)** is one that continually adjusts its stroke as brakes wear to ensure that braking efficiency is not compromised.

Slack adjusters are levers; make sure that the specified clevis pin-to-camshaft dimension is always observed. A small difference in slack adjuster effective length can make a big difference in the resulting brake torque. Each type of automatic slack adjuster requires a slightly different installation procedure, so the OEM instructions should always be observed. **Figure 31–36** and **Figure 31–37** show cutaway and exploded views of two different types of automatic slack adjusters.

INSTALLATION OF SLACK ADJUSTERS

Installation of a new or rebuilt slack adjuster requires several steps, as outlined in the following sections.

Installation Preparation

1. Check the brake camshaft or powershaft, bushings, and seals for wear and corrosion. If necessary, replace the camshaft, powershaft, bushings, or seals. Make sure that the brake operates smoothly. Turn the camshaft or powershaft by hand. If the camshaft or powershaft does not turn easily, inspect and repair the camshaft, powershaft, bushings, or seals as required.
2. Check the return spring in the air chamber to make sure that the spring has enough tension. Apply the service brake and the spring brake several times. Make sure that the return spring quickly and completely retracts the pushrod. If necessary, replace the return spring or the air chamber.
3. The new slack adjuster must be the same length as the old one. Consult the OEM service literature to determine the correct length of slack adjuster.
4. Make sure that the brake is completely released during the installation and adjustment procedures, except when the directions indicate that the brake must be applied. If the brake is not completely released when the clevis and the slack adjuster are installed and adjusted, the slack adjuster will not adjust the brake correctly. **Figure 31–38** shows a typical slack adjuster installation.

FIGURE 31–36 Cutaway view of a typical slack adjuster

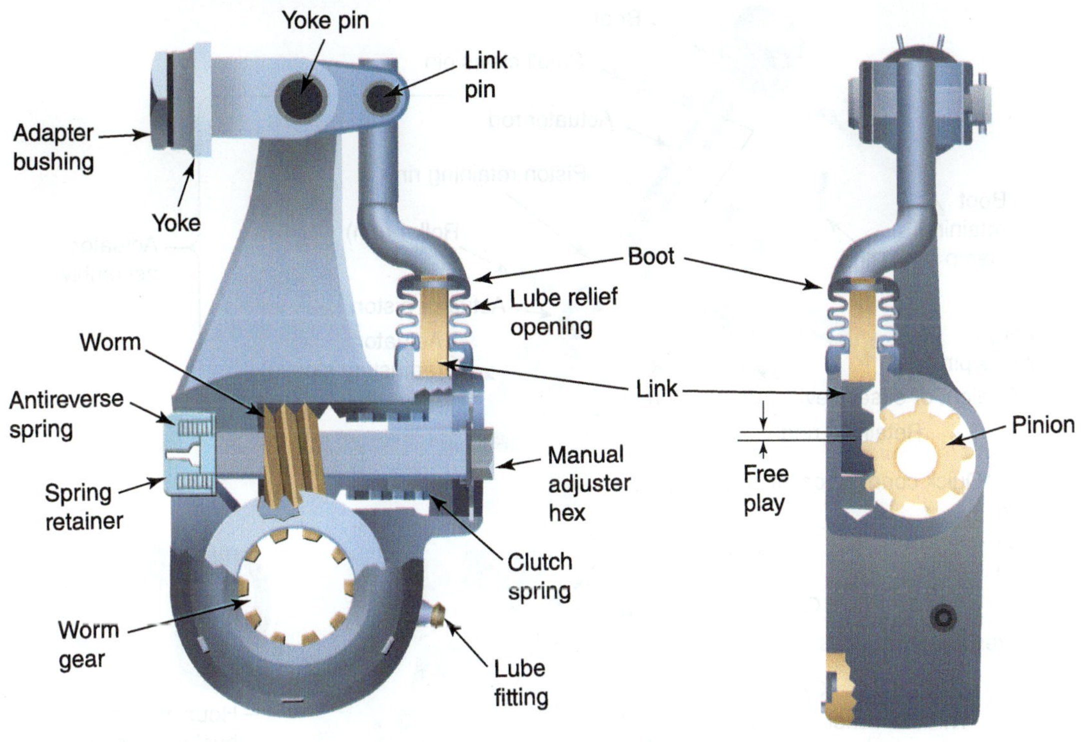

Clevis Installation and Adjustment

Some slack adjuster manufacturers supply installation templates to enable the truck technician to correctly install the adjuster. The templates available can be used with the slack adjuster to check clearances or to ease installation on drum or disc brake foundations. A typical installation template is shown in **Figure 31–39**.

Always use the template(s) prescribed by the manufacturer. If templates are not available, refer to the OEM service literature for specific and detailed installation procedures. The adjustment and installation procedures given here are based on the use of templates.

1. Remove the clevis from the new slack adjuster.
2. Install the new clevis on the pushrod. Do not tighten the jam nut against the clevis.
3. Measure the length of the slack adjuster with the template, as shown in Figure 31–39. The marks by the holes in the small end of this template indicate the length of the slack adjuster.
4. Use the template to install the clevis in the correct position. First, put the large clevis pin through the large holes in the template and the clevis. Select the hole in the template that matches the length of the slack adjuster. Hold the selected hole on the center of the camshaft or powershaft. Look through the slot in the template. The small hole in the clevis must be completely visible. If necessary, adjust the position of the clevis on the pushrod until the small hole in the clevis is completely visible through the slot in the template. At least ½ inch (12.5 mm) of thread engagement must be between the clevis and the pushrod. Also, the pushrod must not extend through the clevis more than ⅛ inch (3 mm) (see Figure 31–32). If necessary, cut the pushrod or install a new pushrod or a new air chamber.
5. Tighten the jam nut against the clevis to hold the clevis in the correct position. Tighten it to the manufacturer's torque specification.

CAUTION

The clevis must be installed in the correct position on the pushrod, or the slack adjuster will not adjust the brake correctly.

Installing the Adjuster

To install a new or repaired slack adjuster, do the following:

1. Lubricate the splines on the slack adjuster gear and the splines on the camshaft or the powershaft with antiseize compound.
2. Install the slack adjuster on the camshaft or powershaft.
3. Install spacing washers and the snapring until a maximum clearance of 0.060 inch (1.5 mm) exists between the washer and the snapring.

FIGURE 31–37 Exploded view of a slack adjuster

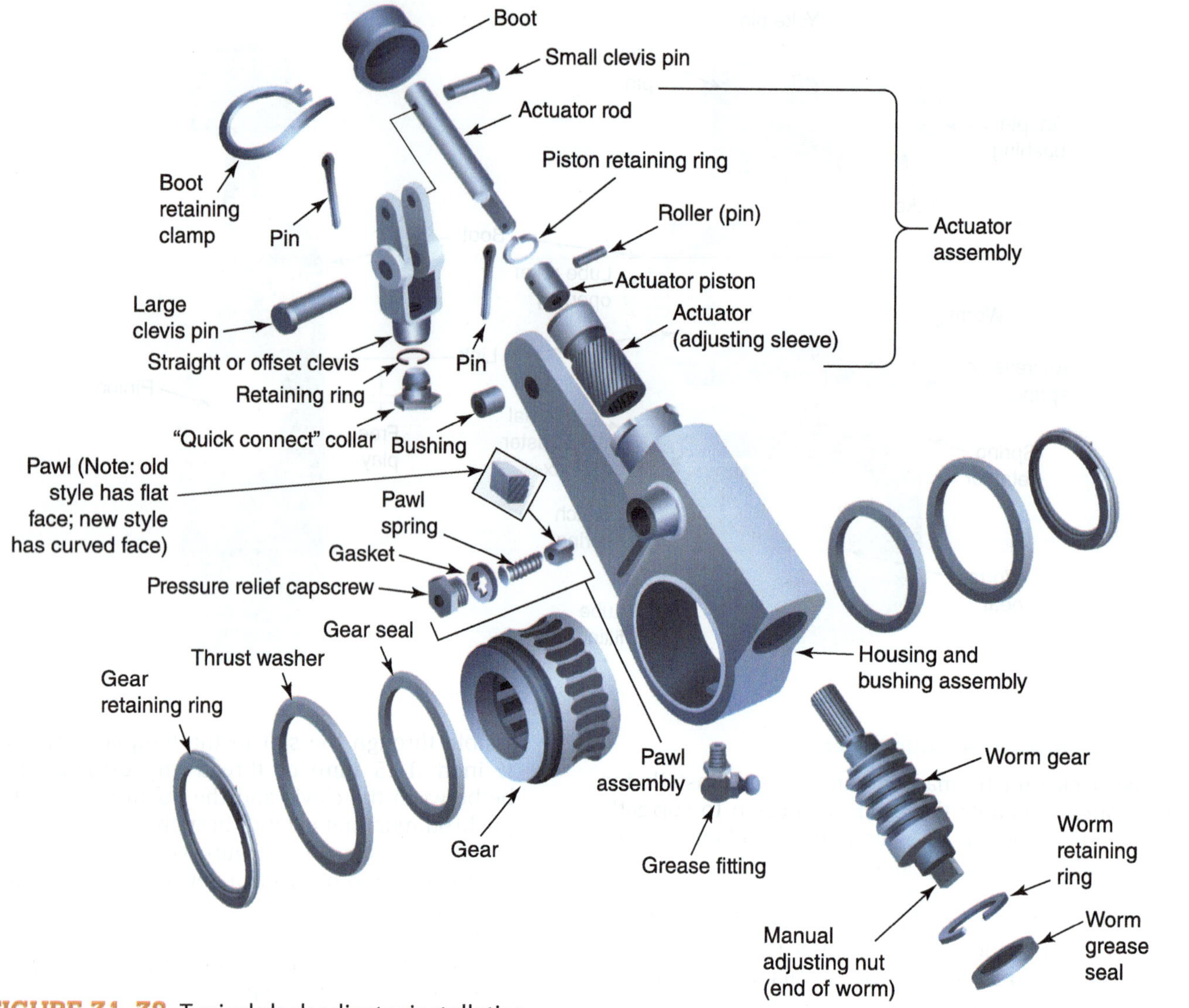

FIGURE 31–38 Typical slack adjuster installation on an S-cam foundation brake

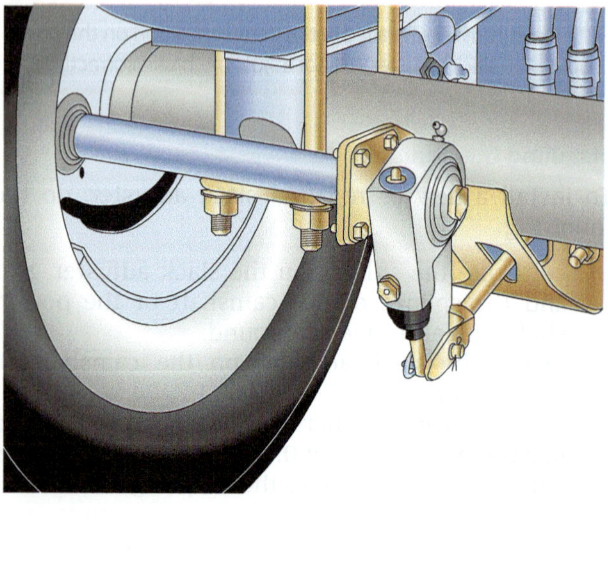

4. Remove or release the pawl assembly from the slack adjuster.
5. Use a wrench to turn the manual adjusting nut to align the hole in the arm of the slack adjuster with the large hole in the clevis, as shown in **Figure 31–40**.
6. Rotate the slack adjuster by hand the same distance as the maximum stroke of the chamber. Make sure that no obstructions will prevent the slack adjuster from rotating when the brakes are applied. Check the OEM specifications to locate the maximum stroke of each size of brake chamber. The adjusted chamber stroke will always be shorter than the maximum stroke.
7. Install both clevis pins through the template, slack adjuster, and clevis. Check again to make sure that the clevis is installed in the correct position. Adjust the clevis if necessary.

FIGURE 31–39 Measuring freestroke using a template

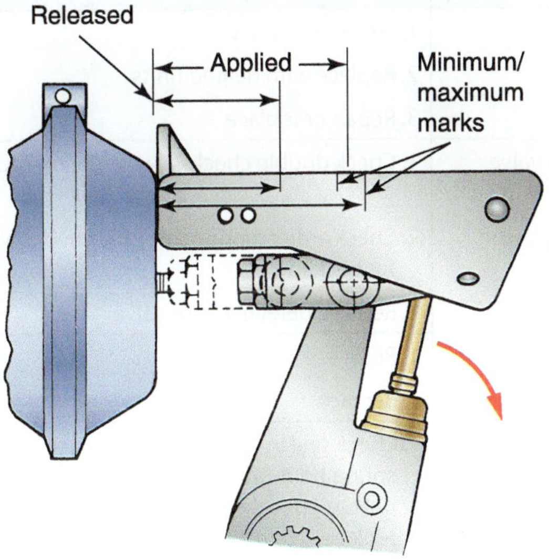

FIGURE 31–40 Use a wrench to turn the adjustment nut to align the arm of the slack adjuster with the large hole in the clevis.

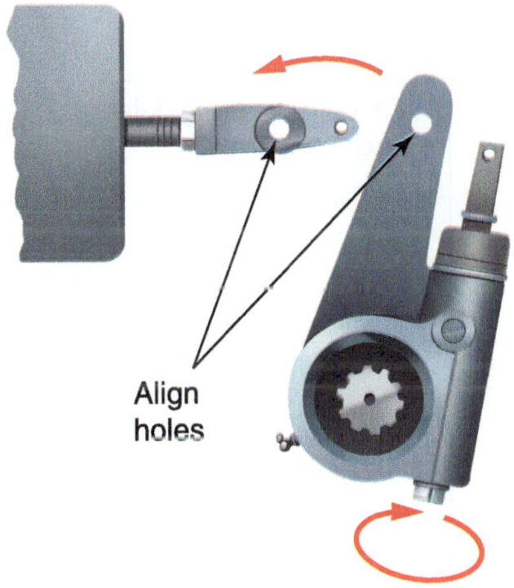

8. Remove the template. Apply antiseize compound to the two clevis pins. Install the clevis pins and install cotter pins to hold the clevis pins in place.
9. Adjust the brakes. Brake adjustment is covered in some detail later in this chapter.

SHOP TALK

The adjusting pawl assembly can be on either side of the housing or on the front of the housing. Make sure that the pawl assembly can be removed after the slack adjuster is installed. The pawl assembly must sometimes be removed when the slack adjuster is serviced.

CAUTION

If the pawl is not removed, the teeth will be damaged when the manual adjusting nut is turned.

31.8 TRAILER BRAKE SYSTEM

Because the trailer brake system is managed entirely from the tractor unit, it is impossible to develop a proper understanding of the trailer brake system without first understanding the tractor brake system. Trailer service technician specialists must ensure that they understand the principles of the tractor braking system before attempting to work on the trailer brakes.

COMPLIANCE ISSUES RELATED TO THE TRAILER BRAKE SYSTEM

Because maintaining a fully functioning trailer brake system that gives the driver full control of brake applications is critical to ensuring that a vehicle can stop as required, a number of regulations address potential deficiencies and are enforced by most jurisdictions. The regulations place certain degraded brake conditions into more severe categories than other conditions.

OOS Conditions of the Trailer Brake System

OOS conditions are the most serious vehicle safety defects. The presence of these conditions normally means that a vehicle is detained until repairs are made, and sometimes a fine is levied. The following OOS criteria are defined by the CVSA and used by vehicle safety enforcement officers across North America:

1. Discharge the trailer supply line; fully apply the tractor service brakes, and ensure that no air discharges from the trailer service line.
2. Fully charge the trailer air system and then disconnect the trailer supply line. The trailer brakes must immediately apply.
3. Audible air leak(s) at brake chamber.
4. Any nonmanufactured hole(s), missing cage bolt rubber plugs, or crack(s) in spring brake chamber(s).
5. Insecure air reservoir.
6. Brake hose that bulges or swells under pressure.
7. Audible leak(s) at any brake hose.
8. Improperly joined or spliced brake hose.
9. Brake hose that is cracked, broken, or crimped.
10. Brake tubing with an audible leak at other than a proper connection.
11. Brake tubing that is cracked, damaged by heat, broken, or crimped.

Air System Valve Troubleshooting Guide

Table 31–2 identifies some common brake valve failure problems and remedies.

TABLE 31–2 Air Valve Troubleshooting Guide

Valves	Symptom	Remedy
Drain valves—automatic	1. Will not drain 2. Will not drain in cold weather 3. Leaks/malfunctions	1. Repair or replace 2. Replace with heated units 3. Repair or replace
Foot valve	1. Leaks at exhaust with trailer hand valve applied 2. Leaks at exhaust with all brakes released 3. Leaks at exhaust with foot brake applied	1. Check double check valve; repair or replace 2. Check anticompound double check valve for backflow 3. Replace defective foot valve
Quick-release valves	1. Leaks (when used in service brake system) 2. Leaks (when used in spring brake system)	1. Replace 2. Replace
Relay valves	1. Leaks at exhaust port with all brakes released 2. Leaks at exhaust port with service brakes applied	1. Check seal in spring brake for backflow of spring hold-off pressure through service port to open exhaust on valve; repair or replace 2. Repair or replace the exhaust valve, which is not seating properly
System park control valve (yellow diamond)	1. Leaks at exhaust port 2. Parking brake will not release 3. Parking brake will not apply	1. Replace 2. Check for full system pressure delivery through valve 3. Replace if it will not release pressure
Trailer supply valve (red octagonal)	1. Leaks at exhaust port 2. Driver may override automatic trailer brakes when tractor air is below 20 psi (1.4 bar) 3. Will not apply trailer immediately when pulled	1. Check for trailer back-leakage; replace if no back-leakage occurs 2. Replace if TP system is automatic and two-line; repair or replace stoplight switch if TP system is automatic and three-line 3. Replace if it will not exhaust; repair or replace stoplight switch
Tractor park valve (blue round) TPV (nonautomatic type)	Leaks at exhaust port 1. Will not respond to trailer charge valve 2. Supply line to trailer with slow bleed down does not shut off tractor air and vents between 45 and 20 psi (3.1 and 1.4 bar). 3. Leaks at exhaust port or tractor service (back through hand or foot valve) or trailer supply	Repair or replace 1. Repair or replace 2. Check trailer supply valve; it should pop to let TPV vent trailer supply; repair or replace 3. Repair or replace

31.9 FOUNDATION BRAKE SERVICE

A brake job refers to the overhaul of the foundation brake assemblies on the tractor and trailer. This is one of the most common service practices in a truck service facility. Although it is thought of as a simple operation, you should never forget that the safety of the entire rig is dependent on it being performed properly.

DISASSEMBLY AND REASSEMBLY OF S-CAM FOUNDATION BRAKE ASSEMBLIES

To disassemble a typical S-cam foundation brake assembly with open anchor shoes, the following procedure should be observed after removing the wheel assembly:

1. Push down on the lower brake shoe. Pull on the cam roller retainer clip to remove the bottom cam roller, as shown in **Figure 31–41**.

FIGURE 31–41 Push down on the lower brake shoe and pull on the roller retainer clip to remove the lower cam roller.

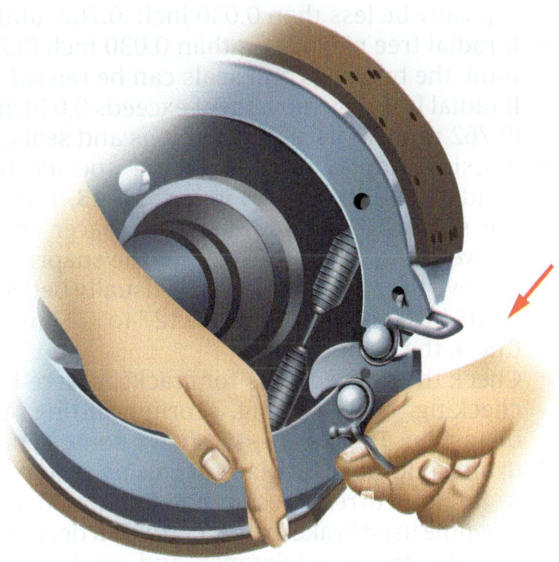

2. Lift the top brake shoe and pull on the cam roller retainer clip to remove the top cam roller.
3. Lift the lower shoe to release the tension on the brake shoe return spring and remove it.
4. Rotate and drop the lower shoe to release the tension on the brake shoe retainer springs, as shown in **Figure 31–42**.
5. Remove the shoe retainer springs and remove both brake shoes.
6. Remove the anchor pins, which often seize in the spider bore and a pin press should be used to remove them. Avoid using heat and sledge hammers to remove anchor pins because they inevitably damage the spider.

FIGURE 31–42 Rotate the bottom shoe to release the tension on both retaining springs.

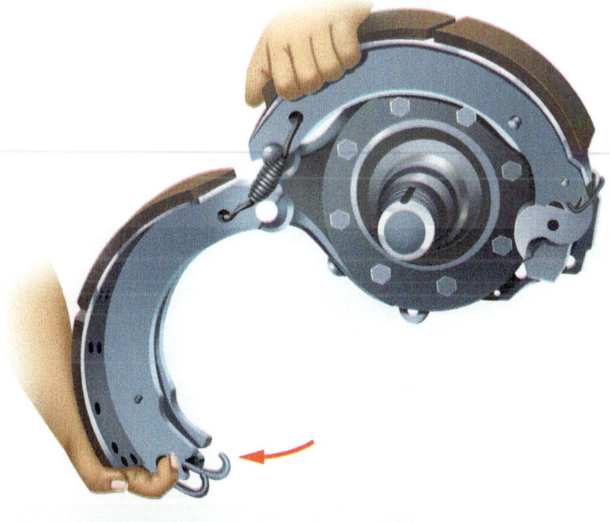

WARNING

Never use compressed air to clean brake dust off a foundation brake assembly. Brake dust may contain asbestos and other harmful microparticles that can endanger health.

Prepare an S-Cam Foundation Brake for Reassembly

Cleaning:

- Use soap and water to clean nonmetal components.
- Some of the heavier components, such as the S-cam, can be cleaned in hot soak tanks.
- Use cleaning solvents to clean the remainder of the components.
- Dry components immediately after cleaning with clean wipers or compressed air.

Corrosion Protection

- If components are assembled immediately after cleaning, lightly lubricate them with grease.
- Avoid getting any grease on friction surfaces such as cam profiles, rollers, and friction faces.
- If components are to be stored after cleaning, apply a corrosion-preventive material. Store components in a special paper or other material that prevents corrosion.

Inspect Components

It is important to inspect all components carefully before assembly, checking them for wear or damage. Repair or replace them as required.

- Check the spider anchor pin bores for oversizing and cracks. Replace damaged spiders and anchor pin bushings. Tighten spider fasteners to manufacturer's recommended torque, as shown in **Figure 31–43**.
- Check the camshaft bracket for broken welds, cracks, and correct alignment. Replace a damaged bracket.
- Check anchor pins for corrosion and wear. Replace damaged anchor pins.
- Check brake shoes for rust, expanded rivet holes, broken welds, and correct alignment. Replace a shoe with any of the listed conditions.
- The distance from the center of the anchor pin hole to the center of the roller must not exceed 12.779 inches (324.6 mm). If it does, the shoe is spread and must be replaced; do not attempt to re-arc.
- Check the S-camshaft for cracks, wear, brinnelling, and corrosion. Check the S-cam profile, bearing journals, and splines. Replace

FIGURE 31–43 Tighten the spider fasteners to the OEM-specified torque.

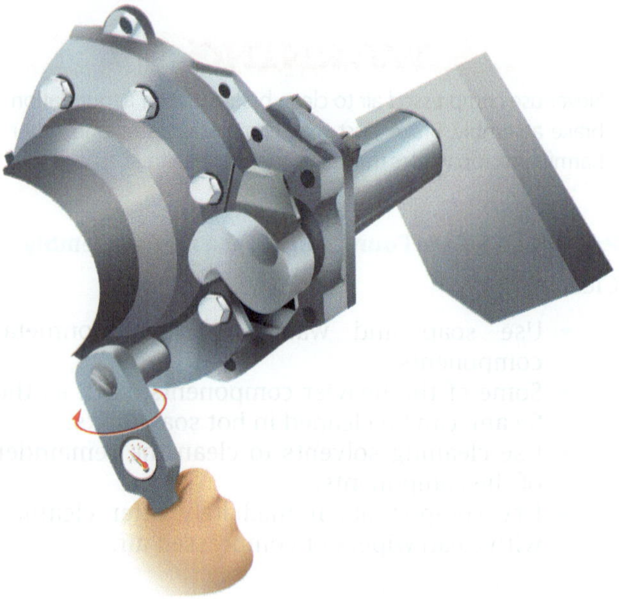

damaged camshafts. Consult the OEM service manual to check the S-camshaft and its bushings.

CAUTION

On 16.5-inch (420 mm) brake shoe foundation assemblies using 1-inch (25 mm) anchor pins, the spider anchor pin bore must not exceed 1.009 inches (25.6 mm).

The following procedure is typical:

- Inspect the cam-to-bushing radial free play and axial endplay. Radial free play should be tested, as shown in **Figure 31–44**, and should typically be less than 0.030 inch (0.762 mm).
- If radial free play is less than 0.030 inch (0.762 mm), the bushings and seals can be reused.
- If radial free play movement exceeds 0.030 inch (0.762 mm), replace the bushings and seals.
- If axial endplay is outside the specification window, as shown in **Figure 31–45**, remove the snapring and add/subtract spacer shims between the slack adjuster and the snapring to achieve the specified free play, usually between 0.005 inch and 0.050 inch (0.127 to 1.27 mm).
- Check the drums.
- Check the brake drums for cracks, severe heat checking, heat spotting, scoring, pitting, and distortion. **Figure 31–46** shows some typical drum conditions. Replace drums as required.
- It is not current recommended practice to machine used brake drums because it decreases the strength, heat tolerance, and service life of the drum.
- Measure the inside diameter of the drum in several locations with a drum caliper or internal micrometer. Check for eccentricity, bell-mouthing, and oversizing. Replace the drum if the diameter exceeds any specifications supplied by the drum manufacturer. A typical 16.5-inch (420 mm) drum has a maximum machining tolerance of 0.090 inch (2.3 mm) and a maximum service diameter of 0.120 inch (3 mm). This means that the maximum diameter that the drum can be machined to and

FIGURE 31–44 Testing radial free play of an S-camshaft

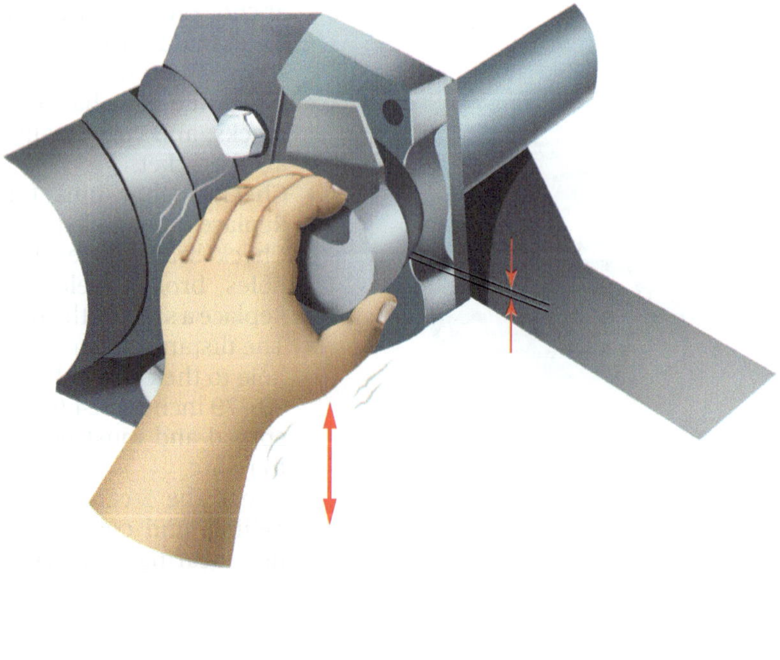

FIGURE 31–45 Measuring S-camshaft endplay

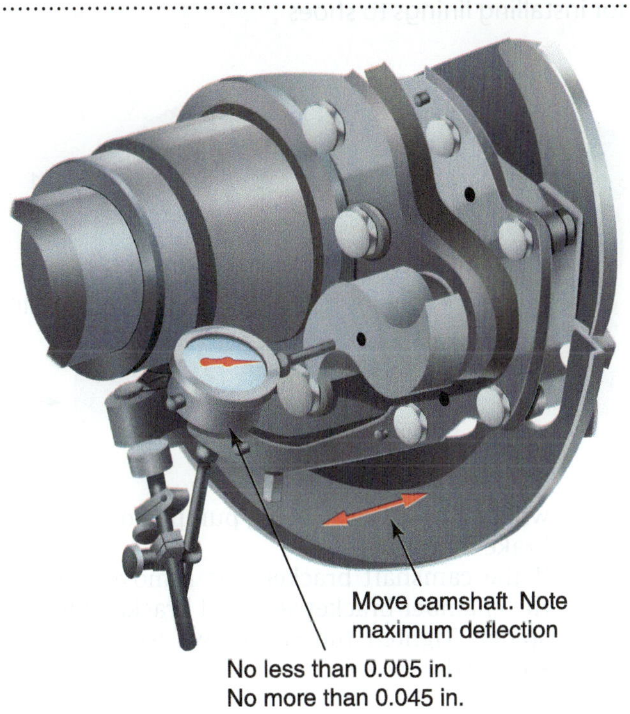

returned to service is 0.090 inch (2.3 mm). Any measurement exceeding 0.120 inch (3 mm) means that the drum has to be scrapped.

- Check dust shields for rust and distortion. Repair or replace damaged shields as necessary.
- Install the drums.
- Ensure that the correct fastening hardware is used when assembling wheel hardware to brake drums and that the correct torque values are observed.

CAUTION

Cast drums are machined in manufacture. However, when stored, especially for long periods, they can deform and lose their concentricity. It is, therefore, a good practice to machine brand new drums to ensure that they are perfectly concentric with the wheel. **Figure 31–47** shows a typical heavy-duty drum lathe.

Linings

If the shoes are to be refaced with new friction linings, ensure that the linings are correctly installed (**Figure 31–48**) and that the correct rivet/torque sequence is observed (**Figure 31–49**).

SHOP TALK

Modern friction facings on brake shoes are much harder than the asbestos base linings used on drum brakes a generation ago. They have a longer service life. The result is that the life of the linings can be the same as the life of the drums in some applications. It is almost impossible to machine/turn down the effects of severe heat checking on a brake drum because the drum becomes so hardened. Cast brake drums are low in cost when compared with the labor cost required to turn a drum. Many operations routinely replace brake drums today when the brakes are relined.

FIGURE 31–46 Drum wear conditions

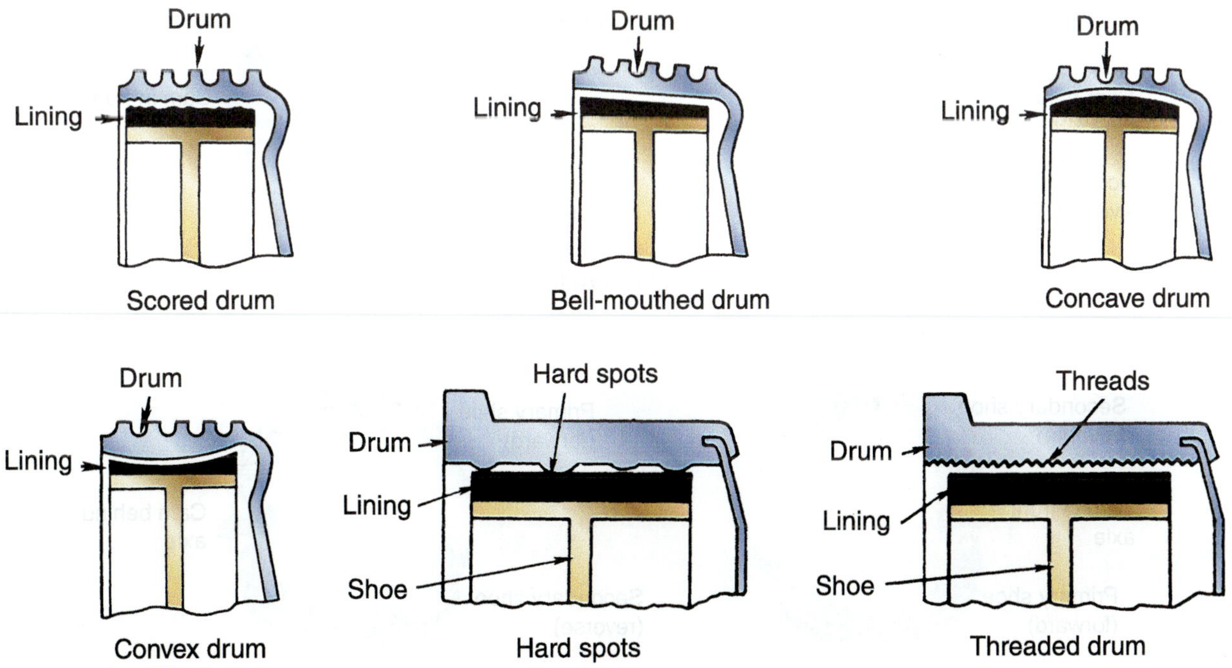

FIGURE 31–47 Heavy-duty drum lathe

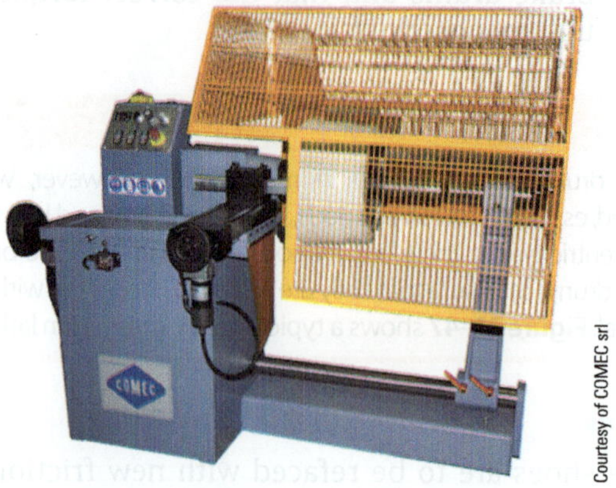

Reassembly of an S-Cam Foundation Brake

It is recommended that new camshaft bushings be installed when installing a new camshaft.

- Tighten all of the spider bolts to the correct torque.
- Use a seal driver to install new camshaft seals and new bushings in the cast spider and camshaft bracket. If equipped with a stamped spider, install both bushings into the bracket.
- Install the seals with the seal lips toward the slack adjuster. Failure to observe this practice will result in grease being pumped out into the brake linings.
- If the camshaft bracket was removed, install the chamber bracket seal and bracket onto the spider. Tighten the capscrews to the torque specified.

FIGURE 31–49 Typical rivet installation sequence for installing linings to shoes

Install the S-Camshaft

- Install the cam head thrust washer onto the camshaft. Apply grease to the camshaft bushings and journals.

FIGURE 31–48 Correct installation for replacement combination linings

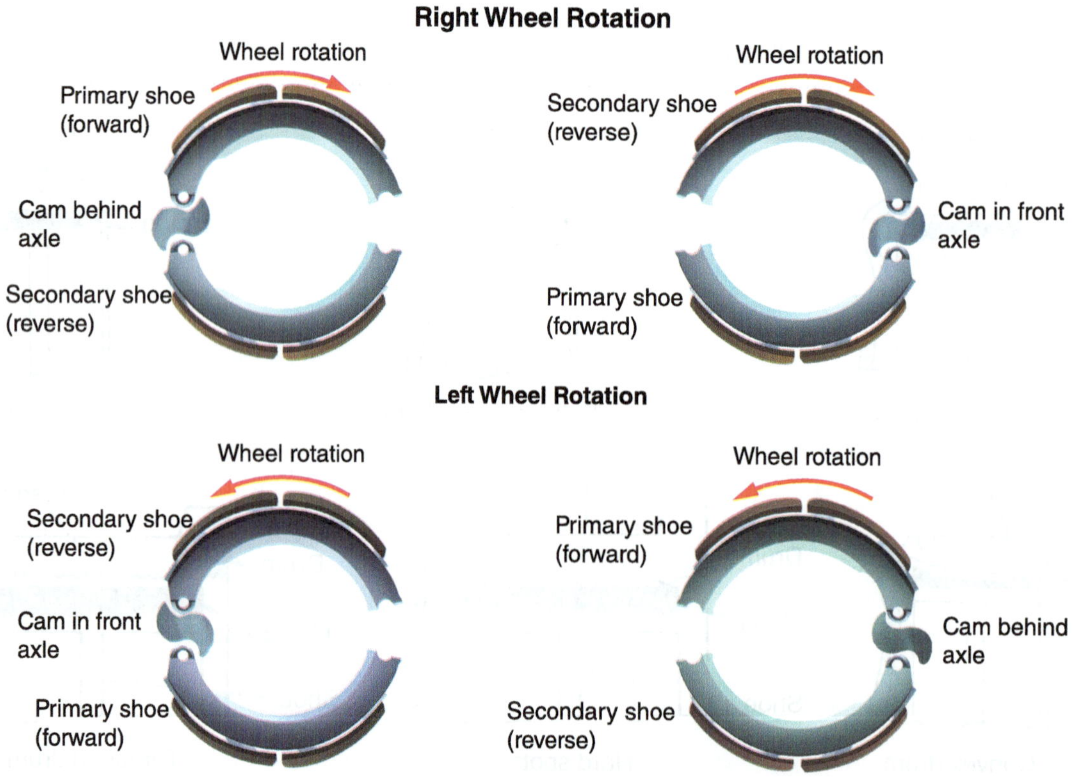

FIGURE 31–50 Install S-camshaft.

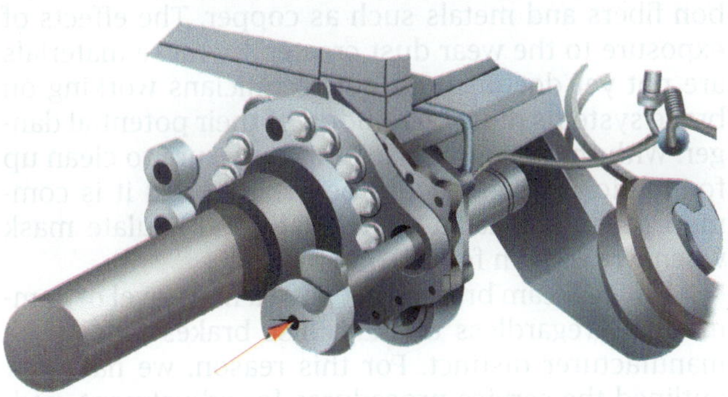

- Install the camshaft through the spider and bracket as shown in **Figure 31–50** and confirm that the camshaft turns freely by hand.

Installing Manual Slack Adjusters

- Apply the service brake and spring brake several times. Check that the chamber return spring retracts the pushrod quickly and completely. If necessary, replace the service chamber.
- Ensure that the new slack adjuster is the same length as the one being replaced.
- If the brake has a spring brake, cage the spring to completely release the brake. No air pressure must remain in the service half of the air chamber.
- Apply antiseize compound to the slack adjuster and cam splines.
- Install the slack adjuster onto the camshaft.
- If necessary, install spacer washers and the snapring to set the OEM-specified endplay clearance of 0.005 to 0.50 inches (0.127 to 12.7 mm).
- Install the clevis onto the pushrod. Do not tighten the jam nut against the clevis.
- Set the clevis position (turn clockwise or counterclockwise) so that a 90-degree angle is achieved when it is in the fully applied position (full service application). Apply antiseize compound to the clevis pin(s). Install the clevis pin(s) through the clevis and the slack adjuster.

SHOP TALK

When an axle wheel seal fails, oil saturates the brake friction linings, which rapidly causes them to glaze, lowering their coefficient of friction or aggressiveness. When this occurs, pressure or steam washing removes only the surface oil, so the linings should be replaced. Also replace the linings on the other end of the axle to maintain brake torque balance over the axle.

Installing Automatic Slack Adjusters

A number of different brands of slack adjusters are on the market, and each requires a different setup procedure. Slack adjusters are described in Chapter 28 as well as earlier in this chapter. Ensure that the OEM instructions are properly observed when fitting automatic slack adjusters to S-camshafts. The consequence of improperly installing automatic slack adjusters is slack adjusters that do not auto-adjust.

SHOP TALK

Take some time to ensure that automatic slack adjusters are properly set up on installation. It takes much longer to properly adjust automatic slack adjusters than manual slack adjusters.

Installing Brake Shoes

- Use a mandrel to install new anchor pin spider bushings if the old ones were removed. Apply high temperature grease to the anchor pins and install them into the brake spider anchor pin bores.
- Place the upper brake shoe into position on the top anchor pin. Hold the lower brake shoe on the lower anchor pin. Install new brake shoe retaining springs, ensuring that the clip eyes face inward.
- Rotate the lower brake shoe forward. Install a new brake shoe return spring with the open end of the spring hooks toward the camshaft, as shown in **Figure 31–51**.
- Pull each brake shoe away from the cam to permit enough space to install the cam roller and cam roller retainer. Press the *ears* of the

FIGURE 31–51 Rotate the lower brake shoe forward and install the brake shoe return spring.

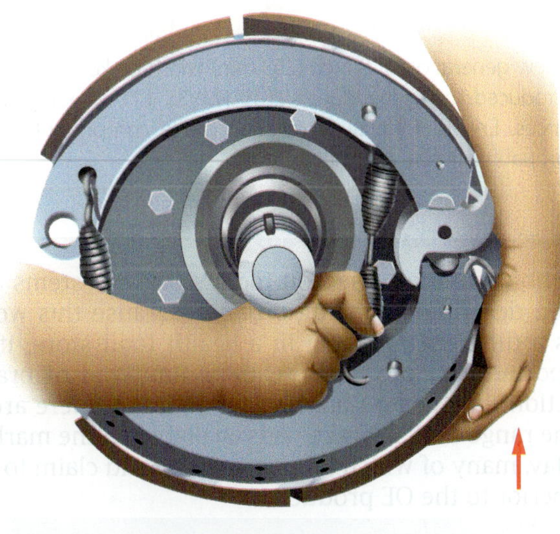

FIGURE 31–52 Press the ears of the retainer together to permit the roller to fit between the brake shoe webs.

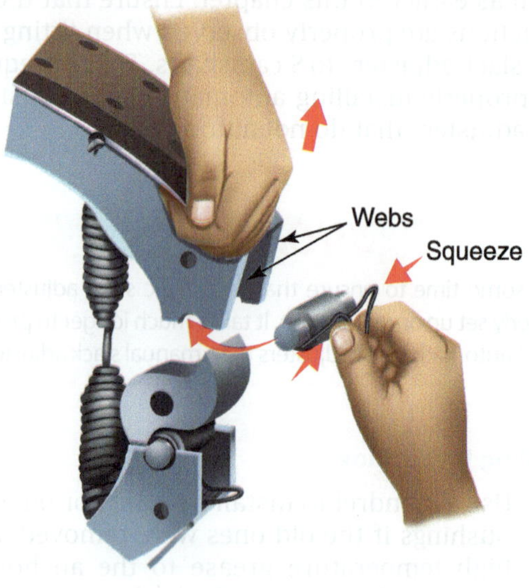

retainer to permit it to fit between the brake shoe webs, as shown in **Figure 31–52**.

- Push the cam roller retainer into the brake shoe until its ears lock in the shoe web holes.
- Lubricate the brake components that require lubrication. Avoid lubricating those components that rely on friction to function properly, such as the S-cam profiles, cam roller faces, and friction faces.

SHOP TALK

Avoid lubricating the cam roller face. Lubricate the cam roller bearings. These are designed to grab at their faces and rotate on their bearings in operation.

CAUTION

A new generation of 8-inch (203 mm) width brake shoes was introduced to meet reduced 2010 FMVSS 121 stopping distances. Ensure that all foundation brake components are compatible before assembly.

Replacement Disc Brake Pads

Friction materials used on today's brake systems are often described as being *organic*. Because this word has a different meaning in a health food store, it is important to note that breathing any kind of brake friction dust can be harmful. In addition, there are a wide range of disc brake pads available on the market today, many of which are more costly and claim to be superior to the OE product.

Although today's brake friction materials contain little or no asbestos, they may be composed of carbon fibers and metals such as copper. The effects of exposure to the wear dust created by these materials are not yet documented but technicians working on brake systems must be mindful of their potential danger. With this in mind, use of air pressure to clean up foundation brakes should be avoided and it is common sense to wear a good quality particulate mask when working on foundation brakes.

Unlike S-cam brakes that have a high level of commonality regardless of OEM, disc brakes tend to be manufacturer distinct. For this reason, we have not outlined the service procedures for adjustment, pad, and rotor replacement. Further information on each system can be sourced directly from the manufacturer. Bendix continues to make excellent quality service and training information available in the form of free downloads so they are recommended:

- Bendix: bendix.com (manual: BRIP–0300)
- Haldex: haldex.com
- Meritor: meritorwabco.com

Express and Extreme Brake Service

Express Brake™ and Extreme Brake™ shoes have been around for a generation and have undergone intensive proofing in applications such as garbage packers and transit buses. Their objective is to reduce the cost, time required, and frequency of foundation brake jobs. A key feature is a stainless steel brake shoe that is designed to last the life of the vehicle. This stainless steel shoe table provides the following advantages:

- Shoe arc is geometrically consistent with the radius of the brake drum. This eliminates flexing of the brake shoe and brake lining. Crowning is a major cause of friction material cracks.
- Elimination of friction crown increases the surface contact area with the drum, especially under light- and medium-duty braking. This enables effective braking while maintaining lower drum temperatures, meaning that drum life is extended.
- Shoe table is designed to last the life of the vehicle.
- Lining life is greatly extended.
- Relines can take place on the wheel end without removing the shoes.
- Rust jacking (development of corrosion under the friction lining) is eliminated.
- Shoe core hassles are eliminated.

Given the emphasis in decreasing costs in today's trucking maintenance environment, this type of foundation brake is well positioned to gain in popularity.

Extreme Brake Construction

The Extreme Brake stainless steel shoes (see **Figure 31–53**) fit to the wheel-end spider anchor

FIGURE 31–53 Extreme Brake shoe: Note the location of the wear indicator. When this is worn flush with the remaining lining surface, the brake lining should be replaced.

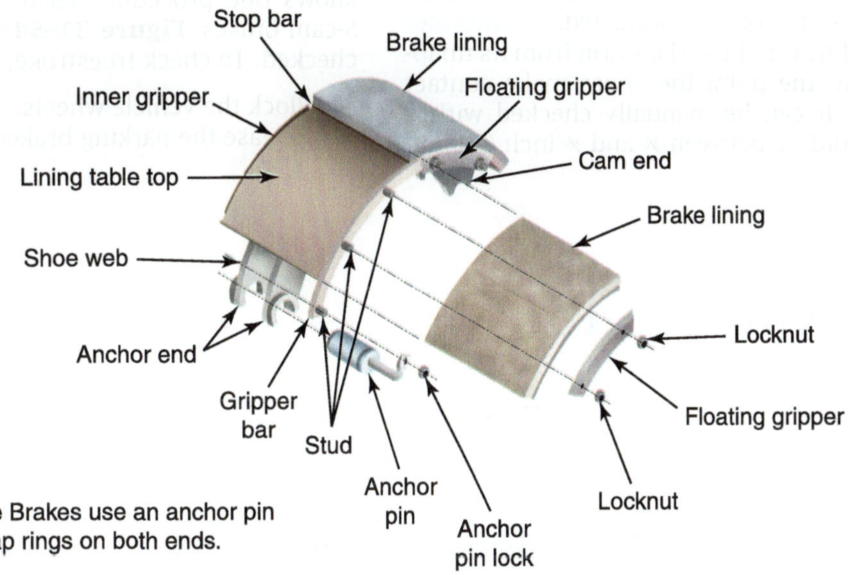

pins in the same way that any other S-cam shoes are designed. The linings are installed in nontapered blocks, two per shoe. The pair of linings on each shoe is retained by removable gripper bars (clamps) on either side of the shoe and a lip on each end. The gripper bars and the fastening hardware are manufactured of stainless steel.

Extreme Lining Replacement

To replace a set of Extreme Brake linings, the wheel assembly should be pulled from the wheel end. The following sequence may be used to replace the friction linings on the shoe:

1. Remove the stainless steel locknuts on each side of the gripper bars.
2. Separate the gripper bars from the stainless steel brake shoe assembly.
3. Pull the used friction pads away from the shoe. Replacement friction facings are made for Extreme Brakes by multiple suppliers.

Extreme Brake manufactures a version of its brake known as Express Brake, which can be replaced on the wheel end without removing the wheel assembly. This replacement can be made by simply removing the backing plates on the foundation assembly. However, it should be noted that this variation cannot be fitted to all wheel ends.

31.10 BRAKE ADJUSTMENT

With the introduction of automatic slack adjusters, the need for frequent brake adjustments on trucks has been reduced. However, it is essential that frequent evaluation of the brake adjustment on trucks and trailers be performed both from a performance (driver) and shop inspection point of view. Brake adjustment determines how well the vehicle brake system will function whether it be under service, parking, or emergency braking. Technicians should ensure that they understand exactly how the hold-off and service circuits function as outlined in Chapter 28. Before beginning the **hold-off circuit** should be actuated. The hold-off circuit is the parking brake release circuit and it requires that the **hold-off diaphragm** in every spring brake chamber on the rig be charged with air pressure. At this point, the brakes on the rig are released; so if you are to perform testing, ensure that the wheels on at least two axles are chocked. Under this condition, when the service brakes are applied by the tractor treadle valve, all of the **service diaphragms** have air pressure acting on them. In other words, when a moving tractor/trailer combination is braked, air pressure is directed into both the hold-off and service chambers on the rig. The first step in verifying brake performance is to check the pushrod stroke on each chamber throughout the truck or tractor/trailer combination.

SHOP TALK

Out-of-adjustment brakes affect all vehicle braking, including emergency braking. Never take chances with brake adjustment. Out-of-adjustment brakes can have lethal consequences, and every technician has a stake in the vehicles on which he or she works.

CHECK FREESTROKE

When the foundation brakes are in good condition, check the length of freestroke at each brake to determine that brakes are correctly adjusted. **Freestroke** is the amount of travel of the slack arm from its unapplied position to the point the shoes make contact with the drum. It can be manually checked with a prybar and should be between ⅜ and ⅝ inch (10 and 16 mm). Less than ⅜ inch (10 mm) can result in brake drag while more than ⅝ inch (16 mm) means that the brakes are out of adjustment. **Photo Sequence 15** shows one procedure used to check freestroke on S-cam brakes. **Figure 31–54** shows freestroke being checked. To check freestroke, proceed as follows:

1. Block the vehicle wheels.
2. Release the parking brakes.

Photo Sequence 15

CHECK FREESTROKE ON S-CAM FOUNDATION BRAKES

P 15–1 Block wheels. If adjusting the brakes on a tractor/trailer combination, block one of the drive axle duals on the tractor and one trailer axle.

P 15–2 Release the parking brakes on the rig by depressing both the System Park and Trailer Supply dash valves.

P 15–3 Ensure that the system air pressure is above the governor cut-out pressure (around 100 psi [7 bar] is ideal) after the brakes have been released.

P 15–4 If the rig has visual stroke indicators, get somebody in the cab to make a full service brake application and hold. Inspect each pushrod stroke indicator to confirm the correct pushrod stroke.

P 15–5 Correct travel when service brakes are applied.

P 15–6 Excessive pushrod travel; brakes require adjustment.

P 15-7 If the pushrods on the rig do not have visual stroke indicators, release the parking brakes and then check each brake for pushrod travel using a heel bar.

P 15-8 If a pushrod travels between ⅜ and ⅝ inch (10 and 16 mm), the brake has the correct amount of free travel and does not require adjustment.

FIGURE 31–54 Measuring freestroke without a template

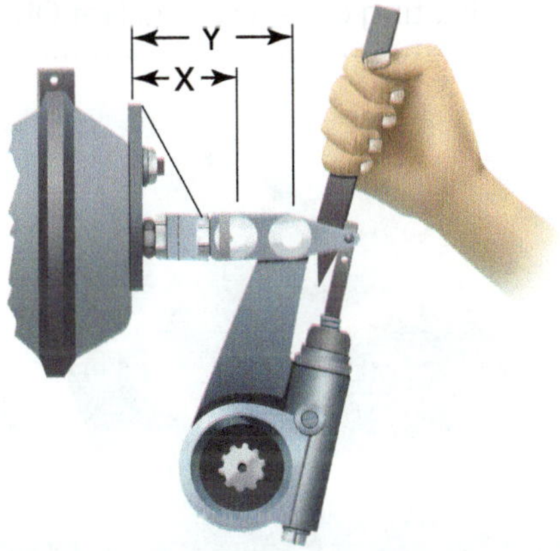

Drum brake: Y minus X must = 5/8 in. – 3/4 in.
In service disc brake: Y minus X must = 3/4 in. – 1 in.
Initial disc brake: Y minus X must = 7/8 in. – 1 1/8 in.

3. Extend each pushrod using a heel bar and measure the distance traveled. Each pushrod should move between ⅜ and ⅝ inch (10 and 16 mm).
4. Apply a full service brake application (85 psi [6 bar]) to ensure that the brakes fully actuate and release.

APPLIED STROKE

In most cases, when freestroke is correct, the applied stroke should fall within the adjustment limit. However, do not assume that this is true in all cases. **Applied stroke** is the slack adjuster travel measured at the clevis pin from its unapplied position to fully applied, meaning an 85 psi (6 bar) service application pressure. Note that all commercial vehicles manufactured since 2007 are required to be equipped with stroke indicators. Stroke indicators show the applied brake stroke and you can check this to specification by consulting **Table 31-3**, which specifies the adjustment limits for S-cam brakes. Within specification freestroke does not guarantee that applied stroke will fall within the adjustment limit when the

TABLE 31–3 Adjustment Limits for Clamp-Type Chambers

Type	Effective Area (sq. in.)	Outside Diameter (in.)	Maximum Stroke at Which Brakes Should Be Readjusted
6	6	4½	1¼
9	9	5¼	1⅜
12	12	5 11/16	1⅜
16	16	6⅜	1¾
20	20	6 11/32	1¾
24	24	7 7/32	1¾
30	30	8 3/32	2
36	36	9	2¼

FIGURE 31–55 Adjuster mechanism properly set on one type of slack adjuster.

foundation brakes are in poor condition. In addition, the adjustment mechanism on the slack adjuster must be properly set; check the manufacturer instructions because not all slack adjusters are set in the same way. **Figure 31–55** shows adjuster set position on one type of slack adjuster.

SHOP TALK

Brake freestroke is typically specified to be between ⅜ and ⅝ inch (10 and 16 mm) regardless of chamber size.

The actual means used to mark the brake chamber pushrods limit the accuracy of this method of checking pushrod stroke. This limitation is sometimes used to challenge OOS brake-related charges, and some operators prefer not to leave reference marks on pushrods. However, in practice, having any errors produced through this practice probably works out in favor of carriers more often than not. **Figure 31–56**, **Figure 31–57**, and **Figure 31–58** show a brake stroke indicator indicating three status conditions.

ALTERNATE BRAKE ADJUSTMENT INSPECTION METHODS

Several methods may be used for checking brake adjustment with varying degrees of reliability. The following methods have proven to be consistently reliable, providing the foundation brake condition is satisfactory. Table 31–3 lists the adjustment limits for clamp-type chambers.

Visual Stroke Indicators

Well-designed visual stroke indicators confirm correct adjustment at a glance and permit the technician to check brake adjustment without getting under the vehicle or having another person's assistance. Properly fitted visual stroke indicators also help to prevent unnecessary brake readjustment. To determine applied stroke using visual stroke indicators:

FIGURE 31–56 A brake stroke indicator with the brakes not applied

FIGURE 31–57 A brake stroke indicator with the brakes applied indicating excess stroke: Not OK.

FIGURE 31–58 A brake stroke indicator with the brakes applied indicating within specified stroke: OK.

1. Block wheels.
2. Release parking brakes.
3. Establish reservoir pressure between 90 and 100 psi (6 and 7 bar).
4. Make a full service brake application and hold.
5. With service brakes applied, use the visual indicators to confirm correct pushrod strokes.

Wheel-Down, Wheel-Up Brake Adjustment

At the time of writing, truck technicians are required to know the procedure for adjusting manual slack adjuster brakes in both the wheel-down and wheel-up positions. Although the number of trucks on our roads with manual slack adjusters has diminished in recent years, knowing this procedure remains required knowledge in certification testing. **Photo Sequence 16** shows the wheel-down, manual slack adjuster procedure and **Photo Sequence 17** shows the wheel-up manual slack adjuster procedure.

Photo Sequence 16

WHEEL-DOWN BRAKE ADJUSTMENT PROCEDURE ON A TRACTOR/TRAILER

P 16–1 Block wheels. If adjusting the brakes on a tractor/trailer combination, block one of the drive axle duals on the tractor and one trailer axle. Note that only manual slack adjusters should be adjusted using this method.

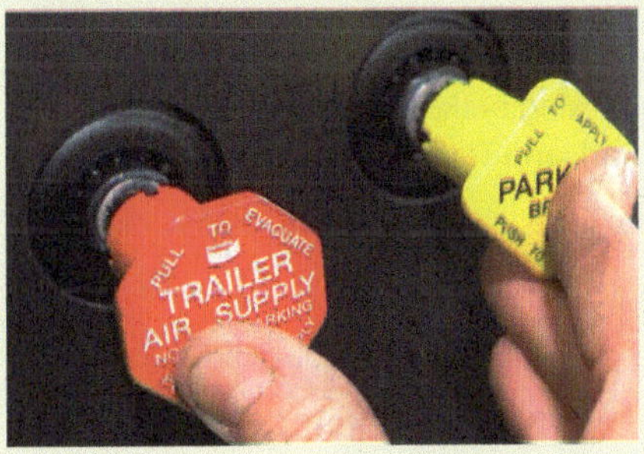

P 16–2 Release the tractor and trailer parking brakes by depressing the System Park and Trailer Supply valves. Ensure that the system air pressure is above the cut-in pressure after the parking brakes have been released. Determine whether the brakes require adjustment by performing the "Check Freestroke" procedure outlined in Photo Sequence 15.

P 16–3 To adjust a manual slack adjuster using the wheel-down method, select the correct-sized combination wrench (9/16 inch [14.3 mm]) and use the closed end to depress the locking collar and engage the adjusting nut.

P 16–4 Now rotate the adjusting nut so that the S-camshaft is rotated in the direction required to actuate the brake. Turn until the brake is fully applied (wrench will not turn further), then back off adjusting nut sufficiently to obtain 1/2 inch (12.5 mm) free pushrod travel. After the adjusting nut has been backed off (typically, this will require 1/3 to 1/2 of a turn), check for free play by once again levering between 3/8 and 5/8 inch (10 and 16 mm), preferably closer to 1/2 inch (12.5 mm).

P 16–5 This brake can now be regarded as being properly adjusted. Ensure that the locking collar engages fully over the adjusting screw.

P 16–6 Proceed to adjust the remaining slack adjusters on the rig in the same manner.

Photo Sequence 17

WHEEL-UP BRAKE ADJUSTMENT PROCEDURE ON A TRACTOR/TRAILER

P 17-1 Wheel-up brake adjustments of manual slack adjusters take longer than wheel-down adjustments but it ensures a more accurate adjustment and can help identify other foundation brake problems. First, block the wheels. If adjusting the brakes on a tractor/trailer combination, block one of the drive axle duals on the tractor and one trailer axle.

P 17-2 Release the tractor and trailer parking brakes by depressing the System Park and Trailer Supply valves.

P 17-3 Ensure that the system air pressure is above the cut-in pressure after the parking brakes have been released.

P 17-4 When performing a wheel-up adjustment, the brakes over every wheel on the unit are adjusted regardless of freestroke. First, use a hydraulic jack to raise one wheel off the floor.

P 17-5 Install a safety stand.

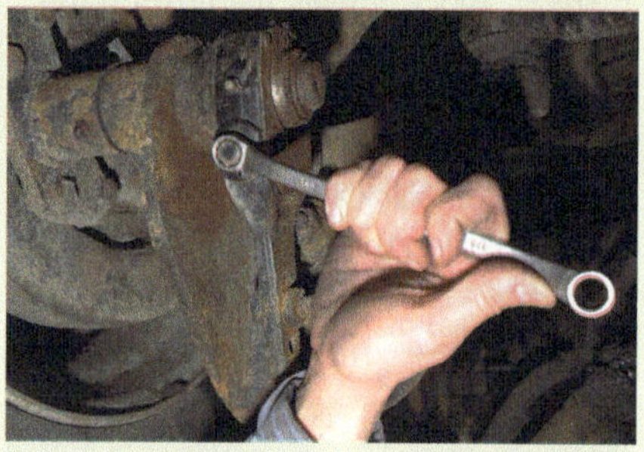

P 17-6 To adjust a manual slack adjuster, select the correct-sized combination wrench (9/16 inch [14.3 mm]) and use the closed end to depress the locking collar and to engage the adjusting nut.

P 17-7 Now rotate the adjusting nut so that the S-camshaft is rotated in the direction required to actuate the brake. Turn until the brake is fully applied (wrench will not turn further).

P 17-8 Now attempt to rotate the wheel.

P 17–9 Next, gradually back off the slack adjuster while attempting to rotate the wheel. Back off the slack adjuster until the wheel can be rotated through a full revolution with minimal but definite drag.

P 17–10 This brake can now be regarded as being properly adjusted. Ensure that the locking collar fully engages over the adjusting screw.

P 17–11 Proceed to adjust the remaining slack adjusters on the rig in the same manner.

E-Stroke Monitoring

Some OEMs use brake E-Stroke monitoring. E-Stroke is designed to alert the driver when brake stroke exceeds a preset value. The preset values are typically 2 inches (50 mm) for a 30/30 standard chamber, 2.25 inches (57 mm) for a long stroke chamber, and 3 inches (75 mm) for super long stroke units used by Navistar and Freightliner. E-Stroke uses a magnetic sensor on the brake chamber actuation rod to measure stroke and messages the chassis data bus when excessive stroke is measured.

Tech Tip

It takes two persons to properly verify applied brake stroke on an air brake–equipped tractor/trailer combination: one to apply the service brakes, the other to inspect the stroke at each wheel. If you have any trouble convincing your boss of this fact, remind him or her that out-of-adjustment brakes account for more CVSA OOS infractions than any other single item.

Adjusting Automatic Slack Adjusters

The Technology and Maintenance Council (TMC) considers it bad practice to undertake repeated adjustments to automatic slack adjusters. When automatic slack adjusters require frequent adjusting, the correct practice is to determine the cause. This may be a defective slack adjuster or a problem in the foundation brake assembly. Some fleets prohibit their techs to manually adjust automatic slack adjusters, requiring them to check the foundation brakes and replace the slack adjusters if necessary.

SHOP TALK

All it takes is two wheels on a rig with excessive brake stroke for a CVSA OOS infraction to occur. Carefully check freestroke and applied brake stroke at each maintenance inspection.

31.11 GENERAL BRAKE VALVE TROUBLESHOOTING

This section takes a look at some air brake system complaints and the remedies required to correct them. Remember, however, that it is not entirely comprehensive and is intended only as a guide.

COMPLAINT: AIR LEAKS

When static leakage occurs, the following conditions should be determined:

1. Check for air system contamination, such as indications of oil in the air circuit. Remember that trace quantities of oil in the air circuit are normal. However, evidence of oil at the exhaust of the brake system valves and in the dash gauges indicates severe oil contamination, and the compressor should be checked.
2. Check for back-feeds from spring brake valves, TPVs, and trailer (if connected). See **Figure 31–59**.
3. Check for leaks at all air line connections and critical valves. Use a soap-and-water solution or an audio air leak detector.

FIGURE 31–59 Sectional view of a typical spring brake control valve

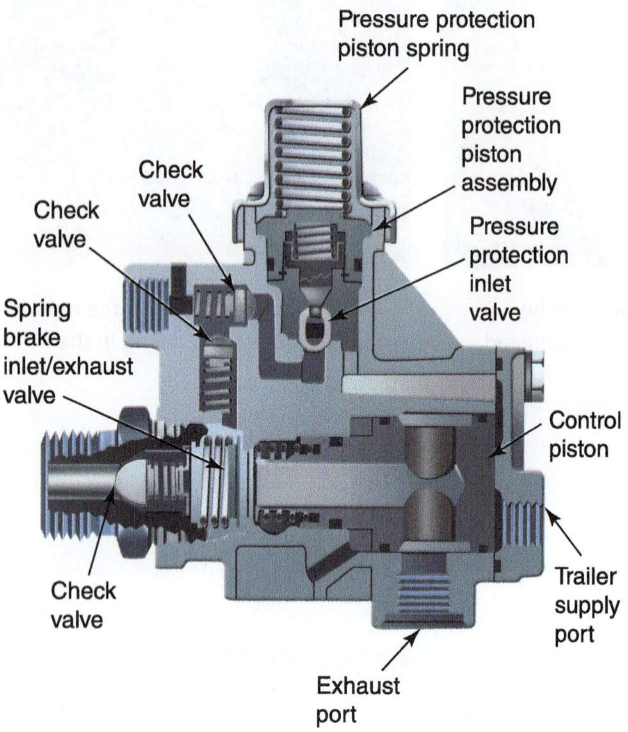

When dynamic leakage occurs, the following conditions should be determined:

1. Check for air system leaks with the system at governed air pressure and the engine off. Chock the wheels and release the parking brakes. Circle the vehicle and listen for leaks. Apply the trailer service brakes and check for leaks.
2. Have someone apply the tractor service brakes and circle the tractor to check for leaks.
3. Inspect the performance of the double check valve located between the trailer control (hand valve) and foot valve. The actual location of this valve can be difficult to determine; use the OEM pneumatic schematic. If this is the cause, leakage will occur when the hand brake is applied and the foot valve released, or vice versa.

COMPLAINT: OVERSENSITIVE SERVICE BRAKES

Check foundation brakes and wheel assembly for the following conditions:

1. Incorrect slack adjuster arm length (check camshaft-to-clevis pin length).
2. Brake chamber clevis is connected to the wrong hole in the slack adjuster (changes leverage).
3. Chamber/slack alignment binding.
4. Inspect friction material.
5. Inspect condition of foundation brake components.
6. Cam rotation counter to wheel rotation. (S-camshafts are directional; it is not unheard of to have them improperly installed.)
7. Inspect tire tread condition and tire mismatch (tread design, flat spotting, standard or low profile).

Check for air system freeze-ups (sticks then breaks free).

1. Inspect the relay valve, quick-release valve, and other brake valves for proper operation.
2. Check the air lines for internal failure causing pressure checking.

Brake valve maintenance items:

1. Plunger/treadle or pedal sticking
   - Treadle or pedal hinge pin is not lubricated.
   - Plunger and roller are not lubricated.
   - Ensure proper treadle with the roller.
   - Plunger cocked in the mounting plate.
   - Boot torn, dirt accumulated around the plunger.
2. Lubrication
   - Too much (damages rubber spring and/or boot).
   - Too little—plunger sticks.
3. Installation
   - Body or mounting plate distortion because of uneven floorboard or firewall mounting surface.
   - Inspect other air system components for malfunction.

COMPLAINT: BRAKES WILL NOT FULLY APPLY

1. Brakes out of adjustment: excessive freestroke (excess slack adjuster travel).
2. Automatic slack adjuster malfunction (underadjusting or seized).
3. Air line or component freeze-up.
4. Air line internal failure causing restriction.
5. Relay valve malfunction caused by freeze-up or piston seizure.
6. Defective foundation brake components.
7. Foot valve pedal movement is restricted: cab floor obstruction (tools, carpet, gravel, debris).
8. Improper chamber/slack adjuster alignment:
   - Pushrod is binding on the chamber nonpressure plate.
   - Pushrod is too long/short.
9. No air pressure in one brake circuit.

10. Air lines are improperly connected to the foot valve, relay valve, quick-release valve, or spring brake valve.
11. Foot valve plunger/pedal is sticking.

COMPLAINT: BRAKES FAIL TO RELEASE COMPLETELY

1. Parking brake is applied and will not release.
2. Condition of foundation brake components—springs, shoes, bushings (elliptical/squeezed), or bent spider—causing anchor misalignment.
3. Air lines are pinched, improperly connected.
4. Check foot valve stop adjustment.
5. System contamination is causing oil swelling of rubber valve components.
6. Relay valve is contaminated or relay piston is sticking.
7. Quick-release or ratio valve exhaust is restricted.
8. Chamber/slack adjuster is mechanically binding (seized clevis pins).
9. Automatic slack adjuster is malfunctioning—overadjusting.
10. Release keyhole cap is missing, resulting in a hold-off chamber full of dirt.
11. Frozen supply lines.
12. Unit has remained parked for an extended period: linings seized or frozen to drum.

31.12 SERVICE PROCEDURES ON AIR DISC BRAKES

Disc brakes are becoming more frequently used, primarily because they offer the superior braking (the operating principles of air disc brakes are discussed in Chapter 28) required of air brake–equipped vehicles manufactured since 2010. The service procedures used on disc brakes vary some what between OEMs because the actuating mechanisms differ. As with any procedure on foundation brakes, the OEM service literature should be consulted, but we have outlined some typical steps. **Figure 31–60** shows an exploded view of a common type of truck disc brake assembly.

FIGURE 31–60 Exploded view of an air disc brake assembly

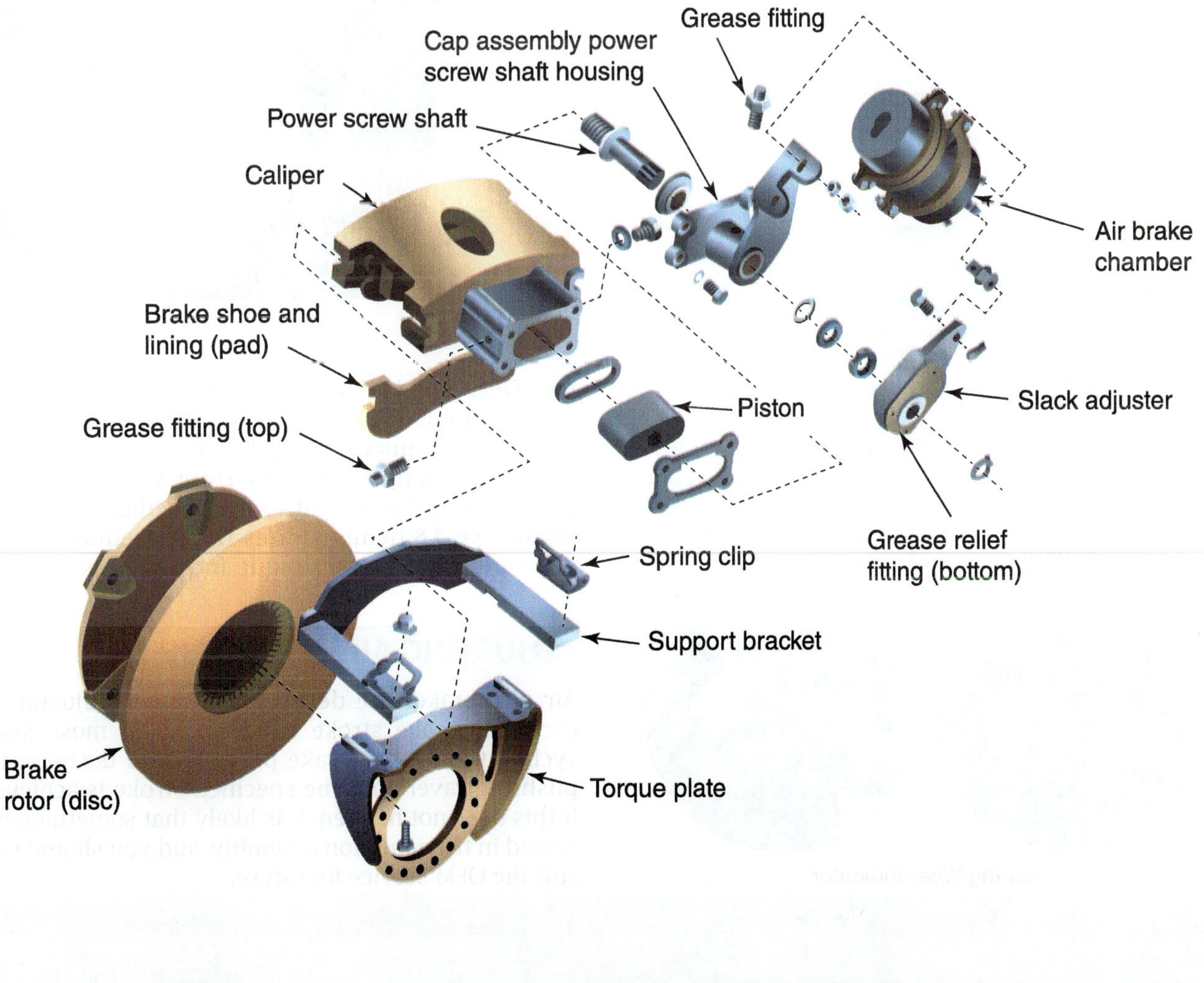

CHECKING LINING WEAR

Lining wear should be inspected by checking the caliper inboard bosses, as shown in **Figure 31–61**. Note the lining wear indicator and the wear dimensions.

LUBRICATING CALIPERS

Disc brake calipers must be lubricated at required intervals. **Figure 31–62** shows the location of the caliper grease fitting and the caliper pressure relief fittings. Ensure that the caliper is not overgreased.

MEASURING ROTORS

Rotors should be inspected and lateral runout measured each time the brakes are serviced. **Figure 31–63** shows the runout being measured. The runout specifications should be checked in the OEM service specifications. After indicating rotor runout, the rotor thickness should be measured with a micrometer. The minimum thickness is usually stamped onto the rotor so you can compare your measurement to the specification. **Photo Sequence 18** demonstrates the procedure used to measure rotor lateral runout and thickness.

FIGURE 31–61 Checking disc brake lining wear

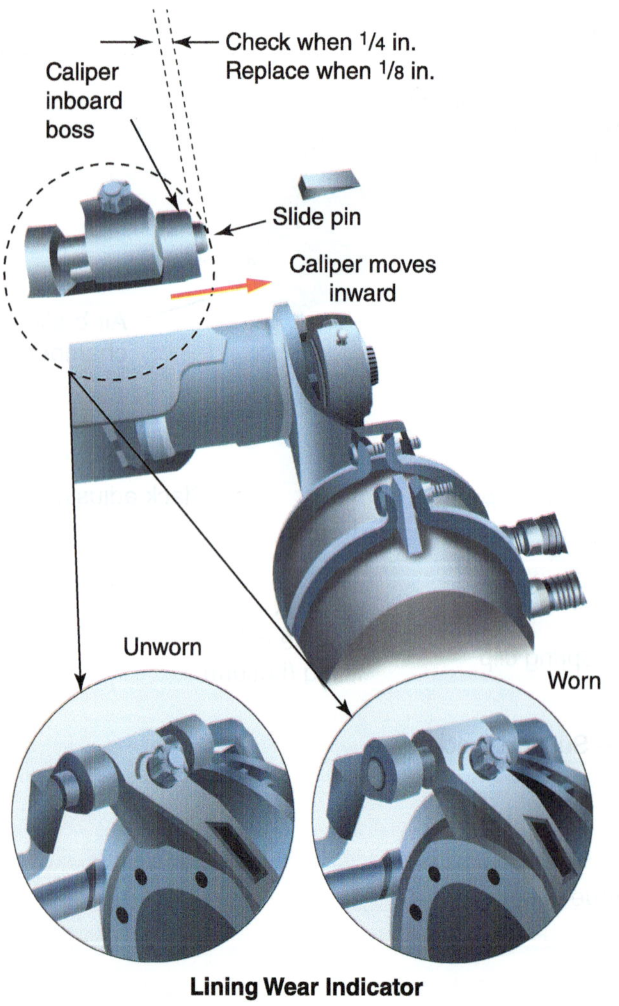

FIGURE 31–62 Lubricating disc brake calipers

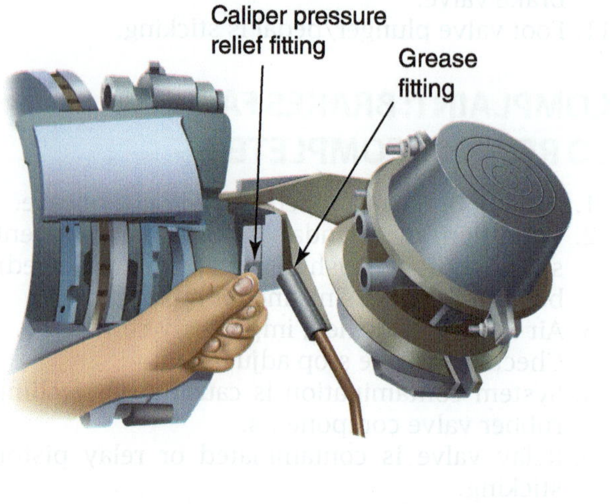

FIGURE 31–63 Checking rotor radial runout

ADJUSTING AIR DISC BRAKES

Air disc brakes are designed to be self-adjusting. If excessive brake stroke is observed, in most cases, cycling the service brake pedal should decrease the pushrod travel until the specified stroke is achieved. If this does not happen, it is likely that something has seized in the actuation assembly, and you should consult the OEM service literature.

Photo Sequence 18

MEASURING A BRAKE ROTOR

P 18-1 With the front wheel removed and the vehicle chassis supported on safety stands, adjust the front-wheel bearings on the side where the rotor is being measured.

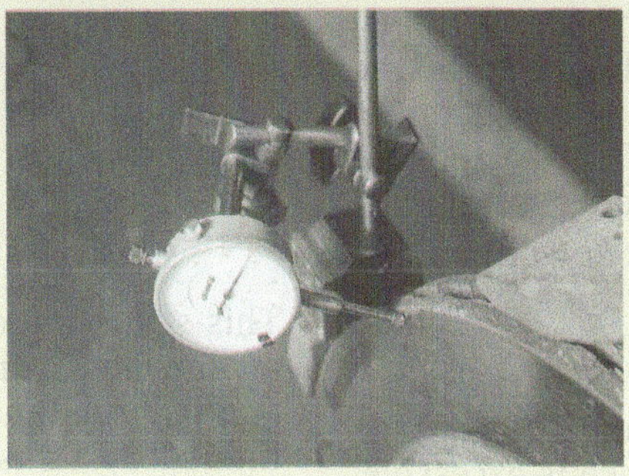

P 18-2 Mount a dial indicator so the plunger contacts the rotor surface approximately 1 inch (25 mm) from the outer edge of the rotor.

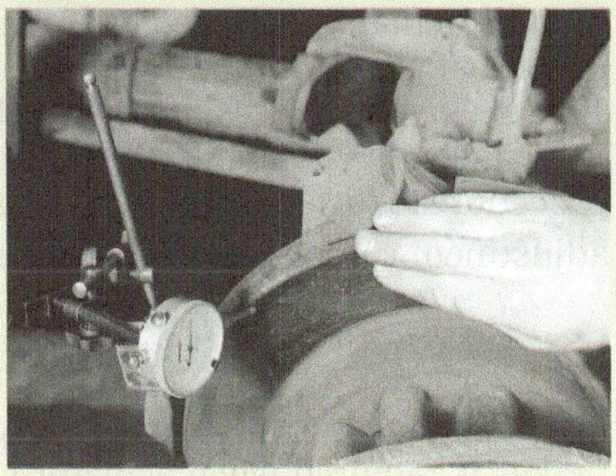

P 18-3 Turn the rotor one revolution and observe the runout on the dial indicator.

P 18-4 Compare the indicated runout on the dial indicator to the runout specification in the service manual. If runout is more than specified, machine or replace the rotor.

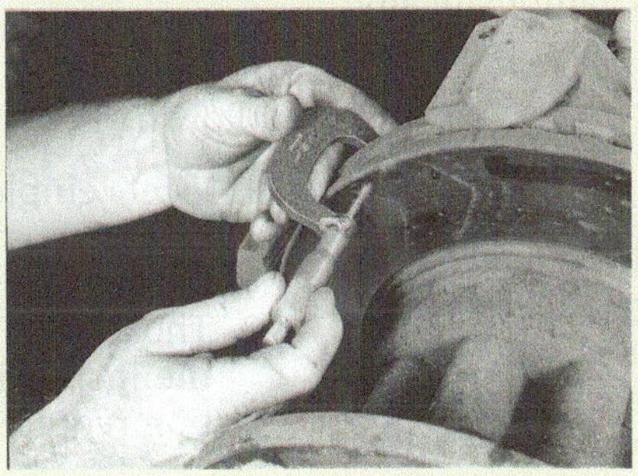

P 18-5 Use a micrometer to measure the rotor thickness approximately 1 inch (25 mm) from the outer edge of the rotor.

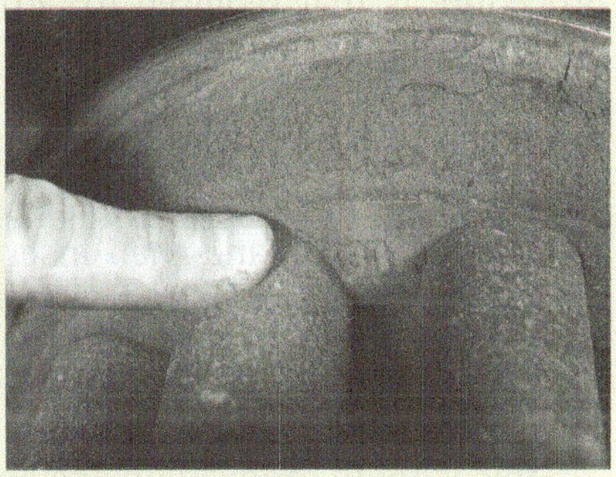

P 18-6 Compare the rotor thickness to the minimum discard thickness stamped on the rotor. If the rotor thickness is equal to or less than the discard thickness, replace the rotor.

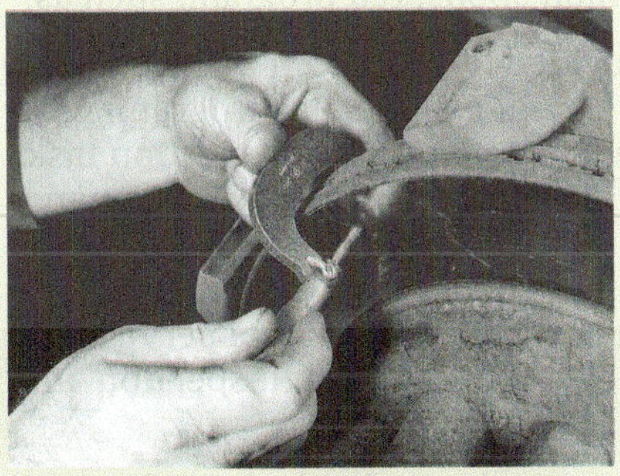

P 18-7 Use the micrometer to measure the rotor thickness at 12 locations around the rotor.

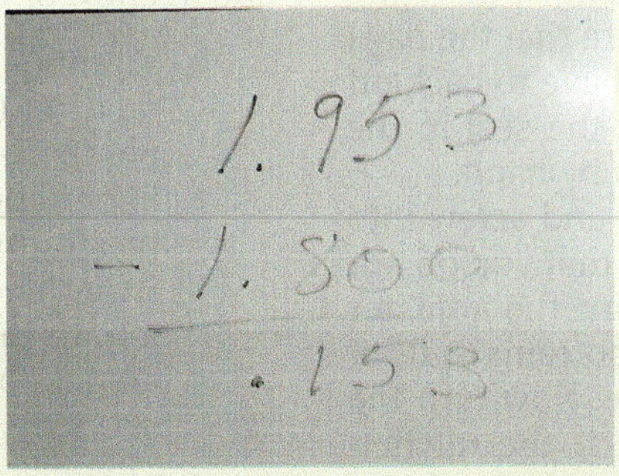

P 18-8 Subtract the minimum and maximum readings to determine the rotor thickness variation.

P 18-9 Compare the rotor thickness variation reading to the specifications in the service manual. If the rotor thickness variation exceeds this specification, machine the rotor.

31.13 SERVICING WEDGE BRAKES

The operating principles of air-actuated wedge brakes are discussed in Chapter 28. Despite their advantages of self-adjustment and superior braking mechanics (double-actuated), air wedge brakes have never been used that much in the trucking industry since the 1980s. One of the reasons is that the self-adjustment mechanism is vulnerable to seizure. Once seized, the procedure to unseize the adjusters and reset the brake adjustment can result in unnecessary downtime.

WEDGE BRAKE ADJUSTMENT

Brake adjustment of automatically adjusting wedge brakes is required when actuator stroke is excessive. In most cases, the adjusting star wheel is seized, meaning that you will often have to disassemble the plunger housing assembly to unseize and lubricate the components. Figure 28–32 shows the wedge and plunger housing assembly, which is the subject of the adjustment procedure outlined here. The following procedure is that required to "adjust" automatically adjusting, double-actuated, wedge brakes, assuming final assembly of the foundation brake or a situation in which the brakes have seized:

1. Lubricate the adjusting plunger and sleeve with chassis grease and then insert the plunger into the plunger housing. The plunger keyway slot should be aligned with the guide hole. Next, assemble the adjusting pawl, spring and threaded cap assembly into the plunger housing. Note that this is tooth-grooved on one end and has a chamfer on the other. Grease the teeth and insert into the plunger housing guide hole, grooved end first with the chamfer facing toward the brake shoe. This allows the pawl grooves to engage with the adjusting plunger grooves.
2. Turn the threaded cap a couple of full turns and assemble the plunger seal and plunger, inserting it in the plunger housing. Make sure that the pawl and sleeve are in mesh. You may have to back off the pawl cap to ensure this. Seat the seal in the plunger and snug up the pawl cap by hand.
3. Grease the adjusting bolt threads and screw into the adjusting sleeve, making sure that you do not pinch the seal in the threads. Turn the adjuster screw into the sleeve until it bottoms out. It should produce a ratcheting noise. Now turn the adjusting screw out three turns. If the ratcheting noise ceases, this indicates that the teeth are properly meshed.
4. Torque the pawl cap to specifications. Coat the adjusting bolt with chassis grease. Work it through the seal hole and turn it in just short of the seal. Do not bottom it out.
5. Pull the bolt outward slightly against the seal to ensure entry of the seal lip into the sleeve groove and then push the bolt back into the plunger.
6. Now set the drum-to-shoe clearance on both shoes to specification by rotating the adjusting star wheel. This should be performed using equal-sized feeler gauges at either side of the brake shoe. Set for equal drag on either side.

31.14 BRAKE CERTIFICATION, INSPECTION, AND TESTING

This section addresses the qualifications required of technicians working on brake systems, inspection, and testing. Brake systems are one of the most scrutinized systems on a truck, and they are routinely inspected by state and provincial transportation safety enforcement personnel.

CERTIFICATION

The truck and bus safety Regulatory Reform Act established the minimum training requirements and qualifications for employees responsible for maintaining and inspecting truck brakes and brake systems. The following fleet personnel need to be certified:

1. Fleet employees responsible for reinspecting, repairing, maintaining, or adjusting the brakes on commercial motor vehicles (CMVs)
2. Those fleet employees who are responsible for the inspection of CMVs to meet periodic inspection requirements

It is not essential that the person who does the work be certified unless he or she is also responsible for ensuring that the task is done properly. The person who is responsible for the service and inspection under this rule is called a brake inspector.

Training can be the completion of an apprenticeship or training program of a state, a province, a federal agency, or a union-sponsored program, such as:

- Training provided by a brake or vehicle manufacturer (or similar commercial program) relating to brake maintenance and inspection tasks
- Experience in performing brake maintenance programs
- Experience in a commercial garage, fleet leasing, or similar facility with brake service and inspection like the assigned inspection and service

Qualification means that individuals understand the brake service or inspection tasks to be done and can do them. They must know and have mastered the methods, procedures, tools, and equipment used and be able to perform the service or inspection based on individual experience or training, or both.

INSPECTION AND TESTING

To check brake systems, it is usually necessary for the inspector to crawl under the vehicle and visually look for defective brake equipment, signs of leaking wheel seals, oily linings, and so on, and measure the slack adjuster pushrod travel. However, jurisdictions and large fleets increasingly are using faster and more accurate means of determining brake performance. These include roller brake dynamometers, plate and pad testers, and sideslip test plates. Many fleets are also specifying brake valve actuators such as the Bendix BVA-85. This valve is specifically designed to enable a driver to perform a pretrip brake inspection unassisted. It is designed to apply an 85 psi (586 kPa) service brake application while the parking brakes are released. It allows the driver to exit the cab and check a tractor/trailer combination for:

- Service application air leaks
- Applied brake stroke lengths

Roller Brake Dynamometers

These units have the appearance of a chassis dynamometer. In minutes, these instruments can gather all sorts of performance data on vehicle brakes, including adjustment, lining and drum condition, and brake force under panic braking conditions. They can also spot a dragging brake and check brake timing and balance throughout a tractor and trailer combination. In addition to providing better data about brake performance and safety, computer-generated results should help the shop correct brake defects and solve braking problems.

Unlike chassis dynamometers that spin wheels at highway speeds and are used to track down engine power complaints, roller brake dynamometers turn the vehicle wheels at less than 1 mph (1.6 kph). It only takes two to three wheel revolutions to assess the total braking picture. The dynamometer rollers have enough torque to slowly turn the wheels even when 40 psi (2.8 bar) or more service air pressure is applied by the driver.

Plate-Type Brake Testers

These recently introduced dynamic brake testers are the fastest and most accurate means of testing brake. **Plate-type brake testers** use precision load cells mounted in flat pads; brake forces and dynamic weight and chassis brake balance can be precisely assessed, analyzed, and printed out in a 60-second test. An example is the Hunter B400T shown in **Figure 31–64**. This type of tester eliminates human error in evaluating brake performance and is becoming popular with enforcement agencies and large fleets.

In both the roller brake dynamometer and plate-type brake testers, computer software program plots brake torque curves on a per wheel basis. An automated test cycle calculates braking force measured at the tire tread surfaces required to drive the wheels under gradually

FIGURE 31–64 Tractor trailer combination on B400T plate-type brake tester

Hunter Engineering Company, Inc.

increasing brake-apply conditions. Data is picked up simultaneously from the left and right wheels by sensors on the load cells and input to the computer. If brake force is unequal, it will be spotted easily in the printouts and graphs as part of the pass/fail analysis.

Each brake on a full semi-rig can be evaluated. When the information is graphed, two graphed data lines are produced. The lower line represents the minimum standard required to pass the test. The upper line represents the actual performance profile of the brakes being tested. If the actual performance line drops down (indicating low brake force) and crosses the pass line, it means that the brake system has failed the test. A screen from the service brake test is shown in **Figure 31–65**. This indicates raw brake force data

FIGURE 31–65 Raw brake force analysis screen from Hunter B400T software

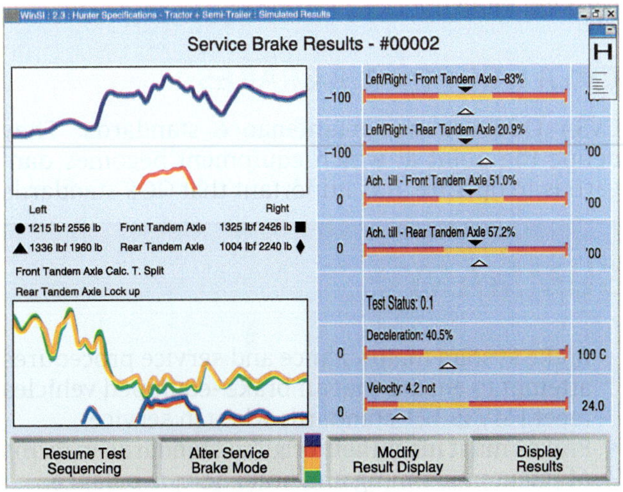

drawn off a tandem tractor with the upper graph indicating steer axle and the lower, drive axle wheel forces. Green line graphs indicate acceptable performance, whereas red indicates a problem. You can learn more about plate-type testers by accessing the Hunter Web site at http://www.hunter.com. Most plate-type testers are also designed to evaluate **sideslip**. Sideslip is any type of tire scuff condition caused by incorrect toe or excessive tandem scrub angle.

31.15 COMMERCIAL VEHICLE STOPPING DISTANCES

In Chapter 28 we stated that the current stopping distance that an air brake–equipped tractor/trailer combination was specified by FMVSS 121 to be within 250 feet (76 m). This stopping distance has to be achieved with the rig loaded, unloaded, and from any speed as mandated by the **National Highway Traffic Safety Administration (NHTSA)**. This requirement was enacted in 2010 and it actually does no more than putting tractor/trailer combinations on par with other commercial vehicles, and makes a fully loaded tractor/trailer combination capable of stopping from 60 mph (97 kph) in almost the same distance as that required of the family automobile at 216 feet (66 m).

HOW BRAKES HAVE CHANGED

Many medium-duty linehaul applications were able to meet the 2010 stopping distance standards simply by beefing up the steer axle brakes. All of the brake system manufacturers have endorsed the NPRM because the technology to achieve shorter stopping distances has existed for some time. The bottom line is that a fully loaded tractor/trailer combination entirely equipped with air disc brakes can be braked as efficiently as an automobile. Foundation brakes over the next decade will change and air disc brakes will continue to grow to become a major market force in the trucking industry.

CVSA OOS FOR AIR BRAKES

CVSA OOS are not maintenance standards. They define the point at which equipment becomes dangerous to operate. It is important that OOS standards are not confused with appropriate maintenance standards. The following are some examples of current brake OOS:

Service Brakes

- Absence of braking action on any axle required to have brakes.
- Missing or broken mechanical components, including shoes, lining pads, springs, anchor pins, spiders, cam rollers, pushrods, and air chamber mounting bolts.
- Loose brake components, including air chambers, spiders, and camshaft support brackets.
- Audible air leak at the brake chamber.
- Readjustment limits. The maximum stroke at which brakes should be readjusted is provided in Table 31–3. Any brake that is ¼ inch (6 mm) or more past the readjustment limit or any two brakes less than ¼ inch (6 mm) beyond the readjustment limit is cause for citation. Stroke should be measured with the engine off and reservoir pressure of 80 to 90 psi (5.5 to 6.2 bar) with the brakes fully applied.

Parking Brakes

Absence of braking when the parking brakes are applied. This includes driveline hand-controlled parking brakes.

Brake Drums and Rotors

- Any external crack or cracks that open upon brake application.
- Any portion of the drum or rotor missing or likely to fall away.

Brake Hose

- Hose with any damage extending through the outer reinforcement ply.
- Bulge or swelling when air pressure is applied.
- Any audible leaks.
- Two hoses improperly joined (such as a splice made by sliding the hose ends over a piece of tubing and clamping the hose to the tube).
- Air hose cracked, broken, or crimped.

Brake Tubing

- Any audible leak.
- Tubing cracked, damaged by heat, broken, or crimped.

SUMMARY

- Brake system maintenance and service procedures attempt to ensure that air brake–equipped vehicles meet FMVSS 121 standards when in service.
- Brakes must meet their original standards in terms of pneumatic timing and brake torque balance.
- Pneumatic timing refers to the management of the air in the system to ensure that the air reaches each actuator at the correct time so that the brakes over one wheel are not applied before or after those on another wheel.

- Brake torque balance refers to the requirement that the same amount of mechanical force be delivered to the wheels on each axle and that the brakes over all the axles on a vehicle work together to bring the vehicle to a stop.
- Disc brake assemblies are increasingly used in truck applications despite higher costs because of their superior brake performance. Service procedures required on disc brake assemblies require you to be able to measure lining wear, lubricate calipers, and measure rotor runout.
- A "brake job" refers to the process of inspecting and replacing brake shoes, linings, and brake drums. It is one of the most frequently performed maintenance tasks in a truck shop.
- Minimum training requirements and qualifications are necessary for employees who inspect, repair, maintain, adjust, or periodically inspect the brakes on commercial motor vehicles (CMVs).
- Roller brake dynamometers and plate testers are used in brake performance testing.

REVIEW QUESTIONS

1. Which of the following is the best practice when instructed to block the wheels on a tractor/trailer combination?
   a. Place a pair of wedged wheel chocks on both sides of one steering axle wheel.
   b. Place a pair of wedged wheel chocks on one tractor steering axle wheel and one trailer axle wheel.
   c. Place a pair of wedged wheel chocks on one tractor drive axle wheel and one trailer axle wheel.
   d. Place a pair of wedged wheel chocks on one trailer axle wheel.
2. What is the required amount of slack adjuster free travel on a properly adjusted, manual S-cam brake?
   a. less than ¼ inch (6 mm)
   b. ¼ to ½ inch (6 to 12.5 mm)
   c. ⅜ to ⅝ inch (10 to 16 mm)
   d. ¾ to 1 inch (19 to 25 mm)
3. Which of the following foundation brake components are lubricated with chassis grease, either for component life or as part of a lube job?
   a. S-cam profile
   b. S-cam roller face
   c. S-cam bushings
   d. brake shoe friction face
4. A 16½-inch (420 mm) brake drum specified to have a machine limit of 0.090 inch (2.3 mm) over and a maximum dimension of 0.120 inch (3 mm) over is measured at 0.098 inch (2.5 mm) over at a brake overhaul. Technician A says that the drum can be turned on a drum lathe, providing that the maximum dimension of 0.120 inch (3 mm) is observed. Technician B says that the drum can be legally returned to service without being turned. Who is correct?
   a. Technician A only
   b. Technician B only
   c. both A and B
   d. neither A nor B
5. A tandem-axle trailer equipped with S-cam brakes has all four slack adjusters originally set with a 6-inch (150 mm) dimension from the clevis pin to the camshaft. One slack adjuster is replaced, and because of a different hole arrangement in the new slack adjuster, it is set at 6½ inches (165 mm). Which of the following is more likely to result?
   a. One of the four wheels will lock up ahead of the others under severe braking.
   b. One of the four wheels will lock up behind the others under severe braking.
   c. One of the service brake chambers will require more air pressure than the others.
   d. One of the service brake chambers will require less air pressure than the others.
6. When adjusting a manual slack adjuster on an S-cam foundation brake, Technician A says that the lock collar should be retracted and the adjusting nut always rotated clockwise to decrease free travel. Technician B says that free travel can be reduced only when the adjusting nut is rotated counterclockwise. Who is correct?
   a. Technician A only
   b. Technician B only
   c. both A and B
   d. neither A nor B
7. A set of brake linings has become glazed because of oil saturation by a failed wheel seal. Technician A says that pressure washing the foundation brake components should be sufficient to remove the effects of the oil saturation. Technician B says that the best practice is to replace the brake linings on both wheels on the affected axle. Who is correct?
   a. Technician A only
   b. Technician B only
   c. both A and B
   d. neither A nor B
8. When adjusting S-cam brakes on a trailer equipped with manual slack adjusters, which of the following must be true?
   a. The tractor parking brakes must be applied.
   b. The trailer service brakes must be applied.
   c. The trailer parking brakes must be released.
   d. The tractor service brakes must be applied.

9. An original equipment 45-degree elbow on a line exiting a relay valve supplying a service chamber is replaced by a 90-degree elbow. Technician A says that this should have no effect on brake performance because it is the service application pressure that determines braking force. Technician B says that the use of the 90-degree elbow increases flow restriction and can affect pneumatic timing. Who is correct?
   a. Technician A only
   b. Technician B only
   c. both A and B
   d. neither A nor B
10. Technician A says that it is recommended practice to disassemble and replace the service diaphragm in a spring brake assembly to minimize the cost of replacing the complete spring chamber. Technician B says that it is no longer possible to change the hold-off diaphragms in most spring brake assemblies because the band clamp cannot be unbolted. Who is correct?
   a. Technician A only
   b. Technician B only
   c. both A and B
   d. neither A nor B
11. What is the main function of a hold-off diaphragm?
   a. It converts air pressure to mechanical service brake pressure.
   b. It prevents compounding of the foundation brakes.
   c. It prevents spring brake force from acting on the pushrod.
   d. It converts spring pressure into pneumatic pressure.
12. When an S-cam foundation brake slack adjuster is in the fully applied position, ideally the slack adjuster-to-pushrod angle should be:
   a. less than 90 degrees
   b. exactly 90 degrees
   c. more than 90 degrees
   d. does not matter
13. A tri-axle trailer is equipped with five S-cam slack adjusters correctly set at an effective length of 6 inches (150 mm) and one incorrectly installed at 5½ inches (140 mm). How much less brake torque will be applied at the brake with the incorrectly set slack adjuster when the service application pressure is 100 psi (7 bar) on a 30-square-inch (194 square cm) chamber?
   a. 150 lb-in. (17 N·m)
   b. 1,500 lb-in. (170 N·m)
   c. 15,000 lb-in. (1700 N·m)
   d. 150,000 lb-in. (17,000 N·m)
14. A leaking axle seal has saturated the brake shoe linings and caused them to glaze. Technician A says that the shoes should be immersed in solvent before being returned to service. Technician B says that if the brake shoes on both sides of the axle are not replaced, brake torque balance could be affected. Who is correct?
   a. Technician A only
   b. Technician B only
   c. both A and B
   d. neither A nor B
15. Technician A says that it is good practice to machine a new drum before putting it into service because it trues the drum to the wheel. Technician B says that most heat discoloration and checking is easy to remove on used drums when brakes are serviced by skimming the drum on a lathe. Who is correct?
   a. Technician A only
   b. Technician B only
   c. both A and B
   d. neither A nor B
16. When a safety pop-off valve continuously unloads, which of the following should be performed first?
   a. The governor should be replaced.
   b. The compressor should be replaced.
   c. The unloader air circuit should be checked.
   d. The primary circuit pressure should be checked.
17. What is the usual freestroke travel window of a properly adjusted, automatic S-cam brake slack adjuster?
   a. marginal preload
   b. zero
   c. ⅜ to ¾ inch (9.5 to 19 mm)
   d. 1 to 1½ inches (25 to 38 mm)
18. What is the usual specified *maximum* S-cam endplay permitted when installing a slack adjuster?
   a. none
   b. 0.030 inch (0.76 mm)
   c. 0.050 inch (1.27 mm)
   d. 0.120 inch (3.05 mm)
19. What is the adjustment stroke limit for a 24-Series brake chamber?
   a. 1 inch (25 mm)
   b. 1.75 inches (44 mm)
   c. 2 inches (50 mm)
   d. 3 inches (75 mm)
20. What is the adjustment stroke limit for a standard 30-Series brake chamber?
   a. 1.5 inches (38 mm)
   b. 2 inches (50 mm)
   c. 2.5 inches (64 mm)
   d. 3 inches (75 mm)

32

Prerequisites: Chapters 1, 2, and 3

VEHICLE CHASSIS FRAME

OBJECTIVES

After reading this chapter, you should be able to:

- Describe the construction of a chassis frame of a heavy-duty truck.
- Define the terms *yield strength*, *section modulus* (SM), and *resist bend moment* (RBM).
- List the materials from which frame rails are made and describe the characteristics of each.
- Explain the elements of frame construction.
- Describe the different ways that frame damage can occur as a result of impact and overloading.
- Perform some basic chassis frame alignment checks and project a frame-to-floor diagram.
- Describe the various categories of frame damage, including diamond, twist, sidesway, sag, and bow.
- Explain how the chassis frame, side rails, and cross-members can be repaired without altering the frame dynamics.
- Specify the appropriate welding methods for repairing tempered and mild steel frame members.
- List some guidelines to follow when using frame repair hardware.

KEY TERMS

applied moment
bending moment
brine
calcium chloride
C-channel
cross-members
galvanic corrosion
Machinery's Handbook
magnesium chloride
rails
resist bend moment (RBM)
section modulus (SM)
sodium chloride
web
yield strength

INTRODUCTION

The chassis frame is the backbone of all heavy-duty vehicles. The main body of the frame on a highway tractor is shaped like a ladder. In fact, we commonly refer to a heavy-duty truck frame as a ladder frame. Although a ladder's function is far different, its two main components—rails and steps—can be compared to truck frame **rails** and **cross-members**. Cross-members are the braces that hold a pair of frame rails together. The cross sections of the rails resemble a "C" or an "I," and as they are increased in size and strength, the duty rating of the ladder or truck frame is upgraded. Cross-members are required to be stronger as the anticipated workload increases.

All frames are engineered and built to provide the versatility, durability, and performance demanded by the varied uses of heavy-duty vehicles. These varied uses require that the frames used should be rigid enough to resist bending and twisting forces, have sufficient strength to carry anticipated loads, provide resistance to fatigue from repeated flexing, and have the durability to absorb the vibration and shocks encountered in daily operation.

ROLE OF THE FRAME

The frame supports the cab, engine, transmission, axles, and the various other chassis components. The cross-members control axial rotation (another way of saying *twist*) and longitudinal motion of the rails. They reduce torsional stress transmitted from one rail to the other. Cross-members also are used for vehicle component mounting and protecting the wires, hoses, and tubing that are routed from one side of the vehicle to the other. The front cross-members may be of a drop style (**Figure 32–1A**) or a straight-through type (**Figure 32–1B**).

FRAME DESIGNS

The drop-style cross-member provides for a low center of gravity, resulting in better vehicle stability. This design also allows for larger radiators to be used and improves service access to the engine and related components. In addition, drop cross-members can make cutting or notching of the frame rails unnecessary. In an area to the rear of the cab, on some heavy-duty truck models, a deep section of frame can be used to provide extra strength at points of high stress. However, when lower cost and greater ruggedness are required—for example, off-highway vehicles—straight-through cross-members are more often used.

FIGURE 32–1 Typical heavy-duty tractor chassis frames: (A) drop style; and (B) straight type

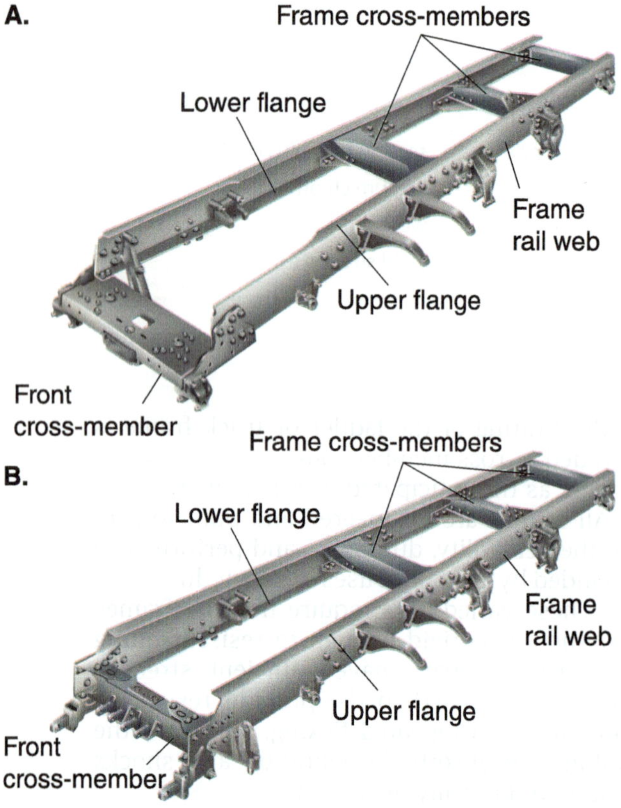

FIGURE 32–2 C-channel rail section

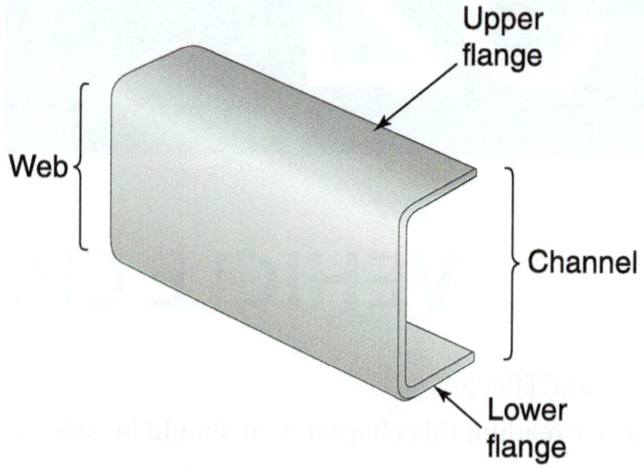

Although I-beams are used in some extra heavy-duty applications such as crane carrier chassis, most frame rails used on Class 6, 7, and 8 trucks are of the **C-channel** rail type. A C-channel frame is shown in **Figure 32–2**. A C-channel has an upper flange, lower flange, and web (the surface between the flanges). Of note is the radius in the bend between the web and flanges. By using a radius bend, stress concentrations are avoided.

A C-channel frame rail accepts cross-members and body mounts easily because they can be fitted between the rail flanges. Each truck manufacturer uses C-channel rails with cross-members of its own design. Cross-members may be I-beam, C-section, tubular, boxed, or other shapes and may be bolted or welded together.

Trailer Frames

Most trailer frames are constructed using very different principles from those used in the tractors that haul them. A majority of trailer frames use a unibody or monocoque principle that shares more with current automobile frame design. Unibody frames are studied in Chapter 33.

32.1 BASIC FRAME TERMS

Before undertaking any frame servicing, the technician should understand the following terms and definitions used to describe frame characteristics and strength. Their use is found throughout this chapter.

- **Yield Strength.** **Yield strength** is the highest stress that a material can sustain without permanent deformation, and it is expressed in pounds per square inch (psi). Light and medium trucks may use rails made of steel alloys with a yield strength of 30,000 to 80,000 psi (207 to 552 MPa). Class 8 trucks use 110,000 psi (7,60 MPa), heat-treated steel. Material yield strength is determined through laboratory testing and the higher the figure, the stronger the material.

- **Section Modulus.** **Section modulus (SM)** factors the shape and size of the rail but not the material. In other words, it is not related to the frame rail material, but it does help define its flexibility. The deeper the "web" (the vertical portion of the rail) and the wider the flange (the horizontal portion attached perpendicular to the web), the greater the rail SM. We will talk about SM in more detail a little later.
- **Resist Bend Moment.** This term is used to accurately express frame rail strength. **Resist bend moment (RBM)** is calculated by multiplying the SM of the rail by the yield strength of the material. This term is universally used in evaluating frame rail strength. The higher the RBM, the stronger the frame. This figure could be in the millions of pounds, and minimum numbers will sometimes be specified by builders of extra heavy-duty equipment, such as wreckers, crane carriers, or roll-on/roll-off bodies.
- **Area.** The area is the total cross section of frame rail (includes all applicable elements) in square inches.
- **Applied Moment.** The term **applied moment** means the specific or focal point that a given load is applied to a frame.
- **Bending Moment.** The term **bending moment** is used to describe the point at which a material deflects when a force is applied to it; this load is distributed across a given section of the frame material. The load applied tends to deflect the frame where the load concentration is greatest. The effect of maximum bending moment on the frame will vary because of different body configurations. For instance, depending on how it is loaded, a straight-body truck will transfer this applied moment in a different manner when compared to a dump truck. You can see this demonstrated in **Figure 32–3**.
- **Safety Factor (Margin of Safety).** This is the amount of load that can safely be absorbed by and through the vehicle chassis frame members. This safety factor (SF) is the relationship of the applied moment to the RBM. If the applied moment and the RBM are the same, then the safety factor is classified as 1. This safety factor is necessary because the applied moment is generally calculated based on a stationary or vehicle static condition.

FIGURE 32–3 Frame applied bending moment locations on (A) straight truck; (B) dump truck; and (C) tractor.

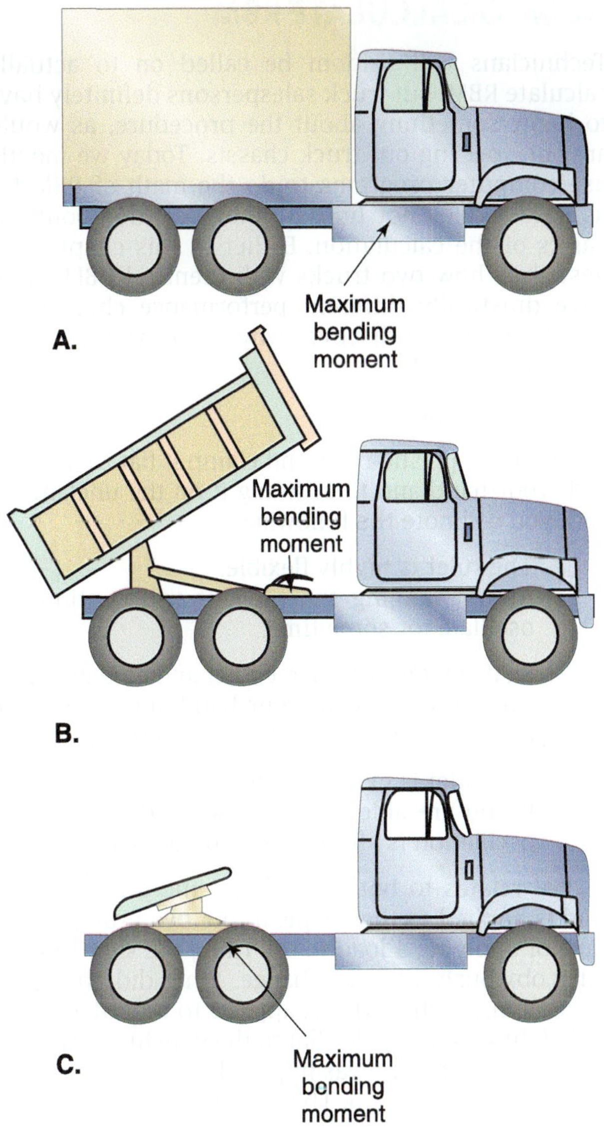

SHOP TALK

Two truck frames with identical RBM can perform very differently. RBM is calculated by factoring SM with yield strength. If two trucks have identical RBM but one is spec'd with a high SM but uses a lower yield strength material, it will be more rigid than a frame with high yield strength but low SM.

32.2 FRAME THEORY

If you are working on trucks today, you should know something about frame theory. It can help you to avoid creating problems when drilling holes in frame members, adjusting frame length, as well as either reinforcing or attaching auxiliary frame components. We begin by taking a look at RBM.

HOW TO CALCULATE RBM

Technicians will seldom be called on to actually calculate RBM, but truck salespersons definitely have to know something about the procedure, as would anyone spec'ing out truck chassis. Today we mostly use computer programs to do the math of RBM for us, but it does not hurt to know a little about the basics of the calculation. Earlier in this chapter, we described how two trucks with identical RBM could have drastically different performance characteristics. Now we will discover why. First, we will take a closer look at SM.

Section Modulus (SM)

If you take a plastic ruler and clamp it flat to a bench with your hand and then twang it on the unclamped end, you will note the following:

- The ruler is highly flexible.
- After twanging the ruler, it will continue to oscillate for some time.

Next, if you take that same ruler and hold it at right angles to the bench with your hand and attempt to twang it, you will note that it behaves very differently:

- The ruler is extremely rigid.
- If you are able to even slightly deflect it, any oscillation is almost instantly damped.

SM relates to both shape and the way in which load is applied to the shape. In the two examples we gave of applying a load to the ruler, the shape of the ruler obviously did not change. What did change is the way in which load was applied to the ruler. When its height was low (ruler flat to the bench), it was flexible. When its height was high (ruler at right angles to the bench), it was rigid. The height factor in a beam is often called the **web**.

A frame rail can be likened to a ruler when it is subjected to loads. Because of this, when RBM is calculated, height has to be emphasized over width. Every shape that can be subjected to load has a different SM formula. The SM calculation for most of the shapes we use in frame construction is provided in a reference book titled ***Machinery's Handbook***. Most truck original equipment manufacturers (OEMs) have spec'ing software designed to make this calculation easier. We are going to focus on SM calculation for the most common shape used in truck frames, the C-channel. **Figure 32–4** shows a typical section of a C-channel frame rail, and we are going to use this to guide us through a SM calculation.

In the formula shown in Figure 32–4, the letters "B(b)" and "D(d)" are used to represent *breadth* (width)

FIGURE 32–4 Frame section modulus formula

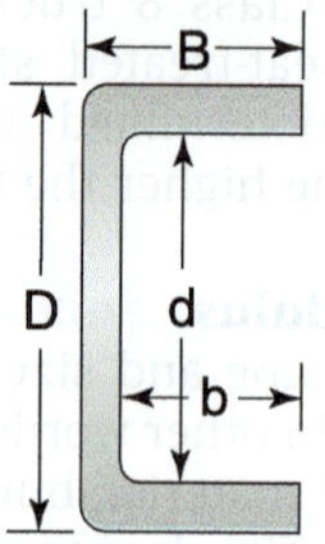

$$\text{Section Modulus} = \frac{BD^2}{6} - \frac{bd^3}{6D}$$

and *depth* (height). Uppercase letters indicate an outside dimension and lowercase letters are used to indicate an inside dimension. Therefore, the SM formula looks like this:

$$SM = \frac{(B \times D^3) - (b \times d^3)}{6D}$$

Remember how height had to be emphasized when we applied load to the ruler; this emphasis on height in the formula is achieved by cubing height (D) in the equation. Now we can use a real-life example to calculate SM on a 10-inch by ¼-inch (250 mm by 6 mm) C-channel rail. The dimensions in this formula are for demonstration purposes only and use standard measurements. *Machinery's Handbook* (pages 234 to 243 of the 29th edition) should be consulted for the metric system formulae.

$$SM = \frac{(3'' \times 10^{3''}) - (2.75'' \times 9.5^{3''})}{6 \times 10''}$$

$$SM = \frac{(3 \times 1000) - (2.75 \times 875)}{60}$$

$$SM = \frac{3000 - 2357}{60}$$

$$SM = \frac{643}{60}$$

$$SM = 10.7$$

The SM value calculated here is a measure of a frame's twistability. The higher the SM, the less flexible a frame tends to be.

SHOP TALK

Before the introduction of tempered steel frame rails with yield strengths up to 110,000 psi, (760 MPa) salespersons spec'd truck frames on the basis of RBM only. Today, tempered steel frame rails mean that high RBM can be achieved with low SM. This is great for highway trucks that required lightweight frames with plenty of flexibility, but it proves to be disadvantageous in applications where high rigidity is required. A fully extended dump truck box requires frame rigidity to control twist at the tail pivot. Twist can result in a raised dump box toppling the chassis. Dump trucks must meet both a specified RBM and have sufficient SM (rigidity) to function safely.

Material thickness usually increases as the intended duty for the truck becomes more severe. On-highway tractors generally use ¼-inch (6 mm) steel frame rail thickness, whereas an on-/off-road dump truck or mixer chassis may have 5⁄16- or ⅜-inch (7.5 or 10 mm)-thick rails, often with reinforcing members. Multiple frame rails are also commonly used to increase SM. The depth or web of the rail also increases with duty severity. Highway tractor rails will usually be 10¼ inches (267 mm) deep, but dump truck rails might be 13½ inches (340 mm) deep or more.

Section Modulus Problems

A frame is a dynamic chassis element. That is, it is designed to flex and move in operation. The extent to which it can flex is defined by its SM. When we do something to change SM in a frame or section of frame, problems can result. For instance, if we were to install a HIAB hoist on a straight truck chassis between the cab and a flatbed deck, the brackets used to mount the hoist to the frame would cause a significant increase in SM over a section of the frame. The increase in SM can set up stress points on either side of the hoist mount brackets that can result in frame fracture failures. These fracture failures are not usually caused by overloading the hoist; they are a result of the fact that the frame has been made extremely rigid in one section, which interferes with normal frame oscillations.

Most hollow tubular section members have high SM combined with light weight; this is why they are used in racecar chassis. Any type of cylindrical tank has high SM, and this must be accounted for when mounting it to a truck or trailer chassis. A tanker trailer will often use the tank itself as the primary frame member, and devices such as the suspension and upper coupler assemblies are fastened to the tank by means of subframes.

Because of its high rigidity, a straight truck tanker must be mounted to the chassis frame rails in such a manner that allows the frame rails to move independently from the tank. Hardwood strips mounted on the rail flanges is one method of accomplishing this. Failure to permit this independent movement of tank and frame rails will result in a hard ride and premature failure of frame chassis or the tank.

Yield Strength

Yield strength is a measure of the material strength of the frame. By definition, yield strength is the maximum stress that a material can withstand before it is permanently deformed. Yield strength is factored with SM to calculate RBM. Two identically shaped steel frame members can have very different material strengths. Heavy-duty chassis are usually manufactured with frame rails of either steel or aluminum alloy. Each material must be handled in a specific manner to ensure maximum service life. The frame material must be clearly identified before attempting repair or modification.

CAUTION

Steel frame rails that appear similar to the eye can have drastically different yield strengths. Always identify the frame material before attempting any repairs.

Steel Frames

Trucks are manufactured with frame rails of mild steel (36,000 [248 MPa] psi yield strength), high-strength low-alloy (HSLA) steel (50,000 psi [345 MPa]) yield strength), or heat-treated steel (110,000 psi [760 MPa] yield strength). Each type has different repair procedures. Steel frame material can be determined by inspecting the frame or referencing the factory line ticket, the chassis data bus, and the sales data book.

There are several methods of identifying heat-treated (110,000 psi [760 MPa] yield strength) frame rails, the most common of which is a stencil marking on the inside middle or rear section of the rail or a stencil mark on one of the cross-members. The stencil data indicates information such as whether the rail is heat treated (aka tempered) and may caution that rail flanges must not be drilled or welded. In this case, the welding of additional brackets or cross-members or the welding of reinforcements to the rails must be avoided. This may cause failure because of frame strength reduction at welds and holes.

Most truck frames carry a decal warning against welding or torching the rails. This is because high temperatures can destroy the heat treatment of the steel, which we call tempering, thereby weakening the material. For instance, if you remove the heat treatment from a tempered truck frame rail specified at 110,000 psi (760 MPa) yield strength, the yield strength is reduced by approximately half. Welding or torching is permitted at the ends of the frame rails and for crack repair when the correct techniques are used.

During assembly, frame rails are custom drilled to accept specific cross-member locations and the attachment of engine mounts, suspensions, and the like. Hole locations are predetermined, based on the intended application of the truck and, therefore, the types of componentry to be attached to the frame. Most builders prefer not to drill extra holes because these weaken the rails.

Aluminum Alloy Frames

It is relatively easy to distinguish aluminum frames from steel frames on the basis of their greater web and flange thickness and nonmagnetic properties. Frame material can be identified by placing a magnet near the frame. The magnet will be attracted to a steel frame. The aluminum frame will not attract the magnet.

Aluminum alloy frames weigh less than their steel counterparts but are not as strong as hardened steel frames. The most common aluminum frames are medium-strength magnesium-silicon-aluminum alloy (37,000 psi [255 MPa] yield strength) and high-yield strength copper-aluminum alloy (80,000 psi [552 MPa] yield strength). Both of these alloys are heat treated. They are much lighter in weight but also both more expensive and susceptible to abrasion, corrosion, and fatigue wear.

Repair procedures for the medium-strength aluminum frame and the high-strength aluminum frame are different. Medium-strength aluminum frames and high-strength aluminum frames cannot be distinguished from one another by appearance. To identify the type of aluminum, consult the factory line ticket or service literature. Sometimes you will have to contact the OEM. **Table 32–1** compares some typical truck frame materials.

SHOP TALK

Although chassis frames are very strong, care must be taken when lifting or moving them to avoid anything that may scratch, cut, or damage an exposed frame assembly. Cushion all chain hoists or cable slings with a section of heavy hose. If the frame rail is raised with a jack, place a block of wood between the jack and the frame rail.

RBM Equation

RBM is a measure of ultimate frame strength. Just remember that RBM tells you nothing about the flexibility so technicians should remember that SM is equally important in assessing frame performance. RBM is calculated as follows:

$$\text{RBM} = \text{SM} \times \text{yield strength}$$

TABLE 32–1 Frame Material Strength Comparisons

Frame Material	Yield Strength in psi
Mild steel (used in light duty only)	28,000–38,000
High-tensile steel	50,000–55,000
Tempered, medium alloy steel	110,000
Magnesium-silicon-aluminum alloy	37,000–50,000
High-yield, copper-aluminum alloy	80,000

So if we use the SM data we calculated a little earlier in this section and apply it to a frame member with a rated yield strength of 110,000 psi (760 MPa), the calculation using the standard measurement system would look like this:

$$\text{RBM} = 10.7 \times 110{,}000 \text{ psi}$$
$$\text{RBM} = 1{,}177{,}000 \text{ inch-pounds}$$

Machinery's Handbook (pages 234 to 243 of the 29th edition) should be consulted for the metric system formulae. RBM calculations are used mostly for comparative purposes in the trucking industry. They are especially important when spec'ing trucks for load and performance and when making any type of adjustment to frame length.

FRAME LOADING

When a C-channel frame member is subjected to a load from above, it places the upper flange under compression (squeezes it) and the lower flange under tension (stretches it). This concept is shown in **Figure 32–5.**

When the same frame member is subjected to road shock from below, the reverse occurs with the lower flange subjected to compressive stress and the upper flange to tensional stress. In both instances, the center of the web is subjected to almost no stress, so we refer to this as the neutral fiber or axis. This is illustrated in **Figure 32–6.**

If you were to ream three perfectly round holes in a vertical line of a C-channel frame member, one through the top flange, one through the lower flange, and one through the neutral fiber, and then load the frame through that section from above, the following would occur:

- The upper hole would elongate inward.
- The hole through the neutral fiber would remain perfectly round.
- The lower hole would elongate outward.

For this reason, OEMs stress that whenever holes have to be drilled in frame rails, they should be positioned as close to the neutral fiber as is realistically possible. They also recommend that scribe marks be made as shallow as possible and that they be entirely removed after drilling. **Figure 32–7** shows the right and wrong ways of drilling holes in frame rails.

FRAME CUTTING AND REPAIR GUIDELINES

Whenever heat is applied to a frame rail in the form of cutting or welding, the temper is affected in hardened steel frame rails. However, there are ways to limit the reduction in material strength when you have to make modifications or repairs to frames.

When frames have to be cut and/or welded, it is advisable to make the cut at an angle of either

FIGURE 32–5 C-channel frame under load

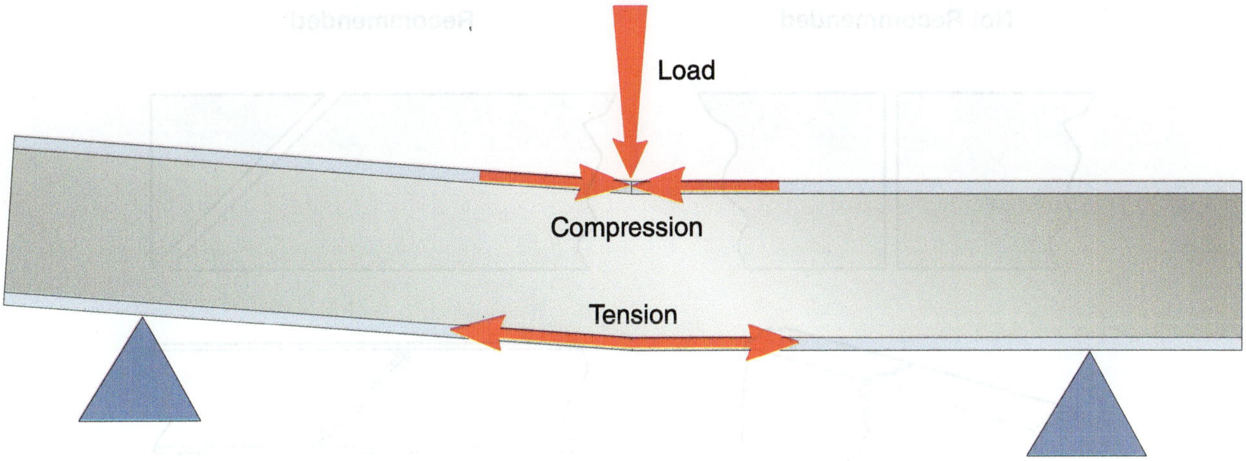

FIGURE 32–6 Neutral fiber on C-channel rail

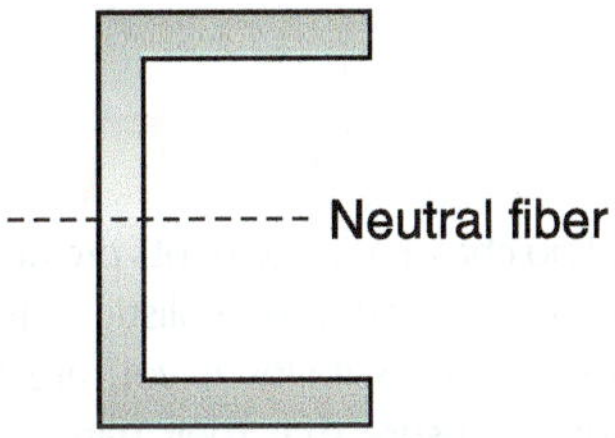

60 degrees (usually recommended) or 45 degrees. In this way, the RBM through any given vertical section will be minimally affected. **Figure 32–8** illustrates these basic guidelines. We address repair methodology in more detail later in the chapter.

32.3 FRAME CONSTRUCTION

Frames may be single, double, or triple construction. Additionally, most frames are available with either inside or partial inside channel reinforcements or outside reinforcements. The reinforcements are used to provide a greater RBM and SM over a section of the frame. Original equipment reinforcements are engineered so seldom provide aftermarket problems. Repair reinforcements are a different story and can often initiate a succession of secondary failures.

CROSS-MEMBERS

Cross-members are designed to connect the frame rails. They provide rigidity and strength, along with

FIGURE 32–7 Drilled hole placement on C-channel rails

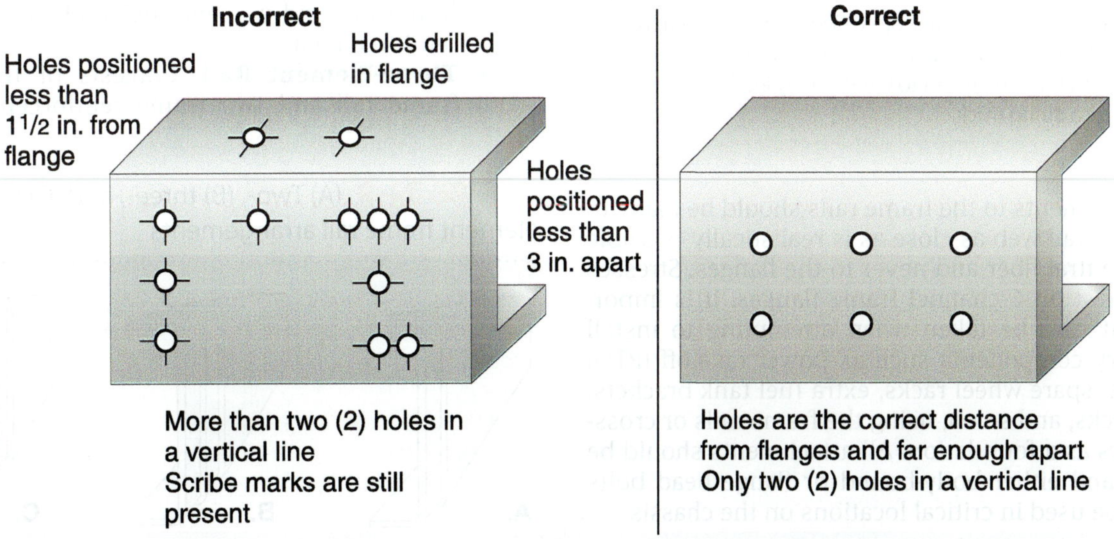

FIGURE 32–8 Cutting and welding tempered frame rails

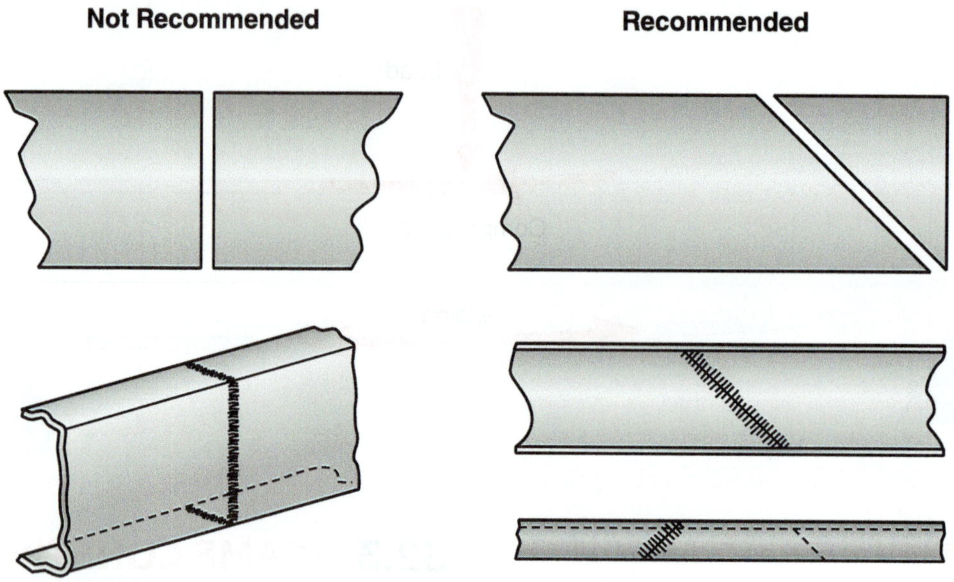

sufficient flexibility to withstand the twisting and bending stresses encountered when operating on uneven terrain. Stamped C-section is a standard type of cross-member. Conventional highway applications use a plate-type cross-member in the first and second positions behind the cab. Stamped C-section cross-members have keystone-shaped gussets for attaching the cross-members to the frame. This design resists weaving or out-of-square (diamonding) frames. Gussets are welded or bolted to the cross-member and then bolted to the frame rail web. Exceptions to this method of frame rail web attachment may occur at the rearmost cross-member.

SHOP TALK

The extreme front and extreme rear of a frame rail are subjected to less stress during normal operation than the sections in between. A notable exception would be the stress the rear of the frame rails are subjected to when a fully loaded dump box is raised on a dump truck.

Attachments to the frame rails should be made to the frame rail web as close as is realistically possible to the neutral fiber and never to the flanges. Stresses are highest on C-channel frame flanges. It is important that care be taken when attempting to install accessory components such as power take-off (PTO) brackets, spare wheel racks, extra fuel tank brackets, chain racks, and so on, using the frame rails or cross-members as a foundation. All attachments should be bolted, and either body-bound or flange head bolts should be used in critical locations on the chassis.

CAUTION

When reassembling chassis components previously assembled with Huck™ fasteners, it is often unrealistic to install new Huck fasteners because of accessibility. If you are replacing Huck fasteners with bolts, ensure that their hardness is consistent with the original fasteners. This will usually, but not always, be equivalent to an SAE Grade 8 fastener.

MULTIELEMENT RAILS

As we have previously described in this chapter, the common frame on heavy-duty highway trucks is the single C-channel configuration. Some other common channel frame configurations are shown in **Figure 32–9**.

- **Two-Element Rail.** Consists of the main frame rail and a single inside channel frame reinforcement.
- **Three-Element Rail.** Consists of the main frame rail and two frame reinforcements: a

FIGURE 32–9 (A) Two-, (B) three-, and (C) four-element frame rail arrangements

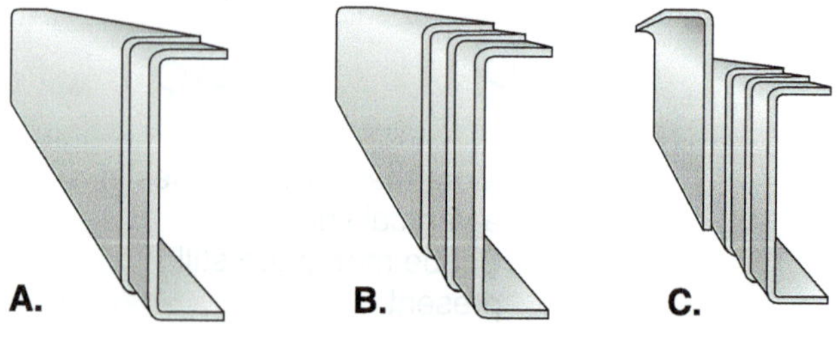

FIGURE 32–10 Alternate frame designs

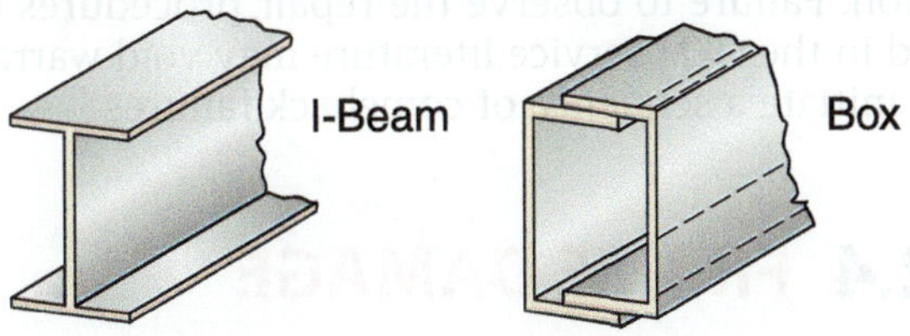

single inside channel and a single outside channel frame reinforcement.

- **Four-Element Rail.** Consists of the main frame and three-frame reinforcements: a single inside channel, a single outside channel, and a single inverted "L" outside frame reinforcement.

Two-frame construction designs—the I-beam and the box—are used in some vocational trucks and often in trailers (**Figure 32–10**).

REINFORCEMENT

Adding reinforcing members to the frame rails increases both their SM and RBM values but also adds weight. The weight per foot (305 mm) of rail can range from below 15 pounds to 40 pounds (7 to 18 kg) and more, depending on the type of reinforcement. These take various forms. For example, inserts may themselves be C-channels, right angles, or inverted or upright Ls clamped within the main rails. L-type reinforcements are shown in **Figure 32–11**. Reinforcing also can be added to the outside of frame rails by using L-channels or nonflanged fishplates. A fishplate is a flat piece of plate reinforcement bolted either outside or inside the frame channel.

FIGURE 32–11 L-type reinforcements

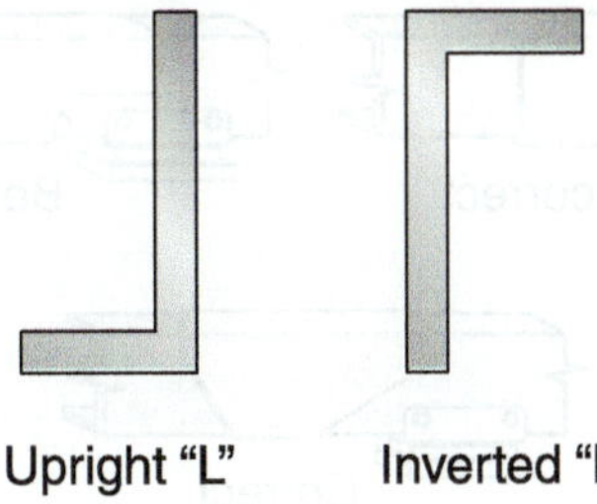

SHOP TALK

When any type of frame reinforcement is added, straight cut fishplates, L-sections, and C-channels should be avoided because this creates a sudden increase in SM. This sudden increase in SM can cause frame failures immediately adjacent to the reinforced section. **Figure 32–12** and **Figure 32–13** show some good and bad practice reinforcement guidelines.

FIGURE 32–12 Fishplate and reinforcement guidelines

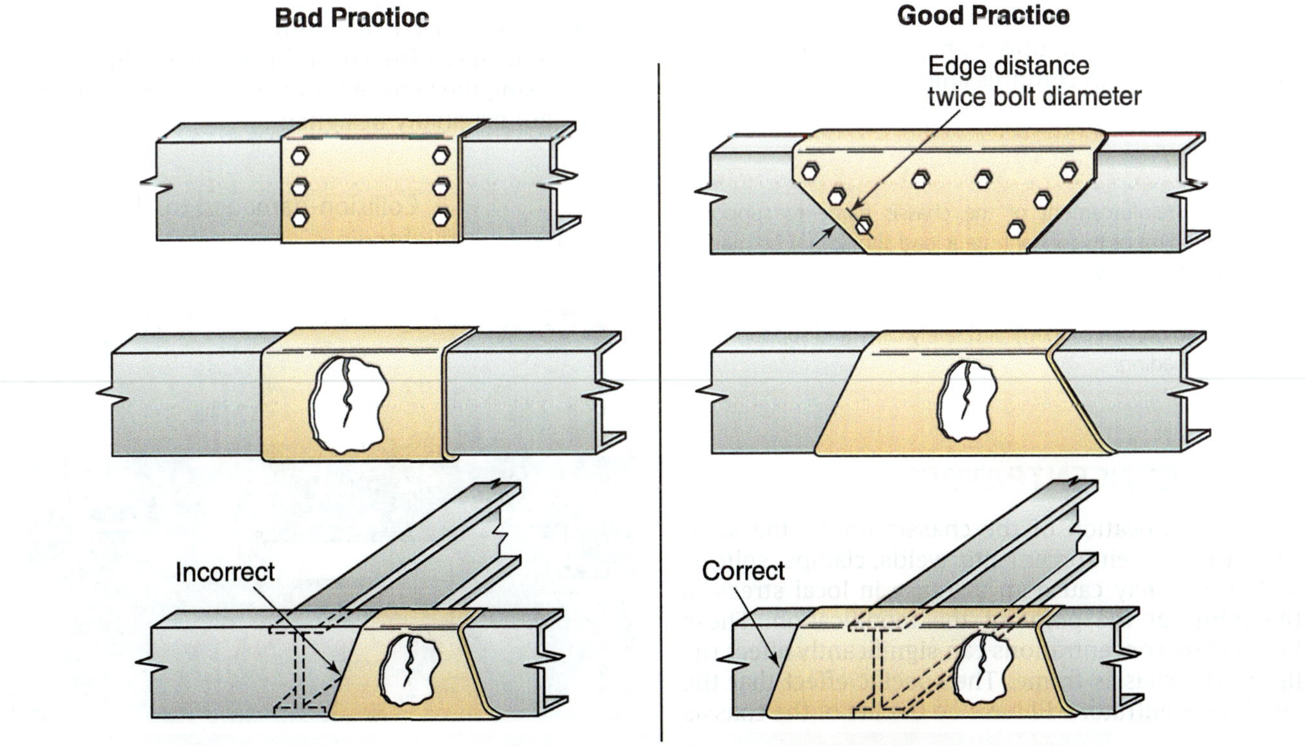

FIGURE 32–13 Reinforcement geometry

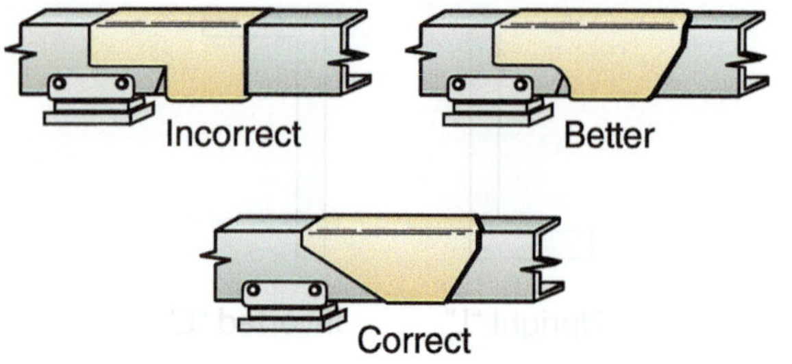

Some OEMs incorporate deep-section or belly extensions to the rails where stress is greatest—toward the middle of the frame, between the axles. This is where weight is suspended without direct support and bowing of the frame is possible. Although it is a more costly alternative, a deep-section frame can add to RBM without the weight penalty incurred by inserts and outside channels or fishplates.

Webs and gussets are added where cross-members join the main rails, and these are often incorporated in the cross-member's design. In severe duty trucks, extra webbing and gusseting can sometimes extend down much of the frame's length.

Reinforcing channels and plates are usually added at points where stress is concentrated. Examples are beneath the tractor fifth wheel and at suspension mounting points, especially for add-on tag or pusher axles or where lift gates are installed. For mounting of certain truck bodies, subframes consisting of shallow C-channels or hardwood strips are attached to the top of the frame rails. Plates or U-bolts join the subframe to the rails and if U-bolts are used, spacers can be wedged between the upper and lower frame flanges to keep them from compressing.

CAUTION

Additional reinforcement of the chassis frame to support greater loading or to concentrate a load should not be made until it has been fully verified that all other vehicle components, such as the brake system, steering system, suspension system, and so on, can properly and safely carry and support the increased loading.

STRESS CONCENTRATORS

In any modification of the chassis frame, the addition of holes, reinforcements, welds, clamps, splices, and so on, may cause an increase in local stress in the frame at the point of the modification. These local stress concentrations can significantly affect the life of the chassis frame. The specific effect that the stress concentrator will have on the life of the chassis frame is influenced by its location, the frequency and severity of the loading, and the type of stress concentration. Failure to observe the repair procedures outlined in the OEM service literature may void warranty and initiate a sequence of comeback failures.

32.4 FRAME DAMAGE

Damage to the chassis frame generally occurs as a result of impact damage to the vehicle, such as the vehicle being involved in a collision or an overturn (**Figure 32–14**). Such damage may often be repaired by straightening and reinforcing the frame, by replacing the frame side-members, or by repairing the damaged area and reinforcing the frame side-member.

SHOP TALK

Frame straightening should be performed only by a qualified frame alignment facility. Because impact-damaged frames are corrected by specialty technicians, this type of frame servicing is not covered in this book.

Damage to the chassis frame, such as a crack in the frame side-member or cross-member not associated with impact damage may be an indication of overloading the vehicle. Overloading can be caused by the following:

- Exceeding either the gross vehicle weight rating (GVWR) or the gross axle weight rating (GAWR) (i.e., loading the frame beyond its design capacity)
- Uneven load distribution
- Improper fifth wheel settings (see Chapter 34)
- Using the vehicle for purposes for which it was not originally designed

FIGURE 32–14 Collision-damaged truck being prepared for straightening

Courtesy of Bee Line

- The use of special equipment for which the frame was not designed
- Improper modification of the frame

CORROSION

Frame damage also may be caused by corrosion caused by the contact between dissimilar metals. If aluminum and steel, for example, are allowed to come into direct contact, **galvanic corrosion** can eat away both materials. Aluminum is anodic with respect to steel and will corrode in its presence. Corrosion of the aluminum frame member will reduce the load-carrying capacity of the frame member and may eventually lead to failure of the frame.

In order to prevent galvanic corrosion, isolation techniques such as nonconductive- or barrier-type spacers or sealers must be used. It is recommended that a sealer be painted onto the surface of both the aluminum frame and the steel reinforcement. Steel bolts passing through the aluminum frame and steel reinforcement, as well as the washers under the head of the bolts and nuts, should be sealed.

Road Salt and Brine Solutions

Galvanic corrosion can also be caused by road salt and brine solutions. Salt and salt solutions continue to be the most common de-icers used on roads in the northern states and Canada because less corrosive chemicals tend to be more costly. The most basic road de-icer is **sodium chloride** in its crystallized form, otherwise known as rock salt. The disadvantage of rock salt as a melting agent is that it freezes at 0°F (–18°C). However, it is probably the least corrosive of de-icing agents in current road use.

More recently, liquid salts known as **brine** have been widely used on roads. One advantage of using a brine solution is that can be cut with **calcium chloride** and **magnesium chloride**, both of which have lower freezing points than sodium chloride but are said to be more caustic. This means that the road de-icer used on any given day of winter can be formulated for the actual temperature conditions predicted by cutting the solution with more costly magnesium chloride and calcium chloride.

Liquid brine applied to roads helps prevent snow and ice from bonding to the road surface, which is a major advantage. However, when sprayed by wheels upward onto the frame and undercarriage the liquid is much more penetrative and has contributed to more rapid corrosion failure. Because brine is so invasive, there has been an increase in corrosion problems especially notable in aluminum van trailers.

Although there are some more natural alternatives to salt-based de-icers such as urea-compounds, these tend to be more costly. In this respect it is important to note that rock and most other salts are *natural* products, so the label does not mean that the products are any less corrosive than existing de-icers. **Figure 32–15** shows the result of rust damage on a steel truck frame rail.

FIGURE 32–15 Effects of rust corrosion on a steel tractor frame rail

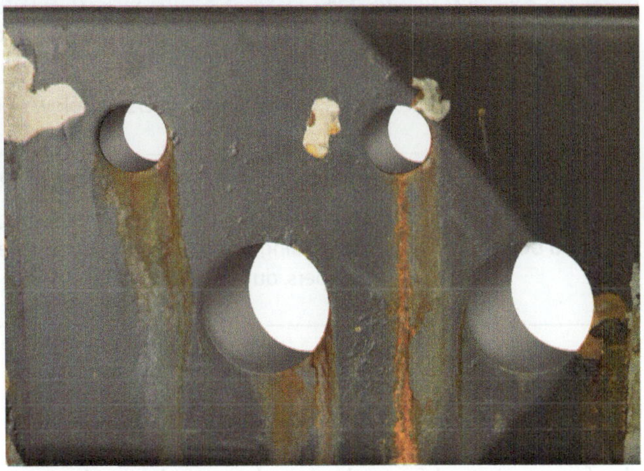

FRAME DAMAGE CATEGORIES

Full-frame damage in trucks and trailers (as opposed to localized damage in the form of fractures) can generally be categorized as follows:

- Diamond
- Twist
- Sidesway
- Sag and bow

When frame damage has been caused by a collision, it sometimes can result in more than one category of damage. **Table 32–2** shows frame damage causes and corrective actions.

Diamond

Diamond damage occurs when one frame rail has moved ahead of the other. It is a characteristic of an offset collision but also can be caused by towing abuse such as when a vehicle is ditched and hauled out with chains attached to just one side. Both frame rails usually remain parallel to each other.

Twist

Twist damage occurs when frame rails are twisted off a level plane in relation to each other. The condition is usually caused by an accident, aggressive docking, or front-end abuse such as snowplow operation. Twist damage can occur either with or without frame rail deformation. If both frame rails measure true, the problem can be confined to the cross-members.

TABLE 32–2 Frame Damage Categories

Damage Category	Causes	Corrective Action
Diamond	Offset collision/towing using one side of the frame as the anchor/snowplow	Usually frame rails remain undamaged; straighten frame and cross-members.
Twist	Collision/docking abuse/snowplow operation	Frame rails may or may not be damaged; straighten frame and cross-members if necessary.
Sidesway	Sideswipe collision or snowplowing abuse	Straighten frame replacing rails and cross-members if required.
Sag and bow	Overloading, uneven loading, weakened frame members, dump box abuse	Sag: replace frame rails. Bow: straighten frame replacing whatever parts are required.

Sidesway

Sidesway is indicated by a banana-shaped curve in the frame viewed from the top that results in the rear wheels not tracking with the front wheels. It is usually caused by a sideswipe-type collision but also can result from snowplow operation that can unequally load and shock-load the frame members.

Sag and Bow

Sag and bow frame failure are bend conditions: Sag is downward bend, bow is upward bend. Sag is usually the result of overloading but also can result when frame rails are weakened in anyway such as by loss of temper, overheating the frame, and drilling too many holes in the frame rails. Bow causes an upward bend arc of the frame and is caused by operating a dump truck with the box up and loaded, snowplow operation, and unequal loading of the frame.

Buckling and Mashing

Buckling and mashing is wrinkling of C-channel flanges that accompany different categories of frame damage. If this type of damage is severe, it is not realistic to expect frame repair facilities to correct the problem. Most OEM dealerships will replace frame rails and cross-members rather than attempt repairs even if the damage is minimal.

Frame Rail Repair

When truck frame rails become damaged, repairs should be undertaken by a specialty frame repair shop. The specialty frame shop will have the equipment to effect repairs if required and the expertise to determine whether component repair or replacement is required. After major repairs are made to a frame, a truck chassis should be aligned on a computerized frame deck.

32.5 FRAME ALIGNMENT

Both tractor and trailer frames should be properly aligned to ensure the best handling characteristics for over-the-road running as well as city operations. A bent frame can affect the directional tracking of a moving vehicle. It can throw a driveline out of alignment and set up vibrations throughout the entire vehicle. More seriously, it can damage the powertrain and cause rapid tire wear and steering problems. If a vehicle has been involved in an accident or has been overloaded, it is recommended that the frame be checked for proper alignment using the methods described later in this chapter.

No special tools are required to make an initial assessment of a frame but the job can obviously be done more quickly by a specialist frame alignment shop using the kind of equipment shown in **Figure 32–16**. For the average truck shop, an initial frame assessment can be made using a plumb bob, a chalk line, tape measure, and paper.

Figure 32–17 demonstrates the remedial principles use to correct the major categories of frame damage on a truck chassis mounted to an alignment and repair deck.

PROJECTING A FRAME DIAGRAM

When diagnosing misalignment conditions, and before disassembling components, be sure that the following items have been checked and are satisfactory:

- Tire inflation.
- Front-end alignment.
- No visual frame damage or bent axle housings.
- Proper wheel and tire balance.
- Tires and rims must be of the proper size and type with no mismatching.
- On disc wheel assemblies, the wheel discs should be the same on all wheels.
- Move the vehicle to a level floor, neutralize the suspension (see Chapter 26), and ensure that the front wheels are tracked as straight as possible.

Frame alignment can be easily checked by transferring vehicle reference points to a level floor with a

FIGURE 32–16 Features of a heavy-duty truck frame straightening deck

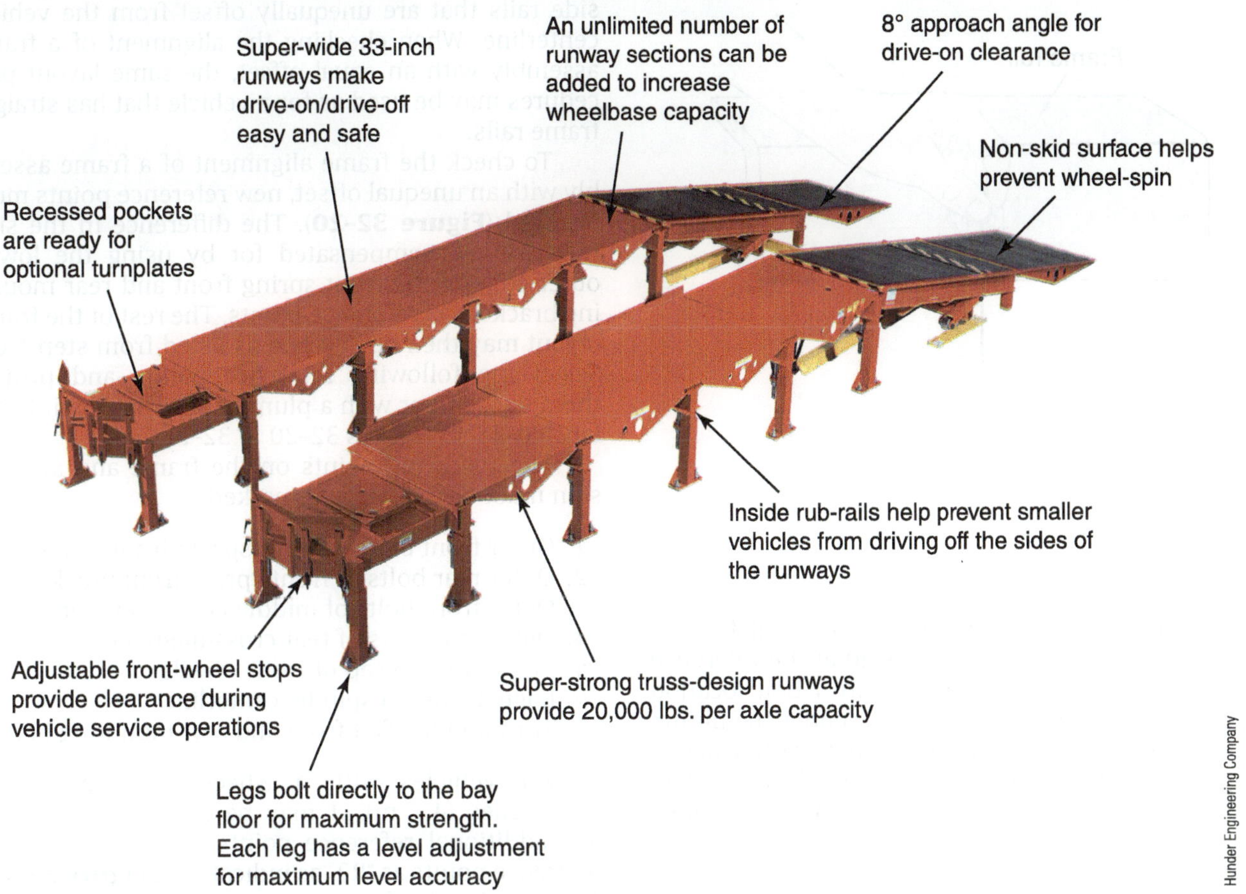

FIGURE 32–17 Four most common types of frame damage.

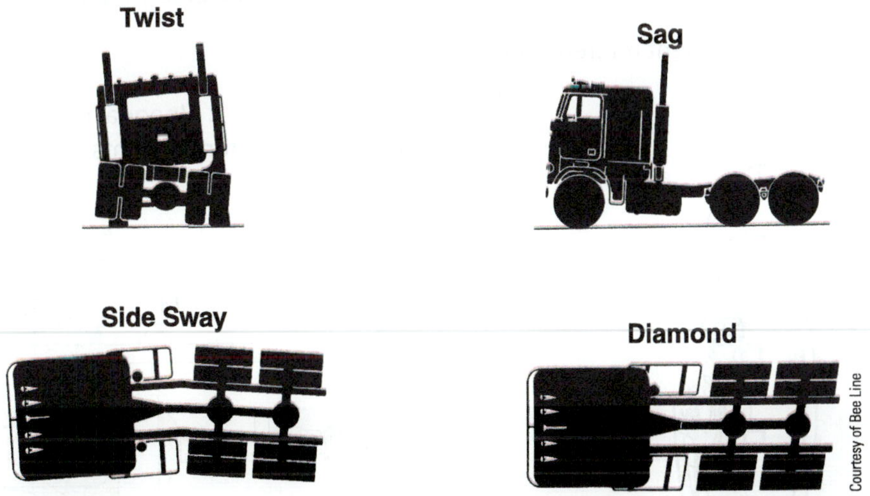

plumb bob. This procedure is known as projecting a frame diagram. You begin by hanging the plumb bob string through the trunnion and wheel hub centerlines and from the center of specified frame rail bolts, as shown in **Figure 32–18**. Do not run the plumb bob string over the nut because this will give an incorrect reading. The bolt location also must be consistent from side to side.

FIGURE 32–18 Method of hanging a plumb bob

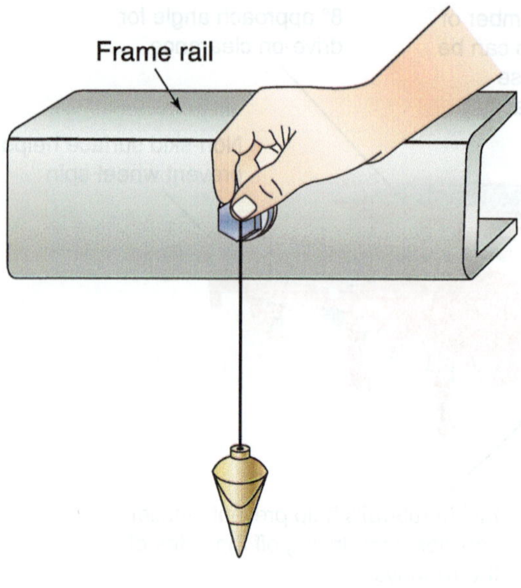

Use the exact same bolt on each rail. If for any reason a plumb bob cannot be used at the reference points indicated, use the next nearest bolt location on both sides. Place a piece of tape under the plumb bob and accurately mark the location with a pencil (**Figure 32–19**). Reference points must be accurately located and marked if the check is to be dependable; consequently, this is a two-person operation.

FRAME LAYOUT

Three basic types of side rails are found on most heavy-duty vehicles. The first type has straight side rails. Another type has side rails that are equally offset from the vehicle centerline. The third type has side rails that are unequally offset from the vehicle centerline. When checking the alignment of a frame assembly with an equal offset, the same layout procedures may be used as for a vehicle that has straight frame rails.

FIGURE 32–19 Marking alignment reference point onto the floor

To check the frame alignment of a frame assembly with an unequal offset, new reference points must be used (**Figure 32–20**). The difference in the side rails will be compensated for by using the lower, outer edge of the front spring front and rear mounting brackets as reference points. The rest of the frame layout may then be checked as listed from step 4 on. Select the following reference points and project them to the floor with a plumb bob. These points are labeled "M" in Figures 32–20 to 32–25.

The following points on the frame and suspension must be located and marked:

1. Outer front bolts of front spring front bracket
2. Outer rear bolts of front spring front bracket
3. Outer front bolts of middle cross-member
4. Outer rear bolts of rear cross-member
5. Center of hub cap of front wheels
6. Bogie trunnion spindle centerline
7. Trunnion bracket front and rear mounting bolts

For vehicles with a wheelbase greater than 184 inches (4.7 m), it may be necessary to establish additional reference points. Identify additional reference points as "R2" as indicated in **Figure 32–21**. Any additional reference points must be located at a common position on both side rails. Every precaution must be taken to ensure that any new reference points are in exact relationship; otherwise, the alignment check will be adversely affected.

FIGURE 32–20 Unequal offset side rail frame alignment reference points

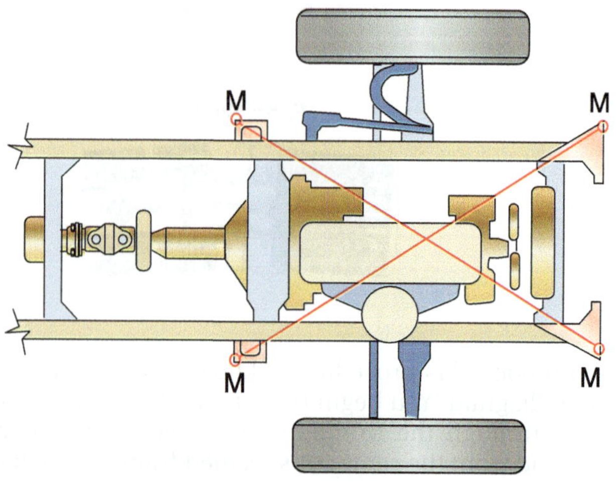

FIGURE 32–21 Straight side rail frame alignment reference points

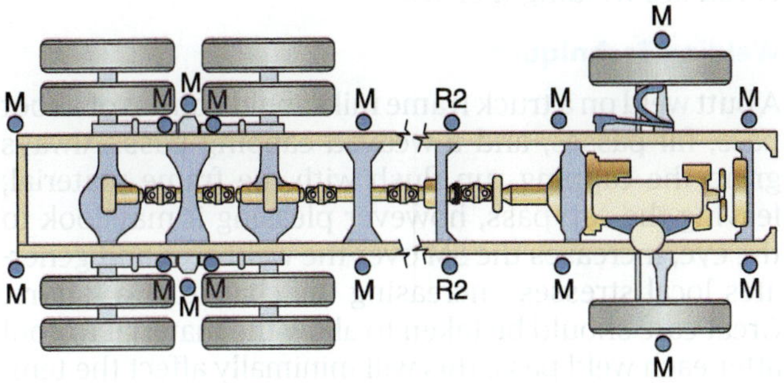

FIGURE 32–22 Connecting alignment reference points

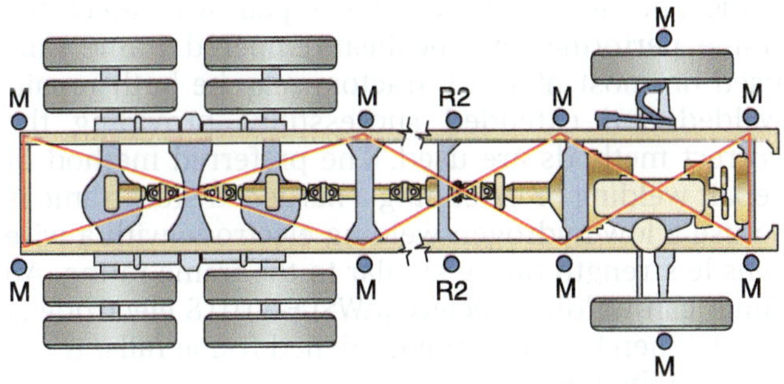

Roll the vehicle away from the points, projected to and marked on the floor. Next connect the points, as shown in **Figure 32–22**. Connecting lines can be marked by locating the ends of a chalk line on the reference points on the floor and snapping it.

Then proceed as follows:

- Bisect the line between the points described in step 1, locating the centerline (CL) at the front of the vehicle.
- Bisect the line between the points described in step 4, locating the CL at the rear of the vehicle.
- With a chalk line, connect the forward and rear CL points as shown in **Figure 32–23**.

FIGURE 32–23 Locating chassis centerline

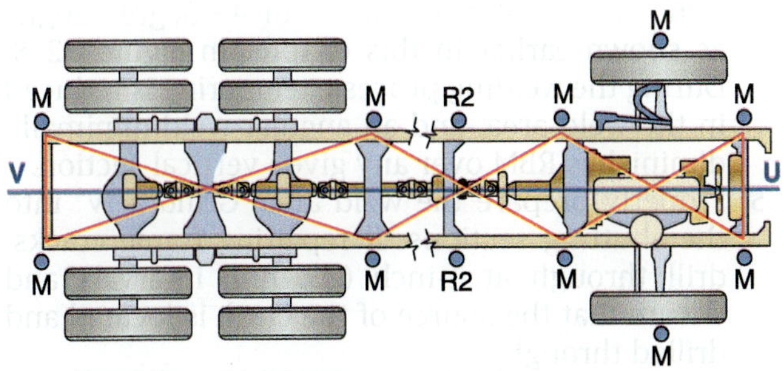

FIGURE 32–24 Checking centerline intersections

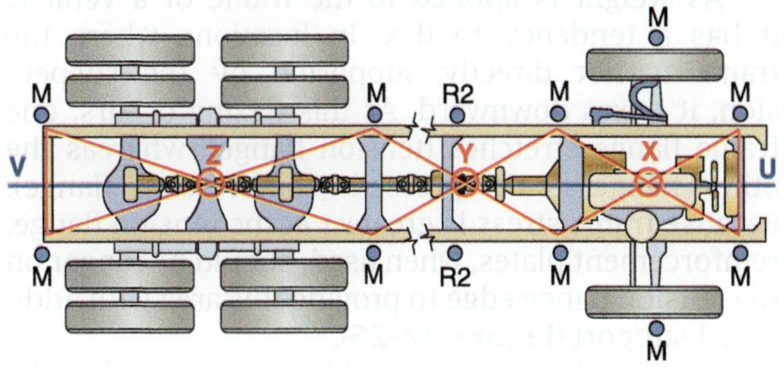

FIGURE 32–25 Comparing lengths of pair diagonals

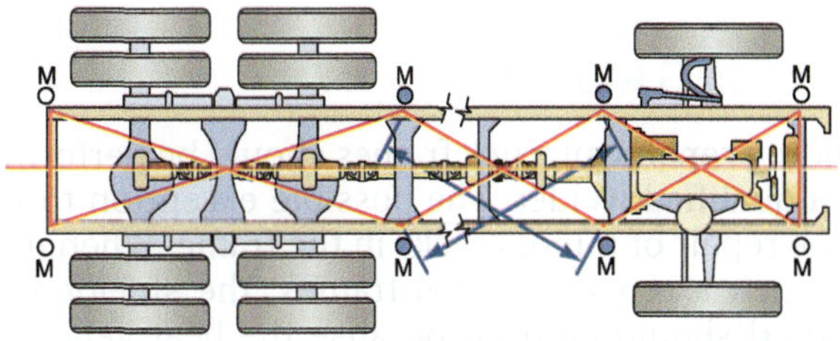

- Next, establish three intersecting points, identified as "X" in **Figure 32–24**. If the frame is true, the CL running through the three intersections X should come within ¼ inch (6 mm) of each other. Variations greater than a ¼ inch (6 mm) indicate misalignment in that area. The damaged area may be located by comparing lengths of the pairs of diagonals and noting the variation at the intersection of the CL. The lengths of each set of diagonals should not vary more than ¼ inch (6 mm), and they should intersect within ¼ inch (6 mm) of the CL, as shown in **Figure 32–25**.

32.6 REPAIR OF FRAME, SIDE RAILS, AND CROSS-MEMBERS

Neither truck OEMs nor the companies that manufacture frame side rails recommend the repair of heat-treated rails. Replacement can be considered to be the best procedure. However, side rails and cross-members are often field repaired by owners and operators to minimize downtime and expense.

To effectively repair frame side rails and cross-members you have to have some knowledge of the factors that cause frame damage in the first place. Damage is primarily caused by high stresses and usually by a local concentration of high stress. Factors that produce localized stress include abrupt change in

SM, incorrect bolt patterns, notches, cracks, improper load concentrations, and loose bolts.

As weight is applied to the frame of a vehicle, it has a tendency to flex. In locations where the frame is not directly supported by the suspension, it flexes downward. As this flexing occurs, one frame flange stretches (tension flange), whereas the other flange is compressed (compression flange). Because frame stress is greatest at the tension flange, reinforcement plates, when used, should be longer on the tension flange edge to provide this area with additional support (**Figure 32–26**).

An effective repair should compensate for the factors that caused the original damage and not introduce any factors that would produce new stress concentrations. The following are certain precautions that must be observed when making the following frame repairs.

WELDING REPAIRS

Welding repairs on truck frames should be performed by a qualified welder. The possible exception to this is the repair of minor cracks in the frame. When making such a repair on steel frames, the shielded arc method should be used because the heat generated during welding is localized and overheating is minimized. Additional advantages are that the finished weld can be ground flush and drilled as necessary.

CAUTION

Cutting and welding in a truck service facility requires strict observance of personal, shop, and equipment safety standards. Make sure that the basics of welding personal safety (see Chapter 2) are understood, fire hazards are minimized (see Chapter 2), and precautions are taken to avoid arcing damage to chassis powertrain and electrical components (this chapter).

Welding Methods

Recommended methods for crack repair of steel frame members are shielded metal arc welding (SMAW); gas metal arc welding (GMAW), also known as metal inert gas (MIG) welding; gas tungsten arc welding (GTAW), also known as tungsten inert gas (TIG) welding; or flux cored arc welding (FCAW).

FIGURE 32–26 Always make reinforcement plates longer than the tension flange edge.

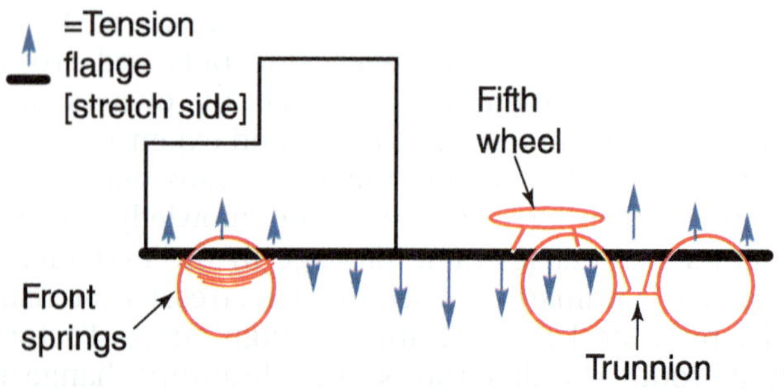

Welding Techniques

A butt weld on a truck frame rail should consist of a root pass, fill passes, and a weaved capping pass. Always grind the capping run flush with the frame material; leaving the cap pass, however pleasing it may look to the eye, increases the SM over the weld area and generates local stresses, increasing the chance of a failure. Great care should be taken to allow the material to cool after each weld pass; this will minimally affect the temper and reduce the chances of weld crystallization.

Welding Procedure

Welds on truck frames are common practice, but they should be performed by a skilled welder with some knowledge of how the repair will affect the frame performance. The heat-tempered frame rails used on most highway tractors may be both repair-welded and extended successfully, providing the correct methods are used. The preferred method of repair welding or extending a hardened steel frame is to use a low-hydrogen welding electrode with a wire tensile strength rating similar to the frame rating. An American Welding Society (AWS) E-11018 electrode is ideal. When butt welding hardened frame rails, try to observe the following:

1. Use the appropriate personal and shop safety equipment during the operation.
2. Multiple passes of a ⅛-inch (3.2 mm) arc welding electrode with a cool-down period between each pass are preferred to using a MIG technique. MIG welding is much faster. However, even if the correct tensile strength welding wire is available, the tendency is to use excess fill during a pass, increasing the chances of overheating the frame material and causing crystallization.
3. Aluminum alloy frames are best welded using a TIG welding process. Ensure that a filler wire compatible with the base material is used. MIG welding can be used, but, again, ensure that the base material is not overheated and that the filler wire is compatible with the base material.
4. When extending both heat-treated alloy steel and aluminum frame rails, ensure that added sections are cut at either a 60- or 45-degree angle, as shown earlier in this chapter in Figure 32–8. During the welding process, tempering is reduced in the weld area, and an angular weld minimally diminishes RBM over any given vertical section.
5. Properly prepare the weld area. Grind a "V" into the abutting sections. If repairing frame cracks, drill through at 1-inch (25 mm) intervals and ensure that the source of the crack is located and drilled through.

6. When the weld has been completed, grind away any profile apparent on either side of the repair so it is flush and preferably not visible. Weld profile increases SM through the weld area, and this can form localized stress and create another failure.

CAUTION

Most frame welds fail not because of a deficiency in the welding technique, but because of a lack of knowledge of frame repair methods on the part of the welder. A hardened steel frame weld that fails by cracking cleanly through the center of the weld profile often does so because the incorrect filler wire (electrode) has been used. A weld that fails by cracking clean to the sides of the weld profile often does so due to crystallization caused by overheating. Crystallization usually means that the welding procedure has been performed too rapidly.

SHOP TALK

Always identify a frame material before attempting to weld it. Most tractors use tempered frame rails with a yield strength of 110,000 psi (760 MPa) and require the use of specialty electrodes. Most trailers are fabricated of mild steel components and can be repair-welded with any general-purpose, mild steel electrodes or filler wires.

Never weld aluminum frames with general-purpose aluminum welding wire. The welds may appear to be sound, but aluminum welding wire is satisfactory only for welding nearly pure aluminums and not the aluminum alloys used in frame rails. When repair welding or extending mild steel frames, the MIG welding process is preferable. Mild steels will not crystallize and because they are not hardened, there is no tempering to destroy. This means that angular weld joints are not required through frame sections.

Figure 32–27 shows a typical frame crack. To repair a crack of this type, first "V" out the crack,

FIGURE 32–27 Preparing a rail crack for welding

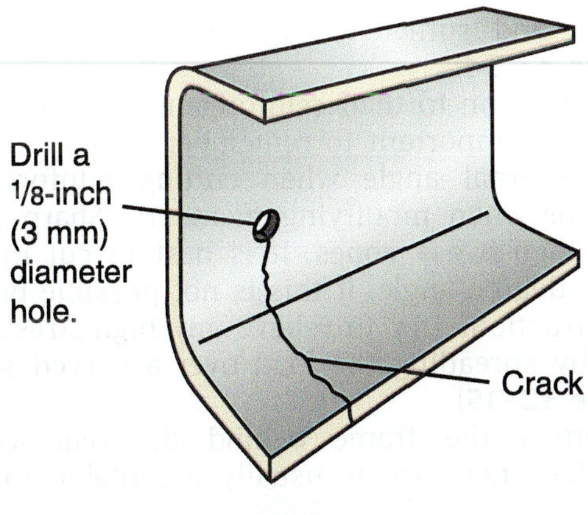

FIGURE 32–28 Grinding a V-groove

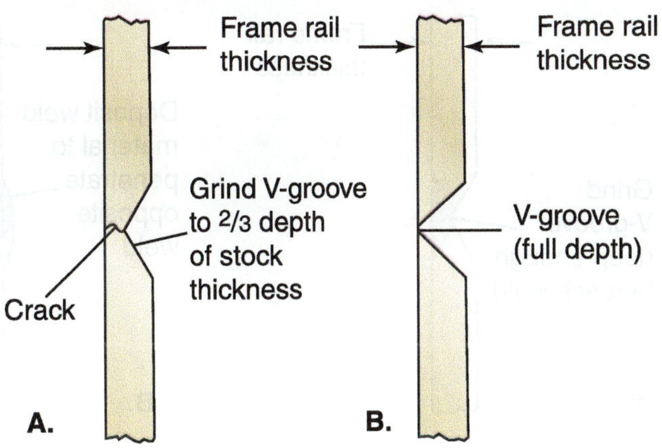

FIGURE 32–29 The first part of making a two-side frame rail weld repair. Note how the weld cap is ground flush with the rail.

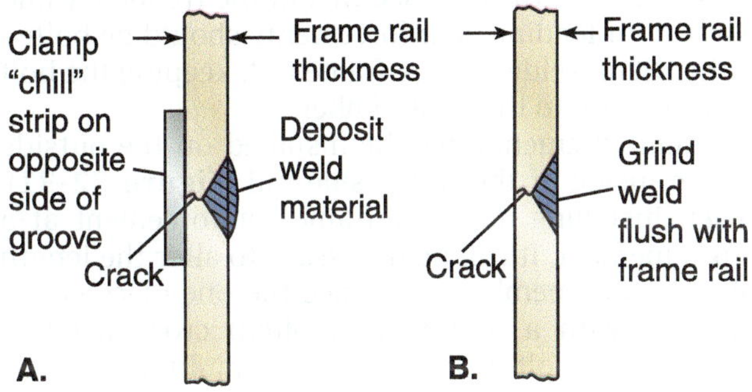

preferably from the inboard side as shown in **Figure 32–28**. Next, weld in the "V" groove as shown in **Figure 32–29**. Then "V" out the crack from the outboard side as shown in **Figure 32–30A** and weld in the "V." If the repair is being performed on a hardened steel frame rail, clamp a chill plate to the opposite side (see **Figure 32–30B**). This will help prevent overheating and crystallization. Finally, grind the weld profile flush (see **Figure 32–30C**) so the SM is not increased through the weld section. This repair method should make any reinforcement unnecessary.

SHOP TALK

Avoid reinforcing weld repairs as a matter of procedure. A frame like your backbone should be dynamic, that is, designed to flex. Think of the loss of mobility in a person who has a couple of vertebrae fused following a back injury. Reinforcing increases SM by increasing rigidity. This can generate a failure elsewhere in the frame, usually right beside your reinforcement.

FIGURE 32–30 Finishing a two-side weld on a frame rail

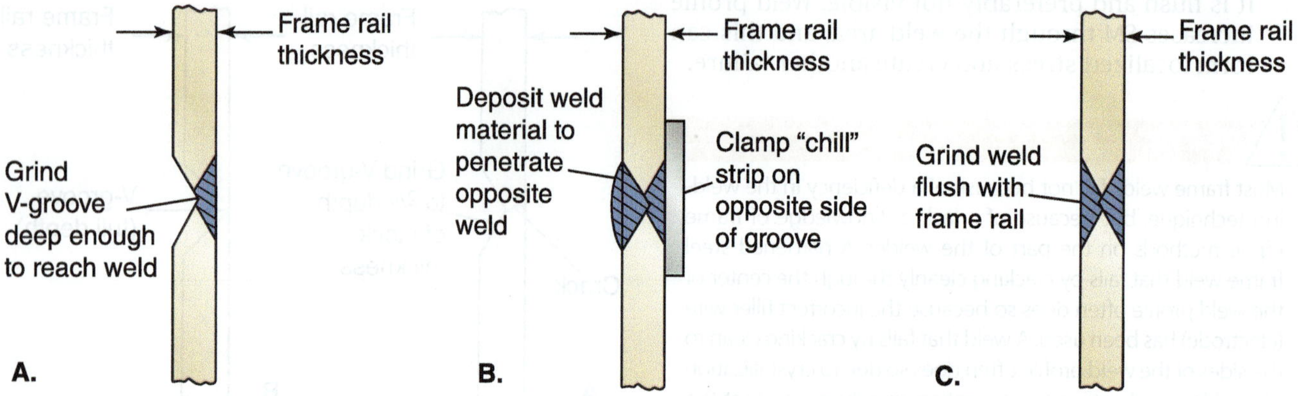

Reinforcements

Reinforcements must be made with suitable channel stock. The reinforcements should be made of a material matching the yield strength of the frame. The length of reinforcement is dependent on the type of service the truck is used in and the frequency and severity of loading. Reinforcements should be bolted rather than welded to the frame rail, keeping the bolt pattern close to the neutral fiber.

Reinforcements may be installed on the outside or the inside of the frame side rail (**Figure 32–31**). When installing an inner frame reinforcement at a cross-member, it will be necessary to alter the length of the cross-member or replace the one-piece cross-member with a multipiece (bolted) cross-member. Alteration of bolt hole locations in a multipiece cross-member may be required to suit the application.

Should it be necessary to cut the reinforcement channel, do so at a 45-degree angle. Cutting at an angle distributes the change in SM (you do not want this to be abrupt) over a greater area than a cut made at right angles to the frame (**Figure 32–32**). Reinforcement plates must be long enough to extend beyond the critical area so that the ends can be cut on an angle instead of square across the frame section.

FIGURE 32–31 Position of a frame reinforcement

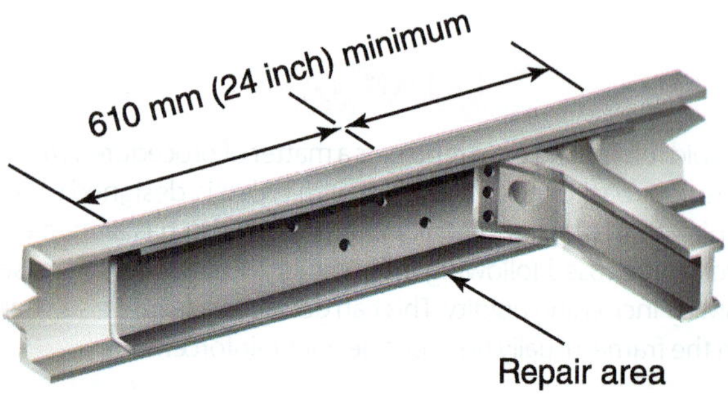

FIGURE 32–32 Reinforcement plate shaping. The angled sides prevent sudden increase/decrease in section modulus.

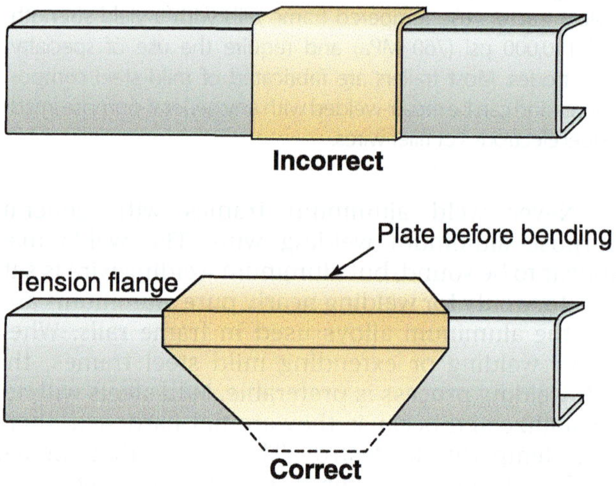

Avoid any section gaps caused by a reinforcement plate stopping short of the ends of adjacent brackets or cross-member gussets (**Figure 32–33**). Also avoid several holes in direct vertical alignment or holes too close together. A staggered bolt pattern with good spacing and sufficient edge distance is desirable (**Figure 32–34**).

In addition to their reinforcement recommendations, it is important to remember to never leave a sharp internal angle when cutting reinforcement plates or when modifying members. Sharp angles create high-stress zones. It is best to cut material at a 45-degree angle. If this is not possible because of obstructions, try to relieve any high-stress pressures by spreading the load over a curved section (**Figure 32–35**).

Cutting the frame behind the rear axle to alter the wheelbase is usually acceptable to most

FIGURE 32–33 Reinforcement plate location

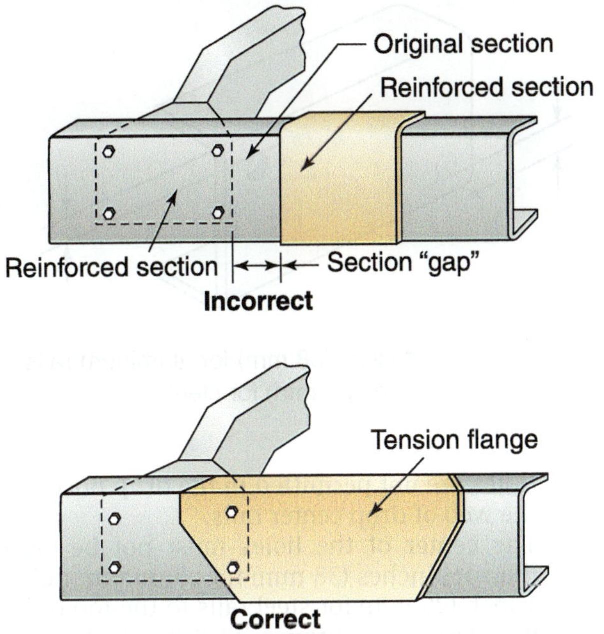

FIGURE 32–34 Reinforcement bolt spacing

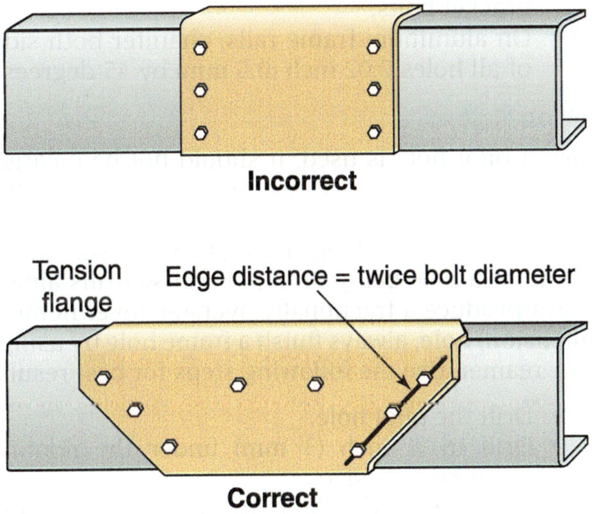

manufacturers. Consult the manufacturer regarding whether reinforcement is required when extending a frame in this manner.

SHOP TALK

When cutting a frame, use a pencil or soapstone to make all lines, points, or other marks. Try to avoid the use of a scriber or tool that will scratch the surface of the frame rail. Use a machinist's square to project all points from the webs to the upper flanges and to measure inboard from the outside face of the frame rails.

FIGURE 32–35 Reinforcement plate angle

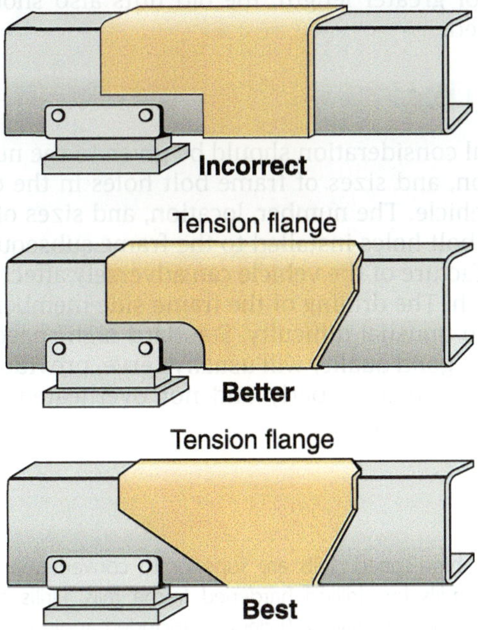

BOLTS AND TORQUE SPECIFICATIONS

Most frames today are assembled with Huck fasteners, bolts, and nuts. Others are riveted. Bolts always must be used when attaching a reinforcement. Rivets can be replaced by bolts when the frame is repaired and reinforced. Details on working with bolts and rivets can be found in Chapter 33.

In bolted joints, the majority of the load is transferred by the clamping force between the members of the joint. Bolts must be properly torqued to develop and maintain the desired clamping force. Loose or improperly torqued bolts can lead to failure of the joint. The bolts and nuts should be inspected periodically to ensure that proper torque is maintained.

Bolts and Huck fasteners must be always replaced by fasteners that meet OEM specifications. These may be SAE Grade 5 body-bound (machined with an interference-fit shank), SAE Grade 5, or SAE Grade 8. Reference Chapter 3 to make sure you understand how fasteners are identified. It is poor practice to replace Grade 5 fasteners with Grade 8 because of the different flex performance of each grade. Ensure that the nuts also meet OEM specification. Never consider replacing body-bound bolts with standard shank bolts because the result will be excess movement at the fastener shank. Body-bound bolts are machined with an interference shank designed to limit movement. Holes that are enlarged or irregularly worn may be reamed to accept the next larger bolt diameter.

If frame components are aluminum, flange-head nuts and bolts should be used, or hardened flat washers must be used next to the aluminum under both the bolt head and nut. If modification or repair requires the replacement of existing bolts with new bolts or bolts of greater length, the old nuts also should be replaced.

DRILLING

Careful consideration should be given to the number, location, and sizes of frame bolt holes in the design of a vehicle. The number, location, and sizes of additional bolt holes installed to the frame subsequent to manufacture of the vehicle can adversely affect frame strength. The drilling of the frame side-member presents no unusual difficulty. Standard high-speed steel drills of good quality will usually serve, provided they are sharpened properly and not overheated during sharpening or use.

SHOP TALK

Cobalt high-speed drills are superior to conventional high-speed drills for drilling hardened frame rails. Drills should be sharpened to give 150 degrees included angle with 7 to 15 degrees lip clearance.

As in any other drilling operation (see Chapter 3), when drilling the frame, sufficient pressure must be applied to the drill bit to maintain continuous cutting. The drill point should frequently be quenched with cutting oil to help cool the drill. Avoid letting a drill bit turn in the work without cutting. This may result in overheating and eventually damage the drill bit. The drill must be held steady during the drilling operation. Avoid wobble or change of drill angle during drilling.

Drilling Guidelines

The following are general guidelines that should be followed when drilling holes in a heavy-duty chassis frame:

- Never drill holes into the flange sections of the frame rails.
- Use existing holes whenever possible.
- Holes should be located as close to the center or neutral axis of the side-member as possible.
- Maintain a minimum of ¾ inch (19 mm) of material between holes.
- There should not be more than three holes located on a vertical line.
- Bolt holes should be no larger than is required for the size of the bolt being used.
- If reinforcements are used, avoid drilling holes closer than 2 inches (50 mm) from the ends of the reinforcement.
- Bolts must be checked periodically to ensure that the proper torque and clamping force is maintained.

FIGURE 32–36 Proper drilling distances

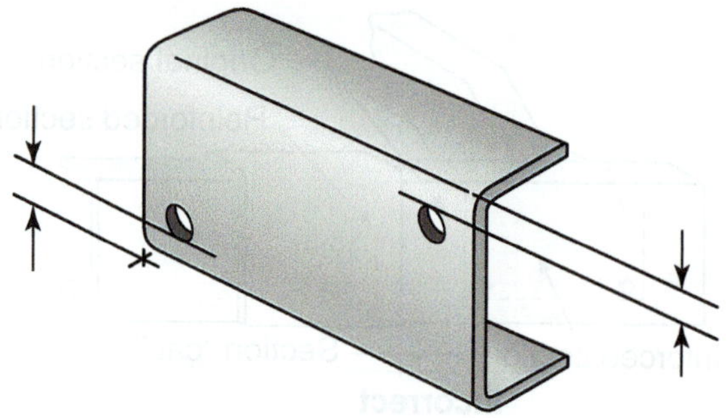

- Holes are not permitted in the drop portion of the web of drop center rails.
- The center of the holes must not be closer than 1½ inches (38 mm) for aluminum rails or 1 inch (25 mm) for steel rails to the top or bottom flange face (**Figure 32–36**) or be less than 3 inches (75 mm) apart.
- Bolt holes must not be larger than those existing in the frame, such as for suspension bracket bolts.
- On aluminum frame rails, chamfer both sides of all holes 0.02 inch (0.5 mm) by 45 degrees.

Drilling Method

When a pilot hole is used, it should not be enlarged in successive stages because rapid wear of drill bits will result. A further step to extend drill bit life is to refrain from breaking through to full size at the bottom of a hole being drilled. Because drills almost always produce a fractionally oversize (over nominal dimension) hole, always finish a frame hole by using a taper reamer. Use the following steps for best results:

- Drill the pilot hole.
- Drill to ⅛ inch (3 mm) under the nominal required hole size.
- Taper-ream the hole to the exact nominal required hole size.

AFTERMARKET MODIFICATIONS

Specialized equipment (such as a dump box, tank, hoist lift, winch, or derrick) is often added to a vehicle by the distributor, installer, dealer, or the fleet shop itself. The addition or installation of this equipment on the vehicle can significantly affect the loading of the chassis frame (**Figure 32–37**). In some cases, it may be necessary to reinforce the frame. Care must be exercised to ensure that the GVWR and/or the GAWR are not exceeded. Installation of this type of equipment may involve state and federal requirements

FIGURE 32–37 Side rail stress concentrations

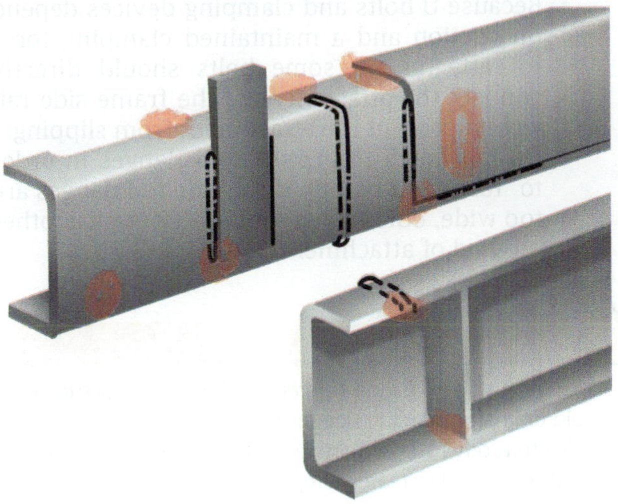

FIGURE 32–38 A channel section side rail loaded as a beam from above.

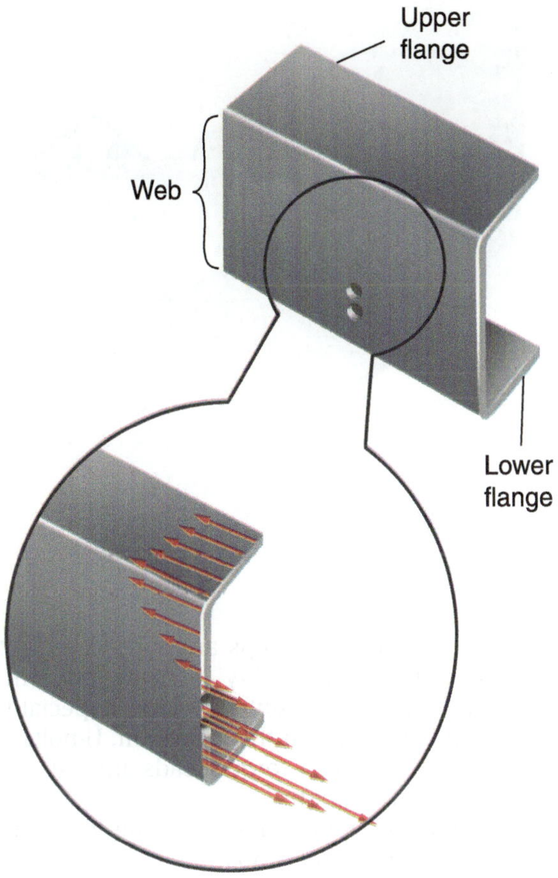

because it can affect vehicle certification for noise emissions, exhaust emissions, brake requirements, lighting systems requirements, and so on. Specialized equipment installers are responsible for the safety and durability of their product, and that Department of Transportation requirements are met.

Attaching Equipment to Frames

The three common methods of attaching specialized equipment to the frame chassis are bolts, U-bolt and clamp attachments, and welding. Bear in mind that anything you attach to a frame rail is going to significantly alter the frame dynamics.

Bolt Attachments

When the holes used for fasteners are in the least critical areas possible, bolting is preferred for their versatility and strength. **Figure 32–38** shows a channel section frame side rail loaded as a beam from above.

This type of loading stretches the lower half of the section (produces tension) and compresses the upper half of the section (produces compression), as indicated by the arrows. If the side rail section was without holes, the highest concentrations of stress (tension and compression) would be in the lower and upper flanges of the rail.

The uppermost hole is located about one-third of the web depth up from the lower flange. Stress (tension) at the bottom of this hole would be approximately equal to the stress in the lower flange. Any hole located less than this distance (for example, the hole located nearer to the lower flange in the illustration) will have significantly higher stress than will the flange. The reason for the difference in stress among the holes is that the upper hole is located closer to the neutral axis of the rail, or the area where the forces of tension and compression are lower. The following recommendations must be considered whenever equipment is to be bolted to the chassis frame:

- Use existing holes where possible. When holes must be drilled, they should be located no closer to the top and bottom side rail flanges than the existing holes placed by the factory.
- Avoid drilling holes in any area of the side rail web except the central one-third area. (The web is the surface area between the flanges.) Depending on loading, the top or the bottom of the side rail can be the tension side. Before drilling holes in the side rails, obtain approval from the OEM regional service representative.
- In all cases, avoid drilling holes in the side rail flanges.

U-Bolt and Clamp Attachment

Generally, clamping devices, such as that shown in **Figure 32–39**, are the least time-consuming and least expensive methods of attachment. U-bolts avoid frame drilling, which is especially helpful when the vehicle is equipped with a heat-treated frame.

FIGURE 32–39 Side rail clamp

But because U-bolts or clamps are not locked to the frame, there is the possibility of heavy equipment moving or being jarred out of place, especially if grease or oil is present on the frame rail. U-bolts and clamps cannot hold high-torque loads and can cause local stress concentrations.

The following guidelines must be considered when using U-bolts and clamps:

- Care should be used to block (with spacers) the side rail channel to prevent collapse of the flanges when the U-bolts are tightened. Steel blocks are preferred because wood may contract and drop out.
- The blocks used to prevent collapse of the side rail flanges should not interfere with plumbing or wiring routed along the frame rails. In addition, the blocks should not be welded to the side rail flanges.
- Because U-bolts and clamping devices depend on friction and a maintained clamping force for attachment, some bolts should directly connect the attachment to the frame side rail web to prevent the attachment from slipping.
- Do not notch frame side rail flanges in order to "force" a U-bolt fit. If the side rail flanges are too wide, obtain a larger U-bolt or use another method of attachment.

CAUTION

Notching a side rail flange severely weakens the structure and could ultimately result in side rail fracture. With complete side rail fracture, other chassis parts could be damaged, with consequent substantial increases in repairs.

Weld Attachment

There are very few cases in which welding on a heat-treated side rail is allowable. Direct welding of the side rail flanges and web should be avoided. To avoid direct welding, equipment brackets should be welded on a separate reinforcement, and the reinforcement should be bolted to the side rail. It is recommended that before any direct welding to the side rails is performed, the manufacturer be consulted. This step could keep the warranty in effect.

After a straightening job has been completed, the technician should check the repaired vehicle carefully. This check should include an inspection of each rail to make sure that there are no buckles or wrinkles. Rear housings or trailer axles must be set at a 90-degree angle to the CL of the chassis. The rear end should be checked to make sure that it is square with the chassis. Reinforcing sections should fit snugly in the old channels. Finally, a road test can help determine whether a frame was properly straightened.

SUMMARY

- The chassis frame is the backbone of all heavy-duty trucks.
- A truck frame is a dynamic component. It is designed to flex when subjected to vehicle loading and road forces. The extent to which it can flex defines the type of operation that the truck is suited to.
- The frame supports the cab, hood, and powertrain components, along with the body and payload.
- The two main components of a ladder-type frame are the two longitudinal members, which are generally referred to as *rails.*
- Ultimate frame strength is measured for comparative purposes by resist bend moment (RBM).
- RBM is factored by section modulus and yield strength.
- Section modulus (SM) concerns the shape of frame beams. High SM produces a more rigid frame. Low SM produces higher flexibility.
- Hardened steel frame rails are formed from high-strength alloy steel, quenched and tempered (heat-treated) to a minimum yield strength of 110,000 psi (760 MPa).

- Low SM frames provide greater flexibility and, therefore, more ride forgiveness. High SM provides the kind of rigidity and twist resistance required by a dump truck application.
- Most frames are available with either inside or partial inside channel reinforcements or outside reinforcements.
- Reinforcements are used to increase rigidity (SM) and provide a greater RBM than can be obtained by using a single mainframe rail.
- Cross-members are designed to connect the frame rails. They provide rigidity and strength, along with sufficient flexibility to withstand twisting and bending stresses encountered when operating on uneven terrain.
- Attachments to the frame rails should be made to the frame rail web and never to the flanges because stresses are highest in the flange areas that are subjected to tensional and compressional loads.
- The most common frame rail used on heavy-duty trucks is the C-channel design. Multiple frame rails can increase SM and RBM.
- A bent frame can decrease the control a driver has over a vehicle during an emergency and increase the chances of an accident occurring.
- Frame damage can be generally categorized as diamond, twist, sidesway, sag, and bow.
- Frame reinforcement can be channel, fishplate, or L-sectioned, but they should be of the same grade and thickness of steel as that of the frame. They should also be designed so that sudden changes in SM are avoided.
- Repair and frame extension welds should be performed by a qualified welder using the correct technique and alloy filler.

REVIEW QUESTIONS

1. Which of the following is a function of the cross-members of the chassis frame?
   a. to control axial rotation and longitudinal motion of the rails
   b. to protect wires and tubing that are routed from one side of the vehicle to the other
   c. to reduce torsional stress transmitted from one rail to another
   d. all of the above
2. The longitudinal components of a truck chassis ladder frame are called:
   a. cross-members
   b. walking beams
   c. rails
   d. gussets
3. The highest stress that a material can stand without permanent deformation or damage is known as:
   a. yield strength.
   b. resist bend moment.
   c. applied moment.
   d. bending moment.
4. When a load is applied to a C-channel frame rail from above, which of the following statements should be true?
   a. The upper flange is under tensional stress.
   b. The neutral fiber is under compressional stress.
   c. The lower flange is under tensional stress.
   d. The lower flange is under compressional stress.
5. Which of the following frame materials has the highest yield strength?
   a. mild steel
   b. high-strength, untempered alloy steel
   c. medium-strength, magnesium-silicon-aluminum alloy
   d. high-strength, copper-aluminum alloy
6. When a frame is said to have excessive flexibility and twist, which of the following would do most to repair the problem?
   a. increasing the resist bend moment
   b. decreasing the section modulus
   c. decreasing the resist bend moment
   d. increasing the section modulus
7. Which of the following causes galvanic corrosion between frame members?
   a. unevenly distributed loads
   b. dissimilar metals in direct contact
   c. overtorquing suspension brackets
   d. lack of lubricant on suspension fasteners
8. Technician A says that most highway tractors use tempered steel, C-channel frame rails. Technician B says that some aluminum alloy truck frame rails can have up to twice the yield strength of mild steel. Who is correct?
   a. Technician A only
   b. Technician B only
   c. both A and B
   d. neither A nor B
9. When projecting a frame diagram to a floor when making a frame alignment check, what tool is used to transfer the reference points on the truck chassis to the floor?
   a. a plumb bob
   b. a chalk line
   c. a measuring tape
   d. a bazooka
10. What procedure do most manufacturers recommend when a frame rail is damaged?
    a. repaired by welding
    b. repaired by fishplating
    c. reinforced with an inverted L-section
    d. replacement

11. What is the correct procedure and the material used to weld a hardened truck frame rail extension into position?
    a. MIG with general-purpose wire
    b. TIG with high alloy steel rod
    c. electric arc with AWS E-11018 electrode
    d. electric arc with AWS E-6011 electrode
12. What are the factors of resist bend moment (RBM)?
    a. tensile strength and frame length
    b. yield strength and section modulus
    c. section modulus and tensile strength
    d. load applied plus total frame length
13. When C-channel stock is used for a frame reinforcement, which of the following must be true?
    a. It should be welded directly to the frame rail.
    b. It must be of the same dimensions as the original frame rail.
    c. It must have the same yield strength as the original frame rail.
    d. It should be installed on only the outside of the frame rail.
14. When extending a frame, Technician A uses a 60-degree angle butt joint. After welding a frame extension butt joint, Technician B grinds the weld cap run flush with the frame. Who is correct?
    a. Technician A only
    b. Technician B only
    c. both A and B
    d. neither A nor B
15. Which of the following methods would do most to ensure that a frame hole was exactly ¾ inch (19 mm) in size?
    a. finish with a ¾-inch (19 mm) taper reamer
    b. finish with a ¾-inch (19 mm) drill bit
    c. blow through with a ¾-inch (19 mm), carbon-arc electrode
    d. use a hand-driven auger drill
16. Technician A says that when fifth wheel brackets are bolted to frame rails, the section modulus over that section of the frame is increased. Technician B says that installing inverted L-section reinforcements is preferred to L-section reinforcements because the section modulus is not affected. Who is correct?
    a. Technician A only
    b. Technician B only
    c. both A and B
    d. neither A nor B
17. In bolting equipment to the chassis frame, Technician A uses existing holes whenever possible. Technician B avoids drilling holes in the frame neutral fiber. Who is correct?
    a. Technician A only
    b. Technician B only
    c. both A and B
    d. neither A nor B
18. When assembling frame and suspension components, Technician A upgrades all SAE Grade 5 fasteners to Grade 8. Technician B says that replacing SAE Grade 8–rated Huck fasteners with Grade 8 nuts and bolts is acceptable. Who is correct?
    a. Technician A only
    b. Technician B only
    c. both A and B
    d. neither A nor B
19. Technician A says that dump trucks usually require a lower section modulus than highway trucks. Technician B says that tempered frame rails allow highway tractors to achieve a higher specified RBM with lighter frame rails. Who is correct?
    a. Technician A only
    b. Technician B only
    c. both A and B
    d. neither A nor B
20. A straight truck is involved in an offset collision. Which of the following types of frame damage would most likely result?
    a. bow
    b. twist
    c. sag
    d. diamond

33

Prerequisites: Chapters 1, 2, and 3

HEAVY-DUTY TRUCK TRAILERS

OBJECTIVES

After reading this chapter, you should be able to:

- Describe what is meant by semitrailers and full-trailers.
- Identify the various different tractor/trailer and train combinations.
- Describe what is meant by full frame, unibody, and monocoque.
- Explain the various types of hitching mechanisms used and their effects on tractor/trailer or train designation.
- Describe the design characteristics of the dry van, reefer, flatbed, tanker, and other types of highway trailer.
- Explain the operating principles of a reefer trailer.
- Identify the changes required for EPA tier IV (2013) compliant reefer engines.
- List the refrigerants currently used by reefer trailers.
- Outline some common trailer maintenance practices.
- Use a MUTT to diagnose trailer problems.

KEY TERMS

A-train
bridge formula
brine
B-train
conspicuity tape
containers
converter dolly
cryogenic systems
C-train
D-train
fresh air exchange
full-trailer
Global Warming Potential (GWP)
hubdometer
landing gear
lead trailer
mobile universal trailer tester (MUTT)
monocoque
multimodal transportation
pintle hook
power line carrier (PLC)
pup trailer
reefers
rivet
semitrailer
sodium chloride
tanker
tire pressure inflation system (TPIS)
unibody
upper coupler
van
vent door

INTRODUCTION

A trailer is any kind of load-carrying vehicle required to be towed by another vehicle. Trailers have no means of motive power so they are towed. A trailer may be towed directly by being coupled to a highway tractor, or indirectly by being coupled to another trailer, itself towed by a tractor. **Figure 33–1** shows some of the many types of trailers used on our roads.

Because trailers are designed to carry a load, they are often designed specifically for the load they are to haul. In addition, they are usually designed to carry a maximum amount of load while also meeting the federal, state, and provincial regulations of the jurisdictions in which they will be operating. These regulations limit maximum vehicle weight, weight over axle, unit length, and total vehicle length. The maximum allowable width of a tractor/trailer combination in North America is 102 inches (2.6 m). In most

FIGURE 33–1 Types of trailers in use today

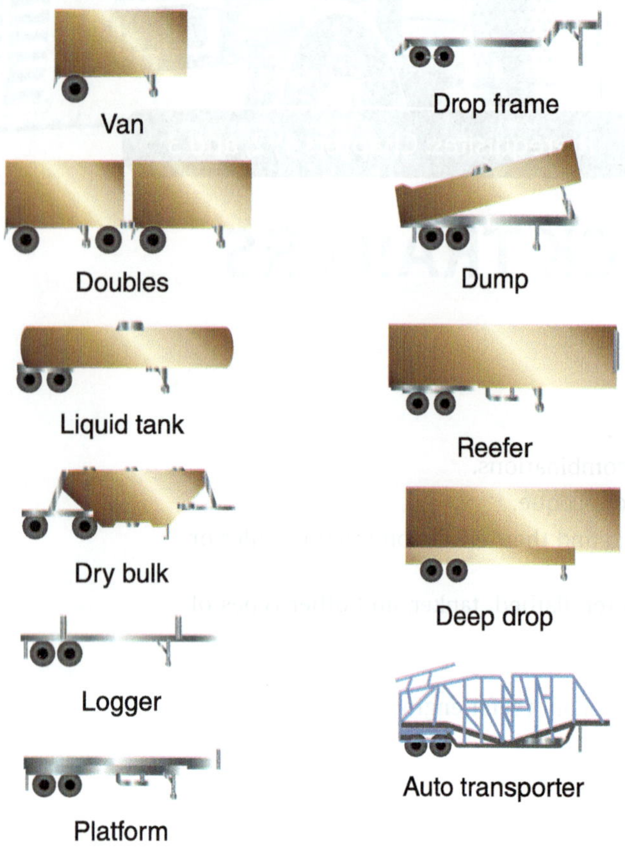

jurisdictions, special permits must be obtained on a per-trip basis if a greater width is required. Individual semitrailers are generally limited to a maximum 57 feet (17.4 m) in length but this may be less in some jurisdictions. In recent years, 53-foot (16 m) trailers have become the norm but many jurisdictions prohibit their use, for example, in New York state, 53-foot (16 m) trailers are permitted on interstates but not within most city limits. Maximum total and over-axle weights vary by jurisdiction, and the fines for load infractions can be considerable. In fact, most fleet operators use software to figure out whether they are in compliance for each jurisdiction where they are in operation.

EVOLUTION OF TRAILERS

Because of the increase in the cost of fuel in recent years and issues such as carbon emissions that directly relate to fuel economy, legislators are taking a serious look at what types of trailers to permit on our roads. Among those things being considered are:

- Increased maximum trailer lengths
- Increased maximum trailer weight loads
- Mandatory aerodynamic features
- Low rolling resistance tire and wheel assemblies

You can expect some changes to take place in most jurisdictions, especially if legislators get serious about implementing carbon emission taxes. In return, truckers will expect and will probably receive some concessions in terms of maximum weight limits and total vehicle lengths.

SHOP TALK

Trailers are usually expected to outlast the tractors that haul them by several times. Van trailers had a potential lifespan of up to 25 years, whereas the life cycle of some types of specialty trailers may be even longer. However, the recent acceptance of 53-foot (16 m) trailers by most jurisdictions has led to a rapid renewal of the national trailer fleet, so that for the first time in many years the average age of trailer fleets nationwide is lower than the average age of the tractor fleets.

33.1 TRACTOR/TRAILER COMBINATIONS

What you see running up and down our highways depends to some extent on where you live on this continent. At this moment in time, state and provincial legislation determines the weight, length, and coupling configuration allowed on the highway. If you reside on the East coast, heavier trailer loads and longer total vehicle lengths are becoming accepted as they have been for some time in western and northern jurisdictions.

SEMITRAILERS

The term **semitrailer** refers to any trailer required to be partially supported by another vehicle. The six wheels on a highway tractor convert into 10-tires (assuming there are duals on the drive axles), which in trucker lingo is *10-wheels.* A 10-wheel highway tractor is designed to haul a semitrailer. The typical semitrailer has two axles and four wheels; assuming that the four wheels are fitted with duals, this converts into eight tires. Add all this up, and you have the *18-wheeler*, a common sight on our roads as shown in **Figure 33–2**. It is often simply referred to as a tractor/trailer combination. The number actually refers to the number of tires on the vehicle rather than the number of wheels. For this reason we can describe an 18-wheeler as follows:

- Five axles: three axles on tractor, two axles on trailer
- Ten wheels: six wheels on tractor, four wheels on trailer
- Eighteen tires: ten tires on tractor, eight tires on trailer

All jurisdictions in North America permit the standard semitrailer combination as we have described it.

FIGURE 33–2 A tractor and semitrailer combination

FIGURE 33–4 Full-trailer

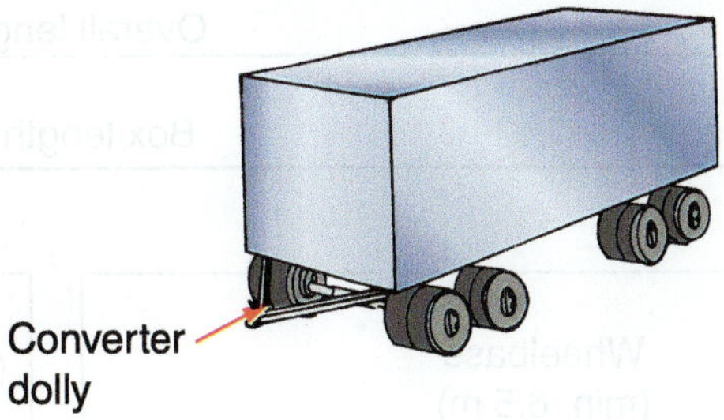

In any semitrailer combination, a percentage of the weight of the trailer is supported on the tractor fifth wheel. The fifth wheel, therefore, acts as both a coupling and weight-supporting device.

Semitraier Trains

A semitrailer can also be coupled to another trailer. You should note that this coupling arrangement is only permitted in a limited number of jurisdictions. Where it is allowed, this type of highway train consists of a tractor and two semitrailers. It is known as a **B-train**. In the B-train configuration, a tractor is coupled to a lead semitrailer by means of a fifth wheel, and a **pup trailer** is coupled to the **lead trailer**, also by means of a fifth wheel. A B-train double is shown in **Figure 33–3**.

A majority of trailers in use on our highways can be described as semitrailers. The semitrailer is always equipped with **landing gear** or dolly legs. Dolly legs are retracted when a semitrailer is coupled and lowered when de-coupled. In other words, they permit the trailer to be self-supporting when parked and stationary. The semitrailer has the ability to couple quickly to any tractor with a fifth wheel and an air brake system.

FULL-TRAILERS

The term **full-trailer** describes any trailer that fully supports its own weight. In other words, none of its weight is supported by the tow unit. **Figure 33–4** shows a full-trailer. A full-trailer could be hauled by a straight truck, or it could be part of another type of train configuration known as an **A-train**. An A-train is a highway train in which a semitrailer is coupled to a tractor by a fifth wheel, and a full pup trailer is coupled to the semitrailer by means of either a

FIGURE 33–3 B-train double combination

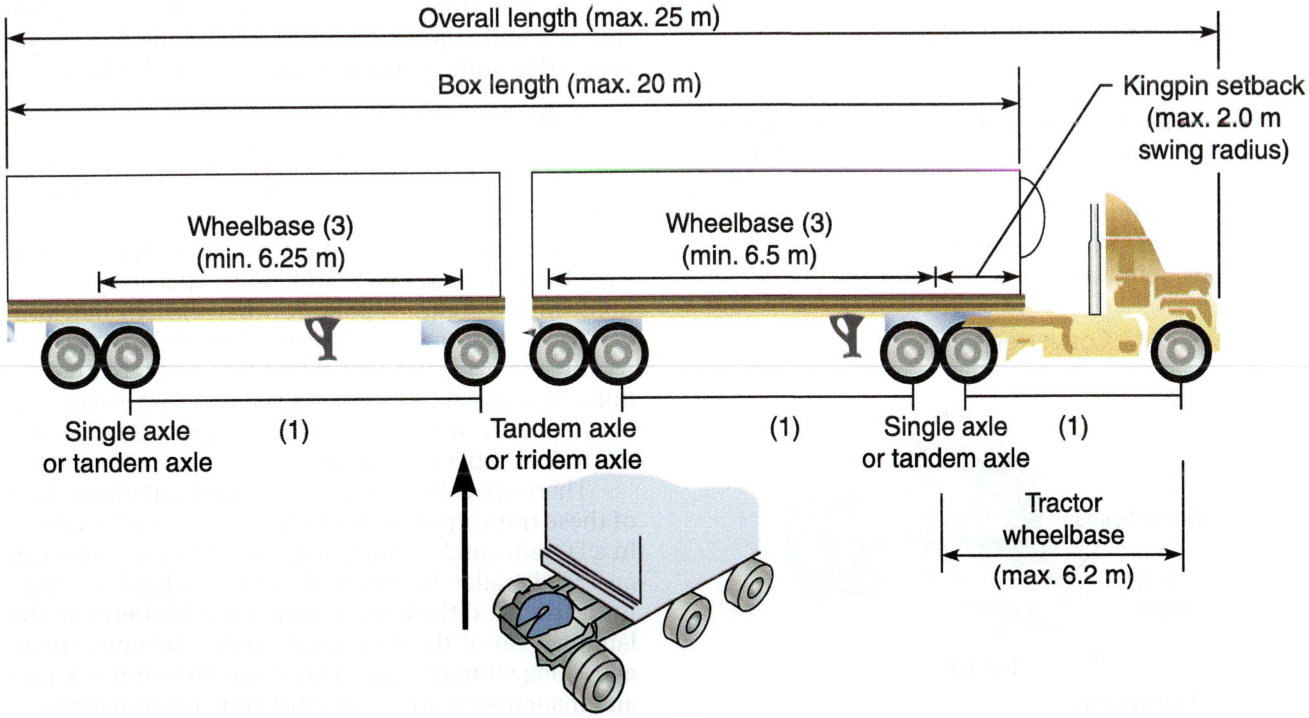

FIGURE 33–5 A-train double combination

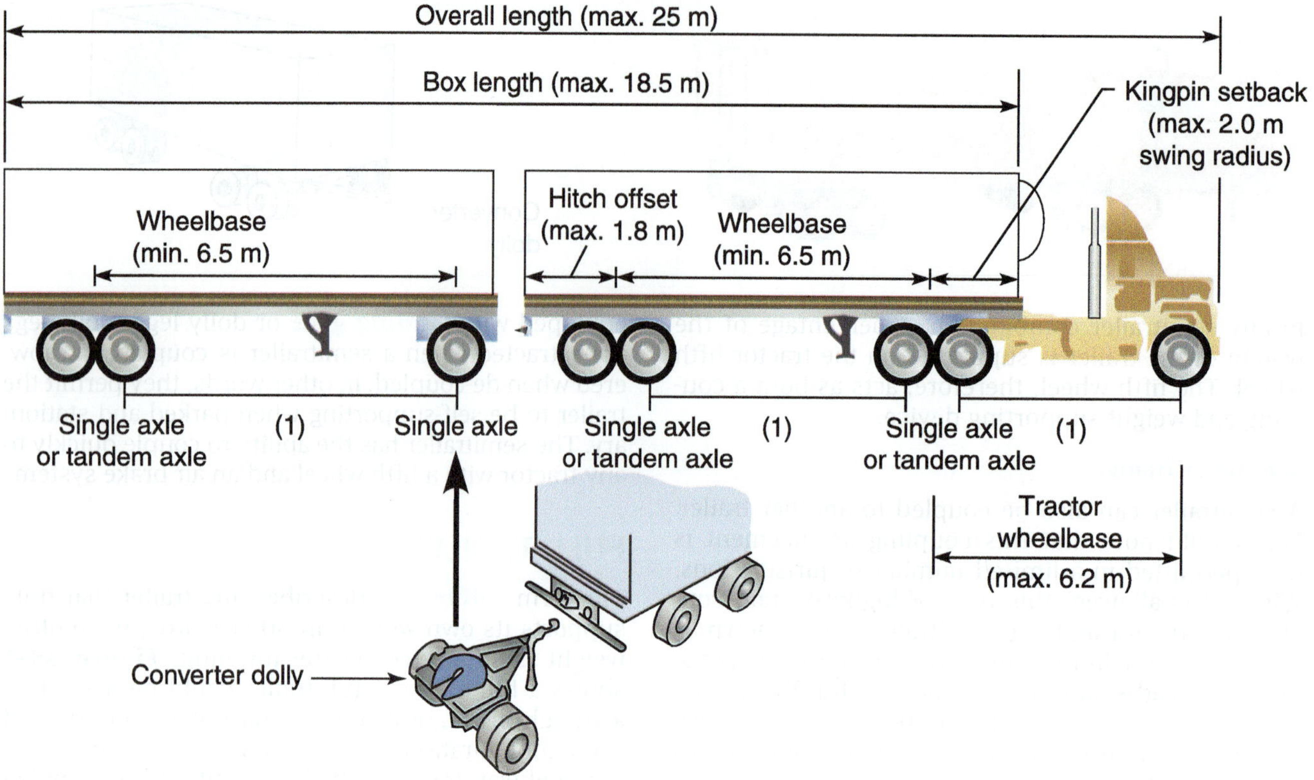

pintle and hook assembly or ball and hitch assembly. **Figure 33–5** shows a typical A-train double combination. There is fairly wide acceptance of A-train combinations throughout most states, Canada, and Mexico, but when two 53-foot (16 m) trailers are coupled in a train, there are many restrictions.

Converter Dollies

Any semitrailer can be converted to a full-trailer by coupling it to a **converter dolly**. A converter dolly consists of an axle, wheels, and frame, supporting a fifth wheel. **Figure 33–6** shows a typical converter dolly. The fifth wheel couples to the **upper coupler** on the pup semitrailer, and a tow bar with a tow eye is used to couple the converter dolly to a **pintle hook** on the lead trailer. An upper coupler is built into the trailer frame and consists of a steel plate and a kingpin that engages to the tractor fifth wheel. A pintle hook is the drawbar assembly used to couple some trailers. Both upper couplers and pintle hooks are covered in some detail in Chapter 34 of this book.

FIGURE 33–6 Converter dolly

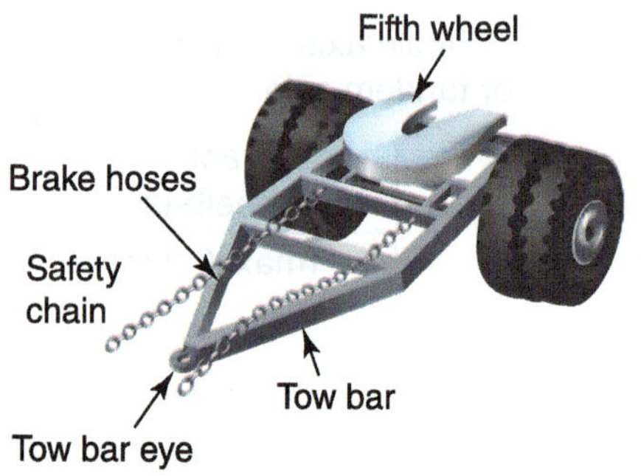

Other Truck Train Configurations

C-trains are less common but also used on our highways. Like the B-train, the C-train couples a tractor with two semitrailers. The lead semitrailer is coupled to the tractor by the usual fifth wheel coupling. However, the pup trailer is mounted to a special converter dolly with a rigid, double drawbar system that prevents any rotation about the drawbar hitch points. Any rotation that occurs will take place at the converter dolly fifth wheel. This results in a unit that is easier to back up than either A- or B-trains. **Figure 33–7** shows a C-train double configuration.

There are other types of trains in limited use. One of these train alternatives is the **D-train** configuration. In a D-train combination, articulation between the cab and lead trailer is removed; a short wheelbase tractor is used and the tractor after-frame fits between the landing gear of the trailer, converting the unit into an extra long straight truck. This is said to improve safety and maneuverability over other train combinations.

FIGURE 33–7 C-train double configuration. Note the drawbar arrangement.

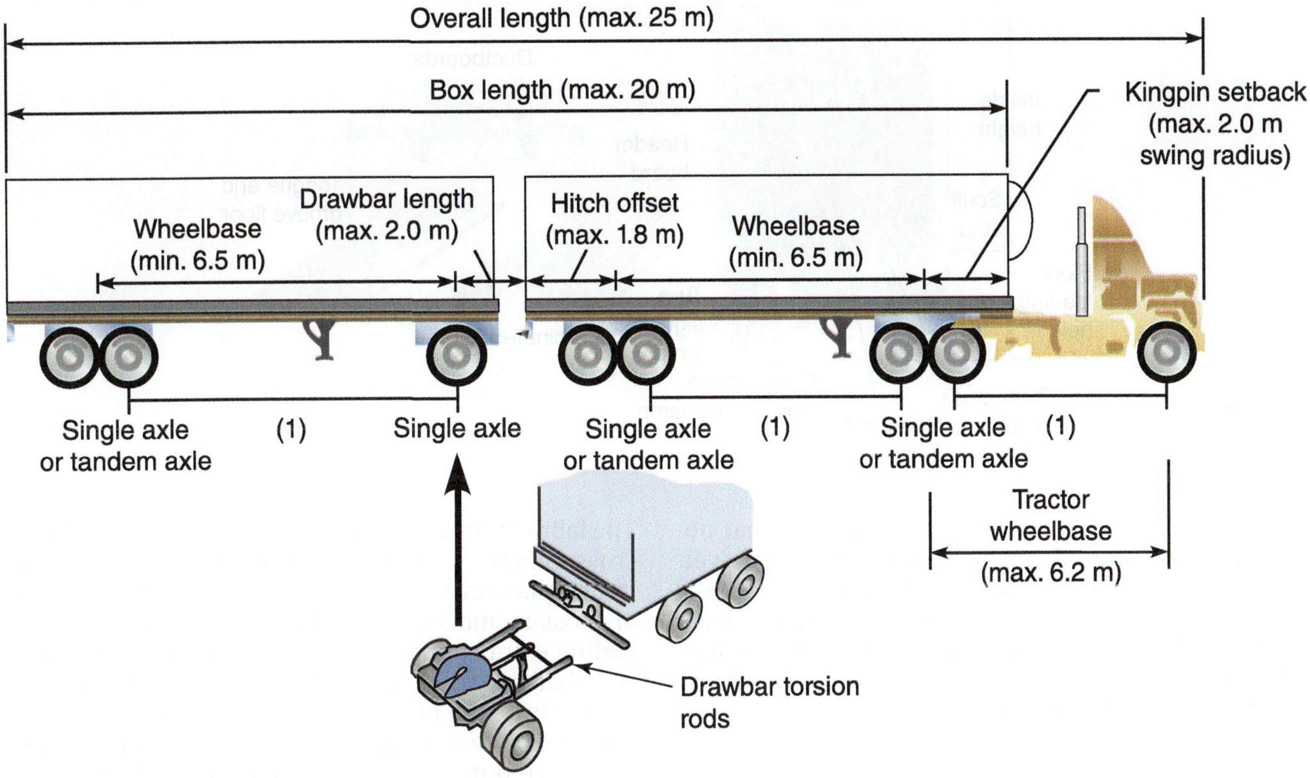

TRAILER FRAMES

Trailers use a variety of different frames. Some are built on a full frame platform such as the gooseneck platform trailer, as shown in **Figure 33–8**. This ladder-type frame has many of the same characteristics as the frame on a highway tractor or truck. However, the common construction of a **van** trailer is the **unibody** or **monocoque** design. Monocoque design is based on the shell design of a chicken's egg. The chicken's egg is unique in that it can be dropped to the bottom of the deepest ocean, withstanding enormous pressure

FIGURE 33–8 Gooseneck platform trailer with a ladder frame

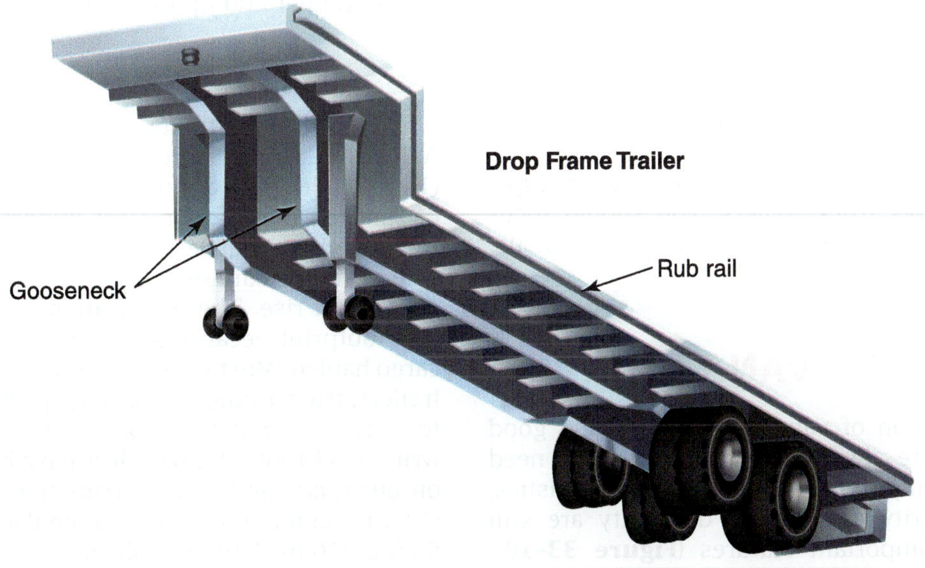

FIGURE 33–9 Unibody, van-type trailer

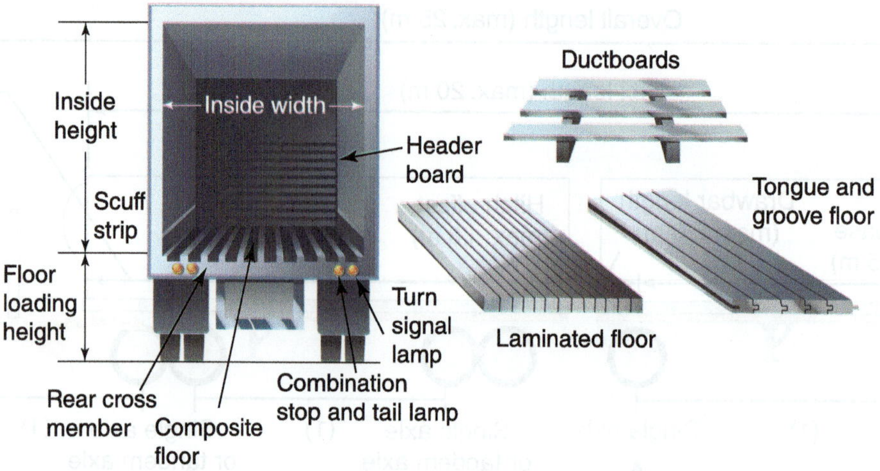

without collapsing in on itself. This is a feat that no human-manufactured device has yet achieved; yet the shell of the egg can be broken with the slightest tap of a spoon. In the unibody or monocoque frame, all the shell components form the frame; compromise any single shell component, for instance a van door, and the frame becomes weakened.

Unibody Van

The monocoque construction of a unibody van means that the van floor, roof, bulkhead, doors, and sidewalls collectively compose the frame assembly. Damage to any shell component results in a reduction of frame strength. Think of how fragile the chicken's egg shell is once it has been broken. **Figure 33–9** shows a typical unibody, semitrailer van construction.

TRAILER SYSTEMS

Trailers share many of the systems with the trucks used to haul them. For instance, there are few differences between the braking systems on truck and trailers, so to properly understand trailer brake operation, you should understand the contents of Chapters 28 through 31. The same would apply to electrical principles, axles, frames, suspensions, and coupling systems, so if you are in a program of study that specifically addresses truck trailers, you should focus on the chapters outlined in the prerequisites at the beginning of this chapter.

33.2 FREIGHT VAN TRAILERS

In the construction of vans, light weight and good aerodynamics are always sought. Despite the need for these two important van trailer characteristics, structural integrity and overall durability are still the two most important features (**Figure 33–10**). The fabrication of the subframe to include steel I-beam cross-member construction has become common in trailer manufacture. Some variables in the spacing of cross-members can be found, with most designs adding extra support in the rear. Although aluminum sheathed plywood-lined sidewalls are common, fiberglass-reinforced plywood (FRP) panels are also widely used. FRP is heavier than aluminum and usually more expensive but has high impact strength, good corrosion resistance, and a smooth finish. It is also easy to repair when damaged.

There are two ways to build lightweight trailers without sacrificing strength. One is by using lighter, more expensive materials such as aluminum and high-strength steel in posts and cross-members (**Figure 33–11**). The other is to give up some cargo space. If the ultimate in interior width and height is not needed, wall posts and floor cross-members can have a deeper section. Aluminum sheet and post vans can be lighter than FRP vans, but post thickness plus a plywood liner add up to more wall thickness.

VAN DIMENSIONS

Why are superthin trailer walls sometimes important? A 1-inch (25 mm) increase in the width of a 53-foot (16 m) trailer adds almost 40 cubic feet (1.13 cubic m) of space to the interior. More important, it can make the difference between getting two pallets side by side or not. Maximum trailer weights and lengths are on a rapid rise. The justification is the issue of carbon footprint because larger trailers represent more cargo hauled. Most jurisdictions are permitting longer trailers, truck trains, and heavier payloads in an effort to reduce overall fuel consumption. At the time of writing, 53-foot (16 m) trailers have become common on our roads and some jurisdictions permit 57 foot (17.4 m) units. In addition, some fleets are operating 53-foot (16 m) A-train doubles.

FIGURE 33–10 Frame construction of a heavy-duty unibody trailer

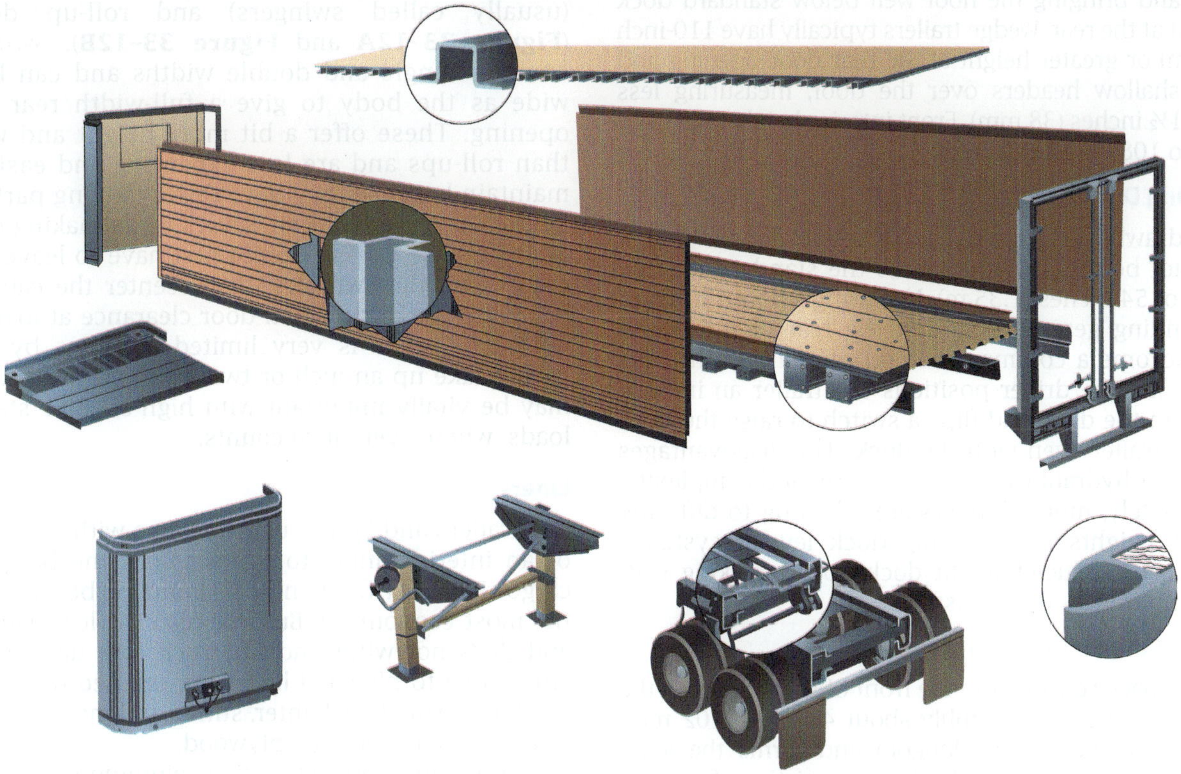

FIGURE 33–11 Typical aluminum dry van tandem axle, the most common trailer on our highways

Extending Length

Extending trailer length is perhaps the least costly way to gain cargo space. Fifty-seven-foot (17.4 m) trailers are rapidly becoming accepted by many jurisdictions across North America. On a plywood-lined sheet and post van, one way to increase van length would be to special-order an FRP front wall that met Department of Transportation (DOT) standards.

Downward Extension

Perhaps the most interesting direction for trailer expansion is down. There is no regulation against lowering the trailer floor. It makes sense to keep the roof at the legal 13 feet 6 inches (4.1 m) and stretch the design downward, because that is where the greatest cargo space can be made. Assuming a completely flat floor for ease of loading, lower limits are determined by tractor fifth wheel height at the front and trailer floor-to-tire clearance in the rear. Low-profile tires have helped at both ends, with fifth wheel heights coming down from the 48-inch (1.2 m) standard to as low as 44 or even 42 inches (1.1 and 1 m, respectively) on tractors with low-profile fifth wheel mounts and lowered frames.

Wedge Design

Further gains also can be made without tractor modifications. For fleets with standard-height tractors, trailer cargo area can be increased with a wedge design.

Such trailers have floors that slope downward from front to rear, minimizing clearance over the trailer tires and bringing the floor well below standard dock height at the rear. Wedge trailers typically have 110-inch (2.8 m) or greater height at the rear door opening and ultra-shallow headers over the door, measuring less than 1½ inches (38 mm). Front interior height is usually 108 to 108½ inches (2.7 m).

Loading Dock Height

One drawback of lowering the trailer floor is that it will not be the same height as the standard loading dock of 54 inches (1.35 m). Mounting a power hydraulic landing gear assembly at the rear of the trailer has become a common method of dealing with this problem. The driver positions the trailer an inch or so from the dock and flips a switch to raise the floor of the trailer even with the dock. The disadvantages of power hydraulics are cost, weight, and complexity. Fortunately, more shippers are adapting to differing trailer heights by installing dock leveler systems, ramps, or reduced-height dock floors, making self-leveling trailers unnecessary.

Upper Coupler Considerations

A real space consumer in the front of semitrailers is the upper coupler, an assembly about 4-inches (102 mm) thick that anchors the kingpin and forms the load-bearing surface of the fifth wheel. Consisting of top and bottom plates and internal bracing, the upper coupler assembly needs a certain amount of vertical space for structural strength. But careful design—and use of high-strength steel—can bring its height down, gaining space inside the trailer. Another space-saving technique is leaving the upper plate of the coupler exposed. This means butting the wood floor to the rear of the coupler, which in turn necessitates the use of special cross-members in the landing gear area. This modification alone can save a cubic foot of space and is usually available with both standard and low-profile couplers.

Trailer Roof

A sturdy watertight roof is mandatory to ensure dry cargo. Some trailers are available with one-piece aluminum sheet roofs. They eliminate seams where short sections are riveted together—joints that are prone to leak after a couple of years. Roof bows that protrude down from the ceiling only ½ inch (12.5 mm) need to be 6-inches (152 mm) wide for strength; they can be made lighter if allowed to take more interior height. Finally, rear door frames can be made stronger with a deep header that cuts 4 or 6 inches (102 or 152 mm) of headroom.

If the van trailer is operated in severe winter conditions, additional roof bows are often added at the rear of the trailer where snow is most likely to collect. Often, snow will slide off a building or terminal roof and land on the rear of the trailer when it is backed up to a loading dock, so the roof has to be tough enough to sustain this type of impact.

Van Doors

There are basically two types of rear doors: hinged (usually called swingers) and roll-up design (**Figure 33–12A** and **Figure 33–12B**). Swingers come in single and double widths and can be as wide as the body to give a full-width rear door opening. These offer a bit more height and width than roll-ups and are less expensive and easier to maintain because they have fewer moving parts.

Roll-ups are real time-savers when making deliveries because drivers do not first have to leave their cabs, open the swingers, then reenter the cab, and back into the dock. Often door clearance at loading/unloading areas is very limited. Roll-ups by their nature take up an inch or two of inside height. This may be vitally important with high-cube or stacked loads, where every inch counts.

Liners

Van trailers and bodies usually come with some kind of an interior lining to protect both the body and cargo. Plywood frequently is standard because it is the most economical. But it is also the least durable and does not withstand lift truck fork damage. An alternative to plywood liners are pressed-wood-fiber products, which are lighter, stiffer, and more resistant to water migration than plywood.

Although more expensive, aluminum, steel, or plastic composite materials are options, as are scuff plates. An alternative to conventional scuff plates is the bonding of additional layers of fiberglass to an FRP trailer wall when the panel is manufactured. Up to three layers can be specified to any desired height with resulting cost and weight less than that of conventional scuff liners.

Flooring

Van trailer and body floors are truly one of the critical areas that can lead to high-maintenance expense. Floors are commonly hardwood, steel, or aluminum with thicknesses from 1¼ to 1½ inches (31 to 38 mm). Some vans contain oak-laminate flooring secured with three 5⁄16-inch (8 mm) screws rather than the traditional two screws per board. In this way, it maintains a tighter floor over the years. Oak-laminate floorboards are used because they are extremely durable and capable of withstanding considerable forklift truck and skid abuse. They are either ship-lapped or tongue-and-groove fitted. In a unibody trailer, oak-laminate flooring provides considerable frame structural strength.

Cross-member spacing, especially at the rear of the trailer or body, can create problems when subjected to extensive lift truck traffic. A popular procedure is for 6-inch (152 mm) cross-member spacing, rather than the more common 12 inches (305 mm), over the rearmost 4 feet (1.2 m) of the trailer/body. On aluminum trailers, some fleets specify upright posts

FIGURE 33–12 Van doors: (A) swing doors; and (B) roll-up or overhead door

on 16-inch (406 mm) centers (rather than 24 inches [610 mm]) from the landing gear back to improve structural strength.

Other Van-Type Trailers

Freight vans are probably the most widely used trailer model in operation. In addition to high-cube and FRP designs, there are several others. **Figure 33–13** illustrates four of the more common designs:

A. **Open Top Van Trailer**. This trailer has no fixed roof but can be easily covered with a nylon or canvas tarpaulin to allow overhead loading of such items as bar steel, wood chips, refuse, and so on. Cubic capacity is similar to a dry freight van, and when a waterproof tarp is used, vehicles can even double as a dry freight van.

B. **Curtain-Sided Freight Van Trailer**. These trailers feature the easy-loading convenience of flatbeds with all the weatherproof benefits of enclosed vans. The curtains allow for full side access on both sides as well as the rear.

C. **Electronic/Furniture Van Trailers**. These trailers—originally designed for the electronic industry but now popular with furniture movers—have a drop floor starting behind the trailer landing gear, allowing for greater internal cargo capacity (approximately 3,000 cubic feet [85 cubic m]) when the maximum drop is used. They usually have one to three side doors, plus the one in the rear. Either swinging or sliding doors may be used.

D. **Doubles**. Because of the initiative to decrease the fuel consumed per ton hauled, many jurisdictions today accept trailer doubles and 53-foot (16 m) doubles. Trailer doubles greatly increase the cubic feet of the cargo that can be hauled by a single tractor and while carbon footprint reduction may have been the trigger to allow them, fleets were quick to capitalize on potential for increased profits. Doubles can use A , B , or C train couplings. A few North American jurisdictions may allow triples on certain routes or highways.

CONTAINERS

The popularity of **multimodal transportation** systems has risen exponentially in recent years. Multimodal transportation means **containers**. Once loaded, a container can be easily transported using trucks, trains, ships, and even aircraft. The contents of a container are more secure for multiple load and unload events. The popularity of containers as a mode of transportation has resulted in trailer systems designed specifically for their haulage. The recent acceptance of 53-foot (16 m) trailers on some highways has led to the introduction of the same length containers.

The biggest challenge in container haulage is keeping track of the millions of containers worldwide. Today many containers use solar-powered, electronic

FIGURE 33–13 Types of cargo trailers: (A) car hauler rig; (B) intermodal containers on flatbed trailer; and (C) integrated intermodal hauler

tracking devices; these, in conjunction with global positioning satellite (GPS) technology, simplify locating containers.

33.3 REFRIGERATED AND INSULATED VANS

Modern refrigerated trailers or insulated vans are dry freight vans that are completely insulated and have a refrigerator unit, usually located at the front or nose of the trailer (**Figure 33–14**). These trailer units are called **reefers**. Depending on outside conditions, reefer units must cool and heat to keep cargoes at proper temperatures. To reduce expensive spoilage reefers must be completely reliable, be fuel efficient, and require little maintenance.

REEFER CONSTRUCTION

The majority of insulated van trailer bodies are constructed of aluminum posts and sheets with 2 to 4 inches (50 to 100 mm) of solid foam insulation on the inside, covered by an aluminum or fiberglass inner liner (**Figure 33–15**). This arrangement works well in most cases but hot spots can be created by

FIGURE 33–14 Reefer unit mounted on a van trailer bulkhead

FIGURE 33–15 Construction features of a refrigerated van

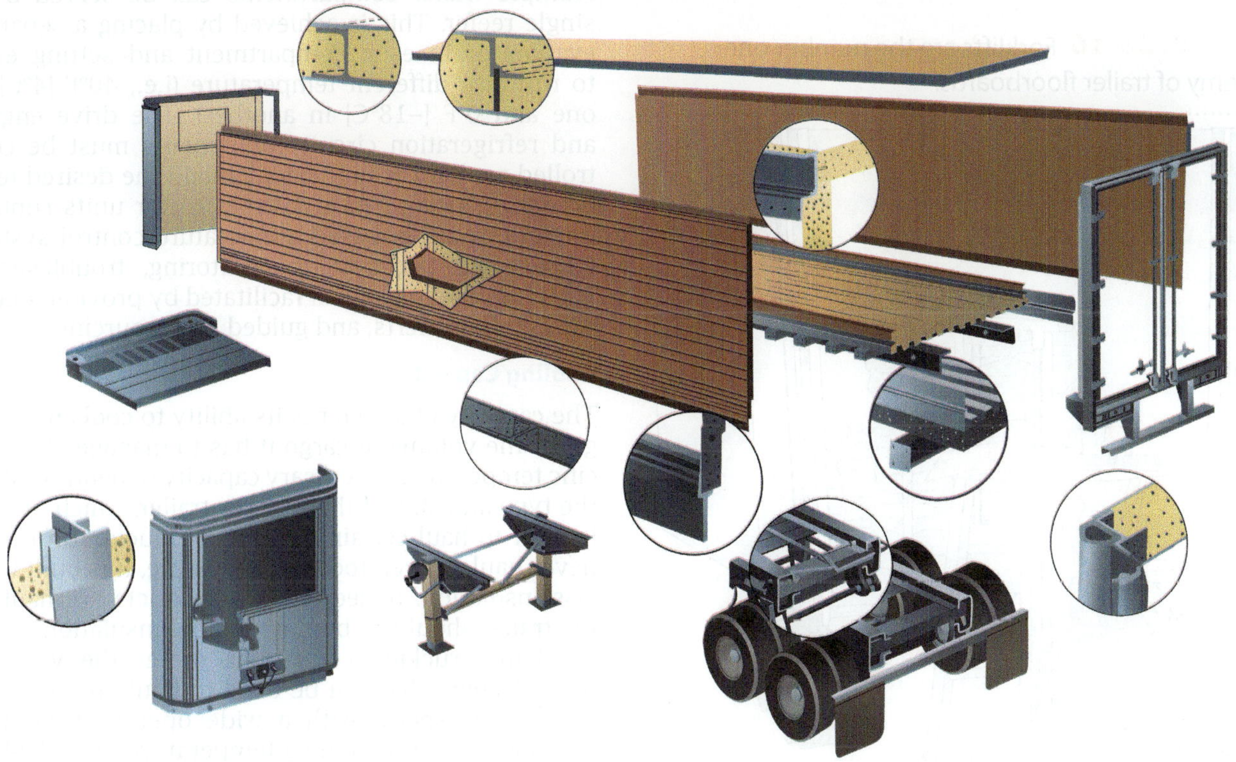

the vertical wall posts spaced every 12 to 24 inches (305 to 610 mm) along the length of the trailer. A large number of posts can compromise the insulation value of the trailer because metal is such a good heat conductor and creates a thermal short circuit.

To overcome this problem, reefer trailer manufacturers have experimented with composite fiber posts in place of aluminum. Because composite fibers are poor conductors of heat, 2-inch (50 mm) walls can insulate as effectively as 2½-inch (62.5 mm) walls using metal posts.

Most refrigerated trailers are equipped with floors made of extruded aluminum, although wood floors are still sometimes used. Aluminum is easy to clean, is not affected by moisture, and can be designed to aid air circulation. Duct-type floors provide channels 1- or 2-inch (25 or 50 mm) deep for return air, but T-bar floors can flow substantially more air volume.

T-Bar Floors

A problem with T-bar floors is their vulnerability to forklift damage (**Figure 33–16**). There is an option—a hybrid floor duct-T. Hybrid floor duct-T floors sustain abuse better than just a T-bar floor and permits better airflow than a duct floor. To minimize forklift abuse, most trailer manufacturers offer heavier gauge extrusions or reinforced floors with extra aluminum inside the hollow portions. This is usually in addition to the extra floor thickness that is standard in the rear of most trailers, where forklift trucks land off loading ramps. Under the floor is foam insulation, and most trailers have a protective sub-pan of fiberglass (the lightest-weight choice), aluminum, or plywood. A ¼-inch (6 mm) plywood floor can add over 100 pounds (45 kg) to the trailer's weight.

Vent Doors and Fresh Air Exchange

Current reefer trailers are required to be fitted with **fresh air exchange** systems that replace the energy wasteful **vent doors** found in older systems. This functions somewhat similarly to a house fresh air exchange system. The heart of a trailer fresh air exchange system is a heat exchanger; refrigerated air within the trailer is exhausted through a heat exchanger core that acts to cool incoming warm but fresh air.

REEFER OPERATION

The refrigerator unit functions on most of the basic principles as the heavy-duty air-conditioning system described in Chapter 35. The reefer unit (**Figure 33–17**) has a condenser/radiator and compressor and is usually driven by its own diesel engine. The reefer unit draws heat out of the air in the cargo compartment. Cooled air is then circulated in the compartment to maintain cargo at the desired temperature. If the reefer must heat the load, the airflow cycle is reversed. The diesel engines used in reefer units are usually electronically managed in current systems, as are the climate control circuits.

FIGURE 33–16 Forklifts are the number one enemy of trailer floorboards.

FIGURE 33–17 Interior of a trailer reefer unit

MULTIPLE COOLING ZONES

Multiple trailer compartments can be served by a single reefer. This is achieved by placing a separate evaporator in each compartment and setting each to run at a different temperature (i.e., 40°F [4°C] in one and 0°F [–18°C] in another). The drive engine and refrigeration circuit components must be controlled so that the unit can maintain the desired temperature. Because most current reefer units contain a microprocessor-based temperature control system (**Figure 33–18**), system monitoring, troubleshooting, and servicing are all facilitated by providing fault codes, alarm alerts, and guided fault sourcing.

Cooling Capacity

The capacity of a reefer is its ability to cool and heat, given the volume of cargo it has to manage at a specific temperature. Necessary capacity depends first on the type and size of the body or trailer, which in turn is built to haul certain types of products. If a trailer never hauls frozen foods, for example, it is built with less insulation; if deep-frozen products are hauled, the trailer should be built with more insulation.

Some trucking companies prefer the versatility of trailers that can be used to haul anything, so reefers are spec'd with a wide operating window and may be maintaining a temperature of 40°F (4°C) one day and –20°F (–29°C) the next. Increasing the

FIGURE 33–18 Microprocessor control console on a reefer unit

trailer cargo area increases the contained volume that has to be cooled or heated, requiring a reefer of higher capacity.

Capacity is usually measured in the British thermal units (Btu) or kilojoules that the reefer can dissipate over a period in the same way that a home air-conditioning unit is spec'd. Reefer ratings are usually rated based on an external temperature (ambient temperature) of 100°F (40°C). The three interior temperatures commonly required are 35°F (1.5°C), 0°F (–18°C), and –20°F (–29°C) because these correlate with food handling standards, but newer electronically controlled units can be set at a full range of temperature within these windows. Rated Btu cooling capacity runs between 14,000 and 25,000 Btu (14,770 and 26,376 kilojoule).

A highway 53-foot (16 m) high-cube reefer trailer with thin walls (and 1½ to 2 inches [38 to 50 mm] of insulation) might require a 25,000 Btu (26,376 kilojoule) reefer unit to maintain lower temperatures. A 45-foot (13.7 m) local distribution trailer with 2½ to 4 inches (62.5 to 100 mm) of sidewall insulation may require less, typically 14,000 to 16,000 Btu (14,770 to 16,880 kilojoule).

Transport refrigeration units are built to standards set by the American Refrigeration Institute. Part of the capacity rating equipment is the ability of a reefer to pull down the temperature of the load on startup. In hot summer conditions, this requires efficiency and high capacity.

Reefer Engines

January 1, 2013 heralded the effective compliance date for the implementation of EPA Tier 4 emissions on engines down to 8 horsepower (6 kW). As usual there is an interim provision to allow manufacturers to phase in product, but the 1980s technology used on small engines until 2013 has changed. EPA-compliant engines for trailer auxiliary power units (APUs) and reefers use diesel particulate filters (DPFs), oxidation catalytic converters, exhaust gas recirculation (EGR), turbochargers, common rail (CR) fuel systems, and computer controls. Each engine OEM may use a different range of these emissions reduction devices to achieve compliancy.

Servicing and repair of this new generation of engines falls outside the coverage of this textbook, but refrigeration technicians should note that both specialized tooling and training is required. Technicians certified to work on older hydromechanical reefer units will have to update their training. All current Thermo King (thermoking.com) reefer power units over 25 hp (19 kW) are EPA compliant and option truck stop, hotel-load electrical power hook up.

Reefer Refrigerants

The following refrigerants may be found in trailer reefer systems:

- **R-22.** Unlike CFC-based refrigerant R-12, R-500, and R-502, HCFC refrigerant R-22 is a more ozone-friendly compound. R-22 will remain available for use in service and replacement until at least 2020 under current legislative guidelines. R-22 is easy to service and recycle and is compatible with current compressors and oils. It has one of the lowest **Global Warming Potential (GWP)** ratings of any commercial refrigerant presently on the market. It is also considered a good transition alternative, but it does not do as good a job as high-CFC refrigerants in maintaining box temperatures below 0°F (–18°C).
- **R-134a.** This was the first low-ozone depletion potential (ODP) refrigerant available in product quantities. It was developed to replace R-12 and is intended for medium- and low-capacity application. It does not produce sufficient capacity for frozen cargoes in typical truck and trailer applications. However, R-134a does meet the capacity requirements of the refrigerated container industry. HCFC refrigerant R-134a in truck and trailer refrigeration is limited to chilling applications or as a component in blended refrigerants. If accidentally released, R-134a has over 1,400 times the global warming potential of a similar weight of CO_2. The proposed replacement for R-134a is R-1234yf.
- **HP-81.** This is a blended refrigerant comprising R-22, R-125, and propane; therefore, it is included in the same phase-out schedule as HCFC refrigerant R-22. Unlike other blends, it exhibits a slightly higher capacity than CFC refrigerant R-502 and has a good performance coefficient.
- **HP-82.** Like HP-81, this is also a blend of refrigerants R-22, R-125, and propane, although in different proportions. Further application testing is necessary before it can be determined whether HP-81 and HP-80 are acceptable alternatives.
- **HP-62.** HP-62 is a low-ODP, all-HFC blended refrigerant. It has performance characteristics comparable to R-502 at slightly higher pressures. It also has a higher GWP than R-22. Although full developmental testing on HP-62 is not yet complete, HP-62 may be a potential candidate for long-term replacement of R-502.
- **R-422D.** This is the most likely replacement for R-22 used in commercial mobile applications, and because it has similar properties to R-22, it can be regarded as a "drop-in" replacement for current systems. It is already being used in some reefer systems.
- **R1234yf.** This is the likely replacement for R134a in automobile A/C systems but is less likely to be used in commercial reefer systems due to high cost. If accidentally released, R1234yf has the GWP equivalent to four times a similar mass of CO_2.

- **Propane.** Used extensively in Mexico (and in most other places in the world), this has effective thermodynamic properties for use as reefer refrigerant and is low in cost. It is illegal for use as a commercial vehicle mobile refrigerant in the United States and Canada so its use is generally limited to refrigerators in mobile homes. Technicians should be aware of explosion hazards when working with propane but there are many who consider propane the most environmental friendly of currently available refrigerants.

Cryogenics

An alternative to trailer refrigeration systems is a **cryogenic system**. By using cryogenics to cool air, the system eliminates the use of both HCFCs and the diesel engine. It usually consists of a stainless steel, double-walled, vacuum-insulated cryogen tank and a small propane or alcohol tank mounted under the trailer. A heat exchanger, an evaporator unit, and various control devices are mounted on the front end of the trailer just as they are in a normal reefer unit but that varies according to which refrigerant is used. Either liquid nitrogen (N_2) or liquid carbon dioxide (CO_2) is stored in the tank. This is known as the cryogenic liquid. It is routed through an evaporator, boils into a gas, enters a heat exchange coil, and then drives a pair of cryomotors. The motors power blower fans and drive an alternator to charge the system battery. After a second pass through the heat exchanger, the gas is vented to the atmosphere. A propane burner can be used to supply heat energy to the heat exchanger when needed.

Cryogenics is not a new technology—systems were built and put into operation in the early 1970s. The system is much simpler than conventional reefer units, weighs less, and has a lower operating cost over the life of the unit. Because of the problems with mobile refrigerants cryogenics is being seriously looked at. A number of units using CO_2 and N_2 in liquid state are in trial operation. Most industry experts feel that if cryogenic systems are widely adopted, CO_2 will be the medium used. Some of the thermodynamic principles that enable cryogenics to function are covered in Chapter 35.

CO_2 Cryogenics

CO_2 has had a bad rap in our press recently because of issues impacting global warming. Natural CO_2 makes up 0.038 percent of earth's atmosphere and is mixed into the air we breathe. Our lungs expel CO_2 every time we exhale. Chemically it has a GWP rating of 1 versus 3,300 for an HCFC. In a cryogenic refrigeration system, CO_2 is both the fuel and the refrigerant. After the CO_2 boils in the cryogenic evaporator it discharges to atmosphere. It does not enter the cargo space. In current trailer reefers it is used at pressures of 150 psi (1,035 kPa) or less and is stored in tanks at around 250 psi (1,724 kPa). Typically the reefer has to carry 500 to 1,000 lb (227 to 452 kg) of CO_2 in a tank. The CO_2 is drawn in liquid form from the bottom of the tank to be routed through a controlled flow area on the way to being boiled to a vapor in the evaporator.

Advantages of CO_2 Cryogenics

Cryogenic refrigeration systems have a very rapid temperature pull-down rate, so performance can generally be regarded as superior to most current competing technologies. The CO_2 used by the system is not produced (as it is when an HC fuel is burned); it is recycled at a current cost ranging between 3 to 16 cents per pound. The cost variability depends mainly on how far away from the production facility it is purchased. Another advantage of cryogenic refrigeration is that it produces almost no noise, typically around 55 dB. This noise level compares with a value of around 72 to 75 dB for other reefers and approximately 65 dB produced by a luxury car. So what is the big hold back? Before we can make large-scale use of cryogenic reefer system a CO_2 storage and distribution infrastructure will be required. This will take some time and probably legislation to establish.

33.4 SPECIALTY TRAILER DESIGNS

In addition to vans, there are numerous other trailer types and designs. One of the more common types is the platform or flatbed trailer. Available in lengths of 42, 45, 48, 50, 53, and 57 feet (12.8, 13.7, 14.6 15.2, 16.2, and 17.4 m), these trailers are used to transport cargo that does not need protection from the elements or is physically too large to fit into a van-type trailer.

FLATBED TRAILERS

There are two basic designs of flatbed trailer: the straight platform (**Figure 33–19A**) and the low-bed trailer (**Figure 33–19B**). The platforms are designed in all-steel, aluminum, and steel/aluminum and are constructed as shown in **Figure 33–20**. Low-beds can be manufactured in lightweight, high-tensile, tempered steels (similar to that used on highway tractors) and be capable of carrying much heavier loads (up to 100 tons) than the platform design. The number of axles and tires used in each design depends on the rated load it has been built to haul.

Dump Trailers

Dump trailers are primarily used to haul aggregates such as sand, gravel, and stones but can also be used to transport a variety of solid commodities that range from soil to slag from steel mills. Some examples of

FIGURE 33–19 (A) Flatbed steel hauler; and (B) drop-nose, quad-axle, low-bed trailer

dump trailers are shown in **Figure 33–21**. Dump trailers can be manufactured of steel or aluminum and many use replaceable liners, especially when subjected to the kind of abrasive wear that occurs in aggregate haulage. The size varies according to the payloads required to be transported, and whether the units have to be operated on highways or exclusively off-highway. Some common designs are:

A. *Frame-type bathtub dumps* are a rugged design built in lengths generally from 21 to 53 feet (6.4 to 16 m). They are engineered for heavy-duty operations.

B. *Slant front bathtub dumps* are available in frame-type and frameless design. They feature an exposed hoist for easy maintenance.

C. *Rock dumps* are engineered to handle the toughest hauling jobs. The illustrated dump trailer is a heavy-duty 100,000 psi (690 MPa) yield steel and is available in two sizes: 27-foot (8 m), 20.8-cubic-yard (16 cubic m) capacity, and 30-foot (9 m), 23-cubic-yard (17.5 cubic m) capacity.

D. *Bottom hopper dump* semitrailers are generally engineered to haul up to 20 cubic yards (15.3 cubic m) of sand, dirt, or aggregates.

E. *Bottom hopper doubles* are usually available in steel or aluminum/steel. Doubles will discharge their loads in 3 to 4 seconds. This type of belly dump is required in some operations such as extreme terrain conditions for safety reasons.

F. *Side dump* trailers discharge their load from the side.

FIGURE 33–20 Construction features of a platform trailer

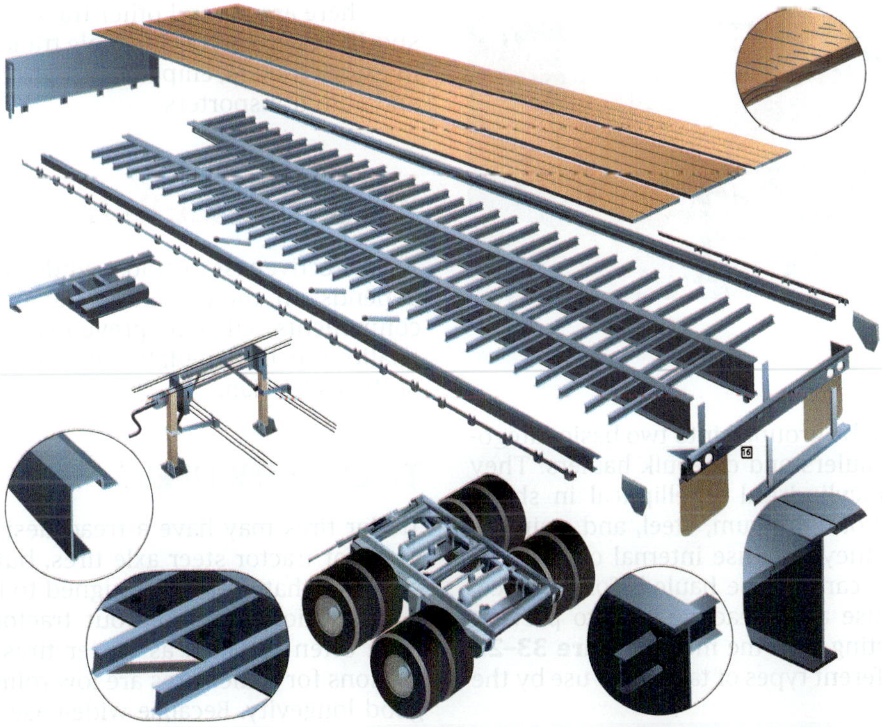

FIGURE 33–21 Selection of dump trailers: (A) B-train double dump trailer; (B) roll-off dump trailer; and (C) quad-axle dump trailer

TANKER TRAILERS

Tanker trailers can be grouped into two basic categories: liquid (wet) haulers and dry-bulk haulers. They are usually either cylindrical or elliptical in shape and manufactured of aluminum, steel, and stainless steel. In addition, they may use internal coatings or liners that suit the cargo to be hauled. For instance, acid tankers will use a nonreactive liner to prevent the acid from reacting with the metal. **Figure 33–22** illustrates four different types of tankers in use by the trucking industry.

A. *Hopper pressure dry-bulk trailers* are used to haul commodities such as dry cement, flour, and talcum powder. Most can be pressurized to facilitate loading and unloading. In most cases, the transfer medium is compressed air produced by a tractor power take-off (PTO)–driven compressor.
B. *Portable bulk storage units* are used as flexible storage containers, allowing job site mobility. The unit shown in **Figure 33–22B** has its own self-powered compressor for pneumatic discharge.
C. *Petroleum/chemical liquid tankers*, a common sight on our roads, can haul liquid commodities such as petroleum products, chemicals, milk, and vegetable oils under atmospheric pressure. Pressurized tankers can haul liquid oxygen, propane, acids, butane, and hydrogen. Petroleum tank/trailers are engineered to comply with DOT MC-306-AL specifications. Each is usually built for a specific application. For instance, to facilitate cleaning, a milk truck is required to have no internal bulkheads.
D. *Hot materials tank trailers* are insulated so that they can haul materials that would be damaged or frozen during transport; liquid asphalt has to be hauled in a hot materials tank trailer. The insulated steel tanks are designed to handle hot liquid materials up to 500°F (260°C).
E. *Live floor trailers* use a walking floor to discharge load. They are used in agriculture, construction, and refuse applications. A live floor trailer typically uses a conveyer belt system turned by drive rollers, but reciprocating floor slat system also exist. **Figure 33–23** shows a walking floor trailer that uses a conveyer belt to distribute load.

There are several other trailer types designed for specific uses, such as log/pole trailers, refuse haulers, livestock trailers, chip haulers, grain and fruit trailers, and auto transporters.

33.5 TRAILER UNDERSIDES

The effectiveness of most undertrailer components depends on three factors: the proper selection of components, efficient preventive maintenance, and quality repairs. The latter two are the responsibility of the technician.

TIRES AND WHEELS

Trailer tires may have a tread design similar to that used for tractor steer axle tires, but today it is more common that tires are designed to be trailer-specific. A generation ago, worn-out tractor steer axle tires were often recycled as trailer tires. Primary considerations for trailer tires are low rolling resistance and good longevity. Because wide-base singles lower the

FIGURE 33–22 Tank trailers: (A) dry bulk; (B) quad-axle petroleum; (C) food grade tanker used to haul liquids for human consumption; and (D) asphalt tanker

rolling resistance, they are gaining in popularity. The reasons for trailer-specific tires are as follows:

1. Longer life and more even wear due to the three-groove design and low drag, shallow tread depth.
2. Lower carbon footprint (better fuel economy) due to light weight and low rolling resistance tread geometry than standard rib tires.
3. Federal excise tax savings are possible due to lighter weight—up to 17 pounds (8 kg) lighter than comparable truck radials.
4. Engineered with retreadability in mind, casings are suitable for multiple retreads.

The maintenance requirements for tractor and trailer tires are similar so you should reference Chapter 27, which deals with tire technology in more detail. Tires are primarily defined by mounting type and tire (tubeless or tube-type), and both cast spoke, stud-piloted, and hub-piloted mounts are used.

Tire Inflation Systems

Trailers today are increasingly being fitted with automatic **tire pressure inflation systems (TPIS)**. This feature is especially desirable in trailers because they tend to suffer from more service neglect that straight trucks and tractors. Tire inflation systems are covered in detail in Chapter 27.

Trailer Brakes

S-cam 16½-inch × 7-inch (420 mm × 178 mm) brakes continue to be universal on highway trailers but some special application trailers are spec'd with disc brakes. Less common today are wedge brakes. Wedge brakes have better retarding efficiency than drum brakes (but not as good as disc brakes) but tend to be more costly to maintain, especially on trailers that are not in continuous service. Tractor/trailer brakes are essentially the same within size, brand, and gross axle weight rating (GAWR). Brake shoes

FIGURE 33–23 Trailer with a walking floor consisting of a conveyer belt on drive rollers

and linings, springs, anchor pins and bushings, rollers and retainers, camshaft hardware, and slack adjusters are similar on both tractors and trailers. The service requirements are also much the same for all brakes. Air brakes are studied in some detail in Chapters 28 and 31. Antilock brake systems (ABS) have been required on trailers since 1998 and these are studied in Chapter 30.

SHOP TALK

For optimum brake performance and safety, both wheel ends of each axle must have the same type of lining and drum equipment. If the trailer has tandem axles, both axles must also have the same type of wheel equipment. The vehicle brake-lining thickness must be the same on each brake shoe and on each side of the axle. It is a DOT requirement that when brake linings wear to ¼ inch (6 mm) in thickness or less at their thinnest point, they should be replaced.

Figure 33–24 illustrates multitrailer axle and tire arrangement. Such an axle arrangement would be used by a trailer hauling heavy loads in a jurisdiction where weight-over-axle limits were low.

FIGURE 33–24 A B-train petroleum tanker rig

Trailer Multiplexing

From the moment trailers started using ABS brakes, it became desirable for the ABS electronics to communicate with the truck data bus. When an ABS failure occurs, the failure should be broadcast to the driver in some way; this could be by using a light on the trailer mounted within the scope of the driver rear view mirror or by using a primitive method of multiplexing known as **power line carrier (PLC)**. PLC communications have been used to network trailers and tractors since 1985 but became mandatory in 2001.

PLC enables communications transactions to take place on a nondedicated communication wire. Trailer ABS requires that a warning light be illuminated in the dash of the truck in the event of a trailer ABS malfunction. Because all the wires on a standard ATA seven-pin connector between truck and trailer were already dedicated, PLC technology was used to convert a communication signal to a radio frequency (RF) signal and then superimpose it the 12-volt auxiliary power wire. A transducer converts the RF signal back to a signal that the receiver electronic control units (ECUs) can use. These signals use J1587/J1708 or J1939 hardware and protocols and are diagnosed accordingly. A full treatment of multiplexing is provided in Chapter 12.

Trailer Suspension

The suspension system of a trailer depends on the application of the vehicle. The type of cargo to be hauled, highest expected gross weight, variability of weights to be hauled, turning frequency, degree of stability required, and application (i.e., highway, city, or vocational) each play a role in determining which suspension would work most effectively.

A suspension that is too light can bottom out under jounce and result in impact damage. A rigid suspension can cause fatigue damage in addition to a rough ride for the driver and cargo. A suspension that functions ideally when subjected to full load may be too unforgiving when the unit is unloaded so the percentage of time that a trailer is to be operated unloaded is also a factor when spec'ing suspensions. Government regulations dictate some design parameters of trailer suspensions, for instance, axle load, axle spacing, and the number of axles in a group.

In recent years, air ride suspensions have become more common than steel leaf-spring suspensions in trailers. The advantage of air ride suspensions is that they are adaptive to varying load and road surface conditions. Air suspension on trailers functions similarly to those on tractors. This subject is covered in detail in Chapter 26. Other trailer suspension choices such as solid rubber mount, walking-beam, and single-point suspensions—usually have relatively narrow applications.

Bridge Formula

Trailers are often equipped with slide tandem suspensions. These increase flexibility and maneuverability and can help meet weight-over-axle limits. Ability to move the bogie on a trailer alters the **bridge formula**. In a simple bridge beam supported at two points, the positioning of the points will determine how much load is supported at each point: Keep the weight the same but move the support points, and the percentage of load supported at each point will shift. **Figure 33–25** demonstrates how the bridge formula can be altered by moving the position of the trailer bogie centerline. From this, we can say:

- When the bogies are positioned in the rearmost position on the trailer, the tractor is supporting the maximum percentage of the load.
- As the bogies are moved forward, the percentage of load supported by the tractor fifth wheel is reduced.
- If it were possible to slide the trailer bogies to the exact midpoint of the trailer, the load supported by the tractor would be reduced to zero; the entire load would be supported by the trailer frame.

FIGURE 33–25 Bridge formula: effect of sliding bogie position. (A) Tractor supports higher percentage of load; (B) as bogie is moved forward, load is removed from tractor.

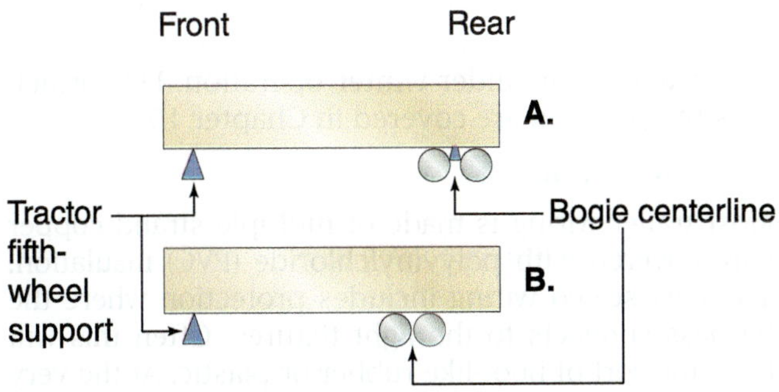

Carefully inspect the slider rail locking pins because these may be vulnerable to damage if they fail to fully retract when the tandem is released on the rails. Typically, there are four locking pins. When inspecting these units, look for full-length slide pads on the slider subframe.

SHOP TALK

One of the weight-savers now being used as an option by some trailer original equipment manufacturers (OEMs) is fiber composite springs (see Chapter 26).They can save up to 50 pounds (23 kg) per spring and can significantly outlast steel springs. Although the initial cost is around four times that of steel springs, they can last the life of the trailer, so they are ultimately an economical choice.

Landing Gear

A variety of landing gear is used on trailers depending on the weight required to be supported and the frequency to which it will be used. Landing gear types include telescoping, folding, round leg, square leg, and hydraulic. In addition, landing gear can be mounted inboard or outboard on a trailer. A typical hand-crank, gear-action landing gear is shown in **Figure 33–26**.

Most landing gear damage is caused by driver and yard jockey abuse, most of it occurring during coupling and uncoupling. Landing gear is designed to support the weight of the trailer from above and not to sustain the effects of a trailer being rammed by a tractor. Cranking the legs up too far can damage the gearing in the cranking gearbox, whereas jamming or not fully engaging the high/low-range cranking gear can fracture gear teeth. **Figure 33–27** shows a cutaway of a two-speed gear housing used on a hand-crank mechanical landing gear unit.

FIGURE 33–26 Typical trailer landing gear

FIGURE 33–27 Two-speed gearbox on a typical trailer landing gear unit

FIGURE 33–28 The worm gear that actuates the legs on hand-crank, trailer landing gear unit

Figure 33–28 shows the worm gear that is actuated by the hand-crank gearbox to raise and lower the landing legs. Trailer landing gear is commonly referred to as dolly legs.

General maintenance of the landing gear requires lubricating both legs using the zerk fittings in the legs at least every 6 months. The gear case should be lubricated by means of the zerk fittings in the gearbox. Specialty landing gear such as hydraulically actuated units should be serviced using OEM procedures.

Spec'ing Landing Gear

Landing gear must be properly spec'd out. There are four different travel heights used in North America and failure to install the appropriate travel height can result in costly trailer damage, especially in the case of replacing a single damaged unit. If requested to replace a single landing gear leg with one of similar appearance, ensure that the parts' numbers are checked for compatibility.

Lighting

Trailer lighting regulations, like those of the truck tractor requirements, are set forth in Federal Motor Vehicle Safety Standard No. 108. The lamp positioning on a van trailer that meets this standard is shown in **Figure 33–29**.

The main concerns for technicians in servicing the electrical system are the repair or replacement of defective wiring, light housings, and bulbs. Most modern trailers use a no-splice sealed wiring system, as shown in **Figure 33–30**. This, combined with the widespread adoption of light-emitting diode (LED) lighting, has greatly reduced the day-to-day electrical service problems caused by trailers.

LED lights significantly outlast incandescent bulbs and require a fraction of the electrical power to operate. The only real disadvantage of LED lighting also happens to be one of its greatest advantages: almost all of the electrical energy supplied to an LED is converted to light energy (good) but because none is converted to heat, lamp units can become obscured by ice and snow under winter operation. LED principles of operation are covered in Chapter 10.

Wiring Requirements

Most trailer wiring is made of multiple strand copper wire covered with polyvinylchloride (PVC) insulation. The best sealed wiring includes protection where the harness connects to the light fixtures. Often this will be some sort of boot-like rubber or plastic. At the very

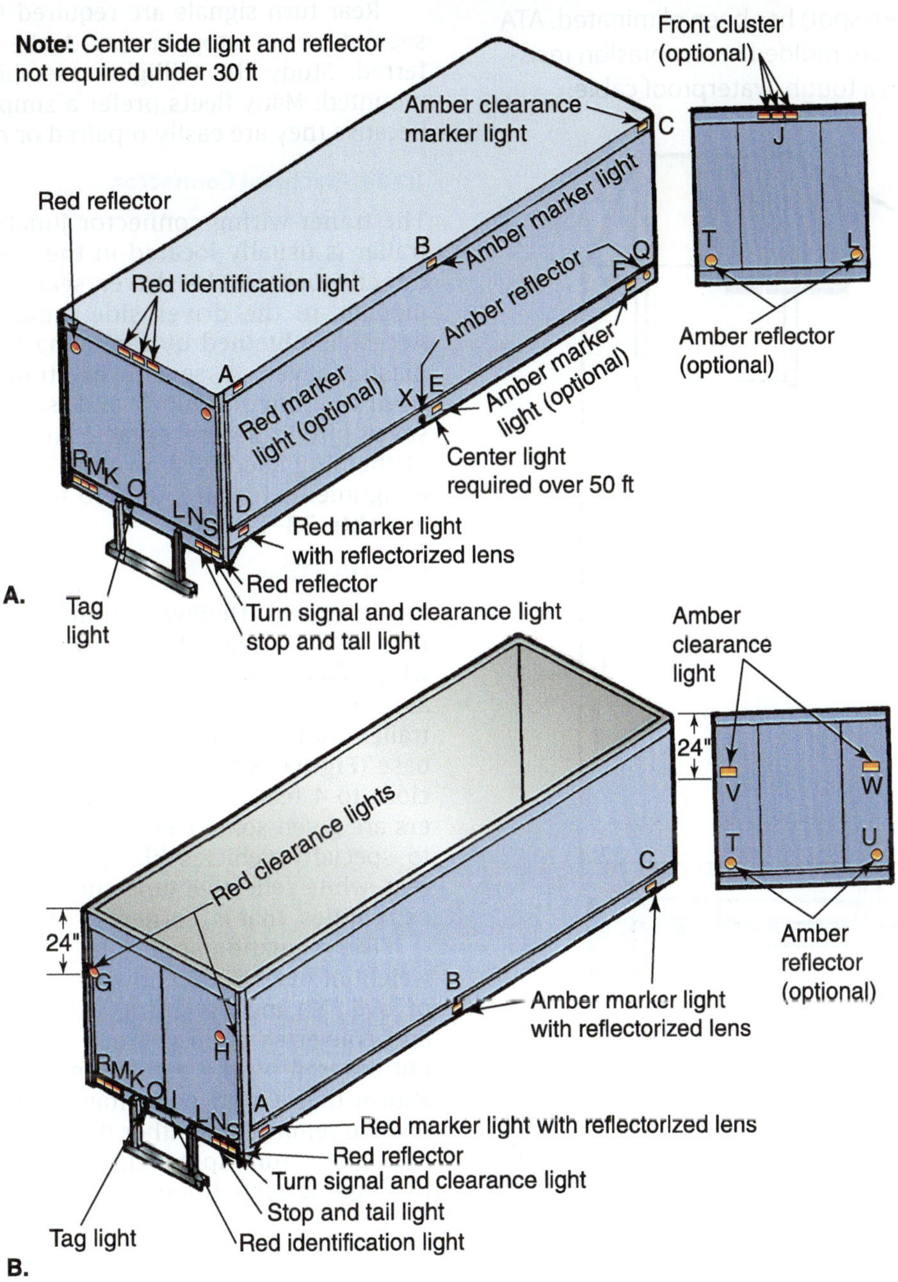

FIGURE 33–29 Lighting location on van trailers: (A) all vans except open-top; and (B) open-top vans.

least, it should include heat-shrunk coverings and *never* just plastic electrical tape or aluminum crimp couplers.

When replacing defective wiring, the factors that determine wire size are the ampere load and distance to the most distant lamp. Follow the lines or bars to arrive at the proper size of wire for this application. For a 12-volt electrical system, a 60-foot (18 m) distance combined with an 8-amp load would require a 12-gauge (3 mm) wire size. If the distance was 80 feet (24 m) or more, the technician should use 10-gauge (5 mm) wire. The Technology and Maintenance Council (TMC) of the American Trucking Association (ATA) recommends that electrical wire in trailers should never be less than 18 gauge (1 mm).

In recent years, the plug-in type of connector has become more popular because it simplifies replacement of components. Trailer harnesses using plug-in connections are generally modular, meaning sections of wiring come in set lengths and couple with other sections and lights to make up the complete system. Male-pin connectors with two or three pins are standard in lighting connectors. Sealed systems use waterproof versions of these connectors.

FIGURE 33–30 Sealed trailer wiring harness. There are no splices for corrosion to attack, and the junction box (a problem spot) has been eliminated. ATA color-coded wires are molded in an abrasion resistant jacket to form a tough, waterproof cable.

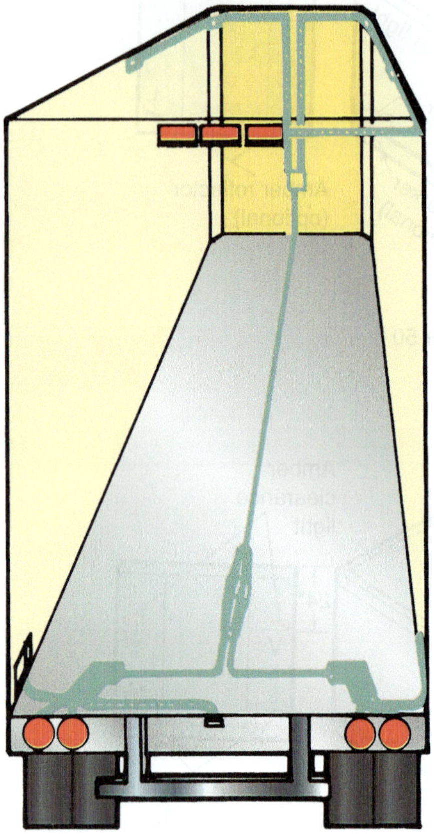

Lamp Unit Construction

Trailer lamp housings should be made of materials resistant to moisture and salt and sealed against contaminants. To lengthen the life of LED and incandescent bulb light units, the fixture should be of a shock-resistant design. Clearance and marker lamps may require protective metal brackets surrounding the lens to guard from impacts.

In a basic lamp fixture, the housing can be manufactured of plastic, aluminum, or steel. Plastic, nylon, or other nonconducting materials are used to insulate the base of the bulb from the housing. A ferrous metal housing will be more likely to rust than the nonferrous and, although nylon and plastic materials do not corrode, they are more susceptible to impact damage than steel.

Lenses are usually made of plastic in appropriate colors, meaning amber or red. Polycarbonate lens compounds can be used in applications more prone to impacts that can crack or break the lens. The slightest crack breaches the light unit sealing capability and allows moisture intrusion, resulting in corrosion, shorting, and costly failures.

Rear turn signals are required to be on a circuit separate from stop lamps. Amber lenses are preferred. Study the taillight assembly and how it is mounted. Many fleets prefer a simple bolt-on design because they are easily repaired or replaced.

Trailer Electrical Connector

The trailer wiring connector junction block for the trailer is usually located in the tractor cab, usually directly behind the driver seat, or in cab sleeper models, in the driver side luggage compartment. Access is obtained by removing the junction block plastic cover. A seven-wire trailer cord extends from the junction block and is routed through the seven-pin trailer receptacle and plug. A diagram of the plug used is shown in Figure 10–18. The pin assignments for an SAE J560 trailer plug are shown in **Table 33–1**.

Reflective Tape

The National Highway Traffic Safety Administration (NHTSA) regulations now require that red and white alternating striped reflective tape, also known as **conspicuity tape**, must be applied to the full trailer width at the rear of the trailer, near the van base (**Figure 33–31**). The tape height should be as close to 4 feet (1.2 m) as possible. But trailer builders are given some leeway, especially when it comes to specialty vehicles like auto transporters. At the rear, white reflective tape should outline the vehicle extremities, that is, corners or curves.

The regulation applies to all trailers with a gross weight of over 10,000 pounds (4,536 kg) and a width of over 80 inches (2 m). Exempt are pole trailers and converter dollies. Truck tractors or bodies are not covered by the regulation because studies have shown that a high percentage of rear-end collisions involve semitrailers rather than straight trucks. Headlight glare from approaching vehicles causes reflective tape to glow brightly at oblique as well as right

TABLE 33–1 Tractor/Trailer Plug Pin Assignments

Wire Color	Light/Signal Circuit
White	Ground return to towing vehicle
Black	Clearance, side marker, and identification lights
Yellow	Left-hand turn signal and hazard warning signal
Red	Stoplights (and ABS devices)
Green	Right-hand turn signal and hazard warning signal
Brown	Tail and license plate lights
Blue	Auxiliary circuit and power line carrier signals

FIGURE 33–31 Conspicuity marking: location of reflective tape on various trailers: (A) tanker curbside; (B) three alternate rear views; (C) rear view (left to right), platform, swing doors, and roll-up door; and (D) sideview layout

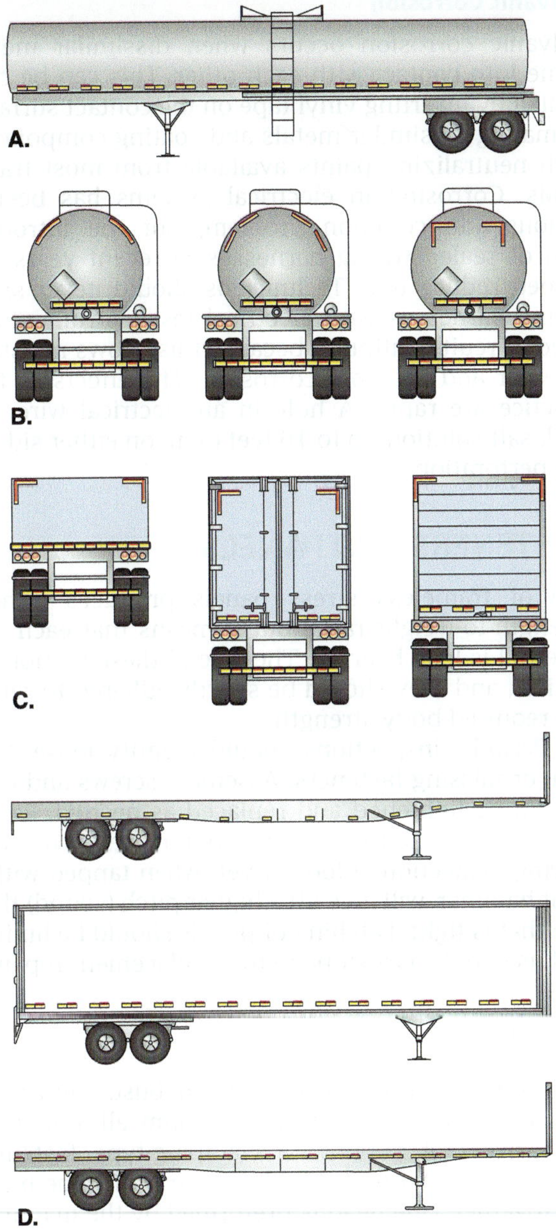

angles, making the vehicle conspicuous and less likely to be hit by an inattentive driver.

Reflective tape needs only to be applied to a clean surface to adhere. Once installed, it requires no more maintenance than occasional washing or wiping. The tape functions by reflecting headlight glare so it consumes no electrical power.

33.6 TRAILER MAINTENANCE

The maintenance of the various system components—brakes, wheel bearings, tires, suspension, and lighting systems—are discussed in earlier chapters of the book and preventive maintenance (PM) programs are described in Chapter 4. However, the following trailer-specific maintenance should be performed.

ROOF MAINTENANCE

Periodic inspection of the roof will reveal any defects before they can damage a load. Damaged areas should be cleaned and resealed with liquid neoprene, caulking, or application-specific compounds for sealing of roof leaks.

DOORS AND DOOR FRAMES

Door seals should be repaired or replaced with new doors. Bonding cement and fiberglass compounds can temporarily repair leakage. Check doors' frames for squareness. When a door frame on a unibody trailer is out-of-square, it can be an indication of more serious structural integrity problems.

SHOP TALK

The entire body shell in a unibody van trailer is the frame and if any body component is damaged the structural integrity of the frame is at risk. A slightly damaged floor or out-of-square door frame can be an indication of a much more serious problem. Take a look at the explanation of how a monocoque frame functions earlier in this chapter.

Aluminum Panels

To ensure long life and appearance of aluminum panels, a regular cleaning program should be established. The trailer should be washed with clean water to flush off surface dirt and reduce the skin temperature (in hot weather) to reduce streaking. An aluminum cleaning solution (select the OEM-recommended solution; an aluminum skin may be oxidized or nonoxidized) should be applied, usually by power washer or soak brush. The cleaning solution should be soluble and self-rinsing. Rinse off the solution, leaving no deposits.

Power washing may be necessary when equipment is infrequently washed or is covered in mud, road tar, oil, or road dirt. Take care not to indent skins by the force of the power washer jet. Improper power wash solutions may result in permanent metal discoloration and increase susceptibility to corrosion. Before power washing, check with local regulations to be sure that it is permissible. For maximum protection, aluminum panels should be waxed to protect the finish, enabling dirt and spot corrosion to be removed more readily. The most economical application requires spraying the surface with a light wax designed for this type of application.

CAUTION

Modern power wash equipment can produce high-impact water jets. Take care when using it on trailer panels and tanker skins because it can indent or cause abrasive damage.

LUBRICATION

Scheduled lubrication is key to operating a problem-free trailer fleet. There is no special skill to lubing trailer chassis, but routines should be adhered to and the recommended lubricants used. Most trailer lubrication schedules should be based on **hubdometer** (a wheel hub located odometer) mileage. The advantage of using hubdometer mileage is that it tracks mileage accomplished by unit. Lube schedules should be undertaken by calendar intervals on low mileage trailers not used regularly.

The critical lubrication points that require inspection and/or service on a trailer are:

- Suspension: covered in Chapter 26
- Wheel end: covered in Chapter 27
- Landing gear: dolly legs and gear case
- Foundation brakes: S-camshaft and slack adjusters covered in Chapter 31

CORROSION PROTECTION

Corrosion is caused primarily by environmental elements and road salt reacting with metals. Paint, besides improving the appearance of transportation equipment, also helps seal metals from the effects of corrosion. Where paint surfaces have been compromised, make a practice of touching up with the appropriate color paint. Surface preparation is a key to ensuring that paint will last. Paint cannot adhere to surfaces that are oxidized or have been coated with an oily film. Glass bead, shot, or sandblasting can be used to prepare larger areas in need of painting. Undersides of some trailers are coated with oxidation inhibitor, usually a coating that never completely hardens and self-heals when punctured. This is important when running vehicles in the north and Canada, where roads are salted and brined during the winter.

Brine Solutions

More recently, liquified salts known as **brine** have been widely used on roads and because liquid **sodium chloride** salt solution can easily mixed with other salt solutions that can lower its freezing point, it has become popular. Liquid brine applied to roads stops snow and ice from bonding to the road surface, but when the liquid is sprayed by wheels upward onto the frame and undercarriage it acts like penetrating fluid. Brine solutions have provided major problems for trailers, especially those with aluminum frames and panels. There is a more detailed description of brine corrosion in Chapter 32.

Galvanic Corrosion

Galvanic corrosion occurs when dissimilar metals come into contact with each other. This can be controlled by inserting vinyl tape on the contact surfaces of mating dissimilar metals and coating components with neutralizing paints available from most trailer OEMs. Corrosion in electrical systems has been a serious and common problem, but the introduction of sealed wiring harnesses in recent years has helped reduce this. Technicians should never spike wiring harnesses with electrical test instruments to check circuit continuity because this allows moisture to enter and promote corrosion. The effects of this practice are rapid. A hole in an electrical wire can wick salt solution up to 10 feet (3 m) on either side of the perforation.

FASTENERS AND PANELS

Use of frameless stress panels produces a high strength-to-weight ratio, but it means that each van sidewall is load bearing. The size of these sections is critical, and they should be strictly adhered to retain the required body strength.

Periodic inspections should identify loose, broken, or missing fasteners. Assembly screws and nuts should be tightened and replaced as needed. Rivets should be checked and repaired or replaced as needed. During inspection, a loose rivet, when tapped with a light hammer, will give off a higher pitch (sound) than one that is tight. Patching of panels should be limited and used only as a stopgap to a replacement repair.

RIVETING

Body skin panels on the tractor cab, buses, and some trailers are fastened using aluminum alloy or mild steel rivets. A **rivet** is a clamping-type fastening device and must have two heads to clamp the material together. One head is preformed by the manufacturer and is referred to as the manufactured head. The other end of the rivet is formed after the rivet has been driven through the material to be clamped. This head is referred to as the bucked head and is shaped when driven against a bucking bar.

A rivet set connects to an air-driven rivet gun. When the proper size rivet set is positioned on the rivet, the rivet gun delivers high-frequency hammer blows that form the bucked head when backed up against a bucking bar (**Figure 33–32**). The size of the rivet set must be the same size as the rivet.

FIGURE 33–32 How a buck rivet holds

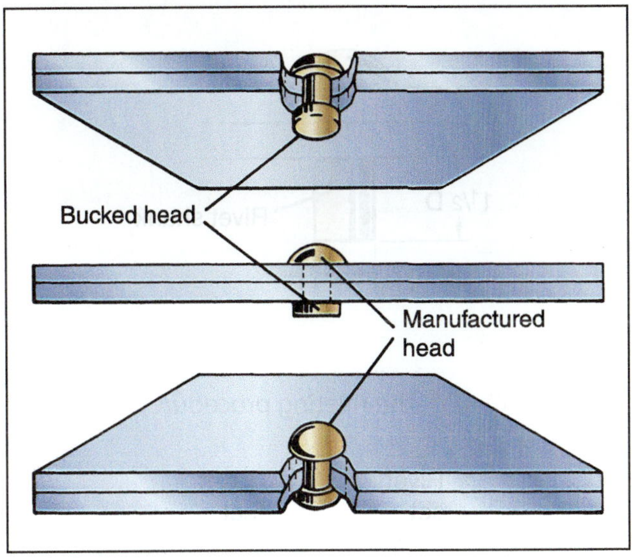

FIGURE 33–33 Brazier and flush type rivet heads

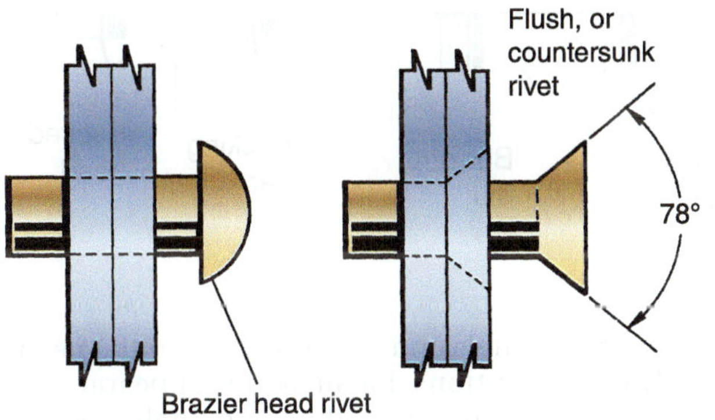

An incorrect size will damage either the rivet or the skin panels. Too large a rivet set will flatten the rivet head and mar the skin panels.

Types of Rivets

There are two methods of clamping trailer and cab panels depending on the type of rivet: brazier head and flush type (**Figure 33–33**). The rivet set used to drive the brazier head rivet is cupped to fit the convex head of the rivet. The *flush type* gets its name from the angle between the shoulder of the rivet. The rivet requires the use of a "mushroom" flush-type rivet set.

The rivet set is the link that transmits the hammer force from the rivet gun to the rivet head. The rivet set end must precisely fit the manufactured head of the rivet (**Figure 33–34**). If they do not, damage to the rivet or the material being fastened can occur and the result will appear unprofessional. A rivet set has the nominal size of the rivet it is designed to buck stamped on its shank.

FIGURE 33–34 (A) Straight rivet set; and (B) offset rivet set

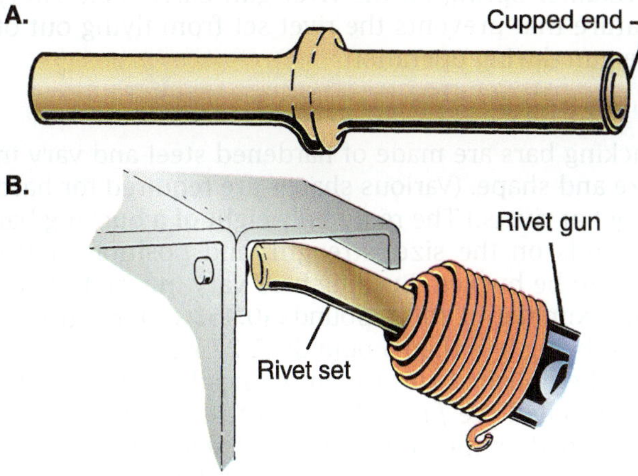

Rivet Gun

The air-driven or pneumatic rivet gun (**Figure 33–35**) provides the hammering force required for riveting. Rivet guns are made in varying sizes. Each rivet gun size has a defined range of rivet sizes that it can drive. The gun sizes are identified as follows: 2X, 3X, 4X, 5X, 9X, and so on. The larger the number, the larger the size rivet it can buck. In general, rivets $\frac{3}{16}$ inch (4.8 mm) in diameter or less can be driven with any one of the following gun sizes: 2X, 3X, or 4X.

FIGURE 33–35 A typical pneumatic rivet gun

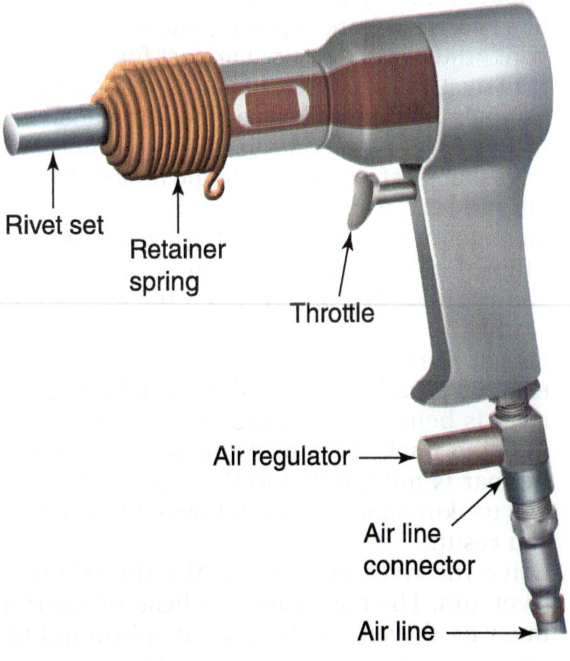

The air connector attaches the air line to the rivet gun. An air regulator adjusts the volume of air pressure to the desired impact power of the rivet gun. A retainer spring on the rivet gun barrel is a safety feature that prevents the rivet set from flying out of the gun during operation.

Bucking Bars

Bucking bars are made of hardened steel and vary in size and shape. (Various shapes are required for hard to get at rivets.) The required weight of a bucking bar depends on the size, strength, and position of the rivet to be bucked. Bucking bars vary in weight from approximately 1 to 15 pounds (0.453 to 6.8 kg) with most being less than 5 pounds (2.27 kg).

The polished surface on the head of the bucking bar is called the *face*. The face of the bucking bar is positioned on the protruding rivet shank and parallel to the panels being fastened when the rivet is ready to be "shot." This forms the bucked head of the rivet. When using a rivet gun, be sure to follow the manufacturer's instructions to the letter. Always wear hearing protection.

GUIDELINES FOR RIVETING

Rivets must be driven and bucked properly. Failure to do so results in damage to cab or trailer skin panels or rivet(s). Failure of one rivet in a row of rivets places additional clamping stress on the remaining rivets, reducing rivet strength and causing possible rivet failure or skin tears.

Riveting should be performed as follows:

1. Connect the air line to the rivet gun. Adjust the air regulator to obtain the desired impact force specified by the rivet gun manufacturer. Test the impact force by pressing the rivet set against a block of wood and opening the throttle of the gun. Adjust to the desired impact force.
2. Drill rivet holes to the specified size. Place the rivet through the drilled holes of the skin panels to be fastened. As a general rule, the projection of the rivet shank before driving should be about 1½ inch (38 mm) "D" wide, with "D" being the unbucked diameter of the rivet shank (**Figure 33–36**).
3. Align the rivet, rivet gun, and bucking bar as shown in **Figure 33–37**. Keep the line of application pressure perpendicular to the rivet head. Ensure that the face of the bucking bar is kept exactly parallel to the skin panels while the rivet is being shot because this will be formed into the bucked head. If the face of the bucking bar is not parallel to the skin panels, damage to skin panels and a deformed bucked head can result.
4. Drive the rivet by opening the throttle of the rivet gun. The manufactured head of the rivets must be driven carefully to avoid dimpling of the skin panels and "eye-browing" of the heads. These conditions blemish the appearance of the vehicle because they are highly visible after painting.
5. Inspect the rivet. The bucked head of the rivet should be 1½ "D" wide, and ½ "D" thick, with "D" being the original diameter of the rivet shank (**Figure 33–38**).

FIGURE 33–36 The correct rivet shank length: D = rivet diameter.

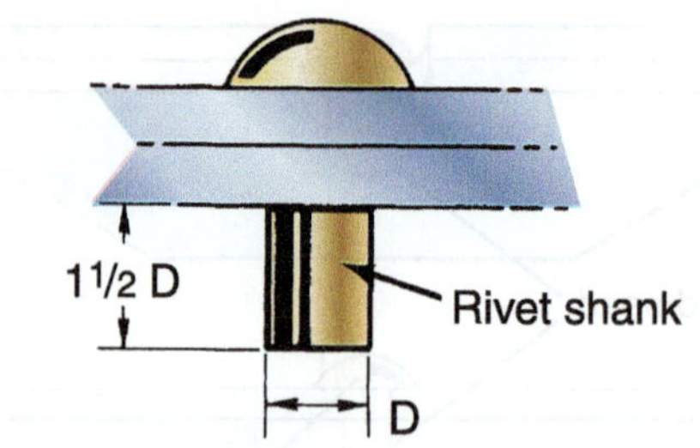

FIGURE 33–37 The riveting procedure

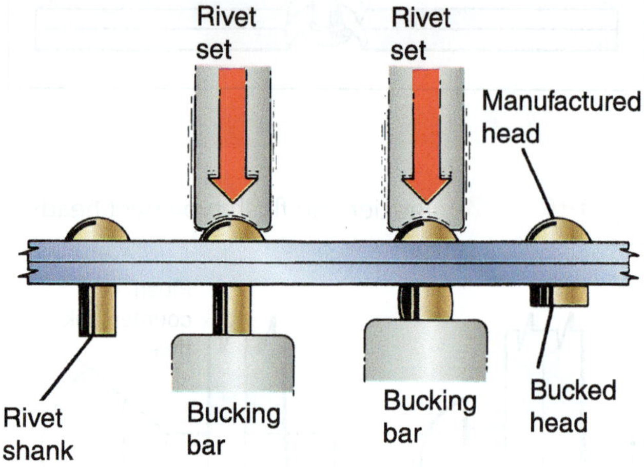

FIGURE 33–38 Correct bucked head sizes: Note that "D" equals the diameter of the rivet.

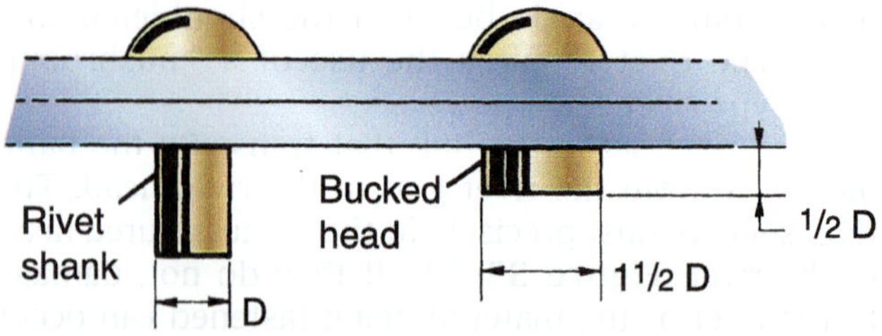

FIGURE 33–39 Removing a buck rivet

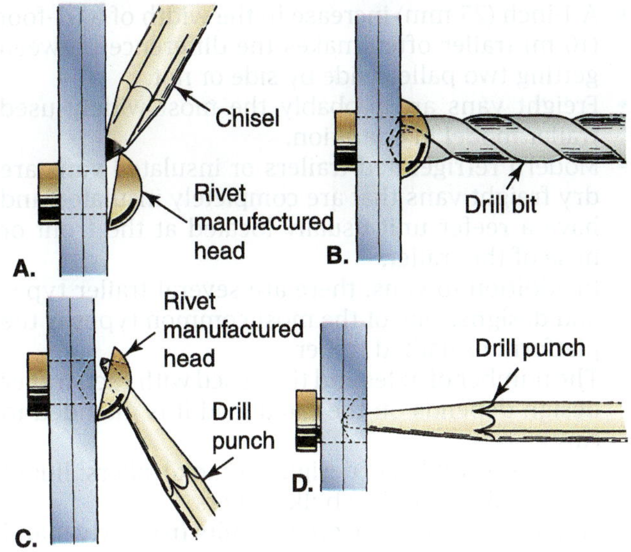

Removing Rivets

As shown in **Figure 33–39**, rivets can be removed in one of the following three ways:

1. Using a cold chisel, cut off the manufactured head of the rivet. See **Figure 33–39A**. Take care not to damage the surrounding material.
2. Using a drill bit that has the same diameter as the rivet shank to be removed. Drill into the exact center of the manufactured head to the depth shown in **Figure 33–39B**.
3. Using a drill punch, pry off the rivet head. See **Figure 33–39C**. Then drive out the shank with the drill punch (**Figure 33–39D**).

33.7 MOBILE UNIVERSAL TRAILER TESTER

Although many service facilities and fleets have homemade trailer electric and air system testers, a better test instrument is the microprocessor-controlled **mobile universal trailer tester (MUTT)**. The unit is designed to quickly test the electrical and pneumatic circuits on an uncoupled semitrailer. MUTT is equipped with a remote control. This means that the tester can be connected at the front of the trailer while the technician verifies turn light and stoplight operation while standing at the rear. **Figure 33–40** shows a MUTT.

LOCATING A SHORT

When a short is located in the electrical circuit, the microprocessor alerts the technician and enters pulsar mode. In pulsar mode, a short pulse of current is unloaded into the shorted circuit every 3 seconds, enabling the technician to source the problem without inflicting further damage to the circuit. Once the short has been repaired, the circuit remains powered up and the alert ceases.

FIGURE 33–40 Mobile universal trailer tester (MUTT)

Courtesy of Innovative Products of America

AIR CIRCUIT TROUBLESHOOTING

The remote control feature allows the troubleshooting technician to quickly locate air leaks. The technician can go under the vehicle and briefly activate the trailer supply and service circuits to help locate a leak. MUTT can also perform leak-down tests of the supply and service circuits. Air ride suspensions can also be checked and serviced by one technician while the trailer is uncoupled.

SUMMARY

- A semitrailer is a trailer in which a portion of the trailer weight is supported by the tow unit—either a tractor or a lead trailer.
- A full-trailer is any trailer in which the trailer weight is self-supported and the unit is towed by a tractor or lead trailer by means of a hook and drawbar assembly.
- A semitrailer can be converted to a full-trailer by means of a converter dolly.
- A train is a combination of a tractor and multiple trailers.
- An A-train double consists of a tractor, semi-lead trailer, and full-pup trailer.
- A B-train double consists of a tractor, semi-lead trailer, and semi-pup trailer.
- A C-train double uses a rigid double drawbar and converter dolly to attach the pup trailer.
- Many van trailers use a unibody or monocoque frame.
- A unibody frame is essentially a shell in which all of the shell components play a role in the frame dynamic, including the floor, side panels, bulkhead, doors, and roof.
- The van trailer is the most common truck trailer on North American roads.
- There are two ways to obtain light weight without sacrificing strength. One is by using lighter, more expensive materials such as aluminum and high-strength steel in posts and cross-members. The other is to use wall posts and floor cross-members that have a deeper section.
- A 1-inch (25 mm) increase in the width of a 53-foot (16 m) trailer often makes the difference between getting two pallets side by side or not.
- Freight vans are probably the most widely used trailer model in operation.
- Modern refrigerated trailers or insulated vans are dry freight vans that are completely insulated and have a reefer unit usually located at the front or nose of the trailer.
- In addition to vans, there are several trailer types and designs. One of the most common types is the platform or flatbed trailer.
- The number of axles and tires used with each trailer design depends on the rated load it is intended to carry.
- There are two basic designs of tank trailers: liquid (wet) haulers and dry-bulk haulers.
- Landing gear damage often results from driver and yard jockey abuse, especially during coupling and uncoupling.
- Trailer wheel and tire designs are generally interchangeable between tractors and trailers.
- Trailer slide tandems allow the load bridge formula to be altered; they increase flexibility, maneuverability, and ability to meet weight laws.
- Trailers must be serviced on regular preventive maintenance schedules. Service schedules may be based on hubdometer mileage or in applications that are used intermittently by calendar.
- The MUTT trailer tester station is useful for identifying both electrical and pneumatic problems on a trailer without the need to couple it to a tractor.

REVIEW QUESTIONS

1. What is the standard loading dock height used in trucking operations?
   a. 44 inches (1.11 m)
   b. 48 inches (1.22 m)
   c. 54 inches (1.35 m)
   d. 60 inches (1.52 m)
2. Technician A says that a damaged floor in a unibody van trailer should be immediately repaired because the floor forms part of the frame. Technician B says that an out-of-square door frame on a unibody van trailer can indicate problems with the structural integrity of the vehicle. Who is correct?
   a. Technician A only
   b. Technician B only
   c. both A and B
   d. neither A nor B
3. What is the function of a hubdometer?
   a. Wheel end.
   b. it lubricates wheel bearings.
   c. it signals wheel hub temperatures.
   d. Odometer.
4. Which of the following is a measure of the cooling capacity of a reefer?
   a. the efficiency rating
   b. the ambient temperature
   c. the Btu capacity
   d. the refrigeration rating
5. What is the major difference between a straight platform and a low-bed trailer?
   a. the degree of protection from the weather
   b. the height of the load that can be carried
   c. the fuel economy
   d. the number of axles and tires
6. How many pivot points are there in an A-train combination?
   a. one
   b. two
   c. three
   d. four

7. What is the difference between trailer-specific tires and tractor tires?
   a. Trailer tires have a shallower tread.
   b. Trailer tires are lighter in weight.
   c. Trailer tires cannot be retreaded.
   d. both a and b
8. How many pivot points are there in a B-train combination?
   a. one
   b. two
   c. three
   d. four
9. What wire color code does power line carrier multiplexing communications use?
   a. white
   b. green
   c. blue
   d. brown
10. What type of trailer design is most commonly seen on our roads?
    a. freight van
    b. flatbed trailer
    c. dump trailer
    d. tank trailer
11. What procedure tends to cause most landing gear damage?
    a. failure to observe lubrication schedules
    b. aggressive coupling practices
    c. cranking dolly legs up in the wrong gear ratio
    d. driving with the dolly legs down
12. Technician A says that in an ATA seven-wire trailer cord, the color green is used for marker lights. Technician B says that the color white is used for chassis ground. Who is correct?
    a. Technician A only
    b. Technician B only
    c. both A and B
    d. neither A nor B
13. Which type of refrigerant could be legally used in some trailer reefer systems but not in a tractor system?
    a. R-134a
    b. R-22
    c. R-12
    d. propane
14. What is the usual maximum permitted width for a trailer operated on the highway?
    a. 102 inches (2.6 m)
    b. 185 inches (4.7 m)
    c. 13 feet (4 m)
    d. 15 feet (4.6 m)
15. Which of the following must be used to couple a semitrailer to a tractor?
    a. a converter dolly
    b. a fifth wheel
    c. a pintle hook
    d. a drawbar
16. Which of the following devices will convert a semitrailer into a full-trailer?
    a. a lift axle
    b. a fifth wheel
    c. a pintle hook
    d. a converter dolly
17. Which of the following train configurations would be made up of two semitrailers?
    a. any double trailer configuration
    b. an A-train
    c. a B-train
    d. a C-train
18. Which of the following train configurations would have to be equipped with a pintle hook?
    a. any double configuration
    b. an A-train
    c. a B-train
    d. a C-train
19. Which of the following terms would best describe the frame used on a typical van-type trailer?
    a. full frame
    b. ladder frame
    c. truss frame
    d. unibody frame
20. Which component on a trailer upper coupler locks to the jaws of the fifth wheel on the tow vehicle?
    a. the kingpin
    b. the deck plate
    c. the scuff plate
    d. the secondary lock

34

Prerequisite: Chapter 33

FIFTH WHEELS AND COUPLING SYSTEMS

OBJECTIVES

After reading this chapter, you should be able to:

- Describe the different types of fifth wheels used on tractors.
- Outline the operating principles of the Holland, Fontaine, and JOST fifth wheels.
- Understand the importance of correctly locating the fifth wheel on the tractor.
- Describe the locking principles of each type of fifth wheel.
- Outline the procedure required to couple and uncouple a fifth wheel.
- Inspect and service the common types of fifth wheels.
- Describe the procedure required to overhaul an SAF-Holland fifth wheel.
- Identify the overhaul procedure required of some common fifth wheels.
- Define high hitch and outline what is required to avoid it.
- Outline the function of the kingpin and upper coupler assembly.
- Describe the operating principles of pintle hooks, ball hitches, and draw bars.
- Inspect and service a pintle hook assembly.
- Identify OOS factors of HD coupling devices.

KEY TERMS

ball hitch	electronic lock indicator (ELI)	kingpin	snubber
bolster plate	fifth wheel	pintle eye	throat
compensating fifth wheel	fully oscillating fifth wheel	pintle hook	turntable
converter dolly	high hitch	secondary lock	upper coupler
drawbar	horn	semi-oscillating fifth wheel	
drawbar eye		sliding fifth wheel	

INTRODUCTION

The **fifth wheel** is a coupling device. Invented by Charles Martin in 1911, the term *fifth wheel* originates from the days of the horse-drawn carriage. Outside of North America, a fifth wheel is often known as a **turntable**.

The purpose of a fifth wheel is to connect a tractive unit to a towed unit. The tractive unit is normally a tractor, but in the case of multiple trailer train, a fifth wheel also can be located on a lead trailer. The fifth wheel permits articulation between the tractive and the towed units. This articulation, or pivoting, is essential to steer a tractor and semitrailer or tractor and multiple trailer combinations properly on the highway. The fifth wheel also must support some of the weight of a semitrailer. A full-trailer is one in which the entire weight of the trailer is supported on its own axles. Most highway trailers are classified as semitrailers because the weight of the front portion of the trailer is supported by the tractor. In cases of multiple trailer combinations, semitrailer weight is supported by the lead trailer. A full description of tractor/trailer combinations and truck trains is provided in Chapter 33 on trailers.

A pintle hook assembly is one means of coupling a full trailer to a tractive unit. It consists of a pintle lockable jaw on a perpendicular plane to which a horizontal drawbar or pintle eye connects. Another means of coupling a full trailer to a tractive unit is by using a ball hitch assembly. The ball hitch consists of a shank and ball on a perpendicular plane to which a horizontal drawbar eye connects. Both systems are described in detail later in this chapter.

FIFTH WHEEL TERMINOLOGY

A fifth wheel consists of a wheel-shaped deck plate usually designed to tilt or oscillate on mounting pins. The assembly is bolted to the frame of the tractive unit. A sector is cut away in the fifth wheel plate (sometimes called a **throat**) to permit a trailer **kingpin** to engage with locking jaws in the center of the fifth wheel. The trailer kingpin is mounted in the trailer **upper coupler** assembly. The upper coupler is fabricated into the trailer frame and contains the kingpin and **bolster plate**. The bolster plate seats directly to the fifth wheel plate when coupled and supports some of the weight of the trailer.

The kingpin is a hardened steel cylindrical flanged stub, and it protrudes downward from the upper coupler plate. As a tractor is backed into the trailer, the trailer bolster plate first contacts the fifth wheel plate. The kingpin then is guided through a sectored recess, the throat, until it contacts the jaw mechanism. The fifth wheel jaw assembly is spring-loaded to enclose the kingpin. Most fifth wheels have primary and **secondary locks** to maximize coupling safety. The flange at the bottom of the kingpin stub prevents a vertical separation from occurring. Some fifth wheels are equipped with rubber block snubbers to help dampen trailer bump.

FIFTH WHEEL LOCKS

There are three general categories of primary fifth wheel locks:

- Trigger-activated (no-slack) Fontaine style
- Kingpin-activated (Euro-style) JOST style
- Double-jaw: jaws spread by kingpin to release, kingpin and spring locked

Later in this chapter, we will take a look at all three types in closer detail.

PINTLE HOOKS

Another method of attaching a trailer to either a truck or lead trailer is by using a **pintle hook**. A drawbar eye engages the pintle hook assembly. The drawbar extends from the **converter dolly** (integral axle, frame, fifth wheel, coupling unit); in all cases in which this coupling method is used, the unit being towed would be classified as a full-trailer. A disadvantage of pintle hook and drawbar coupling is that the number of pivot points is increased. A full-trailer coupled to a truck by a pintle hook would be capable of articulating both at the pintle hook and at the converter dolly.

SHOP TALK

Most truck and trailer coupling mechanisms are simple in design and operating principles. However, because of the extreme consequences of a trailer breakaway, care must be practiced when doing any work on coupling devices. Always consult original equipment manufacturer (OEM) service literature and ask questions when you are unsure of a procedure.

34.1 TYPES OF FIFTH WHEELS

Fifth wheels in use in North America are currently manufactured by three companies:

- SAF Holland
- Fontaine
- JOST

ConMet Simplex (formerly American Steel Foundries [ASF]) was purchased by Holland Hitch in 2005 shortly after which SAF merged with Holland Hitch. The ConMet Simplex range of fifth wheels has been absorbed into the SAF Holland line of fifth wheels and some of them have ceased production. JOST is a German company that has been competing in North America for a generation during which they have gained increased market share. The locking mechanisms on fifth wheels use one of the three general principles identified earlier in this chapter but each has its own small distinctions.

Technicians should understand exactly how each type of fifth wheel lock operates before attempting service and overhaul procedures for some pretty obvious safety reasons. Although many different types of special application fifth wheels exist, by far the most common type is the **semi-oscillating fifth wheel**. A semi-oscillating fifth wheel pivots only on a longitudinal plane.

SEMI-OSCILLATING FIFTH WHEELS

The semi-oscillating fifth wheel articulates about an axis perpendicular to the vehicle centerline. That is, it pivots fore and aft. **Figure 34–1** shows a typical

FIGURE 34–1 Angle-mount bracket semi-oscillating fifth wheel

semi-oscillating fifth wheel. The fifth wheel plate can be a:

- Cast steel plate
- Forged steel plate
- Pressed steel plate
- Forged aluminum alloy plate

The fifth wheel plate has a pair of mounting bosses on either side. The plate pivots on a pair of saddle brackets fitted on a base plate or frame brackets. Pins are inserted through bosses on the fifth wheel plate and the saddle brackets. These pins are bushed and permit the required articulation. The base plate is either rigidly fitted to brackets that in turn are bolted to the frame rails, as shown in **Figure 34–2**, or to a slide plate that permits fore-and-aft positioning of the fifth wheel.

Most fifth wheels are fitted to the tractor in a stationary mounting position. Stationary-mounted fifth wheels are used where the axle loading, trailer kingpin location, and vehicle combination are consistent throughout the fleet for which the tractor is specified.

SLIDING FIFTH WHEELS

A **sliding fifth wheel** assembly in most cases uses a semi-oscillating fifth wheel that can slide fore and aft on the tractor frame. This changes the point at which the tractor is loaded with the trailer weight. Sliding fifth wheels provide the tractor with greater flexibility. Because different jurisdictions have different weight-over-axle regulations, highway trucks can meet a wider range of operational requirements using sliding fifth wheels. Sliding fifth wheels also tend to be fitted to tractors when resale value is critical.

Sliding fifth wheels permit relocation of the point at which the tractor axles are loaded with the weight of the trailer. This allows them to accommodate trailers with different kingpin settings and also enables total vehicle combination lengths to be altered. **Figure 34–3** shows a sliding fifth wheel.

Locking the Slider

Sliding fifth wheels must be locked into a set position before they are used to haul a trailer. The fifth wheel base plate mechanically locks to toothed rails on the slide plate. Two methods are used to release the slide locks. A manual release mechanism is used in applications that do not have to be adjusted often. An air release mechanism is used in applications that require frequent adjustment.

Another major reason for selecting sliding fifth wheels is driver comfort. Most tractors ride more smoothly when the fifth wheel is located closer to the centerline of the bogie or rear axle. When the axles are not overloaded, the driver can extend the fifth wheel for maximum comfort.

FIGURE 34–2 Sliding fifth wheel with a driver controlled, pneumatic release

FIGURE 34–3 Lock cylinder on a sliding fifth wheel.

COMPENSATING AND FULLY OSCILLATING

A **compensating fifth wheel** is designed to articulate on two planes. That is, it pivots fore and aft and side to side. This type of fifth wheel is used in applications in which trailer torque (twist) is critical. In trailers with high rigidity, such as many tankers, trailer torque can be transmitted to a tractor or lead trailer frame and cause handling and premature failure problems. Compensating fifth wheels have made possible the use of aluminum alloy as a frame material in tanker B-trains by minimizing trailer torque transfer effect.

Fore-and-aft pivoting occurs in the same manner as in the common semi-oscillating fifth wheel. Side-to-side pivoting occurs by having trunnion shoes ride in concave tracks. The Holland Kompensator fifth wheel is used in applications in which the trailer center of gravity is located above the fifth wheel top plate. **Figure 34–4** shows the most common type of compensating fifth wheel, and **Figure 34–5** shows its operating principle. Compensating fifth wheels may have an optional lockout that eliminates side-to-side oscillation when engaged. **Figure 34–6** shows a view of the cradle that enables side-to-side oscillation and the lockout used on a Holland Kompensator fifth wheel.

Fully oscillating fifth wheels are used in applications in which the trailer center of gravity is located below the fifth wheel top plate, such as in a gooseneck, low-bed trailer. Full oscillation means that there is articulation on all planes, such as would be found in a ball joint and socket. This is desirable in rough terrain operation, in which trailer torque transfer effect can be considerable and can damage both tractor and trailer frames.

SPECIALTY FIFTH WHEELS

A number of special applications exist for fifth wheels. In most cases, the locking mechanics of specialty fifth wheels are no different from any other type of fifth wheel. A popular specialty fifth wheel is the elevating fifth wheel. Elevating fifth wheels are used on yard shunt tractors to enable trailer hookups and

FIGURE 34–4 Holland Kompensator fifth wheel

Courtesy of SAF-HOLLAND, Inc.

FIGURE 34–5 Holland Kompensator operating principle

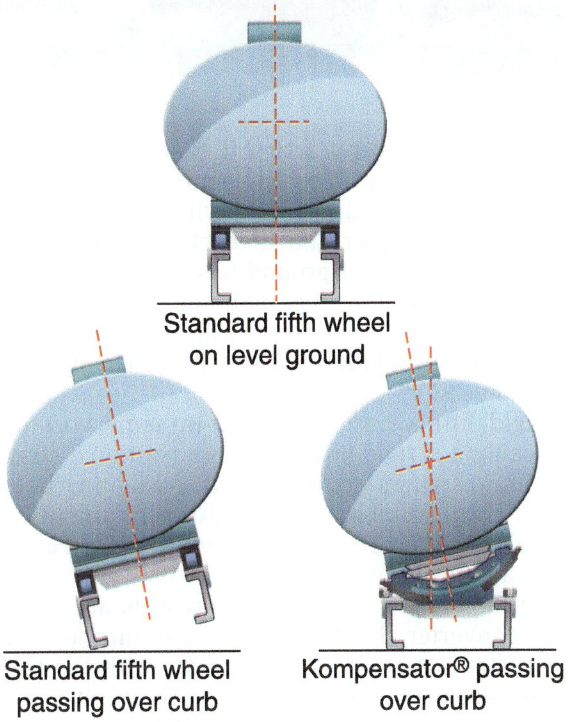

FIGURE 34–6 Holland Kompensator cradle view and lockout mechanism

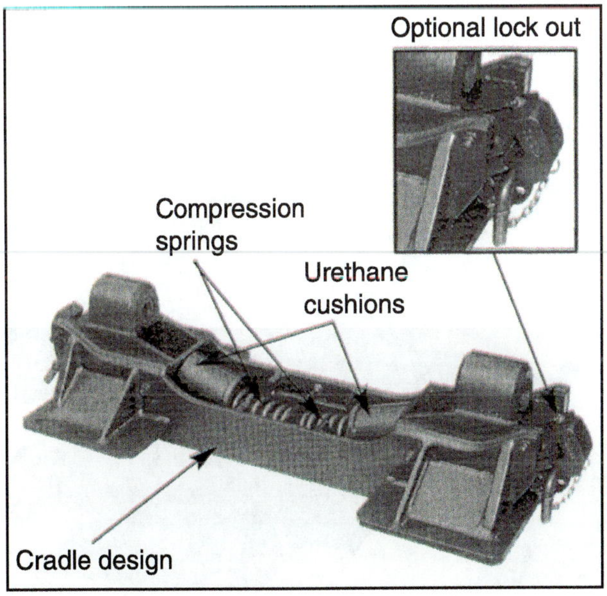

Courtesy of SAF-HOLLAND, Inc.

FIGURE 34–7 Air-actuated elevating fifth wheel

movement without retracting the landing gear. They also can be used for unloading operations that require the trailer to be tilted. The fifth wheel is first engaged to the upper coupler and then elevated so that the trailer dolly legs clear ground level by a safe margin. Elevation is achieved either hydraulically or pneumatically. **Figure 34–7** shows a typical air-actuated elevating fifth wheel, and **Figure 34–8** shows a hydraulically actuated version.

Rigid fifth wheels are used in some vocational applications. A rigid fifth wheel is an option on frameless dump trailers equipped with oscillating bolster plates. This results in the articulation taking place at the trailer rather than the fifth wheel. **Figure 34–9** shows a typical rigid fifth wheel.

Turntable or stabilized fifth wheels are used in some converter dolly assemblies. A turntable fifth wheel allows the top plate to rotate with the trailer bolster plate and allows use of a four-point support on the fifth wheel deck. **Figure 34–10** shows a typical turntable fifth wheel.

FIGURE 34–8 Hydraulically actuated elevating fifth wheel

FIGURE 34–9 Rigid-type fifth wheel

FIGURE 34–10 Underside of Holland FW35 (B-lock) fifth wheel fitted with ELI (electronic lock indicator) that can display lock status to a dash driver display.

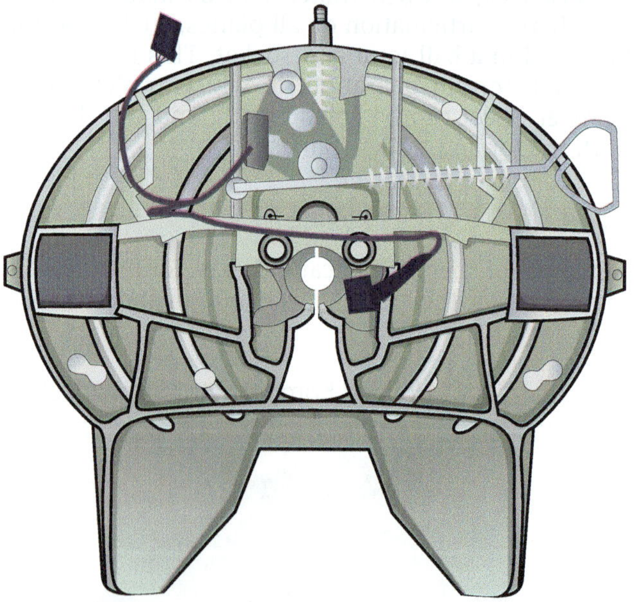

34.2 FIFTH WHEEL HEIGHT AND LOCATION

Most trailers are designed with an intended 47-inch (1.19 m) bolster plate height. This, plus the fact that any trailers operated on North American roads cannot exceed a height of 13 feet 6 inches (4.11 m), determines the plate height of fifth wheels. When it comes to locating the fifth wheel on the tractor frame, more options exist. These options have to be considered because they will determine how the axles of the tractor are loaded with the weight of the trailer.

WEIGHT-OVER-AXLE VARIABLES

The further forward a fifth wheel is positioned on the tractor frame, the higher the percentage of weight on the tractor front axle. As the fifth wheel is moved rearward, weight is shifted off the front axle and onto the rear tractor axles. On a single-drive axle tractor, the center of the fifth wheel must be located ahead of the axle centerline. On tandem-drive axle tractors, the center of the fifth wheel must be located ahead of the bogie centerline. This is an important consideration when installing sliding fifth wheels; the rearmost possible position should never be behind the centerline of the bogies.

Determining the appropriate location of a fixed fifth wheel is the function of the person responsible for determining the vehicle's specifications. The positioning of sliding fifth wheels has been the source of thousands of axle overload tickets. Dispatchers should call the shots when it comes to instructions for locating a sliding fifth wheel, and this type of decision has to be made when factoring in the trailer weight and permissible weight-over-axle in the jurisdictions in which the unit is to be operated.

SHOP TALK

The fifth wheel longitudinal location on the tractor affects the bridge formula, that is, how the load is distributed over the tractor and the trailer axles. Double-check that the fifth wheel is correctly located when installing. Incorrectly placed fifth wheels can result in weight-over-axle infractions and severely imbalanced dynamic braking.

34.3 PRINCIPLES OF FIFTH WHEEL OPERATION

Because the manufacturers of fifth wheels used on highway trucks each use distinct types of locking mechanisms, the truck technician should take nothing for granted when working on them. Having some understanding of how each type operates helps, and some of the more popular types are introduced in this section; however, always consult OEM service literature when working on fifth wheels.

CAUTION

Installing a jaw set into a fifth wheel is a simple, everyday shop procedure. However, because of the serious consequences that result from a fifth wheel separation, it is essential that you carefully observe the OEM service literature during assembly and testing. Double-check every step in the procedure. Test the operation of a completed fifth wheel rebuild by coupling and uncoupling several times to a trailer.

HOLLAND A-TYPE

The Holland A-type locking mechanism uses a single swinging lock jaw and plunger. This is one of the better-known locking mechanisms in the trucking industry; it is found on most heavy duty spec SAF Holland fifth wheels and is illustrated in **Figure 34–11**

FIGURE 34–11 Holland A-type locking mechanism

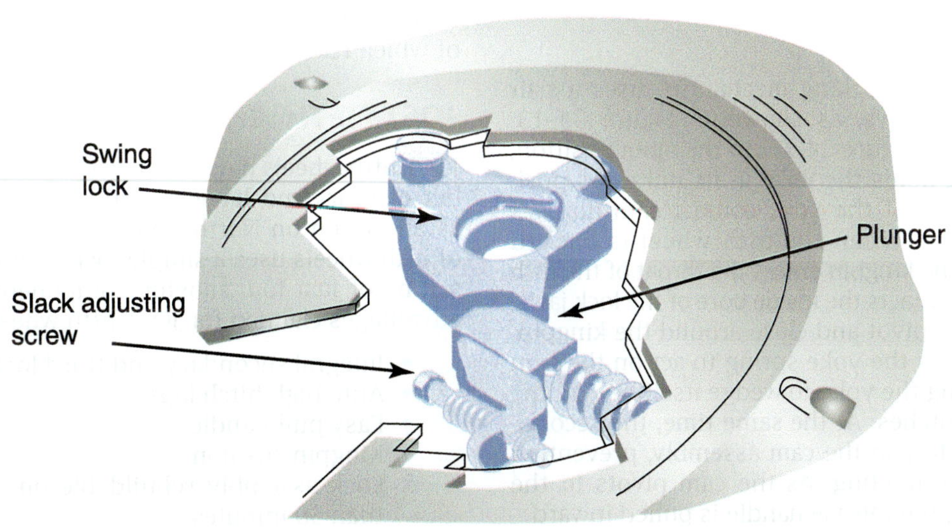

FIGURE 34–12 Underside view of a fifth wheel with a Holland A-type lock

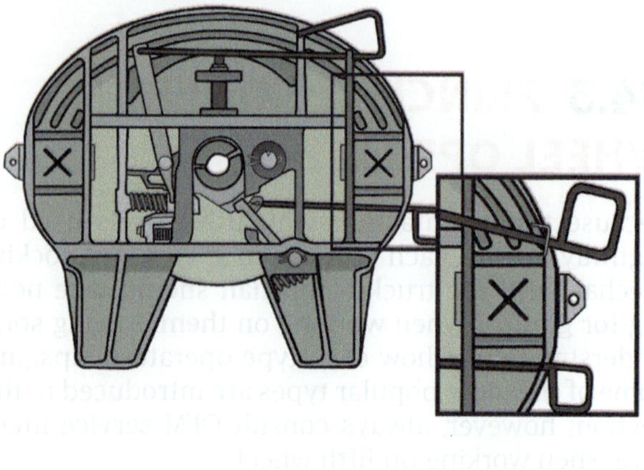

and **Figure 34–12**. It is used on SAF Holland FW0070, FW0100, FW0165, and FW2080 fifth wheels.

A-Operation

When the kingpin enters the throat of the fifth wheel and exerts pressure on the swinging lock, it pivots, enclosing the kingpin. This action causes a spring-loaded plunger to jam into a concave recess in the swinging lock. Slack caused by mechanical wear can be adjusted by means of an adjusting screw in the throat of the fifth wheel, which acts on the jam plunger.

HOLLAND B-TYPE

This type of fifth wheel uses a pair of opposed jaws. **Figure 34–13** shows both an exploded and assembled view of a Holland fifth wheel using a B-type locking mechanism, one of the most commonly used in trucking. It is used on the popular SAF Holland FW35 and FW31 series of fifth wheels. **Figure 34–14** and **Figure 34–15** illustrate the coupling and uncoupling procedure of B-type locks described in the fore-going section.

Coupling

The Holland B-type locking mechanism uses a pair of pivoting lock jaws, as shown in Figure 34–13, Figure 34–14, and Figure 34–15. In the open position, the yoke spring holds the yoke in its unlocked position, and the feet of the yoke contact cam notches on the lock jaws. When the fifth wheel is backed under a trailer, the kingpin enters the throat of the fifth wheel. When it contacts the inside bore of the lock jaws, it forces them to pivot and close around the kingpin. This action causes the yoke spring to act on the cam assembly, forcing the yoke to wedge itself behind the lock jaw cam notches. At the same time, the secondary lock pivots to jam the cam assembly, preventing the yoke from retracting. As the cam pivots to the locked position, the release handle is pulled inward.

Release

To release the fifth wheel lock, the release handle is pulled outward. This simultaneously swings the secondary lock outward and forces the cam around on its pivot, compressing the yoke spring and pulling the yoke away from its wedged position behind the cam notches on the locking jaws. As the truck is pulled away from the trailer, the trailer kingpin spreads the lock jaws, forcing them into their fully retracted position.

Holland Simplex Fifth Wheels

Holland Simplex fifth wheels use cast steel top plates. These fifth wheels incorporate a J-shaped rod handle, which makes it easier for drivers to release the unit. The locking jaw has undergone a redesign since the original version and the new jaw assembly can be retrofit to Simplex I and Simplex II fifth wheels. A major advantage of the Simplex fifth wheel is the small number of parts used in the locking assembly. The locking mechanism used on Simplex fifth wheels is known as Touchloc® and is shown in **Figure 34–16**. The locking mechanism on this fifth wheel can be replaced without removing the fifth wheel from the truck chassis. **Figure 34–17** and **Figure 34–18** show a couple of views of the SAF Holland XA-S1 (Simplex I) fifth wheel.

FONTAINE NO-SLACK® II FIFTH WHEEL

Fontaine fifth wheels use pressed steel top plates. Fontaine fifth wheels tend to be lighter than their equivalently rated competitor versions, which is an important consideration when weight is a factor. **Figure 34–19** shows an exploded view of a Fontaine No-Slack® II fifth wheel. Three models are manufactured: standard duty, heavy duty, and extra heavy duty. The standard duty has an improved locking mechanism. Fontaine fifth wheels use a rear jaw that is engaged to the trailer kingpin by a wedge. The function of the wedge is to prevent slack, and suppress trailer bump, along with providing the system is automatic self-adjusting feature. **Figure 34–20** shows an underview of two types of Fontaine fifth wheels, one of which features an air release.

JOST FIFTH WHEELS

JOST fifth wheels have been used in Europe for a number of years and have recently been made available as an option on North American trucks. This family of fifth wheels uses a simple locking mechanism consisting of just four moving components. Among the advantages claimed for JOST fifth wheels are:

- Integral secondary and third lock
- Anti–high hitch lugs
- Easy pull handle
- Kingpin cushion
- Lock assembly rebuildable on tractor in less than 30 minutes

FIGURE 34–13 Holland B-type locking mechanism: (A) exploded view; (B) front view; and (C) assembled view

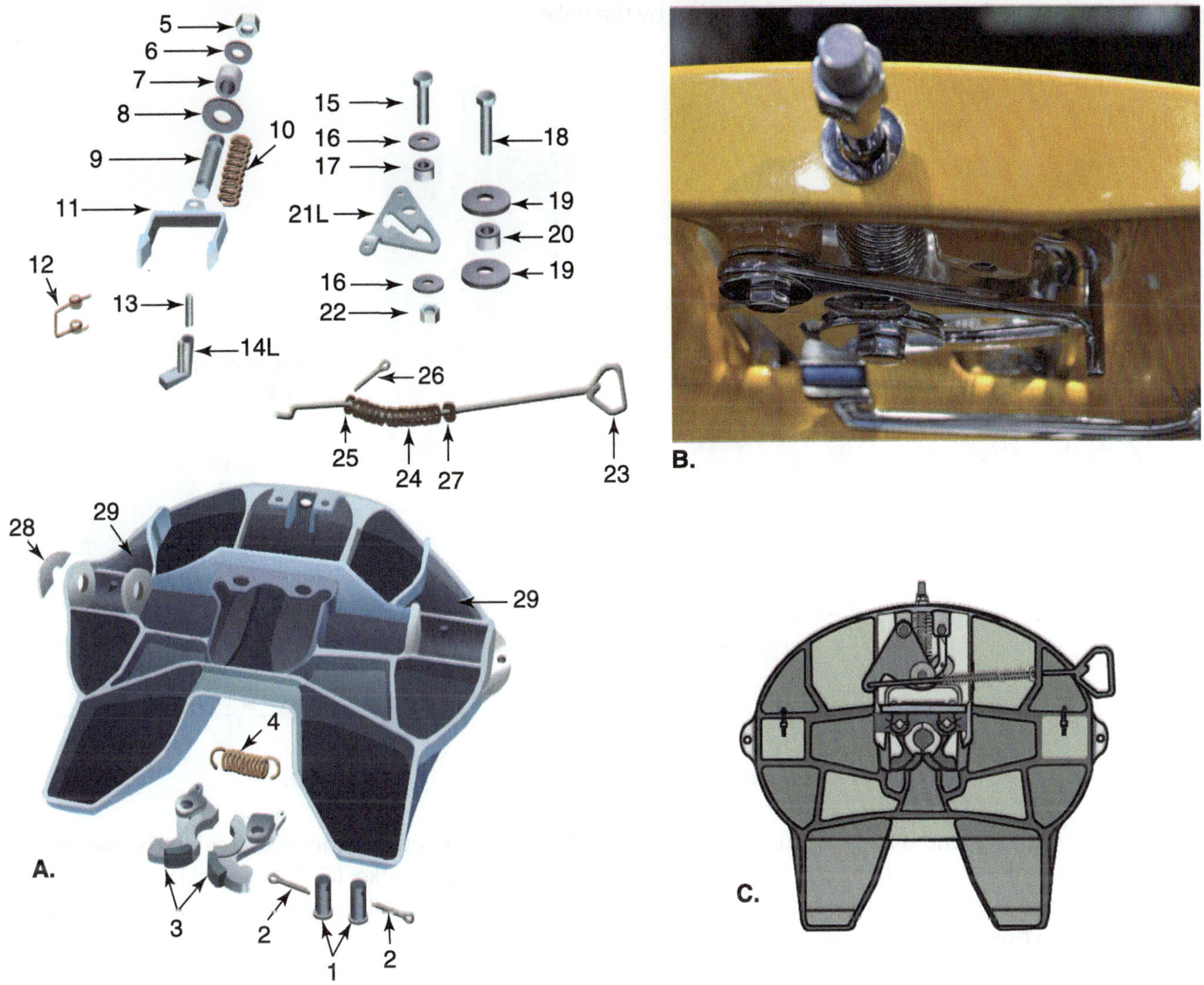

1. Lock pin
2. Cotter pin (1/4 in. × 2 in.)
3. Lock jaw set
4. Extension spring
5. Locknut (3/4 in.-16)
6. Washer (1 1/2 in. O.D. × 13/16 in. I.D.)
7. Rubber washer
8. Lock adjustment tag
9. Yoke shaft
10. Compression spring
11. Yoke
12. Torsion spring
13. Roll pin (1/2 in. × 2 3/4 in.)

14L. Secondary lock (left-hand)

15. Hex head cap screw (1/2 in.-20 × 1 3/4 in.)
16. Washer (1 3/4 in. O.D. × 9/16 in. I.D.)
17. Roller (1/2 in. I.D.)
18. Hex head cap screw (5/8 in.-18 × 1 3/4 in.)
19. Washer (2 5/8 in. × 21/32 in.)
20. Roller (5/8 in. I.D.)

21L. Cam plate (left-hand)

22. Locknut (1/2 in.-20)
23. Release handle (universal for both left and right)
24. Compression spring
25. Washer (1 1/16 in. O.D. × 5/8 in. I.D.)
26. Cotter pin (1/8 in. × 1 1/4 in.)
27. Washer (1 3/8 in. O.D. × 9/16 in. I.D.)
28. Operation tag
29. Grease zerk

FIGURE 34–14 Holland B-lock coupling action: (A) Jaws open with yoke fully retracted; (B) kingpin enters the fifth wheel throat contacting the jaws; (C) kingpin forces jaws to pivot to their closed position; and (D) fifth wheel jaws jammed into a locked closed position by the yoke.

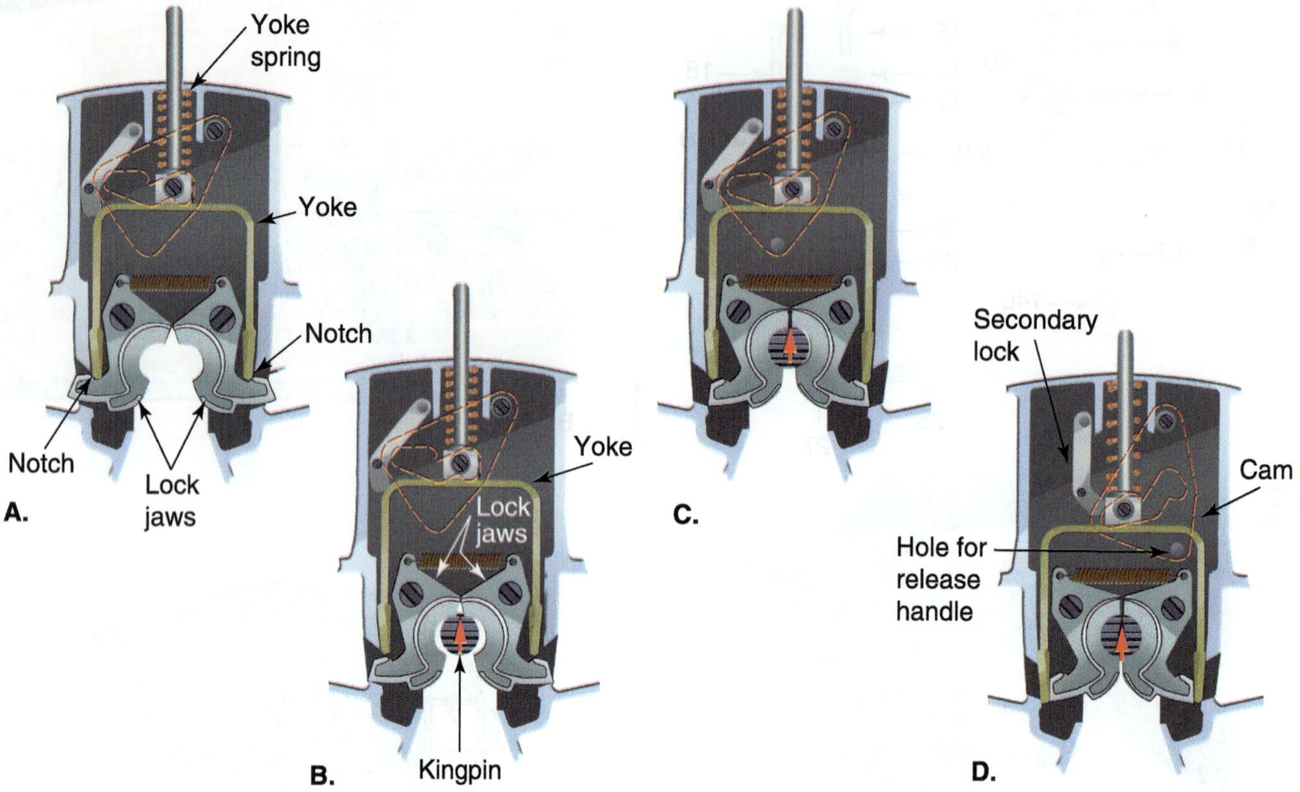

FIGURE 34–15 Holland B-lock uncoupling action: (A) Release handle retracts the yoke freeing the pair of jaws; (B) tractor moves forward to pull fifth wheel jaws away from kingpin yoke; (C) fifth wheel jaws are spread open as they move away from the stationary kingpin; and (D) fifth wheel is fully uncoupled leaving the jaws in their fully open position.

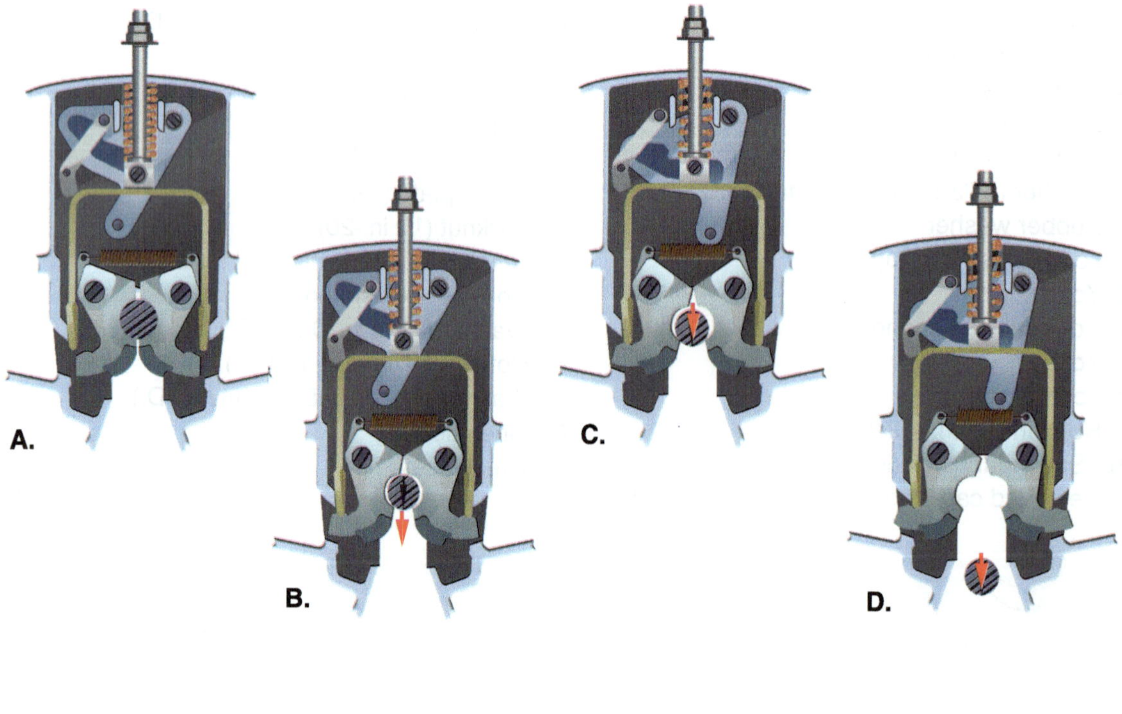

FIGURE 34–16 SAF Holland XA-SI Simplex II fifth wheel components

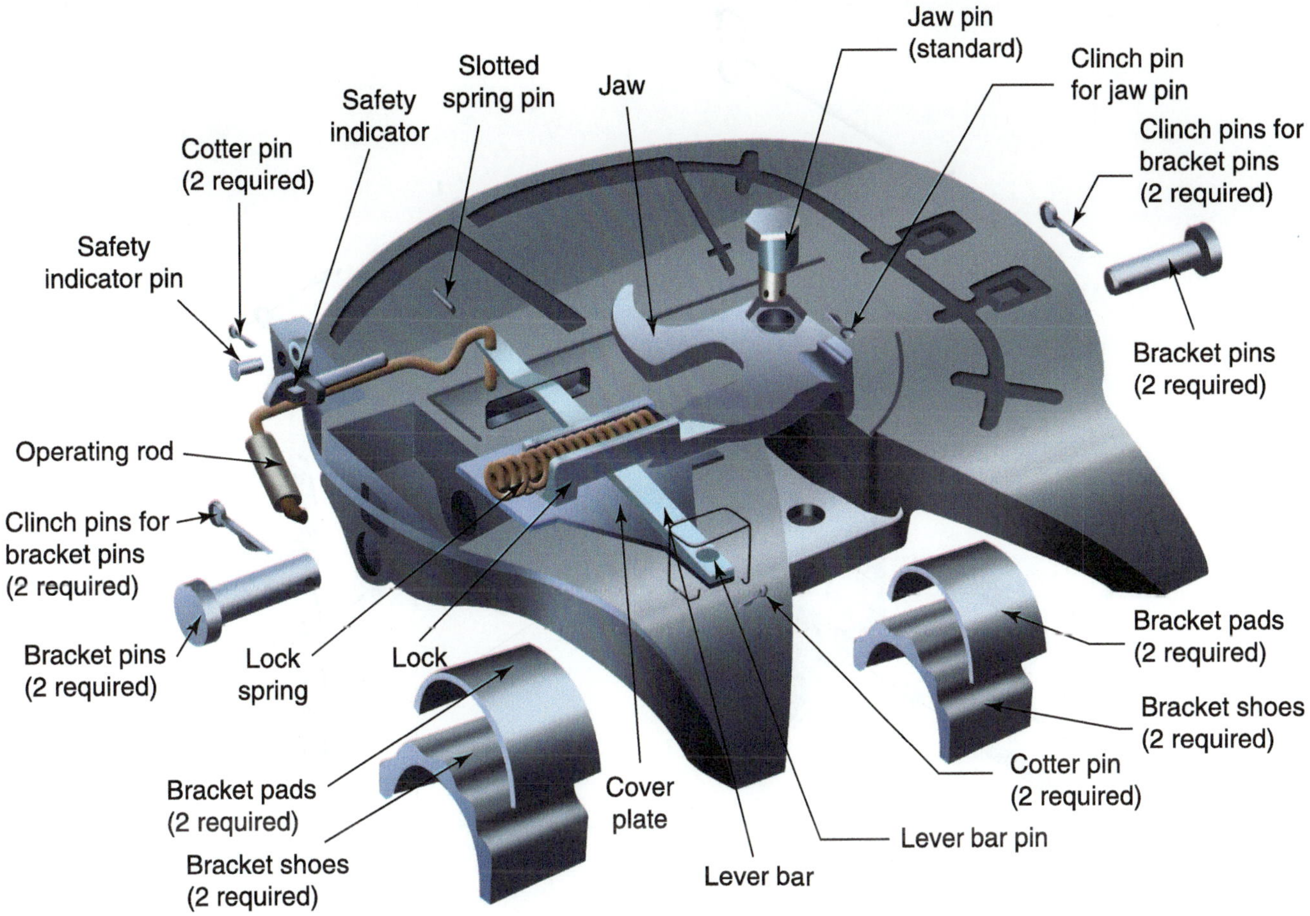

FIGURE 34–17 Underside of a SAF Holland no lube, aluminum turntable fifth wheel

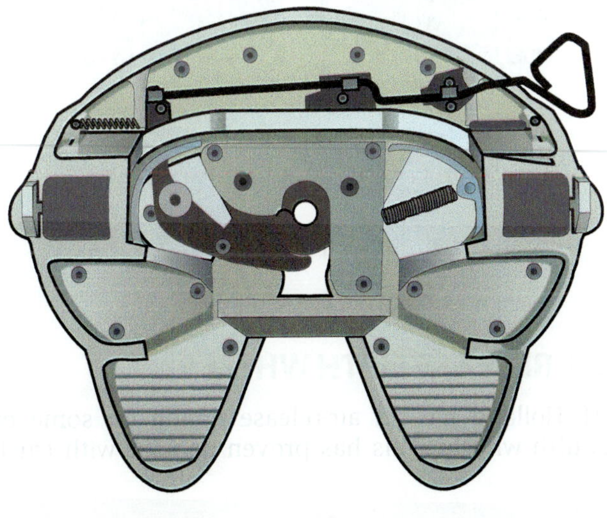

FIGURE 34–18 SAF Holland FWSI (formerly Simplex I) fifth wheel dimensions

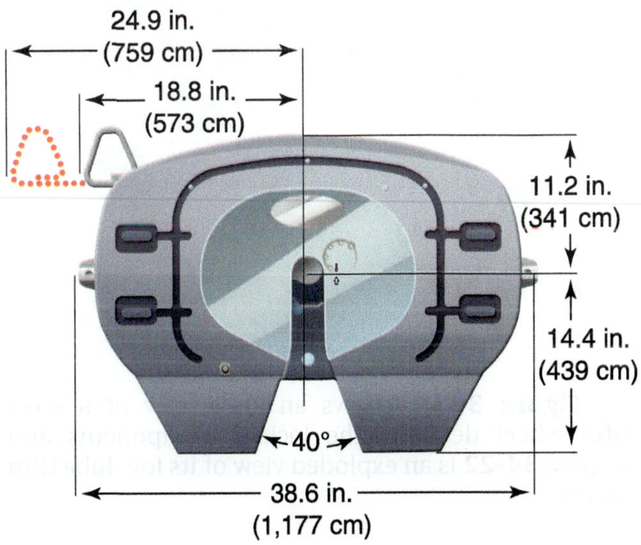

FIGURE 34–19 Exploded view of Fontaine No-Slack® II fifth wheel

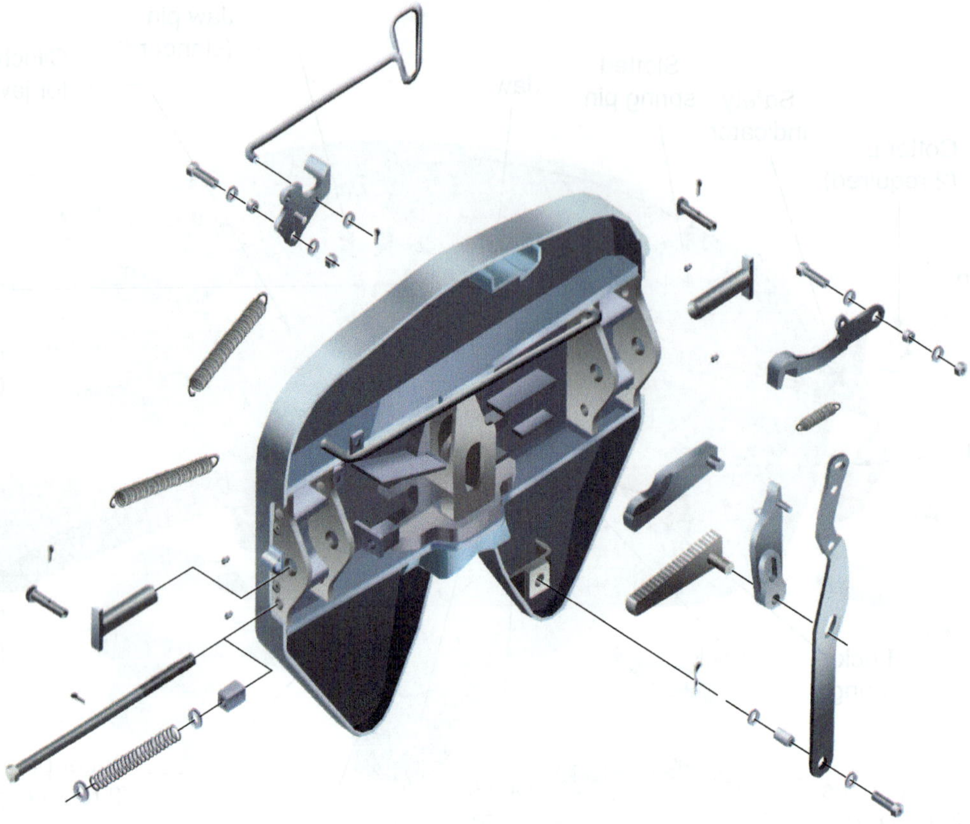

FIGURE 34–20 (A) Fontaine No-Slack® II and (B) Fontaine 5092 air-release fifth wheel

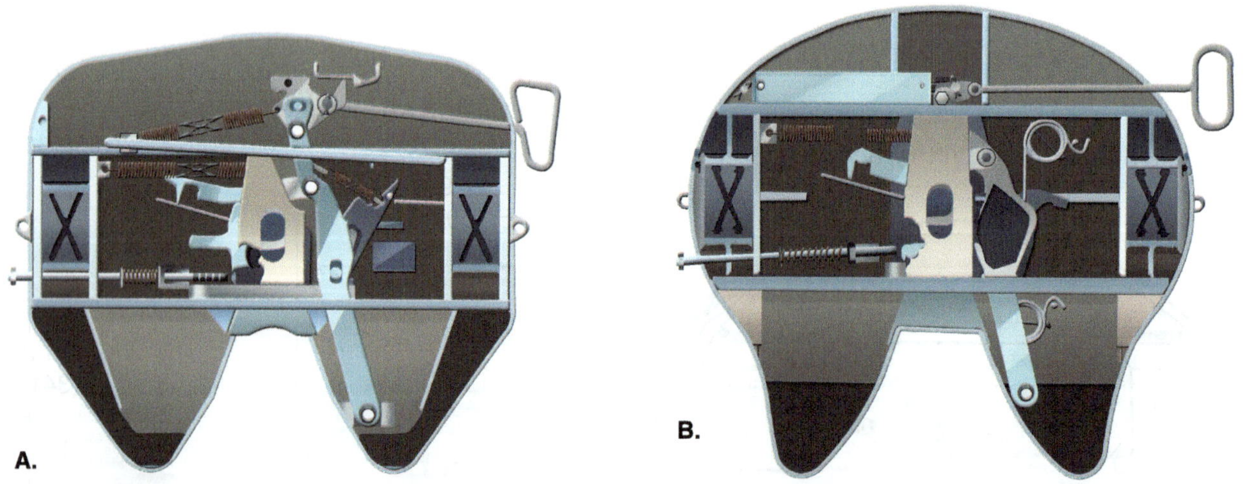

Figure 34–21 shows an underview of a JOST fifth wheel detailing the locking components and **Figure 34–22** is an exploded view of its low-lube fifth wheel.

AIR-RELEASE FIFTH WHEELS

SAF Holland has an air-release option on some of its fifth wheels. This has proven popular with truck

FIGURE 34–21 Locking mechanism used on a JOST fifth wheel

drivers, who can uncouple a trailer without having to get under the unit and pull on a release lever. The system uses a simple pneumatic cylinder to effect the release after an in-cab button is depressed. **Figure 34–23** shows an underview of a Holland FW35 fifth wheel equipped with pneumatic release.

Electronic Lock Indicator

An **electronic lock indicator (ELI)** is an optional safety feature that can be spec'd to a fifth wheel package. It safeguards against **high hitch** and improper lock engagement conditions. High hitch is a dangerous condition in which the fifth wheel locking mechanism engages but the trailer kingpin sits on top of the jaws. The ELI consists of a fifth wheel located sensor and cab dash located display lights capable of signaling three status conditions:

- Ready to couple
- Proper couple
- Improper couple

Figure 34–24 shows ELI fitted to an SAF Holland FW35 fifth wheel.

34.4 FIFTH WHEEL MAINTENANCE

Each type of fifth wheel requires slightly different maintenance procedures; therefore, you should consult the manufacturer's recommendations. This section covers some SAF Holland–recommended service procedures.

MAINTENANCE INTERVAL

This section covers some typical procedures. Many fifth wheels today are sold on the premise that they are low or zero maintenance so technicians should consult OEM service literature before servicing fifth wheels. With a normal maintenance fifth wheel, the first maintenance should be performed within the first 1,000 to 1,500 miles (1,600 to 2,500 km). After this initial service, periodic maintenance at intervals of 30,000 miles (50,000 km) should ensure trouble-free fifth wheel operation. **Figure 34–25**, **Figure 34–26**, **Figure 34–27**, **Figure 34–28**, and **Figure 34–29**

FIGURE 34–22 Exploded view of a JOST low-lube fifth wheel

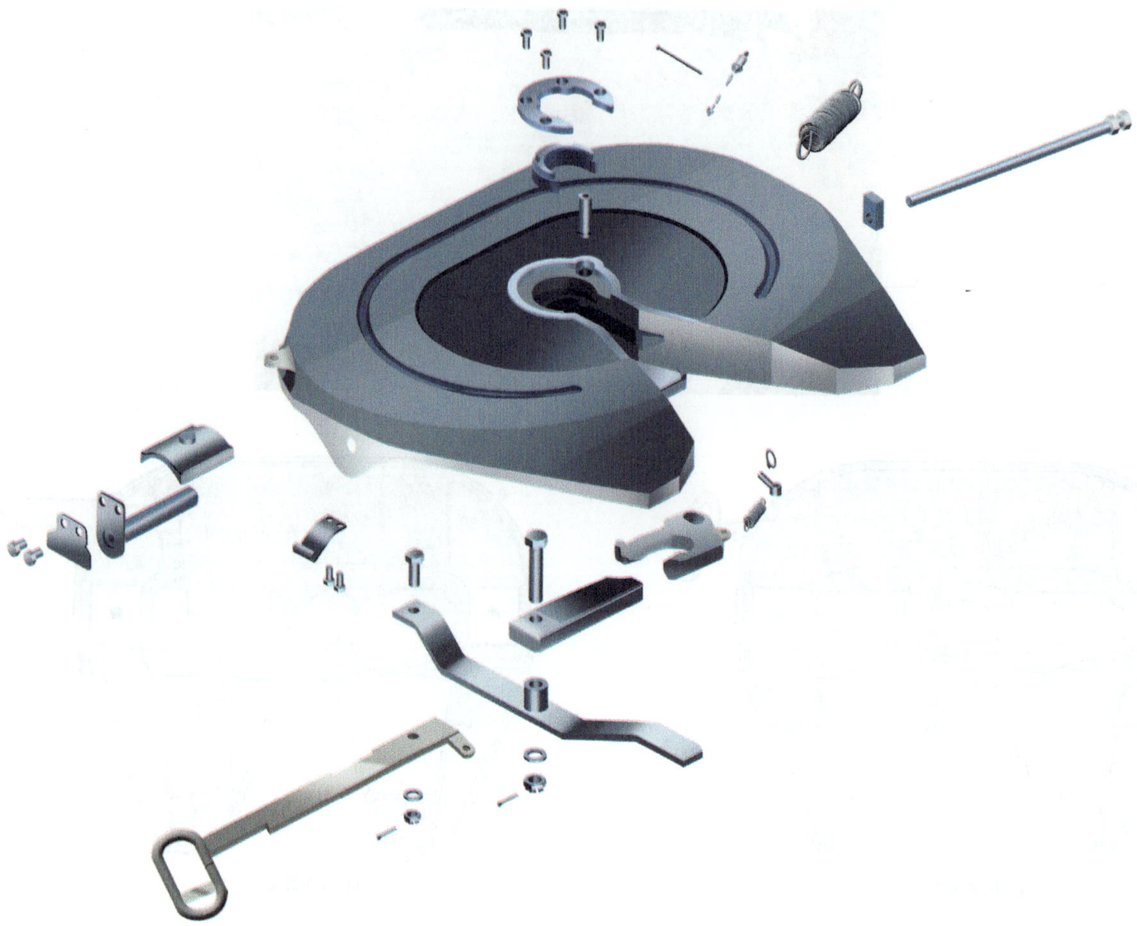

outline some of the recommended maintenance procedures by SAF Holland.

MAINTENANCE PROCEDURE

SAF Holland recommends the following fifth wheel maintenance procedure:

1. Observe fifth wheel operation when coupled to a trailer, checking for slack.
2. Uncouple the fifth wheel from the trailer.
3. Pressure wash or steam clean.
4. Inspect the top plate and brackets for cracks. Replace all damaged, loose, or missing parts.
5. Check the torque on all the fifth wheel mounting bolts.
6. Visually inspect the fifth wheel and mounting plate welds.
7. Adjust and check the fifth wheel operation using a lock tester.
8. Lubricate the fifth wheel.

HOLLAND FIFTH WHEEL ADJUSTMENT

Before attempting to adjust any fifth wheel mechanism, ensure that you have correctly identified it. Apart from the four locking mechanisms cited in this chapter, there are a number of others you will encounter, especially if you are working on older equipment.

A-Type Locking Mechanism

1. Using a Holland lock tester, close the fifth wheel.
2. Tighten the ½-inch (13 mm) Allen head adjustment screw (clockwise) until tight.
3. Loosen the adjustment screw by turning it counterclockwise 1½ (38 mm) turns.
4. Check the fifth wheel operation by using the lock tester several times.

B-Type Locking Mechanism

1. Close the lock jaws and insert the Holland TF-0237 (**Figure 34–30**) 2-inch (5 cm) diameter plug.
2. Check the fit of the plug and turn the adjusting nut counterclockwise to tighten.

FIGURE 34–23 SAF Holland fifth wheel inboard sliding mount equipped with air-release cylinder.

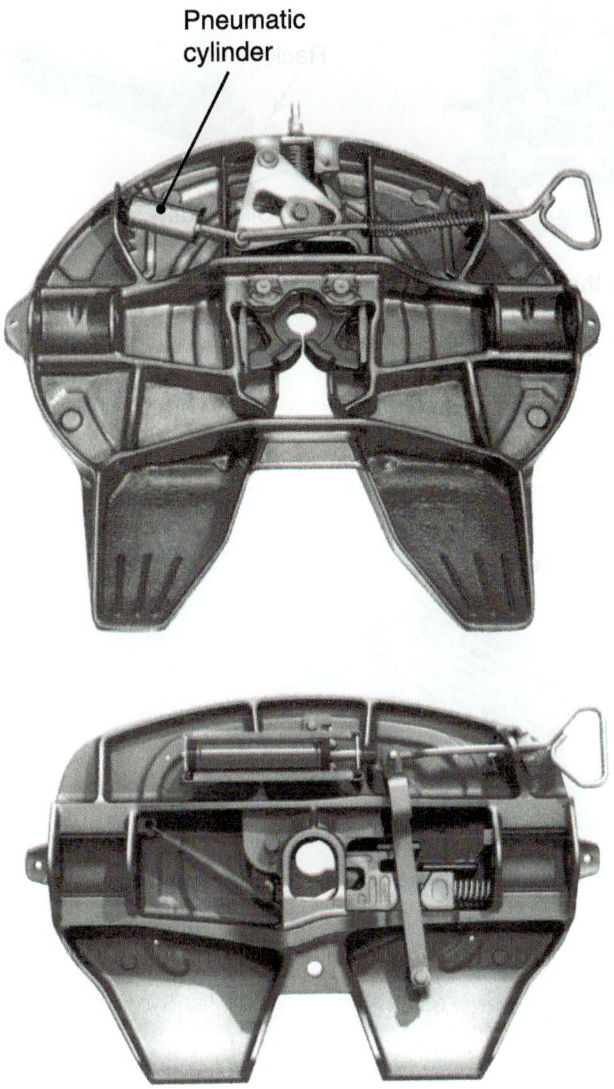

Courtesy of SAF-HOL_AND, Inc.

FIGURE 34–24 Electronic lock indicator on a FW35 fifth wheel

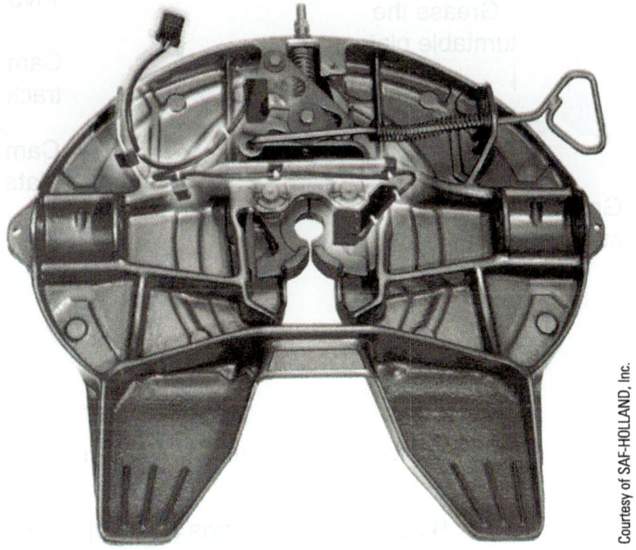

Courtesy of SAF-HOLLAND, Inc.

3. When correctly adjusted, the plug is snug in the lock jaws, but it can be rotated.
4. Check the fifth wheel operation by using the lock tester several times.

Sliding Fifth Wheel Plungers

1. Loosen the locknut and back out the adjustment bolt ½ inch (13 mm).
2. Disengage and engage plungers.
3. Tighten the adjustment bolt (clockwise) until the bolt contacts the slider rack. Turn the bolt an additional ½ turn.
4. Tighten the locknut securely.
5. Check the plunger operation by operating the release mechanism several times.

Routine Service Lubrication

1. Apply water-resistant, lithium-based grease to the trailer contact surface of the fifth wheel top plate.
2. Apply grease to the bearing surface of the support bracket through the grease fittings on the side of the fifth wheel plate. Do not overgrease, because this ends up on the fifth wheel deck plate. The top plate should be tilted back slightly to relieve weight on the bracket when applying the grease.
3. Spray an oil and diesel fuel mixture (80% engine oil, 20% diesel fuel) on the rack and bracket slide path of sliding fifth wheels.

Lubrication after Cleaning

1. Apply a light grease to all moving parts, or spray them with diesel oil.
2. Perform the routine service lubrication procedure.

FIFTH WHEEL SERVICE TOOLS

Figure 34–30 shows some of the basic tools required to test and adjust fifth wheels and kingpins. The kingpin lock testers are available in the SAE standard kingpin dimensions and are used to test fifth wheel locking action in the shop. The C-clamp is used to compress the fifth wheel springs when disassembling and rebuilding fifth wheels. The kingpin gauge or template is used to verify that the kingpin dimensions are to specification; it will test both 2-and 3.5-inch kingpins (50 and 90 mm) (Aus 2175: 50, 75, and 90 mm). The kingpin plug shown in Figure 34–30 is used both to test and reassemble fifth wheels.

FIGURE 34–25 Holland's as-needed lubrication guide

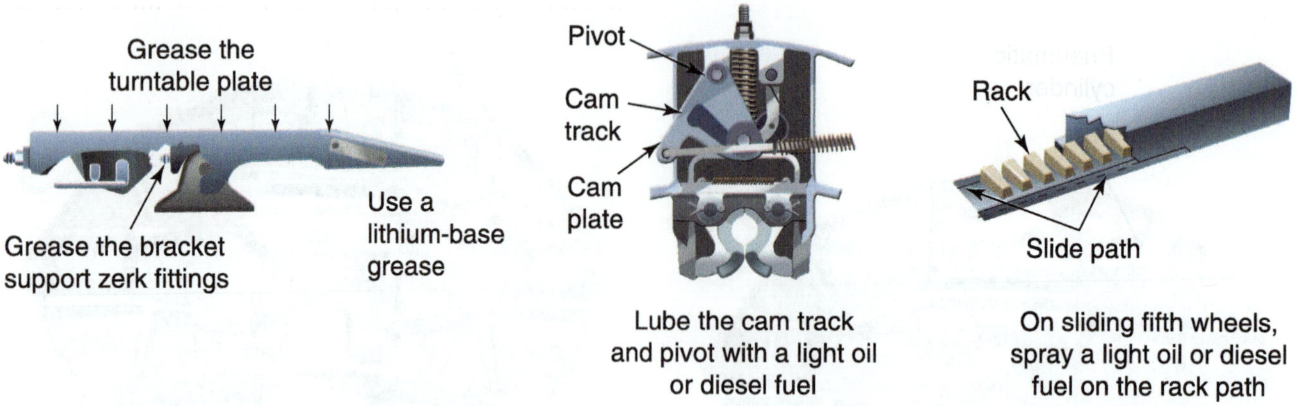

FIGURE 34–26 Holland general fifth wheel inspection

FIGURE 34–27 Inspection of locking mechanism

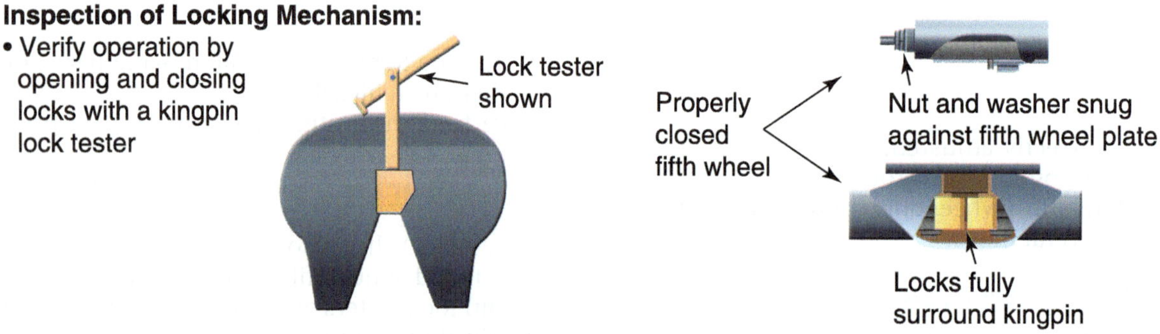

34.5 COUPLING

Each type of fifth wheel uses a different locking mechanism, so although the procedure for coupling each type is more or less the same, the inspection procedure varies. Because the fifth wheel is such a critical safety component on a tractor/trailer combination, be sure to ask to be shown what to look for to ensure that the unit is properly coupled. Properly coupled means that the jaws are fully engaged, the locks are secure, and a high hitch condition does not exist.

FIGURE 34–28 Holland adjustment of locking mechanism

Adjustment of Locking Mechanism:

1. Close jaws using a lock tester.
2. Rotate rubber bushing located between the adjustment nut and casting.
3. If the bushing is tight, rotate the nut on yoke shank *counterclockwise* until it is snug, but still can be rotated.
4. Verify proper adjustment by locking and unlocking using the lock tester.

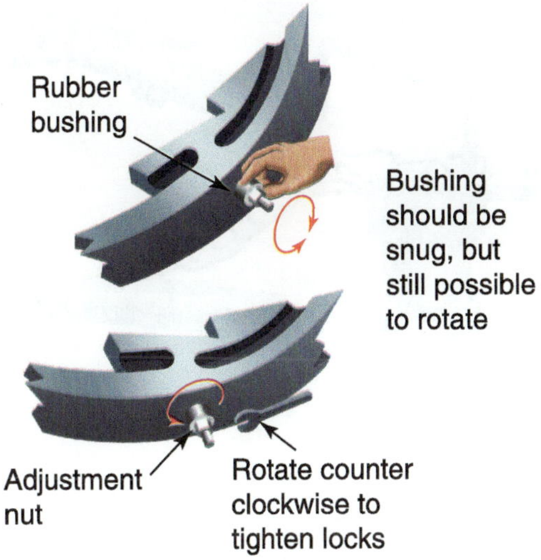

FIGURE 34–29 Adjustment of fifth wheel slide mechanism

Adjusting the Slide Mechanism:

1. Loosen the locknut and turn the adjustment bolt counterclockwise.
2. Disengage and engage the locking plungers. Verify that plungers seat properly when engaged.
3. Next tighten adjustment bolt until it contacts the rack.
4. Turn the adjustment bolt clockwise an additional 1/2 turn, then tighten the locknut securely.

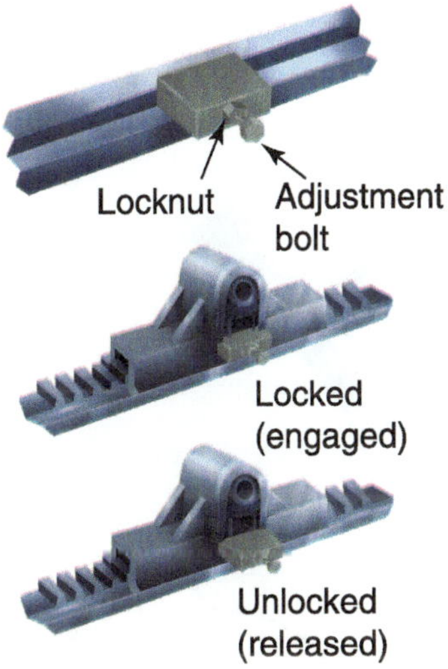

CAUTION

Proper adjustment of the sliding bracket locking plungers must be performed at installation and checked at regular intervals. The specified adjustment is required for proper operation, load transfer, and load distribution.

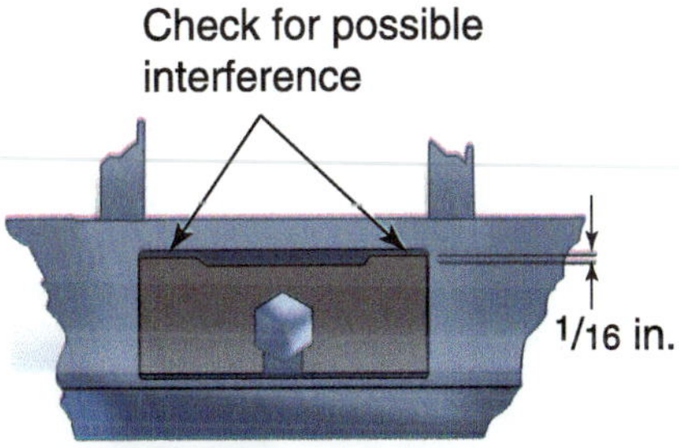

5. If the plungers do not release sufficiently to allow the fifth wheel to slide:
   - Check the air cylinder operation and replace if necessary.
   - Check the plunger adjustment as outlined above.
   - If a plunger is binding on the plunger pocket, remove the plunger using a spring compressor.
   - Grind the top edges of the plunger 1/16 inch as shown. Reinstall and adjust the plungers as outlined above.
6. If the locking plungers are too loose:
   - Check plunger adjustment as outlined above.
   - Check the plunger springs for proper tension and replace if necessary.
   - Check for plunger wear and replace if necessary.

After inspection and adjustment, lubricate all moving parts with a light, oil or diesel fuel.

FIGURE 34–30 Holland fifth wheel and kingpin service tools

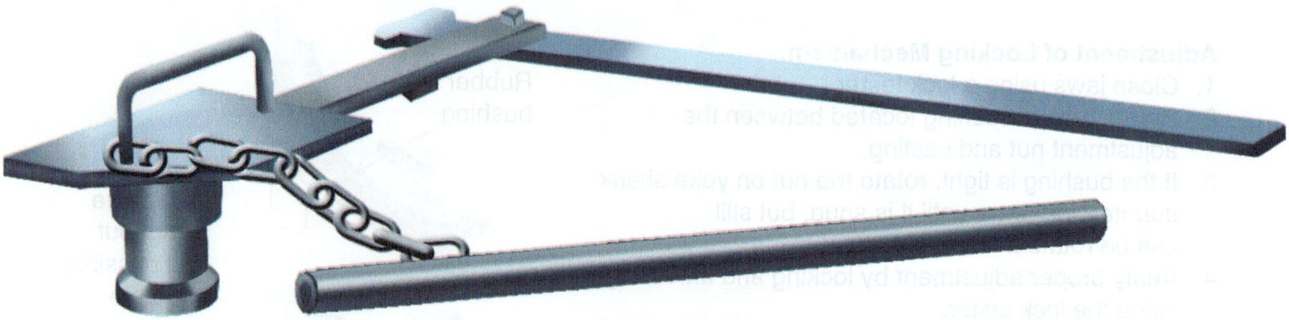

TF-TLN-1000 and TF-TLN-1500
The Holland 1000 (2-inch/50 mm) and 1500 (3.5-inch/90 mm) kingpin lock testers are designed for checking fifth wheel locking action and lock adjustment.

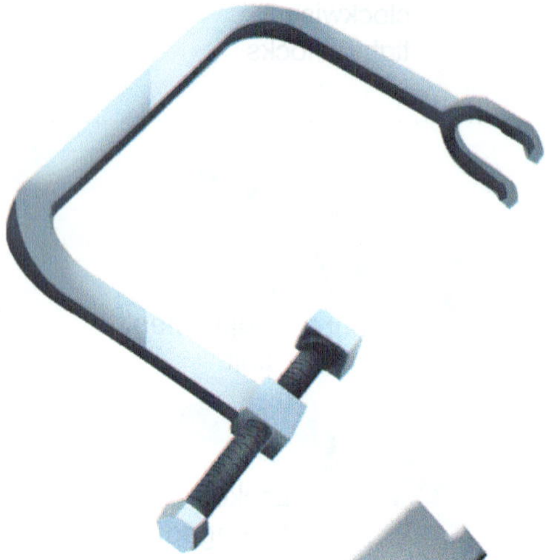

TF-TLN-2500
Designed for removing and installing springs, retainers, and pins on the slide release mechanisms of Holland sliding fifth wheels. Operated with a 5/8 inch (16 mm) socket and ratchet wrench.

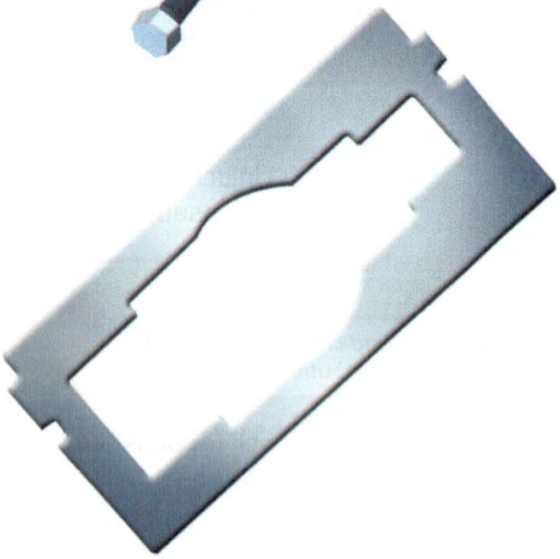

TF-0110
A kingpin gauge quickly identifies an undersized condition and need of replacement of 2 inch and 3 1/2 inch (50 and 90 mm) SAE kingpins. It can also be used to check kingpin length, straightness, and surface deformation of the upper coupler.

TF-0237
The TF-0237 is a 2 inch (50 mm) plug used to check and adjust the kingpin jaws of fifth wheels; it can also be used to assist in lock installation and adjustment.

PRELIMINARY CHECKS

Before attempting to couple a trailer to a tractor fifth wheel you should perform the following preliminary checks:

1. Ensure there is no snow, ice, mud, or other debris build up in the fifth wheel throat.
2. Check the fifth wheel deck plate for clean lubricant. Remove any contamination. Inadequate fifth wheel deck lubrication can result in hard and erratic steering.
3. Check that the trailer is on a level surface.
4. Ensure that both the trailer and tractor air suspensions are properly inflated. Never inflate a trailer suspension after coupling the tractor to the trailer; this practice destroys the fifth wheel locks.
5. Make sure that the trailer parking brakes are fully applied.

COUPLING PROCEDURE

1. Check that the fifth wheel locks are not engaged and that the jaws are in an open position.
2. Align the tractor with the trailer so that the fifth wheel throat is aimed at the trailer kingpin.
3. Carefully and slowly back the tractor up to the trailer. The trailer bolster plate should be just slightly lower than the fifth wheel top plate, causing the tractor to lift the trailer. Back up the tractor until solid resistance is felt. All fifth wheels are designed so that when the kingpin fully enters the throat, the jaws engage, and the locks are activated.
4. Next, ensuring that the trailer brakes remain fully applied, nudge the tractor forward until resistance is felt. The parking brakes of the trailer should prevent forward movement, and the fifth wheel should remain engaged.
5. Exit the tractor and use a flashlight to verify that the fifth wheel jaws are fully engaged, the locks are applied, and the bolster plate is in full contact with the fifth wheel plate. This final step is critical and must be performed even when an ELI indicates the unit is coupled.

UNCOUPLING

Release levers may pull out linearly or slide axially. If uncoupling a fifth wheel with which you are unfamiliar, ask someone about the correct procedure. The following is a general procedure:

1. Park the tractor/trailer combination on a level surface and ensure that parking brakes are applied. If the trailer is loaded, ensure that the ground surface can support the landing gear pads.
2. Gently back into the trailer to unload pressure on the jaw locks. Chock the trailer wheels then drop the landing gear. Raise it enough to relieve some of the trailer weight from the tractor.
3. Disconnect the air and electrical connections from the trailer. Insert the gladhands and trailer plug in the dummy connectors on the tractor and ensure that hoses and the electrical cord are safe from contact with the drive shafts.
4. Release the release lever. Some fifth wheels have secondary locks that must be released before the main release lever is moved. Others have a two-lever release mechanism.
5. Release the tractor brakes and drive the tractor a short distance forward, separating the two units but ensuring that the trailer bolster plate remains over the fifth wheel plate.
6. Check that the landing gear is supporting the trailer.
7. Pull tractor away from trailer.

Hard to Uncouple Problems

When there is difficulty in uncoupling a fifth wheel to a trailer, many problems may be related to the locking mechanism. When this occurs, the following are possible causes and should be checked:

- Trailer weight is loaded onto the lock.
- Trailer air suspension has been deflated.
- Improper fifth wheel jaw lock adjustment.
- Worn kingpin.
- Wamaged upper coupler bolster plate.

Difficulty in uncoupling a tractor from a trailer is good reason to fully inspect the fifth wheel lock mechanism along with the trailer upper coupler. Never take any risks with coupling devices because the consequences can be fatal.

SHOP TALK

Do not attempt to inspect a fifth wheel without removing the grease. The best way to do this is to first scrape off as much as possible, and then apply some solvent before hot pressure washing. Check the turntable plate, throat, and saddle/pivot assembly for cracks. Check the dynamic action of the locking mechanism and the secondary lock integrity. Check all the mounting fasteners and welds.

Photo Sequence 19 shows a typical coupling procedure. Once again, students are reminded that the coupling procedure required by driver license certification testing may differ because this does not account for default to park brake systems. Driver testing may require that the trailer be pneumatically coupled before mechanically engaging the tractor fifth wheel with the trailer upper coupler.

Photo Sequence 19

TRACTOR TO TRAILER COUPLING PROCEDURE

P 19–1 Begin by inspecting the fifth wheel and ensuring its jaws are open. The fifth wheel to be used in this photo sequence is a JOST; this one has been power-washed to demonstrate exactly what the jaws should look like in the open position.

P 19–2 Check that the trailer parking brakes are applied. Chocking the wheels of the trailer as shown will provide an extra measure of security to the coupling process.

P 19–3 Align the tractor with the trailer and back up, stopping just before the trailer bolster plate contacts the fifth wheel. Exit the truck and check the tractor to trailer height. In this image the trailer is low and must be raised.

P 19–4 To elevate the trailer to the appropriate height, crank the landing gear until the bolster plate just clears the fifth wheel plate deck.

P 19–5 Next connect the trailer air supply and trailer brake gladhands. Charge the trailer air circuit to system pressure.

P 19–6 Select the lowest reverse gear ratio in the tractor, apply the trailer service brakes, and back the tractor under the trailer bolster plate until you feel the fifth wheel engage; this will usually be indicated by an audible clunk as the fifth wheel lock(s) engages. Select a low forward gear and nudge the tractor forward with the trailer service brakes applied, to ensure the fifth wheel jaw(s) are fully engaged.

P 19–7 Exit the tractor. First, check that the fifth wheel release lever is in the engaged position. On this JOST fifth wheel, the release lever snaps from its outboard release position to an inboard engaged position. In the image, the hand indicates the release lever in its inboard position.

P 19–8 Next, crawl under the trailer and from behind the fifth wheel, verify that the jaw(s) is properly engaged around the kingpin and that the unit is not high-hitched; there should be no clearance between the bolster plate and the fifth wheel deck.

P 19–9 Finally, inspect the coupling from in front of the fifth wheel to ensure zero clearance between the trailer bolster plate and fifth wheel deck. If this checks out OK, you can raise the trailer landing gear, remove the wheel chocks, and prepare to move the trailer.

34.6 MOUNTING FIFTH WHEELS

The method of mounting fifth wheels on the frame is determined by the design of the fifth wheel, the suspension trunnion, and suspension to frame-mounting brackets. The most popular mounts are:

- **Single-Angle Outboard Mount.** This type of mount has the fifth wheel deck plate welded or bolted to a base plate. The base plate is welded or bolted to a pair of angle iron brackets to be fitted on either side of the tractor frame rails. The angle iron brackets are outboard mounted by bolting them to the truck frame rails using SAE Grade 8 bolts. After fitting the brackets, the fifth wheel base plate can then be welded to them in uniform stitches using AWS E7018 electrodes or a metal inert gas (MIG) equivalent. **Figure 34–31A** shows a single-angle outboard mount, which is probably the most commonly used mounting arrangement in the industry.
- **Double-Angle Inboard-Outboard Mount.** This type of mount is used when a fifth wheel assembly is specified with no mounting plate. Frame-mounting supports are used on both the inside and outside of the frame rail. A spacer plate or steel spacers are used on the inside frame mounting support to keep the mounting surfaces level. The forward edges of the inside and outside angles must be staggered a minimum of 6 inches (150 mm). **Figure 34–31B** shows a double inboard-outboard angle mount used in heavy-duty applications.

FIGURE 34–31 (A) Single-angle outboard mount; and (B) double-angle inboard-outboard mount

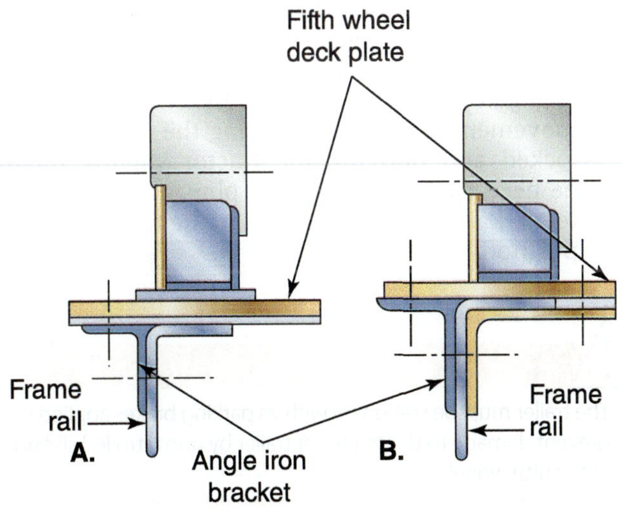

The following considerations are based on one manufacturer's recommendations. Before fitting a fifth wheel, check the manufacturer's requirements, which can vary with fifth wheel rating and vehicle application. **Figure 34–32** shows a properly mounted fifth wheel.

- Angle iron sections used to mount a fifth wheel must be at least 5⁄16-inch (8 mm) ASTM-A36 carbon steel; 3⁄8-inch (9.5 mm) angle iron section also can be used. The angle iron should be at least 3 × 3½ inches (75 × 90 mm) and, at a minimum, 36 inches (91 cm) long. It is good practice to cut a radius on the edges of the mounting angle that contact the tractor frame. Ensure that no bolt hole is located closer than 1 inch (2.5 cm) to the edge of the angle iron. When recesses are cut into the angle iron to avoid suspension brackets, ensure that there is both sufficient clearance and a radius cut.
- Each section must be secured to the frame rails by at least five 5⁄8-inch (16 mm) diameter Grade 8 bolts. Space the bolts 8 inches (20 cm) apart and secure them with Grade C locknuts. Nuts must be torqued to manufacturer's specifications. Mounting angle bolts preferably should be staggered with respect to the horizontal centerline along the length of the mounting angles to avoid weakening the frame. Fifth wheel support brackets are never welded to the truck frame rails.
- The fifth wheel deck plate should be welded to the angle iron frame brackets *after* they have been mounted to the frame rails. Use stitch welding runs; an example would be 5-inch (13 cm) runs interspersed with 5-inch (13 cm) gaps.
- For stationary fifth wheel plates, the angle iron bracket must extend at least 18 inches (46 cm) ahead of the fifth wheel's pivot point and at least 12 inches (30 cm) to the rear. For sliding fifth wheels, the angle iron sections must be long enough that these minimum distances are achieved when the fifth wheel is moved.

FIGURE 34–32 A properly mounted fifth wheel

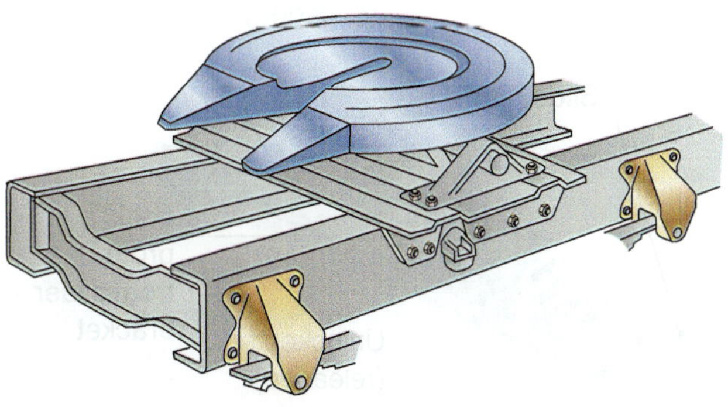

CAUTION

Installation welding on fifth wheel assemblies should be performed by a qualified welder. Improper welding procedures can result in a trailer separation—the consequences of this can be fatal!

34.7 SLIDING FIFTH WHEELS

Sliding fifth wheels increase the versatility of the tractor, but it is important to understand how tractor weight over axle is affected. Locating a sliding fifth wheel to maximize driver comfort often can result in weight-over-axle fines. Large fleets use dispatcher software to calculate the fifth wheel position; this software should be used to avoid fines. The following procedure is typical when adjusting the position of a sliding fifth wheel.

SLIDING PROCEDURE

The fifth wheel should be fully engaged to the trailer for this operation.

1. The tractor and trailer should be parked in a straight line on level ground. Engage the trailer parking brakes.
2. Release the sliding locking plungers. For an air slide release (**Figure 34–33**), put the cab control valve into the unlocked position. For a manual slide release, pull the release lever (**Figure 34–34**).
3. Visually check that both plungers have disengaged from the rail teeth. If the locking plungers are jammed in the rack teeth, try lowering the landing gear to relieve pressure on the plungers. This should allow the fifth wheel to slide easier.
4. Release the tractor brakes and drive the tractor forward or backward slowly to position the fifth wheel.
5. After sliding the fifth wheel to the desired setting, engage the slide locking plungers. For an air slide release, put the control valve in the lock position to engage the plungers to the rack teeth. For a manual slide release, trip the release arm to allow the plungers to engage with the rack teeth.
6. Visually check to see that both plungers are fully engaged to the rail teeth so that no fore-and-aft movement is possible. Leaving the trailer brakes locked and nudging the tractor slightly might be necessary to engage the plungers in the rack teeth. Raise the landing gear to the fully retracted position.

FIGURE 34–33 Air slide release plunger position

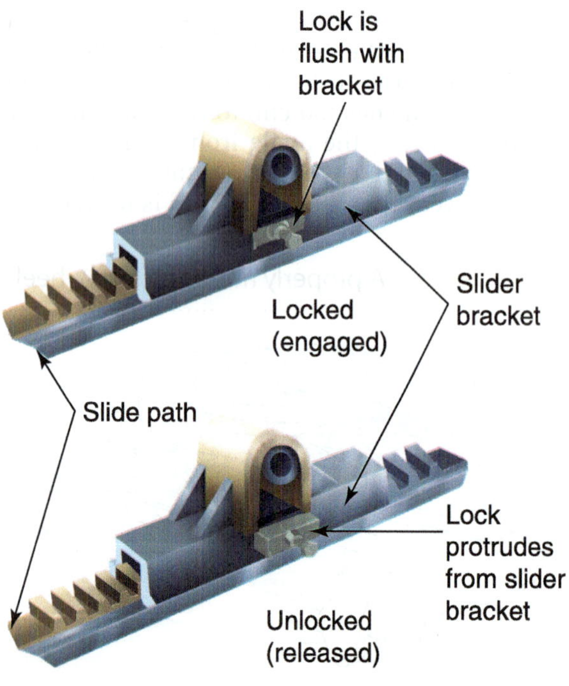

FIGURE 34–34 Manual slide release plunger position

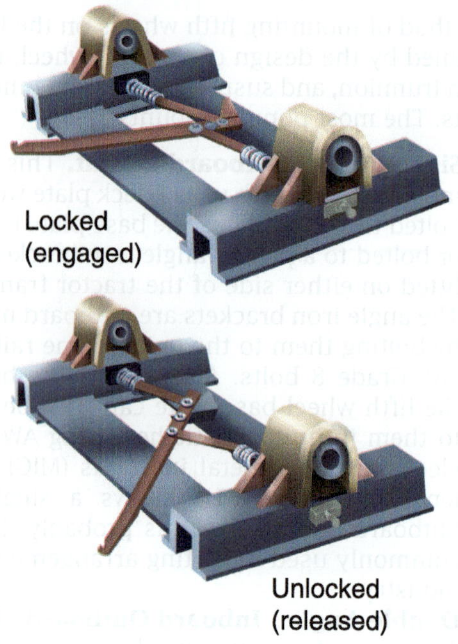

CAUTION

The trailer must be stationary with its parking brakes applied to prevent damage to the tractor or trailer by uncontrolled sliding of the fifth wheel.

CAUTION

Do not operate the vehicle if the plungers are not fully engaged and the landing gear fully retracted because damage to the tractor, trailer, and landing gear can occur.

SLIDING MECHANISM INSPECTION AND ADJUSTMENT

When adjusting the locking plungers:

1. Loosen the locknut and turn adjusting bolt out (counterclockwise).
2. Disengage and engage the locking plungers. Check that the plungers are securely seated without binding.
3. Turn adjusting bolt in (clockwise) until it contacts the rack. Turn adjusting bolt an additional one-half turn and then tighten the locknut securely.

SHOP TALK

On a sliding model fifth wheel, after locating it in the desired position, visually inspect the locking plungers to be sure that they are properly engaged. Also, set the trailer brakes and rock the tractor fore and aft. Do not attempt to shift a sliding fifth wheel when a trailer is in motion!

When locking plungers will not release to permit sliding of the fifth wheel:

1. Check the air cylinder for proper operation and replace if necessary.
2. Check plunger adjustment.
3. If the adjusted plunger binds on the pocket, grind the top plunger edges 1/16 inch (1.6 mm), reinstall, and adjust as described. Use a spring compressor to remove and reinstall the plunger.

CAUTION

Proper adjustment of the sliding bracket locking plungers must be performed at installation and maintained at regular intervals by use of the adjusting bolts provided on both sides. Proper adjustment is required for proper operation and for proper load transfer and distribution.

WARNING

Wear safety glasses when grinding the top plunger edges.

When locking plungers are too loose:

1. Check plunger adjustment.
2. Check the plunger springs for proper compression. Replace if necessary.
3. Check for plunger wear and replace if necessary. Use a spring compressor to remove and reinstall the plungers. Adjust the plungers as described.

CAUTION

Rebuilding of fifth wheels is a common shop practice. Although it is a simple procedure, each type of fifth wheel uses distinct locking mechanisms. Always use the manufacturer procedure included in every rebuild kit. A fifth wheel failure can have fatal consequences!

34.8 KINGPINS AND UPPER COUPLERS

The kingpin stub is mounted on the trailer upper coupler assembly. It is designed to securely engage with the fifth wheel locking mechanism and to permit articulation so that a tractor/trailer or tractor/train combination can turn corners. There are two SAE standard kingpin sizes, 2 inches (50 mm) and 3.5 inches (90 mm). In the U.S. and Canada highway applications, the 2-inch (50 mm) kingpin is more or less universal. The larger 3.5-inch (90 mm) kingpin is found in military applications and in South America. Offshore, the standard kingpin dimensions are 2, 3, and 3.5 inches (50, 75 mm, and 90 mm). Kingpins are required to have an SAE-specified Brinell surface hardness that will produce a yield strength of 115,000 psi (793 MPa) and a tensile strength of 150,000 psi (1,034 MPa).

Most kingpins in the United States and Canada are welded to the upper coupler assembly, but both bolt-on and removable fasteners can also be used. Welding should be performed in accordance with American welding standards (AWS). This generally means welding with an AWS E7018 electrode in a single rotational direction with stitched breaks in the weld runs. An E7018 electrode is used because it has a tensile strength in between the hardened kingpin and the softer upper coupler material.

CAUTION

Always consult the manufacturer for the correct procedure when welding kingpins to upper coupler assemblies. Welding diagrams often are provided with replacement kingpins.

When kingpins are bolted into the upper coupler assembly, SAE Grade 8 bolts, usually ¾-inch (19 mm) diameter, should be used. Tapered, removable kingpins are used in applications in which there may be a possibility of damaging an exposed kingpin when the trailer is not coupled to a tractor. Folding gooseneck trailers are an example of this. A locking dowel pin

FIGURE 34–35 SAE 2- and 3.5-inch (50 and 90 mm) kingpins and specifications. This series of kingpins are forged from 8630 steel with a Brinnell hardness rating of 302-363.

2 in. SAE Kingpin - KP-T-809 Series

Bolster Plt. Thickness	Holes
.25"	No
.31"	No
.38"	No
.50"	No
.25"	8 equally spaced .53" holes on 6.75" diameter for plug welding
.31"	
.38"	
.50"	

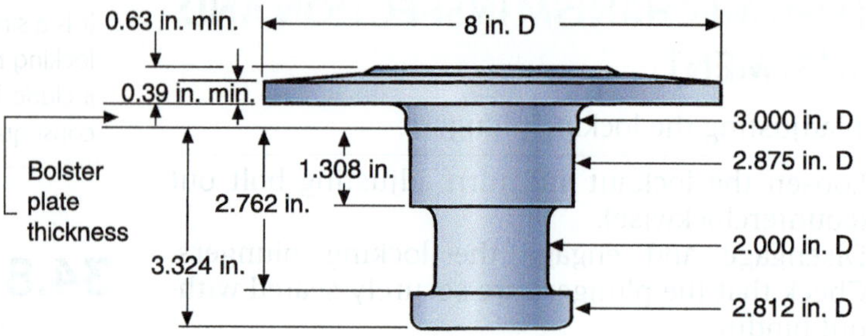

2 in. SAE Kingpin - KP-T-880 Series

Bolster Plt. Thickness	A	B	Holes
.25"	2.88"	2.12"	No
.31"	2.88"	2.12"	No
.38"	2.88"	2.12"	No
.50"	2.88"	2.12"	No
.25"	2.0"	1.5"	No
.31"	2.0"	1.5"	No
.38"	2.0"	1.5"	No
.50"	2.0"	1.5"	No

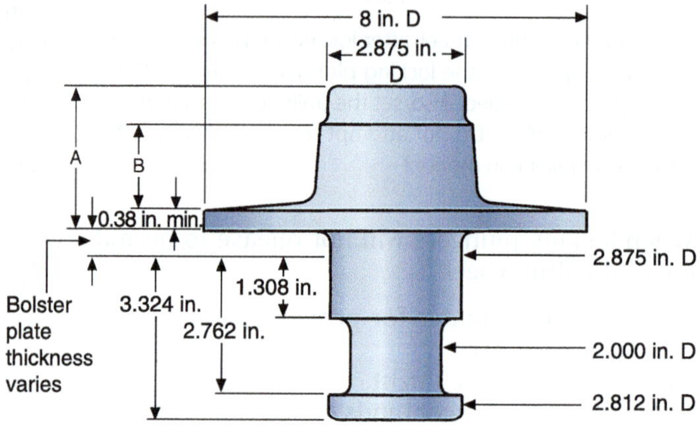

3.5 in. SAE Kingpin - KP-T-847 Series

Bolster Plt. Thickness	Holes
.38"	4 equally spaced 1.25" holes on 8.5" diameter for plug welding
.50"	

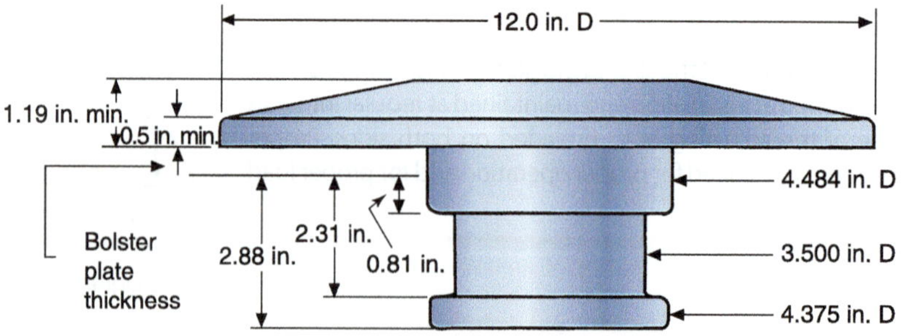

both retains and prevents the rotation of removable-type kingpins. **Figure 34–35** shows some examples of different types of kingpins and their required dimensions.

CHECKING KINGPIN DIMENSIONS

Trailer kingpins should be cleaned and inspected for wear, integrity to the upper coupler, and cracks routinely. Most fleets use a template or gauge to check a kingpin for wear; the template slides over the kingpin. A typical kingpin template capable of measuring both SAE 2- and 3.5-inch (50 and 90 mm) kingpins is shown as TF-0110 in Figure 34–30. Standard procedure is to replace a worn or damaged kingpin.

Kingpin Repair

Manufacturers of kingpins will tell you that when a kingpin is worn out, it should be replaced. Kingpins are measured with a test tool such as the Holland TF-0110 (see Figure 34–30) and can be considered dangerous when undersized. As a rule, wear in excess of 0.125 or a gouge deeper than 0.120 renders one OOS.

The problem with replacing a kingpin lies not in the cost of the kingpin itself but in the fact that the upper coupler has to be rebuilt on many trailers. Mobile repair of kingpins has become accepted by some operators. This type of repair is performed onsite and claims to meet SAE J-133 and J-140 standards. Although mobile repair of kingpins may be controversial, because of the cost and time savings, it has become a fact of life in the trucking industry.

UPPER COUPLERS

Upper coupler assemblies are built into the trailer frame assembly. The lower plate is known as the bolster plate. The kingpin is located through the center of the bolster plate. The bolster plate sits on the tractor fifth wheel plate when coupled and, therefore, provides the weight bearing and pivot surface. Bolster plates come in 0.25-, 0.31-, 0.38-, or 0.50-inch (6, 8, 10, and 12 mm) sizes. Because the weight of the bolster plate is considerable in itself, the lightest plate with a suitable margin of error is usually specified.

Fifth wheels and upper couplers are designed to work together to ensure that the trailer load is evenly distributed to the tractor. Upper couplers manufactured from higher tensile strength steels may reduce the upper coupler total weight, but they also tend to be more flexible. Greater upper coupler flexibility increases the deflection, which can shorten the service life of the assembly. **Figure 34–36** shows both the appearance of a properly loaded fifth wheel and what can occur if the vertical load is concentrated in the center of the fifth wheel assembly. Center loading causes the fifth wheel to act like a beam, flexing with the movement of the vehicle over the road, and it eventually can result in failure of the fifth wheel and upper coupler.

Welding Upper Couplers

The welding procedure followed during the upper coupler assembly, especially when welding too rapidly on lighter duty bolster plates, can result in an upper coupler that has been distorted into ripples. The wavy appearance of the bolster plate results in inadequate surface contact to the fifth wheel and high unit pressures where contact does occur. The result is galling to the fifth wheel and rapid wear of the bolster plate. Observe the manufacturer's welding instructions; stitch welding is usually recommended with plenty of cool-down between runs. E-7018 electrodes are usually recommended. Never weld a kingpin into an upper coupler using a continuous run; this can create stress risers and induce failures. **Figure 34–37** shows a ripple-distorted upper coupler.

Bowed Upper Coupler

A bowed upper coupler is caused by overloading the trailer or by an upper coupler that has been fabricated of gauge materials that are too light. A bowed bolster plate raises the kingpin to a higher than specified height, which can make it difficult to release a fifth wheel. Generally, the maximum permissible amount of bolster plate recess bow is $\frac{1}{16}$ inch (1.6 mm). A recess bow condition can be difficult to detect if the trailer is unloaded, because the deflection may take place only when the trailer

FIGURE 34–36 Proper and center-loaded upper couplers

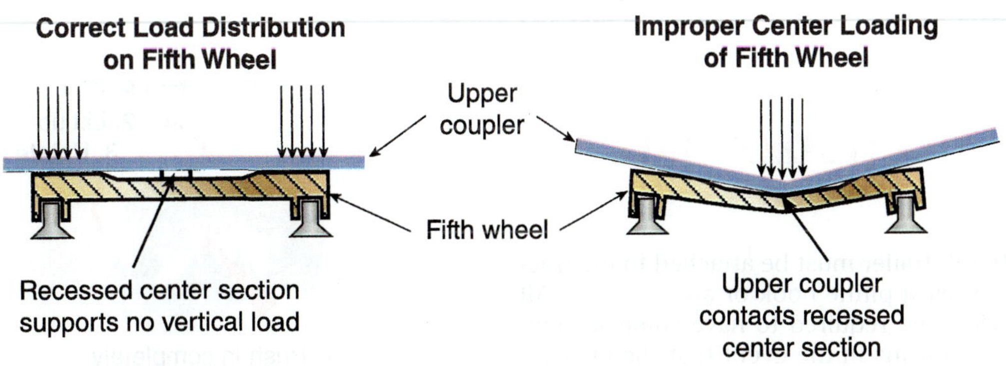

FIGURE 34–37 Ripple-distorted upper coupler

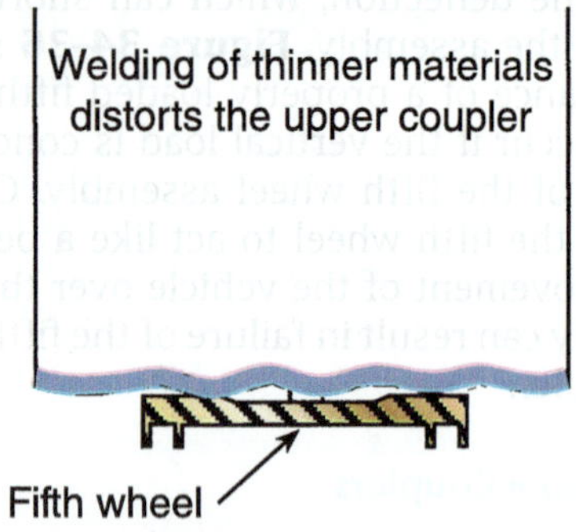

FIGURE 34–38 Bowed upper coupler

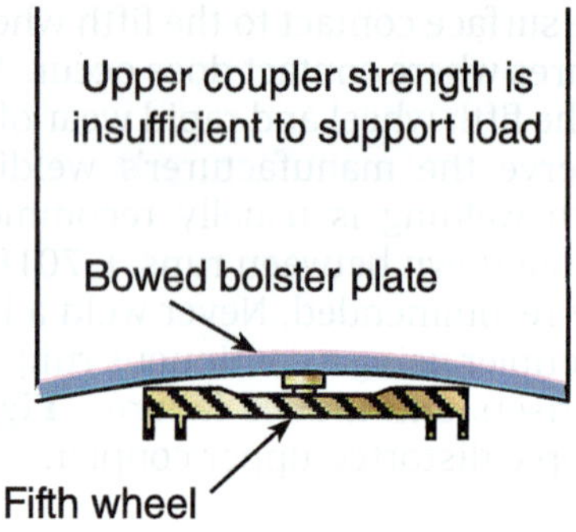

is fully loaded, but it can cause fifth wheel chatter, which should alert the driver to the condition. If the bolster plate is permanently deformed, a straightedge can locate the problem. **Figure 34–38** shows a bowed upper coupler.

Lube Plates

Fifth wheel lube plates come with the promise of less contamination of the grease on exposed fifth wheel turntable plates. However, practice suggests that they should not be used in applications where tractors and trailers are routinely coupled and uncoupled. One problem with fifth wheel lube plates is they shorten the effective length of the kingpin and this can lead to problems. If a lube plate is to be used on a rig that tends to remain coupled, it is recommended that one made by a fifth wheel manufacturer be used.

34.9 PINTLE HOOKS/ BALL HITCHES

A heavy-duty full-trailer must be attached to the tractive unit by either a pintle hook or a **ball hitch**. All coupling devices are required to have some secondary securement means in the event that the primary coupling device fails; the secondary securement usually consists of appropriately rated chain.

PINTLE HOOKS

Pintle hook hitches are more common than ball hitches on heavy-duty highway applications. A pintle hook assembly consists of a perpendicular **horn** (the "hook") and **drawbar eye** or **pintle eye**. The standard design used is the Jeep hook, a name derived from its application on U.S. military Jeeps. **Figure 34–39** shows a typical pintle hook assembly and the means used to lock it. Heavy-duty pintle hook assemblies are entirely manufactured out of tempered alloy steels and they should be bolted, not welded, to the trailer tongue unless otherwise instructed.

The function of a pintle hook is to engage to a drawbar eye, which itself is bolted to the tongue or drawbar assembly of the trailer. Pintle hooks are available in many different ratings determined by the type of application (off-highway, rough terrain operation requires high strength) and the weights of the trailer to be hauled. **Figure 34–40** shows a pintle hook engaged to a drawbar eye. Note how the primary lock is secured by the jam cog shown in the upper left side of the image.

Snubber Unit

Many pintle hooks are mounted to a pneumatic or hydraulic buffer unit known as a **snubber**. Peumatic snubbers tend to be more common in highway applications, while hydraulic snubbers tend to be used in off-road units. The function of a snubber is to minimize the effect of trailer bumping on braking. When the tow unit brakes are even slightly out of phase with the trailer, trailer bumping can cause handling

FIGURE 34–39 Typical pintle hook and locking assembly

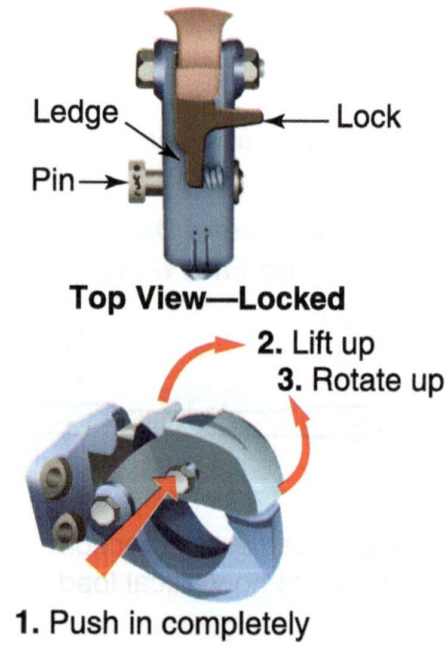

FIGURE 34–40 Heavy-duty pintle hook assembly

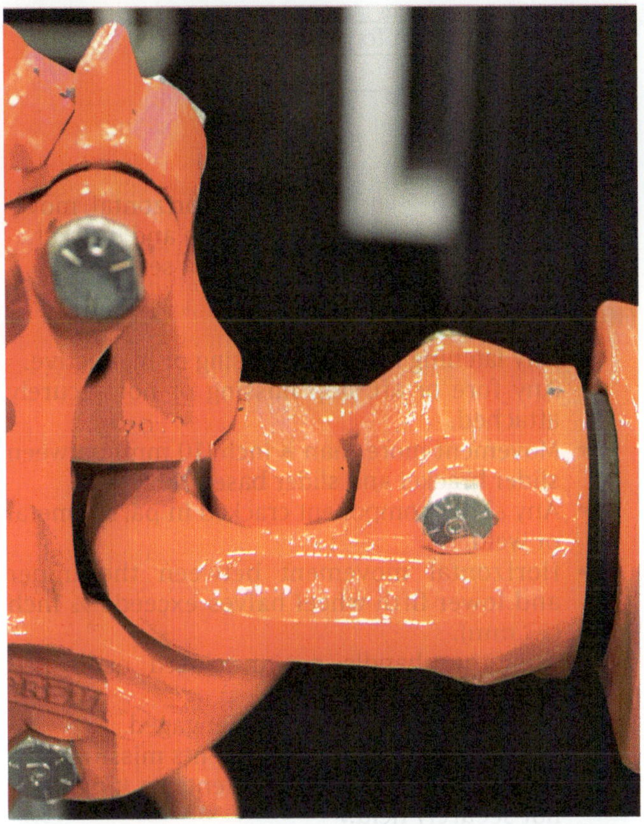

problems and even damage tractor drivetrains. **Figure 34–41** shows an on-highway pintle hook equipped with a pneumatic snubber.

Drawbars

Drawbars are manufactured out of tempered alloy steel with the pintle eye that engages to the pintle hook. They are available in a number of load ratings—again, dependent on application—tongue weight, and total trailer weight. Drawbar eyes are normally fastened to the trailer tongue assembly by means of SAE Grade 8 bolts at the shank. Both rectangular and square shanks are used. Two- and 3-inch (50 and 75 mm) eye diameters are used, with 2.5- and 3-inch diameters (64 and 75 mm) (Aus 2175/NZ/Asia 70 and 90 mm) being the most common. **Figure 34–42** shows two methods of attaching drawbar eyes to a converter dolly tongue.

FIGURE 34–41 Locked heavy-duty pintle hook with a pneumatic snubber

BALL HITCHES

Although not so common in heavy-duty applications, another means of coupling a full trailer to a tow unit is to use a ball hitch. A heavy-duty ball hitch assembly functions differently than the commonly used ball hitches used in automobile and light truck applications, because it is designed to engage with a standard pintle eye. All of the components in the hitch assembly are manufactured from tempered alloy steels and should not be heated. **Figure 34–43** shows a heavy-duty ball hitch locked and pin secured.

Pintle Hook Maintenance

The following maintenance and inspection procedure is recommended by SAF Holland for rigid-mount air-cushioned pintle hooks used on double and triple trailer trains. It can also be used to inspect straight trucks (including dump trucks) that haul rigid tongue trailers. The maintenance procedure should be performed every 30,000 miles (50,000 km) or 3 months, whichever comes first:

1. Use a pintle hook gauge to check for wear on the horn using a pintle hook gauge. The pintle hook should be replaced when wear exceeds ⅛ inch (3 mm).
2. Visually inspect the pintle hook for damage in the form of nicks, gouges, deformation, or cracks. The pintle hook should be replaced if any damage is present that could affect the safe use of the pintle hook.
3. Check the pintle hook plate mounting fasteners for specified torque. Inspect the mounting surface on the truck; cracks or indications of metal fatigue require repair or replacement.
4. Clean and inspect the air cushion system for proper operation; excessively worn, damaged or missing parts should be replaced.
5. Lube the latch pivots with a light oil lubricant. It is important to wipe away excess oil so that it does not attract sand and dirt.
6. Check the plunger adjustment. The plunger should firmly grip a 1.25-inch (32 mm) round bar when the air chamber is energized. When de-energized, a 1.63-inch (42 mm) diameter drawbar should be easily uncoupled from the pintle hook. If the plunger is out of adjustment, refer to the

FIGURE 34–42 Two methods of mounting a drawbar to a tongue

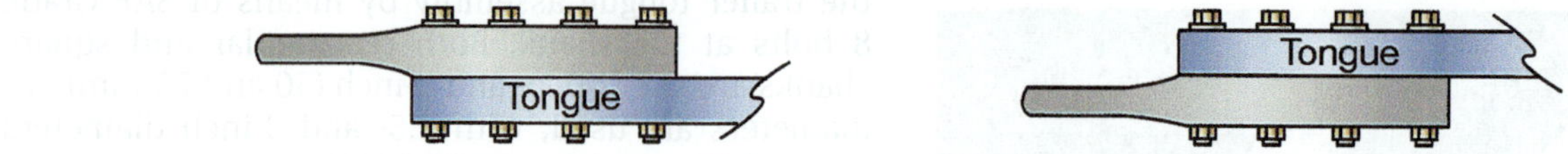

FIGURE 34–43 Heavy-duty ball hitch engaged to a pintle hook.

manufacturer's service literature for the proper adjustment procedure.

7. With the pintle latch closed, measure the clearance between the hook and the latch while lifting up on the latch. The latch should be replaced when the clearance exceeds 3/8 inch (10 mm).
8. Never apply any kind of lubricant to the horn of the pintle hook. This attracts sand and dirt, which is abrasive and accelerates the wear of the pintle and drawbar eye.

34.10 COUPLING DEVICE CVSA OOS

Commercial Vehicle Safety Alliance (CVSA) out-of-service (OOS) are not maintenance standards. They define the point at which highway equipment becomes dangerous to operate. OOS are used to issue citations during safety inspections. The following OOS apply to tractor/trailer coupling devices:

Fifth Wheels

- Mounting to frame fasteners missing or ineffective.
- Movement between mounting components.
- Mounting angle iron cracked or broken.
- Any welds or parent metal cracked.
- More than 3/8-inch (10 mm) horizontal movement between pivot bracket pin and bracket.
- Pivot bracket pin missing or not secured.

Fifth Wheel Sliders

- Any latching fasteners missing or ineffective.
- Any fore or aft stop missing or not securely attached.
- Movement more than 3/8 inch (10 mm) between slider bracket and slider base.
- Any slider component cracked in parent metal or weld.
- Horizontal movement between the upper and lower fifth wheel halves exceeds ½ inch (13 mm).
- Operating handle not in closed or locked position.
- Kingpin cannot be properly engaged.
- Locking mechanism components missing, broken, or deformed to the extent the kingpin is not securely held.
- Separation between upper and lower coupler, allowing light to show through from side to side.
- Cracks in the fifth wheel plate.

Exceptions: Cracks in fifth wheel approach ramps and casting shrinkage cracks in the ribs of the body of a cast fifth wheel.

Pintle Hooks

- Missing or ineffective mounting to frame fasteners.
- Mounting surface cracks extending from point of attachment (cracks in the frame at mounting bolt holes).
- Loose mounting plate.
- Cracked frame cross-member on which pintle hook is mounted.
- Cracks anywhere in pintle hook assembly.
- Any welded repairs to the pintle hook.
- Any part of the horn section reduced by more than 20 percent.
- Latch insecure.

Drawbar Assembly

- Any cracks in mounting attachment welds.
- Any missing or ineffective fasteners.
- Any cracks in drawbar.
- Any part of the eye reduced by more than 20 percent.

- Ineffective latching mechanism drawbar/towbar tongue.
- Missing or ineffective drawbar stop.
- Movement of more than ¼ inch (6 mm) between slider and housing.
- Any leaking, air or hydraulic cylinders, hoses, or chambers.
- Movement exceeding ¼ inch (6 mm) between subframe and drawbar at point of attachment.

SUMMARY

- The purposes of the fifth wheel are both to permit the truck/tractor to articulate around the trailer kingpin when turning corners and to support a percentage of the weight of a semitrailer.
- Fifth wheel plates may be manufactured of cast steel, forged steel, pressed steel, and forged aluminum alloy.
- The semi-oscillating fifth wheel is the most common highway tractor fifth wheel.
- A semi-oscillating fifth wheel articulates or oscillates about an axis perpendicular to the vehicle centerline, that is, on a fore-aft plane.
- Compensating fifth wheels are designed to provide both front-to-rear and side-to-side oscillation between the tractor and semitrailer. Side-to-side oscillation occurs below the fifth wheel plate surface and lessens the effect of trailer torque or twist acting on the frame of the tractor or lead trailer.
- Fully oscillating fifth wheels articulate on all planes and are used with gooseneck, low-bed trailers, especially those operated on rough, off-highway terrain.
- Elevating fifth wheels are used on yard shunt tractors.
- They may be equipped with either air or hydraulic lift mechanisms.
- Many fifth wheels are built with a sliding mechanism that permits longitudinal adjustment placement of the fifth wheel on the tractor frame. This allows the point at which the weight of the trailer is supported by the tractor to be altered according to load and legislative requirements.
- Relocation of a sliding fifth wheel alters the weight-over-axle distribution on a tractor when coupled to a trailer. As a fifth wheel is moved forward, the percentage of weight carried by the tractor steer axle increases.
- Some fifth wheels are manufactured with zero maintenance and a no-lube requirement.
- Some OEMs offer in-cab, pneumatic coupling and uncoupling. Electronic lock indicators display on the dash the coupling status of the fifth wheel.
- Fifth wheel mountings may be plate mounted or angle-on-frame mounted.
- The fifth wheel should be inspected every 30,000 miles (50,000 km) or 3 months, whichever comes first.
- Proper lubrication of the fifth wheel includes keeping water-resistant, lithium-based grease applied to the fifth wheel plate, lubricating all moving parts with light oil, and applying chassis grease to the grease zerks at the pivot points and the cam.
- Kingpins are usually welded to the trailer upper coupler unit and are used to couple a trailer to the fifth wheel.
- Kingpins come in two SAE standard sizes, 2 inches (50 mm) and 3.5 inches (90 mm). They are manufactured in special alloy steels and usually are welded to the upper coupler assembly. Bolted and removable kingpins also are used.
- The 2-inch (50 mm) kingpin is standard in on-highway applications while the 3.5-inch (90 mm) kingpin is more common in off-highway and offshore applications.
- A pintle hook and drawbar assembly is used to couple a full-trailer to a truck or lead trailer.
- Like fifth wheels, pintle and ball hook assemblies are a critical safety item and must be inspected. Most pintle hooks use a snubber device to dampen trailer bump.

REVIEW QUESTIONS

1. Which type of fifth wheel is most commonly used on highway tractors?
   a. fully oscillating
   b. semi-oscillating
   c. stabilized
   d. elevating
2. Which of the following is an advantage of a sliding fifth wheel?
   a. Weight-over-axle on the tractor can be altered.
   b. Weight-over-axle on the trailer can be altered.
   c. Trailer bridge formula can be altered.
   d. all of the above

3. Which type of fifth wheel permits articulation at any angle for use with gooseneck trailers operated on rough terrain?
   a. fully oscillating
   b. semi-oscillating
   c. elevating
   d. all of the above
4. How many fifth wheels would a standard B-train combination rig have in total?
   a. one
   b. two
   c. one and a pintle hook
   d. two and a pintle hook
5. Which type of fifth wheel uses a locking mechanism with a pair of transversely opposed lock jaws?
   a. a Holland A lock
   b. a Holland B lock
   c. a Fontaine
   d. all fifth wheels
6. Technician A uses a double-angle inboard-outboard mount when mounting a fifth wheel assembly that has no mounting plate. Technician B welds fifth wheel deck plate assemblies to the frame mounting angle iron. Who is correct?
   a. Technician A only
   b. Technician B only
   c. both A and B
   d. neither A nor B
7. SAF Holland recommends that a fifth wheel should be inspected every:
   a. 50,000 miles (80,000 km).
   b. 40,000 miles (64,000 km).
   c. 30,000 miles (50,000 km).
   d. 20,000 miles (32,000 km).
8. When lubricating the fifth wheel, Technician A lubricates all moving parts with water-resistant, lithium-based grease. Technician B lubricates the rails on a sliding fifth wheel with a mixture of oil and diesel fuel. Who is correct?
   a. Technician A only
   b. Technician B only
   c. both A and B
   d. neither A nor B
9. When using angle iron to mount fifth wheel brackets to a tractor frame, which of the following methods is recommended?
   a. welding to frame using E-7014 electrodes
   b. welding to frame using E-7018 electrodes
   c. using at least three SAE Grade 5 bolts on either side
   d. using at least five SAE Grade 8 bolts on either side
10. What is the maximum horizontal movement permitted at the fifth wheel before it can be considered CVSA OOS?
    a. ⅜ inch (9.5 mm)
    b. ¼ inch (6 mm)
    c. ⅛ inch (3 mm)
    d. ¹⁄₁₆ inch (1.6 mm)
11. To ensure that the tractor and trailer are properly coupled by the fifth wheel, check by:
    a. pulling the tractor forward with the trailer brakes set.
    b. measuring clearance between the fifth wheel plate and the trailer upper coupler.
    c. listening for an audible snap as the jaws close.
    d. observing that the release lever is in the engaged position.
12. Which is the most common method of attaching a kingpin to a trailer upper coupler assembly?
    a. SAE Grade 8 fasteners
    b. SAE Grade 5 fasteners
    c. welding
    d. lockpin
13. Trailer weight is transferred to the fifth wheel plate by which of the following?
    a. the bolster plate
    b. the kingpin
    c. the pintle hook
    d. the drawbar
14. Which type of fifth wheel is designed to minimize the effect of trailer twist on the tractor frame?
    a. semi-oscillating
    b. rigid
    c. no-tilt
    d. compensating
15. The air-release slide mechanism is preferred over the manual release when:
    a. slide length is short.
    b. slide length is long.
    c. frequent adjustment is necessary.
    d. safety is a factor.
16. Fifth wheel height is defined as:
    a. the distance from fifth wheel plate to the roof of the trailer when coupled.
    b. the distance from ground level to the fifth wheel top plate.
    c. the distance from ground level to the base of the fifth wheel.
    d. the distance from the base of the fifth wheel to the fifth wheel top plate.
17. What is the intended operating height of the bolster plate on most trailer upper couplers in highway use?
    a. 42 inches (107 cm)
    b. 47 inches (120 cm)
    c. 50 inches (127 cm)
    d. 60 inches (152 cm)
18. What is the normal maximum vehicle height on a tractor/trailer combination used on our highways?
    a. 144 inches (3.6 m)
    b. 12 feet 6 inches (3.8 m)
    c. 13 feet (4 m)
    d. 13 feet 6 inches (4.1 m)

19. Which of the following dimensions is equivalent to a standard SAE kingpin size?
    a. 1.5 inches (3.8 cm)
    b. 2.5 inches (6.3 cm)
    c. 3.5 inches (8.9 cm)
    d. 4.5 inches (11.4 cm)
20. Technician A says that when a sliding fifth wheel is moved forward on a tractor, the percentage of total vehicle weight supported by the trailer is reduced. Technician B says that when a tractor sliding fifth wheel is moved rearward, weight is removed from the tractor steer axle. Who is correct?
    a. Technician A only
    b. Technician B only
    c. both A and B
    d. neither A nor B

35

Prerequisites: Chapters 1, 2, 3, 6, and 13

HEAVY-DUTY HEATING, VENTILATION, AND AIR CONDITIONING SYSTEMS

OBJECTIVES

After reading this chapter, you should be able to:

- Understand the basic theory of heavy-duty truck air conditioning systems.
- Outline the requirements of the Clean Air Act that apply to a heavy-duty truck air conditioning system.
- List the five major components of a heavy-duty air conditioning system and describe how each works in the operation of the system.
- Explain how the thermostatic expansion valve or orifice tube controls the flow of refrigerant to the evaporator.
- Identify the refrigerants used in heavy-duty truck air conditioning systems.
- Describe the function of the main components in a typical heavy-duty air conditioning system.
- Recognize the environmental and personal safety precautions that must be observed when working on air conditioning systems.
- Identify air conditioning testing and service equipment.
- Test an air conditioning system for refrigerant leaks.
- Outline the procedure required to service a heavy-duty air conditioning system.
- Perform some simple diagnosis of air conditioning system malfunctions.
- Outline the advantages of connecting air conditioning management electronics to the chassis data bus and explain how to access the system.
- Explain how PWM-actuated controls are used to manage refrigerant circuit switches and move air doors in electronically managed HVAC systems.
- Describe how virtual switches are used in current HVAC systems.
- Identify the key J1939 source addresses (SAs) used to manage HVAC.
- Explain how a truck cab ventilation system operates.
- Describe the role a liquid-cooled heating system plays in a truck cab heating system.
- Describe some types of auxiliary heating and power units.

KEY TERMS

accumulator
air conditioning protection and diagnostics system (APADS)
air conditioning protection unit (ACPU)
block heater
blower motor/fan assembly
BlueCool
British thermal unit (Btu)
Clean Air Act
compressor
condenser
coolant heater
diagnostic trouble code (DTC)

electronic leak detector
electronic service tool (EST)
evaporator
evaporator thermostat
expansion valve
fault mode indicator (FMI)
fuel-fired heater
Global Warming Potential (GWP)
greenhouse gas (GHG)
heater control valve HVAC control module
hydrochlorofluorocarbons (HCFCs)
latent heat
latent heat of condensation
latent heat of vaporization
manifold gauge set
ozone depletion potential (ODP)
pulse width modulation (PWM)
ram air
receiver/dryer
refrigerant
scroll compressor
shorepower
sight glass
signal (detect) and actuation module (SAM)
source address
subcooling
supplementary coolant heater
swashplate compressor
thermistor
thermodynamics
vane compressor

Note: The objective of this chapter is to outline the principles of operation and repair fundamentals of heating, ventilation, and air conditioning systems (HVAC) so some of the mathematics of metric thermodynamics have not been included. However, it should help students to understand upfront that a Btu is almost exactly equivalent to the metric kilojoule; exact conversion values are provided later in the chapter.

INTRODUCTION

Heating, ventilation, and air conditioning (HVAC) systems are designed to keep the cab, or cab and bunk, in a truck at a comfortable temperature. The air conditioning system also helps clean (condition) the air in the cab by removing dust, pollen, smoke, and moisture. This chapter introduces the basics of air conditioning, followed by some of the repair and diagnostic techniques required of the service technician. Our focus is on truck-type air conditioning systems, but most of the principles outlined here also apply to trailer refrigeration and nonmobile refrigeration systems. Most current truck HVAC systems make some use of electronics to manage cab comfort but the core principles of operation have changed little over the years.

THERMODYNAMICS

To understand air conditioning theory, it is important for the service technician to have a basic understanding of how heat behaves. The behavior of heat is a branch of physics known as **thermodynamics**. An air conditioning system uses some very basic thermodynamic principles to remove heat from the cab of a truck and dissipate or lose it to the atmosphere outside the truck (**Figure 35–1**). The temperature inside the cab is reduced by removing heat faster than it comes in from the sun, the ventilation system, the engine (through the firewall), and the road pavement.

FIGURE 35–1 How air is "conditioned" in the cab

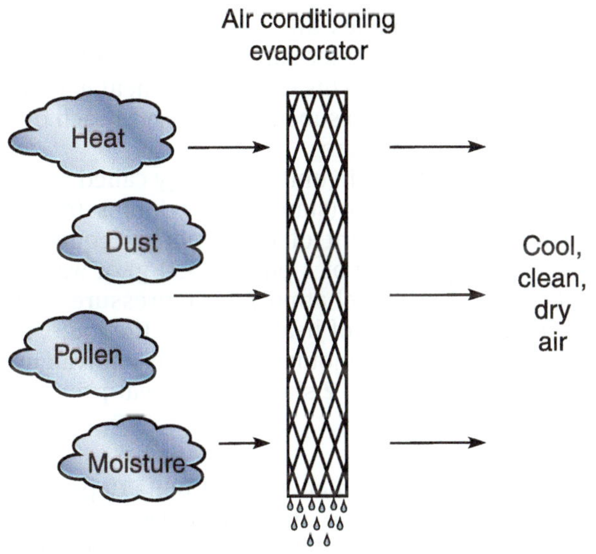

35.1 PRINCIPLES OF REFRIGERATION

Air conditioning and refrigeration systems manage basic thermodynamic principles to produce a more comfortable climate within an enclosed area. To understand how an air conditioning system works, we first have to know something about the states of matter, heat flow, and something called latent heat.

STATES OF MATTER

The three states of matter are solid, liquid, and vapor. Water can be easily observed in all three of these states:

- Solid water is known as ice.
- Liquid water is known as water.
- Vaporized water is known as steam.

The temperature of the water determines which of these three states it is in.

Absolute Heat, Heat Movement, and Measurement

Cold is a relative term. We use it to mean that one substance or atmospheric condition has less heat than another. All matter on Earth contains some heat. Any matter with a complete absence of heat would be at a temperature of –459°F (–273°C) which is known as absolute zero. At this temperature, all molecular movement in matter ceases. Heat is a form of energy. Any substance at a temperature higher than absolute zero has some heat energy.

Heat always moves away from the heat source. In other words, a warmer substance surrenders heat to cooler substances. This exchange of heat generally will continue until two substances close to each other are at equal temperatures.

Heat is measured in units of energy called **British thermal units (Btu)** (standard) or joules (metric). One Btu is the amount of heat energy required to raise the temperature of 1 pound (454 g) of water by 1°F (–17°C) at sea level atmospheric pressure. In the metric system, a unit of heat energy is the joule, normally expressed in kilojoules (kJ); a kJ is equivalent to 1.055 Btus. **Figure 35–2** shows some heat principles.

FIGURE 35–2 Principles of heat quantity, heat measurement, and latent heat

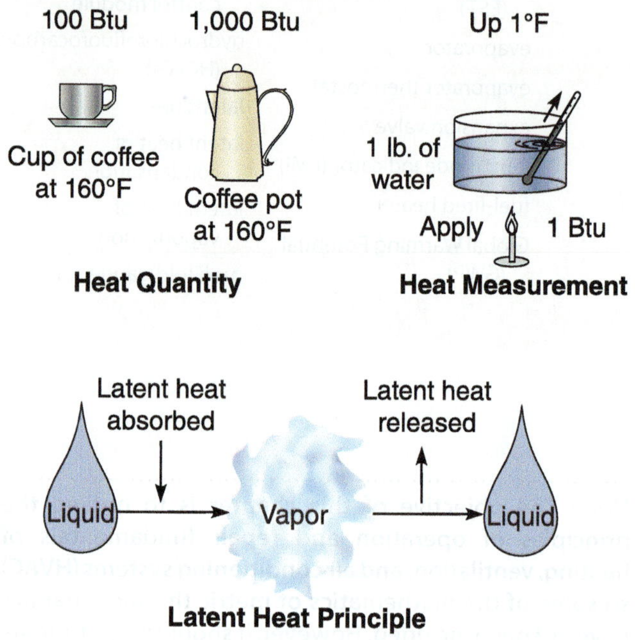

Pressure and Heat

The temperature at which a liquid boils depends on the pressure acting on the liquid. Decreasing the pressure lowers the boil point. Increasing the pressure raises the boil point. We know that atmospheric pressure is lower at high altitudes, which means that water will boil at a lower temperature on a mountaintop. We also know that coolant in a truck cooling system is maintained under pressure. This increases the boil point.

In an air conditioning system, the temperature of **refrigerant**, the cooling medium in the system, is controlled by changing the pressure acting on it. This also will determine the state of the refrigerant. In an air conditioning (A/C) system, we are concerned only with the refrigerant in its liquid and vapor states.

Latent Heat

Whenever a substance changes state, it either releases or consumes heat energy. **Latent heat** is the amount of heat necessary to change a substance from one state to another. When a liquid boils to form a gas, it consumes heat energy. The opposite occurs when a gas condenses to a liquid—when it releases heat. For instance, to convert 1 pound (454 g) of liquid water into steam at 212°F (100°C), 970 Btu (1,023 kJ) must be added.

The heat that we add to the water to change its state without increasing its temperature is known as the **latent heat of vaporization**. When we reverse the change of state, that is, change the water vapor or steam back into liquid, 970 Btu (1,023 kJ) of heat energy are given up or released as the condensation takes place. This is known as the **latent heat of condensation**. Note also that the temperature at which water freezes and at which ice melts are identical. When water freezes, heat energy is released, and when ice melts, heat energy is consumed. Figure 35–2 demonstrates the principle of latent heat.

The principle of latent heat is the basis of all A/C systems. The refrigerants used in A/C systems are chosen for their low boil temperatures, capability to change their state readily, and minimal impact on the environment.

SUMMARY OF AIR CONDITIONING PRINCIPLES

Remember these A/C principles:

- Heat always moves from a warmer area to a cooler area.

- When liquids are heated and evaporate to a vapor state, heat is absorbed.
- When a gas condenses from a vapor to a liquid state, heat is released.

As we go through this chapter, the components of a typical truck A/C system will be studied in some detail. Students should attempt to make the association between HVAC hardware and the thermodynamic principles that enable them to function. **Figure 35–3** shows the location of these components in a typical highway tractor, but we will begin by taking a look at refrigerant.

35.2 REFRIGERANT

The function of a refrigerant is to absorb heat from the air in the cab and transfer it to the atmosphere outside the cab. The refrigerants used in truck A/C systems are industrially manufactured chemicals of some complexity. Refrigerants in mobile systems are controversial and constantly undergoing change. Most refrigerants in use today have some **ozone depletion potential (ODP)** so there are plans to phase them out. However, in this chapter our focus will be on the refrigerants currently in use and we

FIGURE 35–3 Location of components of a typical heavy-duty truck heating and an air conditioning system

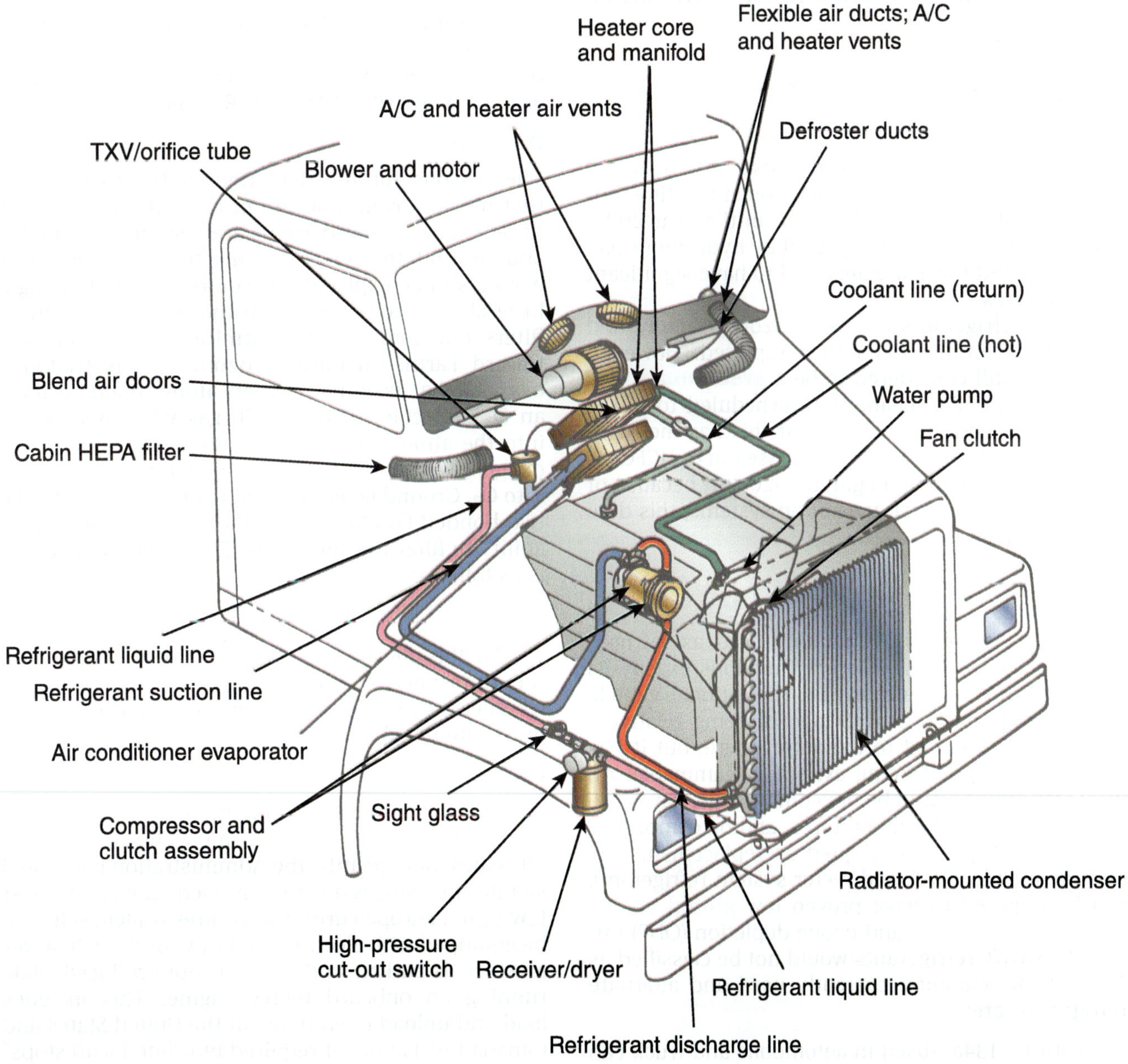

will avoid speculation on what might replace them as much as possible.

CHLOROFLUOROCARBON REFRIGERANTS

Until 1995, the refrigerant used in a truck cab A/C system was known as R-12. This is classified as a chlorofluorocarbon (CFC). R-12 refrigerant boils at –22°F (–30°C), and for years, it was considered an ideal mobile vehicle A/C refrigerant. However, because it is a substance of some toxicity as well as ozone depleting, its use has in theory become strictly controlled under the guidelines of the federal **Clean Air Act**. For whatever reasons, R-12 refrigerant continues to be available today and there is a diminishing but still significant number of truck R-12 HVAC systems on our roads.

HYDROCHLOROFLUOROCARBONS (HCFCs)

Truck and reefer cooling system use **hydrochlorofluorocarbons (HCFCs)** at this moment in time but these are scheduled for phase out in the year 2020. They will be probably be replaced by hydrofluoroolefins (HFOs) but these are also said to have significant disadvantages.

HCFC refrigerants are more ozone friendly and less toxic than the CFCs they replaced. However, HCFCs are still considered to be a **greenhouse gas (GHG)** and the Europeans were scheduled to phase out HCFCs beginning in 2011 but this has met with problems. In North America the phase-out of HCFCs is currently scheduled to begin in 2020, but because of problems with the first-generation systems, this date may be delayed.

ALTERNATIVE REFRIGERANTS

In Mexico, propane is commonly used as refrigerant in reefer systems, where it functions exactly as it does in the refrigerator in a recreational vehicle (RV). Propane has the advantage of being relatively unharmful if released to the atmosphere, but it has the disadvantage of being extremely flammable. An ideal refrigerant would be noncombustible, have convenient freeze/boil temperatures, and be completely nontoxic. Refrigerants adopted as replacements for existing mobile HVAC and reefer system refrigerants will be required to have proven low **global warming potential (GWP)** and ozone depletion (ODP) ratings. Low GWP refrigerants would not be classified as GHGs. Some common currently used and alternate refrigerants are:

- HCFC 134a—used in automobile and truck cab HVAC systems scheduled for phase out in 2020.
- HCFC 22—used in trailer reefer systems: possible phase-out in 2020.
- HCFC 40a—used in trailer reefer systems: possible phase-out in 2020
- HFC 422d—drop in replacement for HCFC 22 with a zero ODP rating but with a much higher GWP than HCFC 22.
- HFO 1234yf—currently used in some auto and truck A/C systems and rated as having 335 times less ODP than HCFC 134a and a GWP four times greater than CO_2. The advantage is that it is almost a drop-in replacement for HCFC 134yf, the disadvantages are that it is more flammable and costly.
- R 744 or CO_2—currently used as an industrial refrigerant and could be used in mobile applications but not a drop-in replacement.

For purposes of study in this chapter, we will refer to refrigerants using the prefix R—rather than HCFC or CFC. In other words, HCFC-134a will be referenced as R-134a and HFO-1234yf as R-1234yf.

Ozone Depletion

The reasons for the depletion of the ozone layer that has occurred at a rate of around 4 percent per decade since the mid-1970s are chemically complex and beyond the scope of this textbook. However, even a simple explanation requires understanding a little about ozone. Ozone in the upper atmosphere filters out some of the harmful sunlight directed toward Earth. Chemically, ozone (O_3) is triatomic oxygen, that is, three oxygen atoms bond to form an O_3 molecule. Chlorine (Cl) gas when discharged into the atmosphere attacks and leaches one of the oxygen (O_2) atoms from the O_3 bond, converting it into O_2. Ground level "oxygen" is diatomic O_2; that is, two bonded O_2 atoms form the molecule. O_2 has no ability to filter the sun's rays. The depletion reaction is as follows:

$$Cl + O_3 = ClO + O_2$$

All this means that any kind of Cl dump into the atmosphere is O_3 depleting. It does not matter whether the Cl is sourced from vehicle HVAC systems or a swimming pool.

Ozone Strategies

In both Europe and California there has been a more focused response to global warming and environmental issues, but recently, the Administration has tabled significant changes due to be implemented in the next few years. Europe currently requires trailer reefers to be equipped with electrical standby for use whenever a vehicle is parked so that CO_2 is not produced while running an onboard reefer engine. This includes load and unload operations. In the United States and Canada this is not yet required but "hotel load stops" providing **shorepower** to rigs are being established at truck stops and weigh stations on major highways, coast to coast. Shorepower required to run trailer reefer systems is supplied at 230 or 460 V-AC.

Other control measures include protecting the integrity of existing systems. To minimize refrigerant losses, flared fittings have been phased out in favor of brazed unions, reducing the incidence of leaks. Also under consideration are the use of hybrids and semi-hermetic compressors. Cryogenic trailer refrigeration systems have progressed beyond the trial phase and most users are happy with their performance. They are not costly, but despite the fact that their ODP and GHG ratings are favorable, insurance companies have regarded them with suspicion making them more expensive to operate. Cryogenic reefers are introduced in Chapter 33 under the heading of "Cryogenics."

Refrigerant Oils

HCFCs require the use of two refrigerant specific oils:

- Polyalkylene glycol (known as PAG)
- Polyolester (known as POE or ester)

Many original equipment manufacturers (OEMs) specify PAGs in heavy-duty truck A/C units using R-134a. Ester oil is hygroscopic (it absorbs moisture) and tends to be used in aftermarket lubricants rather than in PAG.

REFRIGERANT CONTAMINATION

In appearance, R-134a systems look similar to R-12 systems. However, refrigerants should never be mixed in a system; to emphasize this fact, it has been made illegal. Guidelines issued by the Environmental Protection Agency (EPA) specify that all refrigerants must be approved under the Significant New Alternatives Policy (SNAP). Only R-134a meets current SNAP guidelines for small-volume mobile refrigerants, but with a phase-out scheduled for 2020, other mobile refrigerants such as R-1234yf are under trial.

Any refrigerant introduced into an R-134a system other than R-134a itself would be considered a contaminant to the system. The correct refrigerant should always be used when servicing A/C systems. Introducing nonapproved refrigerants into a system can cause both performance and safety problems. Besides the effect of damaging the service equipment, contaminated refrigerant can cause some confusing problems when troubleshooting a system.

For example, with only a 10 percent contamination of R-134a introduced to an R-12 system, the vapor pressure at 100°F (38°C) would change from approximately 115 to 135 psi (7.9 to 9.3 bar). At any given temperature, the operating pressure would be much higher than specified, yet evaporator cooling would be inefficient when compared to the system operating with uncontaminated refrigerant. **Figure 35–4** shows a pressure-to-temperature comparison of R-12 and R-134a.

The service equipment required to diagnose and recharge an A/C system is specific to the refrigerant, so R-12 equipment should not be used to recover or recharge an R-134a system. In practice, you would have to go out of your way to accomplish this, because R-12 and R-134a systems are equipped with different access service fittings. Older R-12 vacuum pumps will not bring the system to a low enough pressure to boil off moisture, so more powerful vacuum pumps (typically ¾–1 hp / 0.56 to 0.75 kW) are used to evacuate R-134a systems.

Refrigerant can be tested before connecting the service equipment. This is done with a refrigerant identifier. These testers identify flammable refrigerants, operate with either R-12 or R-134a, indicate the percentage of the base refrigerant, and indicate that the refrigerant is at an acceptable level (98% pure). This type of device helps protect shop equipment and refrigerant supplies. **Figure 35–5** shows a typical refrigerant identifier kit. It should be noted that, in Mexico, propane is widely used as refrigerant, and if propane is drawn into a recovery station, it will disable it.

SHOP TALK

Refrigerant containers for R-12 and R-134a are color-coded. R-12 containers are white and R-134a containers are light blue and clearly marked. In the United States and Canada R-134a containers use ½-inch 16 acme threads (29° degree thread angle), which cannot be connected to an R-12 gauge set or recovery machine.

REFRIGERANT RECOVERY

The Clean Air Act, passed in 1992, has resulted in major changes in industrial, domestic, and vehicle A/C systems. The Clean Air Act regulates the following:

- It establishes a requirement to recover all ozone-depleting refrigerants. This means that they can never be vented to the atmosphere. Refrigerants must be recovered during the servicing and disposal of A/C or refrigeration equipment.
- It requires persons servicing A/C and refrigeration equipment to observe practices that reduce refrigerant emissions.
- It establishes equipment, off-site reclaimer, and technician certification programs.
- It includes sales restrictions (that may include licensing) on refrigerants regardless of container size.
- It mandates repair of leaks based on the annual leak rates of equipment.

FIGURE 35–4 Temperature/pressure relationship of refrigerant

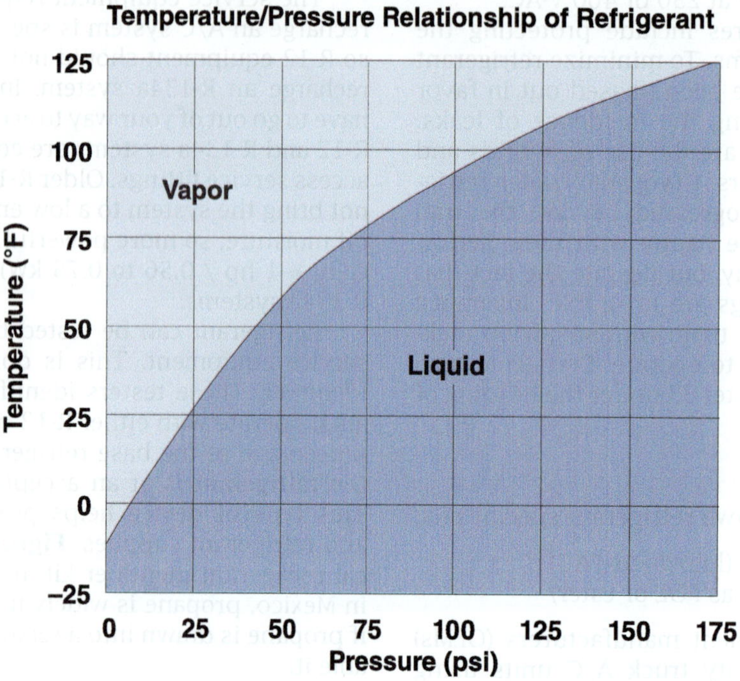

R-12 Temperature/Pressure Chart

Temp (°F)	Press (psig)	Temp (°F)	Press (psig)
16	18.36	100	117.16
18	19.68	102	120.86
20	21.04	104	124.63
22	22.44	106	129.48
24	23.88	108	133.11
26	25.36	110	136.41
28	26.88	112	140.49
30	28.45	114	144.66
32	30.06	116	148.91
34	31.72	118	153.24
36	33.42	120	157.65
38	36.07	122	162.15
40	36.97	124	166.73
45	41.68	126	171.40
50	46.70	128	176.16
55	52.05	130	181.01
60	57.74	135	193.52
65	63.78	140	206.62
70	70.19	145	220.30
75	76.99	150	234.61
80	84.17	155	249.54
85	91.77	160	265.12
90	99.79	165	281.37
95	108.25	170	298.30

R-134a Temperature/Pressure Chart

Temp (°F)	Press (psig)	Temp (°F)	Press (psig)
16	15.33	100	124.27
18	16.66	102	128.58
20	18.03	104	132.98
22	19.45	106	137.48
24	20.92	108	142.08
26	22.43	110	146.79
28	24.00	112	151.59
30	25.62	114	156.51
32	27.29	116	161.53
34	29.01	118	166.66
36	30.79	120	171.89
38	32.63	122	177.24
40	34.53	124	182.70
45	39.52	126	188.27
50	44.90	128	193.96
55	50.69	130	199.76
60	56.90	135	214.78
65	63.55	140	230.54
70	70.67	145	247.08
75	78.27	150	264.40
80	88.38	155	282.53
85	95.01	160	301.49
90	104.19	165	321.29
95	113.94	170	341.96

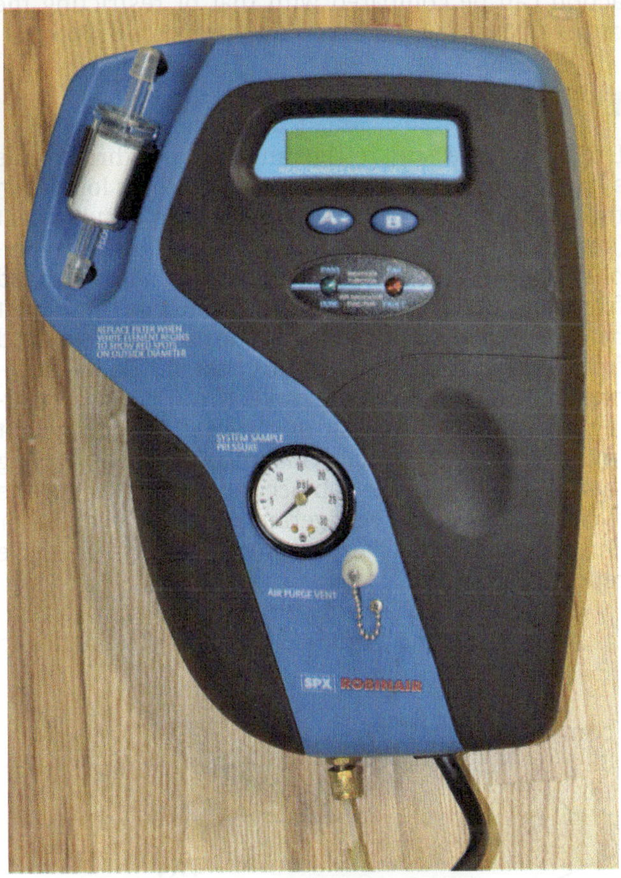

FIGURE 35–5 Refrigerant identifier kit

- It requires that ozone-depleting compounds be removed before the disposal of the appliances, and that all A/C and refrigeration equipment, with the exception of small appliances, be provided with a service fitting that would allow for the recovery of the refrigerant.

AIR CONDITIONING CERTIFICATION

To work on vehicle A/C systems, technicians must be certified. Certification can be obtained by successfully completing a federally approved technician certification program. Three training programs in particular are recognized by the federal government: the Mobile Air Conditioning Society (MACS), the International Mobile Air Conditioning Association (IMACA), and the National Institute of Automotive Service Excellence (ASE). The EPA has reserved the right to require recertification if and when "environmentally significant changes in equipment for motor vehicle air conditioning systems occur." The EPA also has the authority to revoke certification if a technician fails to demonstrate the ability to properly use the equipment. All recovery/recycling equipment must be Underwriters Laboratory (UL) certified to meet EPA and SAE standards.

35.3 THE REFRIGERATION CYCLE

All A/C systems use a refrigerant medium and cycle it continuously through a closed, pressurized system. The cycle consists of four stages: compression, condensation, expansion, and evaporation. As the refrigerant is pumped through this cycle, heat is removed from a hot area (the cab) and released to the atmosphere outside the vehicle. **Figure 35–6** shows a refrigeration cycle. Note that it is divided into a high side and low side. High side and low side are used to describe the pressures of the refrigerant in each side of the system.

COMPRESSOR STAGE

To understand the refrigeration cycle, it is normal to begin at the compressor. The function of the compressor is to change a low-pressure refrigerant gas to a high-pressure gas. Some compressors operate on a principle similar to that of the air compressor used in a truck air system, but other types of pumps also are used. Refrigerant as a low-pressure gas is sucked into the compressor on the downstroke of the piston. The gas is then compressed on the piston upstroke. The process of compressing the refrigerant gas creates heat just as an air compressor heats up the air it compresses. This is known as the heat of compression.

FIGURE 35–6 The refrigeration cycle

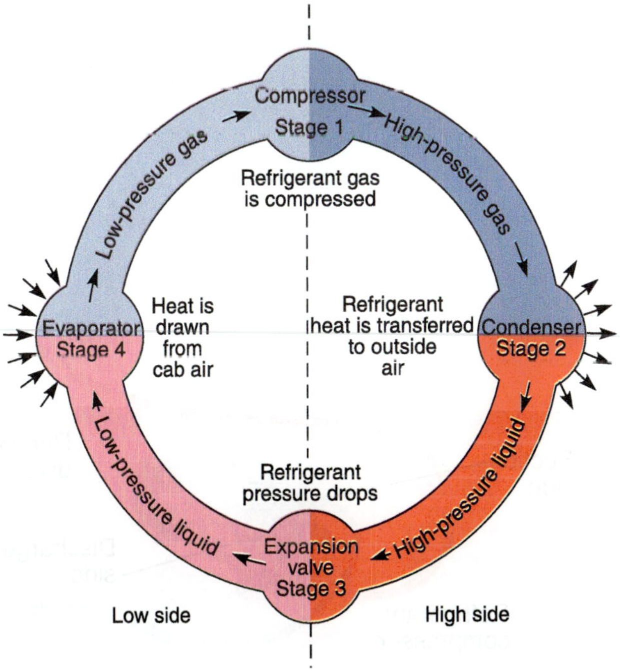

CONDENSER STAGE

High-pressure gas exits the compressor and is forced through the circuit to the **condenser**. The condenser is a heat exchanger. It is usually located in the vehicle airflow so it is subject to the **ram air** effect (that is, the airflow created by vehicle movement) or air movement created by the engine cooling fan. Heat moves from the refrigerant to the outside air. Cooling of the refrigerant causes it to condense. This results in the refrigerant changing to a liquid state. Therefore, cooler high-pressure liquid exits the condenser.

EXPANSION VALVE STAGE

The high-pressure liquid exiting the condenser is then forced through the system to the **expansion valve**. The expansion valve restricts the flow of the refrigerant, which has the effect of lowering its pressure as it exits the valve. Because it provides the restriction that enables the high pressure to be achieved upstream and reduces pressure downstream, it is the dividing point in the system for high/low pressure. The expansion valve allows the expansion of the liquid refrigerant to form a fine, atomized mist, which will allow it to absorb heat and vaporize more easily. Its action can be compared with that of restricting the outlet of a garden hose, creating a mist of droplets.

EVAPORATOR STAGE

Liquid, low-pressure refrigerant is moved through the system to the evaporator. The evaporator is located in the cab of the vehicle. Its function is to remove heat (supplied by the blower) from the cab. This is accomplished when the state of the refrigerant is changed from that of a liquid to a gas. Because of the latent heat of vaporization, when a substance is changed from a liquid to a gas, it absorbs heat. It is in the evaporator stage that heat is removed from the recirculating cab air. It is this heat that causes the refrigerant in the evaporator to boil. This means that the cooling that takes place in the cab occurs in this stage of the cycle. Low-pressure refrigerant exits the evaporator and is returned to the compressor where the cycle repeats itself.

Figure 35–7 diagrams the components in a typical A/C circuit, and **Figure 35–8** uses schematics to

FIGURE 35–7 Heavy-duty truck air conditioning system components

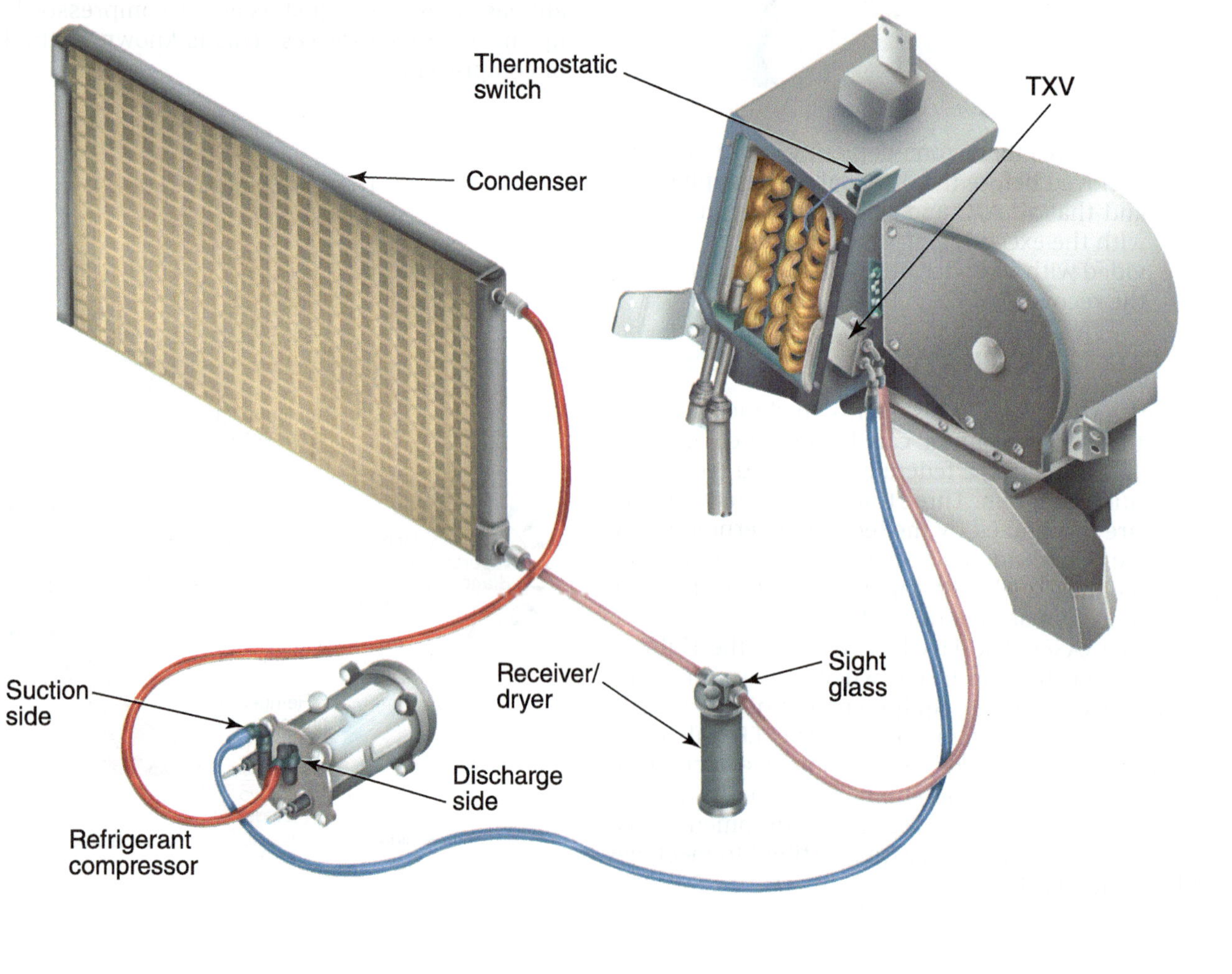

FIGURE 35–8 Refrigerant flow cycle

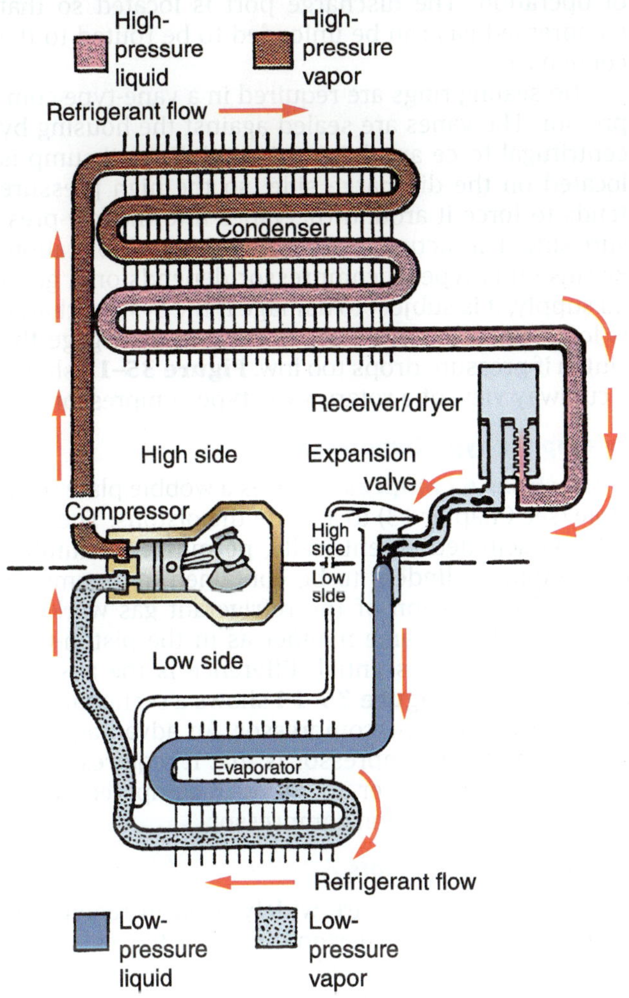

show the same thing. Be sure to note the direction of flow and the state of the refrigerant in each.

35.4 AIR CONDITIONING SYSTEM COMPONENTS

Now we will take a closer look at the components found in a heavy-duty A/C system. The major components are:

- Compressor
- Condenser
- Receiver/dryer or accumulator
- Expansion valve or orifice tube
- Evaporator

However, to put together an A/C system, other components are also required. This section takes a look at both the major and minor components that make an A/C system operate.

THE A/C COMPRESSOR

The **compressor** is the heart of the heavy-duty A/C system. It is one of the components that separate the high-pressure and low-pressure sides of the system (the other is the orifice tube or TXV). It is responsible for the movement of the refrigerant through the circuit. The primary purpose of the A/C compressor is to draw low-pressure vapor from the evaporator and compress it into a higher temperature, high-pressure vapor. This results in the refrigerant having a higher temperature than the surrounding air. The expansion valve located downstream from the compressor discharge provides the restriction required to define the high-pressure side. The compressor also circulates the refrigerant in the system by creating a vacuum on its inlet side and pressure on its outlet side. The compressor is located in the engine compartment and is driven by engine power. Numerous types of compressors are in use today, but they generally can be grouped into the following three types.

Piston-Type Compressors

The piston type of compressor has a crankshaft and its pistons are arranged in an in-line, axial, radial, or V configuration. They operate like a small two-stroke cycle engine. The pistons are actuated by a crankshaft. An intake stroke takes place on piston downstroke, and a compression stroke takes place on piston upstroke. On the intake stroke, low-pressure refrigerant vapor from the evaporator is drawn into the compressor cylinder through reed valves. These one-way valves are designed to permit gas flow in one direction only and are sealed to prevent reverse flow by pressure. The inlet reed valve permits refrigerant vapor to be drawn into the compressor cylinder on piston downstroke, as shown in **Figure 35–9**.

When the piston reverses and is driven upward, the intake reed valves seat, sealing the compressor cylinder. During the upward or compression stroke, the vaporized refrigerant is compressed, as shown in **Figure 35–10**. This increases both the pressure and the temperature of the refrigerant gas without adding heat. Outlet or discharge reed valves then open, allowing the high-pressure, high-temperature refrigerant to move to the condenser. The outlet reed valves represent the beginning of the high side of the system. Reed valves are made of spring steel, which can be weakened or broken if improper charging procedures are used. **Figure 35–11** shows a cutaway view of a piston-type compressor.

Rotary Vane Compressor

A **vane compressor** consists of a rotor with several vanes driven within a rotor housing. As the compressor shaft rotates, the vanes are moved into the rotor housing by centrifugal force, which forms chambers. As the refrigerant is drawn through suction ports

FIGURE 35–9 Compressor piston downstroke creates vacuum which opens the inlet reed valve drawing in low-pressure varporized refrigerant.

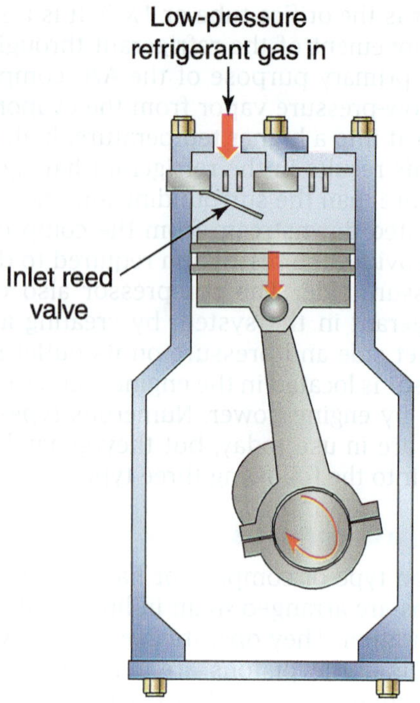

FIGURE 35–10 Compressor piston upstroke closed inlet reed valve and opens exhaust valve to discharge high-pressure/-temperature vapor.

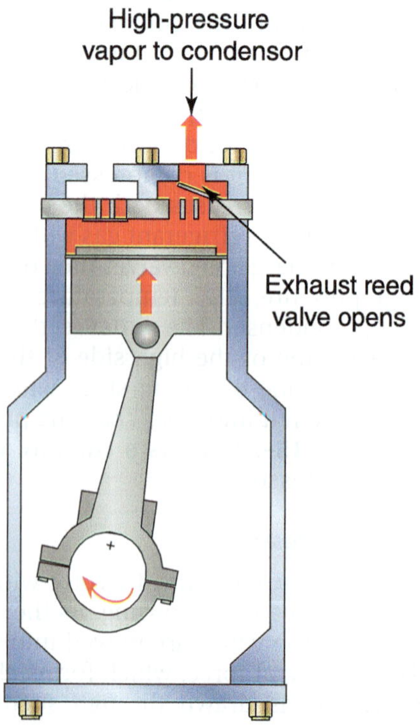

into these chambers, the chamber volume reduces as the rotor turns. **Figure 35–12** shows this principle of operation. The discharge port is located so that compressed gas can be unloaded to be routed to the condenser.

No sealing rings are required in a vane-type compressor. The vanes are sealed against the housing by centrifugal force and lubricating oil. The oil sump is located on the discharge side, so the high pressure tends to force it around the vanes into the low-pressure side. This action ensures continuous lubrication. Because this type of compressor depends on a good oil supply, it is subject to damage if the system charge is lost. A protection device is used to disengage the clutch if pressure drops too low. **Figure 35–13** shows a cutaway view of a rotary vane-type compressor.

Swashplate-Type Compressors

A **swashplate compressor** uses a wobble plate principle (see Chapter 13) to actuate the pistons. The cam rotor is actuated eccentrically, permitting a number of pumping cylinders to be contained in a compact space. Compression of the refrigerant gas works in pretty much the same manner as in the piston-type compressor. The essential difference is the absence of a crankshaft. **Figure 35–14** shows a sectional view of a swashplate-type compressor. An advantage of a swashplate-type compressor is that it requires about one-half of the power to drive it compared to an equivalent-sized piston compressor.

Scroll-Type Compressors

Scroll compressors are widely used in reefer A/C systems, notably by Thermo King, which uses what it calls its hermetically sealed scroll compressor. A scroll compressor consists of a pair of scrolls, one of which is stationary; the nonstationary scroll orbits the stationary scroll, trapping and compressing refrigerant gas. This produces pockets of gas in various stages of compression beginning at induction (low presure) and ending at discharge (high pressure). The result is a simultaneous induction-to-discharge cycle that requires less drive energy and reduces maintenance interventions. This type of compressor is sometimes known a *hermetic scroll compressor*.

COMPRESSOR DRIVES

Almost every compressor is equipped with an electromagnetic clutch as part of the compressor pulley assembly. This clutch is designed to engage the pulley to the compressor shaft when the clutch coil is energized. When the clutch coil is energized, engine power is transmitted to the compressor drive shaft, putting the compressor into effective cycle. When the clutch coil is not energized, the compressor pulley freewheels, and the compressor is not rotated. The clutch is therefore engaged by an electromagnetic field and disengaged by springs when the field collapses.

FIGURE 35–11 Cutaway view of a piston-type compressor

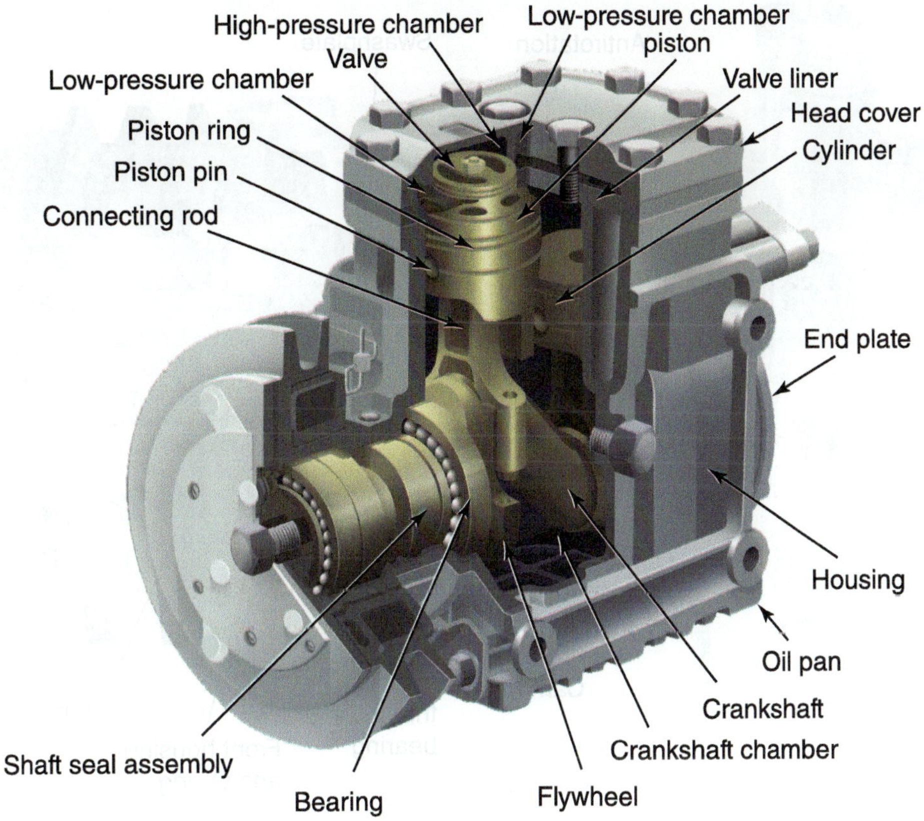

FIGURE 35–12 Operating principle of a rotary vane compressor

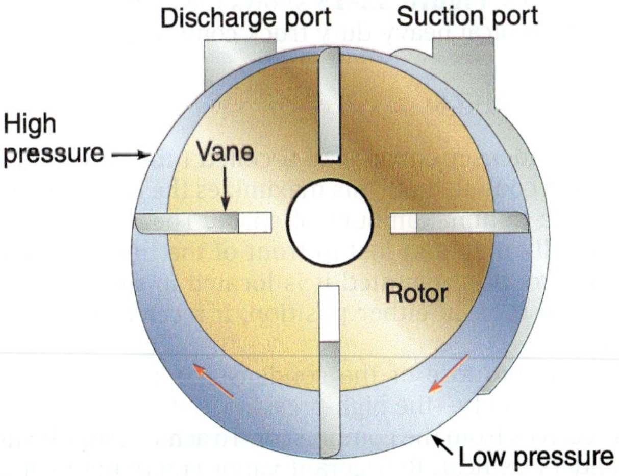

The compressor effective cycles are managed by an electrical circuit, which usually includes a thermostat. In most current truck A/C systems, electronics are used to help manage the A/C cycles.

Two types of electromagnetic clutches are used. Older A/C systems used a rotating coil-type clutch.

FIGURE 35–13 Cutaway view of a typical rotary vane compressor

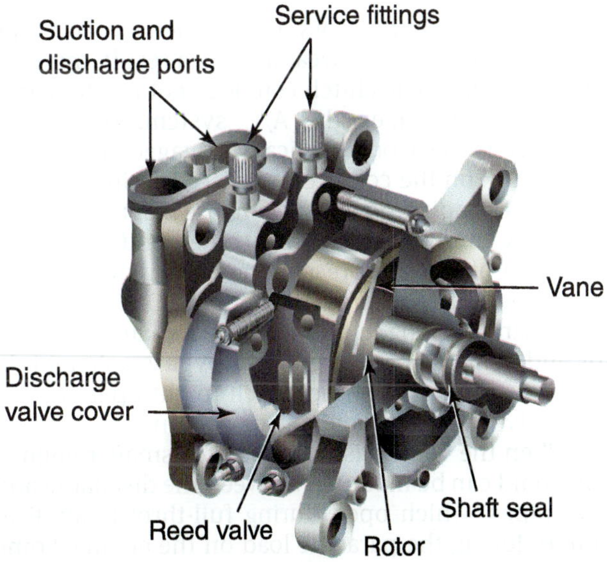

The magnetic coil, which engaged or disengaged the compressor to the drive pulley, was mounted within the pulley and rotated with it. Electrical connections

FIGURE 35–14 Cutaway view of a swashplate compressor

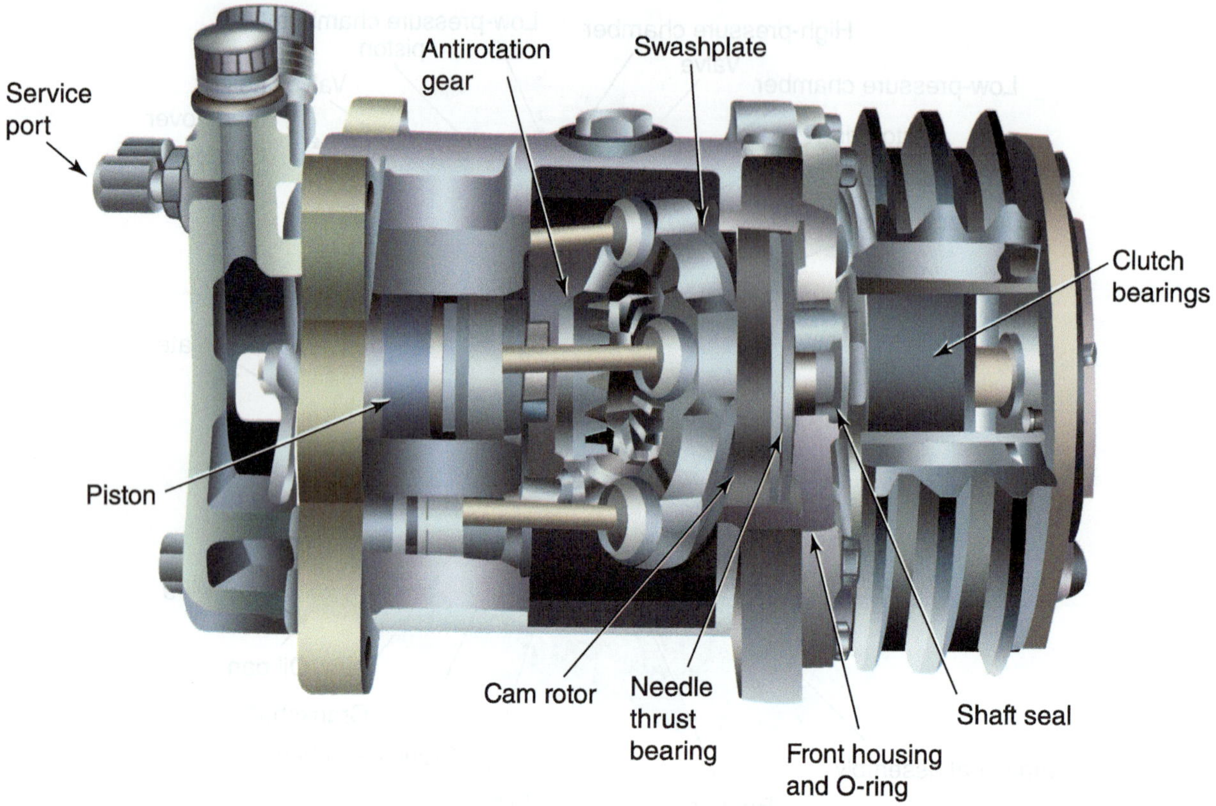

for the clutch operation are made through a stationary brush assembly and rotating slip rings, which are part of the field coil assembly.

Stationary Coil Clutches

More recent compressors use stationary coil-type clutches. These reduce wear and, as a result, increase the service life. The clutch coil does not rotate. When the driver energizes the A/C system, the pulley assembly is electromagnetically engaged to the stationary coil on the compressor body, locking it to the clutch. Depending on the system, the magnetic clutch is usually thermostatically controlled and cycles the operation of the compressor, depending on evaporator temperature, to prevent freezing. One OEM uses a constant turning unit on a swashplate compressor and controls pressure with a bypass solenoid. In some system designs, the clutch might operate continuously when the system is turned on.

When the vehicle is powered by a smaller engine, a solenoid can be mounted between the discharge and inlet pipes, which open during full-throttle application, reducing the parasitic load on the engine. Some truck A/C compressors use a dry reservoir module (DRM) to cut out the compressor to reduce load on the engine at times when full engine power is required. With stationary coil design, service is not usually necessary except for an occasional check on the electrical connections. Some A/C systems, using a two-wire, isolated ground circuit, use a diode to suppress voltage spikes. **Figure 35–15** shows the clutch assembly from a typical heavy-duty truck compressor.

CONDENSER

The condenser consists of a coiled tube surrounded by thin cooling fins. This maximizes the heat transfer, using a minimal amount of space. The condenser is normally mounted just in front of the truck radiator, but if rooftop mounted it is located in the center of the cab roof. In either position, it receives plenty of ram air.

The purpose of the condenser is to condense—that is, liquefy—the high-pressure, hot refrigerant gas it receives from the compressor. To achieve this, it surrenders its heat. Refrigerant vapor enters the inlet at the top of the condenser, and as the hot vapor passes down through the condenser coils, heat is transferred from the hot refrigerant to the cooler air flowing through the condenser coils and fins. This process causes the refrigerant to change state from a high-pressure hot vapor to a high-pressure warm liquid.

FIGURE 35–15 Compressor clutch assembly

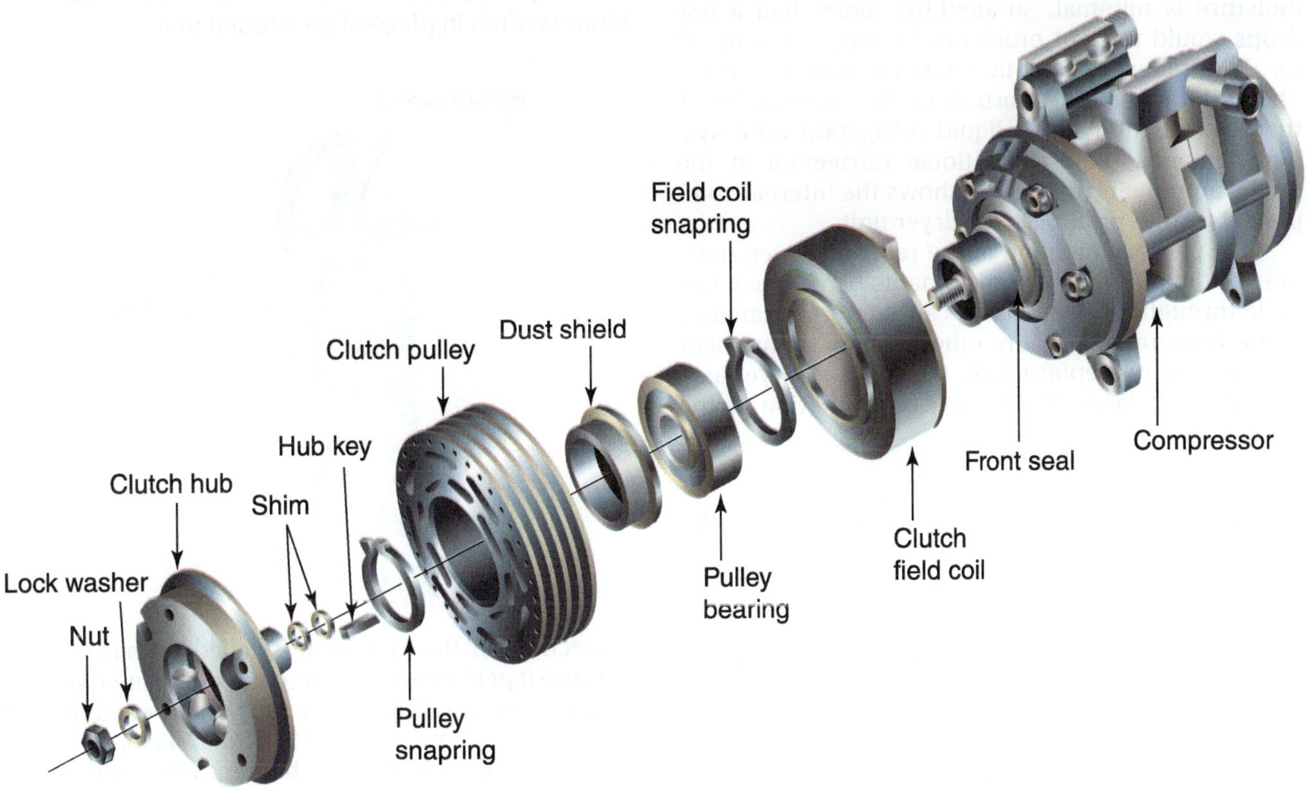

The high-pressure warm liquid flows from the outlet at the bottom of the condenser and is routed to the receiver/dryer or orifice tube, depending on the system. **Figure 35–16** shows a typical condenser. In an A/C system operating under an average heat load, the condenser will have a combination of hot refrigerant vapor (including saturated vapor) in the upper two-thirds of its coils, with the lower third of the coils containing the warm condensed liquid refrigerant.

RECEIVER/DRYER

The **receiver/dryer** performs three important functions in the refrigeration cycle:

- It filters particles from the liquid refrigerant.
- It removes moisture from the liquid refrigerant.
- It stores a small portion of the liquid refrigerant until it is needed in the system evaporator(s) to meet high cooling demand.

High-pressure liquid refrigerant flows from the condenser to the receiver/dryer. To protect the system from possible moisture-generated acid corrosion, a drying agent, or desiccant (moisture-absorbing material such as silica, aluminum, or silica gel), absorbs

FIGURE 35–16 Typical condenser. Restricted airflow (dirt/bent fins) results in poor heat dissipation.

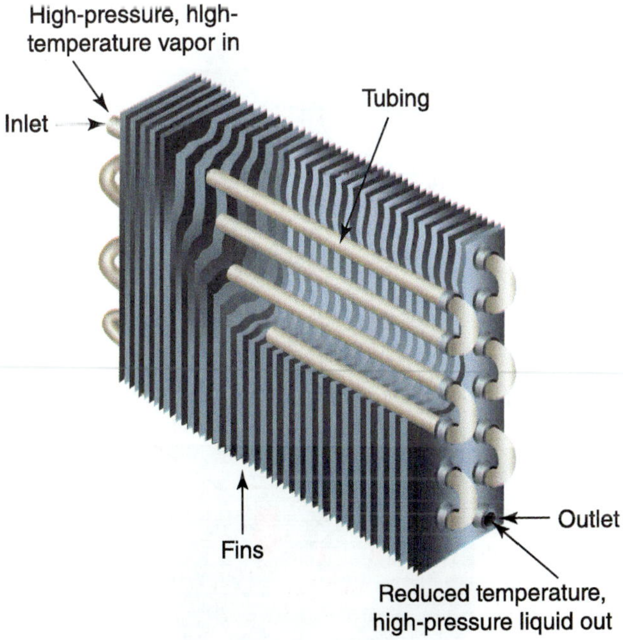

and separates any trace moisture contained in the liquid refrigerant. The ability of desiccant to retain moisture is minimal, so anything more than a few drops would present problems. Leaving it uncapped for 5 minutes results in moisture contamination. The cylinder or tank portion of the receiver/dryer then collects and stores liquid refrigerant until system demands require additional refrigerant at the evaporator(s). **Figure 35–17** shows the internal components of a typical receiver/dryer unit.

The primary function of the receiver/dryer, however, is the removal of contaminants before they can cycle through the system and cause internal damage. If the compressor or any other system component fails or requires replacement, the receiver/dryer also should be replaced to prevent contamination of the system.

Binary Switch

A binary pressure switch can be located on the receiver/dryer. It provides low- and high-pressure protection for the system. It is called a binary switch because it performs two functions with a single switch. When the high-side pressure falls to a predetermined value 15 to below 30 psi (103 to 207 kPa), electrical contacts in the binary switch open. When these contacts open, the compressor clutch will disengage and refrigeration cycle will stop. Electrical contacts in the binary switch also open when the high-side pressure rises above 300 to 350 psi (2068 to 2413 kPa), once again disengaging the compressor clutch and stopping the refrigeration cycle.

More recent electronically managed truck A/C systems use a virtual binary switch. The overall operating principle changes little, but instead of an integral switch, a pressure sensor or transducer signals the HVAC controller which manages the compressor clutch cycling. **Figure 35–18** shows the location of the binary switch on a typical receiver/dryer.

Trinary Switch

Some heavy-duty trucks are equipped with a trinary pressure switch instead of a binary pressure switch. The trinary switch is mounted on the receiver/dryer in a similar fashion to the binary switch. The trinary pressure switch performs three functions to monitor and control pressure inside the A/C system. The low- and high-range functions are the same as on the binary switch system.

1. A low-range pressure function prevents compressor operation when the refrigerant charge has been lost or when ambient temperature is too cold.
2. A midrange pressure function activates the engine fan clutch as system pressure reaches midrange. (On some A/C systems, an electrically powered

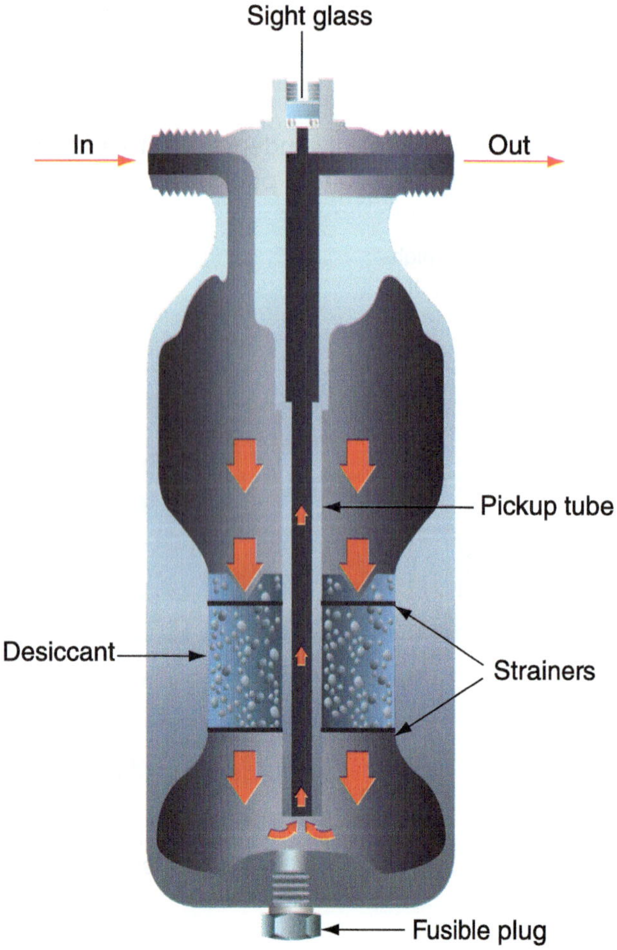

FIGURE 35–17 Internal components of a receiver/dryer unit

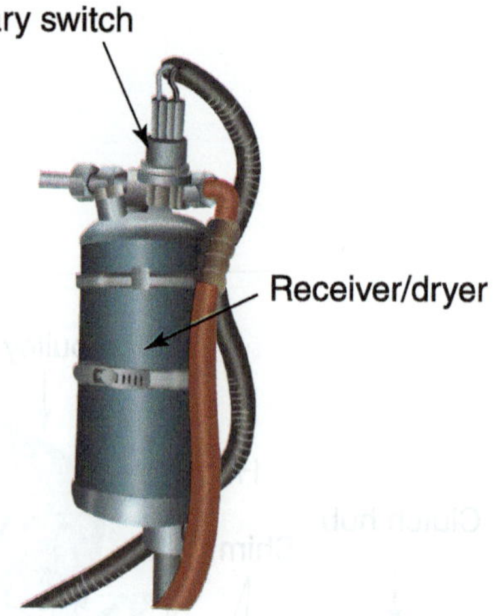

FIGURE 35–18 Location of a binary switch: Some contemporary ECU-controlled units use a virtual binary switch in place of an integral unit.

condenser cooling fan is activated instead.) This increases airflow to the condenser and stabilizes or lowers system operating pressures. The switch cycles on and off to maintain operating pressures.
3. A high-range pressure function turns off the compressor if the system pressure is too high.

All three switch functions reset when the proper system pressures are restored. Some electronically managed truck A/C systems use a virtual trinary switch. The overall operating principle changes little, but instead of an integral switch, a pressure sensor or transducer signals the HVAC controller which manages compressor clutch cycling.

Pressure Relief Valve

A pressure relief valve is also often located on the receiver/dryer. It provides an added high-pressure relief feature if a failure were to develop in the high-pressure cutoff switch. The pressure relief valve is designed to pop off when the refrigerant pressure exceeds a preset maximum safe pressure value. Pop-off pressure can often be as high as 400 psi (2758 kPa).

Fusible Plug

Another protective device is a fusible plug. A fusible plug is usually used in place of a pressure relief valve. The fusible plug will melt when the refrigerant temperature reaches approximately 230°F (110°C). When this happens, the refrigerant is discharged from the system. Unlike the pressure relief valve, when the fusible plug melts, the unit must be replaced. The pressure relief valve will close when the excess pressure is exhausted. An example of a fusible plug is shown in **Figure 35–19**.

Accumulator

An **accumulator** performs three functions:

1. It filters particles from the refrigerant.
2. It removes moisture from the refrigerant.
3. It separates refrigerant into vapor and liquid and allows only vapor to flow to the compressor from the evaporator, preventing compressor damage (a liquid cannot be compressed).

Two of these functions are similar to those of the receiver/dryer. The important difference is the location of the accumulator. An accumulator is located at the outlet of the evaporator, whereas the receiver/dryer is on the inlet side. Systems that use an accumulator operate on what is described as a flooded evaporator system.

The receiver/dryer and the accumulator are often neglected when the A/C system is serviced or repaired. Failure to replace either can lead to poor system performance or failure of other system parts. It is recommended that these components be replaced whenever the system has lost refrigerant or has been open to the atmosphere for any length of time. More important, they should be replaced whenever a compressor has malfunctioned and contaminated the system with metal, aluminum, or plastic particles. **Figure 35–20** shows the operating principle of a typical accumulator.

In addition to receiver/dryers and accumulators, some systems use in-line filters and filter dryers. These provide some supplementary drying and filtering capacity. An in-line filter installed upstream from a replacement orifice tube will prevent repetitive plugging and subsequent compressor damage caused by poor lubricant flow through the orifice tube.

FIGURE 35–19 Fusible plug: This design is not self seal following a high-pressure incident.

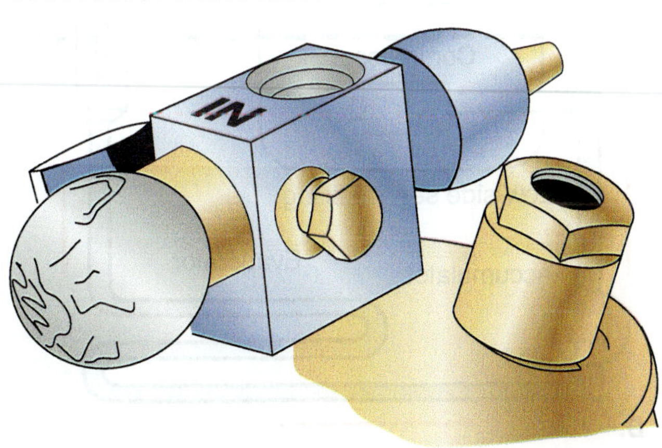

FIGURE 35–20 Typical accumulator: located inline on low-pressure side to protect the compressor from hydraulic lock.

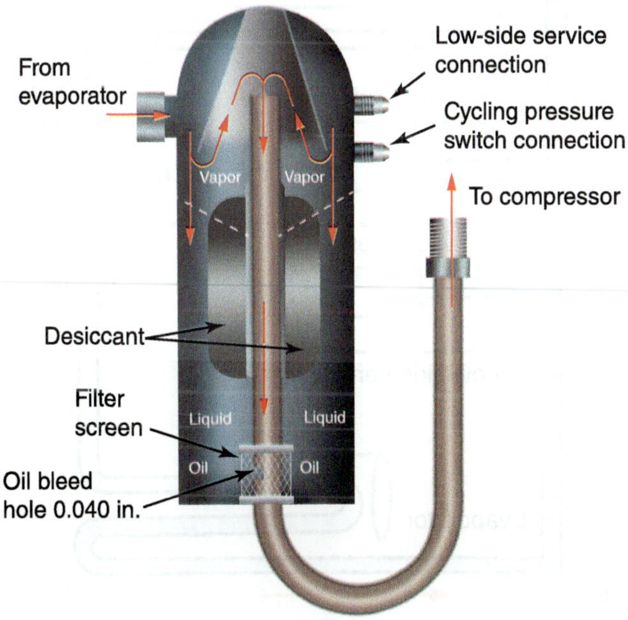

THERMOSTATIC EXPANSION VALVE/ ORIFICE TUBE

A thermostatic expansion valve or orifice tube is used to divide the high-pressure and low-pressure sides of the refrigerant cycle. The refrigerant flow to the evaporator must be controlled to obtain maximum cooling and to ensure complete evaporation of the liquid refrigerant within the evaporator. Two devices are used to accomplish this—either a thermostatic expansion valve (TXV) or a fixed orifice tube (**Figure 35–21**). One of these devices must be incorporated into the system.

- The TXV or fixed orifice tube both function to provide a pressure drop in the circuit, which changes high-pressure refrigerant liquid into a low-pressure liquid.
- The TXV additionally regulates the amount of refrigerant entering the evaporator to match the cooling demand.

Thermostatic Expansion Valve (TXV)

This valve is mounted downstream from the receiver/dryer before the inlet to the evaporator. It separates the high-pressure side of the system from the low-pressure side. A TXV functions to regulate refrigerant flow to the evaporator. Excess refrigerant flow through the evaporator can cause evaporator flooding and compressor damage because of sludging liquid refrigerant. In operation, the TXV regulates the flow to the evaporator by balancing inlet flow to the outlet temperature.

Both external and internal equalized thermostatic expansion-type valves are used in heavy-duty A/C systems. The only difference between the two valves is that an externally equalized TXV uses an equalizer line connected to the evaporator outlet line as a means of sensing evaporator outlet pressure. Internally equalized TXVs sense evaporator inlet pressure through an internal equalizer passage.

Attached to the top of an external pressure-type expansion valve is an externally insulated capillary tube with a temperature measuring device; this can be a measuring bulb or a thermistor. When a measuring bulb is used, it is clamped to the outlet (suction) pipe of the evaporator. The sensing bulb and tube are filled with gas (typically, the refrigerant used in the system) that expands and contracts according to the temperature surrounding the bulb. The expansion and contraction of the bulb gas exerts pressure on an internal diaphragm, which moves the metering needle valve against the closing spring pressure. The temperature bulb senses the temperature of the refrigerant leaving the evaporator.

The externally equalized expansion valve has a second capillary tube attached under the valve

FIGURE 35–21 (A) Expansion valve (receiver/dryer) system; and (B) orifice tube (accumulator system).

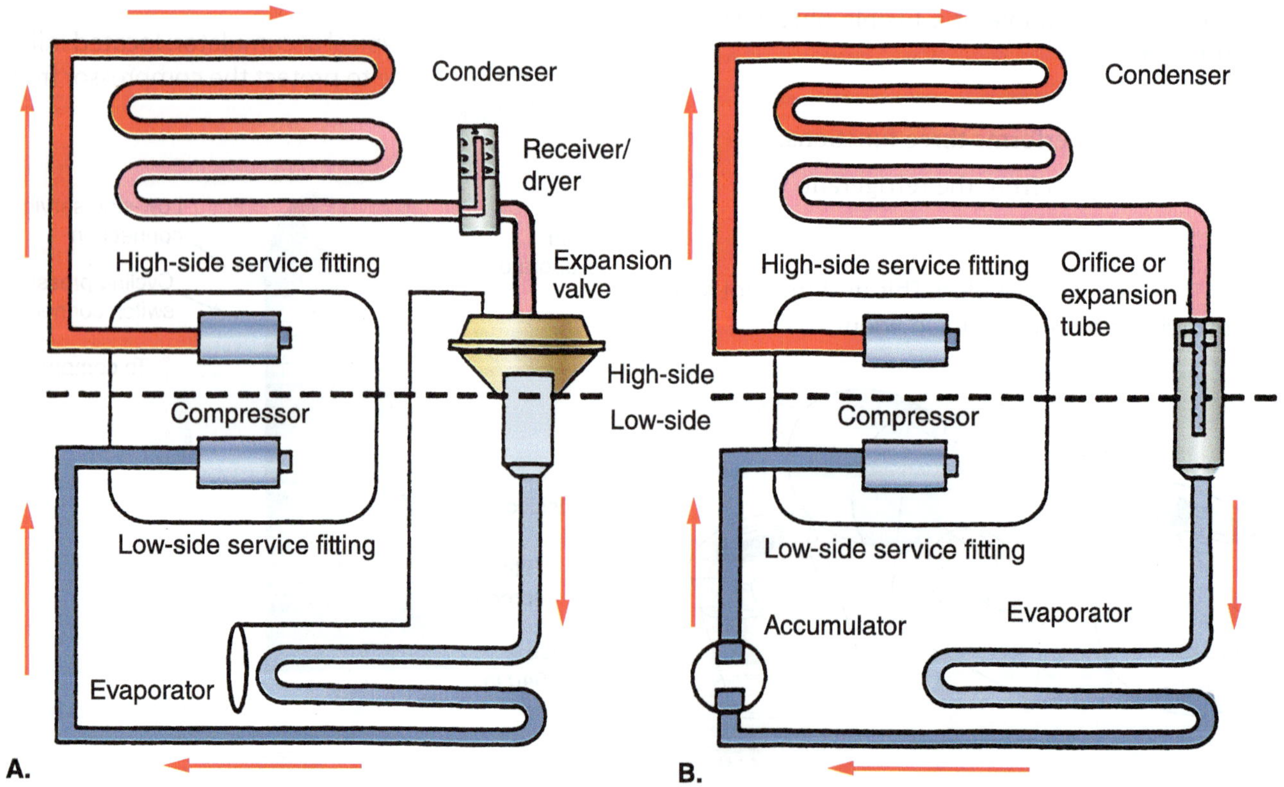

FIGURE 35–22 Older style external TXV used a feeler bulb to determine evaporator outlet temperature.

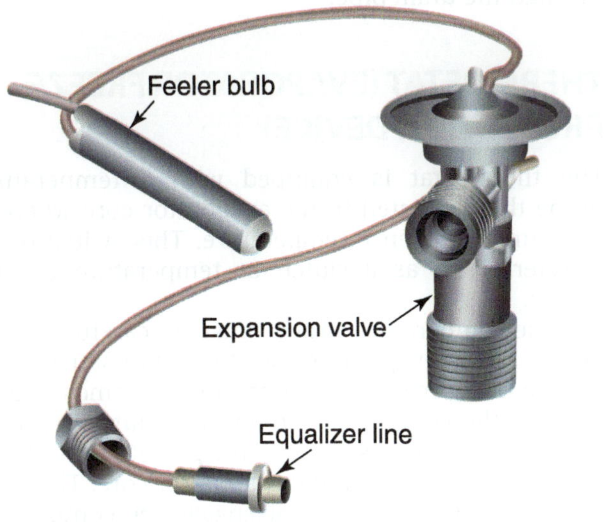

diaphragm. It is connected to the evaporator outlet (suction) pipe. This is an equalizer line that senses suction pressure. Together, they regulate the amount of refrigerant entering the evaporator. **Figure 35–22** shows an external pressure-type expansion valve.

In recent years, the block-type internally equalized TXV has become popular. It is less vulnerable to accidental damage because there is no exposed capillary tube or external connections. Inside the internal-type expansion valve is a sealed sensing tube that responds to changes in the temperature of the vapor flowing out of the evaporator by varying the opening of the metering orifice. As the vapor becomes hotter, the sensing tube expands, and the metering orifice opens further. This allows more liquid refrigerant into the evaporator to meet a higher cooling demand. Likewise, the sensing tube contracts, and the metering orifice becomes smaller as the vapor becomes cooler. This restricts the liquid refrigerant flowing into the evaporator. **Figure 35–23** shows the refrigerant flow through an internally equalized-type expansion valve.

Orifice Expansion Tube

Like the TXV, the orifice expansion tube is the dividing point between the high- and low pressure sections of the refrigerant cycle. However, its metering or flow rate control does not depend on comparing evaporator pressure and temperature. It is a fixed orifice, so it simply defines a flow area. This means that the device has no moving parts. The flow rate is determined by the size of the orifice, the pressure difference across the orifice, and by **subcooling**. *Subcooling* is the term used to describe the additional cooling of the refrigerant in the bottom of the condenser after it has changed from vapor to liquid. The flow rate through

FIGURE 35–23 Refrigerant flow through an internally equalized-type expansion valve

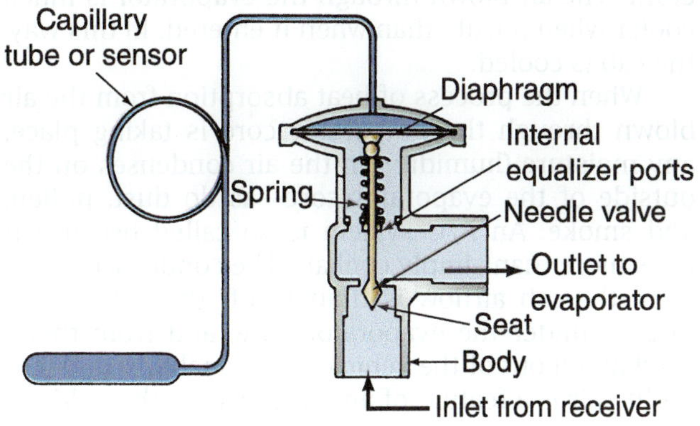

the orifice is more sensitive to subcooling than to pressure difference. Some of the latest orifice tubes used with an R-134a system use a spring-loaded orifice check valve; this opens up flow when pressure becomes too high and eliminates the cracking noise of some R-134a systems. **Figure 35–24** shows an orifice expansion tube.

EVAPORATOR

The **evaporator**, like the condenser, is a heat exchanger. It consists of a refrigerant coil mounted to a series of thin cooling fins. The idea is to provide a maximum amount of heat transfer in a minimum amount of space. The evaporator is located in the cab, frequently beneath the cab dashboard or instrument panel. The heat transfer function of the evaporator is to draw heat out of the cab air.

The evaporator receives the low-pressure, low-temperature liquid refrigerant from the TXV or orifice tube in the form of an atomized spray. This atomized refrigerant is then evaporated (boiled) by the heat of the inlet air blown through the evaporator. That means its state is changed from a liquid to a gas. Heat from the evaporator core surface area is surrendered because of the latent heat of vaporization.

FIGURE 35–24 Orifice expansion tube

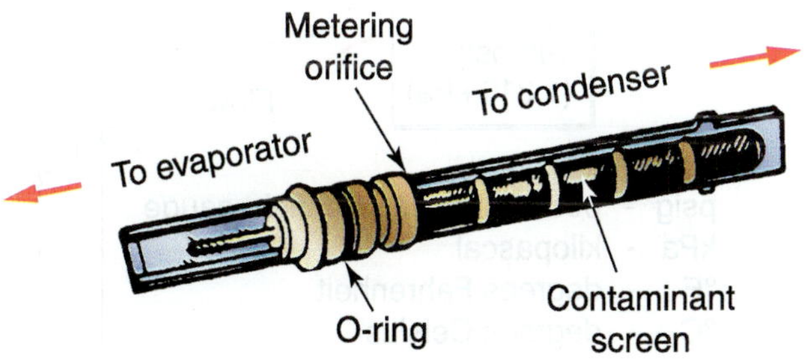

Actually, it is the heat of the cab air blown through the evaporator that causes the refrigerant to boil. Heat from this air is surrendered to the vaporizing refrigerant. The air blown through the evaporator is much cooler when it exits than when it entered. In this way, the cab is cooled.

When the process of heat absorption from the air blown through the evaporator core is taking place, any moisture (humidity) in the air condenses on the outside of the evaporator core—as do dust, pollen, and smoke. An A/C system is so called because it does more than simply cool air. The condensed water from the cab airflow is drained off through a tube located under the evaporator core, and from there, discharged out of the vehicle. The capability to dehumidify air is a feature of any A/C system that adds to the operator's comfort. It also can be used as a means of controlling fogging of the vehicle windows, and some OEMs engage the A/C compressor in defrost mode. **Figure 35-25** shows how an evaporator functions as the refrigerant changes state.

Evaporator cores should be inspected occasionally. Accumulation of dirt over time can plug the cooling fins or restrict the drain. Under high humidity conditions, the catch basin may overflow onto the cab floor. Plugging of the cooling fins can cause an airflow restriction that can reduce evaporator cooling efficiency, but such failures are not common given the protected location of the evaporator. A simple test is to pour water into the air intake at the windshield and observe whether it drains out onto the ground through the drain pipe.

THERMOSTAT (EVAPORATOR FREEZE PROTECTION DEVICE)

The thermostat is equipped with a temperature probe that is routed to the evaporator core where it is mounted to sense temperature. This switch often is referred to as a clutch or temperature cycling switch.

The temperature probe must be carefully positioned in the evaporator core to ensure satisfactory system performance. A switch in the thermostat acts to cycle the compressor on and off (clutch engaged or disengaged) in response to the preset temperature sensed by the probe in the evaporator core. The thermostatic switch opens to disengage the compressor clutch when it senses that evaporator temperature has dropped to a predetermined low, usually in the region of 34°F to 37°F (1°C to 3°C). This prevents evaporator freeze-up that could be caused by moisture condensing on the evaporator fins. **Figure 35-26** shows

FIGURE 35-25 Tube and fin-type evaporator core: Note that the refrigerant changes temperature only after it changes state from a liquid to a vapor.

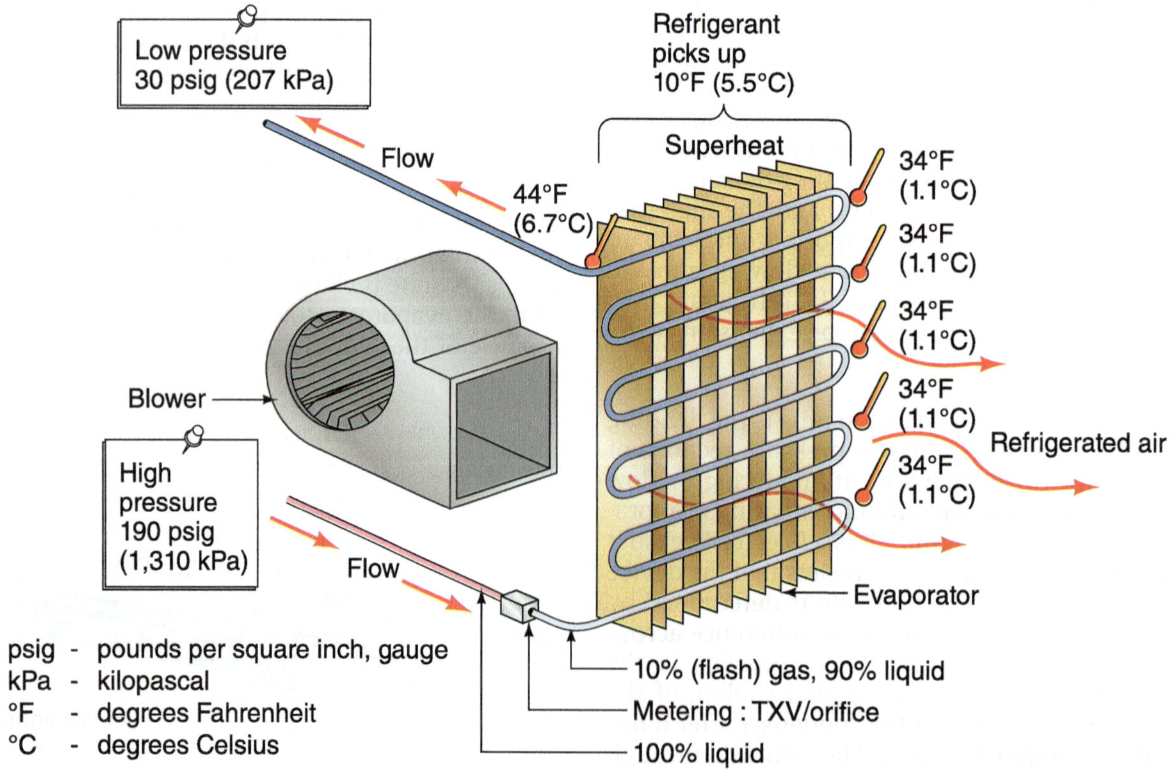

FIGURE 35–26 (A) Infrared temperature gun used to measure evaporator temperature; and (B) display window on an infrared temperature gun.

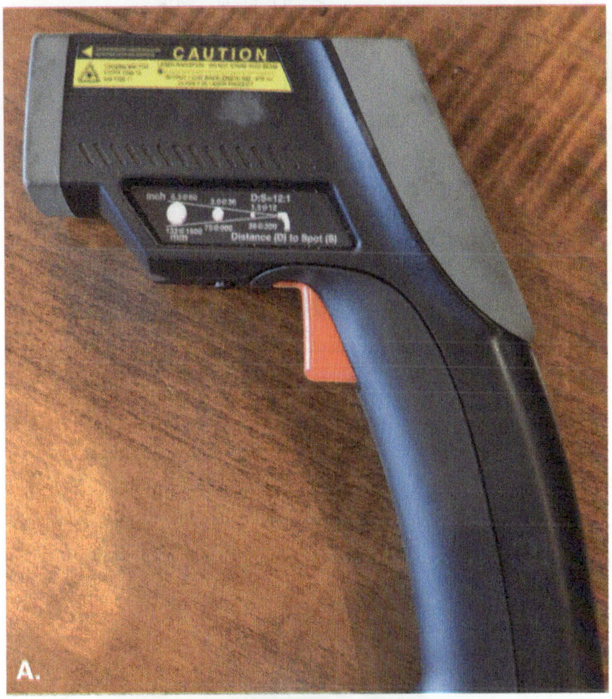

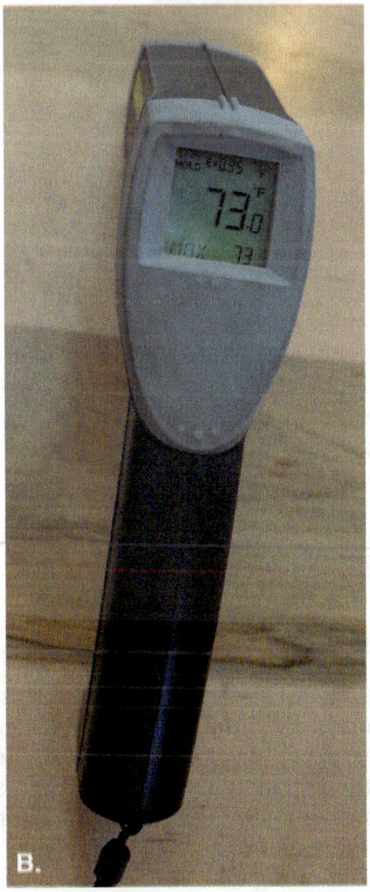

the location of the temperature probe in an evaporator core.

Although the blower motor and fan continue operation, the refrigeration cycle ceases whenever the compressor is disengaged by its drive clutch. This means that refrigerant is no longer pumped through the system. When this occurs, evaporation continues to take place, but the evaporator temperature is permitted to rise. When sufficient temperature rise has taken place, the thermostatic switch again closes, causing the compressor clutch to engage, and the refrigeration cycle resumes.

REFRIGERATION LINES, HOSES, AND COUPLERS

All the major components of the system have inlet and outlet connections that accommodate either flare or O-ring fittings. The refrigerant lines that connect between these units are made up of special refrigerant hose or tubing with flare or O-ring fittings at each end. The hose or tube end of the fitting is constructed with sealing beads to accommodate a hose or tube clamp connection. Typical flare and O-ring fittings are shown in **Figure 35–27**.

The implementation of the Clean Air Act and EPA regulations on refrigerant systems and handling has resulted in specified acceptable leakage rates of refrigerant. These rules have changed the design of refrigerant fittings and hose/fitting construction techniques (**Figure 35–28**). All new refrigerant hose is constructed to minimize refrigerant leakage rates. Because R-134a has a smaller molecular structure than the previously used R-12, it can leak through the standard double braid refrigerant hose used for many years in vehicle A/C applications.

Current refrigerant hose is constructed with a barrier plastic inner layer in the hose to prevent refrigerant leakage through the hose wall. Examples of this are shown in **Figure 35–29**. These new barrier hoses are compatible with both R-12 and R-134a refrigerants. In order to assemble the new hose, new-style bead lock fittings and special crimping tools are required, as shown in **Figure 35–30**.

Refrigerant lines can be classified as follows:

- **Suction Lines.** These lines, which are located between the outlet side of the evaporator and the inlet (suction) side of the compressor, carry the low-pressure, low-temperature refrigerant vapor to the compressor. Suction lines can be identified when the system is running because they are cold to the touch. Low-pressure lines are the larger diameter hoses in the system because they have to handle the same volume but at a lower pressure.
- **Discharge Lines.** Beginning at the discharge outlet on the compressor, the discharge or

FIGURE 35–27 Flare, O-ring, and quick fit connectors: ISO fittings such as these are sensitive to over-torque.

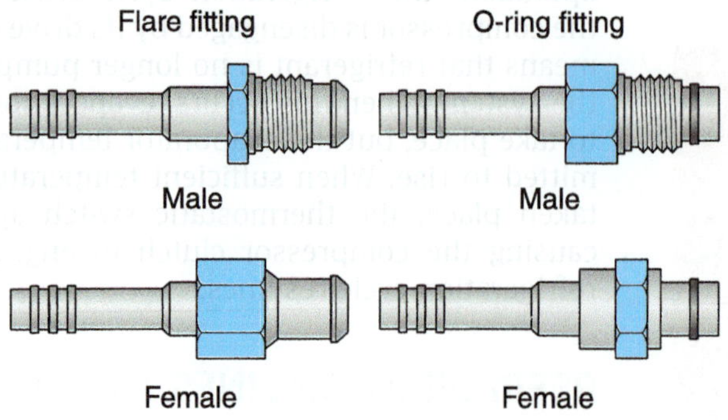

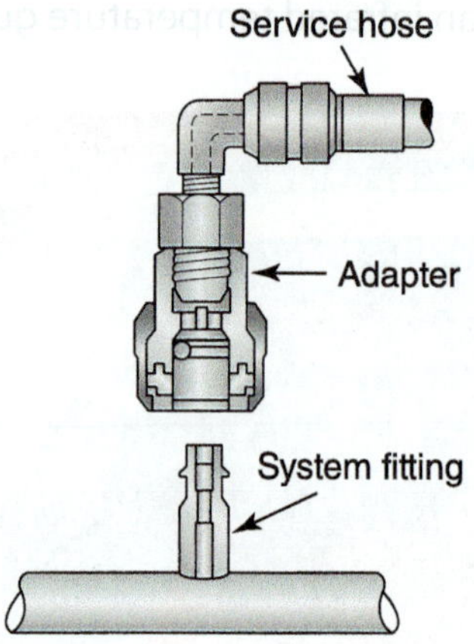

FIGURE 35–28 Hose clamp shell crimp fittings will not seal the new style barrier hose.

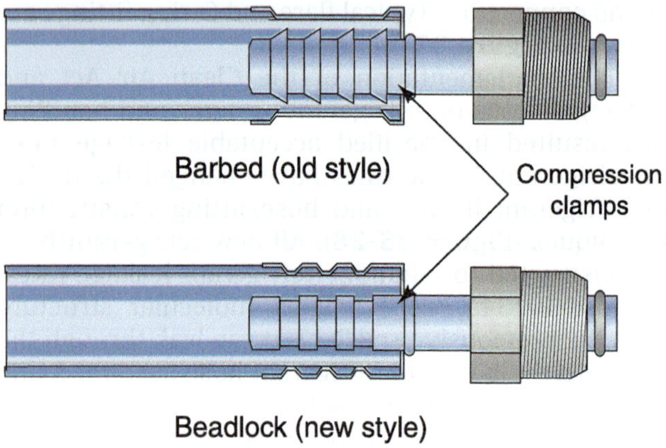

FIGURE 35–30 Special crimping tools

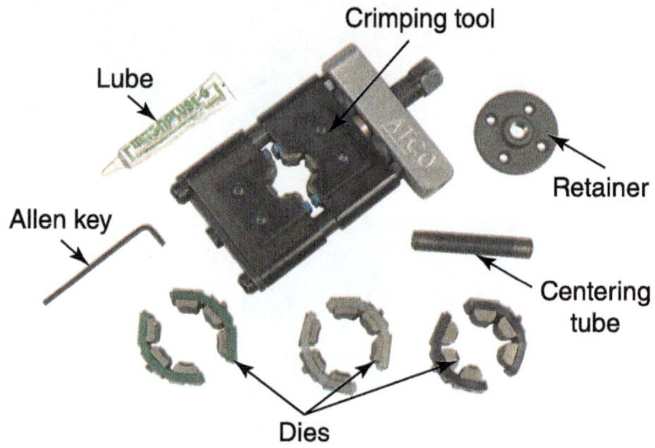

FIGURE 35–29 New hoses are designed to prevent refrigerant gas loss through the hose.

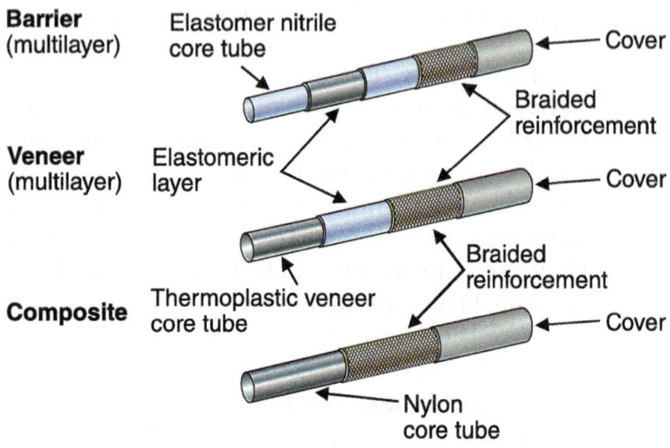

high-pressure lines connect the compressor to the condenser, the condenser to the receiver/dryer, and the receiver/dryer to the inlet side of the expansion valve. Through these lines, the refrigerant travels in its path from a gas state (compressor outlet) to a liquid state (condenser outlet) and, from there, to the inlet side of the expansion valve. Discharge lines are warm to the touch and smaller in diameter than suction lines.

SIGHT GLASS

The **sight glass** allows the technician to see the flow of refrigerant in the lines. It can be located either on the receiver/dryer or inline between the receiver/dryer and the expansion valve or tube. Its functions all relate to diagnosing the refrigerant condition but sight glass observations tend to mislead more than help troubleshooting.

CAUTION

In current vehicle A/C systems, sight glass readings are not regarded as reliable and most manufacturers do not include sight glass observations as valid troubleshooting.

To perform a sight glass check of the refrigerant, run the vehicle engine, open the windows and doors, and set the A/C controls for maximum cooling with the blower on its highest speed. The system should be run for about 5 minutes, so make sure that it is in a well-ventilated area or connected to an exhaust gas extraction system. Always take care when checking the sight glass when the engine is running.

Figure 35–31 shows a typical sight glass and how to interpret some conditions but remember the earlier caution about the inaccuracy of the conditions observed.

- Oil-streaking indicates that the system is empty.
- Bubbles or foaming indicate that the refrigerant is low. Note that some retrofit systems (to R-134a) can show bubbles on a properly charged and functional system.
- Clouding indicates desiccant breakdown with possible contamination throughout the circuit.
- Clear with no bubbles indicates that the refrigerant level is correct with no contamination or completely empty.

Note: Bubbles normally can be observed when the system is started up.

FIGURE 35–31 Sight glass conditions

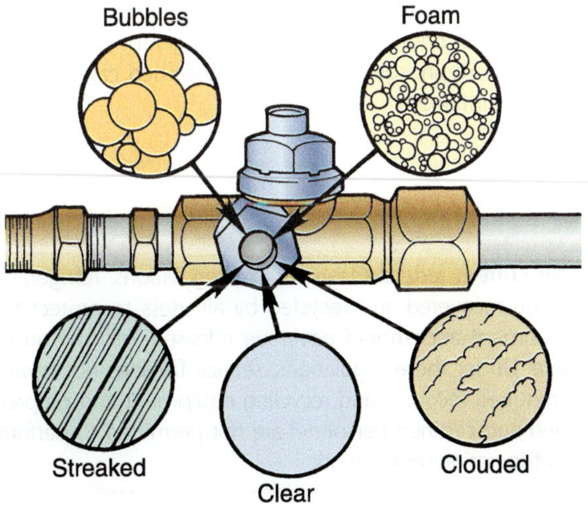

IN-LINE DRYER

Located between the receiver/dryer and TXV or tube, an in-line dryer absorbs any moisture that gets by the receiver/dryer. It also helps to prevent TXV or tube freeze-up.

MUFFLER

Most heavy-duty truck A/C systems now have a muffler installed in the system. The muffler is usually located on the discharge side of the compressor. It also may be located on the suction side. Its function is to reduce the characteristic pumping noise produced by the compressor. Mufflers should be installed into the system correctly, usually with an inlet connection at the top and outlet connection at the bottom. This is done to minimize collection of oil in the unit.

BLOWER MOTOR AND FAN

The **blower motor/fan assembly** is located in the evaporator housing. Its purpose is to blow air through the evaporator and maximize its efficiency as a heat exchanger. The blower draws warm air from the cab, forces it through the coils and fins of the evaporator, and blows the cooled, dehumidified air into the cab. The blower motor is controlled by a fan switch. In most cases, the blower motor/fan assembly also provides warm air to heat the cab in cold weather. The A/C system is designed to cycle only when the blower motor is in the on position; insufficient heat at the evaporator would cause it to freeze in humid conditions. **Figure 35–32** shows the routing of air through the blower motor.

ENGINE COOLING FAN

The engine cooling system plays an important role in the operation of an air conditioner system. The A/C condenser is usually located in front of the engine cooling system radiator. At highway speeds, the ram

FIGURE 35–32 Blower motor increases the airflow in the cab.

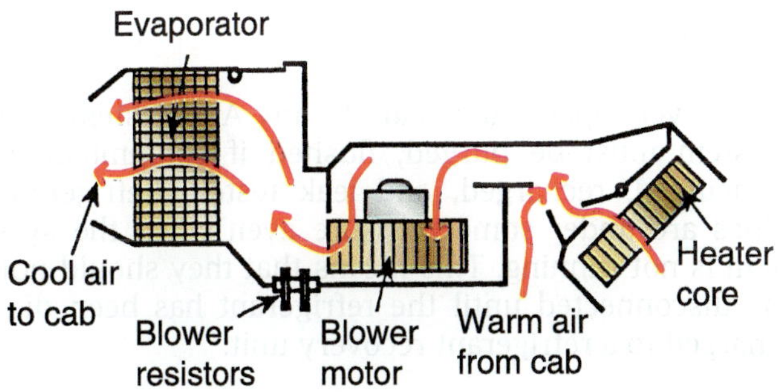

FIGURE 35–33 Electrically driven fan

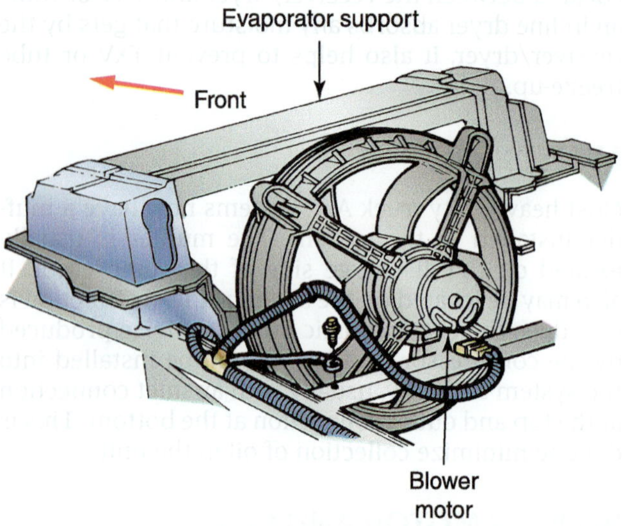

air through the condenser and radiator usually is sufficient for the operation of the A/C system. However, under high heat and low engine and road speeds, the condenser requires additional air, which is delivered by a multiplex controlled engine fan.

In some recent truck engine applications, the conventional belt-driven, water-pump-mounted engine coolant fan has been replaced with an electrically driven fan that is not connected mechanically to the engine coolant pump. The 12-volt, motor-driven fan can be controlled by an engine coolant temperature switch, a sensor (thermostat), and/or the air conditioner controller module. **Figure 35–33** shows an electrically driven fan.

SHOP TALK

Remember that just one drop of water added to refrigerant will lead to corrosion and refrigerant breakdown. Corrosive hydrochloric acid can be produced in the older R-12 system when trace drops of water are added. Also, the smallest amounts of air in the refrigerant system can start chemical reactions that result in system malfunctions.

35.5 SAFETY PRECAUTIONS

Whenever repairs are made to any A/C system, the system must be purged, flushed if contaminated, evacuated, recharged, and leak tested. Refrigerant lines are under some pressure even when the system is not running. This means that they should not be disconnected until the refrigerant has been discharged to a refrigerant recovery unit.

Refrigerants are safe when handled properly. Safety goggles and nonleather gloves should always be worn when testing and servicing A/C systems when discharging, purging, flushing, evacuating, charging, and leak testing the system. When refrigerant contacts leather, the leather can stick to skin.

WARNING

Refrigerant should never come into contact with skin or eyes. Liquid refrigerant, when exposed to the air, quickly evaporates and will almost instantly freeze skin or eye tissue. Serious injury or blindness could result.

Refrigerant splashed in the eyes should be treated as follows:

1. Flush eyes with cold water to raise the temperature above freezing point.
2. Avoid rubbing eyes.
3. If available, apply an antiseptic mineral oil to the affected area. This will form a protective film over the eyeball to reduce the possibility of infection. Then rinse with a weak boric acid solution.
4. Call an eye specialist or doctor immediately and receive medical treatment as soon as possible.

Treatment for refrigerant splashed on the skin should be the same as for frostbite. Gently pour cool water on the area but do not rub the skin. Keep the skin warm with layers of soft, sterile cloth. Call a doctor right away. Even though refrigerant does not burn, when it comes into contact with extreme heat or flame, poisonous phosgene gas is created. This gas also is produced when an open flame leak detector is used. Phosgene fumes have an acrid (bitter) smell.

WARNING

Avoid working in areas where refrigerant may come into contact with an open flame or any burning material, such as a cigarette. When it contacts extreme heat, refrigerant breaks down into poisonous phosgene gas, which, if breathed, causes severe respiratory irritation.

CAUTION

Under current federal Clean Air Act regulations, refrigerants must be recovered and recycled by all users to protect the environment and should never be released into the atmosphere. Under these regulations, service facilities not having the required recovery and recycling equipment and properly trained and certified personnel are not permitted to perform any refrigeration service work.

Refrigerant Storage

Because of its very low boiling point, refrigerant has to be stored under pressure in containers. These containers should never be exposed to temperatures higher than 125°F (52°C). Refrigerant cans should not be left in direct sunlight.

35.6 PERFORMANCE TESTING AN AIR CONDITIONING SYSTEM

This section contains a short introduction to A/C system diagnosis. Because there are differences in each OEM system, this is just a general set of guidelines, and the best approach to testing and troubleshooting a system is to use the OEM service literature. When testing and troubleshooting electronically managed A/C systems, it is important to observe the OEM diagnostic routines which usually require an **electronic service tool (EST)** and appropriate software.

PRELIMINARY CHECKS

Before performance testing an A/C system, perform the following checks:

1. Check that the refrigerant compressor drive belt is not damaged and that it is correctly tensioned. Check the compressor mountings for tightness.
2. Using a thickness gauge, check the compressor clutch clearance to specification.
3. Visually check for broken, burst, or cut refrigerant hoses. Also check for loose fittings on each coupling.
4. Check for road debris buildup on the condenser coil fins. If necessary, this can be cleaned using an air pressure nozzle, whisk brush, and soapy water. Avoid bending the cooling fins.
5. Check the sight glass for refrigerant moisture content and state of charge. The air conditioner should be on when checking the sight glass. In fact, it is best if the system has been running for some time.
6. If there is insufficient cab airflow, check that leaves or other debris have not entered the fresh air ducts under the windshield. If debris has entered, it can clog the evaporator fins. Also, ensure that the ducts are connected to the dash louvers and that the air control flaps are moving properly.

OPERATIONAL TESTING

The following is a guide to conducting a performance test of a truck A/C system.

Cooling Check

Under conditions of extreme heat and high humidity, it will take longer to cool down the cab, especially if the fresh air vents are left open. To make a quick check to determine whether the air conditioner is cooling sufficiently:

1. Set the engine speed at 1,500 rpm or the OEM-governed rpm level. On some computer-controlled engines, this may be less than 1,500 rpm when the vehicle is stationary.
2. Close the doors and windows.
3. Start the engine. If the unit is in a service shop, install an exhaust extraction pipe.
4. Turn on the air conditioner; set the thermostat and blow controls at maximum. Set the vent ducting to recirculate the air.
5. Ensure that the auxiliary A/C, if equipped, is off.
6. Run the system for at least 10 minutes, or, in conditions of extreme heat, for at least 15 minutes.
7. Insert an A/C thermometer into the louver or air duct closest to the evaporator so that it is fully exposed to the airflow, as shown in **Figure 35–34**.
8. Compare the measured temperature readings to the specifications in the temperature/pressure table in the OEM service literature. Higher than specified temperatures indicate that the system is not cooling enough. Some of the possible causes are identified later in this chapter.

Odor Diagnosis

Under some operating conditions, A/C systems may produce a musty odor, usually at startup in hot weather. This type of odor is usually temporary, and, providing it disappears after a short period of operation, should not be of concern. If it persists, it can be an indication of microbial growth on the evaporator core. This requires the removal of the evaporator core and its thorough cleaning with a disinfectant. Some other odor diagnoses are shown in **Table 35–1**.

FIGURE 35–34 Checking outlet air temperature: Test should performed with engine at high idle rpm.

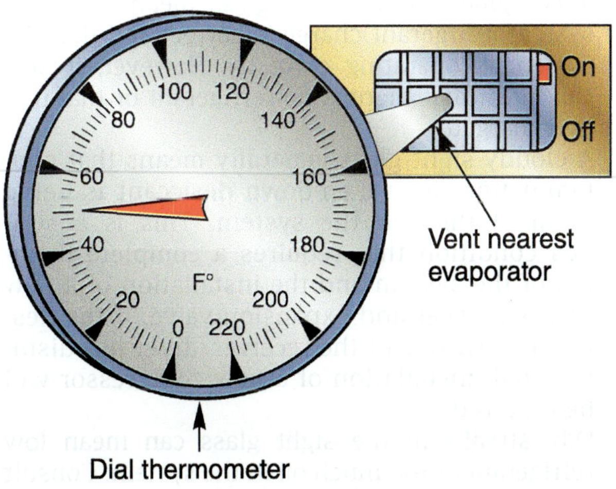

TABLE 35–1 Odor Diagnoses

Problem	Possible Cause	Repair
Musty odor	1. External water leaks 2. Plugged evaporator drain 3. Molds and mildew	1. Seal cab body 2. Unplug drain 3. Clean evaporator
Coolant odor	1. Antifreeze solution leaking at heater core or hoses	1. Replace or repair heater core 2. Tighten hose clamps or replace heater hose
Refrigerant leak	1. Refrigerant oil leaking at evaporator core	1. Replace evaporator core

Refrigerant Check

Look for these four conditions when checking refrigerant in the sight glass but bear in mind throughout that R-134a systems often produce visible bubbles when operating properly:

1. Bubbles normally show only when the system is first started up with R-12 systems. These should disappear within a minute or two. If these bubbles continue to be seen at the sight glass when the system is operating, the refrigerant charge is low and should be recharged. The occasional bubble can be observed during normal A/C operation, particularly when the compressor clutch cycles on. Bubbles may appear in the sight glass in an R-134a system when the system is operating normally.
2. A completely clear sight glass means that the system refrigerant charge is correct and that the system is operating correctly. However, it also can mean that there is no refrigerant or that it is overcharged.
3. A cloudy sight glass generally means that contamination or broken-down desiccant is being pumped through the system. This is a serious condition that requires a complete flushing of the system and the installation of a new receiver/dryer and expansion valve. If the desiccant cartridge of the receiver/dryer has disintegrated, installation of a new compressor will be required.
4. Oily streaks in the sight glass can mean low refrigerant or too much oil in the system. Consult Figure 35–31 to analyze sight glass conditions.

LEAK TESTING A SYSTEM

Testing an A/C system for leaks is an important phase of the troubleshooting process. Over time the refrigerant charge in all A/C systems becomes depleted. In systems that are in good condition, refrigerant losses of up to ½ pound (0.227 kg) per year are considered normal. Higher loss rates indicate the need to locate and repair the leaks. **Figure 35–35** shows some typical potential leak sources.

Leaks are most often found at the compressor hose connections and at the various fittings and joints in the system. Refrigerant can be lost through hose permeation. Leaks also can be traced to pinholes in the evaporator caused by acid etching in R-12 systems. Acids form when water and refrigerant mix. Perform a visual inspection by looking for signs of A/C lubricant leakage and damage or corrosion in the lines, hoses, and components. Suspected leaks can be confirmed by using the following leak detection methods:

1. **Soapsuds Solution.** A solution of soap and water can be sprayed onto a suspected fitting or component. If the leak is large enough, it will cause bubbles to form in the soapy solution at the point of the leak. This method is used when no other leak detection equipment is available.
2. **Fluid Leak Detectors.** Commercial liquid leak detectors perform the same task as the soapsuds solution, but they can locate smaller leaks. The fluid leak detector is applied to the area to be tested. If a leak is present, the liquid will form

FIGURE 35–35 Possible refrigerant leak locations

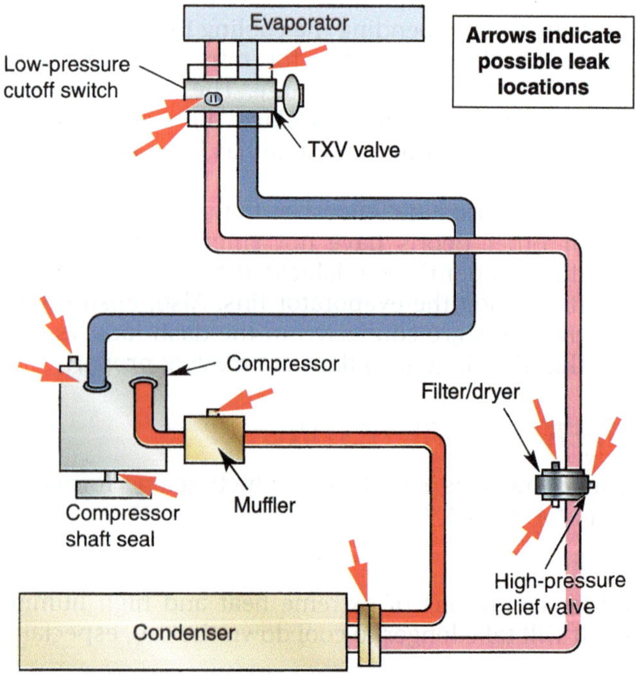

clusters of bubbles or large bubbles (depending on the size of the leak) around the source. A fluid leak detector is useful for locating leaks in a confined space in which the sensitivity of an electronic leak detector makes it an impractical instrument.

3. **Flame-Type Propane Leak Detector.** These are not commonly used today because they were designed to test CFC systems. A halide leak detector is less messy than a soapsuds solution and can locate small leaks. It is actually a small torch. When the flame comes in contact with refrigerant, the flame color changes. Propane leak detectors are a potential fire hazard and poisonous phosgene gas is produced when refrigerant contacts the flame.
4. **Ultraviolet (UV) Leak Detectors.** Sometimes known as black light detectors, UV leak detectors are simple devices capable of identifying small leaks in an A/C system. Systems equipped with this phosphor dye/black light detection system have an identification label on the compressor or receiver/dryer. The phosphor dye also may be introduced to the low side of an A/C system in a specific small quantity. Once the phosphor dye is in the system, shining a standard black light over a leak will produce a bright yellow trace at the leak point. This method of leak detection is used for hard-to-find or pinpoint leaks, especially cold or hot leaks. **Figure 35–36** shows a black light detector kit.
5. **Electronic Leak Detectors.** Any shop regularly performing A/C service work should be equipped with an **electronic leak detector**. It is the most sensitive of the leak detection instruments—so sensitive, in fact, that it can make sourcing a leak difficult, especially in a confined space. Electronic leak detectors are small handheld devices with a flexible probe used to seek out refrigerant leaks. When a leak is detected, a flashing light or buzzer signals the technician. The area should be free of oil, grease, and residual refrigerant before beginning the detection process. Suspected leakage areas should be cleaned using soap and water, not a solvent. A detected leak should be an active flow leak, not a residual condition caused by refrigerant trapped under an oil film. Electronic leak detectors are usually specific to the refrigerant they are designed to test. Most current electronic leak detectors are of the heated diode type and will test only R-134a refrigerant. **Figure 35–37** shows a typical handheld electronic leak detector kit.

FIGURE 35–36 Handheld leak detector (phosphorescent) leak detector kit

CAUTION

Never pressure or leak test R-134a service equipment or vehicle A/C systems with compressed air. Some mixtures of air and R-134a have been shown to be combustible at elevated pressures.

FIGURE 35–37 Black light (UV) leak detection gun

SHOP TALK

Because of their sensitivity, electronic-type detectors must be correctly calibrated before each use to detect the lowest permissible leak rate of the component being checked. Trace the refrigerant system in a continuous path so that no potential leaks are missed. If a leak is found, always continue to test the remainder of the system. At each area checked, the probe should be moved around the location at a rate no more than 1 to 2 inches/second (25 to 50 mm) and no more than ¼ inch (6 mm) from the surface completely around the position. Slower and closer movement of the probe greatly improves the likelihood of finding a leak. It also helps to place the probe lower than a suspected leak because the refrigerant is heavier than air.

ELECTRICAL SYSTEM CHECK

The electrical system should be checked periodically to prevent the truck A/C system from failing unexpectedly. Perform the usual battery checks and follow the wire routing through to the A/C electrical circuit. Because truck A/C systems contain various switches, they can fail completely or intermittently because of corrosion. Check for high resistance at switches and connections using a digital multimeter (DMM). Computer-managed climate control systems have voltage sensitivity parameters, which are usually set at a low of 11 volts and a high of 16 volts. This helps protect the system against the consequences of low-voltage operation and high-voltage spikes.

35.7 AIR CONDITIONING SERVICE EQUIPMENT

Several specially designed pieces of equipment are required to perform test procedures. They include a manifold gauge set, service valves, vacuum pumps, a charging station, a charging cylinder, and recovery/recycling systems. Although both R-12 and R-134a use basically the same types of testing equipment, they are not interchangeable. R-134a systems and equipment have metric connectors, but R-12 use English connections.

MANIFOLD GAUGE SETS

A **manifold gauge set** is used to recycle, recharge, and diagnose A/C systems. The left-side gauge of a standard gauge set is blue and known as the low-pressure gauge. The dial is graduated from 1 to 150 psi (7 kPa to 1,034 kPa) with a cushion up to 250 psi (1,724 kPa) in 1-psi (7 kPa) increments, and, in the opposite direction, in inches of mercury (Hg) vacuum from 0 to 30 (0 to 762 mm/Hg). This gauge is used to check pressure on the low-pressure side of the refrigeration system. The right-side gauge is red and is graduated from 0 to 500 psi (0 to 3,447 kPa) in 10-psi (69 kPa) graduations. This is the high-side gauge, and it is used for checking pressure on the high-pressure side of the system.

The center manifold fitting is common to both the low and the high side and is for evacuating or adding refrigerant to the system. When this fitting is not being used, it should be capped. A test hose connected to the fitting directly under the low-side gauge is used to connect the low side of the test manifold to the low side of the system, and a similar connection is found on the high side. **Figure 35–38** shows a manifold gauge set used with a R-134a systems.

The manifold gauge set is designed to access the refrigeration circuit and control refrigerant flow when adding or removing refrigerant. When the manifold test set is connected into the system, pressure is registered on both of the gauges at all times. During testing, both the low- and high-side hand valves should be in the closed position (turned inward until the valve is seated).

FIGURE 35–38 R-134a manifold gauge set

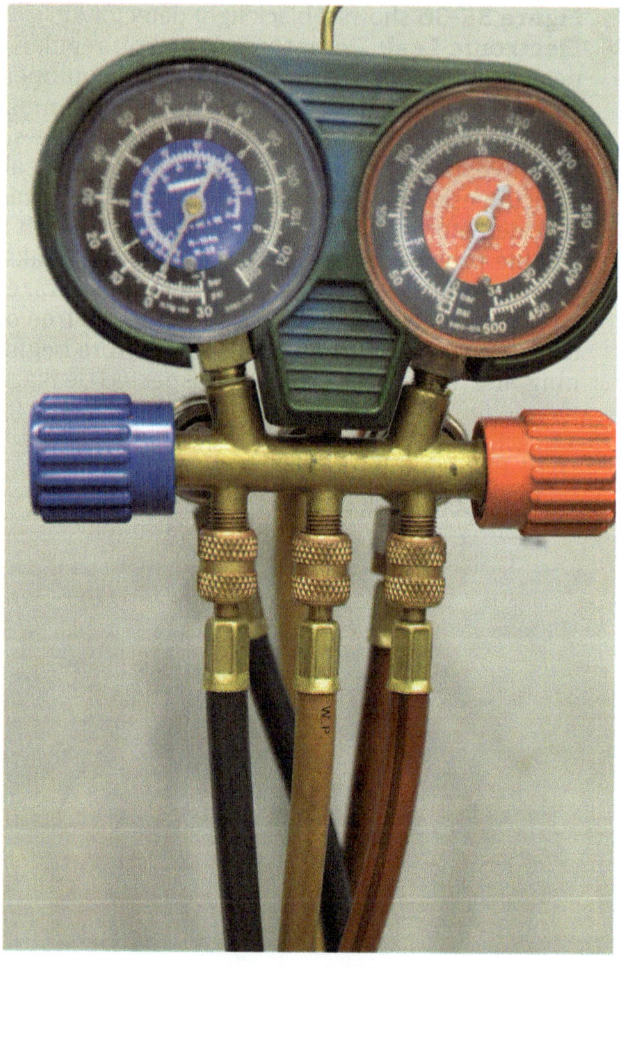

FIGURE 35–39 Manifold gauge set with key components identified

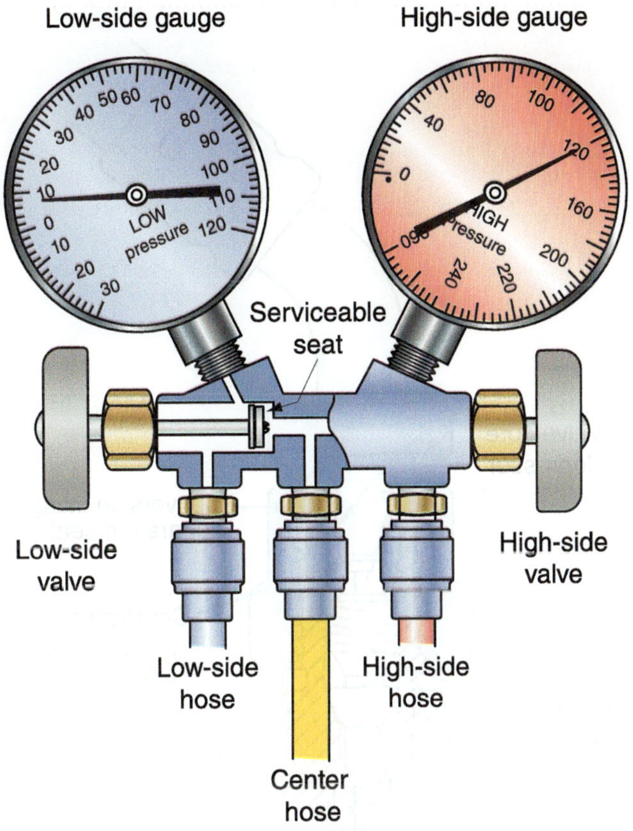

Refrigerant will then flow around the valve stem to the high- and low-side gauges to display the operating pressures on each side of the refrigeration circuit. The hand valves isolate the low and high side from the central portion of the manifold. When the gauges are connected to the gauge fittings with the refrigeration system charged, the gauge lines should always be purged. Purging is done by cracking each valve on the gauge set to allow the pressure of the refrigerant in the refrigeration system to force the air to escape through the center gauge line. Failure to purge lines can result in air or other contaminants entering the refrigeration system. **Figure 35–39** shows a manifold gauge set with color-coded lines.

CAUTION

Never open the high-side hand valve with the system operating and a refrigerant source at the center hose connection. This will cause refrigerant to exit the A/C system under high pressure into the source container, which could cause it to burst. The only time both hand valves should be open is when evacuating the system.

SHOP TALK

Atmospheric pressure reduces as altitude increases. The gauge pressure must be corrected using an OEM correction chart when checking system low-side pressures.

Service Valves

System service valves incorporate a service gauge port for manifold gauge set connections and are provided on the low and high sides of some A/C systems. When making gauge connections, the gauge lines should be purged by cracking the charging valve and bleeding a small amount of refrigerant, followed by immediate connection of the lines. There are two types of service valves used: stem type and Schrader.

Stem-Type Service Valve

The stem-type valve is sometimes used on two-cylinder reciprocating piston-type compressors. Access to the high- and low-pressure sides of the system is provided through service valves mounted on the compressor head. The low-pressure valve is mounted at the inlet to the compressor, and the high-pressure valve is mounted at the compressor outlet. Both of these valves can be used to isolate the rest of the A/C system from the compressor when the compressor is being serviced. These valves have a stem under a cap with the hose connection directly opposite. A special wrench is used to open the valve to one of the three following positions:

- **Back seated.** The normal operating position with the valve stem rotated counterclockwise to seat the rear valve face and to seal off the service gauge port.
- **Midposition.** The test position with the valve stem turned clockwise (inward) 1½ to 2 turns to connect the service gauge port into the system, allowing gauge readings to be taken with the system operating. (A service gauge hose must be connected with the valve completely back seated.)
- **Front seated.** The test position with the valve stem rotated clockwise to seat the front valve face and to isolate the compressor from the system. This position allows the compressor to be serviced without discharging the entire system. The front-seated position is for service only and is never used when the A/C is operating.

Note the valve positions as shown in **Figure 35–40**.

Schrader Type

A Schrader valve resembles a tire valve (**Figure 35–41**). Schrader valves usually are located in the high-pressure line (from compressor to condenser) and in the low-pressure line (from evaporator to compressor) to permit checking of the high side and low side

FIGURE 35–40 Service valve positions: typically used in bus applications so the compressor can be removed without evacuating the system.

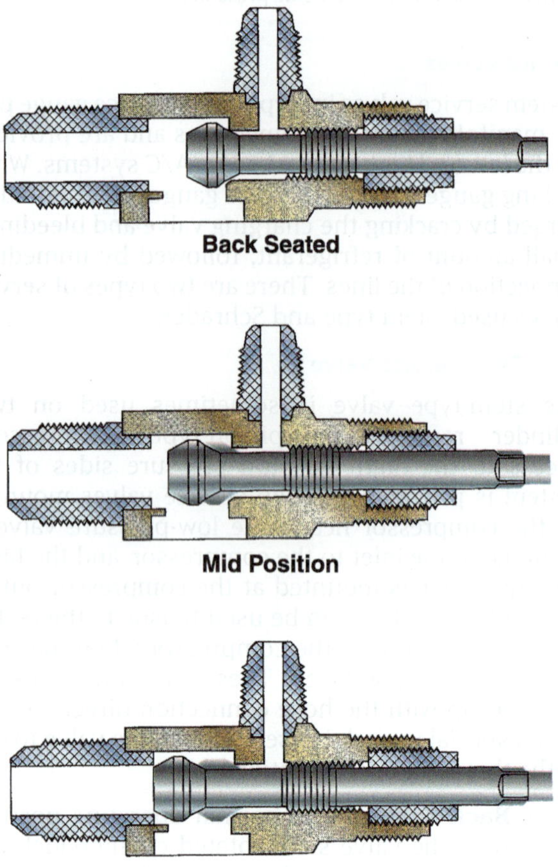

FIGURE 35–41 Schrader valve with a hose pin depressor: Some R134a systems use a hand-actuated knob to depress Schrader valve core.

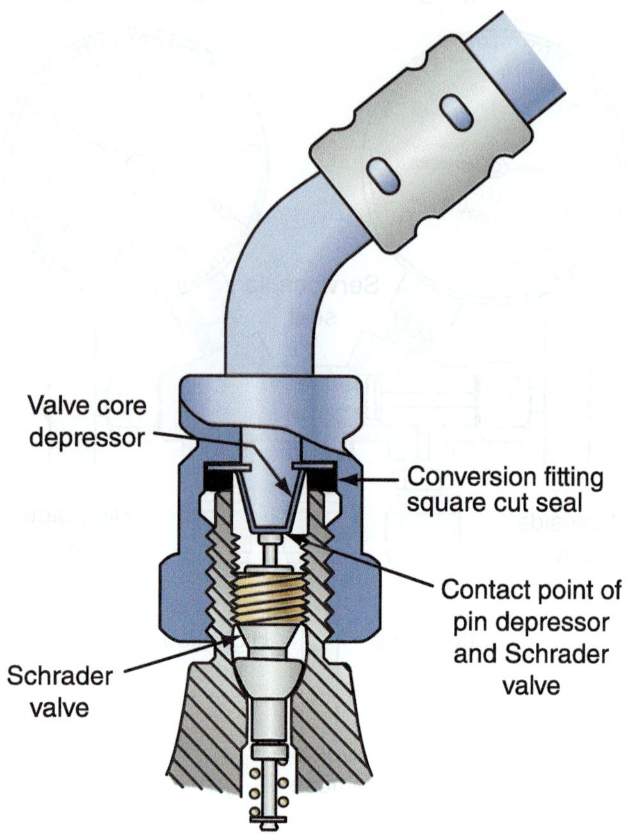

of the system. All test hoses have a Schrader valve core depressor in them. As the hose is threaded onto the service port, the pin in the center of the valve is depressed, allowing refrigerant to flow to the manifold gauge set. When the hose is removed, the valve closes and seats automatically.

SHOP TALK

In R-12 systems, the Schrader fitting on the high-pressure side is smaller than on the low-pressure side, and special adapters are necessary to connect the high-side service hose into the system. The difference in fitting sizes is to prevent mixing up the high- and low-pressure sides of the system when attaching the gauge set. The different size fittings were introduced to prevent disposable 1 lb (0.45 kg) cans of refrigerant from being connected to the high side of the system, causing them to explode. After disconnecting the gauge lines, check the valve areas to be sure that the service valves are correctly seated and that the Schrader valves are not leaking. In R-134a systems, the size as well as the type of connection is different; that is, the larger connection is used for the high side.

MANIFOLD SERVICE HOST SETS

Manifold service hose sets consist of three color-coded (usually blue for the low side, red for the high side, and yellow for the auxiliary) 8-foot (2.4 m) hose sets (**Figure 35–42**). Separate sets are used for R-134a and R-12 refrigerants. They are high-strength, nylon barrier/low permeation-type sets that meet SAE J2196 standards. They have a burst pressure rating of 2,100 psi (14.5 MPa) and a working pressure rating of 470 psi (3.2 MPa).

Vacuum Pump

Any air or moisture that is left inside an A/C system, even in trace quantities, can reduce the system's efficiency and eventually can lead to major problems, such as compressor failure. Air causes excessive pressures within the system, limiting the refrigerant's capability to change its state from gas to liquid within the refrigeration cycle. Water moisture can cause freeze-up at the expansion valve. This can either restrict or completely block refrigerant flow. These problems result in reduced or no cooling.

FIGURE 35–42 Manifold service hose sets

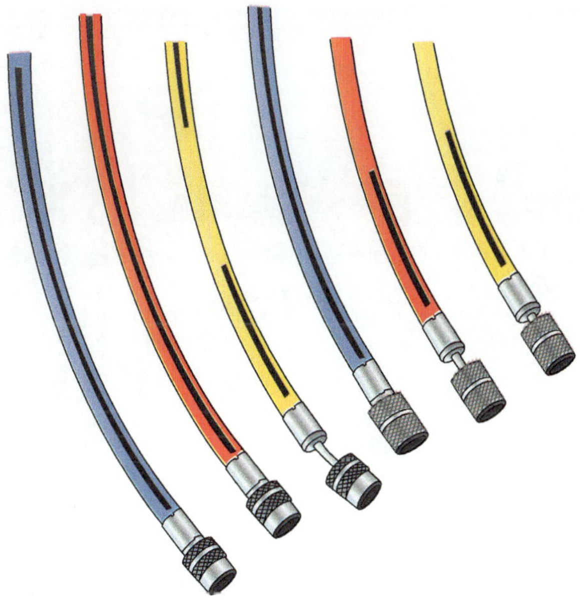

Additionally, moisture can combine with R-12 refrigerant to form corrosive hydrochloric acid, which can destroy the internal components of the A/C system.

The function of the vacuum pump is to evacuate the A/C system—that is, to remove air and moisture from the system. There are several types of vacuum pumps, but the rotary vane type is most common in heavy-duty truck A/C work. The idea is to draw all air and moisture out of the system. The system pressure is reduced until any moisture in the system vaporizes and is eliminated, along with the exhausted air. An R-134a system requires at least a ¾-hp (0.54 kW) vacuum pump to create the 29.9 inches (760 mm) of Hg vacuum required to evacuate the system. **Figure 35–43** shows the operation of a vacuum pump in an A/C system, and **Figure 35–44** shows a typical rotary vane-type vacuum pump.

FIGURE 35–43 A vacuum pump creates and holds a vacuum allowing moisture to be boiled and evacuated from system.

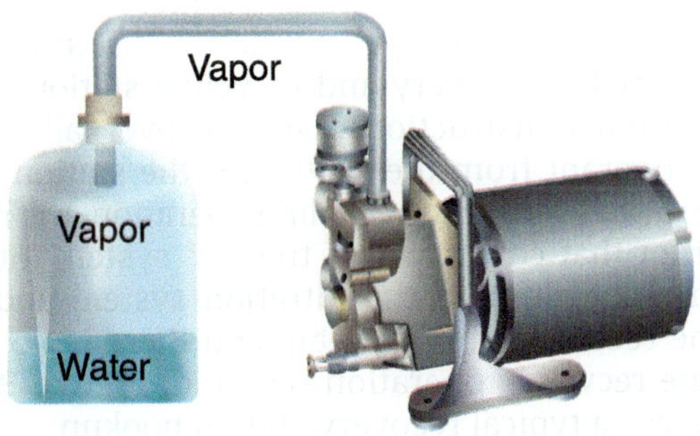

FIGURE 35–44 Rotary vane-type vacuum pump

Courtesy of Robinair, SPX Corporation

SHOP TALK

A vacuum pump is unable to remove moisture in liquid state from a system. It lowers the system pressure and, therefore, the boiling point of liquid moisture. It then removes the vaporized moisture. This is why it is necessary to pull a vacuum on a system for an extended period to ensure that no moisture is left in it.

Thermistor Vacuum Gauge

The electronic thermistor vacuum gauge is designed to work with the vacuum pump to measure the last, most critical inch of Hg vacuum during evacuation. It constantly monitors and visually indicates the vacuum level so that a technician will know for sure when a system is entirely free of air and moisture. The thermistor vacuum gauge is capable of high accuracy and is calibrated in microns (millionths of a meter).

Refrigerant Management Centers

The refrigerant management center recovers, recycles, and recharges an A/C system. For the service facility specializing in A/C service work, refrigerant management is a requirement. The stations are refrigerant specific, so generally an R-134a station is used only for systems using that refrigerant. Such an arrangement not only ensures clean, single-pass, recycled refrigerant but also provides the accuracy of recharging required by current "critical charge" A/C systems.

Most refrigerant management centers permit the technician to switch the refrigeration source, allowing the system to be charged from either a recycled or a virgin refrigerant tank. The manifold gauges are built into a refrigerant management center, allowing the center to be coupled in series with the vehicle system. Some machines permit the operator to divide the center into a recovery/recycling station and a separate charging station. This provides flexibility for busy shops.

Electronic Service Tool

Some OEM truck A/C systems use full authority electronic controls that require the technician to connect an electronic service tool (EST) to the HVAC controller either directly or to the chassis data bus. The EST should be loaded with the OEM software and the diagnostic routines must be undertaken is sequence.

35.8 AIR CONDITIONING SERVICE PROCEDURES

All A/C service procedures should be performed referencing the OEM's service literature and its recommended procedure. Some general guidelines are provided here so that you can develop a general understanding of each procedure. System servicing normally comprises:

- System recovering and recycling
- System flushing
- Compressor oil level checks
- Evacuation
- System recharging

CHARGING CYLINDER OR CHARGING STATION

A charging cylinder is designed to meter out a specific amount of a refrigerant by weight. Compensation for temperature variations is accomplished by reading the pressure gauge of the cylinder. It is essential that this temperature compensation be performed.

When recharging an A/C system, often the pressure in the system reaches a point at which it is equal to the pressure in the cylinder from which the system is being charged. To get more refrigerant into a system to complete the charge, some heat must be applied to the cylinder, or the refrigerant gas can be drawn into the low side with the system running. Some recharging cylinders, such as that shown in **Figure 35–45**, have heaters that eliminate the problem caused by the equalization of pressure between the cylinder and the system being recharged.

All refrigerant must be recovered from the system before repair or replacement of any component (except for compressors with stem-type service valves). This procedure is accomplished through the use of a manifold gauge set or a recovery/recycling station, which makes it possible to control the rate of refrigerant recovered, thus minimizing the loss of oil from the system. Two types of recovery stations are shown in **Figure 35–46**, and the components are shown in **Figure 35–47**.

FIGURE 35–45 R-134A recharging cylinder

The recovery procedure is performed as follows:

1. Remove the caps from the compressor's suction and discharge service valves (**Figure 35–48**).
2. Wearing protective goggles and nonleather gloves, couple the manifold gauge set to the valves by performing the following sequence:
   - Close both manifold valves.
   - Connect the manifold high-side hose to the compressor's discharge service valve.
   - Connect the manifold low-side hose to the compressor suction service valve.
   - Connect the manifold's center hose to an approved recovery and recycling station.
3. Follow the recovery and recycling station manufacturer's instructions, and recover all of the refrigerant from the system. If the station has a one-pass recovery/recycling operation, the refrigerant that comes out of the A/C system into the machine through the filtration system and into the recovery tank is clean, and there is no separate recycling operation required. **Figure 35–49** shows a typical recovery station hookup.

4. Measure oil in the station's oil recovery cup. The compressor will have to be refilled with the same quantity of new refrigerant oil. If the system has become contaminated, all of the compressor oil must be replaced with clean oil after purging and flushing the system. If the system is heavily contaminated with desiccant or grit, replace the compressor, expansion valve, and receiver/dryer.

FIGURE 35–46 (A)Typical refrigerant recovery and recharging station. (B)Technician display window on a refrigerant recovery and recharging station.

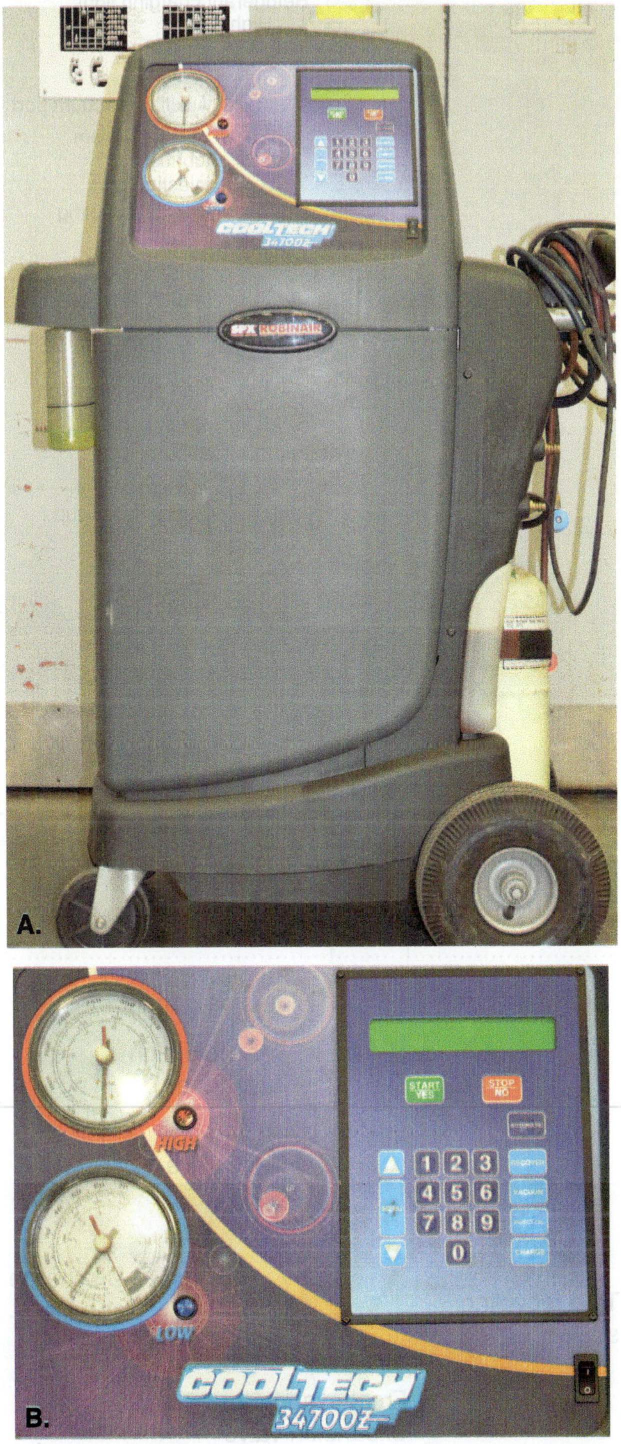

If the station does not have a one-pass recovery/recycling feature, the recycling can be performed in a typical straight recycling machine as follows:

1. The liquid valve on the fill tank should be fully opened at this time. Make sure that the hose is properly connected to the port.
2. Push the toggle switch to the recycle position.
3. Momentarily depress the cycle start button (2 to 3 seconds). The system begins its recycling process. Liquid refrigerant should be seen flowing through the moisture indicator glass at this time.
4. Leave the unit in the recycle mode until the moisture indicator turns from yellow to green. This process will take anywhere from 1 to 6 hours, depending on the amount of moisture contamination in the refrigerant.
5. The color sample dot indicates the minimum level of dryness. When the moisture indicator is darker green than the sample, put the rocker switch from the recycle position to the end cycle position. The unit will then shut off completely after a few seconds. It is normal for the machine to vibrate as it works in a vacuum just before shutting off.
6. Close the liquid valve on the tank, disconnect all of the hoses on the tank, and remove the tank from the unit.
7. Apply the magnetic label to the tank to indicate that it has been recycled and is ready to use.

PURGING AND FLUSHING

Purging removes air, moisture, traces of refrigerant, and loose dirt from the A/C circuit, forcing inert, moisture-free gas through the parts. Flushing is more extreme and removes heavy contamination, such as gritty oil, larger particle dirt buildup, metal, or plastics, which cannot be removed by purging with gas. Flushing is performed by forcing liquid refrigerant through the system, allowing it to pick up the contaminants and flush them out.

The decision as to whether to purge or flush a system depends on the extent of contamination. Most routine servicing and recharging does not require the system to be flushed. Flushing with refrigerant requires a recovery and recycling station. It should be performed in the opposite direction of refrigerant flow; in other words, it is back-flushed.

Note that some OEM systems should not be flushed because of the type of condenser tubes used. Verify that it is okay to flush a system before doing so.

Purging Procedure

Dry nitrogen gas is usually recommended for purging, but safe, inert, moisture-free gases, such as carbon dioxide or argon, may be substituted. A pressure regulator is required to regulate between 0 and 200 psi (0 to 1.38 MPa). Commercial cylinders of nitrogen are stored at pressures exceeding 2,000 psi (13.8 MPa); this pressure must be reduced to 200 psi (1.38 MPa) for purging. (See **Figure 35–50** and **Figure 35–51**.)

FIGURE 35–47 (A) Components of the recovery and recycle portion of a refrigerant center; and (B) recharging portion.

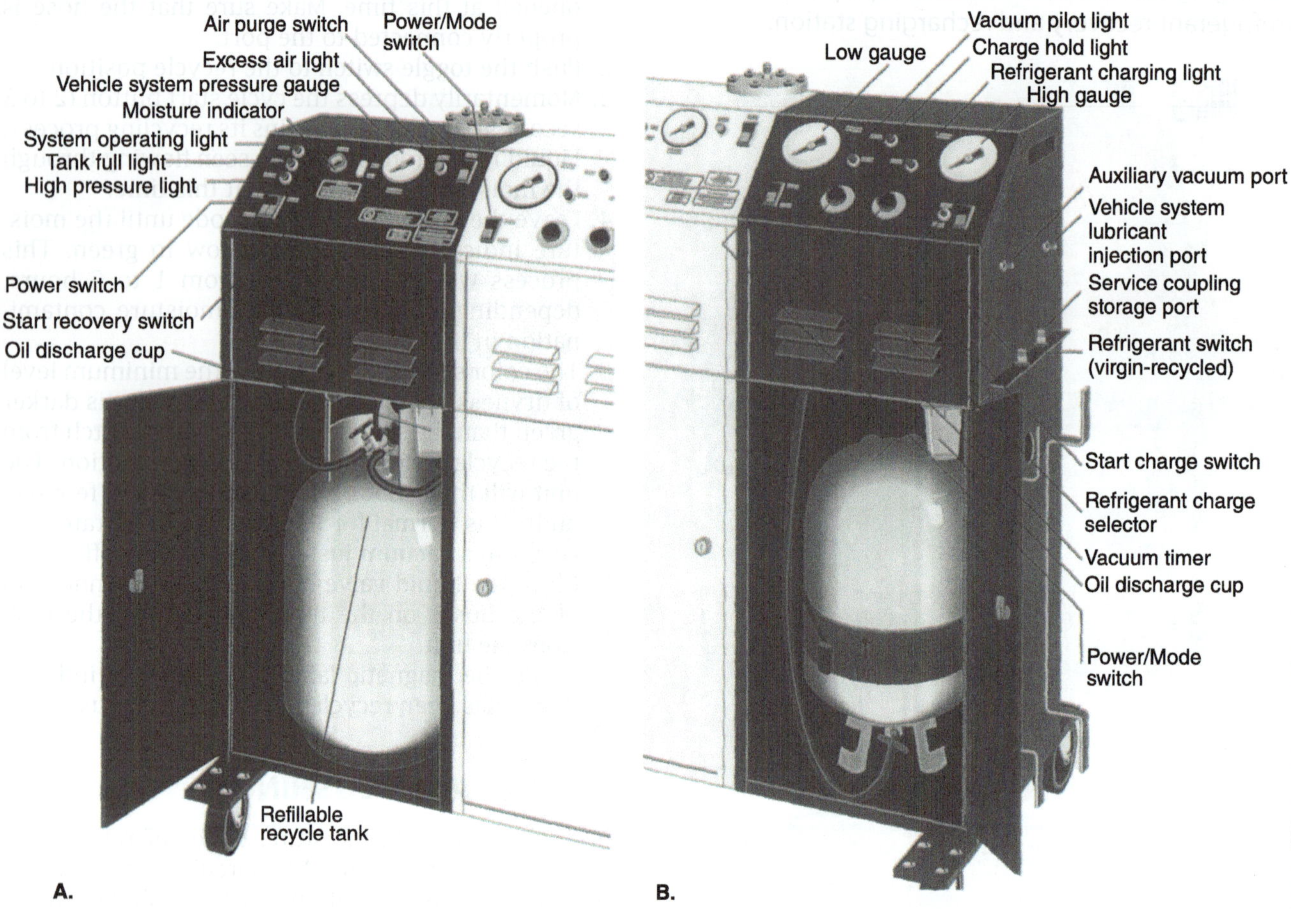

FIGURE 35–48 Discharge and suction service valves

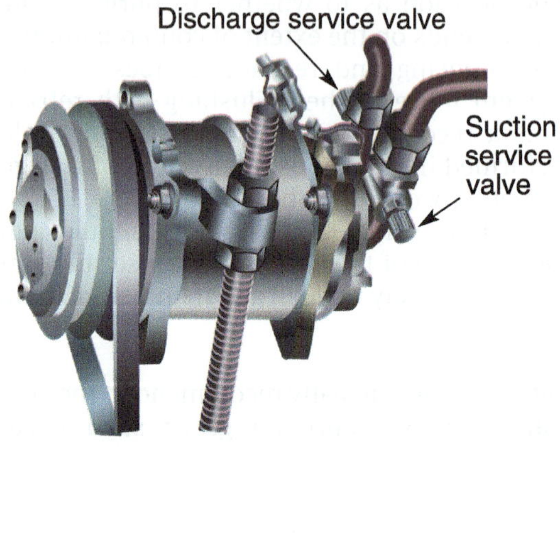

FIGURE 35–49 Recovery hookup

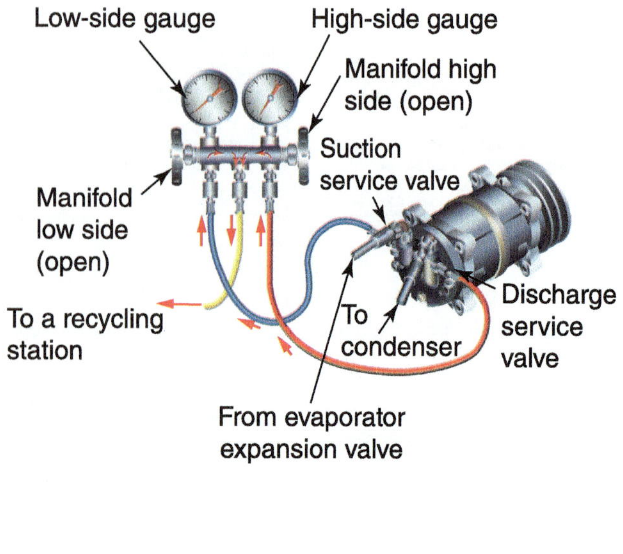

FIGURE 35–50 Arrows show purge and flush points and the direction of flow

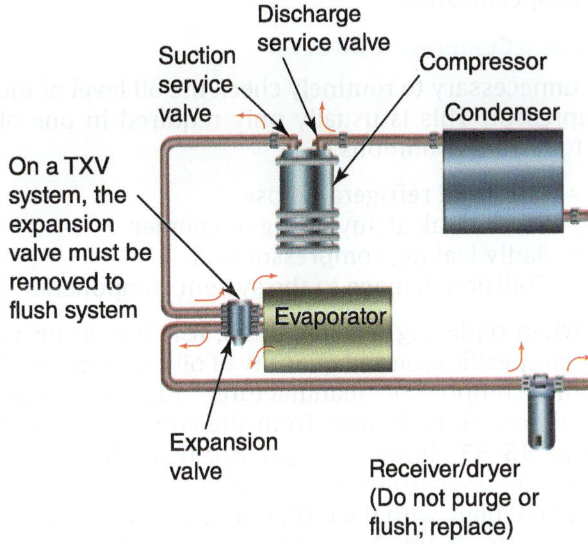

1. Recover the refrigerant from the A/C system.
2. Disconnect both ends of the line or component being purged. Seal the rest of the system by capping.
3. Make sure that the nitrogen cylinder control valve, the purging control valve, and the supply line valve are all closed.
4. Connect the supply line valve to the outlet end of the component or line.
5. Connect the drain line to the inlet end of the component or line.
6. Place the outlet of the drain line into a recycling system container.
7. Adjust the gas regulator to 200 psi (1.38 MPa). Open the nitrogen cylinder control valve and the purging control valve. Then slowly open the supply line valve. Check the drain line for gas flow.
8. Let the nitrogen flow at 200 psi (1.38 MPa) for 25 to 30 seconds.
9. Lower the pressure to around 4 psi (27.6 kPa) and let it flow for 1 to 2 minutes. If the line or

FIGURE 35–51 Purging station: After a liquid flush has been performed using isopropyl alcohol, the flushing process should be completed using nitrogen gas.

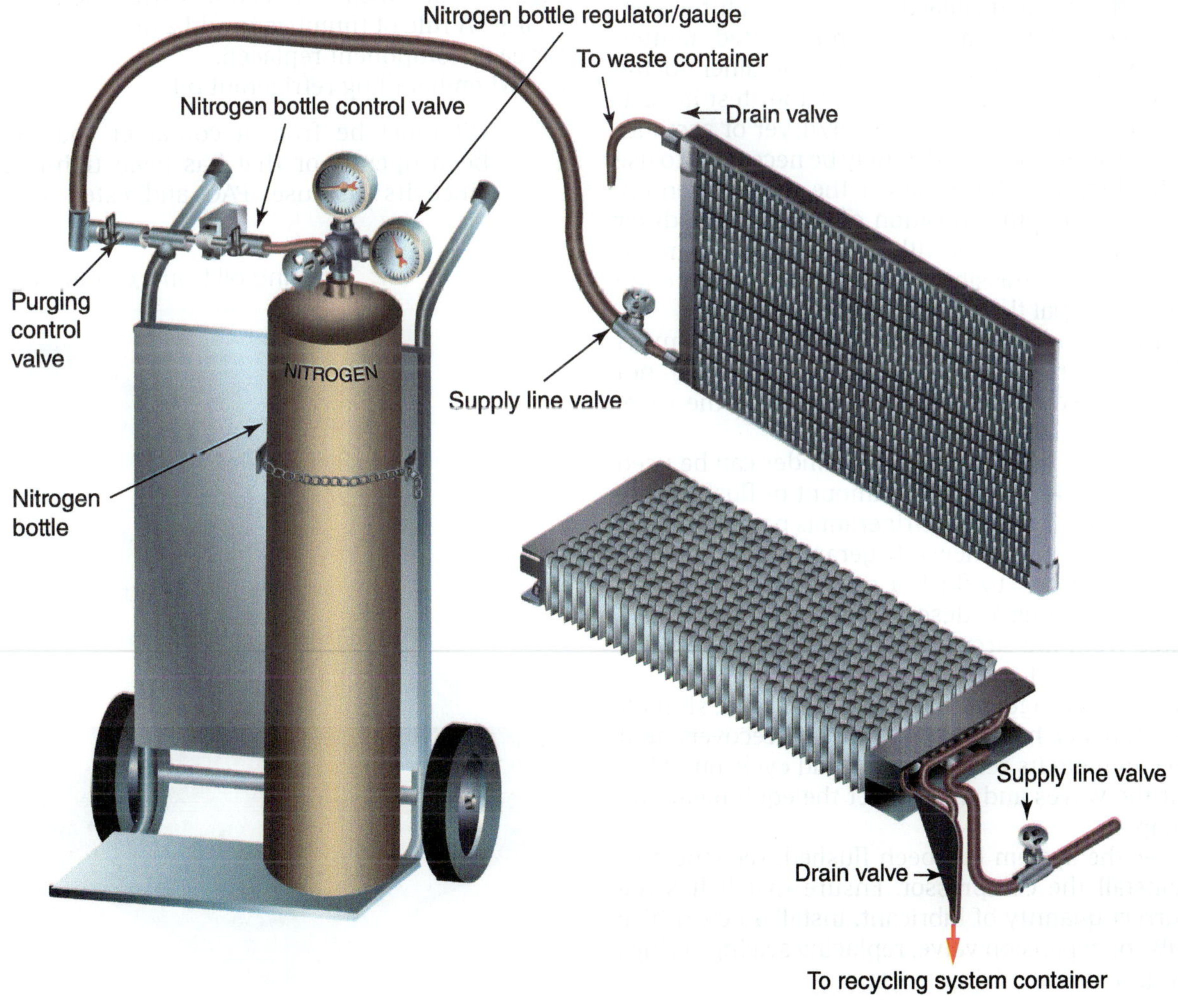

component was visibly wet, allow the gas to flow until no trace of refrigerant oil or other matter is flowing from the drain tube.

10. Close the nitrogen cylinder control valve and the purging control valve first and then close the supply line valve.
11. Disconnect the supply line valve and the drain line. Tightly cap both ends of the component or line.

Flushing Procedure

Flushing should be performed with the specified refrigerant required in the system. In older systems, flushing agents were used, but these are no longer approved. To flush a typical A/C system, proceed as follows:

1. Recover the refrigerant from the A/C system.
2. The compressor, receiver/dryer, accumulator, or control valves cannot be flushed. To isolate these components from the system, install bypass/interconnect flush fittings. If just one component is to be flushed, only the hoses needed to isolate that component need to be disconnected.
3. Remove the expansion valve or orifice tube. Clean the expansion valve and discard the orifice tube.
4. Typically, a system should be flushed with at least twice the system capacity to remove all the contaminated lubricant and contaminated matter. Connect the refrigerant (flush) container to the system with an approved refrigerant test hose at the point at which the receiver/dryer or accumulator was disconnected. It may be necessary to use a flushing adapter. Connect the recovery equipment to the other position of the receiver/dryer or accumulator lines with the proper adapter and hose. It is permissible to flush in either direction of the normal flow in the circuit.
5. With the connectors coupled as described, invert the refrigerant flush container so that liquid, not gas, is used as a flushing agent. Rotate the valve all the way open.
6. A charging scale, station, or cylinder can be used to ensure that the proper amount of flushing liquid is used. Recycled refrigerant is perfect for system flushing, and new refrigerant is not required.
7. The procedure to flush a specific component is much the same as described except that it is isolated from the circuit. Remember, never vent the refrigerant into the atmosphere.
8. After the refrigerant flush has been used, shut off the canister hand valve. Allow the recovery unit to complete its recovery time and cycle out. Shut off the valves and disconnect the equipment and adapters.
9. After the system has been flushed, reconnect or reinstall the compressor. Ensure that it has the correct quantity of lubricant. Install a new orifice tube or expansion valve, replacing sealing O-rings or gaskets.
10. Install a new receiver/dryer and/or accumulator and reassemble using new O-rings. Finish by evacuating the system and recharging the system to specification.

Checking Compressor Oil Level

It is unnecessary to routinely check the oil level of the compressor. This is usually only required in one of the following conditions:

- Ruptured refrigerant hose
- Severe leak at any fitting or coupler
- Badly leaking compressor seal
- Collision damage to the system components

When replacing refrigerant oil, it is important to use the specific type and quantity of oil recommended by the compressor manufacturer. **Figure 35–52** shows how oil is drained from the compressor, and **Figure 35–53** shows a compressor being filled with oil. If there is a surplus of oil in the system, too much oil will circulate with the refrigerant, causing the cooling capacity of the A/C system to be reduced. Too little oil will result in poor lubrication of the compressor. When recovering the refrigerant from the system or replacing system parts, add only enough oil to replace that which has been lost with the refrigerant charge. A rule of thumb is to add 2 ounces (57 g) of oil for each component replaced.

When handling refrigerant oil:

- Oil must be from a container that has not been opened or that has been tightly sealed since its last use. PAG and ester lubricants

FIGURE 35–52 Draining oil from compressor

FIGURE 35–53 Filling compressor with oil: This should be injected into the discharge side.

are hygroscopic (absorb moisture) and are quickly contaminated when left open to the atmosphere.

- The oil should be free of contaminants.
- Never mix any refrigerant lubricants with other types/viscosities of lubricant.
- Tubing, funnels, or other equipment used to transfer the oil should be clean and dry.

During the refrigeration cycle, some of the oil circulates with the refrigerant in the system. When checking the amount of oil in the system, first run the compressor to return the oil to the compressor. The following is a typical procedure for adding oil to a compressor:

1. Turn on the air conditioner (compressor clutch must be engaged) and set the cab blower fan at high speed.
2. Run the compressor for at least 20 minutes between 800 and 1,200 rpm.
3. Shut down the engine and remove the compressor from the engine.
4. Remove the compressor sump drain plug to drain the compressor oil to a clean container. Rotate the drive plate several times by hand to extract the remaining oil through the discharge-side port. About 1 ounce (28.3 g) of oil will remain in the compressor, and this cannot be removed without disassembling it.
5. Measure the drained oil with a measuring cylinder/graduated vial.
6. Visually check the oil for contamination. Unlike engine oil, compressor oil contains no added cleaning agent. Even after extended use, the oil should not become cloudy, as long as there is nothing wrong with the compressor. Inspect the drained oil for any of the following conditions:
   - Cloudiness or oxidized or moisture-contaminated refrigerant. Severely oxidized refrigerant can cause gumming and clogged filters and passages.
   - Color change to red, which indicates varnishing, the result of exposure to air.
   - Presence of foreign particles (metal powder, etc.) in the oil.
7. Install the drain plug and a new O-ring lightly coated with clean compressor oil. Tighten the plug to the specified torque.
8. Fill the compressor with the specified quantity of oil through the suction-side hose fitting.

EVACUATING THE SYSTEM

The refrigerant in the system must be recovered before evacuation. After discharging the system, the low-pressure gauge hose remains connected to the low-pressure test fitting, and both the high- and low-pressure manifold gauge valves should be closed. **Figure 35–54** shows the evacuator hookup.

Proceed as follows:

1. Connect the high-pressure gauge hose to the high-side service port.
2. Connect the center manifold gauge hose to the vacuum pump.
3. With the manifold gauge set and the vacuum pump properly connected (see Figure 35–54), begin evacuation by opening the high- and low-side gauge valves with the vacuum pump operating.
4. Operate the vacuum pump for 30 minutes after the low-side gauge reaches 29.9 inches (760 mm) Hg for R-134 a systems.
5. If the system has been exposed to air or moisture (in other words, left open), it should be partially charged and evacuated again.
6. Check the system for leaks by closing off both the high- and low-side gauge valves and by turning off the vacuum pump. There should be no more than 2 inches (50 mm) of vacuum loss in 5 minutes. A greater loss than this indicates a leak in the system.
7. If a leak is indicated, check all the system connections for specified tightness.
8. Crack open the refrigerant container valve. Open the low-pressure valve and allow the refrigerant to charge into the system. Close both the refrigerant container and the low-pressure valve and check the system for leaks. Correct leaks. Repeat the procedure until all leaks are eliminated.
9. Purge or reclaim refrigerant from the system and reevacuate. Do not discharge refrigerant through the vacuum pump.
10. Allow the vacuum pump to run for 20 minutes at a vacuum of 29.9 inches (760 mm) Hg. Close both manifold gauge valves and then turn off the vacuum pump. The system is now ready for recharging.

FIGURE 35–54 Evacuation hookup

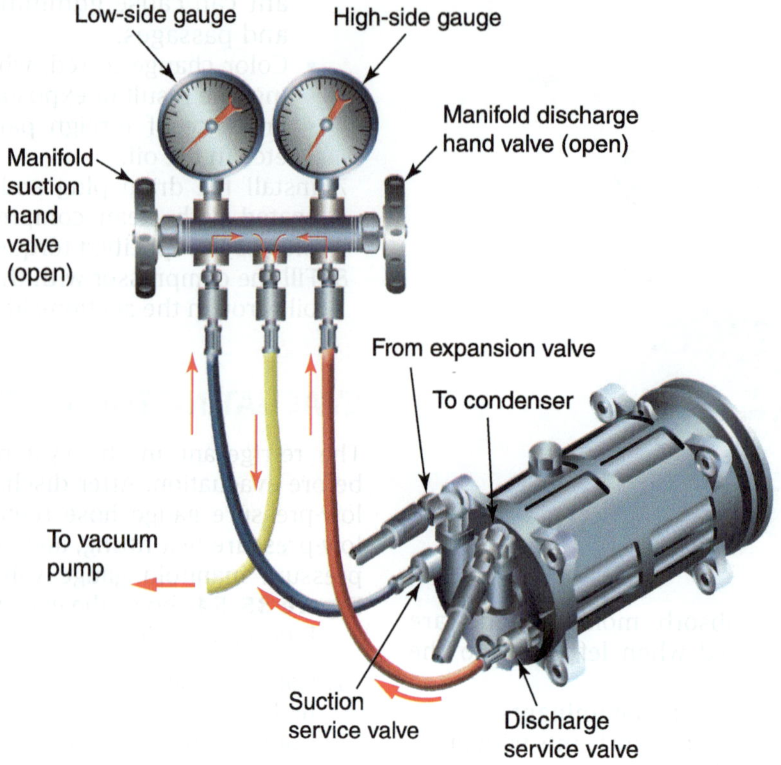

FULL RECHARGE OF AN A/C SYSTEM

Again, it is always important to follow the manufacturer's instructions. The general procedure is as follows:

1. Ensure that the service valves are closed and then connect the service couplings to the high and low ports of the vehicle A/C system. Open both couplings.
2. Set the station power switch to evacuation recharge.
3. Select the refrigerant source switch to virgin or recycled refrigerant. It is usually recommended that recycled refrigerant be selected.
4. Set the desired charge quantity on the station selector.
5. Press the Start Charge button to begin charging. A charging light will illuminate while charging refrigerant.
6. When charging is complete, the charge selector will read 0, and the charging light will go off.
7. Check High-Low pressure gauge readings and system output temperature to ensure that the system is performing within the OEM specifications.
8. Close both couplings, disconnect the station from the vehicle, and reinstall the storage plugs.
9. Set the station power switch to recover and press the Start Recovery button. This will recover the refrigerant vapor left in the hoses.
10. Close the valves on the recycling tank, the valve on the virgin tank, and all hose valves.

TOP-OFF RECHARGE OF AN A/C SYSTEM

Topping-off recharge of an A/C system is not a currently acceptable practice, because it cannot be known how much refrigerant or how much moisture is in a system (**Figure 35–55**). The correct service procedure is to evacuate, purge, and recharge the system with the correct weight of refrigerant (**Figure 35–56**).

SHOP TALK

The importance of the specified charge cannot be stressed enough. The efficient operation of the A/C system greatly depends on the correct amount of refrigerant in the system. A low charge will result in inadequate cooling under high heat loads because of a lack of reserve refrigerant and can cause the clutch cycling switch to cycle faster than normal. An overcharge can cause inadequate cooling because of a high liquid refrigerant level in the condenser. Refrigerant controls will not operate properly, and compressor damage can result. In general, an overcharge of refrigerant will cause higher-than-normal gauge.

FIGURE 35–55 Recharging hookup

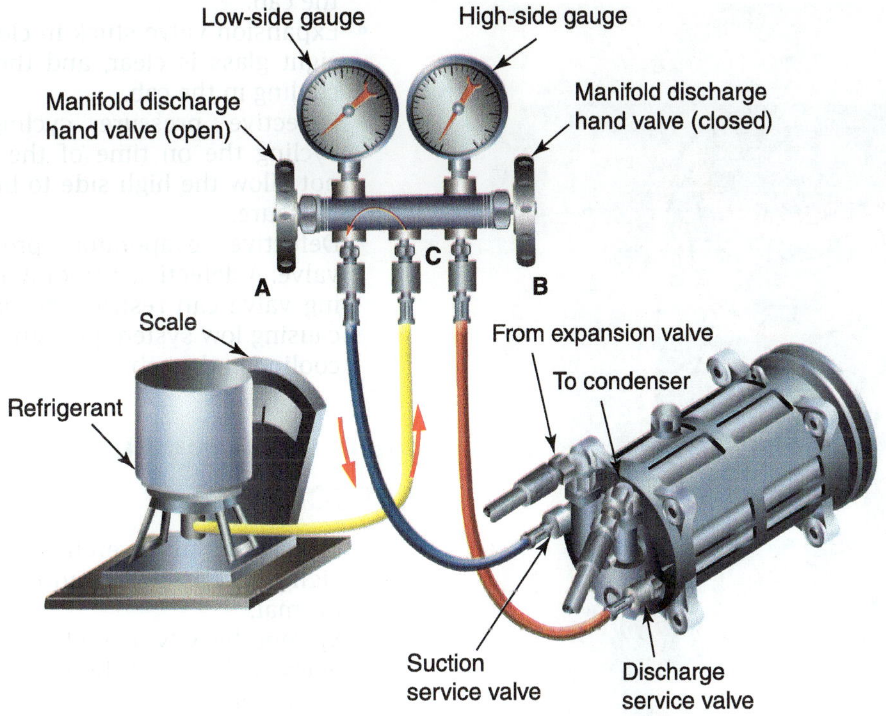

CAUTION

Accumulator systems do not have a sight glass. This is because bubbles are always present in the liquid line, even with a full charge. If the refrigerant is added until the bubbles are gone, serious damage and injury can result from overcharging the system.

35.9 COMMON AIR CONDITIONING PROBLEMS

This section takes a brief look at some A/C problems and their causes. Remember, this is not a comprehensive list; it is just a guide to some typical problems and what may cause them. Use the schematic in **Figure 35–57** to locate the system components, and consult **Figure 35–58** for the performance testing hookups. **Figure 35–59** shows an A/C troubleshooting flow chart used by Freightliner.

NORMAL PRESSURES AND INSUFFICIENT CAB COOLING

- Coolant control valve at the heater core is not completely closed. When tested, high-side and low-side pressures are normal and the sight glass is clear, but it is simply not cool enough in the cab.
- Excessive outside air infiltration. Damaged bunk or door seals or open vent doors on a cab or bunk can make it impossible for the A/C system to keep up with the heat load leaking into the cab and bunk.
- Blend air door/gate in the wrong position. The only time that the A/C system and the heating system should be working simultaneously is in demist or defog mode (defrost position).
- Recirculate/vent door (80/20 door) is jamming in the wrong position. In the max position, there should be a minimum of vent airflow.

HIGH- AND LOW-SIDE PRESSURES ARE LOW

- The system is low on refrigerant charge. High-side and low-side pressures are low; bubbles can be observed in the sight glass; and there is not enough cooling in the cab.
- Insufficient airflow through the evaporator core. Blower motor speed is low or in the wrong direction; the filter is dirty; dirt is clogging

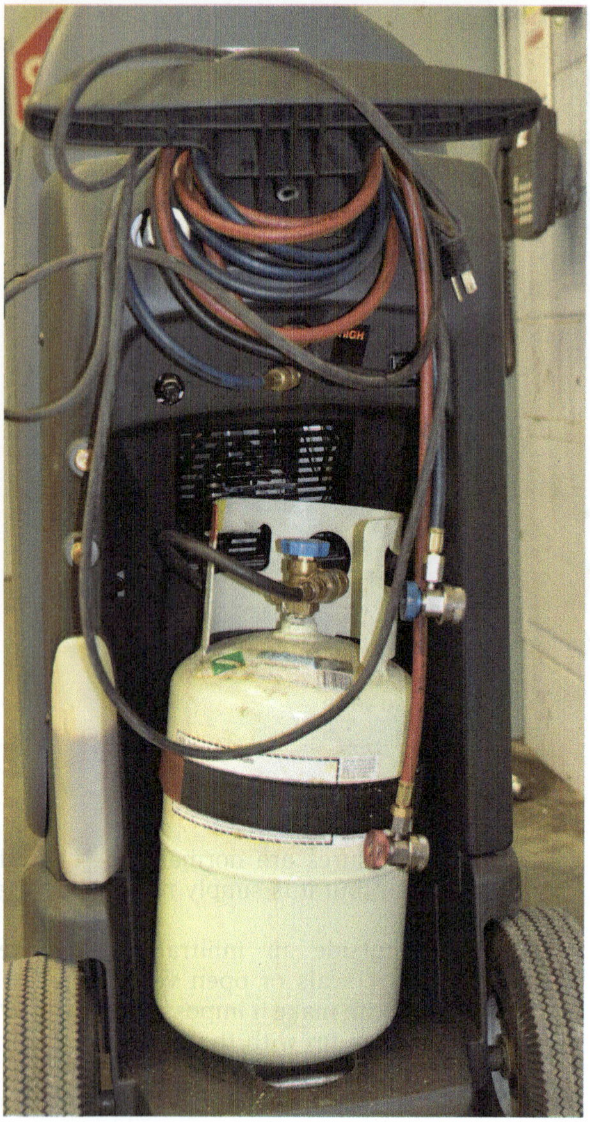

FIGURE 35–56 Weighing a refrigerant canister

the evaporator fins; or evaporator is frozen because of ice buildup.

- Moisture in system. System alternates between normal and poor performance because of moisture freezing at expansion valve and obstructing refrigerant flow. High-side and low-side pressures cycle between normal and low as moisture freezes and thaws at expansion valve.
- Obstructed expansion valve. The sight glass is clear, and there is no cooling in the cab.
- Obstructed orifice tube. Some but not enough cooling in the passenger compartment of the vehicle.
- Obstructed receiver/dryer. The sight glass is clear, and there is not enough cooling in the cab.
- Expansion valve stuck in closed position. The sight glass is clear, and there is not enough cooling in the cab.
- Defective pressure cycling switch. Short cycling the on time of the compressor does not allow the high side to build up to normal pressure.
- Defective evaporator pressure regulating valve. A defective evaporator pressure regulating valve can restrict the flow of refrigerant, causing low system pressures and not enough cooling in the cab.

HIGH- AND LOW-SIDE PRESSURES ARE TOO HIGH

- System slightly overcharged with refrigerant. Sight glass is clear, and cooling in the cab is normal.
- System heavily overcharged with oil. Sight glass is clear, and there is not enough cooling in the cab.
- Large amount of air in system. Bubbles are in the sight glass, and there is not enough cooling in the cab.
- Expansion valve stuck in open position. The sight glass is clear, and there may not be enough cooling in the cab.
- Poor thermal contact of sensing bulb of TXV with outlet of evaporator. The sight glass is clear, and there may not be enough cooling in the cab.
- Insufficient airflow through condenser core. Sight glass is clear, and there is not enough cooling in the cab.

HIGH-SIDE PRESSURE IS LOW AND LOW-SIDE PRESSURE IS HIGH

- Damaged compressor. May be abnormal compressor noise—the sight glass is clear, and there is not enough cooling in the cab.
- Improperly positioned sensing bulb of TXV. Poor thermal contact between the sensing bulb and the outlet of the evaporator can cause the TXV to remain in the full open position, providing too little restriction for normal system function.
- Improperly positioned capillary tube of thermostatic switch. If the capillary tube of the thermostatic switch is not properly installed (inserted too far) in the evaporator core, the

FIGURE 35–57 Truck A/C schematic

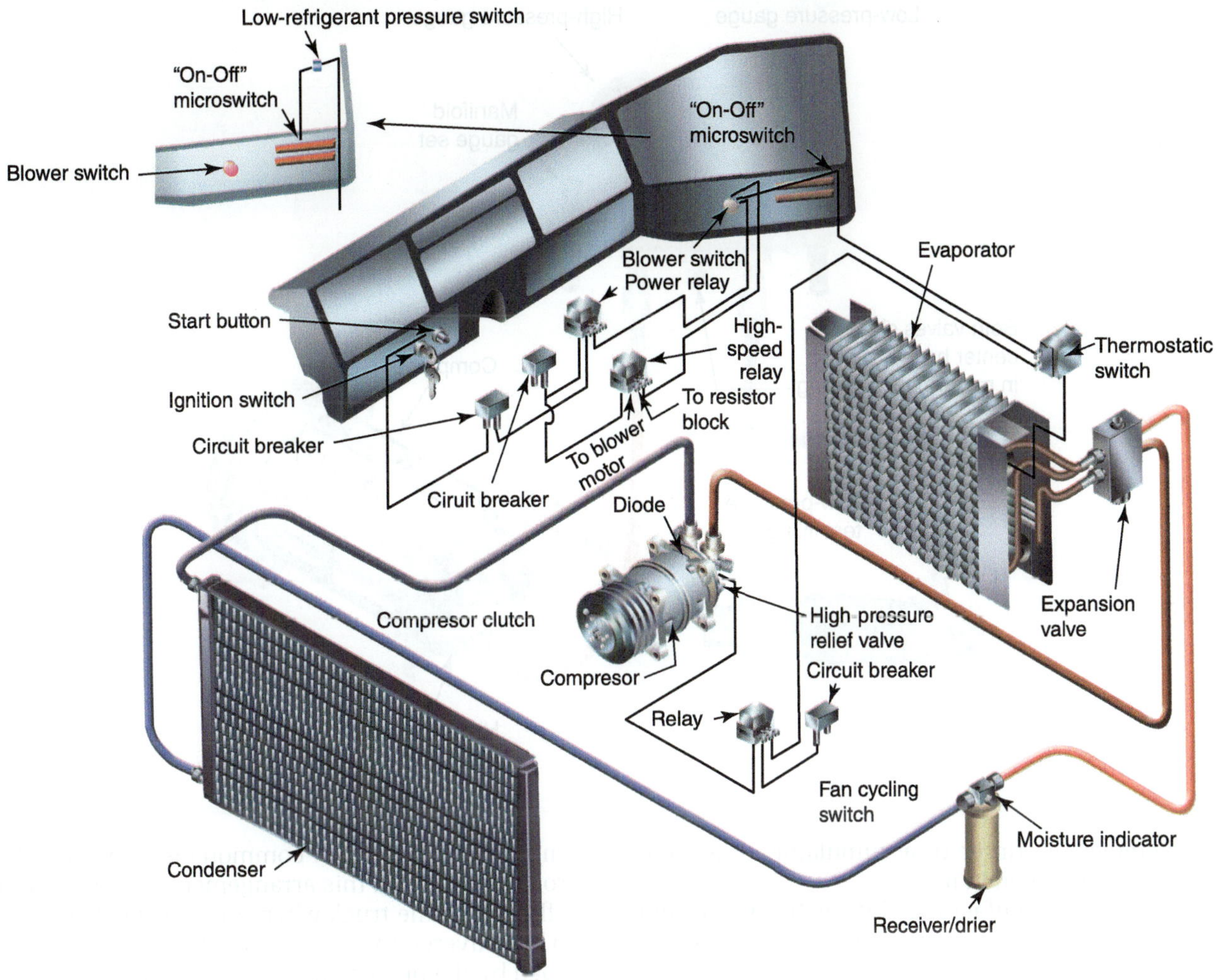

compressor can cut out too soon, not allowing high-side pressure to build or low-side pressure to be pulled down.

35.10 CONVERTING AN R-12 SYSTEM TO AN R-134A SYSTEM

In theory, R-12 should not be available today, meaning that every retrofit should have taken place. In reality R-12 is still readily available. This means that there are trucks running R-12 systems. It also means that shops continue the process of retrofitting systems to comply with the MACS standards. Some of the systems are being aftermarket retrofit to blend systems; some have flammable ingredients; and some can compromise system performance. The technician should practice caution when retrofitting or working on A/C systems that are already retrofit. Personal safety and compliance with federal legislation are both concerns.

RETROFIT PROCEDURE

Always follow the OEM procedure when retrofitting an A/C system. There are no absolute rules to observe, and the following list is only a set of guidelines:

- Use a refrigerant identifier to ensure that the system contains only R-12, not a blend.
- Performance test the system, checking for leaks, pressures, and temperatures.
- Use a recycling station to remove all the R-12 from the system.
- Remove the compressor and drain the oil. Record the amount of oil drained from the compressor.
- Remove the TXV or orifice tube. If it is contaminated, flush the condenser.

FIGURE 35–58 Performance testing

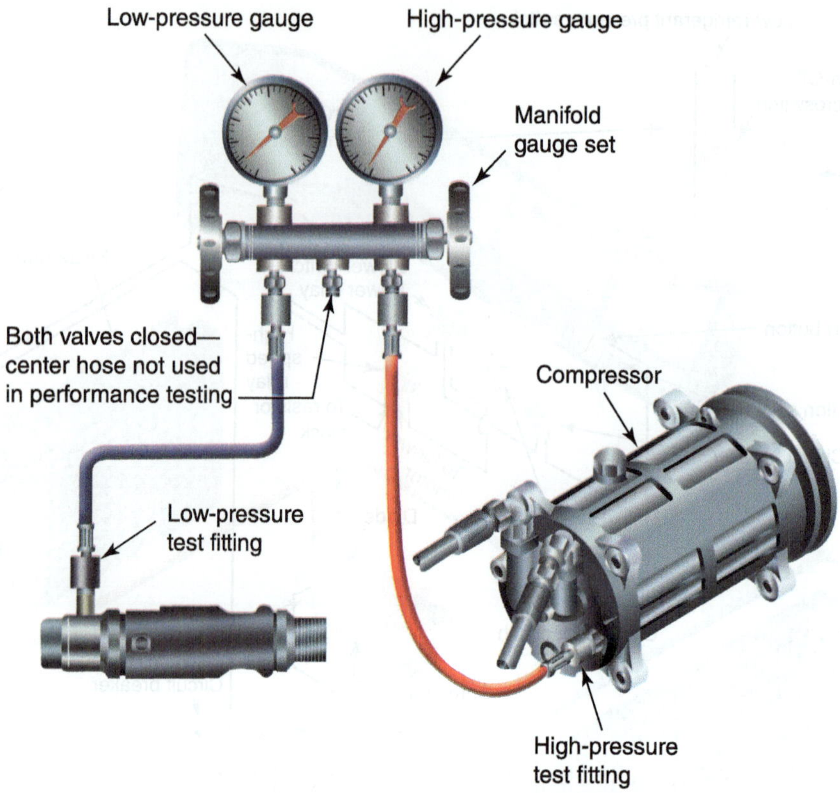

- Remove the dryer or accumulator; drain and measure the oil in it.
- Use the manufacturer kit instructions and remove all components such as hoses, seals, and gaskets that will have to be replaced.
- Install an R-134a-compatible oil into the system, ensuring that the quantity put in equals the quantity removed.
- Install a new dryer or accumulator.
- Install conversion fittings and a high-pressure cutoff switch.
- Remove the R-12 label(s) and install R-134 conversion labels.
- Connect an R-134a recovery station to the system and evacuate for 30 minutes.
- Charge the system to 80 percent capacity.
- Performance test the system and compare with system performance prior to the retrofit.
- Leak test the system.

35.11 CAB VENTILATING AND HEATING SYSTEMS

The ventilating system on most vehicles is designed to supply outside air to the cab through vents. Several systems are used to route vent air into the cab compartment. The most common method is the flow-through system. In this arrangement, a supply of ram air flows into the truck when it is moving. Outside air can be delivered to the cab either by ram air or ram air helped by the heater/air conditioner blower.

AIRFLOW

Most in-cab air movement is accomplished using the ventilation or blower fan whether ram air, the heater, or A/C is used to ventilate the cab. The fan is usually of a squirrel cage design (**Figure 35–60**) and is located behind the dashboard. As the squirrel cage blower rotates, it produces suction on the intake and pressure at its outlet. When the fan motor is energized by using the temperature controls on the dashboard, air is moved through the cab compartment. Many trucks are equipped with air-actuated cylinders to open and close vent gates, which determine how air is sourced and routed through the cab and bunk compartment.

Bunk coolers are used in some applications. These use cells fitted with liquid that freezes at relatively high temperatures. The idea is that the cell liquid is chilled to freezing point by the truck air conditioner when the truck goes down the road, so that when air is blown over it after the A/C system has been shut down, the air cools the sleeper berth.

FIGURE 35–59 Troubleshooting flowchart

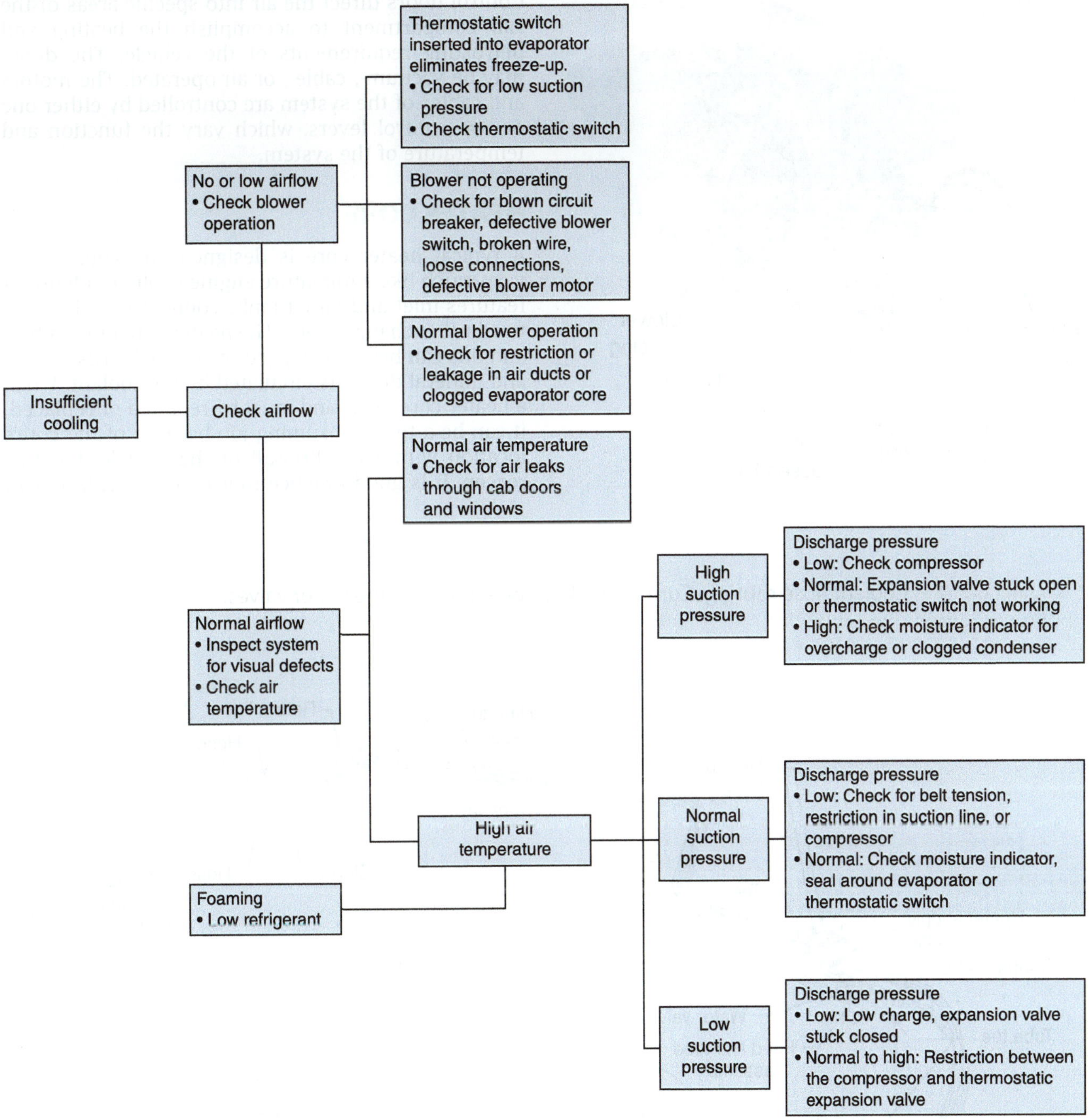

35.12 LIQUID-COOLED HEATING SYSTEM

As in any automobile, a truck's cab heating system is built into the engine cooling circuit. Hot engine coolant is routed through a heat exchanger in the cab, and when cool cab air is blown through this heat exchanger, warm air exits. As the hot engine coolant circulates through the heater core, heat is transferred from the coolant to the tubes and fins of the core. Air blown through the core by the blower motor and the fan then picks up the heat from the surfaces of the core and transfers it into the cab compartment

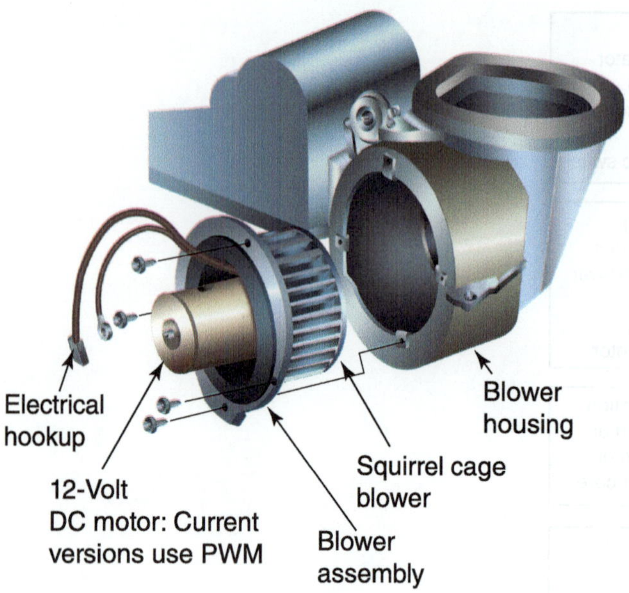

FIGURE 35–60 Typical squirrel cage fan

of the truck. The circulation of the engine coolant is achieved by the engine coolant pump (**Figure 35–61**). Control doors direct the air into specific areas of the cab compartment to accomplish the heating and defrosting requirements of the vehicle. The doors may be vacuum-, cable-, or air-operated. The motors and cables of the system are controlled by either one or two control levers, which vary the function and temperature of the system.

HEATER CORE

A typical heater core is designed and constructed very much like a miniature engine cooling radiator. It features inlet and outlet tanks connected by headers to a heat exchanger core. The heater core tank, tubes, and fins can become clogged over time by rust, scale, and mineral deposits circulated by the coolant. When a heater core leaks and must be repaired or replaced, it can be a time-consuming job because of the core's location within the firewall of the vehicle. For this reason, it is good practice to leak test a replacement

FIGURE 35–61 Coolant hose routing: Current trucks have eliminated the water valves.

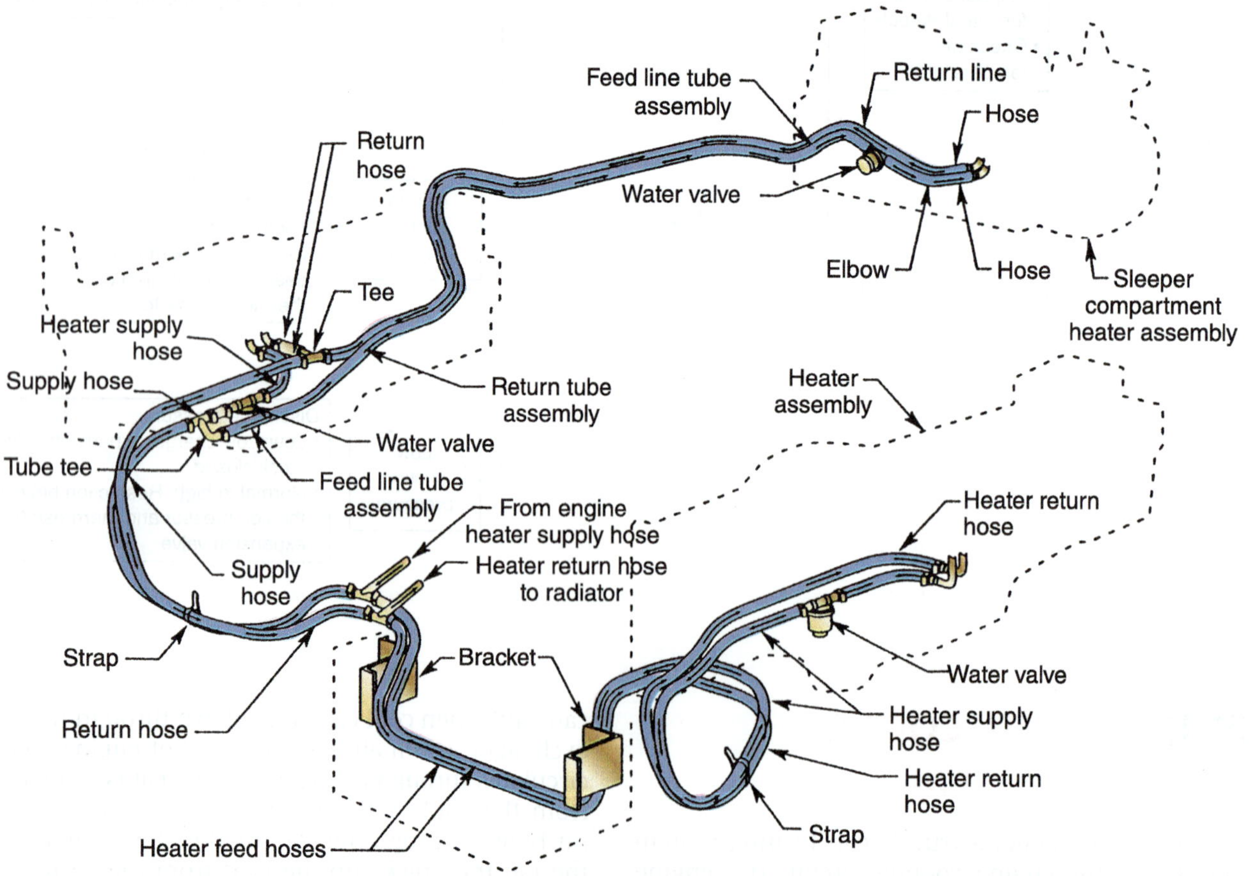

heater core before installation. Flushing the cooling system and seasonal replacement of the coolant can extend the life of all the heat exchangers in the system.

HEATER CONTROL VALVE

A **heater control valve**, also known as a coolant flow valve, controls the flow of coolant into the heater core from the engine. In a closed position, the valve allows no flow of hot coolant to the heater core, allowing it to remain cool. In a fully open position, the valve allows hot coolant to circulate through the heater core, providing maximum heating. Settings between these values permit the heating intensity to be controlled.

THERMOSTAT

The thermostat that helps to regulate the engine coolant temperature plays a role in providing the cab with heat. A malfunctioning thermostat can cause the engine to overheat or not reach normal operating temperature, and either of these conditions will impact the cab heater performance.

COMBINATION HVAC CORE

The combination air conditioner/liquid-cooled heater is a compact unit that heats, cools, and dehumidifies the cab and defrosts the windshield. The heater and air conditioner cab unit is usually mounted below (**Figure 35–62**) and to the right of the instrument panel. This unit contains the heater core and water flow control valve (**Figure 35–63**) for the heater system and the TXV, evaporator core, and cold control switch for the air conditioner. A variable-speed blower motor is mounted in the top of the unit. The blower motor forces air past both coils and supplies heated or cooled air to the ductwork behind the dash panel. A typical control panel is shown in **Figure 35–64**.

Sleeper Compartment Heater and Air Conditioner

Sleeper or bunk heater and air conditioner units (or sleeper boxes) include the heater core, evaporator coil, blower, and control valves. These units are usually mounted in the baggage compartment. The sleeper air conditioner is dependent on the cab's A/C system sharing the same refrigerant and compressor. The bunk heater is plumbed directly to the engine, usually but not always independently of the cab heater.

FIGURE 35–62 Heater core and air duct locations in a cab

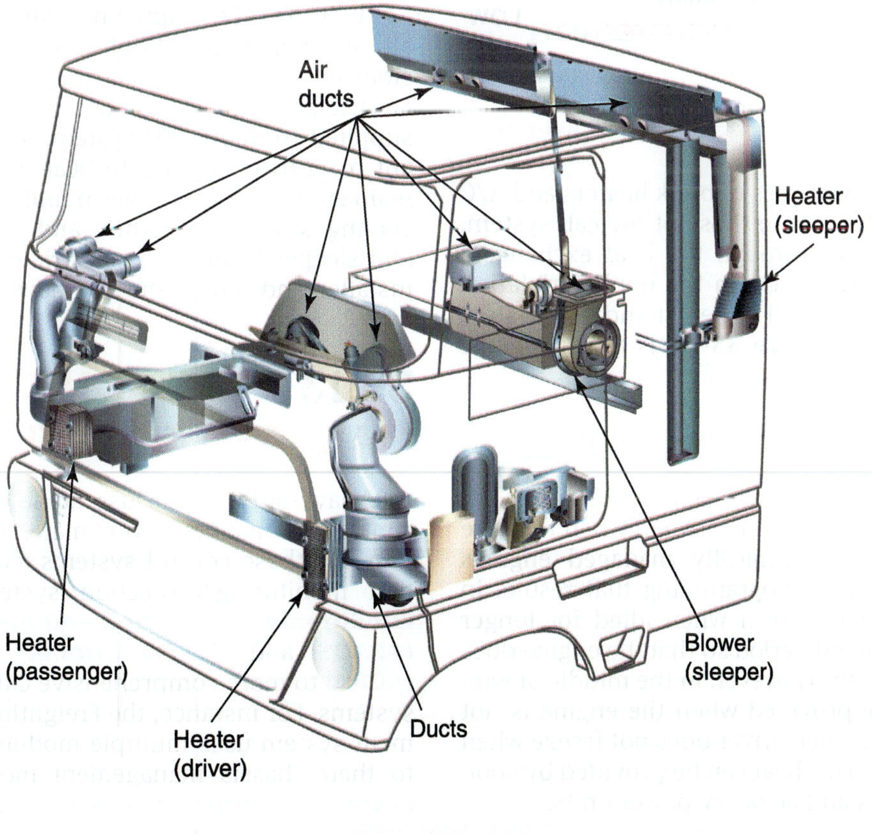

FIGURE 35–63 Cable-controlled coolant valve used to control the flow of coolant through the heater core. Current units have replaced this with movable air-directional doors.

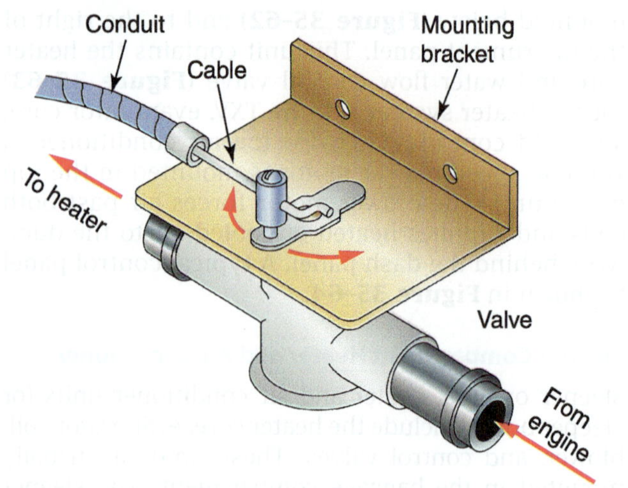

FIGURE 35–64 Control panel and airflow

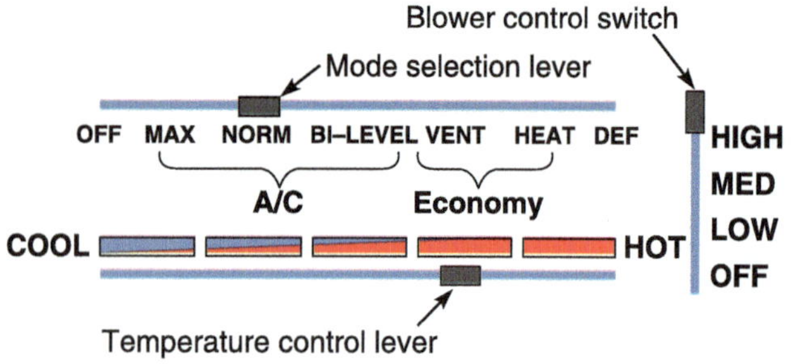

The operating principles of bunk heaters and A/C evaporators are identical to those of the cab system. The blower forces air through the heat exchangers and from there to the ducts in the bunk ventilation circuit and to the sleeper. The heater and A/C system schematic shown in **Figure 35–65** is typical of one found in a heavy-duty truck.

SUPPLEMENTARY HEATING DEVICES

Idling engines can greatly shorten engine life. Also, idling engines tend to cool faster than when they are shut down. Many electronically managed engines today have idle defeat programming that results in the engine being shut down when idled for longer than a preprogrammed period so that the engine does not idle unnecessarily. However, in the middle of winter, heat has to be provided when the engine is not running so that the truck driver does not freeze when asleep in his bunk. This heat can be provided by coolant heating devices and auxiliary power units.

Supplementary Coolant Heater

A **supplementary coolant heater** is a thermostat-controlled, diesel fuel heating device, the function of which is to maintain the engine coolant at a predetermined value. This is usually known as a diesel **fuel-fired heater**. The coolant is maintained at a temperature adequate to provide the truck cab and bunk with sufficient heat during periods of engine shutdown in winter conditions. Most have thermostats operating in conjunction with electronic controls to cycle the system to maintain the required temperature. The unit operates exactly like a kerosene-fired space heater, except that diesel fuel is used as the fuel and the flame acts on a heat exchanger through which coolant is flowed. **Figure 35–66** shows a typical diesel fuel-fired coolant heater. A more primitive type of **coolant heater** consists of an electrically heated element (usually A/C mains supplied) inserted into the cylinder block water jacket. This type of heater is also known as a **block heater**.

Auxiliary Power Units (APUs)

A separate on-board engine and power generator can produce 2,000 to 4,000 watts or more. These small diesel-fueled units consume one-third to one-sixth of the fuel used by the main engine when idling. This type of power source also makes it possible to cool the cab in hot weather using a 120-volt air conditioner or by driving a separate A/C compressor with the APU's engine. One OEM offers a fuel cell–type APU.

With an APU supplying standard main voltage, all other on-board bunk accessories, from televisions and lighting to microwave ovens, can be used. Electric-powered coolant and fuel heaters also can be supplied from an APU system. Some APUs use coolant from their engines to heat the cab, sleeper, and main engine. This involves plumbing the two engines' cooling systems together, and, of course, the cab and sleeper heaters. An electric block heater can be installed and run off the APU power supply.

35.13 ELECTRONICALLY MANAGED CLIMATE CONTROL

All heavy-duty trucks use some level of electronic controls to manage cab climate. The terms used to describe these control systems vary from basic system monitoring/protection systems such as **air conditioning protection and diagnostics system (APADS)** and **air conditioning protection unit (ACPU)** to more comprehensive climate management systems. For instance, the Freightliner HVAC management system uses multiple modules that are bussed to their chassis management module known as a **system actuation module (SAM)**.

FIGURE 35–65 Typical sleeper heater and A/C box: More recent units have replaced the solenoid valve with a PWM-modulated air valve.

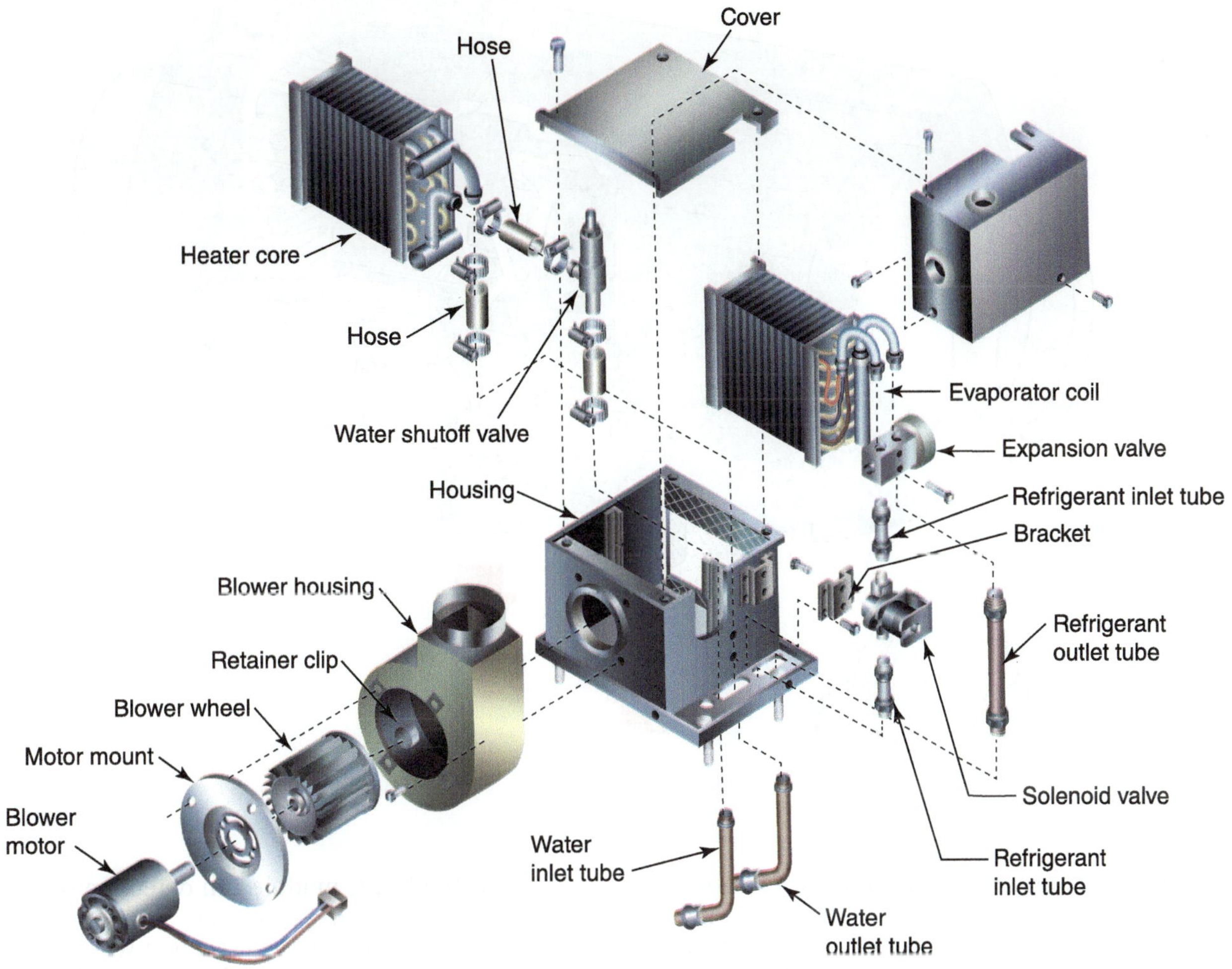

The advantage of using electronic controls is that the HVAC can maintain a specific temperature automatically inside the cab and sleeper compartments with zone variations, and offer a measure of protection to the system components. To maintain a selected temperature, sensors send signals to a computerized control unit that controls the compressor effective cycle, heater valve, blower, and vent door operation.

A typical electronic control system might contain input components, such as a coolant temperature sensor, in-truck temperature sensor, outside temperature sensor, high-side temperature switch, low-side temperature switch, low-pressure switch, vehicle speed sensor (to measure ram air effect), sun-load sensor, and others. Outputs include the management of clutch cycling and other system switching functions such as virtual binary and trinary switches. The HVAC controller is used both to control the A/C system performance and to help in diagnosing A/C system problems. Current A/C control systems are networked to the vehicle data bus and can be accessed by electronic service tools (ESTs).

COMPUTER-CONTROLLED HVAC COMPONENTS

System components include the **HVAC control module**, which performs logic processing of the system inputs and switches the outputs. It also includes an input and output circuit. The HVAC control module masters A/C operation and tracks diagnostic data. It is usually located somewhere in the engine compartment rather than in the vehicle cab.

System Inputs

The input circuit to the HVAC control module necessarily consists of a signal produced from the thermostat and the pressure switches.

- The HVAC thermostat switch is the primary control unit. It is connected in series with the

FIGURE 35–66 Diesel fuel-fired heater

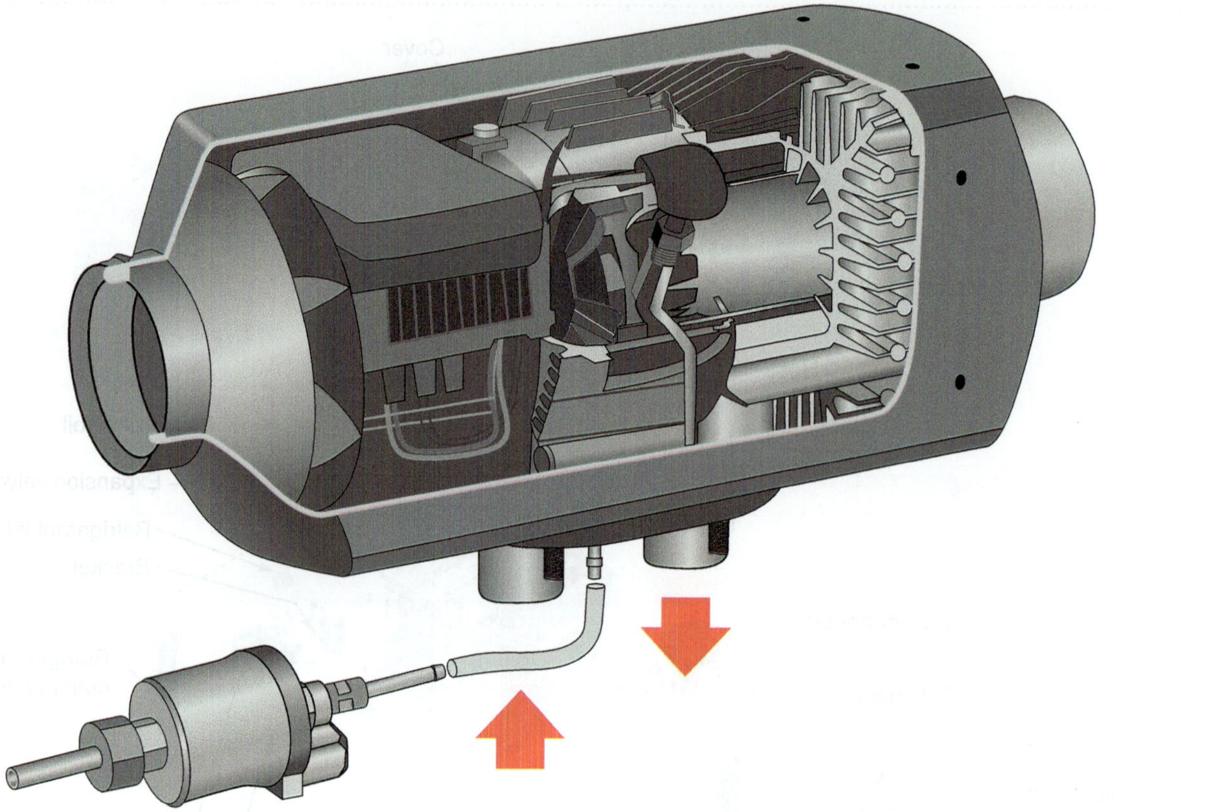

main A/C on/off switch and the **evaporator thermostat**. The HVAC control module will cycle the A/C compressor on and off in response to the thermostat input data. Typically, the thermostat contacts close when the evaporator temperature exceeds 38°F (3°C) on the rise, and open when the temperature decreases below about 32°F (0°C) on the fall. The thermostat is, therefore, used to turn off the effective cycle of the compressor whenever frost begins to form on the evaporator.

- Pressure switches are critical inputs to the HVAC control module. One high-side pressure switch and one low-side pressure switch are used to signal high- and low-side pressure values to the HVAC control module. Both switches are installed into the refrigerant lines. These switches monitor A/C circuit pressures. Both switches are designed to close at low pressure and open as pressure increases. The high-pressure switch typically opens at 300 psi (2,068 kPa) on the rise and closes at 260 psi (1,793 kPa) on the fall. The low-pressure switch opens at 34 psi (234 kPa) on the rise and closes at 10 psi (69 kPa) on the fall. Both switches have a 2,490-ohm resistor in parallel with the contacts so that the HVAC module can detect an open-circuit (broken wire) condition. Static pressure in a properly charged A/C system will cause the low-pressure switch to be open (2,490-ohms resistance) and the high-pressure switch to be closed (0-ohm resistance). However, a closed low-pressure switch would usually indicate a partial or total loss of refrigerant. At very low ambient temperatures, the A/C system pressures can sometimes drop low enough to close the low-pressure switch, even when the refrigerant charge is adequate. It is important to note that the A/C system will not operate when this happens.

HVAC Module

The HVAC module is a microcomputer and in today's modern systems, it is seldom a single, stand-alone unit. The primary controller receives inputs from the HVAC input circuit and off the chassis data bus (see Chapter 12). The inputs are processed along with programmed system instructions to produce outputs. Truck HVAC controllers are usually located in the engine compartment rather than in the cab. They can be read off the chassis data bus using the chassis data connectors and software. The HVAC controller occupies the J939 **source address (SA)** 25/MID 190 address on the data bus and communication with the engine controller (SA 00/MID 128) is required in even the most basic system. The primary HVAC controller may be networked via J1939 to a body controller or chassis management module but in addition, there can be proprietary bus connections to zone (driver side/passenger side/bunk) climate management modules.

The chassis management module occupies the SA 33 address on the bus, but it can be known by names such as body controller or Freightliner's **signal (detect) (and) actuation module (SAM)**. Freightliner favor the use of the acronym SAM in describing modules and in their system you will find there is a SAM-cab, SAM-front, an auxiliary control unit (ACU) and front control unit (FCU), but this makes system description more complicated than it actually is. The term "module" is often used to describe a smart switch by some OEMs. A full description of microprocessors processing cycles and multiplexing is provided in Chapter 6 and Chapter 12.

HVAC Module System Outputs

Two types of output are generated. The first category of outputs comprises those that help optimize system performance and manage the cycling of system components.

The other category of outputs comprises those that protect the system against damage.

The primary HVAC module outputs are request signals to the engine ECM the switch controls for the refrigerant compressor clutch, and the circuit refrigerant and air management controls.

- Although the engine fan is actually controlled by the engine ECM, the HVAC module can provide a fan-request signal input so that other systems, such as A/C, may request fan operation. Whether the fan will actually turn on or not depends on the engine processing cycle at the moment the fan-request signal is received.
- The operation of the refrigerant compressor clutch is directly controlled by the HVAC module. It is usually a 12V-DC switched output of the module.
- Airflow through the heat exchangers in a full-authority HVAC circuit is managed by **pulse width modulation (PWM)** –controlled air doors. PWM signals may be sourced from either the HVAC controller or a chassis module or SAM. The air doors route airflow (warm/cool/fresh/recirculated) through the system.
- In a system in which key controls such as the binary and trinary switches, have been replaced by virtual devices, electronic sensors signal pressure and temperature values to the HVAC controller, which will then effect switching using PWM signals to actuators.

The HVAC module will switch the compressor clutch as an outcome of its own processing cycle. This means that it will be controlled largely as a result of the thermostat input signals from the system electronic circuit. The module can discontinue operation of the compressor clutch if it detects a short- or open-circuit condition.

A/C Clutch Coil

The clutch coil typically has a nominal resistance of 3.5 ohms. The HVAC module calculates the actual coil resistance by making current and voltage measurements. If the calculated resistance is less than about 2 ohms or greater than about 12 ohms, the HVAC module will de-energize the clutch coil circuit. It will make another attempt to energize the clutch coil after a predetermined period (10–30 seconds). Then if the coil resistance is still out of limits, the circuit will be latched off until the next ignition cycle. When this occurs, a fault code is logged in the HVAC module that can be accessed by a data connector and the appropriate EST. **Figure 35–67** shows the location of some of the components in an electronically managed A/C system.

Electronically Managed HVAC Communications

The first generation of HVAC control electronics was a stand-alone system. Today, a typical HVAC electronic control system is multiplexed to the vehicle data, permitting the sharing of hardware with the other vehicle electronic systems. In the event that an HVAC module becomes disconnected from the data bus, the HVAC electronics will assume default values for these

FIGURE 35–67 Components of a typical electronically managed climate control system using a system actuation module (SAM).

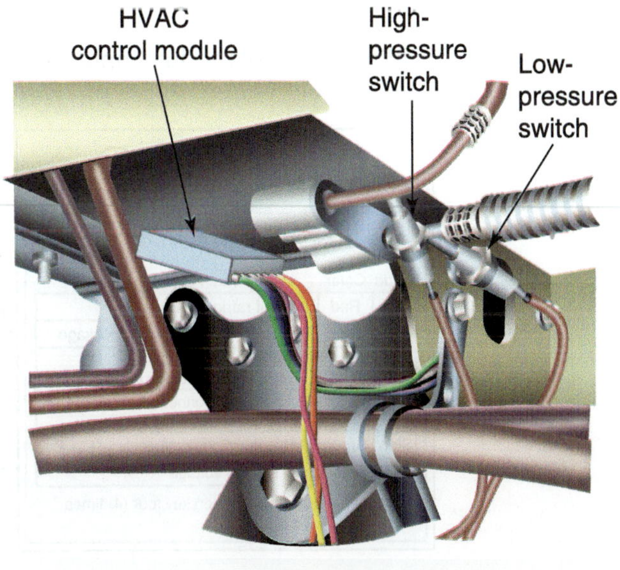

TABLE 35–2 Parameter Defaults

Missing Parameter	Default Value
Ambient air temperature	48°F (9°C)
Road speed	32 mph (51 km/h)
Engine speed	600 rpm
Park brake status	Off
Air pressure	Must 60 psi minimum

parameters. This permits the HVAC system to continue operating. **Table 35–2** lists parameter defaults.

A/C Control Logic

Figure 35–68 shows some of a typical HVAC controller inputs and outputs. The HVAC control module uses the following sequence of logic rules to control the A/C system:

1. A/C start delay. The A/C compressor cannot be driven under effective cycle for the first 15 seconds after the ignition key is switched on and the engine is running at more than 300 rpm.
2. After the ignition has been switched on and the engine has been running at more than 300 rpm for 15 seconds, the A/C compressor engages for 15 seconds. At this point, the following conditions must be satisfied:
   - The high- and low-pressure switches do not indicate out-of-limits or fault conditions.
   - No low air pressure message is being received from the vehicle data bus.
   - Battery voltage sensed by the HVAC module is within specified parameters; typically this means greater than 11 volts and less than 16 volts.
   - There are no logged fault codes that would disable the A/C operation. These faults are low-pressure, open clutch circuit, and shorted clutch circuit.
3. For basic fan operation, the HVAC module initiates a fan request to the engine ECM when the high-side pressure exceeds 300 psig (2,068 kPa), opening the high-pressure switch. The engine ECM should (and normally will) then respond by turning the fan on immediately. When the high-pressure condition ceases to exist and the high-pressure switch closes, the HVAC module fan-request to the ECM remains active for a short period and then ceases. However, the actual length of time the fan runs depends on the engine ECM programming.
4. The HVAC module switches the A/C clutch on and off in response to the thermostat input signal subject to the following conditions:
   - Data read from the data bus to the engine indicates that engine speed is greater than 300 rpm.
   - The on/off cycle frequency of the clutch must exceed 15 seconds.
   - The low-pressure switch remains open.
   - If the high-pressure switch opens, the A/C clutch will be de-energized after a short lag.
   - Locally sensed battery voltage exceeds 11 volts.
   - The A/C clutch will be disengaged if an overcurrent or undercurrent condition in the clutch coil is sensed by the HVAC module. If this happens, the HVAC control module will wait for a predetermined period and then attempt to energize the clutch once again. If the over- or undercurrent condition still exists, the HVAC control module will disable the system until the next ignition cycle.

Because most current HVAC controllers are multiplexed, not only can they broadcast system status and requests on the data bus, they also can pick up requests from other electronic systems networked

FIGURE 35–68 APADS control module inputs and outputs

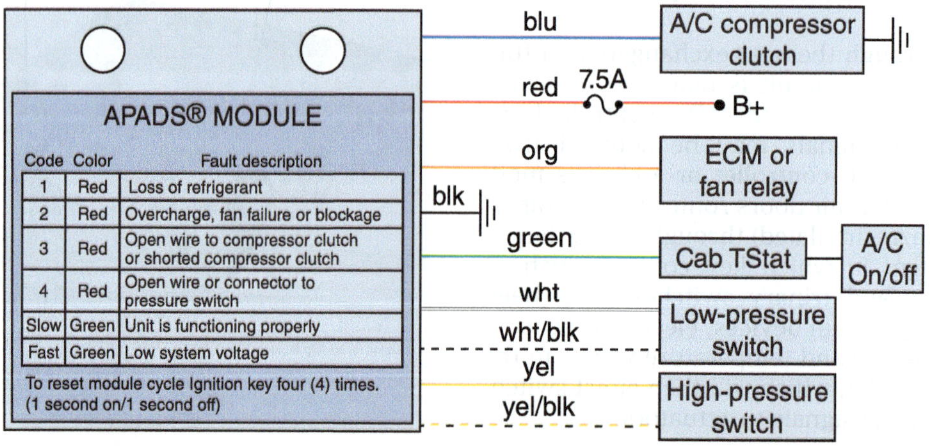

to the bus. For instance, when the engine controller receives an input request for full power acceleration, the engine ECM broadcasts a message "packet" on the data bus, and the HVAC controller responds by putting the A/C compressor clutch in dropout mode. This removes the parasitic load the compressor exacts on the engine for the duration of the full power request from the engine ECM.

Diagnostic Trouble Codes

The method of reading the HVAC system fault/trouble codes is the same as that used for all other truck electronic systems. This process is fully explained in Chapters 11 and 12. First, connect an EST to the dash data connector. Download the entry menu to the EST screen and select the SA (25) or MID (190). Then scroll through the system and to identify the flagged **diagnostic trouble code (DTC)** and **fault mode indicator (FMI)**. Locating HVAC system electronic problems is relatively simple and follows the same rules as other electronic troubleshooting. In at least one OEM system, the diagnostic routine is interactive and requires proprietary software and the ability to enter test data.

FIGURE 35–69 BlueCool truck heat exchanger

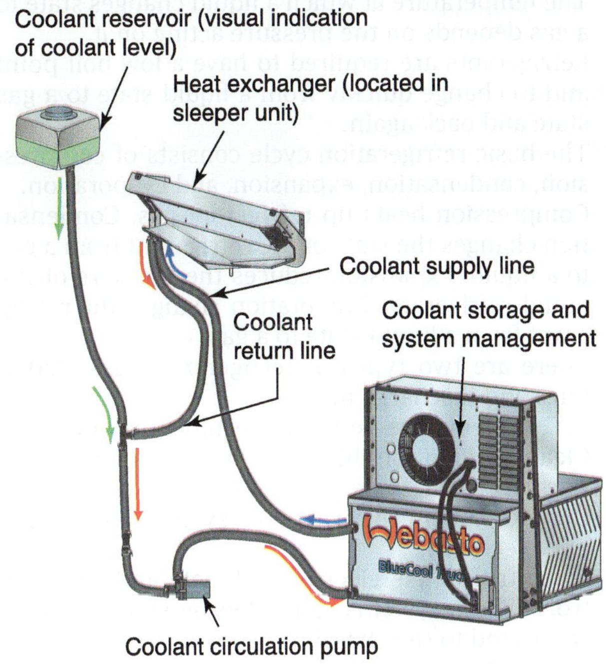

35.14 BLUECOOL TRUCK THERMAL STORAGE SYSTEM

Now that anti-idling restrictions exist in many jurisdictions, keeping a cab cool usually presents more of a challenge to the power capabilities of a highway truck than keeping it warm. Webasto's **BlueCool** truck thermal storage system has recently gained some traction in the marketplace as an environmentally friendly method of keeping a cab and bunk cool when a truck is parked. The system is rated to maintain a bunk temperature of between 68°F and 78°F (20°C and 25°C) at an outside temperature of 95°F (35°C) with a modest consumption of on-board electricity and without consuming any fuel.

BLUECOOL OPERATION

BlueCool uses a pair of aluminum coil circuits surrounded by a graphite matrix storage core contained in insulation. One coil circuit carries refrigerant gas (currently R-134a) and the other, a 50/50 mixture of ethylene glycol (EG) and water. The system uses an icebox principle. The graphite matrix absorbs water in liquid both liquid and frozen states. The storage coil is chilled down to subzero temperatures while the truck is operated on the road. When the engine is shut down, the coil circuit with the EG and water mix is used to circulate the liquid through a bunk heat exchanger. This cooling loop absorbs heat from the sleeper compartment to maintain comfortable temperatures for up to 10 hours. **Figure 35–69** shows the heat exchanger and bunk controls used by BlueCool.

SYSTEM SPECIFICATIONS

Webasto BlueCool recommends that the OEM specified four-battery bank system be used along with an alternator rated at 30 amps over the original factory specification. The cooling system is rated at 17,000 Btus (17,578 kJ). Electrical power consumption ranges from 3.5 to 10 amps depending on how the system is being operated and ambient temperatures. Advantages of BlueCool include zero fuel consumption during cooling cycles, almost silent operation during engine-off running, and an effective cooling cycle long enough for a night's sleep. The system is entirely emissions free during cooling cycles.

SUMMARY

- Heavy-duty heating and air conditioning (A/C) systems are designed primarily to keep the cab comfortable despite the outside weather.
- Liquids absorb heat when changed from a liquid state to a gas.
- Gases release heat when changed from a gaseous state to a liquid.

- Heat always moves from a hotter area to a cooler area.
- The temperature at which a liquid changes state to a gas depends on the pressure acting on it.
- Refrigerants are required to have a low boil point and to change quickly from a liquid state to a gas state and back again.
- The basic refrigeration cycle consists of compression, condensation, expansion, and evaporation.
- Compression heats up refrigerant gas. Condensation changes the state of the refrigerant from a gas to a liquid. Expansion reduces the pressure of the liquid refrigerant. Evaporation changes the refrigerant from a liquid state to a gas.
- There are two types of refrigerant lines: suction lines and discharge lines.
- A/C systems are sensitive to moisture and dirt. Clean working conditions are extremely important.
- The electrical system must be checked periodically to prevent the truck A/C system from failing unexpectedly.
- Some trucks use electronic climate and A/C controls. These systems can either be stand-alone or connected to the chassis data bus.
- Current A/C system controllers are networked to the chassis data bus. This enables them to communicate with other SAs/MIDs to optimize system operation.
- When the A/C controller is networked to the chassis data bus, the system is accessed by the J1939 connector. This allows the system to be read by any software that can interpret SAs/MIDs, SPNs/PIDs, SIDs and FMIs. To perform operations other than simply read the system, proprietary software is usually required.
- In some more recent truck A/C systems, virtual sensor/switches and PWM-controlled air doors have replaced mechanical and electromechanical controls of refrigerant and coolant routing through heat exchangers.
- The HVAC controller can respond to a request off the data bus to put the A/C compressor into dropout mode. An example would be a temporary request for full power from the engine ECM, when it would be desirable to eliminate as many parasitic loads as possible.
- BlueCool uses an icebox principle to cool truck bunks during engine shutdown at zero fuel consumption and with no emissions.
- The HVAC controller occupies the J1939 SA 25 on the bus. It is often designed to operate in conjunction with the body controller (SA 33).
- Some HVAC electronic systems have additional modules that use proprietary bus connections to the HVAC controller.
- The use of shorepower electrical grid supply is beginning to influence HVAC design on trucks, driven by anti-idling legislation and fuel consumption reduction.

REVIEW QUESTIONS

1. The amount of heat required to raise the temperature of 1 pound of water by 10°F is:
   a. 1 Btu.
   b. 10 Btus.
   c. 20 Btus.
   d. 100 Btus.
2. Which of the following is true in a properly functioning refrigeration system?
   a. A liquid changes to a solid and back to a liquid.
   b. A liquid changes to a gas and back to a liquid.
   c. Refrigerant gas reaches a temperature of absolute zero.
   d. Heat flows from a colder object to a warmer one.
3. Which component in an air conditioning system is responsible for removing heat from the circulating refrigerant?
   a. the compressor
   b. the evaporator
   c. the receiver/dryer or accumulator
   d. the condenser
4. Which of the following best describes the function of the TXV?
   a. It separates the high-pressure side of the system from the low-pressure side.
   b. It regulates refrigerant flow to the evaporator.
   c. It balances the inlet flow to the outlet temperature.
   d. all of the above
5. The two main refrigerant lines are known as the:
   a. high-pressure line and low-pressure line.
   b. suction line and discharge line.
   c. compressor line and condenser line.
   d. inlet line and outlet line.
6. Technician A wears safety goggles and rubber gloves when discharging the A/C system. Technician B wears safety goggles and leather gloves when charging and leak testing the system. Who is correct?
   a. Technician A only
   b. Technician B only
   c. both A and B
   d. neither A nor B

7. When checking an A/C sight glass for the moisture content of the refrigerant, Technician A performs the check with the air conditioner running immediately after starting the engine. Technician B says that some observable bubbles in the sight glass at system startup are normal. Who is correct?
   a. Technician A only
   b. Technician B only
   c. both A and B
   d. neither A nor B
8. Which of the following procedures may improve HVAC system performance?
   a. cleaning the outer surfaces of the condenser with modulated compressed air and soap
   b. repairing leaks in the evaporator coil
   c. improving the performance of the engine cooling system
   d. all of the above
9. Which method of detecting a leak in the A/C system would be most sensitive when attempting to locate a very small leak?
   a. application of a soapy water solution
   b. use of an electronic leak detector
   c. use of a flame-type propane leak detector
   d. use of dye and a black light detector
10. Which piece of equipment is used to meter out a desired amount of specific refrigerant by weight?
    a. manifold service hose set
    b. Schrader service valve
    c. recharging cylinder
    d. vacuum pump
11. Purging and flushing an A/C system are alike because both procedures:
    a. use dry nitrogen gas.
    b. use clean, dry liquid refrigerant
    c. remove contaminants.
    d. require the installation of a new receiver/dryer or accumulator.
12. Technician A uses refrigerant from an opened container that he knows has been tightly sealed since immediately after its previous use. Technician B says that recycled refrigerant known to be uncontaminated can be reused on recharging an A/C system. Who is correct?
    a. Technician A only
    b. Technician B only
    c. both A and B
    d. neither A nor B
13. When evacuating an R-134a system, how long should the vacuum pump be run after the low-pressure manifold gauge pulls 29.9 inches (760 mm) Hg vacuum?
    a. 5 minutes
    b. 10 minutes
    c. 30 minutes
    d. 50 minutes
14. Which of the following is the likely result of a slightly low charge of refrigerant in a truck A/C system?
    a. compressor damage
    b. higher-than-normal gauge readings
    c. noisy compressor operation
    d. faster-than-normal clutch switch cycling
15. Which of the following conditions could cause the low-side pressure to read high and the high-side pressure to read low?
    a. system overcharged with refrigerant
    b. insufficient airflow through the condenser core
    c. insufficient airflow through the evaporator core
    d. damaged compressor
16. Which of the following is a typical input of an electronically managed A/C system?
    a. the clutch switch
    b. the high-pressure switch
    c. the fan-request switch
    d. the blower control switch
17. Which of the following components provides the HVAC control module with temperature data?
    a. the thermostat
    b. the data bus
    c. the data connector
    d. the low-temperature switch
18. The expansion valve in an A/C circuit sticks in the open position. What is the likely result when the system is tested?
    a. low high-side pressure
    b. low low-side pressure
    c. high high-side pressure
    d. high low- and high-side pressure
19. Which of the following components is not a heat exchanger?
    a. an evaporator
    b. a condenser
    c. a radiator
    d. a compressor
20. When using generic software to access the vehicle data bus, what SA/MID specifically identifies the HVAC control system?
    a. SA 00/MID 128
    b. SA 03/MID 130
    c. SA 14/MID 140
    d. SA 25/MID 190

APPENDIX A

ACRONYMS

(Most of the terms listed here are explained in the Glossary.)

A	ampere
AAE	accident avoidance electronics
ABD	auto baud detect
ABDC	auto baud detection capable
ABS	antilock brake system
AC	alternating current
A/C	air conditioning
ACB	active cruise with braking system
ACCU	air-conditioning control unit
ACL	automatic chassis lubrication
ACom	Bendix ABS software
ACPU	air-conditioning protection unit
ACU	air bag control unit
A/D converter	analog to digital converter
ADR	Australian Design Rule
AGM	absorbed glass mat (battery)
ALCL	assembly-line communications link
ALDL	assembly-line data link
alt	alternator
AME	automated maintenance environment
AMT	automated manual transmission
ANSI	American National Standards Institute
APADS	air-conditioning protection and diagnostics system
APT	American pipe thread
ARV	antilock relay valve
ASA	automatic slack adjuster
ASC	automatic stability control
ASE	Automotive Service Excellence (see NIASE)
ASIC	application-specific integrated circuit
ASME	American Society of Mechanical Engineers
ASTM	American Society for Testing and Materials
At	ampere turns
AT	automatic transmission
ATA	American Trucking Association
ATEC	Allison Transmission Electronic Controls
Atm	unit of atmospheric pressure
ATV	all-terrain vehicle
AVOM	analog volt-ohmmeter
AWD	all-wheel drive
AWG	American wire gauge
AWS	American welding standard
BASIC	Behavioral Analysis Safety Improvement Category
BAT	battery
BHD	bulkhead
BHM	bulkhead module (FTL multiplex)
BMIS	battery management information system
BOW	brake overuse warning
BSS	blind spot system
Btu	British thermal unit
CA	communication adapter (RP 1210)
CAA	Clean Air Act
CAD	computer-assisted design
CAM	computer-assisted manufacturing/machining
CAN	control area network (multiplex data bus)
CAN 2.0	control area network 2.0 (CAN-C and J1939 data bus platform)
CB	circuit breaker
CBBs	counteract balance beads (see IBA)
CBM	condition-based maintenance
cc	cubic centimeter
CC	cruise control
CCA	cold cranking amps
CCFOT	cycling clutch, fixed orifice tube (A/C)
CCR	cold cranking rating
CCW	counterclockwise
CD	compact disc
CDL	Cat data link
CD-ROM	compact disc, read-only memory

CEC	Commercial Electronic Control (Allison)
CEL	check engine light
CFC	chlorofluorocarbon
cfm	cubic feet per minute
CHM	chassis module (FTL multiplex)
cm	centimeter
CRMS	commercial mobile radio service (V2I comm)
CMS	collision mitigation system
CMV	commercial motor vehicle
CMVSS	Canadian Motor Vehicle Safety Standard
CN	carbon neutral
CNG	compressed natural gas
COE	cab-over-engine
CPP	clutch pedal position
CPU	central processing unit
CRT	cathode ray tube (display monitor)
CSA	Canadian Standards Association
CSA	Compliance Safety Accountability (FMCSA 2010)
CVSA	Commercial Vehicle Safety Alliance
CW	clockwise
CWS	collision warning system
CX28	Caterpillar on-highway automatic transmission
CX31	Caterpillar medium-duty highway automatic transmission
CX35	Caterpillar heavy-duty highway automatic transmission
DBV	data bus viewer (Allison Transmission software)
DC	direct current
DCL	data communication link
DD	digital display
DDD	dash digital display
DDL	diagnostic data link
DDS	driveline disengagement switch
DDR	diagnostic data reader
DDT	digital diagnostic tool
DDU	digital display unit
DEF	diesel exhaust fluid
DER	Department of Environmental Resources
DIN	Deutsches Institut für Normung (German standards)
DIU	driver interface unit (Wingman)
DLC	datalink connector
DLC	diagnostic link connector
DM	datalink mechanical clutch module (Eaton AutoShift)
DMM	digital multimeter
DNS	do not shift
DOC	diagnostic optimized connection (Allison WTEC software)
DOS	disk operating system
DOT	Department of Transportation
DPDT	double pole, double throw (switch)
DPF	diesel particulate filter
DRM	dryer reservoir module
DT	Detroit Transmission
DTC	diagnostic trouble code
DTM	diagnostic trouble mode
DVOM	digital volt-ohmmeter
ECB	electronic circuit breaker
ECM	electronic (engine) control module
ECPC	electronic clutch pressure control (Cat AT)
ECU	electronic control unit
EDI	electronic data interchange
EENG	electronic engine
EEPROM	electronically erasable PROM
EFPA	electronic foot pedal assembly
ELC	extended-life coolant
ELI	electronic lock indicator (fifth wheel)
EMF	electromotive force (voltage)
EMI	electromagnetic interference
EOBR	electronic on-board recorder
EPA	Environmental Protection Agency
EPDM	ethylene, propylene, diene, monomer (belts)
ESN	engine serial number
EST	electronic service tool
ESV	experimental safety vehicle
ETR	evaporator temperature regulator (A/C)
EUI	electronic unit injector
EVS	electronically variable steering
EVT	electric vehicle technology
FET	field effect transistor
FCW	forward collision warning
FCWS	forward collision warning system
FDA	following distance alert (Wingman)
FHWA	Federal Highway Administration
FM	frequency modulation
FMCSA	Federal Motor Carrier Safety Administration
FMI	fault mode indicator
FMVSS	Federal Motor Vehicle Safety Standard
FPGA	field-programmable gate arrays (infotainment processors)
fps	feet per second
FRP	fiberglass reinforced panel (trailers)
FTL	Freightliner truck
FTP	Federal test procedure
FW	firewall
GAWR	gross axle weight rating
GCW	gross combination weight
GEM	greenhouse gas emission model
GHG	greenhouse gas
Gnd	ground
gpm	gallons per minute
GPS	global positioning satellite

GVW	gross vehicle weight
GWP	Global Warming Potential (refrigerant rating)
HAS	headway alert system
HFC	hydrofluorocarbon (refrigerant)
Hg	mercury
HH	handheld
HH-EST	handheld electronic service tool
HOS	hours of service
HTR	heater
H_2O	water
HUD	heads-up display
HVAC	heating, ventilation, and air conditioning
IA	impact alert (Wingman)
IBA	internal balancing agent (see CBBs)
IC	integrated circuit
ICU	instrument cluster unit
id	inside diameter in inches
IFI	Industrial Fastener Institute
IND	Indicator
IR	internal resistance
ISO	International Standards Organization
ITS	Intelligent transportation system (telematics)
JIT	just-in-time (production logistics)
K	kilo (1,000)
KAM	keep alive memory (NV-RAM)
km	kilometer
km/h	kilometers per hour
KOEO	key-on, engine-off
KP	kingpin
kPa	kilopascal
KPI	kingpin inclination
LCD	liquid crystal display
LDW	lane departure warning (system)
LED	light-emitting diode
LIN	local interconnect network (proprietary bus)
LLVW	lightly loaded vehicle weight
LNG	liquefied natural gas
LPV	load proportioning valve
m	meter
MCU	microprocessor control unit
MHz	megahertz
MID	message identifier
MIG	metal inert gas (welding)
MIL	malfunction indicator light
mm	millimeter
mmf	magnetomotive force
MPa	megapascal
mph	miles per hour
MSDS	material safety data sheet
MTIS	Meritor tire inflation system
MTIS	mobile tire inflation system
MUTT	mobile universal trailer tester
NA	not applicable
NACAT	North American Conference of Automotive Teachers
NATEF	National Automotive Technician Education Foundation
NEG	negative
NG	natural gas
NHTSA	National Highway Traffic Safety Administration
NIASE	National Institute for Automotive Service Excellence
NIOSH	National Institute for Occupational Safety and Health
NLGI	National Lubricants and Grease Institute
N·m	Newton-meter
NOx	oxides of nitrogen
NPN	negative-positive-negative (transistor doping)
NPT	National Pipe Thread
NTC	negative temperature coefficient
NV-RAM	nonvolatile random access memory
OBD	on-board diagnostics
O_2	ambient (diatomic) oxygen
OC	occurrence count
od	outside diameter
OD	overdrive
ODP	ozone depletion potential
OE	original equipment
OEM	original equipment manufacturer
OOS	out-of-service
OS	operating system (as in Windows or Linux OS)
OSHA	Occupational Safety and Health Administration
OVRD	override
Pa	pascals
Pb	lead
PBM	performance-based maintenance
PBM	predictive-based maintenance
PC	personal computer
PCM	powertrain control module (automotive/light truck)
PDA	personal digital assistant (handheld computer)
PDI	predelivery inspection
PDM	power distribution module
PG	parameter group
PG	pulse generator
PGN	parameter group number (J1939)
PID	parameter identifier
PLC	power line carrier
PLC	programmable logic controller
PM	preventive maintenance
PNP	positive-negative-positive (transistor doping)
POS	positive
POT	potentiometer
ppm	parts per million

PR	ply rating (tires)
PRD	pressure release device (NG)
PROM	programmable read-only memory
psi	pounds per square inch
psia	pounds per square inch absolute
psig	pounds per square inch gauge
PSV	public service vehicle
PTC	positive temperature coefficient
PTO	power take-off
PTT	professional technician (Volvo-Mack software)
PW	pulse width
PWM	pulse width modulation
PWR	power
QR	quick response (code)
R	resistance
R&D	research and development
RAM	random access memory
RBM	resist bend moment (frames)
RCR	reserve capacity rating (batteries)
RCRA	Resource Conservation and Recovery Act
RDU	remote diagnostic unit (Bendix)
RFI	radio frequency interference
RH	right hand
RLY	relay
rms	root mean square (machining)
ROI	return on investment
ROM	read-only memory
RP	recommended practice (TMC)
rpm	revolutions per minute
RSC	roll stability control
RSG	road speed governing
RSS	road speed sensor
RST	remote sensing technology (alternators)
RTV	room temperature vulcanization
R-12	CFC base refrigerant
RV	recreational vehicle
SA	source address (J1939)
SAE	Society of Automotive Engineers (industry standards)
SAI	steering axis inclination
SAM	signal (detect) (and) actuation module
SCA	supplemental coolant additive
SCI	serial communications interface
SCR	selective catalytic reduction
SCR	silicone controlled rectifier
SEL	stop engine light
SERV	service, emergency, relay valve (trailers)
SF	safety factor
SI	Système International d'Unités (metric system standards)
SID	subsystem identifier
SIR	supplemental inflatable restraint
SM	section modulus (frames)
SMS	Safety Management System (CSA)
SNSR	sensor
SOA	stationary object alert
SOC	state of charge
SOL	solenoid
SOP	standard operating procedure
SPN	suspect parameter number (J1939)
SRC	source (J1939)
SRI	service reminder indicator
SRS	supplemental restraint system
STV	suction throttling valve (A/C)
SUC	shift up control
TAN	total acid number (lubricant)
TBN	total base number (lubricant)
TC	torque converter
TCS	traction control system
TDS	total dissolved solids (coolant)
TERM	terminal
TEV	thermostatic expansion valve (A/C)
TIG	tungsten inert gas (welding)
TIR	total indicated runout
TMC	Technology and Maintenance Council (division of ATA)
TP	throttle position
TPCS	tire pressure control system (Spicer Dana)
tpi	threads per inch
TPMS	tire pressure monitoring system
TPS	throttle position sensor
TPV	tractor protection valve
TRA	Tire and Rim Association
TRU	transportation refrigeration unit
TSB	technical service bulletin
TTS	transmission tailshaft speed
UHS	unitized hub system (Spicer Dana)
U-joint	universal joint
UL	Underwriters Laboratories
ULEV	ultra low emissions vehicle
UNC	Unified National Coarse (threads)
UNF	Unified National Fine (threads)
USB	universal serial bus
V	volt
V2I	vehicle to infrastructure (telematics)
V2V	vehicle to vehicle (telematics)
V-AC	volts/alternating current
V-Bat	battery/system voltage
VC	vehicle clearance
V/C	vehicle body clearance
VD	voltage drop
V-DC	volts/direct current
VFD	vacuum fluorescent display
VI	viscosity index
VIM	vehicle interface module
VIN	vehicle identification number
V-MAC	Mack Trucks chassis controller module
VMRS	vehicle maintenance reporting standards

VOM	volt-ohmmeter
VORAD	vehicle on-board radar
VRE	vehicle retarder enable
V-Ref	reference voltage
VRV	vacuum regulator valve
VSS	vehicle speed sensor
VTF	vacuum tube fluorescent
W	watt
WHMIS	Workplace hazardous materials information system
WIF	water-in-fuel (sensor)
WIPS	wireless inspection processing system
WRI	wireless roadside inspection
WT	World Transmission (Allison)
WTEC	World Transmission Electronic Controls (Allison)
X1	Michelin wide-base tire
ZEV	zero emissions vehicle

APPENDIX B

TRUCK NOMINAL CLASS DESIGNATIONS

CLASS 2 1 to 10,000 lb GVW	CLASS 3 10,001 to 14,000 lb GVW	CLASS 4 14,001 to 16,000 lb GVW	CLASS 5 16,001 to 19,500 lb GVW	CLASS 6 19,501 to 26,000 lb GVW	CLASS 7 26,001 to 33,000 lb GVW	CLASS 8 33,001 lb and over	TRAILER Weight: not specified
Milk/Bread	Milk/Bread	Conventional Van	Rack	Tow	Home Fuel	Fuel	Van
Utility Van	Compact Van	Large Walk-In	Large Walk-in	Furniture	Trash	Dump	Doubles
Pick-up	Walk-in		Bucket	Stake	Fire Engine	Cement	Liquid Tank
Compact Van			Tree Specialist	Coe Van	Sightseeing	Reefer	Dry Bulk
Walk-in			Bottled Gas	School	Transit	Tandem-axle Van	Logger
Multipurpose				Single-axle Van		Intercity	Platform
Compartment Pick-up				Bottler	GCW to 65,000	GVC to 140,000	Drop Frame
Mini				Low-profile Coe	Medium Conventional	Low Profile Tandem Coe	Dump
					High-profile Coe	Heavy Conventional	Reefer
						Heavy-Tandem Conventional	Deep Drop
						Coe Sleeper	Auto Transporter

APPENDIX C

BRAKE SHOE IDENTIFICATION CHART

FOR 16$^{1}/_{2}$ (420 mm) DRUM DIAMETER BRAKES

EATON

Nominal Shoe Thickness 0.209 in./5.3 mm

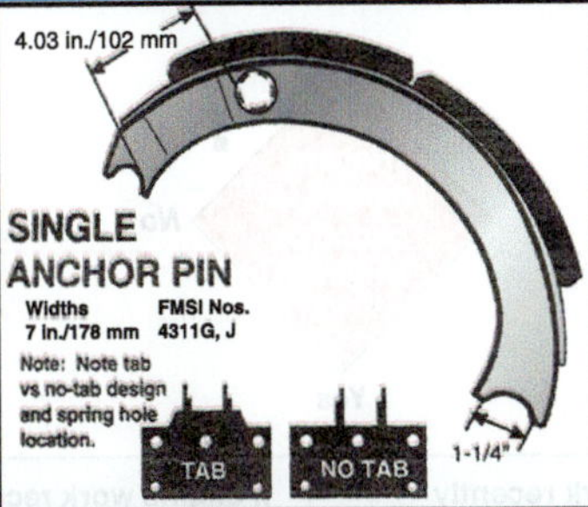

SINGLE ANCHOR PIN

Widths	FMSI Nos.
7 in./178 mm	4311G, J

Note: Note tab vs no-tab design and spring hole location.

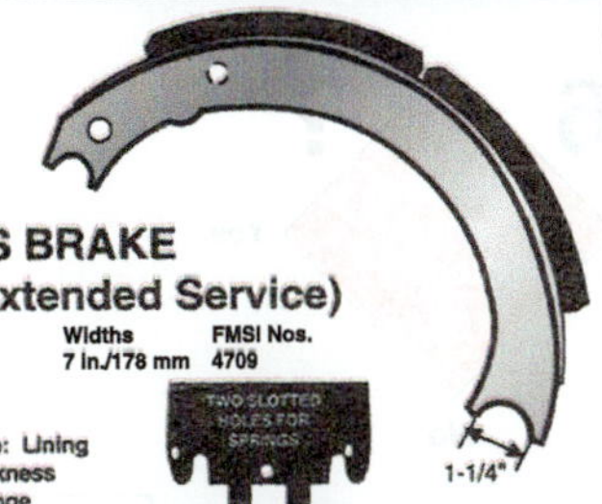

ES BRAKE (Extended Service)

Widths	FMSI Nos.
7 in./178 mm	4709

Note: Lining thickness change

0.179 in./4.55 mm

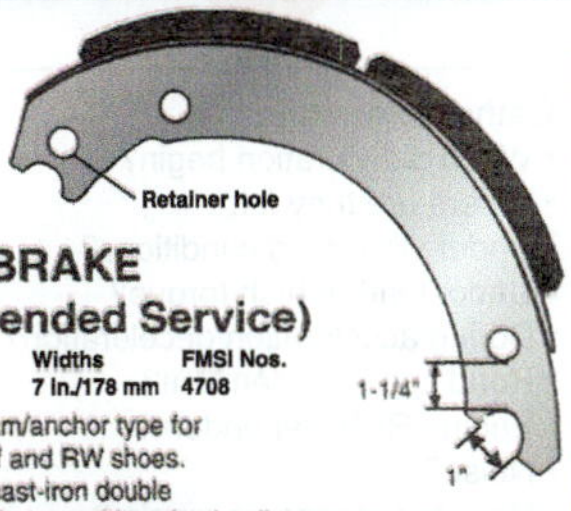

ES BRAKE (Extended Service)

Widths	FMSI Nos.
7 in./178 mm	4708

Note: Cam/anchor type for Fruehauf and RW shoes.
Also: A cast-iron double anchor pin type with revised radius.
Lining thickness change.

ROCKWELL

Nominal Shoe Thickness 0.179 in./4.55 mm

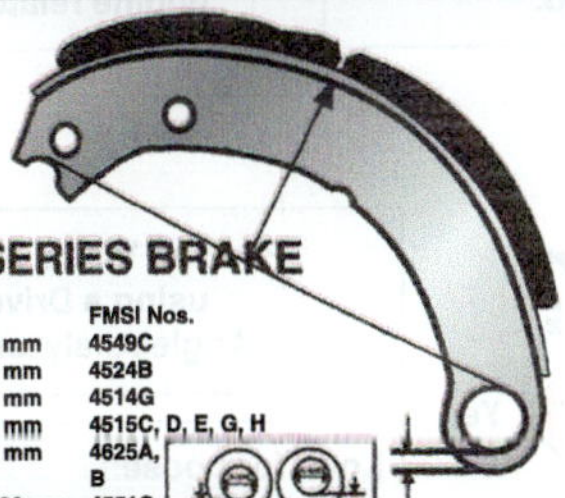

"P" SERIES BRAKE

Widths	FMSI Nos.
4 in./100 mm	4549C
5 in./125 mm	4524B
6 in./150 mm	4514G
7 in./178 mm	4515C, D, E, G, H
7 in./178 mm	4625A, B
8$^{5}/_{8}$ in./220 mm	4551C, D, E
10 in./254 mm	4549C/4514G

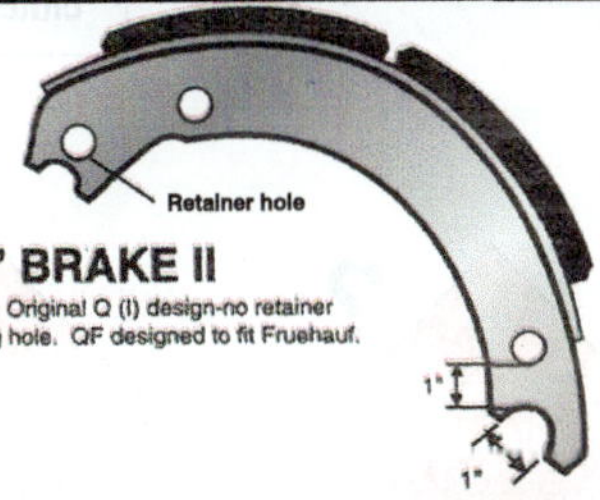

"Q" BRAKE II

Note: Original Q (I) design-no retainer spring hole. QF designed to fit Fruehauf.

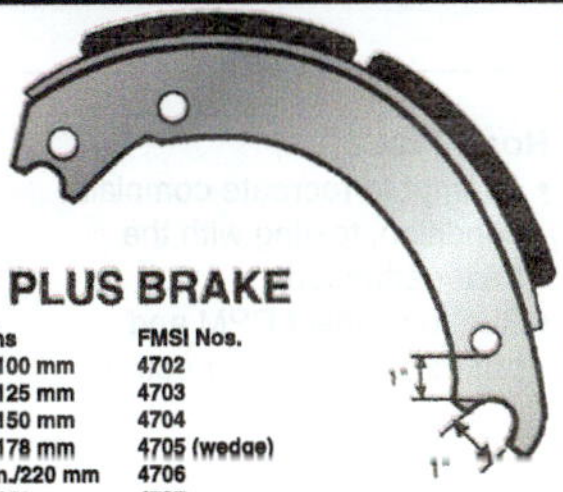

"Q" PLUS BRAKE

Widths	FMSI Nos.
4 in./100 mm	4702
5 in./125 mm	4703
6 in./150 mm	4704
7 in./178 mm	4705 (wedge)
8$^{5}/_{8}$ in./220 mm	4706
7 in./178 mm	4707

Note: Lining thickness change

FRUEHAUF

Nominal Shoe Thickness 0.179 in./4.55 mm

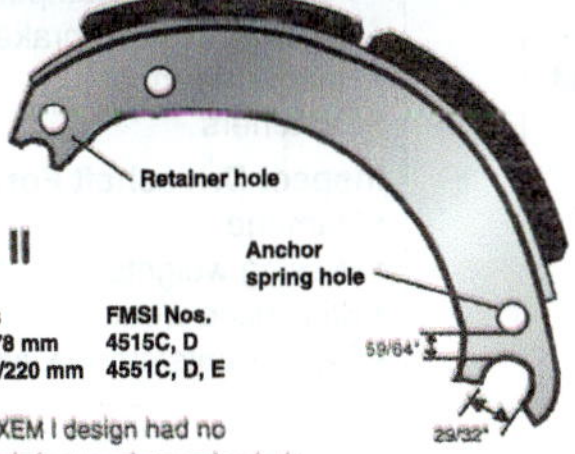

XEM II

Widths	FMSI Nos.
7 in./178 mm	4515C, D
8$^{5}/_{8}$ in./220 mm	4551C, D, E

Note: XEM I design had no retainer hole or anchor spring hole.

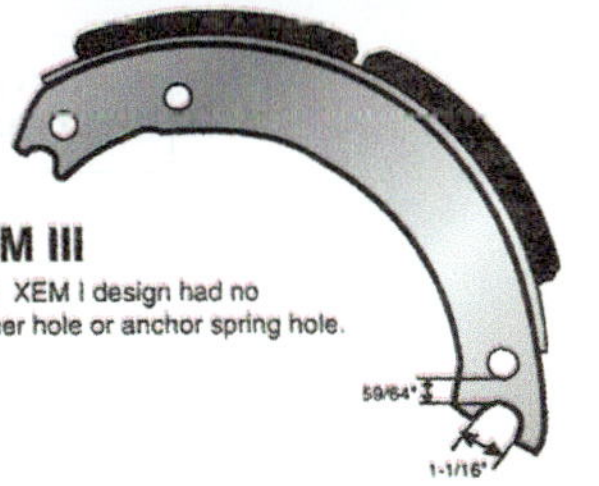

XEM III

Note: XEM I design had no retainer hole or anchor spring hole.

DANA

Nominal Shoe Thickness 0.179 in./4.55 mm

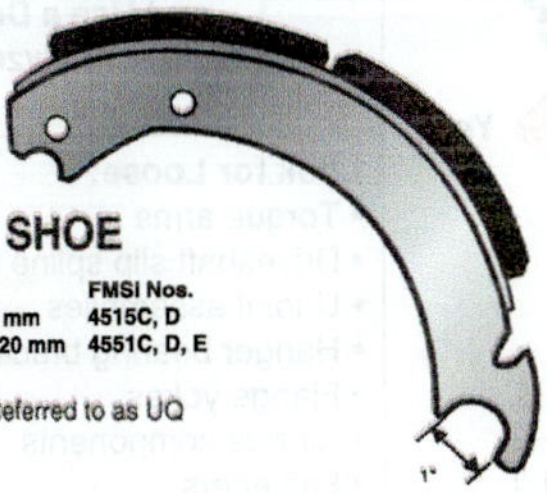

"FC" SHOE

Widths	FMSI Nos.
7 in./178 mm	4515C, D
8$^{5}/_{8}$ in./220 mm	4551C, D, E

Note: Referred to as UQ

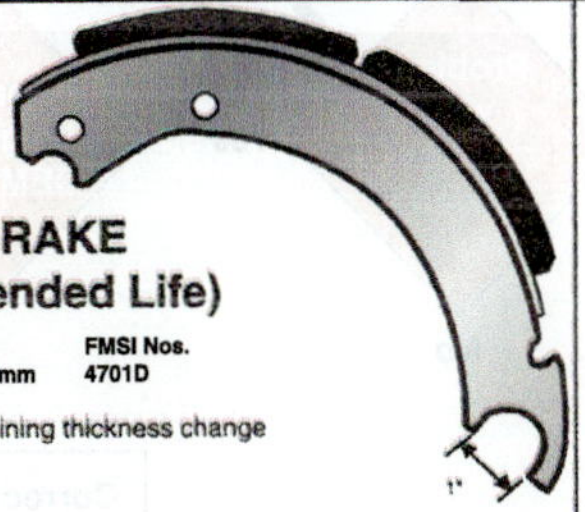

XL BRAKE (Extended Life)

Widths	FMSI Nos.
7 in./178 mm	4701D

Note: Lining thickness change

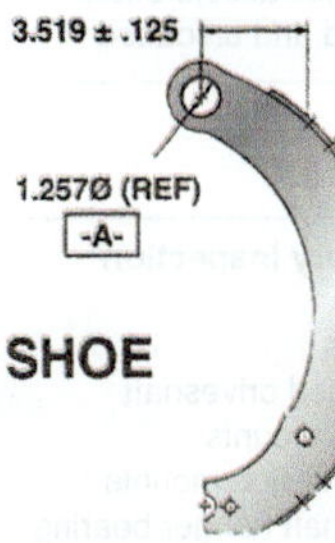

FT SHOE

APPENDIX D

VIBRATION DIAGNOSIS

Start

1

Gather Information
- When did vibration begin?
- Where are they felt?
- Under what road conditions?
- Under load or high torque?
- During acceleration/deceleration?
- Road speed dependent?
- Engine RPM dependent?
- Noise?
- Recent suspension repairs?
- Lube condition and level?

2

Road Test
- Attempt to recreate complaint condition, testing with the trailer attached
- Run to suspect RPM and put transmission in neutral

3

Simulate Conditions
- Road speed
- Engine RPM
- Gear ratio
- Coast
- Under full acceleration
- Loaded and unloaded

4

Record Test Data
- Road speed
- Engine RPM
- Gear ratio
- Coast
- Under full acceleration
- Loaded and unloaded

5

Stationary Inspection
- Tires
- Wheels
- Damaged driveshaft
- Engine mounts
- Transmission mounts
- Driveshaft hanger bearing

6

Vibrates when stationary? Yes / No

Recent repairs to clutch or engine? No / Yes

If clutch work recently done, problem could be clutch related.

If engine work recently done, problem could be engine related.

Road speed related? Yes / No

Does ride height meet OEM specs? Yes / No

Correct per OEM procedures.

Perform Visual Inspection using a Driveline Angle Analyzer (DAA)

Look for Loose:
- U-joints
- Bearing straps
- Flange yokes
- Yoke-mounted damper
- Driveline parking brake
- Hanger bearing
- Fasteners

Inspect Driveshaft For:
- Damage
- Missing weights
- Slip spline wear
- Cab mounts/air ride system

Problem with trailer loaded? Yes / No

Does ride height meet OEM specs? Yes / No

Correct per OEM procedures.

Perform Visual Inspection and Use a Driveline Angle Analyzer (DAA)

Look for Loose:
- Torque arms
- Driveshaft slip spline play
- U-joint assemblies
- Hanger bearing brackets
- Flange yokes
- Air ride components
- Fasteners
- Cab mounts

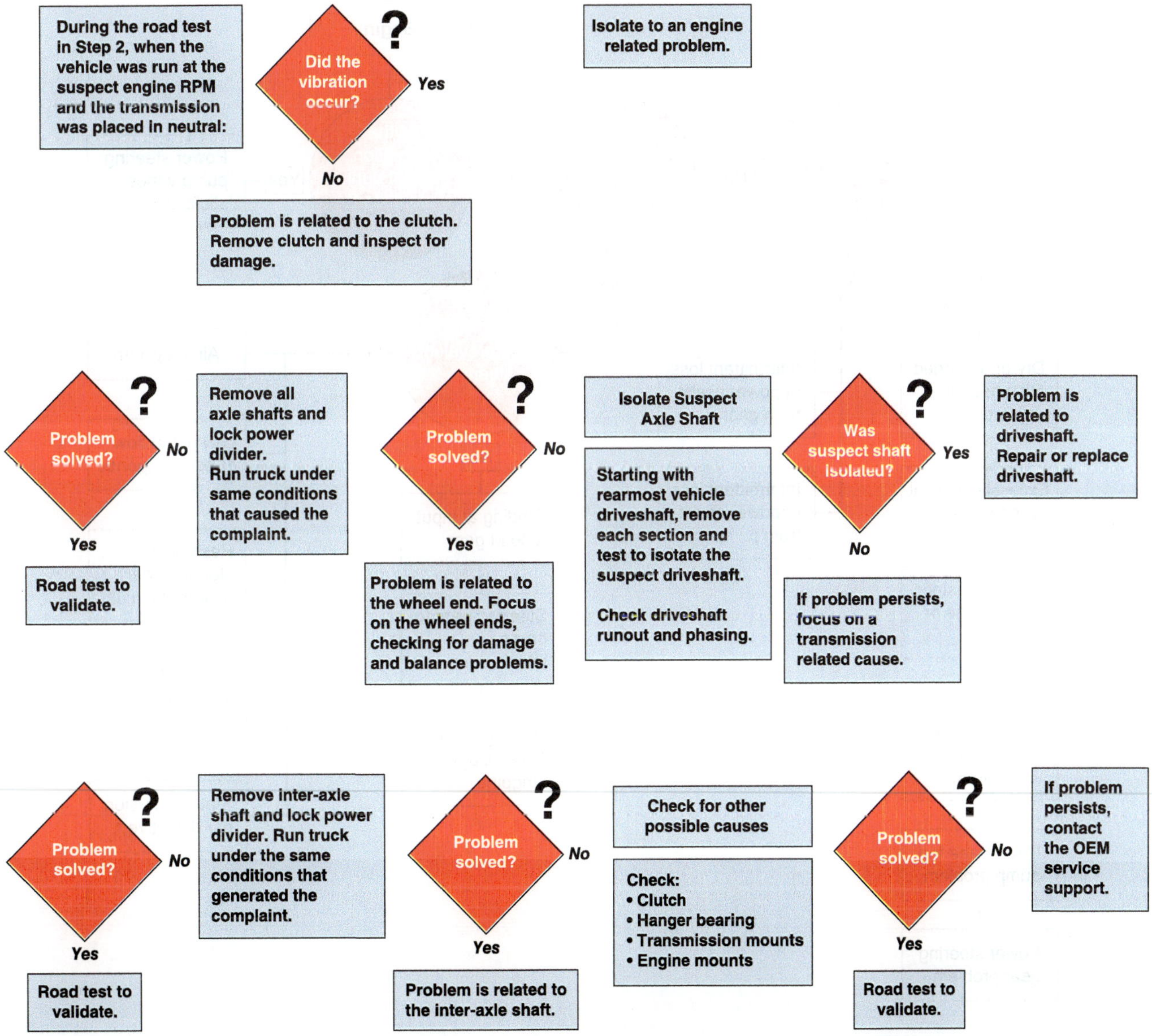
During the road test in Step 2, when the vehicle was run at the suspect engine RPM and the transmission was placed in neutral:
Did the vibration occur?
?
Yes
Isolate to an engine related problem.
No
Problem is related to the clutch. Remove clutch and inspect for damage.
Problem solved?
?
No
Yes
Road test to validate.
Remove all axle shafts and lock power divider. Run truck under same conditions that caused the complaint.
Problem solved?
?
No
Yes
Problem is related to the wheel end. Focus on the wheel ends, checking for damage and balance problems.
Isolate Suspect Axle Shaft
Starting with rearmost vehicle driveshaft, remove each section and test to isolate the suspect driveshaft.
Check driveshaft runout and phasing.
Was suspect shaft isolated?
?
Yes
Problem is related to driveshaft. Repair or replace driveshaft.
No
If problem persists, focus on a transmission related cause.
Problem solved?
?
No
Yes
Road test to validate.
Remove inter-axle shaft and lock power divider. Run truck under the same conditions that generated the complaint.
Problem solved?
?
No
Yes
Problem is related to the inter-axle shaft.
Check for other possible causes
Check:
• Clutch
• Hanger bearing
• Transmission mounts
• Engine mounts
Problem solved?
?
No
If problem persists, contact the OEM service support.
Yes
Road test to validate.

APPENDIX E

HARD STEERING TROUBLESHOOTING FLOW CHART

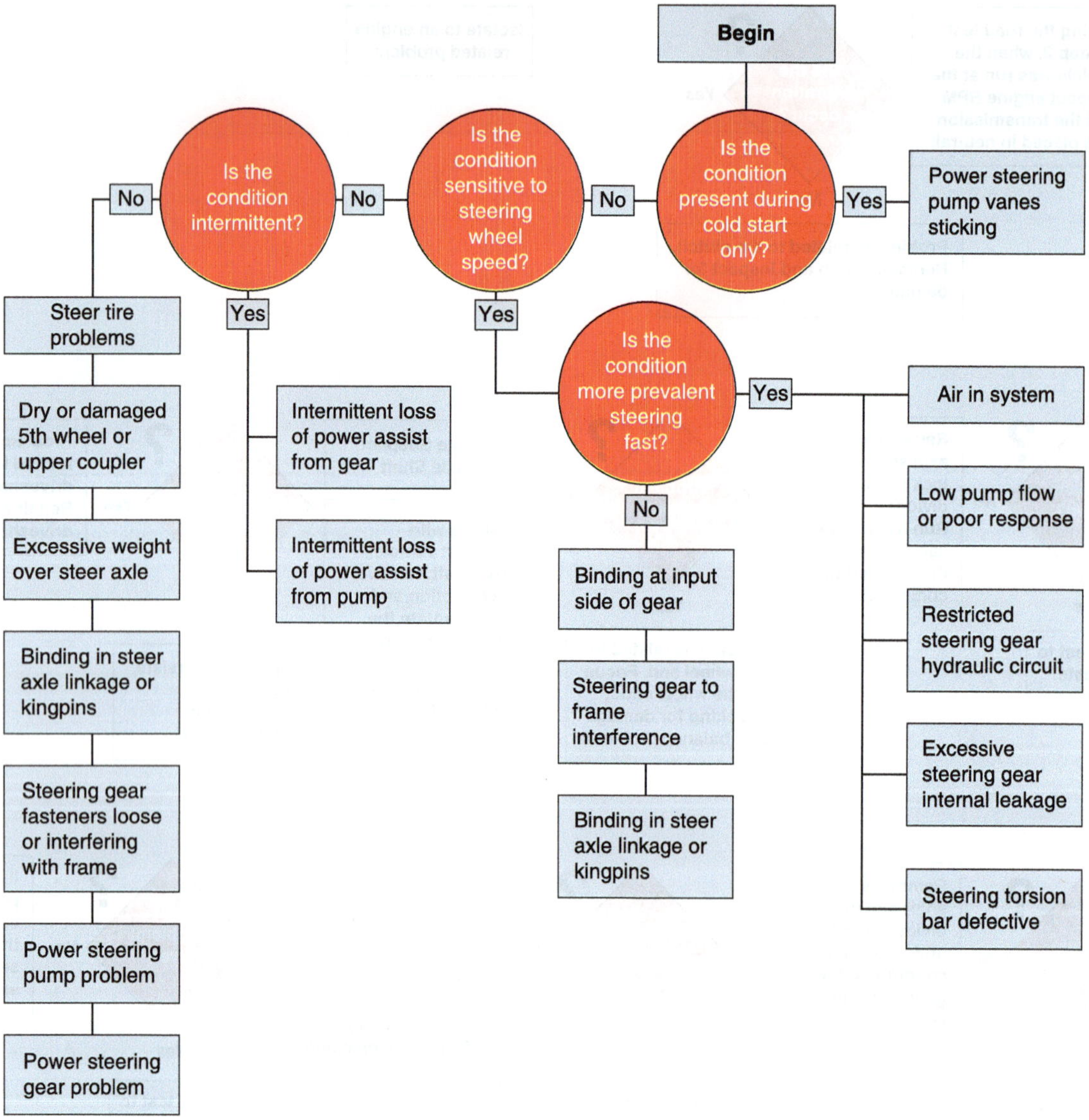

APPENDIX F

DUAL CIRCUIT BRAKE SYSTEM TROUBLESHOOTING

Test 1

GOVERNOR CUT-OUT / LOW PRESSURE WARNING / PRESSURE BUILD-UP

Vehicle Parked, Wheel Chocked	**OK**	**NOT OK**
1. Drain all reservoir to 0 psi	☐	☐
2. Start engine (run at fast idle): Low pressure warning should be on. Note: On some vehicles with anti-lock, warning light will also come on momentarily when ignition is turned on.	☐	☐
3. Low pressure warning: Dash warning light should go off at or above 60 psi.	☐	☐
4. Build up time: Pressure should build from 85 to 100 psi within 25 seconds.	☐	☐
5. Governor cut-out: Cuts out at correct pressure. Check manufacturer's recommendations; usually between 100-130 psi.	☐	☐
6. Governor cut-in: Reduce service air pressure to governor cut-in. The difference between cut-in and cut-out pressure must not exceed 25 psi.	☐	☐

COMPLETE NECESSARY REPAIRS BEFORE PROCEEDING TO TEST 2; SEE CHECKLIST 1 FOR COMMON CORRECTIONS.

Check List 1

If the low pressure warning light or buzzer doesn't come on:		
1. Check wiring	☐	☐
2. Check bulb	☐	☐
3. Repair or replace the buzzer, bulb, or low pressure warning switch(es)	☐	☐
If governor cut-out is higher or lower than specified by the vehicle manual:		
1. Adjust the governor using a gauge of known accuracy.	☐	☐
2. Repair or replace governor as necessary after being sure compressor unloader mechanism is operating correctly.	☐	☐
If low pressure warning occurs below 60 psi:		
1. Check dash gauge with master pressure gauge.	☐	☐
2. Repair or replace a defective low pressure indicator.	☐	☐
If build up time exceeds 25 seconds:		
1. Test the compressor air filter and clean or replace.	☐	☐
2. Check for restricted inlet line if compressor does not have a filter. Replace if necessary.	☐	☐
3. Check compressor discharge port and line for carbon build-up. Replace if necessary.	☐	☐
4 With system charged and governor compressor in unloaded mode, listen at the compressor inlet for leakage. If leakage is evident, apply a small quantity of oil around unloader pistons. If no leakage is evident, then leakage is through the compressor discharge valves.	☐	☐
5. Check the compressor drive for slippage.	☐	☐

RETEST TO CHECK OUT ALL ITEMS REPAIRED OR REPLACED

Test 2

LEAKAGE AT RESERVOIR AIR SUPPLY

GOVERNOR CUT-OUT PRESSURE, ENGINE OFF PARKING BRAKES APPLIED	OK	NOT OK
1. Allow pressure to stabilize for 1 minute	☐	☐
2. Observe the dash gauge pressures for 2 minutes and note any pressure drop		
A. Pressure Drop: Single Vehicle (A 2 psi drop within 2 minutes is permissible for either service reservoir)	☐	☐
B. Pressure Drop: Tractor/Trailer (A 6 psi drop within 2 minutes is permissible for either service reservoir)	☐	☐
C. Pressure Drop: Tractor/2 Trailers (An 8 psi drop within 2 minutes is permissible for either service reservoir)	☐	☐

COMPLETE ALL NECESSARY REPAIRS BEFORE PROCEEDING TO TEST 3; SEE CHECK LIST 2 FOR COMMON CORRECTIONS.

Check List 2

EXCESSIVE LEAKAGE IN THE SUPPLY SIDE OF THE PNEUMATIC SYSTEM, COULD BE CAUSED BY:

NOTE: A leak detector or soap solution will aid in locating the faulty component

1. Supply lines and fittings (tighten)	☐	☐
2. Low pressure indicator(s)	☐	☐
3. Relay valves (antilock modulators)	☐	☐
4. Relay valve	☐	☐
5. Dual circuit foot brake valve	☐	☐
6. Trailer control valve	☐	☐
7. Park control valve	☐	☐
8. Tractor protection valve	☐	☐
9. Spring brake actuators	☐	☐
10. Safety pop-off valve in supply reservoir	☐	☐
11. Governor	☐	☐
12. Compressor discharge valves	☐	☐

***RETEST TO CHECK OUT ALL ITEMS REPAIRED OR REPLACED**

Test 3

LEAKAGE SERVICE AIR DELIVERY

GOVERNOR CUT-OUT PRESSURE, ENGINE OFF PARKING BRAKES RELEASED		
1. Hold a service brake application	☐	☐
2. Allow pressure to stabilize for 1 minute; then begin timing for 2 minutes while observing the dash gauges		
A. Pressure Drop: Single Vehicle (A 4 psi drop within 2 minutes is permissible for either service reservoir)	☐	☐
B. Pressure Drop: Tractor/Trailer (A 6 psi drop within 2 minutes is permissible for either service reservoir)	☐	☐
C. Pressure Drop: Tractor/2 Trailers (An 8 psi drop within 2 minutes is permissible for either service reservoir)	☐	☐
3. Check brake chamber push rod travel (Refer to chart below for allowable tolerances)	☐	☐

Brake Chamber Size	Maximum Stroke Before Readjustment
12	$1\frac{1}{4}$"
16	$1\frac{3}{4}$"
20	$1\frac{3}{4}$"
24	$1\frac{3}{4}$"
30	2"

4. Check the angle formed between the brake chamber push rod and slack adjuster arm. (Should be 90° in the fully applied position)	☐	☐

COMPLETE ALL NECESSARY REPAIRS BEFORE PROCEEDING TO TEST 4; SEE CHECKLIST 3 FOR COMMON CORRECTIONS

Check List 3

Leakage in the service side of the pneumatic system, could be caused by:	**OK**	**NOT OK**
NOTE: A leak detector or soap solution will aid in locating the defective component		
1. Service lines and fittings (tighten)	☐	☐
2. Trailer control valve	☐	☐
3. Stoplight switch	☐	☐
4. Brake chamber diaphragms	☐	☐
5. Tractor protection valve	☐	☐
6. Relay valves (antilock modulators)	☐	☐
7. Service brake valve	☐	☐
8. Front axle ratio valve (optional)	☐	☐
9. Inverting relay spring brake control valve (optional) straight trucks and busses	☐	☐
10. Double check valve.	☐	☐

***RETEST TO CHECK OUT ALL ITEMS REPAIRED OR REPLACED**

Test 4

MANUAL EMERGENCY SYSTEM

GOVERNOR CUT-OUT PRESSURE, ENGINE IDLING 600-900 RPM

FOR STRAIGHT TRUCKS, BUSES, AND BOBTAIL TRACTOR:		
1. Manually operate the park control valve and note that parking brakes apply and release promptly.	☐	☐
FOR TRACTOR/TRAILER COMBINATIONS:		
1. Manually operate tractor protection control valve (trailer supply valve). The trailer brakes should apply and release promptly as control button is pulled out and pushed in.	☐	☐
2. Manually operate system park control (usually yellow diamond button) and check that all parking brakes (tractor and trailer) apply promptly.	☐	☐

COMPLETE ALL NECESSARY REPAIRS BEFORE PROCEEDING TO TEST 5; SEE CHECKLIST 4 FOR COMMON CORRECTIONS

Check List 4

If sluggish performance is noted in either test, check for:		
1. Dented or kinked lines	☐	☐
2. Improperly installed hose fitting	☐	☐
3. A defective relay emergency valve	☐	☐
4. A defective modulator(s)	☐	☐
If the trailer brakes do not actuate and the trailer supply line remains charged, check the:		
1. Tractor protection control	☐	☐
2. Trailer spring brake valve	☐	☐

***RETEST TO CHECK OUT ALL ITEMS REPAIRED OR REPLACED**

Test 5

AUTOMATIC EMERGENCY SYSTEM

GOVERNOR CUT-OUT PRESSURE, ENGINE OFF	OK	NOT OK
1. Drain secondary circuit reservoir to 0 psi.		
A. Primary circuit reservoir should not lose pressure	☐	☐
B. On combination vehicles, the trailer air system should remain charged	☐	☐
C. Tractor and trailer brakes should not apply automatically	☐	☐
2. With no air pressure in the secondary circuit reservoir make a brake application.		
A. Rear axle brakes should apply and release	☐	☐
B. On combination vehicles the trailer brakes should also apply and release	☐	☐
C. The stop lamps should light	☐	☐
3. Slowly drain the primary circuit reservoir pressure.		
A. Spring brake push pull valve should pop out between 35 and 45 psi.	☐	☐
B. Tractor protection valve should close between 45 and 20 psi and trailer supply hose should be exhausted	☐	☐
C. Trailer brakes should apply after tractor protection closes	☐	☐
4. Close drain cocks, recharge system, and drain rear axle reservoir to 0 psi.		
A. Secondary circuit reservoir should not lose pressure	☐	☐
B. On combination vehicles the trailer air system should remain charged	☐	☐
5. With no air pressure in the primary circuit reservoir, make a brake application.		
A. Front axle brakes should apply and release	☐	☐
B. On combination vehicles the trailer brakes should also apply and release	☐	☐
C. If the vehicle is equipped with an inverting relay spring brake control valve, the rear axle brakes should also apply and release		

Check List 5

If the vehicle fails the previous tests outlined, then check the following:

1. Fittings	☐	☐
2. Kinked hose or tubing	☐	☐
3. Single check valves	☐	☐
4. Double check valves	☐	☐
5. Tractor protection valve	☐	☐
6. Tractor protection control valve	☐	☐
7. Parking control valve	☐	☐
8. Relay valves (antilock modulators)	☐	☐
9. Trailer spring brake control valve	☐	☐
10. Inverting relay spring brake control valve (optional) straight trucks and buses	☐	☐

***RETEST TO CHECK OUT ALL ITEMS REPAIRED OR REPLACED**

APPENDIX G

DRIVELINE CONFIGURATIONS

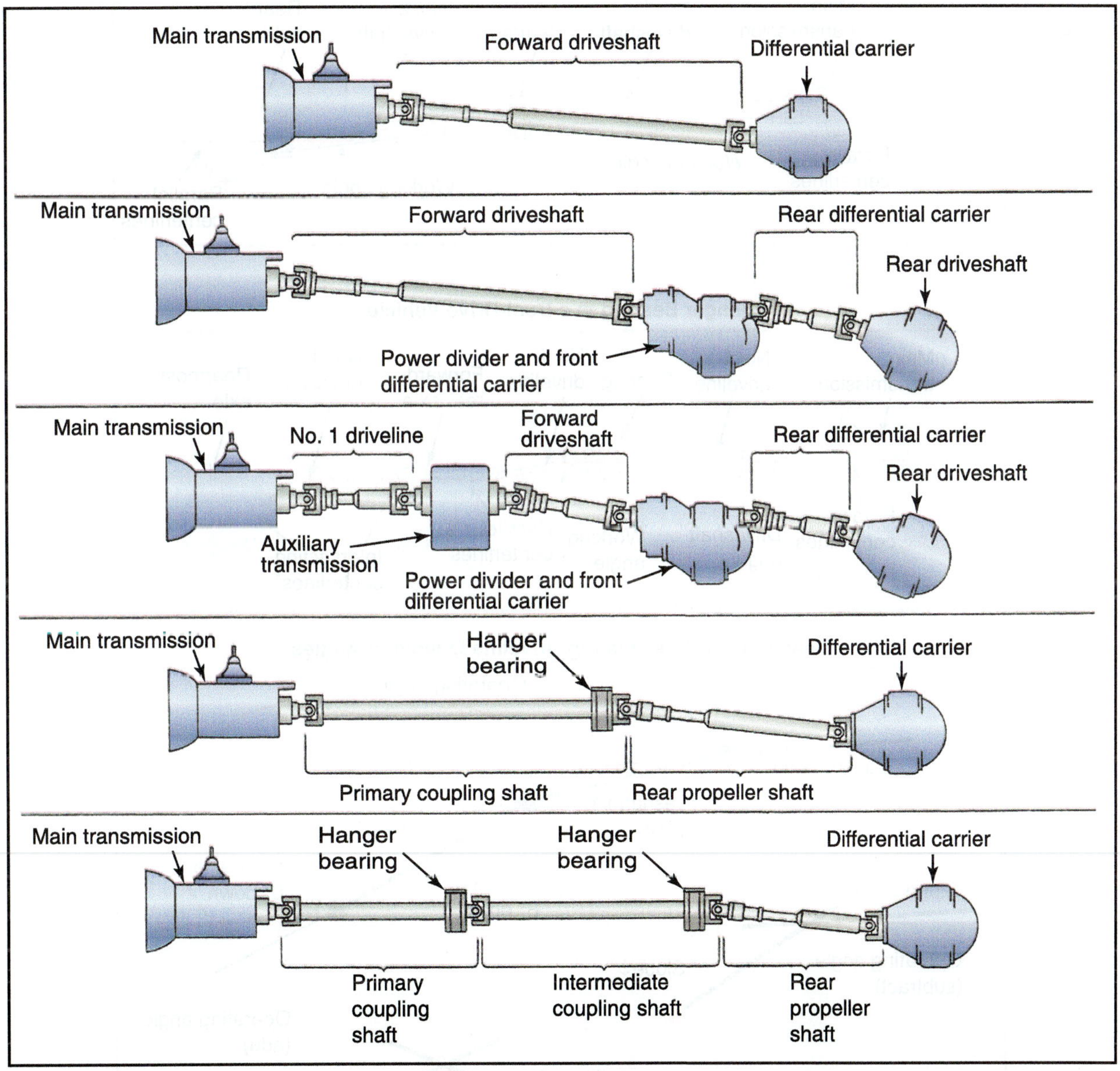

APPENDIX H

MEASURING DRIVELINE ANGLES

Hanger Bearing in a Single-Drive Vehicle

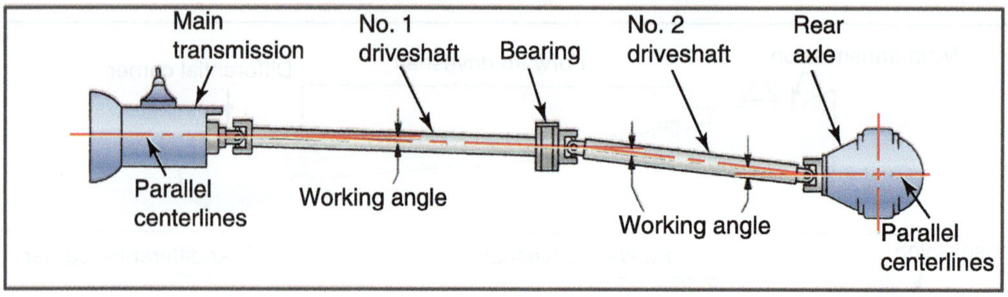

Hanger Bearing in a Dual-Drive Vehicle

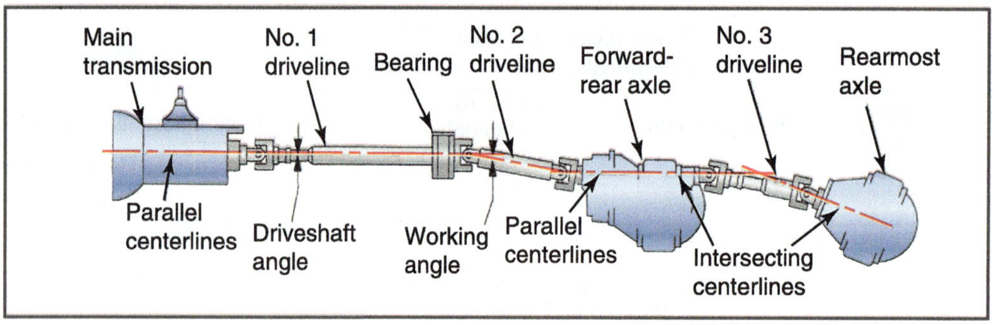

Measuring and Calculating Driveline Operating Angles

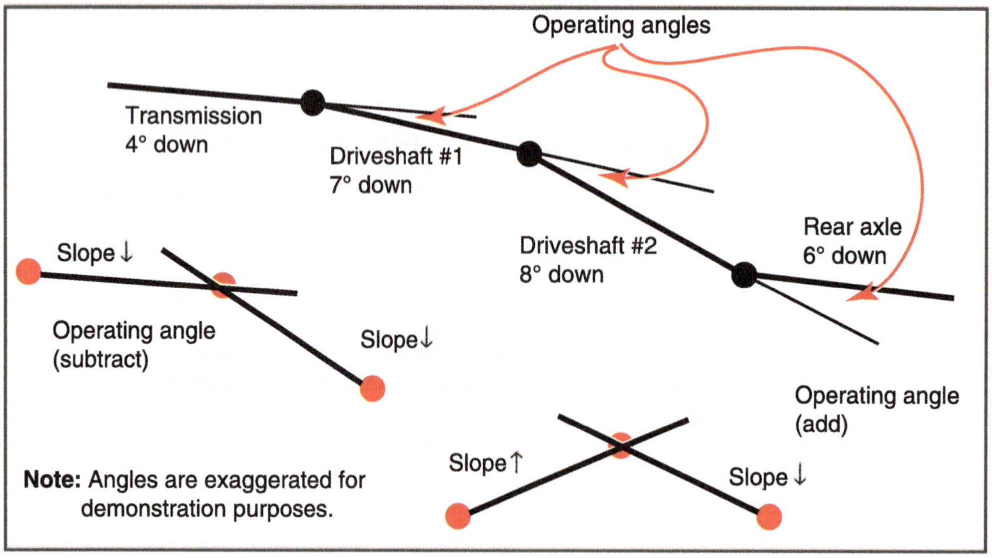

APPENDIX I

VEHICLE CONFIGURATIONS

COE (Cab over Engine)

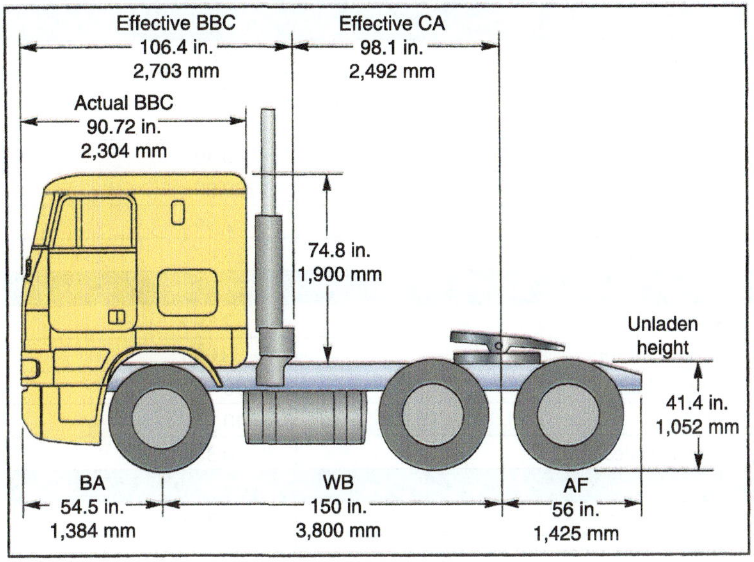

AF = distance between centerline of 5th wheel and rear bumper
BA = distance between front bumper and front wheel centerline
BBC = distance between front bumper and back of cab
BC = body clearance
CA = distance between back of cab and centerline of rear axle(s)
WB = wheelbase

CONV (Convontional)

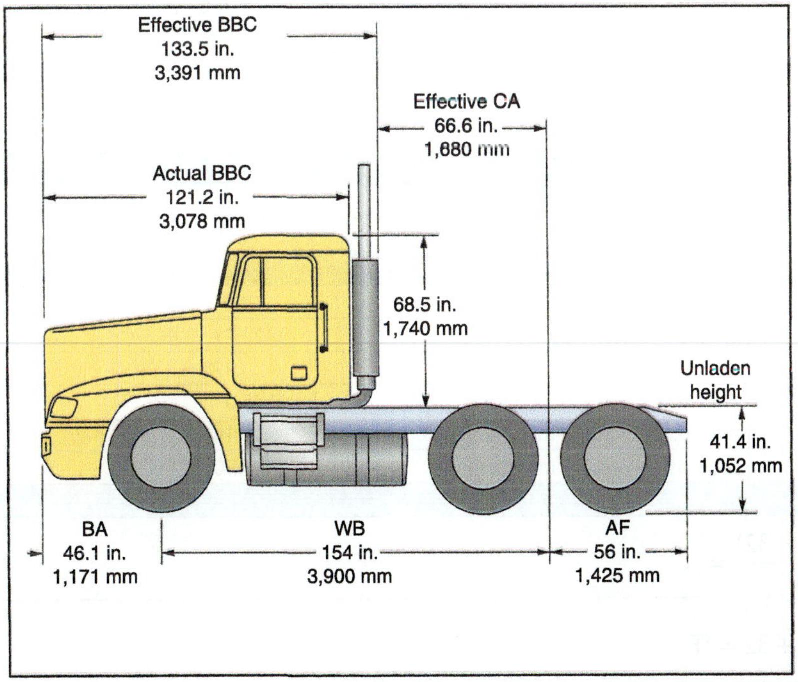

APPENDIX J

QUICK CONVERSION FORMULA

From	To	Multiply by
Distance		
miles	kilometers	1.61
kilometers	miles	0.62
meters	feet	3.28
feet	meters	0.30
kilometers	feet	3,289.80
inches	centimeters	2.54
inches	millimeters	25.4
Area		
square miles	square kilometers	2.590
square feet	square meters	0.093
acres	square feet	43,560
square miles	acres	640
square kilometers	square miles	0.386
Volume		
U.S. gallons	liters	3.785
U.S. gallons	Imperial gallons	0.833
liters	U.S. gallons	0.264
liters	Imperial gallons	0.220
Imperial gallons	liters	4.546
cubic feet	cubic meters	0.028
cubic meters	cubic feet	35.310
cubic inches	cubic centimeters	16.387
Mass		
pounds	kilograms	0.454
kilograms	pounds	2.203
Pressure		
pounds per square inch (psi)	inches of Hg	2.042
pounds per square inch (psi)	inches of water	27.71
pounds per square inch (psi)	kilo-Pascals (kPa)	6.891
megapascals (MPa)	pounds per square inch (psi)	145.0
kilopascals (kPa)	pounds per square inch (psi)	0.145
bar	pounds per square inch (psi)	14.50
Temperature Conversion		
Degrees Fahrenheit to Celsius: $\frac{5 \times (°F - 32)}{9} = °C$		
Degrees Celsius to Fahrenheit: $\frac{9 \times °C}{5} + 32 = °F$		

APPENDIX K

J1939 MESSAGE DECODER

SRC = ______ = ____________________   PGN = ______ = __________________

Byte 1	Byte 2	Byte 3	Byte 4	Byte 5	Byte 6	Byte 7	Byte 8
8 7 6 5 4 3 2 1	8 7 6 5 4 3 2 1	8 7 6 5 4 3 2 1	8 7 6 5 4 3 2 1	8 7 6 5 4 3 2 1	8 7 6 5 4 3 2 1	8 7 6 5 4 3 2 1	8 7 6 5 4 3 2 1

Byte 9	Byte 10	Byte 11	Byte 12	Byte 13	Byte 14	Byte 15	Byte 16
8 7 6 5 4 3 2 1	8 7 6 5 4 3 2 1	8 7 6 5 4 3 2 1	8 7 6 5 4 3 2 1	8 7 6 5 4 3 2 1	8 7 6 5 4 3 2 1	8 7 6 5 4 3 2 1	8 7 6 5 4 3 2 1

Byte 17	Byte 18	Byte 19	Byte 20	Byte 21	Byte 22	Byte 23	Byte 24
8 7 6 5 4 3 2 1	8 7 6 5 4 3 2 1	8 7 6 5 4 3 2 1	8 7 6 5 4 3 2 1	8 7 6 5 4 3 2 1	8 7 6 5 4 3 2 1	8 7 6 5 4 3 2 1	8 7 6 5 4 3 2 1

Byte 25	Byte 26	Byte 27	Byte 28	Byte 29	Byte 30	Byte 31	Byte 32
8 7 6 5 4 3 2 1	8 7 6 5 4 3 2 1	8 7 6 5 4 3 2 1	8 7 6 5 4 3 2 1	8 7 6 5 4 3 2 1	8 7 6 5 4 3 2 1	8 7 6 5 4 3 2 1	8 7 6 5 4 3 2 1

If you want to try populating the above fields, use this University of Tulsa link to get you started: http://tucrrc.utulsa.edu/J1939Database.html

APPENDIX L

TIRE CODE INTERPRETATION AND TREAD PATTERN DESIGNATIONS

Tire

Michelin uses specific numbers or letters to identify different types of tread patterns or casing construction.

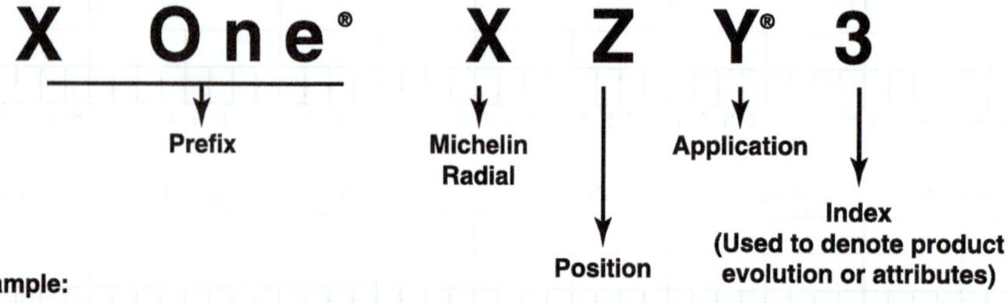

For example:

X =	MICHELIN RADIAL		
Prefix	X One”	=	Single wide tire replacing 2 traditional duals
	X COACH	=	Highway coach
	X MULTIWAY	=	Regional
	X WORKS	=	On/off road
Position	D	=	Drive
	T	=	Trailer
	Z	=	All position
	F	=	Front (steer)
Application Market Segments	A	=	**Highway Applications** • Linehaul • Highway coach
	E	=	**Regional Applications** • Public utility trucks • School bus • Food distribution • Petroleum delivery • Less-than-load (LTL) • Auto carriers • Courier and delivery service
	U	=	**Urban** • Transit buses • Garbage trucks • Fire trucks • Pick-up and delivery trucks
	Y	=	**80% On-Road Use, 20% Off-Road Use** • Construction and mining • Forestry and logging • Oil field
	L	=	**80% Off-Road Use, 20% On-Road Use** • Rock trucks • Forestry and logging • Oil field
Index	⊛	=	Antichip/cut-resistant compound
	HT	=	High torque
	Energy	=	Fuel efficient
	M/S	=	Mud and snow
	S	=	Severe service

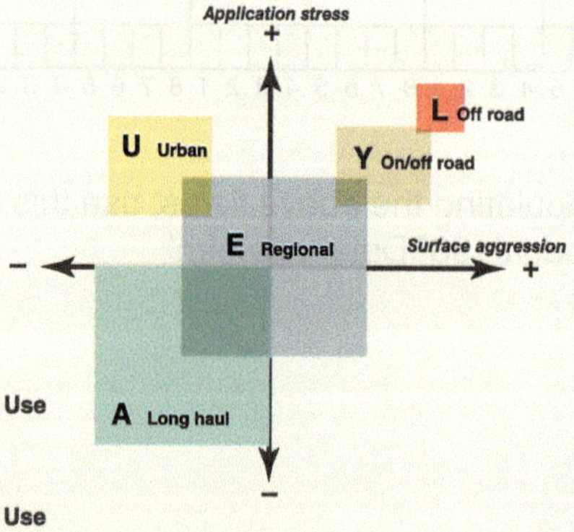

GLOSSARY

absolute pressure a pressure scale with its zero point at absolute vacuum, expressed as pounds per square inch absolute (psia).

absorbed glass mat (AGM) battery type of lead-acid battery in which the electrolyte is absorbed into a very thin boron silicate fiberglass mat. Used in preference to a flooded cell battery because it is less susceptible to vibration and does not have to be oriented in an upright position. Its advantages are zero spill and deep cycling capability.

AC alternating current.

AC conductance testing battery testing instrumentation that pulses AC at a single (usually below 100 Hz) or multiple frequency through a battery circuit. Used by some fleets to assess battery performance in place of more time-consuming testing that requires much higher levels of technician training.

A/C control module the ECU used in a truck climate control system.

accumulator (A/C) filters particles and removes moisture from the refrigerant and separates the refrigerant into vapor and liquid, allowing only vapor to flow to the compressor from the evaporator.

accumulator (hydraulics) a device used to store potential energy; may refer to a pressure storage device in a hydraulic circuit or electrical energy storage devices.

A-circuit voltage regulator located on the ground side of the rotor and battery, voltage is picked up by the field circuit inside the alternator.

Ackerman arm steering component that connects a tie-rod end to the steering knuckle.

Ackerman geometry the front-end setup that allows a vehicle to steer so that the tires track freely during a turn. During a turn, the inboard wheel on a steer axle has to track a tighter circle than the outer wheel. Ackerman geometry is also known as toe-out during turns and it allows the inner and outer wheels to turn at different angles so that both wheels can roll through a turn without tire scrubbing.

ACom software PC-based software using RP-1210 network standards that can be used to read, diagnose, and analyze Bendix ABS. The software also can perform system performance.

acronym a word made up of the initial letters of other words.

active codes fault codes that are read as current by system logic. Active codes cannot be erased until the cause has been repaired and altered their status to historic or inactive. Active codes are assigned an FMI that is broadcast on the vehicle data bus, allowing it to be read by ESTs networked to the vehicle data backbone.

active cruise with braking (ACB) a feature of Eaton Wingman radar electronics that allows the system to be programmed to message requests on the data bus to actuate engine and service braking.

active suspension in trucks this usually refers to an air suspension that has the ability to adapt to changing load and road conditions. Air suspensions can be described as active. See *adaptive suspension.*

actuator a device that produces mechanical motion in response to an electrical, air, or hydraulic signal.

adapter ring used by some two-plate clutches when the clutch pack is installed on a flat-faced flywheel. The adapter ring is bolted between the clutch cover and flywheel.

adaptive logic soft logic processing; processing algorithms in which hard (nonvariable) processing outcomes are replaced by fuzzy logic.

adaptive suspension a suspension designed to adjust to load and road conditions. On trucks this usually means an air suspension system. Adaptive suspensions can be either nonelectronic or electronically controlled. The term *active suspension* is also used.

ADEM Caterpillar's "Advanced Diesel Engine Management" system and module. The ADEM module is used to house the ECU that manages the CX family of on-highway vocational transmissions.

adjustable pliers tools with a multipositional slip joint that permits a wide spread of jaw dimensions.

adjustable wrench a wrench with one fixed and one movable jaw, enabling it to span a wide range of dimensions. The position of the movable jaw is determined by rotating a helical adjusting worm.

adjusting ring used on early-generation Allison automatic transmissions to alter shift speeds. Changing the position of the adjusting rings determines the retaining force of the valve springs in the valve body.

aeration the blending of a gas into a liquid such as the mixing of air into oil by the churning action of rotating gears.

aftercooler can be used to describe a variety of heat exchangers used on engines, transmissions, and air brake systems. An aftercooler on an air brake system is used to condense water from moisture-saturated air leaving the compressor by cooling it.

AGM battery see *absorbed glass mat batteries.*

air bag a supplemental restraint system (SRS) device. In current trucks, SRS air bags may be located in the steering column or side mounted on the outboard side of the driver and passenger seats.

air bag a term used to describe an air spring in a truck suspension or a supplementary restraint device.

air bag control unit (ACU) a control module that uses multiplex-sourced data to generate the deployment of air bag(s) in the event of a collision or rollover.

air brakes pneumatically actuated brakes used on most heavy-duty trucks. These brakes use air pressure to brake the vehicle by means of diaphragms, levers, S-cams, wedges, and rotor assemblies.

air-conditioning protection and diagnostics system (APADS) an electronically controlled HVAC system. The controller is located at the SA 25/MID 190 address on the data bus.

air-conditioning protection unit (ACPU) an electronically monitored HVAC system is multiplexed with an address at SA 25/MID 190 on the data bus.

air dryer a device usually located downstream from the compressor designed to remove moisture from the air system. Air dryers use two principles to eliminate air from an enclosed brake circuit, cooling/condensation and dessicant media.

air dryer reservoir module (ADRM) see *dryer reservoir module.*

air filter a positive filtration device used to remove moisture and contaminants from a pneumatic system or engine.

air filter/regulator an integral filter regulator used on many standard shift truck transmissions.

air governor air pressure sensing and signaling device used to manage truck air brake system compressor effective cycles. May be stand alone, mounted to a compressor, or integrated into an air *dryer reservoir module (DRM).*

air-over-hydraulic brakes a brake system that uses an air control circuit to actuate a hydraulic slave circuit.

air-over-hydraulic intensifier converts control circuit air pressure (from the brake treadle valve) into hydraulic slave circuit pressure to actuate the foundation brakes.

air ratchet wrench a light-duty pneumatic wrench usually with a ⅜-inch (9.5 mm) drive used for repetitive high-speed operations.

air spring see *air bag.*

air spring suspension any of a number of different types of pneumatic suspensions.

Airtek Hendrickson front air suspension system used on its Steertek front axle. This integrates air spring technology with either steel or composite leaf springs.

ALDL connector see *J1962.*

algorithm software term that describes a programmed sequence of operating events; the rules and conditions for producing a processing outcome from a computer. This is also known by some OEMs as *mapping.*

Allen wrench hex-shaped wrenches designed to fit into recessed hex Allen screws.

alloy mixture of metallic elements to create a specific metal; for example, steel is an alloy of iron and carbon.

all-wheel drive (AWD) a term used to describe a vehicle drive configuration used to drive more than one axle in pickup trucks and automobiles but usually describes a three-axle drive system in heavy-duty trucks.

alternating current (AC) electrical current flow that alternates in direction usually produced by rotating a coil in a magnetic field.

amboid gearset a gear arrangement in which the axis of the drive pinion gear is located above the radial centerline of the crown gear it drives.

American National Standards Institute (ANSI) body that standardizes industrial quality and performance standards.

American Society for Mechanical Engineers (ASME) body that sets quality control and material standards.

American Society for Testing Materials (ASTM) American body that regulates material manufacturing and performance standards.

American Trucking Association (ATA) organization with a broad spectrum of representation responsible for standards in the U.S., Canadian, and Mexican trucking industries.

ampere a unit of electrical current flow, usually abbreviated to amp. One ampere is equivalent to 6.28 billion, billion electrons passing a given point per second.

ampere-hour rating amount of current that a fully charged battery can feed through a circuit before the cell voltage drops to 1.75 volts. In a typical 6-cell, 12-volt battery, this would be equal to a battery voltage of 10.5 volts.

analog using physical variables to represent numbers such as voltage or length.

analog-to-digital converter (A/D converter) converts analog voltage values to a digital format before delivery to a computer processing cycle.

analog voltage use of voltage signals that are allowed to range within a window of values.

analog-volt-ohmmeter (AVOM) multipurpose electrical test meter; usually not suited for use on digital circuits.

AND gate a logic or hard channel option gate with two switches. For the gate to flow current, both switches must be closed.

anneal to remove *temper* (heat treatment hardening) of a metal. To anneal hardened steel you heat to cherry red, then slow cool. To anneal copper, you heat it to a green glow and then douse with water.

annulus the largest gear member of a simple planetary (epicyclical) gearset located outside of the planetary gears.

anode positive electrode.

anticompounding valve prevents simultaneous air and spring forces from being applied to the foundation brakes in an air brake system.

anticorrosion agents solutions or additives that inhibit chemical corrosion.

antilock braking system (ABS) an electronically managed hydraulic or pneumatic air brake system that senses wheel speeds and manages application pressures to prevent wheel lockup.

antilock braking system (ABS) module the controller module used in an ABS. It receives inputs from wheel speed sensors and manages the brake cycles by outputting signals to a modulator unit.

antilock relay valve (ARV) replaces a standard relay valve in a pneumatic ABS; the actuator (output) that charges/dumps application pressure to the service brakes.

antirattle springs used to reduce wear and smooth disengagement between the intermediate plate and drive pins in a heavy-duty clutch pack.

antirust agents solutions or additives that inhibit oxidation of irons and steels.

applied moment chassis frame term meaning the focal point of an applied load.

applied stroke a foundation brake term referencing slack adjuster travel measured at the clevis pin from its unapplied position to applied with at least 85 psi service application pressure. Applied stroke is tested at DOT roadside inspections.

arbor press a manually actuated (lever) press used to apply linear force usually on a vertical plane.

armature the moving component within a starter motor or solenoid. It rotates in a starter motor and moves linearly in a solenoid.

articulate the movement permitted between two components connected by a pivot. For example, a tractor/semitrailer articulates at the fifth wheel connection.

aspect ratio a tire term calculated by dividing the section height by the section width.

ATA connector see *data connector.*

atm unit of atmospheric pressure in the standard measurement system; 1 atm is equivalent to 14.7 psi. This is close but not exactly equivalent to 1 bar using metric conversion.

atmospheric pressure ambient pressure at sea level equivalent to 14.7 psia, 101.3 kPa, 1 atm, or 1.01 bar.

A-train a two-trailer combination in which a dolly (converter or turntable) is connected by a pintle hook and drawbar. The resulting connection has two pivot points, meaning that there are a total of three pivot points in the tractor/trailer combination.

audit trail means of recording electronic events such as diagnostic and fault codes into computer memory for subsequent analysis. Audit trails can be open or covert.

Australian Design Rule (ADR) legislation that covers vehicle certification standards for brakes, emission, and safety standards. ADRs 35/01 to 05 cover commercial vehicle brake standards, ADRs 38/00 to 03, trailer standards.

AutoClutch™ the centrifugal clutch pack used by the Eaton UltraShift AMT.

automated maintenance environment (AME) software-driven maintenance scheduling and management of fleets that makes use of predictive and performance-based maintenance strategies. The objective is reduce unnecessary PM. Used extensively by the military.

automated manual transmission (AMT) an electronically controlled clutch/transmission system that uses a standard mechanical transmission as a platform. Some use a three-pedal (third pedal is clutch) system, whereas others entirely dispense with the clutch pedal. In automated transmissions, either all or most driveline clutching is computer controlled making use of centrifugal, hydraulic, or air-actuated clutches.

automated transmission see *automated manual transmission.*

Automatic gear shift (AGS) Mercedes Benz manufactured, automated transmission using a two-pedal system.

automatic moisture ejector an automatic drain valve used on air reservoirs. Also known as a spitter valve.

automatic slack adjuster (ASA) auto-adjusting lever that connects an air brake chamber actuation rod to an S-camshaft that is required in all current vehicles.

AutoShift a Eaton Fuller Roadranger family of electronically automated standard transmissions: phased out in North America but widely used offshore.

AutoShift finger the shift spade used to select and effect shifts in the shift rail housing.

auxiliary filter any of a number of different types of add-on filters used in hydraulic circuits. Often installed on an Allison transmission after a catastrophic failure to remove particulate debris.

axes plural of *axis.*

axial play see *axial runout.*

axial runout wobble from an axial plane of a rotor, shaft, or wheel. Using the example of a wheel, axial runout is wobble as viewed from the front or rear of the vehicle.

axis point about which a shaft, rotor, or component rotates.

axis of rotation see *axis.*

axle a rod or bar on which wheels turn or the housing that supports wheel assemblies. Also used to describe the shaft that transmits drive torque to the drive wheels.

axle range interlock used to prevent axle ratio shifting when interaxle differential action is locked out.

axle shafts used to transmit drive torque between the differential carrier assembly and the drive wheels. Also known as *half shafts* and *driveshafts.*

axle shims thin wedges installed under leaf springs that alter the tilt (cant) angle of the axle. Used to define caster angle on steer axles and U-joint working angles on drive axles.

backdrive testing steering system performance by reversing output and input. The steering gear is rotated by hand pressure acting on the Pitman arm.

backfill pressure a regulator and a relief valve designed to maintain low pressure in the clutch supply passages of some fully automatic transmissions. Consists of a simple spring-loaded relief valve.

backlash the gapping or play between mating gear teeth usually measured with thickness gauges or a dial indicator.

balance beads see *counteract balance beads (CBBs).*

ball hitch a trailer tow system consisting of a shank and ball on a perpendicular plane to which a horizontal drawbar eye connects.

bandwidth volume or capacity of a data transmission medium; the number of packets that can be pumped down a channel or multiplex backbone.

banjo housing differential axle carrier housing, so named because of its banjo shape.

bar metric unit of pressure measurement exactly equivalent to 100 kPa and roughly equivalent to one unit of atmospheric pressure (101.3 kPa).

Battery Council International (BCI) in conjunction with the Society of Automotive Engineers (SAE), it defines and sets the standards for rating batteries.

battery terminals tapered posts or threaded terminal connections on batteries, one positive and one negative.

baud see *baud rate.* Baud is by definition a rate (of data transmission) but almost always in the trucking industry we use the term *baud rate.*

baud rate speed at which data can be transmitted on a data bus, which is important because both the broadcast and receiving devices must use the same transmission speed.

bazooka device for setting trailer axle tracking alignment. A bazooka is fitted to the trailer upper coupler by clamping to the kingpin. Axle extenders can then be fitted to the trailer axles and measurements taken from the bazooka to the axle extensions. Less used today than laser and photo alignment systems.

B-circuit voltage regulator positioned on the feed side to the alternator with battery voltage fed through the regulator to the field circuit, which is then grounded in the alternator.

bead usually a steel wire coil that forms the anchor for tire plies or the rim attachment for a tire or pneumatic spring.

Behavioral Analysis Safety Improvement Categories (BASIC) the foundation of CSA 2010, specifically the seven fields of the SMS outlined by the FMCSA.
bell housing the coupling hub on a transmission that flange couples it to the flywheel housing on the engine and encloses the clutch pack. In automobiles, the term *clutch housing* is more commonly used. In trucks, bell housing-to-flywheel housing mating geometry is standardized by a number of SAE standards allowing a range of transmission to engine options.
bellows a folded, telescoping device that accommodates some movement and flex. Examples would be a bellows-type air spring and the rubber boots that cover flex joints on pivoting components.
bench grinder a rotary abrasive grinder used to dress tools and clean up torch cuts on steel.
bending moment the point at which a material deflects when a force is applied to it.
berm an angled or canted surface beside a roadway.
Bernoulli's Principle states that the sum of the kinetic and potential energy in a moving, incompressible fluid will remain the same. Therefore, in a fluid moving in a steady state in through a circuit, if the potential energy decreases, the kinetic energy must increase so that the sum of the two remains the same.
bevel slope, slant, or angled edge.
bias the definition of this word in electricity simply being voltage potential at input referenced to ground. In electronics, the reference to establish the bias value does not have to be 0 volt. In anything but electricity, bias means directed or leaning toward.
bias belted a bias ply tire that has reinforcing belts under the tread section.
bias ply a tire construction geometry in which the plies are laid either diagonally or at an angle of 35 degrees.
binary something that is represented in two states. *Bits* are the building blocks of the *binary system.*
binary system data representation in which alphanumeric, operating states and logic data can be represented by sequences and stacks of 0s and 1s. For instance, in binary arithmetic, each decimal number is represented by its binary equivalent.
bipolar having two poles or opposite states. A bipolar transistor operates like a switched diode.
bipolar transistor a transistor that operates like a switched diode. It contains base, emitter, and collector terminals and can have NPN or PNP biasing.
bit binary digit with the ability to represent one of two values, usually ON or OFF which equates to an electrical status of voltage-high or voltage-low.
blade fuse a fuse with a pair of contact terminals that are inserted into a female box connector. Also known as *spade fuse.*
blind spot system (BSS) designed to detect typical driver blind spots adjacent to the vehicle and trigger an alert. Typically, BSS alert systems use microwave sensors and are usually integrated into the *headway alert system (HAS)* electronics.
blink codes a means of displaying vehicle fault codes without the use of an EST. Codes are blinked or flashed using dash display lamps. Also known as *flash codes.*
block to mechanically block a wheel, preventing it from moving using wedge-shaped wheel chocks. The term *chock* is also used.
block diagnosis chart a common troubleshooting reference chart listing symptoms, causes, and remedies usually arranged in columns.
block heater an engine coolant heater consisting of an electric element inserted into the cylinder block water jacket; usually AC mains powered and thermostat cycled.
blower fan electrically driven fan used to circulate air in HVAC systems.
blower motor/fan assembly located in the evaporator housing, it blows air through the evaporator, maximizing its efficiency as a heat exchanger.
blowgun a trigger-actuated air nozzle used to clean and dry components.
BlueCool an engine-off heat exchanger unit that uses an icebox principle. It chills the cooling medium down to subzero temperatures while the truck engine is operating. When the engine is shut down, coolant is circulated through a cab heat exchanger to draw heat out of the bunk compartment.
bobtail to run a highway tractor unit without a trailer.
bobtail proportioning valve used in truck air brake systems to reduce service application air to the rear drive axles when bobtailed, reducing lockup tendency when there is no load over the rear of the tractor.
bobtailing a tractor running without a trailer.
body control module (BCM) see *body controller.*
body controller a module that manages chassis and sometimes powertrain functions on a vehicle. The term is more applicable to automobile electronics but some truck OEMs use this as a bus traffic command module to manage events that require actions from multiple controllers on the data bus. Depending on the OEM, this may or may not be synonymous with *chassis controller* or *chassis control module (CCM).*
body roll leaning of a vehicle when cornering; a consideration for truck and trailer suspension design and loading.
bogie tandem-axle undercarriage and associated suspension linkage.
bolster plate term for the support pad on a trailer upper coupler, also known as an upper coupler pad. Bolster plates are normally rigid, but in certain specialty applications such as trailer dumps they may be designed to oscillate.
bottoming occurs when the teeth of one gear contact the lowest point of its mating gear.
bottoming tap a finishing tap used to cut untapered threads as far as possible into a blind hole.
bow damage frame damage condition in which the frame bends in an upward arc. It is caused by operating a dump truck with the box up and loaded, snowplow operation, and unequal loading of the frame.
box-end wrench a hex or double hex wrench designed to completely enclose a bolt or nut and evenly apply torque to it.
Boyle's Law states that for a given, confined quantity of gas, pressure is inversely related to the volume so that as one value goes up the other has to go down.
brake chamber air brake application device in which the potential energy of compressed air is converted to mechanical linear force. At its simplest, it consists of a housing, diaphragm, push-plate, and pushrod. Compressed air acting on the diaphragm forces the push-plate and pushrod to move mechanically, actuating the foundation brakes.
brake dyno term used to describe brake test pads. It consists of drive-on weight and sideslip transducers plus the software for analyzing the data. An example would be the Hunter B400T with WinSI software.

brake fade caused by a number of factors mostly relating to reduction in the coefficient of friction of brake friction materials. Thermal expansion of brake drums under prolonged braking tends to enlarge drum diameters, meaning that the actuation mechanisms such as slack adjusters no longer function at optimum mechanical efficiency, which makes brake fade more of a problem in drum brake systems.

Brake-Link a generic handheld electronic service tool (EST) designed to read, analyze, and program heavy truck brake system electronics.

brake pad friction face and plate assembly used in disc brakes. The brake pads are fitted to calipers that brake by squeezing a rotor.

brake shoe friction face and lever assembly used to apply braking pressure to a brake drum in any drum brake system.

brake spring pliers wide jaw pliers used for the specific purpose of removing and installing brake shoe springs.

brake torque balance achieved when the amount of mechanical braking effort applied to each wheel (the force applied by each pair of shoes to the drums) under braking is balanced.

brake warning light refers to a number of different brake malfunction lights; usually low pressure triggered in air brake systems. See *pressure differential valve* for hydraulic brake systems.

breakout box an electrical/electronic diagnostic tool that can be T-connected into a multiple circuit for purposes of accessing circuit conditions with a DMM when the circuits are live.

breakout T an electrical/electronic diagnostic tool that can be T-connected into a single circuit for purposes of accessing circuit conditions with a DMM when the circuit is live.

breather a device that allows a circuit to be vented to atmosphere; used on reservoirs, axles, etc.

bridge formula used to determine how a truck and semitrailer support their respective loads. Generally, if a trailer sliding bogie is moved forward, the percentage of load supported by the tractor decreases. Bridge formula is also used to determine how tractor axles are loaded when the fifth wheel position is altered. Moving a sliding fifth wheel forward takes load off the drive bogies and transfers it to the steer axle.

brine a salt solution often sprayed on roads as a de-icer.

Brinell hardness a test for rating material hardness that forces a carbide ball of specified diameter into the test material under a specified load. The Brinell hardness number is calculated by dividing the applied load in kilograms by the surface area of the resulting impression in square millimeters.

brinelling minor indentations on the shoulder or in the valley of a bearing race, usually caused by improper installation.

British thermal unit (Btu) unit of calorific value or heat energy. Specifically the amount of heat required to raise the temperature of 1 pound of water, through 1°F.

broken-back driveline driveshaft geometry in which the U-joint working angles are equal but the yokes are not parallel.

brushes stationary devices used to conduct current to or from a rotor in electrical components such as alternators and motors.

B-train a tractor/trailer double configuration in which the tractor is coupled to a lead semitrailer by means of a fifth wheel, and a pup trailer is coupled to the lead trailer, also by means of a fifth wheel.

bus to connect in groups; usually refers to electrical bus bars.

bus arbitration in serial bus multiplexing, the method used to negotiate message priority.

bus systems term used to describe data highways or multiplex networks.

bus topology in multiplexing, it means that no single controller networked to the data bus has more priority than another; in other words, no ECM/ECU is in charge.

butt splice the end-to-end joining of two wires in a circuit by soldering or using a connector.

bypass an optional or secondary route for fluid flow in a circuit.

byte an arrangement of 8 bits. Each bit can express one of two states such as 0 or 1, high or low voltage, etc. One byte can, therefore, be used to express up to 256 different states.

cab-over-engine (COE) a truck or tractor that locates the driver cabin in a forward position directly over the engine. There is no hood.

cage bolt threaded rod with a pair of lugs used to mechanically compress a spring brake assembly spring for installation and removal.

calcium chloride a salt mixture of calcium and chlorine. In solution, it is more corrosive than sodium chloride but it has low freeze point of –62°F (–52°C). Usually mixed into sodium chloride brine to lower its freeze point.

calipers a disc brake component used to effect squeezing force on a rotor. On a bicycle, calipers are actuated mechanically; on trucks, by hydraulic or pneumatic pressure. Calipers can be fixed, floating, or sliding. Also a measuring instrument used to reckon inside or outside diameter.

cam an eccentric or oval-shaped component.

cam bolt an adjusting bolt and eccentric that when rotated forces components to change position.

cam brakes see *S-cam brakes*.

cam ring an eccentric liner, usually stationary, within which mobile pump members rotate.

camber the attitude of a wheel assembly when viewed head-on from the front of the vehicle. A wheel that leans outward at the top (away from the vehicle) has positive camber; one that leans inward at the top has negative camber.

camber roll a change in camber that occurs when a suspension is loaded on one side during a cornering maneuver.

CAN 2.0 see *control area network*. The platform used as the data backbone for CAN-C (auto/light duty) and J1939 (truck/heavy equipment) serial data bus communications.

Canadian Federal Motor Vehicle Safety Standards (CMVSS) the Canadian equivalent of FMVSS; usually but not always identical to the U.S. federal standard concerning highway trucks.

cant to angle or tilt.

capacitor electrical charge storage device. Also known as a condenser.

capacity test (battery) see *load test*.

carbon pile a high-current dynamic battery testing instrument. A high-current path is established through a stack of clamped carbon plates and voltage measured during a timed test run. See *load test*.

carrier the center component of a planetary gearset; a disc to which the planetary pinions are attached.

cartridge fuse consists of a resistor element contained in a glass tube. When a current overload occurs, the element melts, opening the circuit.

case hardening surface hardening treatment in which carbon and/or nitrogen are molecularly absorbed into the outer skin/layer of a steel using heat.

cast iron molten iron that is poured into a cast; several different processes are used but most result in a hard but brittle, uniform shape.

cast steel alloyed steels that are molten and cast to shape without being hot rolled or forged; tend to be more fracture proof than cast irons.

caster the forward or rearward tilt of the kingpin centerline to vertical when viewed from the side. Zero caster indicates that the kingpin is located at true vertical. Positive caster occurs when the kingpin tilts rearward (leans behind) and negative caster occurs when the kingpin tilts forward.

cathode negative electrode.

cavitation bubble collapse erosion in hydraulic or engine coolant circuits.

C-channel the typical frame rail geometry used in heavy-duty trucks. It consists of an upper flange, lower flange, and web. A radius bend is used between the web and flanges.

CEEMAT early generation, not too successful effort by Eaton Rockwell at an automated transmission system; acronym for *converter enhanced, electronic managed, automated transmission.*

center bearing see *center support bearing.*

center support bearing a hanger bearing used to suspend the first section of a driveshaft in a double driveshaft assembly. It is required on truck driveshafts when the distance between the transmission and drive carrier exceeds 70 inches (178 cm) and consists of a rolling bearing element supported by a rubber insulator and bracket. Also known as *hanger bearing.*

centerbolt the bolt used to clamp a multileaf spring pack together and load it under tension to help define its self-dampening characteristics. Self-dampening is created by interleaf friction.

centerline steering condition of having the steering wheel centered when the vehicle is traveling a straight-ahead track; an alignment objective.

central processing unit (CPU) solid-state device used to manage the processing cycle in a computer. It performs arithmetic and logic computations, arranges data in the processing cycle by fetch and carry from data banks, and generates output commands. It is the "brain" of the processing cycle.

centrifugal clutch a clutch pack that uses centrifugal force to engage as engine rpm is spooled up and disengages as it spools down; uses a simple ball-and-ramp engagement hub. An example is the Eaton *AutoClutch* used with the UltraShift AMT.

ceramic fuse not so common fuse found in some offshore vehicles. It consists of a ceramic insulator with a metal strip along one side.

characteristic impedance a measure of how a conductor appears (regardless of length) to a high-frequency signal; for instance, a truck J1939 data backbone has a characteristic of 120 ohms. It cannot be measured with an ohmmeter.

charging system consists of batteries, alternator, voltage regulator, associated wiring, and the electrical loads of the vehicle. The charging system provides the current requirements of the vehicle in operation and recharges the batteries when necessary.

Charles' Law states that the volume occupied by a fixed quantity of gas is directly proportional to its absolute temperature if the pressure remains constant.

chase process of cleaning or restoring existing threads using a tap or die: *thread chasing.*

chassis the base frame of a vehicle but the term is often used to describe the complete vehicle.

chassis controller a module that manages chassis functions on a vehicle. Some OEMs use this as a bus traffic command module to manage events that require actions from two or more controllers on the data bus. Depending on the OEM, this may or may not be synonymous with *body controller* or *body control module (BCM) or signal (detect and) actuation module (SAM).*

chassis control module (CCM) see *chassis controller.*

check valve uses a spring-loaded poppet that seats and unseats. It is designed to permit flow in one direction and to close to prevent flow in the other.

cherry picker a portable shop hoist device consisting of a hydraulically raised lift arm or boom.

chip semiconductor substrate onto which active and passive elements of an electronic circuit can be photo-infused.

chisel impact tool; hardened steel bar dressed with a cutting edge and driven by a hammer.

chlorofluorocarbon (CFC) chlorine base refrigerant used in R-12 and other older refrigerants. It contains chlorine dioxide that damages the ozone by leaching an oxygen atom from the ozone (O_3) molecule, converting it to a diatomic or ambient oxygen (O_2) molecule.

chock to mechanically block a wheel, preventing it from moving using wedge-shaped wheel chocks. The term *block* is also used.

chopper wheel widely used slang term for a toothed reluctor used in a shaft speed sensor. Also known as an exciter wheel, *pulse generator,* tone wheel, etc.

circuit breaker circuit protection device that trips when a specific current value is exceeded. Two types are primarily used on trucks. SAE #1 is cycling and automatically resets. SAE #2 is noncycling, meaning that the circuit must be opened to reset. Two other types are used in limited applications. SAE #3 trips on a current overload at which time a reset button extends; The breaker remains open until manually reset. SAE#4 is similar to #3 but allows manual open and reset after tripping.

Clean Air Act legislation impacting on highway A/C systems. Introduced in 1992 and modified since, it staged the progressive elimination of chlorofluorocarbons from refrigerants.

clearance lights lighting on trucks and trailers used to display the dimensions of the unit for purposes of conspicuity.

client anything in a computer processing cycle or multiplex data transaction that can be described as having a need.

climbing a gear problem caused by excessive wear in intermeshing gears or bearings. Climbing occurs when intermeshing gears move apart sufficiently for one gear to climb over the apex of its mate.

clock speed the speed at which a CPU performs in the processing cycle of a computer system.

closed-center a hydraulic circuit in which the pump is not unloaded to the reservoir when in the center or neutral position.

closed circuit any electrical circuit that is switched to continuity; often used to describe an electrically *energized* circuit.
closed-loop a hydraulic system in which oil is circulated from the pump to the motor and back to the pump without being returned to the reservoir. Used in some hydrostatic drive systems.
cloud point the temperature at which some of the fractions in a diesel fuel begin to wax (solidify) and the fuel takes on an opaque appearance. This can cause problems in the fuel subsystem by plugging filters, resulting in engine shutdown. Most diesel fuel conditioners contain cloud point and pour point depressants.
clutch device used to break torque flow between an engine and transmission and to apply/release application force in other units.
clutch actuator a means of disengaging a mechanical clutch using pneumatics or hydraulics to enable electronic clutch control.
clutch brake circular disc with friction surfaces on both sides mounted on the transmission input shaft; its purpose is to slow or stop the input shaft during initial shifts to prevent gear clash.
clutch brake squeeze the final inch or so of pedal travel as a clutch pedal is depressed toward the floor. The "squeeze" references the clamping of the clutch brake by the clutch release bearing.
clutch free pedal the free downward travel of a clutch pedal during which the clutch release linkage takes up the ⅛ inch (3 mm) of free travel between the release forks and the clutch release bearing. After clutch free pedal has been taken up, the clutch begins to release as the pedal continues to be depressed. Clutch free pedal is correctly set at between 1½ and 2 inches (3.8 and 5 cm).
clutch free travel clutch free travel is the approximately ⅛ inch (3 mm) of clearance between the clutch release forks and the clutch release bearing when the clutch pedal is in the fully up position. When clutch free travel is properly set at ⅛ inch (3 mm), clutch free pedal should be between 1½ and 2 inches (3.8 and 5 cm).
clutch housing the coupling hub on a transmission that flange couples it to the flywheel housing and encloses the clutch pack. In trucks, the term *bell housing* is more commonly used.
clutch pack apply/release mechanisms that typically consists of clutch plates, friction discs, and a pressure plate. In a mechanical clutch pack, the pressure plate is mechanically applied. As a component within an automatic transmission, a pressure plate uses tabs that mate with grooves in the clutch drum and is applied hydraulically or electrically.
clutch release travel see *clutch free travel.*
clutch switch see *driveline disengagement switch (DDS).*
coefficient of friction describes the aggressiveness of friction surfaces; for instance, ice would have a low coefficient of friction, whereas asphalt would be higher. Used to grade brake friction faces.
cold cranking amperes (CCA) the primary method of rating a vehicle battery. It specifies the current load that a battery is capable of delivering for 30 seconds at a temperature of 0°F (−18°C) without dropping the voltage of an individual cell below 1.2 or 7.2 volts for a six-cell battery across the terminals.
cold-rolled metal (usually steel) manufacturing process that reduces hot-rolled stock by compressing it using rolls to a desired thickness.
collision mitigation system (CMS) accident avoidance electronics using radar, microwave, and/or video to sense potential accidents and subsequently alert driver, and using chassis data bus to signal the engine brake and service brakes to make an evasive response. An example is *Wingman.*
collision warning system (CWS) accident avoidance electronics using radar and microwave to alert the driver that an evasive response is required. An example is *VORAD.*
combination air/leaf spring suspension a suspension that combines both air bags and leaf springs to provide good loaded and unloaded performance characteristics.
combination air/torsion spring suspension a sway bar, a torsion spring, and an air suspension used on some Paccar Kenworth trucks. Effective at minimizing wheel hop.
combination pliers slip joint pliers with multiple jaw opening sizes.
combination rubber/torsion spring suspension a solid rubber spring and torsion rod suspension known to maximize wheel-to-road contact and minimize wheel hop.
combination valve used in a hydraulic brake system, it contains the safety switch, metering, and proportioning valves in a single valve body and usually is located close to the master cylinder. Also may refer to other multiple function valves such as the service, emergency, relay valve (SERV) used on air brake systems.
combination wrench a wrench with an open end on one side and box end on the other, both of the same nominal dimension.
come-alongs a cable or chain hook, hand-actuated, linear ratcheting device used for lifting or applying separation pressure.
come-backs the result of a repair that fails shortly after the repair has been made. Usually means that work has to be undertaken without further charge to the customer. Also known as *policy adjustment.*
Commercial Electronic Control (CEC) Allison second-generation, partial authority, electronically controlled transmission. This name replaced the ATEC acronym (Allison Transmission Electronic Control) but today it can be regarded as effectively obsolete.
commercial mobile radio service (CMRS) the vehicle-to-infrastructure (V2I) protocol used by Drivewyze in the U.S. and Canada for commercial truck weigh station bypass.
Commercial Vehicle Safety Alliance (CVSA) defines out-of-service (OOS) criteria used by vehicle safety enforcement officers across North America (United States, Mexico, and Canada) to standardize truck safety standards. It is a nonprofit organization that publishes OOS criteria and inspection forms online and on DVD that can be obtained from http://www.cvsa.org.
communications adapter (CA) a serial link used to transduce bus protocols (J1587 and J1939) to a Windows environment. Standardized by RP 1210. CAs should be auto baud detection capable (ABDC): a non-ABDC CA can cripple a data bus.
commutator conductors connected to armature windings.
compensating fifth wheel a fifth wheel designed to articulate fore and aft, and side to side. Used to limit trailer torque effect transfer; the Holland Kompensator would be an example.
Compliance Safety Accountability (CSA) the FMCSA *safety management system (SMS)* launched in

December 2010 that requires logging and transparency in safety-related citations to state enforcement agencies and consumers. Often known as *CSA 2010*.

compound angle occurs when a driveshaft is angled on both horizontal and vertical planes.

compressor any of a number of mechanically driven devices used to compress a gas. In trucks, compressors are used in air brake systems and A/C systems.

computer any of a number of different types of electronic processing devices with a CPU, input circuit, data retention, and output switching.

concentric two or more components that share a common axis.

condense to change the state of a gas to a liquid.

condenser an electrical charge storage device also known as a capacitor. In an A/C circuit, it is a heat exchanger located in the vehicle airflow and subject to the ram air. Cooling of the refrigerant causes it to condense to a liquid state. This results in the refrigerant changing to a liquid state.

condition-based maintenance (CBM) a term used to describe maintenance planning in which preventive maintenance is minimized and in which maintenance intervention is timed so that parts are replaced or reconditioned just before a failure occurs. A CBM program usually makes use of prognostic analyses and software managed maintenance.

conductance the conducting ability of an electrical component, in DC equal to the *reciprocal* of resistance.

conductance testing battery test instrumentation that pulses AC through a battery circuit at a single frequency (commonly below 100 hertz) to perform the test and verify battery performance. Multifrequency systems are more accurate but require more complex data interpretation software.

conductor any material that permits electrical current to flow; any element with less than four electrons in its valence shell.

conduit channel or pipe used to route fluids or a protective covering used on electrical wiring circuits.

conicity a tire irregularity that causes a tire to take the shape of a cone when loaded; it may generate lateral force.

ConMet PreSet hub a preset hub assembly that eliminates the need for the technician to adjust wheel bearing. A ConMet hub is installed and then torqued to 300 lb-ft.

conspicuity to make noticeable; current trucks and trailers are required to display conspicuity tape to maximize driver awareness of a transport vehicle.

conspicuity tape nighttime reflective tape required on trucks, trailers, school buses, and other highway equipment to maximize the conspicuity on the highway.

constant mesh gears positioned to remain in permanent mesh, not engaged and disengaged by shifting action.

constant rate spring any spring that has a constant deflection rate.

contact to touch.

contact area a term sometimes used to describe the tread surface of a tire that actually contacts the road. See *footprint*.

containers the portable storage vessels used by intermodal transportation systems. Containers are designed for transportation by truck, rail, ship, and aircraft.

continuity used to describe the status of an electrical connection or circuit with little or no resistance to current flow. Checked by using an ohmmeter.

control area network (CAN) data bus systems used in vehicle applications. Serial data buses used in automobile (CAN-C) and truck (J1939) data backbones. CAN 2.0 defines the architecture and network protocols of the current J1939 powertrain data bus.

control circuit used in reference to both cranking systems and air brake systems. In a cranking system, it is the low-current switching circuit that connects the ignition switch with the magnetic switch. In an air brake system, it defines low-volume pilot air signals that are used to activate high-volume slave circuits.

control main pressure a product of the transmission main pressure it uses a regulator to dampen pressure fluctuations and provide smooth, consistent valve actuation.

controlled traction differential gearing that uses a friction plate assembly to transfer drive torque from a slipping wheel (sensed by axle shaft spinout) to its opposite.

controller widely used term for control module. Anything on the data bus that can be assigned source address (SA) or message identifier (MID) status, meaning that it has some processing and networking capability.

conventional theory (of current flow) asserts that current flows from positive to negative. Most automotive wiring schematics, terminology, and troubleshooting procedures are based on conventional theory. Its opposite is *electron theory*.

converter dolly a single axle coupling assembly with a frame, drawbar, and fifth wheel that converts a semitrailer into a full-trailer.

coolant flow valve manages the flow of coolant from the engine cooling circuit to the cab HVAC system. Also known as a *heater control valve*.

coolant heaters used to warm coolant during shutdown in cold weather to enable easier startups and consistent cab temperatures.

coolant hydrometer used to measure coolant specific gravity to determine the degree of antifreeze protection.

cooler any of a number of different types of heat exchangers.

cords the strands of wire or fabric that form the plies and belt in a tire.

corrosive a substance capable of chemically attacking another.

counteract balance beads (CBBs) Teflon-coated glass beads placed inside tires used to dynamically balance axle-end assemblies. Uses centrifugal force to locate the light side of the interior of a tire and counterbalance it.

counterbalance valve a valve that resists flow in one direction while permitting free flow in the other. Often connected to the outlet of a double-acting ram to support weight and prevent sudden, uncontrolled dropping.

counter-electromotive force (CEMF) induced voltage acting in opposition to battery voltage in electric motors. When a conductor cuts across a magnetic field, a voltage that increases with frequency is induced in the conductor. CEMF produced by cranking motors at higher speeds gives the motor a desirable self-regulating characteristic.

counter-rotating torque converter torque converter used in the Voith DIWA transmission that doubles as a driveline retarder.

coupler something that links two components or circuits such as the fittings used in hydraulic circuits; another example of couplers would be *gladhands*. In most cases, couplers have a male and female component.

coupling point occurs in a torque converter when the turbine is rotating at within 10 percent of the impeller speed.
courtesy switch out-of-date term used to describe a headlight *dimmer switch.* Using a dimmer switch when night driving is no longer a courtesy but a legal requirement in most jurisdictions today.
covert literally secret or "undercover" but used in vehicle computer systems to log data events that cannot be readily accessed using standard ESTs and software.
crank to rotate an engine with the objective of starting it.
cranking circuit the chassis electrical circuit that supplies the high-current load required to crank an engine. Also known as *starting circuit.*
cranking motor electric motor that converts electrical energy from the cranking circuit into the mechanical energy required to start the engine. Also known as *starter motor.*
crash sensor produces a signal corresponding with the rate of deceleration in an SRS system; in series with the crash sensor.
cross the cross-shaped trunnion used in a U-joint assembly. Also known as a spider. Also known as a *journal cross.*
cross-members the transverse braces used to connect a pair of frame rails in a ladder type frame assembly.
crosstube transfers steering motion from the steering gear side of the chassis to the opposite side by means of an adjustable rod, forcing the steering knuckles to act in unison. Also called a *tie-rod.*
crown and pinion the matched gear set used in a final drive carrier to convert longitudinal drive torque into transverse drive torque required to turn the drive wheels.
crown area the area of a tire across the tread between the belt areas.
crown gear a large disc-shaped gear usually driven at right angles by a smaller pinion gear. It alters the torque path direction in a final drive carrier by 90 degrees, so that a longitudinal drive shaft can deliver transverse torque to the driving wheels.
crowning a slight curvature machined from end to end in some gear teeth; diminishes concentrated contact loads on teeth.
cryogenic systems reefer units that use the thermodynamic properties of certain liquids to cool. A cryogenic liquid passes through an evaporator, turns into a gas, enters a heat exchange coil, and drives a pair of cryo-motors. Heat is transferred to the cryogenic medium.
CSA 2010 see *Compliance Safety Accountability.*
C-train a train combination of a tractor and two semi-trailers. The lead semitrailer is coupled to the tractor by the usual fifth wheel coupling, but the pup is coupled to a special converter dolly with a rigid, double drawbar system that prevents rotation about the drawbar hitch points. Any rotation that occurs will take place at the converter dolly fifth wheel, producing a double unit that is easier to back up than either A- or B-trains.
curb weight vehicle weight minus its load.
cure the vulcanization process consisting of heat and pressure used to repair tire rubber. Can also be used in the context of the chemical *curing* of sealants.
current flow the quantitative flow of electrons through a closed electrical circuit, measured in *amperes.*
current transformer a DMM accessory device used to measure AC. It reckons AC current flow by transducing 1 amp to 1 mA, enabling the DMM ammeter to safely measure it.
curtain wall van a trailer with canvas wall side that enables easy access to the load.
cushion block usually a solid rubber block used as a primary spring in a rubber block suspension or a *jounce block* in other types of suspensions.
CVSA L1 inspection a 37-step inspection designed to identify OOS defects on a commercial vehicle.
cycling a continuing repetition of events. Repeating a duty cycle.
cycling circuit breaker an SAE type-1 circuit breaker specified to a specific amperage rating, which, if exceeded, heats an armature that then opens the circuit at contact points. When the armature cools, it closes contact points and circuit flow resumes.
damp to buffer or insulate
dampen synonymous with *damp.*
dampened discs clutch friction discs with built-in coaxial springs used to dampen driveline oscillation and shock loads. Many of today's clutches use dampened discs due to the run slow, gear fast drivelines used with high-torque engines.
damping the act of buffering or insulating, having the ability to suppress oscillations or vibration.
Darlington pair an arrangement of two transistors in which one uses very low current to switch a second with a higher current. A solid-state device used in the drivers of chassis system ECMs and ECUs.
dart steering condition in which steering track changes independent of steering wheel inputs; potentially dangerous.
dash control valves any dash-mounted valves, but most often used to refer to air brake system valves such system park and trailer supply valves.
dash digital display (DDD) any of a number of dash-located, digital display units.
dash size the inside diameter of hydraulic hose expressed in 1⁄16-inch (1.6 mm) increments; each dash number indicates 1⁄16 inch (1.6 mm), so a #4 dash hose would be equivalent to 4⁄16 inch (6 mm) or 1⁄4 inch (6 mm).
data electronic information.
data bus multiplex backbone consisting of a twisted wire pair.
data bus communications any type of multiplex transactions.
data connector the means of connecting an electronic service tool (EST) to a chassis data bus. Data bus connectors can be six-pin (J1587/1708) or the newer nine pin (J1939) Deutsch black and green connectors. All U.S.-built trucks have a data connector located on the left side of the steering column, either under the dash or to the left side of the driver's seat.
data bus viewer (DBV) the Allison transmission electronic diagnostic software. Requires a subscription.
data frame a data tag consisting of 100 to 150 bits for transmission to the bus. Each tag codes a message for sequencing transmission to the data bus and also serves to limit the time it consumes on the bus.
data link connector (DLC) the means of making a connection between an EST and a chassis data bus. Examples are the current J1939 black and green 9-pin connectors.
data links means of connecting electronic systems and computers for networking. For instance, to read a chassis electronic system such as the transmission electronics, it has to be accessed by connecting an EST to the chassis data link. Current trucks use a J1939 9-pin black or green (pot 2014) connector.

data processing the act of computing. Often used to describe the processing cycle of a computer from input, fetch-and-carry of data, arithmetic and logic computation, algorithm negotiation, to driving the outcomes.

daylight running lights (DRLs) high-intensity, forward-directed lamps now required by many jurisdictions to increase vehicle *conspicuity*.

dead axle a term used to describe any axle that does not have gearing to drive the vehicle. Lift axles, trailer axles, and steering axles fall into this category.

deadline to take a truck or trailer out-of-service (OOS) because of a safety or running defect. The CVSA defines most truck chassis OOS standards.

deburring to remove sharp or rough edges from a cut.

decelerometer measures the change in angle of a chassis when it is subjected to aggressive braking. This angle is also known as *load transfer, dynamic pitch*.

dedicated contract cartage (DCC) a trucking operation with logistics designed to a specific shipper's requirements. Often tagged into just-in-time (JIT) factory production logistics.

deep cycling the rapid, continuous, full discharging and recharging of batteries.

deflector see *slinger*.

delta a diode bridge arrangement in which three diodes are connected in a triangular arrangement used to rectify AC to DC.

Department of Transportation (DOT) federal government agency that oversees all matters related to highway and transportation safety. Oversees NHTSA and is available at http://www.dot.gov.

desiccant a moisture removal agent. Most air dryers contain desiccant media that may have to be replaced when it becomes depleted or saturated with oil.

detergent oil-soluble substance that holds contaminants in suspension in lube oils and water-based solutions. May be synthetic or petroleum sourced.

detergent additives refer to a range of *detergents* used in some lubricants and hydraulic oils.

Deutsch connector a manufacturer of sealed/weatherproof electronic connectors; these include the SAE standard data bus connectors used to access the data backbone on trucks today. Several of types of Deutsch data bus connector are used: six-pin (J1587/1708), black nine-pin (pre-2014 J1939) and green nine-pin (post-2014 J1939).

diagnostic data reader (DDR) usually refers to electronic service tools (ESTs) used to access a truck chassis data bus; may be a PC (laptop or portable), handheld computer electronic service tool, or phone-based device with appropriate software.

diagnostic flow chart a sequential troubleshooting chart used in OEM service literature and diagnostic software; usually outlines a step-by-step approach to sourcing a problem.

diagnostic link connector (DLC) variation of *data link connector* (preferred).

diagnostic optimized connection (DOC) Allison WTEC read and diagnostic software used in conjunction with HH-ESTs.

diagnostic trouble code (DTC) fault codes. The term is used to categorize faults off a CAN data bus such as the heavy-duty J1939 and the automotive CAN-C. A J1939 DTC incorporates a SPN, SID, FMI, and OC.

diagonal cutting pliers pliers with sharpened jaws more commonly known as *sidecutters*.

dial caliper a mechanical measuring caliper with a rotary display scale.

dial indicator an accurate linear measuring tool actuated by a plunger with a rotary display. Used to measure endplay, runout, recess depth, and so on.

diameter straight line extending across the center of a circle from circumference to circumference. The diameter of a circle is two times its radius.

diamond damage a frame damage category in which one frame rail is positioned further ahead than the other; often caused by offset collision.

Diamond Logic™ the term Navistar International Truck Corporation uses to describe its multiplex truck chassis systems.

die a hand- or machine-actuated tool used for cutting external threads.

die cast a steel casting process that uses metallic dies to produce a form.

dielectric something that limits or blocks current flow. Nonconducting dielectric material is used between poles in a *capacitor* and dielectric grease can be used to protect exposed terminals.

diesel emission fluid (DEF) term used to describe the aqueous urea required by diesel engine selective catalytic reduction (SCR) systems.

diesel exhaust fluid see *diesel emission fluid*.

diff lock commonly used slang term for power divider or differential lockout.

differential drive axle gearing that transmits driveline torque to axle shafts and the drive wheels, plus allows the opposing wheels to rotate at different speeds for cornering and traction.

differential carrier the drive axle assembly that houses the differential gearing.

differential lockout driver-controlled means of locking out differential action to improve traction on a slippery surface.

digital caliper a mechanical measuring caliper with a digital readout. Most possess some processing capability that can be used to convert standard to metric values and vice versa.

digital display (DD) any type of electronic display hardware, including those using CRT, LED, and plasma display media. May be portable, stationary, or on vehicle.

digital display unit (DDU) can refer to any of a number of different types of electronic display units, but most often those that are LED based. Some OEMs use this term to refer to handheld portable devices such as NEXIQ ESTs.

digital multimeter (DMM) a multiple mode electrical measurement instrument capable of measuring AC voltage, frequency, DC, ohms, amperage, and other values. The term *DMM* tends to be used in truck technology; the same instrument is more commonly referred as a digital volt ohmmeter (DVOM) in the auto area.

digital volt-ohmmeter (DVOM) a digital multimeter. This term tends to be used more by automotive OEMs.

dimethyl ether (DME) an alternate fuel option in Volvo trucks. DME is chemically mostly methane and it is produced by biomass fermentation. It has been used as aerosol propellant for many years.

dimmer switch a two-pole switch that switches headlamps between high and low beam positions. The dimmer switch can be floor or stalk switch located. Used to be known as a *courtesy switch*, but today its use is legally required in night driving in most jurisdictions.

diode a two-terminal, solid-state, semiconductor device that allows current flow in one direction only. Acts like a sort of electrical one-way check valve.

direct drive a transmission ratio option in which the input and output shafts rotate at identical speeds, that is, engine speed.

directional control valve a valve that functions to direct or block flow through selected channels or circuits.

directional stability the ability of a vehicle to maintain a straight-ahead tracking line.

disc brakes vehicle brake system in which squeeze pressure is applied to a rotor by a caliper. Both pneumatic and hydraulic disc brakes are used on trucks.

dispatch sheet work schedule.

displacement the volume of fluid that can pass through a motor, pump, or cylinder in a single cycle or stroke.

DM (data link mechanical) clutch the dry clutch pack module used on Eaton's two-pedal system AutoShift transmissions. It is a centrifugal clutch.

dog clutch another term for *sliding clutch.*

dog-tracking a misalignment condition in which the rear wheels of a truck run on a different track line than the front steering wheels, creating a *thrustline* irregularity. More often seen on tractor-semi combinations in which the trailer travels at an angle to the tractor; usually caused by bogie-to-kingpin misalignment.

dolly trailer landing gear or caster wheel hoist device.

dolly axle see *converter dolly.*

dolly legs trailer landing gear system; consists of gear-cranked, retractable legs reinforced by fixed gusset arms.

doping in solid-state electronics, the process by which semiconductor crystals are given a positive (P) or negative (N) bias that determines how they function in a circuit.

Doppler effect frequency shift analysis used to locate and determine relative velocity of moving objects; the principle used in radar collision warning systems.

Doppler radar any type of radar that uses Doppler effect to locate and determine relative velocity using microwave or light waves.

double-acting cylinder a cylinder in which fluid force can be applied in either direction; an example would be a COE cab lift ram.

double check valve used when a component or circuit must receive or must be controlled by the higher of two possible sources of air pressure. The valve will always bias the higher source pressure.

double hex term used to describe a 12-point box-end or socket wrench. It is used to engage a 6-point fastener head; that is, one with six flats, at 30-degree increments of rotation.

double-reduction axle a differential carrier gear set that uses a pair of gear sets for greater speed reduction and increased drive wheel torque. Often used in severe service applications such as cement mixers, construction dump trucks, etc.

downshift to drop down a gear ratio.

drag link a connecting linkage between the steering gear Pitman arm and the steering control arm. It is sometimes adjustable.

drawbar the device used to connect a trailer tongue to either a pintle hook or a ball hitch.

drawbar eye the coupling eye that engages to either a pintle hook or the shank below a heavy-duty ball hitch assembly.

drawing another way of saying *tempering.*

drayage truck a shunt or haulage truck used inside a port authority. In California, it is subject to distinct emissions standards.

drive gear any gear that drives another.

drive shaft used to transmit drive torque between the differential carrier assembly and the drive wheels. A better term is *axle shaft* because it avoids confusion with the term *driveshaft*. Also known as *half shaft.*

driveline the drive transmission components in a chassis extending from the flywheel to the drive tires.

driveline disengagement switch (DDS) a normally closed switch located at the clutch pedal; used to verify that the clutch is fully engaged. When a DDS switch opens, the engine ECM will abort cruise, PTO mode, and engine braking plus reset the idle shutdown timer. Also known as a *clutch switch.*

driven gear a gear that is driven by another.

driver display unit (DDU) a dash-mounted, driver information display system that includes the hardware, software, and visual display screen.

driver interface unit (DIU) the Wingman CMS that displays visual and audio alerts to the driver.

driver's manual set of operating instructions usually provided with a vehicle when new and placed in the glove box. In a truck, this may include such information as shifting instructions, how to interpret a driver digital display, and some basic service procedures.

driveshaft an assembly of one or more propeller shafts (hollow tubes) connected by universal joints (U-joints) to driven driveline components. In a typical tandem drive highway tractor, driveshafts are used to transmit torque from the transmission and forward differential carrier and between the two differential carriers.

driveshaft length distance between the outermost U-joint centers of a propeller shaft when in the installed position.

drivetrain term used to describe the components that move a vehicle down the highway from the engine, transmission, propeller shaft(s), drive axle carrier, drive axle shafts, and drive wheel assemblies. The engine is the prime mover of the drivetrain and the role of the transmission is to manage the output ratios.

Drivewyze an *intelligent transportation system* (ITS) telematic subscription service that uses CMRS to provide bypasses to trucks through state and provincial highway weigh stations. The vehicle CSA record, credentials, and weight are verified using WRI technology and communi cation to the driver can be via EOBR or cell phone.

dropbox term used to describe a transfer case; an auxiliary transmission used to split drive torque from the main transmission between rear and forward drive axles.

Drop-7 an Allison WT transfer case. See *dropbox.*

drum brakes a brake system in which a rotating drum fixed to the wheel assembly turns around stationary friction-faced shoes. Braking action is achieved when the shoes are forced into the drum inside diameter. The common foundation brake system used on trucks using S-cam actuation of the shoes.

dryer reservoir module (DRM) an integrated reservoir, dryer, governor, and pressure protection pack assembly used in some truck air supply circuits.

dry seal a dynamic seal that seals grease (dry) rather than oil, which requires a *wet seal.*

D-train a multitrailer combination in which articulation between the tractor and lead trailer is removed, whereas that between the lead and pup remains.

dual-circuit application valve the *foot valve* used with FMVSS 121–compliant service brakes; masters service applications from primary and secondary circuits. See *treadle valve.*
dual-circuit brakes highway vehicle brake system with full system redundancy in the event of a failure in one brake circuit. Mandatory under FMVSS legislation. Characterized by primary and secondary circuits in air brake systems and by a tandem master cylinder in hydraulic brake systems.
dual-speed axle see *two-speed axle.*
duty cycle usually describes the ON time of a device. This could be measured in milliseconds in a computer-controlled device or as a percentage representing the maximum amount of total ON time over a specified period in equipment such as a welding station.
dynamic moving; the opposite of static.
dynamic balance the balance status of a moving object. For instance, it is common to *dynamically balance* a truck tire and wheel assembly.
E-500 lube a long-life, synthetic gear lubricant recommended by Eaton for use in its transmissions.
ear one of the two projections of a drive yoke on a driveline propeller shaft.
earth term used to describe the negative side of an electrical circuit in Australia, New Zealand, and the United Kingdom. Synonymous with *ground.*
Easy Steer Plus see *unitized axle.*
eddy current small current with a circular dynamic produced in the iron core of an armature of electric motors. Eddy currents produce heat; they can be reduced by using laminated cores.
electric retarder a type of truck/heavy-equipment driveline retarder that uses an electromotive principle. When energized, electromagnets load retarding drag on a rotor.
electric shock-resistant (ESR) footwear specialty footwear required when working around high-voltage potential circuits. Identified by an orange omega on a white triangle.
electricity the science of electrical energy.
electrochemical impedance spectroscopy (EIS) battery testing instrument that evaluates the electrochemical characteristics of a battery by applying alternating current (AC) at varying frequencies and measuring the current response of the battery cell. Frequencies range from 100 microhertz (mHz) to 100 kilohertz (kHz). Known to be more reliable than conductance testing.
electrocution death by electrical overload to the body, usually by heart failure.
electrohydraulic control module mounted at the base of an Allison WT transmission, it effects the results of the TCM processing into action by hydraulically actuating the servo devices.
electrohydraulic valve body mounted at the base of an Allison CEC transmission, it effects the results of ECU processing into action by hydraulically actuating the servo devices.
electrolyte the electrically conductive substance in a voltaic cell that interacts between electrodes (anode and cathode). The electrolytic media in a standard lead-acid battery is a solution of 34 percent sulfuric acid and 66 percent pure (distilled) water.
electromagnetic interference (EMI) low-level radiation (such as that emitted from electrical power lines, vehicle radar) that can interfere with signals on data buses unless it is suppressed.
electromechanical ignition switch a starter control circuit switch that requires the turning of a key or pushing of a button to close the circuit and initiate cranking.
electromechanical switch any type of electrical switch that requires mechanical movement to open or close a circuit.
electromotive force (EMF) electrical pressure or potential usually measured in *voltage.*
electron negatively charged, subatomic particle orbiting the nucleus of every atom.
electron theory asserts that current flow in a circuit is oriented to the movement of electrons through the circuit; that is, negatively charged electrons will flow from a negative to a more positive location in a closed circuit. Used in some electronic schematics. Its opposite is *conventional theory.*
electronic brake system (EBS) an air brake EBS replaces the FMVSS 121 required pneumatic brake control circuit with electronic controls and electrical actuators. Although widely used offshore, when EBS is used in North America, it must have full redundancy to a pneumatic control circuit.
electronic clutch pressure control (ECPC) the external solenoids in a Caterpillar CX transmission that manage the clutch hydraulics.
electronic control module (ECM) the module that contains the processing and output driver capability of an electronically controlled system: the system controller. Depending on the OEM, ECM can be synonymous with ECU. The SAE recommends the use of the term *ECM* to refer to the engine controller and *ECU* to refer to other chassis system controllers.
electronic control unit (ECU) the module that contains the processing and output driver capability of an electronically controlled system: the system controller. Depending on the OEM, ECU can be synonymous with ECM. The SAE recommends the use of the term *ECU* to refer to all chassis system controllers except for the engine controller.
electronic leak detector handheld devices with a flexible probe used to seek out refrigerant leaks. When a leak is detected, a flashing light or buzzer signals the technician.
electronic lock indicator (ELI) an SAF-Holland fifth wheel sensor and dash status indicator that alerts the driver on hitch status.
electronic on-board recorders (EOBRs) data loggers that can store and broadcast data from the vehicle data bus and operational details; read using telematics or by QR scanning at weigh stations and depots.
electronic service tools (ESTs) any of a number of instruments used to work on vehicle electronics. Can be used to refer to a DMM but more commonly used to describe the instruments that connect to the vehicle data bus such as scan tools, HH-ESTs, or PCs.
Electronic Technician (ET) the Caterpillar diagnostic software used for their engine and transmission electronics.
electronically controlled automatic transmission (ECAT) term used to describe fully automatic, computer-controlled transmissions. Fully automatic transmissions are differentiated from AMTs by their near zero spin loss during a ratio shift.
electronically erasable programmable read-only memory (EEPROM) a magnetically retained (in current truck systems) memory component in a computer

system that complements ROM and PROM. In truck and heavy-equipment chassis computers, EEPROM is used to log customer and proprietary data programming, audit trails, fault codes, and other write-to-self capability.

electronically variable steering (EVS) computer-controlled steering provides improved directional stability, steering effort proportional to driving speed, reduced dry park effort at the steering wheel, and reduced system power consumption.

electronics the science of electron behavior.

electropneumatic usually refers to an electrically controlled pilot valve; used extensively on truck chassis. Examples would be an ABS modulator and suspension dump valves.

end yoke term used to describe a half round, output yoke of a transmission that connects to a drive shaft yoke.

endplay linear play in a shaft, wheel, or rotor. The slang term for this, *slop*, describes the condition well. Some OEMs use the term *axial slop* to describe endplay.

engine stall point the rpm at which an engine is overloaded to the extent that it begins to either lug or stall; a factor in performance testing of automatic transmissions.

Engine Synchroshift (ESS) an early-generation ZF Meritor automated transmission; a three-pedal system introduced in 1997 and no longer manufactured.

EPDM current-generation belts are made of EPDM (ethyl ene, propylene, diene, monomer); this material does not crack like the older belts so when inspected, they should be measured for wear.

epicyclic gearset another way of saying *planetary gearset*; refers to orbital gears (planet gears) moving around the circumference of a sun gear. See *planetary gearset*;

equalizer beam one of a pair of longitudinal beams used in a tandem equalizing beam suspension.

equalizing beam suspension a suspension used on bogies to lower the center of gravity of the axle load; there are two types: the leaf spring type and the solid rubber spring type.

Ethernet a multiplexing format with higher speed and bandwith than current CAN 2.0 platforms such as J1939 and CAN-C. Being considered as replacement for J1939.

eutectic mixture of constituents in proportions that are designed to both melt and solidify at a specific constant temperature, that is lower than melting point of any of the constituents. Term is used in welding and heat-sensitive fusible plugs.

evaporate to boil or vaporize a liquid. The change of state that takes place as a substance is changed from a liquid to a gas.

evaporator a heat exchanger used in a refrigeration or A/C circuit; used to boil refrigerant within the circuit.

evaporator thermostat A/C system thermostat. Contacts close when the evaporator temperature exceeds 38°F (3°C) on the rise, and open when the temperature decreases below about 32°F (0°C) on the fall. It turns off the effective cycle of the compressor whenever frost begins to form on the evaporator.

exciter ring preferred term by one OEM for the toothed reluctor used in a wheel speed sensor. See *pulse generator*.

exciter wheel slang for a toothed reluctor used in a shaft speed sensor. See *pulse generator*.

expansion valve restricts the flow of the refrigerant in an A/C circuit, which lowers its pressure as it exits the valve.

external ground a chassis electrical component that is supplied by an insulated positive wire/cable with its electrical circuit completed by being bolted to chassis ground.

extractor any of a number of different types of damaged stud or fastener removal tools; both internal and external types exist.

failure mode indicator (FMI) the universal numeric coding of electronic faults on heavy-duty data buses. Currently there are thirty FMIs to identify failures and imminent failures of components and circuits. Synonymous with *fault mode indicator*.

fairing describes the aerodynamic panels used on some vehicles to minimize airflow resistance.

false brinelling a manufacturing characteristic that can surface polish bearing races without creating any damage. Technicians should learn to recognize this to avoid replacing bearings that are functionally sound.

farad a unit of capacitance.

fast adaptive mode high-speed processing algorithms; Allison uses this term to describe an operating condition in which large changes are made to initial shift conditions to adjust for major system tolerances such as solenoid-to-solenoid, main pressure, and clutch-to-clutch variations. A new or newly rebuilt transmission should be programmed to operate in fast adaptive mode.

fatigue literally *tired* but used to describe components that have been worn out by old age.

fatigue failures catastrophic failures that occur due to old age or excessive wear.

fault code a condition failure code logged to computer memory and often broadcast onto the vehicle data bus. Classified as active or historic/inactive, fault codes are assigned one of twelve failure mode indicators (FMIs) for purposes of diagnosing system conditions and recording to audit trails.

Federal Motor Carrier Safety Administration (FMCSA) body responsible for the governance and enforcement of all transportation safety standards in North America.

Federal Motor Vehicle Safety Standard No. 105 (FMVSS 105) legally mandatory brake standards that all highway hydraulic brake systems must comply with; requires that brakes be dual circuit and have the ability to park and perform an emergency stop mechanically.

Federal Motor Vehicle Safety Standard 108 (FMVSS 108) legally mandatory truck and trailer lighting standards that any truck running on a U.S. highway must comply with.

Federal Motor Vehicle Safety Standard No. 121 (FMVSS 121) legally mandatory air brake standards that all highway trucks must comply with; defines stopping distances, equipment performance, and dual application circuit, among other things.

Federal Motor Vehicle Safety Standards (FMVSS) legally mandatory standards used for all highway equipment; noncompliance will preclude certification of new equipment and is an offense in in-service equipment.

feeler gauge a metal blade precisely finished to a defined thickness; used for measuring clearance and tolerance dimensions. Usually found in sets in which thickness graduates by 0.001-inch (0.025 mm) increments. Also known as *thickness gauges*.

ferrule a compression ring used in many fluid power couplings to create a seal between the line and coupling nut. Ferrules used in truck pneumatic and hydraulic circuits can be made out of malleable plastic, brass, copper, or steels. Many ferrules deform to conform with the fitting when torqued and should not be reused.

fiber composites manufacturing process that produces lightweight components using composites such as carbon and fiberglass in a weaved process that gives them high strength. Used for such varied components as body panels, dash fascia, and suspension springs.

fiber optics the transmission of light pulses through fiberglass strands or other transparent solid mediums.

field coils used in electric motors and other electrical devices to flow current through to establish an electromagnetic field.

field effect transistor (FET) an electronic relay switch of which there are two types: N-channel and P-channel. A *gate* (terminal) is used to alter channel status. Channel allows current to flow or not to flow from the *source* terminal to the *drain* terminal. Used on many current truck electronic systems because of their ability to allow a very small current to control a high current.

field programmable gate arrays (FPGAs) programmable logic devices used for processing applications in which data volumes are low, but the requirement to reprogram due to changing hardware may be high. Used in truck information display/infotainment systems.

fifth wheel the lower coupler assembly used on most highway tractors and lead trailers; typically consists of a pivot plate and locking jaws that engage to the kingpin upper coupler on a trailer.

filament the resistive element in an incandescent bulb: When current is flowed through the filament, electrical energy is converted to both heat and light energy. Usually made out of tungsten and shielded by inert gas.

fillet the radius formed between a pair of perpendicular planes. Bolt caps use a fillet between the cap and shank. When two pieces of flat steel stock are welded at right angles to each other, the result is a fillet weld.

final drive carrier a drive axle assembly that contains gearing to alter the direction of the torque path by 90 degrees and incorporates differential gearing. Also known as a *differential carrier.*

fixed caliper a disc brake design in which the caliper is bolted in place and does not move when the brakes are applied; pistons on both sides of the rotor sandwich the rotating disc to apply the brakes.

flammable capable of burning; means the same as *inflammable.*

flange yoke a half or full round yoke that attaches a drive shaft to a companion flange.

flaring capable of being spread outward; for instance, a *flaring* tool is used to expand tubing.

flash codes a means of displaying vehicle fault codes without the use of an EST. Codes are blinked or flashed using dash display lamps. Also known as *blink codes.*

flex area the area of a tire that deforms when the tire runs through its footprint.

flex disc see *flexplate.*

flexplate used as a drive intermediary between the engine crankshaft and the torque converter used with automatic transmission systems on trucks; has dampening characteristics.

FlexRay a multiplexing format with higher speed and bandwidth than CAN 2.0 platforms such as J1939 and CAN-C. Currently used in Europe but as yet has no traction in the United States. An unlikely replacement for J1939.

flinger see *slinger.*

floating caliper a disc brake design in which the caliper is a one-piece casting that uses single-piston actuation, which allows the assembly to act like a clamp. The inboard hydraulic piston forces the inboard friction pad against the rotor, moving the floating caliper inboard and in doing so clamps the outboard friction pad into the rotor. The result is equal force applied on either side of the rotor.

flooded cell battery a *lead-acid battery* that uses liquid electrolyte.

flow continuous movement of a *fluid* within or outside of a circuit, but the term also can be used to describe the movement of electrons in an electrical circuit.

flow control valve used in hydraulic circuits to regulate the speed of actuators. Functions by defining a flow area and is rated by operating pressure and capacity. It can be used in meter-in or meter-out applications. Two general types are used: bleed-off and fixed orifice.

fluid a substance that has fluidity; that is, it lacks definite shape and both flows readily and yields to the slightest pressure. Most liquids and gases are fluids.

FMVSS 121 legislation that defines minimum standards for all highway operated air brake equipment.

foot-pounds (ft.-lb) although technically incorrect, this is often used to express torque values in place of *pound-feet.* See *pound-feet.*

foot valve the service brake application valve used in a FMVSS 121–compliant system.

footprint the critical contact interface between a tire and the road surface it contacts. Also used to describe the ground uplink/downlink geographic signal area of a satellite communications system; for instance, a truck can move through multiple communication footprints on a cross-continental trip.

force push or pull effort measured in units of weight. Force represents the ability to do work but this ability does not necessarily result in any work done; for instance, you can push all you like on a tractor/trailer combination but it is unlikely to move the vehicle. The formula for force (F) is calculated by multiplying pressure (P) by the area (A) it acts on. $F = P \times A$.

forge an open or closed furnace used for super heating metals, or the process used to form a metal by heating, then shaping it using presses or hammers.

forward collision warning system (FCWS) a radar or video driver alert system designed to warn the driver of a potential collision event.

foundation brakes the braking components downstream from the actuator that mechanically effect braking on a wheel. In a truck S-cam brake assembly, the foundation brakes consist of a slack adjuster, an S-camshaft, brake shoes, and a drum.

frame angle the angle formed between a horizontal line and one drawn parallel to the frame.

frame machine a mechanism for performing frame repairs; usually consists of a rigid, heavy steel deck to which a chassis is attached by means of chains.

franchise a privately owned dealership with exclusive rights for marketing and servicing an OEM product in a given area.

franchised dealership a repair shop with a franchise contract with usually one OEM. The contract usually guarantees some exclusivity over truck and parts sales over a defined geographic area.

free pedal the amount of clutch pedal free travel measured in pedal travel distance in the cab. Free pedal is always directly related to free travel measured at the release fork. Typical specs for clutch *free pedal* would be 1⅛ to 2 inches (28 to 50 mm). This should produce around ⅛ inch (3 mm) of free travel measured at the release forks.

FreedomLine a ZF Meritor automated transmission that uses a two-pedal system; currently available in 12- to 16-speed ratios.

freestroke the amount of travel of a slack adjuster arm at the clevis pin from its unapplied position to the point the shoes make contact with the drum. It can be manually checked with a prybar and should be between ⅜ and ⅝ inch (9.5 and 16 mm).

freight efficiency term that is replacing *fuel economy* by many in trucking because it factors the volume and weight of cargo hauled to fuel consumed, rather than the miles-per-gallon equation of vehicle fuel economy.

frequency number of repetitions of a process or series of events within a specific time frame.

fresh air exchange a reefer heat-exchanger system that replaces the energy-wasteful vent doors used on reefer vans.

fretting a bearing or journal failure characteristic in which the contact surface picks up and then wears on the original manufacturing machining pattern. Also used to describe the rubbing wear that results when two clamped components expand and contract during heating and cooling at different rates.

friction discs the driven discs in a clutch pack. Most heavy-duty truck chassis use double friction discs. These would be clamped between the clutch pressure plate/intermediate plate/flywheel when the clutch was engaged.

fuel-fired heater a supplementary coolant heater used to maintain coolant at a temperature adequate to provide the truck cab and bunk with sufficient heat during periods of engine shutdown. Most are diesel-fueled; some older versions were gasoline-fired.

full cap retread tire in which the shoulder area along with the tread area is applied with new rubber; has the appearance of a new tire.

full floating a component that performs independently of the subsystem that contains it; for instance, a full-floating axle shaft simply imparts drive torque to a wheel assembly and does not support vehicle weight in any way.

full-trailer a trailer that is fully self-supporting and towed by a pintle hook and drawbar.

full-wave rectification occurs when a diode bridge is used in an alternator to convert alternating current (AC) to direct current (DC).

fully oscillating fifth wheel a fifth wheel assembly that allows oscillation on any plane such as a ball joint would.

fuse an electrical circuit protection device designed to overheat and permanently fail when subjected to a current overload. Fuses are rated to the maximum amperage they can sustain without failing.

fuzzy logic ECM or computer processing outcomes that depend on multiple inputs, operating conditions, and operating system commands; processing using soft algorithms. Opposite to closed loop processing.

galling a metal-to-metal wear failure characteristic caused by lack of lubricant or extreme overload.

galvanic describes a device that produces a charge differential by chemical action. A galvanic device requires two electrodes—one positive and one negative—plus some electrolyte to interact between the electrodes. The lead-acid battery is a galvanic device, as are some types of O_2 sensors.

galvanic corrosion a type of corrosion caused by the contact between dissimilar metals such as aluminum and steel.

gateway see *gateway module*.

gateway module a protocol transducer (translator); either a stand-alone module or integrated into another module. Can be known as a *transducer module*. Permits communications between modules or buses that use different communication protocols.

gear any of a number of different types of toothed or spiraled shafts designed to either impart or receive torque from another.

gear pitch the number of teeth per given unit of pitch diameter in a gear.

gear select motor the motor used in automated standard transmissions to move the shift finger and make gear ratio selections.

gear set two or more gears that intermesh.

gel cell term used to describe a gelled electrolyte battery. A type of lead-acid battery that uses a thixotropic (a solid or gel, that liquefies under certain conditions) electrolyte, which liquefies when stirred or shaken but returns to the gelled state when left at rest.

geofence a virtual perimeter drawn around a geographic area that defines a boundary, frequently used in telematics. In a truck application, a geofence can be dynamic such as a radius around a moving vehicle, or static, to define a zone.

geofencing term used by state and provincial safety enforcement agencies when telematics are used to perform "wireless" safety and HOS inspections within a jurisdiction.

geometric centerline in alignment and frame technology, the exact midpoint of the front axle and that of the rear axle(s).

geometry use in technical service literature to mean shape.

geotriggered inspection a WRI alert that can be telematically broadcast to a truck when it crosses a jurisdictional geofence.

gladhands pneumatic couplers used to connect tractor air and brake systems with that on trailers. When coupled, a pair of gladhands has the appearance of a handshake.

Glidecote blue-colored, wear resistant coating applied to Spicer drive shaft slip splines.

global PGN a *parameter group number* message broadcast to all the controllers (SAs) on a J1939 data bus.

global positioning satellite (GPS) the telecommunications satellites used for trilaterated positioning of vehicles for purposes of navigation.

Global Warming Potential (GWP) a rating system for noxious gases. For truck and reefer applications, it rates refrigerants according to their environmental damage potential. For instance, CO_2 has a GWP of 1 and R-12 a GWP rating of 8,500.

governor air pressure sensing and signaling device used to manage truck air brake system compressor effective cycles. May be stand alone, mounted, to a compressor, or integrated into an air *dryer reservoir module (DRM)*.

gradation a calibration index or line used when a single line is referred to; for instance, a single line on a micrometer scale. See *graduations*.

graduations calibration indices or lines usually used when multiple lines are referenced. Also refers to the incremental steps of a measuring instrument display scale.

graphical user interface (GUI) the menu-driven display screen used in Windows-driven software.

grease nipple see *zerk fitting.*

grease zerk see *zerk fitting.*

greenhouse gas (GHG) as far as vehicle technology is concerned, GHGs primarily refer to the CO_2 produced by combusting HC fuels. GHGs are said to play a significant role in the global warming by insulating the stratosphere.

groove the recess between two lands or raised portions in profile; for example, the tread *groove* in a tire.

gross vehicle weight (GVW) the total weight of a fully equipped vehicle and its load. GVW defines much of the specified equipment on a commercial vehicle, including engine, transmission, brakes, axles, suspension, frame, etc.

ground commonly refers to anything on the battery negative side of a load on a vehicle chassis.The term *earth* is used in English speaking countries outside of North America. See *ground circuit.*

ground circuit the negative or noninsulated side of a chassis, negative ground electrical circuit. When testing voltage drop on the ground circuit of a cranking system, total voltage drop should not exceed 0.2 volt in most 12-volt systems.

ground strap used by technicians when working on electronic circuits; grounds electrostatic charge accumulation to prevent it from being discharged into sensitive electronic circuits.

hacksaw a bowed frame saw with a replaceable blade used for cutting metals.

half shaft see *drive shaft* or *axle shaft.*

half-wave rectification occurs when only half of an alternating current (AC) wave is rectified to direct current (DC).

Hall effect see *Hall effect principle.*

Hall effect principle used to signal shaft speed and linear position using a rotating or linear shutter that alternately blocks/exposes a magnetic field from a semiconductor sensor, producing a digital signal. Considered more accurate and less failure prone than *inductive pulse generators* when shaft speeds are higher.

Hall effect probe a DMM accessory used to reckon either AC or DC amperage by transducing 1 amp to 1 mA so it can be read by typical handheld DMMs.

halogen infrared (HIR) headlamp technology that uses a multilayer, thin film bulb that allows light to pass through the bulb glass while reflecting some infrared energy back to a tungsten filament. This redirecting of heat energy back to the tungsten filament increases bulb life and intensity.

halogen lamp see *halogen light.*

halogen light common headlamp constructed with a small quartz/glass envelope within which is a filament surrounded by halogen gas. The bulb unit is contained within a metal reflector and lens assembly.

hand tap a male thread cutting tool that cuts, finishes, or cleans up female threads in bores.

handheld (HH) any of a number of different handheld electronic devices such as battery analyzers, labscopes, and electronic service tools (ESTs).

handheld electronic service tool (HH-EST) refers to a range of electronic system read and diagnostic tools. A common example would be a ProLink iQ.

handshake often used to mean any type of electronic connection, but the term also refers to how electronic data is transmitted between two or more processors. Most often a *handshake* connection describes a physical rather than a wireless connection.

hand-threading die a female thread cutting tool that cuts, finishes, or cleans up threads on screws, bolts, or shafts.

hanger bearing the center support bearing used to suspend the first section of the drive shaft in a double drive shaft assembly; used on truck drivelines when the distance between the transmission and drive carrier exceeds 70 inches (178 mm) or so. Consists of a sealed rolling element bearing that is supported by a rubber insulator and mounted in a bracket assembly. Also known as *center support bearing.*

hard-wire term used to describe switching by using analog voltage transmitted through a dedicated wire to a device.

harness the clustering of vehicle wiring circuits, enclosing them in a loom.

hazardous materials general term used to cover fluids and materials that pose a threat to human or animal health.

head the vertical height of a column of liquid above a given point expressed in linear units (feet). Head can be used to express gauge pressure. Head pressure is equal to the height of a liquid in a column, multiplied by the density of the liquid.

head pressure pressure that is equal to the height of a liquid in a column, multiplied by the density of the liquid. See *head.*

heads-up display (HUD) a technology that superimposes data within the driver's field of vision onto the windshield. Not used a lot in trucks.

headway alert system (HAS) a driver alert system used to signal driver if a safe following distance is not being observed; monitored by radar or video.

heater control valve manages the flow of coolant from the engine cooling circuit to the cab HVAC system. Also known as a *coolant flow valve.*

heavy-duty on-board diagnostics (HD-OBD) legally-required, open access to bus DTCs for all emissions-related problems. Initiated in 2010 (each OEM was required to produce one chassis line that was HD-OBD enabled) and scheduled for universal implementation in 2013, but this has been delayed indefinitely.

heavy-duty truck usually denotes a truck with a GVW of 26,001 lb (11,794 kg) or greater.

height control valve a simple lever-actuated valve used on an air suspension system to automatically maintain chassis ride height at a preset dimension; provides the suspension with its adaptive capability. Also known as a *leveling valve.*

helical something that spirals or is scroll shaped. For instance, the threads on a screw are helical.

hermetic completely air tight, sealed. Often used to refer to a component or system that is both gas and pressure sealed.

hermetic scroll compressor a compact, energy efficient compressor used in A/C and reefer systems.

hertz unit of wave frequency equivalent to one cycle per second.

hex term used to describe a six-point fastener head or a socket or wrench used to hold or apply torque to the fastener: from *hexagonal.* Any bolt head with six flats is a hex head. A hex head bolt can be gripped on six angular planes by a hex wrench and twelve angular planes by a *double-hex* wrench.

hexadecimal numerical notation system that uses 16 as a base rather than 10.

hexadecimal code J1939 messages use hexadecimal code. To do this 16 numeric symbols must be used: 0 to 9 are used for the first nine digits and the letters A through F are used to represent 10 to 16.

hexagonal a six-sided object.

high hitch a potentially dangerous tractor-to-trailer coupling condition in which the fifth wheel locking mechanism engages but the trailer kingpin sits on top of the jaws.

high-intensity discharge (HID) xenon headlamp technology that produces light using an arc between electrodes through a gas (xenon) consisting of an arc source, electrodes, and a ballast module.

hinged pawl switch see *toggle switch.*

historic codes vehicle electronic fault codes that have been logged to memory but are no longer active. Historic codes usually can be erased from the code files but are usually detained in the audit trails. Also known as *inactive codes.*

hoist any of a number of different types of lifting devices, including A-frames, chain falls, cherry pickers, scissor jacks, and cranes.

hold-off chamber the rear chamber in a spring brake assembly that compresses the park spring when charged with air. Consists of a chamber, diaphragm, and powerful coil spring.

hold-off circuit the air brake circuit used to charge air to the spring brake chambers to compress the park spring. The hold-off circuit must be charged with air before a vehicle with spring brakes can move.

hold-off diaphragm the parking brake pancake in an air brake spring brake chamber; must be actuated to release the parking brakes.

horizontal parallel or level with the horizon.

horn term used to describe the vertical hook that engages to the drawbar eye in a pintle hook assembly.

hotel load describes trucks/trailers fitted with an alternate mains electrical supply capability when parked and connected to the grid. Hotel load feed is used to run air conditioning, heating, lighting, and entertainments systems. Also referred to as *shorepower.*

hotel load grid many truck stops and DOT inspection stations along Interstates in the United States and Canada are equipped with hotel load electrical supply. The hotel load grid can support on-vehicle AC-electrical equipment when parked and plugged in, saving fuel and the environment. Shorepower can be at 110, 230, and 460 V-AC depending on the vehicle requirements.

HotSync data transfer software from a handheld device to PC.

hours of service (HOS) legal logging of driver on-duty hours; complicated to interpret so reference CVSA HOS 2011 final ruling for a precise definition.

H-Series Certification ASE certification for transit bus technicians.

hub the wheel assembly that houses the wheel bearing assembly and runs on the axle shaft spindle race(s).

hubdometer a wheel hub located odometer used to track and display mileage; especially useful on trailers to schedule service intervals.

Huck fastener a commonly used production line fastener used in place of a nut and bolt that ensures more accurate clamping pressure. Uses air-over-hydraulic pressure to establish stud clamping force before locking with a crimp-type fastener. Often robotically installed on an assembly line. Less commonly used in service shops because of the bulk and awkwardness of the installation tooling, so Huck fasteners are often replaced with nuts and bolts.

hybrid drive motive drive torque produced by two (usual) or more drive sources. In truck and urban bus applications, hybrid drive systems use a small diesel engine to power electric or hydraulic motors.

hybrid electric drive refers to a powertrain that can option drive torque from either a diesel/NG fueled engine, or electric motor(s) driven by battery/capacitor banks.

hybrid electric vehicle (HEV) term used to refer to most parallel hybrid drive electric vehicles in which the power source is a diesel/NG engine that drives a generator/motor supplying electric motive drive and battery/capacitor banks.

hybrid hydraulic drive refers to a powertrain that can option drive torque from either a diesel/NG fueled engine, or hydraulic motor(s) driven by accumulators or engine-driven pumps. Should not be confused with *hydrostatic drive.*

hydraulic brakes a brake system using hydraulic fluid in a closed circuit and a foot pedal actuator (often power assisted) to apply braking effort to the foundation brakes.

hydraulic launch assist (HLA) powertrain in which the primary power source is a diesel or NG engine, but integrated with a transfer case which can generate and receive accumulated hydraulic pressure. The energy storage device in the HLA powertrain is a nitrogen gas-loaded accumulator

hydraulic power booster in a hydraulic brake system, power-assists the master cylinder to reduce the pedal effort and travel.

hydraulics the science of using fluid power circuits that use liquids, usually specially formulated oils, in confined circuits to transmit power or multiply human or machine effort.

hydrochlorofluorocarbons (HCFCs) current category of refrigerants used in truck and reefer systems today. However, HCFCs are ozone depleting so their use is being phased out in Europe beginning in 2011. North America is scheduled to follow but no schedule was in place at the time of writing.

hydrodynamics the science of forces acting on, or exerted by, liquids, usually, but not always, in confined circuits.

hydrometer an instrument used to measure the specific gravity of a liquid, usually coolant or battery electrolyte.

hydrostatic drive drivetrain in which the main engine drives a hydraulic pump(s) which in turn powers hydraulic motors. Used in off-highway trucks and equipment.

hydrostatics the study of fluid behavior and pressure exerted by liquids at rest.

hygroscopic capable of absorbing moisture; a characteristic of some brake fluids. This is an advantage for hydraulic brake fluid because it inhibits water droplets that can cause corrosion, freeze, or boil.

hypoid gear set a gear arrangement in which the axis of the drive pinion gear is located below the radial centerline of the crown gear it drives. Hypoid gearing is often used in truck differential carrier gearing because its spiral bevel gear geometry allows the absorption of driving torque to be spread over a greater number of teeth, meaning that it is both tougher and quieter.

impact gun term used to describe the air-driven, percussive action tools commonly used in shops; usually available in ½-, ¾-, and 1-inch (1.3, 2, and 2.5 cm) drive sizes and used in conjunction with *impact sockets.*

impact sockets heavier wall, softer steel sockets designed for use with impact wrenches.

impact wrench another term for *impact gun*; any of various types of hammer wrenches usually pneumatically driven although electric versions exist. Impact wrenches drive *impact sockets.* The common size impact wrench used by truck technicians is a ½-inch (1.3 cm) drive lug.

impedance the combined opposition to current flow represented by circuit resistance and reactance.

impeller the driving member of a torque converter torus; the impeller is mechanically connected to the engine crankshaft and should rotate at the same speed.

inactive codes vehicle electronic fault codes that have been logged to memory but are no longer active. Inactive codes can usually be erased from the code files but are retained in the audit trails. Also known as *historic codes.*

inclinometer used to make angular measurements on driveline components and drivelines; may use a liquid level, laser, or electronic operating principle.

included angle in alignment technology, refers to the sum of camber and KPI. In general, refers to the sum of angles in a geometric condition.

indem axle a term sometimes used in place of *tridem* in reference to a triaxle trailer.

individual toe angle formed by the running plane of one wheel versus that of the vehicle centerline.

inductive amp clamp a DMM accessory used to reckon either AC or DC amperage by transducing 1 amp to 1 mA so it can be read by typical handheld DMMs. Also known as a *Hall effect probe.*

inductive pulse generator a shaft speed or position sensor that produces analog AC voltage signal by using a rotating reluctor wheel (aka chopper wheel, pulse wheel) that cuts through a permanent magnetic field; the computer that receives the signal measures the AC frequency and correlates it to rpm. The principle used in wheel speed, transmission tailshaft speed, crankshaft speed, and many other types of shaft speed sensors.

Industrial Fastener Institute (IFI) regulatory body for setting fastener standards; manufacturers of fasteners are required to be registered with IFI.

inertia brake on an automated standard transmission, this solenoid actuated brake helps stop the countershafts on initial engagements when the clutch is either out of adjustment or defective.

inflammable capable of being *inflamed* or burned.

infotainment system term used to describe the driver interface monitor that displays vehicle running data, live video imaging (camera), GPS, satellite, Bluetooth, audio, and data bus readout. In trucks at this moment in time, the emphasis is on information rather than entertainment.

inline fuse a series circuit fuse often used to protect a system subcircuit.

in-phase usually used to mean properly timed or properly balanced. A driveshaft that is *in-phase* would be assembled balanced and with the yokes aligned.

input circuit components and wiring that feed data into a computer, sometimes known as a sensor circuit. Input circuits are composed of command switches and sensors (a TPS, for example) and monitoring switches and sensors (a road speed sensor, for example).

input retarder a driveline brake located between the torque converter housing and the main transmission housing. Input retarders use a paddle wheel design with a vaned rotor mounted between stator vanes in the retarder housing; retardation occurs when oil is directed into the retarder housing.

installation template a cardboard, plastic, or metal plate used to guide the assembly or setup of a component.

instrument cluster unit (ICU) the dash display module usually networked with the chassis data bus.

insulated circuit the positive side of a chassis, negative ground electrical circuit. When testing voltage drop on the insulated circuit of a cranking system, total voltage drop should not exceed 0.5 volt in most 12-volt systems.

insulator a material that resists the flow of electrons when subjected to voltage. Glass and rubber are good insulators. An element with more than four electrons in its valence shell.

integrated circuit a solid-state circuit infused on semiconductor material.

Intelli-check™ a Delco Remy diagnostic instrument designed to rapidly test charging circuit in heavy-duty trucks.

interaxle differential differential gearing used between the drive axles on a tandem drive, highway tractor; divides torque between the two drive axles. Also known as a *power divider.*

intelligent transportation system (ITS) refers to the use of telematics to provide wireless roadside inspections (WRI) using EOBRs and subscription services such as Drivewyze to speed up inspections and reduce weigh scale stops.

interaxle differential lock driver-actuated means of locking out interaxle differential gearing action to temporarily gain better traction.

interleaf friction the dampening principle behind a multileaf spring pack. Because individual leaves are clamped to each other, they have to first overcome friction between the leaves before deflection movement can take place.

intermediary something that fits between two other components or systems; for instance, a U-joint acts as an *intermediary* between a pair of drive shaft yokes, and a transducer module acts as *an intermediary* between two computer systems that use different operating languages.

intermediate plate pins see *separator pins.*

internal balancing agent (IBA) term used to generically refer to in-tire balance beads that can correct axle-end balance conditions. See *counteract balancing beads (CBBs)*

internal ground a chassis electrical component that is supplied by a positive insulated wire/cable and grounded by a dedicated ground wire/cable as opposed to being simply clamped to chassis ground.

internal magnetic solenoid switch (IMSS) a magnetic switch integrated into a cranking motor. Uses a one wire design. More corrosion resistant than external type.

Internet a global network of networks; a communications system.

intersect the crossing point of a pair of lines.

inversion valve a normally open air brake valve used for emergency applications of tractor brakes for a one-time stop during a total system failure. Hold-off air is dumped in relation to application pressure so that spring chamber force can effect emergency service braking. Incorporated in some modulating valves.

ion an atom with either an excess or a deficiency of electrons in the valence; a negatively or positively charged particle.

ISO 9141 specifies the international requirements for data interchange between control modules networked to a chassis data bus.

J1321 the SAE truck fuel consumption test procedure that is being increasingly referenced to validate technologies said to improve fuel economy. The test attempts to minimize the variability that occurs when comparing fuel economy data.

J1587 the software multiplexing protocols used by first-generation truck data bus backbones; retained as auxiliary bus in the J1939 data bus up to 2010.

J1708 the hardware standards required to drive a J1587 data bus; built into most OEM J1939 data buses until 2010.

J1850 hardware and software multiplexing protocols required for CAN 2.0 compliant automotive and light-duty truck data buses up until 2010. Replaced by CAN-C.

J1962 the standard that covers the standard automobile 16-pin data connector required by OBD-II. Used by Volvo Mack to connect to their post-2014 J1939 data bus. Also known as an *assembly line data link (ALDL).*

J1939 hardware and software multiplexing data bus protocols required for medium- and heavy-duty truck, and most off-road equipement; built on CAN 2.0 architecture using a serial, twisted wire pair.

J1939 data bus the physical neural network of a CAN 2.0 network bus using J1939 protocols consisting of a twisted two-wire pair, SA stubs, and two terminating resistors. Used as the primary powertrain bus even in systems using multiple auxiliary buses. Theoretically capable of speeds of up to 1 Mb/s but usually maxed out at 500 Kb/s in current applications.

jack a hydraulic or air-over-hydraulic vertical lifting device.

jackknife caused when the rear wheels of a tractor on a tractor/semitrailer rig lock. The tractor begins to yaw and the forward momentum of the trailer amplifies the swing, making it difficult to steer out of the condition.

Joint Industry Committee (JIC) organization that sets compatibility standards for hydraulic circuits.

jounce the most compressed condition of a spring. Literally *bounce.*

jounce blocks solid rubber blocks used to prevent suspension slam when the vehicle springs are fully jounced. Jounce blocks can be located on the frame or internally within air bag pedestals.

jounce travel travel distance between a suspension in neutral condition and when springs are maximally compressed.

journal a bearing surface area on a shaft or trunnion cross.

journal cross the core component of a U-joint consisting of four equally spaced trunnions on the same plane.

jumper cables large gauge cables used to *jump start* vehicles.

jumper wires means used to bypass a portion of an electrical circuit using a length of wire and alligator clips at either end.

jump start the practice of using a second vehicle or a generator to start a vehicle with dead batteries.

jumpout a condition caused when a sliding clutch is forced out of engagement.

just-in-time (JIT) delivery assembly plant procedure that uses trucking logistics to time the delivery of subcomponents so that they can be phased into a production line without having to be stored as inventory.

kilopascal (kPa) metric unit of pressure used mostly in engineering in North America. One hundred kPa is equivalent to 14.56 psi.

kinetic balance balance status of any moving component. In wheel technology, the balance condition of a wheel related to force generated through a vertical plane.

kinetic energy the energy of motion.

kingpin (1) the linkage pin used in a steering knuckle that permits the assembly to pivot, or the tempered steel, coupling pin mounted on a trailer upper coupler that engages with fifth wheel jaws; (2) the lugged pin built into a trailer upper coupler assembly that engages to jaws in a fifth wheel assembly, permitting articulation at the fifth wheel. Forged from middle alloy steels and tempered. In North America, a kingpin is usually welded into the upper coupler; elsewhere, it is more often bolted.

kingpin inclination (KPI) the inclination of the steering axis to a vertical plane that is the equivalent of steering axis inclination (SAI) in automotive front ends.

Kirchhoff's Law(s) two laws referring to electrical circuit behavior although mostly the second law is referenced. First law: The current flowing into a circuit (or component) must equal the current flowing out. Second law: Voltage drops in proportion to the resistance in a circuit and the sum of the voltage drops in a circuit must equal the voltage applied to it.

kPa kilopascal. Metric unit of pressure measurement. 101.3 kPa = 14.7 psi = 1 atm. One unit of *bar* is equal to 100 kPa.

Kwik-Adjust used in some Eaton adjustable clutches in place of a lockstrap to hold clutch adjusting ring position. The Kwik-Adjust mechanism is turned using a wrench.

Kwik-Konnect a two-piece clutch brake that can be installed onto a transmission input (clutch) shaft without removing the transmission from the chassis.

labscope a modern handheld variation of an oscilloscope; an electronic instrument in which the relationship between two variables can be graphically represented. In motive power technology, the two variables are usually time (frequency) and voltage.

ladder bridge a conjoined ladder of resistors (usually five) integrated into a smart or *ladder switch.*

ladder switch a type of "smart" switch so named because it contains a ladder of resistors, usually five per switch, known as a ladder bridge. The processor that receives data from the ladder switches on the data bus has a library of resistor values that are used to interpret switch status and commands.

laminar flow streamlined flow through a hydraulic circuit. A hydraulic circuit should be designed to minimize friction caused by sharp angles and internal obstructions.

land raised surface beside a groove such as is found on a spool valve.

landing gear the retractable legs used to support a semitrailer when not coupled to a tractor; most have gear-actuated, retractable legs but hydraulic and lock-pin systems are also available.

landing gear assembly the retractable legs used to support a semitrailer when it is not coupled to a tractor. Also known as *dolly legs.*

lane departure warning (LDW) system a video (camera) lane monitoring technology used as part of an accident avoidance electronics. When the camera senses lane departure, an audio and visual warning is broadcast.

laser alignment instrument or test bed that uses laser light beams to align truck and trailer frames, suspension, and wheel assemblies.

latent heat the amount of heat energy required to change a substance from one state to another: a key principle of refrigeration physics. See also *latent heat of condensation* and *latent heat of vaporization.*

latent heat of condensation the amount of heat surrendered (released) by a substance when it changes state from a gas to a liquid; one of the key principles of refrigeration systems.

latent heat of vaporization the amount of heat consumed by a substance when it changes state from a liquid to a gas (boils); one of the key principles of refrigeration systems.

lateral side-to-side or transverse movement.

lateral runout axial wobble of a rotor, wheel, or other rotating component.

launch phase term used to describe the rpm between *touch point* and full clamping in a two-pedal, electronically controlled clutch module.

lead tendency of a moving vehicle to want to track away from the steered (intended) course.

lead trailer the trailer coupled to the tractor in a double or multiple trailer combination. Usually used in reference to an A-, B-, C-, or D-train combination.

lead-acid battery the standard vehicle battery consisting of lead-acid cells arranged in series and a sulfuric acid electrolyte solution. Also known as a *flooded cell* or *wet cell* battery. Most trucks use more than one battery.

leaf spring any of a number of types of steel springs consisting of single or clamped multiple plates used in truck and trailer suspensions.

less-than-truckload (LTL) trucking operation using logistics to collate smaller cargoes in terminals and then bulk shipping first to another major terminal before breaking up the load up for customer delivery. Works a little like a courier service.

leveling valve a simple lever-actuated valve used on an air suspension system to automatically maintain chassis ride height at a preset dimension; provides the suspension with its adaptive capability. Also known as a *height control valve.*

lift axle a dead axle that can be raised and lowered using air pressure.

light beam alignment suspension/chassis alignment instruments that use projected light beams and templates.

light-emitting diode (LED) diode that converts electrical current directly into light (photons) and, therefore, is highly efficient because there are no heat losses. Used increasingly in vehicle lighting systems because of low-current draw and excellent longevity.

lightly loaded vehicle weight (LLVW) describes a vehicle in its unloaded or light loaded state. LLVW is used for certification purposes to ensure that components such as the suspension and brakes meet legal performance criteria when unloaded as well as loaded.

linear actuator device that converts hydraulic energy into linear motion such as COE cab lift cylinders or dump box rams.

linehaul truck transports that operate terminal to terminal. Refers to a truck that operates mainly on major highways.

linehaul miles used by truck chassis and engine miles to reckon longevity. To convert linehaul miles to engine hours, divide by 50.

liner the stationary external pump member used in gear and vane pumps also known as a *cam ring.*

linkage sets of rods, levers, pivots, and bell-cranks used to effect mechanical movement. Examples would be clutch *linkage*, accelerator *linkage*, and COE gearshift *linkage.*

lip-type seal a shaft to hub, dynamic seal that has the seal medium, usually leather or synthetic rubbers, spring-loaded to ride on a wiper ring or wiper race, which is located on the shaft. Commonly used as wet wheel seals.

Li-On battery a type of *lithium-based battery.*

Li-Po battery a type of *lithium-based battery.*

lithium-based batteries high electrochemical potential batteries capable of producing per-cell potentials ranging between 2 and 5 volts. Currently under fast-track development and field testing, their storage and power density make them ideal HEV batteries.

lithium-ion (Li-On) a type of *lithium-based battery* with high-energy density used in transit and inner city HEVs.

lithium-ion polymer (Li-Po) a variant of the lithium battery becoming popular in commercial HEVs.

live axle used to describe an axle that contains drive gearing; functionally opposite to a *dead axle.*

load dump a vehicle electrical circuit condition that occurs when a major electrical current consumer is opened, causing a high-voltage spike.

load proportioning valve (LPV) used on trucks equipped with hydraulic brake systems to sense load on rear axle(s) and proportion brake application pressures accordingly.

load range in tire terminology, the weight limitations of the tire.

load test (battery) a common battery capacity test performed by connecting a carbon pile load tester across the battery or battery bank terminals. Also known as a *capacity test.*

load transfer dynamic pitch the suspension dive angle of a chassis when subjected to aggressive braking. Measured with a decelerometer, it is one indicator of brake torque balance.

local interconnect network (LIN) a lower cost, low-speed bus used to connect a nonmission critical node (proprietary bus) to a system controller (which may itself be networked to J1939).

locking pliers better known as Vise-Grips™.

locking tang a terminal lock used in many types of sealed electronic connector assemblies.

lockstrap the removable locking tang that holds the clutch ring in position; it has to be removed and the clutch ring rotated to make a clutch adjustment.

lockup torque converter a means of locking the impeller and turbine so that they rotate together and eliminate the 10 percent slippage that occurs during torque converter coupling phase.

logic power Eaton term used to refer to the electric feed required to power-up its automated transmission ECUs.

logistics the organization of tracking, moving, storing, and delivering cargo in a trucking operation.
lorry a British and Australian/NZ term for any truck. When the term is used here, it refers to a flatbed truck or railway car.
low-maintenance batteries conventional lead-acid batteries that require a lower level of maintenance.
lug any type of ear-shaped clamp.
lumen light energy emitted through a straight vector angle from one candela per second.
luminosity brightness. Means of rating the brightness of different types of light unit in lumens.
Machinery's Handbook a machinist and technician data handbook containing thousands of formulae and shop reference data. Published by Industrial Press, New York.
machinist's rule a steel ruler used for shop linear measurements and may act as a straightedge.
mag switch slang term for *magnetic switch.*
MagiKey one type of generic *communication adapter (CA)* or serial transducer using TMC RP 1210 protocols that enables the serial link between ESTs with a truck chassis data bus.
magnesium chloride a naturally occurring salt that can be added to sodium chloride brine to lower its freeze point. Its use is controversial because of its capacity to leach into surface water and chlorinate it to the extent that it can inhibit plant growth.
magnetic base indicator a dial indicator on a magnetic base assembly, commonly used in truck shops.
magnetic switch the starter circuit relay that enables the low-current starter control circuit to switch the high-current cranking circuit.
magnetomotive force (mmf) the magnetizing force created by flowing electrical current through a coil.
main memory in a computer system, random access memory (RAM). Main memory is electronically retained and is a major factor (along with clock speed) in determining computing potential.
main pressure the system pressure used to manage the hydraulic circuits in an automatic transmission.
main pressure regulator means of defining main pressure in Allison automatic transmissions. A spool-type valve held in the upward position by spring force: Main pressure acts on the top of the main pressure regulator spool and forces it downward against spring pressure and when this pressure is high enough to drive the spool downward enough, main pressure is spilled, reducing pressure. Used to regulate system pressure in an automatic transmission.
main shaft term used to describe the transmission shaft through which output torque is primarily routed. Term is used in both standard and automatic transmissions but the role played by the main shaft in each differs.
maintenance logs either hard copy or electronic records used to record maintenance events. Mandatory under CSA.
maintenance manual service publication that outlines maintenance procedures and service intervals. Often has to be accessed online or using CD-ROM format.
manifold gauge set the shop station used to recycle, recharge, and diagnose A/C systems.
manometer a single tube bent into a U-shape and mounted to a linear measuring scale. It may be filled to the zero on the calibration scale with either water (H_2O) or mercury (Hg), depending on the pressure range to be measured.
mapping in terms of a vehicle computer processing cycle, the sets of rules and conditions used to produce an output action. Most truck OEMs tend to use the term *algorithm* in preference to mapping.
master cylinder in a hydraulic brake system, the dual circuit application control valve.
material safety data sheet (MSDS) identity label required of all potentially hazardous materials used in the workplace by Right-to-Know Law. WMIS training must be provided to all employees to teach them how to interpret MSDS.
mechanic's gloves artificial fiber tight fitting gloves that provide good tactile sense and a measure of protection when working with light duty components.
memory banks means of optically, and magnetically storing memory within a computer system. Usually does not refer to RAM, typical are ROM, and EEPROM.
Meritor tire inflation system (MTIS) a mobile tire inflation system that monitors tire pressure and uses chassis system air pressure to inflate tires while mobile.
message identifier (MID) identifies an on-vehicle circuit by numeric code on the J1587 data bus. Each MID with an address on the J1587 data bus has at least some data processing capacity. Although the acronym is specific to J1587, some OEMs continue to use it in their current J1939 training.
meter resolution a specification of how fine a measurement can be made by a DMM; digits and counts are used to define meter resolution.
metering the precise measuring of fluid flow.
metering valve used on vehicles equipped with hydraulic front disc and rear drum brakes to improve brake balance; holds off application pressure to front brakes until pressure builds in the hydraulic circuit to move the rear shoes into contact with the brake drums.
Metri-Pack connectors a type of weather-sealed electronic connector used extensively in truck electronic circuits.
micrometer a precise measuring instrument typically consisting of a thimble, barrel, and anvil; used to make id, od, and thickness measurements.
mike the act of making a measurement using a micrometer. Also, the commonly used slang term for *micrometer.*
millimeter metric unit of measurement equivalent to 0.039 inch (1 mm).
minute unit of angular measurement used in alignment instrumentation equivalent to 160 (one-sixtieth) of a degree.
mobile tire inflation system (MTIS) monitors tire pressure and uses chassis system air pressure to inflate tires when mobile.
mobile universal trailer tester (MUTT) a portable device equipped with a trailer electrical supply and air lines used to diagnose trailer electrical and air brake problems.
model year (MY) the EPA official vehicle/engine year that determines what emissions and safety equipment is required on a commercial truck or trailer.
modem telecommunications device that translates digital signals to analog for transmission.
modular describes a major component that is composed of modules. This can make troubleshooting and repair easier because a defective module within a system is identified in troubleshooting and then removed and replaced, eliminating the need to rebuild an entire component. For instance, an Allison WT transmission uses a *modular* construction.
modulate to regulate.

modulating valve a valve that regulates or proportions. In air brake systems, may have *inversion valve* features.
modulator a device that regulates. Term is used to describe the output circuit of an ABS containing the electropneumatic valves.
module term often used to describe the hardware within which a computer system is housed. Also, one of a number of distinct but interrelated units that together make up a component. See also *modular.*
monocoque frame design in which all the shell components form the frame; many automobiles and most van trailers use a monocoque principle. This means that the van floor, roof, bulkhead, doors, and sidewalls collectively compose the frame assembly. Also known as *unibody* design.
MPa megapascal. Metric unit of pressure measurement: 1 million pascals. 1 MPa = 145 psi.
multimodal transportation refers to the transportation of containers that may be moved by truck, railway, ships, and even aircraft.
multiple-disc clutches braking and power transfer devices used in automatic transmissions. They use a series of hollow friction discs to transmit torque or apply braking force.
multiplexing used to describe data bus systems in which two or more controllers communicate with each other; reduces system hard wire and components while optimizing chassis performance.
multiprotocol cartridge (MPC) a second-generation Pro-Link cartridge that connects to the Pro-Link head and uses OEM system specific data cards to access vehicle electronics.
National Automotive Technicians Education Foundation (NATEF) organization that sets standards and certifies secondary and postsecondary auto and truck training programs. Closely allied to ASE.
National Highway Traffic Safety Administration (NHTSA) federal government agency managed by the *Department of Transportation (DOT)* that sets standards and effects safety inspections across the United States. Its Web site is http://www.nhtsa.dot.gov.
National Institute for Automotive Service Excellence (ASE) body that defines certification standards and manages certification testing for the motive power trades, including truck, auto, school bus, transit bus, auto machinists, etc.
natural gas (NG) widely used fuel in inner-city transit vehicles (especially on the West Coast) and becoming more common in truck applications due to low cost. Chemically it is composed primarily of methane.
needle nose pliers a pliers with slim tapered jaws used for electrical work and grasping small objects.
needle rollers the cylindrical roller bearings used in a bearing cup.
negative temperature coefficient (NTC) a term used to describe the electrical characteristic of a conductor in which resistance decreases with temperature rise.
network to communicate, usually in an organized manner. When the term is used in truck electronics, it refers to data bus information transactions on the vehicle and the handshake and wireless means used to connect ESTs to the chassis bus.
neural network term used to describe a complete vehicle multiplex circuit.
neutral safety switch a safety switch located in the control circuit of a cranking system designed to prevent engine cranking when an automatic transmission is in gear.
NEXIQ Brake-Link a user-friendly, generic handheld EST designed to read, diagnose, analyze, and reprogram electronic brake systems.
nickel-metal-hydride (NiMH) see *NiMH battery.*
NiMH battery commonly used battery in commercial HEVs. It uses a potassium hydroxide electrolyte along with a nickel oxide anode and metal hydride cathode.
nitriding a surface-hardening treatment. See *case-hardening.*
nitrogen-filled tires used by some fleets to extend tire life. The tire is inflated with relatively pure nitrogen eliminating the oxygen and moisture that can shorten tire life. Nitrogen-filled tires are also resistant to changes in inflation pressure caused by temperature.
noncycling circuit breaker an SAE type-2 circuit breaker. If specified amperage rating is exceeded, it opens the circuit and remains open until circuit flow is switched off, allowing the armature to cool. May interupt the circuit electronically, in which case the term *virtual* is used.
nonparallel driveline one in which the working angles of the U-joints of the drive shaft are equal but the companion flanges and/or yokes are not parallel. Also known as a *broken-back driveline.*
nose the front of a trailer.
NOT gate any circuit whose outcome is in the *on* or *one* state until the gate switch is in the *on* state at which point the outcome is in the *off* state.
no-tilt fifth wheel a fifth wheel in which the articulation can be locked out, designed to couple to a trailer with an articulating upper coupler.
Occupational Safety and Health Administration (OSHA) federal body that sets standards and monitors employee health and safety in the workplace.
occurrence count (OC) in J1939, the number of incidences of an event such as the logging of a DTC.
offset any plane that differs from another. In alignment technology, the lateral displacement of a wheel or axle relative to a centerline.
ohm unit of measurement of electrical resistance.
Ohm's Law a fundamental law of electrical circuit behavior that states that current, voltage, and resistance work together in a mathematical relationship.
open an electrical term meaning that there is no path for current flow.
open-center a hydraulic circuit in which pump delivery fluid is allowed to recirculate freely to the reservoir in the "center" (neutral) position.
open circuit an electrical circuit that is not energized, that is, no current is flowing. The opposite to a *closed circuit.*
open-end wrench a wrench with parallel jaws in the shape of a Y; contacts a hex nut on two opposing flats only at a given moment.
OR gate a multiple input circuit whose output is in the *on* or *one* state when any of the inputs is in the *on* state.
ORFS fittings O-ring face seal hydraulic coupling fittings that provide a triple sealing faces that withstand vibration better.
original equipment manufacturer (OEM) used to describe a truck chassis manufacturer or manufacturers of major chassis subcomponents such as engines, transmissions, or brakes.
O-ring face seal (ORFS) hydraulic fittings with triple sealing faces. See *ORFS fittings.*

oscillate something that moves or cycles with rhythmic regularity. See *oscillation.* Can be used to refer to back-and-forth movement.
oscillation the act of oscillating. We used this word to describe the natural dynamics of frame rails, the vibrations of drivelines, wheel and tire dynamics, frames, and electrical wave pulses.The wave (nodes and antinodes) frequency of a vibration measured in *hertz.*
out-of-phase term used to describe components that are mechanically out of time or balance with each other.
out-of-round not *concentric; eccentric.*
out-of-service (OOS) to withdraw a truck or trailer from operation because of a safety or running defect. A citation standard, not a maintenance standard. Also known as *deadline.*
output circuit the devices driven by a computer. In a truck computer system this would mean V-Ref, conditioned power-up voltage supply, networking links, and all of the actuator devices.
output devices components controlled by a computer that effect the results of processing. The display screen and printer on a PC system and the modulators on an ABS are all classified as output devices.
output driver the switching apparatus used in most vehicle computer systems used to convert logic processing into actions. Output drivers send electrical signals to actuators.
output retarder hydraulic driveline brakes mounted on the rear a transmission, designed to apply retarding force directly to the driveline.
oval not circular; egg-shaped.
ovality describes something that is not circular; egg-shaped. A measurement of *ovality* is an indication of eccentricity in a supposedly concentric component.
overcrank protection (OCP) protects a cranking motor from thermal damage by inhibiting cranking under adverse conditions. An electromechanical or ECM-managed switch integrated into the starter circuit.
overdrive any drivetrain arrangement in which an output speed exceeds input speed. For instance, most transmissions have overdrive ratio options, meaning that transmission tailshaft speed exceeds engine crankshaft speed.
overinflation in tire technology, describes a tire inflated with a pressure that exceeds specification.
overrunning clutch a drive-protection mechanism that prevents a drive member from being driven; for instance, after a starter motor has started the engine, the overrunning clutch prevents the engine from driving the starter and damaging it.
oversteer an overresponse to steering input that produces vehicle yaw or lateral tracking off the intended turning radius. In other words, a vehicle that has a tendency to turn sharper than the driver intends.
owner-operator (O/O) a term applied to a "for-hire" truck owner and driver. Sometimes known as a *broker* (slang).
oxidation the result of oxygen reacting with a substance such as a metal or oil. Rust is the result of oxidizing iron.
oxidation inhibitors any of a number of ingredients mixed into lubricants that slow high-temperature oxidation of metals.
ozone triatomic oxygen (O_3); an oxygen molecule consisting of three bonded oxygen atoms. The ozone layer in the Earth's stratosphere helps protect the atmosphere from harmful solar ultraviolet radiation.
ozone depletion the chemical reduction of ozone (O_3) to diatomic oxygen (O_2) in the Earth's stratosphere; occurs when a gaseous substance leaches an oxygen atom from the ozone molecule, reducing it to O_2.
ozone depletion potential (ODP) rating of atmospheric gases according to their ability to damage the ozone layer. For instance, CFCs would be rated at a higher ODP than HCFCs.
packer a garbage compactor truck.
packet any data message broadcast on a data bus; may be node-to-node (SA to SA) or occur when a ladder smart switch resistance changes, indicating a change in status.
parabolic in the shape of a parabola; that is, an open plane curve. The cone-shaped headlamp reflector assembly is *parabolic.*
parallel lines or planes that are exactly side by side; a pair of beams that have exact spacing between them.
parallel circuit in electricity, any circuit that has two or more paths for current flow.
parallelism having exact parallelity; for instance, the contact faces on a disc brake rotor would be required to meet a parallelism specification.
parallel-joint driveline in driveline terminology, a drive shaft construction in which all of the companion flanges and yokes are parallel with the working angles of the U-joints set to being equal and opposite.
parallelogram steering linkage steering geometry in which if all pivot points were connected by lines, the lines would be parallel.
parameter group (PG) in J1939, a group of parameters belonging to the same subsystem stream and sharing the same message transmission rate. Each PG has a message length of 8 bytes (minimum and most common) up to 1,785 bytes (maximum).
parameter group number (PGN) the unique numeric value assigned to a PG in J1939.
parameter identifiers (PIDs) primary subsystems and/or components within a MID circuit used in its diagnostic architecture. All OEMs use common PIDs so a 108 PID indicates barometric pressure on all truck chassis.
parity the odd or even quality of a bit (binary). The options are none, odd, or even.
park-on-air an air brake system that uses service chamber application pressure to apply the brakes when in park mode. If the vehicle is FMVSS 121 compliant, should the air pressure bleed down, it does so in both hold-off and service circuits meaning that the spring brakes take over to mechanically park the vehicle.
parts requisition a soft (electronic) or hard (on paper) order for parts within a company or for purchase outside.
Pascal's Law states that pressure applied to a confined liquid is transmitted undiminished in all directions and acts with equal force on all equal areas at right angles to those areas.
performance-based maintenance (PBM) maintenance programming based on analysis and performance of components and systems with the objective of reducing PM frequency.
perpendicular at right angles to.
personal computer (PC) any of a number of different brands of small computers built to drive Windows software; the standard computers used in the business world, homes, and in the truck industry.
personal digital assistant (PDA) term used to describe any of a number of different types of handheld computers

used to read, reprogram, and diagnose vehicle electronic systems. Many PDAs today are telephone based.

personal protective equipment (PPE) equipment and clothing designed for personal protection; includes safety glasses, safety footwear, hearing protection, gloves, and coveralls.

phase angle relative rotational position of each yoke ear on a drive shaft.

Phillips screwdriver a screwdriver tip with four radial prongs in the shape of a cross designed to engage with a similarly shaped recess in a screw.

photo alignment use of wheel-mounted instruments to project light beams onto charts, scales, and sensors to measure toe, caster, and camber. Noted for high resolution and accuracy.

photon a quantum (unit) of light energy.

photonic semiconductor semiconductor that either emits or detects light.

pickup and delivery (P&D) lighter duty trucks used over short distances to pick up and drop off loads to customers; sometimes operated in conjunction with a linehaul or LTL operator.

piggyback assembly a self-contained spring brake and service chamber flange fitted to the existing service chamber. Designed as a replacement repair for a spring brake assembly with a failed spring brake section.

pillow block a rubber insulator or spring used in a bearing clamp or suspension component.

pilot any of a number of devices that receives a signal and then effects an action; can also be known as a relay.

pilot-operated relief valve consists of a main relief valve through which the main pressure line passes and a smaller relief valve that acts as a trigger to control the main relief valve. It is used to handle large volumes of fluid with little pressure differential.

pinhole injection caused by minute high-pressure hydraulic leaks capable of penetrating skin and gloves, often unbeknownst to the injured person; any technician working around hydraulic circuits should be aware of this type of injury and what to if it is suspected.

pinion a smaller, usually stub-shaped gear used to drive a larger gear in an arrangement such as *crown and pinion* gearing in final drive carriers or the starter motor pinion that engages with the flywheel ring gear.

pintle eye the coupling eye on a trailer drawbar that engages with a pintle hook.

pintle hook the standard coupling mechanism used in truck full-trailers. The function of a pintle hook is to engage to the drawbar eye, which itself is bolted to the tongue or drawbar assembly of the trailer.

pitch term meaning *angle* but also used to describe the number of *threads per inch (TPI)* used on a fastener. The pitch of one thread is the distance a nut would advance on a bolt through one full turn.

Pitman arm a steering linkage arm that connects a steering gear to a drag link. A Pitman arm is usually splined to the steering gear sector shaft. In a typical steering gear, the Pitman arm converts the rotary motion of the sector shaft into linear motion at the drag link plus adds leverage.

pitting surface irregularities that usual result from corrosion or abrasive action.

planetary double-reduction axle a drive axle using two reductions: The first reduction takes place at the pinion and crown gearset, and the second reduction is achieved by a planetary gearset.

planetary gearset a gearset consisting of three elements: a sun gear, planetary carrier (containing three or more planetary gears), and a ring gear (annulus). The three separate elements intermesh, and the hold/release status of each determines the output ratios. Also known as an *epicyclic gearset.*

planetary pinion gears the multiple pinion gears, each capable of rotating on their own axes, that are subcomponents of a planetary carrier assembly. The planetary carrier assembly forms the large gear in a planetary (epicyclic) gearset.

plate-type brake testers dynamic brake testers that use precision load cells mounted in flat pads; computer analysis of brake forces, dynamic weight, and chassis brake balance. An example is the Hunter B400T unit.

pliers gripping tool with a pair of jaws mounted on a pivot. There are numerous different types, including lineman, needle nose, and slip-joint.

plow-out results when the tractor steer axle wheels lock, making the tractor difficult to steer. Experienced drivers can usually steer out of plow-out and the incidence has been greatly reduced since the introduction of mandatory ABS.

ply a layer of rubber-coated parallel cords on a tire.

ply rating a method of rating ply strength. It does not necessarily relate to the actual number of plies used in tire construction.

pneumatic in truck technology this term is used mainly to describe chassis air under pressure, as in *pneumatic brakes* or *pneumatic suspension.*

pneumatic brakes air brakes.

pneumatic suspension an air suspension.

pneumatic timing generally refers to the application and release times, combined with synchronization in a truck air brake system.

pneumatics a branch of fluid power physics dealing with pressure and gas dynamics.

polarity the state of electrical charge differential, either positive or negative, of the electrical charge in a circuit or device.

policy adjustment the act of writing off a repair charge usually because a repair or component involved in a repair has failed.

pop-off valve see *safety pop-off valve.*

Posi-Drive™ screwdriver a screwdriver similar in appearance to a Phillips but with a blunter tip that is used a lot in robotic assembly.

positive displacement a characteristic of a pump or motor that has its inlet positively sealed from its outlet so that no fluid can recirculate within the component; that is, what goes into the pump per cycle is what comes out per cycle.

positive filtration term used to describe a filter in which all of the fluid being filtered is forced through the filtration media before it exits.

positive temperature coefficient (PTC) a term used to describe the electrical characteristic of a conductor in which resistance increases with temperature rise. For instance, copper wire has a PTC characteristic.

potentiometer a three-terminal voltage divider. The terminals are identified as supply voltage, signal voltage, and ground. The position of the slider on the device resistive track determines how supply voltage is divided between the signal and ground paths.

pound-foot/feet (lb-ft.) unit of torque measurement: force (lb) by distance (ft.). Often (incorrectly) expressed as *foot-pounds (ft.-lb).*

pounds per square inch (psi) standard unit of pressure used in North America. It is the force acting over a sectional square inch. One psi is equivalent to 6.894 kPa.

pour point the temperature at which diesel fuel begins to solidify to the extent it will not flow properly. Most diesel fuel conditioners contain a pour point depressant that lowers the fuel pour point temperature. It is seldom the cause of a running failure because the engine will have shut down long before the temperature this low is achieved. See *cloud point.*

power the rate of performing work.

power divider interaxle differential gearing used in tandem drive, highway tractors to split torque between the two drive axles. Also known as *interaxle differential.*

power line carrier (PLC) term used to describe communication transactions delivered through the auxiliary (blue) power line in a seven-wire tractor-trailer electrical connection. Signals are converted to radio frequencies for the transaction and subsequently decoded by the receiver ECU. The means used to enable trailer ABS to communicate with the tractor electronics.

power module usually refers to a device that modulates electrical power to an ECM/ECU or electronically controlled circuit. May or may not have processing capability. The term *power distribution module (PDU)* is often used.

power steering on highway vehicles, a system that *assists* manually input, steering effort. In off-highway equipment, may refer to fully hydraulic or drive-by-wire steering.

power steering analyzer a specialty tool for testing steering system hydraulics consisting of hoses, quick-release couplers, a flowmeter, pressure gauge, and flow control valve.

power synchronizer speeds the rotating of main box gearing for smoother automatic downshift and slows main box gearing for smoother automatic upshifts.

power take-off (PTO) any of a number of different means of gear-driving auxiliary apparatus such as hydraulic pumps, compressors, and cement mixer cylinders using engine power. A PTO can be mounted to the engine, transmission (most common), transfer case, or elsewhere on the vehicle drivetrain.

powerflow the mechanical flow path that conducts motive power through a system or component. For instance, the powerflow through a transmission in a specific gear ratio would be the sequential routing of the torque flow path through those gears that are actually engaged.

powershift when used in truck and heavy equipment applications, it denotes the ability to shift under full throttle demand without interrupting driveline torque and incurring spin losses.

powertrain consists of the engine, clutch, transmission, drive shafts, differential carriers, drive axles, and drive wheels on a vehicle.

preadjusted axle a low-maintenance dead axle using synthetic lube and identified by a color-coded (blue) vent cap.

predictive-based maintenance (PBM) maintenance programming based on analysis and performance of components and systems designed to reduce PM. Includes predictions on component life cycles with replacement or maintenance based on average life cycles.

preload a predetermined force applied to a component, such as a bearing on assembly, usually with the objective of preventing unwanted (end) play in operation.

preset hubs hub assemblies with both the bearings and seals already installed. To install on the wheel end, the technician installs the self-piloting hub and torques the assembly in place, eliminating the need for bearing adjustment.

press a usually stationary shop tool capable of applying linear force. Frames may be manually actuated (such as an *arbor press*), hydraulically, or air-over-hydraulically actuated; some shop presses can exert up to 50 tons of linear force.

pressure the force applied over a unit area. Usually expressed in pounds per square inch (psi) or kilopascals (kPa).

pressure differential the difference in pressure values between two points of a circuit or component.

pressure differential valve also known as brake warning light; used in hydraulic brake systems to indicate to the driver when one-half (primary or secondary circuit) of the system is not functioning. Consists of a cylinder inside of which is a spool valve.

pressure drop describes pressure collapse or reduction in a circuit.

pressure protection valve a normally closed, pressure sensitive control valve. Some are externally adjustable. Can be used to delay the charging of a circuit or reservoir or to isolate a circuit or reservoir in the event of a pressure drop-off.

pressure reducing valve used in any part of a circuit that requires a constant set pressure lower than system pressure.

pressure relief hole the grease escape hole in the slip spline assembly of a drive shaft.

pressure relief valve a pressure-limiting device that trips at a predetermined pressure to define circuit pressure or act as a safety device. A *safety pop-off valve* is one type of pressure relief valve.

pressure rise describes a sequence of rising pressures in a circuit or pump element.

preventive maintenance (PM) scheduled maintenance routines designed to preempt on-the-road failures and consisting of a series of inspections and service procedures.

primary circuit in brake systems, refers to the lead circuit in the system; in hydraulic brake systems on straight trucks this is usually the front brake circuit, whereas on large trucks with air brakes, it refers to the brakes over the tandem drive wheels.

primary modulation term used by Allison to describe the amount of ON time of an actuation signal.

prime mover the initial mechanical source of motive power. In the context of a hydraulic circuit, the means used to drive the hydraulic pump would be properly the prime mover, although the term is sometimes used to describe the pump itself.

printed circuit an integral circuit board constructed from thin sheets of plastic onto which conductive material such as copper is laid. The conductive material is then etched by acid to create conductive paths on the board.

priority valve used as a protective device on hydraulic (automatic transmissions) and pneumatic circuits; ensures that the control system upstream from a valve or subcircuit has sufficient pressure to function.

prognostics the science of analyzing component and system life cycles and intervening to repair or replace components before they fail. Developed by the U.S. military, prognostic algorithms are now being applied to maintenance strategy in the trucking field.
programmable logic controller (PLC) microcomputer with multiple input and output paths that easily adapt functionality when updated. Because they may possess high processing density they are commonly used to process video (such as in an LDW system) inputs.
programmable read-only memory (PROM) chip or card retained memory that in a truck chassis computing system is used to qualify ROM data to a specific application; in older systems a PROM chip was often removable from the ECM motherboard because it was the only way in which new/altered data could be programmed to a system.
Pro-Link brand name of a range of generic handheld ESTs used to access vehicle data buses for purposes of reading, effecting customer data programming, and some diagnostics. They vary in capability but they are generally limited to basic levels of readout and troubleshooting.
Pro-Link GRAPHIQ a generic handheld EST capable of reading truck data buses and performing other limited operations.
Pro-Link iQ a generic handheld EST capable of reading truck data buses and performing other operations that vary according to how well NEXIQ is supported by the OEM with software.
propeller shaft see *driveshaft.*
proportioning valve a hydraulic brake valve used on vehicles equipped with front disc and rear drum brake. Its function is to reduce application pressure to the rear drum brakes to prevent premature lockup under severe braking.
protocol networking language and rules.
protons positively charged submatter of atoms located in the nucleus.
protractor a semicircular instrument used to measure angles.
psi pounds per square inch. Standard measure of pressure values used in North America. One psi is equivalent to 6.9 kPa (metric measure of pressure). Also known as gauge pressure.
psia pounds per square inch absolute. Measure of pressure values used in North America that factors in atmospheric pressure of 14.7 psi.
psig pounds per square inch gauge. Gauge pressure is equivalent to *psi*, not *psia.*
P-type semiconductor a semiconductor crystal that has been doped positive.
pull a steering characteristic that causes a vehicle to track off its intended course.
pull circuit in COE cab lift hydraulics, the circuit that brings the cab from a fully tilted position up and over center to lower it.
pull-type clutch any clutch that pulls the release bearing toward the transmission to release it.
pulse generator the principle used for many shaft and wheel speed sensors on vehicles. A toothed reluctor cuts through a magnetic field to generate an AC voltage sent as a signal to a control module. The module reads the signal *frequency* and correlates this to rpm. Also known as a chopper wheel, exciter ring, tone wheel.
pulse wheel the rotating member of a *pulse generator.*
pulse width (PW) see *pulse width modulation.*
pulse width modulation (PWM) constant frequency digital signal in which on/off time can be varied to modulate duty cycle.
pump any of a number of different devices used to move or displace fluids.
pump/impeller assembly the input or drive member of a torque converter that receives torque from the engine and converts this into hydraulic potential to drive the powertrain.
punch an impact tool used to drive out pins, rivets, or shafts, or to scribe/mark components for identification.
pup trailer a term used to describe the second trailer in a double combination train. A pup trailer is coupled to the lead trailer. This term is used in describing A-, B-, C-, or D-train combination trains.
purge to eliminate or expel.
purge tank located upstream from the supply tank in some air brake systems it acts to help reduce the chance of pumping contaminants such as oil through the air system.
push circuit in COE cab lift hydraulics, the circuit that raises the cab up and over center to a fully tilted position.
pusher axle a truck dead axle located in front of a drive axle that may or may not be a lift axle.
push-type clutch any clutch that is disengaged by forcing the release bearing toward the clutch plate where it acts on release fingers.
quench the rapid cooling of a heated metal by immersion in liquid or gases. In most cases (but not all, copper is an exception), this increases hardness.
quick-connect coupler see *quick-release coupler.*
quick-release coupler any of a number of different types of hydraulic or pneumatic connectors with spring-loaded seal couplings designed for fast connect/disconnect.
quick-release valve an air circuit valve that acts like a T fitting when air is applied to it from a source, but it exhausts air at the valve when source air is removed; the objective is to speed release times. Used on various air brake and other vehicle pneumatic circuits.
race a bearing surface runway for balls or rollers. Also used to describe the contact surface of a seal on a shaft such as that of a wheel seal on an axle stub.
rack and pinion steering gear consists of a vertically mounted pinion connected to the steering wheel and a horizontally mounted rack connected to the Ackerman arms at either steer axle wheel. Used in trucks since 2007.
radial extending from the center of an axis or center of a circle; for example, spokes on a wheel.
radial play eccentric movement or runout in a rotating component. See *radial runout* and *radius.*
radial runout the wobble of a turning wheel as viewed from the side of the vehicle; used to describe one type of runout on rotors, shafts, and wheels.
radial tire a tire with circumferential belts in which the plies run at right angles to the beads.
radius a straight line extending from the center of a circle to its circumference. The radius of a circle is half its diameter.
rail used to describe a solid bar, tubular passage, or channel. Could be a component on which another is mechanically guided or through which a fluid is routed. Frame rails are the longitudinal members of a truck ladder frame.

rail select motor a term used to describe the motor that effects transverse movements of the shift finger in an automated standard transmission; determines which shift rail/bar will be selected during a shift sequence.
ram a hydraulic (or pneumatic) linear actuator; examples would be a dump box or COE cab lift rams.
ram air air forced into the engine or passenger compartment when a vehicle moves down a highway. Ram air effect is zero when the vehicle is stationary and increases with vehicle speed. Many truck heat exchangers rely on ram air effect.
random access memory (RAM) electronically retained main memory in a computer system.
ratio valve an air brake valve used on most pre-ABS front axle brakes of highway tractors to gradually proportion application pressures to the front brakes.
ratiometric analog category of sensors using a V-Ref input and output a proportion of that voltage as a signal in "ratio" with change in status.
reactance the nonresistive opposition to current flow in a circuit created by capacitance or inductance. Important in data bus signals because it can result in the current being out of phase with the voltage that pushes it through the circuit, so it must be held at a minimum.
reactive something that alters its physical or chemical status after contact with another substance.
reactivity the property of being *reactive*.
reader/programmer a handheld EST used to read data off a truck data bus and perform limited programming and diagnostic procedures. The NEXIQ family of ESTs is an example of reader/programmers. Used to describe a scan tool with programming capabilities; the term *scan tool* tends to be used by auto technicians to describe the same instrument.
read-only memory (ROM) the magnetically retained (current vehicles), more permanent memory used by a computer control system. Usually the master program for system management is logged in ROM.
rear-axle departure offset amount in inches from the midpoint of a steer axle (or kingpin on a trailer) where the projected thrustline intersects.
rebound the reactive response of a spring after being compressed. It kicks back. Suspensions attempt to limit jounce and rebound cycles by using damping devices such as a shock absorbers.
recall bulletin OEM notices requiring service work or component replacement when a vehicle defect is identified.
receiver/dryer performs three functions on liquid refrigerant in the refrigeration cycle: filters particles, removes moisture, and stores a small amount of the liquid refrigerant until needed in the system evaporator.
reciprocal inversely correspondent: When driving one value up, another is driven downward as a result. For instance, in an electrical circuit, current flow (amps) and circuit resistance (ohms) have a *reciprocal* relationship.
recirculating ball steering gear a steering gear assembly made up of a wormshaft, recirculating balls, and a sector shaft.
recombinant battery a *gel cell* battery.
Recommended Practice (RP) a TMC industry recommended maintenance or engineering standard. RPs cover almost every truck and trailer system and are usually observed by OEMs.
redundancy term means *superfluous* but is also used to describe the fail-safe requirement for an electronically managed system such as ABS to revert to nonelectronic function in the event of electrical/electronic failure. Another example would be the requirement of a power-assisted steering system to function manually in the event of power-assist circuit failure.
reefer a refrigeration trailer.
reference voltage (V-Ref) ECU or ECM conditioned voltage (usually 5 volts DC) supplied to various sensors (such as thermistors, potentiometers) in vehicle electronic circuits. In most cases, a portion of the voltage is returned to indicate the status of sensor.
refractive index the amount a light ray is slowed when it passes through a solution in comparison with pure water. The higher the density of the solution, the more the ray is slowed.
refractometer a device used to measure the refractive index of a solution such as battery electrolyte or antifreeze solution.
refrigerant the cooling media used in air-conditioning and trailer reefer systems; a liquid capable of vaporizing at low temperatures such as ammonia, propane, R-12, R-134a, R-1234yf, etc.
relay any of a number of devices that receives a signal and then effects an action. In vehicle electrical circuits, relays are used to allow a small current circuit to control a large current circuit such as in a starter relay. Also known as a *pilot*.
relay valve in air brake systems, a valve that receives a control signal and then uses a local supply of air to send out to actuator devices. Most relay valves replicate the signal pressure valve to the out circuit, but some are designed to amplify or modulate the signal.
release bearing usually refers to the clutch release mechanism actuated by clutch pedal movement.
reluctance resistance to the continuance of magnetic lines of force.
reluctor a term used to describe a number of devices that use magnetism and motion to produce voltage.
reluctor wheel a toothed rotor that is driven through a stationary magnetic field to produce an AC voltage; used as shaft and wheel speed sensors.
remote diagnostic unit (RDU) handheld EST designed specifically for use on Bendix ABS.
remote sensing used on alternators to compensate for the voltage drop that occurs between the alternator and battery bank. A second sensing wire on the alternator reads voltage at the batteries, thereby compensating for voltage drop and reduces battery charge times especially in cold weather. ECM managed in some applications.
remote sensing technology (RST) Remy term for alternator *remote sensing*.
repair order the means of communicating a maintenance or repair task to the technician responsible for the repair via hard (paper) or soft (electronic) methods. In the event of a disputed bill, the work order assumes a legal status and its language will be interpreted by lawyers. Technicians must be aware of the consequences of anything they write on a repair order. Also known as a *work order*.
reserve capacity rating the amount of time a vehicle can be driven with its headlights on in the event of a total charging system failure. The current output will vary with the type of vehicle electrical system but the critical low-voltage value is 10.5 volts.
reservoir any of a number of different types of storage chamber, both pressurized and nonpressurized, used to

store fluids such as hydraulic oil or the compressed air in the air system.

residual pressure valve a master cylinder-located (usually) valve designed to retain a predetermined pressure in hydraulic brake lines when the brakes are released. Typically set at values between 7 and 14 psi.

resist bend moment (RBM) calculated by multiplying section modulus (geometric shape) by material yield strength; used as a measure of truck frame strength.

resistance in electricity, opposition to current flow measured in ohms.

Resource Conservation and Recovery Act (RCRA) legislation governing the proper storage and disposal of hazardous waste.

retard means *to brake.*

retarder any of a number of different types of auxiliary brake system used on trucks, buses, and heavy equipment.

retarding action braking force.

retread a tire that has been reconditioned by having a new rubber tread cured to it to extend its usable life. Truck tires today are designed to be retreaded several times during their lifecycle.

retrofit to modify a system to adapt to changes made after its manufacture.

returnability in steering technology, the tendency of the steer axle wheels to return to a straight-ahead track.

rheostat a two-terminal variable resistor incorporating a resistive track and slider, resulting in a single path for electrical current flow; the mechanical position of the slider will determine how much current flow is choked down by the resistor.

Right-to-Know Law the Hazard Communication federal legislation administered by OSHA. It requires that any company that produces or uses hazardous chemicals or substances must inform its employees and customers of the danger potential. Designed to ensure transparency of handling hazards in the workplace.

rigid anything that cannot flex or bend.

rigid disc a term used to describe a clutch friction disc plate with no torsional buffering. Provides positive engagement but little torsional forgiveness.

rim component on which a tire is mounted.

ring gear spur gear such as that used on a flywheel for engine cranking, or the outer member of a planetary gearset.

rivet any of a number of different types of one-off use fasteners used to clamp plate. A *bucked rivet* has an oval crown head on one side and after insertion in a hole through both plates, the head on the other side is formed by a pneumatic impact bucking bar.

road crown the curvature of a road from center to curb.

road feel a term that denotes feedback of steering inputs as they are transmitted from the road to the steering wheel.

road isolation the ability of a steering system or suspension to isolate road surface irregularities from the chassis.

road shock forces transmitted to the chassis from the road to the suspension and steering wheel.

Rockwell hardness test a method of rating the hardness of a metal by measuring the depth of penetration with an accurately controlled blow by a diamond cone-shaped penetrator. A variation of the *Brinell hardness* test.

roll stability control (RSC) an electronic add-on to ABS/ATC that monitors vehicle center of gravity and wheel speed and uses the chassis data bus to limit engine torque, apply engine/driveline retarders, and modulate brakes when a potential rollover condition is detected.

roller clutch a clutching assembly with a movable inner race, rollers, accordion application springs, and an outer race. Around the id of the outer race are cam-shaped pockets in which the rollers and accordion springs are located.

RollTek a rollover protection SRS used in trucks. Current RollTek electronics retracts air seats, locks seat belts, and deploys outboard seat side air bags.

roll-up door a sectioned or slatted van door on runner rails that is opened vertically and assisted by spring force.

root the deepest point of a V-notch created in the groove of a thread or between the teeth of a gear. In most threads and gears, the root is radiused (rounded).

root mean square (rms) a term used to mean averaging when performing machining operations and also used to describe AC voltage measurement by giving it an equivalent DC value.

rotary actuator device that converts fluid flow into incremental rotary motion as opposed to continuous cycle rotary motion.

rotary flow fluid movement such as that produced by a paddle wheel in water, or propeller in air, that results in centrifugal force. See *rotary oil flow.*

rotary oil flow in a torque converter, this is created by centrifugal force as the assembly rotates.

rotate to turn or spin.

rotation the act of rotating. Gears, shafts, wheels, and rotors are all subject to *rotation.*

rotor any of a number of different components that rotate. In air brakes, the rotor turns with the wheel and has braking force applied to it, whereas in an alternator, the rotor provides the magnetic fields required to create current flow.

RP 1210 the TMC standard protocol for serial *communication adapters (CAs)* used to enable a connection between a chassis data bus and an EST. RP 1210 generations use alpha codes (a, b, c); the objective was to standardize CAs but some OEMs disregarded the protocol.

runout any kind of wobble in a rotating component. Runout may be axial or radial. Using the example of a wheel on a vehicle, axial runout would be wobble as viewed from the front or rear of the vehicle, and radial runout would be wobble viewed from the side of the vehicle.

SAE J1587 the common software *protocols* used on a J1708 medium/heavy truck data bus. Accessed by connecting to the J1708 data connector, a six-pin Deutsch connector.

SAE J1708 the common hardware regulations and standards used on a J1587 medium-/heavy-truck data bus. The standard six-pin Deutsch connector to the data bus is covered by this standard.

SAE J1850 the CAN 2.0 data bus used by auto and light-truck OEMs up until 2010.

SAE J1939 covers both the hardware and software *protocols* used on the current CAN 2.0 platform for medium-/heavy-truck data bus. A J1939 truck data bus is accessed by connecting to an EST to a nine-pin Deutsch black (pre-2014) or green (post) connector. Post-2014 Volvo-Mack applications require a J1962 connector.

SAE J1962 see *J1962.*

safety factor (SF) a frame term meaning the margin of safety engineered into a frame beyond its rated load.

Safety Management System (SMS) the *FMCSA*-driven CSA seven *BASIC* tracking fields.

safety pop-off valve a term used to describe the safety pressure relief valve used on truck air brake systems. Most pop-off valves are set to trip at 150 psi and are located on the supply (wet) tank in the system.

safety pressure relief valve a term used to describe the pressure relief valve used on truck air brake systems more often know as a *pop-off valve*. Most pop-off valves are set to trip at 150 psi and are located on the supply (wet) tank in the system.

safing sensor produces a signal when a threshold rate of deceleration is achieved indicating a collision, in series with the crash sensor in a typical SRS.

sag damage a frame damage condition in which the frame bends downward. Usually the result of overloading but can also result when frame rails are weakened by loss of temper, overheating the frame, and drilling too many holes.

S-cam brakes drum brake system in which the shoes are driven into the drum by rotating an S-shaped cam, spreading the shoes.

scan tool a term that is primarily used to describe handheld ESTs used to read automobile electronic circuits; initially, scan tools could only read electronic systems but the term has stuck and is applied to ESTs that can read, reprogram, and diagnose auto electronic circuits.

scavenging pump usually refers to an auxiliary-type pump often used to back up a primary oil pump to prevent oil starvation under some operating conditions. Some Allison transmissions use a *scavenging pump*.

screw a threaded fastener.

screw pitch specification of the number of *threads per inch* (tpi) on a threaded fastener.

screw pitch gauge a toothed template gauge used for measuring *screw pitch* or threads per inch (tpi).

scroll compressor a type of A/C and reefer compressor consisting of a pair of intermeshing scrolls, one of which is stationary, around which the other orbits. The result is to produce a continuous, high-efficiency intake to discharge operating cycle,

scrub radius the radius formed at the road surface between the wheel centerline and steering axis.

secondary circuit in brake systems refers to the following circuit in the system usually actuated by the primary circuit. In hydraulic brake systems on straight trucks this is usually the rear brake circuit, whereas on large trucks with air brakes, it refers to the brakes on the steering axle wheels.

secondary lock any of a number of different types of back-up locking device; an example is the secondary locking device used on most fifth wheels, which is either manually or automatically engaged.

secondary modulation Allison term used to describe the control of current flow to an output device. Also known as *submodulation*.

section height the tread center-to-bead plane on a tire.

section modulus (SM) in frame technology, the mathematical coding of the geometric shape of a frame member used to calculate resist bend moment. Each frame shape, for instance, radiused C-channel, has a section modulus formula from which you can calculate the SM by entering the actual dimension data used.

section width the sidewall-to-sidewall measurement on a tire.

self-adjusting any mechanical devices subject to wear that have some kind of automatic adjustment mechanism; examples are self-adjusting slack adjusters and clutches.

self-adjusting clutch a clutch assembly that "automatically" adjusts as friction faces on clutch discs wear. Many current clutches are self-adjusting.

semiconductor material that can act as conductors or insulators depending on how its structure is arranged; elements that have four electrons in the valence shell.

semioscillating fifth wheel the standard fifth wheel that articulates on pivot pins only in a fore and aft direction.

semitrailer any trailer that has part of its weight supported by the towing vehicle. Most highway tractors haul semitrailers coupled by fifth wheels that support a percentage of the trailer weight depending on how the axles are arranged.

sensing voltage the signal voltage (system voltage/V-Bat) used by an alternator to manage charge cycles.

sensor energy conversion device used to signal a condition or state. Sensors are used to signal speed, pressure, temperature, etc.

separator pins used in some twin friction disc clutches to hold an equal distance between each friction disc and the intermediate plate as the clutch is released. Also known as *intermediate plate pins*.

serial communications interface (SCI) communication protocols that permit bus SAs/MIDs to multiplex with SAE J1939.

series circuit an electrical circuit that has only one path for current flow.

series-parallel circuit an electrical circuit with both series and parallel branches.

server in a computer processing cycle or multiplex transaction, the fulfilling of a *client* need is provided by a server.

service brake priority a trailer air brake circuit in which air is routed first to the trailer reservoirs when the trailer supply valve (on the tractor) is activated. This means that air pressure must exceed 60 psi in the reservoirs before the hold-off circuit will release.

service bulletin see *technical service bulletin* (TSB).

service diaphragm the rubber pancake on which service brake pressure acts to apply the brakes on an air brake–equipped vehicle.

service emergency relay valve (SERV) trailer multifunction valve used to modulate hold-off pressure and manage service and emergency foundation braking.

service information system (SIS) a generic term used to refer to OEM online service information software. Data hub based. Access is by Internet and subscription controlled.

service literature a term used to describe most kinds of electronic accessed service instructions and the databases that retain them; replaces hard copy service manuals and TSBs. Most OEMs no longer make hard (paper) copy versions of service literature available.

service manual a hard (paper) copy book that outlines service repair and troubleshooting procedures. Today, most service manuals are electronic due to lower cost and ease of updating; consequently, the term *service literature* is more often used.

ServicePro a Freightliner umbrella software program driven from its Portland, Oregon, data hub. Once the program is loaded onto a PC, it enables online and vehicle communication interchange; it incorporates ServiceLink, ServiceLit, Diagnostic Routines, plus links to its major component supplier data hubs.

servo a component actuated by another. One type of servo is a hydraulic slave piston that is actuated by a control valve.

servo brake braking action that occurs when the movement of one shoe is governed by an input from the other. Used in some hydraulic brake systems. When force is applied to the ends of the brake shoes, the primary shoe reacts first and grabs onto the brake drum, forcing the secondary shoe into the drum with a wedge-action. Servo action increases in proportion to the rotational speed of the drum.

setback angle formed between the vehicle centerline and a line perpendicular to the front axle.

shear force exerted at a right angle to the center line of a component. For instance, scissors apply a shear force to whatever they cut. A bolt is subject to shear forces when the surfaces being clamped slide. Double shear strength is the force a bolt must resist to avoid being cut into three separate pieces (slide action from both clamped sections); it usually exceeds tensile strength by 175 percent.

shells electron orbital paths in an atom.

shift bar sliding shaft to which shift yokes are clamped. When the bar is moved by the gearshift lever spade, the bar slides in bushing supports, moving the shift yoke to effect a gear change. Also known as a *shift rail.*

shift bar housing the manifold on top of a standard transmission that houses the shift rails, usually three, to which shift yokes are connected; the shift tower assembly mounts on top of the shift bar housing. Also known as *shift rail housing.*

shift forks the fork-shaped components that connect shift rails with the sliding clutches in a standard transmission. Also known as *shift yokes.*

shift point adjustment the means of adjusting the shift rpm in Allison transmissions. Adjustment of the shift valves is performed using a valve ring adjusting tool to set shift valve spring tension.

shift rail sliding shaft to which shift yokes are clamped. When the rail is moved by the gearshift lever spade, the rail slides in bushing supports, moving the shift yoke to effect a gear change. Also known as a *shift bar.*

shift rail housing the manifold on top of a standard transmission that houses the shift rails, usually three, to which shift yokes are connected; the shift tower assembly mounts on top of the shift rail housing.

shift yokes the fork-shaped components that connect shift rails with the sliding clutches in a standard transmission. Also known as *shift forks.*

shims gaskets or spacers of a specific thickness used for setting such things as endplay, preload, or tolerances. May be steel plate or fiber.

shimmy front-end oscillation or shake transmitted from steer wheels up to the steering wheels; usually indicative of serious steering geometry problem.

shock absorbers hydraulic devices used to dampen vehicle spring oscillations.

shock failure catastrophic failures that occur due to the sudden application of excessive force. For instance, when a spinning wheel suddenly bites into the road surface following a spinout, the drivetrain is shock-loaded and can cause a *shock failure.*

shorepower term used to describe the means used to connect a truck/trailer combination to AC mains electrical power to supply *hotel load* electricity. A shorepower or hotel load grid exists on many U.S. and Canadian interstate routes. It is referred to 110, 230, and 460 V-AC mains feeds.

short circuit an undesirable current flow path between two electrical wires or components in an electrical circuit. Shorts often occur when wiring insulation wears, exposing bare wire. Shorts may occur within a circuit or to ground.

sidecutter pliers with cutting edges on offset jaws used to strip/cut electrical wiring, steel wire, and cotter pins.

sideslip any type of wheel tracking irregularity caused by incorrect toe or tandem tire scuff.

sideslip tester a load cell plate tester capable of evaluating sideslip. An example is the Hunter SS100T unit.

sidesway damage a frame damage category in which the frame becomes banana shaped with the front wheel not tracking in line with the rear. Usually caused by a side-swipe collision.

sidewall the portion of a tire between the tread and bead.

sight glass allows a technician to see the flow of a liquid in a line (such as fuel or refrigerant) for diagnostic purposes.

signal (detect and) actuation module (SAM) electronic controller that manages chassis functions and bus traffic on a vehicle, specifically on Daimler trucks, a body control module.

single-acting cylinder a hydraulic cylinder designed to produce thrust in one direction only; returned by gravity or spring force.

single-reduction axle drive axle gearing that uses only one stage of reduction gearing achieved by the pinion and crown gear set ratio.

sipes in tire terminology, the tiny notches along the edges of grooves within the tread blocks and ribs. They help relieve rolling stress that initiates river erosion wear during wet pavement operation.

slave cylinders see *wheel cylinders.*

slave valve any of a number of types of air or hydraulic actuators.

sliding caliper a disc brake design similar to a floating caliper with exception that it uses machined surfaces in place of pins for inboard-outboard movement.

sliding clutch actuated by a shift yoke, used to engage a gear on a shaft. Also known as a *dog clutch.*

sliding fifth wheel a fifth wheel mounted on toothed rails, permitting forward and rearward movement; this directly affects weight over the tractor axles by altering the bridge formula and the overall length of a tractor/trailer combination.

sliding vane pump a commonly used fluid pump consisting of a slotted rotor that turns within a pump *liner* or *cam ring.* The rotor is milled with radial slots in which vanes (blades/plates) float, being thrust out by centrifugal force into the liner to form the pump chambers when the rotor turns.

slinger a pressed metal ring used as seal.

slip on a drive shaft, the total normal linear travel permitted at a slip spline.

slip joint the splined section of a drive shaft or steering column. A slip joint is designed to accommodate variations in shaft length but no axial play.

slip rings stationary devices used to conduct current to an alternator rotor.

slip splines used on a drive shaft (propeller shaft) stub and yoke assembly to permit variations in length caused by frame and suspension oscillations while transmitting drive torque.

slip yoke a yoke that accommodates linear movement; the forward yoke in a drive shaft with female splines to which male splines in the main drive shaft tube are inserted to form a *slip joint.*

slipout occurs in transmission gearing, usually when pulling under full power or decelerating with load momentum driving forward movement. Worn sliding clutch teeth walk out of engagement with the gear.

slop widely used slang term for endplay. See *endplay.*

slow adaptive mode slow-speed programming adjustments. Allison uses this term to describe processing adjustments that make minor changes in initial shift conditions to respond to minor system changes.

slug see *slug volume.*

slug volume a term used to describe the fluid volume of a pump chamber or engine cylinder swept by the plunger/piston in one cycle.

smart ignition switch a starter control circuit switch that signals a request to crank message to the engine controller.

smart switch usually refers to a *ladder switch* but also may refer to either a *PLC* or switch with independent processing capability.

SmartShift proprietary Mercedes-Benz AGS shift management software.

SmartWay an EPA initiative that provides loans and funds to help truck operators reduce their carbon footprint by making vehicles more fuel efficient. Includes strategies for replacing older vehicles, aerodynamics, and rolling drag reduction.

snapring a retaining split ring with radial spring pressure; snaprings are usually fitted to grooves and are used to retain shims, springs, bearings, and other components.

snapshot test diagnostic test used by a range of ESTs that can capture frames of data from one to multiple parameters at preset time intervals. A snapshot test can be assigned a set trigger such as the logging of an FMI, or it can be triggered manually. This test is especially useful in isolating intermittent faults in electronic circuits.

snubber a device that damps movement. Some fifth wheels are equipped with rubber block snubbers to damp trailer bump and pintle hook assemblies can be equipped with pneumatic (on-highway) or hydraulic (off-highway) snubbers.

socket a female cavity to which a pin or shaft is inserted. Electrical connectors are equipped with sockets to which terminal pins are inserted to make a connection.

socket wrench an enclosed cylindrical wrench forged with a hex or double hex opening designed to engage with a bolt or nut on one end and a square drive lug recess on the other. Socket wrenches are designed with a full range of standard and metric nominal sizes and are further classified by the size of the drive lug required to turn it; ¼ -, ⅜ -, ½ -, ¾ - and 1-inch (6, 9, 13, 19, and 25 mm) drive sizes are typical.

sodium chloride chemical name for rock salt that is applied to roads during winter in crystallized and liquid brine states. In solid or liquid state, it has no melting effect when temperatures fall below 0°F (−18°C). Under extreme cold conditions, as brine it can be cut with other salts to reduce its freeze point.

software computer programming instructions that are used by the CPU to organize the processing cycle and produce outcomes.

solenoid an electromagnetic device consisting of a coil and armature designed to perform mechanical work when energized. Used in just about every electrical circuit on modern vehicles, whether it be in the starter circuit, transmission, or door locks.

solid rubber suspension a suspension with solid rubber spring blocks used in some vocational trucks; provides a tough suspension best suited in applications that always run loaded.

solid state a semiconductor circuit or device.

solvents substances that dissolve other substances.

source address (SA) a controller with an address on a J1939 vehicle data bus. An SA is designated a unique address on the bus with a numeric code. An SA is the equivalent of a MID on J1587.

spade fuse see *blade fuse.*

spalling surface fatigue that occurs when chips, scales, or flakes of surface metal separate because of fatigue rather than wear.

spec a commonly used slang term meaning to *detail*, as in *to spec a transmission.* It is drawn from the word *specification.*

specialty service shops service facilities that limit work to one or two specialty areas. An example would be a *spring shop* that limits its practice to repairing suspensions.

specific gravity (SG) the weight of a liquid or solid versus that of the same volume of water. A hydrometer is used to make SG measurements.

specific PGN a *parameter group number* message targeted at one specific controller (SA) on a J1939 data bus.

specification an assembly, engineering, or technical detail or value.

speed sensor used to refer to a shaft speed sensor using either an inductive pulse generation or Hall effect principle to signal an ECM or ECU.

spider the axle hub mounting to which truck foundation brakes are mounted. Also a slang term for a U-joint cross or trunnion.

spigot a pilot bearing, usually a straight roller bearing mounted on the head of a shaft. Term is also used to describe a dump valve/tap assembly used on air tanks.

spike a sudden surge in hydraulic pressure or electrical voltage. Also a slang term meaning the tractor steering column mounted, trailer service brake valve.

spindle the shaft on which a wheel assembly rotates.

spinout occurs when differential action allows one drive wheel on a slippery surface to spin while the other remains stationary and receives no driving torque. On tandem drive trucks, because of the additional differential action between the drive axles, it is possible to turn a spinning wheel at four times its normal speed. This can take out the interaxle differential or carrier differential. Spinout is a driver abuse failure of differential carriers.

spinout failures a tandem drive axle failure in which one of the four drive wheels loses traction and receives all driveline torque. Spinout conditions may produce wheel speeds four times greater than specified. Spinout failures result from shock loading when a spinning wheel's motion is suddenly arrested or by lubricant throw-off from rotating components.

spiral bevel gear helical gear set arrangement that has a drive pinion mesh with a crown gear at the radial centerline of the gear.

spitter valve a term used to describe an automatic drain valve used on air tank reservoirs.

splined yokes used on vehicle propeller shafts to positionally engage a stub shaft to a yoke end.

splines radial lands that either protrude from or are milled into a shaft to permit it to engage with female splines for the purpose of transmitting torque.

split guide ring located in both the impeller and turbine sections of a torque converter, a split guide ring suppresses fluid turbulence, allowing more efficient operation.

split-coefficient braking a braking condition under which the wheels on one side of the vehicle have no traction and the others have good traction. ABS can maneuver split-coefficient braking more effectively than non-ABS.

splitter transmission standard transmission in which main box gearing is compounded by an auxiliary section.

splitter valve the shift stick located air shift valve used to range shift an auxiliary section on a compounded transmission.

spontaneous combustion process by which a combustible material autoignites.

spool valve commonly used hydraulic valve in which a sliding spool within a valve body is used to open and close hydraulic subcircuits.

sprag a primitive retarding mechanism inserted into wheel spokes to brake a vehicle; more commonly used to mean *stop*.

spring an elastic device used to store or suppress mechanical force. Springs are used in vehicle suspensions where they act to support the chassis and buffer it from road forces. Some examples of springs are single leaf, multiple leaf, torsion bar, rubber block, and air filled.

spring brake priority a trailer air brake circuit in which air is routed first to the hold-off circuit when the trailer supply valve (on the tractor) is activated. Only after air pressure in the hold-off circuit exceeds 60 psi is air routed to the trailer reservoirs.

spring brake valve manages air routing to spring brake chambers and usually regulates the hold-off air.

spring brakes double-chamber air brake actuators. The forward chamber actuates the service brakes while the rear or hold-off chamber holds off a powerful spring when charged with air. Spring brakes have a mechanical park-by-spring feature and are used on most FMVSS 121 systems.

spring pack a multileaf steel spring assembly with plates clamped together with a center-bolt to form an elliptical shape.

sprocket a wheel with toothed cogs designed to engage with a chain or mesh with a similarly shaped device.

SRC means *source* of a DTC in J1939; the source usually being the SA that has produced the code.

S-Series Certification ASE certification tests for school bus technicians.

stability control electronics (SCE) an ABS add-on that detects and attempts to correct conditions such as spin-out, yaw, rollover, drift-out, and jackknife.

stalk switch a multifunctional turn signal switch. Stalk switches can be either electromechanical (hard wired in and out) or smart (broadcast signals to data bus).

stall stationary or the point at which a moving or rotating object ceases to move. The act of shutting down a system or component.

stall test a test performed on vehicles with automatic transmissions to verify the performance of the engine and transmission, usually to determine if one is at fault.

stall torque ratio a rating of the ability of a torque converter to multiply input torque (from the engine) to the turbine shaft (to the transmission). A stall torque ratio of 2:1 would mean that the torque converter could double the input torque from the engine.

stand pipe usually describes a combined breathing and flowcheck device. On a U-joint assembly, a stand pipe is used to prevent the reverse flow of hot liquid grease during operation.

starter circuit the chassis electrical circuit that supplies the high-current load required to crank an engine. Also known as *cranking circuit*.

starter motor electric motor that converts electrical energy from the cranking circuit into the mechanical energy required to start the engine. Also known as *cranking motor*. Although not common, air-actuated starter motors are also used on trucks.

starter relay the magnetic switch that when energized by the starter control circuit closes the high-current cranking circuit. Also known as *magnetic switch* or *mag switch*.

starting safety switch prevents vehicles with automatic transmissions from being started in gear. The switch closes only when the range selector is in park or neutral.

static balance balance at rest. When a tire is statically balanced, it has no tendency to rotate by itself on its axis regardless of its position.

static electricity capacitive charge caused by electrostatic induction. Results in high voltages and arcing. May damage components and cause fires.

stator (1) a stationary element in a rotating component; (2) the stationary member of an alternator into which electrical current is induced by the rotor; (3) intermediary in a torque converter torus located between the impeller and turbine; redirects oil from the turbine back to the impeller.

stator assembly the reaction member of a torque converter used to multiply torque that is supported on a free wheel roller race splined to the front support assembly.

steering arm see *steering control arm*.

steering axis inclination (SAI) an automotive term that is the equivalent of *kingpin inclination* in truck front-end geometry. Angle formed by a line drawn through the steering axis (kingpin axis) and true vertical.

steering box see *steering gear*.

steering control arm the forged link that connects a drag link to the steering knuckle. Also known as a *steering arm* and a *control arm*.

steering gear steering control gearing of a chassis steering system. Steering input is from a driver-controlled steering column, whereas output is through a sector shaft. Also known as a *steering box*.

steering shaft a tube or rod that connects a steering wheel to the steering gear.

Steertek a combination air/leaf spring front steer axle manufactured by Hendrickson. The leaf springs are either steel or fiber composite construction.
step-down transformer a transformer that induces a lower voltage in its secondary coil by using a lower number of windings than in the primary coil.
step-up transformer a transformer that induces a higher voltage in its secondary coil by using a larger number of windings than in the primary coil.
stop bits the final element in the transmission of a character on a data bus.
straight truck a nonarticulating truck with its cargo body mounted onto its chassis.
stranded wire electrical wire made up of many thin gauge wires twisted together to form a single conductor.
stub shaft the propeller shaft member machined with male splines.
stud remover tool consisting of hardened, knurled eccentric jaws that grip the shank of a stud; used both to remove and install studs.
strut a tubular- or rod-type support between two components.
subcooling term used to describe the additional cooling of A/C refrigerant in the bottom of the condenser after it has changed from vapor to liquid below its freeze point. In physics, this is known as *supercooling*.
submodulation Allison term used to describe the control of current flow to an output device. Also known as *secondary modulation*.
subsystem identifiers (SIDs) the major electronic subsystems that fall within J1587 MID module management in a medium-/heavy-truck electronic system.
suction circuit a term often used to describe the hydraulic line that connects a pump inlet to the reservoir. Lower than atmospheric pressure is created at the pump inlet, which results in atmospheric pressure acting on reservoir oil pushing it through the circuit.
sulfation occurs when sulfate is allowed to build up on battery plates. It causes two problems: First, it lowers electrolyte specific gravity, increasing the danger of freezing in cold weather, and, second, it reduces the reserve power required to crank the engine.
sump a reservoir.
sun gear the central or axis gear in a planetary gearset. It may be optioned to rotate or not rotate, depending on the gear arrangement and desired output ratios.
supercapacitor electrical potential storage device that physically separate the positive and negative charges, holding each in a manner that can be compared to static electrical buildup. Used in conjunction with battery banks in some hybrid electric vehicles. Also known as *ultracapacitors*.
supercool to cool a substance below its freeze point without solidification or crystallization.
SuperTech a nationwide technician rodeo run by the TMC in their annual fall convention. Candidates compete in written and hands-on stations designed to identify technical excellence in the trucking industry.
supplemental restraint system (SRS) general term for vehicle human protection systems in the event of an accident. Current trucks may be equipped with steering column mounted air bags and rollover protection systems such as *RollTek*.
supplementary coolant heater thermostat-controlled diesel fuel heating device used to maintain engine coolant at a predetermined value during periods of engine shutdown.
supply tank in a vehicle air brake circuited the supply tank is the first reservoir in the circuit to receive air from the air compressor; the supply tank feeds the primary, secondary, and auxiliary air tanks in the circuit.
SureShift ZF Meritor three-pedal automated transmission that uses joystick initiated shifts. The clutch pedal is only used to break driveline torque at stall (starting and stopping).
suspect parameter number (SPN) components and subcircuits within a J1939 controller with an SA. The equivalent of PIDs on a J1587 data bus.
suspect PG (SPG) in J1939, an SPG is assigned to each parameter or component. It is used for diagnostic purposes to identify abnormal controller conditions.
suspension means by which axles are attached to the vehicle frame; supports the frame and dampens it from road shock using springs (leaf, air, torsion, rubber, etc).
suspension height distance between a specified point on a vehicle chassis and the road surface.
swage to groove, shape, or reduce alter the taper in materials.
swaging the process of forming, grooving, and bending materials, or alter taper in a bore.
swashplate a stationary canted (angled) plate used in an axial piston pump on which pistons in a rotating cylinder barrel ride; as the pump cylinder rotates, the pistons reciprocate in the barrel bores.
swashplate compressor a type of A/C compressor that uses a *swashplate* pumping principle.
swashplate pump a fluid power pump that uses a *swashplate* pumping principle.
swing see *trailer swing*.
swingers commonly used term for a pair of trailer van doors that are hinged on each side of the van and close to the center.
switch a control device, at its simplest, responsible for managing a circuit on or off. Used to control in electrical and fluid (air/hydraulic) circuits on trucks.
switch status the signal state of a switch, using the example of a simple toggle switch its status could be either on or off.
synergize see *synergy*.
synergy cooperation: components or subsystems organized to function at maximum efficiency. Verb: synergize. Used in truck electronics to describe how nonrelated components are organized to produce optimum efficiencies.
system a group of interrelated or interconnected components that work together to produce a common objective. Systems can be subdivided into subsystems.
System Manager one of two Eaton automated transmission ECUs used in its early generation AMTs. The System Manager interfaced with the truck chassis data bus and mastered the management of the automated transmission system. No longer used in current Eaton AMTs.
system pressure usually refers to the range of pneumatic circuit pressures between cut-out and cut-in pressure values in an FMVSS 121–compliant air brake system. Can refer to other fluid power circuit pressures such as those in transmission hydraulics and engine lube circuits.
tachometer an instrument used to indicate shaft speed.
tag axle a truck dead axle located behind a drive axle; may or may not be a lift axle.

tailshaft the output shaft of a transmission.

tandem two; a pair of.

tandem drive the typical two rear-axle drive system used in a three-axle highway tractor. The front axle on a typical highway tractor is the steer axle, while the tandem rear axles drive the truck.

tandem drive axle one of a pair of drive axles used in tandem drive highway trucks.

tandem lateral offset occurs when the geometric centerline of a set of tandem axles does not cross the midpoint of the other vehicle or combination vehicle axles.

tandem scrub angle formed by the intersection of horizontal lines drawn through each rear axle in a tandem when total toe and offset is zero.

tang release tool a pin or collet used to separate terminals in any of a number of different types of electronic connector assemblies.

tank any of a number of different types of reservoir on a vehicle such as the air tank, fuel tank, etc.

tanker a truck or trailer whose primary function is to transport a liquid load such as petroleum product, milk, liquid chemicals, acids, food products.

tap a hand- or machine-actuated tool used to cut internal threads.

taper reduce in size or thickness at one end of a cylinder or shaft.

TC10 Tractor Series (TS) an Allison fully automatic 10-speed transmission introduced in 2010 for linehaul trucks that consists of an input module (torque converter), twin-shaft gearbox module, and a planetary output module. Designed to be powershifted with no spin-loss at ratio change.

technical service bulletin (TSB) service literature either online or in hard copy that updates or modifies a repair procedure.

Technology and Maintenance Council (TMC) a division of the American Trucking Association (ATA) responsible for setting maintenance and equipment standards in the U.S. trucking industry. Formerly (before 2005) known as *The Maintenance Council.*

Tee fitting a fluid circuit valve usually set up with a single source but distributes to two branches of the circuit; often T-shaped.

temper the heat-treating process used to harden (or in some cases soften) materials. Term is usually applied to heat treatments of middle and high alloy steels. Most highway trucks used *tempered steel* frames. To temper most carbon steels, the temperature is raised to a specific value and then rapidly cooled under controlled temperature drop. Temper can also be applied to alloyed aluminums.

tempered steel any heat-treated steel. For instance, most highway trucks use middle alloy, tempered steel frame rails with a *yield strength* of 110,000 psi. If this steel was to be annealed (that is, remove the temper by heating followed by gradual cooling), the frame yield strength would be less than 60,000 psi.

tempering the process used to temper a metal. Also known as *drawing.* See *temper.*

ten wheeler truck driver term for the typical three-axle configuration of a highway transport tractor; more commonly known as a *tandem drive* tractor.

tensile strength the highest stress that a material can withstand without separating; expressed in psi. In steels, tensile strength typically exceeds *yield strength* by about 10 percent.

tension load applied along the longitudinal centerline of a rod, bolt, or frame member.

terminal a begin, an end, or a connection point in an electrical circuit.

terminating resistor used to prevent the twisted-wire pair used in a data bus from attracting signal interference by acting like a giant antenna. A current data bus uses two terminating resistors that test at 120 ohms. If one is missing, the data bus may function erratically, but if both are missing, no effective networking can take place.

The Maintenance Council (TMC) name changed in 2005. See *Technology and Maintenance Council.* The acronym remains unchanged.

thermalert a warning device integrated into Meritor MTIS consisting of a eutectic plug designed to melt at 280°F (138°C) to warn of an overheat condition in the wheel end. When the thermalert melts a very loud whistling noise is produced by air leakage.

thermistor a temperature sensing device used in various electronic circuits in a truck chassis. Consists of a variable resistor, the variance of which correlates with temperature. When supplied with a reference voltage (V-Ref), some of the supply voltage is returned as a signal from which temperature can be calculated. Most truck electrical circuits use negative temperature coefficient (NTC) thermistors in which resistance drops as temperature increases, meaning that the signal voltage increases with temperature rise.

thermocouple temperature measuring device consisting of two conductive wires each made of a dissimilar material and joined at the *sensing end.* When subjected to temperature rise, a small voltage is produced that may be read at the *read end* by a millivoltmeter. As temperature increases, so does the voltage.

thermodynamics a branch of physics that deals with the behavior of heat; its principles are used to explain the operation of heat exchangers and A/C operation.

thermoplastic hose clamps heat-sensitive dynamic tension hose clamps that maintain a constant tension within a wide band of temperatures.

thermostat any of a number of different types of heat sensing switches; a thermostat usually changes the status of a circuit or system when a specific temperature value is achieved. Found on starters and other electrical components, also used in an engine cooling system where it is used to option coolant circulation to the radiator.

thickness gauge a metal blade precisely finished to a defined thickness; used for measuring clearance and tolerance dimensions. Usually found in sets in which thickness graduates by 0.001-inch increments. Also known as *feeler gauges.*

thread chasing process of cleaning or restoring existing threads using a tap or die.

thread length formula for calculating the thread length on a capscrew: twice the diameter of the capscrew plus ¼ inch (6 mm) for bolts up to 6 inches (15 cm) in length.

threads per inch (tpi) the means used to specify standard thread geometry. The tpi defines the *screw pitch.* Literally, the number of threads cut to a linear inch on a fastener.

three-pedal system used to describe an automated transmission system that retains a clutch pedal (the *third* pedal) that has to be driver actuated under conditions of driveline stall or electively.

three-speed axle a tandem drive axle arrangement in which each of the two drive axles are two-speed units. The third speed is achieved by running one axle in the high-speed range and the other in the low-speed range and letting the interaxle differential balance the differential.
three-way splice a multiple splice wiring repair. When making multiple splice wiring repairs, each repair joint should be staggered.
threshold a lower limit or value. Commonly used in vehicle computer logic maps.
throat any kind of inlet passage such as the entry guide channel to the jaws of a fifth wheel.
throttle position sensor (TPS) usually a potentiometer or Hall effect sensor used to signal linear travel of an accelerator pedal. Potentiometer-type TPS (contact wiper and resistor) were common until 2007, whereas Hall effect TPS (noncontact) are used in more current equipment.
throw output circuits in a switch; a two *throw* switch has two output circuits.
throwout bearing another term for *release bearing*, the linkage-actuated disengagement mechanism used in both push- and pull-type clutches.
thrust force, usually linearly applied.
thrust angle angle formed by the vehicle thrustline and the vehicle geometric centerline. When a thrust angle exists, the result is dog-tracking.
thrust washer a spacer of specified size designed to define or limit shaft endplay. Also used in U-joint bearing cups to prevent galling of the cross trunnions.
thrustline vehicle directional tracking line. Properly set axles should track perpendicular to the vehicle centerline. In other words, the rear wheels should track directly behind the front wheels when the vehicle is moving straight ahead, resulting in a directionally true thrustline. If this were not the case, some dog-tracking tendency would result.
thyristor a three-terminal solid-state switch. It differs from a transistor in that a thyristor is only capable of switching; that is, it has either an ON or OFF status.
tie-rod transfers steering motion from the steering gear side of the chassis to the opposite side by means of an adjustable rod, forcing the steering knuckles to act in unison. Also called a *crosstube*.
tie-rod assembly adjustable rod assembly used in a parallelogram steering linkage. See *tie-rod*.
time the act of phasing two mechanical components so that they interact or sequence properly. Many gears have to be *timed* to each other, as does a fuel system to an engine.
time guide used to compute compensation to dealerships by OEMs for warranty work, often identifying the exact amount of shop time required to complete a specific repair.
timing the procedure used to *time* interacting mechanical components. See *timing marks*.
timing marks means of marking gear set teeth before installation so that they can be placed in proper mesh in the geartrain.
tire pressure control system (TPCS) a chassis tire inflation and monitoring system. May use dual equalization or on-board air for inflation. Provides driver alert in the event of rapid deflation. Examples are Cat's Eye and Crossfire.
tire pressure inflation system (TPIS) an automatic tire pressure management and inflation system that uses chassis system air pressure to maintain a specified tire pressure.
tire pressure monitoring system (TPMS) wireless tire pressure (and sometimes temperature) monitoring system. Examples are SmartWave, SmarTire, Tire Sentry, and Pressure Pro.
toe a steering geometry dimension that specifies the distance between the extreme front and extreme rear of a pair of tires on either end of an axle.
toe-in when the extreme front of a pair of tires on an axle points inward toward the vehicle.
toe-out when the extreme front of a pair of tires on an axle points outward away from the vehicle.
toggle switch a simple two-state switch that either opens (off) or closes (on) an electrical circuit.
tone wheel see *pulse wheel* and *pulse generator*.
tooth contact pattern the intermesh contact surfaces between a pair of mating gears; usually no-load tested using machinist's (Prussian) blue or other semifluid test media.
torque twisting force. Measured in pound-feet (lb-ft.) of applied force.
torque converter a fluid coupling consisting an impeller, stator, and turbine used to transmit driving torque from an engine to transmission; it provides continuously variable drive ratios and can multiply torque.
torque-limiting clutch brake a clutch brake designed to slip when subject to a predetermined torque load, protecting the clutch brake from destruction.
torque multiplication occurs during the vortex flow phase of torque converter operation.
torque proportional differential Mack Trucks power divider that can vary the distribution of torque between the two drive axles of a tandem drive set. Known for its superior ability to contain potential spinout conditions.
torque rod used variously on suspensions to limit torsional effects; many torque rods are integrally adjustable (threads) or adjustable using shims.
torque wrench a wrench used to define final twist force of threaded fasteners on installation; the final twist force of a fastener has a correlation with clamping force so most fastener specifications are provided as torque values.
Torricelli's tube Torricelli discovered the concept of atmospheric pressure by inverting a tube filled with mercury into a bowl of the liquid, then observing that column of mercury in the tube fell until atmospheric pressure acting on the surface balanced against the vacuum created in the tube. His experiment proved that at sea level, vacuum in the column would support 29.92 inches (76 cm) of mercury.
torsion twisting, especially when one end of a beam or body is held.
torsion bar suspension a spring system consisting of spring steel bars loaded under torsion (twist). Not common but has been used on some trailer suspensions, some racecar chassis, and a lot of older Chrysler products.
torsional failures catastrophic failures that result from excessive torque applied to a shaft or beam.
torus the subcomponents of a torque converter consisting of an *impeller*, a *stator*, and a *turbine*.
Torx® fastener a screw with a head with six-prong recess. Often used to secure headlight assemblies, mirrors, and brackets. Can sustain greater torque and also be robotically installed.

Torx® screwdriver manufactured with a six-prong tip used to remove and install Torx fasteners.
total acid number (TAN) a measure of the relative acidity of a lubricant. An abrupt rise in the TAN would indicate imminent breakdown caused by overheating.
total base number (TBN) a measure of the reserve alkalinity of a lubricant, specifically its ability to neutralize harmful acids created by either combustion or overheating.
total base number (TBN) additives lube oil acidity reducers. Designed to minimize acidity in lube oils as they age.
touch point term used to describe the engine rpm at which a two-pedal system dry clutch begins to engage. *Touch point* initiates the launch phase.
toxic poisonous.
toxicity having poisonous characteristics.
track the directional path of a moving vehicle.
tracking the interrelated paths of the steer axle and other axles on a vehicle as it moves.
tractor a heavy-duty highway truck designed for towing trailers. Also any of a number of different types of farm and heavy equipment utility vehicles.
tractor protection valve air system valve designed to isolate the tractor air supply from the trailer air circuits when tractor system pressure drops to a specified value (usually between 25 and 45 psi).
trailer any of a number of different types of nonpowered, load-carrying vehicles towed by another vehicle.
trailer application valve a trolley valve (aka spike or brokerbrake) designed to apply trailer service brakes in isolation from the tractor service brakes; may be air or electric.
trailer swing the result of trailer wheels locking, causing the trailer to swing around the tractor, which remains in a straight-ahead direction. Should be differentiated from *jackknife*, which is a more common occurrence.
transducer device that converts one signal (or energy) format to another; for instance, it converts a pressure measurement to an electrical signal.
transducer module a data bus translator; for instance, early electronic transmissions required a transducer module so they could exchange data with the electronic engine controller. Not often required on today's truck systems due to advanced multiplexing technology. Sometimes known as a *gateway module*.
transfer case an auxiliary transmission used to split drive torque from the main transmission between rear and forward drive axles. Also known as a *dropbox*.
transformer an electrical device used to increase or decrease voltage by using a pair of electromagnetically connected coils. Current flowed through the primary coil induces current in a secondary coil. The number of windings in each coil will determine the input/output voltage and current characteristics.
transformer rectifier (TR) 12/24 alternator an older alternator that managesthe electrical system at 12 volts system voltage and at 24 volts for cranking. Switching takes place in the transformer rectifier module that forms the endplate of the alternator.
transistor a three-terminal semiconductor used to switch and amplify electronic circuits.
transmission control module (TCM) a transmission controller that occupies the SA 03 address on a J1939 data bus.
transmission control unit (TCU) term used by Mercedes Benz to describe the control unit used on its AGS automated transmission.
transmission controller one of two ECUs used to control an early generation Eaton AMT. The unit had some limited processing capability but was mainly responsible for managing the switching requirements of the system.
transverse acting in a crosswise direction; for example, an axle on a truck is *transverse* mounted. Used to describe components (transverse torque rod), positioning (transverse mounted engine), or forces acting on a chassis.
TranSynd a synthetic heavy-duty automatic transmission oil recommended by Allison, which states that it extends oil change intervals by 300 percent.
tread the geometric profile of a tire, specifically the lug and groove geometry manufactured into a tire. Its function is to provide traction and sustain wear.
tread grooves recesses between the tire *tread lugs* or ribs.
tread lugs the raised block elements in tire tread geometry.
tread wear indicator ridges manufactured in tire tread grooves that become flush with the tread land to indicate a worn-out tire.
tread width distance between the centerlines of wheels on the same axle, or on dual wheel assemblies, the distance between the dual centerpoints on each side of the axle.
treadle valve a dual-circuit service brake actuation valve that meters air from the primary and secondary service reservoirs to the service lines and brake chambers. Also known as a *foot valve*.
tree diagnosis chart a graphic schematic used to guide a troubleshooting procedure by guiding the diagnosis step by step; may be paper or software driven. Some current truck systems use interactive diagnostic trees that essentially check each step before proceeding to the next.
triage the process of making a rapid assessment of a condition before determining a course of action. It is not uncommon today for large truck service facilities to have a triage bay in which a technician connects to the data bus for a status assessment prior to routing the unit into a service bay.
triage tool usually an HH-EST used to read the status of controllers with an address on the data bus.
triangulation see *trilateration*.
tridem axle this term is generally, but not exclusively, used to describe a three-axle semitrailer.
trilateration the geometric principle used by vehicle GPS to identify position. Based on having at least three references of known distance and calculating the point at which circles around those reference points intersect. See Chapter 6 for a full explanation.
trip in technology, this term is used to mean to open. For example, a safety valve is *tripped* when its specified pressure is achieved, or a circuit breaker is *tripped* when its specified current is exceeded.
trunnion in U-joints, the bearing races of a cross or spider assembly. Also the suspension subframe assembly used on some tandem drive highway tractors.
truth table a chart showing all of the outcomes possible from a set of inputs; used to depict processing or switch outcomes.
T-Series Certification ASE certification tests for medium-/heavy-duty truck technicians.
tungsten filament the luminescent element in a light bulb; resists current flow through the bulb circuit, heating it to a white-hot temperature and emitting both

heat and visible light in a reaction known as incandescence.

turbine the output (driven) member of a torque converter splined to the input member of an automatic transmission. In fluid power arrangements, turbines are driven by impellers in a fluid coupling. Turbine and transmission input shaft speed should be equal.

turning angle used to describe the difference in turning angle in each front wheel during a turn; more commonly called *Ackerman angle.*

turntable term used to describe a fifth wheel coupling device outside of North America.

twist damage frame damage in which the rails are twisted off a level plane in relation to each other. Usually caused by an accident, aggressive docking, or front-end abuse such as snowplow operation and can occur either with or without frame rail deformation.

twisted-wire pair used as the data backbone of a multiplex system. The wires are twisted to minimize EMI susceptibility. In the J1939 data backbone, the positive wire is color coded yellow and the negative wire is color coded green. High bus wires are additionally protected by foil shielding.

two-pedal system an automated transmission using a clutch but no clutch pedal. In two-pedal systems, clutch operation is controlled electronically and actuated by air or hydraulically.

two-piece clutch brake usually refers to a replacement clutch brake that can be fitted to a transmission input shaft without removing the transmission. An example is the Spicer KwikKonnect.

two-speed axle a final drive axle with two output ratios controlled by the driver using an electric or pneumatic switch. Also known as a *dual range axle.*

two-way check valve an air system shuttle valve that is supplied with air from two sources and has a single outlet. The valve is designed to shuttle and bias airflow from the higher of the two sources to the outlet.

U-bolt a U-shaped steel bolt threaded on each end used to clamp components to a beam or shaft. Multileaf spring packs are clamped to axles with U-joints.

U-joint see *universal joint.*

ultracapacitors electrical potential storage device that physically separates the positive and negative charges, holding each in a manner that can be compared to static electrical buildup. Used in conjunction with battery banks in some hybrid electric vehicles. Also known as *supercapacitors.*

UltraShift an Eaton Fuller two-pedal automated transmission, introduced as a 10-speed ratio but planned for a broader range of ratio options in the near future.

UltraShift Highway Value (HV) an Eaton AMT with claims of increased longevity.

underinflation a tire that has less than the minimum specified pressure.

understeer an underresponse to steering input often caused by steering tire slip at high speeds. In other words, a condition in which a vehicle turns a little less than the driver intends.

undertread the rubber between the base of the tread groove and the top belt.

unibody frame design in which all the shell components form the frame. Many automobiles and most van trailers use a unibody principle, meaning that the van floor, roof, bulkhead, doors, and sidewalls collectively compose the frame assembly. Also known as *monocoque* design.

Unified National Coarse (UNC) (American) standard coarse thread pitch fastener. Previously known as United States Standard (U.S.S) thread.

Unified National Fine (UNF) (American) standard fine thread pitch fastener. Previously known as S.A.E. thread.

unitized axle a unitized steer axle such as the Meritor Easy Steer that incorporates an integral steering knuckle and brake spider assembly or a low maintenance dead axle. A unitized axle has no add-lube window and no field serviceable parts.

Unitized Hub System (UHS) a hub assembly that is factory filled with synthetic grease and eliminates the need for bearing adjustment and seal replacement when removed from the axle.

unitized seal a wheel seal in which a static seal is made at both the axle hub installation bore and the axle spindle; the dynamic portion of the seal is self-contained within the seal itself so it turns within itself.

universal joint an assembly consisting of a trunnion cross and bearings that connects a pair of yokes for purposes of transmitting drive torque on an angled plane.

unsprung weight the weight of a vehicle not supported by the suspension. Ideally, this weight is kept as low as possible because unsprung weight forces react through the suspension.

uplink ground-to-satellite telecommunications hardware.

upper coupler the kingpin deck plate on a trailer that rides on the fifth wheel on a tractor or lead trailer. Built into the subframe of the trailer. Kingpin may be welded or bolted (riveted in much older trailers) into the upper coupler assembly.

upshift to shift up a gear ratio.

Urge-to-Move an Eaton/Dana wet clutch system used in conjunction with its AutoShift transmission; the system provides continuous 45 lb-ft. of torque to the transmission when the vehicle is stationary and the shift selector is in drive.

vacuum a perfect vacuum refers to an absence of matter and, therefore, an absence of pressure. The term is often used to refer to a partial vacuum, which means any pressure condition below atmospheric.

valence see *valence shell.*

valence shell outermost orbital shell in an atom. The number of electrons in the valence plays a role in defining the electrical and physical characteristics of an element.

validity list a listing or database of technical service bulletins (TSBs) indicating which TSBs are in effect.

valve body and governor test fixture use to bench test Allison valve body assemblies. The valve body mates to a test fixture manifold and is used to hydraulically replicate the performance of the transmission.

valve ring adjusting tool a special tool required to adjust the valve rings in Allison automatic transmissions.

van any of a number of different types of rectangular box-shaped vehicles designed to enclose the load they haul, which may be integral with the truck or a trailer. A majority of the trailers we see on our roads today are multipurpose, semitrailer vans that are built on a unibody frame.

vane a blade-shaped projection from a rotor designed to act on or be acted on by fluid movement.

vane compressor an A/C compressor that uses a *vane pump* operating principle.

vane pump a fluid pump consisting of a slotted rotor that turns within a pump *liner* or *cam ring.* The rotor is

milled with radial slots in which vanes (blades/plates) float, being thrust out by centrifugal force into the liner to form pump chambers when the rotor turns.
vapor a gas. One of the three states of matter, the other two being solid and liquid.
vaporization to change state from a liquid to a gas. The result of boiling a liquid. To evaporate.
variable pitch stator used in some off-highway torque converters to permit a larger torque range.
variable-rate spring any number of different types of springs or spring packs designed to alter their deflection rate as they are loaded. Used in truck suspensions to provide a soft spring ride when unloaded and at the same time accommodate heavy weights loaded.
vehicle body clearance (V/C) distance from the inside of a vehicle tire to the outside of the vehicle frame or suspension.
vehicle identification number (VIN) a seventeen-character (alphanumeric) identification assigned to every truck when manufactured. Information contained in a VIN includes OEM, plant of manufacture, date of manufacture, GVW, cab codes, engine type, etc.
vehicle interface module (VIM) Allison module that handshakes data exchange (both ways) between WT electronics and the chassis data bus using SAE J1939 communication protocols and hardware.
vehicle onboard radar (VORAD) see *VORAD*.
vehicle retarder auxiliary braking device used to supplement the service brakes; in most cases these convert chassis kinetic energy into hydraulic or electrical energy and are located in the driveline. Distinct from engine or transmission retarders.
velocity speed.
vent door the means used to introduce outside air into reefer vans prior to the introduction of *fresh air exchange* systems that use a heat exchanger.
Vernier scale a movable scale used on measuring instruments by which fractional readings can be made by divisions of an adjacent graduated scale. Used on micrometers to measure dimensions smaller than one thousandth of an inch.
vertical exactly upright or plumb.
vibration rhythmic or arrhythmic movement or oscillation. In vehicle technology, usually refers to an undesirable condition; for example, *driveline vibration*.
virtual breaker see *virtual circuit protection*.
virtual circuit protection logic controlled circuit protection in which current overloaded circuits can be opened in the same way electromechanical protected circuits are without any physical damage or event taking place.
virtual fuse see *virtual circuit protection*.
viscosity the sheer resistance of a liquid often used to mean oil thickness or resistance to flow.
Vise Grips™ locking pliers manufactured with various different types of jaw design. A very commonly used multipurpose tool.
vocational medium- and heavy-duty trucks used for specific work applications rather than linehaul. Examples would be dump trucks, milk tankers, fire trucks, cement trucks, cranes, fuel tankers, etc.
Voith DIWA an electronically controlled automatic transmission used in some transit buses. The unit features a pair of planetary gearsets to produce four forward speeds and a unique centrally located *counter-rotating torque converter* that doubles as a driveline retarder.
volt unit of electrical pressure, electromotive force, or charge differential. The pressure required to force a current of 1 amp through a resistance of 1 ohm.
voltage electrical pressure, electromotive force (EMF), or electrical potential measured in volts.
voltage drop decrease in voltage measured after current passes through a resistance. Also known as *voltage loss*. See definition for the second of *Kirchhoff's Law(s)*.
voltage limiter any of a number devices used in chassis electrical systems to condition chassis voltage to smooth voltage spikes and current surges. Used in most computer control systems (5 volts is common), dash electronic circuits, and in headlamp circuits to increase bulb life, not allowing voltage to exceed 12.6 volts.
voltage loss see *voltage drop*.
voltage regulator controls voltage in a circuit either electromechanically (obsolete) or electronically.
VORAD vehicle on-board Doppler radar system classified as a *collision warning system (CWS)*. A forward motion detection and monitoring system designed to trigger driver alerts when a possible imminent collision is detected. Recently replaced by Bendix *Wingman* which can generate accident avoidance interventions.
vortex eddy force caused by spiraling fluid. A truck moving down a highway creates a *vortex* behind it as it cuts through the air. Term is used to describe both gas and liquid dynamics.
vortex flow eddy force created by rotational movement. Both a whirlpool and a tornado produce vortex flow, as do the dynamics of a torque converter. See *vortex oil flow*.
vortex oil flow circular oil flow in a torque converter that occurs as oil is forced from the impeller to the turbine, then back to the impeller during operation.
V-Ref reference voltage. ECU or ECM conditioned voltage (usually 5 volts DC) supplied to various sensors (such as thermistors, potentiometers) in truck electronic circuits. In most cases, a portion of the voltage is returned to indicate the status of sensor.
waddle slang term used to describe lateral vehicle movement usually accounted for by a wheel or tire problem.
walking beam a pivoting, tandem-axle suspension beam that maintains axle spacing and alignment while maximizing tire-to-road contact.
wander tendency of a vehicle to drift from side to side off its tracking course.
water-in-fuel sensor (WIF) a common input circuit sensor that distinguishes between the different dielectric properties of water and fuel.
watt a unit of power measurement. In electricity, 1 watt is the power required to accomplish work at the rate of joule per second. One watt of power is consumed when 1 amp of DC is moved through a resistance of 1 ohm in 1 second.
Watt's Law used to calculate power consumed: power (P) equals (=) voltage (V) multiplied (×) by current (A): $P = V \times A$
wear compensator a clutch wear compensation device located in the clutch cover. Consists of an actuator arm, a retainer, and worm gear. When the clutch is actuated, the actuator arm rotates the worm gear, which turns the adjusting ring within the clutch cover, removing slack between the pressure plate and the driven discs.
web term used to describe the height factor of a frame rail used for section modulus calculation.
wedge brakes an air foundation brake system in which an air chamber forces a wedge between brake shoes

forcing them into a brake drum to effect retarding force. Wedge brakes are usually actuated on both sides (unlike S-cam brakes) and so have higher theoretical braking efficiencies.

welch plug a plug fitted into the female end of a slip spline assembly to seal the grease and stub travel cavity.

wet cell battery term commonly used to describe a *lead-acid battery* using a sulfuric acid solution as electrolyte.

wet clutch a fluid clutch system used in Eaton/Dana Urge-to-Move systems in conjunction with some automated transmissions. Wet clutches were also used by Mack Trucks until the 1970s in some vocational applications.

wet seal describes a seal used to seal a liquid medium (oil) rather than grease, which would require a *dry seal.*

wet tank commonly used slang term for an air brake system supply tank. See *supply tank.*

wheel assembly typically consists tires mounted on rims, wheel spacers, hubs, wheel bearings, and an axle spindle.

wheel balance the state of balance of a wheel assembly, not just the tire/rim. In trucks wheel balance is critical because of the high mass of rotating components other than the tires. Usually tested on dynamically on chassis.

wheel chocks wedge-shaped blocks jammed under truck wheels to prevent the rig from moving during brake service.

wheel cylinder a hydraulic piston(s)-type valve used to spread brake shoes into a brake drum when actuated by the master cylinder.

wheel dolly single or dual wheel assembly hoist on rollers. The device can be rolled under a wheel, which is supported by a pair of forks when raised.

wheel end term used to describe the axle spindle and bearing used to mount a wheel assembly to an axle. Wheel end procedure refers to the means used to adjust the bearing endplay or mount preset hub assemblies onto the axle.

wheel hop wheel or axle oscillation that results in irregular tire-to-road contact. Caused by imbalance in the rotating mass (weight) of the wheel assembly.

wheel shimmy steering oscillations either side of the intended steering track.

wheel speed sensor reluctor-type electromagnetic device used to signal wheel speed data to an antilock controller. Consists of a permanent magnetic and toothed reluctor. As the reluctor wheel cuts through a magnetic field, AC is generated and a frequency signal produced. See *pulse generator.*

wheelbase dimension of a tandem drive axle truck measured between the center of the front axle and the centerline of the tandem bogie. In a two-axle vehicle, it is the distance between the centerline of the front and rear axles.

wide-base singles single truck tires designed to replace sets of duals on drive and trailer axles. The current generation of these tires is designed to replace duals without incurring a weight reduction penalty.

wide-base tires references a range of single-tire designs intended to replace duals on highway commercial vehicles. Current versions of wide-base tires may have the same load-carrying capacity as duals with the advantage of a lower rolling resistance.

WinAlign Hunter Engineering Windows driven software used with its current generation of alignment equipment.

windings any of a number of different types of coils used in electromagnetic components, including solenoids, starter motors, alternators, etc.

wind-up axle axial rotation suspension reactions to aggressive braking or acceleration. Wind-up has a clock spring-type effect.

Wingman an active cruise and braking system (ACB). Wingman is an electronic *collision mitigation system (CMS)* that uses Doppler radar to detect the relative velocity of objects/vehicles in front of a bumper-mounted antenna and microwaves to sense lateral objects. The Wingman ECU has an address on the chassis data bus and can broadcast mitigation requests to the engine and brake management modules.

wireless theoretically any type of data transaction that takes place without a physical connection between the units exchanging the information; however, in truck technology, the term usually applies only to the microwave truck-to-terminal sensor exchanges that take place at a terminal refueling island, not to satellite or GPS transactions.

work when effort or force produces an observable result (movement), work results. In a hydraulic circuit, this means moving a load. To produce work in a hydraulic circuit, there must be flow. Work is measured in units of force multiplied by distance, for example, pound-feet.

work order the means of communicating a maintenance or repair task to the technician responsible for the repair: can be hard copy or electronic. In the event of a disputed bill, the work order assumes a legal status and its will be interpreted by lawyers. Technicians must be aware of the consequences of anything they input onto a work order. Also known as a *repair order.*

Workplace Hazardous Materials Information System (WHMIS) Under Right-to-Know Law, all employers are required to provide awareness training to every employee to recognize potentially dangerous materials that might be encountered in his specific workplace.

World Transmission (WT) Allison full authority automatic transmission.

World Transmission Electronic Controls (WTEC) the control electronics used in Allison WT transmissions. The software version is usually identified by a Roman numeral, for instance, WTEC III.

worm gear spiraled single-groove shaft gear.

wrecker a tow truck.

wrench tool for turning or holding fastener heads (nuts and bolts). The width of the jaws or span determines its nominal size.

WTEC III (Allison) World Transmission Electronic Controls version 3. The software version is identified by a Roman numeral.

WTEC controller the Allison World Transmission ECU module.

wye a diode bridge arrangement in which three diodes are connected in a "Y" arrangement with each diode extending from a neutral junction; used to rectify AC to DC.

xenon headlamps also known as *high-intensity discharge* (HID) headlamps, xenon technology produces light using an arc between electrodes through a gas (xenon) consisting of an arc source, electrodes, and ballast module.

XVision a Bendix night vision system that uses infrared technology to assist driver visibility when it is dark. The XVision display is dash mounted and uses infrared thermal imaging to create a virtual image of the road ahead.

yaw vehicle lateral tracking off the intended straight-ahead or turning radius; a vehicle traveling sideways in a skid would be an extreme case of yaw.

yield strength the highest stress a material can withstand without permanent deformation expressed in psi. In steels, yield strength is typically about 10 percent less than *tensile strength*.

yoke any of a number of different fork-shaped components. Shift yokes (on shift rails) fit to dog clutches in a transmission to move them in and out of engagement. A pair of drive shaft yokes couple to each other using a U-joint as an intermediary.

yoke sleeve kit a driveline yoke repair kit consisting of an interference fit sleeve that is installed over the seal race.

zener diode specialty diode designed to reverse-bias conduct only after a specific voltage value is attained. Used in electronic voltage regulators.

zerk see *zerk fitting*.

zerk fitting a standard grease nipple. Its orifice is sealed with a spring-loaded ball that unseats when pressurized grease is applied to it.

INDEX

A

Absolute pressure, 368
Absolute viscosity, 392
Absorbed glass mat (AGM) batteries, 200
Accident analysis audit trails, 184
Accident and damage control response, 184
Accident avoidance systems, 184–185
A/C clutch coil, 1237
A/C compressors, 1199–1200
AC conductance testing, 209
Accumulators, 372, 373, 631, 1205
A/C data logic, 1238–1239
Acetylene cylinders, 33
A-circuit voltage regulators, 234
Ackerman arms, 790
Ackerman geometry, 799–800
Ack field, 349
AC leakage test, 239
ACom diagnostics, 324
ACom software, 1040
Acronyms, 1242–1246
Active codes, 315, 317, 617–618
Active cruise with braking (ACB), 185, 328
Actuators, 380–381
Adapter rings, 405
Adaptive logic, 645
Adaptive suspension, 847–848
Additives, 109
ADEM software, 683
Adhesives, 84
Adjustable pliers, 54
Adjustable wrenches, 51
Adjusting rings, 403, 595
Aftercoolers, 943
Aftermarket modifications, 1126–1028
Air and oil pressure gauge failure, 283
Air bag, 337
Air bag control unit (ACU), 337
Air bag deployment principle, 338
Air bag SRS, 337–338
Air brakes, 9–10, 117–118
    adjustment, 1092–1097
    assessing performance, 1057–1063
    brake adjustment, 1092–1097
    certification, inspection, and testing, 1102–1103
    commercial vehicle stopping distances, 1103–1104
    disc brake service procedures, 1099–1100
    foot valve preventive maintenance, 1069–1071
    foundation brake service, 1085–1092
    general brake valve troubleshooting, 1097–1099
    maintenance and safety, 1053–1056
    parking/emergency circuits, 1076–1081
    safety, 1053
    service application circuit, 1069–1076
    slack adjusters, 1081–1084
    subsystems, 935–937
    supply circuit service, 1063–1069
    system components, 964–981
    system operational checks, 1062–1063
    test equipment, 1056–1057
    wedge brake servicing, 1102
Air compressors, 937–939, 1063–1065
Air conditioning. *See* Heating, ventilation, and air conditioning systems (HVAC)
Air-conditioning protection and diagnostics system (APADS), 1234
Air conditioning protection unit (ACPU), 1234
Air-conditioning systems
    certification, 1197
    full recharge of, 1226
    performance testing, 1213–1216
    problems, 1227–1229
    protection and diagnostic system (APADS), 1234
    protection unit (ACPU), 1234
    purging and flushing, 1221–1225
Air control disassembly, 532–533
Air cranking motors, 985
Air disc brakes, 962–964, 1100
Air drills, 59
Air dryers, 943–944, 1065–1069
Air filter/regulator assembly, 473
Air filter restriction gauge, 117
Airflow, 1230
Air leaks, 527, 1098
Air-over-hydraulic brake systems, 1002–1005
Air pressure gauges, 283, 981
Air ratchet wrenches, 59
Air regulator problems, 527
Air-release fifth wheels, 1170–1171
Air springs, 847
Air spring suspensions
    adaptive suspension, 847–848
    components, 849–852
    servicing, 862–867
Air start systems, 985–986
Air supply circuits, 937–947
Air supply tests, 532
Air tanks, 948, 1055
Alarm systems, 286
Alcohol evaporators, 945
Algorithms, 347
Alignment systems,802–803
Allen wrenches, 51–52
Allison DOC
    connecting DOC to transmission, 634
    DTC and diagnostic information, 634–635
    navigation, 634
    pull-down menus, 634
Allison 1 Electronic Controls (ATEC), 629
Allison hydromechanical transmissions, 578
Allison TC10
    clutches, 667–669
    gearbox module, 667
    hydraulic schematics, 669–670
    input module, 665–667
    maintenance, 681–682
    output module, 667
    powerflows, 670–681
    transmission configuration, 665
    wiring schematics, 682
Alternating current (AC), 138–139
Alternative battery technology, 219–222
Alternators
    bench test, 245
    brushless, 233
    charging system failures and testing, 235–240
    construction of, 228–231
    disassembly, 240–245
    full-field testing, 237
    mountings, 246
    operation, 231–235
    output testing, 237
    reassembly, 245
    residual magnetism and, 245
    33/34 SI specifications, 245–246
    three-phase, 233

Alternators (*Continued*)
transformer rectifier (TR) 12/24 alternators, 234–235
voltage regulators, 233–234
Aluminum alloy frames, 1112
Aluminum bores, 1006
Amboid gear sets, 729
American Trucking Association (ATA), 92
Technology and Maintenance Council (TMC), 19, 100, 181, 201, 921
Ammeters, 283
Ampere, 130
Ampere-hour ratings, 201
Amperes, 130
Amplification, 163–164, 166
Analog integrated circuits, 169
Analog-to-digital converter (ADC), 180
AND gates, 170
Angle spring clutches, 418
Annulus, 558
Anodes, 135, 161, 193
ANSI symbols, 392, 393
Antenna assembly (AA), 186, 187
Anticorrosion agents, 109
Antifreeze, 116, 117
Antilock brake systems (ABS), 184, 1023
add-ons, 1044–1046
automatic traction control (ATC) systems, 1035
channels, 1033–1034
components, 1026–1033
configurations, 1016, 1033–1034
control logic, 1028–1029
diagnostics, 1037–1040
electronic braking systems, 1047–1049
input circuit components, 1026–1027
operation, 1018
operation basics, 1024–1026
output circuit components, 1029–1033
service procedure, 1040–1044
trailer ABS, 1035–1037
trailer ABS warning lights, 1031–1032
voltage check, 1040
Antilock braking system (ABS) module, 1024
Antirattle springs, 405
Antirollover electronics (ARE), 184, 346
Antirust agents, 109
Application pressures, proportioning and timing, 982
Applied moment, 1109
Applied stroke, 1094–1095
Arbitration /Priority fields, 349
Arc welding station, 35
Armatures, 148, 253, 254–255
Armature segments, 256
Armature windings, 255
Asbestos, 36, 425
AS/NZ 2210.3 footwear, 25
Aspect ratios, 888
ASTM Class 75–approved footwear, 25
Atm (atmosphere), 367
Atmospheric pressure, 366–367
Atomic shells, 129
Atomic structure and electron movement, 128–132
Atomic theory, magnetism and, 136
A-train, 1133
Attraction/repulsion, 131
A-type locking mechanism, 1165–1166
AutoClutch, 408
Autodetect, 645–646
Automated maintenance environment (AME), 89
condition-based maintenance algorithm and, 91
cost effectiveness of, 92
managing, 91–92
Automated manual transmissions (AMTs), 8, 449, 603. *See also* Automatic transmissions; Transmissions
advantages of, 603
clutch actuators, 619–621
components, 449
fault detection and data links, 617–619
vs. mechanical, 449
UltraShift, 620
Automatic chassis lube systems (ACLS), 112–114
Automatic downshifts, 580–581
Automatic emergency system tests, 1061
Automatic Gear Shift (AGS), 603
Automatic slack adjuster (ASA), 1081
Automatic traction control (ATC) indicator lights, 1032
Automatic traction control (ATC) systems, 184, 1035
Automatic transmission fluids (ATF), 111
Automatic transmissions, 7, 557. *See also* Automated manual transmissions (AMTs); Transmissions
Automatic upshifts, 580
AutoShift, 603
Auxiliary electronic systems, 183–184
Auxiliary filters, 590
Auxiliary gear sections, 470–486, 514
Auxiliary power units (APUs), 1234
AW Clutch fluid, 443
Axial piston pumps, 376
Axial tire runout, 881
Axis of rotation, 451
Axle adjustments, 870
Axle alignment, 867–870
laser, 867–868
straightedge and tram method, 868–870
Axle classifications, 2–5
Axle-end balance, 909
Axle-end vibration, 908–909
Axle lube service, 753
Axle offset, 801
Axle range interlock, 743, 745–746
Axles, 8
drive axles, 724–734
drive axles shaft configurations, 747
nondriving axles, 747–749
power dividers, 734–740
two-speed drive axles, 740–747
Axle service and repair
carrier, axle and gearing identification, 757
cleaning and inspection, 756–757
drive axle lubrication, 753–756
failure analysis, 757–774
Axle shafts, 723
Axle shims, 716
Axle speed variables, 737

B

Backdrive, 820
Backfill pressure, 631
Backfire, 34
Backlash, 451
Baffles, 372
Balanced vane pumps, 375
Balancing coil, 285
Ball hitch, 1184, 1185
Ball joints, 790
Ball joint separator, 59
Ball screw assemblies, 608–609
Bandwidth, 345
Banjo housings, 725, 750, 754
Bar (barometric pressure), 367
Bar coding, 41
Base terminals, 163
Batteries, 139
alternative technologies, 219–222
charge/discharge cycles, 195–198
charging, 210–216
conductance testing, 216–219
construction, 193–194
corrosion, 201
fast charging, 211
heaters, 120
maintenance, 201–205
maintenance procedure, 202–204
management systems, 223–224
operating principles, 193–200
ratings, 200–201
servicing, 204
sourcing a drain, 293
storage and recycling, 215–216
terminals, 251
testing, 205–210, 216–219, 220
test procedure, Midtronics EXP, 216–218
voltage, 290
winterizing, 204–205
Battery banks, 201
Battery cables, 218
Battery charging
fast charging, 211
guide, 210, 211

instructions for, 211–212
safety, 212
slow charging, 210–211
Battery Council International (BCI), 200
Battery-drain test, 292–293
Battery management information system (BMIS), 224
Baud, 345
B-circuit voltage regulators, 234
Beam splitters, 167
Bearing assemblies, 699
Bearing fatigue, 509
Bearing plate U-joint replacement, 707–709
Bearings, 509–510
Behavioral Analysis Safety Improvement Categories (BASIC), 100
Bell housing removal, 514
Belts and tensioners, 110
Bench grinder, 61
Bending moment, 1109
Bendix XVision, 275
Bernoulli's principle, 370–371
Between zero and full pedal travel, 176
Bias, 160–161, 162
Bias ply tires, 884
Binary switches, 1204
Binary system, 170, 171–173
Bipolar, 164
Bipolar transistors, 164–165
Bits, 171
Blast charging, 214
Blind-spot display (BSD), 187
Blind-spot sensors (BSS), 186, 187
Blink codes, 302, 1038–1039
Block diagnosis chart, 71, 74f3–64
Block heaters, 1234
Block shifting, 461–463
Blower motor/fan assembly, 1211
Blowgun, 60
BlueCool truck thermal storage systems, 1239
Bobtailing, 951
Bobtail proportioning valves, 951, 970
Body control module (BCM), 358
Bogie, 842
Bokodes, 41
Bolster plates, 1161
Bolt and torque specification, 1125–1126
Bolt attachments, 1127
Booster master cylinder, 993–994
Boot-up, 321
Bottoming, 451
Box-end wrenches, 51
Brake certification, inspection, and testing, 1102–1103
Brake chambers, 932, 972–979
Brake drums, 916, 961, 1008–1009
Brake fade, 961
Brake feel, 965
Brake fluids, 990
Brake imbalances, 983–984
Brake lines and hoses, 1012–1013
Brake-Link, 316
Brake pads, 932
Brake shoes, 932, 959–960
Brake spring pliers, 1008
Brake systems, 9–10
air brakes, 932–935
air brake subsystems, 935–937
air brake system components, 964–981
air start systems, 985–986
air supply circuits, 937–947
balance, 981–985
bleeding, 1014–1015
compatibility issues, 983
do not fully apply, 1098–1099
dual circuits, troubleshooting, 1252
fail to release completely, 1099
foundation brakes, 958–964
hydraulic ABS, 1015–1018
hydraulic brakes, 932
imbalances, 983–984
oversensitive, 1098
primary circuits, 947–950
secondary circuits, 950–954
slow application/release, 1074
trailer circuits, 954–958
will not apply/release, 1074
Brake torque balance, 982, 983, 1055
Brake valves, 948, 950, 957, 1097–1099
Breakout boxes, 309
Breakout Ts, 309
Breathers, 590
Bridge formula, 1149
Brine, 1117, 1154
Brinelling, 509, 709–710
Brine solutions, 1117, 1154
British thermal units (Btu), 1192
Broken-back drivelines, 701
Brushes, 229–230, 239, 253
Brushless alternators, 233
B-train, 1133
B type locking mechanism, 1166
Bucking bars, 1156
Buckling and mashing, 1118
Buffing wheels, 61
Bump caps, 26
Bus arbitration, 348
Bushing installer, 59
Bushing removal, 59
Bus systems, 347, 348
Bus topology, 348
Bus troubleshooting, 633
Butt splice, 334
Bytes, 172

C

Cab air suspension systems, 872–874
Cables, 251–252
Cab-lift positions, 108
Cab-over-engine (COE), 26, 119, 466
Cab ventilating and heating systems, 1230
Cage bolt, 1053
Caging spring brakes, 976–977
Calcium chloride, 1117
Calipers, 1000, 1100
Camber, 792, 794–795, 890
Cam ring, 375
CAN 2.0, 347
Capacitance, 149–151, 209
Capacitors, 149–150
functions, 150
types, 150
Capacity test, 208
Carbon-arc cutting, 35
Carburizing flame, 34
Carriers, 558, 559
Caster, 793–796, 890
Caster shims, 795–796
Cast iron, 1006
Cast spoke wheels, 879–881
Caterpillar CX truck transmissions
CX calibration, 688–692
CX construction, 683–684
CX electronics, 687
CX hydraulic circuits, 687–688
definitions, 556–557
ECPCs, 684
external component location, 692
hydraulic retarders, 581
hydraulic systems, 573–581
inspection and maintenance, 585–593
main pressure and tap locations, 686–687
maintenance, 686–687
operational temperature limits, 686
power take-offs (PTOs), 599–600
shift pad, 687
shift point adjustments, 594–599
standard equipment, 683
troubleshooting, 599
WTEC electronics, 643–646
WT hydraulic circuits, 649–650
WT modular construction, 635–636
WT output circuits, 646–650
WT troubleshooting, 663–665
Caterpillar Electronic Technician (ET), 324
Caterpillar Service Information System (SIS), 70, 323
Cathodes, 135, 161, 193
CBM. *See* Condition-based maintenance (CBM); Performance-based maintenance (PBM); Predictive-based maintenance (PBM)
C-channel, 1108
CEL (check engine light), 302
Center support bearings, 700–701
Central gateway (CGW), 358
Central processing unit (CPU), 179, 186–187, 320
Centrifugal clutches, 408
Ceramic linings, 433
Certification, 14
Chalmers suspension, 846–847
Channeling, 111

Characteristic impedance, 350
Charge cycle, 196
Charge differential, 134
Charging circuit testing, 239
Charging cylinders, 1120–1121
Charging stations, 1120–1121
Charging systems, 203, 227–246
  alternator construction, 228–231
  alternator operation, 231–235
  failures and testing, 235–240
Charging voltage, 290–292
Chassis and shop electrical safety, 31–33
Chassis data buses, 360
Chassis electrical circuits, 270–294
  auxiliary electrical equipment, 279–289
  rapid check of truck electrical circuit, 290–294
Chassis frames, 10–11
  alignment, 1118–1121
  construction, 1113–1116
  designs, 1108
  frame, side rails, and cross-members repair, 1121–1128
  frame loading, 1112
  theory of, 1109–1113
  trailer frames, 1108
Chassis grease, 203
Chassis lube systems (ACLS), automatic, 112–115
Chassis lubricants, 111–112, 201
Chassis management module (CMM), 358
Chassis vibration diagnosis, 713–719
Chassis wiring and connectors, 32
Check valves, 380, 933, 972
Chemical tests, 117
Chips, 169
Chisels, 48, 59
Chlorofluorocarbon refrigerants, 1194
Chopper wheels, 178, 1026
Circuit analysis, 289
Circuit board trainers, 169
Circuit breakers, 143, 287–288
Circuit components, 142–143
Circuit protection devices, 143, 287–288
Circuit symbols and diagrams, 334–335
Clamping force, 79
Clean Air Act, 1194
Clean idle certification, 124
Clearance-sensing automatic slack adjusters, 980
Clear codes, 317
Clevis pin installation, 1082
Clients, 345
Climate control, 1234–1239
Climate control, electronically managed, 1234–1239
Climbing, 451
Clock speeds, 347
Closed-center hydraulic systems, 371
Closed-center system, 371
Closed circuits, 142
Closed loop logic, 346–347
Closed-loop system, 371
Close ratio (CR), 570
Clothing, 25
Cloud point, 120
Clutch actuators, 401, 619–621
Clutch adjustment mechanisms, 405
Clutch brakes, 408–409, 425
Clutch brake settings, 425
Clutch brake squeeze, 417
Clutch circuits, 579–580
Clutch discs, 433
Clutches, 7, 401
  adjustment, 416–422
  components of, 401–412
  driven members, 405–407
  driving members, 402–405
  heat, 413–415
  installation, 435–439
  linkages, 410–411
  lubrication, 415–416
  maintenance, 415–422
  with perpendicular springs, 422
  release mechanisms, 407–408
  servicing, 425–443
  troubleshooting, 413–415
Clutch fluid, 619
Clutch free pedal, 416
Clutch function, 401
Clutch inspection, 428–435
Clutch linkages, 410–411, 423
Clutch pack assembly, 411
Clutch packs, 411, 562
Clutch pressure plates, 403
Clutch release fork and shaft inspection, 431–432
Clutch removal, 425–428
Clutch yokes, 510
CO_2 cryogenics, 1144
Coefficient of friction, 960
Coil circuits, 288–289
Coils, 151
Coil-type display gauges, 285
Cold checks, 586–587
Cold cranking amperes (CCA), 200
Cold startup, 587
Collar shift, 458–461
Collector terminals, 163
Collision mitigation systems (CMS), 184, 185–186, 328
Collision warning systems (CWS), 185–186, 328–329
Color-coded supply hoses, 956
Combination air/leaf spring suspensions, 839
Combination air/torsion spring suspension, 847
Combination HVAC core, 1233–1234
Combination pliers, 53–54
Combination rubber/torsion spring suspension, 846
Combination valves, 998
Combination wrenches, 51
Come-backs, 14
Commercial Carrier Journal, 19
Commercial Electronic Control (CEC), 629
Commercial Vehicle Safety Alliance (CVSA), 9, 96, 1053
  OOS air brakes, 1104
  OOS air supply system, 1069
  OOS coupling devices, 1186–1187
  OOS criteria and inspections, 99–105
  OOS for wheels and rims, 884
  OOS service brake system, 1074–1076
  OOS standards, 875–876
  OOS steering axle tires, 914
  OOS steering criteria, 833–834
  OOS tires used on other than the steering axle of a power unit, 914
  OOS trailer brake system, 1084
Commercial vehicle stopping distances, 1103–1104
Communications adapter, 300–301, 358
Commutators, 253, 255
Compensating fifth wheels, 1163
Compliance, Safety, Accountability (CSA), 100
Compound angles, 702
Compounding ratios, 449
Compressed air, 1053
Compression fittings, 389
Compressors, 932, 1197, 1199–1202
Computerized system management, 173
Computers, 41–43, 299
Condensers, 1198, 1202–1203
Condition-based maintenance (CBM), 89
Conductance, 133, 209–210, 216–219
Conductance testing, 209–210, 216–219
Conductors, 128, 133–134, 143
Cone shifting, 461–463
6 × 4 versus 6 × 2 versus 4 × 2 configurations, 3–4
Conflict alerts and intervention, 184
ConMet PreSet hub, 924
Connector dummies, 309
Connectors, 384–391
Conspicuity tape, 1152
Constant mesh, 451
Constant rate springs, 839–840
Containers, 1139–1140
Content-based addressing, 348
Continuity, 305
Control circuits, 258–259
Control circuit testing, 263–264
Controlled traction, 728
Controlled traction differential, 731–732
Controller area networks (CAN), 347
Controllers, 173, 298, 344, 1027
Control main pressure, 631
Control main regulator, 631
Conventional theory of current flow, 128, 134, 135

Conversion formula, 1256
Converter cooler circuit, 577
Converter dollies, 1134, 1161
Converter housing modules, 637
Coolant heaters, 1234
Coolant hoses, 110
Coolant hydrometers, 116–117
Coolant leakages, 590
Coolant temperature gauges, 282
Corrosion, 35, 201, 1086, 1117–1118, 1154
Cost tracking, 43
Counteract balancing beads (CBB), 909, 911
Counterbalance valves, 378
Counter-electromotive force (CEMF), 256
Counter-rotating torque converters, 692–693
Countershaft, 502
Couplers, 386
Coupling guidelines, 388, 1174–1178
Coupling point, 541
Coupling systems
  kingpins, 1181–1184
  pintle hooks, 1184
  upper couplers, 1183–1186
Courier fleets, 12
C3 pressure switch tests, 664
CPU. *See* Central processing unit (CPU)
Crack pressures, 968, 983
Cranking circuit components, 250–252
Cranking circuit testing, 260–266
Cranking current testing, 263
Cranking systems
  circuit components, 250–252
  circuit testing, 260–266
  starter motors, 252–260
  starter rebuild, 266–267
Cranking voltage, 290–292
Crankshaft endplay, 430
Crash sensors, 337
CR (close ratio), 570–574
CR low (first) gear, 572
CR neutral, 572
Cross-members, 1107, 1113–1114
Cross tube, 790
Crown gears, 724
CR reverse gear, 573
CR second through fifth gear, 573
Cryogenic systems, 1144
CSA 2010, 100
CSA Class 1–approved footwear, 25
C-trains, 1132
Cummins QuickServe Online (QSOL), 70
Cupping/scalloping, 890–891
Current, 142
Current flow, 134–135, 255–256
Current routing, 257
Current transformers, 307
Curtain-sided freight van trailers, 1139
Cushion and torsion suspension systems, 845–847
Customer data programming, 318
CVSA L1 inspection, 99
CWS/CMS cab, 188
Cyclic redundancy check (CRC) field, 349
Cycling circuit breakers, 287–288

D

Dampened discs, 406
Darlington, Sidney, 166
Darlington pairs, 166
Dash components, 281–283
Dash control, 951
Dash control valves, 933, 952–953
Dash design, 280
Dash lights, 301–302
Dash size, 384
Dash warning indicator lights, 1030–1031
Data bits, 171
Data bus communications, 344
Data buses, 325, 344, 356–362
Data bus viewer (DBV), 633
Data connector hardware, 315
Data connectors, 300, 344
Data display unit (DDU), 281
Data field, 349
Data frames, 348
Data inputs, 173–182
Data link connector (DLC), 300
Data outputs, 180–181
Data processing, 178–179
Data retention, 182–183
Daylight running lights (DRLs), 271
DC generators, 149
DC motors, 148–149
Dead axles, 747, 748–749
Deadline, 99
Deadlining, 99
Deburring, 48
Decelerometer, 1056
Dedicated contract cartage (DCC), 12
Deep cycling, 197–198, 204
Deep ratio (DR), 570
Deep-reduction transmissions, 485–486
Degree programs, 20
Delta shape, 230
Desiccants, 943
Detent mechanism, 617
Detergent additives, 109
Detroit Diesel Diagnostic Link (DDDL), 70
Detroit Diesel Power Service Literature (PSL), 323
Deutsch connectors, 300, 332–334
Diagnostic codes, accessing, 665
Diagnostic data, 283
Diagnostic link connector (DLC), 300, 351
Diagnostic Optimized Connection (DOC), 324, 634
Diagnostic routines, 324
Diagnostic software, 324
Diagnostic tools, 314–315
Diagnostic trees, 326
Diagnostic trouble codes (DTCs), 325, 327, 360, 634, 1239
Diagonal cutting pliers, 54
Dial caliper, 62–63
Diamond damage, 1117
Diamond Logic™, 355–356
Dielectric, 149
Diesel coolant additive (DCA), 117
Diesel electric parallel drive, 6
Diesel electric series drive, 6
Diesel engines, 5–6, 109–110
Diesel exhaust fluid (DEF), 124
Diesel fuel
  cloud point, 120
  pour point, 120
  storage, 120–121
  tank configuration, 121
Diesel hydraulic parallel drive, 6
Diesel hydraulic series drive, 6
Diesel particulate filter (DPF), 124
Diesel Progress, 19
Differential bearing preload
  axle tracking, 782
  check/adjust crown gear backlash, 780
  crown gear runout check, 779–780
  pinion and crown teeth contact adjustment, 780–781
  thrust screw adjustment, 782
Differential carriers, 123, 724–725, 750, 756–757
Differential gearing types, 728–729
Differential lockout, 728
Differential operation, 725–726
Differentials, 8, 723
Digital display (DD), 644
Digital integrated circuits, 169
Digital measuring tools, 68–69
Digital multimeters (DMM), 167, 299, 302
  accuracy, 305
  breakout boxes, 309
  breakout Ts, 309
  features, 308
  measuring current, 307–308
  measuring resistance, 305–307
  measuring voltage, 305
  operating principles, 305
  resolution, 303
  testing semiconductors, 309
  typical tests, 308–309
Digital signals, 158
Digital-to-analog converter (DAC), 180
Dimethyl ether, 5
Diode bias, 162
Diode holder, 230
Diodes, 161–163, 230, 307
Diode trios, 234, 238, 242
Diode types, 162–163
Dipsticks, 371
Direct current (DC), 138, 307
Direct drive, 451, 561–562
Directional control valves, 380
Direction of rotation, 450
Directly operated relief valves, 379

Disc-and-plate synchronizers, 464–465
Disc brakes, 962–964, 1000–1002
Discharge cycle, 196
Disc wheels, 881–882
Dispatch sheet, 41
DM (datalink mechanical) clutch, 411
DNS lights, 664
DOC help fields, 634
Dog-tracking, 801
Door clearance, 1138
Doping, 160–161
Doppler, Christian, 185
Doppler effect, 185
DOS, 321, 322
DOT codes, 888
Double-acting cylinders, 381, 382
Double check valves, 947
Double diaphragm (DD-3) safety actuator, 979
Double-poppet quick connectors, 390
Double-reduction axles, 732–734
Double-rotating ball quick connectors, 390–391
Double trailers, 1139
Double-wire braid hoses, 386
Downshift fallback, 617
Downshifting, 461, 581
Downshifts, automatic, 580–581
Downshift sequence, 473
Downward extension, 1137
Drag link, 788–789
Drawbar eye, 1184
Drawbars, 1185
Drain plugs, 372
DR (deep ratio), 570–571
Drilling, 1126
Drive and coast sides of gear teeth, 451
Drive axle lubrication, 739, 753–756
Drive axles, 724–734, 761
Drive axle shaft configurations, 747
Drive gears, 450
Driveline angles, measuring, 1254
Driveline balancing, 713–717
Driveline configurations, 1253
Driveline geometry, 701
Driveline phasing, 703
Driveline vibration diagnosis, 713–719
Driven gears, 450
Driver communications, 89–90
Driver display unit (DDU), 187
Driver inspections, 92
Driver interface unit (DIU), 187, 328
Driver seats, 874
Driver's manuals, 71
Driver's report form, 92, 95
Drive shaft assemblies
  chassis vibration diagnosis, 713–719
  construction of, 697–703
  function of, 696–697
  inspection, 704–705
  lubrication of, 705–707
  U-joint replacement, 707–713
Driveshaft runout, 716
Driveshafts, 8, 696
Drivetrain, 448
DR low (first) gear, 570–572
DR neutral, 570
Drop-7, 641
Dropbox, 641
Dropbox modules, 641–642
DR reverse gear, 572
DR second through fifth gears, 572
Drum brakes, 932, 999
Dry element/pleated paper air filter, 1064
Dryer reservoir module (DRM), 944
D-train, 1134
Dual-circuit application valves, 937
Dual circuit brake systems troubleshooting, 1252
Dual-circuit master cylinder construction, 993
Dual in-line package (DIP), 170
Dual range axle shift systems
  axle range interlock systems, 745–746
  axle shift operation, 744
  interaxle differential lockout system, 744–745
  standard system, 743–744
Dual wheels, 884
Dummy gladhands, 956
Dump trailers, 1144–1145
Duty cycles, 158
Dynamic balancing, 713–714, 909–910

E

Ear protection, 25
Eaton DM clutch installation, 442
EC AT J1939, 633
ECM-managed control circuit, 264
ECM networking, 347
Edge codes, 961
Edison, Thomas, 128
Eighteen-speed transmissions, 478–482
Electric arc welding, 35
Electronic Technician (ET), 324, 633
Electrical circuit, rapid checking of, 290–294
Electrical circuits, 141–147
Electrical retarders, 10
Electrical system values, 293
Electrical temperature and fuel display failure, 283–284
Electric circuits, 141
Electricity, 127–128, 132, 139–141
Electric motors, 148–151
Electric shifters, 610, 615
Electric shock resistant (ESR) footwear, 25
Electrochemical impedance spectroscopy (EIS), 209, 323
Electrocution, 25
Electrohydraulic control modules, 643
Electrohydraulic valve bodies, 631
Electrolytes, 193, 194–195
  discharge effect on, 195
Electrolytic-type capacitors, 150
Electromagnetic field force, 138
Electromagnetic induction, 140
Electromagnetic interference (EMI), 349–350
Electromagnetic switches, 152
Electromagnetism, 136–138
Electro-mechanical ignition switch, 259
Electromotive force (EMF), 128, 134
Electronic alignment systems, 802–805
Electronically controlled, automatic transmissions (ECATs), 629
  advantages of, 630
  Allison family, 629–630
  caterpillar family, 630
  clutches, 631
  electrics and electronics, 632–633
  hydraulics, 631
  WT torque paths and range management, 650–658
Electronically erasable programmable read-only memory (EEPROM), 180, 183
Electronically variable steering (EVS), 831–832
Electronic and load-sensing steering, 831–833
Electronic automatic transmissions. *See* Automated manual transmissions (AMTs); Automatic transmission; Transmissions
Electronic braking systems (EBS), 1047–1049
Electronic circuits, 298–299
Electronic clutch pressure control (ECPC), 684
Electronic control module (ECM), 159
Electronic controls, 11
Electronic control unit (ECU), 166, 298, 603, 630, 1003, 1024, 1027–1029
  construction and networking, 173, 347
  functions of, 180
  processing cycle, 181
Electronic data interchange (EDI), 41
Electronic engine management, 157
Electronic/furniture van trailers, 1139
Electronic leak detectors, 1215
Electronic lock indicator (ELI), 1171
Electronic noise, 159
Electronics, 128, 157
Electronic Service Analyst (ESA), 324
Electronic service tools (ESTs), 41, 298, 301, 325, 1025, 1213
Electronic signals, 157–159
Electronic system controller (ESC), 355
Electronic systems, 119, 1216
  accident avoidance, 184–185
  auxiliary electronic systems, 183–184

collision warning systems (CWS), 185
flash or blink codes, 302
handheld diagnostic instruments, 313–318
high-voltage hybrid circuits, 294
multifunctional handheld diagnostic tools, 313
splicing guidelines, 334–335
supplemental restraint systems (SRS), 335–340
troubleshooting, 325
using digital multimeters, 302–312
wiring, connectors, and terminal repair, 329–334
Electron movement, 134
Electron pump, 228
Electrons, 128, 157
Electron theory of current flow, 134, 135
Electropneumatic valves, 1049
Emergency telephone numbers, 28
Emissions control equipment, 124
Emitter terminals, 163
End frame assembly, 230–231
End of frame field, 349
End yoke, 8
Energy, 370
Engine coolant, 115–117, 122
Engine cooling fans, 1211–1212
Engine/electronic control modules (ECMs), 173
Engine/electronic control units (ECUs), 173
Engine oils, 108–109, 122
Engine speed (RPM) shift point adjustments, 594–595
Engine stall point, 593
English/Metric toggle, 317
Environmental Protection Agency (EPA), 1195
EPDM (ethylene, propylene, diene monomer), 110
Epicyclic gearset, 557
Equalizing beam suspensions, 844–846, 858–862
E-stroke monitoring, 1095–1097
Ether starting systems, 118–119
automatic system, 118–119
electronic system, 119
Evaporators, 1207–1208
Evaporator thermostats, 1208, 1236
Exciter rings, 1026–1027
Exhaust backfill regulator and a relief valves, 631
Exhaust pipe cutters, 59
Expansion valves, 1198
Express Brake™ shoes, servicing, 1091
External-gear motors, 382–383
External-gear pumps, 373, 374
External ground, 264
Extractor, 82
Extreme Brake™ shoe servicing, 1091
Extreme pressure (EP) additives, 111
Eye protection, 24

F

Fabric braid hoses, 385
84 factor, 195
Fail relay coil, 1030
Failure analysis, 508
Failure mode indicators (FMIs), 327, 360
Fallback modes, 617
False brinnelling, 710
Faraday, Michael, 128
Faraday's Law of Induction, 148
Farads, 149
Fast adaptive mode, 645
Fast charging, 211
Fasteners
elasticity of, 79
extractors, 82
failures, 81–84
frame fasteners, 78
grades and classes, 76–77
overtorque, 81
replacement of, 80–81
re-use of, 81
rivets, 84
sealants, 84
stud removers, 82
thread lock compounds, 84
thread repair, 81, 83
tightening, 79
washers and lock washers, 80
FastMaint software, 106
Fast reverse, 561
Fatigue failures, 758–759
Fault codes, 317, 617
Fault isolation procedures, 618
Fault mode indicator (FMI), 1239
Federal inspection regulations, 100
Federal Motor Carrier Safety Administration (FMCSA), 100
Federal Motor Carrier Safety Regulations, 92
Federal Motor Vehicle Safety Standard (FMVSS), 40
Federal Motor Vehicle Safety Standard 105 (FMVSS 105), 990
Federal Motor Vehicle Safety Standard 108 (FMVSS 108), 271
Federal Motor Vehicle Safety Standard 121 (FMVSS 121), 933, 985, 1047, 1054, 1056
Feedback problems, 294
Feeler gauge, 69
Ferrule fittings, 389
Ferrules, 389
Fiber optics, 167
Field circuits, 234
Field coils, 253, 254
Field effect transistors (FETs), 164, 289, 354
Field Service Modifications, 71
Fifth-gear powerflow, 460
Fifth wheel locks, 1161
Fifth wheels, 11, 1160
coupling, 1174–1178
height and location, 1164–1165
maintenance, 1171–1174
mounting, 1179
operating principles, 1165–1171
sliding, 1180–1181
uncoupling, 1177
Filament, 271
Files, 55
Filler caps, 371
Filter changes, 587–589
Filters, 167
Final drive axle configurations, 727–728
Final drive carriers, 724
Fire classes, 29
Fire extinguishers, 30
Fire safety, 29–30
First gear, 458, 566
Five-speed fallback, 617
Five-speed transmission powerflow, 569–573
Fixed caliper, 1001
Fixed-displacement pumps, 373, 376–377
Fixed-value capacitors, 150
Flammable, 28
Flammable substances, 28, 35
Flange, 8
Flange yokes, 701, 708
Flared fitting tube couplers, 389
Flareless fittings tube couplers, 389
Flaring, 48
Flashback, 34
Flash codes, 302
Flatbed trailers, 1144–1145
Fleet management software, 123–124
Fleet operations, 11–12
Fleet personnel certification, 1102
Fleet shops, 12–13
Flex discs, 536
Flexplates, 536–537
Floating caliper, 1001–1002
Flooded cell batteries, 193
Flooring, 1138–1139
Flow, 369–371
Flow control valves, 379–380
Flow restrictions, 370
Fluid power, 365
Fluids, 122 123, 365, 391–392
Fluorescent lights, 271
Flush type rivets, 1155
Flux lines, 136
Flywheels, 402
housing face runout, 430
housing pilot runout, 429–430
outer surface runout, 429
Following distance alert (FOA), 328
Fontaine No-Slack® II fifth wheel, 1166
Footprint, 884
Foot valve circuit failures, 966
Foot valves, 935, 948, 950, 964–966, 1069–1072
Footwear, 25
Force, 367
Forward drive axle, 737
Forward regulator valve (four-speed), 577

Forward regulator valves, 577
Foundation brakes, 958–964, 1085–1092
Four-channel systems, 1033–1034
Four-speed transmission powerflow, 565–569
Fourth-gear powerflow, 460, 569
Frame alignment, 1118–1121
Frame damage categories, 1117–1118
Frame diagrams, 1018–1120
Frame fasteners, 78–79
Frame layout, 1120–1121
Frame loading, 1112
Frame machine, 61
Frame rail repair, 1118
Franchised dealerships, 13
Franklin, Benjamin, 128
FreedomLine, 603
Free pedal, 415, 416–417
Free pedal travel, 423
Freestroke, 1093, 1094
Free travel, 410
Freight efficiency, 4
Freightliner Daimler AccessFreightliner, 70
Freightliner EZ-wiring, 324
Freightliner ServiceLink (ServiceLit), 323, 324
Freight van trailers, 1139
Frequency, 158–159
Fresh air exchange, 1141
Fretting, 510
Friction, 79
Frictional electricity, 139
Friction discs, 401
Friction facings, 406
Front axle wheel end service, 916–921
Front brake proportioning, 951
Front-end alignment, 791–801
Front support/charging pump modules, 637
Fuel, 121
Fuel-fired heaters, 1234
Fuel gauges, 284
Fuel heaters, 121
Fuel level gauge, 283
Fuel pressure gauge, 283
Fuel/water separators, 121
Full cut pressure relief, 825–824
Full-fielding, 236
Full-fielding internally regulated alternators, 238
Full floating axles, 747
Full-round end yokes, 712–713
Full trailers, 4, 1133–1134
Full-wave rectification, 232–233
Fully oscillating fifth wheels, 1163
Fuses, 287
Fusible plugs, 1205

G

Galling, 709
Galvanic, 193
Galvanic corrosion, 1117
Gaseous-fueled vehicles, 31
Gas-loaded accumulators, 372
Gassing, 196–197
Gateway, 346
Gateway module, 182
Gauges, 371
Gear and bearing pullers, 55–56
Gear design, 451–453
Gear display, 617
Gear lubricants, 110–111
Gear lubricants classification, 111
Gear motors, 382–383
Gear mounting, 455
Gear pitch, 451
Gear pumps, 373–374
Gear range powerflows, 565
Gear ratios, 453–454
Gears, transmission, inspection of, 510
Gear select motor, 609
Gear slipout, 503
Gear types, 455–456
gel cell, 199
Gelled cell batteries, 199–200
General service repair information, 75–76
Generator charging, 214
Generators, 149
Geroter pumps, 374
Gilbert, William, 128
Gladhands, 98, 937, 955–956
Glidecoat, 698
Global PGN, 359
Global positioning satellites (GPS), 183, 184
Global Warming Potential (GWP), 1194
Gloves, 25
Governor cut-out tests, 1058–1059
Governors, 589, 940–942, 1058
Graphical user interface (GUI), 70
Gray iron components, 510
Grease nipples, 698
Greenhouse gas (GHG), 1194
Green 2014 J1939 connector, 357
Grinding wheels, 61
Gross vehicle weight (GVW), 2
Ground, 144
Ground circuits, 250
Grounded circuits, 144
Ground straps, 326

H

Hacksaw, 47
Hair, 26
Haldex codes, 1041–1042
Half-round end yoke, 708
Half-round yoke assembly replacement, 708
Half-wave rectification, 232
Hall effect, 177, 615
Hall effect fluid level sensors, 178
Hall effect principle, 615
Hall effect sensors, 177
Halogen infrared (HIR) headlamps, 272
Halogen lamps, 272
Hammers, 46, 59
Handheld (HH), 299
Handheld diagnostic instruments, 313–318
Handshake, 322
Hand tap, 55
Hand-threading die, 55
Hand tools
  chisels and punches, 48
  files, 55
  gear and bearing pullers, 55–56
  hammers, 46–47
  pliers, 53–55
  saws and knives, 47–48
  screwdrivers, 49–51
  taps and dies, 55
  wrenches, 51–53
Hanger bearings, 700, 707
Hard disk drives, 320
Hard drive, 320
Hard shifting, 504
Hard steering troubleshooting, 1251
Harnesses, 632
Hazardous materials, 35–36
  handling and disposal of, 36, 40
  laws regulating, 36
  and WHMIS, 35
Head, 367
Headlamp voltage, managed, 274
Headlight fixtures, 275–278
Head pressure, 367
Heads-up display (HUD), 281
Heater control valves, 1233
Heater cores, 1232–1233
Heating, ventilation, and air conditioning systems (HVAC)
  air-conditioning service, 1216–1220
  air-conditioning service procedures, 1220–1226
  BlueCool truck thermal storage system, 1239–1240
  cab ventilating and heating systems, 1230
  common air conditioning problems, 1227–1229
  converting R-12 system to R-134A system, 1229–1230
  electronically managed climate control, 1234–1239
  liquid-cooled heating systems, 1231–1234
  performance testing A/C, 1213–1216
  refrigerants, 1193–1196
  refrigeration, 1191–1199
  refrigeration cycles, 1197–1199
  safety precautions, 1212–1213
Heating/air conditioning (HVAC), 11
Heating devices, supplementary, 1234
Heating systems, liquid-cooled, 1231–1234
Heavy bus, 350
Heavy-duty brake fluids, 1005–1006
Heavy-duty onboard diagnostics (HD-OBD), 299
Heavy-duty truck, 2, 5–11
Height control valves, 849–850

Helical, 51
Helical gears, 455–456
Hertz, 908
Hertz, H. R., 158
Hex, 45
Hexadecimal code, 359–360
HID optics, 273–274
High hitches, 1171
High-intensity discharge (HID) lamps, 272
High/low-range shift systems, 473–478
High oil levels, 585–586
High-range, direct powerflow, 480
High-range, overdrive powerflow, 480–481
High-resistance circuits, 144
High-voltage hybrid electrical vehicle (HEV) circuits, 294
Historic codes, 302, 617
Hold-off chamber, 974–975
Hold-off circuits, 951–952, 1092
Hold-off diaphragm, 1092
Holland A-type locking mechanism, 1165–1166, 1172
Holland B-type locking mechanism, 1166, 1172–1173
Holland Simplex fifth wheels, 1166
Horn, 1184
Hose fittings, 386–387
Hoses, 384–386
Hot checks, 587
Hotel loads, 139, 294
Hotel load grid, 11, 294
Hourmeters, 285
Hours-of-service (HOS), 99
H-Series Certification, 15
Hubdometers, 1154
Hub-piloted wheels, 882
Huck fastener, 79
HVAC control module, 1235
HVAC module, 1236–1237
Hybrid drive, 6, 30–31
Hybrid electric drive, 621–622
Hybrid electric vehicles (HEVs), 30–31, 221
Hybrid hydraulic drive, 623–624
Hybrid-hydraulic drive vehicles, 31
Hybrid transmissions, 621–624
Hydraulic brakes, 10, 932
  air-over-hydraulic brake systems, 1002–1005
  disc brakes, 1000–1002
  drum brakes, 998–1000
  fluids, 990
  master cylinders, 992–998
  service procedures, 1005–1015
  system components, 990–1002
Hydraulic circuits, 366, 651–658
Hydraulic control unit (HCU), 993
Hydraulic cylinders, 381
Hydraulic disc brakes, 1000–1002
Hydraulic fluids, 391–392, 587
Hydraulic hoses, 384–386
Hydraulic jacks, 372
Hydraulic launch assist (HLA), 623
Hydraulic levers, 368–369
Hydraulic motors, 382–384
Hydraulic power boosters, 993
Hydraulic presses, 27–28
Hydraulic retarders, 10
Hydraulics
  actuators, 380–381
  components, 371–372
  conductors and connectors, 384–391
  fluids, 391–392
  fundamentals, 366–371
  maintenance, 392–398
  motors, 382–384
  pumps, 372–380
  safe practice, 392, 394
  symbols and schematics, 392
Hydrochlorofluorocarbons (HCFCs), 1194
Hydrodynamics, 366
Hydromechanical automatic transmissions, 556
Hydrometers, 116, 195, 196, 206
Hydrostatics, 366
Hygroscopic, 990
Hypoid gear set, 728–729
Hypoid pinion gears, 728

I

Idler gears, 455
Ignition coils, 151–152
Ignition switches, 252, 259
IMPACT, 323
Impact alert (IA), 323
Impact sockets, 52
Impact wrenches, 52, 58–60
Impedance, 350–351
Impellers, 537
Inactive codes, 302, 317, 617–618
Inclinometers, 715, 755
Independent post lift systems, 61
Independent truck service shops, 13
Indirect current readings, 307
Inductance, 209
Induction pulse generators, 178, 179
Inductive amp clamps, 307–308
Industrial Fastener Institute (IFI), 77
Inertia brakes, 614
Information processing cycle, 173
Infotainment systems, 189
Inhibitor valves, 579
In-line dryers, 1211
In-phase, 703
In-place fallback, 617
Input circuits, 173, 643, 1026–1027
Input conditioning, 179, 180
Input retarders, 581
Input shafts, 432, 511–512
Instrument cluster unit (ICU), 280
Instrument panels and gauges, 280–283
Insulated circuits, 250
Insulators, 133–134
Intake adapters, 1064
Intake filters, 372, 1064
Integral hydrometers, 206–207
Integrated circuits (I/Cs), 169–170
Integrated diagnostics, 349
Intelli-check, 239–240
Interaxle differential lockout, 738–739, 744–746
Interaxle differentials, 728, 735
Interleaf friction, 839
Intermediate plates, 405, 433–434
Internal balancing agents (IBAs), 909
Internal connection checks, 284
Internal engine brake, 10
Internal-gear motors, 383
Internal-gear pumps, 373–374
Internal ground, 264
Internal magnetic solenoid switch (IMSS), 259
Internal Solenoid, 257
International Mobile Air Conditioning Association (IMACA), 1197
International Standards Organization (ISO), 392
Inversion valves, 952, 969–970
Inverter gates, 171
Ions, 130–131
ISO 32, 392
ISO 9141, 347–349, 648
Isolated hybrid electrical circuits, 193
Isolation transformers, 151
ISO viscosity grades, 392
ISO wiring codes, 335, 337

J

Jackknife, 983, 1045
J1708 data connector, 300
J1939, 630
J1939 backbone, 347, 349–352
J1939 connector, 300, 301, 356–358
J1962 connector, 300, 357
J1587/1708, 346
J1587/1708 connectors, 356
J1939, 347
J1939 data bus, 604
JDR 050, 316
JDR100 CANbus, 316
Jewelry, 26
Job classifications, 14–19
John Deere StellarSupport, 323
JOST fifth wheels, 1166
Jounce, 838, 908
Jounce blocks, 838
Journal cross, 698
Joystick, 321
J standards, 181
Jumpout, 503
Jump-starting, 212, 214–215
Junction FETs (JFETs), 165
Junction-type semiconductors, 165
Just-in-time (JIT) delivery, 12

K

K-baud, 345
Keyboard, 321
Kilopascals (kPa), 366
Kinematic viscosity, 392

Kingpins, 11, 789, 1161
Kingpin inclination (KPI), 797
Kingpins, 11, 789, 1161, 1181–1184
Kirchhoff's Law, 147
Kirchhoff's Law of Current, 147
Kirchhoff's Law of Voltage Drops, 147
Knuckle pins, 789
Kwik-Adjust, 405, 420
Kwik-Konnect, 409

L

Laboratory sample analysis, 117
Lab scopes, 324–325
Ladder switches, 353
Laminar flow, 371
Lamp maintenance, 275
Landing gear, 1133, 1149–1150
Landing gear assembly, 5, 1133, 1149–1150
Lands, 380
Lane departure warning (LDW) system, 185, 346
Laser alignment, 802–803, 867
Latching solenoids, 632
Latent heat, 1191, 1192
Latent heat of condensation, 1192
Latent heat of vaporization, 1192
Launch phase, 412, 620
LDW. *See* Lane departure warning (LDW) system
Lead-acid batteries, 139, 140, 193, 195
Lead trailer, 1133
Leaf springs, 838
Leaf spring suspensions
  applications, 840
  bogle suspensions, 842
  combination air/leaf spring suspensions, 839
  constant rate springs, 839–840
  fiber composite leaf springs, 844–845
  front suspension, 840–842
  rear suspension, single axle, 841–842
  rear suspension, tandem axle, 842
  with shock absorbers, 843–844
  spring pack clamping force, 839
  spring pack principles, 838
  spring pack self-dampening, 839
  with torque rods, 844
  variable rate springs, 840
  walking beams, 840
Leak prevention, 395–396
LED headlights, 274
LED indicators, 1038
Legal requirements, 90
Lens, 167, 273
Less-than-truckload (LTL) carriers, 12
Leveling valves, 849
Lifting and carrying, 26
Lifts and hoists, 61–62
Light beam alignment, 802–803
Light bulb operation, 271
Light bus, 350
Light-emitting diodes (LEDs), 163, 274
Lighting systems, 271–279
Lightly loaded vehicle weight (LLVW), 933
Limit washers, 510
Limp home mode, 665
Linear actuators, 380
Linear-type Hall effect sensors, 178, 179
Linehaul, 12
Liners, 1138
Lip-type seals, 916, 926
Liquid-cooled heating systems, 1231–1234
Lithium-based batteries, 222
Lithium-ion polymer (Li-Po), 222
Lithium-ion (Li-On), 222
Lithium-ion polymer (Li-Po), 222
Litmus tests, 117
Live axles, 723
Load, 141, 143
Load distribution, 892
Load dump, 234
Loading dock height, 1138
Load proportioning valves (LPVs), 998
Load-sensing power steering, 832–833
Load test, 208–209
Load transfer, 997
Local area networks (LANs), 322
Locking pliers, 54
Locking tang, 330
Lockstrap, 405
Lockup circuits, 579
Lockup clutches, 544–545, 551
Lockup torque converters, 544–545
Lock washers, 80
Logic, 346–346
Logistics, 12
Long-stroke chambers, 978
Loss of potentiometer ground, 177
Low-maintenance batteries, 198
Low oil levels, 585–587
Low pressure indicators, 947
Low-profile tires, 888–889
Low-range, direct powerflow, 479
Low-range, overdrive powerflow, 479–480
Lube jobs, 121–123
  brake stroke and lighting, 123
  checking fluids, 122–123
  greasing chassis, 123
Lube pressure management, 631
Lubricants, 80, 108–115, 495, 753–754
  additives, 109
  belts and tensioners, 110
  chassis, 111–112
  coolant hoses, 110
  diesel engine oil classifications, 109–110
  engine oils, 108–109
  gear, 110–111
  mineral-based, 495
  synthetic-based, 495
  trailer, 115, 116
Lubrication, 415, 494–497, 663–664, 705–708, 739–740, 753–756, 1154, 1173
Lubrication failures, 760–761
Luminosity, 272

M

Machinery's Handbook, 1110
Machinist's rule, 62
Mack Trucks Electrical Information System (EIS), 70, 323
Mack Trucks Premium Tech Tool (PTT), 324
Magnesium chloride, 1117
Magnetic switches, 152–153
Magnetism, 135–136
Magnetomotive force (mmf), 138
Main housing modules, 638–639
Main memory, 182
Main modulator solenoid, 631
Main pressure, 631
Main pressure circuit, 574–577, 649
Main pressure regulators, 631, 649
Main relief valve, 631
Mains electrical equipment, 32–33
Mainshaft, 456, 503
Main shaft modules, 640
Maintenance, 71, 88–98, 106–107, 275, 392–398. *See also* Preventive maintenance inspection (PMI)
  practices, 89
  problems, 90
  software, 106
Maintenance-free batteries, 198
Maintenance logs, 71
Maintenance manuals, 71
Manifold gauge set, 1216–1218
Manifold service host sets, 1218–1220
Manometers, 368
Manual chassis lubricating system, 112
Manual parking/emergency system test, 1061
Manual selector linkage inspection, 591–592
Manual steering gears
  centering steering system, 821
  recirculating ball, 816–817
  troubleshooting, 817–822
  worm and sector, 816, 817
  worm shaft preloads, 818–819
Manual transmissions. *See* Automated manual transmissions (AMTs); Automatic transmissions; Transmissions
Manufacturer's service literature, 70–76
Mapping, 347
Margin of safety, 1109
Master cylinders, 989, 992–998, 1006
Master Diagnostics (MD), 324
Master Technician certification, 17
Material safety data sheets (MSDS), 36–39
Matter, states of, 1192
Maximum accelerator pedal angle (full travel), 176
Maximum forward reduction, 559
Maximum overdrive, 560

Measuring tools
  bolt and nut dimension gauges, 69
  dial calipers, 62–63
  digital, 68–69
  feeler gauges, 69
  machinist's rule, 62
  micrometers, 63–68
  screw pitch gauges, 69
  thickness gauges, 69
  torque wrenches, 70
Mechanical engine coolant temperature display, 285
Mechanical modulator controls, 592
Mechanical transmissions, 449. *See also* Automated manual transmissions (AMTs);
Mechanic's gloves, 25
Memory banks, 179
Memory chips, 320
Memory management, 179–180
Mercedes-Benz AGS, 603
Meritor tire inflation system (MTIS), 912
Meritor Toolbox Software, 324
Message bit encoding, 349
Message identifiers (MID), 326–327, 358
Message tagging, 348–349
Metal inert gas (MIG), 35
Metallo-ceramic facings, 406
Metal-oxide semiconductor field effect transistors (MOSFETs), 165
Metal-oxide semiconductors, 165
Metal particle in oil, 590
Meter resolution, 303
Metering valves, 997, 1007
Metric fasteners, 78
Micrometers
  digital display outside, 66
  inch-graduated outside, 64–65
  inside micrometer, 66–68
  metric outside, 65
  outside micrometer with Vernier scales, 65
Microprocessors, 173
Midtronics conductance testing, 210
Midtronics EXP, 216–218, 313
MIL (malfunction indicator light), 301
Minimum forward reduction, 559–560
Misalignment, 889
Mobile Air Conditioning Society (MACS), 1197
Mobile universal trailer tester (MUTT), 1157
Model year (MY), 124, 357
Modems, 321
Modular construction, 635–636
Modulating valve, 952
Modulation, 159
Modulator, 1024
Modulator assemblies, 1018
Modulator pressure circuits, 578
Modules, 635
Moisture ejectors, 946
Molded organic linings, 433
Molecules, 131–132
Monitor, 317, 321
Monocoque design, 1135
Motherboards, 320
Motor Carrier Act of 1980, 11
Motorized cab circuits, 285–286
Mouse, 321
Mufflers, 1211
Multi-element rails, 1114–1115
Multifunctional handheld diagnostic tools, 313
Multimeters, 299
Multimodal containers, 4
Multimodal transportation, 4, 1139
Multipiece Rim/Wheel Matching Chart, 897
Multiple branch parallel circuits, 146
Multiple-disc clutches, 562
Multiplexing, 11, 173, 181–182
  basics, 345–353
  clients and servers, 345
  standardization of, 349
Multiplex switching, 353–354
Multiplex transactions, 354–356
Multiprotocol cartridge (MPC), 314
Mushroom flush type rivet set, 1155
Mutual inductance, 151

N

National Automotive Technicians Education Foundation (NATEF), 19
National Highway Traffic Safety Administration (NHTSA), 10, 97, 933, 1104
National Institute for Automotive Service Excellence (ASE), 14, 1197
National standard dryseal taper pipe thread (NPTF), 389
National standard tapered thread (NPT), 389
Natural gas, 5
Navistar International OnCommand Service Information System (ISIS), 70, 323
Navistar Master Diagnostics (MD), 324
Navistar Service Information (NSI), 323
Needle nose pliers, 54
Negative temperature coefficient (NTC), 133, 173
Negative temperature coefficient-type thermistor, 174
Networking, 322, 356
Neural networks, 344
Neutral safety switches, 259
NEXIQ Brake-Link, 1039
Nickel-metal-hydride (NiMH) batteries, 221–222
Night vision systems, 274–275
NiMH batteries, 221–222
Nitrogen-filled tires, 908
Noise, 25, 159
Noise-cancelling headphones, 26
No-load tests, 265–266
Nominal class designations, 1247
Noncycling circuit breakers, 288
Nondriving axles
  dead axles, 748–749
  steering axles, 747–748
  trailer axles, 749
Nonlatching solenoids, 632
Nonparallel drivelines, 701
Non-servo drum brakes, 999
Nonsynchronized transmission operation, 461
Nonvolatile RAM (NV-RAM), 182
Normal high (NH) solenoids, 633
Normally closed (NC) solenoids, 646
Normally open (NO) solenoids, 646
Normal low (NL) solenoids, 633
No-spin locking differential, 729–731
NOT gates, 171
NPN transistors, 164
N-type semiconductors, 161

O

Occupational Safety and Health Administration (OSHA), 36, 425
Occurrence rate (OC), 360
Odometers, 283, 285
OEM diagnostic software, 324
OEM local area networks (LANs), 322
OEM software, 322, 324
Off-going ratio tests, 664
Ohm, Georg Simon, 141
Ohms, 141, 302
Ohm's Law, 143, 144–145, 305
Oil changes, 495–497, 587–589
Oil contamination, 589–591
oil filter, 496
Oil filter and pump circuit, 574
Oil flow, 541
Oil leaks, 498–500, 503
Oil-level gauges, 371
Oil level readings, 586
Oil pressure gauges, 282
Oil pumps, 565
Oils, engine, 108–109
  2017, 110
  gasoline-fueled vehicles, 110
  natural gas, 110
  used, 110
On-board chassis lubricating systems, 112
Oncoming ratio tests, 664
One-piece dash design, 280
One-speed fallback, 617
One-way check valves, 972
On-off solenoids, 632
Open-center hydraulic systems, 371
Open-center systems, 371
Open circuits, 142, 144, 305
Open-circuit voltage test, 207, 208
Open-end wrenches, 51
Open loop (fuzzy) logic, 347
Open or shorted winding check, 284
Open top van trailer, 1139
Operating angles, 715–716
Operating systems (OS), 321
Optical components, 167
Optical disc reader/writer, 321
Optical light spectrum, 166–167
ORFS (O-ring face seal) fittings, 389

Organic facings, 406
OR gates, 170
Orifice expansion tubes, 1207
Original equipment manufacturer (OEM), 2, 133
O-ring fittings, 389
Oscillation, 838, 908
Outlet lines, 371
Out-of-phase, 703
Out-of-service (OOS), 9, 99–100, 1069. *See also* Commercial Vehicle Safety Alliance (CVSA)
  CVSA OOS, 99–105
  in-house OOS, 99
Output circuits, 180
Output modules, 635
Output retarder modules, 640
Output retarders, 581
Output yokes, 513
Ovality, 717
Overcharging, 203, 235–236
Overcrank protection (OCP), 258
Overdrive, 449, 560
Overinflation, 890
Overrunning clutches, 258, 542–544
Oversteer, 800, 1046
Overtorque, 81
Overvoltage protection, 234
Owner-operator (O/O), 12
Oxidation inhibitors, 109
Oxidizing flame, 34
Oxyacetylene equipment, 33–34
  flame adjustment, 34
  precautions, 34
  torches, 34
Oxygen cylinders, 33–34
Ozone depletion potential (ODP), 1193

P

Paccar Electronic Service Analyst (ESA), 324
Paccar ServiceNet, 70, 323
Packets, 345
Parabolic, 273
Parallel circuits, 143–144
Parallel-joint drivelines, 701
Parallel transmissions, 347
Parameter group (PG), 359
Parameter group number (PGN), 359
Parameter identifiers (PIDs), 327
Parking brakes, 1003
Parking/emergency circuits, 951, 1076–1081
Park-on-air, 957
Partial authority electronics, 603
Parts inventory control, 41
Parts literature, 71
Parts managers, 18
PartsPro, 324
Parts requisition, 41
Pascal, Blaise, 366
Pascal's Law, 366
Passwords, changing, 318
Path, 141
Pedal height, 423
Pencil inspections, 90
Pending codes, 317
Pentavalence, 160
Performance-based maintenance (PBM), 89
Peripherals, 321
Personal computers (PC), 299
  connecting to truck data bus, 322–323
  hardware, 319
  peripherals, 321
  software, 321–322
  system housing, 319–321
Personal safety, 26–27
Phase angles, 703
Phillips screwdriver, 49
Phone-based apps, 322
Photo alignment, 802
Photo diodes, 163
Photoelectricity, 141
Photonic semiconductors, 166–167
Photons, 166
Pickup and delivery (P&D), 12
Pico Scopemeter, 324, 362
Piezo actuators, 153
Piezoelectricity, 141
Piezoelectric sensors, 141
Piezo-resistive pressure sensors, 174
Piggyback assembly, 1078
Pilot bearing, 428
Pilot bearing bore runout, 430
Pilot bearing replacement, 428
Pilot-operated relief valves, 379, 380
Pilot valves, 380
Pinhole injection, 31, 395
Pinhole injection injury, 395
Pinion bearing preload
  check/adjust pinion cage shim pack, 777–778
  differential rolling resistance, 776–777
  press-fit method adjustment, 774–775
  yoke method of adjustment, 775–776
Pinion clearance, adjusting, 267
Pinions, 724
Pin synchronizers, 463–464
Pintle eye, 1184
Pintle hooks, 1134, 1161, 1184–1185
Pipe couplers, 389
Pipes and tubes, 389–391
Piston brakes, 978–979
Piston motors, axial, 384
Piston pumps, 375–378
Piston-type compressors, 1199
Piston-type cylinders, 381
Pitman arms, 788
Pitting, 710
Plain synchronizers, 465
Planetary double-reduction axles, 727
Planetary gearsets, 259–260, 557–564, 637
Planetary pinion gears, 558
Plasma-cutting, 35
Plate-type brake testers, 1103
PLCs. *See* Programmable logic controllers (PLCs)
Pliers
  adjustable, 54
  combination, 53–54
  diagonal cutting, 54
  locking, 54
  needle nose, 54
  safety, 54–55
  sidecutters, 54
  Vise Grips™, 54
Plow-out, 984
Plunger pumps, 376
PMI. *See* Preventive maintenance inspection (PMI)
Pneumatic actuator, 621
Pneumatic balance, 984
Pneumatic circuits, 366
Pneumatic timing, 984, 1054
PNP transistors, 163
Polarized gladhands, 956
Policy adjustments, 14
Polyurethane sponge strainers, 1064
Pop-off valves, 946
Ports, 310
Posi-Drive™ screwdrivers, 50
Position sensors, 609–610
Positive temperature coefficient (PTC), 133, 173
Potential, 227–228
Potential difference (PD), 134
Potentiometers, 174–175
Pour point, 120
Power, 6, 147
Power-assist steering systems
  hydraulic operations, 824–826
  load-sensing power steering, 832–833
  power steering gears, 824
  power steering pumps, 822–824
  troubleshooting, 826–829
Power dividers, 728, 735–739
Powerflow, 457–458, 479–482, 565–573
Power line carriers (PLCs), 346, 1032–1033, 1148
Power management, 613
Power rectifier diodes, 162
Powershift, 630
Power sources, 141, 142
Power steering analyzer, 828
Power steering systems, 826–829
Power supply filters, 150
Power take-off (PTO), 378, 487–491, 718–720
Power tools
  air drills, 59
  air ratchet wrench, 59
  air tool safety, 60
  bench grinders, 61
  blowguns, 60
  impact wrenches, 58–60
  lifts and hoists, 61–62
  presses, 62
  safety, 57–58

torque measuring wheel nut gun, 60
Powertrain, 2
P1 planetary gearset module, 640
P2 planetary gearset module, 640
Prebalance inspections, 910–911
Predictive-based maintenance (PBM), 89
Premium Tech Tool (PTT), 324
Preset hub assemblies, 924
Preset hubs, 879
Presses, 62
Press-fit method adjustment, 774–775
Pressure, 366–368, 1192
Pressure-control valves, 378–379
Pressure differential valves, 997, 1007
Pressure drop, 370
Pressure plate and cover assembly, 432
Pressure protection valves, 947
Pressure reducing valves, 947
Pressure relief valves, 1205
Pressure scales, 367–368
Pressure springs and levers, 403–405
Pressure switches, 632
Pretests, 618
Pretrip inspection, 96
Pretrip inspection procedure, 96–98
Preventive maintenance, 41, 89, 92
Preventive maintenance forms, 92, 93–94
Preventive maintenance inspection (PMI), 40, 89, 92, 497–501. *See also* Maintenance
Preventive maintenance scheduling, 100–107
Priestley, Joseph, 128
Primary circuits, 947–950
Primary coil, 151
Primary modulation, 646
Primary tanks, 948
Prime mover, 370
Printers, 321
Priorities, assigning, 348
Priority valves, 579
Prior learning assessment recognition (PLAR), 20
Processors, 320
Professional advancement, 19–20
Professional associations, 19
Prognostics, 89, 635
Programmable logic controllers (PLCs), 185, 353
Programmable read-only memory (PROM), 180, 183
Programming, 183
Program software, 322
Pro-Link 9000, 313
Pro-Link GRAPHIQ, 316
Pro-Link iQ, 315–316, 1039
PROM. *See* Programmable read-only memory (PROM)
Propane, 1194
Propeller shafts, 696
Proportioning and timing application pressures, 982
Proportioning valves, 997, 1007
Protocols, 345
Protons, 129, 157
Protractors, 755
P-type semiconductors, 160, 161
P-type silicon, 160
Pull circuit, 106
Pull-type clutches, 407–408, 417
Pulse generators, 1027
Pulse radar, 185
Pulse wheels, 177, 1027
Pulse width modulation (PWM), 158, 646, 1237
Pulse width (PW), 158
Pump drives, 378
Pumps
  fixed-displacement, 373
  gear, 373–374
  performance of, 378
  piston, 375–378
  vane, 375
  variable-displacement, 373
Punches, 48
Pup trailers, 1133
Purge tanks, 945
Push circuit, 106
Pusher axles, 748
Push-type clutches, 407, 417
Pyrometer, 283

Q

Quick-release couplers, 390–391
Quick-release gladhands, 956
Quick-release valves, 948, 968, 1074

R

Rack and pinion steering, 829–831
Radar vs. video, 185
Radial piston pumps, 377–378
Radial tires, 884
Rails, 1107, 1114
Rail select motor, 609
Ram air effect, 1198
Rams, 381
Random access memory (RAM), 180, 182, 320
Range control valve problems, 531–532
Range cylinders, 475, 531
Range/shift tests, 664
Range valves, 475–476, 611
Range verification, 664
Ratiometric analog, 173
Ratios and shifting, 449
Ratio valves, 969
Reactive, 35
Reader/programmer, 313
Read-only ESTs, 316
Read-only memory (ROM), 180, 182–183
Rear cover module, 640
Rear drive axle, 737
Rear seal replacement, 500–501
Rebound, 838, 908
Recall bulletins, 71
Receiver/dryer, 1203–1205
Recombinant batteries, 199
Recommended Practices, 19
Record keeping, 90
Record-keeping requirements, federal, 105–106
Rectifier bridge, 230
Reduction-gearing starter motors, 259
Redundancy, 933, 1003, 1018, 1023
Reefers, 1140
  construction, 1140–1141
  engines, 1143
  multiple cooling zones, 1142–1144
  operations, 1141
  refrigerants, 1143–1144
Reflective tape, 1152–1153
Reflectors, 167, 273
Refractive index, 195
Refractometers, 117, 195, 207
Refrigerant management centers, 1219–1220
Refrigerants, 1192, 1193–1197
  alternate, 1194
  contamination, 1195
  recovery, 1195–1197
Refrigerated and insulated vans, 1140–1144
Refrigeration cycle, 1197–1199
Refrigeration lines, hoses, and couplers, 1209–1210
Regrooving, 893–894
Regulator circuit testing, 239
Reinforcement, 1115–1116, 1115–1116
Relay-emergency valves, 957
Relays, 288–289
Relay testing, 292
Relay valves, 948, 950, 967–968, 1072–1073
Release bearings, 407
Relief valves, 378
Reluctance, 130
Reluctance-type generators, 149
Reluctors, 178
Reluctor-type generators, 149
Reluctor wheels, 1026
Remote air solenoid module, 355–356
Remote diagnostic units (RDUs), 1039
Remote power module, 355
Remote sensing, 233–234
Remote sensing technology (RMT), 234
Renting and leasing, 12
Repair order, 18, 90–92
Replace/repair analysis, 41
Reprogram Parameters field, 318
Reserve capacity rating, 201
Reservoir air supply leakage test, 1059–1061
Reservoir drain valves, 1055
Reservoirs, 371
Residual pressure valve, 993
Resistance, 133–134, 142
Resist bend moment (RBM), 1109
Resistor-capacitor circuits (R-C circuits), 151
Resistors, terminating, 350

Resolution, 303
Resource Conservation and Recovery Act (RCRA), 36
Restricted needle movement, 283–284
Retarder controls, 640–641
Retarder electronic controls, 641
Retarder operation, 641
Retarders, 10, 635
Retarding modes, 640
Retreads, 885
Return lines, 371
Reverse gear, 458, 569
Reverse inhibiting, 581
Reverse switches, 612–613
Rheostats, 175–177
Ride analysis, 874–876
Ride height adjustment, 863
Right-hand rule, 137
Right-to-Know Law, 36
Rigid discs, 405
Rims, 879
Ring gears, 558
Rivet, 84, 1154–1156
Rivet guns, 1155–1156
Roller brake dynamometers, 1103
Roller clutches, 542
Rollovers, 1045
Roll stability control (RSC), 872
RollTek, 339
RollTek rollover protection systems, 184, 339–340
Root mean square (RMS) value, 305
Rotary actuators, 380
Rotary control valves, 825
Rotary flow, 538
Rotary-type Hall effect sensors, 177–178
Rotary valves, 380
Rotary vane compressors, 1199–1200
Rotating-cam pumps, 377–378
Rotating clutch modules, 638
Rotating cam radial piston pump, 377–378
Rotating-piston pump, 378
Rotation, 450
Rotochambers, 978
Rotor, 228, 229
Rotor-gear pumps, 374
Rotors, 228, 1011–1012, 1100
RP1210, 300, 358
Running repairs, 13
Runout, 716

S

SAE J1321, 4
SAE J1587, 181, 326
SAE J1708, 181, 326
SAE J1939, 181, 326, 347, 349, 617
SAE J1939 standard, 181
SAE J standards, 181, 298–299
SAE type-2 circuit breakers, 288
SAE viscosity grades, 392
Safety
- battery charging, 212
- brakes systems and, 1053
- chisels and punches, 48i
- clutch servicing, 425
- gear, 24–26
- hammers, 46–47
- hydraulics, 392, 394
- lifts and hoists, 61–62
- pliers and cutter, 54
- power tools, 57–58
- refrigerants, 1212–1213
- saw and knife, 48
- shop tool, 35
- tire and rim, 894–895
- towing precautions, 504
- U-joint removal precautions, 708–709
- wrenches, 53

Safety factor, 1109
Safety Management System (SMS), 100
Safety pressure relief valves, 946
Safing sensors, 337–338
Sag and bow frame failure, 1118
Saws and knives, 47–48
S-cam foundation brakes, 958–962, 1085–1092, 1094
S-camshafts, 958, 1086
Scanners, 321
Scan tool, 316
Scavenging pumps, 641
School buses, 17
Schrader valves, 1217–1218
Scopemeter, 324, 362
Scope testing, 240
Screwdrivers
- Phillips head, 49
- Posi-Drive™, 50
- standard tip, 49
- Torx®, 50

Screw pitch, 69
Screw pitch gauge, 69
Scroll compressors, 1200
Sealants, 84
Seal removal and inspection, 925–926
Seat belts, 97
Secondary circuits, 950–954
Secondary coil, 151
Secondary locks, 1161
Secondary modulation, 646
Secondary tanks, 950–951
Second-gear powerflow, 460, 566–567
Section height, 888
Section modulus (SM), 1109–1111
Section width, 888
Segments, 148
Selective catalytic reduction, 124
Selector valve (four-speed), 577
Self-adjusting clutches, 405
Self-diagnostics, 1039–1040
Self-education, 19
Self-employment, 17–18
SEL (stop engine light), 302
Semiconductors, 133, 157, 159–161, 167–168, 309–313
- bias, 160–161
- doping, 160–161
- testing diodes, 167–168
- testing transistors, 168–169

Semi-floating axles, 747
Semi-oscillating fifth wheel, 1161–1162
Semi-trailers, 4, 1132–1133
Sensing voltage, 233
Sensor and computer alignment systems, 803–805
Separator pins, 405
Sequential troubleshooting, 326
Serial communications interface (SCI), 644, 648–649
Serial data transmission, 347
Series circuits, 143
Series-parallel circuits, 144, 146–147
Servers, 345
ServiceAdvisor, 324
Service brake application, 935
Service brake priority, 956–957
Service brakes, 958, 1074–1076
Service bulletins, 71
Service diaphragm replacement, 1078
Service hour meter, 283
Service information system (SIS), 70–71, 323–324
Service lights, 617
ServiceLink, 324
ServiceLit, 324
Service managers, 18–19
Service manuals, 70
ServiceNet, 323
ServicePro, 324
Service technicians, 14–17
Service valves, 1217
Service writers, 18
Servo brakes, 999
Servos, 376
Servo-type drum brakes, 999–1000
Shift bar housings, 466, 510
Shift fork or yoke, 456
Shift levers, 615–617
Shift point adjustments, 594–599
Shift rail housings, 512–513
Shift selectors, 643–644
Shift towers, 466, 511
Shift tower valve tests, 532
Shock absorber chisel, 59
Shock absorbers, 843
Shock failures, 758
Shop records, 40–43
Shop supervisors, 18
Shorepower, 139, 1194
Short circuits, 144, 294
Short to ground, 144
Shutdown systems, 286
Sidecutters, 54
Sideslip, 1103
Sidesway damage, 1118
Sight glass, 1210–1211
Signal detect and actuation modules (SAM), 358, 1237
Signals, 159
Silicon, 159
Silicon-controlled rectifiers (SCRs), 166
Simpson, Howard W., 557
Simpson geartrain, 557
Single-acting cylinders, 381

Single-phase main, 32
Single-reduction axles, 727, 729–734
Single-wire braid hoses, 386
Sipes, 885
Six-channel systems, 1033
Six-/four-channel systems, 1033
Slack adjusters, 979–980, 1081–1082, 1090–1091, 1097
Slave cylinders, 993
Slave piston, 376
Slave valves, 473, 475, 527
Sleeper compartment heater and air conditioner, 1233
Sleeve and poppet quick-release couplers, 390
Sliding caliper, 1002
Sliding fifth wheels, 1162, 1173, 1180–1181
Sliding gear shift, 456–457
Sliding seal couplers, 390
Slip joints, 698
Slipout, 503
Slip rings, 229–230
Slip splines, 697, 706
Slow adaptive mode, 645
Slow charging, 210–211
Slow overdrive, 560
Slow reverse, 561
Small signal diodes, 162
Smart ignition switch, 259
Smart switches, 270, 353–354
Snapring U-joint replacement, 708
Snapshot tests, 326
Snubber, 1184
Society of Automotive Engineers (SAE), 108, 181, 200
Socket joints, 790
Socket wrenches, 52
Sodium chloride, 1117, 1154
Software diagnostics, 71
Solar cells, 167
Soldering, 286–287
Solenoid regulator valves, 650
Solenoids, 151, 152–153, 256–258, 632–633, 646, 650
Solenoid windings, 256
Solenoid testing, 292
Solid rubber suspension, 845–846
Solid state, 157
Solid-state devices, 157
Solo clutch setup, 440–443
Solvents, 28
Source address (SA), 326–327, 349, 358, 633, 1237
Sourcing problems, 236
Spalling, 709
Specialty fifth wheels, 1163–1164
Specialty service shops, 13
Specialty technicians, 17
Specialty trailer designs, 1144–1146
Specific gravity (hydrometer) test, 205–206
Specific gravity (SG), 194, 195
Specific PGN, 359
Spec'ing landing gear, 1148
Speed of rotation, 450
Speedometers, 283, 285
Speedometer shift point adjustment, 595
Speed sensors, 606, 612, 644
Speed vs. torque, 454–455
Spiders, 698, 959
Spigot bearing, 729
Spike suppressants, 151
Spill response, 28
Spinout, 726, 737–738
Spinout failures, 737–738, 754
Spiral bevel gears, 729
Spiral-wire hoses, 386
Splined yokes, 8
Splines, 510–511
Split-coefficient stops, 1024
Split guide rings, 541–544
Split shifting, 476–477
Splitter cylinder operation, 477–478, 532
Splitter shift problems, 532
Spontaneous combustion, 28
Spool valves, 380
Spring brake chambers, 974
Spring brake control valves, 970
Spring brake priority, 957
Spring brakes, 951–952, 1077 1078, 1079–1080
Spring-loaded accumulators, 372
Spring packs, 838
Spur gears, 455–456
Square waves, 158
S-Series Certification, 15
Stability control electronics (SCE), 184, 1044–1045
Stability management, 1045
Stalk switch, 278–279
Stall tests, 592–593
Stall torque ratio, 545–546
Standard transmissions, 449
Star network structure, 346
Starter circuit, 249
Starter circuit testing, 264–265
Starter motors, 249, 252–260
  construction, 253–255
  operation, 255–256
Starter rebuild, 266–267
Starter relays, 252, 259
Starter relay testing, 261–263
Starter solenoids, 256–258
Starting aids, 118
Starting circuit components, 249–250
Start of frame field, 349
State of charge test, 205–207
Static charge, 31
Static discharge, 31
  and computers, 31–32
Static electricity, 128, 139–140
Static wheel balancing, 909
Stationary object alert (SOA), 328
Stator assembly, 542
Stators, 229, 230, 538, 542–544, 552–553
Stator windings, 149, 238–239
Steel frames, 1111
Steering and alignment
  Commercial Vehicle Safety Alliance (CVSA) OOS steering criteria, 833–834
  components of, 786–791
  electronic alignment systems, 802–805
  electronic and load-sensing steering, 831–833
  front-end alignment, 791–801
  manual steering gears, 816–822
  power-assist steering systems, 822–826
  rack and pinion steering, 829–831
  steering axle inspection, 805–816
  steering system components, 786–791
  troubleshooting power steering systems, 826–829
Steering axle inspection
  adjusting knuckle vertical play, 809–810
  repairs, 810–815
  steering knuckles, 807
  tie-rod inspection, 808–809
  unitized steer axles, 816
  wheel bearing inspection, 809
Steering columns, 786–787
Steering control arms, 789
Steering gears, 788
Steering knuckles, 789–790
Steering systems, 9
Steering wheels, 786
Stem-type service valves, 1217
Step-down transformers, 151
Step-up transformer, 151
Stop lamp switches, 972
Stopping distances, 933, 1103–1104
Straight truck, 2
Stress concentrators, 1116
Stroke-sensing automatic slack adjusters, 980
Stud-piloted wheels, 881–882
Stud removers, 82
Subcooling, 1207
Submodulation, 646
Subsystem identifiers (SIDs), 327, 358
Sulfation, 198, 204
Sump and purge valves, 944
Sun gears, 558
Supercapacitors, 222
SuperTech, 20
Supplemental coolant additives (SCA), 117
Supplemental restraint systems (SRS), 184, 335–337, 346
Supplementary coolant heater, 1234
Supplier guides and catalogs, 75
Supply circuit service, 1063–1069
Supply circuit valves, 947
Supply tanks, 933, 945–947
Suspect parameter group (SPG), 359
Suspect parameter number (SPN), 327, 359
Suspension alignment, 867–872
Suspensions, 9

Suspension systems
air spring, 847–852
alignment of, 867–872
cab air suspension and driver seats, 872–874
cushion and torsion, 845–847
leaf spring, 838–845
ride analysis, 874–876
roll stability control (RSC), 872
rough ride diagnosis, 853–854
servicing air suspensions, 862–867
servicing equalizing beams, 858–862
servicing mechanical systems, 852–858
Swaging, 48
Swashplate, 376
Swashplate-type compressors, 1200
Switched circuits, 288–289
Switches, 143, 178, 353–354
Switch packs, 355
Symbols and schematics, 392
Symptom-driven diagnostics, 618
Synchronized shifting, 461–464
Synchronizers, 510
Synthetic gear lubes, 111
System administration, 324
System actuation module (SAM), 1234
System Manager, 591
System park, 953
System pressure, 942
System redundancy, 1003, 1018

T

Tachometers, 282–283, 285
Tag axles, 748
Tags, 348
Tandem drive axles, 727
Tang release tool, 331
Tankers, 1146
Tanker trailers, 1146
Tapered punches, 59
Tapered roller wheel bearings, 915
Target identifier, 349
T-bar floors, 1141
T/C exterior, 538
TCM-managed PTO mode, 649
TC10 Tractor Series (TS), 629
Technical service bulletins (TSB), 71
Technology and Maintenance Council (TMC), 19, 100, 921
Temperature, 133, 195, 234
Terminals, 135, 289
Terminating resistors, 350
Test lead spoons, 289, 309
Thermistor resistance check, 283
Thermistors, 173–174, 309
Thermistor vacuum gauge, 1219
Thermocouples, 141
Thermodynamics, 1191
Thermoelectricity, 140–141
Thermoplastic hose clamps, 110
Thermostatic expansion valve (TXV), 1206–1207
Thermostats, 258, 1208–1209
Thickness gauges, 69
Third-gear powerflow, 460, 567–569
Thirteen-speed transmissions, 482–486
Thirteen-speed, underdrive transmissions, 485
Thomson, J. J., 157
Thread locking compound, 84
Thread repair, 81–82
Three-pedal systems, 401, 603
Three-phase alternators, 233
Three-speed differentials, 741
Three-way splice, 334
Throats, 1161
Throttle position sensor (TPS), 175–177, 345, 632, 644
Throwout bearings, 407
Thyristors, 166
Tie-rod assemblies, 790–791
Time guides, 71
Timing, 511
Tire pressure control system (TPCS), 912
Tire pressure inflation system (TPIS), 912, 1147
Tire pressure monitoring system (TPMS), 912
Tires. *See* Wheels and tires
Tire service accidents, 895
Tire size ranges, 1040
Tire-to-rim hardware, 883
TMC SuperTech competition, 19–20
TMC wheel bearing adjustment, 921–924
Toe, 791
Toe-in, 791
Toe-out, 791
Tone wheels, 177, 1027
Tooth contact patterns, 780–782
Topology, 348
Torches and tips, 34
Torque, 6, 79, 382, 454–455
Torque converters, 7, 636
construction, 536–540
maintenance and service, 546–554
operating principles, 535–536, 540–546
Torque-limiting clutch brakes, 409–410
Torque multiplication, 538, 542
Torque proportional differentials, 747
Torque wrenches, 69
Torricelli's tube, 367
Torsional failures, 758
Torsion bar suspension, 846
Torus, 536
Torx® screwdriver, 50
Total base number (TBN) additives, 109
Total pedal travel, 423
Touchpad, 321
Touch points, 412, 620
Towing precautions, 504
Toxic, 35
Trackball, 321
Tractor, 2
Tractor park, 954
Tractor protection (TP) valves, 937, 971, 1074
Tractor/semi-trailer dimensions, 4–5
Trade magazines, 76
Trailer ABS warning lights, 1031–1032
Trailer air dryers, 944
Trailer air supply, 954
Trailer application valves, 950, 966–967, 1074
Trailer brake system, 956–957, 1084
Trailer brake valves, 957
Trailer circuit connector, 279–280
Trailer circuits, 954–958
Trailer doubles, 1139
Trailer downward extension, 1137
Trailer frames, 1108, 1135–1136
Trailer hookups, 1056
Trailer length, 1137
Trailer inspection guide, 107
Trailer lubrication, 115, 116
Trailer primary circuit brake signal, 950
Trailer roofs, 1138
Trailers, 4, 115, 1108
air circuit troubleshooting, 1157
aluminum panels, 1153–1154
brakes, 1147–1148
corrosion, 1154
doors and door frames, 1153
electrical connectors, 1152
fasteners and panels, 1154
freight van trailers, 1139
lamp unit construction, 1152
lighting, 1150
lubrication, 1154
maintenance, 1153–1157
mobile universal trailer tester (MUTT), 1157
multiplexing, 1148
refrigerated and insulated vans, 1140–1144
riveting, 1154–1157
roof maintenance, 1153
secondary circuit brake signals, 951
service brakes, 958
service signals, 954–955
specialty trailer designs, 1144–1146
spring and service priority systems, 953
supply hoses, 955
supply valves, 953
suspension, 1148–1149
swing, 983–984
tires and wheels, 1146
tractor/trailer combinations, 1132–1136
undersides, 1146–1152
wiring requirements, 1150–1151
Trailer stability assist, 184
Trailer swing, 983–984
Training literature, 75
Transfer cases, 486–490
Transfer gear modules, 641–642

Transformer rectifier (TR) 12/24
alternators, 234–235
Transformers, 151–152
TransHealth monitor, 634
Transistors, 157, 163–166
Transistor types, 164
Transit buses, 17
Transmission Controller, 603
Transmission control module (TCM), 635, 636
Transmission oil/fluid, 123, 535
Transmissions, 7–8, 557. *See also* Automated manual transmissions (AMTs); Automatic transmissions; Manual transmissions
air leaks, 527
air regulator problems, 527
air shift system troubleshooting, 527–532
auxiliary gear sections, 470–486, 513–514
bearings, 509–510
compounding ratios, 449
deep-reduction, 485–486
disassembly of, 508
eighteen-speed, 478–482
gears, 450–451
gear slipout, 503–504
geartrain configurations, 456
general overhaul guidelines, 511
inspection of, 508–509
lubrication, 494–497, 618–619
main and auxiliary gearing, 449–456
main case reassembly, 508–512
mechanical shift mechanisms, 456–466
noise, 501–502
oil checks, 585
oil leaks, 498–500, 503
overhauls, 507–527
preventive maintenance inspections, 497–501
reinstallation of, 439–440
removal of, 425–428, 507–508
role of, 448–449
servicing, 494–533
shift housing assembly, 466–468
step-by-step overhauls, 511–527
thirteen-speed, 482–486
timing of, 511
towing precautions, 504
transfer cases, 486–491
troubleshooting, 501–506
twin countershaft, 469–470
Transmission tailshaft sensor, 180
Transport Topics, 19
TranSynd, 587, 663
Treadle valves, 937, 964
Tree diagnosis chart, 71, 73f3-63
Triage, 91
Triage tool, 325
Triangulation, 184
Trilateration, 183–184
Trimmer regulator valves, 578
Trimmer valves, 578–579
Trinary switches, 1204–1205
Truck Classifications, 2–5
Trunnions, 698
Truth tables, 170
T-Series Certification, 15
Tube couplers, 389
Tube cutting tools, 48
Tungsten filament, 271
Tungsten inert gas (TIG), 35
Turbines, 537, 551–552
Turn sensor assembly (TSA), 186, 188
Turn signal switches, 279
Turntable, 1160
Twin countershaft transmissions, 469–470
Twist damage, 1117
Twisted axles, 796
Twisted-wire geometry, 350
Twisted-wire pairs, 345
Two-channel systems, 1034
Two-pedal systems, 401, 603
Two-piece clutch brake, 409
Two-speed drive axles, 740–747
Two-speed tandem axles, 741
Two-way check valves, 933, 972
Type 1 virtual fusing, 356
Type 2 virtual fusing, 356

U

U-bolt and clamp attachment, 1127–1128
U-joints. *See* Universal joints (U-joints)
Ultracapacitors (UCs), 222
Ultracapacitors vs. battery discharge rates, 222–223
UltraShift, 603, 604–605, 620–621
UltraShift AW, 620–621
UltraShift AW3, drive coupler for, 442–443
UltraShift components
ball screw assemblies, 608–609
base transmission, 605–606
components, 604–613
Gen3 fault code index, 618
position sensors, 609–610
power management, 613
range valves, 611
reverse switch, 612–613
speed sensors, 612
transmission automation components, 606–608
UltraShift DM, 620
UltraShift DM clutch operation, 620
Unbalanced vane pumps, 375
Uncoupling, 1177
Undercharging, 236
Underinflation, 889
Understeer, 800, 1046
Unibody design, 1135
Unibody van, 1136
Unified National Coarse (UNC), 78
Unified National Fine (UNF), 78
Unitized axles, 816
Unitized Hub Systems (UHS), 924
Unitized seals, 916, 926
Universal joint and tie-rod tool, 59
Universal joints (U-joints), 8, 698–699
drive shaft installation, 711–713
failure analysis, 709–710
lubrication, 705
reassembly, 710–711
removal precautions, 708–709
replacement of, 707–713
working angles, 701–703
yoke inspection, 710
Unsprung weight, 838, 908
Upper couplers, 1134, 1138, 1161, 1183–1184
Upshifting, 461
Upshift sequences, 472–473
Urge-to-Move, 411, 620–621
U.S. Department of Labor
Bureau of Apprenticeships and Training(www.dol.gov/dol/topic/training), 19
USB flash drives, 320, 321

V

Vacuum, 367
Vacuum pumps, 1218
Valence, 129
Valve body and governor test fixture, 597
Valve ring adjusting tools, 595
Valves
directional control, 380
flow control, 379–380
pressure-control, 378–379
Van dimensions, 1136–1139
Van doors, 1138
Vane compressor, 1207
Vane motors, 383
Vane pumps, 375
Vane-type cylinders, 381
Van trailers, 1135, 1139
Variable bleed solenoids (VBS), 632
Variable capacitance (pressure) sensors, 174, 175
Variable-displacement pumps, 373, 376
Variable pitch stators, 544
Variable rate springs, 840
V-Cads, 324
Vehicle and driver performance analysis, 43
Vehicle automation components, 615
Vehicle configurations, 1255
Vehicle harnesses, 617
Vehicle history, 43
Vehicle identification numbers (VIN), 40
Vehicle interface module (VIM), 643, 644
Vehicle Maintenance Reporting Standards (VRMS), 41
Vehicle navigation systems, 183–184
Vehicle Onboard RADar (VORAD), 185
antenna assembly (AA), 186
blind-spot display (BSD), 187
blind-spot sensors (BSS), 186

Vehicle Onboard RADar (VORAD) (*Continued*)
central processing unit (CPU), 186–187
detection electronics, 185–186
driver display unit (DDU), 187
turn sensor assembly (TSA), 186
Vehicle retarders, 10
Vent doors, 1141
Ventilation, 28. *See also* Heating, ventilation, and air conditioning systems (HVAC)
Vibration diagnosis, 1249–1250
Video, 188–189
Virtual breakers, 288
Virtual circuit protection, 288
Virtual fuses, 287, 288, 356
Viscosity, 108, 391–392
Viscosity index (VI), 392
Vise Grips™, 54
Visual stroke indicators, 1095
Vocational trucks, 3
Vocational trucking, 12
Volta, Alessandro, 128
Voltage, 128, 134, 141–142
Voltage checks, 284
Voltage dividers, 175–177
Voltage drop test, 292
Voltage limiter, 283
Voltage regulators, 227, 228, 233–234
Voltmeters, 283, 284–285, 290
Volts, 302
Volvo IMPACT, 70, 323
Volvo V-Cads Pro, 324
VORAD. *See* Vehicle Onboard RADar (VORAD)
Vortex flow, 538

W

Wafers, 160
Wake up transaction, 354
Walking beams, 840
Warranty information, 43
Washers, 80
Waterless engine coolant (WEC), 117
Watt, 147
Watt, James, 147
Wavelength, 166
Wear compensators, 405, 420–421
Wear sensors, 1002
Weatherproof connectors, 329–332
Web, 1110
Wedge brake systems, 962, 1102
Wedge trailer design, 1137–1138
Weight-loaded accumulators, 372
Weight-over-axle variables, 1165
Weld attachment, 1028
Welding and cutting, 27, 287
Welding repairs, 1122–1125
Wet cell batteries, 193
Wet clutches, 411, 620
Wet tanks, 945
Wheel balance, 908
Wheel bearings, 755, 915
Wheel cylinders, 989, 998–999, 1007–1008
Wheel dolly, 916
Wheel-down, wheel-up brake adjustments, 1095
Wheel ends, 9, 123, 879, 925
Wheel hop, 908
Wheel imbalance, 909
Wheels and tires, 9
automated tire service stations, 897
chassis tire pressure management, 912–914
disc wheel installation, 902–903
dual tire matching, 905–908
rib designs, 887
rim inspection, 895–897
speed sensors, 1016, 1026
spoke wheel installation, 901–902
tire, rim, and wheel service, 894–908
tire construction, 884–885
tire maintenance, 889–894
tire rotation and replacement, 892–894
tires, 884–889
tire size, 887
trailer, 1148
tubeless-type radial tires, 905
tube-type radial tires, 905
wheel balancing, 908–911
wheel bearing adjustment, 921–925
wheel end conclusion, 927–930
wheel hubs, bearings, and seals, 914–921
wheels and rims, 879–884
wheel seals, 915–916, 925–927
Wheel speed sensors, 1026
Wide-base singles, 882
Wide-base tires, 886–887
Wide-base wheels, 882–883
WinAlign HD software, 802
Windings, 229, 230, 255, 256
Windows, 322
Windshield wipers, 286
Wingman, 185, 328
Winterizing
air systems, 117–118
battery, 204–205
cab comfort, 119–120
coolant, 115–117
diesel fuel, 120–121
electrical system, 119–120
fuel/water separators, 121
general tips, 121
starting aids, 118–119
Wire wheel brushes, 61
Wobble plates, 376
Work, 370
Work area safety, 28–31
Working angles, 701–703
Work order generation, 43
Work orders, 18, 41–42, 90–92
Workplace Hazardous Materials Information System (WHMIS), 36, 316
World Transmission (WT), 629
Wraparound dash design, 280
Wrenches, 51
adjustable, 51
air ratchet, 59
Allen, 51–52
box-end, 51
combination, 51
open-end, 51
safety, 53
WT diagnostic codes, 635, 636
WT digital display (DD), 644
WTEC controller, 635
WTEC electronics, 643
WTEC TCM, 645
WT hydraulic circuits, 649–650
WT identification, 636, 637
WT modular construction, 635–636
gearbox modules, 637–640
input modules, 636–637
output modules, 640–641
transmission control module, 636
WT output circuits, 643–649
WT powerflows, 658–663
WT servicing, 663–665
WT torque paths and range management, 650
WT troubleshooting, 663–665
Wye shape, 230

X

Xenon headlamps, 272, 273
XVision, 275

Y

Yaw, 801, 1045
Yield strength, 1108–1109, 1111
Yokes, 697, 717–718
Yoke sleeve kits, 500

Z

Zener, Clarence, 162
Zener diodes, 162–163, 234
Zerk fittings, 122, 698
Zero accelerator pedal angle, 176